Clinical Biomechanics in Human Locomotion

Gait and Pathomechanical Principles

Clinical Biomechanics in Human Locomotion
Gait and Pathomechanical Principles

Andrew Horwood
Visiting Fellow, Centre for Biomechanics and Rehabilitation Technologies,
Staffordshire University

With contributions from

Nachiappan Chockalingam
Director, Centre for Biomechanics and Rehabilitation Technologies,
Staffordshire University

Academic Press is an imprint of Elsevier
125 London Wall, London EC2Y 5AS, United Kingdom
525 B Street, Suite 1650, San Diego, CA 92101, United States
50 Hampshire Street, 5th Floor, Cambridge, MA 02139, United States
The Boulevard, Langford Lane, Kidlington, Oxford OX5 1GB, United Kingdom

Copyright © 2023 Elsevier Inc. All rights reserved.

No part of this publication may be reproduced or transmitted in any form or by any means, electronic or mechanical, including photocopying, recording, or any information storage and retrieval system, without permission in writing from the publisher. Details on how to seek permission, further information about the Publisher's permissions policies and our arrangements with organizations such as the Copyright Clearance Center and the Copyright Licensing Agency, can be found at our website: www.elsevier.com/permissions.

This book and the individual contributions contained in it are protected under copyright by the Publisher (other than as may be noted herein).

Notices

Knowledge and best practice in this field are constantly changing. As new research and experience broaden our understanding, changes in research methods, professional practices, or medical treatment may become necessary.

Practitioners and researchers must always rely on their own experience and knowledge in evaluating and using any information, methods, compounds, or experiments described herein. In using such information or methods they should be mindful of their own safety and the safety of others, including parties for whom they have a professional responsibility.

To the fullest extent of the law, neither the Publisher nor the authors, contributors, or editors, assume any liability for any injury and/or damage to persons or property as a matter of products liability, negligence or otherwise, or from any use or operation of any methods, products, instructions, or ideas contained in the material herein.

ISBN 978-0-443-15860-5

For information on all Academic Press publications
visit our website at https://www.elsevier.com/books-and-journals

Publisher: Mara E Conner
Acquisitions Editor: Carrie L. Bolger
Editorial Project Manager: Emily Thomson
Production Project Manager: Kamesh Ramajogi
Cover Designer: Christian J. Bilbow

Typeset by STRAIVE, India

Dedication

For my family, friends, colleagues, students, and all my past patients who have taught me so much.

Contents

About the authors — xiii
Foreword — xv
Preface — xvii
Acknowledgements — xix
Abbreviations — xxi
Introduction — xxiii

1. Understanding human gait

Chapter introduction — 1
1.1 Gait principles — 1
   1.1.1 Introduction — 1
   1.1.2 Gait energetics — 2
   1.1.3 Dividing the body segments to explain gait — 8
   1.1.4 Gait motion and description — 10
   1.1.5 Walking gait phases — 12
   1.1.6 Challenges of upright posture in locomotion — 19
   1.1.7 Human gait models — 22
   1.1.8 Describing gait: Rancho Los Amigos divisions — 34
   1.1.9 Additions and modifications to the Rancho Los Amigos divisions — 43
   1.1.10 Section summary — 48
1.2 Principles of gait analysis data — 49
   1.2.1 Introduction — 49
   1.2.2 Stability and ground reaction force — 49
   1.2.3 Measurements and interpretation of GRF — 56
   1.2.4 Vertical GRF components in walking — 59
   1.2.5 Anterior–posterior and medial-lateral GRF components in walking — 63
   1.2.6 Spatiotemporal parameters in gait — 69
   1.2.7 Measuring joint segment motions — 73
   1.2.8 Measuring and interpreting pressure — 74
   1.2.9 Variability in gait — 77
   1.2.10 Section summary — 80
1.3 Muscle function related to joint motion in gait — 80
   1.3.1 Introduction — 80
   1.3.2 Principles of muscle action in gait — 81
   1.3.3 Primary muscle function during walking gait — 84
   1.3.4 Soft tissue compliance and stiffening of the lower limb — 92
   1.3.5 The effects on walking of terrain, velocity, and gradient — 94
   1.3.6 Muscle activation and dysfunction effects on gait kinematics — 98
   1.3.7 Section summary — 105
1.4 Running gait — 105
   1.4.1 Introduction — 105
   1.4.2 Running energetics — 106
   1.4.3 The running gait cycle — 107
   1.4.4 Running models: Work and power phases in running — 110
   1.4.5 Impact in running: With consideration to walking — 112
   1.4.6 Shock attenuation in the lower limb — 117
   1.4.7 Muscle activity in running — 121
   1.4.8 Spine, pelvis, and arm motion in running — 128
   1.4.9 Running patterns — 132
   1.4.10 Running differences through gender and age — 140
   1.4.11 Foot type and footwear effects on running — 143
   1.4.12 The effects of running terrain — 152
   1.4.13 Section summary — 155
1.5 Variance in gait — 156
   1.5.1 Introduction — 156
   1.5.2 Gender and other morphological differences in walking gait — 157
   1.5.3 Foot function variance in gait — 160

	1.5.4	Joint hypermobility in gait	170
	1.5.5	Gait in pregnancy	173
	1.5.6	Paediatric gait	175
	1.5.7	Ageing and aged-like gait	188
	1.5.8	Leg length discrepancy/inequality	191
	1.5.9	The effects of footwear on gait	195
	1.5.10	Gait in lower limb amputees	196
	1.5.11	Section summary	200
1.6	**Gait in disease**	200
	1.6.1	Introduction	200
	1.6.2	Gait in cerebral palsy	201
	1.6.3	Gait in musculoskeletal disease	203
	1.6.4	Gait in neurological disease	208
	1.6.5	Gait in peripheral vascular disease	212
	1.6.6	Gait in diabetes (mellitus)	213
	1.6.7	Section summary	215
Chapter summary			217
References			217

2. Locomotive functional units

Chapter introduction — 243

2.1 **Soft and hard tissue as functional units** — 244
- 2.1.1 Introduction — 244
- 2.1.2 Principles of tensegrity and biotensegrity revisited — 244
- 2.1.3 Maintenance of biotensegrity structures — 248
- 2.1.4 Principles of core stability — 253
- 2.1.5 Muscle's role in stability–mobility — 254
- 2.1.6 Principles of articular motion and stability — 256
- 2.1.7 Concepts of muscle joint relationships — 257
- 2.1.8 Concepts of form and force closure — 260
- 2.1.9 Concepts of joint packing, congruency, and neutral — 263
- 2.1.10 The skeletal frame — 265
- 2.1.11 Section summary — 265

2.2 **Functional unit of the lumbar spine and pelvis** — 266
- 2.2.1 Introduction — 266
- 2.2.2 The kinematic role of the lumbar spine and pelvis — 266
- 2.2.3 The spine and pelvis as a biotensegrity structure — 270
- 2.2.4 Functional anatomy of the lumbar spine and pelvis — 273
- 2.2.5 Passive soft tissues of the lumbar spine and pelvis — 286
- 2.2.6 Functional joint axes and load distribution of the lumbar spine and pelvis — 289
- 2.2.7 Muscle action at the lumbar spine and pelvis — 291
- 2.2.8 Adaptation and pathology in the lumbar spine and pelvis — 301
- 2.2.9 Section summary — 304

2.3 **Functional unit of the hip** — 304
- 2.3.1 Introduction — 304
- 2.3.2 The kinematic role of the hip — 305
- 2.3.3 The hip as a biotensegrity structure — 305
- 2.3.4 Osseous topography of the hip — 307
- 2.3.5 Passive soft tissues of the hip — 315
- 2.3.6 The hip in lever systems — 317
- 2.3.7 Muscle action at the hip — 323
- 2.3.8 Adaptation and pathology in the hip — 333
- 2.3.9 Section summary — 334

2.4 **Functional unit of the knee** — 334
- 2.4.1 Introduction — 334
- 2.4.2 The kinematic role of the knee — 335
- 2.4.3 The knee as a biotensegrity structure — 337
- 2.4.4 Osseous topography and articular structures of the knee — 339
- 2.4.5 Passive soft tissues of the knee — 345
- 2.4.6 Anatomy of the anterior knee: Patellofemoral joint — 358
- 2.4.7 The functional joint axes of the knee in lever systems — 362
- 2.4.8 Muscle action at the knee in gait — 366
- 2.4.9 Adaptation and pathology in the knee — 375
- 2.4.10 Section summary — 380

2.5 **Functional unit of the ankle** — 380
- 2.5.1 Introduction — 380
- 2.5.2 The kinematic role of the ankle — 381
- 2.5.3 The ankle as a biotensegrity structure — 384
- 2.5.4 Osseous topography of the ankle — 385
- 2.5.5 Passive soft tissues of the ankle — 387

2.5.6	Functional axes of the ankle	395	
2.5.7	The ankle in lever systems	398	
2.5.8	Muscle action at the ankle	405	
2.5.9	Extensor muscles at the ankle	407	
2.5.10	Primary flexor muscles at the ankle	411	
2.5.11	The other plantarflexors of the ankle	426	
2.5.12	Leg to foot rotations around the ankle–subtalar complex and lower limb	435	
2.5.13	Adaptation and pathology in the ankle	437	
2.5.14	Section summary	439	

Chapter summary 439
References 440

3. The foot as a functional unit of gait

Chapter introduction 459
3.1 The foot's material properties 459
- 3.1.1 Introduction 459
- 3.1.2 The foot as a biotensegrity structure 462
- 3.1.3 The mechanical role of the foot 463
- 3.1.4 The role of the foot vault 464
- 3.1.5 The foot as an adjustable multi-tied viscoelastic asymmetrical expanded conical vault 474
- 3.1.6 Span distance and curvature effects on vault stiffness 480
- 3.1.7 Stiffness and pes planus, pes cavus, and the 'normal' foot vault 487
- 3.1.8 Mechanical constraints on the foot's role 489
- 3.1.9 Section summary 493

3.2 The foot's role in gait 494
- 3.2.1 Introduction 494
- 3.2.2 The function and events of heel-toe walking 494
- 3.2.3 Achieving heel contact compliance 499
- 3.2.4 Heel fat pad in compliance 501
- 3.2.5 Muscle action in energy dissipation through the foot 502
- 3.2.6 Energy dissipation through wobbling mass in the lower leg 505
- 3.2.7 The skeletal frame vault in foot compliance 506
- 3.2.8 Transformation from compliance to stiffness 507
- 3.2.9 Terminal stance stiffening mechanisms 512
- 3.2.10 The influence of foot shape and gait speed on compliance and stiffening events 519
- 3.2.11 The foot in modulating viscous spring-damping in running 521
- 3.2.12 Section summary 524

3.3 The foot's functional units 525
- 3.3.1 Introduction 525
- 3.3.2 The role of cutaneous soft tissues and plantar fat pads of the foot 525
- 3.3.3 The role of the passive elastic elements of the foot 528
- 3.3.4 The role and anatomy of the plantar aponeurosis 531
- 3.3.5 The function and anatomy of the muscles of the foot 538
- 3.3.6 The role of the skeletal frame 551
- 3.3.7 Division of the skeletal frame: Medial and lateral columns 551
- 3.3.8 The rearfoot as a functional unit 556
- 3.3.9 The midfoot as a functional unit 571
- 3.3.10 The tarsometatarsal joints and intermetatarsal joints as functional units 584
- 3.3.11 The role of the metatarsal base to head orientations 593
- 3.3.12 The metatarsophalangeal joints and digits as a functional unit 593
- 3.3.13 MTP joints as a stabilising fulcrum 604
- 3.3.14 Section summary 611

Chapter summary 613
References 614

4. Pathology through the principles of biomechanics

Chapter introduction 627
4.1 Understanding the pathological risk 628
- 4.1.1 Introduction 628
- 4.1.2 Patient clerking 628
- 4.1.3 Pathomechanical summation and tissue threshold 631
- 4.1.4 Activity influence on tissue threshold 633
- 4.1.5 Alignment and morphology in tissue threshold 633

- 4.1.6 Instantaneous joint axis location 635
- 4.1.7 Features of gait (gait determinants) 636
- 4.1.8 Soft tissue injuries 639
- 4.1.9 Bone and articular injuries 640
- 4.1.10 Disease-induced pathomechanics 645
- 4.1.11 Principles of free-body diagrams 649
- 4.1.12 Section summary 651
- **4.2 Pathomechanics of the hip** 652
  - 4.2.1 Introduction 652
  - 4.2.2 Principles of free-body diagrams of the hip 653
  - 4.2.3 Sagittal plane free-body diagram of the hip 654
  - 4.2.4 The hip in loading response kinematics 655
  - 4.2.5 The hip in terminal stance 659
  - 4.2.6 Frontal plane free-body diagrams of the hip 664
  - 4.2.7 Transverse plane free-body diagrams of the hip 668
  - 4.2.8 Expanding complexity of hip pathomechanics with free-body diagrams 670
  - 4.2.9 Section summary 671
- **4.3 Pathomechanics of the knee** 672
  - 4.3.1 Introduction 672
  - 4.3.2 Free-body diagrams of the knee 672
  - 4.3.3 Sagittal plane free-body diagrams of the knee 672
  - 4.3.4 Knee muscle pathomechanics in the sagittal plane 676
  - 4.3.5 Pathomechanics in anterior and posterior translations of the knee 678
  - 4.3.6 Frontal plane pathomechanics of the tibiofemoral joint 679
  - 4.3.7 Transverse plane pathomechanics of the tibiofemoral joint 685
  - 4.3.8 Free-body diagrams of patellofemoral joint pathomechanics 691
  - 4.3.9 Section summary 702
- **4.4 Pathomechanics of the ankle and rearfoot** 702
  - 4.4.1 Introduction 702
  - 4.4.2 Sagittal plane free-body diagrams of the rearfoot complex 703
  - 4.4.3 Pathomechanics of weight transference and acceleration 711
  - 4.4.4 Achilles tendinopathy 713
  - 4.4.5 Sagittal plane rearfoot motion and Achilles pathomechanics 717
  - 4.4.6 Achilles tendinopathy in forefoot strikes 724
  - 4.4.7 Achilles ruptures, posterior calcaneal avulsion fractures, and Sever's disease 725
  - 4.4.8 Frontal and transverse plane rearfoot relationships in pathomechanics 727
  - 4.4.9 Ankle sprain pathomechanics 733
  - 4.4.10 Frontal plane Achilles pathomechanics 739
  - 4.4.11 Other tendon injuries around the rearfoot 743
  - 4.4.12 Section summary 746
- **4.5 Pathomechanics of the foot** 747
  - 4.5.1 Introduction 747
  - 4.5.2 Contact to loading response pathomechanics in the foot 748
  - 4.5.3 Early midstance pathomechanics 757
  - 4.5.4 Late midstance pathomechanics 761
  - 4.5.5 Terminal stance phase pathomechanics 765
  - 4.5.6 Tibialis posterior dysfunction 773
  - 4.5.7 The MTP joint fulcrum in pathomechanics 776
  - 4.5.8 MTP joint compliance and stiffness in pathomechanics 786
  - 4.5.9 Soft tissue pathomechanics in the MTP joint fulcrum stability 790
  - 4.5.10 Toe deformities as part of MTP joint pathomechanics 796
  - 4.5.11 Pathomechanics at the 1st MTP joint 800
  - 4.5.12 Pathomechanics of 1st MTP sesamoids 805
  - 4.5.13 Hallux abducto valgus 811
  - 4.5.14 Metatarsal fatigue fractures 822
  - 4.5.15 Plantar fasciopathy and its pathomechanics 827
  - 4.5.16 Section summary 837
- **4.6 Pathomechanics of running** 838
  - 4.6.1 Introduction 838
  - 4.6.2 Overuse principles 838

4.6.3	Assessing the running patient's pathomechanics	840	4.6.8 Tibial fatigue fractures	857
			4.6.9 Medial tibial stress syndrome	863
4.6.4	Foot strike position free-body diagrams	843	4.6.10 Section summary	868
			Chapter summary	868
4.6.5	Frontal plane effects of foot strike positions	849	**References**	869
4.6.6	Running acceleration pathomechanics	852	Appendix	887
4.6.7	Pathomechanics of fascia in shin symptoms of runners	852	Glossary: Gait and pathomechanical principles	889
			Index	901

About the authors

Andrew Horwood has more than 30 years of frontline clinical experience working in biomechanics and the practical and therapeutic interpretation of gait analysis within the clinical environment. He has made a substantial contribution to clinical biomechanics postgraduate provision at Staffordshire University, currently coleading a course titled 'Origins and Principles of Clinical Biomechanics'. He is also a visiting fellow at Staffordshire University. For more than 20 years, he has been a product design consultant for HealthyStep Ltd., working on the development of foot orthoses and other lower limb healthcare products and innovations.

Nachiappan Chockalingam is the Director of the Centre for Biomechanics and Rehabilitation Technologies at Staffordshire University and Fellow of the Institute of Physics and Engineering in Medicine. He is Affiliate Professor at the University of Malta and Visiting Professor at Sri Ramachandra University, India. He is Associate Editor for the journals *Footwear Science*, *Prosthetics and Orthotics International*, and *Rehabilitation for Musculoskeletal Conditions*. He has published numerous peer-reviewed papers and reviews for various global grant-awarding bodies.

Foreword

Biomechanics is a discipline that applies the principles of mechanics to the study of biological systems. For many years, the study of biomechanics was primarily associated with sports science, which still is a major focus for the discipline. However, another focus that has had a profound effect on the discipline is the application of biomechanics in the clinical setting. While most books on biomechanics use examples of examining sports techniques, a few clinical applications are often presented as well. It is now necessary to present a complete book dedicated to the application of biomechanics for clinicians with particular emphasis on the principles of human locomotion and pathology that links to gait dysfunctions.

This new book, *Clinical Biomechanics in Human Locomotion: Gait and Pathomechanical Principles,* is a follow-up to the previous book titled *Clinical Biomechanics in Human Locomotion: Origins and Principles*. This book is primarily concerned with human gait and its pathomechanics. Some of the topics covered in this book include explaining the underlying mechanical principles of efficient human gait, discussing how gait can be evaluated clinically in the context of variability as 'normal', and exploring running gait (of its variable types) as a different form of human locomotion with its own mechanical concepts of efficiency. The text approaches variability in gait through human morphological variation, age, and health status, which includes the amputee. It reviews anatomy acting as functional units across the lower limb that achieve gait by providing a safe biomechanical environment and shows how recent research in foot biomechanics has clearly indicated the importance of the foot as a viscoelastic structure. The human foot is able to energy-dissipate through increased compliance and also has the capacity to reduce this compliance for the application of acceleration power. Importantly, it provides information on how to approach the clerking of a patient to integrate a diagnosis into any locomotor dysfunction and/or other scenarios, which explains the pathomechanical process that lies at the heart of symptoms and pathology. It ends by debating the pathomechanical processes that may help explain the development of many of the common pathologies that compromise patient mobility.

The book has been authored and edited by two very prominent, well-known individuals in the clinical arena who are considered experts in the field of clinical biomechanics. Andrew Horwood is a prominent, practical clinician who dedicated his career to using the principles of biomechanics to improve patient outcomes. To this end, he has explored and studied materials and biological sciences, particularly anatomy, physiology, and evolution to understand the challenges of treating patients as individuals that are expressing a particular pathology. Andrew has given many lectures presenting some of the newest and most controversial research to clinical professionals to help drive more coherent and effective treatments in locomotive dysfunctions. However, he will always say that listening and reading about the ideas and research of colleagues across the biological sciences that support clinical biomechanics, knowingly or inadvertently, is far more important than speaking. Despite this argument, he is a visiting lecturer in clinical biomechanics at Staffordshire University.

Professor Nachiappan (Nachi) Chockalingam leads the clinical biomechanics team at Staffordshire University. He has an internationally recognised applied research profile and a strong track record of enterprise activities. Nachi endeavours to develop world-leading translational research and create a wide-reaching enterprise culture that has a demonstrable socio-economic impact. I have known Nachi for more than 25 years and he is highly published and widely sought after as a speaker at national and international meetings. He is recognised for his ability to work across various disciplines across the academia, industry, and clinical sectors, which compounds his capacity to engage in discourse in the hope of effective, innovative change.

Nachi and Andrew have worked together for well over 20 years. The material presented in this book should act as a guide to the reader in the assessment of the current thought on clinical biomechanics and will help the clinician in understanding the biomechanical principles of human locomotion and its pathomechanics. I suggest to the reader that this book will bring increased awareness of the importance of biomechanics in the clinical field.

Joseph Hamill
Professor Emeritus
University of Massachusetts Amherst

Preface

All truly great thoughts are conceived by walking

Friedrich Nietzsche (1889)

Walking is man's best medicine

Hippocrates (c. 460–c. 370)

The resultant biomechanics of human gait are grounded within the principles of physics and those sciences that underpin biology. Bioengineering brings material, structural, and tissue properties together to explain how the human body is able to manage stresses and internal strains that derive from an inherently unstable orthograde posture during plantigrade bipedalism. Such knowledge allows clinicians to appreciate how and why musculoskeletal pathology develops only under specific circumstances. Human physiological systems support and maintain the health of the locomotive system, while the uniquely evolved biomechanics of human gait, in turn, maintains human physiological homeostasis. Indeed, the effects of active molecules released by muscles during activity (myokines such as interleukin-6) indicate that locomotion is a fundamental part of human health.

The interaction of genetics, epigenetics, developmental biology, and physiology under the influence of locomotive biomechanics and metabolic energetics drives evolution. The interplay of these biological principles within the laws of motion are subjects that have been discussed in this book's companion text *Clinical Biomechanics of Human Locomotion: Origins and Principles*. Such biological pressures on survival are essential in understanding the locomotive biomechanics of modern humans. The clinician must be aware that biomechanical and physiological principles, coupled to genetics/epigenetics, drive individual and population musculoskeletal morphology and tissue properties. Differing lifestyles and environments that focus forces into the anatomy during growth and development explain much of the large diversity of modern human form. This is why there is no such thing as a single 'normal' human phenotype.

Despite early use of plantigrade bipedalism within the hominin family, hominin bipedal locomotive biomechanics has never been of one type among individuals or across time. Different ancient human populations have lived in different environments and utilised terrestrial bipedalism to different extents, requiring different levels of gait adaptability to survive. Those humans that took up more running, either as a result or a driver (or both) of anatomical changes, adapted morphologically to permit a much wider range of terrestrial locomotive speeds. This has proved to be the most successful evolutionary option when compared to the extinct more arboreal hominin species, for of all the hominins that evolved, *Homo sapiens sapiens* is the only one that remains. Yet the morphological plasticity expressed by the hominin musculoskeletal system under biomechanical forces during growth ensures that all modern humans remain adaptable in form, and thus variable among individuals in their locomotive capabilities.

Thus, locomotive biomechanics are highly variable among modern humans. Age and health status are the biggest drivers of biomechanical variance during gait between individuals. However, lifestyle, including gait speed adaptability established particularly during the years of growth, will influence anatomical development. This in turn will influence the expected biomechanics sustained during walking or running. Thus, humans demonstrate high levels of morphological and anatomical variability that make concepts of highly specific lower limb kinematic 'normality' during locomotion erroneous. Humans require the ability to dissipate collision energies to prevent tissue damage during stance phase. However, they also gain mechanical/metabolic advantage if some safe levels of locomotive energies can be stored to add power to their acceleration at the end of stance phase. The terrain any stance phase operates upon (including footwear) is highly influential on what each step requires from the lower limb for safe translation of the body mass above it. Thus, adaptability, not conformity of motion, is associated with health.

Individuals have many locomotive options in how they translate from A to B. Each option has energetic costs that change the biomechanics within the tissues of the lower limb. These forces are then expressed by the motions of the joints and the body posture that maintain balance during transit. What is required to make terrestrial human locomotion safe and what causes pathology within a context of high locomotive and morphological variability must be understood if clinical management of gait-disturbing or gait-induced pathologies are to be managed successfully. It is essential for clinicians to be aware of this, and this book attempts to explore these relationships.

Acknowledgements

I am indebted to Professor Nachiappan Chockalingam for his support, belief, and advice in directing and editing this work. I am also grateful to the staff at Elsevier for their guidance and support of this project. Many thanks go to Rory Markham at HealthyStep Ltd who has had the challenging task of turning my draft sketches into meaningful illustrations. These images have the difficult, yet important, task of helping explain biomechanical principles at play within lower limb anatomy during gait and in dysfunction. I am also most grateful for the assistance I received from Marion McLure in improving the photographs and plates of Edweard Muybridge. A special thank you goes to my very dedicated proofreader and inspector of all things referenced, the amazing William Eric Lee. Any remaining mistakes are my own. I acknowledge the support from all the staff at HealthyStep (Sensograph) Ltd, particularly Tim Hall. Finally, knowledge is founded on enquiry and the friendship with clever people. To the many researchers I reference within this text and the many friends and students who have guided and listened to my multiple ideas within the field of clinical locomotive biomechanics, I am forever grateful.

Abbreviations

2D	two dimensional
3D	three dimensional
AHF	arch height flexibility. Used to establish intrinsic foot stiffness levels
AHI	arch height index. Used to establish the height of the medial foot vault
ANS	autonomic nervous system
A-P	used in the context of radiographs taken in the anterior–posterior (in the foot, dorsal to plantar) view
cm/s	centimetres per second
CNS	central nervous system
CP	cerebral palsy
CoGRF	centre of ground reaction force. An alternative term used in this text for the centre of pressure when used to indicate the location of the centre of GRF forces
CoM	centre of mass
CoP	centre of pressure generated by a pressure plate
CT	computed tomography
DJD	degenerative joint disease (usually in the context of osteoarthritis)
DOMS	delayed onset muscle soreness. Pain that may derive more from the fascial tissues than muscle fibres
EMG	electromyography
EVA	ethyl vinyl acetate
F1	the initial impact peak vertical force expressed on a force–time curve at $\sim 1.3\times$ body weight
F2	the depth of the trough of the vertical force expressed on a force–time curve during midstance of walking gait at $\sim 0.7\times$ body weight
F3	the peak of acceleration vertical force expressed on a force–time curve at $\sim 1.2\times$ body weight
F4	the initial peak of anterior–posterior horizontal force expressed on a force–time curve
F5	the terminal peak of anterior–posterior horizontal force expressed on a force–time curve
FPI	Foot Posture Index. A clinical technique used to semiquantify foot posture in static stance
GJH	general joint hypermobility where the affected individual is asymptomatic
GPa	gigapascal(s). An SI unit of pressure and used for recording Young's modulus
GRF	ground reaction force. The force derived from the interaction of body mass with the ground as an expression of Newton's third law
h	hour
HAT	head arms and trunk. Used to designate these body segments as one segment traveling over the lower limbs
HAV	hallux abducto valgus (hallux valgus). Deformity of the 1st toe (hallux) often and erroneously referred to as a bunion
Hz	hertz
IPJ or IP joint	interphalangeal joint
J	Joule. 1 Joule is equal to the work done by a force of one Newton acting over 1 metre
JHS	joint hypermobility syndrome. Symptomatic joint hypermobility
km/h	kilometres per hour
kPA	kilopascal $= 1000\,Pa$
L	lumbar vertebra (usually followed by a number to denote which particular lumbar vertebra)
LBP	low back pain
LLA	lateral long arch of the foot
LLD	leg length discrepancy
LMN	lower motor neurone
m^2	meters squared
min	minute(s)

MLA	medial longitudinal arch, found on the medial side of the foot vault
mm	millimetre(s)
MPa	megapascal(s), a SI unit of pressure and for recording Young's modulus
MPE	metatarsus primus elevatus. A 1st metatarsal with a low declination angle compared to its companions. The effect is to elevate its metatarsal head and metatarsophalangeal joint relatively to the other lesser metatarsals
MRI	magnetic resonance imaging
ms	millisecond (a thousandth of a second)
m/s	metres per second
m/s^2	metres per second squared
MTP joint	metatarsophalangeal joint
MTSS	medial tibial stress syndrome
mV	millivolt(s) (a thousandth of a volt). Unit of electrical potential, electric potential difference (voltage), and electrical force
N	Newton(s). Measurement of force defined as 1 kg m/s^2
nm	nanometres. A unit of measure of one-billionth of a metre
NM/rad	Newton metre per radian. Used to express torsional stiffness (torque per unit deflection)
OA	osteoarthritis
Pa	pascal. The SI unit of pressure used to quantify internal pressures, stress, Young's modulus, and ultimate tensile strength
Q angle	quadriceps angle formed by the angle from the anterior–superior iliac spine and the middle of the patella
RA	rheumatoid arthritis
RCSP	relaxed calcaneal stance position. The position of the resting foot in static stance
s	second
S	sacral vertebra (usually followed by a number to denote which particular fused sacral segment)
SD	standard deviation
SI joint	sacroiliac joint
SOAP	symptoms, objective observations, analysis, plan of action (treatment planning). An approach to clerking a patient and planning treatment
SOAPIER	as above with intervention, evaluation, and re-evaluation
VM	vastus medialis
VML	vastus medialis longus
VMO	vastus medialis obliquus
VL	vastus lateralis
UMN	upper motor neurone
ym	yoctometre. One-septillionth of a metre. The smallest SI unit of length
μm	micrometre or micron. A scientific (SI) unit of length equal to one thousandth of a mm. The human eye can see objects from around 25 μm in size

Introduction

Nachiappan Chockalingam

This text, the second volume of *Clinical Biomechanics in Human Locomotion: Gait and Pathomechanical Principles*, focuses on the use of biomechanical principles to understand and analyse normal and abnormal human movement, specifically gait. The first volume provided a foundational understanding of biomechanics, while this volume delves into the application of these principles in the context of musculoskeletal pathologies. Biomechanics involves the study of the mechanical principles that govern biological systems, such as the musculoskeletal system in the human body. Pathomechanics, on the other hand, involves applying these biomechanical principles to understand pathological conditions. During human locomotion, various functional units in the body work together in a structured manner. However, if the stress placed on these units exceeds their mechanical limits, it can lead to damage and dysfunction. This mechanical failure of a single functional unit may eventually cause failure of the entire system's biomechanical properties and result in locomotive failure. The clinician must use biomechanical principles to identify the specific functional issues and uncover the origin of dysfunction through history-taking, physical assessment, and gait analysis.

Structured assessment of human locomotion or gait analysis can help in understanding the pathomechanics of the musculoskeletal system by providing a detailed assessment of an individual's gait pattern and the mechanical forces acting upon the body during movement. This information can be used to identify abnormal or problematic gait patterns that may be contributing to musculoskeletal injuries or dysfunctions. Gait analysis can be used to assess an individual's movement patterns during locomotion, including walking and running. It involves the use of techniques such as video analysis, force and pressure sensors, and other motion capture technologies to measure and analyse various kinematic and kinetic variables such as joint angles, joint moments, and ground reaction forces. The assessment of these variables can help identify abnormalities in joint range of motion, muscle strength, and muscle activation patterns that may contribute to gait deviations and cause excessive strain on certain tissues. Chapter 1 introduces gait analysis as a tool used to assess patient locomotion and understand the nature of dysfunction and pathology. It emphasises the importance of distinguishing between the effects of pathology and the origins of pathology. Gait analysis can be used to track the progression of a pathological condition over time and assess the effectiveness of therapeutic interventions.

The mechanical properties of neuromuscular tissues play a role in human locomotion, which can be affected by factors such as age, underlying diseases, and lower limb amputations. Effective gait involves the successful management of energy dissipation and power generation without abnormal stress concentrations that could lead to injury. The information provided through structured gait analysis can also be used to inform treatment strategies, such as prescribing specific exercises or providing orthotic support. Gait analysis is a critical skill in clinical biomechanics that requires a deep understanding of the subject. It involves the ability to describe phases and events during gait, distinguish normal from abnormal function in different client groups, and use the data gathered on gait to improve mobility and prevent injury. The goal of clinical gait analysis is to identify the cause of pathology and dysfunction and use this information to develop a treatment plan to improve patient mobility and health. There are expected patterns of kinematics and kinetics that should be seen during gait assessment, but it is important to note that absolute 'textbook normals' should not always be expected, even in individuals with optimal health and musculoskeletal function. This is because each individual has a preferred gait pattern. However, significant asymmetry in limb function can indicate issues with locomotion. To effectively assess gait, a practitioner must know how to link symptoms and pathology to the gait assessment and should consider factors such as terrain, footwear, and specific activity involved in the patient's symptoms. Clinical gait analysis should be conducted with a knowledge base in biomechanics and a healthy scepticism of the limitations of the information gathered.

Muscle action is often described based on the movement that occurs when a muscle is pulled at its proximal end on its distal segment via its tendon on a distal limb that is free to move. However, this does not always correspond well to living muscle function, especially in the lower limb during the stance phase of gait. To better understand muscle action, it is important to appreciate that muscles have attachments rather than origins and insertions. These attachment points to bone are made through osteotendinous junctions and consist of a mixture of tendon, ligament, aponeurosis, and joint capsule linked to the bone. All connective tissues around a muscle are part of a continuum and muscle cell contraction pulls on them, tightening and expanding them, and increasing regional soft tissue tone. The movement produced by torques around joints during gait is primarily resisted or slowed by muscle activity, with most muscles primarily functioning to prevent or slow movement rather than to produce it. Understanding the role of muscles as resistors of movement is important for correctly interpreting gait data and developing treatment plans to improve mobility and prevent injury.

As outlined within the first volume, the human body is a complex, mobile structure that uses lever arms to move and is controlled by the nervous system. It is important to view the body as more than just a series of levers, fulcrums, and forces, and to consider the role of soft tissues in structural integrity and reactivity during locomotion. Soft tissues have high fluid content and demonstrate viscoelastic, non-Newtonian, and non-Hookean properties. They are effective at resisting and maintaining sustained tensile force and generate force through contractile elements, but also have passive tensile properties. Connective tissues involved in locomotion have passive elements that rely on their mechanical abilities to bear and resist stress and active muscle elements that generate force and control stresses applied to the system. It is important to consider the complex interplay between these elements or in other words functional units in understanding gait and developing treatment plans to improve mobility and prevent injury.

Functional units exist at various levels of the body, including the whole body, limb segment, and tissue and cellular levels. These units rely on the ability to distribute stresses throughout the body and reduce torsion and shear to a minimum. However, any impediment in movement that affects the ability to adjust the shape and structural properties can pose a risk to functional integrity. Biotensegrity offers a solution for managing high stresses within tolerance, but failure in biotensegrity can cause structural property limits to change significantly and negatively impact health and mobility. Chapter 2 discusses these functional units in the lower limb, from the lumbosacral joint to the ankle, and how they work together to achieve lower limb function. This information is important in understanding and treating lower limb dysfunction. Dysfunction in one functional unit can lead to kinematic changes in other areas of the lower limb, and it is important to consider the interconnectedness of these functional units when attempting to identify the cause of pathomechanics.

The human foot is a complex and unique structure that plays a significant role in locomotion and postural stability. It can dissipate and store energy, provide a stable base of support, and contribute to the sensorimotor system that helps maintain balance. The foot is also vulnerable to factors such as ageing and environmental mismatch, and changes in its functional and material properties can result from diseases of the nervous, endocrine, and cardiovascular systems. There is still much to learn about the foot and how it achieves its functions and how its dysfunctions are reflected in the rest of the body. Chapter 3 discusses the foot, a particularly complex functional unit with a variable degree of freedom and a nonlinear evolutionary history. Its mechanical properties are controlled by the relationships between its various anatomical structures, and it has evolved to provide energy-dissipating, braking, and acceleration properties. The function and dysfunction of the foot have been the subject of debate due to its numerous variations and distinct anatomical vault profiles as well as the debate surrounding the flexibility or freedom of motion of certain foot bones and the indistinct nature of some pedal structures. In addition, there has been disagreement over whether one type of foot vault profile is superior to others. It is important to consider the foot's overall mechanical functions, which involve the loading, absorption, and transmission of stress and energy to aid in braking and acceleration during gait when studying the foot.

There are many factors that can contribute to tissue stress breaches in the lower limb, including sex differences, genetic factors, and age-related changes in tissue properties. The mechanical demands of different activities can also play a role in the development of musculoskeletal disorders. To understand and treat these conditions, it is important to apply biomechanical principles and use tools such as gait analysis to identify the specific mechanical issues involved. By understanding the origins of pathology, clinicians can develop targeted treatment approaches that address the underlying issues and improve outcomes for patients. Chapter 4 outlines the injuries and dysfunctions in the musculoskeletal system that can be caused by a variety of factors, including physiological changes, genetic or developmental weaknesses, and trauma. These injuries or dysfunctions can involve any combination of bones, joints, ligaments, fascia, muscle tendons, and nerve supply, and can be caused by a single traumatic event or by the accumulation of fatigue loading. Factors such as healed tissue injuries and ageing can also affect the strength of tissues. While describing pathomechanics as the study of how mechanical principles can be applied to the understanding of pathological conditions in the lower limb locomotive system, the chapter highlights the importance of gathering information about the patient's health, identifying and understanding the pathology, proposing a diagnosis, and applying principles of biomechanics to explain the pathomechanics leading to the pathology. A deep understanding of various subjects, including energetics, the laws of motion, tissue material properties, physiology, developmental processes, evolutionary medicine, locomotive biomechanics, and functional anatomy, is necessary to understand the patient's form of locomotion and choose the appropriate therapeutic intervention. The ultimate goal of pathomechanics is to maintain patient mobility and improve quality of life. It is a crucial aspect of clinical practice, as it helps identify the root cause of dysfunction and inform treatment approaches.

Gait analysis has seen significant advancements in recent years, with the development of various technologies such as optoelectronic motion capture systems and wearable devices. These technologies allow for the precise measurement and analysis of various gait parameters, including kinematics, kinetics, and electromyography, which can provide valuable insights into the underlying musculoskeletal pathologies and inform effective clinical management strategies. One of the main contributions of optoelectronic gait analysis systems is their ability to provide objective and quantifiable data on gait patterns. This can be particularly useful in the clinical setting, where subjectivity and bias can often influence observations and interpretations of gait. By providing objective data, optoelectronic gait analysis systems can help inform clinical decision-making and guide the development of effective treatment strategies. These data can be used to help clinicians understand the underlying pathomechanics of musculoskeletal conditions, such as muscle imbalances or joint abnormalities, and inform treatment decisions. Optoelectronic gait analysis systems can also be used to evaluate the effectiveness of different interventions, such as orthotic devices or physical therapy exercises, and track progress over time. These systems can capture high-speed, three-dimensional movement data, enable detailed analysis of complex gait patterns, and can identify subtle abnormalities that may not be evident during a clinical examination. Some examples of the latest developments in gait analysis technologies include the use of machine learning algorithms to analyse gait data, the integration of virtual reality into gait rehabilitation protocols, and the development of portable gait analysis systems for use in community settings. Overall, optoelectronic gait analysis systems have played a significant role in the development and advancement of gait analysis in clinical practice. They have helped clinicians better understand the pathomechanics of musculoskeletal conditions and develop more effective treatment approaches.

At this point, I am honoured to pay tribute to my dear friend Tom Shannon, who made significant and seminal contributions to the field of clinical gait analysis. He dedicated countless hours to researching and studying the intricacies of human movement and posture with a view to developing technologies to assess them. He developed innovative techniques and technologies that have allowed us to better understand and analyse gait, and his contributions have led to improved treatments and therapies. Tom's dedication to his work has inspired many others to follow in his footsteps, and his legacy will undoubtedly continue to shape the field of clinical gait analysis for years to come. I am grateful to have had the opportunity to work alongside Tom and learn from his expertise. He was truly remarkable, and I am proud to call him my friend.

This book *Clinical Biomechanics in Human Locomotion: Gait and Pathomechanical Principles* and the other companion text *Clinical Biomechanics in Human Locomotion: Origins and Principles* work well together, providing a comprehensive understanding of clinical biomechanics for pregraduate students, postgraduate researchers, and established practitioners. These publications will be valuable resources for universities and other higher education institutions that specialise in teaching clinical biomechanics and practitioners who utilise clinical biomechanics in their treatment approaches.

Chapter 1

Understanding human gait

Chapter introduction

Patient locomotive abilities are commonly assessed through gait analysis. Clinical gait analysis can be used to understand dysfunction and the nature of pathology, both for establishing the cause and the effects of a patient's complaint. It is important that effects of the pathology are not confused with the origins of the pathology. For example, tibialis posterior muscle and/or tendon dysfunction can be confused by its effects with its cause. This muscle is a strong rearfoot invertor and foot vault stabiliser. Its dysfunction can increase ankle-rearfoot eversion and vault lowering during gait, leading the clinician to believe that the large eversion moments and the amount of foot pronation seen associated with the pathology are the cause of the problem, when they are in fact actually the result of the dysfunction with the origin of the pathology lying elsewhere associated to previous pathomechanical events.

Clinical gait analysis is an important tool not only to link pathology to locomotive dysfunction, but also for assessing the nature of the dysfunction and the possible risk of further symptoms and pathology that has or could result from gait abnormalities. Gait analysis is also an excellent indicator of successful treatment intervention (or not), as gait quality should improve with therapeutic interventions. If quantitative data have been recorded at the start of intervention, further gait data collection during and after intervention can be invaluable in monitoring progress beyond patient-reported symptom relief. In athletes, gait analysis can also be used to help establish the cause of a reduction in performance, rather than just being used to understand the origin and/or results of pathology on locomotion.

Human locomotion is not of one type, with the obvious primary distinction being between walking and running. Gait also changes with age and underlying diseases or lower limb amputations. Running itself is also of more than one type, with the requirements of a jogger or sprinter compared to an endurance cross-country runner being quite different. It is thus important to assess clients in the appropriate locomotive task they wish to perform, which can be increasingly difficult in the most extreme distance and rough terrain athletes and hobbyists. Knowing what is relatively 'normal' for each client group prevents inappropriate action or failure to intervene when necessary.

Effective gait of all types occurs through the successful management of energy dissipation during braking events and the generation of power for acceleration, without abnormal stress concentrations that could risk injury. The mechanical properties of the neuromuscular tissues as previously discussed within the companion text 'Clinical biomechanics in Human Locomotion: Origins and Principles', now take on their functional and biomechanical roles in how humans achieve locomotion, and these will be considered throughout this chapter as the events of gait are discussed.

1.1 Gait principles

1.1.1 Introduction

There are expected patterns of kinematics and kinetics that should be seen during gait assessment. Small asymmetries in gait can reflect natural functional differences between limbs, possibly in part due to laterality of the brain with a tendency to use one limb more in preference to another to achieve a functional task, rather than indicating something necessarily insidious (Sadeghi et al., 2000). Absolute 'textbook normals' should not be expected every time even in individuals with optimal health and musculoskeletal function. This is because the evolutionary past and the developmental morphogenesis of individuals, with all of their complex traits lying within human populations and between populations, make the concept of an 'absolute normal' in walking or running gait an impossibility. Each individual has a preferred walking and running gait at a certain pace, and the clinician must be able to perceive when a patient is diverting from their preferred pattern. Clues such as significant amounts of asymmetry in limb function are often indications of locomotive issues. However, the most fundamental part of gait assessment is to be able to link the symptoms and pathology to the gait assessment. It is bad practice to make the symptoms and/or pathology fit the gait.

To be able to assess gait clinically, the practitioner must have the knowledge to systematically review the events, the force generations, and the stress–strain relationships that can link any presenting pathology to the kinetic or kinematic data gathered. Without quantitative biomechanical data, the clinician is left with a simple kinesiological approach of observing gait, which with the right foundational knowledge can still prove highly effective in management of the patient. That being said, this approach can itself be greatly enhanced by the use of simple slow-motion filming.

It must also be remembered that clinical gait analysis is just that: analysis of gait in a clinic with all the artefacts that this can bring. The clinician must evaluate the usual terrain conditions, footwear, and specific activity involved in the patient's symptoms/pathology in a coherent manner in order for an appropriate intervention to be achieved. The principles of biomechanics should provide a knowledge base to allow some 'reflective acquisition' of data with a healthy scepticism as to the information's limitations.

1.1.2 Gait energetics

Human plantigrade bipedal gait is a uniquely novel hominin evolutionary form of locomotion that is relatively 'new' on Earth compared to other forms of animal bipedal locomotion. It is a form of gait that has evolved for a species relatively recently specialised for terrestrial walking, and it has a surprisingly high energetic efficiency. Moreover, it is a smooth process that ideally has no 'sudden' brakes in translation within the motion of the trunk and particularly the head. As a mode of locomotion, human walking is far more efficient than chimpanzee knuckle-walking and covers nearly twice as much distance for the same number of calories (Sockol et al., 2007; Pontzer et al., 2014). Improved walking economy provides humans with a powerful evolutionary advantage, enabling increased range of resource exploitation without high locomotive energy costs. This acquisition of higher levels of energy without significant increases in energy expenditure is most important at the lower levels of physical activity in the mode of transport that human bodies primarily work in (Pontzer et al., 2016), for all creatures do far less high-level activity than low-level activity.

Humans have exceptional walking endurance and even high running endurance when compared to other primates, with long-distance running probably playing a part in later human evolution over the last 2 million years (Pontzer et al., 2010) (Fig. 1.1.2a). The extra energy obtained by improved locomotive energetics can allow more calories to be expended on reproductive and child-raising techniques, longer lifespans, and bigger brains (Pontzer, 2017). The relationship between energy expenditure and amount of exercise is modulated so less exponential energy use occurs (Pontzer et al., 2016). The metabolic cost of walking is defined as the energy cost per unit of distance. This is usually measured by consumption of oxygen and production of carbon dioxide during exercise. It is even possible to work out whether the primary source of the energy is coming from carbohydrate, fat, or protein.

Advances in understanding human locomotive biomechanics have not only led to effective clinical assessment techniques but have also given an understanding that the very early hominin bipedalism conferred little energetic advantage (Pontzer, 2017). In more recent hominin species, such as probable dwarf species *Homo floresiensis* with its long feet, short hind limbs, and upper limbs built for climbing, the expendable energy for large brains was not there. Equally, *Homo naledi* from only 230,000 years ago had a relatively inferior hindlimb (lower limb) bipedal locomotive repertoire compared to modern humans, more similar to that of early *Homo erectus* (another fairly small-brained hominin). It did not develop all the highly derived bipedal lower limb features seen in later *H. erectus* and its derived species nor the increased brain size that followed the lower limb changes (Pontzer, 2017) (Fig. 1.1.2b).

When taking steps, longer legs can smooth out the up and down trajectories of the body's centre of mass (CoM) as it rises and falls with gravity and can greatly increase walking speed. The functional limb length changes through joint flexions and extensions during gait aid such gait smoothness. This is important, as smooth gait CoM oscillations from long legs reduce muscle volume activation per step per metre (Pontzer, 2017). Joint morphology also affects the mechanical advantage of muscles through subtle adaptations that smooth CoM oscillations, thereby improving the exchange of potential and kinetic energy aided by the elastic energy stored and released from tendons and ligaments. These combined effects reduce muscle activation and therefore energy expenditure through the mechanics of the leg length, body posture, and gait (Pontzer, 2017). Humans have long legs that should extend during gait to greatly reduce the cost of transport, developing a U-shaped curve when cost of transport is plotted against speed, thus creating a minimum cost of transport speed during walking (Pontzer, 2017). The overall energetic cost of running in humans is only marginally better than that of chimpanzees because in running, humans use a more flexed hip and knee posture for impact energy dissipation requiring increasing muscular effort and therefore creating greater energy costs (Pontzer, 2017).

Arm swing involves both active and passive elements and has beneficial effects on the energetic costs of gait and the recovery of balance after perturbations, as well as being an integral part of human bipedal gait (Meyns et al., 2013; de Graaf et al., 2019). Arm swing also reduces transverse trunk rotations, vertical angular moments, and vertical ground reaction moments during gait (Meyns et al., 2013; de Graaf et al., 2019), although the arms' swing effects in running gait energetics may be smaller or even negative (Hamner et al., 2010). It is reported that in walking, increasing arm swing

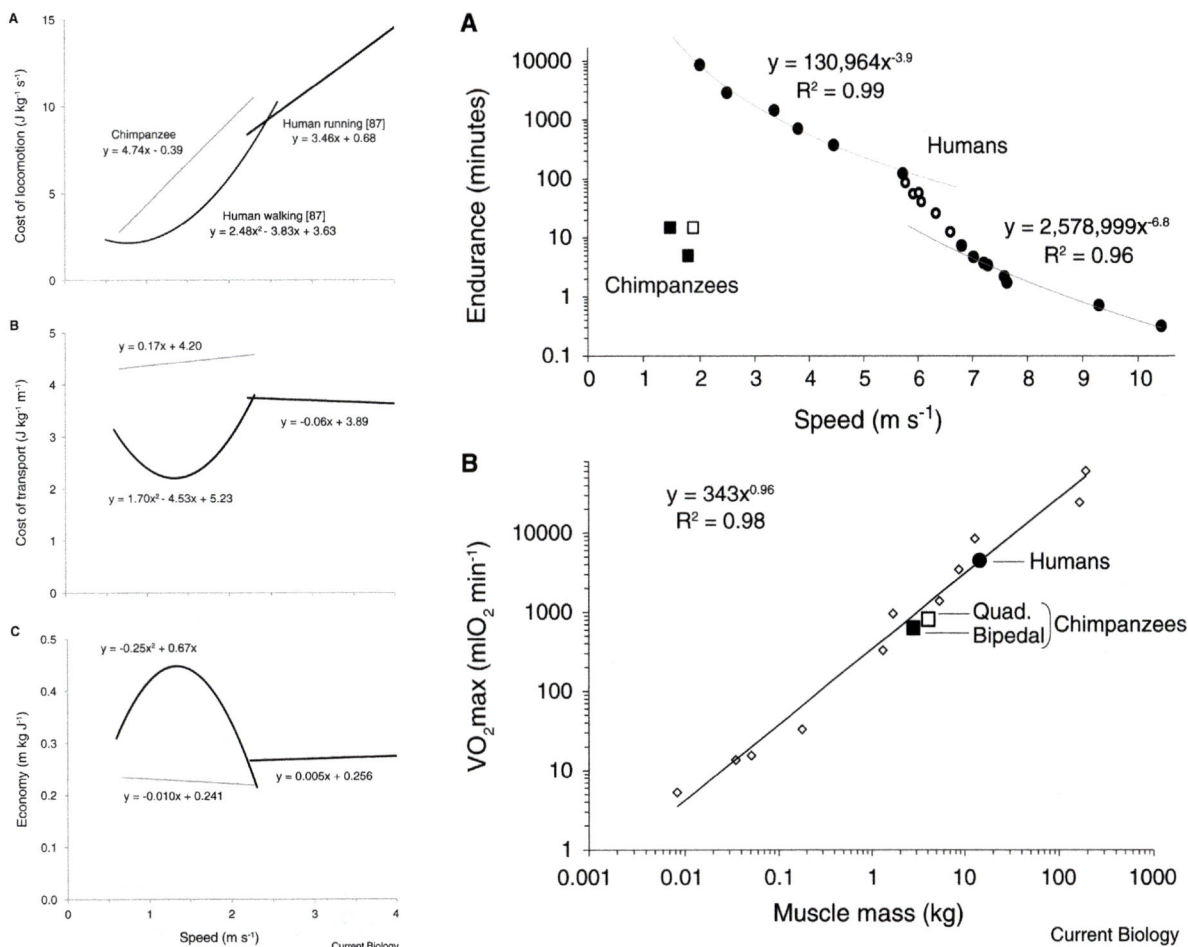

FIG. 1.1.2a Locomotive costs and economies are calculated around the amount of energy used to move a kg of mass. Graphs on the left demonstrate locomotive and transport costs (A and B) and economies (C) of chimpanzee and human bipedal walking and running *(left)*. Mass-specific costs of locomotion (J kg^{-1}s^{-1}) increase curvilinearly during walking and linearly during running. Chimpanzee locomotive costs increase linearly with speed, with walking and running costs similar. Locomotor economy (mkg/J) is the mass-specific distance travelled per unit of energy expended, expressed as the inverse cost of transport. Graphs on the right demonstrate endurance (A) and VO$_{2max}$ (B). The data show a decline in endurance with speed for elite human runners *(circles)*, which is pronounced near speeds of ~6 m/s *(open circles)* and declining steeply over this speed, reflecting anaerobic respiration. Maximum chimpanzee economy *(filled squares* for bipedal and *open* for quadrupedal) is much reduced compared to human running. VO$_{2max}$ strongly correlates to muscle volume across mammals. The term efficiency in locomotion should be reserved for comparing the mechanical work done to move the centre of mass (CoM) and limbs compared to the amount of metabolic energy expended. *(From Pontzer, H., 2017. Economy and endurance in human evolution. Curr. Biol. 27 (12), R613–R621.)*

amplitude continues to reduce vertical angular and vertical ground reaction moments. However, it increases energy transport costs (de Graaf et al., 2019) which may explain both the more rapid arm swings' advantages and their higher costs during running.

In walking, the metabolic costs have been estimated to be broken into ~28% used in body weight support in the earlier stance phases of gait, ~48% to generate the heel lift and forefoot ground reaction force (GRF) in terminal stance, ~10% for the limb in swing, and ~6% for the stabilisation of posture laterally (Gottschall and Kram, 2005). The remaining ~8% is argued to come from respiratory and cardiac activities of gait (Gottschall and Kram, 2005). These distributions of metabolic costs give indications as to the 'difficulty' of tasks within gait, and where gains from the utilisation of gravitational forces improve mechanical efficiency. The 28% metabolic cost used in body weight support during the bulk of stance is in part, from the metabolic costs of braking (negative acceleration) and the energy dissipation muscle activity that is required for this task. The process of generating acceleration in the stance limb is, however, the most expensive, hence the 48% cost of terminal stance (Fig. 1.1.2c). The low cost of swing phase, which pulls the CoM forward to aid weight transference to the next step through centrifugal forces, becomes obviously advantageous when we consider the costs involved in generating acceleration through the stance limb. Indeed, studies have shown reductions of up to 42% in metabolic costs that can be achieved through assisting trunk CoM horizontal progression during gait, with only a 3% benefit if the swing phase foot is

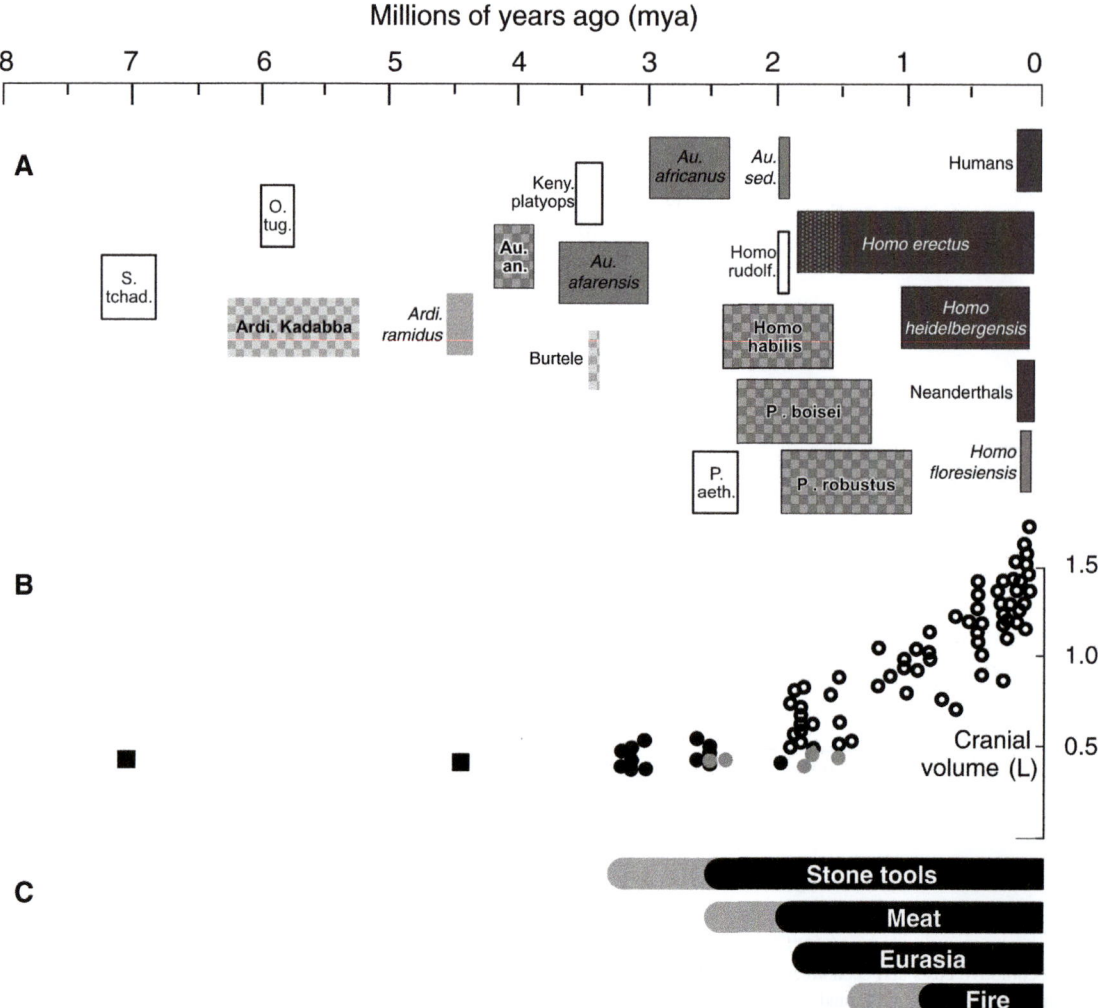

FIG. 1.1.2b Hominin species and an overview of their inferred locomotive capabilities over time, with *Adripithecus* species thought as having poor terrestrial economy and endurance during locomotion. In (A), best gait economy and higher endurance is represented by dark boxes, seen in most *Homo* species with earlier *H. erectus* having less endurance than later forms. Mid-grey boxes indicate high economy but low endurance, seen in *Australopithecus* (particularly *Au. afarensis*) *Paranthropus* species and *H. floresiensis*. Light-grey boxes indicate low economy and endurance of gait, as seen with *Ardipithecus* species and the unknown Burtele foot fossil. Hatched boxes indicate economy and endurance based on more limited evidence and white boxes mean evidence is too limited to estimate. Evolution has honed energetics of bipedal gait, with big improvements from ~2 million-years-ago corresponding with increases in cranial volume, as seen in (B), with *Homo* sp. shown with enclosed white circles, *Australopithecus* as filled black circles, *Paranthropus* by grey circles, and other species by squares. Finally, in (C), the earliest proposed evidence *(grey)* followed by widespread evidence *(black)* for stone tool use, and in the butchery of game animals for meat, human arrival in Eurasia, and the use of controlled fire. *(From Pontzer, H., 2017. Economy and endurance in human evolution. Curr. Biol. 27 (12), R613–R621.)*

assisted (Gottschall and Kram, 2005). Events in gait such as swing and stance phase can be energetically altered by step width and swing phase hip circumduction (Shorter et al., 2017) with implications to the fatigue resistance of individuals, especially when combined with a poor physiological health status.

Energetic collision strategy in gait

Two strategies are used to reduce locomotive costs in human bipedal gait. The first is having legs that simulate elastic springs to create a glancing collision, as used in human running. The second is a sequencing of forces through the leg during a redirection phase from a heel strike impact through to a push-off (acceleration) phase and vice versa, as is seen in human walking (Ruina et al., 2005). In the energetics of a single glancing collision such as occurs in a foot-flat or forefoot running style, the CoM's impact can occur angled behind the limb and always does so during faster running.

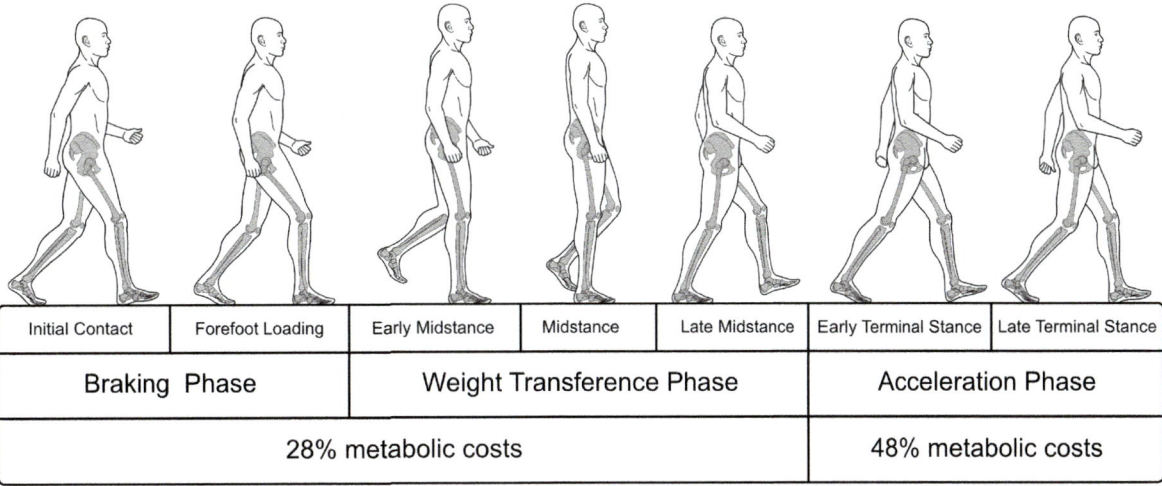

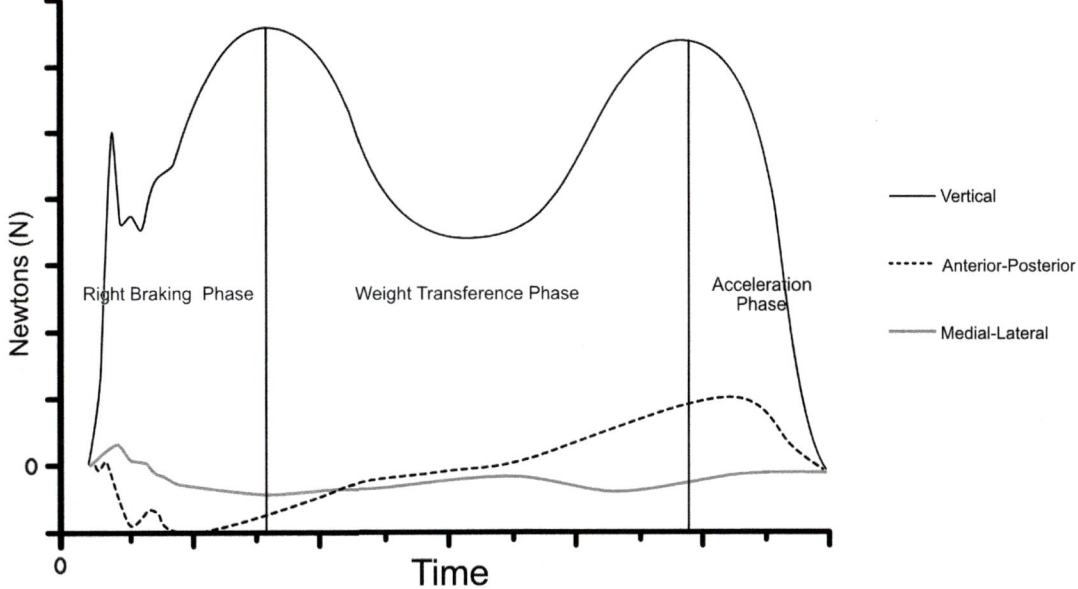

FIG. 1.1.2c The metabolic costs of walking with reference to the stance phase of the gait cycle of the left lower limb and an archetype walking-generated force-time curve. It is clear that acceleration is the most energetically expensive component of walking. *(Data on metabolic costs from Gottschall, J.S., Kram, R., 2005. Energy cost and muscular activity required for the leg swing during walking. J. Appl. Physiol. 99 (1), 23–30. Permission www.healthystep.co.uk.)*

In the energy-absorbing part of the collision, the CoM gets closer to the ground and kinetic energy decreases through eccentric muscle lengthening (energy-buffering effects) around joint flexions until maximal leg compression occurs, after which the distance between the CoM and the impact point then starts to get larger again through elastic recoil (Ruina et al., 2005). This recoil or deflection occurs at a different angle through the leg to the contact angle, tipping over the foot at the ankle and moving the CoM anterior to the leg. The deflection recoil results from the energy stored within the soft tissue's elastic stretch and is also generated by muscle activity causing joint extension (and ankle plantar-flexion) to make up for the lost energy dissipated through the tissues at impact. Without the muscle energy generating phase, running in particular would be very much a 'one-impact wonder' event through loss of potential recoil energy, with no energy left for the collision of acceleration. The response to a glancing collision in gait can be exclusively energy-absorbing (plastic) with a reduction in the anterior drive after the impact, a pseudo-elastic collision with some upward and forward drive, or an exclusively energy-generating collision (highly elastic) where a strong upward and forward force vector can be generated (Fig. 1.1.2d).

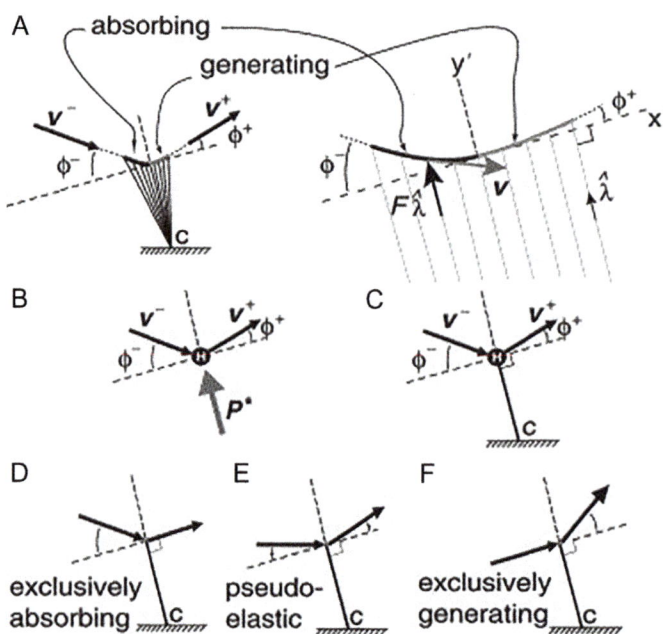

FIG. 1.1.2d A schematic of the glancing collision model of running gait energetics, involving an energy-absorbing and energy-generating phase. Collision occurs on a single leg at a net deflection angle (A). The force (F) comes from the leg at a direction approximated as a constant λ during the collision. The net impulse of one leg (B) on the mass m is **P**. The same collision (C) is shown without the details of the collision's interaction. Three types of collisions can occur: (D) perfectly plastic, (E) pseudo-elastic, and (F) exclusively generating elastic collision. *(From Ruina, A., Bertram, J.E.A., Srinivasan, M., 2005. A collisional model of the energetic cost of support work qualitatively explains leg sequencing in walking and galloping, pseudo-elastic leg behaviour in running and the walk-to-run transition. J. Theor. Biol. 237 (2), 170–192.)*

Understanding collision energetics returns events to the concepts of the coefficient of restitution (see Chapter 2 of the companion text 'Clinical Biomechanics in Human Locomotion: Origins and Principles', Section 2.3.4) and whether an impact involves materials that are plastic (viscous) or elastic, and whether or not and by how much, energy is dissipated or stored within that material. A collision against the ground causes a redirection of energy from down to back up (a bounce), with the final upward speed slower than the downward approach due to energy loss. The more plastic the material, the less the bounce. Yet in gait, a glancing, perfectly plastic collision can appear superficially elastic because, despite the energy lost, the lower limbs have a tipping element (at the ankle and aided by the hip) that changes the angle of deflection through motion. Consequently, the CoM above can accelerate forward with the appropriate ankle angle change (Ruina et al., 2005). Just consider an athlete taking a pole vault with the long, flexible pole angled (directed) in front of him/her during the run up and then pivoting over the pole until it is situated behind (Fig. 1.1.2e). This indicates how important limb angle at heel lift is to gait efficiency.

However, plastic collisions are four times more metabolically expensive than elastic collisions, as to produce recoil power, they need to make up for the energy lost through dissipation. An energy-generating elastic collision is also metabolically expensive in terms of muscle activity to create the necessary stiffness, and such a collision creates higher energy that could be potentially tissue-damaging. For a pseudo-elastic collision, the positive work required to generate acceleration energy through the deflection angle after the impact is proportional to the energy lost at the initial contact angle (Ruina et al., 2005). Pseudo-elastic collisions minimise energetic costs, and even a pseudo-elastic collision without any elastic energy recovery is still more energetically beneficial than the other collision types (Ruina et al., 2005).

The role of heel strike may well be to reduce the peak collision forces, as the collision becomes a two-impact event (heel impact then forefoot impact) rather than one. Dividing the collision into two smaller collisions halves the energetic cost of braking, which is also influenced by the changing impact angle that occurs between heel strike and forefoot loading (Ruina et al., 2005). An event of two-smaller impacts is ideal for walking when a two-stage sequence of motion occurs with a time gap of a significant midstance period that changes the limb angle before acceleration is required. Even in running, heel strike followed by forefoot impact may offer benefits by spreading collision energy over a longer period, particularly

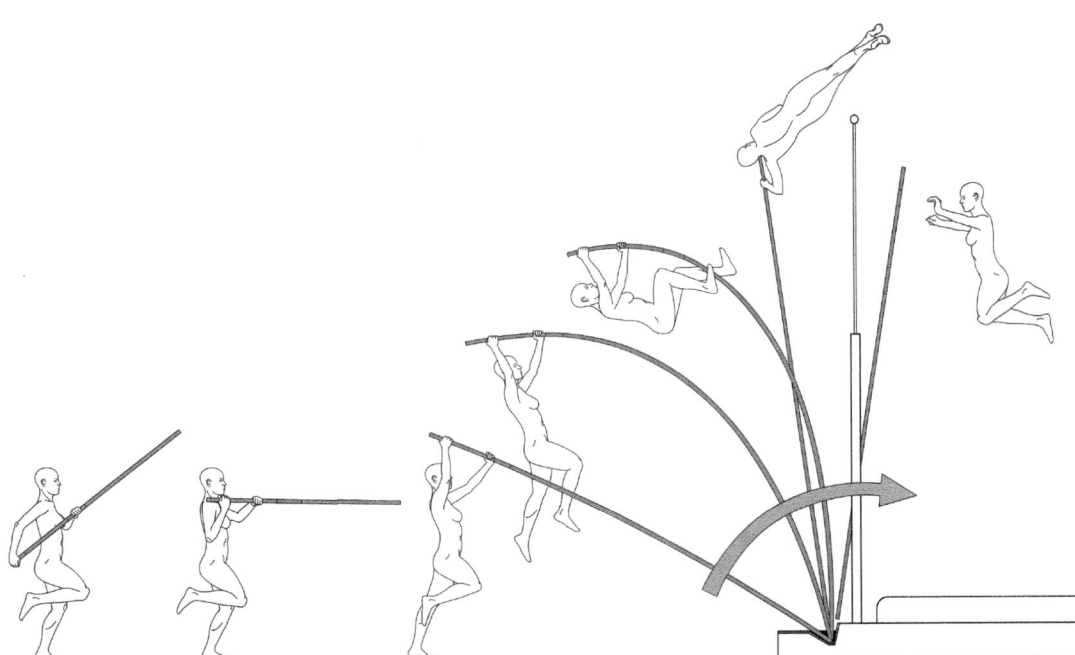

FIG. 1.1.2e A pole vaulter enacts a loading and off-loading collision event much like the lower limb. An accelerated run-up creates a collision of the pole with the ground that drives momentum to flex the pole, using its deflective and elastic properties to lift the athlete up and over the pole. It is the changing pole angle within the indentation known as the pole vaulter's 'box' that allows the athlete to rotate anteriorly over the pole. If the pole does not rotate within the box to change the angle of fall, then the athlete would fall back the way they came. Thus, the pole vault allows the CoM of the athlete to progress forward, despite recoil energy being lost through the pole. Losing energy through pole motion is less important than making sure the athlete ends up facing the right direction. Gait is quite similar in this respect, with the foot and ankle acting as the pole rotating within the box. *(Permission www.healthystep.co.uk.)*

at lower running speeds. Thus, forefoot running does not decrease the collision forces by avoiding initial heel contact but focuses them into one big and longer lasting collision event, rather than two shorter events with a distinct initial heel impact followed by forefoot collision.

Walking collisions

Walking involves the swing limb's velocity before a collision and then the velocity of the CoM following the collision on a low trajectory, just as is also seen in heel strike running. The difference is that the CoM trajectory in walking also involves a high point in the middle of stance as opposed to a middle stance low point as found in running. Walking has also been likened to rolling on a rimless wheel, a polygon, and/or even a concave polygon (Ruina et al., 2005). The leg acts as a pivot point which, once rolled over, falls to the next spoke (of the rimless wheel). The foot, having a distinct heel and forefoot, has two points of contact per foot, like two separate spokes. It can pitch from heel contact point (like a wheel spoke) to forefoot contact point (another spoke) through ankle plantarflexion after heel contact, and then via ankle dorsiflexion to transfer weight forward over the foot, finally using ankle plantarflexion and extension at the metatarsophalangeal (MTP) joints to roll over the forefoot to the next heel (spoke) contact, following the opposite limb's swing forward. These events are like pitching from one rimless wheel spoke to the next by tilting on the ankle and MTP joints, alternating from rearfoot to forefoot and then to the opposite foot's heel. From this rearfoot, it rolls on to its forefoot and then back to the initial foot's heel, and so on.

Indeed, if an ankle is unable to move, the foot can still make a rolling motion if modelled as a concave polygon (Ruina et al., 2005), perhaps giving insight into the development of a convex plantar foot surface as observed in 'rocker bottom foot' deformity that can be developed with restricted ankle joint motion (Fig. 1.1.2f). Restricted or lost ankle joint dorsiflexion may explain the development of the 'rocker bottom foot' profile as seen in neuropathic Charcot feet, most often associated with diabetic neuropathy and loss of ankle joint dorsiflexion range of motion. When ankles and/or MTP joints are nonfunctional, an effective intervention is often a rocker-bottomed shoe, which effectively turns the foot into a convex rolling surface to transfer the CoM from the heel to the forefoot and then on to the next heel strike.

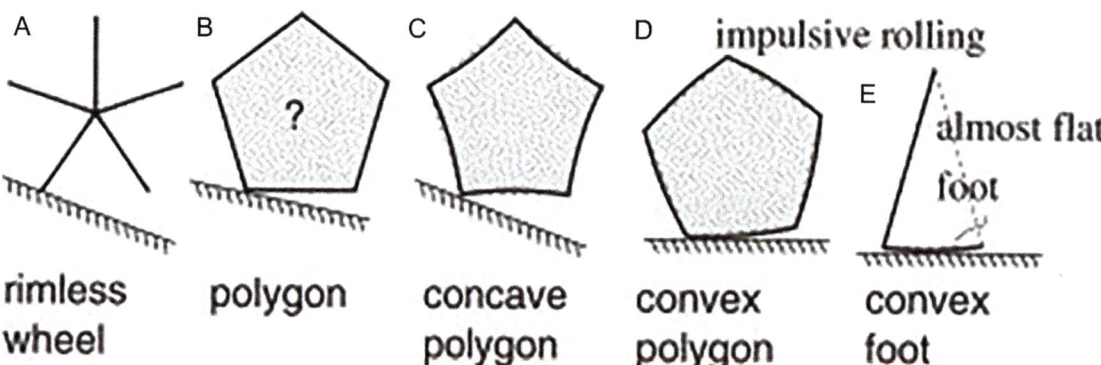

FIG. 1.1.2f The relationship between walking and rolling as explained by Ruina et al. (2005). The rimless wheel (A) or rolling polygon (B) helps to visualise walking as a rolling event with distinct points of collision. The slightly concave polygon (C) helps to understand the difference a distinct foot vault adds to this rolling action, creating respective impulse cycles, acting more like the rimless wheel spokes. A convex polygon (D) creates impulses but is unable to provide energy dissipation. An almost flat foot with a rigid ankle (E) acts more like an impulsive polygon, providing step-to-step impulse generation, but it cannot absorb energy. Such flat foot and rigid ankle situations as seen in Charcot feet associated with certain neuropathies are a serious threat to locomotive health and thus help reveal why ankle joint motion is literally 'pivotal' to locomotive health. *(From Ruina, A., Bertram, J.E.A., Srinivasan, M., 2005. A collisional model of the energetic cost of support work qualitatively explains leg sequencing in walking and galloping, pseudo-elastic leg behaviour in running and the walk-to-run transition. J. Theor. Biol. 237 (2), 170–192.)*

Midfoot break causes excessive midfoot dorsiflexion that functionally presents an 'acceleration only' convex plantar surface, thereby reducing the need for ankle joint dorsiflexion during late midstance. Midfoot dorsiflexion at heel lift often causes a lateral midfoot collision with the ground, rather than a pure forefoot collision as part of the development of an acceleration forefoot GRF at terminal stance. This midfoot collision compromises both the ability to generate acceleration power from the plantarflexors and the extension range of the MTP joints. Thus, acceleration energy is lost with a midfoot break, because it reduces ankle plantarflexion power and the size of the acceleration forefoot collision.

The collisional cost of walking can be reduced by preceding the absorbing events of heel strike with an acceleration from ankle plantarflexion generated by the opposite foot from the previous step. This derives as a consequence of ankle-powered heel lift, which is a generative (muscle-induced) collision event on the opposite forefoot (Ruina et al., 2005), remembering that the forefoot is driven against the ground during heel lift through ankle plantarflexion. This plantarflexion push mechanism onto the next foot reduces collisional energetic costs by a factor of four (Ruina et al., 2005). Thus, active walking is a two-collision event, with an energy-generating push-off collision of one foot, followed immediately by an energy-absorbing heel collision of the other. This energetically results in a two-leg walking collision acting in the manner of a single-leg pseudo-elastic collision with a rolling forward passively under gravity assisted through centrifugal forces from the swing limb while receiving an extra 'shove' on the CoM from behind via heel lifting ankle plantarflexor power (Fig. 1.1.2g).

1.1.3 Dividing the body segments to explain gait

Head, arms, and trunk in gait

The elements consisting of the head, arms, and trunk, referred to as *HAT* (which also includes the pelvis), together constitute 70% of body mass. These body segments are argued not to 'directly' involve themselves in gait but instead act as a passenger unit, being translated forward during gait by the lower limbs. In reality, the upper and lower limbs have a neural coupling between them during gait (Weersink et al., 2021). The pelvis can couple its movements to lower limb motions at higher gait speeds (Bruijn et al., 2008) and also requires the rectus abdominis to become active for trunk stability (Runge et al., 1999). The CoM of this passenger unit is positioned just anterior to the 10th thoracic vertebra within a healthy adult male's body, and in an average man of 184 cm height, this is around 33 cm above the hip joint (Perry, 1992: p. 20). The balance of the passenger unit is therefore dependent upon the instantaneous alignment of the underlying lower limbs to move the base of support below the HAT's momentary CoM position, which is also influenced by trunk muscle activity. The muscle action within the neck, trunk, and arms maintains a stable vertebral alignment with minimal postural change during relaxed gait. Stability of HAT in the upright position is therefore determined by functional balance between the body's CoM, muscle action within HAT, and the position of the locomotor segments below, with the acceleration of the head significantly more attenuated and tightly controlled than that of the trunk (Kavanagh et al., 2005) (Fig. 1.1.3).

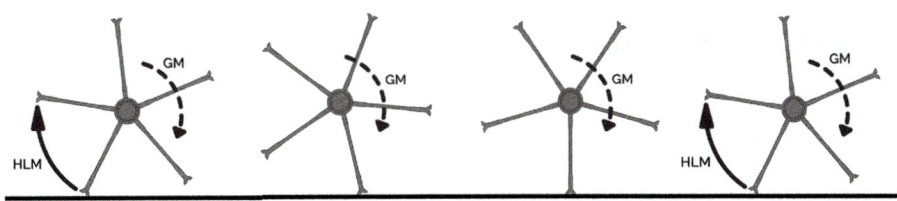

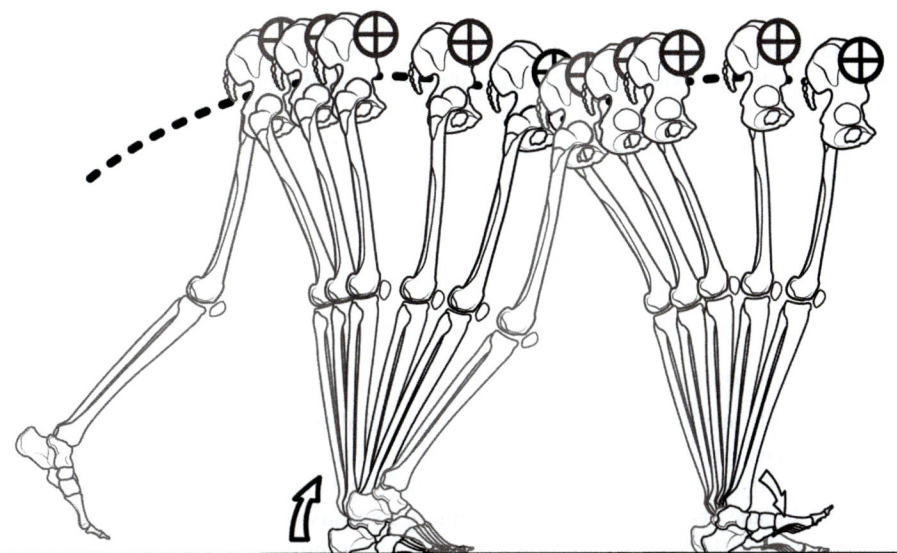

FIG. 1.1.2g The rimless wheel model modified from Ruina et al. (2005), expressed throughout the gait cycle *(upper image)*. The heel lift momentum (HLM) increases gait momentum (GM). GM is driven by gravity and swing limb centrifugal forces. This heel lift-induced forefoot collision concept can be applied to the gravity-induced fall of the body's CoM and the centrifugal 'pull' on the CoM by the swing limb *(lower image)*. Although a powerful ankle plantarflexion moment is not essential to move to the next step, its presence vastly improves gait energetics, giving an extra 'shove' to accelerate the CoM forward. This excellent addition to bipedalism explains the rather 'odd' large muscle mass in the posterior lower limb of triceps surae, found in humans. Such large muscle masses existing so low down is rather unique among large terrestrial mammals. *(Permission www.healthystep.co.uk.)*

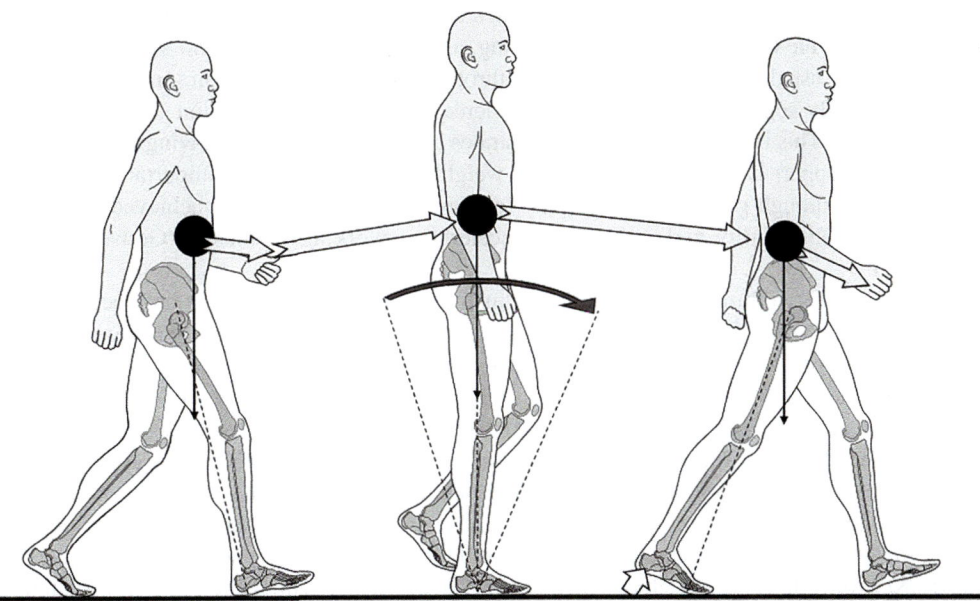

FIG. 1.1.3 The HAT segment *(grey body areas)* is considered to be the passenger segment composed of head, arms, and trunk, including the pelvis, that provides little assistance to locomotion. However, it is the part of the body that needs to be moved from location to location. In reality, the trunk does adjust to perturbations and the pelvis 'swaps' into the locomotive system *(indicated by skeleton)* at higher walking velocities. The body's CoM *(black circle)* starts walking stance phase behind the lower limb of the locomotive segments. It is then carried up and over the lower limb to reach a high point. This is achieved through the arch or rotation at the ankle created by the leg length, which is adjusted by hip and knee flexion-extension angles. From the middle of mid-stance, under the influence of gravity, the CoM is carried anteriorly and downwards in front of the support limb which is lengthening via hip and knee extension. At heel lift, ankle flexion (plantarflexion) functionally lengthens the limb again, further enhancing gait speed. As the next limb loads, a new cycle of HAT carriage begins from behind a newly supporting limb. Thus, the path of the HAT segment and the CoM is a curvilinear, one of up and down over locomotive limbs. The limbs assist in preventing sudden changes in accelerations during gait or from perturbations from reaching the head by adjusting flexions and extensions. CoM oscillations are thus kept smooth. Within the HAT, muscles help moderate any acceleration changes that reach it, further helping to protect the head from velocity changes. *(Permission www.healthystep.co.uk.)*

Locomotor segments and the pelvis in gait

The locomotor system of the lower limb is separated from the passenger unit. The lower limb can also be subdivided into the hip, thigh, leg (shank), and foot. With increasing gait speed, the pelvis can functionally swap from the HAT to being a part of the lower limb. The locomotor segment is constantly changing its posture and alignment to achieve gait, hopefully doing so in an energy-efficient manner. Locomotion thus involves movements at multisegmental levels and interactions with the body mass of the passenger segments above. Each support limb acts as a base over which the HAT passenger can advance while maintaining stability. Using the stable base of the foot, HAT can effectively rotate from behind, over, and beyond the support limb to achieve forward progression, while the other limb is also advancing from behind the HAT to in front of it, ready to receive the falling HAT CoM for the next step. Gait therefore becomes controlled, pivoting and falling under the laws of motion over the foot during stance. The swing phase is also extremely important, for the centrifugal forces and momentum it creates draws the body's CoM forward, encouraging the anterior fall of HAT as it is vaulting over the foot.

1.1.4 Gait motion and description

The laws of motion in the gait cycle

From the point of view of simple mechanics, human bipedal gait is all about the translation of the body's HAT CoM from one point to another. This creates curvilinear motion of the CoM above the lower limb through angular rotations achieved by the joints operating primarily around their z-axis lying in the transverse and frontal planes, with motion primarily occurring along the sagittal plane (x-axis plane). These primary motions are flexion and extension, or in the foot and its parts, plantarflexion and dorsiflexion. Such primary sagittal plane motions are refined by more subtle adjustment activity around the transverse and frontal planes, with all these motions occurring under the influence of gravity, muscle power, and connective soft tissue guiding. Studies have demonstrated that raising gravitational forces heightens the work required but also increases the range of possible walking speeds, allowing us to walk at faster speeds before being directed into running (Cavagna et al., 2000). Gravity is extremely helpful in creating pendular transfers of energy and mass during the stance phase of walking, providing human ambulation with mechanical advantage and reduced metabolic costs (Cavagna et al., 2000). Thus, if humans can raise their CoM, gravity will help to bring it back down in late stance phase, adding momentum to gait.

During walking and running, metabolic energy is consumed to generate the forces that support the body weight against gravity and perform work on the CoM in braking, accelerating, and moving the swing limb's CoM (Gottschall and Kram, 2005). Metabolic costs are usually measured in humans through oxygen consumption, and studies have shown that increasing the loads on body segments such as the feet increases oxygen consumption (Soule and Goldman, 1969; Gottschall and Kram, 2005). This is important, as it demonstrates that there is a cost to moving the mass of limbs in swing as well as costs in single support balance. Thus, the extra weight of footwear can be important. The ability to create a horizontally directed force through gait is critical for any forward translation, which is achieved through a combination of the effects of stance limb stability and swing limb acceleration working in unison (Gottschall and Kram, 2003, 2005; Ellis et al., 2014).

Humans normally use only two limbs for locomotion. Thus, translation of the body can only occur by at least one foot leaving the ground and being swung forward. In human walking and/or running gait, there is a *stance phase* where at least one lower limb is in contact with the ground and a *swing phase*, where one lower limb is not in contact with the ground. The swing limb is translating its own CoM forward, while it also influences motion of the rest of the body's CoM. The stance limb is used to pivot the CoM of the overlying HAT segment forward over the ground, translating the HAT's CoM but not that of its own foot. The whole stance limb only becomes concerned with its own CoM's motion alone at the end of terminal stance in a period known as *preswing*, although the lower limb segments' CoM of the stance limb are always moving. The stance limb is subjected to far more 'interesting' biomechanical events than the swing limb, as it is involved with a continuous collision of the body's weight through the foot and limb into the support surface. This consequently subjects it to forces generated by Newton's third law which entails that every action has an equal and opposite reaction, and Newton's second law which promulgates that mass time acceleration results in a quantity of force being involved. The swing limb is primarily working by Newton's first law in that the body must create momentum through force generation to overcome inertia of the swing limb in its acceleration and also in its braking before impact. This inertia is overcome through muscle contraction. Despite the greater series of stressful events taking place in the stance limb, the swing limb is no less important to gait for it provides limb advancement and influences the forces being driven into the stance foot, functioning as it does as an inverted pendulum during walking (Fig. 1.1.4a).

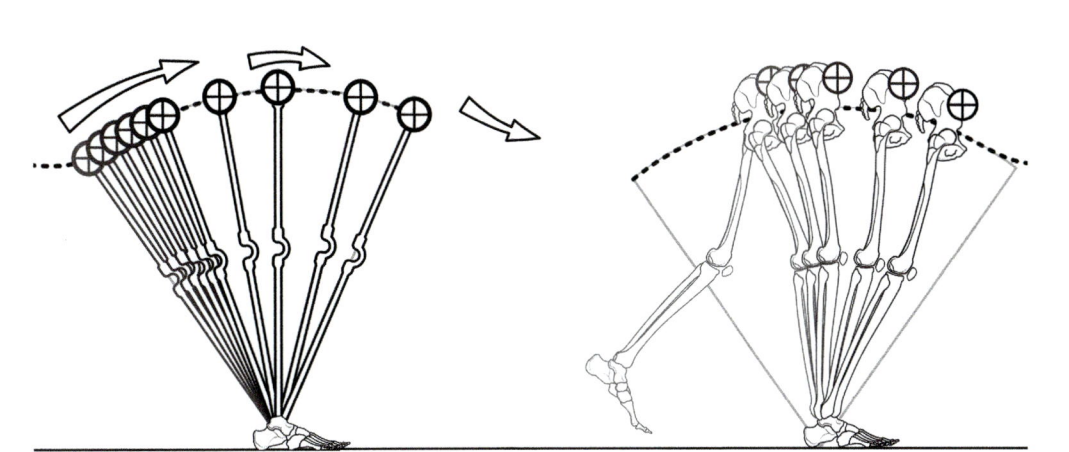

FIG. 1.1.4a The energetic genius of human walking is that the stance limb through single support (midstance phase) uses the lower limb much like an inverted pendulum, pivoting around the ankle *(left)*. This not only changes the direction of the limb from its angle at contact, so that acceleration occurs on a backward-positioned leg, but it also means that during late stance, the fall of the CoM occurs under gravitational forces. Thus, early midstance involves muscle activity (metabolic-induced kinetic energy) that raises the trunk over the foot, but once raised high and loaded with gravitational potential energy, it can fall under free kinetic energy, only requiring muscular control of the descent. *(Permission www.healthystep.co.uk.)*

The swing limb, under the influence of muscle activity and gravitational forces, generates part of the momentum that creates acceleration force on the HAT, pulling the body's CoM forward during late swing phase. This centrifugal momentum helps move the stance limb into dorsiflexion before helping to induce heel lift for the start of terminal stance. Swing is achieved by powerful proximal limb and pelvic muscles, pulling the swing limbs' mass forward in front of the HAT CoM, mainly through iliacus. Thus, the momentum of the swing limb, assisted by gravity, pulls the body's CoM along behind it through the pull of centrifugal force from a pendulum-like effect, as opposed to an inverted pendulum effect as seen in the stance limb (Fig. 1.1.4b).

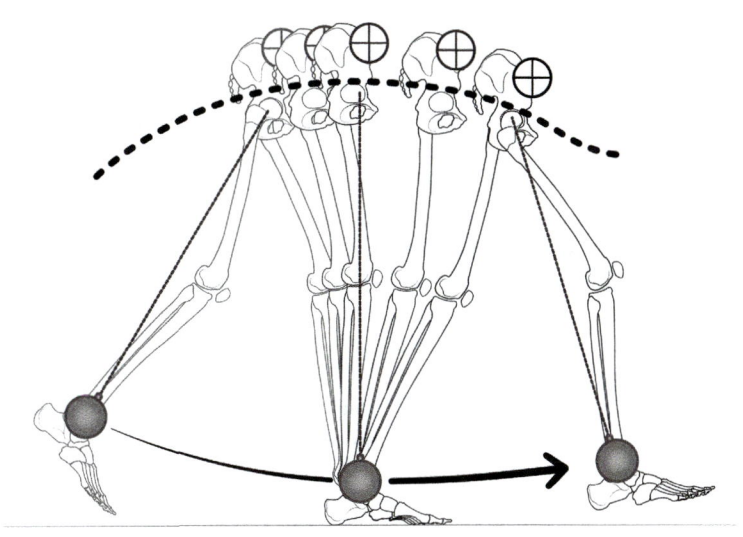

FIG. 1.1.4b The swing phase function is much like a pendulum, rotating at the hip to advance the limb in front of the HAT segment and stance limb, creating a long stride length. The momentum of the swing limb's mass under hip flexor and knee extensor activity during late swing phase pulls the body's CoM along behind it. This aids the anterior progression of the CoM, which is already falling forward under gravity. Long legs enhance this effect and pull the CoM so far forward that it reduces body mass loaded on the support limb, reaching a point where heel lift under plantarflexor power can be achieved easily. *(Permission www.healthystep.co.uk.)*

Gait-type definitions

In walking, there are periods at the end and the beginning of stance where both feet make contact with the ground, and these are known as *double-limb contact phases*. This phenomenon is a distinct characteristic of walking. However, in running (unlike walking), both feet leave the ground at the end of stance and just before the initial contact periods of gait. This means that there is also a *'float'* or *'aerial phase'* as well as a stance and swing phase, but no double-limb contact (Fig. 1.1.4c). This

12 Clinical biomechanics in human locomotion

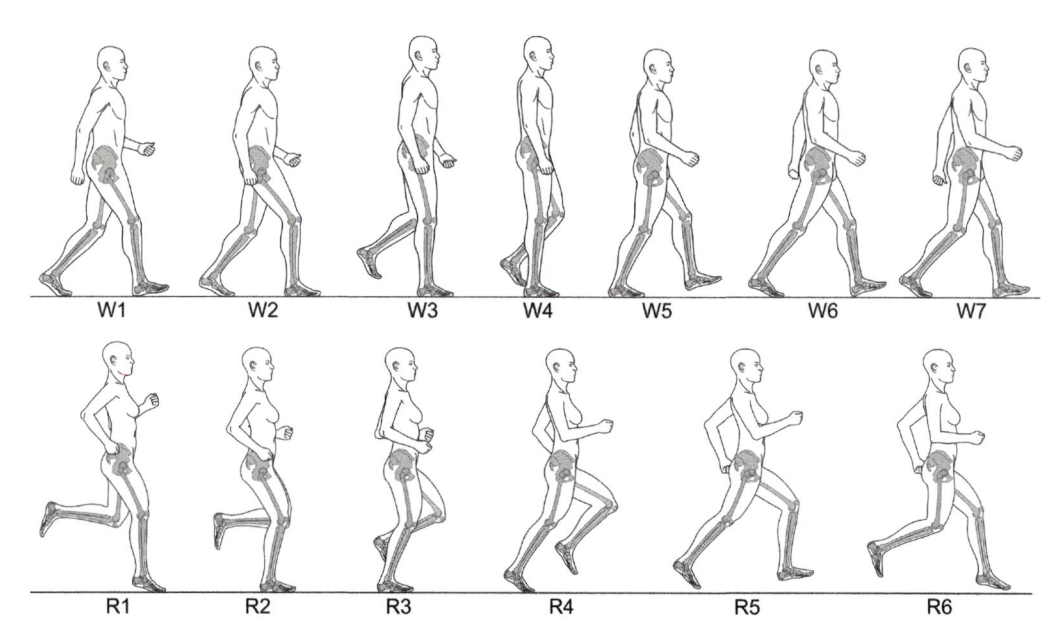

FIG. 1.1.4c Walking and running (forefoot strike shown) are compared. In walking, there are two distinct phases of double-limb contact. These occur during loading (W1–W2) and when off-loading the stance limb (W6–W7). The single-limb support phase (W3–W5) involves the transfer of the CoM over the foot. Potential energy is achieved during walking by creating an extending limb throughout early single-limb support (W3–W4 – early midstance) to raise the CoM to its highest point (W4). In running, whether involving heel strike or forefoot strike (as demonstrated here), loading and off-loading of the foot become tasks of single-limb support with body mass transfer being part of the braking (R1–R3) and acceleration (R3–R4) phases, rather than body mass transfer occurring as a distinct phase of its own. Body mass is also advanced through an aerial phase (R5–R6), with the body's CoM being launched into the air without any limb support to gain potential energy from a fall under gravity afterwards. In running, CoM potential energy is achieved expensively through the high metabolically generated acceleration that produces the aerial phase CoM high point. *(Permission www.healthystep.co.uk.)*

is a defining characteristic of running, whereas jogging is considered to have no double-limb contact and no significant float phase. Hopping also has a float phase which can occur from the same stance foot to the same stance foot or the opposite foot but lacks the rhythmic action seen in running of single support to double swing phase. Jumping is motion from double stance to double aerial and double swing phase. No matter what the mode, in all locomotive gait situations, there must be both a period where at least one foot makes contact with the ground and a swing phase where the limbs advance beyond the HAT CoM. Obviously, if both feet remain on the ground, humans are in static stance.

Human bipedal gait can therefore be neatly divided into a swing phase of nonlimb contact and a stance phase of limb contact. The percentage breakdown of these phases varies with respect to gait speed. In walking, the percentage single-limb support increases with speed as double-limb support decreases, while in running the percentage length of the float phase increases with speed and the length of the stance phase decreases. Therefore, when later discussing normal percentage lengths for periods of the gait cycle, the activity and gait speed will be seen to be highly relevant to the results expected of temporal and spatial gait parameters. There is no set 'normal' within human gait, save that gait should be energetically efficient and asymptomatic, and involve controlled torques through the lower limb and trunk in all three orthogonal planes. Patients will choose a preferred gait speed for a type of locomotion as long as there are no external pressures, and this preferred speed will reflect the health and age of the patient (Fig. 1.1.4d).

1.1.5 Walking gait phases

Swing phase
The swing phase represents around 40% of the walking gait cycle of a limb, depending on gait speed. Swing phase is about generating movement in the limb against the forces of gravity, air resistance, and the resistance of the CoM of each limb segment being moved without any GRF or other external force interventions. To move the lower limb anteriorly, angular moments need to be created at the hip, knee, and ankle to swing the limb forward ready for the next foot contact with the ground. The mass being moved is that of each lower limb segment combined below the hip, which includes any footwear or clothing. Muscle contractions need to overcome the limb segments' resistance resulting from mass, gravity, and inertia, so metabolic energy is required to generate these angular moments (Gottschall and Kram, 2005).

Understanding human gait **Chapter** | 1   13

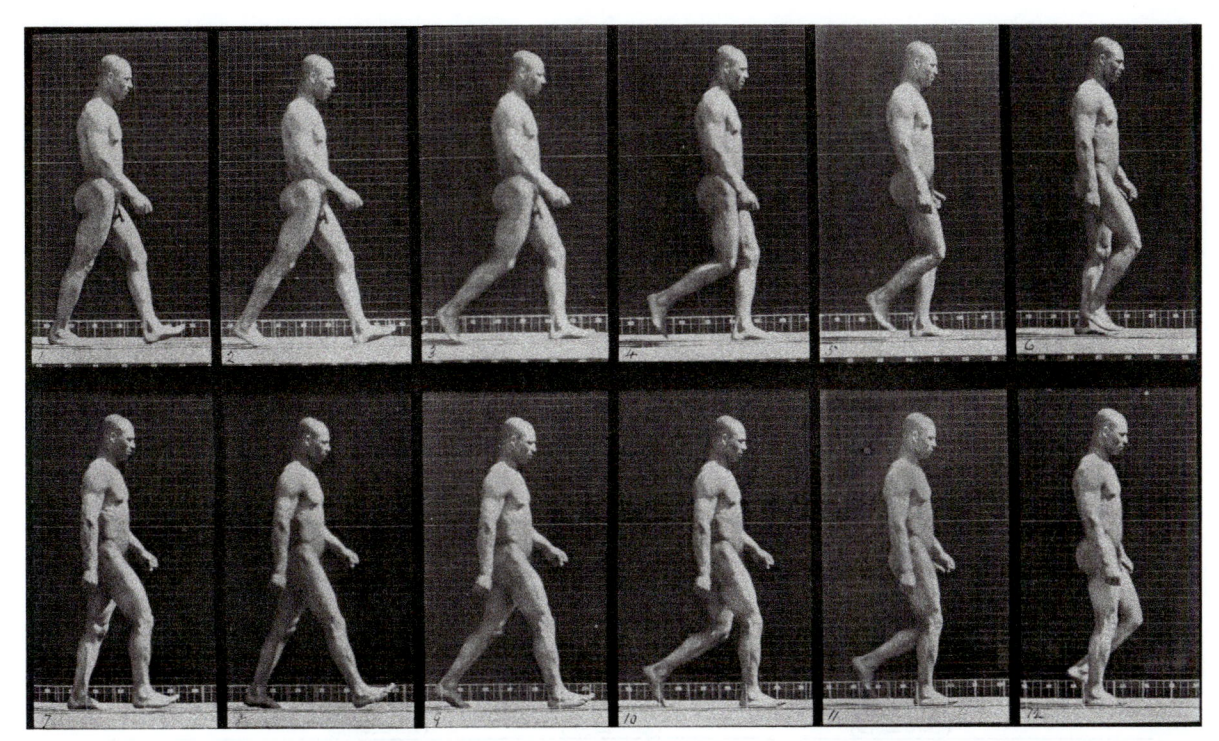

FIG. 1.1.4d The use of a thicker horizontal white line in the middle of a gridded backdrop clearly shows the falls and rising of human anatomy (and thus the CoM of the body) in relation to this line. The CoM falls during right limb midstance and left loading response *(left three upper images)* and also during the late midstance of the left leg and loading response of the right *(left three lower images)*. Compare this to the rise upwards after loading response is completed, during early midstance of the left limb *(right upper three images)* and of the right limb (right lower three images). It was the genius of Eadweard Muybridge in his early photographic assessment of human gait to use a grid backdrop so that postural changes could be more easily visualised. Clinican biomechanists often use a gridded background to video gait analysis to this day.

The metabolic cost of swing primarily comes from the muscular actions needed to initiate and propagate swing which is, in total, around 10% of the net metabolic cost of walking (Gottschall and Kram, 2005). These are perhaps surprisingly small for there are huge energetic benefits to swing phase. From midswing into terminal swing, centrifugal force is generated in the swing limb both from knee extension and hip flexion that helps to pull the CoM of the body forward (Fig. 1.1.5a), assisting gravitational forces to improve CoM transfer over the stance limb (Gottschall and Kram, 2005). The power generated at heel lift has a direct effect on the efficiency of swing initiation, but heel lift does not act as a primary 'accelerator' of the CoM's forward progression. Therefore, the terms '*propulsion*' and '*push-off*' during gait that are used to describe the end of stance phase need to be used with caution and on the understanding that the power generated through the ankle plantarflexors does not itself actually act as the primary driver of CoM progression (Honeine et al., 2013). Heel lift raises the rearfoot and the trailing lower limb to angle the stance limb for the efficient transfer of the CoM to the next step. At the same time, the CoM of the trunk is 'pulled' and 'tipped' forward by gravitational forces and the swing limb's centrifugal force to create the fall forward of the CoM at the end of stance. In addition, heel lift power adds some extra momentum to opposite limb's initial contact (Ruina et al., 2005) (Fig. 1.1.5b).

Swing initiation results from a number of contributory sources including the activity of gluteus maximus, the ankle plantarflexors (particularly medial gastrocnemius for knee flexion), and the ankle extensors such as tibialis anterior. However, the primary muscular initiation of swing is dominated by the hip flexors such as iliopsoas (iliacus in particular) (Gottschall and Kram, 2005). Joint angles in the lower limb at the beginning of swing are around 7° extension at the hip, 39° flexion at the knee, and 13° plantarflexion for the ankle (Barrett et al., 2007). The hip flexors have to move the CoM of the thigh, leg, and foot, around the hip joint's instantaneous axis through flexion movements. The limb's CoM changes its location, moving closer to the hip as the hip flexes out of extension at toe-off and as the knee becomes more flexed during early to midswing. It continues to then move forward with knee extension and continued hip flexion during late swing. This time, however, the CoM moves away from the hip axis. As the CoM comes closer to the hip during midswing, it makes the effort of moving the limb at the hip easier as the resistance comes closer to the axis (see Fig. 1.1.5a). In early swing, more effort is needed to get the hip flexing because the limb's CoM (resistance) is further from the axis. This is why a good

14 Clinical biomechanics in human locomotion

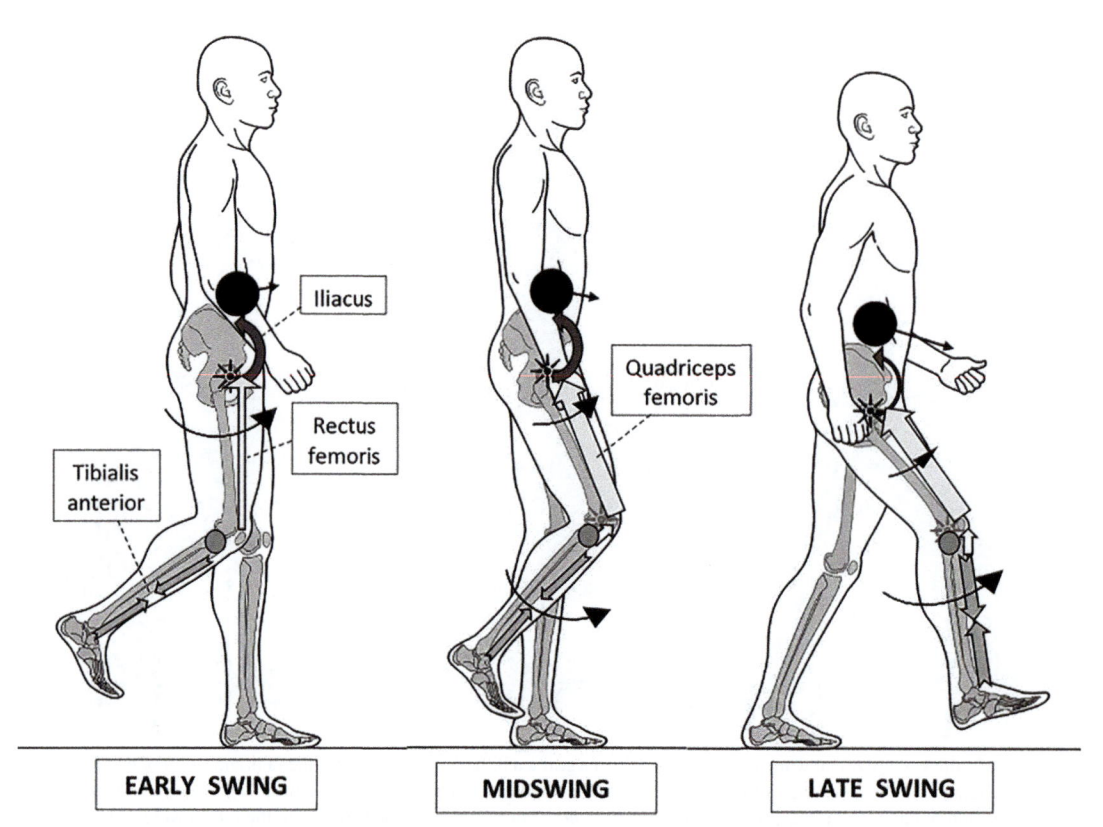

FIG. 1.1.5a Early swing involves hip flexion primarily through iliacus (part of iliopsoas) acting as the primary accelerator of swing limb's CoM *(grey circle)*, aiding by some rectus femoris activity. The limb's CoM moves closer to the hip during early swing phase. Tibialis anterior reduces the residual ankle plantarflexion angle at the end of stance phase to aid foot ground clearance and then holds the foot via near-isometric contraction. The knee is maintained in flexion for the same reason. Midswing begins as the hip swings the thigh under and then anterior to the hip, clearing the lower limb from the ground. Knee extension then begins under full quadriceps femoris activity. Thus, the lower limb's CoM come closest to the hip at midswing and then moves away anteriorly under further hip flexion and knee extension. Tibialis anterior remains isometrically active. During late swing, the knee extends rapidly under increasing quadriceps activity, creating increasing momentum on the body's CoM *(black circle)*, thus allowing iliacus to reduce its activity. The lower limb's CoM is moving away from the hip, placing a centrifugal drag on the HAT CoM. This couples to the gravitational pull on the body's CoM increasing forward momentum. Tibialis anterior activity increases just prior to loading response in preparation of heel strike-induced plantarflexion. *(Permission www.healthystep.co.uk.)*

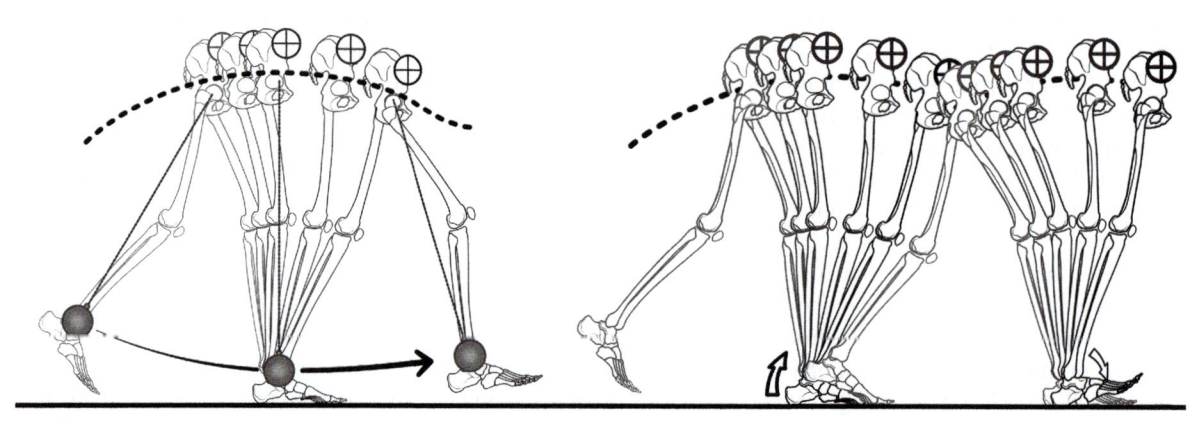

FIG. 1.1.5b The interaction between the swing limb and stance limb is complex to illustrate. When considering the pendular action of the right swing limb, the supporting left foot remains flat to the ground, even as the CoM moves anterior to the ankle and initiates loading forces to the forefoot *(left image)*. Late swing of the right foot is adding anterior momentum to the CoM supported on top of the left limb. Thus, the CoM's anterior fall forward occurs under swing-induced accelerated momentum that increases the ankle dorsiflexion moment on the left foot during late midstance. The resulting Achilles tendon stretched-induced energy storage increases as the CoM is drawn anteriorly by swing, reducing force loading on the left stance foot's heel. Although forces rise through acceleration on the forefoot, the forces on the rearfoot reduce sufficiently to allow the potential energy stored within the Achilles tendon stretch to lift the heel *(right image)*. This terminates ankle dorsiflexion and the swing phase centrifugal 'pull' almost simultaneously, so that heel contact on one foot rapidly follows heel lift of the other. Thus, ankle plantarflexor muscle power causes the swing limb to drop to the ground, starting a new stance phase on the right foot. Whereas the swing limb adds momentum to the anterior progression of the CoM to aid heel lift, the plantarflexion power of heel lift adds momentum into the contact phase of the next step. A powering of weight transference from step to step is achieved through cycles of swing and heel lift, with the timing of these events being everything to energetics. *(Permission www.healthystep.co.uk.)*

powerful extra 'push' from the ankle power in terminal stance phase is important to gait energetics, not just for initial contact momentum on the opposite foot but also to help initiate its own swing. In late swing, the linear momentum generated on the limb's CoM from the hip flexor's power helps pull the leg forward, using the principles of centripetal–centrifugal forces. As the knee extends during late swing, the CoM (resistance) moves away from the hip joint axis, increasing the centrifugal force in drawing the shank forward in front of the rest of the lower limb. In so doing, it increases the centrifugal draw on the HAT's CoM over the support limb. For this process to work, swing limb muscular action must be appropriate and of sufficient power (Fig. 1.1.5c).

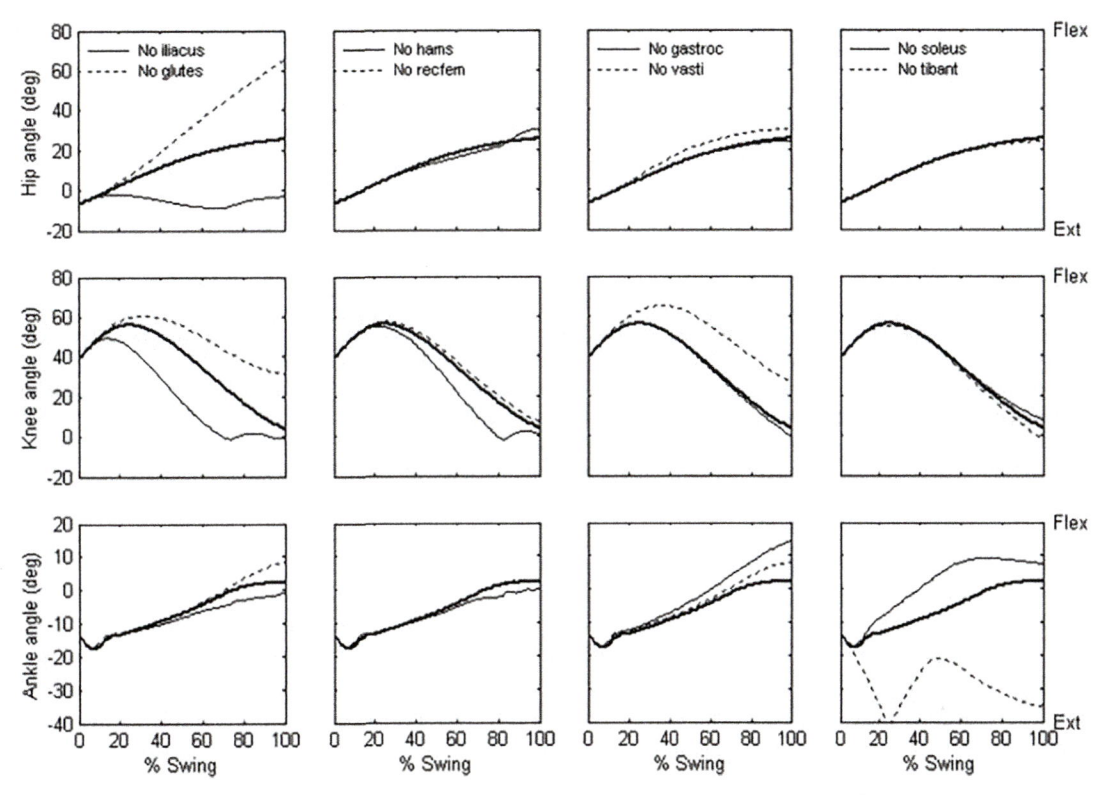

FIG. 1.1.5c Joint angle changes throughout the swing phase are under the control of the swing phase muscles. Computer-modelling simulations can be used to indicate the primary influencing swing muscles during walking on hip, knee, and ankle joint angles through the effects of losing their action. The thick lines in the graphs represent the 'normal' optimal joint angle paths during swing, with hip flexion, knee flexion, and ankle dorsiflexion (despite being extension) all defined as positive rotations. Loss of iliacus and gluteus maximus has dramatic effects on the hip and knee angles, requiring both to provide balancing forces that set angles correctly. The effects of loss of strength in hamstrings and rectus femoris, in comparison, are small. The vasti muscles of the quadriceps along with the hamstrings primarily balance knee angles. Loss of tibialis anterior has dramatic effects on the ability to reduce ankle plantarflexion angles, but loss of gastrocnemius, vasti quadriceps muscles, and particularly soleus restraint can cause the foot to dorsiflex during swing from unopposed tibialis anterior activity at the ankle. *(From Barrett, R.S., Besier, T.F., Lloyd, D.G., 2007. Individual muscle contributions to the swing phase of gait: An EMG-based forward dynamics modelling approach. Simul. Model. Pract. Theory 15 (9), 1146–1155.)*

The mass moving around the knee joint axis consists of the leg and foot and any unnatural additions such as footwear. The knee needs to flex during early swing through to midswing so as to shorten the length of the leg, aiding ground clearance as the foot is still plantarflexed. This action helps avoid tripping. The flexion of the knee reaches around 60–65°, which is important at midswing as the support limb's hip undergoes a varus tilt, which is resisted by the hip abductors on the stance side. Once midswing is reached, the swing knee starts to extend, reaching its most extended swing phase position just before heel contact. Linear momentum generated by the quadriceps femoris assisting hip flexion and extending the knee during the earlier stages of swing phase is so effective that little effort is required to complete knee extension. However, this motion needs resisting. The hip extensors (i.e. gluteus maximus, posterior hip abductors, and adductor magnus) and the knee flexors (hamstrings) need to activate eccentrically in order to slow this knee extension momentum. However, it has been suggested that hamstring activity, at least in faster running, is isometric (Van Hooren and Bosch, 2017a,b) resulting in near-isometric behaviour as the attachment points increase in distance, despite isometric nonlength-changing muscle fibre contraction. Strong restraining activity from these muscles in controlling the limb's CoM anterior momentum can actually drag the heel backwards through elastic recoil. This pulls the heel backwards into the ground during heel contact, causing an effect on the GRF of impact known as *claw back*, which will be discussed shortly (Fig. 1.1.5d).

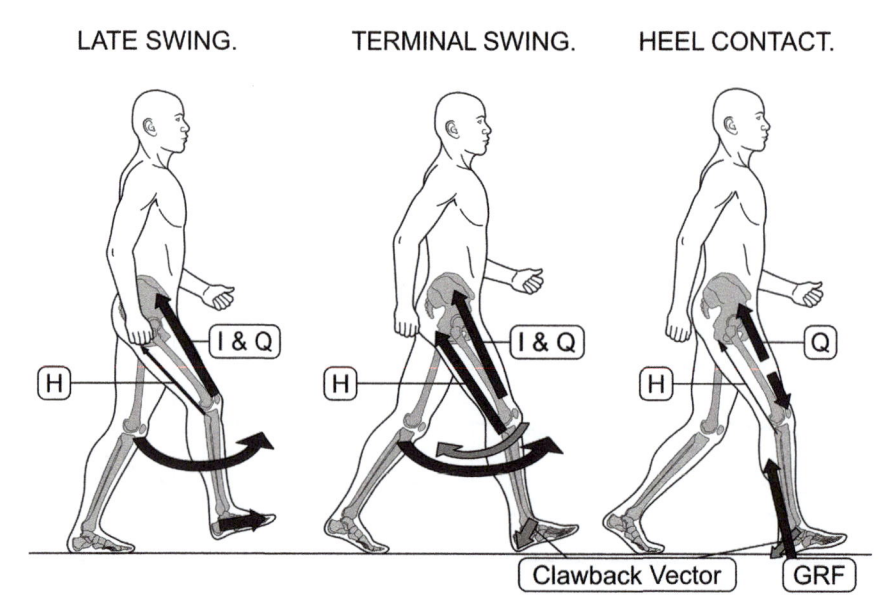

FIG. 1.1.5d During late swing phase, concentric activity of iliacus (I) and the quadriceps (Q) gives anterior momentum to the swing limb and foot *(black arrows)*. The hamstrings (H) start to activate eccentrically or isometrically so that their tendons can start to slow the swing limb momentum, acting as a brake. At terminal swing, hamstring activity peaks, exerting a significant knee extension brake that elastically stretches the hamstring tendons, creating elastic recoil that can pull the swing limb backwards *(grey arrow)*. This can pull the foot back towards the ground. If this occurs, it creates a posteriorly directed element to the GRF at heel strike, known as *claw back*, despite hamstring activity rapidly decreasing at initial contact. This event is far more likely at faster gait speeds where hamstring activation is *stronger* and more rapid, such as in fast walking or heel strike running. *(Permission www.healthystep.co.uk.)*

The ankle is flexed (plantarflexed) at the toe-off phase of gait resulting from heel lifting events during terminal stance. This ankle flexion needs to be reduced in order to aid ground clearance as the limb advances forward over the ground, although the ankle does not usually become extended (dorsiflexed) during the swing phase. After the plantarflexion angle is sufficiently reduced, the ankle remains in a less plantarflexed or neutral attitude throughout the swing phase, mainly due to the isometric activity of tibialis anterior. Failure of the ability to control this position is commonly called '*foot drop*' and results from tibialis anterior weakness or flaccid paralysis. This causes issues with ease of ground clearance of the swing foot, necessitating changes in other joint angles in the swing limb and trunk posture during swing phase, particularly in lifting the swing side hip through hip abduction, known clinically as a *hip hitch*. This usually requires the trunk to sway to the support limb side (Fig. 1.1.5e).

In relation to translation, during swing, the HAT translates less compared to the swing limb's motion, the thigh moves less than the lower leg, and the swing foot's segments move the most. Although most swing motion is generated through muscle effort via concentric contraction from agonistic muscles (e.g. iliacus, the quadriceps muscles, and tibialis anterior) that are regulated by antagonist control (e.g. gluteus maximus, hamstrings, soleus), the segments of the lower limb make up a relatively small mass to move compared to the trunk. The heaviest segment is the thigh which uses the big proximal hip muscles in order to move it, the leg, and the foot through the smallest distance, while the lesser but still substantial quadriceps muscles of the thigh move the mass of the leg and foot larger distances. The much smaller anterior leg muscle group only serve to decrease the plantarflexed position of the foot, to hold a very small foot mass in position. It is the large muscles above the knee that generate the momentum to move the foot a relatively long way. Overall, this results in fairly low-level effort generation throughout swing, especially during walking, where high rates of joint acceleration are not normally required. There are metabolic costs to swing's muscular actions, particularly in initiating and propagating swing, but these are only ~10% of the net metabolic cost of walking (Gottschall and Kram, 2005). Centrifugal forces of the swing limb help reduce metabolic costs of CoM transfer over the stance limb, but the power generated at heel lift has a direct effect on the efficiency of swing phase.

Stance phase

Stance phase involves interaction directly between the lower limb (usually through the foot) and the support surface. The stance phase occurs over approximately 60% of the walking gait cycle. Every force put into the ground by motion of the body's CoM is matched by an equal and opposite reactive force from the ground, the GRF. Stance phase in walking is

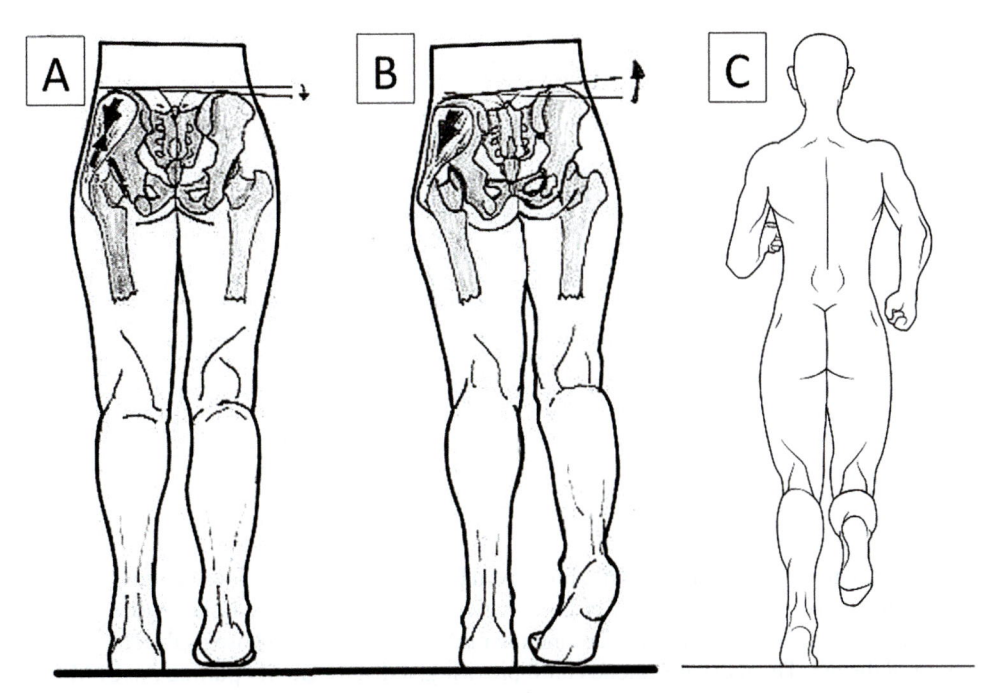

FIG. 1.1.5e Swing phase in walking is aided by contralateral hip abductor isometric contraction that resists hip adduction moments, tilting the swing limb side of the pelvis downward (A). Thus, although the swing side pelvis 'dips', the degree is small, requiring only limited knee flexion and ankle plantarflexion angle reduction to permit the swing limb to pass under the body. If the swing limb's ability to flex hip and knee, and/or ankle joint extension is compromised for ground clearance, the support limb's hip may need to abduct, causing the swing side to 'hitch' upwards, forming a valgus hip angle (B). This raising of the pelvis on the swing limb side requires hip abductor concentric contraction on the support limb side but often also necessitates the trunk tilting towards the support limb. The size of this trunk tilt created by the hip and lumbar spine often indicates the degree of dysfunction in shortening capability of the swing limb for ground clearance during midswing. 'Foot drop' with reduction or loss of tibialis anterior function is one of the most common causes. In running (C), kinematics is somewhat different because the stance limb's knee and hip are significantly more flexed at the midswing point, requiring extensive swing limb knee flexion for enhanced ground clearance. Despite a functionally shorter swing limb during running, hip abductor stabilisation of the body's CoM is even more important as there is no double-limb stance phase where CoM is balanced between the two limbs. *(Permission www.healthystep.co.uk.)*

usually initiated with a collision or impact of the heel with the ground, but variation exists as a consequence of technique (such as in running), terrain issues, and/or pathological situations. As a result, it is better to speak of *initial contact* rather than 'heel strike' as the starting point of stance phase. Runners, in particular, display considerable variation in initial contact location, and indeed, the same runner can change the contact point during a run over time or with changes in speed. In walking, a heel strike is expected and considered normal.

The support limb, through articulations at the hip, knee, ankle, and MTP joints, uses angular momentum to pivot the CoM of the HAT over the foot, following the progression of the swing limb. The HAT's CoM will move anteriorly during the support phase to end up somewhere behind the contact of the next footstep, but in front of the support limb. The support limb must be able to control fluctuations in motion of the HAT's CoM and the changes in the GRFs that result from these fluctuations. In the early loading response phase of stance, the lower limb must dissipate impact forces from accelerating the CoM into the ground anteriorly by losing energy through compliance at joints (motion), connective soft tissues stretching and vibration oscillations, and through muscle-tendon buffering. During single-limb support or midstance phase, the lower limb must achieve stability while permitting the HAT to pivot over it. Finally, effort must be created by muscle contraction and released from stretched connective tissues that result from interactions with the ground. Together, muscles and connective tissues stiffen the foot and ankle to form a stable acceleration platform. Release of energy from stretched connective tissues such as tendons, ligaments, and fascia allows the support limb to accelerate anteriorly throughout heel lift via the stiffened foot against the ground. However, it appears that the energy is not released by the connective tissues in one go but is dictated by where the loads are highest on the foot. Proximal rearfoot ligaments seem to release energy first as they start shortening after lengthening, before more distal structures start to shorten (Welte et al., 2021). However, the important plantarflexor power from the ankle plantarflexor tendons should only be applied to gait to induce and maintain heel lift acceleration. Thus, cycles of lower limb and foot compliance and stiffening periods are created (Fig. 1.1.5f).

18 Clinical biomechanics in human locomotion

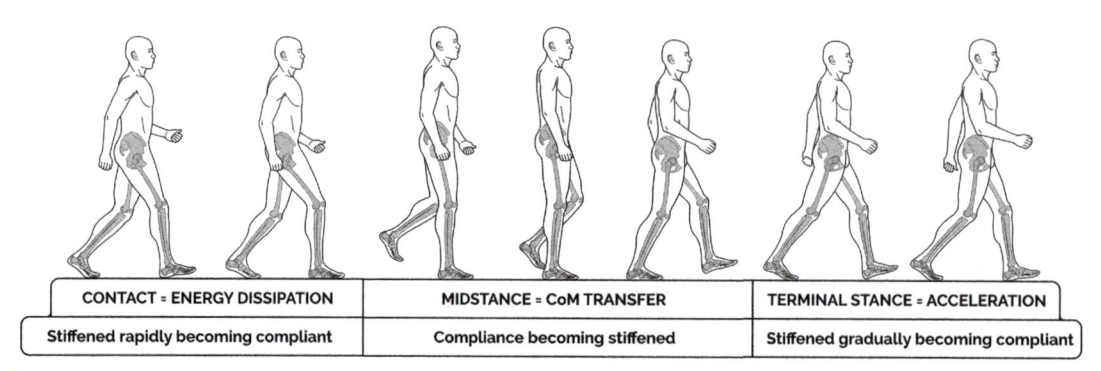

CONTACT = ENERGY DISSIPATION	MIDSTANCE = CoM TRANSFER	TERMINAL STANCE = ACCELERATION
Stiffened rapidly becoming compliant	Compliance becoming stiffened	Stiffened gradually becoming compliant

FIG. 1.1.5f Walking stance phase is broken down into three distinct actions over three phases. Contact phase involves deceleration, shock attenuation, and energy dissipation. Single-limb support or midstance involves weight transference of the body's CoM from a position posterior to the limb to anterior to it. Finally, terminal stance acceleration is required to move the loading forces from the trailing limb to the next step. Energy dissipation at impact requires limb and foot compliance, while stiffness is necessary for acceleration. The weight transference period involves increasing levels of stiffness while coming out of compliance, yet it is important to stress that all periods involve some flexibility to provide the ability to dissipate energy from overstressing tissues. *(Permission www.healthystep.co.uk.)*

The flexions and extensions achieved during stance are driven by the resultant force interactions (i.e. GRF) in relation to joint axes and are restrained and controlled primarily by energy-efficient eccentric muscle contractions. The thigh commences stance in hip flexion, which decreases until around the middle of the stance phase as the HAT's CoM passes more anteriorly and in front of the hip. Then, the hip extends until preswing, aided by gravity pulling the CoM forward, although the rate of hip extension slows during terminal stance. Muscles that resist hip flexion are only active until hip extension begins, with gravity taking over their hip extension role during late midstance. The knee, which should not ever reach full extension, starts stance in its most extended position (at around 4–8° flexed) and then flexes until the end of loading response. This allows the knee to act as a semi-compliant shock absorber through muscle-tendon buffering. Knee flexion reduces from the end of loading response into early single-limb support as the knee begins to actively extend in order to help the whole lower limb lengthen, so as to gain height for the CoM. The trunk's CoM increase in height gains gravitational potential energy (what goes up must come down). The knee continues to slightly extend at a slower rate during late midstance until the advent of heel lift, being prevented from extending by gastrocnemius activity. After heel lift, the knee flexes under the elastic recoil power of gastrocnemius, actually aiding heel lift. It then continues to flex into the terminal stance phase under increasing ankle plantarflexion forces, thereby aiding hip flexion into swing phase. These angular momentums generated in stance are easily understood when considered in relation to the location of the GRF (Fig. 1.1.5g).

A	B	C	D
HIP, KNEE, ANKLE FLEXION MOMENTS	REDUCING HIP, KNEE, ANKLE FLEXION MOMENTS	HIP, KNEE, ANKLE EXTENSION MOMENTS	ANKLE PLANTARFLEXION POWER [flexion] INDUCES KNEE AND HIP FLEXION MOMENTS

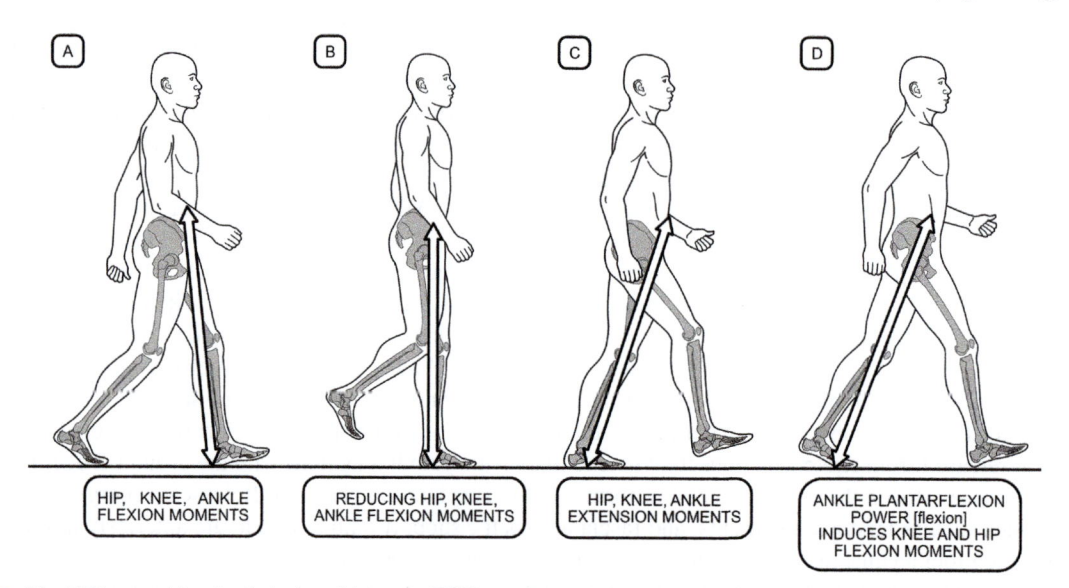

FIG. 1.1.5g The GRF vector driven by the body weight vector (BWV—represented together by the white double-headed arrow), particularly in the sagittal plane, plays a significant part in understanding the large joint motions of the lower limb during gait. In walking, the GRF at heel strike lies behind the ankle and knee, but in front of the hip (A), setting up hip, knee, and ankle flexion moments. These are resisted by hip, knee, and ankle extensor muscles. In early midstance (B), the GRF is moving closer to the joint axis, reducing its effect on joints through distance and velocity changes in angular momentum reducing flexion moments. At the beginning of late midstance (C), the GRF/BWV vector starts to lie behind the hip, but in front of the knee and ankle, creating increasing angular momentum that extends each joint. This angular momentum must be restrained by muscle action, primarily through triceps surae. At heel lift (D), the GRF/BWV vector is largely generated by ankle plantarflexor power from the triceps surae and Achilles elastic recoil, flexing the ankle. This causes the knee and hip to flex too. The hip flexion is accelerated then by iliopsoas at preswing to initiate lift of the limb into swing phase. *(Permission www.healthystep.co.uk.)*

In stance, the thigh moves more than the leg segment, but the whole foot does not actually translate its position until stance is complete, even though there is plenty of movement within the foot and around the ankle. With heel lift, the rearfoot and midfoot translate forward under ankle and midfoot plantarflexion, but the forefoot does not significantly start to translate anteriorly until just before toe-off as the metatarsals off-load. The rate of the different segments' translation is thus reversed from that of the swing phase. The mechanical role of stance therefore lies in moving the 'passenger' segment above the hip anteriorly, whereas swing does the same for the lower limb's CoM. The midstance phase operates like an inverted double-hinged pendulum to move the passenger CoM, whereas the swing limb operates as a double-hinged pendulum to move the lower limb's CoM and helps initiate acceleration of the passenger's CoM to the next step.

1.1.6 Challenges of upright posture in locomotion

Hominin evolution has taken measures to ensure that the CoM lies over the pelvis and lower limbs in bipedal stance and gait. This is, in part, an ontological process that occurs under the influence of the acquisition of bipedalism in infancy that helps set the sacral angles to the lumbar spine and between the ilia (Tardieu et al., 2013). Human gait sets up torques in all three orthogonal planes which permit limb and trunk motions that change with locomotion velocity. These all act as potentially destabilising events to an erect posture through altering the position of the CoM. Sagittal plane torques dominate, for this is the plane of direction of translation for the whole human body. However, frontal plane and transverse plane torques are also significant and occur from trunk to feet. Sometimes, these motions occur with *in-phase relationships* where segments move together in the same rotational directions with similar amounts of rotations. They can also occur *out of phase*, with rotations being very different in degrees and even in direction between segments. These relationships can also change with gait velocity. The upper trunk or thorax rotates together with the pelvis at slow-walking velocities, but this starts to change at around 2.8 km/h with the thorax and pelvis then rotating in opposite directions, with this difference increasing with velocity (Bruijn et al., 2008). Thus, the pelvis starts to couple its transverse plane motions with the leg as gait velocity increases but couples to the thorax at slow-walking pace (Figs 1.1.6a and 1.1.6b). Important as these frontal and transverse plane motions are during gait, the foundations of gait begin in the sagittal plane and it is therefore helpful to learn the sagittal plane influences on gait stability and motion first.

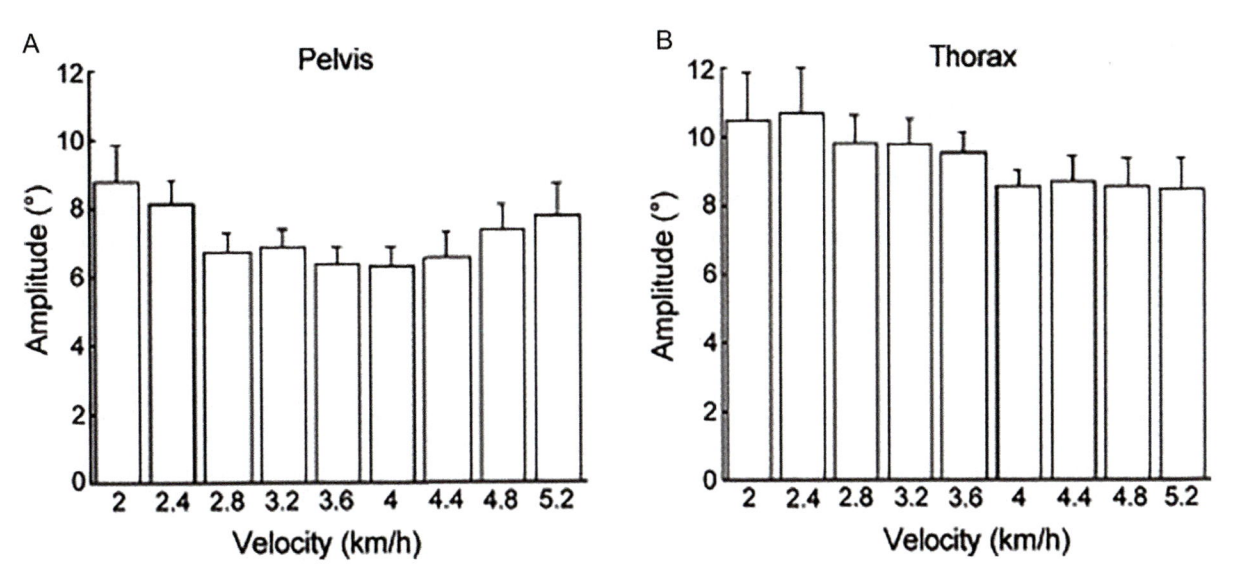

FIG. 1.1.6a Data from nine healthy young adult males, looking at the effects of gait speed on swing limb-induced thorax and pelvic rotations during walking to assess angular momentum of swing. Step length increases with gait velocity, from 0.35 m to 1.37 m. Pelvic rotational amplitude (A) decreases for velocities up to 3.6 km/h and increases again from 4.0 km/h onwards. Thoracic rotational amplitude (B) appears to decrease with increasing velocity, but the effect is not found to be significant. *(From Bruijn, S.M., Meijer, O.G., van Dieën, J.H., Kingma, I., Lamoth, C.J.C., 2008. Coordination of leg swing, thorax rotations, and pelvis rotations during gait: the organisation of total body angular momentum. Gait Posture 27 (3), 455–462.)*

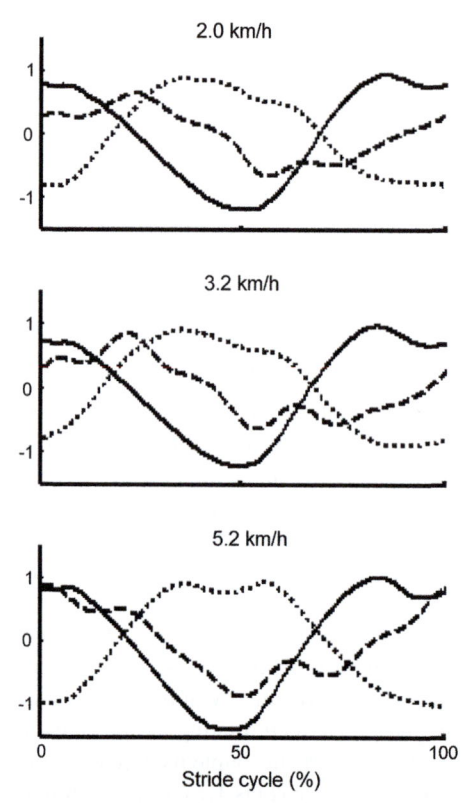

FIG. 1.1.6b Averaged normalised pelvis *(dashed line)* and thorax *(dotted line)* rotations, and forward positions of the right lateral femoral epicondyle *(solid line)* at three velocities of a young adult male. Positive values of thorax and pelvis are rotations to the left of the right side (counterclockwise), and positive values of the femur are forward motion. Although there is variability, patterns across similar subjects produced similar results. *(From Bruijn, S.M., Meijer, O.G., van Dieën, J.H., Kingma, I., Lamoth, C.J.C., 2008. Coordination of leg swing, thorax rotations, and pelvis rotations during gait: the organisation of total body angular momentum. Gait Posture 27 (3), 455–462.)*

In the sagittal plane, human bipedal walking gait torques can be viewed as a vaulting event with the CoM passing over a relatively stiff supporting lower limb acting as an inverted pendulum, while the swing limb acts as a second true pendulum, dragging the CoM of the trunk along behind it as the pendulum swings forward. Walk too fast or overstride, and gravity will fail to keep the support limb fixed stably to the ground. When the mechanics of humans is compared to nonspecialist terrestrial bipedal species such as ducks, the capacity to drive the human swing limb pendulum seems to be a more specialised feature than the support limb (Usherwood et al., 2008). Ducks, including the more terrestrially specialised Indian runner ducks that have a higher CoM posture than other ducks, use a passive swing limb, unable to assist the CoM progression through centrifugal forces (Fig. 1.1.6c). Humans, with increasing locomotive speed, increase their swing limb activity, driving the pendular swing well above the natural pendular frequency of the walking speed, indicating a highly active swing limb that adds momentum to the CoM (Usherwood et al., 2008).

FIG. 1.1.6c The Indian runner duck is a terrestrial specialist compared to other ducks. It has developed a more vertical posture moving its CoM up higher and positioning it closer to lie over its 'webbed' base of support. These are changes perfected in humans, but not in runner ducks. Also, as runner ducks lack long, heavy legs that can be accelerated under hip flexor and knee extensor power passing through large arcs of motion, the duck's segmental limb mass cannot be utilised to add momentum to the anterior progression of the CoM. It appears that human swing phase is a novel energetic addition to bipedal gait. This is a result of relatively large lower leg muscle mass (plantarflexors of the calf) being positioned low down on long legs.

The torques generated within the body and lower limb between segments during gait present a mechanical challenge to maintaining an upright posture within the HAT segment above. This is because of the high human CoM position compared to the small base of support. Walking therefore involves the active management of a large number of articular torques, a process requiring precise neural and mechanical control. A key variable to providing appropriate neural control is the location of the GRF vector which is responsible for the changes in angular momentum force around the lower limb joints in relation to the CoM's position within the trunk. In the sagittal plane, the centre of the GRF (CoGRF) is applied initially in front of the CoM during initial foot contact to then move relatively posteriorly throughout the early gait cycle, eventually passing under the CoM during single-leg support and finally behind the CoM as the trunk passes over the foot when the stance period of gait starts to terminate. The alignment of these GRF vectors, generated over time, crosses each other at a certain point that has been designated the *divergent point* (Gruben and Boehm, 2012a; Vielemeyer et al., 2019). The divergent point sits higher within the trunk than the CoM. This means that the CoM of the trunk does not align directly with the GRF, except at the middle of stance phase during midstance when the GRF is directed vertically up into the CoM (Fig. 1.1.6d).

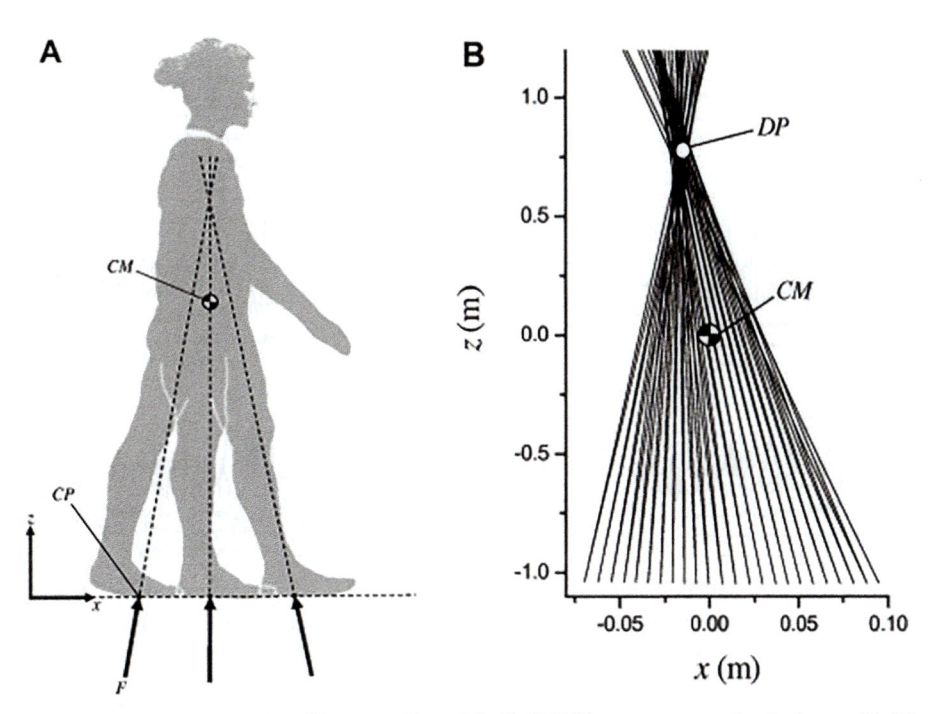

FIG. 1.1.6d The divergent point (DP) is the location of the vector lines of the CoGRF (*F arrows* representing the force, with *CP* representing the GRF location) during the gait cycle (A). The CoGRF vector lines pass above the CoM (*CM* on diagram), only crossing it during the middle of the midstance phase (A and B), creating force vectors pushing posteriorly in early stance and then anteriorly in late stance. This helps to stabilise the trunk around the lower limb torques of locomotion. *(From Gruben, K.G., Boehm, W.L., 2012. Force direction pattern stabilises sagittal plane mechanics of human walking. Hum. Movement Sci. 31 (3), 649–659.)*

By having the divergent point of GRF vectors directed higher than the CoM helps to create a longer moment arm torque that assists in preventing the trunk from tipping forward on initial contact of the foot with the ground. This sudden impact tends to drive the trunk forward into hip flexion. The higher divergent point blocks this. A longer moment arm also prevents the trunk tipping backwards (extending on the hip) when propelling forward on the trailing foot during terminal stance (Gruben and Boehm, 2012a). Thus, the orientation of GRF generation during gait becomes a stabilising mechanism of the trunk (HAT segment), reducing the neuromuscular demand and muscle activity requirements of the body.

Maintaining erect posture in stance and gait seems to involve a mixture of joint strategies. In testing the stance posture balance of normal subjects on a flat support surface subjected to backward ramp displacements, these mixed strategies were recorded and found to relate to the velocity of displacement (Runge et al., 1999). When the velocity of postural perturbations is variably raised across subjects, increased hip torques are always present with ankle torques occurring during greater unevenness of the support surface (Runge et al., 1999). However, the maintenance of erect posture and balance during perturbations that occur during locomotion involve muscles from the neck to the leg, with both hip and ankle accelerations controlling posture (Runge et al., 1999) (Fig. 1.1.6e). Usually, combinations of trunk, hip, and ankle muscles maintain postural stability, with hip strategy dominating, but individuals may start to favour one group of muscles over another with age or disease, which means co-ordination around joint stability may become disturbed. It is likely that foot activity is also frequently at play in postural control, not least through its generation of sensory information but also through muscular activity and the placement of the GRF vector (Gruben and Boehm, 2012b).

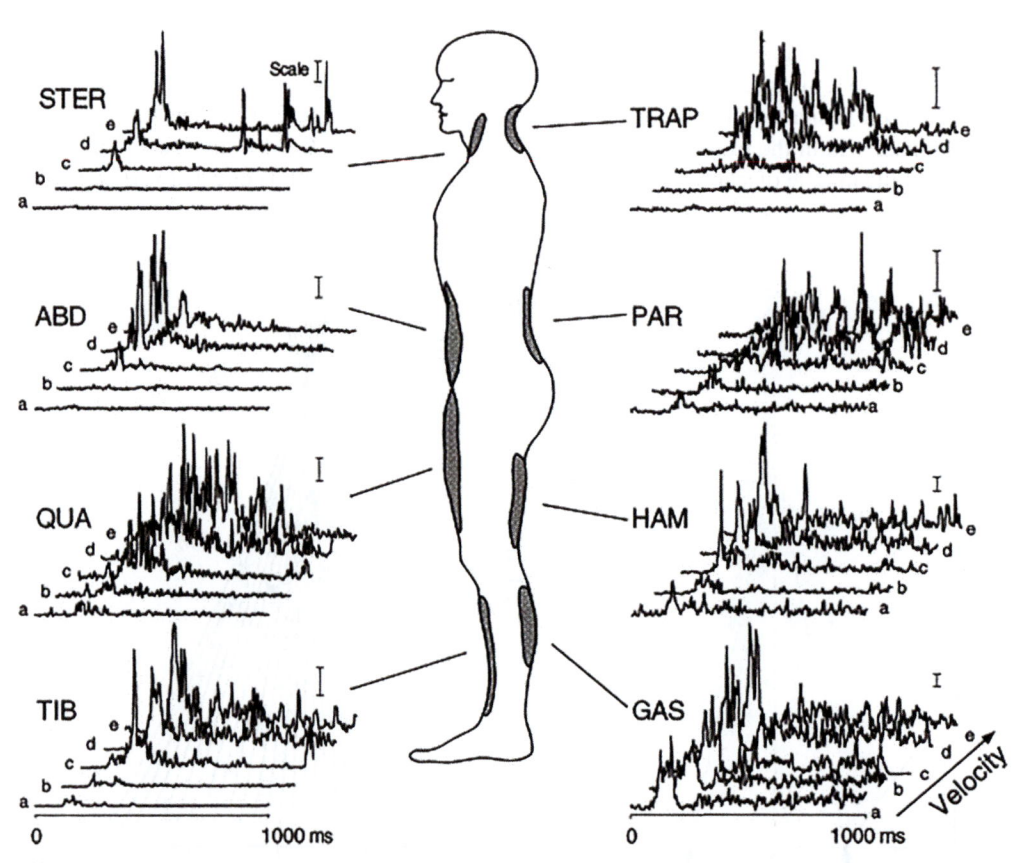

FIG. 1.1.6e EMG activity of postural stability muscles taken from a subject exposed to backward translations on a platform they were stood upon under five changing velocities. These velocities were (a) 15, (b) 20, (c) 25, (d) 32, and (e) 40 cm/s. At lower velocities (a, b), muscle activity on the posterior aspect of the body is from gastrocnemius, hamstrings, and parasinal muscles (GAS, HAM, PAR) and starts with gastrocnemius. At higher velocities of posterior displacement (c, d, e), activity in the upper body muscles on the anterior aspect come into play (rectus abdominis (ABD), quadriceps (QUA), trapezius (TRAP), and sternocleidomastoid (STER)), with rectus abdominis and gastrocnemius the earliest to respond. Such information gives insight into how the body can maintain postural balance under perturbations during gait and activities, with perturbation size altering the location and extent of muscular response. *(From Runge, C.F., Shupert, C.L., Horak, F.B., Zajac, F.E., 1999. Ankle and hip postural strategies defined by joint torques. Gait Posture 10 (2), 161–170.)*

1.1.7 Human gait models

Gait models help to explain the fact that human walking is energetically very efficient, whereas running is not. The modelling of gait as collision events that redirect vertical velocity components from downwards at the beginning of stance to upwards at the end of stance, while at the same time also turning this latter event into an acceleration forward, is extremely important. Running is explained through the legs acting as elastic springs. Walking is explained as sequencing leg forces during a redirection phase of push off, followed by a new heel strike (Ruina et al., 2005). Such modelling explains why two sequenced collisions per given stride length and velocity are usually more energy efficient than is a single collision of the lower limb, as seen in forefoot running. It also serves to explain why pseudo-elastic collisions are more effective than plastic-like collisions, where absorbed energy must then be made up for with muscle action (Ruina et al., 2005). Thus, walking gait requiring a double support phase using sequenced collisions that help power the subsequent collision helps explain why at slower speeds, walking is more efficient than running. Other models than the one discussed under gait energetics (Section 1.1.2) can be used to assist in understanding how walking and running differ in energetics related to velocity of locomotion, giving clinicians more insight into human locomotive techniques.

Human inverted pendular walking gait model

A fundamental principle of human walking gait is the ability to maintain a high CoM over a narrow base of support consisting of one (single stance) or both limbs (double stance) positioned under the body, using each limb in turn to allow a rise and fall of the CoM to maximise energetics. Movement of the CoM is an important descriptor of pathological gait, with CoM displacement to the pelvis being helpful in understanding gait symmetry. Recording lateral and possibly fore and aft deviations throughout gait is easier than vertical deviations, and it must be remembered that all motions relate in part to the patent's height (Gutierrez-Farewik et al., 2006; Carpentier et al., 2017). GRF vectors being directed above the CoM help in stabilising the trunk's CoM around anterior–posterior-directed trunk torques via the divergent point of the GRF vectors (Gruben and Boehm, 2012a). However, the CoM still needs to rise and fall in order to pass over the stance limb during walking. This rising and falling of the CoM during walking occurs through a process similar to pivoting a pole over a pivot point, otherwise referred to as vaulting (like a pole vault), which allows cycles of kinetic energy and potential energy to be used. This interchange of kinetic and potential energy from the CoM vaulting over the foot uses minimal muscle activity.

The CoM is allowed to fall forwards under gravitational pull, creating kinetic energy, only to be stopped from continuing to fall by the opposite limb's contact. This is followed by a process of raising the CoM back upwards (higher) again on the newly contacting limb and then allowing it to repeat the fall forward to the original limb's initial contact. These inverted pendular motions equate to the braking phase of initial heel contact and forefoot loading, the initial CoM raising pivot of early midstance, and finally the late midstance fall forward, pivoting at the ankle to complete the vault. This is then followed by ankle power-generated heel lift and, finally, foot-off to complete the step after the pivot over the foot is completed. This powered heel lift initiates the same process in the opposite foot. Thus, a sequence is created of a pole vault from one limb leading immediately to the next limb vault during walking (Fig. 1.1.7a).

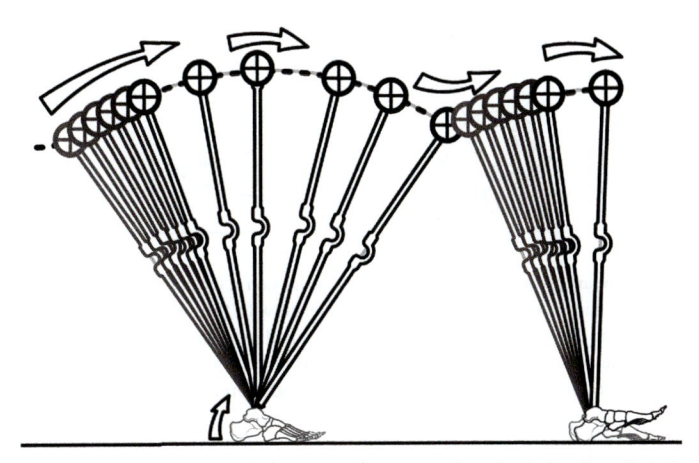

FIG. 1.1.7a The double inverted pendulum that sees rises and falls in the CoM *(crossed circle)* during the gait cycle, with the CoM being pitched up and over the lower limb, functioning like a pole rotating at the ankle. However, to initiate the next cycle of vaulting over the pole, something needs to tip the CoM onto the next 'pole-like' lower limb. Ankle plantarflexion-powered heel lift and swing limb centrifugal forces achieve this. *(Permission www.healthystep.co.uk.)*

The raising of the CoM in early midstance is initiated through contraction of the hip and knee musculature to extend the lower limb, creating a rising effect on a stiffening and extending limb, thus gaining height as the CoM vaults upwards. The momentum from the opposite swing phase limb will help drag the CoM of the body behind it after midswing, tipping the CoM forward from its point of maximum height as it moves into late swing and swing phase knee extension. The result is to regain potential energy from increased CoM height ready for the next fall under gravity instigated by the momentum of the swing limb. The muscular energetic costs of raising the CoM in the early parts of stance phase is significant (Neptune et al., 2004a), but the CoM fall is largely free kinetic energy.

It is this cyclical generation of potential and kinetic energy around a primary axis (in the ankle joint) that gives the impression of an inverted pendulum rising and falling like the pendulum of an old wind-up clock, with the spring impersonating the energy from muscular contraction and connective tissue elastic recoil. In human gait, the weight of the pendulum is the HAT CoM above the hip, so there is a secondary pivot point occurring at the hip as well as another pivot point at the knee (Fig. 1.1.7b). These hip and knee rotations are normally relatively small compared to the ankle, but they are significant for smoothing CoM oscillations and overall CoM accelerations, making the pendulum a two-sectioned rod with three pivot points. The swing limb through its own three-pivot pendulum also influences the stance limb's inverted pendulum by adding momentum to its fall forward, although it is an additional weight to the initial rise of the CoM.

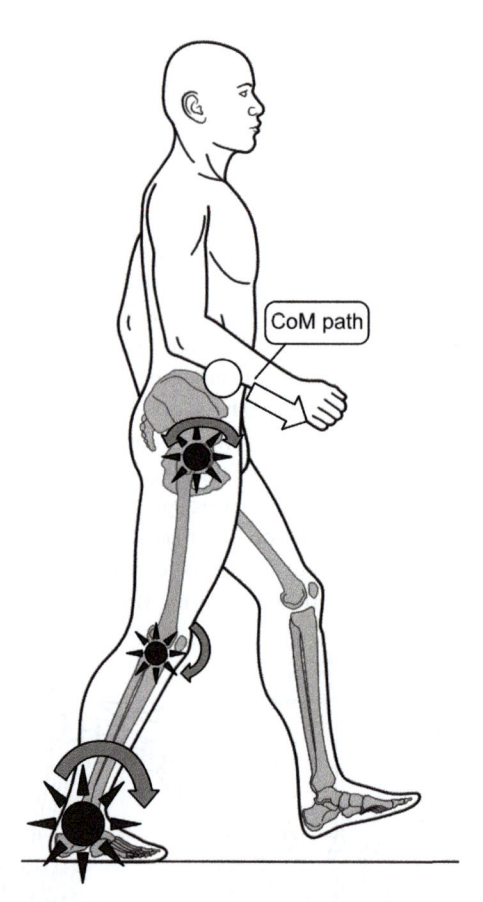

FIG. 1.1.7b The inverted pendulum has a primary pivot axis point of sagittal plane motion at the ankle. The ankle is the primary mover of the CoM through rotations of dorsiflexion (ankle strategy). The hip joint also has significant influence with flexion and extension adjusting the path of the CoM during midstance (hip strategy). The knee should offer the smallest influence using only small extension and flexion moments to help fine-tune the path of the CoM. However, knee extension in early midstance is very important for gaining CoM height for the inverted pendular fall of late midstance. Should motion become excessive or limited at any of these joints, normal inverted vault mechanics and energetics will be disrupted. *(Permission www.healthystep.co.uk.)*

What are the potential benefits of this rather peculiar gait style? Up-and-down motions of the CoM are not unique to humans. Indeed, most erect-postured animals demonstrate it, especially at higher speeds. Modelling bipedal walking as an inverted pendulum has been successfully predictive in accounting for the relative maximum walking speeds in both ducks and humans (Usherwood et al., 2008). The inverted pendulum model during the stance (vaulting) phase of gait has been proposed since the advent of force plate measurements as being both a descriptive and energetic method of explaining the forward motion of human gait (Usherwood et al., 2012).

Inverted pendulum-like gait, efficiently moving the CoM of the body, may be obtained with varying motions of individual body segments. Moreover, it has been reported that in populations of women who carry head-supported loads of up to 20% of their body weight, they do not increase their external mechanical work while walking so loaded (Heglund et al., 1995). This suggests that the position of the CoM, not its size, is important, for heads so loaded still have their CoM within their normal base of support (Fig. 1.1.7c). Furthermore, low-level movement impairment, as a consequence of simple orthopaedic pathologies, does not necessarily seem to cause significant disturbance of the mechanical recovery of energy during walking gait in moving the body's CoM, suggesting that walking gait can easily adapt to minor issues (Detrembleur et al., 2000). However, central nervous system pathology impacts greater disturbance to the energetics of the pendulum gait (Detrembleur et al., 2000). It would therefore seem that walking gait is adaptable to increased loads and minor locomotive changes, but not if neurological control is disturbed.

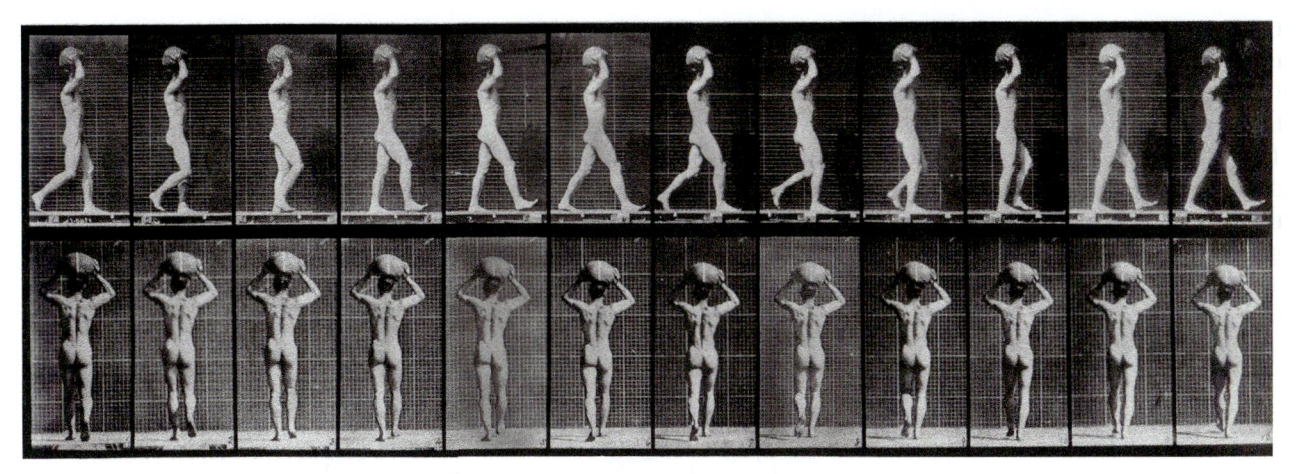

FIG. 1.1.7c The position of the CoM over the supporting limbs seems to be more important than its size during bipedal hominin gait. If a weight is carried in the vertical line above the normal CoM, centred above the pelvis, it does not increase the external mechanical work required for gait. In these images from the Eadweard Muybridge collection, kinematics of the lower limbs appears identical to those seen normally, even with the loss of upper limb counter-swings, during gait.

Nashner and McCollum (1985) proposed two discrete strategies around inverted pendulum motions which could be used separately, or combined through nervous system control, to produce an adaptable regulation of CoM motion. This involved the *'ankle strategy'* that moves the whole body's CoM as a single inverted pendulum through an ankle fulcrum torque, and a *'hip strategy'* that moves the body by a double-segment inverted pendulum with counterphase motion at the hip and ankle (see Fig. 1.1.7e). It was suggested that the hip double-segment system would only be used when body motion at the ankle occurred on compliant surfaces or on a limited support surface which compromised the ankle strategy torque and perhaps also with increased velocity of perturbations in posture (Runge et al., 1999). Slow velocity perturbations in posture are characterised by ankle activity and activity in the gastrocnemius, hamstrings, and paraspinal muscles, with little knee or hip angle changes, although this appears to be variably so between individuals. However, even when slow walking, hip torques and knee and hip motion are not zero, suggesting that to a lesser extent, a mixed strategy of inverted pendulum motion is always utilised in gait and posture maintenance (Runge et al., 1999). Even when the pendulum is primarily moved at the ankle, the upper body segments need to be stabilised as many lower limb muscles act across more than one joint. Hip torques are accompanied by bursts of activity in rectus abdominis with increasing velocity, suggesting that upper body motion is used to restore equilibrium in HAT CoM motion rather than being a true 'passive passenger' during walking gait (Runge et al., 1999).

The faster walking becomes or the less predictable the support surface, the more that hip torque and hip and knee flexion or extension adjustments play a role. When a less hip activity strategy is adopted, the destabilising forces of gravity move the CoM anteriorly throughout the pendulum's fall, and this is countered using plantarflexion ankle muscle-powered restraint against ankle dorsiflexion. The plantarflexor muscle torque pulls the leg backwards, slowing anterior progression of the tibia. In so doing, it drives the trunk to receive a flexion moment at the hip through tipping the trunk forward, unless opposed. This requires proximal hip and trunk muscle activity in order to keep the trunk stable and erect at the hip (Runge et al., 1999). The primary plantarflexion torque is provided by the triceps surae through the Achilles tendon. However, the stabilising torque generated is quite limited compared to the moment of inertia generated by the fall of the CoM about the ankle, else the heel would tend to lift from the support surface before the vaulting cycle was complete. This is a common issue seen clinically that can result from plantarflexor tightness or overactivity.

Despite this, relatively large torques are still required to produce small corrections of the CoM trajectory, even when using the ankle strategy alone (Runge et al., 1999; Umberger, 2010). It has been reported that only in slow translation of the CoM are ankle strategies with little hip torque used in providing enough stability to maintain upright alignment, and that mixed hip-ankle strategies are the more usual (Runge et al., 1999). The magnitude of the hip torques is normally much smaller than the ankle torques in the mixed strategy. Furthermore, the hip torque works with gravity, not against it, as occurs at the ankle.

The result is effectively to divide the body into two, creating a double-segment inverted pendulum. The hip flexors propelling the trunk forward initiate forward rotation of the upper body, while sternocleidomastoid neck muscle activity prevents passive whiplash of the head backwards as the body climbs to the top of the vault, loading potential energy. Gluteus maximus lifts the CoM and extends the hip out of its flexed position during this period. Quadriceps activity assists by straightening the knee, which allows the leg below the hip and above the ankle to mainly move as a single segment. From the CoM high point, the vault over the foot is completed by gravity providing the downward vector, because the trunk leans forward and the natural fall rotates the hip into extension (without hip extensor muscle action), bringing the body's CoM back into equilibrium sagittally as it falls. However, this is only if the whole lower limb can rotate backwards (Runge et al., 1999). Therefore, the inverted pendulum has two important angles to control: that of the ankle to the leg and the other at the thigh via the hip to the trunk. The inverted pendulum action is completed with heel lift. The moment of inertia of the CoM of the body decreases as the trunk falls downward and forward towards the next foot contact. This allows the ankle torque to generate high plantarflexion power and acceleration by stopping eccentric contraction of the ankle plantarflexors to initiate heel lift through the elastic recoil of the Achilles (possibly aided to a small degree through other plantarflexor tendons) (Fig. 1.1.7d).

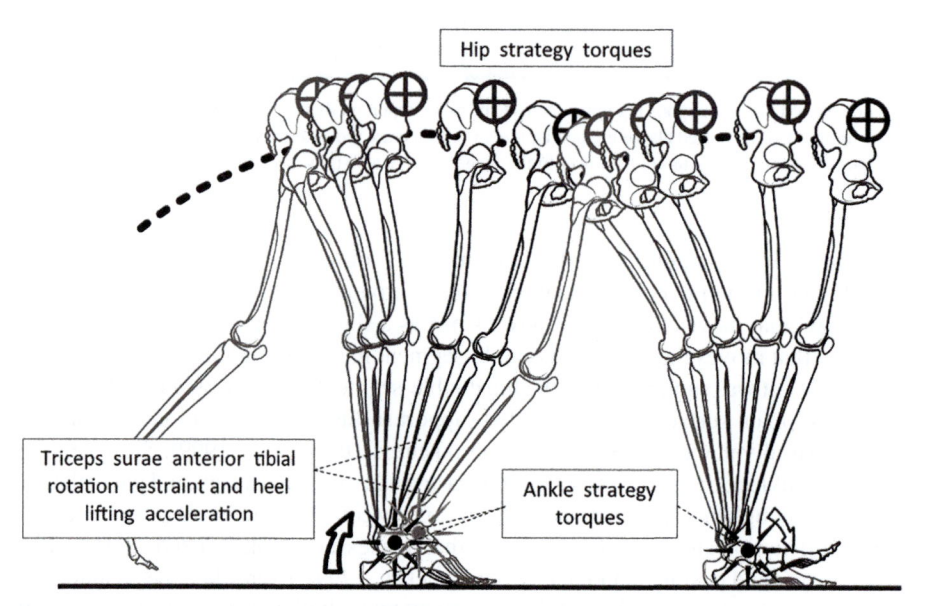

FIG. 1.1.7d Cycles of vaulting through the inverted pendulum concept are brought to a close by heel lift, induced by increasing plantarflexor power. This plantarflexion power-induced heel lift is a result of the inverted pendulum being slowed by the triceps surae braking anterior tibial rotation during the second part of vaulting, as the CoM accelerates forward under gravity. These are motions using ankle strategy to move body mass forward. The trunk adjusts its posture via hip strategy, but hip motion and power should not lead the accelerations of the CoM of the HAT segment. As the CoM moves anteriorly, GRF on the heel reduces while tensional forces within the Achilles rise. A point is reached where the Achilles forces are greater than the GRF maintaining heel contact. Once that point is reached, the heel 'pops-up' and the Achilles plantarflexion recoil power is released at the ankle onto the foot. This not only ends the vaulting on that limb, but the power adds momentum to the CoM, helping to initiate the next vaulting cycle. Thus, ankle and hip strategies are utilised, but ankle torques should dominate. *(Permission www.healthystep.co.uk.)*

Energetics of inverted pendular gait

The mechanical work and energy used to achieve step-to-step transition is a major determinant of the metabolic cost of walking (Donelan et al., 2002). Theoretical collisional analysis demonstrates that some means of powering impulses (forces changing momentum) are less costly than others (Kuo, 2002; Ruina et al., 2005). These models suggest stiff-limbed vaulting combined with powering and energy-dissipative impulses, predicted from a purely collisional approach, are energetically optimal at walking speeds (Srinivasan and Ruina, 2006). These impulses are applied to humans predominantly through impact-generated ankle plantarflexions opposed by the anterior leg muscles (tibialis anterior primarily) at the beginning of stance phase and then powered by the calf muscles' plantarflexion power at the end of stance phase (Usherwood et al., 2012). These actions greatly contribute to the generation of vertical GRF and are consistent with theoretically economical gaits (Anderson and Pandy, 2003; Liu et al., 2006; Usherwood et al., 2012).

However, these models do not explain the plantigrade foot posture alone or why the ankle needs to extend (dorsiflex) during the midstance phase to maintain a plantigrade foot position for so long during stance, as it is neither a theoretical requirement nor the only way to produce the classical M-shaped vertical GRF time curve during gait. Such M-shaped curves are also seen in ostrich walking gaits, human tip-toe walking, and high-heeled footwear walking (Usherwood et al., 2012), see Fig. 1.1.7e.

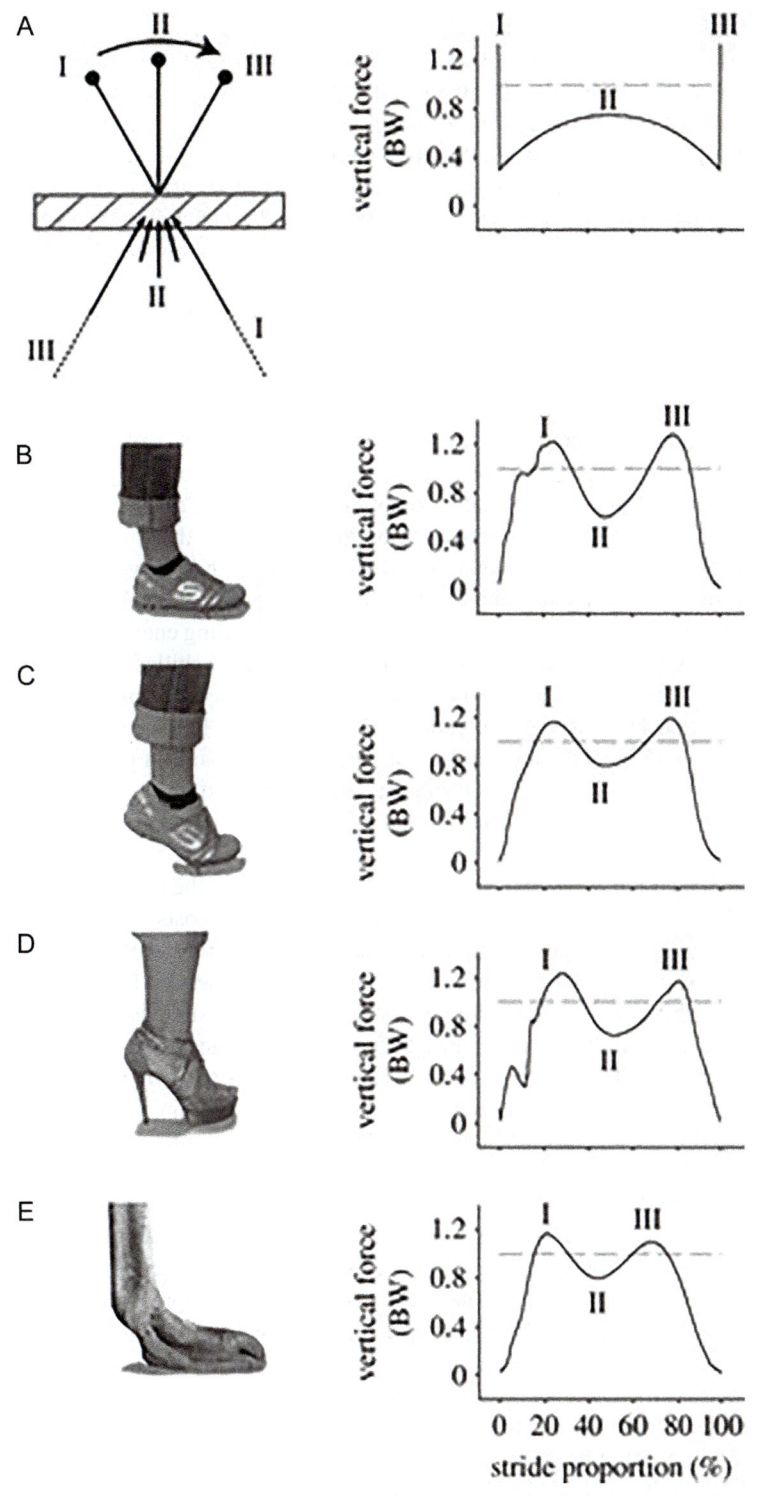

FIG. 1.1.7e Vertical force recordings on force-time curves in human walking demonstrate an 'M'-shaped double peak. This is generated by initial contact impact peaks (A) vector I, the midstance vaulting that elevates and accelerates the CoM away from the ground, reducing vertical forces, seen in vector II, and finally the increased forces of acceleration seen in vector III. Plantigrade, flat shoe gait demonstrates the peaks and troughs associated with these vectors (B). However, toe-walking produces a similar pattern, as does walking in high heels, although differences in troughs to peaks tend to be lower (C and D). Even the digitigrade biped, the ostrich, demonstrates similar vertical vectors when walking (E). However, only plantigrade bipedalism can use large ranges of energy-efficient eccentric muscle contractions. *(From Usherwood, J.R., Channon, A.J., Myatt, J.P., Rankin, J.W., Hubel, T.Y., 2012. The human foot and heel-sole-toe walking strategy: a mechanism enabling an inverted pendular gait with low isometric muscle force? J. Roy. Soc. Interface 9 (75), 2396–2402.)*

The plantigrade foot in inverted pendular gait

To understand the plantigrade foot posture or the need for the foot to be prone (or pronated) during the stance phase of gait, it is necessary to consider the effects of muscle action and the need for the foot to offer compliance and sufficient surface contact during loading and midstance. The human foot has a heel jutting out behind the ankle and a long midfoot and forefoot with shortened toes lying in front of the ankle. The foot has developed a heel-to-toe walking pattern that requires both energetically optimal powering and a support strategy of vaulting. This strategy involves impulsive inverted pendular walking and costly force generation from muscles which can be reduced by changing the mechanical advantage between the muscle action and the GRF, so as to utilise eccentric rather than concentric activity. This allows relatively small muscle forces to interact with the GRF, giving humans high muscular mechanical advantage (Biewener et al., 2004; Cunningham et al., 2010). Low metabolic cost, high force eccentric muscle contraction during late midstance is therefore the benefit of having pronated supporting feet.

Isometric activity might seem to have low energetic costs, but if you stand with your knees flexed for any time, you will find it is hard work. In addition, the behaviour of cartilage also has to be factored in. Cartilage is a biological material that quickly develops deficient mechanical properties under fixed loads. The cost of walking with limited motion and flexed hips and knees in a chimpanzee-like manner increases the energy cost of locomotion by 50%, making limb posture highly relevant to the metabolic costs of hominid terrestrial locomotion (Carey and Crompton, 2005; Pontzer et al., 2009a).

Although it must be appreciated that the foot moves smoothly throughout the stance phase of gait, there are distinct periods of the foot fulfilling specific roles. In considering the inverted pendulum-type action, the stance phase initiates with a loss of mechanical energy from the fall of the CoM until loading response is complete. In other bipedal species that hop or otherwise land on the toes (including forefoot human running), the loading energy from the fall of the CoM is rapidly stored in structures, such as the Achilles tendon, as elastic energy. The human Achilles is not significantly loaded in walking heel strike contact phase. The energy must therefore be dissipated through mechanisms such as the viscoelastic properties of the heel's and forefoot's plantar fat pads, muscle-tendon buffering via hip, knee, and ankle extensors, tissue resonance, tibialis anterior plantarflexion braking, and via controlled proximal joint flexions at the hip and knee. The human tibialis anterior has a greater volume than its counterpart in other apes (Usherwood et al., 2012). Studies suggest that tibialis anterior acts primarily as an energy absorber, contributing negatively to the vertical impulse (Neptune et al., 2008; McGowan et al., 2010).

Following initial energy dissipation upon loading the inverted pendulum of the limb onto the ground, the vaulting phase of moving the CoM over the foot begins and involves significant energetic costs (Neptune et al., 2004a). The inverted pendulum process commences after forefoot loading (another energy dissipating process) and now requires energy input to drive the arm of the inverted pendulum upwards so as to give the CoM height. However, once reached, the walking vaulting mechanism is remarkably effective at achieving a low metabolic cost. This mechanical efficiency accounts for the top walking speeds of humans and the reduced vertical forces during midstance, due in part to the centrifugal force of the mass moving in the arc of radius of the leg length (Usherwood et al., 2008). Significant work is done in muscles during the vaulting period of the gait cycle (Umberger, 2010), possibly to maintain stability in all body planes of the trunk and lower limbs, but also to prevent excessive knee flexion during early stance phase (Fig. 1.1.7f) or knee hyperextension after absolute midstance (Fig. 1.1.7g).

The foot remains plantigrade and prone to the support surface during the vaulting phase, allowing the GRF to lie close to the ankle joint axis. The external length of the GRF angular momentum arm to the ankle joint axis of rotation moves forward, decreasing in length to zero, applying decreasing angular momentum forces. By being plantigrade in early midstance, the foot is able to act optimally energetically, neither having to dissipate nor generate high energy. This removes the requirement for local foot muscles to be loaded extensively throughout this period. The consequence is that the foot tends towards an increasingly prone profile throughout early midstance as it is compressed between the body mass above and the GRFs below, without much active muscle resistance, but with gradually increasing passive resistance from rearfoot plantar ligaments (short and long plantar ligaments and the spring ligament). Only in later midstance is heightened muscle activity for increasing foot vault and forefoot stiffness required.

As the limb vaults over the foot, the CoM moves anteriorly and eventually is situated in front of the ankle joint. The posterior calf muscles are now required to initiate a degree of negative work in order to slow the fall of the CoM and support the body, ensuring its 'controlled' forward progression (Neptune et al., 2001). This causes the gastrocnemius and soleus of the calf to start to generate considerable and important vertical and horizontal components of GRF (Neptune et al., 2001). This requirement on the ankle plantarflexors increases as the dorsiflexion angular moment arm lengthens to producing around 12°–22° of dorsiflexion during stance (Weir and Chockalingam, 2007). This dorsiflexion moment is resisted and controlled by eccentric calf muscle activity, allowing time for the swing limb to progress to a position anterior to the HAT segment, becoming situated so as to act as a stable support to bear load at the next step. This anterior vaulting process loads the energy-storing Achilles tendon with potential energy which it can then elastically release at heel lift, creating significant ankle power (Figs 1.1.7g and 1.1.7h). Moreover, the stance foot will benefit if it can passively stiffen

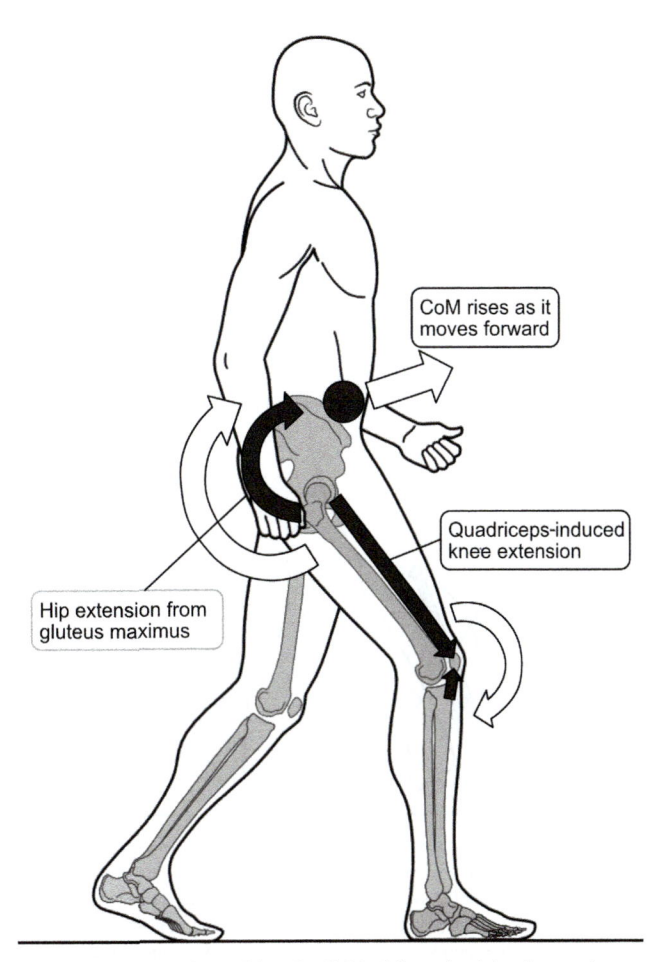

FIG. 1.1.7f At the start of early midstance, vaulting involves raising the CoM while maintaining its anterior progression. This is achieved through hip extension, reducing flexion angles primarily through gluteus maximus and knee extension via the quadriceps muscles reducing the knee flexion angle. The foot is still compliant at this period following its activity in energy dissipation at forefoot loading. Thus, the foot is lengthening and widening to increase its prone posturing on the support surface. As the limb enters single support, the foot is increasing its surface contact area, reducing peak pressure, a process that continues at a decreasing rate throughout the vaulting phase. *(Permission www.healthystep.co.uk.)*

during this late midstance period. As the foot becomes more prone and lengthens (Stolwijk et al., 2014), it becomes stiffer (Bjelopetrovich and Barrios, 2016; Takabayashi et al., 2020) through strain-stiffening properties as would be expected from a structure composed of viscoelastic materials. This also means that elastic energy is stored within the soft connective tissues of the foot as it stiffens. As the CoM vaults to a position over the forefoot, the plantar intrinsic and extrinsic muscles become increasingly active in foot and ankle stiffening (Ferris et al., 1995; Kokubo et al., 2012; Farris et al., 2019, 2020), assisting the multiple passive stiffening structural mechanisms of the foot such as the transverse forefoot vault (Venkadesan et al., 2020).

Heel lift energetics

As the foot moves into the final stages of its vault over the ground, generation of a positive impulse and of positive work is required to power the heel off the ground. The foot is now being loaded with a GRF moving towards the MTP joints anterior to the ankle. This creates an external angular moment arm stabilised through the stiffened midfoot that allows the foot to become an almost horizontal beam lying behind the MTP joint fulcrum. The triceps surae is thought to develop (near-) isometric contraction just prior to heel lift, creating increased elastic stretch within the Achilles tendon (Maganaris and Paul, 2002). This provides the opportunity for highly efficient power generation from the triceps surae (Lichtwark and Watson, 2007; Umberger and Rubenson, 2011).

There is a limited amount of energy that can be stored within the Achilles tendon during inverted pendular ankle extension (dorsiflexion), as excess elastic energy could cause premature ankle plantarflexion and early heel lift, creating an energetically costly deviation from the ideal pendular vaulting gait. Inverted pendular mechanics also predicts that the energy stored within the Achilles will reduce with gait speed and step length (Neptune et al., 2004a). The inverted pendulum model also predicts that at top walking speeds, the GRF available to generate Achilles tendon loads falls. This is consistent

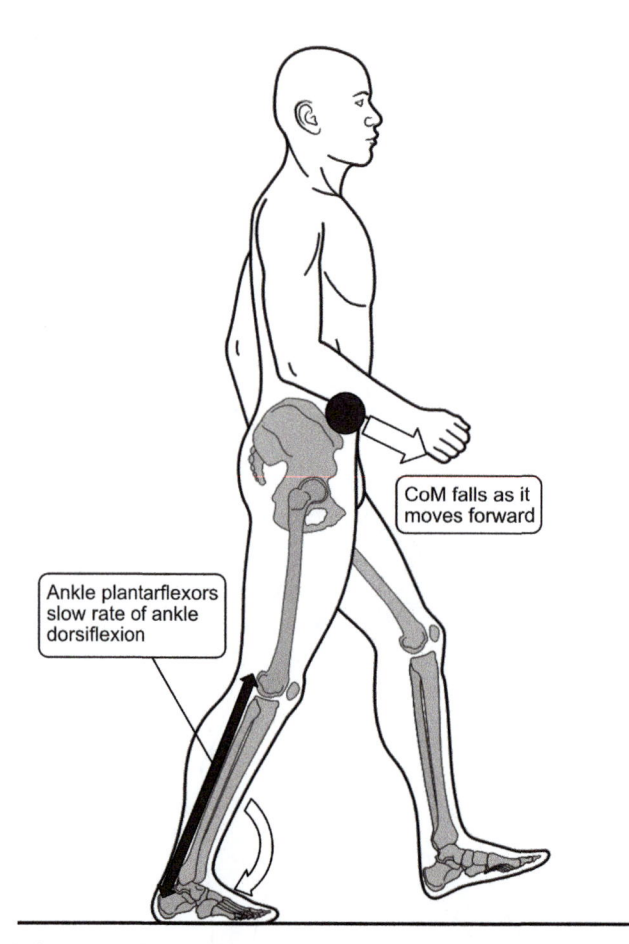

FIG. 1.1.7g During late midstance, the inverted pendulum vaulting action is under the influence of gravity, requiring muscle restraint on the anterior fall of the tibia and lower limb. Eccentric triceps surae action (probably with some minor assistance from the other ankle plantarflexors) moderates the rate of ankle dorsiflexion, prevents knee hyperextension, and gives time for the swing limb to increase its anterior progression ready for weight acceptance. Ankle dorsiflexion also drives a deflection sagging moment into the foot helping to lengthen and widen the foot, stiffening its connective tissue components. Reducing rearfoot loading through the heel will reach a point where the energy stored within the Achilles will cause the heel to lift, hopefully at a moment when the swing limb is ready for ground contact and the foot is adequately stiffened through passive and active components. *(Permission www.healthystep.co.uk.)*

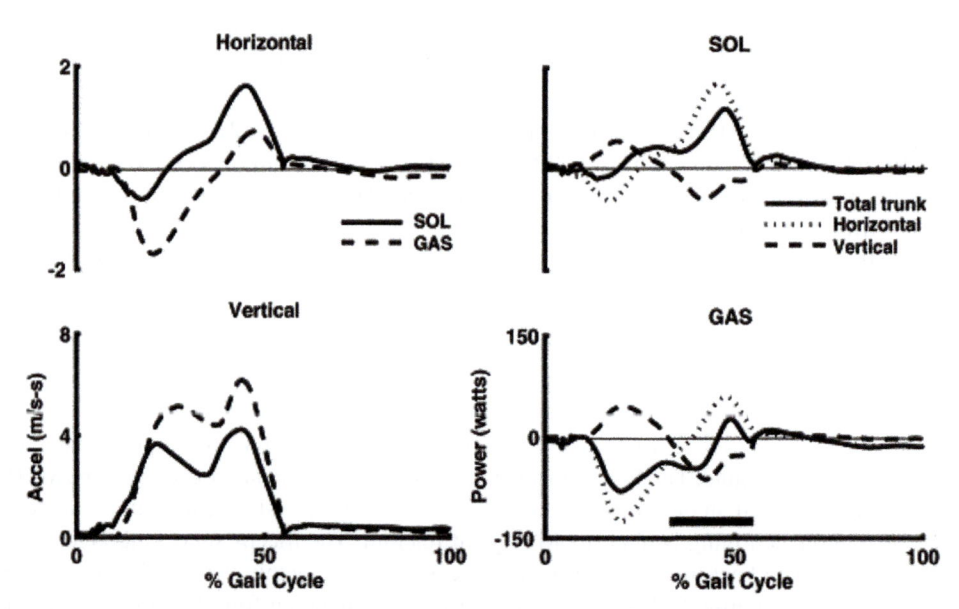

FIG. 1.1.7h Left graphs show soleus and gastrocnemius-induced accelerations of the trunk modelled in the horizontal (positive = forward) and vertical (positive = upward) directions during a gait cycle. On the right, the mechanical power on the trunk from soleus and gastrocnemius is modelled. Horizontal power is the rate of change of horizontal kinetic and potential energy. Vertical power is the time rate of change of the vertical kinetic and potential energy. Total power to or from the trunk (total trunk) is the sum of the horizontal, vertical, and anterior/posterior tilting power. Such anterior/posterior tilting power is very small and not shown. The horizontal bar indicates when the trunk is moving downward. *(From Neptune, R.R., Kautz, S.A., Zajac, F.E., 2001. Contributions of the individual ankle plantar flexors to support, forward progression and swing initiation during walking. J. Biomech. 34 (11), 1387–1398.)*

Understanding human gait **Chapter | 1** 31

with more complex gait modelling, which predicts muscle fibre work increases relative to the tendon work at speeds of 1.2–2.0 m/s (Neptune et al., 2004a). Furthermore, this explains why there is a point where it is better to run than to walk. Mechanically efficient bipedal terrestrial walking gaits are impulse-generating by nature, but the physiology of muscle efficiency and force-velocity relationships precludes the generation of exceedingly high forces.

Much of heel lift can be brought about by the linear momentum of the anteriorly progressing CoM of the HAT and the swing limb's centrifugal pull, as heel lift should occur just before opposite foot contact and body-CoM loading (Fig. 1.1.7i). Therefore, the muscular forces necessary to induce heel lift in ideal human gait are not as high as might be expected. The point of angular rotation (fulcrum) of the foot moves primarily towards the medial MTP joints, achieved through a midfoot eversion and plantarflexion shift of the centre of pressure (CoGRF) within the foot that moves it towards the medial forefoot (Holowka et al., 2017). Once the opposite foot loads with the HAT CoM, the forces around the terminal stance ankle dramatically reduce along with the GRF.

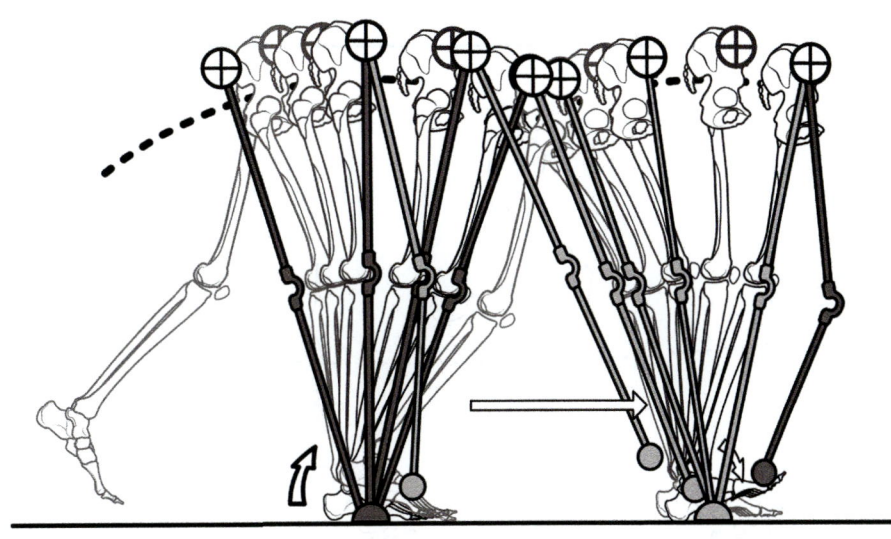

FIG. 1.1.7i The inverted pendulum is not working in isolation to transfer body mass forward. While the CoM is being raised, the swing limb is starting to accelerate under the influence of the hip-flexing contraction of iliopsoas. As the swing limb accelerates past the stance limb at midswing, the swing leg starts to add increasing momentum via centrifugal forces (aided by quadriceps knee extension) to accelerate the anterior fall of the CoM. This is enough to significantly influence the heel's offloading rate, helping to initiate heel lift. Heel lift, in turn, stops contralateral swing phase by 'dropping' it to the ground, turning the swing limb into a new inverted pendulum to vault over. In so doing, heel lift momentum is added to initial contact forces. *(Permission www.healthystep.co.uk.)*

Because the plantigrade human foot has a protruding heel behind the ankle, it does not have a tendon system that acts as a near-obligate distal spring such as that found in the bipedal ostrich. The functional equivalent to the human ankle in the ostrich is the tarsometatarso-phalangeal joints (which obviously lacks a calcaneal posterior projection), and the tendons passing these joints act as springs that enable rapid motion over uneven terrain (Usherwood et al., 2012). Humans have evolved for specialised walking, whereas ostriches have become specialised for running, and this is readily indicated by comparing each species' top speeds. Walking quadrupeds, on the other hand, do not need long-prone human foot profiles. This is because their torqueing muscles do not require to be loaded by body weight throughout the stance phase, as they do not have a single-limb support phase and therefore do not need large feet. However, the size of human feet is limited by the need to rapidly accelerate the foot through swing phase for the next step. Constructing a foot too large requires greater energy to create the momentum to accelerate and then further power to decelerate its inertia, risking swing phase becoming energetically too costly.

Spring-mass model and double pendulum gait

Another way of modelling gait is to bring in elements of compliance and stiffness rather than just stiff beam-like elements as proposed in the inverted pendulum model. The spring-mass model views the lower limb as a linear spring-like structure with a mass situated above it. On loading mass into the spring on ground contact, energy can be stored as potential energy via elasticity to recoil from during the mass-offloading stage of gait. Such a system reduces the energetic needs of locomotion and, like the inverted pendulum, utilises gravitational forces for mechanical advantage. However, this requires metabolic activity in the muscles to 'set' spring-tension in any bi-articular or multi-articular muscle-tendon spring that they create (Fig. 1.1.7j).

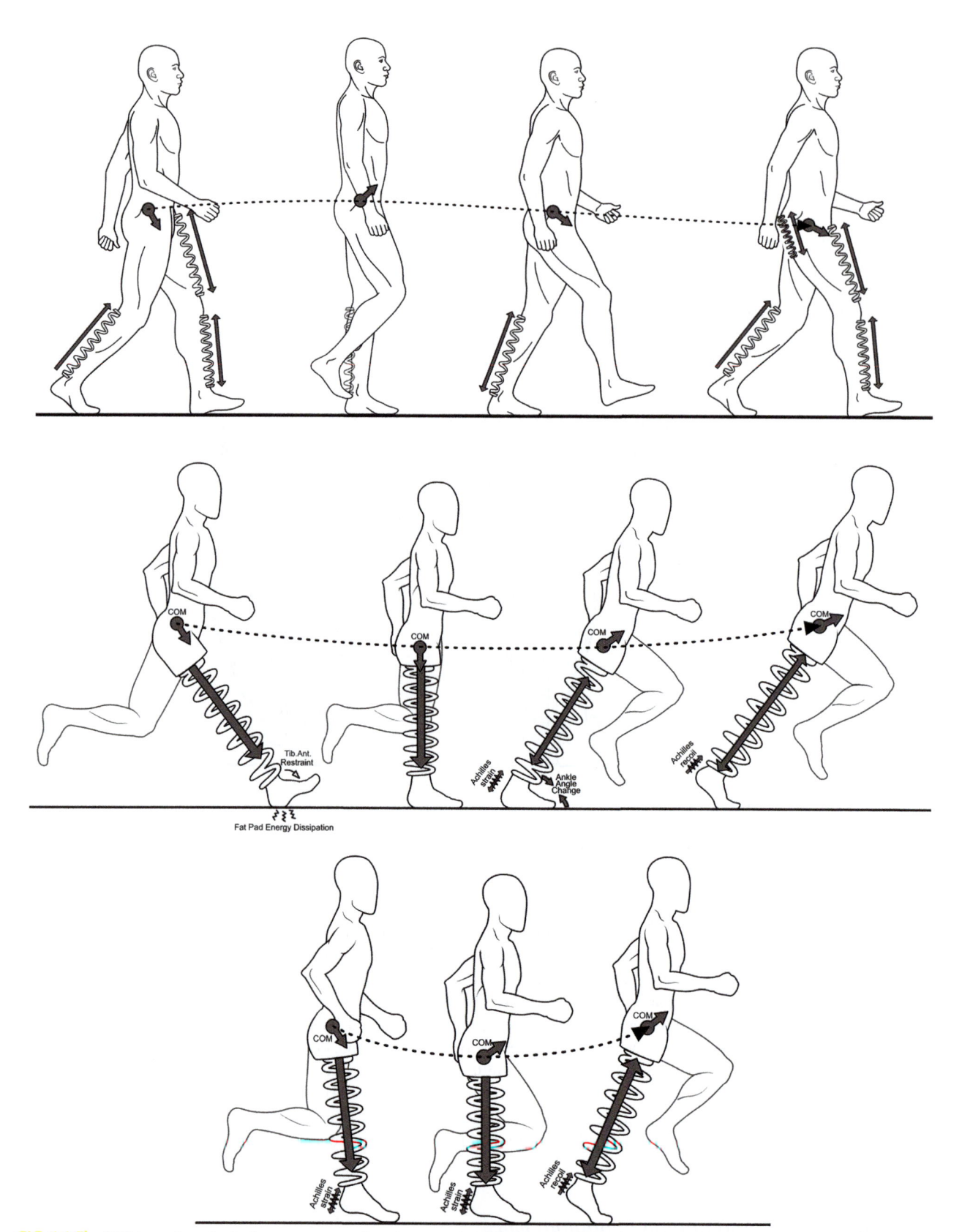

FIG. 1.1.7j Walking and running both use mass over springs, but somewhat differently. In walking *(upper images)*, springs are used to dissipate energy at initial contact through quadriceps femoris at the knee, tibialis anterior at the ankle, and gluteus maximus at the hip (shown on right limb). During midstance, the triceps surae and Achilles act as a lengthening spring, storing energy to be released at heel lift as elastic recoil. The CoM reaches its highest position during the middle of single-limb support. In running *(lower images)*, the whole lower limb becomes a shortening shock-absorbing spring on loading and a lengthening offloading spring of recoiling energy, without a distinct phase of energy storage before energy release. As a result, the CoM reaches its lowest point during the middle of stance phase, with the 'spring-like' limb reaching its most compressed (flexed) posture. This mechanism occurs regardless of whether running's initial contact is via the heel or forefoot. *(Permission www.healthystep.co.uk.)*

This model is particularly important in understanding running because the inverted pendulum model does not explain running gait, with its CoM low point being in the middle of stance phase (Fig. 1.1.7k). However, during walking as much as in running, compliant elements within the lower limb are important to energetics (Iida et al., 2008). The spring-mass model of gait assumes that the entire lower limb is representing a linear spring, although it must be appreciated that this 'spring' limb consists of three segments, namely, the thigh, lower leg, and foot. Not only does the spring-mass model recognise the relatively low activity within the swing muscles, but unlike the inverted pendulum model, it also recognises the spring-like capabilities of the many lower limb bi-articular (multi-articular) muscles (Iida et al., 2008).

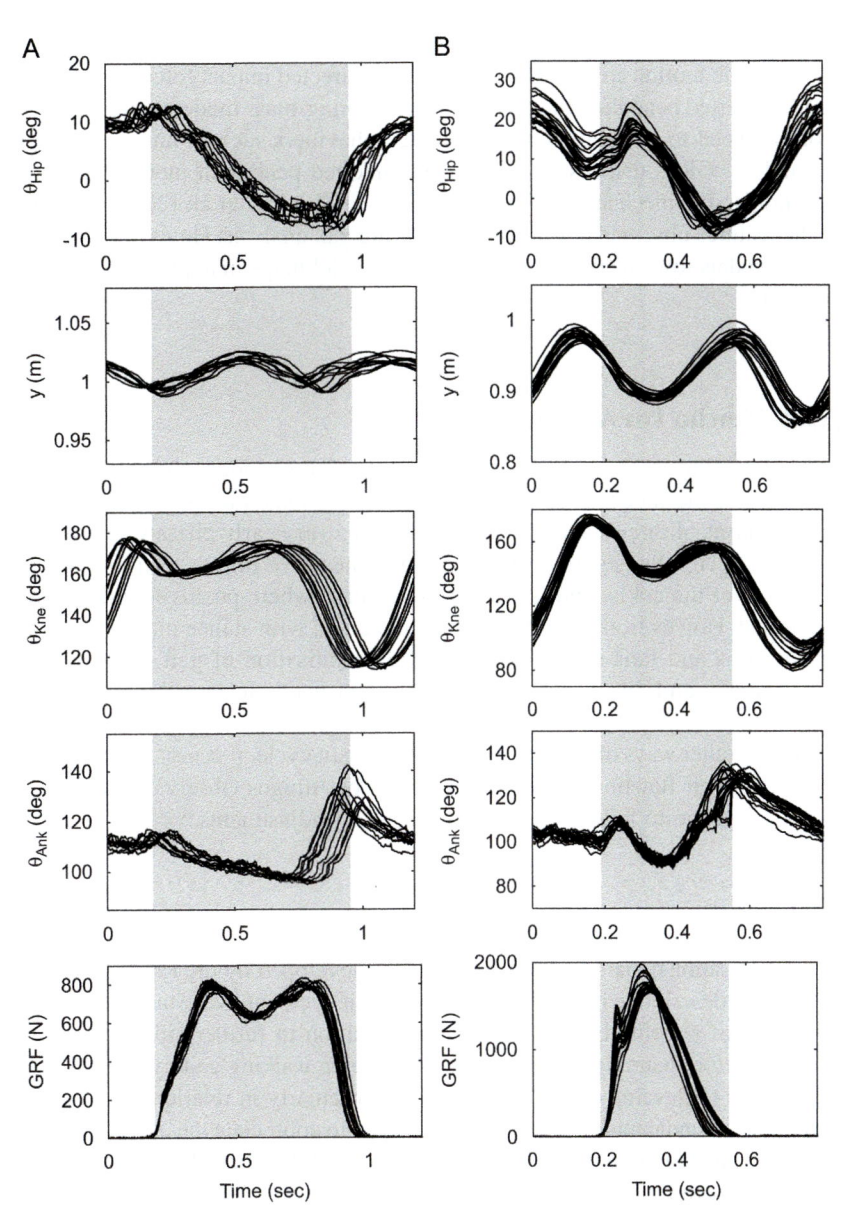

FIG. 1.1.7k Time-series trajectories of human locomotion from 15 steps: (A) walking and (B) running, both at 2 m/s. The stance phase is indicated by the grey areas. From top to bottom, trajectories of the hip joint angles, vertical movements of the body (y), knee joint angles, ankle joint angles, and vertical GRFs are demonstrated. Note the higher initial knee joint flexion angles, greater vertical body displacements, higher vertical GRF forces, and loss of the M-shaped force-time curve in running. *(From Iida, F., Rummel, J., Seyfarth, A., 2008. Bipedal walking and running with spring-like biarticular muscles. J. Biomech. 41 (3), 656–667.)*

The model of the inverted pendulum during the stance phase of walking and the spring-mass model of gait can help explain principles inherent in human locomotion, which have also been successfully applied to robotic walking (Iida et al., 2008). Both models tend to concentrate on the limb's interaction with the GRF and CoM trajectories of the stance limb. Yet interaction with the swing limb is equally important, needing to achieve correct foot placement for the next step which also directly affects the CoM trajectory and influences the GRF of the stance limb (Sharbafi et al., 2017). Therefore, modelling of human gait for completeness needs to consider a double pendulum effect of both legs as well as a segmented spring-mass model working with bi- and multi-articular muscle-tendon springs. Expanding the model can bring together the effects of gravity and active and passive elements of myofascial structures within the stance limb, with the gravitational influence exerted on the stance limb by the swing limb via the active and passive effects of its myofascial structures.

Features seen in gait such as 'claw back' GRF (seen on horizontal GRF data), where the heel is pulled backward into the ground at initial contact, can be explained through the elastic recoil effect of the intrinsic spring of the hamstring tendons after braking knee extension. This creates a backward movement of the swing limb in terminal swing phase (Sharbafi et al., 2017). Such backward swing limb motion creates a lower anteriorly directed impact force by decreasing the anterior acceleration into the ground. The inverted pendulum model of gait and a spring-mass model without a swing limb do not explain such features. Models of gait need to elucidate human walking behaviours such as muscle forces generated, swing and stance limb kinematics, and the GRFs that result. The double inverted pendulum model that considers the bi-articular behaviour of the myofascial passive mechanical effects proposed by Sharbafi et al. (2017) presents a simplified way of accounting for many of the features observed during human walking gait analysis. However, working through the inverted pendulum and spring-mass models towards the double pendulum model helps in approaching the complexities of human gait in a gradualistic manner.

1.1.8 Describing gait: Rancho Los Amigos divisions

Many different ways of delineating the human gait cycle have been developed, each depending on features that are seen as the primary elements of the model used to describe gait. For example, the inverted pendular model tends to view human walking gait in terms of three stance period phases: (i) a very early phase where loss of mechanical energy from the descent of the CoM occurs by negative work creating a negative impulse, (ii) a vault phase where the CoM passes from behind to in front of the ankle, and (iii) a very late phase where positive work creates a positive impulse, driving the heel off the ground. However, this model is only concerned with stance phase mechanics, and although it is excellent at explaining the rises and falls of the CoM and the optimisation of gait energetics in walking, it fails to describe all the primary features and events of the gait cycle and needs to be replaced by the spring-mass model for running.

Therefore, before exploring other ways of looking at the walking gait cycle, it is best to learn to describe gait as a series of 'landmark' events related to their functional role. The Rancho Los Amigos Observational Gait Analysis System is an effective clinically applicable system to follow, assisted by just a few adjustments.

Rancho Los Amigos system of gait description

The Rancho Los Amigos Observational Gait Analysis System was developed by the Rancho Los Amigos Hospital Committee (Perry, 1992: p. 10). This descriptive instrument breaks the gait cycle into very manageable segments that delineate events thoroughly, but it perhaps benefits from the occasional addition to further aid clarity to the locomotive events (Fig. 1.1.8a). A consideration before learning (or relearning) the human walking gait cycle is to try and avoid a tendency of studying the process in order to develop a view of 'normal', particularly in relation to muscle activity. This view of 'normal' is usually based on what fundamentals that we need to know to appreciate the cycle. However, the living situation is more complex in nature, as human anatomical diversity prevents absolute 'normals' from existing. We have to accept a degree of variability of gait between intra-individual steps due to the activity of the sensorimotor system's feedforward-feedback mechanism, before immediately considering abnormality.

This means that muscle activity is, in reality, different to a degree with every step. Lack of consistency is beneficial as long as such variabilities are subtle and kept at a relatively controlled level. Large variability and lack of variability are linked to

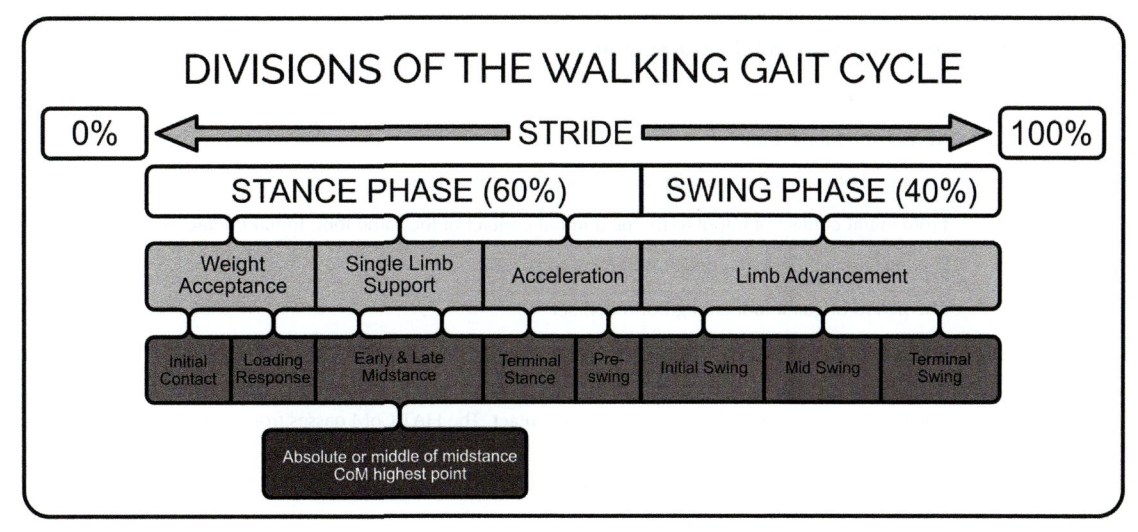

FIG. 1.1.8a The divisions of the gait cycle over a stride, based on the excellent clinical system proposed by Jacquelin Perry through descriptive mapping via the Rancho Los Amigos Observational Gait Analysis System. A few additions have been included to assist further in understanding gait functions. *(Permission www.healthystep.co.uk.)*

pathology, particularly neurological diseases. Variations should be seen within a central theme of 'desirable' plantigrade bipedal human gait. The more patients deviate from the 'normal pattern' of gait motion, the more likely they are to develop problems. Each patient has their own preferred gait style and speed. Recognising patients who are straying from their preferred gait style and speed or have an unusual gait pattern so mechanically and energetically challenging that they are at risk of pathology, is part of the skill of being a clinical biomechanist. By first appreciating the 'desirable' gait pattern, the skill of identifying the issues should follow with a bit of experience. Never be afraid to consult an experienced colleague when in doubt.

The human gait cycle involves a number of important functions mechanically, anatomically, and physiologically. Descriptions of the Rancho Los Amigos Observational Gait Analysis System tend to focus more on one 'functional role' of either braking, translation, or acceleration rather than attempting to combine all energetic and musculoskeletal events, thus avoiding some complexity. Knowledge of different gait models is helpful in understanding the multiple levels of activity occurring during gait that also includes the lower limb's role in assisting its own cardiovascular function through the musculoskeletal calf and foot venous pumps.

Rancho Los Amigos divisions of the walking gait cycle

Bipedal walking gait cycles consist of repetitive sequences of limb motion while maintaining erect stance stability. The limbs and feet are multisegmented structures that are used to propel the HAT segment, said to represent the 'passenger' part of the body being moved by the locomotor lower limbs (Perry, 1992: p. 20). However, this passenger is not actually fully passive (Runge et al., 1999; Bruijn et al., 2008). The lumbopelvic-hip muscles and joints play an important role in locomotion, with pelvic and trunk motions occurring in all three body planes. The lower limbs are considered the primary locomotive generators that go through a limb contact stance phase and an unloaded swing phase. Walking thus involves the reciprocal ground contact of two-limb support and then one-limb support in sequence.

The act of moving the mass from one limb to the next is known as a *step*. The distance between a foot contact and the opposite foot contact is the *step length*. One complete sequence of limb motion from one point in a step to the same point in the next step of the same foot is referred to as a *gait cycle*. The complete sequence from the step of one foot to the next step of the same foot is also known as a *stride*, and the distance between these steps of the same foot is labelled the *stride length*. Giving the gait cycle an absolute starting point just makes it easier to study and describe, but that point could start anywhere within the cycle. It could be argued that, energetically, heel lift would be a good starting point. However, the point of first contact with the ground is an ideal starting point and this is known as *initial contact*. This is usually assigned 0% of the gait cycle as well as being the start of stance phase. Therefore, 100% of the gait cycle is the period from initial contact of the foot to the initial contact of the same foot, when 0% starts again, but for the next stride (Table 1.1.8).

TABLE 1.1.8 The events of a gait cycle and their description.

Event	Description
Step	Moving the foot a distance
Step length	Distance between a foot contact and the opposite foot contact
Gait cycle	From initial contact of a foot to the next initial contact of the same foot. Initial contact datum point of gait
Stride/Stride length	A complete gait cycle/the length of the complete gait cycle
Stance phase	The period of gait a foot is in ground contact
Double stance phase	The period of stance when both feet are in ground contact. The period when the HAT CoM is passed to the leading limb
Single stance phase	The period when only one foot is in ground contact. The HAT CoM passes from behind to in front of the ankle during this period
Swing phase	The period when the lower limb is progressed forward of the HAT without any ground contact

The gait cycle has two obviously distinct phases: (i) support phase where the limb is in contact with the ground, known as the stance phase, and (ii) a period where the limb is being held above the ground, advancing to the point where it can act as the support for the next step, referred to as the swing phase. The implications to the laws of motion and the roles of each stage of gait have been discussed in previous chapters. Stance phase begins with initial contact and ends when the last part of the foot leaves the ground, which is usually the hallux and/or 2nd digit, in a period called preswing. Therefore, stance ideally closes with toe-off, and swing phase starts with preswing/toe-off and closes with initial contact (Fig. 1.1.8b). The stance phase of walking ideally lasts around 60% of the walking gait cycle, while the swing phase around 40% (Perry, 1992: p. 5). However, this is absolutely dependent on gait speed. Most testing in research and within the clinic is undertaken at preferred gait speed, and every patient has their own preferred gait speed. If this is faster, stance percentage will be shorter, and vice versa. It should be noted that older patients and patients carrying physiological health issues tend to walk slower, with consequently longer stance phases.

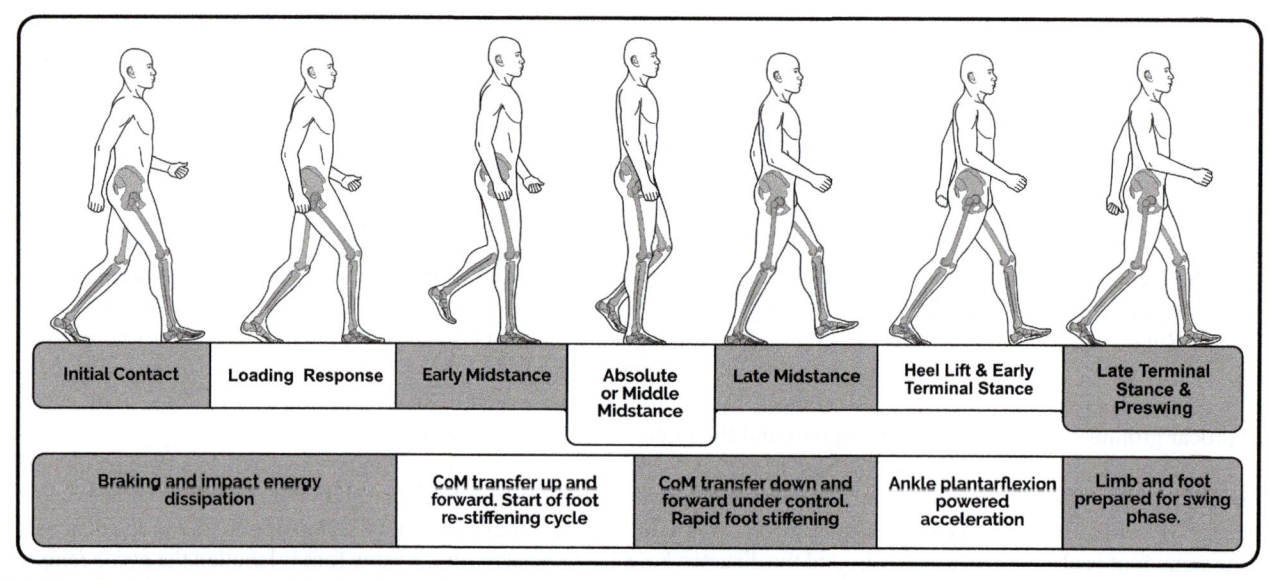

Initial Contact	Loading Response	Early Midstance	Absolute or Middle Midstance	Late Midstance	Heel Lift & Early Terminal Stance	Late Terminal Stance & Preswing
Braking and impact energy dissipation		CoM transfer up and forward. Start of foot re-stiffening cycle		CoM transfer down and forward under control. Rapid foot stiffening	Ankle plantarflexion powered acceleration	Limb and foot prepared for swing phase.

FIG. 1.1.8b The stance phase of the right foot starting with initial contact as 0% (usually on the heel) and ending at toe-off as 100% of stance phase. The loading phase involves decelerating braking and energy dissipation. Early midstance involves raising the CoM to a midstance high point, followed by a controlled fall of the CoM forward during late midstance. Heel lift initiates acceleration while toe-off completes the transfer of all loads onto the next step. *(Permission www.healthystep.co.uk.)*

Stance phase is divided into three distinct intervals, starting with a period where both feet are in contact with the ground as the body mass is transferred to the next foot making a new initial contact with the trailing foot heading towards preswing. Thus, the trailing foot is completing its stance phase, while the initial contact foot begins its stance phase. For the initial contact foot, this phase is termed *initial double stance* or sometimes *loading response double-limb support*. Double-limb support is not a good term, as only for the briefest of moments are both limbs actually 'supporting' the body mass as it is being quickly transferred to the contacting foot just after its initial contact. Therefore, the term should be avoided (Perry, 1992: p. 2), better replaced with double-limb stance. At the end of initial double stance, the opposite foot lifts off the ground and the stance foot now enters a period of ground contact on its own. This is known as single-limb support or single stance. The entire body weight is now bearing down on the support limb and its foot. *Terminal double stance* begins with initial contact of the opposite foot and continues until the original stance phase limb is lifted from the ground. The duration of the double and single stance periods tells us much about the limb's capability to provide stability. Unstable limbs create shorter single-limb support periods and longer double stance periods.

Based roughly on a 60% stance phase of gait, both initial and terminal double stance phases take up 10% each and the remaining 40% of stance is taken up by the single support phase (Perry, 1992: p. 5). However, the precise timings of the gait cycle vary with walking velocity, and at around a comfortable pace of 80 m/min, the timings are usually 62% stance and 38% swing. At slower velocities, the difference between the stance and swing phases becomes greater, with the swing phase percentage decreasing and stance phase increasing. At slower velocities, however, the double stance phase percentage increases and single support shortens in a curvilinear pattern (Perry, 1992: p. 6). As walking velocities increase, the double stance phases shorten as a percentage. When double stance phases are omitted, we have entered into running mode (Perry, 1992: p. 6) and the stance phase continues to shorten as swing phase percentage gradually becomes longer than stance. A phase where neither foot is in contact with the ground develops. This is known as the *float* or *aerial phase*. It is explored further within the exposition on running gait in Section 1.4.

Initial contact in walking is expected to occur on the heel, yet this does not always happen, especially where there is abnormality. Initial contact via the heel is known as heel strike, but the term 'initial contact' describes first contact from any part of the foot's anatomy which is more common in pathological gait patterns such as those found in cerebral palsy. However, regardless of the initial contact point, certain tasks are still required from each gait phase.

Weight acceptance (contact and loading response): 0%–10%

The weight acceptance phase is divided into *initial contact* and *loading response*. Initial contact lasts from 0% at the start of gait to around 2% of the gait cycle and brings about initial double stance. The initial contacting limb will hopefully strike on the heel to initiate a rolling action around the heel, with angular momentum occurring through the ankle joint's axis primarily with motion of plantarflexion in the sagittal plane. Loading response phase lasts to around 10% of the gait cycle, providing a period of shock attenuation, braking (negative acceleration), and weightbearing stability. The forefoot is considered to pivot around a heel rocker, although the angular momentum occurs through the ankle-subtalar joint complex via plantarflexion moments (and translations) of the talus and calcaneus, until the forefoot makes contact with the support surface. This occurs from ~3% to 10% of the gait cycle. This weight acceptance phase ends with the opposite foot having left the ground and therefore the double stance phase also comes to an end during this weight acceptance period. In relation to the inverted pendulum model, this phase is the placing of the pendulum stably to the support surface and is an energy dissipation phase. Energy dissipation is achieved primarily through hip and knee flexion-resisting muscles (gluteal muscles and the quadriceps), tibialis anterior resisting ankle plantarflexion, and by the plantar fat pads and the foot becoming increasingly compliant and prone at forefoot loading (Fig. 1.1.8c).

Single-limb support (mainly midstance): 10%–50%

Gait now enters single-limb support which lasts from around 10% to 50% of the gait cycle. It is divided into two phases, the early midstance (Fig. 1.1.8d) and late midstance (Fig. 1.1.8e), but usually includes the first part of the terminal stance phase at initial heel lift. This is because heel lift should occur while still in single-limb support, to help initiate ground contact of the other foot. Midstance continues until the body weight has become aligned over the forefoot and is equivalent to the vaulting phase of weight transfer discussed under the inverted pendular gait model. Progression of the body weight is considered to occur around a stationary foot through an 'ankle joint rocker', the inferior pivot of the inverted pendulum. This is the ankle joint's motion of dorsiflexion or extension in the sagittal plane. Stability is maintained within the limb and trunk above the ankle. It lasts to around 40% of the gait cycle. The CoM transfer of midstance is considered to end at heel lift, as does the inverted pendulum, but the foot is still in single-limb support at the moment of heel lift. However, heel lift is the first act of stance phase CoM transference or acceleration.

Weight transference/acceleration (terminal stance): 50%–60%

Terminal stance starts from the end of midstance, beginning with heel lift and continuing through the opposite foot's initial contact and its loading response to its own preswing phase. It is initiated when the CoM moves beyond the base of support of

FIG. 1.1.8c The contact phase is a braking event that involves energy dissipation through primarily quadriceps muscle-tendon buffering at the knee via flexion resistance. Gluteus maximus, the posterior segments of gluteus medius and minimus, and the hip adductors provide muscle-tendon buffering at the hip, resisting hip flexion. Tibialis anterior does the same at the ankle, resisting ankle plantarflexion. The foot utilises its plantar fat pads (heel first) and the windlass mechanism via digital extension, to construct itself into an elastic structure which it can then reverses to provide compliance at forefoot loading. At the end of loading response, the relatively compliant foot is in full ground contact and the CoM is at its lowest point during the walking cycle. *(Permission www.healthystep.co.uk.)*

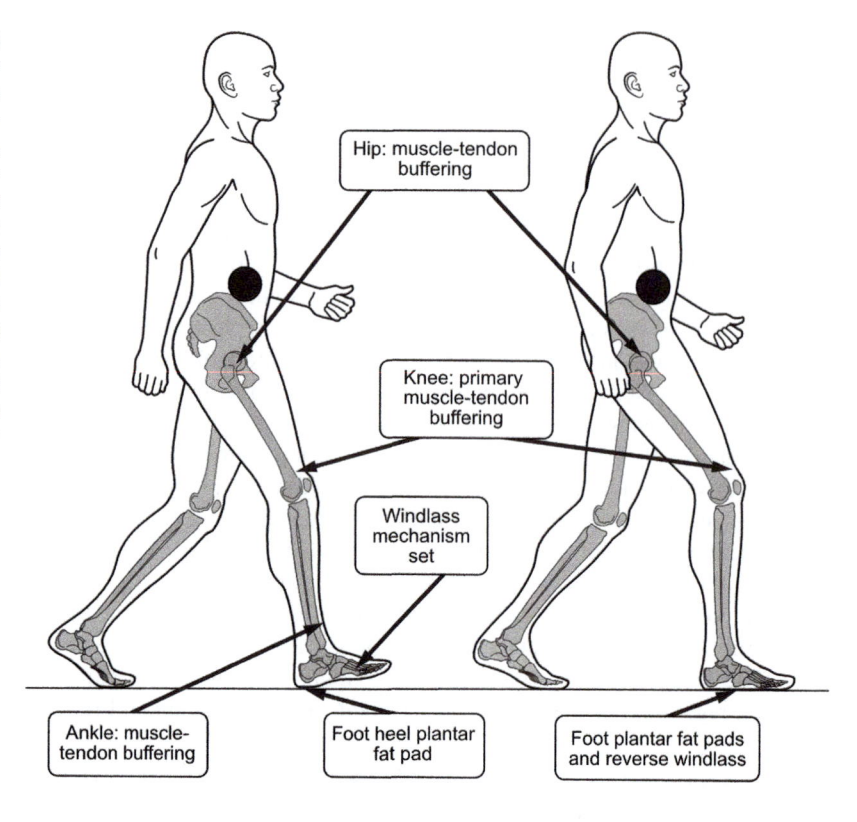

FIG. 1.1.8d Early midstance is a period of CoM *(black circle)* raising through active hip and knee extension out of flexion that also causes the ankle to extend (dorsiflex) out of plantarflexion. This involves positive work. The foot starts this period at its most compliant, lengthening, widening, and reducing the foot vault height most rapidly. Towards the end of this period, triceps surae activity starts in preparation for late midstance. *(Permission www.healthystep.co.uk.)*

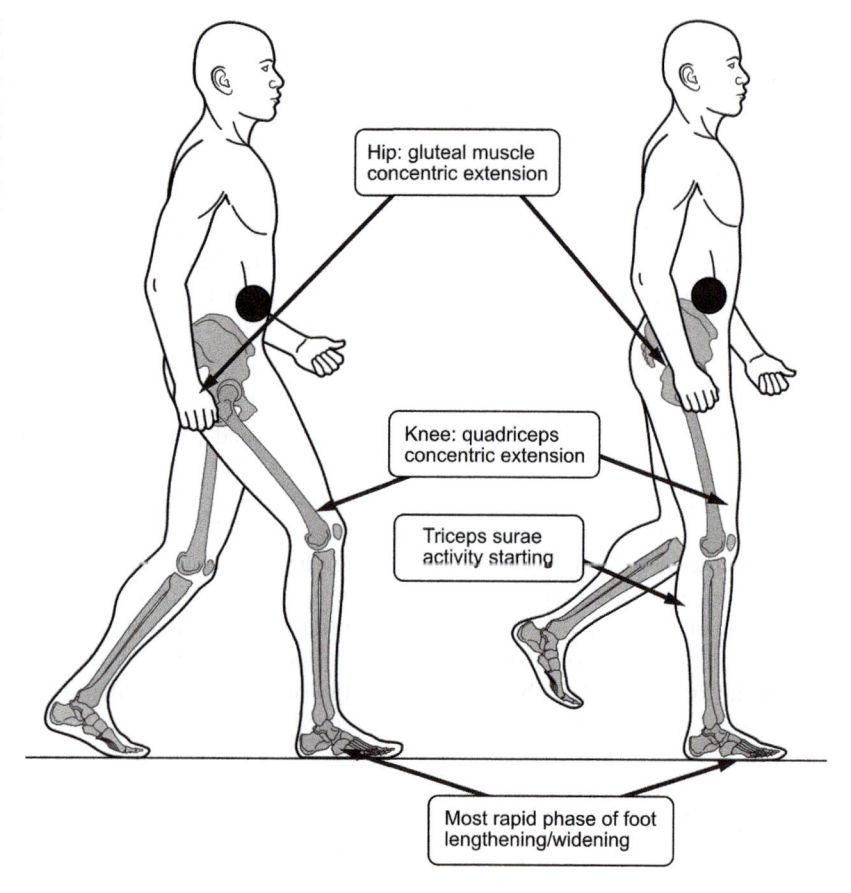

Understanding human gait **Chapter | 1** 39

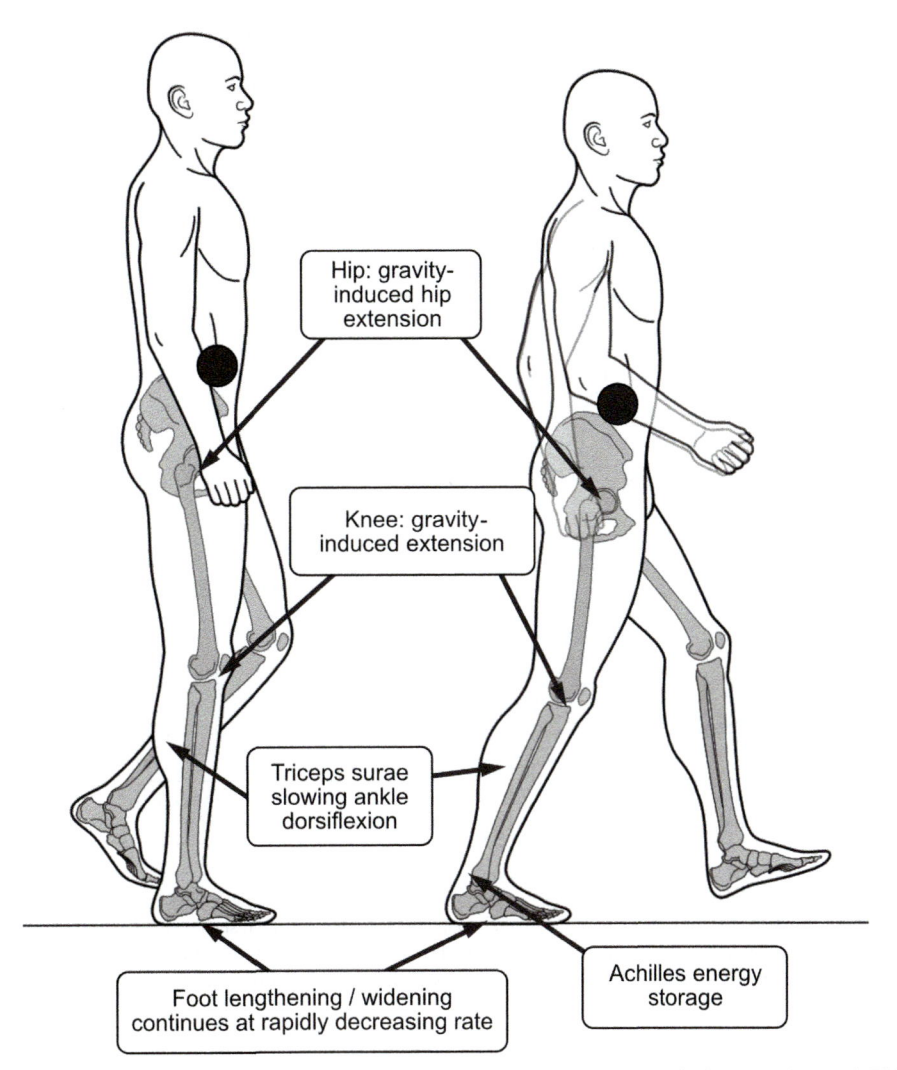

FIG. 1.1.8e Late midstance is a period of anteriorly falling CoM *(black circle)* from the high point of 'absolute midstance'. This fall is controlled by triceps surae, providing negative work very efficiently under eccentric contraction. Hip extension occurs under gravitational forces, while gastrocnemius prevents full knee extension during those same activities. The ankle dorsiflexion that occurs during this phase utilises triceps surae braking, not only to control the forward progression of the CoM but also to store potential 'stretch' energy within the Achilles tendon. *(Permission www.healthystep.co.uk.)*

the stance foot, drawn forward by centrifugal forces of the swing limb. This decreases rearfoot/heel loading forces below the energy stored within the Achilles under ankle dorsiflexion. At the start of the terminal stance phase, the body's CoM has progressed beyond the forefoot, destabilising the body's balance by straying outside of the base of support. The passenger segment 'falls' (and is pulled) forward to the next foot's base of support. It lasts between ~50% and 60% of the gait cycle. Although angular momentum occurs through the ankle via plantarflexion moments produced primarily by triceps surae-generated plantarflexion power, heel lift is considered to occur around a 'forefoot rocker' through MTP joint extension via the MTP joints offering an axis of rotation in the sagittal plane. Digital extension (dorsiflexion) takes place under near-isometric contraction of the digital flexors as the heel lifts, but it is the foot that extends upon the MTP joints and the digits that remain on the support surface, which requires ankle and midfoot plantarflexion.

Towards the end of the terminal stance phase, the limb starts preparatory posturing for limb advancement. This phase is referred to as *preswing*. It begins with opposite foot initial contact and ends with its own toe-off. Preswing has also been referred to as *weight release* or *weight transfer*. Abrupt transfer of body weight occurs onto the opposite foot, but the foot in preswing takes no active part in this event. The terminal stance foot is now unloaded by body weight and so prepares for the rapid demands of swing phase, altering muscle activity to aid ground clearance. The terminal stance part of the gait cycle is largely outside the inverted pendulum model, which is more concerned with moving the CoM over the foot, not about moving it to the next foot. However, the inverted pendulum does provide some of the power of weight transference through midstance ankle dorsiflexion that loads the Achilles with elastic stretch-induced energy, but the swing limb's centrifugal influence and gravity seem to play the most important parts in moving body weight to the next step (Fig. 1.1.8f).

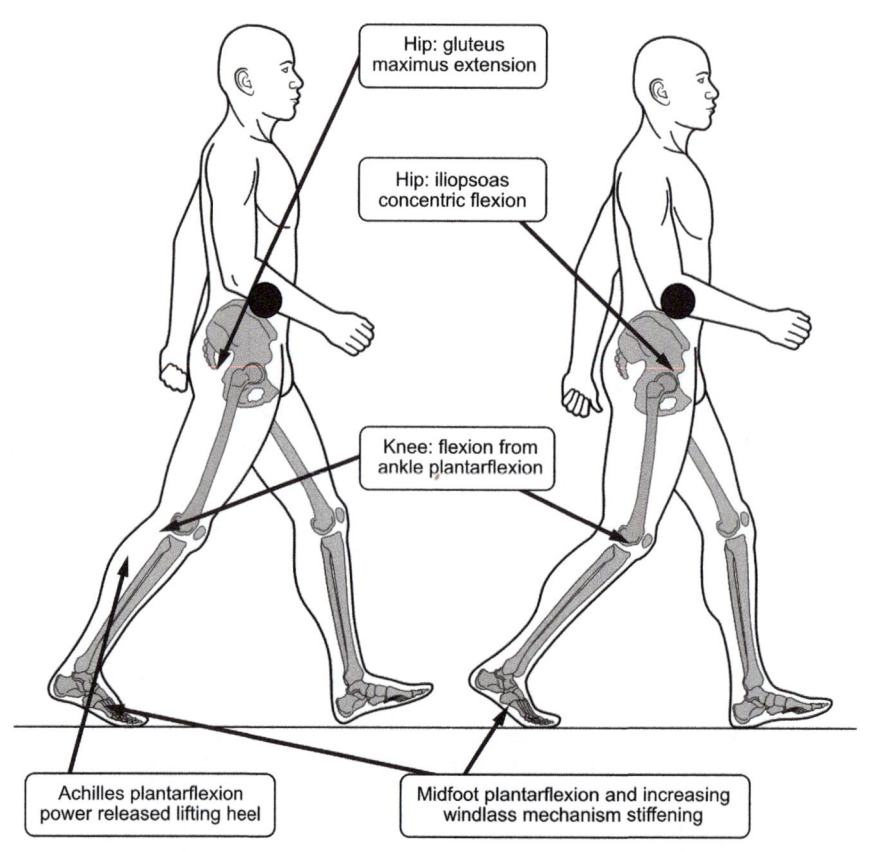

FIG. 1.1.8f At heel lift into early terminal stance, the plantarflexion power, primarily from Achilles tendon energy storage, is released. This tips the CoM onto the next step, adding momentum to the next contact phase. The midfoot and ankle plantarflex, sending the knee into increased flexion, while the MTP joints extend, gradually tightening the windlass mechanism of the plantar aponeurosis. In late terminal stance, the plantarflexion power has been used up. CoM forward progression is maintained through gluteus maximus activity until iliopsoas activity starts to flex the hip as the gait cycle enters preswing. The foot becomes more compliant while its flexor muscle activity largely ceases during toe-off as tibialis anterior and digital extensor activity initiates. *(Permission www.healthystep.co.uk.)*

Limb advancement (swing phase): 60%–100%

Limb advancement through the swing phase has its starting point in preswing. There are extensive changes in muscle activity in the lower limbs at this stage of gait (Gottschall and Kram, 2005; Barrett et al., 2007). Preswing ends at toe-off and followed by initial swing, beginning at around 60% of the gait cycle and ending at around 73% when the swing foot has reached the level of the midstance supporting limb. During this period, the foot needs to be maintained clear of the ground and advancing from a trailing position behind the trunk. It is a continuation of preswing muscle activity of hip flexion, knee flexion angle maintenance, and reduced ankle plantarflexion. Once level with the opposite foot, it has entered *midswing* or the early part of late swing which lasts from ~73% to 87% of the gait cycle. It terminates with continued hip flexion and rapid knee extension, reaching the point where the tibia is vertical to the ground on a limb advanced in front of the body.

Terminal swing starts at around 87% and completes the gait cycle, ending at initial contact. Limb advancement is accomplished as the leg and foot move ahead of the thigh. Muscle activity changes in preparation for the beginning of stance ready for the impulses associated with the foot-ground collision. The hamstrings are used to brake hip and knee flexion before contact (Fig. 1.1.8g).

The swing phase is just as important as stance phase. It plays an important role in moving the CoM of the trunk forward through the linear momentum it sets up in terminal stance via the centrifugal forces of flexing the hip and extending the knee in terminal swing (Gottschall and Kram, 2005). The swing phase also requires important energetic optimum components of gait, such as stride length and step width. The step width selected results in changes in energetic costs to locomotion (Shorter et al., 2017). The foot's motion in swing usually follows a fairly straight path with only minimal ground clearance

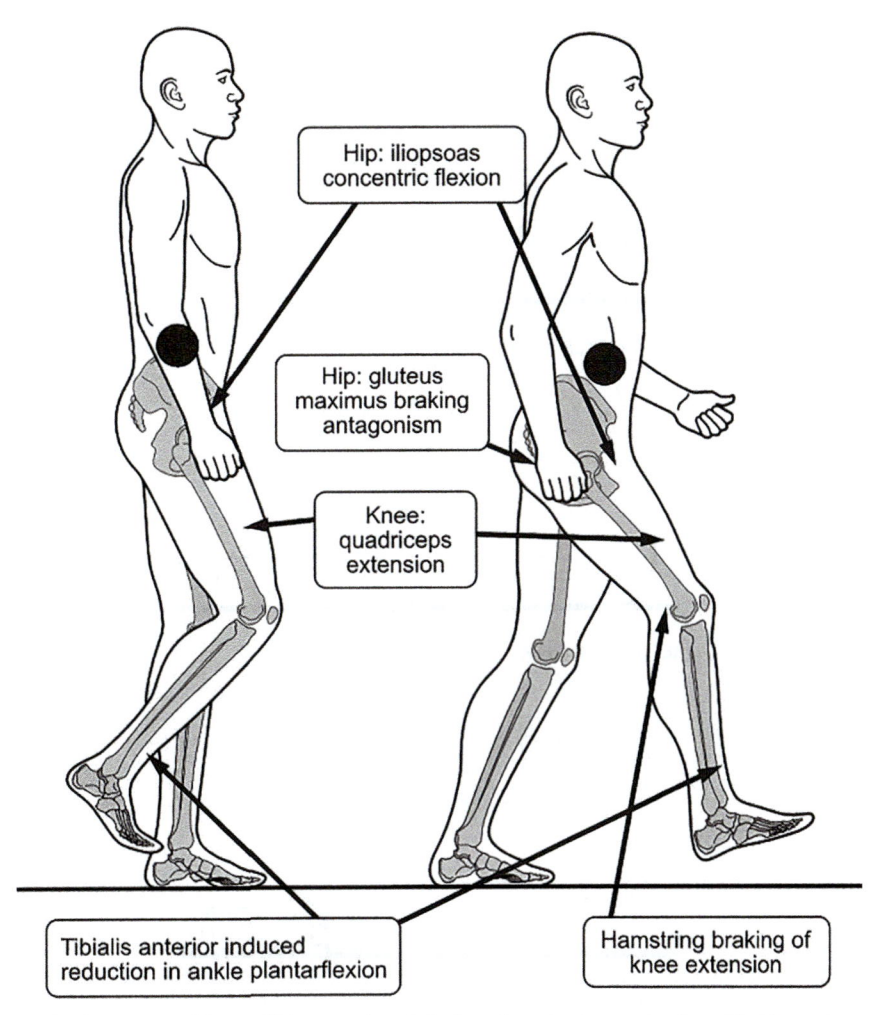

FIG. 1.1.8g Preswing termination occurs with toe-off, but muscle activity in early swing is an extension of that begun in preswing. Tibialis anterior reduces the toe-off ankle plantarflexion angle and iliopsoas continues hip flexion. At midswing, the CoM is back at its highest point over the stance limb and the hip well into flexion, with the knee starting to extend under quadriceps concentric contraction. Quadriceps activity advances the leg in front of the knee and hip during late swing phase. Tibialis anterior maintains the ankle angle assisted by peroneus tertius, if present. The swing limb must be decelerated before contact using gluteus maximus and the hamstrings to antagonise iliopsoas and the quadriceps. Tibialis anterior activity starts to peak in readiness for resisting the ankle plantarflexion moment at heel contact. *(Permission www.healthystep.co.uk.)*

in walking (particularly when compared to running). The legs should take paths mirroring that of the opposing foot, separated by just enough distance to allow movement past each other without collision.

Three-point foot rockers during the stance phase

The Rancho Los Amigos Observational Gait Analysis System proposes that the movements through the stance phase of gait in the sagittal plane be considered around a number of rockers (Perry, 1992: p. 33). The ease of the descriptions has proved very popular, but they perhaps risk overlooking some of the complexities of actual foot motion. Despite this, it is worth considering these rockers briefly, as they serve to simplify the main motions of the foot during gait.

The initial contact (as long as it is on the heel) through to the loading response of the weight acceptance phase is described as using a point of rotation in the heel that acts as a *heel rocker*. Once the forefoot has been loaded and single-limb support phase entered, the point of sagittal plane angular momentum is reported as having moved to the ankle joint axis during midstance, creating the '*ankle rocker*'. Finally, as the heel is lifted from the ground during terminal stance, the axis of sagittal rotation moves to the forefoot to become the *forefoot rocker*. This allows the stance phase to be broken down into three distinctive rockers producing three motions and tasks during the stance phase (Fig. 1.1.8h).

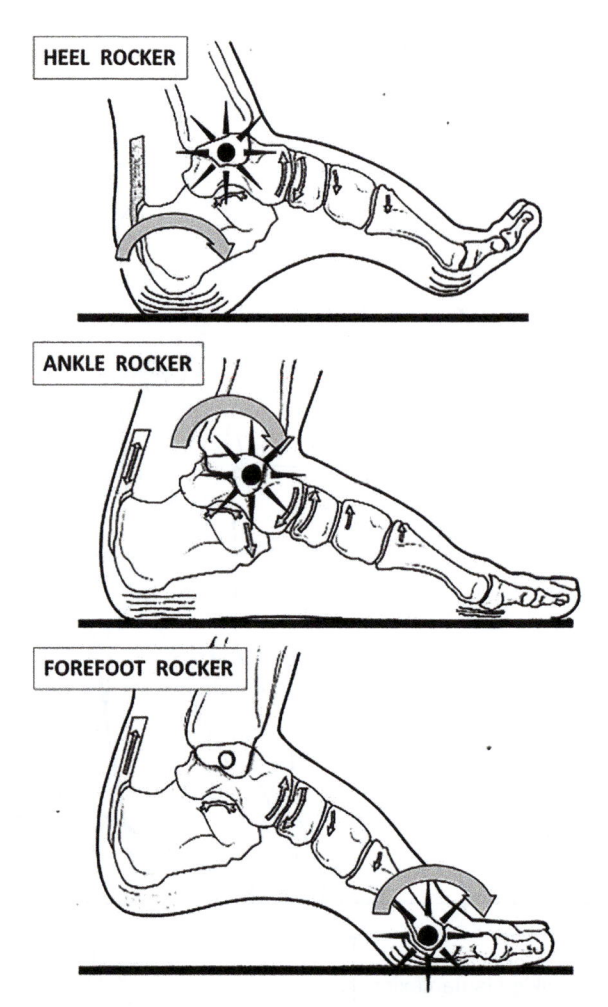

FIG. 1.1.8h The three rockers described for heel contact to loading response (heel rocker), for midstance (ankle rocker), and for terminal stance (forefoot rocker). The simplicity is attractive, but the reality is far more complex. The heel rocker involves plantarflexion at the ankle and subtalar joint (together as the rearfoot articular unit) and heel fat pad displacement. The ankle rocker does primarily involve ankle dorsiflexion but is aided by midfoot dorsiflexion during vault sagging deflection, and some anterior gliding and plantarflexion of the subtalar joint. The forefoot rocker involves MTP joint extensions, but the foot also rotates around this axis through ankle and midfoot plantarflexion motions around their own joint axes, while the forefoot plantar fat pad and other soft tissues experience deforming and energy-dissipating shear strains. *(Permission www.healthystep.co.uk.)*

The problem with this approach is that the sagittal plane motion of the heel rocker is primarily arising from the ankle joint axis, supplemented by gliding and rotations within the subtalar joint and shear within the heel's soft tissues. The heel pad itself is compliant, more so at slower loading rates, so it does not offer a predictable plantar surface to rotate around. The heel rocker motion is therefore actually combined ankle-subtalar joint rotation and heel pad deformation, driven by moments of rearfoot flexion (plantarflexion). Thus, the motion of the heel rocker is derived from combined rearfoot motion.

During midstance, the lower limb rotates over the foot primarily through motion at the ankle joint. Thus, 'ankle rocker' as a term is more appropriate than 'heel rocker'. However, the foot is not 'static' in that it offers compliance and changes its profile as the foot vault is flattening throughout midstance via midfoot dorsiflexion (Holowka et al., 2017). This midfoot flattening of the vault effectively allows continued plantarflexion in the whole rearfoot relative to the forefoot, beyond that of motion around the ankle rocker dorsiflexion alone. This midfoot dorsiflexion can be an important contributor to anterior progress of the body's CoM over the foot as well as helping to dissipate ankle dorsiflexion energies. Thus, the foot vault is playing an important part in influencing the location of the ankle joint's instantaneous axis, which should move anteriorly with increasing plantarflexion of the rearfoot, while permitting ankle dorsiflexion. As the loads on the rearfoot decrease, this rearfoot plantarflexion across the midtarsal articulation should decrease, allowing the proximal plantar ligaments to start to elastically recoil, returning some dorsiflexion energies back into the proximal vault contour.

Those individuals, who demonstrate midfoot break (excessive midfoot dorsiflexion) at the lesser tarsus and tarsometatarsal joints during late midstance, effectively use a 'midfoot rocker' motion as a supplement to the ankle rocker. This midfoot 'rocker' allows the heel to lift while large parts of the more distal midfoot remain in ground contact. Usually, the lateral distal midfoot remains initially weightbearing at heel lift. This not only reduces the influence of the forefoot rocker in terminal stance by reducing MTP joint extension, but also decreases the demands on both the ankle 'rocker' dorsiflexion during midstance and the ankle plantarflexion at and after heel lift.

The forefoot rocker that is described as being initiated at heel lift is primarily occurring through MTP joint extension and coupled to ankle and midfoot plantarflexion after heel lift. Digital extension occurs as the toes are held against the ground by the near-isometric action of the digital flexors, interossei, and lumbricals. The rest of the foot rotates into plantarflexion at the ankle and midfoot behind the MTP joints. This forefoot rocker therefore also requires sagittal plane motion through the ankle and the midfoot as both actively plantarflex. They do so under the influence of the forces generated initially by the triceps surae through its connective tissue recoil (in particular the Achilles tendon) and then through the linear momentum of the advancement of the body onto the next foot generating a midfoot plantarflexion moment (Holowka et al., 2017) assisted through tibialis posterior and peroneus longus activity/elastic tendon-recoil stiffening the foot (Kokubo et al., 2012). Thus, the forefoot rocker point primarily occurs at the MTP joints but also involves shear through the forefoot plantar pad soft tissues and rotational torques around the ankle and midtarsal joints. There are synovial bursae and fibrocartilagenous plates under the lesser metatarsal heads and consistent sesamoids under the 1st MTP joint to aid this function of metatarsal shear and extension rotation.

The concepts of the foot acting as a three-point axis system of heel, ankle, and forefoot rocker probably have its place at a very simple level. In clinical biomechanics, its role as a model of what happens during gait is limited and does not represent the full complexities or variations in the generation of angular momentum in the sagittal plane. However, viewing gait motions via these simple rockers does draw attention to the existence of obvious variances such as midfoot break.

1.1.9 Additions and modifications to the Rancho Los Amigos divisions

A few additions as well as recognition of the limitations of the Rancho Los Amigos System help to highlight some of the key changes in function that occur during the midstance phase of single-limb support. The descriptions and terms discussed here greatly help in understanding the relationship of gait tasks to the generation of force-time curves developed from force plates. Force-time curves can be used to gauge the magnitude and direction of the resultant GRF generated by the body weight interacting with the ground. There are distinct rises and falls in the amount of force generated through impulses and changes in the direction of that force being applied back into the body, all of which help in appreciating what force the body is applying to the ground and what reactive forces are being applied to it (Fig. 1.1.9a).

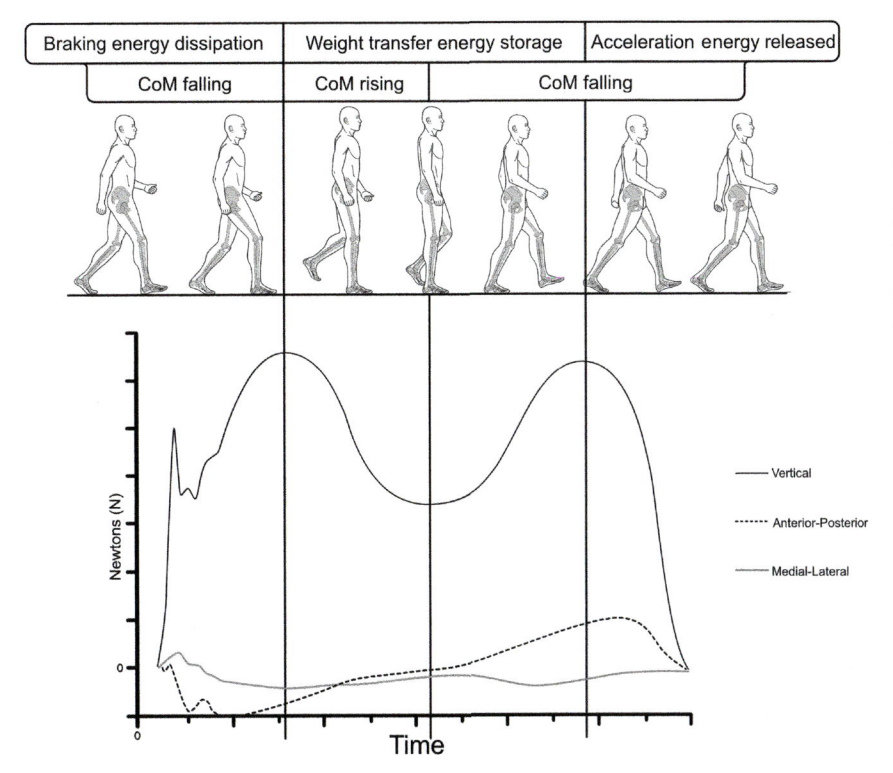

FIG. 1.1.9a The primary functions of walking gait are related to a force-time curve, showing how impulses (changes in force over time) are influenced by changing events. The initial negative acceleration (or deceleration) correlates with the falling CoM and ends with a distinct high peak of vertical force at forefoot loading. This is followed by a period of decreasing vertical force as the CoM is raised over the leg and foot as part of weight transference. At the CoM high point, vertical forces reduce to their lowest and the GRF changes its orientation in the horizontal plane from a posterior negative direction to starting to develop a positive anterior direction (although positive and negative can be set up in the reverse directions on a force plate). Although weight transference continues, the CoM is now falling forward again. Energy is increasingly stored within the soft tissues, and with the aid of muscle-generated energy, a high impulse is generated to accelerate the CoM onto the next step through heel lift. The CoM falls to the next step, reducing vertical forces as the cycle begins again on the other foot. *(Permission www.healthystep.co.uk.)*

44 Clinical biomechanics in human locomotion

The earlier part of the gait cycle in humans is concerned with dissipating energy produced by the impact between the lower limb and the ground. The later part is concerned with generating power around the ankle so as to accelerate the CoM of the body and leg forward. The middle part requires a change from energy dissipation compliance capabilities to stiffer elastic properties. Although the concept is straightforward, the events to achieve this are complex. Nevertheless, in essence, there is a natural role division between parts of the gait cycle in energy dissipation and those of energy generation for acceleration; distinct points of change primarily from one task to another. These processes are accompanied by a CoM falling phase followed by a CoM rising period. Once the CoM has reached its highest point, it starts to fall again producing acceleration of the CoM. During contact and loading response, the CoM of the body is accelerating into the ground while the foot with the lower limb is dissipating energy through its anatomy, achieving 'controlled' braking/cushioning compliance of the limb and foot. During the early part of midstance, the CoM starts to be hauled back up through hip, knee, and ankle extension moments, lengthening the lower limb. This is a process that strains the foot's passive structures, starting the process of passing through its linear elasticity towards stretch-stiffening. Once the CoM has reached its highest point, the CoM achieves its greatest potential energy and starts to fall again, releasing kinetic energy and accelerating the CoM. During the fall of the CoM, the foot starts to lose any remaining compliance and begins to stiffen up rapidly under strain rate-dependent and stress-stiffening behaviours of the connective tissue and increasing muscle action under the foot and posterior to the leg. This prepares the foot to become a semi-stiffened beam as part of a class two lever system for heel lift. If successful, it permits power generation through the action of triceps surae to be applied to a stiffer vault and forefoot to create a stable lever and acceleration platform. This now stiffer foot, combined with the ankle power, creates a significant GRF vertically and horizontally at the forefoot in terminal stance (Fig. 1.1.9b). The forefoot GRF is necessary to achieve stability for weight transference and aids CoM acceleration into the next step.

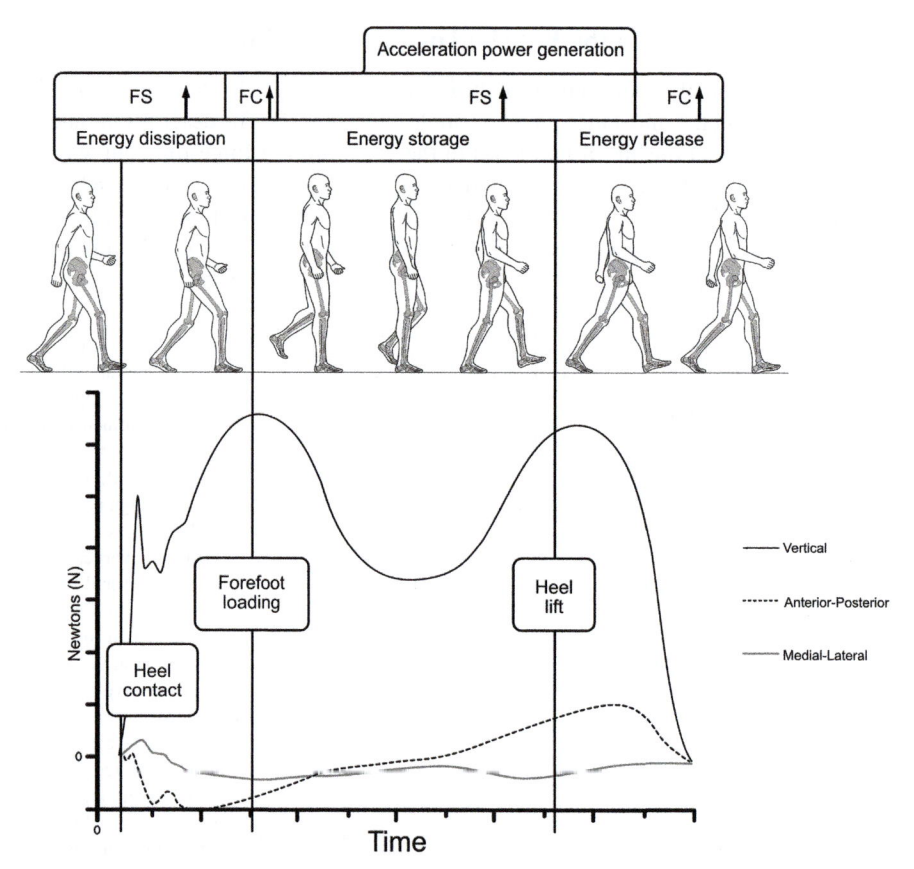

FIG. 1.1.9b The foot undergoes cycles where it is increasing its stiffness (FS↑) or increasing its compliance (FC↑), as well as periods where the lower limb is dissipating, storing, or releasing energy through the soft tissues and joints. These events can be related to the impulses produced on a force-time curve during walking gait, with peak events such as heel contact initiating energy dissipation, forefoot loading completion starting a cycle of gradual energy storage, and heel lift releasing energy. Muscles are generating energy and increasing stiffness for acceleration. *(Permission www.healthystep.co.uk.)*

From the point of view of appreciating the tasks occurring in the single support phase of gait, it is proposed here that the period from the end of loading response to the CoM high point be termed *early midstance*, the period after the CoM high point be termed *late midstance*, and the high point of the CoM at the middle of midstance be termed *absolute midstance*. Ideally, absolute midstance should occur as follows: (i) the opposite limb is entering midswing, and (ii) the trunk is positioned such that the passenger HAT CoM sits over or extremely close to the hip joint axis and over or slightly posterior to the ankle axis in the sagittal plane, with the ankle still in slight plantarflexion. In such a situation, the angular momentum across the joints (collectively) in the lower limb should be at its least during absolute midstance. The knee, which should never fully extend, should have the CoM just behind its axis at a low flexion angle. Furthermore, the knee will also be subject to a very low flexion angular momentum in this position. This means that the CoM, at absolute midstance, sits over the lower limb at its longest stance length (excluding the foot). From here, the HAT's CoM will move anterior to the hip and ankle, helping to extend them, and soon after, move anterior to the knee to apply an extension moment on it via the GRF vector (Fig. 1.1.9c).

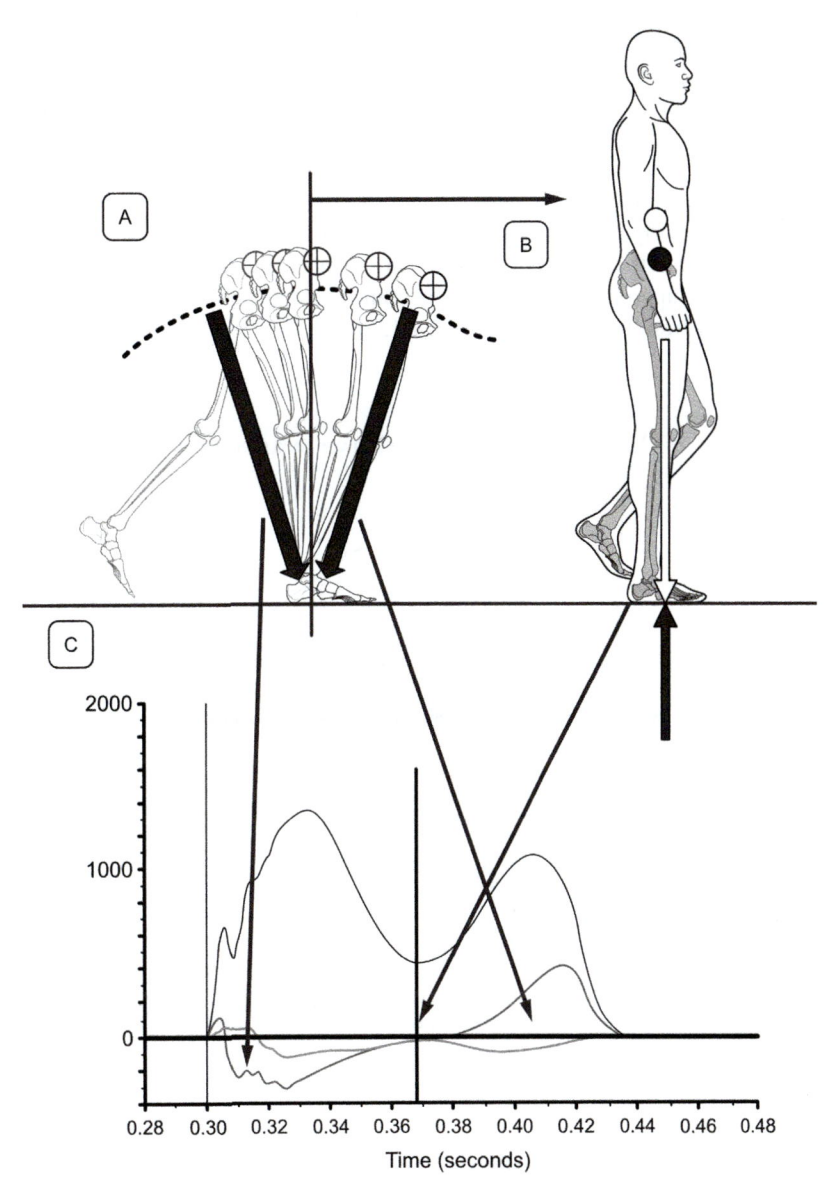

FIG. 1.1.9c Absolute midstance during walking is at the middle of midstance where the arc of the inverted pendulum reaches its highest point. Before absolute midstance, the CoM is rising and the body weight vectors are directed anteriorly, while the GRF is directed posteriorly (A). After absolute midstance, the CoM is falling forward causing the body weight vector to be directed posteriorly with the resulting GRF being directed anteriorly. At absolute midstance, the CoM (black circle) creates a body weight vector and a GRF vector aligned with the divergent point (white circle) (B). The differences seen before and after absolute midstance are shown clearly on the anterior–posterior horizontal forces of a clinical example of a force-time curve, with absolute midstance being the point of crossover between the anterior and posterior GRF that usually aligns close to the lowest vertical GRF indicated by the vertical line (C). Thus, such data will inform where the absolute midstance occurred and whether the change in horizontal force direction correlates with vertical force expectations. *(Permission www.healthystep.co.uk.)*

The great advantage of constructing the midstance as two phases with one distinct transition point is that describing and noting perturbations in gait motion becomes easier by comparing it to ideal motion, position, and events. Ideally, the midswing event should occur as the opposite limb is passing through its absolute midstance point of the highest CoM position. This trunk-limb-extended high point will aid ground clearance of the swing limb. It should also be noted that the arc of motion of the CoM should be upwards during all events associated with early midstance, and that maximum potential energy is achieved at absolute midstance when the CoM can go no higher without rising onto the forefoot (energetically a disastrous thing to do).

Following absolute midstance, the CoM should be on a descending curvilinear path, increasing GRF and positive acceleration inferiorly and anteriorly until heel lift. This utilisation of the kinetic energy released from the potential energy of the CoM high point is controlled primarily by triceps surae eccentric action. It assists in keeping the energy moving in a linear direction of travel, as required from such muscles as tibialis posterior and the peroneal muscles operating to avoid excessive medial-lateral, eversion-inversion, or abduction-adduction excursions across the foot and ankle.

During early midstance, the energy required to maintain ankle stability is very different to that of late midstance. During early midstance, the GRF lies behind the ankle, as it was during weight acceptance. Until the forefoot is fully loaded at the end of forefoot loading, the GRF requires the ankle to plantarflex via angular momentum, which was primarily resisted by tibialis anterior activity. Once weight acceptance and stability through loading response is complete, the GRF plantarflexing angular momentum is blocked by the forefoot's contact with the ground, so little energy is required to further stabilise the ankle. As early midstance progresses, the angular momentum decreases as the CoGRF moves closer to the axis of the ankle joint. At absolute midstance, the CoGRF passes under the ankle axis with the CoM directly above. The instantaneous ankle joint axis moves anteriorly within the ankle once it starts to dorsiflex during late midstance as triceps surae activity increases, tracking anteriorly as the CoGRF moves forward with and beyond it. This effectively reduces the force of angular momentum around the ankle appropriately at the correct time to aid muscular stability energetics by lengthening the triceps surae internal moment arm from the ankle axis (Fig. 1.1.9d).

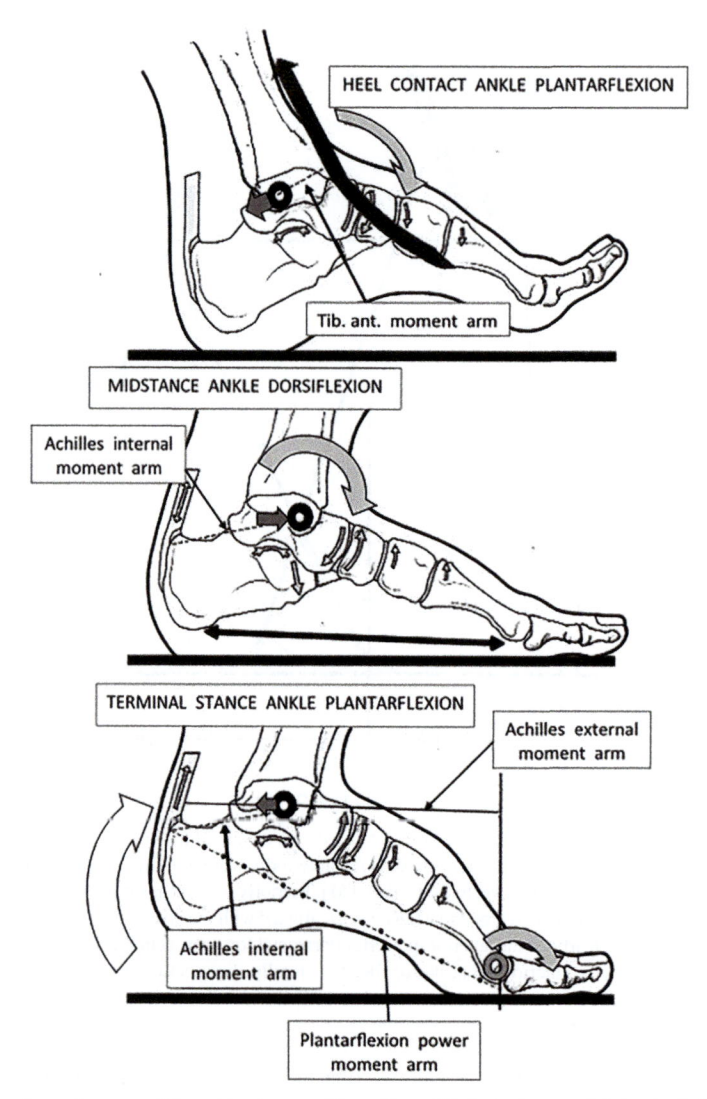

FIG. 1.1.9d The moment arms of muscles around the ankle change with motion of its instantaneous joint axis. At heel contact, the GRF on the heel induces ankle plantarflexion, resisted by tibialis anterior (Tib. ant.). As the tibialis anterior lengthens through eccentric contraction against plantarflexion, the ankle's instantaneous axis moves posteriorly. This lengthens the moment arm to the ankle axis from the tibialis anterior tendon passing through

(Continued)

After absolute midstance, the ankle starts to be subjected to increasing angular momentum of extension (dorsiflexion) as the CoGRF follows the HAT's CoM to lie increasingly anterior to the ankle and over the forefoot, thereby generating increasing dorsiflexion moments at the ankle. To counteract this dorsiflexion force at the ankle, the triceps surae becomes increasingly eccentrically active and the Achilles tensioned to prevent the CoM descending and moving anterior too quickly. This permits time for the swing limb to advance to a terminal swing position suitable for receiving CoM loading. Moreover, absolute midstance point is also highly significant in the behaviour of muscular activity in the foot, ankle, and leg, with distinctly different muscular roles occurring before and after absolute midstance. The muscles become increasingly more active after absolute midstance. An inability to achieve sufficient control of dorsiflexion and adequately control the anterior velocity of the advancing CoM during late midstance can have serious impact on the efficiency of the gait cycle, resulting in, for example, a mistimed early heel lift before the opposite foot is appropriately positioned ready for weight transference.

The concept of foot compliance and foot stiffening also benefits from the midstance phase of gait to be deconstructed into separate parts. During loading response and into early midstance, it is advantageous for the foot to be compliant in order to help dissipate impact and loading forces through deforming strain on the foot. By providing vault flexibility, the foot can help decelerate the fall of the CoM by creating a braking effect through the generation of functional 'crumple-zones' in conjunction with the flexion of the hip and knee above. This phase has no need to store energy elastically within the foot and ankle as this would tend to recoil the foot out of its plantigrade pronated attitude of early stance. Pronation, which improves with vault flexibility, provides contact stability with the ground by increasing surface contact and helping to spread out forces and pressures throughout the plantar tissues. Thus, increasing pronation and compliance should be a positive feature of early midstance, and this is evidenced by low stiffness at initial foot vault loading and increased lengthening rates compared to those during late stance, influenced by the percentage body weight loading forces (Stolwijk et al., 2014; Bjelopetrovich and Barrios, 2016; Takabayashi et al., 2020). Thus, in early midstance, the foot is compliant and less stiff than during later periods of stance. By lengthening the foot easily during early stance, it moves the connective tissues through their linear elastic phase towards their stress-stiffening properties of more rapid stiffening underlying increasing strain-rate, before the end of midstance.

During late midstance, the foot needs to become increasingly stiff in readiness to behave in the manner of a beam to lever the heel off the ground at the end of midstance. As a consequence, the intrinsic and extrinsic muscles supplying the midfoot and forefoot become increasingly active from absolute midstance onwards, increasing their muscular tone and particularly influencing medial-lateral stiffness across the forefoot (Farris et al., 2019, 2020). Despite this, the foot continues to increase its pronated attitude until heel lift (Hunt et al., 2001). This increased late midstance pronation causes the passive tissues to become increasingly stiffer at faster rates, the flatter the foot becomes (Stolwijk et al., 2014; Bjelopetrovich and Barrios, 2016; Takabayashi et al., 2020). This occurs via the stress-stiffening viscoelastic properties of the foot's tissues. Thus, with connective tissue stiffness increased, muscle contractile stiffening can be moderated. Energetically, this is a great technique for making the foot semi-rigid.

As the pedal viscoelastic materials lengthen, they pass through their linear elastic phase and start to behave in a non-linear elastic fashion, stiffening rapidly under small increases in load. This is the expected result of highly tensioning viscoelastic tissues, and it results in a passive stiffening event coupled to foot lengthening. The faster the loading, the stiffer the foot behaves through time-dependent properties of connective tissue. Energetics is not enhanced if muscle action alone is used to stiffen the foot because muscles use metabolic energy, although muscle activity is still required to achieve adequate and adaptable stiffness properties. However, it is far more efficient to control the extent of passive vault-flattening by primarily using muscles for near-isometric control only. Thus, the muscles are protecting the passive connective tissues while also at the same time increasing stiffness within their tendons' own connective tissue, whereas the passive stiffening effects

---

FIG. 1.1.9d, CONT'D the extensor retinaculum, thus improving muscle power efficiency. During midstance, ankle dorsiflexion is resisted by the triceps surae through the Achilles via its internal moment arm. With increasing dorsiflexion, the joint axis moves anteriorly, lengthening the internal moment arm of the Achilles from its insertion. This improves the ability to restrain the rate of anterior displacement of the CoM and CoGRF via braking anterior tibial rotation. This anterior axis positioning also means at heel lift the Achilles internal moment arm is maximised by length to the ankle axis. However, as the fulcrum during heel lift should now be at the 1st and 2nd MTP joints, the internal moment arm to the ankle axis also needs to be considered in relation to the ability to maintain a long external moment arm between the Achilles effort, the ankle joint axis, and the MTP joints used for extension. These axis-to-muscle adjustments help maximise the plantarflexion power-induced forefoot GRFs during acceleration via a plantarflexion power moment arm between the Achilles calcaneal attachment and the medial MTP joints acting as the fulcrum. This maintains good energetics for concurrent ankle plantarflexion with MTP joint extension. With further ankle plantarflexion, the axis moves posteriorly again as the ankle power decreases. The external moment arm from the Achilles enthesis to the MTP joint fulcrum will shorten via midfoot plantarflexion that occurs during terminal stance. The overall moment arm length between the Achilles insertion and the MTP joints is essential for mechanical efficiency at heel lift, and it is best positioned for energetics when the longer medial metatarsal lengths are utilised. *(Permission www.healthystep.co.uk.)*

of the other connective tissues occur with less metabolic energy input as the vault flattens down. The intrinsic and extrinsic muscles such as the toe flexors, tibialis posterior, and peroneus longus can engage before heel lift to help create stiffened elastic tendons suitable for storing energy for elastic recoil (Ferris et al., 1995; Kokubo et al., 2012; Farris et al., 2019). This released stored tendon energy starts to elastically recover the foot vault's profile after heel lift, while maintaining a functionally stiffened forefoot platform from which to drive off from. Stiffening events actuated by soft tissue properties and vault profiles in late midstance are likely crucial for ankle power application to the forefoot during terminal stance. However, the foot is never 'rigid', but in degrees of stiffness. Indeed, too much stiffening across the forefoot would prevent MTP joint extension at heel lift. Interestingly, there are indications that elastic recovery in proximal rearfoot ligaments starts first, with more distal connective tissue structures shortening later (Welte et al., 2021). This is consistent with the location of the CoGRF and peak loads throughout the stance phase (Fig. 1.1.9e).

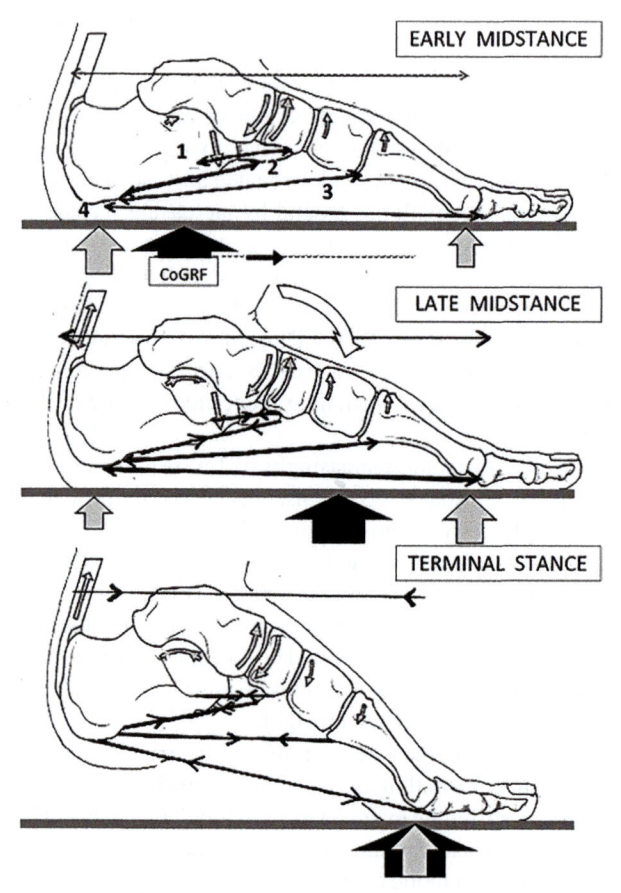

FIG. 1.1.9e A diagram to show the primary plantar ligaments' stretch-recoil cycles occurring in the stance phase. During early midstance, the foot lengthens *(upper image)* most rapidly, stretching the spring ligament (1), short plantar ligament (2), long plantar ligament (3), and plantar aponeurosis (4) through rearfoot and forefoot loading forces *(grey arrows)*. At this stage, the centre of ground reaction forces (CoGRFs) is still under the heel *(black arrow)*. Thus, loading forces are presently higher within the rearfoot than the forefoot. In late midstance *(middle image)*, loading forces are higher in the forefoot as the CoGRF moves anteriorly, while the vault is subjected to ankle dorsiflexion-induced sagging deflection moments *(white arrow)*, causing the distal vault to undergo increasing strain and midfoot dorsiflexion. The foot lengthens and lowers its vault to its maximum at the point of heel lift, but the rate of lengthening slows, while stiffness raises exponentially. Reducing rearfoot loading allows proximal ligaments such as the short plantar ligament and parts of the spring ligament to recoil, with the long plantar ligament and remaining parts of the spring ligament starting to recoil before heel lift. It is suspected that the deep transverse metatarsal ligament is continuing to be stretched as forefoot loads increase and maximise at heel lift *(lower image)*, helping to increase forefoot stiffness. In terminal stance, plantar ligaments of the vault shorten through elastic recoil, although the plantar aponeurosis continues to lengthen initially after heel lift until digital extension angles become greater. Both extrinsic and intrinsic muscles are essential components of foot-shortening and stiffness levels throughout late midstance and terminal stance. *(Permission www.healthystep.co.uk.)*

1.1.10 Section summary

The ability to assign a functional role to stages of gait as well as being able to describe and name phases of gait is a skill fundamental to clinical gait analysis. It allows for pathology to be assigned to a phase, a functional role, and a cause and/or effect of gait, as well as being able to describe dysfunction. It is likely that gait disruption will initially occur in one phase only, or during a period within a gait phase that may merely change the functional ability of the lower limb to achieve the particular functional role that occurs within that phase. Thus, energy dissipation (shock absorption) may be inadequate or the ability to easily accelerate may be compromised, rather than problems existing in both mechanisms. By knowing what part of the anatomy should be doing what task and at which time means that a patient's gait data can be compared to an expected timeline of events and functions.

Locomotion at its core is a simple act of moving from A to B. However, the management of forces, the conservation of energy, and the intrinsic mechanisms involved in achieving this complicates the picture. The features expressed by a patient's gait will indicate the efficiency of their gait, and this will reflect the metabolic costs on their muscles during travel.

1.2 Principles of gait analysis data

1.2.1 Introduction

Interpretation of gait data is an important clinical assessment skill and an essential principle in understanding the origins of pathology and the dysfunctional effects of pathology on gait. Technologies to achieve the gathering of kinetic and kinematic data have gradually become more accurate, and most importantly, far cheaper, albeit some of the best kinematic systems such as 3D video joint tracking and force plates remain outside most clinical budgets. This is due to not only the expense of the equipment but also because of the dedicated space required. However, systems that record more basic kinematic data such as gait speed, step time, stance phase percentages, and/or kinetic pressure data are now very reasonably priced and can usually be easily moved and set up as required, even in a relatively limited space, avoiding the need for dedicated space allocation.

The clinical interpretation of gait data gathered can be quite different from that collected for research but requires reflection and caution when using the data alone to decide on treatments and other interventions. Gait technologies can only provide data on what they record. They do not register a diagnosis nor propose a treatment plan. They can support a diagnosis and highlight particular areas of dysfunction, but in the context of the high variability in human anatomy, each patient should not be expected to achieve 'perfect normality' of gait data. Indeed, one of the most common clinical mistakes the author has observed is for gait analysis to be taken while a patient is in pain during locomotion. Pain alters gait (antalgic manoeuvres such as limping) and all that is consequently recorded is the patient trying to avoid loading the injured tissues. Little light is thrown on the origin of the injury.

Where clinical gait analysis is particularly effective is in assessing improvements instituted through treatment interventions, as improved gait function is associated with successful patient management in many conditions. Gait technologies can also give an idea as to how well a structure has healed. For example, a weak triceps surae or Achilles tendon dysfunction will depress the peak GRF generated on the forefoot at heel lift, resulting in a suppressed impulse peak on a force-time curve or lower pressure peaks on the forefoot. On recovery of normal ankle plantarflexion, improved forefoot GRF should be reestablished, an effect seen on a force plate force-time curve or even on pressure analysis-derived force-time curves. In this context and many others, comparing data from the symptomatic side with that from the asymptomatic side can be particularly helpful in establishing what should be normal for a patient, although of course, the quality of the asymptomatic limb must be taken into consideration.

1.2.2 Stability and ground reaction force

Nothing makes sense in gait without understanding the principles of GRF interacting with CoM motion. Other important principles underpinning gait efficiency and motions are all underscored by the interaction between the support surface and the body's mass. Lower limbs, particularly the feet, are usually the interface between the ground and the body. Where shoes, orthoses, or prostheses lie between the body and the ground, they will also have an influence on the GRF-CoM relationship to the benefit or ill of the individual's gait and postural stability.

CoM influence on stability

Within the body as a collective and also in each body segment is a CoM individualised for each human that gravity pulls upon (Rabuffetti and Baroni, 1999). Gravity wishes to pull the CoM towards the support surface. For gait to occur efficiently, gravity needs to be restrained and utilised to the body's energetic advantage. Each body segment is artificially divided for the convenience of understanding gait. The pelvis finds itself squeezed between the passenger and locomotive segments, despite the fact that it takes part in influencing both regions, although the degree it moves with one or the other depends on gait speed. The swing limb, which is part of the locomotor system, also has an influence on the mass distribution of the passenger segment while it is off the ground during its swing. It does this by its weight drawing the CoM to the side of the body that the swing limb is on. Yet, it also plays a primary role in locomotion by moving the HAT's CoM forward during late swing through centrifugal forces.

During gait, the CoM of the limb segments and the HAT of the passenger unit are moving in all three orthogonal planes, up-down, side-to-side, and forwards (hopefully not backwards). These motions cause 'oscillations' of the CoM. In walking, the CoM normally rises vertically and anteriorly from the end of loading response during the early midstance period of the support limb while the other (contralateral) limb is in early swing. The CoM starts to fall following absolute midstance, during contralateral midswing, and on throughout late midstance and terminal stance into preswing and the start of initial swing, just as the contralateral foot reaches the end of loading response (Crowe et al., 1993, 1996; Crowe and Samson, 1997). Extension of the lower limb during early midstance out of the more flexed posture at the end of loading response functionally lengthens the lower limb, raising the CoM upwards. Afterwards, this gained height enables the trunk through the hip and pelvis to fall forward, extending the hip and ankle behind it and thereby reducing the CoM's height. This motion creates vertical oscillations of the CoM which should be relatively symmetrical left to right when walking on flat surfaces without perturbations in gait (Crowe et al., 1993, 1996; Crowe and Samson, 1997) (Fig. 1.2.2a).

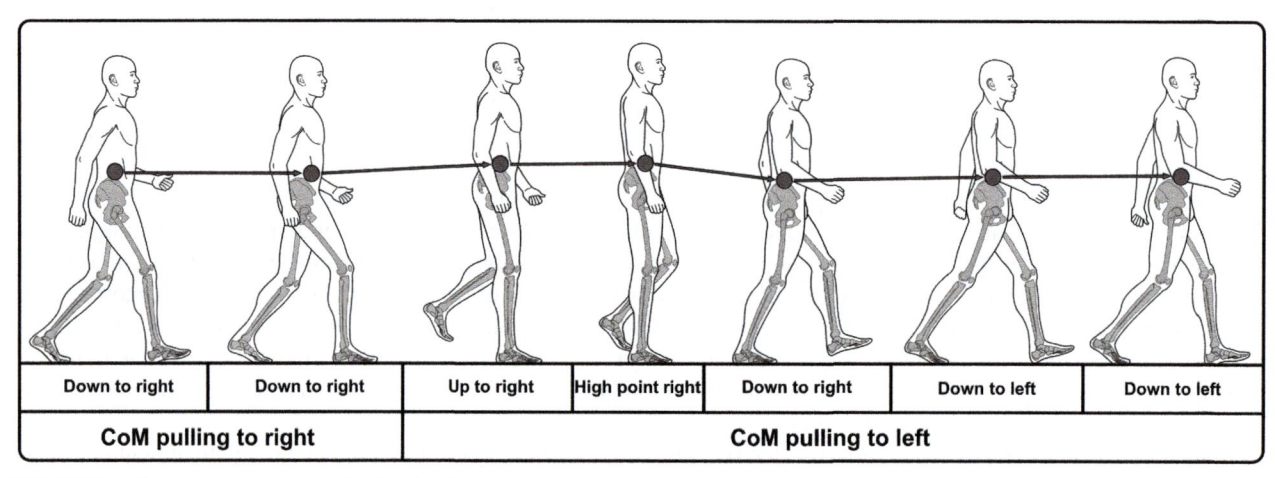

FIG. 1.2.2a Expected movements of the CoM during the right stance phase. Initially, the CoM is transferred across to the right foot as the left foot completes its terminal stance and the body weight loads to the right. Once the left foot enters swing phase, its mass adds to the trunk's CoM drawing it towards the left. Muscles such as the hip abductors limit the CoM's drift to the left, but the left swing limb remains a constant draw to the left side, despite the right leg supporting the CoM. Once the left foot approaches ground contact, the body mass is allowed to move towards the left limb, taking the CoM with it towards the right medial forefoot before transfer to the left foot at its initial contact. This CoM motion across the right foot medially is assisted by extrinsic foot muscles. The situation now reverses as the right limb enters swing and the left limb takes support responsibility for the CoM. *(Permission www.healthystep.co.uk.)*

The CoM oscillations do not just occur in the vertical direction. They also move from left to right, moving towards the stance limb on loading and during most of single support, and then moving back towards the contralateral foot in the direction of the medial forefoot during late midstance and terminal stance to facilitate CoM loading of the opposite limb (Crowe et al., 1993, 1996; Crowe and Samson, 1997). As the CoM position is influenced by the placement of the mass of the body, carrying objects influences CoM positions and movements. Carrying a bag on the left shoulder can shift CoM oscillations more to the left (Crowe and Samson, 1997) and will thus require postural corrections to try and equalise CoM distribution back to the right. The degree of oscillations of the CoM in gait is therefore important to stability as well as energetics. Furthermore, CoM motion is under intense neurological control, particularly in regard to changes in postural balance such as occur during gait initiation (Chastan et al., 2010) resulting in the consequent risk of falls when such controlling mechanisms are inhibited. Moreover, each body segment will have its own oscillation patterns during gait as well as the whole passenger segment and its parts, such as the head or pelvis (Brodie et al., 2015). To appreciate the role of CoM in gait, it is easiest to start with the effects of static standing on the CoM's position.

Statics and ground reaction force relationship

Standing upright in a bipedal posture is far harder to achieve than standing upright in a quadrupedal posture. Consider a four-legged table compared to a two-legged table. Each body segment is being pulled to the ground by gravity, with only the two feet in direct contact with the ground. Thus, the feet at least are not going to fall any further (although the height of the vault of the foot can). To remain erect, it is necessary to restrain each body segment against gravity. This restrained stability can be passive or active. Having a tail helps to balance the CoM posteriorly to anteriorly, but humans have lost theirs, which means anterior–posterior oscillations must be restrained.

Consider someone standing with both feet on the ground. The mass times gravity produces body weight force directed into the ground, which is matched by an equal but opposite force pushing back into the feet. This is the GRF. If the individual stays absolutely static (a hard thing to do), then the forces under the feet will not change. In gait research, the position of these resultant GRFs is termed the centre of pressure (CoP), although they are nothing to do with pressure. Thus, to avoid confusion, in this text, we have referred to these as the CoGRF. When standing on both feet, each foot will generate its own CoGRF. These CoGRFs have a quantity in Newtons and most importantly, being vectors, they have direction. Altering the feet's position or the position of the body over the feet will move the CoGRF, changing the directions and magnitudes of the GRF under each foot. Sway forward and the CoGRFs will move forward towards the forefeet. Sway to the left and GRF will increase in the left foot and decrease in the right, and so forth. In reality, even when we think we are standing still, these CoGRFs are moving constantly as there are tiny sways caused by, among other things, breathing and heartbeat (Conforto et al., 2001; Hodges et al., 2002).

Passive stability is achieved by having the CoM of each body segment stacked above the ones below, down to the pedal weightbearing surfaces. Bipedal creatures such as humans are very top-heavy, with around 70% of the mass located above the lower limbs. Hominin bipeds are unable to use tails to balance any anterior–posterior displacement, so the trunk cannot be angled in the manner of a bipedal dinosaur but must be positioned vertically above the limbs. The CoM-supporting lower limbs are also multisegment points of rotation, and each sits on contoured joint surfaces that tend to rotate to facilitate motion. The ligament-supported lower limb skeleton is built for mobility, with most joint surfaces being rounded. This profoundly narrows the range of positions in which the body can be stable when standing (Figs 1.2.2b and 1.2.2c).

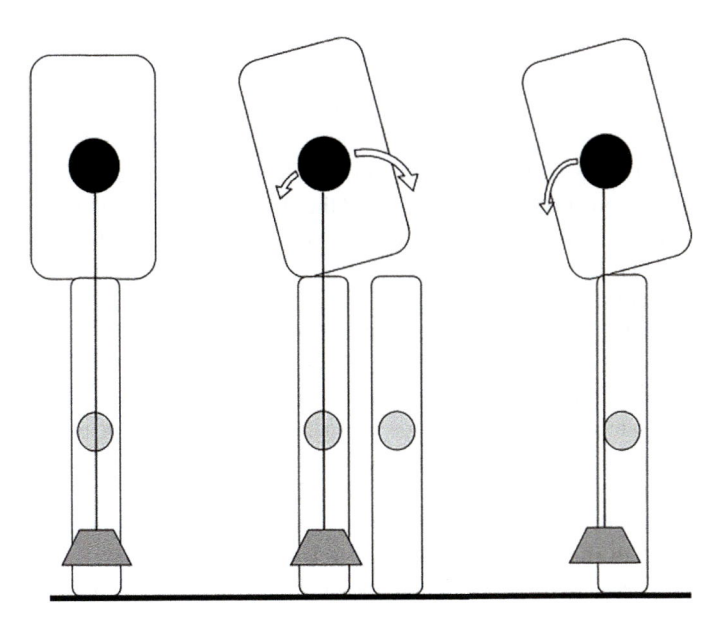

FIG. 1.2.2b The stability of the centre of mass (CoM) in stance and gait is critical and easily appreciated with blocks. As long as the CoM in the top block lies above the CoM of the bottom column, stability is assured. When the CoM lies above a wide column or two columns, displacement of the CoM can be considerable, with gravitational forces still wishing to pull the top block's CoM back towards stability above the supporting columns' CoM. However, displacement of the CoM on a single column is far easier and can initiate falls. In the stance phase of gait, double-limb stance offers better CoM stability, but in single-limb support, higher CoM instability may risk falls. Such instability induces the inverted pendulum mechanics that allows the CoM to move forward. Thus, excessive CoM stability will block motion, while too little risks uncontrollable instability. *(Permission www.healthystep.co.uk.)*

For humans to achieve balanced standing stance, three forces must be controlled. These are the forces from the position of the CoM, the passive connective tissue tension forces (fascia and ligaments), and the muscular activity through tendons and other muscular attachments. The hip and knee, despite having rounded joint surfaces, are able to use tensions from large ligaments to provide postural assistance as and when necessary, with subtle muscle contractions via feedback mechanisms acting together to provide stability. A danger seen in some (usually hypermobile) individuals is that knee hyperextension offers better passive stability than the usual near-extension knee position, because the ligaments are too flexible to provide tensioned stability. The ankle in stance needs to be positioned centrally above the axis of the subtalar joint, which usually has some small yet significant frontal plane flexibility. The posterior heel is closer to the ankle than the forefoot or toes and the calcaneus only loads through the posterior tubercle, although the extensive heel fat pad helps increase the heel's loaded surface area, increasing congruency with the support surface.

FIG. 1.2.2c Stacking thin columns on top of each other greatly increases the chances of CoM displacement from the vertical line of support, causing instability. The size of the base of support becomes very important when upper columns are going to pivot under load. Having a wider structure at the base creates enhanced stability, explaining why the foot offers the ability to spread, widen, and lengthen during stance. As instability of the lower limb column mostly occurs within the sagittal plane, having a posteriorly projecting heel and a long anteriorly projecting midfoot and forefoot can assist CoM stability within the limbs and trunk. The costs of transporting the foot during swing limits the size the support base can be. *(Permission www.healthystep.co.uk.)*

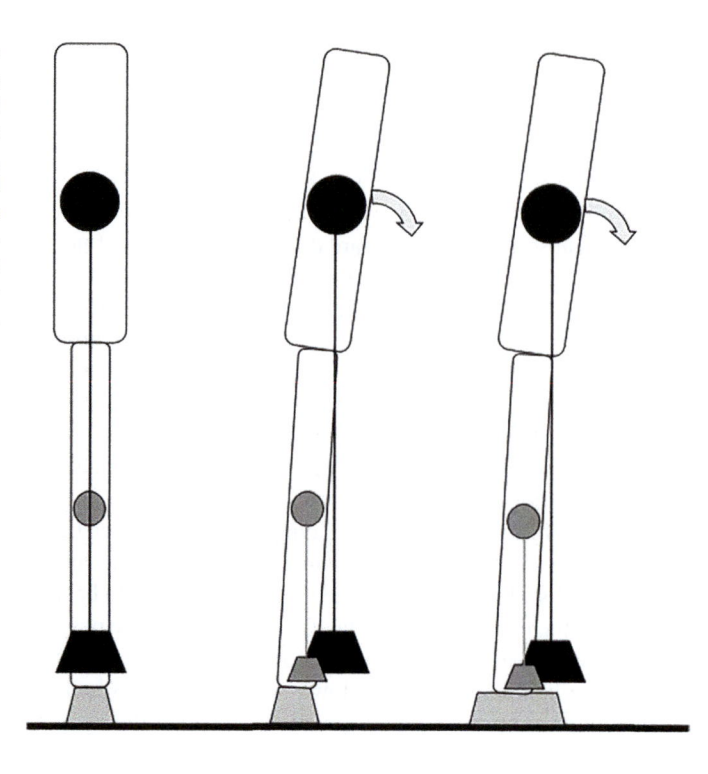

It is theoretically possible that stance could be attained without muscle function if the CoM of the HAT (variably assigned to the anterior margin of the eleventh thoracic vertebra on average) sat precisely above the axis of the hip, knee, and ankle-subtalar complex joints. This would avoid angular rotations within the sagittal plane. However, such stability is lacking as none of these joints are usually restrained in this precise position during stance, and the slightest sway unbalances the segments anyway. The body's CoM vertical vector is always situated in front of the spine in healthy populations (Gangnet et al., 2003; Schwab et al., 2006; Le Huec et al., 2011). It has been calculated, on average, to extend vertically down from the centre of the head, passing 1 cm in front of the 4th lumbar vertebral body, 0.6 cm posterior to the hip, and then passing anterior to the knee and continuing down to 5 cm in front of the ankle with around 2 cm of deviation (Perry, 1992: p. 26). This indicates some significant variation in the resting stance location of the CoGRF within individual's feet. The anterior position of the GRF under the foot during static stance requires a small amount of ankle dorsiflexion if knee extension is to be limited to below zero. In those who demonstrate knee hyperextension when standing, a neutral position for the ankle is possible. However, CoM displacement in gait helps create the angular moments that provide the instability necessary for locomotion, although the correct CoM positioning over joints is critical to prevent falls (Fig. 1.2.2d).

In the frontal plane, balance is maintained in a base of support determined by the width marked out by the distance between the lateral margins of the feet. The tendency for the feet to slightly abduct outwards towards the forefoot increases the base of support anteriorly compared to the heels. Obviously, moving the feet apart increases the frontal plane stability of standing (Fig. 1.2.2e). Indeed, this is a strategy that humans employ when standing on an unstable surface, which is why tightrope walking is such a challenge. Normal healthy individuals use combinations of frontal and sagittal plane positioning to maintain upright stability, countering subtle instability caused by motions such as cardiac dynamics. Thus, there is a constant shift in body weight between the feet in quiet standing.

The relationship between CoGRF and posture can be influenced by postural dysfunction. Back pain has a direct influence on different postural strategies (Moseley and Hodges, 2005; Brumagne et al., 2008), changing the position of the CoGRF location and postural sway when compared to healthy populations, and seemingly limiting their ability to respond to perturbations (Henry et al., 2006). Subjects with low back pain tend to have their CoGRF more posteriorly positioned (Mientjes and Frank, 1999; Brumagne et al., 2008). It has also been shown that the position and velocity in the sagittal plane of the CoGRF is in association with the angulation of the spinal curvature in the cervical, thoracic, and lumbar regions (Boulet et al., 2016). It has been noted that older individuals are not able to adapt their postural control mechanisms to stabilise their CoGRF position (Dault et al., 2003). Furthermore, it has also been suggested that by increasing thoracic

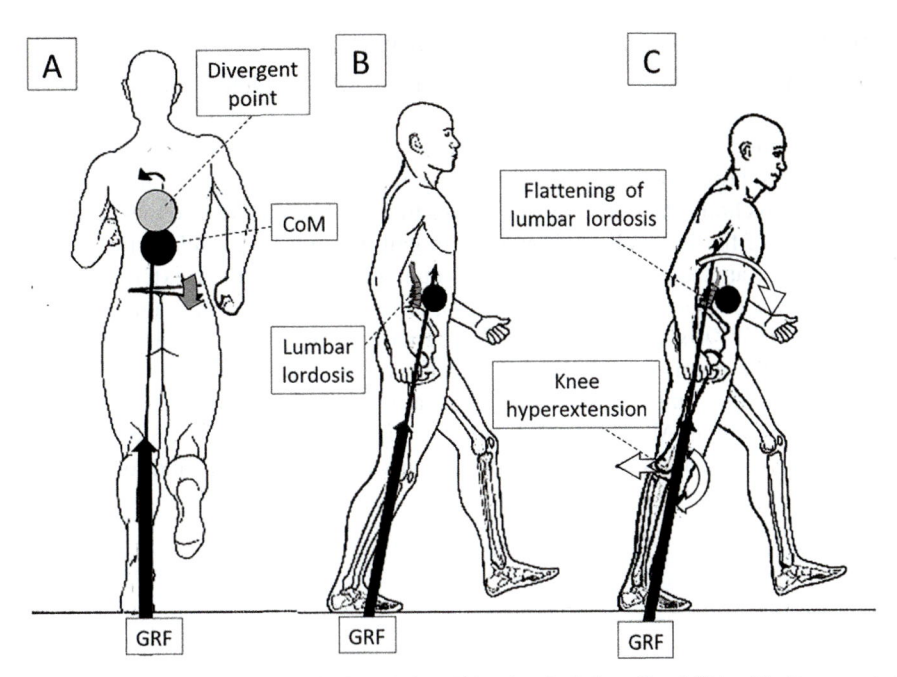

FIG. 1.2.2d In single-limb stance, the CoM's position in the frontal plane (A) and sagittal plane (B and C) is critical to promote the correct joint motion and balance. In the frontal plane, the hip varus tilt caused by the CoM being medial to the support limb risks destabilising the CoM towards the swing limb side. Hip abductor activity displacing the CoM towards the support limb side prevents this and keeps the CoM nearer the GRF vector. These processes are also required in single-limb support standing. In the sagittal plane during gait, the CoM should lie close to and above the GRF vector, although the GRF vector in reality seems to be angled to a point slightly above the CoM that stabilises the trunk's flexion/extension motions (the divergent point). Certain postural and anatomical features such as the lumbar lordosis and the maintenance of slight knee flexion during midstance help to keep the CoM near to the GRF vector. Orthopaedic postural issues (C) such as knee hyperextension and loss of lumbar lordosis can cause the CoM to displace anteriorly during single-limb support, causing gait to destabilise and altering the flexion/extension moments across the hip, knee, and ankle. *(Permission www.healthystep.co.uk.)*

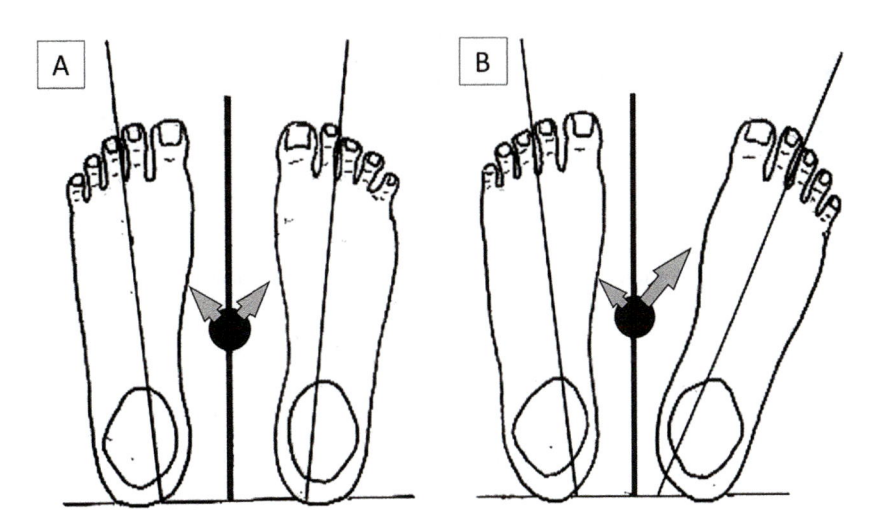

FIG. 1.2.2e In quiet standing, adult feet usually position roughly 9–10 cm apart at an angle of around 7° out-toe or external rotation (A). In this position, both feet provide stabilising resistance to the CoM with a slightly wider base of support anteriorly. If one foot is more abducted (right foot in B), then the CoM can displace anteriorly to the abducted foot side more easily. In gait, such foot abductions may permit more drift of the CoM to the affected limb during loading response, resulting in more work being required to move the CoM to the opposite side in terminal stance. In-toe angles can present the opposite challenge, blocking the anterior movement of the CoM in the affected (adducted foot) stance limb. *(Permission www.healthystep.co.uk.)*

kyphosis and decreasing lumbar lordosis (a common feature of older adults), aged humans may be able to stabilise the CoGRF anteriorly and limit its velocity of movement during stance (Boulet et al., 2016). Widening the feet away from each other and abducting the feet with external limb rotation would also help in increasing stability. When the CoM falls outside the base of support, the trunk starts to fall. Gait is about deliberately forcing the passenger segment to fall, albeit in a controlled manner.

Principles of maintaining stability in stance phase of gait

Upright bipedal posture is maintained by the force of gravity and the GRF, which needs to produce a sufficient vertical component with appropriate aiming to produce a torque about the whole-body CoM (Gruben and Boehm, 2012b). The GRF must produce a vertical magnitude that averages body weight, as well as a vertical component and point of application to produce a torque around the CoM that yields an average angular acceleration of zero, with the nervous system having the flexibility to choose how to control muscles to modulate the GRF to meet these criteria (Gruben and Boehm, 2012b). The stability of the limb's base of support in gait depends on muscle action, the support surface, the position and stability of HAT, and any external forces encountered. The primary external force is the GRF which identifies the *body mass vector* position (Perry 1992: p. 24), representing body mass under the influence of gravity. Hence, '*body weight vector*' (*BWV*) is a more descriptive alternative. The BWV is the vector of force that generates the GRF, and the CoGRF is the precise and averaged location of the GRF centred to one point. Humans are very top-heavy and their multi-articular limbs must change their alignment to accommodate the support surface below and the oscillations of the CoM above. Alignment of the body weight is a dominant feature of stability. In gait analysis, the CoGRF helps tell us about the location of the BWV of the trunk through the limb and foot using Newton's third law (Fig. 1.2.2f).

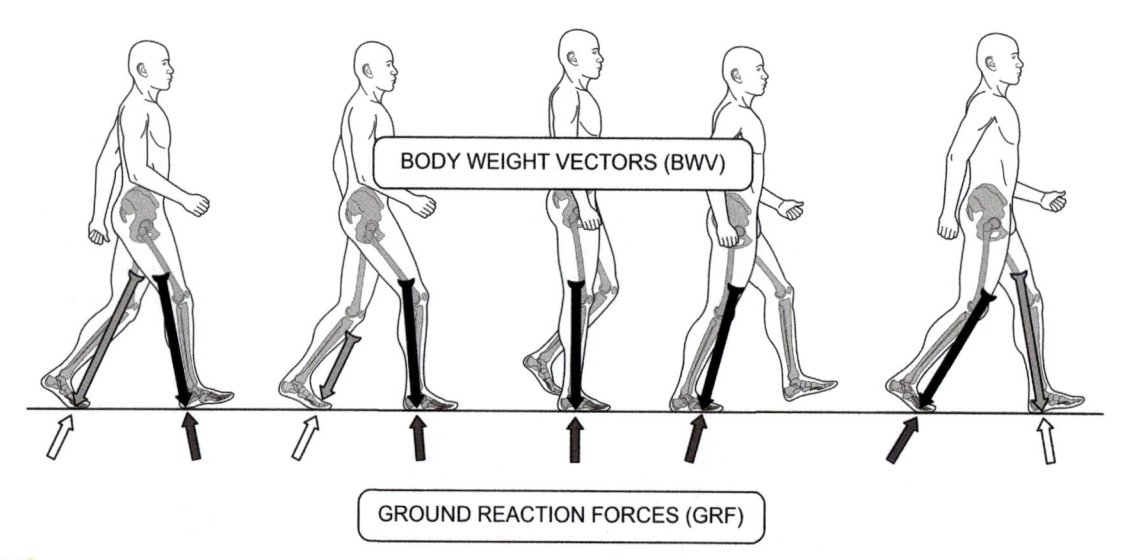

FIG. 1.2.2f The relationship between the direction of the body weight vector (BWV-downward arrows) and the ground reaction force (GRF-upwards arrows) through Newton's third law of every action results in an equal but opposite reaction, underpinning much of the biomechanics of the stance limb. Moving from the left side of the image, the right foot contact creates a BWV angled vertically down and anteriorly, creating an equal but opposite GRF angled vertically up and posteriorly into the limb. On the left foot, the vertical force is rapidly decreasing but the anteriorly directed GRF remains high as the left foot BWV pushes backwards against the ground. Initially, the BWV becomes more vertical and less angled anteriorly through early midstance until forces align vertically without any anterior or posterior element. Only after this point does the BWV direct posteriorly through the stance limb. As a result, the GRF is initially angled posteriorly and then slowly changes to become increasingly angled anteriorly. Vertical GRF decreases in the middle of midstance as the CoM rises and then increases under the pull of gravitational fall until heel lift, when weight starts to transfer to the contralateral (in this case, *left*) foot and limb. *(Permission www.healthystep.co.uk.)*

The BWV is applied to the ground through the leg to the foot. Thus, the angulation of the lower limb and foot to the ground dictates the direction of the GRF vector. This is why the GRF vector does not perfectly align to the body's CoM but rather to the 'divergent point' which lies above the CoM, the exception being at the middle of midstance (absolute midstance) when the CoM lies directly above the GRF vector (Gruben and Boehm, 2012a). GRF vectors align to a point in the trunk which should be angled to the divergent point in such a way that it helps resist angular moments from destabilising the trunk (Gruben and Boehm, 2012a). GRFs are therefore critical to understanding gait stability (Gruben and Boehm, 2012a,b) (Figs 1.2.2g and 1.2.2h).

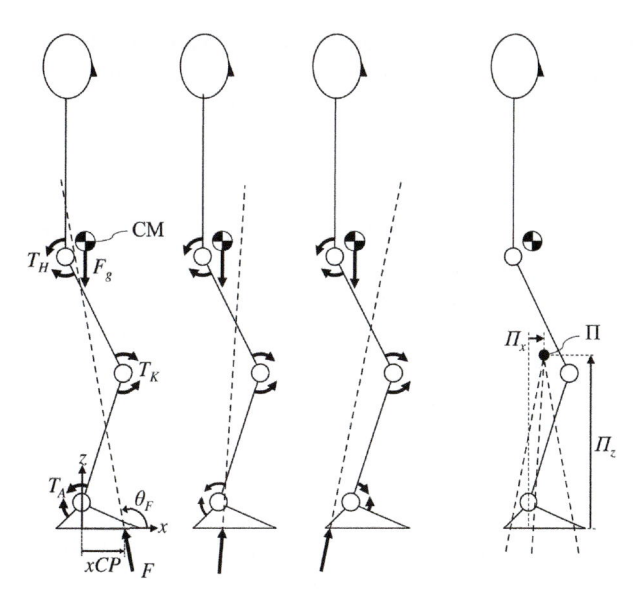

FIG. 1.2.2g Equations of motion show that varying the ankle torque (T_A) results in a specific pattern variation between the instantaneous GRF (F) and the centre of pressure (xCP), which we have referred to in this text as the CoGRF. The three figures on the left show the CoGRF line of action relative to a person with the same posture and hip and knee joint torques, but with a different ankle torque and CoGRF (xCP). Overlaying the three GRF lines of action demonstrates the intersection at Π in the right figure. *(From Gruben, K.G., Boehm, W.L., 2012. Mechanical interaction of center of pressure and force direction in the upright human. J. Biomech. 45 (9), 1661–1665.)*

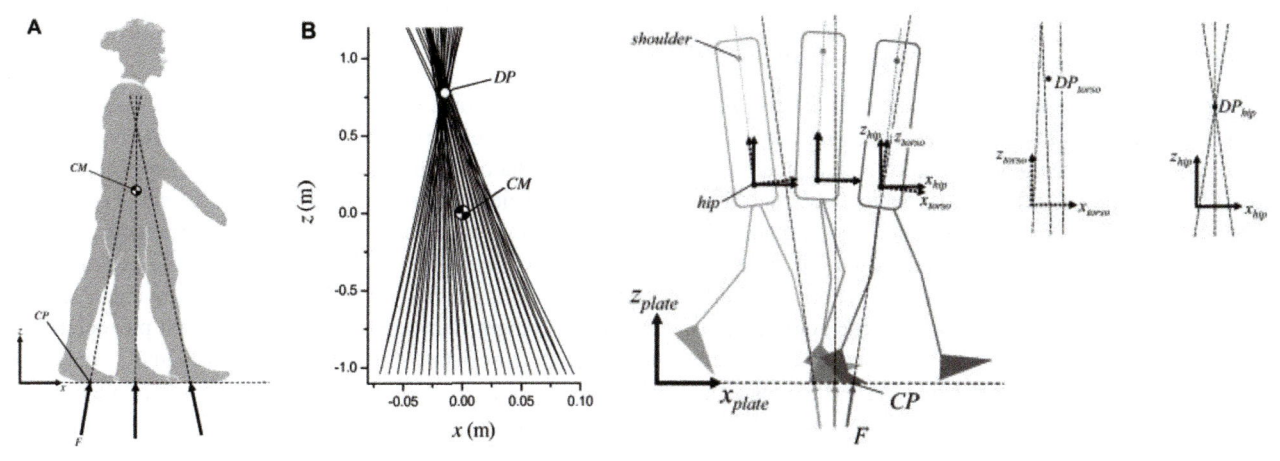

FIG. 1.2.2h The GRF creates torques across the limbs into the trunk to a divergent point (DP) above the centre of mass (CM) that helps stabilise gait (A and B). The divergent point is not fixed because the trunk and hip segments move differently within an individual, so that the projected GRF vector (F) from the CoGRF (CP) results in a unique divergent point with respect to each moment of time during gait *(right images)*. The trunk's pitch is exaggerated in these images, but relative phasing is accurate to the point of the gait cycle, with leaning forward at the start of single-limb support and backwards at the end. The arrival of two GRF forces in double-limb stance should increase stability. *(From Gruben, K.G., Boehm, W.L., 2012. Force direction pattern stabilises sagittal plane mechanics of human walking. Hum. Movement Sci. 31 (3), 649–659.)*

Recording GRF

GRF is recorded in gait analysis by using force plates. The force plate records forces that are generated in 3D, such that a resultant GRF can be produced by combining the vectors from each orthogonal direction of force. As this GRF vector is equal but opposite to the BWV, the direction of the BWV can also be ascertained from the angle of the GRF vector. The individual elements of the force can be broken down further into elements of vertical forces, anterior–posterior (or horizontal) forces, and medial-lateral (or transverse)-directed forces. Under the influence of gravity, vertical forces dominate. Where such forces are applied to the foot depends on its alignment to the support surface at any given moment of time, influencing which, when, and how muscles need to activate in order to maintain stability. This means that gait parameters such as step length, stride length, and base of support are important, as they influence the orientation of the GRF vector, resulting in the location of the divergent point. In the clinical absence of equipment to record the location of the GRF vector, understanding the GRF principles become key to gait interpretation. It is therefore helpful to understand how force plates generate force-time curves for the understanding of events during gait.

Muscle activity occurs in both the locomotor and passenger segments. However, the intensities and functions of these activities are quite different. The passenger HAT unit is largely responsible for its own postural integrity and its demands are kept to an energetic minimum, with events such as arm swing seeming to have little effect on overall energy costs (Perry, 1992: p. 20). The most important contribution the passenger unit makes to gait is to have the CoM alignment positioned to maximise energetics, ensuring appropriate orthogonal displacement of the CoM above the lower limbs. CoM displacement has a profound effect on muscle action and its efficiency. The CoM of the trunk should therefore be positioned in such a way as to ensure that most stance phase muscular activity is eccentric (low energy/high force), utilising the principles of inverted pendulum mechanics to achieve cycles of stance phase kinetic and potential energy generation under gravitational forces.

The foot is the only human structure evolved specifically to manage direct interface GRF interactions. It has evolved to provide an energy-dissipating phase (a deliberate attempt to shed energy) when it first contacts the ground and then to change primarily into an energy-storage structure, before it releases stored potential energy as kinetic energy in order to help propel from the ground during terminal stance. This translates as there being a deceleration, braking, or force reduction event during contact and loading, and then an acceleration or force generation phase before the end of stance. Every force the body creates against the ground during these decelerations and accelerations is equal to the force of the ground pushing back against the body. Force plates record this interaction as force-time curves and are able to reveal the direction of GRF vectors. The interaction between the anatomy and the ground explains much that is observed in human gait and pathology.

1.2.3 Measurements and interpretation of GRF

Force plates/platforms can identify and quantify forces in all orthogonal directions. Under stress (force over cross-sectional area), materials deform. Force can be measured by using materials that have electrical sensing properties that constitute a part of an electric circuit that can deform under load, thereby altering conductivity in a predictable manner. Deformation of these elements causes geometric and structural alterations that change the electrical resistance. The same effect is achieved by stretching a thin copper wire in an electric cable. The thinner it becomes, the harder it is to pass current through it, altering the voltage in an electronic circuit. *Piezoelectric* or *piezoresistive* elements are used in force plates and they are very sensitive to force deformation. Their electrical resistance and voltage are directly related, allowing calculations of the force from the voltage change in the circuit. Although the underlying principles are different between piezoelectric and piezoresistive effects, when connected to a charge amplifier, the piezoresistive concept remains the same as the piezoelectric in that deformation leads to a measurable electrical effect directly related to the amount of force applied. The result of this technology means that kinetic data of the GRF can be recorded.

The applied force in gait continually changes its magnitude. Thus, any quantitative way of assessing this requires equipment with dynamic response capabilities and a transducer, which must always produce the same voltage at the same force in Newtons. These devices need to be sensitive enough to detect small force changes and act at a sufficient range of forces that occur during gait. Hysteresis effects need to be avoided or else the loading force will produce a different voltage than the same force during offloading. The effects of vibration force resonance must also be avoided as structures resonate if the external vibration is at or above the natural frequency of the structure, and this will corrupt data.

Both types of force platforms need to be installed flush to the ground. The piezoelectric force platform is the more sensitive, allowing for a larger range of force measurements. This technology is synonymous with the Kistler force platform. The basis of signal generation in these systems is through deformation of piezoelectric crystals such as quartz, and these generate an electric current as they deform. The co-ordinate system is determined in the platform by the alignment of the crystals within it. Piezoelectric platforms are able to measure frequency content up to 1000 Hz in all three directions. The disadvantage is that such systems experience signal drift and need to be reset following each test. Although this is a simple task, this signal drift may cause problems. Strain-gauge platforms are adequate for walking studies and for clinical use during nonsporting activity assessments. They can measure up to 1000 Hz vertically, but only around 300 Hz horizontally for anterior–posterior and medial-lateral forces, hence, the limitations in running research tests where horizontal forces are higher. They work on the deformation principle that objects change shape when loaded with force. Strain gauges contain materials which, upon distorting, produce resistance. By measuring such resistance changes, the strain can be measured. The signals produced are small and need amplification, either within the platform or externally.

Good force plates measure GRF in three dimensions: vertically, along the length, and across the width of the plate. The various models use a different axis terminology for different directions of force to create a plate reference system. It is worth checking which letter is used in the plate's reference system, although mention must be made that it should be fairly obvious which direction of force you are looking at on force-time curves once you know what to expect from the data (Fig. 1.2.3a).

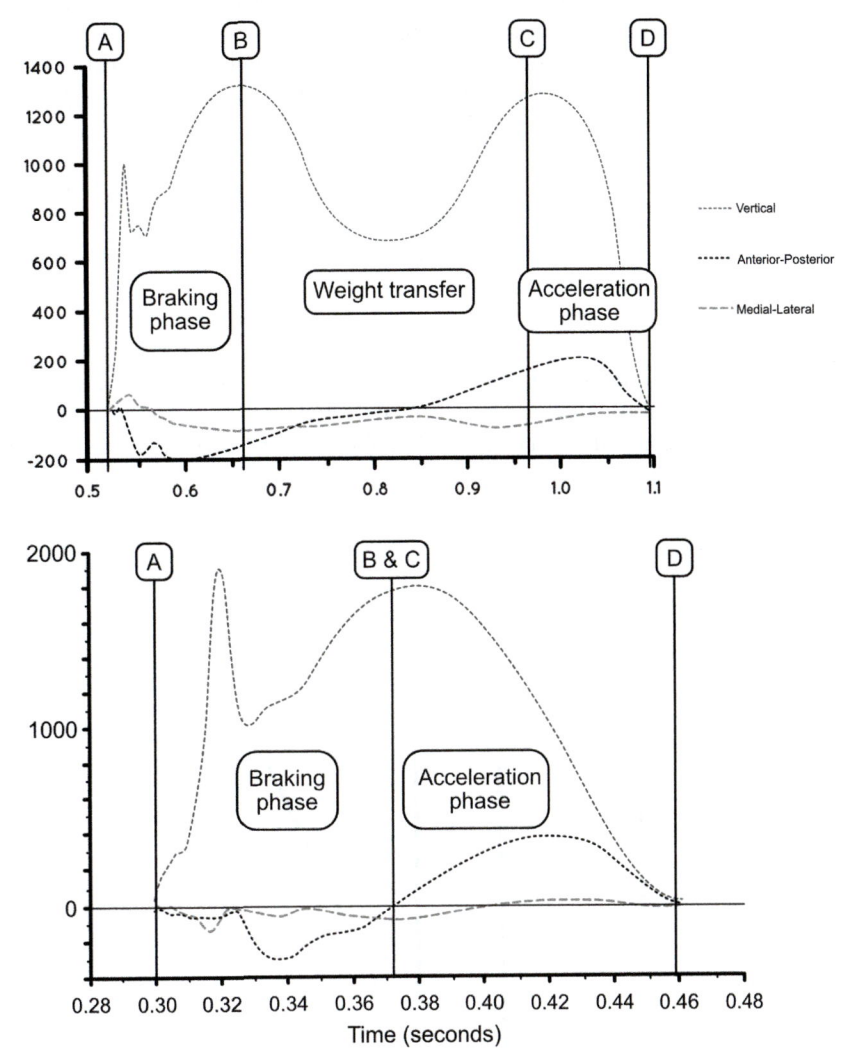

FIG. 1.2.3a The force-time curves of walking *(upper image)* and heel strike running *(lower image)* are compared. Walking and heel strike running have a similar heel strike transient early in the braking phase, a period that lasts from A to B. Both gait styles also demonstrate an acceleration phase ending with toe-off (D). Where walking has a distinct impulse reducing weight transference phase lasting from B to C, running gait braking and weight transference blend from one to the other (line at B and C—lower image). In running, braking and acceleration both involve the ankle angle changing to redirect the CoM over the limb from a position posterior to it to one anterior to it. The changeover from braking to acceleration during running is indicated by the horizontal impulse, demonstrated here where the negative posterior impulses change to anterior positive impulses. *(Permission www.healthystep.co.uk.)*

Force plates only record one resultant force vector regardless of how many objects are applying force to different locations on the plate at the same time. The GRF recorded is therefore numerically and physically equal to the forces that are applied at a given moment of time to the plate. Because the GRF recorded usually arises from a distribution of forces from areas of, for example, the plantar contact points of the foot, the location of the GRF calculated is often referred to as the CoP. This is the centre of all the GRFs gathered together at an averaged location and direction, much as the CoM is the location where all the surrounding mass is brought to a centre point. As stated earlier, the CoP recorded from a force plate has nothing to do with pressure (force divided by area) and is therefore probably not a good term to employ. Unfortunately, it has become ingrained in the biomechanical/gait analysis lexicon, so that replacing the term is difficult. It can cause confusion when learning about gait and gait analysis technologies. The CoP associated with force plates actually represents the centre of the ground reaction forces (CoGRF) to produce a combined GRF vector. As such, in this text, CoGRF will be continued to be used over CoP in learning about force plates, and CoP will only be employed when discussing pressure data generation as the averaged point of pressures.

58 Clinical biomechanics in human locomotion

The CoGRF constantly moves across the plate's surface during stance phase of gait, within the part(s) of the foot or shoe that are in ground contact. In a gait laboratory, it is desirable to perform a 3D kinematic analysis at the same time (if possible), so correlations can be made between the body/limb positions and the CoGRF and the vectors it creates. Thus, kinetic data can be compared to kinematic data. The GRF can be broken up into components depending on the direction of forces that are applied. GRF will be applied in vertical, medial-lateral, and anterior–posterior directions. Force platforms work on the principle that objects vibrate at a particular frequency during collisions. We will soon examine how this is also true of humans. The exact frequency of vibration depends on the mass and the dimensions of the object, and force platforms are no exception to this rule. To work for humans, it is important that the natural frequency of a force plate's vibration is higher than the frequency of the content of the signal of interest, that is, of human gait. Sporting tasks such as running and jumping produce higher frequencies than walking.

Caution needs to be exercised at the initial contact and terminal stance phases of gait. This is because the equations used to calculate the CoGRF co-ordinates can be affected by any slippage in the anterior–posterior and medial-lateral directions that occur at the very beginning and end of the stance phase, when vertical forces are small. That aside, the force-time curves constructed by a force plate reveal the amounts and directions of GRF generated by BWVs. Forces are applied in all three orthogonal directions, but the largest forces are directed vertically from the falls and rises of the body mass into and away from the ground. Remember: force is equal to mass times acceleration. Thus, during stance the greatest impulses are driven by changes in acceleration occurring in the vertical direction. The horizontal forces in the anterior–posterior direction are next largest. They result from the CoM being driven forward from behind the foot in the first half of stance and then change into a posterior direction when the CoM is ahead of the foot (see Figs 1.2.2f and 1.2.3b). Medial-lateral-directed forces are smallest

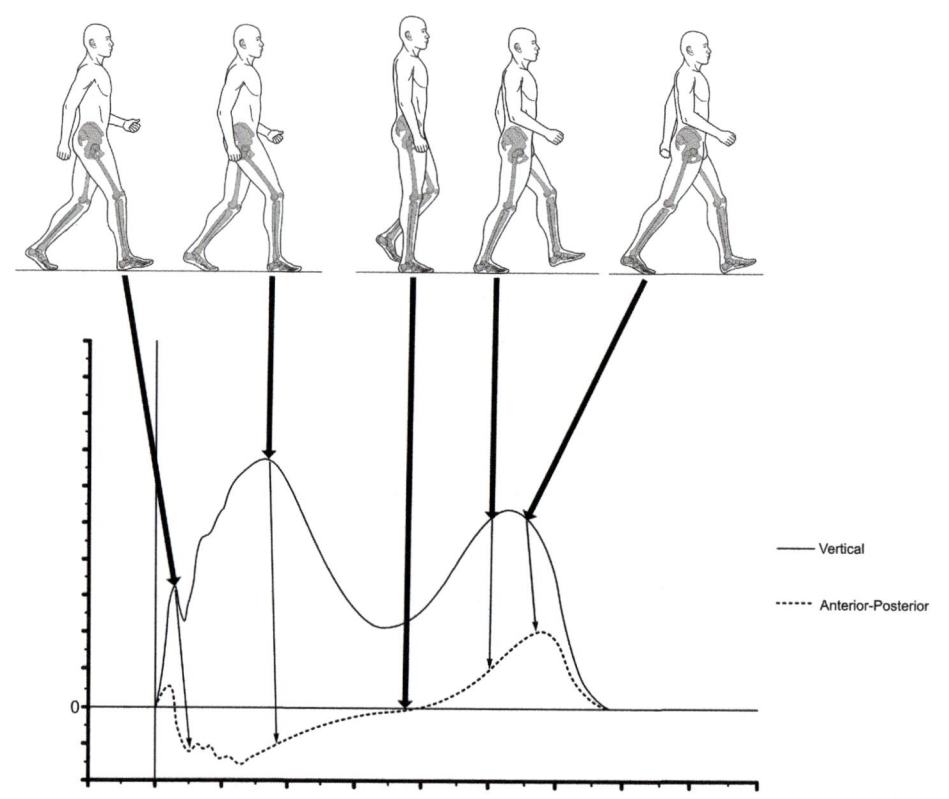

FIG. 1.2.3b The vertical element of the BWV *(indicated by the thick arrow)* and horizontal anterior–posterior element of the BWV *(indicated by the thin arrow)* are located to points of interest during walking stance on a force-time curve. This demonstrates the relationship of the BWV to the GRFs (vertical GRF = *solid line*, anterior–posterior GRF = *dashed line*). The heel strike transient creates a clear spike of vertical force at the same time as generating a posteriorly directed GRF against the BWV angled anteriorly (a negative reading in this example). Initially, there can be an anteriorly directed GRF from a posteriorly directed vector (an initial positive reading) if the heel is pulled back into the ground under hamstring activity at heel strike (known as 'claw back'), as can be seen in this example. Forefoot loading creates the highest early peak of vertical force early in stance, but the BWV is now angled more vertically so that the posterior horizontal GRF vector is decreasing (less negative force). This continues during early midstance. At the absolute midstance, the vertical forces are lowest, as the CoM has been rising away from the ground over the leg as an inverted pendulum. The CoM now being directly above the lower limb means a pure vertical force is created, so there is no anterior–posterior force vector (zero force). During late midstance, the vertical force rises with the CoM's accelerated anterior fall, which also angles the BWV posteriorly into the ground generating an anteriorly directed GRF (here shown as a positive force). Shortly after heel lift, the vertical GRF starts to fall as weight is transferred to the next step, but initially, the anterior GRF increases as the foot accelerates anteriorly over the forefoot, sending a powerful GRF posteriorly against the ground. *(Permission www.healthystep.co.uk.)*

and are dictated by which foot is on the ground, assigning a positive or negative force differently for left and right feet (in the examples shown in this text, the GRF is expressing a positive reading for both lateral and anteriorly directed GRF forces). Overall, medial-lateral forces are a more variable. Usually, they initially involve a small lateral-directed GRF (medial BWV) followed by sustained medial loading of the forces (laterally directed BWV) that later reduces during terminal stance. This reflects the foot motion shifting the CoGRF (and the BWV) medially, when attempting to drive from the medial forefoot, with some small degree of variability in each pattern between steps. Efficient heel-toe gait generates a predictable vertical and anterior–posterior force pattern during gait, and it is important to be able to recognise this pattern and what it means (Fig. 1.2.3c).

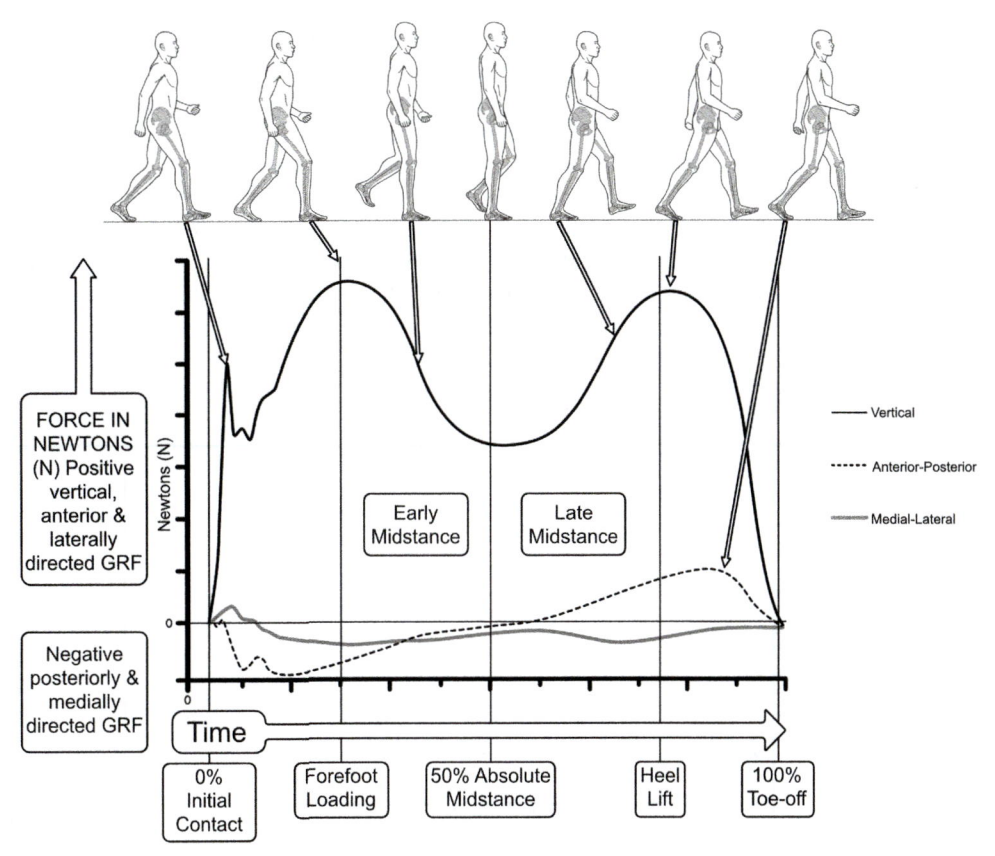

FIG. 1.2.3c The force-time curve of walking stance phase compared to events during the gait cycle. Note that vertical forces can never be negative, as no vertical force is zero. However, force plates are set up to define anterior–posterior and medial-lateral as either negative or positive forces. Thus, it is very important that for consistency of data, subjects always walk over the plate in the same direction. In this and all images generated for this text, examples are shown with posteriorly directed GRF vectors as negative and anterior GRF vectors as positive, reflecting anteriorly and posteriorly directed BWVs, respectively. Medial and lateral-directed forces are more problematic, as whichever way the force plate is set to read positive and negative, left and right feet will read the opposite way. In this text, lateral GRFs are indicated as positive and medial as negative *(shown here as the solid line)*. Should our theoretical subject have been asked to walk over the force plate with the other foot, a mirror image (relatively) of positive and negative forces would be expected in the medial-lateral forces. *(Permission www.healthystep.co.uk.)*

Clinically, due to expense and the need for a dedicated area and a permanent fixation in the floor, force platforms are rarely available. However, the information they provide is extremely useful for understanding gait. Therefore, every clinician involved in biomechanics and intervening in lower limb gait during treatment should at least know how to read a force-time graph and know what forces should be generated in gait, even if they rarely ever get a chance to use the equipment. The peaks and troughs of forces generated by force platforms are readily discernible from force-time graphs and are given *F* numbers to help describe events.

1.2.4 Vertical GRF components in walking

First peak in GRF (F1)

When reading a force-time graph, the vertical component is by far the largest force during gait. This is due to the vertical oscillations of the CoM and the effects of gravitational forces. During normal human walking gait, the vertical components of GRF on the force-time graph can be divided into four sections by looking at the peaks and troughs occurring over time. Usually, the heel makes contact first followed by the loading response that allows weight acceptance and forefoot weight-bearing. The increasing mass and the acceleration of the CoM towards the ground through the loading limb give rise to an increasing force during this period. This creates a slope of escalating vertical force from initial contact to a first peak of vertical force (known as the *F1 peak*). This rise in force is occurring throughout the initial double stance phase, where significant weight is initially still loaded on the contralateral trailing forefoot, but with increasing load arriving on the leading foot's heel. The F1 peak ends with loading response, where the leading foot is fully loaded to the floor through the heel and forefoot (often called *foot flat*), and the trailing foot has offloaded and is progressing into swing. This initial peak should be ~1.2 to 1.4 times body weight during walking.

The escalating F1 slope usually has a marked interruption within it. An initial heel strike usually creates a very distinct peak followed by a short trough before the F1 peak develops fully. This is usually more noticeable when tests are performed barefoot, as the cushioning in the shoes often 'dampens' the effect. This brief vertical force spike (inflection point) is known as the *heel strike transient*. It is followed by a momentary trough as the next acceleration of the body after heel strike takes place through ankle plantarflexion, lowering the forefoot to the ground under the restraining activity of tibialis anterior. This forefoot plantarflexion occurs above the ground, so it is unable to generate a GRF initially and is therefore an unrecorded force event. Thus, there is a drop in the vertical GRF, as the forces generated do not interact with the ground in a vertical direction until the forefoot makes contact shortly after. Following this forefoot contact, the (F1) peak then starts to rise again to develop fully as the CoM falls forward, creating increasing forces that interact with the ground (Fig. 1.2.4a).

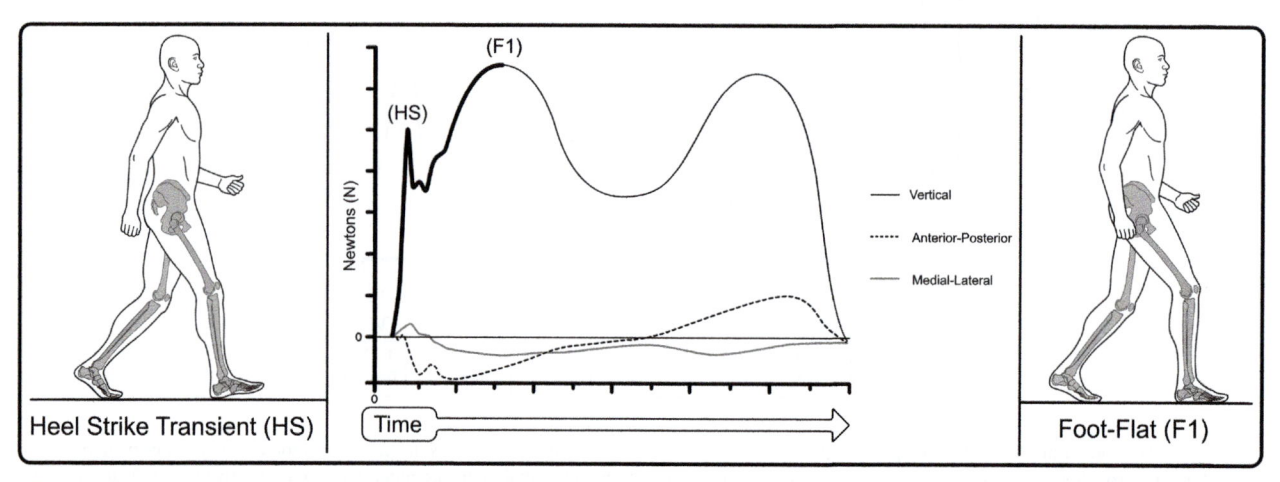

FIG. 1.2.4a Initial contact in walking normally begins with a heel strike transient and the F1 peak on the vertical force-time curve. These are shown with images of gait that this period covers. The F1 peak is reached at the end of loading response with foot flat, as weight completely offloads the opposite foot as it completes its preswing. Thus, the F1 peak should signify the start of single-limb support. Heel strike transients are usually most significant when barefoot, rather than during shod gait. *(Permission www.healthystep.co.uk.)*

First peak (F1) to trough (F2)

As the body's CoM starts to progress over the single support foot following the point at which the F1 peak is reached, the knee and hip start to extend, sending the CoM upwards. There is thus an acceleration of the CoM, but it is against gravity. This reduces the GRF because the acceleration into the ground of the CoM is decreasing (negative acceleration) as the CoM is moving away from the ground. This is similar to the effect seen on bathroom scales when you swing your arms upwards rapidly, thus creating a rapid weight loss. For some twenty years, the author has taught this principle through the concept of driving over a humpback bridge, replicating rises and falls of the CoM via the motion of the vehicle. As you rise up the slope, you get a feeling of lightness until you reach the actual top, whereupon the car now has potential energy from gravity to roll down the other side. As the CoM moves away from the ground with limb extension, it creates a decreasing vertical force slope on the force-time trace known as the *F2 trough*. This reflects a drop in downward vertical GRF. The lowest point of this trough of the vertical GRF should be when the CoM has reached its highest point, an absolute distance point of change in weight transference of the CoM. This is usually when the knee has reached near-extension, the ankle is in slight dorsiflexion, and the hip is neither extended nor flexed, but with the CoM lying over or very near to the hip joint axis. This is the point of maximum CoM height and has been referred to as absolute midstance within this text. The maximum depth of the trough should therefore equate to ~50% of the stance phase, and it represents the peak of lower limb potential energy storage through CoM height gain (Fig. 1.2.4b).

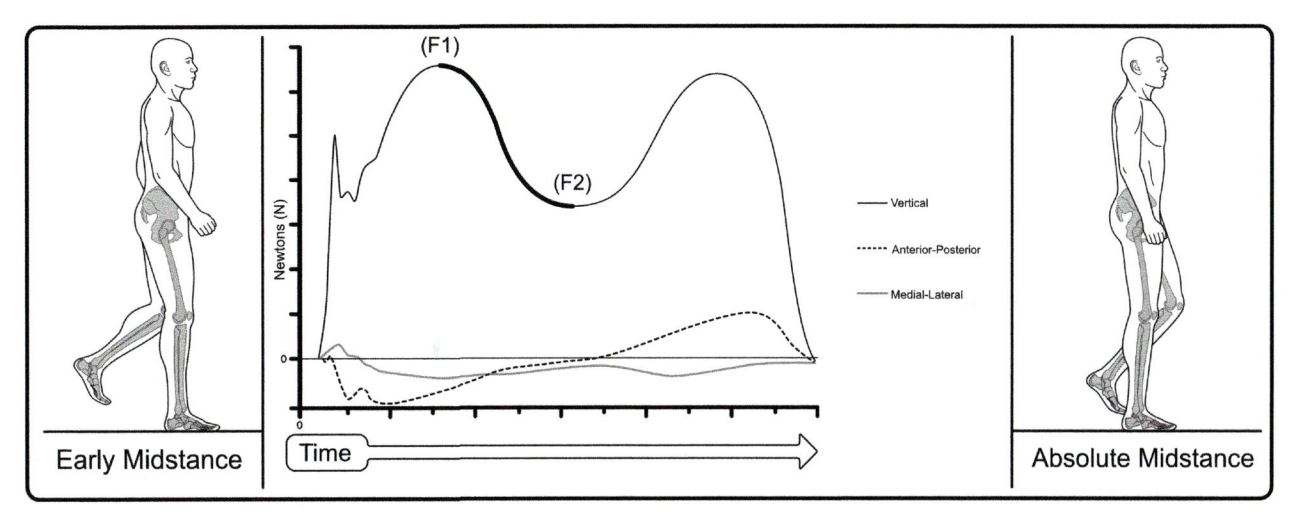

FIG. 1.2.4b Early midstance starts at the F1 peak that signifies the beginning of single-limb support. The limb extends under extensor muscle contraction, accelerating the CoM away from the ground, reducing the vertical component of the GRF. This process ends with the CoM at its highest point during gait around the middle of midstance, representing an absolute midstance point where the vertical component is lowest. This is the F2 trough. *(Permission www.healthystep.co.uk.)*

This low point of the trough should be around 0.7 times the body mass and can therefore be used to assess how well an individual moves over their stance limb to gain height for their CoM. This has big implications to the potential energy stored during gait and reflects on gait energetics. If the trough is deeper than usual, then the CoM could be rising too high. Conversely, if the trough is shallow, then the CoM may not be rising high enough. What goes up must come down, and the fall of the CoM is essential to gait energetics, as is explained by the inverted pendulum model of gait. Too much height requires more muscle energy to gain that height and potentially increases the acceleration of the CoM's fall. A low shallow trough may be related to not generating adequate potential energy. These troughs can also reflect walking speed with higher peaks and deeper troughs with faster walking, and lower peaks and shallower troughs with slow walking, reflecting changes in CoM oscillations that result directly from gait speed. The best peaks and troughs for energetics tend to relate to preferred walking speed, creating lower F1 peaks compared to fast walking (lower force requiring braking) but similar F3 peaks (high acceleration forces) to fast walking (Stolwijk et al., 2014) (Fig. 1.2.4c).

Trough (F2) to second peak (F3)

The trough starts to decrease in depth as the vertical forces increase after absolute midstance, as at this point, the CoM of the body starts to fall forward, increasing the vertical GRF component again. This anterior fall of the CoM should initially be controlled by the activity of the ankle plantarflexor muscles, moderating angular extension (dorsiflexion) moments at the ankle and preventing knee hyperextension through gastrocnemius activity around the knee. A point is reached where the anterior fall of the CoM initiates heel lift, assisted by the contraction and resultant stiffness of the posterior calf ankle plantarflexors through the Achilles tendon. Both the acceleration of the CoM downwards, added to by the active 'pushing' of the plantarflexor force into the ground, causes the generation of an increasing vertical GRF that creates a second peak of vertical force known as the *F3 peak*. The peak of this force should be 1.2 times body weight, roughly matching that of the first F1 peak. A low peak could be related to poor force/power generation from weak ankle plantarflexors. An uneven early peak might result from lifting the heel too soon. This may initially result in a blockage in the fall of the CoM downward, as the early heel lift initially raises the body's CoM before a further fall towards the terminal swing limb can occur (Fig. 1.2.4d).

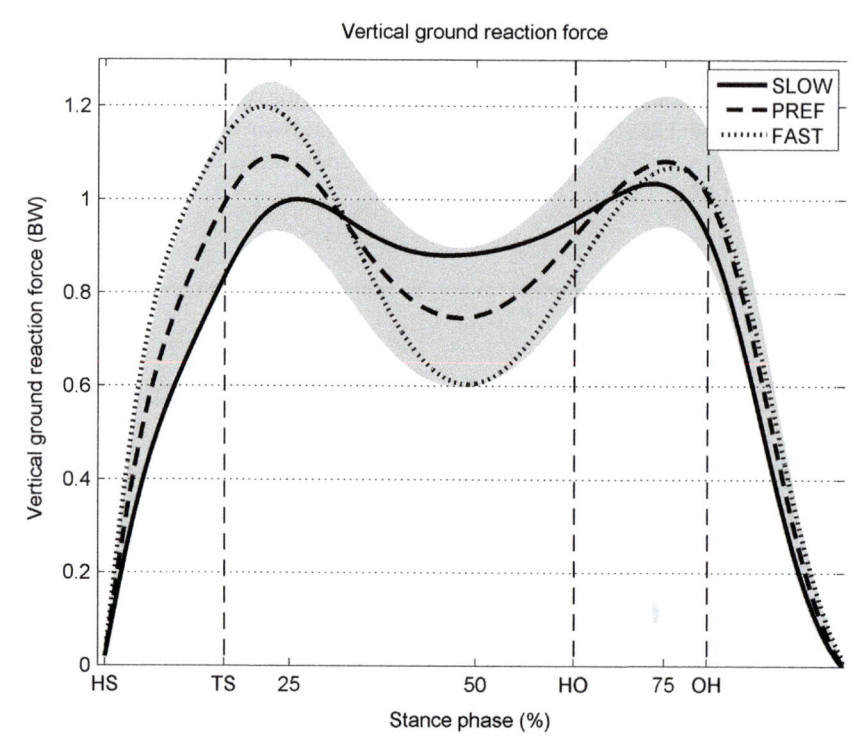

FIG. 1.2.4c The F2 trough depth is related to gait speed, being shallower at slower gait speeds. Note the depressed F1 peak moves further away from the heel strike (HS) and the toe strike (TS – foot flat) during slow gait. On the other hand, vertical GRFs do not reduce to the same extent during midstance, as CoM acceleration away from the support surface is lower. The final F3 peak of acceleration is also depressed and moves closer to heel lift (heel off [HO]) when gait speed is reduced. Faster walking produces a higher F1 peak that requires greater negative work to manage impact energy but produces an F3 peak of acceleration very similar to the one produced at the preferred walking speed. Thus, the effects on vertical forces seem most advantageous to preferred healthy walking speeds. *(From Stolwijk, N.M., Koenraadt, K.L.M., Louwerens, J.W.K., Grim, D., Duysens, J., Keijsers, N.L.W., 2014. Foot lengthening and shortening during gait: A parameter to investigate foot function? Gait Posture 39 (2), 773–777.)*

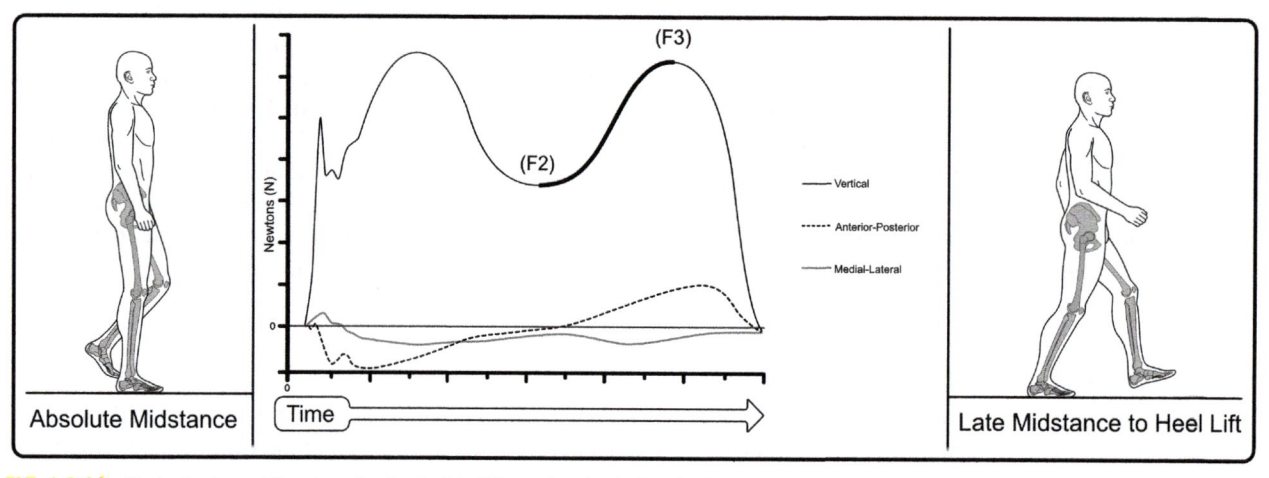

FIG. 1.2.4d From absolute midstance at the depth of the F2 trough to just before the F3 peak reveals a period of increasing vertical forces. This creates an up-slope from the F2 trough to another peak of vertical force associated with heel lift and acceleration, known as the F3 peak. The F3 peak is associated with positive work during gait and reflects the size of the GRF generated by ankle plantarflexion power around heel lift. *(Permission www.healthystep.co.uk.)*

The initial rise in the upslope towards the F3 peak represents a period of storing potential energy which can then be released as part of the peak of force at heel lift. During ankle extension (dorsiflexion) in late midstance, the foot undergoes increasing vault depression and elongation (Stolwijk et al., 2014) which increasingly stiffens the foot under rising percentage body mass force (Bjelopetrovich and Barrios, 2016; Takabayashi et al., 2020). This is aided by intrinsic muscle

Understanding human gait **Chapter | 1** 63

activity (Farris et al., 2019, 2020). By heel lift, the connective tissues should reach an appropriate level of stiffness and stored potential energy in their elastic range, ready to be released once rearfoot unloading starts. The vault should be at its lowest and stiffest at heel lift, with the forefoot stretched out, stiffened, and stable (Hunt et al., 2001; Venkadesan et al., 2020). Thus, the potential energy of the CoM's acceleration can be converted into kinetic energy in order to reduce metabolic costs. As the heel lifts and the forefoot is pressed into the ground, the F3 peak should reach its maximum GRF.

F3 to toe-off

The last part of the vertical force-time graph sees the forces drop away from the F3 peak to nothing, as the load is transferred to the contralateral foot following heel lift. During the period from terminal stance to the preswing terminal double-limb support phase, vertical force rapidly decreases as loading forces move to the next step (Fig. 1.2.4e). It is therefore important that the loading and offloading peaks are viewed in relation to the F1 peak of the contralateral foot. A poor F3 peak and offload may allude to a poor heel strike transient and F1 peak on the opposite foot.

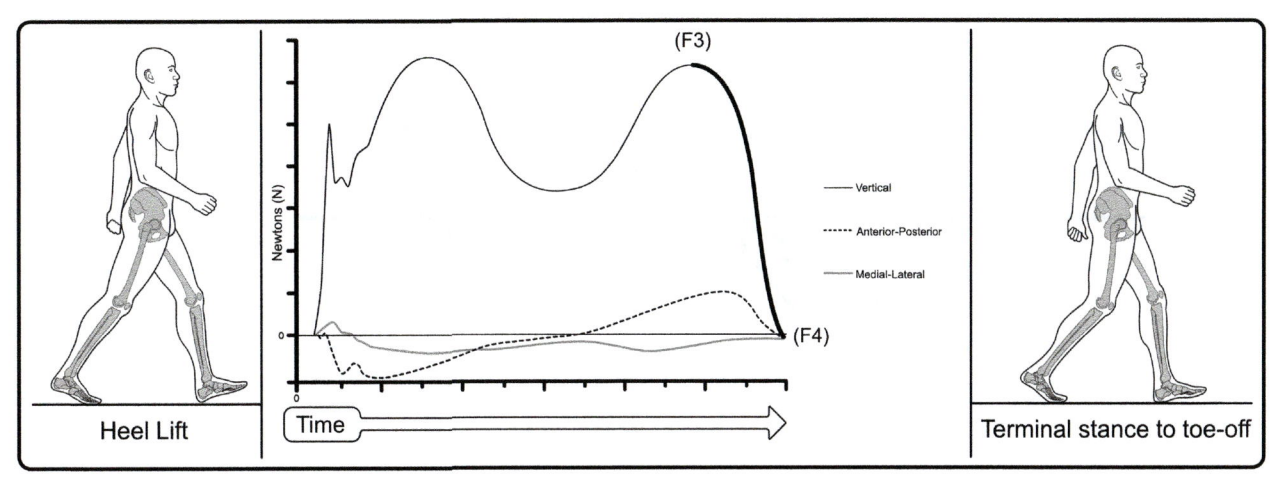

FIG. 1.2.4e The F3 peak of heel lift is followed by decreasing vertical forces that occur during the completion of terminal stance. Weight should be quickly transferred to the contralateral loading heel, as terminal stance muscle-generated forces of acceleration start decreasing around the ipsilateral foot and ankle. Just before vertical forces cease, the foot enters preswing, with toe-off ending all stance's GRFs. *(Permission www.healthystep.co.uk.)*

1.2.5 Anterior–posterior and medial-lateral GRF components in walking

Anterior–posterior components including claw back

The anterior–posterior direction of the GRF is also very important when considering the gait cycle's horizontal components. This GRF cycle is considered in four sections with a crossover point in the middle, which should again equate to ~50% of stance and the absolute midstance point. An occasional feature of the anterior–posterior force-time graph is *claw back*. Claw back is the term for an initial posterior-directed force from the heel (so an anterior-directed GRF) that can occur with initial heel strike but is not always present, especially when footwear is worn. It is caused by the heel being brought backwards into the ground at the end of swing phase, possibly through hamstring braking elastic recoil (see Fig. 1.1.5d). This is a force that can be exaggerated in military-style marching, as the swing limb is deliberately brought back into the ground, resulting in a rapid, often more dramatic heel strike transient due to the resulting impact (Fig. 1.2.5a).

Heel strike to posterior peak (F4)

After claw back (if present), heel contact normally creates a posterior-directed GRF as a result of the anterior direction of the BWV driving down and forward through the heel from the fall of the CoM at initial contact. The beginning of this period involves the loading part of double-limb stance. Every action has an equal but opposite reaction, so an anterior-directed force from the BWV causes a posterior-directed GRF, just as a vertical force downward causes a GRF upwards. On the force-time graph, a fall in vertical force usually appears as a trough dipping towards zero. Not so in horizontal forces, because with anterior–posterior forces, zero is in the middle. Zero now represents a pure vertical component, with no anterior–posterior driving element to the force. Posterior-directed forces are usually set up as negatives down below zero

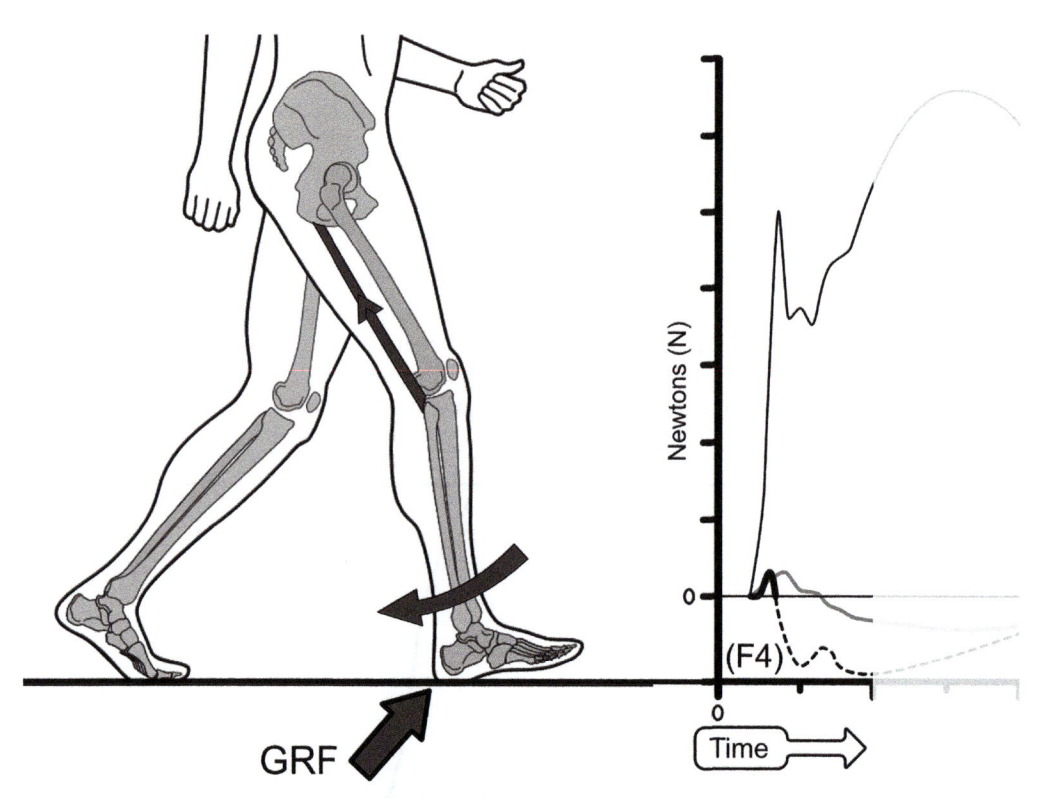

FIG. 1.2.5a Claw back *(dark black line)* is a small anteriorly directed GRF on a force-time curve in the horizontal plane during a period when posterior GRFs are expected in reaction to the body mass being driven through the leg anteriorly. The reason for its presence is the backward pull on the heel that occurs in terminal swing, probably from hamstring tendon elastic recoil from its braking action on knee extension during late swing. It is more likely seen in faster walking and running heel strikes, especially when barefoot. *(Permission www.healthystep.co.uk.)*

and anterior as positive rising above zero. However, if set up as positive forces, then posterior-directed forces will present as a positive peak and anterior-directed GRF as a negative peak. The size of the posterior-directed peak is affected by walking speed and confidence of stability on the support surface. This is because shorter strides tend to decrease the anterior displacement of the body's CoM during steps. The peak of posterior force occurs in loading response as the forefoot becomes weightbearing and is known as the F4 peak (Fig. 1.2.5b).

Posterior peak (F4) to crossover point

As the CoM starts to move over the support limb after loading response, the posterior F4 peak starts to reduce (see Figs 1.2.5b and 1.2.5e). This is concurrent with the CoM being raised through hip and knee extension motion, moving anteriorly, and aligning the limb more vertically. When the CoM has gained maximum height above the limb, there is still a small posterior-directed force. However, shortly after this, the force from the body should be directly and vertically downwards, reaching zero on the force-time curve as there is no horizontal component This is the crossover point and should occur at ~55% of the stance phase, correlating to just after the absolute high point of CoM during midstance.

Crossover to anterior peak (F5)

Following the point of no horizontal or pure vertical component of the GRF, with neither posterior nor anterior-directed forces present through the foot, the CoM of the body begins to move anteriorly to the ankle with increasing forefoot forces angled backwards into the ground from the BWV. These are driven by the fall of the CoM. This equates to an anterior-directed GRF from a posterior-directed BWV which continues to increase throughout late midstance. As the heel lifts from the ground, forces increase backwards due to the action of the ankle and midfoot plantarflexors as they briefly also increase vertical force components before they start to drop. This produces the anterior-directed horizontal peak force known as the F5 peak. The size of this peak is influenced by walking speed and is usually around 0.2 times body weight. A small F5 anterior GRF peak indicates a poor anterior drive through the foot, regardless of the size of the F3 peak on the vertical GRF (Fig. 1.2.5c).

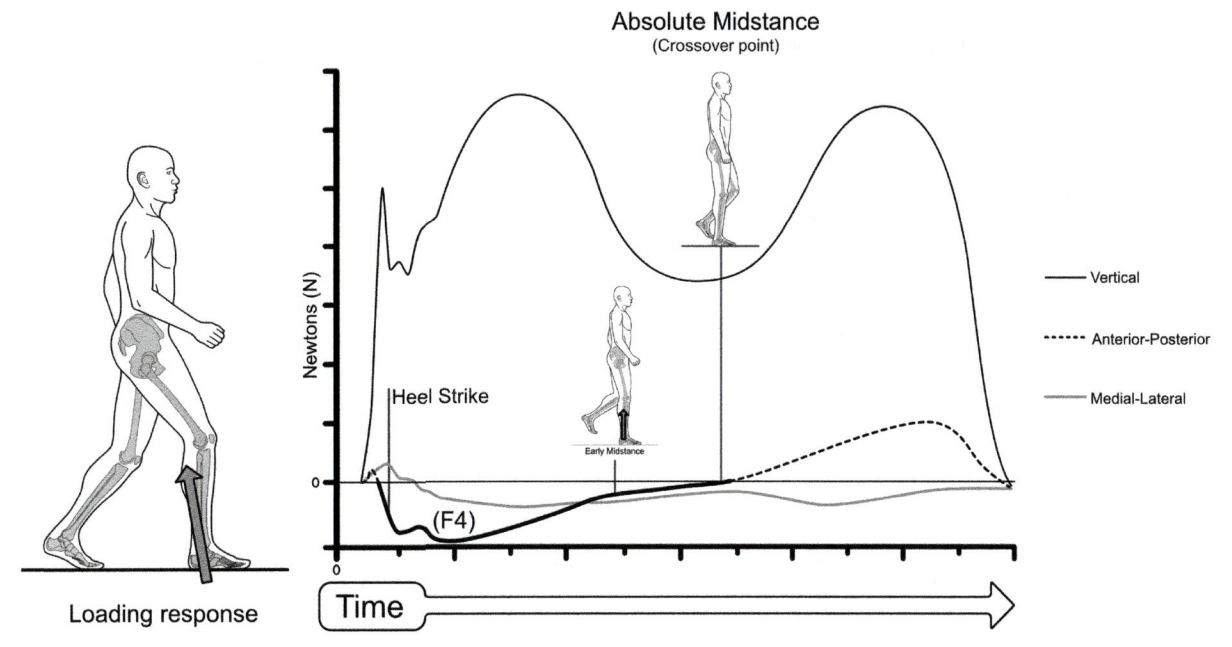

FIG. 1.2.5b The posteriorly directed GRF highlighted with the dark black line forms the F4 peak. F4 is a negative downward peak force (or upward peak if posterior forces are made positive) caused by horizontal components of the GRF in the anterior–posterior direction. It indicates the anterior drive of the BWV against the braking events within the lower limb and braking effects of the GRF. Thus, faster gait should equate to greater F4 forces that require more braking negative work to limit them. The posterior-directed forces should decrease through early midstance as the body becomes more vertically aligned to the lower limb, until forces become purely vertical. This is the crossover point of absolute midstance and usually occurs just after the F2 trough is at its lowest. *(Permission www.healthystep.co.uk.)*

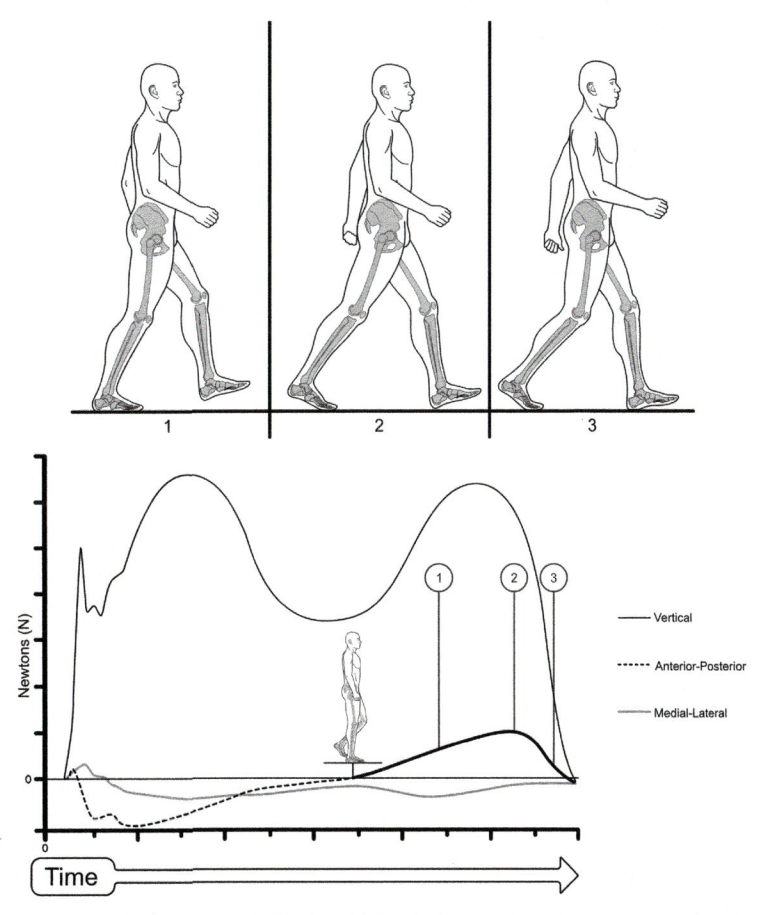

FIG. 1.2.5c The F5 peak demonstrated and explained. Just after the middle of midstance, there is a crossover point where the horizontal GRF becomes directed anteriorly from posteriorly. This is because the body is now driving forces back against the ground (BWV). This anterior GRF increases throughout late midstance (1). At heel lift (2) under ankle and midfoot plantarflexion, the anteriorly directed GRF creates the F5 peak. Following the F5 peak, horizontal (and vertical) forces decrease due to the transfer of weight to the opposite foot under the decreasing power of ankle and midfoot plantarflexion (3). *(Permission www.healthystep.co.uk.)*

F5 peak to toe-off

This is the double support phase of preswing, so the force is now transferring to the contralateral foot. This results in the anterior force production decreasing. The amount of time it takes to offload this trailing foot influences how well the contralateral foot loads under the momentum provided from heel lift.

The effect of arm swing on vertical and horizontal forces

Arm swing in walking influences vertical angular momentum and vertical GRFs (Meyns et al., 2013; de Graaf et al., 2019). When you consider that the left arm is swung back at left heel strike and vice versa, this makes perfect sense. The arm swing's centrifugal force on the trunk will pull the CoM backwards at heel strike, acting as an extra braking moment on the CoM as it falls into ground contact. Increasing arm swing is reported to heighten this effect, decreasing the vertical angular and ground reaction moments further during walking (de Graaf et al., 2019). It is possible that pulling the arm backwards during the initial contact of the lower limb on the same side may also influence initial horizontal forces too. If so, then the forward arm swing during single support may also influence the horizontal force increases of terminal stance too, although the forward swing may be of more assistance to the centrifugal forces of the contralateral side's swing limb. Arm backswing at heel strike may also influence the development of claw back, possibly helping to explain the link between increased walking speed to the development of claw back beyond just hamstring elastic recoil. It may also help restrain the swing limbs centrifugal effects on momentum on the ipsilateral side.

Medial-lateral GRF components

Medial-lateral components produce the smallest elements of GRF and can be split into two sections. Medial forces are usually between 0.05 and 0.1 times body weight and lateral forces usually less. Initially, at heel strike, there is usually a 'lateral thrust' while loading due to the foot working as a compliant adaptive structure. It is moving from a slight position of inversion in the rearfoot at initial contact to a less inverted or everted posture. As it does so, the foot becomes increasingly prone to the support surface. This means that the foot is driving a force medially, but of course, the GRF force drives back laterally. This continues throughout loading as the compliant foot increases its surface contact. After this loading response, the GRF is usually directed medially as the body begins to move over the stance limb, loading more of the lateral side of the foot on the ground due to the vault profile being higher medially. These medial-lateral forces tend to fluctuate in intensity but are usually decreasing towards heel lift. Decreasing medial forces and sometimes small lateral-directed GRF tend to be observed during final push-off as the CoGRF moves towards the medial side of the forefoot and hallux. This probably reflects tibialis posterior, flexor hallucis longus, and peroneal muscle interaction that causes midfoot adduction and eversion as the foot approaches and enters terminal stance. Not only is the medial-lateral component the smaller of the GRFs but it is also the most variable between steps and individuals, and yet it can influence the effects of loading forces across the knee and ankle. Deviations in these GRFs can result in uneven frontal plane compression across these joints during angular moments of flexion and extension. Thus, this is a component of GRF that should not be ignored, despite the relatively small size of it (Fig. 1.2.5d).

Pedotti diagrams

From the study of GRF in the orthogonal planes, it is clear that the GRF can move medially or laterally and posteriorly or anteriorly independent of the size of the vertical component. Usually, the CoGRF moves in a smooth steady pattern from a most posterior-directed force at heel strike (with the BWV driving anteriorly) to a most anterior-directed force during forefoot 'propulsion' (with the BWV driving posteriorly). This is accompanied by a small amount of variable medial-lateral-directed deviation during anterior movement of the CoGRF, while vertical forces rise and fall significantly, all dependent on gait speed and stride length (Fig. 1.2.5e).

These movements of the CoGRF combined with the vertical element of the GRF can be used to plot a *Pedotti diagram*. The Pedotti diagram (also known as a butterfly diagram) allows the magnitude and direction of the force and its changes during gait from heel strike to toe-off to be plotted in 3D, giving a series of changing resultant GRFs during gait. Plotting a Pedotti diagram is dominated by the vertical and horizontal data generated from a force platform (as they are the largest), beginning from heel strike and terminating at the end of the stance phase at toe-off. The resultant GRF forces throughout the gait cycle can be then related to limb positions in relation to the joint axes, especially in real time if the force platform can be used in conjunction with a 3D kinematic joint motion recording system. The direction of the resultant GRF forces at any

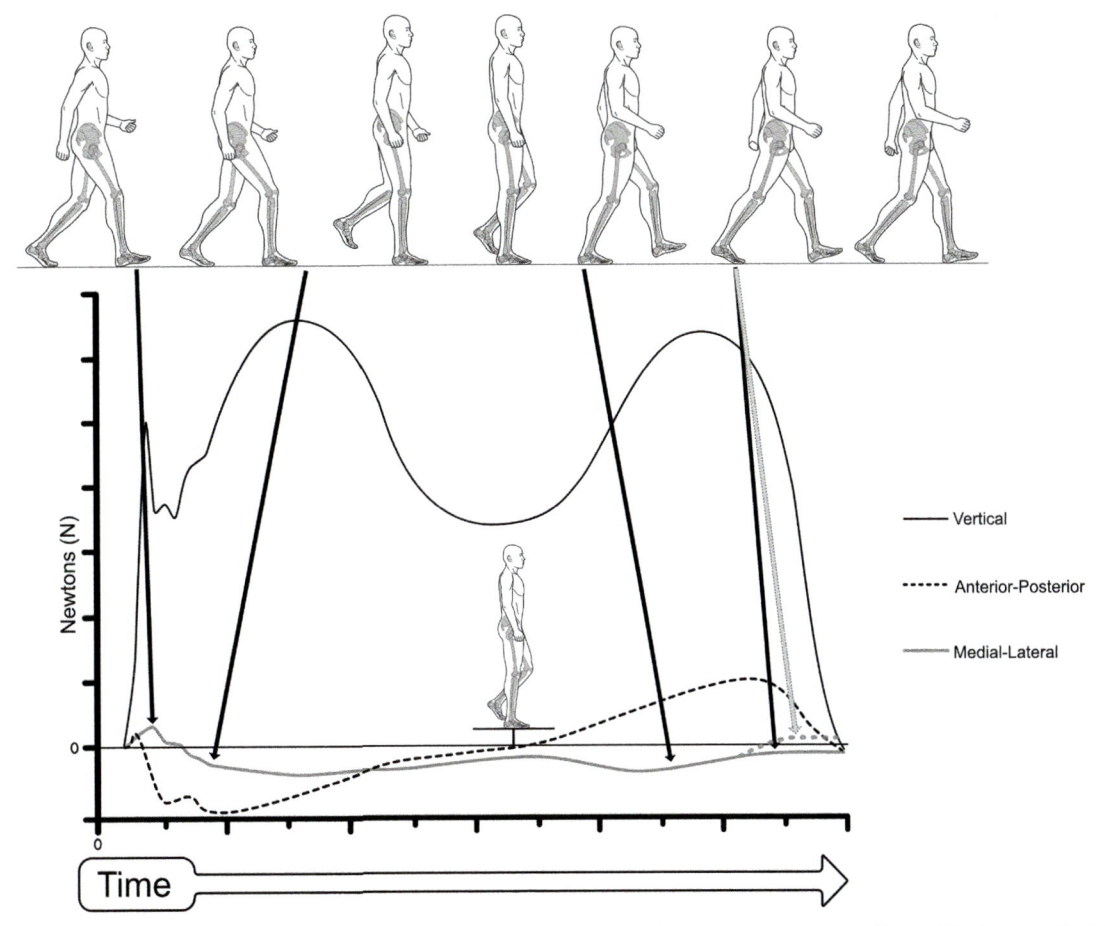

FIG. 1.2.5d Medial-lateral GRFs are less predictable than other GRF vectors. Lateral GRF vectors are represented by positive forces on the right foot in this example and would be reversed for a left foot. Most commonly, there is an initial lateral-directed vector from the foot that is usually moving medially through rearfoot eversion moments *(first black arrow)*. This medial shift of the BWV through the foot creates equal and opposite lateral-directed GRF. During forefoot loading and single-limb stance, GRFs are usually directed medially from more lateral loading of the foot which offers a larger supporting ground contact area *(from second to third black arrows)*. In late midstance, the medial GRF decreases through the activity of the peroneal muscles, a situation that continues into terminal stance *(fourth black arrow)*. Many individuals create a small lateral GRF from a medial BWV applied via the foot during early terminal stance (medial force reduction represented with solid grey line and a more pronounced lateral GRF indicated with dashed grey line), but this is highly variable and neither the finding of one or other GRF path in isolation seems significant to any pathological risk. *(Permission www.healthystep.co.uk.)*

particular moment in relation to any particular joint axis and instantaneous body posture can help explain muscle activity through the need to develop decelerations or accelerations of motion via eccentric or concentric muscular activity. Pedotti diagrams can be constructed quite easily from force-time data without the need for specialist knowledge in programming or specific software (Kambhampati, 2007) (Fig. 1.2.5f).

Comparing Pedotti diagrams to actual kinematics in gait is very useful. For example, GRF generated behind the axis of rotation of the knee will create flexion moments at the knee, while those anterior to the knee will produce extension moments across the knee. GRF in front of the hip will want to flex the hip, while those behind will likewise want to extend it. The size of the resultant GRF dictates how much muscle force must be produced to resist the amount of flexion or extension being generated across the joints affected. Other considerations are the position of the CoM of the body and the limb segment in relation to the GRF. In swing phase, only the CoM of the body segment/limb is of interest because there is no resultant GRF to worry about. All forces are internal from muscle generation and soft tissue restraint. Knowing the principles behind these resultant GRFs helps to explain the kinematic events seen during gait. It may seem strange to start considering gait analysis from GRFs, but in fact, it is difficult to explain the events in gait without first appreciating the generation and influence of GRF.

FIG. 1.2.5e A Pedotti diagram compared to kinematic postures during the stance phase of walking. Pedotti diagrams give a 3D sense to force vectors, although when displayed on a page in a book, this is rather reduced to 2D-combined vertical (y-axis) and anterior–posterior vectors. Despite this, the vertical and horizontal effects of heel strike and claw back are easily identified, while the fall and rise of vertical and horizontal elements to the GRF vectors can be easily visualised to the kinematic events during locomotion. *(Permission www.healthystep.co.uk.)*

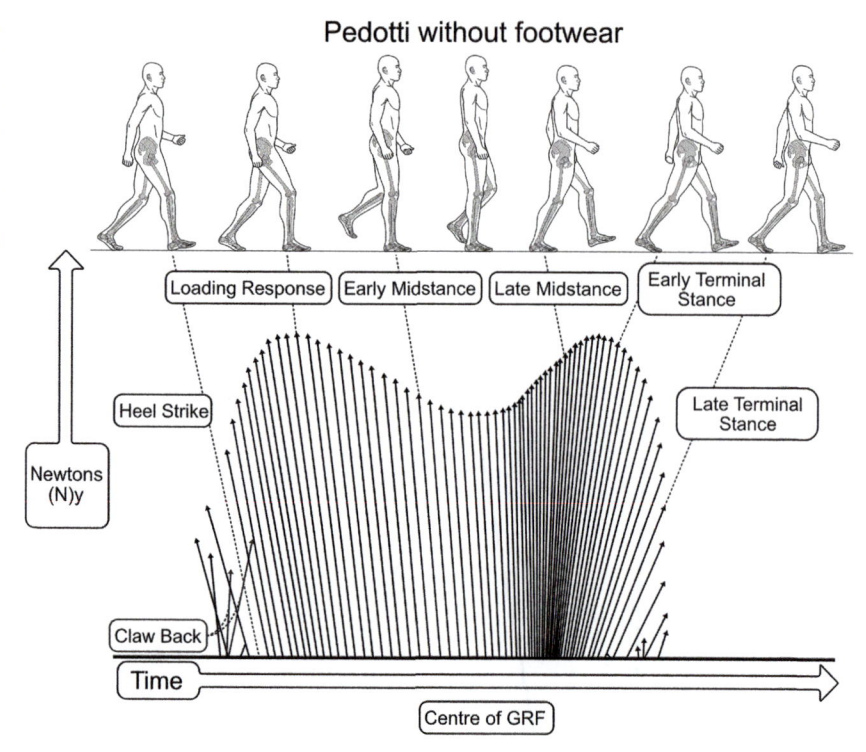

FIG. 1.2.5f Pedotti diagrams constructed using simple 'Microsoft Excel' spreadsheet and charts. The top image shows a Pedotti diagram constructed using such techniques from a shod subject and the lower image, from an unshod subject. Note the higher GRFs in footwear compared to barefoot, particularly during braking. Note also the use of the term 'centre of pressure' for the centre of the ground reaction force (CoGRF) which is standard practice in force plate analysis. *(From Kambhampati, S.B.S., 2007. Constructing a Pedotti diagram using Excel charts. J. Biomech. 40 (16), 3748–3750.)*

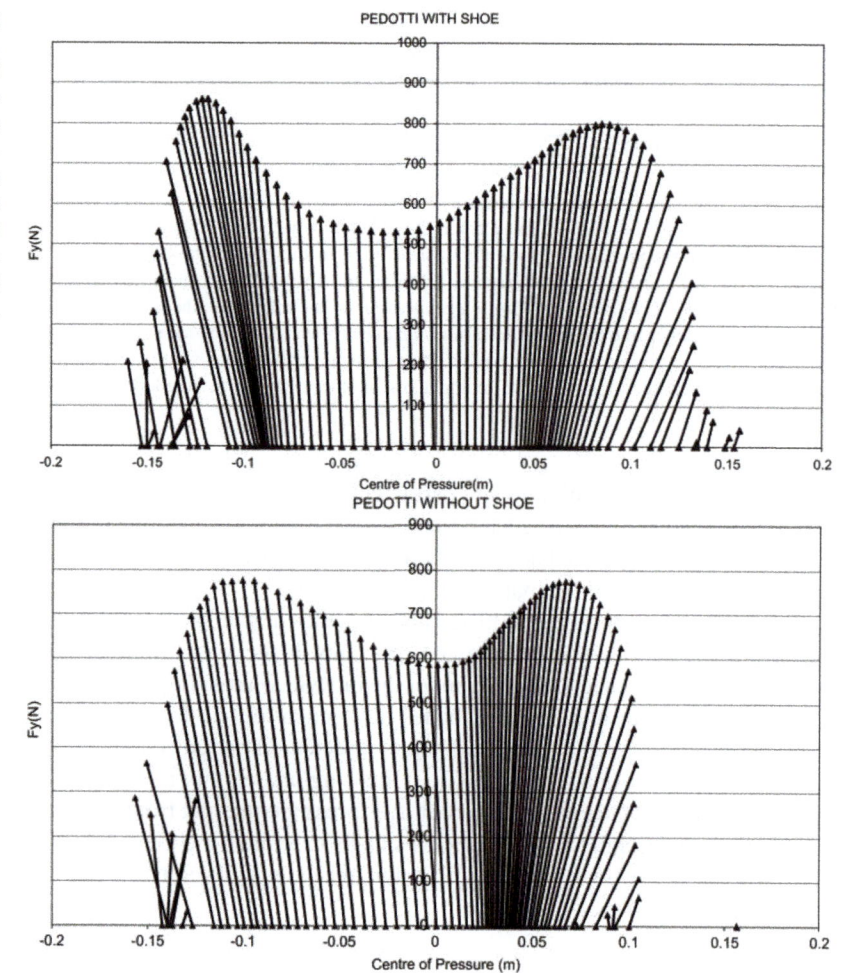

1.2.6 Spatiotemporal parameters in gait

The variability of spatiotemporal parameters characterises the reliability of lower limb motion during gait from stride to stride. If variability is low, then the subject is producing parameters across the strides with only slight deviations in each step. This indicates that gait is rhythmical and automatic. If variability is high, it suggests unpredictability and flexibility within the lower limb motions producing gait. Both situations have their advantages. On uneven ground or with gait perturbations, the ability to create variability-flexibility in gait is advantageous whereas on flat, even ground without perturbations, low variability is suggestive of energetic equilibrium. High or low gait variability is therefore not good or bad but must be appropriate to need. When people walk outside of their self-selected walking speed, variability of spatiotemporal parameters and joint angular motions tend to increase (Jordan et al., 2007; Kang and Dingwell, 2008a,b; Kiss, 2010).

Spatiotemporal parameters and their variability are aids to indicate the appropriate functioning of the lower limbs in gait, and both high and low variability can be associated with pathology. When increased variability of spatiotemporal parameters is associated with decreased joint angular parameters, they may indicate that a subject is not able to reproduce limb coordination from stride to stride. This may predict gait instability and a risk of fall (Beauchet et al., 2007). Kinematic variability therefore represents an index for stability (Hausdorff et al., 2001).

Spatial parameters of gait

The spatial parameters of gait analysis are *step length*, *angle of gait*, and *base width*, also referred to as *step width* or *base of gait* (Richards and Thewlis, 2008). Step length is defined as the distance between two consecutive initial contacts by different feet. Stride length is the distance between two initial contacts by the same foot. Angle of gait or foot angle is considered as being the angle that the foot is orientated in relative to the line of progression. Finally, base or step width is defined as the medial-lateral distance between the centre of the heels of both feet from each other during gait (Richards and Thewlis, 2008). Cadence and average velocity can be calculated from these spatial parameters (Table 1.2.6a).

TABLE 1.2.6a Spatial parameters of gait.

Spatial parameter	Description
Step length	Distance between two consecutive heel strikes
Stride length	Distance between two consecutive heel strikes by the same foot
Angle of gait	Angle of foot orientation from line of progression
Base of gait (gait width)	Medial-lateral distance between the centre of each heel during gait

Humans have subject-specific preferred step lengths, stride lengths, and angle and base of gait widths that affect gait energetics (Holt et al., 1995; Browning and Kram, 2005; Browning et al., 2006; Shorter et al., 2017). The net rate of metabolic expenditure during gait is linked to step width through swing limb circumduction. The width or base of gait should be narrow enough to avoid excessive redirection of the body's CoM trajectory, but wide enough to avoid the swing limb having to rotate around the stance limb causing excess circumduction at the hip (Shorter et al., 2017). Step width also influences the position of the GRF relative to the lower limb joints, for wider bases of support move the GRF more laterally to the trunk and narrow bases of support more medially under the trunk. A narrow base of gait will tend to increase varus moments across the joints from a more medial GRF relative to the limb, while a wide step width will tend to increase valgus moments and decrease varus moments from a more laterally positioned GRF to the limb. In running, a narrower base of support is necessary, reflecting the loss of double-limb contact that requires each single-limb supporting contact to occur more directly below the CoM (see Figs 1.2.6a and 1.2.6b).

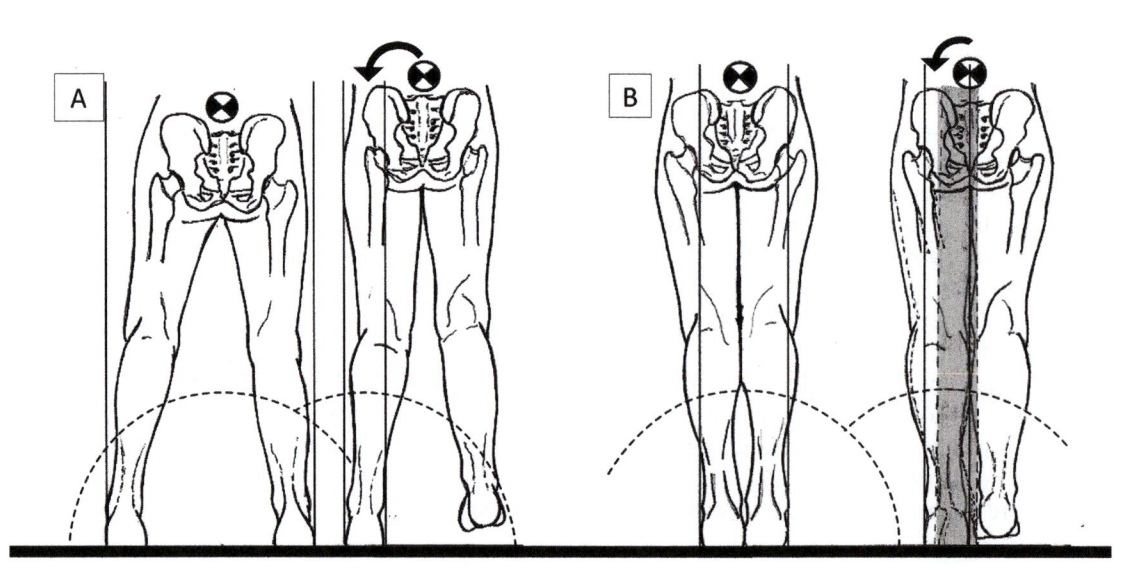

FIG. 1.2.6a The base of gait is important in understanding how easy it is for a patient to support their CoM deviations during locomotion. A wide base of gait (A) allows the CoM to move considerably from left to right without drifting outside of the base of support. In static stance and double-limb stance, a wide base gives advantage, but in single-limb support, the CoM may become too medially positioned to the base of support, making single-limb support stability more problematic. Wider bases of support are therefore less of an issue during slower gaits when double stance periods are longer. A narrow base of support (B) only gives a limited base of support to the CoM in static stance and for double-limb support. However, it positions the base of support more under the CoM during single-limb support of gait. Thus, narrower bases of support offer advantages in faster gait when single support is longer, or as in running, when there is no double-limb support. However, if the base of support increasingly narrows (grey area and hashed limb outline in B), the base of support may be compromised laterally or the stance limb may block the swing limb. In relation to the base of support in gait analysis, considerations to the locomotive style and postural stability are important. *(Permission www.healthystep.co.uk.)*

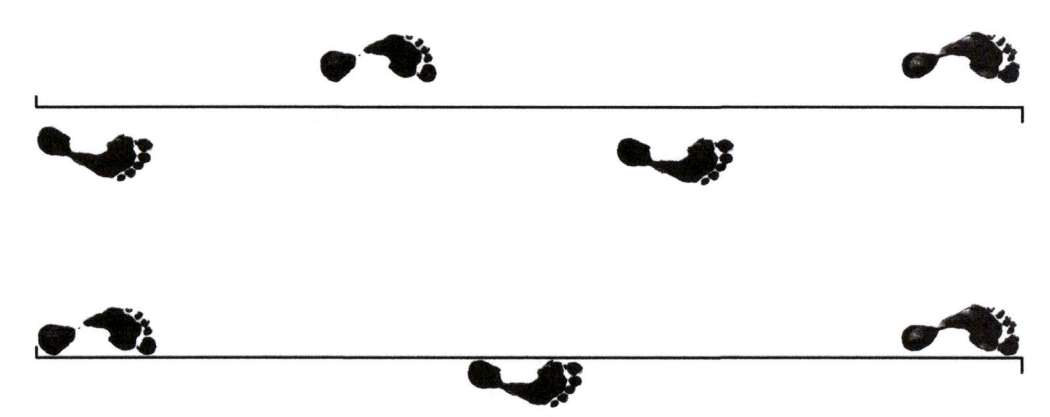

FIG. 1.2.6b The expected base of support should reflect the mode of locomotion used. In walking *(upper image)*, the feet should lie a few centimetres from the line of progression, although variability between individuals should be expected, not least due to different leg lengths. Each foot is positioned a little distance to the left and right of the line of progression to aid in the transfer of the CoM from right to left foot's support limb ready for each single-limb support phase without necessitating the limb being positioned directly under the body. Thus, the CoM moves slightly from left to right following the support limb or in positioning ready to transfer weight to the next support limb during terminal double-limb stance. In running *(lower image)*, not only is the stride length increased but the foot is also positioned closer to or even flush with the line of progression. This reflects the need to reduce left-to-right deviations of the CoM, for there is no double-limb stance for easy weight transfer from one limb to the next. Instead, the CoM is passed from one single supporting limb to the next through a hiatus period in weightbearing (the aerial phase). This requires the support limb to tuck further under the body and closer to the line of progression at contact, reflecting changes in hip joint angulation necessary for running compared to walking. *(Permission www.healthystep.co.uk.)*

The angle of gait plays an important role in aligning the medial MTP joint axis of digital extension to the line of progression of the limb so that sagittal plane motions in the hip, knee, ankle, and medial MTP joints largely align. With the lower leg axis and foot aligned longitudinally, the mechanical properties of the ankle-foot become analogous to passive linear springs creating net near-zero work across a range of walking speeds (Hedrick et al., 2019). Outside of this alignment, energetics may be compromised (Fig. 1.2.6c).

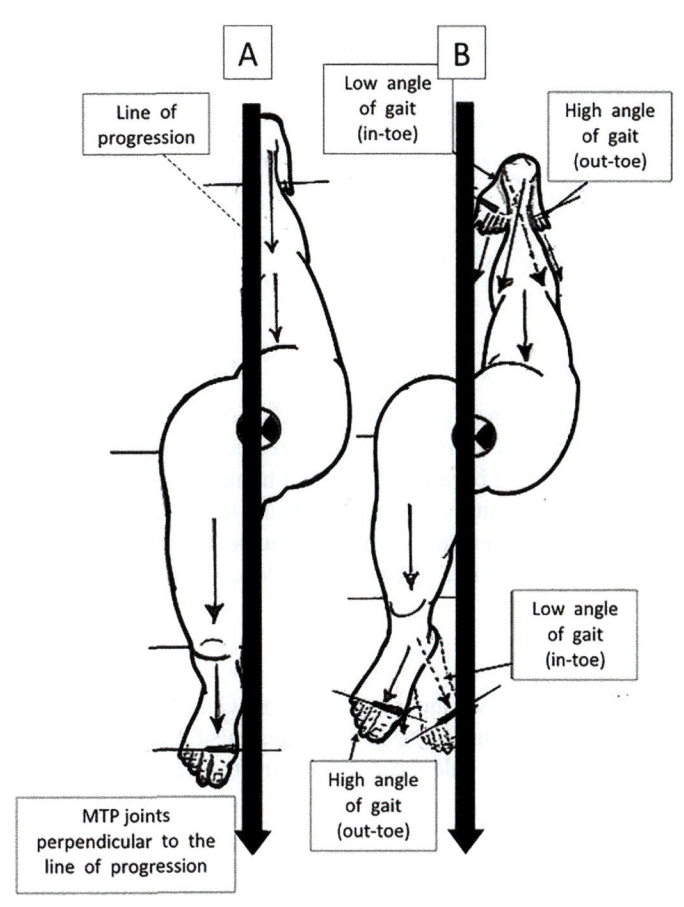

FIG. 1.2.6c The angle of gait should ideally reflect the position of the medial MTP joint fulcrum aligned to the axes of the hip, knee, and ankle in the sagittal plane and near perpendicular to the line of progression *(thick black arrow)*. Together, the joints should facilitate sagittal motion across a range of gait speeds to permit the Achilles to act as a passive linear spring with the ankle and foot (A). Excessive angles of gait (highly abducted or out-toe position) cause the medial MTP joint fulcrum to angle away from the line of progression, while low angles (adducted or in-toe positions) can cause the medial MTP joint axis to angle away from the line of progression (B). Both situations can disturb the proximal joints' efficiency by changing muscular lines of action, preventing structures such as the Achilles from generating acceleration power along the line of progression to act with the foot as a linear spring. In-toe gait often causes the substitution of the medial MTP joint fulcrum to one utilising the lateral MTP joints. This has been termed low gear propulsion, reflecting less mechanical efficiency associated with its shorter moment arm to the Achilles attachment. *(Permission www.healthystep.co.uk.)*

Temporal parameters of gait

Contact times between the foot and the ground can be used to calculate step times and stride times during gait. The amount of time that a single leg compared to two legs supports the body gives the times for the single and double support phases of gait, with swing phase time matching single support phase for the opposite limb. This, and other temporal information allows symmetry of gait to be assessed further (Table 1.2.6b).

TABLE 1.2.6b Temporal parameters of gait.

Temporal parameter	Description (% at comfortable walking speed)
Step time	Time between two consecutive heel strikes. Normally, 50% of stride time is gait cycle-symmetrical
Stride time	Time between two consecutive heel strikes of the same leg; 100% of the gait cycle for that leg
Single support	Time body is supported by one leg only. Approximately 40% of gait cycle and 68% of stance phase
Double support	Time body supported by both legs. Consists of two phases, both lasting approximately 10% of gait cycle and 16% of stance phase each. These periods occur during loading response and late terminal stance, and preswing, respectively
Swing time	Time taken for the leg to swing forward. Always the same time as opposite single support phase, at approximately 40% of gait cycle
Total support	Time body is supported by one leg in a complete cycle and includes the leg in single and double-limb support. Approximately 60% of gait cycle

Although the percentages of the time spent in single and double support are useful, they do vary with gait speed. Double-limb stance increases while single-limb support decreases as gait speed decreases until the point is reached when we do not actually move, whereupon true double support becomes 100% and humans stand statically. As gait speed increases, the reverse is true, with double stance decreasing until we start jogging and then running when there is no double-limb stance at all. Double stance is the parameter that can get race walkers into trouble, as they must have a double-limb stance phase to be walking. These temporal parameters can be used to calculate cadence and velocity.

Cadence, velocity, and symmetry

Cadence is the number of steps per minute or over another given time. Gait velocity is computed by calculating the step length multiplied by the steps per minute (cadence) and then dividing this by the number 60 (seconds of a minute). Step symmetry between the left and right can be easily calculated by dividing the step length of the time of the left foot by that of the right.

Gait speed

Gait speed has been referred to as the sixth determinant or vital sign of morbidity and mortality (Fritz and Lusardi, 2009). Gait speed affects not only the percentage times of the temporal parameters but also changes the percentage time of the CoGRF's progression occurring under any part of the foot, decreasing with increased gait speed in the midfoot but increasing its time under the forefoot during terminal stance and preswing periods (Chiu et al., 2013). Walking speed changes result in both kinematic and kinetic changes (Hanlon and Anderson, 2006). Preferred walking speeds in healthy populations are around 1.05–1.43 m/s (Hanlon and Anderson, 2006). Most peak sagittal plane joint angles demonstrate a significant correlation to gait speed, but with a low predictability between subjects (Lelas et al., 2003; Hanlon and Anderson, 2006). Small changes of 0–10° seem to occur for around a 0.5-m/s change of speed in children (van der Linden et al., 2002), but a reduction in gait speed from 2 to 1 m/s is reported to lead to average absolute changes of between 2° and 11° within the sagittal plane in adults (Lelas et al., 2003). Variability of both spatiotemporal parameters and angular joint motions increase if subjects are taken out of their preferred walking speed (Jordan et al., 2007; Kang and Dingwell, 2008a,b; Kiss, 2010). Knee flexion angle during stance phase significantly correlates to gait speed and may relate to impact and loading force energy dissipation (Hanlon and Anderson, 2006).

Lower limb kinematics are influenced by gait speed, changing joint angles during gait phases (Hanlon and Anderson, 2006). Pathology often reduces gait speed. This makes comparing pathological gait kinematics to healthy gait 'normals' problematic, not least because of the high variation in normal joint angles between individuals at the same walking speed. Such knowledge highlights the need to examine kinematics and kinetics at consistent gait speeds on the same patients between examinations. Caution must be used when using joint angles to compare to a 'normal human model', as patients have their own normal joint angles at different gait speeds. One consideration during treatment of pathology is that a patient usually begins to feel more comfortable at a higher preferred gait speed with successful therapeutic intervention. Thus, increasing speed with symptom relief may be used as a sign of improvement in gait.

Gait speed has a great influence on forces and impulses during gait as it affects acceleration and vertical displacement of the CoM. Faster walking causes increased vertical displacement of the CoM under greater accelerations. This results in higher F1 and F3 peaks and a deeper F2 trough as well as increasing F4 and F5 peaks in the horizontal plane (Fig. 1.2.6d). Slower walking lowers the height of the peaks and reduces the depth of the midstance trough. Speed also increases the elasticity/stiffness of the lower limbs and feet (Kim and Park, 2011; Stolwijk et al., 2014), creating more spring-like behaviour and showing that the lower limb and foot demonstrate time-dependent properties as expected of viscoelastic structures. Slower gait speed should induce greater tissue compliance. If tissue fails to respond to greater strain under lower but longer stresses by increasing compliance, then longer loading times on stiffer structures could create far more issues than high peaks in forces applied over short periods of time.

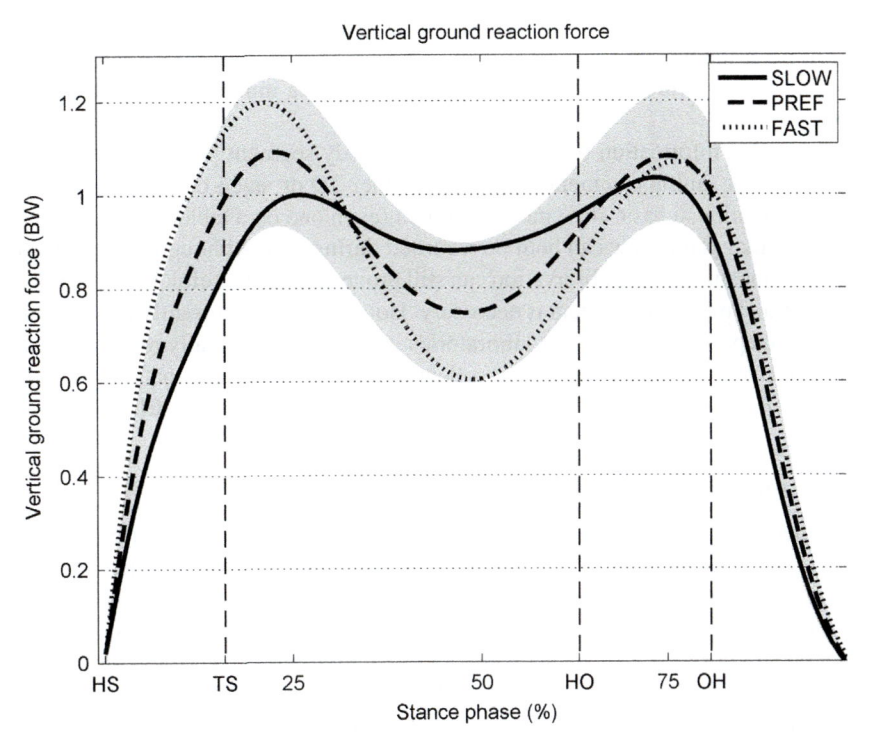

FIG. 1.2.6d The data on vertical GRF forces generated at different walking speeds provided by Stolwijk et al. (2014) are worth looking at again in consideration to temporal gait parameters. Average vertical GRF forces are affected by walking speed, with slow and fast speeds distinctly different to that of the preferred (PREF) walking condition. The *grey area* indicates standard deviation for the preferred walking speeds across a number of subjects, which, as expected, show subject variability. However, it appears that preferred walking speeds for most individuals offer the benefits of reduced initial contact forces compared to fast walking but shows the similar benefits of high acceleration collision vertical forces of fast walking, which are lost during slow walking. In disease, preferred gait speed is often slowed and stride length reduced, which reduces loading GRF forces. As a result, tissues should become more compliant and thus increasingly enabled to dissipate energy via longer loading through increased strain. Tissues unable to alter their mechanical response, such as the permanently stiffened tissues associated with diabetes, will remain at risk at slower walking speeds as although peak forces are reduced, loading time is extended on more brittle tissues. *(From Stolwijk, N.M., Koenraadt, K.L.M., Louwerens, J.W.K., Grim, D., Duysens, J., Keijsers, N.L.W., 2014. Foot lengthening and shortening during gait: A parameter to investigate foot function? Gait Posture 39 (2), 773–777.)*

1.2.7 Measuring joint segment motions

Goniometers, potentiometers, and electrogoniometers can be used to give information on joint angles. Goniometers allow a quick measurement of static joint angles but are of little use in dynamic situations. They either consist of a hinged ruler type, a fluid-filled gravity type, or an electronic angle finder type. The latter can be zeroed to any start position, whereas the fluid-filled is always zeroed by gravity.

Potentiometers measure changes in linear or angular displacement through changes in voltage output. They allow one plane of motion to be measured, for example, extension-flexion within the sagittal plane. They are inexpensive, robust, and allow simple kinematic data to be collected. Measuring more than one joint becomes difficult as they can encumber the patient. Electrogoniometers consist of thin pieces of wire that are sensitive to bending which changes the output voltage. The advantage is that they can be sensitive to triaxial motion, although tests that are biaxial or uniaxial (measuring movement in just one or two planes) are more common. They are relatively inexpensive, accurate, and unobtrusive when fitted on the patient. However, errors can occur through motion within the soft tissues around the joint or if the device has not been correctly placed in relation to the plane being measured, crossing two planes rather than one plane. In such a situation, clinicians/researchers can unknowingly record motion from two planes rather than one, and this is referred to as *'cross talk'*. These devices have finite accuracy due to the limits of strain properties within the wire.

Accelerometers, as the name suggests, measure acceleration. They consist of force transducers designed to measure the reaction force associated with a given acceleration, such as those experienced in the lower limb between segments. However, the data they produce will be dependent on where they have been positioned on the limb. Thus, the more distally they are placed, the more acceleration that will occur during swing phase and the less during stance. They are very good at providing information on shock attenuation and both acceleration and deceleration of body segments but do not give direct information on joint angles. Furthermore, they break easily, and measuring multiple segments requires multiple accelerometers, increasing cost.

By far the best way to gather information on joint motion and body segment position is through movement analysis systems using a single camera or multiple cameras to develop either 2D or, even better, 3D kinematic information. For simple clinical evaluation of motion, a basic slow-motion camera (even one on a tablet or phone) can prove most helpful, especially if a frontal and sagittal plane view can both be recorded during gait. Although measurements of joint angles on such systems are highly questionable, simple observation can still prove clinically useful. For research purposes, a multi-3D camera requiring calibration and marker placement is necessary. However, this is currently usually outside the possibilities of most clinical situations, except in designated gait laboratories where they are usually reserved for more complex cases. Details on such systems are best investigated through excellent specific texts such as that edited by Jim Richards (2008): *Biomechanics in Clinic and Research: An Interactive Teaching and Learning Course*, Churchill Livingstone, Elsevier.

1.2.8 Measuring and interpreting pressure

Pressure data can be obtained using a device which can measure the pressures under the foot. Some comprise insoles full of sensors that fit inside shoes while others consist of a plate or walk-over mat. Pressure mats tend to use capacitive sensor or resistive sensor technology and can sample between 40 Hz and upwards of 500 Hz. Caution with data is needed in pressure data interpretation, for the accuracy is dependent on the number of force-measuring sensors (or cells) within the system. These usually vary from one to four sensors per square cm, although high-quality systems can have 15 sensors per square cm. The fewer the sensors, the larger they are, and as each cell registers the peak force within it, that peak force is recorded and applied to the whole area the sensor covers. The more cells there are, the more specific the forces recorded on the plantar foot. Most clinical systems can only take one step at a time, but mats are increasing in size with more systems now being able to capture multiple-step data. This obviates the need of the patient having to target the mat, which can pose a considerable problem. The other dilemma for the clinician is whether or not to test barefoot or shod. The former does not give direct information about the foot, whereas the latter is not the environment that most feet work in.

In-shoe pressure systems are able to measure the interface between either the foot and the shoe or the foot and an orthosis within the shoe. The difficulty with both is that the curvature of the foot/shoe and/or orthosis interface can distort the sensors, and often transducers need to be fixed to the ankle as they cannot be accommodated within the shoe. This can sometimes influence the gait data as a result. It must be noted that mats and in-shoe pressure systems record quite different data and not just because of the presence of the shoe (Chevalier et al., 2010). Some of the reasons for this are due to the way that the foot strikes the plate which will be approaching perpendicular with a mat (as long as the floor is flat) but is variably angled when using an in-shoe device depending on the foot and shoe angle to the ground at initial contact. This means that you cannot use data comparatively using the two systems (Chevalier et al., 2010).

Pressure systems need to be calibrated, although the calibration techniques vary widely between manufacturers. Techniques vary from standing on the sensors and inputting the patient's weight to using a pressurised bladder. For research purposes, calibration is far more important than in the clinical situation where approximate values are usually more acceptable, and although less accurate, systems that use the patient's body weight are usually quicker to calibrate. Pressure is force over area (a scalar rather than a vector) as the force recorded has no direction. It is just a quantity within an area. Pressure should be measured in Newtons per metres squared (N/m^2) or kilopascals (kPa). Measuring pressure under the plantar foot is useful, not least because it can be associated with pain and pathology from tissue damage. If compressive pressure on the tissues causes mechanical failure through breach of the tissue yield or the point of tissue threshold, then tissue failure occurs. High compression on blood vessels may cause temporary ischaemia, which if highly repetitive, may result in tissue weakening due to poor nutrition and metabolism over time. Thus, repetitive high-stress-induced ischaemia can reduce the mechanical properties of the tissues, setting up a pathological cycle. Ulceration in diabetics, who may already be vascularly compromised from glycation effects on blood vessels, make such lesions particularly vulnerable to high pressure (particularly the shear associated with pressure) because their tissues are less compliant under load to distribute strain throughout the surrounding tissues. As such, they are less able to spread forces to reduce pressure and shear strains within local tissues.

Peak pressures and loading rate

Recording average pressures over the entire foot is not particularly useful, as such data will be influenced by contact area and the weight and gait speed of the patient. Average pressures in specific areas of the foot are of more use, although of more clinical interest is establishing pressures that are causing or could cause pathology. It would be helpful if an absolute quantity of pressure was associated with tissue damage in a human foot. Although it is true to say this exists, the actual point of tissue failure in each patient will be at a quite different pressure quantity and will also be different for a specific location within each foot. Tissue mechanical properties including their yield point are dependent on many factors taking in age and the presence of disease, and it must be appreciated that certain areas within the plantar foot may be more affected by ageing and disease than others. This means that failure of plantar tissues, such as cutaneous tissues under pressure, may vary considerably within the same foot. In part, this may reflect the ability of the anatomy under and around the pressure to dissipate and spread the forces applied more evenly across the tissues.

In normal, healthy, young populations, plantar pressures tend to be around 200–500 kPa during walking. However, in neuropathic diabetics, these can be as high as 1000–3000 kPa, probably through an inability to 'spread' forces throughout surrounding tissues. Recording peak pressures and their locations in the foot can therefore be helpful in aiding to understand pathology. Another useful pressure measurement is the average force in Newtons under different areas of the foot, although caution here is needed as the size of the sensor cells to the size of the area chosen will influence the results. This is because the whole sensor reports the peak pressure, even if the peak pressure only occurred within a tiny area of that sensor. Pressure data can be used to attempt to replicate a force-time curve, better referred to as a *peak pressure–time curve* as such forces are gathered without regard to their direction and will be influenced by the sensor cell sizes. As forces recorded from pressure cannot be given direction, the impulse force directions cannot be identified. This renders the data's usefulness questionable, as horizontal and vertical forces cannot be differentiated.

Pressures that peak over a short period of time are less likely to damage tissues than if a more moderate pressure persists over an area for a sustained time. The best clinical approach to assessing these parameters is to measure the peak or average pressure over time in a specific area under the foot, picking a particular landmark (such as under a metatarsal head) that is prone to ulceration. This then gives a single value of the pressure–time effects during gait on that particular spot. Such measurements are known as *pressure–time integrals* or *peak pressure–time integrals*, depending on whether average or peak pressures are investigated.

Centre of pressure progression

The CoP is an instantaneous point of summation of all the pressures, representing a centre point between every pressure at a given moment. It is calculated from the forces recorded over the area of the plantar surface of the foot in contact with the ground. Standing on a pressure mat with both feet is likely to produce a CoP somewhere between the two feet. Hopefully, if weight is distributed evenly, then it will lie equidistant between the feet, and if the weight is even on the heels and forefeet, then somewhere halfway down the length of each foot. Ideally, the CoP will not actually be under either foot, but between them. Lean forward and the CoP moves forward towards the forefeet. Lean backward and the pressure increases on the heels, moving the CoP posteriorly. Slant to the left and the CoP moves to the left foot. Stand on one foot and the CoP will move under that foot, and if the forefoot and heel are evenly taking pressure, then the CoP will lie somewhere under the midfoot. This is true even if there is nowhere within the midfoot that is actually in contact with the ground.

The CoP progression is the trajectory or path of the CoP running under the foot during gait. When the heel strikes on a pressure plate, a CoP is generated from all the pressure inputs emanating from the heel. As the heel contact area increasingly loads anteriorly, the CoP moves forward. This anterior tracking continues as the forefoot and/or lateral column loads the pressure plate. The pressure mat or insole allows a CoP trace to be positioned on a computerised image of the foot. It is created by pressure applied at any given moment and is a complete trajectory of the path of the CoP over time through the foot. Ideally, the CoP should always move anteriorly as the body mass moves forward over the foot, with subtle medial-lateral deviations dictated by the areas of the foot generating pressure on the mat and movements of the CoM of the body above. On larger pressure walkways that are able to record double support, both the forefoot of one foot and the heel of the next step influence the location of the CoP. At the moment when pressure is increasing more on the leading foot's heel than is on the trailing foot's forefoot, the CoP will start to transfer across to the next footstep, passing anteriorly through the area between the feet. When a number of footsteps are viewed together, the CoP progression will trace across from one foot to the other as well as through each foot (Fig. 1.2.8a).

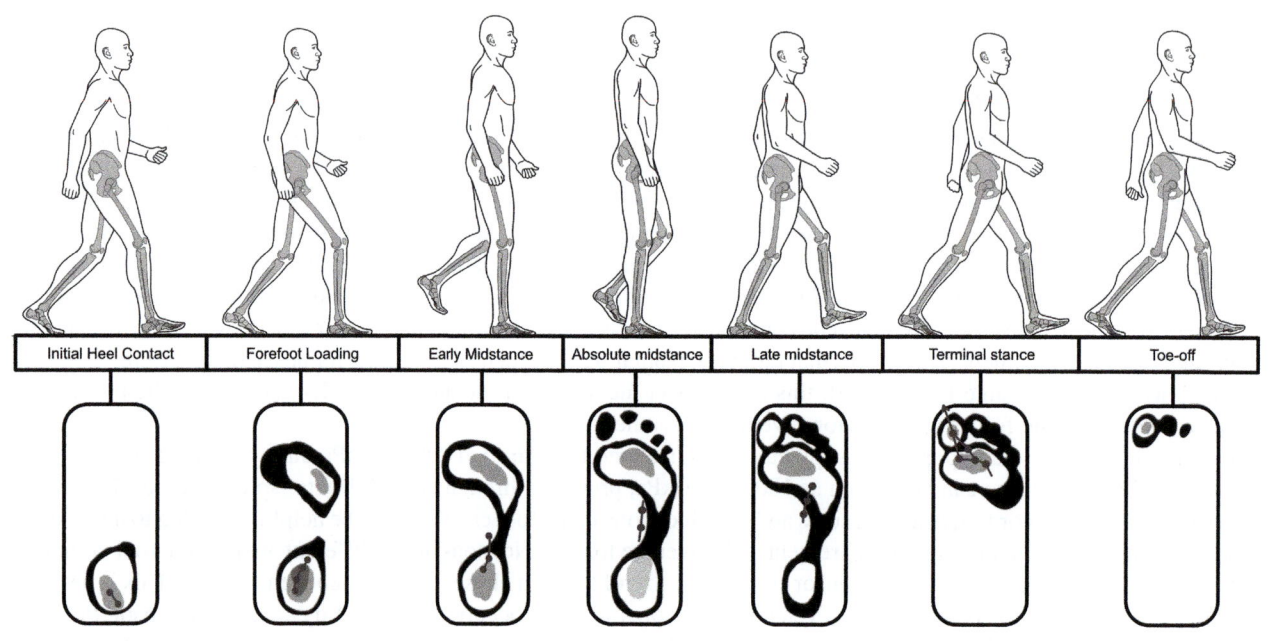

FIG. 1.2.8a Schematic of plantar pressure data (without values) through the stance phase of the right foot, with black indicating low pressures, white the next lowest, and mid-grey the highest pressures. Pressure data will be highly variable between individuals due to variance in foot shape, vault profile, body mass, and spatiotemporal parameters. Trying to suggest pressure data does not meet a 'normal' expectation and is problematic as 'normals' are debatable. Trying to predict kinematic lower limb and trunk motion from only 2D information from the plantar contact area of the foot is also impossible. However, here, simple schematics are submitted to give indications of 'expected' pressure patterns presented to the phases of gait with extreme caution. It does not represent an absolute 'normal' that patients should meet. *(Permission www.healthystep.co.uk.)*

The CoP progression pathway is heavily influenced by the contact areas of the foot because the area loaded, as well as the amount of force applied, are important factors for pressure analysis. For example, if the forefoot increasingly loads more medially, then the CoP will move anteriorly and medially. In walking gait after heel loading, if the first forefoot contact occurs medially, then the CoP will move anteriorly and medially from the heel. If the forefoot first loads along the lateral side as it does more commonly, then the CoP will move anteriorly and laterally from the heel. If the centre of the forefoot loads, then the CoP will just move anteriorly through to the middle of the forefoot. Now consider a foot that initially loads at the heel and then begins to load the proximal lateral column of the foot before starting to load the medial forefoot. The CoP will advance anteriorly in the heel, gradually drifting along a slightly lateral and anterior trajectory, and then more rapidly proceeding anteriorly and shift more medially when the medial forefoot contacts. Thus, whatever part of the of the foot loads next, and how much pressure is produced, influences the tracking of the CoP (Fig. 1.2.8b).

In a low vaulted anatomical pes planus foot, much of the foot vault loads, creating pressures over a larger surface area of the foot than would occur in a higher profiled cavoid foot. Lower profiled feet, especially hypermobile planus feet, tend to have lower peak pressures as they demonstrate a more diffuse contact area compared to high-vaulted feet. In low-vaulted feet, anterior CoP progression tends to be on a more medial trajectory depending on the amount of the plantar vault surface that is contacting the pressure mat medially. However, if only the lateral column of the midfoot is loading with pressures, then the CoP will track laterally and anteriorly during midstance, even if the foot appears to be hyperpronating by lowering the vault extensively. A cavoid foot that does not have any lateral column loading on the mat may develop a medially tracking CoP, particularly if associated with high medial forefoot loading pressures, despite hypopronation, because of a lack of any lateral column pressure. For differences in pressure data between foot types, see Fig. 1.5.3d. In feet with a midfoot break usually associated with planus feet, as the heel lifts, the anterior advancement of the CoP will slow its progression within the midfoot as high pressure often develops on the plantar base of the 5th metatarsal area (see Fig. 1.5.3f). The CoP can only reflect the contact areas of the foot over a time

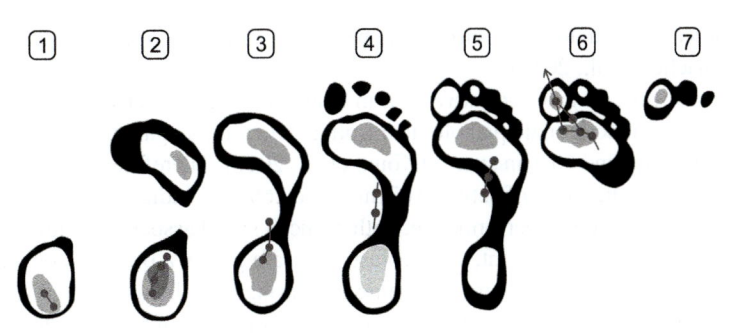

FIG. 1.2.8b A schematic of the centre of pressure (CoP) pathway throughout gait *(dark-grey dots and lines)* related to loaded areas during gait, with black representing lowest pressure areas, light grey the highest pressures areas, and white signifying middle values. Following heel strike (1), CoP will tend to move a little medially at first, reflecting rearfoot eversion moments that increase medial heel loads. What happens next is dictated by forefoot loading (2). Most commonly, the lateral forefoot loads initially, moving the CoP anteriorly and laterally on the heel. Contact areas increase as forefoot loading pressures start to deform the forefoot into the support surface. This occurs variably, reflecting the foot vault profile and soft tissue flexibility. In early midstance (3), the vault continues to depress rapidly. This increases the surface contact area that usually includes the soft tissues of the lateral midfoot. The CoP starts to track under the vault where there is no actual contact, unless the vault has large areas of plantar contact. The CoP path continues anteriorly more rapidly during midstance until the CoP moves into the forefoot (4 and 5). Some increasing toe pressure may occur, but this is highly variable. During late midstance, pressure and surface contact area decreases on the heel. The CoP moves under the central or central-lateral forefoot, but after heel lift (6), it will often move more medially, usually under the 2nd metatarsal head area. It then passes towards the hallux as weight starts to transfer to the opposite foot, taking the CoP to the space between the feet. Some individuals may generate a CoP point anterior to the 1st metatarsal head prior to the hallux (see alternative CoP path). By preswing and toe-off (7), the CoP has transferred to the opposite heel, but if using a single-foot loading platform rather than a walkway, the CoP point will remain under the hallux until toe-off. *(Permission www.healthystep.co.uk.)*

course. CoP pathways are thus a construct of the weightbearing surface area and peak force values in each pressure sensor cell's area based on the foot's vault profile, the amount it spreads into the support surface under load (as a response of its tissue flexibility), and the number and size of the sensors loaded at any moment. Loss of anterior motion and large deviations in the CoP need to be accounted for, but 'a normal CoM path and pattern' should not be the expectation of pressure assessment. In summary, the interpretation of CoP progression must be used as a support to kinematic data, based on the clinical findings rather than trying to fit the CoP progression into a targeted 'normal' at all costs. There will be a reason for each patient's CoP progression path, and the clinician needs to find it and associate or disassociate it from the pathology.

It is worth repeating that, rather confusingly, the term 'centre of pressure' can also mean the position of force coming out of the floor when recording data from a force plate (Richards, 2008). What this essentially signifies is the primary location of the GRF or 'centre' for ground reaction force (CoGRF) generation, which has been discussed earlier in Section 1.2.3. Force is a vector, for it has direction and is not affected by area. The CoGRF does not provide an indication of pressure. Pressure mats can only identify peak forces within an area. When the CoGRF is generated by force plates, it can still move anteriorly-posteriorly or medially-laterally like the CoP generated on a pressure mat. However, CoGRF medial-lateral deviations are generally much more subtle, as they are not influenced directly by the contact area of the foot in the way that CoP is.

1.2.9 Variability in gait

Gait variability describes the variation in movement patterns between repetitive steps and cycles (Stergiou et al., 2006). Variability is an integral part of motor tasks, as movement arises from the integration of a multitude of central and peripheral neuromuscular systems that receive, process, and transmit information to plan and execute an appropriate action (König et al., 2016). These neuromotor processes, which include those of the sensorimotor system, result in low levels of variability during repeated movements. This is perfectly normal, healthy, and advantageous (König et al., 2016). Variability in movement can be adapted flexibly by subjects, depending on the goal of the motor task (Wilson et al., 2008; Wu et al., 2014; Pekny et al., 2015). There is limited evidence to support a 'U-shaped' relationship between motor performance and variability as opposed to the traditional linear relationship, where both extremely high levels of variability and excessively low levels of variability are disadvantageous to motor performance (König et al., 2016). Movement tasks are likely to be executed at an optimum between sufficient accuracy and minimum control costs, as being more controlled in motion uses more energy (König et al., 2016). The more skilled at a task an individual is, the more that task can be performed in a range of variability of motions that use less energy, whereas someone learning a task will use more energy as they work to attempt to keep variability to a minimum.

This suggests that each person develops an optimal amount (or window) of variability for day-to-day tasks such as walking or standing, and that people with a disturbance of their sensorimotor system will operate outside of this window. Therefore, on gathering quantitative gait data such as step and stride length over any duration, clinicians should expect to see subtle variations in kinematic and kinetic data between steps, even of the same foot at the same gait speed during the same test. However, an increase in motor variability can be associated with ageing and neuromuscular pathologies (Hausdorff et al., 1998; Hamacher et al., 2011).

Research suggests that stride time in gait and sway area in stance have an upper threshold of variability of 2.6% and 265 mm^2, respectively, which helps discriminate pathological from asymptomatic gait and stance performance by an overall accuracy of between 80% in walking and 60% in stance (König et al., 2016). Therefore, a variability in stride time between 1.1% and 2.6% in gait and a sway area in stance of between 67 mm^2 and 265 mm^2 indicates a healthy neuromotor status (König et al., 2016). Gait speed is a clear indicator of performance, with reduced speed associated with falls, co-morbidity, and cognitive decline (Al-Yahya et al., 2011). Independent, secure, fast, and safe gait is a primary consideration in the elderly and those affected by neurological disorders, with locomotion being increasingly challenged when locomotive tasks are complicated by concurrent tasks. Dual-task-related changes in spatiotemporal parameters of gait include decreases in speed, cadence, and stride length, along with increases in stride time and stride time variability (Al-Yahya et al., 2011).

Influence of terrain on walking gait analysis parameters and kinematics

Another influence on gait that causes variability is the terrain upon which locomotion takes place. In daily activities, those who are walking or running on uneven cross-country ground will be demonstrating a larger variance in their gait parameters that reflects the greater uncertainty of each step. In the environment of a clinic, the primary variance in terrain is often between the flat clinic floor and a treadmill. Temporal parameters on treadmill and overground walking have proved to be consistent. However, with treadmill walking, it has been demonstrated that there is an increased cadence and amount of knee flexion, as well as a decreased stride time and step length (Strathy et al., 1983), with differences in lower limb joint temporal and angular kinematics compared to overground walking (Lee and Hidler, 2008; Chockalingam et al., 2012). The pelvis is reported to show lower pelvic obliquity on treadmills, with pelvic rotations also different, although gender-induced differences between subjects still produce larger differences than does the treadmill (Chockalingam et al., 2012). Despite similar kinematics in the lower limb, muscle activation patterns have also proven to be different when walking overground compared to walking on treadmills (Lee and Hidler, 2008) (Figs 1.2.9a–1.2.9c). The solution to gathering more

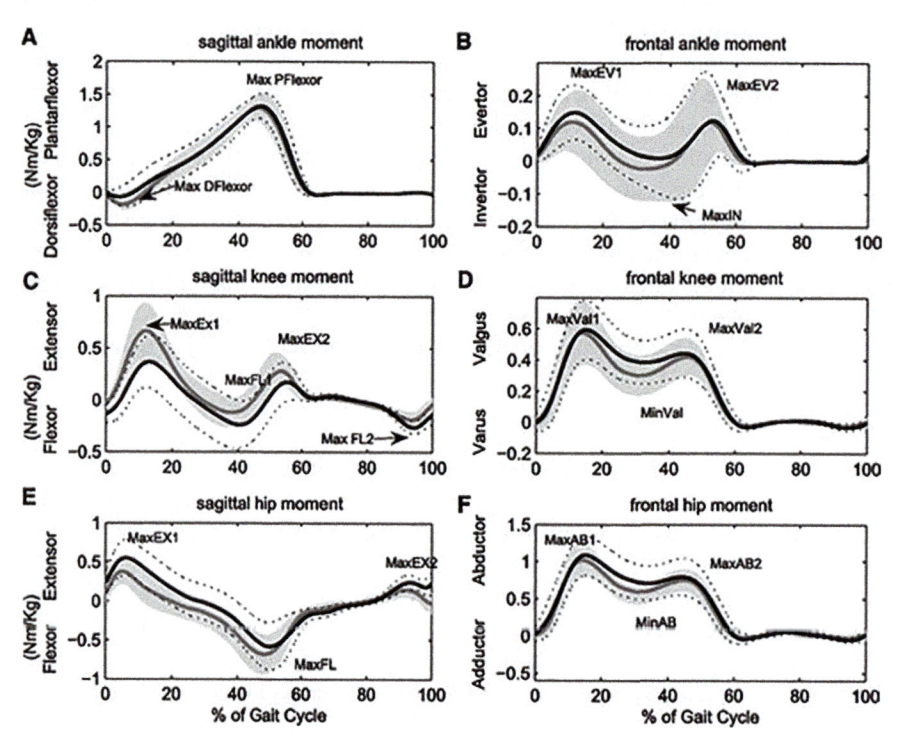

FIG. 1.2.9a Average kinetic joint moment data from a study by Lee and Hidler (2008) during gait in overground compared to treadmill walking (left). Joint moments of the ankle, knee, and hip joints are from the sagittal (A, C, E) and frontal (B, D, F) planes, taken from nine male and ten female healthy adults aged between 18 and 70 years. No age-related differences in this study were found, probably because the older subjects were described as being very fit and active. The grey line indicates mean ± 1 standard deviations for overground trials while the solid black line and two dashed lines represent the mean ± 1 standard deviation for treadmill data. Maximum dorsiflexion (Max DFlexor), maximum eversion at early stance (MaxEV1), and maximum inversion (MaxIN) are noticeably different at the ankle. Maximum plantarflexion (MaxPFlexor) seems unaffected by the treadmill. Knee extension moments (MaxEX1 and 2) and knee flexion moments (MaxFL1 and 2) also indicate a more flexed knee posture on treadmills. At the hip stance phase, maximum extensor moments (MaxEX1) are higher, but maximum flexion moments (MaxFL) are lower on the treadmill, with more flexion in terminal swing in overground walking. Maximum hip abduction moments (MaxAB1 and 2) are similar, but minimum (MinAB) moments are reduced on treadmill walking. *(From Lee, S.J., Hidler, J., 2008. Biomechanics of overground vs treadmill walking in healthy individuals. J. Appl. Physiol. 104 (3), 747–755.)*

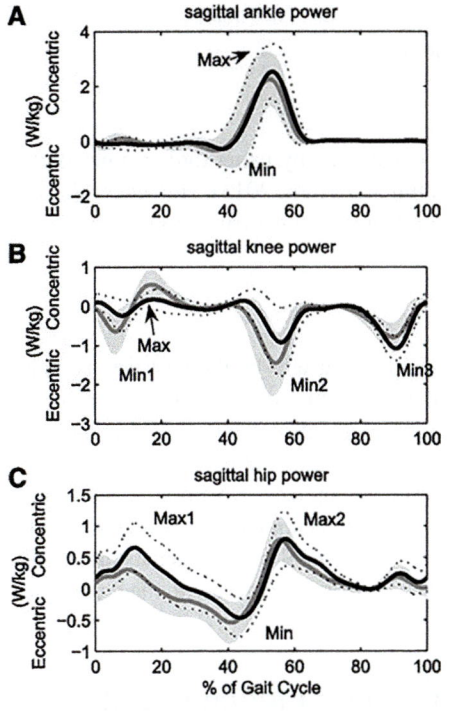

FIG. 1.2.9b Average ankle (A), knee (B), and hip (C) joint power from the study by Lee and Hidler (2008). The grey line indicates mean ± 1 standard deviations for overground trials, whereas the solid black line and two dashed lines represent the mean ± 1 standard deviation for treadmill data. Ankle plantar-flexion power is raised on the treadmill, but knee powers are depressed. The hip expresses more concentric power in early and middle stance, and again in early swing. *(From Lee, S.J., Hidler, J., 2008. Biomechanics of overground vs treadmill walking in healthy individuals. J. Appl. Physiol. 104 (3), 747–755.)*

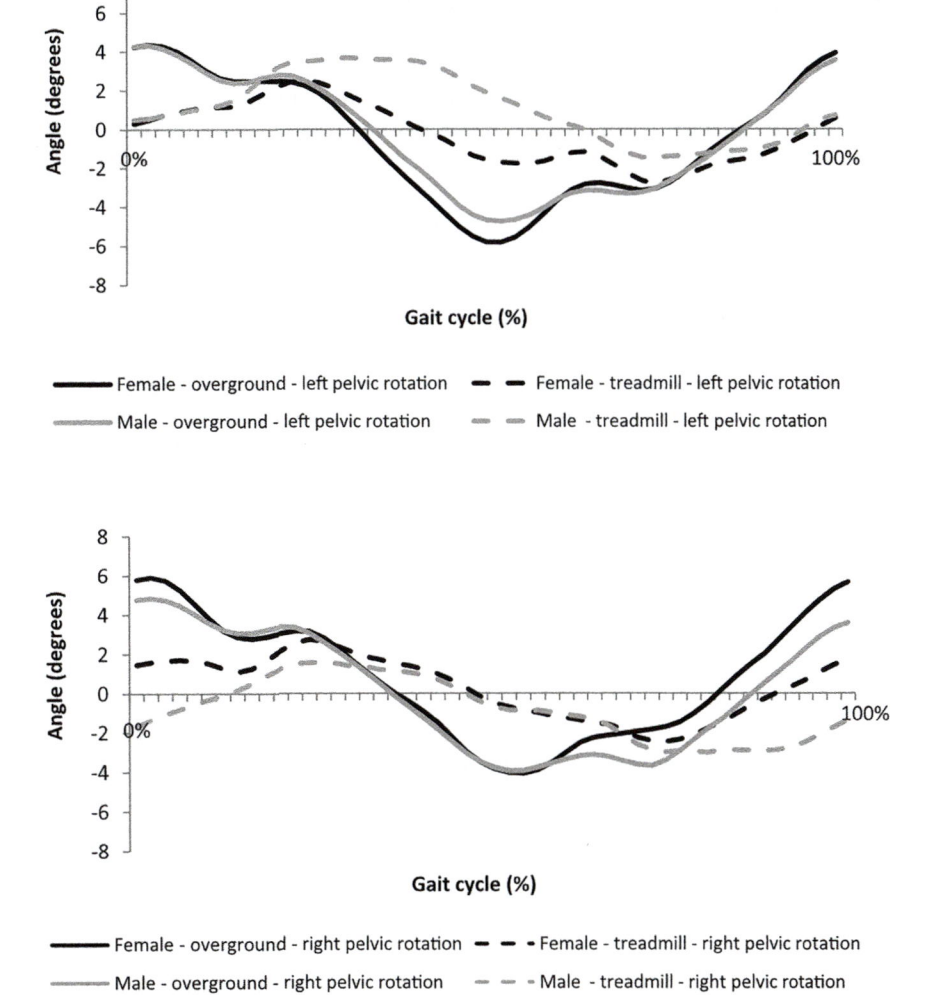

FIG. 1.2.9c Gait patterns reported for men's and women's pelvic rotations at the hip when walking overground compared to walking on a treadmill. Upper data for the left limb and lower data for the right limb. *(From Chockalingam, N., Chatterley, F., Healy, A.C., Greenhalgh, A., Branthwaite, H.R., 2012. Comparison of pelvic complex kinematics during treadmill and overground walking. Arch. Phys. Med. Rehabil. 93 (12), 2302–2308.)*

representative data from treadmills may be achieved by using longer times on the treadmill of over 6 min before starting gait-data capture (Matsas et al., 2000; Cronin and Finni, 2013). However, analysis of gait between assessments should be performed over the same terrain to avoid differences occurring from a change in the terrain alone.

1.2.10 Section summary

A primary clinical skill is the ability to decide what will be the best way of approaching the gait analysis of a particular patient. This reflects the techniques available and the ability of the equipment to provide relevant data. It is important therefore that the clinician be familiar with the data provision of equipment and the significance of the data they gather. Primary clinical errors include the interpretation of data that results from pathology being assumed to be instead, the cause of that pathology. Gait analysis does not give a diagnosis. No single piece of gait analysis equipment can provide all the kinematic and kinetic data necessary for all situations nor can they usually report on real-life locomotion outside the clinic or laboratory. Thus, it is essential therefore that clinical biomechanists are familiar with a range of gait assessment techniques and how to read them. They should become most expert with the techniques they use while also being fully aware of the limitations their equipment presents.

1.3 Muscle function related to joint motion in gait
1.3.1 Introduction

Without muscle function, gait of any type cannot happen. Muscular force during locomotion depends on an animal's speed of gait and size which underlie the energy demands that power locomotion (Biewener et al., 2004). Thus, it is an important requisite to understand the principles of muscle activity in gait. Although it is necessary to know the basic normal muscle patterns of activity in gait, learning them 'parrot fashion' or by rote in association with gait phases without comprehending what they do when and why will not equip the clinician with an understanding as to the variations seen in normal and abnormal changing situations. So-called 'normal' muscle action in healthy individuals is actually variable from one step to the next. The reasons for this lie in the complex interaction between muscle activity and the feedforward-feedback mechanisms of the sensorimotor system. Changes in limb posture affect muscle activity by altering the mechanical advantage around the GRF by presenting different lines of action and muscle lengths at activation.

The mechanical principles underpinning joint angular momentum and the need to control the rise and fall of the CoM is a far better way to appreciate the functional role of muscles. Combining kinematic analysis with GRF data and EMG recordings of muscle activity, motion, and forces within the lower limb can be applied to specific or generalised muscle activity. By knowing the location of GRF force vectors around joints, muscle activity can be anticipated by producing the forces necessary to create, control, or resist motion. The generation of Pedotti diagrams from GRFs are extremely helpful in appreciating the direction of these force vectors. However, in their absence, the clinician may need to imagine them being applied to their patient. The simplest way to approach this is to consider where the GRF is in relation to a joint and each moving segment's CoM in all the orthogonal planes. As humans are primarily using angular momentum of extensions and flexions around their joint axes, it is import to observe whether the GRF is being generated in front of or behind the joint. If the stance foot is in front of the hip, then GRF flexes the hip. If the GRF on the foot is behind the hip, then GRF extends the hip. Depending on whether the movement is desirable or not, muscle action will be needed to prevent or control the motion. This means that the clinician can mine information about patients from just being able to record their gait in slow motion within the sagittal plane. The more planes that gait is recorded in, the more muscle actions can be predicted in detail.

The primary agonist and antagonist muscles of joint moments are always of primary consideration. For example, the braking of hip flexion by gluteus maximus and knee extension by the hamstrings against the hip flexor iliopsoas and knee extensor activity of the quadriceps just prior to initial contact are key events. The ankle plantarflexor activity at late midstance, restraining CoM anterior progression, and the amount of dorsiflexion that occurs during this period are key muscular events prior to heel lift. Other muscles should be considered as assistors and generators of the subtle control of flexion and extension during gait through any transverse and frontal plane perturbations. However, this does not mean that these 'lesser' nonsagittal plane-dominant muscles are not important for limb function, a situation much appreciated once tibialis posterior dysfunction has been observed. Muscle function in relation to anatomical position and mechanical laws explains normal muscle activity, and this helps the clinician remember which muscles are active when and during what action.

As most muscles are bi-articular or multi-articular, they are able to compensate for muscular deficiencies over two or more joints. The hamstrings can potentially compensate for weak hip extensors, secondary to their primary knee flexion role. Gastrocnemius can compensate for soleus weakness, whereas tibialis posterior, the peroneal muscles, and the long digital flexors of the foot are able to compensate for a weak triceps surae through their secondary ankle plantarflexion role. However, the resulting joint kinematics that these compensatory muscles generate will be different.

As a point of caution, sometimes the triceps surae can be referred to as an ankle extensor in gait research. This is because with hip and knee extensor activity, the lower limb lengthens, which is what happens with the addition of plantarflexion at the ankle. This can cause some confusion. Therefore, in this text, such descriptions are translated into 'ankle plantarflexion' for consistency.

1.3.2 Principles of muscle action in gait

Energetics

Changes in energetic cost of muscle force generation underlie most of the changes in energy associated with speed and load-carrying during human walking (Griffin et al., 2003). Studies comparing quadrupeds and bipeds of similar size demonstrate similar metabolic costs related to relative size of limb muscles and the volume of muscles that are recruited to support the body weight during walking locomotion (Roberts et al., 1998). However, the transport costs of human running exceed that of quadrupeds of similar size (Biewener et al., 2004). During walking, muscles have to generate forces that overcome the inertia and gravitational forces of the body and the limb segments' CoMs that generate joint moments, and both counteract and generate GRF to achieve locomotion. The most expensive part of walking gait is the terminal stance phase, when the lower limb assists acceleration to the next step (Gottschall and Kram, 2003, 2005).

Limb posture changes during gait alter mechanical advantage of the GRF and influences the effective line of action of any muscle forces produced at a particular joint. Changing gait speed profoundly influences the effective mechanical advantage through force-velocity relationships as faster contractions risks the production of less force. From walking to running, mechanical advantage is reported to decrease by 68% within the knee extensors, 18% at the hip extensors, and by 23% in the ankle plantarflexors (Biewener et al., 2004). During walking, the knee joint operates at around 154–176° (near the 180° of full extension) within the stance phase, but it is more flexed at around 134–164° in running. This causes a five-fold increase in quadriceps impulse (force over time) required to support the knee during running stance which in itself is associated with an approximate five-fold increase in the GRF moment around the knee (Biewener et al., 2004). However, the hip extensor impulse actually decreases in running, whereas the ankle plantarflexors show little change from walking to running. It is through the knee joint that the resultant transport costs of running are considerably higher (Biewener et al., 2004). This indicates how joint postural alignment can significantly affect the metabolic cost of locomotion. The primary site of muscle accelerator power within the lower limb is the triceps surae through the Achilles tendon, which demonstrates significant changes in both the loading burden and the muscle fascicle-to-tendon tissue lengths in running when compared to walking. This shifts the burden to the Achilles tendon from the associated muscle (Ishikawa et al., 2007).

Force vectors and angular momentum in joints

Joints experience angular momentum or moments of force from inertia, gravity, and the GRF. Gravity and inertia drive GRFs which together are the forces that create moments that wish to rotate joints. GRFs that initiate flexion of a joint in the sagittal plane are referred to as *external flexion moments*, whereas those initiating extension are labelled *external extension moments*. This also works in the frontal plane where adduction and abduction moments are also known as *external varus* (adduction) moments and *external valgus* (abduction) *moments* in the hip and knee or *external inversion moments* and *external eversion moments* within the foot.

Moments, purely in the transverse plane, are not normally generated by GRF but by the internal muscle contraction moments. However, in multiplanar joints, transverse plane angular momentums are often a coupling consequence of other primary joint rotations. The most common uses of multiplanar angular momentum in the context of gait are those referred to by the terms '*supination moments*' and '*pronation moments*' in the foot, particularly in relation to the rearfoot's ankle-subtalar joint complex. Yet when investigated further, in much of the research literature, these terms are often actually being equated to one plane of motion only, usually that of frontal plane inversion with supination and frontal plane eversion with pronation. Such frontal plane moments should be correctly designated as inversion moments and eversion moments.

82 Clinical biomechanics in human locomotion

The muscle action opposing these external moments is termed an *internal moment* of force that are usually opposite to the external force. An external flexion moment is opposed by an internal extension moment, and so on. If the external and internal moments are equal, then the joint will not move, and static stability is created. If the internal moment is less than the external moment, then the external moment wins out so that motion is driven in the direction of the external force. Yet, the overall motion is decelerated by the internal moment, and this reflects the difference between the amounts of the two forces. If the internal moment generated by the muscle is greater than the external moment, then the joint moves in the direction of the internal moment but slowed and restricted under external resistance. Obviously, in swing phase, the situation is different in that the GRF is not involved. Only the weight (gravity effects and air resistance) and inertia of the segments being moved are creating the external force that the internal moment must overcome (Fig. 1.3.2a).

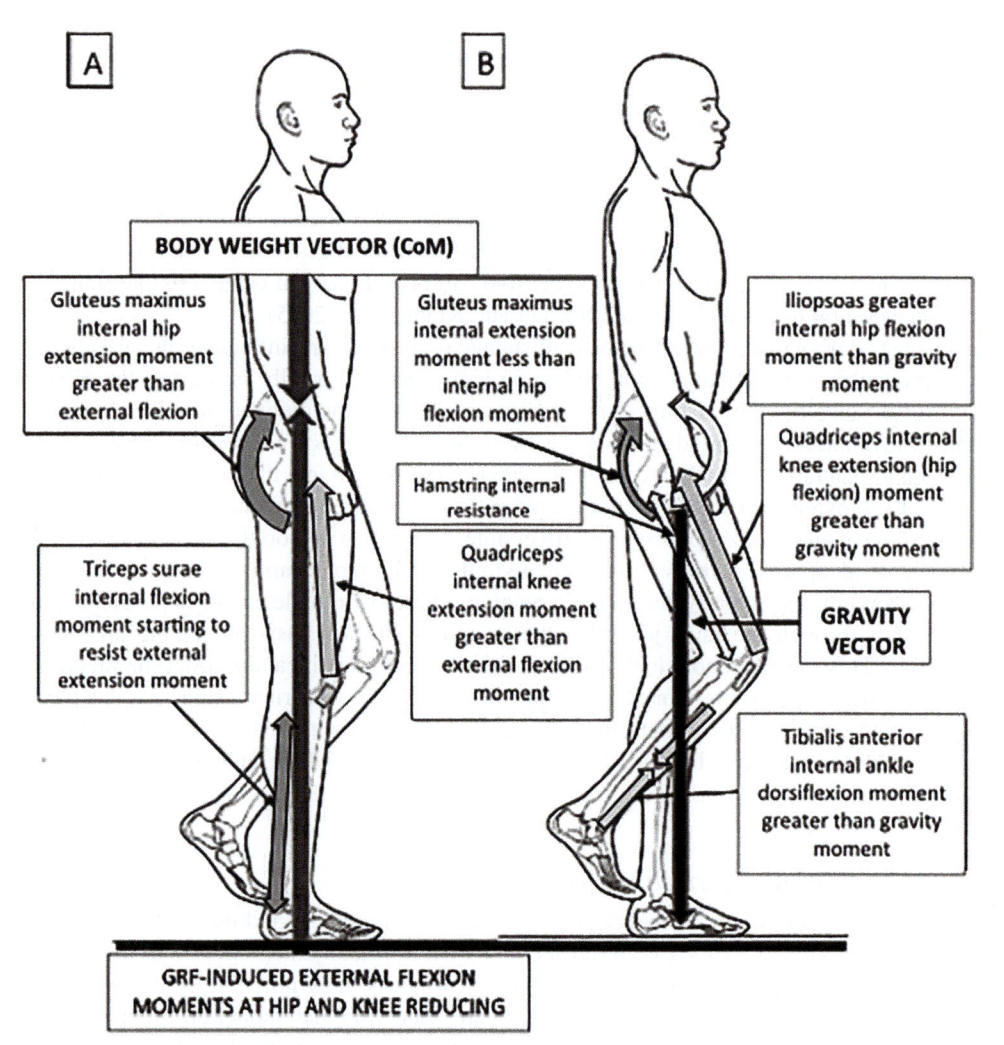

FIG. 1.3.2a Lower limb motion is a combination of internal and external moments. Approaching absolute midstance (A), the right stance limb's gluteus maximus and quadriceps are creating internal extension moments at the hip and knee respectively. These internal moments are greater than external flexion moments (which are reducing rapidly in the middle of midstance) generated from the body weight vector under gravity and the GRF. The triceps surae starts to generate internal ankle flexion (plantarflexion) moments as the GRF slips in front of the ankle to generate ankle extension (dorsiflexion) moments. At midswing (B), the only external force on the right swing limb is gravity, which is trying to draw all limb segments to the vertical. The iliopsoas creates internal hip flexion moments that must be controlled (antagonised) by internal moments generated in muscles such as gluteus maximus and the hamstrings that are producing internal hip extension moments. The quadriceps aid hip flexion through rectus femoris but primarily create an internal knee extension moment. These moments will be antagonised by hamstring-induced internal knee flexion and hip extension moments during late swing. Tibialis anterior uses internal ankle dorsiflexion moments, sufficient to resist the effects of gravity that try to draw the ankle into plantarflexion. *(Permission www.healthystep.co.uk.)*

Simple examples of these situations are the GRF in the single-support limb creating an external varus (or adduction) moment at the hip during single-limb stance. Unopposed, the CoM of the trunk will cause it to rotate on the hip into adduction, creating a severe varus tilt of the pelvis. This is opposed by the internal hip abduction moment from gluteus medius, minimus, parts of gluteus maximus, and tensor fasciae latae, which keep varus pelvic tilt to a minimum. If the internal moment was greater than the trunk segment moment, then the hip would abduct causing valgus tilt of the pelvis or *hip hitch* (see Fig. 1.1.5e). If the internal and external forces are even, then balance of internal abduction moments and external adduction moments is achieved and the trunk maintains its position over the hip within the frontal plane, which is actually a little excessive for gait stability needs. Hence, a little varus tilt occurs at the stance limb hip through the resulting forces from the internal and external moments (Fig. 1.3.2b(A)).

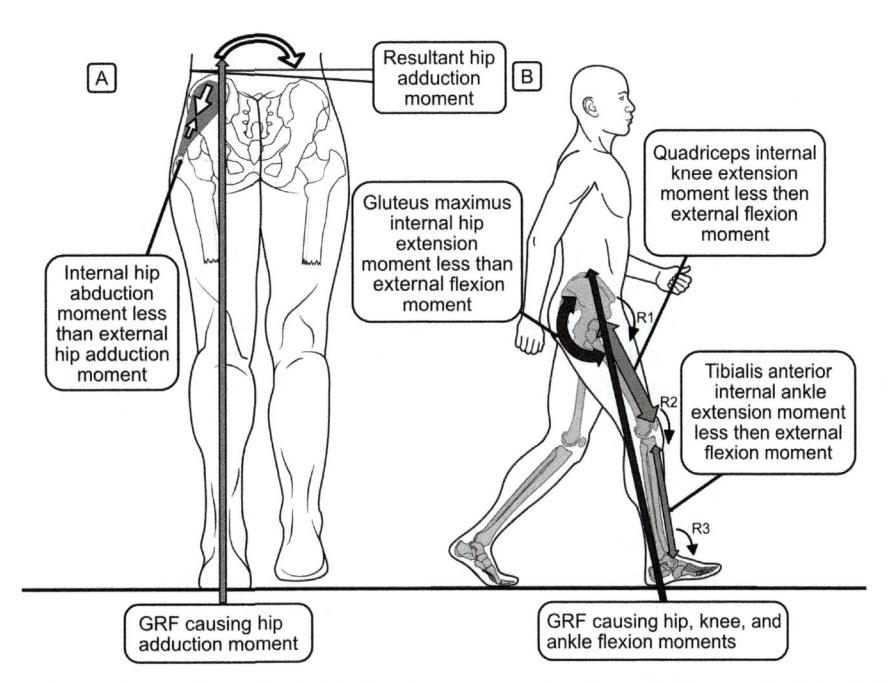

FIG. 1.3.2b Often, internal moments are used to moderate rather than stop or reverse external moments. For example, during single-limb support (A), the GRF sets up an adduction moment at the hip because the GRF and CoM lie medial to the hip in the frontal plane. Hip abductor muscles resist this external moment, but the internal forces are less than the external, being just sufficient for need. Thus, the hip still adducts a little under the final resultant moment. At heel strike (B), hip, ankle, and especially knee flexion results from the external GRF moment on the heel of the foot. This motion is used to allow the large muscles to 'shock-absorb' using their muscle-tendon buffering action. Therefore, although the hip, knee, and ankle extensors create internal moments, they are less than the external moments, allowing resultant moments of hip (R1), knee (R2), and ankle (R3) flexion under eccentric muscle control. In the case of tibialis anterior, the resultant ankle flexion brings about necessary forefoot contact, but in a controlled manner to reduce forefoot impact velocity. Thus, externally derived moments can be exploited usefully for efficient gait biomechanics. *(Permission www.healthystep.co.uk.)*

During weight acceptance, the GRF creates an external flexion moment on the hip which is opposed by the internal extension moment of the gluteus maximus. Initially, the external moment is greater than the internal moment, but this quickly reverses during early midstance. Thus, throughout early midstance, hip flexion reduces under the internal extension moment. At initial contact, the GRF gives rise to an external flexion (plantarflexion) moment on the foot from heel strike. This is opposed by a smaller internal extension moment from tibialis anterior (assisted to a limited extent by the long toe extensors) that allows the forefoot to achieve ground contact through a resultant decelerated plantarflexion moment (internal and external moments combined together). Without this smaller internal extensor moment, the foot would slap onto the floor under rapid acceleration, but if excessive, the forefoot would not load to the support surface during contact phase. Thus, the size of the internal moment generated by tibialis anterior serves to control but not stop the acceleration rate of the external flexion moment of the ankle (Fig. 1.3.2b(B)).

Another complication to muscle function in gait is the neural coupling of upper and lower limb muscle activation as a result of them sharing common subcortical and cortical drivers that co-ordinate rhythmic four-limb gait patterns (Weersink et al., 2021). Such mechanisms may relate back to our evolutionary quadrupedal ancestors, but it is more than that in that within modern humans as it reflects recruitments of neuronal support that can optimise cyclical movement patterns of the lower limbs (Weersink et al., 2021). Upper limb muscle activity seems to influence lower limb activity more than it does vice versa, and such coordination of limb motion may target rehabilitation programmes, especially for the neurologically impaired (Weersink et al., 2021).

Clinically, making limb and joint motion observations in gait analysis can be very useful, even if just using slow-motion video. If muscles are unable to fulfil their role in acceleration, stability, or deceleration of articular moments, then obvious changes will be seen in joint motions. If the forces need to be quantified, then it is necessary that the external moments be quantified first, whereupon the use of inverse dynamic theory is required. For most clinical situations, the observational effects will suffice, as either the wrong motion or a joint's insufficient acceleration/deceleration will indicate that there is an issue with muscle function within the region under investigation.

1.3.3 Primary muscle function during walking gait

Major human lower limb muscle activity occurs during walking gait with an ability to adjust and recruit contractions in response to the environmental need through neuromuscular control. Such variability aside, there are expected muscle activation patterns occurring in walking on level surfaces that are predictable for healthy human gait. Therefore, it is important that clinical biomechanists be very familiar with the expected sequence of muscular activity, especially in relation to the primary flexion-extension muscles that provide the vast bulk of locomotive forces. Classic anatomy texts tend to describe muscle action from the point of view of unresisted motion caused by the pulling of a tendon on a corpse, revealing the moment of force applied to a static unloaded joint. While this is somewhat applicable to the lower limb during swing phase motion, it does not reflect the action of most muscles in gait, particularly during the stance phase. Indeed, most important stance phase muscle activity is eccentric and resisting motion induced by GRF vectors, with only early midstance hip and knee extension significantly contradicting this. Gait research using EMG recordings and the analysis of force vectors and 3D kinematics has increased the understanding of both normal and pathological gaits, despite the complexity of bi-articular/multi-articular joint influences from muscles.

Most studies use surface EMG analysis, which has provided plenty of research data on muscles that are easily accessible through the skin. However, deeper muscles that lie under other muscles need the insertion of fine-wire electrodes (receptors) into them, which is harder to gain ethical committee approval for. Thus, data on such muscles as tibialis posterior and vastus intermedius are harder to obtain, as they sit deep to other muscles. In the case of vastus intermedius, there appears to be considerable interference, known as '*cross-talk*', with the EMG data taken over rectus femoris, often with activity assigned to rectus femoris actually coming from vastus intermedius (Nene et al., 2004).

Gait muscle activation is most easily considered in the sagittal plane, where GRF vectors in the vertical and anterior–posterior direction drive angular moments of flexion and extension in the primary joints of the lower limb. Vertical forces largely drive compressive stress while horizontal anterior–posterior forces largely create shear stress. The reality is more complex because GRF vectors are 3D, and even joints structured mainly for extension and flexion such as the ankle and knee can still rotate in the frontal and transverse planes. However, for ease of understanding, let us imagine an exclusive world of lower limb sagittal plane biomechanics so that basic muscle function concepts can be more easily understood. Such a world allows us to approach muscular contributions within the stance limb's support and motion by looking at the length of time of muscle activation compared to the vertical force exerted by the ground. Using such an approach during walking to model GRFs generated from heel strike necessitates that the ankle dorsiflexors are slowing the plantarflexion moment set up from a foot-to-ground impact behind the ankle. This will continue until the forefoot contacts the ground and the ankle can plantarflex no further. After this 'foot flat' posture forms at the end of loading response and just before contralateral toe-off, limb support is primarily provided by gluteus maximus, the vasti muscles of the quadriceps, and the posterior fibres of gluteus medius and minimus working to moderate flexion at the hip and knee (Anderson and Pandy, 2003) (Fig. 1.3.3a). It is the braking action of these muscles together with gravity that produces the F1 peak GRFs on the force-time curve.

Early midstance support is provided mainly by gluteus maximus, gluteus medius, and minimus of the hip abductors, and the vasti muscles of the quadriceps, plus the influence of gravity on the HAT's CoM. This occurs during the downward slope to the F2 trough of the force-time curve. However, after absolute midstance, the ankle plantarflexors increasingly become involved, providing nearly all late stance phase support and inducing the upward slope of the F3 peak of the force-time curve (Anderson and Pandy, 2003). During walking, medial gastrocnemius muscle fascicles shorten during the first 15% of stance phase due to sudden ankle plantarflexion following heel strike. Thereafter, both the muscle fascicles of medial gastrocnemius and the Achilles tendon lengthen slowly, controlling midstance ankle dorsiflexion with soleus until ~70% of stance. This is followed by muscle fascicle and tendon fibre shortening in the last 70%–100% of stance (Ishikawa et al., 2007), during the acceleration at and after heel lift. Centrifugal forces from the opposite swing limb that

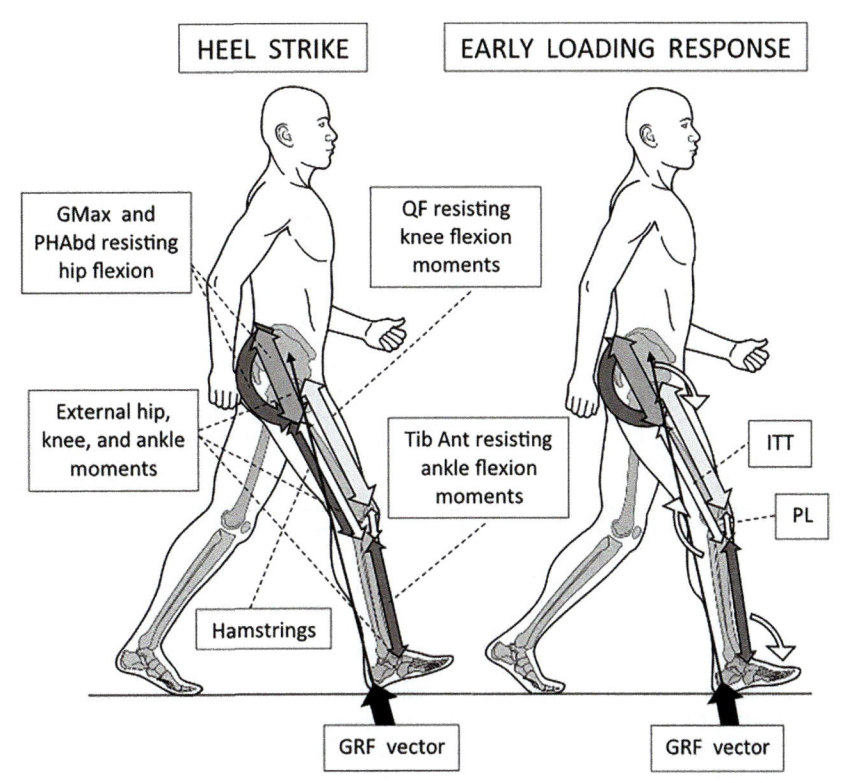

FIG. 1.3.3a Muscle activity at initial contact involves the braking of acceleration from the fall of the CoM of the HAT segment under gravity. Iliopsoas should already have become inactive during late swing. However, its antagonistic hamstrings and gluteus maximus (GMax) that have been resisting hip and knee extension during swing are still active. Hamstring activity is maintained only at initial heel contact to help stabilise knee posture. However, this rapidly ceases. A small amount of gastrocnemius activity also occurs at initial contact, probably as an assistor to stability. The main activity comes from those muscles that resist the external hip, knee, and ankle flexion moments that are brought about by the GRF vector working with the positioning of the CoM. Hip flexion is resisted by GMax aided by the posterior hip adductors (PHAbd). Some of their power helps stabilise knee posture via the iliotibial tract (ITT). The quadriceps femoris (QF) resists knee flexion via the patellar ligament (PL). Tibialis anterior (Tib Ant) resists ankle flexion until foot flat. At foot flat, external hip, knee, and ankle flexion moments start to be overcome by muscle-induced internal hip and knee extension moments and thus create internally induced ankle extension moments. *(Permission www.healthystep.co.uk.)*

provide much of late stance acceleration act to decrease the vertical GRF slightly during terminal stance by pulling the CoM forward. However, they have little influence on the overall vertical forces of stance, whereas the skeleton provides resistance to gravitational forces of around half the body weight throughout the gait cycle (Anderson and Pandy, 2003). Thus, the muscles and the skeletal structures interacting with gravity together provide the whole stability of the body, changing their influences throughout locomotive events (Figs 1.3.3b–1.3.3f).

Muscle function and activity during gait changes with gait speed due to alterations in the timing and the extent of joint angles that occur. Further to this, some muscles such as gastrocnemius can act as both a knee flexor and an extensor during gait (Fox and Delp, 2010). In supporting upright leg posture against gravity, the knee angle stays constant or extends upon muscle contraction. However, once the knee starts to extend, the gastrocnemius flexes the knee, reducing knee support with gravity. This is a situation that is likened to Lombard's paradox and has been discussed in Chapter 4 of the companion text 'Clinical Biomechanics in Human Locomotion: Origins and Principles', Section 4.1.10. Gastrocnemius generates a plantarflexion moment that induces knee extension acceleration. At the same time, gastrocnemius also induces a knee flexion moment that creates knee flexion acceleration with its resultant action being dependent on the ankle and knee moment arms, body position, and foot contact angle (Fox and Delp, 2010) (Fig. 1.3.3g). Hence, during midstance with the foot in dorsiflexion, the gastrocnemius acts as a restrictor of knee extension. Moreover, as it induces ankle plantarflexion at heel lift, it weakly extends the knee, thereby resisting any excess knee flexion that might occur under the influence of the GRF and gravity on the body. This results in soleus and gastrocnemius muscle moments having opposite effects on the knee's acceleration during terminal stance and preswing, despite the fact that both muscles are active ankle plantarflexors until 52% of the gait cycle, delivering energy all the way into preswing even though they deactivate early in terminal stance (Fox and Delp, 2010).

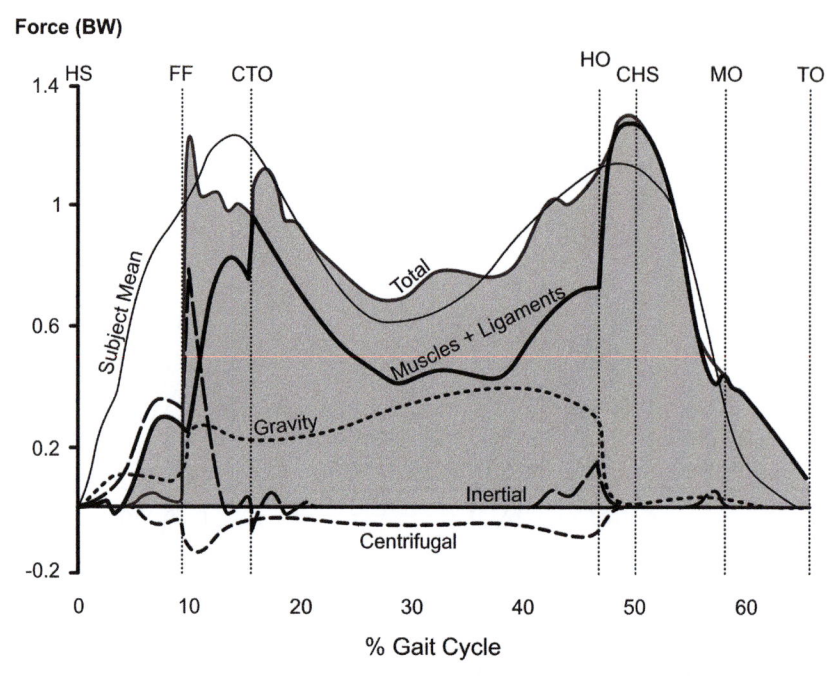

FIG. 1.3.3b Contributions of inertial forces (inertial), centrifugal forces (centrifugal), the resistance to gravity provided by the skeleton and joints (gravity), and the muscles and ligaments (muscles + ligaments) for the body's support during normal gait. The *thick line* represents vertical GRF and the thin black line signifies total force from summing inertial, centrifugal, gravity, and muscles and ligaments force together. The mean vertical GRF is shown as a dotted line. Kinematic events are during stance phase to 100% of the full gait cycle. Events are defined as HS = heel strike; FF = foot flat; CTO = contralateral toe-off; HO = heel-off (heel lift); CHS = contralateral heel strike; MO = metatarsal-off; TO = toe-off; BW = body weight. *(Adapted from Anderson, F.C., Pandy, M.G., 2003. Individual muscle contributions to support in normal walking. Gait Posture 17 (2), 159–169.)*

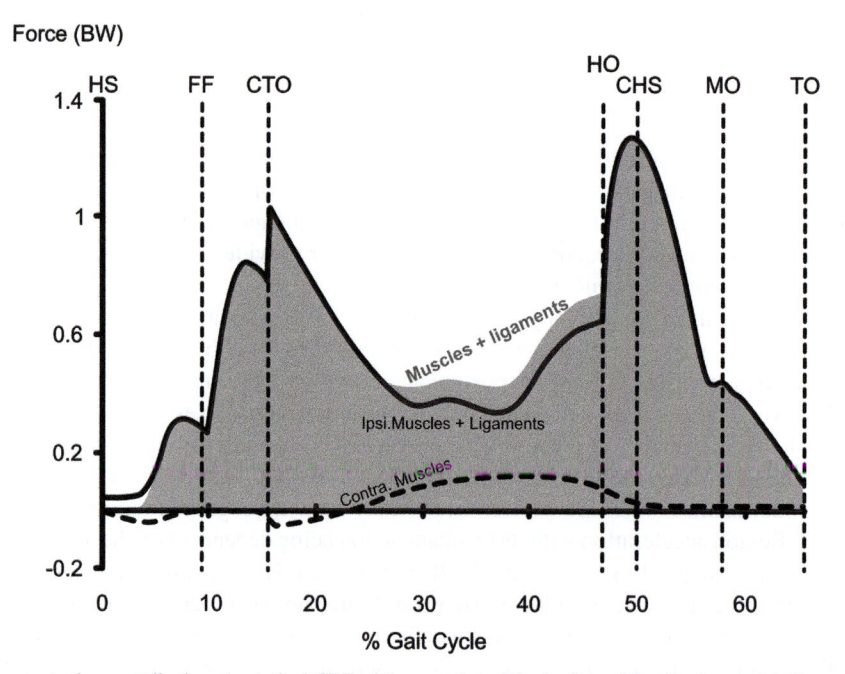

FIG. 1.3.3c A graph demonstrating contributions to vertical GRF of the muscles of the ipsilateral leg (ipsi. muscle + ligaments) and muscles of the contralateral leg (contra. muscles). The *shaded region* represents the total contribution of all the muscles and ligaments to GRF. The curve defined by this region is that labelled muscles+ligaments as in Fig. 1.3.3b. For other symbols of gait events, also see Fig. 1.3.3b. *(Adapted from Anderson, F.C., Pandy, M.G., 2003. Individual muscle contributions to support in normal walking. Gait Posture 17 (2), 159–169.)*

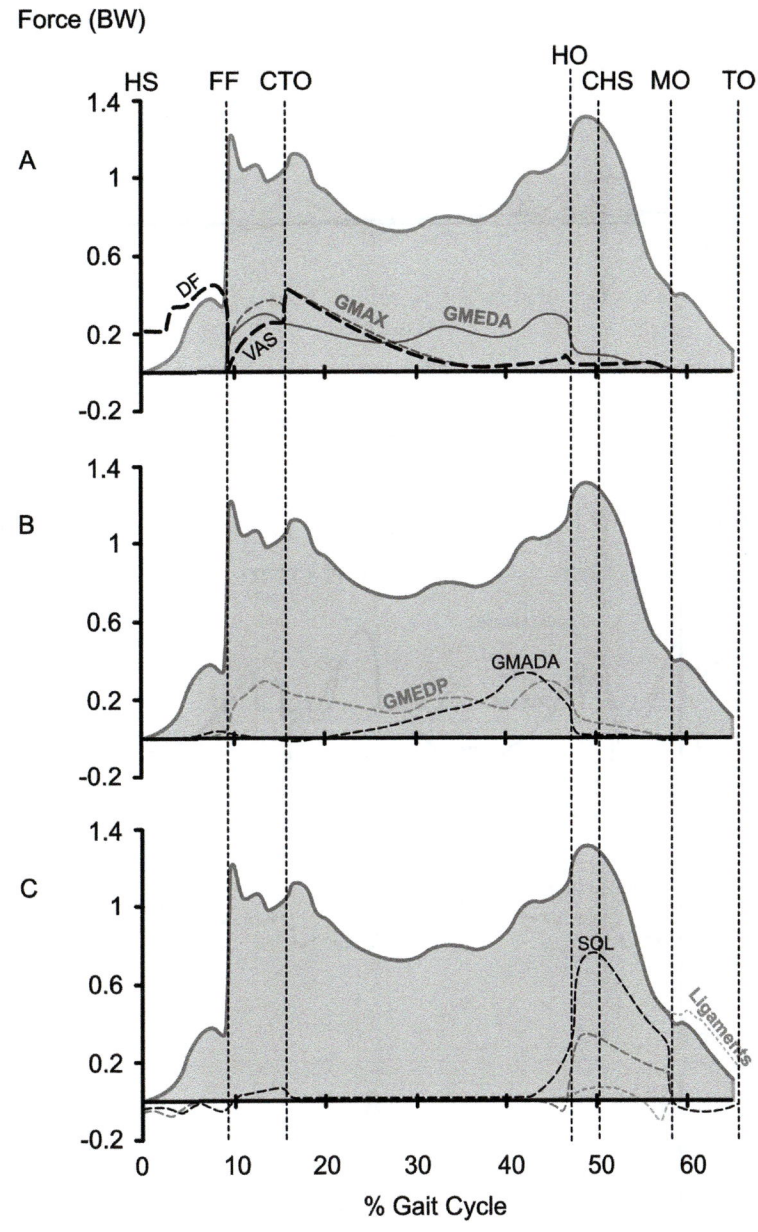

FIG. 1.3.3d Muscles contributing most significantly to body support of vertical forces during stance from modelling. Major contributors to contact phase are shown in (A), midstance (B), and terminal stance (C). Muscles are defined as: DF = ankle dorsiflexors; GMAX = medial and lateral portions of gluteus maximus; VAS = vasti quadriceps muscles; GMEDA = anterior gluteus medius/minimus; GMEDP = posterior gluteus medius/minimus; SOL = soleus; GAS = gastrocnemius; and ligaments = all ligaments. Gait event symbols are presented as in Fig. 1.3.3b. *(Adapted from Anderson, F.C., Pandy, M.G., 2003. Individual muscle contributions to support in normal walking. Gait Posture 17 (2), 159–169.)*

Stance phase muscle activity

The two most important events in the walking stance phase are the deceleration impulse at initial contact/loading and the acceleration impulse at terminal stance phase, with changes in steady-state walking velocity increasing the functional demands (Peterson et al., 2011). The braking (deceleration) impulse seems to have a greater relationship with speed than the acceleration impulse. Braking speed changes are associated with positive changes in hip and knee extensor moments and ankle plantarflexion impulse moments. Acceleration speed change in terminal stance is associated with knee flexor and ankle plantarflexor moments (Peterson et al., 2011). During walking, braking or deceleration actions on the foot are primarily provided by tibialis anterior (Wang and Gutierrez-Farewik, 2011), tibialis posterior, and peroneus longus (Konow

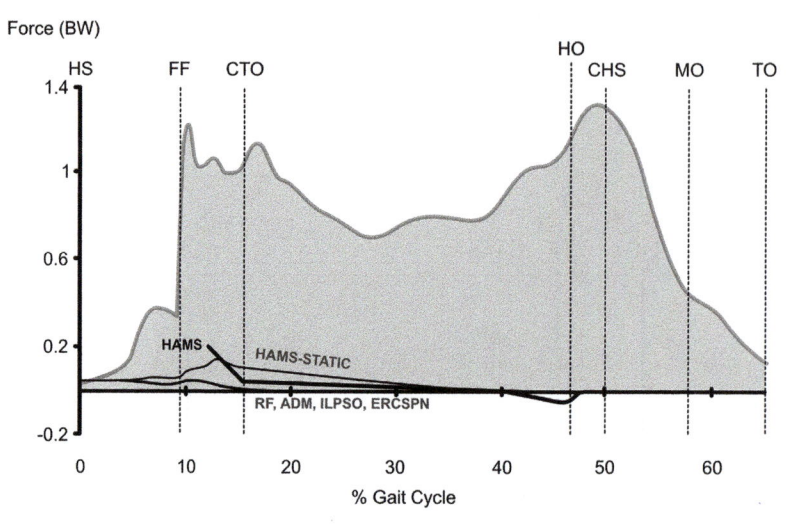

FIG. 1.3.3e Some muscles seem to contribute little to stance phase support, and then only at initial contact despite generating large forces. These include adductor magnus (ADM), iliopsoas (ILPSO), and erector spinae (ERCSPN). The hamstrings (HAMS) and rectus femoris (RF) also provide little support during stance. For gait event symbols, see Fig. 1.3.3b. *(Adapted from Anderson, F.C., Pandy, M.G., 2003. Individual muscle contributions to support in normal walking. Gait Posture 17 (2), 159–169.)*

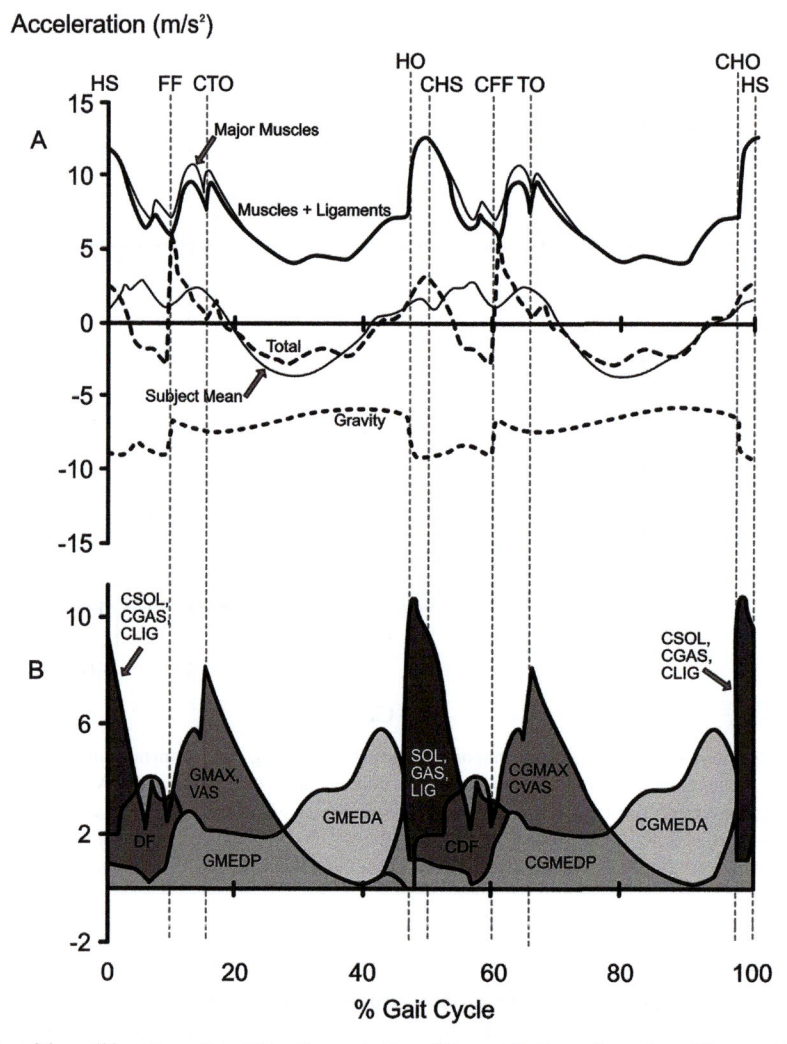

FIG. 1.3.3f Using modelling of the walking stance phase (A), a demonstration of the contributions of muscle and ligament torques (muscles + ligaments – *thick black line*) and the resistance to gravity provided by the skeleton (gravity – *black interrupted line*) to the net vertical acceleration of the CoM (total – *black dashed line*) is shown. The major muscles (major muscles – *thin black line*) represent the total input of the major contributors to support within this model. These are the ankle dorsiflexors, the quadriceps muscles, gluteus medius, soleus, and gastrocnemius. The *lower black line* is the mean vertical acceleration of the CoM of the subjects used in the modelling (subject mean). The periods of activity of the major muscles are illustrated in (B) and show gluteus maximus (GMAX), the ankle dorsiflexors (DF), the vasti muscles of the quadriceps (VAS), gluteus medius anterior (GMEDA) and posterior (GMEDP) sections, soleus (SOL), and gastrocnemius (GAS). Contributions of individual contralateral limb muscles to the net vertical acceleration of the CoM from modelling are also illustrated in (B). These are shown as the contralateral dorsiflexors (CDF), soleus (CSOL), gastrocnemius (CGAS), gluteus maximus (CGMax), the vasti (CVAS), anterior and posterior portions of gluteus medius (CGMEDA and CGMEDP), and contralateral limb ligaments (CLIG). All other symbols of the gait cycle are as explained in Fig. 1.3.3b. *(Adapted from Anderson, F.C., Pandy, M.G., 2003. Individual muscle contributions to support in normal walking. Gait Posture 17 (2), 159–169.)*

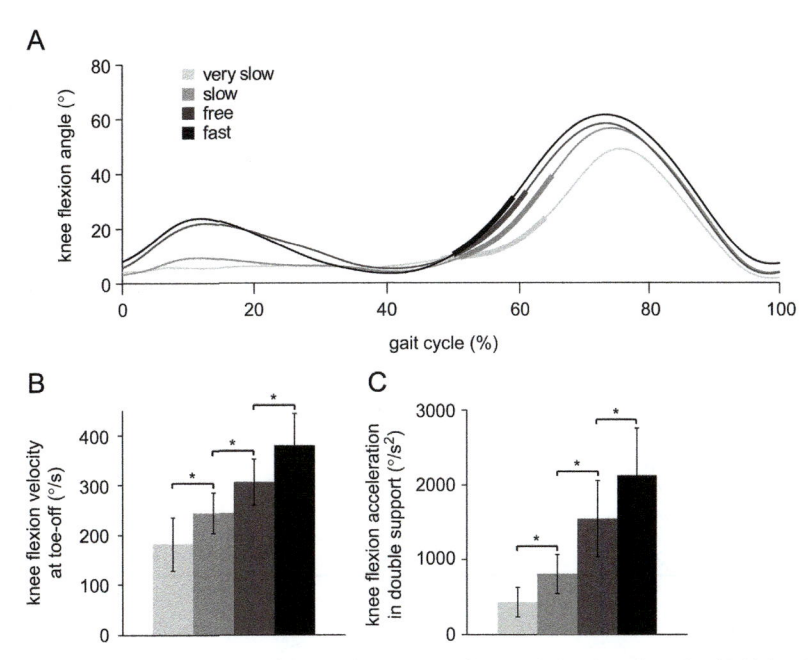

FIG. 1.3.3g Results from walking kinematics reveal that the knee flexion angle changes during the gait cycle with higher flexion angles in early stance around loading, being at their highest during swing. The flexion angles in gait change with different walking speeds, being greater and occurring earlier in stance and swing, and for longer during swing at faster walking speeds (A). Knee flexor velocity increases with walking speed (B) as does the knee flexion acceleration (C) over the double support phase. *(From Fox, M.D., Delp, S.L., 2010. Contributions of muscles and passive dynamics to swing initiation over a range of walking speeds. J. Biomech. 43 (8), 1450–1455.)*

et al., 2012; Murley et al., 2014). For the lower limb, activity in the quadriceps seems central in providing braking action through the moderation and restriction of knee flexion moments (Liu et al., 2006, 2008; Sasaki and Neptune, 2006; Peterson et al., 2011; Ellis et al., 2014) along with supplementary activity from the hamstrings, gluteus maximus, and gluteus medius (Ellis et al., 2014).

Heel lift and ankle power that generates the terminal stance GRF peak is driven primarily by activity within gastrocnemius and gluteus maximus, with assistance by soleus, gluteus medius, and possibly the hamstrings (Gottschall and Kram, 2003; Ellis et al., 2014). There is some disagreement as to the definition of primary and secondary actions of leg muscles, such that contradictions as to the significance of these secondary muscles is found in the literature (Michel and Do, 2002; Neptune et al., 2004b, 2008; Liu et al., 2006, 2008; Sasaki and Neptune, 2006; McGowan et al., 2009), but it appears that the hamstrings contribute little if any power to the GRF vector of terminal stance (Ellis et al., 2014).

Medial gastrocnemius and soleus are not active beyond around 52% of the whole gait cycle, indicating that they are not active in the middle or late parts of terminal stance (Gottschall and Kram, 2005; Fox and Delp, 2010; Honeine et al., 2013). Yet, the energy these muscles deliver into late terminal stance and even into preswing is enough for them to act as assistors in initiating preswing knee flexion (Fox and Delp, 2010), and if they are deficient, gait will require hip flexor compensation (Nadeau et al., 1999; Zmitrewicz et al., 2007). Fascicle and tendon shortening are reported to occur throughout terminal stance in the medial gastrocnemius and Achilles tendon, indicating that triceps surae continues to apply elastic recoil energy to the system at heel lift and throughout terminal stance (Ishikawa et al., 2007). The ability to induce a significant GRF impulse during terminal stance has been shown to help produce a stable gait (Collins et al., 2005), but the 'propulsive' force generating the anterior displacement of the CoM is generated by swing leg oscillations and the trunk's linear momentum trailing ahead of the terminal stance limb (Michel and Do, 2002). It is likely that tibialis posterior and peroneus longus function play an important supporting role in maximising ankle power generation during the middle of terminal stance through their tendon elastic recoil, as part of the overall foot-stiffening mechanism (Fig. 1.3.3h).

Preswing and swing phase muscle activity

The key muscle activities during swing provide movement of the limb against gravitational forces. For the muscles, important events start for swing activation at the preswing-to-swing transition. The swing phase muscular activity starts during preswing in the final moments of stance, when GRF vectors are promoting knee flexion. Muscles are then required for the maintenance of swing limb posture through to midswing as the hip moves into flexion. In addition, muscle activity from knee extensors under antagonsitic control is finally required during terminal swing to advance the lower leg and foot forward through knee extension.

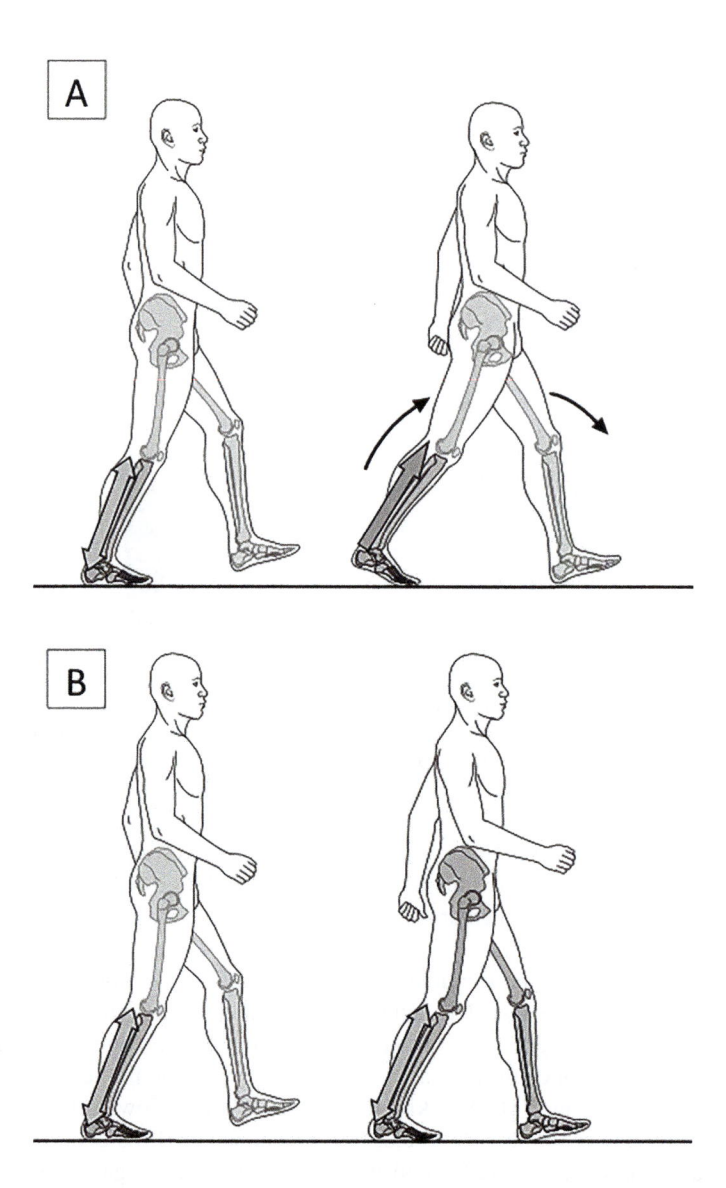

FIG. 1.3.3h Timing of heel lift is everything when adding ankle plantarflexion power to the momentum of the swing limb for the swing limb's heel strike and loading response. Energy loaded into the ankle plantarflexors, primarily the triceps surae-Achilles complex through fibre lengthening, needs to be released to provide momentum into the next footstep at the right momentum during gait. In walking, heel lift (shown for right leg) should occur briefly before opposite left heel contact when the heel is ~1 cm above the ground, allowing the energy released from the right calf to add momentum to weight transfer (A). A premature heel lift means more power must be used in overcoming higher resistance forces, because body weight on the heel will be higher as the CoM will not have moved as far anterior as it should have for heel lift. However, the fall of the swing limb will then be from a greater height, adding more acceleration into heel strike. Heel strike occurring before heel lift (B) means ankle plantarflexor-powered momentum at contact has been lost. If a foot is still flat to the support surface at opposite heel contact, then plantarflexor power cannot be released from muscle-tendon fibre shortening for acceleration during opposite foot contact. If the midfoot and forefoot are too flexible, then acceleration power will be absorbed excessively within the foot (as seen in midfoot break) reducing the power available, despite a normal timing of heel lift. *(Permission www.healthystep.co.uk.)*

The hip flexors play a highly significant role in swing at both its initiation and during its propagation, with activation starting during preswing by acceleration of the knee into flexion during the double stance period of terminal stance (Fox and Delp, 2010). This appears to be accomplished primarily through iliacus and possibly some psoas activity (as iliopsoas), with assistance from rectus femoris (Goldberg et al., 2004; Neptune et al., 2008; Fox and Delp, 2010). There is a peak of hip flexion velocity around toe-off. This is essential in achieving peak knee flexion during swing and is associated with a knee flexion velocity, which increases with greater walking speed (Neptune et al., 2008; Fox and Delp, 2010). Gastrocnemius plays a minor role in aiding preswing knee flexion with the ankle dorsiflexors, while the other ankle plantarflexors, the quadriceps, and the hip extensors and abductors (gluteals) decelerate and stabilise the knee flexion moment during preswing (Neptune et al., 2001; Goldberg et al., 2004; Fox and Delp, 2010). Knee flexion in preswing and early swing phase is aided by the release of the energy stored within the myofascia during heel lift and late stance (Sharbafi et al., 2017). The ankle dorsiflexors (primarily tibialis anterior) are also active during swing, beginning activity during preswing at the end of stance phase by aiding toe-off. Tibialis anterior activity, at about 35% of its total strength, provides an intrinsic moment that reduces the plantarflexion angle of the ankle left at the end of terminal stance (Byrne et al., 2007) (Fig. 1.3.3i).

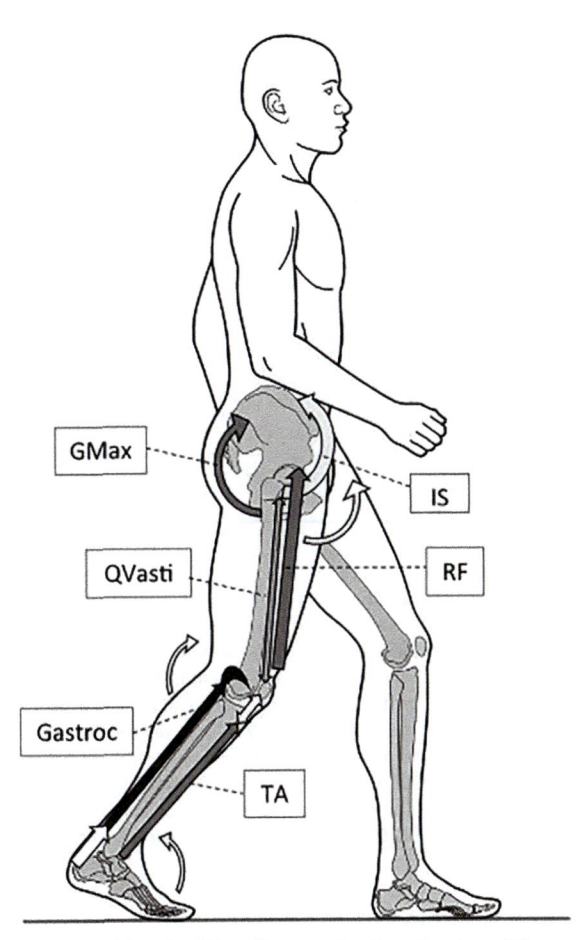

FIG. 1.3.3i Preswing involves a number of muscles aiding foot lift and/or preparing for swing phase. Primary muscle activity involves hip flexors such as iliopsoas (IS) and ankle extensors such as tibialis anterior (TA), aided by digital extensors that lift the lower limb, forefoot, and digits from the ground *(white rotation arrows)*. Gluteus maximus (GMax) briefly adds some momentum to late terminal stance through hip extension prior to the limb's CoM being picked up by IS. Knee flexion is influenced by gastrocnemius (Gastroc) shortening coupled to ankle plantarflexion angles. The quadriceps vasti (QVasti) muscles limit this knee flexion, but rectus femoris (RF) also aids hip flexion. *(Permission www.healthystep.co.uk.)*

Preswing kinematics is also dependent on muscle action in the contralateral limb (Fox and Delp, 2010). The net effect of the contralateral leg muscles is to contribute to knee acceleration during preswing knee flexion through their action on the pelvis via activity of the stance hip extensors and the posterior fibres of the hip abductors (Fox and Delp, 2010). Stance limb knee extension through the quadriceps increases knee flexion moments on the preswing limb (Fox and Delp, 2010). There is also variable but limited influence from the back muscles and their residual forces.

During early swing phase, rectus femoris activity reduces the knee flexion acceleration from foot lift-off by around 30% of swing phase (Sharbafi et al., 2017). Vastus lateralis is active during late swing along with tibialis anterior in preparation for braking forces required for initial ground contact knee and foot flexion moments (Gottschall and Kram, 2005). They reduce the residual plantarflexion angle at the end of terminal stance, thereby maintaining a reduced ankle plantarflexion angle throughout swing via low EMG isometric activity. Tibialis anterior activity increases again at late terminal stance in preparation for initial contact, usually providing around 45% of its total strength (Byrne et al., 2007). It does so in conjunction with the other ankle dorsiflexors and abductor hallucis assisting in toe extension prior to contact (Kelly et al., 2015), helping to initiate the windlass mechanism within the foot that increases its stiffness across the vault (Fig. 1.3.3j).

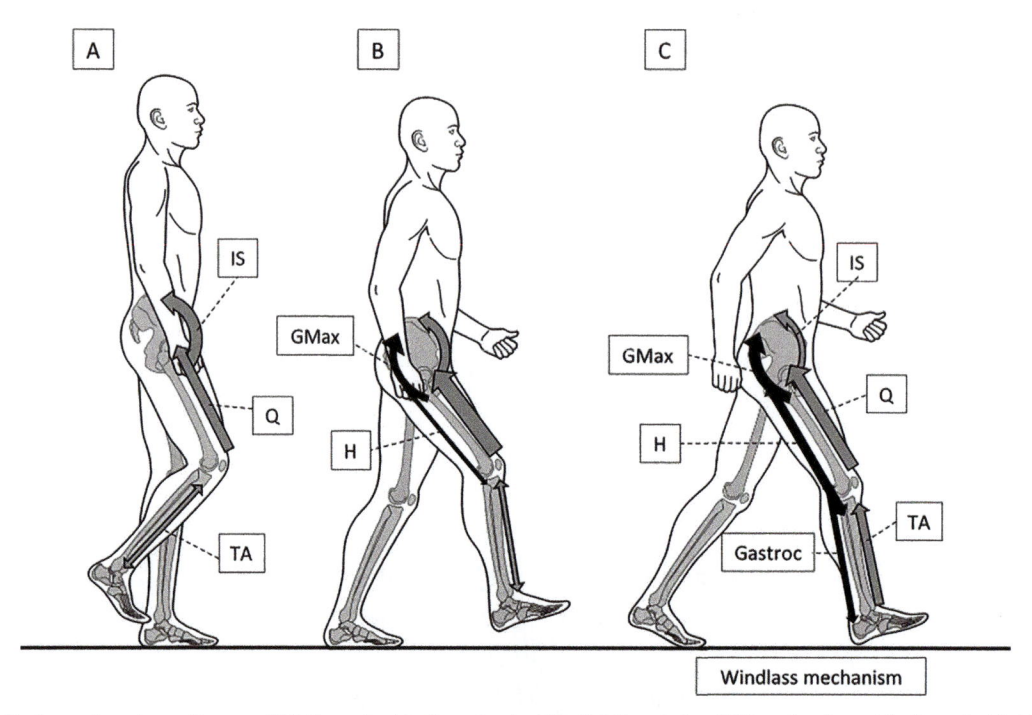

FIG. 1.3.3j During swing phase, iliopsoas (IS) flexes the hip throughout while tibialis anterior (TA), acting isometrically and aided by the digital extensors and peroneus tertius, holds the foot away from significant plantarflexion angles at the ankle. Events start to alter after midswing (A) with a change in quadriceps (Q) activity that initiates knee extension, with rectus femoris continuing to assist hip flexion. In late swing (B), the knee extension and hip flexion moments need to be antagonised by the hamstring muscles (H) and gluteus maximus (GMax) to prevent excessive swing limb momentum and centrifugal forces overflexing the hip and overextending the knee, thus exaggerating stride length. IS's activity reduces during late swing to assist braking of the swing limb. In terminal swing (C), antagonistic muscle action (black muscle vectors) also includes gastrocnemius (Gastroc) that helps the hamstrings slow and limit knee extension. Ankle plantarflexor activity can also moderate increasing tibialis anterior activity, reducing the ankle plantarflexion angle before heel strike. Tibialis anterior and the digital extensors also increase toe extension forces to start to raise the foot vault height via the windlass mechanism through plantar aponeurosis tightening. Tibialis posterior and peroneus longus also begin to activate to assist in increasing vault stiffness just before heel contact, while IS finally switches off before heel strike. *(Permission www.healthystep.co.uk.)*

1.3.4 Soft tissue compliance and stiffening of the lower limb

Muscle plays an essential role in changing the mechanical nature of the lower limb, which in turn affects the body's ability to support and redirect its vertical velocity away from purely up-and-down collision responses. The degree of compliance or stiffness within the lower limb produced through muscle activity and joint motion affects the oscillatory (rise and fall) motion of the CoM in gait and thereby influences the metabolic costs of CoM translations (Holt et al., 2003; Kim and Park, 2011). Achieving the appropriate leg stiffness during each phase of gait maximises the energy dissipation and elastic energy stored by the end of the single support phase before terminal stance. This ensures the momentum change within the stance limb, as it enters terminal double-limb stance, achieves maximum forward propulsion (Kim and Park, 2011). Therefore, increasing leg stiffness with gait speed will beneficially increase the propulsion energy achieved through myofascial passive and active stiffening mechanisms, allowing humans to use spring-like mechanics and thus increase kinetic energy from stored potential elastic energy (Kim and Park, 2011).

Muscle activity reflects which of the two strategies of locomotion are being utilised to reduce locomotive costs. Having legs that simulate elastic springs to create a glancing collision (as used in human running) requires greater muscle stiffening. Walking techniques use a sequencing of forces through the leg during a redirection phase and a 'push' forward prior to heel contact collision on the contralateral foot (Ruina et al., 2005). Energetically, it can be argued that the gait cycle is thus better modelled to start at heel lift rather than heel contact.

The counterbalance to beneficially storing elastic energy is that muscle-stiffening activity increases metabolic energy consumption. Thus, the more active the process of stiffening for elasticity, the more energy consumed. Faster walking leads to potentially more metabolic input for more elastic energy output, whereas walking slowly leads to less metabolic input but results in less elastic energy output. It seems likely that a balance exists where metabolic energy input and elastic output can maximise energetics, influencing the preferred walking speed of an individual. If soft tissues become excessively stiff such as in diabetes, then more compliant walking strategies should be expected such as achieved by a slower gait speed, hopefully finding a new energetic balance. However, it appears such changes from reducing gait speed in neuropathic diabetics may not reap the benefits expected (Fernando et al., 2016). Weakening muscles in old age are often associated with stiffer lower limbs and feet, and this may help to explain the slower walking speed associated with poor health and ageing that are dictating the changes in preferred walking speeds observed in later adult life from those developed during early adulthood. It is the ability to adapt compliance and stiffening mechanisms to the prevailing mechanical advantage that allows faster preferred walking speeds in younger individuals and determines the point of change from walking fast to running.

Modelling the lower limb as a damped compliant structure to produce pseudo-elastic collisions suggests that CoM oscillations during single-limb support benefit from the resonance characteristics of having a spring-like leg (Kim and Park, 2011), something not available on a stiff, unforgiving structure. A lower limb also has to achieve a state of enhanced elastic energy storage through myofascial tension by the end of the single support phase, which is produced by osseous compression and appropriate joint motions. This deformation of the lower limb as a structure is excellent for dissipating potentially harmful energy while also storing some of this energy. The lower limb, including the foot, should then achieve its stiffest properties just prior to terminal stance, because the CoM's forward momentum requires this alteration within the foot structures to create an acceleration lever (Kim and Park, 2011).

Oscillations of the CoM are achieved through gait kinematics resulting from muscle activity that produces negative and positive work (Donelan et al., 2002). Muscle activity is required to make up for any energy lost from energy dissipation mechanisms, with feedforward-feedback sensorimotor systems adjusting the appropriate muscle activity in order to maintain the compliance-stiffness levels required for a particular limb manoeuvre at any moment (Kim and Park, 2011). This means that the spring-like behaviour of the lower limb cannot be a passive event alone but is a result of the muscular activity around the hip, knee, and ankle, across the foot and particularly around the MTP joints in generating appropriate stiffness by reducing osseous motions (Riley et al., 2001; Holt et al., 2003; Zeni and Higginson, 2009; Zeni and Higginson, 2011). It should be expected that the faster humans walk and the greater their step length, then the greater will be the resultant muscle activity (Donelan et al., 2002) as the limb attempts to increase stiffness. This may be a limitation on the walking gait speed that is possible in humans. Such mechanisms may also have implications in the reduced walking speeds observed in older humans and in the presence of pathology (Schrack et al., 2012, 2016), reflecting perhaps a reduced capacity to induce appropriate limb stiffness for good energetic performance (Fig. 1.3.4).

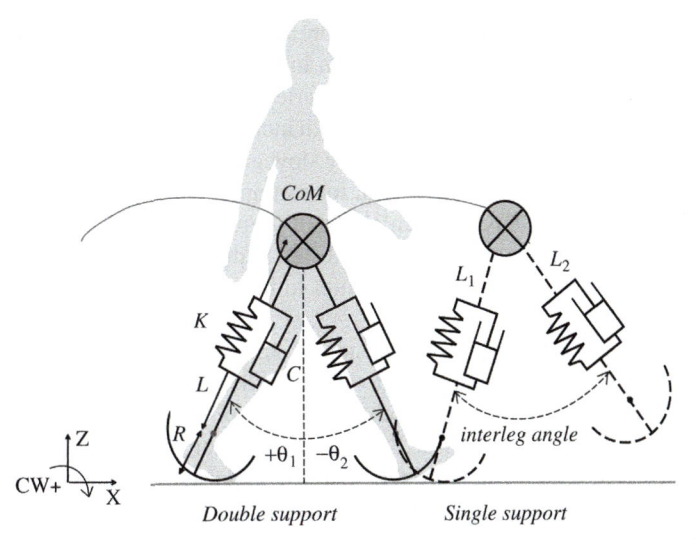

FIG. 1.3.4 Kim and Park (2011) modelled the biomechanics of walking as a damped compliant structure of two massless compliant legs with curved feet. While simplistic, results implied that the CoM momentum change during double-limb support required maximum forward propulsion, and that increasing leg stiffness with speed beneficially increased acceleration energy. This fits nicely into what is known about human gait, both in respect to the whole lower limb and the foot. The faster humans move, the stiffer their lower limbs should become. Failure to adapt stiffness levels to match gait speed biomechanics may underpin many locomotive pathologies. *(From Kim, S., Park, S., 2011. Leg stiffness increases with speed to modulate gait frequency and propulsion energy. J. Biomech. 44 (7), 1253–1258.)*

1.3.5 The effects on walking of terrain, velocity, and gradient

Although the main muscle activities in the different stages of gait have been indicated, it is important to realise that both speed and gradients in terrain produce different muscle activity, just as they change the effects of joint motion. These events are interdependent. As already discussed, terrain affects gait kinematics (including gait speed) which makes clinical testing using treadmills a gait examination exercise to approach with caution, as it is known to alter gait mechanics (Lee and Hidler, 2008; Hollman et al., 2016; Lu et al., 2017; Hutchinson et al., 2021). Gait-instrumented treadmills are reported to increase the velocity, or the backward, medial, and lateral CoGRF velocities, when walking at preferred walking speeds compared to walking on the ground, probably as a result of shear forces from the belt sliding (Hutchinson et al., 2021). Both treadmill-induced kinematic and kinetic changes reflect and cause alterations in muscle activity, particularly in tibialis anterior during stance phase, and the hamstrings, vastus medialis, and adductor longus when in swing phase (Lee and Hidler, 2008). Caution is therefore required from the clinician who only performs gait analysis only on a treadmill when interpreting data. Being aware of the fact that a patient may change gait speed due to (i) where they are assessed, (ii) their mood, (iii) their physiological health, and (iv) the type of footwear they are wearing is important when comparing gait tests, especially on different days.

Leg stiffness increases with walking speed, and particularly during running (Geyer et al., 2006; Kim and Park, 2011). Compliance of the leg is defined by the ratio of the GRF to the CoM displacement during gait, but stiffness determines the dynamics as it reflects the elastic energy stored in the system during the single support phase (Kim and Park, 2011). Modelling the lower limb in walking gait as a damped compliant leg suggests that oscillations in the CoM during single support take advantage of the resonance characteristics of the spring-like structures within the leg, and 'spring-like' elastic storage is maximised at the end of single support phase (Kim and Park, 2011). This correlates with the findings that the foot is lowest, longest, and much stiffer at the end of single support at a time when it is subjected to the greatest percentage body mass (Hunt et al., 2001; Stolwijk et al., 2014; Bjelopetrovich and Barrios, 2016; Takabayashi et al., 2020), supporting the principle that the foot has now entered its viscoelastic stress-stiffened zone. The faster the loading rate, the stiffer the foot behaves through its time-dependent viscoelastic properties, creating the perfect mechanical relationship to the requirement of speed to stiffness.

This implies that the maximum forward 'propulsion' of the CoM from the support limb will occur with increased leg and foot stiffness. Such stiffness is thus achieved through increased muscle tension/activity and passive structures entering their stress-stiffened range, influenced by the loading rate across the joints of the lower limb. This presents possible clinical explanations for dysfunction, in that patients may be unable to match appropriate stiffness to their gait speed, either through muscle or passive tissue structural dysfunctions, including those of ligament or other joint motion-changing pathologies.

Muscle activity changes with walking speed, as the majority of lower limb muscles increase their EMG amplitude on raising their walking speed and reduce it with a slower-than-preferred walking pace (Hof et al., 2002; den Otter et al., 2004; Cappellini et al., 2006; Murley et al., 2014). Activation in certain muscles such as peroneus longus and tibialis posterior also tends to occur slightly earlier in faster walking (Murley et al., 2014), a trend consistent with running muscle activation. GRF profile fluctuations become more consistent with increases in walking speed (Wuehr et al., 2014). This is likely linked to the increased muscle EMG amplitudes. At very slow gait speeds, it is thought that greater muscle activity is required to provide medial-lateral postural stability with, for example, increased EMG amplitude in peroneus longus during stance and rectus femoris in swing having been reported (den Otter et al., 2004). However, a combination of decreased step length, increased step frequency, and increased step width seems to be utilised as the strategies of choice to cope with medial-lateral balance perturbations in an attempt to increase stability and decrease the risk of falls (Hak et al., 2012).

For medial gastrocnemius, tibialis anterior, tibialis posterior, and peroneus longus, it has been demonstrated that walking speed affects activation patterns. Furthermore, it has also been demonstrated that tibialis anterior, tibialis posterior, and peroneus longus display temporal phase-dependent changes (Murley et al., 2014). In these latter three muscles, the change was unique to either contact or midstance phase. Increased activation of these muscles in contact phase may relate to increasing the stiffening of the plantar foot vault through pretensioning prior to forefoot loading. This could help energy dissipation through the reversed windlass mechanism of the plantar aponeurosis, an effect that can increase forefoot compliance upon loading (Stainsby, 1997; Green and Briggs, 2013; Stolwijk et al., 2014). Increased speed requires heightened effort from the tibialis anterior to decelerate the increased plantarflexion moment and increase energy dissipation capacity during loading response (Neptune et al., 2008; McGowan et al., 2010). On the other hand, increased and earlier tibialis posterior and peroneus longus activation during midstance reflects an earlier need for midfoot dorsiflexion bending moment resistance in order to prevent midfoot break at heel lift (Murley et al., 2014) (Fig. 1.3.5a).

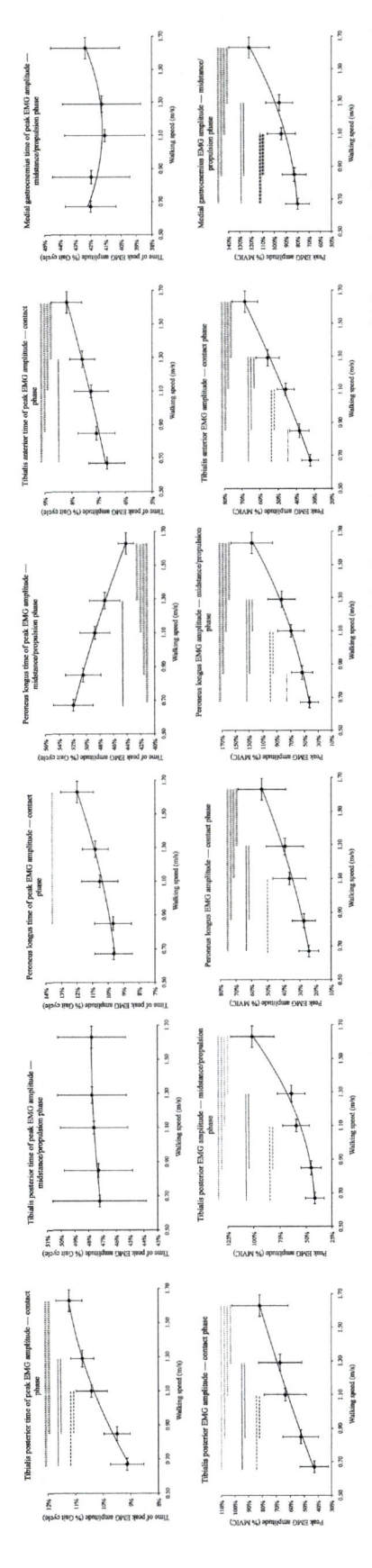

FIG. 1.3.5a Scatter plots and regression lines for the effects of different speeds on EMG parameters *(from left to right)* of tibialis posterior, peroneus longus, tibialis anterior, and medial gastrocnemius during stance. *Solid vertical and horizontal bars* represent 95% confidence intervals, with broken horizontal lines indicating significant differences between walking speeds. *(From Murley, G.S., Menz, H. B., Landorf, K.B., 2014. Electromyographic patterns of tibialis posterior and related muscles when walking at different speeds. Gait Posture 39 (4), 1080–1085.)*

Compared to level walking, additional muscle action is required to raise or lower the CoM during uphill and downhill walking, respectively. This requires changes in muscle activation of larger magnitudes, which increase with the raising of walking velocity (Franz and Kram, 2012). Young adults increase their ankle power around heel lift on inclined treadmills (Franz et al., 2014). This is probably to prevent any backward fall or to accommodate a delay in forward progression of the CoM in late midstance due to a reduction in the anterior pendular fall of the CoM. In uphill walking stance phase, the hip and knee extensors and the ankle plantarflexor muscles progressively increase their activity in response to increasing inclines at all walking speeds compared with level walking. This is statistically significant at all inclines for gastrocnemius medialis (Franz and Kram, 2012). At gradients of over 3°, significant changes compared with level walking speeds are recorded in gluteus maximus, biceps femoris, vastus medialis, and soleus, and at gradients over 6° for rectus femoris (Franz and Kram, 2012). Increased walking speed heightens these muscle activities further, except for rectus femoris (Franz and Kram, 2012) (Fig. 1.3.5b).

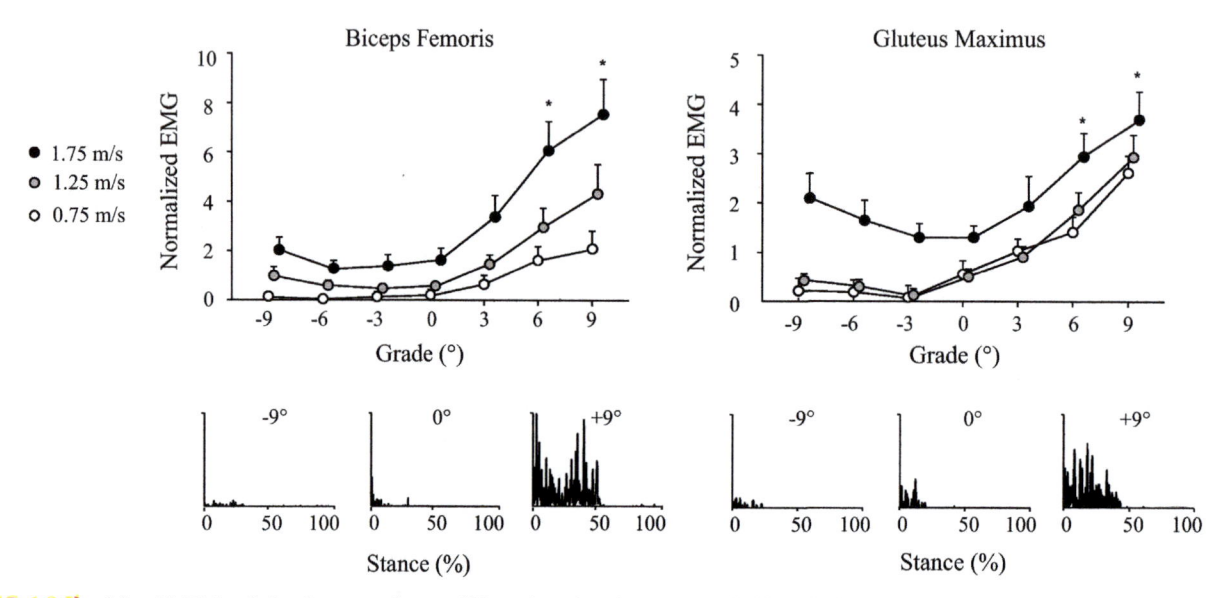

FIG. 1.3.5b Mean EMG signals for gluteus maximus and biceps femoris acting to extend the hip in the first half of stance, normalised to the mean activity during level walking at 1.25 m/s. EMG data are also provided for a single stride of a subject walking at 1.25 m/s on level, inclined (+9°), and declined (−9°) surfaces. Compared to level walking, both muscles increase activity at inclines over 3° across walking speeds but do not seem to be altered on declines. *(From Franz, J.R., Kram, R., 2012. The effects of grade and speed on leg muscle activations during walking. Gait Posture. 35 (1), 143–147.)*

On treadmill declinations, walking at all speeds, only stance phase knee extensor muscle activity is reported to increase at grades steeper than 3° for rectus femoris and 6° for vastus medialis (Franz and Kram, 2012). In contrast, gastrocnemius medialis and soleus reduce their activity when walking downhill, but gluteus maximus and biceps femoris show no changes from level walking (Franz and Kram, 2012). Declination gradients over 3° reduce stance time, and all declination angles give rise to progressively faster strides (Franz and Kram, 2012) (Figs 1.3.5c and 1.3.5d).

The influence of age on inclined/declined surfaces

Changes in muscle function from walking on surface inclinations are affected by age. However, age seems to have little effect on declination angle muscle activation patterns (Franz and Kram, 2013a), whereas there is a significant reduction in peak perpendicular GRF during downhill walking in older adults (Franz and Kram, 2013b). Step length and speed reduce with declination. Further to that, cadence increases in older age and with increasing declination and a shift from anterior–posterior attenuation to medial-lateral attenuation occurs, becoming the dominant focus of postural control (Scaglioni-Solano and Aragón-Vargas, 2015) (Fig. 1.3.5e).

Older people exhibit a smaller increased medial gastrocnemius activity with steeper inclines on treadmills than do younger adults, and a disproportionate recruitment of hip muscles with gluteus maximus activity approaching maximum isometric capacity (Franz and Kram, 2013a). At 9° inclination, older adults are reported to average 73% of their maximum voluntary contraction, as opposed to only 33% in young adults (Franz and Kram, 2013a). Older adults also exhibit diminished ankle joint kinetics but larger hip joint kinetics during heel lift going into terminal stance, with more work being generated by the muscles crossing the knee during single support (Franz and Kram, 2013b; Franz et al., 2014). The

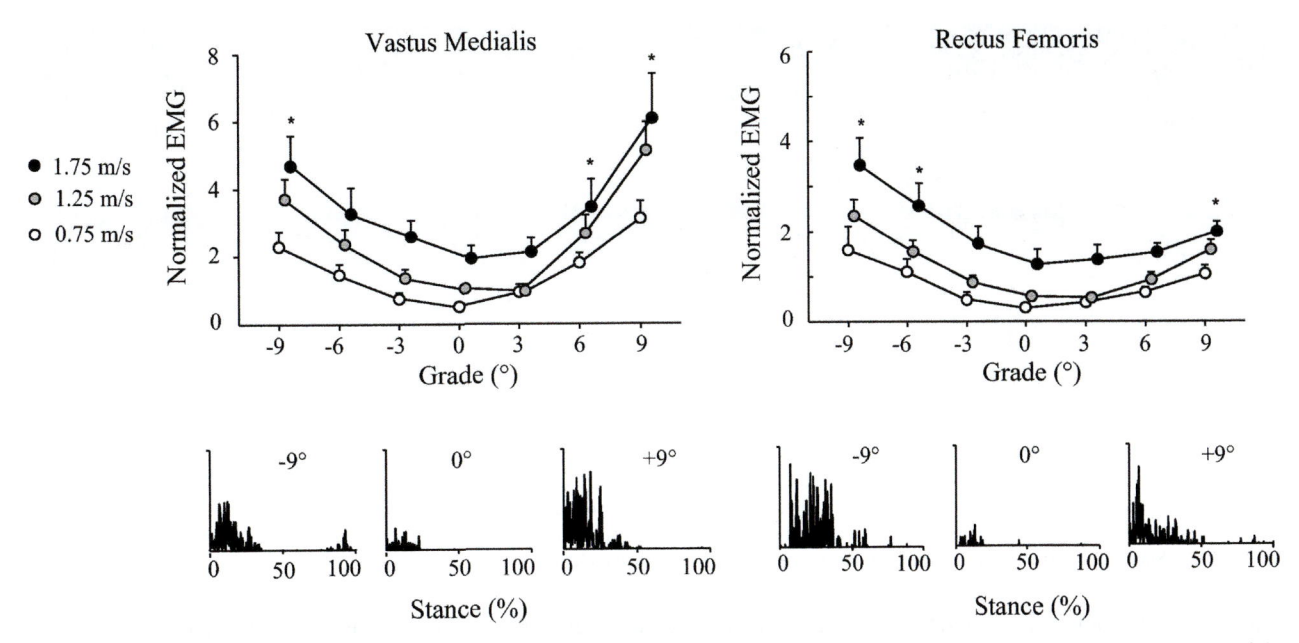

FIG. 1.3.5c Mean EMG signals for vastus medialis and rectus femoris/vastus intermedius during the first half of stance, normalised to the mean activity during level walking at 1.25 m/s. EMG data are also provided for a single stride of a subject walking at 1.25 m/s on level, inclined (+9°), and declined (−9°) surfaces. Compared to level walking, muscle activity increased on inclines greater than 3° for vastus medialis and 6° for rectus femoris/vastus intermedius. When walking on a decline greater than 3°, rectus femoris/vastus intermedius activity increased, and at 6°, vastus medialis activity increased. *(From Franz, J.R., Kram, R., 2012. The effects of grade and speed on leg muscle activations during walking. Gait Posture. 35 (1), 143–147.)*

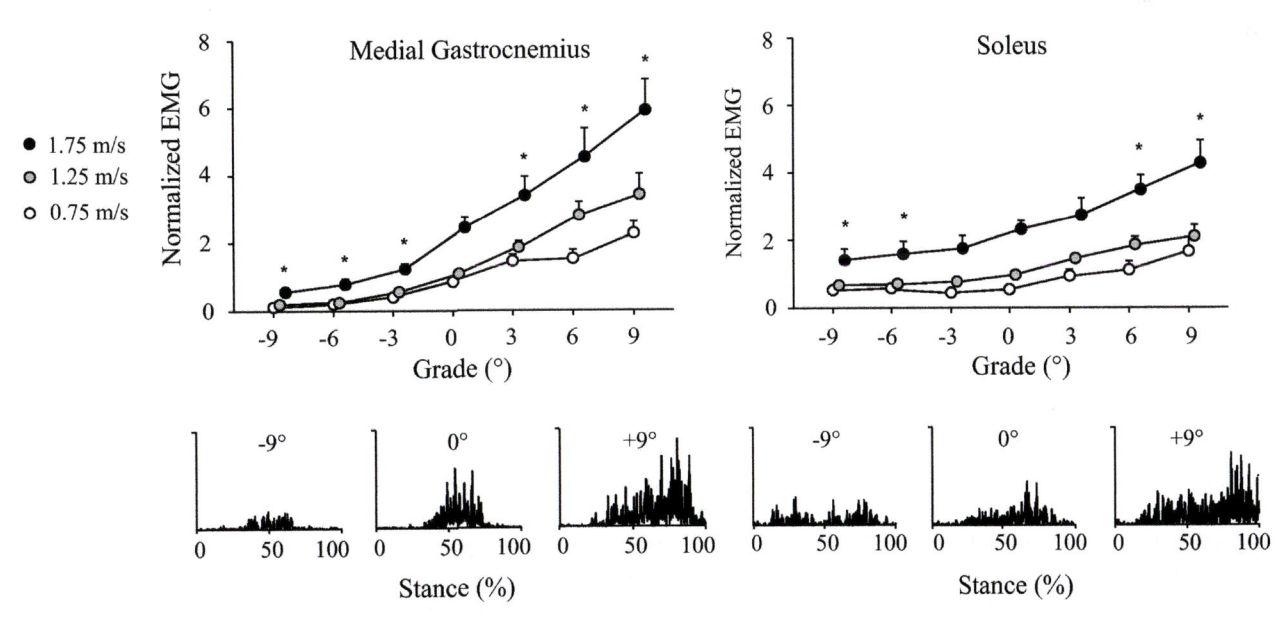

FIG. 1.3.5d Mean EMG signals for medial gastrocnemius and soleus during the second half of stance, normalised to mean activity during level walking at 1.25 m/s. EMG data are also provided for a single stride of a subject walking at 1.25 m/s on level, inclined (+9°), and declined (−9°) surfaces. Compared to level walking, muscle activity in medial gastrocnemius increased on all inclines and increased on inclines greater than 6° for soleus. When walking on a decline, medial gastrocnemius activity increased and soleus activity increased on declines greater than 6°. *(From Franz, J.R., Kram, R., 2012. The effects of grade and speed on leg muscle activations during walking. Gait Posture. 35 (1), 143–147.)*

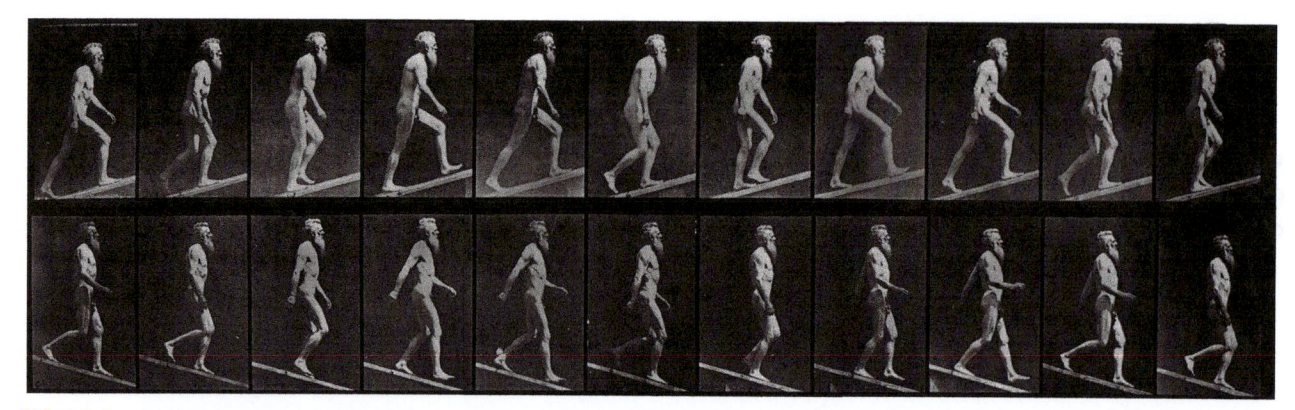

FIG. 1.3.5e Walking on inclines and declines will alter kinematics and muscle activation patterns. This is a result of changing postural balance required due to alterations in the divergent point from GRF interactions and repositioning of the HAT CoM. The CoM moves more anteriorly over the lower limbs on inclines and more posteriorly on declines. These gait adaptations can be seen in these sagittal plane photographs from the Eadweard Muybridge collection. Incline postural changes and muscle activity are altered in older individuals compared to the young-adult. Clinicians must be aware that the age of the patient will affect the adaptations required to maintain stability with terrain perturbations during gait that can result in an instability risk not present in level walking. Such information is particularly helpful in assessing falls risk in the elderly.

GRF produced around heel lift and the ankle power are reduced in older adults both in level and inclined walking (Franz and Kram, 2013b). This indicates that trying to preserve ankle power generation in older adults increases in importance when they encounter an incline.

Final muscular considerations

Clinical gait analysis invariably takes place on flat, even surfaces. However, the patient's real world will regularly involve changes in inclines and declines as well as left and right cambers. This will alter gait, biomechanics, and muscle activation patterns variably throughout the day. Only from good patient clerking can perturbations and terrain factors be equated into understanding the patient's problem. The relationship between reducing ankle power and the compensatory hip mechanisms is an important factor in gait changes with age, as they can influence muscle activation and potential injury locations. It is reported that 60% of falls occur during locomotion in older adults (Cali and Kiel, 1995) and perturbations that require gait changes become an increasing challenge, especially with visual disturbance (Dhital et al., 2010; Reed-Jones et al., 2013). Being aware of how patients cope with changing terrain environments during daily activities can give insight into increasing muscular dysfunction in more challenging situations than flat terrain gait assessments (Fig. 1.3.5f).

1.3.6 Muscle activation and dysfunction effects on gait kinematics

Gait requires agility and balance to achieve good energetics. Disuse, ageing, and disease can lead to loss of muscle strength and affect walking as well as other activities (van der Krogt et al., 2012). Weakening of different muscles can give rise to compensation mechanisms. Overall, human gait is remarkably robust to weakness of up to a 40% loss of strength within muscles, but it is far more vulnerable to the small amounts of weakening within certain muscles more than in others. Gait appears most resistant to the weakening of hip and knee extensors, but most sensitive to weakness of the ankle plantar-flexors, hip abductors, and hip flexors (van der Krogt et al., 2012). This suggests that clinicians should be most alert to loss of strength within these particular muscle groups.

Weakness in a muscle can result in increased activation of the affected muscle and/or those muscles that are compensatory for the weak muscle (van der Krogt et al., 2012). Compensatory activation is costly to gait energetics, generating unbalanced joint moments that require further compensations in order to stabilise the new compensatory motion(s) created (van der Krogt et al., 2012). Energetics therefore decrease as secondary muscle activations increase as a result of the weakness of a primary muscle, with the situation worsening when increasing numbers of muscle weaknesses become involved. It is therefore important to understand the interrelationships of muscles for certain roles in gait so as to appreciate where compensations are likely for weakness of a muscle within a functional chain. This should help target treatment protocols.

Modelling of joint motion through muscle activation models can help explain the effects of muscle weakness and dysfunction during gait, but they are only as good as the muscles they include within the model. Most muscle activation models have been based on surface EMG data and they use rectus femoris activity gained from surface EMG. This is affected by

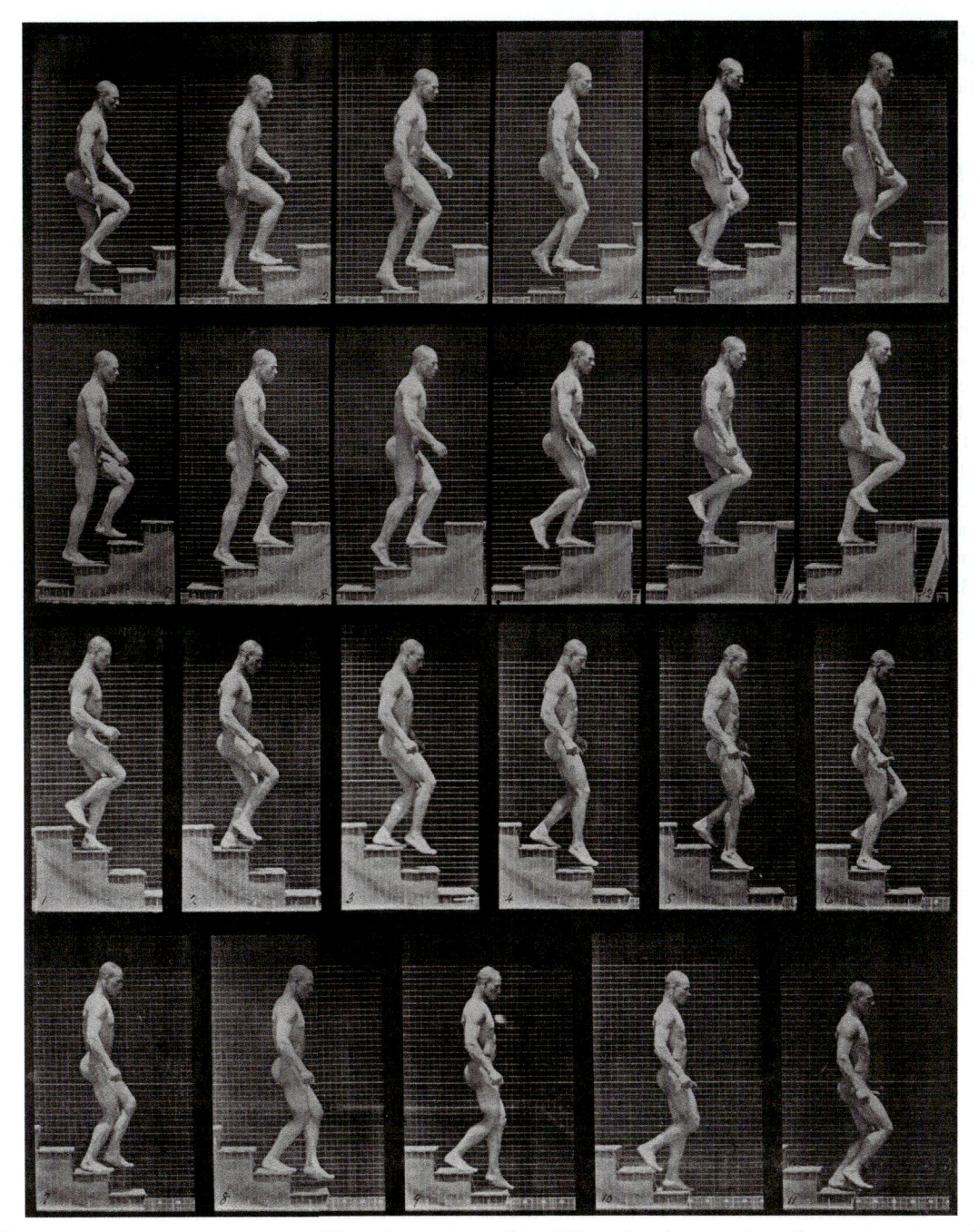

FIG. 1.3.5f Stair ascent and descent also present a different locomotive condition. This requires gluteus maximus, iliopsoas, hamstring, and quadriceps muscle cooperation in managing hip and knee flexions and extensions, primarily around raising or descending the CoM rather than the horizontal translation of it. Assessment of stair locomotion can reveal instability and/or be the source of anterior knee pain through dysfunction in muscular activity not identified in level gait assessment.

vastus intermedius cross-talk, so that rectus femoris loading phase activity should be read as including vastus intermedius activity. Thus, the stance-to-swing transition activity of rectus femoris may actually derive from both muscles (Nene et al., 2004). For ease of understanding, rectus femoris/vastus intermedius are used together here.

Muscle activity around joints can influence motion at more distant joints and also within the contralateral limb, having consequences in seemingly unrelated joint motions. Postural control in gait is reported to involve more coactivation of muscles in older compared to younger individuals and is further reported as being worse in individuals with greater loss of postural control (Nagai et al., 2011). This is likely to make the impact of an individual muscle weakness more significant to the elderly due to potentially greater 'knock-on' effects. The effects of other muscle dysfunctions that cause inappropriate timings in activity or muscle tightness are also likely to have significant effects, for the relationship between the muscles and the kinematic events they influence are complex.

Stance phase muscle function and dysfunction kinematics

Muscles work as myofascial teams or chains to achieve tasks, rather than functionally isolated force-generating units. The resistance of ankle dorsiflexion during midstance and the commencement of plantarflexion motion at heel lift is initiated by activity in the gastrocnemius and soleus utilising energy storage within the Achilles tendon. However, motion around the ankle during the whole of the gait cycle is also heavily influenced by tibialis anterior, gluteus maximus, gluteus medius, biceps femoris, and semitendinosus during stance phase, and also gluteus maximus and semitendinosus within the swing limb (Jonkers et al., 2003a,b). Removal of any of these muscle forces results in a failure to effectively initiate heel lift. Normal heel lift is also controlled through the antagonistic activity of vastus medialis, vastus lateralis, iliacus, and adductor longus, and removal of their activity causes an excessive rate of heel lift (Jonkers et al., 2003a).

Control of the midstance dorsiflexion moment at the ankle over the foot is restrained by soleus, gastrocnemius, adductor longus, and contralateral iliacus and gluteus maximus, with loss of their action leading to excessive midstance ankle dorsiflexion prior to heel lift (Jonkers et al., 2003a). Stance activity in vastus medialis, vastus lateralis, tibialis anterior, biceps femoris, and swing activity in the contralateral gluteus maximus and semitendinosus all play a role in this 'ankle rocker' of the latter midstance phase of gait. Removal of their activity deceases the ankle dorsiflexion moment substantially (Jonkers et al., 2003a) (Fig. 1.3.6a).

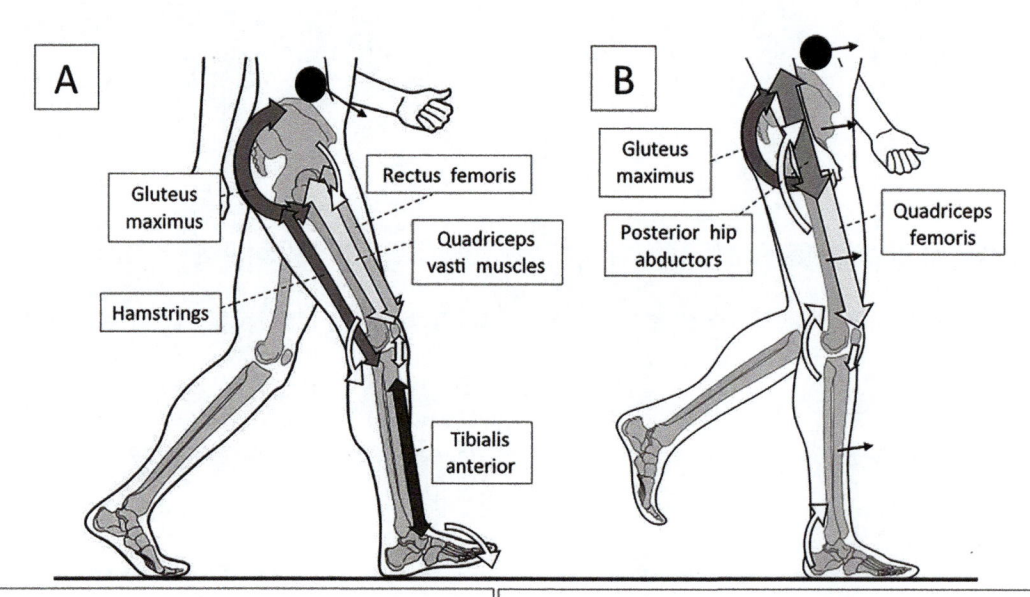

FIG. 1.3.6a In walking, initial contact-to-loading response (A) extensively utilises a team of internal muscle-generated forces to dissipate energy and stabilise the foot on the support surface against external forces try to drop the CoM and flex the lower limb. Thus, the lower limb can act as an energy dissipating spring. During early stance phase (B), stance limb forces are shifted to the muscles that are extending the limb to raise the body's CoM, creating and using previous momentum to drive the pelvis, femur, and tibia forward. This functional limb lengthening provides the space for the swing limb to advance forward beneath the body without detrimentally displacing the CoM. Knee extensor activity within the quadriceps also influences hip extension rate, despite more direct hip control coming from gluteus maximus and posterior hip abductor activity. Failures in any part of these muscular teams achieving their combined task will significantly compromise gait. *(Permission www.healthystep.co.uk.)*

Stance knee motion/stability is controlled via limitation of its extension through iliacus, adductor longus, tibialis anterior, and semitendinosus, and removal of the action of these muscles causes premature knee extension during contact and early midstance. The agonists of stance knee extension are vastus medialis, vastus lateralis, rectus femoris/vastus intermedius, gluteus maximus, and gastrocnemius. Loss of any of these can lead to exaggerated knee flexion during midstance, although this dysfunction does not seem to influence the contralateral swing limb (Jonkers et al., 2003a). Gastrocnemius with soleus plays an important controlling role in ankle dorsiflexion during late midstance, and without soleus for knee flexion in acceleration (Fig. 1.3.6b). However, gastrocnemius can act as a knee flexor or extensor at different stages of the gait cycle, being overall an assistor of knee extension/flexion stability.

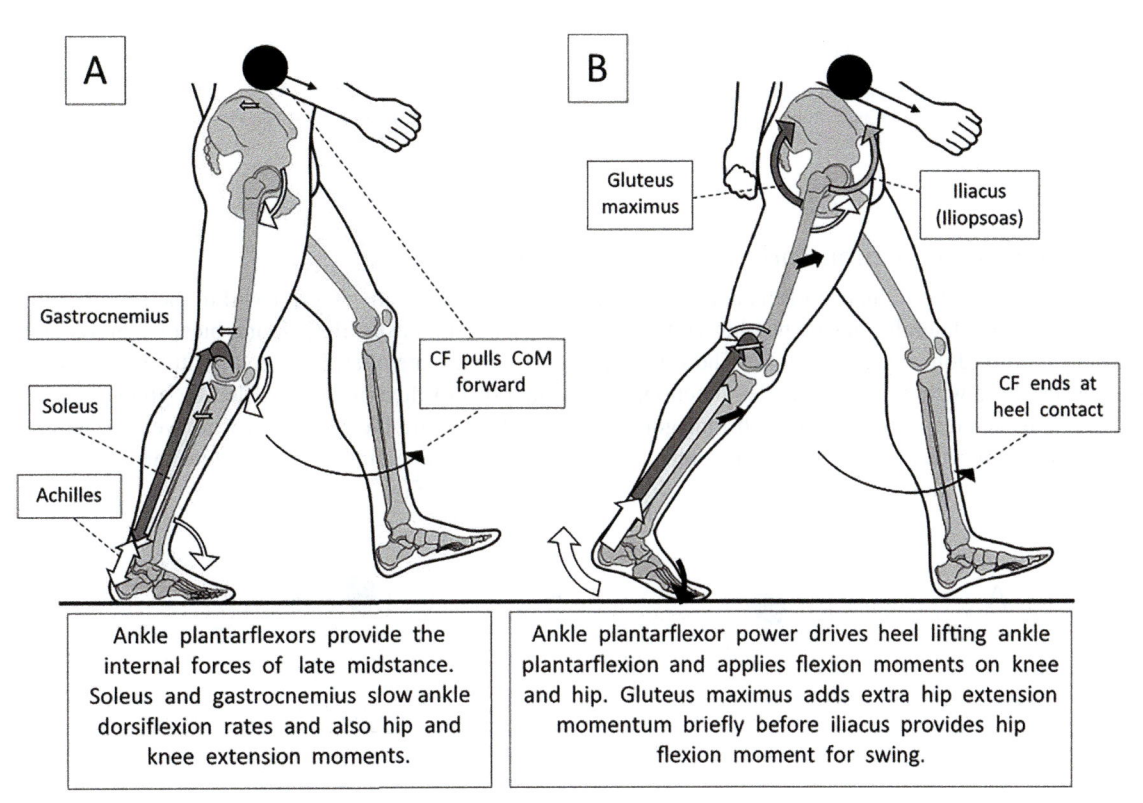

FIG. 1.3.6b During late midstance phase (A), limiting knee and hip extension becomes a priority, as extension moments are driven from GRFs that lie anterior to the hip and knee joints. The fall of the body's CoM is controlled via the ankle plantarflexors restricting hip extension slowing ankle dorsiflexion accelerations (small white arrows of braking). Gastrocnemius is also able to concurrently resist knee extension during late midstance. Centrifugal force (CF), induced by the swing limb's iliacus and quadriceps, influence the power necessary for CoM restraint within the stance limb. Thus, swing limb's gluteus maximus and hamstring braking are also influential. In terminal stance (B), plantarflexor tendons' elastic recoil induces ankle plantarflexion and heel lift, flexing the knee (via gastrocnemius) and hip above as these forces drive the tibia and femur upwards and forwards. The CF is brought to a sudden stop once heel contact occurs, ending most of the plantarflexors' influence on CoM acceleration which is instead aided by a little hip extension momentum provided by gluteus maximus. Iliacus then starts to accelerate hip flexion during preswing and, in so doing, causes further knee flexion into early swing. (Permission www.healthystep.co.uk.)

Stance phase hip extension from its flexed position at contact is produced by the action of vastus medialis, vastus lateralis, rectus femoris/vastus intermedius, gluteus maximus, sartorius, gluteus medius, tibialis anterior, and the contralateral swing limb's activity of gluteus maximus and semitendinosus (Jonkers et al., 2003a). Loss of these muscles reduces stance limb hip extension during late midstance and terminal stance. The hip extension rate during stance is influenced by adductor longus, gastrocnemius, iliacus, semitendinosus, and rectus femoris/vastus intermedius activity within the stance limb. This action is helped by the activity of adductor longus and iliacus in the swing limb (Jonkers et al., 2003a). Loss of any of these muscles can result in a shorter stride length (Jonkers et al., 2003a). An overview of muscle function within the lower limb during gait is presented in Table 1.3.6.

TABLE 1.3.6 An overview of muscle function of individual stance phase limb muscles in the control of joints in the kinematic chain.

Muscle	Foot rotation	Stance ankle dorsi-flexion	Stance knee extension	Stance hip extension	Swing hip flexion	Swing knee flexion/extension	Swing ankle dorsi-flexion
Stance G	Induces	Restrains	Induces	Restrains	Induces	Induces	Induces (via knee flexion)
Stance S	Induces	Restrains	Induces	Restrains	Induces	Induces	
Stance BF	Induces	Induces		Induces	Restrains	Restrains	
Stance GMX	Induces	Restrains	Induces	Induces	Restrains	Restrains	

G, gastrocnemius; S, soleus; BF, biceps femoris; GMX, gluteus maximus.
Data from Jonkers, I., Stewart, C., Spaepen, A., 2003. The complementary role of the plantarflexors, hamstrings and gluteus maximus in the control of stance limb stability during gait. Gait Posture 17 (3), 264–272.

Swing phase muscle function and dysfunction kinematics

The swing hip is under the influence of the adductor longus, iliacus, and tibialis anterior within the swing limb and the vastus medialis, vastus lateralis, gastrocnemius, adductor longus, semitendinosus, iliacus, and biceps femoris on the contralateral stance limb (Jonkers et al., 2003a). The variations in the normal rate of swing (i.e. amplitude of the swing limb) decrease substantially without these muscles functioning. The rate of hip flexion is controlled by gluteus maximus and semitendinosus resistance in the swing limb (antagonistically to iliacus and the quadriceps), as well as being affected by the stance limb's biceps femoris, gluteus maximus, tibialis anterior, semitendinosus, and gluteus medius. Without their action, excessive hip flexion occurs during swing (Jonkers et al., 2003a) (Fig. 1.3.6c).

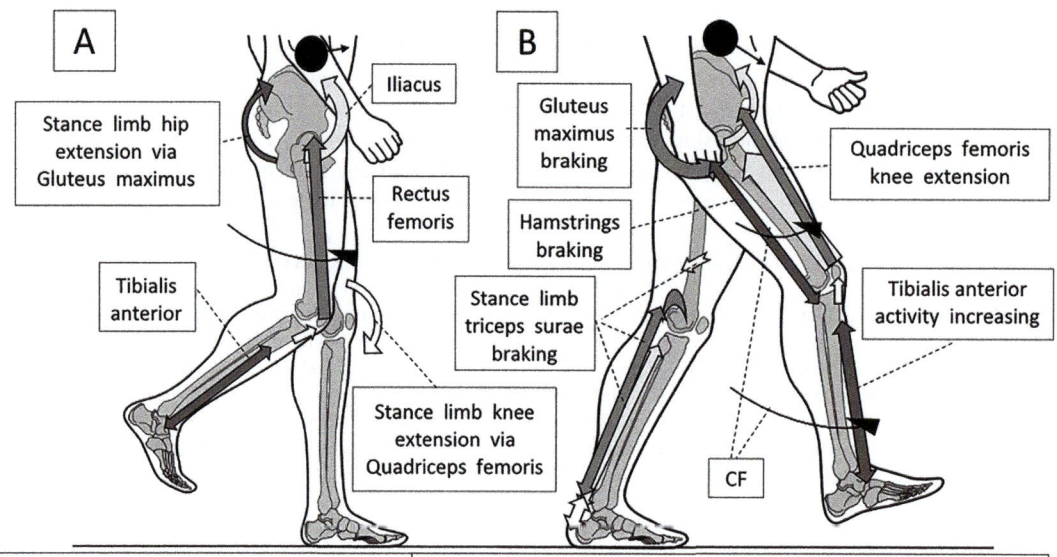

Iliacus and adductor longus hip flexing activity and tibialis anterior ankle dorsiflexing activity helps shorten the swing limb. Stance limb gluteus maximus and quadriceps must concurrently lengthen the support limb.	Iliacus hip flexion is aided by rectus femoris. All the quadriceps muscles extend the knee from mid swing. Together they generate the centrifugal forces that draw the CoM anteriorly. This action is braked by the hamstrings and gluteus maximus in the swing limb and the ankle plantarflexors of the stance limb.

FIG. 1.3.6c Early swing phase muscle activity (A) achieves the task of shortening the swing limb and while translating it anterior to the stance limb. This requires important muscular activity that stabilises the stance limb as well as flexing the hip and knee and extending the ankle (out of plantarflexion) within the swing limb. Late swing phase muscle activity (B) requires the limb to extend at the knee to achieve an adequate stride length and to generate important swing limb centrifugal forces (CF) while doing so. These forces draw and accelerate the body's CoM anteriorly behind the swing limb, restrained by the swing limb's gluteus maximus and hamstrings and the stance limb ankle plantarflexors *(small white arrow)*. Thus, the stance limb's plantarflexor restraint influences how powerful these forces can become without destabilising the stance limb. By creating heel lift before the swing limb's heel contact effectively lengthens the stance limb, allowing a longer, more accelerated stride length. Through decreasing muscle power, such dynamic action around the start of double-limb contact can be lost in the 'unfit' and elderly. *(Permission www.healthystep.co.uk.)*

Swing limb knee extension motion is under the resisting influence of adductor longus, iliacus, biceps femoris, and tibialis anterior within the swing limb and biceps femoris, gluteus maximus, tibialis anterior, semitendinosus, and gluteus medius in the contralateral stance limb, all of which control the rate of knee extension during the terminal swing phase (Jonkers et al., 2003a). Loss of action of any of these muscles can lead to excessive knee extension before initial contact. Gluteus maximus and semitendinosus within the swing limb, and vastus medialis, vastus lateralis, gastrocnemius, soleus, adductor longus, and iliacus within the stance limb, all contribute to knee extension during swing. Their loss would produce insufficient swing knee extension prior to initial contact (Jonkers et al., 2003a).

Ankle motion within the swing limb is controlled primarily by tibialis anterior and semitendinosus, with the loss of their action causing a substantial decrease in ankle joint dorsiflexion (Jonkers et al., 2003a). The rate of ankle dorsiflexion is limited by swing limb soleus and gluteus maximus activity and assisted by stance limb vastus medialis, vastus lateralis, and gastrocnemius activity. Loss of the activity of these muscles would cause exaggerated ankle dorsiflexion (Jonkers et al., 2003a).

Muscle unit function and dysfunction

Muscles, as functional chains, achieve joint kinematics in each orthogonal plane, permitting muscles to compensate for dysfunction. Certain muscles can better compensate for dysfunction occurring in others. The majority of compensations are performed by synergistically acting muscles. Thus, in the sagittal plane, biceps femoris can compensate for stance hip extension in the presence of a weak gluteus maximus (Jonkers et al., 2003b). As gluteus maximus and the ankle plantarflexors both influence distal stance limb joints such as the ankle, a compensation relationship can be established between these muscles. Increased gastrocnemius activity, combined with decreased tibialis anterior activity, can compensate for decreased soleus activity at the ankle (Jonkers et al., 2003b). At the ankle, gastrocnemius and biceps femoris can both influence foot plantarflexion and produce a similar action to gastrocnemius or soleus and gluteus maximus (Jonkers et al., 2003b). This is supported by the findings that gastrocnemius and gluteus maximus provide acceleration power at heel lift, assisting the primary ankle power provided by soleus (Ellis et al., 2014). Thus, there is a potential relationship between increased hip extensor function in the presence of reduced ankle plantarflexor power and more ankle plantarflexor power generated with decreased hip extensor function (Fig. 1.3.6d).

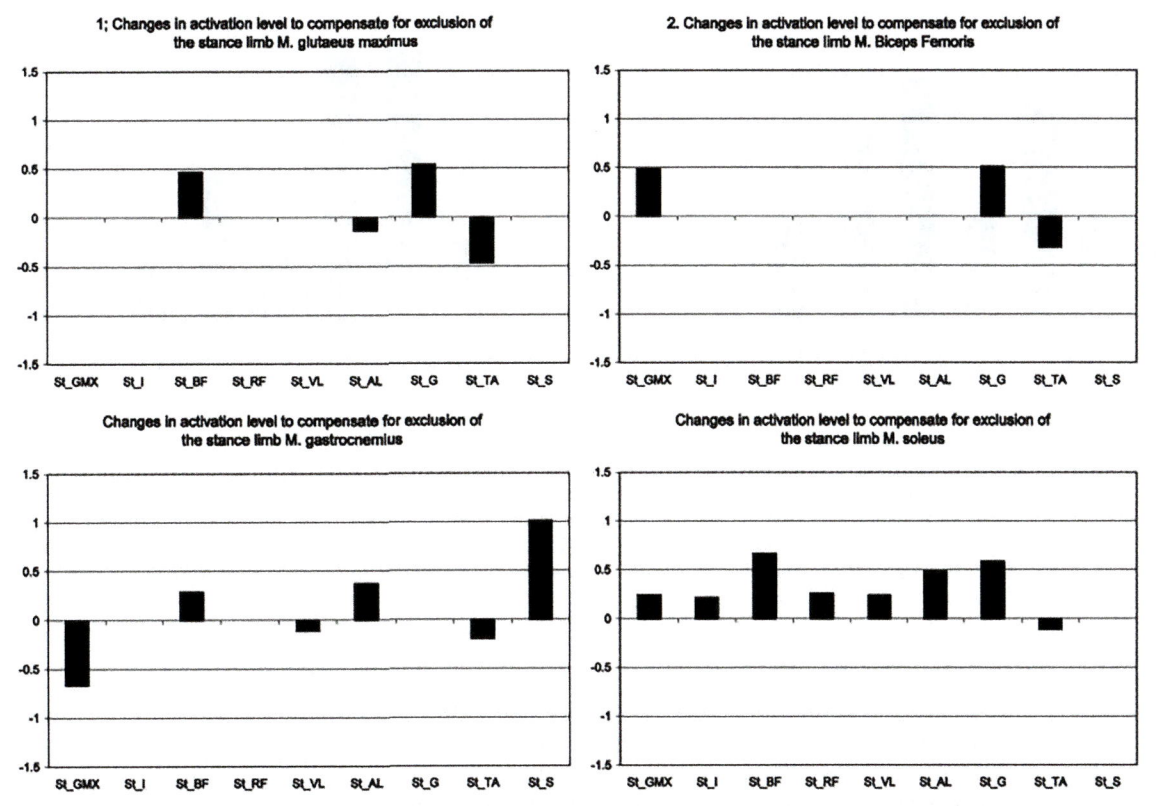

FIG. 1.3.6d Through the modelling of eleven lower limb muscles, changing activation patterns can be assessed in stance (St_). The model used here includes gluteus maximus (GMX), iliacus (I), gluteus medius (GM), adductor longus (AL), biceps femoris (BF), semitendinosus (ST), rectus femoris (RF), vastus lateralis and medialis (V), tibialis anterior (TA), gastrocnemius (G), and soleus (S). Demonstrated are changes in activation levels of modelled stance limb muscles when compensating from loss of certain muscles: (1) gluteus maximus, (2) biceps femoris, (3) gastrocnemius, and (4) soleus. Loss of gluteus maximus results in increased biceps femoris and gastrocnemius activity to restore control over the hip and knee, while decreased tibialis anterior and adductor longus activity assists this. *(From Jonkers, I., Stewart, C., Spaepen, A., 2003. The complementary role of the plantarflexors, hamstrings, and gluteus maximus in the control of stance limb stability during gait. Gait Posture 17 (3), 264–272.)*

Therefore, compensation through synergistic muscle action is complex and may require loss of control at one joint to achieve control at another (Jonkers et al., 2003b). If a single isolated ankle plantarflexor is weak, then the other ankle plantarflexors can compensate. However, this will require increased activity within the hamstrings to restore function in the more proximal joints which the ankle plantarflexors influence (Jonkers et al., 2003a). If there are also deficits in the synergists' capacity to achieve normal joint function, then satisfactory compensation cannot exist. Stance limb stability requires a subtle interplay between muscle activations such that weakness in a single muscle is unlikely to be compensated by increasing or decreasing activity in another single muscle (Jonkers et al., 2003b) (Fig. 1.3.6e). Lower limb joint kinematics in gait result from a multidimensional, well-co-ordinated process of muscle action balanced at different joints under neuro-sensorimotor control, which therefore makes modelling normal and abnormal gait due to muscle activity variations complex and very difficult.

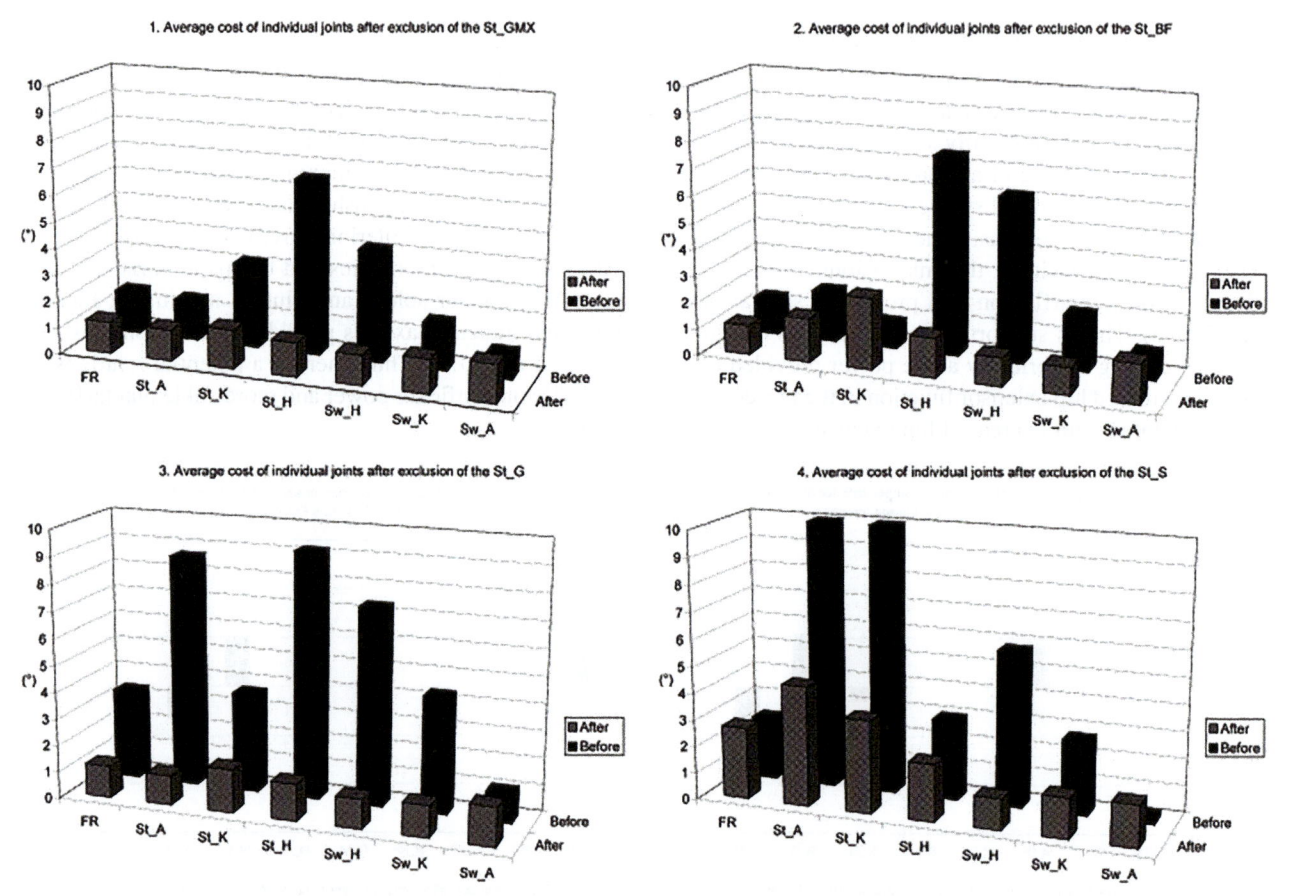

FIG. 1.3.6e An overview of the success of the compensatory changes in muscle activation after exclusion of gluteus maximus (GMX), biceps femoris (BF), gastrocnemius (G), and soleus (S) from the modelling of stance (St_) and swing phase (Sw_) of flexion/extension of the hip (H), knee (K), and ankle (A), and the degree of foot rotation (FR). Losing certain muscles is harder to compensate for and often results in altered joint kinematics and gait control. Biceps femoris compensation is the hardest to compensate for, with good hip extension easily regained by other muscles, but without concurrent knee control. Whereas loss of gastrocnemius can be largely compensated for, loss of soleus compensations lacks knee and ankle stance phase control. *(From Jonkers, I., Stewart, C., Spaepen, A., 2003. The complementary role of the plantarflexors, hamstrings, and gluteus maximus in the control of stance limb stability during gait. Gait Posture 17 (3), 264–272.)*

Each patient's morphology and anatomy will be different, as will be their differences in muscle weakness and subtle muscle firing times, making clinical assessment challenging. Being familiar, at least, with primary sagittal plane muscle activity combined with an awareness of other significant muscle contributions to the areas of the lower limb that are highlighted as being functionally problematic or symptomatic will give a sense of direction to any clinical gait analysis. Gait analysis can then be focused to the pathology/gait complaint of the patient. A knowledge of muscular relationships and the effects of loss of muscle function is part of the process of making clinical gait analysis more patient-focused.

1.3.7 Section summary

Muscles are the internal force providers of support, stability, and power in gait. They heavily influence the structural properties of the lower limb, increasing compliance or stiffness to reflect and manage the locomotive demands from collision effects in the stance limb. Furthermore, they generate the forces within the swing limb that advance it to the next step, which induce the centrifugal forces that draw the CoM along behind it.

Muscles work in teams of prime movers (agonists), synergists, and stabilising and opposing antagonists. By working together, they achieve the functional tasks of GRF management, CoM support, and both displacement and energy management to create efficient energetic locomotion without injury. These tasks are complex for walking and just as complex, but driven more towards their maximum capabilities, during running.

1.4 Running gait
1.4.1 Introduction

Running is the absence of a double-limb stance period with the CoM's lowest point being in the middle of stance phase. Running has long been of interest in gait analysis, both for sport performance and injury assessment. The innovator in photography and movement analysis, Eadweard Muybridge, was hired by Leland Stanford in 1872 to resolve the question as to whether or not all the legs of a horse leave the ground when galloping. Muybridge's studies during the late 19th century, using early sequenced cinematography, included the examination of gait in many animals including humans, and to this day, his images still remain a valuable resource for artists and students of motion alike. Interestingly, his collection of 19th-century photographs of barefoot runners clearly demonstrates a variety of initial strike positions of the foot during running, a subject that caused some debate as recently as the early 21st century. It has now been confirmed that barefoot runners use a variety of foot strike positions as indicated by Muybridge's photographs over 100 years before (Miller and Hamill, 2015) (Fig. 1.4.1a).

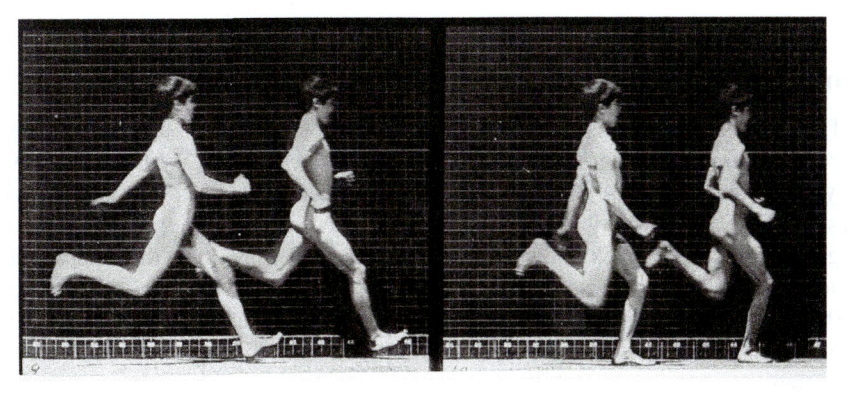

FIG. 1.4.1a Heel strike during running is not an unnatural form of running only expressed when wearing footwear. Running speed has the biggest influence on whether or not a runner rearfoot strikes. Habitual barefoot populations use heel strike running at faster speeds than habitually shod runners when made to run barefoot, suggesting that better rearfoot adaptation occurs with rearfoot running in feet 'developed' alongside barefoot walking. Barefoot heel strikes are clearly indicated in these 19th-century images from Eadweard Muybridge's collection, taken at a time when specific running shoes were very simple and quite a luxury. It is unlikely that these teenagers had ever run wearing specific sports shoes.

Humans are not the greatest runners, but they can outrun some animals including the elephant which can travel at 25 km/h. However, as elephants always have at least one foot on the ground, in essence, they are actually walking very fast rather than running. Humans can also match animals like the dromedary, which has a top speed of around 35.3 km/h over short distances. Other large bipeds such as kangaroos are much faster, and regularly travel at around 50–60 km/h. However, they are bipedal hoppers rather than runners. The human male and female 100 m records at the time of writing are 9.58 s (Usain Bolt) and 10.49 s (Florence Griffith Joyner), respectively. This gives maximum running speeds of around 37 km/h in humans. Compare this to the horse at around 88 km/h, the lion at around 80 km/h, or the fastest land animal, the cheetah, at between 109.4 and 120.7 km/h when at top speed. Other bipedal animals with single-limb support, such as the ostrich, reach around 64 km/h. Top speeds are, of course, short unsustainable speeds, and elite sprint runners as a group of humans are highly specialised and can show anatomical adaptations (Lee and Piazza, 2009; Foster et al., 2021).

A better appreciation of human running ability is to consider average speeds over longer distances. Average running speeds are affected by gender, age, height, weight, terrain, weather, and nutrition/hydration. Elite long-distance human runners can be said to maintain speeds of around 19 km/h. This suggests that humans have evolved more as specialised endurance runners, being able to maintain much longer periods of running than most other animals (Bramble and Lieberman, 2004; Lieberman and Bramble, 2007; Lieberman et al., 2009). The biomechanics of human running is distinctly different to walking biomechanics despite the primary locomotive tasks remaining the same (Fig. 1.4.1b).

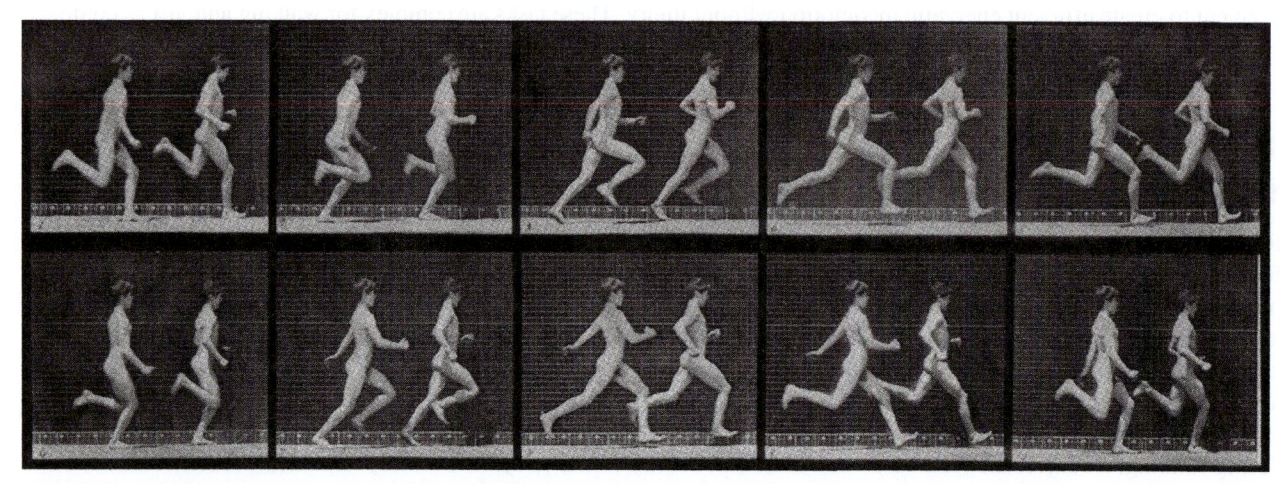

FIG. 1.4.1b Running has its stance and swing phases but lacks double-limb stance. It also provides a period where neither limb is weightbearing, quite unlike walking. Yet, key tasks of impact energy dissipation and braking, HAT CoM transfer over the stance limb, and acceleration, all remain. Heel strike becomes a more optional event dependent on speed, weight, and the use of and previous exposure to footwear.

1.4.2 Running energetics

The metabolic cost of running is expressed as *running economy* and is defined as the total oxygen consumption per kg of mass of the runner per minute during running at submaximal, steady-state velocity (Costill and Winrow, 1970). This can be measured through oxygen consumption (VO_2) or the amount of carbon dioxide used. Through physiology testing, it can also be ascertained as to what is the primary energy source being used, for example, carbohydrate, protein, or fat. The influences on running economy are complex and subject-specific, but all CoM movements diverging from the running direction are likely to significantly affect running energetics, especially those in an exaggerated vertical direction (Gullstrand et al., 2009). During running, CoM movements are usually minimal in the frontal plane (Thorstensson et al., 1984), although any exaggerated movement of the CoM other than in the sagittal plane is likely to be highly influential on running efficiency.

Human running energetics are not the most cost-effective. Transport energy costs in running compared to walking are 50%–80% greater (Farley and McMahon, 1992). This reflects changes in lower limb posture and the loss of mechanical advantage gained from vaulting over the stance foot in the manner of an inverted pendulum. During running, the CoM is at its lowest point during the middle of stance phase, which in walking is when the CoM is at its highest. The resulting increase in muscle action resulting from the loss of the CoM potential energy-storing vaulting mechanism has been considered as being the main reason for the increased metabolic costs (Taylor, 1985; Kram and Taylor, 1990; Griffin et al., 2003). All animals achieve running with the same biological materials as humans, but with quite dissimilar biomechanical techniques. Animals running at high speed, whether they are gallopers, hoppers, or bipedal runners, use their limbs like a spring to allow their bodies to 'bounce' along the ground (Ruina et al., 2005). Human running can be simulated using a spring-loaded inverted pendulum model, which predicts some aspects of running gait such as self-stabilisation. However, adding a spring-mass-damper system to the spring-loaded inverted pendulum allows the effects of energy dissipation through a wobbling mass and reveals that this passive form of energy dissipation is important for running stability (Masters and Challis, 2021).

Humans, like most large animals, save energy needed for running through the use of elastic tissues within their lower limbs to provide cycles of kinetic and potential energy (Ker et al., 1987). The kinetic energy from the braking contact phase is stored as elastic strain and then released during the acceleration phase by elastic recoil. This involves elastic structures such as tendons along with other connective tissues of the distal leg and feet (Ker et al., 1987). However, the reality of human collision biomechanics is that little elastic energy is stored to significantly influence acceleration energetics (Ruina et al., 2005). The Achilles tendon moment arm to the ankle joint axis (also called the internal moment arm) seems particularly important for Achilles energy storage, affecting its spring-like behaviour (Foster et al., 2021). Short Achilles tendon moment arms seem to increase elastic energy storage within this primary energy storage tendon of acceleration. This may explain the relationship between calcaneal length and running economy but also expose a complex relationship between speed, kinematics, lower limb geometry, and foot strike patterns that may all influence Achilles tendon performance (Foster et al., 2021).

What remains particularly interesting and unique about the hominin running style is the plantigrade bipedal nature of running. It gives a number of options in regard to the initial contact position of the foot to the ground, be it the heel, total foot, or forefoot. Each initial contact position alters the biomechanics of the running gait as it changes the location of GRF moments in relation to different joint postures. Throughout stance, the soft tissues have damping effects due to the dissipative work of deformation that removes much of the limb's ability to store energy. When running at 5 m/s, soft tissue energy dissipation within the lower limb is calculated to account for $\sim$30% of the metabolic cost of running alone (Riddick and Kuo, 2016). The foot also absorbs far more energy than it can store, suggesting that it also works as a viscous spring-damper (Kelly et al., 2018). Such negative work needs to be offset by active positive work else gradual deceleration would be the result. Thus, running needs high metabolic input as energy dissipation is high, probably because the energy created by human running collisions is potentially tissue-damaging and it is safer to make up the energy lost rather than expose the body to high storage of impact energy.

Walking gait posture does not change significantly at different speeds and the overall kinematics stay similar, with the only options being walking slower or faster. Furthermore, most people also walk fairly consistently at a preferred walking speed. In running, there is a far wider option of locomotive velocities available, which are roughly broken into jogging, running, and sprinting. Increases in running speed cause changes in lower limb kinetics and kinematics (Mann et al., 1986; Cavanagh, 1987; Arampatzis et al., 1999, 2000; Mercer et al., 2002; Karamanidis et al., 2003). Such changes include, for example, foot strike position. This means that when examining runners in clinic, it is important to assess them running both in the style and upon the terrain they are used to at a self-selected speed, although the actual speed need probably not be fixed (Queen et al., 2006). Those that run cross-country present a particular clinical assessment challenge, which may require some compromises in assessment.

1.4.3 The running gait cycle

The complete running gait cycle is, like walking, constrained within a stride. It commences with initial contact and ends at the next initial contact. The primary difference between walking and running is that there are no double-limb stance phases during running, the GRF is at its maximum during the middle of midstance (rather than at its minimum, as in walking), and there is a period without any ground contact known as the *float phase*, more usually termed the *aerial phase* (Gazendam and Hof, 2007). There are two 'aerial phases', one for each leg. They occur just after stance on the ipsilateral limb, the first aerial phase being referred to as *early aerial phase* which lasts until the contralateral foot contacts. The other occurs after ipsilateral swing phase has completed at the end of the contralateral limb's stance phase as it enters its early aerial phase and is known as the *late aerial phase*. Thus, late aerial phase occurs at the same time as the contralateral limb's early aerial phase. In running, stance only tends to occupy $\sim$40% of the gait cycle and swing/aerial phase $\sim$60%, reflecting a reversal of walking stance and swing times. 'Jogging', as opposed to running, has been defined as having a maximum GRF at midstance, no double-limb support but lacking an aerial phase (Gazendam and Hof, 2007) (Fig. 1.4.3a).

The stance phase of running is divided into an initial contact referred to as a *collision* or *braking* phase, a midpoint of stance that produces the largest GRF and lowest CoM height, and finally, a *rebound*, *propulsive*, or *acceleration* phase. This acceleration phase is about regaining height for the CoM from its midstance low point to achieve a potential energy high point during the aerial phase, thus gaining CoM gravitational acceleration before going into the next step. This is the only way to raise the CoM during running because running has maximum limb flexion in the middle of stance. Thus, the limb extension seen in walking during early midstance, gaining CoM height as part of the foot-vaulting motion, is not possible. In fact, the big energetic difference between running and walking is that in walking, the CoM should be at its highest during absolute midstance, whereas in running, it should be at its lowest at $\sim$50% of the stance phase (Fig. 1.4.3b).

In running, the CoM falls from the aerial phase's high point shortly before initial contact, travelling downwards and anteriorly until contact. At initial contact, this acceleration energy needs to be managed through the lower limb via braking activity and energy dissipation. The early stance phase is therefore a 'braking collision phase' continuing until the CoM reaches its lowest point at midstance through limb flexion. This flexion provides muscle-tendon energy buffering around

108 Clinical biomechanics in human locomotion

A. WALKING GAIT CYCLE

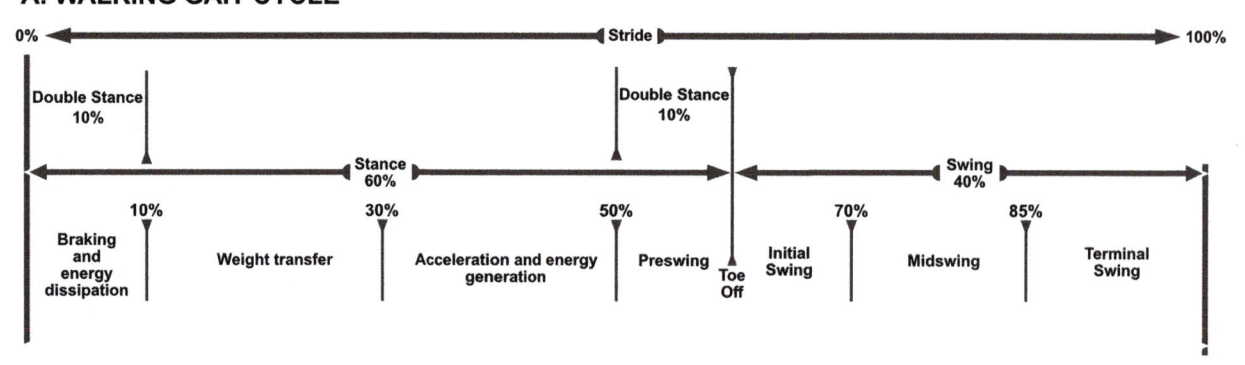

B. RUNNING GAIT CYCLE

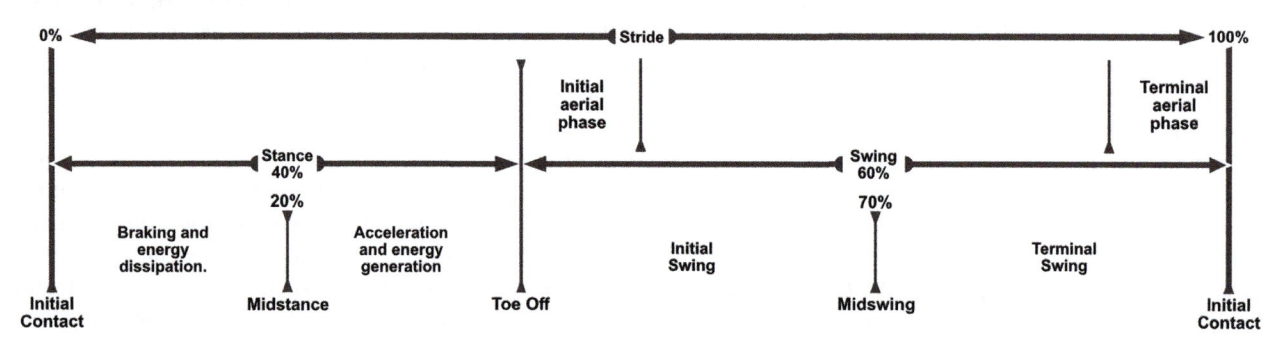

C. JOGGING GAIT CYCLE

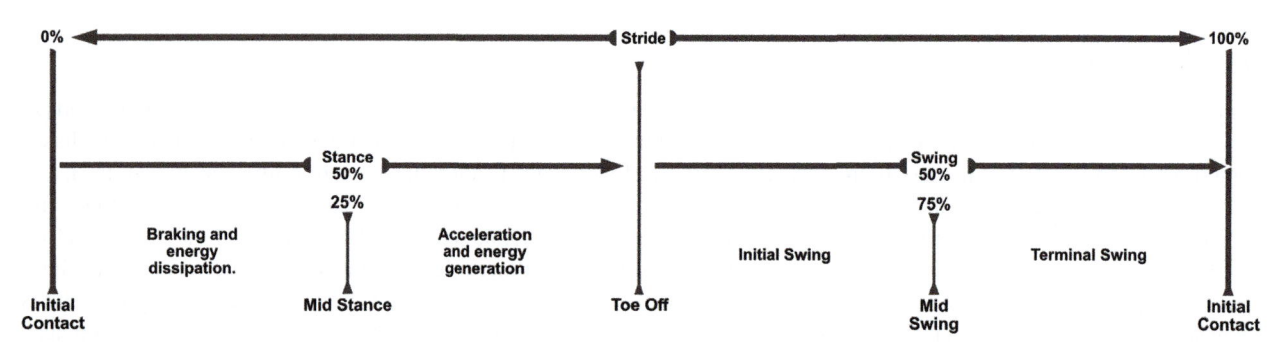

FIG. 1.4.3a Comparisons of the events and approximate percentage timings of the gait cycles for walking (A), running (B), and jogging (C). Note the absence of the limbs being in double stance during running and jogging, and the presence of aerial phases in running alone. However, key functions of energy dissipation and CoM braking during contact phase and a distinct energy-generating acceleration phase remains a part of all human locomotion. *(Permission www.healthystep.co.uk.)*

the hip, knee, and ankle. It is followed by the CoM rising upwards and accelerating anteriorly from midstance through to late stance, creating the acceleration phase via limb extension under limb extensor muscle power (which includes ankle flexion, i.e. plantarflexion) to drive into the early aerial phase to reset the CoM to high again. Generating large horizontal GRF during acceleration is thus desirable for running energetics (Fig. 1.4.3c).

In the frontal plane, running requires that the base of support is brought under the trunk during stance phase. This increases the adduction angle at the hip, necessitating increased hip abduction muscle moments during stance to resist the increased adduction moments set up by the GRF lying more medially to the hip. This, combined with the increased flexion angles of the lower limb joints, is one of the biggest kinematic change from walking, while kinetic changes are a consequence of increasing velocity on force production (Fig. 1.4.3d).

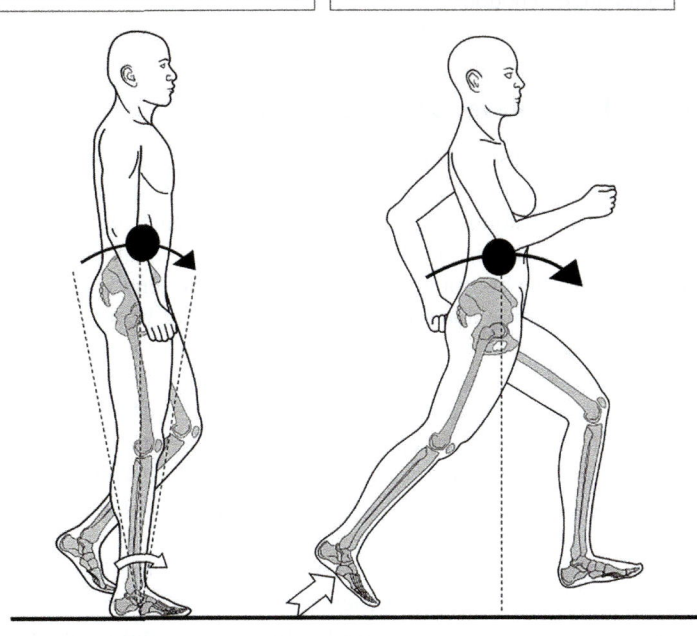

FIG. 1.4.3b To maximise energetics, walking and running utilise the interactions between CoM motion and gravity very differently. In walking, muscular effort brings about limb extension to achieve a CoM high point at absolute midstance. The CoM can then fall forward under the influence of gravity on an arc of rotation defined by the (functional) length of the lower limb throughout late midstance. In running, the CoM is given height by using the muscle power generated during the acceleration phase of late stance, reaching a CoM high point in the aerial phase. It is desirable that the acceleration power of running creates a strong horizontal component so that the CoM's fall is as anteriorly (horizontally) directed as much as possible for maximising energetics. *(Permission www.healthystep.co.uk.)*

FIG. 1.4.3c Following acceleration off the stance foot, toe-off is followed by aerial phase as is clearly seen in this runner from the Eadweard Muybridge collection. Note the change in trunk position higher above the horizontal line set within the background. Aerial phase is unique to running and occurs while the opposite lower limb *(right leg)* is in its late and terminal swing, or late aerial phase. Thus, the body experiences two aerial phases during a running stride, one from the ipsilateral limb and one in late swing from the contralateral limb. The advantages are obvious, as the CoM gains height for the addition of gravitational acceleration before initial contact. The lack of any ground contact means that horizontal acceleration is only impeded by air resistance alone and not in addition to ground contact frictional resistance. Note ankle plantarflexion of the right foot in the far-right image, likely in readiness for forefoot initial contact.

FIG. 1.4.3d Frontal plane running posture requires biomechanics to be set to maintain the CoM over the single support limb of stance phase, while the CoM is moved over the limb. This is a task made more complex than in walking, as running lacks double-limb stance phases. The images from the Eadweard Muybridge collection show a variety of the normal variable methods utilised during a run, such as increasing the hip adduction angle to place the limb more directly under the body than would be normally seen in walking and also sometimes swaying the body mass towards the support limb. The left runner demonstrates both of these techniques more dramatically than does the runner on the right. These techniques, coupled to strong hip abductor activity on the stance limb, provide variables of limb alignment and placement within every running step. Only persistent hip drop through excessive hip adduction moments or sways of the CoM to one particular limb on every step should raise concern.

1.4.4 Running models: Work and power phases in running

Analysis of CoM trajectories and GRF suggests that elastic storage of energy is extremely important to running efficiency (Dickinson et al., 2000; Morin et al., 2007). This has led to the development of spring-mass models of 'bouncing onto the ground' as a theoretical framework for examining running dynamics (Hamner et al., 2010) which consist of a single 'massless' linear 'leg spring', because the leg is in contact with the ground. Although this model also works to explain some of the features of walking through the bi-articular properties of muscle-tendon units (Iida et al., 2008), it is the lack of an inverted pendulum action during running that requires it to be viewed as being mechanically different. The primary parameter of the linear leg spring is stiffness, which is the ratio of the maximal force on the spring in relation to the maximum leg compression in the middle of stance phase.

Lower leg vertical stiffness is used to describe the vertical motion of the CoM during contact (McMahon and Cheng, 1990; Farley and González, 1996). Although the lower limb in running has been mathematically modelled as a glancing collision event on a pseudo-elastic structure (Ruina et al., 2005), muscle activity can produce an adjustable stiffness with the spring being stiffest when the CoM reaches its lowest point during the middle of the stance phase (McMahon and Cheng, 1990; Farley and González, 1996) (Fig. 1.4.4a). The mechanical variability influencing this spring–mass behaviour includes the numerous neuromuscular and biomechanical systems that characterise locomotion, all of which can be potentially disturbed by injury and fatigue (Morin et al., 2011a). Increases in body mass are also likely to affect the stiffness of the spring-mass system, as it has been found that artificially loading runners with more weight increases leg stiffness (Silder et al., 2015).

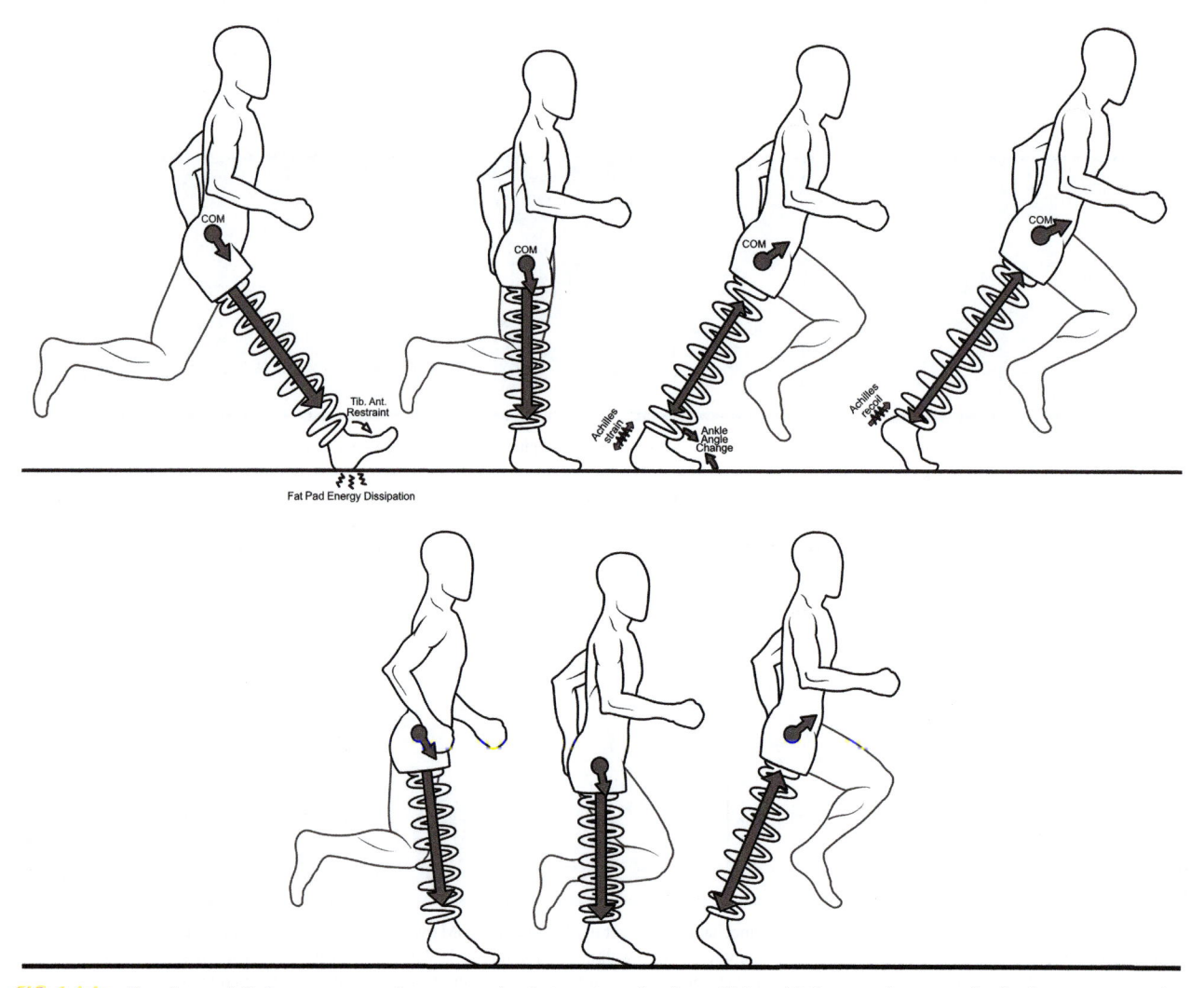

FIG. 1.4.4a Running modelled as a mass operating over a spring that creates a glancing collision with the ground on a pseudo-elastic structure, creating an acceleration 'bounce'. The model works fairly well for running, both with an initial heel or forefoot strike, with the key to success being that the spring is stiff enough to act as both shock absorber, energy storer, and provider of elastic recoil. The recoil bounce-off is at a different angle to that which it collided the ground with initially. The foot acts as a viscous spring-damper, regardless of whether it heel strikes on the heel fat pad or forefoot strikes via the forefoot fat pad, helping the limb dissipate initial contact energy. However, humans dissipate far more energy than they store. *(Permission www.healthystep.co.uk.)*

The running lower limb is required to dissipate more impact energy than it can store to prevent injury, requiring much muscle-generated power to make up for the lost energy. It has also been found that in running, the foot always works as a viscous spring-damper, with greater energy being dissipated than is being stored, and a bigger deficit in the difference between dissipated and stored energy with increases in running speed (Kelly et al., 2018). Therefore, the lower limb 'spring' in human running at contact 'loses' far more energy during its initial collision and braking than it does in storing and releasing for the late stance rebound. The forces that are generated from muscle activity around the vertical displacement of the CoM within the spring-mass system during the stance phase of running, increase horizontal acceleration of the CoM (Romanov and Fletcher, 2007). Gravity causes this horizontal torque as the runner's CoM moves in front of the support foot during late terminal stance, just prior to the aerial phase when the extensor muscles of the knee, hip, and ankle are no longer active. This suggests that significant elastic recoil comes through energy-storage in tendons, regaining much of the CoM's height lost during stance through connective tissue recoil (Romanov and Fletcher, 2007).

Running work and power in considering energetics of gait

Thus, at 'steady' running speed, there is a brief small burst of positive work at initial ground contact (0%–2.5% of a stride) followed by a mass-spring system stance phase (McMahon and Cheng, 1990). This consists of one burst of negative work at 2.5%–15% of a stride, termed the '*collision*', followed by one large burst of positive work (15%–34% stride), termed the 'rebound' or acceleration, that arises from energy storage within soft tissues, primarily energy-storage tendons (McMahon and Cheng, 1990). This is followed by the aerial phase at 34%–50% and then swing, from 50% to 100% of the running stride. The leg joints perform positive and negative work with each stride, with muscle activity accounting for nearly all the metabolic energy expenditure of running (Riddick and Kuo, 2016). Additional energy dissipation occurs through deformation in passive soft tissues such as the heel fat pad, intervertebral discs of the spine, joint cartilage, and the structures of the foot, of which some of the energy applied to these tissues can be returned for locomotion (Ker et al., 1987, 1988; Riddick and Kuo, 2016).

Power is required to lift the CoM vertically against gravity with positive acceleration. Kinetic energy is supplied from stored energy and muscle metabolic energy input that reflects the CoM's height changes during running, and the amount of stretch or compression of the anatomic elastic components (Gullstrand et al., 2009). Elite distance runners demonstrate slightly reduced vertical displacement of their CoM compared to nonelite distance runners (Williams and Cavanagh, 1987) and it has been noted that total vertical impulse, which is proportional to the change in vertical velocity of the CoM, correlates negatively with running economy (Heise and Martin, 2001). Thus, moving the CoM vertically is metabolically expensive during running and so it is better to keep it fluctuating within a lower range. Controlling vertical displacement of the CoM is important to achieve effective horizontal CoM displacement. Yet, significant potential energy is gained through the fall of the CoM supplementary to that gained through lower limb elasticity during stance. An energy balance therefore exists between that gained from CoM displacement and the need to limit the CoM displacement, which costs energy.

It is reported that over one-third of the total metabolic cost of running is used to generate horizontal propulsive forces, with most of the rest used in generating vertical forces (Chang and Kram, 1999). Generating horizontal forces per unit force is more expensive than generating vertical forces, and generating horizontal propulsive forces is much more expensive than generating horizontal braking forces (Chang and Kram, 1999). Horizontal forces are particularly relevant to running speed, as these are the forces that slow the direction of travel at initial contact braking impacts and drive the body forward for acceleration collisions. However, dividing the force generation into planes is less relevant than considering the metabolic costs of the resulting forces of running, as it is the net resultant force generated with the ground that affects the net muscle moment of force required at each joint (Chang and Kram, 1999).

Alterations in limb posture change the effective mechanical advantage of the GRF along with the effective mechanical advantage of limb muscle-force production (Biewener et al., 2004). This change in posture underlies many of the differences in the energy costs of transport from walking to running in different animals (Taylor et al., 1980; Taylor, 1985; Kram and Taylor, 1990). The key facts are that humans have high energy costs in running compared to walking and the changes in posture that occur in the transition to running from walking affect the force-generating requirements, playing a substantial role in energy use (Farley and McMahon, 1992; Griffin et al., 2003; Biewener et al., 2004). During running, the knee is held in a more flexed position associated with the limb functioning as a spring (Farley and McMahon, 1992; Lee and Farley, 1998). The knee converts impact energy into stored elastic energy, assisted by other lower limb joint tendons such as the ankle tendons and those within the foot (Ker et al., 1987). This is different to the vaulting action seen during walking where there is a distinct and relatively long initial compliance phase that dissipates impact energy through tissue viscoelasticity.

Such a shift in action requires a substantial increase in force magnitude and total extensor muscle impulse at the knee in running to generate a GRF from which to drive forward from.

Walking to running and running to walk transitions

Limb muscle mechanical advantage is defined as the ratio of the weighted mean agonist muscle moment arm to the moment arm of the GRF. It has been used to calculate the mechanical efficiency of muscle function during running (Biewener et al., 2004). With increasing gait speed, the maximum GRF joint moments increase steadily at the hip, sharply rise at the interface point of walk-to-run at the knee and yet remain fairly constant at the ankle (Biewener et al., 2004). At the ankle, the inertial and gravitational moments of the limb segments and the body mass are small compared to the GRF moments (Biewener et al., 2004).

The subconscious, spontaneous change from walking to running at gradually increasing speed has long puzzled researchers, with several mechanism proposed to explain why it happens (Voigt and Hansen, 2021). The puzzle is increased when the observation that the run-to-walk transition (out of running) tends to occur at a different, lower locomotive speed than the walk-to-run transition, although this expresses some inter-individual variability. There is mounting evidence that human locomotion involves behaviour of a dynamic nonlinear complex system based around fixed structural anthropometrical dimensions such as body weight, segment mass distribution, muscular properties (such as force-velocity relationships), nerve conduction speed, synapse delays, and vascular properties (Voigt and Hansen, 2021). Voigt and Hansen (2021) have proposed that human motion depends on three functional entities, each consisting of several subsystems. These are the peripheral and central nervous systems and their subsystems, energy turnover (energetics) from respiratory, circulatory, and biochemical metabolism, and from the musculoskeletal system (Fig. 1.4.4b). These determinants organise the dynamic behaviour of the body at a subconscious level, which can of course be overridden by a conscious decision to run or walk. Stride rate and length may play a key role in determining the point that locomotive technique changes, but the whole process within each individual possibly relates to the development during growth and life of each phenotype in the long term and the dynamic processes of self-organisation of each system to neurological organisation and central pattern generators (areas of learned movement programmes) within the central nervous system during acute locomotive events (Voigt and Hansen, 2021) (Fig. 1.4.4c).

Locomotion form	Self-selected walking	Fast walking	Walk-to-run transition	Slow running	Self-selected running
Stride rate	Self-selected 60 strides min^{-1}		70 strides min^{-1}		Self-selected 80 strides min^{-1}
Locomotor drive					

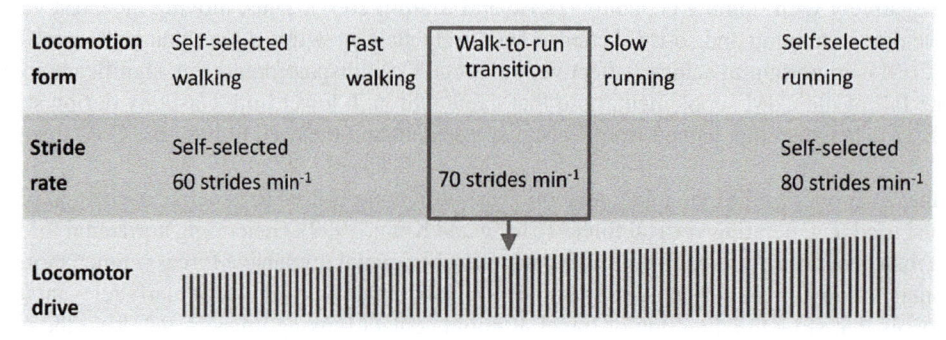

FIG. 1.4.4b A schematic to illustrate the relationship between locomotive drive (the conscious or subconscious needs for locomotion), stride rate, and locomotive form. Locomotive drive can be considered a fundamental control parameter and stride rate can be considered a reflection of locomotive drive that at a certain point, initiates the walk-to-run transition. The speed of gait and the need for postural stability are probably the main factors affecting this. Walk-to-run transitions are reported to occur on treadmill tests when stride rate is midway between self-selected stride rate (preferred) walking speed and preferred running speed. *(From Voigt, M., Hansen, E.A., 2021. The puzzle of the walk-to-run transition in humans. Gait Posture 86, 319–326.)*

1.4.5 Impact in running: With consideration to walking

Impact occurs in both walking and running, but the energy produced from impact is much higher in running. Impact energy is considered a minor pathological risk in walking, at least within healthy tissues, as initial collision negative acceleration (deceleration) forces are relatively low. During running, particularly in fatigued states, impact energy and its dissipation mechanisms become a significant potential source of injury due to the higher rates of negative acceleration engendered during collision events. However, energy dissipation, especially during oscillations within the soft tissues acting as a wobbling mass, may be part of a stabilisation mechanism used in running (Masters and Challis, 2021).

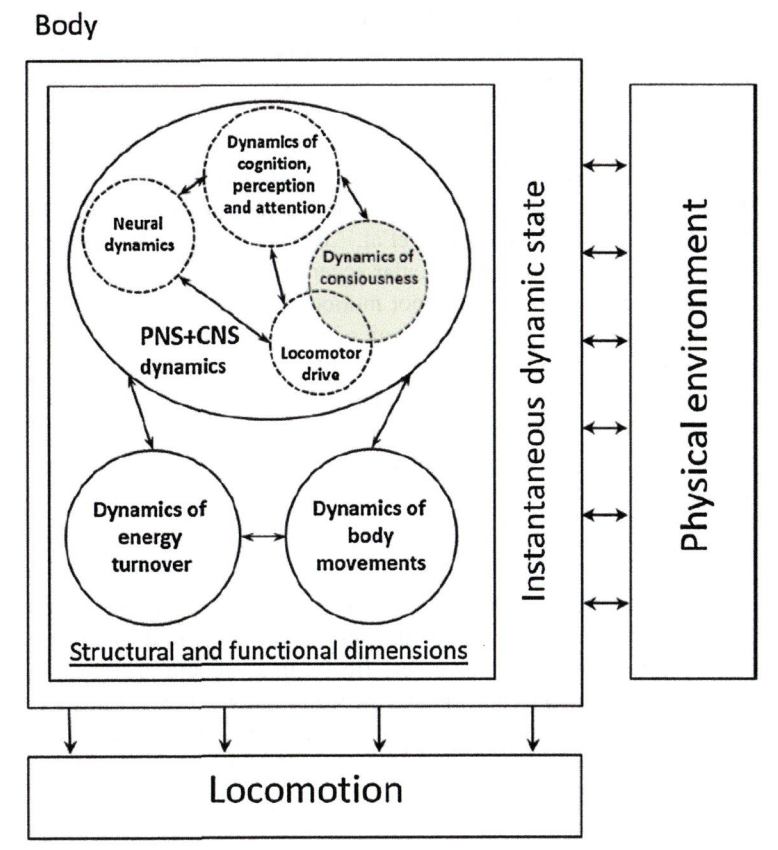

FIG. 1.4.4c A schematic of the dynamic interaction between the peripheral and central nervous systems, energetics, and biomechanics. The net result of the combined dynamic interactions helps determine the type of gait that best fits the instantaneous dynamic needs of locomotion. If gait characteristics of environmental and body demands change gradually, then the instantaneous dynamic state of the body may reach a critically low level of dynamic stability, which initiates a change in the type of locomotion that best regains optimal stability. However, conscious control can almost entirely overrule the self-organised dynamic control of locomotion (subconscious locomotive drives). *(From Voigt, M., Hansen, E.A., 2021. The puzzle of the walk-to-run transition in humans. Gait Posture 86, 319–326.)*

Mechanics of impact revisited

Velocity is a vector of displacement over time. Acceleration is change of velocity over time and is directly proportional to the force and inversely proportional to the mass of the object. Acceleration occurs in the direction of the net force and is positive when it increases or negative when it decreases. This means that a braking automobile has negative acceleration (decelerating) but still has velocity and therefore momentum. It also explains why accidents can still be nasty when vehicles collide while braking. The product of inertia or mass and velocity is momentum. Momentum is a vector, as anything with velocity always has direction. As a simple concept, heavy items have greater momentum than light items at the same velocity, but light items can have greater momentum at higher velocities than slow-moving heavy ones. Moreover, it is easier to get lighter items to move faster than heavier ones and easier to stop lighter ones moving than heavier ones. The greater the velocity, the harder it is to stop any object. Momentum is linear when moving in a straight line, curvilinear if in an arc of motion, and angular if occurring around an axis. It is therefore of interest to point out that when moving from one location to another, humans try and achieve efficient linear motion by producing curvilinear motion of the CoM and angular motion in body segments at joints. This sets up a number of types of momentums within each body segment as discussed within Chapter 1 of the companion text "Clinical Biomechanics in Human Locomotion: Origins and Principles'.

Changes in momentum are known as impulses. These are actions, collisions, or other events that change momentum. Impulses that increase momentum are known as positive impulses, while those that reduce momentum are referred to as negative impulses. Losing mass reduces momentum, but most changes in momentum occur as a result of changing velocity. Increasing running speed increases human body momentum, creating a positive impulse. Slowing down creates a negative impulse, reducing momentum. Muscle activity to increase or decrease motion through internal or external forces together produce impulses. Collision with the ground by the foot directly or via footwear can produce large impulses. In swing, the foot's small mass rotating on a long moment arm at the knee and hip achieves a relatively high velocity compared to the rest of the lower limb and HAT segment, but it must bear the full body weight in single support. As running has no double-limb stance, its initial contact braking and terminal stance acceleration lower limb impulses are more dramatic than those of walking, not least because they occur at higher velocities.

Thus, in walking, the most dramatic gait impulses are at heel strike and heel lift when the foot experiences rapid changes in velocity. Contact phase is therefore a negative impulse braking event, slowing the fall of the CoM. In terminal stance, the ankle plantarflexors primarily generate ankle power that drives the forefoot into the ground, raising the heel to power a positive impulse in driving the CoM forward (Usherwood et al., 2012). At heel strike, an impulse is created by ground contact, stopping the heel's forward momentum and muscular eccentric contractions. That of tibialis anterior slows the angular flexion momentum of the forefoot until the forefoot motion is stopped by ground collision. Indeed, heel strike followed by forefoot strike separates the initial collision energies into two events, allowing reduced peak energy loading within walking that also involves these collisions occurring in double-limb stance, reducing energy loading on each limb (Ruina et al., 2005). Running sustains a dramatic negative impulse at contact on one limb. However, the peak vertical impulse occurs at midstance when the CoM of the body has accelerated to its lowest point before starting to accelerate back up. Thus, in running, peak impulses occur at the changeover from braking to anterior and upward acceleration, regardless of foot strike position (Fig. 1.4.5a).

Impact amplitude and cushioning

Amplitude is the maximum extent of a vibration or oscillation measured from the position of equilibrium. Amplitude is most frequently associated with sound, where it represents the degree of change in the atmospheric pressure caused by a sound wave. Sound waves travel fast as gas atoms move very freely. In a vacuum, there are no atoms, and therefore no vibration of atoms can occur. This results in no sound. Hence, the well-known 1970s tag line for the science fiction horror film, *Alien*: '*In space no one can hear you scream*'. The collisions of the foot and the ground cause sound, as anyone trying to creep or stomp about can testify. The larger the amplitude, the louder the sound. Solids and liquids also experience amplitude as on a molecular level due to the presence of thermal energy, atoms are permanently deviating from their position of equilibrium, even at rest. Sound waves travel through solids and liquids by vibrating or oscillating their atoms, travelling more easily in liquids than solids as liquid atoms move more freely than those of solids.

An impact is an intense energy-loading event. The consequences of an impact are dependent on the details of the transport of the energy and the momentum after impact which redistributes the energy (remembering that energy cannot be destroyed). This energy and momentum redistribution is largely achieved though nonlinear wave propagation, the most dramatic form being commonly known as a *shock wave*. The amplitude of the impact is dependent on the properties of the materials involved in the impact, such as the material strength, viscosity, elasticity, and thermodynamic properties. Any solid structure will respond to transient disturbances through the propagation of stress waves, whether they be slight impulses or dramatic impacts.

A body subjected to a modest impulse loading will propagate as an *elastic pressure wave* or *pulse* at a velocity determined by the material. Although all body tissue is viscoelastic, soft tissue will respond differently to bone through variance in stiffness that effects wave propagation; a phenomenon also dependent on the water content within the tissues. It is these principles of how easily shock waves pass through tissues that are utilised in diagnostic ultrasonography to identify different tissues and pathologies via the difference in stiffness and/or water content within tissues. The velocity of the elastic pulse is determined by the elastic stiffness (or amount of compliance). The shock waves during an impact vibrates the tissues, causing frequency oscillations (measured in hertz [Hz]), the effects of which are dependent on material properties. Bone oscillations occur at largely different amplitudes to soft tissues. Contracted muscle is harder to oscillate than relaxed muscle, as the degree of stiffness influences the ease of sustaining an oscillation. Thus, depending on different tissue stiffnesses at impact, certain tissues will vibrate more freely than others at lower amplitudes. By contracting muscle locally, the human body can modify the effects of impact 'shock waves' requiring different amplitudes to vibrate the tissues, but primarily by damping the amount of vibration. The soft tissue oscillations themselves help to dampen impact energy, particularly during running, and help to stabilise bipedal locomotion (Schmitt and Günther, 2011; Khassetarash et al., 2015; Masters and Challis, 2021).

Cushioning

Cushioning is defined as a reduction of the amplitude of the vertical GRF during impact (Nigg et al., 1995). Therefore, cushioning materials used in orthoses or footwear should be attempting to achieve a reduction in the amplitude of the impact. In heel strike running, peak impact forces in shod runners are up to around three times body weight at

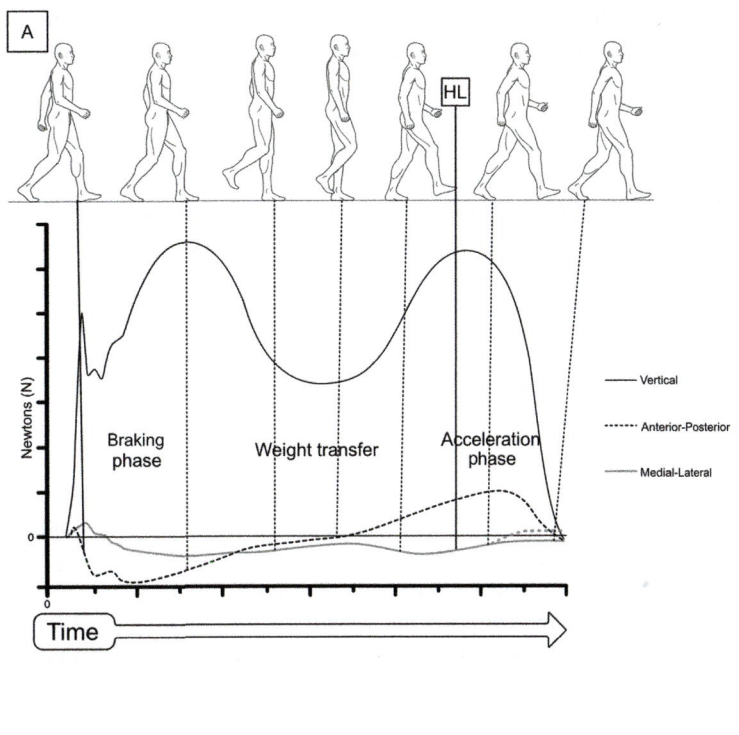

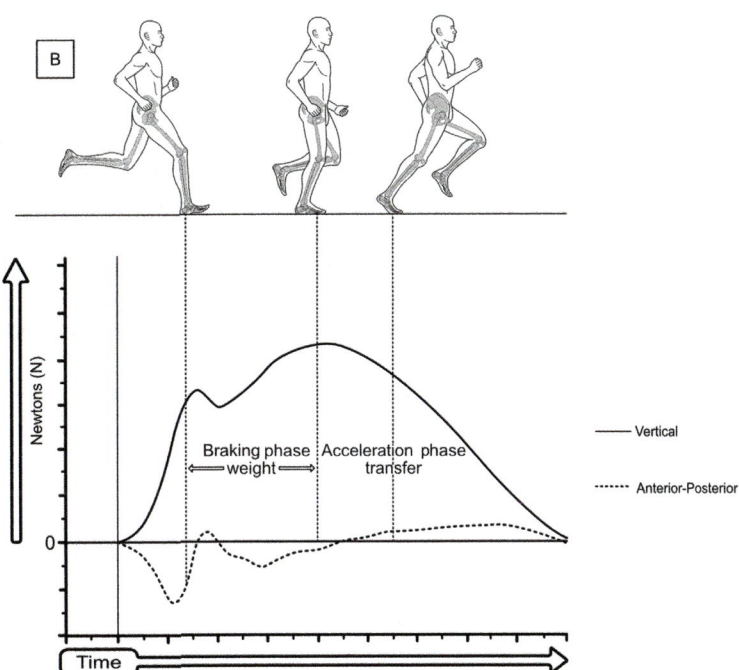

FIG. 1.4.5a Walking and running are different in their management of impact-braking and acceleration collisions, as demonstrated via force-time curves. In walking (A), the heel and forefoot collisions are separated. These impacts are followed by a reduction in forces over time (impulses) when weight is transferred up, over, and down from the rearfoot to the forefoot, as the ankle rotates the CoM anteriorly. Once positioned anteriorly, the impulse rises to a peak of acceleration around heel lift (HL) before weight is transferred to the opposite foot, rapidly decreasing the impulse. In heel strike running (B), the heel and forefoot still collide in separate events, yet under higher acceleration. This requires greater negative work to manage than walking. The forefoot loading impulse blends into the continued braking impulse to end at midstance. This is because vertical forces continue to rise with the fall of the CoM. However, the peak of this second impulse is the start of acceleration as the CoM begins to rise upwards under ankle plantarflexion power. This decreases, as body mass is accelerated away from the support surface under swing limb centrifugal forces towards an aerial phase, where neither limb is weightbearing. Weight transfer in running is a continuous, indistinct event concurrent within braking and acceleration. *(Permission www.healthystep.co.uk.)*

10–30 ms (thousandths of a second). However, because the body has the capacity to modify its own shock attenuation through changing muscle activation patterns and joint positions in response to surface conditions, the results of applying or not applying cushioning are variable (Wakeling et al., 2002). Constructing the ultimate universal cushioning material, orthosis, or shoe is probably impossible. Adding or subtracting cushioning will have individual energetic effects on the patient (Nigg et al., 2003), not least through adding extra mass into the footwear. However, without investigating running economy, the energetic effects may be clinically unclear. In barefoot running, the impact peaks occur at between 5 and 10 ms after initial contact (Nigg, 2000). With sports shoes, the peak impact forces are delayed between 10 and 25 ms and they last up to 50 ms longer (Nigg, 2000). Nevertheless, despite a change in the loading rate when comparing barefoot conditions, running shoes do not change the peak vertical forces of running (Yan et al., 2013). Walking in firm-soled casual shoes does not affect either the vertical peak forces or the loading rate. However, walking in a more cushioned sports shoe does reduce the peak force, but also raises the initial impulse transient (Yan et al., 2013).

Impact implications in running

Both stride length and stride frequency increase with running speed, with stride length and shock attenuation increasing by around 17% and 20% per m/s increase of speed (Mercer et al., 2002). Shock attenuation correlates strongly with stride length, but only moderately with stride frequency across running speeds, whereas shock attenuation increases linearly with running speed (Mercer et al., 2002). Hence, to reduce impact shock, increasing running step frequency is better than increasing stride length. Shock attenuation is due to heightened leg peak impact accelerations. The stride length generated during running has a profound effect on running kinematic changes as well as on the resulting kinetics (Mercer et al., 2002). In addition, stride length is implicated in running injury rates (Bramah et al., 2018), not just because of shock attenuation but also because of the changes in posture and the resultant angular moments that increasing stride length causes.

Anything that affects segmental velocity at the instant before foot-ground collision such as running speed, stride length, stride frequency, and joint orientation will determine the amount of change in the momentum of the foot and leg at initial contact, and thereby the rate and magnitude of the impact force and shock (Gruber et al., 2014). Rearfoot strike running achieves greater impact attenuation throughout the body than does forefoot running (Delgado et al., 2013; Gruber et al., 2014). This is because rearfoot strike is able to absorb lower frequency impacts than forefoot running. Passive mechanisms of shock attenuation are responsible for attenuating higher frequency components of impact and are achieved by shoe deformation (if present), the heel fat pad, ligaments, bone, articular cartilage, and the oscillations (vibrations) of the soft tissue compartments (Chu et al., 1986; Schmitt and Günther, 2011; Khassetarash et al., 2015). Active muscular shock attenuation mechanisms rely on the active responses to impact and occur later in stance, attenuating the lower frequency components of impact forces and include reducing moment loading velocity by eccentric muscle contraction, muscle-tendon buffering, increased muscle activation to limit soft tissue oscillations, adjustments in joint stiffness, and changing limb segment geometry.

The soft tissues of the lower limb are at the root of the transmission and dampening of impact energy during running (Pain and Challis, 2006; Schmitt and Günther, 2011; Riddick and Kuo, 2016) and are able to modulate peak joint forces (Liu and Nigg, 2000; Wakeling and Nigg, 2001) and dissipate mechanical energy (Riddick and Kuo, 2016). This occurs because momentum energy is transferred between bodies during collision, which explains footprints in soft ground and the importance of crumple zones within vehicle design to save lives. Because muscles must actively offset any net energy dissipation within the body, the soft tissue deformations occurring probably affect the metabolic cost of walking as well as running.

Soft tissues perform both negative and positive work during different phases of stance, yielding substantial net negative work and having an increasing role at faster running speeds due to greater impacts (Riddick and Kuo, 2016). Vibration frequencies of soft tissue have been reported to range between 3 and 55 Hz in the lower limb, with maximum wobbling mass (soft tissue mass) excursions relative to bone ranging from 3 mm up to a 4-cm high for the body's CoM vibration excursion (Schmitt and Günther, 2011). More energy seems to be dissipated by wobbling soft tissue mass excursions in the horizontal rather than vertical direction, with less energy dissipated within the lower leg than in the thigh because of higher soft tissue masses in the proximal lower limb (Schmitt and Günther, 2011). The soft tissue energy dissipation during running must be offset (made back up) by equal amounts of muscle activity and could account for 29% of the metabolic cost of running economy (Riddick and Kuo, 2016). This means that soft tissue energy dissipation is quite considerable (Fig. 1.4.5b).

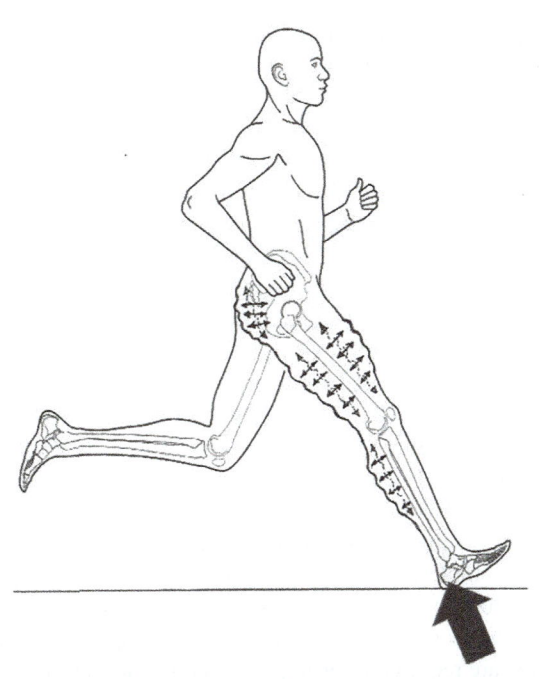

FIG. 1.4.5b Soft tissues in the lower limb can act as energy dissipators by vibrating via shock waves through the muscles, connective tissues, and skin. Such events are most noticeable where soft tissue masses are large (buttocks, thigh, and calf). Tissues will displace superiorly and inferiorly on impact, but horizontal displacement seems more extensive and of greater importance to energy dissipation. This is thus able to convert vertical forces into horizontal deformations that absorb energy in a different plane to that of the peak impact force. Heel strikes at high accelerations seem to produce most dramatic soft tissue oscillations associated with GRFs that produce high initial impulse transients. By having an unusually large muscle mass positioned distally (within the calf) for a terrestrial running animal gives humans a quite unique distal impact dissipation option. *(Permission www.healthystep.co.uk.)*

It is argued that the energetic cost of separating significant wobbling masses from the skeleton may be overcompensated for, both by avoiding the metabolic costs of active impact reduction and decreasing loads on the passive skeletal structures (Schmitt and Günther, 2011). This possibly creates an environment for running-specific pathologies such as medial tibial stress syndrome where fatigue of active muscle energy dissipation is surpassed, muscles become less stiffened, and the reliance on passive soft tissue wobbling mass dissipation is increased. This shakes (via shock waves) the soft tissues away from the bone, excessively tensioning the fibrous attachments of the muscle and fascia cruris to the medial tibial periosteum. It is unlikely that bone is damaged easily during impact, as within reasonable biological limits, impacts are known to be beneficial and necessary for the structural strengthening of bones. Torsion and tension in bone are far more likely to initiate bone pathology than compression, although stress (fatigue) fractures within the lower limb may be precipitated by the failure of active and passive energy-dissipating soft tissue mechanisms in protecting bone from torsion and tension strains. Beyond the risk of injury, impact may also be important in stabilising motion (Masters and Challis, 2021), including by providing mechanical signals for the sensorimotor system.

1.4.6 Shock attenuation in the lower limb

Shock attenuation is the process of absorbing impact energy. The body has both active and passive systems for gait-induced ground collision management for the reduction of shock wave amplitude from the foot to the head and to prevent impact energy reaching the delicate brain through excessive head accelerations (Derrick et al., 1998). Humans have high water content and this is influenced by age, fitness, and gender (fatty tissue contains less water). Structures with high water content are known as *wobbling structures*, which oscillate easily when you hit them (Gruber et al., 1998). This oscillation leads to a vibration load on the structure, but the presence of a passive wobbling mass has a role in maintaining stability during running as well as helping to dissipate energy (Masters and Challis, 2021). To get biological tissues to resonate during impact, the frequency of the impact must be close to the natural frequency of the tissue's resonance. Gait's impact force frequency is around 10–20 Hz, whereas the resonance frequencies of the lower limb's soft tissues are between 10 and 50 Hz (Wakeling et al., 2003). This is therefore a perfect match for gait impact to initiate tissue resonance.

Shock attenuation requirement is affected by velocity of gait which is a product of stride length and stride frequency, although only stride length seems to increase the need for shock attenuation (Mercer et al., 2003). Thus, lots of small impacts appear safer than large ones. Energy-induced tissue shock wave vibration is a necessary and positive stimulus on the healthy maintenance stimulation of the biological response within musculoskeletal tissue. However, in excess, it can be detrimental. Internal vibration forces can become excessive and focused locally within tissues without sufficient time to recover, resulting in tissue damage. Prolonged vibration exposure is known to cause reduction in motor nerve firing rates, reduced muscle contraction force generation, decreased nerve conduction velocity, attenuated sensory perception, and reduced peripheral circulation (Wakeling and Nigg, 2001). Thus, the body must manage but not avoid tissue resonance on a risk-to-benefit basis. Power-trained athletes, such as weightlifters, who have higher amounts of fatiguable fast-twitch fibres can induce higher impact vibration frequencies than nonpower-trained athletes, which may make it harder for them to manage their soft tissue oscillations during prolonged exercise (Chen et al., 2021). This may make some power-trained athletic participants more vulnerable to impact-related pathology during any distance running fitness training.

Muscle activity can increase the damping coefficient, making it harder for tissues to vibrate by increasing stiffness, a task achieved by anticipatory activation (Wakeling et al., 2003). As previously discussed within Chapter 4 of the companion text "Clinical Biomechanics in Human Locomotion: Origins and Principles', Section 4.3, the sensorimotor system relies on a feedforward mechanism to position joints and activate muscles appropriately. This is because of the time it takes to generate a muscle action in response to sensory signals via a feedback mechanism through the nerves. It takes around 100ms for the sensorimotor system to complete the task of sending muscle activation information, while running impact peaks occur at 5–10ms when barefoot, 10–25ms when in shoes, and at 50ms during walking. The only solution is an anticipatory feedforward mechanism of muscle contraction to achieve the appropriate damping effect of tissue vibration (Wakeling et al., 2003). If adjustments are required after impact in anticipation of the next collision, then feedback information gathered via the muscle spindles and joint proprioceptors during the present impact can create changes in skeletal stiffness and damping coefficients through the brain's central pattern generator before the next step (Ivanenko et al., 2000).

Preactivating muscles also give them a role in absorbing and dissipating impact energy. Tibialis anterior is an energy-dissipating specialist postural muscle, having a particularly prominent role in creating a negative component that reduces the vertical impulse component from heel strike impacts (Neptune et al., 2008; McGowan et al., 2010). Using muscles as energy dissipators through muscle-tendon buffering means that energy can be 'lost' without loading more delicate structures within the body. Unlike large-straining passive connective tissue-based dissipative structures, muscles are capable of many cycles without creep and allow some adjustable control of how the energy is dissipated. Muscle-associated fascia may also be involved in energy dissipation, suggested by common symptoms of anterior shin pain being increased during impact activities associated with pathology in both extensor tendons (Herod et al., 2016) and fascia (Stecco et al., 2014). The anterior muscular compartment is particularly vulnerable because it demonstrates stiffer fascia and is supplied with less fatigue-resistant tendons than the flexor muscle compartments.

The soft tissues of the lower limb are at the root of the transmission and damping of impact energy during running (Pain and Challis, 2006; Schmitt and Günther, 2011; Khassetarash et al., 2015) and are able to modulate peak joint forces (Liu and Nigg, 2000; Wakeling and Nigg, 2001). The soft tissue energy dissipation capabilities of impact forces can be divided into those that actively dissipate energy, that is, the muscles and tendons active during impact and those that passively dissipate energy through their resonance frequencies and passive stretching, which in the lower limb range from 3 to 55 Hz (Schmitt and Günther, 2011). As previously stated, more energy seems to be dissipated by wobbling soft tissue mass excursions in the horizontal rather than vertical direction, and in the thigh more so than the shank (lower leg/calf) (Schmitt and Günther, 2011). Therefore, soft tissues perform both positive and negative work during the stance phase, yielding substantial net negative work at impact. However, their role increases within running at a faster pace due to rising impact forces with speed (Riddick and Kuo, 2016).

Joint alignment and stiffness in impact dissipation

Biological shock attenuation can be achieved by adjusting joint stiffness to change compliance, with the greatest effects resulting from knee flexion angles (Wakeling et al., 2003; Podraza and White, 2010). Joint stiffness is achieved through intrinsic muscular properties such as temporal summation, spatial summation, length-tension relationships, and force-velocity relationships under the control of feedforward-feedback mechanisms of the sensorimotor system. Therefore, joint shock attenuation is a joint-to-muscle interaction. All joints of the axial skeleton play a role in impact energy absorption on loading, but the primary joint is the knee which uses mechanisms that are applicable to all joints.

Joint flexion angles during running have been linked to decreased initial vertical impact forces in forefoot strike runners compared to rearfoot strike runners. In the former, the knee and ankle are in increased flexion angles, improving the ability to use large muscle masses such as gluteus maximus, the quadriceps, and triceps surae to apply energy buffering (Pohl and Buckley, 2008). Variability in kinetic chain motions also seems important in avoiding running-related injuries (Hamill et al., 1999; Heiderscheit et al., 2002). Yet, there is a tendency to increase coupling relationships during running compared to walking, but variably so between individuals (Pohl et al., 2007). However, joint coupling relationships tend to decrease with fatigue (Seneli et al., 2021), which suggests that there exists a 'preferred range of coupling' that is variable to each runner and presents a relationship that is vulnerable to fatigue.

GRFs decrease as knee extensor muscle moments and knee flexion angles increase, while higher GRFs occur when the knee is more extended (Podraza and White, 2010). Knee flexion allows the soft tissues around the knee to provide energy dissipation through mechanisms such as muscle-tendon buffering via changes in muscle stiffness. With knee extension, such soft tissue protection is limited, and this explains features of gait such as increased knee extension at initial contact when walking speeds are slow, and then becoming more flexed at faster walking and running speeds. At slower gait speeds, GRFs are lessened as acceleration is lower, thereby reducing impact energy generation, while soft tissues and the foot are more compliant. For example, gluteus maximus activity begins earlier and hamstring activity decreases in slow walking, allowing the extensors to extend the knee under less restraint (Hanlon and Anderson, 2006). As impact energy increases, the dissipation of energies around a joint becomes more significant to the well-being of the locomotive system. It is worth noting that several pathologies have been linked to an inability to flex the knee appropriately to provide the capacity to absorb impact energy through its soft tissues (primarily the myofascial structures) that changes in joint angles can allow. In chronic low back pain, gait is associated with an increase in knee extension unrelated to gait speed (Müller et al., 2015), while 'noncontact' anterior cruciate ligament injuries are associated with near-full knee extension positions in gait (Podraza and White, 2010).

The swing phase of running may also play a part in determining the initial impact force. Smaller impact forces are reported to be associated with larger downward accelerations of the foot prior to impact, which are found with a higher positioned foot and a decreased downward velocity of the lower leg at midswing (Schmitz et al., 2014). Thus, increased hip flexor activity during swing to alter midswing kinematics will decrease the leg's impact velocity, reducing the downward momentum of the leg to ultimately produce a smaller initial loading impact force (Schmitz et al., 2014).

Impact dissipation by specialist passive soft tissues

Muscle activity through changing axial skeletal stiffness and muscle-tendon buffering properties has the greatest effect on energy dissipation. Yet, passive tissues play a role that is proportionally more significant during lower impact stresses of walking, than during the higher ones of running. Passive impact energy dissipation begins with spreading force over larger areas (reducing peak pressures) of the plantar surface of the foot. The cutaneous tissues and plantar fat pads play the same dissipation role during walking as in running, although the areas of the plantar foot initially contacted and fully loaded can vary between individuals and alter via running speeds. The viscoelastic properties of the plantar soft tissues mean that their stiffness or compliance is affected by the loading rate, so the plantar tissues behave more stiffly the faster they are loaded. On the heel, the skin and plantar fat pad make up what is referred to as the *corpus adiposum* and together, these structures work as a shock absorber to dampen the GRFs of heel strike.

The skin's high elasticity helps prevent shear tearing, while the heel fat pad absorbs the bulk of the vertical impact shock. Yet, the skin's stiffness could also have direct effects on heel pressures. Modelling suggests that increased skin stiffness causes a slight rise in skin stress, while softening decreases peak plantar pressures (Gu et al., 2010). The heel pad thickness is reported as being between 12.5 and 24.5mm in several studies using different imaging techniques (Naemi et al., 2016a). The heel pad adipocytes (fat cells) exist within a dense network of fibrous collagenous septa enclosing the fat cells in a honeycomb-like arrangement with the septa preventing free movement of these cells, but allowing lateral spread (Naemi et al., 2016a; Lin et al., 2017). Its chambered nature gives the fat pad a solid connective tissue phase and a fat-fluid phase, allowing it to behave as a visco-poroelastic structure. The presence of vascular pressures from the arterial and venous vessels in the heel also seems to have just a little influence on the heel pad's mechanical properties (Aerts et al., 1995; Gefen et al., 2001; Weijers et al., 2005). Thus, poor peripheral circulation could compromise heel fat pad shock attenuation directly through altered biomechanics, as well as indirectly through ischaemic tissue effects that can atrophy skin and increase skin fragility.

The heel fat pad consists of two layers: a superficial microchamber layer and a deep macrochamber layer. The thinner superficial microchambers protect the fat pad from excess bulging during loading, thereby giving shape to the heel pad (Hsu et al., 2007). The microchamber layer contains mainly elastic fibres within its walls, giving the microchamber different mechanical properties from the macrochamber (Hsu et al., 2007, 2009; Lin et al., 2017). The macrochamber layer is composed of sparser fibre-adipose structures of roughly equal amounts of elastic and collagen fibres within the septa (connective tissue walls) and is a much thicker layer lying beneath the microchamber layer (Naemi et al., 2016b). Under compression, the heel pad initially expands easily because of the initial low stiffness of the solid phase. However, as the tension within the collagen increases with loading of the fat-fluid phase, the heel pad's stiffness increases in the direction of the load. This produces the classic nonlinear force-deformation of the stiffening behaviour of the heel pad under load (Natali et al., 2012). The stiffness of the microchamber layer is greater by a factor of ten compared to the macrochamber layer, but it also receives less strain compared to the macrochamber (Natali et al., 2012). Furthermore, the macrochamber appears to be responsible for larger deformations and cushioning at impact, while the microchamber layer is responsible for restricting tissue splay, thereby protecting the macrochambers beneath (Natali et al., 2012) (Fig. 1.4.6).

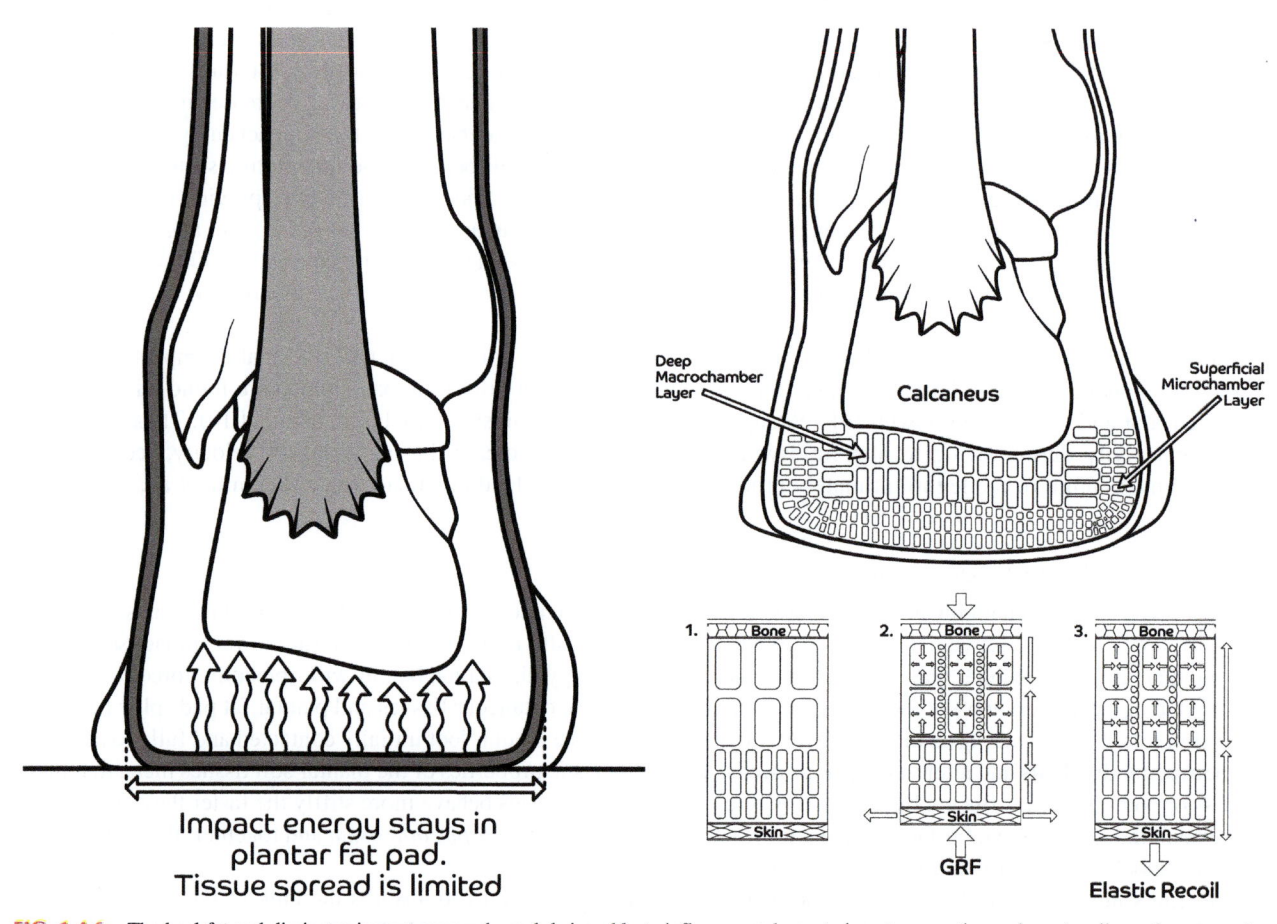

FIG. 1.4.6 The heel fat pad dissipates impact energy through being able to influence and constrain cutaneous tissue shear, tensile, and compression forces. Through its deformation, heel surface area increases. This means that weightbearing pressure is spread over a larger area, decreasing peak pressures on the rearfoot's anatomy. By having a heel fat pad constructed of superficial, more elastic microchambers over less stiff macrochambers, shear stress can be absorbed more through elastic strain in the skin and microchambers, while compressive stresses can be dissipated away from the rearfoot's osseous and articular structures by the macrochambers. On offloading, connective tissues can recoil back into shape, utilising and dissipating the energy absorbed under load to reshape themselves for the impact and loading of the next step. *(Permission www.healthystep.co.uk.)*

On impact and loading, the heel pad acts as a shock absorber and resists compressive loading (Naemi et al., 2016a,b). It also lowers pressure as the soft tissues continue to deform, helping to increase surface area and lower peak pressures until heel lift. As the fat pad is a semi-liquid structure, it demonstrates the hydrostatic properties of fluid (Naemi et al., 2016a,b). Moreover, decreases in water/fluid content may be responsible for the finding that peripheral arterial disease influences the frequency responses of soft tissues to GRFs (McGrath et al., 2012). The mechanical properties of the fat pad are heterogeneous in nature due to the presence of two distinct chambered structures within it. Local heel pad stiffness is at its highest nearest the skin and decreases with depth towards the calcaneus (Lin et al., 2017). Healthy heel pad in vitro quasistatic tests have demonstrated hyperelastic and strain rate behaviours, exhibiting stiffer responses at higher strains and higher strain rates (Ledoux and Blevins, 2007; Grigoriadis et al., 2017). Grigoriadis et al. (2017) reported that on compression, up to 50% strain occurred under forces from 369 to 616 N, with a mean maximum compression force prior to failure at around 6.52 kN. Furthermore, a decrease in the rate of change in strain was noted at higher stresses suggesting that at higher strain rates, a limit in strain is achieved. The Grigoriadis et al. (2017) study also suggested stiffer mechanical behaviours than had been previously reported within in vitro studies. High individual variability exists across specimens, but one of the key features of the healthy heel fat pad is that it unloads like a spring and returns the stored energy from impact to reinstate shape ready for the next impact (Grigoriadis et al., 2017).

Even though the forefoot plantar tissues are not usually first to impact during walking, they are nevertheless the recipient of a significant impact following heel strike at forefoot loading. However, there are times and activities when the forefoot tissues receive the primary impact, such as in forefoot running techniques. This is, of course, usually associated with faster running speeds. From low to high impacts on the forefoot, the elastic modulus in young adults significantly increases from around 300 kPa to about 500 kPa and the energy dissipation ratio also increases from 30% to 60%, from low to high impact velocities (Hsu et al., 2005).

Impact and pathology

There are several mechanisms by which the human's ability to shock attenuate can be disturbed. It has been proposed that repetitive impact forces during activity are not in themselves important as a mechanism of injury, yet they may result in changes in muscle activity necessary to protect the soft tissues from vibration and oscillation that then leads to injury (Nigg and Wakeling, 2001). However, there is no direct association between impact peaks and sports injuries within healthy populations (Gruber et al., 1998; Nigg and Wakeling, 2001). Indeed, rather than peak impact forces being a problem, overworking muscles to prevent tissue vibration through the muscles adjusting body compliance and damping coefficients might lead to fatigue, which in turn may cause issues. The result of fatigued muscles could lead to injury within the musculotendinous unit, or injury could result through the fatigued muscles' inability to protect other soft tissues from excessive oscillation. However, it is also possible that the kinematic mechanisms used to avoid excess vibration could result in excessive soft tissue stiffness that then prevents soft tissue oscillation dissipating energy, leading to soft tissue and bone injuries.

The evidence for an injury link between vertical GRF and impact shock has been highlighted in some studies, but not in others, and remains unclear (Gruber et al., 2014). The impact shock frequency content between forefoot and rearfoot strikers are not identical, suggesting the primary mechanisms for attenuation are anatomically different between the initial contact techniques. It is more likely that the type of injury sustained may differ with different footfall patterns (Gruber et al., 2014). Rearfoot running results in greater peak tibial accelerations and impacts of higher frequency (9–20 Hz) than does forefoot running, with greater ankle compliance in forefoot running contributing to lower rates of tibial acceleration after ground contact (Gruber et al., 2014). This may delay the time to vertical GRF impact peak. This delay in timing probably explains why an initial impact transient peak is not visible on a force plate during forefoot running (Gruber et al., 2014). Head accelerations are not affected by the foot strike position, as the body adapts to maintain head stability. However, with forefoot running, the head and whole body sustain a greater vertical oscillation frequency, probably because of the shortened contact time available to reverse the CoM's downward velocity after impact. This may contribute to greater vertical GRF peaks than rearfoot contact running that can divide braking into two events (Ruina et al., 2005; Gruber et al., 2014).

1.4.7 Muscle activity in running

Muscles in running

Many muscles display similar EMG profiles, but with higher amplitudes in running compared to those during walking. The most noticeable exceptions are those of the calf muscle group, which demonstrate a much greater early stance phase activity rather than just higher activity levels during running than to those observed during late stance of walking (Gazendam and Hof, 2007; Ishikawa et al., 2007). In running, there is early medial gastrocnemius muscle fascicle stretching at 0%–10% of stance during initial contact followed by muscle fascicle shortening within early stance, while the Achilles tendon fibres lengthen (Fig. 1.4.7a). The Achilles fibres only shorten later during acceleration (Ishikawa et al., 2007). This is quite unlike the combined muscle and tendon lengthening seen during midstance and the combined shortening of acceleration seen in walking (Ishikawa et al., 2007) and moves the burden from the ankle plantarflexors more to the Achilles during running, as a result.

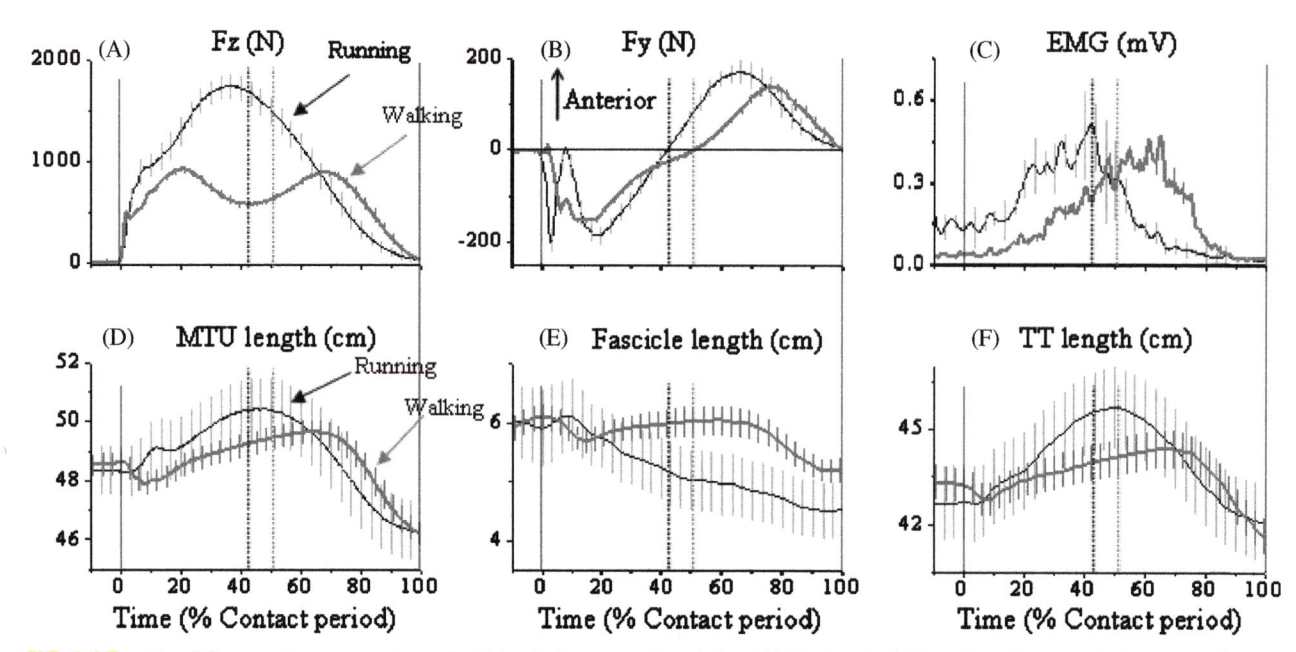

FIG. 1.4.7a The differences between running and walking during stance (from 0% to 100%) of vertical (A) and anterior–posterior horizontal (B) GRF related to EMG activities of medial gastrocnemius (C), the length of the muscle tendon unit (D), muscle fascicle length (E), and tendon length (F). *(From Ishikawa, M., Pakaslahti, J., Komi, P.V., 2007. Medial gastrocnemius muscle behaviour during human running and walking. Gait Posture 25 (3), 380–384.)*

When running, muscles work to develop forces that propel the body forward, while also supporting body weight within the stance limb. Except for the hip muscles, the muscles of the lower limb during running are used in GRF generation, with relatively little energy used to counteract limb segment inertia and gravity (Biewener et al., 2004). In walking, the hip and knee muscle moments are greater than the ankle moments during the first half of stance, and ankle moments are greatest in the latter half when the knee and hip muscle moments are at their least. During running, the knee moments are largest throughout gait with smaller hip and ankle moments throughout stance maintaining the impact burden on the knee but also shifting the mechanical acceleration burden from the ankle to the knee (Biewener et al., 2004). Peak GRF moments at the knee are about five-fold higher in running than during walking and the hip GRF moment is about two-fold higher. However, the ankle experiences little change in GRF moments from heel strike walking (Biewener et al., 2004). It can be expected to be somewhat different for running forefoot strikers, who increase braking forces on the ankle muscles.

The running knee's heightened moments come from increased knee flexion in the stance phase of running compared to walking, which increases the GRF moment arm on the knee (Biewener et al., 2004). The knee is usually flexed only slightly during the stance phase of walking gait and spends most of its time extending out of initial contact's flexion peak until absolute midstance. For the rest of midstance, it is subjected to passive extension resisted by gastrocnemius-induced flexion moments until terminal stance. The hip operates through a similar range of motion in both walking and running, whereas the ankle is more flexed (less dorsiflexed during stance) during running than it is when walking (Biewener et al., 2004). Changing muscle activity influences the kinetics through altering the loading rate and impulse peaks. When conducting instrumented gait analysis on 48 subjects running at 3.3 m/s on a treadmill, Schmitz et al. (2014) found that a smaller impact peak on the stance limb was associated with a larger downward acceleration of the foot, a higher positioned foot, and a decreased distal leg downward velocity at midswing. Moreover, a lower loading rate was associated with a higher positioned thigh at midswing (Schmitz et al., 2014). Therefore, the kinematic changes seen as a result of muscle activity relate back to the stresses the body undergoes during running, even when the changes are associated with the swing limb.

In running, the increase in the knee GRF moment reported in young adult males is associated with a big decrease (by around 68%) in the knee muscles' mechanical advantage. This is due to the more flexed knee position (Biewener et al., 2004). In contrast, the hip and ankle during running have been reported to increase their muscular mechanical advantage by 18% and 23%, respectively (Biewener et al., 2004). Partly due to the changes in muscular mechanical advantages, the muscle impulses during stance limb support when normalised to the GRF impulse, are found to be increasing at the knee, decreasing at the hip, and remaining unchanged at the ankle, compared to walking. In contrast, 9% of the total extensor muscle impulse is generated at the knee during walking and 39% of the total extensor impulse is generated at the knee when running, nearly matching that of the calf muscles (Biewener et al., 2004) (Fig. 1.4.7b). Human running therefore increases stress and injury potential within the knee.

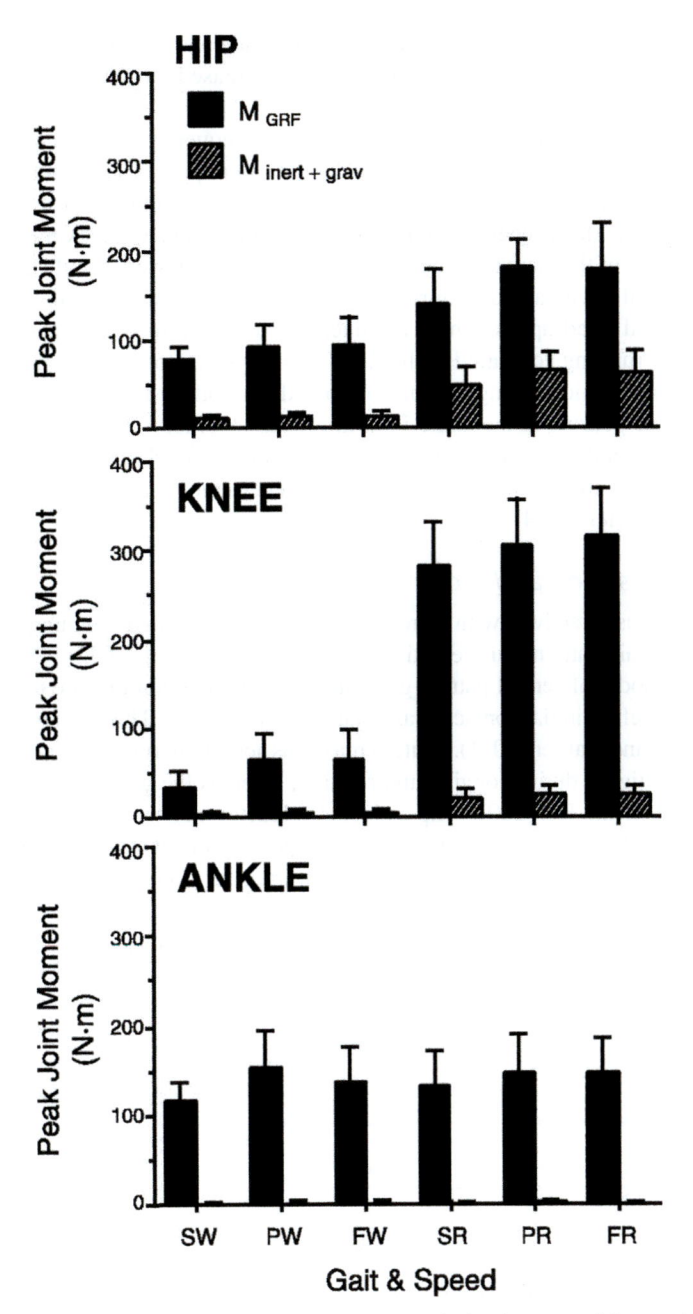

FIG. 1.4.7b Histograms of peak joint moments of GRF (M_{GRF}) and inertial and gravitational moments ($M_{inert+grav}$) versus relative speed and gait style using four subjects to pool information. Locomotion involves slow walking (SW), preferred walking speed (PW), fast walking (FW), slow running (SR), preferred running speed (PR), and fast running (FR). *(From Biewener, A.A., Farley, C.T., Roberts, T.J., Temaner, M., 2004. Muscle mechanical advantage of human walking and running: implications for energy cost. J. Appl. Physiol. 97 (6), 2266–2274.)*

Mean fascicle lengths among hip, knee, and ankle muscles are different, being longest in the hip extensors and shortest within the ankle flexors (Biewener et al., 2004). Changes in muscle force-generation requirements result in a significant shift in active muscle volumes of actively recruited fibres when changing from walking to running, with an average change in volume of the lower limb estimated at being ~2.5-fold (Biewener et al., 2004). In walking, the hip generates most volume change. However, during running, the knee generates around 49% of the volume change (Biewener et al., 2004). A number of studies have reported different estimates on whether or not the ankle or hip produces the next most active volume of muscle change due to variance in the techniques used for these estimations (Griffin et al., 2003; Biewener et al., 2004). In all three main lower limb joints, the volume of active muscle needed to generate inertia and gravitational-resisting moments increases with increasing gait speeds, being greatest at the hip and least at the ankle. In running, the increased muscle action due to inertia and gravity was five-fold higher at the hip and knee than during walking (Biewener et al., 2004). Compared to the GRF moments alone, inertia and gravity moments increased muscle activity by an additional 6% during walking, but by 13% when running (Biewener et al., 2004).

Energetics of running is also likely affected by the increasing stride frequency requirement of faster contracting muscle fibres to generate forces faster and to be able to shorten their length more quickly at higher velocities (Kram and Taylor, 1990). More muscle work is done to move the body per unit of distance during running than when walking (Heglund et al., 1982; Griffin et al., 2003). Although elastic energy storage provides some of this work (Ker et al., 1987), it appears that the quadriceps contributes a major part of the increased energy costs of running (Biewener et al., 2004). Because of the decrease in the knee's muscular mechanical advantage during running, the average human muscular mechanical advantage is less than the hindlimb of similar-sized quadrupeds, which are unchanged at different gait speeds. This explains the greater energy costs observed in human running compared with that of quadrupedal trotting and galloping. Humans demonstrate a 38% greater cost in energy consumption during running compared with other size-matched mammals. The erect bipedal posture and the change in muscle gearing at the ankle joint improves locomotor economy during walking (Rodman and McHenry, 1980) but also increases endurance ability in running (Carrier, 1984). However, bipedalism incurs greater energy transport costs during running than walking, suggesting that running energetic costs were not a selective factor in the human evolution of erect bipedalism (Carrier, 1984).

Muscle transition with gait speed and technique

There is not a gradual change in muscle activity with increasing speed from walking to running, but a distinct difference from walking gait pattern to running gait pattern muscle activity (Gazendam and Hof, 2007). Changes in muscle action from walking to running support the model of central pattern generators within the brain that are programming muscle processes for preactivation to the gait style before initial contact (Gazendam and Hof, 2007). This is learnt automatic behaviour that can be consciously overridden (Voigt and Hansen, 2021). Gluteus maximus activity in running is similar to that of walking, but the EMG amplitude is considerably higher during jogging and running, whereas the gluteus medius EMG amplitude increases linearly with gait speed (Gazendam and Hof, 2007). Hip adductor (adductor magnus particularly) activity has peaks at initial contact and toe-off during walking. However, when running, activity peaks occur at midstance, midswing, and terminal swing, although in slower running and jogging, these peaks are lower and irregular (Gazendam and Hof, 2007) (Fig. 1.4.7c).

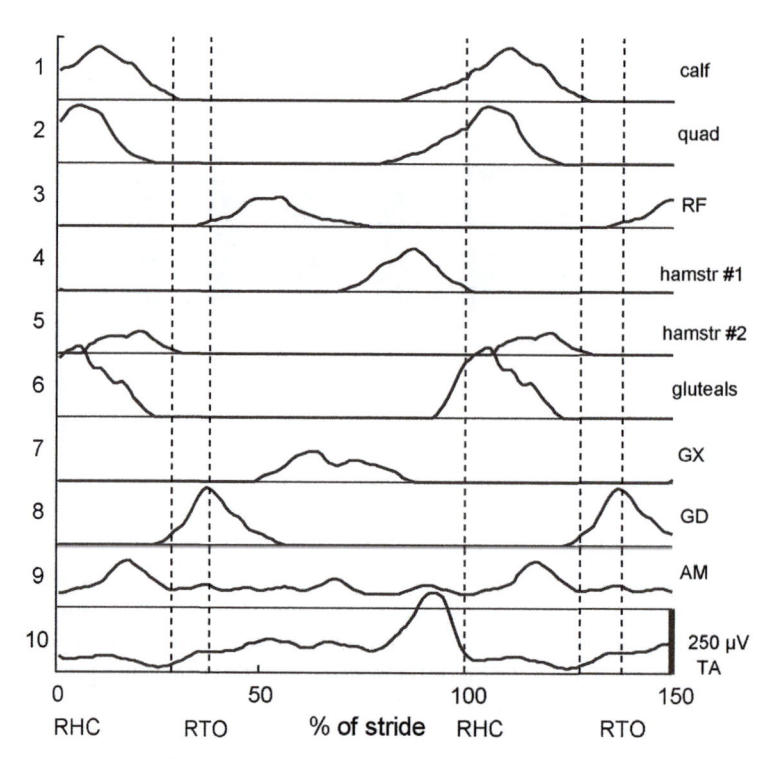

FIG. 1.4.7c The basic patterns of running EMG. The vertical dashed lines show running heel contact (RHC – 0% and 100%) and the range of running toe-off (RTO), which is between 37% at 2.25 m/s and 28% at 4.5 m/s. Functional groups are as follows: calf which includes triceps surae and peroneus longus; quadriceps (quad) including vastus medialis, lateralis, and rectus femoris (RF, no doubt also shows activity from vastus intermedius); hamstr #1 representing biceps femoris; hamstr #2 which includes semimembranosus and semitendinosus; gluteals of gluteus maximus (GX) and gluteus medius (GD); adductor magnus (AM); and tibialis anterior (TA). Patterns are significantly different to walking and jogging. Notice the gluteal group has a first peak of activity together during loading and a second peak of activity of GX in midswing and GD at the transition from stance to swing. *(From Gazendam, M.G.J., Hof, A.L., 2007. Averaged EMG profiles in jogging and running at different speeds. Gait Posture 25(4), 604–614.)*

The quadriceps activity in running has vastus medialis EMG amplitude not changing with speed, vastus lateralis showing a minor increase, but rectus femoris/vastus intermedius demonstrating a substantial ~two-fold increase (Gazendam and Hof, 2007). The minor peak of quadriceps activity associated with knee extension during early midstance seen during walking is missing in running, and the profile when jogging is different to both running and walking (Gazendam and Hof, 2007). In jogging, the period of quadriceps activity during loading is longer, but peak activity is lower than that of running (Gazendam and Hof, 2007). The hamstrings have two peaks of activity similar to walking, at initial contact and terminal swing, with jogging being identical to walking but with higher activity peaks (Gazendam and Hof, 2007).

In running, the calf muscles demonstrate an initial peak before the start of stance, aligned to that of the quadriceps, but occurring a little later. Peroneal and soleal activity is constant throughout stance. The two gastrocnemius muscles increase amplitude significantly with speed (Gazendam and Hof, 2007). Jogging has a calf muscle activity profile between that of walking and running, but like running, it is continuous during stance (Gazendam and Hof, 2007). Tibialis anterior activates during preswing and continues throughout swing to end its activity after heel contact with its duration of activity shorter than that seen during walking. However, its peak before heel contact is earlier in running compared to walking (Gazendam and Hof, 2007). Jogging lacks the prominent peak of tibialis anterior activity seen in running, but jogging speed initiates earlier activity (Gazendam and Hof, 2007). Forefoot running dramatically reduces tibialis anterior activity during contact (Rooney and Derrick, 2013).

Muscle coactivation

Muscular coactivation is the simultaneous contraction of muscles paired around a joint that helps stabilise articular motion during locomotion, controlling stiffness levels within the leg that influence the spring-mass system (Hortobágyi and DeVita, 2000). Muscle coactivation is probably linked to running economy, as longer coactivation of the lower limb extensor muscles (and ankle flexors) has been associated with greater oxygen consumption (Heise et al., 2008; Moore et al., 2014). As speed increases, coactivation of the distal leg flexors becomes shorter in duration (Moore et al., 2014). Thigh coactivation is thought to predominantly act during loading phase helping influence knee flexion angles, for without knee coactivation of the quadriceps and the hamstrings, the leg would collapse. The distal leg coactivations help stabilise the lower limb and ankle during loading, including the anterior sagittal rotation of the tibia (Mann et al., 1986). It is argued that extensor-flexor coactivation of rectus femoris/vastus intermedius and the lateral gastrocnemius stabilises multiple joints and transfers mechanical energy from proximal to distal joints (Heise et al., 2008). Conversely, flexor-extensor activation of biceps femoris and tibialis anterior increases lower limb impact energy absorption strategies and contributes to overall leg stiffness (Moore et al., 2014).

As running speed increases, coactivation in the distal leg decreases with a reduction in coactivation of leg flexor muscles (and ankle dorsiflexors), but proximal muscle and extensor muscle (including the ankle plantarflexors) coactivation increases, ramping up oxygen consumption (Moore et al., 2014). The relative coactivation of the extensors and the proximal muscles remains unchanged across running speeds (Harris et al., 2003; Moore et al., 2014). Activation of tibialis anterior decreases with increasing running speed (Moore et al., 2014), probably because increasing the running speed is also associated with a reduction of the ankle dorsiflexion angle at initial contact (touchdown angle) and finally replaced by forefoot strike running. This negates the need to eccentrically slow the forefoot's descent to the ground (Novacheck, 1998). In forefoot running, tibialis anterior demonstrates far less activity (Rooney and Derrick, 2013). Gastrocnemius remains active over similar periods, regardless of the speed being utilised for energy transfer. Rectus femoris (vastus intermedius) does not proportionally decrease activity, with decrease in stance time being associated with running speed (Moore et al., 2014). Medial gastrocnemius muscle fascicles are reported to lengthen rapidly at initial contact for the first 10% of running gait, but thereafter shorten throughout the rest of stance. However, the Achilles tendon tissues lengthen throughout running stance until braking is complete and acceleration begins, when they shorten (Ishikawa et al., 2007). This shortening of the muscle fascicles while the tendon stretches and then recoils, shifts much of the running braking and acceleration burden to the elastic properties of the Achilles.

By providing joint stability and possibly minimising injury through muscle coactivation, running muscle activity sustains a significant metabolic cost, as the force generated by an agonist will be greater if the antagonist activates at the same time. This is true of walking as well as running (Hortobágyi et al., 2011; Moore et al., 2014). However, the stability necessary through increased coactivation in running may be related to underlying balance control, as improving an individual's dynamic postural control has been shown to reduce the amount of coactivation during balance tasks (Nagai et al., 2012).

To establish the likely activity of muscles during standard heel-toe running techniques, Hamner et al. (2010) developed a model that included 92 muscle-tendon actuators representing 76 muscles of the lower extremities, torso, and arms. They applied this to the running data from 3D motion analysis of a healthy male subject of 183 cm height and 65.9 kg weight. The following data on muscle activation patterns at a self-selected speed of running at 3.96 m/s was generated, giving some indication of muscular activity and functional role during distance running technique (Figs 1.4.7d–1.4.7f).

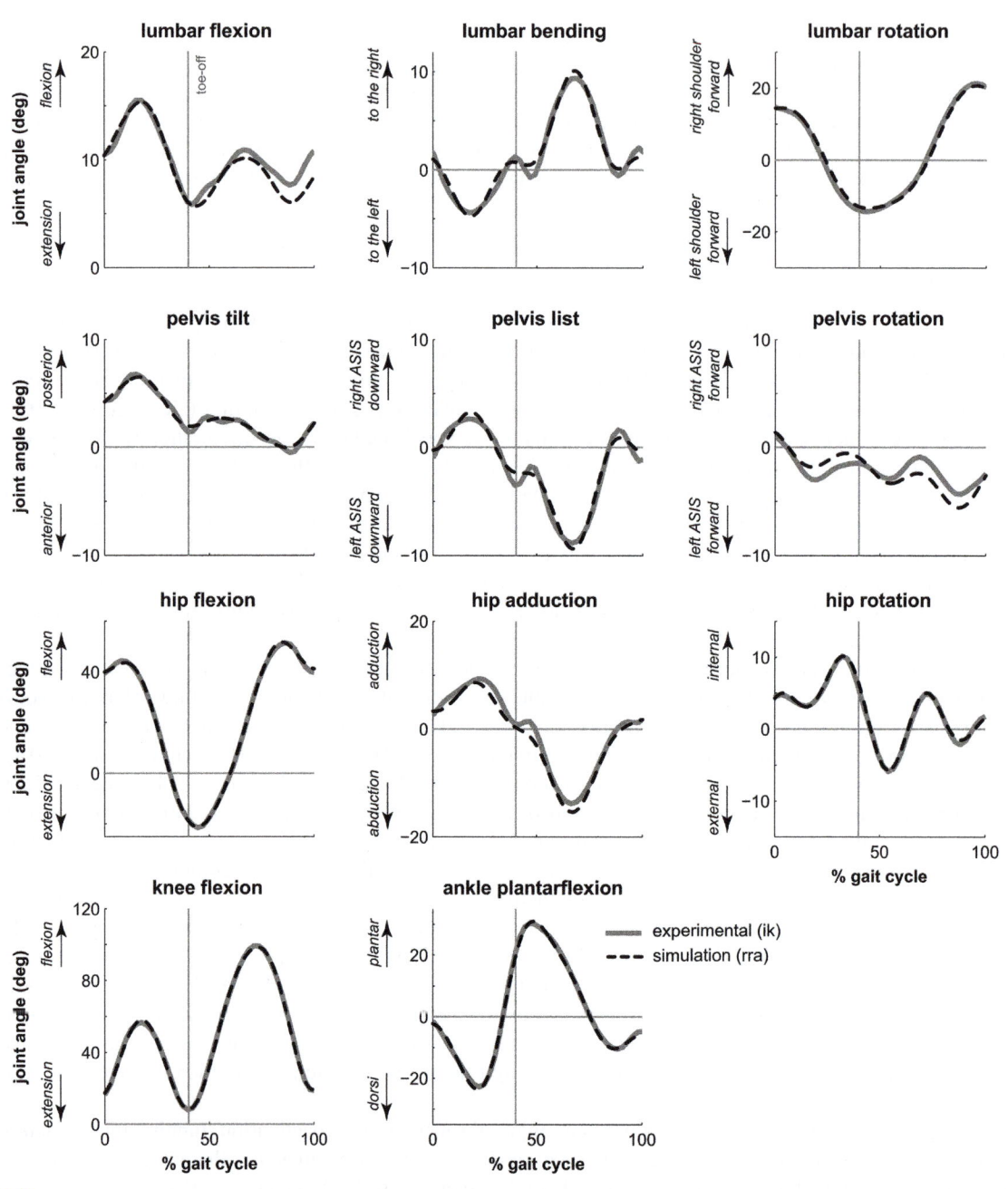

FIG. 1.4.7d Kinematics of the back, pelvis, and lower extremity during running. The *grey line* represents experimental joint angles calculated by inverse kinematics and the *dashed line* represents simulated joint angles produced through computed muscle control. Toe-off is indicated by the vertical line at 40% of the gait cycle. *(From Hamner, S.R., Seth, A., Delp, S.L., 2010. Muscle contributions to propulsion and support during running. J. Biomech. 43 (14), 2709–2716.)*

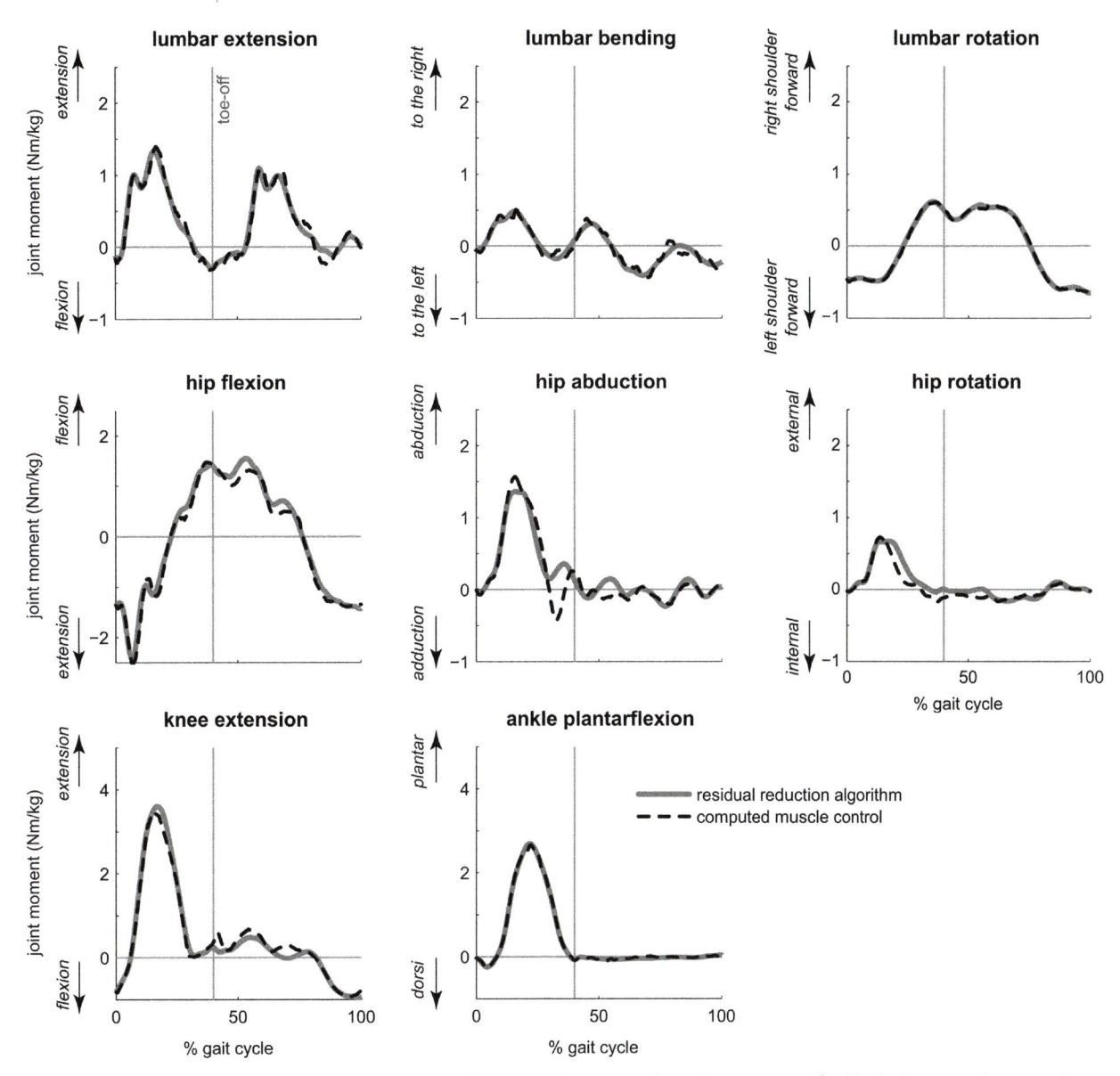

FIG. 1.4.7e Moments about the lumbar spine and lower limbs compared during running. These are normalised by body mass and computed using a residual reduction algorithm *(solid line)* and summing the moments generated by muscle forces *(dashed line)*. Toe-off is shown at 40% of the cycle. *(From Hamner, S.R., Seth, A., Delp, S.L., 2010. Muscle contributions to propulsion and support during running. J. Biomech. 43 (14), 2709–2716.)*

Tendons in running energetics

Metabolic energy consumption increases with running speed. It is at its lowest when tendons reach 2%–3% strain and when muscles are developing maximum isometric force (Uchida et al., 2016). However, which muscles require greater levels of stiffness or compliance is variable. For example, when running at 2 m/s, the soleus consumes less energy with high tendon compliance. This is because the strain it permits allows the muscles to operate near-isometrically during the stance phase, while allowing elastic 'give' in the compliant tendon to allow tensioned motion around the ankle (Uchida et al., 2016). The gastrocnemii consume less energy with higher stiffness because restraining the gastrocnemii muscle fibres allows them to operate closer to their optimal lengths for generating power during stance (Uchida et al., 2016). Thus, restrained motion around the ankle is achieved through the soleus part of the Achilles tendon. However, power is generated from the medial and lateral gastrocnemius muscles.

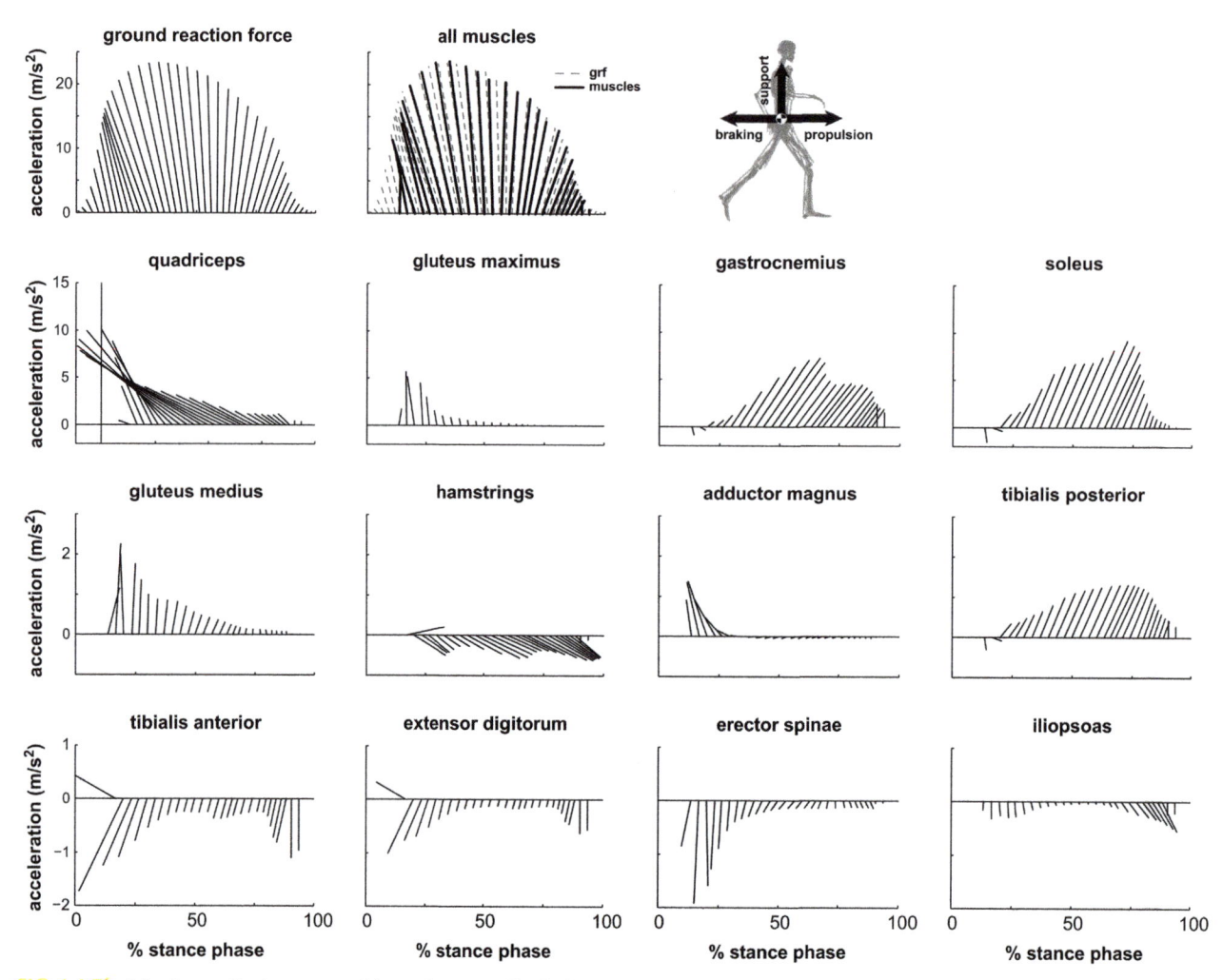

FIG. 1.4.7f Muscle contributions to propulsion and support of body CoM during the stance phase of running defined to 40% of the gait cycle. Each ray is the resultant vector of the vertical and horizontal anterior–posterior accelerations. *(From Hamner, S.R., Seth, A., Delp, S.L., 2010. Muscle contributions to propulsion and support during running. J. Biomech. 43 (14), 2709–2716.)*

1.4.8 Spine, pelvis, and arm motion in running

Movements of the pelvis, lumbar spine, and thoracic spine influence the CoM's position during gait. Arm motion during running functions to counterbalance the rotational angular momentum of the swing legs (Hamner et al., 2010; Arellano and Kram, 2014), which means that the coordination patterns between the spine and the pelvis facilitates an angular momentum balance (Pontzer et al., 2009b). Anterior–posterior accelerations of the CoM are minimised, leading to an *antiphase* (opposite motions) co-ordinated pattern in the sagittal plane (Preece et al., 2016). Posterior pelvic tilt occurs during early stance accompanied by a flexed thorax. However, this is reversed in late stance as the pelvis moves into anterior tilt with extension of the thorax (Preece et al., 2016).

This anterior pelvic tilt in late stance facilitates femoral inclination at toe-off, thereby extending stride length and shifting the CoM more anteriorly, creating a more anteriorly directed GRF to facilitate forward momentum (Novacheck, 1998; Preece et al., 2016). The lumbar extensors are active at initial foot contact and early stance in order to limit forward flexion of the trunk as energy is absorbed within the lower extremity while the CoM decelerates in the first half of stance. During this deceleration, the thorax moves into forward flexion with a small corresponding posterior pelvic tilt as the gluteus maximus acts to initiate hip extension from its flexed loading position. As the lower limb accelerates the CoM through a lengthening stride in the second half of stance, muscle control decelerates thorax extension, most likely via the oblique abdominal muscles which are active later in stance (Saunders et al., 2005) (Fig. 1.4.8a).

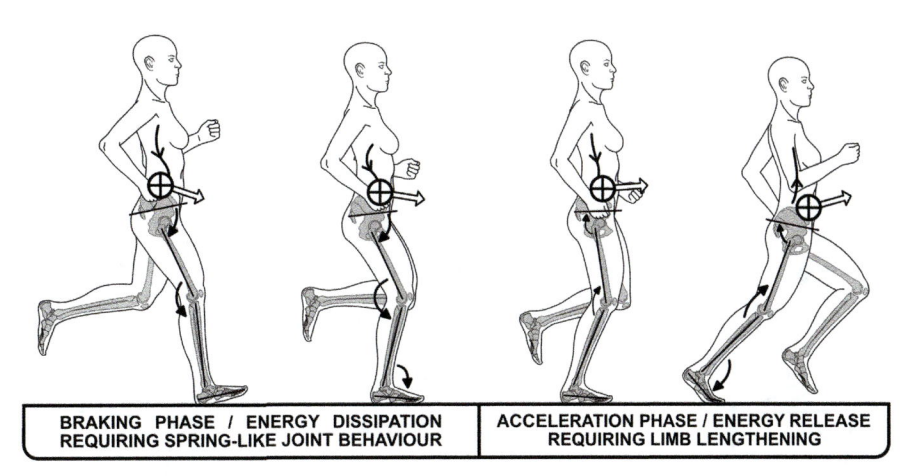

FIG. 1.4.8a During running stance phase (here shown with heel strike), two distinct tasks are performed requiring very different joint postures and muscular activities to decelerate the CoM as it falls and then to send it back up towards the aerial phase during acceleration. The posture of increased lumbar flexion, pelvic posterior rotation, thigh-hip flexion, and lower limb knee flexion permits energy dissipation via flexion motions, absorbing the energy associated with the fall of the CoM. This includes ankle dorsiflexion, a motion also necessary to change the 'push-off' angle. Extension motions that lengthen the limb permit energy recoil and that includes ankle plantarflexion, which is a flexion motion. Because so much negative work is done in braking, much more extensive muscle force generation is required for running acceleration compared to that of walking. *(Permission www.healthystep.co.uk.)*

Frontal plane CoM velocity changes are small (Preece et al., 2016). In early stance, the pelvis is laterally tilted (away) from the stance limb producing a hip varus angle, resulting in a position of the CoM more directly over the foot. This creates a moment around the base of support facilitating transition towards the contralateral foot (Preece et al., 2016). From midstance onwards, the pelvis lifts on the contralateral side during swing until it reaches its maximum position of height at ipsilateral toe-off (Preece et al., 2016). The thorax motion is precisely co-ordinated with this pelvic kinematic pattern in order to minimise the medial-lateral acceleration of the CoM, requiring a smaller range of motion from the thorax (Preece et al., 2016). Gluteus medius is active prior to foot contact and throughout most of the stance phase of running (Gazendam and Hof, 2007; Willson et al., 2012). Gluteus medius is initially likely to be controlling the downward acceleration of the CoM following foot contact and then being used to lift the pelvis on the contralateral side, resisting the varus pelvic tilt. This is consistent with the frontal plane motion occurring at the lumbo-pelvic junction (Preece et al., 2016). In the later stages of stance, the lumbar spine laterally flexes towards the contralateral limb relative to the pelvis, possibly assisted by the contralateral oblique abdominal muscles, which are active at this time (Saunders et al., 2005) (Fig. 1.4.8b).

Transverse plane rotations consist of the pelvis rotating slightly towards the stance limb in early stance, after which it rotates away from the stance limb by around 10° (Schache et al., 2002; MacWilliams et al., 2014; Preece et al., 2016) which has been suggested to reduce horizontal plane braking of the CoM (Novacheck, 1998; Schache et al., 2002). The rotation away from the stance limb may decrease stride length, although the effect is minimal (1%–2%) and probably has little metabolic influence (Preece et al., 2016). The transverse plane rotations of the pelvis may be a secondary consequence of gluteus maximus activity, which is active throughout most of running stance phase (Gazendam and Hof, 2007; Willson et al., 2012). The result is that external hip rotation rotates the pelvis away (posteriorly) from the stance limb, directing the swing limb side of the pelvis to rotate anteriorly (Preece et al., 2016).

Pelvic rotations follow thorax motion, minimising muscle work to passively drive arm swing. From midswing until early stance, the thorax moves from a rotated position to a neutral position relative to the pelvis, probably as a result of stored elastic energy within the connective tissues (Preece et al., 2016). Around midstance, the abdominal muscles become active (Saunders et al., 2005) to limit extension of the trunk in the sagittal plane and actively rotate the thorax relative to the pelvis. This results in greater thorax motion compared to the pelvis (MacWilliams et al., 2014; Preece et al., 2016). Such motion also links to the swinging of the upper limbs during running (Fig. 1.4.8c).

The role of arm swing in running

Humans naturally swing their arms as they run as well as when walking and this has been proposed to reduce energetic costs (Meyns et al., 2013). This linkage between the upper and lower limbs is a neural coupling that influences muscular activity between the limbs, with the upper limbs in walking influencing lower limb muscle activity more than the lower limbs do the upper limbs (Weersink et al., 2021). Shoulder muscles deltoideus anterior and posterior and the ipsilateral thigh muscles

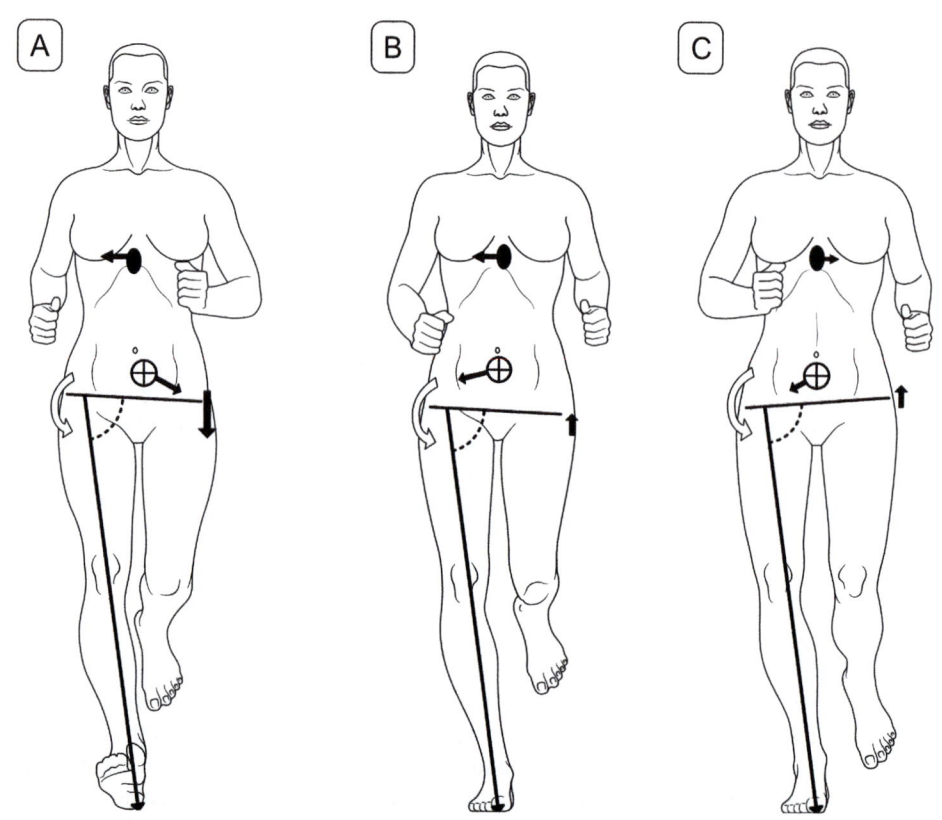

FIG. 1.4.8b Running stance phase frontal plane stability involves correctly positioning the CoM above the support limb. Rotations occur around the lumbosacral junction and hip joints with minor motions of spinal side-flexions. At initial contact (A), the single support limb is brought under the body by strong hip adductor activity. This increases the hip varus (adduction) angle *(dashed line)* to support the CoM, which is initially moving downwards and away from the support limb's hip to lie centrally above the pelvis. This creates a potentially destabilising hip varus moment *(thick black arrow)* that must be resisted by the posterior hip abductors *(white arrow)*, mainly by gluteus medius. Trunk (thorax) side-flexion to the support limb can also aid in managing this varus moment by displacing the upper thorax CoM (small black spot) towards the support limb. During the middle of running stance (B), while still providing some braking, the hip abductors provide forces that fully resist the varus moment, preventing hip drop on the swing side and starting to pull the contralateral side of the pelvis upwards. This creates a small valgus (abduction) moment on the stance hip. By side-flexing the trunk towards the support limb, the effort required within the abductors is reduced. The overall effect is to create more space for the swing limb to pass beneath the body, when the support-limb knee and hip are at their maximal flexion angle. In late stance at the start of acceleration (C), the pelvis may remain in slight valgus under the control of all the hip abductors, but the trunk is allowed to side-flex back towards the swing limb, straightening the trunk. Increasing centrifugal forces from the swing limb will start to reduce the loading forces on the stance limb (seen clearly on a running force-time curve), decreasing the loads that the hip abductors are now resisting. Thus, initial contact and braking hip varus moment control is particularly challenging for human running but are less demanding during acceleration. Less physically prepared runners are vulnerable to inadequate control of hip varus moments. Posturing of the trunk is highly variable, even between steps, reflecting the adaptability of hip varus/valgus moments and degrees of trunk side-flexions in controlling events. *(Permission www.healthystep.co.uk.)*

biceps femoris and rectus femoris seem to show the strongest intermuscular coherence in a diagonal fashion (Weersink et al., 2021). Intermuscular coherence between lower and upper limb muscle activity varies in gait phases and between ipsilateral left and right upper to lower limb relationships, but distal lower limb muscle relationships to the shoulder muscles are less pronounced (Weersink et al., 2021). How much the thoracolumbar and abdominal fascia is playing a role is not yet clear, and how these relationships play out in running is not yet known.

Hamner et al. (2010) suggested the primary function of arm swing is to counterbalance the angular momentum created by the swing leg about the vertical axis to achieve a net vertical angular momentum that fluctuates with a low magnitude, around zero. It has also been speculated that arm swing helps maintain posture and balance and assist GRFs to lift the runner, by aiding the increase of 'bounce-off' from the ground (Hinrichs et al., 1987). However, it appears arm swing contributes less than 1% of the peak CoM acceleration (Hamner et al., 2010). Although there are several techniques to arm swing, the common 'back and forth with slight torso crossover' action is thought to reduce side-to-side motion of the body's CoM during running (Hinrichs et al., 1987). Without arm swing, there appears to be increased amplitude of both shoulder

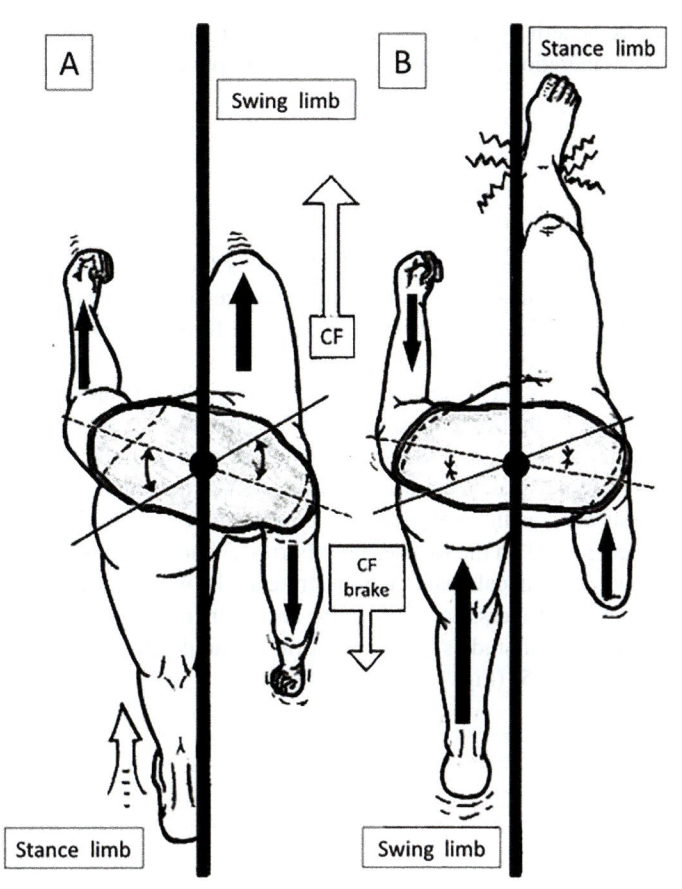

FIG. 1.4.8c In running, because there is no double-limb stance phase, transverse plane pelvic and shoulder rotation patterns are different to those of walking. As the foot completes acceleration phase (A), the swing limb is not ready for weightbearing as the lower leg is only starting to extend from under the thigh. Thus, it is only starting to add significant centrifugal force (CF) to acceleration. At this stage, the swing side pelvis is being rotated forward, while the ipsilateral upper limb and shoulder is in backward swing and rotation. The upper limb back-swing may act as a mild brake on the swing limb's CF. The shoulder girdle is highlighted grey and its bisection (anterior to posterior) as the thin dashed line, while the pelvic girdle below has its bisection marked as a thin, solid black line. Counterrotations increase the angles between these upper and lower segment bisections. The ipsilateral upper limb is rotating anteriorly, setting up a strong diagonal tension across the posterior spinal fascia between it and the swing limb. The stance limb providing plantarflexion power is fixed to the ground but has an anteriorly swinging upper limb and rotating shoulder while the pelvis is, in consequence, rotated posteriorly. This causes tensioning of the abdominal fascia diagonally from the posteriorly positioned stance limb and the anteriorly rotated swing side arm. This contralateral increasing angular rotation around a lumbosacral axis *(black spot)* along the line of progression *(thick black line)* completes its cycle through the aerial phase, ending at heel strike (B). After initial contact of the next step, the rotations start to reverse, with upper and lower limbs reducing their bisectional angles until midstance. After this, left and right sides start to produce opposite rotations, dictated by their stance and swing phases. Evidence is supporting a neural link relationship with shoulder muscle activity and biceps femoris and rectus femoris muscle activity, with more limited associations to distal muscles like soleus. Strengthening of the upper body for powerful shoulder rotations will benefit running training programmes, especially for faster running speeds. *(Permission www.healthystep.co.uk.)*

and pelvic rotation, suggesting that arm swing reduces torso rotation as a counterbalance strategy to oppose the rotational movements from leg swing (Arellano and Kram, 2014).

Metabolic benefits to arm swing have proved hard to establish, with some studies suggesting no benefit (Tseh et al., 2008; Pontzer et al., 2009b) and others between 4% and 8% benefit in running economy (Egbuonu et al., 1990; Arellano and Kram, 2011, 2012). A study by Arellano and Kram (2014) sought to resolve the question by looking at different positions in which the arms are restricted in order to prevent them swinging, while recording oxygen consumption and carbon dioxide production. Compared to natural arm swing, it was found that holding the arms together behind the back, across the chest, and behind the head, all increased metabolic power demands. Curiously, it was also noted that there was a shift to more carbohydrate and less fat utilisation when running without arm swing, which might reflect the fact that oxidising carbohydrates rather than fats yields more energy, reflecting a need for greater energy without arm swing (Arellano and Kram, 2014). However, there is some variability as two runners in Arellano and Kram's (2014) study used less energy with their arms held behind their backs. It appears that the metabolic cost of holding the elbows flexed, the most common running posture for the arms, is offset by the benefit from reduced metabolic control of torso rotation. Arm swing in sprinting might serve a different purpose, where acceleration and maximising power output is more important than maximising metabolic energy costs.

1.4.9 Running patterns

Foot strike position

Foot strike position can occur in three distinct locations: on the heel (rearfoot strike), with the foot-flat (midfoot strike), or on the forefoot (forefoot strike). Foot-striking has been shown to occur over a spectrum, making strict classification difficult (Altman and Davis, 2012; Forrester and Townend, 2015). An extreme form of forefoot striker has also been identified, being referred to collectively as '*toe runners*' (Nunns et al., 2013).

The definition of each strike position is classified by the strike index, which is the CoGRF location at touchdown on a force plate measured as a percentage along the long axis of the foot from heel to toe (Cavanagh and Lafortune, 1980). Locations from 0% to 33% indicate rearfoot strike, 34%–67% midfoot strike, and 68%–100% forefoot strike. However, not every patient can have access to a force plate examination. Thus, foot-strike angle in relation to the ground, measured in the sagittal plane at touchdown, has become the preferred metric as it can be captured on 3D motion or high-speed frame-capture video using surface markers placed on the foot or shoe (Forrester and Townend, 2015). The angle in stance is used as a subtraction from the touchdown angle to calculate the foot strike angle, with 0° being a foot-flat touchdown and angles close to this being considered as midfoot strikers. Larger negative angles are regarded as forefoot strikers and larger positive angles as rearfoot strikers, but a number of studies have used different angles to classify these foot strikers (Altman and Davis, 2012; Forrester and Townend, 2015) (Fig. 1.4.9a).

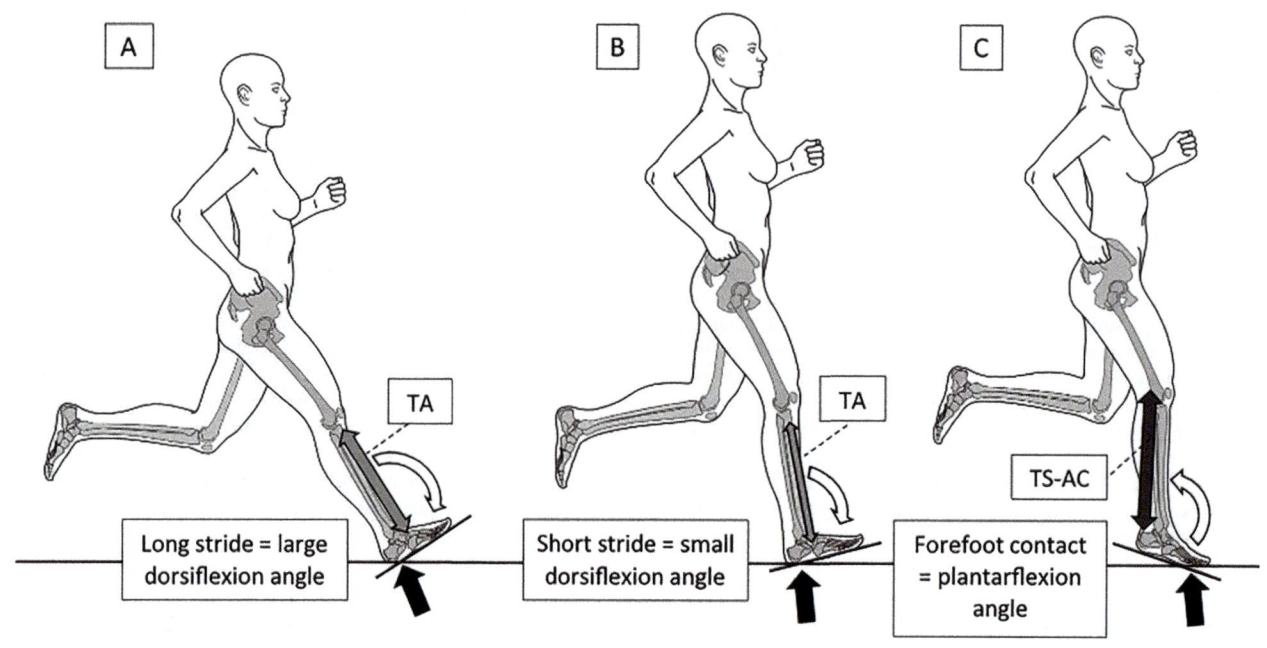

FIG. 1.4.9a Foot strike angles usually relate to three primary factors: stride length and strike position (which positions the initial GRF), which in turn usually couples to running speed. A long stride length with heel strike (A) tends to result in a high dorsiflexion angle of the plantar foot (or sole of the shoe) with the support surface. This sets up a high plantarflexion moment *(white arrow)* that must be resisted by tibialis anterior (TA). Shortening the stride length (B) will tend to reduce the dorsiflexion angle, reducing the plantarflexion moment that must be resisted. Forefoot strikes (C) present a plantarflexion angle, setting up a dorsiflexion moment *(black arrow)* which must be resisted by the ankle plantarflexors, primarily the triceps surae-Achilles complex (TS-AC). The size of this plantarflexion angle and running speed will dictate the size of the dorsiflexion (plantarflexion) moment that must be resisted. Experienced runners tend to have a high cadence rate (lots of strides) as a result of shorter stride lengths, which helps reduce the size of joint moments throughout initial contact. *(Permission www.healthystep.co.uk.)*

Faster running speeds increase forefoot strike rates in both shod and barefoot runners (Hasegawa et al., 2007; Hayes and Caplan, 2012; Hatala et al., 2013; Forrester and Townend, 2015). Forrester and Townend reported that at a running velocity below 5 m/s, rearfoot strikers made up around 70%, midfoot strikers 24%, and forefoot strikers 6%. At faster velocities, the rate of rearfoot strikers fell and increased towards midfoot strikers, although only around 15% of the rearfoot strikers converted to midfoot or forefoot strike (Forrester and Townend, 2015). These results are fairly consistent with other studies

(Nigg et al., 1987; Keller et al., 1996; Hayes and Caplan, 2012). This indicates that with faster speeds, rearfoot strikers are most likely to alter their strike position. Different terrains including midsole material on shoes as well as velocity, also influences foot strike position on an individual basis (Nigg et al., 1987; Tam et al., 2014). Shod runners with different strike positions do not seem to demonstrate differences in contact time or stride length (Nunns et al., 2013), whereas running velocity increases stride frequency and stride length and decreases contact time (Forrester and Townend, 2015). As running velocity increases, the thigh angle becomes more horizontal, dominating the changes over the smaller increase in knee flexion and the negligible change at the ankle. This leads to foot touchdown positioning further anterior to the body (Forrester and Townend, 2015).

Forefoot strike is associated with a flatter foot placement (Divert et al., 2005a,b) and a greater ankle plantarflexion (flexion) and knee flexion angle at impact (Pohl and Buckley, 2008) that together may distribute impact forces over a greater plantar surface area than just the heel alone. Forefoot strike permits the use of the plantar apoeurosis via the reversed windlass mechanism during loading as a significant 'shock absorber' to assist the forefoot plantar fat pad (Caravaggi et al., 2009; Tam et al., 2014). The shift to an impact anterior to the ankle creates an eccentric loading on the triceps surae to resist the dorsiflexion moment, increasing work at the ankle and decreasing that on the knee (Arendse et al., 2004). Forefoot running seems to reduce average and peak loading rates (Yong et al., 2018), yet kinematic data suggest that forefoot and toe strikers may require a stiffer leg than rearfoot and midfoot strikers, which may increase strain on the plantarflexors and Achilles tendon (Almonroeder et al., 2013; Nunns et al., 2013) and extend its loading time (Kernozek et al., 2018). Forefoot strike thus increases the loads on the Achilles and loads the plantar aponeurosis earlier, both acting as shock attenuators. It is thought that in forefoot runners, increased joint stiffness at the ankle will limit the downward velocity of the heel at forefoot strike (Hamill et al., 2011), helping to prevent excessive strain during prolonged loading (Fig. 1.4.9b). Habitual forefoot running does not seem to result in a possibly expected greater adaptive cross-sectional area of the Achilles, despite potentially having greater loading strains upon it (Kernozek et al., 2018).

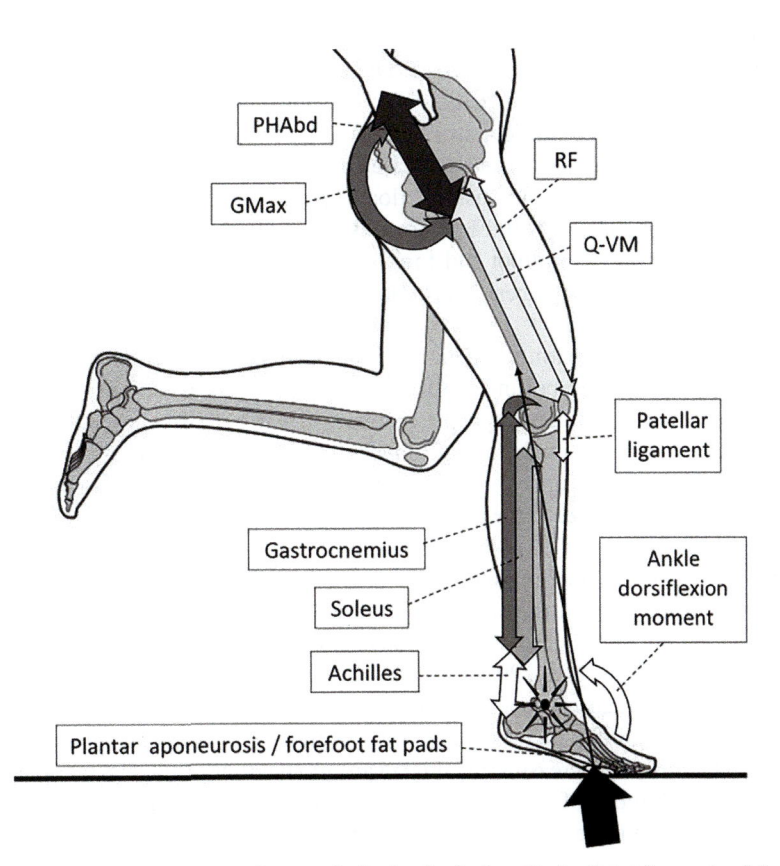

FIG. 1.4.9b Muscle-tendon buffering is an important part of energy dissipation that is altered in forefoot strike running. Like heel strikers, forefoot strike runners primarily use proportionately higher knee flexion moments to dissipate impact than during walking. Restraint of flexion acceleration and muscle-tendon buffering is achieved by the quadriceps including rectus femoris (RF), but mainly via the vasti muscles (Q-VM). The hip in all running styles continues to assist energy dissipation by utilising hip extensor gluteus maximus (GMax) as well as posterior hip abductors (PHAbd) and adductors (not shown). However, in forefoot strike running, as demonstrated here, the forefoot plantar fat pads are now loaded at initial contact, receiving high compression pressures and shears that require plantar aponeurotic tightening and vault stiffening to improve their energy dissipation mechanics. The burden of ankle energy dissipation now falls to the Achilles tendon stretching and triceps surae muscle-tendon buffering mechanisms under the ankle dorsiflexion moment of forefoot strike. The more plantarflexed the ankle angle at impact, the greater the plantarflexor muscle-tendon buffering likely required. *(Permission www.healthystep.co.uk.)*

Coupling relationships are also different with foot strike position. Forefoot strike shows increased variability in the medial articulations of the foot during loading and midstance, while rearfoot strikers do so in the lateral foot articulations (Seneli et al., 2021). Despite the difference in the kinematics of the various strike positions, joint angles for the different running conditions are the same in later stance phase from approximately 60% of the stance phase until toe-off (Wager and Challis, 2016). This suggests that acceleration biomechanics remain much the same for all the different foot strike positions.

Studies have shown that kinematic features associated with heel strike such as greater heel vertical velocity, greater lower leg angle, decreased knee flexion angle, and greater stride length (Laughton et al., 2003) result in greater tibial acceleration at impact (Lafortune et al., 1996; Derrick et al., 1998; Mercer et al., 2002; Lieberman et al., 2010; Potthast et al., 2010; Edwards et al., 2012). Forrester and Townend (2015) reported that midfoot strikers had a relatively high stride frequency and lower stride lengths relative to forefoot strikers. Given the relationship between stride length and impact loading, midfoot strikers may adopt a reduced impact loading technique compared to rearfoot strikers and forefoot strikers (Forrester and Townend, 2015). Midfoot strikers seem to modulate the changes of increased velocity by reducing their foot strike angle (Forrester and Townend, 2015).

Limb positions during terminal swing will influence the velocity of the foot and leg just prior to impact. This will directly affect the force and rate of deceleration of the braking foot and leg motion at initial contact and directly affect the peak impact force magnitude. With the ankle in a more plantarflexed angle during forefoot/midfoot contact, the ankle has greater compliance (Laughton et al., 2003; Hamill et al., 2014) and probably contributes to a lower rate of anterior tibial deceleration after ground contact than it does at heel strike, delaying the vertical peak impact (Gruber et al., 2014). Although the tibial accelerations are different in rearfoot and forefoot strikers, peak head accelerations do not change (Gruber et al., 2014) as the body adapts to varying impact situations in order to maintain head stability and so protect the brain (Lafortune et al., 1996; Derrick et al., 1998).

Advantages of strike positions

Runners who habitually rearfoot strike have been recorded as having a higher overuse injury rate than those who mostly forefoot strike (Laughton et al., 2003; Daoud et al., 2012; Diebal et al., 2012; Nunns et al., 2013). Forefoot strike also reduces tibialis anterior activity (Rooney and Derrick, 2013). This may suggest some benefit to changing strike pattern. Converting runners from rearfoot strike to forefoot strike is associated with lower average and peak loading rates (Shih et al., 2013; Boyer et al., 2014; Yong et al., 2018) and decreases in peak hip adduction angle (Yong et al., 2018). However, changing rearfoot strikers to forefoot striking may risk overloading the Achilles tendon by increasing the loading time on the Achilles during stance from around 25% to 100% (Kernozek et al., 2018). Indeed, the overall evidence is lacking that changing from rearfoot running strikes to midfoot and forefoot running strikes improves running economy and reduces impact forces and injury rates (Hamill and Gruber, 2017). However, running kinetics and kinematics are altered by using different strike positions (Fig. 1.4.9c).

If rearfoot striking is potentially more harmful, then a question needs to be asked, namely, why is it so common, even among habitual barefoot runners? A study reporting on the 2017 IAAF World Championships marathon found that 54% of men and 67% of women were rearfoot strikers, and the number of rearfoot strikers tended to increase with increasing running distance (Hanley et al., 2019). Although 75% of runners maintained their starting strike pattern, the top four male finishers were rearfoot strikers. It therefore appears that a rearfoot strike does not disadvantage endurance running performance (Hanley et al., 2019). This may relate to the metabolic cost of running. Using modelling, Miller and Hamill (2015) found that rearfoot strike was predicted by 57% of the cost of running and was optimal for whole-body energy expenditure and peak joint forces. In barefoot modelling, rearfoot strike was optimal for 55% of the functions.

This suggests that rearfoot strike is the most versatile footfall pattern for achieving the greatest number of goals during locomotion. It may also, by dividing impact into two collision events rather than one, spread peak impact forces across the foot and over a longer time frame (Ruina et al., 2005). Other foot strike patterns tend to be optimal for avoiding localised fatigue in muscle and minimising average joint contact forces (Miller and Hamill, 2015). Additional factors may also play a significant role. It has been suggested that increased hip flexor activity that alters midswing kinematics may ultimately decrease the leg's velocity at collision, decreasing the downward momentum of the leg and thus creating a smaller force at impact (Schmitz et al., 2014). Moreover, forefoot or rearfoot strike running does not seem to change plantar aponeurosis strains or elastic energy storage within the foot during stance phase (Wager and Challis, 2016). This suggests that strike position has little effect on terminal stance acceleration mechanics, reflected by the data that the acceleration phase is not biomechanically different for the variations in strike patterns during running. The fact that running efficiency is based on the horizontal force generated during the latter stages of stance, where joint angles are consistent across running patterns, means that strike patterns primarily influence braking mechanics, not acceleration (Wager and Challis, 2016).

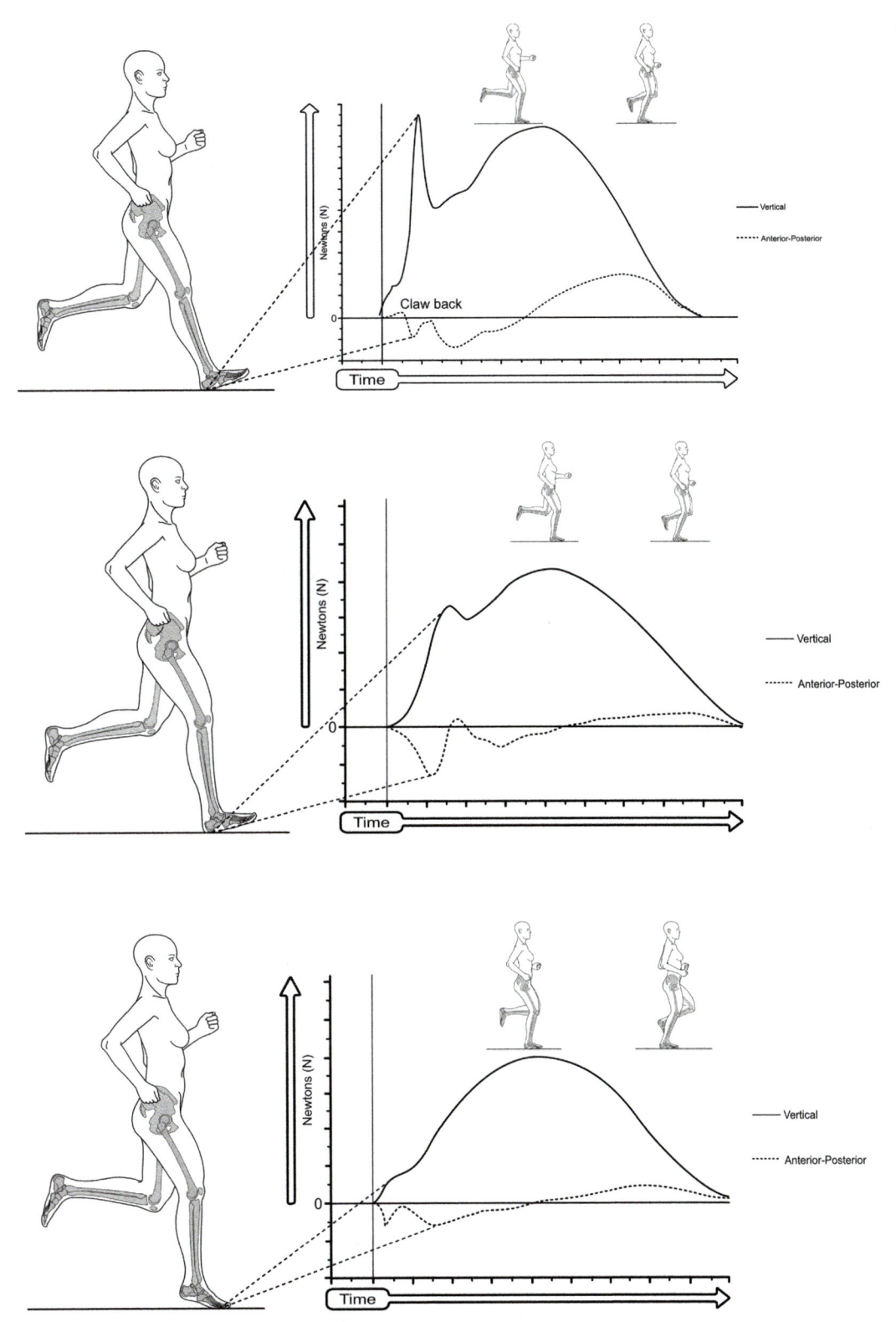

FIG. 1.4.9c Vertical force-time curves are different for each type of foot strike position, although the forces generated are still dependent on the runner's weight and accelerations. The loading rates of forces and the impulses generated are very distinct in heel strikers *(upper two images)*. They generate initial impulse transients from their separate rearfoot contacts before the forefoot collision impulses that are continuous with the braking impulses caused by the continued drop of the CoM. Shorter stride lengths *(middle image)* create reduced ankle dorsiflexion angles tending to reduce the size of the initial impulse transient spike at heel strike. Forefoot strikes *(lower image)* do not produce an initial separate vertical impulse peak, as their impact collision is more continuous with the whole braking event that involves the fall of the CoM increasing vertical forces. Anterior–posterior forces are also different with long strides creating marked heel strike impulses *(upper image)* and generate more posterior-directed GRF at initial contact (often with claw back posterior GRF). Less dorsiflexion-angled heel strikes *(middle image)* and forefoot strikes *(lower image)* generate more variability in these horizontal forces during braking. Vertical and horizontal acceleration forces are much the same for each strike position, reflecting the same acceleration biomechanics of all running styles. *(Permission www.healthystep.co.uk.)*

Sprinting

Sprinting is about power and acceleration rather than running economy. Sprinters are forefoot runners capable of developing rapid and high maximum force, especially in a horizontal direction, while also minimising horizontal braking forces (Harland and Steele, 1997). These requirements are applicable to any rapid acceleration during sport, but rapid acceleration achieves its zenith in sprint races. The ability to accelerate rapidly during the first few strides of a 100-m race separates elite sprinters from merely good ones and is the result of the relationship between GRF impulses and joint kinematics (Hunter et al., 2005). Maximum velocity is reached between 40 and 70 m after the starting blocks and results from an ability to rapidly achieve maximum velocity (Morin et al., 2015).

Running speed is limited by a mechanical interaction between the stance time and the stride length created during swing phase. This is limited by the minimum stance time needed to apply a mass-specific force necessary to drive forward upon, rather than the size of the force created against the ground (Weyand et al., 2010). Increasing from intermediate to top speed requires applying greater GRFs, using shorter periods of stance phase, and repositioning the swing limb more rapidly, thereby reducing the time in aerial phase. Speed in humans is therefore limited by the contact time and the vertical impulse, decreasing to the minimums that provide just enough time to reposition the swing limb for the next step. This requires a swing and aerial time of around 0.35 s at top speed (Weyand et al., 2010).

There appears to be no correlation between running acceleration and the size of the vertical GRF or braking force, with only the anteriorly accelerating horizontal GRF or impulse explaining around 75% of the performance variability in sprinting athletes (Hunter et al., 2005; Kawamori et al., 2013; Morin et al., 2015). To increase speed effectively, a runner can brake less and accelerate more, but the swing phase speed is also a significant limiting factor. This suggests iliopsoas and quadriceps power is very important to running speed. With running speed, the accelerating ankle plantarflexion moment increases significantly more than the knee extension moment, increasing loading on the plantarflexor muscles (Petersen et al., 2014). This means that with increasing speed, the posterior structures within the lower leg are more likely to be vulnerable to injury (Petersen et al., 2014).

Yet speed is not just about how much power is generated around the ankle and knee for acceleration. Human running gait speed is also limited by the mechanical interdependence between the stance and swing phase, as the vertical forces and impulses required to attain speed are dependent on the rate at which the limb can be repositioned (Weyand et al., 2009). Longer aerial/swing time slows repositioning advancement of the limb and increases the GRFs required to elevate the body's CoM in late stance. Shorter swing times effectively reverse the effect, increasing speed (Weyand et al., 2010). It has been noted that minimum swing times (20% shorter than typical values) substantially reduce the vertical forces and impulses required to attain the same sprint speed (Weyand et al., 2009). It seems that muscles are operating at their functional limits during the stance phase of sprinting, but not at their functional limits during swing phase (Weyand et al., 2000). It has been found that the fastest human runners have predominantly fast-twitch muscle fibres that contract and generate force more rapidly (Weyand et al., 2010) and have greater muscle fascicle length (Kumagai et al., 2000; Abe et al., 2001; Seynnes et al., 2007). However, such individuals do not swing their limb any more rapidly at faster running speeds than those with less fast-twitch fibre dominance. Swing time is thus an important limiting factor of human speed (Weyand et al., 2000).

In stance phase mechanics, faster runners apply appreciably greater mass-specific GRFs during a shorter stance phase time than slower runners. (Weyand et al., 2000, 2010). The time of stance phase in sprinting seems too brief to allow the limb muscles to generate their maximum force, with forces peaking around halfway through the stance phase (Weyand et al., 2010). Thus, sprinting stance phase is too short to allow maximum muscle force generation, despite lower limb muscles being activated well before ground contact (Harridge et al., 1996, 1998; Korhonen et al., 2006). The time it takes for a nerve impulse to generate activity within knee or ankle extensors is 81 and 120 ms, respectively, in young men (Harridge et al., 1996). This is roughly twice as long as the stance phase of sprinting. Even if the muscles could activate immediately, the time taken to reach maximum contraction would only allow maximum isometric force of 46% and 22%, respectively (Harridge et al., 1996). Maximum speed in running's 'bouncing', glancing collision gait is a trade-off between the magnitude of the GRF and the step frequency that can be obtained at progressively faster speeds (Weyand et al., 2010).

Thus, constraints on running speed are contact phase time lengths and minimum swing phase time lengths. Fast-twitch muscle fibres that confer faster contractile kinetics found in sprinters do not confer appreciably faster reductions in the aerial or swing phase periods at top speed. Nevertheless, the minimum time for completing the swing process is 343 ± 6 ms (Weyand et al., 2010). Furthermore, the accompanying aerial times required determine the GRF and impulses utilised at any speed. The greater the aerial/swing time required by an individual gait, the greater the slope of any force-speed relationship and the greater the force required to attain top speed (Weyand et al., 2010). The other constraint is the lower limit to the periods of foot-ground contact (i.e. stance phase) during which force can be applied to the ground. This varies between individuals in accordance with the rate of contractile speeds of muscle fibres in the limbs, with faster subjects able to apply greater mass-specific GRF during shorter periods of stance (Weyand et al., 2010). The problem is that slower muscle fibre contraction economises the forces produced during walking and standing and confers greater tendon and bone safety margins of tissue stress (Weyand et al., 2000). Thus, sprinters are more prone to acute injury associated with their tissue and morphological adaptations.

Longer limbs also increase running speed in humans, with an increase of 10cm calculated to increase top speeds by around 9% (Weyand et al., 2010). However, with humans having a small base of support, this could risk increasing the rate of falls from intrinsic instability. These functional trade-offs are part of human evolution but have controversially been manipulated in the design of athletic lower limb prostheses. The muscles of running animals, like cheetahs, function with little difference to humans. Nevertheless, such animals have adapted to develop gait mechanics that prolong their ground contact through pronounced spine bending that increases the ground contact times, despite relatively shorter limbs, allowing much faster running speeds. This is an option that is not available to bipedal humans (Weyand et al., 2010).

Despite human anatomical limitations, elite human sprinters are a morphologically interesting group demonstrating a number of adaptations that improve their performance over nonsprinters. Sprinters display shorter posterior calcaneal tubercles, shorter lower leg segments compared to thigh segments, and longer toes than do nonsprinters (Lee and Piazza, 2009), particularly the 2nd toe, with both the 1st and 2nd metatarsals reported to be longer for elite 400-m sprinters (Tomita et al., 2018) (Fig. 1.4.9d). The Achilles tendon moment arm is around 25% shorter on average than in nonsprinters and the muscular fascicles are 11% longer (Lee and Piazza, 2009). The shorter ankle internal plantarflexor moment arm of the Achilles and the longer toes and metatarsals permit greater generation of anterior impulses, increasing the 'gear ratio' of the foot, maintaining muscle fibre length, and reducing peak fibre shortening velocity (Lee and Piazza, 2009). This is because the Achilles tendon (internal) moment arm is shorter, so that for a little change in muscle fibre or tendon fibre length, proportionally more rotation is achieved at the ankle than with a longer Achilles tendon moment arm, and with greater elastic energy storage (Foster et al., 2021).

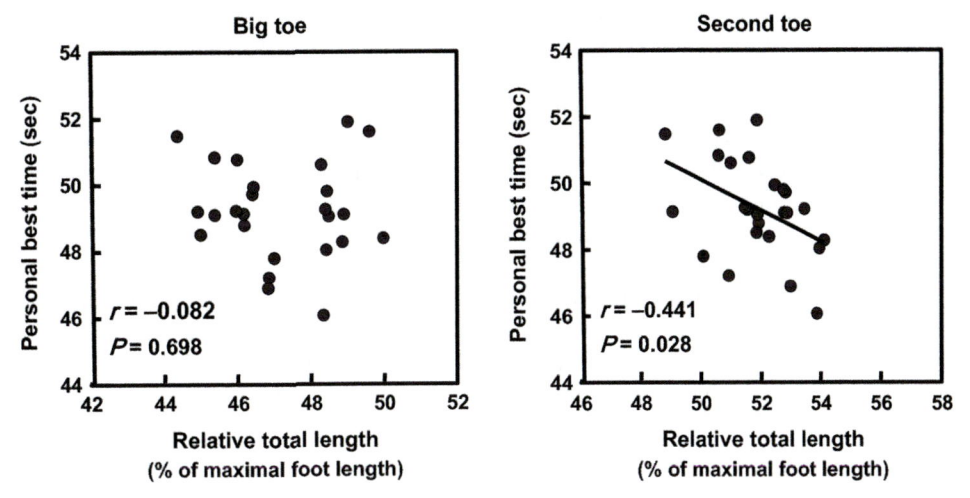

FIG. 1.4.9d The relationship between the relative length compared to the whole foot length and the total length of forefoot bones of the hallux and second toe (including the metatarsals and toe bones) plotted with the personal best 400-m sprint time of athletes. The length of the forefoot bones around the second toe seems most significant. *(From Tomita, D., Suga, T., Tanaka, T., Ueno, H., Miyake, Y., Otsuka, M., et al., 2018. A pilot study on the importance of forefoot bone length in male 400-m sprinters: is there a key morphological factor for superior long sprint performance? BMC Res. Notes 11, 583.)*

Heel length has not been correlated with increased oxygen consumption, suggesting that the energetics are not compromised as a result of the shorter internal Achilles tendon moment arm (van Werkhoven and Piazza, 2017). This is important, as the mechanical benefits of more motion for less fibre length change could be lost through decreased mechanical efficiency of the reduced effort to the MTP joint fulcrum length. This is a problem likely solved by the longer metatarsals found in sprinters that can maintain the external Achilles moment arm length. Thus, subjects with shorter heels experience larger Achilles tendon forces without an increase in metabolic cost (van Werkhoven and Piazza, 2017) and possibly decreasing metabolic cost as a result of the smaller internal moment arm (Scholz et al., 2008). The short posterior calcaneal tubercle length also creates elastic energy storage benefits within the Achilles, and as a consequence, requires less net work during running and sprinting gaits, reducing the cost of locomotion (Foster et al., 2021). This shortened Achilles tendon moment arm in running seems to have a greater effect than the tendon's cross-sectional area, making rearfoot allometry a significant factor in elastic energy storage of the Achilles. However, the moment arm length does not correlate with spring-like tendon behaviour or metabolic costs in walking (Foster et al., 2021). This is probably because the tissue loading rates are much slower during walking, altering stress-stiffening behaviour of the tendon. However, benefits are still seen with jogging (4.44 m/s) but the influence of faster loading rates on tendon spring-like behaviour may explain why calcaneal shortening is seen associated with sprinters more than long-distance runners. Having longer toes prolongs the period of ground contact, giving more time to generate forward acceleration from propulsive GRFs (Lee and Piazza, 2009); that is, as long as stiffness across the MTP joints is adequate to prevent too much energy dissipation via toe extension.

Elite technique sees greater magnitudes of relative propulsion impulses being generated with hip extension velocity, and possibly lower braking impulses associated with a smaller stride length (touchdown distance) (Hunter et al., 2005). Overall, faster athletes are those that demonstrate the highest values of GRF within the horizontal net impulses that assist acceleration (Morin et al., 2015). Thus, utilisation of energy for anterior displacement minimising vertical displacement is a key sprinting trait. At present (2021), the exceptional athlete, Usain Bolt, holds the 100-m sprint record at 9.58 s. He is tall (1.96 m) and had a higher body mass (96 kg) compared to previous record holders who averaged 1.81 m in height and 77 kg in body mass, allowing a step length much longer than competitors, that enabled him to run 100 m in fewer steps (41 compared to 45) than previous record holders (Beneke and Taylor, 2010). This means that he broke records and won races using slower but fewer steps through principles of force-velocity relationships and heat release for increased power development via his exceptional stature and longer legs that reduced the combined time in ground contact (Beneke and Taylor, 2010).

Distance running

Distance running requires good economy to delay fatigue and then adapt the running style to cope with the fatigued condition. Long-distance or endurance running can be equated with running slowly (obviously a relative term) over a sustained distance, considered to be at least 3 km, and is usually associated with heel-toe running. Ultra-distance running is considered to involve distances above that covered by a marathon (42.2 km). It has been reported that the positive work from the ankle plantarflexors is three times greater than from the knee extensors when humans are slow-running (Winter, 1983), with more power being generated proximally as speed increases (Novacheck, 1998). Increases in muscle-tendon complex stiffness within the plantarflexors would seem beneficial to long-distance running. It appears that passive and active muscle stiffness is kept higher in regular long-distance runners compared to untrained controls. However, the tendon stiffness is not changed (Kubo et al., 2015), so it appears that humans adapt the contractile component part of the muscle-tendon complex to regular-distance running.

Ultra-distance running is an exceptionally low energy cost form of running with low exercise intensity, which is disturbed by any musculotendinous and osteoarticular damage (Millet et al., 2012). Increasing running distance increases the probability of a fatigue fracture in bone by between 4% and 10% (Edwards et al., 2009). The running speed of individuals involved in distance running depends both on the objectives of the runner (keeping fit and/or racing) and the distance they run.

As distance in running increases, fatigue effects become an increasing risk. Clinically, it is important to consider the runner's physiological and musculoskeletal capacity to resist fatigue relative to the event they are active in. Failure to gradually develop running distance, technique, and fitness is a common issue that leads to overuse injury, while certain ultra-endurance events will lead to episodes of fatigue that can be either managed by the athlete or not. An inability to adapt to fatigued running potentially leads to injury. Running speed declines regularly with running time, but very noticeably in elite ultra-distance runners at between 8 and 16 h which then remains constant as the body adapts to a new fatigued running technique (Martin et al., 2010).

As of 2021, the male marathon record stands at 2:01:39 by Eliud Kipchoge with the female marathon record standing at 2:14:04 by Brigid Kosgei. Distance running, including marathons, produces elite runners with certain anatomical, morphological, and physiological characteristics (Joyner et al., 2011), yet this is not true of ultra-distance running where there is more diversity among elite endurance runners (Millet et al., 2012). The running economy (based on oxygen consumption) of elite marathon runners is outstanding, and most top male world marathon runners have a height of 170 ± 6 cm and a mass of 56 ± 5 kg, providing a small trunk size to leg length which has favourable effects on oxygen consumption (Joyner et al., 2011). In addition to this, most elite distance runners have had exposure to high altitudes and significant physical activity in early life. However, there also appear to be some genetic influences that go towards creating the genotype/phenotype of a marathon runner (Joyner et al., 2011).

Running economy is important to distance running and depends on mechanical work interplaying with elastic energy, stride frequency, and anthropometric factors (Millet et al., 2012). Slender legs help the lower internal work required to run (Lucia et al., 2006). Long lower limbs, low body mass, shorter calcaneal tubercles (for sprinters, who also require longer trunks for attachment of larger proximal lower limb muscle), long Achilles tendons, low body fat, and a high percentage of type I muscle fibres have all been reported as improving running economy (Millet et al., 2012). Flexibility levels also seem to affect running economy, probably because stiffer musculotendinous structures facilitate better elastic energy storage and recovery. However, flexibility training seems to attenuate exercise-induced muscle damage, limits back pain, and limits the work done by 'bouncing' viscera during running (Millet et al., 2012). It is likely that the capacity to provide some 'middle ground' flexibility in long-distance runners is desirable in limiting tissue damage and preserving running economy.

Freely chosen stride frequencies in distance runners are usually economical, but manipulation of stride frequency can benefit some. A 10% increase in stride frequency has been suggested to reduce fatigue (stress) fracture risk by between 3% and 6%, (Edwards et al., 2009). This is because the fatigue properties of bone are inversely related to the strain magnitude so that many smaller impacts over a long distance are better than fewer but larger impacts. Thus, decreasing stride length probably protects bone by reducing peak loading forces (Heiderscheit et al., 2011). By increasing stride frequency and decreasing aerial/swing time, loading rate and peak forces can be reduced, limiting injury risk. However, this could be detrimental through the increase of running energetics by 6% (Heiderscheit et al., 2011; Millet et al., 2012). In changing a running technique to avoid injury, it is often necessary to warn the patient that performance may initially deteriorate. Nonetheless, it appears that decreasing stride length and increasing frequency seem generally protective due to the reduced tissue stresses during distance running (Edwards et al., 2009; Bramah et al., 2018).

Running economy correlates significantly not only with vertical oscillations of CoM, but also with other biomechanical variables such as stride frequency, stride length, balance time, elbow motion, internal knee moments, foot strike ankle angles, and EMG activity of the quadriceps (Tartaruga et al., 2012). This means that alterations in running technique can change running economy and improve performance (Tartaruga et al., 2012), as well as reducing both hard and soft tissue injury risk (Bramah et al., 2018; Yong et al., 2018). Long-distance running has a high injury risk, the resolution of which often requires changes in training technique (Lun et al., 2004). Training errors are also regularly at fault and it is thought that 65% of chronic injuries in distance running are due to too high a training mileage, rapid increases in mileage, increased training intensity, and/or sudden changes in terrain (Lohman et al., 2011). Recording a patient's approach to training is just as important as assessing their shoes and running techniques.

Effects of fatigue in running

Fatigue changes running mechanics and spring-mass behaviour, decreasing the amplitude of vertical oscillations of the CoM at each step and increasing step frequency, resulting in impairment of muscle function and the structure of running mechanics (Millet, 2011; Morin et al., 2011a,b). These are also features found in older runners (Karamanidis and Arampatzis, 2005; Cavagna et al., 2008) who also reduce overall impact forces during the braking phase (Morin et al., 2011b). Fatigue can be a result of poor fitness levels but is the expected result of running marathons or greater distances. Ultramarathons are often run within challenging environments such as mountains. It is challenging for homeostasis, energetics, and the musculoskeletal system, as ultramarathons often involve running for over 24 h causing sleep deprivation. This therefore creates excellent populations of athletes with which to assess fatigue effects on human activity. Interestingly, sleep deprivation does not seem to have an effect on running pattern during a single 24-h period (Martin et al., 2010).

Studies have looked at the changes in performance in ultra-distance endurance running due to physiological, neuromuscular (Martin et al., 2010), and spring-mass characteristics (Morin et al., 2011a,b). These fatigue changes cause lower peak GRF and an increase in step frequency due to a decreased contact time (Morin et al., 2011a,b). It has also been noted that the vertical displacement of the CoM decreases and the changes in functional leg length throughout gait become lower. This results in higher CoM positioning of around 10% found at midstance and a greater leg stiffness value than that encountered in nonfatigued running (Morin et al., 2011a,b). This suggests energy dissipation by muscle-tendon buffering is compromised. The longer the durations of running and the more extreme the terrain, the more reduced the swing/aerial timings become compared to stance (Morin et al., 2011b; Millet et al., 2012).

Muscle fatigue itself is an exercise-related decrease in maximal voluntary force or power of a muscle group associated with an increase in perceived effort to achieve the desired force (Martin et al., 2010). This decline can potentially occur at all levels of the motor pathway from the brain to the skeletal muscles, and this is often assessed by determining whether the mechanism responsible for fatigue arises from the *peripheral system* from muscle or the *central system* from the nervous system (Enoka and Stuart, 1992; Martin et al., 2010).

Central system fatigue appears to be a method of protecting the peripheral system from a loss of homeostasis to prevent injury. This is responsible for neuromuscular fatigue and is the main mechanism that underlies fatigue processes (Martin et al., 2010). Large central activation deficits are reported, especially affecting knee extensor muscles (Martin et al., 2010). It would seem that the central nervous system (CNS) is primarily responsible for limiting exercise duration (Martin et al., 2010). Fatigue of the stretch-shortening cycle reflexes increase impact peak forces and reduce post-impact forces (Komi, 2000). For fatigued distance runners, this is a particular problem as fatigued subjects must perform greater work during the push-off acceleration phases in order to maintain a constant speed. This requires greater effort that in turn risks even faster rates of fatigue.

Muscle strength loss seems to relate to running duration, with around 30% loss in strength in the knee extensors after 8.5 h of running, increasing to 40% over 24 h (Martin et al., 2010). This suggests a nonlinear relationship of strength loss to exercise duration, with force loss dramatically increasing in the first 2–5 h, then slowly starting to plateau afterwards (Millet and Lepers, 2004). A decrease of 11%–18% of ankle plantarflexor maximum voluntary contraction has been reported from between 1 h and 30 min up to 2 h and 30 min of flat running (Petersen et al., 2007; Racinais et al., 2007; Saldanha et al., 2008), increasing to 30% at longer durations (Martin et al., 2010). This is consistent with a nonlinear relationship and suggests that the plantarflexors are more fatigue-resistant than the knee flexors (Petersen et al., 2007; Martin et al., 2010). Muscle fatigue may be influenced by terrain and the background training of the runner's technique, as large inter-individual variability has been identified (Martin et al., 2010).

The features of reduced vertical oscillations of the CoM, vertical displacement of the spring-mass system, and increased step frequency seen in fatigue are also observed in elderly runners. This could be associated with a 'safer' running style from smaller CoM oscillations (Morin et al., 2011a,b). This seems to fit with reports that the best ultra-distance runners present with significantly higher decreases in swing/aerial time and peak forces, despite overall faster run times when fatigued compared to average ultra-distance runners (Morin et al., 2011b). Although ultramarathon-type running is a relatively rare activity, the fatigue resistance of patients is of concern to gait analysis as the kinetic and kinematic events causing pathology may only be seen once the athlete is fatigued. The changes in gait and/or the failure to achieve a safer running style when fatigued may lie at the heart of an injury. Symptoms developing only after a prolonged period of activity are likely to indicate such scenarios.

Sadly, it is often clinically impractical to view the patient fatigued. But having an idea of likely consequences of fatigue is extremely helpful in planning treatment programmes. For example, it is reported that navicular height and the semiquantitative clinical assessment tool known as the Foot Posture Index (FPI) changes, indicating that there is more and persistent foot 'pronation' after a half-marathon (Cowley and Marsden, 2013; Fukano and Iso, 2016). Furthermore, foot shape changes caused by fatigue seem to take over a week to recover (Fukano et al., 2018). An increase in rearfoot eversion has been reported more in pes planus feet than in pes cavus feet when fatigued (Sinclair et al., 2017). Fatigued low vault profile runners have also been reported to increase pressure under the medial three metatarsal heads and reduce peak pressures under the lateral two, whereas the opposite metatarsal pressure changes are reported in high-foot-vault runners (Anbarian and Esmaeili, 2016). These findings contradict those found at 45 min of running which reported that the more pronated foot types tended to become less pronated with decreased plantar pressures (Bravo-Aguilar et al., 2016). It seems that running time affects foot function, possibly initiating initial increased muscle activity and stiffness for foot protection and improved function which then decreases over time with fatigue, causing biomechanical and alignment/posture changes.

1.4.10 Running differences through gender and age

Male and female runners

Differences in injury rates suggest different biomechanics in running between the genders. Females have a higher incidence of overuse injuries in running, being twice as likely to suffer some conditions, particularly over the age of 50 years (Taunton et al., 2002). Female runners have demonstrated higher incidences of patellofemoral pain syndrome and injuries, iliotibial band syndrome symptoms, knee meniscal injuries, knee osteoarthritis, Achilles tendinopathy, and plantar fasciitis compared to males (Taunton et al., 2002). Females seem particularly more prone to patellofemoral pain, up to 25% more likely to occur among them than in males (Boling et al., 2010).

Female recreational runners appear to demonstrate greater pelvic tilt (Nigg et al., 2012) and exhibit significantly different lower extremity mechanics in the frontal and transverse planes at the hip and knee compared to male recreational runners (Ferber et al., 2003; Chumanov et al., 2008; Nigg et al., 2012; Sinclair et al., 2012). Generally, women have a larger hip width to femoral length ratio (rather than a wider pelvis) that risks greater hip adduction moments (Horton and Hall, 1989; Willson et al., 2012). This leads to a greater angulation of the femur and a resultant functional genu (knee) valgum alignment during stance in females. Women have also been shown to have a greater degree of internal hip rotation than men (Simoneau et al., 1998). Together, this is likely to explain the greater Q-angle that is documented among women (Horton and Hall, 1989; Hsu et al., 1990; Livingston, 1998). An increased Q-angle is associated with increased lateral patellar contact forces (Mizuno et al., 2001) and will increase hip adduction moments, possibly playing a role in the heightened incidence of patellofemoral disorders found in women (Messier et al., 1991; Almeida et al., 1999) (Figs 1.4.10a and 1.4.10b). However, stride length may also be linked to the condition in both genders (Bramah et al., 2018).

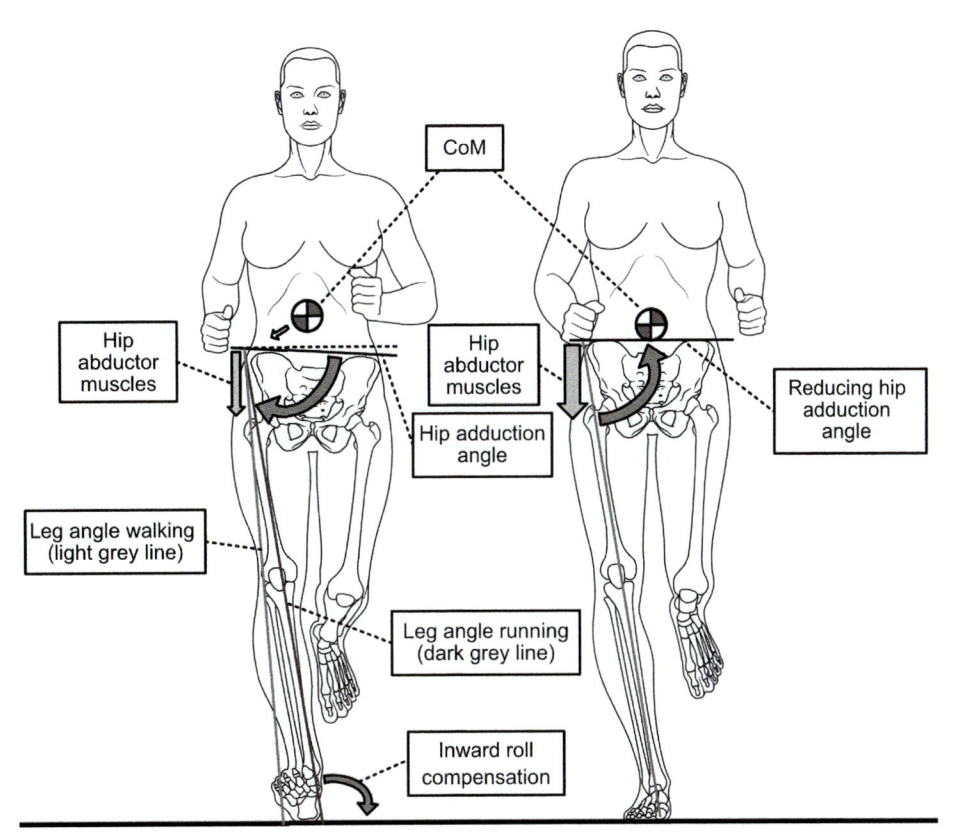

FIG. 1.4.10a Because of the need to place the single support limb closer to lying directly below the CoM (narrowing the base of support), in running the limb is adducted to a greater angle at the hip compared to that required during walking gaits, which have periods of double-limb stance phase. This change in limb angle requires greater hip adductor activity during late swing and at initial contact. This requires hip adductors to be antagonised by the hip abductors, particularly during early stance. Thus, running is subjected to a greater hip adduction moment forming a greater quadriceps (Q) angle during initial contact than those seen of walking. Posterior hip abductor muscles must work harder to help move the CoM towards the support limb hip during contact and loading response *(left image)* to limit the degree of pelvic varus tilt. Once loading is complete *(right image)*, as the CoM moves anteriorly, the CoM moment arm in the sagittal plane shortens, allowing better hip abductor mechanical efficiency to position the CoM over the support-limb and reduce the hip adduction angle and varus pelvic tilt. Usually, before acceleration starts, the pelvis becomes level and the CoM allowed to move a little away from the support limb. Thus, weakness of hip abductors among runners is most problematic during the earliest periods of stance phase. *(Permission www.healthystep.co.uk.)*

It is reported that even though the kinematics are similar between genders, females exhibit more knee valgus in the frontal plane throughout walking and running stance (Kerrigan et al., 1998a; Malinzak et al., 2001; Ferber et al., 2003). Yet, caution is required, because such data are based on mean runners. It must be noted, however, that some female runners will demonstrate more male traits than some male runners do, for there is morphological crossover between males and females. This may explain occasional contradictions observed in the literature on gender differences, especially where subject sample numbers are low. What is probably true is that hip and knee morphology is more extremely expressed in women, and this predisposes them to different kinematic patterns seen in both walking and running. Through most of stance in the frontal plane, knee and hip moments are similar to males, although the hip in females often demonstrates a greater adduction angle and peak adduction velocity at initial contact (Ferber et al., 2003; Chumanov et al., 2008; Willson et al., 2012). Females also absorb greater energy within the hip in the first half of stance than do males (Ferber et al., 2003; Chumanov et al., 2008). The knee also generally demonstrates a greater peak abduction angle throughout stance in females

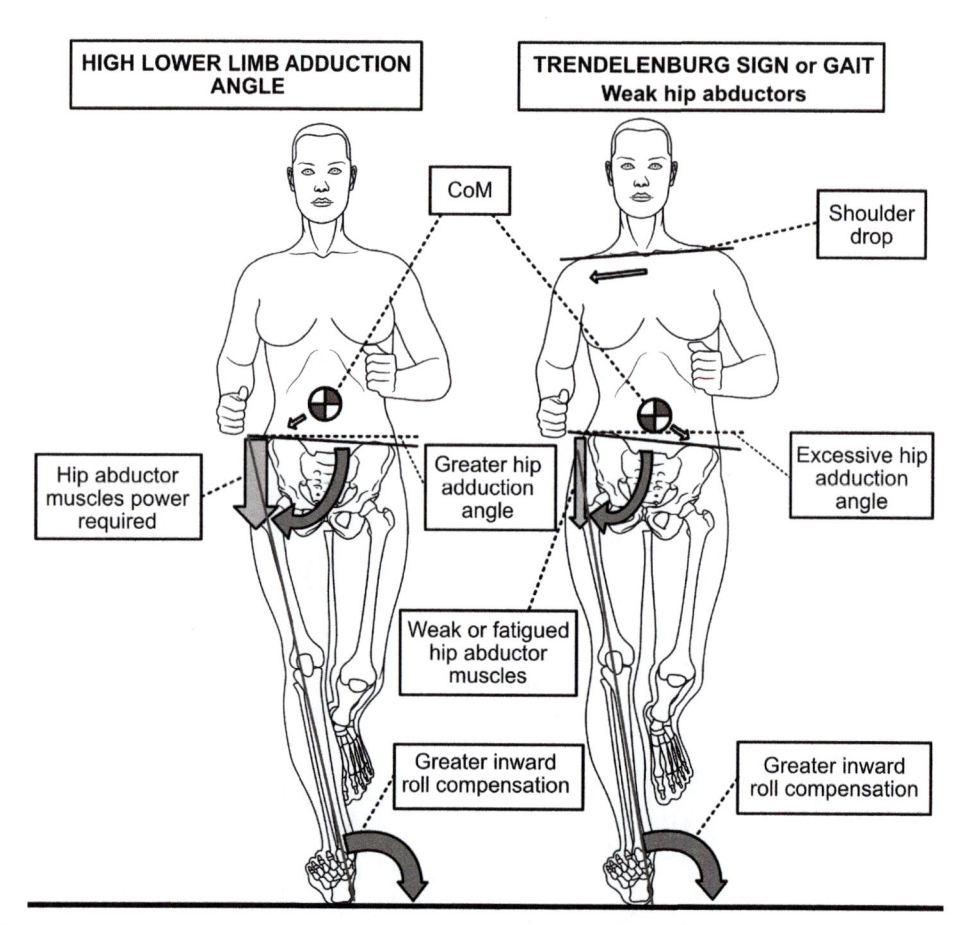

142 Clinical biomechanics in human locomotion

FIG. 1.4.10b The more adducted the lower limb becomes at initial contact, the greater the hip adduction angle and varus pelvic tilt that needs to be restrained by the hip abductors. Females with, on average, shorter lower limb lengths compared to pelvic widths are more prone to creating higher hip adduction angles during running *(left image)*. As a result, they are more vulnerable to weakness or fatigue within the hip abductors destabilising pelvic tilt angle and CoM's position, known as a Trendelenburg gait, that causes swing limb hip drop *(right image)*. This tends to also cause shoulder drop on the stance limb side. Both increased lower limb varus angles and hip abductor insufficiency to the varus moment will alter patellofemoral joint mechanics through functional changes in the quadriceps or 'Q' angle. *(Permission www.healthystep.co.uk.)*

(Ferber et al., 2003; Chumanov et al., 2008). It appears that female runners generally exhibit significantly more hip frontal plane negative work. This can be attributed to greater hip adduction angular velocity, creating greater eccentric demands on the hip abductors than in males (Kerrigan et al., 1998a; Ferber et al., 2003). Thus, it is possible that the greater hip adduction apparent at heel strike and the larger peak vertical GRFs seen in females compared to males, results in greater knee abduction angles and decreased knee internal rotation excursions (Ferber et al., 2003; Willson et al., 2012).

In the transverse plane, female runners exhibit similar movement to males, but tend to display greater hip internal rotation at heel strike and knee external rotation throughout stance, absorbing greater hip and knee energy than men (Ferber et al., 2003). There is no significant difference in temporal parameters and no difference in stance duration between the genders (Ferber et al., 2003). Female sagittal plane knee moments are similar to males with no significant differences in peak hip or knee flexion angles, negative work, peak flexion velocity, or peak extensor movement between the sexes (Ferber et al., 2003). Ferber et al. (2003) reported females as using slightly greater hip flexion and producing a greater hip extensor moment throughout most of stance phase. Conversely, Sinclair et al. (2012) associated males with greater hip flexion.

As joint angulations and motions can be different between the genders in running, we should also expect differences in muscle function. However, this does not mean that we should expect weaker muscles in proportion to body size in one gender more than in the other. Several studies have demonstrated a low association between hip strength and hip kinematics (Dierks et al., 2008; Willson and Davis, 2009) and that strengthening the hip does not influence hip adduction moments (Earl and Hoch, 2011; Ferber et al., 2011; Willy and Davis, 2011). However, other evidence exists that hip abductor weakness in females increases hip adduction angles (Heinert et al., 2008). The differences may lie in the magnitude of hip muscle activity. In females, there is an increased peak and average of gluteus maximus activation compared to males, but gluteus medius activity is the same (Chumanov et al., 2008; Willson et al., 2012). Males with patellofemoral pain run with greater peak knee external adduction moments and less hip adduction moments than do female runners with patellofemoral pain, but also with greater knee adduction moments than do asymptomatic male runners (Willy et al., 2012). It has also been noted that greater gluteus maximus activation occurs in female runners with patellofemoral pain compared to those without (Souza and Powers, 2009).

It is possible that greater gluteus maximus activation in females leads to an earlier onset of fatigue in this muscle during running, resulting in aberrant hip and knee joint kinematics consistent with the hypothesis of Thijs et al. (2011). This hypothesis suggests that poor joint dynamic control due to fatigue results in increased hip adduction velocity over the course of a long run, something noted earlier by Dierks et al. (2008, 2010). However, it is possible that gluteus maximus fatigue may also alter stride length, something that also seems to correlate with patellofemoral pain (Bramah et al., 2018). Yet, this model of dynamic joint stability loss also fits in with the model of increased injury risk in soft tissues as a result of the loss of joint stability (Gabriel et al., 2008).

Differences have also been observed in the foot with greater eversion of the rearfoot (ankle-subtalar joint) in females (Sinclair et al., 2012). It has further been reported that mature female runners (aged between 40 and 60 years) tend to demonstrate greater ranges of rearfoot eversion, knee internal rotation, and knee external adduction moments when compared to younger female runners (Lilley et al., 2013). Although both males and females use the knee as their primary 'shock-absorbing' joint, females have been shown to use the ankle joint and its musculature more in moderating impact forces and absorbing energy, while males are reported to use the larger hip joint musculature (Decker et al., 2003; Kernozek et al., 2005; McClean et al., 2007). This may have a profound effect on injury risk locations when related to aberrant or repetitive impact events during running.

Older runners

The main kinematic feature of older runners (>65 years) is that they use less knee flexion than young adult runners, something that becomes more distinct between runners the larger the age gap (Fukuchi and Duarte, 2008; Nigg et al., 2012). This suggests that knee flexion gradually reduces with age. Age seems to affect the most dominant movements of running rather than the more subtle motions and affects this among runners more than does gender differences (Nigg et al., 2012). Younger runners display greater ranges of motion for flexion at the knee, greater ankle dorsiflexion, and greater vertical displacement of the pelvis than older runners (Nigg et al., 2012).

The primary kinetic changes of older runners are reduced CoM vertical oscillations and reduced peak GRFs (Karamanidis and Arampatzis, 2005; Cavagna et al., 2008). Vertical displacement of the CoM at each step attains a maximum during intermediate speeds but this oscillation is lower in older subjects at around 7.5 cm at approximately 7.5 km/h compared to 10 cm at 10 km/h among the young (Cavagna et al., 2008). Muscular force is reduced in old age, causing changes in the symmetry of the 'rebound' with the muscular push exerted during midstance at the CoM low point. This results in an acceleration upwards, smaller than the acceleration downwards from gravity. The result is a reduced displacement of the CoM during running, with a lower upward acceleration and a reduced duration of aerial phase (Cavagna et al., 2008). As a consequence, less elastic energy is stored, a higher step frequency is necessary, less external work is required to maintain CoM motion, and greater internal work is required to accelerate the limbs relative to the CoM (Cavagna et al., 2008). The net result is that total work increases more steeply in older subjects with speed than in the young, with many of the changes reflecting those encountered with fatigued running (Cavagna et al., 2008).

1.4.11 Foot type and footwear effects on running

Foot types: Compliance and stiffness

Foot types and joint malalignments and aberrations in running kinematics have been associated with increased injury risk and different injury patterns (Simkin et al., 1989; Kaufman et al., 1999; Williams et al., 2001a). While foot posture in running seems to cause changes in lower limb kinetics and kinematics, a precise relationship to lower limb injury risk has not been made (Tong and Kong, 2013; Hollander et al., 2019). One consistent problem is the varying ways in which foot vault profile is determined (Hollander et al., 2019). Despite lack of clarity, foot vault profile has been reported to be associated with an increased injury risk (Kaufman et al., 1999) and a divergence in the type of injury. Low-vaulted planus feet are associated with a higher number of soft tissue injuries tending to be located more medially and more to the knee, while high-vaulted cavoid feet have been associated with more osseous and laterally located injuries in the femur and tibia

(Williams et al., 2001a). As cavoid feet are associated with a stiffer foot type and planus feet with a more mobile foot type (Zifchock et al., 2006), it is possible that the origins of the difference lie in the ability to generate changes in foot compliance and stiffness levels adequately within the appropriate phases during gait. This may be observed in healthy young individuals with different foot flexibility levels, where more flexible feet have higher rearfoot and midfoot impulses than in those with stiff feet who demonstrate higher forefoot impulses (Cen et al., 2020). Thus, it presents the possibility that stiffer feet work harder to dissipate impact and midstance energies, while flexible feet work harder to develop acceleration power. This stiffness difference is further reflected in the knee joint, with cavoid feet associated with significantly stiffer knees than planus feet (Williams et al., 2001b; Powell et al., 2016). It has been further suggested that exaggerated or insufficient stiffness throughout the lower extremity represents a predisposition to injury (Williams et al., 2001b, 2004; Butler et al., 2003). Stiff planus feet and mobile cavoid feet would therefore be expected to demonstrate different injury patterns from their stiff and mobile counterparts due to variance in running biomechanics.

The inequalities between high-vaulted and low-vaulted feet are dominant in the acceleration phase and this probably pertains to the ability of the foot to function as a semi-stiffened beam. There is an association between *arch height index* and foot stiffness with higher arch-associated scores linked with greater rigidity and low arch scores with greater foot mobility (Zifchock et al., 2006), changing the peak impulses generated throughout the foot during running (Cen et al., 2020). Impulses are different in both walking and running in stiff feet compared to mobile feet, with the stiff feet of young healthy adults demonstrating lower heel and midfoot impulses but higher forefoot impulses than age-matched mobile feet (Cen et al., 2020). The change in forefoot impulses is possibly due to the rigid foot being more likely to offer a more easily generated stiffer beam for propulsion than a mobile foot, resulting in a more efficient application of ankle plantarflexion force via the external Achilles moment arm to generate heel lift. See Fig. 1.5.3e for more data on the walking and running effects of differences in foot flexibility from Cen et al.'s (2020) research. However, the potentially reduced range of motion offered by a rigid foot likely causes disadvantages compared to a foot with greater mobility that can adequately generate appropriate stiffness as required during locomotion. Muscle contraction creates greater stiffness and elasticity within a limb when needed, but it cannot create flexibility within a region that is restricted by inflexible connective tissues and articular constraint or joint disease.

The low-vaulted profile, when associated with increased mobility, will have difficulty generating a stiff beam to efficiently apply ankle power to effect heel lift, requiring a greater muscular output. The low-vaulted foot type may also struggle to create forefoot GRF through stable midfoot plantarflexion and adduction. This will reduce the stiffness-induced stability that allows efficient acceleration from the forefoot, reducing the midfoot plantarflexion range of motion required for acceleration (Barnes et al., 2011; Holowka et al., 2017). However, greater foot compliance associated with more mobile feet may allow easier energy dissipation, permitting larger heel impulses to be generated at impact without injury risk (Cen et al., 2020). Thus, consideration should be taken to the initial foot strike position (rearfoot, midfoot, or forefoot) and to how significant the effects might be to the individual depending on foot profile and its (associated) flexibility.

Injury patterns across individuals result from a difference in the altered transmission of force through the kinetic chain in response to impact and power generation patterns (Kaufman et al., 1999; Williams et al., 2001a,b, 2004; Powell et al., 2011, 2012a). Work values in the lower limb are higher during propulsion compared to braking phases, both in walking (DeVita et al., 2007) and running (Heiderscheit et al., 2011). In running, the greatest difference in work values occurs at the ankle joint, with the positive values being two- to three-fold greater than the negative braking values (Heiderscheit et al., 2011). Mobile flat feet likely provide a less efficient 'beam function' in the class two lever system that drives heel lift acceleration, making it harder for muscle power to be converted into forefoot horizontal GRFs. This would increase the active soft tissue demands to stiffen the foot in order to aid propulsion. Stiff cavoid feet are likely to lack sufficient compliance to dissipate braking energy at impact but could be in a more advantageous position at propulsion over the mobile planus foot. It is important for the clinician to be aware that assessment of the degree of foot flexibility is likely to be more important than the nonweightbearing vault profile alone.

Dynamic joint stiffness quantifies the passive and active tissue-loading force attenuation during the braking phase of stance (Gabriel et al., 2008; Lin et al., 2011). Dynamic stiffness values are higher in running than in walking (Gabriel et al., 2008; Powell et al., 2014) and appear higher in barefoot running than in shod running (Lin et al., 2011; Powell et al., 2014). High-vaulted feet in running exhibit significantly greater dynamic joint stiffness values than their low-vaulted comparisons (Powell et al., 2014). High-vaulted feet demonstrate stiffer foot vaults and also absorb energy over a shorter time frame, using a smaller range of motion through multiple planes (Williams et al., 2001b, 2004; Butler et al., 2006; Powell et al., 2011). Low-vaulted runners produce significantly greater loading rates due a significantly earlier peak vertical GRF during running (Williams et al., 2004).

Adult high-vaulted runners reportedly exhibit significantly less net ankle work and use smaller ranges of ankle motion in the sagittal and frontal planes (eversion excursion), particularly at the acceleration portion of the stance phase, than do low-vaulted runners (Heiderscheit et al., 2011; Powell et al., 2014). This is probably a consequence of there being less mobility within the midfoot of stiff cavoid feet, resulting in less motion possible around the midfoot and rearfoot. The net ankle work will however be affected by running velocity, being higher at greater speed and thereby requiring greater mechanical output from the lower extremity (Heiderscheit et al., 2011; Powell et al., 2014). It has been reported that when running, lower-vaulted feet demonstrate smaller forefoot abduction excursions and abduction velocity during early stance than do high-vaulted feet, and thus may exhibit less available ranges of motion through which their forefeet can pass before reaching their end range of joint motions (Barnes et al., 2011). Leg stiffness seems also to be affected by foot vault profile when both running barefoot and shod, with high-vaulted foot runners demonstrating greater overall leg stiffness (Williams et al., 2004; Butler et al., 2006; Powell et al., 2017).

Powell et al. (2014) used female barefoot runners to assess high and low vault profile feet and noticed that all participants running at 2.90m/s used midfoot or forefoot-loading strategies. In such loading postures, the foot should be immediately loaded as a stiff beam structure in a lever arm defined as the horizontal distance in the sagittal plane between the CoGRF and the axis of the ankle joint's rotation, which is positioned further proximally the more the foot is plantarflexed at foot strike. Calculations by Powell et al. (2014) showed that the high-vaulted feet had longer lever arms at loading than did the low-vaulted runners. It can be postulated that on loading, mobile feet's increased midfoot mobility buckles the beam and functionally shortens the lever arm of the foot, requiring greater muscular work to achieve a similar mechanical output (Fig. 1.4.11a). This might require mobile low-vaulted feet to produce work over a longer duration than would a stiffer high-vaulted foot in achieving the same mechanical result during propulsion (Powell et al., 2014). Female runners may be particularly vulnerable as they tend to adopt an ankle-based strategy during athletic tasks (Kernozek et al., 2005; McClean et al., 2007), possibly exaggerating the effect of mobility and vault profile around the ankle.

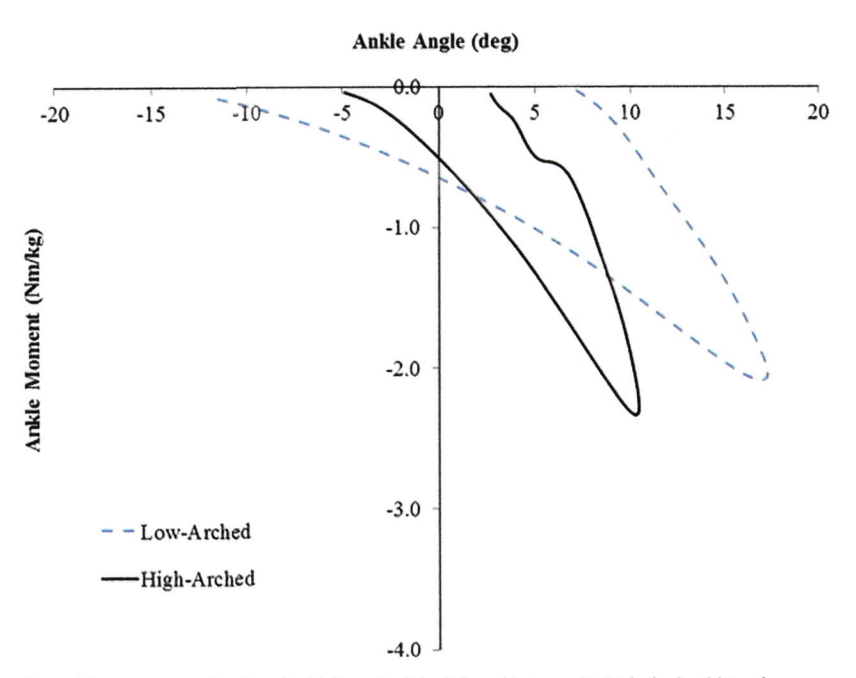

FIG. 1.4.11a Representative ankle moment-angle plots for high-vaulted *(solid)* and low-vaulted *(dashed)* athletes from a sample barefoot running trial during stance phase braking. High-vaulted feet have significantly greater dynamic ankle stiffness than low-vaulted feet, evidenced here by the steeper slope during braking of stance. *(From Powell, D.W., Williams, D.S.B. 3rd, Windsor, B., Butler, R.J., Zhang, S., 2014. Ankle work and dynamic joint stiffness in high- compared to low-arched athletes during a barefoot running task. Hum. Movement Sci. 34, 147–156.)*

Association of vault profile and subtalar joint or rearfoot (ankle-subtalar) eversion is not conclusive, with some studies reporting no correlation (Barnes et al., 2011; Eslami et al., 2014) while others finding correlations (Williams et al., 2001b; Lee et al., 2010; Powell et al., 2011; Sinclair et al., 2017). Eslami et al. (2014) noted an association with navicular drop and peak ankle eversion, but Nigg et al. (1993) found no correlation with foot vault height and rearfoot eversion during shod running. High-vaulted runners possibly use a higher abduction excursion and velocity (Barnes et al., 2011) and more midfoot-forefoot eversion than do low-vaulted runners (Powell et al., 2011). Evidence has not been found for a foot vault link to tibial acceleration differences (Butler et al., 2006) or tibial shock variables (Barnes et al., 2011). After prolonged shod running, tibial shock is reported as being lower in high-vaulted feet in cushioned shoes (Butler et al., 2007), although this effect was not recorded using running sandals (Barnes et al., 2011). Prolonged running to induce fatigue has also brought about changes in metatarsal pressures, increasing medially in low-vaulted feet and laterally in high-vaulted feet (Anbarian and Esmaeili, 2016). Using pressure analysis techniques, it has been reported that centre of pressure (CoP) displacement when barefoot is more lateral in low-foot-vaulted runners than in high-foot-vaulted runners. High foot-vaulted runners have reduced CoP displacement during the forefoot contact phase (De Cock et al., 2008), whereas no difference has been found in force-time integrals using in-shoe pressure analysis (Chuckpaiwong et al., 2008).

It has been observed that high-vaulted foot runners use less ankle plantarflexion at initial contact and less peak dorsiflexion during stance, moving through a smaller range of ankle motion in the sagittal and frontal planes when running than do low-vaulted foot runners during their stance phase (Williams et al., 2001b, 2004; Butler et al., 2006; Powell et al., 2011, 2014). The reduced ankle motion may indicate reduced flexibility within the foot/ankle anatomy, as high-vaulted feet are also associated with greater vault stiffness on vertical loading (Zifchock et al., 2006). There are also indications that high-vaulted feet have less frontal plane motion (Powell et al., 2012b) resulting in a combined inability to absorb energy in the frontal and sagittal planes and increased force magnitudes and loading rates at the ankle (Powell et al., 2014) (Fig. 1.4.11b). Interestingly, the correlation between foot vault profile and sagittal plane ankle motion was not noted during running in children aged 7–14 years, and no significant correlation was found between foot vault profile running biomechanics (Hollander et al., 2018). Children are more likely to have greater foot mobility and smaller impulses than those found in adults, which may explain these findings. Results possibly suggest that foot mobility and body mass are important in how the foot and ankle behaves in response to mobility-rigidity (stiffness), more so than foot vault profile alone.

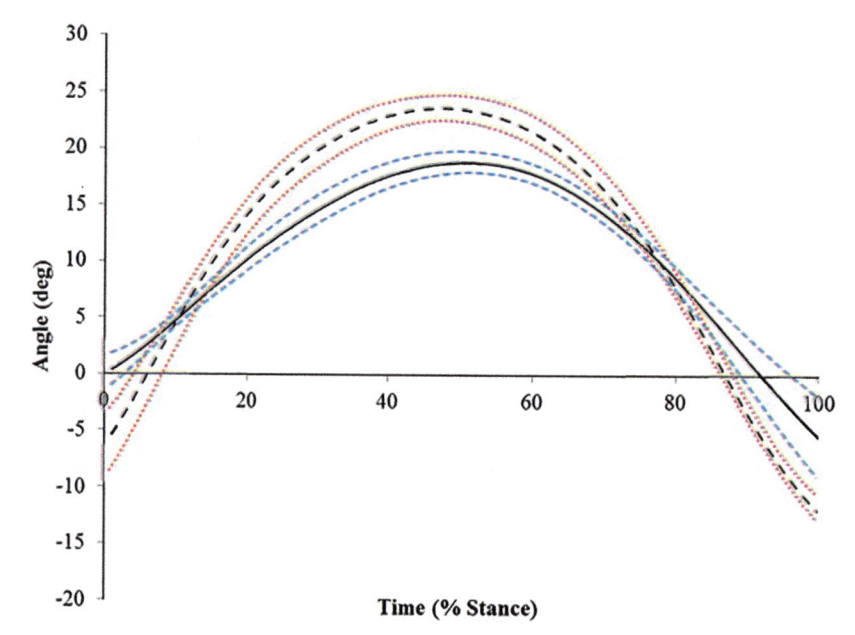

FIG. 1.4.11b Sagittal plane ankle joint angles (+ve scores = dorsiflexion/extension) from representative groups of high-vaulted *(solid line)* and low-vaulted *(thick dashed line)* athletes during barefoot running, with standard deviations shown by fine dashed lines. High-vaulted athletes have less dorsiflexed ankles throughout stance, while low-vaulted athletes exhibit greater peak dorsiflexion and dorsiflexion excursion at their ankles. *(From Powell, D. W., Williams, D.S.B. 3rd, Windsor, B., Butler, R.J., Zhang, S., 2014. Ankle work and dynamic joint stiffness in high- compared to low-arched athletes during a barefoot running task. Hum. Movement Sci. 34, 147–156.)*

Injury rates associated with foot vault profile

Variance in injury location and type is recorded between high and low-vaulted feet in runners (Sullivan et al., 1984; Simkin et al., 1989; Kaufman et al., 1999; Williams et al., 2001a), albeit the rate of injury seems unaffected (Tong and Kong, 2013; Hollander et al., 2019). Stress fractures have been reported more in athletes with high-vaulted feet (Simkin et al., 1989; Kaufman et al., 1999; Williams et al., 2001a), although Sullivan et al. (1984) and Matheson et al. (1987) reported high rates of stress fractures in low-vaulted feet. Modelling the foot has indicated that the amount of energy stored within the foot is affected by the calcaneal inclination angle which under vertical load is reduced in both low and high angles, rising markedly from the extremes to the intermediate angles (Simkin and Leichter, 1990). Coupled with the subtle changes in kinematics with different foot types, this may explain some of the concentrations of stress in particular tissues indicated by increased types of injury at specific sites associated with the different foot vault profiles. Overall, there appears to be a dilemma in that although there seem to be changes in rearfoot joint kinematics, ankle kinetics, lower limb stiffness, and injury locations associated with foot vault profile, the rate of injuries seems largely unaffected by the vault profile itself (Tong and Kong, 2013; Hollander et al., 2019).

Having perfectly 'middle ground' vault profiles does not make a subject immune to running injuries. This is because there are so many biomechanical factors at play during running that foot shape influences on running can be surpassed by far more significant issues, making foot profile one small piece of a larger pathomechanical puzzle. The capacity for avoiding injury in running depends on intrinsic flexibility or stiffness within the foot and lower limb as a whole. Therefore, the ability to create compliance and stiffness appropriately for the shoe worn or the terrain conditions, the running velocity, the foot-strike position, the individual's gender, age, physiological health, and fatigue status while running, could each be contributing factors.

The ability to control and implement the appropriate compliance and stiffness pattern would be necessary, regardless of the foot vault profile. This gives the foot's profile only the capacity to direct the location of the tissue most likely to be overstressed during system compromises, just like any other body alignment that influences biomechanics. Such multiple alignment variations create high variability in individuals being studied, making foot vault profile only a small part of the whole picture and making absolute comments on one feature, that of vault profile, impossible.

Footwear biomechanics in running

Most recreational runners use shoes manufactured for the purpose of running. Yet the human foot did not evolve for footwear, and the modern running shoe was only developed in the 1970s (Lieberman et al., 2010). At the time athletic shoes designed for running entered the market, they attempted to influence foot motion and reduce shock amplification. The idea behind this was that injury rates associated with running would drop. Prices of running shoes increased and so did their weight. The laws of angular momentum state that moving a mass further from the axis of rotation takes more energy to initiate and stop motion. The size of the distal limb segment reflects this, with more muscle mass closer to an axis of rotation being bulked towards the trunk. Having a relatively large plantar surface such as is found on the plantigrade human foot places a disproportionate cost on transportation during swing phase. The torque required to accelerate the mass of the foot is proportional to its moment of inertia, so the mass of the foot is therefore proportional to the increased muscle power required to move it. This results in more metabolic power consumption to achieve motion. The mass of a shoe in sports performance and fatigue is thus important.

An argument therefore developed that running barefoot may be more effective at preventing injury than the technical running shoe that might interfere with natural biomechanics and sensorimotor function, using the principles of environmental mismatch. This subject has caused considerable debate as to whether or not running barefoot will reduce injury rates (Tam et al., 2014). It is certainly more natural for the foot not to be enclosed in footwear. However, the suitability of a foot to walk or run barefoot has a number of factors influencing it. The most important factor is whether or not the foot has been habituated to the barefoot environment. This means whether, during childhood and growth, the feet have been sufficiently exposed to walking and running bare. If shoes have been utilised in growth and development, then mismatch of the anatomy may occur in the barefoot situation. Adult feet suddenly exposed to barefoot running from a previously shod life could suffer, unless time is taken to adjust to the new situation. It is also an environmental mismatch for humans to live in cold climates, so running barefoot in the snow at −5°C is more likely to result in pathology than if wearing shoes.

In principle, there is no reason to consider barefoot running undesirable in an environment that is suitable, but both unshod and shod situations have their advantages and disadvantages. Large numbers of research studies on running shoe biomechanics drives and indeed is driven by a huge and wealthy industry. There has been considerable investigation into comparing barefoot running with shod running. This is of interest to the clinician, not least because patients may arrive as barefoot runners, or they might be thinking of taking it up, or indeed, the clinician may wish to evaluate the patient running barefoot to gather biomechanics data. Therefore, the differences in the two conditions and the expected biomechanics of the two conditions should and must be understood by the clinician.

Running economy can be enhanced by running barefoot, as the mass of footwear increases the energy required to run. Running in a cushioned shoe triggers changes in muscle function that cause the foot vault to behave more stiffly (Kelly et al., 2016). Increased peak flexor digitorum brevis and abductor hallucis activity has been recorded, suggesting that changes in neuromuscular output wearing soft shoes result in altered vault compression and thus the ability to recoil from vault profile adjustments (Kelly et al., 2016). It has been proposed that being barefoot alters strike position of the foot. Heel strike has been associated with shoes and forefoot or midfoot strikes when barefoot (Lieberman et al., 2010; Lohman et al., 2011). Hatala et al. (2013) found that heel striking was common among habitually barefoot runners, with 72% landing on their heels at a preferred running speed. It was further found that more barefoot runners struck on the forefoot or midfoot at faster speeds, and even then, 40% remained heel strikers (Hatala et al., 2013). However, what remains unclear is how much the changes in biomechanics result from being barefoot, or reflect the changes resulting purely from speed. Every 100 g in shoe weight increases metabolic cost by around 1% (Franz et al., 2012). The biggest effect of running barefoot, at least initially for most runners, resides in the shift to forefoot and midfoot contact from heel strike (Lohman et al., 2011; Hall et al., 2013) with a modest reduction in step duration (De Wit et al., 2000; Divert et al., 2008; Squadrone and Gallozzi, 2009; Cronin and Finni, 2013).

Barefoot running seems to be associated with reduced peak GRF, increased foot and ankle plantarflexion angles, and increased knee flexion at ground contact compared to shod. However, the loading rate of force seems more associated with the strike position rather than the presence or absence of a shoe. A forefoot strike reduces the initial loading rate, but a barefoot rearfoot strike pattern increases the loading rate compared to shod (Hall et al., 2013). If barefoot gait results in a forefoot strike, then the loss of the heel strike transient looks impressive on a force-time curve, albeit the overall force applied to the body remains unchanged. It is important to point out that although peak forces and loading rates may be changed from barefoot to shod, the actual forces generated are a result of mass times acceleration. Thus, claims that running barefoot reduces 'force' are erroneous, except for the fact that the lack of a shoe can slightly reduce mass. The overall force is only a result of the mass of the patient and the running pace. The most significant effect of barefoot running appears to be a reduction in energy absorption at the knee, which occurs regardless of the foot strike position (Hashish et al., 2016).

Initial transition from shoe to barefoot seems to result in lower dynamic stability (particularly in acceleration), higher cadence, higher aerial and swing phase time, a shorter contact time, and a lower total vertical displacement and impulse after 2 min of barefoot running with a foot strike index increasing towards the forefoot (Ekizos et al., 2017). Lieberman et al. (2010) found that some habitually shod individuals, when running barefoot, experienced greater impact peaks than when they ran shod, possibly failing to adjust their running style. Habitual shod runners who transition to barefoot and continue to heel strike experience loading rates significantly higher than when shod. However, shod runners who midfoot strike do not demonstrate significant changes (Hashish et al., 2016). Yet interestingly, an increase in average vertical loading rates has also been noted when runners transition from habitually shod to barefoot (Mei et al., 2015), suggesting that some time is required to change braking and acceleration strategies following removal of footwear.

Triceps surae fascicle lengths have also been compared both in running shod and barefoot with no statistical difference being noted, suggesting that although various biomechanical parameters change, they are not reflected in the mechanical behaviour of triceps surae (Cronin and Finni, 2013) (Fig. 1.4.11c). However, this may only be true if the contact strike position in shod and barefoot running remains the same as they did in the study by Cronin and Finni (2013). It would appear that adaptations to barefoot running are very individualistic and that biomechanical changes in running are not predictable but need assessment to avoid injury during any transitioning from the shod to the barefoot state.

Data seems to suggest that going barefoot alone will not decrease injury risk, as shod runners transitioning to barefoot running demonstrate a variety of foot strikes and lower extremity dynamics (Hashish et al., 2016). There is evidence that barefoot running might increase the risk of fatigue fractures within the foot through increased forefoot strikes (Tam et al., 2014). Interestingly, by changing average and peak loading rates, forefoot strike may reduce the risk of tibial stress fractures (Yong et al., 2018). Transitioning to barefoot from shod running is likely to change knee and ankle moments resulting from alterations in strike position (Hashish et al., 2016). Both midfoot and forefoot strike that are more common in barefoot running may increase Achilles tendinopathy risks, as such strikes increase ankle energy absorption (Hashish et al., 2016). It may be that barefoot running is a skill that is not instinctively acquired in a habitually shod population and may require substantial practice for the body and running posture to adapt to it (Tam et al., 2014; Mei et al., 2015; Hashish et al., 2016).

Because of the perceived benefits of barefoot running, footwear designers have tried to develop minimalist or barefoot 'inspired' footwear. Cynically, one has to point out that wearing a shoe to replicate the barefoot condition appears oxymoronic. These shoes offer little if any intrinsic support but are used instead to prevent abrasion and other acute traumas. They are divided into *minimalist shoes* with a traditional toe box shape and *barefoot shoes*, with individual toe pockets built into the shoe. The kinematic and kinetic variables of running barefoot in both habitually barefoot and habitually shod runners are significantly different to shod running, including the use of both minimalist or barefoot 'inspired' shoes (Squadrone and Gallozzi, 2009; Bonacci et al., 2013; Hein and Grau, 2014). Therefore, rather than seeing minimalist and barefoot shoes as an alternative to being barefoot, they should be seen as another running shoe choice. Such minimalist/barefoot shoes produce less dorsiflexion at the ankle at initial contact and reduce stride lengths compared to standard shoes. Nevertheless, they do not alter the foot strike angle (Bonacci et al., 2013; Mann et al., 2015).

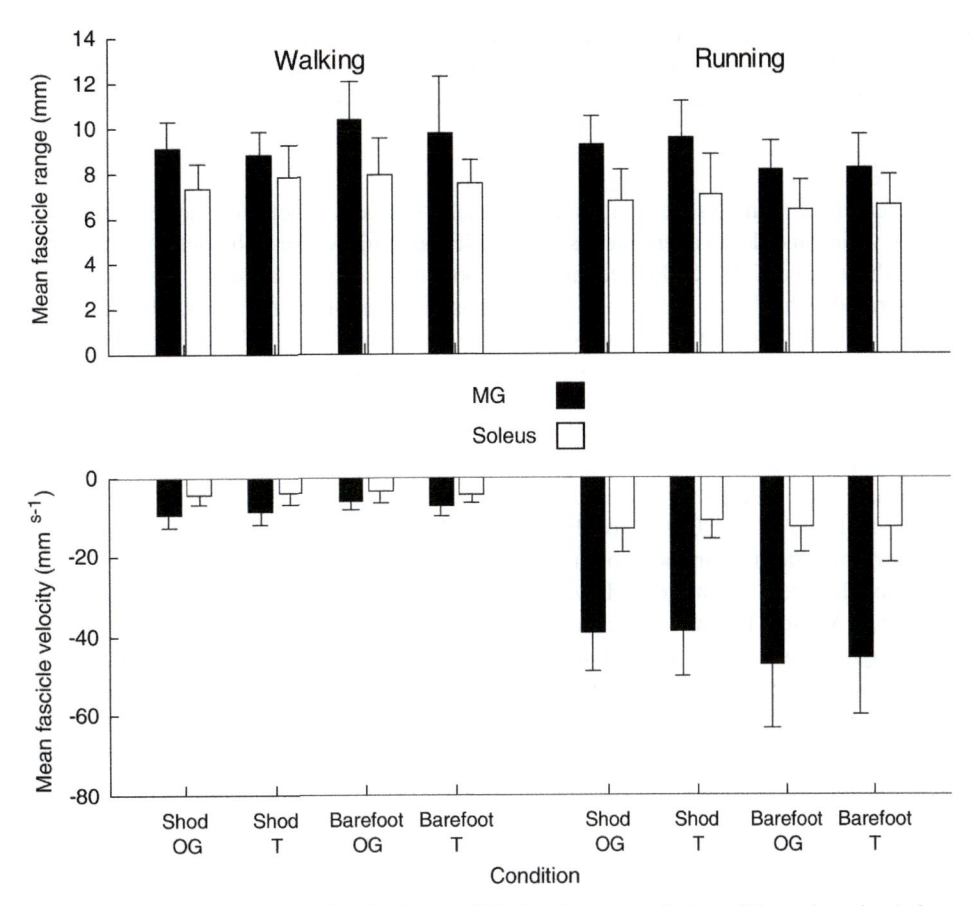

FIG. 1.4.11c Being barefoot or shod, muscle fascicle lengths change within the triceps surae during walking and running (soleus and medial gastrocnemius (MG) being studied here), whether tested on treadmills (T) or overground (OG). The mean range of fascicle lengths *(top graph)* change during gait, determined as maximum-minimum lengths. Mean fascicle velocity of length change is also shown *(lower graph)*. There appears to be no statistical difference. Barefoot results in consistently shorter step duration than when shod, and soleus consistently exhibits smaller and lower velocity length changes compared to medial gastrocnemius. *(From Cronin, N.J., Finni, T., 2013. Treadmill versus overground and barefoot versus shod comparisons of triceps surae fascicle behaviour in human walking and running. Gait Posture 38 (3), 528–533.)*

The effects produced are affected by the amount of cushioning such shoes contain in that any heel lift or cushioning enables runners to land with a more dorsiflexed ankle but do not seem to alter other kinetics or kinematics when compared to using standard shoes (Bonacci et al., 2013). Minimalist shoes also increase loads at the MTP and ankle joints (Firminger and Edwards, 2016) and have been reported to increase an injury risk of Achilles tendinopathy, plantar fasciitis, and metatarsal stress fractures (Salzler et al., 2012; Cauthon et al., 2013). Interestingly, it has been reported that partially minimalist shoes with a little soft 'cushioning' are dramatically associated with a higher increase in rate of injury than are barefoot shoes, although both are reported as being associated with a significant increase in injury rates over standard neutral running shoes (Ryan et al., 2013). It appears the adaptations that occur with barefoot running involve sensory feedback and that the origins of this are within the deep and subcutaneous proprioceptors (Thompson and Hoffman, 2017). Thus, putting any form of footwear on is possibly going to interfere with the sensorimotor system.

Running footwear design

Traditional running shoes developed since the 1970s have tried to use technology to reduce injury rates through a number of design features. These include the primary features of midsole hardness and cushioning, midsole stiffness, and motion control to try and influence kinematics and reduce injury. Running shoes change the biomechanics of running. Impact force peaks in barefoot heel strike running occur at between 5 and 10 ms after final contact, whereas in running shoes, they are delayed to between 10 and 25 ms (Nigg, 2000). Much of this is related to the midsole hardness or amount of cushioning. Although vertical loading rates have been shown to increase with increasing shoe hardness, peak vertical impact forces seem to have no correlation or a negative correlation with shoe hardness (Wright et al., 1998). It appears the body accounts for impact by muscle activation, altering limb stiffness prior to contact through sensorimotor feedforward and feedback mechanisms in response to the loading rate and the material or surface hardness that the foot encounters. This makes human response to material hardness subject specific (Wakeling et al., 2002), as is the energetic response (Nigg et al., 2003).

Midsole cushioning

Midsole hardness is a feature of running shoes, with softer midsoles sold as a benefit for shock attenuation. It is a footwear feature suggested to be varied in order to alter kinetics, kinematics, and injury rates (Nigg et al., 2012). The pinnacle of the cushioned approach is the use of a highly cushioned midsole in both the rearfoot and forefoot, with no increase in the height of the heel from the forefoot. This height difference is known as the *drop* of the footwear. The midsole can be 2.5 times thicker than that found in standard neutral shoes and is commonly known as a *maximal* or *extreme cushioning* running shoe, as opposed to the minimalist shoe with little or no cushioning. However, indications are that maximalist cushioned running shoes do not actually reduce running injuries. Despite no difference being noted in running economy over standard cushioned shoes (Mercer et al., 2018), studies have reported increased impact forces and loading rates when running in maximal cushioned versus standard cushioned neutral running shoes, suggesting that the maximal cushioned shoe might even increase injury risk (Chan et al., 2018; Kulmala et al., 2018; Pollard et al., 2018). It appears that running in these shoes increases leg stiffness (Kulmala et al., 2018), probably as a result of adjustments in muscular activity actioned through the sensorimotor system for softer terrain locomotion as suggested by Nigg and Wakeling (2001).

The effects of midsole hardness are influenced by gender and age but are also subject-independent regardless of these factors (Nigg et al., 2012). Midsole hardness affects components of motion during running in a subject-specific way, predominantly within the sagittal plane at the knee and ankle (particularly the ankle) rather than frontal plane motions at the hip and knee (Nigg et al., 2012). The sagittal plane changes were found to be of the same magnitude as those for ageing, suggesting that altering the midsole hardness may be a way to compensate for the changes that occur with age (Nigg et al., 2012). It has been reported that soft midsoles tend to decrease the range of motion at the hip and knee, but increase ankle dorsiflexion compared to hard midsoles, although again, the results are variable between individuals (Nigg et al., 2012). The problem is that although there are different effects from an individual running on hard, medium, or soft midsoles, the effects depend on that individual. This makes it impossible to accurately advise runners on one midsole density over another without testing them for biomechanical and metabolic data.

Midsole stiffness

Midsole stiffness is the resistance to bending or twisting of the shoe and gives the midsole elasticity. Stiffened shoes have been reported to be able to enhance running economy (oxygen consumption) during submaximal running (Roy and Stefanyshyn, 2006), but more recent research suggests this is not necessarily so (Beck et al., 2020). They create slightly smaller maximum rearfoot eversion excursions (Clarke et al., 1983; Dugan and Bhat, 2005) and less tibial rotation compared to softer midsoles (Dugan and Bhat, 2005). Despite reduced metabolic costs, no significant difference has been noted in lower limb muscle activation (Roy and Stefanyshyn, 2006; Beck et al., 2020). However, stiffness of the shoe across the forefoot that restricts MTP joint dorsiflexion may be counterproductive to efficiency, effectively becoming too stiff to allow any MTP joint fulcrum activation (Oh and Park, 2017). It is thought that the elasticity of stiffened shoes enhances resistive torque during MTP joint motion, reducing the negative work performed during dorsiflexion of the MTP joints, which is resisted by the digital flexors' muscle-tendon components (Stefanyshyn and Nigg, 2000; Roy and Stefanyshyn, 2006; Willwacher et al., 2013). This reduces MTP joint muscle-tendon effort (Oh and Park, 2017), and the elastic restoring force from bending the shoe can then assist forward propulsion during the limb extension on forward acceleration (Willwacher et al., 2013). However, totally blocking any pivoting across the MTP joint region may be counterproductive for acceleration by preventing an appropriate length being created for the plantarflexor power from the external Achilles moment arm. This decreases the acceleration GRF by disturbing dorsiflexion (extension) motion at the MTP joints (Willwacher et al., 2014; Oh and Park, 2017).

With stiffer midsoles, it has been noted that two effects can occur: the first is that the torque at the ankle joint increases to compensate for the longer moment arm, and secondly, stance phase duration is lengthened without significant changes in ankle torque (Willwacher et al., 2014). To maintain the steady-state take-off velocity of the CoM despite the reduction in acceleration GRF, a linear impulse must be maintained by increasing the duration of the propulsion (Oh and Park, 2017). Restriction of natural MTP joint extension (dorsiflexion) caused by stiffened midsoles and the joint torques it initiates in the lower limb serves as a determinant of running energetics. There is a gearing effect between the ankle and MTP joints that gives rise to a mechanical advantage such that greater MTP joint dorsiflexion results in greater transmission of ankle joint torque in developing GRF at propulsion. However, high levels of stiffness will allow the MTP joint extension moment to dissipate (lose) acceleration energy. Thus, desirable MTP joint-resisted dorsiflexion stiffness levels likely exist on an individual-based 'Goldilocks zone', where ankle torque transmission increases without preventing any loss of the ability to apply the GRF to the forefoot. High-stiffness shoes risk losing the mechanical efficiency driven by this gearing. Therefore, shoes or other wearable equipment should avoid significantly restricting natural MTP joint motion (Oh and Park, 2017). These findings suggest that complementary changes in lower limb torques are necessary to maintain steady running in stiffer midsole shoes. Furthermore, because different muscle groups are involved in joint torque changes, the metabolic effects on individuals are likely to vary (Oh and Park, 2017), not least because forefoot stiffness varies among individuals (Oleson et al., 2005; Willwacher et al., 2014). Thus, for each individual, a level of midsole stiffness may exist to improve energetics that reflects the natural stiffness of the MTP joints themselves (Fig. 1.4.11d). Indeed, very flexible MTP joints may benefit from stiffer running shoes across the MTP joint area. Presently, much research effort is being put into this subject to see if athletic performance of elite and average runners can be enhanced through these principles, with such stiffer 'performance' shoes perhaps being better for faster more elite runners (Beck et al., 2020; Day and Hahn, 2020; Ortega et al., 2021).

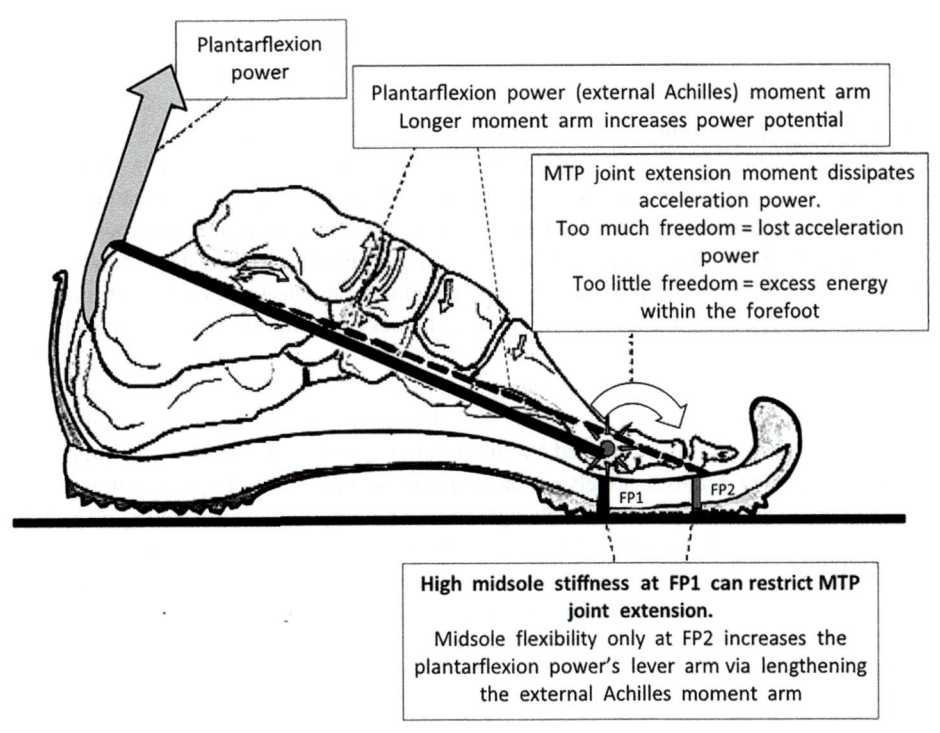

FIG. 1.4.11d Each individual likely has a 'Goldilocks zone' of acceleration energetics/biomechanics at heel lift that reflects factors such as foot flexibility levels, plantarflexor strength, and morphologies such as the Achilles' external and internal moment arm lengths, specific limb length ratios, and foot profiles. Running shoes can take advantage of this if correctly designed and fitted to the individual. Midsole flexibility levels around the forefoot and where that maximal flexibility lies across the shoe may have consequences to the Achilles external moment arm in creating plantarflexion power at the metatarsophalangeal (MTP) joint fulcrum *(star)*. This moment arm is often referred to as the plantarflexion moment arm *(thick black line)*. Shoe forefoot stiffness will also influence how freely the MTP joints can express extension moments around the MTP joints at heel lift. A rigid midsole situated at flexion point 1 (FP1) or with the footwear's flexion point positioned too distally at FP2, can both increase the plantarflexion moment arm length *(thick dashed line)*. Such a plantarflexion moment arm can create greater acceleration power application. However, this will also result in higher stresses on the Achilles by requiring greater ankle plantarflexor power, possibly leading to earlier plantarflexor muscle fatigue and an injury risk for the triceps surae-Achilles complex. It may also create increasing internal stresses within the MTP joints and forefoot soft tissues as well as altering midfoot and ankle plantarflexion moments during acceleration. Flexible shoes may result in more acceleration energy loss via increased MTP joint extension moments. This requires greater internal foot stiffening via the muscles to compensate and provide that safe 'Goldilocks zone' for the plantarflexion moment arm. *(Permission www.healthystep.co.uk.)*

Motion control and stability

'Motion control' and 'stability' are features described within running shoes designed to improve performance through the control of frontal plane motion, usually eversion. Both features demonstrate a rigid plastic reinforcement through the shank of the midsole, stiffening the shoe to prevent bending, except across the forefoot. Motion control shoes are the stiffest and also the heaviest running footwear. They have reinforced material running from the heel through to the forefoot along the shoe medially. Stability shoes have support under the medial foot vault area of the shoe. Both are usually sold as shoes that limit 'pronation' or 'overpronation' (an interesting statement that will need some explanation later), with stability shoes being used for moderate pronation and motion-control shoes for 'severe pronators' (whatever that means). Shoes lacking motion-control are usually referred to as neutral shoes. Motion control shoes have been reported to decrease peak internal tibial rotations (Rose et al., 2011), particularly in low vault-profiled feet (Butler et al., 2007), reduce internal knee rotation, and reduce rearfoot eversion in female runners (Lilley et al., 2013).

In a large randomised controlled trial, overall injury risk among participants who used motion control shoes seemed to be lower than those wearing standard shoes, but only in those runners with more pronated (low-vaulted?) foot postures (Malisoux et al., 2016). The reason for this is likely tied in with the changes noted in more prone feet becoming further pronated with fatigue (Cowley and Marsden, 2013; Fukano and Iso, 2016; Sinclair et al., 2017). On a cautionary note, a smaller randomised controlled trial of female runners found a higher injury rate, and that visual analogue pain scales immediately after running were higher with motion control shoes compared to neutral and stability shoes while taking part in a 13-week half-marathon training programme (Ryan et al., 2011).

Running footwear and foot vaults

Variable running shoe types applied to different foot vaults seem to have different effects on GRFs, with low-vaulted runners having lower instantaneous loading rates in motion control shoes and high-vaulted runners displaying a lower instantaneous loading rate in cushioned shoes (Butler et al., 2006). Shoes may also influence spatiotemporal parameters in foot vault types which are not present when running barefoot (Williams et al., 2004; Hernández-Gervilla et al., 2016; Hollander et al., 2018). This suggests that barefoot changes override the effects of foot vault types on spatiotemporal variables. It appears that fitting the right shoe to the right runner is important, but foot posture alone might not be the only criterion by which this should be achieved. Shoe design, shoe weight-to-patient weight ratio, terrain, training experience, patient morphology, and the distance run will all affect the biomechanics of locomotion, but as yet, some features of running shoes are not so well researched (Sun et al., 2020).

The biomechanical situation with running shoes is therefore difficult to simply unravel. Shoes do change kinetics, kinematics, and energetics. However, the response to any particular shoe is very subject-specific and greatly influenced by the strike position of the foot, which in itself is influenced by speed, fatigue, and terrain. In over 50 years of running shoe technology and research, the injury rate remains relatively unchanged (Lopes et al., 2012). Repetitive impact forces from running may not be directly important in causing injury, but the changes in muscle activity (both in intensity and timing) that attempt to minimise tissue vibration resonance may be linked (Nigg and Wakeling, 2001). Furthermore, this is something that may similarly be connected to kinematic changes in joint positions associated with running style and fatigue. There does not appear to be an injury risk-free running shoe or running style, but it is possible that some situations are more likely to provoke certain injuries than others. Despite much time spent on research, altering running shoe designs alone seems to have failed in decreasing injury rates, even if they have possibly prevented them from increasing (Tam et al., 2014; Sun et al., 2020).

1.4.12 The effects of running terrain

Runners can perform their activity on a number of different terrains such as treadmill, track, road, and country. This poses a dilemma for assessment of running style, as there are differences in biomechanical events between running terrains. The clinician must realise that road and country running can both involve significant hill running, with cross-country likely to produce the most gait perturbations. The constant need to adjust running style changes kinematics and can help to avoid overuse injury by varying motions more throughout each gait cycle. Such 'less sure' ground will tend to decrease stride length, which may have protective benefits from overuse injury. To counterbalance this, 'less sure' ground is more likely to result in traumatic events such as ankle sprains due to failure of the feedforward mechanisms to predict an unseen change in surface. Distance track running places the runner on a surface designed for running, yet unavoidably, it involves the need to run around corners repeatedly in the same direction, usually anticlockwise. This initiates asymmetry in each limbs' biomechanics that may alter anatomy between them, such as plantar soft tissue stiffnesses (Shiotani et al., 2021).

Road running involves running on artificially hard surfaces that tend to encourage longer stride lengths due to surface consistency, potentially increasing the risk of fatigue and overuse injury. Treadmill running provides the most consistent postural running style and is usually the most practical way in which to assess a patient's running style clinically. This is a sprung surface and if treadmills are not the usual running terrain of the patient, then it is important to know how such a surface might alter the running biomechanics from that which was related to the cause of their injury. The use of a treadmill for assessment may not be appropriate if the runner is not familiar to treadmill running, unless time can be given to practice before assessment. This is also true for walking (Meyer et al., 2019), although clinical space for walking assessment is far easier to find than clinical space for running assessment. In running, slightly shorter stride times and higher cadences are reported on the treadmill, although mean differences appear small (Riley et al., 2008; Cronin and Finni, 2013). However, again, a period of acclimatisation for running assessment of 6–8 min is recommended (Lavcanska et al., 2005).

On rough terrains, stability becomes an increasingly important part of biomechanics and potential injury. From an evolutionary standpoint, it is probably the most important terrain to study. Yet, such terrains seem less likely to produce overuse injuries, probably because they initiate increased movement and muscle action variability. Overuse injuries are the result of repeating events. Obstacles and inclines are frequently encountered in the natural environment, the conditions in which human running evolution occurred. Due to changes in surface mechanical properties, the sensorimotor system is far more active in adjusting the locomotive system's compliance and stiffness than on consistently flat surfaces. This is because of the challenge in the latencies of nerve impulses and the dynamic nature of running. In human endurance running, the stance phase lasts around 200 ms, which is only slightly longer that the proprioceptive feedback time of 70–100 ms or the visual feedback delay of 150–200 ms (van Beers et al., 2002). Runners must therefore use anticipatory strategies to control strike position (Dhawale et al., 2019). When feedforward mechanisms fail to correctly predict perturbations in terrain, injury can result from stumbling or *orientation instability* and/or from a failure in the ability to maintain a steady controlled speed, known as *translational instability*.

Energetics during running is affected negatively by uneven terrain, with energetic costs reported to increase by around 5% over flat terrain running (Voloshina and Ferris, 2015) (Fig. 1.4.12a). About half of this energetic increase is explainable from the changes in positive and negative work of increased up and down leg movement for steps over objects. This is consistent with the results from studies comparing walking on flat and uneven terrain which increases net metabolic expenditure by around 28% (Voloshina et al., 2013). Mean muscle activity is reported to increase on uneven terrain for three muscles within the thigh: vastus medialis (7%), rectus femoris (20%), and semimembranosus (19%). Other muscles did not show an increase in activity or increases in muscle coactivation, but they all demonstrated an increased activity variability between strides (Voloshina and Ferris, 2015). Variability is also seen in force-time curves on uneven terrain. Peak maximum force was not statistically different on different terrains, but the peak did increase by around 17%, while the vertical GRF variability more than tripled when running on uneven terrain (Voloshina and Ferris, 2015). It was also noted that body posture became slightly more trunk-crouched and limb-flexed at heel strike resulting in a functionally shorter limb when running on uneven terrains, with around a 15% decrease in maximal change in leg length and the leg being held more stiffly (Voloshina and Ferris, 2015). This probably explains the increase in peak forces reported.

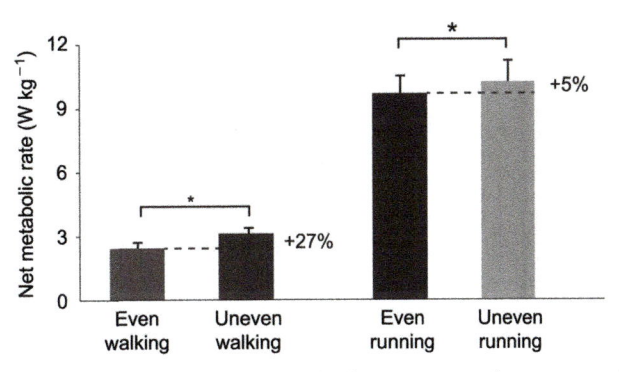

FIG. 1.4.12a The difference in net metabolic effects for walking and running between even and uneven ground. *(From Voloshina, A.S., Ferris, D.P., 2015. Biomechanics and energetics of running on uneven terrain. J. Exp. Biol. 218 (5), 711–719.)*

Reassuringly, for clinicians forced to assess rough terrain runners on flat clinic facilities, joint kinematics and kinetics change little on uneven terrain compared to flat surfaces with the most significant changes occurring at the ankle. On uneven terrain, mean joint angles in the sagittal plane show only slightly higher peak flexion angles in the hip and knee at midstance. This maintains a longer leg (probably aiding swing ground clearance) while the ankle shows a slightly decreased range of motion. Yet, subjects tested seem to maintain a similar heel strike footfall pattern between surfaces (Voloshina and Ferris, 2015). The reduced range of ankle motion is suggested to be a result of landing with the foot less dorsiflexed to the support surface on uneven terrain, with a decrease in ankle range of motion of approximately 14% at midstance (Voloshina and Ferris, 2015). Changes in joint power of 29% and 23%, respectively, were only recorded at the knee and ankle in midstance, but ankle power significantly decreased by 5% prior to acceleration (Voloshina and Ferris, 2015). No difference between terrains was noted in mean step width, length, height, or timing. However, as well as limb stiffness and altered force-time curves, gait variability also increased, including the joint angles with hip and knee variability doubling and ankle movement changing by around 60% (Voloshina and Ferris, 2015) (Figs 1.4.12b and 1.4.12c).

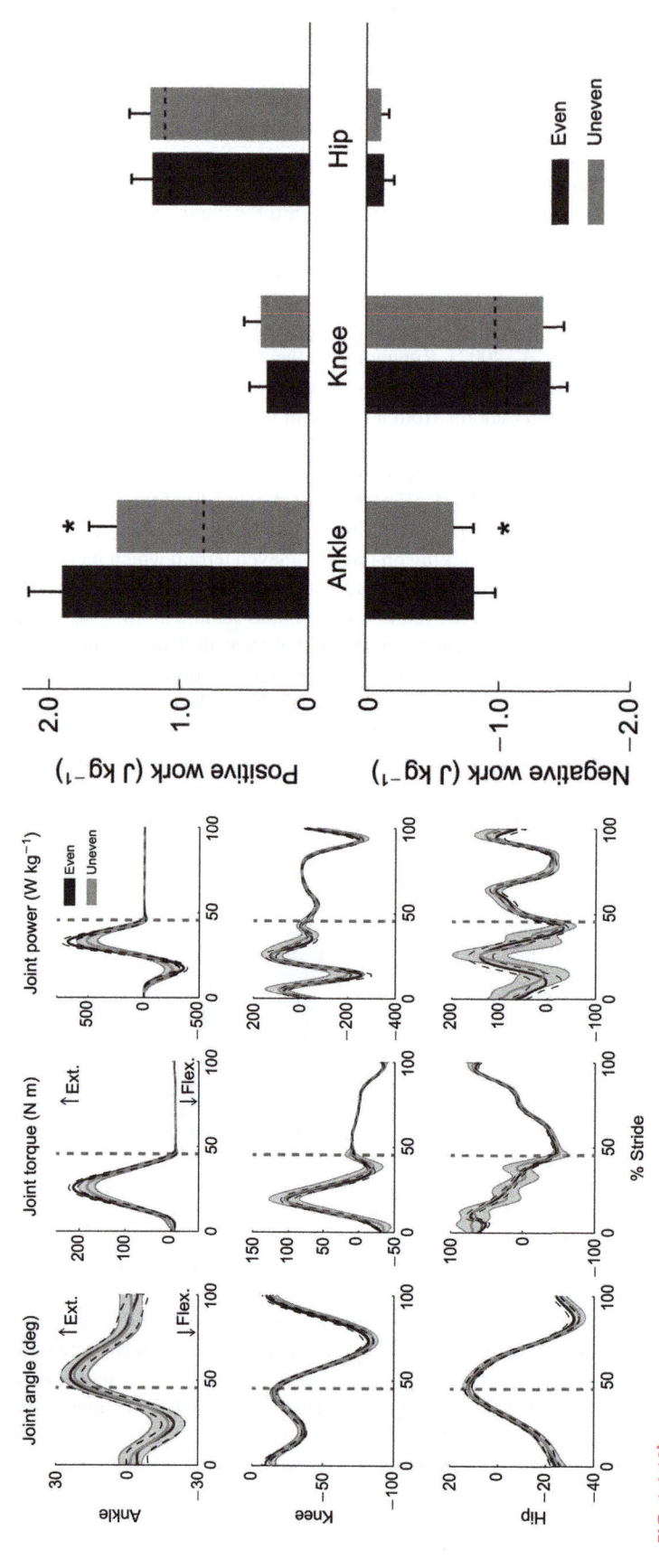

FIG. 1.4.12b Joint angles, torques, and power versus stride time for running on even and uneven terrain. *Solid lines* plot the mean trajectories for ankle, knee, and hip against percentage stride time, with shaded areas denoting the mean standard deviations for uneven conditions and dashed lines for even terrain. The *dashed vertical lines* represent toe-off, negative scores signify flexion, and positive scores extension (dorsiflexion). Ankle, knee, and hip work per stride are compared on the right images. *Dashed lines* indicate net work, *error bars* denote standard deviation, and asterisks a statistically significant difference. *(From Voloshina, A.S., Ferris, D.P., 2015. Biomechanics and energetics of running on uneven terrain. J. Exp. Biol. 218 (5), 711–719.)*

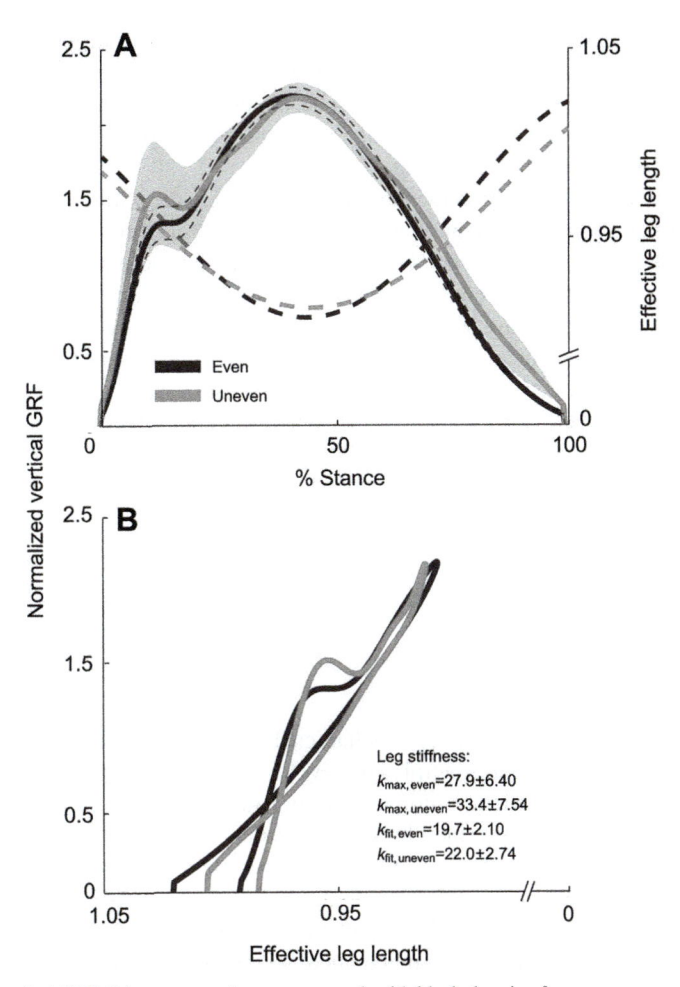

FIG. 1.4.12c The difference in vertical GRF (A) on even and uneven ground, with black showing forces on even ground and grey, uneven. Shaded area denotes the standard deviations across test subjects. Normalised vertical GRF is plotted against normalised effective leg length to calculate leg stiffness (B). Mean stiffness values have been calculated: K_{max} equals the maximum force divided by the maximum leg length displacement and K_{fit} is the slope of the linear fit to the leg stiffness curve. *(From Voloshina, A.S., Ferris, D.P., 2015. Biomechanics and energetics of running on uneven terrain. J. Exp. Biol. 218 (5), 711–719.)*

The lower limb joint compensations during running on uneven terrain are very different to those reported for walking (Voloshina et al., 2013; Voloshina and Ferris, 2015), which reflect the difference in techniques between the inverted pendulum of walking and the spring-mass system of running. The decrease in ankle work seen in uneven terrain running likely reflects the fact that distal joints rely on high-gain proprioceptive feedback as they are first to encounter perturbations (Daley and Biewener, 2006), whereas more proximal muscles have to rely increasingly on feedforward anticipatory control. The reduction in ankle motion is likely to reduce the energy stored within the Achilles tendon and, hence, reduce ankle power. This is also likely to diminish the risk of Achilles injury. Humans tend to stiffen their joints when presented with unfamiliar tasks and not surprisingly uneven terrain produces a similar reaction and induces increased lower limb stiffness. This probably increases peak GRF through landing with a higher contact force, but with the foot placed flatter to the support surface, and this angulation at contact reduces tibial accelerations and the energy dissipation required through tibialis anterior (Voloshina and Ferris, 2015).

1.4.13 Section summary

Running is a far more diverse activity than walking, having its own distinct gait phases and energetics with each style of running changing the biomechanics dramatically. Certain anatomies and morphologies benefit one running style over another where shorter distal legs (not whole legs) and calcanei associated with larger proximal muscle bulks benefit sprinters, and long distal legs with a small, short trunk and low body mass benefit distance running. Ultra-distance runners gain from adapting their running style and physiology to a state of fatigue rather than demonstrating a particular morphology. Gender and age also influence running style through morphology and physiology differences, with aged running demonstrating features associated with fatigued running.

Foot morphology and footwear significantly influence running biomechanics. In the case of the foot vault, the effects seem linked to the ability to provide appropriate compliance and stiffness levels throughout the running phases to dissipate energy. Foot morphology and stiffness will influence energy production to generate efficient GRFs as part of an overall lower limb ability to control these features. The effect of shoes seems to relate primarily to their weight and their influence on the foot strike position, with barefoot running being a viable option if the foot is strong enough to adapt to the changes in loading rate that occurs. Such foot-related factors may relate to the overall morphogenesis of the foot during growth such that those who have always run barefoot may be best left that way, unless shoes are necessary for environmental protection or are carefully selected for performance. Stiffness across the forefoot seems important for performance and if carefully chosen for the runner, seems to enhance performance particularly in elite distance runners. Shoe selection for performance and injury protection is a science-based art, which must be understood in the context that there will never be a universal shoe that is best for every runner.

1.5 Variance in gait
1.5.1 Introduction

The concept of 'normal' is a reoccurring problem in biomechanics. Evolution needs anatomical and genetic variation and developmental plasticity in order to work. The principles of developmental biology, genetics, and epigenetics all play a part in each patient's anatomy and each patient's unique anatomy influences the biomechanical processes that occur during gait. Although fundamental mechanical principles underpin energetically efficient human gait, the subtle nuances of each individual can only be interpreted in the context of their specific anatomy in regard to locomotive efficiency. Most of an individual's musculoskeletal variation such as tendon and muscle variation will remain unknown clinically, which brings us to the 'known unknowns' of clinical biomechanics. However, there can be little doubt that certain morphological and anatomical variances give individuals mechanical advantages and disadvantages during certain tasks and specific activities.

Thus, not all gait should be identical between individuals, and different conditions will change gait. Whether the clinician feels that aspects of gait recorded in the clinic are 'abnormal' and potentially 'pathological' to the subject may reflect more the prejudices of the clinician rather than reflecting the actual science. An ability to appreciate that each patient has a unique 'normal' comfortable gait which adjusts to the demands of living is very important. The concept of an operational comfort zone such as preferred walking speed is important in the complete picture of locomotion and the physiology of the body, set as it is by developmental events and health during the life of the patient thus far. Clinicians aligning themselves with models of gait that have 'absolute expectations' as to what kinematic-kinetic events must occur at certain phases and prescribing 'therapeutic correction' accordingly will probably do patients little good in the long run.

There are anatomical variances among lower limb morphology that increase the likelihood of specific pathologies because of changes in loading patterns in muscles and/or joints. If loads are tending to be concentrated in some anatomical areas or structures, then they are likely to be reduced within others. This would make pathology less likely in the less loaded areas, while increasing pathological risks in others. However, such relationships are not quite as clear-cut as the foregoing statement suggests. This is because life events can still cause injury in areas not normally subjected to high loads, and the effect of the biological response means that tissues in areas under consistently higher loads tend to get stronger and those in areas under less stress have reduced strength. Changing the loading patterns because of perceived increased stress by the clinician may end up increasing load in weaker structures, while reducing the loading in structures that have adapted to higher loads. As always, it is whether the encountered stresses in life breach the mechanical properties of the loaded tissues that indicates whether or not changes need to be made clinically. The key to clinical gait assessment is the ability to link pathology to gait, either as a cause of abnormal parameters or the result of the patient's abnormality. On reaching a conclusion on the pathomechanics, the clinician requires the ability to guide the appropriate interventions in order to achieve resolution or improvement of the mechanical stresses accordingly.

Age is a normal expected determinant of gait, for children are not small adults and they deviate most when initially starting independent locomotion and developing towards an adult gait. Thus, those that assess childhood gait need to be familiar with the expected changes during growth. Old age also results in expected changes through neuromyofascial changes with gait-related age not always reflecting chronological age, unlike changes expected during childhood. Treating normal childhood or elderly gait because it is not like that of a young adult paves the way to a clinical court case.

Footwear use also influences locomotion. The more complex the construction and design of the shoe, the more it will influence gait and posture. Short-term use and variability of styles in daily living is probably important in preventing postural and gait changes becoming fixed through the biological response to footwear-altered stress. Muscular atrophy may also occur due to disuse from reduced ranges of motion caused by features such as higher heels that can decrease the amount of dorsiflexion at the end of midstance. The further a shoe takes the foot away from the barefoot state, the greater the risk of abnormal changes. Gait analysis performed in footwear must consider the influence of the shoes being worn and likewise during barefoot tests must consider the influence on gait and posture of the footwear usually worn.

The use of a prosthetic limb will also result in a gait outside that of the expected values for the possession of two healthy limbs, not least because of reduced sensorimotor input. However, in general, the success of the prosthesis is how well it permits gait to mimic that of healthy populations. In athletic activity, prosthetic assessment becomes a particularly skilled task. Prosthetic joints will similarly cause gait changes from normal healthy gait and hopefully from the pathological presurgical gait, which makes preoperative locomotive assessment particularly helpful in assessing postoperative mobility outcomes quantitatively.

In this section, we consider the many types of gait that clinicians expect to see as a result of health variation, growth, and age. Sadly, not all can be considered within this text. Students and clinicians should seek out further research on their path to specialism in the biomechanics and gait of their specific patient groups. This section is an attempt to instruct the clinical biomechanist that 'normal gait' parameters are set by many factors that can still be 'healthy', even if set outside the classic expectations. If the patient's anatomy also veers away from the 'classic norms', they become more at risk of aberrant events. However, they may still function perfectly well within their anatomy, developmental stage, and age group without increased risk of pathology. Sometimes, these anatomical variances are associated with or present similarly to diseases and pathology, and it thus becomes necessary for the clinician to unpick the origins of the gait variation in order to understand the nature of the events they are investigating clinically.

1.5.2 Gender and other morphological differences in walking gait

Before discussing the differences in gait between males and females, it is important to stress, as discussed in gender differences in running, that the variances are largely brought about by morphological differences. These morphological variances are not unique to one gender or the other, but tend to gravitate more towards one, and may also be influenced by ethnicity. Clinicians should never presume a 'normal' for a patient based on their birth gender, but they must be aware that generalised differences exist based on that sex. Many pathologies such as osteoarthritis show no significant gender difference, while other lower extremity pathologies are more common in females than males (Gabriel et al., 2008).

The average male foot is generally longer and wider than the average female foot, while the female foot is generally narrower and higher in height than the male foot (Luo et al., 2009). Obviously, such dimensional morphological differences affect the data collected during gait analysis, influencing quantitative data such as pressure directly. Males, on average, have a larger angle of CoP progression away from the line of progression of the body during contact, midstance, and propulsion than females generally (Chiu et al., 2013), which possibly reflects their wider foot profile.

Studies have suggested that biomechanical stability is more challenged in females than in males due to less active muscle stiffness (Blackburn et al., 2004, 2006). Females seem to exhibit less dynamic ankle joint stiffness than males during the absorption of mechanical energies throughout midstance and during the generation of energy at late terminal stance acceleration (Gabriel et al., 2008). Dynamic joint stiffness is used as an indicator of joint stability (Hansen et al., 2004; Gabriel et al., 2008) and is associated with gait performance and injury risk (Butler et al., 2003). Increased dynamic joint stiffness is also associated with increased gait velocity and increased gait economy. However, in excess, it is possibly associated with increased bone injury rates, although when insufficient, soft tissue injury (Gabriel et al., 2008). Bone injuries are associated with greater peak forces and loading rates (Williams et al., 2001b, 2004), factors that also correlate with joint stiffness (Butler et al., 2003). However, females are more prone to hormonal-induced bone physiology changes with ageing, altering the potential tissue injury risk compared to younger females and males via altered mechanical tissue properties.

Effects of height

Taller people have a greater leg length which enables them to walk faster (Hof, 1996; Zijlstra et al., 1996). In the normal clinical situation, this matters little, but in monitoring children that grow, it can impact upon differences in gait data collected over time when reviewing from previous clinical consultations. Differences in height can also give rise to issues when comparing data across individuals during studies. Optimal walking speeds should correspond to the same *Froude number*, which is a dimensionless number used to normalise walking speed irrespective of geometric size (Leurs et al., 2011). The Froude number is calculated from speed of locomotion, acceleration due to gravity, and the length of the leg and is useful for comparing the walking of animals of different sizes with similar geometries. This means that it is excellent for comparing humans of different heights (Leurs et al., 2011). Interestingly, Froude number calculations have shown that quadrupeds and humans change from walking to running or trotting at the same Froude number (Leurs et al., 2011).

Optimal walking speed is when the recovery of energy due to the inverted pendulum mechanism is maximal and metabolic cost is minimal. Adults, children, and humans with dwarfism (including pygmies) all have optimal walking speeds that correspond to the Froude number 0.25. It is likely that limb segment (thigh, shank, and foot) lengths are optimised for their role in the energetics and kinematics of walking (Leurs et al., 2011). In a clinical situation, encouraging a patient to walk at their preferred speed should be enough for clinical accuracy. The clinician just needs to be aware that the taller the patient and the longer their legs, the longer their stride and the faster their step time should be. In addition, in those patients that have grown, changes in spatiotemporal gait parameters may reflect the height change and not the treatment.

The effects of lower limb alignment

Lower limb alignment in gait refers to the limb position during functional loading within stance phase, rather than at a single moment in time. This is not to suggest that a change in alignment at a single moment during the gait cycle is insignificant. In fact, it can be quite the contrary and highly significant to the stability capability at a joint. Limb alignments that relate to skeletal positions are part of the processes of development and health that link to adaptations of lifestyle and environments. Yet, these limb alignments may change over time with pathologies such as osteoarthritis. The alignment of individual segments to each other affects the direction and magnitude of the joint moments in each plane of motion and the forces of compression and tension when moments are changed in directions not normally associated with large joint motions. For the hip, knee, ankle, and MTP joints, the primary joint moments are in the sagittal plane (flexion/extension), and these joints are capable of handling high sagittal plane forces through adaptation of muscle activity. However, they are more limited in this ability to protect the joints from high or repeated moderate frontal and transverse plane forces.

The angulation of the knee in the frontal plane during gait can influence whether or not the knee articular surfaces are loaded relatively evenly, thereby keeping stresses spread appropriately within the knee. However, alignments that put the knee in a more varus or valgus orientation can change the intensity of the adduction or abduction moments in the knee respectively. For example, coxa valga and genu or tibial varum will increase the varus alignment of the knee, increasing knee adduction (varus) moments (Fig. 1.5.2). Knee adduction moments in the frontal plane are associated with the development and progression of osteoarthritis of the medial compartment of the knee (Chehab et al., 2014; Mahmoudian et al., 2016; O'Connell et al., 2016). The association of pain levels with this adduction moment changes with the extent of the disease (Henriksen et al., 2012; Hall et al., 2017).

It is impossible to cover all the nuances of the kinetic and kinematic variations that can occur in combinations of alignment variations within this text, but the principles can be explained using specific joint examples. However, it must be appreciated that events occurring due to one set of joint alignments will be influenced by alignments above and below the joint the clinician is interested in. The most significant kinetics and kinematics caused by alignment issues will be those judged to be influencing the pathology. Therefore, the clinician is establishing the *pathomechanics* within gait rather than seeking everything that deviates away from what they perceive as normal. An important part of this assessment is the foot's alignment, as the foot is the final interface with the ground, and everything above is influenced by what happens to the foot during its ground contact. The foot should be the first structure able to compensate for any terrain perturbations.

This brings us to a group of three alignments: the distinct foot 'types' known as *pes planus* (low foot vault profile) and *pes cavus* (high foot vault profile), and the *rectus foot* (of middling foot vault profile) that blends between the two distinct foot types, with the border between these three types having not yet been clearly established quantitatively. Attempts have been made through the clinical semiquantitative FPI (Redmond et al., 2006) and via quantitative techniques such as the arch index and navicular drop (McPoil et al., 2016; Fraser et al., 2017). Radiographical attempts using osseous alignment angles and contours of joints have also been used. The semiquantitative technique of the FPI is presently a commonly used clinical and research technique used in trying to determine foot type, creating scores of 'pronation'. The assessment of gait

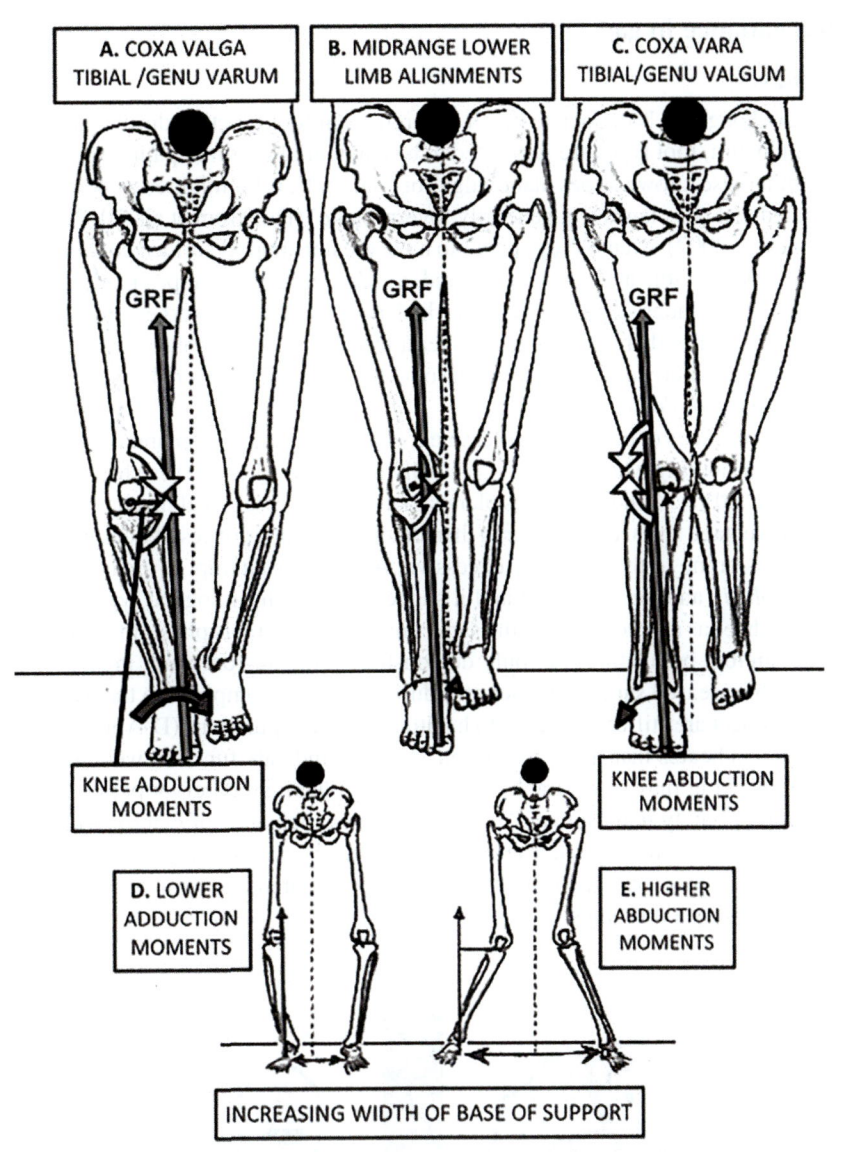

FIG. 1.5.2 Femoral and tibial alignments are important for the generation of knee moments during gait. If angles of frontal plane osseous alignments are in excess, they may provoke pathology. Frontal plane torques at the knee result from not only the osseous alignments of the limbs, but also the hip varus angle during contact and single-limb support. With midrange femoral neck-shaft angles, the foot usually contacts medially to the hip axis as a result of adducting the lower limb under the body to support the CoM during stance (B). This creates small net hip and knee adduction (varus) moments generated by the GRF. Such forces primarily load the medial compartment of the knee, because usually limb adduction also brings the GRF medial to the knee. The foot requires a small rearfoot eversion moment to bring the plantar foot flat to the support surface from the slightly inverted limb contact position. In coxa valga combined with tibial or knee joint (genu) varum alignment (A), the GRF becomes more medially orientated to the knee, increasing the knee adduction moment. The greater inverted limb angulation that this causes increases rearfoot eversion required to bring the foot flat to the support surface. In coxa vara and tibial/genu valgum alignments (C), the GRF force is applied from a foot contact position more lateral to the medial knee compartment. This usually maintains the hip adduction moment but offloads the medial knee compartment by creating an abduction moment that increases lateral knee compartment loading. Less eversion moments will be required to bring the foot flat. However, if the GRF shifts too far medially on the foot as a result, it will induce an inversion moment on the rearfoot at heel contact and may invert the forefoot in loading response that can increase distal vault flattening. The base of support (lower images) is also an important concurrent factor that repositions the GRF. Narrowing the base of support increases knee varus moments, while widening it decreases external varus moments but increases valgus knee moments (D and E), factors which must also be considered in respect to the individual's fixed osseous lower limb alignments. *(Permission www.healthystep.co.uk.)*

kinematics and kinetics are affected by foot anatomical variants, but they often involve multiple alignment relationships between foot segments and are influenced by muscle function and connective tissue compliance. What must also be considered is that each foot has its own internal anatomical variants that make each foot unique, even when assigned clinically to this or that foot type. To attempt to give some coherence to the problem, each foot type's gait function is best considered separately from other lower limb issues that they may be associated with.

1.5.3 Foot function variance in gait

Classifying pes planus

Variable foot structures are thought to be associated with differences in foot function during gait (Buldt et al., 2013, 2018; Hillstrom et al., 2013) and in turn, foot types are suggested to be a potential origin of pathology (Ledoux et al., 2003; Levy et al., 2006). Pes planus, otherwise known as pes planovalgus or 'flat foot', is highlighted as being a source of many pathologies and it is often repeated that pes planus feet generally 'overpronate' (Hillstrom et al., 2013). However, the evidence is not definitive (Michelson et al., 2002), even if they appear to have greater support surface contact and look more prone. Pes planus can be divided into both flexible and rigid groups and, as such, it seems that pes planus feet are far from homogeneous. The pes planus foot type includes several variations which may have led to overgeneralisations of the significance to pathology of some forms. The author therefore suggests at least three forms of pes planus exist: *anatomical*, *hypermobile*, and *acquired*. These represent distinctly different feet with some functional crossover. The acquired pes planus can further be divided into *functional* and *pathological*.

One of the most common mistakes of pes planus assessment is to correlate all pes planus foot types with excessive, over, or hyperpronation, without considering what these terms actually mean and whether these feet match any present definition (Horwood and Chockalingam, 2017). Although all pes planus foot types represent feet that have a lower vault profile and, as a consequence, a greater percentage surface contact area than other foot vault profiles, it does not mean that during gait, the amount that the foot becomes prone to the ground is excessive. A previous attempt at creating a taxonomy of pes planus divided 'flatfoot' into *pathologic* and *physiological* and it suggested that physiological types were associated with loose joints, obesity, and wearing shoes as a child, while at the same time stating that this physiological form was a normal variant (Staheli, 1999). Habitual barefoot populations not only demonstrate a higher foot vault in static stance but also a wider, more prone foot posture with increased surface contact area during walking compared to the habitually shod, although not so during running, when their feet are stiffer compared to habitually shod populations (D'Août et al., 2009; Hollander et al., 2017; Mei et al., 2020). This indicates that static assessment does not indicate function well and that more and adaptable pronation of the foot when walking may be desirable. It is likely that habitual barefoot populations demonstrate stronger feet that can adapt more appropriately to terrain and the biomechanical demands of gait, providing a stiffer foot for static stance and running, and something more responsive and compliant during walking (Figs 1.5.3a and 1.5.3b). Thus, Staheli's (1999) classification possibly needs some revision and although not stipulated, the term 'loose joints' used within this classification may mean hypermobility syndromes.

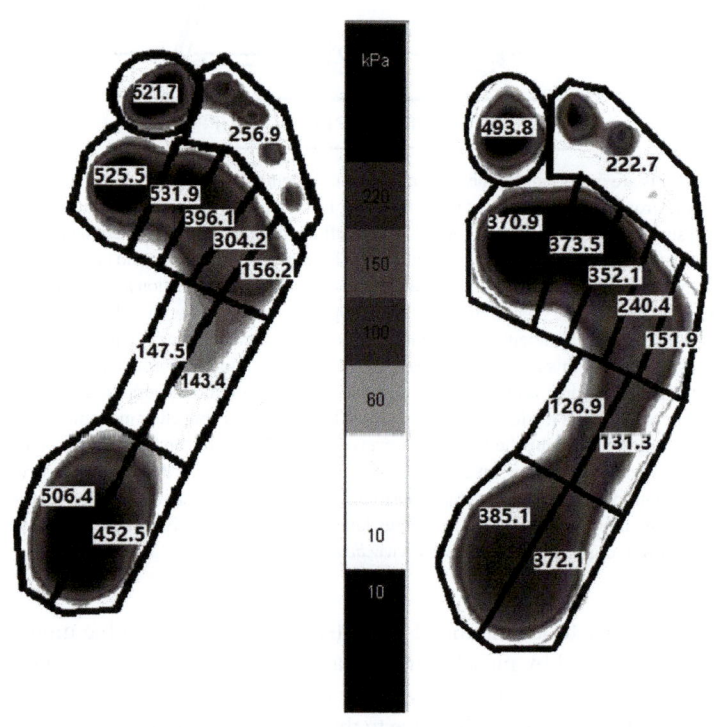

FIG. 1.5.3a Pressure data of mean peak pressures (kPa) taken during walking from habitually shod age-matched young male subjects *(left)* and habitually barefoot male subjects *(right)*, compared. The increased contact surface area and associated decreased peak pressures within habitually barefoot population suggests more foot compliance under loads during walking. *(From Mei, Q., Gu, Y., Xiang, L., Yu, P., Gao, Z., Shim, V., et al., 2020. Foot shape and plantar pressure relationships in shod and barefoot populations. Biomech. Model. Mechanobiol. 19 (4), 1211–1224.)*

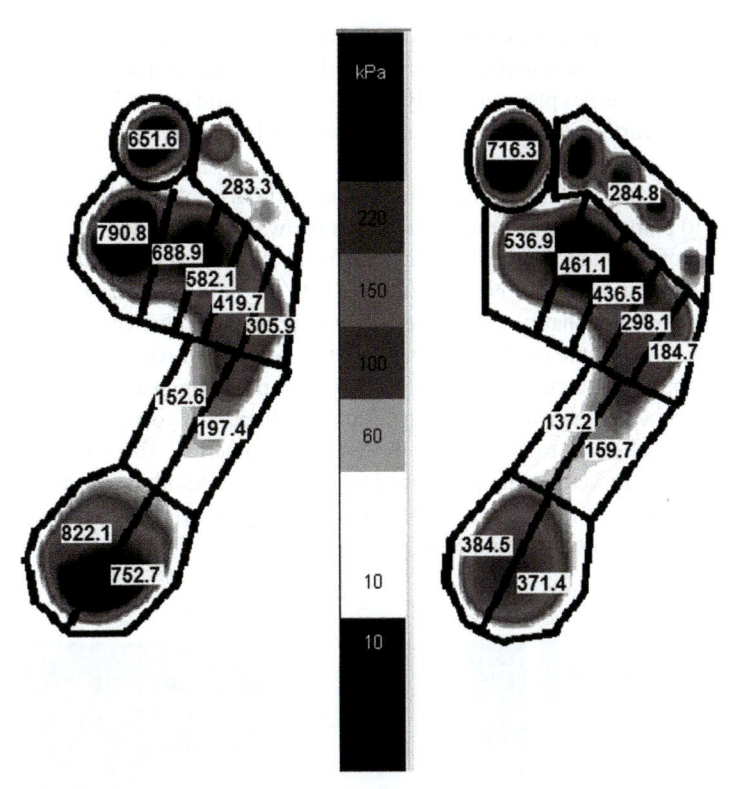

FIG. 1.5.3b Pressure data of mean peak pressures (kPa) taken during running in habitually shod *(left)* and habitually barefoot *(right)* young males. This data appears starkly different to walking data shown in Fig. 1.5.3a, as the habitual barefoot population display a reduced contact surface area, greater hallux pressures, reduced heel peak pressures, and more widely spread forefoot peak pressures. When considered together, the pressure data sets from walking and running may indicate a greater ability to modulate foot stiffness properties and muscle powers appropriately to specific gait requirements in habitual barefoot populations. *(From Mei, Q., Gu, Y., Xiang, L., Yu, P., Gao, Z., Shim, V., et al., 2020. Foot shape and plantar pressure relationships in shod and barefoot populations. Biomech. Model. Mechanobiol. 19 (4), 1211–1224.)*

The anatomical pes planus is a foot phenotype where the profile of the foot is constructed with a low vault, present on both nonweightbearing and weightbearing, and that usually becomes variably lower in profile on weightbearing. This foot type is not always associated with joint hypermobility or pathology resulting in 'hyperpronation', as defined by Horwood and Chockalingam (2017). It should be considered only as a functional anatomical variant, but one that is possibly more associated with certain ethnic groups and growing up habitually shod (Hollander et al., 2017).

The mobile pes planus is a commonly associated clinical feature of hypermobility associated with a significant Beighton scale score within the Villefranche criteria (Beighton et al., 1998). These feet can be low, normal, or even high vault-profiled on nonweightbearing. However, on weightbearing, the vault profile will change dramatically, often resulting in a 'pes planus-looking' foot with a high FPI score as a result of body weight-induced pronation, and during gait will often generate a midfoot break during heel lift (see Figs 1.5.3f and 1.5.4a). Finally, the acquired pes planus is a foot that becomes hyperpronated on weightbearing due to dysfunction of the foot's anatomy and/or its extrinsic soft tissue support muscles. This can be due to general foot weakness, which can be referred to as *functional pes planus* (otherwise termed a hyperpronated foot as the amount of pronation becomes excessive to need), or *pathological pes planus* which is a result of specific pathologies such as tibialis posterior dysfunction. Pathological pes planus feet also often demonstrate highly abnormal and often asymmetrical levels of foot pronation between feet. In time, these feet may become fixed into a planus profile, even though initially, the planus profile may only be noted on weightbearing. A risk to developing these pathologies that creates severe foot vault depression may be associated with certain footwear use as an adult, and too much supportive footwear use, especially during growth (Hollander et al., 2017).

The foot should only be considered to be hyperpronated if the demands placed upon it by kinetic and kinematic events to provide efficient energetic gait and tissue strains for energy dissipation during locomotion are exceeded 'pronation-wise' by the foot (Horwood and Chockalingam, 2017). Thus, they become excessively prone on the support surface. Feet should have a lower vault profile in static stance and during dynamic function than when nonweightbearing, flattening more rapidly at the start of midstance and becoming lowest just before heel lift (Hunt et al., 2001; Stolwijk et al., 2014); events that are part of a necessary foot stiffening process (Bjelopetrovich and Barrios, 2016; Takabayashi et al., 2020). Hyperpronation or similarly used terms must be reserved for stance phase events that permit the foot to become more prone to the support surface than is demanded for efficient gait, for safe terrain traversing, or for perturbation management. Feet or parts of the feet may only become hyperpronated during certain phases of gait, and although hyperpronated when in these phases, they should not be referred to as pes planus feet. Hyperpronation should not be used as a synonym of pes planus, and as such, pes planus as a term needs further clarification. Pes planus should be reserved for feet that appear low profiled in the vault when nonweightbearing.

Tibialis posterior dysfunction is the most common serious acute form of adult acquired flat foot occurring as a result of tendon degeneration leading to pathological acquired pes planus (Edwards et al., 2008). Tibialis posterior creates forces used in controlling the stability of the vault, establishing foot stiffness, and has an important role in generating the terminal stance acceleration GRF (Kokubo et al., 2012; Neville et al., 2013). Its failure results in hyperpronation. Modelling by Wong et al. (2018) clearly demonstrated that tibialis posterior dysfunction will change the vault profile and its mechanics during gait, but they only stated that all the pathomechanical factors associated with it potentially contributed to the progress of pes planus. It is thus important to clarify that tibialis posterior pathology causes the development of acquired pes planus rather than it developing as a proven result of having pes planus. In tibialis posterior dysfunction, the rate and amount of pronation will be greater than it was prior to the onset of the tendon dysfunction. This new foot profile can be classified as an acquired pathological pes planus, producing higher clinical pronation scores within the FPI and a lower vault height than it would have had before the pathology. Unresolved, acquired pes planus feet can become fixed and the patient will usually be aware of a change in their foot profile over time (Fig. 1.5.3c).

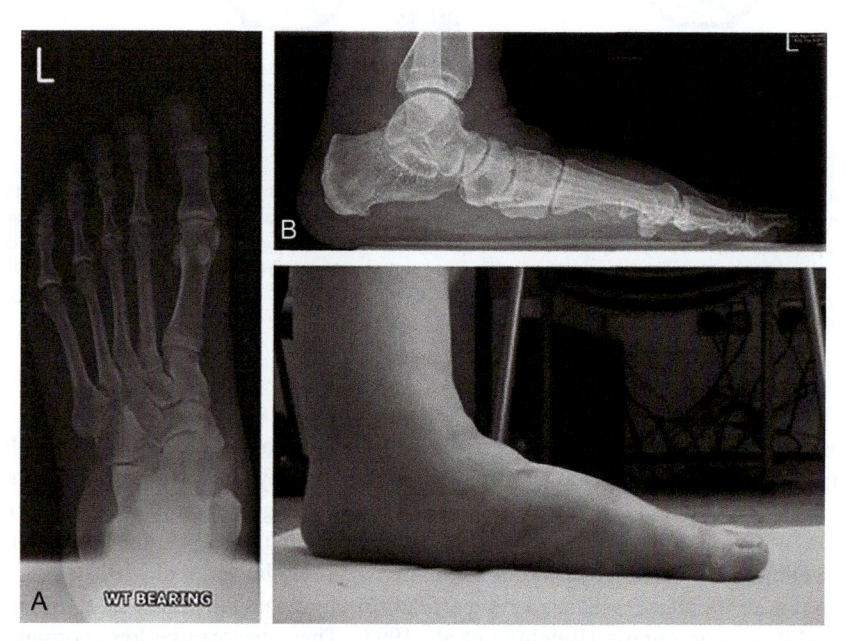

FIG. 1.5.3c Tibialis posterior dysfunction is the commonest form of acute acquired flat foot and clearly demonstrates hyperpronation of the foot. Changes in foot profile in stance are seen on antero-posterior (AP) (A) and lateral (B) radiographic views of the affected foot, with 'uncovering' of the talar head by abduction of the navicular at the talonavicular articulation on AP views and lack of a distinct medial vault on the lateral view. However, clinical observation of vault asymmetry between the more prone affected foot and the unaffected foot, combined with a history of changing foot profile and tibialis posterior muscle strength assessment, are equally diagnostic. *(From Edwards, M.R., Jack, C., Singh, S.K., 2008. Tibialis posterior dysfunction. Curr. Orthop. 22 (3), 185–192.)*

The anatomical pes planus, on the other hand, will also score high on the FPI and in other clinical tests, but it may not actually be hyperpronated, and thus still functions normally, even generating efficient acceleration GRF in terminal stance. In anatomical pes planus feet, the patients will have noted that their foot vault profile has been low since they can remember. These distinctions are important to recognise before assessing gait.

Therefore, acquired pes planus feet, where the feet have changed shape due to dysfunction and/or pathology, are not the same as anatomical pes planus feet which have developed as part of the developmental landscape of the individual. The hypermobile foot represents a different functional group from pathological or anatomical pes planus. A clue to the overlap of hypermobility and pes planus is indicated in a report by Francis et al. (1987) on tarsal tunnel syndrome where eleven reported cases all had pes planus feet, but generalised joint hypermobility was noted in nine of the subjects. Yet all too often, the three pes planus foot types are indiscriminately thrown together by clinicians and researchers, complicating the information on how pes planus feet should function asymptomatically and normally for their anatomical variant phenotype. Anatomical pes planus can be argued as being a common human foot anatomical/morphological variation, in the same way that having a variation in the number or arrangement of tendon slips can be viewed. Yes, it has an effect on function but is not a necessary cause of pathology if it functions within locomotion efficiently. To classify pes planus feet with a normal functional tibialis posterior together with feet that have a dysfunctional tibialis posterior or lax ligaments would be erroneous, as their function, kinematics, and kinetics are going to be different.

Pes planus has been linked to pathologies and dysfunctions such as tibialis posterior dysfunction (Song et al., 1996; Hillstrom et al., 2013), hallux abducto valgus (HAV) or hallux valgus (King and Toolan, 2004; Hillstrom et al., 2013), and osteoarthritis of the 1st MTP joint, that is, degenerative hallux limitus/hallux rigidus (Hillstrom et al., 2013). These links have become ingrained into the clinical mindset in such a profound way that they have almost become mantras of association among clinicians. However, much of the research fails to clearly identify the origin of the test subjects' pes planus, potentially assigning an erroneous statistical pathological link. Pathology and deformity have other causes. For example, Kilmartin and Wallace (1992) and Saragas and Becker (1995) found no association with juvenile HAV and vault height. Hutton and Dhanendran (1979) reported reduced pressure under the hallux in HAV which is quite different to the high plantar hallux pressure reported in association with pes planus (Ledoux and Hillstrom, 2002).

Often, in research, pes planus feet are divided into symptomatic and asymptomatic rather than anatomical, hypermobile, and acquired, which leaves all research somewhat ambiguous as to which type of pes planus is being studied. Painful gait is always different and dependent on where the pain is (Coulthard et al., 2002, 2003; Henriksen et al., 2007, 2010), and gait can change when pain is relieved (Han et al., 2014). Research has reported on asymptomatic 'normal vault' profile feet, asymptomatic pes planus, and symptomatic pes planus in children. It has found the greatest difference in gait parameters between symptomatic and asymptomatic feet, regardless of vault profile (Hösl et al., 2014). This clearly demonstrates the difference between symptomatic and asymptomatic. To explore the gait deviations found in anatomical pes planus, it becomes necessary to rely on research that uses separation of asymptomatic pes planus from symptomatic pes planus and hope that the 'right' subjects were included in each study. It is important to be aware of this problem. Pes planus is usually defined as presenting with an everted rearfoot, a low foot vault profile, and/or a high FPI score in relaxed calcaneal stance position (RCSP), that is, the position the foot takes up during relaxed standing. Some studies have only required one of these features in classifying a foot as pes planus.

Pes planus effects on gait

Younger populations are useful to study, as secondary pathologies have less chance of becoming established and the patient may only have come to the attention of the clinician from the foot's appearance rather than its symptoms. One study of adolescents clearly indicated that midfoot kinematics in adolescent pes planus feet were different from age-matched controls, demonstrating increased midfoot dorsiflexion, eversion, and abduction with less overall midfoot motion during the gait cycle (Caravaggi et al., 2018). In many ways but on a lesser scale, this is motion that has been reported for bipedal foot kinematics among chimpanzees when compared to humans (Holowka et al., 2017). Additional information from symptomatic pes planus adults reports more dorsiflexion (extension) occurring at the 1st tarsometatarsal joint and eversion at both the talonavicular and talocalcaneal joints compared to controls, but total rotations within the foot during gait were only larger at the 1st tarsometatarsal joint (Kido et al., 2013). Whether symptomatic flat feet evert more across the rearfoot or rearfoot eversion itself is a precipitating factor in the development of foot symptoms remains unclear. However, increased medial tarsometatarsal dorsal instability would seem significant in altering normal midfoot acceleration kinematics.

Differences in kinetic data from plantar pressure distribution have reported that subhallucal forces are greater in pes planus feet compared to controls (Ledoux and Hillstrom, 2002). Asymptomatic pes planus defined using the FPI demonstrates a different pressure distribution during gait, with reduced peak pressures under the 4th and 5th toes being the greatest difference (Buldt et al., 2018). It has been suggested that pes planus feet demonstrate higher peak pressures, higher pressure–time integrals, higher maximum forces, higher force-time integrals, and greater contact areas predominantly within the medial arch, the central forefoot, and the hallux, but these variables were lower in the lateral medial forefoot (Buldt et al., 2018). Overall, the effects on pressure of pes planus seem moderate compared to those of average-vaulted feet, with the biggest differences occurring between pes cavus (high-vaulted feet) and pes planus feet (Buldt et al., 2018) (Fig. 1.5.3d).

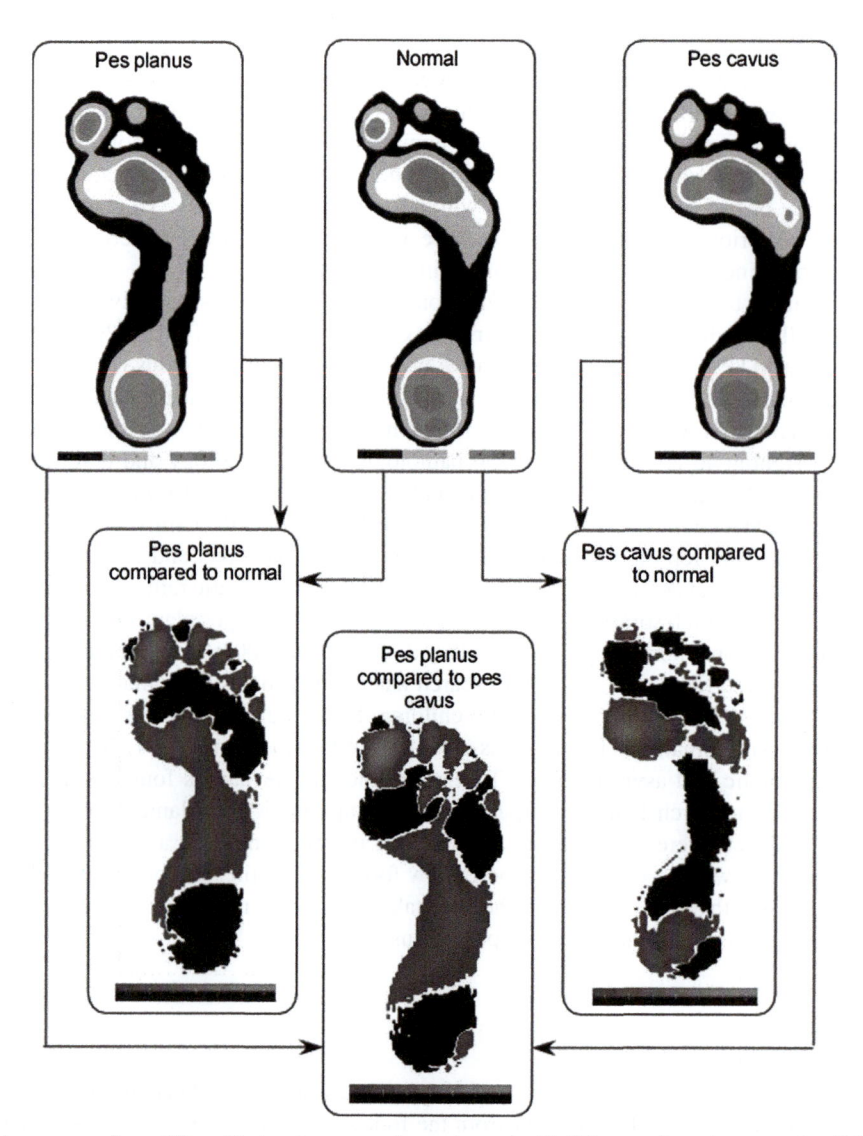

FIG. 1.5.3d Peak plantar pressures from different foot posture groups *(upper images)* with differences in peak pressure for all foot posture groups compared *(lower images)*. The upper images demonstrate reducing relative surface contact areas from pes planus to cavus, consistent with a less prone plantar profile. However, peak pressures tend to occur on the central heel and forefoot in all foot types, with higher peaks under the hallux in planus feet and under the 1st metatarsophalangeal (MTP) joint and heel in cavus feet. In the lower images, higher peak plantar pressure areas are compared to other foot profile types and are indicated by the grey areas, with reduced pressure areas compared to other types indicated in black. *(From Buldt, A.K., Forghany, S., Landorf, K.B., Levinger, P., Murley, G.S., Menz, H.B., 2018. Foot posture is associated with plantar pressure during gait: A comparison of normal, planus and cavus feet. Gait Posture 62, 235–240.)*

A systematic review by Buldt et al. (2013) found that research data had been complicated by a high heterogenicity in the methodology of studies carried out on pes planus kinematics. They highlighted the fact that there is no agreed method of foot type classification. Yet, Buldt et al. (2013) were also able to report some evidence for increased motion within the pes planus foot throughout stance phase, with the most significant differences occurring after heel lift during terminal stance and preswing. The rearfoot was evidenced for differences in frontal plane motion, with increased peak eversion and increased and prolonged forefoot motion within the transverse plane. Interestingly, these associations were stronger in studies that reviewed a number of gait kinematics rather than those that looked at calcaneal alignment in the frontal plane alone (Buldt et al., 2013).

Cen et al. (2020) approached foot types rather differently, classifying feet by their vault stiffness, using changes in vault profile from sitting to standing. They found that gait impulses in both running and walking were different between stiffer and more mobile vaults. With some caution, such findings may be applied to cavus and planus feet. 'Generally', cavus feet are stiffer than planus feet, and thus we should expect lower heel and midfoot impulses in healthy stiffer cavoid feet than in healthy planus mobile ones, but higher forefoot impulses in the stiffer cavus feet than among the mobile planus feet (Cen et al., 2020). The reason being that stiffer feet offer less shock absorption, requiring energy-dissipating leg muscles to work harder, while mobile feet offer less-stable forefeet for acceleration. However, the amount of foot flexibility to the expected impulses generated may be far more important than the vault (arch) profile alone (Fig. 1.5.3e).

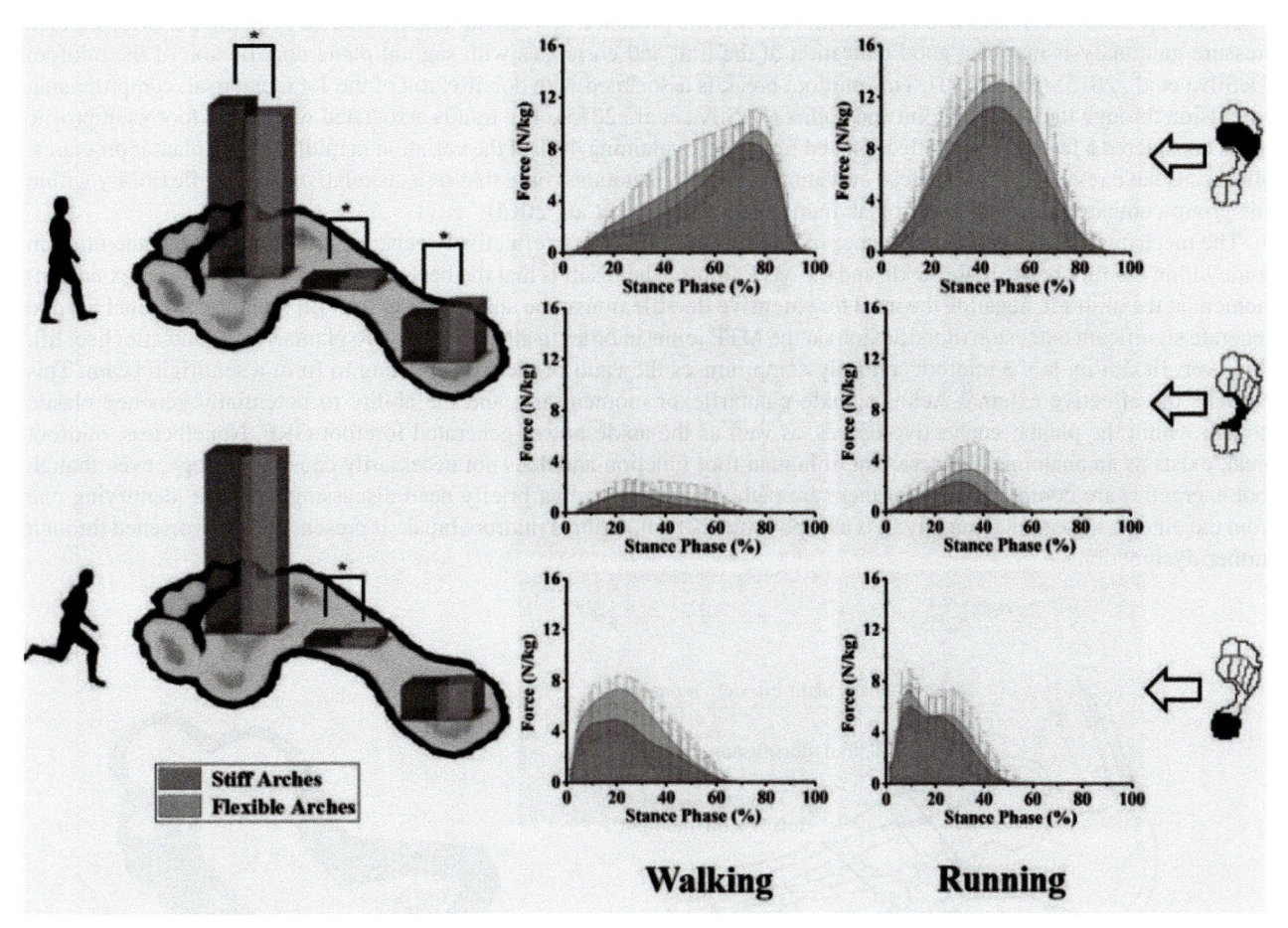

FIG. 1.5.3e The comparison of impulse distribution between different foot vault flexibility during walking and running, may link to foot types through the differences in curvature stiffness between a planus and a cavus foot type. Stiffer healthy feet *(darker grey)* tend to avoid high impulses on the heel and midfoot in both walking and running compared to more mobile feet, which instead are perhaps less able to generate forefoot acceleration forces. If planus feet demonstrate more mobility and cavus feet less mobility, these findings can equate to foot types with caution. *(From Cen, X., Xu, D., Baker, J.S., Gu, Y., 2020. Association of arch stiffness with plantar impulse distribution during walking, running, and gait termination. Int. J. Environ. Res. Public Health 17(6), 2090. https://doi.org/10.3390/ijerph17062090.)*

Clinicians consequently find themselves in a quandary that is best served by associating gait abnormalities to the pathology and not the foot type alone. It is worth being aware that certain kinematics and kinetics are more expected in pes planus and mobile feet compared to cavus and stiff feet.

Midtarsal (midfoot) break during gait

Midtarsal break or midfoot break is the ability to lift the heel from the support surface independently of the rest of the foot. It was first used as a term to describe this ability in nonhuman primates and was thought to be a dichotomous nonoverlapping trait with human feet (DeSilva, 2010; DeSilva et al., 2015). The assumed loss of this ability was suggested from the fossils of *Australopithecus* species and was thought to correlate with the development of a stable midfoot among hominins (DeSilva, 2010; DeSilva and Gill, 2013). With a stable midfoot and a significant foot vault profile, the human foot is able to act as a stable semi-stiffened beam to achieve heel lift around fulcrums at the metatarsal heads. It is now realised by science that a part of the human population maintains midfoot break as a normal part of their gait (DeSilva and Gill, 2013; DeSilva et al., 2015). Clinically, it appears to occur on a continuum, with some midfoot breaks being short and mild, maintaining significant degrees of MTP joint extension, while others appear long, more extensive, and result in little MTP joint terminal stance extension. The midfoot break ability seems to derive primarily from the lateral vault at the cuboid-metatarsal joints and a little from the calcaneocuboid joint, producing a combined ability to dorsiflex within the middle of the foot, with two-thirds of such midfoot dorsiflexion coming from the cuboid-metatarsal joints (DeSilva, 2010).

As heel pressure decreases and is lost with heel lift, the presence of a lateral tarsometatarsal peak plantar pressure on a pressure gait analysis map is a good indication of the trait and correlates with sagittal plane dorsiflexion of the midfoot (DeSilva et al., 2015) (Fig. 1.5.3f). This midfoot break is associated with dorsiflexion of the 1st metatarsal, compromising propulsion through the 1st MTP joint and hallux (DeSilva et al., 2015). The trait is associated with a low foot vault profile and is considered a feature of a highly pronated foot type, explaining 40% of the variation in midfoot peak plantar pressures. MRI scans have revealed a pronounced curvature of the 4th metatarsal base that reflects relative midfoot flexibility within this group compared to nonmidfoot break human feet (DeSilva et al., 2015).

The mechanical/functional consequence of this variable ability is to effectively generate another sagittal plane fulcrum point within the foot between the ankle and the MTP joints. The result is that the heel can lift via a dorsiflexion (extension) moment at the midfoot, negating the need for extensive dorsiflexion at the ankle prior to heel lift and the additional need to generate significant extension (dorsiflexion) at the MTP joints in order to allow the ankle to plantarflex at and after heel lift. However, flexibility at the midfoot seriously compromises the vault of the foot in acting to form a semirigid beam. This shortens the effective external Achilles ankle plantarflexor moment arm and the ability to potentially generate elastic storage within the plantar connective tissues, as well as the ankle power-generated forefoot GRF. Nonetheless, midfoot break exists as an anatomical gait variant of human foot function and does not necessarily cause pathology, even though foot energetics are compromised. Yet there are pathological forms that briefly need discussing, because identifying one from the other is important clinically. It is also possible that a functional midfoot break, if present, can be worsened through further dysfunction.

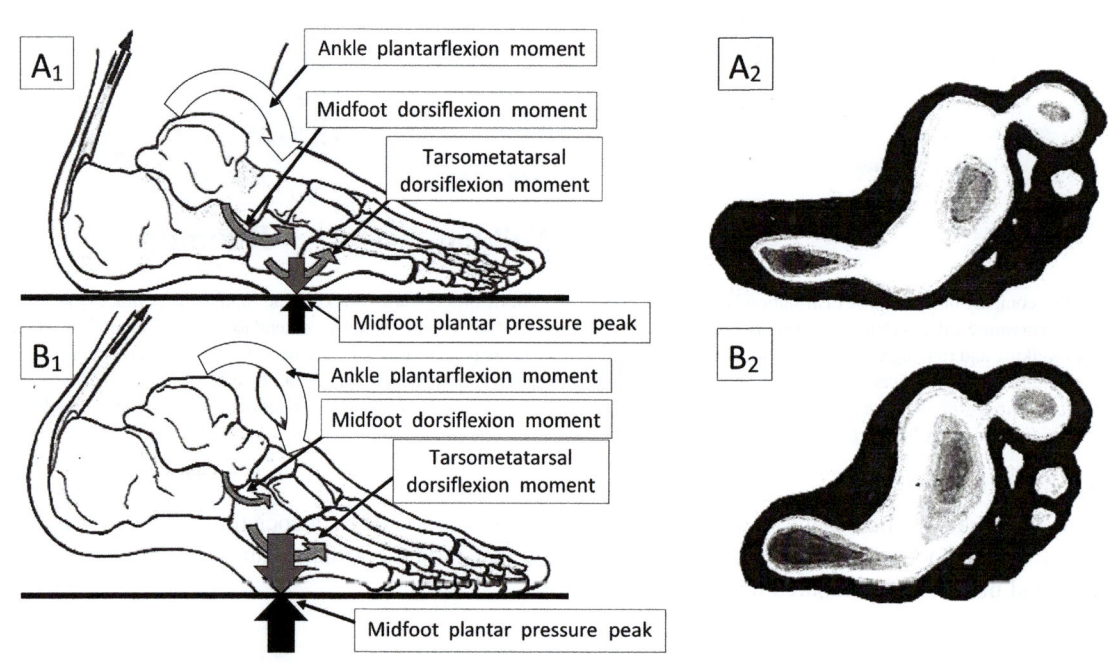

FIG. 1.5.3f Midfoot break does not couple ankle plantarflexion to midfoot plantarflexion and MTP joint extension. Instead, at heel lift (A_1) ankle plantarflexion induces midfoot and tarsometatarsal joint dorsiflexion moments that cause large sagging deflection across the vault. The 5th metatarsal base within the lateral column will start to load to the support surface with its proximal styloid process being forced against the ground, creating a distinct pressure loading peak on pressure data mapping (A_2). As terminal stance progresses (B_1 and B_2), this pressure initially increases as dorsiflexion moments concentrate within the tarsometatarsal joints, especially at the cuboid's articulations. Only as the centre of ground reaction force (CoGRF) starts to move more anteriorly as body weight transfers to the next step, will the peak of pressure under the styloid process and the lateral column generally start to reduce. Then pressure loading moves on to the forefoot alone. This pattern of rising force under the lateral metatarsal base at heel lift seems unique to midfoot break. *(Permission www.healthystep.co.uk.)*

A change in mechanical foot function presents the possibility of acquired midfoot break through restricted ankle joint dorsiflexion during the late midstance phase of gait (Maurer et al., 2014). This can occur if the midfoot anatomy can be compromised into providing a dorsiflexion moment in late stance phase. Such a midfoot break deformity can be due to spasticity or contracture of the ankle plantarflexors, and it is commonly reported in severe foot deformity associated with cerebral palsy (Maurer et al., 2014). Cerebral palsy is a group of permanent neurological disorders that disturb the development of movement and posture, causing activity limitations due to nonprogressive disturbances that occurred in the developing foetal or infant brain (Rosenbaum et al., 2007). Midfoot break deformity is reported as being as high as 30% in cerebral palsy children, with the greatest amount at over 50% in those with quadriplegia (Maurer et al., 2014). The cause is tightness of the triceps surae that pulls the hindfoot into equinus, increasing the forces within the sagittal plane on the foot vault by buckling the midfoot to create either a fixed or flexible vault deformity.

Three types of midfoot break deformity have been reported in cerebral palsy, each with their own distinct kinematic and pressure patterns. The deformity affects all three planes of motion, although the sagittal plane dominates (Maurer et al., 2013, 2014). When compared to normal gait controls, all variations of midfoot break feet in cerebral palsy demonstrate reduced peak ankle dorsiflexion and increased peak midfoot dorsiflexion (Maurer et al., 2013). Cerebral palsy-associated midfoot break has been divided into the following: *flat foot midfoot break* where both the medial and lateral side of the foot remains in support surface contact at heel lift, a *supinated midfoot break* where only the lateral side of the foot remains in support surface contact, and a *pronated midfoot break* where only the medial forefoot remains in ground contact at heel lift (Maurer et al., 2014). It is likely that these various midfoot break patterns exist in the general population in less severe forms and also in those that acquired a midfoot break, making midfoot break heel lift a heterogeneous event. In flat foot midfoot break feet, the pressures are increased evenly between medial and lateral sides in the frontal plane of the forefoot compared to controls, and within the transverse plane, there is increased external forefoot rotation compared to the rearfoot. The pronated midfoot break also demonstrates increased transverse plane external forefoot-to-hindfoot rotation, but in the frontal plane, medial midfoot pressures increase as does peak forefoot eversion. In the supinated midfoot break, lateral midfoot pressures and peak forefoot inversion increase with increased internal forefoot rotation (Maurer et al., 2014). For more on cerebral palsy gait, see Section 1.6.2 and Figs 1.6.2b and 1.6.2c.

Midfoot break should also be evaluated in regard to the known variances of acquired midfoot break deformities. In addition, it would be clinically wise to also consider the acquired midfoot deformity associated with diabetic peripheral neuropathy, known as *Charcot foot*. By what is known of cerebral palsy-related midfoot break deformity that is found associated with restricted ankle dorsiflexion, a better plan of Charcot intervention to prevent and treat this potentially disastrous condition can be considered through improving dorsiflexion flexibility of diabetic patients (Hastings et al., 2016).

Pes cavus in gait

Pes cavus is a high-vaulted foot profile that often has a varus element to the rearfoot, and it is also referred to as *cavovarus* foot. It is usually associated with foot stiffness, but this is not always clinically the case. Pes cavus should be approached just like a planus foot type in that the origin of the foot type must first be established if the nature of the foot's significance during gait is to be understood. Pes cavus can be divided into *neuropathic* and *idiopathic*. There is a strong association of pes cavus with neuropathic disease. Neuropathic pes cavus is often acquired and the patient is aware that their foot shape has changed over time. It is associated with traumatic and genetic neuropathic abnormalities such as damage at birth seen in cerebral palsy, genetically derived neuropathy such as Charcot–Marie–Tooth disease, and also other inherited motor-sensory neuropathies such as Friedreich's ataxia. It can also arise from diabetic peripheral neuropathy, infective neuropathies (e.g. poliomyelitis), and neuropathy including that resulting from diabetes and also following neuropathic damage such as seen after traumatic compartment syndrome (Desai et al., 2010). Pes cavus is also associated with the developmental deformity talipes equinovarus, commonly known as *clubfoot*.

The most common group are idiopathic pes cavus feet, which are often familial to a variable extent and usually bilateral (Desai et al., 2010). The incidence is around 10%–15% of the population with about 60% experiencing symptoms of pain at some point in their lives (Fernández-Seguín et al., 2014). A recommended quick clinical assessment technique for identifying a pes cavus is the 'peek-a-boo' heel sign, with the feet positioned weightbearing straight ahead and facing the clinician. In this position, the medial aspect of the heel can be seen, with the amount of the heel observed indicating the extent of the pes cavus, although a tight Achilles tendon and metatarsus adductus cause false positives and other factors can cause both false positives and negatives (Manoli and Graham, 2018). The FPI can also be effectively used to identify the extent of pes cavus suggested by a negative pronation score. However, neither test indicates the origin of the foot profile. The Enhanced Coleman block test can be used to establish whether plantarflexion of the 1st metatarsal is the primary source of the problem or indicate if there is more to the deformity (Manoli and Graham, 2018) (Fig. 1.5.3g).

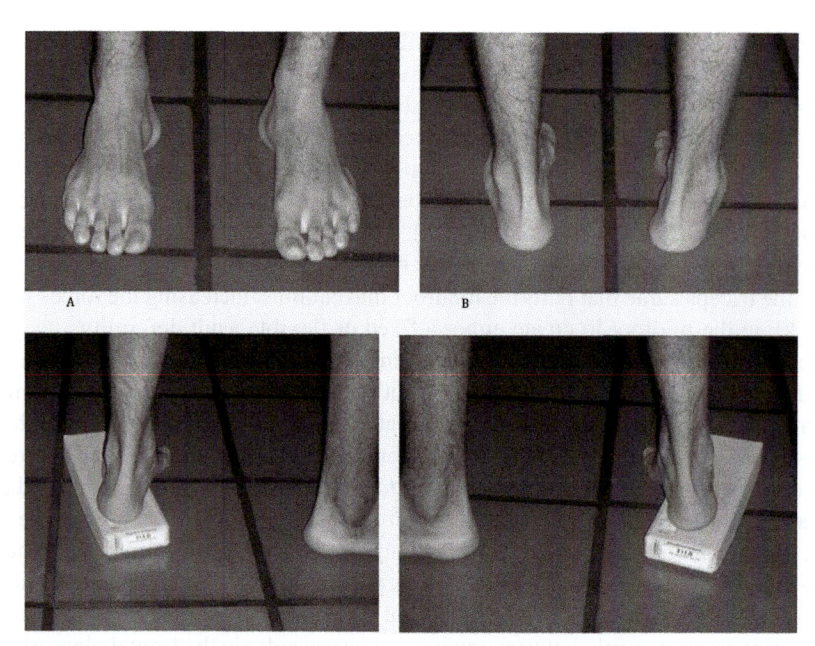

FIG. 1.5.3g Peek-a-boo sign on a 15-year-old male demonstrating a moderate cavus foot (A) with bilateral heel varus also seen from the posterior view, being worse on the right than the left (B). An enhanced Coleman block test on the left foot, when the opposite foot is rotated 90°, resolves the left foot deformity suggesting that an excessive plantarflexed and stiff 1st metatarsal is the source of the problem on the left foot. However, incomplete correction on the right suggests this foot has further problems (C). *(From Manoli, A., Graham, B., 2018. Clinical and new aspects of the subtle cavus foot: A review of an additional twelve year experience. Fuß Sprunggelenk 16 (1), 3–29.)*

Idiopathic pes cavus is also considered to be an intrinsic risk factor for developing injury. This foot type is usually reported to have less mobility than most feet and is thus more susceptible to injuries related to reduced shock attenuation (Williams et al., 2001a,b; Desai et al., 2010). Its relatively reduced surface contact area increases plantar pressures on the heel and forefoot (Burns et al., 2005; Buldt et al., 2018). Pes cavus in athletes is associated with increased stress fractures, ankle instability, impingement syndromes, and tendon disorders (Desai et al., 2010), but as in pes planus, injury relationships of pes cavus are not definitively evidenced (Witvrouw et al., 2001; Lun et al., 2004). This is not to say that any particular injury is a direct result of the foot type a patient presents with, but certain injures do show associations of increased risk with one type of foot over another.

Pes cavus often has a number of strong 'equinus' or plantarflexed elements, meaning structural morphologies that tend to move loads to the forefoot. These include restricted ankle or midfoot dorsiflexion ranges and having alignments of sagittal plane deformity that position distal structures more inferiorly when nonweightbearing. This includes features such as a tight triceps surae, high calcaneal inclination angles, high metatarsal declination angles, and a plantarflexed 1st ray that are all reported central features of pes cavus (Mosca, 2001; Statler and Tullis, 2005). Within the midfoot, there is variability in location where forefoot features of plantarflexion alignment from the rearfoot to the forefoot occur. This can be either at the lesser tarsus or in the tarsometatarsal region (Fig. 1.5.3h). These sagittal plane alignments have an impact on the gait associated with pes cavus through excessively resisting dorsiflexion moments in the ankle-rearfoot or midfoot during gait, although the amount depends on how and when the joints can adapt on loading, and how flexible the foot is. When assessing the triceps surae, the Silfverskiöld test can be useful to isolate gastrocnemius tightness from soleus (Desai et al., 2010).

If the plantarflexion alignments of the forefoot and 1st ray are not accommodated through joint motion or terrain changes (including shoes and foot orthoses) on loading, then not only will the medial forefoot often be the first forefoot contact area after the heel during loading response, but it can also initiate an inversion moment into the rearfoot. This prevents or blocks the rearfoot from meeting its desired loading eversion range during loading response. Depending on the individual foot, the heel can remain inverted on the weightbearing surface on loading and persist throughout midstance until heel lift. If the foot has significantly restricted motion through the midfoot and forefoot, then the foot may continue to remain inverted during terminal stance, forcing a low gear propulsion (Bojsen-Møller, 1979) by accelerating off from the lateral MTP joints. This has poor mechanical implications for the Achilles plantarflexion power but also for the peroneal muscles' mechanics in their attempt to move forces medially across the forefoot during late midstance and terminal stance towards the medial MTP joints (Fig. 1.5.3i).

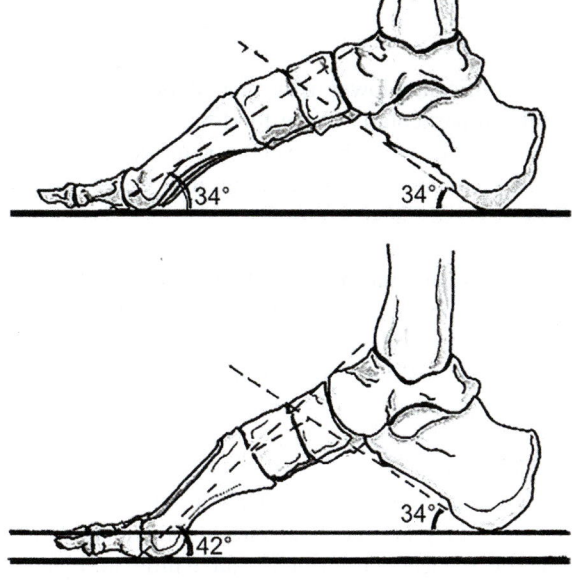

FIG. 1.5.3h The cavus foot can result from a number of postural influences. A high calcaneal inclination angle *(both images)* is common, with angulations over 21° considered to be associated with increasing vault heights and sometimes existing with calcaneal varus. High calcaneal inclination angles necessitate high metatarsal declination angles to bring the metatarsal heads to the support surface. This results in higher resting metatarsophalangeal (MTP) extension angles in stance. Cavus feet can also have excessive metatarsal declination angles due to the joint facet angles at the tarsometatarsal and/or other lesser tarsal articulations *(lower image)*. All metatarsals can be affected or just one, usually the 1st metatarsal. This plantarflexed 1st metatarsal alignment if rigid often tips the rearfoot into inversion to compensate when on hard flat surfaces. These forefoot and/or midfoot 'equinus' deformities may coexist with plantarflexed 1st metatarsals and/or normal or high calcaneal inclination angles. Whether the ankle can compensate for reduced midfoot dorsiflexion by providing increased ankle dorsiflexion angles during late midstance is important as to how well such feet function energetically and biomechanically during locomotion. *(Permission www.healthystep.co.uk.)*

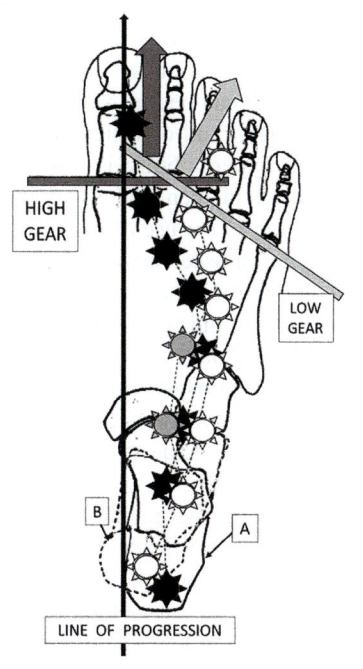

FIG. 1.5.3i During pressure data analysis, mean-vaulted feet (shown by the *solid calcaneal outline*—A) usually demonstrate a centre of pressure line (CoP—black suns) following a path from heel strike that tracks slightly to the medial side (through loading rearfoot eversion) until the start of forefoot loading when the lateral forefoot makes contact. This initial lateral forefoot and/or lateral column pressure loading draws the CoP laterally. Lateral pressures decrease during late midstance as peroneal muscle activity starts to move the midfoot into eversion as tibialis posterior adducts the forefoot, increasing medial forefoot pressures. This places the CoP (and for that matter, the CoGRF) behind the high gear axis *(dark-grey line)* for acceleration *(dark-grey arrow)* aligned to the body's line of progression *(long black arrow)*. Cavus feet (shown by the *dashed calcaneus*—B), usually present with high calcaneal inclination angles and some rearfoot inversion that tends to prevent the medial shift in the cavus foot's CoP (white suns) at heel contact, maintaining increased lateral heel pressure. Although the medial forefoot often loads next due to a plantarflexed 1st metatarsal, the CoP will often tend to lie more laterally during loading response (although this depends on how much pressure develops under the medial forefoot), reflecting the CoGRF position and the higher lateral column loading pressures. However, in more extreme cavoid feet, no part of the lateral column contacts the pressure sensors during midstance, allowing the CoP pathway to track more medially *(grey suns)*, although the foot's CoGRF will still lie more laterally. If the CoGRF is unable to move medially during late midstance from insufficient peroneal midfoot eversion moments, then the foot will tend to accelerate around the low gear axis *(light-grey line)*, sending the foot's line of progression *(light-grey arrow)* lateral to the body's line of progression. *(Permission www.healthystep.co.uk.)*

Plantar pressures are reported to be significantly lower in the medial longitudinal part of the foot vault, with concurrent increased pressures evident at the heel and particularly in the forefoot compared to other foot types (Buldt et al., 2013; Fernández-Seguín et al., 2014). Often, cavoid feet only bear weight on the heel and forefoot due to the plantarflexed positioning of the forefoot and the 1st ray. This reduces the weightbearing surface area of the foot as a consequence of losing any loading of the lateral column's soft tissues (see Fig. 1.5.3e).

Fernández-Seguín et al. (2014) hypothesised that in cavus feet, there is an increased metatarsal verticality (high metatarsal declination angles) which produces a greater transmission of load to the forefoot, especially in the 1st metatarsal head area. All the metatarsals, except the 5th, were observed by Fernández-Seguín et al. (2014) to bear more load than in normal vaulted feet. However, just like normal vaulted feet, the bulk of the load usually fell on the 2nd and 3rd metatarsal areas. The pressure–time integral is higher in cavus feet, and this has been found to correlate with foot pain in different foot types (Burns et al., 2005). Toe pressure was also noted to be significantly reduced in pes cavus, which probably represents the mechanical disadvantage of the digital flexors to bring the toes into the ground via the reversed windlass (Hillstrom et al., 2013; Fernández-Seguín et al., 2014). Claw toes are common in pes cavus, as the relationships of the long flexors and extensors are disturbed (Statler and Tullis, 2005). It is likely that the increased metatarsal declination increases MTP joint extension freedom resulting in excessive extension flexibility across these joints in many cavus feet. If a perfectly healthy cavus foot expressing reduced motion and relative inflexibility across its vault, it should demonstrate reduced heel and midfoot impulses, but higher forefoot impulses than more flexible feet (Cen et al., 2020) and yet it can still function perfectly well without pathology, albeit with some increased inherent vulnerability.

1.5.4 Joint hypermobility in gait

Generalised joint hypermobility (GJH), including joint hypermobility syndrome (JHS), are genetic issues defined as a condition where joint range of motion (RoM) is increased compared to the general population for age, gender, and ethnicity (Hakim and Grahame, 2003; Remvig et al., 2007; Simmonds and Keer, 2007; Grahame, 2009). GJH can be acquired through the maintenance of excessive flexibility through sports training (Gannon and Bird, 1999), inherited as a trait (Dalgleish, 1997; Grahame, 1999), or be linked to connective tissue diseases. Hypermobility takes in a number of conditions including *benign hypermobility syndrome* (which can be far from benign), *Ehlers-Danlos syndrome* (hypermobility form), *Marfan's syndrome*, and *osteogenesis imperfecta*, all of which are heritable genetically linked connective tissue disorders. JHS is the term usually adopted when hypermobility is linked to symptoms (Simmonds and Keer, 2007; Tofts et al., 2009; Smits-Engelsman et al., 2011) and then connected to the underlying connective tissue diseases (Tofts et al., 2009), although there has been variation in the use of JHS. GJH may not be associated with problems, but in some individuals, it predisposes to a range of musculoskeletal symptoms, leading sufferers to seek medical attention. Joint hypermobility is widely prevalent in all communities and has appreciable associations with morbidity and a strong link with chronic pain that is difficult to treat (el-Shahaly and el-Sherif, 1991; Grahame and Hakim, 2008; Grahame, 2009). It is usually diagnosed by clinical assessments (Fig. 1.5.4a).

There are indications that joint hypermobility is a multisystem phenotype that involves the integrity of a brain centre involved in normal and abnormal emotions and physiological responses. This increases vulnerability to neuropsychiatric symptoms (Eccles et al., 2012) as well as joint and soft tissue symptoms and includes carpal/tarsal tunnel syndromes, degenerative joint changes, a high frequency of varicose veins, haemorrhoids, and uterine prolapse (el-Shahaly and el-Sherif, 1991). The connective tissue disorders linked to hypermobility also demonstrate hyperelastic skin, easy bruising, poor wound healing, hernias, and lenticular abnormalities (el-Shahaly and el-Sherif, 1991). However, there is considerable crossover in presentation between familial hypermobility and the full genetic connective tissue disorders (Tofts et al., 2009). There is therefore a benefit in identifying the genetic cause of joint hypermobility. Furthermore, every effort should be made to diagnose the underlying disorder responsible for presentation before or soon after gait analysis in order to assess the long-term implications, as there is no 'clinical' method of confirming familial articular hypermobility from conditions such as Ehlers-Danlos syndrome (Tofts et al., 2009).

Joint hypermobility has been reported as a finding between 3% and 30% of children, being more prevalent in females at ratios between 3:1 and 2:1. However, it is inversely related to age, with the sex difference becoming more noticeable in older children (Smits-Engelsman et al., 2011). In children, GJH is linked to motor delay in infancy, rectal prolapse, juvenile episodic arthritis (arthralgia), 'clicky hips' at birth, congenital hip dislocation, abnormal gait, clumsiness, difficulties in learning, and neuromuscular and motor developmental problems (el-Shahaly and el-Sherif, 1991; Adib et al., 2005). JHS in children seems to overlap to some degree with developmental coordination disorders, so that either JHS or developmental coordination disorder groups of children (Kirby and Davies, 2007) may benefit from gait assessment. Thus, children with GJH and JHS are common attendees for biomechanical assessment associated with symptoms in the lower limbs (Fig. 1.5.4b).

Specific joint laxity	Yes		No
1. Passive apposition of thumb to forearm	☐ Left	☐ Right	☐
2. Passive hyperextension of V-MCP>90°	☐ Left	☐ Right	☐
3. Active hyperextension of elbow >10°	☐ Left	☐ Right	☐
4. Active hyperextension of knee >10°	☐ Left	☐ Right	☐
5. Ability to flex spine placing palns to floor without bending knees	☐ Left	☐ Right	☐

"Each "Yes" is 1 point. A score > 4 out 9 is generally considered an indication of JH. (MCP: metacarpophalangeal).

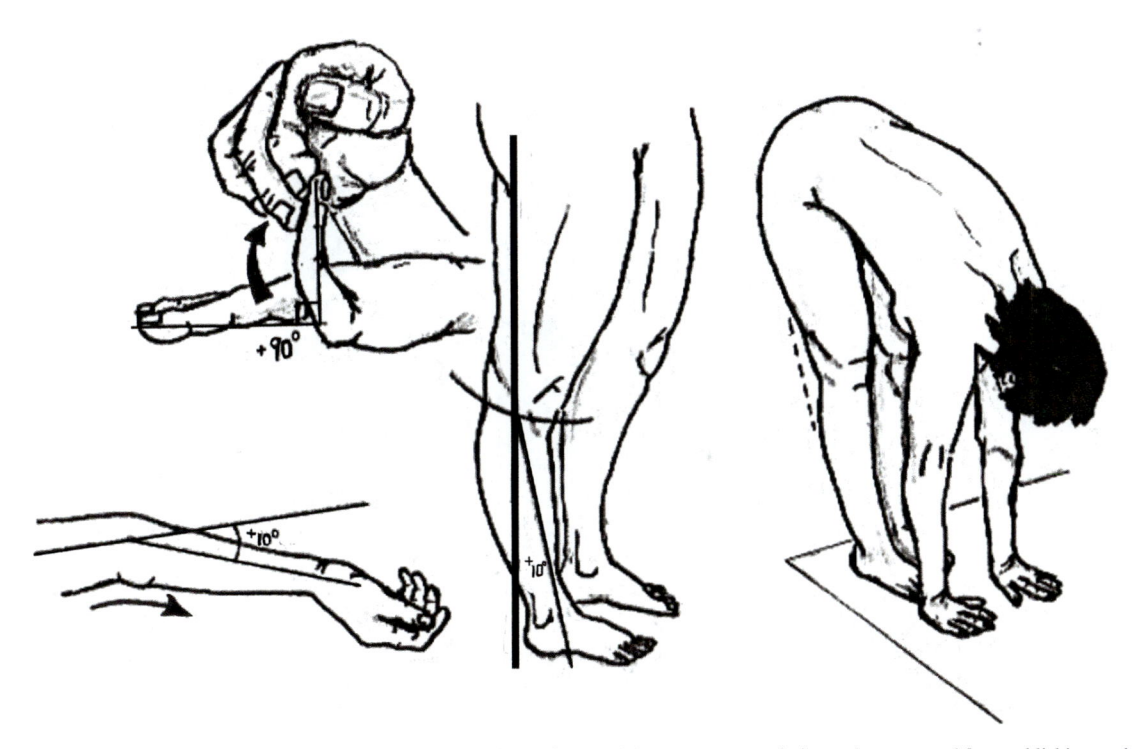

FIG. 1.5.4a The Beighton scale presented as a simple clinical tick box form and the assessment techniques demonstrated for establishing such a score. These involve extending the little finger to 90°, the elbow over 10°, the knee over 10°, and being able to place hands flat on the floor without flexing the knees when standing. Hypermobility in the lower limb is more relevant to gait than in the upper limb, and thus it is also worth looking at changes in the foot profile from nonweightbearing to weightbearing as part of hypermobility assessment. *(Permission www.healthystep.co.uk.)*

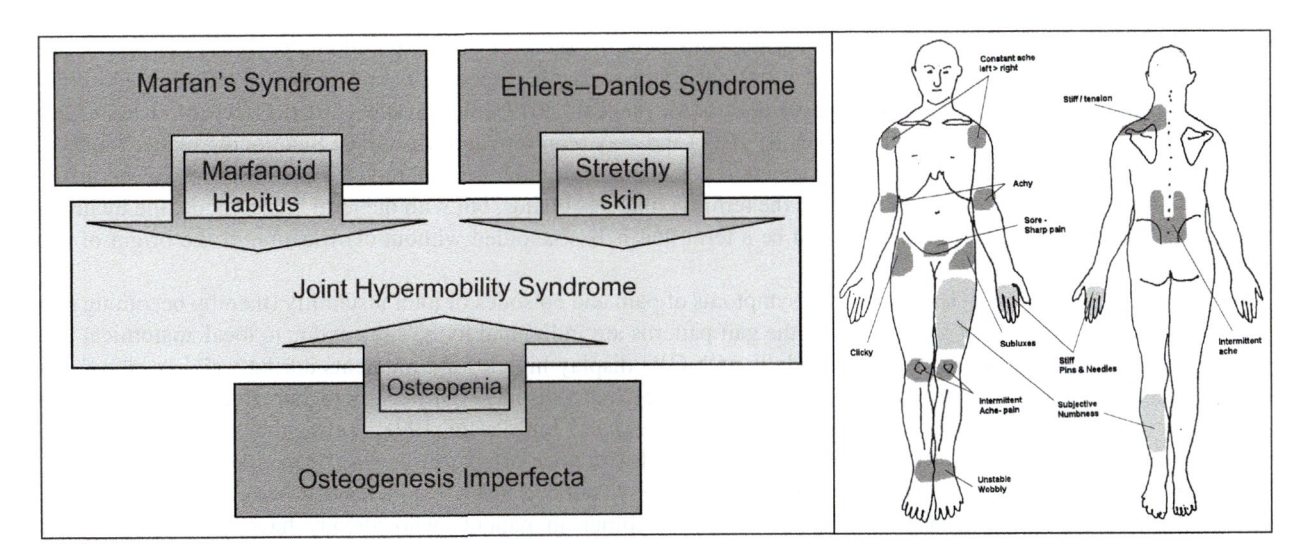

FIG. 1.5.4b A reminder of the interrelationship between heritable connective tissue disorders *(left image)* underscores the need to establish the origins of hypermobility when seen in clinic. Presenting symptoms of joint hypermobility syndrome (JHS) are diffuse *(right image)* and can easily be confused with other musculoskeletal causes. *(From Simmonds, J.V., Keer, R.J., 2007. Hypermobility and the hypermobility syndrome. Manual Ther. 12 (4), 298–309.)*

Diagnosis of all joint hypermobility conditions is achieved through a number of clinical tests, known collectively as the *Beighton score* (see Fig. 1.5.4a). It is positive for hypermobility on a 9-point scale and has been demonstrated as reliable in a number of studies (Hirsch et al., 2007; Smits-Engelsman et al., 2011; Evans et al., 2012) but with recommendations that different scores be used for children, adults, each birth gender, and different ethnicities (Smits-Engelsman et al., 2011). Generally, scores of 6 out of 9 are considered to show hypermobility in children. However, this does not predict future musculoskeletal pain in the short term (El-Metwally et al., 2007) with Smits-Engelsman et al. (2011) reporting 12.3% of children complaining of joint pain and 9.1% complaining of pain after exercise, unrelated to their Beighton scale score. Tests involve assessing the musculoskeletal system's mobility at the little finger, base of the thumb, elbow, knee, and spine. It has been reported that in scores of 4 and over, subjects display decreased proprioception and knee muscle strength (Sahin et al., 2008). GJH is often associated with musculoskeletal and other soft tissue pain, arthralgia, frequent luxations and subluxations, as well as osteoarthritis over time (Hudson et al., 1998; Grahame et al., 2000). Skin flexibility is also a very simple quick assessment tool (Fig. 1.5.4c).

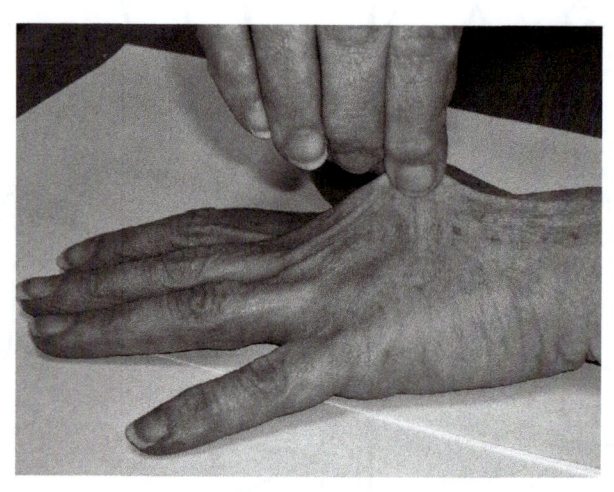

FIG. 1.5.4c Assessing skin flexibility over the hand or foot can be a quick simple clinical test that raises suspicion of hypermobility, but even flexibility noted on shaking an affected patient's hand before the start of a consultation can often be an immediate alert. *(From Simmonds, J.V., Keer, R.J., 2007. Hypermobility and the hypermobility syndrome. Manual Ther. 12 (4), 298–309.)*

The Beighton scale score only uses one test within the lower limb, that of knee extension. This excludes a number of important lower limb joints that can demonstrate hypermobility. In the lower limb, the *Lower Limb Assessment Score* is preferred when assessing the hip, knee, ankle, rearfoot, midfoot, and forefoot ranges of motion, and has been demonstrated to have excellent intra-rater and inter-rater reliability, with a cut-off score of $\geq 7/12$ in children (Ferrari et al., 2005) and $\geq 4/12$ in adults (Johnson et al., 2019) (Fig. 1.5.4d). It also shows excellent specificity and moderate sensitivity, discriminating well between the varying extents of hypermobility (Meyer et al., 2017; Johnson et al., 2019). GJH is often associated with pes planus in children (Smits-Engelsman et al., 2011) and may represent another unique functioning planus foot type other than absolutely mirroring anatomical or acquired pes planus, and therefore should be clinically approached differently. This group is probably represented within the term flexible pes planus, but with the term being vague, the inclusion criteria may also be vague and probably should be a term that is best avoided without clarification of the origin of that flexibility.

The majority of adults with GJH experience symptoms of pain and episodes of joint instability (thereby becoming JHS cases) which can influence their gait. Although the gait patterns are individual to the subject due to local anatomical variance and morphology, it has been noted that adults with GJH display higher joint moments during walking in both the frontal and sagittal planes (Simonsen et al., 2012). Hip and knee joints display significantly higher frontal plane moments in adduction (varus) than do healthy controls by 13% (Schipplein and Andriacchi, 1991; Foroughi et al., 2010; Simonsen et al., 2012). Increased knee adduction moments are associated with knee osteoarthritis (Kerrigan et al., 2001; Gök et al., 2002; Miyazaki et al., 2002; Foroughi et al., 2010). It has been calculated that a 1% increase in knee abduction moment increases the risk of knee osteoarthritis progression by six times in patients who already have knee osteoarthritis

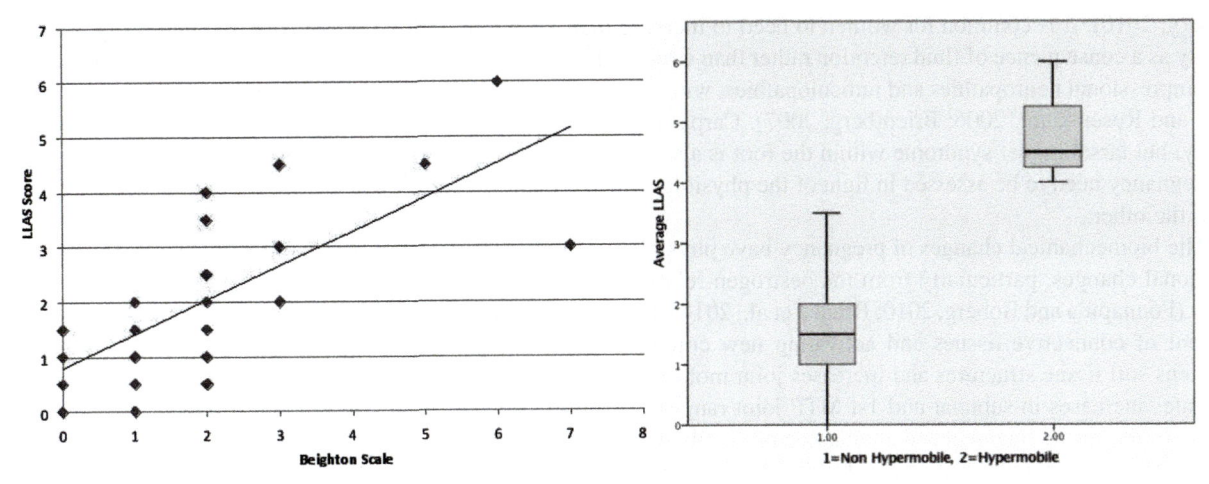

FIG. 1.5.4d A scatter plot of the association between the Beighton scale and the Lower Limb Assessment Score *(left)* and a box plot *(right)* of the median and interquartile ranges of the lower limb assessment score across nonhypermobile and hypermobile groups, as defined by the Beighton scale. As the Beighton scale only has one lower limb assessment test, the Lower Limb Assessment Score is advisable where hypermobility in gait is suspected. *(From Johnson, A.P., Ward, S., Simmonds, J., 2019. The Lower Limb Assessment Score: A valid measure of hypermobility in elite football? Phys. Ther. Sport 37, 86–90.)*

(Miyazaki et al., 2002). The highest compression forces on the knee occur just after heel strike with another peak around 20% into the stance phase, but the second peak is not reported to be statistically higher than those found within healthy controls (Simonsen et al., 2012).

In the sagittal plane, GJH subjects display a moderately higher peak extensor moment at the knee joint, probably as a result of the knee moving into greater flexion angles (Simonsen et al., 2012). The knee is also reported to be more flexed at heel strike in GJH subjects, yet although statistically significant, the changes are small (Simonsen et al., 2012). Walking with a more flexed knee during stance phase has also been reported in cases of knee instability resulting from anterior cruciate ligament ruptures (Alkjær et al., 2003), which may indicate the reason for more flexion in hypermobility subjects. The clinical experience of the author finds midtarsal/midfoot break a relatively common feature within the gait of GJH. If the feet of those with GJH are in a healthy condition, they should produce impulses within them that are likely higher in the rearfoot and midfoot but reduced on the forefoot, as should be expected of flexible feet (Cen et al., 2020). Investigating the impulses across the feet in this subject group may prove interesting.

1.5.5 Gait in pregnancy

Pregnancy is a common condition for women, and it has been estimated that around 80% of women by the age of 44 in the western world will have experienced this state (Anselmo et al., 2017). With some irony, in perceived advanced societies, teenage pregnancy is considered a bad thing, but these problems are associated with social issues rather than physiological ones (Cook and Cameron, 2015). Late teenage years are physiologically a good time for pregnancy for both the mother and child. That aside, a number of musculoskeletal symptoms often result from the anatomical and physiological changes that occur throughout pregnancy due to hormonal and biomechanical changes (Artal et al., 2003; Ponnapula and Boberg, 2010). These can include muscle cramps, soreness, strains, and fatigue associated with modified gait primarily in the second and third trimesters (Foti et al., 2000; Marnach et al., 2003; Borg-Stein et al., 2005; Ponnapula and Boberg, 2010).

In pregnancy, the lower limb is prone to ankle and foot swelling, foot aches and pains, toenail incurvation, nail coarsening, and increased nail growth rates (Ponnapula and Boberg, 2010). Dermatological changes such as hyperpigmentation and pruritus, disruption of eccrine and adrenal gland activity, and increased perspiration and varicosities are common (Ponnapula and Boberg, 2010). Pregnancy triggers a number of vascular changes leading to lower extremity swelling (Ponnapula and Boberg, 2010). It is thought that increasing maternal age is associated with vein valve latency loss and reduced venular vessel tone, and risks telangiectasia, varicosities, and deep vein thrombosis, all of which can be reduced by graded elastic stockings, leg elevation, light exercise, and resting in a left lateral decubitus position (Ponnapula and

Boberg, 2010). It is common for women to need to increase their shoe size during the last trimester and postpartum, primarily as a consequence of fluid retention rather than structural changes (Alvarez et al., 1988). The swelling can also result in compressional neuropathies and radiculopathies, with symptoms of tingling, burning, and numbness in the feet and legs (Sax and Rosenbaum, 2006; Briemberg, 2007). Carpal tunnel syndrome in the hand is a well-known issue during pregnancy, but tarsal tunnel syndrome within the foot is also a risk (Massey and Stolp, 2008). Clinically, common symptoms of pregnancy need to be assessed in light of the physiological and biomechanical effects associating or disassociating one from the other.

The biomechanical changes of pregnancy have physiological links. The foot can develop functional alterations due to hormonal changes, particularly from the oestrogen-related hormone *relaxin*, levels of which peak at around the twelfth week (Ponnapula and Boberg, 2010; Branco et al., 2014). Relaxin triggers the collagenolytic system by increasing the water content of connective tissues and activating new collagen synthesis. The ten-fold surge in relaxin during pregnancy weakens soft tissue structures and increases joint mobility (Ponnapula and Boberg, 2010). The foot is affected and demonstrates increases in subtalar and 1st MTP joint ranges of motion (Alvarez et al., 1988; Ponnapula and Boberg, 2010). Laxity of the spring ligament and attenuation of the tibialis posterior tendon can result in a 1-cm lowering of the talar head and an increase in plantar contact area by 12% in static stance and 8% during gait (Nyska et al., 1997). This probably results functionally in some degree of hyperpronation during late pregnancy, as defined by Horwood and Chockalingam (2017), although Alvarez et al. (1988) felt that the dimension changes were due to fluid retention or an increase in soft tissue spread alone. Generalised foot pain is common, and it has been noted that pelvic and other incapacitating pain in pregnant women has been associated with significantly elevated levels of serum relaxin. Dynamic gait analysis during pregnancy is associated with significantly higher forefoot and midfoot contact times and a 30% increase in pressure as the pregnancy progresses (Anselmo et al., 2017).

Pregnancy causes around a 20% increase in body weight, enough to double the force on a joint during locomotion (Ritchie, 2003; Branco et al., 2014). These changes in mass are associated with morphological changes that cause an anterior displacement of the body's CoM. This, in conjunction with the hormonal change effects on the musculoskeletal system, potentially leads to an inability to maintain postural control, increasing the risk of falls (Moccellin et al., 2015). The postural control changes may result from increased sacroiliac joint and pubic symphysis mobility caused by relaxin levels during early pregnancy (Moccellin et al., 2015). This results both in an increase in the elliptical movement and the amplitude and velocity of the CoGRF displacement during stance phase (Moccellin et al., 2015). The increase in mobility at the sacroiliac ligaments allows an average 4° increase in anterior pelvic tilt, increasing lumbar lordosis, and further displacing the CoM anteriorly, which increases throughout pregnancy (Anselmo et al., 2017). This anterior CoM shift makes controlling angular momentum around joints more difficult and changes gait parameters (Fig. 1.5.5). To accommodate the changes, stride width can increase up to 30%, step time increases, step length decreases, and the stance phase tends to lengthen with longer double-limb stance times (Branco et al., 2014; Anselmo et al., 2017).

The anterior CoM displacement may move the CoGRF some 1.5–2.5 cm more anterior to the ankle joint, resulting in the need for increased dorsiflexion restraining forces from the ankle plantarflexors. However, the common symptom of calf cramping is thought to be more related to mineral deficiency than biomechanics (Anselmo et al., 2017). Gait is also adapted to maximise mediolateral foot stability (Ribeiro et al., 2011). Hip flexion increases are associated with a rise in hip extensor moments during stance phase. The knee flexion angle increases as a result of extension reduction during terminal stance, and both ankle dorsiflexion and plantarflexion decrease during stance phase (Branco et al., 2014). Pregnancy is also associated with a decrease in knee extensor and ankle plantarflexor moments within the sagittal plane and a decrease of pelvic range of motion in both the frontal and transverse planes, possibly to control angular momentum caused by increased inertia of the trunk during late pregnancy. This increasing hip abductor and extensor activity which, combined with increased stretch from the anterior pelvic tilt, may be part of the source of back pain (Branco et al., 2014). Changes in gait generally seem to reflect safety over energetics, with an increase in the step width's base of support, shortening of stride length, and increased double support times; features associated with increased metabolism for gait (Branco et al., 2014).

Postpartum, the effects of pregnancy can lead to continued changes in locomotive biomechanics throughout the rest of life, although most effects such as stride width, pelvic motion, and midfoot pressure changes usually seem to resolve (Anselmo et al., 2017). Although there are contradictions between studies, pregnancy may result in some permanent loss of foot vault height, joint stability, and increases in foot length, postpartum (Anselmo et al., 2017). There are indications that proprioception training techniques may improve postural control during pregnancy (El-shamy et al., 2019), and promoting foot and lower limb strength during and after pregnancy may offer long-term benefits and prevent postural adaptations becoming long-term changes.

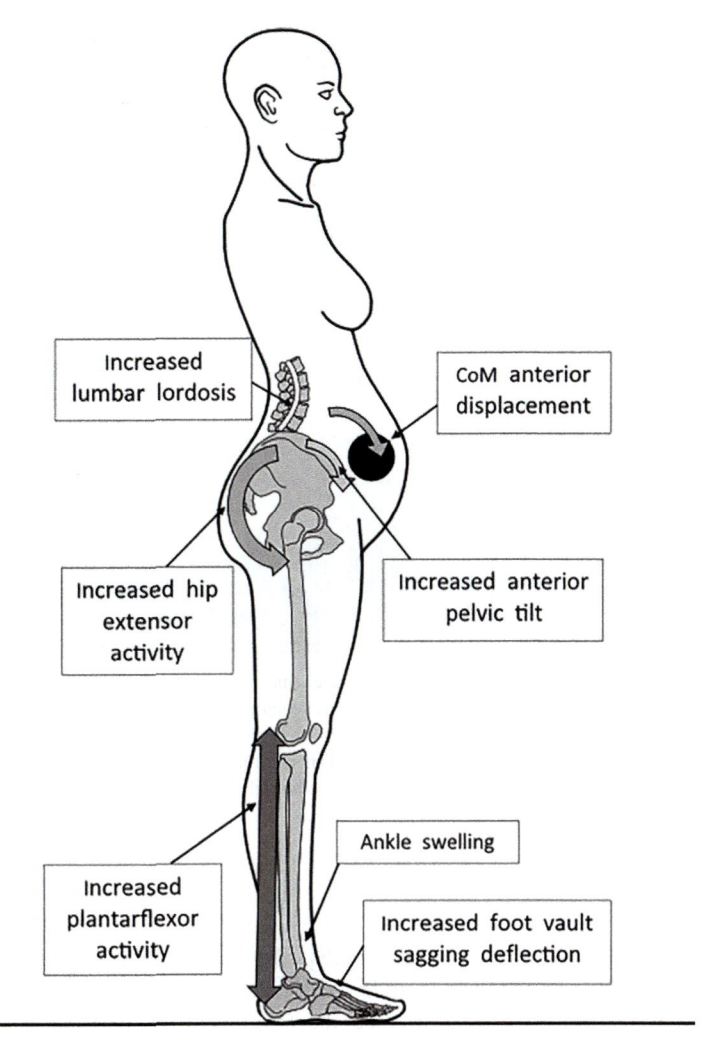

FIG. 1.5.5 A schematic of some of the changes that accompany the anterior shift in the CoM during pregnancy combined with increased connective tissue compliance. The postural changes alter movement, amplitude, and velocity of the centre of ground reaction force (CoGRF) locations during gait, as well as changing kinematics such as widening the base of support, increasing lower limb external rotations, decreasing step length, and increasing double-limb support time. As pregnancy progresses, muscle activity within gluteus maximus and the triceps surae increase as mechanisms to move and restrain the increasingly more anteriorly positioned CoM throughout gait. *(Permission www.healthystep.co.uk.)*

1.5.6 Paediatric gait

The development of gait characteristics

Childhood and adolescence represent anatomical variations in gait from that of adulthood, because neuro-musculoskeletal anatomy is at first very different although with developmental processes becoming increasingly homogeneous to adults over time. Locomotor development in children occurs through gradual multidimensional changes via musculoskeletal growth and maturation of the CNS that gradually alter the biomechanics. Developmentally, this alters the musculoskeletal tissues, setting up a feedback cycle of biomechanics to developmental changes that adjust biomechanics via changing anatomy and altered (improved hopefully) neurological inputs (Fig. 1.5.6a). From reviewing the changes that take place in the formation of adult human gait, an understanding of the role of developmental biology in the context of evolutionary history can be appreciated. This is important for all clinical biomechanists to know and understand, whether they are involved in paediatric cases or not. For example, clinically diagnosed conditions of childhood such as autism are reported to be associated with decreased hip extensor moments, reduced ankle plantarflexion moments and increased dorsiflexion moments in gait compared to age-matched controls. This is possibly due to hypotonia (Calhoun et al., 2011), a condition which may persist into adulthood.

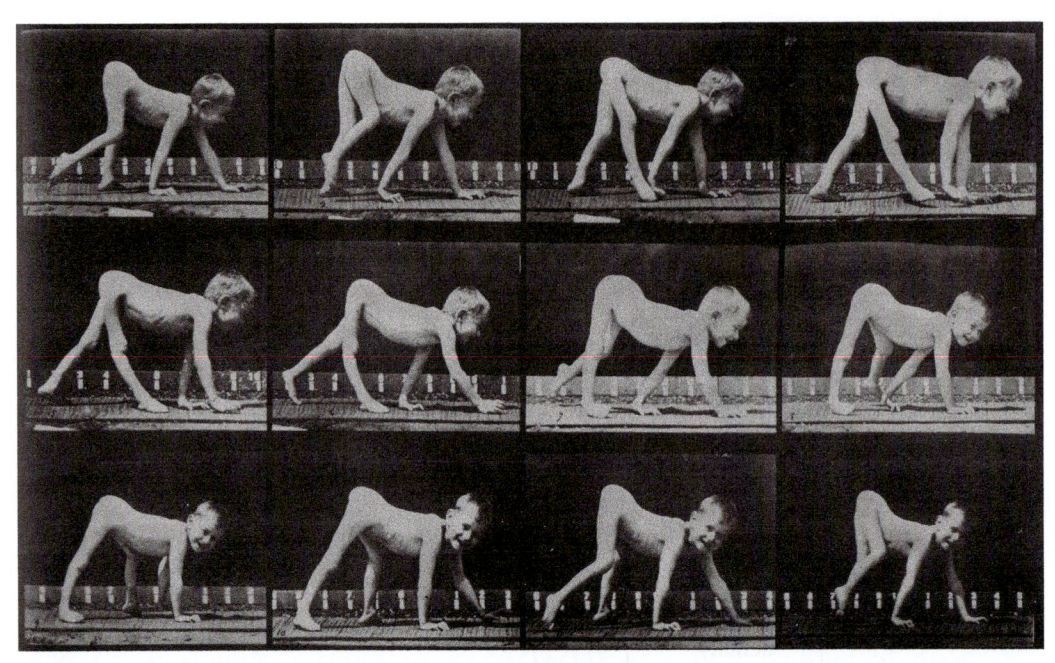

FIG. 1.5.6a The interrelationship between developmental plasticity and biomechanics in the formation of childhood and adult gait patterns is illustrated by these startling images from the Eadweard Muybridge collection. The child has developed habitual quadrupedal gait biomechanics, probably as a result of an autosomal genetic disease altering developmental brain plasticity that seems to reverse gait evolution. This could be due to cerebellar ataxia, mental retardation, and disequilibrium syndrome or most likely, the fascinating Uner Tan syndrome that causes cerebellar malformations. Note the knee hyperextension in the lower (rear) limb's stance phase, the extreme hip flexion angles throughout gait, the loss of lumbar lordosis in late swing and at initial contact, and its increase during acceleration and early swing.

From an infant, sitting to walking takes around nine months. The body weight more than doubles from birth to the infant's first birthday and changes in the anatomy include a 50% increase in limb length by 18 months of age (Price et al., 2018). Foot structure and function also changes to meet the weightbearing challenge, a task that involves important alterations in motor control and coordination (Price et al., 2018). Infants usually go through three phases of developing their locomotion. The first phase uses external supports to transition sideways, known as *cruising*, while the second phase, known as *supported walking*, involves external support to achieve ambulation. Finally, *independent ambulation* allows the infant to walk unaided and permits it to become progressively more mobile with increasing frequency and for longer each day (Adolph et al., 2003). The number of steps occurring in one go is known as a '*bout*', and these bouts increase from usually under four steps at 13 months to mostly over four steps by 19 months of age, with a concurrent reduction in falls (Adolph et al., 2012). It is the repetition of steps and falls, or avoided falls, that teaches the sensorimotor system how to function. However, walking bouts in infants restrict clinical observations as it is hard to see more than five consistent steps in a row, even at age 19 months (Cole et al., 2016) (Fig. 1.5.6b).

The feet change during this period in their shape, structure, and function, converting from an organ of manipulation and reaching towards one that is developed for weightbearing (Galloway and Thelen, 2004). The underlying development of the foot begins in the infant which is characterised by a flat profile, a large plantar contact area, and large levels of subcutaneous fat (Price et al., 2018). It is suggested that 50% of foot length is achieved by 12–18 months of age, which is very high compared to the growth in length still to occur within the rest of the body. The relatively large contact area of the infant foot may help increase stability by enlarging the base of support and thus reducing peak plantar pressures (Price et al., 2018). Short limbs and lower swing phase muscle powers at this age means that the centrifugal forces from swinging a large foot forward should not be a destabilising issue.

Changes in plantar pressure magnitude and distribution suggest significant demands are being placed upon the plantar skin and musculoskeletal structures as gait matures (Hallemans et al., 2003; Bosch et al., 2007). Infants have high plantar surface contact areas and low absolute pressures relative to body mass compared to older children. Between the ages of zero and two months of independent walking, pressures are only around 25%–50% of what can be expected as an adult in proportion to weight (Hallemans et al., 2005; Bosch et al., 2007; Price et al., 2018). Highest pressures occur under the hallux

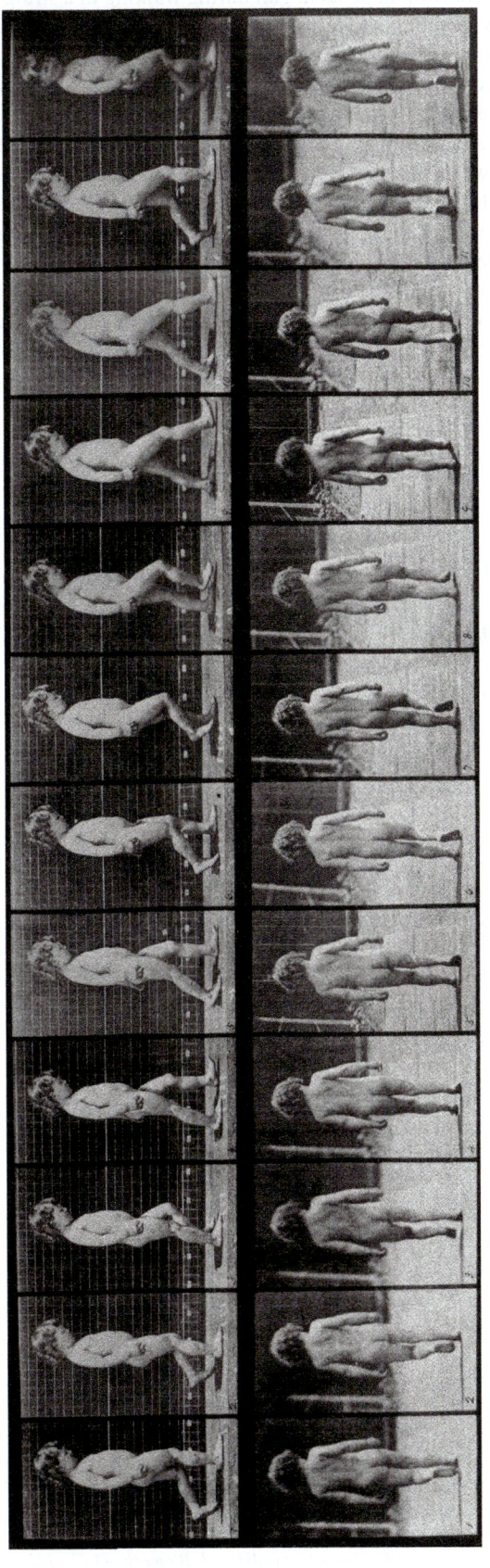

FIG. 1.5.6b Photographs from the Eadweard Muybridge collection recording early development of gait. In childhood, increasing bouts of walking improve motor skills, enhancing confidence in gait. Inverted pendulum walking is reported to be established after only one month of independent walking. Distinct stance and swing phases should quickly develop. Double-limb stance phase is longer in early walkers than those of older children, adolescents, and young adults. The lower limb's nervous system is still immature. The later maturation of lordosis in the lumbar spine and the changing hip neck-to-shaft angle should improve posture and gait over time.

consistently for the first three to eight months of independent gait during walking in a straight line (Hallemans et al., 2003; Price et al., 2018). Increased toe pressures are thought to reflect the need to gain proprioceptive feedback (Price et al., 2018). However, how much the data collected from straight walking represents the usual kinetic forces of life in the real world of an infant is debatable. Yet, the information remains useful within the clinical assessment situation, seemingly indicating that mature plantar pressure distribution is mostly established between the age of five to six years (Hallemans et al., 2003; Bertsch et al., 2004; Bosch et al., 2007; Alvarez et al., 2008). There appears to be no difference in plantar pressure data between boys and girls aged at least between four to seven years (Phethean and Nester, 2012).

GRF pattern maturity, so far, is less clear-cut from the research data but seems to lie somewhere after five years of age before they reach adult-like patterns (Samson et al., 2011). In the youngest walkers, the terminal vertical F3 and horizontal F5 peaks of GRF are almost absent (Hallemans et al., 2006; Samson et al., 2011), but these peaks increase with age (Diop et al., 2005; Samson et al., 2011). Horizontal anterior–posterior forces thus increase with growth and are more adult-like by age five years when medial-lateral forces reduce to also become more adult-like, reflecting a decrease in side-to-side instability (Samson et al., 2011). Takegami (1992) reported that these maximal medial-lateral forces continue to decrease until around age eight years (Fig. 1.5.6c).

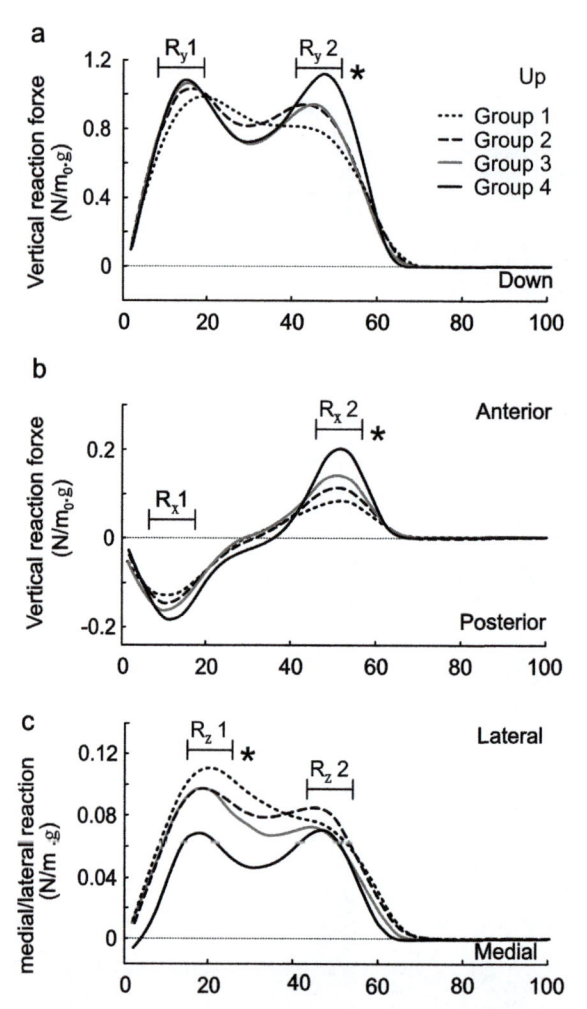

FIG. 1.5.6c Vertical (A), anterior/posterior (B), and medial/lateral (C) GRF forces during gait at different ages. Group 1 are aged two years *(dotted line)*, group 2 are aged three-and-a-half years *(dashed line)*, group 3 are aged five years *(grey line)*, and group 4 are adults *(black line)*. In group 1, the second anterior/posterior peaks are almost nonexistent indicating poor acceleration mechanics, confirmed by the vertical GRF. Children's vertical acceleration GRF remains significantly lower than braking forces but increases with age. *(Adapted from Samson, W., Dohin, B., Desroches, G., Chaverot, J.-L., Dumas, R., Cheze, L., 2011. Foot mechanics during the first six years of independent walking. J. Biomech. 44 (7), 1321–1327.)*

Kinematic changes

Little is known of foot kinematics in infants, but whole-body kinematics have been better investigated (Price et al., 2018). MTP joint motion in children aged two to five years presents as similar to older children and adults, but age has a most significant affect around preswing with increased age causing an increase in MTP joint motion (Samson et al., 2011). It is possible that younger children give priority to stabilisation through their toes rather than via MTP joint extension during propulsion, which is thus suggestive of a more passive MTP joint role (Samson et al., 2011). A greater ankle-rearfoot moment has been noted in younger age groups of around two years of age compared with other older age groups with more experience of walking. Rearfoot motion tends to decrease with age, although the ankle-rearfoot angles are more transient across all ages (Samson et al., 2011). Energy absorption reaches its maximum during early stance in children but occurs by early midstance in adults (Chester et al., 2006; Samson et al., 2009, 2011). These differences can be explained through incomplete foot vault development (Samson et al., 2011) as in the adult foot, the vault drops and the foot lengthens most rapidly in early midstance (Stolwijk et al., 2014), absorbing energy (Fig. 1.5.6d).

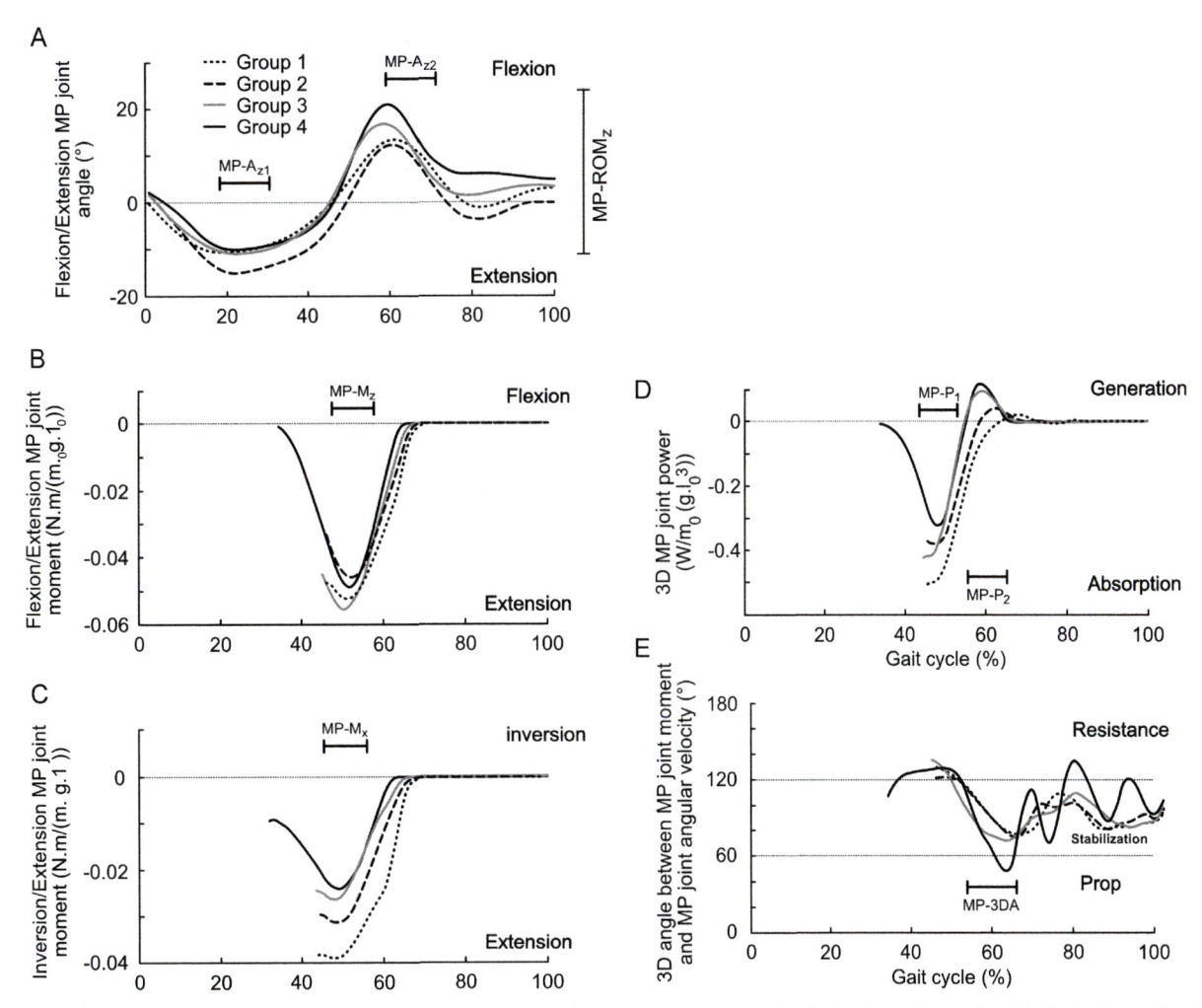

FIG. 1.5.6d Metatarsophalangeal (MTP – here abbreviated to MP) joint motion across ages, with flexion/extension angles (A), flexion/extension joint moments (B), eversion/inversion moments (C), joint power (D), and 3D angle between MTP joint moment and MTP joint angular velocity (E). Group 1 = age two *(dotted line)*, group 2 *(dashed line)* = age three-and-a-half, group 3 = age five *(grey line)*, and group 4 *(black line)* = adult. Maximum differences in range of motion never exceeded 7.1° between groups and differences in MTP joint variables are small. At midstance, maximum absorbed energy is found in two-year-olds with energy generated being lowest in this group at preswing. At preswing, children demonstrate stabilisation while adults demonstrate propulsive acceleration. *(Adapted from Samson, W., Dohin, B., Desroches, G., Chaverot, J.-L., Dumas, R., Cheze, L., 2011. Foot mechanics during the first six years of independent walking. J. Biomech. 44 (7), 1321–1327.)*

Ankle motion and power change during childhood, with ankle power at push-off acceleration doubling by adulthood (Hallemans et al., 2003). This is consistent with noted increases in plantarflexion at initial contact and reduced ankle dorsiflexion during swing in a one-year-old's gait (Price et al., 2018), and the overall reduced range of motion compared to adults, at least up to the age of five years (Samson et al., 2011). Furthermore, infants display increased hip and knee flexion/extension and ankle dorsiflexion ranges of motion (Hallemans et al., 2005, 2006). Children seem to mainly stabilise with the ankle muscles and propel with the hip muscles, while young adults stabilise with the hip muscles and both brake and propel with the ankle muscles (Samson et al., 2009). Ankle and MTP joint mechanics seem to reach relative maturity between three-and-a-half and five years of age (Samson et al., 2011) (Fig. 1.5.6e).

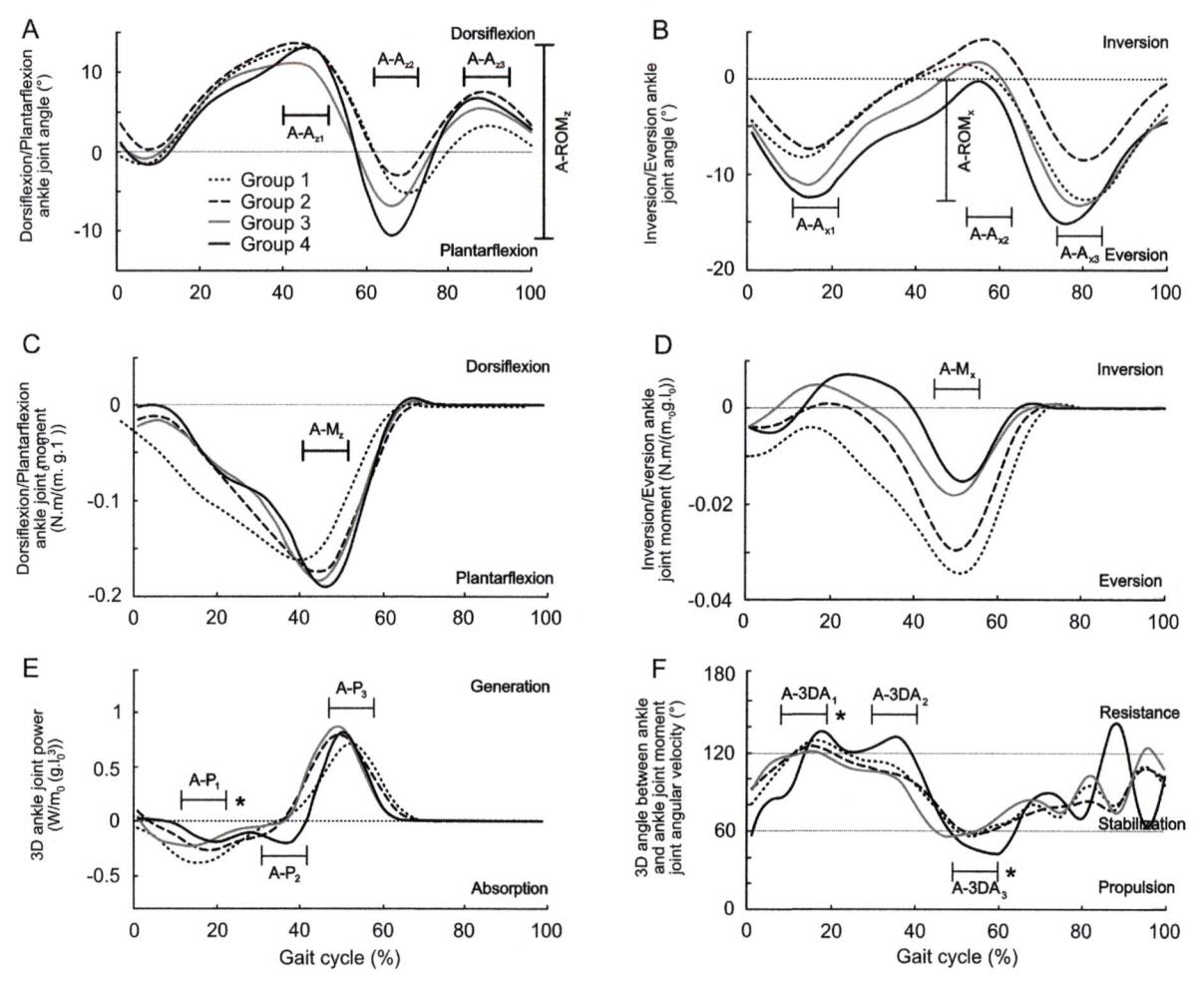

FIG. 1.5.6e Ankle group kinematics at different ages. Ankle joint angles of dorsiflexion/plantarflexion (A) and eversion/inversion (B), dorsiflexion/plantarflexion moments (C), eversion/inversion moments (D), ankle joint power (E), and 3D angle between ankle and ankle joint moment for joint angular velocity (F). * indicates the significant difference when leg lengths are taken into account. Group 1 = age two years *(dotted line)*, group 2 = age three-and-a-half years *(dashed line)*, group 3 = age five years *(grey line)*, and group 4 = adult *(black line)*. Maximum difference in ankle joint angles never exceeded 7.1° in dorsiflexion/plantarflexion and 2.5° in eversion/inversion, but maximum eversion moments tend to decrease with age, while a small inversion moment tends to develop. Energy absorbed in gait changes with age, with maximum absorbed energy appearing to be more important in children during early stance. Joint stabilisation seems to reduce with age. *(Adapted from Samson, W., Dohin, B., Desroches, G., Chaverot, J.-L., Dumas, R., Cheze, L., 2011. Foot mechanics during the first six years of independent walking. J. Biomech. 44 (7), 1321–1327.)*

Trunk oscillations in the sagittal and frontal planes reduce with walking experience during the first one to four months of walking (Assaiante et al., 2000). Three diverse kinematic approaches to trunk accelerations during walking have been noted, although with significant overlap both inter-individually and intra-individually (McCollum et al., 1995). The vertical and lateral movements (amplitude) of the CoM are greater in children before the age of four years, while the forward amplitude remains greater in children before the age of seven years (Dierick et al., 2004). It appears that mature human trunk CoM displacement during gait is a gradual developmental process reflecting neuromotor maturation that is still evolving until at least age seven years (Dierick et al., 2004). Despite this gradual development of the body's CoM displacement/oscillation patterns, the principles of inverted pendulum walking have been reported as becoming apparent by one month after the onset of independent walking, with an increasing prevalence among infants with gait maturation over a period of six months of independent gait (Bisi and Stagni, 2015). When toddlers start to walk, the arms are held high, but as gait matures with age, reciprocal arm swing develops with increasing consistency (Van de Walle et al., 2018). Larger mean shoulder extension angles develop, but axial shoulder rotation remains less than that of adults with full adult-like values not being achieved until ten to fourteen years of age (Van de Walle et al., 2018).

Electromyography (EMG) has been rarely used on infants and children, so a great deal of data on childhood muscle activity remains unknown. Okamoto et al. (2003) studied one child from being a neonate to age seven years. From initial supported to unsupported gait, there was long activation of tibialis anterior and co-contraction of the lower limb's anterior/posterior musculature, but this reduced after six months of independent walking. A larger study started at the onset of walking and three months later found that co-contraction of the lower limb musculature was evident, showing high variability in steps with initial low walking experience (Chang et al., 2006). Longitudinal changes in muscle recruitment strategies have been quantified in infants from six to twelve months of age, with gradual reductions in agonist–antagonist coactivation over time (Price et al., 2018). Tibialis anterior activation during swing in both prewalkers and early walkers is low at around 25%–28%. However, this increases to 63% in those walkers with at least nine months experience of independent gait (Assaiante et al., 2000).

It has also been reported that variability of onset and duration of muscle activation at comfortable walking speeds was no different between four to seven-year-olds and between eight to eleven-year-olds (Detrembleur et al., 1997; Chang et al., 2006). There was greater variability found between seven to nine-year-olds and thirteen to sixteen-year-olds, but only significantly at slower and faster walking speeds, that is, outside of the comfortable preferred walking speed (Tirosh et al., 2013). Kinematic patterns at the ankle appear to reflect maturation of tibialis posterior activation and gastrocnemius antagonist–agonist relationships with experienced walkers around three years of age displaying ankle dorsiflexion during swing but reduced plantarflexion moment and power generation at toe-off compared to adults (Hallemans et al., 2006; Samson et al., 2009, 2011). The ability to generate ankle power at forward progression in walking and running is a marker of gait maturation as well as elastic energy recycling (Lye et al., 2016). This ability is absent until late childhood into early adolescence, suggesting full gait muscle action maturity occurs around this time.

During the first years of walking, it can be appreciated that considerable changes occur in joint kinematics and kinetics (Ivanenko et al., 2005; Chester et al., 2006; Dominici et al., 2007; Chester and Wrigley, 2008; Price et al., 2018). Infant locomotion strategies adapt in manner, achievement, and variability as motor control and physical capacity develops, changing spatiotemporal parameters (Price et al., 2018). There is an increase in step number over distance and the loading pattern across the plantar surface (Price et al., 2018). Initially, early step length is reported to be ~12 cm with an accompanying velocity of ~0.24 cm/s, which increases over the first few months to a step length of ~25 cm and a coincident velocity of ~0.8 cm/s (Chang et al., 2006; Badaly and Adolph, 2008). The step length, in time, can become longer than the leg length, although variability is high among this age group between steps of the same infant (Badaly and Adolph, 2008; Price et al., 2018). In infants, step length is more variable than step width, which is the reverse of healthy adults where step width is more variable than step length (Looper et al., 2006).

Initial foot contacts in very early walkers are highly varied with only ~5% of steps being heel contacts, 60% forefoot contacts, and around 35% being full foot contacts. However, within this group by ~8 weeks of walking, this changes to a dominant 30–58% of initial heel contacts and initial heel contacts become consistent by one year after starting independent gait (Price et al., 2018). With the development of heel-toe gait, pressure distribution changes from the midfoot in younger infants to the heel and forefoot in children (Alvarez et al., 2008). Relative contact time in the midfoot reduces at the start of independent walking from ~76% to 65% of the gait cycle, and the midfoot load reduces from 30% to 20% of the total foot impulse on the plantar foot surface after six months (Bertsch et al., 2004). The development of these changes in the walking pattern has more to do with developmental stages rather than chronological age, and this is suggested to be linked to increased stability during the requirement to load the midfoot and skeletal maturational changes in the medial longitudinal vault (arch) (Price et al., 2018). Along with this shift to initial heel pressure is a shift to more lateral loading with walking experience. This is evidenced by a more lateral deviation of the CoP under the foot and reducing pressures under the hallux (Hallemans et al., 2003).

Maturation of gait

The first step for children in their gait development consists of building up a repertoire of postural strategies. This is followed by the second step wherein they learn to select the most appropriate postural strategy dependent on their ability to anticipate the consequences of the movement in order to maintain efficiency and control balance (Assaiante et al., 2005). The transition from sitting to walking takes around nine months (Price et al., 2018) and it has been suggested that gait maturation occurs by the age of three years (Sutherland et al., 1988). This is justified by the presence of reciprocal arm swing, heel strike, increased walking velocity, step-length, and single support with cadence reduction. However, further gait details engendered through research have revealed a more complex situation. Dusing and Thorpe (2007) found that normalised velocity and step length increased gradually from age one to four years, but such gait features did not stabilise until age five to ten years. Based on the results of their work, Samson et al. (2011) have suggested that foot function maturation during walking gait is 'largely' fully achieved by age five years. Holm et al. (2009) found that there was little change in gait measures from age seven years upwards. Studies on the gait of one to seven-year-olds have reported higher standard deviations in data, as there is a high variability of gait among young children, particularly with speed (Samson et al., 2011, 2013; Van Hamme et al., 2015). Samson et al. (2011,2013) found that different joint biomechanics seem to mature at differing rates, with MTP joint mechanics being more similar to adults by age two years and ankle joint parameters and GRF generation being similar to adults by age four years. Knee and possibly hip kinematics seem to be matured by age six to seven years (Samson et al., 2013). Thevenon et al. (2015) found that all temporal parameters, except double support, differed significantly between six and seven-years-olds compared to nine-year-olds and above, whereas cadence, step time, cycle time, and stance time differed between seven and eight-year-olds. Thevenon et al. (2015) also found that height under growth rising between 110 cm and 130 cm affected temporal parameters and then reached a plateau. Data has also suggested that running propulsion kinetics appear developed by age six years, but that the skill of fast walking requires further neuromuscular maturity (Lye et al., 2016) (Fig. 1.5.6f).

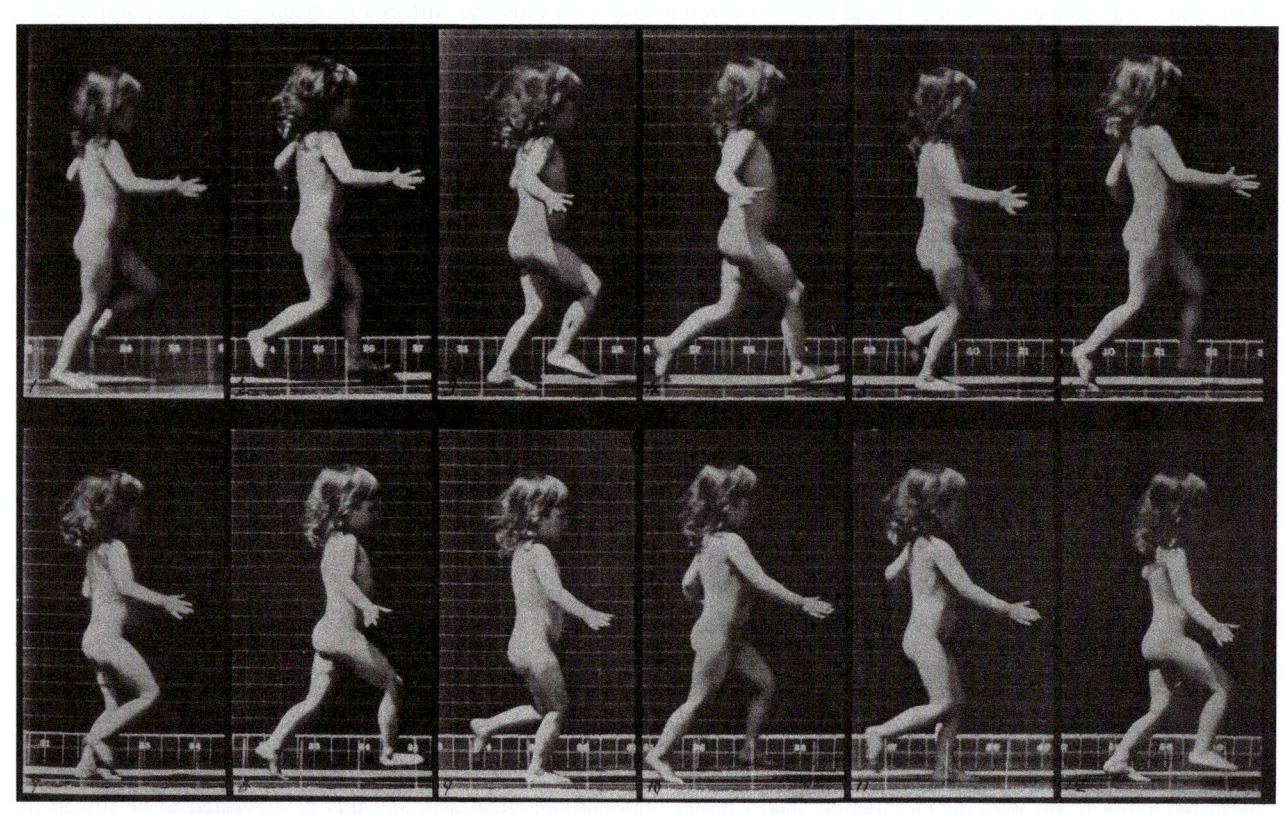

FIG. 1.5.6f Running is another locomotive skill that must be mastered and is largely achieved by age six. These photographs from the Eadweard Muybridge collection clearly show a return to 'arms held high' posture of early walkers through the early stages of running development. It is likely that loss of double-limb stance and the demands of more acceleration power become gait destabilising influences that must be overcome through muscular and postural changes. Yet already, several frames indicate that an aerial phase has started to develop. Faster walking may also be another skill that demands additional neuromuscular maturation beyond that of basic walking, restricting its use to more experienced walkers.

Studies have demonstrated that biomechanical gait parameters are highly influenced by age until mid-childhood, but that sagittal joint kinematics, moments, and power are predominantly influenced by speed of progression rather than age (Van Hamme et al., 2015). Gait speed, step length, base of support, foot progression angle, step time, and stride time increase initially with age, but cadence then reduces as gait becomes relatively stable by age seven years (Sutherland et al., 1988; Dusing and Thorpe, 2007; Lythgo et al., 2009; Thevenon et al., 2015). After age seven years (or 120 cm in height), there seems to be a breakpoint in the rate of increase in temporal parameters such as velocity, step length, and stride length (Moreno-Hernández et al., 2010; Thevenon et al., 2015) which probably reflects a change in coordination abilities. Stance duration and double and single-limb support in children from age five years onwards seem unaffected by age until ~12 years when they change further (Lythgo et al., 2009). Gait in children seems to be highly symmetrical with no significant gender differences (Hillman et al., 2009; Lythgo et al., 2009; Thevenon et al., 2015), suggesting that gender differences in gait only develop sometime after age twelve years.

Adolescent gait

Young humans aged over twelve years demonstrate a reduction in single support phase, but an increase in double-limb stance and overall stance time. This suggests that gait does not fully mature in all aspects until the age of twelve to thirteen years (Peterson et al., 2006; Lythgo et al., 2009). A later maturity of gait is supported by the findings that seven-year-olds produce less peak plantarflexion moments and both less absorption and generation of ankle power during terminal stance than do adults when walking at the same speed (Ganley and Powers, 2005). This indicates that the neuromuscular maturity of the adult has not been reached by age seven years. It occurs later, possibly not being reached until age thirteen years (Lythgo et al., 2009) when gender differences in gait start to develop with different gait variables developing under the influence of maturity at different rates (Fig. 1.5.6g; Table 1.5.6).

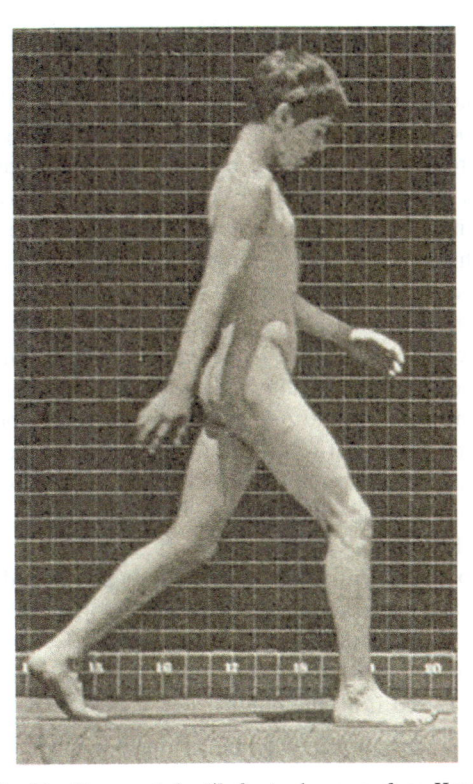

FIG. 1.5.6g Normal adult gait biomechanics should not be expected until after twelve years of age. Here, loading response is completing in the right foot as preswing is occurring within the left foot at the end of double-limb stance phase in an adolescent male from the Eadweard Muybridge collection. Note the CoM is at its lowest point, positioning the chest below the thicker line behind the subject. Even at this age, gait parameters such as double-limb support and overall stance time may still be slightly different compared to mature adults. It is essential that clinicians do not interfere with normal gait development of children and adolescents just because gait data do not match adult expectations.

TABLE 1.5.6 Gait phase timing results from children and young adults revealing early years of adolescence show adult-like walking gait cycles.

Age group	Single support phase (average)	Double support phase (average)	Stance time (average)
Children (5–11 years)	43.2%	13.5%	58.8%
Young adults (12–13 and 19 years)	40.9%	18.6%	59.4%

Data From Lythgo, N., Wilson, C., Galea, M., 2009. Basic gait and symmetry measures for primary school-aged children and young adults whilst walking barefoot and with shoes. Gait Posture 30 (4), 502–506.

Other influences on childhood gait

Children's gait is not only affected by age but also by their weight, rate of motor development, and the presence of footwear and clothing. Obesity has been shown to affect gait in children, possibly as a result of structural changes in the feet causing a lowering of the vault profile (Mickle et al., 2006). It is worth considering again the effect of weight increases, where a 20% increase in body weight reported during pregnancy is enough to double the force on a joint during locomotion (Ritchie, 2003; Branco et al., 2014). Such mass-induced biomechanics are likely to profoundly influence lower limb development in children. Childhood obesity gait changes result in a lower cadence and gait velocity, a longer stance period, and greater step length asymmetry (Hills and Parker, 1992; Thevenon et al., 2015). It also results in a longer time in double support and exhibits a larger base of support (Lythgo et al., 2009) along with altered pressure loading patterns (Dowling et al., 2001, 2004; Mickle et al., 2006; Cousins et al., 2013; Yan et al., 2013). Cousins et al. (2013) found that loading patterns of pressure were also changed in overweight children as well as in those classed as obese, with increased peak and temporal loading of the 5th through to the 2nd metatarsals. However, Phethean and Nester (2012) did not statistically find an effect of body mass on pressure data in four- to seven-year-olds. The thicker plantar fat pad found in obese children does not seem to be an adaptation to increased weight as it only marginally spreads pressure over a larger area, suggesting its presence only reflects the higher body fat levels (Riddiford-Harland et al., 2011).

The kinematic and spatiotemporal characteristics of walking toddlers are reported as being different in slimmer infants, with a decrease in maximum hip adduction in slim children's stance phase, possibly due to less resistance in the frontal plane to sideways movements of the thorax and the head segment (Price et al., 2018). This has implications in the morphogenesis of coxa vara during childhood as the neck-shaft angle at the developing hip joint often initially decreases via developing limb adduction to then increase under the influence of hip abductor muscle forces in a mammalian hip that matures its osseous structures relatively late (Hogervorst and Vereecke, 2014). Hip abductor strength in gait may be inadequate to resist the larger hip adduction moments provoked within overweight/obese children that may in response, develop a more adducted limb posture at the hip to support their larger CoM (see Fig. 1.5.6h). Hip femoral neck-shaft angles decrease from juveniles to set within early adulthood, finally falling somewhere between 120° and 140°. Global means are ~127°, although they are commonly found outside these values and they can decrease again after age 60 (Gilligan et al., 2013). There are many often interlinked factors influencing the development of femoral neck-shaft angles during growth, including climate, lifestyle, and clothing use, although perhaps surprisingly, not gender (Gilligan et al., 2013). Thermal effects may be particularly influential. Cold climate seems to lower population mean angles, but the use of complex warm clothing appears to counteract or buffer this influence, yet these are only some of the factors that are part of a complex process that has set variable means within populations around the world. The femoral neck-shaft angle is therefore shown to be a significantly plastic anatomical feature of human growth, and childhood weight may be another significant influence on setting the angle.

Significant effects are caused by footwear on the gait of habitually shod children as well as their foot morphology (Hollander et al., 2017). Compared to barefoot, shod children increase gait speed, step length, stride length, step time, and stride time. Shoes also reduce the foot progression angle and cadence (Lythgo et al., 2009). Clinically, this creates dilemmas, for the age of the child, its health status and body morphology, and whether clinical gait tests should involve being shod or unshod needs to be considered if all the clinical gait data is to be useful. Problems become particularly apparent if all the initial gait data were taken barefoot, yet outcomes with the intervention of devices such as foot orthoses are assessed with shoes on.

It has been reported that material stiffness of clothing may also affect gait in toddlers. Trousers, tracksuits (flexible material), underwear, and nappies (diapers) are all found to affect basic gait patterns at six to eighteen months of walking experience (Théveniau et al., 2014). This was particularly so with trousers which decreased walking velocity and step length, with lighter tracksuits decreasing step length only compared to underwear (Théveniau et al., 2014). The effect of clothes seems greater on more experienced walking toddlers than earlier walkers and results suggests trousers may actually cause biomechanical constraints that could decrease hip and/or knee ranges of motion by preventing joint freedom (Théveniau et al., 2014). Nappies (diapers) did not affect step length or gait velocity, but they did influenced step width variability (possible through adduction hip constraint) which decreased when nappies were replaced with underwear (Théveniau et al., 2014). This is something worth considering with a trend to now use nappies for longer on children in 'developed' nations in relation to the development of femoral neck-shaft angles. Lower limb clothing factors may have altered the kinematic and kinetic data collected on subjects within gait studies, as the clothing used by subjects is not normally reported on. It is also something to consider and note, as most research is performed with

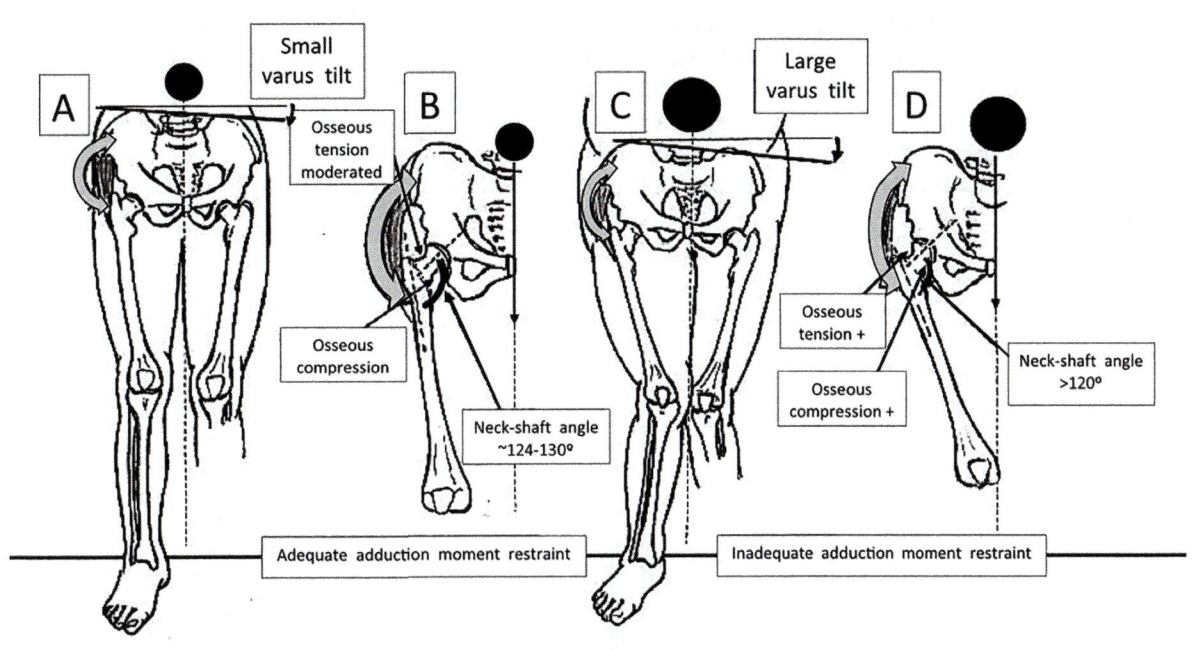

FIG. 1.5.6h   The human femoral neck-shaft angle is set under lifestyle factors and biomechanical forces during childhood. High hip adduction moments during loading and early single-limb support are generated by the CoM's position in relation to the support limb. This influences the femoral neck forces that help set the femoral neck-shaft angle and femoral carrying angle at the knee. The results help positions the quadriceps or Q-angle during gait. Hip adduction moments in stance phase that create varus hip tilt during single support become greater as the body and gait matures during childhood, with forces initially reflecting the small mass of the child on proportionately shorter lower limbs increasing the hip adduction angle necessary to support the CoM. This should provoke the hip abductor muscles to grow and develop in proportion to gradual increases in body mass and adductor moment size generated under locomotive and lifestyle factors that start to set the neck-shaft angle appropriately for adulthood (A). Abductor muscle forces *(grey arrows)* should be proportional to the hip varus moment and should moderate superior neck tensile forces generated creating resistive compression forces superiorly. In immature growing femoral necks, such force balancing helps set the neck-shaft angle appropriately for an efficient hip adduction moment-resisting lever arm (B). If the weight of a child rises dramatically and quickly it becomes excessive for normal osseous development. Hip abduction muscle forces may be inadequate to moderate the immature hip varus tilt sufficiently to maintain postural stability of the CoM, unless the support limb is positioned more medially under the body (C). This necessitates greater hip adductor muscle activity at initial contact, which if combined with inadequate hip abductor strength, will result in increased tensile stresses on the superior femoral neck while compressive forces increase inferiorly (D). This encourages a lower neck-shaft angle. A coxa vara with its higher femoral carrying angle increases the risk of other lower limb alignment changes such as genu valgum. Such lower limb alignment changes associated with childhood obesity may in turn encourage a lower vault profile from higher medial forefoot loading forces under greater overall weight-induced loads on immature anatomy. *(Permission www.healthystep.co.uk.)*

children dressed in normal daily clothing and also often wearing shoes, making some of the information given within this text open to some degree of error, at least when compared to natural human development. In relation to urban environments and how children move in the modern world, such data would be perfectly valid.

Gait in paediatric pes planus

In children, the term 'normal' in relation to foot vault profile is even more problematic than in adults because the foot changes shape with age and is expected to look much flatter at younger ages (Uden et al., 2017). Thus, no age has been concluded to be 'the' age at which a child's foot ceases to develop further (Uden et al., 2017). However, this is hardly surprising as changes in morphology continue throughout life, the foot included. Despite this caveat, pes planus is reported in 33% of all children and is one of the main features of children with hypermobility (Yazgan et al., 2008; Smits-Engelsman et al., 2011). It is also associated with childhood obesity (Mickle et al., 2006). Pes planus is considered to be one of the most commonly seen conditions in paediatric orthopaedic clinics, but it remains unclear as to what age a child may grow a more pronounced foot vault, or even if flat feet should be defined as a pathology (Evans, 2008).

The probable reasons for this, as already debated, are that not all flat feet (pes planus) are in the same condition, and if associated with hypermobility, are probably a more significant finding. It is important to note that clinical assessments such as relaxed calcaneal stance position (RCSP) and radiographic methods have very little ability to predict gait deviance in paediatric pes planus, even though radiographic findings of talometatarsal angle correlate with RCSP and talocalcaneal angle with maximal internal and external rotation angle of the knee during gait (Lee et al., 2009). Despite the presence of a thicker plantar fat pad within the medial midfoot of children making feet look flatter, it has been reported that 'flat feet' are found more frequently in preschool boys than in preschool girls, suggesting that the medial vault profile develops at a slower rate in boys than in girls (Mickle et al., 2008); something worth being aware of in paediatric gait clinics. Higher arch profiles increase in frequency with age, regardless of sex (Fig. 1.5.6i).

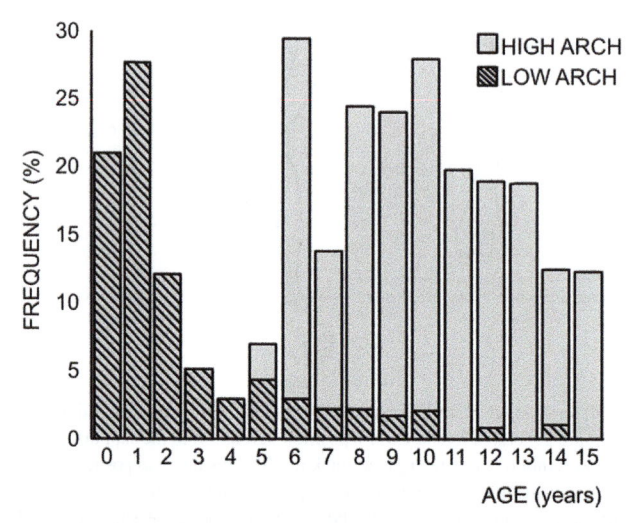

FIG. 1.5.6i Chart of different rates of children's flat feet reported with changes in age from a study by Volpon (1994) using the length of a footprint and evaluating the medial longitudinal arch. The height of the feet's structural vault increases between two- and six-year-olds. However, caution is still required with older children, as the foot is still maturing and feet that are assessed as high or low from their medial longitudinal arch, may still function perfectly well. Why a foot is the shape it is (developmental factors) and whether it can provide adequate cycles of changing compliance and stiffness levels during gait are 'the' key facts to gather in order for the clinician to decide if intervention is warranted or not. Children should have more compliant feet than adults. *(Modified from Volpon, J.B., 1994. Footprint analysis during the growth period. J. Paediatr. Orthop. 14 (1), 83–85.)*

Developing foot vault profiles during childhood may be important to monitor, not so much for their final shape alone but for the level of compliance and stiffness they are able to generate in concert with soft tissue strength and overall foot flexibility. The younger the child, the more mobile the foot should be expected to be. There are kinematic differences in pes planus feet in children and early adolescents compared to controls with less rearfoot dorsiflexion, increased midfoot dorsiflexion and/or ankle-rearfoot eversion motion, and increased forefoot inversion and abduction. Yet, these asymptomatic planus feet are generally reported to be stiffer within this age group (Hösl et al., 2014; Saraswat et al., 2014). Kinetic differences include a reduced ankle plantarflexion moment and reduced ankle and midfoot joint power (Saraswat et al., 2014). This suggests issues with acceleration power around the foot during terminal stance. It has been observed that passive dorsiflexion is greater in asymptomatic compared to symptomatic pes planus children and early adolescents but without significant differences being noted in gait, save that symptomatic planus feet lack positive joint energy in terminal stance (Hösl et al., 2014) (Fig. 1.5.6j). It could well be that the difficulty for pes planus feet, particularly symptomatic ones, lies in the generation of terminal stance power; an argument supported by the lower and reduced impulse of the forefoot found in running and walking with more mobile vaulted feet as has been reported by Cen et al. (2020).

Paediatric equinus

Toe-walking or *equinus gait* is associated with an absence of heel strike in walking and is not unique to children, being seen even in adults with neurological conditions such as cerebral palsy. However, it is far more common in paediatric neurological conditions and also occurs idiopathically in some children without neurological deficit. So, idiopathic toe-walking is usually a diagnosis of exclusion (Pendharkar et al., 2012; Davies et al., 2018). Toe-walking is not initially uncommon but is considered abnormal after age three, with a prevalence of 4.9% by age five-and-a-half years (Davies et al., 2018). Pains in the legs or feet and trips and falls are the most common complaints, with limited passive ankle dorsiflexion that can cause ankle injuries and restrict normal daily activities (Davies et al., 2018). Children with idiopathic toe-walking often have

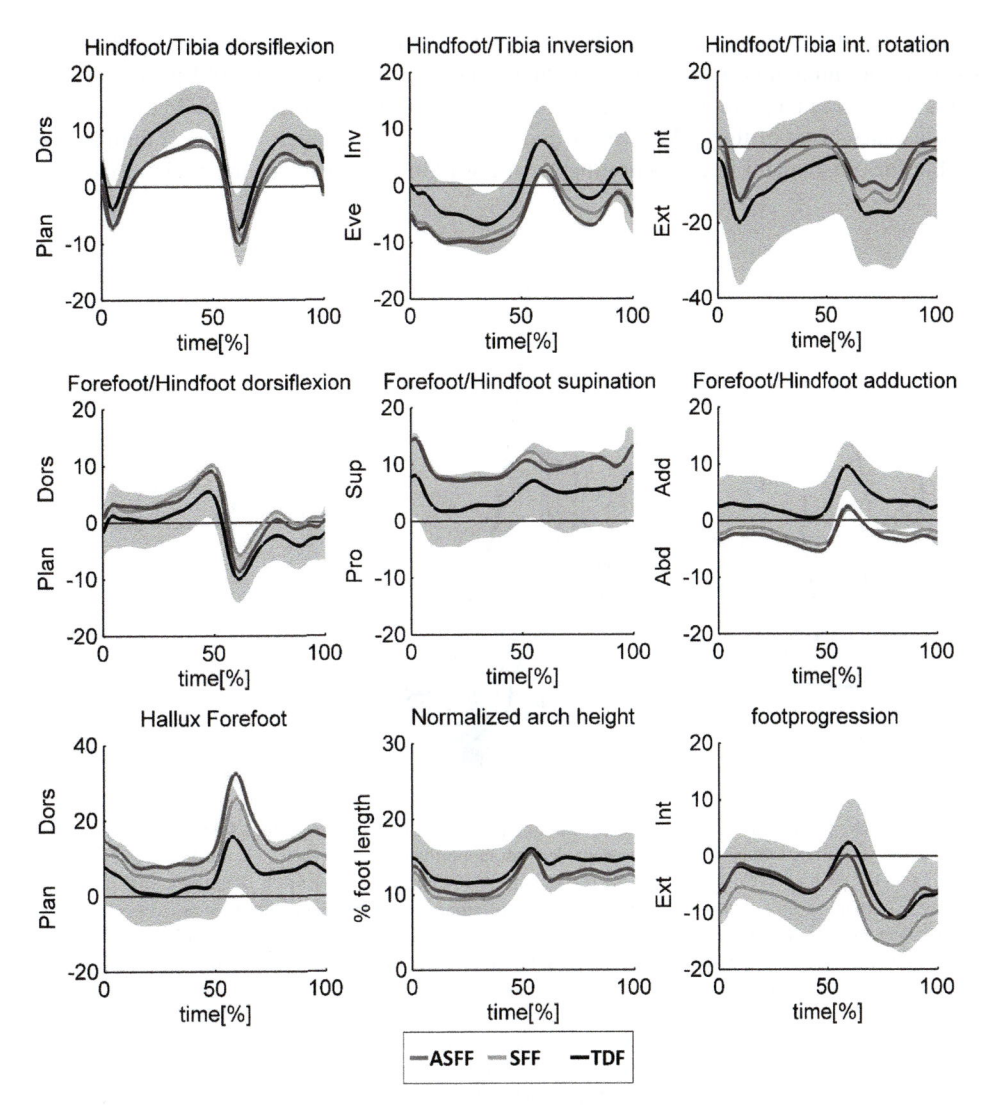

FIG. 1.5.6j Average foot kinematics (°) of children aged between seven years to early adolescence (fourteen years) during the gait cycle for asymptomatic flexible flat feet (ASFF—*mid-grey line*) and symptomatic flexible flat feet (SFF—*light-grey line*) compared to typical developing feet (TDF—*black line*). Shaded area shows ±1 standard deviation across controls. *(Modified from Hösl, M., Böhm, H., Multerer, C., Döderlein, L., 2014. Does excessive flatfoot deformity affect function? A comparison between symptomatic and asymptomatic flatfeet using the Oxford Foot Model. Gait Posture 39 (1), 23–28.)*

histories of developmental delays or neurodevelopmental diagnoses such as autism, language disorders, and sensory processing dysfunctions (Williams et al., 2010a,b, 2014; Pollind et al., 2019).

In simulated equinus gait in healthy children, the ankle plantarflexors become markedly biphasic with triceps surae becoming the primary loading muscle complex at initial contact (Houx et al., 2013). A number of variable postural changes in the limb and trunk occur with equinus gait, with the hips becoming more flexed and the pelvis more anteriorly tilted during loading response in addition to the knee becoming more flexed and the ankle plantarflexed (Houx et al., 2013). During midstance, the knee flexion angles usually increase rather than decrease as normal, but the knee can become hyperextended in some individuals. The knee is also subjected to increased varus moments, while the hip tends to experience less hip adduction moments and the foot becomes more internally rotated to the line of progression (Houx et al., 2013).

The triceps surae-Achilles complex normally becomes a walking energy dissipator as well as a midstance decelerator of the CoM into late midstance and an accelerator during terminal stance. However, with paediatric equinus, this ability is compromised. Ankle plantarflexion power becomes much reduced through power absorption due to the increased ankle plantarflexion angle (Matjačić et al., 2006; Houx et al., 2013). In a plantarflexed angle, the ankle plantarflexors find themselves at a contractile disadvantage, reducing their ability to generate force (Neptune et al., 2007). With an increased ankle plantarflexion angle throughout gait, such individuals tend to walk with their CoM positioned high during stance phase. This requires decreases in hip motion and halves the normal knee extensor moments (Matjačić et al., 2006) resulting in smaller vertical displacements of the CoM (Usherwood et al., 2012). However, because a number of postural change options are available for causing and in compensating for toe-walking, toe-walkers are not homogenous (Fig. 1.5.6k).

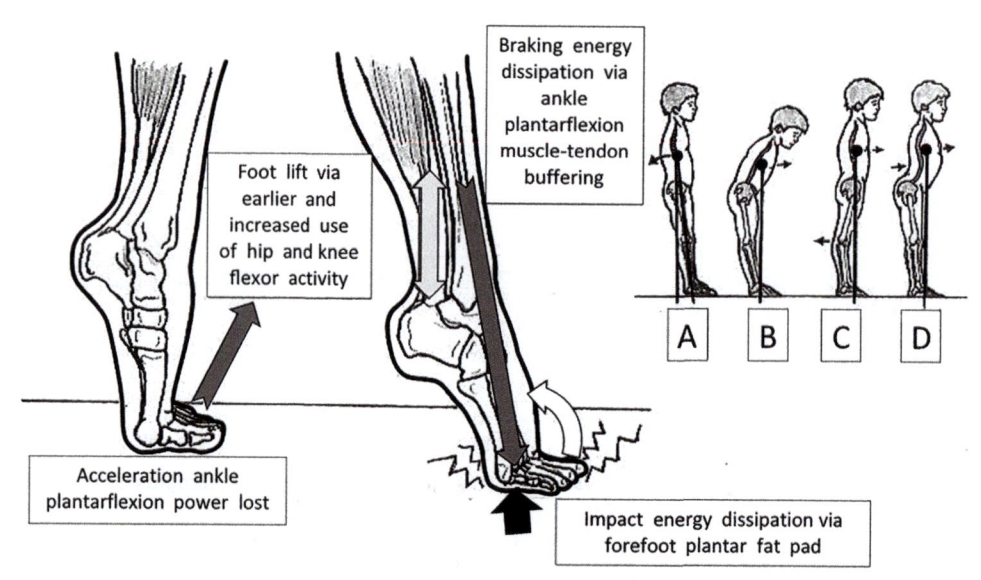

FIG. 1.5.6k Equinus gait (meaning horse-like) refers to forefoot walking and loss of the plantigrade foot. Energy dissipation and acceleration biomechanics are completely altered. The weight transfer phase during single-limb support is intrinsically unstable due to a much smaller base of support, foot contact surface area, and a higher positioned CoM than normal. The foot needs to maintain greater levels of stiffness throughout stance, but as walking involves a lower velocity than running, time-dependent connective tissue stiffening properties cannot aid these processes during walking forefoot gait mechanics. Muscles are thus required to provide the stiffness through vault posture support and contractile forces during all of stance phase. Plantarflexors must provide energy dissipation at impact, yet they are unable to store normal stretch energy via large weight transfer-induced dorsiflexion motions during late midstance for powering acceleration. Foot lift is instead provided by increased and earlier activity within the hip and knee flexors. As anatomy and posture follow biomechanical forces, then if this gait is maintained, hip, knee, and ankle flexors will shorten and tighten their muscle and connective tissue fibres. Tight ankle plantarflexors that then prevent ankle dorsiflexion make providing a prone foot posture difficult without tipping the CoM posteriorly (A). Compensatory hip flexion can draw the CoM anteriorly (B), as can knee hyperextension (genu recurvatum) (C) and increasing the lordotic curvature of the spine (D). Children afflicted with equinus gait will develop combinations of these postural changes to overcome some of their toe-walking and resultant limited ankle dorsiflexion. The longer toe-walking persists, then the more problematic and clinically challenging it is to resolve these functional adaptations. *(Permission www.healthystep.co.uk.)*

Idiopathic equinus may be found in 100% of the steps of some children, while in others it can be variable, transitory, and self-resolving. In persistent idiopathic cases and those found associated with neuropathic pathologies such as cerebral palsy, it creates more of a clinical challenge not least because there tends to be more gait sway, especially in cerebral palsy patients. As toe-walkers grow and develop a larger body mass, persistent toe-walking becomes increasingly more problematic due to instability from a gait with a higher CoM and reduced vertical oscillations that together increase the potential risk of foot and ankle instability. Persistent cases may require therapeutic interventions such as surgery, rehabilitation, and shoe and orthosis management (Davies et al., 2018; Radtke et al., 2018). Idiopathic toe-walkers may benefit from vibration biofeedback from wearable sensor-based insoles (Pollind et al., 2019).

1.5.7 Ageing and aged-like gait

Ageing of gait

Disorders of gait in old age are common and can be physically and socially devastating, leading to severe deterioration in quality of life, mobility, and increased mortality. Changes in gait with ageing are well established in the clinical data and include decreasing comfortable preferred walking speed (gait velocity), step length, step width, and single support time. The increased double support time that results from the latter seems based within physiological changes

(Judge et al., 1996; Kerrigan et al., 1998b; McGibbon, 2003; Callisaya et al., 2010; Herssens et al., 2018), probably from central and peripheral nervous system alterations (Callisaya et al., 2010; Kiss, 2010). Walking and standing require complex control mechanisms from the sensorimotor system to provide the timing for balance and coordination, which start to deteriorate after the sixth decade of life (Aagaard et al., 2010). It is possibly the result of neurological deterioration that affects muscle activity and power that underlies the gait changes of age, rather than through loss of muscle strength due to muscle atrophy (*sarcopenia*). Exercises to strengthen musculature alone appear to fail in improving walking ability in the elderly (Nowalk et al., 2001; Nelson et al., 2004; Kalapotharakos et al., 2005; Hanson et al., 2009; Alfieri et al., 2010; Geirsdottir et al., 2012; Holviala et al., 2012).

Gait variability is significantly higher from around 65 years of age onwards compared to the younger healthy human adult population and is consistent with the slowing of gait (Dubost et al., 2006; Hollman et al., 2007; Kang and Dingwell, 2008a,b). Reduced ankle power around heel lift seems to be a particular problem (Winter et al., 1990; Judge et al., 1996; Prince et al., 1997; Franz and Kram, 2013b), although it appears during testing that older adults have underutilised their plantarflexor muscle power reserves available during walking, which could be enhanced therapeutically (Franz et al., 2014).

The physiological changes underlying gait deterioration probably include neuromuscular-induced reduction in strength and range of motion, diminished vestibular function, loss of visual acuity, and reduced cardiopulmonary function. However, it is difficult to discriminate cause from effect in human physiological behaviours (McGibbon, 2003). Gait changes in disabled or functionally limited elderly populations are similar to those of healthy elderly populations, even though it appears that different neuromuscular adaptations underlie the resulting kinematic and kinetic changes within the lower extremity (McGibbon, 2003). The decline in neuromotor control results in an increased risk of falls and thus, as a result, injury (König et al., 2016).

Separating age changes disassociated from underlying pathology can be challenging, especially in the early stages of disease, as sources of neuromuscular gait changes known as *idiopathic sources* are often overlaid by co-morbidities such as osteoarthritis at specific joints (McGibbon, 2003). However, even completely healthy adults show changes in the metabolic cost of walking with increased age (Malatesta et al., 2003; Mian et al., 2006; Ortega and Farley, 2007). These are probably associated with increased coactivation of antagonistic muscles about the thigh, although such changes do not appear to occur in and around the lower leg (Peterson and Martin, 2010). This suggests that age-related lower extremity neuromuscular adaptation impacts the hip and knee's sagittal plane stability during walking, but not the ankle's (Peterson and Martin, 2010).

Variations in gait measures from one step to the next increase in the elderly and is referred to as *intra-individual gait variability*. This is a better measure for the risk of falls and mobility problems than the general changes of gait speed and step length, although there is an association between such general changes and increased gait variations (Callisaya et al., 2010). Greater age is associated with greater intra-individual variability in step length and step time, independent of chronic disease, and this possibly represents a decline in automated stepping mechanisms from central motor control (Callisaya et al., 2010). Interestingly, a greater association in females has been found in step variation related to age (Callisaya et al., 2010). Step width variability in old age has been reported by several studies (Stolze et al., 2000; Grabiner et al., 2001; Owings and Grabiner, 2004; Woledge et al., 2005; Callisaya et al., 2010) and may represent attempts to stabilise balance (Callisaya et al., 2010). Step length variability and other gait parameters in old age do not appear to be linked to the presence of chronic disease (Callisaya et al., 2010). The important role the CNS plays in controlling rhythmic gait is reflected by gait changes found in diseases of the CNS. As ageing is associated with alterations in brain structure associated with regions that are important for intrinsic automaticity of gait such as the basal ganglia, this may be the cause of some of these age-related changes (Callisaya et al., 2010).

Loss in power in the plantarflexors around the beginning of terminal stance and a simultaneous reduction in knee energy absorption power are particularly significant issues in ageing (Winter et al., 1990; Judge et al., 1996; Prince et al., 1997; Franz and Kram, 2013a). Ankle torque reduction and smaller peak propulsive forces associated with reduced mechanical work from the trailing leg appear with age (Franz and Kram, 2013a). The reduced ankle plantarflexor power impairs swing initiation and CoM progression in late stance during double-limb stance phase, and further affects the ability to stabilise the limb and trunk during initial contact on the leading contact limb of double-limb stance (McGibbon, 2003). Muscle activity and power can still increase in the elderly with walking speed (Kerrigan et al., 1998b; Riley et al., 2001), but the same muscle activity peaks are not achieved as in younger adults (Franz and Kram 2013a; Franz et al., 2014).

Kerrigan et al. (1998b) and Riley et al. (2001) reported that the primary impairment of elderly gait occurs in the hips, with increased activity there rather than at the ankle plantarflexors. However, alteration in either result in the same sort of changes to the gait parameters. If elderly subjects are able to regain ankle plantarflexion power, they may not need more hip flexion power during late stance to adjust for the loss of ankle power (McGibbon, 2003). Hip compensation reported by

Kerrigan et al. (1998b) noted that elderly subjects increase their hip extensor moments at increased walking speeds. However, they display reduced maximal extension compared to young populations and do not increase hip extension at higher walking velocities, which further limits step length (Kerrigan et al., 1998b). It therefore appears that loss of ankle plantarflexion power is a potentially limiting factor in elderly gait and that hip flexion compensation increases with gait velocity as a result, explaining a reduction in gait velocity to avoid higher compensations. However, if ankle plantarflexor strength is maintained, then hip extension range impairment may be a limiting factor on comfortable step length (McGibbon, 2003). Evidence suggests that age-related changes cause a redistribution of muscle moments and power, generally with an increasing output by hip musculature and a decreasing output by the more distal musculature. However, individually, this effect is variable, particularly where concurrent lower limb disability also exists (McGibbon, 2003).

Thus, overall, healthy elders reduce ankle plantarflexion power output compared to the young, and in the lower extremity-disabled elderly compared to the healthy elderly, ankle power is reduced further (McGibbon, 2003). This age-induced neuromuscular functional limitation explains changes in step length and comfortable gait velocity because the reduced power output decreases the body's forward progression through diminished joint momentum to produce forward swing of the limb (McGibbon, 2003). Calf muscle strength is a strong predictor of ankle plantarflexor power and the ability to generate rapid ankle torques (Judge et al., 1996; Thelen et al., 1996). The compensation for this loss of ankle power in the healthy and disabled elderly comes from the hips (McGibbon, 2003). The reason for this is likely to be that the gastrocnemius, soleus, and the gluteal muscles all play a prominent role in trunk stabilisation during stance for controlling the CoM's forward progression, and they are important for limb swing initiation. Consequently, their diminished function requires neuromuscular adaptations within the lower limbs to compensate, with the hips seemingly being the most suited to take on the role of triceps surae (McGibbon, 2003). The overall picture is complicated by the indications that elderly reduced step length and gait velocity are due to limited hip extension and may be caused by hip muscle contracture, giving rise to a loss in ankle plantarflexor power (McGibbon, 2003). Thus, the deterioration may be a cyclical situation between the joints (Fig. 1.5.7).

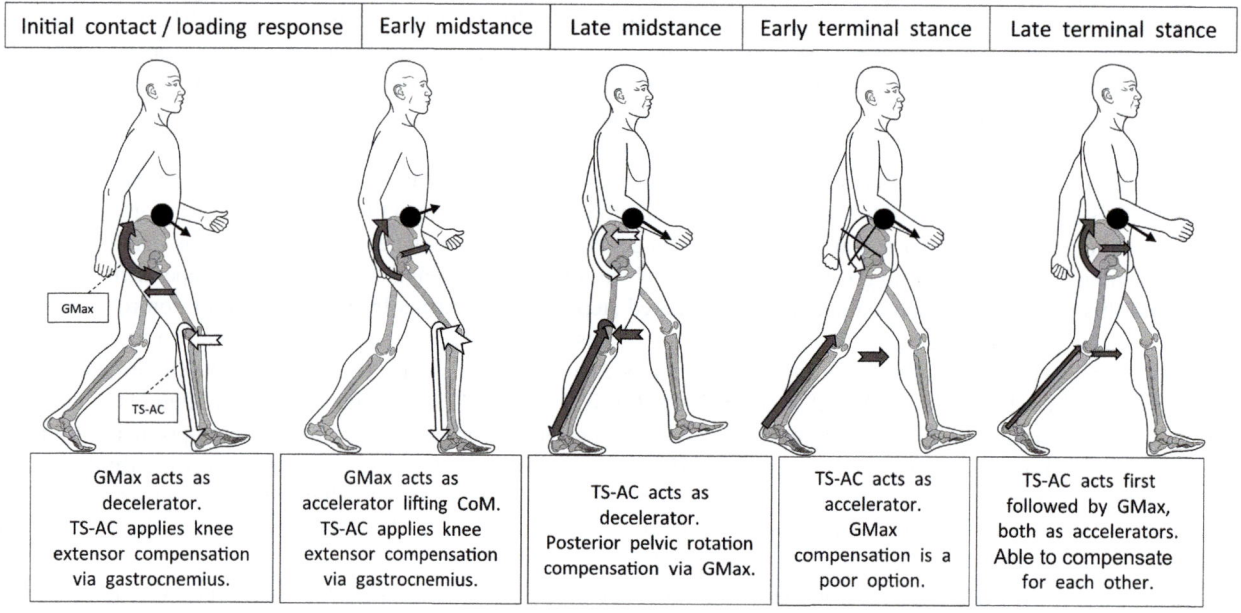

FIG. 1.5.7 The maintenance of a stable velocity for the body's CoM during gait is essential for energetics. The primary responsibility for this within the stance limb, is the interplay of acceleration-controlling activity between gluteus maximus (GMax) during early stance and the triceps surae-Achilles complex (TS-AC) within late stance. These muscles provide either acceleration forces *(dark-grey arrows)* or deceleration forces *(white arrows)* on the CoM. During walking's loading response and early midstance, poor gluteus maximus activity results in excess hip flexion which can be compensated for by increasing the rate of knee extension, thus raising the CoM more via the knee. However, this might also induce a deceleration moment on the CoM via earlier triceps surae-Achilles complex activity, limiting the ankle flexion moment. Within late midstance, triceps-surae provides a brake on CoM acceleration forward over the limb at the ankle. Any triceps surae weakness can be compensated for in part, by gluteus maximus hip extension activity, pulling the CoM backwards at the pelvis. However, at heel lift, gluteus maximus activity is unable to compensate for weak plantarflexion acceleration power as although hip extension could provide for acceleration, the posterior pelvic force necessary for increasing acceleration would posteriorly rotate the pelvis, decelerating the CoM of the trunk. In late terminal stance, as ankle plantarflexion acceleration power is diminishing, gluteus maximus adds a final pulse of acceleration before iliopsoas picks up the mass of the trailing leg during preswing. While dysfunction in gluteus maximus can to some extent be compensated for by triceps surae-Achilles complex activity, loss of ankle plantarflexion strength, particularly at heel lift, cannot be adequately compensated for by gluteus maximus' hip extension activity. Such gait dysfunctions are more frequently found in older age when loss of plantarflexor strength becomes a common issue. *(Permission www.healthystep.co.uk.)*

Changing gait strategy with age

There seems to be a change in neuromuscular patterning among the elderly from the use of a 'trunk-leading' strategy found in a young adult gait to a 'pelvis-leading' strategy in elderly gait, causing a significant increase in lower back energy expenditure (McGibbon and Krebs, 2001). It has been noticed that a striking difference exists between the healthy, able-bodied elderly and those with lower extremity impairment such as knee osteoarthritis, resulting in a significant increase in eccentric hip power in the disabled group (McGibbon, 2003). Increased eccentric hip power will occur if the thigh rotates at a greater rate posteriorly relative to the pelvis. It may be that having stiff hip flexors allows increased eccentric action as a means of storing and releasing energy within passive mechanisms to bring about limb advancement into swing phase (McGibbon, 2003). Pathologies such as knee osteoarthritis may also shift gait to one of less energetic use of hip flexors through contraction of hip flexors thereby changing limb advancement to using eccentric activity instead of concentric contraction (McGibbon, 2003). Osteoarthritis is associated with greater step time variability possible due to pain or decreased joint strength (Callisaya et al., 2010).

Gait disturbance is associated with many diseases during old age that in turn are themselves associated with advancing age. Neurological ageing issues that affect gait changes from average spatiotemporal parameters are not distinctive to specific diseases (Moon et al., 2016). Gait changes are associated with many neurological diseases including small cerebral bleeds, particularly those causing subcortical white matter lesions and reduced cortical thickness in brain regions such as the primary and supplementary motor cortices (de Laat et al., 2010, 2012). Different locations of motor cortical thinning are associated with alterations in different gait parameters such as changes in velocity, stride length, cadence, and stride width (de Laat et al., 2012). Gait changes are also found in Parkinson's disease, multiple sclerosis, Huntington's disease, cerebellar ataxia, and Alzheimer's disease (Moon et al., 2016), as well as diseases that cause deteriorations in mental function (Snijders et al., 2007). There is heterogeneity between neurological diseases, with Huntington's disease (a hyperkinetic movement disorder) seemingly creating the most alterations in gait variability within pathological groups. Parkinson's disease, Alzheimer's disease, and multiple sclerosis (disorders of hypokinetic movement) have much lower variations reflecting the nature of the specific disease processes (Moon et al., 2016). Part of the role of the clinical biomechanist is to identify common changes in gait associated with ageing and then to be able to distinguish them from something more sinister. In the early stages of the disease, this can be challenging. With experience and reading around the subject further, the clinical biomechanist can become more alert to significant changes in gait.

One final consideration for the clinician is that older adults tend to adopt a more cautious gait pattern when wearing socks rather than walking barefoot, something that does not seem to affect the young (Tsai and Lin, 2013). This includes a slower walking speed, decreased stride length, and a reduced CoM minimal velocity of forward progression in single-limb support, indicating that older patients should be cautious about just wearing socks around the house, especially if a balance deficit already exists (Tsai and Lin, 2013). Undoubtedly, the problem is related to age disturbance of the sensorimotor system, particularly via cutaneous receptors associated with plantar surface insensitivity that occurs with ageing (Perry, 2006).

Endurance and fatigue in walking

Endurance is important to gait. It is defined as the maximum amount of time that an individual can sustain a given speed. The physiology behind this ability is not clearly understood. Current models propose that endurance is modulated by the brain via peripheral indicators of fatigue such as muscle oxygenation, lactate production, and energy reserves, and then limits performance through slowing or stopping activity before injury develops (Noakes, 2012). Endurance is strongly dependent on running speed in animals, as slower speeds can be sustained for longer. Energy and oxygen consumption rise with running speed with a steep decline in endurance occurring as oxygen demand approaches the lactate threshold and VO_{2max}, the maximum energy expenditure through aerobic respiration. Endurance falls to below 15 min in humans if the VO_{2max} threshold is exceeded. In mammals, this corresponds to the number of mitochondria within the muscles available to take up oxygen (Weibel et al., 2004). Type I muscle fibres are the most mitochondria-rich and fatigue-resistant.

In healthy populations, failure in endurance is normally only an issue during running, which again indicates how good humans are at walking. Endurance and fatigue issues in patients who only walk should be physiologically and medically investigated further.

1.5.8 Leg length discrepancy/inequality

The role of leg length discrepancy or LLD (also known as *anisomelia*) in gait disturbance and as a predisposing factor in musculoskeletal pathology is a controversial subject. Some clinicians have suggested that LLDs as small as 3–5 mm need intervention, others as large as 20–30 mm before intervention is required (Gurney, 2002). Back pain through pelvic torsion (Cooperstein and Lew, 2009), patellofemoral pain (Carlson and Wilkerson, 2007), increased knee osteoarthritis (Harvey et al., 2010), and increased levels of stress fractures among athletes (Bennell et al., 1996) have all been associated with LLD, but pathology associations with something like an LLD does not mean that the LLD is the cause. A study of 576 cadavers investigated for LLD and osteoarthritic changes within the spine, hip, and knee did not find any link (Liu et al., 2018). LLD is commonly reported as being found to be present in between 40% and 90% of the population, with 20% >9 mm (Gurney, 2002; Knutson, 2005a,b). However, it does not appear to be clinically significant until ≈20 mm (Knutson, 2005a,b). This is hardly surprising, as growth of the two lower extremities to exactly the same length in every human would be a remarkable act of biological symmetry occurring during growth and development. Therefore, it should be expected that human anatomy can easily compensate and manage for any small differences. This ability is something that is also necessary for motion over uneven terrain. True and functional LLDs are also quite diverse in anatomical origin and compensations, many involving just small pelvic malalignments (Fig. 1.5.8a).

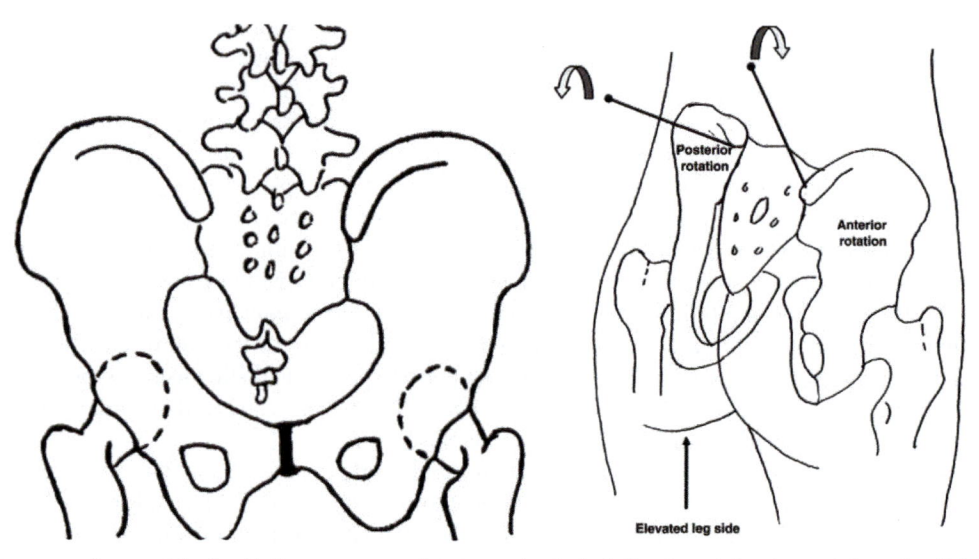

FIG. 1.5.8a The causes of apparent leg length discrepancy can arise at many levels, including the pelvis, where pelvic torsion *(left image)* can cause asymmetry between left to right pelvic landmarks. This gives the impression of differing leg lengths where there may be none. However, 'true' inequality between limb lengths will elevate the pelvis on the longer side which in turn can lead to compensatory rotations between the ilia across the sacroiliac joints. Functional inequalities can cause similar anatomical asymmetry across pelvis and limbs, making clinical static stance identification of true and functional differences problematic. *(From Cooperstein, R., Lew, M., 2009. The relationship between pelvic torsion and anatomical leg length inequality: a review of the literature. J. Chiropractic Med. 8 (3), 107–118.)*

There are two classifications of LLD: *true discrepancy* and *functional discrepancy* (Knutson, 2005a). True discrepancy is when there is an actual anatomical difference between the lengths of the two limbs, between the proximal edge of the femoral head to the distal edge of the tibia, which can be developmental/congenital or acquired as a consequence of trauma, degenerative disease, or surgery such as a joint replacement. Functional LLD is defined as an asymmetry of length as a result of compensation responses such as due to altered limb mechanics due to soft tissue contracture, weak or short muscles, and altered joint alignment or motion in any of the three planes. These differences cannot be examined by diagnostic images alone, but only through careful clinical alignment and posture assessment in stance and/or gait (Khamis and Carmell, 2017a).

It has been suggested that individuals with a long-standing LLD are able to adjust to larger differences in the individual lengths of their lower limbs in comparison to those who have been subjected to an induced difference as can be seen following a fracture or surgery such as a total hip replacement (Gurney, 2002; Maloney and Keeney, 2004; Desai et al., 2013). Post-hip replacement LLDs of less than 1 cm are common and are suggested by some to be well tolerated (Maloney and Keeney, 2004). Yet, LLD can also be a source of general dissatisfaction, back pain, sciatica, and gait disorders (Konyves and Bannister, 2005; Desai et al., 2013). Interestingly, patients whose leg length is perceived to have been increased postoperatively are more likely to experience problems than those who feel that their leg is now shorter, and this perception does not change over time in 86% of cases (Konyves and Bannister, 2005). When hips are replaced, setting the femoral offset position is considered important. This is the distance from the centre of rotation of the femoral head to a line bisecting the long axis of the femur, which is taken from a radiograph. However, CT scans are more accurate (Lecerf et al., 2009). It has been reported that leg length discrepancies of over 5 mm in combined leg length and offset position at 12 months following total hip arthroplasty led to altered gait kinematics with a lower Froude number, less hip range of motion, and a slower walking speed than those with a difference within 5 mm (Renkawitz et al., 2016). Under or overcorrections of over 5 mm were both associated with worsening gait patterns (Renkawitz et al., 2016).

Younger individuals may be better capable in adjusting to such induced LLDs than older people, for older individuals are less able to adapt to any new motor tasks (Gurney, 2002). Individuals who participate in higher activity levels also seem more sensitive to differences, which suggests that the ability to tolerate the difference or not may reflect activity levels. Thus, a number of factors need to be contemplated before considering a particular quantitative amount that may be significant to the individual. If the LLD is thought to be the cause of pathology, then how much LLD is necessary to create significant gait and mechanical perturbations? And furthermore, how can you clinically measure the difference accurately?

A study by Resende et al. (2016) looked at artificially induced mild leg length discrepancy using built-up sandals on one side. Using young subjects with no intrinsic difference greater than 0.5 cm, the study demonstrated that the shorter limbs use biomechanical strategies that enable them to functionally lengthen the shorter limb in gait such as by increased rearfoot plantarflexion and smaller knee and hip flexion angles. The longer limb uses greater ranges of ankle dorsiflexion and more knee and hip flexion that give rise to limb-shortening strategies. During the first 50% of stance phase, these long and short kinematic situations seem reversed (Resende et al., 2016), findings that are in concordance with studies on larger leg length discrepancies of over 2 cm (Song et al., 1997). Needham et al. (2012) compared artificial LLDs composed of high-density ethyl vinyl acetate (EVA) heel lifts of 1, 2, and 3 cm applied directly to the plantar right heel of seven young adult males as opposed to using their natural barefoot without any lift on the left side. These unilateral heel lifts caused on average an increased pelvic obliquity of 2.67°, 5.29°, and 3.95°, respectively. The pelvic lower obliquity angle occurring with a 3-cm heel lift under one foot was considered a result of changes in hip and knee flexion angles on the right leg with the lift (see Fig. 1.5.8b). However, the authors also reported that pelvic and lumbar spine motion was unchanged in the heel lift condition compared to natural barefoot in both the frontal and transverse planes (Needham et al., 2012). Small differences in lumbar spine sagittal plane motion were also noted using 3 cm heel lift, with asymmetries in lumbar flexion during gait of only around 2°. The overall results indicated that artificial LLD alters 3D pelvic and lumbar spine motion along with motion of other joints of the lower limb such as increased knee and hip flexion and ankle dorsiflexion in any combination, and limb motion asymmetry is thus able to accommodate smaller LLDs (Needham et al., 2012). However, heel lifts may create different kinematics compared to raising the entire foot as the midfoot and forefoot can still ground contact through ankle and midfoot plantarflexion angle changes in gait, which in turn will set up new gait biomechanics. As adding a heel lift is a common therapeutic intervention for LLD, this should be considered.

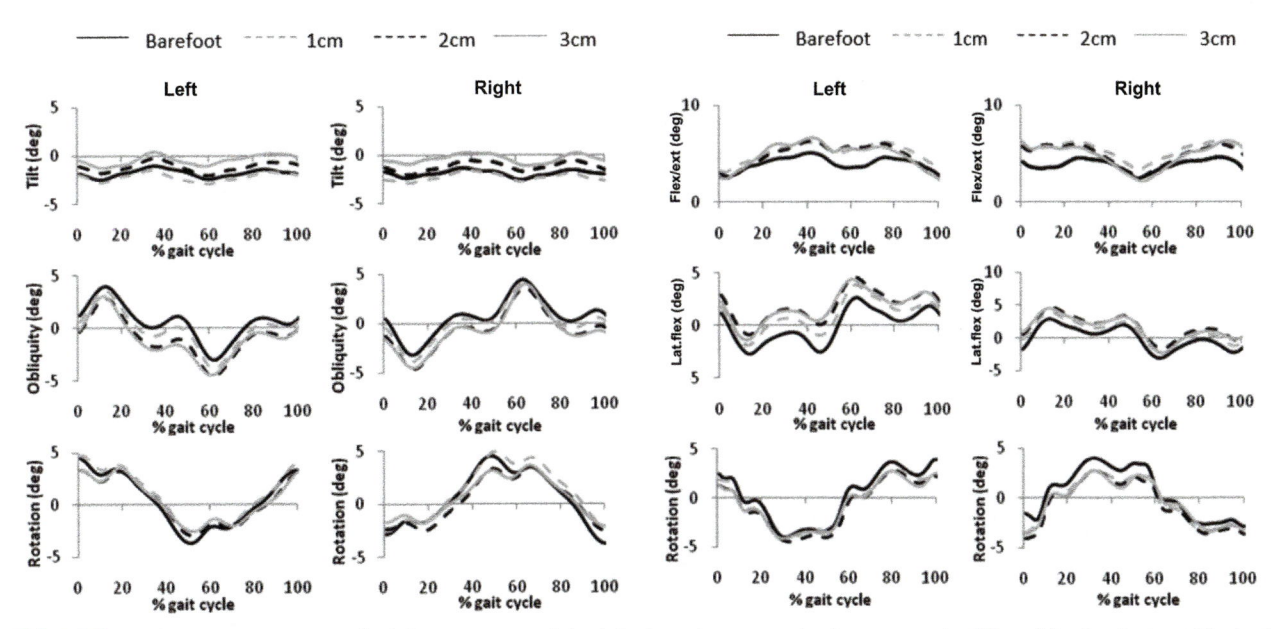

FIG. 1.5.8b Left charts show patterns of pelvic movements and the right charts demonstrate lumbar movements while walking barefoot or with a heel raise (1, 2, or 3 cm) trying to represent a leg length discrepancy. Such an approach of using just a heel lift may only produce changes in functional limb length via altered joint angles that also adjust foot posture, rather than raising the whole limb throughout the stance phase as would a true osseous limb difference. However, the effects are important to consider as clinically, unilateral heel lifts are frequently added to shoes and foot orthoses to balance perceived leg length differences. *(Modified from Needham, R., Chockalingam, N., Dunning, D., Healy, A., Ahmed, E.B., Ward, A., 2012. The effect of leg length discrepancy on pelvis and spine kinematics during gait. Stud. Health Technol. Inform. 176, 104–107. In: Kotwicki, T., Grivas, T.B. (Eds.), Research into Spinal Deformities. 8. IOS Press.)*

On using a large range of full-length sole lifts inside footwear between 5 and 40 mm on seven young healthy individuals, Khamis and Carmeli (2018) reported that LLD causes both kinematic deviations in all planes and changes in dynamic limb length during gait, even at 5-mm differences. It was reported that examining alterations in the dynamic limb length compensations for LLD could be quantified and as reported by others, requires the short leg to functionally increase its length and the longer limb to functionally reduce its length. Shortening of the longer limb occurred in swing significantly with a 10-mm lift or higher being used, with increased hip abduction and external foot rotation being demonstrated. In stance phase, more flexed positions on the longer limb and extended positions in the shorter limb were observed, starting at 5 mm (Khamis and Carmeli, 2018). In artificially induced LLD during loading to absolute midstance, the knee extension moment developed out of a less flexed knee posture and demonstrated a greater ankle plantarflexion moment (Resende et al., 2016). On the longer side, the hip showed an increased flexion angle during the first 50% of stance (Resende et al., 2016). In the frontal plane within the short limb, the rearfoot was more inverted compared to the longer limb side which displayed greater eversion (Resende et al., 2016). This is an expected result, as rearfoot inversion is associated with functional rearfoot lengthening and eversion with shortening (Resende et al., 2015). These lower extremity compensations to correct for the discrepancy by functional changes may alter the kinetics of the tissues and perhaps should not be ignored, even in differences as little as 5 mm (Resende et al., 2015; Khamis and Carmeli, 2018) (Fig. 1.5.8c).

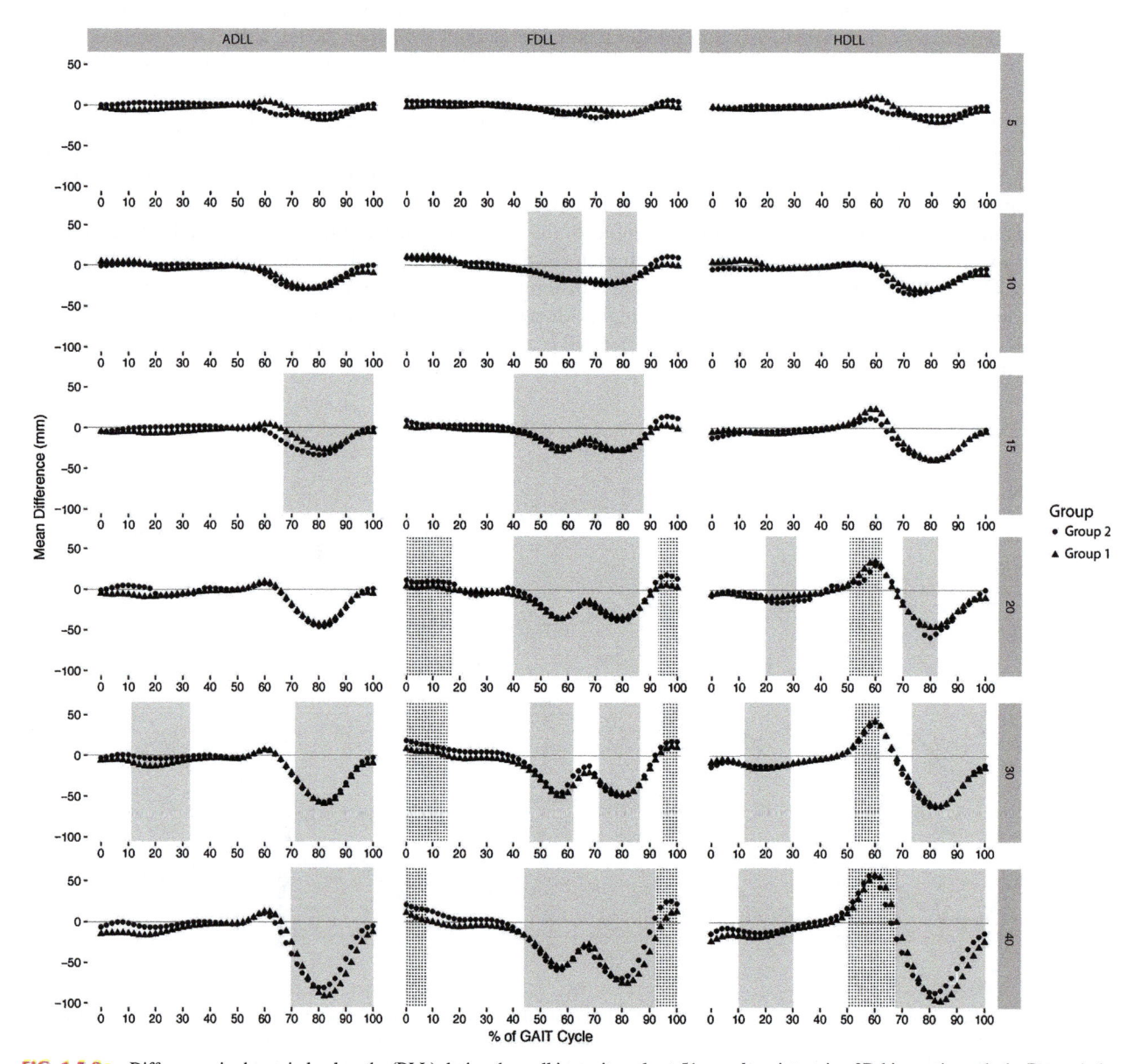

FIG. 1.5.8c Differences in dynamic leg lengths (DLL) during the walking gait cycle at 51 sample points using 3D kinematic analysis. Dynamic leg length was measured from the hip joint centre to the heel marker (HDLL), to the ankle joint centre (ADLL), and to the forefoot marker (FDLL). Group 1 was using foot lifts on the left side and group 2 used foot lifts on the right side, showing very similar results. The shaded areas demonstrate the proportions of gait with significant negative differences and the dotted areas, positive differences. *(From Khamis, S., Carmeli, E., 2018. The effect of simulated leg length discrepancy on lower limb biomechanics during gait. Gait Posture 61, 73–80.)*

In the pelvis, Resende et al. (2016) suggested that the pelvic strategies are less able to compensate, demonstrating increased obliquity of the pelvis on the shorter side throughout stance and increased hip adduction during the first 60% of stance. The longer limb demonstrated increased pelvic obliquity upwards throughout stance and increased knee abduction moment between 10% and 90% of walking stance (Takacs and Hunt, 2012; Resende et al., 2015, 2016). This is possibly supported by Needham et al.'s (2012) finding that hip obliquity is reduced with a 3-cm heel lift. Certainly, the findings are coherent with expectations of functional limb-shortening strategies, but these are artificial and suddenly created differences. Moreover, during in-life walking, the ground is not always even, so these relationships will alter with changes in terrain.

A number of ways are proposed to measure LLD. These include radiography, scanograms, computerised digital radiographs, and computerised tomography (CT) (Khamis and Carmeli, 2017a). Although considered as being the most accurate techniques, they are all expensive, rarely justified clinically, and will also miss functional differences unless performed in carefully positioned stance, which is rarely possible. Clinical techniques involve measuring the distance between two anatomical points on a limb, usually from the anterior superior iliac spine to the medial malleolus. However, research results on the accuracy are mixed (Khamis and Carmeli, 2017a). Indirect clinical methods have used external blocks of known height to level the pelvis heights on left and right side, taking account of hip, knee, and foot position to make sure they are symmetrical. But again, research on reliability is mixed (Khamis and Carmeli, 2017a).

Functional leg length differences are generally established by looking at limb and joint alignment symmetry in relation to pelvis alignment. If pelvic alignment is not level and lower limb joint positions are not symmetrical, it may be found that improving joint positional symmetry may bring the pelvis level, indicating a functional difference. 3D motion gait analysis probably offers the most promise in identifying dynamic leg length discrepancies that significantly influence gait (Khamis and Carmeli, 2017a), but this is a time-consuming and initially expensive form of clinical examination which must be justified. 3D kinematics shows that during swing phase, compensatory strategies seem to reduce any asymmetries between the shorter and longer limbs, but 3D analysis technique can identify kinematic and dynamic lengths of the lower extremities, recognising the contributions of segmental compensations (Khamis and Carmeli, 2017a). Compensation mechanisms should be seen in both the longer and shorter limbs, with greater discrepancies initiating more strategies of compensation, mainly in the sagittal plane. However, some frontal plane compensations at the pelvis, hip, and foot can occur (Khamis and Carmeli, 2017b).

The key question remains. Does LLD, when discovered clinically, influence musculoskeletal disorders and symptoms? If correlations can be made, then the expense of gait analysis and diagnostic imaging may be necessary to confirm a link. Initial detection, by clinical means, needs to establish a potential link between any pathology and the true or functional differences found. Ideally, 3D motion capture looking at dynamic leg length asymmetries seems the best option for assessment. More simple gait analysis multiple-step pressure systems can identify consistent differences in step times or stride lengths between limbs which, combined with the presence of persistent pathology or symptoms that logically correlate to the true or functional discrepancy, may warrant intervention. Any functional LLD is likely to need resolving first before shoes and orthoses are brought in that could 'capture' the functional deformity's posture (Knutson, 2005b). LLD in gait remains a clinical quandary that needs a careful nonbiased approach when approaching treatment, with changes in stance posture alone being possibly insufficient to establish influences on locomotion.

1.5.9 The effects of footwear on gait

Humans evolved with naked feet. However, many humans now grow and develop their gait within footwear. Modern societal demands do not allow humans to live barefoot, and fashion continues to drive much of footwear design. Footwear design encompasses everything from the minimum of a simple toe-thong and sole to a knee-high laced-in boot, a functional protective hiking boot, or a fashion stiletto heel. All the potential effects on gait of various styles of footwear cannot be covered in such a text as this. However, some broad statements on the use of footwear can be made and considered, with arguments developed in regard to environmental mismatch made in Section 6.4.7 of the companion text 'Clinical Biomechanics in Human Locomotion: Origins and Principles' and within this text, on running footwear in Section 1.4.11.

Gait patterns are altered when walking (or running) barefoot compared to when shod. Spatiotemporal parameters change between shod and barefoot, with barefoot resulting in reduced hip range of motion and shorter strides (Shakoor and Block, 2006; Lythgo et al., 2009; van Engelen et al., 2010; Dames and Smith, 2015). The walking barefoot condition elicits lower initial and loading response peak vertical GRFs and yet a higher propulsive GRF, probably as a result of a shorter stride length than demonstrated when walking in sports footwear (Keenan et al., 2011). Walking in sports shoes seems to induce a 6.5% increase in stride length and results in increased GRF in all three orthogonal planes (Keenan et al., 2011). Sports shoes also increase peak knee joint moments, with a 9.7% increase in initial peak varus knee moments and increased hip flexion and extension moments (Keenan et al., 2011). This is something to carefully consider as often, sports shoes are recommended to patients as part of their musculoskeletal pathology management. It might also be necessary to add advisory information about attempting to control the patient's stride length in such shoes.

Regular casual shoes are also known to elevate knee loads in normal individuals and are thought to be related to the increase in walking speed associated with shoes (Kerrigan et al., 2003; Mündermann et al., 2004). However, when walking speed was controlled, normal casual footwear, compared to barefoot in patients with osteoarthritic knees, demonstrated 11.9% lower knee adduction moments. This is a reduction of forces larger than that reported for the effects of lateral-wedged insoles that are used to reduce knee adduction moments therapeutically (Shakoor and Block, 2006). Peak loads at the hip and knee are decreased in such patients when barefoot, and again, stride length, cadence, angle of gait, and joint range of motion also change, suggesting that shoes detrimentally increase loads on the lower extremity through altering gait (Shakoor and Block, 2006). Knowing that shoes will change gait data from barefoot and knowing also that there are a large variety of shoe styles demonstrating different material construction raises the question as to whether gait analysis data taken barefoot can be usefully compared with data taken in shoes, especially so when treatment interventions such as foot orthoses are added to the footwear. The answer really must be no. It cannot. Even if gait kinematics at the hip or knee were only to be investigated, then it would be best to first assess the patient in the shoes they wear, without the orthoses. The research also clearly underlines the need to make the shoe consistent across gait analysis data-gathering.

1.5.10 Gait in lower limb amputees

Prosthetic-wearing amputees are not a homogeneous group, not least because the level of amputation is variable as is the level of physiological health, and some patients are double lower limb amputees. Also, individual response produces significant levels of heterogeneity, even when wearing the same prosthesis (Wong et al., 2016). Limb amputation can be a result of trauma, cancer, or secondary to physiological disease such as diabetes and other vascular diseases. The cardiovascular status and age of the amputee affects the resultant gait, with those suffering amputation due to cardiovascular disease tending to fair less well than those amputated through traumatic injury. Diabetics are around 20 times more likely to sustain an amputation than nondiabetics, and male diabetics are more at risk than females. Although the most common diabetic amputation is for toes, higher heights of the amputation correlate with the diabetics age (van Houtum et al., 1996). Each level of amputation is listed in Fig. 1.5.10a.

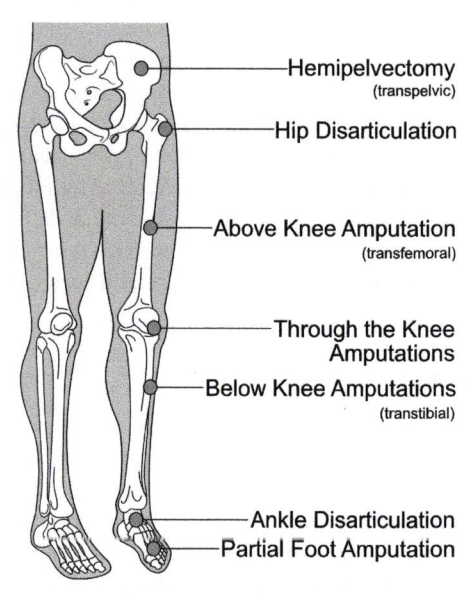

FIG. 1.5.10a The different levels used for amputation and the terms used for such amputations. *(Permission www.healthystep.co.uk.)*

The type of prosthesis prescribed reflects the type of amputation (e.g. digital, partial foot amputations such as at the midfoot, wholefoot, or ankle disarticulations, transtibial below-knee amputation, through-knee disarticulations, transfemoral above-knee amputations, hip disarticulations, and hemipelvectomy). Prosthetic choice should also reflect the level of activity of the patient. Thus, any prostheses should be prescribed using quantitative need and evidence. Oftentimes, however, only empirical knowledge is used. It must be remembered that not only is muscle lost from the segments amputated, but the muscles that supply the amputated segment are also functionally lost, altering muscular forces and lines of action across more proximal articulations during closed chain phases of gait. Compensation strategies are required in the muscular, skeletal/articular, and neurological systems, with some individuals more likely to rely on one form of compensation mechanism more than another. Risks are raised for tissue damage and overload in both the amputation side as well as within the intact limb that tends to have to extend its single-limb support time. This becomes a vulnerable period of gait especially during the prosthetic side single-limb support, when sensorimotor information is much reduced. The prevalence of osteoarthritis is significantly greater for both knee and hip within traumatic leg amputees (Struyf et al., 2009). High-activity young transtibial amputees are known to have an increased risk of knee osteoarthritis in their intact knee, while the amputee knee is less at risk, probably resulting from a change in knee joint contact forces and higher GRF on the intact side (Ding et al., 2021). Transfemoral amputation is associated with an increased risk of hip osteoarthritis in both the amputation and intact limb side (Sagawa et al., 2011).

Gait and physiological analysis are effective ways to quantify both motion and energy expenditure in amputees (Sagawa et al., 2011) and the more that biomechanical data reflects normal age-matched controls, the better the mobility outcomes are likely to be (Fig. 1.5.10b). It is clear from the literature that lower limb amputation leads to asymmetrical gait patterns that risk long-term health and locomotive problems (Clemens et al., 2020). Nevertheless, personalised exercise programmes can reduce falls and improve gait (Schafer et al., 2018), indicating that following the gait data may lead to better treatment-focused plans. Changing muscle function may protect joint overload in both the amputated and intact limb (Ding et al., 2021) (Fig. 1.5.10c).

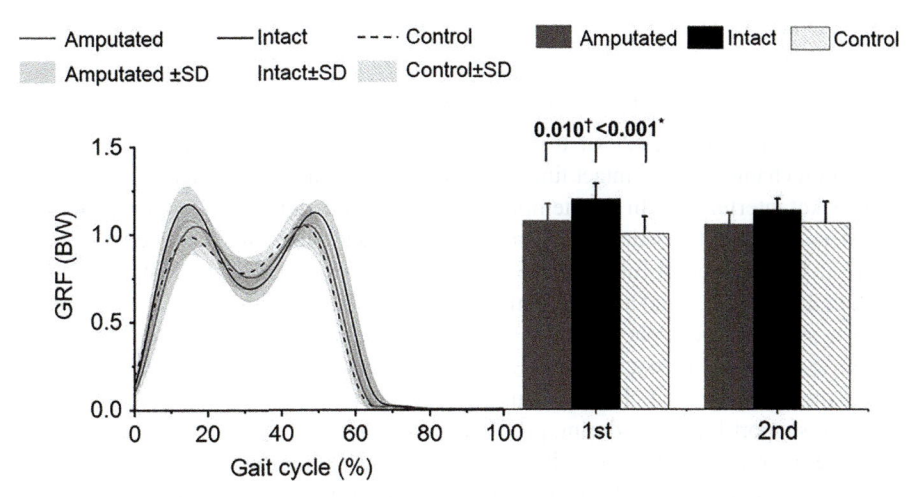

FIG. 1.5.10b GRF peaks showing the difference between controls and highly functioning below-knee amputees and their intact limbs. The intact limb can be clearly seen to be experiencing higher GRFs peaks compared to the amputated limb or those of the control group limbs. *(From Ding, Z., Jarvis, H.L., Bennett, A.N., Baker, R., Bull, A.M.J., 2021. Higher knee contact forces might underlie increased osteoarthritis rates in high functioning amputees: A pilot study. J. Orthop. Res. 39 (4), 850–860.)*

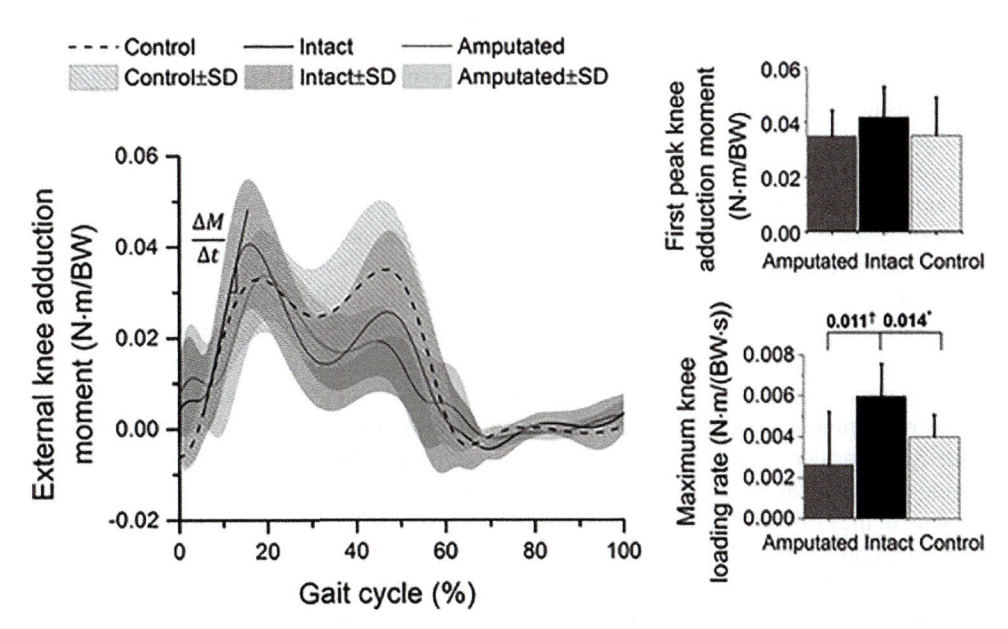

FIG. 1.5.10c External knee adduction moments and maximum knee loading rate (defined as the maximum change in external knee adduction moment per unit of time). The knee's first peak adduction moment and loading rate has become increased in the intact limb and is greater than that found in the knees of normal healthy controls. *(From Ding, Z., Jarvis, H.L., Bennett, A.N., Baker, R., Bull, A.M.J., 2021. Higher knee contact forces might underlie increased osteoarthritis rates in high functioning amputees: A pilot study. J. Orthop. Res. 39 (4), 850–860.)*

Spatiotemporal changes in amputees

Useful parameters to investigate include spatiotemporal parameters such as stance time, stance time ratio, and step time ratio, with most studies indicating that in single limb amputees, the stance time will be longer on the intact limb, contributing to a more asymmetrical gait pattern (Sagawa et al., 2011). The increased loading time on the intact limb may lead to the development of complications in that limb, particularly if it is also physiologically compromised (Sagawa et al., 2011). Better fitting of prosthetic limbs may have outcomes and the use of an appropriate socket may reduce the degree of asymmetry and improve limb control and positioning (Board et al., 2001).

The initial foot-flat or forefoot contact period from the end of loading response into early midstance shows different spatiotemporal parameters that vary with the level of amputation. Normally, forefoot contact occurs by 12%–17% of gait with an external plantarflexion moment normally lasting for around 9% of gait. With a solid foot/ankle prosthesis with a cushioned heel, only the heel makes contact during the first 20%–44.5% of stance, with an absent or limited ability to allow the forefoot to plantarflex to the ground. This is despite an external plantarflexion moment persisting within the first 20% of gait (McNealy and Gard, 2008). Lack of early forefoot ground contact risks instability and alteration of proximal muscle function (Sagawa et al., 2011). Those with transtibial amputation need to maintain knee stabilisation throughout early stance due to a prolonged heel-only contact, unless prosthetic feet are able to provide motion around the ankle to permit forefoot contact (Sagawa et al., 2011). Some prosthetic limb types are able to provide a forefoot contact by 14% of gait.

Changes in joint angles and moments resulting from amputation

Sagittal plane reduction in hip motion is associated with transfemoral amputation, primarily with loss of extension and compensatory pelvic motion changes on the intact limb side to maintain gait speed (Sagawa et al., 2011). This is thought to result from ischium socket interference. In an attempt to maintain a functional step length, intact limb-side hip flexion increases in such amputees compared to able-bodied individuals to twice the angles expected ($4° \pm 1°$ to $8° \pm 5°$). This creates a strategy that risks back symptomatology in the long run (Sagawa et al., 2011). Pelvic frontal plane rotations significantly increase with unilateral transtibial and transfemoral lower limb amputations, with a tendency to lift the pelvis on the amputated side during swing phase; a phenomenon known as *hip hiking* or *hip hitching*. This is believed to be a consequence of being unable to dorsiflex the foot at the ankle through the ankle extensors in order to aid ground clearance. Hip hiking/hitching aids prosthetic foot ground clearance but increases metabolic demand. It usually requires the CoM to be moved further towards the support limb side during prosthetic swing phase.

The greatest difference between able-bodied individuals and those with a lower limb prosthesis is the external knee flexion moment in loading response. In the able-bodied, knee energy dissipation through flexion and muscle-tendon buffering is important to protect the limb from energy dissipation biomechanics, but this ability is significantly reduced in transtibial amputees and often lost in transfemoral amputees (Sagawa et al., 2011). This situation worsens with gait speed, for where the normal limb side can still change its capacity to dissipate energy with changes in gait speed, the prosthetic side cannot. Even when adding a microprocessor to control knee flexion in transfemoral prosthetic devices, the sensation of knee buckling seems to prevent the wearer from improving knee shock attenuation (Sagawa et al., 2011).

Normally, knee extensors (all the quadriceps) control knee flexion moments, but in transtibial amputees, there is a significant increasing in vastus lateralis activity reported with increases in its duration of activity compared to the able-bodied. This is despite negligible demand because of a significantly smaller external flexion moment during stance (Sagawa et al., 2011). Inconsistency between the joint moment and the muscle activity indicates a discrepancy between demand and physiological response among amputees. Transfemoral amputees have an even smaller knee flexion moment. The prosthesis cannot supply quadriceps stabilising muscle activity to the knee, although once again, the design of the prosthetic device influences the size of the knee flexion moment (Sagawa et al., 2011).

Ankle plantarflexion on contact to achieve foot-flat with a prosthetic limb is limited as already discussed, but the degree of forefoot contact that is possible earlier in stance is dependent on prosthetic design. Dynamic prostheses tend to use a blade without an articulating ankle joint, so that functional plantarflexion is due to heel compression only with better ranges of plantarflexion being associated with those designed with an articulated foot (Sagawa et al., 2011). Dorsiflexion in midstance necessary for translation of the CoM over the foot is reduced by some $7°$ with a prosthetic limb, but again this is influenced by the design being dependent on the capacity of the foot/blade to bend (Sagawa et al., 2011). Overall, complete ankle range of motion in the sagittal plane is often increased in amputees on the normal limb side, an adjustment required to assist ground clearance of the prosthetic limb during its swing phase. But again, this is affected by differences in prosthetic design, with possibly less prosthetic ankle motion resulting in greater intact limb ankle motion (Sagawa et al., 2011).

In able-bodied loading response, knee flexion reduces the duration and the amount of ankle plantarflexion required. However, with reduced or absent knee flexion and a lack of ankle joint motion, amputees must spend an increased time rotating on their 'heel rocker' to bring about forefoot contact (Sagawa et al., 2011). On the normal limb side, there is reported to be a sudden increase in the external dorsiflexion moment at around 26% of the gait cycle in an attempt to facilitate toe clearance of the prosthetic limb during swing (Nolan and Lees, 2000). The external dorsiflexion moment at the end of stance in transtibial amputees is only 60%–70% of that found among able-bodied individuals, likely due to the lack of ankle plantarflexor muscles but also possibly due to the functionally shorter 'keel' under the prosthetic foot compared to the natural foot. This shortens the moment arm from the GRF to the ankle joint axis centre (Sagawa et al., 2011). However, the ankle moment on both the prosthetic foot and the intact limb can vary with rehabilitation and the type of prosthesis provided, with a dynamic flexing foot type of prosthesis offering a 15% greater ankle dorsiflexion moment compared to a nondynamic foot prosthesis (Sagawa et al., 2011).

The changes in joint motion during gait due to prosthetic use necessitates changes in joint power. Increased concentric hip power is required on the intact limb during double-limb stance as the prosthetic limb is in its terminal stance. But again, the extent is dependent on how dynamic the prosthetic foot segment is (Sagawa et al., 2011). This increase in concentric hip power seems likely to result from reduced active acceleration power on the prosthetic limb side, utilising the hip muscles to draw the trailing prosthetic limb along behind it.

On the amputation side, increased ankle power is also required prior to toe-off to enhance the anterior and upward drive from preswing into swing that is usually assisted by the ankle power generated at heel lift (Sagawa et al., 2011). The knee's problem is the failure to absorb energy at initial loading, which is reduced in transtibial amputees, but further worsened in individuals with transfemoral amputation. With reduced or no energy dissipation and little energy being stored within the knee muscles to assist hip muscle power, amputees must either compensate by using their hip as the primary shock absorber and energy generator or risk overloading the intact limb's knee (Sagawa et al., 2011).

The ankle also operates in an energy-dissipating, storing, and generation cycle that is interrupted by a prosthetic limb. In such a situation, acceleration from the ankle is most seriously compromised, reaching at best around 20% of the power generated in healthy able-bodied populations (Sagawa et al., 2011). It must be remembered that ankle plantarflexion power is a major source of energy generation during the acceleration of terminal stance, which can only be compensated for to some degree through the hip flexors at preswing. This underlines once again the functional compensatory link between poor ankle power and hip muscle compensation seen in a variety of gait patterns.

Muscular rehabilitation programmes for amputees seem helpful, especially with visual feedback. However, consensus on the best approach to exercise has not been reached (Darter and Wilken, 2011; Highsmith et al., 2016; Wong et al., 2016). Data certainly suggests earlier intervention is more helpful and can reduce the risk of falls from perturbations in gait regardless of the reasons for amputations (Schafer et al., 2018; Schafer and Vanicek, 2021). Gait data, including kinetic data, should be considered in the fitting and alignment of prostheses, with precise alignment being required for each patient. Such specialist 'tuning' of a prosthetic limb requires gait data to allow the patient to most closely reach 'normal' gait expectations, in conjunction with traditional selection factors such as comfort (Jonkergouw et al., 2016).

Effects of bilateral amputation on gait

Although far less researched, gait data exists on bilateral limb amputees and indicate considerable differences when set against normal able-bodied population parameters. In bilateral transtibial amputees with the same dynamic response foot prosthesis attached to both prosthetic limbs, walking is slower with lower cadence and shorter step lengths, wider step widths, and hip hiking/hitching upwards during swing phase (Su et al., 2007). This indicates greater hip abductor function during single support. Walking at a comparable speed to healthy nonamputee controls reveals reduced ankle dorsiflexion and knee flexion in stance phase, and also reduced peak ankle flexor moment and ankle power in late stance. This results in an increase in hip power to compensate for poor ankle power (Su et al., 2007, 2008; McNealy and Gard, 2008).

Such ankle-to-hip changes in power is reported to result in an increase in energy expenditure of 40%–120% or more per distance than in normal controls (Su et al., 2007), along with an exaggerated lateral trunk sway in gait that may expose amputees to reduced walking stability (Major et al., 2013). Yet again, the importance of calf muscle power in human gait is revealed. There is little difference in gait, except perhaps in methods of ground clearance, between those double-amputees that have resulted from trauma and those resulting from peripheral vascular disease. This cannot be accounted for through self-selected gait speed, but trauma-induced amputees do tend to walk faster than those resulting from physiological disease, undoubtedly due to the former's better vascular health before and after amputation (Su et al., 2008). Prosthetic limbs with ankle units have been shown to improve sagittal plane motion through restoration of the midstance fulcrum (McNealy and Gard, 2008; Su et al., 2010).

1.5.11 Section summary

Gait is far from homogeneous in the human population due to natural development of gait, ageing, morphological variance, and transient changes in morphology such as pregnancy. The changes so initiated reflect alterations in limb length, muscle power, neurological development, adjustments to the CoM location with growth, and in the case of pregnancy, dramatic changes in mass over a short period. Changes in connective tissue restraint such as hypermobility will shift the burden of control more to the active elements of muscle activity in order to create stability and guide and restrain motion around joints. Foot and lower limb morphology will change how moments are generated and managed around joints, the levels of compliance and stiffness, and where best to generate fulcrum points within joints to manage CoM transference from one limb to the next. Each variant of morphology will develop new energetics of motion, which as long as health is maintained, will present a functional gait pattern that achieves the goal of locomotion without symptoms of pathology.

However, not all development is equal, and changing lifestyles (particularly during childhood) may prevent normal morphological and neuromuscular development. These changes may, in time, threaten premature failure of the locomotive tissues. Thus, a child developing abnormal weight and/or weak hip abductors could induce coxa varum and genu valgum as a consequence of femoral head and neck development that increases valgus/abduction moments in the knee. A hypermobile pes planus foot with a midfoot break that is present during childhood is likely to be more problematic if hypermobility is maintained with age and if adult weight is increased over time while inactivity increases, resulting in relative muscle weakening. In these sorts of situations, the midfoot dorsiflexion moment of the midfoot break may start to generate forces that overload the plantar foot muscles and ligaments that resist the midfoot buckling moment. The addition of physiological disease and the effects of sarcopenia and osteopenia associated with age may also change a previously stable situation, even in individuals with a formerly fairly average gait and morphology.

Artificial changes in morphology and gait function from footwear use also pose another set of confounding factors to healthy human locomotion, although some of which may enhance locomotion in the presence of morphological, anatomical, and physiological variations and dysfunctions. Artificially induced gait alterations caused through the introduction of prosthetic limbs are essential to keep patients bipedal despite the loss of one or more limbs. Prostheses are limited in their replication of more natural gait function due to the challenges they create by altered muscle function and power. This is in part due to loss of sensory information from the amputated limb or limbs. In those whose amputation lies in physiological morbidity, the picture is more complex as the intact limb may also be compromised in function, making it a poor candidate for the increased functional burden imposed upon it due to the opposite limb's amputation. However, technologies keep progressing to create prosthetic limbs that are more functional for locomotion, and less concern is being considered to just the cosmetic appearance. Indeed, at the extreme end of technology, prosthetic limbs are being used to enhance locomotion performance in the athletic population.

1.6 Gait in disease
1.6.1 Introduction

Clinical gait analysis is an integral part of clinical biomechanics that is used to investigate the consequences or causes of locomotive dysfunction, with the aims of formulating a coherent treatment plan and/or assessing the locomotive results of therapeutic interventions. Depending on the equipment utilised, it can give rough estimations on changes in movement patterns or precise kinematic data of lower limb motion and/or the kinetic data during locomotion. Gait analysis should be a part of the process of locating and isolating aberrant components in the context of injury and pathology. There is still controversy concerning the use of gait analysis in clinical decision-making due the limited training in gait analysis of clinicians, the amount of data collected, the limited validation of techniques, and the reliability and validity of results. Yet, the ability for gait analysis to differentiate diseased patients from healthy controls is clearly shown across the research.

Ideally, clinical gait analysis should be performed in a designated gait laboratory which can provide kinetic, kinematic, and EMG data to provide insights into lower limb function and dysfunction through well-established protocols. This is not always a possibility in the clinical setting such that the clinician is often limited to just observational or at best, using slow-motion video kinesiology without quantifiable data. In the most limited of scenarios, a patient should at least be observed walking in a corridor at their preferred walking speed. Although the information gathered from such a task is limited, it is still better than trying to manage diseases of locomotion through static and nonweightbearing clinical tests and diagnostic images alone, which can be quite limited without some indication of the locomotive environment that such pathological processes are taking place in. This failure is perhaps most clearly seen in the thousands of diabetic, arterial, neuropathic, and venous foot ulcers treated every year, without any attempt to investigate the locomotive processes that these ulcers coexist in. Yet, it is even quite common for orthopaedic patients not to have been assessed dynamically before surgery.

Pain is a common disruptor of normal gait so that during any analysis that induces significant pain within the patient, such information is unlikely to provide anything more useful than data on attempted pain avoidance. Discriminating cause from the effect of pathology will always remain a clinical challenge. Thus, for gait analysis to be effective, the influence of symptoms and disease processes on gait should be understood. Disease presentation and patients with the same disease are not homogeneous, and neither are the gaits they develop. However, certain disease processes result in particular locomotive changes that should, and indeed, need to be identified by the clinician. These changes reflect the effect(s) of the disease process on locomotion rather than the nature of the disease, and they can give an insight into the extent of the disease's influence on the individual. It is not possible to discuss all diseases and their influences on the locomotive system with this text. Nevertheless, an introduction to the concepts of locomotive changes in the presence of key systemic and multisystem diseases is helpful in establishing the link between pathology and biomechanics, such that clinical biomechanics can be seen as an integral part of medicine.

1.6.2 Gait in cerebral palsy

Cerebral palsy (CP) is an umbrella term for nonprogressive neurological developmental dysfunction resulting from brain malformation or damage during early development, with defining characteristics of motor and postural impairment that limits activity and independent living. It is the most common cause of childhood disability, occurring in 2 per 1000 live births (Gómez-Pérez et al., 2019), and is classified according to the level of disability, with levels I and II being unassisted walkers and levels III–V requiring walking aids or wheelchairs (Booth et al., 2018). A single lower limb (hemipelgia) or both lower limbs (diplegia) may be affected, with variable amounts of upper limb involvement. Gait abnormalities are due to abnormal muscle tone and loss of motor control and balance issues, which result in secondary abnormalities that arise during development in response to abnormal locomotive techniques. Together, these issues compound the walking pattern abnormalities, often requiring multilevel surgical interventions involving both bone and soft tissue procedures, physical and occupational therapy, and neurosurgical and pharmacological interventions to reduce hypertonia (Lamberts et al., 2016).

Attempts have been made to construct gait-type classifications based around either anatomical levels of motion and position, primarily in the sagittal plane, or stride length and cadence. However, no single classification can address the full range of gait deviations seen in CP children (Dobson et al., 2007). Needless to say, CP gait is far from homogeneous, and management requires specialist knowledge and experience. For example, midfoot break is common in CP children, particularly in diplegic cases (Gaston et al., 2011; Maurer et al., 2013; Sees and Miller, 2013), and three types of midfoot break have been identified in CP children (Maurer et al., 2014).

Although spatiotemporal and kinematic parameters are more widely used in assessment of gait in CP children, kinetic and EMG studies are less widely reported (Gómez-Pérez et al., 2019). This may be due to the expense and time required for instrumented gait analysis. The most responsive data comes from joint angles over time in the sagittal plane, with further research being required to determine the significance and relevance of frontal and transverse plane data, kinetic data, and EMG data in gait dysfunction (Gómez-Pérez et al., 2019). Ambulatory CP children may present with several common gait patterns due to differing patterns of muscular spasticity. These commonly include a *crouch gait* displaying excessive hip and knee flexion. Continuous knee flexion during gait can result from hamstring and gastrocnemius contracture and can present with a combination of planovalgus feet, heel equinus, talonavicular joint dislocation, midfoot break, and external tibial rotation that all contribute to the crouch gait through lever arm dysfunctions. Limited or absent ankle dorsiflexion and toe-walking (*equinus*) are common, often being associated with varus or inversion deformity which can become fixed if not clinically resolved during early growth and development (Sees and Miller, 2013). Forefoot initial contacts are not uncommon in CP limbs, either unilaterally or bilaterally. In bilateral lower extremity spasticity (*diplegic*), a rigid planovalgus deformity becomes common, and this can be combined with ankle dorsiflexion restriction that facilitates midfoot break (Maurer et al., 2013; Sees and Miller, 2013).

In the transverse plane, in-toe (internal or adduction) rotations in the angle of gait are common. The association between equinus toe-walking and in-toe gait is most likely linked to the extra anterior stability of having the forefoot positioned close to the line of progression of the CoM in this gait style. This type of gait has a small base of support and a particularly high and relatively anteriorly positioned CoM. The lower limb transverse rotations can arise from internal rotations in the femur, tibia, or both, with the final foot angle also being influenced by pelvic/hip alignment. The angle of gait that causes in-toe is a sum of the transverse plane angles throughout the lower limb (Dobson et al., 2007), including the foot (Fig. 1.6.2a).

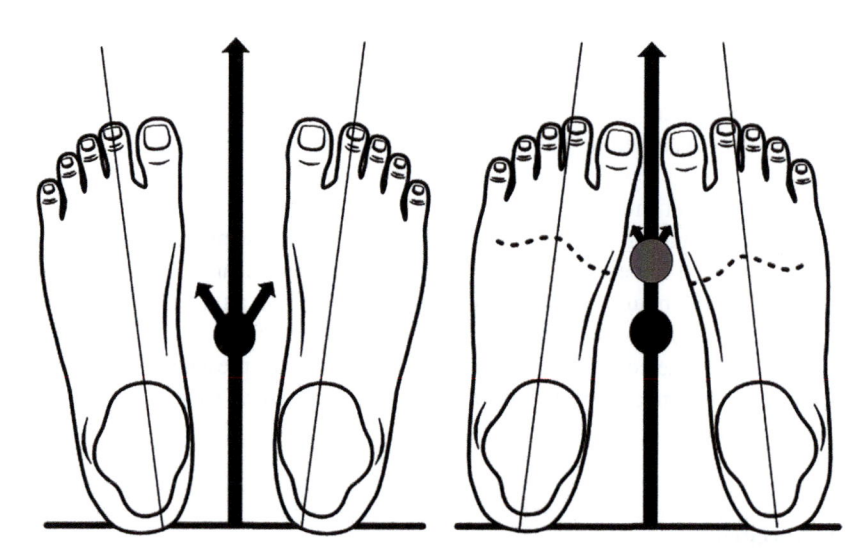

FIG. 1.6.2a Compared to foot abduction angles of slight out-toe as normally demonstrated during gait *(left image)*, an in-toe or adducted foot posture *(right image)* brings the forefoot closer to the line of progression *(thick black arrow)*. The forefoot, thus positioned, can act to partially block the anterior drift of the CoM deviating to the left or right over the foot while it is normally moving anteriorly towards each support limb during double-limb stance phase. Compared to the usual slightly abducted angles of gait, adducted feet offer a far narrower anterior base of support anteriorly. Equinus toe-walking gait only produces weightbearing on the forefoot *(dashed lines)* and thus they move their CoM anteriorly, risking greater instability over a small, narrow base of support. This may itself encourage the development of an anterior drift-blocking, CoM-stabilising in-toed gait. Such in-toe postures increase the difficulty of passing the CoM across the forefoot during terminal stance, helping to balance the need for more anterior CoM displacement stability among toe-walkers which if unrestrained, risks them toppling forward. Midfoot break within equinus gaits may temporarily help increase the weightbearing surface stability during late stance phase. Compared this image to Fig. 1.2.2e on out-toe stance and gait. *(Permission www.healthystep.co.uk.)*

Research shows a correlation between midfoot break, ankle/heel equinus, and internal hip rotations in hemiplegic and diplegic CP patients with transverse plane rotation of the foot related to those of the hip and pelvis (Gaston et al., 2011). When the foot is in closed chain, fixed to the support surface, any transverse external rotation required through deformities such as midfoot break have to be compensated for by the whole limb. This is because of the lower limb's rigidity, and the resultant inefficient ankle plantarflexion to knee extension coupling needs to be modulated by the equinus found in most CP patients (Gaston et al., 2011). When managing and examining CP patients, the effects of the fixed and functional deformities on the lever arms must be considered. Midfoot break is common in cerebral palsy and it results in lever arm dysfunction and weakness for acceleration (Maurer et al., 2013). It is found in several forms (Maurer et al., 2014), undoubtedly each with their own limiting effects on lower limb lever arm function compared to normal acceleration biomechanics (Fig. 1.6.2b).

Indeed, the locomotive postures found in CP have been referred to as *lever arm disease* (Gaston et al., 2011). Gait data gathered around lower limb lever system mechanics can provide biomechanical information that is essential to therapeutic and surgical management. For example, it is reported that subtalar fusion or lateral calcaneal lengthening surgery in pes planovalgus CP feet can increase knee extension in the gait of patients that have higher preoperative ankle joint dorsiflexion ranges in stance and better levels of ankle power. This thus improves lower limb lever arm function without the need for knee surgery (Kadhim and Miller, 2014). The biomechanical capabilities and heterogenous features of gait, such as which type of midfoot break is used, should be foundational in constructing the correct interventions in CP (Fig. 1.6.2c).

During childhood, a common therapeutic aim of those with CP is to improve mobility and walking ability through increasing muscle strength and joint range of motion. This is achieved via addressing impairments in activity and participation through utilising repetitive task-specific movements to reconstruct motor pathways (Booth et al., 2018). Functional gait training should therefore result in increased independence and activity participation. This approach to intervention is more recent. Adult CP sufferers were in the past subjected to stretching and increased mobility interventions alone, without as much regard being given to the functional effects of interventions. Therefore, joint mobility could have been increased surgically, such as through Achilles tendon lengthening, without realising that the restrictions on mobility was the result of muscle weakness and that tendon lengthening thus risked further functional disability.

Around 25%+ of ambulant adults with CP experience a gait decline with age, especially in those who are less independent in their gait. These individuals have bilateral impairment and higher levels of pain and fatigue during locomotion, giving gait analysis a role in presenting data useful for patient planning (Morgan and McGinley, 2014). CP adults are often

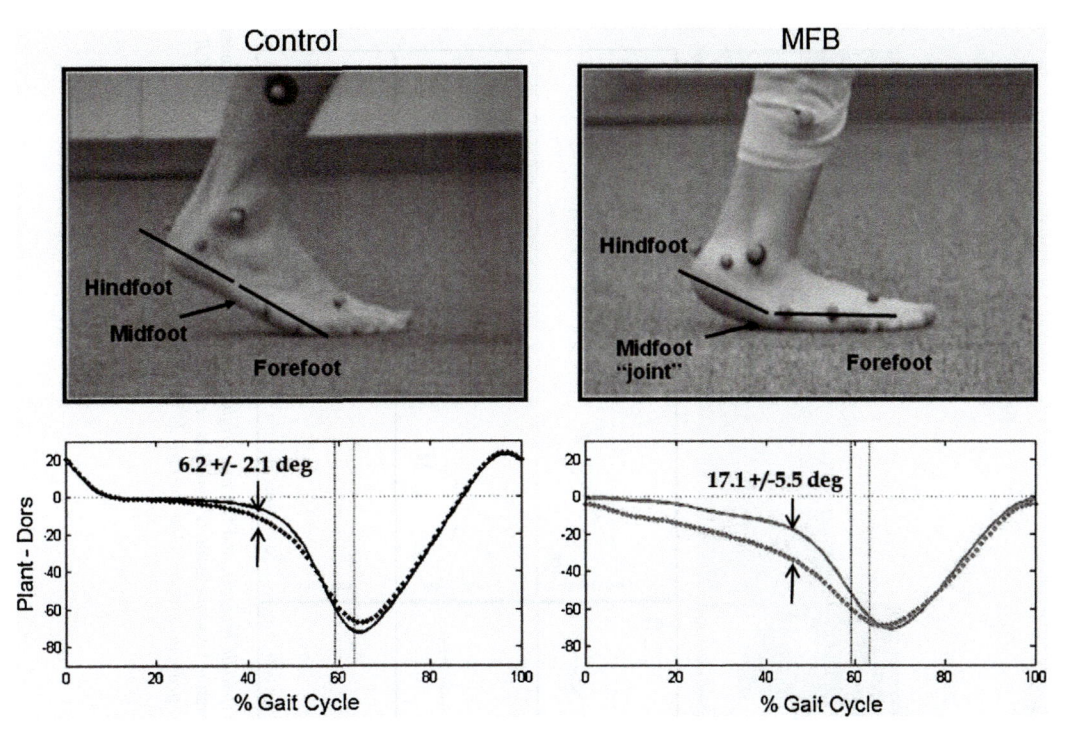

FIG. 1.6.2b Average forefoot *(solid line)* and hindfoot angles *(dotted line)* shown for normal controls on the left and cerebral palsy children with midfoot break on the right. Data clearly shows the difference in the rearfoot angle at heel lift. Children with midfoot break have decreased peak ankle dorsiflexion angles and increased peak midfoot dorsiflexion angles. This, combined with loss of ankle plantarflexion during heel lift, severely compromises the plantarflexion power that can be applied to the forefoot during acceleration. Midfoot break in cerebral palsy feet depowers gait and, combined with in-toe gait and excessive limb flexion, is another important cause of lever arm dysfunction within the lower limb. *(From Maurer, J.D., Ward, V., Mayson, T.A., Davies, K. R., Alvarez, C.M., Beauchamp, R.D., et al., 2013. A kinematic description of dynamic midfoot break in children using a multi-segment foot model. Gait Posture 38 (2), 287–292.)*

relatively neglected in relation to monitoring of their locomotive ability until further dysfunction requires increasing levels of intervention and social care support. This is perhaps a neglected area where clinical gait analysis and planned biomechanical management could play a larger part in the long-term planning of patient mobility needs.

1.6.3 Gait in musculoskeletal disease

Potential causes of musculoskeletal disease are quite vast and can essentially be split, less than neatly, into those arising from a systemic disease origin or biomechanically initiated through mechanical stress overload. However, the crossover of this division is obviously quite considerable. The intention here is just to open up the concept that pathology within the musculoskeletal system will induce changes in gait with a degree of predictability to the joint and the structures affected by the pathology and the compensation mechanisms.

Arthritis

In diseases of the musculoskeletal system such as seronegative and seropositive inflammatory arthropathies, locomotive ability is usually affected early. Although effects of such disease processes are not homogeneous, reflecting the different joints affected and the dysfunction and deformities that result, certain parameters are changed more consistently. These include slower walking speeds, longer double stance time, decreased cadence, decreased stride length, decreased ankle power, and avoidance of more extreme joint positions. These changes are also common indicators of systemic diseases and advanced age. Such spatiotemporal features have been observed in rheumatoid arthritis but with reduced walking velocity also noted in osteoarthritis and gout (Baan et al., 2012; Carroll et al., 2015). In systemic lupus erythematosus, force generation capacity within the lower limbs is reported as being reduced compared to healthy age-matched controls, with concurrent changes in foot posture towards increased pronation along with more reported foot symptoms (Stewart et al., 2020). However, these differences are not observed in psoriatic arthritis and as yet, there is not sufficient data on polymyalgia rheumatica and systemic sclerosis (Carroll et al., 2015).

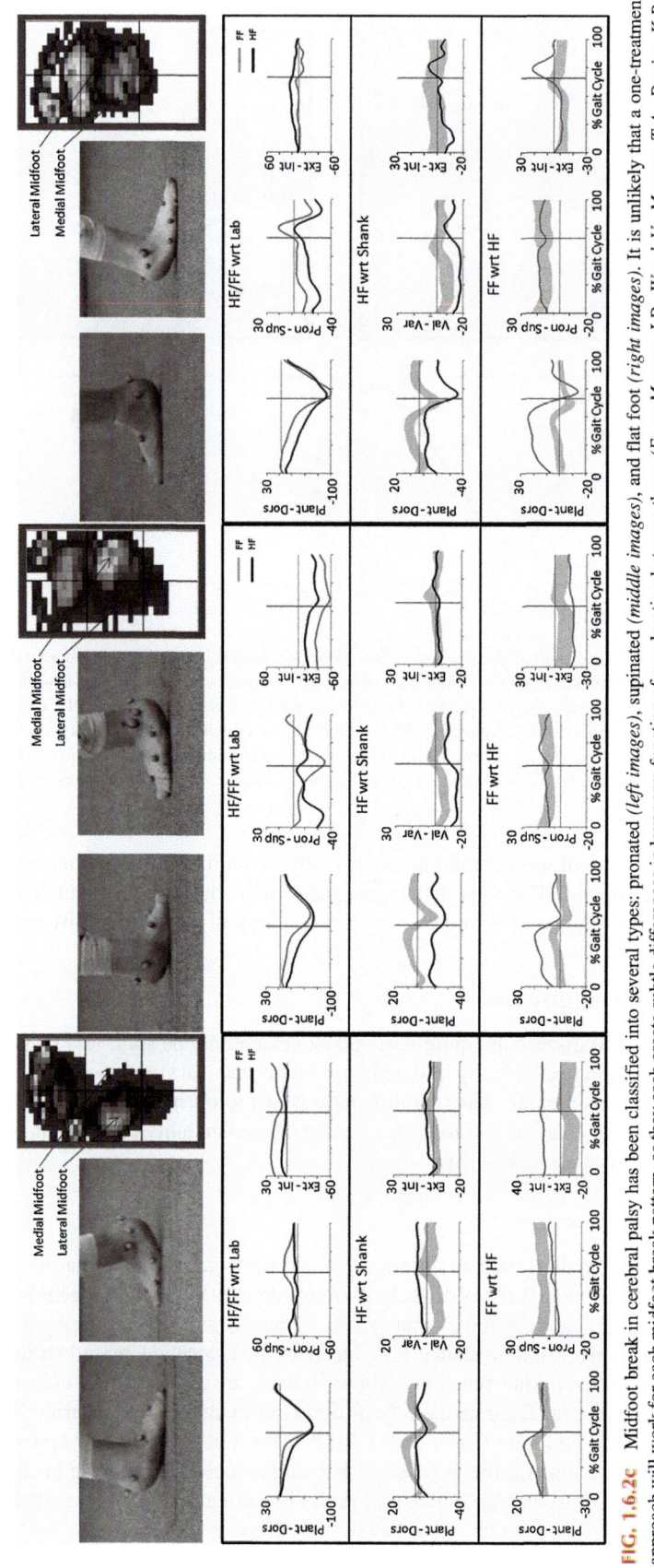

FIG. 1.6.2e Midfoot break in cerebral palsy has been classified into several types: pronated (*left images*), supinated (*middle images*), and flat foot (*right images*). It is unlikely that a one-treatment approach will work for each midfoot break pattern, as they each create subtle differences in lever arm function of acceleration between them. *(From Maurer, J.D., Ward, V., Mayson, T.A., Davies, K.R., Alvarez, C.M., Beauchamp, R.D., et al., 2014. Classification of midfoot break using multi-segment foot kinematics and pedobarography. Gait Posture 39 (1), 1–6.)*

Deformity is more common (and can be extensive) in the feet of rheumatoid arthritis sufferers than in any of the other inflammatory arthropathies, and includes hallux valgus, lesser toe deformities, pes planovalgus, along with other severe rearfoot deformities that change foot kinematics (Woodburn et al., 2002). This can alter gait parameters such as causing decreased speed of heel strike during contact phase, decreased acceleration time of terminal stance, an overall prolonged stance time, shortened stride time, and increased cadence with a reduction in walking speed by 20% compared to controls (Khazzam et al., 2007). The greater the extent of deformity and foot dysfunction involving areas of the forefoot, rearfoot, and midfoot together, the greater the changes in gait parameters. These include reduced walking speed and extended double-stance phase compared to those that are seen in individuals with rheumatoid arthritis changes only affecting either the forefoot or rearfoot alone, producing quite different pressure data that reflects the extent of the deformities (Turner and Woodburn, 2008) (Fig. 1.6.3a).

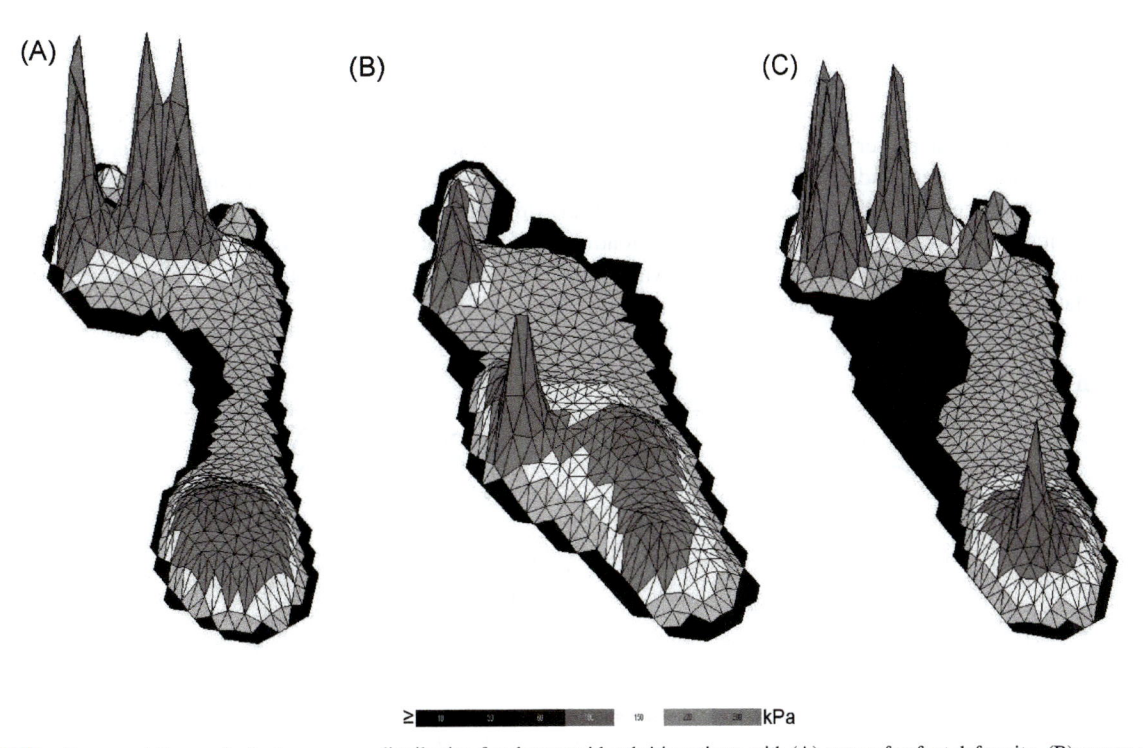

FIG. 1.6.3a Representative peak plantar pressure distribution for rheumatoid arthritis patients with (A) severe forefoot deformity, (B) severe rearfoot deformity, and (C) both forefoot and rearfoot deformity. Whereas forefoot deformity only leads to local changes in forefoot pressures (in this case, raised peak plantar metatarsal head pressures), any rearfoot deformity has a tendency to shift pressure medially, indicating higher levels of foot pronation (hyperpronation?) and suggesting compromise in foot's vault function during gait. *(From Turner, D.E., Woodburn, J., 2008. Characterising the clinical and biomechanical features of severely deformed feet in rheumatoid arthritis. Gait Posture 28 (4), 574–580.)*

Patients change their gait in response to both hip and knee osteoarthritis (Bejek et al., 2006; Elbaz et al., 2012, 2014). Although the origins of osteoarthritis may be insidious, traumatic, or a secondary consequence to diseases such as haemophilia, the resulting loss of mobility and pain at the joint will influence the locomotive strategies of patients regardless of the cause. Spatiotemporal gait parameters demonstrate that hip osteoarthritis sufferers as a group are associated with a slower walking speed and a greater asymmetry between their limbs than are age-matched controls. The self-selected gait speed between individuals is highly heterogeneous, with a mean in hip osteoarthritis sufferers that is 26% ± 20% slower than age-matched controls (Constantinou et al., 2014). Hip joint degeneration significantly worsens the variability of gait data on the affected side, with the nonaffected joints and pelvis compensating through increased flexibility, and by adapting the step variability. This indicates that interventions need to focus on stability and reliability of limb motion with increasing levels of hip arthritis and increased walking speed (Kiss, 2010). Clinical assessment of gait in hip osteoarthritis should be done both at the patient's preferred walking speed and at higher speeds. This enables the clinician to see the effects that changing gait speed may have on gait stability during daily activities. Adjusting the walking speed during clinical assessment may reveal variability in parameters that reflect the severity of the disease (Kiss, 2010). Following total hip replacement surgery, improvements in gait parameters should be expected, but these will not return to a situation comparable with normal controls for at least 12 months following surgery (Bahl et al., 2018) and may not ever return to normal levels of adjustability necessary for gait speed changes.

Knee osteoarthritis is the most common joint disease. Once again, spatiotemporal parameters of gait can be used as indicators of disease severity in this condition, which may outperform radiographic findings and questionnaires on disability (Debi et al., 2011; Elbaz et al., 2012, 2014; Mills et al., 2013a). Knee osteoarthritis increases gait deviations compared to controls, such as greater stride time, increased double stance phase, shorter single-limb support time, shorter step length, slower gait velocity, and decreased cadence, with changes reflecting the severity of disease (Debi et al., 2011; Elbaz et al., 2012; Mills et al., 2013b). It is likely that these changes indicate the knee's inability to bear increased loads and to extend normally from more flexed postures during single-limb stance. Thus, after therapeutic input, a longer single-limb stance time, a longer stride length, and a higher cadence would be indicators that treatment was actually improving function.

Ankle osteoarthritis causes a significant disruption of several spatiotemporal parameters. It affects normal rearfoot triplanar motion, lowers vertical GRFs (especially decreasing the F3 peak in terminal stance associated with reduced ankle power, as well as maximal medial GRF), and also reduces sagittal and transverse ankle joint moments (Valderrabano et al., 2007). Lower rearfoot dorsiflexion angles in midstance also result (Nüesch et al., 2012) and this probably links to the reduction in available ankle power at heel lift. Pressure data changes are also noted with ankle osteoarthritis with maximum peak force and contact areas reduced over the affected foot, while peak rearfoot pressure and toe area pressures are decreased (Horisberger et al., 2009). These changes are probably, in part, indicative of pain avoidance strategies. Together, there is consequential loss of ankle plantarflexion on loading, reduced dorsiflexion in midstance, and reduced plantarflexion and ankle power in terminal stance; findings which are all consistent with the reduced vertical peaks reported from force plate data (Figs 1.6.3b and 1.6.3c).

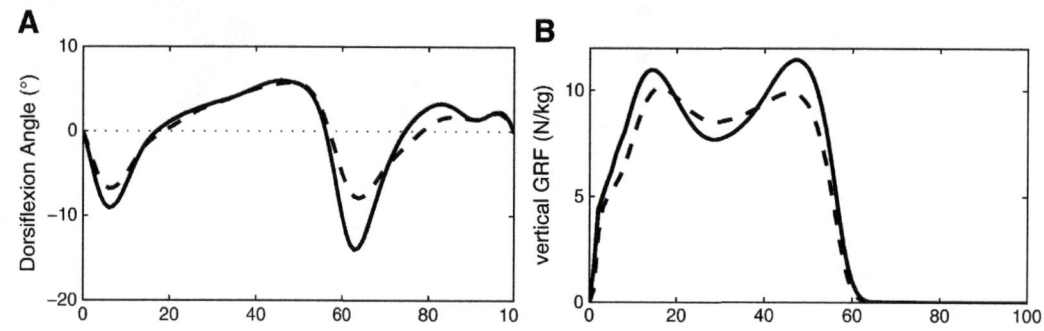

FIG. 1.6.3b The mean hindfoot dorsiflexion angles (A) of the nondominant leg of controls *(solid line)* and affected side of asymmetrical ankle osteoarthritis patients *(dashed line)*. In (B), mean vertical GRFs between groups are compared on a force-time curve. Note lower peaks and shallower troughs in ankle arthritis. *(Modified from Nüesch, C., Valderrabano, V., Huber, C., von Tscharner, V., Pagenstert, G., 2012. Gait patterns of asymmetric ankle osteoarthritis patients. Clin. Biomech. 27 (6), 613–618.)*

Gait data indicates that initially, gait parameters deteriorate following total ankle replacement arthroplasty, and that it may be better to wait 6–12 months to assess gait data following surgery for a better indication of functional outcomes (Valderrabano et al., 2007). In time, patients with ankle arthroplasty should develop a gait pattern more closely resembling normal as compared to a patient who has undergone an ankle arthrodesis, where gait becomes quite abnormal (Singer et al., 2013). Conservative interventions should also aim to change gait data to reflect more normal gait parameters, as well as improve symptoms, if indeed improved function is the aim of intervention.

Osteoarthritis is capable of affecting any joint within the foot, although it is relatively rare in the lesser MTP joints and digits. This is unless it occurs secondary to a history of trauma, a developmental disturbance such as an avascular necrosis (e.g. Freiberg's disease), or an inflammatory arthropathy. Rheumatoid arthritis has a particularly strong association with secondary osteoarthritis of the foot as part of the deformities and functional disabilities rheumatoid arthritis can create (Khazzam et al., 2007; Turner and Woodburn, 2008). Thus, gait changes and expectations of foot osteoarthritis relate to the joint or joints affected, increased dysfunction being likely with increased areas of degeneration of the rearfoot, midfoot, and forefoot. Subtalar joint osteoarthritis is likely to influence stance phase ankle function as part of its unified joint function during stance phase, reducing rearfoot motion. On the other hand, loss of mobility at the talonavicular joint will likely result in loss of subtalar joint motion, midfoot midstance dorsiflexion, and terminal stance phase plantarflexion, thereby reducing forefoot pressure/force generation in terminal stance (Fogel et al., 1982; Fishco and Cornwall, 2004).

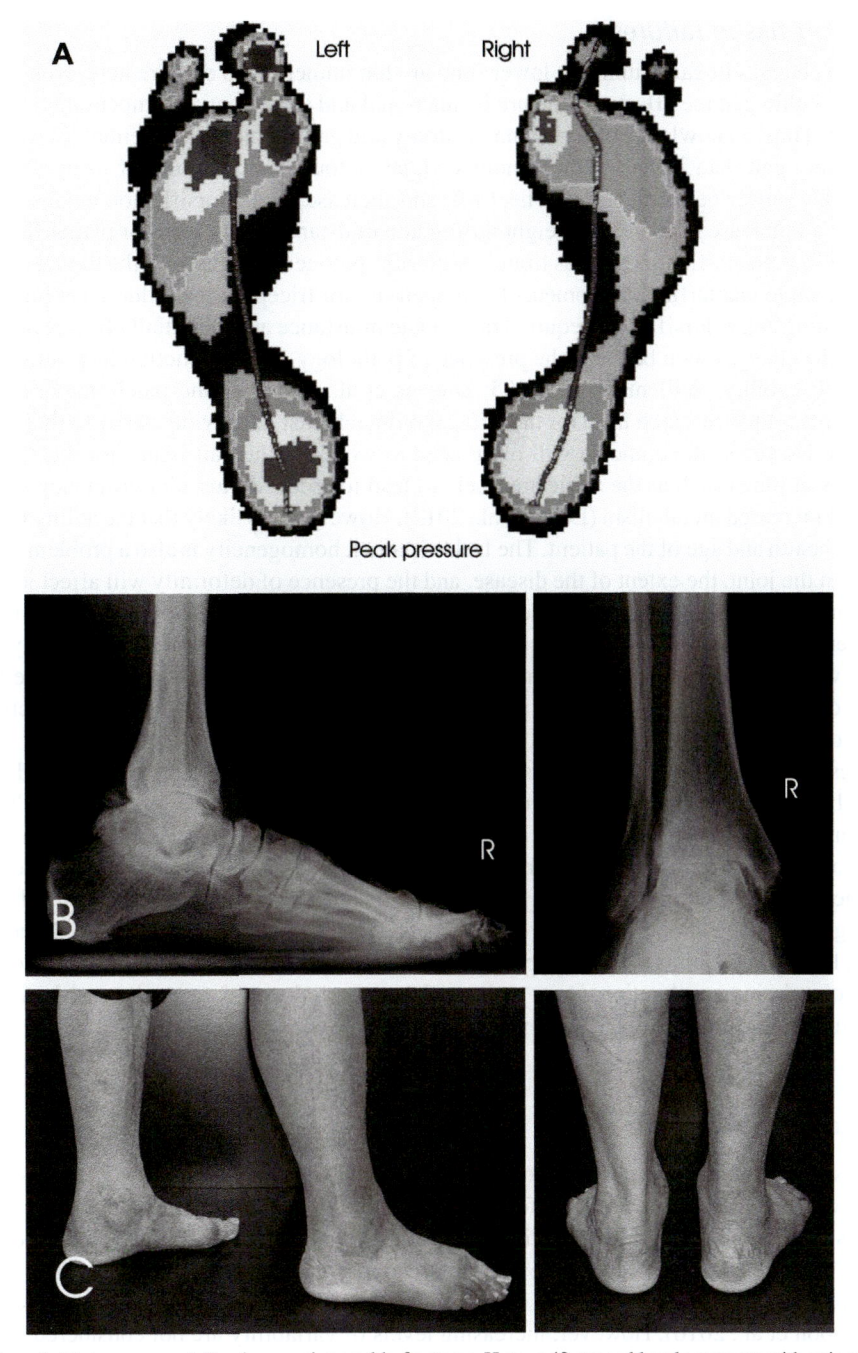

FIG. 1.6.3c Ankle osteoarthritis is common following previous ankle fractures. Here, a 40-year-old male presents with pain and loss of motion due to severe osteoarthritis in the right ankle, 12 years after a malleolar fracture (B and C). Pressure data from pedobarography (A) shows reduced peak plantar pressure in the affected right foot, probably due to protective mechanisms during contact energy dissipation and loss of plantarflexion power during acceleration. *(From Horisberger, M., Hintermann, B., Valderrabano, V., 2009. Alterations of plantar pressure distribution in posttraumatic end-stage ankle osteoarthritis. Clin. Biomech. 24 (3), 303–307.)*

Osteoarthritis of the 1st MTP joint is clinically known as hallux rigidus, although milder forms can be referred to as hallux limitus. One cause can be secondary to inflammatory arthropathies, in particular, gout (Stewart et al., 2016). Here, the primary motion lost is hallux extension, with a more variable loss of plantarflexion. Often, the sesamoid articulations are also involved. It can be a progressive and debilitating condition, often resulting in changes within gait that utilise the lateral forefoot during acceleration at terminal stance to avoid hallux extension, thus decreasing the acceleration forces from ankle power (Brodsky et al., 2007; Callaghan et al., 2011) through the loss of the normal medial MTP extension fulcrum. This is discussed further in Chapter 4, Section 4.5.11.

Musculoskeletal soft tissue failures

Soft tissue dysfunction changes in gait within the lower limb are too numerous to explore here. Nonetheless, gait changes should reflect the time within gait the affected structure is functional and the potential compensatory mechanics that might try to limit dysfunction. Thus, a knowledge of functional anatomy and gait moments is essential for a clinical biomechanist to be able to unravel any gait data in musculotendinous or ligamentous dysfunction. For example, an Achilles tendon rupture will reduce ankle power generation, delay heel lift, and increase ankle dorsiflexion angles in stance before heel lift. It may also induce an increase in foot vault height during late midstance as the assistor plantarflexor muscles attempt to induce heel lift. The foot plantarflexors such as tibialis posterior, peroneus longus, and the digital flexors can stiffen the foot and produce weak ankle plantarflexion moments to compensate for triceps surae failure. Yet together, they also raise the foot vault, thus resisting ankle dorsiflexion required during late midstance and potentially further delaying heel lift. Data has been collected on the changes seen between the presence of pathology and dysfunction in several common conditions such as chronic ankle instability (Willems et al., 2005; Drewes et al., 2009a,b) and patellofemoral pain (Levinger and Gilleard, 2007), requiring a little research time for the clinical biomechanist to provide clarity to their examination of specific pathology groups. However, the clinician will often need to work out the gait issues for themselves.

Changes in mobility at joints such as the ankle are likely to lead to reduced muscular efficiency and an increase in the energetic costs through increased metabolism (Lobet et al., 2012). However, it is likely that the ability to compensate will be affected by the overall health and age of the patient. The lack of patient homogeneity is also a problem in assessing any joint arthritis, as areas within the joint, the extent of the disease, and the presence of deformity will affect joint motion and locomotion patterns that are also a reflection of other morphological features of the individual. Gait changes in knee osteoarthritis will be different in a varus-aligned knee compared to a valgus-aligned knee, so that only groups showing a homogeneous nature with regard to their location and extent of disease should be compared. Despite this, gathering data on locomotive values during gait are useful if targets are set for improved motor performance through clinical interventions, whatever the origins of the locomotive dysfunction (Fig. 1.6.3d).

Prospective studies to identify risk factors of musculoskeletal injury perhaps hold the future for advancing the value of clinical gait analysis, both in preventing injury and guiding rehabilitation. For example, Willems et al.'s (2005) prospective study identified a number of features that seemed to increase ankle inversion. These included a more laterally situated CoP at initial contact, and a more mobile foot type at 1st metatarsal contact, foot-flat (forefoot contact), and heel off, displaying prolonged stance time pronation with more medial pressure and delayed maximum knee flexion. They found that 'resupination' or vault height re-establishment was delayed, and that acceleration occurred less through the medial MTP joints and more through the lateral MTP joints. This, they proposed, could be due to a failure to establish medial forefoot stability around the 1st MTP joint during acceleration. Once an injury has occurred, gait data can be at risk of contamination from protective mechanisms induced by the injury, rather than revealing an initial cause.

1.6.4 Gait in neurological disease

Neuromotor processes are inherently 'busy' processes that create a lot of neurological 'noise', and this produces variability in motion through a combination of feedforward-feedback sensorimotor mechanisms. This is not necessarily a disadvantage as deterministic processes have been identified within movement variability that improve energetics of motion (König et al., 2016). Absolute controlled repeatable motion uses more energy than slightly variable motions that achieve the same repeatable result. As a consequence, low levels of gait parameter variability are considered usual and healthy (König et al., 2016; Moon et al., 2016). However, increasing levels of variability are not considered healthy and are associated both with neurological disease and the extent of the disease (König et al., 2016; Moon et al., 2016).

Evidence is showing that gait variability provides some unique information on the characteristics found in different neurological diseases. Variability in gait is common in individuals with neurological disorders and is associated with an increased risk of falls (Moon et al., 2016). Specific characteristics of gait can be different and distinctive between neurological disorders, which are all consistently abnormal compared to healthy controls (Moon et al., 2016). Spatiotemporal changes such as reduced gait speed seem universal across neurological diseases, but stride-to-stride variations are a more unique marker of disease (Moon et al., 2016). Increased temporal gait variability in stride times and step length (anterior–posterior spatial parameters) are believed to reflect loss of rhythmic gait, while disruption of medial-lateral spatial parameters such as step width variability are associated with diminished balance (Bruijn et al., 2013). An increase in gait variability is found in all neurological diseases investigated compared to healthy controls (König et al., 2016; Moon et al., 2016). Overall, spatiotemporal parameters within neurological disease seem to present with changes in magnitude from normal of 11% to 40% (Crowther et al., 2007).

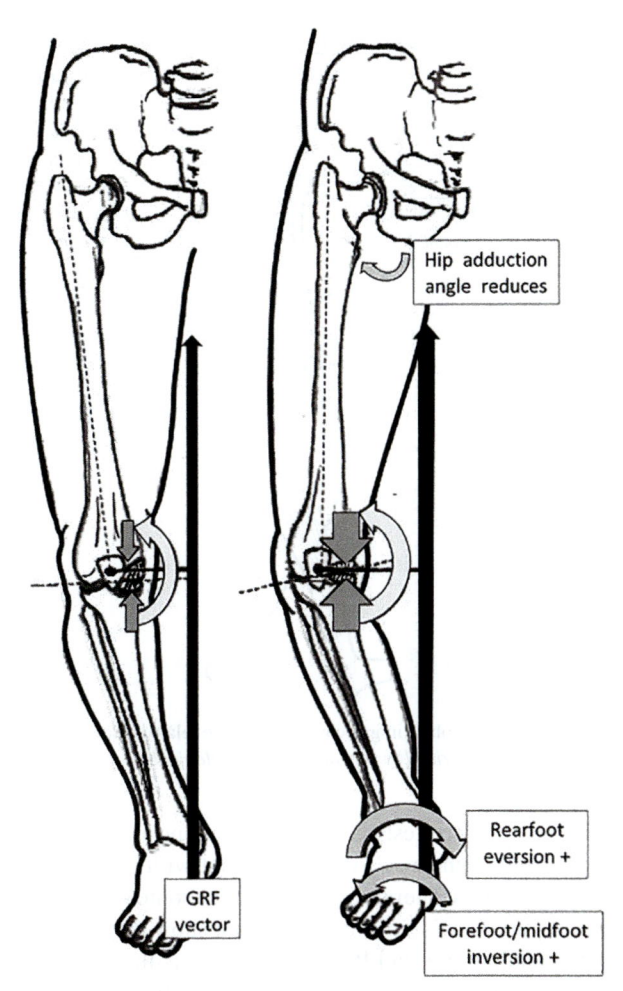

FIG. 1.6.3d Osseous alignments and gait parameters (such as base of support) can influence joint moments that can induce degenerative joint changes that in their turn can alter joint moments further. For example, a high femoral neck-shaft angle coupled to a genu/tibial varum and a narrow base of gait (A) will create a GRF medial to the knee. This causes increased knee adduction moments *(white arrow)* on a large adduction lever arm in the frontal plane, raising compression forces *(dark-grey arrows)* within the medial compartment of the knee. If this in time leads to degenerative changes within the joint and its subchondral bone, then the femoral carrying angle *(dashed lines)* at the knee may change from slight valgus to varus, necessitating a compensatory reduction in the hip adduction angle during stance (B). The increased varus alignment at the knee within this posture increases the adduction moment arm as the knee drifts more lateral to the GRF. Thus, a cycle of increasing medial compartment degeneration and increasing compression forces within the medial knee compartment can be initiated. Such a situation also requires greater rearfoot eversion and midfoot/forefoot inversion to maintain a foot-flat posture on an increasingly inverted lower leg. This, in turn, may change forces detrimentally elsewhere within the soft tissues. In valgus knees, the reverse cycle of increasing abduction moments and tibiofemoral lateral compartment degeneration can occur. Therefore, very different pathomechanics on the knee can cause variable osteoarthritic changes and raise a variety of soft tissue stresses, dependent on internal and external vector alignments during gait. *(Permission www.healthystep.co.uk.)*

Maintaining 'gait rhythm' consistency is a multifactorial process dependent on a number of diverse neurological structures so that damage to any neurological structure can be expected to have unique influences on gait parameter variability. Neurological impairment of the cerebral cortex in the prefrontal, premotor, primary motor, and posterior parietal cortices, along with supplementary motor areas is likely to impair voluntary control of gait (Takakusaki, 2008). Disruption of voluntary control of gait will cause problems with initiation and selection of appropriate movement plans, precise limb control, and efficient and effective postural adjustments (Takakusaki, 2008, 2013) increasing variability during locomotion. Altered emotional behaviours such as the loss of fight-flight reactions and fear of falls may also override automatic gait control, decreasing accuracy during adaptive gait (Takakusaki, 2013; Young et al., 2016). Damage in the basal ganglia disrupts both the voluntary and automatic processes of gait, as the ascending projections excite the cerebral cortex while the descending projections regulate the brainstem activities of locomotor rhythm and postural muscle tone (Takakusaki, 2013). The cerebellum receives information from the cerebral cortex, spinal cord, and vestibular system, so that damage within the cerebellum will cause issues in regulation of timing and the rate and force of muscle activity that maintain gait consistency (Moon et al., 2016). Gait variability may therefore provide disease-specific information (Fig. 1.6.4a).

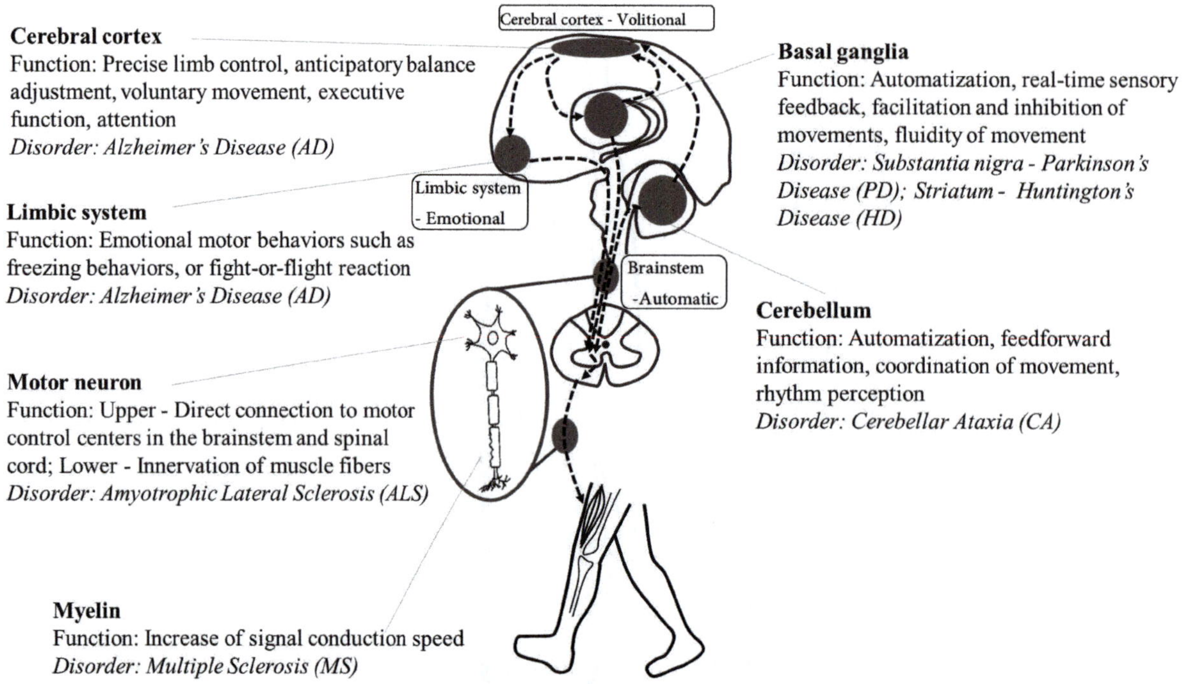

Cerebral cortex
Function: Precise limb control, anticipatory balance adjustment, voluntary movement, executive function, attention
Disorder: Alzheimer's Disease (AD)

Limbic system
Function: Emotional motor behaviors such as freezing behaviors, or fight-or-flight reaction
Disorder: Alzheimer's Disease (AD)

Motor neuron
Function: Upper - Direct connection to motor control centers in the brainstem and spinal cord; Lower - Innervation of muscle fibers
Disorder: Amyotrophic Lateral Sclerosis (ALS)

Myelin
Function: Increase of signal conduction speed
Disorder: Multiple Sclerosis (MS)

Basal ganglia
Function: Automatization, real-time sensory feedback, facilitation and inhibition of movements, fluidity of movement
Disorder: Substantia nigra - Parkinson's Disease (PD); Striatum - Huntington's Disease (HD)

Cerebellum
Function: Automatization, feedforward information, coordination of movement, rhythm perception
Disorder: Cerebellar Ataxia (CA)

FIG. 1.6.4a A schematic of the neurological structures involved in gait control and some related disorders that will alter gait function. *(From Moon, Y., Sung, J., An, R., Hernandez, M.E., Sosnoff, J.J., 2016. Gait variability in people with neurological disorders: A systematic review and meta-analysis. Hum. Movement Sci. 47, 197–208.)*

Huntington's disease is reported to have the greatest gait variations among all groups with neurological disease, whereas Parkinson's disease, multiple sclerosis, and Alzheimer's disease present with the least amount of variation. Furthermore, diseases such as cerebral ataxia and amyotrophic lateral sclerosis present somewhere in the middle, demonstrating heterogeneity between neurological diseases. Parkinson's sufferers as a group show the least variability (Moon et al., 2016). The difference in gait variations between Huntington's and Parkinson's reflects the impairment of different neural pathways involved in gait regulation, despite both diseases affecting the basal ganglia. Huntington's is associated with excessive basal ganglia inhibition resulting in a hyperkinetic disorder (irregular limb and trunk movements). Parkinson's, on the other hand, is a result of the loss of dopamine release in the substantia nigra leading to excessive inhibition of the cortico-basal ganglia loop which causes a hypokinetic disorder of less movement and decreased range of limb motion (Moon et al., 2016).

A specific characteristic of cerebral ataxia is the lack of intra-limb coordination, which will probably result in greater gait variability as a result of arrhythmic reciprocal flexor and extensor muscle activity during walking (Morton and Bastian, 2006). Both Huntington's disease and cerebral ataxia exhibit abnormal limb coordination due to an inability to suppress unnecessary movement. However, whereas Huntington's is compromised in real-time sensory inputted reactive adaptations, cerebral ataxia maintains these functions (Morton and Bastian, 2006; Takakusaki, 2013). Cerebral ataxia may be able to compensate for some coordination deficit through the brainstem or lower motor neurone centres to gain reactive feedback-driven gait adaptations, resulting in lower levels of gait variations than Huntington's (Moon et al., 2016). Tests on populations with multiple sclerosis and Alzheimer's disease have been undertaken in the mild-to-moderate cases of the conditions, which may suppress the levels of gait variability that may be found in the worse stages of the disease. Greater disequilibrium of gait has been reported in the more advanced stages of these diseases (Moon et al., 2016).

Base of support and step width is not reported to be significantly different in neurological groups compared to controls. It has been suggested that such medial-lateral-directed parameters are related to balance, and it is possible that in neurological disease, the ability to retain balance through manipulating medial-lateral parameters in gait remains intact (Moon et al., 2016). This observation in Parkinson's disease might indicate progressive decline in lower extremity muscle tone and postural reflexes that, in turn, could reduce levels of variability in step width in an attempt to maintain a stable CoM in the medial-lateral direction (Rochester et al., 2014). This would actually suggest that loss of control in one dimension might result in tightening of control in another (König et al., 2016). It may be that a limited capacity in adjusting to internal and external perturbations in gait from too low a level of variability may also be problematic, and that both too much variability and too little might be associated with falls (Singh et al., 2012). Alternatively, the lack of difference could simply be a result of the wide confidence interval of the parameters in gait stability, even when gait rhythmicity is lost. Variability of motion in health and disease remains a subject that is in need of much further investigation (Fig. 1.6.4b).

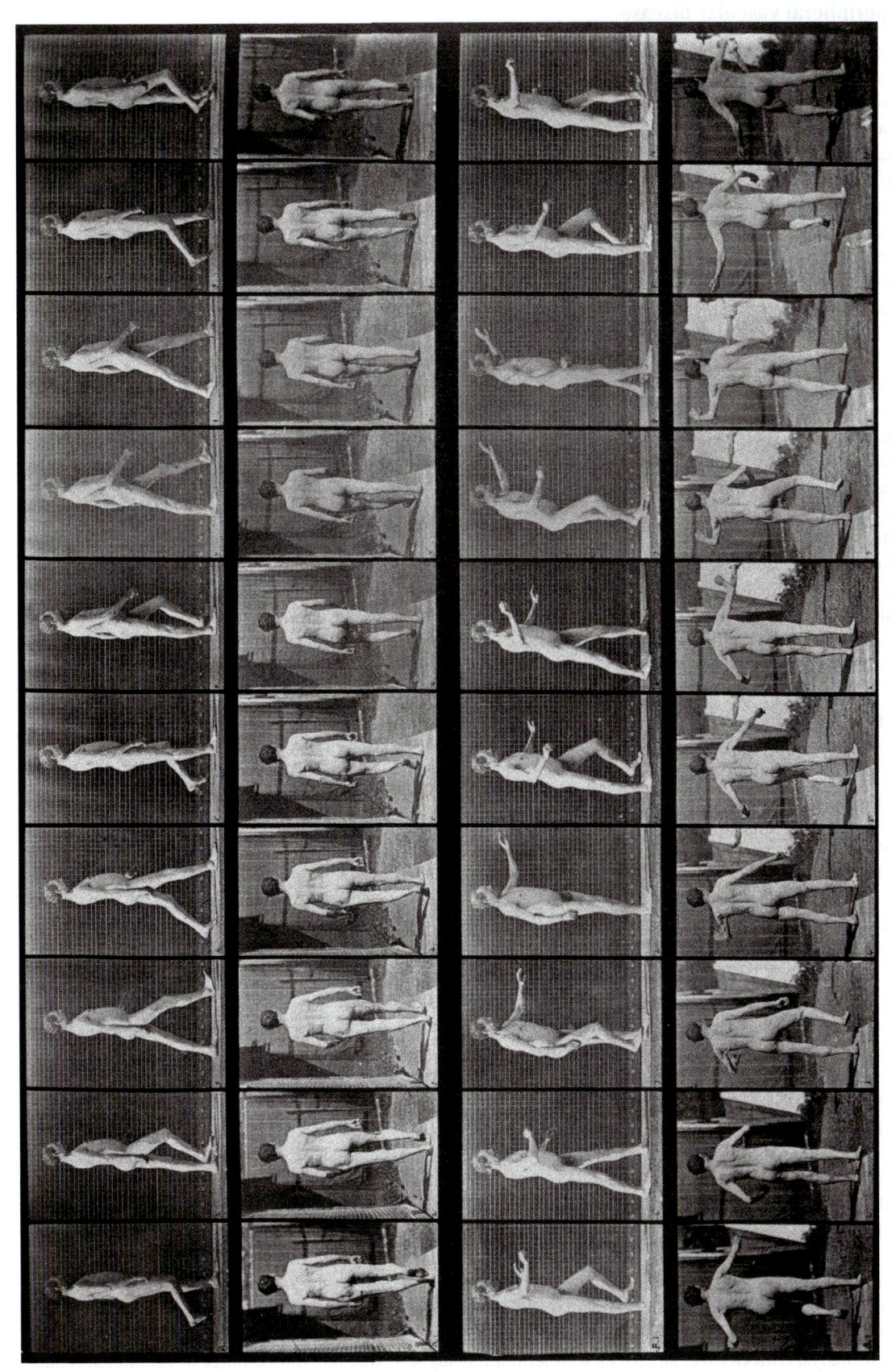

FIG. 1.6.4b The sad deterioration of gait over time as a result of neurological disease captured by Eadweard Muybridge. Sadly, the time frames involved have not been discovered by this author, but the changes in gait of the individual in the lower two lines of images compared to those in the top two lines are clear to see. Note the more subtle disturbance in gait in the top images such as trunk sway to the right limb in right midstance, knee hyperextension in both limbs during late midstance, large pelvic anterior tilt, and exaggerated lumbar lordosis. As the disease progresses, normal upper and lower limb gait strategies are lost as maintaining postural balance overrides energetics.

1.6.5 Gait in peripheral vascular disease

Peripheral arterial disease is a chronic arterial occlusive disease of the lower extremity caused by atherosclerosis and affects around 20%–30% of older patients attending general medical practice (Celis et al., 2009). It commonly causes intermittent claudication, an exercise-induced pain experienced in the calves, thighs, or buttocks that is relieved by rest. Most patients suffering peripheral arterial disease do not complain of intermittent claudication, but instead suffer more subtle signs and symptoms of disease. Walking problems are linked to mitochondrial and neuromotor dysfunction associated with peripheral arterial disease. The neuromotor dysfunction is further exacerbated by the metabolic demand of active leg muscles for oxygenation which are not being satisfied due to the restricted blood flow to the lower extremities (Celis et al., 2009; McGrath et al., 2012). When the resulting oxygenation deficit is great enough, cramping pain results, known as claudication. This occurs with variable onset time depending on the amount of ambulation.

Regardless of the presence of claudication pain, the effects of peripheral arterial disease on gait are similar (Chen et al., 2008; Koutakis et al., 2010; Myers et al., 2011; McGrath et al., 2012). Patients with peripheral arterial disease demonstrate reduced physiological capacity, reduced lower limb mobility, reduced walking performance, reduced lower extremity strength, impaired balance, higher prevalence of falls, and reduced physical activity levels (Crowther et al., 2007; Celis et al., 2009; McGrath et al., 2012).

This reduction in lower limb mobility causes a decreased quality of life as well as changes in spatiotemporal gait parameters such as stride length and gait speed (Daley and Spinks, 2000; Crowther et al., 2007; McGrath et al., 2012). When comparing to controls matched by age, height, BMI, mass, and percentage body fat, sufferers of intermittent claudication/peripheral arterial disease demonstrate significant differences in spatiotemporal parameters, resulting in shorter walking distances (McDermott et al., 2001), slower walking speeds, shorter step lengths, and reduced cadence (Gardner et al., 2001; Crowther et al., 2007; Watson et al., 2011). Intermittent claudication sufferers and those with peripheral arterial disease have a slower walking pace, take longer to complete a gait cycle with a smaller stride length and lower cadence and are reported to spend longer in both single and double-limb support compared to age-matched controls (Gardner et al., 2001; Crowther et al., 2007). Overall, peripheral arterial disease seems to increase the percentage of time spent in double stance phase and increases the time it takes to complete a stride, with reduced stride length and walking speeds. This may relate in part to a higher prevalence of lower limb sarcopenia in peripheral vascular disease (Addison et al., 2018).

Differences in joint kinematics are also noticeable with reductions in joint powers (Crowther et al., 2007; Chen et al., 2008; Celis et al., 2009; Koutakis et al., 2010; Myers et al., 2011), with greater ankle plantarflexion in early stance after initial heel strike. The time to minimum plantarflexion was shorter, but the time to maximum dorsiflexion was longer during the stance phase (Celis et al., 2009) while remaining more plantarflexed during swing phase (Crowther et al., 2007). This indicated significant dysfunction in ankle joint range of motion, suggesting weakness of the posterior compartment (Crowther et al., 2007; Chen et al., 2008; Celis et al., 2009). The peripheral arterial disease population seem to present with a type of 'foot drop' at heel contact resulting from possible nerve damage or muscle myopathy from chronic ischaemia, leading to poor eccentric control by the dorsiflexors such as tibialis anterior (Celis et al., 2009). Involvement of joints more proximal to the ankle may be unaffected by the level of occlusion, as Celis et al. (2009) did not report changes in hip and knee kinematics. Nevertheless, others have reported a reduction in hip and knee motion (Crowther et al., 2007; Chen et al., 2008; Myers et al., 2011). These studies together suggest individual variability with regard to the extent of the effects. Mobility is reduced in the peripheral arterial disease population and produces a more shuffling gait pattern with reduced joint angular displacement, velocities, and accelerations compared to healthy age-matched controls (Crowther et al., 2007). Possible mechanisms of action to changing gait are due to lower limb ischaemia and the myopathy associated with that ischaemia (Crowther et al., 2007; Myers et al., 2011). It is possible that such patients will not respond well to strengthening rehabilitation (King et al., 2012).

Looking at GRFs over force plates during the gait of peripheral arterial disease sufferers, variances with controls are noted in frequency bandwidth in anterior–posterior directions along with more sluggish oscillatory components of the neuromotor system as the body applies force to the ground (McGrath et al., 2012). Anterior–posterior components of GRF are altered with such changes indicating a reduced range of movement frequencies, as patients with peripheral arterial disease propel themselves forward. However, these changes do not affect vertical-directed forces (McGrath et al., 2012). The frequency analysis of vibration (oscillation) of movement determines the range of frequencies associated with all components of the neuromotor system within bones, nerves, muscles, and connective tissue as they interact to produce motion. Reduced bandwidth oscillation frequencies of these movements indicate constrained oscillations in one or more of these structures during anterior–posterior-directed motion. This correlates with the reduced ankle plantarflexion power in late stance reported in kinematic analysis (McGrath et al., 2012). Furthermore, it suggests that the limited range of oscillation relates to the dysfunction associated with peripheral arterial disease-induced myopathy (McGrath et al., 2012). When experiencing claudication pain, higher oscillatory movements were detected in the anterior–posterior direction which research has linked to tremors and instability of movement patterns, suggesting that claudication initiates gait instability (McGrath et al., 2012). Patients with claudication have an 86% increase in ambulatory stumbling and unsteadiness compared to controls, and a 73% higher history level of falls (Gardner and Montgomery, 2001).

It is possible that the reduction in GRF generation due to peripheral vascular disease or other pathologies such as quadriplegia and neuropathies that prevent or reduce GRF have a knock-on effect on venous return (Carvalho and Cliquet Jr, 2005; De Carvalho et al., 2006; Newland et al., 2009; Shiman et al., 2009; Yim et al., 2014). Both GRF and lower limb muscle activity are essential for the foot and calf pumps that return venous blood from the lower limbs (Newland et al., 2009; Horwood, 2019). Thus, the management of venous insufficiency and ulcers may benefit from the use of gait data to plan rehabilitation to improve venous function (Kirsner, 2018). Such concepts link to multisystem diseases such as diabetes, where poor biomechanics may compromise failing physiological systems further.

1.6.6 Gait in diabetes (mellitus)

Diabetes mellitus (diabetes) is a multisystem disease process that in many respects results from the expression of accelerated ageing. For the locomotive system directly, it is the damage done through the deterioration of vascular and neurological tissue combined with a resulting dysfunction in tissue strength, loss of compliance, and loss of tissue damage warning systems that collectively lie at the heart of lower limb pathology. These problems are exacerbated by poor pressure distribution in the cutaneous tissues and overall, limited tissue force-management abilities. These issues arise as a consequence of a loss or reduction of muscle strength, tendon energy storage, tissue compliance controls, and any resultant foot deformity that accompanies the attendant neuromuscular disease processes. This interrelationship of multilayered pathology and dysfunction reaches its locomotive zenith in biomechanical failure with the development of Charcot foot deformity, bringing with it an increased risk of foot and lower limb amputations. Development of deformities and amputations are a failure in patient management from either or both the patient's actions and the clinician's interventions. The proper utilisation of clinical biomechanics has a large part to play in the management of diabetic pathology and improvements in health and mobility. If interventions are early, then increases in exercise levels and dietary controls can reverse the disease process to variable degrees and can change movement pathways to prevent the development of deformity, even should the disease process continue.

Plantar foot ulcer management takes up much time and creates considerable healthcare costs. Plantar soft tissues in diabetics are much stiffer than healthy tissues, reducing their protective role in dissipating shear stresses and thus opening the foot up to the possibility of ulceration (Pai and Ledoux, 2012). Sadly, tissue shear cannot yet be measured directly during gait analysis. However, using careful observation of pressure data and, if available, force and kinematic data of diabetics, can indicate areas likely to be subjected to increased shear and therefore to being more vulnerable. It must be understood that local differences in tissue pathology as a result of the presence of advanced glycosylated end-products probably leads to plantar soft tissues with heterogeneous responses and failure rates under loading, so that a 'Rubicon' of pressure intensity should not be sought to act as a guide for triggering intervention. Patients developing foot ulcers seem to have less cumulative plantar stresses compared with those who do not develop foot ulcers, whereas those that develop greater variability of activity seem more at risk (Wrobel and Najafi, 2010). Thus, trying to reduce foot pressure with footwear or orthoses alone is not the answer, particularly as another problem is the lack of use of protective footwear/insoles in the home by patients. Using thermometry to monitor skin temperature elevations is beneficial in identifying areas of imperceptible inflammation within the skin early, particularly in the neuropathic diabetic devoid of nociceptive feedback. However, damage starts deeper in the underlying tissues at the bone-soft tissue interface (Naemi et al., 2016c), so that early damage and dysfunction can be initially missed.

The production of advanced glycosylation end-products affects all connective tissues containing collagen, including tendon, ligament, bone, fat pad, and muscle, as well as the nervous tissue leading to neuropathic and tissue mechanical dysfunction that disturbs gait on many levels. The whole picture of gait parameters, joint angles, and muscle power must be viewed together in assessing risk of deterioration. Diabetics tend to be slower in gait, take shorter steps and stride lengths with a wider base of support, and demonstrate a longer double support phase with much higher metabolic costs during walking than age-matched controls (Wrobel and Najafi, 2010; Petrovic et al., 2016). Such changes cause deterioration in energetics. These changes may reflect instability induced by peripheral neuropathy, damage to the vestibular, autonomic, and somatic nervous system and may also result from psychological factors of depression. Diabetic-associated depression possibly influences gait speed by up to 50%, although reduced muscle strength explains most of these changes. The changes in step width are more likely to be due to neurological degeneration, reflecting a delay in reaction time and a need for a wider support base (Wrobel and Najafi, 2010). Vertical, anterior–posterior, and mediolateral GRFs, along with maximum foot loading times are increased in diabetic patients compared to controls, and even in those with foot ulcers, the vertical and mediolateral GRFs remain high (Wrobel and Najafi, 2010). Slowing preferred gait speed is also likely a result of changing tissue mechanics.

Diabetics have more limited knee and ankle mobility, with lower plantarflexion moments and power at the ankle than do controls. Reduced ankle mobility and peak plantarflexion moment and power is significantly associated with increased glycated haemoglobin (HbA1c) (Wrobel and Najafi, 2010). Mean plantarflexion peak torques may be as low as 55% compared to matched controls, with a very significant loss in muscle strength and volume at the knee and ankle. It has been noted that there are delays in EMG responses in the thigh and leg compared to controls, with significant EMG response delays in tibialis anterior and vastus lateralis, which are muscles with important shock attenuation properties (Wrobel and Najafi, 2010). Diabetic patients seem to rely more on passive torques, depending more on intramuscular structures for strength and stiffness. However, overall muscle stiffness in diabetics does not appear to be different to healthy controls, which may reflect changes from muscle atrophy being offset by connective tissue collagen cross-linking stiffening (Wrobel and Najafi, 2010). Tendon, on the other hand, certainly does seem to stiffen (Petrovic et al., 2018). Peripheral neuropathy diabetics seem more affected than other diabetics, with a larger loss of joint mobility and with both ankle and 1st MTP joint motion reduced by around 2°–3° in neuropathic diabetics compared to other diabetics. Subtalar joint mobility loss has also been found in the presence of diabetic foot ulcers (Wrobel and Najafi, 2010) (Fig. 1.6.6a).

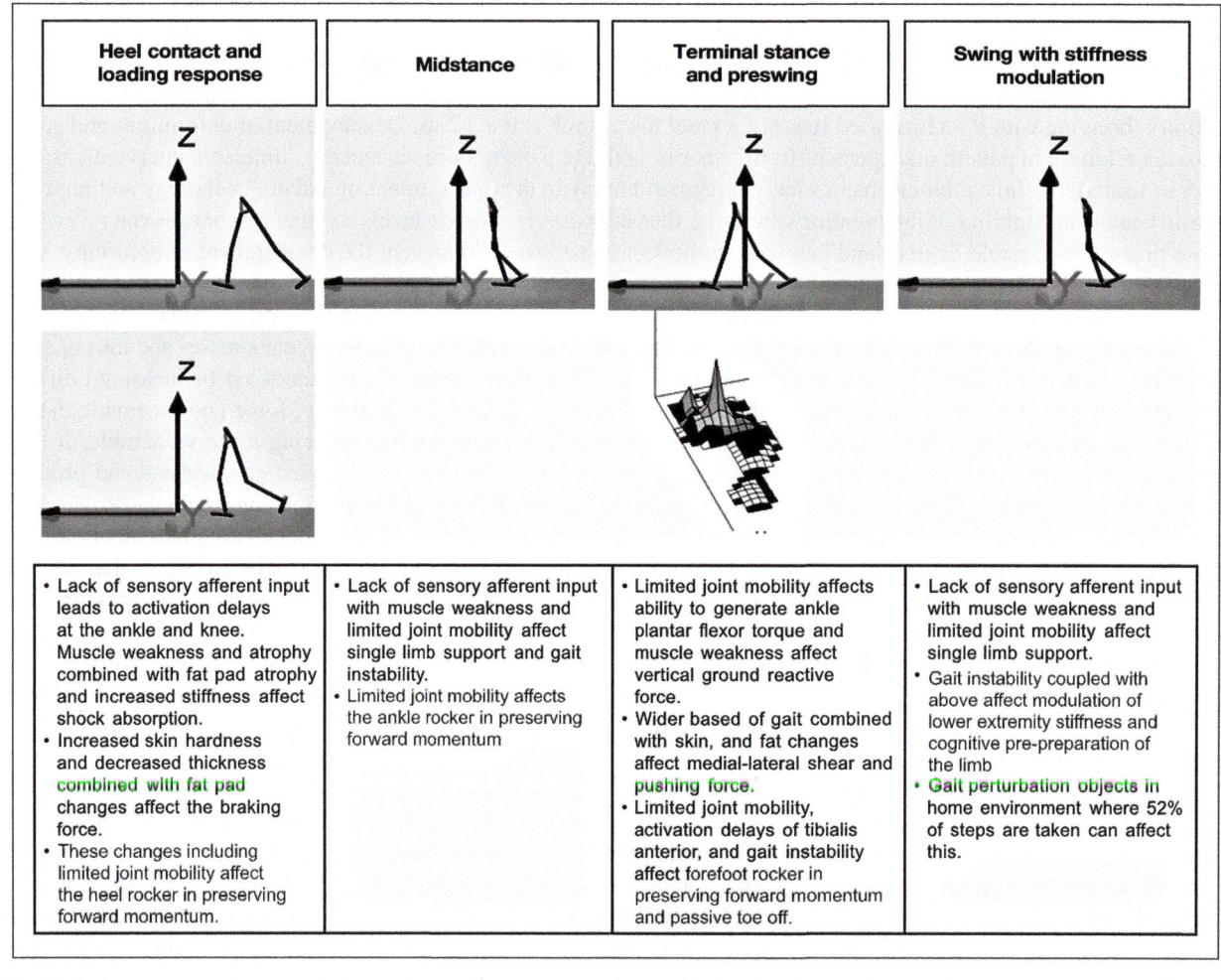

Heel contact and loading response	Midstance	Terminal stance and preswing	Swing with stiffness modulation
• Lack of sensory afferent input leads to activation delays at the ankle and knee. Muscle weakness and atrophy combined with fat pad atrophy and increased stiffness affect shock absorption. • Increased skin hardness and decreased thickness combined with fat pad changes affect the braking force. • These changes including limited joint mobility affect the heel rocker in preserving forward momentum.	• Lack of sensory afferent input with muscle weakness and limited joint mobility affect single limb support and gait instability. • Limited joint mobility affects the ankle rocker in preserving forward momentum	• Limited joint mobility affects ability to generate ankle plantar flexor torque and muscle weakness affect vertical ground reactive force. • Wider based of gait combined with skin, and fat changes affect medial-lateral shear and pushing force. • Limited joint mobility, activation delays of tibialis anterior, and gait instability affect forefoot rocker in preserving forward momentum and passive toe off.	• Lack of sensory afferent input with muscle weakness and limited joint mobility affect single limb support. • Gait instability coupled with above affect modulation of lower extremity stiffness and cognitive pre-preparation of the limb • Gait perturbation objects in home environment where 52% of steps are taken can affect this.

FIG. 1.6.6a Some of the common gait characteristic changes in persons with diabetes that can be seen in the orthogonal planes along with pressure data. The heel rocker equates to loading response, the ankle rocker to midstance, and the forefoot rocker to terminal stance. For a reminder of the heel, ankle, and forefoot rockers described here, see Fig. 1.1.8h in Section 1.1.8. *(Adapted from Wrobel, J.S., Najafi, B., 2010. Diabetic foot biomechanics and gait dysfunction. J. Diabetes Sci. Technol. 4 (4), 833–845.)*

The primary issues for gait of diabetics, particularly those with neuropathy, seem to be that they have significantly smaller moment arms to the ankle joint axis from the Achilles tendon attachment. This moment arm has been referred to as the Achilles internal moment arm (Petrovic et al., 2017), which should increase with anterior displacement of the ankle joint axis with increasing ranges of dorsiflexion. Loss of ankle joint dorsiflexion seen in diabetics may set up a cycle where such a loss limits the ability to displace the GRF further from the ankle before heel lift, shortening the external moment arm distance from the GRF to the ankle axis (Petrovic et al., 2017). Diabetics with neuropathy tend to either generate their acceleration GRF closer to the ankle joint axis or more at an angle to the ankle joint axis, both of which will reduce the external lever arm length. This depowers the ankle joint plantarflexion moment, increases the energetic costs of gait, and significantly lowers lower limb joint work and the muscular demands of walking (Petrovic et al., 2016, 2017). However, this altered leverage around the ankle offers diabetics a novel mechanism to reduce joint moments at the ankle, despite matching the walking speed of nondiabetics (Petrovic et al., 2017) (Fig. 1.6.6b). Nevertheless, this may do nothing to protect tissues from increasing stresses even with decreasing gait speeds, for loss of midstance ankle dorsiflexion creates greater risks, particularly to midfoot structures (Fernando et al., 2016; Maeshige et al., 2021).

Higher metabolic costs during walking make locomotion more difficult and fatigue more likely, resulting in negative actions during gait. Diabetics with and without neuropathy demonstrate higher metabolic costs in their walking, especially those with neuropathy (Petrovic et al., 2016). Possibly, this reflects an increasing cadence to compensate for reducing gait speed from the loss of Achilles tendon energy storage transference into acceleration kinetic plantarflexion power (Petrovic et al., 2018). Tendons and fascia are affected within diabetics, with both the plantar aponeurosis and Achilles tendon demonstrating thickening; the situation worsening with the severity of diabetes. The diabetic Achilles is also able to store less energy than the healthy Achilles because end-products of glycation make the tendon stiffer and thus less able to lengthen elastically, particularly in neuropathic diabetics, taking away the ability to stretch and store energy (Petrovic et al., 2018). The hysteresis of the tendon is also higher (greater energy loss and increased dissipation) in diabetics, particularly with neuropathy, indicating why altered Achilles tendon function is likely to increase metabolic costs (Petrovic et al., 2018). This is because the tendon is less able to store and release energy passively, requiring the muscles to compensate with greater contractile activity. The loss of midstance ankle dorsiflexion range also increases peak plantar pressures, which even simple calf stretching programmes in diabetics seems to be able to reduce (Maeshige et al., 2021).

In addition, diabetics, especially those with neuropathy, are associated with muscle atrophy (including intrinsic foot muscles and tibialis anterior), while the bone tissue density decreases (Wrobel and Najafi, 2010). The plantar skin becomes harder when tested with a durometer particularly at the sites of ulcers, with the skin becoming thicker compared to non-ulcerated areas. This likely increases pressure and shear locally, although initially, before ulceration, plantar tissue thickness reduces (Wrobel and Najafi, 2010; Naemi et al., 2016a,c). Further to this, the plantar fat pads become less thick and demonstrate fibrotic atrophy, losing their shear resistance and cushioning properties and causing significant changes in mechanical properties across the plantar soft tissues, thus reducing normal mechanical protection (Wrobel and Najafi, 2010; Naemi et al., 2016a,c). All this coupled with the loss of joint mobility, perturbation response, and muscle strength, sets up a dangerous situation for locomotive loading. Failure to prevent deformation of the foot through the presence of digital deformity, or the formation of pes cavus through motor neuropathy, or Charcot foot because of total tissue function breakdown (including bone), will lead to further dramatic kinetic and kinematic changes which will exacerbate the dangers of locomotion for the patient. Severe foot damage with neuropathy removing self-protecting mechanisms of pain and a lowered immunity means that ulcers have a high risk of finally leading to infections, extensive spreading tissue damage, and amputations. Yet, focused early rehabilitation and therapeutic interventions targeted by gait data monitoring may prevent much of this process from developing in diabetics.

1.6.7 Section summary

Gait analysis is a clinical skill developed through knowledge and experience, particularly in relation to the equipment that is available to gather information. Being able to identify data that links to specific aspects of a disease or symptoms that are of interest can be particularly difficult, especially early on in a clinician's career. Those new to gait analysis or the utilisation of a new piece of equipment will benefit greatly from working and communicating with colleagues who have more experience.

The value of clinical gait analysis in disease and pathology management is dependent on how that data can be utilised to plan and assess treatments and outcomes. Specific gait data targets can be set if pre- and posttreatment data can be obtained, and this can be most useful in long-term monitoring of mobility. If more prospective data can be gathered on a large range of diseases and pathology, then it is possible that in time, gait analysis will become part of the process of diagnosis, disease severity assessment, and injury prevention, helping to increase the health and mobility of the population.

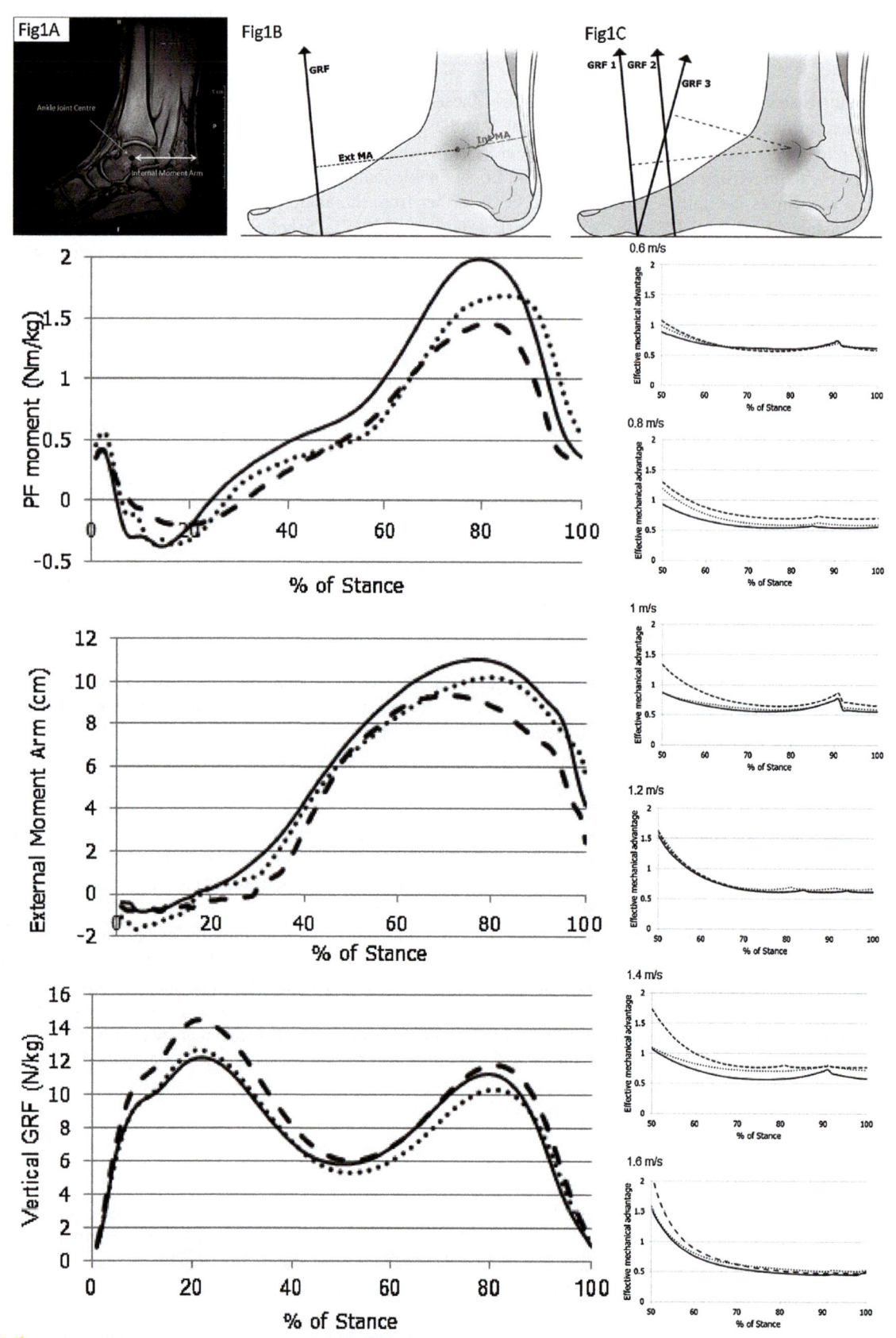

FIG. 1.6.6b The definition of the internal ankle moment arm (Int. MA) derives from the ankle joint axis of rotation (here shown as the joint centre on an MRI) to the Achilles tendon (*top left*: Fig. 1A). The external moment arm (Ext MA) is the distance from the ankle joint axis perpendicular to the resultant GRF vector, which changes during gait (Figs 1B and 1C). GRF 1 gives the longest external moment arm for the Achilles. Graphs demonstrating the ankle plantarflexion (PF) moment, Ext MA, and vertical GRF during walking at 1.4 m/s *(lower left graphs)* in healthy controls *(solid line)* in nonneuropathic diabetics *(dotted line)*, and neuropathic diabetics *(dashed line)* are compared. The mechanical advantage of late midstance and acceleration across the different subjects are compared on the lower right graphs, showing that neuropathic diabetics demonstrate the worse mechanical advantage for the application of plantarflexion power. *(From Petrovic, M., Deschamps, K., Verschueren, S.M., Bowling, F.L., Maganaris, C.N., Boulton, A.J.M., et al., 2017. Altered leverage around the ankle in people with diabetes: A natural strategy to modify the muscular contribution during walking? Gait Posture 57, 85–90.)*

Chapter summary

Gait analysis is a critical and principal skill in clinical biomechanics that necessitates a deep knowledge of the subject that will keep expanding throughout any clinician's career. The ability to describe phases and events during gait is necessary for communication between clinicians. Furthermore, the appreciation of what each event mechanically initiates helps in the understanding of normal function and energetics that can be easily conveyed back to the patient, so that they in turn can appreciate their dysfunction and/or rehabilitation needs to improve performance.

Each client group, from patients with disease to the developing child, injured athlete, or recreational runner, will have their own expectations of locomotive ability. This needs to be considered in relation to performance and expectation for each client group which, as can be seen throughout this chapter, is diverse. The clinician must be able to distinguish healthy and normal from abnormal in each client group, and via the data gathered on gait, advise what action if any is required to improve mobility and prevent injury during locomotion. The 'one-intervention-fits-all' option can never be appropriate, and with increasing complexity of disease and age, treatment options will require careful consideration focused on the dysfunctional features of the locomotion presented to the clinician.

In summary, the art and science of clinical gait analysis is the ability to associate or disassociate the biomechanical data of locomotion from the complaint, and thus identify those features that might be causing pathology and dysfunction from those that are the result of pathology and dysfunction. Finally, the data should permit the planning of therapeutic input so as to try and establish better pain-free locomotive techniques that can improve patient mobility and health, with the possibility that data can later be re-assessed to quantify or qualify improvements.

References

Aagaard, P., Suetta, C., Caserotti, P., Magnusson, S.P., Kjaer, M., 2010. Role of the nervous system in sarcopenia and muscle atrophy with aging: strength training as a countermeasure. Scand. J. Med. Sci. Sports 20 (1), 49–64.

Abe, T., Fukashiro, S., Harada, Y., Kawamoto, K., 2001. Relationship between sprint performance and muscle fascicle length in female sprinters. J. Physiol. Anthropol. Appl. Hum. Sci. 20 (2), 141–147.

Addison, O., Prior, S.J., Kundi, R., Serra, M.C., Katzel, L.I., Gardner, A.W., et al., 2018. Sarcopenia in peripheral arterial disease: prevalence and effect on functional status. Arch. Phys. Med. Rehabil. 99 (4), 623–628.

Adib, N., Davies, K., Grahame, R., Woo, P., Murray, J.K., 2005. Joint hypermobility syndrome in childhood. A not so benign multisystem disorder? Rheumatology 44 (6), 744–750.

Adolph, K.E., Vereijken, B., Shrout, P.E., 2003. What changes in infant walking and why. Child Dev. 74 (2), 475–497.

Adolph, K.E., Cole, W.G., Komati, M., Garciaguirre, J.S., Badaly, D., Lingeman, J.M., et al., 2012. How do you learn to walk? Thousands of steps and dozens of falls per day. Psychol. Sci. 23 (11), 1387–1394.

Aerts, P., Ker, R.F., De Clercq, D., Ilsley, D.W., Alexander RMcN., 1995. The mechanical properties of the human heel pad: a paradox resolved. J. Biomech. 28 (11), 1299–1308.

Alfieri, F.M., Riberto, M., Gatz, L.S., Ribeiro, C.P.C., Lopes, J.A.F., Santarém, J.M., et al., 2010. Functional mobility and balance in community-dwelling elderly submitted to multisensory versus strength exercises. Clin. Interv. Aging 5, 181–185.

Alkjær, T., Simonsen, E.B., Jørgensen, U., Dyhre-Poulsen, P., 2003. Evaluation of the walking pattern in two types of patients with anterior cruciate ligament deficiency: copers and non-copers. Eur. J. Appl. Physiol. 89 (3–4), 301–308.

Almeida, S.A., Trone, D.W., Leone, D.M., Shaffer, R.A., Patheal, S.L., Long, K., 1999. Gender differences in musculoskeletal injury rates: a function of symptom reporting? Med. Sci. Sports Exerc. 31 (12), 1807–1812.

Almonroeder, T., Willson, J.D., Kernozek, T.W., 2013. The effect of foot strike pattern on Achilles tendon load during running. Ann. Biomed. Eng. 41 (8), 1758–1766.

Altman, A.R., Davis, I.S., 2012. A kinematic method for footstrike pattern detection in barefoot and shod runners. Gait Posture 35 (2), 298–300.

Alvarez, R., Stokes, I.A., Asprinio, D.E., Trevino, S., Braun, T., 1988. Dimensional changes of the feet in pregnancy. J. Bone Joint Surg. 70-A (2), 271–274.

Alvarez, C., De Vera, M., Chhina, H., Black, A., 2008. Normative data for the dynamic pedobarographic profiles of children. Gait Posture 28 (2), 309–315.

Al-Yahya, E., Dawes, H., Smith, L., Dennis, A., Howells, K., Cockburn, J., 2011. Cognitive motor interference while walking: a systematic review and meta-analysis. Neurosci. Biobehav. Rev. 35 (3), 715–728.

Anbarian, M., Esmaeili, H., 2016. Effects of running-induced fatigue on plantar pressure distribution in novice runners with different foot types. Gait Posture 48, 52–56.

Anderson, F.C., Pandy, M.G., 2003. Individual muscle contributions to support in normal walking. Gait Posture 17 (2), 159–169.

Anselmo, D.S., Love, E., Tango, D.N., Robinson, L., 2017. Musculoskeletal effects of pregnancy on the lower extremity: a literature review. J. Am. Podiatr. Med. Assoc. 107 (1), 60–64.

Arampatzis, A., Brüggemann, G.-P., Metzler, V., 1999. The effect of speed on leg stiffness and joint kinetics in human running. J. Biomech. 32 (12), 1349–1353.

Arampatzis, A., Knicker, A., Metzler, V., Brüggemann, G.-P., 2000. Mechanical power in running: a comparison of different approaches. J. Biomech. 33 (4), 457–463.
Arellano, C.J., Kram, R., 2011. The effects of step width and arm swing on energetic cost and lateral balance during running. J. Biomech. 44 (7), 1291–1295.
Arellano, C.J., Kram, R., 2012. The energetic cost of maintaining lateral balance during human running. J. Appl. Physiol. 112 (3), 427–434.
Arellano, C.J., Kram, R., 2014. The metabolic cost of human running: is swinging the arms worth it? J. Exp. Biol. 217 (14), 2456–2461.
Arendse, R.E., Noakes, T.D., Azevedo, L.B., Romanov, N., Schwellnus, M.P., Fletcher, G., 2004. Reduced eccentric loading of the knee with the pose running method. Med. Sci. Sports Exerc. 36 (2), 272–277.
Artal, R., O'Toole, M., White, S., 2003. Guidelines of the American College of Obstetricians and Gynecologists for exercise during pregnancy and the postpartum period. Br. J. Sports Med. 37 (1), 6–12.
Assaiante, C., Woollacott, M., Amblard, B., 2000. Development of postural adjustment during gait initiation: kinematic and EMG analysis. J. Mot. Behav. 32 (3), 211–226.
Assaiante, C., Mallau, S., Viel, S., Jover, M., Schmitz, C., 2005. Development of postural control in healthy children: a functional approach. Neural Plasticity 12 (2–3), 109–118.
Baan, H., Dubbeldam, R., Nene, A.V., van de Laar, M.A.F.J., 2012. Gait analysis of the lower limb in patients with rheumatoid arthritis: a systematic review. Semin. Arthritis Rheum. 41 (6), 768–788.
Badaly, D., Adolph, K.E., 2008. Beyond the average: walking infants take steps longer than their leg length. Infant Behav. Dev. 31 (3), 554–558.
Bahl, J.S., Nelson, M.J., Taylor, M., Solomon, L.B., Arnold, J.B., Thewlis, D., 2018. Biomechanical changes and recovery of gait function after total hip arthroplasty for osteoarthritis: a systematic review and meta-analysis. Osteoarthr. Cartil. 26 (7), 847–863.
Barnes, A., Wheat, J., Milner, C.E., 2011. Fore- and rearfoot kinematics in high- and low-arched individuals during running. Foot Ankle Int. 32 (7), 710–716.
Barrett, R.S., Besier, T.F., Lloyd, D.G., 2007. Individual muscle contributions to the swing phase of gait: an EMG-based forward dynamics modelling approach. Simul. Model. Pract. Theory 15 (9), 1146–1155.
Beauchet, O., Allali, G., Berrut, G., Dubost, V., 2007. Letter to the editor: is low lower-limb kinematic variability always an index of stability? Gait Posture 26 (2), 327–328.
Beck, O.N., Golyski, P.R., Sawicki, G.S., 2020. Adding carbon fiber to shoe soles may not improve running economy: a muscle-level explanation. Sci. Rep. 10, 17154. https://doi.org/10.1038/s41598-020-74097-7.
Beighton, P., De Paepe, A., Steinmann, B., Tsipouras, P., Wenstrup, R.J., 1998. Ehlers-Danlos syndromes: revised nosology, Villefranche, 1997. Am. J. Med. Genet. 77 (1), 31–37.
Bejek, Z., Paróczai, R., Illyés, A., Kiss, R.M., 2006. The influence of walking speed on gait parameters in healthy people and in patients with osteoarthritis. Knee Surg. Sports Traumatol. Arthrosc. 14 (7), 612–622.
Beneke, R., Taylor, M.J.D., 2010. What gives Bolt the edge – A.V. Hill knew it already! J. Biomech. 43 (11), 2241–2243.
Bennell, K.L., Malcolm, S.A., Thomas, S.A., Reid, S.J., Brukner, P.D., Ebeling, P.R., et al., 1996. Risk factors for stress fractures in track and field athletes: a twelve-month prospective study. Am. J. Sports Med. 24 (6), 810–818.
Bertsch, C., Unger, H., Winkelmann, W., Rosenbaum, D., 2004. Evaluation of early walking patterns from plantar pressure distribution measurements. First year results of 42 children. Gait Posture 19 (3), 235–242.
Biewener, A.A., Farley, C.T., Roberts, T.J., Temaner, M., 2004. Muscle mechanical advantage of human walking and running: implications for energy cost. J. Appl. Physiol. 97 (6), 2266–2274.
Bisi, M.C., Stagni, R., 2015. Evaluation of toddler different strategies during the first six-months of independent walking: a longitudinal study. Gait Posture 41 (2), 574–579.
Bjelopetrovich, A., Barrios, J.A., 2016. Effects of incremental ambulatory-range loading on arch height index parameters. J. Biomech. 49 (14), 3555–3558.
Blackburn, J.T., Riemann, B.L., Padua, D.A., Guskiewicz, K.M., 2004. Sex comparison of extensibility, passive, and active stiffness of the knee flexors. Clin. Biomech. 19 (1), 36–43.
Blackburn, J.T., Padua, D.A., Weinhold, P.S., Guskiewicz, K.M., 2006. Comparison of triceps surae structural stiffness and material modulus across sex. Clin. Biomech. 21 (2), 159–167.
Board, W.J., Street, G.M., Caspers, C., 2001. A comparison of trans-tibial amputee suction and vacuum socket conditions. Prosthetics Orthot. Int. 25 (3), 202–209.
Bojsen-Møller, F., 1979. Calcaneocuboid joint and stability of the longitudinal arch of the foot at high and low gear push off. J. Anat. 129 (1), 165–176.
Boling, M., Padua, D., Marshall, S., Guskiewicz, K., Pyne, S., Beutler, A., 2010. Gender differences in the incidence and prevalence of patellofemoral pain syndrome. Scand. J. Med. Sci. Sports 20 (5), 725–730.
Bonacci, J., Saunders, P.U., Hicks, A., Rantalainen, T., Vicenzino, B.G.T., Spratford, W., 2013. Running in a minimalist and lightweight shoe is not the same as running barefoot: a biomechanical study. Br. J. Sports Med. 47 (6), 387–392.
Booth, A.T.C., Buizer, A.I., Meyns, P., Oude Lansink, I.L.B., Steenbrink, F., van der Krogt, M.M., 2018. The efficacy of functional gait training in children and young adults with cerebral palsy: a systematic review and meta-analysis. Dev. Med. Child Neurol. 60 (9), 866–883.
Borg-Stein, J., Dugan, S.A., Gruber, J., 2005. Musculoskeletal aspects of pregnancy. Am. J. Phys. Med. Rehabil. 84 (3), 180–192.
Bosch, K., Gerss, J., Rosenbaum, D., 2007. Preliminary normative values for foot loading parameters of the developing child. Gait Posture 26 (2), 238–247.
Boulet, S., Boudot, E., Houel, N., 2016. Relationships between each part of the spinal curves and upright posture using multiple stepwise linear regression analysis. J. Biomech. 49 (7), 1149–1155.

Boyer, E.R., Rooney, B.D., Derrick, T.R., 2014. Rearfoot and midfoot or forefoot impacts in habitually shod runners. Med. Sci. Sports Exerc. 46 (7), 1384–1391.
Bramah, C., Preece, S.J., Gill, N., Herrington, L., 2018. Is there a pathological gait associated with common soft tissue running injuries? Am. J. Sports Med. 46 (12), 3023–3031.
Bramble, D.M., Lieberman, D.E., 2004. Endurance running and the evolution of Homo. Nature 432 (7015), 345–352.
Branco, M., Santos-Rocha, R., Vieira, F., 2014. Biomechanics of gait during pregnancy. Sci. World J. 2014, 527940. https://doi.org/10.1155/2014/527940.
Bravo-Aguilar, M., Gijón-Noguerón, G., Luque-Suarez, A., Abian-Vicen, J., 2016. The influence of running on foot posture and in-shoe plantar pressures. J. Am. Podiatr. Med. Assoc. 106 (2), 109–115.
Briemberg, H.R., 2007. Neuromuscular diseases in pregnancy. Semin. Neurol. 27 (5), 460–466.
Brodie, M.A.D., Beijer, T.R., Canning, C.G., Lord, S.R., 2015. Head and pelvis stride-to-stride oscillations in gait: validation and interpretation of measurements from wearable accelerometers. Physiol. Meas. 36 (5), 857–872.
Brodsky, J.W., Baum, B.S., Pollo, F.E., Mehta, H., 2007. Prospective gait analysis in patients with first metatarsophalangeal joint arthrodesis for hallux rigidus. Foot Ankle Int. 28 (2), 162–165.
Browning, R.C., Kram, R., 2005. Energetic cost of preferred speed of walking in obese vs. normal weight women. Obes. Res. 13 (5), 891–899.
Browning, R.C., Baker, E.A., Herron, J.A., Kram, R., 2006. Effects of obesity and sex on the energetic cost and preferred speed of walking. J. Appl. Physiol. 100 (2), 390–398.
Bruijn, S.M., Meijer, O.G., van Dieën, J.H., Kingma, I., Lamoth, C.J.C., 2008. Coordination of leg swing, thorax rotations, and pelvis rotations during gait: the organisation of total body angular momentum. Gait Posture 27 (3), 455–462.
Bruijn, S.M., Meijer, O.G., Beek, P.J., van Dieën, J.H., 2013. Assessing the stability of human locomotion: a review of current measures. J. R. Soc. Interface 10 (83), 20120999. https://doi.org/10.1098/rsif.2012.0999.
Brumagne, S., Janssens, L., Knapen, S., Claeys, K., Suuden-Johanson, E., 2008. Persons with recurrent low back pain exhibit a rigid postural control strategy. Eur. Spine J. 17 (9), 1177–1184.
Buldt, A.K., Murley, G.S., Butterworth, P., Levinger, P., Menz, H.B., Landorf, K.B., 2013. The relationship between foot posture and lower limb kinematics during walking: a systematic review. Gait Posture 38 (3), 363–372.
Buldt, A.K., Forghany, S., Landorf, K.B., Levinger, P., Murley, G.S., Menz, H.B., 2018. Foot posture is associated with plantar pressure during gait: a comparison of normal, planus and cavus feet. Gait Posture 62, 235–240.
Burns, J., Crosbie, J., Hunt, A., Ouvrier, R., 2005. The effect of pes cavus on foot pain and plantar pressure. Clin. Biomech. 20 (9), 877–882.
Butler, R.J., Crowell, H.P., Davis, I.M., 2003. Lower extremity stiffness: implications for performance and injury. Clin. Biomech. 18 (6), 511–517.
Butler, R.J., Davis, I.S., Hamill, J., 2006. Interaction of arch type and footwear on running mechanics. Am. J. Sports Med. 34 (12), 1998–2005.
Butler, R.J., Hamill, J., Davis, I., 2007. Effect of footwear on high and low arched runners' mechanics during a prolonged run. Gait Posture 26 (2), 219–225.
Byrne, C.A., O'Keeffe, D.T., Donnelly, A.E., Lyons, G.M., 2007. Effect of walking speed changes on tibialis anterior EMG during healthy gait for FES envelope design in drop foot correction. J. Electromyogr. Kinesiol. 17 (5), 605–616.
Calhoun, M., Longworth, M., Chester, V.L., 2011. Gait patterns in children with autism. Clin. Biomech. 26 (2), 200–206.
Cali, C.M., Kiel, D.P., 1995. An epidemiologic study of fall-related fractures among institutionalized older people. J. Am. Geriatr. Soc. 43 (12), 1336–1340.
Callaghan, M.J., Whitehouse, S.J., Baltzopoulos, V., Samarji, R.A., 2011. A comparison of the effects of first metatarsophalangeal joint arthrodesis and hemiarthroplasty on function of foot forces using gait analysis. Foot Ankle Online Jl. 4 (12), 1. https://doi.org/10.3827/faoj.2011.0412.0001.
Callisaya, M.L., Blizzard, L., Schmidt, M.D., McGinley, J.L., Srikanth, V.K., 2010. Ageing and gait variability – a population-based study of older people. Age Ageing 39 (2), 191–197.
Cappellini, G., Ivanenko, Y.P., Poppele, R.E., Lacquaniti, F., 2006. Motor patterns in human walking and running. J. Neurophysiol. 95 (6), 3426–3437.
Caravaggi, P., Pataky, T., Goulermas, J.Y., Savage, R., Crompton, R., 2009. A dynamic model of the windlass mechanism of the foot: evidence for early stance phase preloading of the plantar aponeurosis. J. Exp. Biol. 212 (15), 2491–2499.
Caravaggi, P., Sforza, C., Leardini, A., Portinaro, N., Panou, A., 2018. Effect of plano-valgus foot posture on midfoot kinematics during barefoot walking in an adolescent population. J. Foot Ankle Res. 11, 55. https://doi.org/10.1186/s13047-018-0297-7.
Carey, T.S., Crompton, R.H., 2005. The metabolic costs of 'bent-hip, bent-knee' walking in humans. J. Hum. Evol. 48 (1), 25–44.
Carlson, M., Wilkerson, J., 2007. Are differences in leg length predictive of lateral patello-femoral pain? Physiother. Res. Int. 12 (1), 29–38.
Carpentier, J., Benallegue, M., Laumond, J.-P., 2017. On the centre of mass motion in human walking. Int. J. Autom. Comput. 14 (5), 542–551.
Carrier, D.R., 1984. The energetic paradox of human running and hominid evolution [and comments and reply]. Curr. Anthropol. 25 (4), 483–495.
Carroll, M., Parmar, P., Dalbeth, N., Boocock, M., Rome, K., 2015. Gait characteristics associated with the foot and ankle in inflammatory arthritis: a systematic review and meta-analysis. BMC Musculoskelet. Disord. 16, 134. https://doi.org/10.1186/s12891-015-0596-0.
Carvalho, D.C.L., Cliquet Jr., A., 2005. Response of the arterial blood pressure of quadriplegic patients to treadmill gait training. Braz. J. Med. Biol. Res. 38 (9), 1367–1373.
Cauthon, D.J., Langer, P., Coniglione, T.C., 2013. Minimalist shoe injuries: three case reports. Foot 23 (2-3), 100–103.
Cavagna, G.A., Willems, P.A., Heglund, N.C., 2000. The role of gravity in human walking: pendular energy exchange, external work and optimal speed. J. Physiol. 528 (3), 657–668.
Cavagna, G.A., Legramandi, M.A., Peyré-Tartaruga, L.A., 2008. Old men running: mechanical work and elastic bounce. Proc. R. Soc. B Biol. Sci. 275 (1633), 411–418.
Cavanagh, P.R., 1987. The biomechanics of lower extremity action in distance running. Foot Ankle 7 (4), 197–217.

Cavanagh, P.R., Lafortune, M.A., 1980. Ground reaction forces in distance running. J. Biomech. 13 (5), 397–406.

Celis, R., Pipinos, I.I., Scott-Pandorf, M.M., Myers, S.A., Stergiou, N., Johanning, J.M., 2009. Peripheral arterial disease affects kinematics during walking. J. Vasc. Surg. 49 (1), 127–132.

Cen, X., Xu, D., Baker, J.S., Gu, Y., 2020. Association of arch stiffness with plantar impulse distribution during walking, running, and gait termination. Int. J. Environ. Res. Public Health 17 (6), 2090. https://doi.org/10.3390/ijerph17062090.

Chan, Z.Y.S., Au, I.P.H., Lau, F.O.Y., Ching, E.C.K., Zhang, J.H., Cheung, R.T.H., 2018. Does maximalist footwear lower impact loading during level ground and downhill running? Eur. J. Sport Sci. 18 (8), 1083–1089.

Chang, Y.-H., Kram, R., 1999. Metabolic cost of generating horizontal forces during human running. J. Appl. Physiol. 86 (5), 1657–1662.

Chang, C.-L., Kubo, M., Buzzi, U., Ulrich, B., 2006. Early changes in muscle activation patterns of toddlers during walking. Infant Behav. Dev. 29 (2), 175–188.

Chastan, N., Westby, G.W.M., du Montcel, S.T., Do, M.C., Chong, R.K., Agid, Y., et al., 2010. Influence of sensory inputs and motor demands on the control of the centre of mass velocity during gait initiation in humans. Neurosci. Lett. 469 (3), 400–404.

Chehab, E.F., Favre, J., Erhart-Hledik, J.C., Andriacchi, T.P., 2014. Baseline knee adduction and flexion moments during walking are both associated with 5 year cartilage changes in patients with medial knee osteoarthritis. Osteoarthr. Cartil. 22 (11), 1833–1839.

Chen, S.-J., Pipinos, I., Johanning, J., Radovic, M., Huisinga, J.M., Myers, S.A., et al., 2008. Bilateral claudication results in alterations in the gait biomechanics at the hip and ankle joints. J. Biomech. 41 (11), 2506–2514.

Chen, C.-H., Yang, W.-W., Chen, Y.-P., Chen, V.C.-F., Liu, C., Shiang, T.-Y., 2021. High vibration frequency of soft tissue occurs during gait in power-trained athletes. J. Sports Sci. 39 (4), 439–445.

Chester, V.L., Wrigley, A.T., 2008. The identification of age-related differences in kinetic gait parameters using principal component analysis. Clin. Biomech. 23 (2), 212–220.

Chester, V.L., Tingley, M., Biden, E.N., 2006. A comparison of kinetic gait parameters for 3-13 year olds. Clin. Biomech. 21 (7), 726–732.

Chevalier, T.L., Hodgins, H., Chockalingam, N., 2010. Plantar pressure measurements using an in-shoe system and a pressure platform: a comparison. Gait Posture 31 (3), 397–399.

Chiu, M.-C., Wu, H.-C., Chang, L.-Y., 2013. Gait speed and gender effects on center of pressure progression during normal walking. Gait Posture 37 (1), 43–48.

Chockalingam, N., Chatterley, F., Healy, A.C., Greenhalgh, A., Branthwaite, H.R., 2012. Comparison of pelvic complex kinematics during treadmill and overground walking. Arch. Phys. Med. Rehabil. 93 (12), 2302–2308.

Chu, M.L., Yazdani-Ardakani, S., Gradisar, I.A., Askew, M.J., 1986. An in vitro simulation of impulsive force transmission along the lower skeletal extremity. J. Biomech. 19 (12), 979–981. 983–987.

Chuckpaiwong, B., Nunley, J.A., Mall, N.A., Queen, R.M., 2008. The effect of foot type on in-shoe plantar pressure during walking and running. Gait Posture 28 (3), 405–411.

Chumanov, E.S., Wall-Scheffler, C., Heiderscheit, B.C., 2008. Gender differences in walking and running on level and inclined surfaces. Clin. Biomech. 23 (10), 1260–1268.

Clarke, T.E., Frederick, E.C., Hamill, C.L., 1983. The effects of shoe design parameters on rearfoot control in running. Med. Sci. Sports Exerc. 15 (5), 376–381.

Clemens, S., Kim, K.J., Gailey, R., Kirk-Sanchez, N., Kristal, A., Gaunaurd, I., 2020. Inertial sensor-based measures of gait symmetry and repeatability in people with unilateral lower limb amputation. Clin. Biomech. 72, 102–107.

Cole, W.G., Robinson, S.R., Adolph, K.E., 2016. Bouts of steps: the organization of infant exploration. Dev. Psychobiol. 58 (3), 341–354.

Collins, S., Ruina, A., Tedrake, R., Wisse, M., 2005. Efficient bipedal robots based on passive-dynamic walkers. Science 307 (5712), 1082–1085.

Conforto, S., Schmid, M., Camomilla, V., D'Alessio, T., Cappozzo, A., 2001. Hemodynamics as a possible internal mechanical disturbance to balance. Gait Posture 14 (1), 28–35.

Constantinou, M., Barrett, R., Brown, M., Mills, P., 2014. Spatial-temporal gait characteristics in individuals with hip osteoarthritis: a systematic literature review and meta-analysis. J. Orthop. Sports Phys. Ther. 44 (4), 291–303. B1–B7.

Cook, S.M.C., Cameron, S.T., 2015. Social issues of teenage pregnancy. Obstet. Gynaecol. Reprod. Med. 25 (9), 243–248.

Cooperstein, R., Lew, M., 2009. The relationship between pelvic torsion and anatomical leg length inequality: a review of the literature. J. Chiropract. Med. 8 (3), 107–118.

Costill, D.L., Winrow, E., 1970. A comparison of two middle-aged ultramarathon runners. Res. Quart. 41 (2), 135–139.

Coulthard, P., Pleuvry, B.J., Brewster, M., Wilson, K.L., Macfarlane, T.V., 2002. Gait analysis as an objective measure in a chronic pain model. J. Neurosci. Methods 116 (2), 197–213.

Coulthard, P., Simjee, S.U., Pleuvry, B.J., 2003. Gait analysis as a correlate of pain induced by carrageenan intraplantar injection. J. Neurosci. Methods 128 (1–2), 95–102.

Cousins, S.D., Morrison, S.C., Drechsler, W.I., 2013. Foot loading pattens in normal weight, overweight and obese children aged 7 to 11 years. J. Foot Ankle Res. 6, 36. https://doi.org/10.1186/1757-1146-6-36.

Cowley, E., Marsden, J., 2013. The effects of prolonged running on foot posture: a repeated measures study of half marathon runners using the foot posture index and navicular height. J. Foot Ankle Res. 6, 20. https://doi.org/10.1186/1757-1146-6-20.

Cronin, N.J., Finni, T., 2013. Treadmill versus overground and barefoot versus shod comparisons of triceps surae fascicle behaviour in human walking and running. Gait Posture 38 (3), 528–533.

Crowe, A., Samson, M.M., 1997. 3-D analysis of gait: the effects upon symmetry of carrying a load in one hand. Hum. Mov. Sci. 16 (2-3), 357–365.

Crowe, A., Schiereck, P., de Boer, R., Keessen, W., 1993. Characterization of gait of young adult females by means of body centre of mass oscillations derived from ground reaction forces. Gait Posture 1 (1), 61–68.

Crowe, A., Samson, M.M., Hoitsma, M.J., van Ginkel, A.A., 1996. The influence of walking speed on parameters of gait symmetry determined from ground reaction forces. Hum. Mov. Sci. 15 (3), 347–367.

Crowther, R.G., Spinks, W.L., Leicht, A.S., Quigley, F., Golledge, J., 2007. Relationship between temporal-spatial gait parameters, gait kinematics, walking performance, exercise capacity, and physical activity level in peripheral arterial disease. J. Vasc. Surg. 45 (6), 1172–1178.

Cunningham, C.B., Schilling, N., Anders, C., Carrier, D.R., 2010. The influence of foot posture on the cost of transport in humans. J. Exp. Biol. 213 (5), 790–797.

D'Août, K., Pataky, T.C., De Clercq, D., Aerts, P., 2009. The effects of habitual footwear use: foot shape and function in native barefoot walkers. Footwear Sci. 1 (2), 81–94.

Daley, M.A., Biewener, A.A., 2006. Running over rough terrain reveals limb control for intrinsic stability. Proc. Natl. Acad. Sci. U. S. A. 103 (42), 15681–15686.

Daley, M.J., Spinks, W.L., 2000. Exercise, mobility and aging. Sports Med. 29 (1), 1–12.

Dalgleish, R., 1997. The human type I collagen mutation database. Nucleic Acids Res. 25 (1), 181–187.

Dames, K.D., Smith, J.D., 2015. Effects of load carriage and footwear on spatiotemporal parameters, kinematics, and metabolic cost of walking. Gait Posture 42 (2), 122–126.

Daoud, A.I., Geissler, G.J., Wang, F., Saretsky, J., Daoud, Y.A., Lieberman, D.E., 2012. Foot strike and injury rates in endurance runners: a retrospective study. Med. Sci. Sports Exerc. 44 (7), 1325–1334.

Darter, B.J., Wilken, J.M., 2011. Gait training with virtual reality-based real-time feedback: improving gait performance following transfemoral amputation. Phys. Ther. 91 (9), 1385–1394.

Dault, M.C., de Haart, M., Geurts, A.C.H., Arts, I.M.P., Nienhuis, B., 2003. Effects of visual center of pressure feedback on postural control in young and elderly healthy adults and in stroke patients. Hum. Mov. Sci. 22 (3), 221–236.

Davies, K., Black, A., Hunt, M., Holsti, L., 2018. Long-term gait outcomes following conservative management of idiopathic toe walking. Gait Posture 62, 214–219.

Day, E., Hahn, M., 2020. Optimal footwear longitudinal bending stiffness to improve running economy is speed dependent. Footwear Sci. 12 (1), 3–13.

De Carvalho, D.C.L., Martins, C.L., Cardoso, S.D., Cliquet, A., 2006. Improvement of metabolic and cardiorespiratory responses through treadmill gait training with neuromuscular electrical stimulation in quadriplegic subjects. Artif. Organs 30 (1), 56–63.

De Cock, A., Vanrenterghem, J., Willems, T., Witvrouw, E., De Clercq, D., 2008. The trajectory of the centre of pressure during barefoot running as a potential measure for foot function. Gait Posture 27 (4), 669–675.

de Graaf, M.L., Hubert, J., Houdijk, H., Bruijn, S.M., 2019. Influence of arm swing on cost of transport during walking. Biol. Open 8 (6), bio039263. https://doi.org/10.1242/bio.039263.

de Laat, K.F., van Norden, A.G.W., Gons, R.A.R., van Oudheusden, L.J.B., van Uden, I.W.M., Bloem, B.R., et al., 2010. Gait in elderly with cerebral small vessel disease. Stroke 41 (8), 1652–1658.

de Laat, K.F., Reid, A.T., Grim, D.C., Evans, A.C., Kötter, R., van Norden, A.G.W., et al., 2012. Cortical thickness is associated with gait disturbances in cerebral small vessel disease. NeuroImage 59 (2), 1478–1484.

De Wit, B., De Clercq, D., Aerts, P., 2000. Biomechanical analysis of the stance phase during barefoot and shod running. J. Biomech. 33 (3), 269–278.

Debi, R., Mor, A., Segal, G., Segal, O., Agar, G., Debbi, E., et al., 2011. Correlation between single limb support phase and self-evaluation questionnaires in knee osteoarthritis populations. Disabil. Rehabil. 33 (13–14), 1103–1109.

Decker, M.J., Torry, M.R., Wyland, D.J., Sterett, W.I., Steadman, J.R., 2003. Gender differences in lower extremity kinematics, kinetics and energy absorption during landing. Clin. Biomech. 18 (7), 662–669.

Delgado, T.L., Kubera-Shelton, E., Robb, R.R., Hickman, R., Wallmann, H.W., Dufek, J.S., 2013. Effects of foot strike on low back posture, shock attenuation, and comfort in running. Med. Sci. Sports Exerc. 45 (3), 490–496.

den Otter, A.R., Geurts, A.C.H., Mulder, T., Duysens, J., 2004. Speed related changes in muscle activity from normal to very slow walking speeds. Gait Posture 19 (3), 270–278.

Derrick, T.R., Hamill, J., Caldwell, G.E., 1998. Energy absorption of impacts during running at various stride lengths. Med. Sci. Sports Exerc. 30 (1), 128–135.

Desai, S.N., Grierson, R., Manoli, A., 2010. The cavus foot in athletes: fundamentals of examination and treatment. Oper. Tech. Sports Med. 18 (1), 27–33.

Desai, A.S., Dramis, A., Board, T.N., 2013. Leg length discrepancy after total hip arthroplasty: a review of literature. Curr. Rev. Musculoskelet. Med. 6 (4), 336–341.

DeSilva, J.M., 2010. Revisiting the "midtarsal break". Am. J. Phys. Anthropol. 141 (2), 245–258.

DeSilva, J.M., Gill, S.V., 2013. Brief communication: a midtarsal (midfoot) break in the human foot. Am. J. Phys. Anthropol. 151 (3), 495–499.

DeSilva, J.M., Bonne-Annee, R., Swanson, Z., Gill, C.M., Sobel, M., Uy, J., et al., 2015. Midtarsal break variation in modern humans: functional causes, skeletal correlates, and paleontological implications. Am. J. Phys. Anthropol. 156 (4), 543–552.

Detrembleur, C., Willems, P., Plaghki, L., 1997. Does walking speed influence the time pattern of muscle activation in normal children? Dev. Med. Child Neurol. 39 (12), 803–807.

Detrembleur, C., van den Hecke, A., Dierick, F., 2000. Motion of the body centre of gravity as a summary indicator of the mechanics of human pathological gait. Gait Posture 12 (3), 243–250.

DeVita, P., Helseth, J., Hortobágyi, T., 2007. Muscles do more positive than negative work in human locomotion. J. Exp. Biol. 210 (19), 3361–3373.

Dhawale, N., Mandre, S., Venkadesan, M., 2019. Dynamics and stability of running on rough terrains. R. Soc. Open Sci. 6 (3), 181729. https://doi.org/10.1098/rsos.181729.

Dhital, A., Pey, T., Stanford, M.R., 2010. Visual loss and falls: a review. Eye 24 (9), 1437–1446.

Dickinson, M.H., Farley, C.T., Full, R.J., Koehl, M.A.R., Kram, R., Lehman, S., 2000. How animals move. An integrative view. Science 288 (5463), 100–106.

Diebal, A.R., Gregory, R., Alitz, C., Gerber, J.P., 2012. Forefoot running improves pain and disability associated with chronic exertional compartment syndrome. Am. J. Sports Med. 40 (5), 1060–1067.

Dierick, F., Lefebvre, C., van den Hecke, A., Detrembleur, C., 2004. Development of displacement of centre of mass during independent walking in children. Dev. Med. Child Neurol. 46 (8), 533–539.

Dierks, T.A., Manal, K.T., Hamill, J., Davis, I.S., 2008. Proximal and distal influences on hip and knee kinematics in runners with patellofemoral pain during a prolonged run. J. Orthop. Sports Phys. Ther. 38 (8), 448–456.

Dierks, T.A., Davis, I.S., Hamill, J., 2010. The effects of running in an exerted state on lower extremity kinematics and joint timing. J. Biomech. 43 (15), 2993–2998.

Ding, Z., Jarvis, H.L., Bennett, A.N., Baker, R., Bull, A.M.J., 2021. Higher knee contact forces might underlie increased osteoarthritis rates in high functioning amputees: a pilot study. J. Orthop. Res. 39 (4), 850–860.

Diop, M., Rahmani, A., Belli, A., Gautheron, V., Geyssant, A., Cottalorda, J., 2005. Influence of speed variation and age on ground reaction forces and stride parameters of children's normal gait. Int. J. Sports Med. 26 (8), 682–687.

Divert, C., Baur, H., Mornieux, G., Mayer, F., Belli, A., 2005a. Stiffness adaptations in shod running. J. Appl. Biomech. 21 (4), 311–321.

Divert, C., Mornieux, G., Baur, H., Mayer, F., Belli, A., 2005b. Mechanical comparison of barefoot and shod running. Int. J. Sports Med. 26 (7), 593–598.

Divert, C., Mornieux, G., Freychat, P., Baly, L., Mayer, F., Belli, A., 2008. Barefoot-shod running differences: shoe or mass effect? Int. J. Sports Med. 29 (6), 512–518.

Dobson, F., Morris, M.E., Baker, R., Graham, H.K., 2007. Gait classification in children with cerebral palsy: a systematic review. Gait Posture 25 (1), 140–152.

Dominici, N., Ivanenko, Y.P., Lacquaniti, F., 2007. Control of trajectory in walking toddlers: adaptation to load changes. J. Neurophysiol. 97 (4), 2790–2801.

Donelan, J.M., Kram, R., Kuo, A.D., 2002. Mechanical work for step-to-step transitions is a major determinant of the metabolic cost of human walking. J. Exp. Biol. 205 (23), 3717–3727.

Dowling, A.M., Steel, J.R., Baur, L.A., 2001. Does obesity influence foot structure and plantar pressure patterns in prepubescent children? Int. J. Obes. Relat. Metab. Disord. 25 (6), 845–852.

Dowling, A.M., Steele, J.R., Baur, L.A., 2004. What are the effects of obesity in children on plantar pressure distributions? Int. J. Obes. Relat. Metab. Disord. 28 (11), 1514–1519.

Drewes, L.K., McKeon, P.O., Paolini, G., Riley, P., Kerrigan, D.C., Ingersoll, C.D., et al., 2009a. Altered ankle kinematics and shank-rear-foot coupling in those with chronic ankle instability. J. Sport Rehabil. 18 (3), 375–388.

Drewes, L.K., McKeon, P.O., Kerrigan, D.C., Hertel, J., 2009b. Dorsiflexion deficit during jogging with chronic ankle instability. J. Sci. Med. Sport 12 (6), 685–687.

Dubost, V., Kressig, R.W., Gonthier, R., Herrmann, F.R., Aminian, K., Najafi, B., et al., 2006. Relationships between dual-task related changes in stride velocity and stride time variability in healthy older adults. Hum. Mov. Sci. 25 (3), 372–382.

Dugan, S.A., Bhat, K.P., 2005. Biomechanics and analysis of running gait. Phys. Med. Rehabil. Clin. N. Am. 16 (3), 603–621.

Dusing, S.C., Thorpe, D.E., 2007. A normative sample of temporal and spatial gait parameters in children using the GAITRite electronic walkway. Gait Posture 25 (1), 135–139.

Earl, J.E., Hoch, A.Z., 2011. A proximal strengthening program improves pain, function, and biomechanics in women with patellofemoral pain syndrome. Am. J. Sports Med. 39 (1), 154–163.

Eccles, J.A., Beacher, F.D.C., Gray, M.A., Jones, C.L., Minati, L., Harrison, N.A., et al., 2012. Brain structure and joint hypermobility: relevance to the expression of psychiatric symptoms. Br. J. Psychiatry 200 (6), 508–509.

Edwards, M.R., Jack, C., Singh, S.K., 2008. Tibialis posterior dysfunction. Curr. Orthop. 22 (3), 185–192.

Edwards, W.B., Taylor, D., Rudolphi, T.J., Gillette, J.C., Derrick, T.R., 2009. Effects of stride length and running mileage on a probabilistic stress fracture model. Med. Sci. Sports Exerc. 41 (12), 2177–2184.

Edwards, W.B., Derrick, T.R., Hamill, J., 2012. Musculoskeletal attenuation of impact shock in response to knee angle manipulation. J. Appl. Biomech. 28 (5), 502–510.

Egbuonu, M.E., Cavanagh, P.R., Miller, T.A., 1990. Degradation of running economy through changes in running mechanics. [Abstract No. 100]. Med. Sci. Sports Exerc. 22 (2 Suppl), S17.

Ekizos, A., Santuz, A., Arampatzis, A., 2017. Transition from shod to barefoot alters dynamic stability during running. Gait Posture 56, 31–36.

Elbaz, A., Mor, A., Segal, O., Agar, G., Halperin, N., Haim, A., et al., 2012. Can single limb support objectively assess the functional severity of knee osteoarthritis? Knee 19 (1), 32–35.

Elbaz, A., Mor, A., Segal, G., Debi, R., Shazar, N., Herman, A., 2014. Novel classification of knee osteoarthritis severity based on spatiotemporal gait analysis. Osteoarthr. Cartil. 22 (3), 457–463.

Ellis, R.G., Sumner, B.J., Kram, R., 2014. Muscle contributions to propulsion and braking during walking and running: Insight from external force perturbations. Gait Posture 40 (4), 594–599.

El-Metwally, A., Salminen, J.J., Auvinen, A., Macfarlane, G., Mikkelsson, M., 2007. Risk factors for development of non-specific musculoskeletal pain in preteens and early adolescents: a prospective 1-year follow-up study. BMC Musculoskelet. Disord. 8, 46. https://doi.org/10.1186/1471-2474-8-46.

el-Shahaly, H.A., el-Sherif, A.K., 1991. Is the benign joint hypermobility syndrome benign? Clin. Rheumatol. 10 (3), 302–307.

El-shamy, F.F., Ribeiro, A.P., Abo Gazia, A.A., 2019. Effectiveness of proprioceptive training on dynamic postural balance during pregnancy: a randomized controlled trial. Physiother. Pract. Res. 40 (1), 77–85.

Enoka, R.M., Stuart, D.G., 1992. Neurobiology of muscle fatigue. J. Appl. Physiol. 72 (5), 1631–1648.

Eslami, M., Damavandi, M., Ferber, R., 2014. Association of navicular drop and selected lower-limb biomechanical measures during the stance phase of running. J. Appl. Biomech. 30 (2), 250–254.

Evans, A.M., 2008. The flat-footed child - to treat or not to treat: what is the clinician to do? J. Am. Podiatr. Med. Assoc. 98 (5), 386–393.

Evans, A.M., Rome, K., Peet, L., 2012. The foot posture index, ankle lunge test, Beighton scale and the lower limb assessment score in healthy children: a reliability study. J. Foot Ankle Res. 5, 1. https://doi.org/10.1186/1757-1146-5-1.

Farley, C.T., González, O., 1996. Leg stiffness and stride frequency in human running. J. Biomech. 29 (2), 181–186.

Farley, C.T., McMahon, T.A., 1992. Energetics of walking and running: insights from simulated reduced-gravity experiments. J. Appl. Physiol. 73 (6), 2709–2712.

Farris, D.J., Kelly, L.A., Cresswell, A.G., Lichtwark, G.A., 2019. The functional importance of human foot muscles for bipedal locomotion. Proc. Natl. Acad. Sci. U. S. A. 116 (5), 1645–1650.

Farris, D.J., Birch, J., Kelly, L., 2020. Foot stiffening during the push-off phase of human walking is linked to active muscle contraction, and not the windlass mechanism. J. R. Soc. Interface 17 (168), 20200208. https://doi.org/10.1098/rsif.2020.0208.

Ferber, R., Davis, I.M., Williams 3rd., D.S., 2003. Gender differences in lower extremity mechanics during running. Clin. Biomech. 18 (4), 350–357.

Ferber, R., Kendall, K.D., Farr, L., 2011. Changes in knee biomechanics after a hip-abductor strengthening protocol for runners with patellofemoral pain syndrome. J. Athl. Train. 46 (2), 142–149.

Fernández-Seguín, L.M., Diaz Mancha, J.A., Sánchez Rodríguez, R., Escamilla Martínez, E., Gómez Martín, B., Ramos Ortega, J., 2014. Comparison of plantar pressures and contact area between normal and cavus foot. Gait Posture 39 (2), 789–792.

Fernando, M.E., Crowther, R.G., Lazzarini, P.A., Sangla, K.S., Buttner, P., Golledge, J., 2016. Gait parameters of people with diabetes-related neuropathic plantar foot ulcers. Clin. Biomech. 37, 98–107.

Ferrari, J., Parslow, C., Lim, E., Hayward, A., 2005. Joint hypermobility: the use of a new assessment tool to measure lower limb hypermobility. Clin. Exp. Rheumatol. 23 (3), 413–420.

Ferris, L., Sharkey, N.A., Smith, T.S., Matthews, D.K., 1995. Influence of extrinsic plantar flexors on forefoot loading during heel rise. Foot Ankle Int. 16 (8), 464–473.

Firminger, C.R., Edwards, W.B., 2016. The influence of minimalist footwear and stride length reduction on lower-extremity running mechanics and cumulative loading. J. Sci. Med. Sport 19 (12), 975–979.

Fishco, W.D., Cornwall, M.W., 2004. Gait analysis after talonavicular joint fusion: 2 case reports. J. Foot Ankle Surg. 43 (4), 241–247.

Fogel, G.R., Katoh, Y., Rand, J.A., Chao, E.Y.S., 1982. Talonavicular arthrodesis for isolated arthrosis: 9.5-year results and gait analysis. Foot Ankle 3 (2), 105–113.

Foroughi, N., Smith, R.M., Lange, A.K., Baker, M.K., Fiatarone Singh, M.A., Vanwanseele, B., 2010. Dynamic alignment and its association with knee adduction moment in medial knee osteoarthritis. Knee 17 (3), 210–216.

Forrester, S.E., Townend, J., 2015. The effect of running velocity on footstrike angle – a curve-clustering approach. Gait Posture 41 (1), 26–32.

Foster, A.D., Block, B., Capobianco 3rd, F., Peabody, J.T., Puleo, N.A., Vegas, A., et al., 2021. Shorter heels are linked with greater elastic energy storage in the Achilles tendon. Sci. Rep. 11, 9360. https://doi.org/10.1038/s41598-021-88774-8.

Foti, T., Davids, J.R., Bagley, A., 2000. A biomechanical analysis of gait during pregnancy. J. Bone Joint Surg. 82-A (5), 625–632.

Fox, M.D., Delp, S.L., 2010. Contributions of muscles and passive dynamics to swing initiation over a range of walking speeds. J. Biomech. 43 (8), 1450–1455.

Francis, H., March, L., Terenty, T., Webb, J., 1987. Benign joint hypermobility with neuropathy: documentation and mechanism of tarsal tunnel syndrome. J. Rheumatol. 14 (3), 577–581.

Franz, J.R., Kram, R., 2012. The effects of grade and speed on leg muscle activations during walking. Gait Posture 35 (1), 143–147.

Franz, J.R., Kram, R., 2013a. How does age affect leg muscle activity/coactivity during uphill and downhill walking? Gait Posture 37 (3), 378–384.

Franz, J.R., Kram, R., 2013b. Advanced age affects the individual leg mechanics of level, uphill, and downhill walking. J. Biomech. 46 (3), 535–540.

Franz, J.R., Wierzbinski, C.M., Kram, R., 2012. Metabolic cost of running barefoot versus shod: is lighter better? Med. Sci. Sports Exerc. 44 (8), 1519–1525.

Franz, J.R., Maletis, M., Kram, R., 2014. Real-time feedback enhances forward propulsion during walking in old adults. Clin. Biomech. 29 (1), 68–74.

Fraser, J.J., Koldenhoven, R.M., Saliba, S.A., Hertel, J., 2017. Reliability of ankle-foot morphology, mobility, strength, and motor performance measures. Int. J. Sports Phys. Ther. 12 (7), 1134–1149.

Fritz, S., Lusardi, M., 2009. White paper: "walking speed: the sixth vital sign". J. Geriatr. Phys. Ther. 32 (2), 46–49.

Fukano, M., Iso, S., 2016. Changes in foot shape after long-distance running. J. Funct. Morphol. Kinesiol. 1 (1), 30–38.

Fukano, M., Inami, T., Nakagawa, K., Narita, T., Iso, S., 2018. Foot posture alteration and recovery following a full marathon run. Eur. J. Sport Sci. 18 (10), 1338–1345.

Fukuchi, R.K., Duarte, M., 2008. Comparison of three-dimensional lower extremity running kinematics of young adult and elderly runners. J. Sports Sci. 26 (13), 1447–1454.

Gabriel, R.C., Abrantes, J., Granata, K., Bulas-Cruz, J., Melo-Pinto, P., Filipe, V., 2008. Dynamic joint stiffness of the ankle during walking: gender-related differences. Phys. Ther. Sport 9 (1), 16–24.

Galloway, J.C., Thelen, E., 2004. Feet first: object exploration in young infants. Infant Behav. Dev. 27 (1), 107–112.

Gangnet, N., Pomero, V., Dumas, R., Skalli, W., Vital, J.-M., 2003. Variability of the spine and pelvis location with respect to the gravity line: a three-dimensional stereoradiographic study using a force platform. Surg. Radiol. Anat. 25 (5-6), 424–433.

Ganley, K.J., Powers, C.M., 2005. Gait kinematics and kinetics of 7-year-old children: a comparison to adults using age-specific anthropometric data. Gait Posture 21 (2), 141–145.

Gannon, L.M., Bird, H.A., 1999. The quantification of joint laxity in dancers and gymnasts. J. Sports Sci. 17 (9), 743–750.

Gardner, A.W., Montgomery, P.S., 2001. Impaired balance and higher prevalence of falls in subjects with intermittent claudication. J. Gerontol. A Biol. Sci. Med. Sci. 56 (7), M454–M458.

Gardner, A.W., Forrester, L., Smith, G.V., 2001. Altered gait profile in subjects with peripheral arterial disease. Vasc. Med. 6 (1), 31–34.

Gaston, M.S., Rutz, E., Dreher, T., Brunner, R., 2011. Transverse plane rotation of the foot and transverse hip and pelvic kinematics in diplegic cerebral palsy. Gait Posture 34 (2), 218–221.

Gazendam, M.G.J., Hof, A.L., 2007. Averaged EMG profiles in jogging and running at different speeds. Gait Posture 25 (4), 604–614.

Gefen, A., Megido-Ravid, M., Itzchak, Y., 2001. In vivo biomechanical behavior of the human heel pad during the stance phase of gait. J. Biomech. 34 (12), 1661–1665.

Geirsdottir, O.G., Arnarson, A., Briem, K., Ramel, A., Tomasson, K., Jonsson, P.V., et al., 2012. Physical function predicts improvement in quality of life in elderly Icelanders after 12 weeks of resistance exercise. J. Nutr. Health Aging 16 (1), 62–66.

Geyer, H., Seyfarth, A., Blickhan, R., 2006. Compliant leg behaviour explains basic dynamics of walking and running. Proc. R. Soc. B Biol. Sci. 273 (1603), 2861–2867.

Gilligan, I., Chandraphak, S., Mahakkanukrauh, P., 2013. Femoral neck-shaft angle in humans: variation relating to climate, clothing, lifestyle, sex, age and side. J. Anat. 223 (2), 133–151.

Gök, H., Ergin, S., Yavuzer, G., 2002. Kinetic and kinematic characteristics of gait in patients with medial knee arthrosis. Acta Orthop. Scand. 73 (6), 647–652.

Goldberg, S.R., Anderson, F.C., Pandy, M.G., Delp, S.L., 2004. Muscles that influence knee flexion velocity in double support: implications for stiff-knee gait. J. Biomech. 37 (8), 1189–1196.

Gómez-Pérez, C., Font-Llagunes, J.M., Martori, J.C., Samsó, J.V., 2019. Gait parameters in children with bilateral spastic cerebral palsy: a systematic review of randomized controlled trials. Dev. Med. Child Neurol. 61 (7), 770–782.

Gottschall, J.S., Kram, R., 2003. Energy cost and muscular activity required for propulsion during walking. J. Appl. Physiol. 94 (5), 1766–1772.

Gottschall, J.S., Kram, R., 2005. Energy cost and muscular activity required for the leg swing during walking. J. Appl. Physiol. 99 (1), 23–30.

Grabiner, P.C., Biswas, S.T., Grabiner, M.D., 2001. Age-related changes in spatial and temporal gait variables. Arch. Phys. Med. Rehabil. 82 (1), 31–35.

Grahame, R., 1999. Joint hypermobility and genetic collagen disorders: are they related? Arch. Dis. Child. 80 (2), 188–191.

Grahame, R., 2009. Joint hypermobility syndrome pain. Curr. Pain Headache Rep. 13 (6), 427–433.

Grahame, R., Hakim, A.J., 2008. Hypermobility. Curr. Opin. Rheumatol. 20 (1), 106–110.

Grahame, R., Bird, H.A., Child, A., 2000. The revised (Brighton 1998) criteria for the diagnosis of benign joint hypermobility syndrome (BJHS). J. Rheumatol. 27 (7), 1777–1779.

Green, S.M., Briggs, P.J., 2013. Flexion strength of the toes in the normal foot. An evaluation using magnetic resonance imaging. Foot 23 (4), 115–119.

Griffin, T.M., Roberts, T.J., Kram, R., 2003. Metabolic cost of generating muscular force in human walking: insight from load-carrying and speed experiments. J. Appl. Physiol. 95 (1), 172–183.

Grigoriadis, G., Newell, N., Carpanen, D., Christou, A., Bull, A.M.J., Masouros, S.D., 2017. Material properties of the heel fat pad across strain rates. J. Mech. Behav. Biomed. Mater. 65, 398–407.

Gruben, K.G., Boehm, W.L., 2012a. Force direction pattern stabilizes sagittal plane mechanics of human walking. Hum. Mov. Sci. 31 (3), 649–659.

Gruben, K.G., Boehm, W.L., 2012b. Mechanical interaction of center of pressure and force direction in the upright human. J. Biomech. 45 (9), 1661–1665.

Gruber, K., Ruder, H., Denoth, J., Schneider, K., 1998. A comparative study of impact dynamics: wobbling mass model versus rigid body models. J. Biomech. 31 (5), 439–444.

Gruber, A.H., Boyer, K.A., Derrick, T.R., Hamill, J., 2014. Impact shock frequency components and attenuation in rearfoot and forefoot running. J. Sport Health Sci. 3 (2), 113–121.

Gu, Y., Li, J., Ren, X., Lake, M.J., Zeng, Y., 2010. Heel skin stiffness effect on the hind foot biomechanics during heel strike. Skin Res. Technol. 16 (3), 291–296.

Gullstrand, L., Halvorsen, K., Tinmark, F., Eriksson, M., Nilsson, J., 2009. Measurements of vertical displacement in running, a methodological comparison. Gait Posture 30 (1), 71–75.

Gurney, B., 2002. Leg length discrepancy. Gait Posture 15 (2), 195–206.

Gutierrez-Farewik, E.M., Bartonek, A., Saraste, H., 2006. Comparison and evaluation of two common methods to measure center of mass displacement in three dimensions during gait. Hum. Mov. Sci. 25 (2), 238–256.

Hak, L., Houdijk, H., Steenbrink, F., Mert, A., van der Wurff, P., Beek, P.J., et al., 2012. Speeding up or slowing down?: Gait adaptations to preserve gait stability in response to balance perturbations. Gait Posture 36 (2), 260–264.

Hakim, A., Grahame, R., 2003. Joint hypermobility. Best Pract. Res. Clin. Rheumatol. 17 (6), 989–1004.

Hall, J.P.L., Barton, C., Jones, P.R., Morrissey, D., 2013. The biomechanical differences between barefoot and shod distance running: a systematic review and preliminary meta-analysis. Sports Med. 43 (12), 1335–1353.

Hall, M., Bennell, K.L., Wrigley, T.V., Metcalf, B.R., Campbell, P.K., Kasza, J., et al., 2017. The knee adduction moment and knee osteoarthritis symptoms: relationships according to radiographic disease severity. Osteoarthr. Cartil. 25 (1), 34–41.

Hallemans, A., D'Août, K., De Clercq, D., Aerts, P., 2003. Pressure distribution patterns under the feet of new walkers: the first two months independent walking. Foot Ankle Int. 24 (5), 444–453.

Hallemans, A., De Clercq, D., Otten, B., Aerts, P., 2005. 3D joint dynamics of walking in toddlers: a cross-sectional study spanning the first rapid development phase of walking. Gait Posture 22 (2), 107–118.

Hallemans, A., De Clercq, D., Aerts, P., 2006. Changes in 3D joint dynamics during the first 5 months after the onset of independent walking: a longitudinal follow-up study. Gait Posture 24 (3), 270–279.

Hamacher, D., Singh, N.B., Van Dieën, J.H., Heller, M.O., Taylor, W.R., 2011. Kinematic measures for assessing gait stability in elderly individuals: a systematic review. J. R. Soc. Interface 8 (65), 1682–1698.

Hamill, J., Gruber, A.H., 2017. Is changing footstrike pattern beneficial to runners? J. Sport Health Sci. 6 (2), 146–153.

Hamill, J., van Emmerik, R.E., Heiderscheit, B.C., Li, L., 1999. A dynamical systems approach to lower extremity running injuries. Clin. Biomech. 14 (5), 297–308.

Hamill, J., Russell, E.M., Gruber, A.H., Miller, R., 2011. Impact characteristics in shod and barefoot running. Footwear Sci. 3 (1), 33–40.

Hamill, J., Gruber, A.H., Derrick, T.R., 2014. Lower extremity joint stiffness characteristics during running with different footfall patterns. Eur. J. Sport Sci. 14 (2), 130–136.

Hamner, S.R., Seth, A., Delp, S.L., 2010. Muscle contributions to propulsion and support during running. J. Biomech. 43 (14), 2709–2716.

Han, J.-H., Kim, M.-J., Yang, H.-J., Lee, Y.-J., Sung, Y.-H., 2014. Effects of therapeutic massage on gait and pain after delayed onset muscle soreness. J. Exercise Rehabilit. 10 (2), 136–140.

Hanley, B., Bissas, A., Merlino, S., Gruber, A.H., 2019. Most marathon runners at the 2017 IAAF World Championships were rearfoot strikers, and most did not change footstrike pattern. J. Biomech. 92, 54–60.

Hanlon, M., Anderson, R., 2006. Prediction methods to account for the effect of gait speed on lower limb angular kinematics. Gait Posture 24 (3), 280–287.

Hansen, A.H., Childress, D.S., Miff, S.C., Gard, S.A., Mesplay, K.P., 2004. The human ankle during walking: implications for design of biomimetic ankle prostheses. J. Biomech. 37 (10), 1467–1474.

Hanson, E.D., Srivatsan, S.R., Agrawal, S., Menon, K.S., Delmonico, M.J., Wang, M.Q., et al., 2009. Effects of strength training on physical function: influence of power, strength, and body composition. J. Strength Cond. Res. 23 (9), 2627–2637.

Harland, M.J., Steele, J.R., 1997. Biomechanics of the sprint start. Sports Med. 23 (1), 11–20.

Harridge, S.D., Bottinelli, R., Canepari, M., Pellegrino, M.A., Reggiani, C., Esbjörnsson, M., et al., 1996. Whole-muscle and single-fibre contractile properties and myosin heavy chain isoforms in humans. Arch. Eur. J. Physiol. 432 (5), 913–920.

Harridge, S.D.R., Bottinelli, R., Canepari, M., Pellegrino, C., Reggiani, C., Esbjörnsson, M., et al., 1998. Sprint training, in vitro and in vivo muscle function, and myosin heavy chain expression. J. Appl. Physiol. 84 (2), 442–449.

Harris, C., Debeliso, M., Adams, K.J., 2003. The effects of running speed on the metabolic and mechanical energy costs of running. J. Exercise Physiol. Online 6 (3), 28–37.

Harvey, W.F., Yang, M., Cooke, T.D.V., Segal, N.A., Lane, N., Lewis, C.E., et al., 2010. Association of leg-length inequality with knee osteoarthritis: a cohort study. Ann. Intern. Med. 152 (5), 287–295.

Hasegawa, H., Yamauchi, T., Kraemer, W.J., 2007. Foot strike patterns of runners at the 15-km point during an elite-level half marathon. J. Strength Cond. Res. 21 (3), 888–893.

Hashish, R., Samarawickrame, S.D., Powers, C.M., Salem, G.J., 2016. Lower limb dynamics in shod runners who acutely transition to barefoot running. J. Biomech. 49 (2), 284–288.

Hastings, M.K., Mueller, M.J., Woodburn, J., Strube, M.J., Commean, P., Johnson, J.E., et al., 2016. Acquired midfoot deformity and function in individuals with diabetes and peripheral neuropathy. Clin. Biomech. 32, 261–267.

Hatala, K.G., Dingwall, H.L., Wunderlich, R.E., Richmond, B.G., 2013. Variation in foot strike patterns during running among habitually barefoot populations. PLoS ONE 8 (1), e52548. https://doi.org/10.1371/journal.pone.0052548.

Hausdorff, J.M., Cudkowicz, M.E., Firtion, R., Wei, J.Y., Goldberger, A.L., 1998. Gait variability and basal ganglia disorders: stride-to-stride variations of gait cycle timing in Parkinson's disease and Huntington's disease. Mov. Disord. 13 (3), 428–437.

Hausdorff, J.M., Rios, D.A., Edelberg, H.K., 2001. Gait variability and fall risk in community-living older adults: a 1-year prospective study. Arch. Phys. Med. Rehabil. 82 (8), 1050–1056.

Hayes, P., Caplan, N., 2012. Foot strike patterns and ground contact times during high-calibre middle-distance races. J. Sports Sci. 30 (12), 1275–1283.

Hedrick, E.A., Stanhope, S.J., Takahashi, K.Z., 2019. The foot and ankle structures reveal emergent properties analogous to passive springs during human walking. PLoS ONE 14 (6), e0218047. https://doi.org/10.1371/journal.pone.0218047.

Heglund, N.C., Fedak, M.A., Taylor, C.R., Cavagna, G.A., 1982. Energetics and mechanics of terrestrial locomotion. IV. Total mechanical energy changes as a function of speed and body size in birds and mammals. J. Exp. Biol. 97 (1), 57–66.

Heglund, N.C., Willems, P.A., Penta, M., Cavagna, G.A., 1995. Energy-saving gait mechanics with head-supported loads. Nature 375 (6526), 52–54.

Heiderscheit, B.C., Hamill, J., van Emmerik, R.E.A., 2002. Variability of stride characeristics and joint coordination among individuals with unilateral patellofemoral pain. J. Appl. Biomech. 18 (2), 110–121.

Heiderscheit, B.C., Chumanov, E.S., Michalski, M.P., Wille, C.M., Ryan, M.B., 2011. Effects of step rate manipulation on joint mechanics during running. Med. Sci. Sports Exerc. 43 (2), 296–302.

Hein, T., Grau, S., 2014. Can minimal running shoes imitate barefoot heel-toe running patterns? A comparison of lower leg kinematics. J. Sport Health Sci. 3 (2), 67–73.

Heinert, B.L., Kernozek, T.W., Greany, J.F., Fater, D.C., 2008. Hip abductor weakness and lower extremity kinematics during running. J. Sport Rehabil. 17 (3), 243–256.

Heise, G.D., Martin, P.E., 2001. Are variations in running economy in humans associated with ground reaction force characteristics? Eur. J. Appl. Physiol. 84 (5), 438–442.

Heise, G., Shinohara, M., Binks, L., 2008. Biarticular leg muscles and links to running economy. Int. J. Sports Med. 29 (8), 688–691.

Henriksen, M., Alkjær, T., Lund, H., Simonsen, E.B., Graven-Nielsen, T., Danneskiold-Samsøe, B., et al., 2007. Experimental quadriceps muscle pain impairs knee joint control during walking. J. Appl. Physiol. 103 (1), 132–139.

Henriksen, M., Graven-Nielsen, T., Aaboe, J., Andriacchi, T.P., Bliddal, H., 2010. Gait changes in patients with knee osteoarthritis are replicated by experimental knee pain. Arthritis Care Res. 62 (4), 501–509.

Henriksen, M., Aaboe, J., Bliddal, H., 2012. The relationship between pain and dynamic knee joint loading in knee osteoarthritis varies with radiographic disease severity. A cross sectional study. Knee 19 (4), 392–398.

Henry, S.M., Hitt, J.R., Jones, S.L., Bunn, J.Y., 2006. Decreased limits of stability in response to postural perturbations in subjects with low back pain. Clin. Biomech. 21 (9), 881–892.

Hernández-Gervilla, Ó., Escalona-Marfil, C., Corbi, F., 2016. Relación entre la postura del pie y la cinemática de la carrera: estudio piloto (Correlation between foot posture and running kinematics: a pilot study). Apunts. Medicina de L'Esport. 51 (192), 115–122.

Herod, T.W., Chambers, N.C., Veres, S.P., 2016. Collagen fibrils in functionally distinct tendons have differing structural responses to tendon rupture and fatigue loading. Acta Biomater. 42, 296–307.

Herssens, N., Verbecque, E., Hallemans, A., Vereeck, L., Van Rompaey, V., 2018. Do spatiotemporal parameters and gait variability differ across the lifespan of healthy adults? A systematic review. Gait Posture 64, 181–190.

Highsmith, M.J., Andrews, C.R., Millman, C., Fuller, A., Kahle, J.T., Klenow, T.D., et al., 2016. Gait training interventions for lower extremity amputees: a systematic literature review. Technol Innov 18 (2–3), 99–113.

Hillman, S.J., Stansfield, B.W., Richardson, A.M., Robb, J.E., 2009. Development of temporal and distance parameters of gait in normal children. Gait Posture 29 (1), 81–85.

Hills, A.P., Parker, A.W., 1992. Locomotor characteristics of obese children. Child Care Health Dev. 18 (1), 29–34.

Hillstrom, H.J., Song, J., Kraszewski, A.P., Hafer, J.F., Mootanah, R., Dufour, A.B., et al., 2013. Foot type biomechanics part 1: Structure and function of the asymptomatic foot. Gait Posture 37 (3), 445–451.

Hinrichs, R.N., Cavanagh, P.R., Williams, K.R., 1987. Upper extremity function in running. I: Center of mass and propulsion considerations. Int. J. Sport Biomech. 3 (3), 222–241.

Hirsch, C., Hirsch, M., John, M.T., Bock, J.J., 2007. Reliability of the Beighton Hypermobility Index to determinate the general joint laxity performed by dentists. J. Orofac. Orthop. 68 (5), 342–352.

Hodges, P.W., Gurfinkel, V.S., Brumagne, S., Smith, T.C., Cordo, P.C., 2002. Coexistence of stability and mobility in postural control: evidence from postural compensation for respiration. Exp. Brain Res. 144 (3), 293–302.

Hof, A.L., 1996. Letter to the editor: scaling gait data to body size. Gait Posture 4 (3), 222–223.

Hof, A.L., Elzinga, H., Grimmius, W., Halbertsma, J.P.K., 2002. Speed dependence of averaged EMG profiles in walking. Gait Posture 16 (1), 78–86.

Hogervorst, T., Vereecke, E.E., 2014. Evolution of the human hip. Part 1: the osseous framework. J. Hip Preserv. Surg. 1 (2), 39–45.

Hollander, K., de Villiers, J.E., Sehner, S., Wegscheider, K., Braumann, K.-M., Venter, R., et al., 2017. Growing-up (habitually) barefoot influences the development of foot and arch morphology in children and adolescents. Sci. Rep. 7, 8079. https://doi.org/10.1038/s41598-017-07868-4.

Hollander, K., Stebbins, J., Albertsen, I.M., Hamacher, D., Babin, K., Hacke, C., et al., 2018. Arch index and running biomechanics in children aged 10-14 years. Gait Posture 61, 210–214.

Hollander, K., Zech, A., Rahlf, A.L., Orendurff, M.S., Stebbins, J., Heidt, C., 2019. The relationship between static and dynamic foot posture and running biomechanics: a systematic review and meta-analysis. Gait Posture 72, 109–122.

Hollman, J.H., Kovash, F.M., Kubik, J.J., Linbo, R.A., 2007. Age-related differences in spatiotemporal markers of gait stability during dual task walking. Gait Posture 26 (1), 113–119.

Hollman, J.H., Watkins, M.K., Imhoff, A.C., Braun, C.E., Akervik, K.A., Ness, D.K., 2016. A comparison of variability in spatiotemporal gait parameters between treadmill and overground walking conditions. Gait Posture 43, 204–209.

Holm, I., Tveter, A.T., Fredriksen, P.M., Vøllestad, N., 2009. A normative sample of gait and hopping on one leg parameters in children 7-12 years of age. Gait Posture 29 (2), 317–321.

Holowka, N.B., O'Neill, M.C., Thompson, N.E., Demes, B., 2017. Chimpanzee and human midfoot motion during bipedal walking and the evolution of the longitudinal arch of the foot. J. Hum. Evol. 104, 23–31.

Holt, K.G., Jeng, S.F., Ratcliffe, R., Hamill, J., 1995. Energetic cost and stability during human walking at the preferred stride frequency. J. Mot. Behav. 27 (2), 164–178.

Holt, K.G., Wagenaar, R.C., LaFiandra, M.E., Kubo, M., Obusek, J.P., 2003. Increased musculoskeletal stiffness during load carriage at increasing walking speeds maintains constant vertical excursion of the body center of mass. J. Biomech. 36 (4), 465–471.

Holviala, J., Kraemer, W.J., Sillanpää, E., Karppinen, H., Avela, J., Kauhanen, A., et al., 2012. Effects of strength, endurance and combined training on muscle strength, walking speed and dynamic balance in aging men. Eur. J. Appl. Physiol. 112 (4), 1335–1347.

Honeine, J.-L., Schieppati, M., Gagey, O., Do, M.-C., 2013. The functional role of the triceps surae muscle during human locomotion. PLoS ONE 8 (1), e52943. https://doi.org/10.1371/journal.pone.0052943.

Horisberger, M., Hintermann, B., Valderrabano, V., 2009. Alterations of plantar pressure distribution in posttraumatic end-stage ankle osteoarthritis. Clin. Biomech. 24 (3), 303–307.

Hortobágyi, T., DeVita, P., 2000. Muscle pre- and coactivity during downward stepping are associated with leg stiffness in aging. J. Electromyogr. Kinesiol. 10 (2), 117–126.

Hortobágyi, T., Finch, A., Solnik, S., Rider, P., DeVita, P., 2011. Association between muscle activation and metabolic cost of walking in young and old adults. J. Gerontol. Ser. A Biol. Med. Sci. 66 (5), 541–547.

Horton, M.G., Hall, T.L., 1989. Quadriceps femoris muscle angle: normal values and relationships with gender and selected skeletal measures. Phys. Ther. 69 (11), 897–901.

Horwood, A., 2019. The biomechanical function of the foot pump in venous return from the lower extremity during the human gait cycle: an expansion of the gait model of the foot pump. Med. Hypotheses 129, 109220. https://doi.org/10.1016/j.mehy.2019.05.006.

Horwood, A.M., Chockalingam, N., 2017. Defining excessive, over, or hyper-pronation: a quandary. Foot 31, 49–55.

Hösl, M., Böhm, H., Multerer, C., Döderlein, L., 2014. Does excessive flatfoot deformity affect function? A comparison between symptomatic and asymptomatic flatfeet using the Oxford Foot Model. Gait Posture 39 (1), 23–28.

Houx, L., Lempereur, M., Rémy-Néris, O., Brochard, S., 2013. Threshold of equinus which alters biomechanical gait parameters in children. Gait Posture 38 (4), 582–589.

Hsu, R.W., Himeno, S., Coventry, M.B., Chao, E.Y., 1990. Normal axial alignment of the lower extremity and load-bearing distribution at the knee. Clin. Orthop. Relat. Res. 255 (June), 215–227.

Hsu, C.-C., Tsai, W.-C., Chen, C.P.-C., Shau, Y.-W., Wang, C.-L., Chen, M.J.-L., et al., 2005. Effects of aging on the plantar soft tissue properties under the metatarsal heads at different impact velocities. Ultrasound Med. Biol. 31 (10), 1423–1429.

Hsu, C.-C., Tsai, W.-C., Wang, C.-L., Pao, S.-H., Shau, Y.-W., Chuan, Y.-S., 2007. Microchambers and macrochambers in heel pads: are they functionally different? J. Appl. Physiol. 102 (6), 2227–2231.

Hsu, C.-C., Tsai, W.-C., Hsiao, T.-Y., Tseng, F.-Y., Shau, Y.-W., Wang, C.-L., et al., 2009. Diabetic effects on microchambers and macrochambers tissue properties in human heel pads. Clin. Biomech. 24 (8), 682–686.

Hudson, N., Fitzcharles, M.A., Cohen, M., Starr, M.R., Esdaile, J.M., 1998. The association of soft-tissue rheumatism and hypermobility. Br. J. Rheumatol. 37 (4), 382–386.

Hunt, A.E., Smith, R.M., Torode, M., Keenan, A.-M., 2001. Inter-segment foot motion and ground reaction forces over the stance phase of walking. Clin. Biomech. 16 (7), 592–600.

Hunter, J.P., Marshall, R.N., McNair, P.J., 2005. Relationships between ground reaction force impulse and kinematics of sprint-running acceleration. J. Appl. Biomech. 21 (1), 31–43.

Hutchinson, L.A., De Asha, A.R., Rainbow, M.J., Dickinson, A.W.L., Deluzio, K.J., 2021. A comparison of centre of pressure behaviour and ground reaction force magnitudes when individuals walk overground and on an instrumented treadmill. Gait Posture 83, 174–176.

Hutton, W.C., Dhanendran, M., 1979. A study of the distribution of load under the normal foot during walking. Int. Orthop. 3 (2), 153–157.

Iida, F., Rummel, J., Seyfarth, A., 2008. Bipedal walking and running with spring-like biarticular muscles. J. Biomech. 41 (3), 656–667.

Ishikawa, M., Pakaslahti, J., Komi, P.V., 2007. Medial gastrocnemius muscle behavior during human running and walking. Gait Posture 25 (3), 380–384.

Ivanenko, Y.P., Grasso, R., Lacquaniti, F., 2000. Influence of leg muscle vibration on human walking. J. Neurophysiol. 84 (4), 1737–1747.

Ivanenko, Y.P., Dominici, N., Cappellini, G., Lacquaniti, F., 2005. Kinematics in newly walking toddlers does not depend upon postural stability. J. Neurophysiol. 94 (1), 754–763.

Johnson, A.P., Ward, S., Simmonds, J., 2019. The Lower Limb Assessment Score: a valid measure of hypermobility in elite football? Phys.l Ther. Sport 37, 86–90.

Jonkergouw, N., Prins, M.R., Buis, A.W.P., van der Wurff, P., 2016. The effect of alignment changes on unilateral transtibial amputee's gait: a systematic review. PLoS ONE 11 (12), e0167466. https://doi.org/10.1371/journal.pone.0167466.

Jonkers, I., Stewart, C., Spaepen, A., 2003a. The study of muscle action during single support and swing phase of gait: clinical relevance of forward simulation techniques. Gait Posture 17 (2), 97–105.

Jonkers, I., Stewart, C., Spaepen, A., 2003b. The complementary role of the plantarflexors, hamstrings and gluteus maximus in the control of stance limb stability during gait. Gait Posture 17 (3), 264–272.

Jordan, K., Challis, J.H., Newell, K.M., 2007. Walking speed influences on gait cycle variability. Gait Posture 26 (1), 128–134.

Joyner, M.J., Ruiz, J.R., Lucia, A., 2011. The two-hour marathon: who and when? J. Appl. Physiol. 110 (1), 275–277.

Judge, J.O., Davis 3rd, R.B., Ounpuu, S., 1996. Step length reductions in advanced age: the role of ankle and hip kinetics. J. Gerontol. Ser. A Biol. Med. Sci. 51 (6), M303–M312.

Kadhim, M., Miller, F., 2014. Crouch gait changes after planovalgus foot deformity correction in ambulatory children with cerebral palsy. Gait Posture 39 (2), 793–798.

Kalapotharakos, V.I., Michalopoulos, M., Tokmakidis, S.P., Godolias, G., Gourgoulis, V., 2005. Effects of a heavy and a moderate resistance training on functional performance in older adults. J. Strength Cond. Res. 19 (3), 652–657.

Kambhampati, S.B.S., 2007. Constructing a Pedotti diagram using excel charts. J. Biomech. 40 (16), 3748–3750.

Kang, H.G., Dingwell, J.B., 2008a. Separating the effects of age and walking speed on gait variability. Gait Posture 27 (4), 572–577.

Kang, H.G., Dingwell, J.B., 2008b. Effects of walking speed, strength and range of motion on gait stability in healthy older adults. J. Biomech. 41 (14), 2899–2905.

Karamanidis, K., Arampatzis, A., 2005. Mechanical and morphological properties of different muscle-tendon units in the lower extremity and running mechanics: effect of aging and physical activity. J. Exp. Biol. 208 (20), 3907–3923.

Karamanidis, K., Arampatzis, A., Brüggemann, G.-P., 2003. Symmetry and reproducibility of kinematic parameters during various running techniques. Med. Sci. Sports Exerc. 35 (6), 1009–1016.

Kaufman, K.R., Brodine, S.K., Shaffer, R.A., Johnson, C.W., Cullison, T.R., 1999. The effect of foot structure and range of motion on musculoskeletal overuse injuries. Am. J. Sports Med. 27 (5), 585–593.

Kavanagh, J.J., Morrison, S., Barrett, R.S., 2005. Coordination of head and trunk accelerations during walking. Eur. J. Appl. Physiol. 94 (4), 468–475.

Kawamori, N., Nosaka, K., Newton, R.U., 2013. Relationships between ground reaction impulse and sprint acceleration performance in team sport athletes. J. Strength Cond. Res. 27 (3), 568–573.

Keenan, G.S., Franz, J.R., Dicharry, J., Della Croce, U., Kerrigan, D.C., 2011. Lower limb joint kinetics in walking: the role of industry recommended footwear. Gait Posture 33 (3), 350–355.

Keller, T.S., Weisberger, A.M., Ray, J.L., Hasan, S.S., Shiavi, R.G., Spengler, D.M., 1996. Relationship between vertical ground reaction force and speed during walking, slow jogging, and running. Clin. Biomech. 11 (5), 253–259.

Kelly, L.A., Lichtwark, G., Cresswell, A.G., 2015. Active regulation of longitudinal arch compression and recoil during walking and running. J. R. Soc. Interface 12 (102), 20141076. https://doi.org/10.1098/rsif.2014.1076.

Kelly, L.A., Lichtwark, G.A., Farris, D.J., Cresswell, A., 2016. Shoes alter the spring-like function of the human foot during running. J. R. Soc. Interface 13 (119), 20160174. https://doi.org/10.1098/rsif.2016.0174.

Kelly, L.A., Cresswell, A.G., Farris, D.J., 2018. The energetic behaviour of the human foot across a range of running speeds. Sci. Rep. 8, 10576. https://doi.org/10.1038/s41598-018-28946-1.

Ker, R.F., Bennett, M.B., Bibby, S.R., Kester, R.C., Alexander RMcN., 1987. The spring in the arch of the human foot. Nature 325 (6100), 147–149.

Ker, R.F., Alexander, R.M.C.N., Bennett, M.B., 1988. Why are mammalian tendons so thick? J. Zool. 216 (2), 309–324.

Kernozek, T.W., Torry, M.R., van Hoof, H., Cowley, H., Tanner, S., 2005. Gender differences in frontal and sagittal plane biomechanics during drop landings. Med. Sci. Sports Exerc. 37 (6), 1003–1012.

Kernozek, T.W., Knaus, A., Rademaker, T., Almonroeder, T.G., 2018. The effects of habitual foot strike patterns on Achilles tendon loading in female runners. Gait Posture 66, 283–287.

Kerrigan, D.C., Todd, M.K., Della, C.U., 1998a. Gender differences in joint biomechanics during walking: normative study in young adults. Am. J. Phys. Med. Rehabil. 77 (1), 2–7.

Kerrigan, D.C., Todd, M.K., Della Croce, U., Lipsitz, L.A., Collins, J.J., 1998b. Biomechanical gait alterations independent of speed in the healthy elderly: evidence for specific limiting impairments. Arch. Phys. Med. Rehabil. 79 (3), 317–322.

Kerrigan, D.C., Lelas, J.L., Karvosky, M.E., 2001. Women's shoes and knee osteoarthritis. Lancet 357 (9262), 1097–1098.

Kerrigan, D.C., Karvosky, M.E., Lelas, J.L., Riley, P.O., 2003. Men's shoes and knee joint torques relevant to the development and progression of knee osteoarthritis. J. Rheumatol. 30 (3), 529–533.

Khamis, S., Carmeli, E., 2017a. Letter to the editor: a new concept for measuring leg length discrepancy. J. Orthop. 14 (2), 276–280.

Khamis, S., Carmeli, E., 2017b. Relationship and significance of gait deviations associated with limb length discrepancy: a systematic review. Gait Posture 57, 115–123.

Khamis, S., Carmeli, E., 2018. The effect of simulated leg length discrepancy on lower limb biomechanics during gait. Gait Posture 61, 73–80.

Khassetarash, A., Hassannejad, R., Enders, H., Ettefagh, M.M., 2015. Damping and energy dissipation in soft tissue vibrations during running. J. Biomech. 48 (2), 204–209.

Khazzam, M., Long, J.T., Marks, R.M., Harris, G.F., 2007. Kinematic changes of the foot and ankle in patients with systemic rheumatoid arthritis and forefoot deformity. J. Orthop. Res. 25 (3), 319–329.

Kido, M., Ikoma, K., Imai, K., Tokunaga, D., Inoue, N., Kubo, T., 2013. Load response of the medial longitudinal arch in patients with flatfoot deformity: in vivo 3D study. Clin. Biomech. 28 (5), 568–573.

Kilmartin, T.E., Wallace, W.A., 1992. The significance of pes planus in juvenile hallux valgus. Foot Ankle 13 (2), 53–56.

Kim, S., Park, S., 2011. Leg stiffness increases with speed to modulate gait frequency and propulsion energy. J. Biomech. 44 (7), 1253–1258.

King, D.M., Toolan, B.C., 2004. Associated deformities and hypermobility in hallux valgus: an investigation with weightbearing radiographs. Foot Ankle Int. 25 (4), 251–255.

King, S., Vanicek, N., Mockford, K.A., Coughlin, P.A., 2012. The effect of a 3-month supervised exercise programme on gait parameters of patients with peripheral arterial disease and intermittent claudication. Clin. Biomech. 27 (8), 845–851.

Kirby, A., Davies, R., 2007. Developmental coordination disorder and joint hypermobility syndrome – overlapping disorders? Implications for research and clinical practice. Child Care Health Dev. 33 (5), 513–519.

Kirsner, R.S., 2018. Editorial: exercise for leg ulcers: "Working out" the nature of venous ulcers. J. Am. Med. Assoc. Dermatol. 154 (11), 1257–1259.

Kiss, R.M., 2010. Effect of walking speed and severity of hip osteoarthritis on gait variability. J. Electromyogr. Kinesiol. 20 (6), 1044–1051.

Knutson, G.A., 2005a. Anatomic and functional leg-length inequality: a review and recommendation for clinical decision-making. Part I, anatomic leg-length inequality: prevalence, magnitude, effects and clinical significance. Chiropr. Osteopat. 13, 11. https://doi.org/10.1186/1746-1340-13-11.

Knutson, G.A., 2005b. Anatomic and functional leg-length inequality: A review and recommendation for clinical decision-making. Part II, the functional or unloaded leg-length asymmetry. Chiropr. Osteopat. 13, 12. https://doi.org/10.1186/1746-1340-13-12.

Kokubo, T., Hashimoto, T., Nagura, T., Nakamura, T., Suda, Y., Matsumoto, H., et al., 2012. Effect of the posterior tibial and peroneal longus on the mechanical properties of the foot arch. Foot Ankle Int. 33 (4), 320–325.

Komi, P.V., 2000. Stretch-shortening cycle: a powerful model to study normal and fatigued muscle. J. Biomech. 33 (10), 1197–1206.

König, N., Taylor, W.R., Baumann, C.R., Wenderoth, N., Singh, N.B., 2016. Revealing the quality of movement: a meta-analysis review to quantify the thresholds to pathological variability during standing and walking. Neurosci. Biobehav. Rev. 68, 111–119.

Konow, N., Azizi, E., Roberts, T.J., 2012. Muscle power attenuation by tendon during energy dissipation. Proc. R. Soc. B Biol. Sci. 279 (1731), 1108–1113.

Konyves, A., Bannister, G.C., 2005. The importance of leg length discrepancy after total hip arthroplasty. J. Bone Joint Surg. 87-B (2), 155–157.

Korhonen, M.T., Cristea, A., Alén, M., Häkkinen, K., Sipilä, S., Mero, A., et al., 2006. Aging, muscle fiber type, and contractile function in sprint-trained athletes. J. Appl. Physiol. 101 (3), 906–917.

Koutakis, P., Johanning, J.M., Haynatzki, G.R., Myers, S.A., Stergiou, N., Longo, G.M., et al., 2010. Abnormal joint powers before and after the onset of claudication symptoms. J. Vasc. Surg. 52 (2), 340–347.

Kram, R., Taylor, C.R., 1990. Energetics of running: a new perspective. Nature 336 (6281), 265–267.

Kubo, K., Miyazaki, D., Yamada, K., Yata, H., Shimoju, S., Tsunoda, N., 2015. Passive and active muscle stiffness in plantar flexors of long distance runners. J. Biomech. 48 (10), 1937–1943.

Kulmala, J.-P., Kosonen, J., Nurminen, J., Avela, J., 2018. Running in highly cushioned shoes increases leg stiffness and amplifies impact loading. Sci. Rep. 8, 17496. https://doi.org/10.1038/s41598-018-35980-6.

Kumagai, K., Abe, T., Brechue, W.F., Ryushi, T., Takano, S., Mizuno, M., 2000. Sprint performance is related to muscle fascicle length in male 100-m sprinters. J. Appl. Physiol. 88 (3), 811–816.

Kuo, A.D., 2002. Energetics of actively powered locomotion using the simplest walking model. J. Biomech. Eng. 124 (1), 113–120.

Lafortune, M.A., Lake, M.J., Hennig, E.M., 1996. Differential shock transmission response of the human body to impact severity and lower limb posture. J. Biomech. 29 (12), 1531–1537.

Lamberts, R.P., Burger, M., du Toit, J., Langerak, N.G., 2016. A systematic review of the effects of single-event multilevel surgery on gait parameters in children with spastic cerebral palsy. PLoS ONE 11 (10), e0164686. https://doi.org/10.1371/journal.pone.0164686.

Laughton, C.A., Davis, I.M., Hamill, J., 2003. Effect of strike pattern and orthotic intervention on tibial shock during running. J. Appl. Biomech. 19 (2), 153–168.

Lavcanska, V., Taylor, N.F., Schache, A.G., 2005. Familiarization to treadmill running in young unimpaired adults. Hum. Mov. Sci. 24 (4), 544–557.

Le Huec, J.C., Saddiki, R., Franke, J., Rigal, J., Aunoble, S., 2011. Equilibrium of the human body and the gravity line: the basics. Eur. Spine J. 20 (Suppl 5), 558–563.

Lecerf, G., Fessy, M.H., Philippot, R., Massin, P., Giraud, F., Flecher, X., et al., 2009. Femoral offset: anatomical concept, definition, assessment, implications for preoperative templating and hip arthroplasty. Orthop. Traumatol. Surg. Res. 95 (3), 210–219.

Ledoux, W.R., Blevins, J.J., 2007. The compressive material properties of the plantar soft tissue. J. Biomech. 40 (13), 2975–2981.

Ledoux, W.R., Hillstrom, H.J., 2002. The distributed plantar vertical force of neutrally aligned and pes planus feet. Gait Posture 15 (1), 1–9.

Ledoux, W.R., Shofer, J.B., Ahroni, J.H., Smith, D.G., Sangeorzan, B.J., Boyko, E.J., 2003. Biomechanical differences among pes cavus, neutrally aligned, and pes planus feet in subjects with diabetes. Foot Ankle Int. 24 (11), 845–850.

Lee, C.R., Farley, C.T., 1998. Determinants of the center of mass trajectory in human walking and running. J. Exp. Biol. 201 (21), 2935–2944.

Lee, S.J., Hidler, J., 2008. Biomechanics of overground vs. treadmill walking in healthy individuals. J. Appl. Physiol. 104 (3), 747–755.

Lee, S.S.M., Piazza, S.J., 2009. Built for speed: musculoskeletal structure and sprinting ability. J. Exp. Biol. 212 (22), 3700–3707.

Lee, J.H., Sung, I.Y., Yoo, J.Y., 2009. Clinical or radiologic measurements and 3-D gait analysis in children with pes planus. Pediatr. Int. 51 (2), 201–205.

Lee, S.Y., Hertel, J., Lee, S.C., 2010. Rearfoot eversion has indirect effects on plantar fascia tension by changing the amount of arch collapse. Foot 20 (2-3), 64–70.

Lelas, J.L., Merriman, G.J., Riley, P.O., Kerrigan, D.C., 2003. Predicting peak kinematic and kinetic parameters from gait speed. Gait Posture 17 (2), 106–112.

Leurs, F., Ivanenko, Y.P., Bengoetxea, A., Cebolla, A.-M., Dan, B., Lacquaniti, F., et al., 2011. Optimal walking speed following changes in limb geometry. J. Exp. Biol. 214 (13), 2276–2282.

Levinger, P., Gilleard, W., 2007. Tibia and rearfoot motion and ground reaction forces in subjects with patellofemoral pain syndrome during walking. Gait Posture 25 (1), 2–8.

Levy, J.C., Mizel, M.S., Wilson, L.S., Fox, W., McHale, K., Taylor, D.C., et al., 2006. Incidence of foot and ankle injuries in West Point cadets with pes planus compared to the general cadet population. Foot Ankle Int. 27 (12), 1060–1064.

Lichtwark, G.A., Watson, A.M., 2007. Is Achilles tendon compliance optimised for maximum muscle efficiency during locomotion? J. Biomech. 40 (8), 1768–1775.

Lieberman, D.E., Bramble, D.M., 2007. The evolution of marathon running: capabilities in humans. Sports Med. 37 (4-5), 288–290.

Lieberman, D.E., Bramble, D.M., Raichlen, D.A., Shea, J.J., 2009. Chapter 8: Brains, brawn, and the evolution of human endurance running capabilities. In: Grine, F.E., Fleagle, J.G., Leakey, R.E. (Eds.), The First Humans: Origin and Early Evolution of the Genus Homo. (Vertebrate Paleobiology and Paleoanthropology Series). Springer, Dordrecht, pp. 77–92.

Lieberman, D.E., Venkadesan, M., Werbel, W.A., Daoud, A.I., D'Andrea, S., Davis, I.S., et al., 2010. Foot strike patterns and collision forces in habitually barefoot versus shod runners. Nature 463 (7280), 531–535.

Lilley, K., Stiles, V., Dixon, S., 2013. The influence of motion control shoes on the running gait of mature and young females. Gait Posture 37 (3), 331–335.

Lin, C.-F., Chen, C.-Y., Lin, C.-W., 2011. Dynamic ankle control in athletes with ankle instability during sports maneuvers. Am. J. Sports Med. 39 (9), 2007–2015.

Lin, C.-Y., Chen, P.-Y., Shau, Y.-W., Tai, H.-C., Wang, C.-L., 2017. Spatial-dependent mechanical properties of the heel pad by shear wave elastography. J. Biomech. 53, 191–195.

Liu, W., Nigg, B.M., 2000. A mechanical model to determine the influence of masses and mass distribution on the impact force during running. J. Biomech. 33 (2), 219–224.

Liu, M.Q., Anderson, F.C., Pandy, M.G., Delp, S.L., 2006. Muscles that support the body also modulate forward progression during walking. J. Biomech. 39 (14), 2623–2630.

Liu, M.Q., Anderson, F.C., Schwartz, M.H., Delp, S.L., 2008. Muscle contributions to support and progression over a range of walking speeds. J. Biomech. 41 (15), 3243–3252.

Liu, R.W., Streit, J.J., Weinberg, D.S., Shaw, J.D., LeeVan, E., Cooperman, D.R., 2018. No relationship between mild limb length discrepancy and spine, hip or knee degenerative disease in a large cadaveric collection. Orthop. Traumatol. Surg. Res. 104 (5), 603–607.

Livingston, L.A., 1998. The quadriceps angle: a review of the literature. J. Orthop. Sports Phys. Ther. 28 (2), 105–109.

Lobet, S., Hermans, C., Bastien, G.J., Massaad, F., Detrembleur, C., 2012. Impact of ankle osteoarthritis on the energetics and mechanics of gait: the case of hemophilic arthropathy. Clin. Biomech. 27 (6), 625–631.

Lohman 3rd, E.B., Balan Sackiriyas, K.S., Swen, R.W., 2011. A comparison of the spatiotemporal parameters, kinematics, and biomechanics between shod, unshod, and minimally supported running as compared to walking. Phys. Ther. Sport 12 (4), 151–163.

Looper, J., Wu, J., Angulo Barroso, R., Ulrich, D., Ulrich, B.D., 2006. Changes in step variability of new walkers with typical development and with Down syndrome. J. Mot. Behav. 38 (5), 367–372.

Lopes, A.D., Hespanhol Júnior, L.C., Yeung, S.S., Pena Costa, L.O., 2012. What are the main running-related musculoskeletal injuries? A systematic review. Sports Med. 42 (10), 891–905.

Lu, H.-L., Lu, T.-W., Lin, H.-C., Chan, W.P., 2017. Comparison of the body's center of mass motion relative to center of pressure between treadmill and over-ground walking. Gait Posture 53, 248–253.

Lucia, A., Esteve-Lanao, J., Oliván, J., Gómez-Gallego, F., San Juan, A.F., Santiago, C., et al., 2006. Physiological characteristics of the best Eritrean runners – exceptional running economy. Appl. Physiol. Nutr. Metab. 31 (5), 530–540.

Lun, V., Meeuwisse, W.H., Stergiou, P., Stefanyshyn, D., 2004. Relation between running injury and static lower limb alignment in recreational runners. Br. J. Sports Med. 38 (5), 576–580.

Luo, G., Houston, V.L., Mussman, M., Garbarini, M., Beattie, A.C., Thongpop, C., 2009. Comparison of male and female foot shape. J. Am. Podiatr. Med. Assoc. 99 (5), 383–390.

Lye, J., Parkinson, S., Diamond, N., Downs, J., Morris, S., 2016. Propulsion strategy in the gait of primary school children; the effect of age and speed. Hum. Mov. Sci. 50, 54–61.

Lythgo, N., Wilson, C., Galea, M., 2009. Basic gait and symmetry measures for primary school-aged children and young adults whilst walking barefoot and with shoes. Gait Posture 30 (4), 502–506.

MacWilliams, B.A., Rozumalski, A., Swanson, A.N., Wervey, R., Dykes, D.C., Novacheck, T.F., et al., 2014. Three-dimensional lumbar spine vertebral motion during running using indwelling bone pins. Spine 39 (26), E1560–E1565.

Maeshige, N., Uemura, M., Hirasawa, Y., Yoshikawa, Y., Moriguchi, M., Kawabe, N., et al., 2021. Immediate effects of weight-bearing calf stretching on ankle dorsiflexion range of motion and plantar pressure during gait in patients with diabetes mellitus. Int. J. Low Extrem. Wounds. https://doi.org/10.1177/15347346211031318.

Maganaris, C.N., Paul, J.P., 2002. Tensile properties of the in vivo human gastrocnemius tendon. J. Biomech. 35 (12), 1639–1646.

Mahmoudian, A., van Dieen, J.H., Bruijn, S.M., Baert, I.A.C., Faber, G.S., Luyten, F.P., et al., 2016. Varus thrust in women with early medial knee osteoarthritis and its relation with the external knee adduction moment. Clin. Biomech. 39, 109–114.

Major, M.J., Stine, R.L., Gard, S.A., 2013. The effects of walking speed and prosthetic ankle adapters on upper extremity dynamics and stability-related parameters in bilateral transtibial amputee gait. Gait Posture 38 (4), 858–863.

Malatesta, D., Simar, D., Dauvilliers, Y., Candau, R., Borrani, F., Préfaut, C., et al., 2003. Energy cost of walking and gait instability in healthy 65- and 80-yr-olds. J. Appl. Physiol. 95 (6), 2248–2256.

Malinzak, R.A., Colby, S.M., Kirkendall, D.T., Yu, B., Garrett, W.E., 2001. A comparison of knee joint motion patterns between men and women in selected athletic tasks. Clin. Biomech. 16 (5), 438–445.

Malisoux, L., Chambon, N., Delattre, N., Gueguen, N., Urhausen, A., Theisen, D., 2016. Injury risk in runners using standard or motion control shoes: a randomised controlled trial with participant and assessor blinding. Br. J. Sports Med. 50 (8), 481–487.

Maloney, W.J., Keeney, J.A., 2004. Leg length discrepancy after total hip arthroplasty. J. Arthroplasty 19 (4 Suppl 1), 108–110.

Mann, R.A., Moran, G.T., Dougherty, S.E., 1986. Comparative electromyography of the lower extremity in jogging, running, and sprinting. Am. J. Sports Med. 14 (6), 501–510.

Mann, R., Malisoux, L., Urhausen, A., Statham, A., Meijer, K., Theisen, D., 2015. The effect of shoe type and fatigue on strike index and spatiotemporal parameters of running. Gait Posture 42 (1), 91–95.

Manoli, A., Graham, B., 2018. Clinical and new aspects of the subtle cavus foot: a review of an additional twelve year experience. Fuß Sprunggelenk 16 (1), 3–29.

Marnach, M.L., Ramin, K.D., Ramsey, P.S., Song, S.-W., Stensland, J.J., An, K.-N., 2003. Characterization of the relationship between joint laxity and maternal hormones in pregnancy. Obstet. Gynecol. 101 (2), 331–335.

Martin, V., Kerhervé, H., Messonnier, L.A., Banfi, J.-C., Geyssant, A., Bonnefoy, R., et al., 2010. Central and peripheral contributions to neuromuscular fatigue induced by a 24-h treadmill run. J. Appl. Physiol. 108 (5), 1224–1233.

Massey, E.W., Stolp, K.A., 2008. Peripheral neuropathy in pregnancy. Phys. Med. Rehabil. Clin. N. Am. 19 (1), 149–162.

Masters, S.E., Challis, J.H., 2021. Increasing the stability of the spring loaded inverted pendulum model of running wth a wobbling mass. J. Biomech. 123, 110527.

Matheson, G.O., Clement, D.B., McKenzie, D.C., Taunton, J.E., Lloyd-Smith, D.R., MacIntyre, J.G., 1987. Stress fractures in athletes: a study of 320 cases. Am. J. Sports Med. 15 (1), 46–58.

Matjačić, Z., Olenšek, A., Bajd, T., 2006. Biomechanical characterization and clinical implications of artificially induced toe-walking: differences between pure soleus, pure gastrocnemius and combination of soleus and gastrocnemius contractures. J. Biomech. 39 (2), 255–266.

Matsas, A., Taylor, N., McBurney, H., 2000. Knee joint kinematics from familiarised treadmill walking can be generalised to overground walking in young unimpaired subjects. Gait Posture 11 (1), 46–53.

Maurer, J.D., Ward, V., Mayson, T.A., Davies, K.R., Alvarez, C.M., Beauchamp, R.D., et al., 2013. A kinematic description of dynamic midfoot break in children using a multi-segment foot model. Gait Posture 38 (2), 287–292.

Maurer, J.D., Ward, V., Mayson, T.A., Davies, K.R., Alvarez, C.M., Beauchamp, R.D., et al., 2014. Classification of midfoot break using multi-segment foot kinematics and pedobarography. Gait Posture 39 (1), 1–6.

McClean, S.G., Fellin, R.E., Suedekum, N., Calabrese, G., Passerallo, A., Joy, S., 2007. Impact of fatigue on gender-based high-risk landing strategies. Med. Sci. Sports Exerc. 39 (3), 502–514.

McCollum, G., Holroyd, C., Castelfranco, A.M., 1995. Forms of early walking. J. Theor. Biol. 176 (3), 373–390.

McDermott, M.M., Ohlmiller, S.M., Liu, K., Guralnik, J.M., Martin, G.J., Pearce, W.H., et al., 2001. Gait alterations associated with walking impairment in people with peripheral arterial disease with and without intermittent claudication. J. Am. Geriatr. Soc. 49 (6), 747–754.

McGibbon, C.A., 2003. Toward a better understanding of gait changes with age and disablement: neuromuscular adaptation. Exerc. Sport Sci. Rev. 31 (2), 102–108.

McGibbon, C.A., Krebs, D.E., 2001. Age-related changes in lower trunk coordination and energy transfer during gait. J. Neurophysiol. 85 (5), 1923–1931.

McGowan, C.P., Kram, R., Neptune, R.R., 2009. Modulation of leg muscle function in response to altered demand for body support and forward propulsion during walking. J. Biomech. 42 (7), 850–856.

McGowan, C.P., Neptune, R.R., Clark, D.J., Kautz, S.A., 2010. Modular control of human walking: adaptations to altered mechanical demands. J. Biomech. 43 (3), 412–419.

McGrath, D., Judkins, T.N., Pipinos, I.I., Johanning, J.M., Myers, S.A., 2012. Peripheral arterial disease affects the frequency response of ground reaction forces during walking. Clin. Biomech. 27 (10), 1058–1063.

McMahon, T.A., Cheng, G.C., 1990. The mechanics of running: how does stiffness couple with speed? J. Biomech. 23 (Suppl 1), 65–78.

McNealy, L.L., Gard, S.A., 2008. Effect of prosthetic ankle units on the gait of persons with bilateral trans-femoral amputations. Prosthetics Orthot. Int. 32 (1), 111–126.

McPoil, T.G., Ford, J., Fundaun, J., Gallegos, C., Kinney, A., McMillan, P., et al., 2016. The use of a static measure to predict foot posture at midstance during walking. Foot 28, 47–53.

Mei, Q., Fernandez, J., Fu, W., Feng, N., Gu, Y., 2015. A comparative biomechanical analysis of habitually unshod and shod runners based on a foot morphological difference. Hum. Mov. Sci. 42, 38–53.

Mei, Q., Gu, Y., Xiang, L., Yu, P., Gao, Z., Shim, V., et al., 2020. Foot shape and plantar pressure relationships in shod and barefoot populations. Biomech. Model. Mechanobiol. 19 (4), 1211–1224.

Mercer, J.A., Vance, J., Hreljac, A., Hamill, J., 2002. Relationship between shock attenuation and stride length during running at different velocities. Eur. J. Appl. Physiol. 87 (4-5), 403–408.

Mercer, J.A., Devita, P., Derrick, T.R., Bates, B.T., 2003. Individual effects of stride length and frequency on shock attenuation during running. Med. Sci. Sports Exerc. 35 (2), 307–313.

Mercer, M.A., Stone, T.M., Young, J.C., Mercer, J.A., 2018. Running economy while running in shoes categorized as maximal cushioning. Int. J. Exercise Sci. 11 (2), 1031–1040.

Messier, S.P., Davis, S.E., Curl, W.W., Lowery, R.B., Pack, R.J., 1991. Etiologic factors associated with patellofemoral pain in runners. Med. Sci. Sports Exerc. 23 (9), 1008–1015.

Meyer, K.J., Chan, C., Hopper, L., Nicholson, L.L., 2017. Identifying lower limb specific and generalised joint hypermobility in adults: validation of the Lower Limb Assessment Score. BMC Musculoskelet. Disord. 18, 514. https://doi.org/10.1186/s12891-017-1875-8.

Meyer, C., Killeen, T., Easthope, C.S., Curt, A., Bolliger, M., Linnebank, M., et al., 2019. Familiarization with treadmill walking: how much is enough? Sci. Rep. 9, 5232. https://doi.org/10.1038/s41598-019-41721-0.

Meyns, P., Bruijn, S.M., Duysens, J., 2013. The how and why of arm swing during human walking. Gait Posture 38 (4), 555–562.

Mian, O.S., Thom, J.M., Ardigò, L.P., Narici, M.V., Minetti, A.E., 2006. Metabolic cost, mechanical work, and efficiency during walking in young and older men. Acta Physiol. 186 (2), 127–139.

Michel, V., Do, M.C., 2002. Are stance ankle plantar flexor muscles necessary to generate propulsive force during human gait initiation? Neurosci. Lett. 325 (2), 139–143.

Michelson, J.D., Durant, D.M., McFarland, E., 2002. The injury risk associated with pes planus in athletes. Foot Ankle Int. 23 (7), 629–633.

Mickle, K.J., Steele, J.R., Munro, B.J., 2006. Does excess mass affect plantar pressure in young children? Int. J. Pediatr. Obes. 1 (3), 183–188.

Mickle, K.J., Steele, J.R., Munro, B.J., 2008. Is the foot structure of preschool children moderated by gender? J. Pediatr. Orthop. 28 (5), 593–596.

Mientjes, M.I., Frank, J.S., 1999. Balance in chronic low back pain patients compared to healthy people under various conditions in upright standing. Clin. Biomech. 14 (10), 710–716.

Miller, R.H., Hamill, J., 2015. Optimal footfall patterns for cost minimization in running. J. Biomech. 48 (11), 2858–2864.

Millet, G.Y., 2011. Can neuromuscular fatigue explain running strategies and performance in ultra-marathons? The flush model. Sports Medicine. 41 (6), 489–506.

Millet, G.Y., Lepers, R., 2004. Alterations of neuromuscular function after prolonged running, cycling and skiing exercises. Sports Med. 34 (2), 105–116.

Millet, G.Y., Hoffman, M.D., Morin, J.B., 2012. Sacrificing economy to improve running performance – a reality in the ultramarathon? J. Appl. Physiol. 113 (3), 507–509.

Mills, K., Hunt, M.A., Ferber, R., 2013a. Biomechanical deviations during level walking associated with knee osteoarthritis: a systematic review and meta-analysis. Arthritis Care Res. 65 (10), 1643–1665.

Mills, K., Hettinga, B.A., Pohl, M.B., Ferber, R., 2013b. Between-limb kinematic asymmetry during gait in unilateral and bilateral mild to moderate knee osteoarthritis. Arch. Phys. Med. Rehabil. 94 (11), 2241–2247.

Miyazaki, T., Wada, M., Kawahara, H., Sato, M., Baba, H., Shimada, S., 2002. Dynamic load at baseline can predict radiographic disease progression in medial compartment knee osteoarthritis. Ann. Rheum. Dis. 61 (7), 617–622.

Mizuno, Y., Kumagai, M., Mattessich, S.M., Elias, J.J., Ramrattan, N., Cosgarea, A.J., et al., 2001. Q-angle influences tibiofemoral and patellofemoral kinematics. J. Orthop. Res. 19 (5), 834–840.

Moccellin, A.S., Nora, F.G.S.A., Costa, P.H.L., Driusso, P., 2015. Static postural control assessment during pregnancy. Brazil. J. Motor Behav. 9 (1), 1–9.

Moon, Y., Sung, J., An, R., Hernandez, M.E., Sosnoff, J.J., 2016. Gait variability in people with neurological disorders: a systematic review and meta-analysis. Hum. Mov. Sci. 47, 197–208.

Moore, I.S., Jones, A.M., Dixon, S.J., 2014. Relationship between metabolic cost and muscular coactivation across running speeds. J. Sci. Med. Sport 17 (6), 671–676.

Moreno-Hernández, A., Rodríguez-Reyes, G., Quiñones-Urióstegui, I., Núñez-Carrera, L., Pérez-SanPablo, A.I., 2010. Temporal and spatial gait parameters analysis in non-pathological Mexican children. Gait Posture 32 (1), 78–81.

Morgan, P., McGinley, J., 2014. Gait function and decline in adults with cerebral palsy: a systematic review. Disabil. Rehabil. 36 (1), 1–9.

Morin, J.B., Samozino, P., Zameziati, K., Belli, A., 2007. Effects of altered stride frequency and contact time on leg-spring behavior in human running. J. Biomech. 40 (15), 3341–3348.

Morin, J.-B., Samozino, P., Millet, G.Y., 2011a. Changes in running kinematics, kinetics, and spring-mass behavior over a 24-h run. Med. Sci. Sports Exerc. 43 (5), 829–836.

Morin, J.B., Tomazin, K., Edouard, P., Millet, G.Y., 2011b. Changes in running mechanics and spring-mass behavior induced by a mountain ultra-marathon race. J. Biomech. 44 (6), 1104–1107.

Morin, J.-B., Slawinski, J., Dorel, S., de Villareal, E.S., Couturier, A., Samozino, P., et al., 2015. Acceleration capability in elite sprinters and ground impulse: push more, brake less? J. Biomech. 48 (12), 3149–3154.

Morton, S.M., Bastian, A.J., 2006. Cerebellar contributions to locomotor adaptations during splitbelt treadmill walking. J. Neurosci. 26 (36), 9107–9116.

Mosca, V.S., 2001. The cavus foot. J. Pediatr. Orthop. 21 (4), 423–424.

Moseley, G.L., Hodges, P.W., 2005. Are the changes in postural control associated with low back pain caused by pain interference? Clin. J. Pain 21 (4), 323–329.

Müller, R., Ertelt, T., Blickhan, R., 2015. Low back pain affects trunk as well as lower limb movements during walking and running. J. Biomech. 48 (6), 1009–1014.

Mündermann, A., Dyrby, C.O., Hurwitz, D.E., Sharma, L., Andriacchi, T.P., 2004. Potential strategies to reduce medial compartment loading in patients with knee osteoarthritis of varying severity: reduced walking speed. Arthrit. Rheumat. 50 (4), 1172–1178 (Erratum appears in: Arthritis & Rheumatism. 50 (12), 4073).

Murley, G.S., Menz, H.B., Landorf, K.B., 2014. Electromyographic patterns of tibialis posterior and related muscles when walking at different speeds. Gait Posture 39 (4), 1080–1085.

Myers, S.A., Pipinos, I.I., Johanning, J.M., Stergiou, N., 2011. Gait variability of patients with intermittent claudication is similar before and after the onset of claudication pain. Clin. Biomech. 26 (7), 729–734.

Nadeau, S., Gravel, D., Arsenault, A.B., Bourbonnais, D., 1999. Plantarflexor weakness as a limiting factor of gait speed in stroke subjects and the compensating role of hip flexors. Clin. Biomech. 14 (2), 125–135.

Naemi, R., Chatzistergos, P., Sundar, L., Chockalingam, N., Ramachandran, A., 2016a. Differences in the mechanical characteristics of plantar soft tissue between ulcerated and non-ulcerated foot. J. Diabetes Complicat. 30 (7), 1293–1299.

Naemi, R., Chatzistergos, P.E., Chockalingam, N., 2016b. A mathematical method for quantifying in vivo mechanical behaviour of heel pad under dynamic load. Med. Biol. Eng. Comput. 54 (2-3), 341–350.

Naemi, R., Behforootan, S., Chatzistergos, P., Chockalingam, N., 2016c. Chapter 10: Viscoelasticity in foot-ground interaction. In: El-Amin, M.F. (Ed.), Viscoelastic and Viscoplastic Materials. IntechOpen, pp. 217–243, https://doi.org/10.5772/64170.

Nagai, K., Yamada, M., Uemura, K., Yamada, Y., Ichihashi, N., Tsuboyama, T., 2011. Differences in muscle coactivation during postural control between healthy older and young adults. Arch. Gerontol. Geriatr. 53 (3), 338–343.

Nagai, K., Yamada, M., Tanaka, B., Uemura, K., Mori, S., Aoyama, T., et al., 2012. Effects of balance training on muscle coactivation during postural control in older adults: a randomized controlled trial. J. Gerontol. Ser. A Biol. Med. Sci. 67 (8), 882–889.

Nashner, L.M., McCollum, G., 1985. The organization of human postural movements: a formal basis and experimental synthesis. Behav. Brain Sci. 8 (1), 135–150.

Natali, A.N., Fontanella, C.G., Carniel, E.L., 2012. A numerical model for investigating the mechanics of calcaneal fat pad region. J. Mech. Behav. Biomed. Mater. 5 (1), 216–223.

Needham, R., Chockalingam, N., Dunning, D., Healy, A., Ahmed, E.B., Ward, A., 2012. The effect of leg length discrepancy on pelvis and spine kinematics during gait. Studies in Health Technology and Informatics. 176: 104-107. In: Kotwicki, T., Grivas, T.B. (Eds.), Research into Spinal Deformities. IOS Press, p. 8.

Nelson, M.E., Layne, J.E., Bernstein, M.J., Nuernberger, A., Castaneda, C., Kaliton, D., et al., 2004. The effects of multidimensional home-based exercise on functional performance in elderly people. J. Gerontol. Ser. A Biol. Med. Sci. 59 (2), M154–M160.

Nene, A., Byrne, C., Hermens, H., 2004. Is rectus femoris really a part of quadriceps? Assessment of rectus femoris function during gait in able-bodied adults. Gait Posture 20 (1), 1–13.

Neptune, R.R., Kautz, S.A., Zajac, F.E., 2001. Contributions of the individual ankle plantar flexors to support, forward progression and swing initiation during walking. J. Biomech. 34 (11), 1387–1398.

Neptune, R.R., Zajac, F.E., Kautz, S.A., 2004a. Muscle mechanical work requirements during normal walking: the energetic cost of raising the body's center-of-mass is significant. J. Biomech. 37 (6), 817–825.

Neptune, R.R., Zajac, F.E., Kautz, S.A., 2004b. Muscle force redistributes segmental power for body progression during walking. Gait Posture 19 (2), 194–205.

Neptune, R.R., Burnfield, J.M., Mulroy, S.J., 2007. The neuromuscular demands of toe walking: a forward dynamics simulation analysis. J. Biomech. 40 (6), 1293–1300.

Neptune, R.R., Sasaki, K., Kautz, S.A., 2008. The effect of walking speed on muscle function and mechanical energetics. Gait Posture 28 (1), 135–143.

Neville, C., Flemister, A.S., Houck, J., 2013. Total and distributed plantar loading in subjects with stage II tibialis posterior tendon dysfunction during terminal stance. Foot Ankle Int. 34 (1), 131–139.

Newland, M.R., Patel, A.R., Prieto, L., Boulton, A.J.M., Pacheco, M., Kirsner, R.S., 2009. Neuropathy and gait disturbances in patients with venous disease: a pilot study. Arch. Dermatol. 145 (4), 485–486.

Nigg, B.M., MacIntosh, B.R., 2000. Forces acting on and in the human body. In: Nigg, B.M., Mester, J. (Eds.), Biomechanics and Biology of Movement. Human Kinetics, Champaign, IL, pp. 253–267 (Part III, Chapter 14).

Nigg, B.M., Wakeling, J.M., 2001. Impact forces and muscle tuning: a new paradigm. Exerc. Sport Sci. Rev. 29 (1), 37–41.

Nigg, B.M., Bahlsen, H.A., Luethi, S.M., Stokes, S., 1987. The influence of running velocity and midsole hardness on external impact forces in heel-toe running. J. Biomech. 20 (10), 951–959.

Nigg, B.M., Cole, G.K., Nachbauer, W., 1993. Effects of arch height of the foot on angular motion of the lower extremities in running. J. Biomech. 26 (8), 909–916.

Nigg, B.M., Cole, G.K., Brüggemann, G.-P., 1995. Impact forces during heel-toe running. J. Appl. Biomech. 11 (4), 407–432.

Nigg, B.M., Stefanyshyn, D., Cole, G., Stergiou, P., Miller, J., 2003. The effect of material characteristics of shoe soles on muscle activation and energy aspects during running. J. Biomech. 36 (4), 569–575.

Nigg, B.M., Baltich, J., Maurer, C., Federolf, P., 2012. Shoe midsole hardness, sex and age effects on lower extremity kinematics during running. J. Biomech. 45 (9), 1692–1697.

Noakes, T.D., 2012. Fatigue is a brain-derived emotion that regulates the exercise behavior to ensure the protection of whole body homeostasis. Front. Physiol. 3, 82. https://doi.org/10.3389/fphys.2012.00082.

Nolan, L., Lees, A., 2000. The functional demands of the intact limb during walking for active trans-femoral and trans-tibial amputees. Prosthetics Orthot. Int. 24 (2), 117–125.

Novacheck, T.F., 1998. The biomechanics of running. Gait Posture 7 (1), 77–95.

Nowalk, M.P., Prendergast, J.M., Bayles, C.M., D'Amico, F.J., Colvin, G.C., 2001. A randomized trial of exercise programs among older individuals living in two long-term care facilities: the FallsFREE program. J. Am. Geriatr. Soc. 49 (7), 859–865.

Nüesch, C., Valderrabano, V., Huber, C., von Tscharner, V., Pagenstert, G., 2012. Gait patterns of asymmetric ankle osteoarthritis patients. Clin. Biomech. 27 (6), 613–618.

Nunns, M., House, C., Fallowfield, J., Allsopp, A., Dixon, S., 2013. Biomechanical characteristics of barefoot footstrike modalities. J. Biomech. 46 (15), 2603–2610.

Nyska, M., Sofer, D., Porat, A., Howard, C.B., Levi, A., Meizner, I., 1997. Plantar foot pressures in pregnant women. Isr. J. Med. Sci. 33 (2), 139–146.

O'Connell, M., Farrokhi, S., Fitzgerald, G.K., 2016. The role of knee joint moments and knee impairments on self-reported knee pain during gait in patients with knee osteoarthritis. Clin. Biomech. 31, 40–46.

Oh, K., Park, S., 2017. The bending stiffness of shoes is beneficial to running energetics if it does not disturb the natural MTP joint flexion. J. Biomech. 53, 127–135.

Okamoto, T., Okamoto, K., Andrew, P.D., 2003. Electromyographic developmental changes in one individual from newborn stepping to mature walking. Gait Posture 17 (1), 18–27.

Oleson, M., Adler, D., Goldsmith, P., 2005. A comparison of forefoot stiffness in running and running shoe bending stiffness. J. Biomech. 38 (9), 1886–1894.

Ortega, J.D., Farley, C.T., 2007. Individual limb work does not explain the greater metabolic cost of walking in elderly adults. J. Appl. Physiol. 102 (6), 2266–2273.

Ortega, J.A., Healey, L.A., Swinnen, W., Hoogkamer, W., 2021. Energetics and biomechanics of running footwear with increased longitudinal bending stiffness: a narrative review. Sports Med. 51 (5), 873–894.

Owings, T.M., Grabiner, M.D., 2004. Variability of step kinematics in young and older adults. Gait Posture 20 (1), 26–29.

Pai, S., Ledoux, W.R., 2012. The shear mechanical properties of diabetic and non-diabetic plantar soft tissue. J. Biomech. 45 (2), 364–370.

Pain, M.T.G., Challis, J.H., 2006. The influence of soft tissue movement on ground reaction forces, joint torques and joint reaction forces in drop landings. J. Biomech. 39 (1), 119–124.

Pekny, S.E., Izawa, J., Shadmehr, R., 2015. Reward-dependent modulation of movement variability. J. Neurosci. 35 (9), 4015–4024.

Pendharkar, G., Percival, P., Morgan, D., Lai, D., 2012. Automated method to distinguish toe walking strides from normal strides in the gait of idiopathic toe walking children from heel accelerometry data. Gait Posture 35 (3), 478–482.

Perry, J., 1992. Gait Analysis: Normal and Pathological Function. SLACK Incorporated, Thorofare, NJ.

Perry, S.D., 2006. Evaluation of age-related plantar-surface insensitivity and onset age of advanced insensitivity in older adults using vibratory and touch sensation tests. Neurosci. Lett. 392 (1-2), 62–67.

Petersen, K., Hansen, C.B., Aagaard, P., Madsen, K., 2007. Muscle mechanical characteristics in fatigue and recovery from a marathon race in highly trained runners. Eur. J. Appl. Physiol. 101 (3), 385–396.

Petersen, J., Nielsen, R.O., Rasmussen, S., Sørensen, H., 2014. Comparisons of increases in knee and ankle joint moments following an increase in running speed from 8 to 12 to 16 km h^{-1}. Clin. Biomech. 29 (9), 959–964.

Peterson, D.S., Martin, P.E., 2010. Effects of age and walking speed on coactivation and cost of walking in healthy adults. Gait Posture 31 (3), 355–359.

Peterson, M.L., Christou, E., Rosengren, K.S., 2006. Children achieve adult-like sensory integration during stance at 12-years-old. Gait Posture 23 (4), 455–463.

Peterson, C.L., Kautz, S.A., Neptune, R.R., 2011. Braking and propulsive impulses increase with speed during accelerated and decelerated walking. Gait Posture 33 (4), 562–567.

Petrovic, M., Deschamps, K., Verschueren, S.M., Bowling, F.L., Maganaris, C.N., Boulton, A.J.M., et al., 2016. Is the metabolic cost of walking higher in people with diabetes? J. Appl. Physiol. 120 (1), 55–62.

Petrovic, M., Deschamps, K., Verschueren, S.M., Bowling, F.L., Maganaris, C.N., Boulton, A.J.M., et al., 2017. Altered leverage around the ankle in people with diabetes: a natural strategy to modify the muscular contribution during walking? Gait Posture 57, 85–90.

Petrovic, M., Maganaris, C.N., Deschamps, K., Verschueren, S.M., Bowling, F.L., Boulton, A.J.M., et al., 2018. Altered Achilles tendon function during walking in people with diabetic neuropathy: implications for metabolic energy saving. J. Appl. Physiol. 124 (5), 1333–1340.

Phethean, J., Nester, C., 2012. The influence of body weight, body mass index and gender on plantar pressures: results of a cross-sectional study of healthy children's feet. Gait Posture 36 (2), 287–290.

Podraza, J.T., White, S.C., 2010. Effect of knee flexion angle on ground reaction forces, knee moments and muscle co-contraction during an impact-like deceleration landing: Implications for the non-contact mechanism of ACL injury. Knee 17 (4), 291–295.

Pohl, M.B., Buckley, J.G., 2008. Changes in foot and shank coupling due to alterations in foot strike pattern during running. Clin. Biomech. 23 (3), 334–341.

Pohl, M.B., Messenger, N., Buckley, J.G., 2007. Forefoot, rearfoot and shank coupling: effect of variations in speed and mode of gait. Gait Posture 25 (2), 295–302.

Pollard, C.D., Ter Har, J.A., Hannigan, J.J., Norcross, M.F., 2018. Influence of maximal running shoes on biomechanics before and after a 5K run. Orthop. J. Sports Med. 6 (6), 2325967118775720. https://doi.org/10.1177/2325967118775720.

Pollind, M., Soangra, R., Grant-Beuttler, M., Aminian, A., 2019. Customized wearable sensor-based insoles for gait re-training in idiopathic toe walkers. Biomed. Sci. Instrum. 55 (2), 192–198.

Ponnapula, P., Boberg, J.S., 2010. Lower extremity changes experienced during pregnancy. J. Foot Ankle Surg. 49 (5), 452–458.

Pontzer, H., 2017. Economy and endurance in human evolution. Curr. Biol. 27 (12), R613–R621.

Pontzer, H., Raichlen, D.A., Sockol, M.D., 2009a. The metabolic cost of walking in humans, chimpanzees, and early hominins. J. Hum. Evol. 56 (1), 43–54.

Pontzer, H., Holloway 4th, J.H., Raichlen, D.A., Lieberman, D.E., 2009b. Control and function of arm swing in human walking and running. J. Exp. Biol. 212 (4), 523–534.

Pontzer, H., Rolian, C., Rightmire, G.P., Jashashvili, T., Ponce de León, M.S., Lordkipanidze, D., et al., 2010. Locomotor anatomy and biomechanics of the Dmanisi hominins. J. Hum. Evol. 58 (6), 492–504.

Pontzer, H., Raichlen, D.A., Rodman, P.S., 2014. Bipedal and quadrupedal locomotion in chimpanzees. J. Hum. Evol. 66, 64–82.

Pontzer, H., Durazo-Arvizu, R., Dugas, L.R., Plange-Rhule, J., Bovet, P., Forrester, T.E., et al., 2016. Constrained total energy expenditure and metabolic adaptation to physical activity in adult humans. Curr. Biol. 26 (3), 410–417.

Potthast, W., Brüggemann, G.-P., Lundberg, A., Arndt, A., 2010. The influences of impact interface, muscle activity, and knee angle on impact forces and tibial and femoral accelerations occurring after external impacts. J. Appl. Biomech. 26 (1), 1–9.

Powell, D.W., Long, B., Milner, C.E., Zhang, S., 2011. Frontal plane multi-segment foot kinematics in high- and low-arched females during dynamic loading tasks. Hum. Mov. Sci. 30 (1), 105–114.

Powell, D.W., Long, B., Milner, C.E., Zhang, S., 2012a. Effects of vertical loading on arch characteristics and intersegmental foot motions. J. Appl. Biomech. 28 (2), 165–173.

Powell, D.W., Hanson, N.J., Long, B., Williams 3rd., D.S.B., 2012b. Frontal plane landing mechanics in high-arched compared with low-arched female athletes. Clin. J. Sport Med. 22 (5), 430–435.

Powell, D.W., Williams 3rd, D.S.B., Windsor, B., Butler, R.J., Zhang, S., 2014. Ankle work and dynamic joint stiffness in high- compared to low-arched athletes during a barefoot running task. Hum. Mov. Sci. 34, 147–156.

Powell, D.W., Andrews, S., Stickly, C., Williams, D.S.B., 2016. High- compared to low-arched athletes exhibit smaller knee abduction moments in walking and running. Hum. Mov. Sci. 50, 47–53.

Powell, D.W., Paquette, M.R., Williams 3rd., D.S.B., 2017. Contributions to leg stiffness in high- compared with low-arched athletes. Med. Sci. Sports Exerc. 49 (8), 1662–1667.

Preece, S.J., Mason, D., Bramah, C., 2016. The coordinated movement of the spine and pelvis during running. Hum. Mov. Sci. 45, 110–118.

Price, C., Morrison, S.C., Hashmi, F., Phethean, J., Nester, C., 2018. Biomechanics of the infant foot during the transition to independent walking: a narrative review. Gait Posture 59, 140–146.

Prince, F., Corriveau, H., Hébert, R., Winter, D.A., 1997. Gait in the elderly. Gait Posture 5 (2), 128–135.

Queen, R.M., Gross, M.T., Liu, H.-Y., 2006. Repeatability of lower extremity kinetics and kinematics for standardized and self-selected running speeds. Gait Posture 23 (3), 282–287.

Rabuffetti, M., Baroni, G., 1999. Validation protocol of models for centre of mass estimation. J. Biomech. 32 (6), 609–613.

Racinais, S., Girard, O., Micallef, J.P., Perrey, S., 2007. Failed excitability of spinal motoneurons induced by prolonged running exercise. J. Neurophysiol. 97 (1), 596–603.

Radtke, K., Karch, N., Goede, F., Vaske, B., von Lewinski, G., Noll, Y., et al., 2018. Outcomes of noninvasively treated idiopathic toe walkers. Foot Ankle Special. 12 (1), 54–61.

Redmond, A.C., Crosbie, J., Ouvrier, R.A., 2006. Development and validation of a novel rating system for scoring standing foot posture: the Foot Posture Index. Clin. Biomech. 21 (1), 89–98.

Reed-Jones, R.J., Solis, G.R., Lawson, K.A., Loya, A.M., Cude-Islas, D., Berger, C.S., 2013. Vision and falls: a multidisciplinary review of the contributions of visual impairment to falls among older adults. Maturitas 75 (1), 22–28.

Remvig, L., Jensen, D.V., Ward, R.C., 2007. Are diagnostic criteria for general joint hypermobility and benign joint hypermobility syndrome based on reproducible and valid tests? A review of the literature. J. Rheumatol. 34 (4), 798–803.

Renkawitz, T., Weber, T., Dullien, S., Woerner, M., Dendorfer, S., Grifka, J., et al., 2016. Leg length and offset differences above 5mm after total hip arthroplasty are associated with altered gait kinematics. Gait Posture 49, 196–201.

Resende, R.A., Deluzio, K.J., Kirkwood, R.N., Hassan, E.A., Fonseca, S.T., 2015. Increased unilateral foot pronation affects lower limbs and pelvic biomechanics during walking. Gait Posture 41 (2), 395–401.

Resende, R.A., Kirkwood, R.N., Deluzio, K.J., Cabral, S., Fonseca, S.T., 2016. Biomechanical strategies implemented to compensate for mild leg length discrepancy during gait. Gait Posture 46, 147–153.

Ribeiro, A.P., Trombini-Souza, F., de Camargo Neves Sacco, I., Ruano, R., Zugaib, M., João, S.M.A., 2011. Changes in the plantar pressure distribution during gait throughout gestation. J. Am. Podiatr. Med. Assoc. 101 (5), 415–423.

Richards, J., 2008. Chapter 3: Ground reaction forces, impulse and momentum. In: Richards, J. (Ed.), Biomechanics in Clinic and Research: An Interactive Teaching and Learning Course. Churchill Livingstone Elsevier, Philadelphia, pp. 35–49.

Richards, J., Thewlis, D., 2008. Chapter 4: Motion and joint motion. In: Richards, J. (Ed.), Biomechanics in Clinic and Research: An Interactive Teaching and Learning Course. Churchill Livingstone Elsevier, Philadelphia, pp. 51–65.

Riddick, R.C., Kuo, A.D., 2016. Soft tissues store and return mechanical energy in human running. J. Biomech. 49 (3), 436–441.

Riddiford-Harland, D.L., Steele, J.R., Baur, L.A., 2011. Medial midfoot fat pad thickness and plantar pressures: are these related in children? Int. J. Pediatr. Obes. 6 (3-4), 261–266.

Riley, P.O., Della Croce, U., Kerrigan, D.C., 2001. Effect of age on lower extremity joint moment contributions to gait speed. Gait Posture 14 (3), 264–270.

Riley, P.O., Dicharry, J., Franz, J., Della Croce, U., Wilder, R.P., Kerrigan, D.C., 2008. A kinematics and kinetic comparison of overground and treadmill running. Med. Sci. Sports Exerc. 40 (6), 1093–1100.

Ritchie, J.R., 2003. Orthopedic considerations during pregnancy. Clin. Obstet. Gynecol. 46 (2), 456–466.

Roberts, T.J., Chen, M.S., Taylor, C.R., 1998. Energetics of bipedal running. II. Limb design and running mechanics. J. Exp. Biol. 201 (19), 2753–2762.

Rochester, L., Galna, B., Lord, S., Burn, D., 2014. The nature of dual-task interference during gait in incident Parkinson's disease. Neuroscience 265, 83–94.

Rodman, P.S., McHenry, H.M., 1980. Bioenergetics and the origin of hominid bipedalism. Am. J. Phys. Anthropol. 52 (1), 103–106.

Romanov, N., Fletcher, G., 2007. Runners do not push off the ground but fall forwards via gravitational torque. Sports Biomech. 6 (3), 434–452.

Rooney, B.D., Derrick, T.R., 2013. Joint contact loading in forefoot and rearfoot strike patterns during running. J. Biomech. 46 (13), 2201–2206.

Rose, A., Birch, I., Kuisma, R., 2011. Effect of motion control running shoes compared with neutral shoes on tibial rotation during running. Physiotherapy 97 (3), 250–255.

Rosenbaum, P., Paneth, N., Leviton, A., Goldstein, M., Bax, M., Damiano, D., et al., 2007. A report: the definition and classification of cerebral palsy April 2006. Dev. Med. Child Neurol. 49 (s109), 8–14.

Roy, J.-P.R., Stefanyshyn, D.J., 2006. Shoe midsole longitudinal bending stiffness and running economy, joint energy, and EMG. Med. Sci. Sports Exerc. 38 (3), 562–569.

Ruina, A., Bertram, J.E.A., Srinivasan, M., 2005. A collisional model of the energetic cost of support work qualitatively explains leg sequencing in walking and galloping, pseudo-elastic leg behavior in running and the walk-to-run transition. J. Theor. Biol. 237 (2), 170–192.

Runge, C.F., Shupert, C.L., Horak, F.B., Zajac, F.E., 1999. Ankle and hip postural strategies defined by joint torques. Gait Posture 10 (2), 161–170.

Ryan, M.B., Valiant, G.A., McDonald, K., Taunton, J.E., 2011. The effect of three different levels of footwear stability on pain outcomes in women runners: a randomised control trial. Br. J. Sports Med. 45 (9), 715–721.

Ryan, M., Elashi, M., Newsham-West, R., Taunton, J., 2013. Examining the potential role of minimalist footwear for the prevention of proximal lower-extremity injuries. Footwear Sci. 5 (Suppl 1), S31–S32.

Sadeghi, H., Allard, P., Prince, F., Labelle, H., 2000. Symmetry and limb dominance in able-bodied gait: a review. Gait Posture 12 (1), 34–45.

Sagawa Jr., Y., Turcot, K., Armand, S., Thevenon, A., Vuillerme, N., Watelain, E., 2011. Biomechanics and physiological parameters during gait in lower-limb amputees: a systematic review. Gait Posture 33 (4), 511–526.

Sahin, N., Baskent, A., Ugurlu, H., Berker, E., 2008. Isokinetic evaluation of knee extensor/flexor muscle strength in patients with hypermobility syndrome. Rheumatol. Int. 28 (7), 643–648.

Saldanha, A., Nordlund Ekblom, M.M., Thorstensson, A., 2008. Central fatigue affects plantar flexor strength after prolonged running. Scand. J. Med. Sci. Sports 18 (3), 383–388.

Salzler, S., Bluman, E.M., Noonan, S., Chiodo, C.P., de Asla, R.J., 2012. Injuries observed in minimalist runners. Foot Ankle Int. 33 (4), 262–266.

Samson, W., Desroches, G., Cheze, L., Dumas, R., 2009. 3D joint dynamics analysis of healthy children's gait. J. Biomech. 42 (15), 2447–2453.

Samson, W., Dohin, B., Desroches, G., Chaverot, J.-L., Dumas, R., Cheze, L., 2011. Foot mechanics during the first six years of independent walking. J. Biomech. 44 (7), 1321–1327.

Samson, W., Van Hamme, A., Desroches, G., Dohin, B., Dumas, R., Chèze, L., 2013. Biomechanical maturation of joint dynamics during early childhood: updated conclusions. J. Biomech. 46 (13), 2258–2263.

Saragas, N.P., Becker, P.J., 1995. Comparative radiographic analysis of parameters in feet with and without hallux valgus. Foot Ankle Int. 16 (3), 139–143.

Saraswat, P., MacWilliams, B.A., Davis, R.B., D'Astous, J.L., 2014. Kinematics and kinetics of normal and planovalgus feet during walking. Gait Posture 39 (1), 339–345.

Sasaki, K., Neptune, R.R., 2006. Differences in muscle function during walking and running at the same speed. J. Biomech. 39 (11), 2005–2013.

Saunders, S.W., Schache, A., Rath, D., Hodges, P.W., 2005. Changes in three dimensional lumbo-pelvic kinematics and trunk muscle activity with speed and mode of locomotion. Clin. Biomech. 20 (8), 784–793.

Sax, T.W., Rosenbaum, R.B., 2006. Neuromuscular disorders in pregnancy. Muscle Nerve 34 (5), 559–571.

Scaglioni-Solano, P., Aragón-Vargas, L.F., 2015. Age-related differences when walking downhill on different sloped terrains. Gait Posture 41 (1), 153–158.

Schache, A.G., Blanch, P., Rath, D., Wrigley, T., Bennell, K., 2002. Three-dimensional angular kinematics of the lumbar spine and pelvis during running. Hum. Mov. Sci. 21 (2), 273–293.

Schafer, Z.A., Vanicek, N., 2021. A block randomised controlled trial investigating changes in postural control following a personalised 12-week exercise programme for individuals with lower limb amputation. Gait Posture 84, 198–204.

Schafer, Z.A., Perry, J.L., Vanicek, N., 2018. A personalised exercise programme for individuals with lower limb amputation reduces falls and improves gait biomechanics: a block randomised controlled trail. Gait Posture 63, 282–289.

Schipplein, O.D., Andriacchi, T.P., 1991. Interaction between active and passive knee stabilizers during level walking. J. Orthop. Res. 9 (1), 113–119.

Schmitt, S., Günther, M., 2011. Human leg impact: energy dissipation of wobbling masses. Arch. Appl. Mech. 81 (7), 887–897.

Schmitz, A., Pohl, M.B., Woods, K., Noehren, B., 2014. Variables during swing associated with decreased impact peak and loading rate in running. J. Biomech. 47 (1), 32–38.

Scholz, M.N., Bobbert, M.F., van Soest, A.J., Clark, J.R., van Heerden, J., 2008. Running biomechanics: shorter heels, better economy. J. Exp. Biol. 211 (20), 3266–3271.

Schrack, J.A., Simonsick, E.M., Chaves, P.H.M., Ferrucci, L., 2012. The role of energetic cost in the age-related slowing of gait speed. J. Am. Geriatr. Soc. 60 (10), 1811–1816.

Schrack, J.A., Zipunnikov, V., Simonsick, E.M., Studenski, S., Ferrucci, L., 2016. Rising energetic cost of walking predicts gait speed decline with aging. J. Gerontol. Ser. A Biol. Med. Sci. 71 (7), 947–953.

Schwab, F., Lafage, V., Boyce, R., Skalli, W., Farcy, J.-P., 2006. Gravity line analysis in adult volunteers: age-related correlation with spinal parameters, pelvic parameters, and foot position. Spine 31 (25), E959–E967.

Sees, J.P., Miller, F., 2013. Overview of foot deformity management in children with cerebral palsy. J. Child. Orthop. 7 (5), 373–377.

Seneli, R.M., Beschorner, K.E., O'Connor, K.M., Keenan, K.G., Earl-Boehm, J.E., Cobb, S.C., 2021. Foot joint coupling variability differences between habitual rearfoot and forefoot runners prior to and following an exhaustive run. J. Electromyogr. Kinesiol. 57, 102514. https://doi.org/10.1016/j.jelekin.2021.102514.

Seynnes, O.R., de Boer, M., Narici, M.V., 2007. Early skeletal muscle hypertrophy and architectural changes in response to high-intensity resistance training. J. Appl. Physiol. 102 (1), 368–373.

Shakoor, N., Block, J.A., 2006. Walking barefoot decreases loading on the lower extremity joints in knee osteoarthritis. Arthritis Rheum. 54 (9), 2923–2927.

Sharbafi, M.A., Mohammadi Nejad Rashty, A., Rode, C., Seyfarth, A., 2017. Reconstruction of human swing leg motion with passive biarticular muscle models. Hum. Mov. Sci. 52, 96–107.

Shih, Y., Lin, K.-L., Shiang, T.-Y., 2013. Is the foot striking pattern more important than barefoot or shod conditions in running? Gait Posture 38 (3), 490–494.

Shiman, M.I., Pieper, B., Templin, T.N., Birk, T.J., Patel, A.R., Kirsner, R.S., 2009. Venous ulcers: a reappraisal analyzing the effects of neuropathy, muscle involvement, and range of motion upon gait and calf muscle function. Wound Repair Regen. 17 (2), 147–152.

Shiotani, H., Yamashita, R., Mizokuchi, T., Sado, N., Naito, M., Kawakami, Y., 2021. Track distance runners exhibit bilateral differences in the plantar fascia stiffness. Sci. Rep. 11, 9260. https://doi.org/10.1038/s41598-021-88883-4.

Shorter, K.A., Wu, A., Kuo, A.D., 2017. The high cost of swing leg circumduction during human walking. Gait Posture 54, 265–270.

Silder, A., Besier, T., Delp, S.L., 2015. Running with a load increases leg stiffness. J. Biomech. 48 (6), 1003–1008.

Simkin, A., Leichter, I., 1990. Role of the calcaneal inclination in the energy storage capacity of the human foot – a biomechanical model. Med. Biol. Eng. Comput. 28 (2), 149–152.

Simkin, A., Leichter, I., Giladi, M., Stein, M., Milgrom, C., 1989. Combined effect of foot arch structure and an orthotic device on stress fractures. Foot Ankle 10 (1), 25–29.

Simmonds, J.V., Keer, R.J., 2007. Hypermobility and the hypermobility syndrome. Man. Ther. 12 (4), 298–309.

Simoneau, G.G., Hoenig, K.J., Lepley, J.E., Papanek, P.E., 1998. Influence of hip position and gender on active hip internal and external rotation. J. Orthop. Sports Phys. Ther. 28 (3), 158–164.

Simonsen, E.B., Tegner, H., Alkjær, T., Larsen, P.K., Kristensen, J.H., Jensen, B.R., et al., 2012. Gait analysis of adults with generalised joint hypermobility. Clin. Biomech. 27 (6), 573–577.

Sinclair, J., Greenhalgh, A., Edmundson, C.J., Brooks, D., Hobbs, S.J., 2012. Gender differences in the kinetics and kinematics of distance running: implications for footwear design. Int. J. Sports Sci. Eng. 6 (2), 118–128.

Sinclair, C., Svantesson, U., Sjöström, R., Alricsson, M., 2017. Differences in Pes Planus and Pes Cavus subtalar eversion/inversion before and after prolonged running, using a two-dimensional digital analysis. J. Exercise Rehabilit. 13 (2), 232–239.

Singer, S., Klejman, S., Pinsker, E., Houck, J., Daniels, T., 2013. Ankle arthroplasty and ankle arthrodesis: gait analysis compared with normal controls. J. Bone Joint Surg. 95-A (24), e191 (1-10).

Singh, N.B., König, N., Arampatzis, A., Heller, M.O., Taylor, W.R., 2012. Extreme levels of noise constitute a key neuromuscular deficit in the elderly. PLoS ONE 7 (11), e48449. https://doi.org/10.1371/journal.pone.0048449.

Smits-Engelsman, B., Klerks, M., Kirby, A., 2011. Beighton Score: a valid measure for generalized hypermobility in children. J. Pediatr. 158 (1), 119–123.

Snijders, A.H., van de Warrenburg, B.P., Giladi, N., Bloem, B.R., 2007. Neurological gait disorders in elderly people: clinical approach and classification. Lancet Neurol. 6 (1), 63–74.

Sockol, M.D., Raichlen, D.A., Pontzer, H., 2007. Chimpanzee locomotor energetics and the origin of human bipedalism. Proc. Natl. Acad. Sci. U. S. A. 104 (30), 12265–12269.

Song, J., Hillstrom, H.J., Secord, D., Levitt, J., 1996. Foot type biomechanics: comparison of planus and rectus foot types. J. Am. Podiatr. Med. Assoc. 86 (1), 16–23.

Song, K.M., Halliday, S.E., Little, D.G., 1997. The effect of limb-length discrepancy on gait. J. Bone Joint Surg. 79-A (11), 1690–1698.

Soule, R.G., Goldman, R.F., 1969. Energy cost of loads carried on the head, hands, or feet. J. Appl. Physiol. 27 (5), 687–690.

Souza, R.B., Powers, C.M., 2009. Differences in hip kinematics, muscle strength, and muscle activation between subjects with and without patellofemoral pain. J. Orthop. Sports Phys. Ther. 39 (1), 12–19.

Squadrone, R., Gallozzi, C., 2009. Biomechanical and physiological comparison of barefoot and two shod conditions in experienced barefoot runners. J. Sports Med. Phys. Fitness 49 (1), 6–13.

Srinivasan, M., Ruina, A., 2006. Computer optimization of a minimal biped model discovers walking and running. Nature 439 (7072), 72–75.

Staheli, L.T., 1999. Planovalgus foot deformity. Current status. J. Am. Podiatr. Med. Assoc. 89 (2), 94–99.

Stainsby, G.D., 1997. Pathological anatomy and dynamic effect of the displaced plantar plate and the importance of the integrity of the plantar plate-deep transverse metatarsal ligament tie-bar. Ann. R. Coll. Surg. Engl. 79 (1), 58–68.

Statler, T.K., Tullis, B.L., 2005. Pes cavus. J. Am. Podiatr. Med. Assoc. 95 (1), 42–52.

Stecco, C., Pavan, P., Pachera, P., De Caro, R., Natali, A., 2014. Investigation of the mechanical properties of the human crural fascia and their possible clinical implications. Surg. Radiol. Anat. 36 (1), 25–32.

Stefanyshyn, D.J., Nigg, B.M., 2000. Influence of midsole bending stiffness on joint energy and jump height performance. Med. Sci. Sports Exerc. 32 (2), 471–476.

Stergiou, N., Harbourne, R., Cavanaugh, J., 2006. Optimal movement variability: a new theoretical perspective for neurologic physical therapy. J. Neurol. Phys. Ther. 30 (3), 120–129.

Stewart, S., Dalbeth, N., Vandal, A.C., Rome, K., 2016. The first metatarsophalangeal joint in gout: a systematic review and meta-analysis. BMC Musculoskelet. Disord. 17, 69. https://doi.org/10.1186/s12891-016-0919-9.

Stewart, S., Dalbeth, N., Aiyer, A., Rome, K., 2020. Objectively assessed foot and ankle characteristics in patients with systemic lupus erythematosus: a comparison with age- and sex-matched controls. Arthritis Care Res. 72 (1), 122–130.

Stolwijk, N.M., Koenraadt, K.L.M., Louwerens, J.W.K., Grim, D., Duysens, J., Keijsers, N.L.W., 2014. Foot lengthening and shortening during gait: a parameter to investigate foot function? Gait Posture 39 (2), 773–777.

Stolze, H., Friedrich, H.J., Steinauer, K., Vieregge, P., 2000. Stride parameters in healthy young and old women - measurement variability on a simple walkway. Exp. Aging Res. 26 (2), 159–168.

Strathy, G.M., Chao, E.Y., Laughman, R.K., 1983. Changes in knee function associated with treadmill ambulation. J. Biomech. 16 (7), 517–522.

Struyf, P.A., van Heugten, C.M., Hitters, M.W., Smeets, R.J., 2009. The prevalence of osteoarthritis of the intact hip and kee among traumatic leg amputees. Arch. Phys. Med. Rehabil. 90 (3), 440–446.

Su, P.-F., Gard, S.A., Lipschutz, R.D., Kuiken, T.A., 2007. Gait characteristics of persons with bilateral transtibial amputations. J. Rehabil. Res. Dev. 44 (4), 491–501.

Su, P.-F., Gard, S.A., Lipschutz, R.D., Kuiken, T.A., 2008. Differences in gait characteristics between persons with bilateral transtibial amputations, due to peripheral vascular disease and trauma, and able-bodied ambulators. Arch. Phys. Med. Rehabil. 89 (7), 1386–1394.

Su, P.-F., Gard, S.A., Lipschutz, R.D., Kuiken, T.A., 2010. The effects of increased prosthetic ankle motions on the gait of persons with bilateral transtibial amputations. Am. J. Phys. Med. Rehabil. 89 (1), 34–47.

Sullivan, D., Warren, R.F., Pavlov, H., Kelman, G., 1984. Stress fractures in 51 runners. Clin. Orthop. Relat. Res. 187 (July-August), 188–192.

Sun, X., Lam, W.-K., Zhang, X., Wang, J., Fu, W., 2020. Systematic review of the role of footwear constructions in running biomechanics: implications for running-related injury and performance. J. Sports Sci. Med. 19 (1), 20–37.

Sutherland, D.H., Olshen, R.A., Biden, E.N., Wyatt, M.P., 1988. The Development of Mature Walking. [Clinics in Developmental Medicine. No. 104/105.]. Mac Keith Press, London.

Takabayashi, T., Edama, M., Inai, T., Nakamura, E., Kubo, M., 2020. Effect of gender and load conditions on foot arch height index and flexibility in Japanese youths. J. Foot Ankle Surg. 59 (6), 1144–1147.

Takacs, J., Hunt, M.A., 2012. The effect of contralateral pelvic drop and trunk lean on frontal plane knee biomechanics during single limb standing. J. Biomech. 45 (16), 2791–2796.

Takakusaki, K., 2008. Forebrain control of locomotor behaviors. Brain Res. Rev. 57 (1), 192–198.

Takakusaki, K., 2013. Neurophysiology of gait: from the spinal cord to the frontal lobe. Mov. Disord. 28 (11), 1483–1491.

Takegami, Y., 1992. Wave pattern of ground reaction force of growing children. J. Pediatr. Orthop. 12 (4), 522–526.

Tam, N., Astephen Wilson, J.L., Noakes, T.D., Tucker, R., 2014. Barefoot running: an evaluation of current hypothesis, future research and clinical applications. Br. J. Sports Med. 48 (5), 349–355.

Tardieu, C., Bonneau, N., Hecquet, J., Boulay, C., Marty, C., Legaye, J., et al., 2013. How is sagittal balance acquired during bipedal gait acquisition? Comparison of neonatal and adult pelves in three dimensions. Evolutionary implications. J. Hum. Evol. 65 (2), 209–222.

Tartaruga, M.P., Brisswalter, J., Peyré-Tartaruga, L.A., Ávila, A.O.V., Alberton, C.L., Coertjens, M., et al., 2012. The relationship between running economy and biomechanical variables in distance runners. Res. Q. Exerc. Sport 83 (3), 367–375.

Taunton, J.E., Ryan, M.B., Clement, D.B., McKenzie, D.C., Lloyd-Smith, D.R., Zumbo, B.D., 2002. A retrospective case-control analysis of 2002 running injuries. Br. J. Sports Med. 36 (2), 95–101.

Taylor, C.R., 1985. Force development during sustained locomotion: a determinant of gait, speed and metabolic power. J. Exp. Biol. 115 (1), 253–262.

Taylor, C.R., Heglund, N.C., McMahon, T.A., Looney, T.R., 1980. Energetic cost of generating muscular force during running: a comparison of large and small animals. J. Exp. Biol. 86 (1), 9–18.

Thelen, D.G., Schultz, A.B., Alexander, N.B., Ashton-Miller, J.A., 1996. Effects of age on rapid ankle torque development. J. Gerontol. Ser. A: Biol. Med. Sci. 51 (5), M226–M232.

Théveniau, N., Boisgontier, M.P., Varieras, S., Olivier, I., 2014. The effects of clothes on independent walking in toddlers. Gait Posture 39 (1), 659–661.

Thevenon, A., Gabrielli, F., Lepvrier, J., Faupin, A., Allart, E., Tiffreau, V., et al., 2015. Collection of normative data for spatial and temporal gait parameters in a sample of French children aged between 6 and 12. Ann. Phys. Rehabilit. Med. 58 (3), 139–144.

Thijs, Y., Pattyn, E., Van Tiggelen, D., Rombaut, L., Witvrouw, E., 2011. Is hip muscle weakness a predisposing factor for patellofemoral pain in female novice runners? A prospective study. Am. J. Sports Med. 39 (9), 1877–1882.

Thompson, M.A., Hoffman, K.M., 2017. Superficial plantar cutaneous sensation does not trigger barefoot running adaptations. Gait Posture 57, 305–309.

Thorstensson, A., Nilsson, J., Carlson, H., Zomlefer, M.R., 1984. Trunk movements in human locomotion. Acta Physiol. Scand. 121 (1), 9–22.

Tirosh, O., Sangeux, M., Wong, M., Thomason, P., Graham, H.K., 2013. Walking speed effects on the lower limb electromyographic variability of healthy children aged 7-16 years. J. Electromyogr. Kinesiol. 23 (6), 1451–1459.

Tofts, L.J., Elliott, E.J., Munns, C., Pacey, V., Sillence, D.O., 2009. The differential diagnosis of children with joint hypermobility: a review of the literature. Pediatr. Rheumatol. Online J. 7, 1. https://doi.org/10.1186/1546-0096-7-1.

Tomita, D., Suga, T., Tanaka, T., Ueno, H., Miyake, Y., Otsuka, M., et al., 2018. A pilot study on the importance of forefoot bone length in male 400-m sprinters: is there a key morphological factor for superior long sprint performance? BMC Res. Notes 11, 583. https://doi.org/10.1186/s13104-018-3685-y.

Tong, J.W.K., Kong, P.W., 2013. Association between foot type and lower extremity injuries: systematic literature review with meta-analysis. J. Orthop. Sports Phys. Ther. 43 (10), 700–714.

Tsai, Y.-J., Lin, S.-I., 2013. Older adults adopted more cautious gait patterns when walking in socks than barefoot. Gait Posture 37 (1), 88–92.

Tseh, W., Caputo, J.L., Morgan, D.W., 2008. Influence of gait manipulation on running economy in female distance runners. J. Sports Sci. Med. 7 (1), 91–95.

Turner, D.E., Woodburn, J., 2008. Characterising the clinical and biomechanical features of severely deformed feet in rheumatoid arthritis. Gait Posture 28 (4), 574–580.

Uchida, T.K., Hicks, J.L., Dembia, C.L., Delp, S.L., 2016. Stretching your energetic budget: how tendon compliance affects the metabolic cost of running. PLoS ONE 11 (3), e0150378. https://doi.org/10.1371/journal.pone.0150378.

Uden, H., Scharfbillig, R., Causby, R., 2017. The typically developing paediatric foot: how flat should it be? A systematic review. J. Foot Ankle Res. 10, 37. https://doi.org/10.1186/s13047-017-0218-1.

Umberger, B.R., 2010. Stance and swing costs in human walking. J. R. Soc. Interface 7 (50), 1329–1340.

Umberger, B.R., Rubenson, J., 2011. Understanding muscle energetics in locomotion: new modeling and experimental approaches. Exerc. Sport Sci. Rev. 39 (2), 59–67.

Usherwood, J.R., Szymanek, K.L., Daley, M.A., 2008. Compass gait mechanics account for top walking speeds in ducks and humans. J. Exp. Biol. 211 (23), 3744–3749.

Usherwood, J.R., Channon, A.J., Myatt, J.P., Rankin, J.W., Hubel, T.Y., 2012. The human foot and heel-sole-toe walking strategy: a mechanism enabling an inverted pendular gait with low isometric muscle force? J. R. Soc. Interface 9 (75), 2396–2402.

Valderrabano, V., Nigg, B.M., von Tscharner, V., Stefanyshyn, D.J., Goepfert, B., Hintermann, B., 2007. Gait analysis in ankle osteoarthritis and total ankle replacement. Clin. Biomech. 22 (8), 894–904.

van Beers, R.J., Baraduc, P., Wolpert, D.M., 2002. Role of uncertainty in sensorimotor control. Philos. Trans. Roy. Soc. Lond. B: Biol. Sci. 357 (1424), 1137–1145.

Van de Walle, P., Meyns, P., Desloovere, K., De Rijck, J., Kenis, J., Verbecque, E., et al., 2018. Age-related changes in arm motion during typical gait. Gait Posture 66, 51–57.

van der Krogt, M.M., Delp, S.L., Schwartz, M.H., 2012. How robust is human gait to muscle weakness? Gait Posture 36 (1), 113–119.

van der Linden, M.L., Kerr, A.M., Hazlewood, M.E., Hillman, S.J., Robb, J.E., 2002. Kinematic and kinetic gait characteristics of normal children walking at a range of clinically relevant speeds. J. Pediatr. Orthop. 22 (6), 800–806.

van Engelen, S.J.P.M., Wajer, Q.E., van der Plaat, L.W., Doets, H.C., van Dijk, C.N., Houdijk, H., 2010. Metabolic cost and mechanical work during walking after tibiotalar arthrodesis and the influence of footwear. Clin. Biomech. 25 (8), 809–815.

Van Hamme, A., El Habachi, A., Samson, W., Dumas, R., Chèze, L., Dohin, B., 2015. Gait parameters database for young children: the influences of age and walking speed. Clin. Biomech. 30 (6), 572–577.

Van Hooren, B., Bosch, F., 2017a. Is there really an eccentric action of the hamstrings during the swing phase of high-speed running? Part I: A critical review of the literature. J. Sports Sci. 35 (23), 2313–2321.

Van Hooren, B., Bosch, F., 2017b. Is there really an eccentric action of the hamstrings during the swing phase of high-speed running? Part II: Implications for exercise. J. Sports Sci. 35 (23), 2322–2333.

van Houtum, W.H., Lavery, L.A., Harkless, L.B., 1996. The impact of diabetes-related lower-extremity amputations in the Netherlands. J. Diabetes Complicat. 10 (6), 325–330.

van Werkhoven, H., Piazza, S.J., 2017. Does foot anthropometry predict metabolic cost during running? J. Appl. Biomech. 33 (5), 317–322.

Venkadesan, M., Yawar, A., Eng, C.M., Dias, M.A., Singh, D.K., Tommasini, S.M., et al., 2020. Stiffness of the human foot and evolution of the transverse arch. Nature 579 (7797), 97–100.

Vielemeyer, J., Grießbach, E., Müller, R., 2019. Ground reaction forces intersect above the center of mass even when walking down visible and camouflaged curbs. J. Exp. Biol. 222 (14), jeb204305. https://doi.org/10.1242/jeb.204305.

Voigt, M., Hansen, E.A., 2021. The puzzle of the walk-to-run transition in humans. Gait Posture 86, 319–326.

Voloshina, A.S., Ferris, D.P., 2015. Biomechanics and energetics of running on uneven terrain. J. Exp. Biol. 218 (5), 711–719.

Voloshina, A.S., Kuo, A.D., Daley, M.A., Ferris, D.P., 2013. Biomechanics and energetics of walking on uneven terrain. J. Exp. Biol. 216 (21), 3963–3970.

Wager, J.C., Challis, J.H., 2016. Elastic energy within the human plantar aponeurosis contributes to arch shortening during the push-off phase of running. J. Biomech. 49 (5), 704–709.

Wakeling, J.M., Nigg, B.M., 2001. Modification of soft tissue vibrations in the leg by muscular activity. J. Appl. Physiol. 90 (2), 412–420.

Wakeling, J.M., Pascual, S.A., Nigg, B.M., 2002. Altering muscle activity in the lower extremities by running with different shoes. Med. Sci. Sports Exerc. 34 (9), 1529–1532.

Wakeling, J.M., Liphardt, A.-M., Nigg, B.M., 2003. Muscle activity reduces soft-tissue resonance at heel-strike during walking. J. Biomech. 36 (12), 1761–1769.

Wang, R., Gutierrez-Farewik, E.M., 2011. The effect of subtalar inversion/eversion on the dynamic function of the tibialis anterior, soleus, and gastrocnemius during the stance phase of gait. Gait Posture 34 (1), 29–35.

Watson, N.L., Sutton-Tyrrell, K., Youk, A.O., Boudreau, R.M., Mackey, R.H., Simonsick, E.M., et al., 2011. Arterial stiffness and gait speed in older adults with and without peripheral arterial disease. Am. J. Hypertens. 24 (1), 90–95.

Weersink, J.B., de Jong, B.M., Halliday, D.M., Maurits, N.M., 2021. Intermuscular coherence analysis in older adults reveals that gait-related arm swing drives lower limb muscles via subcortical and cortical pathways. J. Physiol. 599 (8), 2283–2298.

Weibel, E.R., Bacigalupe, L.D., Schmitt, B., Hoppeler, H., 2004. Allometric scaling of maximal metabolic rate in mammals: muscle aerobic capacity as determinant factor. Respir. Physiol. Neurobiol. 140 (2), 115–132.

Weijers, R.E., Kessels, A.G.H., Kemerink, G.J., 2005. The damping properties of the venous plexus of the heel region of the foot during simulated heelstrike. J. Biomech. 38 (12), 2423–2430.

Weir, J., Chockalingam, N., 2007. Ankle joint dorsiflexion: assessment of true values necessary for normal gait. Int. J. Ther. Rehabil. 14 (2), 76–82. https://doi.org/10.12968/ijtr.2007.14.2.23518.

Welte, L., Kelly, L.A., Kessler, S.E., Lieberman, D.E., D'Andrea, S.E., Lichtwark, G.A., et al., 2021. The extensibility of the plantar fascia influences the windlass mechanism during human running. Proc. Roy. Soc. B. Biol. Sci. 288 (1943), 20202095. https://doi.org/10.1098/rspb.2020.2095.

Weyand, P.G., Sternlight, D.B., Bellizzi, M.J., Wright, S., 2000. Faster top running speeds are achieved with greater ground forces not more rapid leg movements. J. Appl. Physiol. 89 (5), 1991–1999.

Weyand, P.G., Bundle, M.W., McGowan, C.P., Grabowski, A., Brown, M.B., Kram, R., et al., 2009. The fastest runner on artificial legs: different limbs, similar function? J. Appl. Physiol. 107 (3), 903–911.

Weyand, P.G., Sandell, R.F., Prime, D.N.L., Bundle, M.W., 2010. The biological limits to running speed are imposed from the ground up. J. Appl. Physiol. 108 (4), 950–961.

Willems, T., Witvrouw, E., Delbaere, K., De Cock, A., De Clercq, D., 2005. Relationship between gait biomechanics and inversion sprains: a prospective study of risk factors. Gait Posture 21 (4), 379–387.

Williams 3rd, D.S., McClay, I.S., Hamill, J., 2001a. Arch structure and injury patterns in runners. Clin. Biomech. 16 (4), 341–347.

Williams 3rd, D.S., McClay, I.S., Hamill, J., Buchanan, T.S., 2001b. Lower extremity kinematic and kinetic differences in runners with high and low arches. J. Appl. Biomech. 17 (2), 153–163.

Williams 3rd, D.S., Davis, I.M., Scholz, J.P., Hamill, J., Buchanan, T.S., 2004. High-arched runners exhibit increased leg stiffness compared to low-arched runners. Gait Posture 19 (3), 263–269.

Williams, K.R., Cavanagh, P.R., 1987. Relationship between distance running mechanics, running economy, and performance. J. Appl. Physiol. 63 (3), 1236–1245.

Williams, C.M., Tinley, P., Curtin, M., 2010a. The Toe Walking Tool: a novel method for assessing idiopathic toe walking children. Gait Posture 32 (4), 508–511.

Williams, C.M., Tinley, P., Curtin, M., 2010b. Idiopathic toe walking and sensory processing dysfunction. J. Foot Ankle Res. 3, 16. https://doi.org/10.1186/1757-1146-3-16.

Williams, C.M., Tinley, P., Curtin, M., Wakefield, S., Nielsen, S., 2014. Is idiopathic toe walking really idiopathic? The motor skills and sensory processing abilities associated with idiopathic toe walking gait. J. Child Neurol. 29 (1), 71–78.

Willson, J.D., Davis, I.S., 2009. Lower extremity strength and mechanics during jumping in women with patellofemoral pain. J. Sport Rehabil. 18 (1), 76–90.

Willson, J.D., Petrowitz, I., Butler, R.J., Kernozek, T.W., 2012. Male and female gluteal muscle activity and lower extremity kinematics during running. Clin. Biomech. 27 (10), 1052–1057.

Willwacher, S., König, M., Potthast, W., Brüggemann, G.-P., 2013. Does specific footwear facilitate energy storage and return at the metatarsophalangeal joint in running? J. Appl. Biomech. 29 (5), 583–592.

Willwacher, S., König, M., Braunstein, B., Goldmann, J.-P., Brüggemann, G.-P., 2014. The gearing function of running shoe longitudinal bending stiffness. Gait Posture 40 (3), 386–390.

Willy, R.W., Davis, I.S., 2011. The effect of a hip-strengthening program on mechanics during running and during a single-leg squat. J. Orthop. Sports Phys. Ther. 41 (9), 625–632.

Willy, R.W., Manal, K.T., Witvrouw, E.E., Davis, I.S., 2012. Are mechanics different between male and female runners with patellofemoral pain? Med. Sci. Sports Exerc. 44 (11), 2165–2171.

Wilson, C., Simpson, S.E., van Emmerik, R.E.A., Hamill, J., 2008. Coordination variability and skill development in expert triple jumpers. Sports Biomech. 7 (1), 2–9.

Winter, D.A., 1983. Moments of force and mechanical power in jogging. J. Biomech. 16 (1), 91–97.

Winter, D.A., Patla, A.E., Frank, J.S., Walt, S.E., 1990. Biomechanical walking pattern changes in the fit and healthy elderly. Phys. Ther. 70 (6), 340–347.

Witvrouw, E., Bellemans, J., Lysens, R., Danneels, L., Cambier, D., 2001. Intrinsic risk factors for the development of patellar tendinitis in an athletic population: a two-year prospective study. Am. J. Sports Med. 29 (2), 190–195.

Woledge, R.C., Birtles, D.B., Newham, D.J., 2005. The variable component of lateral body sway during walking in young and older humans. J. Gerontol. Ser. A Biol. Med. Sci. 60 (11), 1463–1468.

Wong, C.K., Ehrlich, J.E., Ersing, J.C., Maroldi, N.J., Stevenson, C.E., Varca, M.J., 2016. Exercise programs to improve gait performance in people with lower limb amputation: a systematic review. Prosthetics Orthot. Int. 40 (1), 8–17.

Wong, D.W.-C., Wang, Y., Leung, A.K.-L., Yang, M., Zhang, M., 2018. Finite element simulation on posterior tibial tendinopathy: load transfer alteration and implications to the onset of pes planus. Clin. Biomech. 51, 10–16.

Woodburn, J., Helliwell, P.S., Barker, S., 2002. Three-dimensional kinematics at the ankle joint complex in rheumatoid arthritis patients with painful valgus deformity of the rearfoot. Rheumatology 41 (12), 1406–1412.

Wright, I.C., Neptune, R.R., van den Bogert, A.J., Nigg, B.M., 1998. Passive regulation of impact forces in heel-toe running. Clin. Biomech. 13 (7), 521–531.

Wrobel, J.S., Najafi, B., 2010. Diabetic foot biomechanics and gait dysfunction. J. Diabetes Sci. Technol. 4 (4), 833–845.

Wu, H.G., Miyamoto, Y.R., Gonzalez Castro, L.N., Ölveczky, B.P., Smith, M.A., 2014. Temporal structure of motor variability is dynamically regulated and predicts motor learning ability. Nat. Neurosci. 17 (2), 312–321.

Wuehr, M., Pradhan, C., Brandt, T., Jahn, K., Schniepp, R., 2014. Patterns of optimization in single- and inter-leg gait dynamics. Gait Posture 39 (2), 733–738.

Yan, S.-h., Zhang, K., Tan, G.-q., Yang, J., Liu, Z.-c., 2013. Effects of obesity on dynamic plantar pressure distribution in Chinese prepubescent children during walking. Gait Posture 37 (1), 37–42.

Yazgan, P., Geyikli, I., Zeyrek, D., Baktiroglu, L., Kurcer, M.A., 2008. Is joint hypermobility important in prepubertal children? Rheumatol. Int. 28 (5), 445–451.

Yim, E., Vivas, A., Maderal, A., Kirsner, R.S., 2014. Neuropathy and ankle mobility abnormalities in patients with chronic venous disease. J. Am. Med. Assoc. Dermatol. 150 (4), 385–389.

Yong, J.R., Silder, A., Montgomery, K.L., Fredericson, M., Delp, S.L., 2018. Acute changes in foot strike pattern and cadence affect running parameters associated with tibial stress fractures. J. Biomech. 76, 1–7.

Young, W.R., Olonilua, M., Masters, R.S.W., Dimitriadis, S., Williams, A.M., 2016. Examining links between anxiety, reinvestment and walking when talking by older adults during adaptive gait. Exp. Brain Res. 234 (1), 161–172.

Zeni, J.A., Higginson, J.S., 2011. Knee osteoarthritis affects the distribution of joint moments during gait. Knee 18 (3), 156–159.

Zeni Jr., J.A., Higginson, J.S., 2009. Dynamic knee joint stiffness in subjects with a progressive increase in severity of knee osteoarthritis. Clin. Biomech. 24 (4), 366–371.

Zifchock, R.A., Davis, I., Hillstrom, H., Song, J., 2006. The effect of gender, age, and lateral dominance on arch height and arch stiffness. Foot Ankle Int. 27 (5), 367–372.

Zijlstra, W., Prokop, T., Berger, W., 1996. Adaptability of leg movements during normal treadmill walking and split-belt walking in children. Gait Posture 4 (3), 212–221.

Zmitrewicz, R.J., Neptune, R.R., Sasaki, K., 2007. Mechanical energetic contributions from individual muscles and elastic prosthetic feet during symmetric unilateral transtibial amputee walking: a theoretical study. J. Biomech. 40 (8), 1824–1831.

Chapter 2

Locomotive functional units

Chapter introduction

If there is one primary difficulty in understanding kinesiology and biomechanics, it is in the way that anatomy is often taught to pregraduates. Andreas Vesalius (1514–1564) is the figure most responsible for the way anatomy is approached, based on dissection. This is hardly surprising, as all anatomists since have mapped out their subject via dissection of cadavers. The gastrointestinal tract, the eyes, the brain, and the genitourinary tract as such are not a significant problem as descriptions of their anatomy do not influence the concepts underpinning their physiological function. It is when describing the musculoskeletal system that conspicuous issues arise because the tissues are described as having primary distinct proximal (origin) and distal (insertion) attachment points, a convention that overlooks the intricate interconnection of general fascial tissues and their undeniable role in the transmissions of forces.

Every muscle is described by an origin proximally and an insertion distally, and these attachments are more often than not made via a tendon of variable size and length. A description of the muscle as it lies from origin to insertion is given and this is known as the muscle's route, passing as it does from origin to insertion in an ordered way. After delineating the blood and nerve supply, the muscle action is then described as providing that movement which occurs in the cadaver when the muscle is pulled at its proximal end on its distal segment via its tendon on a distal limb that is free to move. This does not always correspond well to living muscle function, especially so in the lower limb during the stance phase of gait. Indeed, it can cause considerable confusion, as often the muscle is providing a reverse restraint of the movement that has been described as its actual action of motion. For example, in gait, the gluteus maximus primarily protects the limb from hip flexion moments in order to keep the trunk and pelvis vertical. It actually becomes largely inactive around the middle of midstance when hip extension is occurring during later stance phase. It is only active briefly at the very end of stance phase to provide a little hip extension power during terminal stance, around preswing. Yet, this muscle is commonly described as a hip extensor rather than a resistor of hip flexion.

To understand muscle action, particularly in the lower limb, it is essential to appreciate that muscles have attachments rather than origins and insertions. These attachment points to bone are either tendinous or fibrous (both proximally and distally), made through osteotendinous junctions and via the tendons themselves, to join to the muscle body via myotendinous junctions (Józsa and Kannus 1997, p. 46). Generally, muscles have two primary attachment areas rather than points. However, many have several proximal or distal attachments. All these myofascial soft tissue attachment areas consist of a mixture of tendon, ligament, aponeurosis, and joint capsule linked to the bone, and are known as entheses (McGonagle et al., 2003). The entheses consist of two types, fibrous and fibrocartilaginous, the latter of which are associated with resisting shear and compression and are linked with muscular attachments through tendons (McGonagle et al., 2003). Beyond these primary osteotendinous attachment points, there are multiple attachments to bone via ligaments, periosteum, aponeurosis, retinaculum, and other fascias that to some extent transmit muscle forces.

However, because all connective tissues are part of a continuum, muscle cell contraction pulls on all the local connective tissues around it, tightening them, expanding them, squeezing them, and generally increasing the regional soft tissue tone. It is only in situations where a segment is free to move distally that the muscle attachments produce such movements as described within anatomy texts. Except at the initiation and early stages of swing phase, most muscle activity primarily prevents or slows the movement generated by torques around joints that have been created from the interplay of GRFs and the body's CoM being acted upon by gravity. Such muscle activity stabilises the limb to support the body throughout stance phase, assisting and correcting the translation of the CoM. Muscle activity also provides energy dissipation through braking effects or supplies additional energy by adding extra power for acceleration as required. This effect is an interplay between all the tissues acting in concert as a functional unit.

Biomechanics is usually represented as a system of bones, muscles, and tendons in a lever system, and this remains an important concept. Yet even without these macro-elements, soft tissues aid in maintaining structural integrity, with a key element to structural stability being rigidity or stiffness (Smit and Strong, 2020). Forces are constantly generated and transmitted throughout tissues. They are generated by external and internal moments requiring changes in shape and thus the enablement of tissues to glide past each other at all hierarchical levels. If forces exceed the ability of the tissues to manage them, they will exceed a *'tissue threshold'* resulting in force-induced tissue harm (Smit and Strong, 2020). It is through the biotensegrity systems of hierarchical structure acting as individual functional units that such risks are minimised, and tissue integrity is maintained. How tissues manage and change forces are still not as yet fully understood, making precise prediction of tissue failure challenging (Smit and Strong, 2020). Nevertheless, understanding integral functional unit anatomy in relation to tensegrity is important in appreciating function and dysfunction. However, in dividing anatomy into functional units, some of the continuity concepts of biotensegrity are necessarily sacrificed.

2.1 Soft and hard tissue as functional units

2.1.1 Introduction

There is a tendency in musculoskeletal medicine and biomechanics to view the human body through the role of bones, joints, ligaments, and muscle forces, with other soft tissues draped around in a minor supportive role. Kinematic joint motion studies have dominated research in the past with many of the models, particularly when explaining foot function, being limited to focusing in on a specific joint such as the subtalar joint or the 1st metatarsophalangeal (MTP) joint in a 'predictable' coupling motion, linking the effects of motion at one joint to others throughout the skeletal chain (Lee, 2001; Petcu and Colda, 2012). This is a limited vision of the locomotive system's function that restricts an understanding of the soft tissues being a major component of structural integrity possessing reactivity during locomotion. The human body is a mobile, flexible, multihinged, omnidirectional structure (Levin, 1995). It uses the simple engineering principles of lever arms to move, but it is so much more than merely a series of levers, fulcrums, and forces, as the nervous system provides the anatomy with predictive and reactive mechanical abilities.

For example, the spine, consisting of short bones rather than long bones, cannot be treated as functioning like a beam within a lever system. Similarly, the foot is a multiarticulated structure that only becomes beam-like in the latter stages of stance phase through increased stiffening, and when compared to a single long bone, can at best be described as a semirigid beam. Due to complexity, the total forces required to maintain stability within a mobile column such as the spine or a variably mobile structure like the foot, while the body's CoM is constantly moving over and around another series of jointed beams functioning as an inverted pendulum, are quite possibly incalculable. This is especially so when considering the ability of the human body to work as an almost all-terrain structure, providing mixed walking, running, climbing, and even swimming abilities.

Locomotive soft tissues have high fluid content and demonstrate viscoelastic, non-Newtonian, and non-Hookean properties (Gatt et al., 2015). The high fibrillar collagen content found in musculoskeletal tissues (except the type II collagen of articular cartilage) makes soft tissues highly effective as resisters and maintainers of sustained tensile force. The musculoskeletal system utilises these tensions to create compressions across articulations and bones. In the case of muscles, tendons, and their associated fascial structures, the tensile forces they interact with are also part of the forces they generate through their contractile elements. Therefore, the connective tissues involved in locomotion should always be considered as having passive elements that rely on their mechanical abilities to bear and resist stress so as to maintain stability (ligaments, aponeuroses, deep fasciae, passive myofascial elements). Added to this are also active muscle (myofascial) elements that generate force and therefore, through their activity, control and influence the stresses applied to the system. It must be remembered that this simple picture is slightly blurred as muscles have passive tensile properties (Herbert and Gandevia, 2019) and fascia demonstrates smooth muscle cell activity that can alter tensional properties (Schleip et al., 2005, 2019).

2.1.2 Principles of tensegrity and biotensegrity revisited

Consider a human hero or heroine in a 'movie' using their body as a beam in order to allow people to cross over a gap bridging some dangerous fall. Their feet lie on one side of the gap and their hands grip tightly on the other side. To allow their companions to cross, the principal character cannot rely on his or her bones to act as a stable loading frame, for there are too many mobile joints in between the feet and the hands. In order for the human bridge to work, the ligaments need to keep the bones restrained and together. Yet, the passive tension on loading ligaments will not be enough to compress the bones together under the forces caused by a human walking over the hero/heroine's back (or even support the hero/heroine's own body weight for that matter). Thus, to prevent the body deflecting under load and falling into the abyss, active muscle tone must be increased to bring about articular compression and turn the body into a rigid or semirigid beam through near-isometric contraction. Under neurological control through the sensorimotor system, appropriate muscle activation can be provided to each area of the body as need demands to manage the load of any person crossing the human bridge, creating support through tension–compression principles. In essence, this simple idea explains the processes of core stability and biotensegrity structures. Indeed, anyone who participates in the 'plank exercise' will know (especially for those of us built un-hero/heroine-like) that even maintaining one's own body weight through the arms and legs takes considerable effort to maintain (Fig. 2.1.2a).

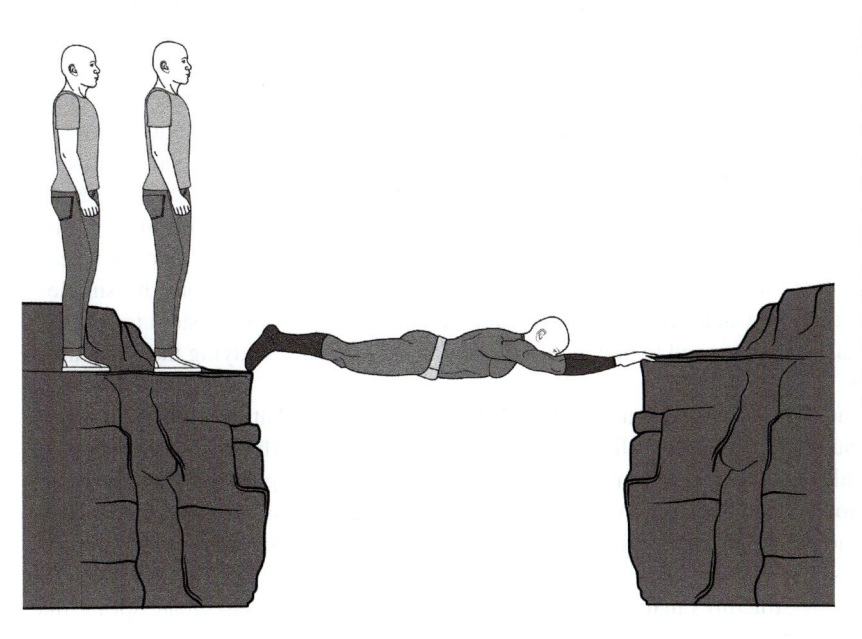

FIG. 2.1.2a A superhero attempts to form a bridge by using their body to span a 'chasm of doom'. A certain amount of muscular tension is required to maintain stability so that joints do not buckle under body weight (upper image). The addition of someone else's weight will require increased muscular tensional forces (lower image) to compress the joints together, maintaining stability and shifting muscular effort to follow the progression of the person crossing. As hip and spinal extension is easier to resist than flexion, logic dictates the bridge is created face down. Muscle stability under changing loads and postures is an essential part of musculoskeletal health and efficiency. *(Permission www.healthystep.co.uk.)*

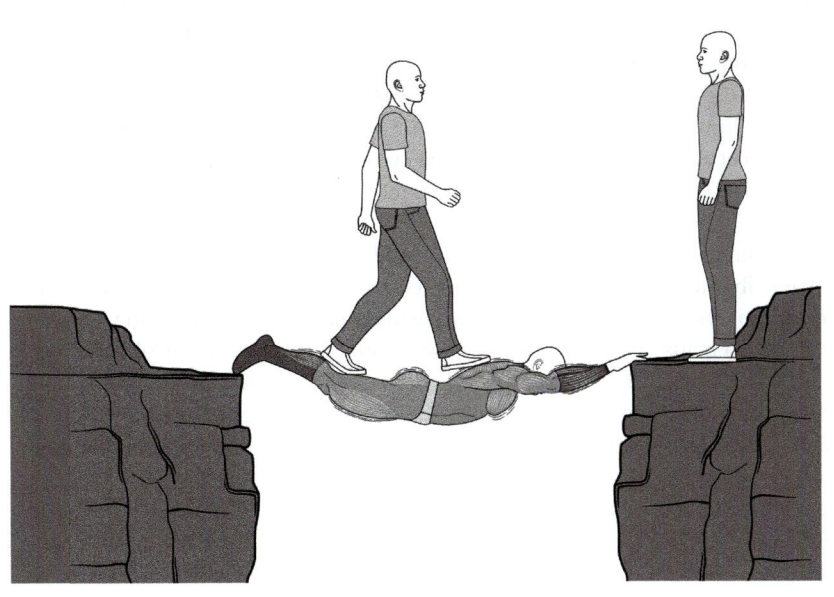

Bending a truss or beam produces deflection across its structure. This is resisted by the tensions and compressions of the material, depending on the direction of deflection. In osseous vertebrate animal bodies, active tension forces are applied via muscle contractions. Passive tensional resistance is provided by the fascia (connective tissue) in all of its forms such as ligaments, tendons, deep fascia, and fascial layers within muscles. Also, the muscles' own intrinsic viscoelastic properties via proteins such as titin, leading up to its macrostructures such as tendon, all provide passive tensional forces. These tensional elements are integral to the body's support (Levin, 1995). However, the truss that the soft tissues maintain is a certain type of truss known as a *tensegrity structure* (Levin, 1995) which has previously been discussed in Chapter 1 of the companion text 'Clinical Biomechanics in Human Locomotion: Origins and Principles', Section 1.1.4 and within Chapter 2, Section 2.4.6 of the same text. Tensegrity is a state of pre-stress imposition that imparts tension to all the tensional elements (tension cables) in stabilising a structure, providing first-order stiffness to all of its infinitesimal mechanisms (Tibert, 2002). Thus, a tensegrity structure is both a set of struts under compression and an arrangement of cables under tension that always balance out in the most energetically efficient configuration (Scarr, 2011). In animals, this equates to bones (struts) and the largely passive ligaments, fascia, and active controllable muscle–tendons (cables). This concept is referred to as biotensegrity.

The tension in tensegrity reduces the object's shape to a minimum and creates an infinite variety of stable shapes. This is made possible through changes in the lengths/arrangements of their compression members and yet requires very little energy to control such shape changes (Scarr, 2011). Each component influences all the others, distributing stresses throughout the system and creating a structure that can react to external forces from any direction without collapsing (Levin, 2002). Using such a system, organisms can move with minimal energy expenditure without losing stiffness or stability (Levin, 2002; Scarr, 2011). The tensional and compressional forces are considered to remain separate, allowing the material properties of the components to be optimised through hierarchies, achieving a significant reduction in mass used in tissue construction (Scarr, 2011). This provides a functional connection at every hierarchical level from the simplest cell-level structures up to the most complex macrostructures of a living organism, allowing the whole system to function as an integrated single unit. This concept has been proposed to better explain human motion than studying the anatomical components alone (Dischiavi et al., 2018).

Tensegrity structures form omnidirectional, gravity-independent, flexibly hinged structures that have nonlinear, non-Newtonian, and non-Hookean properties. Such properties fit perfectly with the known mechanical behaviour of biological materials. Tensegrity structures automatically assume a position of stable equilibrium in a configuration that minimises their stored elastic energy, yet also allow movement with a minimal energy expenditure without loss of stiffness or stability (Scarr, 2011). Biotensegrity structures demonstrate nonlinear viscoelasticity with fluid-like motion that results from the action of all components within the whole system (Scarr, 2011). Different arrangements and numbers of struts form different shapes that permit construction of stable structures that can be twisted into tensegrity helixes to create spring-like elasticity. By so doing, they can form cylindrical walls around central spaces to construct tubular elastic-compressive sheaths through increasing scales of hierarchy (Scarr, 2011). This is similar to having many levels of compression hosiery inside each other in the manner of a Russian nested babushka doll. It has thus been argued that biotensegrity is the fundamental architecture of life (Ingber, 1998). Biotensegrity structures can be found at all hierarchical levels of life (Levin, 1995; Scarr, 2011) and have been associated with cellular cytoskeletons (Ingber, 2008), spider silk (Du et al., 2006), mammalian and avian lungs (Moore et al., 2005; Watson et al., 2007; Weibel, 2008), the cerebral cortex (Van Essen, 1997), the cranium (Scarr, 2008), the fascial matrix (Levin, 1997; Parker and Ingber, 2007), the shoulder (Levin, 1997), the spine (Levin, 2002), and the pelvis (Levin, 2007).

Tensegrity structures form a continuous interconnecting network. Type I collagen fibrils are the predominant tensors in biotensegrity structures and are virtually inextensible under tension (other collagen mechanical properties are less well understood), whereas proteoglycan and glycosaminoglycan amounts tend to increase in tissues under compression. In widespread type I collagen, repeated sequences of amino acids spontaneously form left-handed helixes of procollagen, with three of these helices forming a right-handed helix of tropocollagen (Scarr, 2011). Five tropocollagen molecules then coil in a staggered helical array which lengthens longitudinally. The supplementation of more tropocollagen forms a microfibril (Orgel et al., 2006), with more complex hierarchical arrangements giving rise to fibrils, fibres, and fascicles (Hulmes et al., 1995).

These hierarchies of structure enable mechanical forces to be transferred down to a smaller scale, dissipating such forces to avoid damaging stresses (Gao et al., 2003; Gupta et al., 2006), with the tissues being responsive to the load they receive through mechanotransduction mechanisms (Ingber, 2006). This continues down to the atomic/molecular level where the basic forces of attraction and repulsion automatically balance these stresses in the most energetically efficient configuration (Scarr, 2011). The helical and tensegrity structural systems complement each other and are based on the properties of the icosahedron and tetrahedron, where a chain of tensegrity icosahedra contain the crossed-helical fibres of a tube. Tubes are everywhere in biology, ranging from muscle, tendon, and ligament fascicles to macrostructures such as blood vessels, muscle fascial compartments, tendon sheaths, and long bones. Tubular shapes are inherently stable, yet flexible.

The principles of tensegrity state that tensional and compressive forces are separated into discrete components, always acting in straight lines which results in an absence of shear stresses and bending moments. The presence of curved struts (bones) contradicts this, but the principles are consistent in terms of hierarchy (Scarr, 2011). Curved struts can only remain stable if their molecular structure is strong enough to resist potentially damaging shear stresses that can result in the structure buckling, or if they exist in a tensegrity structure that eliminates such deflection stresses by its nature. This means that curved struts may appear curved on a macroscopic level, but when looked at in more detail, they actually have structural components that handle tension and compression in straight lines. A good example illustrating this is the trabecular arrangements seen in a curved long bone, such as the neck of the femur. The fibre angle within any particular tissue is likely to depend on its functional context, with previous descriptions of 'randomly arranged' collagen fibres arising from misinterpretations of the actual tensegrity alignment (Williams et al., 2005) (Fig. 2.1.2b).

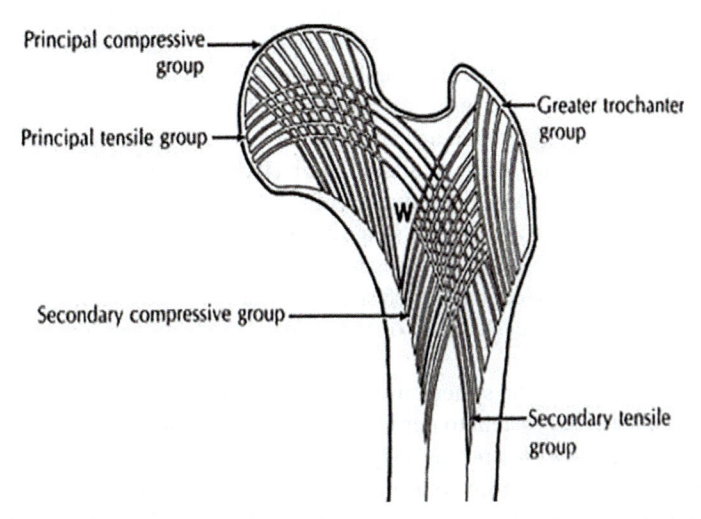

FIG. 2.1.2b The trabecular structures within the human femoral head and neck are not the result of random orientations, but the result of stress-induced responses during bone formation. They are arranged in groups within the femoral head and neck to transfer forces to and from the femoral shaft derived through internal and external sources. The medial side of the femoral shaft is subjected to more compressive stresses, while the lateral side experiences higher tensile strains. Orientations of protein-based fibres such as collagen are highly responsive to repetitive stress loading so that areas of stress in a repeated direction organise fibre arrangements specifically to resist loads. In contrast, areas of multidirectional loading organise fibres in multidirectional arrangements. These orientations are seen variably at different levels throughout the skeleton. Loss of such orientations of trabecular structure in areas of high repetitive stress such as the femoral head and neck can help forewarn of osteoporotic changes when seen on radiographs. *(Image from Sapthagirivasan, V., Anburajan, M., Mahadevan, V., 2013. Bone trabecular analysis of femur radiographs for the assessment of osteoporosis using DWT and DXA. Int. J. Comput. Theory Eng. 5(4), 616–620.)*

Axial compression of the tensegrity helix initiates rotation in the direction that is dictated by the helical angle and the strut orientation, i.e. left or right-handed (known as chirality). Such compression rotation corresponds to a decreasing central diameter. Axial extension of the helix results in expansion, causing a negative Poisson's ratio. This is a common phenomenon in biological tissues (auxetics) and is noted in tendons under low tensional forces (Gatt et al., 2015). Surrounding a helix with another one with an opposite chirality is found in collagen on a microscale and within vertebral discs on a macroscale. Chirality greatly increases resistance to axial compression, as each helical layer counteracts the compressive rotation of the other (Shadwick, 2008). In vascular tubes, the compression-resisting element is provided by the fluid pressure within tensioning walls. The fluid therefore becomes the compression strut (Scarr, 2011), linking the mechanical concept of poroelasticity to biotensegrity. This concept of poroelasticity in tensegrity is important for the understanding of plantar fat pad behaviour as well as blood pressures. The tensegrity helix is therefore an energy-efficient solution to molecular close-packing, and has an inherent ability to form hierarchies. The most widespread structural protein, collagen, consists of several helical levels. Collagen is the major component of extra-cellular matrix which is continuous with fascial tissues, nicely fulfilling the needs of biotensegrity (Scarr, 2011).

The 'concept' of helical tubes within larger helical tubes means that fascial compartments within the trunk and limbs, as well as the concentric rings in intervertebral discs, can be viewed in similar ways, i.e. providing tensile reinforcement (Hukins and Meakin, 2000) where the fluid contained within the compartment or the disc replicates the role of compressive struts in situations where hard structures such as bones are missing. Although there are objections to the idea, it has been suggested that muscle contraction (through an increase in diameter acting against a contained volume of fluid) heightens tensional forces on the fascia in much the same way as do the mechanisms described by Shadwick (2008) for soft-bodied animals, where a contained volume of fluid acts as the struts within a tensegrity structure. In the context of biotensegrity, fascia provides the main component of tension, suspended as it is between the bones under compression, with smaller compartments taking their origins from larger ones and the muscles becoming the controllable motors of the amount of tension provided (Scarr, 2011). In addition, compression-resistance structures in the body such as the fluid enclosed within tissue pores and contracted muscle also potentially act to provide more transient struts, in addition to the permanent struts that the bones provide.

2.1.3 Maintenance of biotensegrity structures

A key element of structural stability, integrity, and motion is the ability to provide stiffness at an appropriate level. The unhelpful term of *quasistiffness* is sometimes used in the literature but until there is a precise definition, some caution to its use is required. From the cellular to the tissue to the structural level, degrees of rigidity or stiffness permit changes in shape to occur without loss of function, an ability that if lost risks harm through excessive or insufficient changes in shape (deformation). The ability to avoid this is achieved through dispersing force over a larger area by stiffening, stretching, and transferring forces throughout different structures that have dissimilar mechanical properties such as skin, bone, fascia, tendon, and adipose tissue (Smit and Strong, 2020). As a consequence of the 'structure of a continuum' within biotensegrity, this system of force handling cooperation between tissues via shape changes and energetic management is possible. Evidence is mounting that muscles should no longer be viewed as independent structures that simply connect one bone to another but rather are part of a network of 'in series' connections which span the entire musculoskeletal system (Dischiavi et al., 2018). Thus, the sum of the whole anatomy is greater than any one of its parts, and any part lost reduces the mechanical sum of the whole.

The support elements of the musculoskeletal system are via compression and tension (even in relaxed states) to variable degrees, thus enabling the musculoskeletal system to permit combinations of mobility, compliance, stability, and rigidity. Such compression under tension in biotensegrity makes body-supported mechanisms act like a truss. A truss is an inherently stable structure with free-moving hinges. There are no beams or levers within a truss, the load being distributed throughout the structure in a triangulated manner. It is worth reminding the reader that triangles are excellent at resisting shear strains. When made of rope or wire, the tensional elements of a truss are unidirectional. In biological structures, the tensional elements are naturally far more complex. Nevertheless, developing a comprehension of the principles underlying what a truss is by recourse to the basic rope/wire model remains useful (Fig. 2.1.3a). In understanding human anatomy, concepts such as beams cannot be wholly discarded, as the long bones and the stiffened foot are still utilised in beam-like ways during locomotion. However, the addition of the concept of biotensegrity brings mechanical concepts more realistically into the musculoskeletal system.

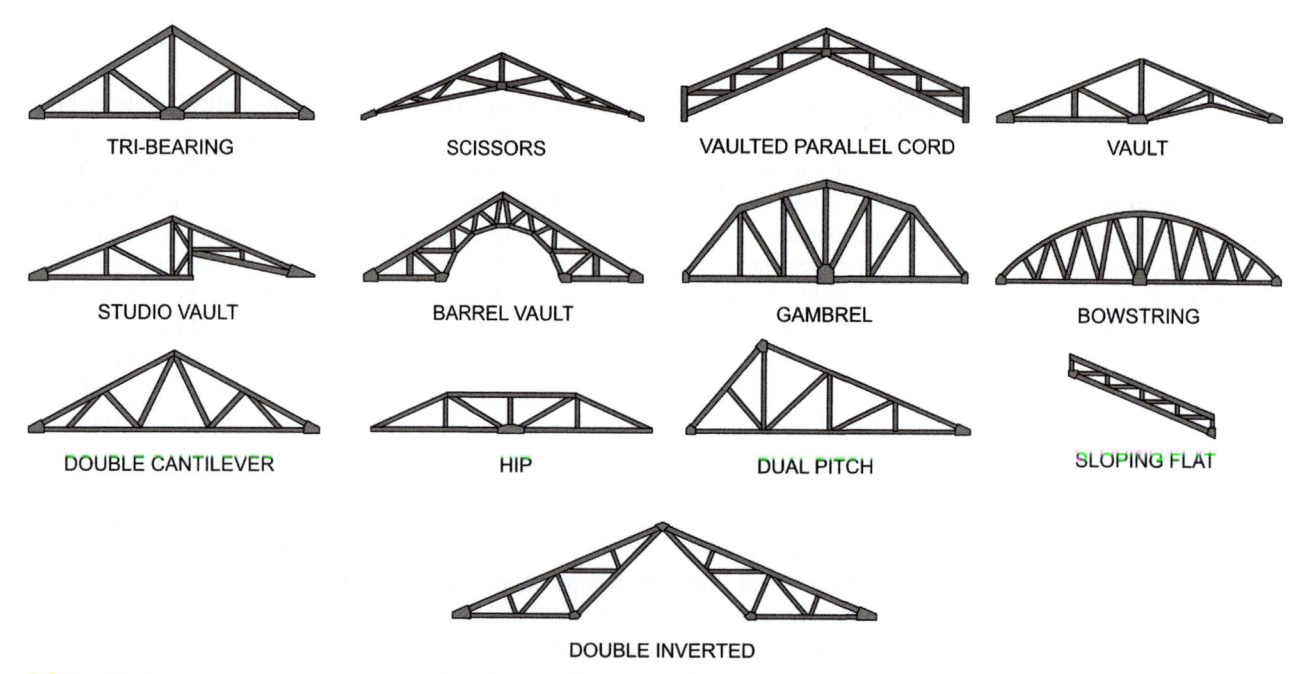

FIG. 2.1.3a Some common types of trusses used in engineering. The truss distributes stress and strain through compression and tension in a triangulated manner of smaller triangles within larger triangles. Triangles are exceptional structures in achieving stability. They are most resistant to shear strain of any shape and are thus the basis of many inherently stable structures. Engineers have long known this, using truss principles for roof vaults, bridges, and beam-like support structures. Using triangles means that material costs can be reduced while structural stability is increased. The use of 'truss principles' in vault construction should be noted particularly when reflecting on the foot by considering the foot vault profile in each orthogonal plane. Muscle vectors also use triangulated forces when stabilising joints. *(Permission www.healthystep.co.uk.)*

In the human body on the macroscale, the compression elements are under tensional forces from the soft tissues. The compression-resistant bones are suspended within the tensegrity structure network. Bones do not normally directly bear force with the ground. For example, in the weightbearing foot, bones are suspended within the soft tissues, lying above the plantar cutaneous tissues, the plantar fat pads, the plantar aponeurosis, the muscular compartmental fasciae, the extensor hood apparatuses, and the ligaments and muscles. When this mechanism fails, as can be seen in severe connective tissue diseases such as rheumatoid arthritis, the metatarsal heads can appear to 'drop' through the soft tissues to lie just under the skin, becoming extremely prominent and thus potential sources of pain, pressure, and dysfunction. In such situations, the bone no longer 'floats' within the support matrix of the soft tissues, now being only covered by cutaneous tissues.

With bones acting as the compressive elements and the soft tissues as the tension elements, stability is created in any position during multiple joint motions. Shortening one soft tissue element has a rippling effect throughout the structure. Movement, in creating a new posture, instantly creates a stable new limb/body shape, which should present low energy costs to maintain. This is true of the spine, as much as it is of any other locomotive structure such as the foot. The foot vault is able to alter shape under loading stress thereby changing its tensions, and yet it is still able to achieve soft tissue 'floating' of the compressional elements (the bones) in the manner dictated by the principles of tensegrity. This supportive tension continues as long as the entire foot structure remains intact under manageable tensile forces. The complication is that the foot needs variable degrees of tensioned stability throughout locomotion if it is to fulfil its compliant energy-dissipation braking role, and then its energy storage and stiffened elastic acceleration role.

Clinical implications of the biotensegrity model

The suspension of the osseous compression components through tensegrity is maintained via the tensional elements of the soft tissue, although soft tissue structures can also take on the role of compressional components if highly stiffened. Tension elements are achieved by two distinct characteristics of the passive fascial and the contractile muscular elements. Either or both of these elements can fail with serious consequences to the ability of the body part to suspend the osseous structures appropriately to reflect the force-induced shape changes. Regardless of the soft tissue tensional elements that fail, the consequences are the same in that the ability to 'float' the bones are reduced or lost. This will cause consequential changes in joint function and stability where compressive struts come together. Without the active elements of fluctuating tension provided by muscle contractions, the ability to resist gravity and to move against or control against gravity is lost. Yet, the bones do not fall on to the inner surface of the skin when muscle activity stops.

Problems with soft tissue function can therefore be simplified into two situations: excessive tension (i.e. excessive stiffness) and insufficient tension (i.e. excessive compliance). If tensions are excessive in 'floating' and 'fixing' the osseous structures, then the ability to deform and change shape under stress in order to dissipate energy and also to store tension-induced elastic energy as kinetic energy, can be restricted by insufficient strain (deformation). The tension forces applied to the compression elements may also be too great. Compressing articular surfaces together excessively will reduce freedom of motion within the joint. In time, such a scenario would cause dystrophic changes among the cartilage of the joints. Over-compression results in increased shear forces between cartilaginous hierarchical levels during motion, risking a breach in tissue threshold forces. When compression is localised to one functional area of a joint, the loss of motion and deformation within one joint area will tend to transfer the need for motion/deformation to neighbouring areas within the same joint or at other joints that can provide similar motion. Furthermore, when tensional excesses are systemic, the whole musculoskeletal system becomes limited in motion and its ability to provide appropriate compliance and thus shape changes are reduced. Such a situation is exhibited by hypomobile patients and is present in those suffering with diseases that increase tissue stiffness, such as diabetes.

Excessive compression–tension forces resulting in inadequate ranges of motion are associated with an increase in intra-articular compression, frictional shear force, and tensional overloading on soft tissue attachment sites such as at the entheses. These over-tensioned areas of anatomy are also more likely susceptible to torsional forces within the bone shafts as the system seeks more motion that is simply not available through the joint and soft tissue motions. This consequently focuses torques towards the more rigid tissues. The classic presentation of these problems would be in the patient with generalised widespread joint motion restrictions (hypomobility) and stiff cavoid feet. Such morphology presents with less motion, impaired shock attenuation, and a resultant increased risk of fatigue fractures (Buldt et al., 2015). Should a tension element such as a ligament fail in a 'stiff' individual and increase the joint range of motion locally, then the ligament failure becomes a mechanism for increased motion where tension restrictions have been lifted. This would tend to focus additional motion and forces to the new site of freedom, potentially exacerbating forces on the surrounding soft tissues that are still restricting motion.

The reverse situation exists in the under-tensioned, excessively compliant individual. Lack of tensional forces leads to an inadequate ability to maintain 'floatation' of the osseous elements within the matrix of soft tissues. There is also a lack of requisite tensional forces to develop sufficient compression forces across the osseous anatomy to adequately stabilise joints to create stiffness. Thus, excessive motion of bones can occur at a joint, changing joint torques in ways that disadvantage mechanical moments and energetics. Failure in the ability to provide stability can occur from both the passive (connective tissue) or active (muscular) tensional elements, or indeed, both together. Therefore, systemic hypermobility, locally lax or damaged ligamentous or fascial tissues, and/or weakened muscles can result in loss of functional biotensegrity, with a consequential loss of joint stability during motion. Essentially, it mirrors a truss where the triangular forming support elements are being lost in sections of the whole structure, thereby permitting shear.

Soft tissue dysfunction in human biotensegrity

Hypotheses have been developed suggesting that myofascial pathways form global interdependent chains throughout their biotensegrity structure. Thus, failure in one muscle or ligament at a particular joint will have immediate consequences to many of the other joint motions that normally and globally accompany a specific motion at that particular joint, and which are directly affected through synergistic activity (Dischiavi et al., 2018). Clinical observations such as 'medial collapse' are thereby consequently explained. Medial collapse can be seen in manoeuvres such as stepping down with weak posterior hip abductors at high angles of hip flexion that can lead to contralateral pelvic drop, femoral internal rotation, knee valgus, tibial internal rotation, and foot pronation (vault-flattening) that can initiate patellofemoral pain on descending stairs (Fig. 2.1.3b). Proximally, trunk posture may also be altered via anterior and posterior abdominal and thoracolumbar myofascial changes, respectively. These proximal structures' own dysfunction could in turn initiate medial collapse of the contralateral lower limb. Body-spanning muscular chains could provide integrated neuromotor input, linked together via viscoelastic myofascial envelopment (Dischiavi et al., 2018). This would be a perfect solution to co-ordinated movement, but it does risk global failure from a single element's dysfunction.

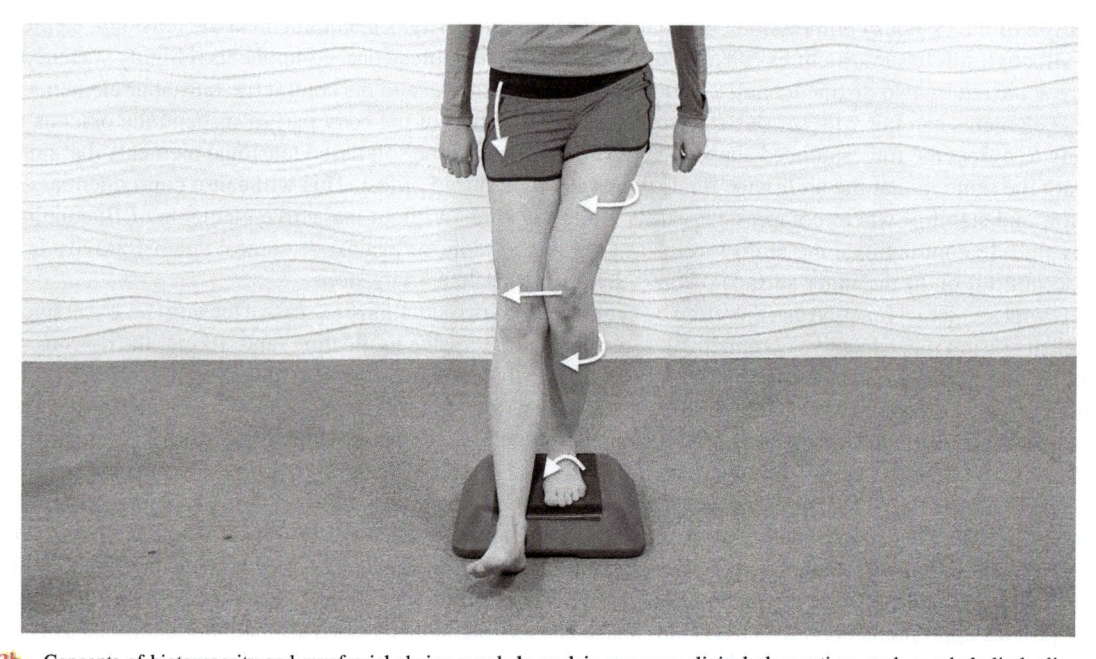

FIG. 2.1.3b Concepts of biotensegrity and myofascial chains may help explain common clinical observations such as whole-limb alignment issues associated with a single muscle weakness. Medial collapse can occur throughout static stance, gait, or just during certain activities such as descending stairs. The origins of such problems may lie in dysfunction somewhere from the pelvis to the foot, for any failure within the chain of structures that maintain correct posture during motion is significant. More severe problems are likely to affect even static stance while more minor dysfunction imbalances only affect particular activities that challenge the postural chain more extensively. *(Image from Dischiavi, S.L., Wright, A.A., Hegedus, E.J., Bleakley, C.M., 2018. Biotensegrity and myofascial chains: A global approach to an integrated kinetic chain. Med. Hypotheses 110, 90–96.)*

Hypermobile individuals such as those with idiopathic hypermobility, Ehlers-Danlos syndrome, or Marfan's syndrome are the classic examples of a loss of effective tensegrity through an inability to adequately produce compressive stabilisation through tensile forces via their passive elements. Such widespread connective dysfunction means that osseous structures are not able to 'float' as easily within tensile–compression constraints resulting in, for example, a much flatter foot vault profile on weightbearing as loads seek connective tissue tensions at greater lengths. There is a tendency throughout the body, in situations of hypermobility, to take joints towards their end-ranges of motion such as with full knee extension and lowered foot vault profile. Increased soft tissue lengthening occasioned by taking a joint to its end-range of motion induces more passive tension throughout the ligaments and fasciae of these individuals, and takes the joint to positions where further motion is difficult. These positions are referred to as *close-packed positions* (Fig. 2.1.3c).

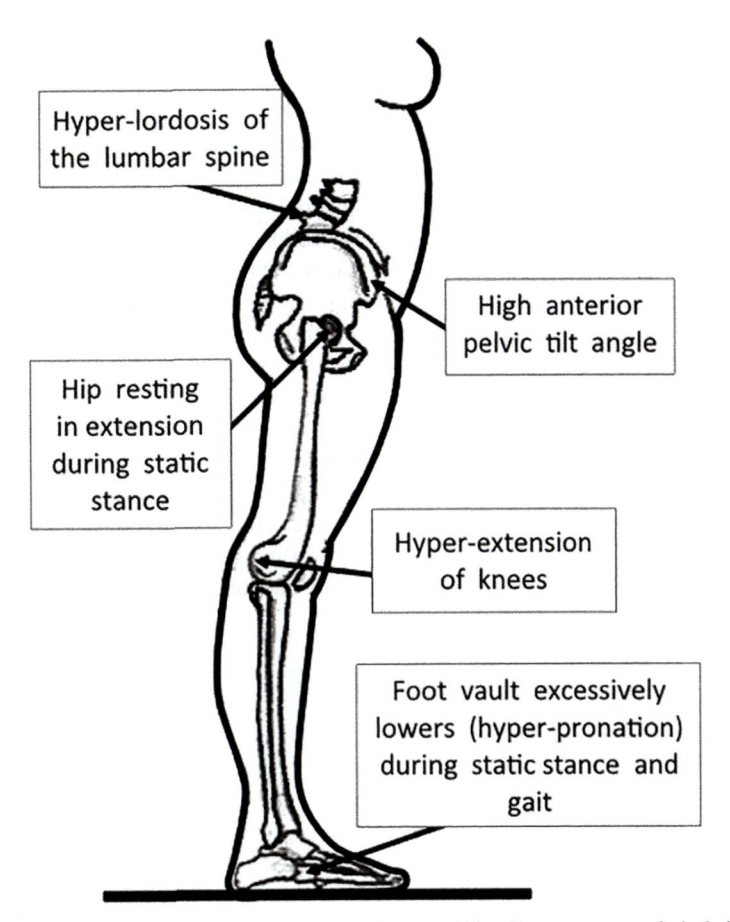

FIG. 2.1.3c Some common lower limb postural changes seen in cases of hypermobility. Features commonly include exaggerated lumbar curvature (hyper-lordosis), excessive pelvic tilt, excess relaxed hip extension positions, knee hyper-extension, and excessively lowered foot vaults (hyper-pronated feet). In hypermobility conditions, the body seeks positions of greater joint stability and reduced mobility, often found towards joint motion end-ranges and in close-packed joint positions (see Chapter 1, Section 1.5.4 for further information on hypermobility during gait). *(Permission www.healthystep.co.uk.)*

Hypermobility is far more common in the young, and many cases should improve with age under the natural maturing of the collagen structures to achieve increasing stiffness. Forces are much lower within children due to their smaller size and mass, so the reduction in tensile elements should be of little concern as long as this improves with growth and the muscles that support them remain strong. In obese children, further increasing body mass presents a pathological risk due to the larger potential forces placed upon the joints. For those individuals who do not significantly improve their soft tissue tension with age, hypermobility presents a medical quandary. Unfortunately, such individuals can end up suffering with musculoskeletal pains that are challenging to manage. In the model of biotensegrity, the approach to passive tensile insufficiency is through trying to increase active tensile-generated forces by improving muscle activation and strength. Strengthening the muscles and teaching the individual how to activate them in order to achieve appropriate tensions, both statically and dynamically, will be necessary for improved postural support and motion. Invariably, this means postural neuromuscular training is necessary.

Most clinical cases of poor tensile force stability involve a local loss of biotensegrity due to the atrophic dysfunction of some soft tissues, soft tissue injuries, and/or functional loss of a tension-producing structure. Loss of one or more structures providing tensile stability within a functional unit results in a loss of compression-stability capacity locally. This can lead to joint instability resulting in excessive joint mobility. However, if the osseous alignments are altered through an inability to maintain positions across joints via appropriate joint compressions, then joint restriction and/or functional immobility may result instead. Such an eventuality can happen if the joint ends up in its close-packed position too soon during motion, or if the osseous structures end up impinging or overly compressing each other during motion. If a joint is unstable or moving inappropriately, the normal positioning and motion of its instantaneous axis of motion can be lost, altering muscle moment arms around the joint. Sometimes, the joint can even get 'stuck' in a restricted close-packed position through a soft tissue failure to stabilise appropriately. This will prevent normal shape changes during motion around the affected joint. It is likely that this principle of lost joint freedom to permit changes in shape during motions underlies the benefits of mobilisation and manipulation of joints. Such clinical manoeuvres can serve to improve motion at whole functional units such as the foot, ankle, or spine.

Examples of changes in osseous alignment and joint dysfunction as a result of the loss of tensile-generated compression abound. A particularly good example indicating the significance of a structure in functionally maintaining biotensegrity is that of tibialis posterior. Although not a particularly large muscle, the tibialis posterior provides a significant role in supporting the biotensegrity structure within the foot. It not only provides tensile–compression forces across the joints it bridges, but it also provides tendon compression directly via its tendon sheath where it runs within the malleolar groove of the tibia and under the groove on the navicular. In these locations, the tensile collagen structure of the tendon is changed to a more cartilaginous form in order to resist the increased compressive shear forces exercised on the tendon material. The tibialis posterior tendon attaches to the plantar navicular (and sometimes the medial cuneiform and sustentaculum tali of the calcaneus) and then runs under the navicular to attach to multiple and variable sites across the plantar surfaces of the lesser tarsus and tarsometatarsal joints. The tibialis posterior therefore exerts a powerful effect by compressing the osseous structures across the vault of the foot in an oblique-transverse, lateromedial, and distoproximal direction, maintaining a 'float' compression stability throughout the vault's motion and its shape changes during the stance phase of gait. It is assisted in this role by multiple passive tensile and active tensile elements, the most significant being the long plantar ligament, short plantar ligament, spring (calcaneonavicular) ligament, and the peroneus longus muscle–tendon. Failure in or hypermobility of any part of this 'biotensegrity team' leads to a loss of vault function and shape change stability, with the osseous structure shifting more towards the ground through loss of the ability to adequately 'float' the skeletal frame under loading (Fig. 2.1.3d).

FIG. 2.1.3d Schematic of the primary anatomical structures of foot tensegrity that allow it to remain stable throughout shape changes and altered levels of flexibility during gait. Tibialis posterior (1) and peroneus longus (2) have particularly significant roles in foot tensegrity. The medial vault's (top image) function is supported by flexor hallucis longus (3), abductor hallucis (4), ligaments of the subtalar joint complex, i.e. the cervical and bifurcate ligaments (A), and the spring (plantar calcaneonavicular) ligament (B) under the talocalcaneonavicular joint. The lateral vault (bottom image) is supported by peroneus longus (2), peroneus brevis (5), the abductor digiti minimi (6), and the strong short and long plantar ligaments (C). Together, adjustable forces drive the talus anteriorly and the more distal bones proximally in a dorsal direction that compresses the vault together, stiffening it. Part of the genius of its construction is the lateral-to-medial and medial-to-lateral transverse connections of tibialis posterior and peroneus longus, respectively (insert). It must be pointed out that equally complex assisting muscles and ligaments (not shown) are all playing important roles in foot biotensegrity through the midfoot and forefoot. *(Permission www.healthystep.co.uk.)*

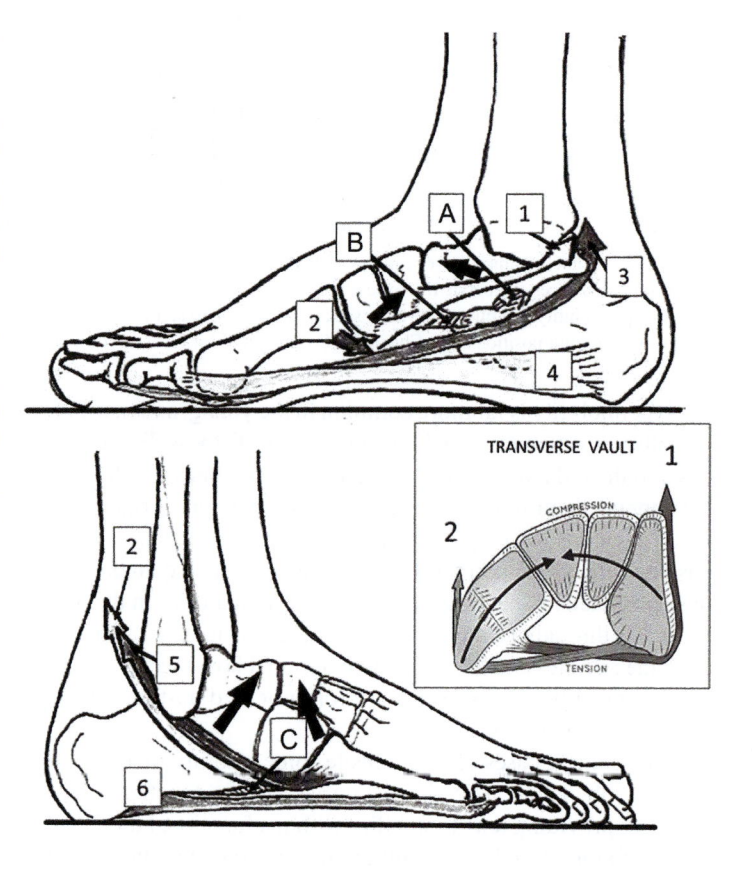

Thus, failure of biotensegrity can occur from either dysfunction in the active or passive components of the tensile elements. Obviously, these principles apply to any joint or functional unit, and each situation requires the clinician to consider the local circumstances of the anatomy injured in order to unravel the primary factor(s) resulting in the loss of the biotensegrity structure. Biotensegrity principles underpin anatomical functional stability during motion.

2.1.4 Principles of core stability

Biotensegrity involves adjusting force–shape relationships to every movement throughout locomotion. Therefore, a balance needs to be maintained between the passive tension–compressive restraints, the active responsive tension–compressive restraints, and the neurological feedforward predictive activation and responsive feedback information. Together, they correctly tune the biotensegrity structure for the appropriate functional task required. This may seem potentially very complex, but consider the metaphor of the hero/heroine bridge cited earlier and the concept can be easily appreciated. The human bridge stiffness is set initially for the 'predicted' mass of the person about to walk over them, but this stiffness would need some ability to adapt to the 'actual' loading stresses when someone starts to walk over them. This capability to partially predict and then fully provide the requisite stability and flexibility to permit controlled motion or postural change fits well with the concept of core stability, something that has long been considered in regard to the spine (Panjabi, 1992a, b), but has also been applied to other aspects of the musculoskeletal system including the foot (McKeon et al., 2015).

Core stability has been defined as the ability to control the position and motion of the trunk over the pelvis to allow for optimum production, transfer, and control of force and motion to the terminal lower limb segments during locomotion (Kibler et al., 2006). Core muscle activity has been stated as being the preprogrammed integration of local single joint and multijoint muscles to provide stability and produce motion. This results in the provision of proximal stability for distal mobility for a proximal-to-distal patterning of force generation and the creation of interactive motions that move and protect the distal joints (Kibler et al., 2006). Although initially promoted as a model of rehabilitation therapy for the pelvis and lower spine in response to back pain, the concepts have since been investigated in relation to foot function (Cobb et al., 2014; McKeon et al., 2015). Whether it is actually possible to train people to preprogramme muscle activity rather than just strengthening them is a debate for elsewhere, bearing in mind the role of the sensorimotor system in integrating appropriate muscle function. Authors have certainly questioned the evidence (Cleland et al., 2002; Koumantakis et al., 2005; Mills et al., 2005; Hibbs et al., 2008; Lederman, 2010; Okada et al., 2011). Yet, here is a fundamental concept where posture and movement require that muscles activate under neurological control, interacting with passive tensional intrinsic forces which affect the ability to maintain appropriate posture and motion. Keeping muscles strong and movement pathways controlled is likely important for locomotive health.

The concepts describing spinal stability published by Panjabi (1992a, b, 2003, 2006) gave an explanation as to how loss of stability impacts upon spinal function and dysfunction. It stressed the interplay between passive and active elements, or subsystems as he termed them. The passive subsystems of joint congruity, ligaments, and passive myofascial structures work in harmony with an active myofascial subsystem under direction from the control subsystem (neural) in the promotion of spinal stability. This model, through the concepts of biotensegrity, can be applied to the whole musculoskeletal system. Hoffman and Gabel (2013) expanded Panjabi's model to differentiate between stability and mobility supported by the passive, active, and neurological subsystems, so that instead of there being three subsystems linking together to bring about locomotive and postural stability, six subsystems of three stability (neural, passive and active) and three mobility subsystems (neural, passive and active) all interlink (Fig. 2.1.4). The central nervous system (CNS) is active in stabilising the body prior to a predicted challenge, and these stabilising mechanisms increase neurological activity proportionately to the increase in functional challenge (Grillner et al., 1978). This results in increased muscular activity occurring throughout the body prior to movement (Bouisset and Zattara, 1981). The passive subsystems rely on their mechanical properties under the loading stress they receive. Postural control and mobility are therefore proposed to be an integrated system within the musculoskeletal system's continuity (Hoffman and Gabel, 2013).

It is proposed that both stability and mobility subsystems work synergistically, determining the quality of motion with autonomous determination of the ratio of stability to mobility according to the task at hand (Hoffman and Gabel, 2013). Thus, archery emphasises relative stability over mobility, while ballroom dancing emphasises mobility over stability, yet the two subsystems are important to both activities. When walking, the stance limb needs stability, the swing limb needs mobility, and the foot during contact requires greater mobility than it does at midstance. The foot in terminal stance requires considerably more stability than mobility but requires significant MTP joint and considerable ankle joint mobility. Therefore, the subsystems function to counterbalance and merge forces in order to create actions of full body motion with stability and mobility occurring in the right place and at the right time (Hoffman and Gabel, 2013).

When considering gait, certain functional units of anatomy (such as the pelvis) are more frequently involved in stability events, whereas others depend on the particular phase they are in at any moment during gait, changing from stability to mobility dominance as required. Stability is dependent on the movement of multiple body segments under appropriate muscle activity. Impaired movement may be apparent because of a loss of adequate synergy between mobility and stability which are dependent on movement in multiple body segments (Hodges et al., 2002). Alterations in the normal activation

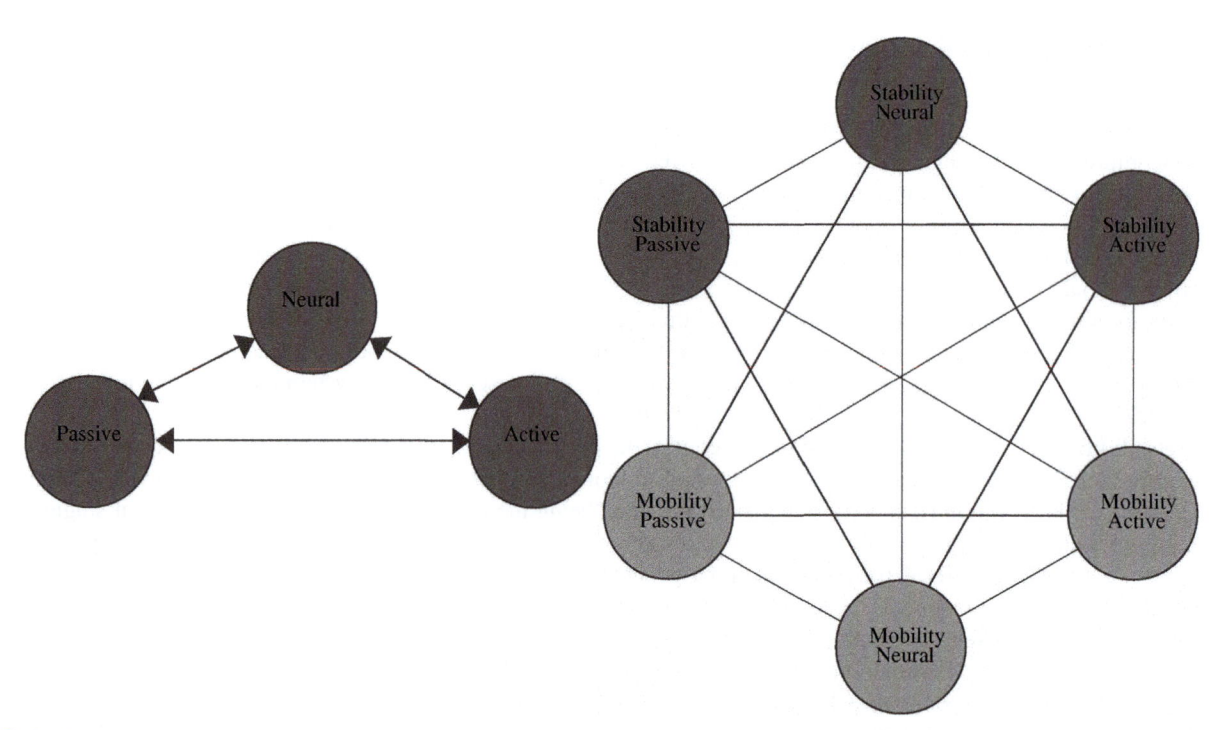

FIG. 2.1.4 Panjabi's subsystems of spinal stability (left) and the six subsystems proposed by Hoffman and Gabel (right) that stress the difference and interaction between stability and mobility. *(Image from Hoffman, J., Gabel, P., 2013. Expanding Panjabi's stability model to express movement: a theoretical model. Med. Hypotheses 80(6), 692–697.)*

timing of activity between stability and mobility muscles can be disrupted by pain through feedforward mechanisms (Hodges and Moseley, 2003) and may even be affected by complex biopsychosocial sources (Tsao et al., 2011). The potential disrupting operations of feedforward stability mechanisms are important to gait as the CNS may initiate stability mechanisms led by activation of the transverse abdominis muscle prior to the initiation of lower limb mobility (Hodges and Richardson, 1997). In the presence of temporary low back pain, it has been reported that the sequence of firing between stability muscles first and then mobility muscles is reversed, so that mobility muscles are activated prior to stability muscles but revert to normal when the pain has resolved (Hodges and Richardson, 1996).

2.1.5 Muscle's role in stability–mobility

Human resting muscle tone or myofascial tone is the passive tension of skeletal muscle that is derived from its intrinsic mechanical viscoelastic properties, independent of CNS activation through structures such as muscle spindles. This provides muscles with a passive tensional element as well as a CNS-initiated active role within which they are usually considered. Therefore, resting/myofascial tone is not the same as the muscle tone achieved during isometric contraction, for nature is frugal in that humans have adapted to erect postural gravitational forces by evolving mechanisms in skeletal muscle to economically enhance stability. The energetic increase from a supine to an erect body posture minimally increases energy costs by around only 7% for long durations, without fatigue (Masi and Hannon, 2008). Resting muscle tone is therefore EMG-silent, not involving any muscle contractile activity, and is derived from the muscle's intrinsic viscoelastic properties (Masi and Hannon, 2008). It therefore must be classified as one of the passive tensile elements that provides stability and mobility as part of the biotensegrity of the human body overcoming external forces, including gravity. Yet, most biotensegrity muscle influence involves active tensile forces, in both motion and stability, formed by the active compression that results.

Muscle fibres are usually considered as consisting of two types: slow twitch (type I or red fibres) and fast twitch (type II or white fibres). Fast twitch fibres can be further categorised as type IIa, a hybrid between slow and fast twitch, and type IIb, the classic fast twitch fibre. A variable distribution of fast and slow twitch fibre type is found in different muscles (Elder et al., 1982; Verbout et al., 1989). Slow twitch fibres are considered to be stability-biased, while fast twitch fibres are associated with bursts of sudden power and are considered to be mobility-biased. Muscles often have one type dominating within them, but slow twitch fibres can convert to fast twitch with ageing and chronic musculoskeletal disorders (Elder et al., 1982; Verbout et al., 1989). It is therefore likely that as we age, some stability at joints is lost. Several attempts have been made to classify muscles into either being stabilisers or movers. However, during movement, muscles share both stability and mobility depending on their functional role at the time (Hoffman and Gabel, 2013).

Muscles are challenged during gait, adapting from one role to another, which is dictated by the speed of gait. Slower muscle contraction will tend to produce more power through the effects of force–velocity relationships. With reduced stabilising ability changes reported in muscles through ageing, such age-induced reduction in gait speed can perhaps be explained as a consequence of the need to increase stability at the cost of reducing gait speed. These changes are likely to affect some muscles more than others. Gastrocnemius is primarily a mobility muscle, whereas soleus is a mixed mobility and stability muscle. During gait, gluteus medius on the stance limb side addresses more stability issues rather than mobility issues, as it stabilises the body's CoM within the frontal plane. However, on the swing limb side, gluteus medius is mildly involved in hip mobility. Some muscles may serve both stability and mobility roles equally, such as tibialis posterior.

Variance in role may select for the mechanical needs of an energy-storage tendon with lots of interfascicle mobility, a positional tendon if the role is more stabilising, or if the role is more hybrid, then a tendon that displays something in-between. It is possible that trying to use one muscle/tendon type for a primary role that is unsuited to its morphology may lie at the biomechanical root of some tissue stress pathology. It is worth noting that muscle behaviour is likely set during development and growth under mechanical loads, particularly in the noncollagenous interfascicular matrix (Zamboulis et al., 2020), so that changes in muscle 'use' during adulthood are likely to create mismatches between a muscle–tendon's mechanical properties and its mechanical use. Age-associated muscle fibre changes may present an unavoidable risk of muscle–tendon functional alterations, with consequences on the anatomical functional unit as well as the muscle–tendon unit itself.

Muscle–tendon complexes

Muscles and tendons are part of the same functional unit that creates motion, energy absorption, and stability around joints. In locomotion, muscle action seldom involves pure (near-)isometric, concentric, or eccentric contraction, but more usually stretch-shortening cycles involving all contraction types. These cycles arise from the repetition of limb segments being exposed to impact and tensile forces, making the muscles either eccentrically contract or near-isometrically contract, followed by them concentrically contracting (Komi, 1984, 2000; Komi and Nicol, 2000) (Fig. 2.1.5). Such stretch-shortening cycles involve reactivation via feedforward mechanisms of the sensorimotor system followed by variable activation from

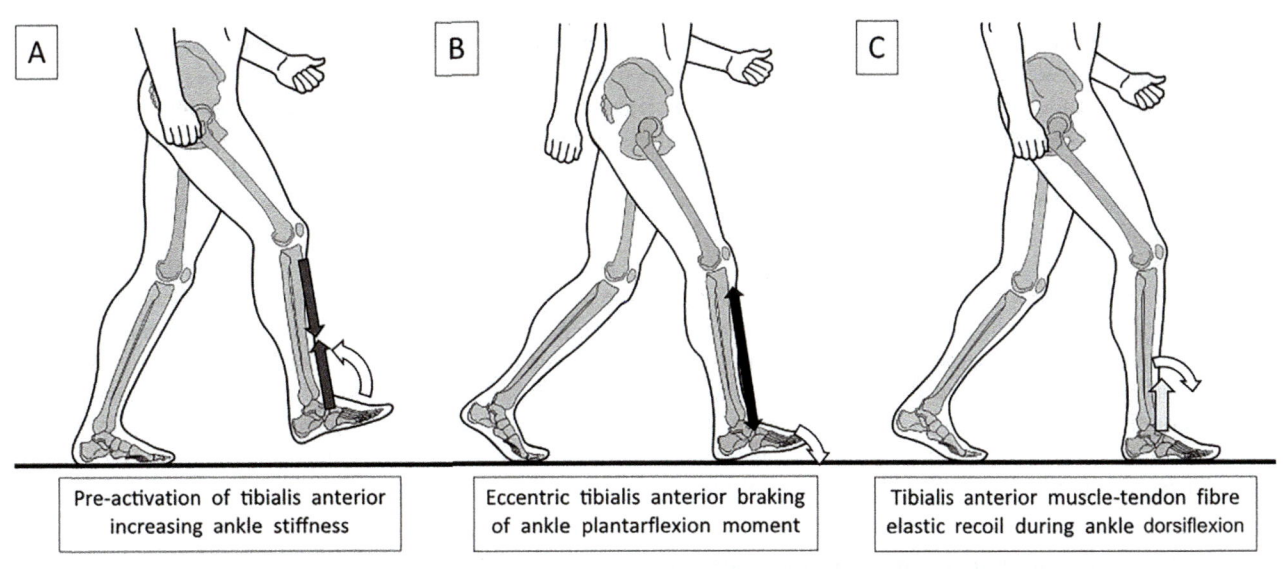

FIG. 2.1.5 The stretch-shortening cycle of tibialis anterior involves muscle preactivation in late terminal stance prior to heel contact (A). This helps to stiffen the anterior ankle via increased muscle tone and may slightly increase the ankle dorsiflexion angle. This is the fibre tightening and/or shortening part of the cycle. However, this muscle activity is more isometric than concentric. Activity before heel strike also helps in stiffening the foot through the windlass mechanism. At heel strike, the GRF-induced ankle plantarflexion is resisted by tibialis anterior via lengthening (thus stretching) under eccentric contraction (B). By lengthening from a preactivated muscle, the tendon can lengthen first before the muscle fibres, helping in producing better muscle–tendon energy buffering. After forefoot loading (C), the muscle and tendon fibres shorten, mainly as a result of connective tissue elastic recoil. Yet, this creates little effect on motion as the ankle is dorsiflexing out of plantarflexion under the activity of the hip and knee extensors as the CoM moves anteriorly, preventing the forefoot from springing back up. Stretch-shortening cycles are generally more important in running, prompting muscle to activate much earlier and create more stability and stiffness than during walking. Preactivation generates larger internal forces and higher muscle tone prior to the muscle force being required. With earlier preactivation in running, more dramatic stretch-shortening cycles occur, helping create more elastic recoil power through the tendons. *(Permission www.healthystep.co.uk.)*

the feedback mechanisms of the sensorimotor system (Komi, 2000). Using muscle fibres under isometric contraction and utilising tendon length changes (near-isometric contraction) seems to produce great energetic benefits (Uchida et al., 2016), and are extremely important in running where there is a great deficit between braking energy lost and the acceleration energy that is put back in.

The mechanical properties of the muscle–tendon complex involve a contractile component, a parallel elastic component, and a series elastic component (Zajac, 1989). The parallel elastic component characterises the passive stiffness of slow stretch in a relaxed muscle (Gajdosik, 2001; Morse, 2011) while stiffness of the series elastic component is characteristic of active conditions, which are thought to play a major role during the stretch-shortening cycle in the storing of elastic energy and the transmission of tension from muscle contraction (Komi, 1984). Stiffness created under high muscle activity may provide an advantage in events that develop high forces (Fouré et al., 2012), but this also risks injury if these events result in a lack of adequate energy absorption during loading (McNair and Stanley, 1996).

Injury within tissues can be caused by a lack of extensibility due to stiffness or by excessive compliance failing to absorb damping energy within the tissues. The musculoskeletal system must be tuned appropriately to the correct settings of stiffness required at a joint at the appropriate time through muscle activation. Many muscle and tendon injuries seem to occur in the midrange, rather than at over-extended joint positions, possibly as a result of a loss of extensibility due to being too stiff (Fouré et al., 2012). Muscle–tendon complex stiffness, which is the ability to resist lengthening, is important in providing elasticity through energy storage within the tissue and to transfer muscular force which is particularly important in athletic activity to create rapid rates of force development (Komi, 2000). Training techniques such as plyometrics (utilising speed and force) increases the elastic series stiffness in muscle–tendon complexes to improve springing activities like jumping (Fouré et al., 2011).

Muscle and tendon stiffness are both influenced by their cross-sectional areas and intrinsic tissue properties, although muscle behaviour is also affected by the state of contraction. Patellar and Achilles tendon cross-sectional areas have been reported as being greater in men than in women (Fouré et al., 2012). Male muscle physiological cross-sectional area, fascicular length, fascicular angle, and the cross-sectional area of the type II fibres are also greater in males (Fouré et al., 2012). Women demonstrate lower stiffness in their in series elastic components within the passive structures (aponeurosis, tendon, etc.) of their foot plantarflexors than do men, but have a higher stiffness in the active muscular part. Moreover, males have more of a global angular joint stiffness (Fouré et al., 2012), whereas linear stiffness appears similar between the sexes (Fouré et al., 2012). Geometric parameters seem to be the primary factor explaining the differences in passive structure stiffness rather than the intrinsic structural properties, which may explain the greater incidence of Achilles tendon ruptures in men (Fouré et al., 2012).

The ability of muscle and tendon to work as a complex is important to the way in which forces and strain rates are managed in muscles, and how quickly the load is applied to the structures that tendons finally attach to. It is worth repeating here that muscle–tendon units lengthen out of phase by first lengthening through tendon fascicle sliding under load before the muscle fascicles lengthen (Roberts and Konow, 2013). This mechanism protects the muscle fibres from high strain rates and peak forces and enables the muscle–tendon complex to dissipate energy, while at the same time allowing the tendon to maintain its capacity to act as an elastic spring returning stored energy to the system (Reeves and Narici, 2003; Roberts and Azizi, 2010; Konow et al., 2012; Roberts and Konow, 2013).

Temporarily storing energy within the tendon reduces the chance of damage because it allows muscle contraction forces to be applied slower, less powerfully, and at a lower force, while not significantly reducing muscle lengthening (Roberts and Konow, 2013). The ability of the muscle–tendon complex to behave in this way permits muscles like tibialis anterior to act as an energy dissipator at heel strike, while the Achilles can dissipate energy on forefoot loading. During forefoot contact loading, the Achilles can store and release energy as an energy-storage tendon to increase power generation for acceleration through it acting as a stiff elastic spring. This ability to store energy after impact is of little value to tibialis anterior during heel contact, as it does not need to generate acceleration power during gait. Hence, the Achilles is an energy-storage tendon, and tibialis anterior has a tendon of the positional type.

2.1.6 Principles of articular motion and stability

For an intrinsically unstable gait like that of humans, appropriate joint motion and stabilisation is extremely important. The stability and mobility of the skeletal structures are directly influenced by the relationship of dynamic and passive tensile elements of anatomy. Osteoligamentous stability throughout the body is variable and more limited where mobility is needed. For example, the osteoligamentous stability of the vertebral spine is barely sufficient to hold the weight of the head (Masi and Hannon, 2008). Muscle force generation is essential to move, stabilise, and/or resist joint motion, but this costs energy and can result in muscle fatigue. The more a joint can be stabilised passively using joint shape and ligament binding/tightening, the easier it is to keep a joint stable and minimise muscle activity. The counterbalance to this is insufficient motion when needed in, for example, within the hip or knee, where restricted motion would result in more steps to move from A to B and less ability to adapt to encountered terrain and activity demands. Thus, the flip side to having stable joints is that they make motion more difficult. An appropriate balance has to therefore become established within the skeletal system, where anatomical regions of stability tend to have stiffer joints and areas of mobility have joints with larger degrees of freedom. These properties are at least in part set during childhood growth under mechanical loading.

The bones themselves can carry and transfer loads as well as acting as a framework for the soft tissues to operate on. In their capacity to carry high compressive loads, long dense compact bones make ideal levers, while the high porosity of short bones and the high concentration of trabecular bone situated around joints within long bones, makes these areas more suitable to act as energy absorbers through their poroviscoelastic properties. This trabecular 'spongy bone' protects articular surfaces under high compressive loads. Yet overall, bones have little bioengineering properties for locomotion without the presence of soft tissues. Equally, soft tissues have limited potential for locomotion without a stiff framework to interact with, and both need neurological inputs that provide the information to appropriately integrate bone and soft tissue to perform the tasks of locomotion. A repeating theme of locomotion is the interaction between neural input, adjustable active force input, and passive force input. These three elements are necessary and thus fully integrated into motion. Joints have passive and active elements to produce stability and mobility under the influence of mechanical loading and feedforward–feedback modulated inputs from the nervous system. The whole joint 'knows what to do' and how to make subtle adjustments from articular and muscle proprioceptors by initiating a chain of events that result in appropriate muscle activation/inhibition. The connective tissues can only passively respond to events.

Muscles can generate force to create motion (mobility) or resist motion (stability) within any segment or any joint that a contractile muscle–tendon unit crosses. This internally derived force produces compression on the joint surfaces during motion or during the stopping of motion which stabilises the joint. This is because the contractile/tensile elements of the muscle belly have a proximal and distal attachment point on both sides of either a single joint or a number of joints. Therefore, when crossing several joints, from one muscle contraction, stability can be provided in one joint whereas at the same time, motion can be provided in one or more other joints, depending on the number crossed. This provides the body with both excellent flexibility and stability during motion. The success or failure of the mechanics of joint motion/stability is dependent on the health of the biological materials constructing the joint, the health of the force-generating elements, and the appropriate positioning of the joints via utilising the sensorimotor system. Failure in any part of static/dynamic control will cause kinematic and kinetic changes that can prove detrimental.

2.1.7 Concepts of muscle joint relationships

Oftentimes, joint motion is only taught through the concept of uni-articular muscle control. For such a prototypical muscle, the proximal attachment is close to the force-generating muscle belly that is attached to the proximal bone through a large attachment site. The distal attachment is found on the other side of the joint consisting of the tendon transitioning into the osteotendinous junction to attach to the distal bone, usually over a smaller site area. Most muscles do not conform to this model, as we shall see. The vastus intermedius of the quadriceps muscle and the popliteus are relatively simple examples of prototypical muscles, although for popliteus, the model is reversed with its long tendon arising proximally to terminate at an extended fibrous distal attachment (Fig. 2.1.7a).

For the sake of simplicity, we will discuss the prototypical muscles such as the above stated vastus intermedius. This muscle has a long fibrous proximal attachment along its muscle belly on and around the anterior surface of the femur and distally attaches via the patellar ligament/tendon (enclosing a large sesamoid bone, the patella) through a relatively small enthesis to the tibial tuberosity. Between its proximal and distal attachments, the joint it passes over is the knee or tibiofemoral joint. When sitting in a chair with the leg free from the ground, activating the vastus intermedius by concentric contraction causes extension of the knee, lifting the resistance mass of the lower leg away from the ground. This is an example of class three lever arm mechanics inducing knee extension moments. Regardless of the type of contraction, activity in vastus intermedius creates compression stability within the knee during motion.

This uncomplicated model demonstrates the simple concept of joint compression stability, although in reality, things are actually more complex. The vastus intermedius muscle bridges two joints in that the patella also has its own articulation with the anterior femur: the patellofemoral joint. It is thus a bi-articular muscle, and when contracting, it is also creating shear and compression across the patellofemoral joint. Vastus intermedius is not usually functioning in isolation but does so with the other quadriceps muscles, i.e. vastus lateralis and vastus medialis which also cross the same two joints, and rectus femoris that also crosses the hip. Vastus lateralis and vastus medialis have proximal attachments on the anterior and posterior (linea aspera) femoral surfaces aligned obliquely to the tibiofemoral and patellofemoral joints. Thus, their lines of

FIG. 2.1.7a Vastus intermedius (left) and popliteus (right) initially seem to present as simple uni-articular muscles. Yet, functionally, vastus intermedius spans the tibiofemoral and the patellofemoral joints and provides part of a collection of force vectors across the knee articulations with the other three quadriceps femoris muscles. Popliteus appears to be an 'upside-down' muscle with its tendinous attachment found proximally. It also links to surrounding connective tissue structures, such as the lateral meniscus, and creates different force vectors in different knee positions. A classic example of a single muscle action over one joint is not easily found within the distal lower limb. *(Permission www.healthystep.co.uk.)*

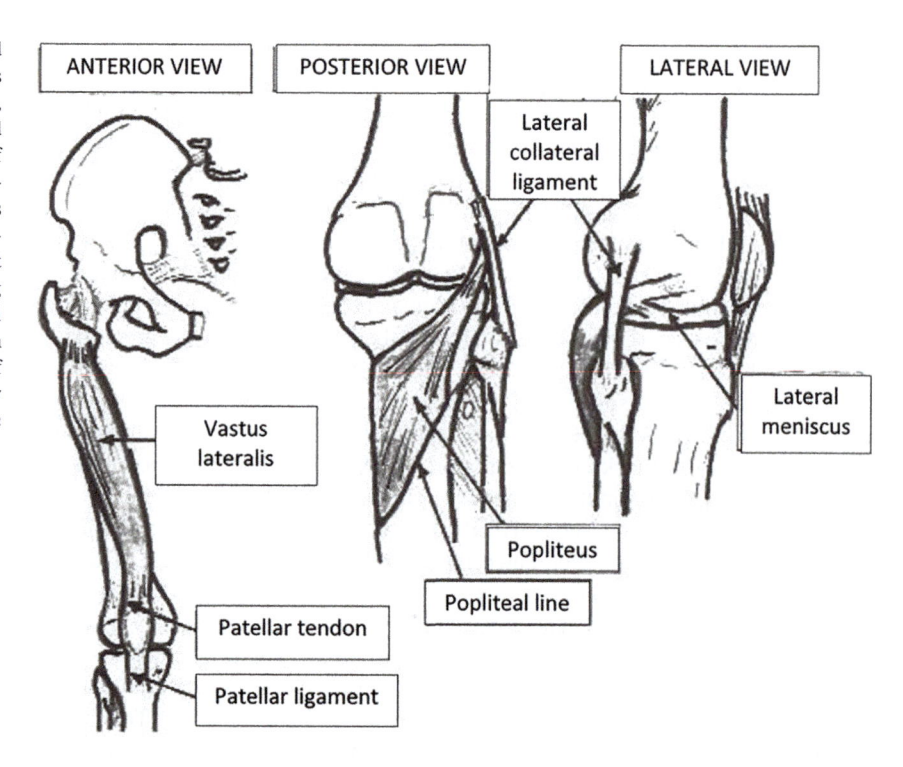

action are oblique, despite sharing a common distal attachment to the tibial tuberosity. This obliquity creates frontal/transverse plane torque within the sagittal plane knee angular momentum and patellar glide of the patellofemoral joint. Vastus medialis also demonstrates a medial section of muscle with fibres oriented more transversely known as *vastus medialis obliquus* (*VMO*), creating two differently angled force vectors into the patella from the single muscle.

The quadriceps action becomes more complex again as we add in the action of rectus femoris. Rectus femoris has two proximal attachment points, one of which attaches to the anterior inferior iliac spine of the ilium and the other arising from a groove above the hip's acetabulum, with some muscle fibres attaching to the hip capsule (Draves, 1986, p. 228). Thus, it becomes a multi-articular muscle in that it crosses three joints: the hip, the patellofemoral joint, and the tibiofemoral joint. Therefore, the quadriceps involve three bi-articular muscles (the vasti) and one multi-articular muscle (rectus femoris). This means that the simple picture of a single muscle action over a single joint is far removed from reality (Fig. 2.1.7b).

Uni-articular muscles are uncommon in the distal regions of the lower limb, but they are more common around the hip. These include gluteus medius and minimus, the obturators, and the gemelli. Piriformis crosses the sacroiliac and hip joints. Gluteus maximus has proximal attachments to the sacrum and coccyx as well as the ilium, and through its attachment and action on the iliotibial tract, it influences both hip and knee moments as does tensor fasciae latae. The hamstrings are also bi- and multi-articular. They attach proximally to the ischial tuberosity of the ischium proximally and cross the hip and tibiofemoral joint. The biceps femoris also crosses the proximal tibiofibular joint. Although hamstring proximal attachments are very close together on the ischial tuberosity, their distal attachments are not. Semimembranosus attaches distally as an expansion to the horizontal groove on the posterior surface of the medial tibial condyle, and via the popliteal ligament and fascia to the soleal line of the tibia. Semitendinosus attaches distally to the anterior medial tibia passing over the pes anserinus bursa, whereas biceps femoris attaches to the fibular head with some slips also attaching to the lateral tibial condyle and crossing the proximal tibiofibular joint. Not only do hamstrings act together in providing knee and hip extension–flexion moments, but they can also create rotation torques within the hip, knee, and tibiofibular joints (Fig. 2.1.7c).

The muscles of the lower leg and foot, with the exception of the interossei, lumbricals, and flexor digiti minimi brevis, are usually multi-articular muscles (see Table 2.1.7). The significance of this is that on contraction, these muscles influence motion-stability across multiple articulations. This is essential as the foot and ankle work together and the foot has to achieve changeable mechanical properties and points of motion. Initially, a state of relative compliance is required of the foot through loading response, allowing soft tissue and osseous motion during the early part of the gait cycle. It then gradually changes to a stiff platform from which to drive forward from. This requires quite different levels of joint mobility and stability at different times and at different locations. If multi-articular muscles did not exist, then each joint in the foot

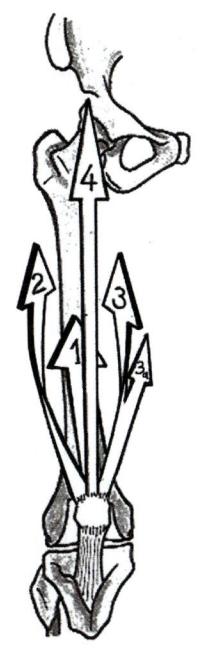

FIG. 2.1.7b Vastus intermedius' force vector (1) is part of a global collection of whole quadriceps femoris-induced vectors from vastus lateralis (2), vastus medialis (3), and rectus femoris (4), which crosses the hip as well as the patellofemoral and tibiofemoral joints. Thus, the quadriceps together can flex the hip and extend the knee, but individually, each muscle has quite different effects. Thus, weakness or tightness in one alters the balance of the force vectors. Note that vastus medialis has a second more medially directed force vector from its more oblique medial fibres (3a). *(Image modified from Kapandji, I.A., 1987. The Physiology of the Joints. Volume 2: The Lower Limb. fifth ed. Churchill Livingstone Inc, New York, NY.)*

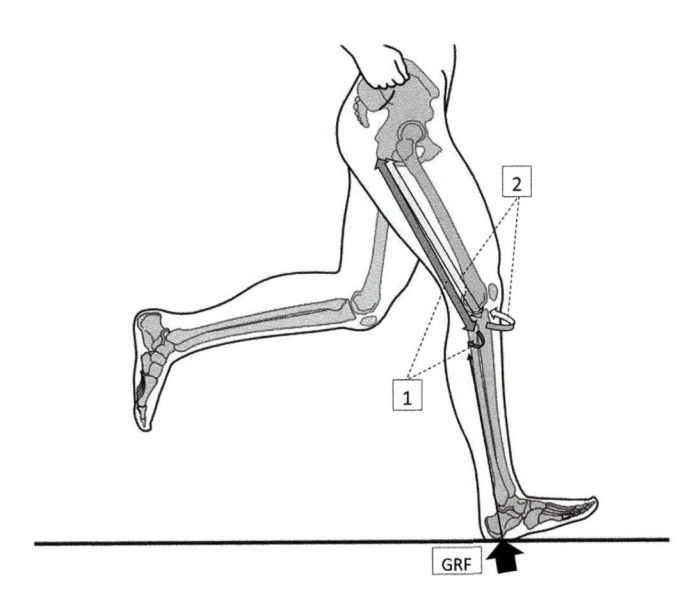

FIG. 2.1.7c Muscles often function in a state of a force-induced 'balancing act'. In late swing phase, before initial contact, the hamstrings brake knee extension and hip flexion, acting in unison. However, biceps femoris (1) through its lateral fibular head attachment can externally rotate the tibiofibular joint, imparting an external rotation vector on the lower leg. This may help tension the interosseous ligament and membrane between the fibula and tibia, helping dissipate torque during impact. The more medially aligned hamstrings, particularly semitendinosus (2) that attaches to the anterior medial tibia, can impart an internal rotation moment to the lower leg. Thus, a balanced force vector medially and laterally can be achieved while resisting knee and hip flexion, keeping the lower leg straight. Alternatively, medial or lateral moments can be allowed to dominate. This adds adaptability between the hamstrings in positioning the limb prior to initial contact, something more likely significant in running and fast walking when hamstring braking action becomes greater. Weakness in one hamstring may upset this adaptability. *(Permission www.healthystep.co.uk.)*

and ankle would need a separate muscle to control stability and motion across itself. By having multi-articular muscles, stability across multiple foot articulations can be achieved by utilising a much smaller number of muscles acting as tie bars in a foot vault behaving as a functional unit.

The concept of muscles crossing a number of joints makes the separation of functional units such as the pelvis, hip, and knee artificial. This is because through bi- and multi-articular muscle action, the stability and motion at one joint is directly influencing other functional units. Yet the ability for each unit in the lower limb chain to operate independently is also possible. Movement of one joint or unit affecting the movement of another joint or limb segment is known as *coupling*.

TABLE 2.1.7 Some examples of intersegmental complexity of muscular influence on joint motion and stability within the lower limb in consideration to a muscle being uni-, bi-, or multi-articular.

Muscle	Functional unit(s)	Joint stability type	Number of joints crossed
Gluteus medius	Pelvis–Hip	Uni-articular	1 Joint
Gluteus maximus	Pelvis–Hip–Knee	Multi-articular	3 Joints
Piriformis	Pelvis–Hip	Bi-articular	2 Joints
Tensor fasciae latae	Pelvis–Hip	Bi-articular	2 Joints
Adductor magnus	Pelvis–Hip	Uni-articular	1 Joint
Adductor longus	Pelvis–Hip	Uni-articular	1 Joint
Sartorius	Hip–Knee	Multi-articular	3 Joints
Rectus femoris	Hip–Knee	Multi-articular	3 Joints
Vastus intermedius	Knee	Bi-articular	2 Joints
Popliteus	Knee	Uni-articular	1 Joint
Gastrocnemius	Knee–Ankle	Multi-articular	3 Joints
Soleus	Ankle	Bi-articular	2 Joints
Tibialis posterior	Ankle–Foot	Multi-articular	10 Joints (variable)
Tibialis anterior	Ankle–Foot	Multi-articular	4–5 Joints (variable)
Flexor digitorum Longus	Ankle–Foot	Multi-articular	21 Joints (variable)
Flexor digitorum Brevis	Foot	Multi-articular	17 Joints
Abductor hallucis	Foot	Multi-articular	4 Joints

The amount of joint coupling seems to be variable both in relation to individuals and different motions. Generally, where more stability is required such as in running as compared to walking, more joint coupling tends to occur (Pohl et al., 2007).

2.1.8 Concepts of form and force closure

The interplay between joint shape and muscle activity is key to joint stability. Some joints primarily use shape to provide stability, while others rely primarily on muscle compression forces. Synovial (diarthrodial) joints are therefore stabilised in one of two ways, with large variation and crossover between the two mechanisms. *Form closure* of a joint is a term used for the passive elements of joint stability such as ligaments, joint capsule, and bone surface shape to fit closely together. *Force closure* of joints is concerned with the contractile elements such as the muscles and their tendons that cross a joint. In truth, no joint is purely one or the other, but one mechanism does tend to dominate and this underpins much of the activity required at a particular functional unit during gait and activity. The general rule is that joints with little motion tend to be stabilised more by form closure, whereas those that require high mobility need to be unrestricted by joint shape and rely on force closure for stability. Those that require high levels of both, such as the hip, have joint morphology lying somewhere between.

The best example of form closure within the lower limb is in the pelvis at the sacroiliac joint, where an articular surface lies between the bones of the ilium and sacrum. There are two of these joints, one on each side of the body. They are broader superiorly and narrower inferiorly. In adults, the ilium is fused to the ischium and the pubis bones. The two pubic bones join together as a fibrocartilaginous union at the pubic symphysis, which normally permits only the slightest rotational motion between the two pubic bones. With the two sides of the pelvis firmly fixed in fibrocartilaginous union anteriorly, the sacrum fits into the reciprocal articular surface posteriorly at the ilium like a keystone in an arch. The sacrum is broader superiorly, narrower inferiorly, and is angled anteriorly, allowing it to sit in near-perfect alignment to the articular surfaces of the ilium. The sacroiliac joint is then compressed together by numerous ligaments such as the iliolumbar, posterior sacroiliac, anterior sacroiliac, and sacrotuberous ligament that crosses the pelvis to the sacrum, assisted by force closure from gluteus maximus activity. This results in the sacrum sitting in stable joint alignment, passively compressed together in a 'sandwich' between the body weight vector (BWV) under the influence of gravity and the resultant GRF generated during gait. Although the sacroiliac joint is synovial and can move by small torques of nutation (anterior rotation) and counternutation (posterior rotation), little muscle effort is required to keep these joints stable, bound together as they are by passive tight ligaments. The sacroiliac joints are about as distinctly 'form closed' as a synovial joint can get. However, in reality, the muscles that cross this joint (in addition to other spinal joints and the hip) such as the erector spinae muscles, psoas major, and piriformis, as well as gluteus maximus, are able to influence joint stability and motion through force closure (Fig. 2.1.8a).

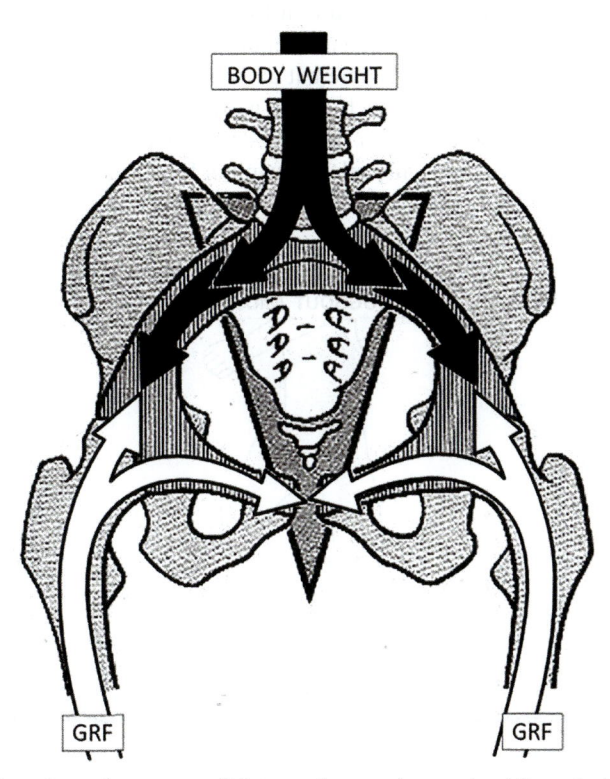

FIG. 2.1.8a A schematic of sacrum form closure from osseous fit between the opposing sacral and iliac articular surfaces at the sacroiliac joints (indicated by sides of the central triangle). These surfaces are wider superiorly, narrower inferiorly, crudely forming the sacrum into a triangular wedge. The articulation is also narrower posteriorly than anteriorly. Although an excellent example of form closure, full 'wedging' under vertical compression can only occur when a human is erect and in double-limb support. It is obvious that force closure from surrounding muscles is still very important to this joint's stability. *(Image adapted from Kapandji, I.A., 1974. The Physiology of the Joints. Volume 3: The Trunk and the Vertebral Column. Churchill Livingstone, Edinburgh.)*

Force closure is the preferred method of joint stability in joints that require large ranges of mobility, such as the MTP joints. The best example within the lower limb is the knee. Here, the congruency (matching shapes of the opposing joints) is extremely poor, with the highly curved condylar surfaces of the femur sitting superior to the relatively flat tibial articular surfaces. Joint stability is enhanced by soft tissue structures such as the cruciate ligaments, collateral ligaments, and menisci, but these are mainly effective in providing stability and congruency around end-range extension and gliding motions, rather than stabilising primary extension–flexion motion. Flexion is limited by the presence of large muscle bulks in the thigh and calf when contacting each other. Humans are able to hold the knee in fixed degrees of flexion through the compression forces that are initiated by activity of the muscles that cross the knee. Force closure is therefore only achievable through muscle activity making the knee joints reliant on strong healthy muscles working in balance to sustain effective stability–mobility relationships. Weakness and/or tightness of knee muscles have a big impact on knee joint function, whereas weakness or tightness around the form closed sacroiliac joint (while far from desirable) does not result in the same degree of disability from instability and mobility issues than it does within the knee. The knee requires high sagittal plane mobility, but very limited transverse and frontal plane mobility. Therefore, the passive restraint of strong ligaments targets the transverse and frontal planes.

As a general rule in lower limb joints, the greater the requirement for mobility, the less form closure will be present and the more a joint is reliant on force closure. The hip needs a large degree of stability but also a high degree of mobility in all planes, albeit mostly in the sagittal plane. The answer for the hip joint is to use a ball-and-socket arrangement that has excellent congruency and is assisted by the tightening of ligaments around the hip with changing motion to achieve ligament-increased form closure. In motion occurring in one direction, some hip ligaments tighten while others loosen in conjunction with changing muscle activity that goes to create stability and/or motion as required.

The ankle also requires stability, with stability being dominant in the frontal and transverse planes where only small ranges of motion are required. It is 'boxed' dorsally between the distal tibial and fibular malleoli which are firmly attached to each other through ligaments, including the large interosseous membrane. The superior articular surface of the talus is therefore congruent within the articular surfaces presented to it by the tibia and fibula, which primarily only permit relatively free sagittal plane flexion and extension motion that is required by the ankle joint during gait. The ankle therefore requires force closure to keep flexion and extension stable but also to remain mobile (Fig. 2.1.8b). This force closure is supplied by multi-articular and bi-articular muscles such as the gastrocnemius, soleus, tibialis posterior, the peroneals, tibialis anterior, and the long toe flexors and extensors. Because transverse and frontal plane form closure is strong in the ankle due to osseous and ligamentous restraint, only relatively minor force closure is provided for frontal and transverse plane motions through (primarily) the tibialis posterior and peroneal muscles (see Fig. 2.1.3c).

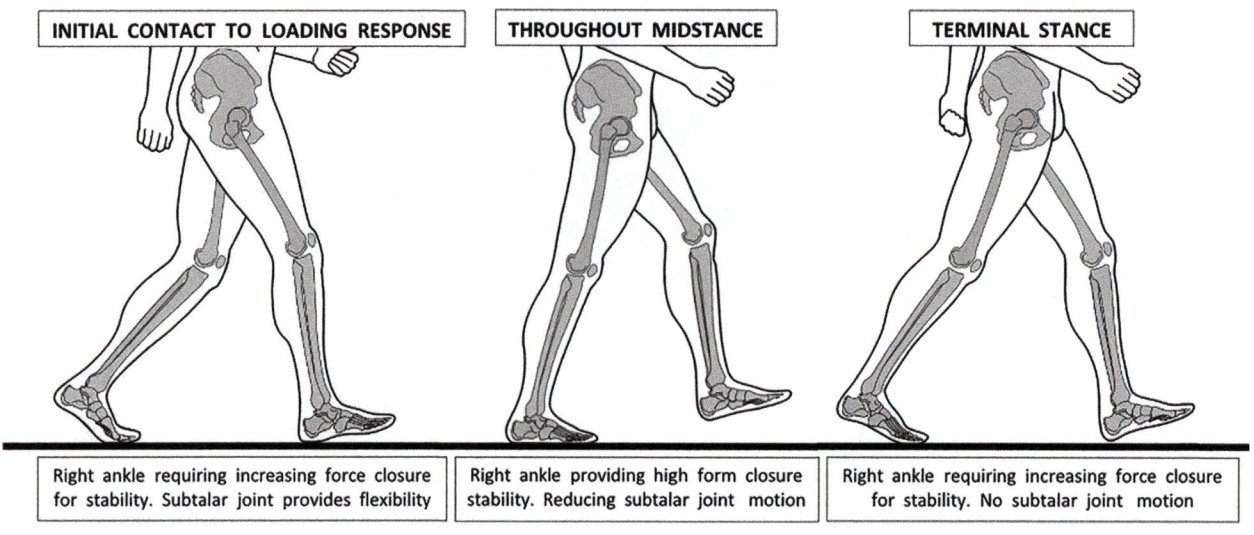

FIG. 2.1.8b Ankle joint form and force closure relationships change during gait and directly influence subtalar joint motion. The subtalar joint only demonstrates motion through ligament flexibility under external loading. Therefore, the calcaneus moves with the talus when the heel is not in ground contact. At initial heel contact, the foot is relatively plantarflexed and ankle form closure is less pronounced. Ligament and muscle activity is thus important in maintaining joint stability. The subtalar joint can provide some frontal and transverse flexibility to reduce ankle torques under externally derived forces (i.e. GRF). As the ankle becomes increasingly dorsiflexed during midstance, form closure improves because the wider anterior trochlear surface of the talus becomes more tightly fitted between the articular surfaces of the tibia and fibula. Subtalar joint motion should also reduce through decreasing talar freedom. After heel lift, with reducing ankle dorsiflexion and then rising plantarflexion, form closure reduces and more force closure for stability is required. When the calcaneus becomes unloaded, subtalar motion is lost. *(Permission www.healthystep.co.uk.)*

The talus is also wider anteriorly at its superior articular surface, so the ankle demonstrates better form closure in dorsiflexion (extension) than it does in plantarflexion (flexion). As the foot is in single-limb support when the ankle is undergoing dorsiflexion during gait, the increased stability associated with stance phase dorsiflexion is desirable. The ankle also utilises a close coupling relationship with the subtalar joint below it, with a distinct change in that relationship between rearfoot weightbearing and nonweightbearing. When the rearfoot is in ground contact, the subtalar joint provides some freedom of motion between calcaneal and talar forces. However, when the rearfoot is nonweightbearing, ligamentous force closure is such that the calcaneus moves with the talus as though the subtalar joint did not exist (Leardini et al., 2001).

The foot presents a quandary to the principles of form and force closure, although this is easily explained through its evolution. The foot is made of many 'flattish' gliding-surfaced joints demonstrating poor form closure, with only the subtalar joint having any significant form closure. The rest of the foot's articulations are shaped to provide mobility, with strong ligamentous joint force closure restrictions coming primarily from the multi-articular long plantar ligament, the uni-articular short plantar ligament, and the bi-articular plantar calcaneonavicular (spring) ligament that all lie on the plantar osseous surfaces. The foot has evolved from a grasping, climbing, arboreal structure requiring high mobility to one that has been co-opted by time into bipedal gait, requiring greater stability. In its present function, it has to achieve two states during gait. It must first function as a fairly compliant structure to dissipate impact and single support stance energy and then conform to the support surface through joint motion and soft tissue flexibility. During stance, it must increasingly become a stiffer structure to provide a firm base to accelerate forward on, converting the foot into a semirigid lever arm by increasing its elastic properties.

Although the foot does not provide a large degree of internal motion (except at the MTP joints), imagine the impact of a foot having only excellent form closure and being stiff throughout gait. With the provision of such characteristics, conditions would be ideal for this foot to perform the role of a stiff structure. However, its ability to create compliance to dissipate loading energies would be severely limited. Equally, if it retained its ancestral mobility, then it would be too compliant to provide a stiff acceleration platform. By using force closure to control the degree of compliance and then create stiffness, the foot is able to perform its two important roles of braking and acceleration while at the same time acting as a base of stability; providing that muscle activity and passive tissue tensioning is appropriate in both its timing and strength. It is surely no coincidence that the primary midfoot joint that requires the most mobility, but also stability, is the talonavicular joint that has an articular surface shaped somewhat similar to the hip, being ball-and-socket-like (Pisani, 1994, 2016).

2.1.9 Concepts of joint packing, congruency, and neutral

Close-packing refers to a position where there is maximum congruency of joint surfaces and the ligaments are at their most taut, with all the other articular positions being referred to as *open-packed* (otherwise known as *loose-packed*). A synovial joint's role is to provide rather than restrict motion, and they are almost frictionless when in good health. Stability is achieved through muscle activity and internal reaction forces which together achieve a state of equilibrium. Ligaments are neither taut nor lax as they express a degree of both in any particular position. Their complex fibre arrangements allow variable areas of tautness in various positions; a rare exception being the posterior portion of the medial collateral knee ligament in flexion (Hull et al., 1996). Close-packed joint positions often bring some ligaments or areas within the ligaments to their peak of tautness, causing an end-range of joint motion where additional motion is not possible without disruption of the joint's anatomical integrity, such as a ligament tear or cartilage damage.

This end-range of joint motion concept makes close-packed positions perilous, as in such positions, a joint cannot be protected from negative work via eccentric muscle contraction which is the usual protector of the musculoskeletal system's integrity (Bull and Amis, 1998). Forces applied into such positions at the end-range of motion can cause bone to compress on cortical areas as well as trabecular bone surrounding the joint. Cortical bone has a higher elastic modulus than trabecular bone and is unable to dissipate the cartilaginous load as would the trabecular bone (Lovejoy, 2007). This makes close-packed positions mechanically undesirable and not a part of normal joint stability. Often, geometric contiguity of joints in such positions is lost, resulting in asymmetrical compression forces developing within the joint surfaces. If prolonged, this can lead to changes in the frictional properties of the cartilage.

Close-packed positions are usually associated with the end-range motion of a joint in a particular direction. Examples of this are knee end-range extension, ankle end-range dorsiflexion, and subtalar end-range eversion. Such close-packed positions are very stable, as motion from these positions is extremely difficult in one direction. However, in that direction, such motion can be potentially injurious and it usually requires significant muscle length changes to move the joint into and out of this position. The disadvantage of such positions is obvious in that increased force further in the direction of the close-packed end-range results in further tissue deformation to the joint structures such as the ligaments, the cartilage, and the subchondral bone. In the case of the knee, in end-range extension, patellofemoral congruency is poor in that the patella's position is only retained by soft tissues alone as no retention mechanism within the bony structural ridges on the femur (femoral groove) has evolved to protect the patella from dislocation in this position (Lovejoy, 2007). Full knee extension is not a position of normal human gait, but it can occur under high knee extension forces or with ligamentous hypermobility.

A close-packed position is not the same as *joint congruency* which represents a position where the joint surfaces are maximally in contact, and loads are most evenly distributed throughout the joint. In joint congruency, articular surfaces are aligned with relatively even contact forces across the joint surfaces without asymmetrical surface deformation. Often, congruency is the natural resting position of a relaxed joint uninfluenced by gravity. Motion in any direction from the congruency position should be relatively easy. Examples of such positions include the knee in near-full extension (but not in full extension), the ankle when it is neither dorsiflexed nor plantarflexed and angled ~90° to the leg, and the subtalar joint where it is neither everted nor inverted. Should force be directed into a joint in any direction when congruent, it should be more easily able to dissipate such energies through free motion.

The difference between neutral position and joint congruency is somewhat open to interpretation and is for some joints much the same position; the two terms having become blurred. Neutral position should be considered to be where the joint is in a position that equates to the anatomical position. This is easy in relation to the hip, as a hip is neither extended nor flexed, internally nor externally rotated, and neither abducted nor adducted while in the anatomical position. In the knee, however, the joint is always described as either extended (near-end-range of extension) or flexed (anything outside that position). The ankle is in its neutral position when neither dorsiflexed (extended) nor plantarflexed (flexed), making neutral position again easy to define. The foot should always be pronated to the ground (plantar surface prone) in the anatomical position, but not maximally prone. In clinical biomechanics, the 'neutral' position of the subtalar joint has caused the most confusion in assessment and treatment. Neutral position within the foot in relation to subtalar joint neutral position was proposed as a position the foot worked best from and near to and was described as the position found at one-third away from maximum eversion and two-thirds away from maximum inversion by Root et al. (1971, p. 38). Root and his co-workers also defined a neutral position for the hip based on ranges of motion (Root et al., 1971, pp. 104–108). The history and debates over this position and the model of foot function associated with this concept can be explored by reading Lee (2001), but the conceptual prominence that the subtalar joint neutral position once held is no longer felt to be relevant to clinical practice (McPoil and Cornwall, 1994, 1996; Jarvis et al., 2017). However, this does not in any way mean that the subtalar joint's motion is not relevant.

The open-packed position describes a position where the joint is least congruent. It can still involve some joint compression at the end-range of the open-packed position, although the areas of contact are usually smaller than in the close-packed position. With minimal joint surface contact, motion in the joint usually increases outside the primary plane of freedom, and the ligaments become the critical restraining mechanism outside of muscle contraction rather than the joint surface. Good examples of this are ankle plantarflexion and subtalar inversion. The more plantarflexed the ankle, the more the ankle can rotate in the transverse and frontal planes. Because the subtalar and ankle joint complexes are so interdependent, a supinated subtalar joint with its own plantarflexion, adduction, and inversion is considered as an open-packed position. This links to increased frontal plane motion within the rearfoot and reduced congruency of the joint surfaces in the ankle and subtalar joints together. The subtalar joint is a stiff joint with little freedom of motion and is well restrained by ligaments, but more so in ankle dorsiflexion and when unloaded by external force (Leardini et al., 2001). Although end-range knee flexion is prevented by soft tissue in the calf and thigh, when in greater knee flexion, the knee can move more freely, i.e. rotate and adduct or abduct with increasing ease within the frontal plane.

Clinical implications of joint stability

Knowing how joint stability is primarily achieved in each joint helps in establishing the mechanism of injury in the clinic. Stable joints such as the sacroiliac joint are far less likely to suffer excess mobility resulting from muscle weakness. However, excessive or uneven stiffness, or flexibility between the two sacroiliac joints across the pelvis may have some implications on function and the generation of symptoms. Joints like the knee and those within the foot are highly reliant on muscle strength to keep them stable, so muscle weakness risks profound effects on function. Finding muscle weakness around joints with a high need for force closure is likely to be clinically very significant. Restrictions through soft tissue dysfunction that either result in limited motions, offer resistance to full motion, or cause changes in planal dominance of motion are again significant findings. A patient with an abducted foot position in gait may develop restricted ankle motion through lack of sagittal plane ankle freedom for use in the line of progression of the foot, which will no longer align to the line of progression of the rest of the lower limb. The problem then requires increased rotation of the rearfoot (eversion) or through the midfoot (inversion) in the frontal plane, as the line of progression is angled across the foot rather than near or along its longitudinal length. An abducted foot in gait can result from whole limb external rotation, derived from, for example, excessive external hip rotation either because of osseous joint shape or hip muscle weakness/tightness. In these cases, knee and ankle torques will also be altered to the line of progression of the trunk.

The implications of articular packing are that joints in close-packed positions are far more likely to sustain end-range joint compression injuries if subjected to sudden perturbations. Examples include osteochondral compression and cartilage injuries. Open-packed positions are far more likely to result in ligament and muscle–tendon injuries as movement is freer if soft tissue failure occurs. Ankle ligament sprains, ruptures, or avulsion fractures are thus more likely to occur in ankle plantarflexion under lower stresses when the ankle has freer motion associated with events such as slipping when walking downstairs, walking on a steep slope, or wearing high heels. A patient demonstrating a hypermobile foot with a more laterally positioned centre of ground reaction force (CoGRF) will likely be more vulnerable to ankle sprains (Willems et al., 2005), as ankle passive restraint is compromised in a more unstable rearfoot position.

These concepts of joint position have their clinical limitations in understanding failure within any lower limb functional unit. Provide enough force and the tissues will fail regardless of their joint position. For example, twisting the lower leg beyond the capacity of the interosseous membrane and associated ligaments to shock absorb torques across the tibiofibular joint and the fibula may fracture bone and tear ligaments in either direction of rotation. Take a joint beyond its end-range in an open-packed position and it will become compressed at certain sites, developing subchondral injuries and fractures. Force a close-packed position joint beyond its end-range limits and not only are subchondral lesions likely, but ligament and even tendon failures and bone fractures (usually induced by compressive-rotational forces) can still occur. The concepts of close-packed and open-packed joints when used in reference to the time of injury can only help in deciding what tissues are most likely to be injured through also appreciating the direction of the force applied. Therefore, establishing what a patient was doing at the time of injury is critical to establishing the most likely mechanism of injury, the tissues at highest risk of being injured, and the pathology most likely to result.

2.1.10 The skeletal frame

The skeletal frame as a unit can be used as a term to describe the osseous, articular, and ligamentous structures that together act as a stiff elastic structure, thus separating the deep passive tensile ligaments and bone structures under compression from the myofascial unit. Despite this separation, it must be appreciated that many connective tissues blur the boundary between the units. The skeletal frame forms a stiff deep core structure that has a changing specific role during the particular tasks of locomotion. The changing mechanical properties and roles the skeletal frame provides are ultimately dependent on the activity of the myofascial tissues. The skeletal frame is the functional target of the myofascial tissues that can alter its shape to change the flexibility within it. Simply stated, it is whether the skeletal frame becomes more stiff or less stiff as an entire structure that influences the balance between mobility and stability. It is the joints that provide mobility, diseases of which can lead to loss of motion through disturbed articular integrity and/or pain.

The skeletal frame of the foot probably most clearly illustrates the usefulness of the term, as its twenty-eight regular bones and their joints provide an extensive range of possible shapes and degrees of flexibility. The foot's skeletal frame can be variably stiffened or softened through changing its shape and ranges of motion. Extending (dorsiflexing) the MTP joints stiffens the foot's skeletal frame and raises its vault through the windlass mechanism, brought about by the plantar aponeurosis being shortened and tightened across the foot's vault. Allowing the toes to move out of this extension gradually increases flexibility across the foot and its vault, enabling the vault to be more easily lowered. Tensing muscles under the vault stiffens the foot, making further vault lowering more difficult. Muscle activity can also raise the vault, causing the foot to behave more stiffly from increased curvature mechanics. Standing on the foot with the skeletal frame loaded medially from heel to toe may cause the medial foot to lengthen more and as a result, cause medial digital extension to become increasingly difficult. Thus, medial foot vault raising also becomes restrained as a result of preventing hallux extension. Therefore, these changes in shape and loadings of the foot's skeletal frame from internally derived forces (muscle activity, connective tissue stress-stiffening) and externally derived forces (GRF) become important in understanding how the foot can modify its shape and stiffness to cope with the transition of roles during locomotion. The shape changes around the skeletal frame are an essential part of modifying the foot's mechanical properties. It also underlies the fact that should these adaptable abilities be lost through skeletal frame dysfunction, then gait disturbance and perturbation can result.

The skeletal frame is therefore an artificial construct that serves as a way of describing tasks to indicate changes in structural mechanical properties that the myofascial soft tissue can create within the bone-joint-ligament core structures of the limb. It represents a more intimate structural unit within a greater functional unit of the biotensegrity structure.

2.1.11 Section summary

Functional units exist on many levels and will be discussed within the concept of anatomical levels and joints that perform functional tasks as a team during gait. However, functional units exist at a whole-body level as well as at a limb segment level, and even down to tissue and cellular levels. Biological structures have developed ingenious methods of distributing stresses throughout the body's anatomy into manageable tension and compression forces via protein-based structures, reducing torsion and shear (with its associated friction) to an absolute minimum. In so doing, biology has been able to keep most life stresses well within tissue stress thresholds (Smit and Strong, 2020).

However, like all functional units, the sum is dependent on the parts. Any impediment in the ability to move so as to adjust shape and structural properties, represents a risk to functional integrity. Although biotensegrity offers a structural solution to managing the preferred high stresses (compression and tension) well within tolerance, failure in biotensegrity can soon cause structural property limits to change significantly and become catastrophic for health and mobility.

2.2 Functional unit of the lumbar spine and pelvis

2.2.1 Introduction

Lower back dysfunction, including the sacroiliac joint and its associated structures, is a common source of lumbopelvic or pelvic girdle pain and disability. It is usually considered a multifactorial problem with considerable physical, psychological, social, and economic impacts that are rarely associated with one pathological cause (Cohen et al., 2013; Buchbinder et al., 2018; Poilliot et al., 2019). The degeneration of ligaments and joints has not proved to be a conclusive source of symptoms in the lower back (Hartvigsen et al., 2018). Trunk muscles function differently in back pain (Hodges and Tucker, 2011; Hug et al., 2014) and the cross-sectional area of the muscles is often reduced (Sions et al., 2014). The thoracolumbar fascia also represents a possible source of pain and dysfunction (Benjamin, 2009; Vleeming et al., 2014).

There is no clear single source of low back pain, but associations are found with altered movement patterns (Hodges and Tucker, 2011). Understanding the spine as a functional unit may provide clinicians with a better tool to approach pathology rather than just a diagnostic image in the majority of cases. Sub-failure of interspinal ligaments, facet joint capsules, and associated mechanoreceptors, as well as large fascial structures such as the thoracolumbar fascia may lie at the heart of understanding mechanical low back pain (Schleip et al., 2007). The human lumbosacral region finds itself as a hub between erect stability of the trunk and lower limb locomotive mobility, and supplies the anatomy that directly crosses this dramatically different functional interface.

2.2.2 The kinematic role of the lumbar spine and pelvis

If the hip is the start of the lower limb, then the pelvic bones which form the superior half of the hip joint are the origin of its musculature and must be duly considered in explaining lower limb locomotion. Yet, if the pelvis plays a fundamental role in lower limb locomotion, then the interaction between pelvic and spinal stability should be considered. Therefore, to avoid expanding one's clinical remit to cover the full human body's anatomy and biomechanics, a focus on the lower limb must arbitrarily start somewhere. With regard to human locomotion, consideration must be given to the fact that the lumbar spine and pelvis represent part of the trunk-passenger segment that not only interfaces with the lower limb but also acts differently to the upper trunk and thorax. This is not to suggest that the thoracic spine should be considered rigid, for it is clearly not and thus should not therefore be modelled as a rigid segment (Needham et al., 2016). However, the consensus of 'normal' spinal mobility continues to remain a subject of debate (Chockalingam and Needham, 2020). From an evolutionary standpoint, the lumbar spine and pelvis have altered profoundly across time to achieve a morphology that copes with plantigrade bipedal locomotion, while the upper limb has evolved separate beneficial anatomical consequences from these changes (mobile manipulative upper limbs) and this must be the justification for the convention of starting the locomotive functional units at the lumbosacral interface.

The spine must fulfil roles of stability and mobility as required. The whole vertebral column is often referred to as being like the mast of a tall sailing ship with the mast of the vertebrae resting on the deck of the pelvis. The shoulders are likened to the mainyard of a ship set transversely through the scapular girdle. At all levels, muscles and ligaments behave like stays (tensional supporting ropes), linking the mast to the pelvic deck. While the vertebral mast is vertical and the shoulders level, a position of symmetry is maintained with equal forces to the left and right side of the spine (Kapandji, 1974, p. 10). It is worth pointing out that such a structure demonstrates triangulated truss tensegrity (Levin, 2002). However, the sailing ship analogy is limited because the spinal 'mast' is curved sagittally and constantly moves with motion (Myers, 2020). When we stand on one leg, the opposite side of the pelvis tends to drop a little (varus tilt), controlled and stabilised by the stance limb hip abductors. However, the pelvis may rise up to counterbalance the pelvic drop by a valgus tilt via increased hip abductor activity. Single-limb support (statically or dynamically) results in a loss of equilibrium of spinal stability, and muscles must be activated in order to maintain postural balance through the extrapyramidal nervous system and full sensorimotor system, adjusting postural tone (Fig. 2.2.2a).

Repetitive destabilisation of one side and then the other side of the spine through alternating single-limb stance phases during gait requires the spine to have the flexibility that the vertebrae as multiple components give. Ligaments provide consistent nonspecific underlying passive tensile stability, with muscles supplying specific stability as required to balance forces generated during activity and motion. The pelvis acts as the stabilising 'hub', sandwiched between the mobility of the lower limbs and the spine. Because of the more restricted morphology of the thoracic spine, mobility is focused to the cervical spine enabling head mobility and to the lumbar spine for pelvic and lower limb mobility. What complicates this relationship vastly is that the spine also acts as a protective conduit for the spinal cord supplying all the information to and from the lower limbs that permit them to function and adapt correctly during activity through the sensorimotor system.

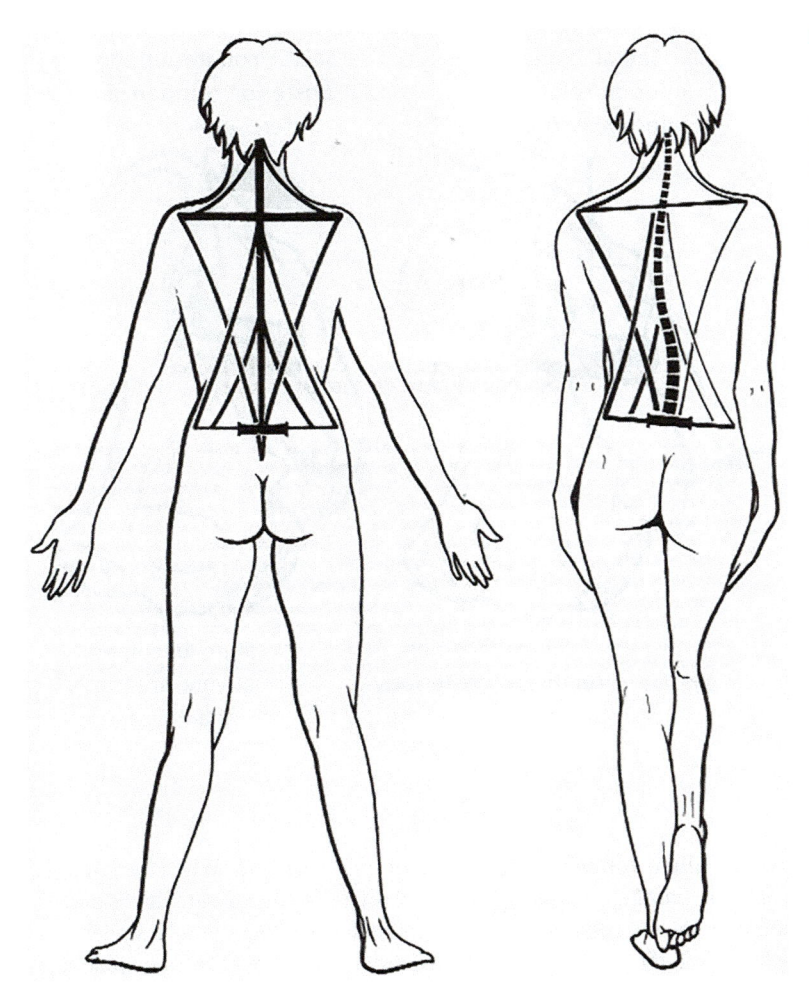

FIG. 2.2.2a Kapandji's images of spinal stability demonstrate principles of triangulated truss tensegrity by using the rigging of a sailing ship. The decking and mast change angulations to each other during locomotion, making the analogy to the mast and rigging problematic. However, the triangulated nature of the support created can be seen, even during single-limb support. *(Image from Kapandji, I.A., 1974. The Physiology of the Joints. Volume 3: The Trunk and the Vertebral Column. Churchill Livingstone, Edinburgh.)*

The pelvis serves as the stable anchor point for ligaments and muscles that both stabilise and provide motion above and below the pelvis in the sagittal, frontal, and transverse planes. With these soft tissues, the pelvis likely helps to dissipate sagittal plane rotation torques set up by flexion and extension of the lower limbs through their asymmetry of motion during gait. This still remains a controversial area in biomechanics and anatomy, but the logic of such an ability is difficult to dismiss. The reason is that sacroiliac (SI) joint motion is complex, generally poorly understood, and is variable between individuals, but it normally seems symmetrical within individuals (Jeong et al., 2018). The fact that motion occurs at this joint there is no doubt, for unilateral fusion creates a 26% decrease in off-axis mobility, but only a 6% reduction within a single plane, possibly explaining why many studies have failed to identify significant motion at the SI joint (Jeong et al., 2018). Further research is therefore warranted. However, with posterior rotation of the pelvis at the SI joint, the lumbar curve is decreased but with anterior rotation of the pelvis, it increases. Hip flexion induces posterior pelvic rotation with SI flexion, whereas hip extension induces anterior rotation with SI joint extension. In gait, opposite motions are being generated in each lower limb, and thus an ability to moderate the torque effects on the lumbar spine would seem highly beneficial. This moderation is thought to occur through SI joint rotations–torques limiting lower limb and pelvic stresses being applied directly to the lumbosacral joint.

Motion of the SI joint has been assigned its own descriptions, as motion largely fails to confirm easily to triplanar motion descriptions. This is due to articular surface shape, the difficulty in locating any instantaneous joint axis, and because motion occurs in part through cartilage deformations. A flexion torque directed on the pelvic bones from the hip during gait and the drive of the CoM of the trunk downwards and forwards induces *nutation* motion at the SI joint. Sacral nutation results in relative anterior rotation and an inferior and posterior gliding of the sacrum on the ilium. During hip extension, a *counternutation* torque is applied to the SI joint, causing the sacrum to rotate relative to the ilium in a posterior direction with superior and anterior gliding (Fig. 2.2.2b). These torques across the SI joint are also likely to generate torque across the

FIG. 2.2.2b Schematic of the motion of the sacrum during nutation (right) and counternutation (left) at the left sacroiliac (SI) joint. The SI joint's articular surfaces are exaggerated in size to help demonstrate motion, with the surfaces becoming increasingly roughened and irregular with age. Small, yet significant motion at the SI joint has now been established in vivo, but the SI joint remains a joint where further research is required to fully understand the extent and role of its complex multiplanar motion. *(Permission www.healthystep.co.uk.)*

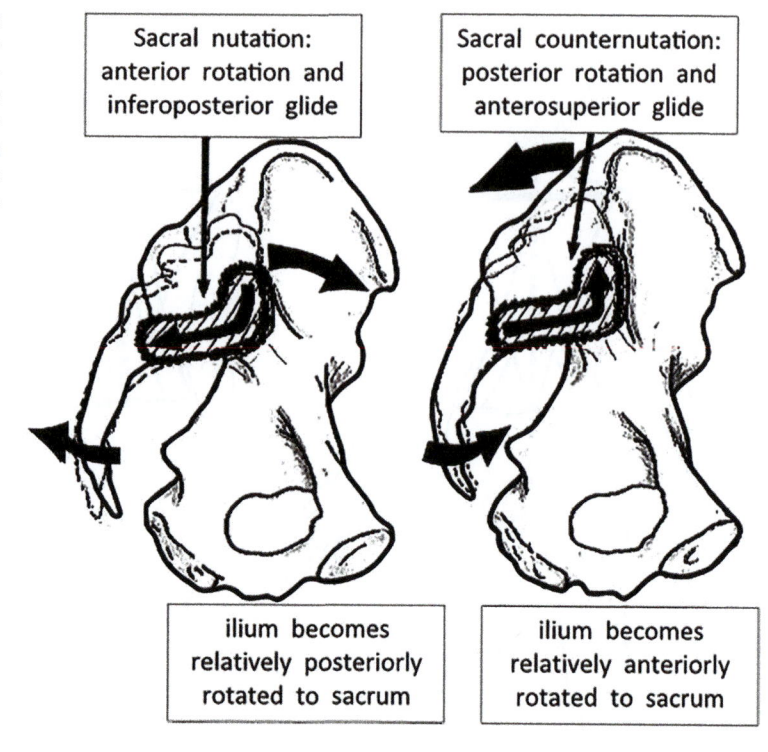

pubic symphysis through the asymmetrical motions of the ilium during the double support phases of gait. When the femur is flexed at the hip during gait, the innominate bones of the pelvis are relatively posteriorly rotated to the sacrum that is undergoing nutation, whereas in hip extension, the innominate bones become anteriorly rotated to the sacrum as it undergoes counternutation (Gibbons, 2007). However, the situation is more complex under GRF-induced motion of the spine and the whole pelvis. The contact GRF vector tends to flex the whole pelvis and trunk on the hip, increasing the net anterior pelvic tilt and increasing the lumbar lordosis. This is resisted by gluteus maximus and the posterior trunk muscles. During late midstance and terminal stance, the GRF vector drives the pelvis and trunk into posterior rotation via hip extension, decreasing the net anterior pelvic tilt and reducing the lumbar lordosis; motion resisted largely by psoas and the abdominal muscles that maintain the lordosis. Thus, nutations and counternutations are occurring at the SI joints within a whole pelvic reference frame in relation to the trunk above and lower limb motions below. For example, SI nutation at contact causes the sacrum to anteriorly rotate and glide inferoposteriorly on an ilium that itself is anteriorly rotating through flexion around the hip, despite the fact the ilium is now becoming relatively posteriorly rotated to the sacrum more than it was before initial contact (Fig. 2.2.2c).

In-phase and antiphase motion

Trunk motion has a significant influence on human locomotion, contributing to step length at velocities higher than 3 km/h. An inability to adapt trunk and pelvic rotations in gait that are associated with diseases such as Parkinson's, and with symptoms of low back and pelvic girdle pain (Bruijn et al., 2008). In the transverse plane, opposite rotations between the pelvis and the thorax–shoulder girdle occur and are known as *antiphase* motions (Fig. 2.2.2d). This is greater at faster walking speeds and is referred to as the '*pelvic step*' (Bruijn et al., 2008). During low walking velocities below 2 km/h, the pelvis and thorax tend to rotate together 'in-phase'. However, with increase in velocities, this changes to antiphase motion between the pelvis and thorax of around 120° (Lamoth et al., 2002; Bruijn et al., 2008). In the frontal plane, trunk and pelvic motions become more tightly co-ordinated and less variable with increased walking speed, while muscles such as the lumbar erector spinae increase their activity (Lamoth et al., 2006a).

The thorax rotates in antiphase with the femur at all velocities of gait, but the pelvis has a large phase difference to the femur only during lower gait velocities. This then decreases in difference from around 2.8 km/h upwards, becoming increasingly in-phase with rotations of the upper lower limb while becoming increasingly out of phase with the thorax

Locomotive functional units **Chapter | 2** 269

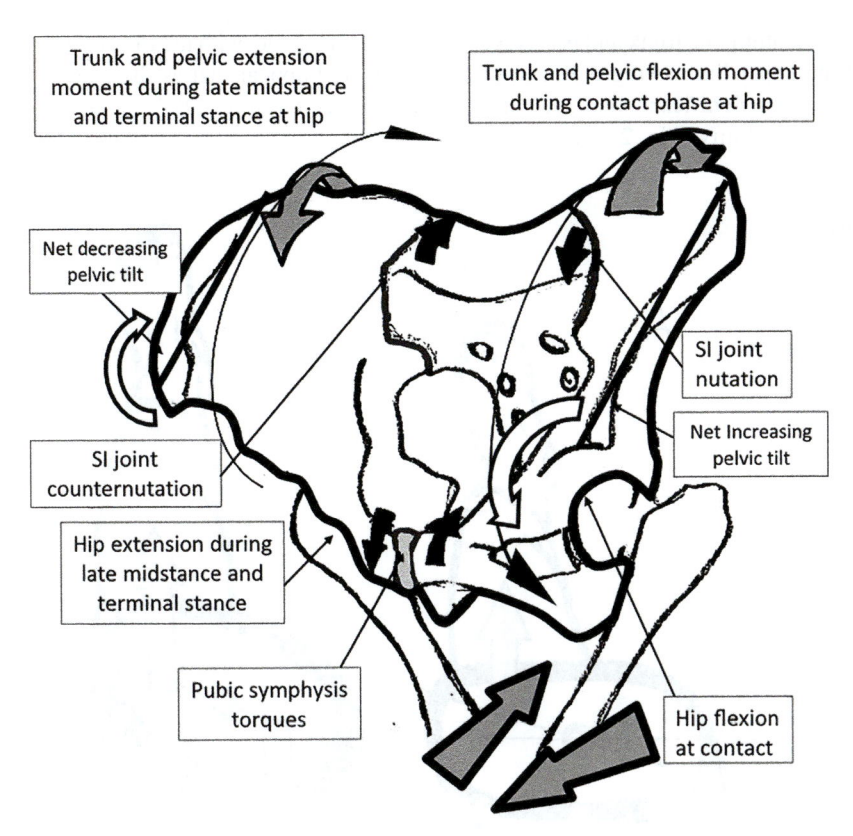

FIG. 2.2.2c Although motions of nutation and counternutation are relatively simple, these motions occur together at opposite SI joints within the complexity of a range of pelvic motions. At contact and loading, the sacrum anteriorly rotates and glides inferiorly and posteriorly on the ilium while the pelvis is flexing at the hip. On the opposite side, with the lower limb in its terminal stance, the sacrum is in counternutation. Thus, the sacrum posteriorly rotates and glides anteriorly and superiorly on the ilium, which itself is extending at the hip as part of the whole pelvis. This means the sacrum also torques within the transverse plane. The SI joint and the pubic symphysis become important moderators of lower limb torques between the two sides of the pelvis, protecting the lower spine from stresses induced through opposite limb motion. *(Diagram modified from Gibbons, S., 2007. Clinical anatomy and function of psoas major and deep sacral gluteus maximus. In: Vleeming, A., Mooney, V., Stoeckart, R. (Eds.), Movement, Stability & Lumbopelvic Pain: Integration of Research and Therapy, second ed. Churchill Livingstone, Elsevier, Edinburgh, UK, pp. 95–102 (Chapter 6).)*

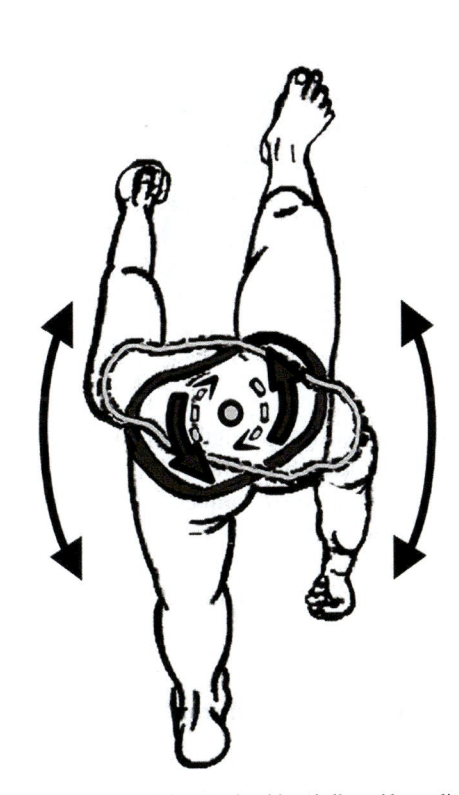

FIG. 2.2.2d Antiphase (out-of-phase) motion occurs between the thorax–shoulder girdle and lower limbs during gait. This means that the ipsilateral arm rotates backwards when the hip flexes forward during lower limb late swing phase, whereas it swings forward on the shoulder as the hip extends during late stance phase. Thus, contralateral upper and lower limbs advance forward together during swing phase, with the extent influenced by gait speed or via conscious intervention. *(Permission www.healthystep.co.uk.)*

(Bruijn et al., 2008). For generating angular momentum, the trunk and pelvis seem to play a minor role, with the lower limbs generating 60% of angular momentum and the upper limb swing only around 25%. It would appear that as the angular momentum of the lower limb increases, it creates a significant rotational 'drag' on the pelvis, causing antiphase motion with the trunk that helps in maintaining postural stability (Fig. 2.2.2e).

FIG. 2.2.2e During slow walking (A), the pelvic transverse plane axis alignment (PTPA) is in-phase with motion of the thorax, while the femur and lower leg axis (LL) move in antiphase with the pelvis. With increase in gait speed (~2.8 km/h +), the pelvis starts to rotate in-phase with the femur and out of phase with the thorax–shoulder girdle (B). The faster the gait speed, the more in-phase transverse plane motion becomes, helping to increase stride length potential through transverse hip rotation. *(Permission www.healthystep.co.uk.)*

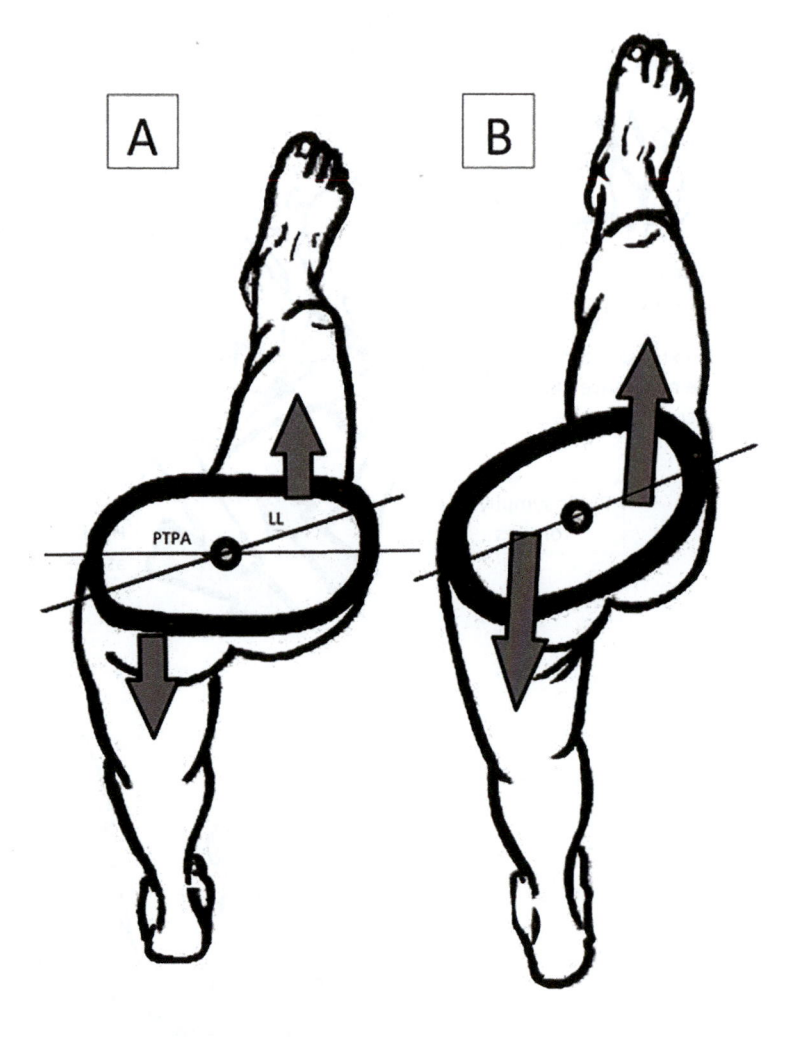

The GRFs generated by the lower limb's increased step length that occur with increased gait velocity may assist trunk stability through the force vector alignment to the divergence point within the thorax (Gruben and Boehm, 2012). This may help to negate the destabilising influence of the transverse torques between the pelvis and thorax. It is possible that the benefits to trunk stability generated by the divergence point's increased forces resulting from raised gait velocity, outweigh the destabilising effects of antiphase motion of the pelvis following the lower limbs' motions. At slower gait speeds, divergent point stability from lower GRFs may be insufficient in stabilising the trunk, requiring in-phase pelvic and trunk motion.

2.2.3 The spine and pelvis as a biotensegrity structure

Due to the continuity of anatomy and the body-wide totality of biotensegrity principles, separation of the lower limb from the rest of the body's regions is artificial and the boundaries between such regions debatable. The lower spine and the pelvis share load transfers between the trunk and lower limbs but have different roles in dynamic motion and stability. The lumbar spine is mobile with osseous form closure restricted to its posterior articulations while anteriorly, the vertebrae are linked by fibrocartilaginous plates. The lumbar spine's stability is highly dependent on the active and passive elements of biotensegrity, with the active elements most important for maintaining an erect posture. The thoracic spine above the lumbar spine has far less mobility. Being aware of the precise nature of regions of overriding stability and mobility is helpful, but human locomotion is best viewed under a holistic or global approach. Muscles work in series with the connective tissues to form viscoelastic myofascial chains across anatomy that can be appreciated through the concept of humans as biotensegrity structures (Dischiavi et al., 2018) (Fig. 2.2.3a).

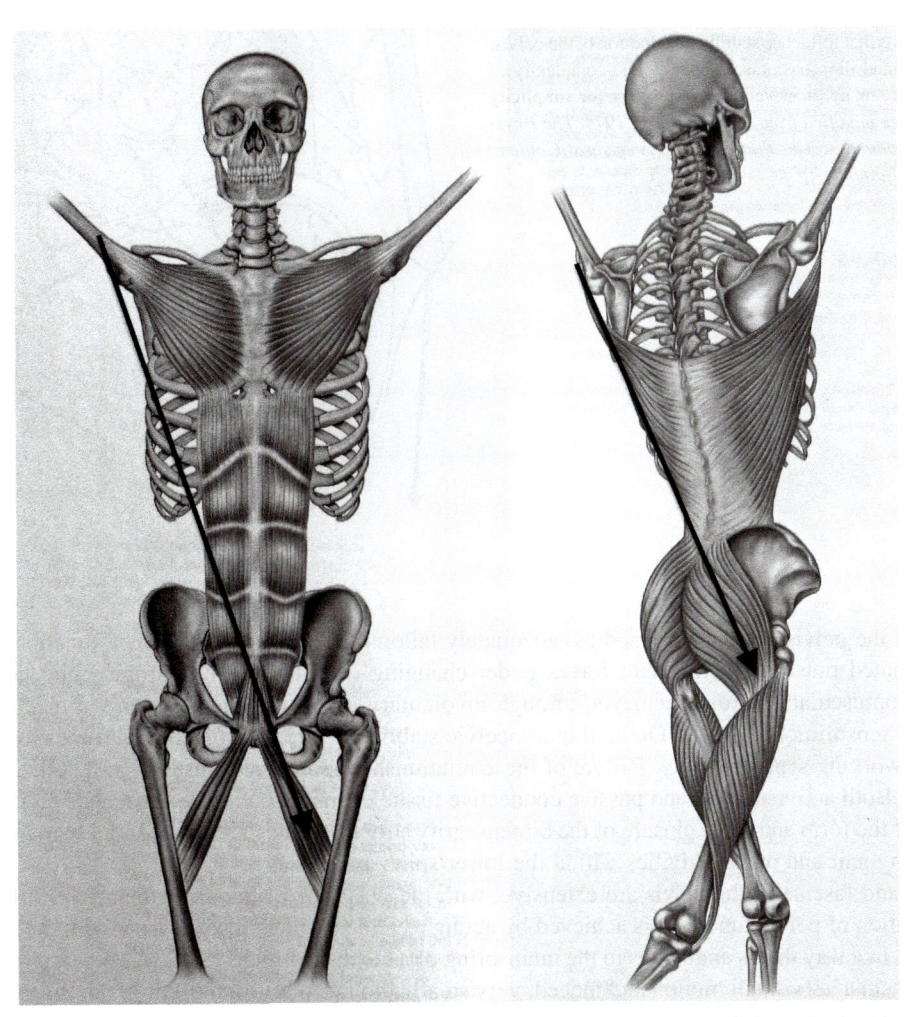

FIG. 2.2.3a Schematics that indication of the complexity of myofascial chains that span from the lower limbs to the shoulders in anterior (left) and posterior (right) views. Such examples indicate the dangers of becoming too reductionist in the approach to anatomy and clearly place the lumbopelvic functional unit at the heart of trunk and lower limb interaction that extends to the upper limbs. *(Images from Myers, T.W., 2014. Anatomy Trains: Myofascial Meridians for Manual and Movement Therapists, third ed. Elsevier, Edinburgh.)*

The pelvis consists of three bones fused together and the sacrum, which is a fused segment of vertebrae, with the coccyx being a fused segment of the residual tail below it. The pelvis demonstrates excellent form closure and passive biotensegrity elements that vastly dominate pelvic stability, with tensile and compressive stresses found across the SI joint (Hammer et al., 2013). Motion is restricted to small amounts at the synovial SI joints and minimal motion at the fibrocartilaginous pubic symphysis, allowing small rotational motions and cartilaginous absorption of torques between the two halves of the pelvis that support each limb (Fig. 2.2.3b).

FIG. 2.2.3b Schematic of the force vectors across the lower lumbar spine and pelvis showing its significant form closure from articular fit. This stability is kept in compression by multiple powerful sacroiliac ligaments (1), including the iliolumbar ligaments superiorly. Furthermore, force closure is provided by muscular force vectors such as from the hip abductors, including the gluteus maximus posteriorly and its linkage to the iliotibial tract (arrow-tipped black line) combined with the other hip abductor muscles. This force closure is represented by force vectors 2 within the image. The hamstrings and hip adductors provide vectors 3, and together with the abductors create the compression vector of the hip (4). However, many abdominal, spinal, and proximal lower limb muscles have attachments that influence stability and motion of the pelvis, in particular, the conjoined iliopsoas muscle. Despite the complexity of the displayed schematic, many other vectors are missing for simplicity of understanding. *(Image modified from Kapandji, I.A., 1974. The Physiology of the Joints. Volume 3: The Trunk and the Vertebral Column. Churchill Livingstone, Edinburgh.)*

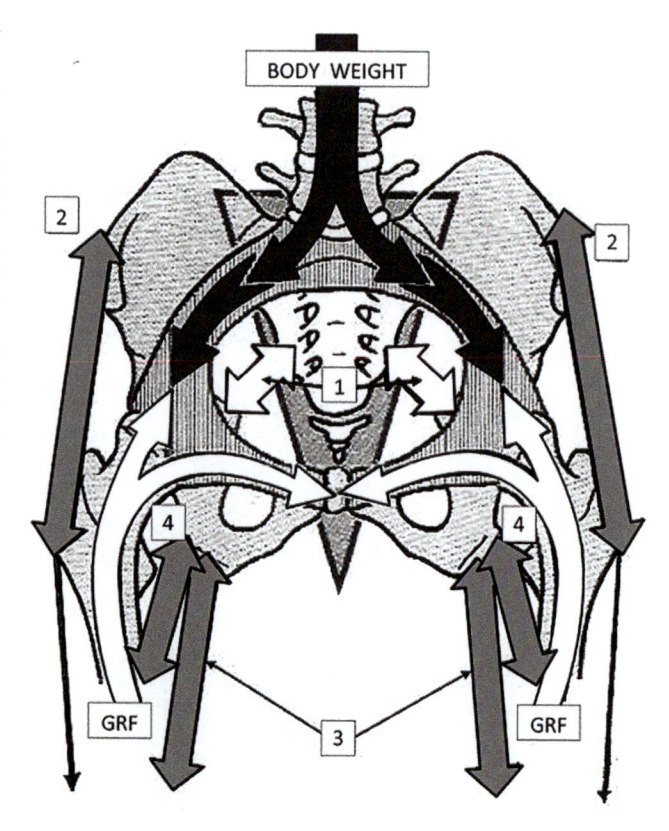

Stability within the pelvis has been defined as 'adequately tailored joint compression, as a function of gravity', supported by co-ordinated muscle and ligament forces under changing conditions and neuromuscular control (Vleeming et al., 2008). Neuromuscular control is achieved through involuntary activation and reaction via feedforward–feedback mechanisms of the sensorimotor system. Optimal lumbopelvic stability is a result of form and force closure under neuromuscular control, working synergistically. Failure of these relationships potentially results in altered kinematics, altered stability, and pain. Both active muscle and passive connective tissue elements comprise the tensile–compressive components that construct the form and force closure of the biotensegrity structure. They are all essential in maintaining form and function both during gait and other activities within the lower spine and pelvis.

The ligaments and fasciae of the pelvis are extensive, with the posterior SI ligaments being the strongest within the body. The stabilisation of pelvic functions is achieved by acting as a hierarchical icosahedron, with the massive posterior SI ligaments acting in a way that is analogous to the mainspring of a clock (Dorman, 1995). A very strong, short spring can store a lot of energy in a very small motion and indeed, very small amounts of motion occur in the SI joints, torquing very strong elastic ligaments. During walking gait, the head and trunk move up and down as a consequence of changes in lower limb joint angles under altering functional limb lengths from ground to pelvis. The gravitational energy is dissipated and stored within the musculoskeletal anatomy, which is released in order to help acceleration in the terminal stance limb as it starts to enter preswing.

Carrying weights behind or in front of the body can displace the body's CoM and restrict its vertical displacements during gait. This detrimental mechanism can be avoided if weights are carried above the trunk, explaining why it is more efficient to carry a weight on the head than using a backpack (Maloiy et al., 1986). The spine and pelvis play an important role in gait, with small spinal muscle contractions and pelvic rotations interacting with the lower limbs. Therefore, immobilisation of the trunk retards gait efficiency, increasing oxygen use (Dorman, 1995).

An extensive network of musculature traverses the pelvis, creating active tensional forces across the biotensegrity continuum of the spine and lower limb. Muscles of the lower spine, abdominal wall, and lower limb arise and course over joints either as bi-articular or often multi-articular structures that can create tensional and compressive forces throughout a diverse range of postural changes. Suboptimal pelvic stability is associated clinically with lumbopelvic, groin, and hamstring pains (Arumugam et al., 2012).

2.2.4 Functional anatomy of the lumbar spine and pelvis

Lumbar vertebrae and the lumbosacral joint

The osseous features of the entire spine and pelvis are complex due to regional specificity and anatomical variation, with more than one hundred articulations in the spine alone (Chockalingam and Needham, 2020). Rather than discussing the whole spine in detail (which can be obtained elsewhere in the many anatomical descriptions), a general review and a more detailed anatomical–functional description of the fourth and fifth lumbar vertebral bodies (L4–L5) along with their articulations in the lumbar spine and the lumbosacral joint (L5–S1) will be primarily discussed here. The spinal articulations are highly specialised structures involving bone, hyaline cartilage, fibrocartilage, and ligaments working with highly organised connective tissue structures known as *intervertebral discs*. The vertebral bodies and intervertebral discs take most of the axial load of the upper body when erect.

The anatomy of the vertebral column essentially establishes two functional pillars. The anterior column consists of vertebral bodies and intervertebral fibrocartilaginous discs that act as a mobile weightbearing pillar. The posterior column consists of two weightbearing pillars formed by the articular processes running on each side of the vertebral arch. Between the vertebral articular processes (otherwise known as zygapophyses) run synovial zygapophyseal or facet joints. Therefore, a vertebra has four facet joints: two superior and two inferior articulations. The anterior and posterior columns functionally link together and act in the manner of a class one lever system in the sagittal plane. The facet joints operate as a fulcrum, working with the compression forces of the intervertebral discs and the anterior spinal ligaments' tensile forces on one side opposed by the tensile forces of the posterior spinal ligaments on the other (Figs 2.2.4a and 2.2.4b).

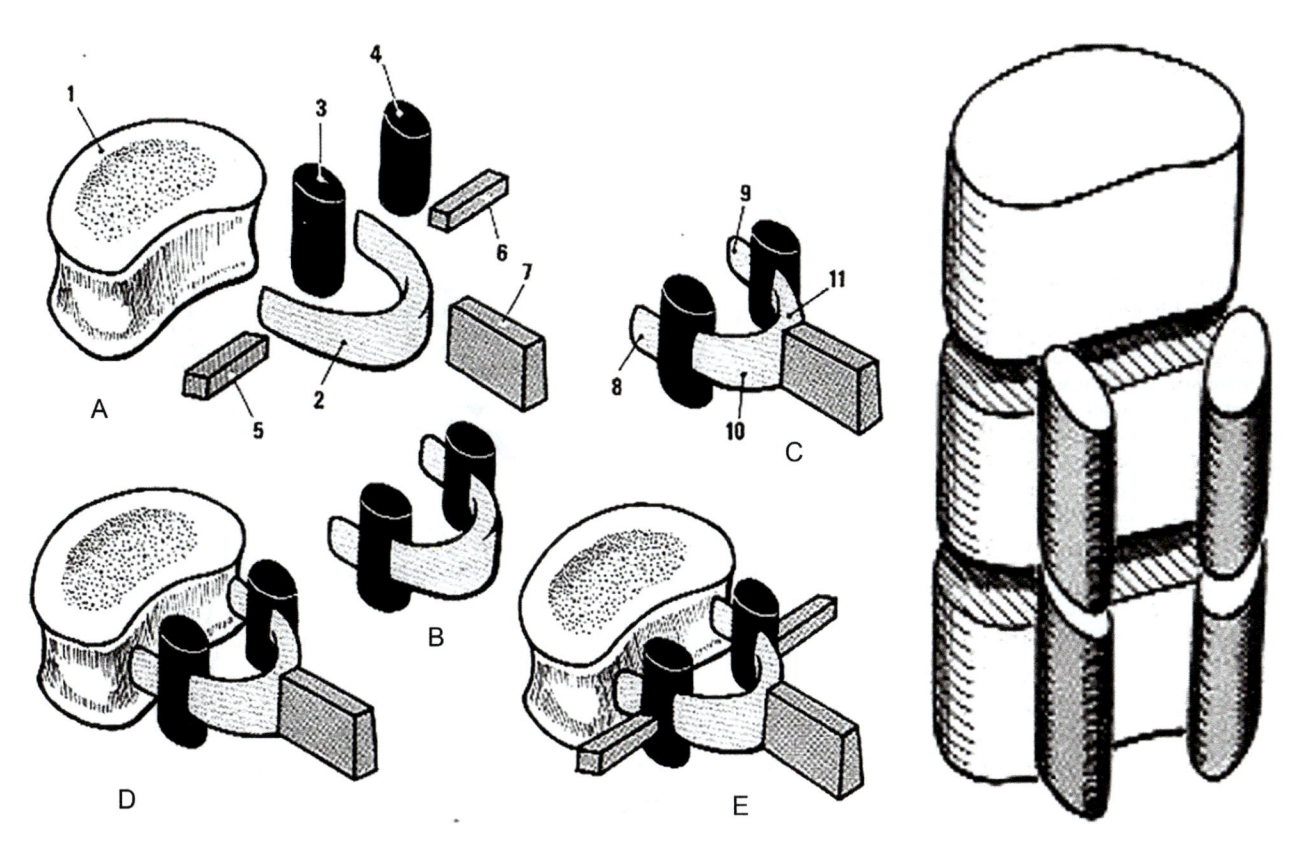

FIG. 2.2.4a A schematic (right) to show that vertebrae can be deconstructed into (A) the vertebral body (1), the vertebral arch (2), the articular processes (3 and 4), the transverse processes (5 and 6), and the spinous process posteriorly (7). The articular processes divide the vertebral arch into the pedicles anteriorly (C—8 and 9) and the laminae posteriorly (10 and 11). Each of these features come together to construct a vertebra (B, D, E). It is the vertebral body through intervertebral discs and the articular processes through synovial joints that form the load-bearing surfaces creating anterior and posterior spinal columns (left image). The other features give osseous integrity between the two load-bearing structures and also places for soft tissue attachment linkage and force closure stability between each vertebra. *(Image from Kapandji, I.A., 1974. The Physiology of the Joints. Volume 3: The Trunk and the Vertebral Column. Churchill Livingstone, Edinburgh.)*

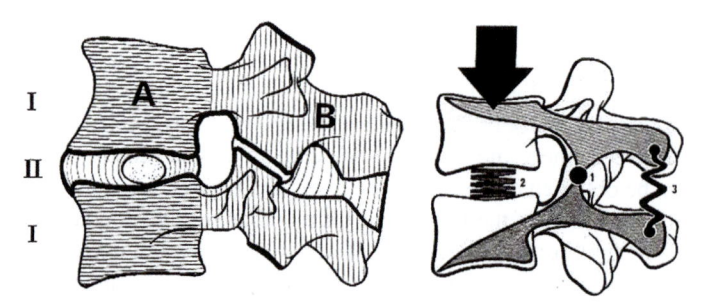

FIG. 2.2.4b Lateral views of two articulating vertebrae (left) help visualise the anterior column or pillar (A) and the combined posterior column constructed from two smaller columns of articular processes (B). The posterior column has a more dynamic role reflected by its articular surfaces. The bodies present a rigid structure (I) and the mobile dynamic border formed by the intervertebral disc, intervertebral foramen, the articular processes, the ligamentum flavum, and interspinous ligament (II and outlined in thicker black line). The interrelationship of loading between the anterior and posterior columns (right) shows that the articular processes acting as a fulcrum (1) can moderate compression loads on the bodies and discs (2) by allowing paravertebral muscles (3) to adjust forces on the bodies and discs through tensional active contraction forces. *(Image from Kapandji, I.A., 1974. The Physiology of the Joints. Volume 3: The Trunk and the Vertebral Column. Churchill Livingstone, Edinburgh.)*

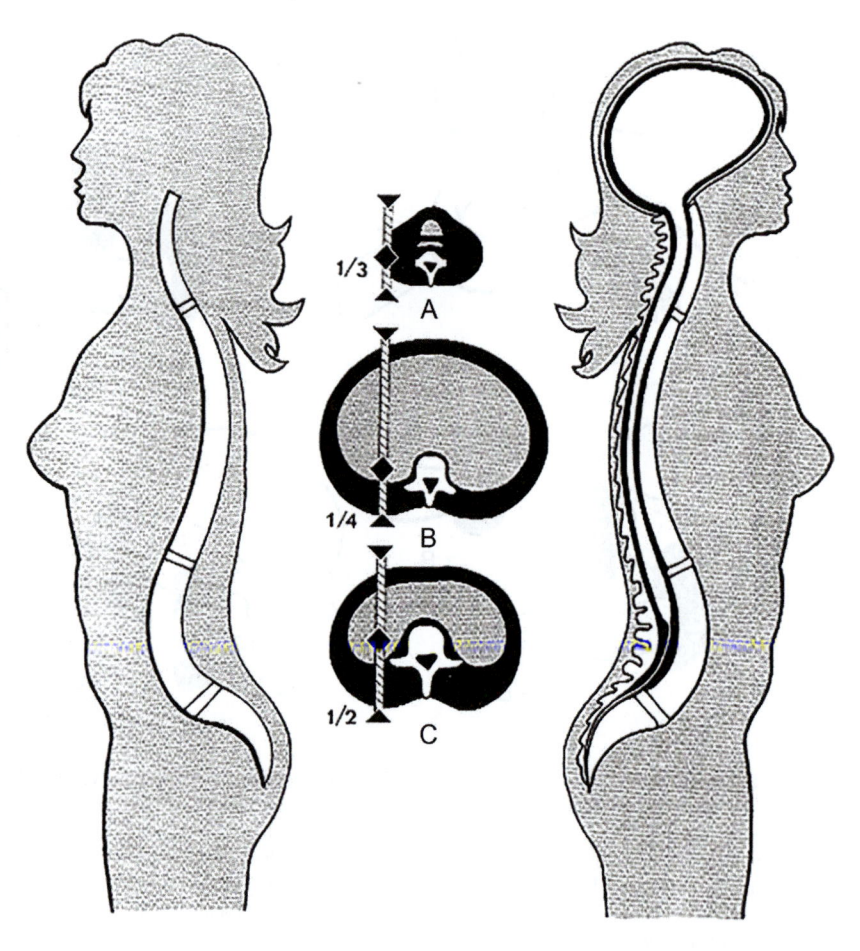

The spine acts as an asymmetrical sinusoidal column, differently positioning the vertebrae in relation to the exterior body surface throughout its length (Fig. 2.2.4c). The vertebra is a short trabecular bone with a dense cortex and spongy medulla. The superior and inferior surfaces form vertebral plateaus that connect to a cartilaginous plate that forms in the epiphysis giving rise to a distinct rim, becoming fixed to the vertebral body by around puberty. During adolescence, epiphysitis can occur within this anatomical site. Ossification of the cartilage rim in later life is known as *Scheuermann's disease*. The fibrocartilaginous intervertebral discs attach to the cartilage ring of the vertebral bodies, superior and inferior to the disc.

FIG. 2.2.4c The position of the spine acting as an asymmetrical sinusoidal column changes its position in relation to the dorsal and ventral surfaces throughout its length. In the thorax, it lies most posteriorly (B) and in the neck more anteriorly (A), with the most anterior placement being in the lumbar region where it becomes almost central (C). The spine also protects the spinal cord within its neural arch, providing a flexible encasing. However, this protection is not absolute, for nerve roots need to pass out between vertebrae and intervertebral discs can herniate to compress the spinal cord or nerve roots. *(Image from Kapandji, I.A., 1974. The Physiology of the Joints. Volume 3: The Trunk and the Vertebral Column. Churchill Livingstone, Edinburgh.)*

Two articular processes (i.e. the zygapophyses) are linked to the vertebral arch (also known as the neural arch). This is formed 'horseshoe-like', enclosing the vertebral canal of the spinal cord, and runs from one side of the posterior aspect of the vertebral body to the other. The lateral bony parts of the vertebral arch anterior to the articular processes are called the *pedicles* and those found posteriorly, the *laminae*. The vertebral arch provides protection for the spinal canal. Arising posteriorly from the laminae lies a large spinous process that divides said laminae into left and right. Laterally, two transverse processes extend from the articular processes. These processes are attachment points for the ligaments of the spine (see Figs 2.2.4a and 2.2.4d).

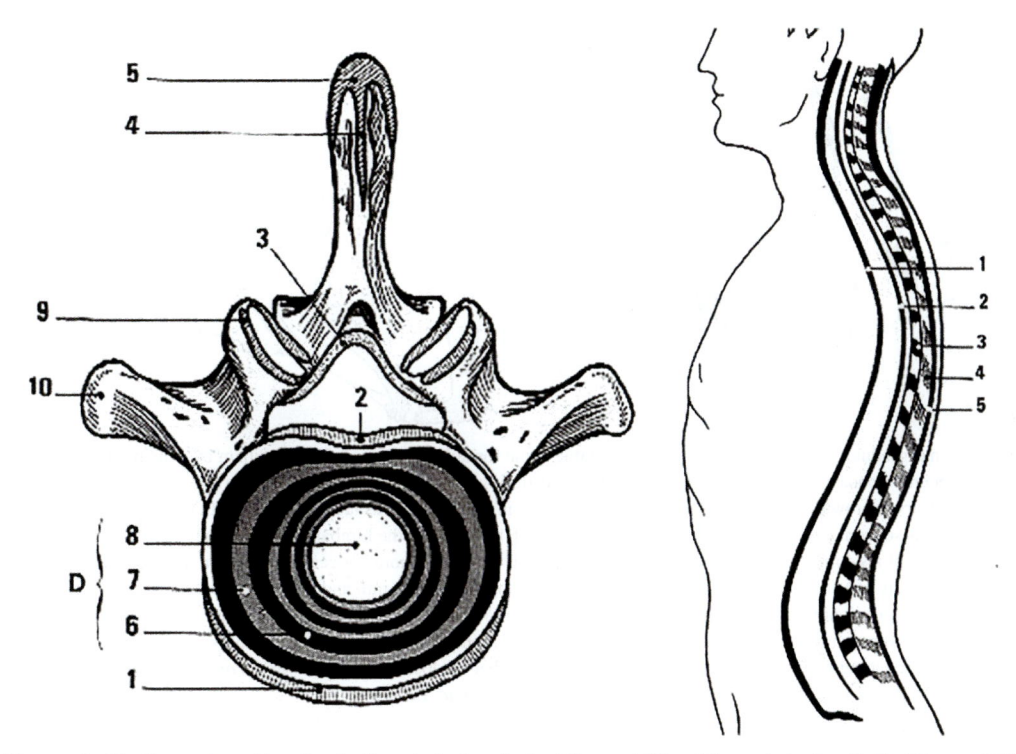

FIG. 2.2.4d The vertebral bodies are bound together by the anterior longitudinal ligament (1) and the posterior longitudinal ligament (2), interlinked by the intervertebral disc (D). The disc is constructed of two primary parts of concentric layers of fibrous connective tissues (e.g. 6 and 7) enclosing a central gelatinous nucleus pulposus (8). The vertebral arch is bound anteriorly by the ligamentum flavum (3). Lying between the spinous processes are the interspinous ligaments (4), which are continuous with the posterior supraspinous ligaments (5) that bind the spinous processes together posteriorly. Two powerful anterior and posterior ligaments strengthen the capsular ligaments of the articular processes (9). The intertransverse ligaments link the transverse processes of the vertebrae together (10). *(Image from Kapandji, I.A., 1974. The Physiology of the Joints. Volume 3: The Trunk and the Vertebral Column. Churchill Livingstone, Edinburgh.)*

The ligaments of the spine are the passive stabilisers that rely on their intrinsic strength and the lever arm through which they act. They consist of the anterior longitudinal ligament running on the anterior surface of the vertebrae and discs and the posterior longitudinal ligament running on the anterior aspect of the vertebral canal on the posterior surface of the vertebral bodies and discs. Many ligaments connect the vertebral arches of the individual vertebrae (Fig. 2.2.4d). The *ligamentum flavum* is a strong ligament attached to the deep surface of the laminae and upper vertebra passing to the superior margin of the next lower vertebra. The *interspinous ligament* runs superiorly–inferiorly between the posterior process's longitudinal ligament of each vertebra and is linked to the *supraspinous ligament*, running posteriorly from each posterior process. The supraspinous ligament is less well defined in the lumbar region. The transverse processes are linked by the *inter-transverse* ligament. The articular processes have anterior and posterior ligaments that strengthen the capsule of the facet joints.

Facet joints (bilaterally situated) supply articular weightbearing surfaces between the vertebrae, other than those provided by the intervertebral discs. In the lumbar spine when standing, they carry between 10% and 20% of compressive loads, but over 50% of anterior shear. In extension, the facet joints increase the weightbearing load, heightening stress on the spinal laminae. The facet joints of the lumbar spine are synovial, lying between the inferior process of one vertebra and the superior articular process of the adjacent vertebra and possess all the anatomical features of synovial joints. Each articular process has around 2–4 mm of hyaline cartilage which resists instantaneous compressive forces and demonstrates the same biomechanical behaviour of all articular cartilage.

The shape of the articular surfaces of the facet joints changes with growth throughout childhood, concurrently with the development of a lordotic lumbar spine that becomes curved and biplanar, with their posterior surfaces orientated in the sagittal plane and their anterior aspects orientated to the frontal plane (Figs 2.2.4e and 2.2.4f). The lordotic shape developed could be a determining factor in the relation between lateral flexion and axial rotation because lumbar extension and lordosis improves anterior shear resistance in the lower spine through increasing articular

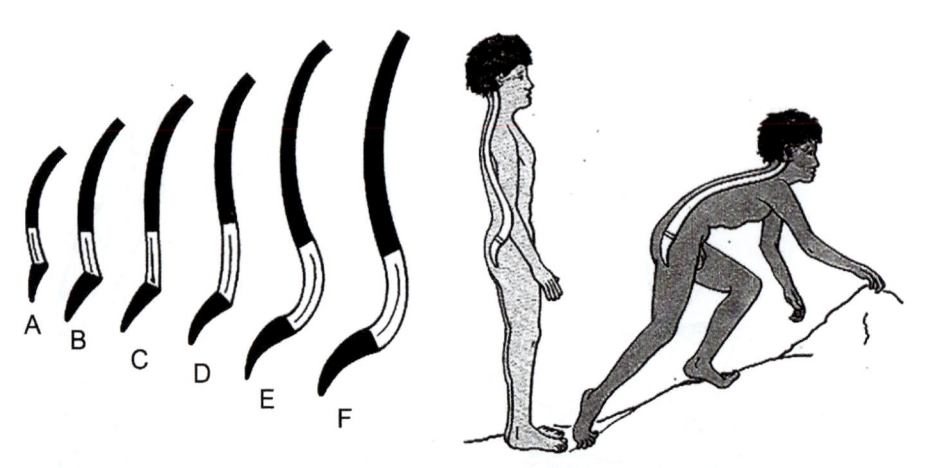

FIG. 2.2.4e Evolution has gradually furnished humans with a suitable spine for tailless plantigrade bipedalism by causing the lumbar spine to initially straighten and then to become concave posteriorly. These changes occur during infancy and childhood, in part as a response to biomechanical forces experienced during early years of growth. This explains population variability in features such as facet joint shape. In the foetus, the whole spine is concave anteriorly (A), becoming less so from around 6 months in utero to newborn (B). However, the lumbar spine becomes straight around the initiation of gait (C). Lumbar lordosis thus develops during the early years of gait development, becoming more noticeable by age three to four (D), conspicuously obvious by age eight (E), and more adult-like around early adolescence (F). Minor changes continue into and during adulthood. Significant periods of gait during childhood likely influence the development of appropriate levels of spinal curvature, with other activities such as clambering and climbing maintaining spinal flexibility and strength through regular shape changes (left). Extended hours of sitting in children may be detrimental to long-term spinal morphology and biomechanics, reducing mobility and potentially decreasing lumbar lordosis. *(Image adapted from Kapandji, I.A., 1974. The Physiology of the Joints. Volume 3: The Trunk and the Vertebral Column. Churchill Livingstone, Edinburgh.)*

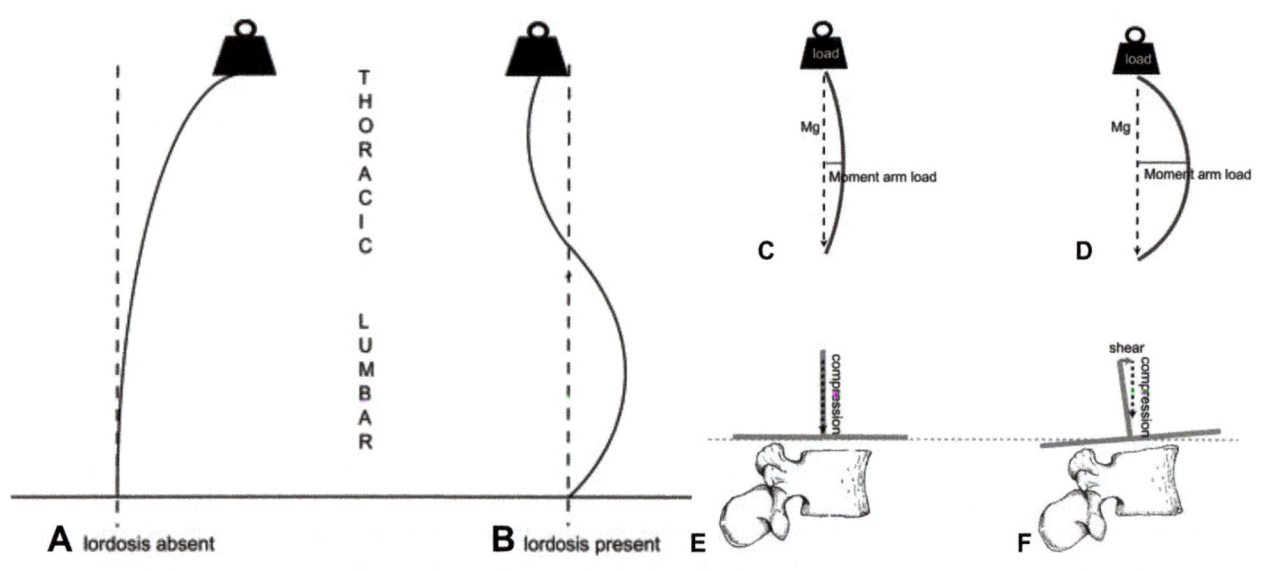

FIG. 2.2.4f Without the double curvatures within the human spine, the trunk would always be in a state of potential flexion (left image A). The presence of lumbar lordosis (B) avoids this problem. Increasing lordotic angles heightens potential shear strains on the lumbar joints. The angled morphology of each lumbar vertebra and the positions of the facet joints use the anterior shear created by the lordotic curvature to create stability within the lumbar spine (right image C–F). *(Images from Whitcome, K.K., 2012. Functional implications of variation in lumbar vertebral count among hominins. J. Hum. Evolut. 62(4), 486–497.)*

compression (Chockalingam and Needham, 2020). This helps to explain hypermobile patient's tendencies to increase their lumbar lordosis in the presence of more flexibility within their ligaments. The facet joint surfaces are convex on the inferior process and concave on the superior process of the articulating vertebra below. However, it is not a true ball-and-socket arrangement. The superior articular process' concavity is functionally increased at the posterior aspect by the ligamentum flavum which serves as an anterior wall of the facet joint capsule (Fig. 2.2.4g). Hypertrophy of the ligamentum flavum can be seen during clinical imaging of advanced degeneration within the facet joints and resulting from spinal stenosis.

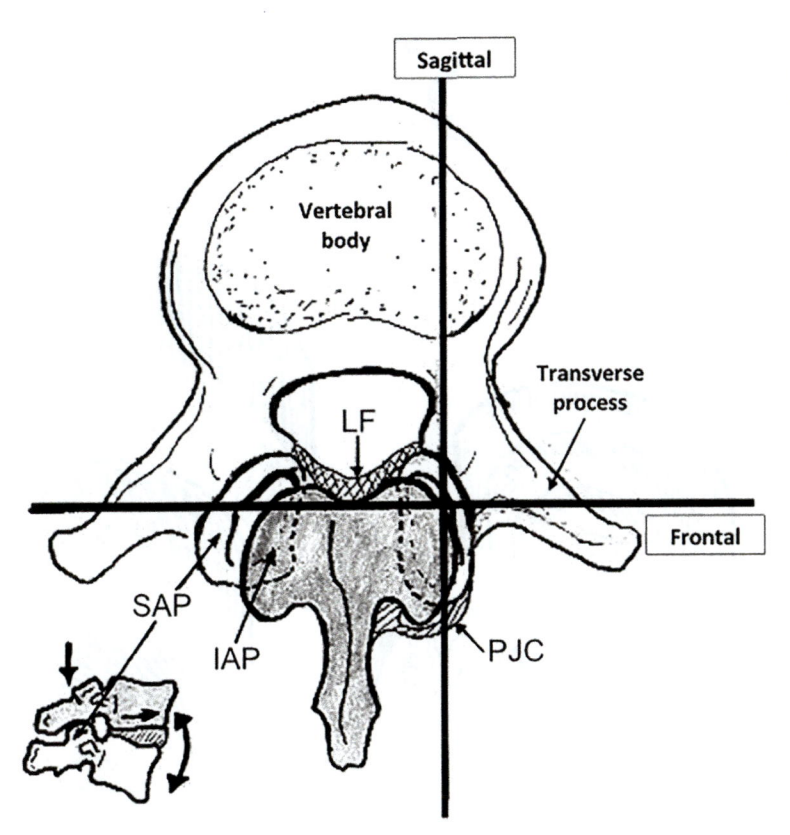

FIG. 2.2.4g The articular facets of the spinous process consist of the convex inferior articular process (IAP) of one vertebra sitting in the relatively concave superior articular process (SAP) of the vertebra below. The anterior aspect of the joint is orientated in the frontal plane, while the posterior aspect is orientated to the sagittal. The posterior joint capsule (PJC) encloses the joint posteriorly. These joints' anterior wall is formed from the lateral aspects of the ligamentum flavum (LF). Lumbar extension (or increased lordosis) increases the compression loading of these joints, providing better stability, particularly against anterior shear (lower left image). Posture that can induce more flexion of the lumbar spine, such as when sitting, can leave the lumbar articular facets less protected. However, excessive compression can cause stress fractures and degeneration of these joints. *(Permission www.healthystep.co.uk.)*

Vertebral bodies and intervertebral discs

The intervertebral discs sit between the vertebral bodies. Their bodies are shaped as solid cylinders but are variable in their form between regions of the spine. Generally, the width and depth of the vertebral bodies increase moving inferiorly from the upper cervical region down to the lower lumbar spine, with the height of the body also following a similar pattern. Exceptions are cervical vertebra body 6 which is normally less high than cervical bodies 5 and 7, and the lumbar vertebrae which are all smaller in height than the second lumbar vertebra (L2). The load-carrying capacity of the vertebral body depends on its size, shape, bone density, and trabecular integrity (Chockalingam and Needham, 2020). The vertebral bodies and discs demonstrate an important functional synergetic interplay in force distribution with the facet joints. The facet joints are responsible for direction and amplitude of movement while the bodies and discs bear the majority of the load (Chockalingam and Needham, 2020). Differences in the orientation of the instantaneous joint axis of the facet joints dictate the motion, with the lumbar joints orientated for the sagittal plane motions of flexion and extension, but not for rotation (Chockalingam and Needham, 2020). On forward bending in the lumbar region, the CoM of the body moves forward. This causes anterior shear and stiffening from compression and tension within the facet joints and increased disc pressure through anterior compression between L4 and L5, stiffening the anterior edge of the disc and producing stabilisation.

FIG. 2.2.4h Schematics of the structure of the intervertebral disc, with the central gel-filled nucleus pulposus (N) and the annulus fibrosus (AF). The annulus fibrosus is made up of concentric layers of collagen fibres within fibrocartilage, which primarily run vertically in the outer fibres, but with increase in obliquity towards the centre (A and B). This reflects the more vertical tensional forces superficially and more transverse tensions centrally. Fibres alternate their direction between layers so that some layers are tensioned whichever direction the spine is twisted. Like articular cartilage, it is the biphasic nature and the fibre orientation and integrity that provides the intervertebral disc with its mechanical properties. *(Image from Kapandji, I. A., 1974. The Physiology of the Joints. Volume 3: The Trunk and the Vertebral Column. Churchill Livingstone, Edinburgh.)*

The intervertebral disc demonstrates the material characteristics of both cartilage and ligament through rotation and translation properties resulting from its deformation behaviour. However, motion is dependent on the geometry of the facet joints, the anatomical variation of which (between individuals) explains the relative differences found within populations (Chockalingam and Needham, 2020). The disc is composed of a fibrous outer layer known as the *annulus fibrosus* which is made up of concentric layers of collagen fibres running more obliquely in the deeper layers and more vertically in the outer layers, alternating in their direction between the layers. Such arrangement provides good elastic recoil when the discs are rotated and stretched in any direction. The central part is referred to as the *nucleus pulposus* and consists of a gelatinous substance that is roughly spherical when put under pressure inside the intact, tough annulus fibrosus (Fig. 2.2.4h). This permits the disc to act as a swivel joint, permitting intervertebral tilting, rotating, and gliding/shearing motion, giving it 6° of freedom. This means that sagittal flexion–extension, side flexion, sagittal and frontal plane gliding, and rotation in the transverse plane are all possible. Each intervertebral joint has only a small range of motion, but when multiple segments act together, sizeable motion can be achieved within the spine. If a joint is restricted, compensatory changes are possible at the other joints. However, if restriction is excessive and prolonged, they can lead to musculoskeletal pathology and symptoms. Measuring and modelling spinal kinematics and kinetics remains a considerable challenge (Chockalingam and Needham, 2020).

When at rest and in erect standing, pressures due to body weight loading within the centre of the nucleus tensions the annulus fibrosus layers. On spinal tension, the annulus fibrosus is put under vertical tension/stretching, pulling in on the nucleus and compressing it from the sides which reduces internal pressures. On vertical compression of the spine, the disc is compressed, raising the internal nucleus pressure which tensions the annulus fibrosus layers. In vertebral extension, the posterior aspect of the nucleus is compressed, increasing nucleus pressure and forcing the gel to flow anteriorly which tensions the annulus fibres anteriorly. This restricts intervertebral extension. In intervertebral flexion, the nucleus is compressed anteriorly, increasing pressure on the posterior nucleus wall which increases tension in the posterior annulus fibrosus layers, restricting further flexion. On side flexion, the side towards the flexion is compressed and the opposite side is tensioned, resisting the direction of side flexion. During axial rotation, because of the alternating orientation of the annulus fibrosus fibres in each layer, different layers become tensioned in one direction of axial loading and the other layers in the opposite direction. However, the outer layers with their vertical orientation are tensioned in both directions. This is the mechanical genius of vertebral disc mechanics (Fig. 2.2.4i).

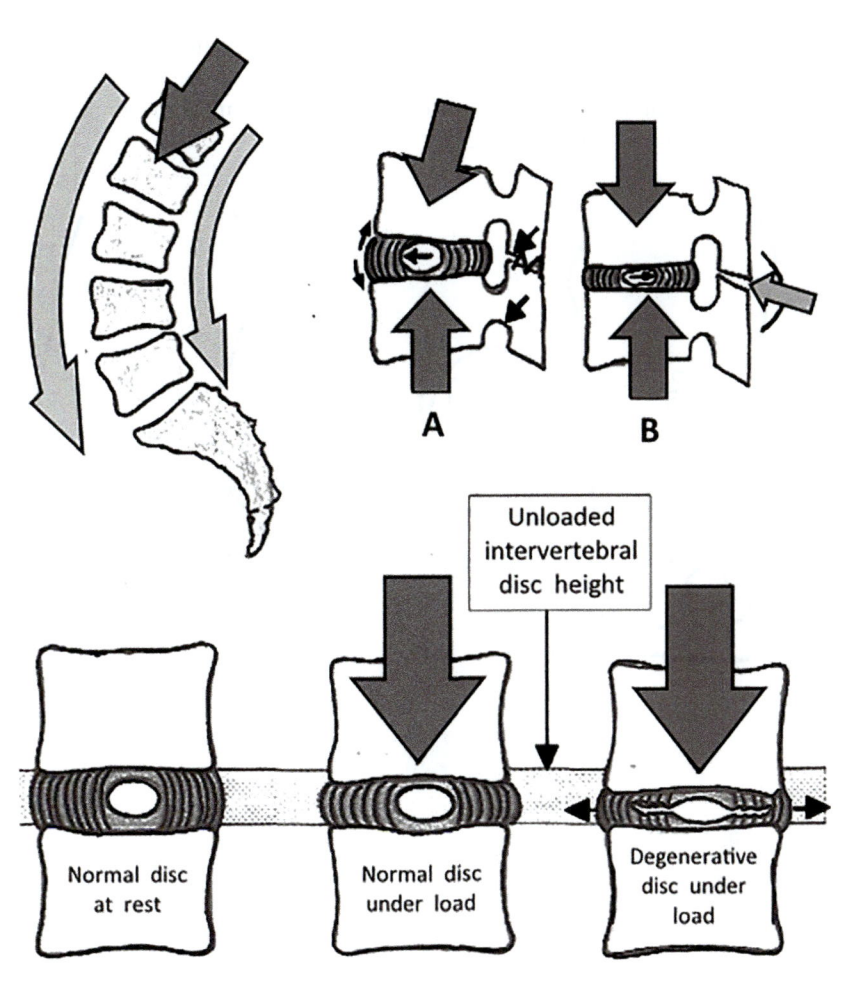

FIG. 2.2.4i The curvature of the lumbar spine results in slightly divergent compression loads from the vertical, creating anterior shear with compression over the vertebral bodies. For better intervertebral disc function within this region, the nucleus pulposus is positioned more anteriorly than elsewhere. Under these compression forces, the nucleus' gel is displaced anteriorly, tensioning and stiffening the anterior annulus fibrosus fibres, while its facet joints are compressed to reduce any anterior shear (A). Spinal flexion and reduction of lumbar lordosis upsets this intrinsic lumbar stability, losing its facet joint shear stability and changing compression vectors within the discs (B). The situation is further complicated when discs become dysfunctional through degeneration, usually as a result of annulus fibrosus failures. It is then that the gel within can become displaced through spinal compression, risking disc herniation and excessive compression forces on the vertebral bodies and facet joints. *(Images modified from Kapandji, I.A., 1974. The Physiology of the Joints. Volume 3: The Trunk and the Vertebral Column. Churchill Livingstone, Edinburgh.)*

The pressure within the nucleus is never at zero. This creates a preloaded state which reduces deformation on loading by pressuring and stretching the fibrous walls of the annulus fibrosus. The nucleus has high water content, but under load, this gradually reduces. Thus, in the morning following sleep, it is at its most pressurised with high water content and is least pressurised by the end of the day, reducing the internal preload state. During constant loading, the fluid slowly flows out in a manner laid down by poroelastic principles, but then refills again on offloading. Therefore, like articular cartilage, prolonged loading is mechanically more detrimental than loading under movement. The nucleus bears around 75% of the force passing through the disc (Kapandji, 1974, p. 32). On asymmetrical loading, the upper vertebral plateau of the disc tilts down on the loaded side, but the annulus fibrosus wall on the opposite side undergoes tension which restricts the vertebral tilt. In regions of the spine with greatest curvature such as the lumbar spine, the disc loading is more asymmetrical. The lumbar spine increases its internal loading as it approaches the sacrum and as such, the nucleus of the disc is offset posteriorly to mechanically adjust for this.

Disease and damage of the intervertebral disc is highly detrimental to spinal mechanics and will have significant influence in regions of greater spinal loads and mobility such as in the lumbar spine. Disc thickness therefore varies throughout the spine and they are thickest in the lumbar region at around 9 mm, although the ratio of thickness to vertebral body height is more mechanically important. The greater the ratio, the greater the mobility at the joint. In the lumbar region, because of the presence of lordotic curvature, loading forces on the disc alters its anatomy in a way that the nucleus becomes more posteriorly positioned, and with more of the annulus fibrous lying anterior to it. The nucleus pulposus is also larger than in other regions.

The 4th–5th (L4–L5) lumbar spine articulation

The degree of frontal plane and sagittal plane orientation of the facet joints varies at the level of the lumbar vertebrae, and asymmetry between the two sides is reported to be around 35% in the lumbar spine, raising questions over the significance of clinical comparisons of movement in individuals between sides. The articular facet joint surfaces fully engage with anterior shear of the superior vertebrae. This anterior shear is greatest at the L4–L5 and L5–S1 levels due to the greater lordotic curve within the upper lumbar spine. This anterior stabilising shear is greater in stance than in modern sitting positions, because lumbar lordotic angles are reduced in sitting positions (Fig. 2.2.4j).

FIG. 2.2.4j The lumbar spine has evolved for erect static and dynamic stance. In modern sitting positions, the lumbar spine angles reduce with the less lordotic curvature resulting and thus preventing efficient joint stability biomechanics. Long hours of sitting in chairs discourages maintenance of the lordotic curve, made worse when leaning forward over desks. Increasing rates of back pain seem to correlate with increased deskwork and reduced weightbearing dynamic work activity within the population. *(Upper image: Adapted from Porterfield, J.A., DeRosa, C., 1998. Mechanical Low Back Pain: Perspectives in Functional Anatomy, second ed. W.B. Saunders, Philadelphia. Lower image: Permission www.healthystep.co.uk.)*

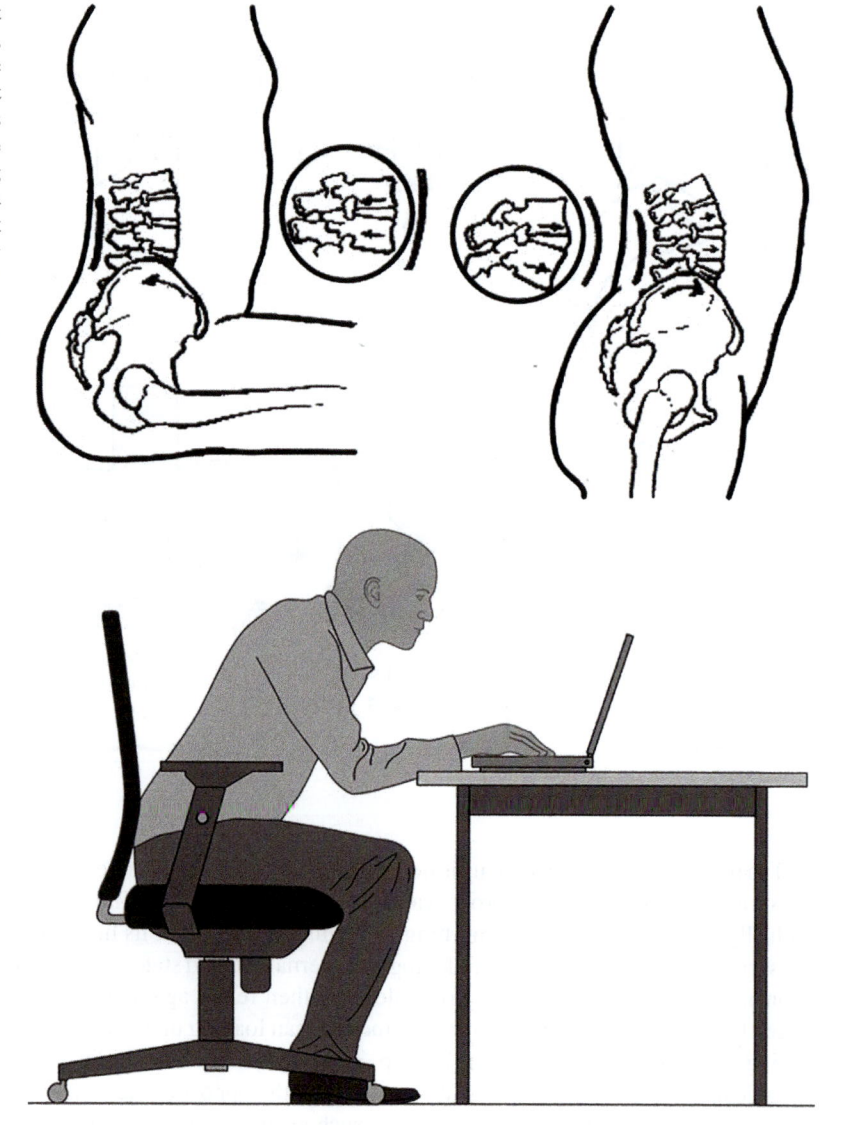

The ligaments and joint capsule fibre orientations as well as the articular surface orientations influence motion within the lumbar spine, with most motion occurring in the sagittal plane and least motion in the transverse plane. Planar dominance of motion across the spine is highly variable, but flexion and extension are greatest across the cervical and lumbar regions of the spine. In the latter region, sagittal plane motion increases inferiorly at the lumbosacral joint of L5–S1, with L4–L5 and L5–S1 making up 40%–50% of the sagittal plane motion of the lumbar spine. In comparison, only around 3–5° of rotation occurs within the transverse plane of the lumbar spine. The lumbar spine is best able to resist compression when flexed at around 60%–80% of its erect standing flexion range, a position that correlates well with good lifting posture.

The lumbosacral joint

The human sacrum is a fusion of five sacral vertebrae perforated by four pairs of *neuroforamina*. The sacrum connects to the coccyx, a residual part of the human tail that now serves little function other than as a point for important ligament attachments. The sacrum and coccyx act as part of the pelvic functional unit rather than as a part of the spine so that from a locomotive perspective, the lumbosacral joint should be considered the end of the mechanical spine. The body of the S1 vertebra articulates with the body of L5 by way of the lumbosacral intervertebral disc, with compression and shear between the articulation also occurring at the lumbosacral facet joint. The plane of orientation of the lumbosacral joint is variable between individuals and often asymmetrical from left to right. Functionally, the lumbosacral joint operates much in the manner of the lumbar articulations. However, at the lumbosacral joint, weight from the trunk is transferred from the mobile spine into the stable pelvic hub before passing to the lower limbs, while forces generated through the lower limb are transferred to the spine above through the pelvis. The joint is force closed by the muscles of the lumbosacral region and the iliolumbar ligament, which often blends with the anterior sacroiliac ligament (see Fig. 2.2.3b).

The angle the sacrum makes in the sagittal plane, known as the *angle of sacral incidence*, presents significant correlations with the lumbar spine curvature that plays an important part in balancing the upper body CoM over the lower limbs (Tardieu et al., 2013). The angle of sacral incidence increases from neonates to adults to become negatively correlated with the sacro-acetabular distance which through growth, helps bring the CoM backwards to centre it over the pelvis and lower limbs in stance. This change in angulation is in co-ordinated response to the acquisition of bipedal gait during infancy (Tardieu et al., 2013) (Figs 2.2.4k and 2.2.4l).

The pelvis and sacroiliac (SI) joint

The pelvis consists of the fused coxal bones of the ilium, ischium, and pubis (the *innominate bone*), and the sacrum. Two sacroiliac (SI) joints lie between the sacrum and the left and right ilia. The pubic symphysis lies between the left and right pubic bones. The pelvis also articulates with the femora through the acetabula at the hip joint surfaces, each of which is constructed from all three coxal bones. The cartilage present at all five articulations of the pelvis therefore influences stresses within the pelvis on loading (Hammer et al., 2013). The pelvic bones are highly trabecular, but covered in a high-density cortical layer of bone forming a low-weight structure able to support high loads (Poilliot et al., 2019). The SI joints between the sacrum and the ilia are suggested to be a symphysis, with some characteristics of a synovial joint being confined to the distal cartilaginous portions on the iliac side of the joint (Puhakka et al., 2004). The SI joint has been referred to as a diarthro-amphiarthrosis, arguably because of its articular and syndesmosis parts (Poilliot et al., 2019). Embryologically, it forms like any other free-moving diarthrodial joint. However, it starts to alter at around ten weeks in utero, and although the sacroiliac joint in children and young adults presents with the characteristics of a diarthrodial joint, older individuals demonstrate stiffer amphiarthrodial characteristics (Poilliot et al., 2019).

On average, male and female pelvic morphology is different, but there is some crossover between individuals. The female pelvis is usually (appears) shorter and wider proportionately to that of males because of shorter (femoral) limb lengths in females. It is also more mobile, with a degree of mobility associated with hormonal changes such as the release of relaxin during pregnancy. The SI joint is more inferiorly orientated in males relative to the anterior superior iliac spine, whereas females have a smaller SI surface area (Poilliot et al., 2019) (Fig. 2.2.4m).

The coxal bones have a number of significant anatomical features that provide important muscular and ligamentous attachments, including the anterior superior and anterior inferior iliac spines, the posterior superior and posterior inferior iliac spines, the iliac crest, the ischial tuberosity, and the pubic tubercles. From the anteromedial body of the pubis, the superior ramus passes posteriorly and superiorly to join the ilium and ischium at the acetabulum of the hip. The inferior pubic ramus courses posteriorly, inferiorly, and laterally to join the ischial ramus inferiorly. The pubis provides attachments for the thigh adductor muscles including gracilis, pectineus, rectus abdominis, both obturator internus and obturator externus around the edges of the obturator membrane, and the pyramidalis and levator ani muscles.

FIG. 2.2.4k The sacrum or sacral plate represents the boundary of the pelvis with the spine. Its inclination angle is called the sacral slope, forming an angle of sacral incidence (upper images). In conjunction with the spinal curves, the setting of this angle should keep the CoM over the base of support of the pelvis (lower image—A). Different morphologies can achieve this including extremes of very large (lower images B—left) and very low sacral incidence (B—right), which require different adjustments within the curves of the spine. *(Images from Tardieu, C., Bonneau, N., Hecquet, J., Boulay, C., Marty, C., Legaye, J., et al., 2013. How is sagittal balance acquired during bipedal gait acquisition? Comparison of neonatal and adult pelves in three dimensions. Evolutionary implications. J. Hum. Evol. 65(2), 209–222.)*

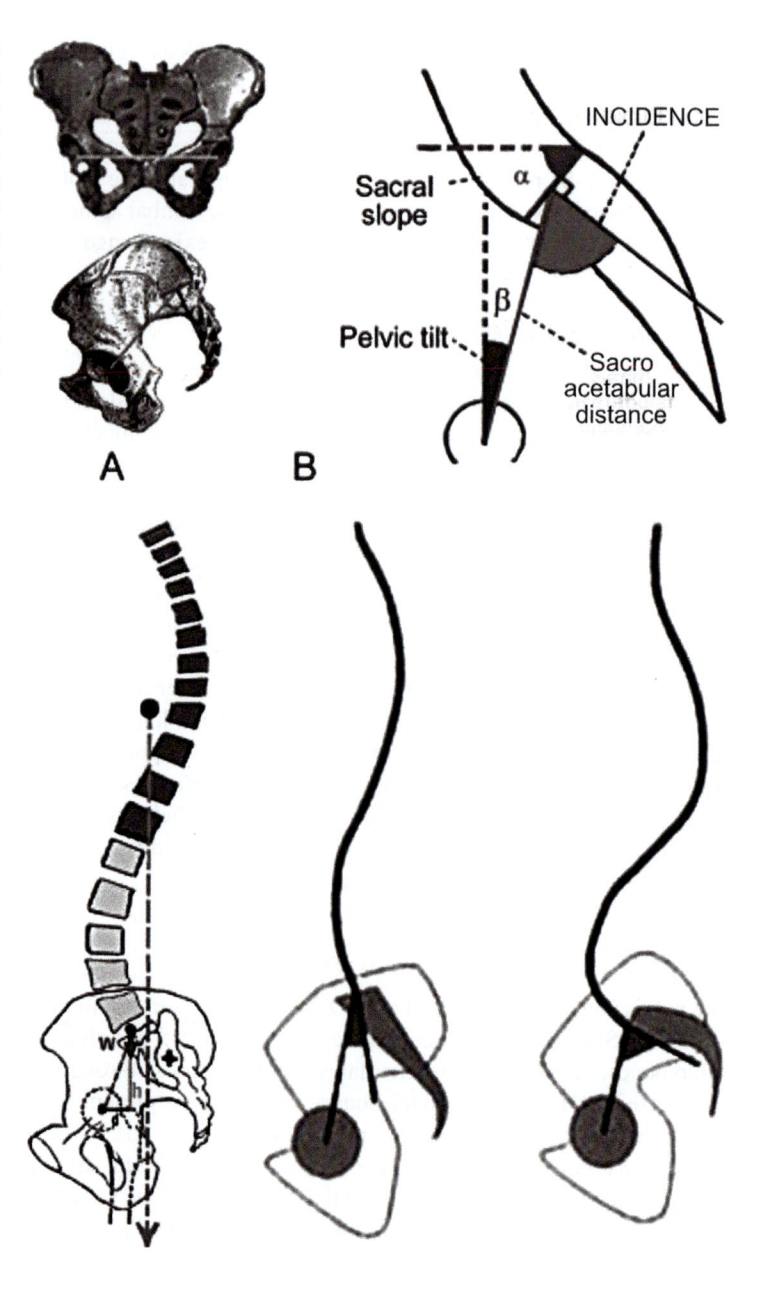

The SI joints through their powerful ligamentous supports serve as a bracing mechanism and shock absorber in gait and jumping activities. They store energy that aids walking efficiency with antigravitational energy stored within their ligaments released appropriately during gait (Dorman, 1995). Over 80% of the articular surface of the sacrum arises from the 1st, 2nd, and 3rd sacral segments (S1, S2, S3) Fig. 2.2.4n). The sacrum is wider superiorly than inferiorly, forming a rough, truncated, inverted pyramid shape. This, combined with the ligaments, gives the joint its excellent form closure. The SI joint contains synovial fluid in its joint cavity (Le Huec et al., 2019). However, the interfacing thick hyaline cartilage is sometimes reported as only found on the sacral side with a more fibrous cartilage being found on the ilium side, whereas other studies have reported cartilage on both sides (Poilliot et al., 2019). The sacral hyaline cartilage can be two to three times thicker than the ilium fibrocartilage, with sacral cartilage being thicker in women (Poilliot et al., 2019). The iliac cartilage changes from being composed of collagen bundles in infancy to more hyaline cartilage over time. However, it is prone to early degenerative changes from age 20 years onwards, possibly linked to the SI joint space decreasing from age 12 years onwards (Poilliot et al., 2019).

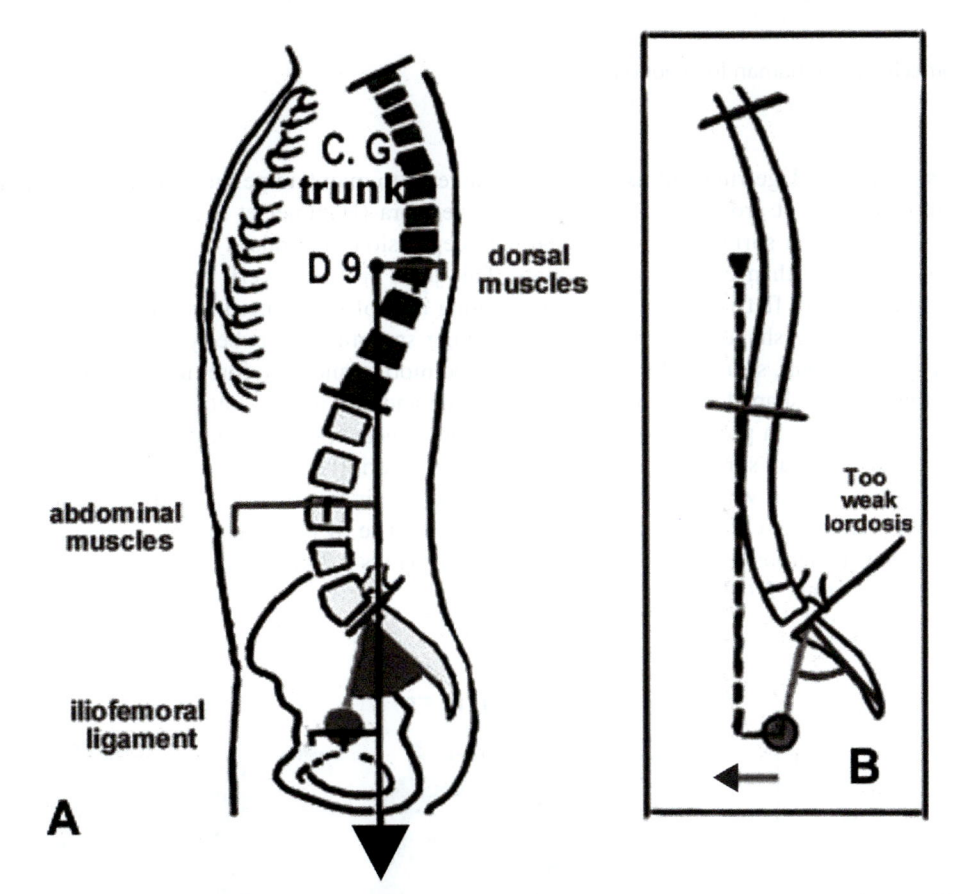

FIG. 2.2.4l Ideally (A), the line of gravity should lie just anterior to the 9th thoracic vertebra and thus anterior to the dorsal curve of the thoracic kyphosis, but posterior to the lumbar lordosis. This gives the abdominal muscles a long lever arm to the lumbar curve and the iliofemoral ligament a longer lever arm to the hip. The position of the GRF-induced divergent point is also possibly at play in helping in setting the line of gravity across the spine. Balance is compromised if lumbar lordosis is inadequate (B) due to a high angle of incidence (over 56°), moving the line of gravity increasingly anterior to the lumbar spine and femoral heads, requiring increased lumbar spine, gluteus maximus, and hamstring muscle activity to counteract gravity. *(Image from Tardieu, C., Bonneau, N., Hecquet, J., Boulay, C., Marty, C., Legaye, J., et al., 2013. How is sagittal balance acquired during bipedal gait acquisition? Comparison of neonatal and adult pelves in three dimensions. Evolutionary implications. J. Hum. Evol. 65(2), 209–222.)*

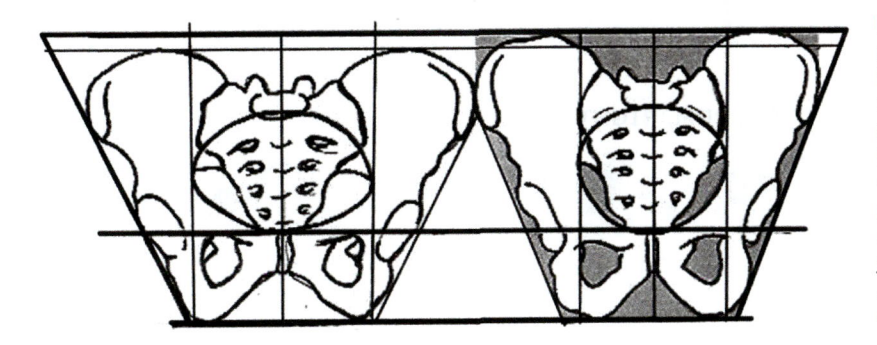

FIG. 2.2.4m Females express different 'average' pelvic morphologies than males. A schematic of the female pelvis (left) presents a broader superior but shorter triangulated shape than that of the male. The pelvic opening or brim tends to be more open that the males. However, there is considerable variation with female-to-male crossover in pelvic morphology. Within the immature pelvis of children, sexual dimorphism is not apparent. *(Image adapted from Kapandji, I.A., 1974. The Physiology of the Joints. Volume 3: The Trunk and the Vertebral Column. Churchill Livingstone, Edinburgh.)*

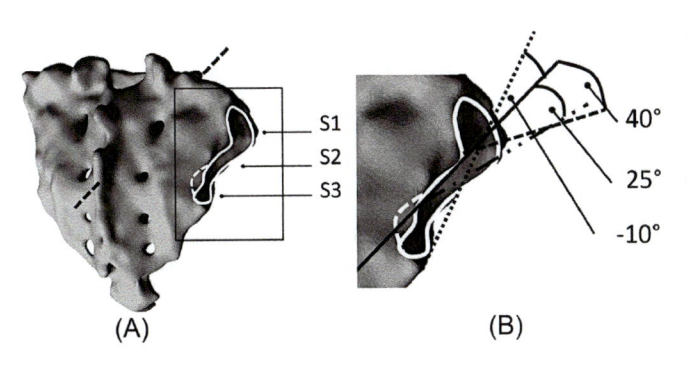

FIG. 2.2.4n Schematic of the sacral articular surface on the right side of the pelvis demonstrating average angles at the S1, S2, and S3 levels of the articular surface. *(Images from Casaroli, G., Bassani, T., Brayda-Bruno, M., Luca, A., Galbusera, F., 2020. What do we know about the biomechanics of the sacroiliac joint and of sacropelvic fixation? A literature review. Med. Eng. Phys. 76, 1–12.)*

The SI joint relies on the cartilage ring and its extensive ligaments to maintain pelvic stability with the instantaneous centre of sacral motion lying at around the level of the S2 vertebra (Hammer et al., 2013).

The adult SI joint's articular surfaces are roughened with depressions and raised ridges that result in a joint with a higher co-efficient of friction than other synovial joints. This joint profile is very unlike that of infants and young children which display smooth flat joints parallel with the long axis of the spine. In response to growth and bipedal gait, the joint surfaces change shape and orientation, remaining smooth until puberty. The adult joint becomes the contradiction of a synovial joint, structured for highly restricted motion and therefore more for stability. The articular surfaces are very roughly 'L'-shaped when viewed from the side and are divided into superior and inferior segments. The sacral side is more convex in profile and the iliac side more concave, but the joint is not orientated to any single body plane. The sacral articular surface is orientated laterally, inferiorly, and posteriorly. The resulting joint can therefore only rotate in a limited way around a transverse plane x-axis to create a type of flexion–extension torque, referred to as sacral nutation and counternutation (see Fig. 2.2.2b). The SI joint shape and arrangement, as well as its supporting anatomy, has numerous reported anatomical variations (Figs 2.2.4o and 2.2.4p), and fat can also be present within the joint (Poilliot et al., 2019; Casaroli et al., 2020).

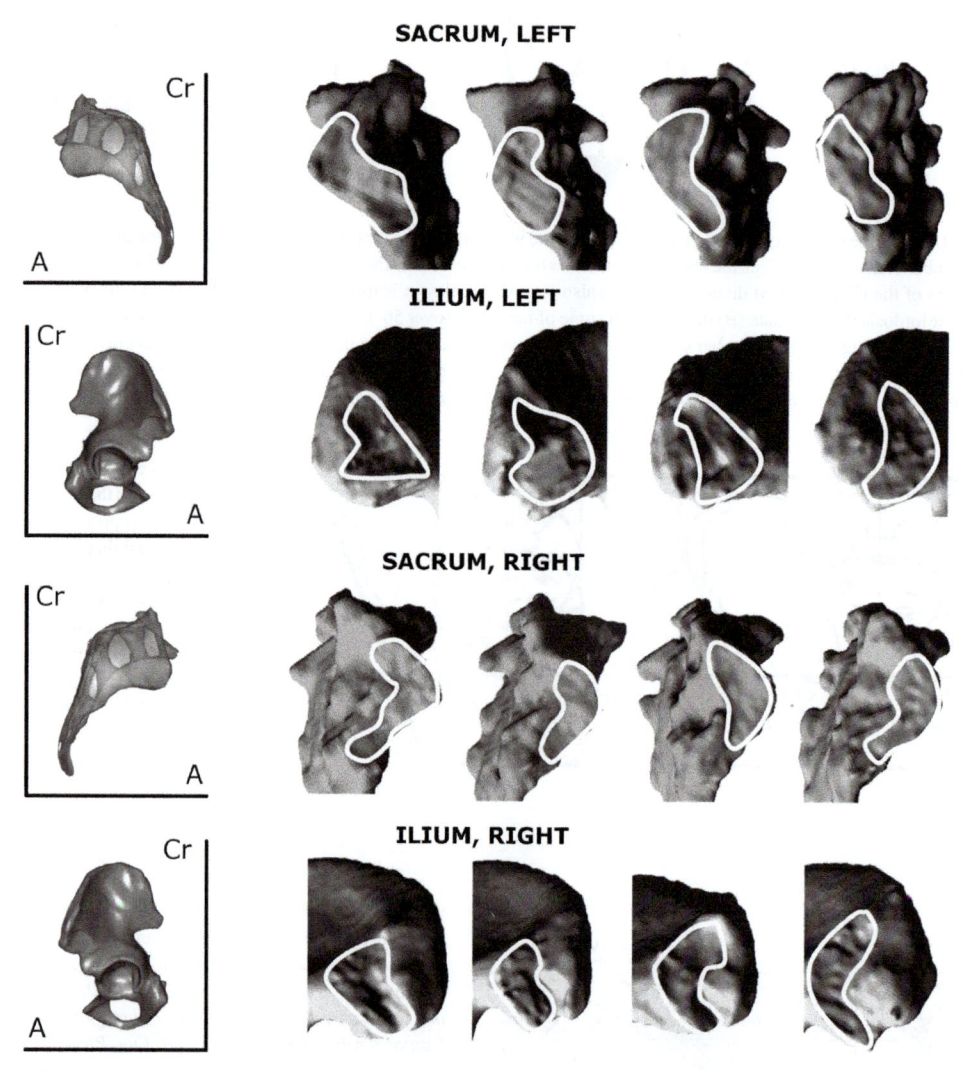

FIG. 2.2.4o Image demonstrating some of the reported differences in articular shape on the opposing sides of the SI joint. Cranial (Cr) or superior and anterior (A) orientations are indicated. These surfaces were reconstructed from radiographic and CT scans of living subjects. *(Images from Casaroli, G., Bassani, T., Brayda-Bruno, M., Luca, A., Galbusera, F., 2020. What do we know about the biomechanics of the sacroiliac joint and of sacropelvic fixation? A literature review. Med. Eng. Phys. 76, 1–12.)*

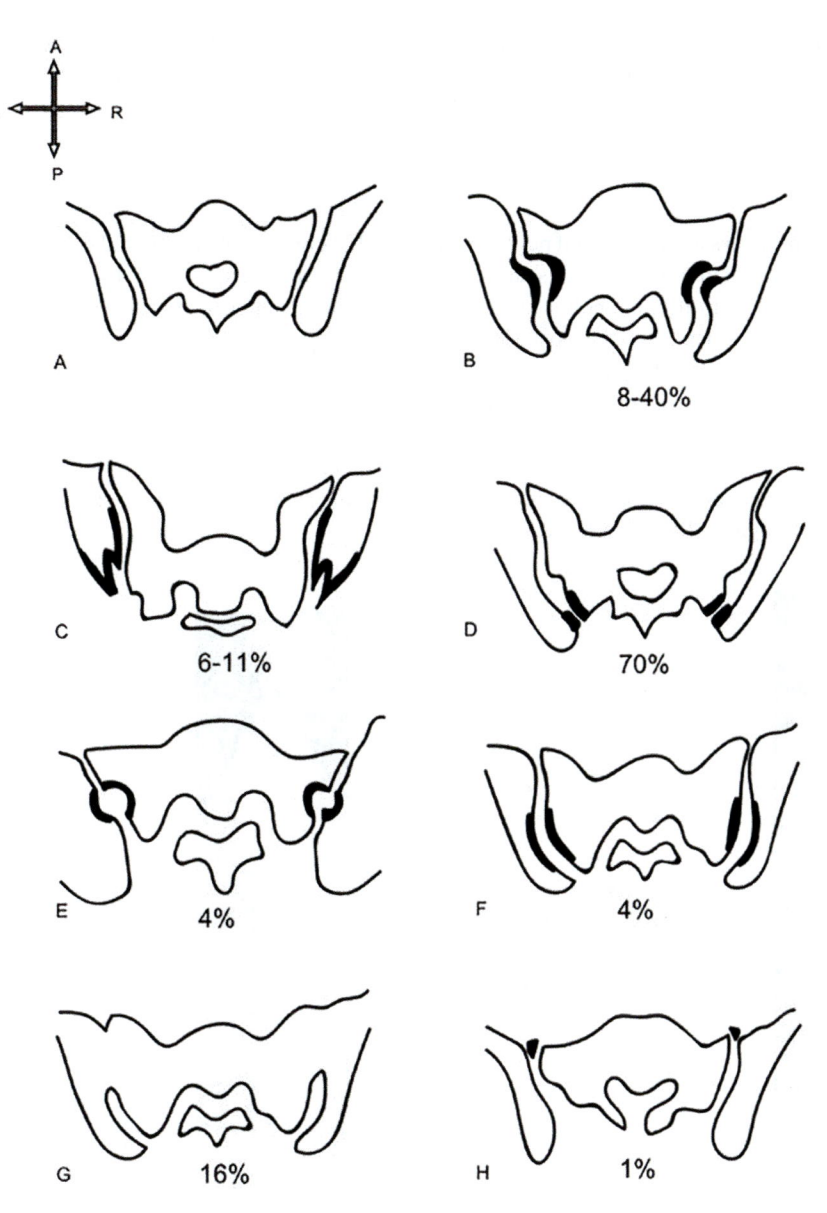

FIG. 2.2.4p A demonstration of some examples of variations in SI joint morphology found between the ilium and the fused S1–S4 sacral vertebrae of the sacrum, with some percentage indication of how frequently they are reported within the literature from specimens studied. (A) shows the most common form, (B) an accessory SI joint, (C) a pronounced iliosacral complex, (D) a bipartite appearance with dysmorphic posterior iliac changes, (E) semicircular defects, (F) crescent articular surfaces, (G) articular fusion (*synostosis*), and (H) separate sacral wing ossification centres. This indicates how variable precise pelvic biomechanics can be. *(Image from Poilliot, A.J., Zwirner, J., Doyle, T., Hammer, N., 2019. A systematic review of the normal sacroiliac joint anatomy and adjacent tissues for pain physicians. Pain Physician 22(4), E247–E274.)*

The powerful ligamentous framework across the joint cavity consists of the anterior and posterior *sacroiliac ligaments* traversing the joint along with the *interosseous ligament*. The joint capsule and ligaments contain a good supply of mechanoreceptors and have extensive innervation, so that sacroiliac-induced pain can mimic that of lumbar origin (Fortin et al., 1994). Sacroiliac maximum rotations and displacements amount to less than 2° and 1 mm, respectively (Casaroli et al., 2020). Larger motions reported are associated with skin marker displacement, but in pathology, joint displacements can become larger (Casaroli et al., 2020).

Opposing the two SI joints anteriorly is the semirigid union of the two pubic bones via the pubic symphysis. The pubic symphysis is a fibrocartilaginous joint unified with a disc of fibrocartilage called the *interpubic disc*, and is stabilised by four pubic ligaments. The joint unites the anterior parts of the innominate bones and on average it is morphologically different between males and females. Motion at the joint is small and probably relates to plasticity within the bone and the join of the fibrocartilage. Cadaveric testing has produced pubic symphysis motions (deformations) of 3.20 mm ± 2.5 mm, the largest of these deformations occurring in the transverse plane, with significantly more mobility being demonstrated in females (Pool-Goudzwaard et al., 2012). In vivo tests have been rare but have reported average translations of 1.1 mm in the sagittal and transverse planes, 2.5 mm of vertical translation, and 0.5° rotation in the frontal plane (Walheim et al., 1984). *Osteitis pubis* is an aseptic inflammatory process involving the pubic symphysis that can form oedema within the surrounding bone (which may also involve the surrounding soft tissues, tendon, and muscles) and is a common cause of groin pain, particularly in elite athletes (Gaudino et al., 2017).

Pelvic tilt angle

Pelvic tilt is one of the pelvic parameters, along with sacral slope and pelvic incidence, that creates coherence of the postural spinal parameters of notably the lumbar lordosis and thoracic kyphosis, and is a relationship that changes from, for example, standing to sitting postures (Lazennec et al., 2013). In life, the pelvis sits at an angle from the anterior superior part of the anterior superior iliac spine to the posterior superior part of the posterior superior iliac spine. Increases or decreases in pelvic tilt or asymmetry in tilts between sides are possibly associated with clinical problems (Herrington, 2011) and may cause biomechanical asymmetry. These tilts are measured clinically using handheld callipers, with surprisingly good intratester reliability. Furthermore, it has been reported that most pelvic tilts are rotated anteriorly, reportedly found in 85% of males and 75% of females at values of around 6–7° (Herrington, 2011). Males are also reported to have slightly smaller levels of asymmetry than females in healthy young populations, but the significance of the asymmetry and the tilt angle in relation to pathology remains unknown (Herrington, 2011) (Fig. 2.2.4q).

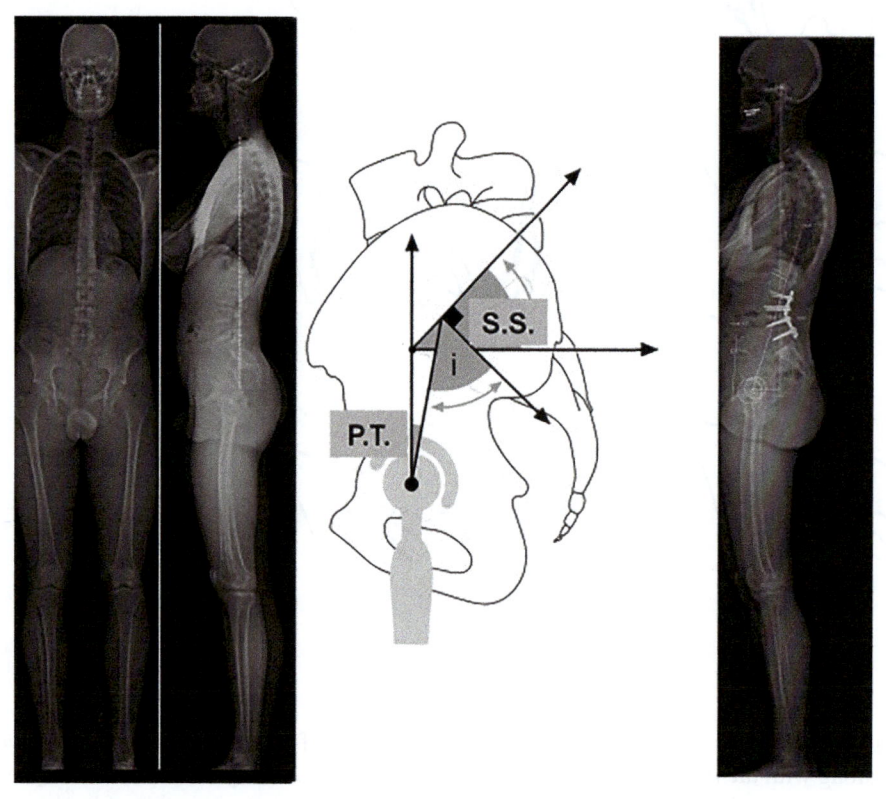

FIG. 2.2.4q The pelvic tilt angle (i) is related to the sacral slope (SS) and the pelvic tilt (PT), as the pelvic incidence angle is equal to the sacral slope plus the pelvic tilt. Thus, pelvic tilt is a mechanism that should help maintain the line of gravity for locomotive efficiency. It is important that during lumbar spinal fusion or other lumbosacral surgery, these relationships are appropriately maintained as much as possible. *(Image from Lazennec, J.Y., Brusson, A., Rousseau, M.A., 2013. Lumbar-pelvic-femoral balance on sitting and standing lateral radiographs. Orthop. Traumatol. Surg. Res. 99(1 Supplement), S87–S103.)*

2.2.5 Passive soft tissues of the lumbar spine and pelvis

Pelvic ligaments

Stability within the pelvis is maintained through the pelvic ring by compressions and tensions found across the SI joint and the pubic symphysis, created by the extensive and stiff network of pelvic ligaments. The ligamentous morphology of the SI joint seems to be gender-dependent (Steinke et al., 2010). Data from finite element modelling suggest that motion at the pubic symphysis and the SI joints decreases with increased stiffness of the ligaments and cartilage, the exceptions being the *sacrotuberous* and *sacrospinous* ligaments where increased stiffness of these structures seems to increase SI joint motion (Hammer et al., 2013). The effects of the interactions between the cartilage of the SI joints and pubic symphysis and the ligaments straddling these joints have specific regional consequences, which are stiffness-dependent (Hammer et al., 2013). Ligamentous structures such as the *obturator membrane*, *inguinal ligament*, and *pubic ligament* are calculated to influence pelvic stability and motion very little (Hammer et al., 2013), and are therefore considered to have little effect on pelvic biomechanics. Although total load on the pelvic ligaments is reduced to around 90% of that produced on standing, sitting tends to concentrate load in the iliolumbar, interosseous, and posterior sacroiliac ligaments, which are suggested to be generators of low back pain (Hammer et al., 2013). Thus, sedentary activities rather than locomotion are more likely to be the source of dysfunction that can eventually lead to disturbed gait.

Intrinsic ligaments

The SI joint is stabilised through intrinsic and extrinsic ligaments. There is limited knowledge on the SI joint ligaments, with discrepancies in descriptions of nomenclature, layers, attachment sites, and topographical relationships between ligaments and nerves, probably reflecting population variances (Poilliot et al., 2019; Casaroli et al., 2020). The intrinsic ligaments are the long posterior sacroiliac ligament, the interosseous ligaments, and the anterior and posterior sacroiliac ligaments, which are statistically larger in males than in females (Steinke et al., 2010). The anterior ligament is a connective tissue thickening consisting of a fan-like spreading of upper, middle, and lower parts arising from the anterior sacral wing of the ilium (Poilliot et al., 2019). The ligament is continuous with the posterior sacroiliac, sacrospinous, and sacrotuberous ligaments (Fig. 2.2.5a).

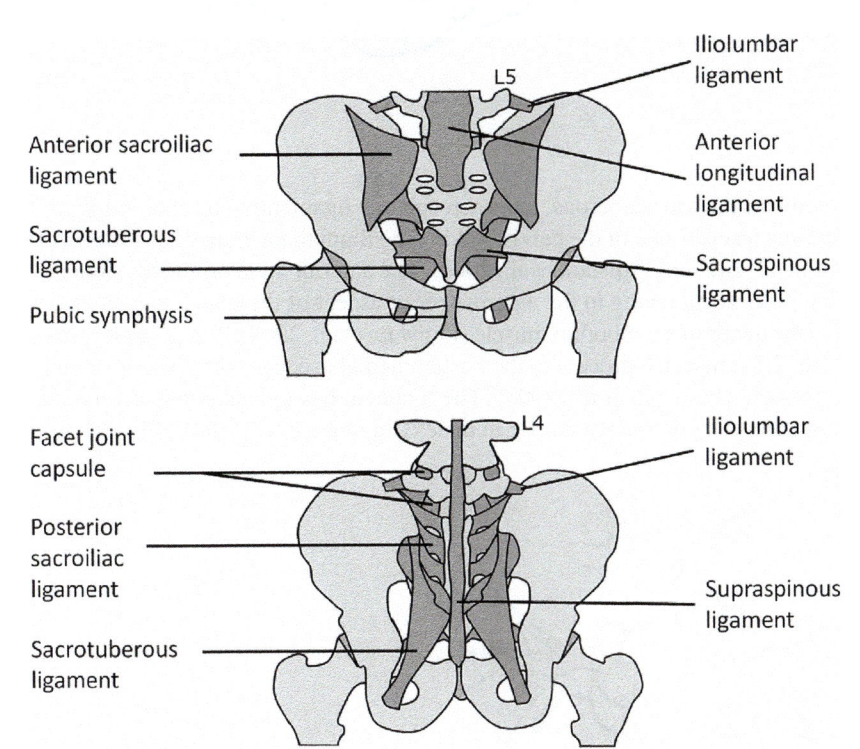

FIG. 2.2.5a Schematic representations of the primary sacroiliac ligaments found in the pelvis that constrain its motion and improve its energy storage and transfer capacity between the lower limb and spine. The anteroposterior view is demonstrated in the upper image and the posteroanterior view in the lower image. *(Images from Casaroli, G., Bassani, T., Brayda-Bruno, M., Luca, A., Galbusera, F., 2020. What do we know about the biomechanics of the sacroiliac joint and of sacropelvic fixation? A literature review. Med. Eng. Phys. 76, 1–12.)*

The posterior ligament is continuous with the long posterior sacroiliac ligament, sacrospinous, sacrotuberous, and interosseous ligament. The long posterior sacroiliac ligament lies deep to the posterior ligament, but there is much mixing of fibres between them (Poilliot et al., 2019). The interosseous ligament of the sacroiliac joint has the largest volume and greatest attachment surfaces of all the sacroiliac joint's ligaments, and they are generally larger in females (Steinke et al., 2010). It fills the syndesmotic joint space attached to the iliac tuberosity, but only in the superior aspect of the joint (Poilliot et al., 2019) (Fig. 2.2.5b).

FIG. 2.2.5b The primary sacroiliac ligaments in the transverse plane viewed superiorly with the posterior aspect to the top of the image. The fused sacrum (S1–S5) is sandwiched between each ilium. The long posterior (or dorsal) sacroiliac ligament (LPSL) runs inferiorly from the ilium to the inferior sacrum. The intrinsic SI joint ligaments consist of the short posterior sacroiliac ligament (SPSL), the interosseous ligaments (IL), and the anterior sacroiliac ligament (ASL), which between them enclose the sacroiliac joint cavity (SIC) creating local SI joint compression (black arrows). Inferiorly, the sacrospinous ligament (SP) binds the sacrum to the inferior posterior iliac spine, while the sacrotuberous ligament (ST) binds below to the ischial tuberosity from the sacrum and coccyx. Nutation and counternutation tension different ligaments, maintaining pelvic stability, elastic energy storage, and energy release throughout gait. This passive ligamentous force closure is supplemented by that of muscles such as those of the abdominal obliques' vectors (AOV), helping to counter some of the posterior ligament forces that are dominant. Loss of abdominal strength thus risks disturbing lumbopelvic biomechanics and can initiate issues such as pubic symphysis symptoms. Posterior and lateral muscles equally have important roles in lumbopelvic stability. Compare this image to Fig. 2.2.3b. *(Permission www.healthystep.co.uk.)*

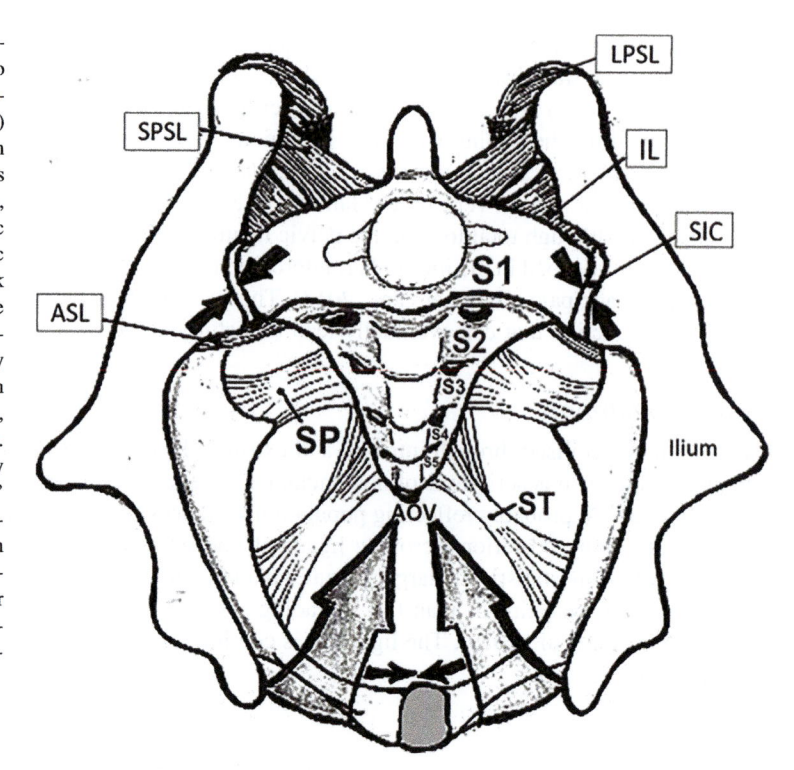

Extrinsic ligaments

The SI joint's extrinsic ligaments are the iliolumbar, the sacrotuberous, and sacrospinous ligaments. The iliolumbar ligaments consist of two bands (anterior and posterior) on each side of the pelvis, linking the ilium to the transverse process of L5 and adding additional stability to the lumbosacral junction (Fujiwara et al., 2000). The anterior band originates from the anterior and lateral aspects of the L5 transverse process, attaching to the anterosuperior aspect of the iliac tuberosity by a broad attachment linked to the anterior fascia of the quadratus lumborum muscle (Fujiwara et al., 2000). The posterior band runs from the posterior and inferior aspects of the L5 transverse process to the posteromedial aspect of the posterior ilium, and is again linked to the quadratus lumborum fascia (Fujiwara et al., 2000). The ligament has two anatomical variants, displays variation between ethnic human groups, and is significantly shorter in males (Fujiwara et al., 2000) (Fig. 2.2.5c).

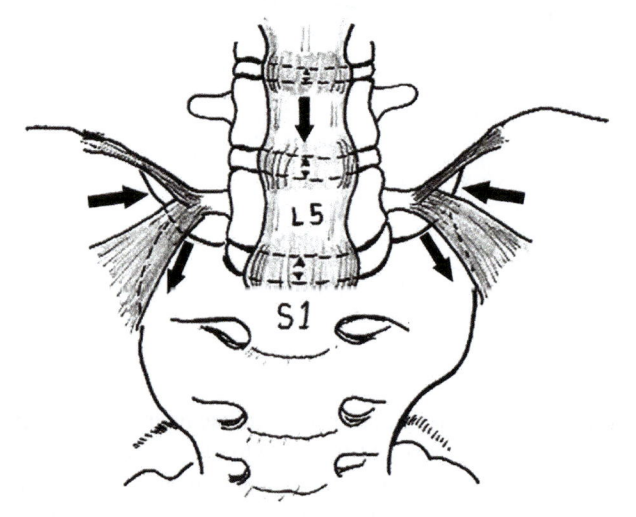

FIG. 2.2.5c Tension and compression forces induce lumbosacral and sacroiliac stability. The iliolumbar ligament creates compressive forces across the lumbosacral and SI joints as well as binding the ilia to the 5th lumbar vertebra. The longitudinal ligaments, with the anterior surface demonstrated here, provide compression through the vertebral bodies, resisted by the properties of the intervertebral discs and the facet joints' articular cartilage. *(Image modified from Porterfield, J.A., DeRosa, C., 1998. Mechanical Low Back Pain: Perspectives in Functional Anatomy, second ed. W.B. Saunders, Philadelphia.)*

The iliolumbar ligament restricts lateral bending in the lumbar spine so that the left-sided ligament restricts right-sided lateral bending and vice versa, but functioning together, they restrict flexion–extension and rotation of the lumbosacral junction (Yamamoto et al., 1990). The iliolumbar ligaments also restrict SI joint mobility, with the inferior band providing the most significant restriction (Pool-Goudzwaard et al., 2003). It is therefore thought that the iliolumbar ligament is important in preventing disc degeneration between L5 and the sacrum (Aihara et al., 2002). It has a rich nerve supply of mechanoreceptors (mainly Pacinian) and plays an important role in the proprioception coordination of the lumbosacral region (Kiter et al., 2010) (Fig. 2.2.5d).

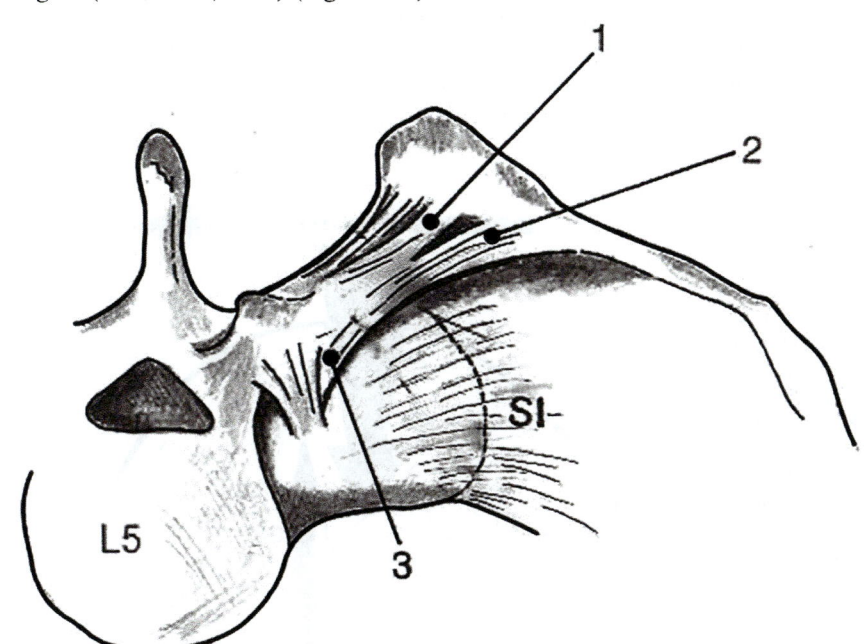

FIG. 2.2.5d The iliolumbar ligament in the transverse plane viewed superiorly with 1: the dorsal band, 2: the ventral band, and 3: the sacroiliac part. The line of the sacroiliac (SI) joint is indicated as is the 5th lumbar vertebra (L5). The iliolumbar ligament blends with the anterior sacroiliac ligament. It can be appreciated from this view that lumbar rotations at the lumbosacral joint will cause these ligaments to tighten. (Image modified from Pool-Goudzwaard, A., Hoek van Dijke, G., Mulder, P., Spoor, C., Snijders, C., Stoeckart, R., 2003. The iliolumbar ligament: its influence on stability of the sacroiliac joint. Clin. Biomech. 18(2), 99–105.)

The sacrospinous ligaments connect the ischial spine of the ischium to the sacrum through the last two sacral segments and the 1st coccygeal segment, blending fibres with the anterior sacroiliac ligament superiorly and the sacrotuberous ligament (Poilliot et al., 2019). The sacrotuberous ligament arises from the posterior inferior iliac spine and the inferior edge of the sacrum and upper coccyx. It shares attachments with the muscles of the erector spinae, piriformis, gluteus maximus, biceps femoris, and the thoracolumbar fascia (Poilliot et al., 2019). These ligaments cause direct force transmission and give rise to axial rotation and translation forces between the bones, limiting sacral nutation rotations on vertical loading (Hammer et al., 2013). These ligaments have a greater lever arm ratio than the interosseous ligament at the sacroiliac joint, despite the interosseous ligament having a larger cross-sectional area. However, they have minimal stabilising effect on the acetabula and pubic symphysis (Steinke et al., 2010; Hammer et al., 2013).

2.2.6 Functional joint axes and load distribution of the lumbar spine and pelvis

The function of the lumbar spine, pelvis, and hip is to accentuate the linkage across the pelvic hub from the lower spine and trunk above to the lower limb below, through all locomotive function. In the spine, the orientation of the facet joints and intervertebral discs contribute to and limit the direction of motion. The discs allow motion in many planes due to their fibre arrangement within the annular rings, whereas the facet joints direct and limit motion. In the lumbar spine, facet joints allow sagittal plane motion but limit motion in other planes. The axis of rotation of each lumbar vertebra runs through the facet joints on both sides of the processes by virtue of an axis within the transverse plane angled to the particular vertebral orientation. This is orientated to any asymmetry between the left and right-sided facet joints.

Normally the axis of transverse rotation in the lumbar spine is vertically orientated and located at the postero-central aspect of the intervertebral disc, but in extreme end-ranges of rotation, the articular surfaces of the facet joints engage on one side, shifting the axis posteriorly (Porterfield and DeRosa, 1998, p. 131). This changes the axial torques occurring at the disc–vertebra interface, altering the location of stresses. Therefore, in end-range rotation, the lumbar spine experiences increased ipsilateral engaged facet compression, increased shear on the intervertebral disc–vertebral body interface, and tension–distraction of the joint capsule on the contralateral side.

The pelvic joints are subject to compression, tension, and shear stresses (Li et al., 2006, 2007; Hammer et al., 2013). When the pelvis is loaded with vertical force through the acetabula, motion occurs at the pubic symphysis giving rise to a widening of around 0.5 mm (Hammer et al., 2013). The sacrum apex rotates superiorly and posteriorly on vertical loading. This is part of sacral nutation which also involves some translation between the iliac and sacral surfaces. This sacral rotation is limited by the sacrospinous and sacrotuberous ligaments via a centre of rotation positioned around the level of the second sacral vertebra through the interosseous ligament (Puhakka et al., 2004; Hammer et al., 2013). This is effectively a point of least resistance to motion rather than a point of motion. Both iliac crests become spread from the sacrum when loaded as compared to the unloaded condition (Hammer et al., 2013). During gait, the ilium moves with the lower limb torques, so that hip flexion causes posterior rotation on the ilium and hip extension causes anterior rotation on the ilium. Therefore, the sacrum nutates and counternutates in opposite directions through the SI joints during gait, with nutation occurring during swing and initial stance phase and counternutation occurring throughout midstance and terminal stance (see Figs 2.2.2b, 2.2.2c, and 2.2.6).

FIG. 2.2.6 A schematic to show the relationship between the BWV and GRF at the sacroiliac (SI) joint, with the pelvis viewed from the right side and its anterior surface (A) facing to the right. The body mass of the trunk compresses the spine into the sacrum helping drive the SI joint inferiorly into form closure, but also rotating the sacrum anteriorly. In static stance, the GRF ascending from the lower limb across the hip tends to posteriorly rotate the ilium, but locomotive posture is dependent on the angle of the lower limb during the stance phase of gait. At initial contact, a strong total pelvic flexion moment is applied by the GRF at the hip, while the BWV through the spine rotates the sacrum anteriorly on the ilium. These forces together cause nutation of the SI joint, which is resisted through tension in ligaments such as the sacrospinous and sacrotuberous ligaments (shown here posteriorly). In static stance, nutation is largely passively resisted. In gait, as the hip passes through flexion and extension the GRF vector is applied variably into the ilium. Hip flexion at initial contact tips the GRF posteriorly, rotating the ilium backwards, helping nutation of the SI joint, and increasing energy storage within the ligaments through rising tensions. As the hip extends throughout late stance phase, the GRF is directed anteriorly, rotating the ilium forward to cause counternutation at the SI joint. This releases the stored energy within the ligaments, aiding motion. Such a mechanism means that the SI joint has energy dissipating and storage capabilities as long as torques create tissue deformation around the joint. *(Permission www.healthystep.co.uk.)*

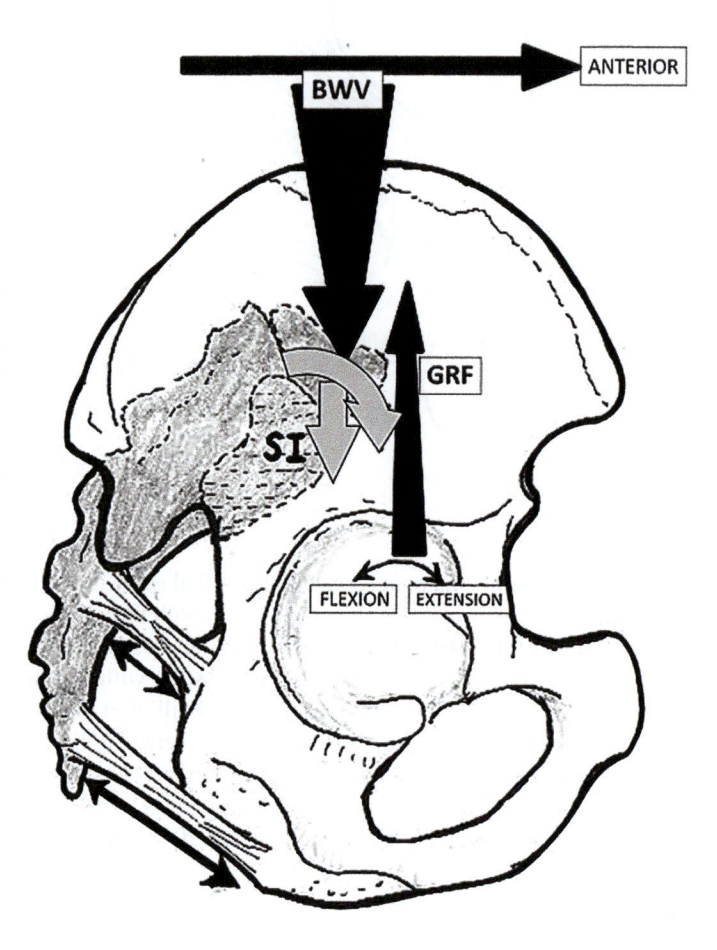

The pubic symphysis experiences tensile stresses developed at the superior and anterior aspects of the pubic arch, with compression developing at the superior and posterior parts of this arch. An 'intermediate' stressed region forms between the compression and tension regions from the bending moments across the joint and the pubic arch (Hammer et al., 2013). The SI joint experiences compressive forces at the anterior articular surface, with tensile stresses occurring posteriorly in the region covered by the ligaments and concentrated to the superior posterior SI joint compartment and the interosseous and posterior sacroiliac ligaments (Hammer et al., 2013). Least stress is applied at the level of the second sacral vertebra, because it is at the centre of SI joint rotation (Hammer et al., 2013).

Increasing cartilage stiffness in the SI joint and its ligaments decreases SI joint motion and changes the stresses expressed at the pubic symphysis and acetabula (Hammer et al., 2013). Stresses in the acetabula in the anterior–posterior sagittal plane on standing and sitting are similar. However, sitting increases vertical stress in the acetabula but decreases transverse plane stress (Hammer et al., 2013).

2.2.7 Muscle action at the lumbar spine and pelvis

There are 29 muscles that have pelvic attachments. Twenty link the pelvis to the femur and the rest link the spine to the pelvis. The muscles of the lower spine, abdominal wall, and pelvis exert a considerable influence on the pelvic ring's stability and that of the sacroiliac joint (Beales et al., 2010; Hu et al., 2010; Park et al., 2010; Arumugam et al., 2012; Jung et al., 2013). The changing trunk coupling torques between muscles in the trunk and the pelvis are complex. Significant force can be generated through the pelvis and into the lumbar spine by various combinations of hip and knee muscle activity. It is easier to stabilise a pelvis when standing than sitting because of this. More than one name is given to many of the muscles of the spine, adding complications to learning the anatomical complexity. For example, erector spinae can be divided into iliocostalis, longissimus, and spinalis, each with their own three sections. As the biomechanical principles of locomotion are the key here, complexity of names and minor specifics of muscles will be avoided as much as possible.

Thoracolumbar fascia

The thoracolumbar fascia is an aponeurosis that forms an extensive and highly organised fascial network, linking powerful lumbopelvic muscles together (Vleeming et al., 2014). It is divided into three layers: anterior, middle, and posterior. The middle and posterior layers are the most mechanically significant and are termed the deep and superficial layers, respectively. The superficial posterior layer is subdivided into two lamellae (Porterfield and DeRosa, 1998, pp. 82–83). It is thick and attached to the spinous processes of the vertebrae and supraspinous ligaments, coursing laterally to cover the erector spinae muscles. At its most lateral aspect, it forms the proximal attachment to the internal abdominal oblique and transversus abdominis muscles of the abdominal wall. The fascia also attaches to the anterior surface of the erector spinae muscles to form the deep fascial layer and attaches to the lumbar spine's transverse processes and intertransverse ligaments. The thoracolumbar fascia attachments to the lumbar spine's spinous and transverse processes serve to compartmentalise the primary muscles of spine extension (Porterfield and DeRosa, 1998, pp. 82–83).

Inferiorly, the thoracolumbar fascia is attached bilaterally to the sacrum and ilium, blending with the fascias of each contralateral gluteus maximus muscle. Superolaterally, the fascia on both sides of the spine blends with the latissimus dorsi muscles. This mechanically links mutually contralateral gluteus maximus and latissimus dorsi muscles through the superficial layer, coupling the action of the muscles across the back. In so doing, the arrangement sets up lines of tension that can store energy and increase compression obliquely across each sacroiliac joint, approximately aligned perpendicular to the joint line. The shoulder rotates and the arm swings in antiphase between the ipsilateral lower limb and pelvis, but in-phase between the contralateral lower limb and pelvis. This in-phase contralateral relationship highlights the need for reciprocal upper and lower limb exercises when working on locomotive strength.

The thoracolumbar fascia, through osseous and muscular attachments, links the lumbar vertebrae to the pelvis by spanning the lumbar, lumbosacral, and SI joints, connecting them through linkage to the gluteus maximus muscles. The thoracolumbar fascia links forces across the hip and shoulder into the lower and upper limbs. Tensions are thus permitted to be generated and altered through the thoracolumbar fascia by changing joint angular alignments and positions. This enables the thoracolumbar fascia to minimise aberrant motion between bony segments and helps in the storage and release of elastic energy. The thoracolumbar fascia also has extensive muscular linkages through the erector spinae muscles, multifidus, gluteus maximus, latissimus dorsi, and the abdominal muscle walls. These muscles are able to generate active tensions within the fascia, creating a stabilising envelope around the trunk-to-pelvis interface. There are co-dependent mechanisms involving balanced tensions through the aponeurotic components of the thoracolumbar fascia that link the deep abdominal and lumbar spinal muscles together (Vleeming et al., 2014).

Latissimus dorsi

Although considered an upper limb adductor and medial rotator, the latissimus dorsi has, through the thoracolumbar fascia, important effects on lumbopelvic mechanics. It attaches to the lower six thoracic spinous processes and all the lumbar and sacral spinous processes throughout the thoracolumbar fascial network to the lilac crest, often via which the latissimus dorsi muscle has direct attachment. Superiorly, the latissimus dorsi attaches to the lesser tubercle and the intertubercular (bicipital) groove of the humerus. It has the longest moment arm of any muscle that influences the posterior trunk, with the position of the humerus influencing tension through the thoracolumbar fascia during arm motion. This makes arm swing position significant to generating thoracolumbar tensions. If the humerus is abducted and flexed such as when lifting or pulling an object towards the body, then latissimus dorsi becomes stretched. This tensions the thoracolumbar fascia further if the pelvis also posteriorly rotates and the gluteus maximus contracts. Contraction of both latissimi muscles pulls on the spinous and transverse process attachments of the thoracolumbar fascia, heightening lumbar resistance to rotation and flexion and thereby increasing stabilisation.

A counterforce occurs through ipsilateral latissimus dorsi contraction, inducing contralateral oblique and transverse abdominal wall muscle contraction and gluteus maximus contraction. The upper and lower limbs always move in antiphase to the ipsilateral side. During slow walking speeds, the trunk and pelvis usually rotate in-phase, but at faster walking speeds and when running, motion of the pelvis follows that of the lower limb while arm swing also increases the antiphase motion between the trunk and pelvis. This mechanism, assisted by the abdominal muscles, increases tension within the thoracolumbar and abdominal wall aponeurosis through opposing rotations. The mechanism stores energy within and then releases it from the fascial collagen fibres through the tensions produced by these counterrotations and the contralateral muscular activity during gait (Fig. 2.2.7a).

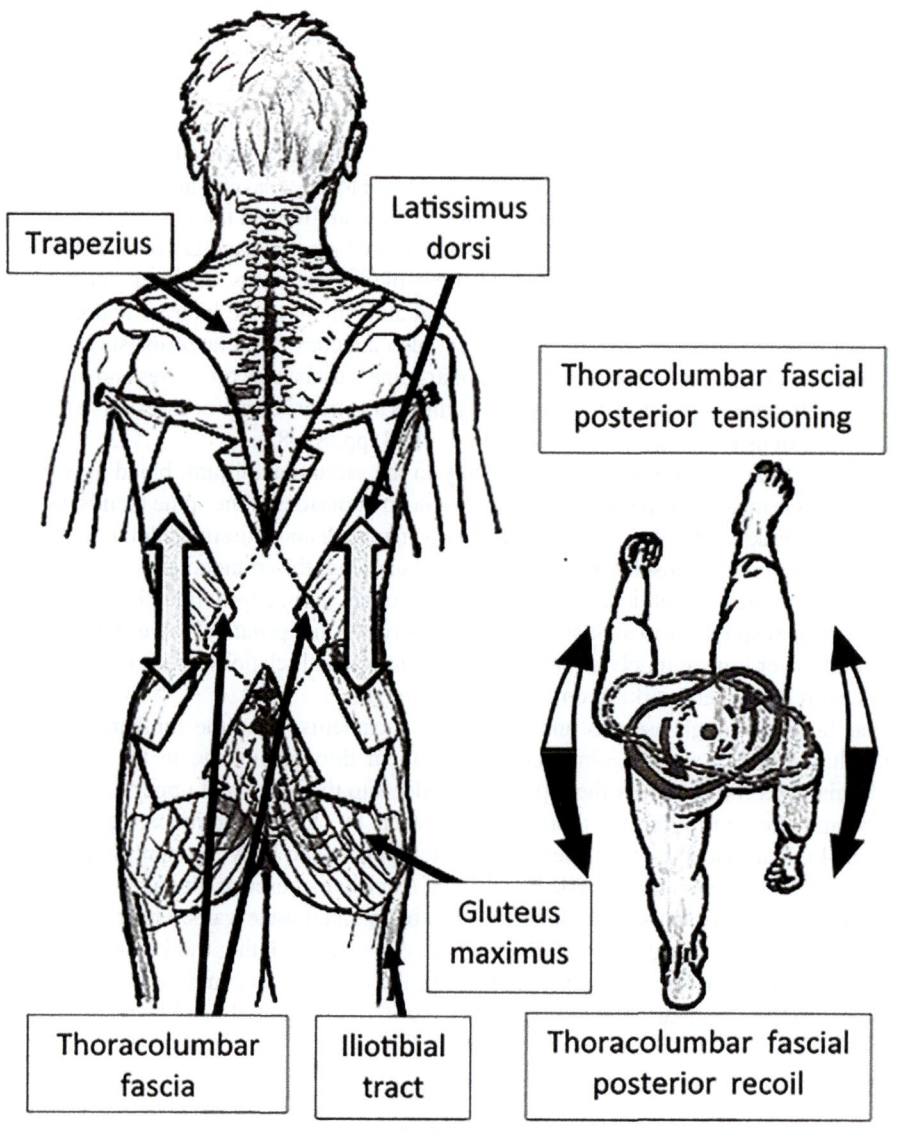

FIG. 2.2.7a Large fascial structures such as the thoracolumbar fascia set up complex biomechanical interrelationships across anatomy. By linking left and right latissimus dorsi, gluteus maximus, and the abdominal wall with bony attachments to the humerus, iliac crest, and spinous processes, the upper and lower limbs and the shoulder and pelvic girdles are biomechanically bound together. Two distinct but related tensional mechanisms are set up dorsally as a result. Antiphase motion between the ipsilateral upper and lower limbs, shoulders, and pelvic girdles during locomotion create tension from left to right. When the arm, shoulder, and contralateral lower leg swing in front of the trunk, the fascia is loaded with potential energy. During arm backswing and hip extension, this energy is released from the thoracolumbar fascia. Thus, left arm works with right leg and vice versa. This energy-storage mechanism increases with faster gait speeds as pelvic motion becomes antiphasic with the trunk. Coupled to this is tension on the ipsilateral side between the limbs, with tension increased as the limbs move away from each other, storing energy and then releasing it as the limbs move closer together in midstance and midswing. Trunk side flexion also increases tension within the fascia on the contralateral side. These posterior rotations and motions across the trunk between the pelvic and shoulder girdles are influenced by and also add influences on, the abdominal wall stresses anteriorly. *(Permission www.healthystep.co.uk.)*

Erector spinae muscle

The erector spinae muscle can be simply divided into superficial and deep divisions. The superficial division is attached by a broad, flat, fibrous aponeurotic tendon to the iliac crest. This erector spinae muscle aponeurosis has a broad medial–lateral expansion from the lumbar and sacral processes to the ilium, with the erector spinae muscle arising from its anterior aspect to run and attach to the lower ribs. The superficial erector spinae muscles do not attach to the lumbar spine directly but have an optimal lever arm for lumbar extension through pulling the thorax posteriorly. The deeper divisions of the erector spinae muscles arise proximally from the ilium above and laterally to the posterior iliac spine, and the deep surface of the erector spinae aponeuroses. Distally, they attach superiorly to the lumbar transverse processes. The muscles twist on themselves, presenting their superolateral muscle fibres to the ilium attachments at their lowest insertion points on the lower lumbar transverse processes. The inferomedial fibres at the ilium have the highest proximal attachments on the upper transverse processes. The deep erector spinae muscles create an inclination of pull posteriorly, giving rise to vertical compression and posterior shear in a triangulated manner, stabilising and resisting anterior shear of the lumbar spine.

Eccentrically, the erector spinae muscles control lumbar flexion although they are electromyographically silent at the end of trunk bending (Kippers and Parker, 1984), probably as the point of stabilisation is relinquished at large flexion angles to other structures. The erector spinae muscles are important in trunk and pelvic coordination and their activity varies with walking velocity (Lamoth et al., 2006a). To adapt in preserving stable efficient gait patterns, the erector spinae muscles display biphasic activity with peak activity around initial foot contact to resist any trunk flexion moments created by the GRF, but it has little activity during swing phase (Lamoth et al., 2006b). The erector spinae are important for trunk stability in both the frontal and sagittal planes and increase their activity with increase in gait velocity (Saunders et al., 2005; van der Hulst et al., 2010). They work in concert with psoas to maintain erect posture in the lumbopelvic region. Hip flexion in swing is associated with iliopsoas, but it seems that only iliacus is involved in significant hip flexion, with psoas assisting the erector spinae muscles to maintain an erect trunk (Gibbons, 2007) (Fig. 2.2.7b).

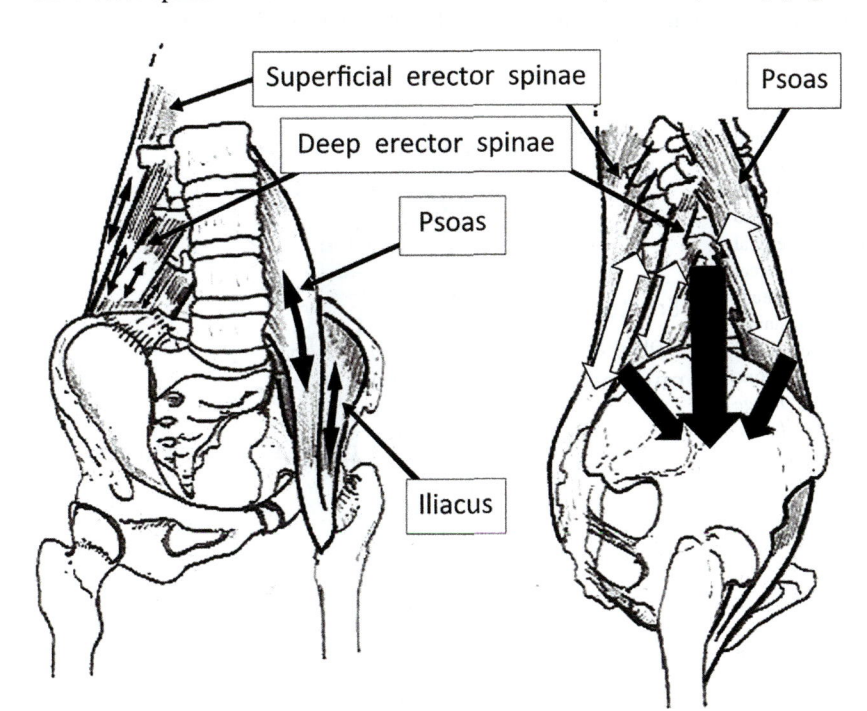

FIG. 2.2.7b Schematics of the erector spinae, psoas, and iliacus are shown contralaterally in the left image for ease of appreciating the anatomy. Note the conjoined tendon of iliopsoas attached to the femur. These muscles are part of the stabilisation mechanism of the lumbosacral and pelvic region. Viewed from the right (right image), the ipsilateral erector spinae and psoas create a resultant force that compresses the lumbar spine into the sacrum and sacrum into the SI joint, providing trunk-to-pelvis stability. Rotations of flexion can also be influenced around the lumbar spine, SI joint, and hip joint. However, iliacus and gluteus maximus have far greater influence on hip motion and stability. *(Permission www.healthystep.co.uk.)*

Multifidus

Multifidus arises distally from the posterior surface of the sacrum, the sacrotuberous ligament, the aponeuroses of the erector spinae muscles, the medial surface of the posterior superior iliac spine, and the posterior sacroiliac ligament. It attaches to the mammillary processes found on the transverse processes of the lumbar vertebrae. The muscle runs bilaterally, inferiorly to superiorly, up the length of the spine, but is most developed within the lumbar region. Multifidus is a large prominent muscle filling the space between the transverse and spinous processes of the vertebrae and lying on the posterior surface of the sacrum. This muscle's largest cross-sectional area lies transversely at the lumbosacral joint, and this diminishes as it becomes more closely attached on either side of the spine progressively as the spinous processes of the vertebrae become smaller superiorly. This is an important muscle for spinal stability that can be functionally assessed through soft tissue ultrasound scanning (Sions et al., 2014).

Multifidus is more effective at creating an extension moment arm over the lumbar spine than elsewhere within the spine, resisting trunk flexion and anterior shear. Like the erector spinae muscles, it also applies compressive vertical stabilising forces between the lumbar vertebrae and across the lumbosacral joint, helping to resist torsional forces and protecting the vertebral discs. A high percentage of type I muscle fibres found in multifidus suggests a stabilising function rather than a dynamic role. Multifidus recruitment is different in sufferers of chronic low back pain (Rantanen et al., 1993) and the histology associated with dysfunction reveals degeneration in its type I fibres and atrophy of its type II fibres (Flicker et al., 1993).

There are a series of small muscles between the lumbopelvic region that connect one intervertebral segment to the next, and they are usually named by the segments that they link. These are the *interspinales* and *intertransversarii* muscles. The former is found on either side of the interspinal ligaments and the latter attached adjacent to the mammillary and transverse processes. They probably fine-tune forces generated by multifidus and provide sensorimotor muscle spindle input.

Quadratus lumborum

The quadratus lumborum is a thin, flat muscle with fibres orientated in several directions lying deep to the thoracolumbar fascia. The muscle can be divided into anatomical components from its different attachments. The *iliotransverse* part runs from the iliac crest to the lumbar transverse processes, the *costotransverse* part passes from the lumbar transverse processes to the lower ribs, and the *iliocostal* from the iliac crest to the lower ribs (Porterfield and DeRosa, 1998, pp. 81–82). On the iliac crest, quadratus lumborum lies deep to erector spinae and its aponeurosis. It has far less cross-sectional area than the erector spinae and has a role in stabilising the pelvis to the spine within the frontal and transverse planes, particularly during actions like side-bending and trunk rotation. Working with the iliolumbar ligament, the erector spinae muscles, psoas, and the hip flexors and extensors, these muscles are likely elements of triplanar stability across the lumbopelvic region (Figs 2.2.7c and 2.2.7d).

FIG. 2.2.7c A simplified schematic of posterior spine, pelvis, and lower limb muscle force vectors. Through their multiple attachments, these force vector relationships provide stability, resist flexion, and supply extension mobility to the region. Erector spinae muscles (represented as black muscles), work with multifidus (white muscles) providing a vertical and extension force component from the spine to the sacrum. Quadratus lumborum (grey muscles) adds further vertical and extension power but with oblique force components to direct forces left and right via the ilium into the lower limbs. Gluteus maximus (striated muscles) adds extra stability to this mechanism when extending the hip. However, when gluteus maximus is not active in later stance phase, this role moves to the hip abductors, shifting the resultant force vector into the single support limb. *(Image adapted from Kapandji, I.A., 1974. The Physiology of the Joints. Volume 3: The Trunk and the Vertebral Column. Churchill Livingstone, Edinburgh.)*

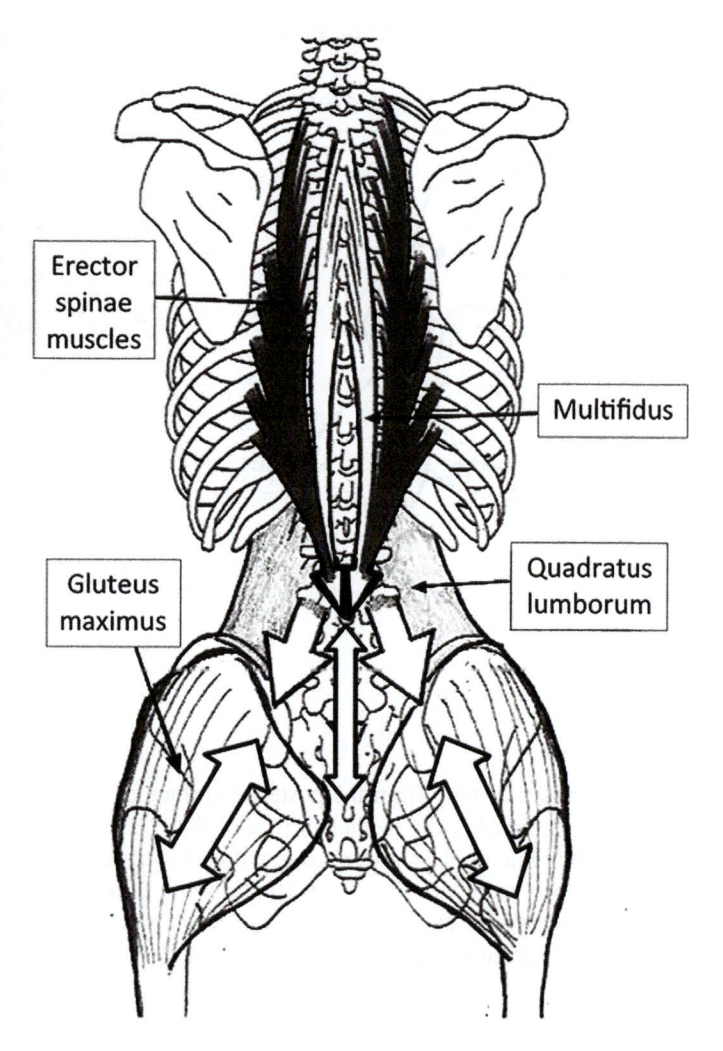

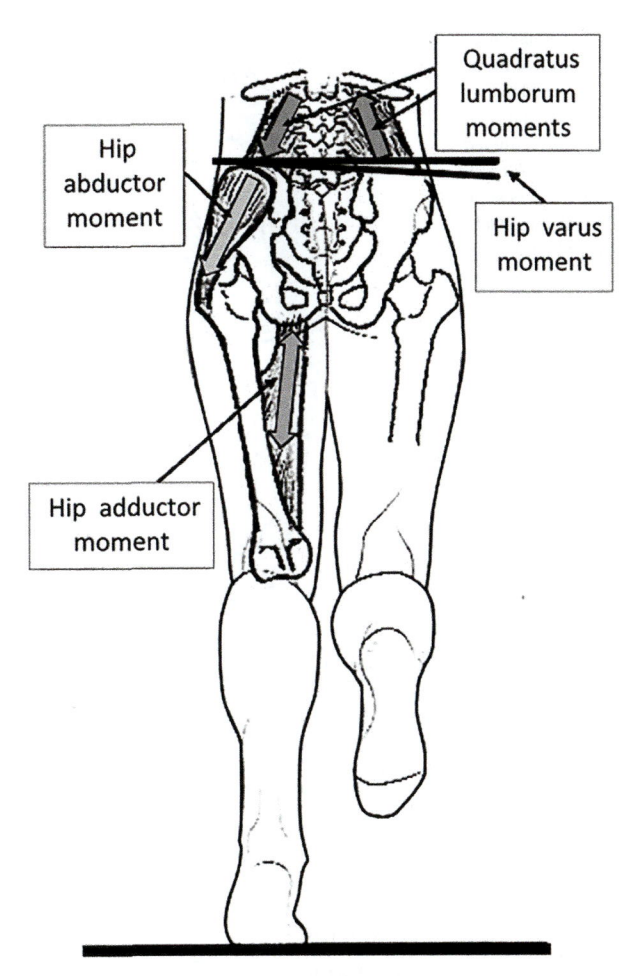

FIG. 2.2.7d In single-limb support, quadratus lumborum's frontal plane role becomes significant in working with the hip abductors and adductors to balance the varus moment tilt on the pelvis from displacement of CoM by the swing limb. This requires quadratus lumborum activity on both sides of the pelvis. With running only having a single-limb support phase, this muscle activity is particularly important to trunk–pelvic–lower limb postural stability in this form of locomotion. Indeed, any frontal plane trunk-to-lower limb motion or perturbation requires quadratus lumborum activity. *(Permission www. healthystep.co.uk.)*

The abdominal aponeurosis and wall muscles

Several layers of superficial fascia lie above the abdominal wall, connected to the superficial fascia of the thigh and perineum. However, the most significant abdominal fascia is the abdominal aponeurosis associated with the external and internal abdominal oblique and transverse abdominis muscles. There is also a strong fascial sheath associated with and enclosing the rectus abdominis muscle that links to the aponeurosis. The formation and arrangement of the fascial sheath is different above and below the umbilicus. Above the umbilicus, the aponeurosis splits at the lateral border of rectus abdominis and tracks anterior and posterior to rectus abdominis, whereas below the umbilicus, the aponeurosis remains double-walled in front of rectus abdominis. The double-walled region is found where maximum anterior shear stress occurs posteriorly within the lumbar spine. It is possible that with the lateral abdominal wall muscles attaching to the double-walled aponeurosis inferiorly, there is additional connective tissue support lying anteriorly to resist any anterior shear strain on the lumbar spine generated by the lordotic lumbar curve.

The aponeurosis is integral to the function of the abdominal wall muscles, anatomically and functionally. The abdominal aponeurosis is attached to the pectoralis major muscle of the chest wall superiorly. Fascial elements cross the midline of the fascia so that abdominal fascia links to abdominal muscles on one side across the pectoralis major's fascia attachments on the contralateral side, as well as to the superficial aspect of the rectus sheath. This produces linkages between the opposing shoulder and pelvic girdles on the anterior aspect of the trunk, much as occurs posteriorly with the thoracolumbar aponeurosis between the gluteus maximus and the contralateral latissimus dorsi (Fig. 2.2.7e).

FIG. 2.2.7e The abdominal fascia binds to and encloses the abdominal muscles, acting as a strong connective tissue restraint that muscles can apply force to. It is tethered tightly at the linea alba centrally (which separates the left and right abdominal muscles and encloses the umbilicus), tethered inferiorly to the inguinal ligament and anterior iliac crest, and links superiorly to pectoralis major on the chest to connect by upper arm attachments. Ipsilateral arm backswing and hip extension increases tension (white arrow), but it is the contralateral arm backswing and hip extension that couple together in gait creating tensions superior and inferior and left to right across the aponeurosis (black arrow). The abdominal aponeurosis works with the thoracolumbar fascia to utilise the antiphasic motion of the shoulder girdle and arm swing against the lower limbs. In gait, posterior arm swing and contralateral hip extension tightens the aponeurosis in an opposite but complementary way to the thoracolumbar fascia. Thus, on each side of the body, either the thoracolumbar fascia or the abdominal aponeurosis should be tightening during locomotion. *(Permission www.healthystep.co.uk.)*

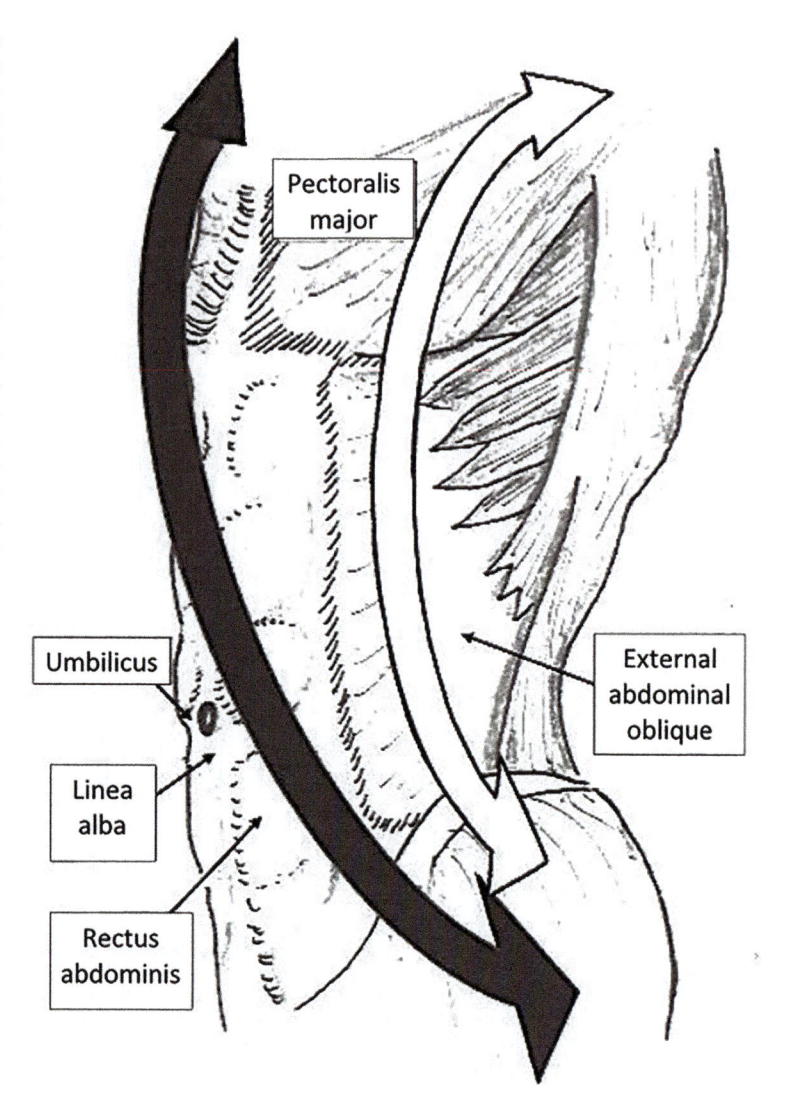

External abdominal oblique

The external abdominal oblique muscle (obliquus externus abdominis) arises from fleshy digitations to the last eight ribs. It extends inferiorly to attach by long fibrous attachments to the iliac crest inferomedially to blend into the abdominal aponeurosis. This in turn is attached to the *linea alba*, a tense thickening (raphe) within the aponeurosis. The external abdominal oblique is the largest and most superficial of the abdominal muscles and at its inferior margin, it forms a thick enrolled band called the *inguinal ligament* which represents the inferior limit of the abdominal aponeurosis.

The external abdominal oblique functions by flexing the lumbar spine, bringing the thorax closer to the pelvis. Thus, flexing the lumbosacral junction through rotating the pelvis posteriorly compresses the abdominal contents towards the spine. This concentric contraction increases tension of the abdominal aponeurosis, rotating the thorax to the contralateral side, or in activation of both external obliques, pulls the aponeurosis latero-posteriorly. Activity in multifidus antagonises the flexion vector of the external abdominal oblique to achieve pure trunk rotation as necessary. The muscle's inferior fibres are medially orientated and positioned to resist anterior and inferior pelvic rotation within the sagittal plane. The muscle is also important to lumbopelvic movement in the transverse plane (Saunders et al., 2005). Activity in the muscle increases with walking velocity (van der Hulst et al., 2010), indicating its importance in preventing the abdomen from shifting the CoM anteriorly.

Internal abdominal oblique

The internal abdominal oblique lies deep to the external muscle and attaches to the lateral part of the inguinal ligament, iliac crest, and the inferior portion of the lateral raphe of the thoracolumbar fascia. From these attachments, the muscle extends superomedially to attach to the cartilaginous border of the last three or four ribs, the abdominal aponeurosis, and the linea alba. The muscle compresses the abdominal contents posteriorly and increases tension in the abdominal aponeurosis. It flexes the lumbar spine and rotates the trunk on the pelvis.

Transversus abdominis

The transversus abdominis muscle lies deep to internal abdominal oblique muscle and attaches to the lateral inferomedial third of the inguinal ligament, the inner lip of the iliac crest, and a shared attachment to the thoracolumbar fascia with the internal abdominal oblique. The fibres of transversus abdominis, as suggested by their name, are more horizontal in orientation than those of the other abdominal muscles. They travel to the linea alba anterior to rectus abdominis and below the umbilicus, helping to reinforce the aponeurosis. The muscle increases tension within the abdominal fascia and on into the thoracolumbar aponeurosis to complete a 'ring' of support around the beltline of the trunk–pelvic interface.

Rectus abdominis

The rectus abdominis extends vertically from the pubic tubercles to attach superiorly to the rib cage on either side of the sternum. The muscle is long and strap-like and divided into intermuscular compartments separated by fascial bands. It is covered by the fascia of the rectus sheath which the muscle is attached to at all levels via fascial bands. This makes rectus femoris largely an interfascial muscle that lies on either side of the linea alba, filling the rectus sheath. The presence of rectus abdominis in the aponeurosis structure produces a mechanical linkage between the tensional forces of the oblique muscles that pull laterally, stabilising the rectus abdominis attachment points (Fig. 2.2.7f).

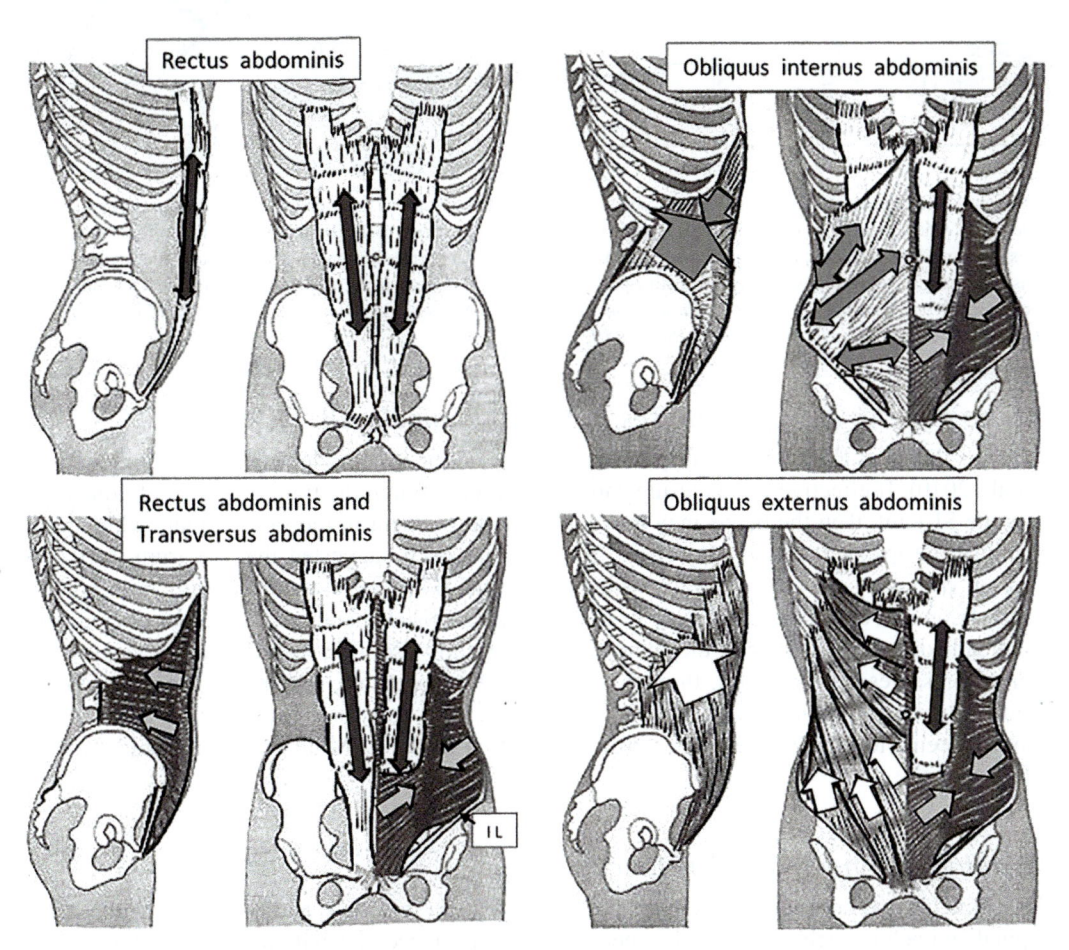

FIG. 2.2.7f The abdominal muscles create a complex network of force vectors that can flex the trunk with the inguinal ligament, which is an important ligamentous attachment. However, trunk flexion is not a desirable motion during gait, so abdominal muscle activity is important for tightening and compressing the abdominal contents to prevent anterior displacement of the CoM and helping to control anterior pelvic tilt. Through this mechanism, erect postural control is synchronised with the posterior muscles so that the fascia anterior and posterior to the spine can be utilised in energy transfer from antiphasic lower limb/pelvic rotations to upper limb/shoulder girdle rotations. *(Images modified from Kapandji, I.A., 1974. The Physiology of the Joints. Volume 3: The Trunk and the Vertebral Column. Churchill Livingstone, Edinburgh.)*

Rectus abdominis can function segmentally rather than just as one long strap. Above the umbilicus, the external abdominal oblique has aponeurotic linkage to the rectus abdominis, but below the umbilicus, all three lateral abdominal muscles pass their aponeurotic expansions anterior to rectus abdominis. Rectus abdominis is important to lumbopelvic motion in the sagittal plane (Saunders et al., 2005) and is possibly the functional antagonist to the erector spinae muscles (van der Hulst et al., 2010). Activity of the muscle heightens with increase in walking velocity and has a strong link to hip function (Neumann, 2010; van der Hulst et al., 2010) (Fig. 2.2.7g).

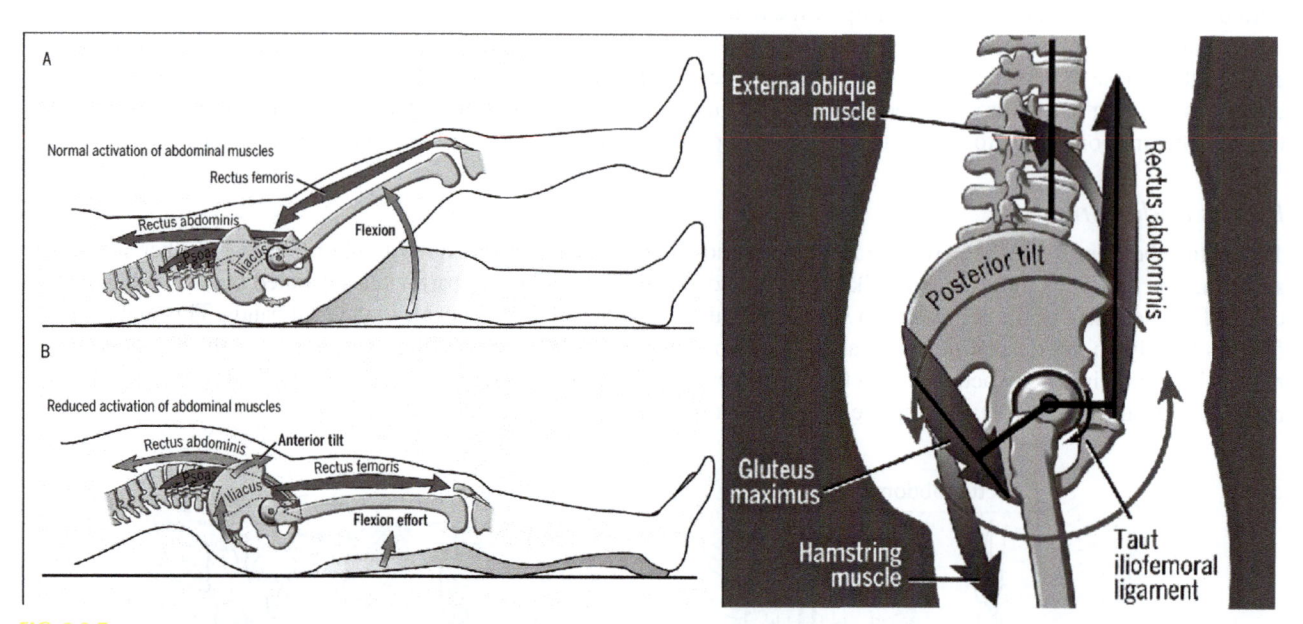

FIG. 2.2.7g The synergistic nature of rectus abdominis as an abdominal muscle to assist psoas function during hip flexion (left image) can be illustrated during a straight leg raise, where their action on stabilising the pelvis prevents anterior rotation of the ilium (A). With abdominal weakness (B), contraction of iliacus produces anterior tilt of the pelvis, increasing lumbar lordosis. This action can further be appreciated when considering how abdominal muscles interact with hamstring and gluteus maximus activity in resisting pelvic flexion (right image), particularly during initial contact and loading response hip flexion moments. *(Images from Neumann, D.A., 2010. Axial skeleton: muscle and joint interactions. In: Kinesiology of the Musculoskeletal System: Foundations for Rehabilitation, 2nd ed. Elsevier, pp. 379–422 (Section III, Chapter 10).)*

Function of the muscles and aponeuroses of the abdominal wall

The flexors such as iliopsoas exert an anteriorly rotating torque on the ilium, which the abdominal wall seems to help restrict (Hu et al., 2010). If effective force closure can be achieved through the transverse and oblique abdominal wall muscles across the SI joint, then the pelvis may move as a single segment in the sagittal plane allowing hip extensor activity to stabilise the ipsilateral ilium (Hu et al., 2010). It has been suggested that the trunk acts like a cylinder within this region providing an anterior wall, while the pelvic floor muscles, diaphragm, and epiglottis convert the thoracic and abdominal chambers filled with air into rigid pressurised chambers. These chambered cylinders could theoretically unload some of the stress on the spine and are certainly useful in respiration and venous return through changing internal pressures. The muscles, through their fascial connections, can increase tension circumferentially through the abdominal and thoracolumbar fascia, helping to create soft tissue stability around the lumbar spine. For locomotion, it seems to be the ability to control the position of the CoM and influence pelvic tilt that aids gait and spinal mechanics.

Hip muscle influence on lumbopelvic function

Many muscles of the hip and knee joints have proximal attachments in the pelvis and can therefore work synergistically to attenuate spinal and pelvic forces, particularly when the foot is in stance. Several powerful hip muscles cross both the hip and sacroiliac joints directly, and in the case of psoas, the lumbar articulations as well. By far the most important of these muscles are iliopsoas and gluteus maximus. However, it is worth considering the smaller influence of some other lower limb muscles first. The tensor fasciae latae and the hip abductors attach to the ilium and cross the hip into the lower limb. They control stance phase pelvic stability, primarily within the frontal plane. Weakness of the hip abductors causes an excessive varus tilt of the pelvis. This is known as a Trendelenburg gait, where the pelvis drops on the contralateral side during the single-limb stance phase, lowering the height for swing limb ground clearance. The large hip abductors and gluteus maximus are innervated by the 4th and 5th lumbar nerve roots. Thus, issues at this nerve root can cause a Trendelenburg gait of large pelvic varus tilt, which is associated with uncontrolled forward bending of the trunk at heel strike as hip flexion resistance is also lost.

The hip adductors similarly cross the hip to the femur distally and laterally from the ischium and pubis. Further to this, a number of deep hip rotators also cross from the pelvis to the femur, deep to gluteus maximus, and include piriformis, obturator internus, obturator externus, the superior and inferior gemelli, and the quadratus femoris. Piriformis attaches to the sacrum and to both the anterior surface of the sacrotuberous ligament and the SI joint capsule to run distally and attach to the trochanteric fossa of the femur. Piriformis contraction increases tension on the SI joint capsule and sacrotuberous ligament, helping to compress and stabilise the sacrum against the ilium (Fig. 2.2.7h). This muscle can occasionally cause symptoms (piriformis syndrome) through entrapment or compression of the sciatic nerve creating pelvic girdle and/or buttock pain/spasm, with referred symptoms along the sciatic nerve route into the back of the thigh that can reach the foot similar to sciatica from other causes.

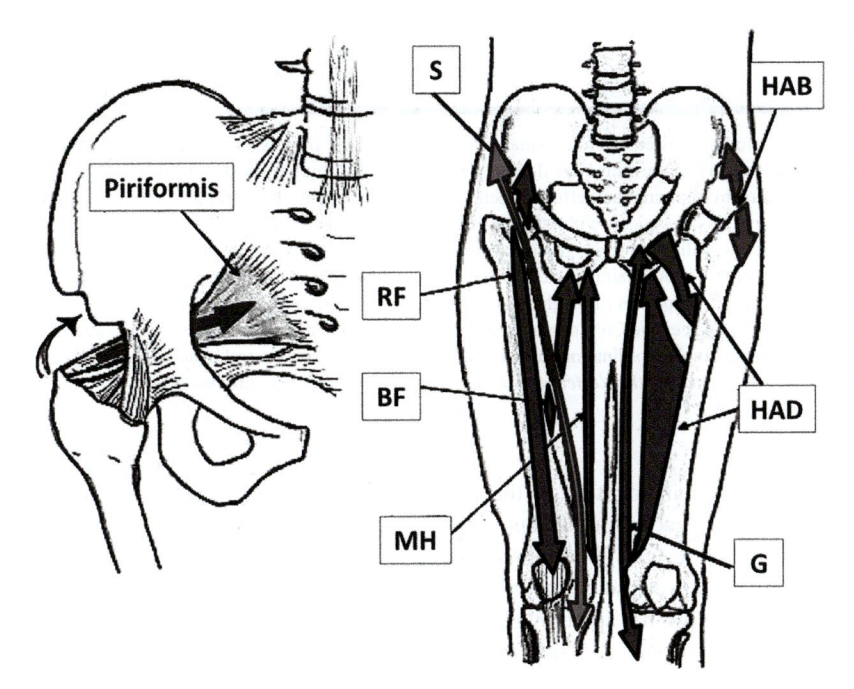

FIG. 2.2.7h Piriformis provides some hip-to-sacrum compression and weak hip external rotation and abduction. A significant number of other small and large muscles aid hip compression and motion particularly in the transverse and frontal planes such as the hip abductors (HAB) and adductors (HAD). Long muscles such as biceps femoris (BF), the medial hamstrings; semimembranosus and semitendinosus (MH), rectus femoris (RF), gracilis (G), and sartorius (S) link the pelvis to the lower leg across the knee. However, they do not provide direct stability within the lumbosacral or sacroiliac region. What they do indicate is that there is interdependency between the lower limb and pelvic posture. *(Permission www.healthystep.co.uk.)*

The quadriceps have a direct effect on the pelvis through rectus femoris which crosses the hip anteriorly and can assist anterior pelvic rotation at the pelvis as a part of hip flexion, increasing extension of the lumbosacral joint. The three hamstrings attach proximally to the ischial tuberosity. Only the short head of biceps femoris fails to cross both the hip and knee joints, but its long head attaches to the sacrotuberous ligament, tensioning it on contraction and pulling the sacrum against the ilium to increase sacroiliac stability. The hamstrings influence pelvic rotation over the hip during forward bending from erect stance, as well being part of a chain of muscles extending from spine to ankle that stabilise the posterior surface of the body (Fig. 2.2.7i). Hamstring tension can constrain pelvic rotation rhythms and perturbations in gait. Other less powerful muscles such as sartorius attach to the anterior superior iliac spine and link distally to the medial proximal tibia, assisting muscles like rectus femoris as an antagonist in stabilising the pelvis during the stance phase of gait and when in static stance.

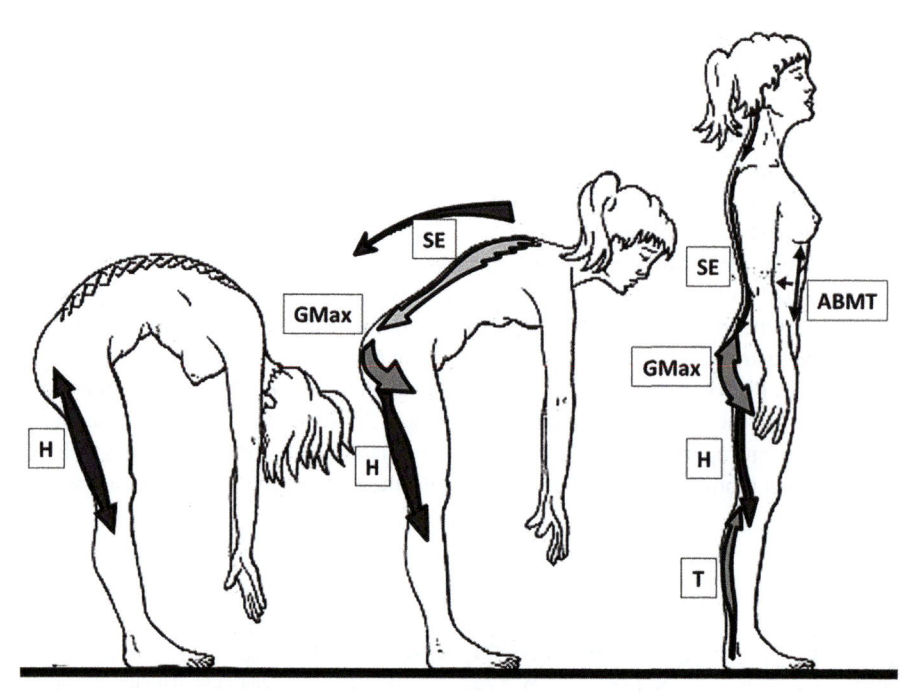

FIG. 2.2.7i The posterior or dorsal muscles in humans have a combined extension role to maintain an erect posture from head to ankle. Many muscles partake in this action. The hamstrings (H) even passively create tension that prevents knee flexion, most noticeable during trunk flexion exercises, potentially limiting the ability to 'touch toes'. Despite being knee flexors, by applying their effort at their proximal attachments, hamstrings help hip extension against resisted knee flexion during trunk extension. They achieve this in concert with activity of gluteus maximus (GMax) and the spine's extensor muscles (SE). In static stance, this assembly of posterior muscles are utilised to resist gravity against trunk flexion, but with a vertical posture, gravity provides vertical compression stability necessitating little muscle activity in maintaining static posture. Thus, hamstrings are little used during static stance and instead, passive muscle tone and posterior connective tissue elastic properties (through collagen crimp behaviour) provide most of the trunk flexion restraint. The abdominal muscle tone (ABMT) and the aponeurosis help keep the CoM over the base of support. Deliberate CoM instability necessary for locomotion dramatically complicates the situation, requiring trunk-to-limb coordination to maintain instantaneous CoM control. *(Image adapted from Kapandji, I.A., 1974. The Physiology of the Joints. Volume 3: The Trunk and the Vertebral Column. Churchill Livingstone, Edinburgh.)*

Iliopsoas

Iliopsoas is comprised of psoas major and iliacus that join together at the level of the inguinal ligament and attach together distally to the greater trochanter of the femur. Each muscle has independent innervations, making selective activation possible at each segment they control (Torry et al., 2006). Psoas major is one of the two psoas muscles, the other being psoas minor. It is an important hip flexor, yet psoas major seems to have little influence on hip flexion on its own (Gibbons, 2007). It attaches proximally to the lumbar spine by segmental attachments and to the SI joint as it crosses it, giving the muscle an important role in lumbopelvic stability as well as hip flexion through its union with iliacus as iliopsoas. However, it seems that psoas major's part in iliopsoas's hip flexion role is quite limited (Gibbons, 2007). Thus, it is Iliacus that is thought to provide a primary hip flexor role and psoas major a lumbar spine–sacroiliac joint stability role (Gibbons, 2007).

Psoas major is a unipennate muscle that has posterior and anterior segments that attach proximally to the lumbar spine by segmental attachments, and also to the SI joint. This gives the muscle an important part to play in lumbopelvic stability (see Fig. 2.2.7b) by creating axial compression on the lumbar spine, resisting shear (Gibbons, 2007). Its line of action is very close to the spine, giving it a limited role in the spinal motions of extension at L1–L3 and small amounts of flexion at L4 and L5 (replicating lordosis alignment). However, it has a further powerful effect on segmental stability (Gibbons, 2007). By also crossing the SI joint as well as the vertebral articulations, psoas major is able to provide a posterior rotation force on the pelvis (Gibbons, 2007). Psoas major is difficult to access through EMG lying as it does so deep in the pelvis, making its precise role difficult to assess. Primarily, it seems to be a provider of lumbar spine and SI joint stability during hip flexion, helping to maintain an erect posture in gait when it may also be important in eccentrically controlling trunk side flexion (Andersson et al., 1997; Gibbons, 2007). Psoas minor is a thin paired muscle lying anterior to psoas major. It attaches proximally to the T12 and L1 vertebrae to attach distally to the iliopubic eminence and pectineal line (pecten pubis) on the superior pubic ramus, to assist psoas major in resisting spinal extension.

Gluteus maximus

Gluteus maximus is the largest muscle within the human body and is powerful in influencing lumbopelvic, hip, and knee function directly and also has a profound influence on gait and locomotion generally. It has large, long fibrous attachments to the iliac crest, gluteal fascia, thoracolumbar fascia, sacrum, coccyx, and sacrotuberous ligament. Distally, it runs to attach inferolaterally to the posterior aspect of the iliotibial tract of the fascia lata and the gluteal tuberosity of the femur. It is its ability to influence thoracolumbar fascia tension along with latissimus dorsi that gives it a significant lumbopelvic stability function as well as its proximal attachments crossing the SI joint that are helping to brace it. The *gluteal raphe* is a fascial thickening that lies over the posterior medial aspect of gluteus maximus. It acts as a strong fascial junction, blending gluteus maximus to multifidus and the erector spinae aponeurosis. The effect of this fascial tissue is to join the three powerful extensors at the posterior aspect of the lumbopelvic region into an 'extensor assemblage' of the lumbar spine, pelvis, and hip.

Contraction of gluteus maximus tensions the sacrotuberous ligament, minimising and stabilising motion between the articular surfaces of the SI joint. The attachment to the iliotibial tract (a thick band of fascia within the thigh) involves the majority of the gluteus maximus's fibres, with the remaining fibres running to the gluteal tuberosity. This means that the most powerful muscle in the human body has extensive fascial attachments that anatomically reflect its complex mechanical function across the lumbopelvic region and much of the lower limb. The iliotibial tract spans both the hip and the knee joints. Thus, gluteus maximus activity increases tension through the fascia that consequently leads to increased hip and knee joint stability. The effects of the muscle depend on the multiple joint orientations that its tensional–compressive forces bridge, essentially from the upper limbs through linkages to the contralateral latissimus dorsi and down to the knee joint.

Gluteus maximus has a synergistic relationship with iliopsoas in controlled hip flexion–extension and pelvic stability. It crosses the SI and hip joints, and via the iliotibial tract bridges and influences the knee. It has three segmental divisions. These are the superficial sacral fibres, deep sacral fibres, and deep ilium fibres, all of which demonstrate different attachment points but share a common nerve supply (Gibbons, 2007). The distal attachments of the superficial fibres join the posterior superior aspect of the iliotibial tract and also the gluteal tuberosity of the femur. The deep sacral fibres cross the SI joint to attach to the posterior pelvic rim lateral to the posterior iliac spine. However, the sacral segment is only present in around two-thirds of individuals (Gibbons, 2007). The deep ilium fibres attach to the gluteal tuberosity of the femur which with the superficial sacral fibres that also cross the hip, suggest that these segments influence hip extension–flexion as well as pelvic stability (Gibbons, 2007) (Fig. 2.2.7j). The deep sacral part of the muscle may have a role in vertical loading of the spine in a superior-to-inferior direction and play an active part in counternutation (Gibbons, 2007). Gluteus maximus also maintains CoM velocity in stance phase, working in a synergistic relationship with triceps surae as discussed previously in Chapter 1, Section 1.3.3.

2.2.8 Adaptation and pathology in the lumbar spine and pelvis

Many common symptoms are related to suboptimal lumbopelvic stability in a diverse range of patients. This is a major health and socioeconomic problem (Vleeming et al., 2008; Arumugam et al., 2012; Buchbinder et al., 2018), albeit the mechanisms are still not fully understood (Cohen et al., 2013; Buchbinder et al., 2018; Poilliot et al., 2019). Low back pain (LBP) has been defined as pain below the 12th rib and above the gluteal fold, and pelvic girdle pain has been defined as symptoms between the posterior iliac crest and the gluteal fold around the SI joint (Vleeming et al., 2008) (Fig. 2.2.8). Pain may radiate into the thigh or the pubic symphysis and usually limits the ability to stand, walk, and/or sit. Aside from the many soft tissue structures that can fail around the spine and pelvis, the facet joints being constructed of hyaline cartilage and subjected to high compressive forces are vulnerable to osteoarthritis, especially in the presence of intervertebral disc degeneration which can alter axial loading forces.

Chronic LBP can cause difficulty during locomotion with slower-than-average walking than in asymptomatic age-matched controls, hinting at poor motor control (Lamoth et al., 2002, 2006a; van der Hulst et al., 2010; Müller et al., 2015). Altered trunk–pelvic biomechanics is associated with the development of nonspecific chronic LBP, both as a response to and cause of LBP (Hodges et al., 2003; Müller et al., 2015). This is highlighted by the increased prevalence of chronic LBP in transtibial amputees who are known to walk kinematically differently to able-bodied individuals (Yoder et al., 2015). Normal walking involves cyclical coupling motions of the limbs, pelvis, trunk, and head that creates stable yet adaptable flexibility that are responsive to changes in locomotion velocity; adjusting to lower limb kinematic changes caused by altering walking speed (Hanlon and Anderson, 2006). Populations who are able-bodied but suffer with low

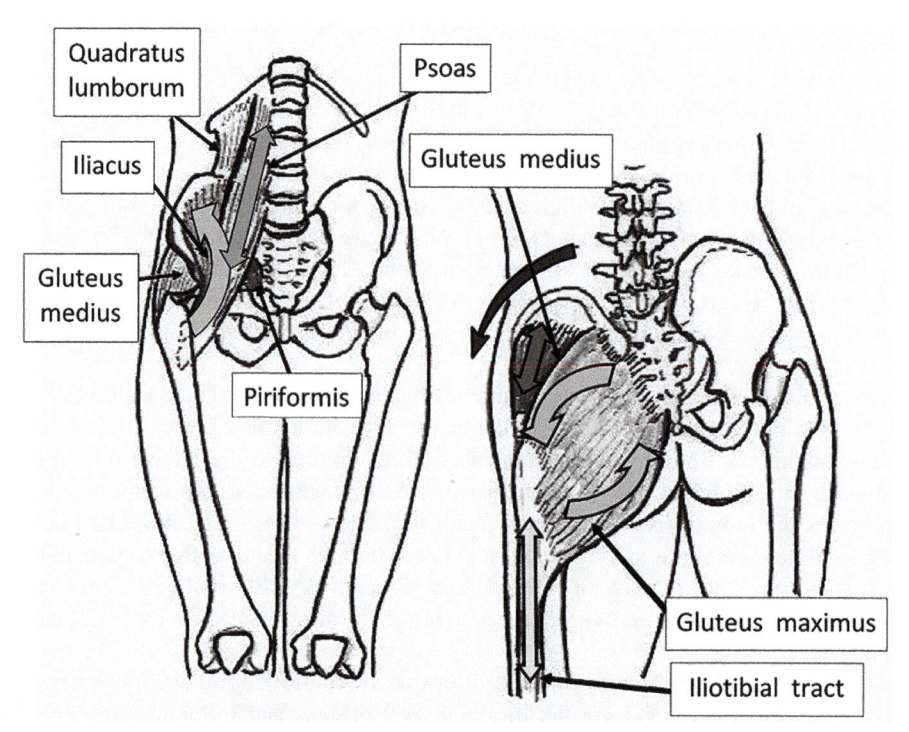

FIG. 2.2.7j Although lumbopelvic relationships to the lower limb are maintained through many muscles and ligaments, two muscles dominate. Iliopsoas (left image), through psoas binding the lumbar spine to the femur and iliacus binding the ilium to the femur, is able to create a mechanism of hip-to-lumbar spine extension resistance against the spinal and hip extensors antagonistically. Yet, in agonistic fashion with the extensors, this combined muscle works with the extensors to create compressive stability across the region via its shared distal tendon and separate proximal attachments. Iliacus primarily controls hip flexion, working antagonistically with gluteus maximus. Gluteus maximus (right image), through its attachments with the sacrum, coccyx, ilium, thoracolumbar fascia, femur, and iliotibial tract, is a powerful extensor of the hip and lumbopelvic region. It compresses the sacrum towards the hip but also sets up linkages to the lower leg across the knee. This creates a relationship between trunk, pelvic, and lower limb biomechanics. Here, it is illustrated creating hip abduction with gluteus medius, tightening the iliotibial tract and putting extension moments between the back and hip. Thus, gluteus maximus' pelvic stabilising role is also part of the hip abduction moment control necessary for single-limb support of an erect plantigrade biped. *(Permission www.healthystep.co.uk.)*

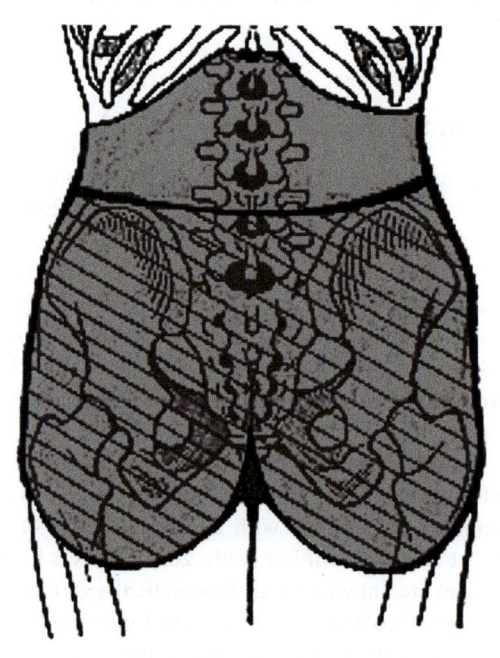

FIG. 2.2.8 Low back pain is considered to be defined by symptoms between the 12th thoracic vertebra (T12) to the gluteal fold (all grey areas). Pelvic girdle pain (hashed grey area) is considered as lying between the iliac crest and the gluteal fold. There is some considerable crossover between the two terms which the clinician needs to be aware of when patient's report 'low back pain'. Much anatomy is contained within, necessitating a more precise diagnosis or origin of symptoms by which to plan biomechanical assessment and intervention. *(Permission www.healthystep.co.uk.)*

chronic LBP tend to demonstrate less variability in their trunk–pelvic axial motions across strides, at the expense of increased variability within frontal plane motion (Seay et al., 2011). Healthy walkers are capable of easily compensating for internal and external perturbations during gait through kinematic variations (Lamoth et al., 2006a; Müller et al., 2015). Reduction in antiphase segment coupling and an increase of in-phase transverse motion between the trunk and pelvis are seen in both walking (Lamoth et al., 2002; Bruijn et al., 2008; Wu et al., 2014) and running with chronic LBP (Seay et al., 2011). It is possible that loss of walking speed seen in chronic LBP results from a loss in the ability to adapt trunk–pelvis coordination to the normal changes seen with increased velocity. Such normal changes alter lower back and pelvic relationships to antiphase transverse plane motion to bring the pelvis back to being in-phase with the femur (Lamoth et al., 2006a; Müller et al., 2015). Controls have been reported as having greater trunk–pelvis antiphase motion than do chronic LBP subjects (Lamoth et al., 2002, 2006a, b).

With increase in gait changes, normal subjects demonstrate alterations in thoracic-pelvic and lumbopelvic coordination within the transverse plane. However, in chronic LBP, this variability of motion is lost, albeit frontal plane intersegmental coordination becomes more variable and less tightly coupled (Lamoth et al., 2006a). Frontal plane variability increases, suggesting that loss of antiphase rotations in the transverse plane requires increased frontal plane variability and loss of the normal global patterns of gait motion. Walking on uneven surfaces changes the situation again with a decrease in transverse plane trunk–pelvic rotations being replaced with increased trunk inclination at initial contact within chronic LBP subjects (Müller et al., 2015).

In healthy subjects, increased walking speed is associated with an increased knee flexion angle at initial contact of gait, with inversely more extension occurring with reduced gait speed. In chronic LBP, the knee joint angle at initial contact is significantly extended and does not relate to walking speed (Müller et al., 2015). A trunk-flexed gait posture that occurs with walking on uneven terrain requires compensation within the lower limb with increased knee flexion (Saha et al., 2008). It seems that chronic LBP patients change to a more trunk-flexed posture and maintain an extended knee at initial foot contact in walking and running (Müller et al., 2015). No difference in chronic LBP subjects has been noted in ankle joint kinematics in either running or walking compared to healthy controls (Müller et al., 2015). In healthy subjects, a more extended knee at initial contact, maintaining the same step length, will result in an increase in vertical forces (Podraza and White, 2010). Despite a more extended knee in chronic LBP subjects, the early vertical contact peak in gait was decreased compared to controls. To achieve this lower initial loading vertical force peak, chronic LBP sufferers need to slow down their gait to compensate for the extended knee impact force and use an earlier onset and higher muscle activation in gait for energy buffering (Müller et al., 2015).

The lumbar erector spinae muscles' activation patterns are also disturbed in LBP, with loss of adaptability in activation at higher gait velocities. Additional erector spinae bursts of activity are noted in chronic LBP subjects, possibly in an attempt to stabilise the spine through increased stiffness (Lamoth et al., 2006a; van der Hulst et al., 2010). Changes occurring in the trunk through superficial abdominal and lumbar muscle activity in chronic LBP sufferers may represent a 'guarding' strategy, which may be provoked at higher walking velocities (van der Hulst et al., 2010). Heightened activity of the trunk muscles increases with gait velocity, but coactivation of the erector spinae muscles (Lamoth et al., 2006b; van der Hulst et al., 2010) and rectus abdominis are notably increased in subjects with chronic LBP compared to asymptomatic control subjects. However, this has not been reported for the external abdominal oblique muscles (van der Hulst et al., 2010). The erector spinae muscles, rectus abdominis, and the external abdominal obliques all increase their activity with increasing gait velocity in symptomatic and asymptomatic subjects. In the case of the erector spinae muscles and rectus abdominis, their increased activation in chronic LBP sufferers may reflect guarding activity to 'control' or limit increased spinal-pelvic antiphase motion (van der Hulst et al., 2010). Studies have also linked increased erector spinae activation to stress and anxiety as well as chronic LBP (van der Hulst et al., 2010).

The muscles of the hip arising from the pelvis that link to the lower limbs will be affected by LBP. In heathy populations, gluteus maximus activity begins earlier in swing phase with decreasing gait speed, while hamstring activity decreases (Müller et al., 2015). Patients with chronic LBP demonstrate gluteus maximus activity during early swing phase with increased hamstring activation at the end of swing phase (Vogt et al., 2003). In healthy subjects, hamstring activity at the end of swing phase prior to initial contact decreases with decreasing gait speed, and the knees become significantly more extended (Hanlon and Anderson, 2006). Chronic LBP subjects have been shown to activate biceps femoris and gluteus maximus earlier in the swing phase and with greater activity than in healthy subjects. This displays a different hamstring adaptation compared to normal slow walking (Vogt et al., 2003).

The SI joint is commonly accepted as a source of many cases of low back and posterior pelvic pain that are thought to be related to pelvic ligament dysfunction (Fortin et al., 2003; Cohen, 2005; Mens et al., 2006; Pel et al., 2008; Poilliot et al., 2019). Arthritis at the SI joints as well as arthritis elsewhere within the locomotive system can significantly affect pelvic motion (Görke et al., 2010). Sitting (so-called *slouching movements*), which so much of the modern human world's population now partake in during both working and leisure, also changes loading stresses within the pelvic articulations. Sitting in chairs increases stress in ligaments such as the iliolumbar, interosseous, and posterior sacroiliac which may in themselves serve as generators of LBP due to their extensive nerve supply (Hammer et al., 2013). Using muscular strengthening techniques or a pelvic compression belt may induce increased force closure on the pelvis through the transverse and oblique abdominal muscles pressing the ilia against the sacrum to relieve LBP symptoms (Hu et al., 2010).

2.2.9 Section summary

The lumbopelvic region represents a complex hub linking trunk-to-lower limb transition of loading body forces and locomotive stresses and motions. Stability through tension-induced compression is a key element to lumbopelvic stability, clearly supporting biotensegrity principles. In-phase and antiphase motions between the trunk and pelvis and the pelvis and lower limb change with gait speed, and these events can be influenced by pathology and symptoms within the region. Changes in gait speed with chronic LBP may be an attempt to decouple pelvic motion from the lower limb, returning pelvic motion to an in-phase relationship with the spine. This LBP-type decoupling of the lower limb-to-pelvis is different to normal slower speed gait decoupling and may lead to a new state of dysfunction during locomotion.

The origins of chronic LBP and pelvic girdle pain remain a source of much debate. Certainly, changes in the daily round of modern life due to increased hours of sitting and less active manual motion may be placing the spine into a new environmental mismatch vulnerability. Assessment of gait may give more data on the results of the pathology rather than uncovering the pathomechanics. In making therapeutic changes to improve gait, a 'healthier' slow gait may be the initial target, while a return to normal gait's adaptable variability may indicate success in treatment programmes as a whole, which could act as targets for therapy for both the patient and clinicians.

2.3 Functional unit of the hip
2.3.1 Introduction

The hip joint allows for large motions between the body's 'passenger segment' and the locomotive lower limb. Although the pelvis, the hip, and the thigh are a continuum of the anatomy both structurally and through principles of biotensegrity, the hip joints allow for large asymmetries of motion between the lower limbs across the pelvis in order to allow the limbs to advance one in front of the other. Alternate hip flexion and extension dictates the lengths of step and stride. Such motions are directly influenced by the sagittal plane mobility of the hip. The hip is a ball-and-socket joint that permits large ranges and combinations of motion, allowing the body and the lower limb to achieve multiple orientations and thereby providing human locomotion with an 'all-terrain' capacity. Healthy hip function permits us to walk, run, jump, climb, and even swim, and our ability to perform any of these locomotor tasks is severely restricted by the absence of multiplanar mobility at the hip. Therefore, the existence of a strong and freely mobile hip joint profoundly affects gait energetics.

The muscles of the hip arise proximally within the pelvis and assist in maintaining both pelvic and hip stability and mobility. However, whereas most pelvic stability comes from significant form closure and strong passive ligamentous tensions, hip stability is heavily reliant on the ability of active muscle–tendon units to achieve force closure activity during motion and stance because of its large freedom of motion. Not surprisingly therefore, there are seventeen muscles that bridge the hip. Gluteus maximus, iliopsoas, piriformis, the three hamstrings, the long head of rectus femoris, gracilis, sartorius, and tensor fasciae latae all cross the hip and the sacroiliac joints and/or the hip and knee joints. The anterior deep muscle fibres of gluteus maximus have extensive distal fascial connections to a thickening of the thigh's deep fascia called the iliotibial tract or band, as well as their attachments to the gluteal tuberosity of the femur. The iliotibial tract also has proximal attachments to the tensor fasciae latae, linking gluteus maximus posteriorly to the tensor fasciae latae anteriorly around the hip's functional unit. In addition, the iliotibial tract also has a linked distal attachment to *Gerdy's tubercle* on the lateral anterior tibia. Therefore, gluteus maximus can influence motion across three major joints: the sacroiliac, hip, and knee, and possibly also motion at the pubic symphysis. Passive stability of the hip comes from a number of important hip-bridging ligaments.

Yet, despite this need for active stability during mobility, the hip joint itself is one of the most congruent joints in the body with an oval femoral head enclosed within a concave oval facet constructed from the union of the ilium, ischium, and pubic bones (along with the acetabular labrum—see Section 2.3.4) known as the acetabulum. The acetabula form the sockets for the ball-like femoral heads to fit into. The hip is thus a joint with excellent force and form closure that allows large freedom of stable motion. The only other lower limb joint remotely similar in form to the hip is the talonavicular joint of the foot which has been referred to as the *coxa pedis*, meaning 'hip of the foot' (Pisani, 1994, 2016). However, this joint is far less form closed than the hip and provides far less mobility, although like the hip, it also offers multiplanar motion.

2.3.2 The kinematic role of the hip

The role of the hip during locomotion is divided into two primary functions. The swing phase role is to allow the lower limb to be advanced as required, passing under the HAT (head, arms, and trunk) segment to a position anterior to the HAT. The hip's stance phase role is to enable the HAT to move over the lower limb from a position posterior to the hip to a position anterior to the hip, while the lower limb is supporting the body weight. Roles change from one hip to the other during gait, although in running during float phase, both limbs are moving below the body at the hip without any ground contact of the lower limbs. Primary hip motions occur in the sagittal plane and involve cycles of flexion and extension moments. Thus, the hip is the fulcrum point that allows the HAT or lower limb to advance around the other segment with the pelvis acting as a significant myofascial hub for stability and motion between the segments.

The hip rotates around 40° within the sagittal plane throughout gait with a maximum flexion angle of between 30° and 35° occurring during late swing phase at ∼85% of the whole gait cycle. Maximum extension is around 10° and is reached before toe-off during preswing (Krebs et al., 1998). The GRF moves progressively from a position anterior to the hip towards a maximal posterior position by the end of stance. The GRF position drives the flexion and extension torques throughout gait during stance phase. The large extension and flexion-generated torques enable an unusual range of extension motion to develop at the hip compared to other animals, which is necessary for efficient human plantigrade bipedalism. This is required because of the lack of a trunk mass-stabilising, counterbalancing tail. Achieving hip joint extension gives humans a large metabolic advantage during relaxed static stance and during midstance of walking, for the erect stance that results allows the line of gravity to fall posterior to the axis of rotation of the femoral head. This causes the hip's capsular ligaments to wind up tautly in erect stance so that relatively little muscle effort is required to maintain relaxed standing for short periods (Neumann, 2010). In the late stance phase of gait, further hip extension brings the CoM to lie anterior to the hip. As the CoM can now fall forward under gravity, it extends the limb without requiring further muscle contraction, restrained only by hip ligament tightening and ankle plantarflexor activity below.

The hip also permits internal and external motion in the transverse plane, adduction and abduction within the frontal plane (i.e. taking up a more varus or valgus position, respectively), and circular circumduction motions involving all three planes. Hip transverse plane internal rotation during gait has been reported to range between around 3° and 12°. It is at its largest angles during loading and early midstance (throughout 13% of gait cycle) and at its lowest towards the end of stance phase at ∼56% of the gait cycle (Uemura et al., 2018). Small external rotations occur during the swing phase (Krebs et al., 1998). Hip adduction in the frontal plane starts in early stance, reaching a maximum of at around 40% of the gait cycle, whereas abduction of 5–7° occurs in early swing (Krebs et al., 1998). This loading adduction moment is caused by the hip adductors helping to stabilise the limb angle under the body to achieve an effective base of support during single-limb support. During stance, GRFs and the HAT's CoM drive further frontal plane hip adduction torques, which are counteracted by activity within the hip abductor muscles. Thus, little hip adductor activity is usually required during stance, but more is required during late swing and initial contact.

Hip mobility gives the lower limb or HAT the opportunity to move in all planes (as required) to maintain postural stability and increase limb placement possibilities, especially in response to perturbations in gait and to extreme terrain adaptation. The hip provides a range of motion second only in magnitude to the shoulder in humans, despite being a weightbearing joint that must provide postural stability.

2.3.3 The hip as a biotensegrity structure

The hip is a ball-and-socket joint surrounded by powerful, balanced muscles and ligaments that provide excellent stability while permitting a large range of motion. The mechanical aims of the hip and pelvis are quite different. The pelvis requires strong stability through limited motion and shape changes. The hip requires stability during demands for high mobility and large shape changes, with the degree of mobility presenting a stability challenge. Because the tissues found at the hip present quite different mechanical properties, deformity occurs across the region in a nonuniform manner which may cause a breach in tolerated stresses unless mechanisms can moderate the stresses more equally. The key element in protecting the tissues at the hip is to provide structural integrity through rigidity, fine-tuned to allow necessary shape changes throughout motion. Biomechanical systems must work to distribute any external or internal force, while maintaining and safeguarding tensional homeostasis, allowing tissues to maintain their mechanical function. Biotensegrity offers the best model to explain how this is achieved in structures such as the hip, where the tissues are able to adapt their behaviour during position changes or movement.

Tensegrity structures are considered to present a solution to combining hard and soft components, and in robotics, they have been successfully used to create human-like lower limbs and joints (Jung et al., 2019). Tensegrity techniques have also been used in creating other types of automaton-like 'biological' locomotions such as in robotic swimming fish

(Chen and Jiang, 2019), which demonstrates how flexible and successful tensegrity principles are in explaining biological anatomy and locomotion.

At the hip, biotensegrity compression elements arise from the osseous pelvis and the femur, with passive tensile elements arising from the dense strong ligaments of the hip that envelop the femoral head and most of the neck. The active tensile elements are the immense and powerful gluteus maximus, iliopsoas, the hamstrings, the hip abductors and adductors, sartorius, and the long head of rectus femoris. These are coated and interwoven with the more superficial tensile elements of the deep fascia and the deep intermuscular fascial septa of the thigh known as the fascia lata along with its dense lateral reinforcement of the iliotibial tract. The hip ligaments change their passive tensile forces throughout motion, thereby altering peak compression force directions across the hip. Muscular function creates further compressive stability assisted by the fascia that generates the primary compression stability across the hip (Fig. 2.3.3). It is biotensegrity structures that permit the HAT to travel over the hip with mobility under compressive joint stability.

FIG. 2.3.3 A schematic of the primary force vectors around the right hip viewed from an oblique anterolateral position. As a biotensegrity structure, the hip offers excellent mobility and stability, utilising form closure from its articular structures including a fibrocartilage ring (acetabular labrum, shown in black). Primary passive restraints are the iliofemoral, ischiofemoral (not shown for clarity), and pubofemoral ligaments. In extension, these ligaments wind around the femoral neck, acting as tensioning restraints (especially the inferior part of the iliofemoral ligament) that can elastically recoil the hip towards flexion after heel lift. Hip muscles are numerous and are here limited to the primary muscles for clarity. Gluteus maximus and tensor fasciae latae acting through the iliotibial tract add lower limb continuity to hip stability as a biotensegrity structure. The hip adductors provide strong medial tension restraint, helping to position the trunk over the base of support ready for single-limb support. Gluteus medius and minimus stabilise the pelvis superiorly over the femur to limit pelvic tilt during single-limb support. Iliacus provides most active hip flexion, but psoas through the iliopsoas tendon provides biotensegrity structural linkage between the lumbar spine and the lower limb. The hamstrings and rectus femoris (not shown) likewise do the same between the pelvis and the lower leg. *(Permission www.healthystep.co.uk.)*

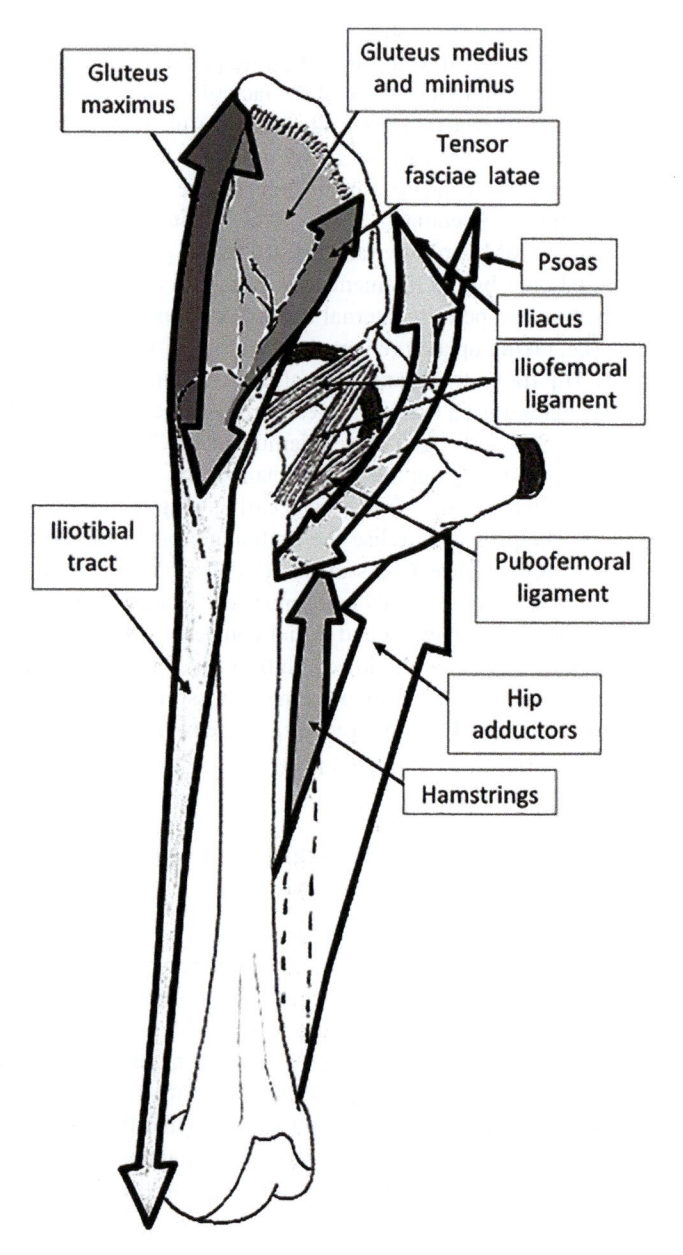

2.3.4 Osseous topography of the hip

The human hip is a multidirectional ball-and-socket joint composed of the femur and the acetabulum. The bony anatomy of the hip begins where the pelvis ends, with the three fused coxal bones that make up the innominate bone forming the lunate/horseshoe-like and incomplete cup-shaped acetabulum (socket) of the hip joint. The incomplete part, situated anteroinferiorly, is known as the *acetabular notch*. The ischium and ilium each make up around 40% of the acetabulum and around 20% by the pubis. Ossification of the acetabulum completes relatively late in development. Eventually, a 'Y'-shaped growth plate is formed between the three bones by around age 10 years. It starts fusing together from the age of 15 years and is completed by age ~23 years (Draves, 1986, p. 42; Byrne et al., 2010). The acetabulum forms the articular floor of the hip socket of this synovial joint, with its lunate-shaped surface covered with articular (hyaline) cartilage. A central area known as the *central inferior acetabular fossa* is filled with a synovial-covered fat pad (pulvinar) and contains the attachment point of a ligament linking the acetabulum to the femoral head, known as the *ligamentum teres* (Byrne et al., 2010).

The ring of bone is incomplete inferiorly at the *acetabular notch* and is widest superiorly where it transmits the most body weight into the superior aspect of the femoral head. The notch, positioned inferiorly, serves as a space for vessels and nerves to pass into the acetabular fossa, femoral head, and ligamentum teres that run between the acetabulum and the femoral head. The notch is bridged by the *transverse acetabular ligament* and the *acetabular labrum*. The acetabular labrum consists of a fibrocartilaginous ring that attaches to the rim of the acetabulum and transverse ligament to deepen the acetabulum and increase contact stability, thereby decreasing the risk of hip dislocation. The femoral head cannot be withdrawn from a hip joint without rupturing the acetabulum labrum (Fig. 2.3.4a). Such a tight fit serves to restrain synovial fluid within the joint, creating negative hip joint pressures (Ferguson et al., 2003).

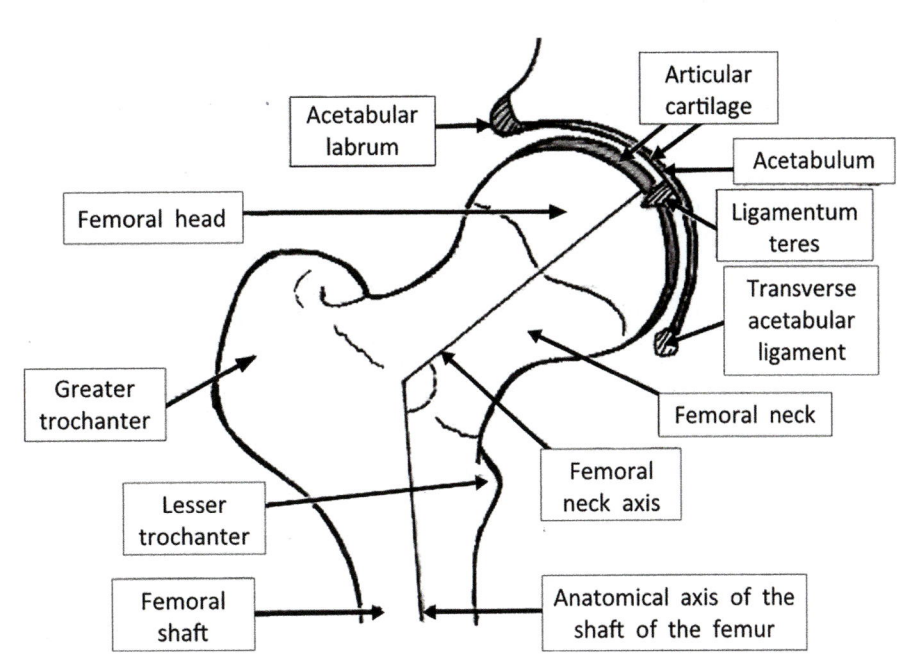

FIG. 2.3.4a A schematic of the primary osseous and articular features of the hip in cross section, with the primary extra-articular ligaments not shown. The femoral neck–shaft angle is illustrated, constructed from the angle between the bisection of the head and shaft. The acetabular labrum forms an enclosing fibrocartilage ring around the articular cartilage of the acetabulum, with the transverse acetabular ligament bridging the notch below the joint under which blood vessels and nerves can pass into the hip. The ligamentum teres connects the femoral head to the acetabulum with a considerable amount of laxity within it to prevent interference in motion. The greater and lesser trochanter form important muscular attachment points that influence the lever arms of the hip muscles via altered femoral positioning during gait. *(Permission www.healthystep.co.uk.)*

The femur is the largest long bone of the human body and therefore represents the longest rigid beam of the lower limb lever system. Strains across the femur are the consequence of both internal muscular and other soft tissue forces and the external GRFs which together cause the femur to accrue more bone mass under physical loading during development. This process gives the bone its distinct trabecular microarchitecture in addition to stimulating the thickness and curvature of its diaphysis (Edwards et al., 2016). The femur has a number of very significant features and prominent landmarks that reflect these developments and the widely different strains generated across it during gait (Edwards et al., 2016). Femoral bending moments occur around an anteromedial axis, with the highest principal tensile strains occurring along the lateral and anterior surfaces and the highest principal compressive strains along the medial and posterior surfaces (Edwards et al., 2016). The curvatures of the femoral shaft reflect such loading strains (anteriorly bowed). These strains change with loading of the hip in gait as muscles activate around the pelvis and femur. They are at their highest during initial limb loading when the hip is most flexed (Figs 2.3.4b and 2.3.4c).

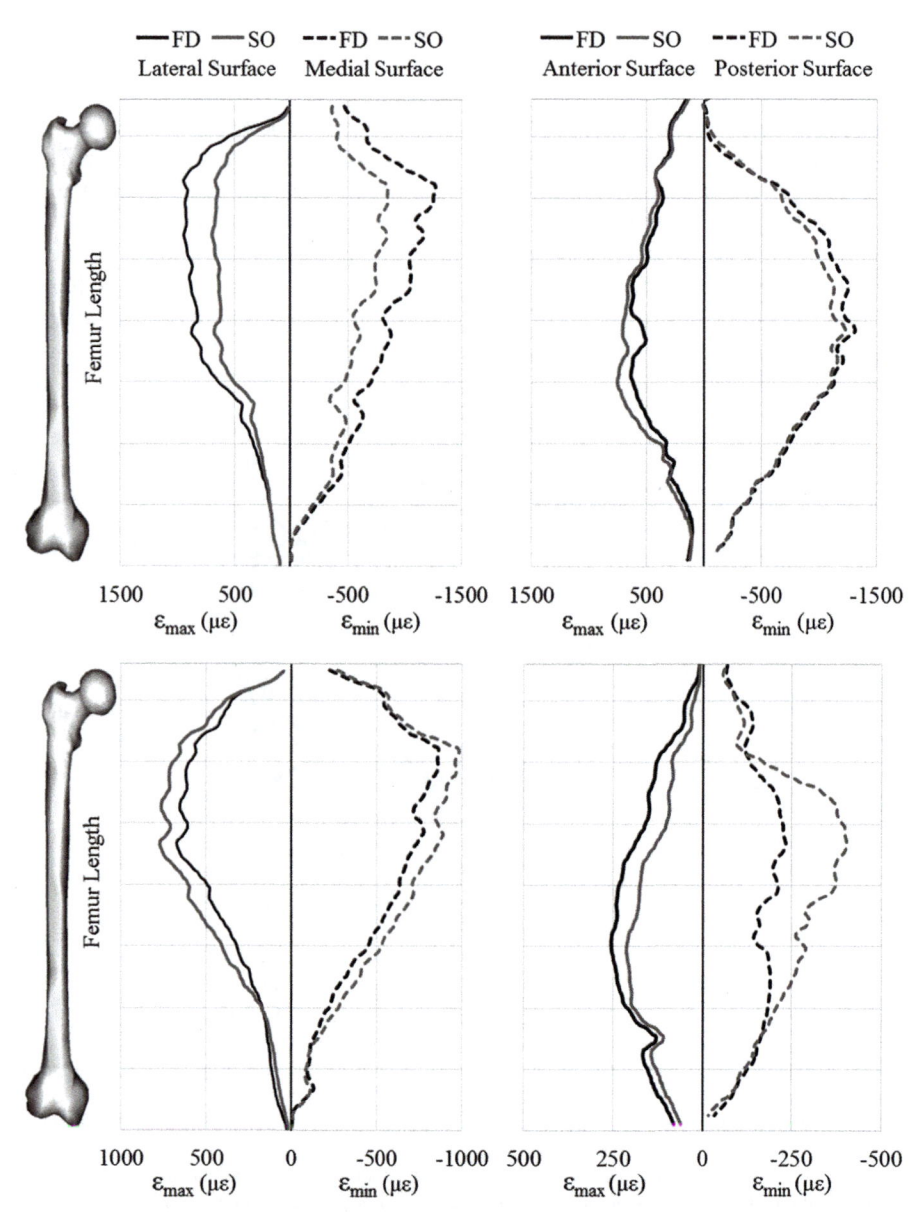

FIG. 2.3.4b Human femurs are put under considerable tensile and compressive stresses during loading response as body weight is transferred across the femur into the lower leg. Using forward dynamics (FD) and static optimisation (SO) when modelling femoral stresses produces similar results that correlate with in vivo strain gauge studies of femoral stresses. However, large differences in strain magnitudes occur in the first half of stance phase between these techniques, where forward dynamics better reflects the gluteal muscle forces put on the femur during initial contact, loading response, and early midstance. Hip joint contact and muscle forces produce femoral bending moments about an anteromedial axis. The largest tensile stresses occur along the lateral surface of the femoral shaft followed by the anterior surface, and the highest compressive forces occur on the medial surface followed by the posterior surface. *(Image from Edwards, W.B., Miller, R.H., Derrick, T.R., 2016. Femoral strain during walking predicted with muscle forces from static and dynamic optimization. J. Biomech. 49(7), 1206–1213.)*

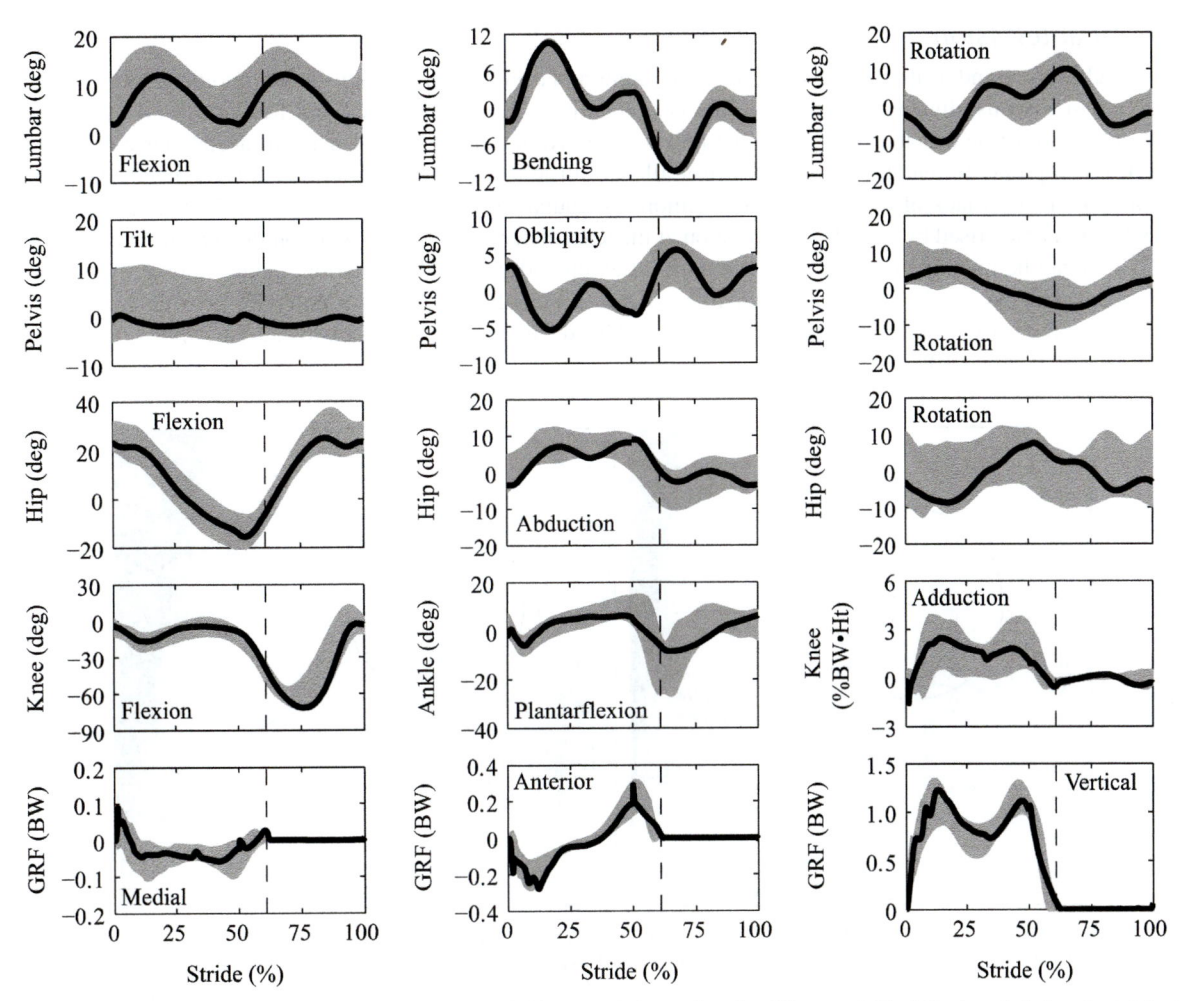

FIG. 2.3.4c Data from a study undertaken on fourteen healthy adult humans during preferred walking speed. The full gait cycle is from heel strike to heel strike (100%) indicating lumbar, pelvic, and lower limb joint angles, and GRF during walking. Shaded areas represent ± 2 standard deviations. *(Image from Edwards, W.B., Miller, R.H., Derrick, T.R., 2016. Femoral strain during walking predicted with muscle forces from static and dynamic optimization. J. Biomech. 49(7), 1206–1213.)*

Proximally, the femur consists of a spherical head and a protruding neck. The head of the femur is 60%–70% covered in articular cartilage except at the *central fovea capitis* which is the attachment site of the ligament at the head of the femur, termed the ligamentum teres femoris (Byrne et al., 2010). The cartilage is thickest where it bears most load, i.e. anteriorly, laterally, and superiorly to the centre (Draves, 1986, p. 57). The femoral head is attached to the shaft by the neck which is variable in length depending on the size of the individual, and narrowest midway down its length (Byrne et al., 2010). Mammalian femoral heads can be classified along a spectrum wherein they are characterised as having either a more restricted motion approximating to a hip stereotype referred to as coxa recta (nonspherical), or approximating that of a coxa rotunda, a hip stereotype demonstrating a round femoral head. Humans reveal a more coxa recta-like femoral head shape than most primates despite having evolved from an ape rotunda head, likely as a result of decreased climbing and increased terrestrial weightbearing activities including running (Hogervorst et al., 2011). For the reasons behind this see Chapter 6, Section 6.2.5, of the companion text 'Clinical Biomechanics in Human Locomotion: Origins and Principles'. The human femoral head and neck are now formed for high vertical loading and running through biomechanically induced evolutionary changes. High vertical loading requires greater hip stability rather than just large freedoms of motion, in a joint that yet still requires mobility including for providing for unusual hip extension ranges.

Acetabular anteversion

The bone architecture and anatomical geometry of the femur and acetabulum are significant to joint stability, yet the hip requires certain osseous alignments to achieve the best stability. These include femoral head–neck offset angles, *acetabulum anteversion* (the acetabular angle of the pelvis to the frontal plane), and sufficient acetabular coverage of the femoral head. Appropriate femoral head–neck offset allows full joint range of motion without impingement of the acetabular labrum. Thus, a lack of offset may cause femoroacetabular impingement (Torry et al., 2006). A *cam deformity* (Fig. 2.3.4d) is characterised by extra bone formation at the anterolateral head–neck junction that results in a nonspherical deformity that can be forced into the acetabulum during flexion and internal hip rotation, increasing the risk of hip joint degeneration (Agricola et al., 2014). Abnormalities of the acetabulum such as osteophytic growths can also cause impingement of the neck, while during growth, the femoral head can move on the neck to cause a *slipped capital femoral epiphysis* (Byrne et al., 2010).

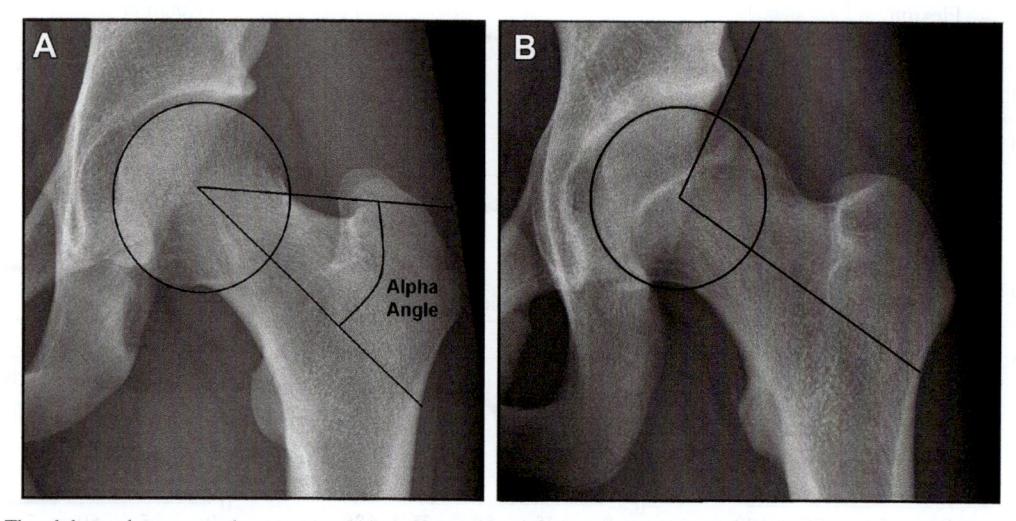

FIG. 2.3.4d The *alpha angle* measures the extent to which the femoral head deviates from a sphere shape. It is measured by drawing a 'best fit' circle around the femoral head, and then a line through the centre of the femoral neck to the centre of the head. From the centre of the head circle, a second line is drawn to the point where the superior surface of the head–neck junction first departs the circle. The angle between the lines is the alpha angle. A 'normal' angle of 41° (left) is compared to one of 98° (right) that represents a cam deformity. Angles over 78° are considered a pathological concern. *(Image from Agricola, R., Waarsing, J.H., Thomas, G.E., Carr, A.J., Reijman, M., Bierma-Zeinstra, S.M.A., et al., 2014. Cam impingement: defining the presence of a cam deformity by the alpha angle: data from the CHECK cohort and Chingford cohort. Osteoarth. Cartil. 22(2), 218–225.)*

The acetabulum is usually positioned to intersect the sagittal plane at around 40° and the transverse plane at around 60°, opening anteriorly and laterally to be angled 45° inferiorly and 15° anteriorly (Torry et al., 2006). Acetabular anteversion is considered normal between 15° and 20° with decreased anteversion being considered to be between 10° and 14° and increased anteversion between 21° and 25° (Torry et al., 2006). The degree of acetabular anteversion influences the amount of internal–external rotation of the hip, with an increase in internal rotation being associated with high acetabular anteversion angles and decreased angles increasing external rotation. Recognising acetabular anteversion gives the clinician an indication of femoral head coverage provided by the acetabulum and can be measured radiographically as the *central edge angle of Wiberg*, defined as the angle between the horizontal line through the centre of the femoral head and a line tangent to the superior and inferior acetabular rims (Torry et al., 2006). The normal centre edge is angled 30°, with a decreased angle of 20° being associated with hip dysplasia and a potential rapid onset of osteoarthritis due to poor acetabular coverage and load transfer across the hip (Torry et al., 2006). However, computerised tomography (CT) also offers an extremely good way to measure the acetabular angles and has given average angles of ~16° in adult males and ~19° in adult females, with average angles of version increasing with age (Klasan et al., 2019). Increased angles, known as retroversion, are low in prevalence in young adults, but around a tenth of the population display significant asymmetry (Klasan et al., 2019) (Fig. 2.3.4e).

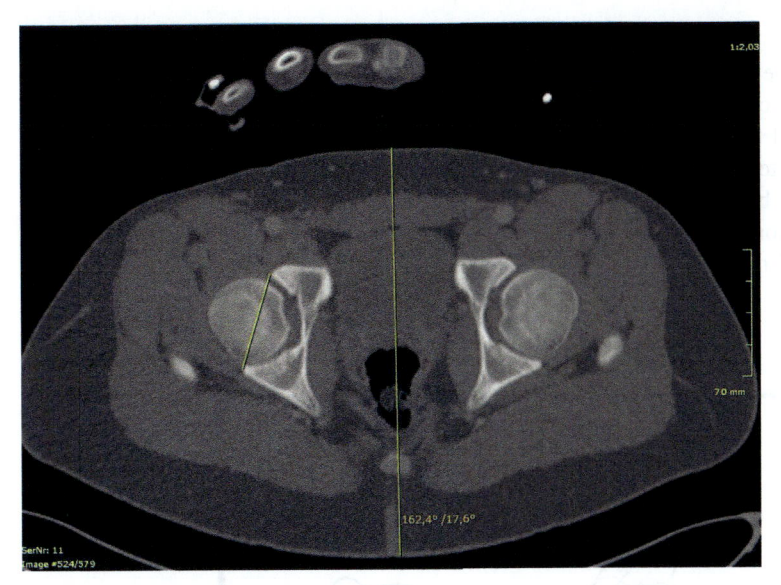

FIG. 2.3.4e Acetabular anteversion angles being measured using computerised tomography (CT) images. The angle of acetabular version is made between the reference line bisecting the body and a line drawn through the most lateral, anterior, and posterior wall of the acetabulum as seen on the left. *(Image from Klasan, A., Neri, T., Sommer, C., Leie, M.A., Dworschak, P., Schofer, M.D., et al., 2019. Analysis of acetabular version: retroversion prevalence, age, side and gender correlations. J. Orthop. Transl. 18, 7–12.)*

Femoral neck–shaft angle

The neck joins the femoral shaft at an angle of 125° ± 5° in adults (Byrne et al., 2010). If the angle exceeds 130°, it causes the shaft to angle outward to the neck creating an alignment variation known as coxa valga, whereas if less than 120°, the shaft angles more inwards from the neck to cause an alignment referred to as coxa vara. This alteration in angulation between the neck and the shaft is significant because it changes the lever arms of the hip abductors to the pelvis (Byrne et al., 2010). This can possibly be influenced during growth through hip abductor strength, as the osseous hip joint is relatively immature during childhood going into puberty, being particularly retarded for a large animal's joint at birth (Hogervorst and Vereecke, 2014) (Fig. 2.3.4f).

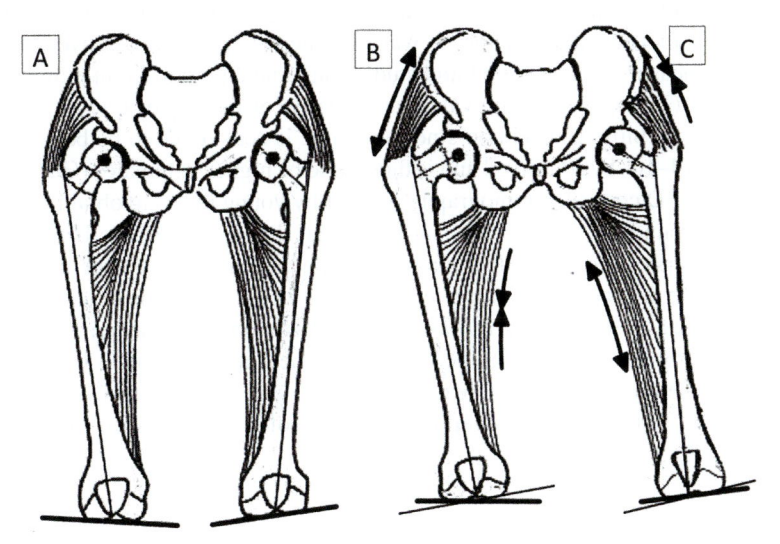

FIG. 2.3.4f The femoral neck–shaft angle should be ~125°, helping set up a small valgus carrying or bicondylar angle (black lines) at the femoral condyle of ~8–14° (A). This keeps the knee closer to being under the CoM during gait. Angles less than 120° are known as coxa vara (B). They position the knee increasingly medially, risking the creation of a varus bicondylar angle (thin black line) unless the bicondylar angle responds during development through the distal femoral growth plate to continue to form a small valgus angle (thick black lines) in compensation. When above 130° (C), they are known as coxa valga and the femoral shaft becomes aligned laterally away from under the body, risking an excessive valgus bicondylar angle (thin black line). Bicondylar angles during development are highly biomechanically responsive to loads that result from the femoral neck–shaft angle, but they should remain slightly valgus whatever the femoral neck–shaft angle. However, the position of the body's CoM more medial or lateral to the knee, can change frontal plane torques across the developing knee creating altered alignments to the tibia below that may induce loading patterns that alter the tibia's development. If femurs are fractured or surgical hip replacements occur, the neck–shaft angle-to-bicondylar angle relationships can be disturbed. *(Permission www.healthystep.co.uk.)*

Femoral anteversion

Femoral anteversion is a term used to describe the proximal angle between the femoral neck and the distal femoral condyles. The femoral condyles are aligned internally compared to the axis of the femoral neck, which itself lies anterior to the transcondylar plane. Anteversion is also referred to as *antetorsion* or *anterotation* (Gulan et al., 2000; Cibulka, 2004; Scorcelletti et al., 2020). Two angles are at play to determine the internal/external rotation at the hip between the pelvis and the femur. The *hip rotation angle* is the angle between the transverse axis of the pelvis and the posterior condylar line running between the posterior aspects of the femoral condyle. *Proximal femoral rotation* angle influences femoral anteversion and represents the internal rotation of the femoral condyles at the knee to the femoral neck's hip joint angle (Uemura et al., 2018) (Fig. 2.3.4g).

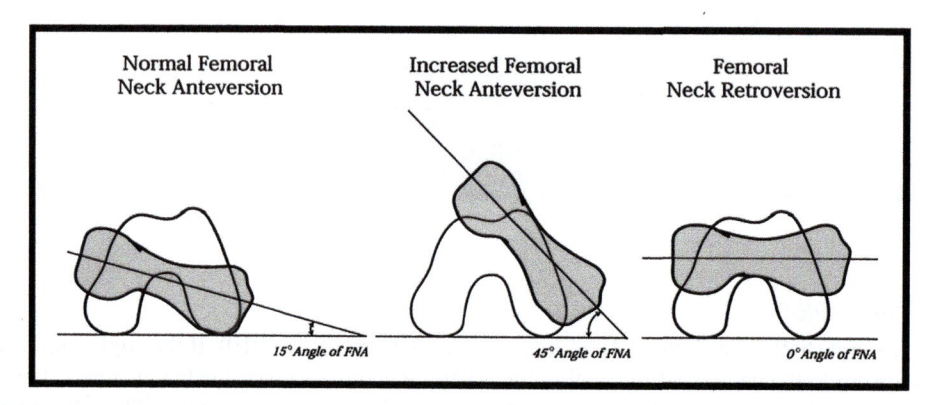

FIG. 2.3.4g Schematic representations of 'normal' and 'abnormal' femoral neck anteversion (FNA) rotations within the right femur. The grey area represents the femoral neck and the white area the distal condylar area. Note that low angles or negative angles are often termed retroversion. *(Image from Cibulka, M.T., 2004. Determination and significance of femoral neck anteversion. Phys. Ther. 84(6), 550–558.)*

Anteversion is variable within the population and with age as the alignment changes during childhood, probably as a result of mechanical loading but also hereditary factors as well. The normal values for anteversion are variable within the literature because protocols of measuring have varied and studies undertaken on different ethnic human groups may also have contributed to that variation (Table 2.3.4). Adult values are usually considered to be 15–20° (Gulan et al., 2000; Byrne et al., 2010; Botser et al., 2012). At birth, anteversion's normal range is considered to be between 30° and 40°, reducing throughout childhood to around 14–16° by adulthood and is thought to be higher in females (Gulan et al., 2000; Torry et al., 2006). Uemura et al. (2018) used 3D computed tomography on healthy young males to obtain measurements and reported femoral anteversion values of around 16° ± 6°. They suggested that this technique has better inter-observer reliability although it records a higher anteversion score than does magnetic resonance imaging (Botser et al., 2012). Some individuals demonstrate a reverse angle where the angulation of the femoral neck faces backwards in relation to the femoral condyles rather than forward. This is known as *retroversion*, *retrotorsion*, or *retrorotation*. These terms can also confusingly be used to describe antetorsion values below those reported as being normal. The sum of the angles of femoral and acetabular anteversion predicts hip instability when summed angles are of 60° or more, while low summed angles of less than 20° predict a low risk of instability (Torry et al., 2006). It has been reported that 38% of hips display low instability index values and 6% demonstrate high instability index values (Torry et al., 2006).

TABLE 2.3.4 Femoral anteversion ranges expected with age reported in: Gulan, G., Matovinović, D., Nemec, B., Rubinić, D., Ravlić-Gulan, J., 2000. Femoral neck anteversion: values, development, measurement, common problems. Colleg. Antropol. 24(2), 521–527. In brackets are those values reported by: Fabry, G., 1997. Normal and abnormal torsional development of the lower extremities. Acta Orthop. Belgica 63(4), 229–232.

Birth	Age 2–4	Age 4–6	Age 6–8	Age 8–10	Age 10–12	Age 12–14	Age 14–16	Adult
36°	33° (31°)	28°	26° (24°)	25°	22°	21°	16° (15°)	15°

Femoral anteversion is an important factor in hip stability during gait and are the result of heredity, foetal development, intrauterine position, and the mechanical forces generated during growth and development (Gulan et al., 2000; Uemura et al., 2018). The degree of anteversion is a femoral anatomical parameter that is believed to influence hip kinematics, especially rotational movements in the axial or transverse plane of the femur (Uemura et al., 2018) (Fig. 2.3.4h). Anteversion is associated with in-toe gait in early childhood and hip and knee osteoarthritis in adulthood (Gulan et al., 2000; Heller et al., 2001), as well as affecting surgical total hip replacement outcomes (Heller et al., 2001).

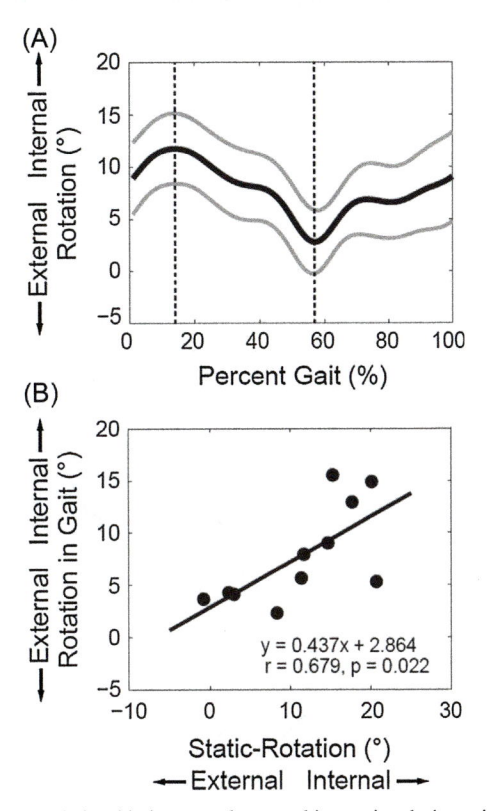

FIG. 2.3.4h Hip rotations during gait (A) and the relationship between the mean hip rotation during gait as measured during standing (B) are reported. Hip rotation—solid black line in (A)—is maximised at 13% of gait and minimised at 56% (vertical dotted lines) during late terminal stance. The grey lines indicate the 95% confidence interval of hip rotation. Note that mean gait rotations seem to positively correlate with hip rotation angle in static stance. This suggests that factors such as anteversion and soft tissue restrictions and strengths which influence static stance hip posture, will likely influence hip rotation during gait. *(Image from Uemura, K., Atkins, P.R., Fiorentino, N.M., Anderson, A.E., 2018. Hip rotation during standing and dynamic activities and the compensatory effect of femoral anteversion: an in-vivo analysis of asymptomatic young adults using three-dimensional computed tomography models and dual fluoroscopy. Gait Posture 61, 276–281.)*

Larger degrees of anteversion may increase proximal bending moments, influencing bone modelling during growth and maturity (Heller et al., 2001). Hips are internally rotated at around 11° when humans are stood in static stance and at around 8° during gait. However, humans with increased magnitudes of femoral anteversion display increased internal hip rotation on standing, although these effects are usually eliminated during gait (Uemura et al., 2018). The hip is normally internally rotated during gait with maximums peaking around early stance phase and minimums around early terminal stance phase. These may increase in cases of larger anteversion angles in order to help improve hip stability created by the co-activation of internal and external hip rotators (Uemura et al., 2018).

Femoral loading patterns

The femur has a concavity/convexity in its superior–inferior axis within the sagittal plane of its shaft, with the convexity facing anteriorly. This gives the appearance of an anteriorly directed deflection to the femoral shaft. The contouring helps to resist the bending moments applied to the femur during stance phase, particularly during initial contact phase of walking when the femur is in its most horizontal posture under body weight loads (Fig. 2.3.4i). In running, this peak horizontal alignment occurs at the end of braking phase. Tensile forces are concentrated to the anterior femur during early walking stance phase (Edwards et al., 2016). This is because the femur acts as a cantilever to the HAT segment, maintaining the CoM's height initially and then starting to lift the CoM of the trunk upwards and forwards during early midstance. Tensile forces are controlled by quadriceps activity, creating compression across the anterior surface with the hamstrings limiting the deflection from the posterior surface at loading response. As the femur starts to take up a more vertical alignment, it undergoes increase in vertical compression, becoming more column-like in its loading. In running, the more horizontal loading is prolonged with hip flexion through to the middle of stance, maintaining higher anterior tensile forces for longer. The anterior surface is subjected to less tensile forces during late stance as the CoM of the trunk moves anterior to the hip throughout midstance.

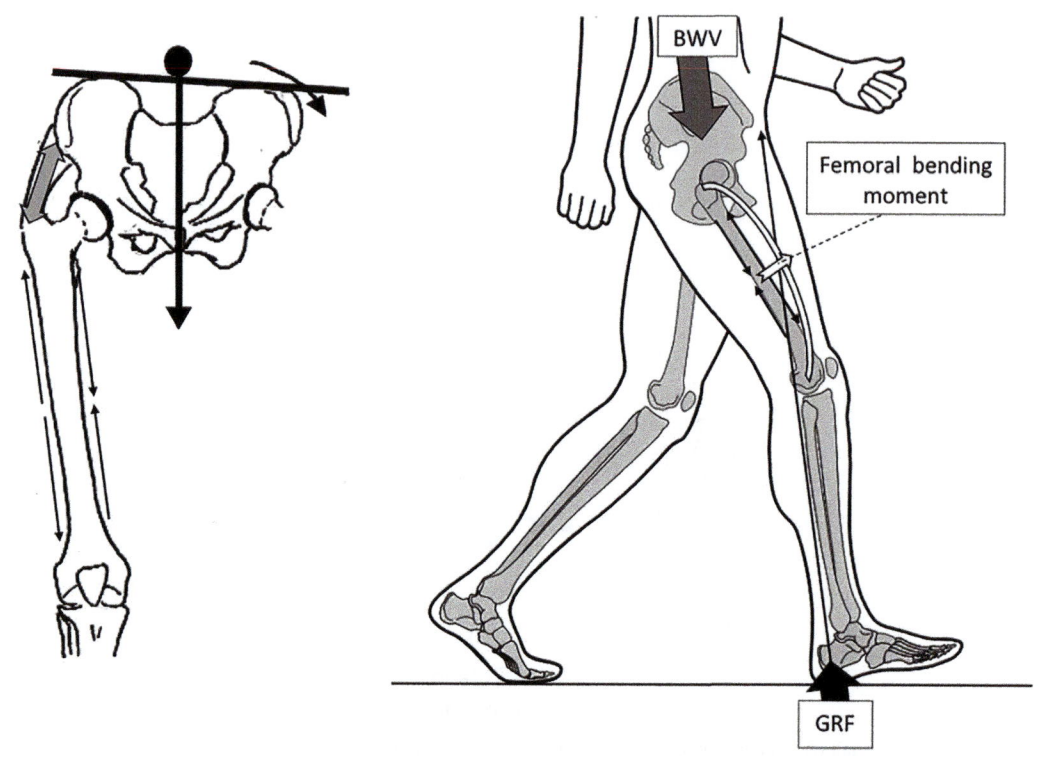

FIG. 2.3.4i Femoral resultant bending moments arise from external moments via the GRF acting with body mass causing frontal plane bending that compresses the medial side and tensions the lateral side of the femur (left). Frontal plane bending moments are most significant during single-limb support under the hip adduction moment, resisted by muscular internal moments such as those from the hip abductors. In the sagittal plane during walking (right), bending moments are most dramatic during contact and loading response. With the body weight vector (BWV) behind the lower limb, the anterior-directed resultant bending moment will tension the anterior femoral surface and compress the posterior. Internal moments from the quadriceps are important in moderating this anterior surface tension as their contraction produces anterior femoral compression. *(Permission www.healthystep.co.uk.)*

During contact and early midstance, gluteus maximus is generating substantial compressive forces across the posterior surface of the femur via the gluteus maximus attachment to the iliotibial tract, giving a lateral orientation to the compression (Edwards et al., 2016). Proximally, this compression from gluteus maximus is directed through the femoral neck. Moreover, the quadriceps have the role of resisting the higher tensile stresses expressed on the anterior and lateral femur during initial loading (Fig. 2.3.4j).

Femoral loading forces are directly attributable to instantaneous muscle activity around the hip during gait and other activities (Duda et al., 1998; Heller et al., 2001). Femoral loads are dominated by large compressive forces, particularly around midstance, with relatively small amounts of shear increasing from the femoral diaphysis through to the metaphyseal region during walking (Heller et al., 2001). Forces in the femur heighten with increasing walking speed (Bergmann et al., 1995) and are related to the relatively large forces generated by the hip abductors (Heller et al., 2001). In stair climbing, peak femoral loading forces move to the diaphyseal part of the femur due to quadriceps loading forces associated with increased hip and knee flexion angles (Heller et al., 2001). Bending moments across the long axis dominate torsional stresses around the long axis and lie maximally within the diaphyseal bone, although the loading magnitude patterns vary slightly between individuals (Heller et al., 2001). General strain ratios between compression and shear stresses and bending and torsional stresses remain similar throughout activity, and the muscles are capable of creating high loading in both the

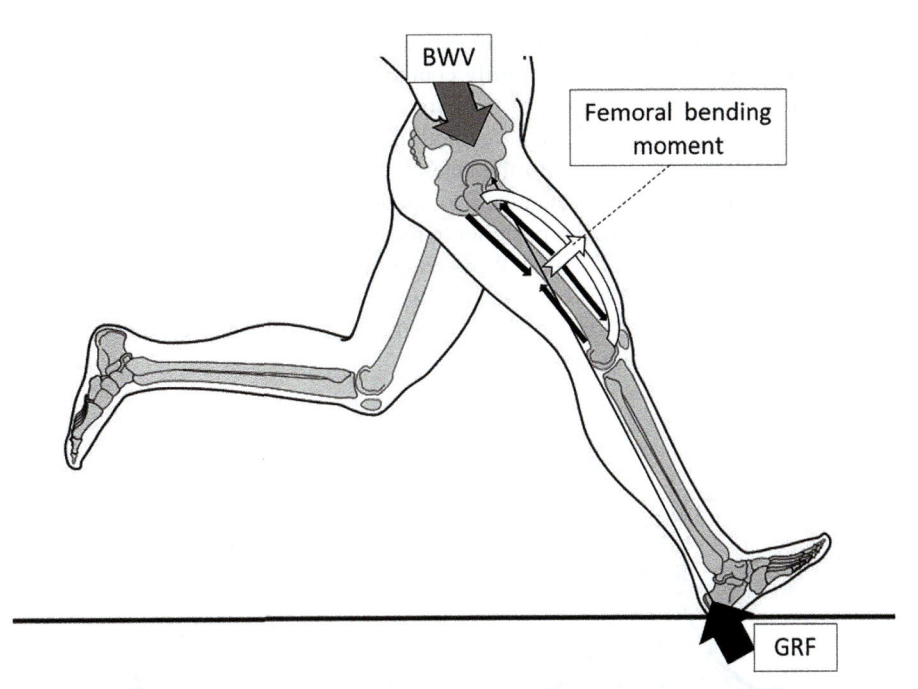

FIG. 2.3.4j In the sagittal plane as indicated here, bending moments are greater in running because the femur becomes more horizontally orientated during early stance phase to the end of the braking phase. The femur is also exposed to higher forces from greater acceleration and a lack of double stance phase. In the frontal plane, greater hip adduction angles associated with running increase the bending moment within the frontal plane. Together, this increases anterior and lateral tension and posterior and medial compression on the femur, especially in those with a long stride length. Correspondingly, the quadriceps and hip abductors become increasingly loaded to protect the femur during running. *(Permission www.healthystep.co.uk.)*

hip joint and the femur as a whole (Heller et al., 2001). Compression through the femur directed into the hip for joint stability is achieved through biotensegrity principles that moderate shear forces and spread the loads across tissues.

2.3.5 Passive soft tissues of the hip

Hip ligaments

The joint capsule of the hip is relatively loose to allow free motion from its neutral position when extended or flexed. The capsule is adapted to facilitate erect posture and lower limb-to-pelvis stability during stance and locomotion. The hip joint capsule is comprised of five ligaments and has firm and continuous attachments to the acetabular rim's periosteum and the femur's intertrochanteric line, but has no posterior osseous attachment (Wagner et al., 2012). The capsule of the hip is fibrous, dense, and strong and encloses all of the femoral head and most of the neck, save some of the posterior areas. It is thickest near its acetabular origin around the posterosuperior and superior regions, and thinnest near its femoral attachments in the posterior and posteroinferior hemi-quadrants (Walters et al., 2014).

Both the *ligamentum orbicularis* and the ligamentum teres femoris are intracapsular. The flat, strap-like ligamentum teres is enclosed in synovial membrane and begins as a medial and lateral band attached to the pubic and ischial parts of the acetabular notch and floor. It runs deep to the transverse ligament with the bands converging as they head towards the femoral head, covered in synovial membrane. This makes it an intraarticular structure, yet it is also extrasynovial (Byrne et al., 2010). Failure of the ligamentum teres increases the hip's range of motion, particularly adduction and some abduction, although the ligament does not contribute significantly to the overall stability of the hip, thus working as a secondary restraint (Byrne et al., 2010).

The ligamentum orbicularis (also known as the *zona orbicularis* or ring ligament) is an annular ligament than encircles the neck of the femur. It plays little part in hip stability (Byrne et al., 2010). However, it does help in preventing hip impingement. The *transverse ligament* is also intracapsular but does not bridge the hip. It merely spans the acetabular notch, joining the fibrocartilaginous acetabular labrum to complete the 'ring' of the acetabulum. Failure of the transverse ligament and thus the acetabular labrum may not significantly affect hip stability (Konrath et al., 1998), although such failure is more likely to result in hip osteoarthritis.

The iliofemoral, pubofemoral, and ischiofemoral ligaments are the important functional ligaments of the hip. Hip capsule function is not homogeneous due to it being partly composed by these discrete ligaments (Hewitt et al., 2002; Telleria et al., 2011). The anterior ligaments consist of the two arms of the iliofemoral ligament which is considerably stronger and stiffer than the posteriorly positioned ischiofemoral ligament (Hewitt et al., 2002). The iliofemoral ligament attaches via its medial and lateral arms proximally to the anterolateral supraacetabular region of the ilium, forming an inverted Y-shape running in a spiral direction to attach to the femoral intertrochanteric line (Byrne et al., 2010; Telleria et al., 2014). The pubofemoral ligament lies posterior to the iliofemoral ligament and anterior to the ischiofemoral ligament without bony attachment (Telleria et al., 2014). Weak points are created where the pubofemoral and ischiofemoral ligaments conjoin the iliofemoral ligament. The iliofemoral ligament is the strongest ligament reported in the body, with a tensile strength greater than 350 N, and is used to keep the pelvis tilted posteriorly in erect stance. It is taut in extension and relaxed in hip flexion (Byrne et al., 2010) (Fig. 2.3.5).

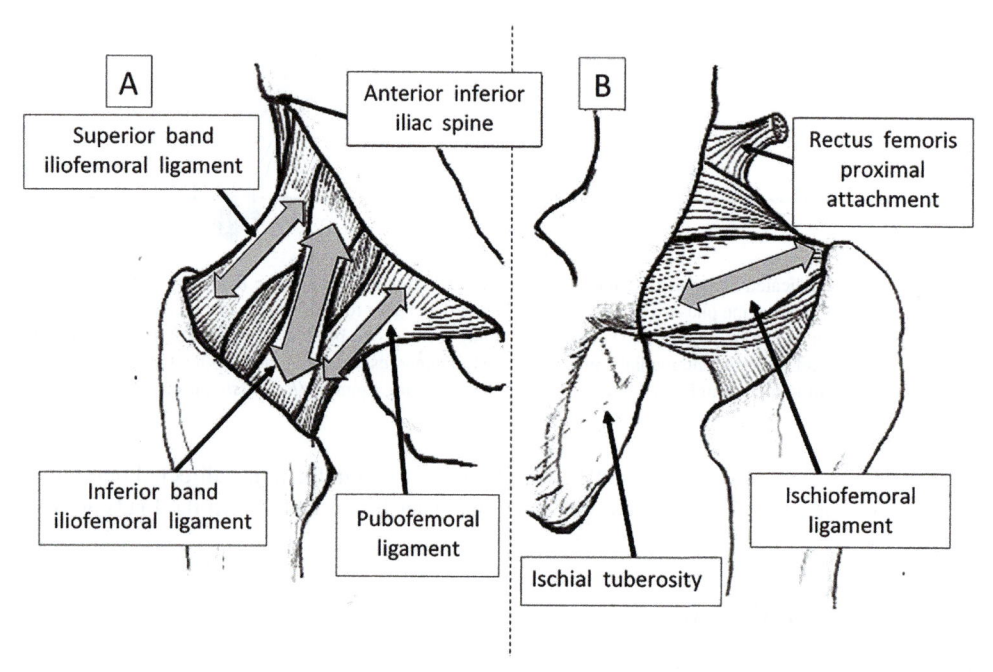

FIG. 2.3.5 The interconnected ligaments of the hip link the pelvis to the medial surface of the femoral trochanter. They play an important passive stability role in support of muscles and assist hip acceleration out of extension. Anteriorly (A), the schematic shows that the iliofemoral ligament has two distinct superior and inferior reinforcements. The superior band connects to the gluteus minimus through an aponeurotic expansion. The pubofemoral ligament attaches proximally to the iliopubic eminence to blend with fibres of the pectineus muscle. Together, these ligaments create a crude connective tissue-shaped letter 'Z' over the anterior femoral neck. A bursa lies between these ligaments and the iliopsoas tendon. Posteriorly (B) lies the weaker ischiofemoral ligament which connects to the tendon of obturator externus muscle. The erect posture of hominin bipedalism has coiled these ligaments around the femoral neck. Hip extension during late stance phase tensions these ligaments, particularly the inferior band of the iliofemoral ligament (grey arrows). The release of their elastic recoil energy starts the acceleration out of hip extension during terminal stance, likely assisting initial iliopsoas activity at preswing. Replacement hip surgery that requires severance of these ligaments may compromise this early hip flexion acceleration force. *(Permission www.healthystep.co.uk.)*

Each ligament acts as a primary restraint of hip internal–external rotation at different degrees of rotation (Torry et al., 2006; van Arkel et al., 2015). The high strains exercised on these hip ligaments correspond to the directional arrangement of their fibres. The strains are highest in the inferior iliofemoral ligament at maximum extension at either 10° or 20° of external rotation with maximal extension, the ischiofemoral ligament at either 10° or 20° of abduction in maximal internal rotation, and the pubofemoral ligament in maximal abduction at 10°, 20°, or 30° of external rotation (Hidaka et al., 2014).

The capsular ligaments play an important part in hip stability (Torry et al., 2006; Myers et al., 2011; van Arkel et al., 2015). The iliofemoral ligament's lateral arm and the ischiofemoral ligament are the primary restraints in two thirds of hip rotated positions (van Arkel et al., 2015), with the iliofemoral ligament limiting external rotation and anterior translation of the femur (Myers et al., 2011). Ligamentum teres acts as a secondary restraint in thigh flexion, adduction, and externally rotated hip positions. The acetabular labrum provides secondary impingement stability/protection at high and low flexion angles, full abduction, and internal rotation, or in extension in any adducted or abducted and externally rotated position

(Myers et al., 2011; van Arkel et al., 2015). Rectus femoris, obturator externus, and gluteus minimus have capsular connections through their tendons, creating dynamic elements to the capsular structures (Walters et al., 2014). These are particularly important in the context of the sensorimotor system. Excessive hip rotation can lead to impingement, high joint wear, and even dislocation (van Arkel et al., 2015). However, owing to the large tensile forces from the hip ligaments, hip dislocation requires extremely high forces to occur, except in children who have a relatively shallow acetabulum compared to adults (Torry et al., 2006). Abnormal function of the iliofemoral ligament is linked to an audible clicking or snapping sound of the hip in motion, known as *coxa saltans* (Torry et al., 2006).

Fascia lata

The hip musculature is enclosed within the fascia lata that forms a fascial sheath within the thigh. It is continuous with the deep fascia of the pelvis and trunk and the fascia crura of the lower leg. Like all deep fascia, the fascia lata provides a mechanically superficial elastic restraint and 'mechanoreceptor hub' to improve contractive efficiency of the region's muscles. It varies in thickness from the hip to the knee, receiving muscular fibres from the gluteus maximus and tensor fasciae latae laterally through an aponeurotic thickening known as the iliotibial tract. This is a connective tissue extension from the gluteus maximus and the tensor fasciae latae muscles that acts as their distal tendinous attachment (Peabody and Bordoni, 2021). The iliotibial tract as a lateral reinforcement within the fascia lata creates a significant lateral force vector from the pelvis across the knee to the proximal leg (see Fig. 2.3.3).

At the gluteus maximus, the fascia lata divides into superior and deep layers. Respectively, these layers traverse deeply and superiorly to reunite at the inferior border of gluteus maximus forming a thick fibrous band. This band is continuous with the iliotibial tract and is also continuous with the deep aspect of the tensor fasciae latae and the lateral part of the hip joint capsule (Peabody and Bordoni, 2021). The iliotibial tract attaches distally to the lateral tibia, fibula, lateral femoral condyle, and parts of the lateral knee ligaments and fascia as is discussed in more detail in Section 2.4.5. The fascia lata has extensive attachments to the tibial and femoral condyles and the fibula, as well as the fascia crura of the leg distally.

The fascia lata is thicker proximally and laterally, especially where the fibrous tissue from gluteus maximus becomes integrated with the iliotibial tract and around the knee to become part of the patellar retinaculum distally. Medially, the fascia lata invests the adductor muscles and receives a fibrous continuation from the biceps femoris, sartorius, and quadriceps femoris muscles (Peabody and Bordoni, 2021). Proximally, it is attached to the sacrum and coccyx. It also has attachments at the ischial tuberosity posteromedially, the iliac crest laterally, the inguinal ligament and superior pubic ramus anteriorly, and the inferior ramus of the pubis and lower border of the sacrotuberous ligament medially (Peabody and Bordoni, 2021).

The fascia lata further reinforces the quadriceps anteriorly with transverse orientated fibres attached to the distal portions of these muscles, except for rectus femoris (Peabody and Bordoni, 2021). Travelling in a distal direction, the deep surface of the fascia lata separates into two intermuscular septa of the thigh that attach to the linea aspera of the femur. This forms a significant fibrous attachment system for the muscles that supply the knee, with the lateral intermuscular septum being the strongest and thickest (Peabody and Bordoni, 2021). This septum extends from the gluteus maximus to the lateral femoral condyle, dividing the vastus lateralis anteriorly from the biceps femoris and giving partial attachment to both muscles (Peabody and Bordoni, 2021). The medial intermuscular septum is less substantial and separates the vastus medialis from the adductor muscles. The fibre orientation of the fascia lata is dominant in a longitudinal direction and is slightly thicker within males (Peabody and Bordoni, 2021).

2.3.6 The hip in lever systems

Instantaneous joint axis of the hip

The hip joint's instantaneous axis is important to locate if both hip and knee joint moments and the capacity of muscles to generate moments and forces around the hip need to be calculated. Locating the hip joint's centre in vivo is difficult without radiographs, CT, or MRI scans which can locate the centre with errors as little as 2mm. However, these are expensive options (Kainz et al., 2015) and crude estimations can be sufficient for clinical needs. In fully erect standing, the body's CoM lies just posterior to the hip joint's axis in the sagittal plane, causing posterior pelvic tilt on the femoral heads. This is opposed by tensile forces caused by a stretching placed on the strong anterior hip joint capsule via the iliofemoral ligament, reducing the muscular input required to maintain static stability (Torry et al., 2006). This mechanism also plays a significant role in trunk postural stability during gait, mechanically assisting muscle function around the hip. It also underlines why it is important to be able to maintain the ability to stand straight.

The line of force and action of the muscles that cross the hip joint are dependent on the axis of the hip and the orientation of the muscles' femoral attachments. Hip motions change the alignment of the femoral attachment points of the muscles in relation to their proximal attachment points across the hip, directly affecting the relative torques generated across the joint. This is because of changes to the moment arm lengths of the muscles (Neumann, 2010). Therefore, muscle actions reported for the hip can change during the femur's arcs of movement during gait (Fig. 2.3.6a).

FIG. 2.3.6a A lateral view schematic of the sagittal plane lines of force within the right hip of several of the hip muscles in association with the joint's axis in erect stance. Extensor muscles are shown with dashed arrows, flexors with solid arrows. The hip axis is indicated by the circle. The internal moment arm of rectus femoris is shown here as the thick black line from the hip axis, but this could be applied to any muscle indicated to give comparable moment arms for each muscle. However, it must be remembered that these moment arms from the hip axis change during hip motions in gait and that larger muscles can generate greater forces based on their cross-sectional areas, fibre arrangements, and speed of contraction. *(Image from Neumann, D.A., 2010. Hip. In: Kinesiology of the Musculoskeletal System: Foundations for Rehabilitation, second ed. Elsevier, pp. 465–521 (Section IV, Chapter 12).)*

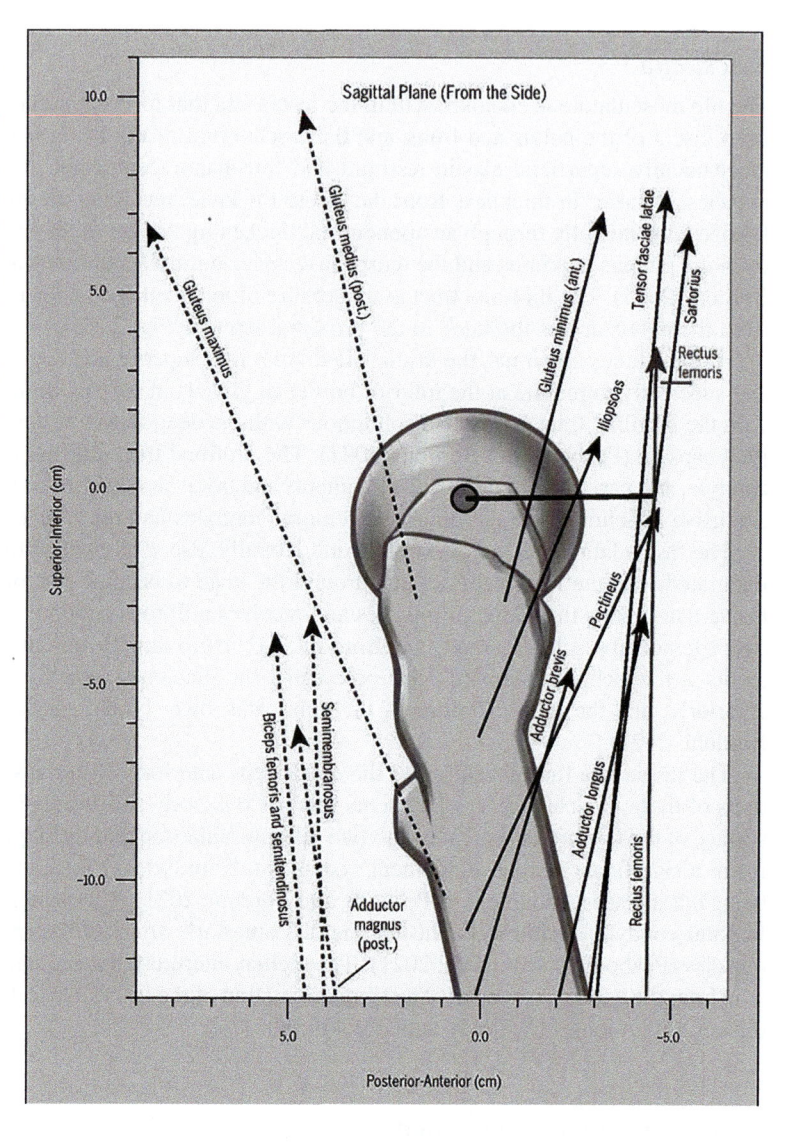

The hip muscles also involve their own complex geometric architecture, which combined with a variable subject-specific joint axis alignment (due to femoral and acetabulum anteversions) makes calculating muscle function around the hip in a specific patient challenging. MRI and CT scanning has enhanced the ability to construct hip muscle force vectors for modelling hip function. However, such modelling only allows for one length of muscle fibre and one moment arm to be estimated for each muscle path, and it poorly represents in vivo force–velocity behaviours of the hip muscles (Torry et al., 2006). The considerable changes in muscle fibre moment arms are dependent on the femoral, pelvic, and lumbar motions combined (Arnold and Delp, 2001).

Lever arms of the hip

The hip fulfils its role in locomotion as two different class three lever systems, one simple class one lever system, and one debatable class two lever system. Its primary function is to provide hip extension and flexion both in stance and during swing. These sagittal plane motions occur through a functional axis within the transverse and frontal planes to provide motion in the sagittal plane. In reality, motions occur within all planes during gait because the hip joint possesses a ball-and-socket arrangement that has an ability to orientate its functional axis instantaneously and easily, allowing flexion to occur with adduction or abduction and internal or external rotation. Thus, only simplistically do the lever systems of the hip work on the basis of providing only sagittal plane or frontal plane motion through class two, class three, and class one lever systems, respectively. Despite this, knowing these core lever arms is still useful in understanding hip function.

Sagittal plane levers

In swing phase, the role of the hip is to move the lower limb from a position posterior to the hip to a position in advance of the hip. In so doing, it must use muscular activity to overcome the inertia and control the momentum of moving the CoM on the lower leg. This CoM consists of the mass of the whole swing limb segments (and any clothing or footwear). The hip is the fulcrum and the lower limb's CoM the resistance, with effort being provided by the distal attachments of the muscles–tendons that cross the hip joint. This results in the effort arm being shorter than the resistance arm, forming the class three lever.

At preswing, iliopsoas starts to flex the hip out of late stance extension, with some assistance from hip ligament elastic recoil and previously generated ankle plantarflexor power and gluteus maximus activity that maintains forward CoM momentum. During early swing phase, the CoM of the lower limb is being brought closer to the fulcrum through knee flexion and reduced ankle plantarflexion. By moving the CoM closer to the hip from a position furthest away at toe-off to midswing means that inertia is reduced for the lower limb. For the same muscle effort, the lower limb can accelerate towards hip flexion from extension, giving each lower limb's CoM greater momentum. After midwing, the lower limb extends through increasing knee extension, moving the CoM away from the hip. This now gives the lower limb's CoM greater momentum, helping to pull the CoM of the body along behind it through centrifugal forces and thereby reducing the effort on the hip flexors and knee extensors to bring the leg into a lengthened position ahead of the body and ready for initial contact.

Shortening the limb through early swing phase by hip and knee flexion is also there to aid ground clearance and fulfils a role of mechanical efficiency. This is a good explanation as to why in running, swing limb knee flexion increases to gain greater ground clearance as at contralateral midstance, the support lower limb length is shortest as opposed to in walking when it is at its longest. The need for greater accelerated hip flexion in running also enhances the centrifugal pull on the CoM from the swing limb during late swing. Because the hip is flexing throughout the swing phase in both walking and running, only the hip flexors provide significant effort to the lever arm, although the extensors such as gluteus maximus (and the hamstrings through their knee attachments) aid in braking acceleration of the lower limb during late swing (Fig. 2.3.6b).

In stance phase, motion of the hip is used to move the HAT's CoM over the stance limb. During walking, this is achieved through inverted pendulum vaulting action, whereas in running, it is achieved through spring–mass function. However, the overall lever principles are the same. The body's CoM is therefore the 'resistance', which includes the opposite swing limb's mass, while the hip flexors and extensors provide effort through their proximal attachments which are all below the CoM. Therefore, the effort arms are shorter than the resistance arms, creating a class three lever system.

During contact phase, the CoM of the body is behind the fulcrum of the hip and must be moved anteriorly. The hip extensors (primarily gluteus maximus) are used to pull the thigh out of hip flexion towards extension. This requires a debatable class two lever that is only present in walking, as during running stance phase, the hip continues to flex until after 50% of stance. In walking, with hip flexion decreasing during early midstance, the CoM of the body moves closer to the hip in the sagittal plane, reducing the flexion moment at the joint and thus reducing the effort required from gluteus maximus to apply its extension moment. Once the CoM lies perfectly vertical over the hip around absolute midstance, the angular momentum is lost and the hip extensors can switch off. Because the CoM remains above the effort of gluteus maximus through its attachments around the iliac crest and posterior gluteal surface of the ilium, the lever system is vertically aligned as a class three lever. However, within the horizontal plane, there is a class two lever system at work. After absolute midstance, the CoM moves anterior to the hip extending it, but this CoM movement is controlled distally through triceps surae activity. Although the hip extends, extensor muscle action at the hip is not required, with hip extension restrained directly by hip ligament tightening and via the anterior hip abductors. Hip extension is thus naturally limited by passive structures and articular form closure, primarily through iliofemoral ligament-induced restraint (Fig. 2.3.6c).

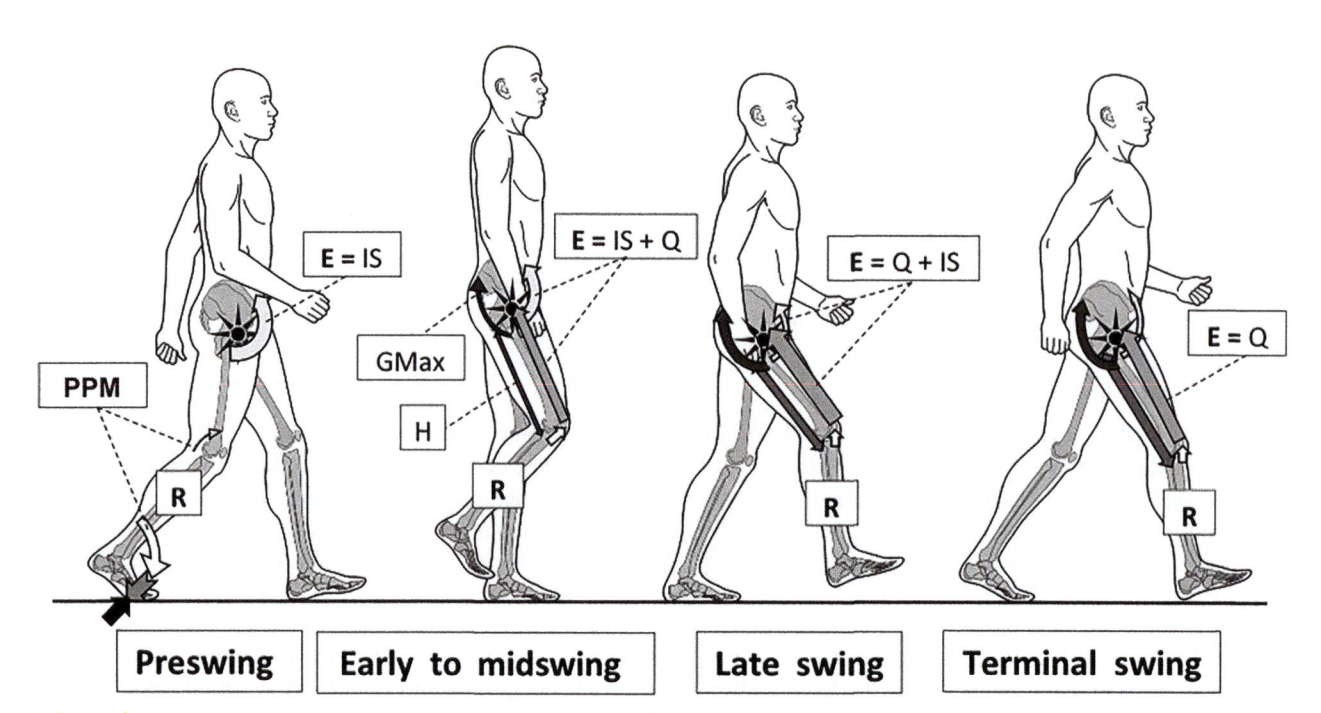

320 Clinical biomechanics in human locomotion

FIG. 2.3.6b A schematic of the hip acting as a class three lever during each stage of swing phase, showing the changes in the lower limb's resistance (R) from varying its CoM location. Iliopsoas (IS) provides the hip flexion effort (E) at the lesser trochanter. The hip's joint axis is the fulcrum (star). Note how the limb's CoM, represented by the resistance, moves closer to the hip's axis of rotation during early swing through knee flexion and reduced ankle plantarflexion when moving the limb under the body. The resistance then increases its distance from the fulcrum in late swing, aided by quadriceps-induced (Q) knee extension, as IS effort reduces. In so doing, limb momentum and centrifugal forces are raised, pulling the body's CoM forward. Preswing hip action is heavily influenced by the ankle plantarflexor power's momentum (PPM) during terminal stance that accelerates the upper body's CoM from the foot, as well as some brief acceleration power from gluteus maximus (GMax) activity at the hip. Gluteus maximus and the hamstrings (H) act mainly as antagonistic stabilisers, becoming more active in late and terminal swing to decelerate hip flexion and knee extension momentum from iliopsoas and the quadriceps swing phase activity. *(Permission www.healthystep.co.uk.)*

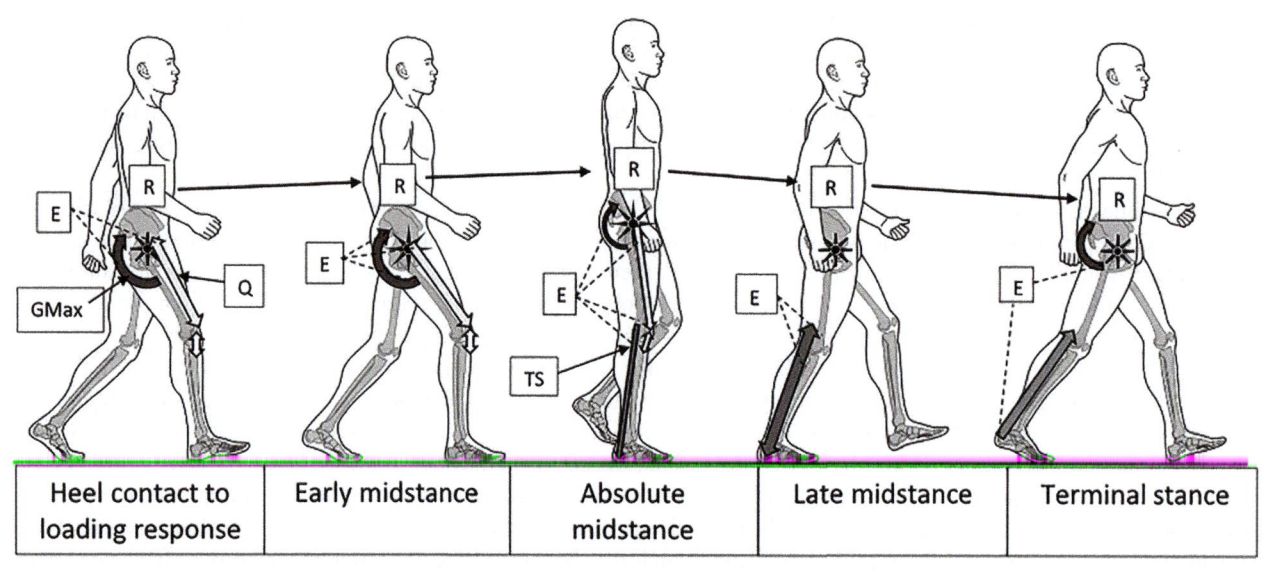

FIG. 2.3.6c In stance phase of gait, the resistance (R) being moved is the CoM of the HAT segment (plus the weight of the swing limb). The fulcrum is the hip joint axis (star) and the effort (E) changes with the muscles that are providing continued forward motion to the CoM. This is being significantly affected by gravity. At initial contact and loading response, gluteus maximus (GMax) and the quadriceps (Q) resist the fall of the body's CoM via hip flexion resistance while helping to manage impact forces through muscle–tendon buffering. Effort is provided at these muscles' proximal attachments. In early midstance, the GMax and Q effort is used to extend the hip and knee, raising the CoM upwards and forwards. At absolute midstance, gravity 'takes over' forward momentum of the resistance under the braking of acceleration by the ankle plantarflexors (mainly triceps surae—TS) acting through their proximal attachments. This activity peaks in late midstance until plantarflexor power stored with the Achilles becomes greater than the forces maintaining heel contact, which then induces a heel lift. This changes the role of triceps surae from brake to accelerator of the R (CoM) via its distal Achilles elastic recoil lifting its distal attachment. Rotation occurs at the ankle that aids hip and knee flexion. Once ankle plantarflexor power starts to decrease, GMax creates some CoM acceleration towards the loading contralateral foot before iliopsoas takes up ipsilateral limb's acceleration of its CoM for swing phase (see Fig. 2.3.6b). *(Permission www.healthystep.co.uk.)*

Frontal plane lever

The hip requires frontal plane stability when in stance phase, particularly during single-limb support. This involves keeping the pelvis level and preventing the trunk from excessively adducting on the hip in the frontal plane under the influence of gravity. Such a mechanism allows the support limb and trunk to maintain its functional length for the swing limb to pass underneath the body's CoM. The CoM should move forward in the sagittal plane without any large deviations in the other planes. Were the pelvis to significantly drop down on the swing limb side (large varus tilt at the hip), then ground clearance space for the swing limb would be decreased and stance limb balance potentially lost. This in turn would require greater hip and knee flexion and foot dorsiflexion on the swing limb. Not only this, but if the CoM of the body fell towards the swing limb and away from the support limb, then single-limb support phase stability could be lost. It is particularly important that left and right-sided hip abductor mechanisms work well during running, as there is only a single-limb support phase during stance for stability. In addition, ground clearance space is already reduced in running due to the spring–mass function of the lower limb that decreases support limb's length to a minimum as the swing limb passes beneath it.

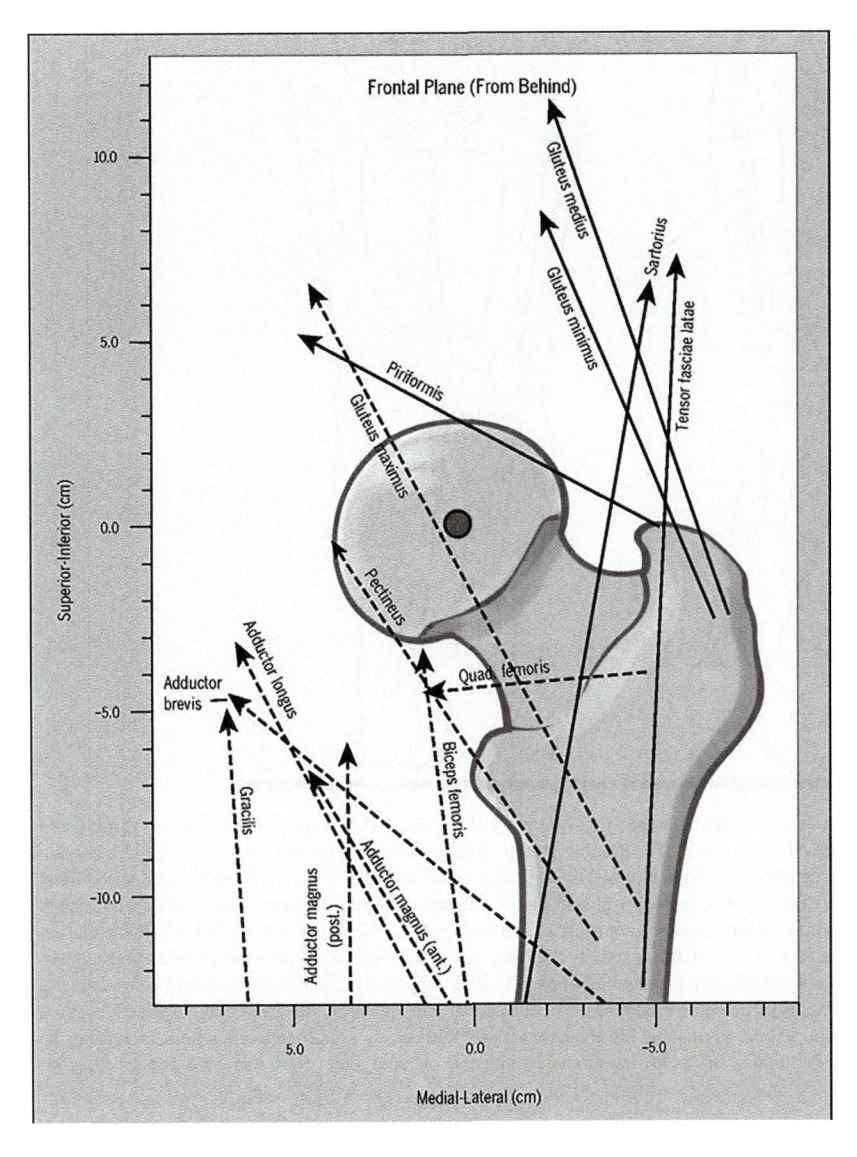

FIG. 2.3.6d A schematic of the posterior view of the right hip to show the frontal plane muscular force vectors. The axis of rotation (circle) is now positioned in the anterior–posterior direction (sagittal plane), so that hip motion is now considered to be in the frontal plane with adduction and abduction. Hip abductors are indicated with solid black lines and adductors by dashed lines. *(Image from Neumann, D.A., 2010. Hip. In: Kinesiology of the Musculoskeletal System: Foundations for Rehabilitation, second ed. Elsevier, pp. 465–521 (Section IV, Chapter 12).)*

The muscles primarily responsible for this hip stabilisation are the hip abductors, i.e. gluteus medius, gluteus minimus, and tensor fasciae latae, and at initial contact and early stance, the superficial parts of gluteus maximus (Fig. 2.3.6d). These muscles provide their effort through their proximal attachments to the posterior lateral surface of the ilium and the deep fascial layer known as the gluteal aponeurosis. Their proximal attachments lie superior and lateral to the hip. The CoM resistance they control in the frontal plane is formed by the weight of the swing limb and the HAT segment, which lies medial and superior to the abductor effort. Therefore, the lever system created is a crude class one lever system with effort and resistance lying on either side of the hip fulcrum point. The effort lies closer to the hip than the resistance, so the resistance arm is longer and has mechanical advantage. In double stance, the hip abductors are active on both limbs, increasing CoM-balanced stability more easily (Fig. 2.3.6e). This is a luxury not afforded to running. The hip abductors are also active in the swing limb, but they are only controlling the mass of the swing limb against hip adduction. Failure of this system can cause an increased hip adduction motion (Trendelenburg gait) where the swing side hip drops downwards during single-limb support. This hip adduction instability tends to cause the body to flex towards the support limb to maintain balance, disrupting normal gait kinematics and kinetics.

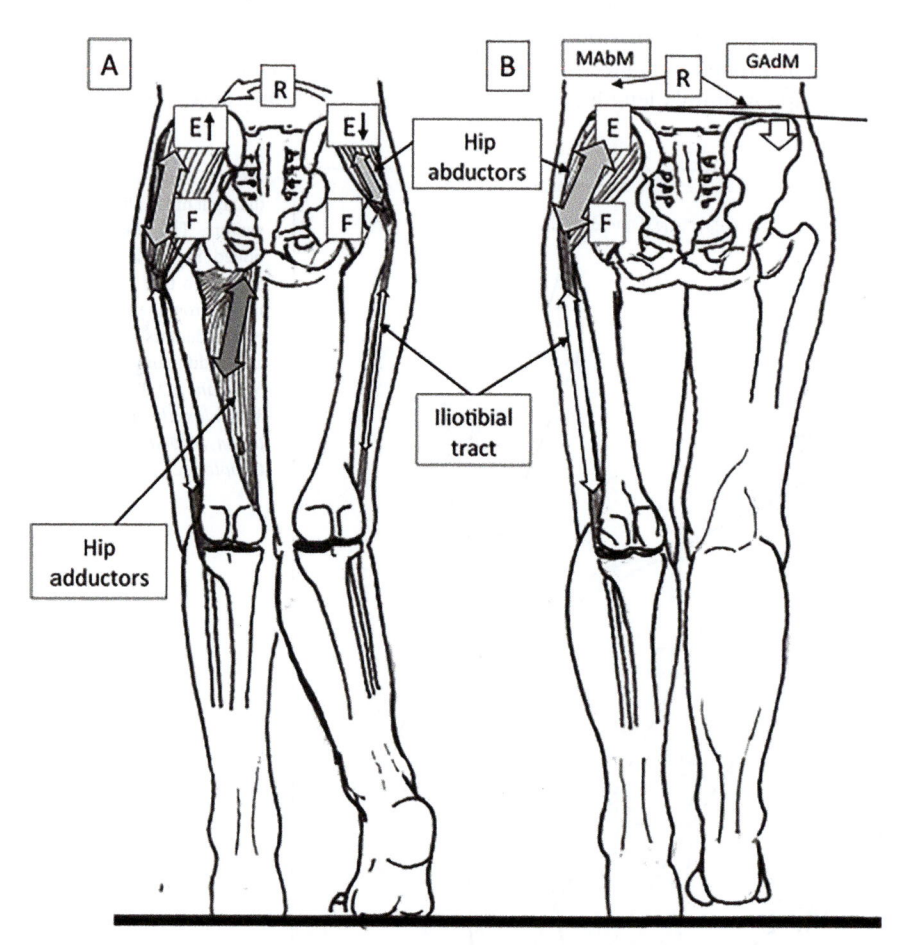

FIG. 2.3.6e A schematic to show that the hip abductor muscles' lever arm in the frontal plane is crudely class one. The proximal attachment effort (E) of these muscles to the ilium is superolateral to the hip acting as the fulcrum (F). The iliotibial tract helps give these muscles a longer lever arm than those that only have attachments to the greater trochanter such as gluteus medius and minimus. The resistance (R) is the CoM of the HAT segment and swing limb, lying superomedial to the hip. In double-limb support (A), the terminal stance leg (right leg shown) has decreasing hip abductor activity (E↓) as the weight of the HAT segment is transferred to the loading left limb under increasing abductor muscular effort (E↑). The offloading limb is primarily using anterior hip abductor activity while the loading limb has increasing posterior hip abductor activity. Hip abductor activity on both limbs makes double-limb support stance phase more stable during the anterior passage of the CoM onto the leading limb. In single-limb support (B), the hip abductors use near-isometric activity to create a muscular abductor moment (MAbM) that is sufficient to minimise the pelvic adductor moments created by gravity-induced adduction moments (GAdM) interacting with the HAT and swing limb's CoM. In running, hip abductor activity initiates on a single supporting limb, increasing the muscular effort necessary to stabilise weight transfer while limiting hip adduction moments under accelerated forces from increased gait speed. *(Permission www.healthystep.co.uk.)*

The ability to stride effectively using both types of lever systems maintains hip stability during gait. This muscular stability is essential to efficient energetics and may also have an important effect on injury avoidance. Failure to maintain appropriate lever systems around the hip may be linked to the development of increased injury risk. Increased hip adduction and overextension of the knee at initial contact, as a consequence of excessive stride length, have both been linked to increased running injuries (Bramah et al., 2018).

2.3.7 Muscle action at the hip

Of the hip muscles, gluteus medius, gluteus maximus, and gluteus minimus provide the majority of support of the CoM during the first 0%–30% of the stance phase of gait. From forefoot loading at the end of loading response (10%–50% of stance phase), gluteus maximus and the posterior parts of gluteus medius and gluteus minimus contribute most significantly to control the vertical GRFs (Anderson and Pandy, 2003). With the addition of the weight of the bones and joints and the action of the other parts of the gluteal muscles, almost all of the vertical forces are explained during midstance by these muscles (Anderson and Pandy, 2003). Only towards the end of midstance do the anterior muscle bundles of gluteus medius and gluteus minimus contribute to support, with iliopsoas developing substantial forces during late terminal stance, moving

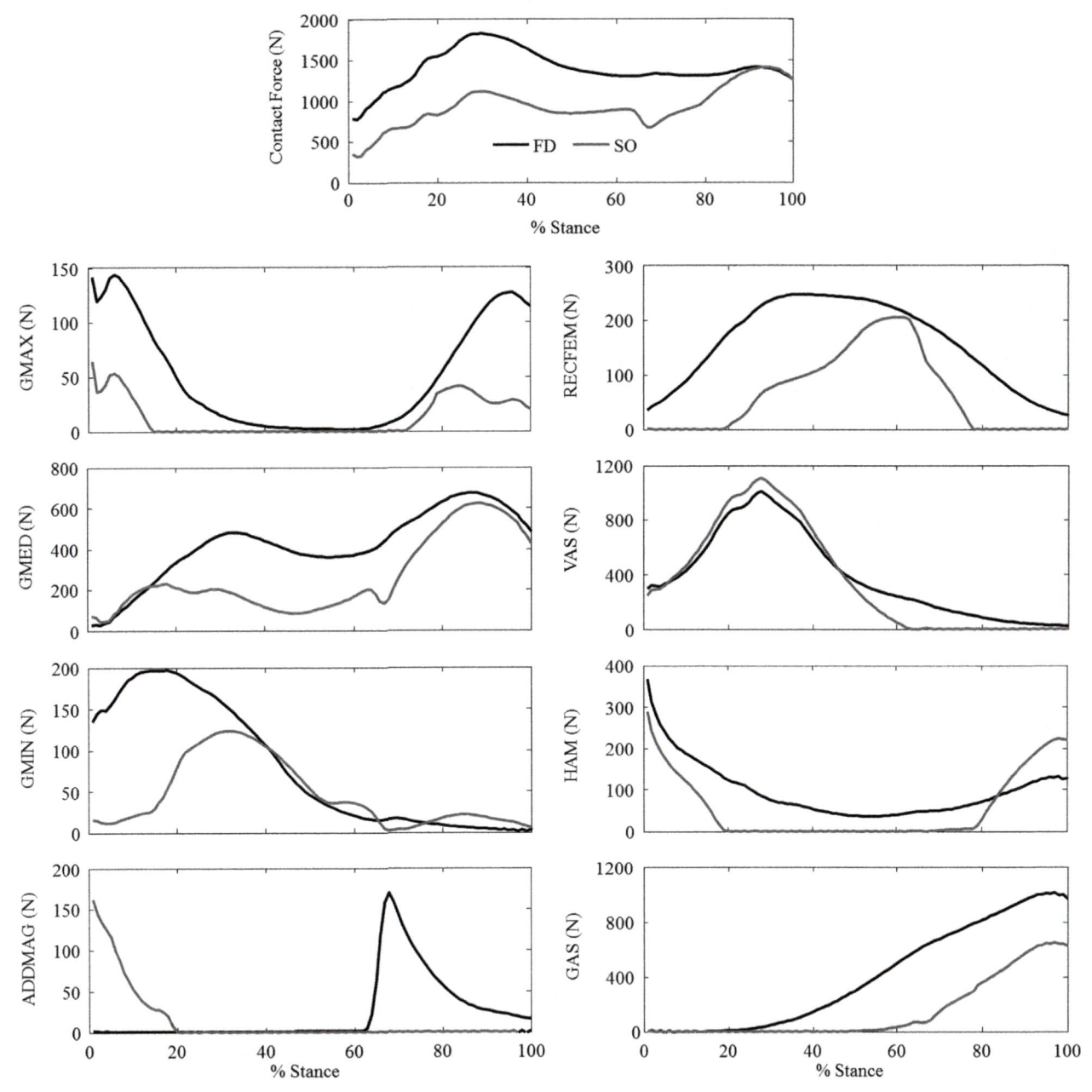

FIG. 2.3.7a The resultant hip joint contact forces and selected muscle forces acting on the femur during stance phase are demonstrated. The hip joint contact force has been calculated as the vector sum of the resultant joint reaction force and forces from the muscles that span the hip. GMAX = gluteus maximus, GMED = gluteus medius, GMIN = gluteus minimus, ADDMAG = adductor magnus, RECFEM = rectus femoris, VAS = vasti quadriceps muscles, HAM = hamstring muscles, GAS = gastrocnemius. *(Image from Edwards, W.B., Miller, R.H., Derrick, T.R., 2016. Femoral strain during walking predicted with muscle forces from static and dynamic optimization. J. Biomech. 49(7), 1206–1213.)*

towards preswing (Anderson and Pandy, 2003). Hip stability is therefore managed through muscle tensions applying compression into the joint. This is initially supplied by the more posterior hip muscles during loading, moving the burden to the more anterior hip muscles as the hip moves from flexion into extension and the CoM of the body moves from posterior to the hip to anterior throughout stance phase (Fig. 2.3.7a).

Hip flexors

The primary hip flexor is iliopsoas which contributes directly to stability and movement of the trunk, pelvis, and lower limb. This is a muscle comprised of psoas major and iliacus that conjoin at the level of the inguinal ligament to collectively gain attachment distally at the greater trochanter of the femur. Each muscle has independent innervations, making selective activation possible to each segment they control (Torry et al., 2006). Psoas major attaches proximally to the lumbar spine by segmental attachments and to the sacroiliac joint as it crosses it, giving the muscle an important role in lumbopelvic stability as well as hip flexion through its union with iliacus as iliopsoas. As already stated, psoas major's role in hip flexion may be quite limited (Gibbons, 2007). Iliacus is thought to provide the primary hip flexor role with psoas major providing more lumbar spine–sacroiliac joint stability with psoas minor (Gibbons, 2007). The hip flexor iliopsoas and the hip extensor gluteus maximus both have lumbopelvic stability effects as well as hip motion and stability roles. This sets up a synergistic and antagonistic relationship between these muscles.

Iliopsoas contributes stability to the femoral head in the acetabulum and with rectus femoris, it reinforces the anterior hip capsular ligaments (Shu and Safran, 2011). Iliacus and gluteus maximus are the most influential muscles during swing phase (Barrett et al., 2007). Loss of iliacus function is modelled to decrease both hip and knee flexion (Piazza and Delp, 1996; Barrett et al., 2007). Iliopsoas displays a marked increase in EMG amplitude in running speeds of over 2 m/s compared to walking (Andersson et al., 1997).

Iliopsoas is assisted in hip flexion by sartorius, rectus femoris, and tensor fasciae latae (Andersson et al., 1997). Sartorius arises from the anterior superior iliac spine of the ilium and runs distally to the pes anserine attachment site on the anteromedial surface of the proximal tibia. It also influences hip abduction and aids in resisting internal rotation. Sartorius has a small physiological cross-sectional area, suggesting fairly limited force production. Rectus femoris demonstrates weak hip flexor activity and is the only quadriceps muscle that crosses the hip. It attaches proximally to the anterior inferior iliac spine and near to the acetabulum and distally via the patella and its ligament to the tibial tuberosity at the anterior knee. It has a moderately sized physiological cross-sectional area compared to the other quadriceps muscles, so its power is limited. The other vastus quadriceps muscles, through their contraction momentum during knee extension in the first 15% of stance, help induce hip extension moments indirectly (Zajac et al., 2003) (Fig. 2.3.7b).

FIG. 2.3.7b The induced hip intersegmental force generated by the quadriceps muscles that does not cross the hip still provides trunk support and hip extension moments that progress the trunk over the limb. Thus, the vastus muscles of the quadriceps decelerate the lower limb (negative signs) but accelerate the CoM (positive signs) through reducing hip flexion in the first ~15% of the walking gait cycle (right image). Rectus femoris, that crosses both the knee and hip, also accelerates the trunk, but during late stance, this is achieved through generating hip flexion from ~45% of the walking gait cycle during terminal stance. *(Images from Zajac, F.E., Neptune, R.R., Kautz, S.A., 2003. Biomechanics and muscle coordination of human walking. Part II. Lessons from dynamical simulations and clinical implications. Gait Posture 17(1), 1–17.)*

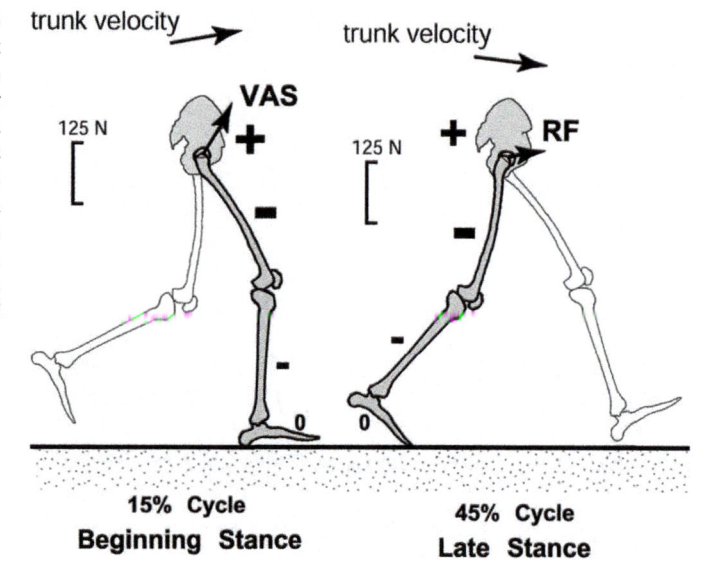

Hip extensors

Hip extension moments peak during preswing at the stance-to-swing transition (DeJong et al., 2020). Yet, hip extensors are not active in the generation of these extension angles. The hip extensors thus operate by decreasing hip flexion moments at contact and reducing hip flexion angles by providing extension moments during early midstance. They are again active in braking and resisting hip flexion moments during swing. The primary concentric hip extensor, i.e. the powerful gluteus maximus, is the largest muscle of the human body and not surprisingly demonstrates a large physiological cross-sectional area. It is also the most superficial of the hip extensors covering the dorsolateral aspect of the sacrum, the posterior ilium, and is itself covered by thoracolumbar fascia (Byrne et al., 2010). Anteriorly, it overlays the posterior gluteus medius which can create difficulty of 'crosstalk' on EMG studies between these muscle (DeJong et al., 2020). Gluteus maximus usually demonstrates three divisions of superficial sacral fibres, deep sacral fibres, and deep ilium fibres with their own distinct attachment points, yet all share a common nerve supply (Gibbons, 2007).

The distal attachments of the superficial sacral fibres are located to the posterior aspect of the iliotibial tract and gluteal tuberosity of the femur, crossing the hip and sacroiliac joint. The deep sacral fibres cross the sacroiliac joint to attach to the posterior rim of the iliac crest lateral to the posterior superior iliac spine and across the posterior iliac surface posterior to the gluteal line, but this specific arrangement is only present in around two-thirds of humans (Gibbons, 2007). The deep ilium fibres gain attachment to the gluteal tuberosity of the femur and again cross the hip. Therefore, the superficial sacral and deep ilium segments of gluteus maximus both influence hip extension–flexion moments. Concentrically, gluteus maximus produces hip extension moments to reduce hip flexion angles and accelerations. However, it also resists and controls hip flexion moments from primarily iliacus during swing phase through eccentric activity (Neptune and McGowan, 2011; Castermans et al., 2014). Gluteus maximus activity is greater in running and climbing than during walking, but by far the greatest activity is recorded in sprinting, suggesting that it has only a limited trunk stabilisation role in running (Bartlett et al., 2014). It is either the large CoM-lifting power as seen in climbing a tree or the dramatic CoM acceleration power for sprinting fast that demands large hip extension power from gluteus maximus.

Gluteus maximus also influences transverse and frontal plane hip rotations through its deep ilium fibres and frontal plane adduction/abduction moments through its superficial fibres, helping to create stability at the hip. In walking gait, it primarily functions as a resistor and reducer of hip flexion and internal hip rotation. It is active from before initial contact (throughout late and terminal swing) through to the middle of midstance while the hip is in stance phase flexion. It is also a very influential muscle in swing phase kinematics in concert with iliacus, affecting both hip and knee motion (Barrett et al., 2007). Just before initial contact, gluteus maximus and the hamstrings activate to decelerate iliacus and quadriceps-induced hip flexion and knee extension. This braking action gives gluteus maximus an important role in energy dissipation during initial contact and has been reported to have the capacity to compensate for quadriceps weakness. This is because both the quadriceps and gluteus maximus decelerate and 'brake' the body's CoM downward progression during initial loading (Thompson et al., 2013).

However, the gluteus maximus' capacity to absorb impact and loading energy is far less than that of the quadriceps, which limits the capacity of gluteus maximus to compensate for quadriceps weakness for use in energy dissipation (Thompson et al., 2013). It has been estimated that for every 1 N of decreased force generation by the quadriceps, gluteus maximus would need to generate 4 N to compensate at ~30% of the gait cycle (Thompson et al., 2013). Gluteus maximus activity may also influence internal tibial rotation rates by providing resistance during loading response (Fig. 2.3.7c). It has been suggested that this is achieved through external rotation torque via its femoral attachments (Preece et al., 2008). However, it may also exert these effects through its iliotibial tract attachments to the tibia, in which case it will affect whole lower limb's internal rotations.

Gluteus maximus activity continues throughout early stance for as long as the GRF vector lies anterior to the hip. As the flexor angular moment decreases through shortening of the GRF's moment arm under reducing hip flexion angles, the gluteus maximus' activity gradually decreases to cease in the middle of midstance only to activate again during late terminal stance phase to aid CoM acceleration to the opposite foot (DeJong et al., 2020). Concentric contractile thickness of the muscle increases around 14% of the gait cycle or ~154 ms before influencing hip kinematics, although its influence on hip kinetics actually occurs earlier, as does its EMG activity. This is consistent with an electromechanical delay between nerve impulses and muscle fibre changes (DeJong et al., 2020). Thus, gluteus maximus activity follows hip flexion moments, causing volumetric increases (thicknesses) in the muscle to occur around 66 ms after the flexion moment initiates (DeJong et al., 2020).

Gluteus maximus and the ankle plantarflexors both slow forward progression of the body's CoM during gait, but throughout different phases. Gluteus maximus works from initial contact through to just before absolute midstance, gradually moving the CoM forwards on reducing hip and knee flexion angles. The ankle plantarflexors take over the role from just before absolute midstance, prior to the ankle starting to dorsiflex, to allow the CoM to advance in front of the leg, falling forward over the ankle. This plantarflexor braking CoM activity continues until heel lift when these muscles are involved in CoM acceleration through generating plantarflexor power via the Achilles elastic recoil (Neptune et al., 2004; Liu et al., 2006). Gluteus maximus also contributes to the CoM acceleration in late terminal stance, along with some active hip extension after heel lift (Neptune et al., 2004; Neumann, 2010; DeJong et al., 2020), effectively taking over the acceleration from the ankle plantarflexors before iliacus 'picks up' the CoM through hip flexion (Fig. 2.3.7d).

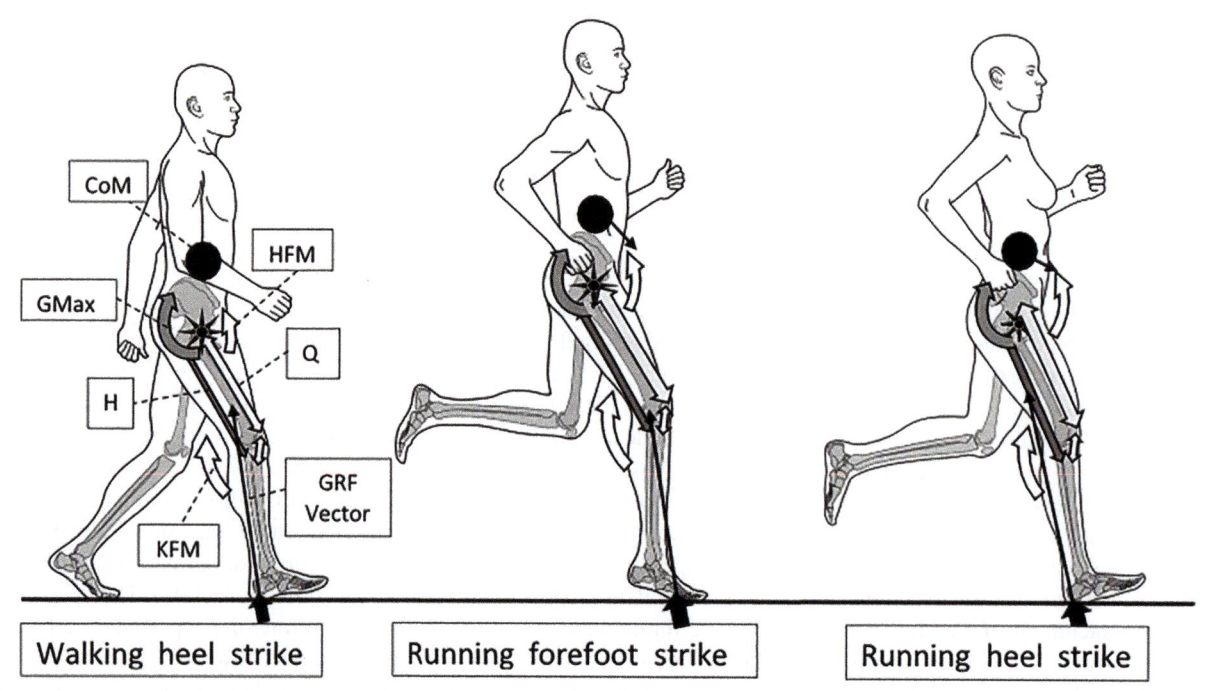

FIG. 2.3.7c Schematics of different types of initial contact and braking events around the hip. In walking, gluteus maximus (GMax) and the quadriceps (Q) act as the primary braking and energy dissipation muscles that control hip and knee flexion moments created by the GRF vector (black arrows). They are initially aided in this role by the hamstrings (H). In walking, the GRF vector is usually relatively close to the hip's joint axis (star) and acceleration forces are relatively low. Thus, the knee flexion moment (KFM) and knee flexion angle should be restrained but will be influenced by the stride length and walking speed. In running, larger accelerations create greater GRF vectors on a single-limb contact, necessitating greater muscular braking activity. In forefoot running, the knee and hip flexion moments will be larger, but the stride is usually short. In longer stride lengths associated with heel strike running, the GRF vector moves further forward from the hip joint axis, increasing the hip and knee flexion moments further. This necessitates that the hamstrings, gluteus maximus, and quadriceps operate at longer, less efficient muscular lengths. Angular moments will increase in the sagittal plane as a result. With a narrower base of support in running, the hip abductors will be at a mechanical disadvantage from internal torques associated with initial contact. *(Permission www.healthystep.co.uk.)*

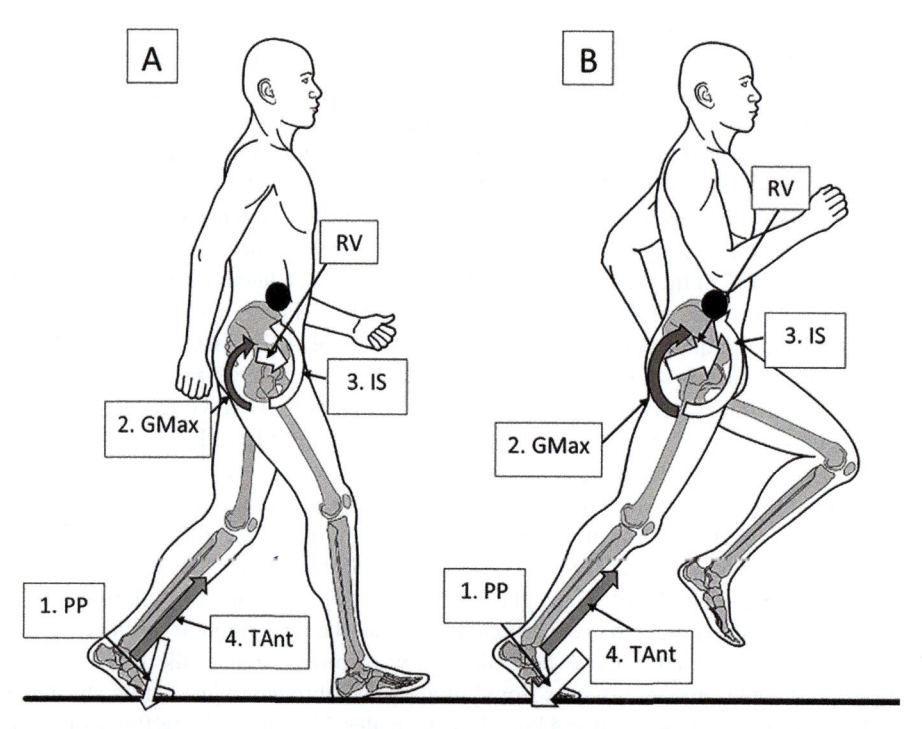

FIG. 2.3.7d Schematics of the acceleration events at the end of terminal stance in (A) walking and (B) running. The acceleration of the CoM at the end of stance phase involves four primary muscles operating in sequence. The triceps surae power, through the Achilles, will first have to generate significant acceleration via ankle plantarflexion power (1. PP). This will need to be much larger in running than during walking. As this ankle plantarflexion power is lost through the release of elastic energy from the Achilles, the gluteus maximus (2. GMax) briefly takes over acceleration, maintaining an anteriorly directed CoM resultant vector (RV—white arrow) which can be 'picked up' by iliopsoas (3. IS) as it starts to flex the hip for swing. Tibialis anterior (4. TAnt) activates to lift the foot away from the support surface as iliopsoas gets swing phase underway via hip flexion, permitting the lower limb to follow the anterior path of the CoM. During running, muscle power must be far greater, for acceleration is faster and the CoM is moved into an aerial phase rather than simply being transferred to the contralateral weightbearing limb. *(Permission www.healthystep.co.uk.)*

Other hip extensors are considered to be the posterior head of adductor magnus, piriformis, and the hamstrings. Both the posterior head of adductor magnus and the hamstring muscle semitendinosus have the greatest moment arm for hip extension in hip flexion, increasing up to 60° of flexion, but with the adductor magnus having a greater cross-sectional area than the hamstrings (Neumann, 2010). Piriformis is orientated laterally from an inner pelvic attachment on the sacrum to an attachment at the apex of the greater trochanter of the femur, where it can aid in flexing the torso at the hip.

Although the hamstrings are primarily considered as knee flexors or resistors of knee extension, they also play a limited role in hip extension (or more correctly resistors of hip flexion) due to their bi-articular orientation across the hip and knee. The hamstrings all arise from the inferior ischial tuberosity but have divergent distal attachments to the tibia and fibula. The hamstrings have a large influence on knee motion in swing and control the velocity of heel contact, but only exert a weak effect directly on hip motion (Barrett et al., 2007). It is worth repeating that although they only cross the knee, the vastus quadriceps muscles, by producing knee extension and decelerating lower limb motion in early stance, generate a significant acceleration on the trunk that creates an important extension moment on the hip (Zajac et al., 2003).

Deep hip rotators

The deep musculature of the hip is functionally less known because the individual muscles are relatively inaccessible to EMG examination. This means their remains some debate over whether these muscles are hip stabilisers, assistors of motion, or only of minimal significance (Giphart et al., 2012; Vaarbakken et al., 2015; McGovern et al., 2017; Parvaresh et al., 2019; Meinders et al., 2021). These muscles are thought to take up the role of 'fine-tuning' hip function, having only short lever arms. Pectineus is reported to have a moment arm less than 9 mm during the stance phase of gait, yet it may be significant enough to cause flexion deformity with internal rotation in overactivity associated with cerebral palsy (Torry et al., 2006). It is moderately active throughout the stance phase, limiting femoral abduction and assisting in the resisting of transverse plane torques. This is because it possesses a small external moment arm if walking is performed with marked internal thigh rotation, whereas it normally has an internal moment arm during erect standing (Arnold and Delp, 2001). Further to this, some minor EMG activity in pectineus is also reported during swing (Torry et al., 2006).

Muscles that are considered to be primary external hip rotators include gluteus maximus, piriformis, obturator internus, gemellus superior, gemellus inferior, and quadratus femoris. The smaller muscles run from medial to lateral across the hip joint, from the pelvis to the greater trochanter. Quadratus femoris is a flattened muscle running between the upper ischial tuberosity to the intertrochanteric crest, situated between the greater and lesser trochanters posteriorly. This is a muscle that externally rotates the hip or resists internal femoral rotation and which may also assist hip adduction. The obturator externus and internus attach from the outer and inner surface of the bone margins of the obturator foramen rim and membrane, respectively. They run to attach almost horizontally with obturator externus attaching to the trochanteric fossa of the posterior medial aspect of the greater trochanter. Obturator internus runs over the posterior ischial surface to create a groove on the ischium where a bursa separates the tendon from the bone before it runs horizontally to the trochanteric fossa on the medial aspect of the greater trochanter in common with the gemelli. The superior gemellus attaches to the ischial spine, while the inferior gemellus attaches at the superior ischial tuberosity and then after running its course, conjoins with the tendon of obturator internus. All act as weak external hip rotators and superior hip stabilisers.

Secondary external rotators are the posterior segments of gluteus medius and gluteus minimus, obturator externus, sartorius, and the long head of biceps femoris. They are not considered as being primary internal hip rotators, but there are a number of secondary internal rotators. These are the anterior segments of gluteus minimus and gluteus medius, adductor longus, adductor brevis, the posterior head of adductor magnus, pectineus, and tensor fasciae latae (Fig. 2.3.7e).

Hip adductors

There are considered to be five primary hip adductors: pectineus, adductor brevis, adductor longus, gracilis, and adductor magnus. The hip adductors are mostly proximally attached to the pubic bone and distally on the femur below the level of the greater trochanter (Robinson et al., 2007; Robertson et al., 2009). Pectineus attaches to the pubis proximally and the pectineal line and linea aspera on the femur distally. By so doing, it contributes to external rotation and hip flexion with adduction. Adductor brevis attaches to the anterior surface of the inferior pubic ramus, inferior to the origin of adductor longus and attaches to the pectineal line and superior medial lip of the linea aspera. Adductor longus attaches proximally and medially to pectineus and distally to pectineus on the linea aspera between adductor brevis and adductor magnus. Because pectineus, adductor brevis, and adductor longus have attachments to the posterior surface of the femur on the linea aspera, they are able to assist external hip rotation or resist internal femoral rotation as well as adduct the hip (Fig. 2.3.7f).

FIG. 2.3.7e A schematic showing the force vectors for transverse plane motion and stability, with external rotators indicated with solid black lines and internal rotators with dashed lines. The hip axis is now orientated vertically to the y-axis of the hip. Once again, it is worth stating that these moment arms from the hip axis change during hip motions during gait. *(Image from Neumann, D.A., 2010. Hip. In: Kinesiology of the Musculoskeletal System: Foundations for Rehabilitation, second ed. Elsevier, pp. 465–521 (Section IV, Chapter 12).)*

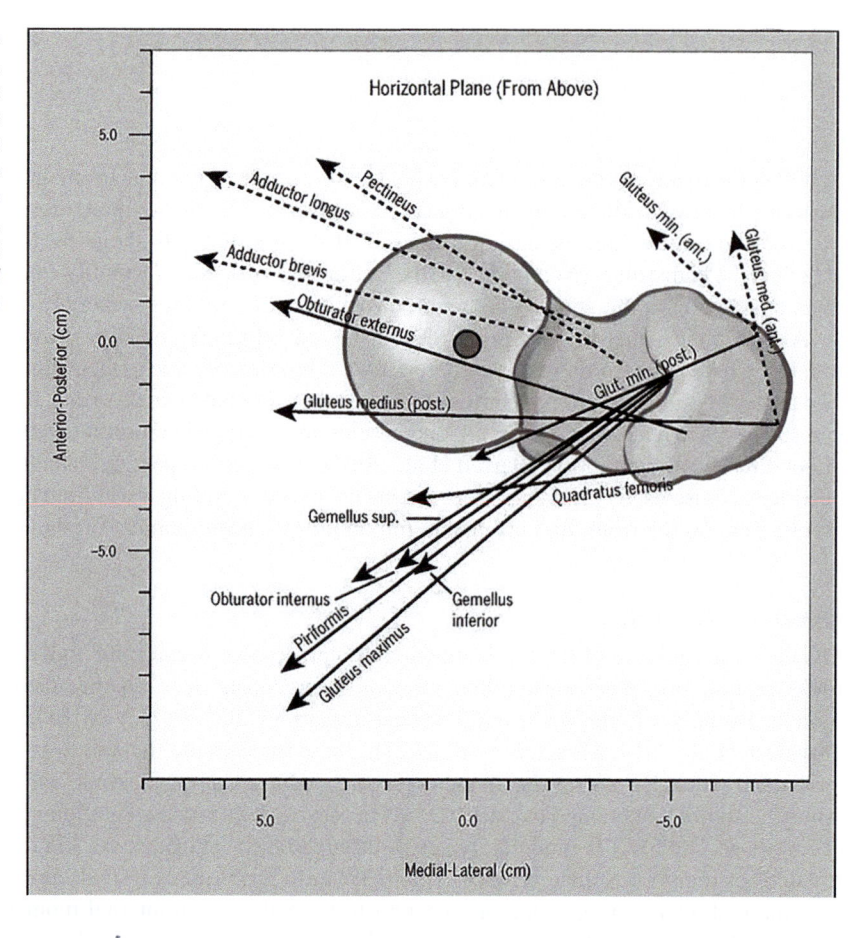

FIG. 2.3.7f A schematic of the adduction moments of the hip around the hip axis (star). Adductor magnus (1) dominates these forces. Its distinct medial and lateral fibres have different proximal and distal attachments with the medial fibres having a moment arm more horizontal, giving more powerful but similar actions to pectineus (4). A more vertical set of fibres create another distinct muscle belly in adductor magnus (1a) running to the adductor tubercle on the medial femoral condyle, forming yet another moment arm for this muscle. Other important adductors are adductor longus (2), adductor brevis (3), and obturator internus, obturator externus, quadratus femoris, and the gemelli (5). Their effect on internal or external rotation depends on the position of the hip's instantaneous axis at any given moment. Biceps femoris (6), semimembranosus (7), and semitendinosus (8) can act as hip adductors in hip abduction angles, but also as hip abductors in extreme positions of adduction, dependent on the moment arms between their distal and proximal attachments in any given position. *(Permission www.healthystep.co.uk.)*

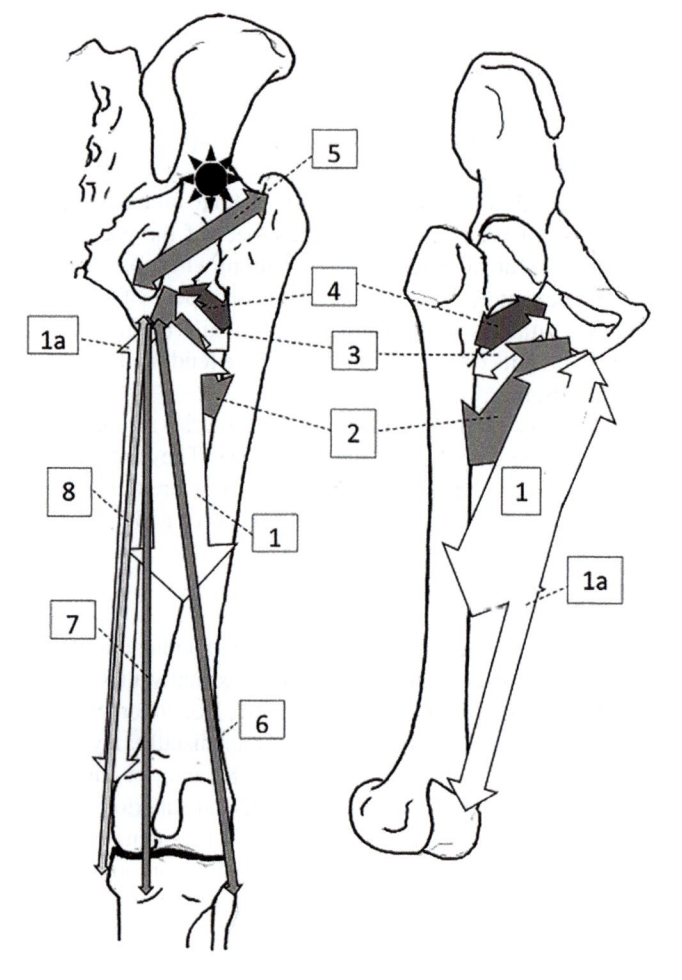

Adductor magnus, a large muscle that arises from the inferior ramus of the ischium and the inferolateral ischial tuberosity, runs to attach almost horizontally into the gluteal tuberosity of the femur. It then courses to a long fibrous enthesis down the medial lip of the linea aspera to the medial supracondylar ridge and on to the adductor tubercle, with a distinct separate tendinous enthesis attachment on the condyle. At its inferior attachments, the muscle fibres run almost vertically to slowly fan out to an almost horizontal fibre orientation proximally at the gluteal tuberosity. Adductor magnus is a powerful hip adductor through its horizontal fibres, but it uses its vertical fibres for hip extension. Gracilis is a thin muscle on the medial border of the thigh, arising on the pubic symphysis and inferior pubic ramus and runs almost vertically to the tubercle behind sartorius' attachment at the pes anserine site, often blending with the sartorius tendon and fascia crura of the leg.

The hip adductors work in relatively 'near-isometric contraction' during stance phase, with their action affected by their alignment relative to the position of the hip joint's axis in relation to their femoral attachments. When stood erect on one leg, the ipsilateral-sided adductors largely remain EMG-silent (Torry et al., 2006). Thus, lower force values from these muscles are reported at hip angles at neutral (0° flexion–extension) but also at 90° of flexion when compared to 45° of flexion. Reduced EMG output from these muscles has overall been correlated with a history of groin strain in athletes (Lovell et al., 2012). During walking gait, the adductors demonstrate different types of phasic activity and a marked difference between the two parts of adductor magnus, with the upper part possessing a pure adductor role that is active throughout the whole gait cycle (Torry et al., 2006). Adductor brevis and longus show triphasic activity with peaks mainly around toe-off at the end of terminal stance (Torry et al., 2006). Secondary hip adductors are the long head of biceps femoris, the posterior fibres of gluteus maximus, quadratus femoris, and obturator externus.

Hip abductors

The hip abductors primarily generate forces to maintain frontal plane stability between the pelvis and lower limb across the hip during the stance phase of gait, producing maximum abduction moments at ~35% of the gait cycle during midstance (DeJong et al., 2020). Hip adduction (varus) moments increase with body mass and gait velocity and occur most significantly from initial contact through to midstance (Rutherford and Hubley-Kozey, 2009). The hip abductor muscles are the resistors of hip adduction rather than being the generators of hip abduction. The most significant adduction moment resistance is generated by gluteus medius and then minimus that lie under the lateral fascia lata. Both tensor fasciae latae and gluteus maximus have important assistor roles via the iliotibial tract (Rutherford and Hubley-Kozey, 2009).

Gluteus medius is proximally attached to almost all of the iliac crest and is shaped like an inverted triangle that attaches distally to the anterolateral aspect of the greater trochanter. It plays a vital role in hip stability during the stance phase of gait as the main hip abductor to stabilise the pelvis on the lower limb (Kumagai et al., 1997; Mickelborough et al., 2004; Bervet et al., 2013). As the primary hip abductor, it is a large flat muscle with a large physiological cross-sectional area orientated in an anterior-to-posterior direction. It is composed of three functional segments of muscle bundles: anterior, middle, and posterior, with potentially independent CNS control to each (Semciw et al., 2013). The posterior segment lies deep to gluteus maximus. All parts of gluteus medius are biphasic (having two bursts of EMG peak activity). However, the anterior portion produces a later peak amplitude during both of these activity bursts, while the posterior and middle segments work in synchronicity (Semciw et al., 2013). Gluteus medius provides femoral head stability through tension-induced compression of the femoral head within the acetabulum. This is probably achieved through the posterior segment which has muscle fascicles arranged parallel to the femoral neck, while the middle section with its larger physiological cross-sectional area has a large moment arm to resist pelvic adduction on the closed chain stance phase lower limb (Semciw et al., 2013).

The anterior fibres are orientated in a relatively vertical direction with a large frontal plane moment arm assisting the middle gluteus medius segment in resisting hip adduction during stance phase. It has an additional role through its later peak activity of helping to reduce anterior hip joint extension forces by resisting hip extension when the hip is already extended (Lewis et al., 2007; Semciw et al., 2013). Gluteus medius may also be recruited to produce a forward contralateral rotation of the pelvis in the transverse plane. Its large physiological cross-sectional area makes it highly suitable for aiding anterior terminal swing foot placement in the contralateral limb through pelvic anterior rotation (Semciw et al., 2013). The anterior segment of gluteus medius is also better suited than the other segments to stabilise the femoral head in mid-to-late stance phase (Semciw et al., 2013). However, it contributes little support during early stance (Anderson and Pandy, 2003) (Fig. 2.3.7g).

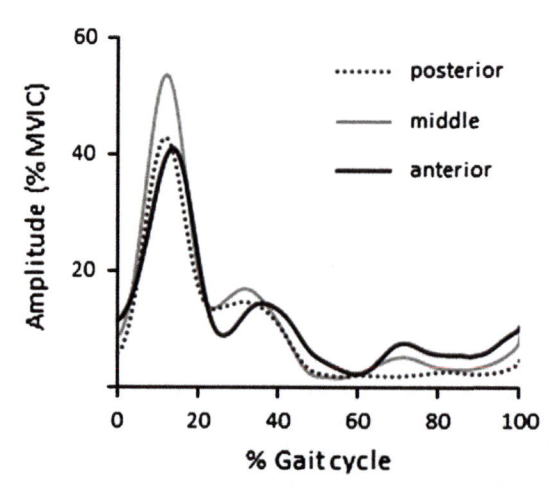

FIG. 2.3.7g EMG assembly averages for the anterior, middle, and posterior segments of gluteus medius during the walking gait cycle. Note the peak activity of the posterior and middle segments in loading response and in early midstance. The anterior segments initiate slightly later, but become more significant compared to the other segments during late single support and into terminal stance. *(Image from Semciw, A.I., Pizzari, T., Murley, G.S., Green, R.A., 2013. Gluteus medius: an intramuscular EMG investigation of anterior, middle and posterior segments during gait. J. Electromyogr. Kinesiol. 23(4), 858–864.)*

Gluteus medius assists as a hip rotator as well as assisting hip adduction (Al-Hayani, 2009; Semciw et al., 2013; Uemura et al., 2018; DeJong et al., 2020). Bursts of activity are noted at initial contact and at midstance, indicating biphasic stance phase activity with the greatest activity occurring during early stance. Around 50% of that activity amplitude is noted during midstance (Hof et al., 2002; Mickelborough et al., 2004; Semciw et al., 2013) creating strong hip abduction and weaker external rotation moments (DeJong et al., 2020). This activity corresponds to the CoM-induced pelvic 'drop' into adduction at the hip observed in the frontal plane during midstance (Chang et al., 2005; Mündermann et al., 2005; Astephen et al., 2008a, b; Semciw et al., 2013). Gluteus medius starts to 'bulk up in thickness' due to contraction volume increases prior to the changing hip motions at ~20% of the gait cycle in the transverse plane and at ~17% in the frontal plane (DeJong et al., 2020).

Gluteus minimus lies deep to the gluteus medius and is attached to the gluteal surface of the ilium. It courses distally to attach to the anterolateral aspect of the greater trochanter, deep to gluteus medius (Fig. 2.3.7h). Its muscle fascicles run parallel to the femoral neck and they are aligned to draw the head of the femur in a superior and medial direction into the acetabulum, creating superior tensile hip compression stability along with gluteus medius (Al-Hayani, 2009; Semciw et al., 2014). These compression forces peak at contralateral toe-off (~18% of gait cycle) as the limb starts single support and again at the end of single support with the contralateral limb's heel strike (~45% of gait cycle) (Semciw et al., 2014). Gluteus minimus is an internal hip rotator as well as an abductor, but can be either a flexor or an extensor depending on which areas (anterior or posterior, respectively) of the muscle activate (Semciw et al., 2014) (Fig. 2.3.7i).

Gluteus minimus has anterior and posterior segments that are capable of independent function (Semciw et al., 2014). The posterior segment has its greatest activity earlier in gait, from initial contact through to 20% of gait. The anterior segment demonstrates peak activity from 20% to 60% of gait, and the two segments seem to be able to function and act independently (Semciw et al., 2014). It appears that the initial burst of gluteus minimus activity stabilises the hip through medial–superior hip joint compression force in early stance from the muscle's posterior segment. This is followed by co-contraction in late midstance to counteract the high anterior hip joint forces that can occur with hip extension during late stance, working in conjunction with iliopsoas (Semciw et al., 2014). This should help protect the iliofemoral ligament and anterior–superior parts of the acetabular labrum. The anterior segment of gluteus minimus is also highly active during internal hip rotation and has a large internal rotation torque potential. With the lower limb fixed to the ground during stance, it is able to provide assistance to rotate the pelvis forward on the contralateral side, albeit this role may arise more from anterior gluteus medius which has the larger cross-sectional area (Al-Hayani, 2009; Semciw et al., 2013, 2014).

Tensor fasciae latae attaches to the anterior part of the iliac crest, inferior to gluteus medius and the anterior superior iliac spine, with strong fascial attachments to the fascia lata overlying it. It runs inferiorly to attach to the anterior fibres of the iliotibial tract. Tensor fasciae latae is a hip abductor, internal rotator, and weak flexor that helps create lateral knee and hip stability. It demonstrates moderate activity in hip flexion, medial rotation, and abduction, acting biphasically during stance phase (Torry et al., 2006). It is also reported to be active during the hip flexion phase of cycling, unlike the glutei (Torry et al., 2006). Piriformis, sartorius, and rectus femoris can at times assist in hip abduction, with their overall effect depending on the hip positioning of the femur in influencing their lever arms.

Locomotive functional units **Chapter | 2** 331

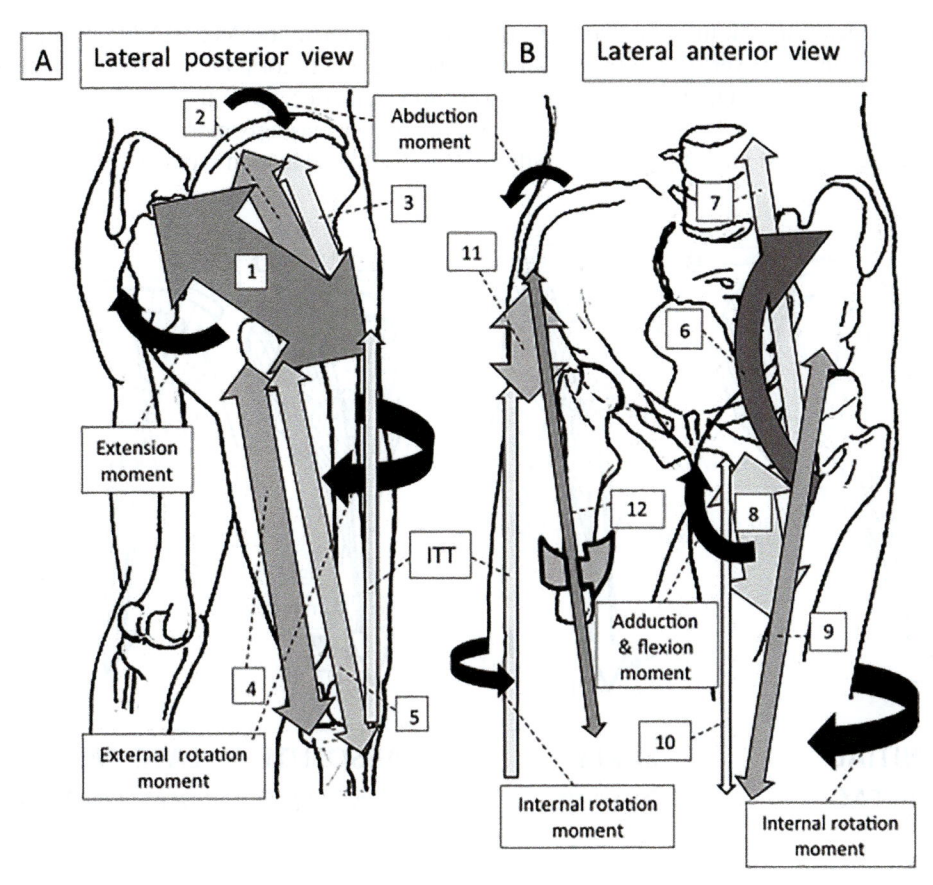

FIG. 2.3.7h Schematics of the hip abductors and surrounding muscle vectors illustrating their complex interactions with the other hip muscles. The posterior muscular area (A) can resist hip moments of internal rotation and hip flexion that are occurring during loading response into early midstance via their extension and external rotation moments. Muscles primarily involved are gluteus maximus (1), the posterior and intermediate fibres of gluteus medius (2), and the posterior fibres of gluteus minimus (3). Multiplanar hip moments can be fine-tuned at initial contact through gluteus maximus' influence on the iliotibial tract (ITT) and the medial (4) and lateral hamstrings (5), while it also maintains hip adduction resistance during hip extension out of flexion. Such resultant moments oppose those of the hip flexor and adductor muscles, that are setting the stance limb's position at initial contact (B—left limb) following previous activity from iliacus (6) and psoas (7) during swing, and the adductors (8), and rectus femoris (9) on loading. The hip adductor's combined resultant force is variably angled through distinct activity within adductor longus, adductor magnus, and pectineus, aided by gracilis (10). The anterior hip abductors consist of the anterior fibres of gluteus medius, gluteus minimus, and tensor fasciae latae (11). These muscle bundles are primarily active during hip extension after absolute midstance (B—right limb). They not only maintain the hip abduction moment during late midstance and terminal stance to resist pelvic varus tilt, but they also produce a significant internal rotation and flexor moment on the hip. This is because their proximal attachments lie anterior to the hip joint's vertical instantaneous axis during late stance phase. These moments can help pelvic anterior rotation during late contralateral swing (adding iliopsoas limb advancement) and assist in resisting hip extension in the ipsilateral limb's late stance phase. Sartorius (12) links the ilium to the tibia and seems to stabilise the knee on an internally rotating pelvis during late stance when the knee is near full extension. *(Permission www.healthystep.co.uk.)*

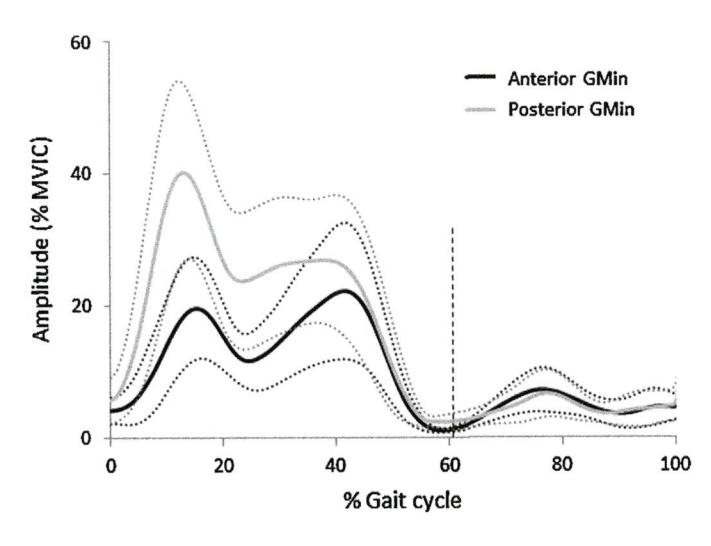

FIG. 2.3.7i Gluteus minimus is a hip abductor and internal rotator aiding pelvic frontal plane stability and anterior pelvic rotation for the contralateral swing side. EMG averages of the anterior and posterior segments are demonstrated with 95% confidence intervals shown by dotted lines. Differences between the muscle segments' activities are likely to relate to the areas where their EMG lines do not overlap the confidence intervals. Data indicate that the anterior segment tends to activate later than the posterior segment. Posterior segment muscle activity peaks in walking gait at late loading response and during early midstance with some anterior segment activity. During late single-limb support to heel lift, both segments are active, but the anterior segment's peak occurs during this period. *(Image from Semciw, A.I., Green, R.A., Murley, G.S., Pizzari, T., 2014. Gluteus minimus: an intramuscular EMG investigation of anterior and posterior segments during gait. Gait Posture 39(2), 822–826.)*

The way the hip abductors appear to work is from posterior-to-anterior activity. As the CoM of the body moves from a position posterior to one anterior over the lower limb and hip, the more posterior portions of gluteus medius and gluteus minimus initially provide peaks of activity. As the CoM moves further anteriorly over the hip, the anterior parts of gluteus medius and gluteus minimus and finally the tensor fasciae latae become increasingly more active. Thus, a 'pass the parcel' situation exists to move the CoM anteriorly while avoiding excessive hip adduction moments (Fig. 2.3.7j).

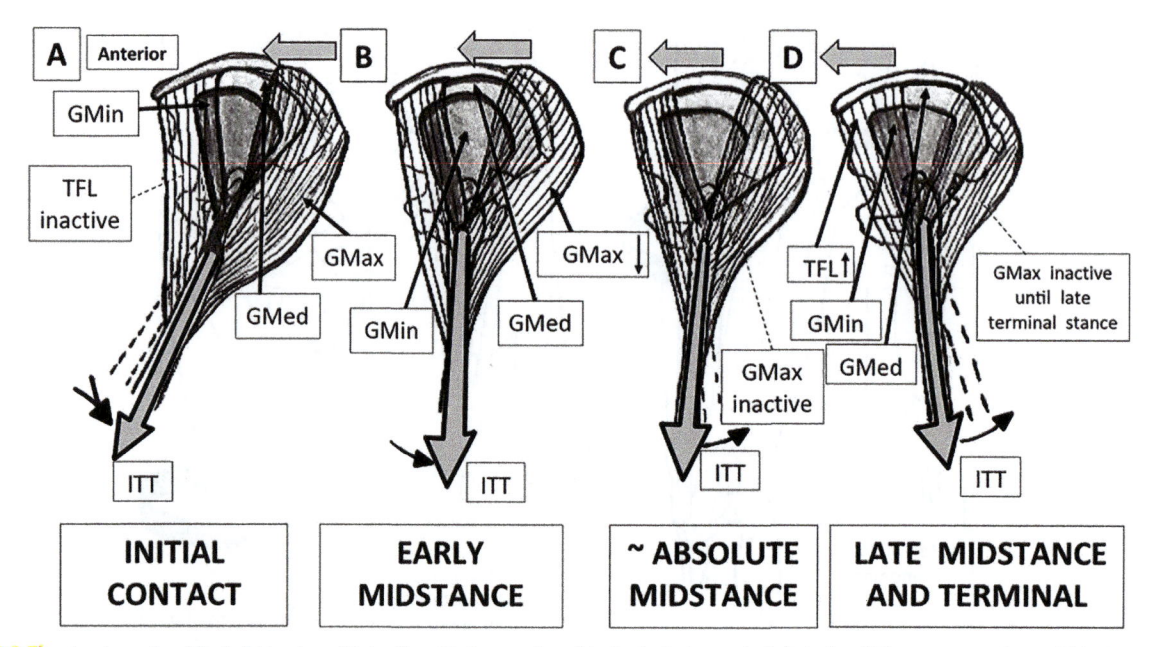

FIG. 2.3.7j A schematic of the left hip viewed laterally with the anterior of the body facing to the left during different stance phases. This demonstrates how the hip abductors can be utilising in the sagittal plane as assistors of hip extension motion while maintaining resistance to hip adduction moments. At initial contact (A), the gluteus maximus bound into the posterior fibres of the iliotibial tract (ITT), the posterior and intermediate fibres of gluteus medius, and the posterior fibres of gluteus minimus are primarily active. Tensor fasciae latae (TFL) is largely inactive at contact and early midstance. At this time, the HAT segment's CoM (represented by the superior grey arrow) is positioned posterior to the hip but moving forward. The hip is under adduction and internal rotation moments during early stance phase that are resisted by the posterior segments of the hip abductors. In early midstance (B), the hip abductor activity continues with gluteus maximus activity decreasing until it falls silent at absolute midstance (C). As the HAT CoM passes anterior to the hip, it requires a change in muscle activation patterns with gravity-induced adduction moments still requiring resistance, but extension moments now needing some braking management. This initiates an increase in anterior hip abductor activity, including tensor fasciae latae (D), with the burden of effort falling more towards the anterior fibres of gluteus medius and minimus. This anterior activity provides internal hip rotation that aids anterior pelvic rotation to increase step length on the swing limb and helps resist hip extension on the stance limb. In this way, the forward passage of the HAT segment's CoM above the hip is moved anteriorly in stages through activity initially in the more posterior muscular parts of the hip abductors during loading response and then towards the anterior muscular components during late midstance and acceleration. *(Permission www.healthystep.co.uk.)*

Hip muscle stabilisation

The muscles of the hip constitute the primary active elements of structural stability. The distribution of 80% of the femoral loading force is directly attributable to instantaneous muscle activity (Duda et al., 1998; Lewis et al., 2010). Forces are dominated by large compressive forces with relatively small amounts of shear, both of which increase from the diaphyseal to the metaphyseal region of the femur during walking (Heller et al., 2001). This effect is related to the relatively large forces generated from the hip abductors (Heller et al., 2001) and conforms to the tension-to-compression stability that reduces shear through biotensegrity principles. The action of the muscles around the hip is affected by their position at activation because of changes in the muscles' lines of action and lever arm lengths and positions in relation to the hip's instantaneous axis of rotation. This alteration in function usually affects the secondary muscles' roles more than those of primary muscles. Adductor longus is a hip flexor at 50° of hip flexion but becomes an extensor at 70° of flexion (Byrne et al., 2010). With hip flexion, lines of action are moved anteriorly, usually increasing the hip flexor effects, whereas extension can move more muscle tendons to lie posterior to the hip axis, increasing the effects of the extensors.

Vertical GRFs tend to peak at just over body weight, but vertical hip forces have been calculated at over four times body weight as a consequence of gravity, inertia, and muscle activity (Stansfield and Nicol, 2002; Lewis et al., 2010). Muscle forces provide stabilisation, which if imbalanced or weak, may result in increased forces within the joint (Bergmann et al.,

2004; Lewis et al., 2009). Anterior hip joint forces are generated by the anterior portion of gluteus medius and iliopsoas (Correa et al., 2010; Lewis et al., 2010). Thus, it could be expected that reduced hip flexor activity would reduce anterior hip joint forces. Nevertheless, reduction in iliopsoas activity during preswing results in substantial synergistic hip flexor activity that increases the joint force (Lewis et al., 2009). This is likely to result in less compression but increased shear force. To reduce anterior hip joint forces, it may be better to increase the ankle power to aid preswing (Lewis and Ferris, 2008), possibly assisted by increased gluteus maximus strength.

Hip joint moments have been linked to knee joint loading, particularly in the frontal plane (Chang et al., 2005; Mündermann et al., 2005; Bennell et al., 2007). Increased hip joint adduction moment has been associated with osteoarthritis of the knee (Chang et al., 2005; Mündermann et al., 2005; Astephen et al., 2008a, b), knee ligament injuries (Russell et al., 2006; Herman et al., 2008), iliotibial tract pathology (Fredericson et al., 2000), and symptoms of anterior knee pain (Brindle et al., 2003). No clear association between hip abductor strength and knee adduction moments has been found (Rutherford et al., 2014). However, peak knee adduction moment increases with an ipsilateral Trendelenburg gait by 25% (Dunphy et al., 2016), suggesting that hip abductor function probably plays an important role in protecting the medial compartment of the knee from varus compression.

2.3.8 Adaptation and pathology in the hip

Hip pathology can affect strength, control, and muscle extensibility, resulting in symptoms and loss of gait energetics. Abnormal hip kinetics and kinematics can result in degenerative changes within the hip joint or its connective tissues. Young athletes are commonly affected by hip and groin pains due to traumatic events. Minor hip injury from repetitive motions commonly leads to chronic pathological conditions, resulting in capsular laxity and labral tears (Torry et al., 2006). Athletes with previous groin injuries demonstrate a significant fall in some adductor muscle EMG outputs (Lovell et al., 2012). In addition, there is also some relatively strong evidence pointing to a link between patellofemoral joint symptoms and a reduction in gluteus medius activity duration in runners (Semciw et al., 2016).

Distribution of pain associated with hip osteoarthritis is usually in the groin and buttock with ~85% of patients with hip dysfunction reporting groin pain. Interestingly, however, around 47% of patients experience pain below the knee (Khan et al., 2004). Anterior hip pain or groin pain may be symptoms that are brought on initially as a consequence of excessive hip extension during gait. Gait models have indicated that increased hip extension moments lead to increased force in the anterior hip (Lewis et al., 2007, 2010), with only a 2° increase in maximum hip extension estimated to result in a 24% increase in anterior hip joint force through muscle activity (Lewis et al., 2010). Anterior hip joint forces reach their peak at near-maximal extension (Lewis et al., 2007, 2009). Thus, reduced hip extension in gait may be a possible protective mechanism for hip pain (Lewis et al., 2010). Increased gluteal and iliopsoas activity can reduce anterior joint force (Lewis et al., 2007) and should therefore be a target for rehabilitation for the protection of the anterior hip.

Hip osteoarthritis (OA) is associated with changes in hip muscle size and function, with decreased hip flexor function resulting in increased hip joint forces (Lewis et al., 2007, 2009; Mendis et al., 2014). This could be a result or cause of hip OA. Hip flexor strength is reported to be greater on the dominant side, but it is also reported as being weaker on the pathological hip's side. However, in acetabular labral tears which are associated with the development of OA, the size of the hip flexors has been reported as being similar to healthy controls, despite reduced strength (Mendis et al., 2014). This suggests that hip muscle atrophy occurs after OA development and may not be a cause of it.

OA in the hip joint has significant effects on gait biomechanics and is a condition usually affecting older adults. A slowing of preferred walking speeds reflects normal decreases associated with age, but individuals with hip OA demonstrate gait speed changes and changes in spatiotemporal parameters occurring on the affected hip side only (Kiss, 2010). Asymmetry in both spatiotemporal parameters and angular joint motions increases if walking speed occurs outside that of OA patients' preferred walking speeds, which possibly explains the increased risk of falls in this group (Kiss, 2010). Variability and asymmetry in spatiotemporal parameters increase with the severity of hip OA, while the variability of hip motion decreases with reduction in hip flexibility, thereby reducing the ability to adapt motion around the hip during gait (Kiss, 2010).

Total hip arthroplasty with insertion of a prosthetic hip as a consequence of OA is a successful surgical procedure used to alleviate hip pain and improve quality of life measures. However, the procedure does not return patients to normal gait mechanics 12 months postoperatively (Beaulieu et al., 2010). It is not unusual to see patients with gait quality issues or gait-induced symptoms in the clinic following hip replacement. Such patients demonstrate less hip abduction moment generation (and thus larger hip adduction moments in stance phase) and decreased sagittal plane motion. They also generate lower power and achieve less energy absorption/dissipation compared to heathy controls, even when the slower walking speeds of hip replacement patients are taken into account (Foucher et al., 2007; Mont et al., 2007; Beaulieu et al., 2010).

Kinematic changes are also seen at the ankle joint with greater peak dorsiflexion angles (Mont et al., 2007; Beaulieu et al., 2010). The unoperated side also demonstrates kinematic changes with reduced hip adduction in transition from single to double stance compared to healthy controls (Beaulieu et al., 2010). The abnormal kinematics could either be a result of pain-avoidance techniques being maintained postsurgically or through adaptations to or apprehensions to the prosthesis. Therefore, clinicians should not expect patients with replaced hips to display normal gait kinematics. Hips following resurfacing surgery show hip abductor and extensor moments and hip kinematics that are more 'normal' in comparison to prosthetic hips (Mont et al., 2007). The insertion of prosthetic hips also risks upsetting angles between the femur and acetabulum, suddenly changing life-developed musculoskeletal relationships between the pelvis and lower limb. This can lead to problems both locally at the hip and/or elsewhere within the lower limb's biomechanical chain.

2.3.9 Section summary

The hip is a functional unit that expresses high mobility and stability in its articulation through highly adaptable muscular tensile components that induce strong compression stability, minimising shear. This indicates the classic biotensegrity features of a mobile joint. The hip is the most proximal segment of the lower limb that utilises primarily class three lever energetics in the sagittal plane to induce large ranges of motion through small ranges of muscle length change excursions. It controls the HAT segment's angle over the support limb during sagittal translation, assisted by a class one lever mechanism provided by the hip abductors. The hip also facilitates the advancement of the lower limb under the pelvis in the sagittal plane during swing, using class three lever energetics. Transverse rotations of the pelvis on the hip help facilitate swing limb anterior translation. Thus, the swing lower limb can initiate centrifugal forces that aid in moving the CoM anteriorly, improving energetics via transverse pelvic and sagittal thigh rotations.

A deep core of passive tensile elements maintains the hip joint within its state of relatively high congruency, with ligaments that do not only relax for hip flexion in order to permit free forward progression of the swing limb, but also contain a tightening mechanism, particularly at the iliofemoral ligament to provide trunk stability throughout hip extension. Passive stability is supplemented by large powerful hip muscular activity throughout its motion by creating strong hip compressive forces particularly during stance phase. This muscular burden of stability is gradually transferred from posterior to anterior as the stance phase progresses. Gluteus maximus and iliacus muscles provide the most significant sagittal plane mobility and stability. The hip abductors provide the most significant frontal and transverse plane stability during stance. They are aided by a large assembly of secondary assistor muscles creating an ability to adjust hip motion and limb and trunk posture and stability, thus adapting to any gait perturbation adjustments as and when required. This allows the hip joint to participate in a greater range of functions, permitting humans to cross complex terrains including the ability to climb, swim, and partake in unusual forms of biological locomotion such as cycling.

2.4 Functional unit of the knee

2.4.1 Introduction

The human knee is anatomically an inherently unstable structure reliant on its soft tissue forces to maintain stability in motion. Its articular surfaces offer poor congruency and form closure, and as with the hip, it operates in an unusually extended position compared to the knees of other animals. Knee stability is achieved primarily by muscle activity and aided by mechanical guiding and restriction from ligaments and special fibrocartilaginous structures called *menisci*. Knee stability is highly reliant on sensorimotor feedforward–feedback relationships from information provided by the soft tissue structures themselves. The knee is sagittal plane-dominant in its motion, with very limited amounts of anterior–posterior translation, transverse rotation, and frontal plane motion under normal circumstances. Motion becomes far more restricted in these planes when the knee is loaded at near-extension via tensile–compression stability during stance phase than it does during a nonweightbearing clinical examination.

The knee comprises two articulations with three distinct articular surfaces. The joints are the tibiofemoral joint, split into its medial and lateral compartments, and the patellofemoral joint. This latter joint is made up of the patella (the largest sesamoid bone in the human body) and the anterior distal part of the femur through their corresponding articular surfaces. Stability of both knee articulations is governed by passive ligaments and active muscular tensional forces creating joint compression. The tibiofemoral joint compresses together through the meniscocapsular fibrocartilage over the local chondral and osseous topography. The knee fits the criteria of a biotensegrity structure, as its soft tissue 'cables' are essential in maintaining stability through a large range of motion. Biotensegrity principles help explain the 'stability-relationships' of muscles and ligaments in creating stability throughout mobility around the knee's instantaneous functional axis (Kim and Park, 2018).

2.4.2 The kinematic role of the knee

The knee has two fundamental roles when operating as simple engineering levers. These are either to move the lower limb below the knee (lower leg and foot) or to allow the body and thigh segments above to move around the knee. Primarily, these motions occur in the sagittal plane of the lower limb. In swing phase, the primary demands are those of knee flexion during early swing, used to gain ground clearance to midswing, followed by knee extension to advance the leg and foot anterior to the knee before ground contact. Knee extension in late swing causes increased momentum on the CoM of the lower leg and foot to aid centrifugal forces to pull the body's CoM along behind it (Fox and Delp, 2010; Yamazaki et al., 2012). Such knee extension moves the whole lower limb's CoM away from the hip joint's axis, helping to increase the centrifugal force by increasing the hip's angular momentum. The hip must flex in the swing limb throughout swing phase to aid this process (Fig. 2.4.2a).

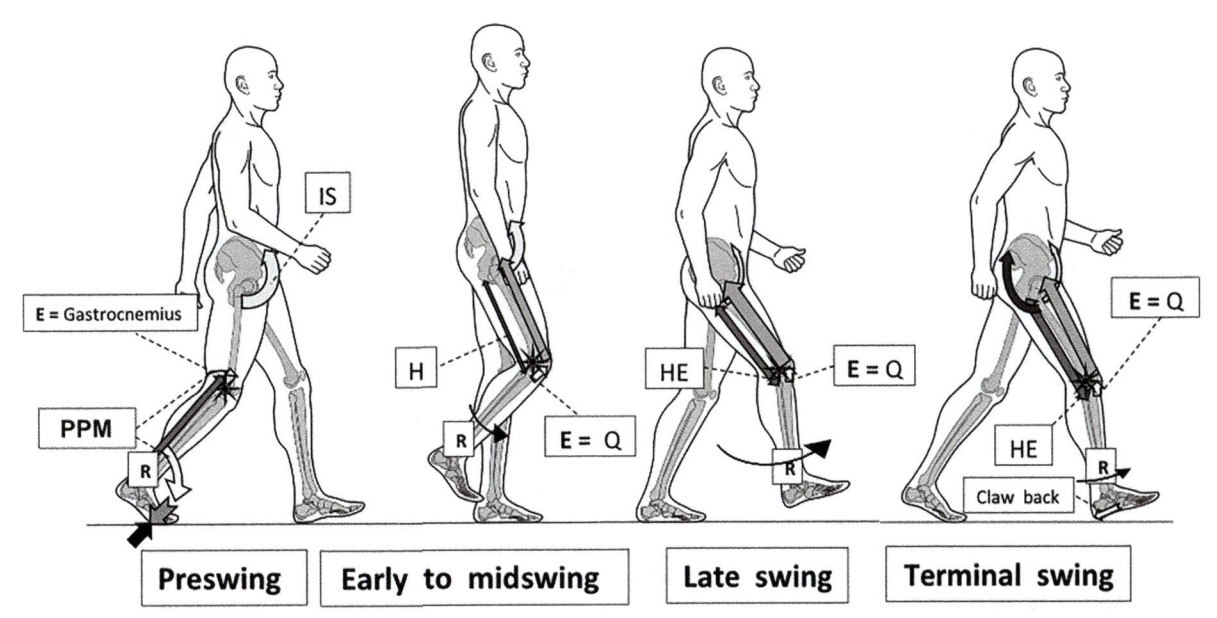

FIG. 2.4.2a The swing phase resistance (R) is the CoM of the lower leg and foot below the knee joint axis, acting as the fulcrum (star). In preswing, knee flexion occurs under the ankle plantarflexor powered moments (PPM) generated by gastrocnemius that bridges the knee, ankle, and rearfoot via the Achilles. This aids iliopsoas (IS) in advancing the lower limb forward. Once in early swing, the knee flexion position can be largely maintained through a little hamstring isometric activity providing weak hamstring effort (HE) at their distal attachments. At midswing, quadriceps activity starts to increase, providing significant effort (QE) at the tibial tuberosity through the patellar ligament, extending the knee. This sets up increasing momentum on the lower leg. As R moves away from the hip axis through knee extension, it imparts increasing centrifugal forces to the swing limb. By terminal swing, significant hamstring effort is required to control the rate of knee extension, so that effort from both the quadriceps and hamstrings occurs below the knee axis, with the quadriceps effort dominant. Just before heel strike, hamstring forces should match that of the quadriceps knee extension effort and the momentum of R, stopping further knee extension. Hamstring effort may now pull the heel backwards into the ground just before initial contact, causing the claw back phenomenon seen on a force–time curve. This probably derives from recoil elasticity within the hamstring's tendons. *(Permission www.healthystep.co.uk.)*

In stance phase, the knee motion occurs under muscle and body weight- compression-induced stability, otherwise postural equilibrium above would be lost. Due to poor form closure of the articular surfaces, the active and passive soft tissue structures of the knee largely limit knee flexion and extension motions through strong force closure. The knee should never become fully extended and should only extend near to 0°, remaining in some degree of flexion. Therefore, the lever system for the knee has to provide for a variation in knee flexion angles through flexion and extension motions.

The knee is subjected to internal extension moments after initial contact to resist external flexion moments, but there are variations, most significantly during running when the knee continues to flex throughout all of early stance phase. The contact GRF is usually slightly posterior to the knee, but sometimes the GRF can pass directly into the knee axis when exposed to particularly long stride lengths. In such circumstances, the knee may even experience external extension moments that must be resisted. This is because very long stride lengths move the heel more anterior to the knee. Initial contacts on shorter stride lengths keep the heel behind the knee, creating GRFs that induce knee flexion. The nearer the GRF is to the knee's instantaneous joint axis, the smaller the GRF moment arm and the easier muscular stabilisation of the knee moments will be. The HAT's CoM acting as the resistance will be posterior to the knee at contact, setting up a flexion torque across the joint. Concurrent hamstring and quadriceps co-activation efforts manage these initial contact GRFs and CoM moment arms around the knee. Shortly after heel contact as loading response progresses, flexion moments 'win out' requiring continued quadriceps activity only, thereby preventing knee over-flexion during early stance phase loading. Knee flexion management itself is an important part of energy dissipation at impact (Fig. 2.4.2b).

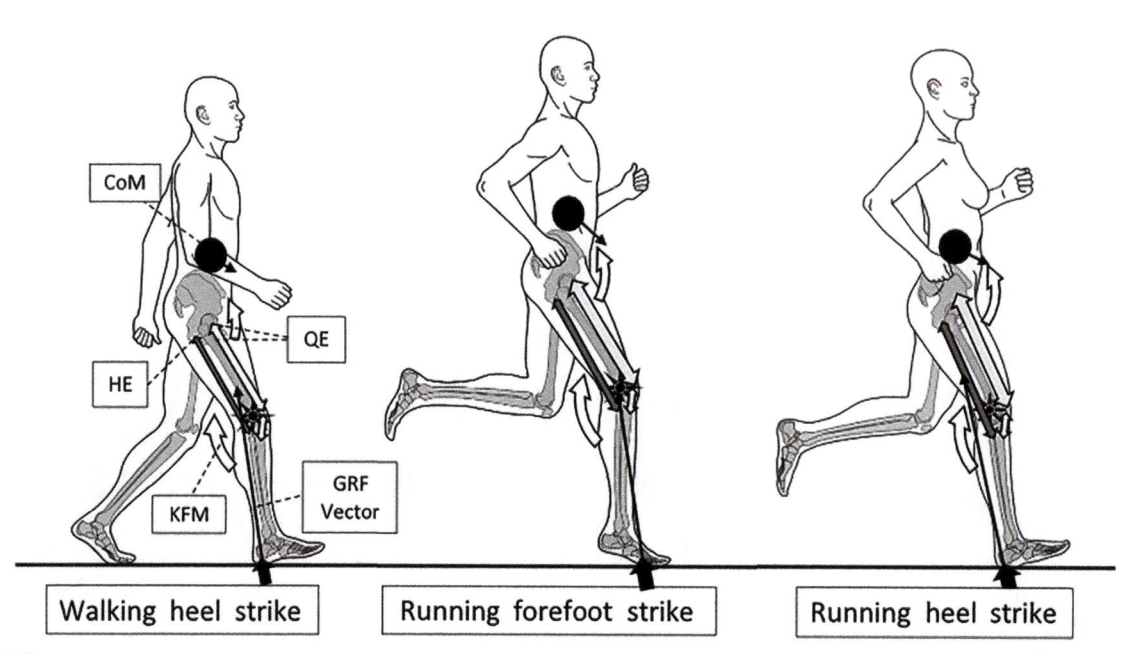

FIG. 2.4.2b Walking heel contacts usually occur with the knee at a small but significant, knee flexion angle. This means that heel contact occurs behind the knee joint axis (fulcrum—star) setting up a knee flexion moment (KFM). The resistance is the HAT CoM, which being behind the knee, is also creating a knee flexion moment. Muscular activity minimises the resultant flexion force. In running, resistance also includes the weight of the contralateral swing limb as initial contact is now a single-limb support event. The primary agonistic muscle is shown with light-grey arrows and the assisting/antagonistic muscle with a dark-grey arrow in each contact situation. In walking, knee flexion moments are resisted by the quadriceps setting up points of effort at their proximal attachments (QE) to the anterior femur and in the case of rectus femoris, on the anteroinferior pelvis. The hamstrings are also active, providing a proximal effort (HE) to the ischial tuberosity. The HE is resisting hip flexion moments more than generating knee flexion, but this force can initially help stabilise the knee, providing an antagonistic effect against the quadriceps. Any running strike positions, including rearfoot strikes, that generate a GRF posterior to the knee axis sets up an external KFM, the size of which is dictated by the knee and foot position at initial contact. However, the very long strides sometimes seen in heel strike running can move the GRF directly into or anterior to the knee joint axis. This risks increased articular compression and/or creates knee extension moments that require increased hamstring activation to protect the knee and help initiate knee flexion. *(Permission www.healthystep.co.uk.)*

During walking, early midstance knee flexion angles and moment arm lengths are reduced through the activity of the quadriceps and gluteus maximus, reducing the flexion moment at the hip. This shortens the distance between the knee's instantaneous joint axis and the body's CoM until the knee sits below the latter at around absolute midstance, reaching its maximally extended position. After absolute midstance and into late midstance, the sagittal moments of the knee are controlled through the ankle plantarflexors as the CoM falls forward under gravity through the inverted pendulum effect of walking. Thus, the knee is consequently subjected to a significant extension moment. This must be resisted by the knee-bridging proximal gastrocnemius attachments and through the anterior hip abductor muscle effects via the iliotibial tract, keeping the knee slightly flexed (Fig. 2.4.2c). Should the knee slip into extension, the iliotibial tract and gastrocnemius are able to become knee extensors, which should be avoided. After heel lift, the knee flexes due to gastrocnemius activity (as long as the knee does not become fully extended) via the Achilles elastic recoil and through the action of iliopsoas at the hip.

Locomotive functional units **Chapter | 2** 337

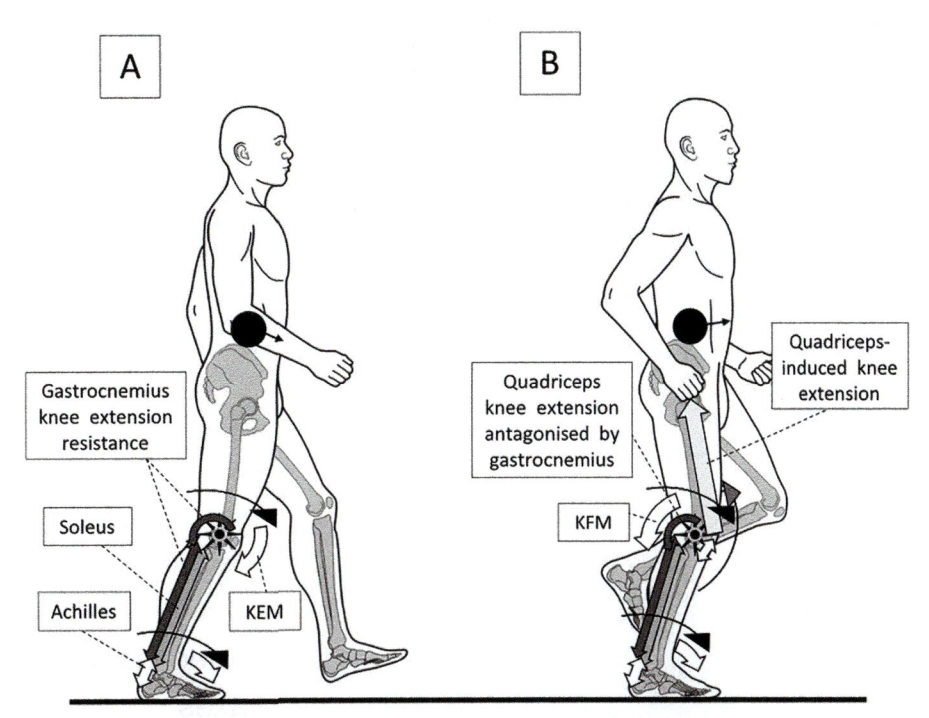

FIG. 2.4.2c The control of the knee during late stance phase is quite different between walking and running. In walking, after absolute midstance (A), the knee extension moment (KEM) is achieved through gravitational forces on the HAT's CoM falling anteriorly. This requires knee flexion moments to prevent excess knee extension by controlling the anterior progression of the CoM over the foot rotating at the ankle. The proximal attachments of gastrocnemius above the knee on the posterior femoral condyles are perfectly placed to provide this restraint as part of the triceps surae complex activity with soleus (S) and the Achilles, that together are restraining CoM acceleration via anterior tibial rotation control. In running (B), there is no such inverted pendulum phase of weight transference. Instead, triceps surae and the quadriceps (Q) are both active throughout stance, with the triceps surae complex storing ankle plantarflexion power for acceleration as it limits ankle dorsiflexion, and the quadriceps muscles braking knee flexion moment (KFM) and then initiating knee extension acceleration. Gastrocnemius is well placed to assist in providing both knee and ankle acceleration power in running, by creating a posterior translation on the femoral condyles as the quadriceps extend the knee. Thus, gastrocnemius can be a resistor or a provider of knee extension power. *(Permission www.healthystep.co.uk.)*

During running, the knee flexes to far greater degrees throughout early stance to a peak flexion angle at the middle of stance, which serves as an important braking and energy dissipation mechanism. After the middle of stance, the knee starts to extend out of flexion to assist acceleration. The knee's mechanical burden in running is far greater than during walking, especially for the quadriceps.

Complex models of knee function have been developed to explain the mechanical features of the knee. Such models have looked at the knees changing muscle moment arms, articular contact points, and ligament kinematics to help understand the knee's mechanical function and the stresses that develop in adult's (Clément et al., 2015; Gasparutto et al., 2015) and children's gait (Barzan et al., 2019). It can be hoped that through these models and comparisons with living subjects, a greater comprehension of human knee kinetics and kinematics will be achieved to advance our understanding in the treatment and prevention of common knee pathologies.

2.4.3 The knee as a biotensegrity structure

Knee biotensegrity components include the tensile active muscles and passive guiding ligaments which compress the articular contact points together around the bony compression struts. The articular surfaces are orientated to impose internal tensions that reduce the freedom of motion and ensure immediate mechanical stability responses throughout the system in both static weightbearing and dynamic motion of the knee (Kim and Park, 2018; Jung et al., 2019; Kim, 2020). Tensional forces transmit themselves over the shortest distance between two points, so that the elastic biotensegrity members are precisely positioned to best withstand the applied stress, offering maximal strength for any given amount of material to create stable equilibrium (Kim, 2020). This means that the orientation of the ligaments and the attachment points of the muscles have evolved so that despite an inherently unstable articular shape, the knee remains stable in all functional orientations throughout stance and swing phase. This is, of course, as long as all the structures remain intact. Kim (2020) takes the biotensegrity model further to argue that the knee's instantaneous axis always aligns very close to the GRF vector to create a 'torque-less' situation, allowing the GRF to be loaded by the whole tensegrity structure of the knee, helping to avoid large magnitudes of angular momentum (Fig. 2.4.3a).

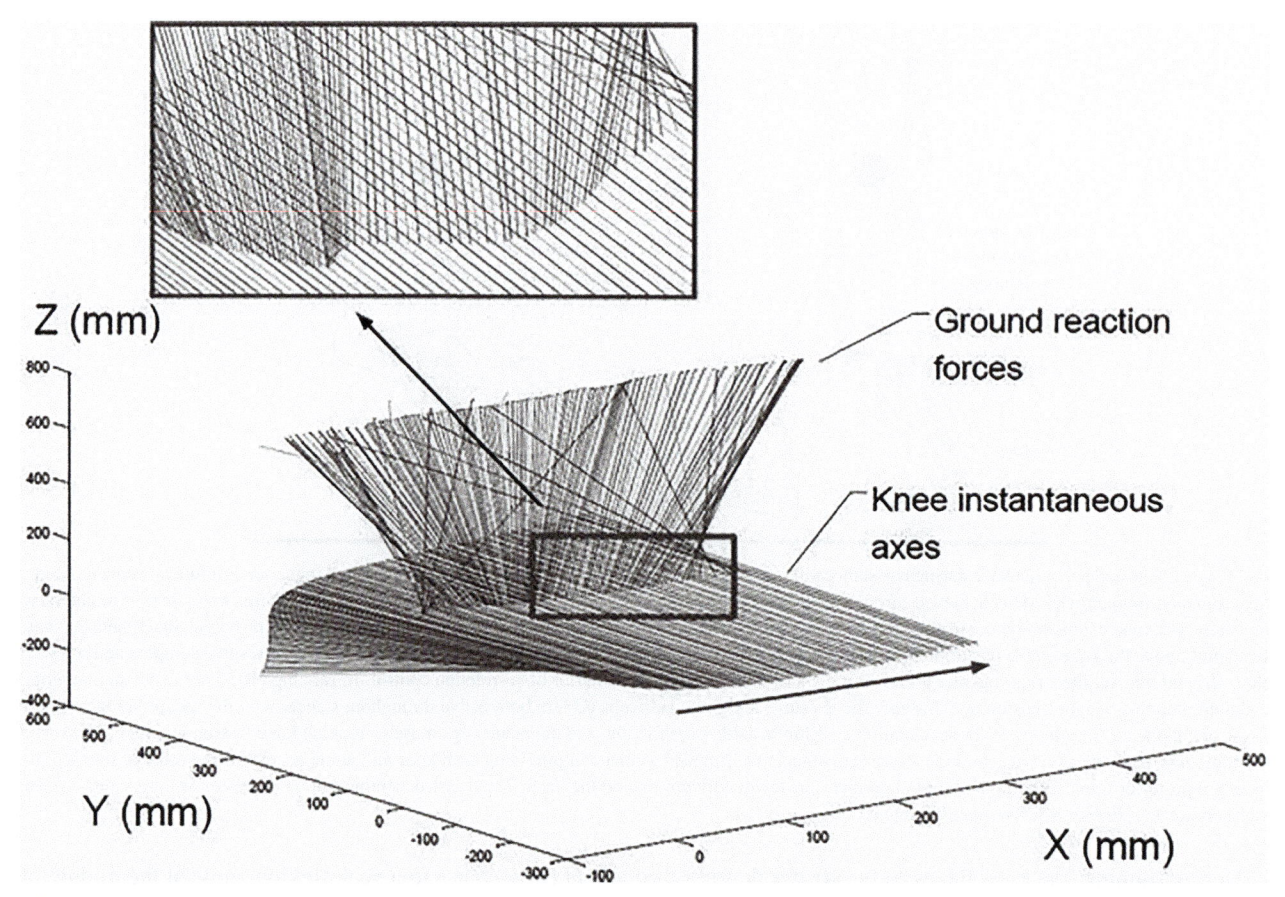

FIG. 2.4.3a In mathematics, an *invariant* is a property of an object which remains unchanged after operations or transformations of certain types are applied to the object. Here, a unique combination of the knee's instantaneous joint axis and GRF is used to demonstrate an invariant, with the positions of the CoGRF in relation to the knee's instantaneous axis occurring during gait over time. When the knee is 'deformed' through motion and its shape changed by the lower leg's relationship to the ground via the GRF, the resulting strain is distributed over the whole structure and not localised to the knee joint. The reaction torque on the knee joint will be zero if the GRF line of action intersects the joint axis allowing the body to exert a large force on the ground without overloading it. A considerable GRF can be exerted on the foot when the GRF vector coincides with the position of the knee joint axis, thus avoiding significant angular momentum. It is argued that this gives a 'sense of support' based on the close correspondence of the vector of the GRF via the location of the CoGRF and the instantaneous joint axis of the knee with fluctuations at the spatial scale of a millimetre. Thus, movement of the CoGRF should couple to the motion of the instantaneous knee joint axis. *Image from Kim, W., 2020. Tibial femoral tunnel for isokinetic graft placement based on a tensegrity model of a knee. In:* Nogueiar J.B.S. (Ed.), *Knee Surgery—Reconstruction and Replacement. IntechOpen. (Chapter 1).*

The knee provides a large range of motion which is largely restrained to one plane of motion, demanding considerable stability within the other planes. The soft tissues of the knee define the stability of the knee joint which is heavily reliant on tensile force closure. Stability of movement is best explained through biotensegrity models (Kim and Park, 2018; Jung et al., 2019; Kim, 2020). Continued tensional forces from the knee's extensive soft tissues provide inherent tension–compression stability in all positions. It is argued that this mechanism dramatically reduces joint torque on the knee by essentially trying to keep the CoGRF as close as possible to the knee's instantaneous joint axis throughout motion to minimise angular momentum (Kim, 2020). Therefore, to understand knee biomechanics, the soft tissues (both active and passive) need to be known and understood as mechanisms of tensional restraint. The importance of these soft tissue structures is underlined by the fact that ligaments and tendons are associated with significant bony prominences or tubercles around the knee, reflecting their strong entheses (Fig. 2.4.3b).

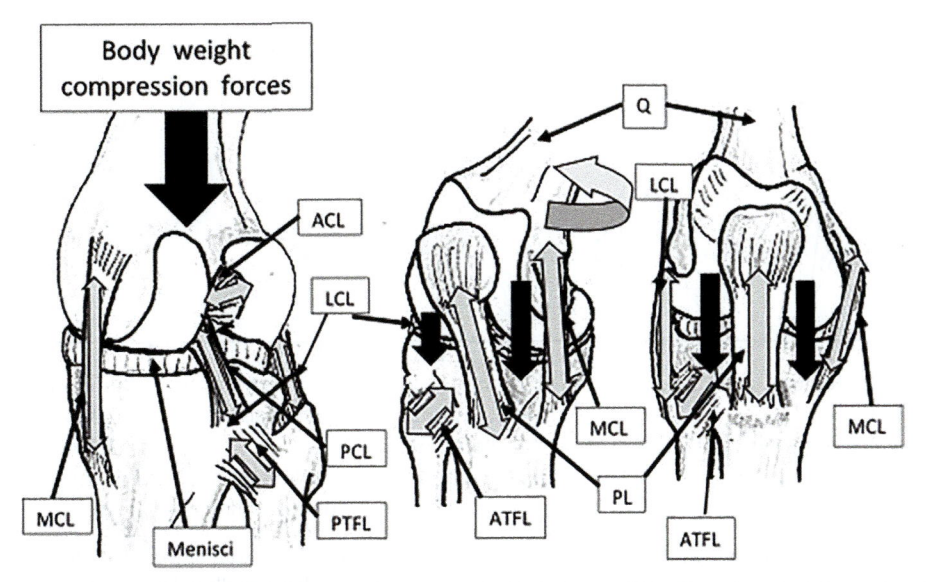

FIG. 2.4.3b The knee as a biotensegrity structure consists of a range of powerful tensile restraining ligament vectors (grey arrows) that maintain compressive stability (black arrows) across the articular surfaces of both the tibiofemoral and patellofemoral joints. These directly related forces occur through the entire range of knee motion, with the ligaments guiding articular movements including any femoral or tibial rotations or translations. Under weight-bearing compression, the menisci provide fibrocartilaginous improvements to articular compression and torque management. The simplistic schematics of the extended knee in posterolateral view (left) and anterior views in angles of knee flexion (right) show the primary passive ligament restraints. The medial collateral ligament (MCL) and lateral collateral ligament (LCL) primarily resist valgus and varus moments, respectively. The anterior cruciate ligament (ACL) and posterior cruciate ligament (PCL) primarily restrict posterior and anterior gliding of the femur over the tibia, respectively, during the stance phase. They also moderate internal and external articular surface torques. Through direct connections to the surrounding muscles and via the sensorimotor system, the active muscular tensegrity components reinforce stability during knee motion. The quadriceps muscles (location Q) provide the best example by acting directly on the patellar ligament (PL) via the patella, creating stabilising anterior articular compression in degrees of knee flexion. The posterior (PTFL) and anterior (ATFL) proximal tibiofibular ligaments via muscular and ligamentous connections help to link the knee to ankle motion and stability via tibiofibular motion. *(Permission www.healthystep.co.uk.)*

The ligaments and muscles work together as a functional unit to achieve healthy knee biomechanics. Their function should be understood in detail, and this can be approached by reviewing the functional anatomy. Failure within each anatomical area is associated with a variety of knee pathologies.

2.4.4 Osseous topography and articular structures of the knee

Knee anatomy reflects its evolutionary past through which it needed to acquire increased knee extension motion. This is a necessity of plantigrade bipedal human gait as without a balancing tail, the CoM of the trunk must be kept over the small bipedal base of support in both single support and double stance phases of gait. Such a morphology makes a significant flexed knee posture a severe disadvantage to energetics. The advantage of larger ranges of hip and increased knee extension ability is that it permits a longer stride and step length. Extension at these joints means an increased CoM height in midstance to gain advantage from gravity in inverted pendulum foot vaulting, and an increased arc of motion in the CoM's fall which can be utilised through ankle dorsiflexion in late midstance to improve inverted pendulum mechanics. The knee is anatomically and functionally divided into two separate joints, the tibiofemoral joint and the patellofemoral joint, and these joints are in close-coupled motion (Fig. 2.4.4a).

The articular surfaces at the knee are composed of the femoral condyles of the distal femur interfacing with the plateaus on the proximal tibia (tibiofemoral joint) and the posterior surface of the patella (at the patellofemoral joint) moving on the anterior femoral condyles. The tibiofemoral joint can be subdivided into two functional units: the medial and lateral compartments of the knee (Fig. 2.4.4b). The tibiofemoral joint is exposed to peak forces during the first half of the stance phase, averaging around 3.9 times body weight (Winby et al., 2009). Left and right tibiofemoral joints are generally morphologically similar in individuals, but this is not always so, with some subjects reported as demonstrating larger intra-individual differences between their knees than are usually found between individuals (Dargel et al., 2009).

HEEL STRIKE - END LOADING RESPONSE	EARLY – ABSOLUTE MIDSTANCE	LATE MIDSTANCE	TERMINAL STANCE
Tibiofemoral joint flexion. Patellofemoral joint inferior glide.	Tibiofemoral joint extension. Patellofemoral joint superior glide.	Tibiofemoral joint stability. Patellofemoral joint stability.	Tibiofemoral joint flexion. Patellofemoral joint inferior glide.

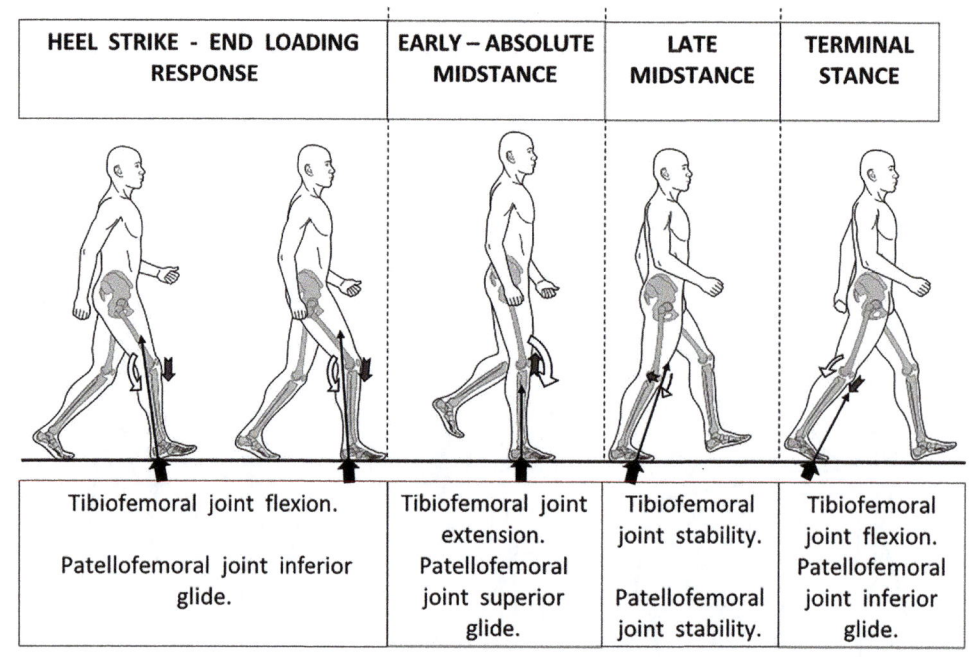

FIG. 2.4.4a The tibiofemoral and patellofemoral joints of the human knee couple their motions together with the patella largely following the tibia's motion via its strong patellar ligament attachment, gliding the patella over the femoral articular surface. Walking usually follows a cycle of initial loading tibiofemoral flexion under the external GRF vector that aligns anterior to the knee. This causes the patella to glide relatively inferiorly within the patellofemoral joint under compression of the flexion resistance of the quadriceps muscles. Motion continues until the foot is fully flat at the end of loading response, whereupon the tibiofemoral joint starts to extend, reducing the knee flexion angle under internal quadriceps moments. This causes the patella to glide relatively superiorly within the patellofemoral joint, continuing until absolute midstance. Once the GRF vector slips anterior to the knee, it applies an external extension moment on the knee which is resisted by gastrocnemius activity. This should lead to a stable knee with little motion. At heel lift, ankle plantarflexion power via gastrocnemius as part of triceps surae initiates a knee flexion moment, added by hip flexion. This causes the patella to glide inferiorly again within the patellofemoral joint. The higher knee flexion angle is maintained into early swing to then begin to return towards knee extension at midswing in readiness for initial contact. However, the knee is now under lower forces from swing phase muscle action. *(Permission www.healthystep.co.uk.)*

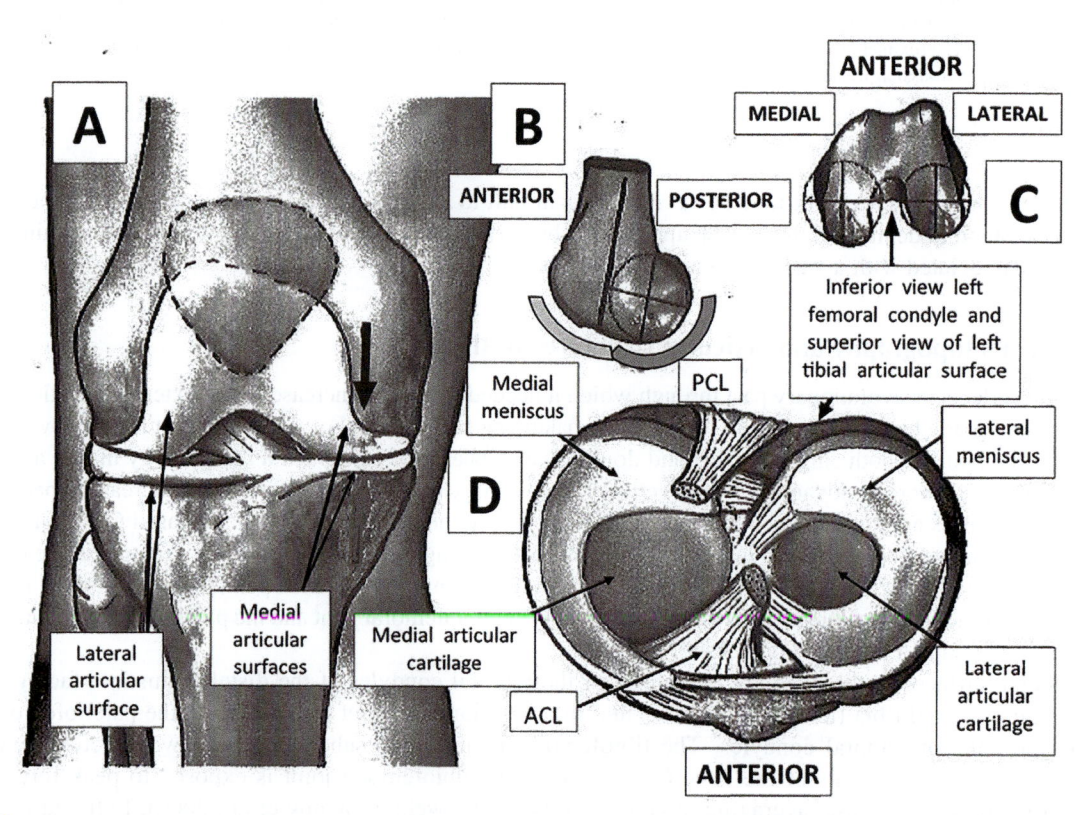

FIG. 2.4.4b Although often considered to be one joint, the tibiofemoral joint consists of distinct medial and lateral articular surfaces (A). The medial articulations tend to be subjected to greater compression loads during gait (thick black arrow) than the lateral surfaces. In the sagittal plane (B), the articular surfaces of the femur are shaped differently, being flatter anteriorly and more rounded posteriorly. This provides for a flatter surface form closure towards extension when the anterior surface is loaded. The rounded posterior surface is loaded more in flexion. The posterior femoral condyles also have widely separated, rounded, and convex articular surfaces (C) which encourage more frontal and transverse motion at high flexion angles. Such divisions of the articular surfaces make describing and locating lesions within the articular surfaces easier. The superior joint surfaces of the tibia (D) are covered in distinct protective fibrocartilage pads lying over the articular cartilage, known as the menisci. These tibial surfaces also contain the distal attachment points for the anterior and posterior cruciate ligaments (ACL and PCL, respectively). *(Permission www.healthystep.co.uk.)*

The medial tibiofemoral joint consists of a convex femoral surface on a concave tibial surface. This geometry gives some limited stability to the knee, improving the dynamic situation for soft tissue injuries of the medial knee after injury (James et al., 2015). However, the medial compartment of the knee is loaded with far more stress than the lateral compartment (Morrison, 1970; Winby et al., 2009; Marouane et al., 2016), calculated at being around 60% of tibiofemoral load during stance (Winby et al., 2009). Thus, it is not surprising that this compartment is some ten times more prone to osteoarthritis (Pollo, 1998). The medial compartment of the knee is highly vulnerable to osteoarthritic changes if exposed to large knee varus (adduction) moments which tend to unload the lateral compartment (Schipplein and Andriacchi, 1991; Winby et al., 2009). Muscular function from the hamstrings, quadriceps, and the medial and lateral parts of gastrocnemius with assistance from the tensor fasciae latae through the iliotibial tract helps increase lateral condyle loading and aids in protecting the medial condyle from over-compression (Winby et al., 2009).

The lateral articular surfaces of the knee also have a convex-on-concave geometry, creating inherent instability (James et al., 2015). As a result, lateral soft tissue knee injuries are less likely to heal following injury and are associated with lateral knee gapping into varus. This risks medial compartment osteoarthritis development and meniscal tears from the compression forces induced by the varus moments (James et al., 2015). The tibial plateau also has an individually variable posterior slope with some consensus that a higher degree of posterior sloping is associated with anterior cruciate ligament (ACL) injuries (Marouane et al., 2016). It appears that during all periods of stance from heel strike to toe-off, an increase in the angulation of posterior tibial sloping on either plateau, or at least on the one most loaded, increases anterior tibial translation and loading force on the posterior lateral bundle of fibres of the ACL (Marouane et al., 2016).

Contact forces over the tibiofemoral joint during the stance phase of gait have crucial biomechanical and pathological consequences, affected by varus and valgus angulations within the knee. In the first 25% of stance phase (5% of gait cycle), the lateral plateau is loaded most significantly (Marouane et al., 2016) and is not usually totally unloaded at any point during stance (Winby et al., 2009). From 25% of stance and onwards throughout the remainder of stance phase, the medial plateau is loaded far more (Marouane et al., 2016). Large excursions of over 17 mm have been calculated to occur on the medial plateau in a mediolateral direction during the stance phase (Marouane et al., 2016) (Figs 2.4.4c and 2.4.4d).

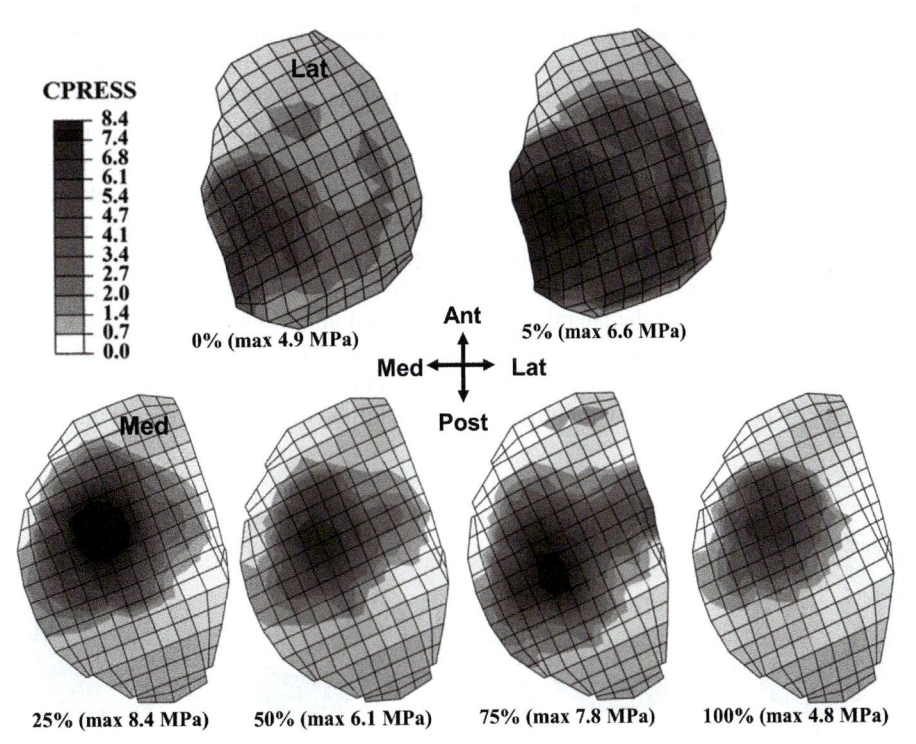

FIG. 2.4.4c Predicted contact pressure distribution at the articular surfaces of the tibial plateaus at different periods of stance phase during gait. From initial contact (0%) to early loading response (5%), the lateral plateau is primarily loaded, but thereafter, the medial plateau is normally sustaining the highest loading forces through to 100% of stance. *(Image adapted from Marouane, H., Shirazi-Adl, A., Adouni, M., 2016. Alterations in knee contact forces and centres in stance phase of gait: a detailed lower extremity musculoskeletal model. J. Biomech. 49(2), 185–192.)*

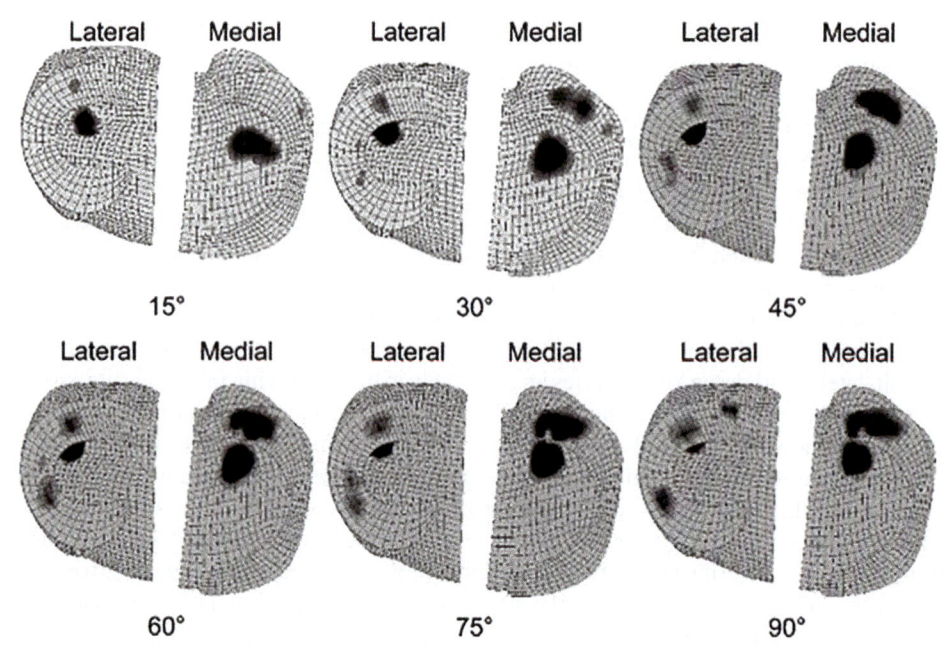

FIG. 2.4.4d Distribution of compressive forces at the articular cartilage of the tibial compartments during knee flexion angles. The contact surface is composed of part of the tibial cartilage uncovered by the menisci and the part covered by the menisci. *(Image from Ramaniraka, N.A., Terrier, A., Theumann, N., Siegrist, O., 2005. Effects of the posterior cruciate ligament reconstruction on the biomechanics of the knee joint: a finite element analysis. Clin. Biomech. 20(4), 434–442.)*

Menisci: Anatomy and mechanical role

The complex fibrocartilage structures known as the menisci increase form closure and have important compressive and torque protective functions within the knee. Their loss increases degenerative changes within the knee joint (McDermott et al., 2008). The menisci are two crescent-shaped fibrocartilaginous structures, one found between each femoral condyle and articular section of the tibial plateau, which play an important role in load-bearing, impact energy dissipation, and secondary stabilisation within the knee (McDermott et al., 2008). They also play roles in joint lubrication and through fluid flow, nutrient distribution (McDermott et al., 2008). In addition, they are part of the sensorimotor system's proprioceptive structures (Jerosch and Prymka, 1996; Al-Dadah et al., 2011; van der Esch et al., 2013) that can be adversely affected by abnormality or injury (Jerosch and Prymka, 1996; van der Esch et al., 2013). Furthermore, sensorimotor deficits can persist after meniscal repair surgery (Karahan et al., 2010; Al-Dadah et al., 2011).

The menisci maintain their optimum load-bearing function by moving with the articular surface, helping to maintain better form closure of the joint at all positions. From full extension to full flexion, the menisci displace posteriorly by 5 mm for the medial meniscus and by 11 mm for the lateral meniscus (Fig. 2.4.4e). Moreover, the anterior meniscal horns move more than do the posterior horns when the knee is unloaded, but less so when loaded (McDermott et al., 2008). During loaded situations, significant peripheral motion occurs with the lateral meniscus' anterior horn showing the greatest motion (Vedi et al., 1999). The lateral meniscal role in joint translational stability is far less significant than the role of the medial

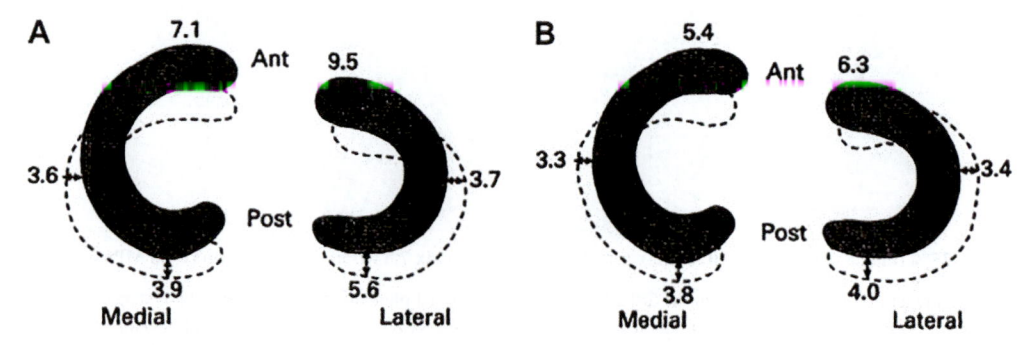

FIG. 2.4.4e Mean movements of the menisci in mm from extension (shaded region) to flexion (dashed region) during weightbearing (A) and in the unloaded knee (B). *(Image from McDermott, I.D., Masouros, S.D., Amis, A.A., 2008. Biomechanics of the menisci of the knee. Curr. Orthop. 22(3), 193–201.)*

meniscus, with less functional effects resulting from lateral meniscal resection than with medial. However, any knee stability lost with meniscal damage only occurs with concurrent loss of the cruciate ligaments (McDermott et al., 2008). The loss of the ACL along with the medial meniscus results in increased anterior tibial loads and anterior tibial translation of around 5mm (Allen et al., 2000). The immobile posterior horn is possibly acting a bit like a 'chock block', resisting posterior translation and giving the medial meniscus a secondary role in knee stability (McDermott et al., 2008).

A number of variable ligaments are associated with the menisci. The *anterior intermeniscal ligament* or *transverse genicular ligament* connects the anterior fibres of the medial and lateral anterior horns, found in around two-thirds of the population (McDermott et al., 2008). Two *meniscofemoral ligaments* are more consistent and are found attaching the posterior horn of the lateral meniscus to the lateral side of the medial condyle of the femur's intercondylar notch (McDermott et al., 2008). These latter ligaments have been reported as providing around 28% of the total posterior draw-resisting force at 90° flexion in intact knees, and 70% in posterior cruciate-deficient knees (McDermott et al., 2008) (Fig. 2.4.4f).

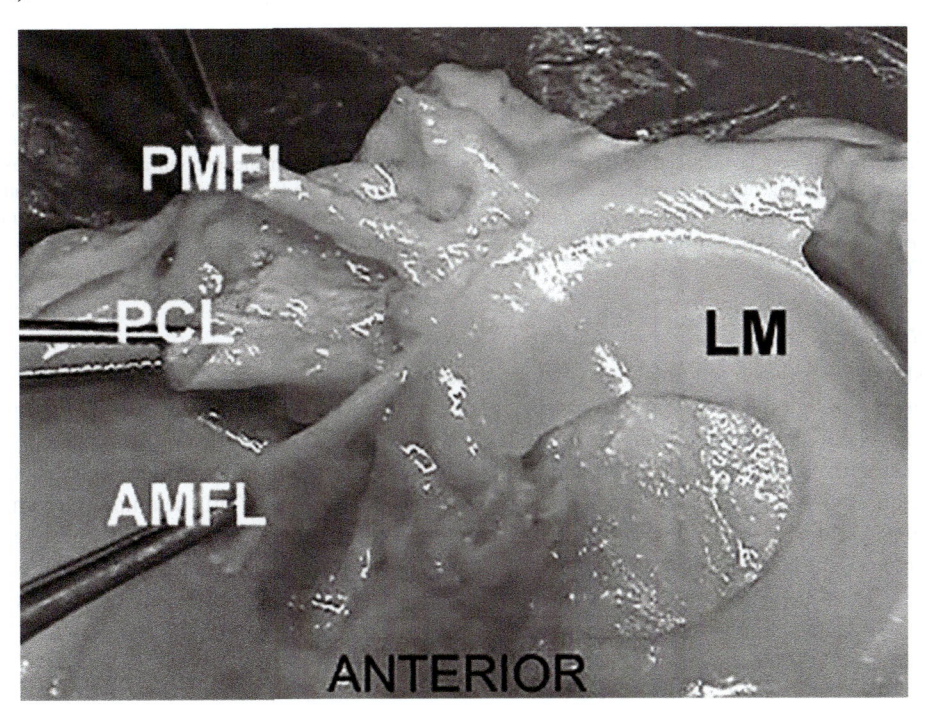

FIG. 2.4.4f Photograph of dissection of the meniscal ligaments, showing the lateral meniscus (LM), posterior cruciate ligament (PCL), anterior meniscofemoral ligament (AMFL), and posterior meniscofemoral ligament (PMFL). *(Image from McDermott, I.D., Masouros, S.D., Amis, A.A., 2008. Biomechanics of the menisci of the knee. Curr. Orthop. 22(3), 193–201.)*

There is structural variation between the superficial and deep layers of the menisci. The menisci are predominantly composed of fibrocartilage (containing type I and II collagen fibres) interposed with cells, an extracellular matrix of proteoglycans and glycoproteins, with a water content of around 74% (McDermott et al., 2008). The superficial layers consist of randomly arranged collagen fibres similar to that of articular hyaline cartilage, creating a low-friction surface between the femur and the tibia during motion (McDermott et al., 2008). The deeper layers consist of an inner third of radially arranged collagen sandwiched between two layers of outer circumferentially arranged collagen fibres, suggesting inner compression resistance and outer tension resistance (McDermott et al., 2008). In places, the radial fibres pass into the outer layers, possibly acting as ties to resist longitudinal splitting (McDermott et al., 2008) (Fig. 2.4.4g).

The material properties of the menisci relate to their macroshape as well as their microstructure of collagen matrix, which overall results in anisotropic properties (McDermott et al., 2008). The elastic modulus in circumferentially arranged collagen segments is substantially higher than that of radially orientated fibre segments. The material properties also vary depending on the depth of the menisci they come from, with the elastic modulus being the highest in the middle zone. It is next highest for the deepest zone, with the superficial zone reported to have a modulus of only around one quarter of the middle zones (McDermott et al., 2008). Meniscal stiffness is affected by location and meniscal thickness, probably due to the amount of collagen fibres and their orientations in specimens tested (Lechner et al., 2000). It is estimated that overall, menisci have a tensile modulus of around 150MPa, with around 50%–75% of that estimated for the ACL (McDermott et al., 2008). Compression properties seem to be less affected by location than by tensile properties, but posterior deeper areas seem to have a higher compressive modulus (McDermott et al., 2008).

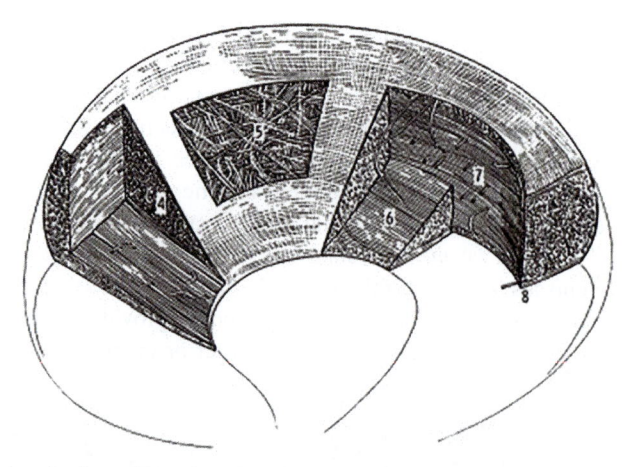

FIG. 2.4.4g Diagramatic representation of collagen fibre orientation within the meniscus that is shown in different regions, from zones 4 to 8. Fibres are arranged radially at the tibial surface in zone 4, whereas in zone 5, they are lying tangential to the surface with the collagen bundles in a woven arrangement. In zone 6, the fibres are orientated in a circumferential manner. The area through the thickness of the meniscus parallel to its edge in zone 7 has circumferential fibres but arranged radially near the tibial surface. The radially disposed fibres 'curl up' from the tibial surface of zone 8 into the substance of the meniscus. The shape and collagen arrangement of the menisci means that they become tensioned under loading conditions, with the radially disposed fibres acting as tie fibres to resist longitudinal splitting, particularly near the osseous surfaces. Strength is also dependent on the nonfibrillar matrix. *(Image from Bullough, P.G., Munuera, L., Murphy, J., Weinstein, A.M., 1970. The strength of the menisci of the knee as it relates to their fine structure. J. Bone Joint Surg. 52-B(3), 564–570.)*

The meniscal role

The menisci transfer force between the tibial and fibular surfaces under compressive restraint of the soft tissues. Circumferential stress passes through the meniscal tissues via the circumferential collagen fibres and insertions, extruding the menisci peripherally under compression from the condyles and plateau articular surfaces (McDermott et al., 2008). This fibrocartilage extrusion is resisted by the firm attachments at the horns through the insertional ligaments, so as the radius increases, the circumference simultaneously increases (McDermott et al., 2008). This sets up a circumferential tension known as 'hoop stress' within the menisci that resists further displacement. Thus, compression leads to tension that resists the compression between the bones by classic tensegrity principles. The menisci are reported to cover 59%–71% of the joint surface contact area. They are also estimated to halve the peak contact pressures between the articular surfaces and their removal decreases the contact area within the joint by 75%, consequently increasing local peak contact forces by around 235% (McDermott et al., 2008) (Figs 2.4.4h–2.4.4j).

FIG. 2.4.4h A stress–relaxation curve demonstrating the behaviour of meniscal tissue compressed to a set value of displacement. Its behaviour is typical of a viscoelastic material. *(Image from McDermott, I.D., Masouros, S.D., Amis, A.A., 2008. Biomechanics of the menisci of the knee. Curr. Orthop. 22(3), 193–201.)*

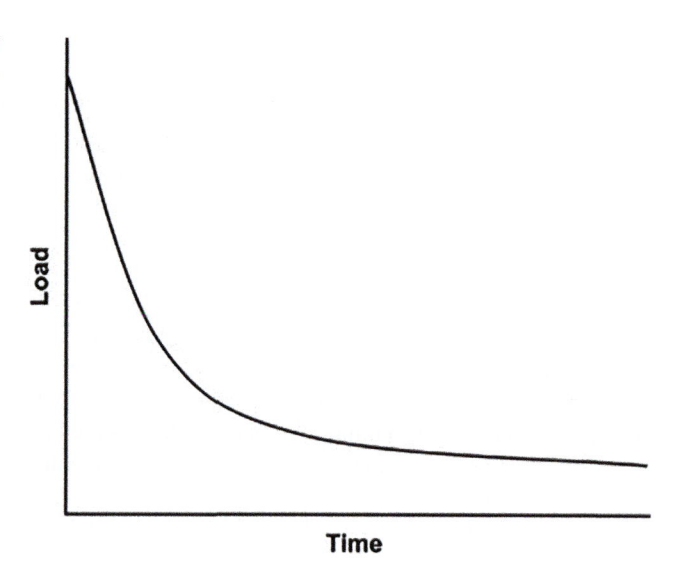

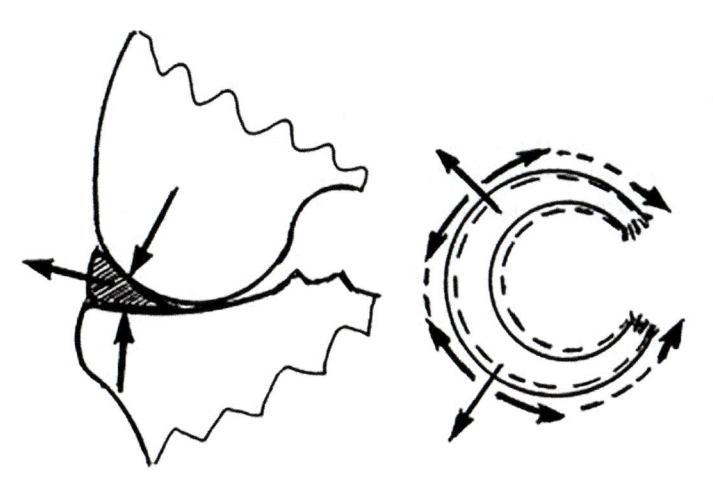

FIG. 2.4.4i Schematic of the conversion of axial load into extruding meniscal 'hoop stress'. *(Image from McDermott, I.D., Masouros, S.D., Amis, A.A., 2008. Biomechanics of the menisci of the knee. Curr. Orthop. 22(3), 193–201.)*

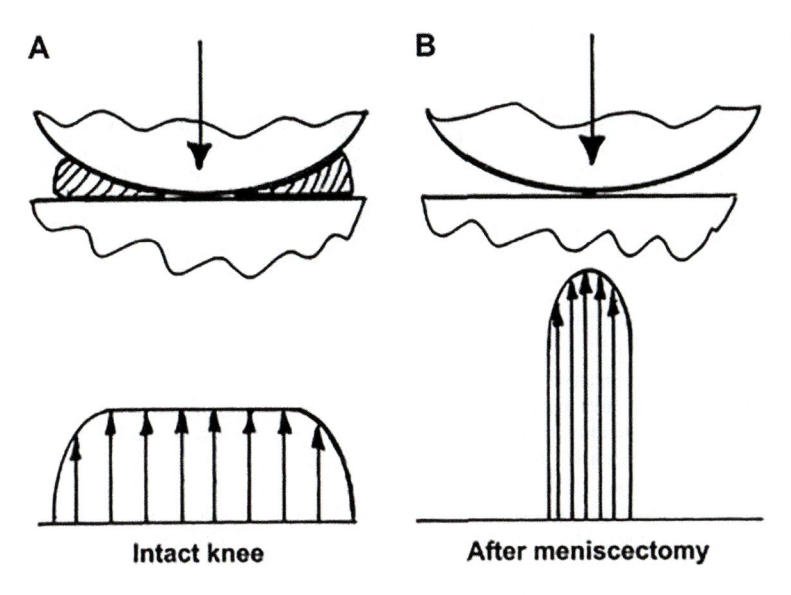

FIG. 2.4.4j Schematic of the normal compressive effects of having menisci (A) compared to the increase in peak local contact pressures after meniscectomy (B). *(Image from McDermott, I.D., Masouros, S.D., Amis, A.A., 2008. Biomechanics of the menisci of the knee. Curr. Orthop. 22(3), 193–201.)*

From the material and structural properties and the effects of having menisci, it can be ascertained that they play an important role in gait and other weightbearing activities, providing impact shock attenuation (energy dissipation) under changing impulses. This is achieved by converting axial vertical forces into 'hoop stresses' within the transverse plane and using the solid phase of the menisci. This is a significant part of the mechanical abilities of collagen and the poroviscoelastic properties of the structure (McDermott et al., 2008). It is estimated that knee shock attenuation is reduced by 20% in the absence of the menisci (Voloshin and Wosk, 1983).

2.4.5 Passive soft tissues of the knee

The ligaments and tendons of the knee have many connections between the passive and active structures, making it quite difficult to clearly define one from the other. Thus, while this section attempts to consider the ligaments separately from the muscles, descriptions of the tendinous involvement in the ligamentous structures of the knee indicate that this is really impossible. 'Passive' ligaments, as a concept, is always somewhat limited, as muscles create forces within their structures both indirectly through changing joint angles and also through the many direct tensional linkages via indirect attachments. The knee is extremely reliant on the force closure that these muscle–tendon–ligament attachments are essential in achieving this.

Cruciate ligaments of the knee: Anatomy and role

The two most significant intrinsic joint ligaments of the knee are the cruciate ligaments which are less directly connected to muscle fibres than most other knee ligaments. Like all ligaments, they play a motion-guiding role as well as acting as a mechanical restraint of knee stability and they also play a significant part in the sensorimotor system (Solomonow and Krogsgaard, 2001; Barba et al., 2015; Nyland et al., 2017). The cruciate ligaments are composed of dense connective tissue connecting the femur to the tibia, and they are widest at their attachments and narrowest in their mid-portions (Woo et al., 2006; Bowman and Sekiya, 2009) (Fig. 2.4.5a).

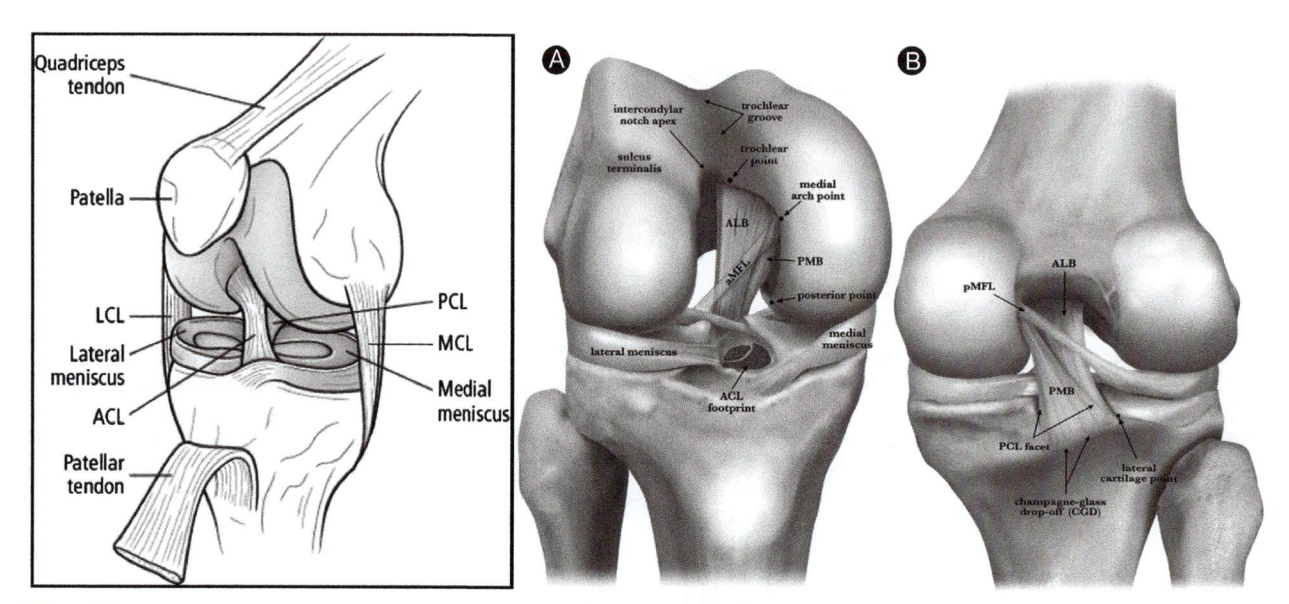

FIG. 2.4.5a A schematic of the anterior (ACL) and posterior (PCL) cruciate ligaments, with the femur and tibia distended apart (left). The ACL attaches proximally to the posteromedial surface of the lateral femoral condylar area to pass inferomedially to the medial aspect of the anterior intercondylar area of the tibia. The stronger and shorter PCL runs proximally and posteriorly from the anterior aspect of the lateral side of the medial femoral intercondylar fossa of the tibia to the attachment of the posterior horns of the menisci. The lateral and medial collateral ligaments (LCL and MCL, respectively) work together with the cruciate ligaments to provide frontal, transverse, and anterior–posterior shear gliding restraint within the knee. The posterior cruciate anatomy is better seen in the right image (A anterior view, B posterior view) where the anterior cruciate ligament is stripped away. More detail such as its distinct anterolateral band (ALB) and posteromedial band (PMB), as well as the anterior meniscofemoral ligament (aMFL) and posterior meniscofemoral ligament (pMFL), is demonstrated. *(Left image from LaBella, C.R., Hennrikus, W., Hewett, T.E., Council on Sports Medicine and Fitness, Section on Orthopaedics, 2014. Anterior cruciate ligament injuries: Diagnosis, treatment, and prevention. Pediatrics 133(5), e1437–e1450. Right image from Barba, D., Barker, L., Chhabra, A., 2015. Anatomy and biomechanics of the posterior cruciate ligament and posterolateral corner. Oper. Tech. Sports Med. 23(4), 256–268.)*

The ACL resists anterior translations and internal rotational loads (Matsumoto et al., 2001; Duthon et al., 2006). It has a mean length in extension of 32 mm and is around 7–12 mm wide (Duthon et al., 2006). It attaches to the lateral femoral condyle within the intercondylar notch posteriorly and to the anterior part of the tibial plateau (Woo et al., 2006). The ACL consists of an anteromedial bundle and a posterolateral bundle. The anteromedial bundle lengthens in flexion while the posterolateral bundle shortens, and the reverse occurs during extension (Gabriel et al., 2004; Duthon et al., 2006; Woo et al., 2006). This makes the ACL ideal for limiting excessive anterior tibial translations and internal rotations (or femoral posterior translations and femoral external rotations on a fixed closed chain tibia) as well as axial and valgus rotations across the knee. Such an ability results from the elastic effects of the complex macrostructure/microstructure that allows the ligament to manage the changing tensile strains set up during knee motion (Gabriel et al., 2004; Duthon et al., 2006; Woo et al., 2006). The anteromedial bundle carries a higher load of anterior tibial translation stress than does the posterolateral bundle in flexion angles greater than 30°, but the posterolateral bundle carries higher loads nearer to knee extension (Gabriel et al., 2004; Woo et al., 2006) (Fig. 2.4.5b). Under combined valgus and internal rotation loads at 25° knee flexion, the bundles share the load evenly (Woo et al., 2006).

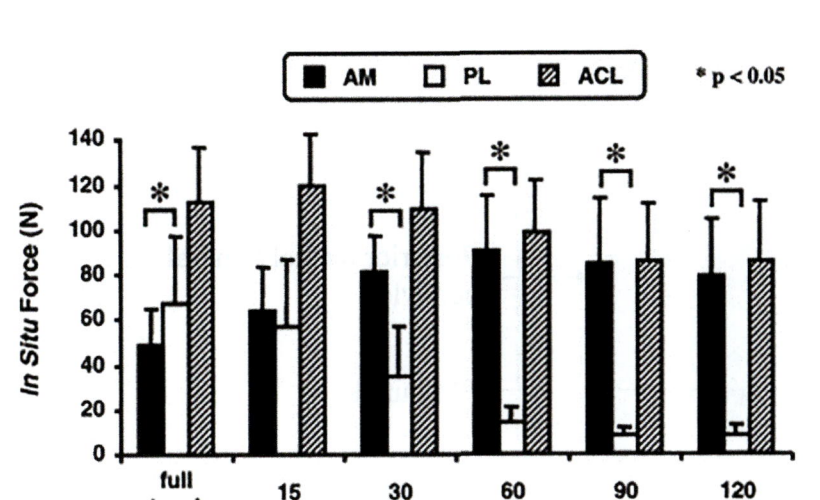

FIG. 2.4.5b A chart of in situ forces in intact anterior cruciate ligaments (ACL) and their anteromedial (AM) and posterolateral (PL) bundles in response to a 134 N anterior tibial load at different knee flexion angles. *(Image from Gabriel, M.T., Wong, E.K., Woo, S.L.-Y., Yagi, M., Debski, R.E., 2004. Distribution of in situ forces in the anterior cruciate ligament in response to rotatory loads. J. Orthop. Res. 22(1), 85–89.)*

The posterior cruciate ligament (PCL) arises from the lateral surface of the medial femoral condyle and passes behind the ACL to attach distally to the posterior tibia. Like the ACL, it is also interarticular, surrounded by a reflection of the posterior synovial membrane medially, laterally, and anteriorly (Bowman and Sekiya, 2009). The PCL attachment is wider at its femoral attachment than at its tibial attachment and has a concave semilunar shape, with fibre bundles arranged anteroposteriorly (Bowman and Sekiya, 2009). The PCL has an average length of 38 mm and an average width of 13 mm (Bowman and Sekiya, 2009). It is composed of two bundles, the anterolateral and posteromedial (Ramaniraka et al., 2005; Bowman and Sekiya, 2009), which are separated at the femoral attachment by a bony ridge. The tibial attachment is located at the intercondylar fossa between the two articular sections of the tibial plateau, with bundles of fibres orientated in a mediolateral fashion (Bowman and Sekiya, 2009). The anterolateral bundle attaches to the superolateral portion of the attachment. The posteromedial bundle attaches to the posterior tibial rim. The bone density of the attachment of the PCL is less than that at the ACL, potentially making surgical graft reattachment of the PCL more complicated, but this also suggests that less stress is placed on the PCL attachment than at the ACL attachment (Bowman and Sekiya, 2009).

The anterolateral bundle is slightly shorter than the posteromedial bundle in all flexion angles, and it increases in length with increase in knee flexion angles (Bowman and Sekiya, 2009). The posteromedial bundle decreases in length at flexion angles between 0° and 45° and then starts to increase slightly from 60° to 120°, with overall length changes far less than those demonstrated by the anterolateral bundle (Bowman and Sekiya, 2009). The posteromedial bundle's fibres orientate vertically in extension and become more horizontal in flexion, whereas the anterolateral bundle's fibres are horizontal in extension and become more vertical in flexion beyond 30°. Fibre orientation changes around 120° of flexion as the posteromedial bundle's fibres become more horizontal than the anterolateral bundle's fibres (Bowman and Sekiya, 2009). Fibre orientation change with motion means that some parts of the PCL are resisting vertical stress while others are resisting horizontal stress in all positions of knee motion. Modelling of tensile stresses inside the PCL suggests an increase in stresses during knee flexion that reach a peak at 90° of flexion, with stresses concentrated to the attachment zones where the ligaments are wider (Ramaniraka et al., 2005) (Fig. 2.4.5c).

The PCL limits posterior tibial translation and is a secondary stabiliser of posterolateral and external tibial rotation and varus loading, with a maximum tensile strength of between 730 and 1627 N (Bowman and Sekiya, 2009; Barba et al., 2015). The primary role of the PCL is to restrain posterolateral translation of the tibia (or anteromedial translation of the femur, thereby resisting posterior knee stresses), limiting translation with increased knee flexion and providing 95% of the stability at 90° and 30° flexion (Bowman and Sekiya, 2009; Barba et al., 2015). The anterolateral bundle restrains 50%–74% of the translation between 40° and 120° flexion, with the posteromedial bundle being responsible for 57% of the posterior translation of the tibia at angles greater than 120° flexion which matches the changes in fibre orientation (Bowman and Sekiya, 2009). The anterolateral bundle is relatively slack in full extension and at its tightest in mid-flexion, while the posteromedial bundle is tight in full extension and becomes progressively lax in mid-flexion before re-tightening at the limits of flexion (Bowman and Sekiya, 2009). The tensile strength of the anterolateral bundle is reported to be higher at 1620 N, compared to 258 N in the posteromedial bundle, due to the larger cross-sectional area of the former (Bowman and Sekiya, 2009). The PCL's overall cross-sectional area is larger than that of the ACL, making it a stronger ligament and raising an interesting

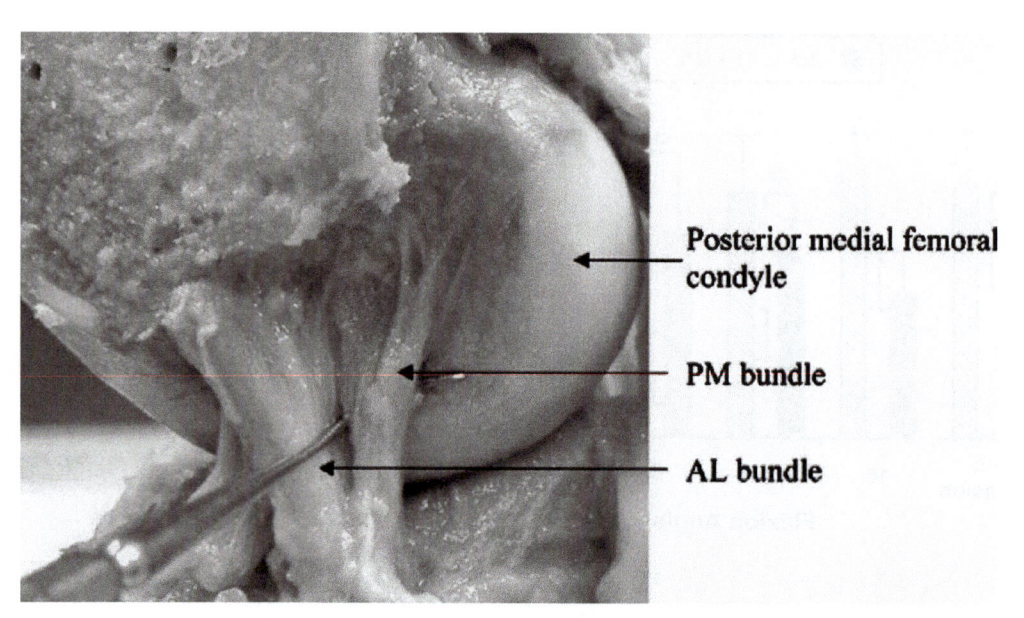

FIG. 2.4.5c A sagittal view photograph of the posterior cruciate ligament in dissection (femoral condyle sectioned) showing the positions of the posteromedial (PM) and anterolateral (AL) bundles. *(Image from Edwards, A., Bull, A.M.J., Amis, A.A., 2007. The attachments of the fibre bundles of the posterior cruciate ligament: an anatomic study. Arthrosc.: J. Arthrosc. Rel. Surg. 23(3), 284–290.)*

question regarding its seemingly weaker bone attachment. Most of the PCL's strength lies in the anterior bundle, reported to be around 730–1627 N by Bowman and Sekiya (2009). This estimate was gathered from cadaveric specimens of aged individuals, so it is likely lower than the mechanical properties to be found in young healthy populations (Fig. 2.4.5d).

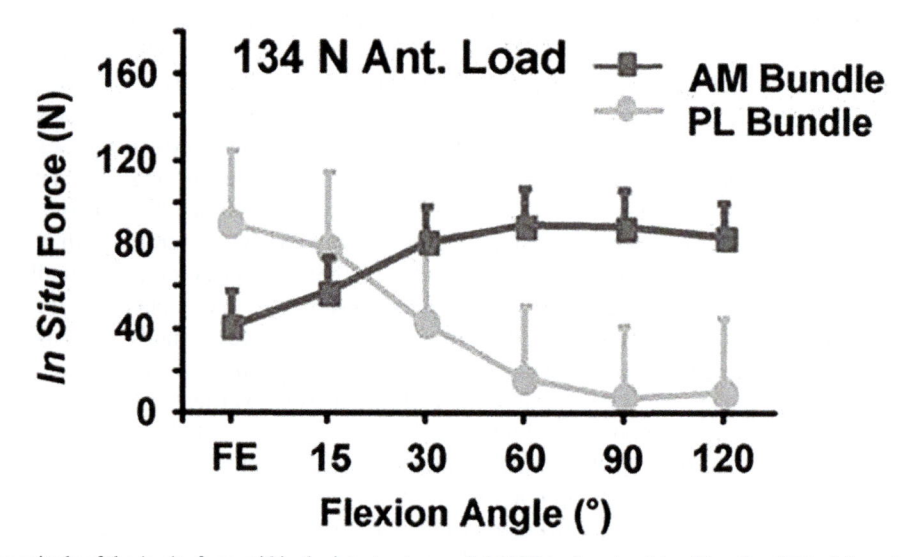

FIG. 2.4.5d The magnitude of the *in situ* force within the intact anteromedial (AM) and posterolateral bundles (PL) of the anterior cruciate ligament in response to 134 N anterior tibial loads at different flexion angles. *(Image from Gabriel, M.T., Wong, E.K., Woo, S.L.-Y., Yagi, M., Debski, R.E., 2004. Distribution of in situ forces in the anterior cruciate ligament in response to rotatory loads. J. Orthop. Res. 22(1), 85–89.)*

As the knee approaches its terminal extension ranges, the PCL's role decreases as the primary translation restraint of the *medial collateral ligament* and posterior joint capsule become the primary stabilisers (Bowman and Sekiya, 2009). The PCL's role in varus and external rotation is secondary to that of other posterior and lateral structures that are known collectively as the *posterolateral corner complex*. This complex consists of the *lateral collateral ligament*, the popliteus tendon, and the *popliteofibular ligament*. However, the PCL plays an important role in the stabilising function of the posterolateral corner complex (Bowman and Sekiya, 2009) while also protecting the knee from excessive compressive stresses (Fig. 2.4.5e).

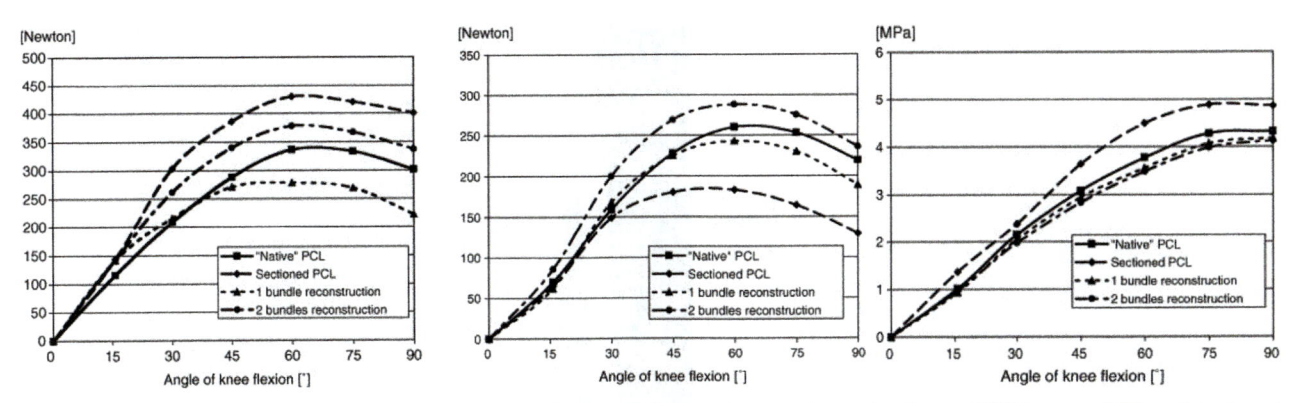

FIG. 2.4.5e Comparing compressive forces in the knee during flexion with intact native posterior cruciate ligament (PCL), resected PCL, and single- and double-graft reconstructed PCL in the medial tibiofemoral compartment (left graph), the lateral tibiofemoral compartment (centre graph), and the patellofemoral joint's compressive stress (right graph). The results suggest that lost PCL function creates excessive compressive forces that can lead to articular degeneration, with reconstruction only partially restoring the biomechanics of knee flexion, with the double-graft reconstruction subjected to the highest tensile stress. *(Image from Ramaniraka, N.A., Terrier, A., Theumann, N., Siegrist, O., 2005. Effects of the posterior cruciate ligament reconstruction on the biomechanics of the knee joint: a finite element analysis. Clin. Biomech. 20(4), 434–442.)*

Posterior knee soft tissues: Anatomy and role

The posterior aspect of the knee, known as the popliteal fossa, is made up of a complicated network of passive and active soft tissue stabilisers. Its core components are the popliteus muscle belly and the series of fascial attachments of the semimembranosus hamstring muscle attaching to the posterior tibia and knee (LaPrade et al., 2007a). Semimembranosus displays multiple attachments in addition to its direct attachment to the posterior tibia, with its distal tibial expansion forming a fascial layer over the popliteus muscle (LaPrade et al., 2007a). These widespread attachments include a tendinous expansion to the *oblique popliteal ligament* (which it contributes fibres to), the coronary ligaments of the medial meniscus, and the posterior joint capsule (Bowman and Sekiya, 2009). The *oblique popliteal ligament* is the largest structure of the posterior knee, forming a broad fascial sheath of approximately 48 mm in length, 9.5 mm wide at its medial attachment, and 16.4 mm wide at its lateral attachment (LaPrade et al., 2007a). It arises laterally from the posterolateral joint capsule and the lateral aspect of the PCL facet on the posterior tibia (LaPrade et al., 2007a; Bowman and Sekiya, 2009). The ligament also attaches to the cartilaginous or osseous fabella sesamoid found at the proximal attachment of the lateral gastrocnemius and the plantaris muscle, without direct attachment to the femur (LaPrade et al., 2007a; Bowman and Sekiya, 2009).

The posterior joint capsule of the knee extends from the posterior femoral condyles to the posterior margin of the tibial plateau, where the distal extent of the PCL becomes continuous with the capsular and periosteal fibres of the tibia (Bowman and Sekiya, 2009). The popliteus muscle connects to the joint capsule through a posterior capsular thickening near the intercondylar notch which is part of the 'posterior popliteus complex' (LaPrade et al., 2007a). The popliteus muscle attaches to the posterior medial tibia, transitioning into a tendon at the lateral third of the popliteal fossa while passing superiorly. The popliteus tendon becomes intracapsular after giving attachment to the *popliteofibular ligament*. It then continues on to the lateral collateral ligament attachment to the popliteal sulcus, distal and anterior to the lateral epicondyle of the femur (Bowman and Sekiya, 2009). As already mentioned, the three structures of the popliteus tendon, the popliteofibular ligament, and the lateral collateral ligament are known as the posterolateral corner complex (LaPrade et al., 2003; Bowman and Sekiya, 2009; Agha, 2017) (Fig. 2.4.5f).

Medial knee soft tissues: Anatomy and role

The medial epicondyle is described as the most anterior and distal prominence of the medial femoral condyle. Here lies the attachment of adductor magnus onto the medial supracondylar line and the adductor tubercle that is found at the distal end of this line (LaPrade et al., 2015). Slightly distal and posterior to the adductor tubercle and proximal and posterior to the medial epicondyle lies the gastrocnemius tubercle. This is the site of attachment of the medial gastrocnemius proximal tendon (LaPrade et al., 2007b, 2015). Apart from the 'passive' ligaments, the medial side of the knee is stabilised by the active tensions of the adductor magnus tendon, the medial hamstring tendons, the proximal medial gastrocnemius tendon, and the vastus medialis muscle via the patella, its retinaculum, and tendon-like ligament. It is the organisation of the attachment sites of these structures to the femoral condyle, medial meniscus, and medial tibia that provides the stability (LaPrade et al., 2015).

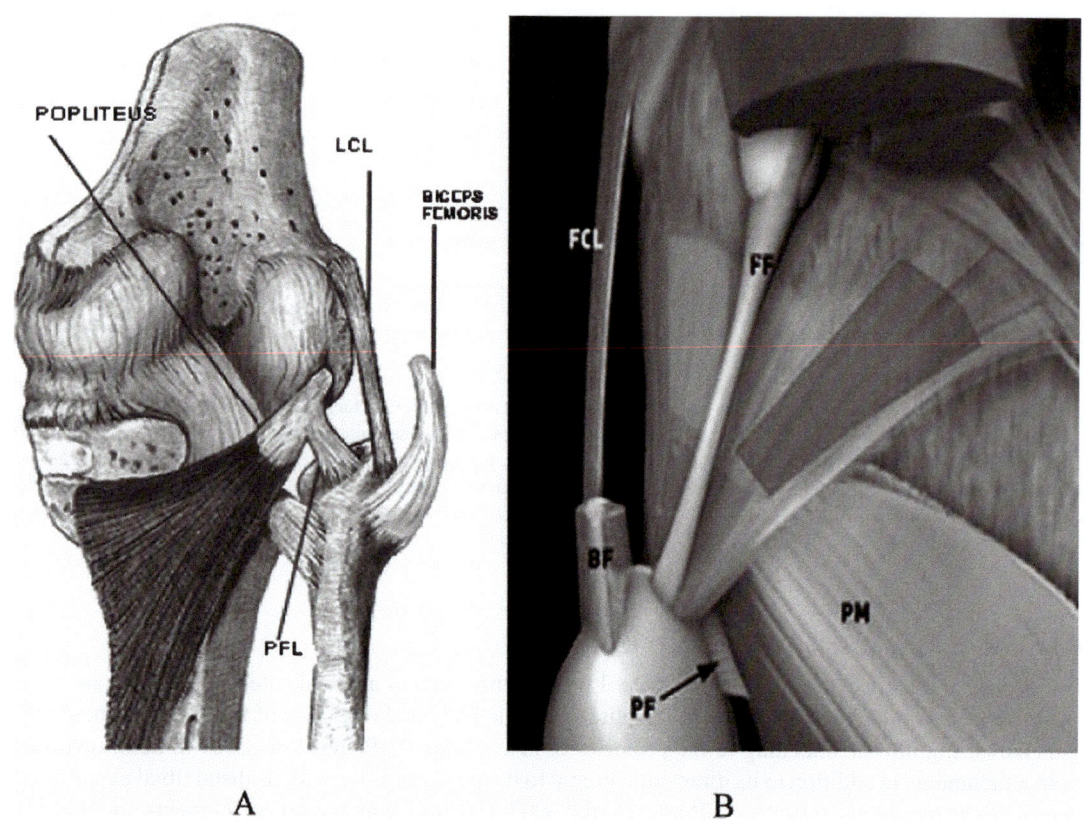

FIG. 2.4.5f An anatomical diagram of the posterior lateral corner complex (A) showing the popliteus and also biceps femoris' distal attachment to the fibular head. The popliteofibular ligament (PFL) and the lateral collateral ligament (LCL) are shown 'tensioned and distracted' for visualisation. A more local schematic (B) with biceps femoris (BF) removed shows the Y-shaped arcuate ligament composed of the medial (dark-grey) and lateral (light-grey) segments attached to the lower (mid-grey) region that then attaches to the fibular head. Other local ligaments such as the fibular collateral ligament (FCL), fabello-fibular (FF) ligament, the popliteofibular (PF) ligament, along with the popliteus muscle (PM) are also indicated. *(Image from Agha, M., 2017. MRI of the posterolateral corner of the knee, please have a look. Alexandria J. Med. 53(3), 261–270.)*

The ligaments primarily responsible for maintaining medial knee stability are the superficial and deep parts of the medial collateral ligament (MCL) and the *posterior oblique ligament* (LaPrade et al., 2015). These demonstrate a consistent attachment pattern (LaPrade et al., 2007b). The *medial patellofemoral ligament* that helps stabilise the patellofemoral joint is also located here (Amis et al., 2003; LaPrade et al., 2007b). The MCL checks valgus instability of the knee by preventing the opening of the medial joint space (Matsumoto et al., 2001). The superficial MCL is approximately 94 mm long and has a proximal attachment on the femoral condyle around 3 mm above and 5 mm proximal to the medial epicondyle (LaPrade et al., 2007b). It is the largest of the medial ligamentous structures (LaPrade et al., 2015). The ligament has two separate tibial attachments, the most distal of which is around 61 mm distal to the tibiofemoral joint (LaPrade et al., 2007b).

The deep MCL consists of portions attaching from the femur to the medial meniscus and from the medial meniscus to the tibia (LaPrade et al., 2007b). It reinforces the anterior aspect of the medial joint capsule, comprised of the *meniscotibial* and *meniscofemoral components* or sections, that attach distally and proximally to the joint line, respectively (LaPrade et al., 2015).

The posterior oblique ligament has a femoral attachment around 8 mm distal and 6 mm posterior to the adductor tubercle, and a distal attachment just anterior to the gastrocnemius tubercle (LaPrade et al., 2007b). This ligament also has a capsular attachment reinforcing the posteromedial joint capsule (LaPrade et al., 2015). The medial patellofemoral ligament attaches from the femur just anterior to and around 4 mm distal to the adductor tubercle (LaPrade et al., 2007b). It is important in restraining the patella in the trochlear groove of the patellofemoral joint, running as it does from the adductor tubercle to the medial patella (LaPrade et al., 2015) (Figs 2.4.5g and 2.4.5h).

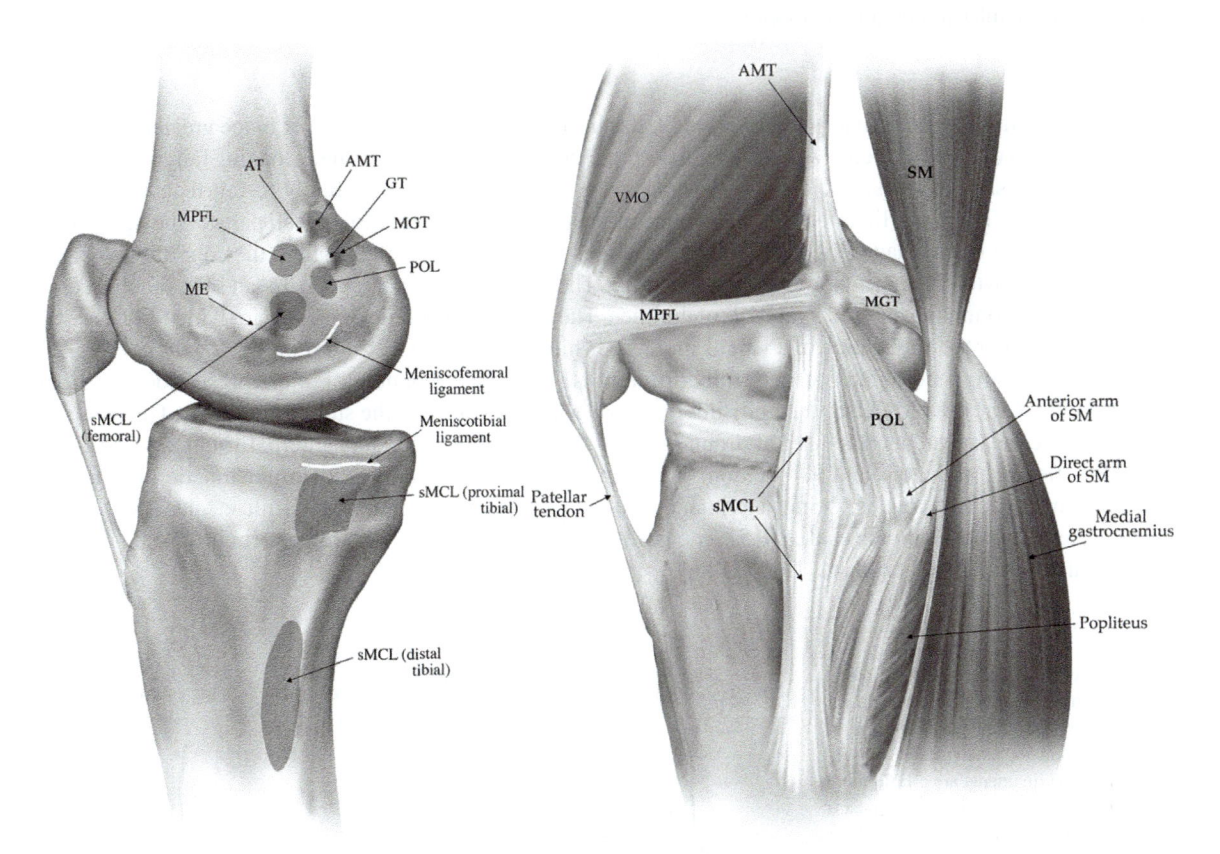

FIG. 2.4.5g Anatomical illustration of the femoral osseous landmarks and attachments of the main medial knee structures (left) and the soft tissues in situ (right). Demonstrated are: Adductor tubercle (AT), gastrocnemius tubercle (GT), medial epicondyle (ME), adductor magnus tendon (AMT), medial gastrocnemius tendon (MGT), superficial medial collateral ligament (sMCL), medial patellofemoral ligament (MPFL), posterior oblique ligament (POL), semimembranosus muscle (SM), and vastus medialis and vastus medialis obliquus (VMO) muscle. *(Images from LaPrade, R.F., Engebretsen, A.H., Ly, T.V., Johansen, S., Wentorf, F.A., Engebretsen, L., 2007. The anatomy of the medial part of the knee. J. Bone Joint Surg. 89-A(9), 2000–2010.)*

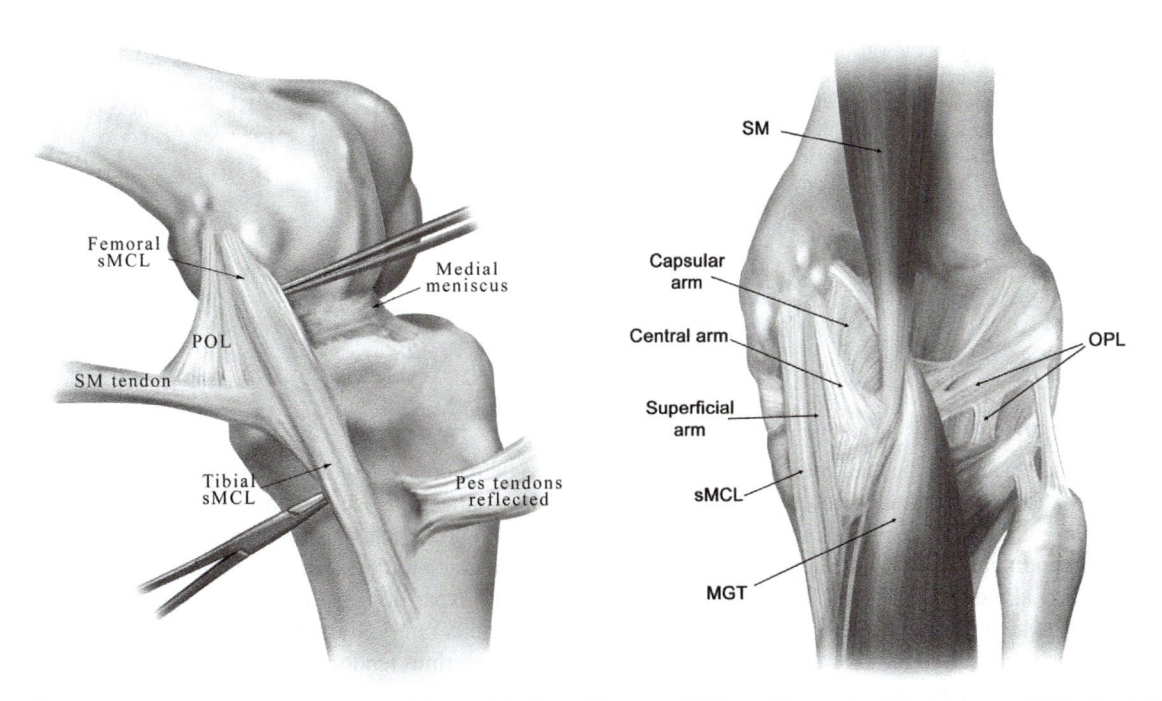

FIG. 2.4.5h Anatomical illustration (left) of the superficial medial collateral ligament (sMCL) and the posterior oblique ligament (POL) of the left knee, with the upper forceps placed under the anterior edge of the femoral portion, and the lower forceps between the proximal and distal tibial attachments of the collateral ligament. The right illustration gives a medioposterior view in the right knee of the three arms (capsular, central, and superficial) of the posterior oblique ligament. It also shows the anatomy of the superficial medial collateral ligament, the oblique popliteal ligament (OPL), the medial gastrocnemius and tendon (MGT), and semimembranosus (SM) muscles and tendons. *(Images from LaPrade, R.F., Engebretsen, A.H., Ly, T.V., Johansen, S., Wentorf, F.A., Engebretsen, L., 2007. The anatomy of the medial part of the knee. J. Bone Joint Surg. 89-A(9), 2000–2010.)*

The mean loading failure of the intact superficial MCL has been reported to be 557 N, over twice greater than that of the POL, with the deep part of the MCL being weaker (Wijdicks et al., 2010a). A much lower mean loading failure was reported in each of the individual slips of the superficial MCL which were some six times more likely to fail than when combined (Wijdicks et al., 2010a). This underlines the fact that ligamentous structural properties rely on all parts of the ligaments being intact, so that the whole biomechanics of ligamentous structures within joints is greater than the sum of each part. Such findings are consistent with the expectations of biotensegrity. The superficial and deep MCLs and the other medial knee stabilisers are the most commonly injured ligaments of the knee. They are usually clinically tested for with the knees extended or flexed at 20–30°, under the application of a valgus moment, which they should restrain if intact (Wijdicks et al., 2010b). However, even valgus stress radiographs cannot fully differentiate the actual ligament(s) injured (LaPrade et al., 2010).

The components of the deep MCL serve stabilising functions similar to those of the superficial MCL. The meniscotibial portion acts as a secondary valgus stabiliser at around 60° of flexion and a secondary internal rotation stabiliser at full extension to 30° flexion, and at around 90° flexion. The meniscofemoral portion has a similar function but has a more significant role in all flexion angles, being a primary stabiliser of internal rotation (Griffith et al., 2009). The deep MCL is the weaker structure (LaPrade et al., 2015). The knee collateral ligaments are therefore primarily knee frontal plane stabilisers, but also serve the other planes (Fig. 2.4.5i).

FIG. 2.4.5i A schematic of the medial collateral ligament (MCL) and lateral collateral ligament (LCL) as the primary restraining mechanisms of valgus and varus moments on the knee, respectively. In flexion, these ligaments are a little more relaxed so as to permit more motion, but this means higher knee flexion angles are more vulnerable to frontal plane insult. Large valgus torques on the knee risk MCL failure while varus torques risk LCL tears and ruptures. These failures can change articular loading patterns on the tibiofemoral joint, with loss of MCL restraint risking increased lateral compartment compression forces and LCL failure promoting increased medial compartment varus compression (vertical black arrows). Collateral ligament failures most commonly occur within combined triplanar torques, the most common being internal and valgus rotations with posterior shear translation of the femur on a closed chain tibia. These events can result in both MCL and anterior cruciate ligament injuries. As muscular activity can prevent large joint torques, strong knee muscles resistant to fatigue will reduce the risk of incidents. *(Permission www.healthystep.co.uk.)*

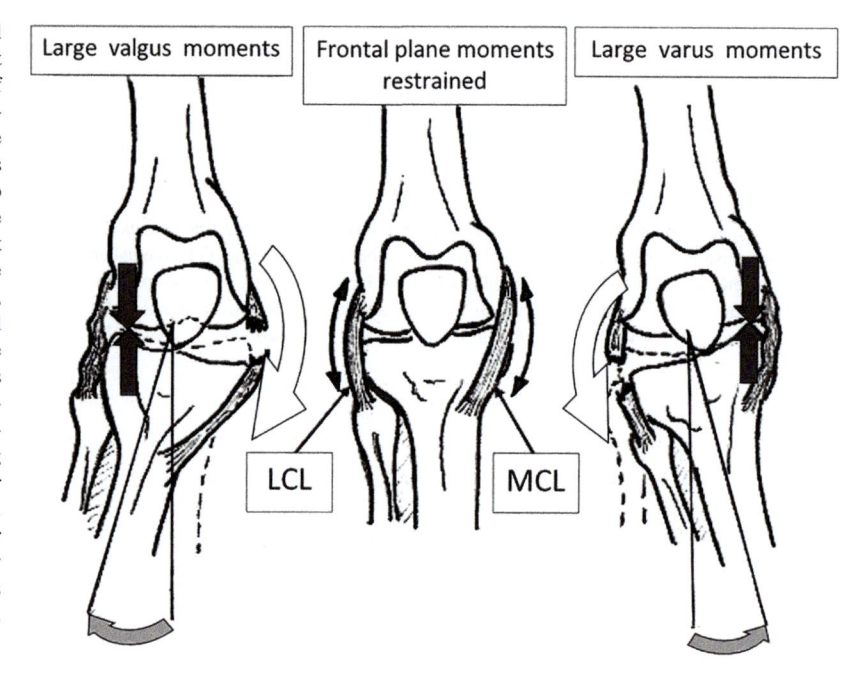

Medial knee muscles

The muscle tendons around the knee are quite interlinked as well as being connected directly to many of the ligaments that have just been discussed. The adductor magnus tendon demonstrates connections with vastus medialis and the medial gastrocnemius through its thick tendinous sheath (LaPrade et al., 2015). Adductor magnus has its primary attachment through its thick long tendinous sheath to the supracondylar line and does not attach to the apex of the adductor tubercle, but just proximal to it (LaPrade et al., 2015). In addition, it also has a fascial expansion that attaches proximally to the medial gastrocnemius tendon and the posteromedial joint capsule. This muscle is rarely injured (LaPrade et al., 2015), probably due to its extensive and large surface attachment area to bone. The pes anserinus tendon complex that comprises the tendons of gracilis, sartorius, and semitendinosus attaches on the anteromedial aspect of the proximal part of the tibia, forming the roof over the pes anserine bursa just before its enthesis attachment to the proximal anterior tibia (LaPrade et al., 2015) (Fig. 2.4.5j).

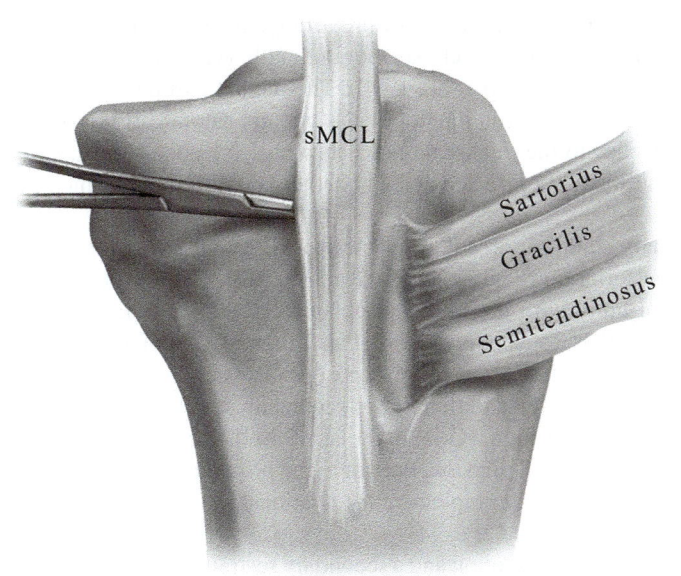

FIG. 2.4.5j An anteromedial view of the left knee to reveal the pes anserine bursa location (smooth area between sMCL and the conjoined tendon complex) relative to the superficial medial collateral ligament (sMCL). The tendon complex is reflected anteriorly to better view its attachment site. The pes anserine bursa lies behind these tendons to protect them from friction on the anterior tibial surface. It is a relatively common site of anterior knee pain symptoms associated with moderate and high levels of activity. *(Image from LaPrade, R.F., Engebretsen, A. H., Ly, T.V., Johansen, S., Wentorf, F.A., Engebretsen, L., 2007. The anatomy of the medial part of the knee. J. Bone Joint Surg. 89-A(9), 2000–2010.)*

The proximal tendon attachment of medial gastrocnemius is located over the posteromedial edge of the medial femoral condyle, just proximal to the gastrocnemius tubercle (see Figs 2.4.5f and 2.4.5g). It also has two other fascial attachments along the lateral aspect of the adductor magnus tendon and the posterior-medial aspect of the posterior oblique ligament's capsular arm (LaPrade et al., 2015). The link between foot plantarflexion and knee flexion–extension has been demonstrated through eccentric contraction of medial gastrocnemius that results in stressing the attachment of the gastrocnemius to the femur (Patterson et al., 2014). As the knee is subjected to extension moments during late stance phase, it can be resisted through gastrocnemius activity via ankle dorsiflexion. The eccentric contractions of gastrocnemius and soleus that control anterior tibial rotation at the ankle increase the stiffness of the Achilles. This helps to induce ankle plantarflexion power for heel lift, with the knee's degree of flexion at heel lift influencing ankle plantarflexion power (Fig. 2.4.5k).

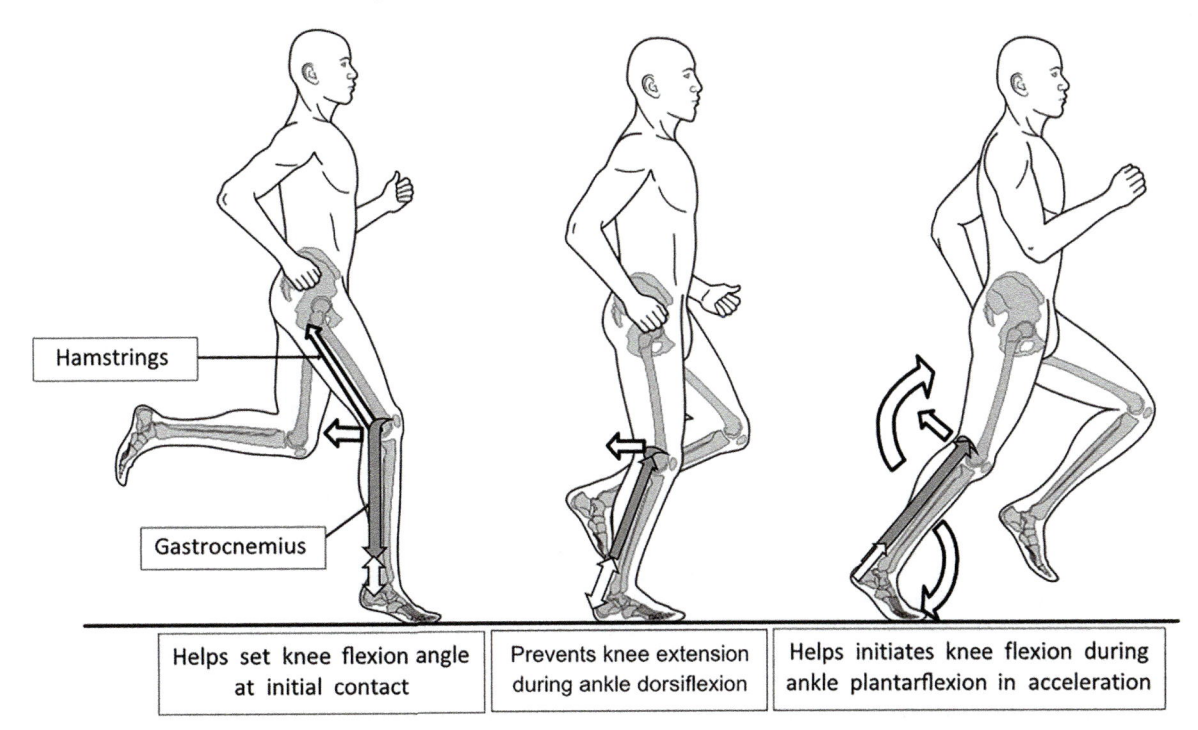

FIG. 2.4.5k Although not always considered as a knee muscle, gastrocnemius through its proximal medial and lateral tendons on the posterior superior surfaces of the femoral condyles plays a significant part in knee biomechanics. It creates a posteriorly directed moment on the femoral condyles that helps restrain anterior femoral translation on the tibia at initial contact and opposes knee extension, promoting knee flexion and resisting femoral posterior translation in the latter stages of stance. Thus, it is an important protecting muscle of ligament forces within the knee. Here, a schematic of its action in forefoot running is shown. It is an important influencer of knee motion both in walking and during all running styles. *(Permission www.healthystep.co.uk.)*

Vastus medialis is a quadriceps muscle that has two separate parts with differing fibre arrangements (Hubbard et al., 1997). Its lateral part is anatomically known as the *vastus medialis longus*, which has longitudinally (vertically) arranged fibres and is reported to be directly contributing to knee extension (Hubbard et al., 1997). The medial part has obliquely orientated muscle fibres and is widely known as vastus medialis obliquus (oblique), but it is not an anatomically separate muscle (Hubbard et al., 1997). Vastus medialis obliquus attaches to the thick tendinous sheath of the adductor magnus tendon as well as joining the patella, helping to increase the medially directed forces on the patella through a medial attachment point (LaPrade et al., 2015).

In essence, the medial soft tissue structures provide primary stabilisation and resistance to valgus knee rotations, stabilisation of external and internal rotations, and anterior and posterior translations (LaPrade et al., 2015). The overall effects of the medial ligaments are realised through their functional unity with the local musculature. The superficial MCL's proximal division has a primary valgus resistance function as well as a secondary external and internal rotation stabilising function in certain degrees of flexion (Griffith et al., 2009). Its distal division serves as a primary stabiliser of internal rotation and at 30° flexion, it also acts as an external rotation stabiliser. In addition, it is a secondary stabiliser to external rotation at near-knee extension and at 20° and 60° degrees of flexion (Griffith et al., 2009). Load response to external–internal torques is also variable to knee flexion angle in the distal division, but not in the proximal division which remains consistent (Griffith et al., 2009). The superficial MCL also has a complementary relationship with the posterior oblique ligament (Griffith et al., 2009; Wijdicks et al., 2010a).

The posterior oblique ligament provides a primary restraint to internal rotation and a secondary restraint to valgus torque and external rotation throughout all knee flexion angles. However, most load occurs around full extension (LaPrade et al., 2015). The ligament also shares loads against posterior and anterior tibial translation and it is an important stabiliser of rotation and valgus stress in cases of isolated MCL injury (LaPrade et al., 2015).

Combined superficial MCL and posterior oblique ligament tears cause chronic valgus instability, underscoring the important load-sharing relationship between these two ligaments (LaPrade et al., 2015). In consideration of anterior–posterior translation, it must be remembered that during stance phase, posterior tibial translation is actually femoral anterior translation, and anterior tibial translation equates to femoral posterior translation as the tibia becomes fixed to the ground through the foot and ankle in closed chain (Fig. 2.4.5l).

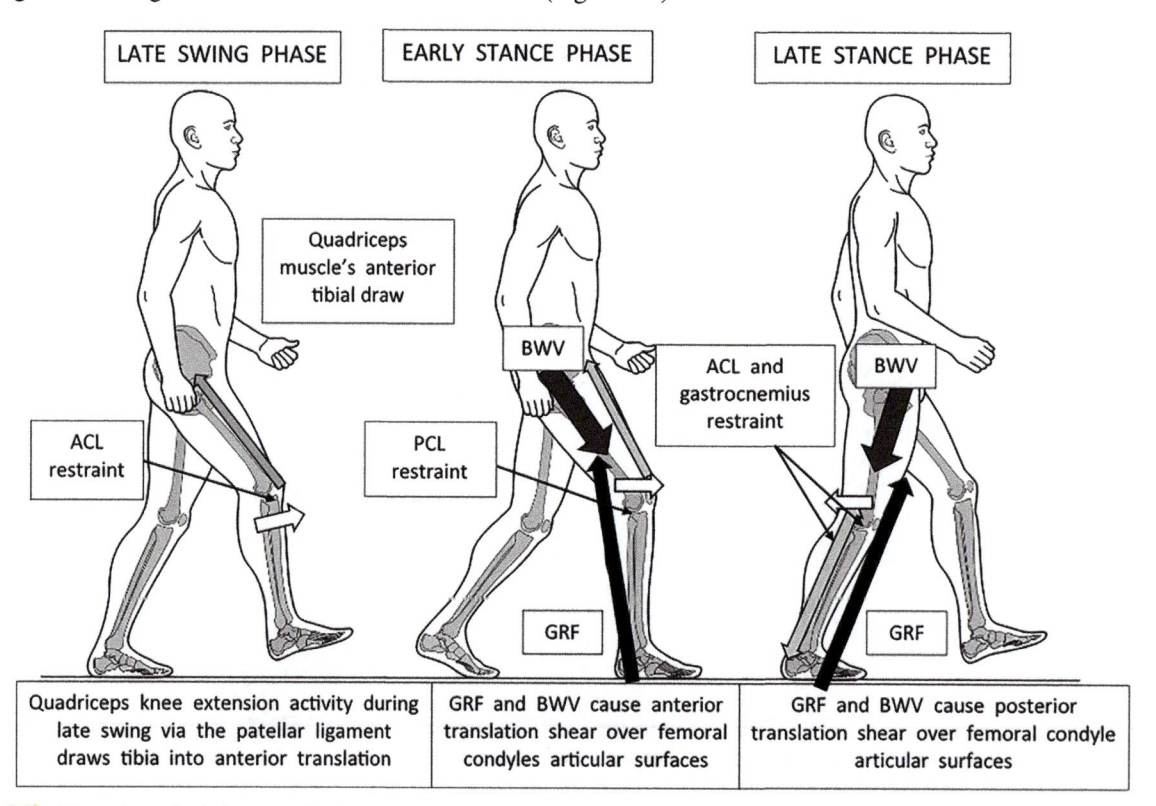

FIG. 2.4.5l The actions of soft tissues on tibial translations within the knee are usually described in open chain. For example, the ACL is often described as restraining anterior tibial translation. Such open chain translations are only true of swing phase. However, larger stresses are experienced within the knee during closed chain activities when the tibia is fixed to the ground via the foot. Thus, during stance phase of gait, perceived anterior tibial translation is in reality posterior femoral translation. Anterior femoral translations are a normal part of initial contact and early stance phase, restrained by the posterior cruciate ligament (PCL). Once the CoM has slipped anterior to the knee during late stance, the femur will be subjected to posterior translation, gliding over its condylar surface in response to GRF and body weight vectors (BWV) until heel lift. This requires ACL restraint of femoral posterior translation on the tibia, assisted by gastrocnemius in preventing excess knee extension. *(Permission www.healthystep.co.uk.)*

Lateral knee soft tissues: Anatomy and role

The lateral aspect of the knee consists of 28 unique structures. It is extensively supported by ligaments, the popliteal muscle as part of the popliteal muscle–tendon–ligament complex, the lateral collateral ligament (LCL), and the posterolateral capsule working together as a functional unit to provide stability (Pasque et al., 2003; James et al., 2015). Stability of the posterolateral aspect of the knee is important in preventing abnormal strains occurring in the ACL and PCL ligaments (Pasque et al., 2003). The primary stabilising structures of the region are the LCL (also known as the fibular collateral ligament), the popliteus tendon, and the popliteal ligament which together resist lateral compartment gapping and rotational instability. In this, they are provided with assistance from the iliotibial tract, biceps femoris, the lateral gastrocnemius proximal tendon, and other local ligaments (James et al., 2015). It appears that lateral knee joint injuries rarely heal and lead to lateral compartment gapping, medial compartment osteoarthritis, and medial meniscal tears (James et al., 2015).

The LCL is a tubular ligament around 66 mm (range 59–77 mm) in length that arises from the lateral femoral condyle, roughly 2 cm above the knee joint line. It proceeds distally and posteriorly above the joint capsule to the head of the fibula with the popliteus tendon coursing beneath it (LaPrade et al., 2003; Davies et al., 2004; Bowman and Sekiya, 2009; James et al., 2015). The *popliteofibular ligament* is postage stamp-sized and is found deep to the arcuate ligament, arising from the posterior portion of the fibular head's styloid process to attach into the popliteus tendon at its musculotendinous junction (Fig. 2.4.5m). Together with the popliteus tendon, it forms an inverted Y-shaped musculotendinous attachment between the tibia, fibula, and lateral femur (Davies et al., 2004). It is potentially a provider of strength to the posterolateral corner of the knee, but is not always present (Davies et al., 2004).

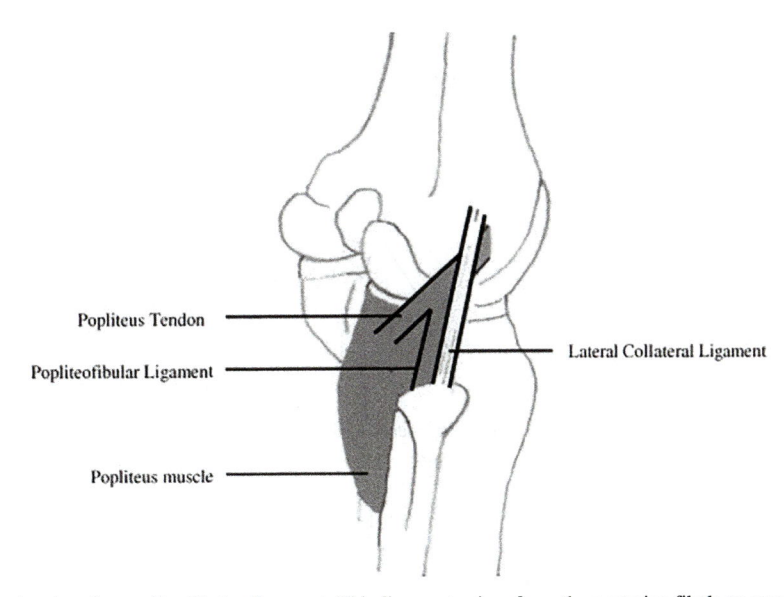

FIG. 2.4.5m A schematic showing the popliteofibular ligament. This ligament arises from the posterior fibula to merge with the popliteus tendon, inserting into the femoral condyle under the lateral collateral ligament. Via popliteus, this structure provides a helpful passive-active posterolateral stabilising restraint that protects the cruciate ligaments by sharing some of their stresses. It is not always present. *(Image from Davies, H., Unwin, A., Aichroth, P., 2004. The posterolateral corner of the knee: Anatomy, biomechanics and management of injuries. Injury 35(1), 68–75.)*

The *arcuate ligament* is Y-shaped, arising from the posterior part of the joint capsule. It runs around the distal surface of the femur to attach distally to the posterior aspect of the fibular head, running over the popliteus muscle and coursing deep to the blood vessels of the inferior-lateral knee. However, its presence is more variable, probably existing in less than 50% of knees (Davies et al., 2004). The *fabello-fibular ligament* runs parallel to the LCL from the fabella, a sesamoid fibrocartilage or bone found within the tendinous head of lateral gastrocnemius. Both this ligament and the fabella are inconsistent, being present in around 78%–80% of knees (Davies et al., 2004), with the sesamoid existing as either a fibrocartilaginous structure or an osseous sesamoid. If present, the ligament runs from the fabella to the head of the fibula, posterior to the biceps femoris tendon (Fig. 2.4.5n).

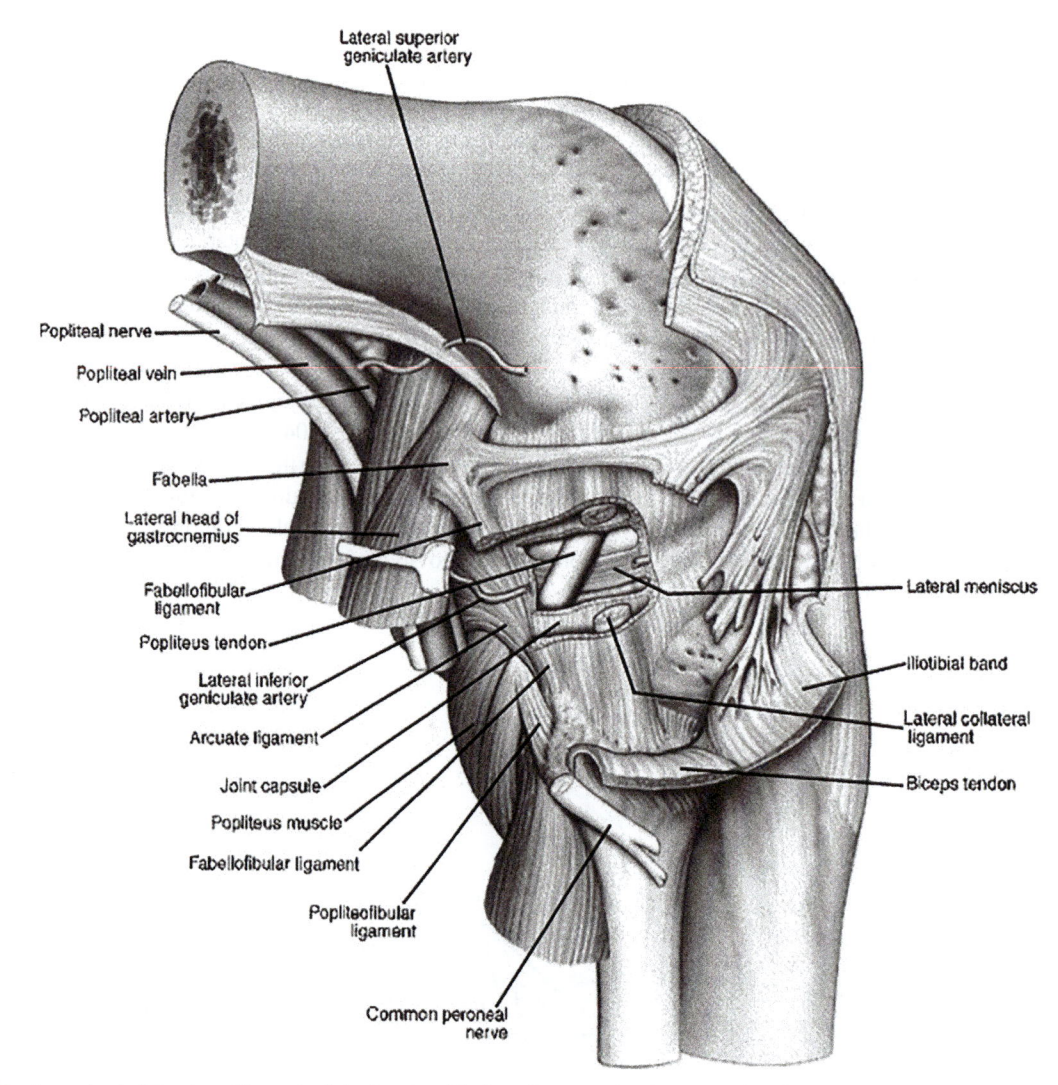

FIG. 2.4.5n An anatomical drawing of the posterolateral corner of the knee demonstrating the complexity of structures within this region. *(Image from Macgillivray, J.D., Warren, R.F., 1999. Treatment of the acute and chronic injuries to the posterolateral and lateral knee. Oper. Tech. Orthop. 9(4), 309–317.)*

The lateral retinaculum of the knee is orientated longitudinally with the knee in extension, and consists of a superficial and obliquely orientated retinacular ligament running from the fascia lata to the patella (Fulkerson and Gossling, 1980). Deep to this lie separate transverse fibrous bands to the patella, an epicondylar band, and a patellotibial band (Fulkerson and Gossling, 1980). The lateral posterior knee is reported as having three distinct soft tissue layers providing functional stability (Davies et al., 2004). The outer layer consists of superficial fascia, the iliotibial tract, and the biceps femoris tendon. The middle layer consists of the patellar retinaculum and the lateral patellofemoral ligament. The deep layer consists of the LCL, arcuate ligament, fabello-fibular ligament, popliteofibular ligament, and the popliteus tendon (Davies et al., 2004). The popliteus tendon and the joint capsule have little ability to control anterolateral rotation in the absence of an intact ACL (Fulkerson and Gossling, 1980).

If the LCL or the lateral deep ligamentous structures are cut, then there is an increase in varus rotation of the knee. However, the severance of the LCL has the most significance (Davies et al., 2004). The severance of the LCL and deep structures creates a slight increase in posterior translation at knee flexion angles and increases coupling of external rotation with posterior force. The combined severance also increases varus rotation greater than LCL severance alone, with greater effects at 30° flexion (Davies et al., 2004). These lateral ligaments in conjunction provide a primary role of resistance against varus moments, with some influence on posterior translation and external knee moments.

PCL failure causes increased posterior translation, increasing from 0° to 90° flexion, with concurrent loss of the coupling of external rotation with posterior force. With concurrent loss of the PCL and the LCL, there is much greater posterior translation, external rotation, and varus motion at all flexion angles, and significantly so at 60° (LaPrade et al., 2002; Davies et al., 2004). Hyperextension with varus force is likely to injure these structures (Davies et al., 2004), indicating that this is the force direction that these ligaments resist collectively. The loss of motion-guiding from any of the knee ligaments could change the knee's instantaneous joint axis in relation to the GRF vector, increasing the angular momentum arm and causing loss of the functional tensegrity-balanced loading system as proposed by Kim (2020).

Iliotibial tract

The iliotibial tract (ITT) is a fascial aponeurosis that is both tendon and ligament-like in functions, playing a role in lateral knee stability (James et al., 2015). It demonstrates a high interconnectedness with the anatomy of the region that impacts on its function (Vieira et al., 2007). At the hip, through its attachments to the gluteus maximus, gluteus medius, and tensor fasciae latae muscles, the ITT acts as a lateral hip stabiliser helping the hip abductors resist hip adduction moments on the stance limb (Noehren et al., 2007). It has four distinct distal attachments to the knee (Moorman and LaPrade, 2005). The primary part is the superficial layer that covers the majority of the lateral aspect of the knee and where it attaches distally to Gerdy's tubercle, i.e. the bony enthesis found on the lateral anterior tibia (Moorman and LaPrade, 2005). Its anterior expansion, known as the *iliopatellar band*, has an important role in patellar tracking (Moorman and LaPrade, 2005). The deeper layer of the ITT attaches to the medial aspect of the superficial layer of the lateral intermuscular septum of the distal femur, linking the tract to the muscular septa of the posterior and anterior thigh laterally. Deep and posterior to this is a capsulo-osseous layer that attaches to the lateral proximal head of gastrocnemius and the short head of biceps femoris, which also runs to attach to Gerdy's tubercle, 1 cm posterior to it (Moorman and LaPrade, 2005).

Linkage of the ITT across the hip and knee means that fascial tension within the tract increases with hip abductor activity and with internal rotation around the knee (Fig. 2.4.5o). Loss of control of these movements is linked to pathology (Noehren et al., 2007; Ferber et al., 2010; Louw and Deary, 2014; Aderem and Louw, 2015). Through this hip–knee linkage, the tensor fasciae latae muscle plays a significant role in countering knee adduction moments, assisting the gastrocnemii,

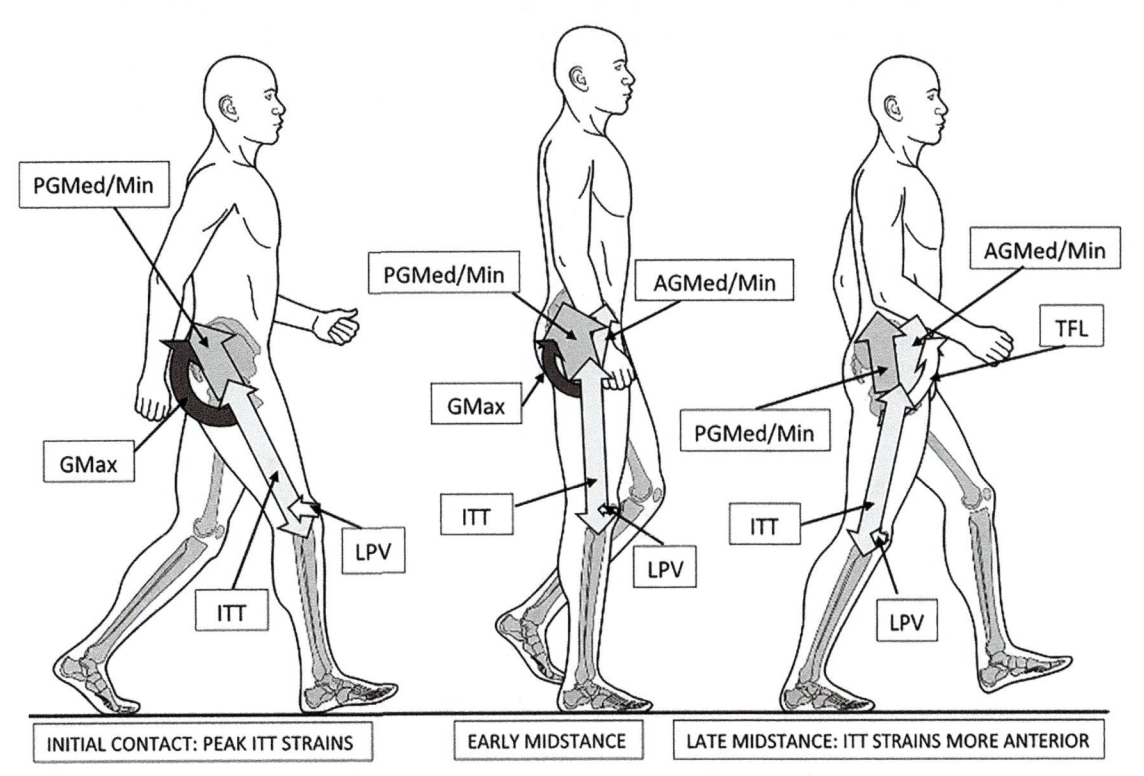

FIG. 2.4.5o A schematic of the role of the iliotibial tract (ITT) as seen in the sagittal plane during stance. During contact phase, gluteus maximus (GMax) has the greatest influence on the ITT, with assisting hip abduction control from the posterior and intermediate muscle fibres of gluteus medius and minimus (PGMed/Min). The ITT thus has a role in restraining knee flexion at initial contact, and then assisting hip extension moments, while helping resist hip adduction moments at the start of single-limb support. Its attachments to the lateral patella also provide a lateral patellar vector (LPV). As midstance progresses, GMax influence decreases and stops while the anterior muscle bundles of gluteus medius and minimus (AGMed/Min) start to become more influential. In late midstance towards heel lift, although the posterior and intermediate hip abductor fibres remain important, it is the anterior muscle bundles and tensor fasciae latae (TFL) that become proportionately more influential on the ITT tensions. TFL produces more anterior ITT forces during terminal stance, aiding knee flexion after heel lift. The ITT deep intermuscular septal connections to the lateral femur provide diffuse attachment areas for the ITT tensions to be applied to the lower limb, but these tensions tend to focus towards the distal attachments to the tibia and surrounding soft tissues. *(Permission www.healthystep.co.uk.)*

the quadriceps, and the hamstrings in this role (Winby et al., 2009). The tensor fasciae latae via the ITT has a relatively large effect on tibiofemoral frontal plane stability, creating up to 25% of lateral compartment loading force (Winby et al., 2009). The ITT linkage to large muscles like gluteus maximus as well as tensor fasciae latae plays a significant role in its ability to create lateral knee stability via tension-induced compression.

The ITT is commonly associated with lateral knee pain in (primarily female) runners and cyclists. This is a condition that was first reported in detail as late as 1975 (Baker et al., 2011). It has been linked to peak hip adduction and internal rotation of the knee, possibly causing increased compression through excessive tension over the lateral femoral condyle (Noehren et al., 2007; Ferber et al., 2010; Baker et al., 2011; Louw and Deary, 2014; Aderem and Louw, 2015). Tract pain in runners has been associated to just after heel strike as the knee is flexed around 20–30° and is exacerbated in downhill runners when the knee experiences a greater flexion excursion than occurs during level running (Noehren et al., 2007). It is also a pathology linked to long stride lengths and increased hip adduction moments during running (Bramah et al., 2018).

2.4.6 Anatomy of the anterior knee: Patellofemoral joint

The anterior aspect of the knee is covered by the patella which is the largest sesamoid bone (a bone within a tendon) present within humans. The patellofemoral joint lies between the anterior femoral condyles at the anterior articular facet and the posterior articular surface of the posterior patella. The joint is supported by the soft tissue tendon structures of the quadriceps muscles, the patellar retinaculum, the patellar tendon/ligament, and the medial patellofemoral ligament. The patellar retinaculum forms a fascial reinforcement that encloses the patella and covers both joints of the knee anteriorly. The fascia, quadriceps tendons, and patellar ligament via the patella can generate appropriate levels of posteriorly directed compression into the femoral condyles to help restrict femoral anterior translation over the tibia during contact and loading response. Together, they act like an anterior elastic sling restraint on the femur. The patellar retinaculum, as part of this mechanism, is most mechanically beneficial when the quadriceps are at reduced mechanical advantage with the knee in low flexion angles near extension. The quadriceps generate increasing compression on the patella at higher flexion angles, with these forces rising and falling in response to muscle activity (Fig. 2.4.6a). The length of the hamstrings also seems to influence patellofemoral joint reaction forces, with shortened hamstring lengths reported to increase forces across the joint, particularly laterally at 60° flexion during squatting tests (Whyte et al., 2010).

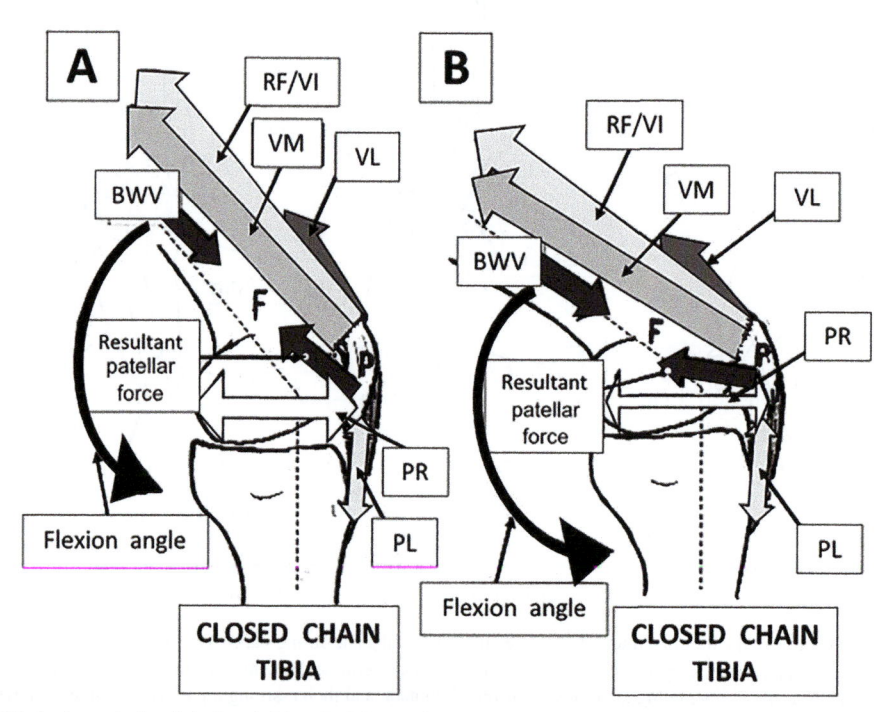

FIG. 2.4.6a A simplified schematic (medial view, left knee) of the patellofemoral joint acting as an anterior constraint. Forces derived from the quadriceps muscles of rectus femoris, vastus intermedius (RF/VI), vastus lateralis (VL), and vastus medialis (VM) are opposed by tensional resistance from within the patellar ligament (PL) and patellar retinaculum (PR). The femoral (F) knee flexion angle on a fixed closed chain tibia dictates the resultant force direction. In moderate degrees of flexion (A), the resultant force compresses the patella (P) into the patellofemoral joint in a superoproximal direction, restricting anterior femoral translational shear from the body weight vector (BWV) and assisting in moderating anterior articular compressions within the tibiofemoral joint. At higher flexion angles (B), femoral anterior shear becomes greater. However, the changing quadriceps forces alter the resultant compression-resistance forces of the patella, matching the more horizontal nature of the translation shear. The patellar retinaculum plays little part in higher knee flexion angles, only increasing its patellofemoral stabilising role towards knee extension. *(Permission www.healthystep.co.uk.)*

The lateral articular facet of the femur is larger than the medial facet, consistent with the prevailing lateral force vector found at the joint (Amis et al., 2003). Patellofemoral joint stability is produced by a complex interplay between active muscle and passive fascial tensions and compressive forces within the joint, which alter with changing knee angles (Ateshian and Hung, 2005; Senavongse and Amis, 2005). The alignment of the femur to the patella and the tibia is significant to the application and direction of joint stresses. The flexion angle is also significant, as increasing flexion heightens joint contact forces and causes a rise in contact area and maximum peak joint pressures (Wünschel et al., 2011). However, the forces from the retinaculum that hold the patella tight to the knee around extension reduce with increased knee flexion (Amis et al., 2003) (Figs 2.4.6b and 2.4.6c).

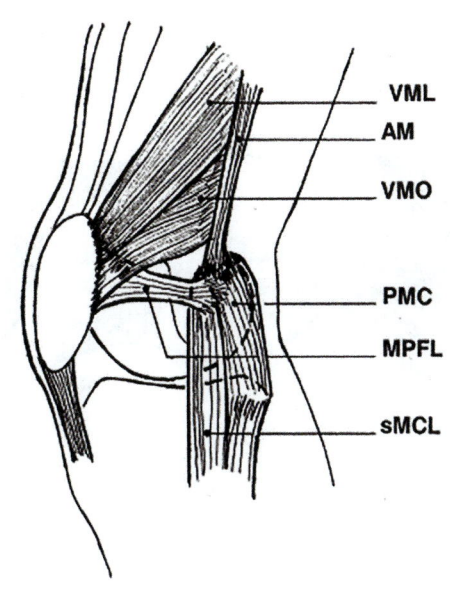

FIG. 2.4.6b The anatomy of the medioanterior knee showing the vertical fibres of the distal adductor magnus (AM), the medial patellofemoral ligament (MPFL), the posteromedial joint capsule (PMC), the superficial band of the medial collateral ligament (sMCL), vastus medialis longus (VML), and vastus medialis obliquus (VMO). *(Image from Amis, A.A., Firer, P., Mountney, J., Senavongse, W., Thomas, N.P., 2003. Anatomy and biomechanics of the medial patellofemoral ligament. Knee 10(3), 215–220.)*

The angular geometry of the joint is important to patellar function and potential dislocation. This is dependent on equality of medial–lateral force vectors around the joint. To understand the biomechanics of the patellofemoral joint, the quadriceps need to be split into different component vectors in relation to the vertical axis of the femur and patella. The rectus femoris and vastus intermedius parts of the quadriceps exert a largely vertical tension on the patella. Both vastus lateralis and vastus medialis longus/obliquus (mainly through its obliquus or VMO fibres) have small, displaced oblique components that diverge from the vertical axis, providing a significant transverse force vector to the joint that instigates medial–lateral stabilisation (Farahmand et al., 1998). There is some evidence to suggest that the VMO part of vastus medialis is more prone to weakness than the other parts of the quadriceps, and this can be associated with symptoms within and around the patellofemoral joint (Powers et al., 1996). The muscles of the knee affect the compressive forces generated within the patellofemoral joint through anterior–posterior and internal–external translational changes between the tibia and femur. This indicates an intrinsic coupled relationship between tibiofemoral and patellofemoral joint kinematics and kinetics (Li et al., 2004). An increase in tibial posterior translation and external rotation or femoral anterior translation and internal rotation is accompanied by increasing patellofemoral contact pressures (Li et al., 2004) (Fig. 2.4.6d).

The resultant tensional force of all the quadriceps should run along the vertical axis of the femur into the patella, resisted by the tensional stresses of the patellar tendon/ligament. These forces are not collinear but diverge at around a 15° angulation, known as the quadriceps angle or '*Q*' *angle*. This creates a functional knee (genu) valgum angle (Amis et al., 2003). Although it was long considered that the wider female pelvis caused a greater incidence of higher Q angles and patellofemoral pain, the pelvis is not actually wider in females. It is their shorter on average limb length that makes it appear so (Wu, 2015). The origins of the Q angle are more likely a result of increasing the hip adduction angle, particularly during running where shorter lower limb's need greater hip adduction angles to bring them under the CoM. The patella has a natural resultant lateral force vector because of the Q angle, which is resisted by a more prominent lateral facet on the femoral trochlear groove surface (Amis et al., 2003). This means that a deficient size in the lateral facet risks patellar lateral

FIG. 2.4.6c A dissection photograph of the human knee viewed distally with removal of the quadriceps muscle at 15° flexion (upper image) and 90° flexion (lower image). Note the changing position of the patella relative to the femoral surface, with increasing flexion angles. Changing articular forces and locations helps hold the patella tight against the femur in higher flexion angles with the patellar retinaculum helping restrain the patella at low knee flexion angles nearing extension. *(Image from Amis, A.A., Firer, P., Mountney, J., Senavongse, W., Thomas, N.P., 2003. Anatomy and biomechanics of the medial patellofemoral ligament. Knee 10(3), 215–220.)*

FIG. 2.4.6d Changes in patellar cartilage compressive stresses during knee flexion at different angles. Variations in quadriceps muscle activity are likely to alter the locations in peak compressive stress (darker areas), which if consisting of a persistent abnormal pattern, could lead to articular degeneration. *(Image from Ramaniraka, N.A., Terrier, A., Theumann, N., Siegrist, O., 2005. Effects of the posterior cruciate ligament reconstruction on the biomechanics of the knee joint: a finite element analysis. Clin. Biomech. 20(4), 434–442.)*

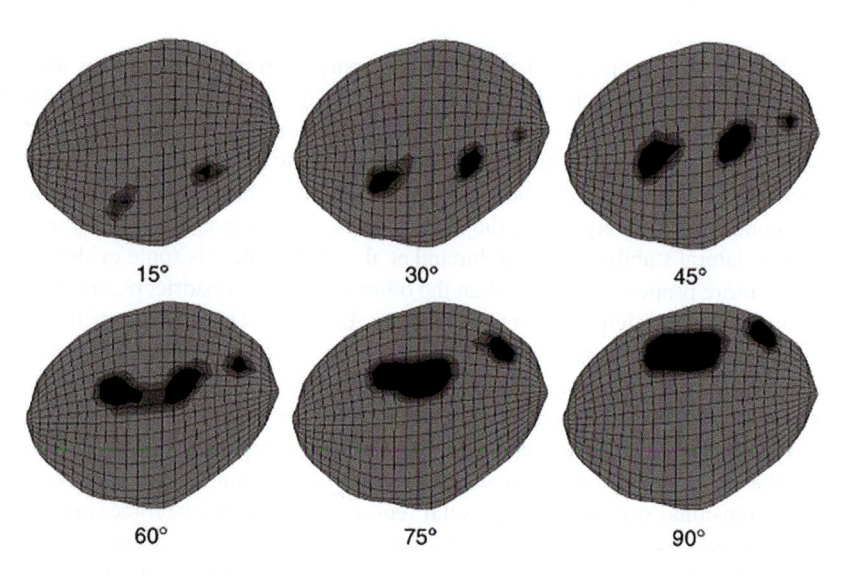

dislocation (Amis et al., 2003). The soft tissue active and passive force vectors need to resist this larger lateral vector to achieve functional efficient patellar tracking in the groove with knee motion. This is achieved through the action of the quadriceps, the patellar retinaculum and tendon, and the medial patellofemoral ligament. The dominant lateral force vector influenced by the overall Q angle requires effective vastus medialis resistance, more so if Q angles are large (Fig. 2.4.6e).

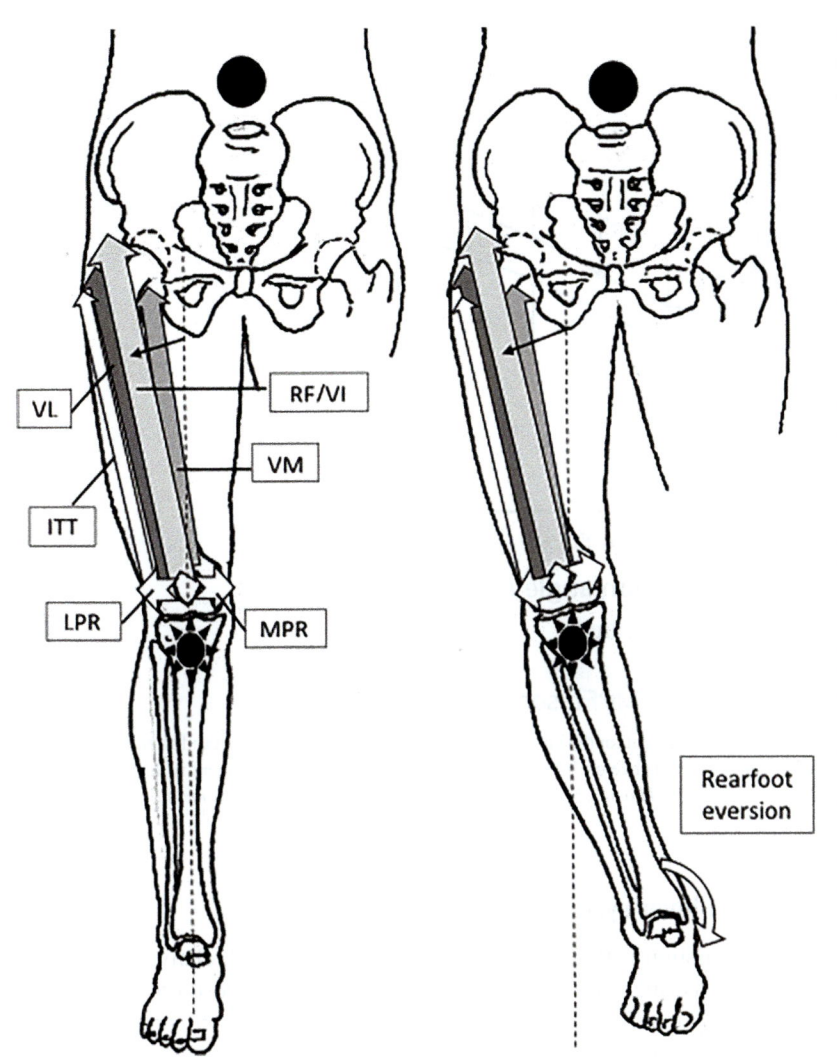

FIG. 2.4.6e A schematic of forces involved in changes of the Q angle. Vectors from rectus femoris and vastus intermedius (RF/VI), vastus lateralis (VL), vastus medialis (VM), iliotibial tract (ITT), and medial and lateral patellar retinaculum (MPR and LPR) are demonstrated. The quadriceps or Q angle is the angle between the quadriceps muscles force (as a single combined vector from all the muscles) and the patella through its tibial tuberosity ligament attachment (black sun). The quadriceps vector is naturally lateralised in humans (left image) and changes its lateral orientation with flexion (reduces) and extension (increases). It is considered high if the angle is over 20° near full knee extension. It can be influenced by femoral neck-to-shaft angles which can cause alignment changes within the femur (coxa vara increases angle, coxa valga reduces it). The Q angle is also 'functionally changeable', with greater hip adduction angles increasing it (right image). Running, which lacks a double-limb stance phase, is particularly notorious for increasing the hip adduction angle to place the foot under the body's CoM. If the hip abductors and quadriceps cannot adapt their force vectors to this change in limb position, the resultant quadriceps vector can become increasingly laterally orientated. This will pull the patella more into the lateral side of the patellofemoral joint and increase stresses on the medial connective tissue restraints such as the medial patellar retinaculum (MPR). High hip adduction angles at contact can present the foot more inverted to the support surface, requiring rearfoot eversion to bring the foot flat. *(Permission www.healthystep.co.uk.)*

The medial patellofemoral ligament is a retinacular band found deep to the deep fascia, connecting the femoral medial epicondyle to the medial edge of the patella (Amis et al., 2003). It is around 55 mm long with a reported wide variation in width, widening at its attachments and fanning out towards the patella from the femur (Amis et al., 2003). In addition, it is surprisingly strong with a mean tensile strength of around 200 N (Amis et al., 2003). The ligament is overlaid by VMO to a variable extent, often having merging fibres with this muscle to provide the primary passive restraint of lateral patellar displacement (Amis et al., 2003). The connection to active VMO resistance obviously potentially enhances the restraint mechanism.

Its capacity to restrain lateral deviation of the patella is greater with less knee flexion, with the patella most likely to displace laterally at around 20° of flexion. It is protected at lesser flexion angles because the retinaculum is at its tightest, pulling the patella into its articular facet around full extension, and only slackening with flexion (Amis et al., 2003). The quadriceps should protect the patella from dislocation at higher flexion angles, so the risk is in the range of flexion where patellar stabilisation is passed from one area of anatomy to the other. The femoral attachment of the patellofemoral ligament is reported to be poorly delineated, convergent with a number of structures including the superficial MCL around the medial femoral epicondyle (Amis et al., 2003). The patellar attachment is wider than the femoral, and occurs either on the medial aspect of the patella for a length of around 20 mm or sometimes along the entire medial patellar edge (Amis et al., 2003). Sectioned medial patellofemoral ligaments in cadaveric knees have resulted in greater lateral patellar displacement during motion compared to intact ligaments as the knee approaches extension (Amis et al., 2003) (Fig. 2.4.6f).

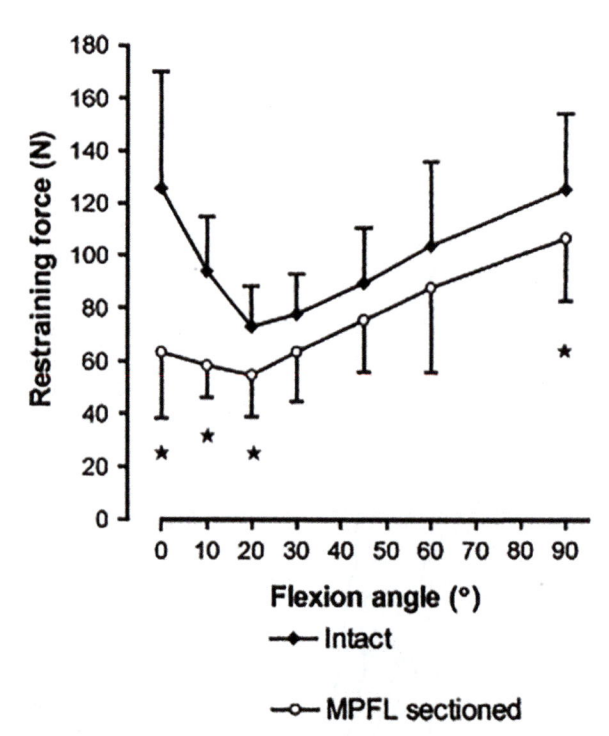

FIG. 2.4.6f Graph of patellar lateral displacement force at 10 mm displacement for the intact knee under 175 N quadriceps tension and after transection of the medial patellofemoral ligament (MPFL sectioned). This ligament's influence on VMO's ability to restrain lateral displacement is thus shown to be significant. *(Image from Amis, A.A., Firer, P., Mountney, J., Senavongse, W., Thomas, N.P., 2003. Anatomy and biomechanics of the medial patellofemoral ligament. Knee 10(3), 215–220.)*

2.4.7 The functional joint axes of the knee in lever systems

During walking's swing phase, the knee has to control motion of the lower leg and foot from an initial relatively extended position at preswing to a maximum flexed position at midswing, and then through to a near-extension position prior to heel strike. In running, initial contact can occur with the knee more flexed, but this is dependent on stride length and foot strike position. The primary knee action of late swing is to extend the knee into terminal swing from its flexed position of midswing. This is achieved through the action of the quadriceps applied via its distal attachments through the patellar tendon/ligament to the tibial tuberosity. The precise timing of this activation within gait is variable and associated with gait speed and technique, i.e. walking or running (Gazendam and Hof, 2007). The lever system utilised is always class three. This is because the extension effort is applied just below the fulcrum of the knee while the CoM of the leg and foot segments is below the effort, creating a longer resistance arm than the effort arm. As the knee extends angular momentum increases on the knee, which in turn heightens the centrifugal force (Fig. 2.4.7a). This is restrained by hamstring activity.

In walking stance phase, the knee continues to function as a class three lever, but the lever system is now inverted compared to swing. Effort is thus applied to the proximal attachment points of the quadriceps around the femoral shaft, and the resistance becomes the CoM of the HAT, the thigh, and the mass of the swing limb. In normal heel strike walking gaits, loading forces are greatest during loading response and during early midstance when the body's CoM lies behind the knee joint. This generates a flexion angular moment that needs to be resisted by the quadriceps. However, at heel contact during very long strides (usually associated with running), the knee can experience a GRF in line with or in front of the knee joint axis. This can cause increased articular compression and it can drive an extension moment onto the loading knee. The size of this moment is dependent on the stride length and will require greater hamstring activation to resist it. It is important that the knee is able to flex after initial contact in order to help dissipate energy through its motion via muscle–tendon buffering, but it must also remain stable on initial loading. The flexion moment from the body's CoM is opposed by the quadriceps and any GRF extension moment is opposed by the hamstrings keeping the knee angulation fairly stable for contact, but with the ability to respond with appropriate flexibility once loading starts (Fig. 2.4.7b).

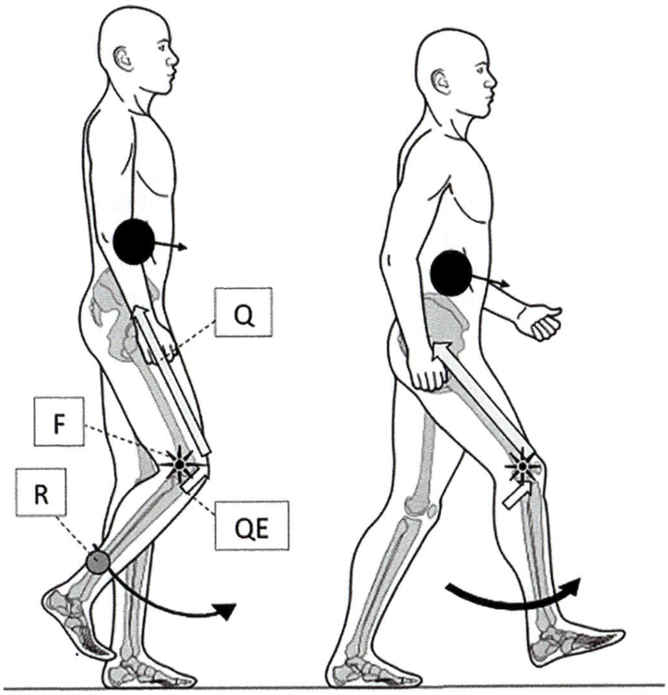

Locomotive functional units **Chapter | 2** 363

FIG. 2.4.7a The knee presents a simple class three lever system in swing phase with the knee joint as the fulcrum (F—star) and the quadriceps (Q) provide the effort (QE) via the patellar ligament's attachment on the tibial tuberosity. The resistance (R) is the CoM of the lower leg and foot below the knee. This mechanism is active from midswing into terminal swing when the contraction forces of the quadriceps knee extension become part of the centrifugal forces accelerating the anterior progression of the body's CoM towards the next step. Rectus femoris is also assisting iliopsoas in providing hip flexion. *(Permission www.healthystep.co.uk.)*

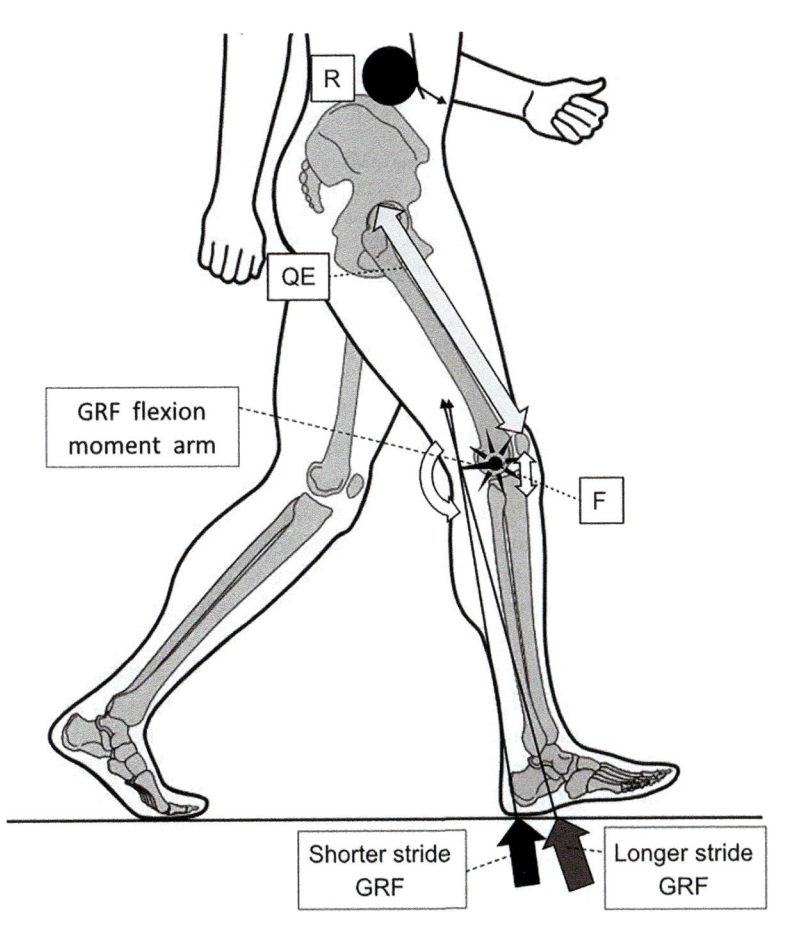

FIG. 2.4.7b The knee's lever arm remains class three at initial contact and loading response. The normal GRF (black arrow) of heel strike on a shorter stride is behind the knee, initiating a flexion moment. This is resisted by the eccentric action of the quadriceps (light grey double-headed arrow). Its internal extension forces provide the quadriceps effort (QE) onto the proximal shaft of the femur that is used against the knee flexion moment, initiating shock absorption through muscle–tendon buffering with the patellar ligament (white double-headed arrow) to dissipate impact energy. QE at this time is controlling the drop of the body's CoM acting as the resistance (R—black circle) via knee and hip flexion restraint. Thus, the fulcrum is the knee (F—star). A longer stride length will shorten the GRF's (dark-grey arrow) knee flexion moment arm, but the CoM will now lie further behind the knee, making control of CoM descent more challenging and weight transfer more difficult. If the GRF lines up directly into the knee joint axis, as can occur on a very long stride, the joint will receive a direct impact compression force and knee flexion energy dissipation will be harder to initiate, although the initial contact joint flexion torque will be smaller. *(Permission www.healthystep.co.uk.)*

The GRF-induced knee flexion moment during contact initially increases, flexing the knee to dissipate energy so that quadriceps activity increases concurrently with the ceasing of any hamstring activity. Once the forefoot loads and early midstance begins, the quadriceps utilise their proximal effort on the femur to draw the knee towards extension. This moves the CoM of the body nearer to the fulcrum (axis) of the knee, decreasing the knee flexion moment which allows the effort required from the quadriceps to decrease. By absolute midstance, the lever arm has completed its task of drawing the knee towards extension, and the quadriceps can reduce activity with the CoM starting to pass anterior to the knee.

In late midstance, the knee largely functions without significant muscle-induced motion as the torque of the inverted pendulum changes to ankle dorsiflexion, with the action of gastrocnemius hopefully preventing excessive knee flexion being developed from the BWV and GRF-induced extension torque (Fig. 2.4.7c). The quadriceps (mainly rectus femoris) become active again at the knee in preswing, helping draw the knee forward with the hip flexors. The lever arm utilised in preswing is still class three with the resistance being the mass of the lower leg and foot as in swing, and the effort returning to the tibial tuberosity of the patellar tendon attachment.

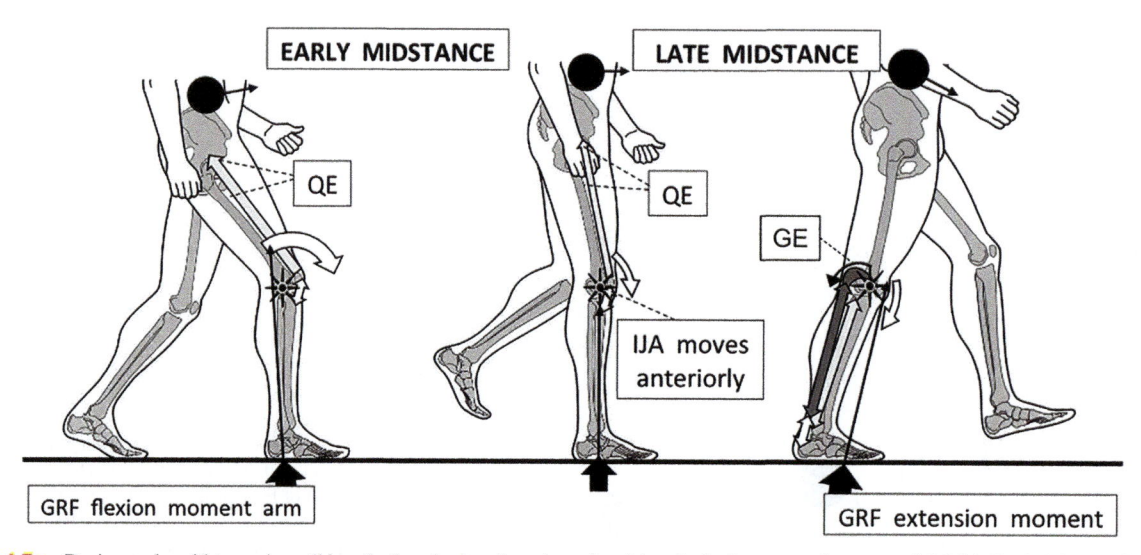

FIG. 2.4.7c During early midstance in walking, the knee's class three lever is raising the body mass resistance or CoM (black circle) up and over the lower limb. This action is aided by the hip extensors extending the hip out of flexion. The concentrically applied quadriceps effort (QE) draws the femur forward around the knee joint fulcrum (star), creating a knee extension moment that raises the CoM upwards and anteriorly. The instantaneous knee joint axis moves anteriorly with this extension. The quadriceps activity gradually decreases as the CoM high point is approached at absolute midstance. In late midstance, the CoM remains the resistance, but the knee is now under accelerating extension moments from the CoM's fall and the increasingly anteriorly directed GRF lying behind the knee. This is resisted by eccentric gastrocnemius effort (GE—dark grey arrows) as part of the triceps surae CoM braking action. As GE is applied very close to F, it can only provide a little but usually sufficient, knee extension resistance. At and after heel lift, effort from gastrocnemius-induced Achilles elastic recoil is sufficient to aid knee flexion (coupled to ankle plantarflexion) that is required for early swing phase. *(Permission www.healthystep.co.uk.)*

In running, events are different as the knee becomes increasingly flexed to the middle of stance phase, rather than extending out of flexion. However, the lever arm events are also shaped by stride length and foot strike position. Thus, heel strikes in running result in an initial knee loading pattern similar to that of walking, but forefoot strikes sustain greater initial knee and ankle flexion moments. However, midstance knee flexion and terminal stance acceleration knee extension in running is fairly consistent, regardless of the initial foot strike position (Fig. 2.4.7d). Increased flexion moments and ranges of motion at the knee during running increase the energetic burden, which explains the increased prevalence of knee symptoms in running compared to walking.

Kim (2020) has argued that the instantaneous axis of the knee joint moves anteriorly at initial contact in such a way as to minimise or avoid joint torques or the angular momentum induced by GRF vectors during gait. Effectively the GRF is kept in line with or extremely close to the joint axis, reducing the angular momentum generated. Knee tension and compression elements during locomotion are argued to balance out through utilising the knee's instantaneous joint axis position in such a way as to result in virtually no work (Kim, 2020). Should such a system be proven, then it reduces the significance of the lever arm system and dramatically reduces the potential influence of the GRF vectors that affect muscle activity. However, the proposal does not totally remove joint torques such that potential flexion and extension torques, although much reduced, still need to be managed through appropriate muscle contraction to control and create appropriate motion. Failure of the tensegrity structure's tension–compression balance, as a result of pathology in the tensioning units such as the ligaments, may explain the associated degenerative changes that occur in the knee through ligament injuries, as angular moments may start to change. In ligament dysfunction, the associated close relationship between the instantaneous joint axis alignment and the GRF vector may be lost, and joint torques consequently greatly increased, damaging tissues.

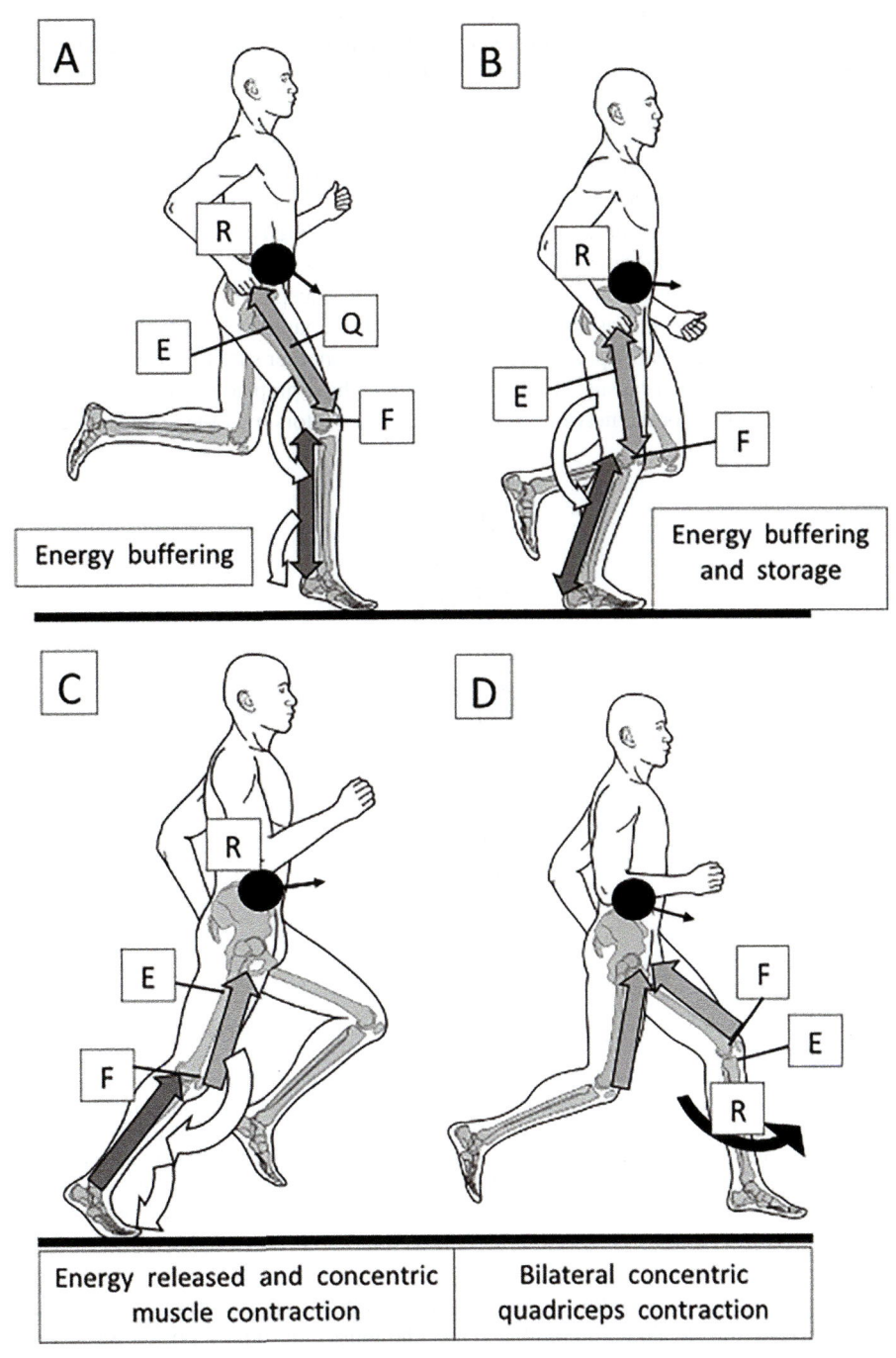

FIG. 2.4.7d Running represents a change in action from walking, but the knee remains the fulcrum (F) in a class three lever system. At initial contact (A), the effort (E) remains via the quadriceps (Q) eccentric contraction force applied to the femur. In forefoot strike, the ankle plantarflexors (dark-grey arrows) can assist in impact energy buffering. Knee flexion is required throughout the braking phase (B), with the HAT's CoM acting as the resistance (R) that must not fall too far inferiorly or lose too much anterior momentum. The end of braking initiates acceleration (C) via knee extension, requiring a concentric quadriceps contraction force to provide the effort for limb extension combined with the ankle plantarflexors' and hip extensors' power to raise the CoM. Late swing phase (D) requires knee extension, but initial aerial phase (right leg) requires knee flexion that is initiated by plantarflexion power. *(Permission www.heathystep.co.uk.)*

2.4.8 Muscle action at the knee in gait

Extensor muscle action at the knee

The quadriceps femoris (quadriceps) is translated as 'four-headed of the femur' and refers to the primary knee extensors. From lateral to medial, they consist of the vastus lateralis, rectus femoris, vastus intermedius, and vastus medialis. The latter muscle has two different areas of fibre arrangements known as vastus medialis longus and vastus medialis obliquus, commonly referred to as the VMO, dividing its muscles force vectors. With the exception of the long head of rectus femoris, the quadriceps are all bi-articular muscles bridging the tibiofemoral and patellofemoral joints via their tendinous attachments to the patella through the patellar ligament (tendon).

The long head of rectus femoris attaches to the anterior inferior iliac spine. Its shorter head attaches from a groove just above the acetabulum of the hip joint capsule. Rectus femoris has weak hip flexor activity as well as having knee extensor activity. Vastus medialis attaches proximally to the lower part of the intertrochanteric line, the spiral line, the linea aspera, and the superior part of the supracondylar ridge of the anterior surface of the femur through a long fibrous enthesis. Vastus intermedius attaches to the superior two thirds of the anterior and lateral femoral shaft and the lateral intermuscular septum, to blend with vastus medialis at the linea aspera. Vastus lateralis is the largest of the quadriceps and attaches through a broad tendon to the intertrochanteric line and the anterior and inferior borders of the greater trochanter, the lip of the linea aspera, and the lateral gluteal tuberosity of the femur (Draves, 1986, pp. 228–229). There is also another small knee extensor muscle attached to the femur, *articularis genu*, that tenses the joint capsule in extension to retract the suprapatellar bursa (Astur et al., 2011; Grob et al., 2017). Together, these muscles should form a network of balanced force vectors over the anterior knee (Fig. 2.4.8a).

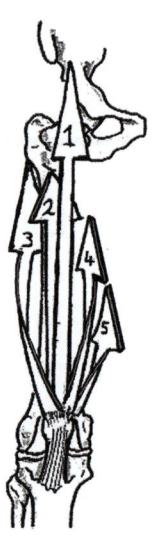

FIG. 2.4.8a The quadriceps muscles represent an interesting set of force vectors at the knee. Through biomechanical investigation, this has become increasingly complex. Rectus femoris (1) has two proximal attachments, one to the anterior inferior iliac spine and the other to the groove just above the acetabulum. This muscle seems to play a more important role in linking swing phase hip flexion to knee extension and has a tendon consistent with a positional role. Thus, stance phase anterior knee joint superior vectors for the patellofemoral and tibiofemoral joints seem to derive from vastus intermedius (2) attached from a long fibrous enthesis to the anterior-lateral femur. Its tendon is an energy-storage tendon and beneath it lies articularis genu (not shown) that lifts the anterior joint capsule and bursa on extension. To each side lie vastus lateralis (3) and medialis (4 and 5) that envelop most of the femur by both of them attaching to the linea aspera on the posterior femur. Vastus lateralis creates the largest single quadriceps by cross-sectional area. Vastus medialis is split into two distinct areas of muscle fibre orientations, vertical (4) and oblique (5), that create different force vectors on the patella and knee. *(Permission www.healthystep.co.uk.)*

Of the quadriceps muscles, vastus intermedius is the hardest to study as it lies deep to rectus femoris, making surface EMG studies of the intermedius muscle alone impossible. Thus, the muscle activity reported for rectus femoris, as previously stated, probably represents combined rectus femoris and vastus intermedius activity, yet it could also just amount to vastus intermedius activity. Using fine wire EMG data taken directly from within these muscles suggests rectus femoris only seems to be active during stance at the stance-to-swing transition stage of the gait cycle and also at terminal stance during faster walking speeds (Nene et al., 2004). Surface EMG exhibits a biphasic patterned burst of activity for rectus femoris at both initial contact and the terminal stance-to-swing transition. Therefore, the biphasic activity seen in the quadriceps at contact and loading activity in early stance is only consistent to vastus lateralis, medialis, and intermedius.

The quadriceps as a whole are most active around terminal swing and initial contact, and also at the stance-to-swing transition during preswing. Using surface EMG at self-selected walking speeds, the central quadriceps (rectus femoris and/or vastus intermedius) are reportedly characterised by different activations within different strides of the same walk (Di Nardo and Fioretti, 2013). Despite these variations, two clear patterns of central quadriceps activity are identified: at the stance-to-swing transition to aid limb advancement into swing and from terminal swing to midstance acting with the medial and lateral quadriceps muscles (Di Nardo and Fioretti, 2013). Thus, the quadriceps accelerate the leg in swing and then resist knee flexion moments during loading and early midstance. The synergistic action of these muscles is to create a knee-straightening extension moment or a flexion resistance moment (Fig. 2.4.8b). However, the contributions between the muscles are not even, with eccentric and isometric recruitment of vastus lateralis being greater than vastus medialis and rectus femoris/vastus intermedius recruitment occurring only at low-to-intermediate intensities (Pincivero et al., 2006).

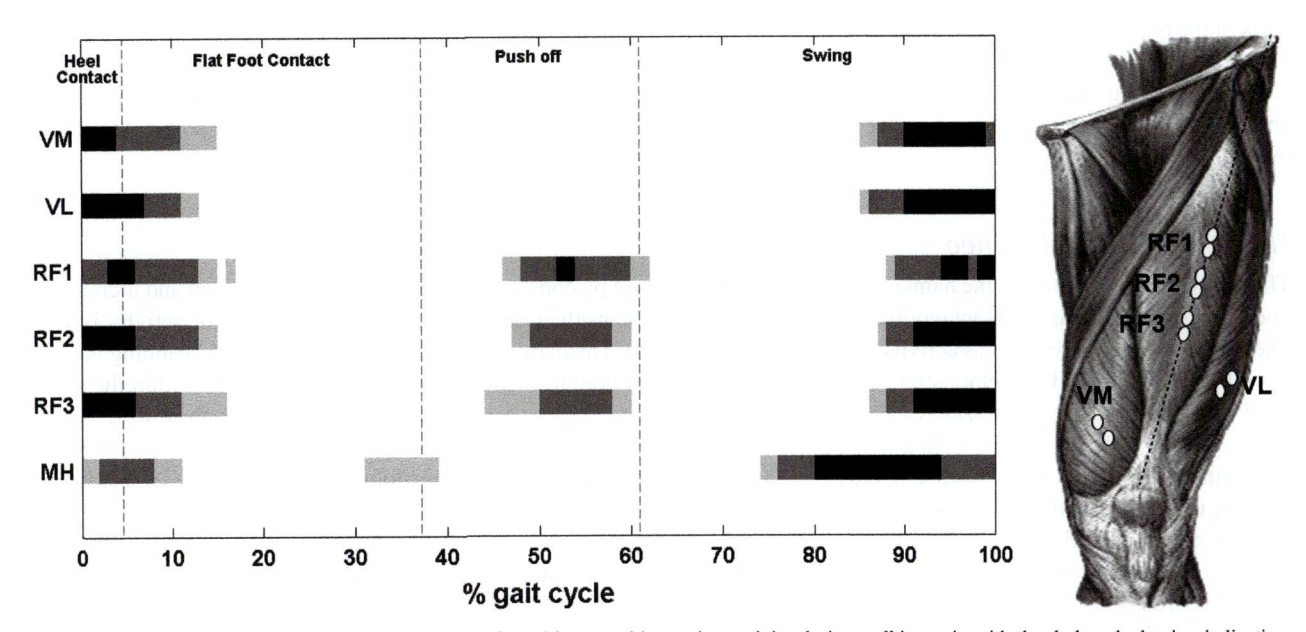

FIG. 2.4.8b Quantised (packaged) average patterns of quadriceps and hamstring activity during walking gait, with the darker shadowing indicating higher levels of EMG amplitude and the light shading representing low levels. These are shown for vastus medialis (VM), vastus lateralis (VL), and the medial hamstrings (MH). Note that flat foot means from loading response (first dashed line) and throughout midstance, while push off refers to acceleration after heel lift (second dashed line) at the start of terminal stance, ending at toe-off (third dashed line). The results for rectus femoris (RF1, 2, 3) depend on the electrode placement (see right image) and likely presents levels of crosstalk from vastus intermedius beneath. *(Image from Di Nardo, F., Fioretti, S., 2013. Statistical analysis of surface electromyographic signal for the assessment of rectus femoris modalities of activation during gait. J. Electromyogr. Kinesiol. 23(1), 56–61.)*

Females are reported to generate greater EMG activity in their quadriceps during concentric knee extension, but at a lower velocity than males. They further demonstrate a significantly greater decrease in vastus lateralis and rectus femoris/vastus intermedius EMG activity during isometric contraction compared to vastus medialis, which reportedly shows no gender difference (Pincivero et al., 2006). These reported variances in quadriceps activities between the genders may be significant to pathology rates within the knee, as changing quadriceps loading patterns alter tibial axial and varus-valgus rotations (Wünschel et al., 2011). Higher internal tibial rotations are associated with higher medial muscle activity. This results in optimising patellar tracking by reducing the Q angle but increases the patellofemoral contact pressures and areas. Increasing vastus lateralis activity induces a reduction in patellofemoral contact forces but also creates a reduced contact area by 30% at 90° of flexion (Wünschel et al., 2011).

Fatigue impairs quadriceps biomechanics and neuromuscular function during gait (Murdock and Hubley-Kozey, 2012). In fatigue, knee extensor torques are reduced and mean power frequencies decreased, knee external rotation angles increase, while the net flexion and external rotation moments decrease (Murdock and Hubley-Kozey, 2012). In fatigue of the quadriceps, rectus femoris/vastus intermedius sustains greater activity reductions than the other muscles (Karlsson et al., 2003; Pincivero et al., 2006). The vastus medialis seems more resistant to fatigue, probably due to the relatively greater number of type I slow twitch muscle fibres it contains (Pincivero et al., 2006). However, following vastus medialis fatigue, low-demand walking knee motion and loading characteristics were altered compared to those observed prior to fatigue in a manner that might leave the knee joint susceptible to a greater risk of joint injury through altered kinematics (Murdock and Hubley-Kozey, 2012).

Patellar ligament or tendon?

The patellar ligament is structurally and functionally more tendon-like than ligament-like (Franchi et al., 2009) and should be viewed as an extension of a combined tendon of all the quadriceps. However, there are distinct morphological differences between the proximal quadriceps tendon region superior to the patella and the distal patellar ligament region of this structure (Hadjicostas et al., 2007). Strains, behaviour, and elastic energy storage have been found to be uneven between the parts of the quadriceps tendon in the attachment to the patella in rabbits (Franchi et al., 2009). The tendinous parts of vastus intermedius respond to greater forces by providing much greater elastic recoil (Franchi et al., 2009). This suggests that the vastus intermedius tendon is an energy-storage tendon. The tendon sections arising from rectus femoris play a more 'ligament-like' positional role of limiting excessive knee flexion (Franchi et al., 2009), which suggests it is a postural muscle–tendon unit.

Flexor muscles at the knee

The primary knee flexors are the hamstrings, muscles that function by constraining anterior tibial translation and increasing knee flexion with an agonist–antagonist relationship between themselves and the quadriceps. Knee flexor activity is also likely assisted by gastrocnemius activity (Mengarelli et al., 2018). The hamstrings are bi-articular muscles spanning the hip as well as the knee and all demonstrate similar EMG activity, although smaller degrees of activity occur within the short head of biceps femoris (Kumazaki et al., 2012). The short head of biceps femoris that arises from the posterior femoral surface on the lateral lip of the linea aspera seems to have a slightly different functional role in early swing from the other hamstrings (Goldberg et al., 2004; Neptune et al., 2008; Fox and Delp, 2010). This is probably because it does not span the hip, yet it is still bi-articular, as the fibres from the long head tendon cross the proximal tibiofibular joint. Thus, the long head of biceps femoris' fibres cross three joints (multi-articular).

The other hamstrings (including the long head of biceps femoris) all arise from a tendinous proximal attachment on the ischial tuberosity (Kumazaki et al., 2012). They cross the hip inferomedially and are normally aligned to adduct and extend the hip as well as providing flexion moments to the knee during gait. Semimembranosus and semitendinosus attach distally to the posterior medial and medial knee structures, respectively (Kumazaki et al., 2012). Being bi-articular muscles, the hamstrings' kinematics and kinetics result in and also from, both the hip and knee positions and from the orientations of the pelvis, femur, tibia, and fibula (Fig. 2.4.8c). Passive tension stress is highest within the hamstrings when the pelvis tilts anteriorly (Nakamura et al., 2016). Hamstrings dominate frontal plane stability of the tibiofemoral joint briefly after heel strike (Winby et al., 2009) and generate maximal flexion torques when the knee is almost fully extended (Kumazaki et al., 2012). Isometric contraction forces of the hamstrings decrease significantly with increased flexion angles (Kumazaki et al., 2012).

Hamstrings are secondary hip extensors as well as primary knee flexors, although their hip stability role in gait seems mainly related to resisting hip extension during the late swing phase and just into the initial contact phase of gait (Ellis et al., 2014). They do not seem to provide much hip extension power during gait, as during terminal stance and preswing, they seem to have only a minor role in hip extension (Gottschall and Kram, 2003; Ellis et al., 2014). However, the short head of biceps femoris is assisting iliopsoas in early swing to provide knee stability with flexion of the hip (Goldberg et al., 2004; Neptune et al., 2008; Fox and Delp, 2010). The action of the hamstrings in mid-to-late swing has long been considered to be that of eccentric braking. However, at least in running, this action of the hamstrings may be more an isometric contraction, while the attachment points proximally and distally move apart (Van Hooren and Bosch, 2017a, b). This would seem like another situation where the term 'near-isometric contraction' could better describe the situation, with passive connective tissue lengthening during isometric muscle fibre contraction (Fig. 2.4.8d).

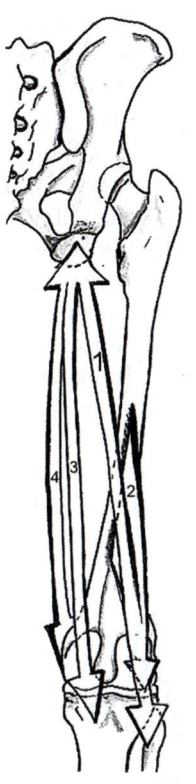

FIG. 2.4.8c Hamstring muscles (except the short head of biceps femoris) are set up to pull the posterior proximal tibia and fibula towards the pelvis and vice versa, extending the hip or flexing the knee. The hamstrings attach proximally to the superior medial quadrant of the ischial tuberosity. Biceps femoris (1) also has a short head (2) with a very direct flexion moment on the knee via its proximal strong attachment to the linea aspera on the posterior femur and lateral intermuscular septum of the thigh. It only blends with the long head near the lateral aspect of the knee, before biceps femoris attaches via three slips to the lateral tibial condyle, the lateral collateral ligament, and the head of the fibula. Semimembranosus (3) attaches distally on the posterior surface of the medial tibial condyle and blends with the oblique popliteal ligament. Semitendinosus (4) attaches through a long slender tendon to the anteromedial aspect of the medial tibial condyle, crossing the medial collateral ligament to join sartorius and gracilis at the pes anserine attachment. A bursa lies between it and the medial collateral ligament and the tibia as it folds around the knee. There are subtle variances in hamstring roles that make their different attachments important. *(Permission www.healthystep.co.uk.)*

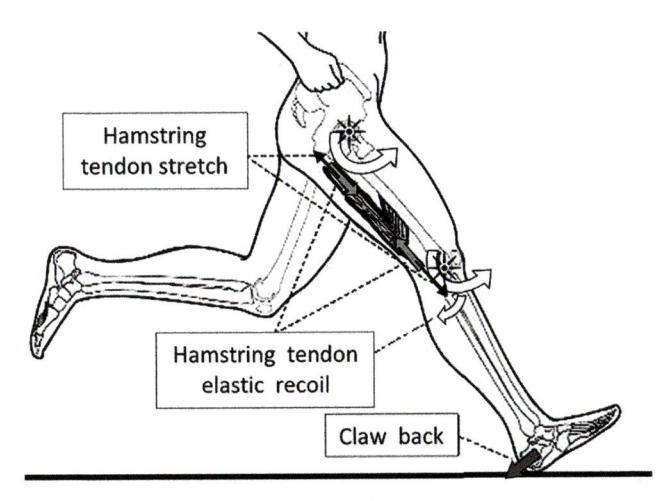

FIG. 2.4.8d The hamstrings make perfect brakes of hip flexion and knee extension in late and terminal swing. They are particularly important during running due to higher swing limb accelerations. Hamstring isometric contraction creates tension forces between the lower leg and pelvis, resisting both swing hip flexion and knee extension acceleration. This stores elastic energy within the proximal and distal hamstring tendons as they are stretched (thin black arrows). It provides ideal elastic recoil energy (grey thick arrows) at initial contact to aid the start of hip extension and knee flexion at the initiation of stance phase. This elastic recoil probably explains the posteriorly directed GRF often seen at heel strike (claw back on a force–time curve) as the heel recoils backwards under the hamstring-induced elasticity at the end of swing phase. This is most pronounced on long stride lengths (as demonstrated here) during faster running. This effect may be important for initiating a knee flexion moment that would otherwise be initially limited on a longer stride. *(Permission www.healthystep.co.uk.)*

Biceps femoris has a secondary role in providing lateral knee stability (James et al., 2015) as well as being a knee flexor. It attaches proximally at two points that divide the muscle into a long and a short head. The long head of biceps femoris attaches proximally to the ischial tuberosity through a common tendon with semitendinosus, and runs distally to form medial and lateral attachments, each with an anterior and a posterior component (Tubbs et al., 2006; Ertelt and Gronwald, 2017). The short head arising from the linea aspera has muscle fibres orientated at around 40° to attach distally as part of the common biceps femoris tendon on the styloid process of the fibular head and on the tibia 1 cm posterior to Gerdy's tubercle, where the ITT also attaches (Moorman and LaPrade, 2005; Ertelt and Gronwald, 2017). The short head muscle also connects to a capsular fascial sheath that attaches to the posterolateral joint capsule of the knee and laterally to the gastrocnemius complex. This union is the *arcuate ligament* (see Fig. 2.4.5n) with the distal edge of this structure being the fabello-fibular ligament (Davies et al., 2004; Moorman and LaPrade, 2005).

The distal biceps femoris attachments cover the lateral aspect of the knee, with primary attachments to the fibular head and the tibia, and with fascial attachments also to the posterior aspect of the fascia of the popliteus muscle tendon. Furthermore, it has an anterior fascial arm to the LCL (Moorman and LaPrade, 2005; Tubbs et al., 2006; Branch and Anz, 2015). The anterior arm is also associated with a bursa over the LCL (Moorman and LaPrade, 2005).

The hamstrings express Lombard's paradox because they can act as both knee flexors and extensors depending on the hip and knee position when the muscles are activated (Ertelt and Gronwald, 2017) (Figs 2.4.8e and 2.4.8f). With hip

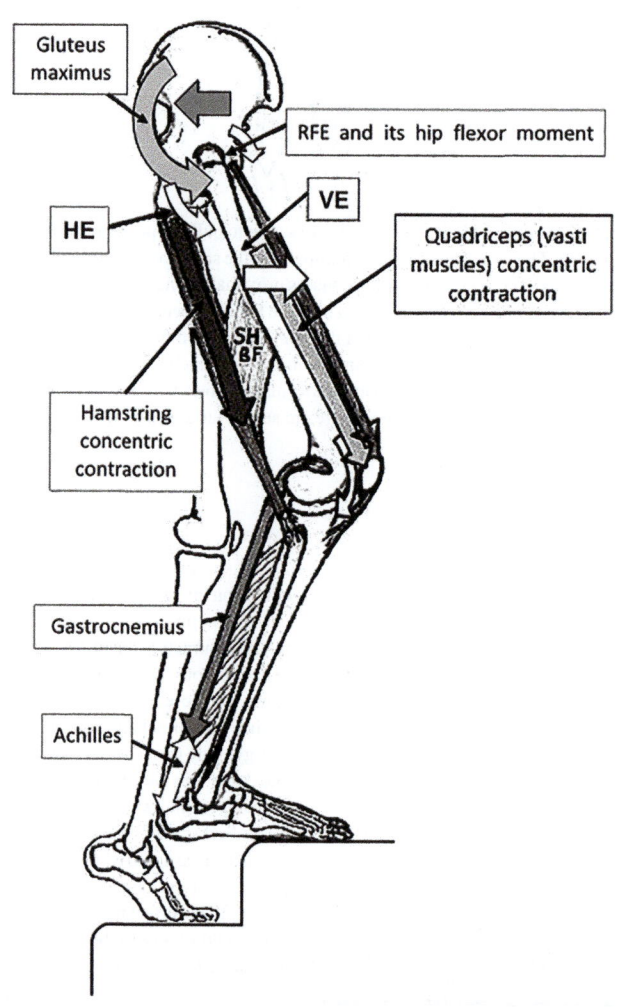

FIG. 2.4.8e Lombard's paradox is only a paradox for traditional anatomy. The hamstrings as knee flexors work together with the quadriceps as knee extensors in activities such as rising from and moving into sitting in addition to stair climbing, as demonstrated in the illustration. With the lower limb in closed chain, the solution to why this happens is simple. Once it is appreciated that these muscles are working at their proximal attachments, then the hamstrings' effort (HE) becomes one of hip extension, pulling the pelvis posteriorly (black arrow) as the vasti quadriceps' effort (VE) produces knee extension, pulling the femur anteriorly (white arrow). Thus, the knee and hip seem to be extended by the knee flexors and extensors through lower limb lengthening, with gluteus maximus assisting at the hip and triceps surae (gastrocnemius) at the knee. Only rectus femoris effort (RFE) is antagonising the hamstrings to produce some hip stabilising flexion moment, while the short head of biceps femoris (SHBF) does the same against knee extension. *(Permission www.healthystep.co.uk.)*

Rectus femoris muscle

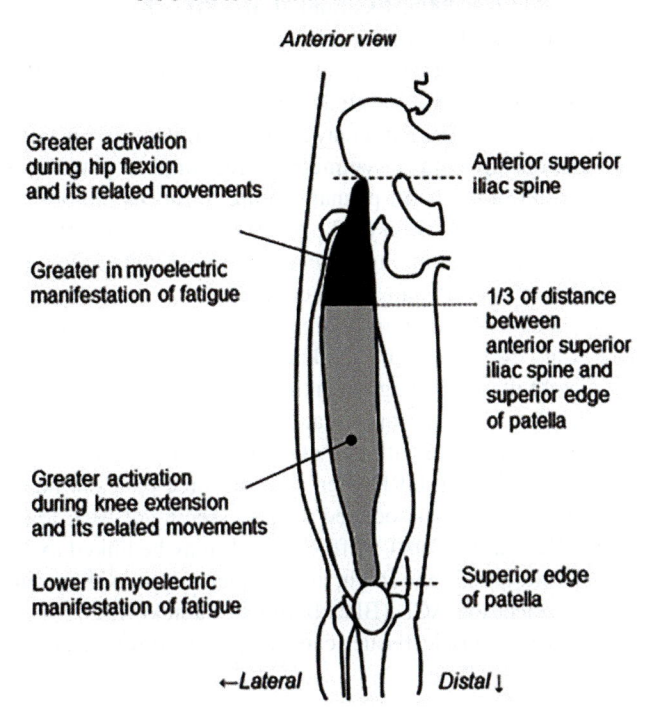

Anterior view

- Greater activation during hip flexion and its related movements
- Greater in myoelectric manifestation of fatigue
- Anterior superior iliac spine
- 1/3 of distance between anterior superior iliac spine and superior edge of patella
- Greater activation during knee extension and its related movements
- Lower in myoelectric manifestation of fatigue
- Superior edge of patella

←Lateral   Distal↓

FIG. 2.4.8f Much of Lombard's paradox can also be explained by appreciating that muscles activate different fibre areas to achieve variance in tasks. Studies using surface EMG have identified region-specific activity within rectus femoris. This demonstrates more proximal activation during hip flexion and greater distal activation during knee extension, and when resisting knee flexion. *(Image from Watanabe, K., Vieira, T.M., Gallina, A., Kouzaki, M., Moritani, T., 2021. Novel insights into biarticular muscle actions gained from high-density electromyogram. Exerc. Sport Sci. Rev. 49(3), 179–187.)*

extension and low knee flexion angles, the hamstrings can become knee extensors when the foot is fixed to the support surface, working with quadriceps such as rectus femoris in tasks such as sitting down and rising from sitting. Such changes in kinematic effects can also be found with human jumping, albeit with some inter-subject variability (Ertelt and Gronwald, 2017). The attachment to the popliteus tendon also suggests coupling mechanisms with this muscle, despite certain aspects of the biomechanics of these muscles remaining unclear (Tubbs et al., 2006; Ertelt and Gronwald, 2017). Biceps femoris is important in influencing the loading forces within the knee during stance and swing (Ertelt and Gronwald, 2017; Sharbafi et al., 2017), as well as influencing tibiofemoral joint loading stresses through laterally directed forces that balance those of the more medially positioned hamstrings (Shelburne et al., 2006; Winby et al., 2009). Furthermore, biceps femoris is exposed to an instantaneous high tensile force during late swing phase in sprinting, increasing the risk of muscle strain (Higashihara et al., 2016). In terminal swing of sprinting, the long head of biceps femoris, semimembranosus, and semitendinosus can reach 12%, 10%, and 9%, respectively, beyond their normal erect stance length (Schache et al., 2012).

Semimembranosus arises from the superior lateral ischial tuberosity of the ischium and runs distally to the knee where it forms five tendinous arms (Beltran et al., 2003). These attachments intertwine with the posterior oblique popliteal ligament in the posterior medial knee, providing stability and acting synergistically with the popliteal muscle. It actively pulls on the posterior horn of the medial meniscus during knee flexion (Beltran et al., 2003). Both biceps femoris and semimembranosus display a hemi-pennate muscle architecture and a fibre length per total muscle shorter than that of semitendinosus (Kumazaki et al., 2012). The semimembranosus is a slanted trapezoid muscle with longer muscle fibres proximally and shorter ones distally (Kumazaki et al., 2012). It sustains the highest passive tension stresses of all the hamstrings (Nakamura et al., 2016), yet seems less at risk of injury than biceps femoris although more so than semitendinosus (Kumazaki et al., 2012).

Semitendinosus is a fusiform muscle with longitudinal fibres intersected by a tendinous septum (Kumazaki et al., 2012). The muscles attaching proximally to the ischial tuberosity pass to the anteromedial aspect of the tibia on the pes anserinus tubercle. Here, they join the tendons of sartorius and gracilis to attach via a fibrocartilaginous enthesis to the anterior medial proximal tibia. Its distal attachment lies posterior to gracilis and sartorius and all three tendons are separated from the MCL and superior tibia by the pes anserine bursa (see Fig. 2.4.5j).

Hamstring injuries are common in sport, especially those involving high running speeds and frequent changes in direction (Woods et al., 2002; Hägglund et al., 2005; Ekstrand et al., 2011; Kumazaki et al., 2012; Valle et al., 2017). Extension of the knee joint increases the risk of muscle injury due to the maximum torques generated within this knee posture (Kumazaki et al., 2012). Injuries mostly affect biceps femoris (Ekstrand et al., 2012; Kumazaki et al., 2012), particularly at the musculotendinous junction (Fiorentino and Blemker, 2014). This could be because biceps femoris lengthens more than the other hamstrings (Schache et al., 2012) and may need to compensate for fatigue in semitendinosus, as both muscles have a close synergistic relationship with a symmetrical recruitment pattern (Schuermans et al., 2014). Its increased injury risk might also be a result of the hemi-pennate nature of biceps femoris that causes greater fibre length changes than in the fusiform semitendinosus (Kumazaki et al., 2012; Schache et al., 2012). Changes in the proximal aponeurosis dimensions of biceps femoris during fast running may concentrate strain to the biceps femoris musculotendinous junction proximally (Silder et al., 2010; Fiorentino and Blemker, 2014).

Injuries to the hamstrings are associated with positions of hip flexion and knee extension (Petersen and Hölmich, 2005; Cohen and Bradley, 2007; Askling et al., 2008), altering the knee torque demands on the hamstrings (Kumazaki et al., 2012). These positions are associated with initial loading when the hamstrings are known to be important for contact limb stability. However, it is also suggested that injury occurs in late swing just prior to initial contact (Woods et al., 2002), possibly associated with the elastic recoil associated with the 'claw back' event. This results from the near-isometric contraction seen in running and the anteriorly directed GRF often seen on force plates at heel strike.

Greater hamstring musculotendinous stiffness is associated with a lessening of ACL loading mechanisms, suggesting a protective relationship (Blackburn and Pamukoff, 2014). Stiffness levels may be linked to the amount of intermuscular fat, lower leg mass, and lesser muscle per unit mass in the thigh, all of which suggests that training to reduce fat mass and to increase hamstring strength may help protect the ACL (Blackburn and Pamukoff, 2014). Previous biceps femoris injury seems to alter in-series stiffness within the muscle tissue, causing larger strains near the proximal musculotendinous junction (Silder et al., 2010).

Popliteus muscle action at the knee

The popliteus muscle arises from a proximal tendon attachment of around 6 mm on the lateral aspect of the femur, near the lateral condyle's articular cartilage. It attaches distally as a broad fibrous attachment to the posterior aspect of the tibia, forming the floor of the popliteal fossa (Jadhav et al., 2014; James et al., 2015). The proximal tendon courses obliquely in a posterior–inferior direction, running deep to the LCL, becoming extra-articular near the popliteal hiatus, and gains anchorage to the lateral meniscus by three *popliteomeniscal fascicles* or bundles (Jadhav et al., 2014; James et al., 2015). From full extension to $\sim$112° of knee flexion angulation, the popliteus tendon lies over the lateral femoral condyle, but on further flexion, it engages into the popliteal sulcus located on the lateral side of the lateral femoral condyle (LaPrade et al., 2003; James et al., 2015). Coursing from beneath the LCL and the tendon of biceps femoris inferomedially, the muscle belly arises from the musculotendinous junction with its muscle fanning out to run and attach distally via a long fibrous enthesis to the tibia at the *soleal* or *popliteal line* (Draves, 1986, p. 266; Jadhav et al., 2014; James et al., 2015). The popliteus muscle is usually described as having superficial fibres that run to the joint capsule and more variable deep fibres that can have lateral meniscal attachments (Jadhav et al., 2014) (Fig. 2.4.8g).

The popliteus muscle is an internal rotator of the knee through the tibia and an external rotator of the femur (resisting internal femoral rotation) when the distal segment is fixed in closed chain (Ullrich et al., 2002; Pasque et al., 2003; James et al., 2015). The popliteus tendon and the popliteofibular ligament have an intimate relationship with the ligament providing a functional attachment of the popliteal tendon to the fibula, as well as popliteus having attachments to the femur and tibia (Pasque et al., 2003). Popliteus induces internal tibial rotation in open chain. This slackens the popliteofibular ligament and opposes external tibial rotation which tenses the popliteofibular ligament (Ullrich et al., 2002). Failure of the popliteofibular ligament seems to have little effect on knee motion alone. This is because it is a secondary stabiliser to the LCL, thus working with the LCL to resist external and varus rotations and posterior translations of the tibia through the knee (Pasque et al., 2003). In stance, with the tibia fixed by GRF through the foot and ankle, popliteus and popliteofibular ligament restraint equates to resisting internal femoral rotations, anterior translations of the femur, and varus moments at the knee. Popliteus does not influence knee varus or anterior–posterior translation in isolation as it primarily creates open chain tibial internal rotation and closed chain femoral external rotation (Pasque et al., 2003). This effect is greatest at knee flexion angles of between 90° and 120° when the tendon lies in its sulcus (Pasque et al., 2003) with the internal and external arcs of the knee rotations increasing from extension to 90° of flexion (Ullrich et al., 2002). Popliteus also plays an important role in stabilising the lateral meniscus and by providing an integral part of the posterolateral corner complex (Ullrich et al., 2002; Jadhav et al., 2014) (Fig. 2.4.8h).

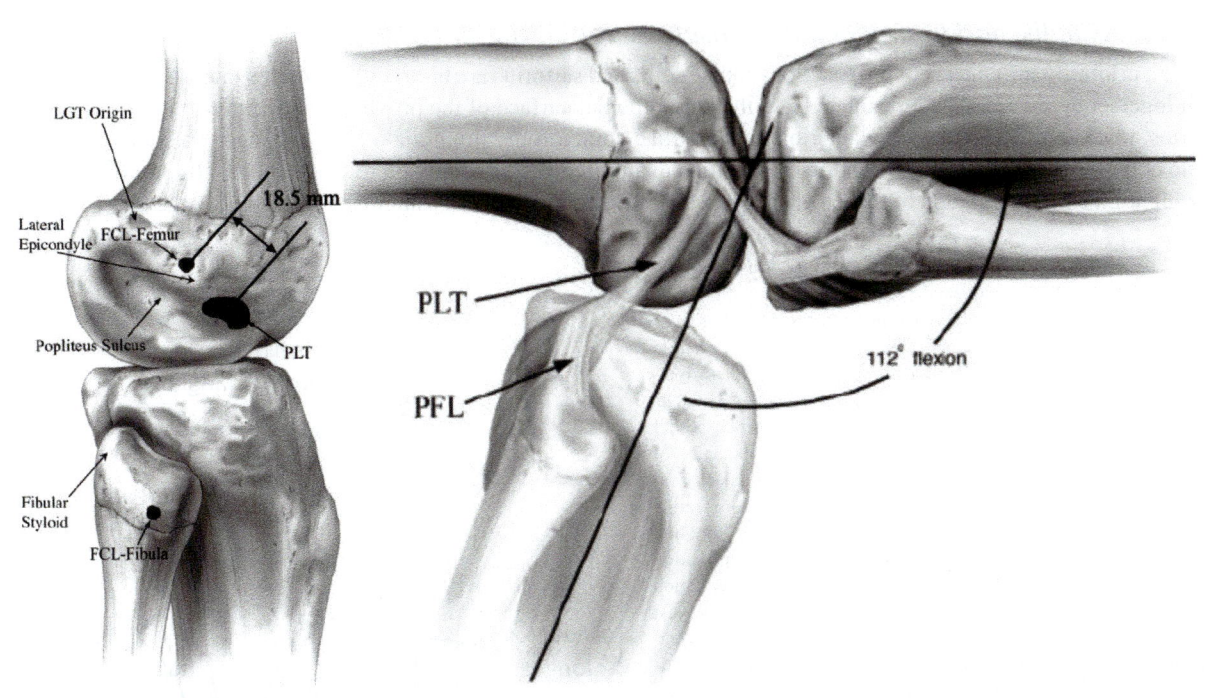

FIG. 2.4.8g An anatomical drawing of the osseous features around the lateral knee that pertain to popliteus (left). Attachments are shown for the lateral (fibular) collateral ligament (FCL), lateral gastrocnemius tendon (LGT), the popliteofibular ligament (PFL), and the popliteus tendon (PLT) at the anterior aspect of the peroneal sulcus. With increase in degrees of flexion after an angle of 112°, the popliteus tendon passes into the sulcus (right). *(Images from LaPrade, R.F., Ly, T.V., Wentorf, F.A., Engebretsen, L., 2003. The posterolateral attachments of the knee: a qualitative and quantitative morphologic analysis of the fibular collateral ligament, popliteus tendon, popliteofibular ligament, and lateral gastrocnemius tendon. Am. J. Sports Med. 31(6), 854–860.)*

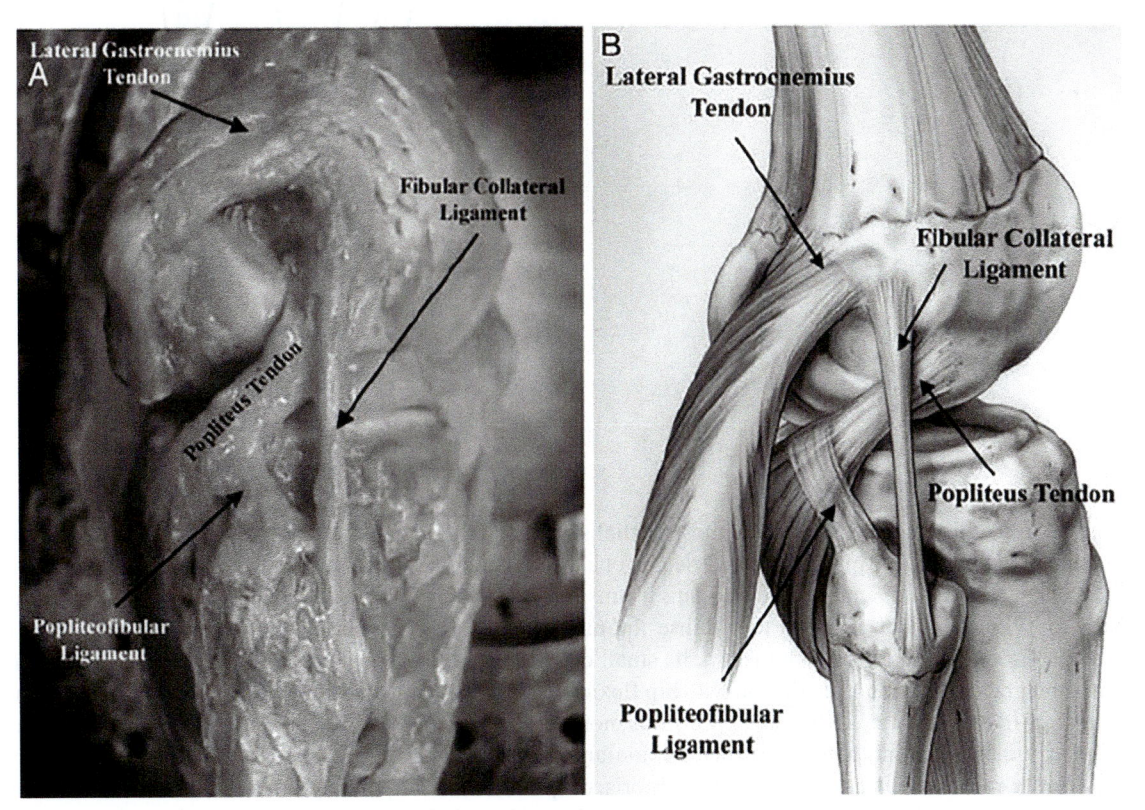

FIG. 2.4.8h A lateral dissection photograph of the anatomy around the popliteus' proximal attachment (A) compared to an anatomical drawing of the same view (B). Its action in the stance phase needs to be visualised with the tibia fixed distally and the femur rotating above. *(Image from LaPrade, R.F., Ly, T.V., Wentorf, F.A., Engebretsen, L., 2003. The posterolateral attachments of the knee: a qualitative and quantitative morphologic analysis of the fibular collateral ligament, popliteus tendon, popliteofibular ligament, and lateral gastrocnemius tendon. Am. J. Sports Med. 31(6), 854–860.)*

Sartorius, gracilis, and gastrocnemius muscle action at the knee

Although primarily influencers of hip function, both gracilis and sartorius are bi-articular muscles that also cross the knee to attach together at the pes anserinus tubercle on the anteromedial surface of the proximal tibia, anterior to semitendinosus. Sartorius attaches proximally to the anterior superior iliac spine on the ilium via a fibrocartilage-tendinous enthesis. Its muscle belly lies superficial to the quadriceps and runs in an inferomedial direction to the knee. Its primary role is in open chain hip joint external rotation of the leg via the hip and closed chain internal rotation of the pelvis on the hip (Fig. 2.4.8i).

FIG. 2.4.8i An anatomical drawing of gracilis and sartorius indicating their force vectors and action among the associated anatomy. Sartorius, lying superficial to the quadriceps muscles, provides the lateral border of the femoral triangle. This is an anatomical region that contains the lateral cutaneous and femoral nerves, the femoral artery and vein, and lymphatics as they pass through it. A prominent pulse is located here. Its superior border or base is the inguinal ligament and its medial border the medial side of adductor longus. Its floor consists medially of adductor longus and laterally of iliopsoas (iliacus and psoas major). Gracilis lies medial to adductor longus, joining sartorius at the pes anserine attachment. Together, they provide closed chain influences on knee and hip varus moments, creating a resultant force that initiates hip flexion and adduction. In closed chain, sartorius provides some internal rotation of the pelvis on the hip and knee. When the knee is flexed, they assist knee flexion, but when the knee is extended, they assist extension stability. Once again, the moment arms of these muscles at the time of activation are critical to their action. *(Permission www.healthystep.co.uk.)*

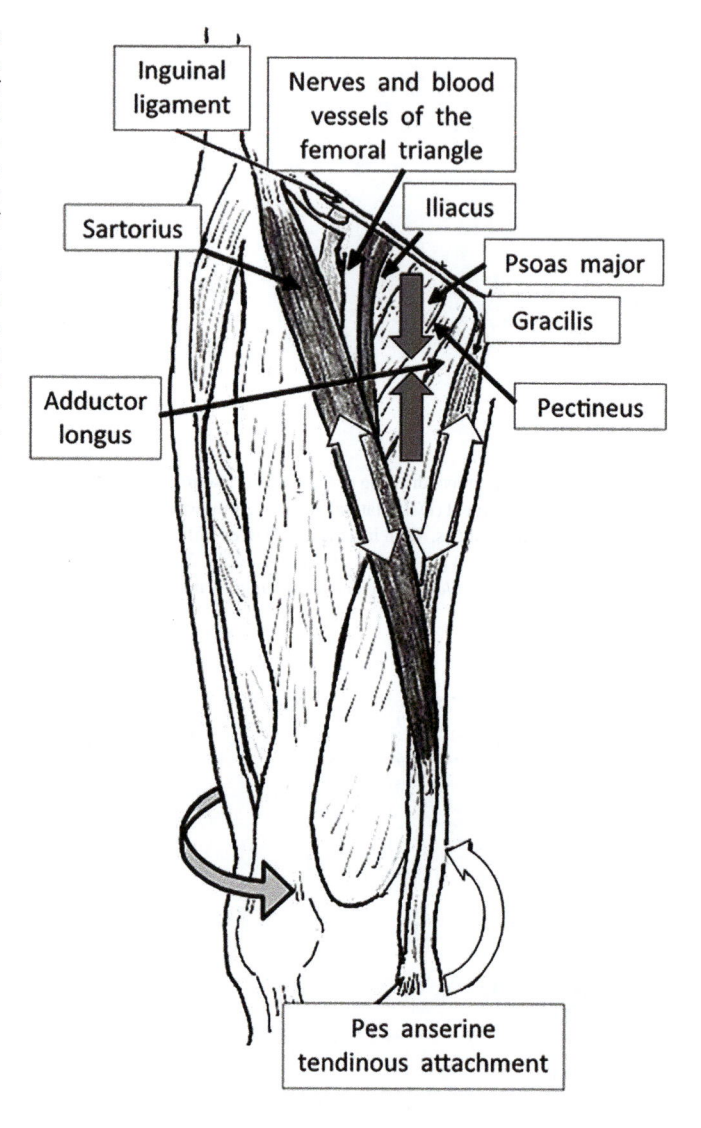

Gracilis is a long, thin muscle found in the most medial superficial thigh, having a long distal tendon that is part intramuscular and part external to the muscle belly (Dziedzic et al., 2018). It arises from the pubic symphysis and pubic inferior ramus at the connection to the ramus of the ischium. It runs inferiorly to the pes anserinus, just posterior to sartorius, and blends partially with its tendon. It helps to stabilise the medial tibiofemoral joint and influences the adduction angles between the pelvis and the thigh through the hip. Its small cross-sectional area suggests only an assistor role during gait.

Sartorius and gracilis provide knee flexion with hip flexion and yet they assist the knee extension torque when the knee is extended. They likely have a small input on knee varus moments and some internal rotation torque in knee flexion angles. Sartorius activity increases with increasing knee flexion angles (Mohamed et al., 2002). This indicates a primary knee flexion role for the muscle as an assistor to biceps femoris, despite its anterior pelvic proximal attachment. Both sartorius and gracilis undergo a significant reduction in their level of hip and knee influence with elongation, but even their effect on knee flexion is relatively minor because both muscles have small cross-sectional areas (Mohamed et al., 2002).

Gastrocnemius is primarily an ankle flexor (plantarflexor), but by crossing the knee and subtalar joints, it also influences these joints too. Gastrocnemius activity is well known to take place through the latter half of midstance, peaking before heel lift, yet it has also been reported to show some activity at heel contact (Di Nardo et al., 2016; Mengarelli et al., 2017, 2018) when an active ankle plantarflexion moment would seem inappropriate. However, it displays co-activation lasting for ~9% of the gait cycle from the start of stance, along with hamstring and quadriceps activity, which could enhance knee stiffness/stability at a time when the knee flexion angle is initially very close to full extension (Mengarelli et al., 2018). Another period of co-activation between gastrocnemius and the quadriceps occasionally seems to occur during late midstance. It appears that the quadriceps become active to adjust (resist) the knee flexion angle (if required) as the lower limb pivots over the ankle when the gastrocnemius is acting as a knee flexor controlling knee extension (Mengarelli et al., 2018). It seems likely that if gastrocnemius is overdoing knee flexion, then some counterbalancing knee extension is required from the quadriceps. There is now clear evidence that the gastrocnemius influences ACL elongation as an antagonist (Klyne et al., 2012; Mokhtarzadeh et al., 2013; Adouni et al., 2016).

Active stabilisation effects of muscles on knee loads

Joint forces are created by a combination of external (GRF) and internal forces, the latter being a result of soft tissue passive and active restraint and active motion. Knee contact forces are therefore influenced by stabilisation of external forces through myofascial tissues. For example, they resist an external varus/adduction moment, but this stability results in increased joint compression loading in the opposite direction (Schipplein and Andriacchi, 1991). Precisely how this relationship occurs in healthy knees is unknown due to the difficulty in recording internal joint contact loads in vivo without risking joint damage. It appears that the bulk of frontal plane stability and contact forces within the knee are provided for by the quadriceps and gastrocnemius (Shelburne et al., 2006).

The external loads on the knee account for around 26%–55% of the medial compartment loads and between −33% and 29% of the lateral compartment loads, although with greater inter-subject variability reported medially (Winby et al., 2009). The minus external loading of the lateral knee means that the lateral compartment can be loaded 100% through internal forces from muscular contributions to counteract the unloading effect of externally generated forces of the GRF (Winby et al., 2009). Internal medial compartment loads are initially generated predominantly by the hamstrings followed by the quadriceps in early stance, with gastrocnemius producing the dominant loads in late stance (Winby et al., 2009). The lateral compartment loads are also generated by the same muscles but with tensor fasciae latae through the ITT making a greater contribution, particularly in mid-to-late stance, although with subject variability reported (Winby et al., 2009). Overall, the tibiofemoral joint force comes primarily from the quadriceps, hamstrings, and gastrocnemius, with some force arising from the tensor fasciae latae and very small loads from sartorius and gracilis (Fig. 2.4.8j).

2.4.9 Adaptation and pathology in the knee

There are a vast number of knee pathologies that reflect the complexity of the anatomy involved. Generally, the long-term consequences of structural and functional failure of the knee are degenerative changes leading to osteoarthritis (OA). This results in joint space narrowing and osteophyte formation through breakdown of cartilage biomechanics and underlying bone. Severity is clinically assessed for through radiographic classification of gross anatomical changes, but such classification poorly correlates to function (Hunter et al., 2013; Barr et al., 2015). Associated soft tissue derangements are assessed using MRI (Hunter et al., 2011). Each individual develops their own unique bony shape variation with degenerative changes in regard to tibial flattening, widening, and femoral condylar squaring (Lynch et al., 2019). Individual cases also present with their own level of pain and functional limitation, regardless of the changes seen on the image (Hunter et al., 2013).

Symptomatic OA knees display osseous expansions larger than asymptomatic controls. Such changes may explain loss of full knee flexion and extension, and the extent of these changes can distinguish symptomatic from asymptomatic knees by around 95% (Lynch et al., 2019). These bony expansions arise from the edges of the tibial and femoral cartilage plates (Neogi et al., 2013; Barr et al., 2015; Bowes et al., 2015; Lynch et al., 2019). Reduced space in the intercondylar notch is due to expansion of bone from the medial and lateral condyles of the femur (Lynch et al., 2019). The changes in the femoral and tibial geometry more than likely alter the tibiofemoral kinematics, including loss of full flexion due to osseous impingement (Lynch et al., 2019), with the function lost being dependent on the location of the degeneration (Figs 2.4.9a and 2.4.9b).

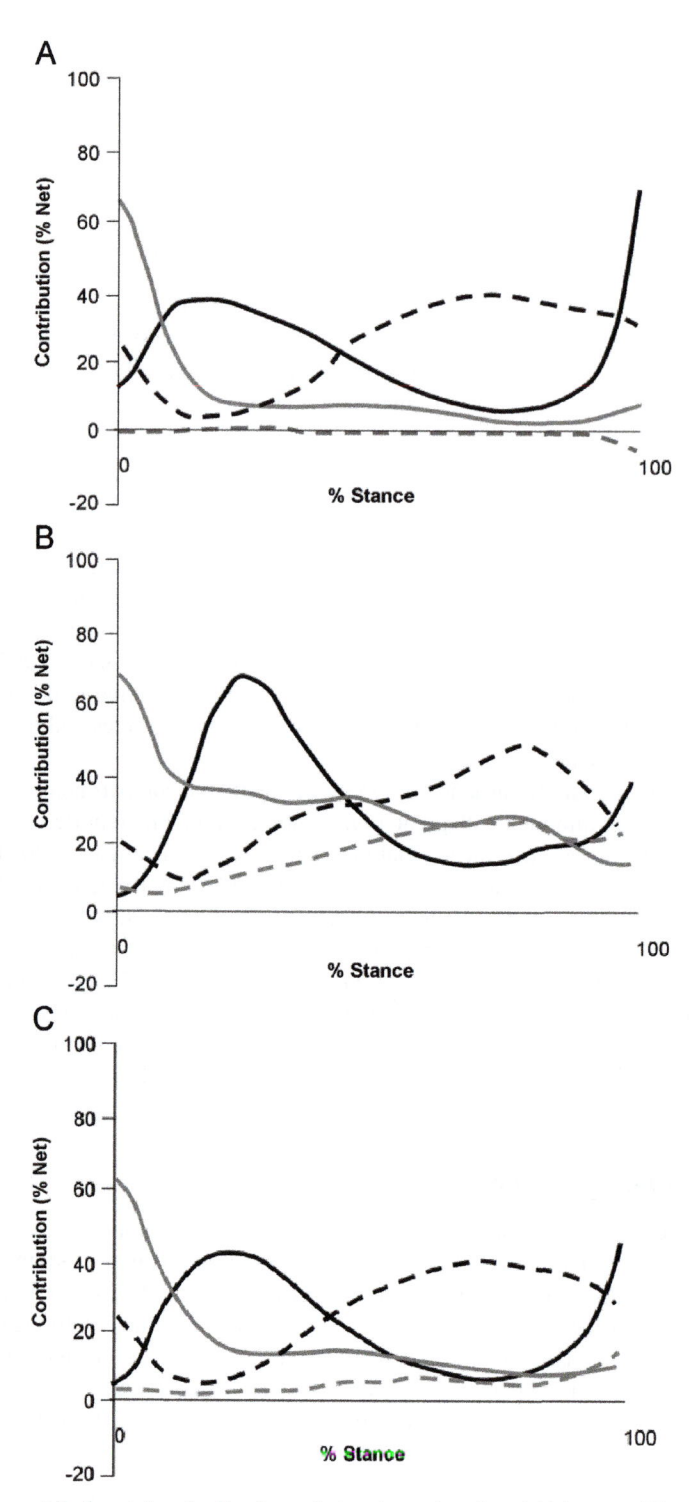

FIG. 2.4.8J Data gathered on the contributions to knee loading forces during stance phase from initial contact 0% to toe-off at 100%. The contributions are calculated for the quadriceps (dark line), the hamstrings (light-grey line), gastrocnemius (dark dashed line), and tensor fasciae latae (light-dashed line). *(Image from Winby, C.R., Lloyd, D.G., Besier, T.F., Kirk, T.B., 2009. Muscle and external load contribution to knee joint contact loads during normal gait. J. Biomech. 42(14), 2294–2300.)*

Knee OA is commonly seen as being a secondary problem that develops following previous soft tissue deficits such as cruciate ligament and meniscal injuries to the knee. It seems that despite the knee joint's poor form closure, when functioning with sufficient force closure, the joint is mechanically sound and well protected from excessive loading until there is a failure of the soft tissue balanced tensile–compression forces. It can be postulated that this leads to failure of the knee as a stable biotensegrity structure, which vastly changes the position of the instantaneous joint axis, moments, and strains across the tissues as a consequence.

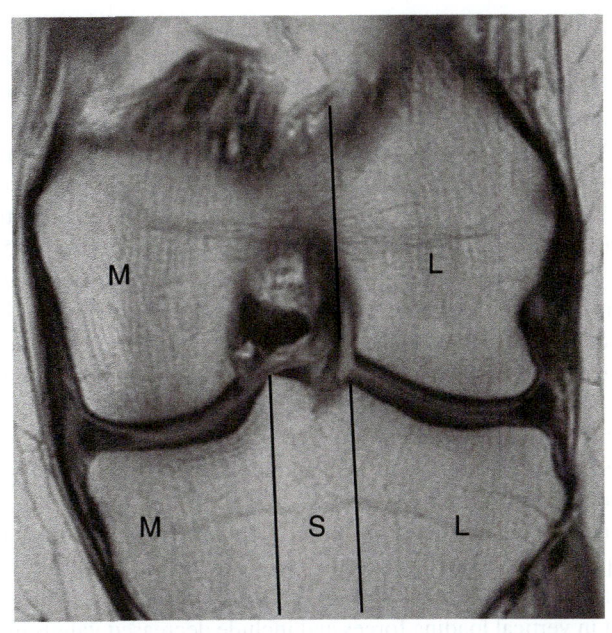

FIG. 2.4.9a Being specific in identifying degenerative changes within the knee articulation can greatly help in understanding the biomechanics that underlie the pathology and aid in planning intervention. An intermediate-weighted magnetic resonance image (MRI) of the tibiofemoral joint in the frontal (coronal) plane is shown delineated into medial (M) and lateral (L) condyles. The tibia is delineated into medial, subspinous (S) and lateral subregions. The intercondylar notch of the femur is considered part of the medial femur. *(Image from Hunter, D.J., Guermazi, A., Lo, G.H., Grainger, A.J., Conaghan, P.G., Boudreau, R.M., et al., 2011. Evolution of semi-quantitative whole joint assessment of knee OA: MOAKS (MRI Osteoarthritis Knee Score). Osteoarthr. Cartil. 19(8), 990–1002.)*

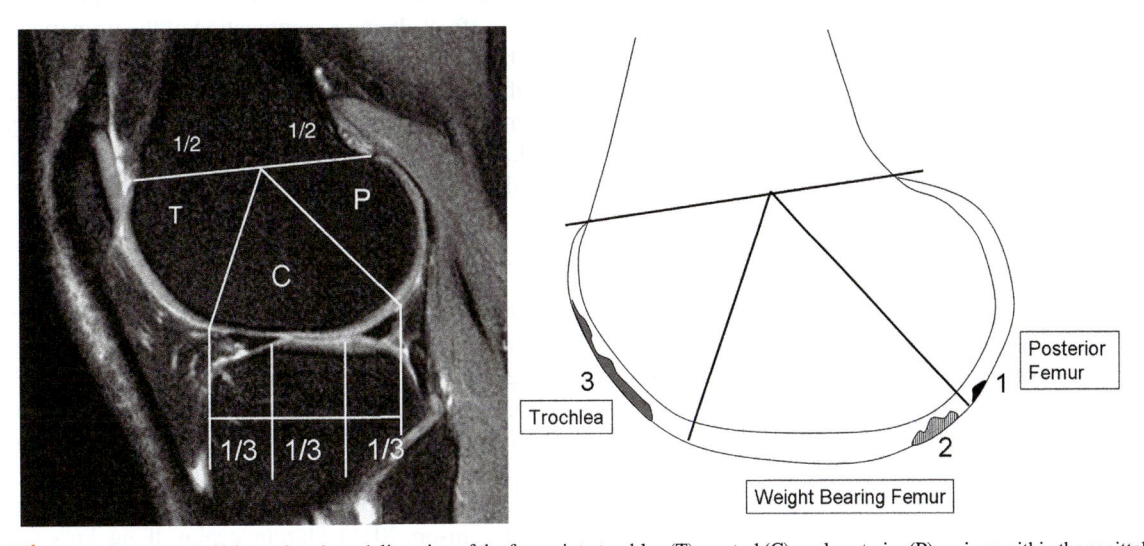

FIG. 2.4.9b An MRI image (left) is used to show delineation of the femur into trochlea (T), central (C), and posterior (P) regions within the sagittal plane. The tibia is divided into thirds to delineate it into anterior, central, and posterior subregions. Degenerative changes identified in each region can then be graded for size of cartilage loss within each region (right). Grade 0 means none is lost, grade 1 <10% lost, grade 2 10%–75% lost, and grade 3 >75% lost. Understanding the specific location and extent of pathology rather than using broad diagnoses such as knee osteoarthritis is important in clinical biomechanics. *(Image from Hunter, D.J., Guermazi, A., Lo, G.H., Grainger, A.J., Conaghan, P.G., Boudreau, R.M., et al., 2011. Evolution of semi-quantitative whole joint assessment of knee OA: MOAKS (MRI Osteoarthritis Knee Score). Osteoarthr. Cartil. 19(8), 990–1002.)*

Ligament failures

Mid-substance cruciate ligament tears cannot heal. Such injuries can manifest as the knee 'giving way', causing considerable disability in multidirectional sports that demand pivoting and cutting manoeuvres (Woo et al., 2006). Loss of ACL integrity has been demonstrated to be associated with bone contusions, MCL injury, ACL ruptures (Yoon et al., 2011), larger areas of cartilage degeneration medially, and varus deformity within knees (Harman et al., 1998; Vasara et al.,

2005; Moschella et al., 2006). Its failure allows greater anterior tibia translation (femoral posterior translation) during locomotion. Loss of the ACL function also results in an increase in internal and valgus rotations at the knee allowing some opening of the medial joint space (Matsumoto et al., 2001). The resultant increased valgus/internal rotation is associated with increased bone contusions across the joint (Yoon et al., 2011).

Although less common than ACL failure, loss of function of the PCL ligament results in increased posterior tibial translation (anterior femoral translations) at all knee flexion angles, but larger magnitudes occur at greater flexion angles of around 90°. PCL deficiency risks increased articular contact pressures and thus degeneration in the medial compartment and the patellofemoral joint (Barba et al., 2015). Clinically, tibial translation ranges from 4.5 to 10 mm of laxity on posterior draw tests with PCL failure. On high flexion radiographs with patients statically squatting, an increase in posterior tibial translation of around 6 mm occurs compared to uninjured contralateral sides (Bowman and Sekiya, 2009). At knee flexion angles of less than 70°, the effect is not as pronounced. It has also been reported that the PCL results in slightly lower compressive forces being generated within the lateral compartment of the tibiofemoral joint and significantly higher forces both in the medial compartment and within the patellofemoral joint, increasing with higher knee flexion angles (Logan et al., 2004; Ramaniraka et al., 2005). This results from an anterior and medial translation of the femoral and tibial contact forces, altering the kinematics at flexion are consistent with an increased risk of medial compartment and patellofemoral OA after PCL ligament tears (Logan et al., 2004; Barba et al., 2015).

However, studies report that despite clinical laxity, PCL-deficient knees demonstrate only small differences in biomechanical and neuromuscular adaptations compared to controls, with strength of the lower limb being unaffected (Bowman and Sekiya, 2009). Yet, the amount of instability caused by PCL failure corresponds to the degree of disability (Iwata et al., 2007). Changes occur during gait in vertical loading forces and include decreased valgus moments in stance and a bilateral reduction in GRF during jumping and high impact activities (Fontboté et al., 2005). The decreased posterior tibial translation restraint of PCL failure disadvantages the quadriceps lever arm, requiring increased quadriceps activity to compensate which leads to increased patellofemoral compression forces (Bowman and Sekiya, 2009). Rotation deficits with PCL deficiency are not significant as long as the posterolateral corner complex structures remain intact, as popliteus can restrain both posterior tibial translation and external rotation instability (Bowman and Sekiya, 2009).

The effects of isolated failure of smaller knee ligaments remain largely unknown, but the effects of combined injuries of knee ligaments have received greater investigation, with the resulting effects being a combination of the primary and secondary stabilising roles of the ligaments being lost. Failure of combinations of ligaments that work synergistically potentially has the worst outcomes, such as a lost PCL along with any of the posterolateral corner complex structures that control posterior tibial translation (Bowman and Sekiya, 2009). Concurrent loss of the ACL and the medial collateral ligament cause a larger valgus instability within the knee (Matsumoto et al., 2001). The result of excessive or inappropriate knee motion is exposure of the articular cartilage to damaging shear and unequal compressive stresses in small loading areas.

Tibiofemoral degenerative joint disease

Whether a primary or secondary problem, knee OA or degenerative joint disease can be particularly disabling, despite the fact that every patient has their unique compensation mechanisms in response to joint pathology. Knee OA causes a decrease in walking velocity compared to controls (Kaufman et al., 2001; Zeni and Higginson, 2011). Joint motion changes that occur are likely due to neuromuscular strategy in response to joint pain and/or muscle weakness rather than changes in walking speed alone (Zeni and Higginson, 2011). Kinematic and kinetic changes have been identified at the hip, knee, and ankle joints. However, they depend on the severity of the degeneration and walking speed (Astephen et al., 2008a). Generally, ankle motion contributions increase, and less hip motion occurs with OA knees at faster walking speeds, but such changes are reflecting each individual's control strategy (Zeni and Higginson, 2011). During level walking, knee OA peak motion is reduced by around 6°, with reduced peak flexion and early stance extension moments occurring compared to normal controls (Kaufman et al., 2001; Astephen et al., 2008a; Heiden et al., 2009). The sequence of knee movements in walking gait of loading knee flexion followed by extension to midstance and ending with knee flexion again remains unchanged, but with the range of motion reduced (Kaufman et al., 2001). Healthy female knees demonstrated greater peak knee flexion, extension, and external rotation moments than those of healthy males during level walking gait. However, in OA knees, the difference in ranges between males and females becomes greater in extension and less in external rotation (Kaufman et al., 2001). Loss of femoral cartilage and articular shape seems more predictive of which knees are most likely to require total knee prosthetic surgery than tibial articular changes. This is probably because the femur receives more load from twice the number of articular surfaces as either the tibia or patella (Barr et al., 2016).

Knee adduction moments are increased in more severe symptomatic OA knees, while the hips above demonstrate decreased peak internal rotation and adduction moments, and the ankles display decreased peak dorsiflexion

(Mündermann et al., 2005; Astephen et al., 2008a). Secondary gait changes in medial compartment OA seem to occur mainly within the frontal plane, but hip moments dominate those derived around the rearfoot. Attempts seem to be made to move the trunk CoM more laterally over the support limb to reduce the varus knee moment, something better achieved in patients with less knee OA, perhaps indicating that they maintain better hip abductor muscle strength than those with more severe degeneration (Mündermann et al., 2005). Patients with medial OA knees make initial contact with their knees more extended and experience a more rapid increase in GRF, indicating that they use a strategy to transfer body weight more rapidly from the contralateral limb to the support limb. However, this seems successful only in patients with less severe OA (Mündermann et al., 2005). OA knee patients also use predominantly lateral muscle activation (biceps femoris, vastus lateralis, gastrocnemius lateralis) during stance, possibly through an attempt to resist the increased knee adduction (varus) moment (Heiden et al., 2009). Biomechanical treatment interventions in OA knees need to address kinematic and kinetic changes of the disease as well as the underlying mechanical failure, such as ligament instability.

Patellofemoral dysfunction

Patellofemoral pain is also common, although clarification of the tissues being either injured or dysfunctional seems less easily established clinically. This results in a clinical tendency to label many cases with the blanket term 'patellofemoral pain syndrome'. From a biomechanical point of view, establishing the nature of the tissue stress overload is important in order to direct the correct intervention (Wünschel et al., 2011). If this is not done, it risks the generic symptom-based descriptor of patellofemoral pain syndrome which is too broad a label by which to understand the pathomechanics. The specific location of pathology within the patellofemoral joint is thus important (Fig. 2.4.9c).

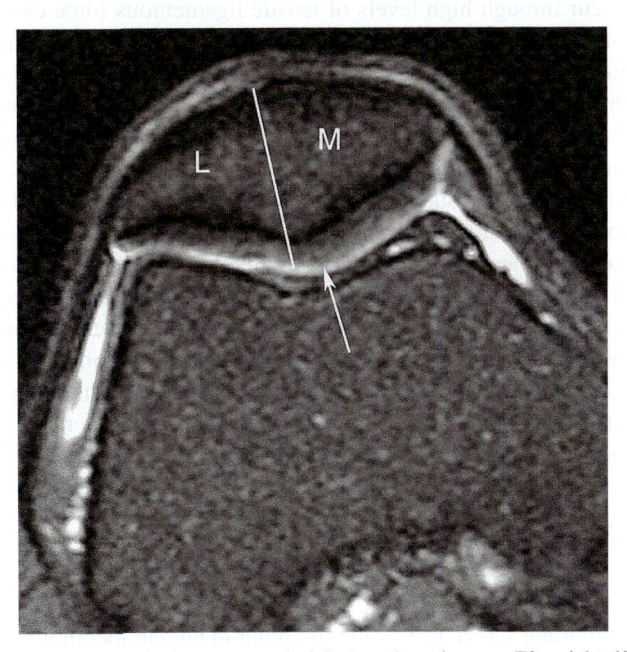

FIG. 2.4.9c The patella is divided into two subregions in the transverse (axial) plane, here shown on T2-weighted MRI. This helps identify the location of degenerative changes within the articular surface of the patellofemoral joint, as forces that damage only the lateral articular surface are likely to be quite different to those that instigate degeneration within the medial articulation. Medial (M) and lateral (L) portions are shown with the patellar crista or apex (arrow) as part of the medial subregion. *(Image from Hunter, D.J., Guermazi, A., Lo, G.H., Grainger, A.J., Conaghan, P.G., Boudreau, R.M., et al., 2011. Evolution of semi-quantitative whole joint assessment of knee OA: MOAKS (MRI Osteoarthritis Knee Score). Osteoarthr. Cartil. 19(8), 990–1002.)*

Cadaveric studies indicate that VMO activity, lateral condyle profile, the medial retinaculum (Senavongse and Amis, 2005), and the patellofemoral ligament (Amis et al., 2003) all play a significant and complex role is stabilising the patella against the inherent lateral force vector. Senavongse and Amis (2005) found that a flat lateral femoral condyle caused a 70% reduction in lateral force vector stability at 30° flexion. Relaxation of VMO tension caused a 30% reduction in the lateral force vector. Around extension, loss of the retinaculae had the largest effect with a 49% reduction in joint stability.

Hip abductor strengthening seems effective in reducing patellofemoral symptoms, suggesting that more proximal soft tissue stabilisations are also at play, despite no change in functional knee frontal plane angles from such interventions (Earl and Hoch, 2011; Ferber et al., 2011). In contrast, vastus medialis weakness is associated with reduced patellofemoral contact pressures, implying that in cases of patellofemoral OA, vastus medialis strengthening may be counterproductive (Csintalan et al., 2002). This shows a complex interaction with varying soft tissue functions and kinematics of the knee affecting the relative stability issues (see Figs 2.4.6a, 2.4.6b, 2.4.6e, and 2.4.8a for some of the sources of the patella's vectors) with no single factor likely to be the cause of all patellofemoral pathology and pain (Lee et al., 2003).

Patellofemoral pain is reported to be more common in active young females (Arendt, 1994; Nakagawa et al., 2012). This has been suggested as being a result of a higher Q angle and increased hip adduction and knee abduction values at all knee flexion angles, than are found in males (Arendt, 1994; Csintalan et al., 2002; Nakagawa et al., 2012). Yet, an increased Q angle does not necessarily lead to patellofemoral symptoms with both increased and decreased Q angles having been associated with patellofemoral pain (Csintalan et al., 2002).

A significant increase in mean contact pressure has been observed in cadaveric female knees compared to male knees at 0° and 30° flexion, and greater changes in joint contact pressures in response to vastus medialis loading at flexion angles of 0°, 30°, and 60° (Csintalan et al., 2002). Females have also been reported to demonstrate diminished hip abductor torques and increased gluteus medius activation compared to males (Nakagawa et al., 2012). Females with patellofemoral symptoms demonstrate lower hip abductor torques compared to both males and asymptomatic females (Nakagawa et al., 2012). These findings may reflect structural differences between male and female knee anatomy, helping to explain the discrepancy in frequency of symptoms at this joint between men and women.

2.4.10 Section summary

Stable knee mobility needs to occur through high levels of tensile ligamentous force closure that focuses motion to the sagittal plane. Yet small levels of translation and internal and external rotations are permitted between the femur and tibia to improve the muscular functional lever arms as they change during motion around the instantaneous knee joint axis. Frontal plane motion needs high levels of constraint to avoid GRF and bodyweight vectors abnormally loading one side of the tibiofemoral joint over the other, and altering the vertical vector on the patella. The patella is a component of anterior constraint on the knee, preventing excessive femoral anterior translation during loading response and utilising both its retinacular and muscular components to facilitate stabilising compression directed posteriorly.

Thus, the knee's complex soft tissues have highly intimate ligament, tendon, and fascial relationships that interlink muscle activity in such a way that most knee ligaments are more directly linked to active contraction than perhaps in other lower limb areas. This heightens the complexity of the knee when modelled as a biotensegrity structure and highlights the significance of soft tissue failure to knee stability. Although large ligamentous structures such as the cruciate ligaments and collateral ligaments present significant structures to injure (structures that have dominated clinical focus), it is likely that smaller structures in the grip of dysfunction still have considerable effects in maintaining biotensegrity stability through mobility at the knee. This is because knee stability is the sum of all the parts, not the result of one single part, and together these parts are influencing the position of the instantaneous joint axis.

2.5 Functional unit of the ankle
2.5.1 Introduction

The ankle is the fundamentally important stance phase fulcrum point in human locomotion, being extensively involved in lower limb mobility for changing angles of the leg above the foot. The ankle joint or talocrural joint has an intimate relationship with the subtalar joint of the rearfoot such that they are functionally indissociable when loaded by external forces (Leardini et al., 2001; Hamel et al., 2004). This functional joint unit is usually described as the *ankle–subtalar complex*. The ankle–subtalar complex has three primary roles in stance phase kinematics. These are to permit the foot to achieve plantigrade contact with the support surface, to enable the leg to pivot over the foot, and to allow plantarflexion power to be generated at the ankle to achieve heel lift in preparation for foot ground clearance after applying a strong acceleration force into the ground through the forefoot. Abilities to adapt to changes in the angulation of the foot to the support surface beneath, make the subtalar joint's capacity to provide some frontal and transverse plane motion helpful in permitting ankle joint sagittal motion without perturbation. The subtalar joint is only a functional joint during weightbearing (Leardini et al., 2001), and then mainly for only the first 25% of stance phase (Hamel et al., 2004).

The ankle demonstrates strong but variable form closure, its form closure stability increasing with dorsiflexion. Thus, greater reliance on active and passive tissues to create force closure is required in plantarflexion angles. It is worth a quick reminder that ankle plantarflexion is a flexion motion and ankle dorsiflexion is an extension, despite dorsiflexion bringing the foot and leg closer together, effectively shortening the lower limb's length. The easiest way to remember which ankle–foot motion is flexion, is to take note of the fact that the muscles that plantarflex the toes crossing behind the ankle are the long digital flexors.

Ankles have extensive medial and lateral ligaments providing strong frontal and transverse plane stability but have very limited sagittal plane restraint. Through the ankle and subtalar joint functioning together, more frontal and transverse plane mobility can be offered. The ankle also offers more freedom of motion in these planes when at plantarflexion angles. Muscular activity around the ankle is thus more important in stabilising the joint during periods of ankle plantarflexion, when muscles with significant frontal plane moment arms, such as the tibialis posterior and peroneal muscles, are active.

As a functional unit, the ankle is separated from the rearfoot in this section to help to discuss the anatomy's relationship to ankle function, divorcing it as much as possible from that of the foot. However, such a task is difficult because of the intimate relationship between the ankle and the rearfoot through the subtalar joint and also because so much of the local anatomy crosses both regions. This dilemma will become particularly apparent when discussing mechanisms of pathology, and so the two anatomical regions will be approached together in Chapter 4 where pathomechanics are dealt with. Inevitably, the rearfoot has to be discussed with reference to the ankle functional unit. Therefore, further anatomical information will be discussed regarding the foot as a functional unit in Chapter 3. The tendinous continuation of most of the ankle's muscles into and across the midfoot makes separating the ankle from the entire foot an artificial construct, a situation which must be suspended from the mind in order to coherently appreciate human locomotion biomechanics.

2.5.2 The kinematic role of the ankle

The ankle is the primary fulcrum for the lower limb's motion during most of the stance phase. The lever arm's beam is constructed from the thigh and lower leg segments that allows the HAT's CoM to vault over the supporting base of the foot via the hip, as discussed in the inverted pendulum model of walking. At initial heel contact, the ankle facilitates foot flexion to allow the forefoot's supporting base to fully contact the ground as the lower limb loads with the full body weight behind it. The ankle then goes on to allow the whole limb to vault over the foot through increasing angles of dorsiflexion taking the body mass anteriorly over it. At the end of the inverted pendular vaulting, powerful muscle-generated plantarflexion moments and reducing heel loads spring the heel off the ground via ankle flexion, helping to accelerate the CoM of the body, leg, and rearfoot anteriorly and superiorly around MTP joint extensions. Finally, the ankle aids foot lift for swing through an extension moment that reduces the plantarflexed angle remaining at the end of terminal stance. Thus, during walking, the ankle undergoes two cycles of plantarflexion each followed by dorsiflexion motion during every stride, with the last phase being a dorsiflexion moment occurring upon entering and ending swing (Fig. 2.5.2a).

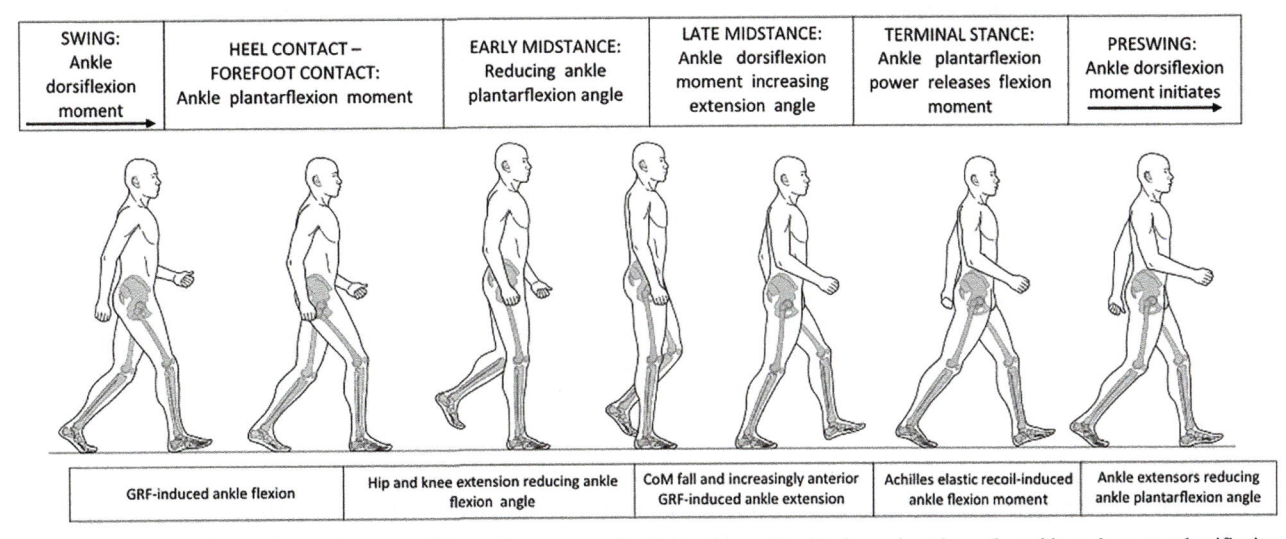

FIG. 2.5.2a The ankle during walking stance phase undergoes a cycle of changing angles. During swing phase, the ankle undergoes a dorsiflexion moment that resists ankle plantarflexion gravitational pull. Heel strike occurs with the foot angled in dorsiflexion to the support surface (but not usually at the ankle) with the GRF behind the ankle joint initiating a plantarflexion moment. The ankle plantarflexes until forefoot loading with its acceleration restrained by the ankle dorsiflexors. Early midstance consists of a period of reducing ankle plantarflexion under hip and knee flexion muscular activity. Late midstance consists of increasing ankle dorsiflexion as a result of the forward fall of the HAT's CoM and the contralateral swing limb's accelerating centrifugal forces. This is restrained by the ankle plantarflexors that stored this muscular energy of restraint within their tendons. As the CoM offloads the heel, the Achilles' elastic recoil is released, causing ankle plantarflexion and heel lift. Ankle plantarflexion continues throughout terminal stance, ending with muscular dorsiflexion moments at preswing that reduce the ankle plantarflexion angle for swing phase. *(Permission www.heathystep.co.uk.)*

During walking swing phase, the ankle recovers from its most plantarflexed position at toe-off, dorsiflexing to a near-ankle neutral position before heel strike. The ankle should not actually become dorsiflexed during any part of swing. Except for the presence of low isometric ankle extensor activity, the ankle carrying the foot is very much a 'passenger' during most of the swing phase. Towards the end of terminal swing, the ankle muscles adjust the foot's posture in preparation for initial heel contact with muscles that bridge the ankle–subtalar complex becoming active. Heel strike running follows a similar ankle motion path to that of walking during stance phase, of contact plantarflexion followed by dorsiflexion, and then acceleration plantarflexion. However, in forefoot running, there is only one extended phase of ankle dorsiflexion during forefoot loading before the start of plantarflexion associated with acceleration. This is followed by dorsiflexion during swing phase to reduce the plantarflexion angle, but in forefoot strike running the ankle is positioned in terminal swing for forefoot contact in a plantarflexed posture. In all forms of running, swing phase dorsiflexion moments occur after toe-off, necessitating greater knee flexion in early swing than walking to give the ankle dorsiflexors time to reduce the take-off plantarflexion angle for midswing when the support limb is functionally shortened.

During the stance phase in all forms of locomotion, the ankle becomes the key fulcrum that changes the impact or collision angle of the leg posterior to the foot during braking, to one angled anterior to the foot during the acceleration phase. The amount of angle change is particularly important in running during the 'spring–mass' function of impact collision and recoil 'bounce-off'. This is because the change in ankle angle is a key part in directing acceleration forces more horizontally than vertically to aid running acceleration (Fig. 2.5.2b).

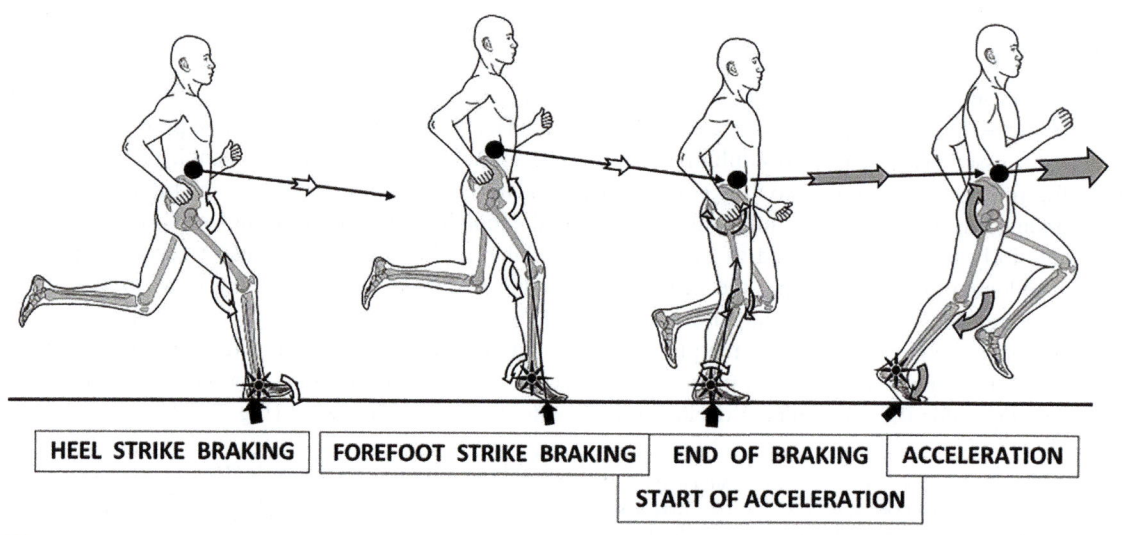

FIG. 2.5.2b Being a single collision and bounce-off event without a distinct separate weight-transfer phase period, the effects of changing ankle angles from loading and braking collision to offloading acceleration collision become more obvious in running than during walking. During heel strike running, the initial loading impact occurs with the ankle angled posteriorly to the foot so the lower limb can act as a brake to acceleration by developing a posteriorly directed GRF vector (thin black arrow) into the limb, with hip, knee, and ankle flexion (white arrows) acting as shock absorbers. This braking phase gradually sees the ankle's posterior angulation disappear as ankle plantarflexion angles are reduced through ankle dorsiflexion motion coupled to knee and hip flexion. Once the ankle becomes angled anteriorly at the braking/acceleration boundary, the application of force derived from the ankle plantarflexors starts to create a forefoot GRF (grey arrow) with a large anterior horizontal element that drives the body forward towards its aerial and swing phase. Although acceleration events are identical in forefoot running to that of heel strike running, initial loading involves a more vertical forefoot impact GRF. This necessitates ankle dorsiflexion dropping the heel towards the ground and requires ankle plantarflexors to aid hip and knee flexors in providing shock attenuation. During forefoot strike running, the impact ankle dorsiflexion moment is continuous with the dorsiflexion motion that is required to angle the lower limb to face anteriorly for acceleration. *(Permission www.healthystep.co.uk.)*

At heel strike in walking and running, the GRF occurs behind the ankle, setting up a rapid plantarflexion moment across the ankle that accelerates the forefoot towards the support surface. Thus, initial ankle motion is controlled through the muscles of the anterior muscle group (particularly tibialis anterior) opposing ankle plantarflexion (Byrne et al., 2007). This anterior muscle group action occurs concurrently with antagonistic gastrocnemius activity (Di Nardo et al., 2016; Mengarelli et al., 2017, 2018), probably assisting in stabilisation of the knee flexion angle with the hamstrings rather than generating ankle plantarflexion (Mengarelli et al., 2018). The plantarflexing resistance activity of tibialis anterior provides significant impact energy dissipation, assisted at forefoot contact by the windlass effect of toe extension and by the agonistic/antagonistic team of tibialis posterior and peroneus longus that together influence an appropriate level of stiffness/compliance within the foot prior to forefoot contact (Kokubo et al., 2012).

During walking midstance, the ankle's role is to change the leg's angle from lying posterior to the foot to lying anterior to the foot. Initially, this is driven by impact phase momentum which gains assistance from contralateral heel-lifting ankle plantarflexion power and aided by quadriceps and gluteus maximus activity, reducing knee and hip flexion, respectively, as they extend the lower limb. This combined action reduces the ankle plantarflexion angle at and after forefoot loading. Around absolute midstance, the ankle angle changes to one of dorsiflexion with further dorsiflexion initiated by the anterior fall of the CoM from its inverted pendulum high point. This sets up anterior acceleration of the tibia at the ankle which is controlled through eccentric negative acceleration and work from the posterior calf muscle group, particularly the superficial muscle group of the triceps surae via its energy-storing Achilles tendon. This continues until heel lift. The ankle motion at terminal stance is a result of ankle plantarflexor-generated power created at the end of midstance and the reducing loads on the rearfoot. Although the acceleration fulcrum at heel lift moves to the MTP joints, the ankle still passes through a considerable range of plantarflexion motion in terminal stance to permit the ankle plantarflexion power to maintain a stabilising forefoot GRF. The arc of rotation of the ankle is reduced throughout MTP joint extension during acceleration by the assistance of midfoot plantarflexion during terminal stance, which shortens the external moment arm of the Achilles in late terminal stance. By so doing, less ankle plantarflexion motion is required to achieve MTP joint extension angles necessary before toe-off.

Terminal stance ankle plantarflexion power has been explained as a mechanism to add energy to facilitate the opposite initial contact momentum, improving double-limb stance energetics (Ruina et al., 2005). This model links the function of the two ankle joints together during double-limb stance (Fig. 2.5.2c). As the support role is transferred to the leading foot, activity shifts to the terminal stance foot's anterior muscle group to start to reduce the ankle plantarflexion angle ready for swing.

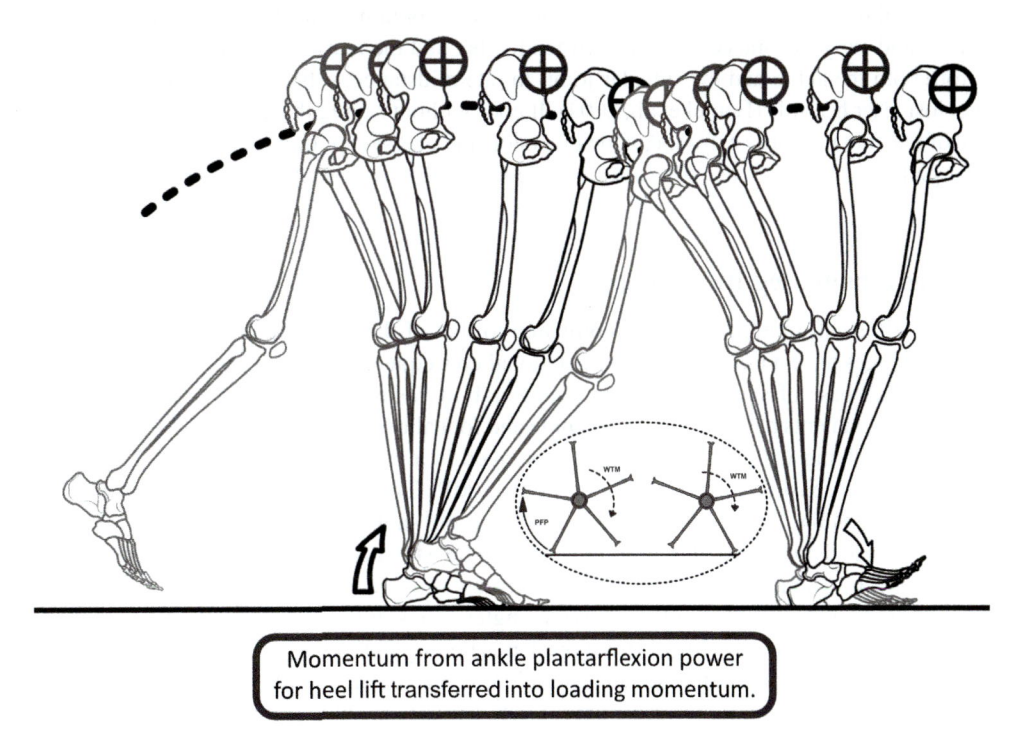

FIG. 2.5.2c The ability of the ankle plantarflexors' power (PFP) to add momentum into the next step is essential for effective gait energetics. The insert image shows the events applied to the wheel model proposed by Ruina et al. (2005) showing how heel lift momentum can assist in weight transfer momentum (WTM) that can persist into early midstance during human locomotion. *(Permission www.healthystep.co.uk.)*

During running, the ankle's initial motion can change depending on the strike position. In heel strike running, the pattern is much the same as walking, with the limb angled posterior to the ankle. However, although the leg vaults over the foot, the knee and hip continue to flex until the middle of stance. The leg is working like a spring–mass in a bounce-like collision. The limb starts to recoil out of flexion after the middle of stance with the ankle now dorsiflexed, angling the limb anterior to the ankle which aids horizontal acceleration. The ankle quickly follows this limb extension motion by plantarflexing under Achilles elastic recoil. In forefoot running, the ankle strikes in plantarflexed posture, impacting the ground via the forefoot.

The limb is usually still positioned posterior to the ankle but angled more vertically than seen in a heel strike. This posture sets up a dorsiflexion moment at the ankle from the GRF on the forefoot. The heel is thus accelerating towards the support surface, resisted by the ankle plantarflexors (Kernozek et al., 2018). The result is that forefoot running reduces tibialis anterior activity (Rooney and Derrick, 2013) but increases triceps-surae/Achilles loading (Kernozek et al., 2018).

The position and direction of the GRF which results from the strike position in walking and running as well as the nature of the terrain, largely dictates how the ankle will initially be loaded, and the motions that will be required for the ankle angle to change to allow the body's CoM to pass over it. Thus, during all phases and types of gait, the ankle and the foot work together to create lower limb motion and stability during stance. The ankle function is dependent on effective foot function, just as the foot is dependent on effective ankle function.

The ankle as part of the rearfoot

One of the confounding factors in appreciating the role of the ankle joint is its intimate link with the talocalcaneal and talocalcaneonavicular joints (known together as the subtalar joint). When the rearfoot is unloaded, the calcaneus moves with ankle motion only. There is no freedom of motion occurring at the subtalar joint between the talus and calcaneus, almost as if such a joint did not exist (Leardini et al., 2001). Once under extrinsic forces (such as GRF), the calcaneus moves independently of the talus through the subtalar joint, but this subtalar motion occurs mostly when the ankle is plantarflexed rather than when dorsiflexed (Leardini et al., 2001; Hamel et al., 2004). Thus, subtalar motion mainly occurs around the first 25% of stance phase during heel contact gaits.

Rearfoot motion in stance can be difficult to locate without this knowledge. When the rearfoot is unloaded such as during terminal stance, all rearfoot motion between the foot and leg should be considered as coming from the ankle. When loaded at heel contact and early midstance, motion between the leg and the rearfoot should be considered as being a combination arising from both the ankle and the subtalar joint. In late midstance, subtalar joint motion becomes increasingly restricted, and then lost after heel-off (Hamel et al., 2004). It goes without saying that subjects will vary as to which joint is the primary source of rearfoot motion, i.e. frontal plane inversion and eversion or transverse plane adduction and abduction. This is likely to be particularly true in the presence of ankle–rearfoot ligament dysfunction.

2.5.3 The ankle as a biotensegrity structure

The ankle has a relatively small motion excursion compared to that of the hip and knee. Yet, being the support base in stance, its own angle changes have a large resultant translational consequence to the segments above it. The foot and ankle represent a relatively small base of support requiring strong stability as the body mass rotates above them. The diverse collection of muscles and ligaments passing around the ankle and foot allows the ankle to act as a concentrated sensorimotor 'hub'. The muscle spindles lying within the muscles are arranged in a variety of directions around the ankle, improving multidirectional sensorimotor sensitivity (Kavounoudias et al., 2001). Thus, the ankle acts as a sensorimotor hub for both the lower limb's posture upon the foot and the foot's position on the support surface.

Together, the ligaments and muscles of the tibiofibular and ankle–subtalar complex give the ankle strong force closure, with increasing form closure stability occurring through increasing joint congruency during dorsiflexion. More force closure stability is required when increasing plantarflexion angles. Thus, muscle activity around the ankle provides most stability when the ankle is plantarflexing at heel strike or plantarflexing during terminal stance, and also when starting to move out of plantarflexed postures. In forefoot loading situations or indeed any increased plantarflexed postures, the ankle is more reliant on force closure. An ability to create appropriate changes in ankle form and force closure is dependent not only on tissue integrity but also on stable and appropriately timed joint motions (Fig. 2.5.3).

Through the concept of biotensegrity, ankle shape change is just as important to soft tissue tensional stability as soft tissue tensional stability is to controlling joint position. The ankle is required to maintain stability throughout weightbearing and its ability to achieve this is highly dependent on the foot's capacity to provide stability. This is because force closure derives from those muscles that arise in the leg to supply the foot (the extrinsic foot muscles) that all crowd around the ankle, helping to compress and brace it as they tension. The principles of biotensegrity and myofascial chains still very much apply to the ankle. However, it is difficult to discuss ankle biotensegrity thoroughly without consideration being given to the foot's biotensegrity which will discussed further in Chapter 3.

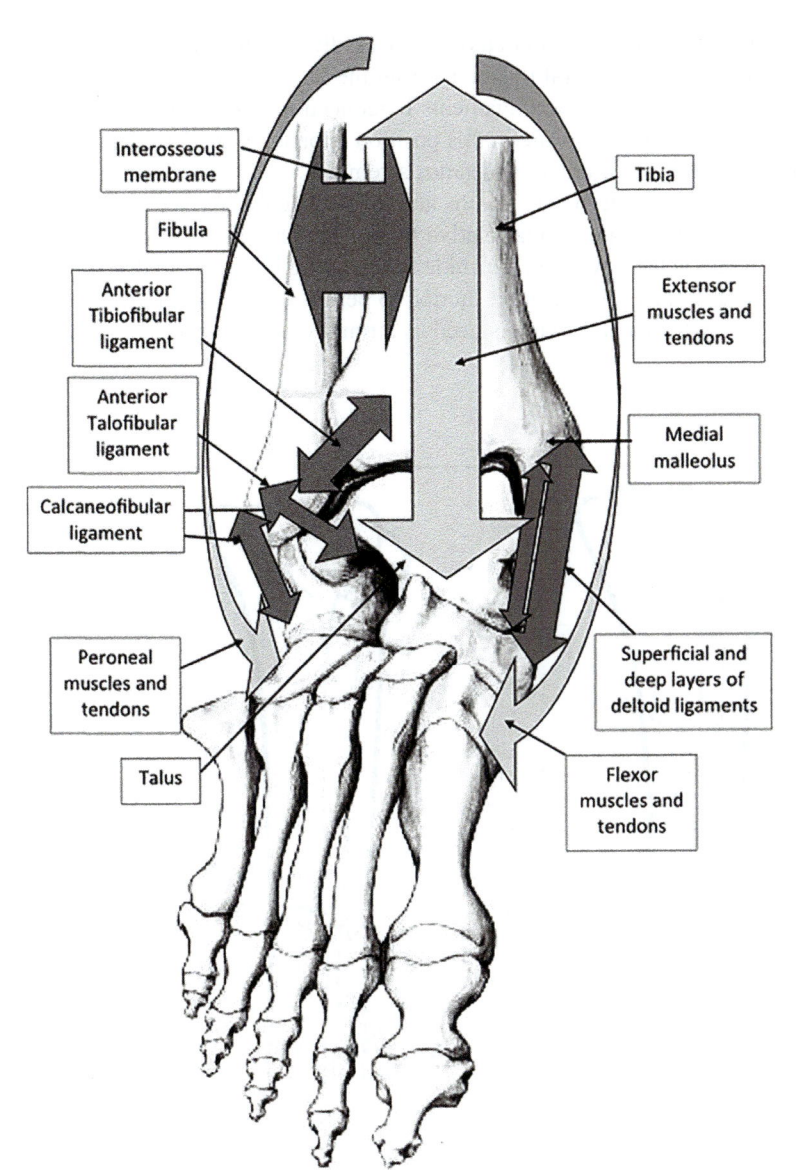

FIG. 2.5.3 A schematic of the anterior, medial, and lateral structures that maintain ankle biotensegrity. The talus is constrained within a mortice with its articular surfaces restrained against the articular facets of the tibia superiorly and medially, and the fibular facet laterally. Passive tension maintains this mortice. This is achieved using the syndesmotic anterior and posterior (not shown) tibiofibular ligaments and the long powerful ligament known as the tibiofibular interosseous membrane. The proximal tibiofibular joint is similarly supplied with superior, anterior, and posterior tibiofibular ligaments. The deltoid ligament, consisting of distinct superficial and deep layers, provides a trapezoid medial restraint to eversion moments. The lateral collateral ligaments of the anterior and posterior (posterior not shown) talofibular ligaments and the calcaneofibular ligament all resist inversion. Active adjustable muscular tensional forces provide compressive stability, with the ankle–foot extensors providing stability against plantarflexion moments and the powerful triceps surae via the Achilles (not shown) acting against moments of dorsiflexion. These posterior muscle–tendon structures also provide the all-important ankle plantarflexor power of acceleration. The long foot flexors of tibialis posterior, flexor hallucis longus, and flexor digitorum longus, and also the peroneal muscles laterally, provide a little dorsiflexion moment resistance and provide limited plantarflexor power. However, they are more important in moderating inversion, eversion, abduction, and adduction moments around the instantaneous joint axes of the ankle–subtalar complex. These muscles are also playing important complementary role in foot stiffness moderations during gait. *(Permission www.healthystep.co.uk.)*

2.5.4 Osseous topography of the ankle

The ankle is constructed from the inferior articulations of the distal tibia (an area often referred to as the *tibial plafond*) and fibula with the dorsal articular or trochlear surface of the talus. There are also two articular surfaces (one proximal and one distal) between the tibia and fibula which are also held together by a strong interosseous membrane along their opposing shafts. The proximal tibiofibular joint is not usually considered an anatomical part of the ankle. However, functionally, through its coupling of motion, it is useful to consider it with the ankle. The ankle, as already mentioned, also has an intimate coupling of motion relationship with the subtalar joint below the talus (Wright et al., 1964; Leardini et al., 2001; Leardini and O'Connor, 2002; Bonnel et al., 2010).

The tibia's motion couples through the ankle joint to the motion of the talus at the subtalar joint, but calcaneal motion directly follows motion of the talus during swing and terminal stance, and largely does so during late midstance. This is because the subtalar joint does not have any true freedom of motion, only some flexibility across the joint needing external force distally to separate talar from calcaneal motion (Leardini et al., 2001).

Like all joints, ankle motion is guided by the osteoarticular surfaces and ligamentous structures through changing force and form closure relationships. The ankle has a strong syndesmotic ligamentous and interosseous membrane complex across the tibia and fibula that prevents osseous separation. This is essential for ankle stability, creating a comparison with a mortice (Tourné et al., 2019). The syndesmotic ligamentous unit is composed of distinct individual ligament components (Golanó et al., 2010; Yuen and Lui, 2017). Medial and lateral collateral ligaments from the tibia and fibula limit frontal, sagittal, and transverse plane motion across the ankle. The role of these ligaments is reduced in ankle dorsiflexion and subtalar eversion due to increased close-packing and osseous form closure in this position (Bonnel et al., 2010). In dorsiflexion, form closure is particularly strong, but only if the tibiofibular syndesmosis remains intact to prevent osseous splaying. The syndesmotic articulations allow the tibiofibular unit to vary widths across the articular surface of the talus through slightly ascending and medially rotating movements of the fibula relative to the tibia during dorsiflexion. This motion helps accommodate talar maximal articular width anteriorly when the ankle is becoming increasingly dorsiflexed (Golanó et al., 2010). Both the proximal and distal tibiofibular joints exhibit motion to permit this particular accommodation during ankle dorsiflexion. This motion between the tibia and fibula is used as a transverse torque shock absorber across the lower leg (Fig. 2.5.4a).

FIG. 2.5.4a A schematic to show the effects on the tibiofibular joints during ankle motion that makes these articulations extremely useful in dissipating torque from the ankle joint. Ankle plantarflexion presents the narrower posterior articular surface of the talus within the ankle mortice. This position slackens the tensions relatively across the proximal and distal tibiofibular joints and the interosseous membrane. The reduced ligament tensions allow the fibula to glide inferiorly and internally rotate, moving closer to the tibia and thus narrowing the ankle mortice. Dorsiflexion of the ankle presents the wider anterior articular surface of the talus requiring the mortice to adapt by widening. This is achieved by the fibula gliding superiorly and external rotation which tightens the ligaments and interosseous membrane between the tibia and fibula. The stretching and elastic recoiling of these connective tissues during ankle motion stores and releases energy through cycles of ankle extension and flexion, dissipating torque energy and improving the ankle's form closure through adjusting articular morphology within the ankle mortice. *(Permission www.healthystep.co.uk.)*

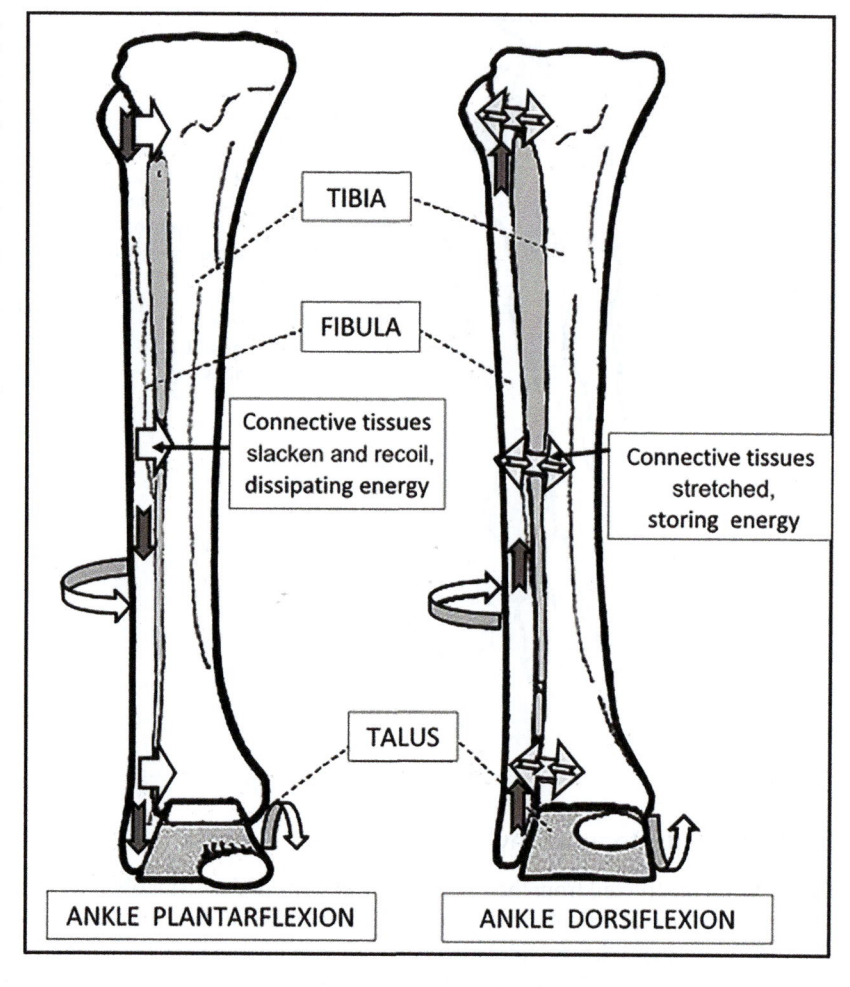

The ankle is often described as a hinge joint and as such, there is a tendency to see it as a simple provider of sagittal plane flexion and extension motion around the foot. However, this is not the reality because of the changing instantaneous centre of rotation that occurs with ankle motion (Leardini and O'Connor, 2002; Bonnel et al., 2010). Ankle motion complexity arises from the movement of the functional axis within the joint's motion that can be somewhat subject-specific. This arises from specific articular and ligamentous morphology in and around the ankle, something that is variable within the population (Bonnel et al., 2010). The human talus morphology presents an asymmetrical truncated cone, the dimensions of which have been classified crudely into three types (Figs 2.5.4b and 2.5.4c).

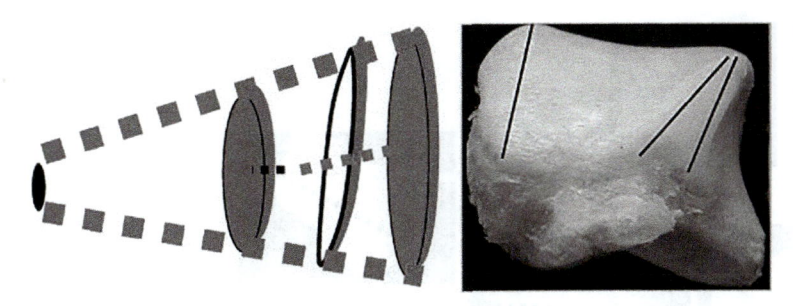

FIG. 2.5.4b Asymmetry in the articular trochlear surface of the talus results from it being shaped as a truncated cone. *(Image from Bonnel, F., Toullec, E., Mabit, C., Tourné, Y., Sofcot, 2010. Chronic ankle instability: biomechanics and pathomechanics of ligaments injury and associated lesions. Orthop. Traumatol. Surg. Res. 96(4), 424–432.)*

RC: radius of curvature

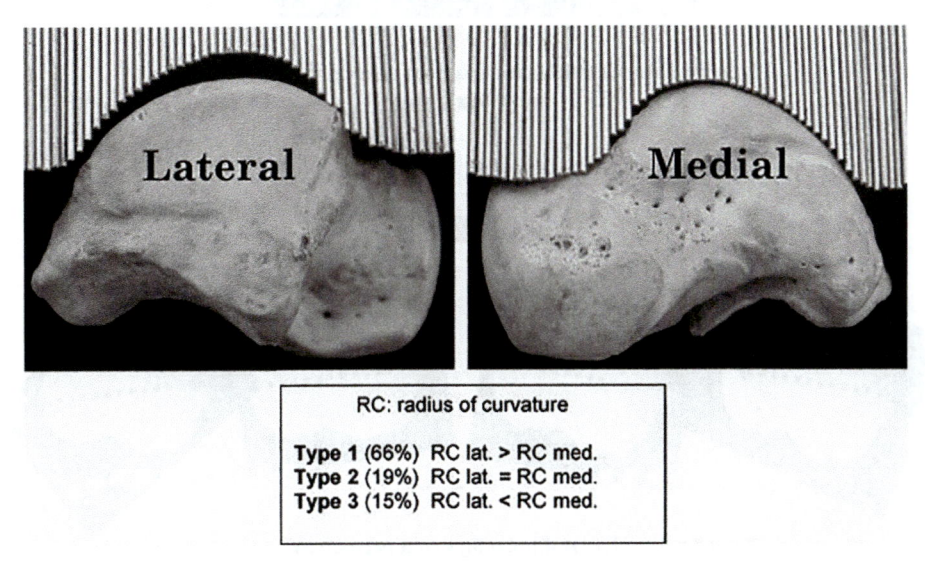

Type 1 (66%) RC lat. > RC med.
Type 2 (19%) RC lat. = RC med.
Type 3 (15%) RC lat. < RC med.

FIG. 2.5.4c The biometrics of the three types of human talar articular surface reported. The photographs show the lateral and medial views of the type 1 talar morphotype with the radius of curvature (RC) of the lateral trochlear surface being greater than that of the medial. Type 2 is considered to have similar curvatures lateral to medial, while type 3 has a larger medial radius of curvature. These different shapes among human tali exist on a continuum and influence ankle motion. *(Image from Bonnel, F., Toullec, E., Mabit, C., Tourné, Y., Sofcot, 2010. Chronic ankle instability: biomechanics and pathomechanics of ligaments injury and associated lesions. Orthop. Traumatol. Surg. Res. 96(4), 424–432.)*

Thus, the ankle joint offers variable motion in all spatial planes (Bonnel et al., 2010). The talus actually provides combined rolling flexion–extension associated with horizontal rotation in the frontal plane and abduction/adduction in the transverse plane through multiaxis changing dynamics (Bonnel et al., 2010). Axis motion is down to the morphology of the trochlear surface of the talus which is shaped as an asymmetrical truncated cone. This is wider anteriorly than posteriorly by 4 ± 2 mm, reducing the joint space between the malleoli anteriorly (Bonnel et al., 2010). Three types of talus have been described based on the type of curvature of the trochlear surface. Type I has a medial radius of curvature smaller than the lateral radius of curvature, and is the most common (Bonnel et al., 2010). In type II, the radii of curvature are equal and in type III, the medial radius of curvature is larger than the lateral and is considered a protective morphotype to the anterior talofibular ligament, helping to reduce the risk of injury to this particular ligament (Bonnel et al., 2010) (Fig. 2.5.4d).

2.5.5 Passive soft tissues of the ankle

Stable ankle kinematics are dependent on ligaments, most significantly those that hold the fibula to the tibia, constraining the talar trochlear surface between them. The tibiofibular syndesmosis ligaments and interosseous membrane play an extremely important role in maintaining osseous form closure. The role of ankle ligaments in joint stability is dependent upon the load and direction of the applied force (Bonnel et al., 2010; Watanabe et al., 2012). Ankle sprains are very common in humans. They provide a significant number of cases seen in emergency departments (Waterman et al., 2010) and account for many sports injuries (Golanó et al., 2010; Hertel and Corbett, 2019), with collision/contact sports being more likely to cause syndesmotic injuries (Tourné et al., 2019). Most ankle sprains occur around frontal plane perturbations, and most

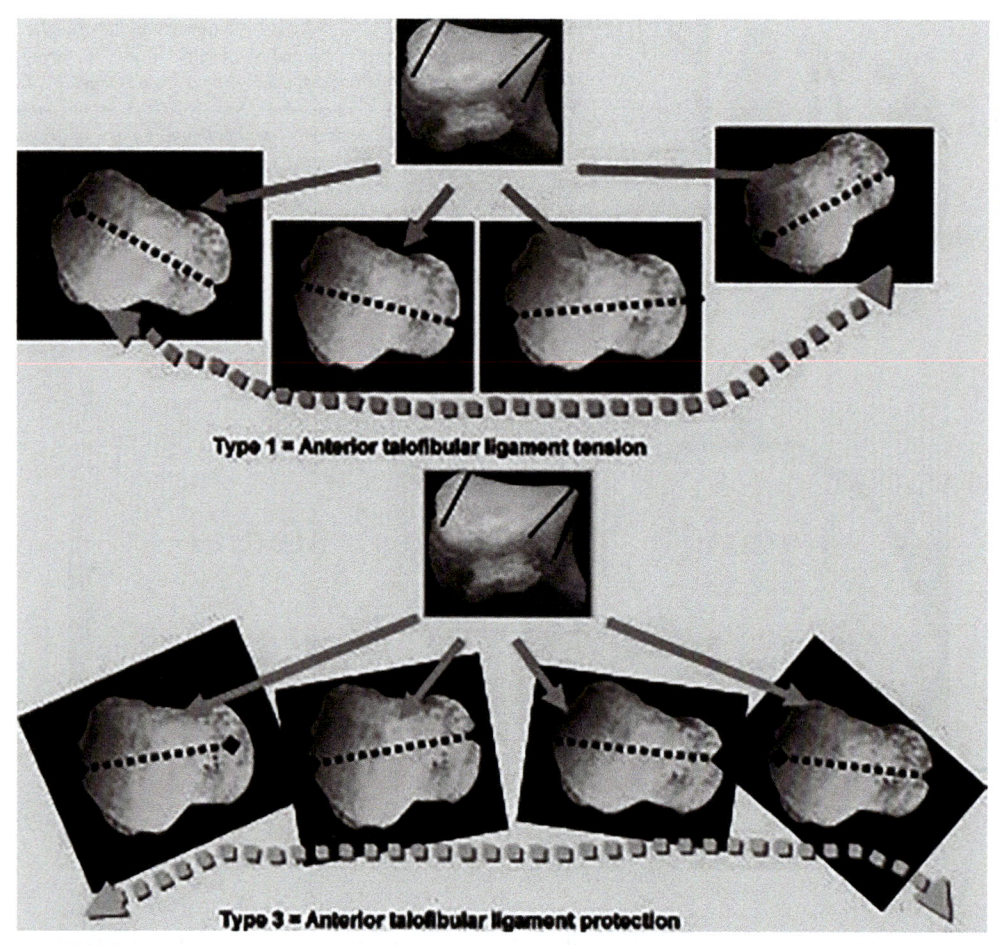

FIG. 2.5.4d Two instantaneous joint axis orientations exist within the ankle. One is orientated bottom to top and from medial to lateral for plantarflexion and the other orientated from top to bottom and medial to lateral for dorsiflexion. Together, these allow for a pure rotation motion and a rolling (rotation and sliding) motion which can associate together. Rolling develops with the motion of the instantaneous joint axis within the talar body. The images represent the different motions of rolling/sliding kinematics of the talus within the ankle associated with varying shapes of the human talus. Type 1 is thought to increase anterior talofibular ligament tensions, potentially increasing injury risk, whereas type 3 may provide some protection to this ligament. However, loading stress peaks are going to be most important in regard to whether any injury occurs or not. *(Image from Bonnel, F., Toullec, E., Mabit, C., Tourné, Y., Sofcot, 2010. Chronic ankle instability: biomechanics and pathomechanics of ligaments injury and associated lesions. Orthop. Traumatol. Surg. Res. 96(4), 424–432.)*

frequently are only collateral ligament stressing in origin. In the sagittal plane, ligamentous stability is syndesmosis ligament-based. These ligaments, during terminal stance plantarflexion, neutralise shearing force by embedding the talus within the mortice with the talus then acting as a keystone, and the convex talar surface-to-concave tibial surface coming into play by providing form closure (Bonnel et al., 2010). Hyper-dorsiflexion and hyper-plantarflexion injuries risk separation of the syndesmotic ligaments (Golanó et al., 2010). The loss of ligamentous ankle stability in the longer term can lead to degenerative joint changes.

Tibiofibular ligaments

Ankle joint form closure is produced by its stabilising articular geometry that sustains physiological loadings of around 700 N, with the ankle joint being most stable in dorsiflexion and least stable when in plantarflexion (Stiehl et al., 1993; Watanabe et al., 2012). Strong force closure is due to the increased joint congruency within ankle dorsiflexion that reduces medial–lateral rotation compared to plantarflexion. The form closure accounts for 100% of translation stability and 60% of external–internal rotation stability (Watanabe et al., 2012), with heightened axial loading increasing stability further (Stiehl et al., 1993; Tochigi et al., 2006). Thus, the ankle joint position dramatically influences the stabilisation effects of the ligaments of the ankle through dorsiflexion and plantarflexion (Li et al., 2019). The ankle ligaments consist of three functional units: lateral collateral ligaments, medial collateral ligaments, and syndesmotic ligaments of the tibiofibular joints (Golanó et al., 2010).

The syndesmotic ligaments hold the fibula and tibia together, both proximally and distally, with proximal and distal anterior and posterior tibiofibular ligaments that join the epiphyses together (Golanó et al., 2010). Throughout the length of the two bones runs an interosseous membrane of strong connective tissue linking the lateral side of the tibia to the medial side of the fibula. This syndesmotic ligament complex ensures stability between the distal tibia and fibula, resisting axial, rotational, and translational forces that attempt to separate the fibula from the tibia (Golanó et al., 2010). However, the ligament complex does permit some external rotation and superior glide with ankle dorsiflexion, while elastically retaining compression stability across the talus (Golanó et al., 2010). The ligament complex functions as an external torque shock absorber coupled to ankle motion (Fig. 2.5.5a). Distally, the ligaments are the anteroinferior tibiofibular ligament, the posteroinferior tibiofibular ligament, and the interosseous tibiofibular ligament which work collectively with the distal interosseous membrane (Golanó et al., 2010). Distal to the interosseous membrane attachment lies the tibiofibular synovial recess which has within its posterior surface a bundle of adipose tissue, referred to as the *fatty synovial fringe*. In plantarflexion, this structure lowers towards the ankle and retracts and rises in dorsiflexion. It is implicated in chronic ankle pain following ankle sprains, causing a condition known as *syndesmotic impingement* (Golanó et al., 2010).

Ankle collateral ligaments

The lateral collateral ligaments of the ankle consist of the anterior talofibular ligament, the calcaneofibular ligament, and the posterior talofibular ligament. All three attach to the fibula and terminate on either the talus or calcaneus, stabilising both the ankle and subtalar joints (Bonnel et al., 2010; Golanó et al., 2010). The anterior tibiofibular ligament is short but widens distally as it passes from the lateral malleolus horizontally, anteriorly, and inferiorly to attach to the neck of the talus. The calcaneofibular ligament is covered by the peroneal tendon sheaths and arises from the highest point of the lateral malleolus, running inferiorly and slightly posteriorly to the lateral side of the calcaneus along a 45° orientation to the lateral malleolar long axis (Bonnel et al., 2010; Golanó et al., 2010). This ligament is able to stabilise the fibular-talocalcaneal unit, assisting the cervical ligament within the subtalar joint (Bonnel et al., 2010). The calcaneofibular ligament is the primary ligamentous restraint of inversion, influencing the ankle mostly in plantarflexion and at the subtalar joint when the ankle is dorsiflexed (Li et al., 2019). The posterior talofibular ligament extends horizontally from the posterior lateral malleolus to the posterior surface of the talus (Fig. 2.5.5b). Anatomical variation exists among these ligaments, with doubling of the anterior ligaments and an accessory ligament between the anterior and middle ligament having been reported (Golanó et al., 2010; Bonnel et al., 2010) (Fig. 2.5.5c).

The medial collateral ligaments or *deltoid ligaments* are comprised of two distinct superficial and deep layers (although precise anatomical descriptions vary in the literature), and are reported as consisting of five or six components (Golanó et al., 2010; Bonnel et al., 2010). The superficial layer is triangular, arising from the highest point of the medial malleolus of the tibia and then fanning out its fibres to run inferiorly. The posterior tibiotalar part is the strongest portion which runs to a tubercle on the medial side of the talus. The tibiocalcaneal section runs to the medial edge of the *sustentaculum tali* (a prominent tubercle of the calcaneus), while the tibionavicular and anterior tibiotalar parts attach to the navicular and spring ligament (Bonnel et al., 2010). The spring ligament, working as part of the deltoid ligament, provides anteromedial stabilisation of the talocalcaneonavicular joint which is the site of multiplanar rotatory mechanisms occurring between the ankle–rearfoot and midfoot (Pisani, 1994). The deep layers of the deltoid ligament are made of strong and fibrous bundles, extending diagonally from the medial malleolus to the entire medial side of the talus situated below the trochlear surface and most anteriorly to the talar neck (Bonnel et al., 2010). The tibialis posterior tendon sheath covers the middle part of the deltoid ligament (Fig. 2.5.5d).

Recent studies have shown that the superior fascicle of the anterior talofibular ligament is a distinct anatomical feature, whereas the inferior parts of this ligament share fascicles with the calcaneofibular ligament, forming a functional unit named the *lateral fibulotalocalcaneal ligament complex*, dividing the anterior talofibular ligament into two functionally distinct structures with the superior fascicles more likely to heal following injury (Vega et al., 2020). This ligament complex is also bound into *Kager's fat pad* found between the Achilles tendon, flexor hallucis longus, and the posterior surface of the calcaneus, creating fascial linkage across much of the posterior anatomy of the ankle (Szaro et al., 2021).

Subtalar ligaments

The function of the medial and lateral collateral ligaments of the ankle joint occurs in concert with the ligaments of the subtalar joint, because most of these ligaments have connections across the subtalar joint to the calcaneus from the tibia and fibula. Thus, it is necessary to discuss them here. They consist of the subtalar interosseous ligament that runs in the *sinus tarsi*, the osseous tunnel formed by the talar roof and the calcaneal floor. The interosseous ligament functions as a central pivot point, providing rotatory stability within the subtalar joint in much the same manner as the cruciate ligaments of the

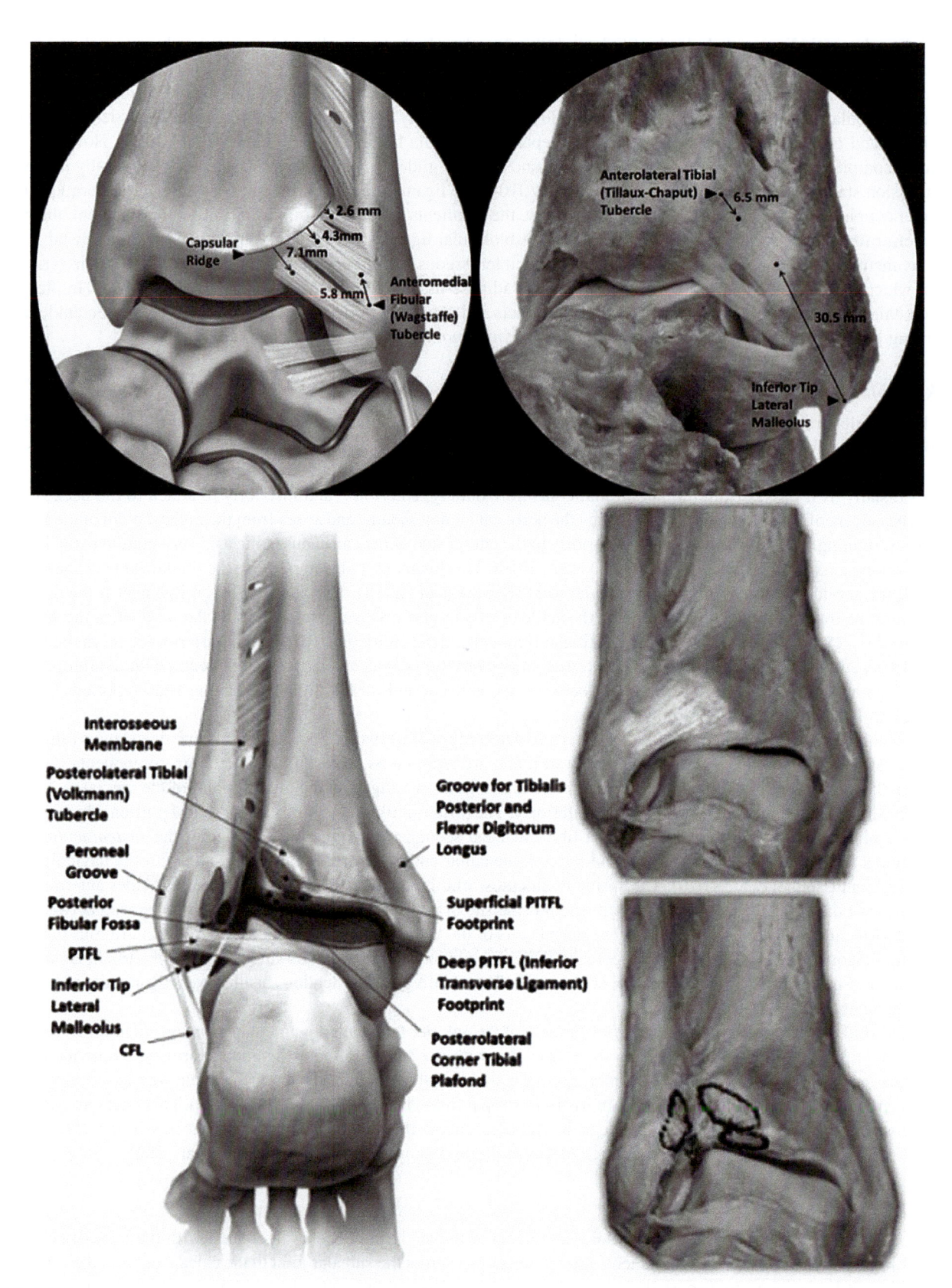

FIG. 2.5.5a The tibiofibular ligaments and the interosseous membrane are essential for tibiofibular stability, which means the ankle relies on their integrity both superiorly at the proximal and inferiorly at the distal tibiofemoral joints. The anterior inferior tibiofibular ligament is trapezoidal (upper images) and consists of three distinct bundles. The posterior inferior tibiofibular ligament (PTFL—lower images) is also trapezoidal, more solid, horizontal, and is longer than the anterior, being wider at its tibial compared to its fibular attachment. *(Images from Tourné, Y., Molinier, F., Andrieu, M., Porta, J., Barbier, G., 2019. Diagnosis and treatment of tibiofibular syndesmosis lesions. Orthop. Traumatol. Surg. Res. 105(8 Supplement), S275–S286.)*

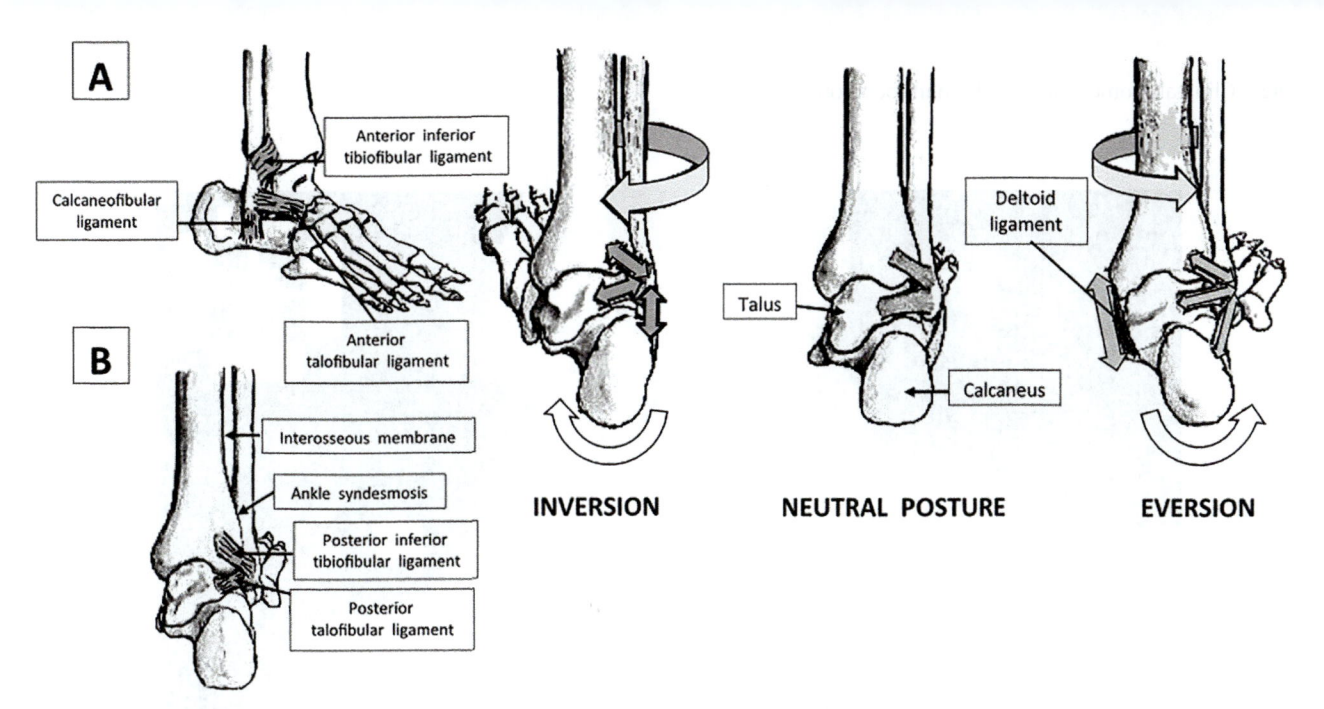

FIG. 2.5.5b Diagrams of the lateral collateral and inferior tibiofibular ligaments viewed anterolaterally (A) and posteriorly (B). Under inversion moments, lateral ligaments receive the dominant tensional stresses while medial (deltoid) ligaments undergo greater stress during ankle eversion. Plantarflexion moments focus stresses on anterior ligaments and dorsiflexion moments on posterior ligaments. However, transverse plane rotational torques adjust the locations and orientations of peak strains. The inferior tibiofibular ligaments resisting tibiofibular separation torques will tend to have more stress posteriorly with external tibial rotation (often accompanying inversion) and more anteriorly during internal tibial rotation (often seen with eversion). *(Images modified from Dubin, J.C., Comeau, D., McClelland, R.I., Dubin, R.A., Ferrel, E., 2011. Lateral and syndesmotic ankle sprain injuries: a narrative literature review. J. Chiropract. Med. 10(3), 204–219.)*

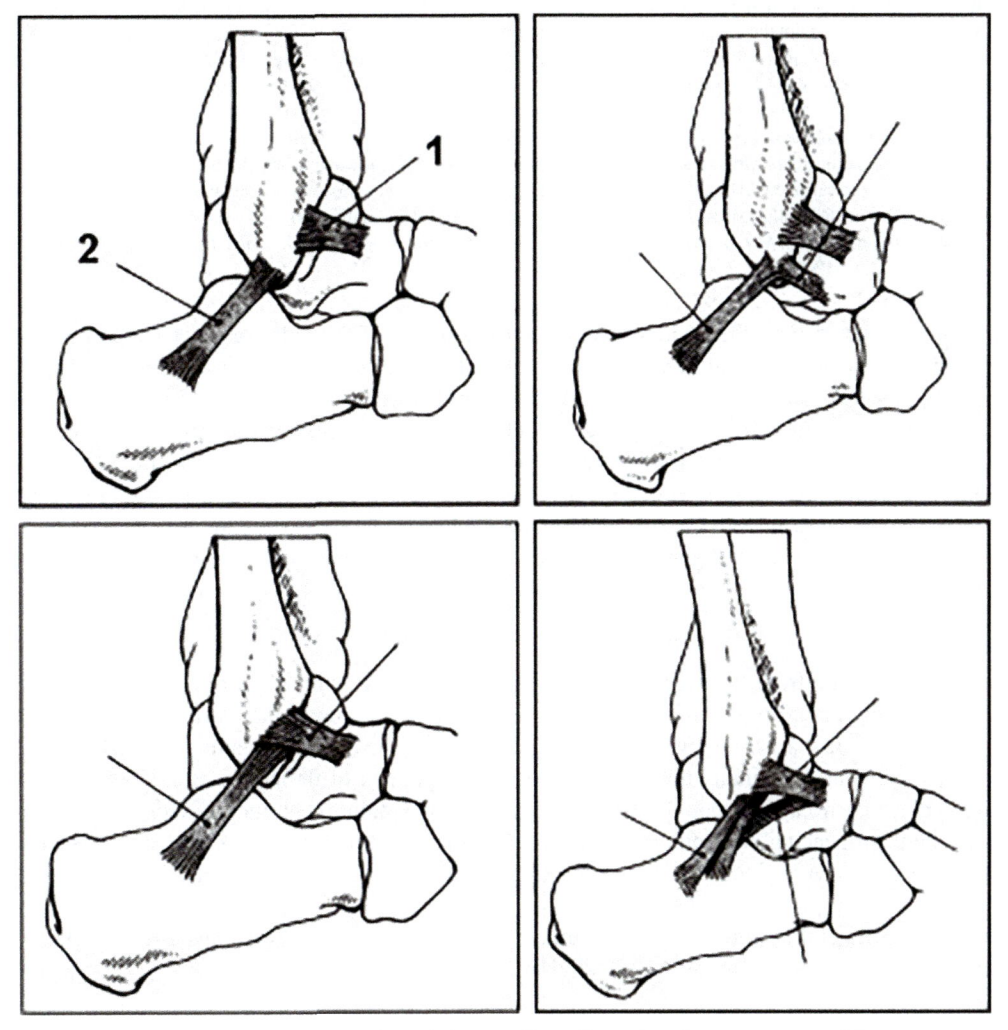

FIG. 2.5.5c Lateral collateral ankle ligaments are not consistent in humans. These illustrations demonstrate some of the common anatomical variations reported for the lateral ankle ligaments. *(Image from Bonnel, F., Toullec, E., Mabit, C., Tourné, Y., Sofcot, 2010. Chronic ankle instability: biomechanics and pathomechanics of ligaments injury and associated lesions. Orthop. Traumatol. Surg. Res. 96(4), 424–432.)*

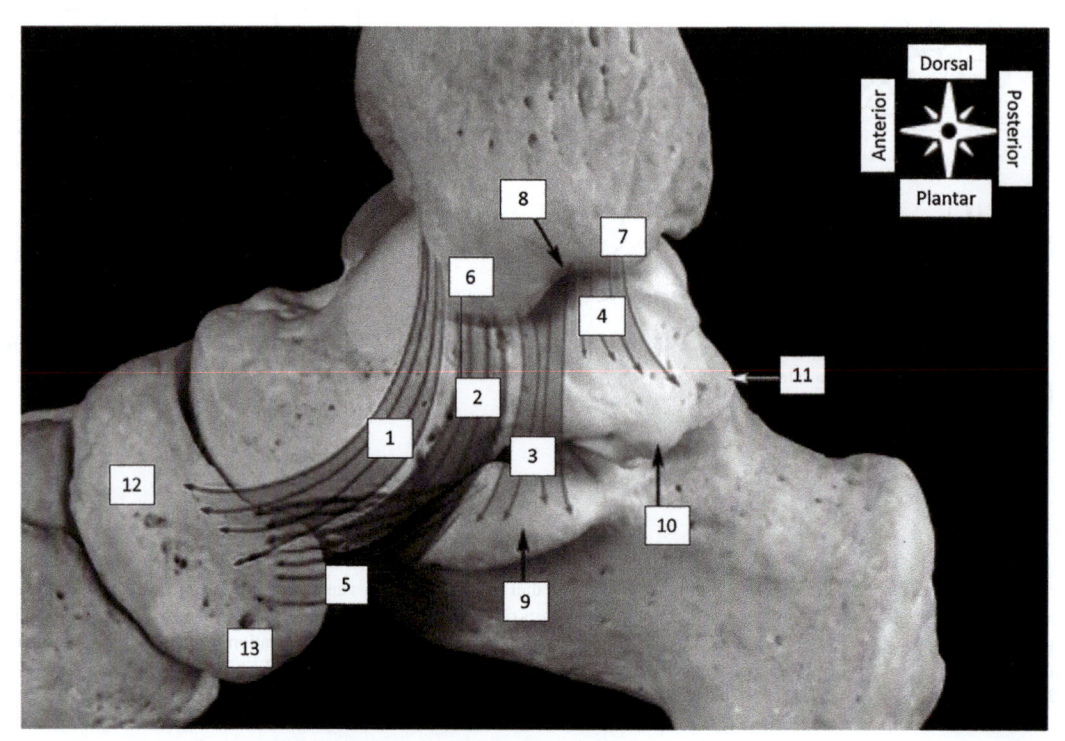

FIG. 2.5.5d The main components of the medial collateral or deltoid ligament consist of the tibionavicular ligament (1), tibiospring ligament (2), tibiocalcaneal ligament (3), deep posterior tibiotalar ligament (4), and the associations with the spring (calcaneonavicular) ligament complex (5). These are superimposed over the osseous anatomy, indicating features of the medial malleolus of the anterior colliculus (6), posterior colliculus (7), and the intercollicular groove (8). The sustentaculum tali (9) of the calcaneus forms part of the important distal attachment for the deltoid complex as does the medial talar process (10) and navicular (12), including its tuberosity (13). The lateral posterior talar process (11) forms the lateral edge of the groove that flexor hallucis longus lies within. The anterior medial collateral ligaments come under greater strain in plantarflexion. Posterior and spring ligament-associated parts of this medial deltoid ligament increase strain under ankle and midfoot dorsiflexion. However, these ligaments are most important for restricting posterior translation of the talus and calcaneus. High weightbearing ankle–rearfoot complex eversion combined with anterior talar and tibial translation moments commonly injures the medial collateral ligaments, subtalar ligaments, and the tibialis posterior tendon. *(Images adapted from Golanó, P., Vega, J., de Leeuw, P.A.J., Malagelada, F., Manzanares, M.C., Götzens, V., et al., 2010. Anatomy of the ankle ligaments: a pictorial essay. Knee Surg. Sports Traumatol. Arthrosc. 18(5), 557–569.)*

knee, with the point of rotation over the middle facet (or proximal part of the anterior facet when these facets are continuous) that lies above sustentaculum tali (Bonnel et al., 2010; Fernández et al., 2020). Ligaments of the subtalar joint lie (variably) medial, posterior, and lateral to the joint, including the intrinsic cervical and interosseous talocalcaneal ligaments, as well as the extrinsic ankle collateral ligaments such as the calcaneofibular and tibiocalcaneal part of the deltoid ligament (Figs 2.5.5e and 2.5.5f). The subtalar joint's centre of rotation should lie in the joint's middle facet, also providing some translation on its helical axis that reflects the elastic nature of these subtalar ligament restraints (Fernández et al., 2020). Loss of these ligament restraints risks excessive mobility, whereas overly stiff ligaments can lead to loss of mobility and energy dissipation across the subtalar joint, risking degenerative changes within the ankle–subtalar complex.

The role of the ankle–rearfoot ligaments

With reduction in form closure from the ankle dorsiflexed position during plantarflexion, potential instability develops through increased freedom of motion. During external forced supination motion of the rearfoot (plantarflexion, inversion, and adduction), stabilisation first arises within the subtalar joint through the cervical ligament, then the lateral fibres of the subtalar interosseous ligament, and finally the anterior talofibular ligament of the ankle (Bonnel et al., 2010). This ligamentous restraining mechanism causes a double joint stabilisation that explains the clinical association of impairment of the ankle and subtalar joint influencing each other (Hertel and Corbett, 2019) (Fig. 2.5.5g). However, when the rearfoot is unloaded, the supination burden is placed on the ankle-bridging ligaments alone, as subtalar joint flexibility and motion is lost.

The ankle's lateral collateral ligaments are the primary restraint of anterior talar translation (posterior tibial translation) and provide secondary talar posterior translation restraint. The lateral ligaments seem to provide 75% of

Locomotive functional units **Chapter | 2** 393

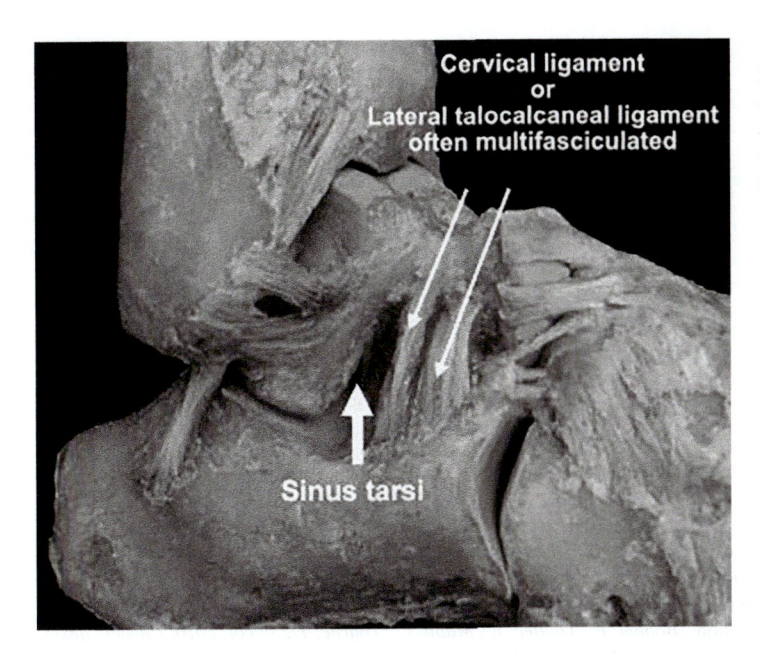

FIG. 2.5.5e The cervical ligament of the subtalar joint is shown in dissection. This ligament acts an anterolateral subtalar lock. *(Image from Bonnel, F., Toullec, E., Mabit, C., Tourné, Y., Sofcot, 2010. Chronic ankle instability: biomechanics and pathomechanics of ligaments injury and associated lesions. Orthop. Traumatol. Surg. Res. 96(4), 424–432.)*

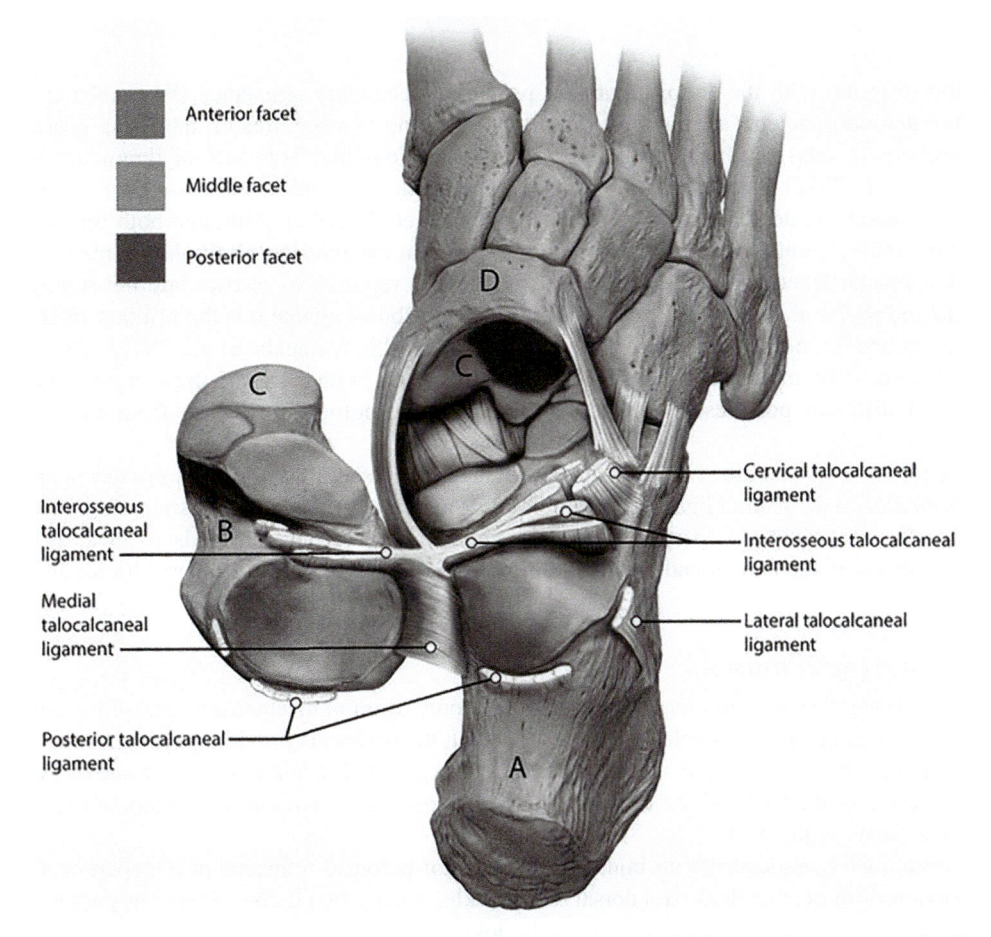

FIG. 2.5.5f An anatomical diagram of the subtalar joint and its ligaments. The ligaments of this joint are an important part of ankle–rearfoot complex stability and mobility. The calcaneus (A) has variable articular surfaces, here shown as three articular surfaces providing a middle facet. The talus (B) has been reflected medially so that the talonavicular facet (C) can be seen. The important cervical and interosseous talocalcaneal ligaments have been transected but can clearly give insight into their orientations that are used for restraining anterior and medial talar displacement on the calcaneus. At heel strike, the GRF subjects the calcaneus to a posteriorly and laterally directed force vector, while the talus is subjected to anterior and medial shear momentum over the calcaneus from the BWV via the ankle and leg. Such strong ligaments that prevent freedom of motion within the subtalar joint are essential in restraining excess anteromedial talar shear under high initial heel contact forces (the heel strike transient on the force–time curve). Their loss is a serious risk to developing ankle–rearfoot and midfoot pathology. *(Image from Fernández, M.P., Hoxha, D., Chan, O., Mordecai, S., Blunn, G.W., Tozzi, G., et al., 2020. Centre of rotation of the human subtalar joint using weight-bearing clinical computed tomography. Scientific Rep. 10, 1035.)*

1. anterior talofibular lig.
2. calcaneofibular lig.
3. cervical lig.
4. interosseous talocalcaneal lig.

Sprain of the calcaneofibular lig.
= integrity of the talocalcaneal unit

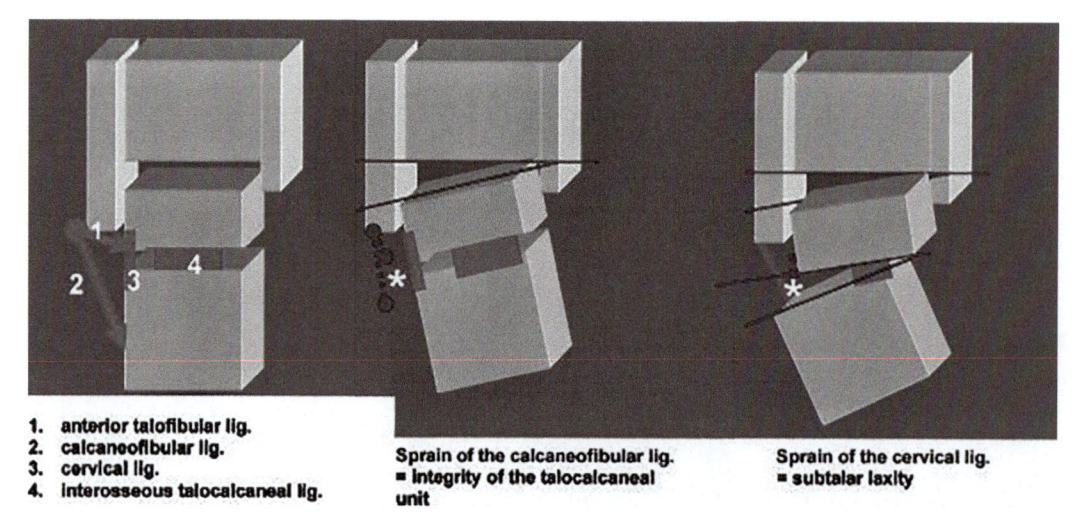

Sprain of the cervical lig.
= subtalar laxity

FIG. 2.5.5g Through the intimate relationship between the ankle and rearfoot, damage to the ligaments within each complex will influence stresses on the others, such that an ankle sprain can lead to midfoot degenerative joint changes through damage to the ankle collateral ligament complex. However, as demonstrated through this simple schematic, calcaneofibular ligament sprains are less likely to cause subtalar instability, whereas a cervical ligament sprain will immediately lead to greater subtalar joint freedom of motion. *(Bonnel, F., Toullec, E., Mabit, C., Tourné, Y., Sofcot, 2010. Chronic ankle instability: biomechanics and pathomechanics of ligaments injury and associated lesions. Orthop. Traumatol. Surg. Res. 96(4), 424–432.)*

anterior translation restraint, with the deltoid ligament providing secondary assistance (Watanabe et al., 2012). The medial collateral or deltoid ligaments are the most important ligamentous structures in stabilising posterior talar translation, which correlates to anterior tibial translation on the talus. Together, both sets of ligaments provide rotatory stability (Watanabe et al., 2012). Rupture of the lateral ligaments in vivo and in vitro has been reported to increase anterior laxity of the ankle and decrease ankle stiffness (Watanabe et al., 2012). Although both the lateral and deltoid ligaments restrict internal–external rotations, it has been reported that the anterior talofibular ligament restricts internal talar rotation in the plantarflexed position. The deltoid ligament is reported to restrict internal rotation in the ankle neutral and dorsiflexed position. For external rotation, the calcaneofibular ligament is the primary restraint in all ankle positions (Rasmussen and Kromann-Andersen, 1983; Rasmussen, 1985; Watanabe et al., 2012). This suggests a synergistic function between the medial and lateral collateral ligaments with parts of these ligaments coming under increasing strains in different postures, undoubtedly aiding sensorimotor information from the ankle's ligament complex.

Differences are reported for peak loads among the ankle ligaments, with the anterior talofibular ligament possessing similar loading properties to the medial ligaments (Butler and Walsh, 2004). The viscoelastic nature of the ligaments is revealed by their stress–relaxation properties that are demonstrated even under loads as low as 5 N, with less stress–relaxation occurring in the medial and anterior talofibular ligaments than in the others (Butler and Walsh, 2004).

Ligaments in concert with muscles

The components of the osteoligamentous morphology are not the only restraint mechanisms controlling ankle motion. The muscles and their fascial components also play a pivotal role, with the peroneals providing active lateral stability and the tibialis posterior primarily providing medial stability (Bonnel et al., 2010). The ankle is also the site of union of a continuation between the fascia of the foot and the fascia of the leg (fascia crura), with several important fascial thickenings known as retinaculae found in the area.

Lateral ankle instability is associated with injury to the superior peroneal retinaculum (Geppert et al., 1993). Three (sometimes four) extensor muscle tendons pass dorsal to the ankle, usually two peroneal tendons pass to the lateral side, and three deep flexor tendons pass medial and posterior to the ankle. However, the most significant tendon to cross the ankle and subtalar joint is the Achilles tendon which is formed by four distinct subtendons. The close coupling relationship of the ankle and subtalar joint explains much of the biomechanics of the Achilles, as each subtendon passing the ankle has its own distinct relationship between ankle and foot motion through tensile–compressive forces that both induce and restrict joint motion.

2.5.6 Functional axes of the ankle

The ankle joint axis location is fundamental to the sagittal plane plantarflexion (flexion) moment arm of the Achilles tendon. It determines the mechanical advantage of triceps surae relative to the GRF during the latter part of stance phase. This internal moment arm length in turn affects the muscle force generation by setting the amount of muscle–tendon shortening required per unit of ankle joint rotation excursion. Where the centre of rotation is at any particular moment during gait will affect the moment arm. Using the static anatomy as a guide is not a good estimation of where the functional axis lies, and usually results in shortened estimations of the moment arm (Wade et al., 2019). Using functional moment arms estimated from ankle kinematics that move with angles of plantarflexion creates larger moment arms than when fixing the axis to a position of the static anatomy such as the malleolar positions (Wade et al., 2019). This is because the ankle's axis is instantaneous and moves.

The centre of rotation's effect on the muscular lever arms is a critical determinant to the mechanics and efficiency of ankle motion during gait. Ankle movement dictates the position of the axis, which moves anteriorly with dorsiflexion and posteriorly with plantarflexion. Motion of the axis directly explains an individual patient's kinematics and kinetics around the ankle. There is reported to be an increase in the functional axis moment arm in loaded ankles compared to unloaded ankles, probably as a result of muscle bulging with contraction that is altering the moment arm which likely increases with increased muscular contraction (Wade et al., 2019).

During ankle plantarflexion, the talus adducts (medially rotates) in the transverse plane, while during dorsiflexion, the talus abducts with most of the transverse plane talar motion produced during movement from ankle neutral into dorsiflexion (Bonnel et al., 2010). Ankle neutral position is more easily defined than the neutral position assigned to the subtalar joint (i.e. neither pronated nor supinated). For the ankle, neutral is a position where the ankle–rearfoot is neither dorsiflexed (extended) nor plantarflexed (flexed) to the tibia or lower leg at the ankle when at 90° to the plantar rearfoot. When the tibia is fixed and the ankle and foot are free to move (in open chain), the talus makes a pronation (dorsiflexion, eversion, abduction) movement during ankle dorsiflexion and a supination (plantarflexion, inversion, and adduction) movement during plantarflexion around an anteroposterior axis and a transverse axis (Bonnel et al., 2010). When the talus is fixed across the subtalar joint (subtalar closed chain), the tibia rotates along an anteroposterior axis which during plantarflexion makes the tibia externally rotate and in dorsiflexion, internally rotate. However, this motion is subject-specific to the radii of curvature shape of the trochlear surface of the talus, changing the location of the rotary stresses around the ankle depending on the talus curvature type. For example, type I talar trochlear surfaces (see Fig. 2.5.4c) permit more internal and inversion rotations (Bonnel et al., 2010). The accompanying tibiofibular syndesmosis' subtle superior–inferior glide and the external–internal rotation that occurs with ankle motion can act as a shock absorber of ankle joint stresses, protecting the malleoli from the risk of torsional stress and associated fatigue fracture (Bonnel et al., 2010).

Ankle joints demonstrate a large variation in motion, both actively and passively (Weiss et al., 1986a, b). Based on the determination of the curvature centres of the medial and lateral edges of the talar trochlear surface, several axes have been described (Bonnel et al., 2010). The axes of the ankle joint do not align with the z-axis, lying as they do in the transverse plane perpendicular to the y-axis alone to provide pure sagittal plane motion (motion along the x-axis). The ankle joint axis is variably angled so that the joint provides motion in all three orthogonal planes. When unloaded, the ankle has been reported as having a greater freedom of motion and does not behave like a simple hinge. However, under loading from GRF and muscular activity in stance, the ankle seems to become more hinge-like and sagittal plane-dominant (Scott and Winter, 1991). The ankle has long been known to have a different functional axis location in dorsiflexion compared to that of plantarflexion (Wright et al., 1964; Leardini and O'Connor, 2002; Bonnel et al., 2010) (Fig. 2.5.6a). The plantarflexion axis is orientated obliquely from bottom to top and from medial to lateral, while the axis for dorsiflexion is orientated from top to bottom and from medial to lateral (Bonnel et al., 2010).

This axis arrangement permits two types of movement: a pure rotational motion and a rolling rotational-sliding motion which can be associated with each other. The rolling motion creates an instantaneous centre of rotation through the central zone on the talar body and dispersal from this zone might reflect joint dysfunction (Bonnel et al., 2010). As the ankle dorsiflexes, its axis of rotation moves anteriorly while in plantarflexion, it moves posteriorly (Leardini and O'Connor, 2002). Changes in the position of the axis alter the lever arm lengths of the muscles found around the ankle. As the ankle joint axis moves, so do the tendons, altering the ability of muscles to produce joint torques to generate or resist motion. This relationship of ankle axis movement means that the Achilles tendon has its greatest mechanical advantage through a longer effort lever arm to the ankle axis (fulcrum) when the ankle is most dorsiflexed, as it is prior to heel lift. However, it requires greater muscle–tendon fibre shortening recoil for the degree of plantarflexion achieved. This situation is assisted during later terminal stance by the midfoot plantarflexing, which shortened the distance between the Achilles to the MTP joint fulcrum (Achilles external moment arm), reducing the ankles arc of rotation and the amount of fibre shortening necessary

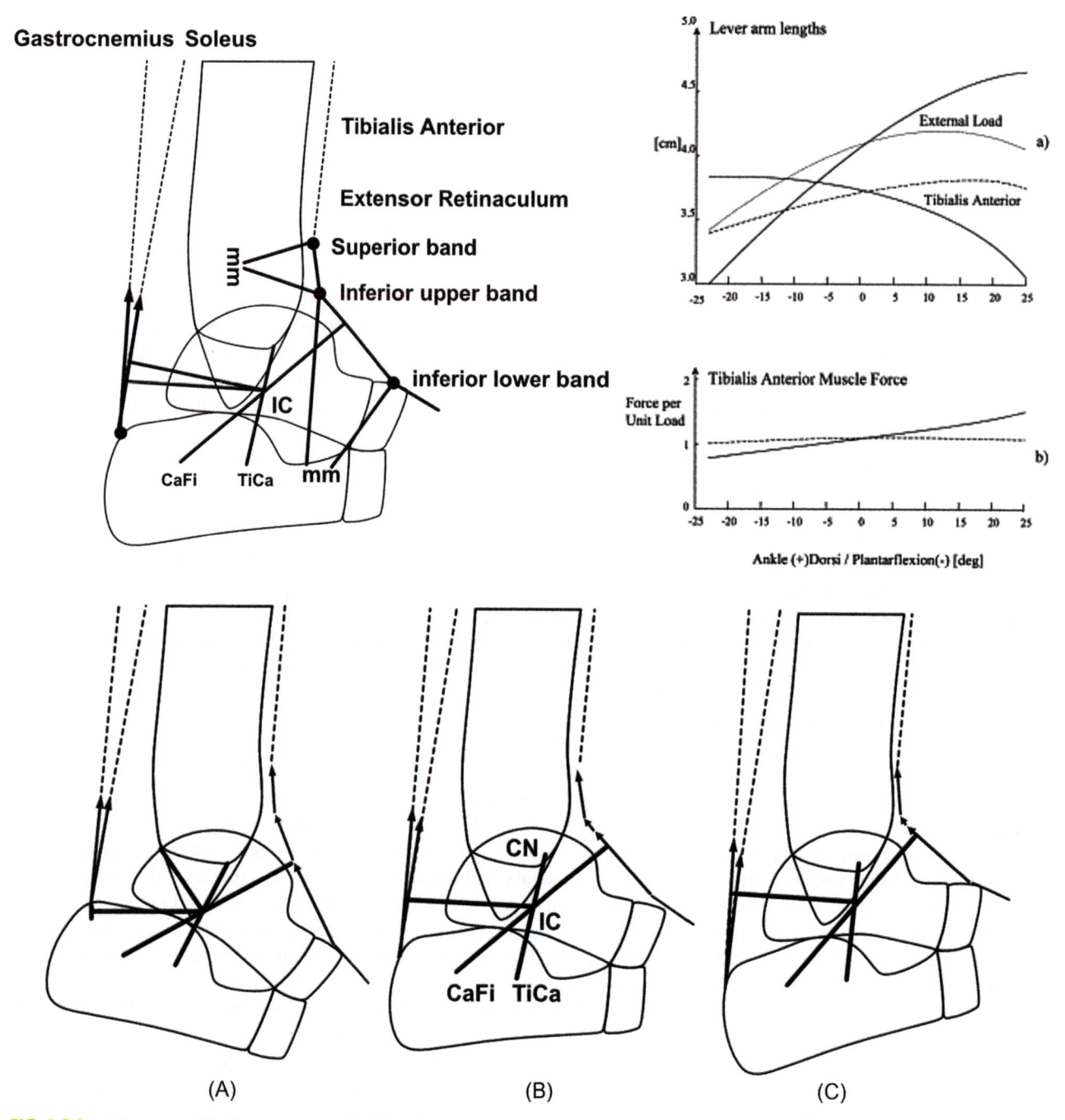

396 Clinical biomechanics in human locomotion

FIG. 2.5.6a Movement of the instantaneous ankle joint axis changes the position of the tendons to the axis, adjusting the lever arm lengths throughout gait. This is positioned so that extensors have mechanical advantage when plantarflexion needs resisting and flexors advantaged when dorsiflexion needs resisting. Ligaments, such as the calcaneofibular (CaFi) and the tibiocalcaneal ligament (TiCa) maintain tensional stability in positioning the instantaneous centre of rotation (IC) and influencing the primary joint contact location (CN) during motion. The charts on the upper right demonstrate these effects on tibialis anterior. *(Images from Leardini, A., O'Connor, J.J., 2002. A model for lever-arm length calculation of the flexor and extensor muscles at the ankle. Gait Posture 15(3), 220–229.)*

for increased angles of plantarflexion. Tibialis anterior, on the other hand, has the most mechanical advantage when the ankle is plantarflexed after heel strike and at the end of terminal stance through a longer lever arm to the posteriorly positioned ankle axis. This is in accord with the efficiency of human ankle biomechanics and energetics for the activity of these muscles (Fig. 2.5.6b).

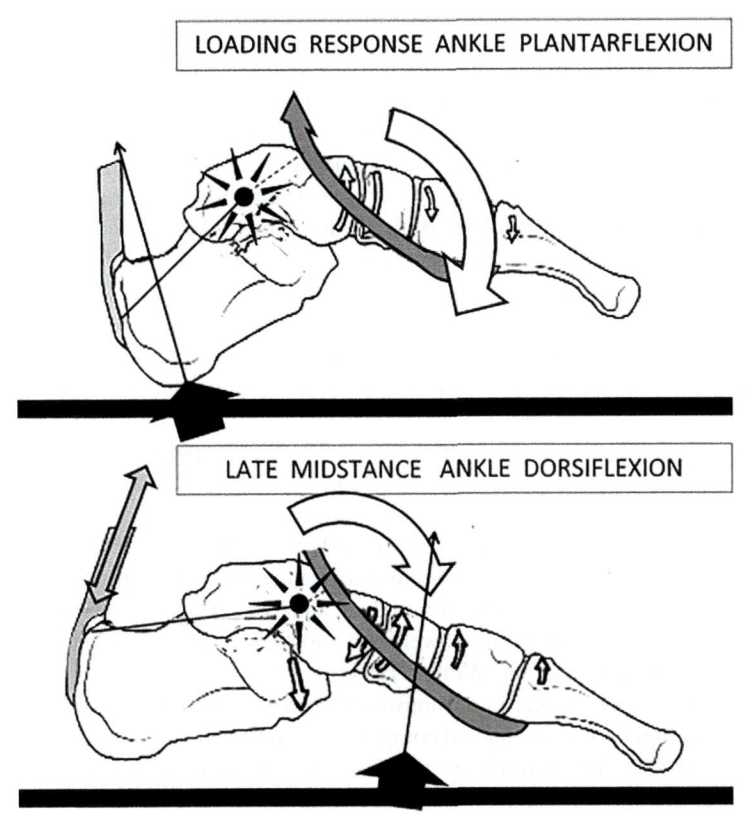

FIG. 2.5.6b A schematic to show the effects of instantaneous ankle joint axis motion during gait on the internal moment arms of tibialis anterior and triceps surae via the Achilles (tibia removed for ease of visualisation). During walking's loading response, ankle plantarflexion moves the instantaneous axis posteriorly. This increases the distance of the tibialis anterior tendon moment arm from the axis (dashed black line), improving its mechanical efficiency. At the same time, the Achilles moment arm (solid black line) is shortened. However, as triceps surae is not required at loading response during heel strike, this is not important. During late midstance as the tibia anteriorly rotates inducing dorsiflexion, the ankle's instantaneous axis moves anteriorly, lengthening and improving the internal moment arm for the Achilles thus making deceleration of anterior tibial rotation easier. Heel lift should thus occur with the longest internal moment arm for the Achilles. As the ankle plantarflexes in terminal stance, the Achilles moment arm gradually shortens as body mass is offloaded to the next step, while tibialis anterior's internal moment arm increases towards preswing, when it becomes active again. Such ankle joint axis motions are important for all ankle plantarflexors and dorsiflexors. (Permission www.healthystep.co.uk.)

The instantaneous length of the muscle lever arm is calculated as the ratio between the instantaneous values of tendon excursion from the joint's axis of rotation. This is effectively the distance a tendon moves from the underlying bone during contractions of the muscle (Leardini and O'Connor, 2002). Therefore, structures that restrain tendon motion and maintain their positions, such as the extensor, flexor, and peroneal retinacular sheaths found at the ankle, are extremely important in influencing the lever arm lengths around the ankle axis as the joint moves. These fascial specialisations, that act as frictionless pulleys tethered to the tibia, fibula, talus, and calcaneus, enclose the tendons to alter the line of force with each change of joint angle while the joint axis also repositions in relation to the muscular torque (see effects on tibialis anterior in Fig. 2.5.6a).

The overall effect of ankle joint axis movement working with soft tissue anatomy mechanics has been calculated to increase the lever arm of tibialis anterior by around 23% in plantarflexion, whereas a fixed axis would decrease the lever arm in plantarflexion by around 11% (Leardini and O'Connor, 2002). The gastrocnemius has a maximum lever arm length at 7° of dorsiflexion whereas a fixed axis would leave a 40% decrease in lever arm length at this angle. Soleus has a maximum lever arm length at ankle joint neutral (0° flexion–extension). However, with a fixed axis, this would be at 7° plantarflexion (Leardini and O'Connor, 2002). By having a moving ankle axis, the muscle force necessary to resist an external load by tibialis anterior in maximum dorsiflexion is 41% larger than with a fixed axis, but 22% less in maximum plantarflexion. For soleus and gastrocnemius, forces to resist external load in maximum dorsiflexion are calculated to be 20% and 21% larger, respectively, in a fixed axis and 4% and 5% less (respectively) with the moving ankle joint axis (Leardini and O'Connor, 2002). Therefore, the movement of the functional instantaneous axis of rotation is pivotal for muscular efficiency of the large torques around the ankle. In clinical situations that block this free ankle motion, gait biomechanics will change. It is therefore of critical mechanical importance that patients have stable but freely moving ankle joints, with significant ranges of motion both in dorsiflexion and plantarflexion.

The ankle joint's triplanar axis is also orientated to the vertical, rising higher as it runs to the medial side. Yet, the GRF vector acting on the foot during stance tends to be initially applied laterally on the heel and then drives the foot medially throughout the stance phase of gait (Scott and Winter, 1991). The raised orientation of the medial side of the joint axis creates a resistance to the force component that acts along the ankle joint axis lateral-to-medial, a force that does not contribute to the ankle joint's angular momentum (Scott and Winter, 1991). It is possible that higher vertical axis orientation is helpful in resisting ankle joint eversion moments, which may act as a subject variable constraint of rearfoot eversion. If so, some patients will have better resistance to foot eversion moments through their ankle morphology than others.

In concert with the subtalar joint

The intimate relationship between the ankle and subtalar joints (the talocalcaneal part in particular) in achieving the functional positioning of the foot to the leg means that these two joints are best considered together as primary providers of all orthogonal plane motions between the leg and rearfoot. It is important to remember that individuals demonstrate significant differences in ranges of motion at these joints in each plane (Scott and Winter, 1991), a finding that makes the establishments of a clinical 'average' normal axis position pointless. However, being aware of any significantly unusual direction of motion within the ankle–subtalar complex is helpful to establish for any patient. Both the ankle and subtalar joints demonstrate a circular arc of movement for their joint axis positions during motion, providing frontal, transverse, and sagittal plane motions, but they also tend to show some out-of-phase motion between the two joints (Scott and Winter, 1991). This often means that when the ankle undergoes loaded negative rotations of plantarflexion, adduction, and inversion, the subtalar joint tends to demonstrate positive rotations of dorsiflexion, abduction, and eversion and vice versa. However, this is not the case in all phases of gait, as when the subtalar joint is unloaded by external force, the calcaneus just follows talar motion.

At initial heel contact, the ankle plantarflexes as the forefoot lowers to the ground. This is followed by positive rotations of dorsiflexion between 10% and 75% of stance which is then followed by rapid plantarflexion during the final 22%–28% of stance during walking acceleration. The talocalcaneal joint moment of the subtalar joint is initially positive (closed chain pronation motions), reaching a peak at around 75% of stance (Scott and Winter, 1991). At forefoot loading, the subtalar joint plantarflexes (distal calcaneus lowers), everts, and abducts. These are the motions of closed chain subtalar pronation. During midstance, as the ankle dorsiflexes, the subtalar joint continues to plantarflex, abduct, and evert. These out-of-phase rotations, that lower the proximal vault, help maintain the leg's posture and the lower limb's joint motions within the sagittal plane along the line of progression throughout stance phase (Scott and Winter, 1991).

An important clinical consideration in examining the ankle and subtalar joints nonweightbearing is that they have more freedom of motion and more variability in their axes of rotation under low manual compressive loads than those associated with gait (Scott and Winter, 1991). This makes assessment of the location of their joint axes and ranges of motion in nonweightbearing relatively meaningless when compared to the motion that actually occurs during gait.

2.5.7 The ankle in lever systems

To help appreciate the action of the muscles around the axes of the ankle, it is helpful to consider the simple engineering principles of the lever arms acting around the joint during gait. The ankle functions as part of a class three lever system from initial loading and throughout midstance as the lower limb pivots around the foot. This is regardless of any initial contact position. However, the foot briefly works as part of a class two lever system when achieving heel lift as it applies the ankle plantarflexion power to the forefoot. The ankle joint itself acts as the fulcrum when functioning as a class three lever system. However, it is not the primary fulcrum for the class two heel lift lever of terminal stance, when the primary rotation occurs at the MTP joints. This fulcrum should ideally form between the 1st and 2nd MTP joints. However, significant ankle plantarflexion motion should still occur during terminal stance to permit elevation and anterior rotation of the now trailing lower limb to initiate and complete double stance phase; motion which is coupled to the MTP joint extension (dorsiflexion) angles. Weight is transferred from the terminal stance limb to the contralateral contact limb as the support of the body efficiently moves from one leg to the next via this mechanism. The leg effectively rotates forward over the foot as the ankle plantarflexes to permit leg rotation and knee flexion.

Without ankle plantarflexion after heel lift, the knee and hip would need to maintain their extension motion during terminal stance, until preswing when rapid knee and hip flexion would need to occur. Loss of ankle plantarflexion in terminal stance takes away gastrocnemius' knee flexion power at this time. During terminal stance phase, the ankle has to transfer its plantarflexion power generated by all the muscles posterior to it into a stabilising forefoot GRF. This requires ankle plantarflexion on a stiffened forefoot so as to achieve anterior progression of the lower limb's CoM, following that of the HAT segment which is falling onto the opposite limb. Midfoot plantarflexion during terminal stance reduces the amount of ankle plantarflexion required during this task and also adds a mechanism of adaptability.

Terminal stance class two lever

The ankle plays a secondary role to the fulcrum at the MTP joints for rotating the trailing limb anteriorly, but it is the joint around which plantarflexion power is applied to the forefoot that drives the lever. The Achilles tendon releases powerful effort energy to the posterior calcaneus that lifts the rearfoot on a midfoot that is acting as a semirigid lever. This tips the foot up and around the MTP joints. This heel lift action shifts the CoM of the trailing lower limb anteriorly and upwards, rotating it forward. This assists the HAT CoM's anterior acceleration on to the next step, adding extra momentum to the opposite initial contact collision (Ruina et al., 2005). Throughout this action, the ankle is thus acting as a hinge joint that allows the leg angle to be adjusted to the foot's MTP joint angle so as to smooth the CoM's passage anteriorly. As the resistance is the mass that lies over the foot from the trailing leg, the resistance has a shorter distance to the MTP joint fulcrums than the effort lying behind the ankle. The effort arm is thus longer than the resistance arm, so a class two lever system forms. Ideally, as terminal stance proceeds, the resistance reduces due to more mass being transferred to the support of the opposite foot, which should now be in double-limb stance phase as the load-supporting foot. Therefore, as a single-limb support event, peak effort is required to initiate heel lift, but decreasing effort is required to maintain the heel-lifting action as soon as the contralateral foot makes ground contact. The amount of effort required is dependent on the load on the rearfoot when heel lift occurs, and that is dependent on how far anteriorly advanced the HAT's CoM has become to the rearfoot in late midstance. This is seen clearly in normal electromyography studies of the triceps surae which demonstrate high activity at the end of late midstance to initiate heel lift, but this activity rapidly switches off after heel lift (Honeine et al., 2013). Achilles elastic recoil energy maintains ankle plantarflexion under reducing power throughout terminal stance (Fig. 2.5.7a).

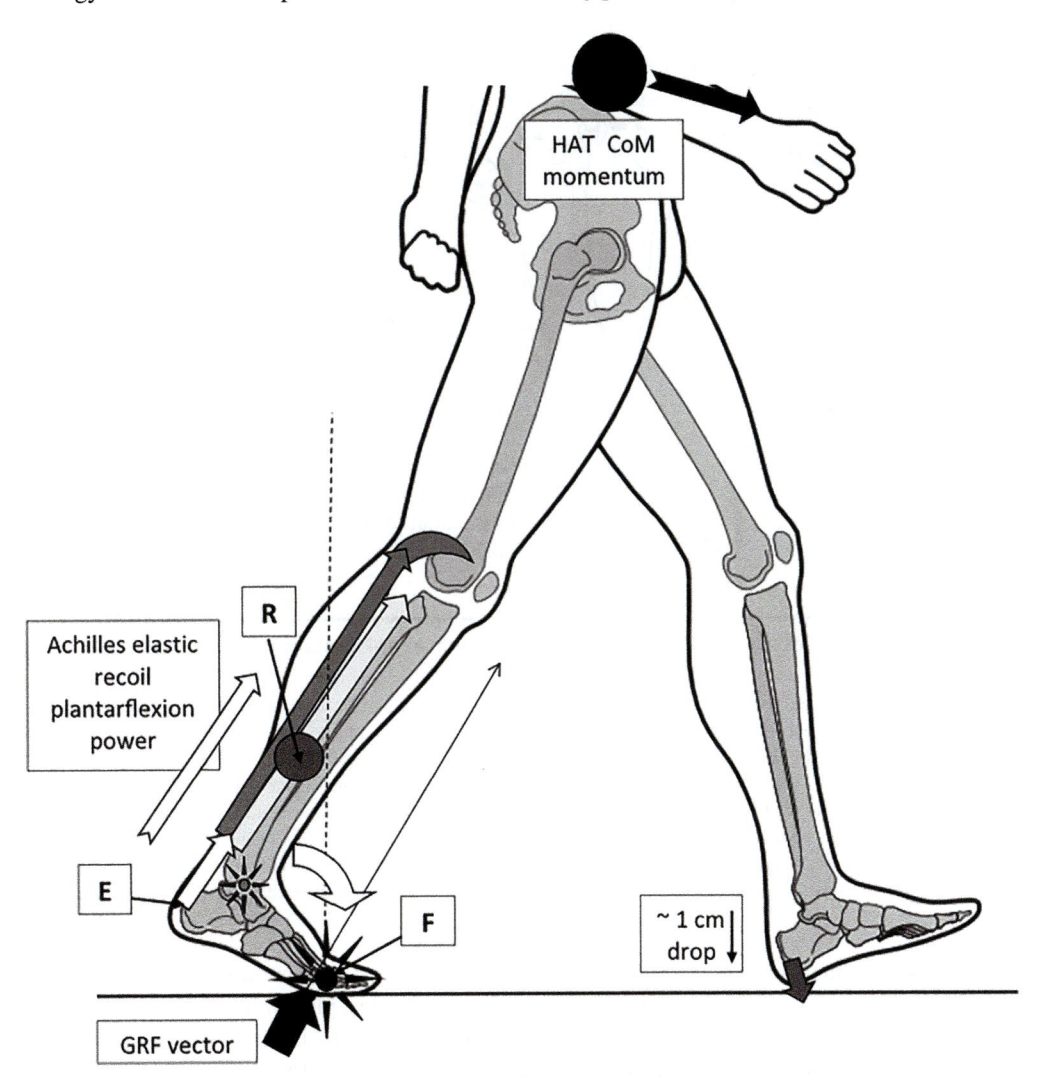

FIG. 2.5.7a Although the power that brings about heel lift derives primarily from the triceps surae muscles, it is the release of energy stored within the Achilles from the midstance eccentric muscle contraction that supplies the effort required to spring the heel off the ground. The heel lift effort raises all of the lower limb mass lying behind the metatarsophalangeal (MTP) joints acting as the fulcrum (F and vertical dashed line). This lower limb mass, consisting of the rearfoot and lower leg, provides the resistance to be moved (R—grey circle). However, plantarflexion-powered heel lift achieves so much more than lifting R. By forcing the foot to plantarflex at the ankle against the ground, the plantarflexion power generates a large GRF angled anteriorly that accelerates the body's CoM forward and downward, assisting gravity. This drops the swing limb around 1 cm to accelerate the rate of contact with the support surface, ending swing phase. Although this class two lever system does not primarily move the body's CoM anteriorly (gravity and centrifugal forces from the swing limb do that), the effort it applies to gait during the terminal stance phase helps 'power' weight transfer onto the opposite limb, adding momentum to the HAT's CoM that can persist into contralateral early stance phase. *(Permission www.healthystep.co.uk.)*

In reality, heel lift and the transfer of the trailing limb's CoM to the next step is achieved by a number of assistor muscles and their tendons' elastic recoil. These muscles are providing only slight assistance to the triceps surae and the Achilles tendon power while also generating midfoot plantarflexion after heel lift and increasing the foot vault contour. Such muscles include peroneus longus and tibialis posterior ((Kokubo et al., 2012; Holowka et al., 2017). As the ankle power is largely used up by late terminal stance, the power of the extrinsic and intrinsic midfoot and toe plantarflexors can continue to generate the forefoot GRF, sending the 'ever-decreasing limb mass' anteriorly to the opposite support limb over a stable forefoot. Like the triceps surae, the extrinsic midfoot plantarflexion power of tibialis posterior and peroneus longus derives from the elastic recoil of their tendons, as their peak muscle activity also ends around the late midstance–terminal stance boundary (Fig. 2.5.7b).

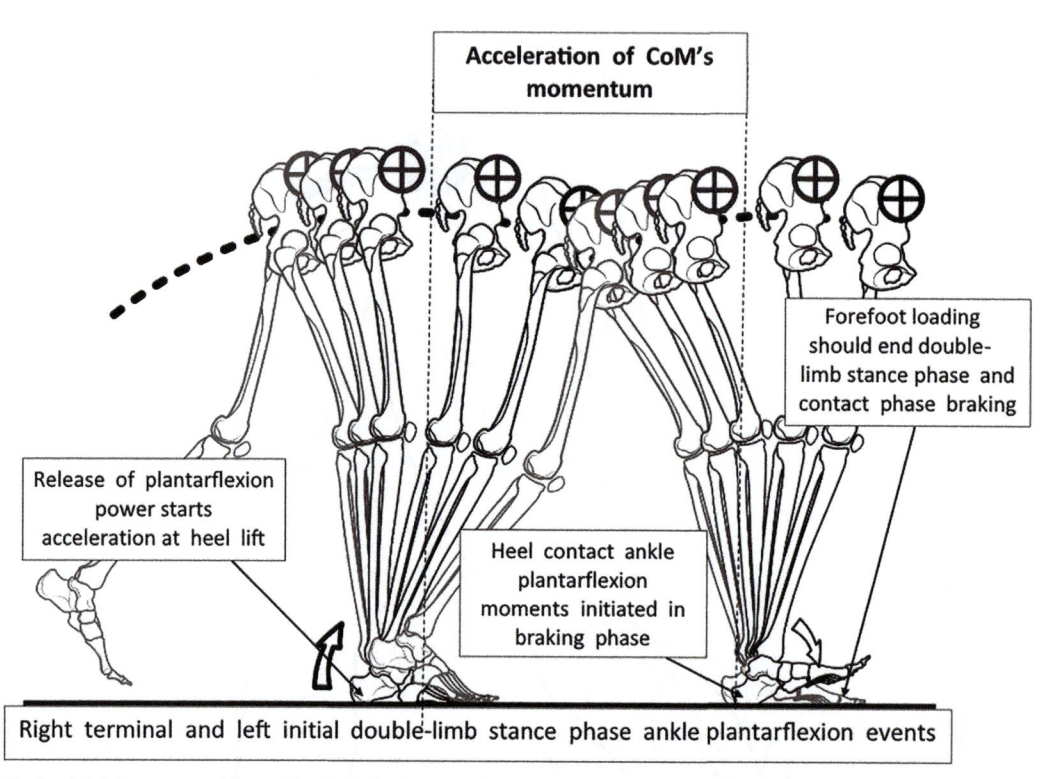

FIG. 2.5.7b Both ankle joints are providing ankle plantarflexion in double-limb stance phase of walking, shown here as a schematic of right heel lift and left heel contact. The ankle undergoing heel lift should initiate the start of double-limb stance phase. Powerful ankle plantarflexion reduces the swing limb's heel height from the ground, initiating its heel contact. This heel contact sets up a GRF that initiates a strong plantarflexion moment on the ankle that the ankle dorsiflexors must decelerate. Ankle plantarflexion power of heel lift thus initiates its own stance limb's terminal double-stance phase, and the momentum it provides to the anterior fall of the CoM should accelerate the start of initial double-limb stance of the contralateral limb coming out of swing. Plantarflexion power from heel lift helps power the ankle plantarflexion moment that brings the forefoot of the initial contact foot to the ground (area between two dashed vertical lines). Thus, a foot's loading response stresses are in part a result of the opposite limb's heel-lifting power. Momentum from heel lift can be sufficient to assist the ascent of the CoM (after its low point of forefoot loading) into early midstance. *(Permission www.healthystep.co.uk.)*

For this power transfer to the forefoot to work effectively, the foot must become a stable semirigid beam from the rearfoot to the MTP joints and many of the assistor ankle plantarflexor muscles, plantar intrinsic muscles, and passive connective tissues are at play in achieving this. These include the plantar aponeurosis through the action of passive digital extension at the MTP joints, particularly as terminal stance progresses. Failure of foot stiffness risks a midfoot break that reduces the lever arm length of ankle plantarflexion power to a much shorter midfoot fulcrum. This, in turn, causes excessive midfoot energy dissipation that removes power from acceleration.

For efficiency of walking, it is also necessary for the contralateral foot and leg to be ready to become the support limb very rapidly after heel lift. Usually, only around a ~1 cm drop is required from the swing limb to initiate its heel contact during opposite heel lift. Therefore, the body's CoM must be well in advance of the contralateral stance foot when the previous step's support limb's terminal stance begins. The long posterior tubercle of the human heel and the dorsiflexion angle of the foot to the ground, help reduce the height of the drop required to make heel contact. If the swing limb is not advanced sufficiently, then the resistance load at heel lift is greatly increased and the effort required from the Achilles is much higher. When considering the foot and ankle action interplay during terminal stance as a class two lever, it can be seen that many events can result in problems. This will be discussed further under pathomechanics of the ankle and foot in Chapter 4 (Fig. 2.5.7c).

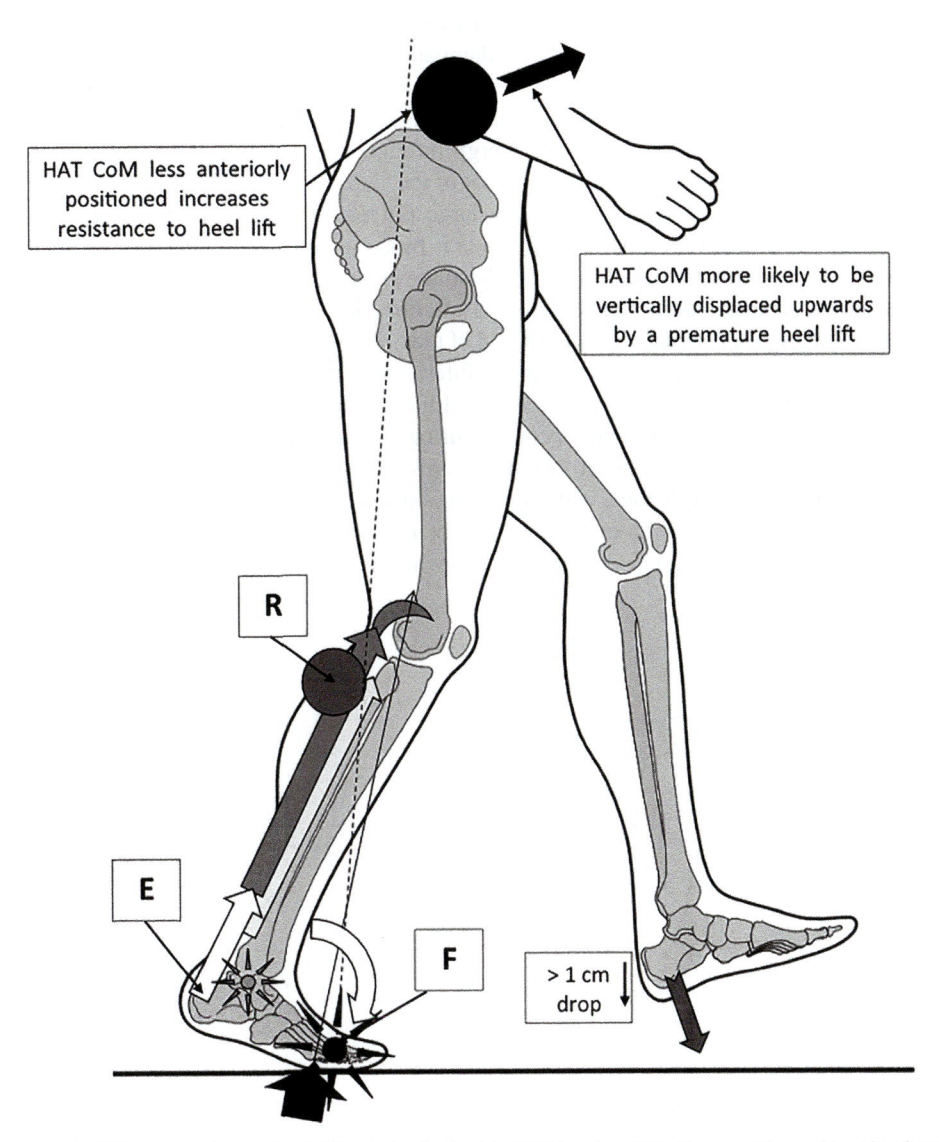

FIG. 2.5.7c A premature heel lift occurs when the heel offloads, but the body's CoM is not sufficiently moved forward to gain advantage of gravitational force and the swing limb has not advanced far enough ahead of the CoM to express its full centrifugal effects on the HAT's CoM. It also occurs before the swing limb is correctly placed for heel contact to provide a suitable stride length. Ankle plantarflexion that starts too early will be causing a premature heel lift, knee flexion, and early metatarsophalangeal extensions. As a result, effort (E) will be applied on a less efficient class two lever with potentially more lower limb mass and therefore greater resistance (R) lying behind the MTP joints' fulcrum (F—dashed line). With the HAT's CoM being positioned less anterior to the foot, there will be less acceleration under gravity acting upon the CoM at heel lift. As a result, greater effort to lift the heel off the ground and to accelerate the CoM anteriorly will be required. Early heel lifting may initially add an upward vertical moment against the gravitational fall of the HAT's CoM, greatly diminishing energetics by increasing vertical CoM oscillation. The GRF will have a smaller horizontal anterior driving component as a result. The earlier the heel lift, the worse the mechanical position the heel lift lever arm finds itself in, and assisting gluteus maximus acceleration power in late terminal stance will be compromised. Tight posterior calf myofascial tissues and/or osseous limitation of ankle dorsiflexion range in late midstance are the most common reasons for a premature heel lift. Stiff cavoid feet with limited vault drop in midstance can also cause similar results by blocking the functional ankle dorsiflexion range. *(Permission www.healthystep.co.uk.)*

The ankle as class three lever systems
Other than at heel lift, the ankle is usually functioning as a class three lever system, both in stance and swing.

Heel contact
During heel contact, the GRF on the posterior heel sets up a plantarflexion (flexion) moment at the ankle, accelerating the forefoot towards the ground. Unrestricted acceleration of the plantarflexion moment would create impact forces on the forefoot that could cause tissue damage through tissue resonance within the lower limb. Thus, a number of mechanisms are utilised within the foot and around the ankle to help 'dampen' the impact energy and decelerate the plantarflexion velocity.

The anterior muscles of the leg (shank) act as an eccentric brake, using the ankle as a fulcrum to decelerate the GRF-induced ankle plantarflexion moment and thus absorb energy through muscle–tendon buffering (Neptune et al., 2008; McGowan et al., 2010). The resistance decelerated is only that of the CoM of the foot beyond the rearfoot, as the heel will be in ground contact. Anterior shin muscles do not need to be large as a result and compared to the posterior calf muscles of the leg, they are not. As tibialis anterior by cross-sectional area is by far the most significant muscle in this braking action (Byrne et al., 2007), so the effort point can be considered to be at the plantar medial aspect of the medial cuneiform and the medial 1st metatarsal base at the distal attachment for tibialis anterior. However, plantarflexion resistance from the dorsiflexors is also applied via the extensor retinaculum. The CoM of the midfoot and forefoot is anterior to this effort point and both effort and resistance lie anterior to the ankle acting as the fulcrum point. As the resistance arm length is greater than the effort, this situation sets up a class three lever that is active until the forefoot becomes a supportive structure at the end of loading response. The more plantarflexed the foot becomes during contact, the more the ankle's functional joint axis is posteriorly positioned, increasing the extensors effort arm towards the end of the motion. Thus, the angle of the foot at heel contact influences the lever arm mechanics, with more initial dorsiflexion ankle angles being associated with longer strides, increasing the tibialis anterior duration of activity and potentially initially reducing its moment arm. This concept explains why the foot at the ankle should not be dorsiflexed before heel contact, and why it tends to remain slightly plantarflexed to the ankle (Fig. 2.5.7d).

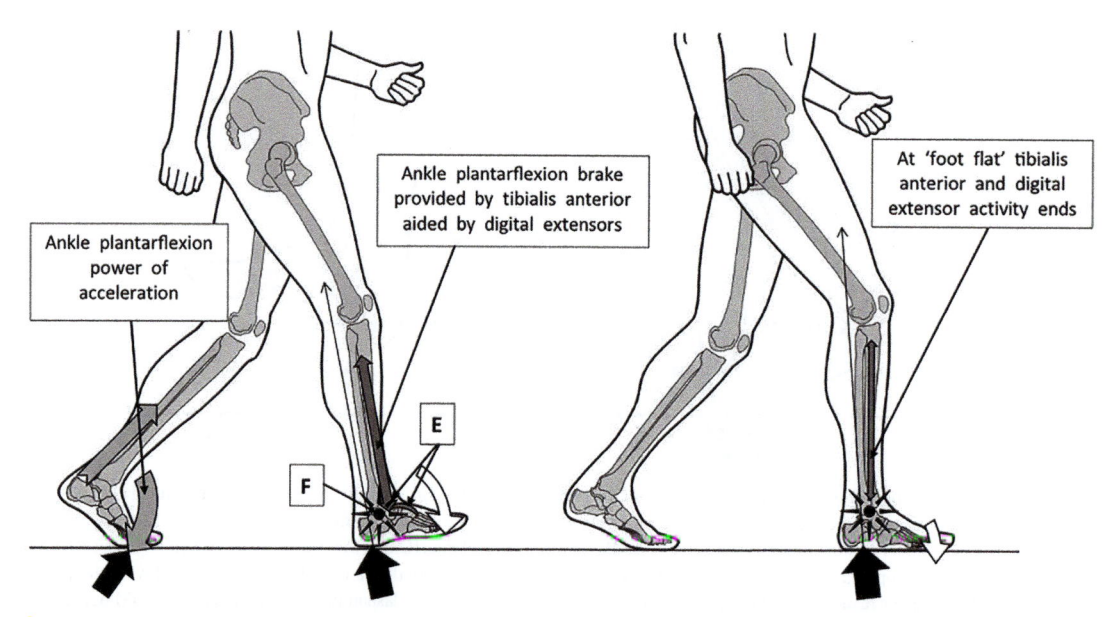

FIG. 2.5.7d The lever arm of decelerating ankle plantarflexion initiated by heel strike posterior to the ankle joint axis is a simple class three. The ankle dorsiflexors' effort (E), dominated by tibialis anterior, eccentrically resists but does not stop the GRF-induced ankle plantarflexion moment. This very efficiently brings the forefoot resistance (R) safely into ground contact. Using the GRF from a heel strike to cause an angular momentum to bring the forefoot to the ground splits impact forces, spreading them over two energy dissipation collisions, with the ankle dorsiflexors using their eccentric activity not only to decelerate forefoot impact but also to impart muscle–tendon energy buffering. Heel lift that plantarflexes the contralateral ankle in its terminal stance initiates and adds energy to heel strike and loading response. Ankle plantarflexion's initial contact role ceases at the end of loading response with the whole foot flat on the support surface and this is when ankle extensor activity largely stops. *(Permission www.healthystep.co.uk.)*

Activity of the tibialis anterior after forefoot contact is no longer required, and this is consistent with its reducing EMG activity. However, the situation is more complex because of the co-activity of the long digital extensor tendons through their dorsiflexion action on the toes via the MTP joints. Nonetheless, this effect is unlikely to be enough to negate the validity of the lever arm which is dominated by tibialis anterior. The extensor digitorum longus has been reported as either being much reduced or even absent on anatomical cadaveric studies (Newton, 2014), yet these abnormalities were found on cadavers that had no 'known' issues with gait during life. Extensor digitorum brevis alone may be sufficient for lesser digital extension during gait. The anterior muscle group seems to be highly variable in relation to the anatomy of the long digital extensors and peroneus tertius. The presence of tibialis anterior and extensor hallucis longus is consistent albeit with anatomical variation occurring within their tendon slips and through the nature of their arrangement of attachments (Al-saggaf, 2003; Anagnostakos et al., 2006; Tezer and Cicekcibasi, 2012). The consistent presence of these muscles highlights their significance compared to other anterior muscles. Therefore, a class three lever system for tibialis anterior as the primary effort point seem to be valid in understanding the normal mechanics and pathomechanics of foot loading around the ankle in heel strike position, but with some appreciation of an assistor role from the long digital extensors.

It can be argued that the fulcrum point is at the plantar heel during heel contact, which is rounded to help facilitate a rotation rather than plantarflexion within the ankle. Much of the 'rotation' around the heel is actually caused by soft tissue deformation rather than true pivoting, whereas the ankle truly rotates around an angular moment. From an engineering standpoint, whichever fulcrum is used (ankle or posterior heel) makes little difference as the distance of the effort to either fulcrum is shorter than the CoM of the foot's resistance arm to the fulcrum.

Forefoot contact

In forefoot contact, the lever system changes in orientation but not in class, which remains three. The forefoot contacts with the ankle more plantarflexed than it is during a heel strike. This generates a GRF anterior to the ankle that gives rise to a dorsiflexion (extension) moment on the foot at the ankle driving the heel towards the support surface. This must be resisted by the triceps surae through the Achilles tendon attachment to the posterior calcaneus. In this situation, the resistance is not the CoM of the foot which is stabilised by the support surface but the CoM of the body (HAT segment and lower limbs) that is now lying just behind and above the ankle joint fulcrum. This is obviously a considerably larger resistance than the forefoot, and therefore, deceleration of the dorsiflexion moment of the ankle needs a large muscle mass to resist it. Through the triceps surae, humans have this power. The change in lever orientation at forefoot contact greatly increases the peak loading stresses on the Achilles tendon and the duration of high stresses on the Achilles during such gaits compared with heel contact gait (Kernozek et al., 2018).

Midstance

During early midstance, the ankle joint reduces its plantarflexed orientation until absolute midstance, with the instantaneous joint axis moving anteriorly with this motion. This is largely a passive motion for the ankle, brought about by hip and knee extension moments extending the leg. Hip and knee flexion angles are thus primarily reduced through gluteus maximus and quadriceps activity, and some continuing momentum brought about by heel lift power from the opposite foot also likely assisting (Ruina et al., 2005). Prior to absolute midstance, the triceps surae starts to activate, but this is more in force-generating 'anticipation' of resisting the large dorsiflexion moment at the ankle occurring after absolute midstance. During this period, the functional ankle axis should be moving further anteriorly as the GRF moves closer and passes anterior to the ankle joint (Leardini and O'Connor, 2002; Bonnel et al., 2010) (Fig. 2.5.7e).

The ankle now enters another class three lever system as the GRF and the HAT's CoM both move anterior to the ankle, setting up a dorsiflexion (extension) moment at the ankle. To prevent the CoM of the body translating too quickly anteriorly and therefore not giving adequate time for the swing limb to be positioned correctly before heel lift, the triceps surae must control the acceleration of the body's CoM via the tibia and femur as it falls forward, vaulting over the foot. The ankle is the fulcrum, the resistance the CoM above the ankle, and the effort is the proximal attachments of the triceps surae on the posterior femoral condyles of the femur for gastrocnemius and the posterior proximal aspect of the tibia and fibula for soleus. The lever system is thus angled and orientated vertically and anteriorly. The CoM of the body is further away from the ankle than are the proximal attachment points of triceps surae providing the effort. Therefore, the resistance arm is longer compared to the effort arm, creating a class three lever system.

As the CoM moves further anteriorly, the GRF on the forefoot increases with the resistance's velocity rising, requiring an increasing level of resistive effort in the triceps surae throughout late midstance. As the ankle becomes more dorsiflexed,

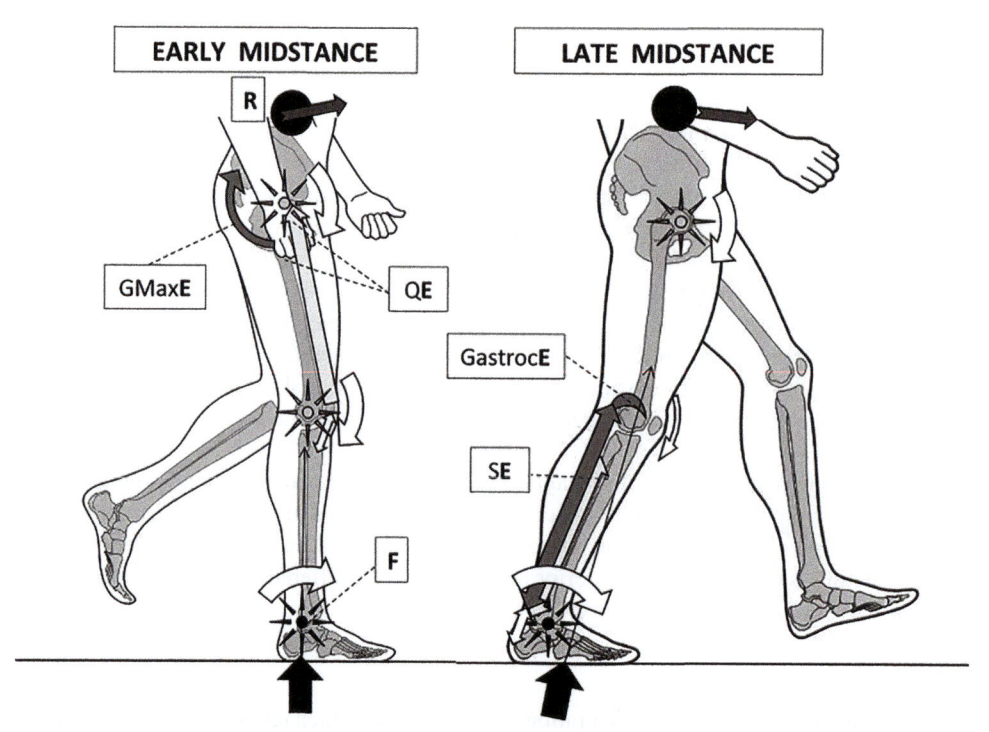

FIG. 2.5.7e Walking's early midstance (left) involves reducing ankle plantarflexion angles around its fulcrum (F—black star), but not through ankle dorsiflexor activity which would not be energetically efficient. Instead, reduced ankle plantarflexion is coupled to decreasing hip flexion angles via effort from gluteus maximus (GMaxE) activity and reducing knee flexion angles through quadriceps muscle effort (QE). Together with the hip abductors and adductors, these muscles help pull the CoM up to its high point over the ankle, reducing ankle plantarflexion angles towards dorsiflexion. It is thus the lever arms operating around the hip and knee fulcrums (hollow-centred stars) that control the ankle moments. However, around absolute midstance, the CoM of HAT slips anterior to the ankle joint axis giving the ankle plantarflexor muscles mechanical advantage to restrain and control the CoM's anterior progression during late midstance (right image). Triceps surae now eccentrically controls acceleration around and against increasing ankle joint dorsiflexion moments brought about by the interplay of the GRF and body weight vector as the body mass falls under gravity. This sets up a class three lever arm with the triceps surae proximal attachments becoming points of effort (GastrocE and SE) that slow the proximal resistance (R) of the HAT CoM. Power for heel lift is being stored within the elastic stretch of the Achilles by this action. Gastrocnemius' attachment to the posterior aspects of the femoral condyles can also resist the extension moment on the knee generated by the fall of R. *(Permission www.healthystep.co.uk.)*

the functional axis of the ankle moves anteriorly, increasing the Achilles tendon's internal moment arm. Rising myofascial tensions in this lever system coupled to the advancement of the resistance of the body's CoM and the swing limb momentum-induced centrifugal force, eventually result in heel lift through elastic recoil power as loads on the rearfoot decrease. This force overcomes the mass behind the MTP joints, ending this class three lever system at the starting of a class two lever system during the initiation of terminal stance.

Swing phase

During the swing phase, the ankle also works as a class three lever primarily through the action of tibialis anterior on the CoM of the whole foot. This means that the CoM of the foot is nearer to the ankle fulcrum than during contact phase, but the mass is larger as it includes the whole foot. However, there is no acceleration moment resulting from a GRF during swing. Therefore, the lever system at play in swing involves a reduction of the plantarflexed angulation at the ankle joint that is present at the end of terminal stance, so that a plantarflexed foot does not get in the way of the ground during swing. If the foot remains plantarflexed, it could be a problem, requiring greater hip and knee flexion to compensate. This is a problem in tibialis anterior dysfunction which causes a plantarflexed swing phase foot condition known as 'foot drop'. This can be caused by a neurological deficiency in nerve supply (common peroneal nerve) to the anterior muscles or by tibialis anterior tendon rupture. By reducing ankle plantarflexion, the normal foot is able to achieve a suitable angulation for heel ground contact at the end of swing. Therefore, how much of a reduction in plantarflexion and whether the reduced plantarflexion angle decreases further or increases before contact depends on the foot strike position required (Fig. 2.5.7f).

| PRESWING | EARLY SWING - LATE SWING | LATE SWING - TERMINAL SWING |

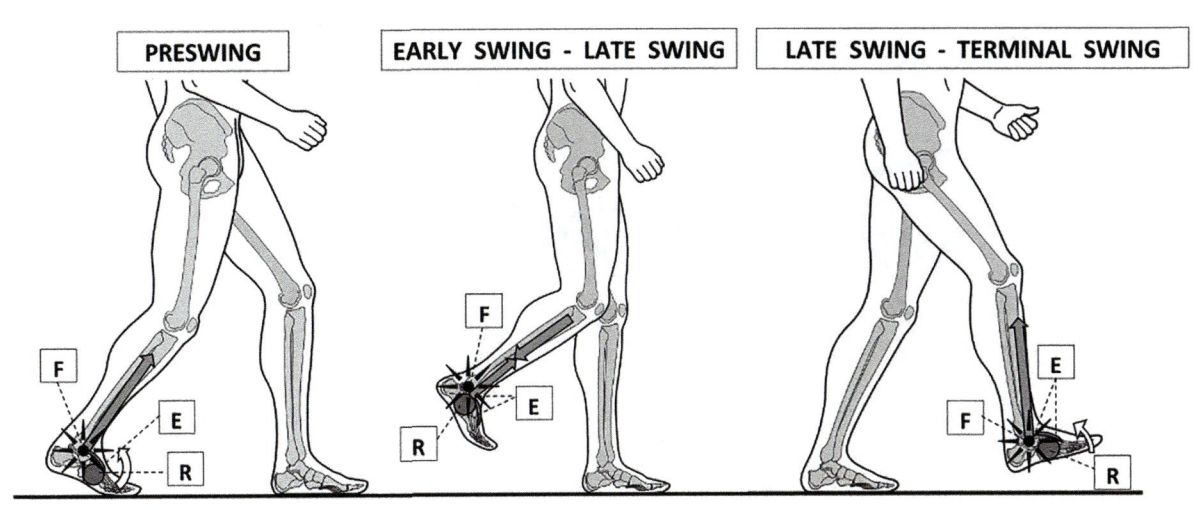

FIG. 2.5.7f Swing phase involves an open chain ankle dorsiflexion moment that lifts the foot's CoM as the resistance (R) by the ankle dorsiflexors around the ankle fulcrum (F). By far the greater part of this force derives from tibialis anterior activity. Thus, its distal attachment to the medial plantar surface of the medial cuneiform and base of the 1st metatarsal and in addition, where its tendon passes under the extensor retinaculum should be considered the points of effort (E). However, there is some inevitable digital extension which aids ground clearance. This class three lever initiates at preswing (left) with concentric contraction reducing the plantarflexion angle of the ankle left over from late terminal stance. This extensor activity also maintains digital extension. As early swing progresses to midswing, isometric dorsiflexor activity holds the foot in a position short of ankle neutral, which is enough when coupled to knee flexion angles to see the swing limb pass the support limb without making ground contact (centre). After midswing and during terminal swing (right), the knee rapidly extends under quadriceps muscular activity. The ankle dorsiflexors start to increase their activity again in preparation for resisting the heel strike plantarflexion moment. This rising activity spike may further decrease the ankle plantarflexion angle and increase digital extension prior to initial heel contact. *(Permission www.healthystep.co.uk.)*

In all locomotive situations of swing phase, the lever system remains the same. The ankle is the fulcrum, the CoM of the whole foot the resistance, and the tibialis anterior distal attachment to the medial cuneiform and 1st metatarsal base is the primary effort. The resistance is small and therefore the effort generated is not great. It is only enough to move the mass of the foot at the ankle. In reality, the other anterior muscles will assist through digital extension, but the consequences of tibialis anterior dysfunction as seen in foot drop firmly indicate that tibialis anterior is providing the primary role (Anagnostakos et al., 2006). Compensatory digital extensor activity in tibialis anterior dysfunction may cause digital extension deformity in consequence of trying to assist ankle dorsiflexion; a mechanism known as *extensor substitution*.

2.5.8 Muscle action at the ankle

The muscles of the lower leg that cross the ankle can be divided anatomically into the anterior, posterior, and lateral groups. The muscles crossing the ankle joint can be functionally divided into the ankle plantarflexors (flexors) consisting of plantarflexors–invertors–internal rotators and plantarflexors–evertors–external rotators, and the dorsiflexors (extensors), consisting of dorsiflexors–invertors–internal rotators and dorsiflexors–evertors–external rotators. Each muscle's action depends on whether the tendons cross anteriorly or posteriorly and medially or laterally to the ankle joints axis (Fig. 2.5.8a). The ankle axis is triplanar in orientation and it moves, making muscle moment arms changeable. In addition, anatomical variation makes each patient's muscular relationship to joint axis rather individual. Another confounding factor in appreciating muscle-induced ankle motion or motion resistance is that the muscles that pass the ankle are all multiarticular muscles.

Soleus crosses just the ankle–subtalar joint complex, although it also influences motion at the proximal as well as the distal tibiofibular joints via ankle motion. Crossing the subtalar joint means rearfoot motions profoundly affect soleus function through altered Achilles stresses. When the subtalar joint is unloaded, soleus-induced motion only occurs at the ankle joint but when the rearfoot is loaded during stance, subtalar joint motion can change the orientation of the soleal part of the Achilles subtendon's. Thus, rearfoot motion is influencing the point of effort applied to the calcaneus through changing subtalar inversion or eversion angles (Zifchock and Piazza, 2004; Lee and Piazza, 2008; Lersch et al., 2012). Tibialis posterior is far more complex, crossing nine joints in total. Thus, its action at the ankle and within the midfoot changes with the positions of its distal and proximal attachment points across the foot and ankle.

The closer the muscle tendons pass to the ankle's functional axis, the weaker their plantarflexion–dorsiflexion role. However, with the instantaneous axis moving anteriorly and posteriorly throughout gait, the moment arms of the muscles

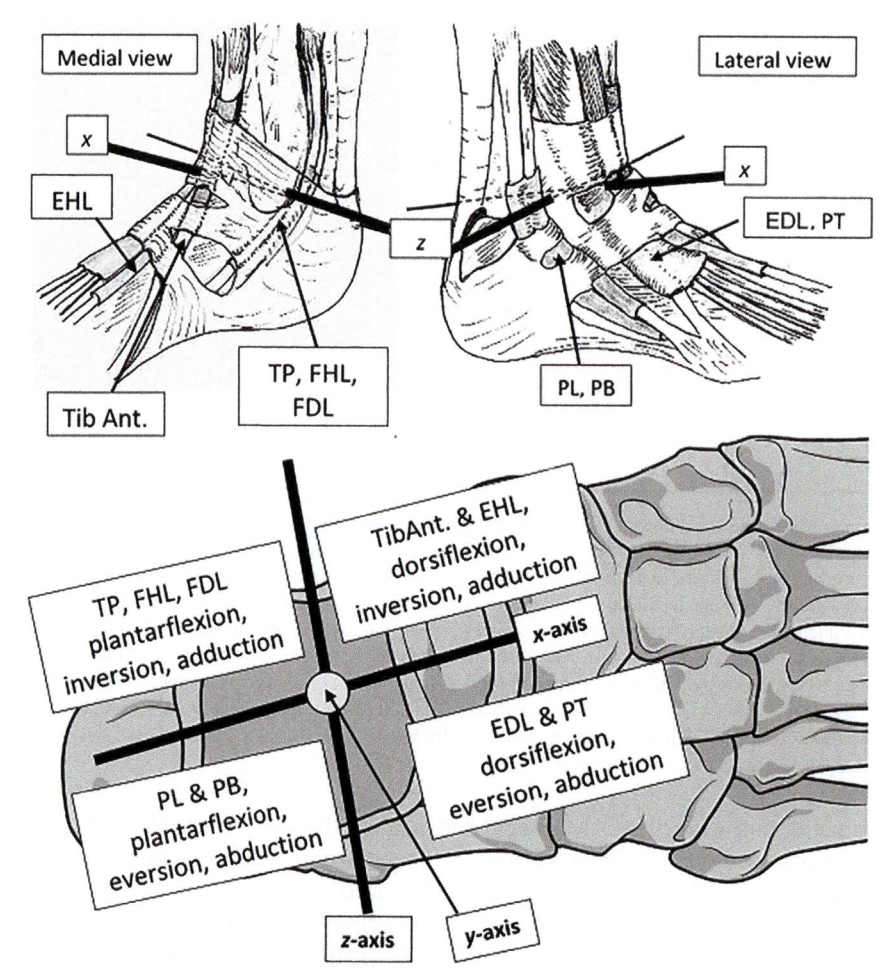

FIG. 2.5.8a The positions of the extrinsic foot muscle tendons as they pass the ankle tightly bound to the osseous structures by retinacular structures, as well their distal attachment points within the foot, influence the moments they apply on the ankle. During muscular contraction, the tendons can directly influence forces around the talus by applying tensions and compressions. The ankle–rearfoot complex provides motion in all three orthogonal planes, with motion dominant around its z-axis for the sagittal plane motion of extension (dorsiflexion) and flexion (plantarflexion). Transverse plane internal/external (adduction/abduction) occurs around the y-axis and frontal plane inversion/eversion around the x-axis. Thus, muscle–tendons can be divided into extensors or flexors, invertors or evertors, and adductors or abductors depending on their location to each rearfoot axis. In stance phase closed chain, they are resistors of the opposite open chain motions that they are associated with. This schematic helps to demonstrate these concepts. Thus, medially, tibialis anterior (TibAnt.) and extensor hallucis longus (EHL) are dorsiflexors, invertors, and adductors resisting plantarflexion, eversion, and abduction. Tibialis posterior (TP), flexor hallucis longus (FHL), and flexor digitorum longus (FDL) are plantarflexors, invertors, and adductors that are also resisting the opposite motions. Laterally, extensor digitorum longus (EDL) and peroneus tertius (PT) are dorsiflexors, evertors, and abductors, with peroneus longus (PL) and peroneus brevis (PB) also being evertors and abductors, but they are plantarflexors, sitting behind the z-axis. Once again, during stance, they oppose the opposite motions. The position of the FHL tendon gives this muscle the most homogenising effect on the muscular force vectors around the ankle during gait, but changes in the instantaneous joint axis position during motion or resulting from pathology can subtly change moment arms. *(Permission www.healthystep.co.uk.)*

change during locomotion. Attachment to the elongated calcaneal posterior tubercle gives the triceps surae (via the Achilles) the longest lever arm to the ankle, producing powerful plantarflexion moments from a large muscle mass with a large cross-sectional area via a tendon that also has a large cross-sectional area. However, the length of the calcaneal posterior tubercle influences the lever arm and how quickly moments can be generated around the ankle from triceps surae (Lee and Piazza, 2009; van Werkhoven and Piazza, 2017). Longer posterior tubercles create longer Achilles moment arms but require more muscle–tendon fibre length change to create degrees of ankle motion. Mechanical efficiency by lever arm length is improved with longer posterior calcaneal tubercles, but potential speeds of joint angular changes are slower. Thus, the mechanical requirements of faster ankle plantarflexion as seen among sprinters, are enhanced by having shorter posterior calcaneal tubercles but long-distance runners and walkers will do better with longer ones.

The deep posterior muscles found in the calf are all weak plantarflexors, having tendons very close to the joint axis. Their influence on the ankle lies more around the transverse and frontal planes as their tendons pass medial to the ankle. They include flexor hallucis longus, flexor digitorum longus, and tibialis posterior. The most medial is the tibialis posterior tendon which has the longest lever arm to produce ankle–subtalar joint inversion in open chain or eversion resistance within ankle–subtalar closed chain (Klein et al., 1996; Imhauser et al., 2004). The lateral muscle group of peroneus longus and peroneus brevis sit laterally and posterior to the fibula, generating ankle-subtalar complex eversion and plantarflexion in rearfoot open chain and inversion and dorsiflexion resistance within rearfoot closed chain.

The instantaneous axis of the ankle increases plantarflexion efficiency of these posterior muscles when the ankle joint is in maximum dorsiflexion as it moves anteriorly, which is also when the subtalar joint provides the least amount of flexibility (Leardini et al., 2001). This results in a tendency to increase the effectiveness of these multi-articular muscles in influencing ankle joint motion rather than creating motion within the rearfoot at the end of midstance. In ankle plantarflexion, the instantaneous axis starts to move posteriorly, decreasing these posterior muscles' influence on the ankle, but opens up freedom to influence plantarflexion moments across the midfoot.

All the muscles that bridge the ankle have an influence on the stresses that occur within the joint (Potthast et al., 2008). The anterior aspect of the ankle joint is predominantly stressed throughout midstance loading dorsiflexion, with overall loads being most influenced through the Achilles tendon forces (Potthast et al., 2008). The smaller muscles that run into the foot primarily influence the loading of the ankle joint across the transverse and frontal planes. Peroneus longus and peroneus brevis activity increases lateral compression, while tibialis posterior, flexor digitorum longus, and flexor hallucis longus act together, stressing and compressing the joint more medially (Potthast et al., 2008). Changes in the joint contact area are most influenced by tibialis posterior with its activity increasing medial joint stress, while the flexor hallucis longus plays a role of homogenising muscle-induced stresses across the joint (Potthast et al., 2008). Thus, these two muscles with those of triceps surae are particularly important for ankle joint biomechanics, although forces via the Achilles dominate (Fig. 2.5.8b).

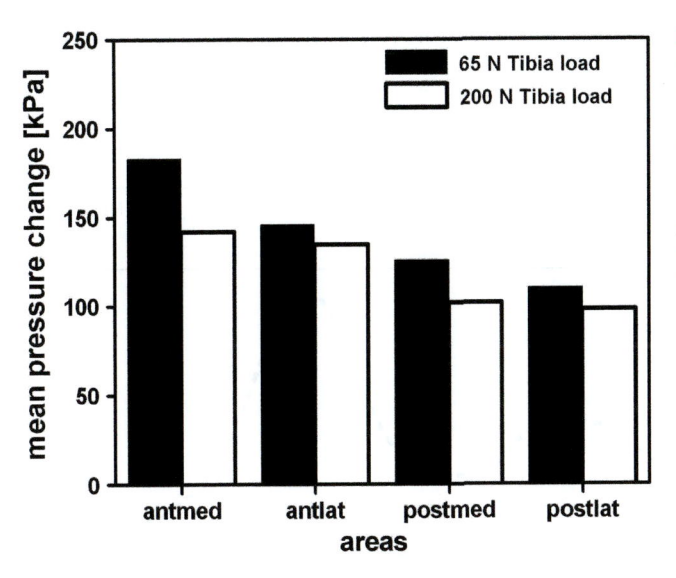

FIG. 2.5.8b Changes in talocrural joint pressures through an Achilles tendon force in two axial lower leg loading conditions of 65 and 200 N are demonstrated. This shows the loading pressures on the anteromedial (antmed), anterolateral (antlat), posteromedial (postmed), and posterolateral (postlat) articular surfaces. These loading pressures from the Achilles forces are not even. Activation of other muscles passing the ankle also have their own effects on the final pressures across the joint. *(Image from Potthast, W., Lersch, C., Segesser, B., Koebke, J., Brüggemann, G.-P., 2008. Intraarticular pressure distribution in the talocrural joint is related to lower leg muscle forces. Clin. Biomech. 23(5), 632–639.)*

Tibialis posterior and the peroneal muscles significantly influence rearfoot and ankle mechanics and pathology. Dysfunction within these muscles is usually associated with 'ankle pain' and instability, as well as profoundly affecting the foot's function. Discussing the foot's extrinsic muscles' primary action necessitates that their function be considered slightly differently within either the functional units of the ankle or the foot, despite their obvious interconnection.

2.5.9 Extensor muscles at the ankle

The main roles of the ankle dorsiflexors or extensors are to reduce the ankle plantarflexion foot angle during swing phase in order to aid ground clearance and reduce forefoot plantarflexion acceleration after heel contact, while at the same time achieving energy dissipation during contact. Ankle dorsiflexors thus only actually dorsiflex the ankle into a position of dorsiflexion during actions such as working a pedal in a car or negotiating uneven and inclined terrains, but do not do so in normal gait. The anterior muscle group of the leg is dominated by tibialis anterior, which is larger in humans than in any other apes (Usherwood et al., 2012). Of the extensors, only tibialis anterior has a fibrocartilaginous enthesis, indicating greater tendon motion and stresses (Frowen and Benjamin, 1995).

The extensor tendons are firmly tethered to the anterior ankle and leg through the extensor retinaculum, which exists as three reinforced fascial thickenings. These consist of a broad, medial-to-lateral coursing superior band of the retinaculum linked to a Y-shaped inferior retinaculum running from a stem to divide into a medial superior and inferior oblique band. Sometimes, a superior lateral band also arises from the inferior extensor retinaculum found in 25% of cases, changing the Y into an X shape (Draves, 1986, pp. 151–152). Both extensor retinaculae have fascial linkages to the periosteum of the calcaneus and talus as well as to the tibia and fibula, and it envelops the extensor tendons to form tendon sheaths (Draves, 1986, pp. 152–153). It is the tethering of the tendons through the sheaths and retinaculae that alters the lever arms and the line of pull as the ankle flexes and extends.

Tibialis anterior is a postural muscle–tendon complex (Birch et al., 2008) which lies under a stiff deep fascia (Stecco et al., 2014). This reflects its role in controlling the position of the free foot during swing and decelerating ankle plantarflexion during heel contact. These actions do not require energy storage to initiate a recoil dorsiflexion. The muscle also plays a prominent role in energy dissipation due to its energy buffering effects achieved through muscle–tendon complex lengthening (Neptune et al., 2008; McGowan et al., 2010; Roberts and Konow, 2013). Despite these two important functional roles, tibialis anterior is not normally subjected to large resistance forces. In swing, it reduces the angle of plantarflexion from around the 10° that is left over at the ankle at the end of terminal stance to aid ground clearance, achieving an angle of around ankle-neutral or an angle that is plantarflexed a little short of neutral, during swing phase (Perry, 1992, p. 57; Byrne et al., 2007). In swing, it merely has to maintain the position of the resistance of the mass of the foot, only utilising around 35% of maximal muscle force with reducing EMG activity until terminal swing when it starts to increase again (Byrne et al., 2007) (Fig. 2.5.9a). Tibialis anterior is probably aided in swing by peroneus tertius (if present) and gastrocnemius lateralis (Jungers et al., 1993; Di Nardo et al., 2013).

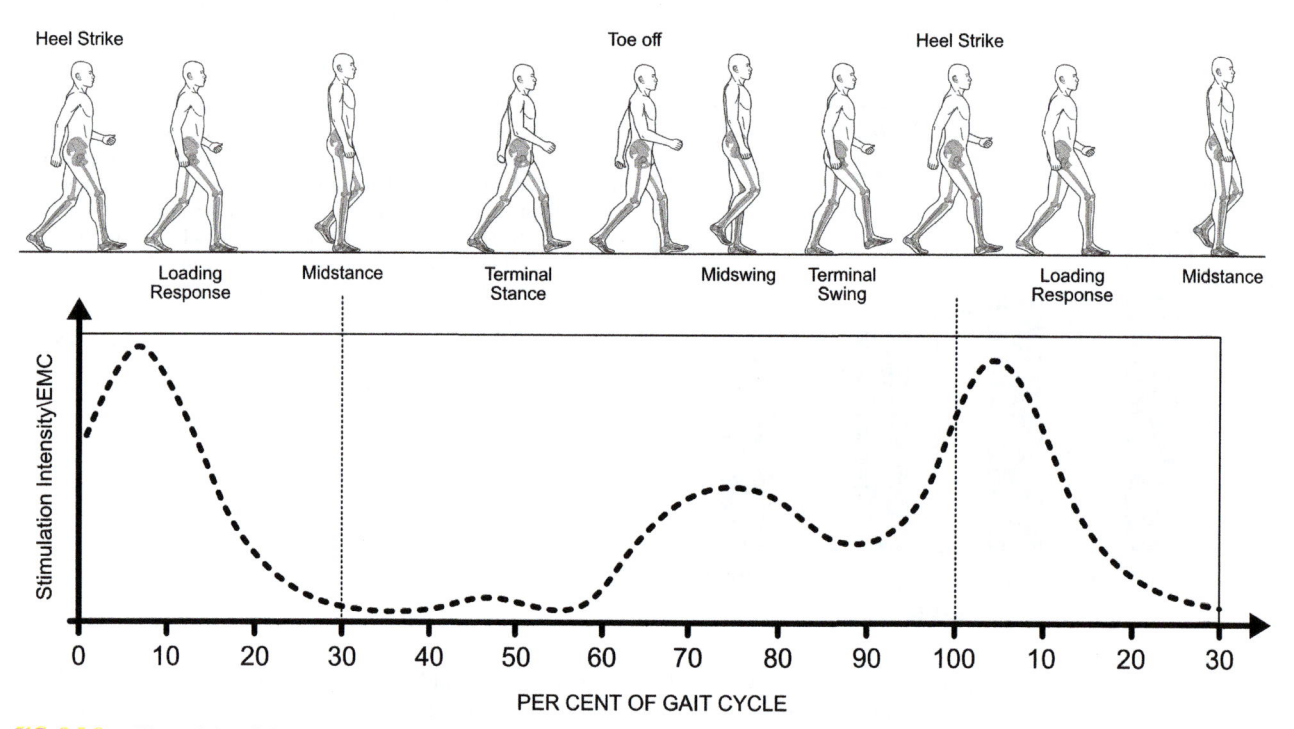

FIG. 2.5.9a The activity of tibialis anterior related to phases of the gait cycle with activity beginning to peak before and during initial contact, reducing through loading response and with little activity again until preswing. EMG data from Byrne et al. (2007). *(Permission www.healthystep.co.uk.)*

Peak activity is usually within two periods: in terminal stance around preswing and then starting again at precontact-to-loading response. It can also show single or (more rarely) triple phasic activity from preswing to loading response (Sutherland, 2001; Di Nardo et al., 2013). Tibialis anterior activity rises before heel contact (Byrne et al., 2007) and allows preactivation stretch-shortening mechanics so that the tibialis anterior is primed for eccentric contraction against ankle plantarflexion prior to heel strike. Initial heel contact usually occurs at an ankle angle between neutral and 3–5° plantarflexed, permitting forefoot plantarflexion of around 7° controlled by 45% of maximal contraction force, after which tibialis anterior activity rapidly decreases (Byrne et al., 2007).

Walking speed affects the bursts of peak tibialis anterior activity and in the slowest walking speeds, peaks occur around toe-off. However, at preferred walking speeds and faster activity, peaks occur around heel contact, showing a clear relationship with the size of the heel impact force (Byrne et al., 2007). In forefoot contact gaits, the activity of tibialis anterior is much reduced (Rooney and Derrick, 2013). Midstance activity is only present in around 15%–45% of strides, and only lasts for 10% of the total gait cycle when present (Agostini et al., 2010; Di Nardo et al., 2013). This midstance activity may represent medial balancing of the foot vault (if necessary) to provide primarily frontal plane stability. Tibialis anterior also plays a role in maintaining erect stance, CoM stability, and is more active in individuals with increased CoM sway and with a narrower base of support in stance when the feet are closer together (Lemos et al., 2015).

Tibialis anterior's muscle–tendon complex has a mechanical advantage in positions of ankle plantarflexion through instantaneous ankle joint axis positioning (Leardini and O'Connor, 2002). As the tendon crosses the ankle axis anteriorly and medially, it has a variable small inversion moment arm on the foot (Klein et al., 1996; Wang and Gutierrez-Farewik, 2013, 2014). With rearfoot inversion, its inversion moment arm tends to increase (Wang and Gutierrez-Farewik, 2013, 2014). Tibialis anterior is a multi-articular structure that crosses the ankle, the talocalcaneonavicular portion of the subtalar and talonavicular joints, the medial cuneonavicular joint, and the 1st metatarsocuneiform joint. Through this route, tibialis anterior activity creates stabilising proximally directed compression forces into the medial column of the foot and the ankle–subtalar complex. Activity before and during contact to loading response give tibialis anterior a role in stiffening the medial side of the foot vault, improving pretensioning of the skeletal frame of the foot prior to and during forefoot loading as an active part of the windlass mechanism.

Tibialis anterior's late terminal stance activity will influence medial column adduction and compressive stability via its preswing activation. The forefoot adduction moment reported by Holowka et al. (2017) during terminal stance may be influenced by both tibialis anterior as well as tibialis posterior. Tibialis anterior has a small inversion moment arm in all rearfoot frontal plane positions, but with rearfoot inversion, its ability to dorsiflex at the ankle is restricted compared to when being in an everted position (Wang and Gutierrez-Farewik, 2011). The action of tibialis anterior in the sagittal plane also helps to initiate contact phase knee flexion by restraining ankle plantarflexion, and this is also more significant when accompanied with rearfoot eversion (Wang and Gutierrez-Farewik, 2011). Tibialis anterior activity is also related to the transition from walking to running, with its heel contact activity reaching maximal levels before initiating running (Malcolm et al., 2009). However, transition will occur at slower speeds if tibialis anterior becomes fatigued (Segers et al., 2007).

The other anterior muscles are far more anatomically variable and occasionally some can be missing (Draves, 1986, pp. 250–254; Newton, 2014; Carlis et al., 2017). They consist of the long digital extensors, i.e. extensor hallucis longus and extensor digitorum longus, and peroneus tertius. These muscles play a supporting role in positioning the foot in swing, aiding in ground clearance, and decelerating ankle plantarflexion. Peroneus tertius muscle is present in 93%–95% of humans, but not seen in other apes. It is probably advantageous for swing energetics, avoiding the need for peroneus longus and brevis recruitment during swing phase; something that is necessary for other primates' bipedal gaits (Jungers et al., 1993; Yammine and Erić, 2017).

Although the long digital flexors aid ankle positioning, their combined mass and cross-sectional area is considerably smaller than tibialis anterior's, suggesting a fairly limited role (Perry 1992, pp. 55–56). Although a relatively rare occurrence (Jain et al., 2015), when tibialis anterior is ruptured, gait presents much as a classic neurological foot drop or 'dropped foot', despite the ability of the assistor muscles to continue to provide some ankle dorsiflexion (Anagnostakos et al., 2006). The lesser extensor muscle lever arms are more lateral to the ankle, so none can assist in the small but important, inversion lever arm motion of tibialis anterior during swing and at initial heel contact. Their assistor action in tibialis anterior dysfunction causes the rearfoot to become more everted during swing and contact phase. The long digital extensors should be seen as primarily functioning with the only dorsal muscle of the foot, *extensor digitorum brevis*. Together, they provide digital extension during swing to aid forefoot ground clearance and through digital extension to pretension the foot vault prior to forefoot contact loading response. This influences foot stiffness at contact (Fig. 2.5.9b).

Extensor pretensioning of the skeletal frame of the foot is probably important in loading response, as the connective tissues around the foot can behave more elastically if stiffened, helping to dissipate energy, especially if stiffness can be relaxed on stress-loading. The extensors, by extending the toes, tighten the plantar aponeurosis through the windlass mechanism, pretensioning the foot prior to forefoot contact (Carlson et al., 2000; Iwanuma et al., 2011). Even then, the ankle angle achieved by the combined action of the ankle extensors seems to have a greater effect on plantar aponeurosis stiffening than does digital extension alone (Wager and Challis, 2016). As the forefoot loads, toe extension is released which allows for further motion and deformation of the skeletal frame's foot vault to permit energy absorption.

This loss of digital extension and relaxation of the plantar aponeurosis on loading has been termed the *reversed windlass* or *reverse windlass* (Stainsby, 1997; Aquino and Payne, 2000; Green and Briggs, 2013; Stolwijk et al., 2014). In high-vaulted cavoid feet, the potentially reduced ability to unextend the digits at the MTP joints due to excessive extension of the toes may be linked to increased foot stiffness and reduction of energy dissipation capability, despite normal tibialis anterior function (Fig. 2.5.9c). This capacity to use plantar aponeurosis mechanics to aid energy dissipation is considered

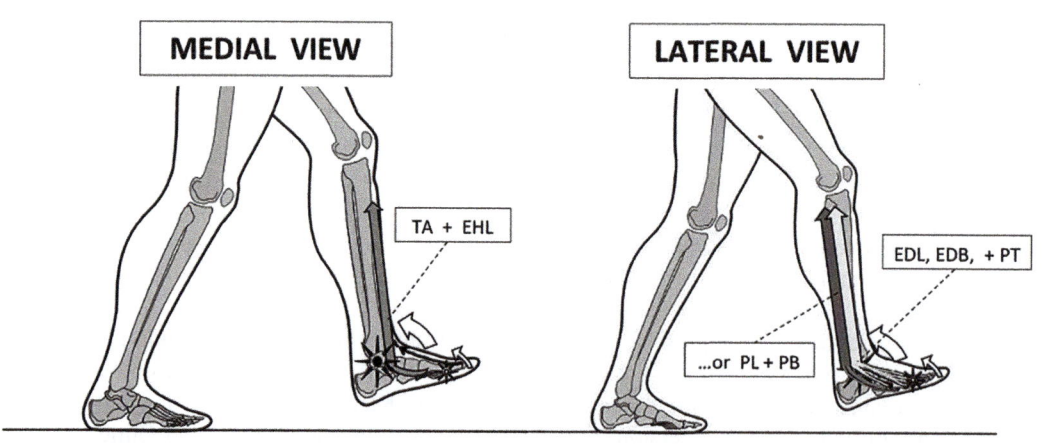

FIG. 2.5.9b Little muscular force is required to resist the mass of the foot around the ankle during swing. The primary moment required is ankle dorsiflexion (white arrow) that can be easily provided by tibialis anterior (TibAnt.) and aided by a little digital extension via extensor hallucis longus (EHL), extensor digitorum longus (EDL), and the more laterally positioned extensor digitorum brevis (EDB). Another uniquely human (among large primates) ankle extensor, peroneus tertius, can also aid swing phase ankle dorsiflexion and increase foot and ankle abduction and eversion. Eversion of the ankle helpfully shortens the swing limb as opposed to inversion that lengthens it. However, this useful muscle is not always present, requiring peroneus longus and brevis to provide for any extra eversion moments in its absence. These muscles are ankle plantarflexors, so if utilised, they require their activity to be offset by more dorsiflexion moments. *(Permission www.healthystep.co.uk.)*

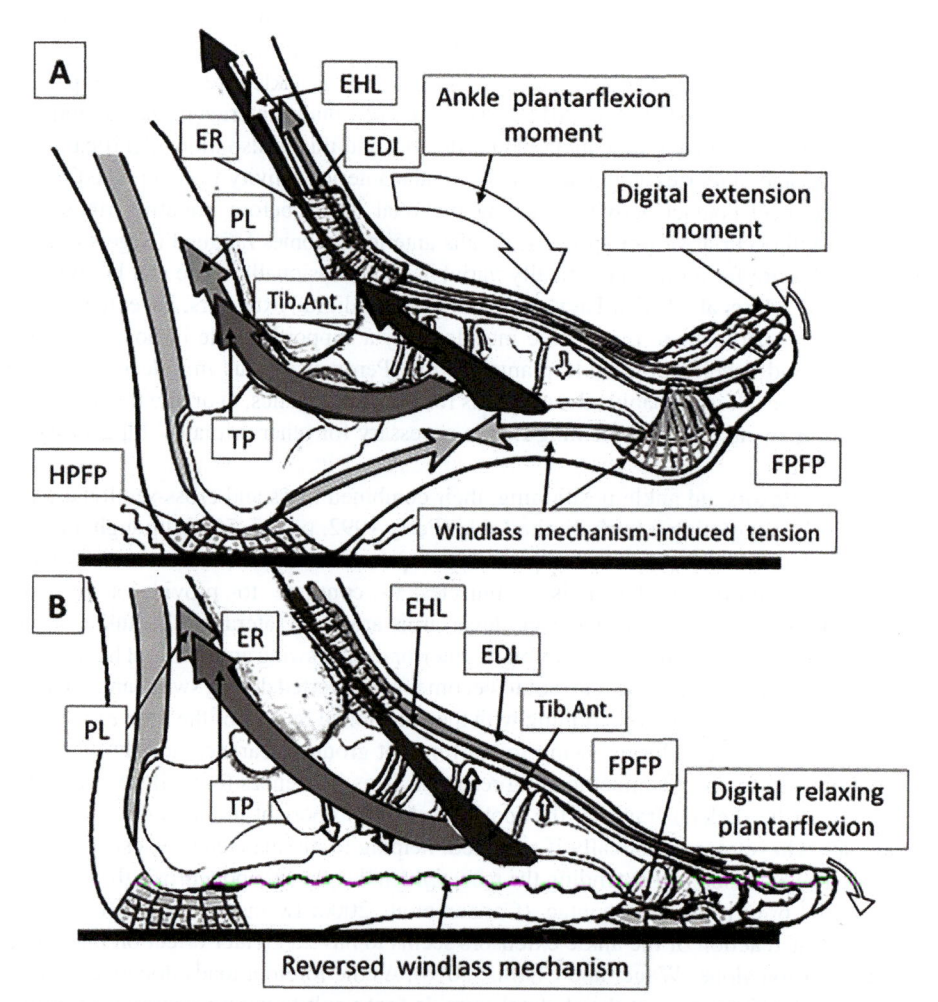

FIG. 2.5.9c Extensor hallucis longus (EHL) and extensor digitorum longus (EDL) have a dorsiflexion moment arm to the ankle axis, and yet they provide minimal influence on ankle plantarflexion moments during loading response (A). They have relatively small muscular cross-sectional areas compared to tibialis anterior (Tib.Ant.) and operate over many joints distal to the ankle, including mobile digital joints. This restricts their ankle effort being applied via the extensor retinaculum (ER). Their effect on the foot and ankle derives more from inducing the windlass mechanism after heel strike on the heel plantar fat pad (HPFP), rather than decelerating foot plantarflexion at the ankle. The windlass mechanism tightens the plantar aponeurosis and its associated forefoot plantar fat pad (FPFP) through digital extension from EHL, EDL, and extensor digitorum brevis (not shown). This tensions and stiffens the plantar foot prior to forefoot contact, aided by activity in tibialis posterior (TP) and peroneus longus (PL) that together plantarflex the midfoot articulations, shortening and raising the foot vault. On forefoot loading (B), the long digital extensors, TP, and PL reduce activity, allowing the toes to rotate out of extension, relaxing the plantar aponeurosis and together increase foot compliance. This allows midfoot articulations to glide into dorsiflexion under load. Thus, they all assist in impact shock attenuation. *(Permission www.healthystep.co.uk.)*

further in the pedal functional unit during Chapter 3. It is interesting to note that anatomical variance among the long digital extensors (both muscles and/or tendon slips) has an intimate relationship to tendon slip variations in the short digital extensor. Extensor digitorum brevis is seemingly able to cover for loss of long extensor tendon slips, underlining the functional interrelationship of these three muscles. Such an ability may explain why the absence of a long extensor muscle has been reported on random cadaver examinations (Newton 2014) rather than through a patient presenting with gait difficulty.

Considering the action of the long extensors of the toes during gait, the presence of compression-resistant sesamoid fibrocartilage on the deep surface of the digital extensor tendons in the digits might seem surprising. As found in the hands, the tendons of the toe extensors form part of the joint capsule of the interphalangeal (IP) joints and press against the proximal phalanx during flexion (Milz et al., 1998). Toe extensor tendons are most tensioned during flexion, a position they rarely find themselves in today, but despite this, they are often under higher stress when flexed compared to those tendons found in the fingers. Compression-resistive type II collagen is present in all the toe extensor tendons but is particularly characteristic of the tendons to the lesser toes, suggesting that the lesser toes undergo greater flexion moments at the IP joints than occur at the hallux (Milz et al., 1998). This is particularly true in climbing when using the toes to grip. Thus, the feature may relate more to the evolutionary past of hominins, although biomechanical forces are usually required to maintain such features. The hallux IP joint is subjected to high extension moments during late terminal stance, forces that are less expressed on lesser toe IP joints.

2.5.10 Primary flexor muscles at the ankle

The high net ankle torque from ankle power indicates the torque-generating capabilities of the plantarflexors (flexors). The peak moment incorporates both the passive and active torques, with the power primarily resulting from the active contraction of the plantarflexors (Salsich and Mueller, 2000). The moments generated around the ankle by the plantarflexors control the anterior displacement of the body's CoM and generate the terminal stance GRF at the forefoot. This GRF provides stability during weight transfer to the next step during heel lift. The plantarflexors of the ankle are more numerous and vastly larger by mass and cross-sectional area than are the extensors and they also include the thickest tendon within the human body, the Achilles.

The triceps surae muscle-Achilles tendon complex

The ankle plantarflexors restrain forward angular momentum of the tibia on the talus during midstance, produce ankle and knee stability during stance, and create heel lift without causing large vertical oscillations in the CoM of the body. They are extremely important to human gait, being able to compensate for hip and knee extensor and flexor deficiencies. Proximal lower limb muscles cannot, in return, compensate appropriately for plantarflexor deficits (Goldberg and Neptune, 2007). The most important ankle plantarflexor is the triceps surae, which controls both ankle dorsiflexion rates and generates ankle plantarflexion motion during stance. Gait speed influences the forces generated within the muscles through force–velocity relationships, with muscles less able to generate power at faster speeds but connective tissue more able to store and release energy. Although the power generated by the triceps surae is fundamental in restraining the forward motion of the CoM throughout late midstance to store acceleration power, it does not generate the primary forces that drives the body's CoM forward (Honeine et al., 2013).

In this sense, the plantarflexors are not really 'push-off/propulsion' muscles. They are instead CoM accelerators that generate a prominent GRF through the foot as the heel lifts and the ankle plantarflexes. This action controls the exchange and flow of potential energy stored within the connective tissues of the posterior leg and plantar foot during midstance that are being released as kinetic energy for acceleration in which the significance of the Achilles tendon as an energy-storing tendon becomes apparent (Honeine et al., 2013). The forefoot GRF created occurs in both vertical and anterior–posterior directions with increasing loading rate resulting in increases in anteriorly–posteriorly peak GRF, but without any increase in muscle activity (Honeine et al., 2013). This relates back to the force–velocity relationship of muscle that can generate less force during faster loading, and the time-dependent properties of connective tissue that behaves more stiffly (elastically) during faster loading. Therefore, more muscle power is generated at slower loading rates in triceps surae (and other ankle flexor muscles), but the Achilles behaves more elastically, storing greater energy when operating at faster loading rates. The faster the gait speed, the more of terminal stance acceleration from the plantarflexors is driven through elastic recoil of the tendons (Fig. 2.5.10a).

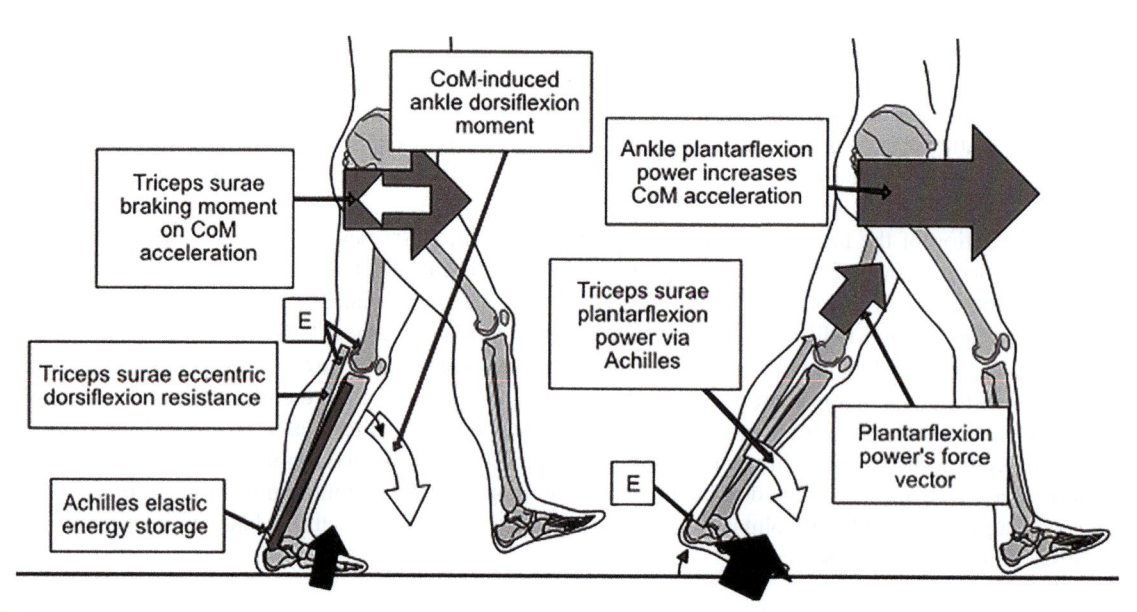

FIG. 2.5.10a The large forefoot GRF generated during terminal stance assists CoM acceleration (large grey arrows). Yet it derives its power from CoM deceleration against gravity during midstance (left image) and releases it at the start of terminal stance (right image). Braking of anterior tibial rotation during late midstance is necessary to control CoM-induced ankle dorsiflexion (white lower rotatory arrow) and this is achieved through eccentric contraction of the triceps surae muscles applying effort (E) via their proximal attachments. This not only slows the acceleration and anterior fall of the CoM (insert white arrow within CoM grey arrow of acceleration) but also stretches the Achilles, loading it with elastic energy. As the GRF (black arrows beneath stance foot) moves anteriorly, it increasingly offloads the heel. Once forces of elastic recoil within the Achilles are greater than forces remaining on the heel, the heel 'springs up' into heel lift and terminal stance begins. Ankle plantarflexion power (white arrow—right image) is now released from the triceps surae-Achilles complex to create a large forefoot GRF with E applied at the Achilles distal attachment. This creates an upward and anterior resultant force from plantarflexion power into the terminal stance limb (dark-grey arrow within limb) that assists the acceleration of the CoM into the next step. The resulting plantarflexion power and ankle motion that is generating the forefoot GRF is also creating limb stability against the ground, which increases its anterior angulation as the ankle plantarflexes over the foot, increasing the horizontal drive. *(Permission www.healthystep.co.uk.)*

The CoM of the body is not pushed forward but drawn forward by centrifugal forces of the swing limb during terminal swing and gravity through the fall of the CoM. The triceps surae supports the CoM as the body travels over the ankle and foot, restraining the CoM of the body from uncontrolled anterior falling velocity, and indirectly controlling step length and walking speed through this control (Honeine et al., 2013). Thus, the triceps surae is responsible for balance control via the braking of the CoM's vertical displacement during walking or running (Cavagna et al., 2000; Honeine et al., 2013). Controlling the body's support in late midstance, the triceps surae can modulate the anterior–posterior distance between the body's CoM and the CoGRF, thereby controlling the walking velocity (Honeine et al., 2013). This is important to triceps surae and Achilles tendon stress generation, as the CoGRF should be transferred significantly into the forefoot before the heel lifts, and not before the swing limb is correctly positioned and its muscles preactivated for initial contact.

At heel lift, the triceps surae through the Achilles tendon is only responsible for lifting the CoM of the rearfoot, midfoot, and the lower limb segment that lies behind the fulcrum point of the MTP joints. This allows the body's CoM to fall around 1 cm in height to the heel of the opposite foot onto the support surface during walking, providing some extra momentum to the CoM's forward progression. Anything that changes the progression of the CoGRF such as triceps surae strength, midfoot break, ankle range of motion, or foot vault profile (high or low, hypo or hypermobility) may cause detrimental energetic changes in the control of the CoM to CoGRF relationship.

The triceps surae component muscles

The triceps surae muscle–tendon complex involves three separate muscles: soleus, gastrocnemius medialis (medial gastrocnemius), and gastrocnemius lateralis (lateral gastrocnemius). These gradually merge via their aponeuroses into the Achilles tendon in which the precise nature of this merger expresses human individual variation within the muscle and the tendon anatomy (Bojsen-Møller et al., 2004; Pękala et al., 2017). The Achilles tendon arises proximally from the superficial and deep (soleus) aponeurosis of the gastrocnemius muscle complex to attach to the posterior calcaneal enthesis, with the tendon fascicles and subtendons undergoing a twist around each other (Edama et al., 2015; Pękala et al., 2017) (Figs 2.5.10b and 2.5.10c). Gastrocnemius and soleus differ in their influence on forward displacement of the CoM during heel lift and yet provide similar contributions to vertical support and CoGRF movement (Francis et al., 2013).

Locomotive functional units **Chapter | 2** 413

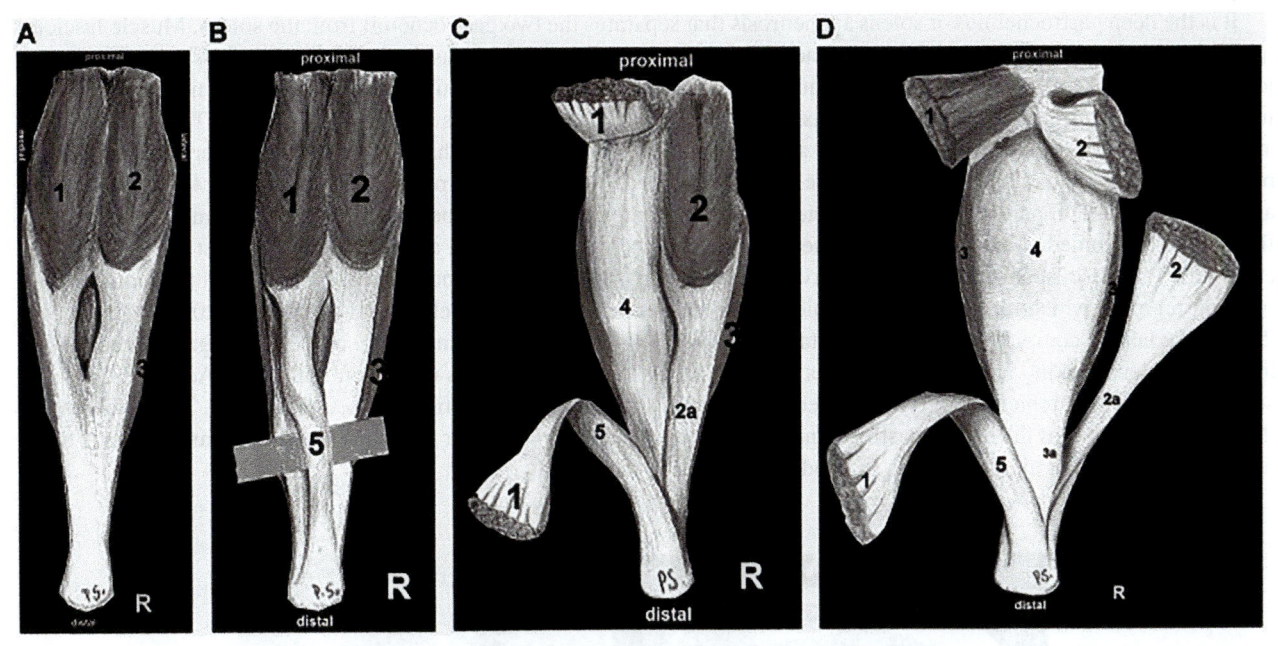

FIG. 2.5.10b Stages of dissection of the triceps surae in posterior view show the medial head (medial gastrocnemius or gastroc. medialis) of gastrocnemius (1), the lateral head (lateral gastrocnemius or gastroc. lateralis) of gastrocnemius (2), and the connective tissue tendon fibres from the lateral head (2a). Soleus (3) and its tendon fibres (3a) lie under the aponeurosis of the soleus (4) lying deep to gastrocnemius. The fascicle from the medial head of gastrocnemius (5) lies the most superficial to the posterior aspect of the Achilles. *(Image from Szaro, P., Witkowski, G., Smigielski, R., Krajewski, P., Ciszek, B., 2009. Fascicles of the adult human Achilles tendon—an anatomical study. Ann. Anat. 191(6), 586–593.)*

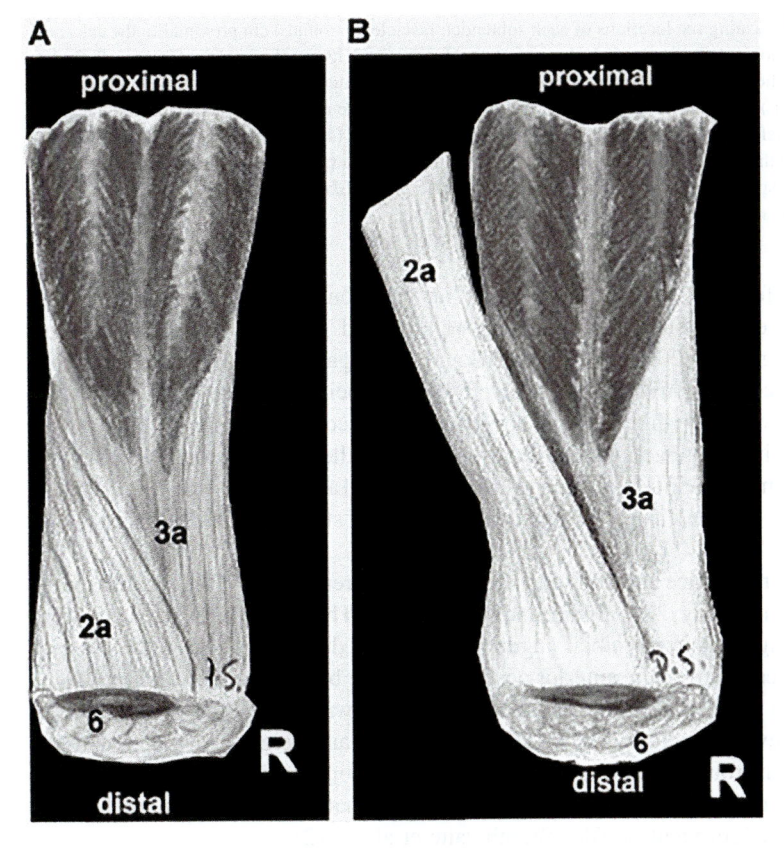

FIG. 2.5.10c Images of dissection from the anterior surface view of the right Achilles tendon clearly demonstrates the twisting of the fibres from the lateral head of gastrocnemius (2a) and soleus (3a). Kager's fat pad (6) sits anteriorly to this part of the Achilles. *(Image from Szaro, P., Witkowski, G., Smigielski, R., Krajewski, P., Ciszek, B., 2009. Fascicles of the adult human Achilles tendon—an anatomical study. Ann. Anat. 191(6), 586–593.)*

It is the deep gastrocnemius or soleus aponeurosis that separates the two gastrocnemii from the soleus. Muscle fascicles of the gastrocnemii (medial and lateral) attach to the aponeurosis with interfascicular connective tissue before finally joining with the Achilles tendon through a musculotendinous junction. The Achilles narrows from the musculotendinous junction but widens again distally towards its osteotendinous calcaneal enthesis (Pękala et al., 2017). The Achilles, as it arises from the triceps surae, twists to form a precise pattern at its distal attachment so that the posterior or superficial portion of the distal attachment arises from the medial bundles of the medial gastrocnemius, the posterior lateral attachment arises from the lateral part of medial gastrocnemius and the anterolateral or deepest areas from lateral gastrocnemius, and finally, the anteromedial and medial attachment arises from the soleus (Szaro et al., 2009; Handsfield et al., 2017; Pękala et al., 2017). These arrangements show individual variability, with three distinct variations in the subtendon degree of twisting reported which may influence the biomechanical properties of each Achilles (Edama et al., 2015; Pękala et al., 2017). The largest component at the distal attachment site (~44.5%) is that emanating from the lateral gastrocnemius followed by soleus making up ~28% (Pękala et al., 2017). Each of these sections of tendon within the Achilles represents subtendons at a hierarchical level lying between full tendon and fascicle (Handsfield et al., 2016, 2017). The twists within the superficial fibres of the medial gastrocnemius are less than those of the deeper lateral gastrocnemius and soleus (Edama et al., 2015; Pękala et al., 2017) (Fig. 2.5.10d and 2.5.10e).

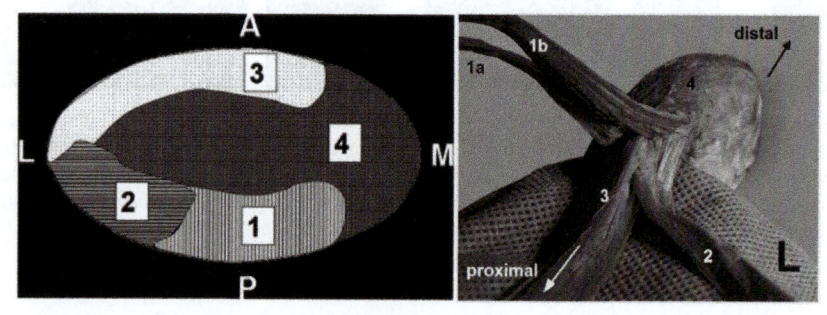

FIG. 2.5.10d Cross section schematic of the left Achilles indicating the locations of each subtendon fascicle at around 1 cm proximal to the calcaneal enthesis (left image) compared to an anatomical dissection photograph directed inferiorly onto a left calcaneus near its attachment (right image). In the left image, fibres from the medial part of medial gastrocnemius (1) lie posteriorly (P) (superficially), while those of its lateral fibres lie posterolaterally (2). The fibres of lateral gastrocnemius (3) supply most of the lateral (L) and anterior (A) aspect of the distal Achilles. Soleus (4) supplies variable medial areas. Distal twisting is indicated in the right dissection image of the left Achilles near its insertion the posterior calcaneus (4), with fibres from the medial part of medial gastrocnemius (1a) and its lateral fibres (1b) occupying the lateral and posterior aspect of the tendon. The lateral gastrocnemius occupies the lateral and anterior (deep) aspect (2) and soleus (3) the anteromedial aspect. *(Image from Szaro, P., Witkowski, G., Smigielski, R., Krajewski, P., Ciszek, B., 2009. Fascicles of the adult human Achilles tendon—an anatomical study. Ann. Anat. 191(6), 586–593.)*

Different parts of the triceps muscle cause differential strain behaviours within the aponeurosis and Achilles tendon. Muscular combined forces are estimated at between 1400 and 2600 N during walking and 3100–5330 N during running (Scott and Winter, 1990; Finni et al., 1998; Giddings et al., 2000). Normal EMG activity of gastrocnemius is reported from the end of loading response to early terminal stance, with stride-to-stride variability also being reported (Di Nardo et al., 2013). The Achilles is the strongest and thickest tendon within humans, constructed to accommodate these high in vivo stresses created via GRF and triceps muscle activity interaction. Despite this, the Achilles is commonly affected by pathology and is the most commonly reported human tendon to rupture (Wren et al., 2001a, b). Understanding Achilles strains is complex due to the four subtendon units and the changes in the concentration of strain within different regions of the Achilles throughout gait (Handsfield et al., 2017).

In walking, triceps surae activity starts in early midstance and continues throughout the remainder of the single support phase of gait (Crenna and Frigo, 1991; Anderson and Pandy, 2001; Franz and Kram, 2012). This early midstance activation is in preparation for CoM deceleration by resisting the anterior tibial angular momentum during ankle dorsiflexion that occurs in late midstance. This ankle dorsiflexion results from the anterior positioning of the body's CoM to the ankle joint axis after absolute midstance. Early single support phase activation occurs even at slower walking speeds when the CoM momentum is reduced (Murley et al., 2014). However, its activity can be affected by foot vault height with soleus activity having been reported in a small study to be greater in around 50% of low-vaulted feet at forefoot loading than in high-arched feet (Branthwaite et al., 2012). The peak of triceps surae activity occurs just before heel lift, but this crescendo of preheel lift activity still demonstrates variability unaffected by foot vault profile (Branthwaite et al., 2012).

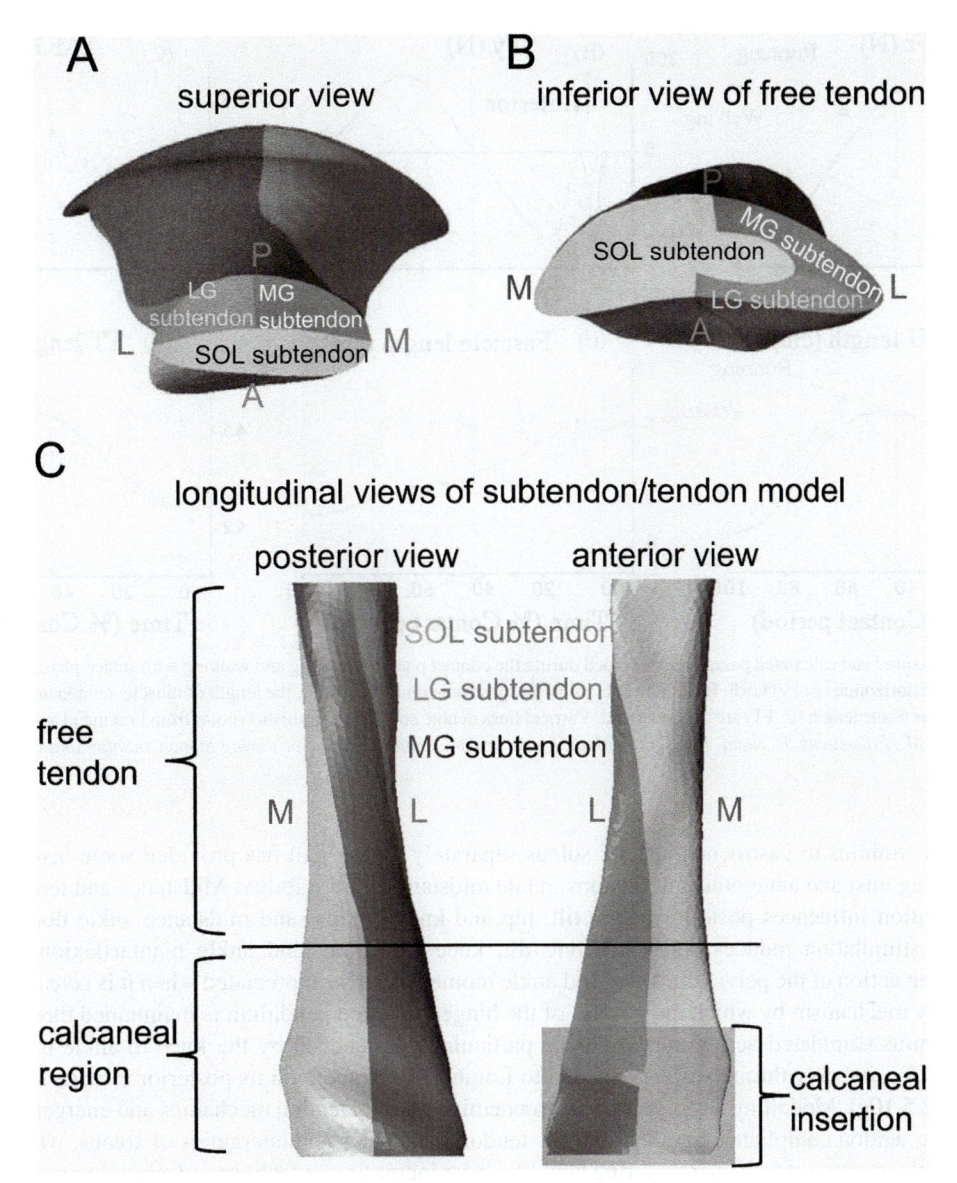

FIG. 2.5.10e The illustrations of the locations of the subtendons as modelled by Handsfield et al. (2017). Areas clearly show the twisting of fibres from the proximal musculotendinous junction to the enthesis, with lateral gastrocnemius (LG), medial gastrocnemius (MG), and soleus (SOL) changing their positions throughout their descent through the Achilles. *(Image from Handsfield, G.G., Inouye, J.M., Slane, L.C., Thelen, D.G., Miller, G.W., Blemker, S.S., 2017. A 3D model of the Achilles tendon to determine the mechanisms underlying nonuniform tendon displacements. J. Biomech. 51, 17–25.)*

In walking, the muscle fascicles of medial gastrocnemius have been found to shorten initially from 0–15% of stance during loading response followed by lengthening of the muscle fascicles and Achilles tendon tissue during 15%–70% of stance (Ishikawa et al., 2007). During midstance, the muscle–tendon unit controls ankle dorsiflexion and tibial anterior rotation as a CoM decelerator. During the final 70%–100% of stance, at heel lift into terminal stance, both the muscle fascicles and tendon fibres shorten (Ishikawa et al., 2007). This is not the same as in running. In running, initial contact is marked by rapid muscle fascicle and tendon lengthening (0%–10%). The muscle fascicles then shorten throughout the rest of ground contact, while the tendon initially lengthens throughout the rest of the braking phase to then shorten during acceleration in the final 70%–100% of stance (Ishikawa et al., 2007) (Fig. 2.5.10f). Thus, running uses greater concentric–isometric contraction within the triceps surae, increasing the Achilles' elastic burden of stretch and recoil to control ankle motion. Hence, running is more vulnerable to triceps surae fatigue than walking which in turn increases Achilles tendon overloading.

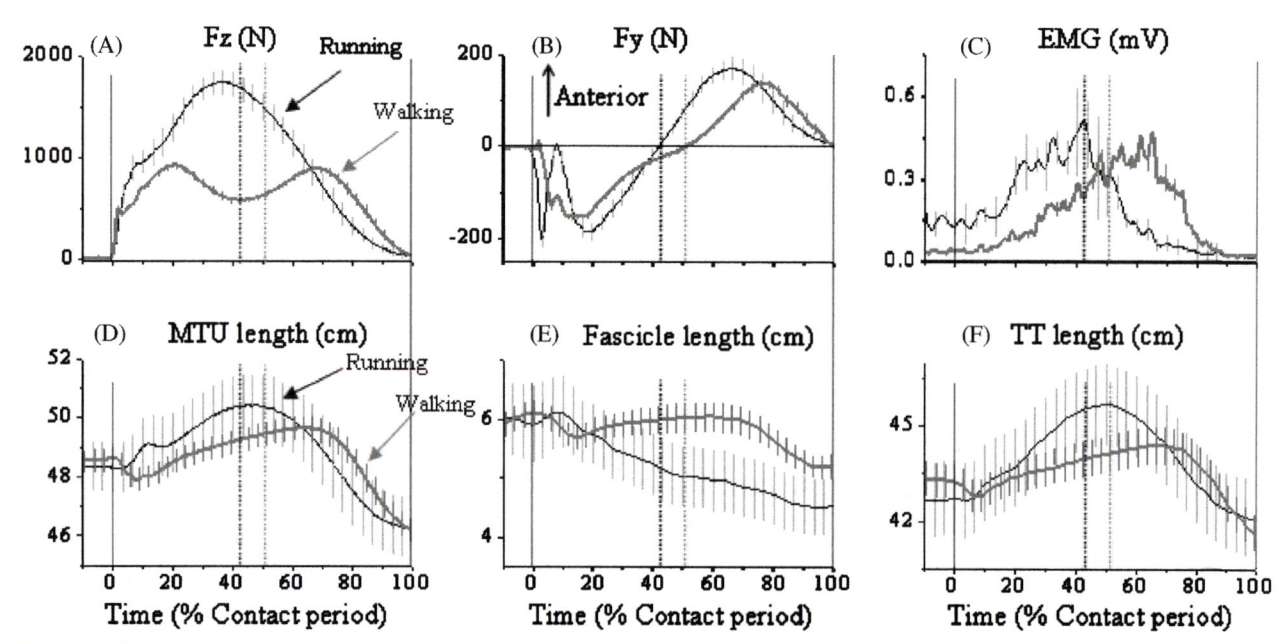

FIG. 2.5.10f The measured and calculated parameters recorded during the contact phase of running and walking with stance phase normalised to 100%. The vertical (A: Fz) and horizontal (B: Fy) GRF, EMG activities of medial gastrocnemius (C: EMG), the length of muscle–tendon unit (D: MTU), fascicle length (E), and tendinous tissue length (F: TT) are demonstrated. Vertical lines denote contact, the transition point from braking to acceleration and toe-off. *(Image from Ishikawa, M., Pakaslahti, J., Komi, P.V., 2007. Medial gastrocnemius muscle behavior during human running and walking. Gait Posture 25(3), 380–384.)*

Applying extra stimulus to gastrocnemius and soleus separately during gait has provided some insight into gastrocnemius and soleus agonist and antagonist interactions in late midstance limb stability. Midstance and terminal stance gastrocnemius stimulation influences posterior pelvic tilt, hip and knee flexion, and midstance ankle dorsiflexion angles. Soleus midstance stimulation induces anterior pelvic tilt, knee extension, and ankle plantarflexion (Lenhart et al., 2014). The opposing action of the pelvic tilt, knee, and ankle moments can be appreciated when it is considered that triceps surae is the primary mechanism by which the posture of the hinged inverted pendulum is maintained throughout late midstance. Gastrocnemius-stimulated action appears to be particularly influenced by the knee-to-ankle moment arm ratio (Lenhart et al., 2014), probably through effects likened to Lombard's paradox via its posterior femoral condyle proximal attachments (Fig. 2.5.10g). Modelling suggests that gastrocnemius muscle–tendon mechanics and energetics are more sensitive to changes in tendon compliance than the muscle–tendon mechanics and energetics of soleus, with gastrocnemius increasingly less able to return stored energy during heel lift under increasing compliance (during slower gait speeds) compared to soleus (Orselli et al., 2017). This probably reflects the increased number of joints crossed by gastrocnemius, and this is important in relation to subtendon loading of the Achilles as stiffness is often compromised with age or injury (Orselli et al., 2017).

The separate gastrocnemius muscle bellies are linked through the gastrocnemius aponeurosis before joining the soleus to form the Achilles tendon. The sliding shear displacement between the gastrocnemius and soleus is affected by the angulation of the knee during ankle motion. Shear displacement of the aponeurosis during ankle plantarflexion has been reported to be 36% more in knee extension than in flexion (Karakuzu et al., 2017). Greater gastrocnemius aponeurosis displacement at maximal force exceeds that of soleus (Bojsen-Møller et al., 2004). However, in knee flexion, soleus displacement was greater with the direction of shear between the muscles' connective tissues differing significantly between the two positions and the range of contractions, likely influencing changes within the local strains of the Achilles subtendons (Maganaris, 2003; Bojsen-Møller et al., 2004). The width of the aponeurosis increases with muscular force production (Scott and Loeb, 1995; Maganaris et al., 2001). Length changes of the aponeurosis are greater when passive than during muscle activity (Scott and Loeb, 1995; Lieber et al., 2000), but the overall aponeurosis displacement relates to the Achilles tendon lengthening deformation (Magnusson et al., 2003).

Regional patterns of recruitment within the muscles change during dynamic tasks, influencing the pattern of strains within the muscles during varying activities (Wakeling, 2009). This includes changes in the way forces are distributed proximally and distally within the muscles (Karakuzu et al., 2017). During submaximal plantarflexion of the ankle, an activation torque occurs in soleus larger than that within gastrocnemius but with transmission of some of this active force

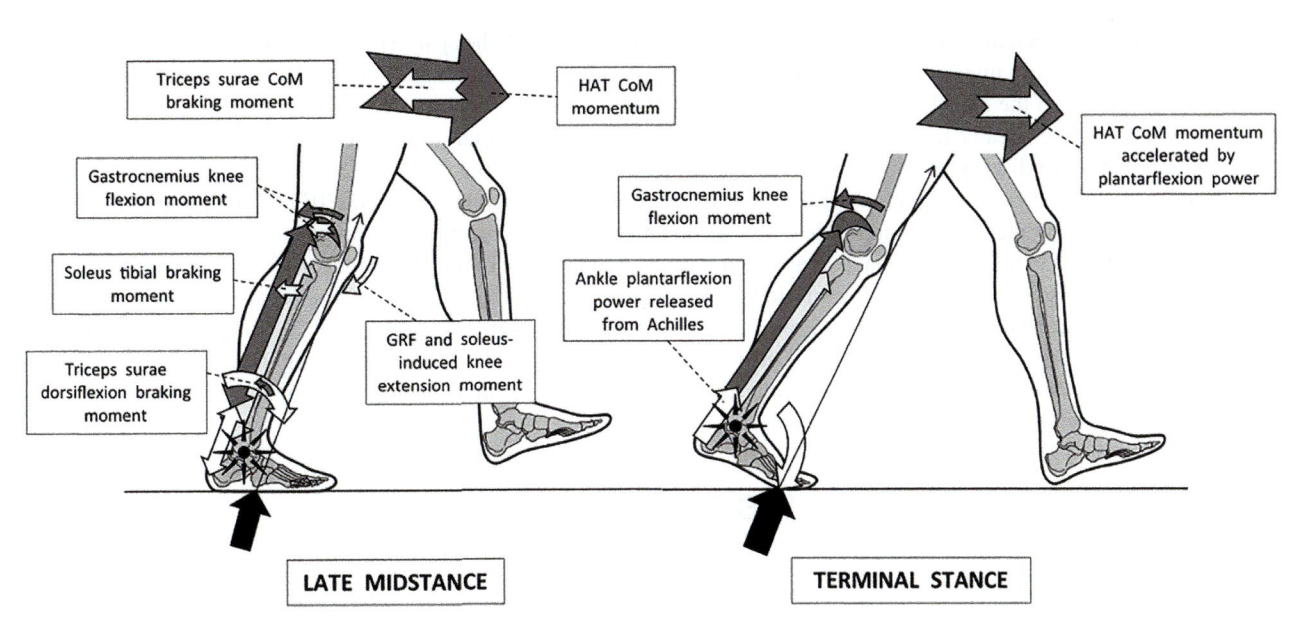

FIG. 2.5.10g Schematics of late midstance (left) and early terminal stance (right) demonstrating how triceps surae functions through generating moments around the ankle joint. The combined action of triceps surae is to brake acceleration of the CoM (dark grey arrow inserted in large white arrow of CoM momentum) as the ankle undergoes dorsiflexion, which in walking occurs during late midstance. Although gastrocnemius and soleus work together to provide this moment of restraint and in so doing load the Achilles with elastic stored energy, they also antagonise each other at the knee. The restraining action on anterior tibial rotation caused by soleus causes the knee to undergo an extension moment as the CoM continues forward. Gastrocnemius, by crossing the knee with its medial and lateral proximal tendons, develops a knee flexion moment during late midstance as it restrains the anterior rotation of the tibia. At heel lift, the Achilles releases its stored plantarflexor power, accelerating the CoM (grey arrow now points in same direction as CoM white arrow). Through the influence of the gastrocnemius moment, ankle plantarflexion induces an increased knee flexion moment. The result is that the knee flexes as the ankle plantarflexes during terminal stance, continuing this action as the hip flexors activate at preswing. *(Permission www.healthystep.co.uk.)*

onto the distal parts of the gastrocnemius fascicles, elevating the loads through the fascial elements (Karakuzu et al., 2017). This causes local lengthening by inter-synergistic myofascial force transmission (Karakuzu et al., 2017), a capability noted in passive (Huijing et al., 2011; Yaman et al., 2013) and active human muscles (Bojsen-Møller et al., 2010).

Two distinct forms of soleus–gastrocnemius junctions are known, the most common of which involves the two muscles contributing collagen fibres directly to the Achilles. The alternative junction is one in which the gastrocnemius aponeurosis inserts more proximally into the underlying soleus aponeurosis, with a long rotation of the separate collagen fibres at the junction (Bojsen-Møller et al., 2004). The muscle and tendon fibres twist as they descend distally, with the medial fibres of gastrocnemius rotating 90° anticlockwise to become posteriorly situated at the calcaneal attachment and the posterior fibres gaining attachment laterally (Bojsen-Møller et al., 2004). Variation also occurs in the anterior–posterior diameter of the Achilles and within the soleus–gastrocnemius junction formation compared to the calcaneal attachment of the Achilles. This latter variation affects the so-called 'free tendon length' that represents the length of tendon with no musculotendinous junction on it (Bleakney et al., 2002; Kader et al., 2002; Magnusson et al., 2003).

Gastrocnemius is known to be primarily active with soleus during late midstance to heel lift, but there are periods of activity outside of this. These include at initial contact, probably influencing loading response's knee flexion angles and stability (Di Nardo et al., 2016; Mengarelli et al., 2017, 2018). Gastrocnemius lateralis' most common activation, found in 30%–50% of strides, consists of two patterns (Di Nardo et al., 2013). The first pattern starts at the end of loading response into early midstance, which only ends after heel lift in terminal stance. However, this can continue into preswing (Agostini et al., 2010; Di Nardo et al., 2013). Another period of activation occurs during mid and terminal swing (Di Nardo et al., 2013). In around 20%–40% of strides, the gastrocnemius lateralis splits the stance phase into two separate activations. This is something that can also occur during swing. However, in 10–50% of strides, the swing phase activity is totally missing (Di Nardo et al., 2013). Stance phase gastrocnemius lateralis activity is always present and fits expectations with reducing the plantarflexion ankle angle to aid CoM lifting through limb extension in early midstance. This is followed by increasing activity during periods resisting ankle dorsiflexion throughout late midstance to finally generating an ankle plantarflexion moment at heel lift. In around 70% of strides, gastrocnemius lateralis is silent during preswing (Di Nardo et al., 2013) suggesting some fine-tuning behaviour in some strides but not all. The explanation of the midswing and preswing activity may relate to frontal plane positioning to aid ground clearance in the former, and to assist foot eversion and adduction in the latter if required.

Half of medial gastrocnemius is made up of type I muscle fibres, although the spatial distribution is unknown (Karakuzu et al., 2017). Low intensity myofascial force is maintained by recruitment of small and intermediate muscle units of type I and II muscle fibres, but the muscle shows heterogeneity throughout (Karakuzu et al., 2017) (Fig. 2.5.10h). Distal parts of the muscle demonstrate greater fatigue resistance characteristics than proximal parts (Gallina et al., 2011), with slower motor units found more distally and faster ones more proximally (Hodson-Tole et al., 2013). At higher intensity plantarflexion contractions, the proximal regions of the muscle are more active (Kinugasa et al., 2011). Submaximal isometric plantarflexion causes greater changes in fascicle pennation angles in distal compared to proximal regions of the gastrocnemius medialis (Rana et al., 2013). Greater changes in fascicle length occur in the passive gastrocnemius medialis muscle during ankle motion rather than during knee motion (Hodson-Tole et al., 2016). During submaximal activation, heterogeneous strain distribution has been noted with extended knee positions (Karakuzu et al., 2017) with lengthening of the proximal ends and shortening of the distal ends of the fascicles (Pamuk et al., 2016). It is also thought that in knee extension, the gastrocnemius aponeurosis is displaced a greater distance in a proximal direction during isometric positions of ankle plantarflexion (Bojsen-Møller et al., 2004).

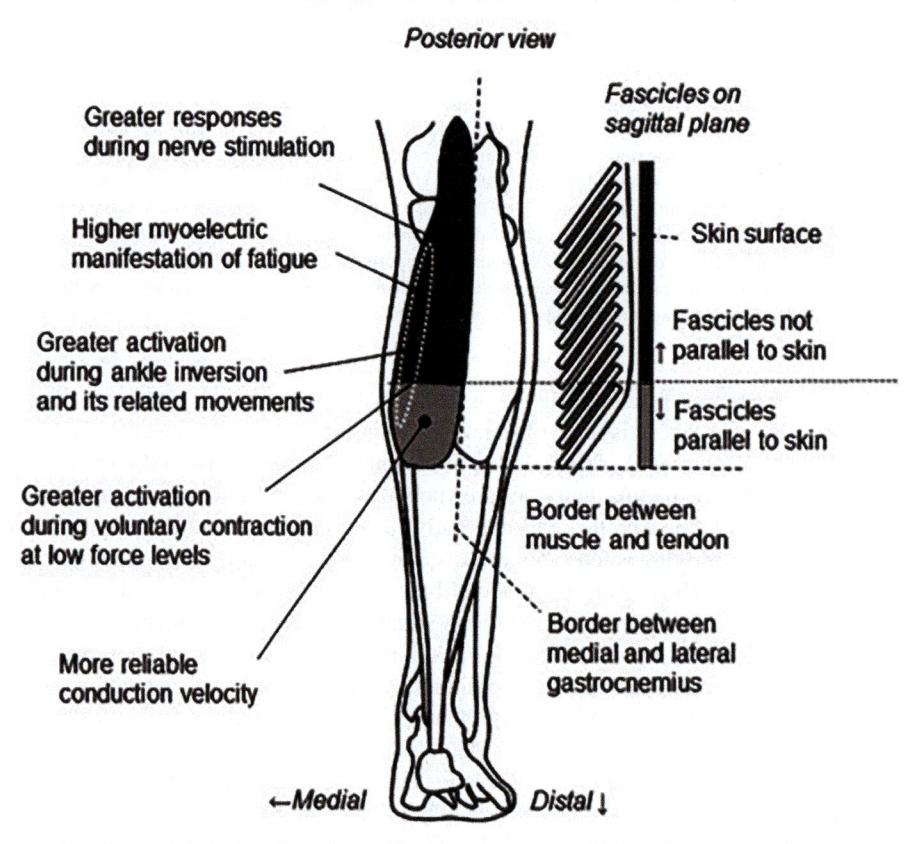

FIG. 2.5.10h Differences have been noted within regions of the medial gastrocnemius muscle through the use of high-density surface electromyography. Alignment of muscle fascicles to the skin also varies in different regions and certain areas seem more vulnerable to fatigue. Note the region associated with activation on ankle inversion (eversion resistance) is particularly medially positioned, providing a longer lever arm through the Achilles to the calcaneus to resist such motion. *(Image from Watanabe, K., Vieira, T.M., Gallina, A., Kouzaki, M., Moritani, T., 2021. Novel insights into biarticular muscle actions gained from high-density electromyogram. Exerc. Sport Sci. Rev. 49(3), 179–187.)*

Achilles tendon

The Achilles is subjected to high strains induced by mechanical loading, and as a high-stress energy-storage tendon, it holds considerable elastic energy which it must release appropriately during gait. It experiences higher in vivo stresses than other human tendons (Ker et al., 1988). Yet interestingly, it seems to have similar mechanical properties to many tendons rather than being materially better adapted to high in vivo stresses through developing differences in its material properties (Wren et al., 2001a). This appears to explain why the Achilles needs to be so thick (Ker et al., 1988) and also to be constructed from four subtendons as this increases mechanical strength through extra layers of structural hierarchy. As in all tendons, the

Achilles demonstrates viscoelastic properties of strain rate dependency with failure stress and strain being 15% higher during faster loading rates, despite there being no significant difference in elastic modulus during different rates of loading (Wren et al., 2001a) (Table 2.5.10). Stiffness increases with loading rate in a linear manner (Theis et al., 2012) so that like all tendons, the Achilles is stiffer, stronger, and more elastic in running and jumping than during walking. Furthermore, the slower the walk, the more compliant and mechanically weaker the tendon.

TABLE 2.5.10 Mean (SD) mechanical properties of the human Achilles tendon from grouped data, with modulus and stress reported in megapascals and failure load in Newtons. The 1 mm s^{-1} rate corresponds with a strain rate of approximately 1% s^{-1}. The 10 mm s^{-1} rate corresponds with a strain rate of approximately 10% s^{-1}.

	Rate 1 mm s^{-1} ($n = 9$)	10 mm s^{-1} ($n = 9$)	Combined ($n = 18$)
Modulus (MPa)	816 (218)	822 (211)	819 (208)
Failure load (N)	4617 (1107)	5579 (1143)	5098 (1199)
Failure stress (MPa)	71 (17)	86 (24)	79 (22)
Failure strain %			
Bone–tendon complex	12.8 (1.7)	161.1 (3.6)	14.5 (3.2)
Tendon substance	7.5 (1.1)	9.9 (1.9)	8.8 (1.9)

Data from Wren, T.A.L., Yerby, S.A., Beaupré, G.S., Carter, D.R., 2001. Mechanical properties of the human Achilles tendon. Clin. Biomech. 16(3), 245–251.

Strain is presented as a percentage change in dimensions such as length or width. Strain passes through the myofascial structures, such as the aponeurosis of gastrocnemius and each subtendon of the Achilles, in variable amounts to reach the enthesis on the calcaneus, and each area has been investigated. On cyclical loading to failure-strain distribution, differences are noted between the proximal and distal tendon with the distal tendon showing higher strain rates prior to failure at over 20% (Wren et al., 2001a). Subregional strain differences have been recorded at the intra-tendinous level in cadavers (Wren et al., 2001a; Lyman et al., 2004; Lersch et al., 2012), with experimentation surprisingly showing that greater displacement occurs in the anterior tendon on ankle dorsiflexion (Handsfield et al., 2017) (Fig. 2.5.10i). This may result from different mechanical properties in different subtendons and possibly indicates that the medial gastrocnemius subtendons, which form the posterior Achilles fibres, are stiffer than the anterior subtendon of soleus (Handsfield et al., 2017). The deep (anterior/ventral) portion of the Achilles undergoes greater compressive strain than the superficial (posterior/dorsal) portion of the Achilles as they are pressed against the posterior calcaneus during ankle dorsiflexion (Lersch et al., 2012). The position of the calcaneus also alters the strain between the subtendons in the frontal plane (Zifchock and Piazza, 2004; Lersch et al., 2012).

The mechanical properties of the Achilles and the triceps surae muscle are therefore influenced by sagittal plane flexion–extension joint angles, as relatively slackened lengths within the muscle and tendon occur at different joint angulations (Hug et al., 2013). Shear elastic modulus is linearly related to muscle stress both isometrically (Bouillard et al., 2011, 2012) and during passive stretching (Maïsetti et al., 2012), and is altered depending on the knee and ankle position (Hug et al., 2013). At 90° knee flexion, the shear elastic modulus remains low in medial gastrocnemius which remains relatively slack until around 10° of ankle dorsiflexion. However, the Achilles elastic modulus is low at around 40° plantarflexion but increases rapidly when passing into dorsiflexion (Hug et al., 2013).

This confirms that joint angulations influence the locations of the peak strains within the triceps surae-Achilles complex. Forces within the medial portion of the Achilles tendon exceed those of the lateral portion when gastrocnemius medialis and soleus are loaded, but lateral forces exceed medial forces when either the gastrocnemius lateralis is loaded or the whole triceps surae is loaded (Arndt et al., 1999). It is clear that muscle activity within the triceps surae directly affects stress–strain within the Achilles and that joint angulation in the sagittal plane also influences strain location within the tendon. However, the position of the calcaneal enthesis in orientation to the muscles also plays a significant role.

A number of important structures help protect the Achilles from compression and friction on the posterior surface of the calcaneus, most of which are intimately related to the enthesis. Kager's fat pad is an adipose structure located between the Achilles and flexor hallucis longus tendon and the posterior calcaneus, participating in lubrication and helping distribute anterior compression stresses away from the Achilles along with the retrocalcaneal bursa. It has connections with the inferior part of the anterior talofibular ligament and calcaneofibular ligament (a region known together as the lateral fibulotalocalcaneal ligament complex), the posterior talofibular ligament, and the flexor and superior peroneal retinaculum, with many fascia crura connections orientated in the frontal plane at the level of the posterior talus (Szaro et al., 2021). Such intermuscular myofascial connections likely influence local muscle and tendon function (Szaro et al., 2021). These complex connections support the existence of a fascial integration system with Kager's

nonuniform motion expected
more displacement in posterior portion due to larger moment arm

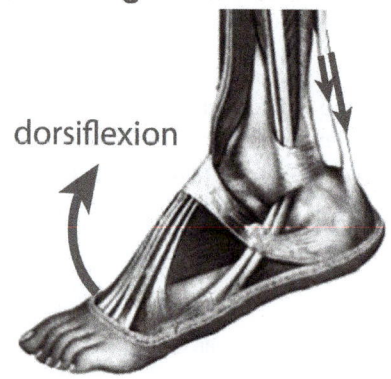

dorsiflexion

nonuniform motion observed
less displacement in posterior portion

dorsiflexion

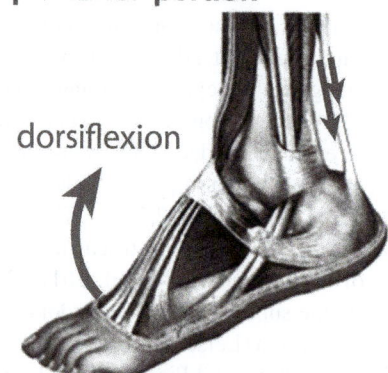

FIG. 2.5.10i Images demonstrating the nonuniform strain displacement on the Achilles during dorsiflexion, with greater anterior fibre displacements occurring compared to those of the posterior fibres, something not expected by pure logic. *(Image from Handsfield, G.G., Inouye, J.M., Slane, L.C., Thelen, D.G., Miller, G.W., Blemker, S.S., 2017. A 3D model of the Achilles tendon to determine the mechanisms underlying nonuniform tendon displacements. J. Biomech. 51, 17–25.)*

fat pad, which may link to dysfunction of the fat pad and its interconnected structures to some Achilles tendinopathies (Szaro et al., 2021) (Fig. 2.5.10j).

Achilles enthesis biomechanics

The Achilles attaches to the posterior surface of the calcaneus with the area proximal and deep to the attachment being occupied by the retrocalcaneal bursa. This is because the anterior surface of the Achilles can be subjected to high compressive stresses against the posterior calcaneus under dorsiflexion angles. During dorsiflexion, the tendon bends and presses into the bursa which flattens between the tendon and bone. Muscular, ligamentous, or joint capsular attachments are termed enthesis organs (Benjamin and McGonagle, 2001) and the enthesis organ of the Achilles tendon includes three fibrocartilage structures: two lining the bone at the bursal walls to protect it from tendon-bone compression and referred to as the periosteal fibrocartilage and one for the tendon-bone junction, known as the sesamoid fibrocartilage (Milz et al., 2002; Robson et al., 2004). The orientation of the fibrocartilage at the Achilles attachment is clinically relevant as MRI techniques demonstrate subtle signal intensity abnormalities on images taken in patients with chronic Achilles pain. These are associated with a number of conditions such as ectopic tendon calcification, calcaneal spurs, retrocalcaneal bursitis, and *Haglund's deformity* (Milz et al., 2002). Vaishya et al. (2016) and Debus et al. (2019) differentiate between Haglund's deformity, which is an exostosis of the posterior superior calcaneus, and Haglund's syndrome, which is the same but includes retrocalcaneal bursitis, calcaneal tendon bursitis, and Achilles tendinosis. Simply stated, Haglund's deformity is a bone stress reaction and enlargement around the enthesis, whereas the syndrome also involves local soft tissue inflammation (Fig. 2.5.10k).

FIG. 2.5.10j The Kager's fat pad region and its fascial communications at the level of the posterior talus with the local tendons shown in black, muscles in mid-grey, and bones in light grey. Upper left image demonstrates cross section of left ankle (with tip of fibula and the talar trochlear surface shown) and lower left image presents sagittal plane schematics. Structure demonstrated are: (1) the lateral fibulotalocalcaneal ligament complex (LFTCL), (2) the posterior talofibular ligament, (3) the superior lamina of the LFTCL, (4) branch of the LFTCL to the node-like structures that link to the fascia crura, (5) the node-like structures, (6) the superior peroneal retinaculum, (7) the connection between the anterior talofibular ligament and the superior peroneal retinaculum, (8) the connection between the node-like structures and the superior peroneal retinaculum, (9) the connection between the lateral part of the Achilles paratenon (10) and the superior peroneal retinaculum, (11) connections between node-like structure and the medial part of the retinaculum, (12) the flexor retinaculum, (13) the connection between the node-like structures and the flexor retinaculum, (14) the muscular septum between the flexor digitorum longus and tibialis posterior, (15) the posterior talofibular ligament, (16) the connection between the posterior talofibular ligament and flexor hallucis longus via the node-like structures, and (17) the direct connections between the posterior talofibular ligament and flexor hallucis longus. Dysfunction of any of these structures via their complex fascial interconnections may influence function among the others, explaining why Achilles tendinopathy may be so multifactorial in origin demonstrating quite different causes between individuals. *(Images from Szaro, P., Polaczek, M., Ciszek, B., 2021. The Kager's fat pad radiological anatomy revised. Surg. Radiol. Anat. 43(1), 79–86.)*

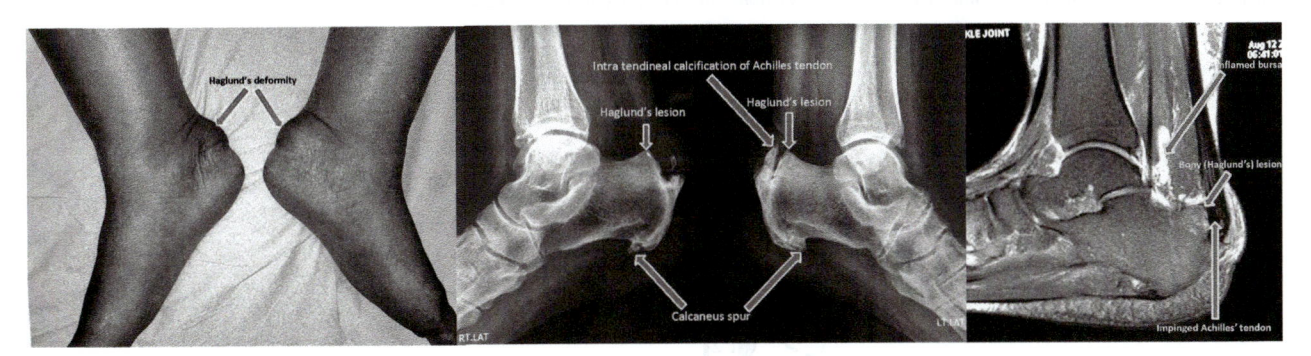

FIG. 2.5.10k From left to right, a clinical photograph, a lateral view radiograph, and an MRI image of Haglund's syndrome. The enlargement of the posterior heel, which is osseous and soft tissue in origin, can clearly be seen in each image. It is the bony prominence that is the Haglund's lesion. Calcifications with the Achilles and plantar aponeurosis are also extensive, indicating that excessive stresses on fibrocartilage and surrounding inflammatory processes have initiated ossification. *(Images from Vaishya, R., Agarwal, A.K., Azizi, A.T., Vijay, V., 2016. Haglund's syndrome: a commonly seen mysterious condition. Cureus 8(10), e820.)*

There appears to be a close correspondence between the position and size of the periosteal and sesamoid fibrocartilage, probably because their functions are interdependent. The superior calcaneal tuberosity, on which the periosteal fibrocartilage is located, acts as a pulley for the distal part of the Achilles tendon (Milz et al., 2002). This requires a corresponding area of tendon to be modified in order to withstand the compression that occurs within this pulley mechanism. The sesamoid fibrocartilage serves such a purpose, dissipating the bending of tendon fibres away from the hard bone tissue interface (Fig. 2.5.10l). This effectively creates another pulley mechanism suggesting two functional pulleys running in series, providing an efficient moment arm for the Achilles (Milz et al., 2002). At neither pulley does significant longitudinal lengthening excursion occur within the Achilles tendon fibres (Fig. 2.5.10m).

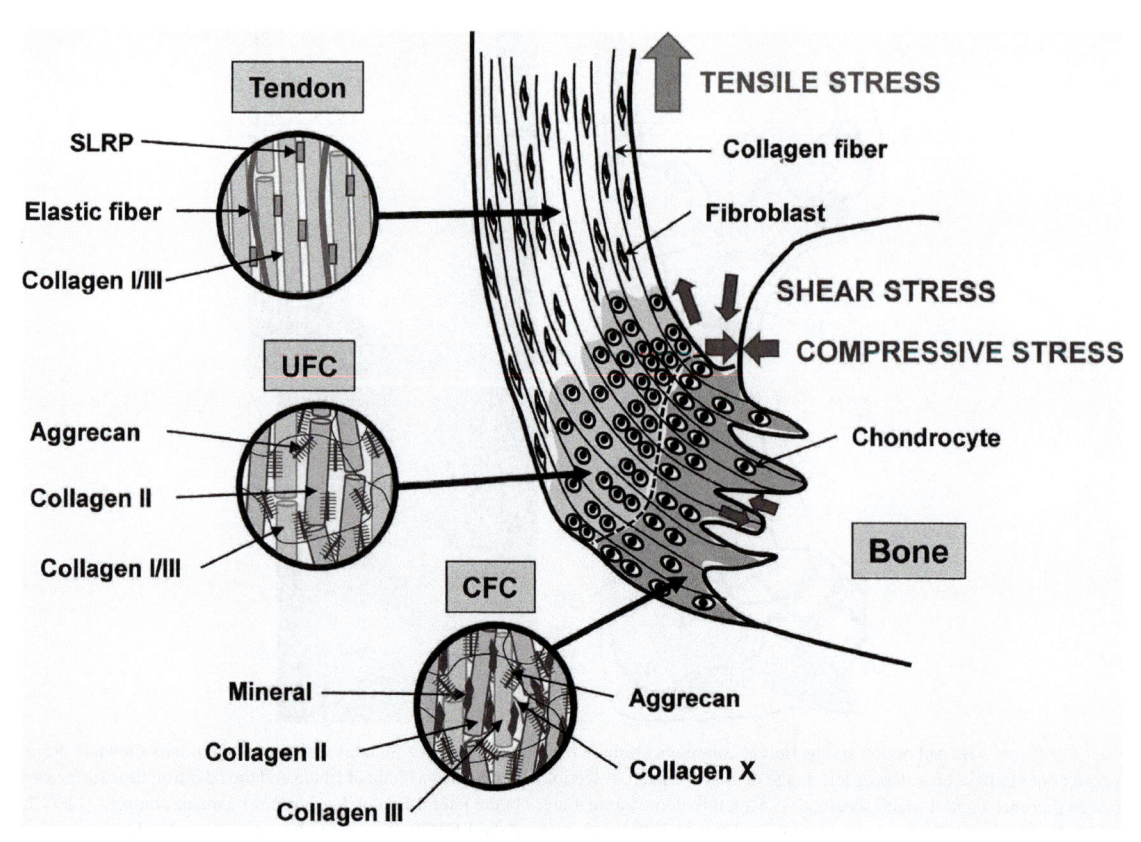

FIG. 2.5.10l A schematic of the enthesis organ, a structure exemplified by the Achilles tendon enthesis. The tendon approaching the enthesis expresses its proteoglycans (SLRP) and elastic and collagen fibres. The enthesis organ consists of the unmineralised fibrocartilage zone (UFC) and the calcified fibrocartilage zone (CFC). Different types of collagen help alter material properties across the enthesis organ to prevent zones of sudden change in mechanical properties. At the tendon-to-bone interface, tendon fibroblasts are replaced by chondrocytes within the fibrocartilage. The enthesis organ is subjected to areas of high shear and compressive stresses, while the tensile stresses are mainly managed through the more flexible tendon fibres. These forces are complex in the Achilles enthesis through its four subtendon enthesis organs and because of changing angles that alter compression forces against the posterior calcaneal tuberosity. The presence of sesamoid cartilage within the Achilles tendon in the area lying over the superior posterior tuberosity, a bursa situated between the tendon and bone, and periosteal fibrocartilage over the posterior calcaneal surface deep to the bursa, should all help to moderate peak shear and compression forces. *(Image from Roffino, S., Camy, C., Foucault-Bertaud, A., Lamy, E., Pithioux, M., Chopard, A., 2021. Negative impact of disuse and unloading on tendon enthesis structure and function. Life Sci. Space Res. 29, 46–52.)*

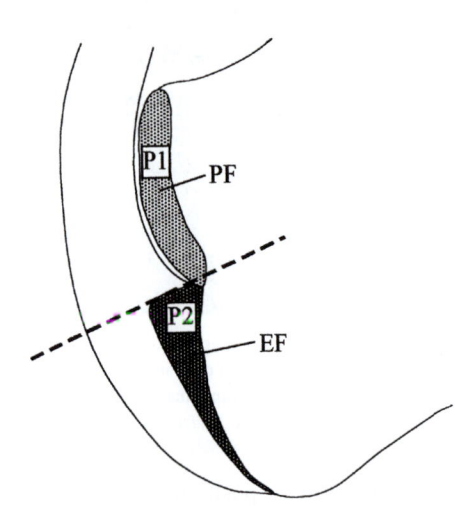

FIG. 2.5.10m A schematic to represent how the periosteal fibrocartilage (PF) and enthesis fibrocartilage (EF) can be viewed as a two-pulley (P1 and P2) system arranged in-series with each other to increase the moment arm of the Achilles tendon at its distal attachment. The periosteal fibrocartilage covers the superior tuberosity of the calcaneus, acting as a pulley below the level of the broken line during dorsiflexion, altering the direction of the collagen fibres within the tendon. *(Image from Milz, S., Rufai, A., Buettner, A., Putz, R., Ralphs, J.R., Benjamin, M., 2002. Three-dimensional reconstructions of the Achilles tendon insertion in man. J. Anat. 200(2), 145–152.)*

The Achilles has a greater attachment proximally to the enthesis than distally which supports its role as a pulley (Milz et al., 2002). Regional variances exist in the quantity of calcified fibrocartilage in different parts of the enthesis, with more calcified tissue presenting more proximally than distally (Milz et al., 2002). The interface between bone and tendon is primarily subjected to tension within the calcified interface, allowing force transfer into the calcaneus. However, the tensile forces of triceps surae do not just transfer into the enthesis via Sharpey's fibres that ensure adhesion of tendons, ligaments, and periosteum to the bone. The 3D orientation of the trabecular spicules in the inferior posterior calcaneal tubercle suggests that tensional forces are relayed through the calcaneus in the direction of the plantar aponeurosis (Milz et al., 2002), thereby allowing force to be transferred through bone to the plantar aponeurosis and via this structure to the forefoot. This would functionally assist and complement the transmission of ankle power at plantarflexion into the forefoot.

Achilles tendon stress contributions

The Achilles tendon and plantar aponeurosis are continuous below the calcaneus in the early years of life, although this continuation reduces with age (Snow et al., 1995). While the continuation is present, the attachment of the Achilles to the calcaneus and the plantar aponeurosis helps couple ankle motion to vault mechanics and forefoot posture. This helps to explain some of the findings of force generation of the triceps surae around the posterior calcaneal tubercle as reported by Zifchock and Piazza (2004) and Lee and Piazza (2008), possibly accounting for some of the common loss in ankle power reported to be occurring with age (Judge et al., 1996; McGibbon, 2003). Indeed, this may help explain the development of the (usually) 'middle-aged' issue of plantar fasciopathy. It is worth noting that increased medial gastrocnemius stiffness has been associated with the presence of plantar fasciopathy (Zhou et al., 2020), with loss of triceps surae power and flexibility being a known problem associated with ageing.

The Achilles tendon demonstrates considerable variability in fascicle strain amplitudes and deviations, probably because of inter-individual variation in calf muscle–tendon complex anatomy (Dalmau-Pastor et al., 2014; Hansen et al., 2017). It has long been noted that about half the population has a soleus-dominant contribution to the Achilles tendon, with a long distal gastrocnemius aponeurosis contribution (Cummins et al., 1946; Bojsen-Møller et al., 2004). However, it has also been reported that around a third of individuals have the muscles contributing equally to the Achilles and no distal gastrocnemius aponeurosis (Cummins et al., 1946), implying differences in fibre length and orientation (Karakuzu et al., 2017). Myofascial connections are complex and have nonlinear (non-Newtonian) properties and variable distributions, varying the loading on different parts of the muscle and tendon in different individuals (Karakuzu et al., 2017). Measurements of human muscle force in tendons indicate that differences in force variation mechanics exist among individuals (Yucesoy, 2010; Yucesoy et al., 2010), and it is important that clinicians are aware of this.

Frontal plane influence on Achilles strain distributions

As already discussed, the ankle and knee flexion–extension angles affect triceps surae and tendon biomechanics (Hug et al., 2013). However, the Achilles also crosses the subtalar joint and thus the ankle–subtalar joint has variable frontal plane influences on the Achilles. The subtalar joint lacks freedom of motion and requires external forces to create flexibility between the talus and the calcaneus. Without these external forces, the calcaneus is coupled to ankle motion as if continuous with the talus (Scott and Winter, 1991; Leardini et al., 2001).

However, alterations in the position of the calcaneus result in changes in the intra-tendinous strains of the Achilles (Lersch et al., 2012) and adjust the primary point of effort on the Achilles (Zifchock and Piazza, 2004; Lee and Piazza, 2008). Medial–lateral, proximal–distal, and dorsal–ventral distributions in tendon strains within the Achilles seem more influenced by the position of the calcaneus and subtalar (whole rearfoot) kinematics than through muscular variations in force generation within the triceps surae (Lersch et al., 2012). This means that the twists within the triceps surae and Achilles muscular and tendon fascicles are of paramount importance to balancing heterogeneous tendon strains in relation to calcaneal position (Lersch et al., 2012) (Fig. 2.5.10n).

Using cadaveric specimens, Lersch et al. (2012) found that increasing calcaneus frontal plane angles (from 0° to 7.5° and then to 15°) increased tensile strain in both inversion and eversion from 0°, but caused ~20% more strain in eversion. Increasing inversion heightens load in the lateral aspect of the tendon and decreases strain in the medial aspect. Increasing eversion resulted in increasing strain in the medial tendon and decreasing strain in the lateral portion. This correlated with the findings of Zifchock and Piazza (2004) and Lee and Piazza (2008) who found that the point of effort on the calcaneus changed from lateral during inversion to medial with eversion. However, Lersch et al. (2012) also found that increasing frontal plane angles and increasing muscle loading forces resulted in further unilateral increases of strain in both inversion and eversion, but more so when angled in inversion. The central portions of the Achilles experienced lower strain differences than the medial and lateral portions throughout the frontal plane changes (Fig. 2.5.10o).

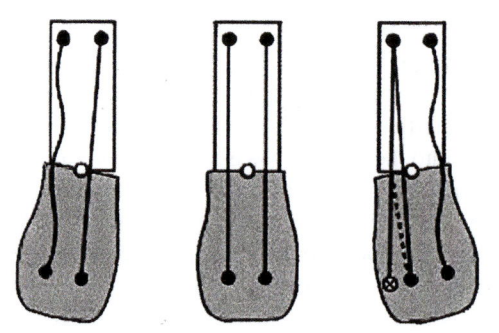

FIG. 2.5.10n A simple schematic of a two-point model of Achilles tendon function applied to the right posterior heel demonstrates the frontal plane advantages of having four subtendons functioning within the Achilles. The position of the ankle–subtalar joint axis within the frontal plane is indicated in the image by the white circle. Medial and lateral sides of the Achilles are indicated by the black lines, with black spots indicting the distal attachment's sides. Inversion (left) causes increased tensile strain in the lateral fibres, but it reduces strain medially while a vertically positioned rearfoot keeps tensions even. Eversion (right image) tensions the medial fibres more than the lateral fibres. Interestingly, experiments by Zifchock and Piazza (2004) suggested that the point of effort during rearfoot eversion moved beyond the medial margins of the Achilles attachment (circle marked by X). Motion of the instantaneous joint axis of the ankle–subtalar joint that moves medially and laterally during gait, even when the subtalar joint may not have freedom of motion, should also be considered. *(Image from Zifchock, R.A., Piazza, S.J., 2004. Investigation of the validity of modeling the Achilles tendon as having a single insertion site. Clin. Biomech. 19(3), 303–307.)*

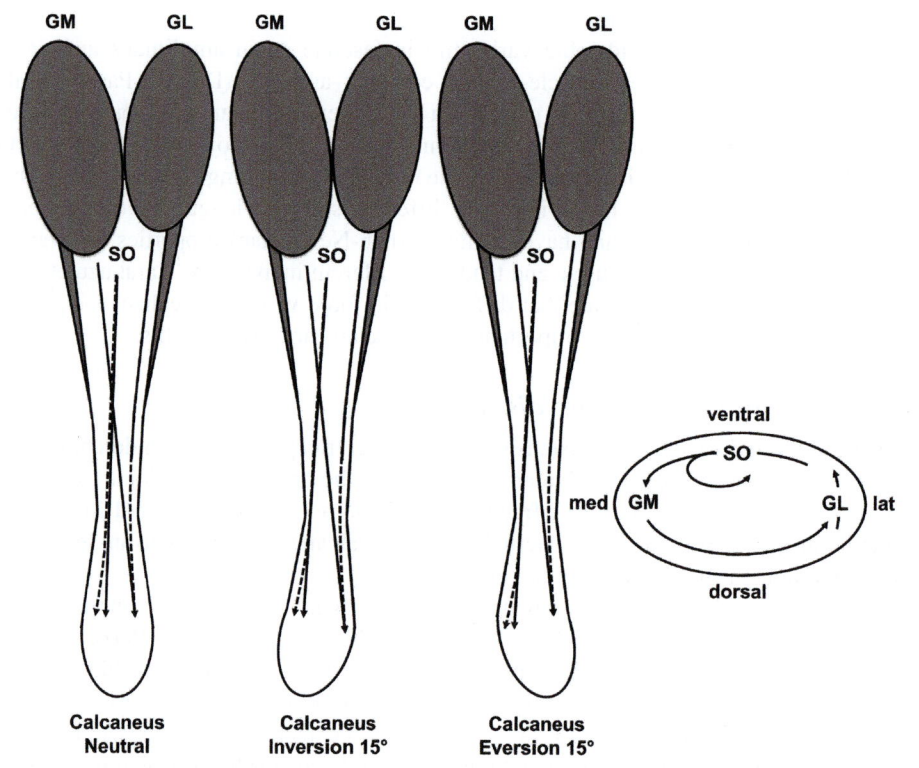

FIG. 2.5.10o Strain behaviour of the right Achilles tendon during 15° inversion and 15° eversion with strain magnitude represented by the length of the arrows, the direction of which indicates the orientation of the fascicles. These are gastrocnemius medialis (GM), gastrocnemius lateralis (GL), and soleus (SO). Dashed lines represent the anterior stresses, while the transverse view shows the fascicle orientation from myotendinous junction to enthesis attachment. *(Image from Lersch, C., Grötsch, A., Segesser, B., Koebke, J., Brüggemann, G.-P., Potthast, W., 2012. Influence of calcaneus angle and muscle forces on strain distribution in the human Achilles tendon. Clin. Biomech. 27(9), 955–961.)*

As already discussed, the deeper (anterior) portions of the Achilles are reported to show consistently larger lengthening displacements than the superficial tendon by around 0.9–2.6 mm (Arndt et al., 2012; Slane and Thelen, 2014; Handsfield et al., 2017), and are under greater compression during dorsiflexion (Lersch et al., 2012). In eversion, the deep or anterior-lateral aspect undergoes greater compression, while the deep-medial portion undergoes greater tensile and compression strains. These strains are considerably less within the superficial areas of the tendon (Lersch et al., 2012). Displacements are smaller in isometric contractions than when the triceps surae is stretched passively, except in the superficial tendon when the knee is flexed at 90° but displacements are usually larger with the knee positioned at 30° flexion (Slane and Thelen, 2014).

Proximal and distal parts of the Achilles are also subjected to variable strains through the changes in calcaneal positions. The superficial aspect of the tendon experiences differences in strains in a proximal–distal direction. Increasing eversion angles are associated with increased compressive strain in the proximal–medial area and tensile strains in the distal–medial superficial tendon (Lersch et al., 2012). Distal–lateral strains are associated with increasing inversion angles that are higher than the proximal–lateral tendon areas, which rise with increasing tendon-loading force (Lersch et al., 2012). Overall, the differences between intra-tendinous strains between medial and lateral aspects did not exceed 1.5% at the superficial aspect or 2.5% in the deep aspects. However, in general, distal strains exceeded proximal strains, with the lateral strains being greater than the medial strains within the proximal Achilles areas and medial strains exceeding the lateral values within the distal tendon (Lersch et al., 2012) (Fig. 2.5.10p). Variation is also noticed between and within specimens, but the Achilles is likely more vulnerable to injury due to the heterogeneous nature of the Achilles strains during various movements on an individual basis, which may be a mechanism worthy of consideration in helping in the development of healing/strain offloading strategies during treatment (Arndt et al., 2012).

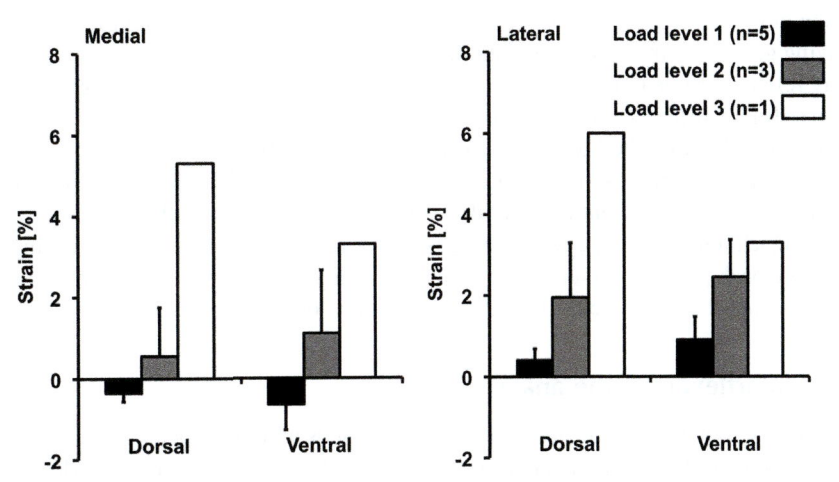

FIG. 2.5.10p Mean dorsal (superficial) and ventral (deep) strains from cadaveric specimens of the medial and lateral aspects of the Achilles tendon under different loading conditions. Triceps surae ratio has been set for soleus, gastrocnemius medialis, and gastrocnemius lateralis at 3:1:1 for loads at level 1 and 2. Level 1 was the more proximal tendon portion near the joining of gastrocnemius and soleus fibres. Level 2 was the mid-portion of the Achilles. At level 3, just proximal to the enthesis, the ratio was set at 6:1:1. Results indicate that medial and lateral and superficial and deep fibres display differences under loading of the Achilles tendon. Distal portions are more strained than the proximal regions. *(Image from Lersch, C., Grötsch, A., Segesser, B., Koebke, J., Brüggemann, G.-P., Potthast, W., 2012. Influence of calcaneus angle and muscle forces on strain distribution in the human Achilles tendon. Clin. Biomech. 27(9), 955–961.)*

Achilles in strength and stiffness

The Achilles and the triceps surae tendon–aponeurosis complex does not seem to adapt its mechanical properties or dimensions in response to increased exercise levels that are sufficient in creating improved cardiovascular performance (Hansen et al., 2003), suggesting that tendon morphology does not respond as easily to increased exercise as does the cardiovascular system. However, habitual runners appear to have a greater cross-sectional area of their Achilles compared to nonrunners (Rosager et al., 2002). Mammalian studies seem to confirm that high stress energy-storing tendons do not adapt their stress–strain properties in response to exercise alone, even though low stress postural tendons seem to do so (Hansen et al., 2003). It has also been reported that animal tendon cross-sectional areas initially reduce with exercise, which is then followed later by hypertrophy (Woo et al., 1982). This suggests a biphasic response of collagen to exercise with a net breakdown response before a net collagen increase, creating a period of possible injury risk on commencing new exercise levels. Because of the narrowing of the tendon as it approaches the enthesis organ of the Achilles, there is a point of stress concentration through reduced cross-sectional area within the triceps surae muscle–tendon unit. The Achilles tendon can rupture, usually occurring within this narrower area, or the enthesis strains can cause a bony avulsion. Both injuries can arise by similar loading mechanisms, with pathology focused to the anatomy associated with the weakest tissue within the individual, i.e. bone or tendon (Wren et al., 2001b).

Ankle plantarflexor stiffness as well as strength seem important as factors contributing to gait characteristics (Salsich and Mueller, 2000). Plantarflexor strength and plantarflexion peak moments are positively correlated (Salsich and Mueller, 2000), but stiffness accounts for around 10% of variability within the plantarflexion peak moment (Salsich and Mueller, 2000). The stiffness of the gastrocnemius and Achilles tendon increases from ankle neutral into dorsiflexion (Aubry et al., 2013; Chino and Takahashi, 2018; Huang et al., 2018; Liu et al., 2018; Taş and Salkin, 2019). Achilles stiffness is increased on average by around 40% of dorsiflexion, although only by 14%–16% within the medial gastrocnemius and yet this stiffness behaviour is quite subject-specific (Taş and Salkin, 2019).

The plantar aponeurosis also stiffens with ankle plantarflexion and toe extension (Huang et al., 2018), as seen in later terminal stance via the windlass mechanism. Plantar aponeurosis shortening during acceleration is only reached during later terminal stance, suggesting that more active mechanisms stiffen the foot prior to and during heel lift (Farris et al., 2020). The whole foot vault has also demonstrated stiffening properties that correlate to ankle dorsiflexion in that increasing foot vault loading and lengthening heightens vault stiffness (Bjelopetrovich and Barrios 2016; Takabayashi et al., 2020); a situation expected of viscoelastic structures during late midstance. Such stiffening-to-length coupling creates a functional link between ankle dorsiflexion during stance phase to the forming of a stiffened ankle–foot complex along with high elastic storage capabilities prior to heel lift. This creates stability and power that is necessary for the start of plantarflexion when coming out of ankle and foot dorsiflexion. This fits in with the energetic relationship reported by Takahashi et al. (2016) of an increased force–velocity relationship and ankle power with a stiffened foot at heel lift, but with increasing metabolic output required from the ankle plantarflexors (primarily soleus) as a result of this stiffness.

Gender also seems to have an effect on the plantarflexor unit stiffness through the Achilles and medial gastrocnemius muscle, with males having greater stiffness (Taş and Salkin, 2019). The difference in stiffness values centres on the medial gastrocnemius and may reflect hormonal differences such as oestrogen interfering with collagen synthesis. However, the whole subject of tendon stiffness difference between males and females does not yet appear to have reached consensus due to the wide differences reported within the literature (Taş and Salkin, 2019).

In patients with diabetes mellitus that is known to cause increased tissue stiffness, the effects on plantarflexion peak moments seem more pronounced and potentially highly pathomechanical, especially for those expressing concurrent peripheral neuropathy (Salsich and Mueller, 2000). There is also a reported significant relationship between improving plantarflexion and reducing peak plantar pressures and the pressure-time integral during gait, making weightbearing daily calf stretches a potentially important treatment to reduce the risk of foot ulcers (Maeshige et al., 2021).

2.5.11 The other plantarflexors of the ankle

The other plantarflexors of the foot at the ankle are located deep to soleus within the deep posterior compartment of the leg and also include the lateral compartment's peroneal muscles. The role of these muscles at the ankle is to collectively act as a tensile brace of frontal and transverse plane movements and stresses (Potthast et al., 2008). From medial to lateral, the posterior deep compartment contains the flexor digitorum longus attached to the posterior surface of the tibia below the popliteus and the intermuscular septum shared with tibialis posterior. Tibialis posterior attaches proximally to the lateral posterior surface of the tibia, the medial posterior surface of the fibula, and the posterior surface of the interosseous membrane between these two bones (Draves, 1986, pp. 267–269). Laterally, flexor hallucis longus attaches to the posterior surface of the fibula, the lower part of the interosseous membrane, and the intermuscular septum from the posterior surface of tibialis posterior, which it largely overlies, medially (Draves, 1986, p. 270).

Inferior to the medial malleolus of the ankle, the tendons of the three muscles cross each other so that the tibialis posterior passes under flexor digitorum longus to run most medial and inferiorly to the malleolus. The flexor hallucis longus tendon travels behind the posterior talar process within a bony groove. It continues distally to a groove under the sustentaculum tali of the calcaneus. The tendon then courses to the hallux, passing between the two sesamoids under the first metatarsal head to attach to the plantar base of the distal phalanx of the hallux through a fibrous enthesis (Frowen and Benjamin, 1995). The tibialis posterior tendon lies deepest, passing around and under the medial malleolus to enter into the foot, and has the longest medial moment arm to the ankle. Its distal attachments are complex, but it primarily attaches to the navicular tuberosity through one tendon slip and to several plantar tarsometatarsal joints (Figs 2.5.11a and 2.5.11b).

Apart from variations in tendon slips (including supplementary slips linking to flexor hallucis longus), flexor digitorum longus usually divides into four slips around the level of the navicular plantar surface (Draves, 1986, p. 267). The posterior lateral side of this tendon is linked to the distal attachment of the quadratus plantae (flexor accessorius) muscle, which is attached proximally via two heads from the plantar surface of the calcaneus body, the lateral of which lies just anterior to the lateral process of the plantar calcaneal tuberosity (Draves, 1986, pp. 294–295). Activation of quadratus plantae straightens the pull of the oblique alignment of the flexor digitorum longus tendon as it enters the plantar foot medially. This likely improves lesser digital flexion force alignment, but also exerts a medial stabilising effect on the lumbrical muscles that are acting on the MTP joints. The four lumbrical muscle bellies bind proximally on the flexor digitorum tendon slips to attach distally to the medial side of the lesser MTP joints where they play an important role in MTP joint stability. The four tendon

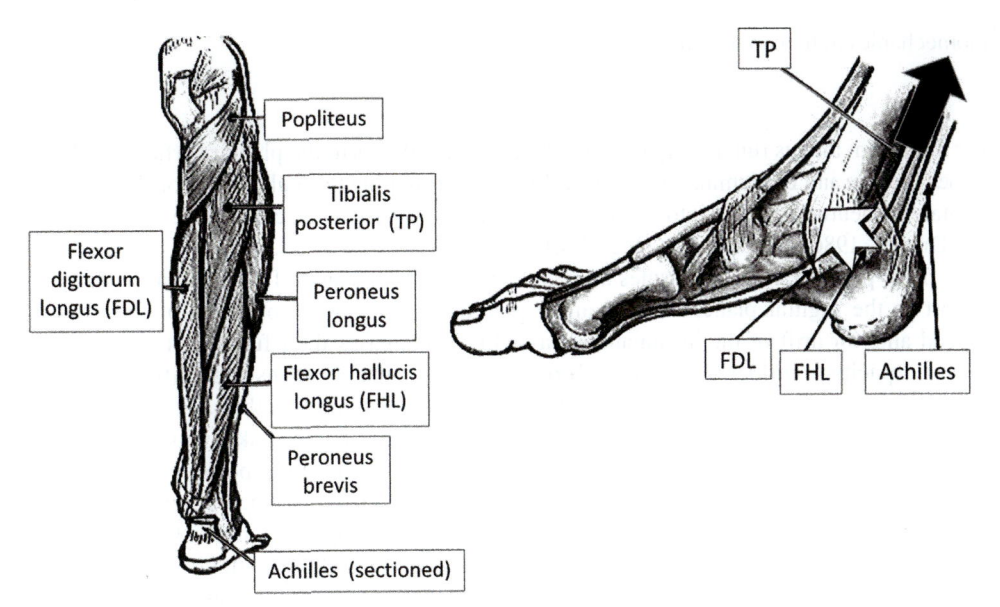

FIG. 2.5.11a The deep muscular compartment of the posterior leg is revealed beneath the triceps surae muscles (gastrocnemius, soleus, and Achilles removed) in the left anatomical drawing of the right leg. With the exception of popliteus, all these muscles can generate ankle plantarflexion moments, as they lie posterior to the ankle joint axis. However, they lie very close to this axis, limiting the plantarflexion influence of their moment arms. Each muscle has a more significant role in frontal and transverse plane stability around the foot and ankle, with more plantarflexing influence on the foot. On contraction (black arrow—right medial view ankle image), their tendons are compressed up against their bone pulleys, behind the malleoli for tibialis posterior (TP) and flexor digitorum longus (FDL), and on the posterior talus for flexor hallucis longus (FHL), providing posterior compression-induced stability to the ankle. This is particularly so during ankle dorsiflexion angles during late midstance. In the right image, the compression vector (white arrow) is shown for the medial muscle group only, pressing the tendons into the tibia. The peroneal tendons act in a similar manner behind the fibula. *(Permission www.healthystep.co.uk.)*

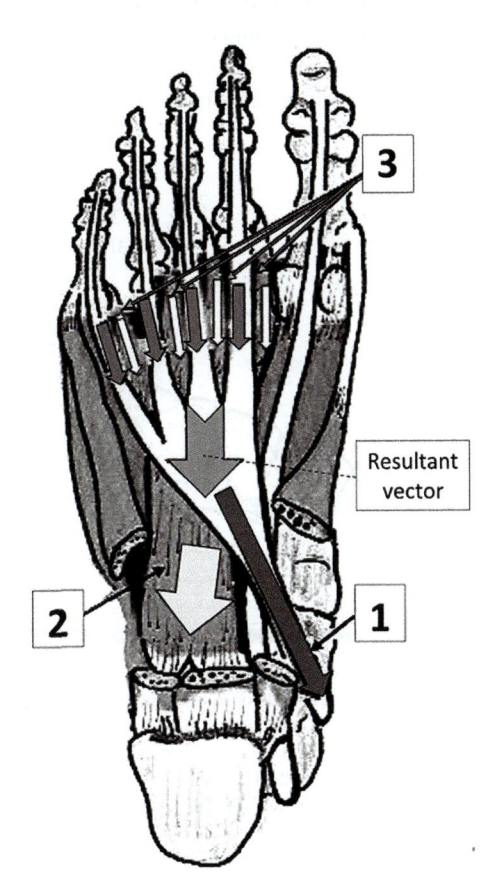

FIG. 2.5.11b The proximal–medial force vector of flexor digitorum longus (1) is adjusted through the force vector of quadratus plantae (2—white arrow) within the foot. Quadratus plantae has strong medial and lateral attachments to the plantar calcaneus under the more superficial plantar muscles (resected here) and is attached distally to the tendon of flexor digitorum longus. This ensures that the tendon slips of the long flexor to the digits pull in a 'purely' proximal direction (small dark-grey arrows), creating a resultant force (large grey arrow) that straightens the action of the lumbricals in the process (3—small white arrows). Lumbrical muscles arise from the flexor digitorum tendon slips. Failure of quadratus plantae risks adduction deformity of the lesser toes. However, the medial stabilisation of the flexor digitorum longus tendon is not lost from the ankle but altered with quadratus plantae dysfunction. *(Permission www.healthystep.co.uk.)*

slips of the flexor digitorum longus run distally to attach by fibrous entheses to the plantar surfaces of the base of the distal phalanx of each toe (Frowen and Benjamin, 1995). They do this after running through the flexor sheaths created by the deep fibres of the plantar aponeurosis bound into a complex of MTP joint regional connective tissues (Bojsen-Møller and Lamoreux, 1979; Draves, 1986, pp. 267–268) (Fig. 2.5.11b).

The tendons of these plantar extrinsic muscles all pass very close to the ankle functional axis. Thus, their ankle lever arms are very short in the sagittal plane, limiting ankle plantarflexion power potentials of these muscles. The ankle dorsiflexion-induced anterior drift of the instantaneous axis likely increases their functional lever arms in dorsiflexion, suggesting that their capacity to influence as plantarflexors is more a capacity to resist dorsiflexion and anterior rotation of the tibia. These assistor plantarflexors will also have their greatest lever arms in the sagittal plane at maximum dorsiflexion just prior to heel lift, and supply just a little of the plantarflexor power at the ankle with triceps surae. Their activity decreases after heel lift as the ankle plantarflexes and the instantaneous axis shifts posteriorly. This suggests that their function shifts away from the ankle joint and more towards the foot as terminal stance progresses.

The assistor ankle plantarflexors all bridge the vault of the foot. This is important in assisting midfoot stiffness prior to heel lift, resisting midfoot dorsiflexion, and helping to achieve midfoot plantarflexion after heel lift. Midfoot plantarflexion during terminal stance shortens the length of the foot, reducing the ankle's plantarflexion arc of rotation when lifting the heel from the ground, which likely helps maintain Achilles recoil power for longer. However, should the triceps surae function be lost through Achilles rupture, then an attempt to utilise these ankle plantarflexion muscles as assistors will only provide very weak ankle power generation around heel lift. The poor ankle plantarflexor mechanics of these flexor muscles cannot compensate for much of the triceps surae and Achilles ankle power if lost. Increased compensation activation of these lesser ankle plantarflexors will tend to induce midfoot plantarflexion, raising the height of the vault during midstance instead of the vault height depressing throughout midstance. An increasing vault height during midstance will then interfere with motion of the ankle joint axis, blocking the anterior tibial tilt that occurs during normal midstance ankle and midfoot dorsiflexion. The talus will thus be restrained by the increased arch height, further reducing the lever arm capacity to generate power for heel lift. The result will be a flexing or a gripping of the toes during late midstance, vault raising, and a very delayed heel lift on an abnormally high-vaulted foot profile, with foot lift-off occurring only after weight transfer is completed to the opposite foot. Heel lift and toe-off will be completed by increased hip muscle power attempting to compensate before toe-off (Fig. 2.5.11c).

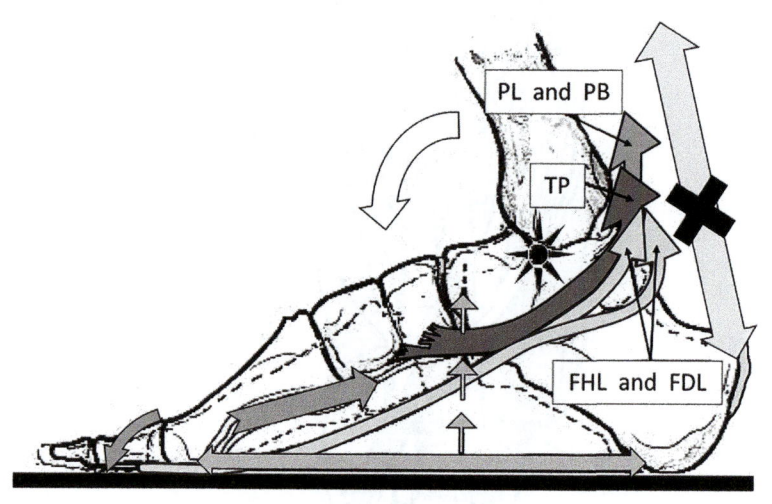

FIG. 2.5.11c If triceps surae's force is lost from the ankle plantarflexion power through significant weakness or full Achilles tendon rupture, then the other plantarflexors can assist. These muscles are tibialis posterior (TP), peroneus longus and brevis (PL and PB), and flexor digitorum longus (FDL). Flexor hallucis longus (FHL) potentially offers a greater plantarflexor lever arm as it runs behind the talus. However, their action together will be quite different because their moment arm length to the instantaneous ankle joint axis (star) is much shorter than that of the Achilles, and their combined muscle cross-sectional area far smaller than that of triceps surae. Thus, the plantarflexion force they apply is much smaller, delaying heel lift and de-powering acceleration during gait. Through their attachments across the midfoot and forefoot, these muscles also raise the foot vault and flex the toes into the support surface. As a result, activating the plantarflexor accessory muscles to induce ankle plantarflexion power will cause the foot's vault to rise, the foot to shorten, and the toes to flex (claw) into the ground during late midstance. This prevents normal anterior tibial rotation and instantaneous joint axis motion that occurs with a falling vault height. Triceps surae plantarflexion power cannot be replicated by any other muscles. *(Permission www.healthystep.co.uk.)*

The position of the flexor muscles to the medial and lateral sides of the ankle–subtalar complex has significance to frontal and transverse plane motions around the rearfoot and ankle, assisting the collateral ankle ligaments. Most important in this are the peroneals and tibialis posterior. To help maintain tendon positions in the medial–lateral directions of motion, the medial flexors are restrained firmly by the flexor retinaculum while the peroneals are held by the peroneal retinaculum, which provide further connective tissue ankle restraint and sensorimotor input.

Tibialis posterior in ankle function

Tibialis posterior is vital for medial stability of the ankle and foot vault during gait (Durrant et al., 2011). Its activity prior to forefoot loading is important in stiffening the foot with peroneus longus before the forefoot collision, setting vault stiffness ready for forefoot contact (Kokubo et al., 2012; Murley et al., 2014). Thus, tibialis posterior helps set the vault height during loading response so that the vault can dissipate energy through permitting increasing compliance during forefoot loading. This deformation lengthening continues rapidly throughout forefoot loading and into initial midstance. It achieves its vault-stiffening effect through being a multi-articular muscle and plays a key role in controlling eversion during rearfoot (ankle–subtalar complex) loading (Ferber and Pohl, 2011).

Its muscle–tendon unit actively lengthens on initial forefoot contact and loading into early midstance while its muscle fascicles are shortening, thus storing and dissipating energy within its elastic myofascial tissues and tendon (Maharaj et al., 2016). Much energy from its muscle–tendon buffering is stored within the tendon throughout the rest of midstance as its muscle activity increases again, with this energy released during late stance acceleration. Late midstance tibialis posterior activity is when fascicle and tendon shortening occur together, generating and storing power for terminal stance despite muscle activation starting to decrease immediately after heel lift (Murley et al., 2014; Maharaj et al., 2016). Late midstance peak tibialis posterior activity and its tendon elastic recoil play a primary role in stiffening the foot throughout the late midstance and early terminal stance phases of gait (Kokubo et al., 2012; Konow et al., 2012; Murley et al., 2014). It is a pivotal muscle for terminal stance midfoot plantarflexion and adduction, creating medial terminal stance stability and increasing foot stiffness prior to and during heel lift (Kokubo et al., 2012; Holowka et al., 2017).

Tibialis posterior has three primary distal tendon slips arising at the level of the navicular. The largest attaches to the navicular tuberosity and often to the plantar surface of the medial cuneiform, with a variable posterior slip being sent to the sustentaculum tali (Draves, 1986, p. 270; Semple et al., 2009). The distal slip continues under the navicular through a groove, radiating distally and laterally to the plantar surfaces of multiple structures variably across many articulations of the vault (Draves, 1986, p. 270; Semple et al., 2009). These attachments commonly include the plantar surface of the intermediate cuneiform, the bases of the 2nd, 3rd, and 4th metatarsals, with additional slips found at the lateral cuneiform, cuboid, and sometimes the 5th metatarsal. The slip to the 4th metatarsal base is reported to be the most consistent and strongest attachment (Draves, 1986, p. 270). All of these slips also blend with and strengthen the plantar ligaments of the vault that they cross (Draves, 1986, p. 270). Together, these distal attachments stabilise the articulations they bridge via tension-induced compression forces, providing stability to the vault as a biotensegrity structure.

Tibialis posterior entheses are fibrocartilaginous, indicating significant attachment motion (Frowen and Benjamin, 1995). Its proximal attachments to the posterior tibia, fibula, and interosseous membrane also give a role in proximal and distal tibiofibular joint motion and stability. Oftentimes, tibialis posterior is described as having a vital role in dynamically stabilising the medial longitudinal arch (Semple et al., 2009). However, in view of its anatomy, this really underestimates its importance in locomotive biomechanics. Here, we will try to consider its role at the ankle. However, this is difficult in isolation from the foot, as tibialis posterior is another example of a primary ankle–foot linkage (Fig. 2.5.11d).

At the ankle, the tibialis posterior is acutely angled around the medial malleolus, subjecting the tendon to significant compression within a pulley mechanism. Here, the tendon expresses a flattening associated with increased fibrocartilage content (Benjamin et al., 1995; Petersen et al., 2004) and as a result, an avascular region (Petersen et al., 2002). The tendon around the medial malleolus is enclosed within a synovial sheath, firmly held by the flexor retinaculum which forms the roof of the *tarsal tunnel*. The tibialis posterior tendon usually has a fibrocartilage sesamoid for compression protection at the navicular tuberosity attachment, which creates a consistently high signal intensity on proton-density-weighted MRIs. However, there can occasionally be a sesamoid bone often referred to as an *os tibiale externum* within this location (Delfaut, et al., 2003). In addition, an accessory bone known as the *os naviculare* (but also commonly and erroneously referred to as *os tibiale externum, os tibiale*, or *naviculare secundarium*) is not uncommon at this site, which may mechanically undermine the medial tibialis posterior attachment's stability, thereby affecting foot and ankle mechanics. A separate bone at the attachment having a fibrotic connection to the normal navicular bone is considered an accessory ossicle, but if found within the tendon with no fibrotic connection, the bone should be considered a sesamoid (Aparisi Gómez et al., 2019). The navicular tendon attachment site itself is associated with widening or flaring of the tendon to upwards of 8 mm from a more proximal width of 4 mm (Delfaut, et al., 2003). This is consistent with a tendon passing through a pulley as it winds around the navicular, medially restraining and compressing it (Fig. 2.5.11e).

FIG. 2.5.11d An anatomical schematic is compared to a photograph of the foot in late midstance, a period where ankle and foot vault mechanics directly influence each other. As the triceps surae works eccentrically to control ankle dorsiflexion by applying braking effort at its proximal attachments, it causes a plantarflexion moment on the distal calcaneus (upper image) that takes the talus into plantarflexion with it. This is restrained in part by the subtalar ligaments. This motion is part of the strong vault-bending moment from ankle dorsiflexion under the body weight vector (BWV) derived from the CoM above. These forces could easily cause excessive sagging deflection and thus create injury throughout the vault's anatomy. The assistor ankle plantarflexors together with the plantar intrinsic muscles provide an active mechanism of tensile bending moment resistance that can also help dissipate the ankle strain energy during vault deflection. Passive connective tissues are also at play (such as the plantar ligaments and aponeurosis) using their viscoelastic properties to stiffen the foot vault as they strain. They can also store and dissipate energy as the foot elongates and widens during single-limb support. Thus, vault deflection is managed by the assistor ankle plantarflexors, but not prevented. Preventing vault deflection disrupts mechanisms of tissue-protecting energy dissipation and storage for use in acceleration, and also limits anterior tibial rotation and anterior displacement of the instantaneous ankle joint axis. *(Permission www.healthystep.co.uk.)*

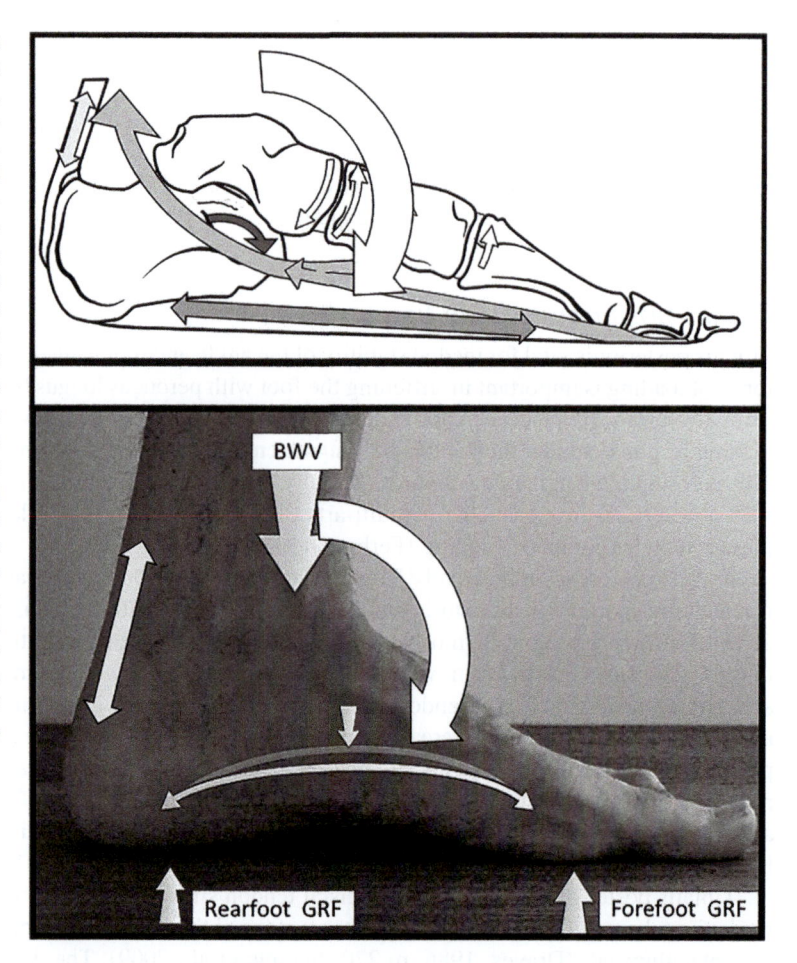

FIG. 2.5.11e Although quite a common asymptomatic anatomical variant, the presence of an accessory navicular bone or a sesamoid within tibialis posterior's tendon (many names are used interchangeably for these bones) can cause loss of some active medial ankle stability and symptoms. This problem seems more common in hypermobile children when connective tissues can present with higher compliance. The tibialis posterior proximal effort is disturbed in the presence of an accessory bone as it has to act through a variably flexible fibrous union from its attachment with the accessory bone to the navicular that it desires to apply effort directly to. Thus, tension must first be applied via the fibrous tissue between the bones (white double arrow). This potentially reduces its effort if significant 'play' occurs between these bones. A sesamoid bone replacing compressive force-dissipating fibrocartilage within the tendon as it passes over and around the navicular can increase compression forces onto the attachment site (small grey arrow). Symptoms are therefore reported with both accessory bone types, especially if the malleolar groove for tibialis posterior is also shallow. A shallow malleolar groove places greater tensile strain directly on the tendon attachment on the navicular rather than through an efficient pulley first. A deeper groove on the medial malleolus spreads loads from tibialis posterior contraction into two compression sites and creates more proximally directed forces onto the attachment at the navicular, redirecting some muscular contraction force into compression onto the posterior malleolar surface. *(Permission www.healthystep.co.uk.)*

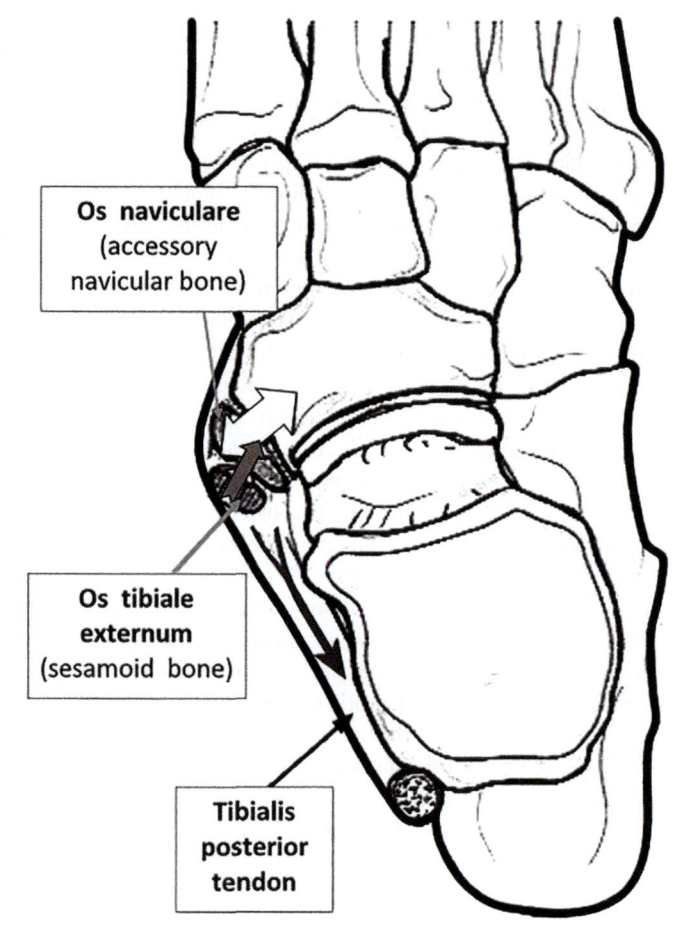

Because of the difficulty in accessing a deep muscle such as tibialis posterior, EMG studies are usually small in size. However, they have demonstrated tibialis posterior activity to be biphasic, with activity occurring in peaks at contact and around the midstance–terminal stance transition. This is characterised by a high variation among healthy adults during walking (Semple et al., 2009). In running, a single-stance phase burst has been reported, but at the fastest running speeds, a second burst during midswing has also been noted (Semple et al., 2009). Changes in EMG peak amplitudes increase systematically during stance phase with increased walking speed, which also changes the timing of the peaks (Murley et al., 2014). In walking, tibialis posterior maintains a two-burst pattern of activity regardless of speed but displays significant gains in peak EMG amplitude with increasing gait speed (Murley et al., 2014). Peak EMG amplitude seems little changed across slow to very slow walking speeds during contact and midstance/propulsion (Murley et al., 2014) (Fig. 2.5.11f). This possibly reflects an increased demand for mediolateral foot stability at slower speeds due to a more mobile vault resulting from reduced stress-stiffening and reduced time-dependent stiffening rates of connective tissues that are normally gained by using more rapid loading and increased tibialis posterior activity.

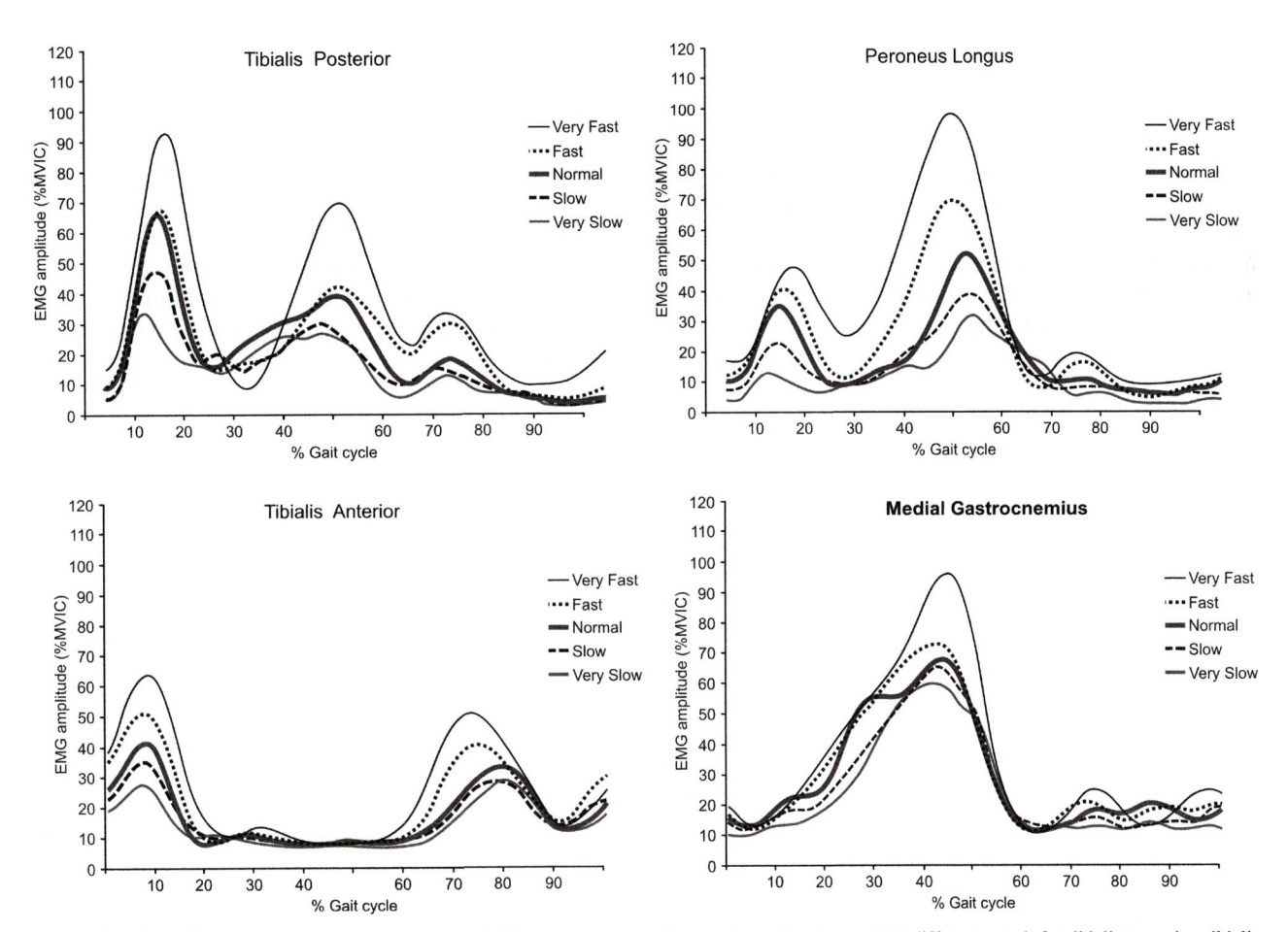

FIG. 2.5.11f EMG ensemble averages for the gait cycle (0% heel contact—100% ipsilateral heel contact) at different speeds for tibialis posterior, tibialis anterior, peroneus longus, and medial gastrocnemius (gastrocnemius medialis). Raised EMG spikes with increase in walking speed reflect the need for greater foot stiffness and ankle plantarflexion power that increases muscular work during faster gait speeds. *(Image from Murley, G.S., Menz, H.B., Landorf, K.B., 2014. Electromyographic patterns of tibialis posterior and related muscles when walking at different speeds. Gait Posture 39(4), 1080–1085.)*

The tibialis posterior is most commonly highlighted as being the primary supinator of the rearfoot (Semple et al., 2009; Kokubo et al., 2012) with a large inversion arm to the subtalar joint (Semple et al., 2009; Durrant et al., 2011) that helps support the medial longitudinal vault (Thordarson et al., 1995; Kitaoka et al., 1997). This is demonstrated most easily by activating tibialis posterior when the foot is nonweightbearing, indicating it is the rearfoot not the subtalar joint that it moves. Yet, stance phase peak amplitude of activity may indicate a far more complex and interesting function within the foot, working as it does as an agonist as well as an antagonist with tibialis anterior and peroneus longus in energy dissipation during contact, foot-stiffening, and driving midfoot motion during midstance and terminal stance. As always, clues to the function of a muscle can be gathered as much from studying dysfunction as it can from studying normal function. Tibialis posterior activity increases energy dissipation rates, reduces osseous displacement, and increases stiffening of the foot when working in concert with peroneus longus. However, it appears to be the primary foot-stiffening muscle when acting alone without peroneus longus (Kokubo et al., 2012).

The kinematics of the foot vault are profoundly affected by loss of tibialis posterior function, and are quite similar to loss of the plantar ligaments (Imhauser et al., 2004). This indicates a strong active/passive functional relationship between the structures in maintaining foot vault function/stiffness. At the ankle, fatigue of tibialis posterior has been reported as disrupting the typical shank-to-foot coupling patterns and results in increased variability of motion during walking (Ferber and Pohl, 2011) and in runners that are classified as 'over-pronators' (Cheung and Ng, 2007).

The peroneals in ankle function

The peroneal muscles lie in the lateral fascial compartment of the leg, with peroneus longus filling most of the superior half of the compartment, and peroneus brevis lying below. Rarely, peroneus brevis and peroneus longus are fused (Draves, 1986, p. 256). Peroneus quartus is an accessory muscle found in 12%–22% of the population. Its tendon lies behind peroneus brevis, arising from the distal fibula to form a variable distal attachment to the calcaneus, cuboid, 5th metatarsal, peroneal tendons, and retinaculum (Wang et al., 2005). The proximal osseous attachment of peroneus longus is to the upper two-thirds of the lateral fibular surface, with fascial attachments to the anterior and posterior intermuscular septa. Peroneus longus remains superior to peroneus brevis in the lateral compartment and becomes tendinous in the middle of the compartment. Peroneus brevis arises from the lateral fibular surface deep to and below peroneus longus, as well as from the inferior parts of the intermuscular septum. It becomes tendinous as it passes with the peroneus longus tendon posterior to the lateral malleolus to lie anterior to the tendon of peroneus longus. Peroneus brevis has a simple distal attachment to the lateral surface of the styloid process of the 5th metatarsal base after passing under the peroneal retinaculum and superiorly to peroneus longus and the peroneal tubercle (trochlea) on the lateral calcaneus (Draves, 1986, p. 258) (Fig. 2.5.11g).

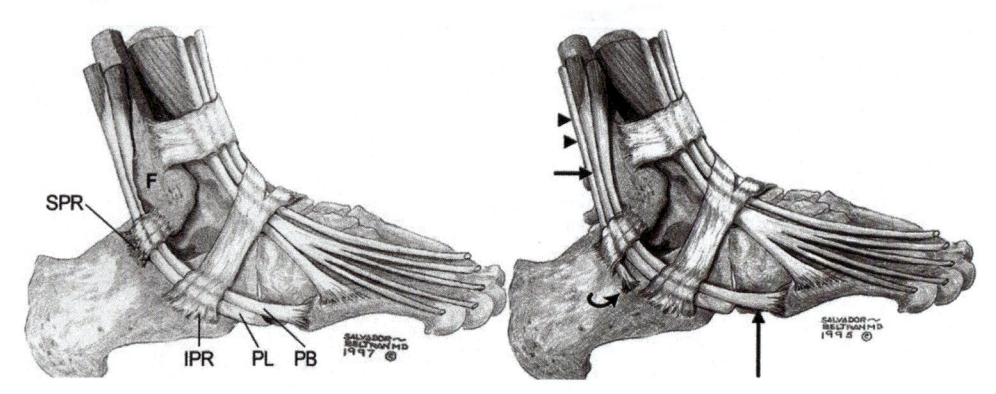

FIG. 2.5.11g An anatomical drawing (left) of the common anatomy of the peroneal muscle group. The peroneal tendons lie posterior to the fibula (F), bound down to the osseous anatomy via the superior peroneal retinaculum (SPR) and inferior peroneal retinaculum (IPR). The peroneus longus tendon (PL) and peroneus brevis tendon (PB) descend together, usually in the same tendon sheath to separate distally with PB attaching to the proximal shaft and styloid process of the 5th metatarsal. Peroneus longus passes beneath the cuboid, which it deeply grooves, to cross the vault of the foot running distomedially to the lateral plantar aspect of the medial cuneiform and 1st metatarsal base. This gives peroneus longus with peroneus brevis an important ankle-supporting role through its tendon and associated retinaculum, but it also provides important foot vault-influencing mechanics. In ankle dorsiflexion (right), the peroneal tendons provide compressive anteromedially directed force closure stability to the ankle. A third peroneal muscle, peroneus quartus, highlighted with black arrows in the right image, is a common variant that can add extra lateral stability and eversion power as long as its presence does not disturb the biomechanics of the other peroneals. *(Image from Wang, X.-T., Rosenberg, Z.S., Mechlin, M.B., Schweitzer, M.E., 2005. Normal variants and diseases of the peroneal tendons and superior peroneal retinaculum: MR imaging features. RadioGraphics 25(3), 587–602 (Courtesy of Salvador Beltran, Albons, Girona, Spain).)*

The peroneals are usually considered mostly in relation to lateral ankle stability, providing an active element to the passive restraint of the lateral collateral ankle ligaments. However, they are also evertors of the foot at the ankle–subtalar complex and weak ankle plantarflexors (Wang et al., 2005). Peroneus longus attaches distally to the plantar surface of the posterior lateral aspect of the medial cuneiform, with occasional slips attaching to bases of the 2nd, 3rd, and 4th metatarsals and the adductor hallucis muscle (Draves, 1986, p. 257). These distal attachments beyond the midfoot tarsal joints suggest a significant role in developing midfoot plantarflexion moments in conjunction with tibialis posterior and also midfoot eversion moments that are reported in foot kinematics during terminal stance (Holowka et al., 2017). Peroneus longus also induces midfoot abduction moments, but these are opposed by tibialis posterior-induced adduction moments (Kokubo et al., 2012). Tension on the tendon produces a slightly plantar but laterally dominant-directed force around the cuboid to produce significant eversion of the medial column (Johnson and Christensen, 1999). All the articulations the muscles cross will be compressed together during peroneal activity that with tibialis posterior, create stability and stiffness of the midfoot. Another 'locking' mechanism within the 1st ray of the medial column through the peroneus longus plantarflexing and everting the medial cuneiform and 1st metatarsal base is also suggested (Johnson and Christensen, 1999). However, it is as agonists of plantarflexion, and as the antagonists of inversion and adduction that the peroneals produce the midfoot-stiffening effects with tibialis posterior that are the key to influencing terminal stance ankle and midfoot stability.

This combined action increases the medial convexity of the foot vault, stiffening the medial column but also effectively plantarflexing the medial side of the midfoot more than the lateral. Clues as to its full action are indicated from peroneus longus activity in cavoid feet where hyperactivity and overdrive of the muscle are suggested to be the cause of the stiffening of the high-vaulted foot profile (Mosca, 2001; Manoli 2nd and Graham, 2005), resulting in inferior impact dissipation properties within the foot. Yet, peroneus longus has been indicated to be a muscle that exhibits a greater compliance influence on the foot than tibialis posterior, but only if tibialis posterior is not coactive (Kokubo et al., 2012).

The long and short peroneals, when active together, provide an important stability moment to the ankle through their resistance to inversion. In this role, they are assisted by lateral parts of the extensor digitorum longus and the lateral collateral ligaments, lying as they do lateral and anterior to the ankle. Extensor digitorum longus offers only weak eversion and is normally only active during swing with peroneus tertius (Jungers et al., 1993). However, together they offer ankle dorsiflexion, something the peroneus longus and peroneus brevis lying behind the lateral malleolus cannot do. The ankle is more vulnerable to lateral instability when plantarflexed because the trochlear surface of the talus is narrower posteriorly than anteriorly, making it less congruent. As a result, more medial–lateral displacement and internal–external rotations occur within ankle plantarflexion than in dorsiflexion (Watanabe et al., 2012). This is something which the extensor evertors anterior to the ankle could help to reduce in plantarflexed lateral ankle instability if activated and strong enough.

Peroneus brevis

Peroneus brevis is multi-articular, crossing the ankle, subtalar, calcaneocuboid, and cuboid–5th metatarsal joint where it attaches to the 5th metatarsal base's styloid process. Apart from providing active lateral ankle stability, it is a rearfoot evertor and abductor of the calcaneocuboid joint. It can provide laterally orientated compression stability through the joints it crosses. Peroneus brevis in the barefoot state shows activity before heel strike suggesting some minor foot 'posture' role for initial contact, but as stance phase progresses, it demonstrates increasing activity with its peak amplitude occurring during terminal stance (Roca-Dols et al., 2018). It has a slower reaction time to ankle inversion moments than does peroneus longus (Konradsen and Ravn, 1991; Karlsson and Andreasson, 1992; Benesch et al., 2000). Greater ankle plantarflexion angles increase the reaction time for both peroneal muscles. This is possibly due to pretensioning of the anterior tibiofibular ligament, provoking increased proprioceptive activity (Benesch et al., 2000). Both peroneal muscles have demonstrated different EMG activity with different types of running shoes compared to barefoot (Roca-Dols et al., 2018). Running shoes tend to reduce activation of peroneus brevis and peroneus longus in the terminal stance phase. However, cushioned and medially wedged running shoes increased peak amplitude of peroneus longus during midstance (Roca-Dols et al., 2018).

Peroneus longus

Peroneus longus EMG activity in stance is very similar to tibialis posterior, as both are synergistic as ankle plantarflexors and agonistic and synergistic in providing medial–lateral foot vault stability and midfoot plantarflexion. Activity remains significant during slow and very slow walking (Murley et al., 2014), probably for providing medial–lateral stability with tibialis posterior across the foot vault but certainly influencing the stiffness and compliance with tibialis posterior through the foot vault (Kokubo et al., 2012). However, in faster walking speeds, tibialis posterior's peak amplitude timing changes little, whereas peroneus longus develops an earlier peak amplitude (Murley, et al., 2014).

It is possible that peroneus longus has a more significant role to play in preventing midfoot break and providing vault stability in midstance and terminal stance, although its activation in isolation of tibialis posterior seems to indicate that it influences energy dissipation and increases compliance (Kokubo et al., 2012). This function would fit the anatomical fact that the peroneus longus passes under the cuboid and runs plantar and medially across the tarsometatarsal joints. It is often reported to plantarflex the 1st metatarsal base as well as being an ankle–subtalar complex plantarflexor and evertor (Kokubo et al., 2012). It may also explain some of the increased foot mobility that occurs across the foot vault with tibialis posterior dysfunction when peroneus longus abduction and eversion activity is left unopposed or un-antagonised.

434 Clinical biomechanics in human locomotion

The peroneus longus and tibialis posterior functional relationship

The peroneus longus, from its anatomical positioning and the timing of its peak activity, suggests a muscle involved in impact dissipation and foot-stiffening. Cadaveric experiments suggest that peroneus longus achieves its function in concert with tibialis posterior (Kokubo et al., 2012), a muscle it usually shares an activation pattern with (Murley et al., 2014). However, results suggest that peroneus longus activity is more important in energy dissipation than it is in foot stiffening, while tibialis posterior's function has little benefit on energy dissipation, with foot-stiffening increased most when tibialis posterior is activated alone (Kokubo et al., 2012) (Fig. 2.5.11h). This may have some influence in cavus foot-stiffening when the foot's increased inverted posture at the ankle may reduce the mechanical advantage of peroneus longus in favour of tibialis posterior.

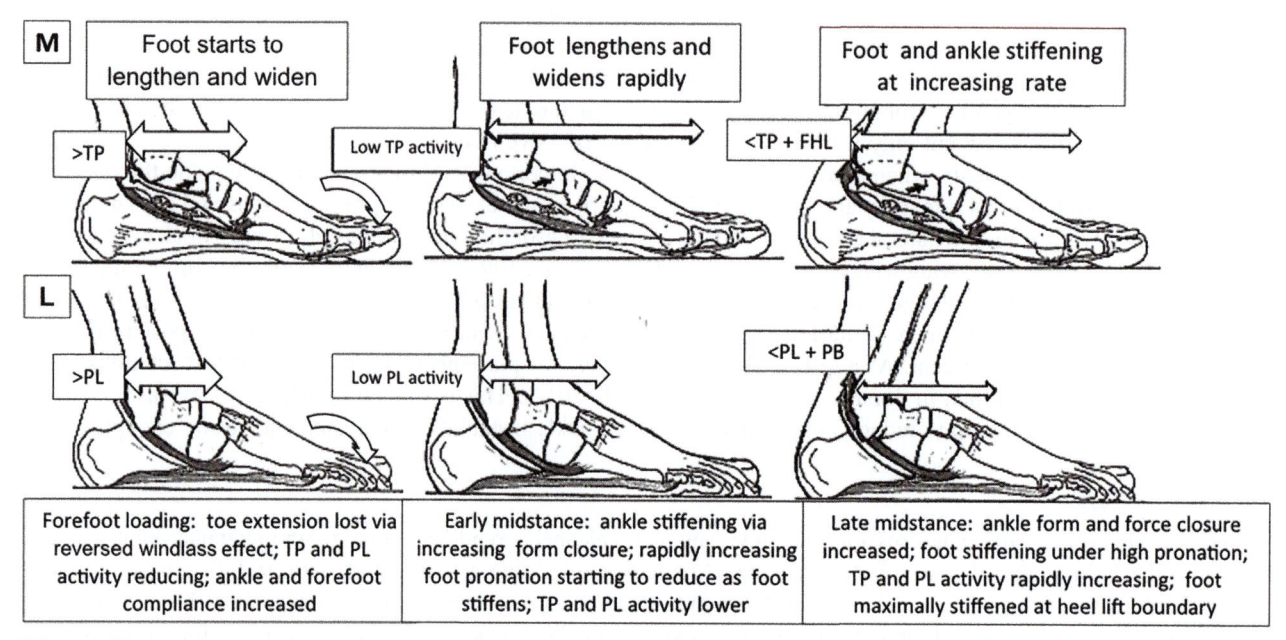

FIG. 2.5.11h Muscle activity over the ankle and into the foot is synergistic for foot and ankle stability during midstance. Here, schematics of the medial view (M) of the left foot (upper images) and lateral view (L) of the right foot (lower images) demonstrate how ankle plantarflexors create frontal, transverse, and sagittal plane stability during stance phase transfer of the CoM over the foot. On forefoot loading, (left) the stiffness created by tibialis posterior (TP—white tendon, medial image) and peroneus longus (PL—white tendon, lateral image) is relaxed through decreasing muscular activity creating increased energy-dissipating foot compliance. The windlass mechanism of loading response is also deactivated. This allows early midstance to become a period of relative compliance around the foot and ankle, allowing the foot to become increasingly prone to the support surface as the ankle plantarflexion angle decreases. The rapid foot elongation and widening (white double-headed arrows) coupled to increasing ankle form closure starts to raise foot stiffness levels by absolute midstance (centre images). During late midstance (right images), the ankle's form closure rapidly increases assisted by rising muscular activity within all the ankle plantarflexors, producing increased force closure. This occurs in concert with increasing plantar intrinsic muscle contraction and connective tissue stress stiffening. At the heel lift boundary, TP is powering foot stiffness and supporting the ankle medially via its tendon's compression onto the medial malleolus, assisted and moderated by flexor hallucis longus (grey tendon, medial image), with the medial vault also supported by abductor hallucis (lower white muscle—medial image). Laterally, PL is now assisted by peroneus brevis (PB—dark-grey tendon, lateral image) to stiffen and evert the lateral column, while their tendons compress against the lateral malleolus. The lateral vault is also supported by abductor digiti minimi (lower grey muscle). Thus, with foot stiffness also resulting from a lower vault height that has stretched and stiffened the connective tissues, the force closed ankle is able to provide enough stability for application of the ankle plantarflexor power to the forefoot at the initiation of terminal stance. *(Permission www.healthystep.co.uk.)*

The peroneus longus and tibialis posterior are muscles that appear to work in unison, with peak activity bursts being demonstrated during contact and loading and around late midstance into terminal stance (Murley et al., 2014). This activity of the two muscles moderating stiffness and compliance and providing energy dissipation around the ankle and within the foot can only occur by working together (Kokubo et al., 2012). Peroneus longus seems to control flexibility and energy dissipation whereas tibialis posterior seems to induce stiffness (Kokubo et al., 2012). A mechanism exists where through moderating activity of one or other of the muscles, the body is able to adjust energy dissipation and stiffness throughout the primary energy dissipation period of contact and the primary foot-stiffening period at the midstance–terminal stance boundary. However, neither peroneus longus nor tibialis posterior have extensive attachments to the hindfoot. Thus, their activity seems to create a relationship across the ankle–subtalar joint complex between the midfoot and the ankle and leg that can couple motion between them. In effect, we have muscles that work together as agonists of midfoot/ankle plantarflexion but work antagonistically in controlling frontal and transverse plane rearfoot and midfoot motions. Only by working together can foot stiffness levels be appropriately managed to lower leg postures.

2.5.12 Leg to foot rotations around the ankle–subtalar complex and lower limb

It has long been hypothesised that the amount of foot pronation has a coupling relationship with rotations of the tibia and femur (Buchbinder et al., 1979; Reischl et al., 1999). It could therefore be expected that peak pronation timing of the foot should be associated with peak rotations of the femur and tibia in gait. This relationship, however, has not been established, and peak foot pronation does seem to relate to peak lower extremity segment rotations (Reischl et al., 1999). In fact, kinematic coupling between the forefoot, rearfoot, and lower leg is weak during walking, especially between rearfoot eversion/inversion and lower leg external/internal rotations (Pohl et al., 2007). However, coupling is much stronger during running, which also shows better sagittal and transverse plane coupling of forefoot to frontal plane rearfoot motion than does walking (Pohl et al., 2007; Pohl and Buckley, 2008).

The problems with assessing rotations within the lower limb during gait are that internal rotation of the knee could be either the result of the tibia internally rotating on a fixed femur, the femur rotating on a fixed tibia, or both segments externally rotating but with the tibia externally rotating less than the femur. The femur also does not consistently follow the tibia (Reischl et al., 1999) for the knee is not a rigid link and is likely capable of absorbing rotatory stresses during stance phase. Tibial rotation would logically link to rearfoot weightbearing motion at the subtalar joint, as movements of the talus (which articulates with high congruency with the tibia) should be expected to couple across the ankle–subtalar joint (Fig. 2.5.12).

Averages created from 30 healthy subjects' kinematics reported an internal tibial rotation of 3.7° at 15% of the gait cycle during loading response, probably via the internal rotation of the talus through the lateral loaded subtalar joint (Reischl et al., 1999). Yet, the peak timing of pronation established through the triplanar motion of the rearfoot, forefoot, and midfoot is reported well beyond initial double-limb support at around 27% of the gait cycle. Peak tibial and femoral rotations occur well before forefoot loading (Reischl et al., 1999). This suggests that midfoot (talonavicular and calcaneocuboid) joints have a minimal coupling relationship with the lower limb above the ankle. Of the 30 subjects examined by Reischl et al. (1999), an average peak external femoral rotation of 2° was reported at 14% of gait, with little difference between males and females. However, they found that 12 of their 30 subjects demonstrated internal rotation of the femur after initial contact, which other studies have also reported (Reischl et al., 1999). It may be that rotations of the proximal segments demonstrate significant subject and step variability.

The relationship of lower limb rotations to the foot at the ankle may lie in the muscles that bridge the joints. Tibialis posterior takes its proximal attachment from the posterior lateral aspect of the tibia and posterior surface of the fibula and the peroneals from the lateral surface of the fibula. This gives the muscles a capacity to influence the distance between distal and proximal attachments through limb rotation. If a coupling relationship between the foot and lower leg via the ankle is being influenced by activity within these muscles then it would be expected that at faster speeds that increase muscle activity and connective tissue stiffness, there would be increased coupling between the foot segments and the lower leg. It is reported that coupling between the forefoot, rearfoot, and lower leg is weakest during walking, but increased during running (Pohl et al., 2007; Pohl and Buckley, 2008). This supports increased connective tissue stiffness resulting from timedependent properties as being the mechanism behind foot-leg motion coupling, assisted by increasing muscle activity.

A study investigating the effects of fatigue of tibialis posterior by Ferber and Pohl (2011) also supports significant coupling of segment motion through muscular activity, rather than just through osseous relationships. At around 30% loss of maximal voluntary isometric contraction force, changes in coupling relationships occurred through changes in coupling angles, and the tibia and forefoot increased their motions relative to the rearfoot. There appears to be only a 1:1 ratio of tibial rotation coupling to rearfoot inversion/eversion during loading response, with greater rearfoot motion registered compared to the tibial motion during the remainder of stance during normal gait (Reischl et al., 1999; DeLeo et al., 2004; Pohl et al., 2007; Ferber and Pohl, 2011). With a fatigued tibialis posterior during early midstance, tibial motion increased relative to the rearfoot as did forefoot sagittal plane motion compared to rearfoot frontal plane motion during all of midstance (Ferber and Pohl, 2011).

Tibialis posterior and peroneus longus do not have direct muscle–tendon attachments to the calcaneus or talus, save occasionally for a tendon slip to the sustentaculum tali of the calcaneus for tibialis posterior (Draves, 1986, p. 270). These muscles do have multiple indirect connective tissue attachments to the rearfoot and direct attachments to the midfoot and metatarsal bases. It seems likely that the loss of normal muscle power output due to fatigue reduces the ability to control motion within the lower leg and forefoot around a rearfoot that these muscles have little direct influence on. Fatigue of tibialis posterior in particular also results in increased joint coupling variability for tibial rotation and forefoot transverse

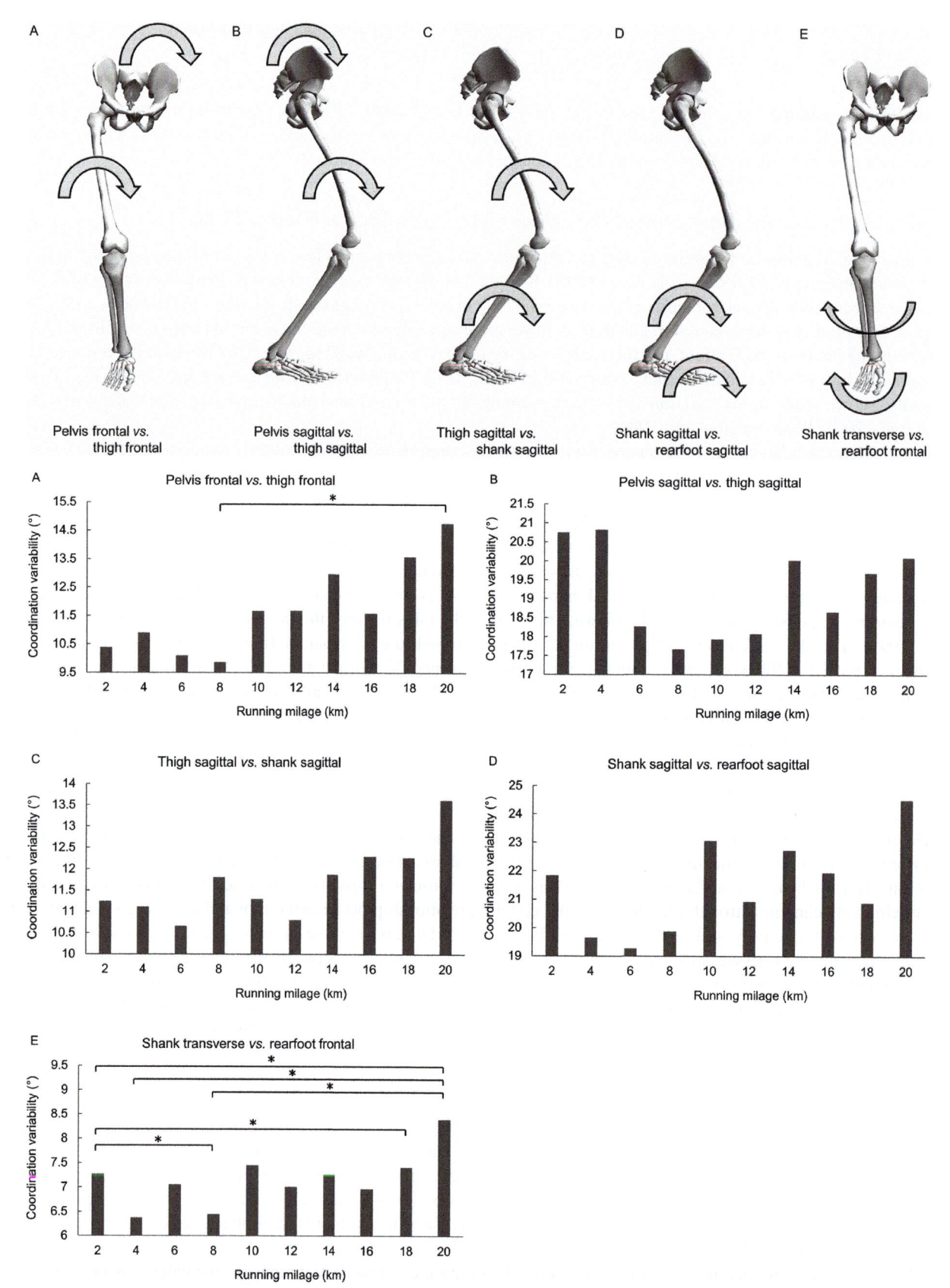

A	B	C	D	E
Pelvis frontal vs. thigh frontal	Pelvis sagittal vs. thigh sagittal	Thigh sagittal vs. shank sagittal	Shank sagittal vs. rearfoot sagittal	Shank transverse vs. rearfoot frontal

FIG. 2.5.12 Coupling occurs between the segments of the lower limb through correlations required for limb motion. In running, most interest has focused on the sagittal and frontal planes proximally (A, B, C, D), but at the ankle, motion between the lower leg and foot is also of particular interest within the transverse plane compared to those of the frontal plane (E). Frontal plane rearfoot motion has been thought to closely couple to lower leg transverse plane internal and external rotations. However, the picture is more complex and can change over time such as during a half-marathon of 20 km. Gait speed seems important to expectations of coupling between segments of the lower limb, with running generally demonstrating closer relationships between segmental motions that walking. *(Images from Chen, T.L.-W., Wong, D.W.-C., Wang, Y., Tan, Q., Lam, W.-K., Zhang, M., 2022. Changes in segment coordination variability and the impacts of the lower limb across running mileages in half marathons: Implications for running injuries. J. Sport Health Sci. 11(1), 67–74.)*

plane abduction/adduction-to-rearfoot frontal plane motion during the midstance phases of gait. In running, studies have reported loss of thigh-to-leg variability in patients with patellofemoral pain syndrome (Hamill et al., 1999) and loss of variability between tibial rotation and rearfoot eversion in those with iliotibial tract/band syndrome (Miller et al., 2008) compared to controls.

Such loss of variability may reflect protective mechanisms to injury. Variability in movement patterns can be both beneficial and harmful, depending on the global or local variability patterns measured (Ferber et al., 2011). For example, Ferber et al. (2011) found that a three-week strengthening protocol in patellofemoral pain syndrome patients increased hip abductor muscle strength and reduced pain, but also reduced variability in stride-to-stride knee joint kinematics without influencing the knee's functional valgum angle. Thus, it is possible that intersegmental variability of motion is good as long as specific joint kinematics remain consistent, while intersegmental fixation may be detrimental to joints by forcing them to alter their biomechanics to create extra intersegmental motion stability.

Miller et al. (2008) found that symptomatic iliotibial tract/band syndrome runners compared to asymptomatic controls had greater coupling variability in knee flexion–extension to forefoot adduction/abduction before running to fatigue level, but less variability of thigh adduction/abduction to foot inversion/eversion after running to fatigue. It was also found that after running, there was less variability in thigh adduction/abduction to tibial rotation but no change in tibial rotation to rearfoot eversion coupling (Miller et al., 2008). All these studies suggest that increased muscle activation, whether by increasing speed, strength, or protective contraction, results in more predictability in intersegmental joint coupling relationships. The effects of activity-induced fatigue may be subject-specific, making one type of injury more likely than another. It seems most probable that an injury changes muscle activation both before and after fatigue as a protective mechanism. Increased coordination variabilities on the frontal and transverse planes of motion may be associated with fatigue, as seen within the duration of half-marathon running, which causes increasing variabilities in the coupling of pelvic frontal plane movement, thigh frontal plane motions, and lower leg transverse plane motions with those of the rearfoot within the frontal plane (Chen et al., 2022).

2.5.13 Adaptation and pathology in the ankle

The ankle is a common site of injury and pathology because of its relative narrow cross-sectional area and significance in maintaining balance between the foot base and lower limb angles during locomotion. The body's CoM, being a long way from the ankle, gives displacement of the CoM a long moment arm by which to create problems. Ankle fractures associated with inversion and eversion events with variable degrees of transverse plane rotations are common, and ligament injuries even more so. Both can lead to the degenerative joint changes of osteoarthritis (OA) through loss of joint congruency (Valderrabano et al., 2006, 2009; Bloch et al., 2015). Loss of function in the subtalar joint also risks degenerative ankle joint changes through its close functional relationship with the ankle (Bloch et al., 2015). Functional instability in the ankle causes unbalanced loading within the ankle joint, resulting in an increased risk of abnormal cartilage loading (Golditz et al., 2014). Given time, this can lead to ankle degenerative changes such as OA, with the latency of this time usually reflecting the severity of injury (Fig. 2.5.13a).

The influence of the ankle soft tissues on the skeletal frame's kinematics can be appreciated by reviewing the data from studies on subjects with an ankle arthrodesis compared to those with normal ankles. Changes have been noted after ankle arthrodesis in both muscle activity, force data, and foot and leg kinematics (Wu et al., 2000). Compared to free ankles, unilateral ankle arthrodesis results in decreased swing phase and increased stance time on the affected limb with increased swing and decreased stance time on the unaffected side (Wu et al., 2000). The peak of the F1 curve is delayed and the F2 trough on the vertical force–time curve is reduced with ankle arthrodesis indicating a greater vertical GRF at midstance, while the F3 peak is lowered (Wu et al., 2000). This indicates that there is a decreased loading rate with loss of free ankle plantarflexion throughout loading response and that the body's CoM elevation is reduced during early midstance, without the freedom of decreasing ankle plantarflexion that usually occurs during the CoM raising manoeuvre. The findings also indicate that during terminal stance, less power is generated around the ankle without freedom of ankle plantarflexion. The anterior–posterior components of GRF (F4 and F5 peaks) are thus significantly reduced (Wu et al., 2000). From this change in kinetics, it can be seen that the ankle is essential for gait energetics as the GRF reflects the ability to absorb, store, and return energy to drive locomotion (Fig. 2.5.13b).

The foot and lower limb kinematics are also altered with loss of the ankle's freedom of motion (Wang et al., 2015; Deleu et al., 2021). Sagittal plane motion is significantly decreased in the rearfoot during gait and there is a loss of early midstance forefoot-to-rearfoot dorsiflexion (Wu et al., 2000). Forefoot sagittal plane and transverse plane motion in the rearfoot, midfoot, and forefoot, develop larger motion and changing kinetic and kinematic patterns (Wu et al., 2000; Wang et al., 2015; Deleu et al., 2021). This suggests that the compensatory motion occurs in the midfoot/forefoot during

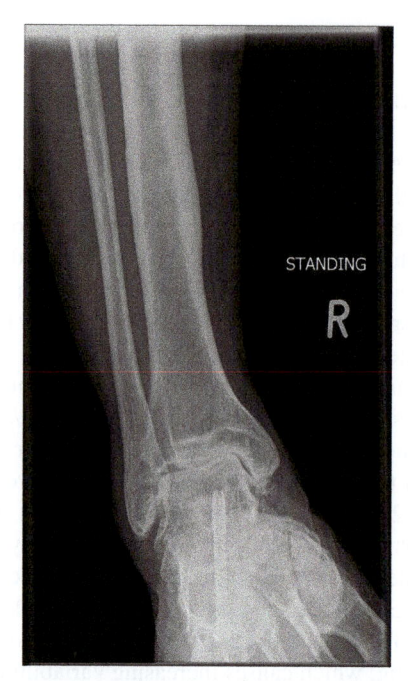

FIG. 2.5.13a Ankle osteoarthritis (OA) is common, which is not helped in the presence of subtalar joint dysfunction. A radiograph of an OA ankle with varus talar tilt is demonstrated. This case is reported to have occurred secondary to previous subtalar fusion. However, ankle and subtalar OA may have their own distinct pathomechanical origins, quite separate from each other. *(Image from Bloch, B., Srinivasan, S., Mangwani, J., 2015. Current concepts in the management of ankle osteoarthritis: a systematic review. J. Foot Ankle Surg. 54(5), 932–939.)*

FIG. 2.5.13b Data presented from force plate studies by Wu et al. (2000) following ankle arthrodesis give some indication to as the kinetic effects of the loss of ankle joint freedom. Average changes in force–time curves of GRF in ten patients with ankle arthrodesis (solid line) compared to normal healthy controls (broken line) are demonstrated. Fore–aft in the top graph refers to anterior–posterior horizontal forces (with posterior GRF, positive), assigned an initial F4, an F5, and an F6 peak, with the latter two equivalent to the F4 and F5 peaks discussed in Chapter 1. T4, T5, and T6 represent the time from initial contact to the force peaks. Such horizontal forces are suppressed by ankle arthrodesis. Medial–lateral forces (centre graph) have been assigned an F7, F8, and F9 peaks with medial GRF positive and T7, T8, and T9 for the time between these peaks. The arthrodesis group show high shear stresses across the foot. Vertical forces (lower graph) have been assigned the usual F1 and F3 peaks and the F2 trough of lowered midstance forces, with the time between them from initial contact shown as T1, T2, and T3, respectively. Midstance forces are higher and acceleration forces suppressed in the presence of ankle arthrodesis. *(Image adapted from Wu, W.-L., Su, F.-C., Cheng, Y.-M., Huang, P.-J., Chou, Y.-L., Chou, C.-K., 2000. Gait analysis after ankle arthrodesis. Gait Posture 11(1), 54–61.)*

Force-plate recordings:

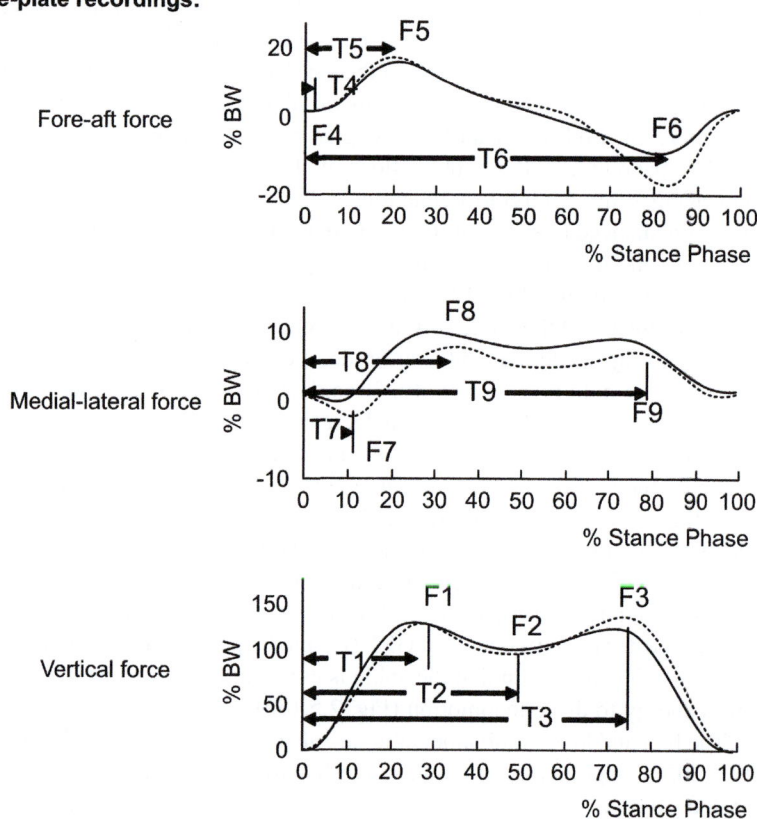

midstance (and possibly the subtalar joint in earlier midstance) and particularly so in terminal stance when the midfoot is known to plantarflex (Holowka et al., 2017). However, the midfoot articulations are less able to provide changes in midstance sagittal plane motion as midfoot dorsiflexion is usually very limited. In pes planus feet that express greater midfoot dorsiflexion during midstance, compensations may be better but in stiff cavoid feet, they are likely to be far more limited (Deleu et al., 2021). This has implications to patients needing ankle arthrodesis, as concurrent midfoot arthritis may severely compromise the ability for the foot to help compensate for the loss of ankle motion (Wang et al., 2015).

The loss of the ankle's freedom of motion has effects on muscular activity, with soleus and the quadriceps shown to change activity considerably (Wu et al., 2000). Indications are that triceps surae activity is much reduced in midstance, the normal graduated rise in activity before heel lift is lost, and peak activity is delayed until well within the terminal stance phase. Quadriceps activity is much disturbed in gait with lost ankle mobility, except during swing, and is delayed during loading response (Wu et al., 2000).

Loss of the anterior muscle group leads to 'foot drop', a serious disability resulting in abnormal gait. The condition can arise from common peroneal nerve neuropathy at the neck of the fibula, anterior horn cell disease, cerebrovascular accidents, lumbar plexopathies, L5 radiculopathy, and sciatic neuropathies (Stewart, 2008). In addition, it can also appear through tibialis anterior tendon rupture (Anagnostakos et al., 2006). Foot drop particularly disrupts the swing phase biomechanics from preswing toe-off through to initial contact, reducing ground clearance and centrifugal draw of the swing limb.

Although the pathomechanical mechanisms remain unclear, physiological–biochemical as well as biomechanical influences seem to be at play in Achilles tendinopathy. Achilles tendinopathy presents complex pathomechanics where sagittal plane and frontal plane angles and forces play a significant role in understanding stress patterns. This is explored further within Chapter 4.

2.5.14 Section summary

The ankle is the distal 'balance hub' that offers significant sagittal plane mobility but limited transverse and frontal plane torques. The limited capacity to increase motion outside of the sagittal plane utilises the close coupling relationship to the subtalar joint, making the ankle–subtalar joint complex a better functional unit to consider as opposed to separating one joint from the other. This combined joint functional unit is able to use form closure effects of the osseous topography to alter freedom/flexibility of motion within both joints. Ankle plantarflexion and a loaded rearfoot increase motion in the ankle and subtalar joints, respectively. With the rearfoot loaded by external forces of the GRF, the subtalar joint displays independent flexibility. However, ankle dorsiflexion not only reduces ankle freedom but also restrains subtalar motion. This means that the clinician can use loading positions to locate the likely source (ankle, subtalar, or both) of motions around the rearfoot during gait.

These important kinematic changes to joint motion include the degree of force closure required. Rearfoot plantarflexion requires the most force closure, especially if the rearfoot is suddenly loaded by an external force in this position as motion within the subtalar joint can then also come into play. With the ankle in plantarflexion, frontal plane moments induced on forefoot loading can only be accommodated through ankle motion at the rearfoot. Flexion and extension of the ankle also changes the instantaneous joint axis position, altering the lever arm moments to the ankle joint from the surrounding muscles. The more posterior positioning of the axis in plantarflexion provides mechanical advantage to the dorsiflexors, and an anterior axis location during ankle dorsiflexion gives the plantarflexors mechanical advantage. This motion relationship tends to increase the mechanical advantage to the appropriate functioning muscles as they achieve peaks of activity during the stance phase of gait.

The ability to create appropriate changes in form and force closure is thus dependent on tissue integrity and free yet stable joint motion. In the concept of biotensegrity, ankle shape changes are just as important to the soft tissues' tensional stability as is the soft tissue tension stability in controlling joint geometry. Thus, ankle dysfunction in its many potential forms can be a clinical challenge to manage and one that can profoundly alter foot and leg kinematics and kinetics.

Chapter summary

The functional units from the lumbosacral joint to the ankle have been discussed and the concepts of tissues and regions working in coherence to achieve lower limb function have been presented. These concepts in clinical biomechanics are critical to planning therapeutic intervention. Problems can take time to unravel, particularly where dysfunction in one functional unit leads to kinematic changes elsewhere within the lower limb. Such problems often need further reflective research, and this chapter should be a useful guide to establishing potential explanations for pathological clinical presentations. The continuity of segments with each other means that the origin(s) of pathomechanics can lie outside the area of symptoms and local tissue stress overload may arise at dysfunctions proximal or distal to the pathology via interconnected kinetic and kinematic changes. Each functional unit must in time be associated with motions in all of the others. This includes the highly complex foot functional unit which is discussed in Chapter 3.

References

Aderem, J., Louw, Q.A., 2015. Biomechanical risk factors associated with iliotibial band syndrome in runners: a systematic review. BMC Musculoskelet. Disord. 16, 356. https://doi.org/10.1186/s12891-015-0808-7.

Adouni, M., Shirazi-Adl, A., Marouane, H., 2016. Role of gastrocnemius activation in knee joint biomechanics: gastrocnemius acts as an ACL antagonist. Comput. Methods Biomech. Biomed. Eng. 19 (4), 376–385.

Agha, M., 2017. MRI of the posterolateral corner of the knee, please have a look. Alexandria J. Med. 53 (3), 261–270.

Agostini, V., Nascimbeni, A., Gaffuri, A., Imazio, P., Benedetti, M.G., Knaflitz, M., 2010. Normative EMG activation patterns of school-age children during gait. Gait Posture 32 (3), 285–289.

Agricola, R., Waarsing, J.H., Thomas, G.E., Carr, A.J., Reijman, M., Bierma-Zeinstra, S.M.A., et al., 2014. Cam impingement: defining the presence of a cam deformity by the alpha angle: data from the CHECK cohort and Chingford cohort. Osteoarthr. Cartil. 22 (2), 218–225.

Aihara, T., Takahashi, K., Ono, Y., Moriya, H., 2002. Does the morphology of the iliolumbar ligament affect lumbosacral disc degeneration? Spine 27 (14), 1499–1503.

Al-Dadah, O., Shepstone, L., Donell, S.T., 2011. Proprioception following partial meniscectomy in stable knees. Knee Surg. Sports Traumatol. Arthrosc. 19 (2), 207–213.

Al-Hayani, A., 2009. The functional anatomy of hip abductors. Folia Morphol. (Warsz) 68 (2), 98–103.

Allen, C.R., Wong, E.K., Livesay, G.A., Sakane, M., Fu, F.H., Woo, S.L., 2000. Importance of the medial meniscus in the anterior cruciate ligament-deficient knee. J. Orthop. Res. 18 (1), 109–115.

Al-saggaf, S., 2003. Variations in the insertion of the extensor hallucis longus muscle. Folia Morphol. (Warsz) 62 (2), 147–155.

Amis, A.A., Firer, P., Mountney, J., Senavongse, W., Thomas, N.P., 2003. Anatomy and biomechanics of the medial patellofemoral ligament. Knee 10 (3), 215–220.

Anagnostakos, K., Bachelier, F., Fürst, O.A., Kelm, J., 2006. Rupture of the anterior tibial tendon: three clinical cases, anatomical study, and literature review. Foot Ankle Int. 27 (5), 330–339.

Anderson, F.C., Pandy, M.G., 2001. Dynamic optimization of human walking. J. Biomech. Eng. 123 (5), 381–390.

Anderson, F.C., Pandy, M.G., 2003. Individual muscle contributions to support in normal walking. Gait Posture 17 (2), 159–169.

Andersson, E.A., Nilsson, J., Thorstensson, A., 1997. Intramuscular EMG from the hip flexor muscles during human locomotion. Acta Physiol. Scand. 161 (3), 361–370.

Aparisi Gómez, M.P., Aparisi, F., Bartoloni, A., Ferrando Fons, M.A., Battista, G., Gugliemi, G., et al., 2019. Anatomical variation in the ankle and foot: from incidental finding to inductor of pathology. Part II. Midfoot and forefoot. Insights Imaging 10, 69. https://doi.org/10.1186/s13244-019-0747-1.

Aquino, A., Payne, C., 2000. The role of the reverse windlass mechanism in foot pathology. Austral. J. Podiatr. Med. 34 (1), 32–34.

Arendt, E.A., 1994. Orthopaedic issues for active and athletic women. Clin. Sports Med. 13 (2), 483–503.

Arndt, A., Brüggemann, G.-P., Koebke, J., Segesser, B., 1999. Asymmetrical loading of the human triceps surae. I. Mediolateral force differences in the Achilles tendon. Foot Ankle Int. 20 (7), 444–449.

Arndt, A., Bengtsson, A.-S., Peolsson, M., Thorstensson, A., Movin, T., 2012. Non-uniform displacement within the Achilles tendon during passive ankle joint motion. Knee Surg. Sports Traumatol. Arthrosc. 20 (9), 1868–1874.

Arnold, A.S., Delp, S.L., 2001. Rotational moment arms of the medial hamstrings and adductors vary with femoral geometry and limb position: implications for the treatment of internally rotated gait. J. Biomech. 34 (4), 437–447.

Arumugam, A., Milosavljevic, S., Woodley, S., Sole, G., 2012. Effects of external pelvic compression on form closure, force closure, and neuromotor control of the lumbopelvic spine—a systematic review. Man. Ther. 17 (4), 275–284.

Askling, C.M., Tengvar, M., Saartok, T., Thorstensson, A., 2008. Proximal hamstring strains of stretching type in different sports: injury situations, clinical and magnetic resonance imaging characteristics, and return to sport. Am. J. Sports Med. 36 (9), 1799–1804.

Astephen, J.L., Deluzio, K.J., Caldwell, G.E., Dunbar, M.J., 2008a. Biomechanical changes at the hip, knee, and ankle joints during gait are associated with knee osteoarthritis severity. J. Orthop. Res. 26 (3), 332–341.

Astephen, J.L., Deluzio, K.J., Caldwell, G.E., Dunbar, M.J., Hubley-Kozey, C.L., 2008b. Gait and neuromuscular pattern changes are associated with differences in knee osteoarthritis severity levels. J. Biomech. 41 (4), 868–876.

Astur, D.C., Oliveira, S.G., Badra, R., Arliani, G.G., Kaleka, C.C., Jalikjian, W., et al., 2011. Updating of the anatomy of the extensor mechanism of the knee using a three-dimensional viewing technique. Rev. Brasil. Ortopedia (Engl. Ed.) 46 (5), 490–494.

Ateshian, G.A., Hung, C.T., 2005. Patellofemoral joint biomechanics and tissue engineering. Clin. Orthop. Relat. Res. 436 (July), 81–90.

Aubry, S., Risson, J.-R., Kastler, A., Barbier-Brion, B., Siliman, G., Runge, M., et al., 2013. Biomechanical properties of the calcaneal tendon in vivo assessed by transient shear wave elastography. Skelet. Radiol. 42 (8), 1143–1150.

Baker, R.L., Souza, R.B., Fredericson, M., 2011. Iliotibial band syndrome: soft tissue and biomechanical factors in evaluation and treatment. PM&R: J. Injury Funct. Rehab. 3 (6), 550–561.

Barba, D., Barker, L., Chhabra, A., 2015. Anatomy and biomechanics of the posterior cruciate ligament and posterolateral corner. Oper. Tech. Sports Med. 23 (4), 256–268.

Barr, A.J., Campbell, T.M., Hopkinson, D., Kingsbury, S.R., Bowes, M.A., Conaghan, P.G., 2015. A systematic review of the relationship between subchondral bone features, pain and structural pathology in peripheral joint osteoarthritis. Arthritis Res. Ther. 17 (1), 228. https://doi.org/10.1186/s13075-015-0735-x.

Barr, A.J., Dube, B., Hensor, E.M.A., Kingsbury, S.R., Peat, G., Bowes, M.A., et al., 2016. The relationship between three-dimensional knee MRI bone shape and total knee replacement—a case control study: data from the Osteoarthritis Initiative. Rheumatology 55 (9), 1585–1593.

Barrett, R.S., Besier, T.F., Lloyd, D.G., 2007. Individual muscle contributions to the swing phase of gait: an EMG-based forward dynamics modelling approach. Simul. Model. Pract. Theory 15 (9), 1146–1155.

Bartlett, J.L., Sumner, B., Ellis, R.G., Kram, R., 2014. Activity and functions of the human gluteal muscles in walking, running, and climbing. Am. J. Phys. Anthropol. 153 (1), 124–131.

Barzan, M., Modenese, L., Carty, C.P., Maine, S., Stockton, C.A., Sancisi, N., et al., 2019. Development and validation of subject-specific pediatric multibody knee kinematic models with ligamentous constraints. J. Biomech. 93, 194–203.

Beales, D.J., O'Sullivan, P.B., Briffa, N.K., 2010. The effects of manual pelvic compression on trunk motor control during an active straight leg raise in chronic pelvic girdle pain subjects. Man. Ther. 15 (2), 190–199.

Beaulieu, M.L., Lamontagne, M., Beaulé, P.E., 2010. Lower limb biomechanics during gait do not return to normal following total hip arthroplasty. Gait Posture 32 (2), 269–273.

Beltran, J., Matityahu, A., Hwang, K., Jbara, M., Maimon, R., Padron, M., et al., 2003. The distal semimembranosus complex: normal MR anatomy, variants, biomechanics and pathology. Skelet. Radiol. 32 (8), 435–445.

Benesch, S., Pütz, W., Rosenbaum, D., Becker, H.-P., 2000. Reliability of peroneal reaction time measurements. Clin. Biomech. 15 (1), 21–28.

Benjamin, M., 2009. The fascia of the limbs and back—a review. J. Anat. 214 (1), 1–18.

Benjamin, M., McGonagle, D., 2001. The anatomical basis for disease localisation in seronegative spondyloarthropathy at entheses and related sites. J. Anat. 199 (5), 503–526.

Benjamin, M., Qin, S., Ralphs, J.R., 1995. Fibrocartilage associated with human tendons and their pulleys. J. Anat. 187 (3), 625–633.

Bennell, K.L., Hunt, M.A., Wrigley, T.V., Hunter, D.J., Hinman, R.S., 2007. The effects of hip muscle strengthening on knee load, pain, and function in people with knee osteoarthritis: a protocol for a randomised, single-blind controlled trial. BMC Musculoskelet. Disord. 8, 121. https://doi.org/10.1186/1471-2474-8-121.

Bergmann, G., Graichen, F., Rohlmann, A., 1995. Is staircase walking a risk for the fixation of hip implants? J. Biomech. 28 (5), 535–553.

Bergmann, G., Graichen, F., Rohlmann, A., 2004. Hip joint contact forces during stumbling. Langenbeck's Arch. Surg. 389 (1), 53–59.

Bervet, K., Bessette, M., Godet, L., Crétual, A., 2013. KeR-EGI, a new index of gait quantification based on electromyography. J. Electromyogr. Kinesiol. 23 (4), 930–937.

Birch, H.L., Worboys, S., Eissa, S., Jackson, B., Strassburg, S., Clegg, P.D., 2008. Matrix metabolism rate differs in functionally distinct tendons. Matrix Biol. 27 (3), 182–189.

Bjelopetrovich, A., Barrios, J.A., 2016. Effects of incremental ambulatory-range loading on arch height index parameters. J. Biomech. 49 (14), 3555–3558.

Blackburn, J.T., Pamukoff, D.N., 2014. Geometric and architectural contributions to hamstring musculotendinous stiffness. Clin. Biomech. 29 (1), 105–110.

Bleakney, R.R., Tallon, C., Wong, J.K., Lim, K.P., Maffulli, N., 2002. Long-term ultrasonographic features of the Achilles tendon after rupture. Clin. J. Sport Med. 12 (5), 273–278.

Bloch, B., Srinivasan, S., Mangwani, J., 2015. Current concepts in the management of ankle osteoarthritis: a systematic review. J. Foot Ankle Surg. 54 (5), 932–939.

Bojsen-Møller, F., Lamoreux, L., 1979. Significance of free-dorsiflexion of the toes in walking. Acta Orthop. Scand. 50 (4), 471–479.

Bojsen-Møller, J., Hansen, P., Aagaard, P., Svantesson, U., Kjaer, M., Magnusson, S.P., 2004. Differential displacement of the human soleus and medial gastrocnemius aponeuroses during isometric plantar flexor contractions in vivo. J. Appl. Physiol. 97 (5), 1908–1914.

Bojsen-Møller, J., Schwartz, S., Kalliokoski, K.K., Finni, T., Magnusson, S.P., 2010. Intermuscular force transmission between human plantarflexor muscles in vivo. J. Appl. Physiol. 109 (6), 1608–1618.

Bonnel, F., Toullec, E., Mabit, C., Tourné, Y., Sofcot, 2010. Chronic ankle instability: biomechanics and pathomechanics of ligaments injury and associated lesions. Orthop. Traumatol. Surg. Res. 96 (4), 424–432.

Botser, I.B., Ozoude, G.C., Martin, D.E., Siddiqi, A.J., Kuppuswami, S., Domb, B.G., 2012. Femoral anteversion in the hip: comparison of measurement by computed tomography, magnetic resonance imaging, and physical examination. Arthrosc.: J. Arthrosc. Rel. Surg 28 (5), 619–627.

Bouillard, K., Nordez, A., Hug, F., 2011. Estimation of individual muscle force using elastography. PLoS One 6 (12), e29261. https://doi.org/10.1371/journal.pone.0029261.

Bouillard, K., Hug, F., Guével, A., Nordez, A., 2012. Shear elastic modulus can be used to estimate an index of individual muscle force during a submaximal isometric fatiguing contraction. J. Appl. Physiol. 113 (9), 1353–1361.

Bouisset, S., Zattara, M., 1981. A sequence of postural movements precedes voluntary movement. Neurosci. Lett. 22 (3), 263–270.

Bowes, M.A., Vincent, G.R., Wolstenholme, C.B., Conaghan, P.G., 2015. A novel method for bone area measurement provides new insights into osteoarthritis and its progression. Ann. Rheum. Dis. 74 (3), 519–525.

Bowman Jr., K.F., Sekiya, J.K., 2009. Anatomy and biomechanics of the posterior cruciate ligament and other ligaments of the knee. Oper. Tech. Sports Med. 17 (3), 126–134.

Bramah, C., Preece, S.J., Gill, N., Herrington, L., 2018. Is there a pathological gait associated with common soft tissue running injuries? Am. J. Sports Med. 46 (12), 3023–3031.

Branch, E.A., Anz, A.W., 2015. Distal insertions of the biceps femoris: a quantitative analysis. Orthop. J. Sports Med. 3 (9). https://doi.org/10.1177/2325967115602255.

Branthwaite, H., Pandyan, A., Chockalingam, N., 2012. Function of the triceps surae muscle group in low and high arched feet: an exploratory study. Foot 22 (2), 56–59.

Brindle, T.J., Mattacola, C., McCrory, J., 2003. Electromyographic changes in the gluteus medius during stair ascent and descent in subjects with anterior knee pain. Knee Surg. Sports Traumatol. Arthrosc. 11 (4), 244–251.

Bruijn, S.M., Meijer, O.G., van Dieën, J.H., Kingma, I., Lamoth, C.J.C., 2008. Coordination of leg swing, thorax rotations, and pelvis rotations during gait: the organisation of total body angular momentum. Gait Posture 27 (3), 455–462.

Buchbinder, M.R., Napora, N.J., Biggs, E.W., 1979. The relationship of abnormal pronation to chondromalacia of the patella in distance runners. J. Am. Podiatry Assoc. 69 (2), 159–162.

Buchbinder, R., van Tulder, M., Öberg, B., Costa, L.M., Woolf, A., Schoene, M., et al., 2018. Low back pain: a call for action. Lancet 391 (10137), 2384–2388.

Buldt, A.K., Levinger, P., Murley, G.S., Menz, H.B., Nester, C.J., Landorf, K.B., 2015. Foot posture is associated with kinematics of the foot during gait: a comparison of normal, planus and cavus feet. Gait Posture 42 (1), 42–48.

Bull, A.M., Amis, A.A., 1998. Knee joint motion: description and measurement. Proc. Inst. Mech. Eng. H J. Eng. Med. 212 (5), 357–372.

Butler, A.M., Walsh, W.R., 2004. Mechanical response of ankle ligaments at low loads. Foot Ankle Int. 25 (1), 8–12.

Byrne, C.A., O'Keeffe, D.T., Donnelly, A.E., Lyons, G.M., 2007. Effect of walking speed changes on tibialis anterior EMG during healthy gait for FES envelope design in drop foot correction. J. Electromyogr. Kinesiol. 17 (5), 605–616.

Byrne, D.P., Mulhall, K.J., Baker, J.F., 2010. Anatomy & biomechanics of the hip. Open Sports Med. J. 4, 51–57.

Carlis, S., Pollack, D., McTeague, S., Khaimov, G., 2017. Peroneus quartus muscle autograft: a novel approach to the repair of a split peroneus brevis tendon tear. J. Am. Podiatr. Med. Assoc. 107 (1), 76–79.

Carlson, R.E., Fleming, L.L., Hutton, W.C., 2000. The biomechanical relationship between the tendoachilles, plantar fascia and metatarsophalangeal joint dorsiflexion angle. Foot Ankle Int. 21 (1), 18–25.

Casaroli, G., Bassani, T., Brayda-Bruno, M., Luca, A., Galbusera, F., 2020. What do we know about the biomechanics of the sacroiliac joint and of sacro-pelvic fixation? A literature review. Med. Eng. Phys. 76, 1–12.

Castermans, T., Duvinage, M., Cheron, G., Dutoit, T., 2014. Towards effective non-invasive brain-computer interfaces dedicated to gait rehabilitation systems. Brain Sci. 4 (1), 1–48.

Cavagna, G.A., Willems, P.A., Heglund, N.C., 2000. The role of gravity in human walking: pendular energy exchange, external work and optimal speed. J. Physiol. 528 (3), 657–668.

Chang, A., Hayes, K., Dunlop, D., Song, J., Hurwitz, D., Cahue, S., et al., 2005. Hip abduction moment and protection against medial tibiofemoral osteoarthritis progression. Arthritis Rheum. 52 (11), 3515–3519.

Chen, B., Jiang, H., 2019. Swimming performance of a tensegrity robotic fish. Soft Robot. 6 (4), 520–531.

Chen, T.L.-W., Wong, D.W.-C., Wang, Y., Tan, Q., Lam, W.-K., Zhang, M., 2022. Changes in segment coordination variability and the impacts of the lower limb across running mileages in half marathons: implications for running injuries. J. Sport Health Sci. 11 (1), 67–74.

Cheung, R.T.H., Ng, G.Y.F., 2007. Efficacy of motion control shoes for reducing excessive rearfoot motion in fatigued runners. Phys. Ther. Sport 8 (2), 75–81.

Chino, K., Takahashi, H., 2018. Association of gastrocnemius muscle stiffness with passive ankle joint stiffness and sex-related difference in the joint stiffness. J. Appl. Biomech. 34 (3), 169–174.

Chockalingam, N., Needham, R., 2020. Spinal biomechanics. In: Moramarco, M., Borysov, M., Ng, S.Y., Weiss, H.-R. (Eds.), Schroth's Textbook of Scoliosis and Other Spinal Deformities. Cambridge Scholars Publishing, Newcastle upon Tyne, UK, pp. 39–67 (Chapter 2).

Cibulka, M.T., 2004. Determination and significance of femoral neck anteversion. Phys. Ther. 84 (6), 550–558.

Cleland, J., Schulte, C., Durall, C., 2002. The role of therapeutic exercise in treating instability-related lumbar spine pain: a systematic review. J. Back Musculosk. Rehab. 16 (2), 105–115.

Clément, J., Dumas, R., Hagemeister, N., de Guise, J.A., 2015. Soft tissue artifact compensation in knee kinematics by multi-body optimization: performance of subject-specific knee joint models. J. Biomech. 48 (14), 3796–3802.

Cobb, S.C., Bazett-Jones, D.M., Joshi, M.N., Earl-Boehm, E., James, C.R., 2014. The relationship among foot posture, core and lower extremity muscle function, and postural stability. J. Athl. Train. 49 (2), 173–180.

Cohen, S.P., 2005. Sacroiliac joint pain: a comprehensive review of anatomy, diagnosis, and treatment. Anesth. Analg. 101 (5), 1440–1453.

Cohen, S., Bradley, J., 2007. Acute proximal hamstring rupture. J. Am. Acad. Orthop. Surg. 15 (6), 350–355.

Cohen, S.P., Chen, Y., Neufeld, N.J., 2013. Sacroiliac joint pain: a comprehensive review of epidemiology, diagnosis and treatment. Expert. Rev. Neurother. 13 (1), 99–116.

Correa, T.A., Crossley, K.M., Kim, H.J., Pandy, M.G., 2010. Contributions of individual muscles to hip joint contact force in normal walking. J. Biomech. 43 (8), 1618–1622.

Crenna, P., Frigo, C., 1991. A motor programme for the initiation of forward-oriented movements in humans. J. Physiol. 437 (1), 635–653.

Csintalan, R.P., Schulz, M.M., Woo, J., McMahon, P.J., Lee, T.Q., 2002. Gender differences in patellofemoral joint biomechanics. Clin. Orthop. Relat. Res. 402 (September), 260–269.

Cummins, E.J., Anson, B.J., Carr, B.W., Wright, R.R., 1946. The structure of the calcaneal tendon (of Achilles) in relation to orthopedic surgery, with additional observations on the plantaris muscle. Surg. Gynecol. Obstet. 83 (1), 107–116.

Dalmau-Pastor, M., Fargues-Polo Jr., B., Casanova-Martínez Jr., D., Vega, J., Golanó, P., 2014. Anatomy of the triceps surae: a pictorial essay. Foot Ankle Clin. 19 (4), 603–635.

Dargel, J., Feiser, J., Gotter, M., Pennig, D., Koebke, J., 2009. Side differences in the anatomy of human knee joints. Knee Surg. Sports Traumatol. Arthrosc. 17 (11), 1368–1376.

Davies, H., Unwin, A., Aichroth, P., 2004. The posterolateral corner of the knee: anatomy, biomechanics and management of injuries. Injury 35 (1), 68–75.

Debus, F., Eberhard, H.-J., Olivieri, M., Peterlein, C.D., 2019. MRI in patients with Haglund's deformity and its influence on therapy. Arch. Orthop. Trauma Surg. 139 (7), 903–906.

DeJong, A.F., Koldenhoven, R.M., Hertel, J., 2020. Cross-correlations between gluteal muscle thickness derived from ultrasound imaging and hip biomechanics during walking gait. J. Electromyogr. Kinesiol. 51, 102406. https://doi.org/10.1016/j.jelekin.2020.102406.

DeLeo, A.T., Dierks, T.A., Ferber, R., Davis, I.S., 2004. Lower extremity joint coupling during running: a current update. Clin. Biomech. 19 (10), 983–991.

Deleu, P.-A., Naaim, A., Chèze, L., Dumas, R., Devos Bevernage, B., Goubau, L., et al., 2021. The effect of ankle and hindfoot malalignment on foot mechanics in patients suffering from post-traumatic ankle osteoarthritis. Clin. Biomech. 81, 105239. https://doi.org/10.1016/j.clinbiomech.2020.105239.

Delfaut, E.M., Demondion, X., Bieganski, A., Cotten, H., Mestdagh, H., Cotten, A., 2003. The fibrocartilaginous sesamoid: a cause of size and signal variation in the normal distal posterior tibial tendon. Eur. Radiol. 13 (12), 2642–2649.

Di Nardo, F., Fioretti, S., 2013. Statistical analysis of surface electromyographic signal for the assessment of rectus femoris modalities of activation during gait. J. Electromyogr. Kinesiol. 23 (1), 56–61.

Di Nardo, F., Ghetti, G., Fioretti, S., 2013. Assessment of the activation modalities of gastrocnemius lateralis and tibialis anterior during gait: a statistical analysis. J. Electromyogr. Kinesiol. 23 (6), 1428–1433.

Di Nardo, F., Mengarelli, A., Burattini, L., Maranesi, E., Agostini, V., Nascimbeni, A., et al., 2016. Normative EMG patterns of ankle muscle co-contractions in school-age children during gait. Gait Posture 46, 161–166.

Dischiavi, S.L., Wright, A.A., Hegedus, E.J., Bleakley, C.M., 2018. Biotensegrity and myofascial chains: a global approach to an integrated kinetic chain. Med. Hypotheses 110, 90–96.

Dorman, T.A., 1995. Elastic energy in the pelvis. Spine: State Art Rev. 9 (2), 365–379.

Draves, D.J., 1986. Anatomy of the Lower Extremity. Williams & Wilkins, Baltimore, MD.

Du, N., Liu, X.Y., Narayanan, J., Li, L., Lim, M.L.M., Li, D., 2006. Design of superior spider silk: from nanostructure to mechanical properties. Biophys. J. 91 (12), 4528–4535.

Duda, G.N., Heller, M., Albinger, J., Schulz, O., Schneider, E., Claes, L., 1998. Influence of muscle forces on femoral strain distribution. J. Biomech. 31 (9), 841–846.

Dunphy, C., Casey, S., Lomond, A., Rutherford, D., 2016. Contralateral pelvic drop during gait increases knee adduction moments of asymptomatic individuals. Hum. Mov. Sci. 49, 27–35.

Durrant, B., Chockalingam, N., Hashmi, F., 2011. Posterior tibial tendon dysfunction: a review. J. Am. Podiatr. Med. Assoc. 101 (2), 176–186.

Duthon, V.B., Barea, C., Abrassart, S., Fasel, J.H., Fritschy, D., Ménétrey, J., 2006. Anatomy of the anterior cruciate ligament. Knee Surg. Sports Traumatol. Arthrosc. 14 (3), 204–213.

Dziedzic, D.W., Bogacka, U., Komarnitki, I., Ciszek, B., 2018. Anatomy and morphometry of the distal gracilis muscle tendon in adults and foetuses. Folia Morphol. (Warsz) 77 (1), 138–143.

Earl, J.E., Hoch, A.Z., 2011. A proximal strengthening program improves pain, function, and biomechanics in women with patellofemoral pain syndrome. Am. J. Sports Med. 39 (1), 154–163.

Edama, M., Kubo, M., Onishi, H., Takabayashi, T., Inai, T., Yokoyama, E., et al., 2015. The twisted structure of the human Achilles tendon. Scand. J. Med. Sci. Sports 25 (5), e497–e503.

Edwards, W.B., Miller, R.H., Derrick, T.R., 2016. Femoral strain during walking predicted with muscle forces from static and dynamic optimization. J. Biomech. 49 (7), 1206–1213.

Ekstrand, J., Hägglund, M., Waldén, M., 2011. Injury incidence and injury patterns in professional football: the UEFA injury study. Br. J. Sports Med. 45 (7), 553–558.

Ekstrand, J., Healy, J.C., Waldén, M., Lee, J.C., English, B., Hägglund, M., 2012. Hamstring muscle injuries in professional football: the correlation of MRI findings with return to play. Br. J. Sports Med. 46 (2), 112–117.

Elder, G.C., Bradbury, K., Roberts, R., 1982. Variability of fiber type distributions within human muscles. J. Appl. Physiol. 53 (6), 1473–1480.

Ellis, R.G., Sumner, B.J., Kram, R., 2014. Muscle contributions to propulsion and braking during walking and running: insight from external force perturbations. Gait Posture 40 (4), 594–599.

Ertelt, T., Gronwald, T., 2017. M. biceps femoris—a wolf in sheep's clothing: the downside of a lower limb injury prevention training. Med. Hypotheses 109, 119–125.

Farahmand, F., Senavongse, W., Amiss, A.A., 1998. Quantitative study of the quadriceps muscles and trochlear groove geometry related to instability of the patellofemoral joint. J. Orthop. Res. 16 (1), 136–143.

Farris, D.J., Birch, J., Kelly, L., 2020. Foot stiffening during the push-off phase of human walking is linked to active muscle contraction, and not the windlass mechanism. J. R. Soc. Interface 17 (168), 20200208. https://doi.org/10.1098/rsif.2020.0208.

Ferber, R., Pohl, M.B., 2011. Changes in joint coupling and variability during walking following tibialis posterior muscle fatigue. J. Foot Ankle Res. 4, 6. https://doi.org/10.1186/1757-1146-4-6.

Ferber, R., Noehren, B., Hamill, J., Davis, I.S., 2010. Competitive female runners with a history of iliotibial band syndrome demonstrate atypical hip and knee kinematics. J. Orthop. Sports Phys. Ther. 40 (2), 52–58.

Ferber, R., Kendall, K.D., Farr, L., 2011. Changes in knee biomechanics after a hip-abductor strengthening protocol for runners with patellofemoral pain syndrome. J. Athl. Train. 46 (2), 142–149.

Ferguson, S.J., Bryant, J.T., Ganz, R., Ito, K., 2003. An in vitro investigation of the acetabular labral seal in hip joint mechanics. J. Biomech. 36 (2), 171–178.

Fernández, M.P., Hoxha, D., Chan, O., Mordecai, S., Blunn, G.W., Tozzi, G., et al., 2020. Centre of rotation of the human subtalar joint using weight-bearing clinical computed tomography. Sci. Rep. 10, 1035. https://doi.org/10.1038/s41598-020-57912-z.

Finni, T., Komi, P.V., Lukkariniemi, J., 1998. Achilles tendon loading during walking: application of a novel optic fiber technique. Eur. J. Appl. Physiol. Occup. Physiol. 77 (3), 289–291.

Fiorentino, N.M., Blemker, S.S., 2014. Musculotendon variability influences tissue strains experienced by the biceps femoris long head muscle during high-speed running. J. Biomech. 47 (13), 3325–3333.

Flicker, P.L., Fleckenstein, J.L., Ferry, K., Payne, J., Ward, C., Mayer, T., et al., 1993. Lumbar muscle usage in chronic low back pain. Magnetic resonance image evaluation. Spine 18 (5), 582–586.

Fontboté, C.A., Sell, T.C., Laudner, K.G., Haemmerle, M., Allen, C.R., Margheritini, F., et al., 2005. Neuromuscular and biomechanical adaptations of patients with isolated deficiency of the posterior cruciate ligament. Am. J. Sports Med. 33 (7), 982–989.

Fortin, J.D., Dwyer, A.P., West, S., Pier, J., 1994. Sacroiliac joint: pain referral maps upon applying a new injection/arthrography technique. Part I. Asymptomatic volunteers. Spine 19 (13), 1475–1482.

Fortin, J.D., Vilensky, J.A., Merkel, G.J., 2003. Can the sacroiliac joint cause sciatica? Pain Physician 6 (3), 269–271.

Foucher, K.C., Hurwitz, D.E., Wimmer, M.A., 2007. Preoperative gait adaptations persist one year after surgery in clinically well-functioning total hip replacement patients. J. Biomech. 40 (15), 3432–3437.

Fouré, A., Nordez, A., McNair, P., Cornu, C., 2011. Effects of plyometric training on both active and passive parts of the plantarflexors series elastic component stiffness of muscle-tendon complex. Eur. J. Appl. Physiol. 111 (3), 539–548.

Fouré, A., Cornu, C., McNair, P.J., Nordez, A., 2012. Gender differences in both active and passive parts of the plantar flexors series elastic component stiffness and geometrical parameters of the muscle-tendon complex. J. Orthop. Res. 30 (5), 707–712.

Fox, M.D., Delp, S.L., 2010. Contributions of muscles and passive dynamics to swing initiation over a range of walking speeds. J. Biomech. 43 (8), 1450–1455.

Franchi, M., Quaranta, M., Macciocca, M., De Pasquale, V., Ottani, V., Ruggeri, A., 2009. Structure relates to elastic recoil and functional role in quadriceps tendon and patellar ligament. Micron 40 (3), 370–377.

Francis, C.A., Lenz, A.L., Lenhart, R.L., Thelen, D.G., 2013. The modulation of forward propulsion, vertical support, and center of pressure by the plantarflexors during human walking. Gait Posture 38 (4), 993–997.

Franz, J.R., Kram, R., 2012. The effects of grade and speed on leg muscle activations during walking. Gait Posture 35 (1), 143–147.

Fredericson, M., Cookingham, C.L., Chaudhari, A.M., Dowdell, B.C., Oestreicher, N., Sahrmann, S.A., 2000. Hip abductor weakness in distance runners with iliotibial band syndrome. Clin. J. Sport Med. 10 (3), 169–175.

Frowen, P., Benjamin, M., 1995. Variations in the quality of uncalcified fibrocartilage at the insertions of the extrinsic calf muscles in the foot. J. Anat. 186 (2), 417–421.

Fujiwara, A., Tamai, K., Yoshida, H., Kurihashi, A., Saotome, K., An, H.S., et al., 2000. Anatomy of the iliolumbar ligament. Clin. Orthop. Relat. Res. 380 (November), 167–172.

Fulkerson, J.P., Gossling, H.R., 1980. Anatomy of the knee joint lateral retinaculum. Clin. Orthop. Relat. Res. 153 (November–December), 183–188.

Gabriel, M.T., Wong, E.K., Woo, S.L.-Y., Yagi, M., Debski, R.E., 2004. Distribution of in situ forces in the anterior cruciate ligament in response to rotatory loads. J. Orthop. Res. 22 (1), 85–89.

Gajdosik, R.L., 2001. Passive extensibility of skeletal muscle: review of the literature with clinical implications. Clin. Biomech. 16 (2), 87–101.

Gallina, A., Merletti, R., Vieira, T.M.M., 2011. Are the myoelectric manifestations of fatigue distributed regionally in the human medial gastrocnemius muscle? J. Electromyogr. Kinesiol. 21 (6), 929–938.

Gao, H., Ji, B., Jager, I.L., Arzt, E., Fratzl, P., 2003. Materials become insensitive to flaws at nanoscale: lessons from nature. Proc. Natl. Acad. Sci. U. S. A. 100 (10), 5597–5600.

Gasparutto, X., Sancisi, N., Jacquelin, E., Parenti-Castelli, V., Dumas, R., 2015. Validation of a multi-body optimization with knee kinematic models including ligament constraints. J. Biomech. 48 (6), 1141–1146.

Gatt, R., Vella Wood, M., Gatt, A., Zarb, F., Formosa, C., Azzopardi, K.M., et al., 2015. Negative Poisson's ratios in tendons: an unexpected mechanical response. Acta Biomater. 24, 201–208.

Gaudino, F., Spira, D., Bangert, Y., Ott, H., Zobel, B.B., Kauczor, H.-U., et al., 2017. *Osteitis pubis* in professional football players: MRI findings and correlation with clinical outcome. Eur. J. Radiol. 94, 46–52.

Gazendam, M.G.J., Hof, A.L., 2007. Averaged EMG profiles in jogging and running at different speeds. Gait Posture 25 (4), 604–614.

Geppert, M.J., Sobel, M., Bohne, W.H.O., 1993. Lateral ankle instability as a cause of superior peroneal retinacular laxity: an anatomic and biomechanical study of cadaveric feet. Foot Ankle 14 (6), 330–334.

Gibbons, S., 2007. Clinical anatomy and function of psoas major and deep sacral gluteus maximus. In: Vleeming, A., Mooney, V., Stoeckart, R. (Eds.), Movement, Stability & Lumbopelvic Pain: Integration of Research and Therapy, second ed. Churchill Livingstone, Elsevier, Edinburgh, UK, pp. 95–102 (Chapter 6).

Giddings, V.L., Beaupré, G.S., Whalen, R.T., Carter, D.R., 2000. Calcaneal loading during walking and running. Med. Sci. Sports Exerc. 32 (3), 627–634.

Giphart, J.E., Stull, J.D., LaPrade, R.F., Wahoff, M.S., Philippon, M.J., 2012. Recruitment and activity of the pectineus and piriformis muscles during hip rehabilitation exercises: an electromyography study. Am. J. Sports Med. 40 (7), 1654–1663.

Golanó, P., Vega, J., de Leeuw, P.A.J., Malagelada, F., Manzanares, M.C., Götzens, V., et al., 2010. Anatomy of the ankle ligaments: a pictorial essay. Knee Surg. Sports Traumatol. Arthrosc. 18 (5), 557–569.

Goldberg, E.J., Neptune, R.R., 2007. Compensatory strategies during normal walking in response to muscle weakness and increased hip joint stiffness. Gait Posture 25 (3), 360–367.

Goldberg, S.R., Anderson, F.C., Pandy, M.G., Delp, S.L., 2004. Muscles that influence knee flexion velocity in double support: implications for stiff-knee gait. J. Biomech. 37 (8), 1189–1196.

Golditz, T., Steib, S., Pfeifer, K., Uder, M., Gelse, K., Janka, R., et al., 2014. Functional ankle instability as a risk factor for osteoarthritis: using T2-mapping to analyze early cartilage degeneration in the ankle joint of young athletes. Osteoarthr. Cartil. 22 (10), 1377–1385.

Görke, U.-J., Günther, H., Nagel, T., Wimmer, M.A., 2010. A large strain material model for soft tissues with functionally graded properties. J. Biomech. Eng. 132 (7), 074502. https://doi.org/10.1115/1.4001312.

Gottschall, J.S., Kram, R., 2003. Energy cost and muscular activity required for propulsion during walking. J. Appl. Physiol. 94 (5), 1766–1772.

Green, S.M., Briggs, P.J., 2013. Flexion strength of the toes in the normal foot. An evaluation using magnetic resonance imaging. Foot 23 (4), 115–119.

Griffith, C.J., LaPrade, R.F., Johansen, S., Armitage, B., Wijdicks, C., Engebretsen, L., 2009. Medial knee injury. Part 1. Static function of the individual components of the main medial knee structures. Am. J. Sports Med. 37 (9), 1762–1770.

Grillner, S., Nilsson, J., Thorstensson, A., 1978. Intra-abdominal pressure changes during natural movements in man. Acta Physiol. Scand. 103 (3), 275–283.

Grob, K., Gilbey, H., Manestar, M., Ackland, T., Kuster, M.S., 2017. The anatomy of the articularis genus muscle and its relation to the extensor apparatus of the knee. JB & JS Open Access 2 (4), e0034. https://doi.org/10.2106/JBJS.OA.17.00034.

Gruben, K.G., Boehm, W.L., 2012. Force direction pattern stabilizes sagittal plane mechanics of human walking. Hum. Mov. Sci. 31 (3), 649–659.

Gulan, G., Matovinović, D., Nemec, B., Rubinić, D., Ravlić-Gulan, J., 2000. Femoral neck anteversion: values, development, measurement, common problems. Colleg. Antropolog. 24 (2), 521–527.

Gupta, H.S., Seto, J., Wagermaier, W., Zaslansky, P., Boesecke, P., Fratzl, P., 2006. Cooperative deformation of mineral and collagen in bone at the nanoscale. Proc. Natl. Acad. Sci. U. S. A. 103 (47), 17741–17746.

Hadjicostas, P.T., Soucacos, P.N., Berger, I., Koleganova, N., Paessler, H.H., 2007. Comparative analysis of the morphologic structure of quadriceps and patellar tendon: a descriptive laboratory study. Arthroscopy: J. Arthrosc. Rel. Surg. 23 (7), 744–750.

Hägglund, M., Waldén, M., Ekstrand, J., 2005. Injury incidence and distribution in elite football—a prospective study of the Danish and the Swedish top divisions. Scand. J. Med. Sci. Sports 15 (1), 21–28.

Hamel, A.J., Sharkey, N.A., Buczek, F.L., Michelson, J., 2004. Relative motions of the tibia, talus, and calcaneus during stance phase of gait: a cadaver study. Gait Posture 20 (2), 147–153.

Hamill, J., van Emmerik, R.E., Heiderscheit, B.C., Li, L., 1999. A dynamical systems approach to lower extremity running injuries. Clin. Biomech. 14 (5), 297–308.

Hammer, N., Steinke, H., Lingslebe, U., Bechmann, I., Josten, C., Slowik, V., et al., 2013. Ligamentous influence in pelvic load distribution. Spine J. 13 (10), 1321–1330.

Handsfield, G.G., Slane, L.C., Screen, H.R.C., 2016. Letter to the editor: Nomenclature of the tendon hierarchy: an overview of inconsistent terminology and a proposed size-based naming scheme with terminology for multi-muscle tendons. J. Biomech. 49 (13), 3122–3124.

Handsfield, G.G., Inouye, J.M., Slane, L.C., Thelen, D.G., Miller, G.W., Blemker, S.S., 2017. A 3D model of the Achilles tendon to determine the mechanisms underlying nonuniform tendon displacements. J. Biomech. 51, 17–25.

Hanlon, M., Anderson, R., 2006. Prediction methods to account for the effect of gait speed on lower limb angular kinematics. Gait Posture 24 (3), 280–287.

Hansen, P., Aagaard, P., Kjaer, M., Larsson, B., Magnusson, S.P., 2003. Effect of habitual running on human Achilles tendon load-deformation properties and cross-sectional area. J. Appl. Physiol. 95 (6), 2375–2380.

Hansen, W., Shim, V.B., Obst, S., Lloyd, D.G., Newsham-West, R., Barrett, R.S., 2017. Achilles tendon stress is more sensitive to subject-specific geometry than subject-specific material properties: a finite element analysis. J. Biomech. 56, 26–31.

Harman, M.K., Markovich, G.D., Banks, S.A., Hodge, W.A., 1998. Wear patterns on tibial plateaus from varus and valgus osteoarthritic knees. Clin. Orthop. Relat. Res. 352 (July), 149–158.

Hartvigsen, J., Hancock, M.J., Kongsted, A., Louw, Q., Ferreira, M.L., Genevay, S., et al., 2018. What low back pain is and why we need to pay attention. Lancet 391 (10137), 2356–2367.

Heiden, T.L., Lloyd, D.G., Ackland, T.R., 2009. Knee joint kinematics, kinetics and muscle co-contraction in knee osteoarthritis patient gait. Clin. Biomech. 24 (10), 833–841.

Heller, M.O., Bergmann, G., Deuretzbacher, G., Claes, L., Haas, N.P., Duda, G.N., 2001. Influence of femoral anteversion on proximal femoral loading: measurement and simulation in four patients. Clin. Biomech. 16 (8), 644–649.

Herbert, R.D., Gandevia, S.C., 2019. The passive mechanical properties of muscle. J. Appl. Physiol. 126 (5), 1442–1444.

Herman, D.C., Weinhold, P.S., Guskiewicz, K.M., Garrett, W.E., Yu, B., Padua, D.A., 2008. The effects of strength training on the lower extremity biomechanics of female recreational athletes during a stop-jump task. Am. J. Sports Med. 36 (4), 733–740.

Herrington, L., 2011. Assessment of the degree of pelvic tilt within a normal asymptomatic population. Man. Ther. 16 (6), 646–648.

Hertel, J., Corbett, R.O., 2019. An update model of chronic ankle instability. J. Athl. Train. 54 (6), 572–588.

Hewitt, J.D., Glisson, R.R., Guilak, F., Vail, T.P., 2002. The mechanical properties of the human hip capsule ligaments. J. Arthroplasty 17 (1), 82–89.

Hibbs, A.E., Thompson, K.G., French, D., Wrigley, A., Spears, I., 2008. Optimizing performance by improving core stability and core strength. Sports Med. 38 (12), 995–1008.

Hidaka, E., Aoki, M., Izumi, T., Suzuki, D., Fujimiya, M., 2014. Ligament strain on the iliofemoral, pubofemoral, and ischiofemoral ligaments in cadaver specimens: biomechanical measurement and anatomical observation. Clin. Anat. 27 (7), 1068–1075.

Higashihara, A., Nagano, Y., Ono, T., Fukubayashi, T., 2016. Relationship between the peak time of hamstring stretch and activation during sprinting. Eur. J. Sport Sci. 16 (1), 36–41.

Hodges, P.W., Moseley, G.L., 2003. Pain and motor control of the lumbopelvic region: effect and possible mechanisms. J. Electromyogr. Kinesiol. 13 (4), 361–370.

Hodges, P.W., Richardson, C.A., 1996. Inefficient muscular stabilization of the lumbar spine associated with low back pain. A motor control evaluation of transversus abdominis. Spine 21 (22), 2640–2650.

Hodges, P.W., Richardson, C.A., 1997. Feedforward contraction of transversus abdominis is not influenced by the direction of arm movement. Exp. Brain Res. 114 (2), 362–370.

Hodges, P.W., Tucker, K., 2011. Moving differently in pain: a new theory to explain the adaptation to pain. Pain 152 (3 Supplement), S90–S98.

Hodges, P.W., Gurfinkel, V.S., Brumagne, S., Smith, T.C., Cordo, P.C., 2002. Coexistence of stability and mobility in postural control: evidence from postural compensation for respiration. Exp. Brain Res. 144 (3), 293–302.

Hodges, P.W., Moseley, G.L., Gabrielsson, A., Gandevia, S.C., 2003. Experimental muscle pain changes feedforward postural responses of the trunk muscles. Exp. Brain Res. 151 (2), 262–271.

Hodson-Tole, E.F., Loram, I.D., Vieira, T.M.M., 2013. Myoelectric activity along human gastrocnemius medialis: different spatial distributions of postural and electrically elicited surface potentials. J. Electromyogr. Kinesiol. 23 (1), 43–50.

Hodson-Tole, E.F., Wakeling, J.M., Dick, T.J.M., 2016. Passive muscle-tendon unit gearing is joint dependent in human medial gastrocnemius. Front. Physiol. 7, 95. https://doi.org/10.3389/fphys.2016.00095.

Hof, A.L., Elzinga, H., Grimmius, W., Halbertsma, J.P.K., 2002. Speed dependence of averaged EMG profiles in walking. Gait Posture 16 (1), 78–86.

Hoffman, J., Gabel, P., 2013. Expanding Panjabi's stability model to express movement: a theoretical model. Med. Hypotheses 80 (6), 692–697.

Hogervorst, T., Vereecke, E.E., 2014. Evolution of the human hip. Part 1. The osseous framework. J. Hip Preserv. Surgery 1 (2), 39–45.

Hogervorst, T., Bouma, H., de Boer, S.F., de Vos, J., 2011. Human hip impingement morphology: an evolutionary explanation. Journal of Bone and Joint. Surgery 93-B (6), 769–776.

Holowka, N.B., O'Neill, M.C., Thompson, N.E., Demes, B., 2017. Chimpanzee and human midfoot motion during bipedal walking and the evolution of the longitudinal arch of the foot. J. Hum. Evol. 104, 23–31.

Honeine, J.-L., Schieppati, M., Gagey, O., Do, M.-C., 2013. The functional role of the triceps surae muscle during human locomotion. PLoS One 8 (1), e52943. https://doi.org/10.1371/journal.pone.0052943.

Hu, H., Meijer, O.G., van Dieën, J.H., Hodges, P.W., Bruijn, S.M., Strijers, R.L., et al., 2010. Muscle activity during the active straight leg raise (ASLR), and the effects of a pelvic belt on the ASLR and on treadmill walking. J. Biomech. 43 (3), 532–539.

Huang, J., Qin, K., Tang, C., Zhu, Y., Klein, C.S., Zhang, Z., et al., 2018. Assessment of passive stiffness of medial and lateral heads of gastrocnemius muscle, Achilles tendon, and plantar fascia at different ankle and knee positions using the MyotonPRO. Med. Sci. Monit. 24, 7570–7576.

Hubbard, J.K., Sampson, H.W., Elledge, J.R., 1997. Prevalence and morphology of the vastus medialis oblique muscle in human cadavers. Anat. Rec. 249 (1), 135–142.

Hug, F., Lacourpaille, L., Maïsetti, O., Nordez, A., 2013. Slack length of gastrocnemius medialis and Achilles tendon occurs at different ankle angles. J. Biomech. 46 (14), 2534–2538.

Hug, F., Hodges, P.W., van den Hoorn, W., Tucker, K., 2014. Between-muscle differences in the adaptation to experimental pain. J. Appl. Physiol. 117 (10), 1132–1140.

Huijing, P.A., Yaman, A., Ozturk, C., Yucesoy, C.A., 2011. Effects of knee joint angle on global and local strains within human triceps surae muscle: MRI analysis indicating in vivo myofascial force transmission between synergistic muscles. Surg. Radiol. Anat. 33 (10), 869–879.

Hukins, D.W.L., Meakin, J.R., 2000. Relationship between structure and mechanical function of the tissues of the intervertebral joint. Am. Zool. 40 (1), 42–52.

Hull, M.L., Berns, G.S., Varma, H., Patterson, H.A., 1996. Strain in the medial collateral ligament of the human knee under single and combined loads. J. Biomech. 29 (2), 199–206.

Hulmes, D.J., Wess, T.J., Prockop, D.J., Fratzl, P., 1995. Radial packing, order, and disorder in collagen fibrils. Biophys. J. 68 (5), 1661–1670.

Hunter, D.J., Guermazi, A., Lo, G.H., Grainger, A.J., Conaghan, P.G., Boudreau, R.M., et al., 2011. Evolution of semi-quantitative whole joint assessment of knee OA: MOAKS (MRI Osteoarthritis Knee Score). Osteoarthr. Cartil. 19 (8), 990–1002.

Hunter, D.J., Guermazi, A., Roemer, F., Zhang, Y., Neogi, T., 2013. Structural correlates of pain in joints with osteoarthritis. Osteoarthr. Cartil. 21 (9), 1170–1178.

Imhauser, C.W., Siegler, S., Abidi, N.A., Frankel, D.Z., 2004. The effect of posterior tibialis tendon dysfunction on the plantar pressure characteristics and the kinematics of the arch and the hindfoot. Clin. Biomech. 19 (2), 161–169.

Ingber, D.E., 1998. The architecture of life. Sci. Am. 278 (1), 48–57.

Ingber, D.E., 2006. Cellular mechanotransduction: putting all the pieces together again. FASEB J. 20 (7), 811–827.

Ingber, D.E., 2008. Tensegrity-based mechanosensing from macro to micro. Prog. Biophys. Mol. Biol. 97 (2–3), 163–179.
Ishikawa, M., Pakaslahti, J., Komi, P.V., 2007. Medial gastrocnemius muscle behavior during human running and walking. Gait Posture 25 (3), 380–384.
Iwanuma, S., Akagi, R., Hashizume, S., Kanehisa, H., Yanai, T., Kawakami, Y., 2011. Triceps surae muscle-tendon unit length changes as a function of ankle joint angles and contraction levels: the effect of foot arch deformation. J. Biomech. 44 (14), 2579–2583.
Iwata, S., Suda, Y., Nagura, T., Matsumoto, H., Otani, T., Toyama, Y., 2007. Posterior instability near extension is related to clinical disability in isolated posterior cruciate ligament deficient patients. Knee Surg. Sports Traumatol. Arthrosc. 15 (4), 343–349.
Jadhav, S.P., More, S.R., Riascos, R.F., Lemos, D.F., Swischuk, L.E., 2014. Comprehensive review of the anatomy, function, and imaging of the popliteus and associated pathologic conditions. Radiographics 34 (2), 496–513.
Jain, K., Asad, M., Joshi, Y., Syed, A., 2015. Tibialis anterior tendon rupture as a complication of first tarsometatarsal joint steroid injection: a case report and review of literature. Foot 25 (3), 179–181.
James, E.W., LaPrade, C.M., LaPrade, R.F., 2015. Anatomy and biomechanics of the lateral side of the knee and surgical implications. Sports Med. Arthrosc. Rev. 23 (1), 2–9.
Jarvis, H.L., Nester, C.J., Bowden, P.D., Jones, R.K., 2017. Challenging the foundations of the clinical model of foot function: further evidence that the Root model assessments fail to appropriately classify foot function. J. Foot Ankle Res. 10, 7. https://doi.org/10.1186/s13047-017-0189-2.
Jeong, J.H., Leasure, J.M., Park, J., 2018. Assessment of the biomechanical changes after sacroiliac joint fusion by application of the 3-dimensional motion analysis technique. World Neurosurg. 117, e538–e543.
Jerosch, J., Prymka, M., 1996. Proprioception and joint stability. Knee Surg. Sports Traumatol. Arthrosc. 4 (3), 171–179.
Johnson, C.H., Christensen, J.C., 1999. Biomechanics of the first ray. Part I. The effects of peroneus longus function: a three-dimensional kinematic study on a cadaver model. J. Foot Ankle Surg. 38 (5), 313–321.
Józsa, L., Kannus, P., 1997. Structure and metabolism of normal tendons. In: Human Tendons: Anatomy, Physiology, and Pathology. Human Kinetics, Champaign, IL, pp. 46–90 (Chapter 2).
Judge, J.O., Davis 3rd, R.B., Ounpuu, S., 1996. Step length reductions in advanced age: the role of ankle and hip kinetics. J. Gerontol. Ser. A Biol. Med. Sci. 51 (6), M303–M312.
Jung, H.-S., Jeon, H.-S., Oh, D.-W., Kwon, O.-Y., 2013. Effect of the pelvic compression belt on the hip extensor activation patterns of sacroiliac joint pain patients during one-leg standing: a pilot study. Man. Ther. 18 (2), 143–148.
Jung, E., Ly, V.T., Buderi, A., Appleton, E., Teodorescu, M., 2019. Design and selection of muscle excitation patterns for modeling a lower extremity joint inspired tensegrity. In: 2019 Third IEEE International Conference on Robotic Computing (IRC). Naples, Italy, pp. 282–287, https://doi.org/10.1109/IRC.2019.00053.
Jungers, W.L., Meldrum, D.J., Stern Jr., J.T., 1993. The functional and evolutionary significance of the human peroneus tertius muscle. J. Hum. Evol. 25 (5), 377–386.
Kader, D., Saxena, A., Movin, T., Maffulli, N., 2002. Achilles tendinopathy: some aspects of basic science and clinical management. Br. J. Sports Med. 36 (4), 239–249.
Kainz, H., Carty, C.P., Modenese, L., Boyd, R.N., Lloyd, D.G., 2015. Estimation of the hip joint centre in human motion analysis: a systematic review. Clin. Biomech. 30 (4), 319–329.
Kapandji, I.A., 1974. The Physiology of the Joints. Volume 3: The Trunk and the Vertebral Column. Churchill Livingstone, Edinburgh.
Karahan, M., Kocaoglu, B., Cabukoglu, C., Akgun, U., Nuran, R., 2010. Effect of partial medial meniscectomy on the proprioceptive function of the knee. Arch. Orthop. Trauma Surg. 130 (3), 427–431.
Karakuzu, A., Pamuk, U., Ozturk, C., Acar, B., Yucesoy, C.A., 2017. Magnetic resonance and diffusion tensor imaging analyses indicate heterogeneous strains along human medial gastrocnemius fascicles caused by submaximal plantar-flexion activity. J. Biomech. 57, 69–78.
Karlsson, J., Andreasson, G.O., 1992. The effect of external ankle support in chronic lateral ankle joint instability: an electromyographic study. Am. J. Sports Med. 20 (3), 257–261.
Karlsson, J.S., Östlund, N., Larsson, B., Gerdle, B., 2003. An estimation of the influence of force decrease on the mean power spectral frequency shift of the EMG during repetitive maximum dynamic knee extensions. J. Electromyogr. Kinesiol. 13 (5), 461–468.
Kaufman, K.R., Hughes, C., Morrey, B.F., Morrey, M., An, K.-N., 2001. Gait characteristics of patients with knee osteoarthritis. J. Biomech. 34 (7), 907–915.
Kavounoudias, A., Roll, R., Roll, J.-P., 2001. Foot sole and ankle muscle inputs contribute jointly to human erect posture regulation. J. Physiol. 532 (3), 869–878.
Ker, R.F., McN, A., Bennett, M.B., 1988. Why are mammalian tendons so thick? J. Zool. 216 (2), 309–324.
Kernozek, T.W., Knaus, A., Rademaker, T., Almonroeder, T.G., 2018. The effects of habitual foot strike patterns on Achilles tendon loading in female runners. Gait Posture 66, 283–287.
Khan, A.M., McLoughlin, E., Giannakas, K., Hutchinson, C., Andrew, J.G., 2004. Hip osteoarthritis: where is the pain? Ann. R. Coll. Surg. Engl. 86 (2), 119–121.
Kibler, W.B., Press, J., Sciascia, A., 2006. The role of core stability in athletic function. Sports Med. 36 (3), 189–198.
Kim, W., 2020. Tibial femoral tunnel for isokinetic graft placement based on a tensegrity model of a knee. In: Nogueira, J.B.S. (Ed.), Knee Surgery—Reconstruction and Replacement. IntechOpen, https://doi.org/10.5772/Intechopen83631 (Chapter 1).
Kim, W., Park, A., 2018. Topological space of the knee tensegrity system. Biomed. J. Scientific Tech. Res. 3 (2), 3116–3118.
Kinugasa, R., Kawakami, Y., Sinha, S., Fukunaga, T., 2011. Unique spatial distribution of in vivo human muscle activation. Exp. Physiol. 96 (9), 938–948.
Kippers, V., Parker, A.W., 1984. Posture related to myoelectric silence of erectores spinae during trunk flexion. Spine 9 (7), 740–745.

Kiss, R.M., 2010. Effect of walking speed and severity of hip osteoarthritis on gait variability. J. Electromyogr. Kinesiol. 20 (6), 1044–1051.

Kitaoka, H.B., Luo, Z.P., An, K.N., 1997. Effect of the posterior tibial tendon on the arch of the foot during simulated weightbearing: biomechanical analysis. Foot Ankle Int. 18 (1), 43–46.

Kiter, E., Karaboyun, T., Tufan, A.C., Acar, K., 2010. Immunohistochemical demonstration of nerve endings in iliolumbar ligament. Spine 35 (4), E101–E104.

Klasan, A., Neri, T., Sommer, C., Leie, M.A., Dworschak, P., Schofer, M.D., et al., 2019. Analysis of acetabular version: retroversion prevalence, age, side and gender correlations. J. Orthop. Transl. 18, 7–12.

Klein, P., Mattys, S., Rooze, M., 1996. Movement arm length variations of selected muscles acting on talocrural and subtalar joints during movement: an in vitro study. J. Biomech. 29 (1), 21–30.

Klyne, D.M., Keays, S.L., Bullock-Saxton, J.E., Newcombe, P.A., 2012. The effect of anterior cruciate ligament rupture on the timing and amplitude of gastrocnemius muscle activation: a study of alterations in EMG measures and their relationship to knee joint stability. J. Electromyogr. Kinesiol. 22 (3), 446–455.

Kokubo, T., Hashimoto, T., Nagura, T., Nakamura, T., Suda, Y., Matsumoto, H., et al., 2012. Effect of the posterior tibial and peroneal longus on the mechanical properties of the foot arch. Foot Ankle Int. 33 (4), 320–325.

Komi, P.V., 1984. Physiological and biomechanical correlates of muscle function: effects of muscle structure and stretch-shortening cycle on force and speed. Exerc. Sport Sci. Rev. 12 (1), 81–121.

Komi, P.V., 2000. Stretch-shortening cycle: a powerful model to study normal and fatigued muscle. J. Biomech. 33 (10), 1197–1206.

Komi, P.V., Nicol, C., 2000. Stretch-shortening cycle fatigue. In: Nigg, B.M., BR, M.I., Mester, J. (Eds.), Biomechanics and Biology of Movement. Human Kinetics, Champaign, IL, pp. 385–404 (Chapter 20).

Konow, N., Azizi, E., Roberts, T.J., 2012. Muscle power attenuation by tendon during energy dissipation. Proc. R. Soc. B Biol. Sci. 279 (1731), 1108–1113.

Konradsen, L., Ravn, J.B., 1991. Prolonged peroneal reaction time in ankle instability. Int. J. Sports Med. 12 (3), 290–292.

Konrath, G.A., Hamel, A.J., Olson, S.A., Bay, B., Sharkey, N.A., 1998. The role of the acetabular labrum and the transverse acetabular ligament in load transmission in the hip. J. Bone Joint Surg. 80-A (12), 1781–1788.

Koumantakis, G.A., Watson, P.J., Oldham, J.A., 2005. Trunk muscle stabilization training plus general exercise versus general exercise only: randomized controlled trial of patients with recurrent low back pain. Phys. Ther. 85 (3), 209–225.

Krebs, D.E., Robbins, C.E., Lavine, L., Mann, R.W., 1998. Hip biomechanics during gait. J. Orthop. Sports Phys. Ther. 28 (1), 51–59.

Kumagai, M., Shiba, N., Higuchi, F., Nishimura, H., Inoue, A., 1997. Functional evaluation of hip abductor muscles with use of magnetic resonance imaging. J. Orthop. Res. 15 (6), 888–893.

Kumazaki, T., Ehara, Y., Sakai, T., 2012. Anatomy and physiology of hamstring injury. Int. J. Sports Med. 33 (12), 950–954.

Lamoth, C.J.C., Beek, P.J., Meijer, O.G., 2002. Pelvis-thorax coordination in the transverse plane during gait. Gait Posture 16 (2), 101–114.

Lamoth, C.J.C., Daffertshofer, A., Meijer, O.G., Beck, P.J., 2006a. How do persons with chronic low back pain speed up and slow down? Trunk-pelvis coordination and lumbar erector spinae activity during gait. Gait Posture 23 (2), 230–239.

Lamoth, C.J.C., Meijer, O.G., Daffertshofer, A., Wuisman, P.I.J.M., Beek, P.J., 2006b. Effects of chronic low back pain on trunk coordination and back muscle activity during walking: changes in motor control. Eur. Spine J. 15 (1), 23–40.

LaPrade, R.F., Muench, C., Wentorf, F., Lewis, J.L., 2002. The effect of injury to the posterolateral structures of the knee on force in a posterior cruciate ligament graft: a biomechanical study. Am. J. Sports Med. 30 (2), 233–238.

LaPrade, R.F., Ly, T.V., Wentorf, F.A., Engebretsen, L., 2003. The posterolateral attachments of the knee: a qualitative and quantitative morphologic analysis of the fibular collateral ligament, popliteus tendon, popliteofibular ligament, and lateral gastrocnemius tendon. Am. J. Sports Med. 31 (6), 854–860.

LaPrade, R.F., Morgan, P.M., Wentorf, F.A., Johansen, S., Engebretsen, L., 2007a. The anatomy of the posterior aspect of the knee: an anatomic study. J. Bone Joint Surg. 89-A (4), 758–764.

LaPrade, R.F., Engebretsen, A.H., Ly, T.V., Johansen, S., Wentorf, F.A., Engebretsen, L., 2007b. The anatomy of the medial part of the knee. J. Bone Joint Surgery 89-A (9), 2000–2010.

LaPrade, R.F., Bernhardson, A.S., Griffith, C.J., Macalena, J.A., Wijdicks, C.A., 2010. Correlation of valgus stress radiographs with medial knee ligament injuries: an in vitro biomechanical study. Am. J. Sports Med. 38 (2), 330–338.

LaPrade, M.D., Kennedy, M.I., Wijdicks, C.A., LaPrade, R.F., 2015. Anatomy and biomechanics of the medial side of the knee and their surgical implications. Sports Med. Arthrosc. Rev. 23 (2), 63–70.

Lazennec, J.Y., Brusson, A., Rousseau, M.A., 2013. Lumbar-pelvic-femoral balance on sitting and standing lateral radiographs. Orthop. Traumatol. Surg. Res. 99 (1 Supplement), S87–S103.

Leardini, A., Stagni, R., O'Connor, J.J., 2001. Mobility of the subtalar joint in the intact ankle complex. J. Biomech. 34 (6), 805–809.

Leardini, A., O'Connor, J.J., 2002. A model for the lever-arm length calculation of the flexor and extensor muscles at the ankle. Gait Posture 15 (3), 220–229.

Lechner, K., Hull, M.L., Howell, S.M., 2000. Is the circumferential tensile modulus within a human medial meniscus affected by the test sample location and cross-sectional area? J. Orthop. Res. 18 (6), 945–951.

Lederman, E., 2010. The myth of core stability. J. Bodyw. Mov. Ther. 14 (1), 84–98.

Lee, W.E., 2001. Podiatric biomechanics: an historical appraisal and discussion of the Root model as a clinical system of approach in the present context of theoretical uncertainty. Clin. Podiatr. Med. Surg. 18 (4), 555–684.

Lee, S.S.M., Piazza, S.J., 2008. Inversion–eversion moment arms of gastrocnemius and tibialis anterior measured in vivo. J. Biomech. 41 (16), 3366–3370.

Lee, S.S.M., Piazza, S.J., 2009. Built for speed: musculoskeletal structure and sprinting ability. J. Exp. Biol. 212 (22), 3700–3707.

Lee, T.Q., Morris, G., Csintalan, R.P., 2003. The influence of tibial and femoral rotation on patellofemoral contact area and pressure. J. Orthop. Sports Phys. Ther. 33 (11), 686–693.

Le Huec, J.C., Tsoupras, A., Leglise, A., Heraudet, P., Celarier, G., Sturresson, B., 2019. The sacro-iliac joint: a potentially painful enigma. Update on the diagnosis and treatment of pain from micro-trauma. Orthop. Traumatol. Surg. Res. 105 (1 Supplement), S31–S42.

Lemos, T., Imbiriba, L.A., Vargas, C.D., Vieira, T.M., 2015. Modulation of tibialis anterior muscle activity changes with upright stance width. J. Electromyogr. Kinesiol. 25 (1), 168–174.

Lenhart, R.L., Francis, C.A., Lenz, A.L., Thelen, D.G., 2014. Empirical evaluation of gastrocnemius and soleus function during walking. J. Biomech. 47 (12), 2969–2974.

Lersch, C., Grötsch, A., Segesser, B., Koebke, J., Brüggemann, G.-P., Potthast, W., 2012. Influence of calcaneus angle and muscle forces on strain distribution in the human Achilles tendon. Clin. Biomech. 27 (9), 955–961.

Levin, S.M., 1995. The importance of soft tissues for structural support of the body. Spine: State Art Rev. 9 (2), 357–363.

Levin, S.M., 1997. Putting the shoulder to the wheel: a new biomechanical model for the shoulder girdle. Biomed. Sci. Instrum. 33, 412–417.

Levin, S.M., 2002. The tensegrity-truss as a model for spine mechanics: biotensegrity. J. Mech. Med. Biol. 2 (3&4), 375–388.

Levin, S.M., 2007. A suspensory system for the sacrum in pelvic mechanics: bio-tensegrity. In: Vleeming, A., Mooney, V., Stoeckart, R. (Eds.), Movement, Stability and Lumbopelvic Pain. Churchill Livingstone, pp. 229–238 (Chapter 15).

Lewis, C.L., Ferris, D.P., 2008. Walking with increased ankle pushoff decreases hip muscle moments. J. Biomech. 41 (10), 2082–2089.

Lewis, C.L., Sahrmann, S.A., Moran, D.W., 2007. Anterior hip joint force increases with hip extension, decreased gluteal force, or decreased iliopsoas force. J. Biomech. 40 (16), 3725–3731.

Lewis, C.L., Sahrmann, S.A., Moran, D.W., 2009. Effect of position and alteration in synergist muscle force contribution on hip forces when performing hip strengthening exercises. Clin. Biomech. 24 (1), 35–42.

Lewis, C.L., Sahrmann, S.A., Moran, D.W., 2010. Effect of hip angle on anterior hip joint force during gait. Gait Posture 32 (4), 603–607.

Li, G., DeFrate, L.E., Zayontz, S., Park, S.E., Gill, T.J., 2004. The effect of tibiofemoral joint kinematics on patellofemoral contact pressures under simulated muscle loads. J. Orthop. Res. 22 (4), 801–806.

Li, Z., Alonso, J.E., Kim, J.-E., Davidson, J.S., Etheridge, B.S., Eberhardt, A.W., 2006. Three-dimensional finite element models of the human pubic symphysis with viscohyperelastic soft tissues. Ann. Biomed. Eng. 34 (9), 1452–1462.

Li, Z., Kim, J.-E., Davidson, J.S., Etheridge, B.S., Alonso, J.E., Eberhardt, A.W., 2007. Biomechanical response of the pubic symphysis in lateral pelvic impacts: a finite element study. J. Biomech. 40 (12), 2758–2766.

Li, L., Gollhofer, A., Lohrer, H., Dorn-Lange, N., Bonsignore, G., Gehring, D., 2019. Function of ankle ligaments for subtalar and talocrural joint stability during an inversion movement—an in vitro study. J. Foot Ankle Res. 12, 16. https://doi.org/10.1186/s13047-019-0330-5.

Lieber, R.L., Leonard, M.E., Brown-Maupin, C.G., 2000. Effects of muscle contraction on the load-strain properties of frog aponeurosis and tendon. Cells Tissues Organs 166 (1), 48–54.

Liu, M.Q., Anderson, F.C., Pandy, M.G., Delp, S.L., 2006. Muscles that support the body also modulate forward progression during walking. J. Biomech. 39 (14), 2623–2630.

Liu, C.L., Li, Y.P., Wang, X.Q., Zhang, Z.J., 2018. Quantifying the stiffness of Achilles tendon: intra- and inter-operator reliability and the effect of ankle joint motion. Med. Sci. Monit. 24, 4876–4881.

Logan, M., Willams, A., Lavelle, J., Gedroyc, W., Freeman, M., 2004. The effect of posterior cruciate ligament deficiency on knee kinematics. Am. J. Sports Med. 32 (8), 1915–1922.

Louw, M., Deary, C., 2014. The biomechanical variables involved in the aetiology of iliotibial band syndrome in distance runners—a systematic review of the literature. Phys. Ther. Sport 15 (1), 64–75.

Lovejoy, C.O., 2007. The natural history of human gait and posture. Part 3. The knee. Gait Posture 25 (3), 325–341.

Lovell, G.A., Blanch, P.D., Barnes, C.J., 2012. EMG of the hip adductor muscles in six clinical examination tests. Phys. Ther. Sport 13 (3), 134–140.

Lyman, J., Weinhold, P.S., Almekinders, L.C., 2004. Strain behavior of the distal Achilles tendon: implications for insertional Achilles tendinopathy. Am. J. Sports Med. 32 (2), 457–461.

Lynch, J.T., Schneider, M.T.Y., Perriman, D.M., Scarvell, J.M., Pickering, M.R., Asikuzzaman, M., et al., 2019. Statistical shape modelling reveals large and distinct subchondral bony differences in osteoarthritic knees. J. Biomech. 93, 177–184.

Maeshige, N., Uemura, M., Hirasawa, Y., Yoshikawa, Y., Moriguchi, M., Kawabe, N., et al., 2021. Immediate effects of weight-bearing calf stretching on ankle dorsiflexion range of motion and plantar pressure during gait in patients with diabetes mellitus. Int. J. Lower Extrem. Wounds: OnlineFirst. https://doi.org/10.1177/15347346211031318.

Maganaris, C.N., 2003. Force-length characteristics of the in vivo human gastrocnemius muscle. Clin. Anat. 16 (3), 215–223.

Maganaris, C.N., Kawakami, Y., Fukunaga, T., 2001. Changes in aponeurotic dimensions upon muscle shortening: in vivo observations in man. J. Anat. 199 (4), 449–456.

Magnusson, S.P., Hansen, P., Aagaard, P., Brønd, J., Dyhre-Poulsen, P., Bojsen-Møller, J., et al., 2003. Differential strain patterns of the human gastrocnemius aponeurosis and free tendon, in vivo. Acta Physiol. Scand. 177 (2), 185–195.

Maharaj, J.N., Cresswell, A.G., Lichtwark, G.A., 2016. The mechanical function of the tibialis posterior muscle and its tendon during locomotion. J. Biomech. 49 (14), 3238–3243.

Maïsetti, O., Hug, F., Bouillard, K., Nordez, A., 2012. Characterization of passive elastic properties of the human medial gastrocnemius muscle belly using supersonic shear imaging. J. Biomech. 45 (6), 978–984.

Malcolm, P., Segers, V., Van Caekenberghe, I., De Clercq, D., 2009. Experimental study of the influence of the *M. tibialis* anterior on the walk-to-run transition by means of a powered ankle-foot exoskeleton. Gait Posture 29 (1), 6–10.

Maloiy, G.M.O., Heglund, N.C., Prager, L.M., Cavagna, G.A., Taylor, C.R., 1986. Energetic cost of carrying loads: have African women discovered an economic way? Nature 319 (6055), 668–669.

Manoli 2nd, A., Graham, B., 2005. The subtle cavus foot, "the underpronator," a review. Foot Ankle Int. 26 (3), 256–263.

Marouane, H., Shirazi-Adl, A., Adouni, M., 2016. Alterations in knee contact forces and centers in stance phase of gait: a detailed lower extremity musculoskeletal model. J. Biomech. 49 (2), 185–192.

Masi, A.T., Hannon, J.C., 2008. Human resting muscle tone (HRMT): narrative introduction and modern concepts. J. Bodyw. Mov. Ther. 12 (4), 320–332.

Matsumoto, H., Suda, Y., Otani, T., Niki, Y., Seedhom, B.B., Fujikawa, K., 2001. Roles of the anterior cruciate ligament and the medial collateral ligament in preventing valgus instability. J. Orthop. Sci. 6 (1), 28–32.

McDermott, I.D., Masouros, S.D., Amis, A.A., 2008. Biomechanics of the menisci of the knee. Curr. Orthop. 22 (3), 193–201.

McGibbon, C.A., 2003. Toward a better understanding of gait changes with age and disablement: neuromuscular adaptation. Exerc. Sport Sci. Rev. 31 (2), 102–108.

McGonagle, D., Marzo-Ortega, H., Benjamin, M., Emery, P., 2003. Report on the second international enthesitis workshop. Arthritis Rheum. 48 (4), 896–905.

McGovern, R.P., Kivlan, B.R., Martin, R.L., 2017. Length change of the short external rotators of the hip in common stretch positions.: a cadaveric study. Int. J. Sports Phys. Ther. 12 (7), 1068–1077.

McGowan, C.P., Neptune, R.R., Clark, D.J., Kautz, S.A., 2010. Modular control of human walking: adaptations to altered mechanical demands. J. Biomech. 43 (3), 412–419.

McKeon, P.O., Hertel, J., Bramble, D., Davis, I., 2015. The foot core system: a new paradigm for understanding intrinsic foot muscle function. Br. J. Sports Med. 49 (5), 290. https://doi.org/10.1136/bjsports-2013-092690.

McNair, P.J., Stanley, S.N., 1996. Effect of passive stretching and jogging on the series elastic muscle stiffness and range of motion of the ankle joint. Br. J. Sports Med. 30 (4), 313–317. Discussion 318.

McPoil, T.G., Cornwall, M.W., 1994. Relationship between neutral subtalar joint position and pattern of rearfoot motion during walking. Foot Ankle Int. 15 (3), 141–145.

McPoil, T.G., Cornwall, M.W., 1996. The relationship between static lower extremity measurements and rearfoot motion during walking. J. Orthop. Sports Phys. Ther. 24 (5), 309–314.

Meinders, E., Pizzolato, C., Goncalves, B.A.M., Lloyd, D.G., Saxby, D.J., Diamond, L.E., 2021. The deep hip muscles are unlikely to stabilize the hip in the sagittal plane during walking: a joint stiffness approach. IEEE Trans. Biomed. Eng. 69 (3), 1133–1140.

Mendis, M.D., Wilson, S.J., Hayes, D.A., Watts, M.C., Hides, J.A., 2014. Hip flexor muscle size, strength and recruitment pattern in patients with acetabular labral tears compared to healthy controls. Man. Ther. 19 (5), 405–410.

Mengarelli, A., Maranesi, E., Burattini, L., Fioretti, S., Di Nardo, F., 2017. Co-contraction activity of ankle muscles during walking: a gender comparison. Biomed. Signal Process. Control 33, 1–9.

Mengarelli, A., Gentili, A., Strazza, A., Burattini, L., Fioretti, S., Di Nardo, F., 2018. Co-activation patterns of gastrocnemius and quadriceps femoris in controlling the knee joint during walking. J. Electromyogr. Kinesiol. 42, 117–122.

Mens, J.M.A., Damen, L., Snijders, C.J., Stam, H.J., 2006. The mechanical effect of a pelvic belt in patients with pregnancy-related pelvic pain. Clin. Biomech. 21 (2), 122–127.

Mickelborough, J., van der Linden, M.L., Tallis, R.C., Ennos, A.R., 2004. Muscle activity during gait initiation in normal elderly people. Gait Posture 19 (1), 50–57.

Miller, R.H., Meardon, S.A., Derrick, T.R., Gillette, J.C., 2008. Continuous relative phase variability during an exhaustive run in runners with a history of iliotibial band syndrome. J. Appl. Biomech. 24 (3), 262–270.

Mills, J.D., Taunton, J.E., Mills, W.A., 2005. The effect of a 10-week training regimen on lumbo-pelvic stability and athletic performance in female athletes: a randomized-controlled trial. Phys. Ther. Sport 6 (2), 60–66.

Milz, S., McNeilly, C., Putz, R., Ralphs, J.R., Benjamin, M., 1998. Fibrocartilages in the extensor tendons of the interphalangeal joints of human toes. Anat. Rec. 252 (2), 264–270.

Milz, S., Rufai, A., Buettner, A., Putz, R., Ralphs, J.R., Benjamin, M., 2002. Three-dimensional reconstructions of the Achilles tendon insertion in man. J. Anat. 200 (2), 145–152.

Mohamed, O., Perry, J., Hislop, H., 2002. Relationship between wire EMG activity, muscle length, and torque of the hamstrings. Clin. Biomech. 17 (8), 569–579.

Mokhtarzadeh, H., Yeow, C.H., Goh, J.C.H., Oetomo, D., Malekipour, F., Lee, P.V.-S., 2013. Contributions of the soleus and gastrocnemius muscles to the anterior cruciate ligament loading during single-leg landing. J. Biomech. 46 (11), 1913–1920.

Mont, M.A., Seyler, T.M., Ragland, P.S., Starr, R., Erhart, J., Bhave, A., 2007. Gait analysis of patients with resurfacing hip arthroplasty compared with hip osteoarthritis and standard total hip arthroplasty. J. Arthroplasty 22 (1), 100–108.

Moore, K.A., Polte, T., Huang, S., Shi, B., Alsberg, E., Sunday, M.E., et al., 2005. Control of basement membrane remodeling and epithelial branching morphogenesis in embryonic lung by Rho and cytoskeletal tension. Dev. Dyn. 232 (2), 268–281.

Moorman 3rd, C.T., LaPrade, R.F., 2005. Anatomy and biomechanics of the posterolateral corner of the knee. J. Knee Surg. 18 (2), 137–145.

Morrison, J.B., 1970. The mechanics of the knee joint in relation to normal walking. J. Biomech. 3 (1), 51–61.

Morse, C.I., 2011. Gender differences in the passive stiffness of the human gastrocnemius muscle during stretch. Eur. J. Appl. Physiol. 111 (9), 2149–2154.

Mosca, V.S., 2001. The cavus foot. J. Pediatr. Orthop. 21 (4), 423–424.

Moschella, D., Blasi, A., Leardini, A., Ensini, A., Catani, F., 2006. Wear patterns on tibial plateau from varus osteoarthritic knees. Clin. Biomech. 21 (2), 152–158.

Müller, R., Ertelt, T., Blickhan, R., 2015. Low back pain affects trunk as well as lower limb movements during walking and running. J. Biomech. 48 (6), 1009–1014.

Mündermann, A., Dyrby, C.O., Andriacchi, T.P., 2005. Secondary gait changes in patients with medial compartment knee osteoarthritis: increased load at the ankle, knee, and hip during walking. Arthritis Rheum. 52 (9), 2835–2844.

Murdock, G.H., Hubley-Kozey, C.L., 2012. Effect of a high intensity quadriceps fatigue protocol on knee joint mechanics and muscle activation during gait in young adults. Eur. J. Appl. Physiol. 112 (2), 439–449.

Murley, G.S., Menz, H.B., Landorf, K.B., 2014. Electromyographic patterns of tibialis posterior and related muscles when walking at different speeds. Gait Posture 39 (4), 1080–1085.

Myers, T.W., 2020. Tension-dependent structures in a stretch-activated system. J. Bodyw. Mov. Ther. 24 (1), 131–133.

Myers, C.A., Register, B.C., Lertwanich, P., Ejnisman, L., Pennington, W.W., Giphart, J.E., et al., 2011. Role of the acetabular labrum and the iliofemoral ligament in hip stability: an in vitro biplane fluoroscopy study. Am. J. Sports Med. 39 (1 Supplement), 85S–91S.

Nakagawa, T.H., Moriya, E.T.U., Maciel, C.D., Serrão, F.V., 2012. Frontal plane biomechanics in males and females with and without patellofemoral pain. Med. Sci. Sports Exerc. 44 (9), 1747–1755.

Nakamura, M., Hasegawa, S., Umegaki, H., Nishishita, S., Kobayashi, T., Fujita, K., et al., 2016. The difference in passive tension applied to the muscles composing the hamstrings—comparison among muscles using ultrasound shear wave elastography. Man. Ther. 24, 1–6.

Needham, R., Naemi, R., Healy, A., Chockalingam, N., 2016. Multi-segment kinematic model to assess three-dimensional movement of the spine and back during gait. Prosthetics Orthot. Int. 40 (5), 624–635.

Nene, A., Byrne, C., Hermens, H., 2004. Is rectus femoris really a part of quadriceps? Assessment of rectus femoris function during gait in able-bodied adults. Gait Posture 20 (1), 1–13.

Neogi, T., Bowes, M.A., Niu, J., De Souza, K.M., Vincent, G.R., Goggins, J., et al., 2013. Magnetic resonance imaging-based three-dimensional bone shape of the knee predicts onset of knee osteoarthritis: data from the Osteoarthritis Initiative. Arthritis Rheum. 65 (8), 2048–2058.

Neptune, R.R., McGowan, C.P., 2011. Muscle contributions to whole-body sagittal plane angular momentum during walking. J. Biomech. 44 (1), 6–12.

Neptune, R.R., Zajac, F.E., Kautz, S.A., 2004. Muscle force redistributes segmental power for body progression during walking. Gait Posture 19 (2), 194–205.

Neptune, R.R., Sasaki, K., Kautz, S.A., 2008. The effect of walking speed on muscle function and mechanical energetics. Gait Posture 28 (1), 135–143.

Neumann, D.A., 2010. Kinesiology of the hip: a focus on muscular actions. J. Orthop. Sports Phys. Ther. 40 (2), 82–94.

Newton, B.W., 2014. Unilateral absence of an extensor digitorum longus muscle and variations of toe tendons. Int. J. Anat. Variat. 7 (1), 42–45.

Noehren, B., Davis, I., Hamill, J., 2007. ASB Clinical Biomechanics Award Winner 2006: prospective study of the biomechanical factors associated with iliotibial band syndrome. Clin. Biomech. 22 (9), 951–956.

Nyland, J., Gamble, C., Franklin, T., Caborn, D.N.M., 2017. Permanent knee sensorimotor system changes following ACL injury and surgery. Knee Surg. Sports Traumatol. Arthrosc. 25 (5), 1461–1474.

Okada, T., Huxel, K.C., Nesser, T.W., 2011. Relationship between core stability, functional movement, and performance. J. Strength Cond. Res. 25 (1), 252–261.

Orgel, J.P.R.O., Irving, T.C., Miller, A., Wess, T.J., 2006. Microfibrillar structure of type I collagen in situ. Proc. Natl. Acad. Sci. U. S. A. 103 (24), 9001–9005.

Orselli, M.I.V., Franz, J.R., Thelen, D.G., 2017. The effects of Achilles tendon compliance on triceps surae mechanics and energetics in walking. J. Biomech. 60, 227–231.

Pamuk, U., Karakuzu, A., Ozturk, C., Acar, B., Yucesoy, C.A., 2016. Combined magnetic resonance and diffusion tensor imaging analyses provide a powerful tool for in vivo assessment of deformation along human muscle fibres. J. Mech. Behav. Biomed. Mater. 63, 207–219.

Panjabi, M.M., 1992a. The stabilizing system of the spine. Part I. Function, dysfunction, adaptation, and enhancement. J. Spinal Disord. 5 (4), 383–389.

Panjabi, M.M., 1992b. The stabilizing system of the spine. Part II. Neutral zone and instability hypothesis. J. Spinal Disord. 5 (4), 390–397.

Panjabi, M.M., 2003. Clinical spinal instability and low back pain. J. Electromyogr. Kinesiol. 13 (4), 371–379.

Panjabi, M.M., 2006. A hypothesis of chronic back pain: ligament subfailure injuries lead to muscle control dysfunction. Eur. Spine J. 15 (5), 668–676.

Park, K.-M., Kim, S.-Y., Oh, D.-W., 2010. Effects of the pelvic compression belt on gluteus medius, quadratus lumborum, and lumbar multifidus activities during side-lying hip abduction. J. Electromyogr. Kinesiol. 20 (6), 1141–1145.

Parker, K.K., Ingber, D.E., 2007. Extracellular matrix, mechanotransduction and structural hierarchies in heart tissue engineering. Philos. Trans. R. Soc. Lond. B: Biol. Sci. 362 (1484), 1267–1279.

Parvaresh, K.C., Chang, C., Patel, A., Lieber, R.L., Ball, S.T., 2019. Architecture of the short external rotator muscles of the hip. BMC Musculoskelet. Disord. 20 (1), 611. https://doi.org/10.1186/s12891-019-2995-0.

Pasque, C., Noyes, F.R., Gibbons, M., Levy, M., Grood, E., 2003. The role of the popliteofibular ligament and the tendon of popliteus in providing stability in the human knee. J. Bone Joint Surg. 85-B (2), 292–298.

Patterson, J.T., Jokl, P., Katz, L.D., Lawrence, D.A., Smitaman, E., 2014. Isolated avulsion fracture at the medial head of the gastrocnemius muscle. Skelet. Radiol. 43 (10), 1491–1494.

Peabody, T., Bordoni, B., 2021. Anatomy, Bony Pelvis and Lower Limb, Fascia Lata. StatPearls [Internet], StatPearls Publishing, Treasure Island (FL). 32491429.

Pękala, P.A., Henry, B.M., Ochala, A., Kopacz, P., Tatoń, G., Młyniec, A., et al., 2017. The twisted structure of the Achilles tendon unraveled: a detailed quantitative and qualitative anatomical investigation. Scand. J. Med. Sci. Sports 27 (12), 1705–1715.

Pel, J.J.M., Spoor, C.W., Goossens, R.H.M., Pool-Goudzwaard, A.L., 2008. Biomechanical model study of pelvic belt influence on muscle and ligament forces. J. Biomech. 41 (9), 1878–1884.

Perry, J., 1992. Gait Analysis: Normal and Pathological Function. SLACK Incorporated, Thorofare, NJ.

Petcu, D., Colda, A., 2012. Foot functioning paradigms. Proc. Roman. Acad. Ser. B 14 (3), 212–217.

Petersen, J., Hölmich, P., 2005. Evidence based prevention of hamstring injuries in sport. Br. J. Sports Med. 39 (6), 319–323.

Petersen, W., Hohmann, G., Stein, V., Tillman, B., 2002. The blood supply of the posterior tibial tendon. J. Bone Joint Surg. 84-B (1), 141–144.

Petersen, W., Hohmann, G., Pufe, T., Tsokos, M., Zantop, T., Paulsen, F., et al., 2004. Structure of the human tibialis posterior tendon. Arch. Orthop. Trauma Surg. 124 (4), 237–242.

Piazza, S.J., Delp, S.L., 1996. The influence of muscles on knee flexion during the swing phase of gait. J. Biomech. 29 (6), 723–733.

Pincivero, D.M., Gandhi, V., Timmons, M.K., Coelho, A.J., 2006. Quadriceps femoris electromyogram during concentric, isometric and eccentric phases of fatiguing dynamic knee extensions. J. Biomech. 39 (2), 246–254.

Pisani, G., 1994. The coxa pedis. Eur. J. Foot Ankle Surg. 1 (2–3), 67–74.

Pisani, G., 2016. "Coxa pedis" today. Foot Ankle Surg. 22 (2), 78–84.

Podraza, J.T., White, S.C., 2010. Effect of knee flexion angle on ground reaction forces, knee moments and muscle co-contraction during an impact-like deceleration landing: implications for the non-contact mechanism of ACL injury. Knee 17 (4), 291–295.

Pohl, M.B., Buckley, J.G., 2008. Changes in foot and shank coupling due to alterations in foot strike pattern during running. Clin. Biomech. 23 (3), 334–341.

Pohl, M.B., Messenger, N., Buckley, J.G., 2007. Forefoot, rearfoot and shank coupling: effect of variations in speed and mode of gait. Gait Posture 25 (2), 295–302.

Poilliot, A.J., Zwirner, J., Doyle, T., Hammer, N., 2019. A systematic review of the normal sacroiliac joint anatomy and adjacent tissues for pain physicians. Pain Physician 22 (4), E247.

Pollo, F.E., 1998. Bracing and heel wedging for unicompartmental osteoarthritis of the knee. Am. J. Knee Surg. 11 (1), 47–50.

Pool-Goudzwaard, A., Hoek van Dijke, G., Mulder, P., Spoor, C., Snijders, C., Stoeckart, R., 2003. The iliolumbar ligament: its influence on stability of the sacroiliac joint. Clin. Biomech. 18 (2), 99–105.

Pool-Goudzwaard, A., Gnat, R., Spoor, K., 2012. Deformation of the innominate bone and mobility of the pubic symphysis during asymmetric moment application to the pelvis. Man. Ther. 17 (1), 66–70.

Porterfield, J.A., DeRosa, C., 1998. Mechanical Low Back Pain: Perspectives in Functional Anatomy, second ed. WB Saunders, Philadelphia.

Potthast, W., Lersch, C., Segesser, B., Koebke, J., Brüggemann, G.-P., 2008. Intraarticular pressure distribution in the talocrural joint is related to lower leg muscle forces. Clin. Biomech. 23 (5), 632–639.

Powers, C.M., Landel, R., Perry, J., 1996. Timing and intensity of vastus muscle activity during functional activities in subjects with and without patellofemoral pain. Phys. Ther. 76 (9), 946–955. discussion 956–967.

Preece, S.J., Graham-Smith, P., Nester, C.J., Howard, D., Hermens, H., Herrington, L., et al., 2008. The influence of gluteus maximus on transverse plane tibial rotation. Gait Posture 27 (4), 616–621.

Puhakka, K.B., Melsen, F., Jurik, A.G., Boel, L.W., Vesterby, A., Egund, N., 2004. MR imaging of the normal sacroiliac joint with correlation to histology. Skelet. Radiol. 33 (1), 15–18.

Ramaniraka, N.A., Terrier, A., Theumann, N., Siegrist, O., 2005. Effects of the posterior cruciate ligament reconstruction on the biomechanics of the knee joint: a finite element analysis. Clin. Biomech. 20 (4), 434–442.

Rana, M., Hamarneh, G., Wakeling, J.M., 2013. 3D fascicle orientations in triceps surae. J. Appl. Physiol. 115 (1), 116–125.

Rantanen, J., Hurme, M., Falck, B., Alaranta, H., Nykvist, F., Lehto, M., et al., 1993. The lumbar multifidus muscle five years after surgery for a lumbar intervertebral disc herniation. Spine 18 (5), 568–574.

Rasmussen, O., 1985. Stability of the ankle joint. Analysis of the function and traumatology of the ankle ligaments. Acta Orthop. Scand. 56 (Supplementum 211), 7–75.

Rasmussen, O., Kromann-Andersen, C., 1983. Experimental ankle injuries. Analysis of the traumatology of the ankle ligaments. Acta Orthop. Scand. 54 (3), 356–362.

Reeves, N.D., Narici, M.V., 2003. Behavior of human muscle fascicles during shortening and lengthening contractions in vivo. J. Appl. Physiol. 95 (3), 1090–1096.

Reischl, S.F., Powers, C.M., Rao, S., Perry, J., 1999. Relationship between foot pronation and rotation of the tibia and femur during walking. Foot Ankle Int. 20 (8), 513–520.

Roberts, T.J., Azizi, E., 2010. The series-elastic shock absorber: tendons attenuate muscle power during eccentric actions. J. Appl. Physiol. 109 (2), 396–404.

Roberts, T.J., Konow, N., 2013. How tendons buffer energy dissipation by muscle. Exerc. Sport Sci. Rev. 41 (4), 186–193.

Robertson, B.A., Barker, P.J., Fahrer, M., Schache, A.G., 2009. The anatomy of the pubic region revisited: implications for the pathogenesis and clinical management of chronic groin pain in athletes. Sports Med. 39 (3), 225–234.

Robinson, P., Salehi, F., Grainger, A., Clemence, M., Schilders, E., O'Connor, P., et al., 2007. Cadaveric and MRI study of the musculotendinous contributions to the capsule of the symphysis pubis. Am. J. Roentgenol. 188 (5), W440–W445.

Robson, M.D., Benjamin, M., Gishen, P., Bydder, G.M., 2004. Magnetic resonance imaging of the Achilles tendon using ultrashort TE (UTE) pulse sequences. Clin. Radiol. 59 (8), 727–735.

Roca-Dols, A., Losa-Iglesias, M.E., Sánchez-Gómez, R., López-López, D., Becerro-de-Bengoa-Vallejo, R., Calvo-Lobo, C., 2018. Electromyography comparison of the effects of various footwear in the activity patterns of the peroneus longus and brevis muscles. J. Mech. Behav. Biomed. Mater. 82, 126–132.

Rooney, B.D., Derrick, T.R., 2013. Joint contact loading in forefoot and rearfoot strike patterns during running. J. Biomech. 46 (13), 2201–2206.

Root, M.L., Orien, W.P., Weed, J.H., Hughes, R.J., 1971. Biomechanical Examination of the Foot: Volume I. Clinical Biomechanics Corporation, Los Angeles.

Rosager, S., Aagaard, P., Dyhre-Poulsen, P., Neergaard, K., Kjaer, M., Magnusson, S.P., 2002. Load-displacement properties of the human triceps surae aponeurosis and tendon in runners and non-runners. Scand. J. Med. Sci. Sports 12 (2), 90–98.

Ruina, A., Bertram, J.E.A., Srinivasan, M., 2005. A collisional model of the energetic cost of support work qualitatively explains leg sequencing in walking and galloping, pseudo-elastic leg behavior in running and the walk-to-run transition. J. Theor. Biol. 237 (2), 170–192.

Russell, K.A., Palmieri, R.M., Zinder, S.M., Ingersoll, C.D., 2006. Sex differences in valgus knee angle during a single-leg drop jump. J. Athl. Train. 41 (2), 166–171.

Rutherford, D.J., Hubley-Kozey, C., 2009. Explaining the hip adduction moment variability during gait: implications for hip abductor strengthening. Clin. Biomech. 24 (3), 267–273.

Rutherford, D.J., Hubley-Kozey, C., Stanish, W., 2014. Hip abductor function in individuals with medial knee osteoarthritis: implications for medial compartment loading during gait. Clin. Biomech. 29 (5), 545–550.

Saha, D., Gard, S., Fatone, S., 2008. The effect of trunk flexion on able-bodied gait. Gait Posture 27 (4), 653–660.

Salsich, G.B., Mueller, M.J., 2000. Effect of plantar flexor muscle stiffness on selected gait characteristics. Gait Posture 11 (3), 207–216.

Saunders, S.W., Schache, A., Rath, D., Hodges, P.W., 2005. Changes in three dimensional lumbo-pelvic kinematics and trunk muscle activity with speed and mode of locomotion. Clin. Biomech. 20 (8), 784–793.

Scarr, G., 2008. A model of the cranial vault as a tensegrity structure, and its significance to normal and abnormal cranial development. Int. J. Osteopathic Med. 11 (3), 80–89.

Scarr, G., 2011. Helical tensegrity as a structural mechanism in human anatomy. Int. J. Osteopathic Med. 14 (1), 24–32.

Schache, A.G., Dorn, T.W., Blanch, P.D., Brown, N.A.T., Pandy, M.G., 2012. Mechanics of the human hamstring muscles during sprinting. Med. Sci. Sports Exerc. 44 (4), 647–658.

Schipplein, O.D., Andriacchi, T.P., 1991. Interaction between active and passive knee stabilizers during level walking. J. Orthop. Res. 9 (1), 113–119.

Schleip, R., Klingler, W., Lehmann-Horn, F., 2005. Active fascial contractility: fascia may be able to contract in a smooth muscle-like manner and thereby influence musculoskeletal dynamics. Med. Hypotheses 65 (2), 273–277.

Schleip, R., Vleeming, A., Lehmann-Horn, F., Klinger, W., 2007. Letter to the Editor concerning "A hypothesis of chronic back pain: ligament subfailure injuries lead to muscle control dysfunction" (M. Panjabi). Eur. Spine J. 16 (10), 1733–1735.

Schleip, R., Gabbiani, G., Wilke, J., Naylor, I., Hinz, B., Zorn, A., et al., 2019. Fascia is able to actively contract and may thereby influence musculoskeletal dynamics: a histochemical and mechanographic investigation. Front. Physiol. 10, 336. https://doi.org/10.3389/fphys.2019.00336.

Schuermans, J., Van Tiggelen, D., Danneels, L., Witvrouw, E., 2014. Biceps femoris and semitendinosus - teammates or competitors? New insights into hamstring injury mechanisms in male football players: a muscle functional MRI study. Br. J. Sports Med. 48 (22), 1599–1606.

Scorcelletti, M., Reeves, N.D., Rittweger, J., Ireland, A., 2020. Femoral anteversion: significance and measurement. J. Anat. 237 (5), 811–826.

Scott, S.H., Winter, D.A., 1990. Internal forces of chronic running injury sites. Med. Sci. Sports Exerc. 22 (3), 357–369.

Scott, S.H., Winter, D.A., 1991. Talocrural and talocalcaneal joint kinematics and kinetics during the stance phase of walking. J. Biomech. 24 (8), 743–752.

Scott, S.H., Loeb, G.E., 1995. Mechanical properties of aponeurosis and tendon of the cat soleus muscle during whole-muscle isometric contractions. J. Morphol. 224 (1), 73–86.

Seay, J.F., Van Emmerik, R.E.A., Hamill, J., 2011. Low back pain status affects pelvis-trunk coordination and variability during walking and running. Clin. Biomech. 26 (6), 572–578.

Segers, V., Lenoir, M., Aerts, P., De Clercq, D., 2007. Influence of M. tibialis anterior fatigue on the walk-to-run and run-to-walk transition in non-steady state locomotion. Gait Posture 25 (4), 639–647.

Semciw, A.I., Pizzari, T., Murley, G.S., Green, R.A., 2013. Gluteus medius: an intramuscular EMG investigation of anterior, middle and posterior segments during gait. J. Electromyogr. Kinesiol. 23 (4), 858–864.

Semciw, A.I., Green, R.A., Murley, G.S., Pizzari, T., 2014. Gluteus minimus: an intramuscular EMG investigation of anterior and posterior segments during gait. Gait Posture 39 (2), 822–826.

Semciw, A., Neate, R., Pizzari, T., 2016. Running related gluteus medius function in health and injury: a systematic review with meta-analysis. J. Electromyogr. Kinesiol. 30, 98–110.

Semple, R., Murley, G.S., Woodburn, J., Turner, D.E., 2009. Tibialis posterior in health and disease: a review of structure and function with specific reference to electromyographic studies. J. Foot Ankle Res. 2, 24. https://doi.org/10.1186/1757-1146-2-24.

Senavongse, W., Amis, A.A., 2005. The effects of articular, retinacular, or muscular deficiencies on patellofemoral joint stability: a biomechanical study in vitro. Journal of Bone and Joint. Surgery 87-B (4), 577–582.

Shadwick, R.E., 2008. Foundations of animal hydraulics: geodesic fibres control the shape of soft bodied animals. J. Exp. Biol. 211 (3), 289–291.

Sharbafi, M.A., Mohammadi Nejad Rashty, A., Rode, C., Seyfarth, A., 2017. Reconstruction of human swing leg motion with passive biarticular muscle models. Hum. Mov. Sci. 52, 96–107.

Shelburne, K.B., Torry, M.R., Pandy, M.G., 2006. Contributions of muscles, ligaments, and the ground-reaction force to tibiofemoral joint loading during normal gait. J. Orthop. Res. 24 (10), 1983–1990.

Shu, B., Safran, M.R., 2011. Hip instability: anatomic and clinical considerations of traumatic and atraumatic instability. Clin. Sports Med. 30 (2), 349–367.

Silder, A., Reeder, S.B., Thelen, D.G., 2010. The influence of prior hamstring injury on lengthening muscle tissue mechanics. J. Biomech. 43 (12), 2254–2260.

Sions, J.M., Velasco, T.O., Teyhen, D.S., Hicks, G.E., 2014. Ultrasound imaging: intraexaminer and interexaminer reliability for multifidus muscle thickness assessment in adults aged 60 to 85 years versus younger adults. J. Orthop. Sports Phys. Ther. 44 (6), 425–434.

Slane, L.C., Thelen, D.G., 2014. Non-uniform displacements within the Achilles tendon observed during passive and eccentric loading. J. Biomech. 47 (12), 2831–2835.

Smit, H.J., Strong, P., 2020. Structural elements of the biomechanical system of soft tissue. Cureus 12 (4), e7895. https://doi.org/10.7759/cureus.7895.

Snow, S.W., Bohne, W.H.O., DiCarlo, E., Chang, V.K., 1995. Anatomy of the Achilles tendon and plantar fascia in relation to the calcaneus in various age groups. Foot Ankle Int. 16 (7), 418–421.

Solomonow, M., Krogsgaard, M., 2001. Sensorimotor control of knee stability. A review. Scand. J. Med. Sci. Sports 11 (2), 64–80.

Stainsby, G.D., 1997. Pathological anatomy and dynamic effect of the displaced plantar plate and the importance of the integrity of the plantar plate-deep transverse metatarsal ligament tie-bar. Ann. R. Coll. Surg. Engl. 79 (1), 58–68.

Stansfield, B.W., Nicol, A.C., 2002. Hip joint contact forces in normal subjects and subjects with total hip protheses: walking and stair and ramp negotiation. Clin. Biomech. 17 (2), 130–139.

Stecco, C., Pavan, P., Pachera, P., De Caro, R., Natali, A., 2014. Investigation of the mechanical properties of the human crural fascia and their possible clinical implications. Surg. Radiol. Anat. 36 (1), 25–32.

Steinke, H., Hammer, N., Slowik, V., Stadler, J., Josten, C., Böhme, J., et al., 2010. Novel insights into the sacroiliac joint ligaments. Spine 35 (3), 257–263.

Stewart, J.D., 2008. Foot drop: where, why and what to do? Pract. Neurol. 8 (3), 158–169.

Stiehl, J.B., Skrade, D.A., Needleman, R.L., Schiedt, K.B., 1993. Effect of axial load and ankle position on ankle stability. J. Orthop. Trauma 7 (1), 72–77.

Stolwijk, N.M., Koenraadt, K.L.M., Louwerens, J.W.K., Grim, D., Duysens, J., Keijsers, N.L.W., 2014. Foot lengthening and shortening during gait: a parameter to investigate foot function? Gait Posture 39 (2), 773–777.

Sutherland, D.H., 2001. The evolution of clinical gait analysis. Part I. Kinesiological EMG. Gait Posture 14 (1), 61–70.

Szaro, P., Witkowski, G., Smigielski, R., Krajewski, P., Ciszek, B., 2009. Fascicles of the adult human Achilles tendon—an anatomical study. Ann. Anat. 191 (6), 586–593.

Szaro, P., Polaczek, M., Ciszek, B., 2021. The Kager's fat pad radiological anatomy revised. Surg. Radiol. Anat. 43 (1), 79–86.

Takabayashi, T., Edama, M., Inai, T., Nakamura, E., Kubo, M., 2020. Effect of gender and load conditions on foot arch height index and flexibility in Japanese youths. J. Foot Ankle Surg. 59 (6), 1144–1147.

Takahashi, K.Z., Gross, M.T., van Werkhoven, H., Piazza, S.J., Sawicki, G.S., 2016. Adding stiffness to the foot modulates soleus force-velocity behaviour during human walking. Sci. Rep. 6, 29870. https://doi.org/10.1038/srep29870.

Tardieu, C., Bonneau, N., Hecquet, J., Boulay, C., Marty, C., Legaye, J., et al., 2013. How is sagittal balance acquired during bipedal gait acquisition? Comparison of neonatal and adult pelves in three dimensions. Evolutionary implications. J. Hum. Evol. 65 (2), 209–222.

Taş, S., Salkin, Y., 2019. An investigation of the sex-related differences in the stiffness of the Achilles tendon and gastrocnemius muscle: inter-observer reliability and inter-day repeatability and the effect of ankle joint motion. Foot 41, 44–50.

Telleria, J.J.M., Lindsey, D.P., Giori, N.J., Safran, M.R., 2011. An anatomic arthroscopic description of the hip capsular ligaments for the hip arthroscopist. Arthrosc.: J. Arthrosc. Rel. Surg. 27 (5), 628–636.

Telleria, J.J.M., Lindsey, D.P., Giori, N.J., Safran, M.R., 2014. A quantitative assessment of the insertional footprints of the hip joint capsular ligaments and their spanning fibers for reconstruction. Clin. Anat. 27 (3), 489–497.

Tezer, M., Cicekcibasi, A.E., 2012. A variation of the extensor hallucis longus muscle (accessory extensor digiti secundus muscle). Anat. Sci. Int. 87 (2), 111–114.

Theis, N., Mohagheghi, A.A., Korff, T., 2012. Method and strain rate dependence of Achilles tendon stiffness. J. Electromyogr. Kinesiol. 22 (6), 947–953.

Thompson, J.A., Chaudhari, A.M.W., Schmitt, L.C., Best, T.M., Siston, R.A., 2013. Gluteus maximus and soleus compensate for simulated quadriceps atrophy and activation failure during walking. J. Biomech. 46 (13), 2165–2172.

Thordarson, D.B., Schmotzer, H., Chon, J., Peters, J., 1995. Dynamic support of the human longitudinal arch. A biomechanical evaluation. Clin. Orthop. Relat. Res. 316 (July), 165–172.

Tibert, G., 2002. Deployable tensegrity structures for space applications. [Doctoral Thesis.] Technical Reports from Royal Institute of Technology, Department of Mechanics, SE-100 44, Stockholm, Sweden. http://www-civ.eng.cam.ac.uk/dsl/publications/TibertDocThesis.pdf.

Tochigi, Y., Rudert, M.J., Saltzman, C.L., Amendola, A., Brown, T.D., 2006. Contribution of articular surface geometry to ankle stabilization. J. Bone Joint Surg. 88-A (12), 2704–2713.

Torry, M.R., Schenker, M.L., Martin, H.D., Hogoboom, D., Philippon, M.J., 2006. Neuromuscular hip biomechanics and pathology in the athlete. Clin. Sports Med. 25 (2), 179–197.

Tourné, Y., Molinier, F., Andrieu, M., Porta, J., Barbier, G., 2019. Diagnosis and treatment of tibiofibular syndesmosis lesions. Orthop. Traumatol. Surg. Res. 105 (8 Supplement), S275–S286.

Tsao, H., Danneels, L.A., Hodges, P.W., 2011. ISSLS prize winner: smudging the motor brain in young adults with recurrent low back pain. Spine 36 (21), 1721–1727.

Tubbs, R.S., Caycedo, F.J., Oakes, W.J., Salter, E.G., 2006. Descriptive anatomy of the insertion of the biceps femoris muscle. Clin. Anat. 19 (6), 517–521.

Uchida, T.K., Hicks, J.L., Dembia, C.L., Delp, S.L., 2016. Stretching your energetic budget: how tendon compliance affects the metabolic cost of running. PLoS One 11 (3), e0150378. https://doi.org/10.1371/journal.pone.0150378.

Uemura, K., Atkins, P.R., Fiorentino, N.M., Anderson, A.E., 2018. Hip rotation during standing and dynamic activities and the compensatory effect of femoral anteversion: an in-vivo analysis of asymptomatic young adults using three-dimensional computed tomography models and dual fluoroscopy. Gait Posture 61, 276–281.

Ullrich, K., Krudwig, W.F., Witzel, U., 2002. Posterolateral aspect and stability of the knee joint. I. Anatomy and function of the popliteus muscle-tendon unit: an anatomical and biomechanical study. Knee Surg. Sports Traumatol. Arthrosc. 10 (2), 86–90.

Usherwood, J.R., Channon, A.J., Myatt, J.P., Rankin, J.W., Hubel, T.Y., 2012. The human foot and heel-sole-toe walking strategy: a mechanism enabling an inverted pendular gait with low isometric muscle force? J. R. Soc. Interface 9 (75), 2396–2402.

Vaarbakken, K., Steen, H., Samuelsen, G., Dahl, H.A., Leergaard, T.B., Stuge, B., 2015. Primary functions of the quadratus femoris and obturator externus muscles indicated from lengths and moment arms measured in mobilized cadavers. Clin. Biomech. 30 (3), 231–237.

Vaishya, R., Agarwal, A.K., Azizi, A.T., Vijay, V., 2016. Haglund's syndrome: a commonly seen mysterious condition. Cureus 8 (10), e820. https://doi.org/10.7759/cureus.820.

Valderrabano, V., Hintermann, B., Horisberger, M., Fung, T.S., 2006. Ligamentous posttraumatic ankle osteoarthritis. Am. J. Sports Med. 34 (4), 612–620.

Valderrabano, V., Horisberger, M., Russell, I., Dougall, H., Hintermann, B., 2009. Etiology of ankle osteoarthritis. Clin. Orthop. Relat. Res. 467 (July), 1800–1806.

Valle, X., Alentorn-Geli, E., Tol, J.L., Hamilton, B., Garrett Jr., W.E., Pruna, R., et al., 2017. Muscle injuries in sports: a new evidenced-informed and expert consensus-based classification with clinical application. Sports Med. 47 (7), 1241–1253.

van Arkel, R.J., Amis, A.A., Cobb, J.P., Jeffers, J.R.T., 2015. The capsular ligaments provide more hip rotational restraint than the acetabular labrum and the ligamentum teres: an experimental study. Bone Joint J. 97-B (4), 484–491.

van der Esch, M., Knoop, J., Hunter, D.J., Klein, J.-P., van der Leeden, M., Knol, D.L., et al., 2013. The association between reduced knee joint proprioception and medial meniscal abnormalities using MRI in knee osteoarthritis: results from the Amsterdam osteoarthritis cohort. Osteoarthr. Cartil. 21 (5), 676–681.

van der Hulst, M., Vollenbroek-Hutten, M.M., Rietman, J.S., Hermens, H.J., 2010. Lumbar and abdominal muscle activity during walking in subjects with chronic low back pain: support of the "guarding" hypothesis? J. Electromyogr. Kinesiol. 20 (1), 31–38.

Van Essen, D.C., 1997. A tension-based theory of morphogenesis and compact wiring in the central nervous system. Nature 385 (6614), 313–318.

Van Hooren, B., Bosch, F., 2017a. Is there really an eccentric action of the hamstrings during the swing phase of high-speed running? Part I. A critical review of the literature. J. Sports Sci. 35 (23), 2313–2321.

Van Hooren, B., Bosch, F., 2017b. Is there really an eccentric action of the hamstrings during the swing phase of high-speed running? Part II. Implications for exercise. J. Sports Sci. 35 (23), 2322–2333.

van Werkhoven, H., Piazza, S.J., 2017. Does foot anthropometry predict metabolic cost during running? J. Appl. Biomech. 33 (5), 317–322.

Vasara, A.I., Jurvelin, J.S., Peterson, L., Kiviranta, I., 2005. Arthroscopic cartilage indentation and cartilage lesions of anterior cruciate ligament-deficient knees. Am. J. Sports Med. 33 (3), 408–414.

Vedi, V., Williams, A., Tennant, S.J., Spouse, E., Hunt, D.M., Gedroyc, W.M.W., 1999. Meniscal movement: an in-vivo study using dynamic MRI. J. Bone Joint Surg. 81-B (1), 37–41.

Vega, J., Malagelada, F., Manzanares Céspedes, M.-C., Dalmau-Pastor, M., 2020. The lateral fibulotalocalcaneal ligament complex: an ankle stabilizing isometric structure. Knee Surg. Sports Traumatol. Arthrosc. 28 (1), 8–17.

Verbout, A.J., Wintzen, A.R., Linthorst, P., 1989. The distribution of slow and fast twitch fibres in the intrinsic lumbar back muscles (ILBM). Clin. Anat. 2 (2), 119–124 (In: Abstracts: Presented at the Annual General Meeting of the British Association of Clinical Anatomists, January 9, 1989, University College, London, UK).

Vieira, E.L.C., Vieira, E.A., da Silva, R.T., dos Santos Berlfein, P.A., Abdalla, R.J., Cohen, M., 2007. An anatomic study of the iliotibial tract. Arthrosc.: J. Arthrosc. Rel. Surg. 23 (3), 269–274.

Vleeming, A., Albert, H.B., Östgaard, H.C., Sturesson, B., Stuge, B., 2008. European guidelines for the diagnosis and treatment of pelvic girdle pain. Eur. Spine J. 17 (6), 794–819.

Vleeming, A., Schuenke, M.D., Danneels, L., Willard, F.H., 2014. The functional coupling of the deep abdominal and paraspinal muscles: the effects of simulated paraspinal muscle contraction on force transfer to the middle and posterior layer of the thoracolumbar fascia. J. Anat. 225 (4), 447–462.

Vogt, L., Pfeifer, K., Banzer, W., 2003. Neuromuscular control of walking with chronic low-back pain. Man. Ther. 8 (1), 21–28.

Voloshin, A.S., Wosk, J., 1983. Shock absorption of meniscectomized and painful knees: a comparative in vivo study. J. Biomed. Eng. 5 (2), 157–161.

Wade, F.E., Lewis, G.S., Piazza, S.J., 2019. Estimates of Achilles tendon moment arm differ when axis of ankle rotation is derived from ankle motion. J. Biomech. 90, 71–77.

Wager, J.C., Challis, J.H., 2016. Elastic energy within the human plantar aponeurosis contributes to arch shortening during the push-off phase of running. J. Biomech. 49 (5), 704–709.

Wagner, F.V., Negrão, J.R., Campos, J., Ward, S.R., Haghighi, P., Trudell, D.J., et al., 2012. Capsular ligaments of the hip: anatomic, histologic, and positional study in cadaveric specimens with MR arthrography. Radiology 263 (1), 189–198.

Wakeling, J.M., 2009. The recruitment of different compartments within a muscle depends on the mechanics of the movement. Biol. Lett. 5 (1), 30–34.

Walheim, G., Olerud, S., Ribbe, T., 1984. Mobility of the pubic symphysis. Measurements by an electromechanical method. Acta Orthop. Scand. 55 (2), 203–208.

Walters, B.L., Cooper, J.H., Rodriguez, J.A., 2014. New findings in hip capsular anatomy: dimensions of capsular thickness and pericapsular contributions. Arthrosc.: J. Arthrosc. Rel. Surg. 30 (10), 1235–1245.

Wang, R., Gutierrez-Farewik, E.M., 2011. The effect of subtalar inversion/eversion on the dynamic function of the tibialis anterior, soleus, and gastrocnemius during the stance phase of gait. Gait Posture 34 (1), 29–35.

Wang, R., Gutierrez-Farewik, E.M., 2013. Compensatory strategies for excessive muscle co-contraction at the ankle. Gait Posture 38 (Supplement 1), S15.

Wang, R., Gutierrez-Farewik, E.M., 2014. Compensatory strategies during walking in response to excessive muscle co-contraction at the ankle joint. Gait Posture 39 (3), 926–932.

Wang, X.-T., Rosenberg, Z.S., Mechlin, M.B., Schweitzer, M.E., 2005. Normal variants and diseases of the peroneal tendons and superior peroneal retinaculum: MR imaging features. Radiographics 25 (3), 587–602.

Wang, Y., Li, Z., Wong, D.W.-C., Zhang, M., 2015. Effects of ankle arthrodesis on biomechanical performance of the entire foot. PLoS One 10 (7), e0134340. https://doi.org/10.1371/journal.pone.0134340.

Watanabe, K., Kitaoka, H.B., Berglund, L.J., Zhao, K.D., Kaufman, K.R., An, K.-N., 2012. The role of ankle ligaments and articular geometry in stabilizing the ankle. Clin. Biomech. 27 (2), 189–195.

Waterman, B.R., Owens, B.D., Davey, S., Zacchilli, M.A., Belmont Jr., P.J., 2010. The epidemiology of ankle sprains in the United States. J. Bone Joint Surg. 92-A (13), 2279–2284.

Watson, R.R., Fu, Z., West, J.B., 2007. Morphometry of the extremely thin pulmonary blood-gas barrier in the chicken lung. Am. J. Physiol. Lung Cell. Mol. Physiol. 292 (3), L769–L777.

Weibel, E.R., 2008. Editorial: how to make an alveolus. Eur. Respir. J. 31 (3), 483–485.

Weiss, P.L., Kearney, R.E., Hunter, I.W., 1986a. Position dependence of ankle joint dynamics. I. Passive mechanics. J. Biomech. 19 (9), 727–735.

Weiss, P.L., Kearney, R.E., Hunter, I.W., 1986b. Position dependence of ankle joint dynamics. II. Active mechanics. J. Biomech. 19 (9), 737–751.

Whyte, E.F., Moran, K., Shortt, C.P., Marshall, B., 2010. The influence of reduced hamstring length on patellofemoral joint stress during squatting in healthy male adults. Gait Posture 31 (1), 47–51.

Wijdicks, C.A., Ewart, D.T., Nuckley, D.J., Johansen, S., Engebretsen, L., LaPrade, R.F., 2010a. Structural properties of the primary medial knee ligaments. Am. J. Sports Med. 38 (8), 1638–1646.

Wijdicks, C.A., Griffith, C.J., Johansen, S., Engebretsen, L., LaPrade, R.F., 2010b. Injuries to the medial collateral ligament and associated medial structures of the knee. J. Bone Joint Surg. 92-A (5), 1266–1280.

Willems, T., Witvrouw, E., Delbaere, K., De Cock, A., De Clercq, D., 2005. Relationship between gait biomechanics and inversion sprains: a prospective study of risk factors. Gait Posture 21 (4), 379–387.

Williams, R.M., Zipfel, W.R., Webb, W.W., 2005. Interpreting second-harmonic generation images of collagen I fibrils. Biophys. J. 88 (2), 1377–1386.

Winby, C.R., Lloyd, D.G., Besier, T.F., Kirk, T.B., 2009. Muscle and external load contribution to knee joint contact loads during normal gait. J. Biomech. 42 (14), 2294–2300.

Woo, S.L., Gomez, M.A., Woo, Y.K., Akeson, W.H., 1982. Mechanical properties of tendons and ligaments. II. The relationships of immobilization and exercise on tissue remodeling. Biorheology 19 (3), 397–408.

Woo, S.L.-Y., Wu, C., Dede, O., Vercillo, F., Noorani, S., 2006. Biomechanics and anterior cruciate ligament reconstruction. J. Orthop. Surg. Res. 1, 2. https://doi.org/10.1186/1749-799x-1-2.

Woods, C., Hawkins, R., Hulse, M., Hodson, A., 2002. The Football Association Medical Research Programme: an audit of injuries in professional football - analysis of preseason injuries. Br. J. Sports Med. 36 (6), 436–441.

Wren, T.A.L., Yerby, S.A., Beaupré, G.S., Carter, D.R., 2001a. Mechanical properties of the human Achilles tendon. Clin. Biomech. 16 (3), 245–251.

Wren, T.A.L., Yerby, S.A., Beaupré, G.S., Carter, D.R., 2001b. Influence of bone mineral density, age, and strain rate on the failure mode of human Achilles tendons. Clin. Biomech. 16 (6), 529–534.

Wright, D.G., Desai, S.M., Henderson, W.H., 1964. Action of the subtalar and ankle-joint complex during the stance phase of walking. J. Bone Joint Surg. 46-A (2), 361–382.

Wu, C.-C., 2015. Does pelvic width influence patellar tracking? A radiological comparison between sexes. Orthop. Traumatol. Surg. Res. 101 (2), 157–161.

Wu, W.-L., Su, F.-C., Cheng, Y.-M., Huang, P.-J., Chou, Y.-L., Chou, C.-K., 2000. Gait analysis after ankle arthrodesis. Gait Posture 11 (1), 54–61.

Wu, W.H., Lin, X.C., Meijer, O.G., Gao, J.T., Hu, H., Prins, M.R., et al., 2014. Effects of experimentally increased trunk stiffness on thorax and pelvis rotations during walking. Hum. Mov. Sci. 33, 194–202.

Wünschel, M., Leichtle, U., Obloh, C., Wülker, N., Müller, O., 2011. The effect of different quadriceps loading patterns on tibiofemoral joint kinematics and patellofemoral contact pressure during simulated partial weight-bearing knee flexion. Knee Surg. Sports Traumatol. Arthrosc. 19 (7), 1099–1106.

Yamamoto, I., Panjabi, M.M., Oxland, T.R., Crisco, J.J., 1990. The role of the iliolumbar ligament in the lumbosacral junction. Spine 15 (11), 1138–1141.

Yaman, A., Ozturk, C., Huijing, P.A., Yucesoy, C.A., 2013. Magnetic resonance imaging assessment of mechanical interactions between human lower leg muscles in vivo. J. Biomech. Eng. 135 (9), 91003. https://doi.org/10.1115/1.4024573.

Yamazaki, N., Ohta, K., Ohgi, Y., 2012. Mechanical energy transfer by internal force during the swing phase of running. Procedia Eng. 34, 772–777.

Yammine, K., Erić, M., 2017. The fibularis (peroneus) tertius muscle in humans: a meta-analysis of anatomical studies with clinical and evolutionary implications. Biomed. Res. Int. 2017, 6021707. https://doi.org/10.1155/2017/6021707.

Yoder, A.J., Petrella, A.J., Silverman, A.K., 2015. Trunk-pelvis motion, joint loads, and muscle forces during walking with a transtibial amputation. Gait Posture 41 (3), 757–762.

Yoon, K.H., Yoo, J.H., Kim, K.-I., 2011. Bone contusion and associated meniscal and medial collateral ligament injury in patients with anterior cruciate ligament rupture. J. Bone Joint Surg. 93-A (16), 1510–1518.

Yucesoy, C.A., 2010. Epimuscular myofascial force transmission implies novel principles for muscular mechanics. Exerc. Sport Sci. Rev. 38 (3), 128–134.

Yucesoy, C.A., Baan, G., Huijing, P.A., 2010. Epimuscular myofascial force transmission occurs in the rat between the deep flexor muscles and their antagonistic muscles. J. Electromyogr. Kinesiol. 20 (1), 118–126.

Yuen, C.P., Lui, T.H., 2017. Distal tibiofibular syndesmosis: anatomy, biomechanics, injury and management. Open Orthop. J. 11, 670–677.

Zajac, F.E., 1989. Muscle and tendon: properties, models, scaling, and application to biomechanics and motor control. Crit. Rev. Biomed. Eng. 17 (4), 359–411.

Zajac, F.E., Neptune, R.R., Kautz, S.A., 2003. Biomechanics and muscle coordination of human walking. Part II. Lessons from dynamical simulations and clinical implications. Gait Posture 17 (1), 1–17.

Zamboulis, D.E., Thorpe, C.T., Ashraf Kharaz, Y., Birch, H.L., Screen, H.R.C., Clegg, P.D., 2020. Postnatal mechanical loading drives adaptation of tissues primarily through modulation of the non-collagenous matrix. elife 9, e58075. https://doi.org/10.7554/eLife.58075.

Zeni, J.A., Higginson, J.S., 2011. Knee osteoarthritis affects the distribution of joint moments during gait. Knee 18 (3), 156–159.

Zhou, J.-P., Yu, J.-F., Feng, Y.-N., Liu, C.-L., Su, P., Shen, S.-H., et al., 2020. Modulation in the elastic properties of gastrocnemius muscle heads in individuals with plantar fasciitis and its relationship with pain. Sci. Rep. 10, 2770. https://doi.org/10.1038/s41598-020-59715-8.

Zifchock, R.A., Piazza, S.J., 2004. Investigation of the validity of modeling the Achilles tendon as having a single insertion site. Clin. Biomech. 19 (3), 303–307.

Chapter 3

The foot as a functional unit of gait

Chapter introduction

Whereas most of the lower limb can be divided by large joints lying between rigid long bones, the foot presents a particularly interesting biotensegrity structure. It is constructed from 28 bones, of which nineteen are long bones, seven are short bones, and two are the constant sesamoid bones situated under the 1st metatarsal head. However, not all sesamoid bones are constant, and it is common for there to be several more sesamoid variants and accessory ossicles within a foot (Nwawka et al., 2013). Furthermore, the intermediate and distal phalanges of the 5th toes are sometimes fused (Draves, 1986: p. 176). Between each foot bone lie joints, ligaments, fascial sheets, and muscles among which a large number of anatomical variations have been reported. The foot is therefore an area of intense and highly variable anatomy.

The mechanical properties of the foot are controlled by alterations in the relationships between the anatomical structures. The foot's function and dysfunction has been mired in debate, not least because of its numerous variations but also because of its distinct anatomical vault profiles. Further debate exists as to whether certain foot bones that are tightly bound together also usually move together, and thus represent areas of either limited flexibility or joints with true freedom of motion. In addition, some pedal anatomical structures can appear as being rather indistinct, meaning that some ligaments and muscles are argued as being one structure rather than several structures. Finally, in relation to foot shape, argument has raged as to whether one type of foot vault profile is functionally 'better' than another. This is the problem when approaching the foot without considering its overall mechanical functions that depend on stress (energy) loading, absorption, and transmission to assist in braking and acceleration during gait.

The foot is a particularly complex functional unit that shares a similar soft tissue and skeletal anatomy to the hand. Moreover, the foot and hand also share an evolutionary history of functional climbing as grasping, manipulating appendages. The human foot consists of many articulations, with variable degrees of freedom and nonlinear evolutionary histories. The hands and the feet are both in the 'primitive' five-digit (pentadactyl) format as is to be found in almost all primates. Compared to many other large terrestrial mammals such as those with hooves or cloven feet, human feet stand out.

If ever the phrase '*if I wanted to get there, I would not start from here*' was applicable to human anatomy, then the foot for terrestrial gait provides a fine example of this. Having evolved initially for the completely different task of arboreal living, there is much within the foot that is at odds with terrestrial locomotion. Yet despite this, the foot has taken advantage of its highly mobile evolutionary past to provide mechanisms of varying compliance and stiffness so as to achieve energy-dissipating, braking, and acceleration properties.

3.1 The foot's material properties

3.1.1 Introduction

The healthy foot and ankle working together in walking produce a near-zero net energetic effect of an energy neutral system, where the negative work put in is nearly equal to the positive work returned (Takahashi et al., 2017; Hedrick et al., 2019). When optimally functioning, the foot and ankle present properties that are analogous to a passive spring. This is a task that is demanding for its anatomy. The foot has extremely poor form closure at most of its joints due to its primate evolution generating an anatomy geared to functional expressions of mobility and as a result, it is now largely dependent on force closure to achieve rigidity and stability. As a biotensegrity structure, it uses extensive passive and active soft tissue tensional elements to maintain stability. The foot as a spring plays an essential role in energy dissipation, energy storage, energy transfer, and energy release, providing important kinetic energy during gait (Ker et al., 1987; Erdemir et al., 2004; Kelly et al., 2015, 2018; McDonald et al., 2016). It achieves multiple changes in its functional role during weightbearing activities through the relaxing and tensing of its biotensegrity soft tissue structures, permitting either greater or lesser motion across its numerous joints. Thus, the foot can be more compliant and thus more easily 'deformed' or it can become stiffer and elastic, resisting shape changes through extrinsic forces as locomotive situations demand.

Changing compliance and stiffness can be caused by muscle activity that alters the foot's shape via internal forces or through external loading forces that both lengthen and widen the foot. Alterations in compliance are achieved through four primary mechanisms. The first of these mechanisms is through muscular activity-induced tensional increases that create compressing forces across multiple joints. As a consequence of the foot's inherent mobility, this changing muscle activity alters its compliance and stiffness properties while the foot remains structurally stable throughout different terrain situations. This variability in muscular activity also occurs in response to different body and lower limb postures. Two other flexibility influencing mechanisms involve connective tissue passive viscoelastic properties. As connective tissue demonstrates stress-stiffening properties, it provides most compliance on initial loading (from the crimp effect) but behaves exponentially stiffer at higher strains. Connective tissues' time-dependent properties are also very important, for the faster they are loaded with stress, the stiffer they behave. Thus, connective tissue is able to generate variable tensions within itself under differing stress loading rates so that compliance or stiffness can be adjusted both passively and through the amount and rate of muscular activity. The final mechanism utilises muscular and connective tissue properties together and involves the foot's unique midfoot vault profile and the consequences of changing that profile. By increasing curvature, foot stiffness can be increased. Muscle activity is able to change the contouring of the vault to create stiffer and higher foot vault profiles. However, the inherent flexibility of the connective tissues that bridge the vault can also influence the degree of midfoot contouring possible under load. This is achieved with motion, altering the foot's shape (e.g. lengthening and thus eventually stiffening the vault) and changing moment arms across varying joint orientations throughout the foot's many articulations.

These mechanisms can work together, enabling muscle activity to raise the vault and also increase the tensioning within connective tissues such that each mechanism can stiffen the foot simultaneously or separately. Thus, these mechanisms can work in opposition so that on vault loading, the curvature can decrease, increasing compliance that results in lengthening and widening of the foot which in turn leads to tensioning and stiffening of the connective tissues. Therefore, increasing compliance in one mechanism can eventually lead to stiffening in another.

The foot also plays pivotal roles in the sensorimotor system, helping to guide posture and muscle activity (Fiolkowski et al., 2005). In addition, it assists the cardiovascular system in that loading forces deform the foot so that it acts like a hydraulic pump to return blood from the limbs during walking or more intense exercise (Horwood, 2019). A complex role requires a complex structure, and in the foot, we definitely have one.

When approaching function, it is worth considering the foot as a number of functional subunits making up a whole. This can be approached in one of two ways. The first way is to divide the foot into a rearfoot, a midfoot, and a forefoot (Carson et al., 2001; Leardini et al., 2007). This is logical when you consider that the CoGRF has to sequentially pass through each of these segments during walking gait. However, these segmental divisions have often been assumed to be rigid units (in modelling) which they most certainly are not, as many local motions can occur within them.

The second approach is by dividing the foot into medial and lateral columns (Fig. 3.1.1). The medial column consists of the talus, the navicular, the three cuneiforms, the three medial metatarsals, and for completeness, should perhaps also include their digital bones. The lateral column consists of the calcaneus, the cuboid, the lateral two metatarsals, and the digital bones. From a functional and evolutionary point of view, this makes sense as these osseous structures within each column have intimate coupling relationships with some distinct soft tissue lines of separation between each column (Bojsen-Møller, 1999). They also seem to have different evolutionary timelines with the lateral column seemingly having evolved to its modern form earlier than the medial column (Kidd and Oxnard, 2005; Zipfel et al., 2009; McNutt et al., 2018). The medial column demonstrates greater mobility and thus may require a greater ability and capacity to be able to modify its compliance and stiffness through muscular, passive soft tissue, and curvature changes than the inherently stiffer lateral column. This might explain the need for a higher medial side to the foot vault.

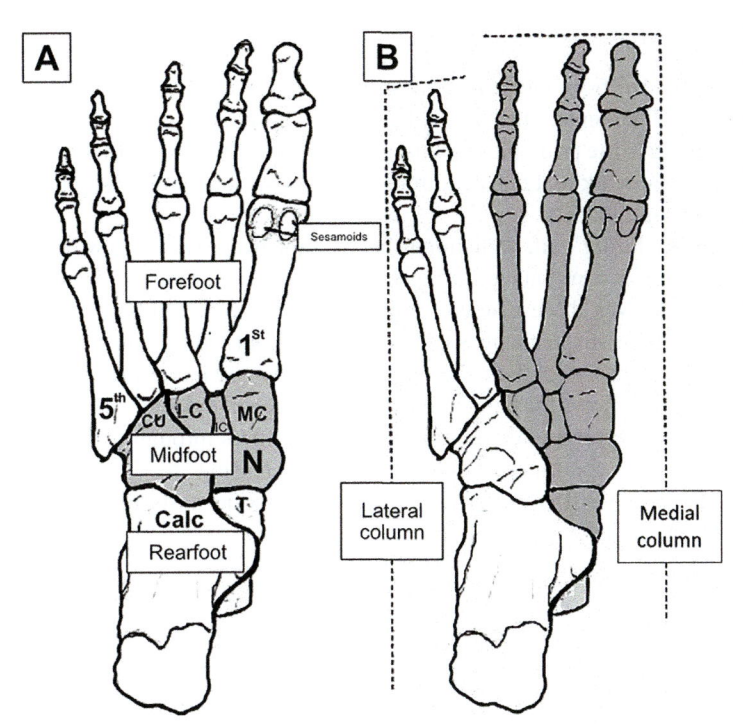

FIG. 3.1.1 The 28 bones of the foot viewed from the plantar surface so that the sesamoids are not missed. The bones can be separated into (A) the rearfoot (calcaneus—Calc and talus—T), the midfoot (grey area: navicular—N, medial cuneiform—MC, intermediate cuneiform—IC, lateral cuneiform—LC, cuboid—Cu) and the forefoot. The forefoot consists of each of the five long metatarsal bones and arguably the phalanges of each toe and the medial and lateral sesamoids under the 1st metatarsal head. In (B), the foot is divided into the medial column (grey area: talus, navicular, the three cuneiforms, the three medial metatarsals, their sesamoids, and for completeness, their digital phalanges) and the lateral column (calcaneus, cuboid, lateral two metatarsals, and their digital phalanges). The foot can be logically argued to be divided into either set of functional units. Such artificial divisions are based on little more than the human desire to separate anatomy for easier comprehension. *(Permission www.healthystep.co.uk.)*

Notwithstanding these segmental divisions, the foot is a complete functional unit. Indeed, it cannot truly be understood until it is approached through its holistic mechanical role during locomotion. The foot has an active osseous suspension system based on biotensegrity principles that protects itself and the rest of the body from two distinct and significant collision events. The first of these pedal collision events stems from the repetitive acts of regenerative braking which can express a variable ability to store and release elastic energy that can be then used to aid acceleration. The acceleration creates a second collision event against the ground through the forefoot. When healthy, the foot can modulate its stiffness and compliance to dampen energy spikes resulting from these collisions while aiding stability and providing the information necessary for sensorimotor system modulation. This creates a structure that can change its mechanical response on demand to match the need of the moment or at least, most of the time it can.

The bones of the foot under normal circumstances should never be directly loaded via the skin alone above the support surface and indeed, normally they are not. The bones 'float' within the soft tissues under tension and resultant compression. The foot is able to provide a semi-rigid propelling lever arm to facilitate the transfer of ankle power to the ground through the cutaneous soft tissues, despite being relatively mobile when at rest and nonweightbearing (Yawar et al., 2017). Stability, compliance, and stiffness must be provided through the foot for locomotion to be efficient and adaptable. The changes in tensional forces that 'float' the bones are responsible for this adaptability.

Despite a long history of scientific exploration into the foot's function, there still remain gaps in our knowledge about how the foot achieves its optimal mechanical function. This is particularly true in understanding the role of the so-called 'arches', or more appropriately, vaults of the feet. The ideal function of the foot involves a capacity for variation and subtle adjustment as and when needed, as well as an ability to compensate for more dramatic perturbations that can disturb mechanical efficiency. This allows the foot to achieve safe terrain manoeuvres and avoid injury. It is through the foot's ability to control its compliance and stiffness, guided by the sensorimotor system's muscular activity commands and intrinsic passive material properties, that allows it to achieve so much.

The foot's resultant material properties are not static quantities, for muscles can modulate stiffness (Holowka et al., 2017; Farris et al., 2019, 2020), and passive connective tissues such as ligaments and the plantar aponeurosis can tighten and stiffen or relatively slacken under length and load adjustments during gait (Hicks, 1954; Ker et al., 1987; Griffin et al., 2015; Venkadesan et al., 2017a,b; Yawar et al., 2017). Vault profiles change due to muscle activity and connective tissue loading and add to curvature mechanics, influencing the compliance and stiffness mix. The foot has to function appropriately to the environmentally induced mechanical variables it encounters. By having a number of variability mechanisms within the extent and timings of the compliance and stiffness levels provided, the human foot represents the ideal responsive organ as long as it remains fit and healthy.

3.1.2 The foot as a biotensegrity structure

The skeletal frame of the foot lies in the deep layers, suspended from the ground by structures such as the plantar aponeurosis, the interconnected muscular fascial septa, and many ligaments, muscles, and tendons. These tissues apply tension-induced compression across the bones and joints, conforming to the principles of tensegrity. This firmly bound structure sits on pads of fatty tissue emmeshed in a connective tissue framework that can act through viscoelastic and poroelastic properties as a variably compressive resistant structure under high loads.

The foot needs to change shape and maintain stability during gait, something that its biotensegrity structure freely allows. It needs to have a variable stiffness that can increase compliance during initial contact and loading response, but under increasing loads, achieve greater stiffness. It achieves these changes both by its passive elements tightening as the foot widens and lengthens and via muscle activity. These shape changes can occur as the foot reduces its curvature on loading under reducing muscle activity with the connective tissues at their shortened length, thereby increasing its compliance. The compliant freedom of the reduced vault curvature allows the connective tissues to stretch under tension so that they can achieve stress-stiffening through their viscoelastic properties. As the foot increases stiffness through connective tissue stretching, particularly across the forefoot, it can combine muscle contraction tension to increase its tension–compression structural stability. By initiating passive stiffening first, muscular metabolic costs are reduced (Fig. 3.1.2).

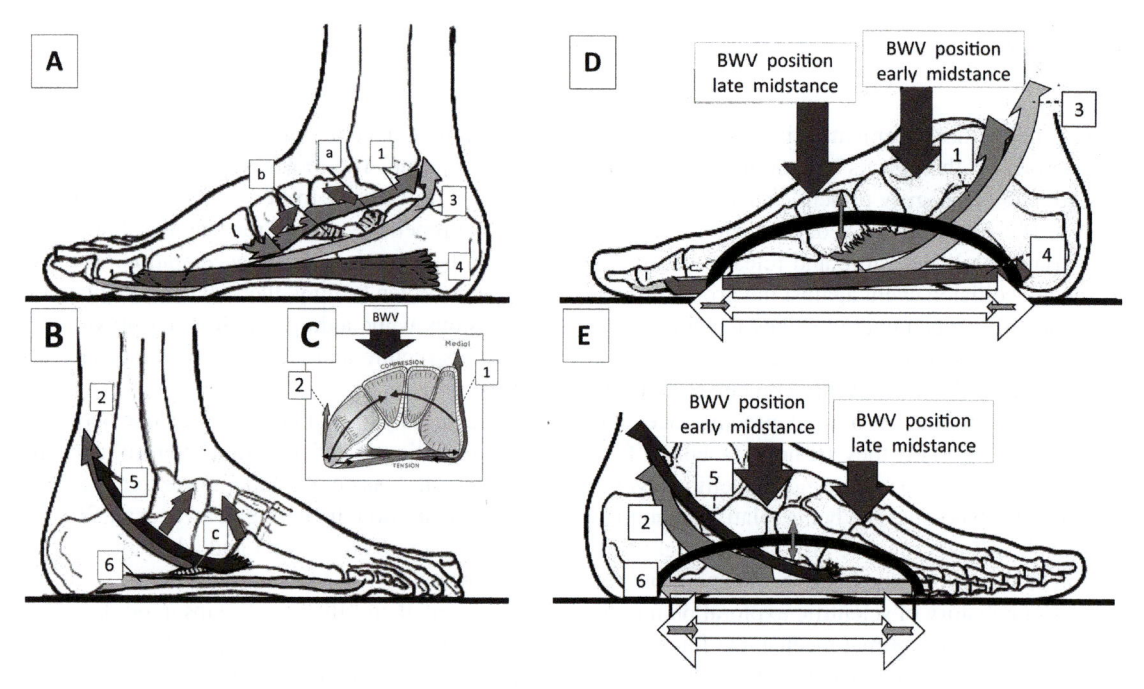

FIG. 3.1.2 The foot is a complex biotensegrity structure, for it has multiple articulations loaded at different times throughout locomotion. This fact makes different loading strategies such as heel or forefoot contact important during gait analysis. Primary anatomy supporting the biotensegrity structure is viewed medially (A) and laterally (B). They include the tibialis posterior tendon (1), peroneus longus tendon (2), flexor hallucis longus tendon (3), abductor hallucis muscle (4), peroneus brevis tendon (5), and the abductor digiti minimi muscle. This is not to underplay the importance of other intrinsic muscles and extrinsic muscle tendons within the foot. Ligaments are also important, represented here by those of the subtalar joint (a), the spring ligament (b), and long plantar ligament (c). In the transverse plane (C—insert), tibialis posterior and peroneus longus tendons 'stirrup-strap' and tension the medial and lateral foot such that under their muscle activation, the lesser tarsus is bound tightly, compressing the midfoot articulations together. This acts against sagging deflections that attempt to widen and flatten the transverse curvature of the foot, helping maintain foot stiffness. The deep transverse metatarsal ligament fulfils a similar binding tension role across the metatarsal heads. During loading in early midstance, the foot's longitudinal plane medially (D) and laterally (E) experiences body weight vectors (BWV) that initially align vertically over more proximal foot structures. These are resisted by ligaments such as the short (not shown) and long plantar and spring ligaments. As midstance progresses, the BWV loads move distally so that more proximal structures such as the short plantar, spring, and subtalar ligaments start to unload. However, the tarsometatarsal and intermetatarsal region's ligaments now become increasingly stretched. Such anatomy is protected and the biotensegrity structure maintained by the application of greater tensile forces from increasing muscle activity during late midstance, stiffening the foot as vault height lowers. The medial column should lower and lengthen more than the stiffer and shorter lateral column which requires much greater force to depress it. *(Permission www.healthystep.co.uk.)*

The foot demonstrates a number of structures that, through biotensegrity principles, utilise soft tissue solid and fluid phases. Thus, the viscoelastic properties of the foot's fat pads work by deforming and elastically recoiling as they load and offload, with the fat inside their chambers acting as an incompressible fluid. Tensegrity structures also make ideal hydraulic pumps as a consequence of their ability to increase compression through tensional–compressive forces. Moreover, the foot appears to have a significant part to play as both an active and passive pump, returning blood through venous return during gait via a bioelastic-hydraulic pumping action (Horwood, 2019). The foot as a biotensegrity structure is therefore able to fulfil locomotive and cardiovascular physiological roles through its biotensegral structural properties.

3.1.3 The mechanical role of the foot

The foot holds a most important site in bipedal locomotion, acting as it does between the support surface and the rest of the lower limb. It provides shock absorption, energy storage, and energy return during gait. Overall, the sum of the foot's energy management alone is net energy dissipation (braking dominates acceleration creating net negative work), fulfilling an increasing initial stance phase energy dissipation demand which increases with body mass (Papachatzis et al., 2020). As a consequence, the lower limb is required to increase positive work to compensate during late midstance-to-terminal stance for the lost energy of impact (Papachatzis et al., 2020). Through its passive and active structures, the foot cyclically dissipates impact energy, supports body weight during CoM transport anteriorly, and converts ankle plantarflexor power into a forefoot stabilising GRF. In fulfilling these central roles, the nature of the foot must change from a compliant structure that can deform and absorb energy to a stiffened elastic structure that returns energy by creating a collision with the support surface. Through greater stiffness, the foot translates more calf muscle plantarflexion power into acceleration which would be lost if the foot deformed significantly due higher compliance throughout the latter stages of stance. During swing, the foot is a passenger, contributing nothing but mass to swing phase (Fig. 3.1.3).

Initial (heel) contact	Loading response	Early midstance	AMS	Late midstance	Terminal stance	Preswing
Forefoot stiffened prior to its impact	Forefoot fat pad and vault energy dissipation	Foot impact energy dissipation complete, foot straining		Energy dissipation and storage in connective tissues	Energy dissipation of ankle dorsiflexion moment, storage of plantarflexion energy	Plantarflexion power application via vault to forefoot, its fat pad and MTP joint energy dissipation	Plantarflexion energy used up
Heel fat pad compliance but foot stiffens	Vault compliant, foot deforms rapidly	Vault deformation slowing as foot starts stiffening		Vault stiffness increasing rapidly, vault profile is lowest	Vault stiffened and its height increasing	Vault high, muscle activity low, compliant foot

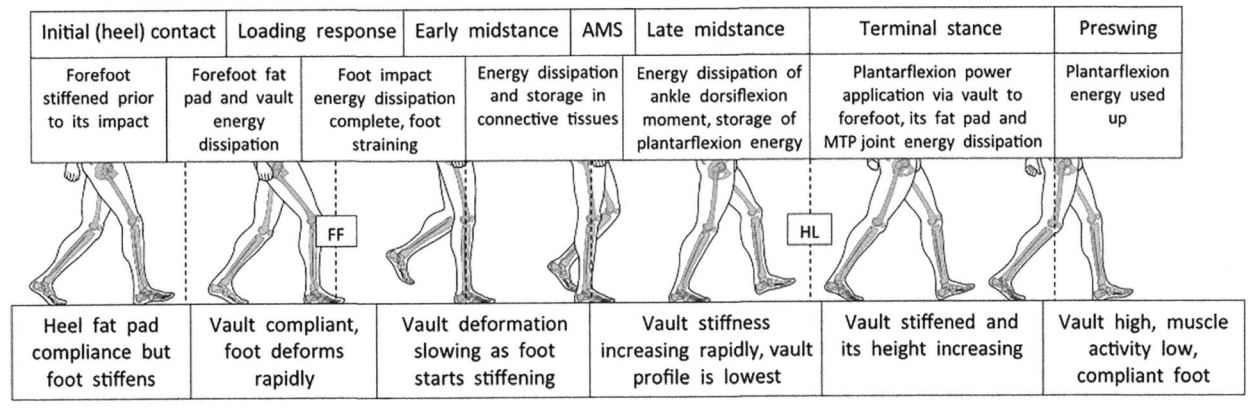

FIG. 3.1.3 The foot and its anatomy have different roles at different times during gait. During initial heel contact walking, the heel fat pad and ankle–rearfoot complex start the energy dissipation process with plantarflexion braking via tibialis anterior, while the vault is stiffened through muscular vault raising and tensioning via the windlass mechanism. At forefoot loading, the windlass mechanism and vault-stiffening muscles relax allowing the vault to reduce its tensions, increasing compliance to allow the forefoot fat pad, foot vault, and toe flexion out of dorsiflexion to assist in dissipation of impact energy. In early midstance, the foot is still comparatively compliant and deforms relatively easily under loading, but a cycle of gradual stiffening is starting by approximately absolute midstance (AMS). During late midstance, continued but slowing, vault lowering maintains energy dissipation via plantar soft tissue stretching (under ankle dorsiflexion moments). The rate of stiffening dramatically increases aided by extensively increasing muscle activation. By heel lift, the foot is at its flattest and is in a semi-rigid state. Ankle plantarflexion power built up during late midstance within the Achilles can now be applied to the forefoot via the stiffened vault, to create a large forefoot GRF. This important GRF peak provides stability and aids momentum into the next footstep. This power is moderated by digital extension and the forefoot fat pad energy dissipation as the ankle and midfoot plantarflex, raising the vault. By preswing, the plantarflexion power from the muscles is used up, tension in the foot decreases, and the foot increases compliance for swing phase with the vault raised. *(Permission www.healthystep.co.uk.)*

Therefore, the foot represents a spring-like elastic interface with the ground, providing a combination of energy dissipation and mechanical storage as an elastic structure. Muscles that cross the foot and ankle have the capacity to act as dampers and motors that dissipate and generate energy, respectively. Despite the profound anatomical complexity involved in achieving this, the foot can be considered as a single elastic body that transmits and receives loads from the support surface and transforms these applied loads into displacements under mechanical laws. The process of creating foot stiffness involves muscle–tendon activity, ligament/fascia viscoelastic properties, vault contour changes, and neurological input and controls, all acting together to provide joint stability.

By appreciating the foot's mechanical constraints and its evolutionary history, an understanding of how and why the foot works in the way it does becomes apparent. These simple yet well-timed cycles of mechanical properties allow the clinician to grasp the individual consequences of anatomical variation and pathology within the foot. The foot is a multi-jointed biotensegrity structure that forms a central inconstant vault profile, restrained by variable elastic soft tissue tensions, and is padded on its weightbearing surfaces by viscoelastic cutaneous tissues and connected to the deep fascia beneath which is reinforced as the plantar aponeurosis. The behaviour of this mechanical structure is controlled through predictive internal force generation which can be modulated in response to external force application through a sensory feedback mechanism. Failure of any part risks dysfunction and injury.

3.1.4 The role of the foot vault

One of the most interesting biotensegrity structures in the human body is the stable yet changeable human foot vault profile. The foot vault's overall stiffness or compliance during gait affects the GRFs/impulses generated by the foot (Cen et al., 2020). It demonstrates stress-stiffening mechanisms of increased stiffness with increasing load (Stolwijk et al., 2014; Bjelopetrovich and Barrios, 2016; Takabayashi et al., 2020) and time-dependent properties such that the faster humans walk, the stiffer their feet behave (Stolwijk et al., 2014). The foot vault is both at its flattest and at its longest at heel lift (Hunt et al., 2001; Stolwijk et al., 2014). Foot lengthening couples with stiffening prior to heel lift. This stiffening that is stabilising the lever arm for the ankle joint plantarflexors allows the feet to apply power to the forefoot, which has beneficial energetic effects on locomotion (Holowka et al., 2017; Honert et al., 2020).

There have been many models proposed to explain mechanical foot function which to varying degrees have been based on either explanation around arch profiles, the motions of specific joints and joint axes (Lee, 2001; Petcu and Colda, 2012), or muscle activity dependent on 'core' stability' principles (McKeon et al., 2015). The twisted *osteoligamentous plate* concept proposes that the metatarsal heads correspond to the anterior part of a plate horizontally orientated, which gradually twists throughout the length of the midfoot to angle the plate vertically at the calcaneus (Sarrafian, 1987; Eslami et al., 2007; Leardini et al., 2014; Araújo et al., 2019). The amount of twist is suggested to determine the profile of the longitudinal and transverse contour of the vault. The twist is maintained by the soft tissues, i.e. the fascia, ligaments, and muscles. Increased twisting of the plate is associated with supination and stiffening of the foot whereas untwisting the osteoligamentous plate is associated with pronation (or foot flattening) and increasing foot compliance (Araújo et al., 2019). This foot model and the very similar *calcaneopedal unit* concept (Ghanem et al., 2019) have attempted to search for an explanation for the source of foot pathology and explain the ability of the foot to change its compliance–stiffness properties and shape, with some degree of association existing between shape change and compliance–stiffness (Fig. 3.1.4a). The concept of the foot stiffening

FIG. 3.1.4a The twisted osteoligamentous concept with the anterior part described as representing the metatarsal heads and the posterior part the calcaneus, although a better description would be the forefoot and rearfoot, respectively. Diagrams (A) and (B) represent the 'relaxed' so-called 'neutral' position, with reduced curvature associated with foot pronation demonstrated in (C) and (D). *(From Araújo, V.L., Souza, T.R., Magalhães, F.A., Santos, T.R.T., Holt, K.G., Fonseca, S.T., 2019. Effects of a foot orthosis inspired by the concept of a twisted osteoligamentous plate on the kinematics of foot-ankle complex during walking: A proof of concept. J. Biomech. 93, 118–125.)*

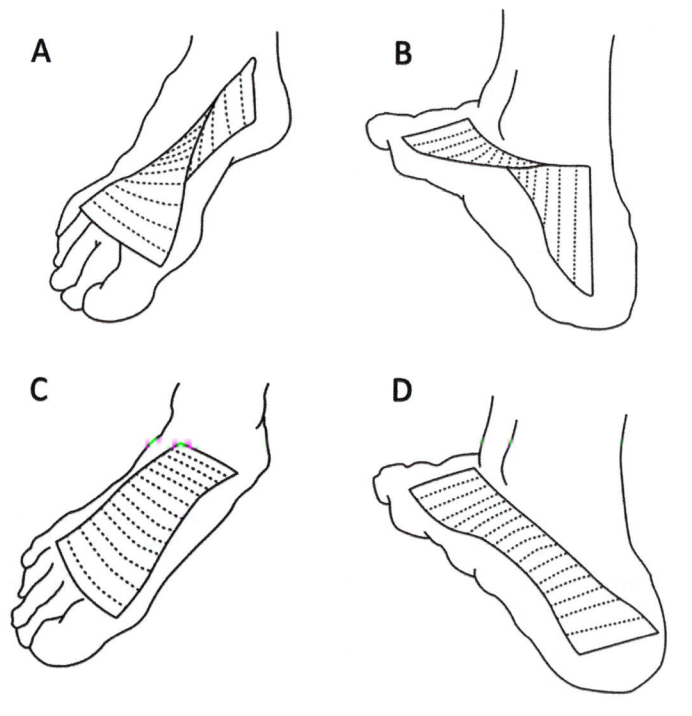

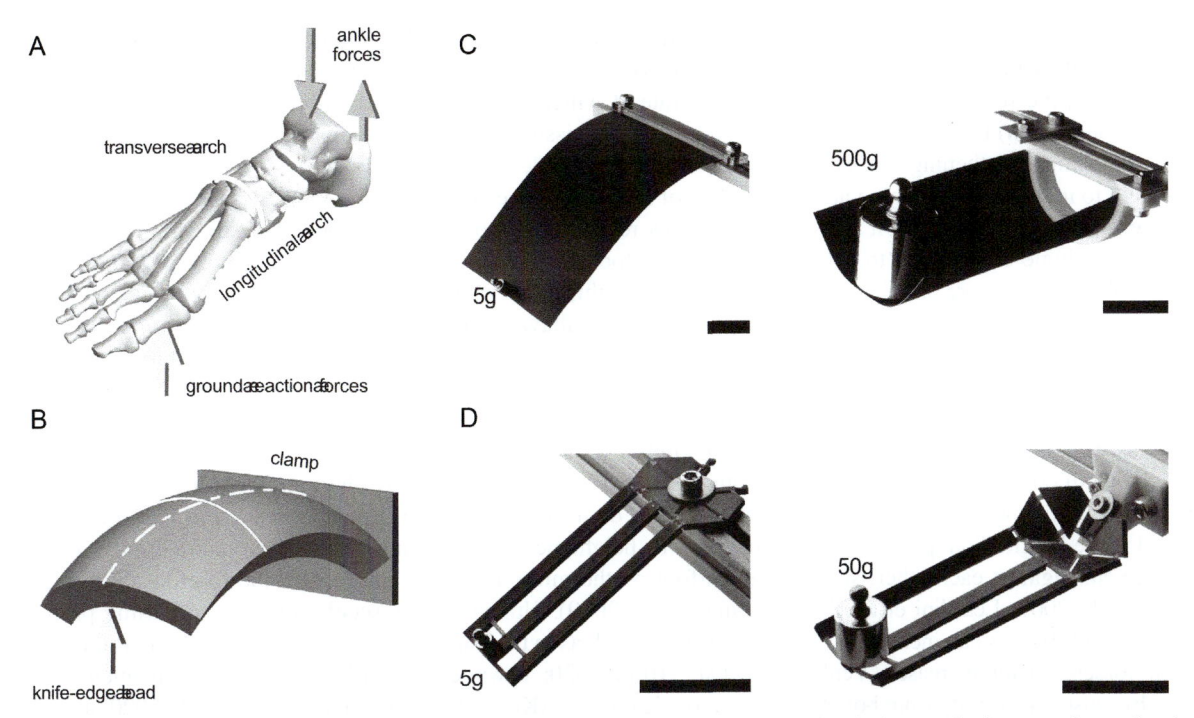

FIG. 3.1.4b Foot stiffness can be altered through its transverse plane curvature. In image (A), the foot is shown as having two primary directions of curvature, commonly referred to as the longitudinal and the transverse arches. If the contoured (vaulted) profile of the foot is mathematically represented as an elastic shell clamped at its proximal end (representing the ankle) with a free knife-edge under load (metatarsal heads), then the bending moments it is subjected to during terminal stance can be calculated under different curvature effects (B). In images (C) and (D), the effects of increasing longitudinal curvature and transverse curvature can be compared to illustrate that stiffness is far more dramatically influenced through transverse curvature. The reason is simple. The foot's span distance is far shorter in this direction so stiffness can be more easily increased. The scale bars in the image represent 5 cm. *(From Venkadesan, M., Dias, M.A., Singh, D.K., Bandi, M.M., Mandre, S., 2017a. Stiffness of the human foot and evolution of the transverse arch. https://arxiv.org/abs/1705.10371.)*

through alterations in curvature has also been proposed by Venkadesan et al. (2017a, 2020), but through changes led within the transverse plane of the foot (Fig. 3.1.4b).

The key to understanding biological viscoelastic materials is through their non-Newtonian properties of stress-stiffening and their loading rate-dependent properties (Naemi et al., 2016). These attributes provide an explanation as to why foot shape alterations result in material property changes of stiffness. Biological structures use curvature of the whole macrostructure and the twisting of collagen-based connective tissues at the microstructural level to increase stiffness (Nguyen, 2010; Nguyen et al., 2017, Yawar et al., 2017). Being able to increase or decrease curvature through twist changes affords biological structures the ability to adjust their stiffness properties through both changing shape and twisting connective tissue at the same time. Yet, stretching connective tissue also causes tissue to express its stress-stiffening behaviour where it stiffens at an exponential rate in response to further lengthening after passing through its linear elastic zone.

Armed with this core knowledge, understanding foot function becomes a process of appreciating the numerous effects of curvature, twist, lengthening (and widening), loading rate, stress-stiffening properties, and fatigue, in addition to other viscoelastic responses to loading. This moves models of foot function away from the slavish adherence to foot 'pronation' equals compliance and foot 'supination' equals stiffness. Instead, consider that the foot vault profile is at its lowest and therefore at its most 'pronated' just prior to heel lift (Hunt et al., 2001). Static tests have demonstrated greater foot stiffness when the foot is more stress-loaded and consequently flattened/lengthened through its stress-stiffening properties of lengthening (Stolwijk et al., 2014; Bjelopetrovich and Barrios, 2016; Takabayashi et al., 2020). Thus, arch stiffness is increasing with greater degrees of 'flatness' (pronation), peaking for a brief moment before heel lift. The foot then attempts to increase its stiffness further at the heel lift boundary and throughout terminal stance by increasing curvature within the vault (Venkadesan et al., 2017b, 2020; Yawar et al., 2017). Should the foot only be able to stiffen through becoming more supinated, then the vault would need to rise and the ankle–rearfoot invert prior to heel lift; something that is not reported as a feature of normal gait (Hunt et al., 2001).

Increasing foot pronation upon loading initially increases compliance through loss of the vault's curvature, enabling the foot to lengthen more easily (Stolwijk et al., 2014). Yet, as a consequence of lengthening and widening, the foot eventually starts initiating increased stiffness within the connective tissues. Altering vault curvatures and the rates of change of curvature within the vault means that viscoelastic properties can be used variably, influenced by active and controllable muscle activity. Thus, the force–velocity relationship (the rate at which actin–myosin cross-bridge formation can occur under speed of contraction) between muscle power and connective tissue stiffness is established in relation to gait speed. Flattening–widening–lengthening connective tissue-stiffening principles support the concept of curvature change in the transverse vault to induce foot stiffening at the beginning of terminal stance, as reported by Venkadesan et al. (2017a, 2020) and Yawar et al. (2017). However, it does raise questions regarding the osteoligamentous plate and similar such concepts serving as a single explanation to explain variations in foot stiffness. Under such models, a more pronated foot should become increasingly more mobile throughout its range, yet this does not appear to be the case during static testing on foot lengthening under load (Bjelopetrovich and Barrios, 2016; Takabayashi et al., 2020). Instead, pronation and vault curvature reduction only seem to demonstrate increased compliance on initial loading of normal healthy feet, which is then followed by gradual stiffening. The foot's musculature is certainly an important part of the overall regulation of foot vault stiffness, joint mobility, and stability (Farris et al., 2019, 2020), not least because it can change the vault's curvature through joint motion as well as alter compression forces across joints.

Twisting of collagen-based structures on both the macro and microscale likely plays an important part in adjusting the foot's viscoelastic mechanical properties as much as they do in stretching them. However, these mechanisms do not seem coupled to the longitudinal plane alone. Stiffness can be heightened by increasing the vault curvature through twisting the foot segments against each other, such as the rearfoot on the midfoot and the midfoot on the forefoot (Sarrafian, 1987; Eslami et al., 2007; Leardini et al., 2014; Araújo et al., 2019). Utilising the viscoelastic stress-stiffening properties of the passive soft tissues can increase structural curvatures such as in the transverse vault (arch) of the foot, as under forefoot loading forces, tension increases (Venkadesan et al., 2017a, 2020; Yawar et al., 2017). This can be supplemented and controlled by muscular activity from both the foot's intrinsic muscles (Kelly et al., 2015; Farris et al., 2019, 2020) as well as the extrinsic muscles that supply the foot, such as the long digital flexors (Ferris et al., 1995), tibialis posterior, and peroneus longus (Kokubo et al., 2012).

The modern human medial column has inherently more mobility potential than the lateral column, as it has more articulations and a greater span distance. A higher medial vault profile is possibly required to give the medial foot greater variable control of compliance and stiffness levels through greater changes in its curvatures, and by also being able to lengthen more than the lateral side. The lateral column is intrinsically stiffer with a much shorter span distance and this suggests that higher levels of adjustable compliance are not required here. This may reflect the lateral column's important role in acting as a stable 'pulley' site for peroneus longus. Such difference in properties required for gait might explain the possible earlier evolution of a modern lateral column and the later development of a more complex medial column (Kidd and Oxnard, 2005; McNutt et al., 2018; DeSilva et al., 2019). The more mobile lateral column that is reportedly to have been present in Australopithecus sediba remains curious, but likely reflect increased foot mobility driven by a return to more climbing activities.

Increased twisting for heightened stiffness is a technique utilised within tendons and other connective tissues as much as it is for changing vault curvature-induced stiffness. It is used as a mechanism to increase triceps surae–Achilles stiffness during late midstance ankle dorsiflexion which increases twisting of its fascicles and subtendons in concert with plantar fascial lengthening and stiffening (Huang et al., 2018). This coupled mechanism occurs throughout late midstance, increasing compression across the skeletal structures via increased myofascial tensions, thus maintaining foot stability through vault-lowering changes that peak just before heel lift. Foot-stiffening mechanisms are increased by the forces generated by muscles active during late midstance (Kelly et al., 2015; Farris et al., 2019, 2020). Late midstance through to the terminal stance boundary can be seen as the period of rapidly increasing stiffness. This is followed by initial terminal stance, a period of further stiffness resulting from elastic recoil and re-curvature of the vault. Terminal stance finally ends with increasing compliance as muscle activity within and into the foot reduces.

Supination or less pronation stiffening

The foot, when less pronated (i.e. less prone to the ground, reducing its plantar surface contact area), is usually referred to as supination. The word 'supination' is not fully appropriate as the plantar foot never becomes actually 'supine' in gait, but rather becomes less 'prone'. This 'supination' motion is linked to structural increases in curvature and the collagen twisting of certain soft tissue structures across the midfoot. Foot supination does not directly link to stress-stiffening viscoelastic properties across the arch, except that the increase in vault curvature causes twists within the connective tissues that bridge the vault during terminal stance. Terminal stance is also associated with digital extension which tensions the digital plantar-linked connective tissues such as the plantar aponeurosis, which also stiffens the foot.

A decrease in foot pronation only occurs after heel lift (Hunt et al., 2001). If supination was required to induce stiffness, then the stiffening mechanism would have to initiate vault profile rises against the body mass-induced midfoot bending moments occurring across the foot, requiring dramatic vault-recontouring muscle activity. If the foot could not stiffen as it becomes lengthened and wider (more prone) under the midstance midfoot dorsiflexion bending moment, then the heel would have to lift on a compliant foot. However, if the connective tissue cannot stiffen adequately and muscles cannot actively compensate, then heel lift may still occur on an inappropriately compliant midfoot, giving some explanation to events such as the midfoot break. The foot clearly demonstrates a difference in its material and structural properties that are both influencing stiffness and compliance during its function. Indeed, there is a positive correlation demonstrated between a lower arch height in gait and joint (connective tissue) laxity, rather than a relationship with the rearfoot angle (Kanatli et al., 2006). This finding appears to show an association between foot vault stiffness and connective tissue-stiffening properties rather than rearfoot-to-forefoot opposed twisting-stiffening before heel lift.

Vault height in foot function

Stiffness across the vault may better reflect the foot's appropriate functional adaptability rather than using arch height measures (Zifchock et al., 2019). Stiffness is calculated by comparing the differences between the foot vault profile during standing and sitting positions normalised to 40% body weight to account for weight difference when sitting (10% body weight per foot) and standing (50% body weight per foot), with larger changes reflecting greater mobility. This has been referred to as *arch height flexibility* or *AHF* (Zifchock et al., 2019). A technique known as the *arch height index* (*AHI*) is used to establish the vault height at 50% of the entire foot length, except in the presence of toe deformities where the 1st metatarsophalangeal (MTP) joint should be used instead (McPoil et al., 2008). Showing good intra-tester and inter-tester reliability, the AHI is taken in sitting and standing positions with the changes recorded to provide the AHF (Butler et al., 2008; Fraser et al., 2017).

The equation for calculating arch height flexibility (AHF) is:
AH = arch height
BW = body weight

$$AHF = \frac{AH_{standing} - AH_{sitting}}{0.4 \times BW} \times 100$$

Dominant feet tend to demonstrate higher vaults than non-dominant feet. However, despite this, no difference in the same asymptomatic individual's foot stiffness has yet been reported. Only a weak association is found directly between vault height and stiffness levels (Zifchock et al., 2006). Young female Japanese feet are reported to have a lower AHI compared to age-matched Japanese males, yet there is no demonstrated difference in AHF between the genders (Takabayashi et al., 2020). Stiffer feet have been reported to show a greater proportion of medial–lateral directed motion during rearfoot strike running compared to that which is found in more compliant feet in young adults (Zifchock et al., 2019) (Fig. 3.1.4c). Foot stiffness influences impulses during walking and running gaits, with healthy stiffer feet demonstrating lower heel and midfoot impulses but greater forefoot impulses when compared to more flexible feet (Cen et al., 2020) (Fig. 3.1.4d).

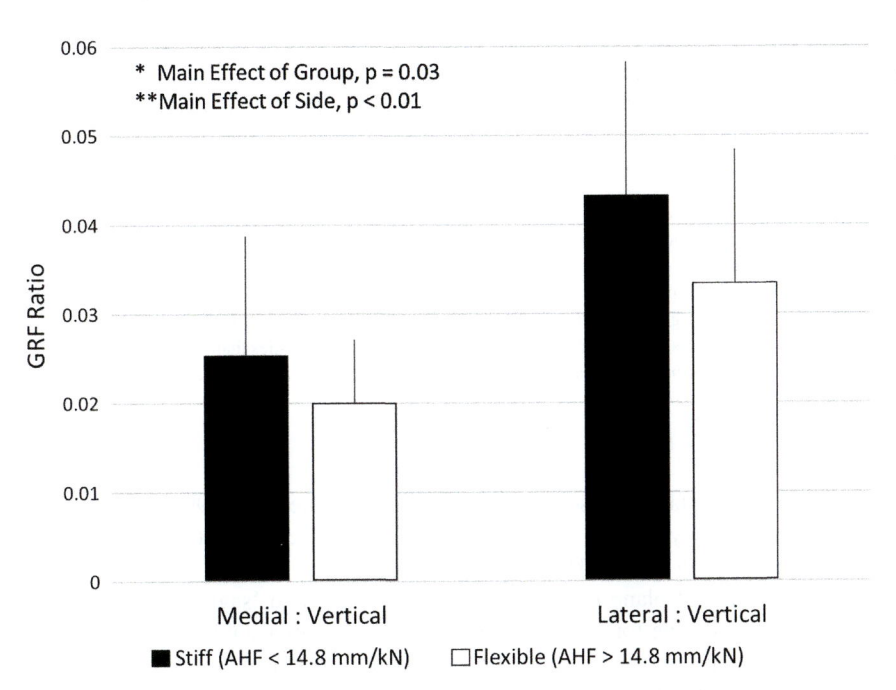

FIG. 3.1.4c Average (± standard deviation) proportional distribution of medial and lateral GRFs for individuals with stiff *(black)* and flexible *(white)* feet during rearfoot strike running, indicating that stiffer feet demonstrate higher medial and lateral vertical forces across the foot although lateral–vertical impulses are higher in both groups. *(From Zifchock, R., Parker, R., Wan, W., Neary, M., Song, J., Hillstrom, H., 2019. The relationship between foot arch flexibility and medial–lateral ground reaction force distribution. Gait Posture 69, 46–49.)*

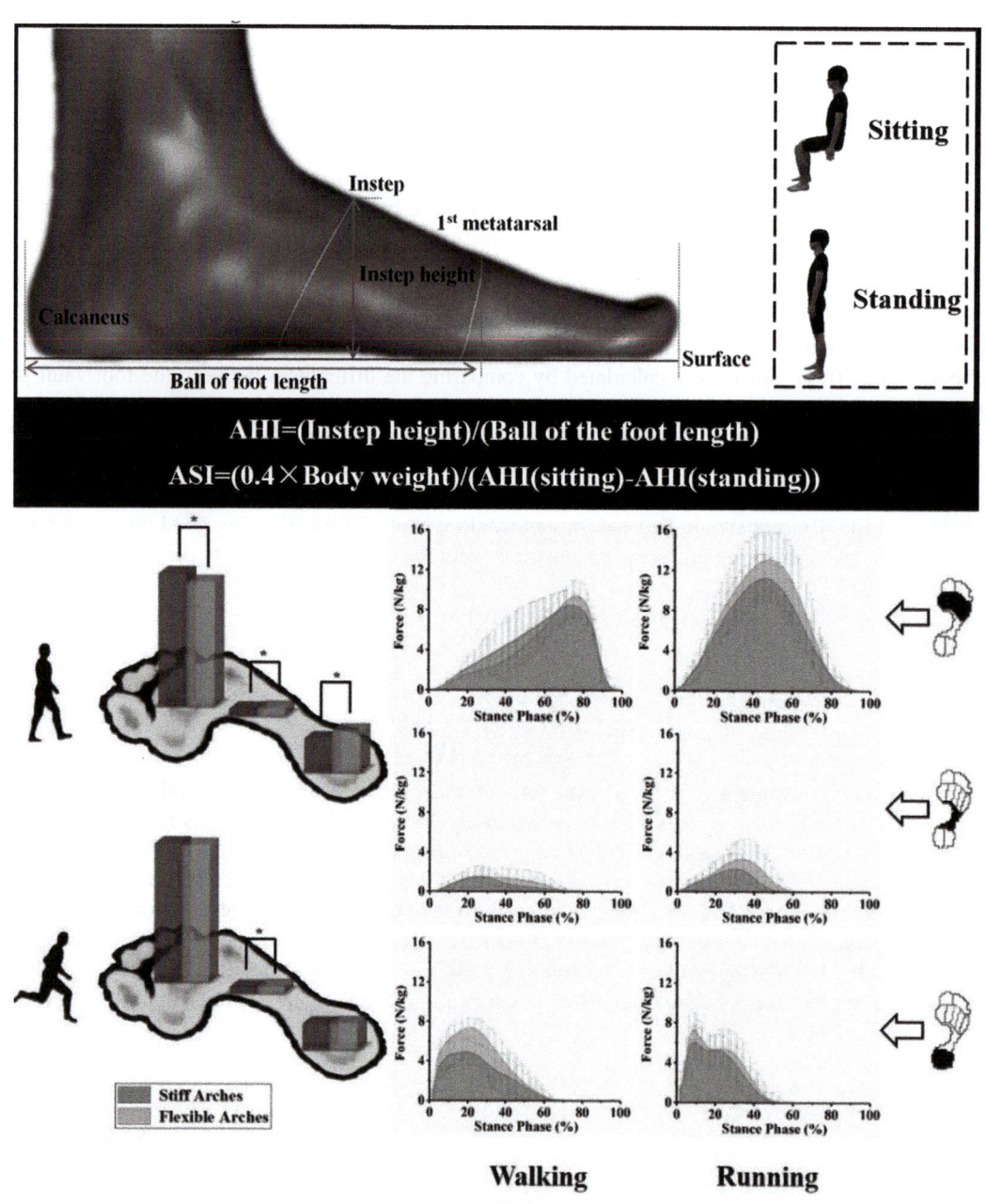

FIG. 3.1.4d The technique for calculating arch height flexibility (AHF) also known as arch stiffness index (ASI: as used in *upper image*) is shown. When comparing varying foot stiffness levels in walking and running, differences in pressure generated by pressure plates along the plantar longitudinal axis reveal something quite interesting *(lower image)*. Stiffer feet *(darker grey* columns and graphs) and more compliant feet *(lighter grey* columns and graphs) in walking and running are shown to present differences between their impulses. Flexible healthy feet present larger impulses in their rearfoot and midfoot but smaller impulses within their forefoot compared to stiffer feet during both walking and running. Flexible feet may be better able to use vault deformation to transfer impulses and dissipate loading energies but may be less able to transfer impulses into acceleration energies than stiffer feet. *(From Cen, X., Xu, D., Baker, J.S., Gu, Y., 2020. Association of arch stiffness with plantar impulse distribution during walking, running, and gait termination. Int. J. Environ. Res. Public Health 17 (6), 2090.)*

Terminal stance stiffness: The foot as a beam

Forefoot forces exceed body weight during the terminal stance phase of walking gait (Hayafune et al., 1999). These forces are a combination of the mass (body weight) being lifted by the foot during heel lift and the power generated from the ankle plantarflexors (as accelerators) that create the heel lifting action. The forces generated across the foot subject the metatarsals, tarsometatarsal, and tarsal joints to large sagittal plane torques that attempt to cause a 'sagging' deflection of the midfoot at heel lift into early terminal stance. Increasing forefoot–midfoot stiffness to form a stable beam during late midstance for heel lift helps to maximise mechanical power generation for acceleration (Takahashi et al., 2016; Pontzer, 2017), as well as limiting midfoot sagging and forefoot widening or splaying.

A foot that can act as a stiff cantilever at heel lift is advantageous energetically. A cantilever is a rigid structural element that is anchored at one end to a support from which it protrudes, much like the forefoot being fixed to the ground by the forefoot GRF at heel lift. A 'rigid' stiff foot presenting a convex dorsal surface and a concave plantar surface would be best contoured to resist sagging at heel lift, making an excellent mechanical cantilever. Increasing the foot vault curvature during terminal stance can thus assists in maintaining stiffness. This requires activity from the intrinsic plantar foot muscles and the extrinsic foot plantarflexors aided by higher connective tissue stiffness within structures such as the long plantar ligament, the plantar aponeurosis, the deep transverse metatarsal ligaments, and the intermuscular septa. These structures can be stiffened passively prior to heel lift by late midstance forces that lengthen and widen the foot, stretching the connective tissues into their stiffer range of their material properties.

Through plantar intrinsic and extrinsic muscle activity, later assisted by the windlass effect, digital extension and the digital flexor muscles create MTP joint stiffness which can be highly effective in increasing vault curvature throughout terminal stance in conjunction with ankle and midfoot plantarflexion (Ferris et al., 1995; Farris et al., 2019, 2020). The plantar soft tissue tension created maintains high plantar tensional forces helping to compress the bones on the concave side of the plantar skeletal frame of the vault together, reducing potential deflection sagging via motions within the foot articulations. In so doing, compression forces on the dorsal aspect of the skeletal frame's articulations are concurrently reduced. Thus, the vault is tied most strongly during heel lift from medial to lateral and from anterior to posterior along the changing plantar concave profiles (Fig. 3.1.4e).

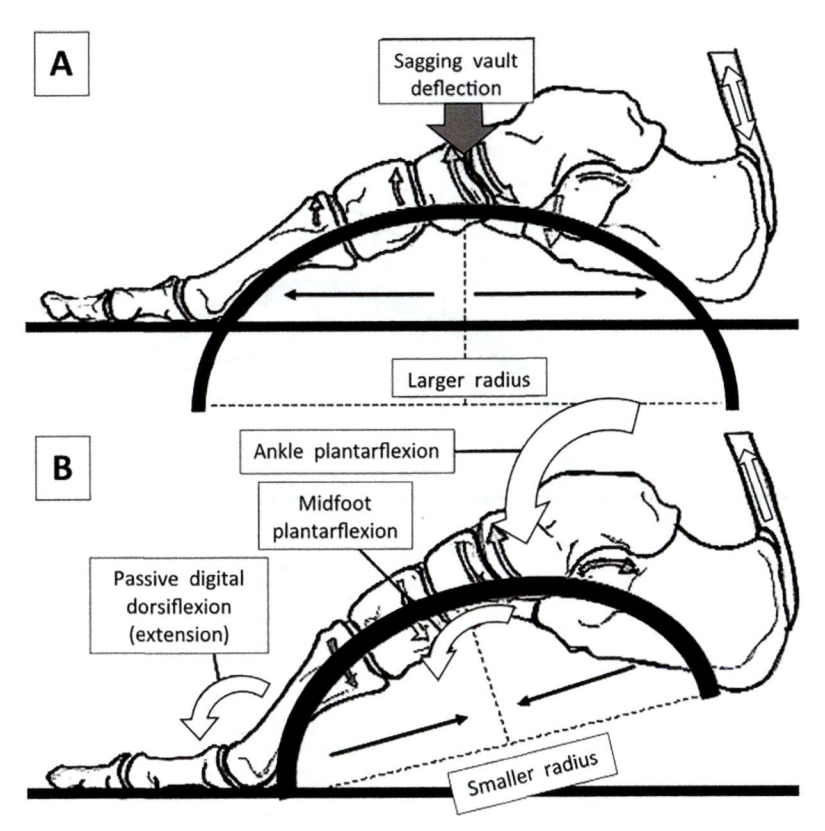

FIG. 3.1.4e During midstance (A), the radius of curvature of the vault is increasing, meaning that stiffness via the mechanical laws of curvature is decreasing. The advantage of this is two-fold: energy can be dissipated through higher vault flexibility and the flexibility lengthens connective tissues, stretch stiffening them and storing elastic energy. After heel lift (B), ankle and midfoot plantarflexion couples to digital extension, actions that together raise the foot vault. This action decreases the radius of the vault's dome which causes increased stiffness. A stiffer foot helps transfer stored plantarflexion muscle power into the forefoot creating a GRF that can be directed anteriorly to aid horizontal elements of acceleration off the grounded foot. The higher curvature longitudinally sets up a strong cantilever beam for lifting the CoM of the trailing limb during terminal stance. However, it is the increased curvature across the smaller span distances within the transverse plane that stiffen the foot from medial to lateral that are most likely necessary for permitting longitudinal changes in vault curvature and stiffness. *(Permission www.healthystep.co.uk.)*

The foot vault and midfoot break

Midfoot break, where the midfoot 'buckles' or sags at heel lift, represents failure of the normal cantilever mechanism. In midfoot break, the midfoot acts as a fulcrum for heel lift, utilising sagittal plane motion primarily at the talonavicular, calcaneocuboid, and particularly the cuboid–metatarsal joints. This is the mechanism that is still seen operating within chimpanzees during normal bipedal foot motion (Holowka et al., 2017; Holowka and Lieberman, 2018). However, it is also a heel-lifting variant present within modern human populations (DeSilva, 2010; DeSilva and Gill, 2013; DeSilva et al., 2015). Chimpanzees have a shorter, less terrestrially specialised triceps surae–Achilles complex (Kuo et al., 2013; Griffin et al., 2015). This short chimpanzee Achilles is best served by a midfoot that can dissipate energy at heel lift and reduce the 'stretching' loading time on the Achilles tendon through using the midfoot as a fulcrum point. This midfoot fulcrum provides a much smaller Achilles external moment arm than that derived from the MTP joints.

Humans with a longer specialised energy storage Achilles tendon can use the longer external moment arm, creating a cantilever running to the medial MTP joints' collective fulcrum in order to apply greater power at the ankle to produce heel lift. Therefore, humans can improve acceleration work ratios and provide superior acceleration mechanics. By having a stiffer foot vault at terminal stance, humans increase the force–velocity behaviour of the foot but the price of this is increased metabolic cost of locomotion through added force demands placed upon the ankle plantarflexors (Takahashi et al., 2016). The use of the medial MTP joints' fulcrum rather than the tips of the toes means that the passive extension buckling effect at the MTP joints helps dissipate acceleration energy, thus reducing ankle power through lower force–velocity behaviour. It is possible that the MTP joints represent an important motional compromise to protect the plantarflexors from further increased metabolic demands that could become excessive. Using the analogy of 'Goldilocks and the three bears', the Goldilocks zone of being 'just right' in terms of safe human plantarflexor stresses for maximum acceleration benefit may reside within the region of the medial MTP joints. The lateral MTP joints and midfoot break may represent reduced acceleration capabilities and yet still provide Achilles protection. Propulsion on fulcrums anterior to the MTP joints may represent a risk to over stressing the Achilles (Fig. 3.1.4f).

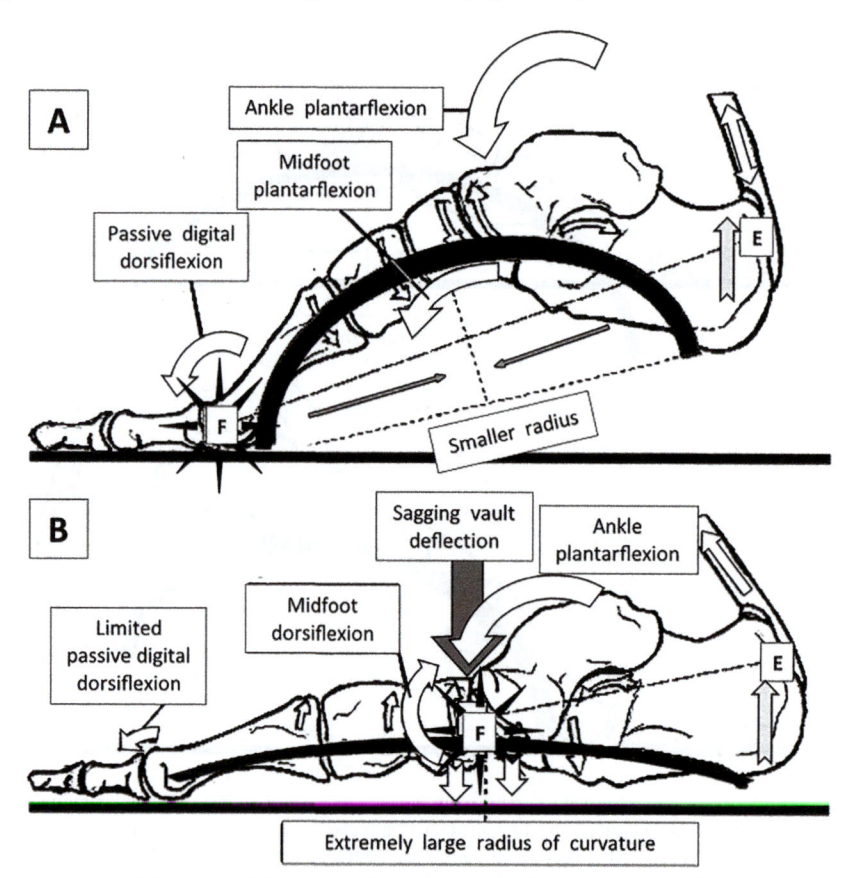

FIG. 3.1.4f When compared to a normal heel lift (A) of increasing vault curvature and the efficient ankle and midfoot plantarflexion moments that generate passive extension around an MTP joint fulcrum (F), midfoot break (B) presents something quite different. Normally, the external moment arm from the Achilles effort (E) to the fulcrum at the medial MTP joints is relatively long (thick dashed line). In midfoot break, the Achilles external lever arm is much shorter, reaching only to the midtarsal joints. Longitudinal vault curvature is lost within a midfoot break, almost presenting the foot as a longitudinally flat sheet. How serious a lack of longitudinal curvature is, depends on how well a transverse curvature across the foot is maintained to provide some stiffening of the foot. Midfoot break ankle plantarflexion power not only provides heel lift but through a midfoot fulcrum also drives a sagging deflection moment through the midfoot, much as if midstance vault lowering has been prolonged. As the rearfoot lifts, the forefoot's metatarsals may initially become increasingly prone to the support surface through midfoot dorsiflexion moments when their bases should be in rotation upwards, away from the support surface under the combined ankle–midfoot plantarflexion moment. As terminal stance progresses and weight is transferred onto the next step, the MTP joints will start to experience some extension, but the application of plantarflexion power through stable metatarsals that demand MTP joint extension (dorsiflexion) from early terminal stance will have been lost. This is not to say that human midfoot break does not serve for terminal stance acceleration, for it does for many healthy humans. It is just a significant compromise in human energetics which causes much of the advantage of a long Achilles to be forfeited. Thus, midfoot break can present a focus for dysfunction in human acceleration biomechanics. *(Permission www.healthystep.co.uk.)*

Arches in the vault

The longitudinal profile of the vault has long been deconstructed into two 'arches' running the length of the foot from the plantar tubercle of the calcaneus to the metatarsal heads. The medial longitudinal 'arch' (MLA) is usually described as having the navicular as its highest point, acting as a keystone within this 'arch'. The MLA runs from the calcaneal inclination angle and up through the body, neck, and head of the talus via the navicular body and then declining down through the medial cuneiform and the declination angle of the 1st metatarsal, terminating at the 1st metatarsal head. The lateral longitudinal 'arch' (LLA) is described as running up through the lateral inclination of the calcaneus to the calcaneocuboid joint, via the cuboid (acting as the keystone), and then finally progressing down the 5th metatarsal declination angle to terminate at its metatarsal head (Fig. 3.1.4g).

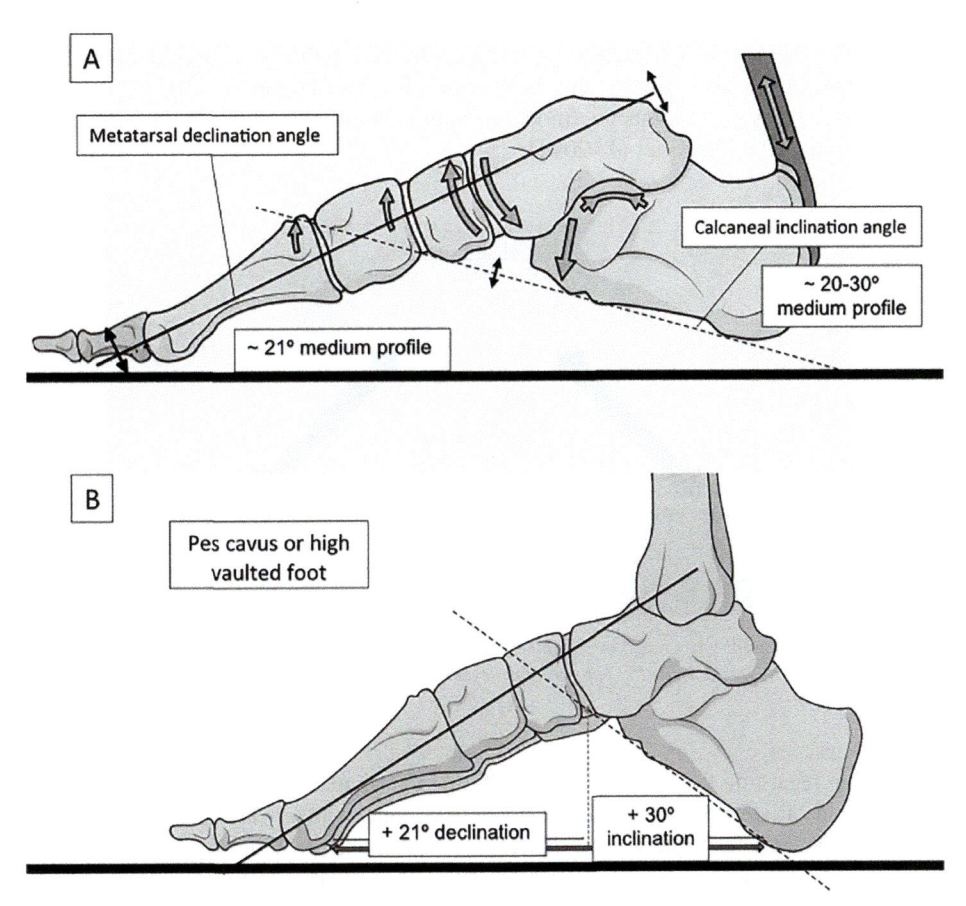

FIG. 3.1.4g The medial longitudinal arch rather than the lateral longitudinal arch (which is usually much lower) is most often the focus for research on measuring and comparing vault (arch) height. The most common technique in orthopaedics is by taking measurements from a lateral view static weightbearing radiograph of the calcaneal inclination angle and the metatarsal declination angle (A). It is very important to remember that these angles change on weightbearing and during gait due to significant bone motion across the foot. The calcaneal inclination angle *(dashed line)* uses the flat weightbearing surface and the angle it makes to the plantar surface of the calcaneus. Angles between 20 and 30° are considered normal. Above this value they are considered cavus, high-vaulted feet (B) and below these values, planus, low-vaulted feet. A similar technique is used for the metatarsal declination angle, sectioning the 1st metatarsal into dorsal and plantar halves from the head to its base *(solid line)*. Values of ~21° are considered normal with higher scores associated with cavus feet and lower scores, planus feet. Although such angulation of anatomy influences the radius of curvature of the vault, how valuable these measurements are in predicting foot behaviour during gait remains in doubt. Such angles are in constant flux during gait and there is some error when taking the measurements. Each metatarsal also has its own distinct and functionally changing declination angle. *(Permission www.healthystep.co.uk.)*

The fact that profile of the LLA is lower than that of the MLA has focused continued research for an explanation as to the evolution of this asymmetry within the foot's shape related to hominin bipedalism (Holowka and Lieberman, 2018; DeSilva et al., 2019). Much research has also been centred on the height of the MLA in order to understand kinematics and pathology within the foot. Attempts to quantify changes in MLA profile commonly use the calcaneal inclination and metatarsal declination angles radiographically, as well as using the AHI, the long arch angle, and navicular drop as a means to classify foot types clinically (Williams and McClay, 2000; Menz and Munteanu, 2005; Menz et al., 2005; Rathleff et al., 2010; McPoil et al., 2016; Fraser et al., 2017). Modelling different foot vault profiles with and without the plantar aponeurosis has indicated that structures like the plantar aponeurosis seem to have mechanical effects that are linked to the profile of the foot. The profile itself affects vault compliance and stiffness levels, with higher arch profiles proving stiffer and the plantar aponeurosis having more influence on such modelling of higher profiled feet (Arangio et al., 1998). This suggests that the contouring of the foot and thus the height of the MLA and LLA is functionally significant.

These foot vault 'arches' have been suggested as being supported by active elements of the plantar intrinsic and extrinsic muscles (Kapandji, 1987: pp. 202–227; McKeon et al., 2015) and the passive elements of the ligaments of the foot, including the plantar aponeurosis (Kim and Voloshin, 1995; Gu and Li, 2012; McKeon et al., 2015). Such soft tissues essentially act as engineering ties of the MLA and LLA. These ties are viscoelastic and/or reactive in nature, altering their stiffness in response to loading rates and their lengthening, influenced by neurological derived muscle activity. The muscular elements of support have led some authors to believe that the best strategy to resolving foot pathology might be to approach it through physical therapy focused rehabilitation based on 'core stability', a system often used in the lumbopelvic region based on the principles of global movers and local stabilisers (McKeon et al., 2015) (Fig. 3.1.4h). There is some evidence supporting core stability as a therapeutic model of approach elsewhere within the body (Sharrock et al., 2011; Huxel Bliven and Anderson, 2013; Stuber et al., 2014).

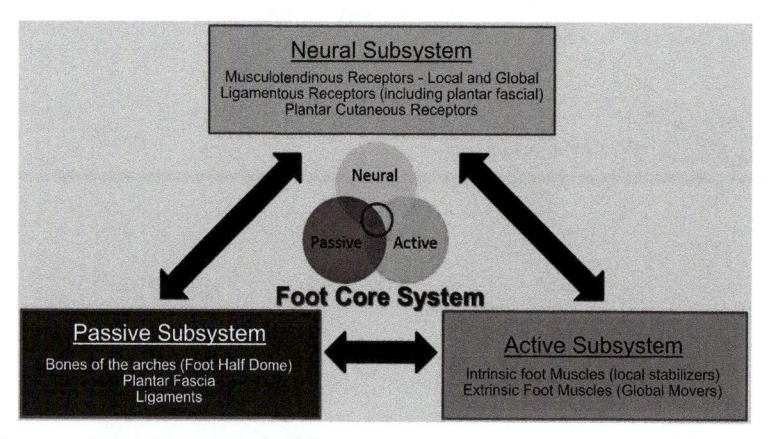

FIG. 3.1.4h The model of foot core stability has much in common with that proposed for stability in the spine by Panjabi (see Fig. 2.1.4). *(From McKeon, P.O., Hertel, J., Bramble, D., Davis, I., 2015. The foot core system: a new paradigm for understanding intrinsic foot muscle function. Br. J. Sports Med. 49 (5), 290.)*

Much has been written regarding the height of the midfoot (in particular, the MLA) and attempts have been made to assign normal and abnormal variants to arch height. But do these variants matter? Due to the effects of curvature mechanics on bending stiffness, the more curved a structure becomes, the stiffer it behaves, hence why modelling the foot with a higher vault profile stiffens its behaviour (Arangio et al., 1998). This means that a high foot vault profile engenders a foot-stiffening effect while a lowering of the foot vault profile gives rise to an increased compliance effect; that is, as long as we disregard any effect this lowering has on the soft tissue mechanics beneath the vault. A high vault with a relatively short moment arm to the midfoot joints and a relatively shorter span distance as seen in a short high-arched foot is more likely to present a stiff midfoot with reduced midfoot bending moments. A low-arched long foot with a relatively long span distance potentially presents a foot that is more prone to larger 'sagging' bending moments, creating greater midfoot deflection and increased compliance. It can now be appreciated that the stiff cavoid and thus relatively shorter foot, can resist midfoot deflection more effectively. Such a stiffer foot is still able to freely bend at the MTP joints, needing a potentially larger range of passive toe extension because of the steeper metatarsal declination angle such a highly curved foot shape offers. As long as the MTP joints move freely, the Achilles lever arm stresses should be well within tissue tolerance. A long planus foot might find it mechanically easier to partially buckle at the midfoot using the midfoot joints as a fulcrum during heel lift because the external moment arm to the MTP joints might be too long to safely load the Achilles. Such a midfoot break requires less MTP joint passive extension throughout terminal stance and protects the Achilles from high stresses, but at the cost of less acceleration power being applied (see Fig. 3.1.4f).

The situation is further complicated by the soft tissues that bridge the foot because tensioned passive tissue properties also affect stiffness, as do active muscle tensions. For instance, a hypomobile and muscularly strong planus foot might resist midfoot bending moments more effectively than a hypermobile, weak cavoid foot. On the other hand, a hypermobile, muscularly weak planus foot is far more prone to midfoot sagging deflection than a hypermobile cavoid foot. The hypermobile planus foot is thus more likely to use a midfoot break than a hypermobile cavus foot, utilising a midfoot fulcrum for heel lift because it is more efficient than trying to use large levels of metabolic muscle activity to restrain the midfoot break. Such a mechanism of heel lift would require little MTP joint passive extension, but the price for this would be an increased difficulty in generating plantarflexion power from the ankle to the forefoot during terminal stance for acceleration. Yet, the benefit would be reduced plantarflexor metabolism and stress. Mobile planus feet should therefore be expected to express greater midfoot dorsiflexion, inversion, and abduction during midstance with less midfoot range of motion overall, as midfoot plantarflexion during terminal stance demonstrated by most feet would be much reduced or absent, due to it no longer being required. This is supported by the literature (Caravaggi et al., 2018). This makes such mobile planus feet more similar kinematically to those reported for the midfoot dorsiflexion motion found within chimpanzee bipedal gait rather than those found in most modern humans (Holowka et al., 2017).

A stiff, strong-muscled planus foot would still have the ability to create a powerful plantarflexion torque, but the resultant cost would be a much greater metabolic demand placed upon the foot and ankle plantarflexors to maintain and utilise MTP joint extension. A stiff planus foot is likely to have greater difficulty in developing midfoot plantarflexion, eversion, and adduction during terminal stance and would have a lesser capacity to adjust its vault curvature properties throughout gait. A stiff, muscularly strong cavoid foot would only be able to use the collective MTP joint fulcrum (never a midfoot break), making passive digital extension very important in the application of ankle plantarflexion power generation to prevent the metabolic cost on the plantarflexors reaching pathologically high levels. A relatively mobile and weak cavoid foot is unlikely to have the option of a midfoot break and yet may be unable to generate adequate ankle power to induce an appropriately timed heel lift during gait (Fig. 3.1.4i).

It can be appreciated that each potential combination of foot vault mechanics scenarios has costs and benefits, and whether the individual has the right combination of foot compliance properties will depend on their lifestyle and activity levels. Such a set of costs and benefits related to a variable foot vault profile and soft tissue mechanics would present evolutionary and developmental processes to the human species with a range of options 'best compromised' to any lifestyle circumstances. Most modern human vault profiles lie somewhere in the middle-'ish' of height and stiffness levels as an 'expected' of a general population compromise. However, the vault of shod populations is seemingly still different to humans that live more naturally in the barefoot condition. Barefoot populations seem to be better able to adapt their compliance and stiffness levels effectively.

There is much anatomical variation in the way connective tissues 'hold' the foot bones together, both between individuals and often between left and right feet of the same individual (Gwani et al., 2017). The foot's shape and dimensions are also affected by racial and regional variations, gender, lifestyle, shoe style, nutritional status, as well as congenital and climatic conditions (Gwani et al., 2017). Both foot length and navicular height change with age, increasing in the growing child and adolescent (Waseda et al., 2014). Thus, there are many reasons underlying population variation in foot vault heights.

Studies on insoles prescribed for 'flat feet' have shown that both insole arch height and material hardness (stiffness) will increase together, behaviours that can also be predicted for the foot. Higher, stiffer foot vaults will alter stresses across the foot, increasing midfoot joint and ligament stresses (mostly at the calcaneocuboid joint) with higher arch heights also increasing ankle joint stresses (Su et al., 2017). The effects brought about through insole stiffness on the foot undoubtedly affect bending moments across the midfoot, ankle power generation at the ankle, and forces generated across the forefoot. It is for the clinician to carefully weigh up whether the changes they make with any insole will stiffen the foot's behaviour beneficially or detrimentally.

The transverse arch of the vault

The vault of the foot has transverse vaulting profiles running medio-laterally across the foot, as well as the more often considered longitudinal arches. This is most frequently discussed as being at the level of the lesser tarsus through the cuneiforms and also through the metatarsals near to their heads, where it has been termed the anterior transverse arch (Kapandji, 1987: p. 224). The transverse arch has been relatively overlooked until more recently, yet it may be far more significant to foot function and evolution than the longitudinal 'arches'. The transverse forefoot profile now seems more evidenced as having significance in initiating changes in the foot stiffness properties required for heel lift and terminal stance, a process that allows more proximal stiffening events to occur effectively afterwards (Yawar et al., 2017; Venkadesan et al., 2017a, 2020).

The changing nature of vault stiffness across 'arches' brings home another important point. The foot requires the capacity to change its degree of compliance and stiffness throughout the stance phase to assist in energy dissipation, surface adaptation, energy storage, and acceleration mechanics. Feet are therefore disadvantaged if they have a 'set' mechanical property throughout the gait cycle. Fixed stiff feet and fixed compliant feet are potentially mechanically destructive as they cannot provide the appropriate nuanced mechanical properties that the changes in the foot's functions demand during gait,

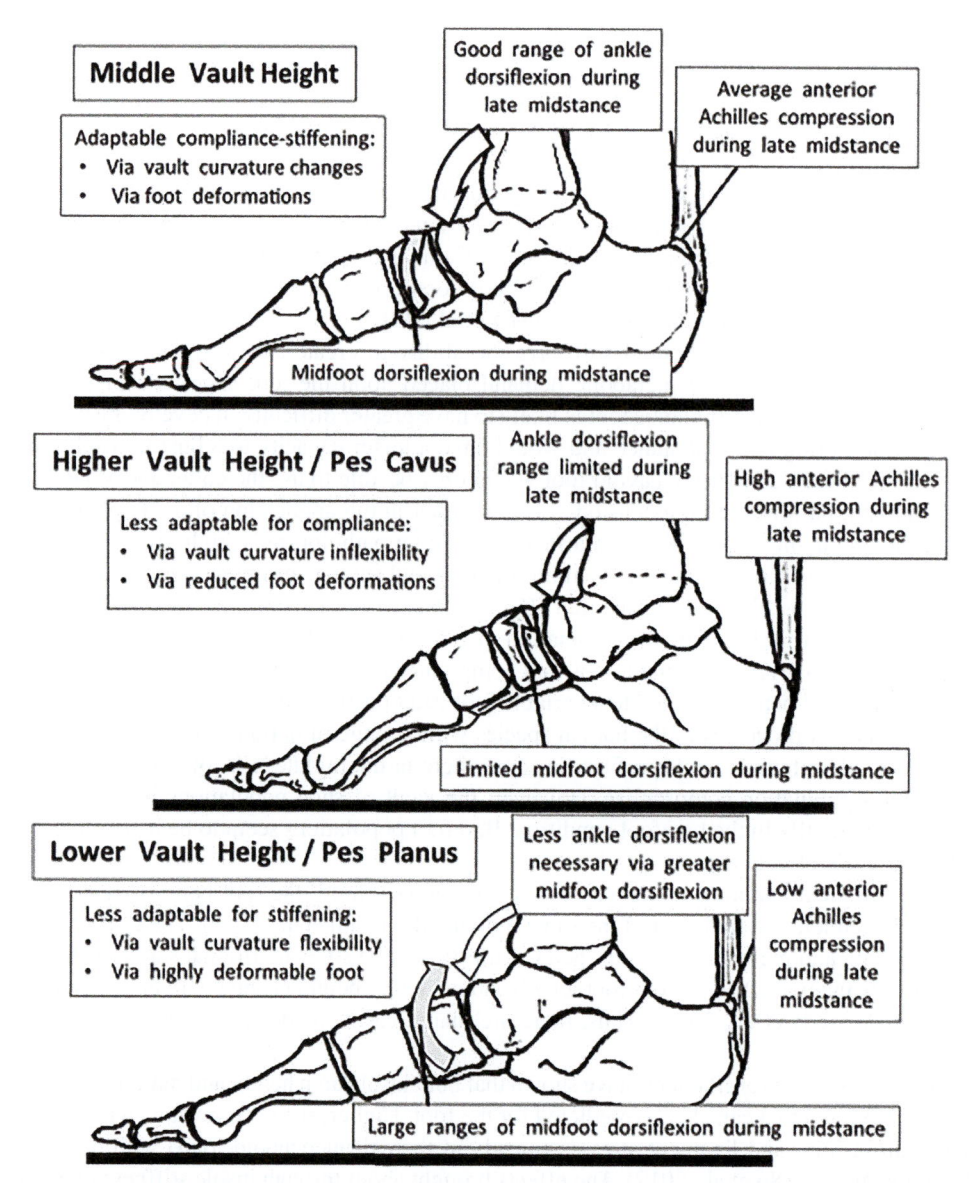

FIG. 3.1.4i Differences in foot vault profile across individuals should be considered as anatomical variances that can influence foot function rather than being distinct abnormalities. Changes in foot vault heights (outside of gradual increases during infancy and childhood) are an important part of clinical history that should not be missed. Vault behaviour is also likened to connective tissue behaviour so that hypermobility and hypomobility are also significant factors in predicting how a particular foot will behave during gait. Expect feet of middle-'ish' profiles to offer a middle range of changing curvature radii, lengths, and widths during stance phase, permitting good compliance and stiffness capabilities if foot strength is adequate. Cavoid feet are likely to express less of these changes via their reduced radius of curvature increasing the inherent stiffness. Planus feet in comparison may still express less vault changes during gait, having started stance with a lower profile, but their large radius of curvature tends to permit more flexibility across the vault resulting in greater foot deformation potential and larger energy dissipation. This makes large forefoot GRFs during acceleration more difficult to generate in mobile pes planus feet compared with other vault types. Structures like the Achilles will also experience different compression stresses during late midstance through altered ankle and midfoot dorsiflexion moments, both of which tend to be restricted in cavus feet. Within planus feet, midfoot dorsiflexion tends to increase at the expense of ankle dorsiflexion range. (Permission www.healthystep.co.uk.)

such as energy dissipation, storage, and application. Trying to drive forward on a soft flexible structure is as problematic as trying to cushion impact forces on something very hard.

3.1.5 The foot as an adjustable multi-tied viscoelastic asymmetrical expanded conical vault

Kapandji (1987: pp. 218–219) discussed the three arches of the foot as being the architectural supports of the 'plantar vault', supports that lie at the corners of an equilateral triangle with weight applied to a keystone situated within the lesser tarsus.

For this structure to maintain stability, the three arches construct a triangular-shaped '*vault*' (Fig. 3.1.5a). On the other hand, a half-dome has also been used to describe the plantar foot vault very effectively (McKenzie, 1955) (Fig. 3.1.5b). However, there is another architectural structure that better fits the irregularity found in the pedal skeletal frame's plantar contour than either of these two descriptions and that is the asymmetrical expanded conical vault. This construction far better models the foot's structure without the necessity of having to assign a 'normal arch height' and it helps explain differences in foot flexibility associated with changing foot profiles. However, an asymmetrical expanded conical foot vault still needs to be considered through the lens of the functional influence of soft tissue ties composed of viscoelastic materials, with some of these ties having the ability to actively and instantly change their tensional properties (muscles).

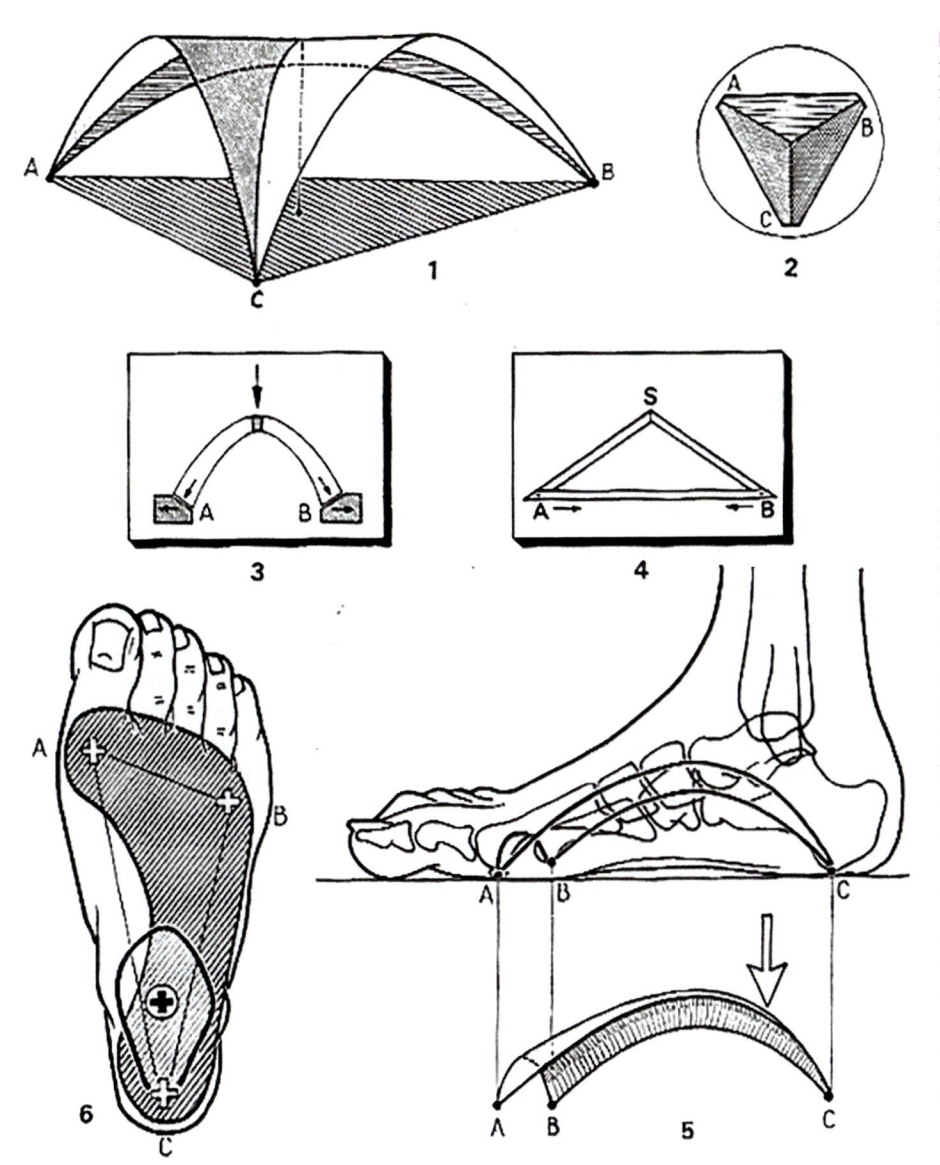

FIG. 3.1.5a The plantar vault as proposed by Kapandji is a vault structure modelled as having three points of support (A, B, and C) to form an equilateral triangle (image 1 and 2). This structure provides 'arches' running between each point of contact (3) with the bones forming rafters and the soft tissues forming tensions between rafters, establishing trusses between the contact points (4). Being triangular, this meets the requirements of biotensegrity and suggests structures that are resistant to shear deformation. When these concepts are applied to the foot via the metatarsal heads and plantar calcaneal tubercle, medial and lateral longitudinal trusses are formed as well as a transverse truss that is continuous from forefoot to rearfoot, narrowing proximally (5 and 6). Issues arise from the fact that the 1st metatarsal head (point A) and the 5th metatarsal head (point B) represent the most mobile metatarsals within the foot, providing relatively poor points of stability on loading. However, the truss profile remains even if the dimensions change on weightbearing between the metatarsals providing stability. Furthermore, the transverse plane truss should be continued to the point where all the metatarsal heads lie in the transverse plane horizontally on the support surface. This is because the deep transverse metatarsal ligament and its associated structures form an important tie-bar, binding metatarsal heads together to the most stable and usually longest 2nd metatarsal, forming yet another truss-like triangular structure lying on its side. (From Kapandji, I.A., 1987. The Physiology of the Joints. Volume 2: The Lower Limb, 5th ed., Churchill Livingstone Inc., New York, NY.)

A vault is a 3D structure creating a 'dome' shape that can be used to support a structure such as a roof. It is made of many combined arches that may run in a number of planes of direction depending on how the vault is constructed. The vault exerts lateral thrusts into the centre of the dome that requires resistance to maintain vault integrity. When 'arch' profiles are symmetrical and complete in all directions, they construct what is referred to as a *cloister* or *domical vault*, a structure seen in many religious buildings such as the Great Synagogue of Rome. A *barrel vault* runs longest in one plane of direction, rather like cutting a barrel in half. The *expanding conical vault* is similar to a barrel-like vault but consists of an 'arched' profile smaller at one end than the other, so that the size of the arch increases in height and width as you pass from the narrow low

FIG. 3.1.5b The foot represented as a half-dome as proposed by McKenzie (1955). Note that the top of the dome is considered to be the trochlear surface of the talus and that the profile of the foot bones is lower than the dome shape proposed. The profile of this dome is also in constant flux during gait and any model of the foot's plantar convex profile has to take this motion into account. *(From McKeon, P.O., Hertel, J., Bramble, D., Davis, I., 2015. The foot core system: a new paradigm for understanding intrinsic foot muscle function. Br. J. Sports Med. 49 (5), 290.)*

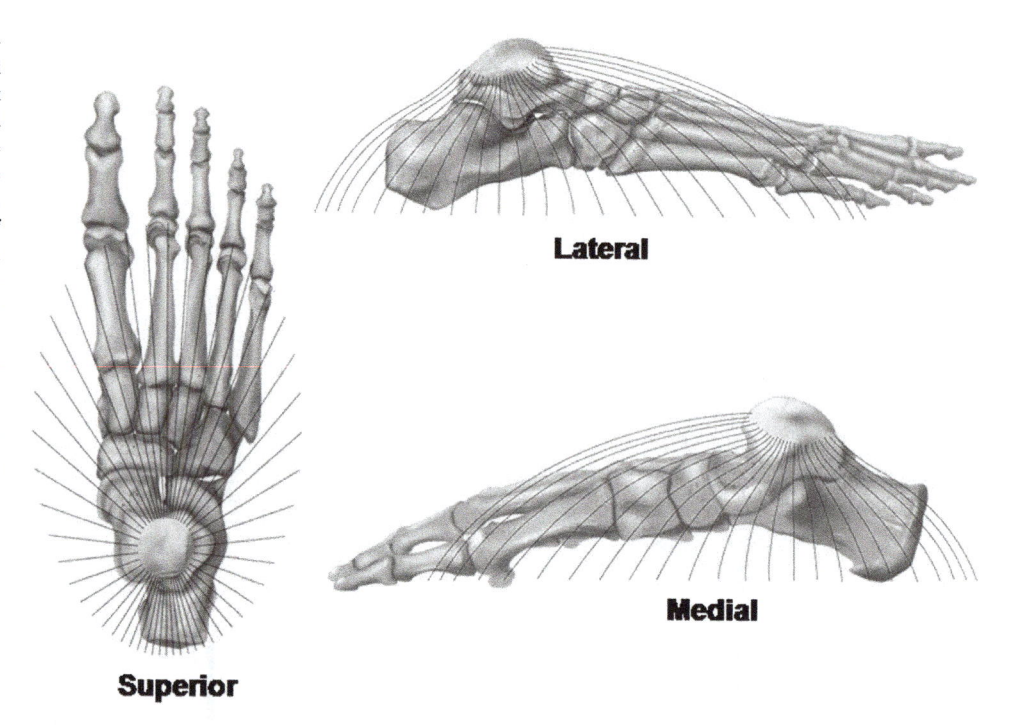

end to the broad high end. The span distance of the vault is therefore greater at the wide end and reduces towards the narrow end, while the curvature angles (radii) change from the narrower to the wider end (Fig. 3.1.5c).

This structure already represents the foot better than trying to apply to it two longitudinal arches and a transverse arch existing together without continuity. The expanding conical vault has a profile that allows a multitude of 'arch profiles' to be achieved in any plane or across many oblique directions, just as is found within the foot. The foot's expanded conical shape also includes asymmetrical profiling where the width and height of the vault is not uniform from medial to lateral, being asymmetrically curved along the supporting edges due in part to metatarsals being of different lengths and declined at different angles. This creates a vault with a large number of potential supporting curves across its profile.

Despite its irregularity, as long as the vault still conforms to the requirements of exerting lateral thrust across any plane of direction throughout its structure, it remains stable. The foot's asymmetrical expanding conical vault uses multi-vault ties to form a tied asymmetrical expanding conical vault where lateral thrust compression of the structure comes from tension via the vault ties. This mechanism is aided by external GRF during stance. The engineering genius of the foot resides in the fact that these soft tissue ties extend in a multitude of directions, running from posterior to anterior and medial to lateral, often obliquely, and at a variety of angles and levels throughout the vault. These ties consist of the ligaments, fascial sheaths, intermuscular septa, extrinsic muscle tendons, and intrinsic muscle–tendon complexes that use tension from stretch and active contraction to compress the skeletal frame together, creating a supported vault structure (Fig. 3.1.5d).

The connective and muscular tissue ties exist at a number of levels within the foot and provide a multi-layered, multi-directional, tensioned support to the skeletal structures above. This is much like a 'through arch' or 'half-through arch' bridge that can have the road or railway line passing across it at any height and still be supported (Koshi and Kottalil, 2016). Indeed, the foot forms a structure much as if a through bridge and a few half-through bridges have been combined together. In this way forces can be spread across the foot and passed from rearfoot to forefoot beneath the skeletal frame, reducing the peak loads on bones. In the foot, the external load (body weight) passes over the 'suspension-arch' rather than through and along the linear road–railway platform beneath the arch as would occur on a bridge. Forces are able to be distributed throughout the skeletal vault that act as the superiorly lying arch as found within such bridges, with the connective and muscular tissues attached below acting as its tension supports. Furthermore, within the foot, the horizontally layered soft tissues make a horizontal 'roadway' that becomes another set of tension cable mechanisms running anterior-to-posterior (and medial-to-lateral). Yet despite the differences in the way forces are loaded upon the foot compared to through-arch bridges, the stability principles are much the same in that tension within any of the structural planes pulls on the cables, increasing tensional strains and thus heightening compressions across the structure. In this way, the lateral thrusts needed to maintain the vault profile are retained and the vault is able to resist any instability caused during

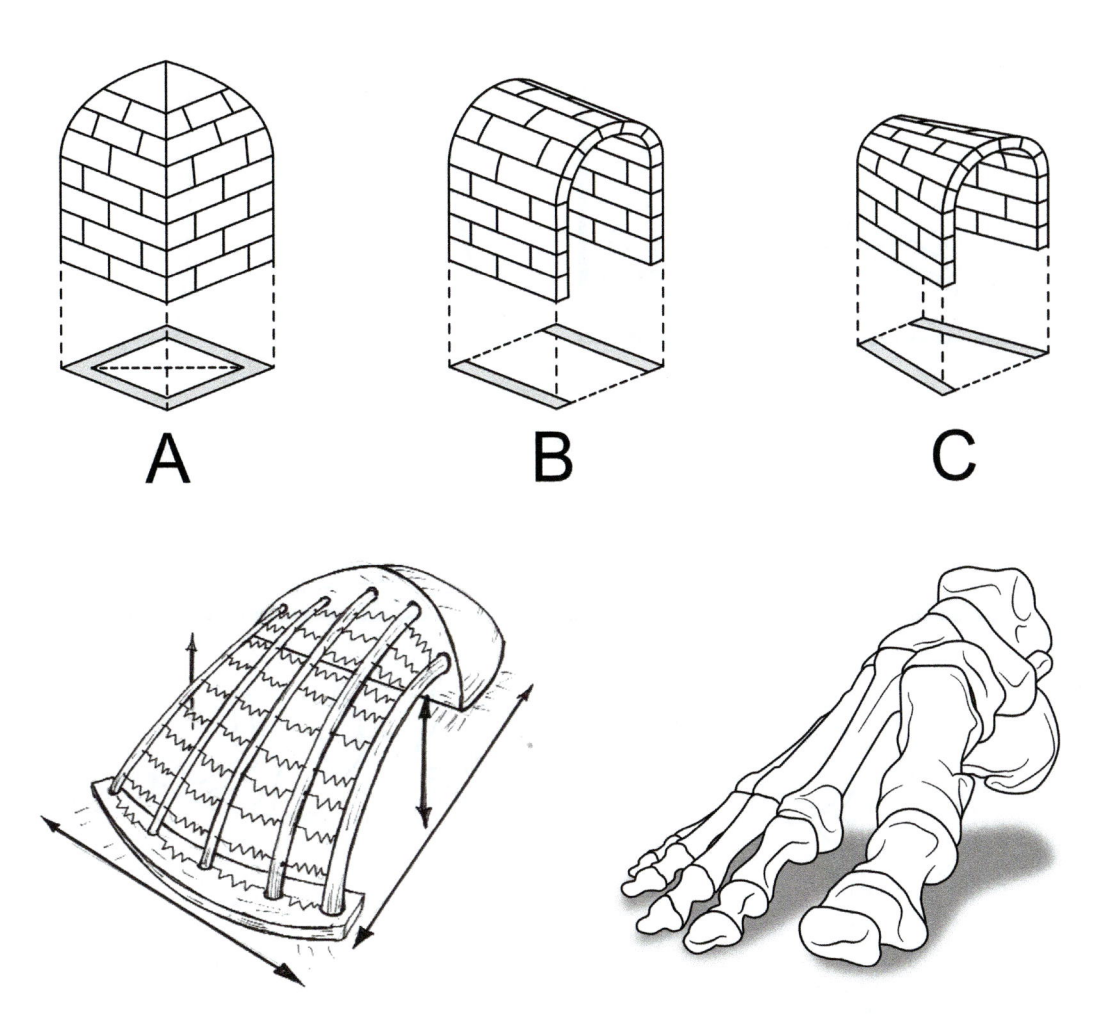

FIG. 3.1.5c Simple engineering schematics of some of the types of vaults found in architecture. The cloister vault (A) is symmetrical in all directions and enclosed. The barrel vault is open at both ends but symmetrical (B), whereas the conical or expanding vault (C) is more representative of the different lengths and heights of the longitudinal nature of the medial and lateral sides of the foot that are asymmetrical. However, the foot vault *(lower images)* supplies asymmetrical lateral thrusts, with the anterior wall formed by metatarsals of different lengths and declination angles tied firmly together by the deep transverse metatarsal ligament that stretches and stiffens when loaded. The foot vault is able to deform under loads and recoil on stress offloading, being constructed with viscoelastic properties. This makes it quite unlike anything used in the construction of a building but more similar in nature to modern popular 'pop-up' tents. *(Permission www.healthystep.co.uk.)*

profile and stress changes. Inferior tensile structures thereby take on a prominent role in maintaining the superior arch made of struts. Of course, the foot is much more deformable than a bridge under load. But be aware. Modern engineering often provides hinges within the structure of the bridge so that it too can deform further under higher loads, reducing the stress concentrations that would occur within the bridge if its structure was fixed (Fig. 3.1.5e).

In the foot, the deep tensile ties include the deep transverse metatarsal ligaments that form a transverse plane forefoot tie-bar and the plantar ligaments. Next come the deeper plantar intrinsic muscles such as the interosseous muscles, adductor hallucis, and flexor hallucis brevis. In the next layer are found intrinsic muscles that associate with the extrinsic tendons. The largest intrinsic muscles are the most superficial. These are abductor hallucis, flexor digitorum brevis, and abductor digiti minimi. Superficial and plantar to these lies the plantar aponeurosis. However, it should be remembered that all parts of the tensile soft tissue network will be either of some greater or lesser significance to the functioning of the whole including even the small dorsal, plantar, and interosseous ligaments of the midfoot. The complex connective tissue arrangement across the forefoot consists of the deep and superficial transverse metatarsal ligaments, extensor apparatus (extensor hood complex), fibrocartilaginous plantar plates, and the metatarsal suspensory ligaments that connect to the tendons and the plantar aponeurosis within the forefoot. Collectively, they form a particularly important tie-bar across the forefoot. Within the midfoot, the complex tarsometatarsal ligaments provide more restraint with tibialis posterior

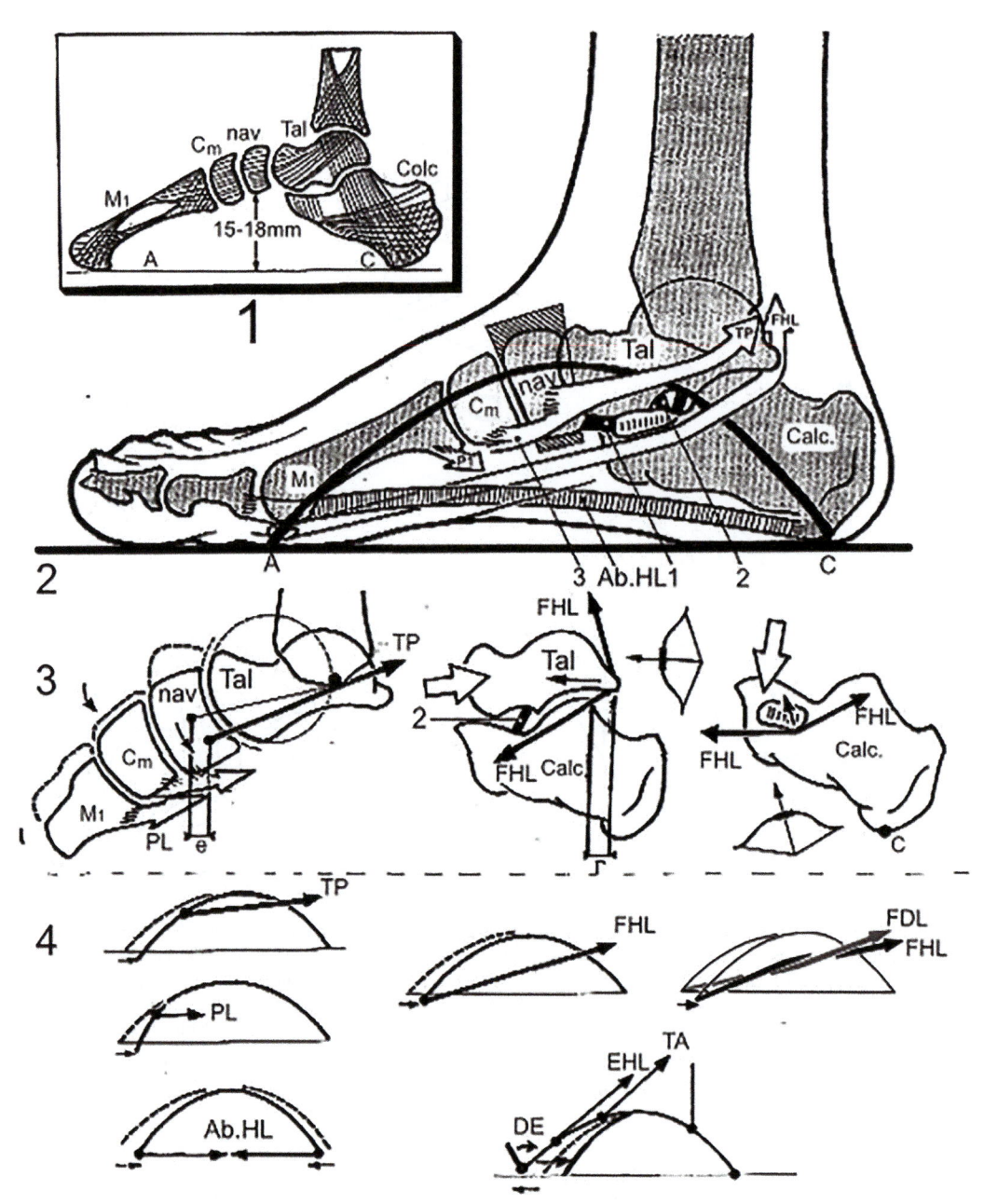

FIG. 3.1.5d Academics such as I. A. Kapandji have long indicated that the foot's plantar contour is maintained via ligamentous and muscular structures during static stance and gait. These soft tissues influence trabecular arrangements within the foot bones as seen in the medial vault (A to C) in image 1 demonstrated within the calcaneus (Calc.), talus (Tal), navicular (nav), medial cuneiform (C_m), and 1st metatarsal (M_1). The plantar foot is packed with anatomical structures that between them create a coherent restraint against loading deformation, shown passing across the vault or dome in the sagittal plane in image 2. Abductor hallucis tethers the medial side (Ab.HL) and many plantar ligaments such as the plantar calcaneonavicular (spring) ligament (1) and the talocalcaneal ligaments (2) within the subtalar joint guide and inform on motion around the vault's articulations. Muscle entheses also play their part in vault control such as the medial attachments of tibialis posterior's tendon (3). Not only do the muscles create or resist motion through the midfoot, they also create compression via their tendon tensions, applying further important stabilising forces into the foot. In image 3, the plantarflexion moments of peroneus longus (PL) and tibialis posterior (TP) are demonstrated across the midfoot. Anterior and superior-directed forces are created by flexor hallucis longus' (FHL) tendon as it slings around the posterior talus and the sustentaculum tali opposing the prevailing forces (white arrows) generated during terminal stance and late midstance, respectively. Image 4 demonstrates the actions of each of the major plantarflexors of the foot vault (including flexor digitorum longus, FDL) that plantarflex the midfoot creating vault-shortening moments, while Ab.HL pulls the vault together medially, raising it. The extensors such as tibialis anterior (TA) and extensor hallucis longus (EHL) can potentially flatten the vault height if their action is not associated with digital extension (DE), which will shorten and heighten the vault via the windlass mechanism. The foot has to utilise structures that have previously evolved for arboreal and manipulative use that have over time been merely 'co-opted' to the role they now perform. Usually, they perform their tasks rather well. *(Adapted from Kapandji, I.A., 1987. The Physiology of the Joints. Volume 2: The Lower Limb, 5th ed., Churchill Livingstone Inc., New York, NY.)*

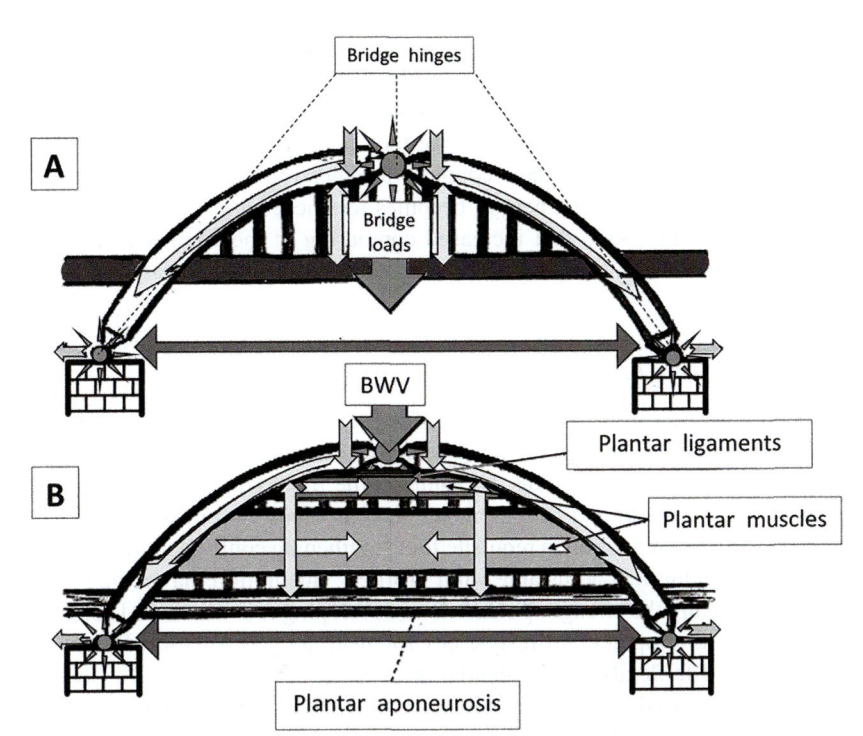

FIG. 3.1.5e A schematic of a three-hinged half-through bridge (A) as a very crude engineering model of how the foot's vault is maintained via ties. The hinges permit some motion on loading and are often used on large modern bridges to avoid fatigue fracture of the structure over time. Such bridges carry loads through their lower beam so that the tensions this causes from the cables compress the ends of the bridge together. The foot experiences most of its loading directed down from the dorsum during midstance, particularly just before heel lift. A better model (yet still too simplistic for the foot) is to combine half-through and through bridges together (B). Consider the beams that carry traffic as further tension cables, representing the plantar ligaments superiorly, the four muscle layers within their muscle compartments within the middle, and the plantar aponeurosis as the lower tie. Any loading forces now tension these cables, compressing the structure together both horizontally and vertically. However, do not just consider the mechanics in the longitudinal direction, for such principles are also at play medial-to-lateral across the foot. *(Permission www.heathystep.co.uk.)*

and peroneus longus, setting up another transverse tie-bar that is rapidly adjustable through associated tendon and muscle tensioning. The short plantar ligament provides a longitudinal tie across the calcaneus to the cuboid laterally and the spring ligament a tie from the calcaneus to the navicular medially. The long plantar ligament extends the longitudinal tie from the calcaneus to the metatarsal bases, and the fact that such a strong stiff plantar ligament is a human feature and not one of other great apes indicates something special about human foot function (DeSilva et al., 2019). Of course, all these ties provide elastic components of restraint that vary their tensions based on the rate at which forces are applied to them. Thus, the vault can be more compliant when walking at a slower pace and more elastic and thus stiffer when walking faster.

The asymmetrical expanding conical vault changes its profile through adjustments in the tensions from the elastic components of the tie-bars. This can occur in multiple directions. Muscles obviously represent vault ties with instantaneously adaptable tensioning capabilities, dependent on their activity. The passive fasciae and ligaments are largely dependent on loading rate and stress-tensioning to alter their tensional properties. Any vertically directed loading of the vault generates increased tie tensions via loads attempting to reduce curvature of the vault through sagging deflection. These are met by passive tensional resistance and the faster or greater the vertical force, the stiffer the ties become and thus the better they are able to resist loading. Active muscle vault ties can change their tension, increasing or decreasing their tensional force to allow, prevent, or attempt to raise the vault profile in a multitude of directions. Again, it is worth noting that fascia also possesses contractile smooth muscle elements, allowing the passive ties to adjust their tone through much slower neurological stimulation (Schleip, 2003a,b; Schleip et al., 2005, 2019; Klingler et al., 2014). This means that foot tone is changeable at the connective tissue material property level through slow sensorimotor 'awareness' control settings.

When the vault is no longer loaded with significant force, it can elastically recoil. Such elastic recoil returns much of the energy put into the vault through the soft tissue's elastic properties returning the foot and its tissues to shape for the next step. Obviously, some is lost to events such as muscle–tendon buffering. The ability to store energy is dependent on the soft tissue stiffness at the time of recoil and the foot vault is able to repeat this process over and over again. It is likely that much of the potential energy stored within the vault ties during the stance phase is utilised mainly to return the foot back into an unloaded shape for the next step. The foot vault is therefore an adjustable, viscoelastic, asymmetrical, expandable conical vault. No static building's 'arch' is ever likely to use one of these. The foot is a dynamic structure, built through evolution and requiring an ability to change its function without loss of stability.

The adjustable, viscoelastic, asymmetrical, expanding conical vault helps in explaining the functional changing compliance and stiffness properties of the foot. The presence of a vault within the foot fulfils the requirement of a biotensegrity structure in that it can move, changing its shape while maintaining stability, and still suspend the skeletal frame within the soft tissues. By changing curvature and its active and passive vault tie tensions, the foot vault can be made variably compliant or stiffened in a multi-planar way, with vault stiffness in any plane coupling to other vault planes as long as the anatomy remains intact. The actual profile of the vault, including its height, becomes less necessary in explaining foot function than does its compliance–stiffening cyclical ability, save that increased contouring of the vault will heighten stiffness and decreased contouring will enhance compliance initially until tissues are stretched into increased stiffness (Arangio et al., 1998; Pini et al., 2016).

Developing an individual multi-tied asymmetrical conical vault

The vault height may have developmental–functional relationships. A high vault profile keeps the soft tissue ties at a shorter length than does a lower vault profile which may limit the loading–stretching lengths of the connective tissues during weightbearing, encouraging them to remain shorter and thereby keeping the vault profile more curved. Environmental factors that reduce the need to lengthen the foot during growth and development, such as shoes with raised heels, may reduce connective tissue strains during development. Such a reduction in requisite foot lengthening could encourage stiffer, less lengthening-responsive connective tissues, encouraging less adjustable vault profiles. This in turn may reduce muscular demand for increasing foot stiffness, thereby reducing the strength potentials of developing foot muscles. Therefore, without changing connective tissue requirements in a developmental lifestyle environment, a cycle of 'form-following-function' is developed, influencing the foot vault's profile flexibility throughout growth.

An excessively low vault profile will likely result in reduced stiffness capability within the foot through altered curvature mechanics (Arangio et al., 1998; Pini et al., 2016), unless adjustability burdens are taken by the foot muscles or the connective tissues become more stiffly set. Planus feet may form due to a lack of foot strength during development as a consequence of less locomotive challenges during growth that require large changes in foot profile. This may include using feet in climbing, for tree climbing is a natural human capability and still used extensively among the young of that other terrestrial great ape, the gorilla. With the foot vault not being challenged through repetitive profile changes during exercise, the connective tissue ties may end up consistently bridging relatively longer spans, encouraging increased flexibility within the connective tissue vault ties with weaker muscles less able to compensate. Increasing activity levels, size, and mass in adulthood or loss of tissue material properties with age could cause developmentally induced planus feet to become problematic.

The viscoelastic nature of the vault ties is important to stiffness. Concepts of physiological laxity (hypermobility) or stiffness (hypomobility) within connective tissue can be accounted for. However, not all vault ties seem to have the same significance in influencing vault height and stiffness. Certain ligaments in the foot, such as the short plantar ligament and the spring (calcaneonavicular) ligament, are known to be essential to midfoot stiffness yet do not seem to be dependent on foot vault height, being equally important stiffeners in human 'arched' feet and the flat-profile feet of monkeys (Ker et al., 1987; Bennett et al., 1989). The strong long plantar ligament is a human trait (DeSilva et al., 2019) and the spring ligament and plantar aponeurosis are more developed among humans and seem to have a stronger linked role to foot vault profile and flexibility (Arangio et al., 1998; Griffin et al., 2015; DeSilva et al., 2019). Only time will tell if modelling the foot as an adjustable multi-tied asymmetrical conical vault proves more successful in unravelling foot biomechanics further.

3.1.6 Span distance and curvature effects on vault stiffness

As is discussed in Chapter 1 of 'Clinical Biomechanics in Human Locomotion: Origins and Principles', Section 1.2.5 and in Chapter 2 of the same text, within Section 2.1.3, span distance, beam stiffness, and beam contouring/curvature determines the bending moment and maximum deflection within a beam. The foot needs to act as a stiffened beam in order to be able to effectively transfer energy and apply power generated by the ankle plantarflexors to the forefoot so as to create a stabilising GRF during terminal stance. This ability aids momentum by facilitating weight transference for the next step. As the human foot is not normally fully prone to the support surface, it has span distances for loads to cross over through a vaulted profile. Mechanically, the foot represents a crude shell beam with a width narrower than its length. As an asymmetrical expanding conical vault, the foot has multiple span distances in a multitude of directions which are continuous and multiplanar. Generally, in the longitudinal anterior–posterior direction, span distances are greater, whereas they are shorter in medial–lateral orientations, being narrower proximally than distally (Fig. 3.1.6a).

Despite all the spans of the foot being constructed from a number of bones compressed together by soft tissue acting as tensioned vault ties, the potential for deflection in the shorter transverse spans is far less than within the longitudinal spans (Venkadesan et al., 2017a, 2020; Yawar et al., 2017). The amount of deflection will reflect the span distance, the curvature,

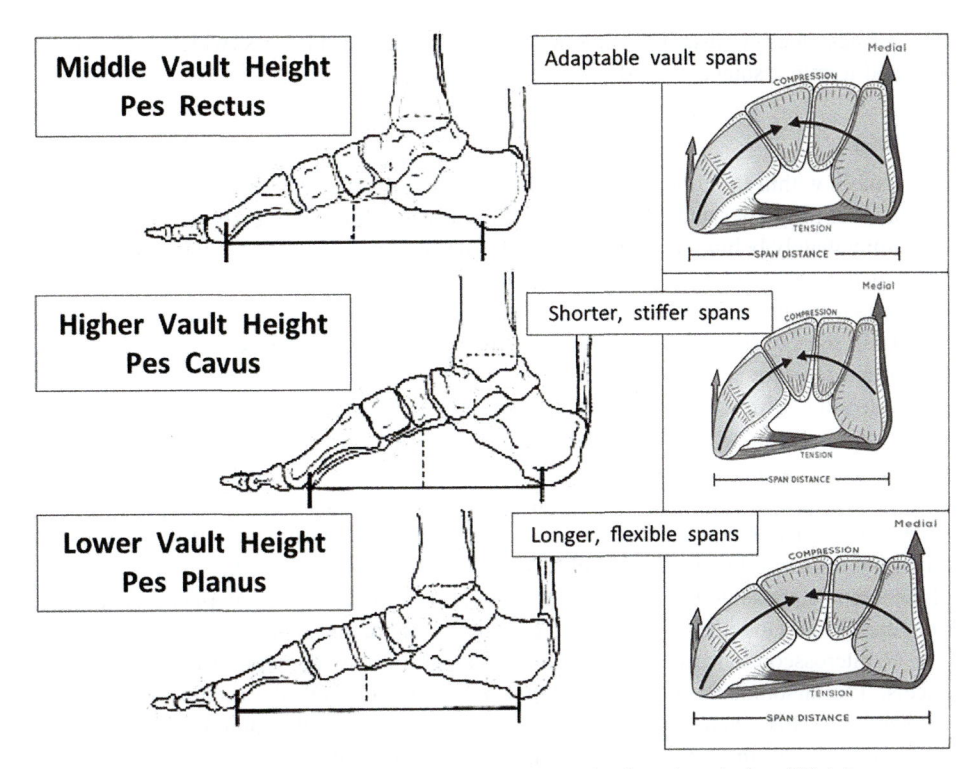

FIG. 3.1.6a The vault's span distances couple stiffness to the radius of curvature variably throughout the foot. This is important, not only to the intrinsic compliance–stiffness relationships but also to how easily it is to adjust compliance across the foot. These principles are at play both in the sagittal plane *(left)* and within the frontal plane *(right)*. The right images demonstrate the transverse vault of the foot at the level of the tarsometatarsal joints. The articular surfaces of the cuneiforms and cuboid are shown as well as some of the distal attachments of tibialis posterior to the medial side of the cuboid and the peroneus longus to the medial cuneiform. Larger, longer, broader feet with lower vaults will tend to behave more flexibly than smaller, shorter, narrower feet with higher vaults which tend to be intrinsically stiffer. Many potential combinations influencing intrinsic foot flexibility and adaptability under loading forces can therefore exist. A general concept is that flatter feet with relatively longer spans, longitudinally and transversely across the foot, will behave more flexibly under load. *(Permission www.healthystep.co.uk.)*

the applied load, and the soft tissue integrity within any foot. Foot vault shape is highly variable, so the potential span distance and curvature is also highly subject specific. In addition, foot muscle strength and connective tissue stiffness are also highly variable. In the case of very 'flat' feet, the cutaneous tissues may load to the support surface all across the plantar foot and in such feet, longitudinal spans are less functionally significant albeit they are still present. Only in feet with loss of osseous integrity such as can occur in Charcot feet can the vault mechanics be totally lost. Moreover, the foot also changes shape during gait. Thus, at different moments within the stance phase of gait, span distances will change and behave differently depending on foot lengthening or shortening (Fig. 3.1.6b). During the midstance phase

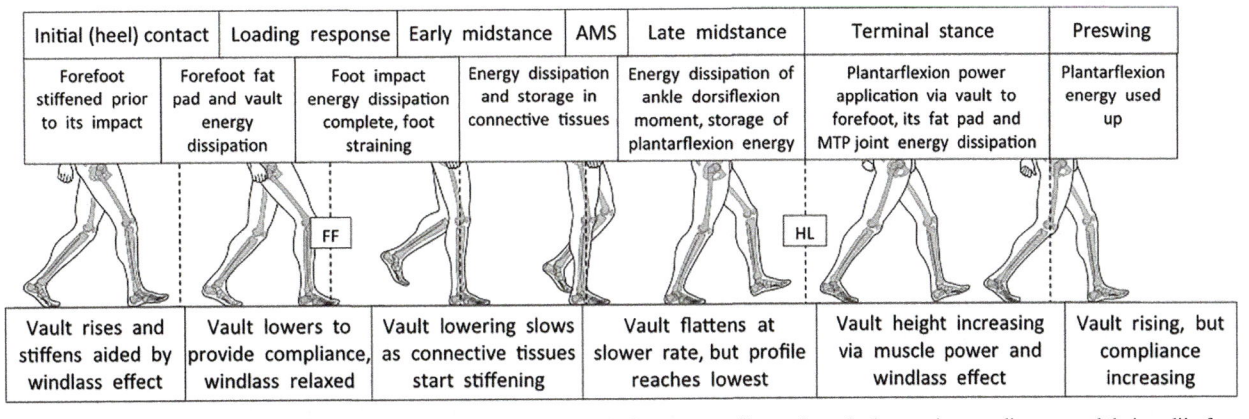

Initial (heel) contact	Loading response	Early midstance	AMS	Late midstance	Terminal stance	Preswing
Forefoot stiffened prior to its impact	Forefoot fat pad and vault energy dissipation	Foot impact energy dissipation complete, foot straining	Energy dissipation and storage in connective tissues	Energy dissipation of ankle dorsiflexion moment, storage of plantarflexion energy	Plantarflexion power application via vault to forefoot, its fat pad and MTP joint energy dissipation	Plantarflexion energy used up
Vault rises and stiffens aided by windlass effect	Vault lowers to provide compliance, windlass relaxed	Vault lowering slows as connective tissues start stiffening	Vault flattens at slower rate, but profile reaches lowest	Vault height increasing via muscle power and windlass effect	Vault rising, but compliance increasing	

FIG. 3.1.6b Vault profiles change during gait, altering their capacity to influence compliance through changes in span distance and their radii of curvature. The foot vault profile also affects connective tissue stress-stiffening processes and the moment arms of muscles, influencing effects beyond the material property changes that result directly from their shape change. Other factors such as an individual's muscular strength and connective tissue behaviour (e.g. hypermobility) across the foot are also at work in setting foot compliance–stiffness levels. *(Permission www.healthystep.co.uk.)*

of gait, vault height stiffness does not usually couple with connective tissue stiffness (which stiffen as the foot flattens) but on forefoot loading and terminal stance, they do couple together.

Transverse vault span

None of the span distances within the foot are bridged by a single biological material or structure. Thus, each expresses variable material behaviour properties and all of them have multiple mobility points (hinges) due to the presence of joints. The transverse span immediately behind the metatarsal heads has no interosseous articulations. The area is instead tightly bound together from medial to lateral by a strong connective tissue tie-bar, in particular through the transversely running deep transverse metatarsal ligament. The interosseous muscles, particularly the dorsal interossei that consistently attach proximally from adjacent sides of the metatarsals, provide a potential active reinforcement mechanism against metatarsal spreading, usually termed forefoot splay. Width tension develops across these intermetatarsal ligaments, resulting from forefoot widening on loading which the interossei can actively help restrict through their transverse plane orientated anatomy. Muscle-induced stiffness across the distal forefoot provides an active support mechanism to supplement the passive ligamentous and fascial binding across the transverse plane curvature that lies between the metatarsals. At the proximal intermetatarsal joints, plantar, dorsal, and interosseous intermetatarsal ligaments bind each metatarsal base together (Draves, 1986: p. 163). All these anatomical structures influence longitudinal stiffness of the foot through coupling actions between the transverse and longitudinal vaults (Venkadesan et al., 2017a, 2020; Yawar et al., 2017) (Fig. 3.1.6c).

At the level of the tarsometatarsal joints, the transverse span is reinforced by dorsal, plantar, and interosseous ligaments binding the intermetatarsal bases together, assisted by indentations within the tarsal row at the 2nd and 4th metatarsal bases. There are, however, no interosseous tarsometatarsal ligaments at the 1st and 5th metatarsal bases, reducing stability at these tarsometatarsal facets. The tarsometatarsal joints together help maintain both the medial–lateral and anterior–posterior plantar midfoot concavities. This plantar midfoot concavity persists through the intercuneiform and cuboid articulations and the navicular–cuboid region under the distal calcaneal area. The midfoot transverse span concavity is maintained actively from the obliquely orientated tendons of tibialis posterior and peroneus longus, with the peroneus longus tendon 1st metatarsal base attachment possibly making up for a lack of an interosseous tarsometatarsal ligament.

Lateral vault span

The lateral span distance is bridged by the calcaneus through its inclination angle, the cuboid, and the 4th and 5th metatarsals via their declination angles. These osseous structures are tethered together anterior-to-posterior by ligaments such as the long and short plantar ligaments, the dorsal calcaneocuboid ligament, and the bifurcate ligament, assisted by other connective fascial tissues and muscles. The peroneus longus tendon underrides the cuboid such that the cuboid's role as a pulley for the peroneus longus helps to 'hold' the cuboid up during stance phase (Fig. 3.1.6d).

Medial vault span

The medial longitudinal span consists of the medial metatarsals, the three cuneiforms, the navicular, the talus, and the calcaneus. With more joints and a longer span, the potential deflection through the medial span joints is much higher than it is laterally. Deflection can be particularly significant at the talonavicular joint which has important dorsiflexion–plantarflexion motion (Wolf et al., 2008; Holowka et al., 2017). It is a ball-and-socket-like joint (Pisani, 1994) that moves in relation to and with the talus by being the anterior portion of the subtalar joint. Other joints of the medial span are capable of gliding inferiorly–superiorly on relatively vertical flat articular surfaces, but these joints are less mobile than the talonavicular joint (Wolf et al., 2008). These articulations are also usually angled to provide a slight dorsal overhang so that plantar gliding is less restricted than dorsal gliding. The medial longitudinal span has its distal edge at the metatarsal heads and its proximal edge at the plantar calcaneal process. Ligamentous support includes the medial parts of the long plantar ligament on the lateral side of the medial span, the dorsal, plantar, and interosseous tarsometatarsal ligaments, the plantar and dorsal cuneonavicular ligaments, and the highly significant plantar calcaneonavicular or spring ligament.

Vault span stability mechanics

Span distance stability across the vault occurs through the plantar concave curvature and via the length changes that influence the stiffness levels required to induce a cantilevered effect to lift the rearfoot at heel lift (Pini et al., 2016). Vault stiffness is affected by vault width, length, curvature, and its active and passive soft tissue vault ties, influenced by the soft tissue generated Young's modulus and Poisson's ratio (Venkadesan et al., 2017a). The living human foot is stiffer than

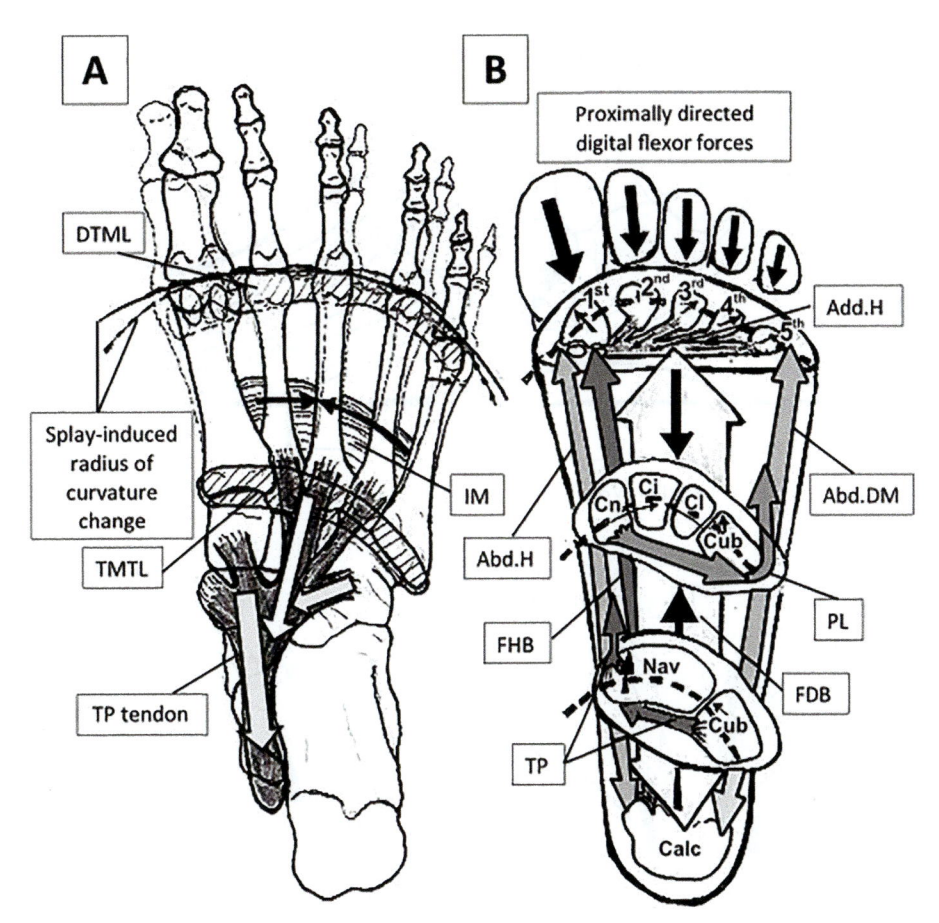

The foot as a functional unit of gait **Chapter | 3** 483

FIG. 3.1.6c Schematics to help explain the interaction between transverse and longitudinal vault stiffening during late midstance. Body mass acceleration over the grounded foot moves the CoGRF towards the forefoot (A), creating forces under which the metatarsals increasingly spread apart (splay, see ghost metatarsal/digital outlines), with motion being greatest at the 1st (medial), 4th and 5th (lateral) metatarsal heads. This reflects strain deformation of the deep transverse metatarsal ligament (DTML) which is variable in its elasticity between each metatarsal. The metatarsal heads, varying as they do in length, create a radius of curvature that increases under splay forces *(solid black line to dashed line)*. Thus, the increasing stretch stiffening of the DTML is coupled to increasing curvature which stiffens the transverse plane of the vault distally. Proximally, the metatarsal bases are restrained together by activity within the interosseous muscles (IM). This is aided via tibialis posterior (TP) activity that provides central tarsometatarsal compression, extensively aided by the tarsometatarsal ligaments (TMTL). (B) The transverse muscle fibres of adductor hallucis (Add.H) restrain the metatarsals splay through its attachments to the DTML, actively adding to distal forefoot stiffening. These events are concurrently occurring with other transverse and longitudinal ligament tensioning events and extensive intrinsic and extrinsic muscle activities across the vault. By increasing stiffness across the forefoot transversely, each muscle is better able to apply their own forces. The extrinsic muscles TP and peroneus longus (PL) also decrease transverse vault radius *(dashed curved lines)* through compression of the lesser tarsal articulations via the cuneiforms (medial Cm, intermediate Ci, and lateral Cl), cuboid (Cub), and navicular (Nav). TP lifts the Nav and Cm while PL is able to compress the vault together in the opposite direction to TP, lifting the Cub and helping maintain lateral vault stability. Activity within flexor hallucis brevis (FHB), flexor digitorum brevis (FDB), abductor hallucis (Abd.H) and Abductor digiti minimi (Abd. DM) can now be more influential in applying forces across the stiffer vault, which is less able to deflect under load, by pulling the calcaneus (Calc) towards the forefoot. This combined vault action aids TP and PL midfoot plantarflexion after heel lift. During vault stiffening and plantarflexion, these plantar muscles are aided by the extrinsic long digital flexors (not shown). The digital flexors create strong proximal forces towards the metatarsal via their digital attachments, pulling the digits into the metatarsal head's articular surface and onto the ground. Rearfoot stiffening in late midstance occurs through ligament tightening, muscular contraction, and restriction of rearfoot motion via changes in ankle/subtalar articular form and force closure during ankle dorsiflexion. *(Permission www.healthystep.co.uk.)*

other primate feet to a degree that can only be accounted for by plantar muscle activity working on the curvature of the plantar surface, with the greatest contribution to stiffness being at terminal stance and coming initially from the transverse vault (Holowka et al., 2017; Venkadesan et al., 2017a). The tarsal joints form an elastic foundation in series within the stiff transverse vault of the foot, assisted by the muscles that can modulate the structural stiffness of the vault (Venkadesan et al., 2017a). These particularly involve the tibialis posterior and peroneus longus assisted by the other plantar extrinsics but stiffening also requires action from the plantar intrinsic muscles (Kokubo et al., 2012; Farris et al., 2019).

Increasing the cantilever stiffness of the foot at heel lift in order to stop deflection would seem harder to achieve through the longitudinal span distance than it would through the transverse span because of the shorter span in the latter. If the

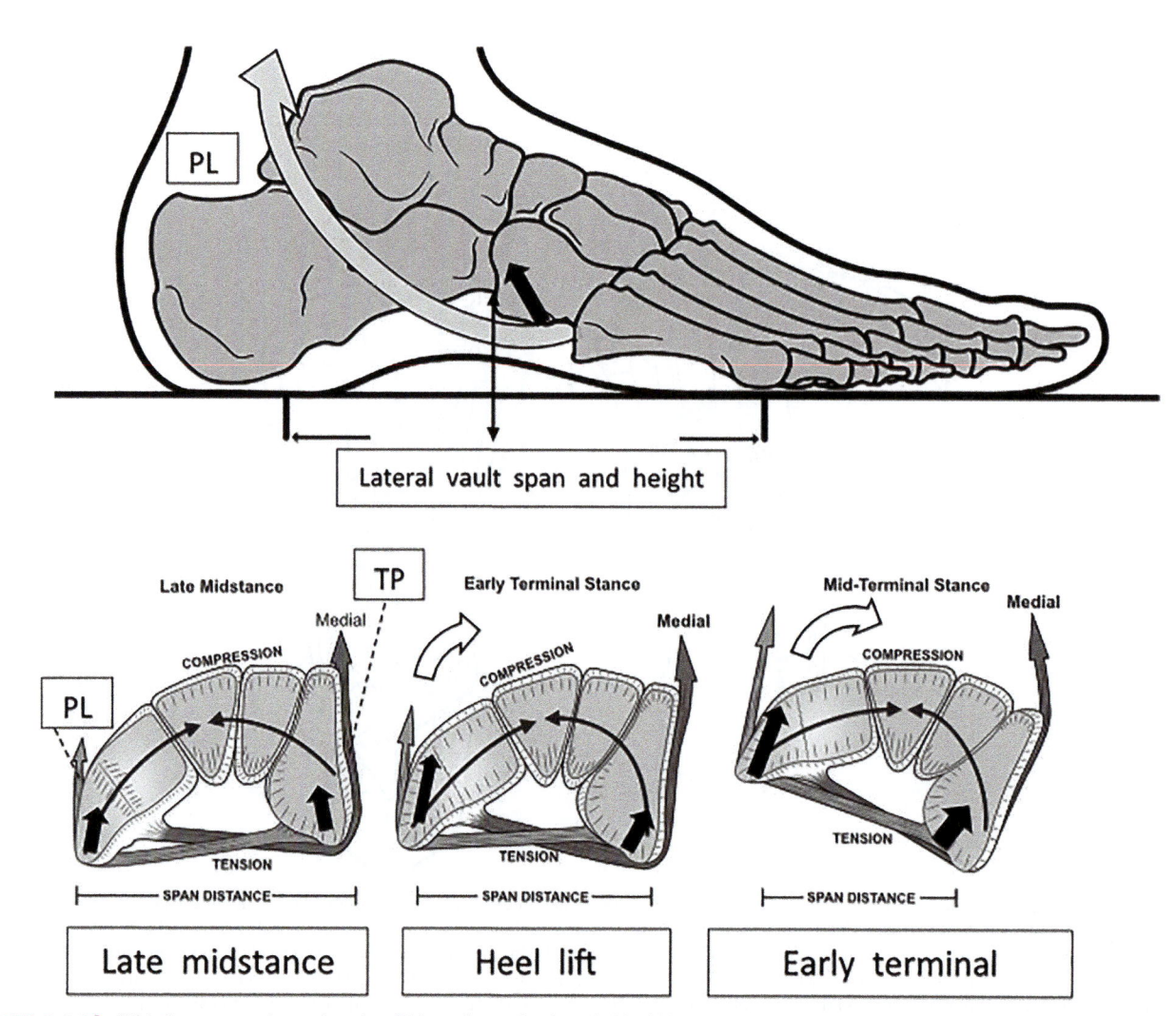

484 Clinical biomechanics in human locomotion

FIG. 3.1.6d With the peroneus longus' tendon (PL) passing under the cuboid within a deep groove on its plantar surface, it can create a lifting force *(black arrow)* under the cuboid while creating sagittal plane compression towards the calcaneus. This stabilises the lateral column proximally. The peroneus longus also provides stability transversely across the vault by compressing the lesser tarsal bones and 1st metatarsal base laterally towards the cuboid, while tibialis posterior (TP) pulls the cuboid and central metatarsal bases medially. This maintains some height for the cuboid within the lateral side of the foot vault. By doing so, it can apply an eversion moment *(white arrow—lower images)* across the stiffened midfoot into the forefoot during terminal stance. At the same time, TP will produce a midfoot adduction moment through its extensive tarsal bone attachments. *(Permission www.healthystep.co.uk.)*

stiffening of the transverse beam could be coupled so that transverse stiffening resulted in longitudinal stiffening, then a far more mechanically efficient mechanism would exist as opposed to trying to stiffen to prevent deflection across the longer longitudinal spans on their own.

Just such a mechanism has been proposed (Venkadesan et al., 2017a, 2020; Yawar et al., 2017) which, aided by active muscle stiffening (Farris et al., 2019, 2020), helps explain the kinematics of midfoot plantarflexion and adduction which is followed by an abrupt increase in pedal flexibility during late terminal stance (Holowka et al., 2017). This increased flexibility at the end of terminal stance correlates well with the findings of sudden deactivation of the tibialis posterior at the end of stance phase (Kitaoka et al., 1997; Murley et al., 2009, 2014). The elastic recoil of the extensive tendon slips of tibialis posterior would provide for early terminal stance stiffness with the recoil energy being fully utilised by late terminal stance. Late terminal stance is a period of decreasing forefoot loading forces, reducing stretching of transverse ligaments before toe-off when little body mass is loaded on the foot. What remaining stiffness is still required within the forefoot is probably largely provided for by the plantar aponeurosis' windlass mechanism.

Transverse plane curvature

Structural curvature is widely found in nature, stiffening thin biological structures (Pini et al., 2016). The foot is a relatively thin structure superior-to-inferior in the context of its osseous skeletal frame across the midfoot and forefoot, being thinner dorsoplantarly when compared to its width and its length. The foot is also a composite material structure composed of laminated beams of bones sandwiched between stiffened soft tissues, particularly plantarly. If curved composite structures are subjected to bending that tends to flatten or compress the composite structure, then interlaminar stresses are generated in the direction of greatest thickness of the composite material, which can cause delamination failure of the material (Nguyen, 2010). Stresses are also affected by the beam curvature of the structure they cross, influencing both radial stresses (towards or away from the central axis) and tangential stresses, i.e. around the circumference (Nguyen, 2010).

In midstance, the foot behaves like an encastre (fixed) beam, bridging a span distance with the heel and forefoot both bearing load. However, at heel lift, only the forefoot is fixed to the ground, hence its change of behaviour to that of a cantilever. Cantilever stiffness greatly increases with small amounts of transverse plane curvature and thus develops internal stresses (Pini et al., 2016). The stiffness of cantilever sheets (thin beams) significantly increases with transverse curvature. The bending stiffness of a cantilevered sheet is different and greater when the cantilever is transversely curved than when it is longitudinally curved (Pini et al., 2016). This has been termed bending asymmetry (Pini et al., 2016). These curvature effects on the foot have been assessed and a conclusion drawn that the transverse arch of the foot is primarily responsible for affecting foot stiffness because it has a small radius and span distance, making it far easier to stiffen through changing its radius by just a little (Venkadesan et al., 2017a, 2020; Yawar et al., 2017).

Transverse stiffening seems to be initiated in the forefoot (see Fig. 3.1.6c) through the increasing tension within the intermetatarsal structures (Yawar et al., 2017; Venkadesan et al., 2020). The foot's asymmetric conical vault profile provides a selection of complex 3D curvatures, giving an extensive adjustability to curvature and undoubtedly using the principles described by Pini et al. (2016). Intermetatarsal motions tensing the interosseous soft tissues can probably alter curvature stiffness on the same principles already described for the bone rays of fish fins by Nguyen et al. (2017). Twisting or stretching soft tissues between fin or metatarsal bones can tighten the connective tissue through rotational motion-inducing curvature changes and thus moderate transverse plane stiffness across the structure, be it fin or forefoot. Thus, metatarsal rotation and intermetatarsal splaying can tighten the intermetatarsal connective tissues so as to increase the transverse plane curvature profile, providing a mechanism distally that can couple with muscular activity proximally leading to longitudinal stiffness across the whole foot vault (Fig. 3.1.6e).

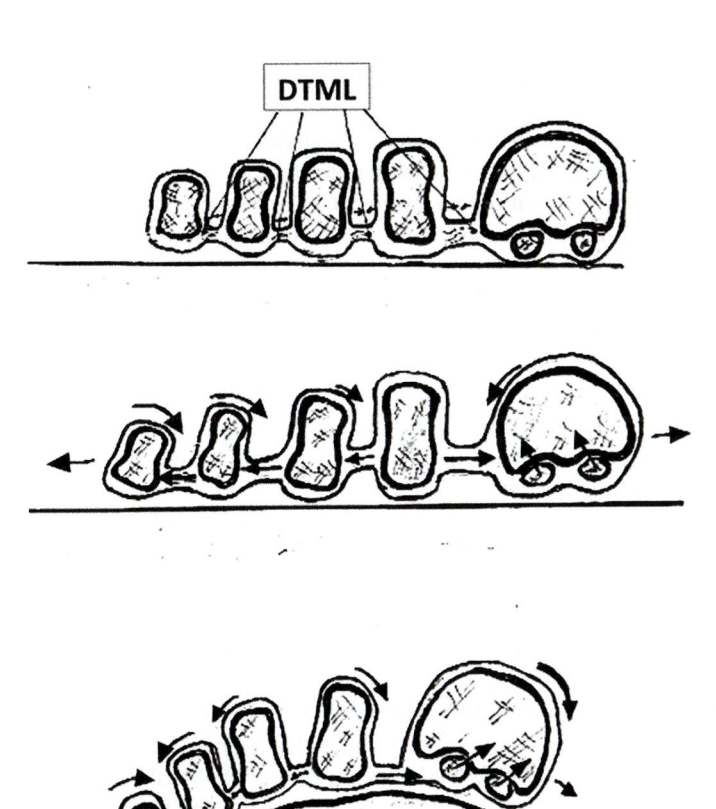

FIG. 3.1.6e The deep transverse metatarsal ligament (DTML) is bound into the many connective tissue structures that surround the metatarsophalangeal (MTP) joints, such as the extensor apparatus and fibrocartilaginous plantar plate. Together these structures maintain a tension tie-bar across the metatarsal heads and necks that influence MTP joint stability. On increasing forefoot loading as the heel gradually offloads on a flat surface during gait, the metatarsals splay apart tensioning the DTML between the metatarsals asymmetrically. This is because the 4th, 5th, and 1st metatarsals have more motion at their tarsometatarsal joints than the 3rd, and especially the 2nd. The 2nd metatarsal base is usually tightly 'fixed' in a mortice-like joint between the cuneiforms and thus moves relatively little on loading. Thus, the intermetatarsal parts of the DTML express variable flexibility to reflect this asymmetry of motion between the metatarsals. Not only does the spreading of the metatarsals tension the DTML but rotations around the metatarsals shaft also tightens and stiffens it. The fact that rotations between metatarsals and spreading distance will tension the DTML means that on uneven ground, the forefoot will stiffen under load, especially if the metatarsals rotate in opposite directions to each other or move up or down in opposite directions within the sagittal plane. This allows the forefoot to stiffen under many terrain conditions. *(Permission www.healthystep.co.uk.)*

It is suggested that 40% of the longitudinal stiffness arises from the structures that cross the transverse foot vault (Venkadesan et al., 2020). As a result, the human foot vault can utilise extensive passive and some active stiffening mechanisms to match environmental changes such as rough uneven terrain surfaces and minor perturbations to the stability of the base of support through alterations in the transverse span distance of the foot. Through this mechanism, postural stability via the principles of biotensegrity can be maintained. The transverse plane seems the most logical 'target' to induce stiffening, as the transverse plane with its small span distance offers an easier structure to create and maintain stiffness within by adjusting its curvature radius.

The function of the viscoelastic asymmetrical conical foot vault

A permanently stiff foot would be able to provide an acceleration beam, easily allowing plantarflexor power-derived energy to cross to the forefoot, but it could not provide for sufficient energy dissipation. A permanently compliant foot would allow for impact energy dissipation but would be unable to transfer energy across itself for acceleration, risking stress overload of the tissues at the midfoot and forefoot. It takes both properties to achieve effective energy dissipation and acceleration. Modelling the foot as a rigid beam results in too great an estimation of ankle power, but also an underestimation of energy dissipation during gait (Farinelli et al., 2019). The foot can transfer from a state of relative stiffness to one of relative loading compliance and then on to a stiffer state for acceleration, and this is an important ability. Indeed, being stiff at initial impact means the release of compliance can be controlled, while having some ability to provide flexibility during acceleration means loading energies within tissues can be moderated.

The foot utilises a cycle of transitional preloading partial stiffening before forefoot impact through vault raising via digital extension and selective muscle activation. This can be reduced during forefoot loading to increase compliance, allowing the vault curvature to reduce and stiffness to be lost with energy dissipation. Throughout midstance, vault curvature reduction, lengthening, and widening causes gradual stiffening via stress-stiffening properties influenced by loading rate, while this strain within the connective tissues dissipates and stores energy. Finally, the foot starts to rapidly stiffen during the late midstance into terminal stance transition, changing the foot's elasticity to aid in acceleration of the CoM as a propelling rigid beam. This requires the forefoot connective tissues to stiffen the forefoot, while the extrinsic and intrinsic muscles contract. Muscle activity and elastic recoil from active and passive soft tissues, including the plantar aponeurosis windlass effect, reset the foot vault during late terminal stance in preparation for swing as the foot moves back towards a state of compliance around toe-off (see Figs 3.1.3 and 3.1.6b).

This changeability should reach its best functional compromise in behaviour as an energy dissipator and stiff beam-like structure when most needed. The foot can then return to being more flexible approaching preswing in order to start to pretension again prior to contact through the windlass mechanism via both digital extension (Caravaggi et al., 2009) and tibialis posterior and peroneus longus activity (Kokubo et al., 2012). The mechanics and the anatomy of this process are discussed in detail later.

Span distances within the osseous structure of the foot permit the foot vault to act effectively as an encastre beam or cantilever when needed using muscle activity, connective tissue stiffening, and/or vault curvature to influence the stiffness levels required under load. The ability to allow deflection deformation through curvature loss-induced mobility within the foot improves energy dissipation and surface contouring at loading to improve stance stability and balance. However, in midstance, the same process induces stiffness through connective tissue stress-stiffening behaviour. The engineering principles of span distances suggest that foot stiffness is most easily modified through the transverse vault profile (Venkadesan et al., 2017a, 2020; Yawar et al., 2017), and that longitudinal vault stiffening occurs in part through coupling to transverse forefoot stiffening (Yawar et al., 2017) and muscular activity (Kokubo et al., 2012; Farris et al., 2019, 2020).

Forefoot metatarsal head spreading apart or splaying, tightens the deep transverse metatarsal ligament and stiffens the transverse arch first, then stiffens the foot proximally through and into the longitudinally spanning vaults within the sagittal plane (Yawar et al., 2017). The longitudinal span distance deflections are restricted during stance via the passive stress-stiffening mechanisms of the myofascial tissues. This explains why most foot lengthening is reported just after forefoot loading as connective tissues pass through their toe (crimp-straightening) and linear elastic stages. The rate of lengthening decreases during late midstance as connective tissues pass out of their linear elastic phase and the plantar muscles increase contraction. This is occurring when most body mass is being loaded on the midfoot to forefoot. The foot vault thus becomes flattest prior to heel lift (Hunt et al., 2001; Stolwijk et al., 2014; Bjelopetrovich and Barrios, 2016; Takabayashi et al., 2020).

This mechanism means that longitudinal span length increases can enhance some foot stiffness through connective tissue viscoelastic stress-stiffening mechanisms. It is likely that similar stress-stiffening mechanisms associated with forefoot splay width increase at the level of the deep intermetatarsal ligaments which are responsible for initiating the enhanced transverse vault stiffness prior to heel lift. These stiffening mechanisms prepare the foot for appropriate loading as a cantilever at heel lift so as to permit ankle plantarflexor power transfer to the forefoot. In so doing, the force–velocity behaviour of muscle contraction and the time-dependent properties of connective tissue can set up compensatory mechanisms for gait velocity. The return of the foot's full vault profile begins just after heel lift via midfoot plantarflexion (Holowka et al., 2017) and through the action of the plantar intrinsic and extrinsic muscles (Ferris et al., 1995; Kokubo et al., 2012; Farris et al., 2019, 2020) until stiffness of the foot can be relaxed. This is in combination with the windlass action via both passive digital extension and plantar aponeurosis tightening (Hicks, 1954; Ker et al., 1987) which becomes more significant a little after heel lift. Therefore, the plantar aponeurosis initially lengthens during heel lift to start shortening a little later (Farris et al., 2020; Welte et al., 2021).

The plantar intrinsic muscles do not provide significant resistance to longitudinal beam deflection. However, they do influence foot stiffness (Farris et al., 2019), probably through actively assisting passive connective tissue stiffness tension-induced joint compressions across the plantar foot. It is possible that much of the energy stored within the foot's soft tissues as potential energy is used to help reset the foot's profile, much like the energy stored within the plantar fat pads resetting their material properties on offloading. Accelerating ankle plantarflexion power is generated primarily through triceps surae-induced muscular energy loaded into and released from the Achilles assisted by elastic recoil from other plantarflexor tendons, rather than through soft tissue recoil from within the foot's passive connective tissues.

3.1.7 Stiffness and pes planus, pes cavus, and the 'normal' foot vault

Rather than approaching certain foot profiles as being abnormal and potentially pathological, it might be more helpful to approach them as variations on a continuum with very flat (uncurved) feet at one end and very high-vaulted, extremely curved feet at the other. In isolation of other factors and in terms of stiffness, the structural curvature behaviour of these feet would be the flatter the vault, the higher the 'compliance potential' and the higher the vault, the higher the 'stiffness potential'. Intimately related to this property is passive soft connective tissue mobility which also has effects on compliance and stiffness locally around joints and globally across whole-body anatomy. The more hypermobile the foot, the greater the compliance while the more hypomobile, the stiffer the foot. Finally, there are the muscles to consider which introduce control and adjustability into the system. The weaker the muscles, the less ability there is to control stiffness levels. The more the muscles are in a state of high passive tone that restricts motion, the less their ability to permit compliance. Muscle strength is required for energy dissipation through energy buffering which will be deficient in weak muscles. Thus, strong muscles functioning well at a variety of lengths are desirable.

Foot vault profiles are therefore an important factor in compliance and stiffness through curvature mechanics within a framework of several factors. How stiff or compliant a foot behaves is dependent on all these factors. Thus, a hypermobile cavoid foot with weak muscles may have more in common mechanically with a hypomobile pes planus with moderate muscle strength than it will with a hypomobile cavus foot with strong muscles. The former two feet may present with more comparable mechanical pathological risks in more similar locations than will the strong hypomobile cavus foot, despite the latter looking more like the hypermobile cavoid foot nonweightbearing.

Feet behave in a biomechanically variable manner between individuals, an observation which is clearly demonstrated by research. Pathology is not consistent when the foot vault is used as the only factor determining foot type. Modelling of vault profiles has indicated that high-vaulted feet are generally subjected to increased stress and strain within the plantar aponeurosis and metatarsals with increased forefoot pressure (probably in part from reduced surface contact area). Conversely, planus feet experience increased stress and strain within the calcaneus, navicular, and cuboid, possibly reflecting different impulse patterns (Sun et al., 2012; Cen et al., 2020). Pathology may relate to loss of normal impulse loading patterns that certain degrees of foot flexibility or stiffness demand. Thus, avoiding high rearfoot vertical collisions would seem wise for rigid feet, and individuals should be expected to develop a gait pattern and an anatomy that reflect this need. Rigid feet are more likely prone to injury through vertical loading than are more compliant feet who will express more horizontal (transverse) plane component forces (Fig. 3.1.7). Midfoot articular stresses and strains are expected to increase the forefoot abduction moments that associate with greater vault compliance, affecting the type and location of injury and possibly helping to explain the differences in 2nd and 3rd metatarsal fractures as reported by Dixon et al. (2019).

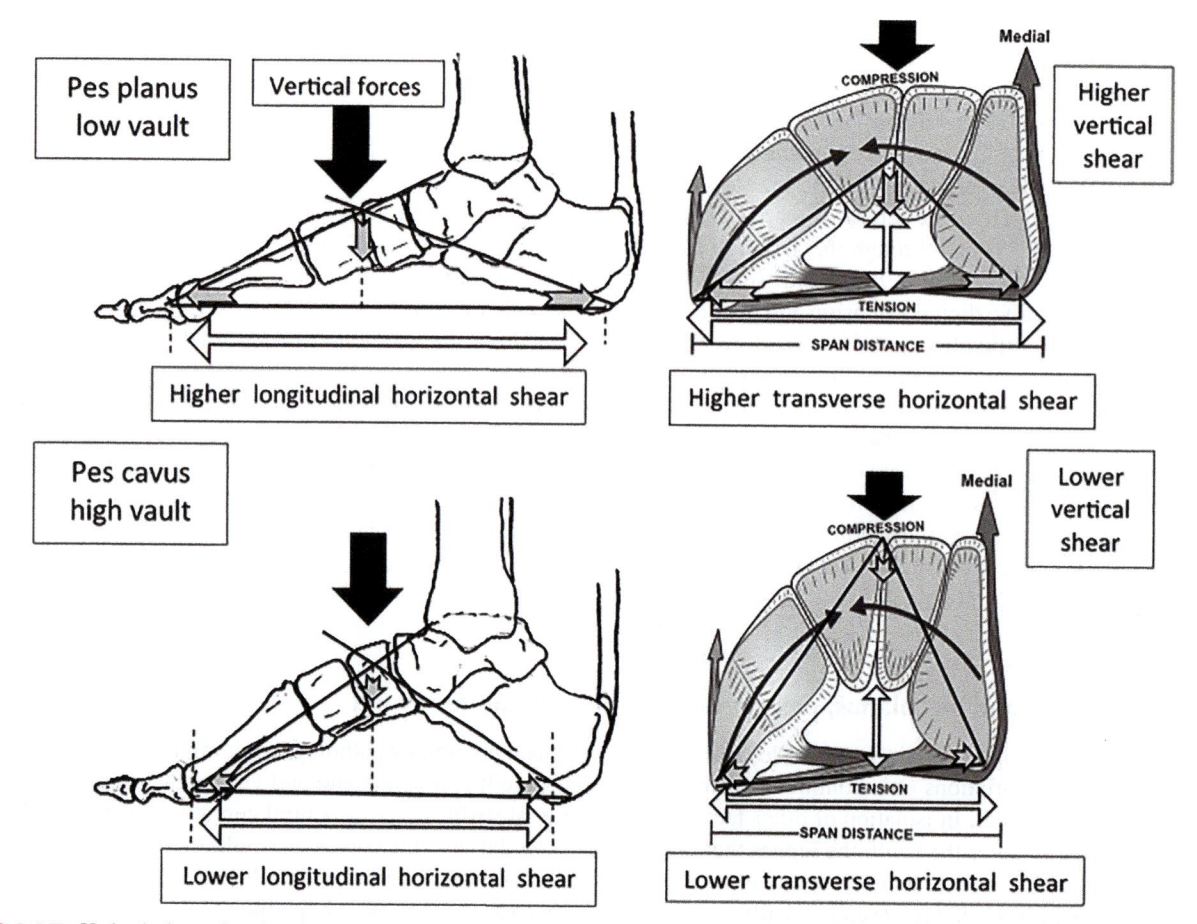

FIG. 3.1.7 Under the laws of engineering and the principles of biotensegrity structures, triangular-shaped components are best at resisting shear strains. They are the strongest shape in nature. They can be seen everywhere in human engineering designs and within biological structures once one becomes aware of this. Planus, low-vaulted feet offer less fully formed triangular structures and can shear more easily under vertical loads, allowing the foot to deform relatively easily. This is good for dissipating vertical forces into shear strain, both horizontally and vertically, but is not so good for storing energy and creating stability within the foot. Stiffer high-vaulted cavoid feet through their high calcaneal inclination and metatarsal declination angles offer more pronounced curves and relatively short spans. These form almost complete equilateral triangular structures both longitudinally and transversely across the foot vault. Such structures resist shear strain very well, which means they also resist vertical compressions as they are unable to deform easily (as long as their tension cables remain tight). Thus, they are poor at dissipating vertical loading energies. Therefore, every foot vault profile is to some degree a compromise to loading, dissipating, and storing energy. Human locomotion should demand both stability and energy dissipation. Thus, a vault avoiding an extreme profile is more desirable, but it will reflect lifestyle demands and soft tissue behaviour (tension cable tightness). Each vault height increases or decreases the risk of certain types of pathology, but nature cannot provide a vault that is capable of avoiding injury in all circumstances. *(Permission www.heathystep.co.uk.)*

Differences in compliance and stiffness properties are going to change the kinematics of gait. For example, during late midstance, ankle joint dorsiflexion stiffens the ankle, Achilles, and plantar aponeurosis (Huang et al., 2018). A hypomobile cavoid foot may experience earlier stress-stiffening of connective tissues and less vault curvature flattening, preventing foot lengthening. This risks an increase in osseous metatarsal and midfoot articulation bending moment stresses as the soft tissues are less able to dissipate loading forces by utilising strain. The premature stiffness of the foot and ankle could also initiate an early heel lift in order to maintain body CoM anterior translation speed. Early heel lift will increase the resistance load during heel lift as the body's CoM will not have progressed as far anteriorly due to the restriction of ankle dorsiflexion. An early heel lift, as a result of its increased resistance, will also heighten the dorsiflexion bending moments on the foot and the plantarflexor power necessary to induce the heel lift and thus increase the forefoot GRF. A cavus foot may also offer a smaller forefoot surface contact area due to the increased metatarsal declination angles found in cavus feet, and such feet may also demonstrate reduced forefoot splay as a result of concurrent hypomobility. Both features will increase peak pressures on the forefoot.

It is the ability to apply, control, and adjust the mechanical properties during foot function that highlights the potential significance of the vault. As a result of the effects on compliance and stiffness via muscle function, connective tissue viscoelastic properties, and vault curvature adjustment changes, the functional behaviour of the foot is explained. This gives an insight into potential pathological mechanisms and dysfunctions due to individual foot strengths and weaknesses within the mechanics of the soft tissues and the osseous alignments. Research has often tried to link pathology to foot vault shape rather than looking at a specific pathology in a specific location and then investigating what factors may be influencing their development. Further investigations may be able to assign the foot vault as being an influence on certain pathological risks by altering the pattern of stance phase cycles of compliance and stiffness needed. However, it is unlikely that foot vault profile alone is important.

One consideration highlighted by looking at foot vault profiles is that midrange-vaulted foot profiles and midrange tissue flexibility combined with good muscle strength present the most easily adjustable 'centre ground' to provide compliance and stiffness adjustability. Developmental processes should direct the foot to develop this 'centre ground' morphology, unless lifestyle requires or causes something a bit different to be developed. Should such environmental 'need' be presented during growth, then the foot has the developmental plasticity to react to lifestyle influences which could include many environmental mismatches such as footwear. This is likely to explain much of the differences noted in the behaviour of habitually shod and unshod feet.

3.1.8 Mechanical constraints on the foot's role

Foot size

The size of the foot is an important mechanical constraint. A large, strong foot might be perceived as an effective way to deal with high force applications and improve stability during stance phase. However, this is not an option for humans as the foot's size is limited by the laws of energetics. Most terrestrial animals actively swing their legs forward under their body because of the energetic benefits this has over any other options such as lizard-like side-to-side body flexions (Kuo, 2007). The laws of angular momentum state that moving a mass further from the axis of rotation takes more energy to initiate and terminate motion through inertial effects. This mechanical law makes the size of the distal limb segments very important as it is desirable to have the CoM of the segment closer to an axis of rotation. Segments descend in size the further they are positioned from the trunk, and muscle mass distally is bulked as near as possible to the joint axis above. Thus, the large posterior muscle bulk within the calf of the lower leg of humans is an unusual feature for a terrestrial animal.

Having a relatively large plantar surface, such as that found on the plantigrade human foot, places a disproportionate transport cost on the human foot during swing phase. The lower leg's large muscle bulk compounds this issue, yet this is essential for human bipedal acceleration mechanics for the ankle plantarflexors are necessarily placed here for acceleration power of heel lift. The torque required to accelerate the mass of the foot is proportional to its moment of inertia. Thus, the mass of the foot is proportional to the increased muscle power required to move it and the metabolic power consumption that results from this motion. Therefore, there is a constraint on the size that feet can be and why the mass of a shoe is important, especially in sports performance and particularly when the athlete is fatigued. In gait models, metabolic power consumption as measured by oxygen consumption is increased by 20% by distributing 5% of body mass to the ankles from the waist (Venkadesan et al., 2017b). The swing phase foot makes up 30% of the swing limb's metabolic transport costs (Doke et al., 2005). Thus, locomotive costs to achieve weight transfer of the foot limit its size.

However, the majority of the foot's metabolic energy consumption is not derived from swing, but from the stance phase in making up for the loss of energy dissipated during surface collisions (impacts). A larger foot would also result in higher collision metabolic costs because it causes the limb to have a higher moment of inertia at impact (Venkadesan et al., 2017b). Higher gait impulses would require the addition of material to increase bone strength and increase the cushioning capabilities. The addition of extra tissue mass may adversely affect energetics and increase the forces involved during impact disproportionately by increasing the inertia of the leg. Making a foot out of materials with high strength and low density is energetically ideal, although whether the biological materials used in the foot are 'optimal' for this role is a fundamental area for future research (Venkadesan et al., 2017b) for they may show that they are simply just the best available.

Neural transmission

Another constraint is linked to the rate of neural impulse transmissions that cause a neural time delay between collision and response in order to set the foot's damping and stiffness properties correctly. As a foot increases in weight, its distal mass effect on momentum limits the ability of the nervous system to control the leg through sensorimotor feedback with resultant implications to impact, terrain irregularities, or other gait perturbations (Venkadesan et al., 2017b). However, the important role the foot plays in acting as a mechanical base and sensory interface with the ground outweighs any benefit of dispensing with a human foot totally by using just a smaller tip or pole-like structure (Venkadesan et al., 2017b). A hoof for a human-like erect biped is therefore not a viable option for both stability and sensory reasons.

The need for adaptable material properties

A foot must be able to modulate its ability to dampen energy through compliance but also possess the ability to stiffen so as to achieve energy return for acceleration. This demand for property changes presents challenges to energy consumption and stability and risks injury (Venkadesan et al., 2017b). The foot's ability to behave compliantly is essential during contact and loading impacts and in so doing prevents high energy-induced tissue injury, particularly during running and jumping. By acting as a compliant energy dissipation brake through structural deformation for the duration of impact, high energy loading time is extended, and thus peak forces are reduced. This is in much the same way as shock absorbers act in a car or how 'crumple zones' in high impact collisions are designed to protect the passengers from harmful energy. The ability to break impact into a two-stage collision of heel strike and then forefoot strike also reduces impact forces being applied all at one time (thus also decreasing the loading rate), requiring compliance in tissues during both events.

In contrast, the foot must be stiff during the acceleration part of stance phase in order to convert the energy generated around the ankle into useful positive work. This accelerates the CoM of the body and helps initiate lift of the trailing limb's mass forward towards the next step. This power is achieved from ankle joint muscle activity of the plantarflexors that have tendons passing around the ankle and into the foot (Kuo et al., 2005; Collins et al., 2015; Takahashi et al., 2017). Without a stiff foot, ankle power could not be applied to the support surface at heel lift. Instead, plantarflexor energy would be absorbed within the foot's compliance as the midfoot deformed (strained) under midfoot dorsiflexion moments and by the forefoot as it was spreading out in excess. Midfoot dorsiflexion energy loss is what occurs during midfoot break, reducing the power available for acceleration, whereas digital extension also dissipates some energy yet maintains an appropriate lever arm length for acceleration power application.

The ability to modulate stiffness is essential and indicates that the foot is an adaptive structure. A passive structure reliant on fixed material properties would be either always compliant, or stiff, or always somewhere compromised in the middle, whereas an adaptive structure is able to achieve multiple tasks. For a structure like a human foot, the term *temporal filter* is used (Venkadesan et al., 2017b). This means that it is a dynamic structure that transforms inputs into outputs over time. The foot also behaves as a *spatial filter* giving it the capacity to mitigate initial collisions, but it is also able to smooth out the effects of CoM deviations resulting from roughness in terrain. The plantar foot is also coated in glabrous skin characterised by a ridged surface (rete ridges). Within it lie Pacinian corpuscle mechanoreceptors sensitive to vibration. This cutaneous tissue sits over the viscoelastic soft tissue padding of the plantar foot. The proprioceptors of the musculoskeletal system are intimately associated with the multitude of muscles, tendons, joint capsules, and ligamentous structures of the foot that provide the neural outputs from it. Through sensorimotor feedback, the foot muscles adjust their elasticity throughout stance by setting up a highly responsive and corrective stiffening mechanism. The concentrated need for musculoskeletal proprioceptors may explain some of the foot's anatomical complexity as each extra ligament and muscle increases the sensorimotor information and adjustment repertoire.

The contact area of the foot and the support surface are influenced by both the foot's shape and the properties of the support surface. If the foot and the ground were both perfectly rigid, then the forces and torques transmitted to the ankle would depend only on the precise shape of the ground and the foot. The foot deforms in response to the nature of the support surface. This means that the amount the foot must deform varies with terrain-induced need and requires spatial filtering. Such an ability is the result of two mechanisms: soft tissue padding and musculoskeletal flexibility which is made possible due to the numerous joints within the foot. The viscoelastic plantar soft tissues filter the ground's unevenness in depth while the musculoskeletal flexibility filters unevenness further by spreading the foot out in width (Venkadesan et al., 2017b), motion associated with vault profile changes. The flexibility within the foot filters out uneven surfaces so that the whole foot conforms to the features of the support surface, allowing it to become changeably more or less prone. This fits with the philosophically justifiable definition of pronation as proposed by Horwood and Chockalingam (2017). Pronation of the foot is an ability that has been preserved in part from our ancestral species, where changes in shape to improve compliance were required for grasping and transport on the highly irregular surfaces of tree branches in arboreal environments.

A foot that has compliant 'softness' at initial contact can act as a shock absorber, but it can also 'mould' to the support surface. Soft cutaneous tissue can adapt to small terrain irregularities such as stones and small pebbles, but the foot must also be flexible enough to adjust to any planar inclines or declines encountered. The tied asymmetrical conical vault of the human foot gives profiles in the frontal, transverse, and sagittal planes that allow it to conform to the presenting support surface via increased pronation. Becoming adequately prone through compliance to match the needs of the kinetic and kinematic events initiated by terrain contact, even if involving high levels of pronation, is not hyper-pronation. This is because it is not in excess of what is needed to perform the task energetically and/or safely under kinetic and kinematic demands (Horwood and Chockalingam, 2017).

Vault adjustability

This adjustable vault profile, in conjunction with the muscular and connective tissue restraints and viscoelastic properties, can therefore flatten to permit compliance. However, pronation can also stiffen the foot profile through passive connective tissue stress–strain properties. The resting skeletal profiling of the foot vault has a role to play in this compliance–stiffening relationship. Stiffening capability allows the asymmetrical conical foot vault to act as a spring-like structure, brought about by an increase in the flattening of the foot's curvature that stretches the connective tissues. Increasing the vault profile and twisting the connective tissues across the foot also increases its elasticity/stiffness through increased curvature stiffness and connective tissue twisting.

Thus, increasing stiffness can be created at both the increased vault height profile end of the range and also at the decreased vault height profile range, with variable degrees of stiffness existing between. Certain key elements are highly important in maintaining the foot's elastic response. These are the transverse metatarsal ligaments through their effect on the transverse contouring of the vault, the muscles and their tendons that span the vault, the plantar fascia (aponeurosis) via its windlass effect, and the longitudinal ligaments and other connective tissues. The transverse metatarsal elements seem to surpass the importance of the longitudinal stiffening elements of the vault, at least for initiating heel lifting stiffness levels (Yawar et al., 2017).

Back to the transverse vault profile in foot stiffening

The adaptability of the foot's shaping during locomotion may centre on factors within the forefoot rather than the rearfoot. The role of the transverse plane profile of the vault in foot stiffness can be appreciated when you consider that it is easier to stiffen a thin elastic object by curving it across a narrower plane than it is for curving it across a longer plane. Curling a bank note transversely drastically increases its stiffness, whereas trying to do so longitudinally still leaves the bank note limp (see Fig. 2.4.2a in Chapter 2 of 'Clinical Biomechanics in Human Locomotion: Origins and Principles'). Thin structures are easier to stretch in-plane than out of plane, because transversely increased curvature couples a restriction to longitudinal bending. Flattening of the transverse curve couples to increased freedom in longitudinal bending.

In the foot, this increased transverse and horizontal curvature equates to a spreading or splaying of the distal metatarsal heads and the coupled change in sagittal freedom in the midfoot due to sagittal (longitudinal) plane vault height lowering. On weightbearing of the forefoot, the transverse vault splays (widens) which places tension on the transversely orientated deep intermetatarsal ligaments, with the amount of forefoot splay in part dictated by the hardness of the support surface. This splay stiffening of the forefoot increases through stress-stiffening principles as the GRF moves anteriorly with the CoM passing over the foot. Metatarsal splaying increases the transverse/horizontal curvature of the forefoot and is estimated through modelling to dramatically increase stiffness in the foot by nearly 100% (Venkadesan et al., 2017a). Thus, it is likely that altering transverse curvatures across the vault profile of feet also significantly influences foot-stiffening capacity. These transverse curvature changes are certainly at play in altering kinetics and kinematics within different foot types such as planus and cavus feet, rather than just the longitudinal profiles. The significance of the overall vault profile once again returns to the effect that foot profile has on the compliance–stiffness parameters of the foot; but transversely, not just longitudinally.

Summary of constraints

The foot is therefore limited in size and weight by energetics. It must start stance with the ability to provide a compliant structure to dissipate impact energy and conform to the terrain. The foot must end midstance as a stiffer structure in order to transmit forces into the ground for acceleration power. It must achieve this through changing from one structural role to another to improve energetics and while changing from one to the other, it is advantageous to be able to store potential energy. This is achieved by deforming the foot vault by lowering it to tension the connective tissue is such a way that it stiffens the foot as stance phase progresses. The stretching of connective tissues become a source of stored elastic energy while also stiffening it. Such stiffening is attained through passive beam deflection mechanics during midstance. Indeed, raising the vault before heel lift to generate stiffness through muscular contraction alone would use far more energy than simply utilising passive stress-stiffening and strain rate principles, which are achieved through vault lowering and lengthening that develop the lowest vault profile around heel lift (Hunt et al., 2001).

During terminal stance, stiffness can be further increased just after heel lift to create a stabilising forefoot GRF that permits ankle muscle power to accelerate the CoM towards the next step. This adds extra energy to the momentum of the next step (Ruina et al., 2005). The foot must achieve these mechanical roles in walking while at the same time permitting an inverted pendulum-type mechanism pivoting over the foot at the ankle (Usherwood et al., 2012). However, during running, the foot is utilised as part of a spring–mass mechanism (Hamner et al., 2010).

During a range of walking speeds, the foot and ankle are able to produce near-zero or negative net mechanical work when the foot is compressed and recoiled along the longitudinal axis of the lower leg (Hedrick et al., 2019). This is achieved through a complex of mechanisms that can change stiffness through the foot's structural, hierarchical, viscoelastic anatomy (Fig. 3.1.8a).

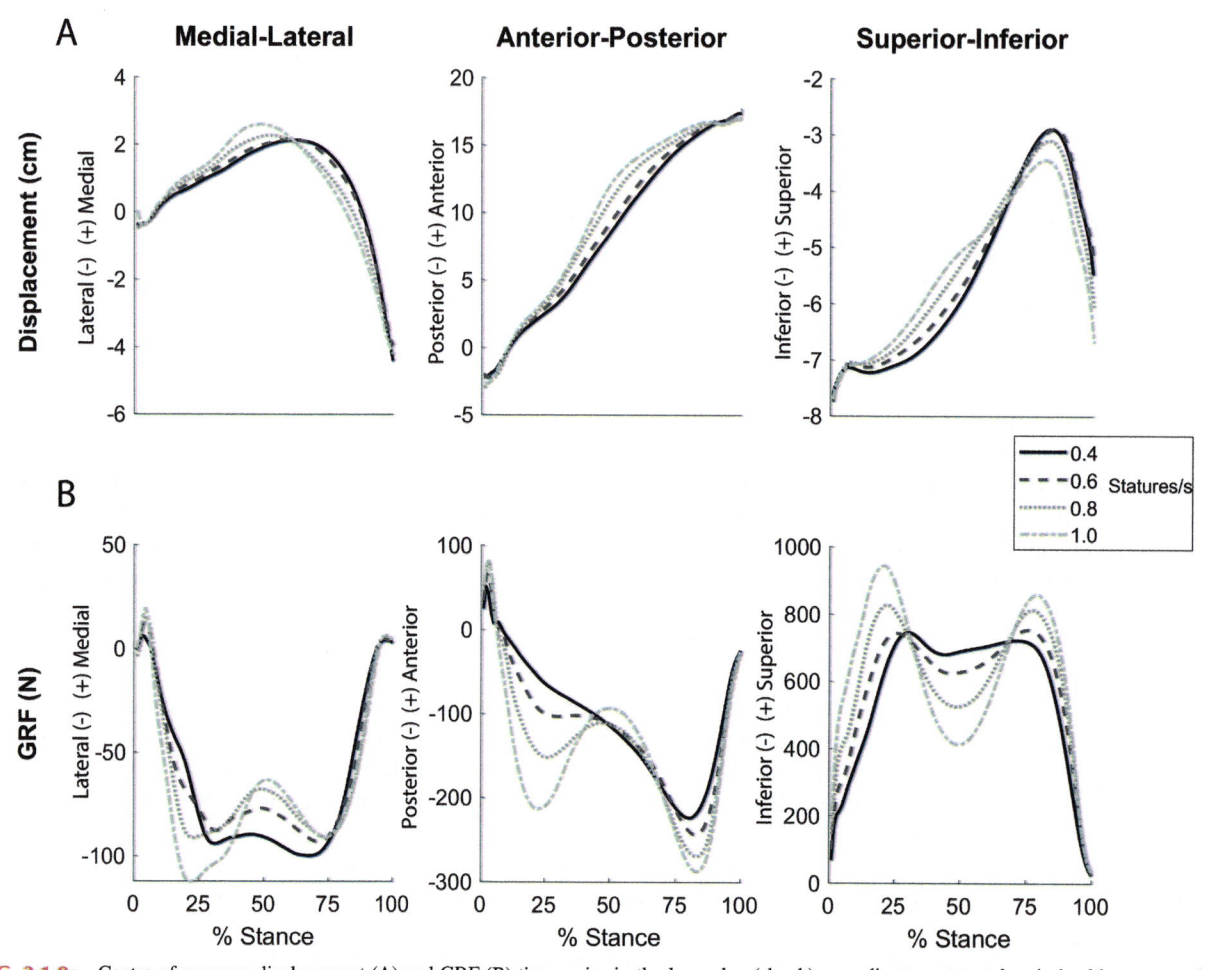

FIG. 3.1.8a Centre of pressure displacement (A) and GRF (B) time series in the lower leg (shank) co-ordinate system taken in healthy young adults during walking barefoot. The different lines represent different speeds, with solid line slowest (~7 m/s–0.4 statures/s), dashed line indicates next slowest (~1.03 m/s–0.6 statures/s), dotted line faster speeds (~1.4 m/s–0.8 statures/s), and dash-dot line fastest (~1.7 m/s–1.0 statures/s). Increasing gait speeds increase foot stiffness (elasticity). Compare these graphs to the vertical GRF at different walking gait speeds reported by Stolwijk et al. (2014) that shows similar results (see Fig. 1.2.4c). *(From Hedrick, E.A., Stanhope, S.J., Takahashi, K.Z., 2019. The foot and ankle structures reveal emergent properties analogous to passive springs during human walking. PLoS ONE 14 (6), e0218047.)*

The primary mechanisms that influence foot function include:

1. Gait speed that influences time-dependent properties. Increasing the loading rate stiffens the foot while decreasing it reduces stiffness. It also alters force–velocity relationships that influence muscle contraction power in that faster loading rates require more rapid contractions that are less able to produce as much muscular force.
2. Stress–strain behaviour through increased connective tissue lengthening. The greater the lengthening of the connective tissues, the stiffer they behave.
3. Increased twisting of connective tissue. The greater the twist on the connective tissue fibres, the stiffer they behave. Thus, a forefoot that can move in the opposite direction to the rearfoot can influence connective tissue twisting, as can tendons that twist under certain joint motions like the Achilles during ankle dorsiflexion.

4. Increasing the radius of curvature of structure. The more curvature (smaller the radius) put into structures, particularly into the foot vault, the stiffer it behaves.
5. Increasing the active component tension (muscle strength) which can also increase connective tissue tension, joint compression forces, or create increased structural curvatures, thus generating further coupled stiffness properties.

By decreasing stiffness, collision energy can be dissipated. By increasing stiffness, energy can be stored. Through the foot's stiffness in the latter stages of stance, muscle-generated ankle power can be efficiently applied for heel lift and add power to acceleration into the next step without significant energy loss from the foot deforming through excess compliance. Sufficient MTP joint extension adds some compliance that moderates the ankle power, avoiding overstrain of the plantarflexor tissues. In this way, foot energetics achieves its negative–positive work balance (Takahashi et al., 2017; Hedrick et al., 2019) (Fig. 3.1.8b). This could not be achieved by a foot expressing a single mechanical property. Thus, a foot offering both compliance and stiffness at appropriate points during gait needs to be considered when designing prosthetic limbs and lower limb orthoses.

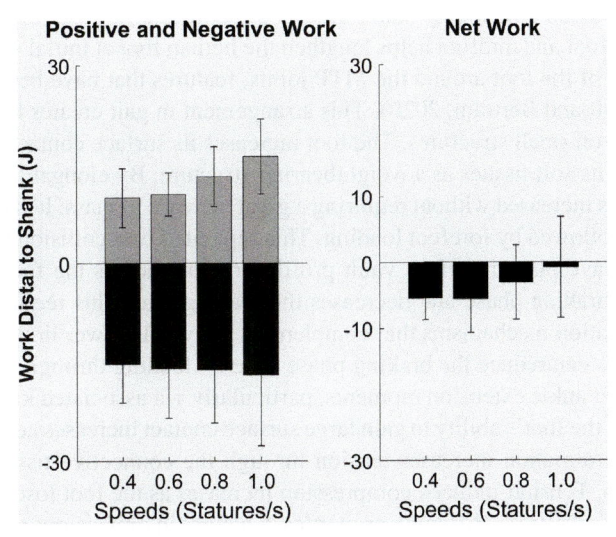

FIG. 3.1.8b Average mechanical work distal to the shank (lower leg) during stance phase of gait and the net work that it provides at different speeds (see caption Fig. 3.1.8a for statures/s speeds). Overall, negative work is slightly higher than positive work. The magnitude of positive work increased with speed, but not the negative work. This is probably a benefit of increasing foot and lower limb stiffness with gait speed. *(From Hedrick, E.A., Stanhope, S.J., Takahashi, K.Z., 2019. The foot and ankle structures reveal emergent properties analogous to passive springs during human walking. PLoS ONE 14 (6), e0218047.)*

3.1.9 Section summary

The human foot is constrained in its size by energetic demands from angular momentum. However, it is required to provide a base of support with the ground while dissipating impact/collision energies and yet also needs to be stiff and elastic enough to provide leverage off the support surface at the end of stance. An adaptable foot vault profile provides a number of benefits in that changes in the foot's curvature, length, and width dimensions that alter passive soft tissue tensions and muscle activity can manipulate the contact surface area as well as its intrinsic compliance and stiffness levels. The foot can stiffen before impact by raising the vault to then use reduction of this stiffness and increasing vault deformation to absorb impact energy. By impacting through a plantigrade heel before the forefoot collision, impact loading forces can be spread across two impact events, greatly extending the loading time and decreasing the loading rate.

Throughout the loading of a more compliant foot, the vault can decrease in height and increase in width, length, and contact area, increasing the base of support as the foot becomes increasingly prone to the support surface. This increase in dimension and skin contact area is likely to enhance proprioceptive information and reduce peak pressures. Through stress-stiffening principles, the viscoelastic properties of connective tissue will enhance foot stiffening as the vault increases its depression under stance phase bending mechanics, enabling stance-induced energies to be dissipated and stored as necessary.

Finally, the foot requires to function as a stiff cantilever to lift the heel around the MTP joints' fulcrum, permitting ankle power to be transmitted into a stabilising GRF. This forefoot stability facilitates an acceleration force that helps transfer the

CoM to the next step and imparts extra momentum into the loading of that next step. The stiffening from the increasingly pronated foot prior to heel lift should assist in pre-stiffening the foot, with increased forefoot loading leading to a transverse plane stiffening mechanism emanating from the deep transverse metatarsal ligament. This stiffening mechanism is enhanced by digital flexor extrinsic and intrinsic muscle activity, with further foot stiffening extrinsic muscle activity arising from peroneus longus and tibialis posterior more proximally. Transverse forefoot stiffening will couple with longitudinal stiffening through such muscle activity, including toe flexor function, and later through the windlass effect via increased digital extension. Getting these processes correctly activated during gait is essential for healthy locomotive energetics with each step requiring slightly different patterns of compliance and stiffness, reflecting the postural and balance needs at any given moment.

3.2 The foot's role in gait

3.2.1 Introduction

Constrained by size, the feet provide a relatively small base of support to a body-CoM that is positioned high over long lower limbs. Having a long rearfoot and midfoot helps lengthen the human foot at initial contact and terminal stance, while short toes allow easy extension of the foot around the MTP joints, features that have been influenced by the evolution of plantigrade gait energetics (Croft and Bertram, 2020). This arrangement in gait creates large vertical forces and displacements resulting in high impacts on small structures. The foot increases its surface contact area with the support surface by utilising its long calcaneus and its soft tissues as a weightbearing structure. By elongating the weightbearing surface posteriorly, the foot's contact area is increased without requiring a great increase in mass. It also allows the foot to impact in the two-phase event of heel strike followed by forefoot loading. This separated two-collision mechanical advantage is assisted through the presence of a concave plantar surface vault profile bridging across the foot. Vault-loading deformation on forefoot contact lengthens the braking phase and decreases the loading rate. This requires the rearfoot and the forefoot to have their own energy dissipation mechanisms that complement the whole lower limb's energy dissipation mechanics. At high running speeds, humans can reduce the braking phase time by loading through the forefoot alone. However, this requires greater knee flexion and ankle extension moments, particularly via associated knee soft tissue energy dissipation.

Once in single-limb support, the foot's ability to gain large surface contact increases adhesion to the support surface by it becoming more prone. This deformation increases tension through the connective tissues to maintain stabilising compression across the articulations. Tension-induced compression increases as the foot loses vault height and the connective tissue elasticity increases through its stress-stiffening properties. Changes in dimensions and associated stiffening increases most rapidly during early midstance (Stolwijk et al., 2014) with static testing indicating that the foot stiffens as loads on the foot increase (Bjelopetrovich and Barrios, 2016; Takabayashi et al., 2020). Connective tissue stiffness peaks before heel lift as the foot becomes maximally prone (Hunt et al., 2001) and this is accompanied by increased stiffening at the ankle through dorsiflexion and Achilles tendon stiffening (Leardini et al., 2001; Aubry et al., 2013; Chino and Takahashi, 2018; Huang et al., 2018; Liu et al., 2018; Taş and Salkin, 2019). The foot and ankle thus become loaded with connective tissue-stored potential energy through increased stiffness/elasticity, thereby also providing high foot stability prior to heel lift.

At heel lift, further forefoot transverse plane stiffening mechanisms initiate a chain of events involving distal forefoot connective tissue stiffness and proximal transverse plane midfoot stiffening from peroneus longus and tibialis posterior (Kokubo et al., 2012; Venkadesan et al., 2017a, 2020; Yawar et al., 2017). This allows the long digital flexors and the intrinsic foot muscles to further stiffen the foot longitudinally via compression (Farris et al., 2019). Eventually, stiffening through the windlass effect of the plantar aponeurosis supplements stability later during terminal stance rather than at heel lift, under increasing digital extension (Farris et al., 2020). The windlass effect likely enhances terminal stance midfoot plantarflexion through assisting the shortening of the vault, increasing vault height to re-establish the foot profile prior to toe-off. The role of the terminal stance foot is to assist ankle plantarflexor power in providing stabilising GRF to accelerate the CoM onto the next step. To achieve this, the foot must attain stiffness in order to prevent excess 'give' across the foot acting as a lever arm at the MTP joint fulcrums. Let us now consider these events in more detail.

3.2.2 The function and events of heel-toe walking

Heel collision, inverted pendulum, followed by heel lift acceleration seen in walking is the most energetically efficient of human gaits. At heel contact, the viscoelastic properties of the heel fat pad assist energy dissipation from the proximal joint muscles that are providing muscle–tendon energy buffering (Sutherland, 2001; Neptune et al., 2008;

McGowan et al., 2010; Di Nardo et al., 2013). Tibialis anterior activity in decelerating ankle plantarflexion following heel contact provides ankle muscle–tendon energy buffering and decelerates both forefoot impact peak force and the loading rate during forefoot loading response. The forefoot plantar fat pad further dissipates energy on forefoot loading. Forefoot impact energy is also dissipated via the effects of pre-tensioning the foot's skeletal structural frame before forefoot impact. The vault is raised and tensioned by the activity of tibialis posterior, peroneus longus (Kokubo et al., 2012), and the plantar aponeurosis windlass mechanism via foot extensor muscle activity (Caravaggi et al., 2009). This tension enables the vault to act as a spring-like shock absorber that reduces its elasticity, increasing its shock attenuation abilities on forefoot loading. As muscle activity decreases after forefoot contact, the reversed windlass effect occurs with digital extension reducing as the toes flex downwards towards the ground. This is accompanied by fairly rapid vault depression and foot elongation into the start of early midstance (Fig. 3.2.2a).

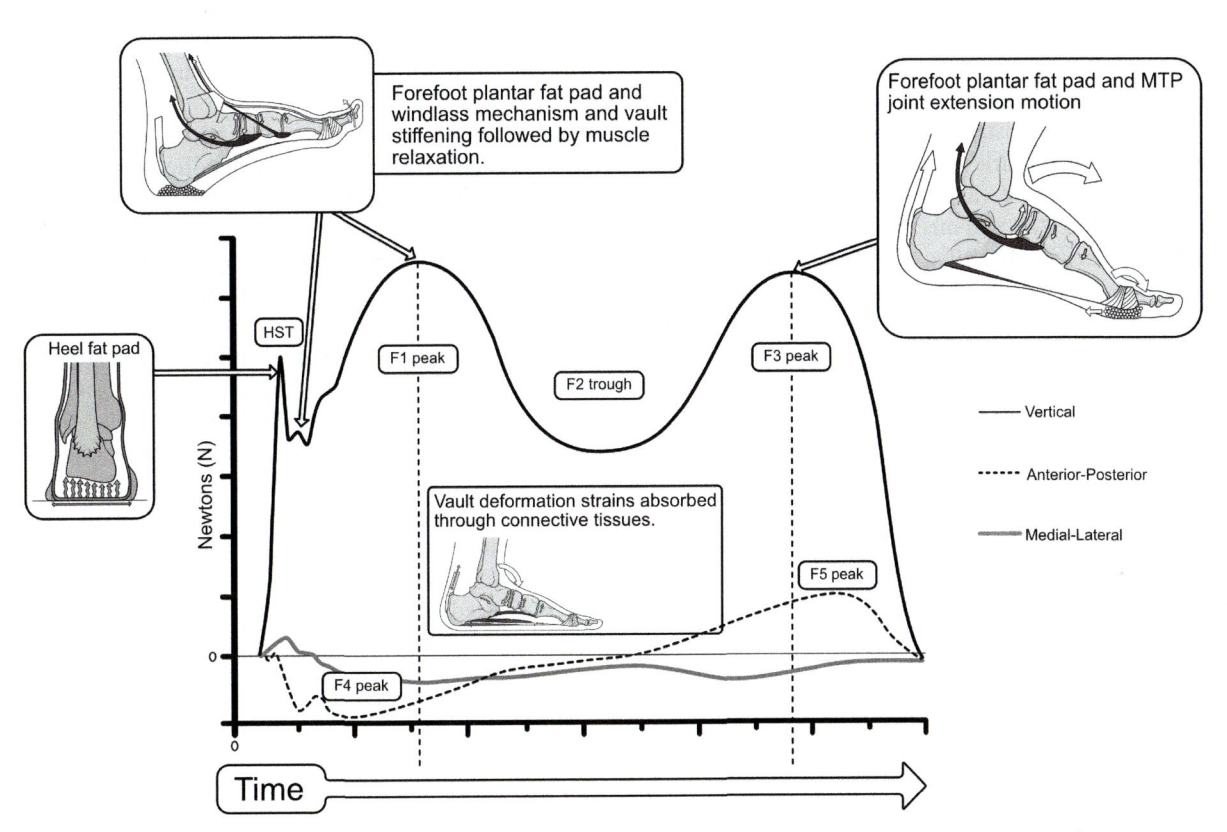

FIG. 3.2.2a Energy dissipation through the foot and ankle involves a number of mechanisms which can be related to events demonstrated on the force–time curve. At heel strike (heel strike transient—HST), the heel's plantar fat pad is active in dissipation with the extensor muscle to decelerate the descent of the foot towards forefoot contact. This braking muscle action also includes tibialis posterior and peroneus longus. Such combined muscle action actively stiffens the proximal vault as the extensors' activity also activates the windlass mechanism, tensioning and stiffening the forefoot fat pad via plantar aponeurosis tightening. On forefoot loading, the digits come out of extension as extensor activity largely ceases and tibialis posterior and peroneus longus activity decrease. This allows increasing levels of foot compliance through the vault and permits relatively free soft tissue straining across the vault and forefoot at and briefly after, the F1 peak. In the foot-loading period of gait, there is considerable net negative work done, aided by more proximal mechanisms of soft tissue resonance oscillations and knee and hip muscle–tendon buffering during flexion. Together, these mechanisms manage F1 and F4 GRF peaks. During midstance, vault compression, lengthening, and widening dissipates energy in all orthogonal planes. Weight-bearing energy is driven into the foot by ankle dorsiflexion moments during late midstance as forces rise out of the F2 trough. Gradually, decreasing energy dissipation is replaced by increasing energy storage under stress-stiffening mechanisms via the vault's connective tissues, including the plantar aponeurosis. Intrinsic and extrinsic muscle activity greatly influences this stiffening. At heel lift, the plantar forefoot fat pads stiffen yet continue to supply significant shear and compression energy dissipation, as does the extension ranges of motion of the metatarsophalangeal (MTP) joints. These mechanisms help prevent tissue stress overload during the F3 and F5 GRF peaks. However, overall, acceleration produces net positive work as muscular work dramatically generates energy while stored energy is released from the connective tissues more than it is being dissipated by them. This combined energy is utilised for acceleration, and some passes into the momentum of the next step. Pathology can upset these energy balances across the foot. *(Permission www.healthystep.co.uk.)*

During loading response, the GRF vector is primarily derived from vertical and posteriorly directed forces that produce the F1 and F4 peaks of the force–time curve. It is the F1 and F4 peaks of force (energy) that are being dissipated by the lower limb and foot at impact. From the end of loading response, vertical and horizontal GRFs are reducing. The need for compliance and energy dissipation are lessened as the lower limb enters single-limb support. GRFs start to fall as the CoGRF moves nearer to the ankle axis, decreasing angular momentum around the ankle, with vertical GRFs decreasing as the body's CoM starts to move away from the support surface through increasing limb extension. During early single support (early midstance), the foot is starting to rapidly decrease its level of compliance under relatively low midstance phase GRFs. The foot should demonstrate rapid changes in length during early midstance (Stolwijk et al., 2014) as it is loaded while compliant so that the connective tissues pass quickly through their toe region into their linear elastic zone and towards their stress-stiffening properties (as seen on a stress–strain curve). This rapid deformation process initiates increasing levels of stiffness within the foot's connective tissues throughout the rest of single-limb support.

By absolute midstance, the CoM of the body reaches its highest point through lower limb extension at the hip and knee. The CoM gains potential energy with this positioning and should lie roughly over the hip joint axis, but slightly posterior to the knee and the ankle axis. The CoGRF should lie roughly under the ankle axis and the GRF should be at its midstance phase lowest. Activation of triceps surae begins just before this point, ready for increasing activity during late midstance that generates powerful dorsiflexion moments across the midfoot (Murley et al., 2014).

After absolute midstance, the CoM begins to fall forward under the braking control of eccentric triceps surae activity (Di Nardo et al., 2013; Murley et al., 2014). Increasing loading through the Achilles tendon by ankle dorsiflexion and the fall of the CoM reduces foot vault height through beam deflection mechanics, increasing the strain in the plantar aponeurosis (Cheung et al., 2006) and other connective tissues. As the CoGRF starts to advance distally from the ankle axis, the foot and ankle increasingly stiffen. Combined passive and active stiffening mechanisms are driven by increasing forces generated by the fall in the CoM, rising triceps surae braking activity, and increasing angular momentum that is coupled to the lengthening distance between the ankle axis and CoGRF.

The ankle stiffens through articular congruency changes (Leardini et al., 2001) and the Achilles tendon stiffens with dorsiflexion through stress–strain properties and fibre twisting during late midstance (Aubry et al., 2013; Chino and Takahashi, 2018; Huang et al., 2018; Liu et al., 2018; Taş and Salkin, 2019). The CoM's fall produces accelerating ankle dorsiflexion moments, increasing forces across the foot vault which causes sagging deflection via a strong midfoot dorsiflexion moment. The distal calcaneus undergoes plantarflexion under increasing proximal vault loading, taking the talus and navicular inferiorly with it and thus inducing talonavicular and (to a lesser extent) calcaneocuboid dorsiflexion.

This vault deflection and forefoot splaying under late midstance loading across the forefoot causes the plantar surface of the foot to become increasingly prone, stretching connective tissues into increasing levels of stiffness. It is important for foot energetics that the sagittal plane deflecting bending moment is resisted, else midfoot break will occur at heel lift and acceleration power lost. However, deflection should not be totally blocked else foot vault energy dissipation cannot protect anatomy. The midfoot stabilisation process should occur under rising forces and increased foot stiffness deflection, which can be used to help dissipate the energies being applied through the foot during single support. Ideally, the increasing levels of elasticity from connective tissue stiffening can then partially store the midstance loading energies in the form of potential energy for acceleration.

There is a proportionate relationship between midfoot dorsiflexion and transverse foot splay where the amount of splay greatly affects the range of midfoot dorsiflexion (Yawar et al., 2017). During late midstance, the non-Hookean properties of the foot are clearly demonstrated, with stiffness increasing disproportionately in relation to dimension changes. Thus, disproportionate stiffness increases are achieved for little change in vault profile. The result is that the foot continues to stiffen while the rate of foot vault profile change decreases. This behaviour is supported by static and dynamic testing (Stolwijk et al., 2014; Bjelopetrovich and Barrios, 2016; Takabayashi et al., 2020).

Throughout late midstance, the ankle acts as a class three lever controlling the vaulting effect of the leg and thigh over the foot associated with the inverted pendulum mechanism (Usherwood et al., 2012). The reduction in vault height possibly aids this vaulting action throughout the late midstance of single-limb support, permitting higher ankle dorsiflexion angles to the support surface derived in part from additional talonavicular plantarflexion associated with navicular drop (Fig. 3.2.2b).

At the end of midstance, the foot vault should be at its stiffest and lowest profile. This allows the foot to fulfil its role as a semi-stiff beam and enables the power generated at the triceps surae to gain mechanical advantage through a class two lever system, utilising a stiff forefoot as a propulsive platform on a mobile MTP joint-formed fulcrum. The action of the triceps surae could change to concentric contraction, but usually the elastic energy storage and stiffness of the Achilles reaches sufficient power levels to overcome the reducing loads on the rearfoot for elastic recoil-induced heel lift. This power is converted to a large anteriorly directed GRF under the forefoot to propel the CoM to the opposite contacting foot (Ruina et al., 2005).

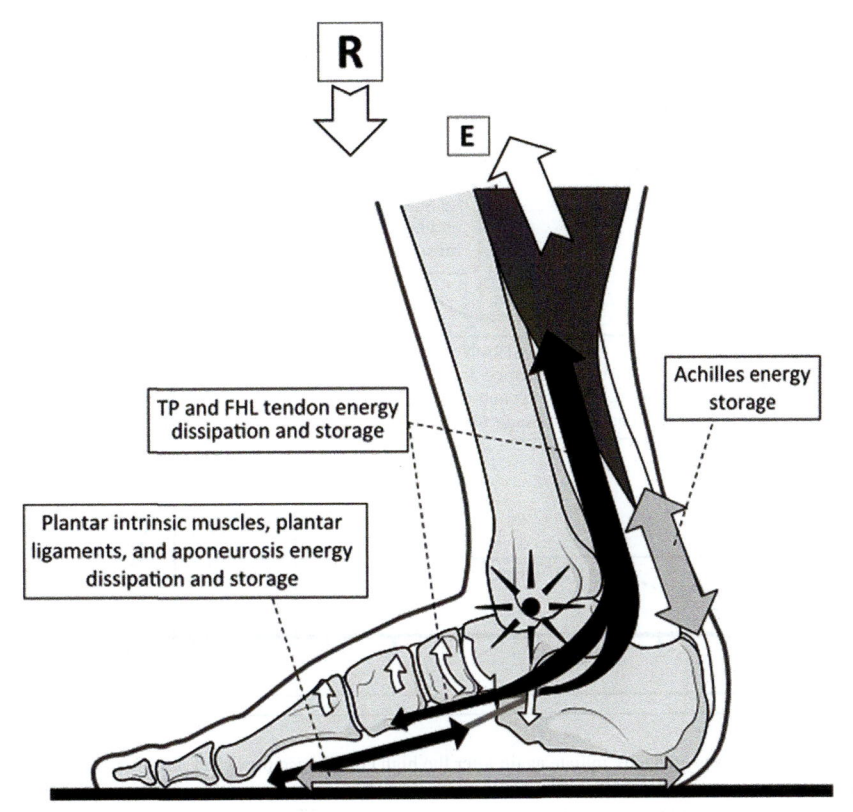

FIG. 3.2.2b The foot, via its vault profile, acts as a weightbearing flexible and superiorly curved beam, taking a supporting base role in a class three lever system rotating at the ankle fulcrum (star) during midstance. The triceps surae complex applies effort (E) at its muscular proximal attachments to the tibia and femur, slowing the acceleration of the CoM of the body which acts as the resistance (R). This effort permits and yet restrains the velocity of the dorsiflexion moments at the ankle, requiring the distal Achilles to be attached to a fixed object (the posterior calcaneus in its closed chain). As the distance between the distal and proximal attachments increases, the Achilles stretches and loads with energy *(dark-grey double-headed arrow)*. With the foot being relatively flexible under the dorsiflexion moment, forces are consequently directed into its vault. The vault is subjected to sagging bending moments, rotating the vault's bones (including the distal calcaneus) in its proximal articulations into plantarflexion and its distal articulations into dorsiflexion. This stretches both plantar ligaments and fascia, as well as the increasingly active plantar intrinsic muscles' tendons *(light-grey double-headed arrows)*. The amount of deflection is influenced by the vault curvature, which at higher curvatures during midstance will behave more stiffly. Obviously, soft tissue tone also influences the amount of deflection. Muscles such as tibialis posterior (TP) and flexor hallucis longus (FHL) *(black arrows)* demonstrate increasing activity in late midstance, which loads energy within their tendons from resisting both ankle dorsiflexion and vault deflection at the same time. Thus, feet with higher vaults and stiffer soft tissues during midstance resist bending moments more effectively, but risk losing the ability to use the deflection for energy dissipation during early midstance and for storage later. The vault deflection stiffens the connective tissue to store energy through the lengthening strains of the tissues' fibres, entering a stress-stiffened zone. This changes the foot into a semi-stiffened beam suitable for acceleration. The foot should reach maximal deflection and stiffening around the heel lift boundary when dorsiflexion moments and Achilles energy storage peak. *(Permission www.healthystep.co.uk.)*

The transverse plane curvature stiffness across the forefoot is increasing throughout late midstance. However, with heel lift, the increased forefoot loading force should induce further forefoot connective tissue stress-stiffening, producing greater forefoot stiffness (Venkadesan et al., 2017a; Yawar et al., 2017). This stiffness seems to be assisted by activity of the intrinsic foot muscles (Kelly et al., 2015; Farris et al., 2019, 2020), tibialis posterior and peroneus longus (Kokubo et al., 2012), and the long flexors (Ferris et al., 1995). The power generated, coupled through the stiff foot from the ankle plantarflexors, is now seen as the F3 and F5 peak GRFs (Fig. 3.2.2c).

Pedal stiffness is assisted in terminal stance by a plantarflexion, adduction, and eversion motion through the midfoot (Kokubo et al., 2012; Holowka et al., 2017) which increases vault curvature. After heel lift, terminal stance events occur under decreasing GRFs as the CoM of the body is transferring to the next step. This is aided by increasing ankle plantarflexion and coupled MTP joint dorsiflexion angles that tip the lower limb towards knee and hip flexion. The changing ankle plantarflexion and MTP joint extension angles will facilitate and couple midfoot plantarflexion in concert with more proximal joint motions such as those within the subtalar joint via its ligamentous elastic recoil. Midfoot plantarflexion should occur with midfoot adduction and eversion allowing the CoGRF to shift more medially to utilise the longer fulcrum

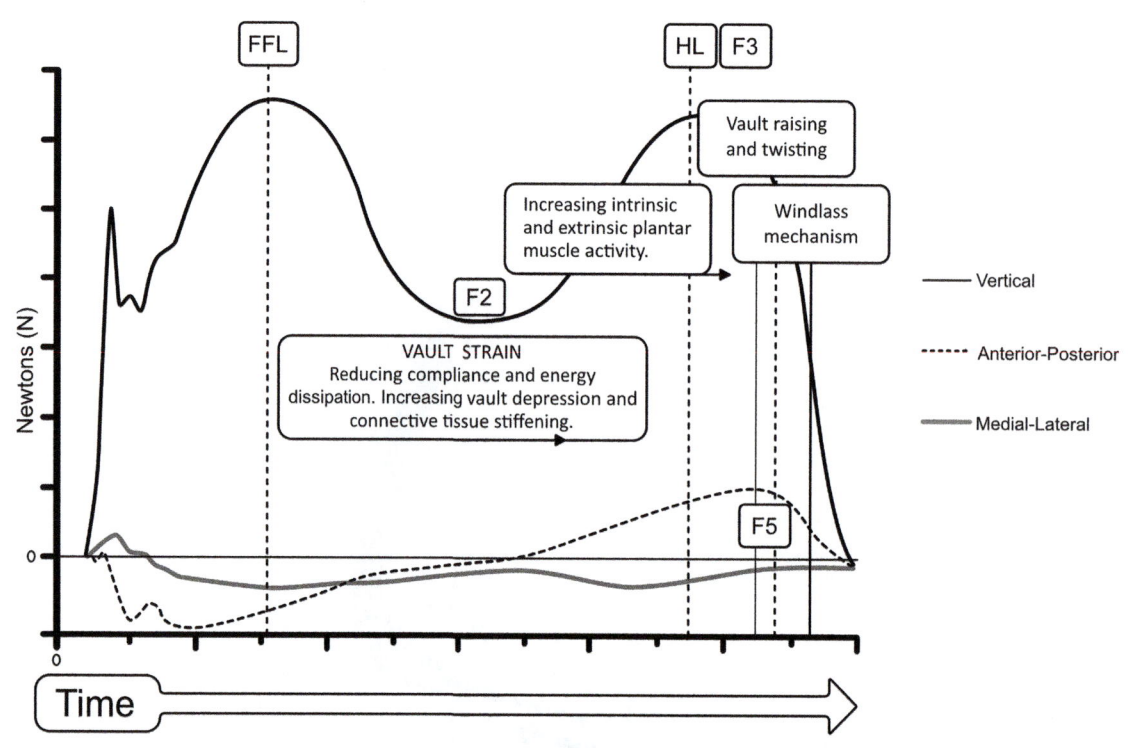

FIG. 3.2.2c Foot-stiffening mechanisms gradually initiate again after the high compliance presented just after forefoot loading (FFL) for energy dissipation and increased surface contact area. As the vault profile undergoes strain-inducing depression, lengthening, and widening, energy from body weight and ankle dorsiflexion forces are dissipated throughout midstance. However, after the F2 trough, energy is increasingly stored within the connective tissues. This energy storage mechanism includes activity via the tendons of the rising intrinsic and extrinsic plantarflexor muscles spanning the vault that increase their activity during late midstance. Thus, under passive and active mechanisms, the foot stores energy, although more proximal ligaments start to release their energy and recoil as the rearfoot increasingly offloads as the CoGRF moves anteriorly. At heel lift (HL), the foot is under high vertical forces of the F3 peak and yet is also at its lowest vault profile. Fortunately, active and passive mechanisms have conspired to turn this 'flattened' foot into a semi-rigid beam that can now only rotate easily at the MTP joints under ankle plantarflexion moments, releasing the plantarflexion power to apply a GRF to stabilise a stiffened forefoot on the ground. Muscle activity largely ceases from the plantarflexors (save those to the digits) soon after the heel lift boundary, but the elastic energy released from storage within the tendons maintains the plantarflexion power that aids vault twisting (adduction/eversion) and midfoot plantarflexion. These motions raise and stiffen the vault, aided by the windlass mechanism that begins to also stiffen the foot as muscle power wanes. This means that the foot maintains adequate stiffness until after the horizontal F5 GRF peak, before compliance increases at preswing. *(Permission www.healthystep.co.uk.)*

axis of the medial MTP joints that lie furthest from the ankle. This enhances the length of the external lever arm to the Achilles, providing the ankle power with a larger effort moment arm (Holowka et al., 2017). In the GRF's medial shift, peroneus longus likely plays a significant role through both midfoot plantarflexion and eversion, aided by peroneus brevis and with tibialis posterior providing the adduction moment.

Passive toe extension, under the stabilising resistance of the plantar intrinsics and extrinsics, enables the plantar aponeurosis to engage its windlass effect to enhance foot stiffness after heel lift during terminal stance (Caravaggi et al., 2010; Farris et al., 2020). The windlass effect will assist midfoot plantarflexion, aiding in re-establishing the vault profile before swing phase. It is likely that stored elastic potential energy within the connective tissues also helps in returning the foot to the 'offloaded' (nonweightbearing) vault profile during terminal stance. However, the assistance of the windlass will provide some foot stiffening when other stiffening mechanisms are reducing, such as from extrinsic muscle activities, as tendon (and other connective tissue) elastic recoil energy has already been largely utilised (Fig. 3.2.2d).

The importance of acceleration stiffness

Foot stiffness in healthy young individuals has been reported to affect the size of the impulse during walking. Asymptomatic, healthy, more flexible feet demonstrate larger impulses within the rearfoot and midfoot than do stiffer feet, whereas flexible feet have a decreased ratio of impulse in the forefoot compared to stiff feet (Cen et al., 2020). This would suggest that gait is utilised to avoid high initial loading impulses in stiffer feet, but these stiffer feet are more able to transfer the ankle power into a forefoot GRF than are flexible feet. Flexible feet may be better able to dissipate impact and midfoot

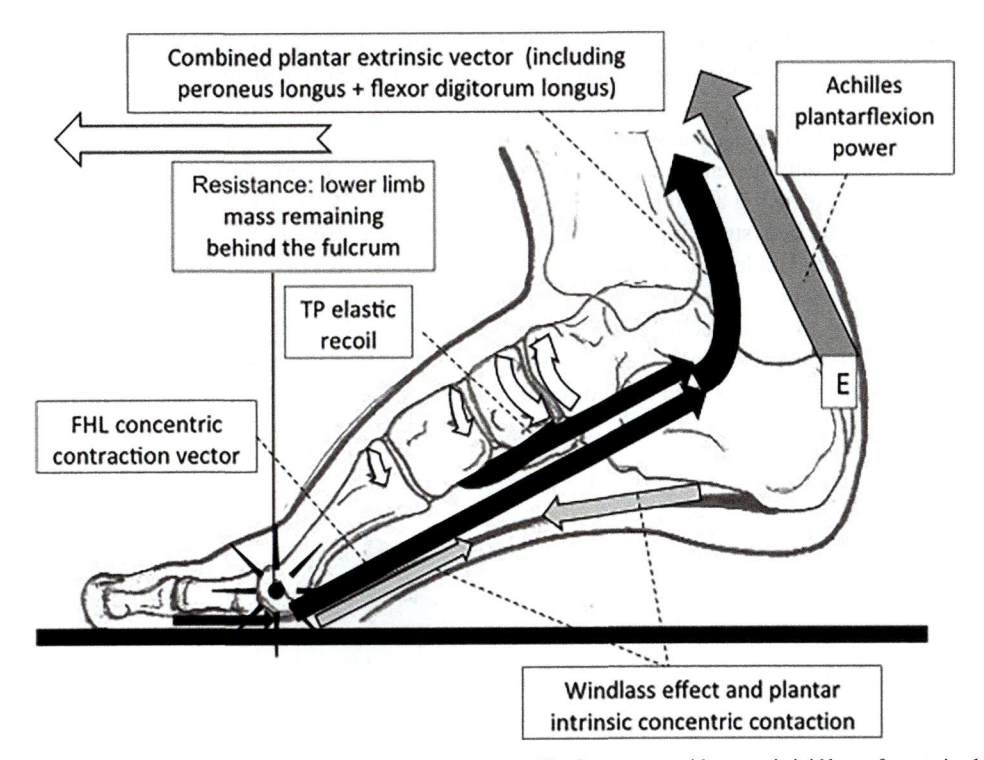

FIG. 3.2.2d At heel lift, acceleration is initiated using a class two lever system. The foot now provides a semi-rigid beam for rotating between extension at the MTP joint fulcrum (star) and ankle joint flexion. The Achilles attachment effort point (E) on the posterior calcaneus releases its plantarflexion power *(dark-grey arrow)* across the vault from the calf to the forefoot. The resistance is the remaining lower limb mass behind the MTP joint fulcrum, as the body's CoM should be anterior to the foot at heel lift and rapidly moving onto the next step's support limb. Any significant instability within the vault when raising the resistance will risk sagging deflection, preventing the foot forming an efficient cantilever to accelerate forward from. Heel lift releases both midstance-derived calf and vault energy to aid acceleration that will also help recontour the vault via midfoot plantarflexion, adduction, and eversion. These latter midfoot motions derive from the elastic tendon recoil of tibialis posterior (TP) and peroneus longus (on lateral side of foot), as well as continuing activity from flexor hallucis longus (FHL) and the other digital flexors. These vectors are represented by the black arrows in the image. The digital extension at the MTP joints soon engages the windlass mechanism, tightening the plantar aponeurosis to help stiffen across the plantar aspect of the foot, but the vault is also assisted in this by continued plantar intrinsic muscle activity driving aponeurotic shortening *(light-grey arrows)*. Twisting and raising the vault maintains a level of forefoot and vault stiffness via a higher curvature as the vault shortens and muscle power decreases with energy lost from the connective tissues. As body mass completes its transfer to the next step, foot compliance can increase ready for swing phase. *(Permission www.healthystep.co.uk.)*

impulses. However, the price is less stability to generate an acceleration impulse. Pathological variations of stiff and flexible feet may demonstrate different impulses, but these data give some early insight into what may be happening in differing phenotypic foot stiffnesses.

The series of events described here appears coherent, fulfilling the energetic efficiency requirements of the inverted pendulum and rimless wheel models, accounting for impact/collision energy dissipation and the elasticity necessary to achieve terminal stance acceleration. These processes explain how the foot should transfer from compliance to stiffness and return to compliance without creating perturbations within the gait cycle. Yet despite over one hundred years of studying foot function mechanisms, some aspects of the foot's mechanics still remain unclear. Further research is therefore required to confirm and investigate the processes discussed here before absolute conclusions can be made.

3.2.3 Achieving heel contact compliance

Stiff tissues cannot absorb sufficient energy on loading and are therefore more likely to become injured as a result (Fouré et al., 2012). During the weight acceptance part of stance phase, the foot is subjected to large forces initiated by the impact between it and the ground, thus being sandwiched between the body weight vector (BWV) and the GRF vector. Vertically and posteriorly directed GRFs are high and rise rapidly as the CoM continues to accelerate forward and down towards the ground throughout loading response. This creates impact energy that has the potential to cause tissue injury through tissue resonance and excessive deformation strains.

A very stiff foot would not be advantageous because stiff tissues resist lengthening and cannot sufficiently deform to dissipate energy; however, a stiffer foot has the ability to act more elastically and store more energy. Therefore, at heel strike, the foot must provide for dissipating energy through compliance, using combinations of biphasic soft tissues that can deform freely under load and muscle action that creates reducible stiffness through joints. The foot with the rest of the lower limb is able to supply muscle–tendon energy buffering during articular motion. High compliance is required during loading response, with GRF peaking at around 12% of the stance phase (Perry, 1992: p. 39) after which compliance can reduce. Energy dissipation in heel strike bipedal gait is achieved by five primary lower limb mechanisms: displacement within the viscoelastic fat pads, ankle joint plantarflexion deceleration, a tensioned deformable lower limb and foot skeletal frame, the leg's soft tissue wobbling mass, and through having muscle–tendon complexes acting as energy buffers. The foot is an integral part acting within these mechanisms, with its fat pads and reducible vault profile playing an important part in impact protection (Qian et al., 2010). Heel strike followed by forefoot loading sets up two distinct phases of energy dissipation compliance within the foot as seen related to GRF peaks in Fig. 3.2.3.

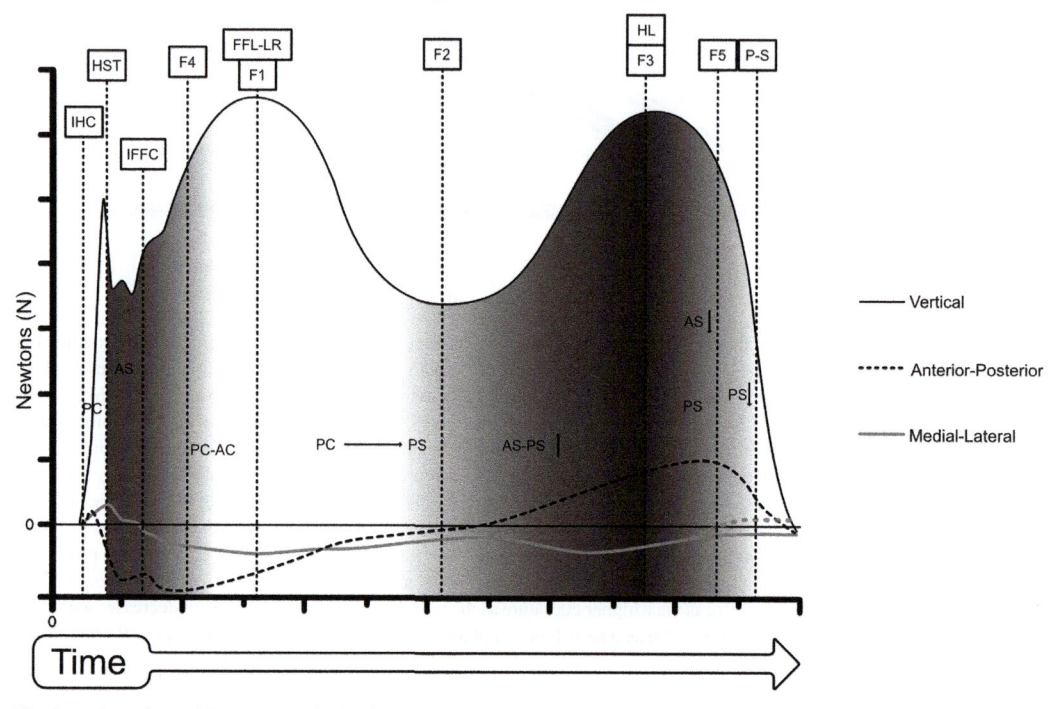

FIG. 3.2.3 The interplay of providing energy-dissipating (shock-absorbing) compliance, cushioning capabilities, energy-storing abilities, and foot-stabilising stiffness sets up a complex pattern during walking stance phase. No period consists of only one ability to provide either stiffness or compliance, but a state of net overall compliance or stiffness exists to manage forces within the foot. However, the logic of the pattern is easier to explain via consideration to the force–time curve. Periods of peak compliance are shown in white and net foot compliance increasing in lightening grey, while periods of increasing and net stiffness are displayed as darkening grey areas. At initial heel contact (IHC), the heel fat pad provides passive compliance (PC) under an inflexible rearfoot during the heel strike transient (HST). The midfoot and forefoot are undergoing a process of active stiffening (AS), ready to act as a forefoot impact shock absorber. This is achieved primarily via tibialis posterior and peroneus longus activity and the windlass mechanism. At initial forefoot contact (IFFC), this foot stiffness is released to permit an increase of both passive and active compliance (PC–AC), allowing the foot to deform across the vault with the forefoot plantar fat pad. This ability provides protection during the high vertical and horizontal forces of the F1 and F4 GRF peaks. Following forefoot loading and loading response (FFL–LR), a relatively compliant foot continues to provide passive compliance (PC) and energy dissipation through vault depression. The rate of flattening and the compliance offered decreases under vault strain, causing some passive stiffening (PS) to initiate as midstance progresses. By the bottom of the F2 trough, energy dissipation capacity is being replaced by energy storage in the stiffening connective tissues. With increase in muscle activation, energy dissipation decreases and active and passive stiffening (AS–PS) rapidly increases towards heel lift (HL) and the F3 GRF peak. AS and PS reach their maximum at the HL boundary. AS starts to reduce in mid-terminal stance before the F5 GRF peak. PS is initially maintained due to elastic tissue recoil primarily from the tendons that continue to induce vault raising and twisting in terminal stance, with the additional benefits of the windlass mechanism from passive digital extension. PS starts to decline in late terminal stance and the foot becomes increasingly compliant for preswing (PS). *(Permission www.healthystep.co.uk.)*

At the end of opposite single-limb support, the swing limb's heel should be ~1 cm above the ground. After and initiated by heel lift on the support foot, the swing foot drops that small height in ~0.02 s causing an abrupt acceleration-assisted impact (Perry, 1992: p. 38). This is observed on force plate data as the heel strike transient. This impact force is dampened by the soft tissues of the heel and lower leg via knee, hip, and ankle muscle activity. Heel contact occurs behind the axis of rotation of the ankle, initiating the plantarflexion torque with around 60% of body weight and requiring a linear relationship in tibialis anterior activity that correlates directly with increasing walking/running speeds (Byrne et al., 2007). The acceleration of this torque is thus resisted and controlled primarily by tibialis anterior, absorbing energy through the buffering activity of its muscle–tendon complex. The knee's torque flexion angle explains around 76% of acceleration variability of the heel contact force (Beschorner and Cham, 2008), underscoring the knee's important impact energy dissipation function. Compliance within the foot's rearfoot skeletal structures is limited to ankle–subtalar complex motion until the forefoot starts to load. In some individuals, parts of the lateral column can begin to load before the forefoot if the lateral vault profile is low.

On forefoot loading, the pre-tensioned foot vault profile can provide energy dissipation through articular motion and soft tissue deformation. Pre-tensioning of the foot vault is caused by digital extension tightening of the plantar aponeurosis (Caravaggi et al., 2009, Qian et al., 2010), tibialis posterior activity, peroneus longus activity (Kokubo et al., 2012), and possibly some plantar intrinsic activity (Farris et al., 2019). By expanding its dimensions on loading, the vault absorbs energy in all three orthogonal planes assisted by the forefoot plantar fat pads. The braking energy absorbed in the forefoot will be that which is not already absorbed by the heel fat pad, hip, knee, and ankle flexion mechanisms. During walking, it is normally the viscoelastic heel fat pad and lower limb soft tissues that encounter impact energy first. In energy dissipation, knee muscle–tendon buffering mechanics express the most important role. However, many other mechanisms are functioning, such as soft tissue oscillations, that are combined with the ability to adjust gait speed and joint angles. These give many options in spreading out and managing impact energy dissipation during human gait.

3.2.4 Heel fat pad in compliance

The heel fat pad and lower limb soft tissues acting as a wobbling mass act in concert to dissipate impact energies around the ankle and rearfoot during heel strikes (Pain and Challis, 2001). The plantar heel fat pad is a peculiar structure consisting of specialised fibro-adipose tissue that dissipates energy on loading. The vertical GRF heel strike transient combined with the posteriorly directed GRFs of the F4 peak, as well as the anteriorly directed horizontal 'claw back' GRF-induced shear (often found associated with higher gait speeds), can each present injury risks to tissues that the heel fat pad can provide protection from (Qian et al., 2010; Wearing et al., 2014). The heel fat pad provides a smoothing of all orthogonal loading forces and eases the distribution of pressure on the heel by increasing surface area on contact, preventing local force concentrations within the rearfoot (Natali et al., 2012). It consists of fibrous compartments that enclose adipose tissue and is divided into superficial and deep layers. The superficial layer forms microchambers of adipose tissues surrounded by collagen/elastin fibrous septal walls, with deeper layers being composed of larger macrochambers (Jahss et al., 1992; Hsu et al., 2007a; Natali et al., 2012). This arrangement of fat and fibrous-walled chambers is essential for the heel fat pad's ability to dissipate impact energy, providing greater initial compliance superficially (Qian et al., 2010; Naemi et al., 2016) (Fig. 3.2.4a).

The collagen fibres within the septa are arranged spirally and fixed either to bone or other septa. The fat is arranged in globules joined by connective tissue and grouped together by bundles of collagen and elastin fibres (Jahss et al., 1992). The central portion chambers are orientated vertically, while in the posterior and lateral regions, they are smaller and transversely orientated (Rome, 1998). The superficial layer is attached to the epidermis and the deep layer is attached to the superficial layer and the calcaneal periosteum. Overall, the tissue is ~12.5–24.5 mm thick (Naemi et al., 2016), but depth is also influenced by body mass index (Rome et al., 2002). The mechanical properties are also slightly influenced by the presence of fluid within the vascular structures of the heel (Aerts et al., 1995; Gefen et al., 2001; Weijers et al., 2005). However, the propulsion of blood present in the heel does not affect the heel fat pad compression rates at loading speeds higher than 0.4 m/s, and even at lower rates does not affect the tissue stiffness more than ~3% (Weijers et al., 2005). The fat pad forms a hyperviscoelastic material that can operate across a wide range of strains and strain rates, providing material properties that influence transmission of loading forces into the lower limb's bones and joints (Grigoriadis et al., 2017).

The heel pads bear repeated loads during gait, with each step's heel strike force being spread over the calcaneus through fat pad deformation (Grigoriadis et al., 2017). In walking barefoot at preferred walking speeds, it has been reported that the plantar heel fat pad deforms with irregular load–deformation curves more compliantly at initial loading than at final loading, as its stiffness under load duration increases (Wearing et al., 2014). This behaviour is expected for the connective tissue that forms its septa. Peak deformation is ~10.3 mm, and around 1.0 J (J) of energy is dissipated at each step (Wearing et al., 2014). The heel fat pad is inhomogeneous and anisotropic, exhibiting nonlinear viscoelastic behaviour due to its biphasic (solid and fluid-fat phases) properties and the different attributes of its two layers (Rome, 1998). Both imaging (Gefen et al., 2001; Tong et al., 2003) and indentation techniques as used for bone or testing material shore scores have been used in vivo to quantify its material properties (Rome et al., 2001; Tong et al., 2003; Erdemir et al., 2006). However, the

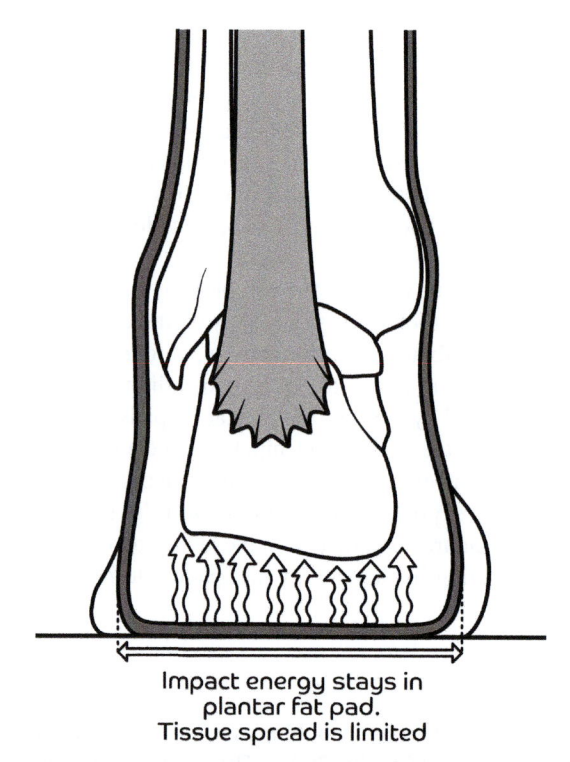

Impact energy stays in plantar fat pad. Tissue spread is limited

FIG. 3.2.4a The heel fat pad is able to deform in all three orthogonal planes, utilising its biphasic (fat-fluid phase and connective tissue-solid septa phase) to provide a structure with variable strain rates. The heel fat pad is able to behave more elastically at higher loading rates and more plastically at lower loading rates, moderating the amount of tissue spread and compression under load. Thus, it protects the rearfoot from shear and compression forces appropriately to the need of impact energy peaks. This is a useful trait for dealing with variable vertical and posteriorly directed GRFs from walking and running. The anterior GRF from claw back can also be managed within the fat pad. Transversely arranged fibres in the posterior and lateral regions are better adapted for shear than the central vertical arranged fibres. The heel fat pad is thus as beneficial to protect the compliant plantar skin from shear as it is to protect the stiff bone deep to it from compression. *(Permission www.healthystep.co.uk.)*

results depend on the location and the size of the indentor being used (Spears and Miller-Young, 2006), making modelling a better option to examine the whole structure (Natali et al., 2012; Grigoriadis et al., 2017).

Fat pad biomechanics

The plantar fat pad exhibits hyperelastic, stress-stiffening, and strain rate-dependent properties, becoming stiffer in response to higher strains at higher rates (Erdemir et al., 2006; Ledoux and Blevins, 2007; Natali et al., 2012; Naemi et al., 2016; Grigoriadis et al., 2017). Thus, at faster loading velocities, the heel fat pad will behave more stiffly. It is the adipose (fat) tissue itself that provides the high compression resistant fluid element (Natali et al., 2012). When compressed slowly, the fat globules expand circumferentially and stretch the fibrous septa of the chambers, restricting the fatty tissue spread. The greatest spread occurs internally within the chambers on initial loading, resulting in increased pressure on the septa leading to raised stiffness/elasticity within them. This stiffness/elasticity wall deformation springs back on off-loading to dissipate any stored energy by restoring the chambers' shape (Grigoriadis et al., 2017). The quicker the septa stress, the more stiffly the plantar fat pad behaves, allowing it to be passively responsive to changing impact loading rates (Fig. 3.2.4b).

3.2.5 Muscle action in energy dissipation through the foot

Leg muscles that cross the ankle to support the foot vault have an important role in energy dissipation through their muscle–tendon buffering activity during contact phase (Roberts and Konow, 2013). Through extensor muscles extending the toes during swing and early loading response, pre-tensioning of the plantar aponeurosis can occur, stiffening the foot vault before forefoot contact. By then allowing the windlass effect to reverse on forefoot loading, compliance is increased (Stainsby, 1997; Green and Briggs, 2013; Stolwijk et al., 2014). This mechanism provides additional energy dissipation through active and passive soft tissues (Carlson et al., 2000; Wager and Challis, 2016). The primary muscles of foot pre-stiffening prior to and during initial contact are tibialis anterior, tibialis posterior, and peroneus longus working in

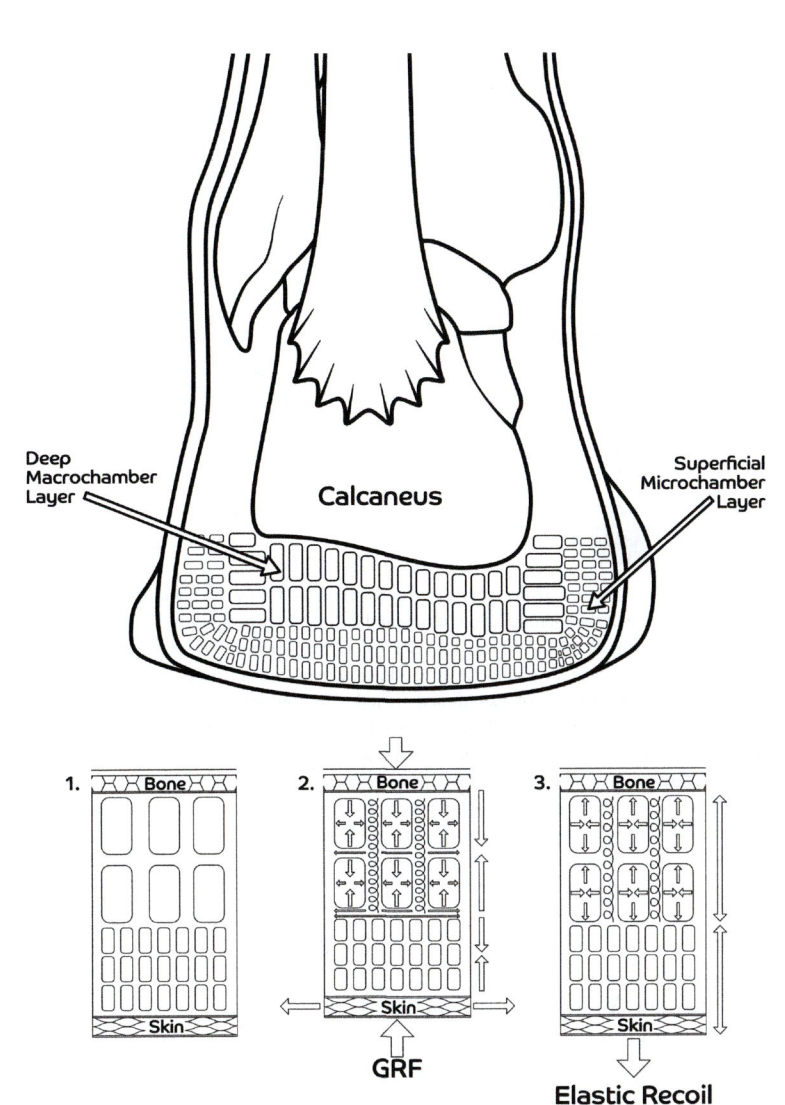

FIG. 3.2.4b The presence of superficial microchambers and deeper macrochambers (upper image and lower image 1) offers an ability to alter structural properties through the fat pad, providing different loading rate behaviours regionally. Transversely orientated fibres within the septa (walls) found to the sides and to the posterior heel help resist greater shear from tissue spread and more posteriorly driven horizontal forces that are present during initial contact. On loading (image 2), skin and chamber septa shear and tension while the fat deforms vertically under compression. This squeezes the fat to tension the chamber septa that are full of collagen and elastin fibres. The connective tissue is stretched due to fat deformation, storing the strain energy within its fibres. On offloading (3), the septa recoil, springing back into their relaxed tension, pushing the compressed fat back into its pre-loaded shape and thereby releasing the impact energy to reform the plantar fat pad for the next step. Only the macrochambers are shown deforming in the image to improve visualisation, but all chambers act on the same principles. The microchamber regions found superficially have proportionally more connective tissue septal content so they behave more elastically, offering greater initial compliance than the deeper microchamber regions. *(Permission www.healthystep.co.uk.)*

unison with the digital extensors (Neptune et al., 2008; McGowan et al., 2010; Kokubo et al., 2012). Abductor hallucis may also assist in setting the degree of digital extension during swing and loading response (Kelly et al., 2015).

Tibialis anterior in energy dissipation

At heel strike, the ankle's axis of rotation moves progressively posteriorly under ankle plantarflexion motion. The lever arm distance from the posterior heel contact point of the GRF to the ankle axis is minimised if the foot is more plantarflexed or less dorsiflexed at heel strike, reducing the angular momentum arm of the GRF while that of tibialis anterior becomes lengthened. This is a useful mechanical trait if foot strike occurs more posteriorly on the heel as a consequence of a longer stride length. The foot usually contacts the ground with the ankle between 0° and 3–5° plantarflexed (Perry, 1992: p. 53; Byrne et al., 2007). If the foot were dorsiflexed, then the moment arm of the GRF to the ankle would be longer due to a more anteriorly positioned ankle axis. A higher dorsiflexion angle of the foot at contact results in the generation of more plantarflexion angular momentum that must be resisted by tibialis anterior. The muscle has a short moment arm when in dorsiflexion angles, so is initially at mechanical disadvantage.

A more dorsiflexed foot at heel strike also extends the plantarflexion moment to bring the forefoot to the ground. The faster the walking speed and the longer the stride length, the larger the plantarflexion moment and the less time there is to decelerate it. Footwear with a prominent posterior flare will exacerbate the effect by lengthening the GRF moment arm. Intensity of tibialis anterior activity (peak amplitude on electromyography) during heel strike to the end of loading response is mainly dependent on walking speed to prevent the accelerated angular plantarflexion moments, with failure to control causing high forefoot accelerated impacts seen as a 'forefoot slap' down (Byrne et al., 2007) (Fig. 3.2.5).

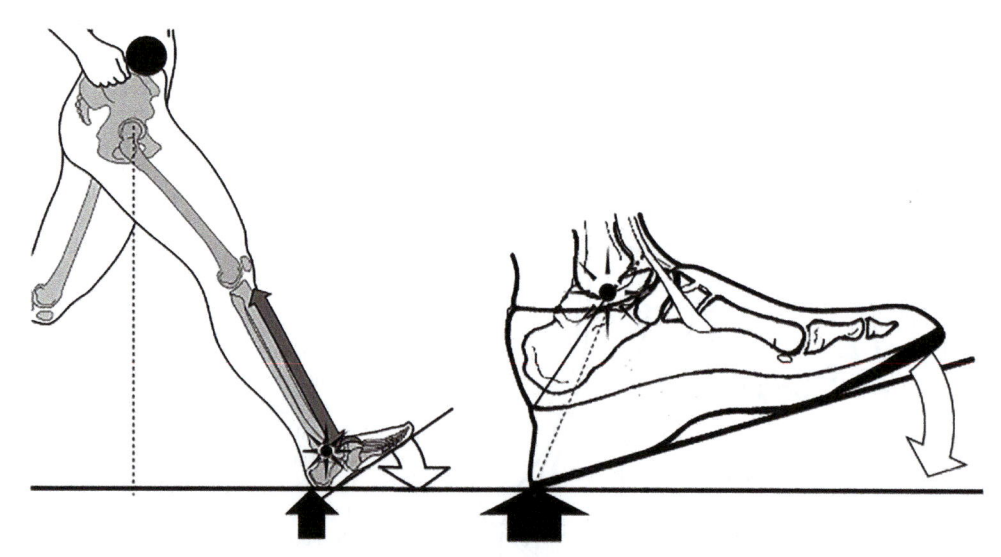

FIG. 3.2.5 Tibialis anterior is an important extrinsic muscle in protecting the forefoot from excessive impact energies following heel strike. Faster gait speeds that involve a heel strike will increase the loading rate and thus the force of the plantarflexion moment that the eccentric activity of tibialis anterior has to resist. Longer stride lengths (A) can also increase the foot dorsiflexion contact angle, moving the GRF *(black arrow)* more posteriorly on the heel to increase the external moment arm to the ankle (B—*solid line*). This results in extending the plantarflexion moment around the ankle axis (star), necessary to bring the forefoot to the ground *(white curved arrows)*. If the tibialis anterior is weak or becomes fatigued, then the forefoot impact velocity will be greater having been insufficiently braked. This will heighten tissue stresses within the forefoot at its collision. Footwear, when present, sets up a new dorsiflexion angle to the ground using the sole of the shoe rather than the plantar foot surface. Footwear structured with a posterior flare, especially with a significant heel lift (as demonstrated in B), can also increase the external ankle plantarflexion lever arm of the GRF *(dashed line)* compared to the natural barefoot one from the posterior heel *(solid line)*. This is also a result of the GRF being moved further posterior from the ankle joint axis. The resistance arm of tibialis anterior remains the same, and thus, its mechanical efficiency is reduced. The most common complaint when ankle plantarflexion moment energies are not well braked and dissipated by tibialis anterior muscle–tendon buffering, is anterior shin aching associated with myofascial inflammation. This is often referred to as (anterior) shin splints. *(Permission www.healthystep.co.uk.)*

Tibialis anterior does not need to store energy during gait. An ability to elastically recoil into ankle dorsiflexion after resisting plantarflexion would be an energetic disaster, causing the forefoot to spring back away from the support surface after forefoot loading. The capacity of the tibialis anterior muscle–tendon unit to dissipate energy relies on the lengthening events within the muscle and tendon that are out of phase. A period of tendon lengthening occurs first, accompanied by minimal muscle fascicle lengthening. This is followed by a period when the tendon stays the same length but the muscle fascicles lengthen instead (Roberts and Konow, 2013). The relaxation of the muscle allows for decay in the muscle force and the elastic tendon recoil without significantly shortening the muscle–tendon complex. This produces an internal shuttling of energy so that strain energy is first loaded to tendon stretch and then later released as the muscle lengthens, dissipating the energy (Roberts and Azizi, 2010). Any elastic recoil is avoided in the tibialis anterior tendon, as after its eccentric loading activity, the muscle–tendon unit is shortened in length through ankle dorsiflexion motion (reducing plantarflexion angle) during early midstance. With the ankle dorsiflexing, the elastic modulus (i.e. the deformation under load) will be decreasing.

This reduction in tibialis anterior elastic modulus has been reported in 50% of subjects in a study by Koo et al. (2014), although in the other 50% of subjects, only the deep tibialis anterior fascicles demonstrated a decrease. In these subjects, the superficial fascicles displayed an increase in elastic modulus, probably in part due to the tight, dense nature of the anterior fascia crura found superficial to the tibialis anterior (Koo et al., 2014). These results suggest that stiffness of the fascia above tibialis anterior directly affects this muscle's energy-dissipating material properties. This could be important for discovering the cause of anterior shin pain in runners. The primary role of tibialis anterior in impact dissipation is to decrease the forefoot acceleration element of forefoot impact force by slowing the plantarflexion moment and dissipating energy through the muscle–tendon complex. The fascia above may play a part in this mechanism.

Assistor muscles in extensor energy dissipation

Important assistor muscles during heel contact foot energy dissipation to tibialis anterior are extensor hallucis longus and extensor digitorum longus, which are also active prior to, during, and after heel contact. These muscles have a longer moment arm to control the forefoot loading rate via the digits, although they are far less powerful (smaller cross-sectional areas). They preload the plantar aponeurosis during initial contact through MTP joint dorsiflexion, increasing the plantar aponeurotic strain, shortening the foot vault length, and thereby heightening stiffness across the vault to improve forefoot energy dissipation potential (Flanigan et al., 2007; Caravaggi et al., 2009; Wager and Challis, 2016). This suggests that the digital extensors help mainly in forefoot energy dissipation rather than acting as important ankle plantarflexion decelerators after heel strike.

The ankle is rarely actually dorsiflexed at initial contact (Perry, 1992: p. 53; Byrne et al., 2007), probably for good reason. Studies on heel-toe running confirm higher plantar aponeurotic strains at initial stance in rearfoot runners compared to forefoot runners who impact the ground with the ankle and foot plantarflexed (McDonald et al., 2016; Wager and Challis, 2016). Forefoot running also demonstrates less tibialis anterior activity (Rooney and Derrick, 2013). To improve rearfoot contact energy dissipation through tibialis anterior and the plantar aponeurosis, heel strike may require high digital extension; something difficult to achieve if the foot becomes dorsiflexed at the ankle. The resulting higher foot vault profile obtained from initiating more of the windlass mechanism may allow further energy to be dissipated within the foot vault's connective tissues via extensor muscle–tendon buffering immediately on forefoot loading through toe flexion motion out of extension. This is because the vault's initial increased stiffness will affect the rate of its deformation and the amount the foot can change shape under external impact stress.

Abductor hallucis and quadratus plantae both display activity prior to heel contact, while quadratus plantae also remains active during the early stance phase (Kelly et al., 2015). This suggests a role for both these muscles by assisting or (in the case of quadratus plantae) antagonising digital extension in stressing the plantar aponeurosis as part of the reversed windlass. Vault-lengthening control through quadratus plantae may prevent an excessive loss of vault height on loading in early stance through tensing the lesser digital flexor tendons. However, studies so far seem to suggest that the vault profile and tibialis anterior play a more significant role in preloading the plantar aponeurosis with strain than does MTP joint dorsiflexion through active digital extensor muscles (Wager and Challis, 2016). The process by which the digits flex out of extension at the MTP joints during loading response in order to bring themselves into ground contact has been termed the reversed or reverse windlass effect (Stainsby, 1997; Aquino and Payne, 2000). The timing of this digital plantarflexion may help influence energy dissipation through the influencing of initial vault deformation rates, assisted by a burst of quadratus plantae activity during early stance acting through its attachments on the flexor digitorum longus tendons (Kelly et al., 2015). This assistor muscle activity influencing plantar aponeurotic mechanics may also aid tibialis posterior and peroneus longus in the setting of stiffness within the foot vault throughout the skeletal frame prior to impact, giving greater deformation adaptability for energy absorption during impact.

3.2.6 Energy dissipation through wobbling mass in the lower leg

The lower leg, particularly the posterior calf, has muscles that control foot motion and vault shape. This muscle bulk can help in energy dissipation by acting as a wobbling mass at impact (Pain and Challis, 2001) as well as through active muscle–tendon buffering. It is known that soft tissues moving independently of the skeleton play a significant role in energy dissipation (Gruber et al., 1998; Pain and Challis, 2001; Khassetarash et al., 2015). The displacement of wobbling masses is only important to impact energy dissipation in the first 10–30 ms of contact phase, but within this period, they seem to have a considerable ability to reduce impact force being passed to the trunk, without having much influence on kinematics (Gruber et al., 1998).

The fact that the human body consists of a wobbling mass rather than rigid body segments means that the body's CoM moves downward more rapidly via soft tissue displacement than if it were rigid and thus only able to move down with increased joint torques (Gruber et al., 1998). When the GRF force is modelled on a wobbling body, the impact force produced is lower than that modelled on a rigid body for the same mass and geometry. If the body were a rigid structure, joint torques required to decelerate impact forces would need to be greater, causing greater collapse of limb and body posture at impact (Gruber et al., 1998). This would create greater stress for muscle–tendon buffering mechanisms.

Measurements in vitro and in vivo of the energy dissipation properties of the heel fat pad have shown considerable differences. When introducing deformable elements and wobbling masses to pendulum-based impact models, heel pad properties tend to better match those found in vivo, helping to explain the quantity of energy lost during impact. This implies that both the heel pad and the soft tissues of the lower leg make a significant contribution to energy dissipation during heel impacts during gait (Pain and Challis, 2001). In modelling undertaken by Pain and Challis (2001), wobbling masses reduced the peak forces on heel pads by 55%, producing similar values to in vivo results reported by Cavanagh et al. (1984) and Aerts et al. (1995). Peak forces modelled on a heel fat pad on a rigid lower leg were 100% higher than when modelled with a wobbling mass lower leg, indicating a significant role for soft tissue resonance on impact forces during collisions.

The maximum excursion of any wobbling mass relative to the underlying rigid structure on modelling was less than 17 mm from its original position (Pain and Challis, 2001). Soft tissue motion during running impacts is reported to be over

40mm (Cappozzo et al., 1996) suggesting a greater amount of tissue excursion in vivo than in modelling. Interestingly, off-centre pendular impacts had a greater effect on the behaviour of wobbling mass models as they induced relative turning moments that produced large amounts of inter-segmental motions (Pain and Challis, 2001). This suggests that rotation in the ankle has a role to play in energy dissipation and that off-centre impacts on the heel may reduce energy dissipation (Pain and Challis, 2001).

The clinical biomechanist should be very interested in the heel contact point during gait and foot stiffness established through arch height flexibility (AHF) investigations. Foot stiffness predicts the level of impulse that should be expected to be generated in normal young populations. Out of the subjects so far discussed, initial heel impact impulses through their increased ability to dissipate energy should be expected to be higher in those individuals who have more flexible feet and lower in those with stiffer feet, as reported by Cen et al. (2020). This is because stiffer feet need to take significant action to moderate impact forces to a greater extent, while more flexible feet do not. Stiff feet with high rearfoot GRF impulses may thus indicate poor energy dissipation and risk failures in impulse adjustment strategies.

3.2.7 The skeletal frame vault in foot compliance

The foot's *skeletal frame* represents the rigid struts of a biotensegrity structure supported by cables of elastic connective tissue that create an important energy-absorbing and storing structure during gait (Qian et al., 2010). This ability to dissipate energy is tied into the initial rapid change in vault height at the end of loading response. On forefoot contact, distal metatarsal loading will cause a spreading of the metatarsal heads, widening the distance between the metatarsals in the transverse/horizontal plane. This process utilises the energy-dissipating elastic properties of the many ligaments tethering the foot together, including the deep transverse metatarsal ligament (Qian et al., 2010). The energy not dissipated through soft tissue deformation within intermetatarsal and tarsal motions will be applied as bending moments on the metatarsal bones. Although the order of metatarsal loading can be quite variable during gait, usually the more mobile lateral metatarsals are loaded first. This lateral metatarsal base mobility can absorb energy through cuboid–metatarsal motions, especially horizontal shear. Spreading of the metatarsals permits energy-absorbing deformation to occur in all three planes rather than just the vertical. Vertical GRF causes the rearfoot–midfoot bones through their articulations to primarily dorsiflex, explaining the rapid initial lengthening that occurs during early loading of the vault (Stolwijk et al., 2014; Bjelopetrovich and Barrios, 2016; Takabayashi et al., 2020).

Widening, elongation, and depression within the vault permit it to act like a shock-absorbing spring. Therefore, the stiffness of the vault's 'spring' and the amount it can deform on loading are important properties in dissipating energy (Fig. 3.2.7). The windlass effect through digital extension that can reverse and the collective increasing and then decreasing activity of tibialis posterior and peroneus longus, are critical to the vault's ability to act as a shock absorber through pre-tensioning and then relaxing the spring. Limited or absent toe extension prior to contact and loading response will reduce the ability of the plantar aponeurosis windlass mechanism to reverse on forefoot loading. Thus, inadequate tibialis posterior activity throughout loading response will reduce muscle-induced vault stiffness before forefoot loading, reducing the energy-dissipating properties of the foot vault due to excessive initial compliance at forefoot loading. Indeed, any failure in the muscles or the connective tissues to provide elastic restraint of the vault (such as the spring, long plantar, or short plantar ligaments) also risks reducing the effectiveness of controlled vault depression. Loss of the vault's capacity to act as an elastic spring for controlling and slowing the loading rate of the forefoot impact's impulse by releasing the foot's compliance properties for energy dissipation, is potentially injurious.

The effect of soft tissue failures to create vault stiffness prior to forefoot contact is much like using worn-out shock absorbers or bottomed-out cushioning foam. Such dysfunction is likely to reduce shock attenuation abilities and increase injury risk from impulse energies. Equally problematic is a reversed windlass that does not reverse on time. A tight tibialis anterior muscle or a particularly stiff vault, due to a high vault and/or connective tissue hypomobility, will prevent effective energy dissipation at impact through the vault not being able to deform sufficiently.

The normal rapid vault 'drop' and its lengthening/widening at the end of loading response and the beginning of midstance should take the connective tissues through their stress–strain toe region, ending the compliance–energy dissipation phase and initiating the beginning of the stress-stiffening process within the connective tissues before late midstance. Vault compliance should occur just in time for the foot to start to develop its single-limb support stability. The energy dissipation mechanism of the vault should protect the feet and lower limbs from excessive collision energies. Foot stiffness that can be established through arch height flexibility (AHF) should help predict the level of impulse generated in normal young healthy populations and give insight as to what kinetics should be expected. Impact loading impulses on the rearfoot are higher in those individuals with more flexible feet, and lower in those with stiffer feet (Cen et al., 2020). This is probably a result of increased negative work in stiffer feet to provide higher levels of energy dissipation through the active components. Thus, stiff feet with high levels of heel GRF forces may indicate energy dissipation failure.

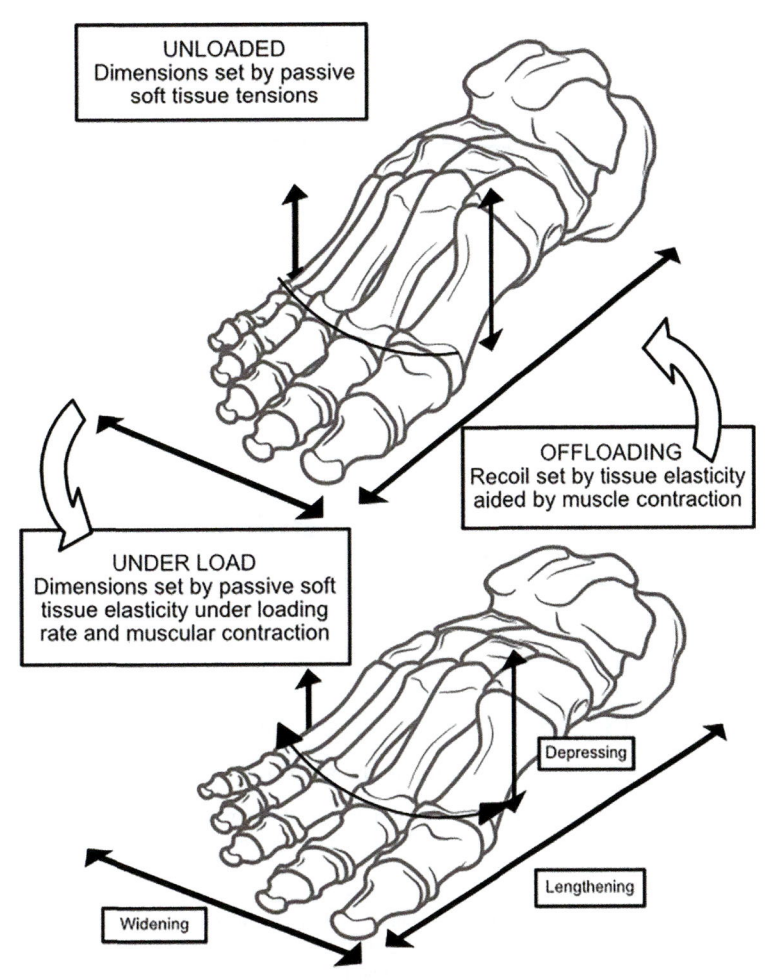

FIG. 3.2.7 The foot vault is a viscoelastic structure that deforms on loading by widening, lengthening, and depressing. The tensions created on weight-bearing stiffen the foot, increasing internal elasticity. In the forefoot, widening decreases the radius of curvature across the metatarsal heads as well as tensioning the deep transverse metatarsal ligament, helping to increase overall stiffness. Increased elasticity via deformation not only provides energy dissipation and storage but also creates internal forces that conspire to 'spring' the vault back into its preloaded profile on offloading. Foot tension is therefore reduced and flexibility increased, much like a relaxed spring. During gait, lengthening, widening, and flattening reach a peak at the heel lift boundary. Elastic recoil starts to occur proximally as peak loads transfer to the forefoot during late midstance, but midfoot and forefoot recoiling only starts to occur following heel lift. Peak-loaded areas across the foot are constantly changing and are variably adapted to each step depending on terrain and the 'events' within a single step. Variability requires information on specific loads coupled to the ability to apply active control of tensions within soft tissues and compressions across articulations. Such a responsive system is supplied by the sensorimotor system, fully integrated into muscle activation. Feet are not just passive mechanical structures but are responsive and adaptive in utilising their connective tissue viscoelasticity properties. *(Permission www.healthystep.co.uk.)*

3.2.8 Transformation from compliance to stiffness

Out of compliance

After forefoot loading, the foot starts to change its role from an energy dissipator to that of a semi-stiffened beam. Stiffness in biology is the degree of resistance offered by tissues in response to forces of lengthening, hence why measuring the foot's length (width) under different loading stresses can indicate its stiffness. The stiffening process in the foot starts just after loading response completes, probably as the toes load to the support surface and plantar aponeurosis preloading is relaxed (Caravaggi et al., 2009). A relatively rapid increase in lengthening of the foot should be expected and is indeed reported at this time (Stolwijk et al., 2014) (Figs 3.2.8a and 3.2.8b). It is possible that the presence of the toes aligning directly distally to the metatarsal heads along the metatarsals' long axes after toe-strike on the ground, influences the lengthening rate of the foot during vault depression by presenting a 'restraint' on the metatarsal heads anterior drift. This would indicate an extra significance to the reversed windlass mechanism being completed by the end of loading response. In hypomobile feet, such a stabilising mechanism would be of less significance, and the need to get the toes to the ground early in stance of less importance to the vault stability than in more mobile feet. Toe deformities, such as hallux abducto valgus and hammer toes, could compromise this anterior restraint.

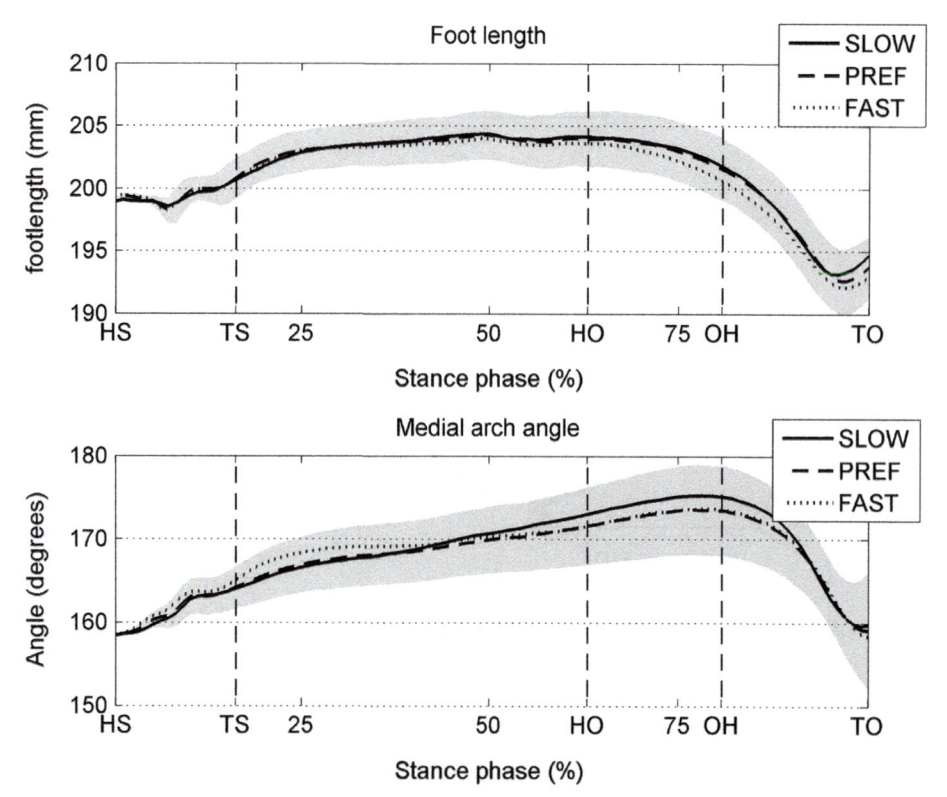

FIG. 3.2.8a The average change in foot length *(upper panel)* and medial arch angle (lower panel) during gait for three different walking speeds (slow, preferred, and fast), with grey area being the standard deviation at preferred walking speeds. At heel strike (HS) for all walking speeds, foot length is relatively shorter than for most of stance phase, decreasing until forefoot loading initiates the reversed windlass action, ending with toe strike (TS) onto the support surface. The foot length increases rapidly during early midstance and then the rate of change slows, ending at heel lift or heel-off (HO). Now the foot length decreases rapidly, especially after the opposite foot's heel contact (OH), continuing until just before toe-off (TO) as the foot becomes more compliant. The medial arch angle (the angle between markers placed at the 2nd MTP joint, the heel, and navicular tuberosity as used in this study) increases throughout stance until loading of the opposite foot initiates the start of vault profile recoil. This demonstrates that vault height is lowest during heel lift at the start of terminal stance. The stiffer foot behaviour during fast walking speed has a limiting effect on foot lengthening and is only significantly different from late midstance into terminal stance. *(From Stolwijk, N.M., Koenraadt, K.L.M., Louwerens, J.W.K., Grim, D., Duysens, J., Keijsers, N.L.W., 2014. Foot lengthening and shortening during gait: A parameter to investigate foot function? Gait Posture 39 (2), 773–777.)*

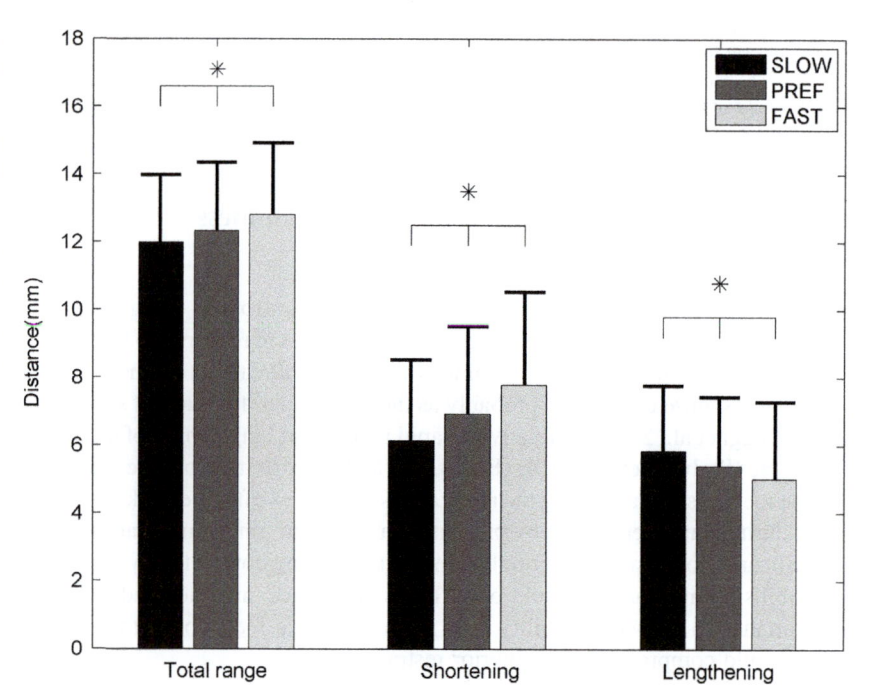

FIG. 3.2.8b Mean and standard deviations for lengthening, shortening, and total lengthening range at three different walking speeds of slow, preferred, and fast. Note that connective tissue behaves more stiffly (elastically) the faster it loads, helping explain the different ranges of length changes. Despite less lengthening, there are greater overall length changes at faster walking speeds. *(From Stolwijk, N.M., Koenraadt, K.L.M., Louwerens, J.W.K., Grim, D., Duysens, J., Keijsers, N.L.W., 2014. Foot lengthening and shortening during gait: A parameter to investigate foot function? Gait Posture 39 (2), 773–777.)*

The vault profile at the start of midstance depresses through lengthening and widening so as to increase its surface contact area as the foot undergoes pronation and the muscle–tendon units within the foot lengthen (Kelly et al., 2015). The rate of depression during early midstance is likely to be controlled by two factors: the stress–strain behaviour of the soft tissue elements of the foot and the rate of rise in the body's CoM that occurs with leg extension during early midstance, a process that also reduces vertical force peaks. The higher the early midstance GRF and the less stiff the foot, the greater the degree of pronation is likely to be. During this phase of gait, there is little active muscular contraction within the foot although on initial loading, quadratus plantae is active (Farris et al., 2019) with tibialis posterior being the most 'variably' active throughout early midstance (Murley et al., 2009; Semple et al., 2009). Tibialis anterior has also been reported as intermittently active during early midstance, but highly variably between steps (Byrne et al., 2007). Most of the soft tissue mechanics is likely dependent on strain rates under load applied to the collagen-dominated ligamentous and fascial components lying plantar to the skeletal frame. These initially pass through their faster-lengthening un-crimping phase before undergoing linear elastic behaviour.

The CoGRF remains behind the ankle joint axis at the start of early midstance, although it is progressing anteriorly. This means that the posteriorly directed GRF is decreasing at the same time that the vertical and horizontal GRFs are also decreasing due to limb extension, moving the CoM upwards. The foot vault is therefore being initially loaded and lengthened under descending forces, passing the connective tissues through their linear elastic phase under a relatively low loading force. This helps set a safety margin to the forces driven into the foot, while it is particularly compliant. Vault profile depression will initially be more extensive, despite the declining forces due to the lengthening effects of passing through the toe region's stress–strain mechanics within the connective tissues. This is quickly followed by vault stabilisation as the connective tissues enter their linear elastic region. GRFs continue to decrease to absolute midstance, helping to reduce the vault's rate of depression as it begins to stiffen. As these midstance forces tend to remain higher at slower walking speeds, the foot vault may be more vulnerable to increased depression strains if muscles do not remain strong under the plantar foot at slower walking speeds. Increasing stiffness as early midstance progresses becomes desirable, for excess vault compliance could cause interference with the CoM's ascent through limb extension lengthening. Despite the reduction in early midstance GRFs, the soft tissues across the vault undergo increasing strain at a pace that is dependent on the loading rate. Faster loading rates cause increased connective tissue stiffness through strain rate-dependent properties, decreasing the amount the foot lengthens and thus influences vault height changes (see Figs 3.2.8b and 3.2.8c).

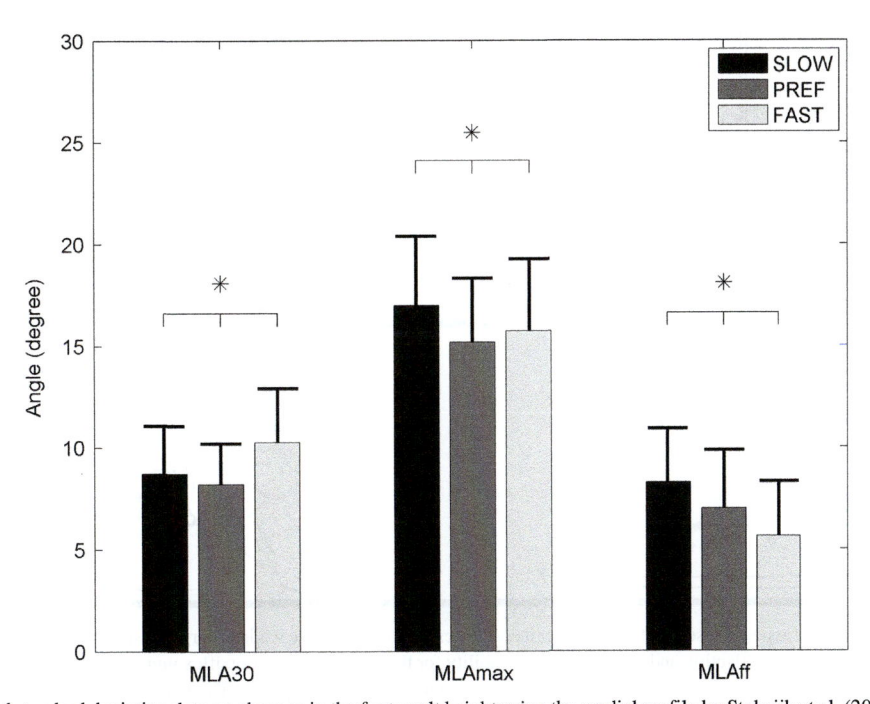

FIG. 3.2.8c Mean and standard deviation data on changes in the foot vault height using the medial profile by Stolwijk et al. (2014) give insight into the pattern of vault profile changes during stance phase. Using a medial long arch (MLA) angle formed by markers placed on the heel, the 2nd MTP joint, and the navicular tuberosity, changes were recorded on VICON (Oxford, UK) 3D motion analysis during three different walking speeds. Changes between heel strike and 30% of the stance cycle (MLA30) and heel strike and the maximum MLA angle (MLA$_{max}$) of stance phase were calculated as an indicator of change in the MLA angle during early stance phase and the total stance phase. The difference between MLA30 and MLA$_{max}$ was calculated to indicate the change in the MLA angle during the foot flat phase (MLA$_{ff}$) that exists throughout midstance. Note that preferred walking speed offers the least maximal changes in the vault profile during stance. *(From Stolwijk, N.M., Koenraadt, K.L.M., Louwerens, J.W.K., Grim, D., Duysens, J., Keijsers, N.L.W., 2014. Foot lengthening and shortening during gait: A parameter to investigate foot function? Gait Posture 39 (2), 773–777.)*

Late midstance increased stiffening rate

Following absolute midstance, the foot starts to stiffen more rapidly with an increase in GRF towards the F3 and F5 peaks caused by potential energy release due to the fall of the CoM from its high point. This results in further vault depression, moving from the linear stress–strain into the stress-stiffening part of connective tissue strain behaviour. The CoGRF becomes situated anterior to the ankle joint axis and the ankle becomes subjected to increasing dorsiflexion moments. The ankle joint axis moves anteriorly with dorsiflexion and the ankle stiffens through an increase in joint congruency (Leardini et al., 2001; Leardini and O'Connor, 2002), increasing the plantarflexor muscles' internal lever arm to improve mechanical efficiency in resisting dorsiflexion. Despite this anterior axis shift, the GRF dorsiflexion lever arm increases as the CoGRF continues to advance anteriorly. Under the influence of the increasing midfoot dorsiflexion (sagging-deflection) moment caused by the CoGRF passing under the midfoot, the vault profile continues to depress, lengthen, and widen, but at a much lower rate because of increased connective tissue stiffening. Increasing triceps surae load through the Achilles controlling the dorsiflexion moment in the ankle increases the strain in the plantar aponeurosis, stiffening it (Cheung et al., 2006) as well as the Achilles tendon. As the load within the ligaments and the other connective tissues increases, they exhibit their non-Hookean strain-stiffening properties (Kitaoka et al., 1994).

The vault will stiffen more rapidly through vault lowering brought about via increasing dorsiflexion moments across the midfoot coupled to increasing Achilles stiffness and anterior GRF forces in late midstance, despite little further connective tissue lengthening. A possible benefit of continued vault depression in this phase of gait, beyond such motion assisting in energy dissipation, is that the vault profile depression allows the ankle joint to dorsiflex unhindered by any significant blocking resulting from the height of the midfoot that could limit anterior-plantar displacement of the talus. This avoids the risk of impinging the ankle dorsiflexion range within the anterior ankle mortice. Vault depression will permit easier limb vaulting over the foot through increasing ankle freedom of motion and a greater functional ankle dorsiflexion angle to the ground. This extra midfoot dorsiflexion motion prevents ankle dorsiflexion being limited by the ankle becoming close-packed too early to permit full limb vaulting before heel lift. This increases the energy that can be stored during Achilles stretching. Without this vault depression, a premature heel lift might be necessary to prevent interference in the anterior progress of the CoM, reducing the energy stored within the Achilles during late midstance (Fig. 3.2.8d).

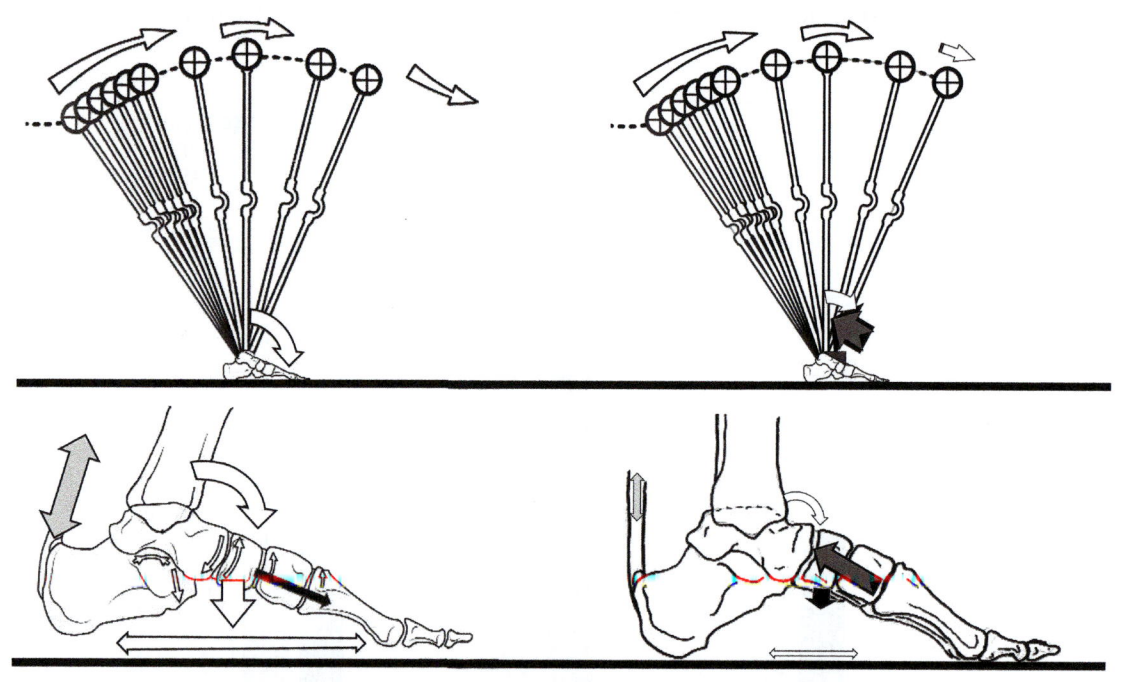

FIG. 3.2.8d Elastic vault depression is essential for normal gait energetics, not least because it adds energy dissipation and storage capabilities to single-limb support as the foot undergoes deflection and lengthening. The ability for the ankle to freely dorsiflex under the restraint of triceps surae requires an adequate range of dorsiflexion to move the swing limb anteriorly on an anteriorly pivoting pelvis. The pelvis' anterior progress is dependent on ankle dorsiflexion range which in turn influences the progression of the swing limb's stride length. By lowering the vault, as seen on the left, the lower limb can pivot freely as the talus is able to plantarflex and glide distally, opening up more functional dorsiflexion. In stiff, high-vaulted feet as demonstrated on the right, a lack of vault depression and lengthening can cause the navicular to distally block the anterior glide of the talus during dorsiflexion. This action is much like a chock under a wheel. Such rearfoot and midfoot dorsiflexion blockage limits the anterior progression of the pelvis, restricting the posterior angulation of the stance limb before heel lift. The limited limb rotation in turn restrains the swing limb's freedom to create its stride length and centrifugal forces, necessitating increased swing limb flexor power at the hip and extensor power at the knee. In turn, loss of dorsiflexion limits energy storage within the Achilles for acceleration. *(Permission www.healthystep.co.uk.)*

Exponential foot vault stiffening is supported by the work of Stolwijk et al. (2014) by decreasing rates of lengthening observed during late midstance. Under stance loads, both Bjelopetrovich and Barrios (2016) and Takabayashi et al. (2020) have found linear and curvilinear trends in the arch height index under increasing load. These studies have demonstrated that most arch deformation occurs in the ambulatory load range for healthy feet. The feet appeared to lengthen by 4 mm (± 1.5) under static loading forces from 10% body weight to 80% body weight, but with no further lengthening occurring from 80% to 120% body weight. Furthermore, dorsal foot height reduced by 5 mm (± 0.4) over the lower weight range and only by 1 mm from 80% to 120% body weight (Bjelopetrovich and Barrios, 2016) (Fig. 3.2.8e). Takabayashi et al. (2020) also found that changes in arch height decreased and stiffness increased nonlinearly with increases from 10%–50%–90% loading conditions. However, caution should be exercised when predicting dynamic arch function from static data. This is because the CoGRF is moving dynamically, applying different peak stress locations over time and the Achilles via the triceps surae is generating considerable power throughout late midstance via dorsiflexion moments across the vault, which eventually initiates heel lift.

As momentum and high Achilles loads are eliminated in static assessments, the actual deformations of the foot are likely to be underestimated compared to dynamic locomotion (Bjelopetrovich and Barrios, 2016). However, increased loading rates should result in increased vault stiffness. The other factor likely to be underestimated is the effect of the dynamic plantar intrinsic and extrinsic foot and ankle muscle activity that starts to become increasingly prominent during late midstance. Yet, the data supports the concept that feet constructed extensively of collagenous soft tissues present an exponential stress-stiffening relationship as expected. This indicates that higher strain-length changes occur at lower loads, and the vault stiffens without significantly increasing length or decreasing height at higher loads. As the foot stiffens, it has the potential to behave more elastically, creating greater overall length changes. It does seem possible that populations of humans may express gender behaviour differences in vault flexibility (Zifchock et al., 2006; Saghazadeh et al., 2015; Bjelopetrovich and Barrios, 2016). However, Takabayashi et al. (2020), looking at a young Japanese population, did not find any vault flexibility gender differences (Figs 3.2.8e and 3.2.8f).

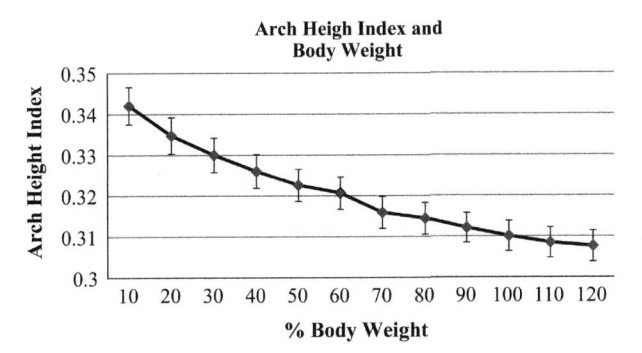

FIG. 3.2.8e Arch height index (AHI) changes vs body weight % indicating a curvilinear relationship between arch height and incremental loading, with the vault behaving more stiffly under increasing load. Female participants in this study had significantly less stiffness than male participants in all loading conditions. *(From Bjelopetrovich, A., Barrios, J.A., 2016. Effects of incremental ambulatory-range loading on arch height index parameters. J. Biomech. 49 (14), 3555–3558.)*

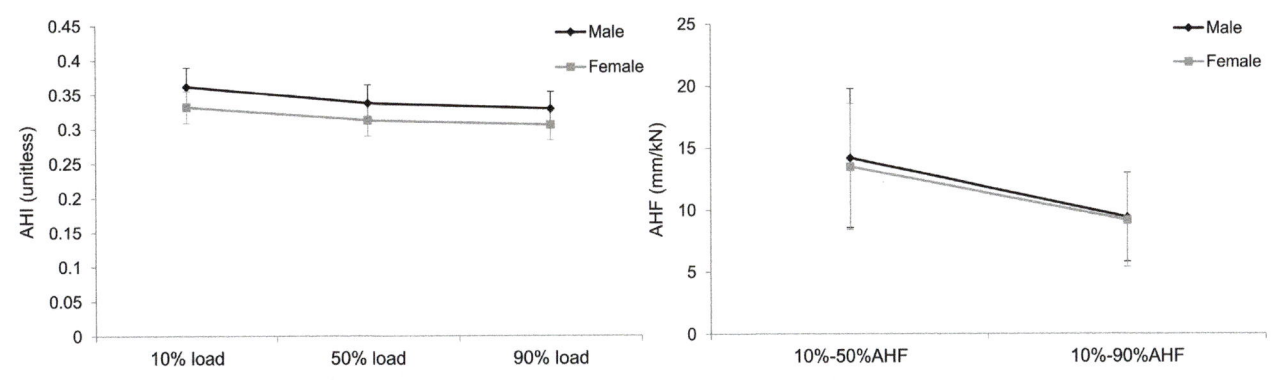

FIG. 3.2.8f Changes in arch height index (AHI) *(left)* of males and females for 10%, 50%, and 90% and arch height flexibility (AHF) *(right)* between 10 and 50% and 10–90% static body weight loading conditions are shown for young Japanese feet (18–24 years of age). This reveals that changes in AHI changes under load are significantly different between genders but AHF is not different between the genders in young Japanese individuals. When investigating AHI, gender, and loading, conditions must be considered. However, AHF should be investigated in consideration to the changes in loading forces as the foot behaves more stiffly under increasing loads. *(From Takabayashi, T., Edama, M., Inai, T., Nakamura, E., Kubo, M., 2020. Effect of gender and load conditions on foot arch height index and flexibility in Japanese youths. J. Foot Ankle Surg. 59 (6), 1144–1147.)*

Stiffening the human foot allows for elastic energy storage and for the creation of a rigid lever beam from which to propel (Bojsen-Møller, 1979; Kuo et al., 2005). It is the vaulted profile that is thought to underlie the ability to stiffen (Yawar et al., 2017), undoubtedly aided by curvature mechanics. However, it is the soft tissue elements, both passive and active, that 'control' the processes that influence vault curvature. A changeable vault profile allows the connective tissues to stress-stiffen. If it did not exist, the adjustable mechanism would be lost. Outside of curvature mechanics, the vault's mechanical attributes are also expressed via collagen properties through hierarchy in the connective tissues and brought about by lowering the vault under increasing loads to cause stress-stiffening effects. The rate of loading the soft tissues through their viscoelastic material properties of increasing stiffness behaviour via their loading rate-dependent properties is also important. Finally, by contracting the active muscular elements of the intrinsic and extrinsic foot muscles, stiffening forces are created through influencing foot profile and creating active tensile stiffness. The plantar intrinsic muscles in late midstance are most likely working in near-isometric contraction with the tendon fascicles sliding and lengthening to permit vault depression (Kelly et al., 2015). Creating active twisting of the vault's curvature also twists collagen fibres; a mechanism that can assist in stiffen them. However, changes in vault profile that create this normally seem solely to occur in terminal stance, as only then can the foot shorten once the forefoot alone has become loaded (Stolwijk et al., 2014).

The passive elements related to connective tissue-stiffening properties and the active mechanisms of force generation are both at play in generating stiffening within the foot. The ability to achieve success in stiffening the foot may result through a change in restricting sagittal plane midfoot dorsiflexion by only around 4° in the human foot, as reported by Holowka et al. (2017). This dorsiflexion resistance to the midfoot midstance phase bending moment prevents midfoot break, which is absent in most humans but is the normal situation seen in chimpanzees. This suggests that most humans can produce enough stiffness through the foot prior to heel lift to prevent midfoot buckling via a midtarsal–tarsometatarsal fulcrum. The primary mechanical stiffening mechanism within the foot at the late midstance–terminal stance boundary may lie within the forefoot's transverse plane stiffness (Venkadesan et al., 2017a,b; Yawar et al., 2017). However, the foot-stiffening process begins long before terminal stance and heel lift with depression of the vault at and after loading response.

3.2.9 Terminal stance stiffening mechanisms

At the midstance–terminal stance boundary, the foot must provide high stiffness in order to permit itself to act as a stable lever to apply ankle power through the forefoot. The medial longitudinal foot arch profile has long been thought of as the source of foot stiffening (Hicks, 1954; Bojsen-Møller, 1979; Ker et al., 1987; Williams and McClay, 2000; Prang, 2016). This hypothesis proposes that the height of the medial longitudinal arch is predictive of foot stiffness when the forefoot is loaded, and is supported by the finding in cadaveric studies that transecting the plantar aponeurosis lowers foot stiffness by around 25% (Ker et al., 1987). In addition, the windlass mechanism was purported to increase foot stiffness during terminal stance due to changes in geometric shape and its non-Hookean material properties (Hicks, 1954; Kitaoka et al., 1994; Gelber et al., 2014). However, these mechanisms cannot fully account for the high stiffness of the foot observed when the forefoot is under high loads or the fact that the windlass mechanism is not a significant part of heel lift foot stiffness (Yawar et al., 2017; Farris et al., 2020).

Living human feet are two to three times stiffer than chimpanzee's feet (Venkadesan et al., 2017b), yet mobility in the midfoot is very similar in both cadaveric human and chimpanzee feet (Bates et al., 2013; Greiner and Ball, 2014). The hallux dorsiflexes relative to the cuneiform twice as much in planus feet due to its increased midfoot mobility under body-weight loading than it does in average-vaulted feet (Kido et al., 2013). This would suggest that average vault-profiled feet are ≈100% stiffer than planus feet, more than could be accounted for by plantar fascia tightening alone (Yawar et al., 2017). The windlass mechanism relies on the diameter of the metatarsal heads and the MTP joints in the forefoot (Hicks, 1954), which are not known to be altered in planus foot profiles compared to other feet. Moreover, radiographs of planus feet in toe extension show that the windlass is engaged in planus subjects (Gelber et al., 2014) and suggests that the mechanism is still available for use in planus feet. However, in heel raise tests where the toes are free to extend, diabetic planus foot subjects were unable to support their body weight on the forefoot (Hastings et al., 2014); a finding that more likely reflects lack of muscular strength rather than loss of the windlass. The lack of a clear pathological link to longitudinal arch morphology also raises questions with regard to the effects of longitudinal foot mechanics on dysfunction (Murphy et al., 2003; Tong and Kong, 2013).

Producing an increased curved vault profile stiffens a structure, and the shortest length curvatures are the most efficient stiffeners of a shell or beam because of effects brought about through the radius of curvature and via span distance. Mechanically, it would seem inefficient to increase the curvature of the longitudinal foot profile to stiffen the foot when a shorter span distance to stiffen is available that could then aid longitudinal stiffening. Thus, the transverse plane curvature is probably more of a significant target for initiating foot-stiffening mechanisms (Venkadesan et al., 2017a,b; Yawar et al., 2017). At heel lift, all weightbearing forces are transferred to the forefoot before being transmitted to the opposite foot. It appears, through modelling, that forefoot spreading that occurs under increasing load has a coupling relationship that underlies the ability of the foot to provide stiffness as a direct consequence of changing transverse vault contouring (Yawar et al., 2017). Therefore, it can come as no surprise that transverse plane curvature is a feature of modern healthy human feet, even within longitudinally flatter profiled ones (Drapeau and Harmon, 2013; Venkadesan et al., 2017b).

Yawar et al. (2017) used tape around the forefoot to limit forefoot splay and stiffen the forefoot. They extended the hallux to record differences in change of midfoot height with and without the tape present and compared this with a control group of tape loosely applied to the forefoot. As a consequence of doing this, they discovered that increasing transverse stiffness by using firm tape to increase the transverse curvature profile of the forefoot enhanced stiffness over the entire foot by 50%. This resulted in significantly reduced changes in arch height with digital extension compared to the no tape and loose tape condition (Yawar et al., 2017). This may explain the findings of Farris et al. (2019) who found that the plantar intrinsic muscles seemed to contribute minimally to longitudinal arch stiffness, yet helped stiffen the distal foot to aid 'push-off' against the ground when walking or running. The stiffening of the foot as it resists forefoot splay suggests that a significant mechanical spring exists across the tightly bound metatarsal heads in the form of the deep transverse metatarsal ligament and associated anatomy (Yawar et al., 2017) (Figs 3.2.9a and 3.2.9b).

It is likely that the orientation of some of the plantar intrinsic muscles restricts forefoot splaying when active, increasing the stiffening effects of the transverse plane vault. The interossei and the lateral muscle belly of adductor hallucis seem particularly well placed to help influence transverse intermetatarsal stability. Muscles such as adductor hallucis, through its two heads, provide oblique and transverse stability across the midfoot and forefoot, respectively. More proximally, transverse stability can arise through tibialis posterior and peroneus longus activity together stiffening the foot from the tarsometatarsal level (Kokubo et al., 2012). These muscles work collectively as an agonist–antagonist in that they moderate the stresses in the agonist and increase the stress in the antagonist tendons (Morales-Orcajo et al., 2017).

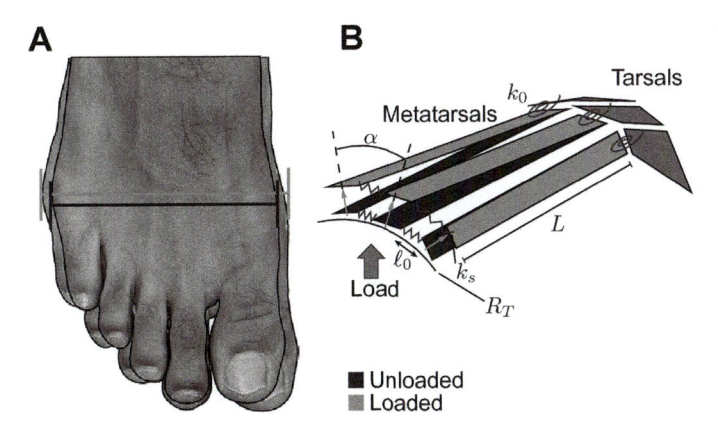

FIG. 3.2.9a Upon loading of the forefoot under body weight, the distal metatarsals splay apart in the transverse plane with the overlaid image (A) expressing the unloaded narrower foot profile. The model (B) presents a simplified three-metatarsal arrangement as torsional springs at their proximal ends (the ligamentous restrained tarsometatarsal joints), with the distal linear springs (the deep transverse metatarsal ligament) resisting the splay to stiffen across the forefoot's transverse plane. *(From Yawar, A., Korpas, L., Lugo-Bolanos, M., Mandre, S., Venkadesan, M., 2017. Contribution of the transverse arch to foot stiffness in humans. file:///C:/Users/user%23/Downloads/Contribution_of_the_transverse_arch_to_foot_stiffn%20(2).pdf.)*

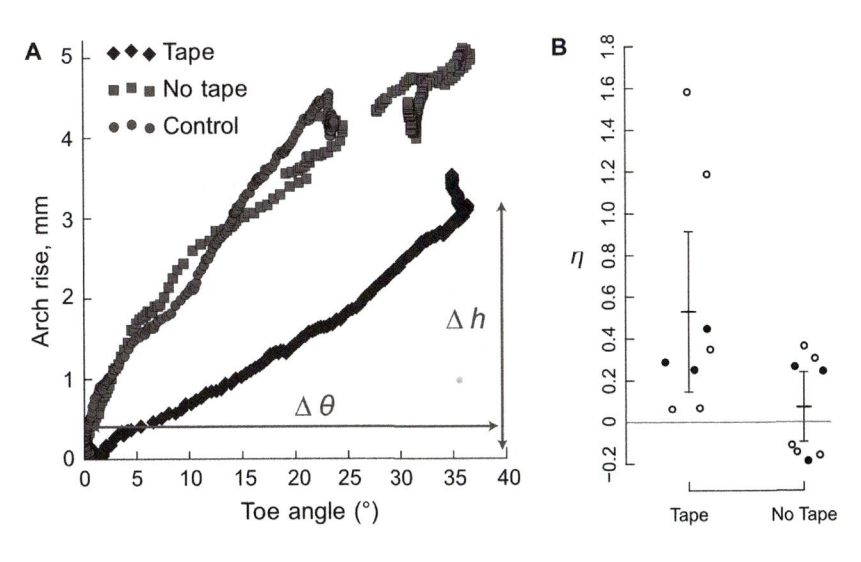

FIG. 3.2.9b The effects of taping the forefoot to bind the metatarsals closer together distally, compared to loosely applied tape or no tape, demonstrate that tightening the forefoot stiffens the metatarsal region. Upon extending the toe (A), the midfoot height (h) increases for all three conditions but least in the pre-stiffened taped condition. (B) shows a summary of the statistics demonstrating an incremental rise in midfoot height per degree of toe extension, with the taped condition illustrated to be stiffer than the other conditions. Solid markers are for male subjects and open for female. *(From Yawar, A., Korpas, L., Lugo-Bolanos, M., Mandre, S., Venkadesan, M., 2017. Contribution of the transverse arch to foot stiffness in humans. file:///C:/Users/user%23/Downloads/Contribution_of_the_transverse_arch_to_foot_stiffn%20(2).pdf.)*

Active forefoot stiffening

Most flexor plantar intrinsic muscles are orientated to create a stabilising, proximally directed force across the metatarsal heads, demonstrating most activity in late midstance and early terminal stance (Farris et al., 2019, 2020). The long extrinsic flexors are also active during this period, certainly with important influences from flexor hallucis longus (Ferris et al., 1995). Through the MTP joint's connective tissues that the intrinsic muscles are intimately bound to, they achieve a stabilising role that has much the same effect as the hole known as a 'box' that pole-vaulters place their pole into before initiating a vault. In employing such an analogy, the lift of the athlete can be likened to lifting the foot's rearfoot and lower leg segments' CoM during terminal stance. The plantar intrinsics are likely assisted by an already stiffened foot, but they and the long extrinsic flexors are also well positioned to proximally compress the phalanges into the metatarsal heads across the MTP joints as the toes start to extend following heel lift. As the plantar aponeurosis windlass mechanism stiffening becomes greater with digital extension, it seems logical that the muscles lead the way in stiffening with the windlass effect assisting later in terminal stance (Farris et al., 2020) as digits become more extended (Fig. 3.2.9c).

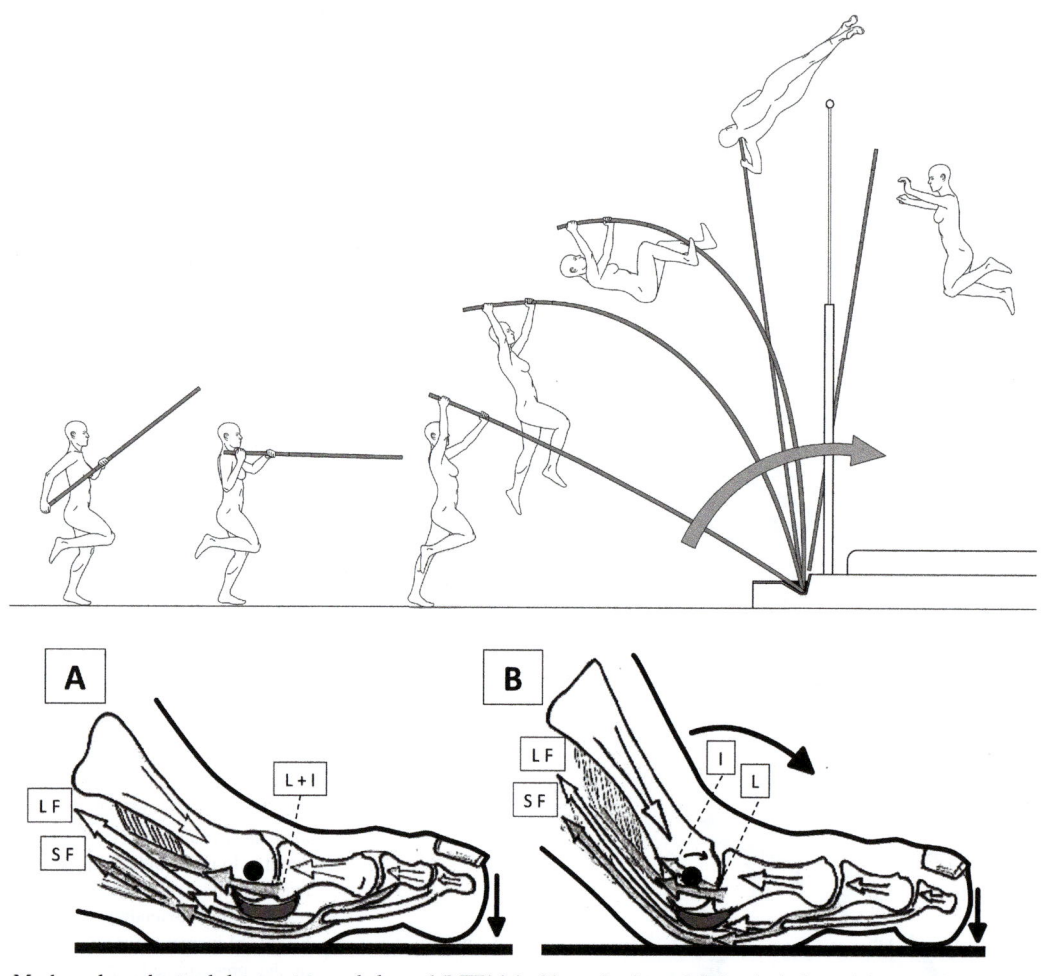

FIG. 3.2.9c Much can be understood about metatarsophalangeal (MTP) joint biomechanics and their role during gait by considering the action of a pole vault. It is essential for a successful pole vault that the tip of the pole is made stable to form the leverage by which the athlete can apply the momentum from the run up into that of rising and rotating over the bar. In late midstance (A), increasing activation of toe-flexing muscles such as the extrinsic long flexors (LF), the intrinsic short flexors (SF), and lumbrical and interosseous muscles (LI) sets up a plantarflexion moment on the digits at the MTP joints. This pulls the digits into the ground, but more importantly, creates a proximally directed force vector into the metatarsal heads which is itself being driven anteroinferiorly into the ground under body weight-induced vault deflection and lengthening. This compression force creates MTP joint stability and reduces shear under the metatarsal heads, much like the box used to insert the pole for the pole vault. Increasing tension from lengthening the plantar aponeurosis during late midstance likely aids this process. As heel lift initiates gait into terminal stance acceleration (B), the compression force from the digital flexors maintains stability while the MTP joint rotates into dorsiflexion as the ankle and midfoot plantarflex behind it. The tightening of the plantar aponeurosis as part of the windlass mechanism is likely an important aid to this process and may compensate for the loss of the interossei's flexion moment during late terminal stance. At the end of terminal stance, the interosseous tendons will slip above the MTP joints' instantaneous joint axis, converting them into toe extensors. This should only occur when there is little loading on the late terminal stance foot, and this may be helpful for preswing toe extension activity ready for toe-off. *(Permission www.healthystep.co.uk.)*

It can be surmised that extra stiffening of the foot at the late midstance–heel lift into terminal stance boundary via muscle contraction is critical in initiating adequate terminal stance stiffness. Thus, a combination of factors that achieve transverse plane stiffening, directly coupled to an increased sagittal plane stiffening restriction of midfoot dorsiflexion, crescendo during and just following heel lift. The long and short digital flexors stabilise the MTP joints and in so doing set up proximally directed forces into the foot while still permitting the digits to extend under ankle plantarflexion moments. This maintains a proximally directed trans-MTP joint forefoot stiffness as the other earlier foot stiffening elements such as the proximal connective tissue stress-stiffening under vault lengthening, disengage. This muscular mechanism can be aided later by the plantar aponeurosis and its associated structures via the windlass mechanism coupled to digital extension during terminal stance. Furthermore, vault curvature also increases under ankle and midfoot plantarflexion and MTP joint dorsiflexion, shortening the foot once terminal stance develops. The anatomy of the plantar aponeurosis is intimately linked to both the superficial and deep transverse metatarsal ligaments, the extensor hood apparatuses, the plantar plates, and the plantar intrinsic muscles that surround the MTP joints; an intimacy that allows stiffening mechanisms to easily couple.

Muscular transverse plane stiffening

The interossei have extensive proximal attachments that lie on adjacent metatarsals, effecting transverse arch stiffening by tethering the adjacent metatarsal shafts together. These are bipennate muscles consisting of three plantar interossei lying between each of the 2nd to 5th metatarsals and four dorsal interossei lying between each of the 1st to 5th metatarsals (Draves, 1986: pp. 301–302). These intrinsic muscles are likely providing either isometric or near-isometric contraction. The stability created by this proximal muscular intermetatarsal linkage will help stabilise the action of the long and short toe flexors, aiding a proximally directed compression force onto the metatarsal heads and thereby preventing distal translation of the metatarsal heads under high forefoot loading forces. This stability at the MTP joints will enhance windlass efficiency. While holding the proximal part of the metatarsals and their bases together while the metatarsal heads splay, the forefoot can create a decreasing radius of curvature from distal to proximal, further enhancing stability up to the tarsometatarsal joints. The weightbearing forefoot can thus increase its width and stability while the vault narrows proximally during terminal stance (Fig. 3.2.9d).

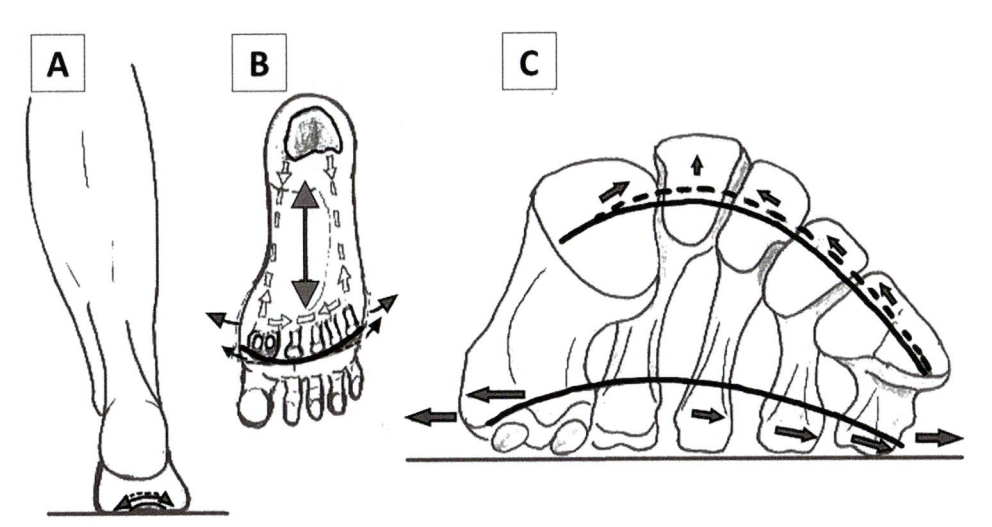

FIG. 3.2.9d Decreasing the radii of curvature and reducing span distances across the foot vault to increase stiffness is a complex process initiated by increasing forefoot loads. It is important to point out that the central metatarsal heads do not lift from the support surface to create changes in the transverse/horizontal radius during terminal stance. The 2nd and 3rd metatarsal heads are very important weightbearing structures. Only the areas proximal to the metatarsal heads undergo a process of reducing radius and span distance (A). The schematic of the plantar view on the right foot (B) attempts to demonstrate that under weightbearing loads, the foot vault lengthens and widens *(solid black arrows and arrows)* resisted by increasing soft tissue tensions across the vault from medial to lateral, transversely, and longitudinally across the foot. This consists of passive connective tissues undergoing stress-stiffening and muscular active forces from contraction creating near-isometric vault motion restraints. This process includes muscles that pull the proximal metatarsal bases together, while the spread of the distal shafts and metatarsal heads are limited via deep transverse metatarsal ligament restraint. The metatarsal splaying, being uneven, can decrease the radius of curvature proximal to distal within the horizontal plane of the metatarsal heads. With the proximal ends of the metatarsal bases pulling together more than the metatarsal heads (C), a smaller radius of curvature is created proximally to distally forming a stabilising 'scoop-like' profile across the transverse vault of the plantar forefoot. This creates a relatively wide but increasingly curved, forefoot surface contact area while the midfoot behind narrows. *(Permission www.healthystep.co.uk.)*

The other intrinsic muscle likely to be directly involved in transverse plane stiffening is the adductor hallucis. It is divided into two separate muscles with one common attachment to the lateral or fibular sesamoid and the lateral plantar aspect of the proximal phalanx of the hallux. The oblique muscle arises from the bases of the plantar–medial aspects of the 2nd, 3rd, and 4th metatarsals, running distally and medially to a distal attachment on the lateral sesamoid and proximal hallucal phalanx. The transverse part of this muscle arises distally from attachments to the plantar plates, plantar metatarsophalangeal ligaments, deep transverse metatarsal ligament, consistently from the 3rd and 4th MTP joints, and variably from the 5th and/or 2nd MTP joint areas (Draves, 1986: pp. 298–299). The orientation of this muscle suggests a significant role in limiting forefoot splay, actively assisting the passive restraint of the deep transverse metatarsal ligament. Its variable activity can add adaptability into forefoot stiffness levels (Fig. 3.2.9e).

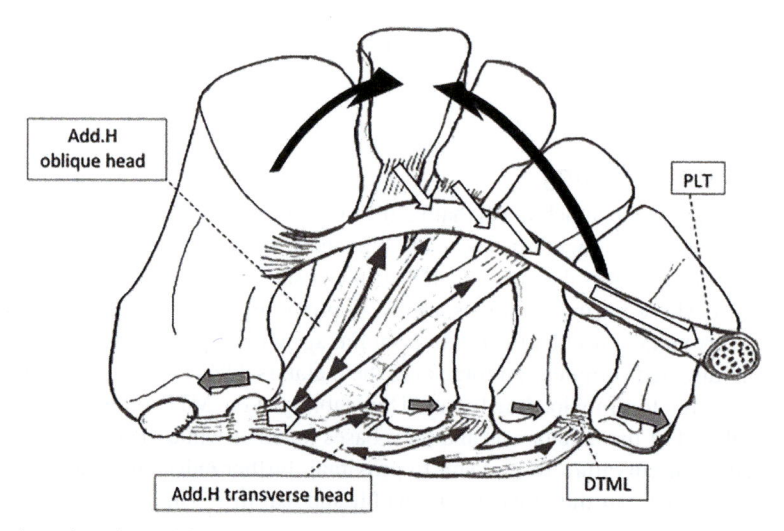

FIG. 3.2.9e Adductor hallucis produces internal force vectors that are important for forefoot transverse vault mechanics, with a significant linked longitudinal action that works in concurrence with the dorsal and plantar interosseous muscles. Through its transverse head (Add.H. transverse head), the muscle provides lateral attachments from the deep transverse metatarsal ligament (DTML) between the 2nd and 3rd, 3rd and 4th, and 4th and 5th metatarsal heads, running muscle fibres to an attachment on the lateral (fibular) sesamoid and its restraining soft tissues. This sets up vectors that can actively and controllably restrain metatarsal head splay both medially and laterally (grey arrows) coupled to DTML tightening. The larger and more powerful oblique belly (Add.H. oblique head) that arises proximally from the bases of the central three metatarsals, lateral cuneiform, and cuboid provides proximal to lateral directed vectors to the sesamoid anatomy. This muscle belly can tension the foot mediolaterally and obliquely in a distal to proximal direction, most likely through near-isometric contraction. There is a close relationship with the peroneus longus tendon (PLT), for Add.H. oblique also has proximal attachments to this important tendon that compresses the midfoot articulations in a similar direction to peroneus longus under the foot. Thus, as the peroneus longus tightens, the proximal attachments of Add.H. oblique stabilise. Together, these muscles therefore improve vault stability. In this way, distal transverse vault mechanisms that stiffen the forefoot are coupled to more proximal transverse and longitudinal vault stiffening through both muscular and ligamentous connections. *(Permission www.healthystep.co.uk.)*

The dorsal and plantar interossei muscles are likely important in stabilising the forefoot in a complimentary manner to adductor hallucis (Fig. 3.2.9f). However, the plantar intrinsic muscles are not acting alone in this transverse plane stiffening. The foot's extrinsic muscles, such as tibialis posterior and peroneus longus, have oblique tendon routes that are orientated to reinforce the transverse vault across the tarsometatarsal and lesser tarsus regions. This is a region of extensive ligament restraints. Tibialis posterior has a primary distal attachment point to the tuberosity on the medial navicular, with a variable extension to the medial cuneiform and calcaneus (Draves, 1986: p. 270) providing an important frontal plane lever arm to resist frontal plane eversion moments of the rearfoot (Imhauser et al., 2004).

However, such medial attachment points are not the only ones. Tibialis posterior's tendon provides a highly significant tendon slip with a lever arm used in maintaining the transverse arch. This tendon slip passes through the groove near the navicular tuberosity to run distally and laterally to attach variably to the plantar surface of the bases of the 2nd, 3rd, and 4th metatarsals as well as the plantar surface of the 2nd cuneiform and cuboid (Draves, 1986: p. 270). The most consistent parts of this tendon's attachments are to the 2nd or intermediate cuneiform and the 4th metatarsal, with variant slips found gaining attachment to the lateral cuneiform, cuboid, and 5th metatarsal (Draves, 1986: p. 270). These attachment slips also blend with and strengthen the plantar ligaments they cross, giving a potential mechanism for coupling transverse and sagittal (longitudinal) stiffening between ligaments and tendons. Tibialis posterior is highly active during late midstance into

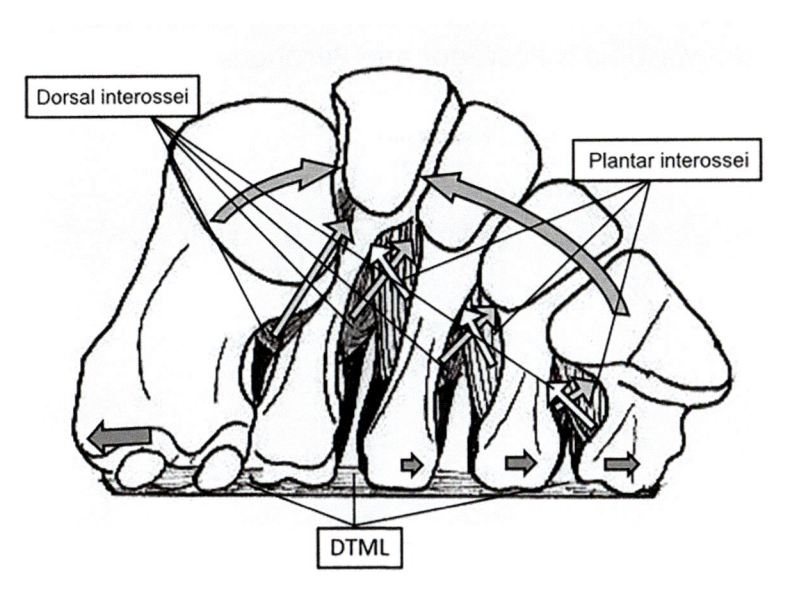

FIG. 3.2.9f The four dorsal interosseous muscles *(dark-grey muscle and vector arrows)* and the three plantar interosseous muscles *(striated muscle and white vector arrows)* have fibrous proximal attachments that lie between the metatarsal shafts. The dorsal interosseous muscles have attachments that cross the intermetatarsal space on opposing sides of all five metatarsals. Together, they provide transverse vectors that draw the metatarsal shafts together proximally. The interossei have tendinous distal attachments onto the bases of the proximal phalanges that can link MTP joint stability to intermetatarsal base compression forces. With powerful tarsometatarsal ligaments proximally and the deep transverse metatarsal ligament (DTML) distally, the forefoot becomes a controllable semi-elastic spring with a hemicone like narrower radius proximally than distally. *(Permission www.healthystep.co.uk.)*

the early terminal stance phase of gait (Murley et al., 2014) where it has considerable influence on foot stiffness (Kokubo et al., 2012).

The tendon of peroneus longus is orientated to bring about transverse plane stability across the lesser tarsal joints and the 1st and 2nd metatarsal base tarsometatarsal articulations. It runs through the peroneal groove (or near to it) on the cuboid, passing plantar to the cuboid to course medially and distally under the distal long plantar ligament that serves as the fibrous part of an osseofibrous canal (Draves, 1986: p. 257). The tendon proceeds across the midfoot to attach distally to the plantar posterolateral aspect of the medial cuneiform and the lateral side of the plantar surface of the base of the 1st metatarsal, with a common slip gaining attachment to the base of the 2nd metatarsal. Occasionally, slips to the 3rd and 4th metatarsals and to the oblique adductor hallucis muscle are found (Draves, 1986: p. 257). The resultant force vector of this muscle is primarily that of a laterally and proximally directed compression across the transverse profile of the foot. If unresisted, the peroneus longus vector also results in eversion and abduction of the foot. This muscle is active in late midstance and terminal stance in conjunction with tibialis posterior, which together increase foot stiffness around heel lift (Kokubo et al., 2012) (Fig. 3.2.9g).

Late terminal stance stiffness

After heel lift, the vault of the foot starts to rise accompanied by foot shortening (Stolwijk et al., 2014). The toes extend and the ankle moves into plantarflexion. During this final phase, the midfoot creates a plantarflexion and adduction motion that enhances ground contact to the medial side of the forefoot. The benefit of this is a longer lever arm length between the 2nd and 1st MTP joints and the ankle axis of rotation and the distal Achilles attachment from which to apply the ankle plantarflexion moment and power (Holowka et al., 2017). Use of the medial MTP joints' fulcrum has been termed *high gear* (Bojsen-Møller, 1979). Midfoot plantarflexion, eversion, and adduction 'twists' the midfoot in its frontal plane into a higher vault profile. Torsion of the midfoot has been defined as rotation of the forefoot relative to and opposite in direction to the rearfoot (Graf et al., 2017) and occurs through the lesser tarsus via the transverse tarsal joint (Chopart's joint), consisting of the talonavicular and calcaneocuboid joints through to the tarsometatarsal joints (Lisfranc's joint). The tarsometatarsal joints consists of the cuboid and all three cuneiform joints with their metatarsal bases. These joints allow for triplanar joint motion. In the frontal plane, they are reported as having 15°–20° eversion and 35°–40° of inversion when the rearfoot is fixed. Rearfoot inversions or eversions during midstance and at heel lift will influence the midfoot's ability to stabilise (Fig. 3.2.9h).

The midfoot joints pass through a range of plantarflexion, eversion, and adduction, maintaining forefoot contact under ankle plantarflexion power (Holowka et al., 2017). While the benefit of midfoot and ankle plantarflexion in maintaining forefoot contact to the support surface is obvious, it is the adduction moment shifting the CoGRF across the forefoot towards the medial MTP joints that further enhances mechanical efficiency of ankle power from the middle of terminal stance (Holowka et al., 2017). The midfoot plantarflexion and adduction reported by Holowka et al. (2017) are probably driven primarily by the actions of peroneus longus and tibialis posterior while they are maintaining foot stiffness, with

Proximal Transverse Vault Control via Tibialis Posterior and Peroneus Longus

A Late Midstance

B Early Terminal Stance

C Mid-Terminal Stance

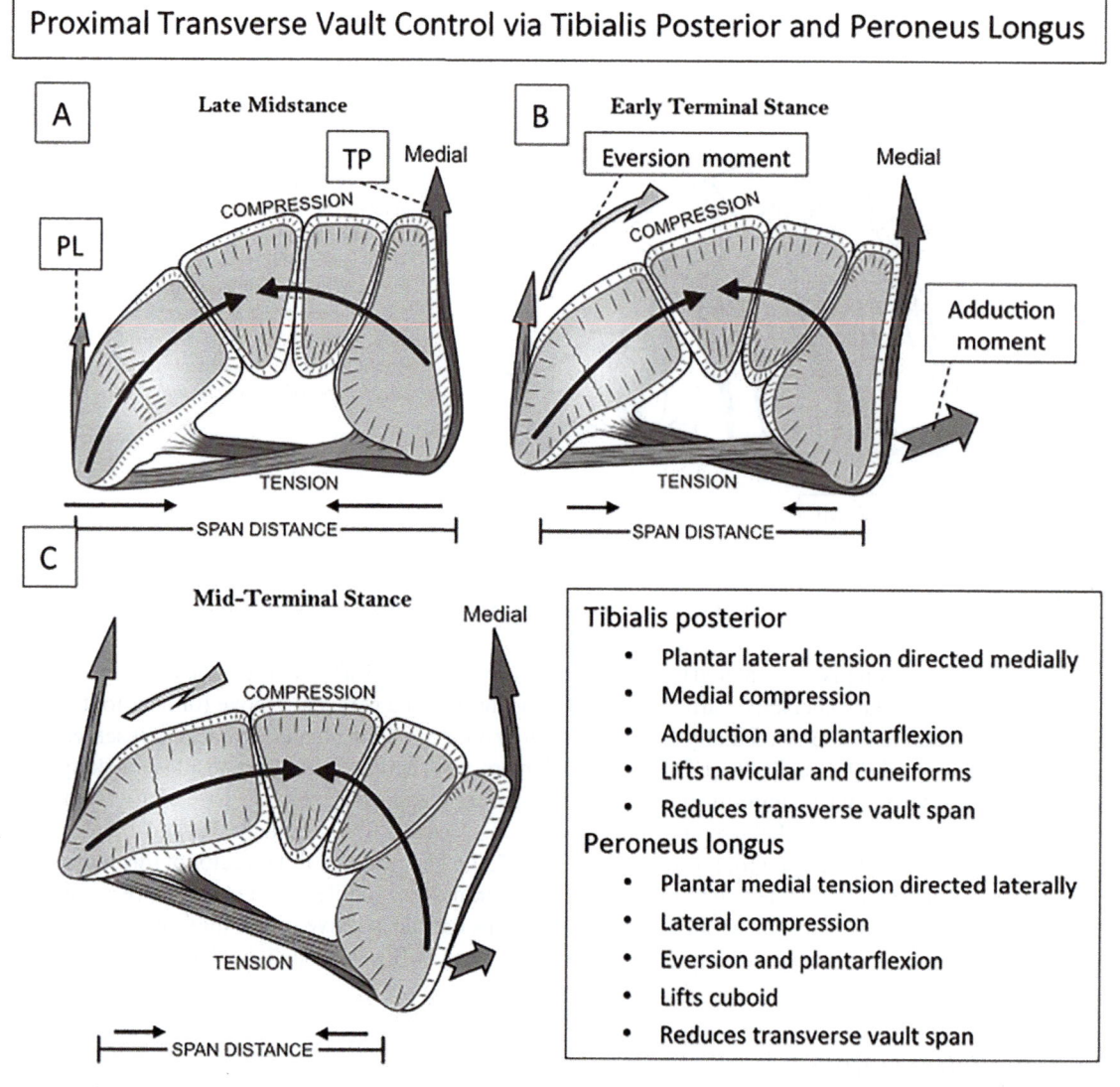

Tibialis posterior
- Plantar lateral tension directed medially
- Medial compression
- Adduction and plantarflexion
- Lifts navicular and cuneiforms
- Reduces transverse vault span

Peroneus longus
- Plantar medial tension directed laterally
- Lateral compression
- Eversion and plantarflexion
- Lifts cuboid
- Reduces transverse vault span

FIG. 3.2.9g Tibialis posterior (TP) and peroneus longus (PL) are agonistic and antagonistic muscles that work synergistically to stabilise and control motion around the midfoot and tarsometatarsal boundary. In loading response, but particularly during late midstance and early terminal stance, these muscles apply force vectors through their tendon tensions across the vault aided by wrapping them around the medial and lateral sides of the vault, that provide high compression stability across the midfoot articulations. This at first limits the widening and then narrows the transverse profile of the foot vault (A). Together, these muscles' tendons act as a tensioned support stirrup across the plantar aspect of the vault and provide midfoot and tarsometatarsal plantarflexion moments resisting vault sagging during midstance. Tibialis posterior also supplies a dominant midfoot adduction moment while peroneus longus (aided by peroneus brevis) generates an eversion moment *(black arrows)* across the midfoot as terminal stance begins (B). As the heel lifts, elastic energy released from their tendons increases the plantarflexion, adduction, and eversion moments that shift the peak weightbearing surface to the medial metatarsals (especially the most stable 2nd metatarsal). By doing so, the foot can utilise the longer external moment arm at the medial MTP joint fulcrum for application of the plantarflexor power for acceleration through terminal stance (C). *(Permission www.healthystep.co.uk.)*

peroneus longus and brevis acting together to create midfoot eversion (Kokubo et al., 2012). Tibialis posterior applies the adduction moment which must overcome the abduction moment created by peroneal activity. The midfoot plantarflexion, eversion, and adduction together 'twist' and compress the rearfoot and the forefoot in opposition to each other, driving the vault back into a higher profile and conforming to the stiffening model of the osteoligamentous plate (Sarrafian, 1987; Eslami et al., 2007; Leardini et al., 2014; Araújo et al., 2019).

This twisting of the midfoot maintains the transverse vault profile and stability across the lesser tarsus, tarsometatarsal joints, and metatarsals, which has previously been reported as 'midfoot locking'. The longitudinal profile of the vault in the sagittal plane increases, coupled to this midfoot twisting and also through the windlass effect. The windlass effect via the plantar aponeurosis only contributes around 25% of the foot's stiffness through the longitudinal arch (Ker et al., 1987;

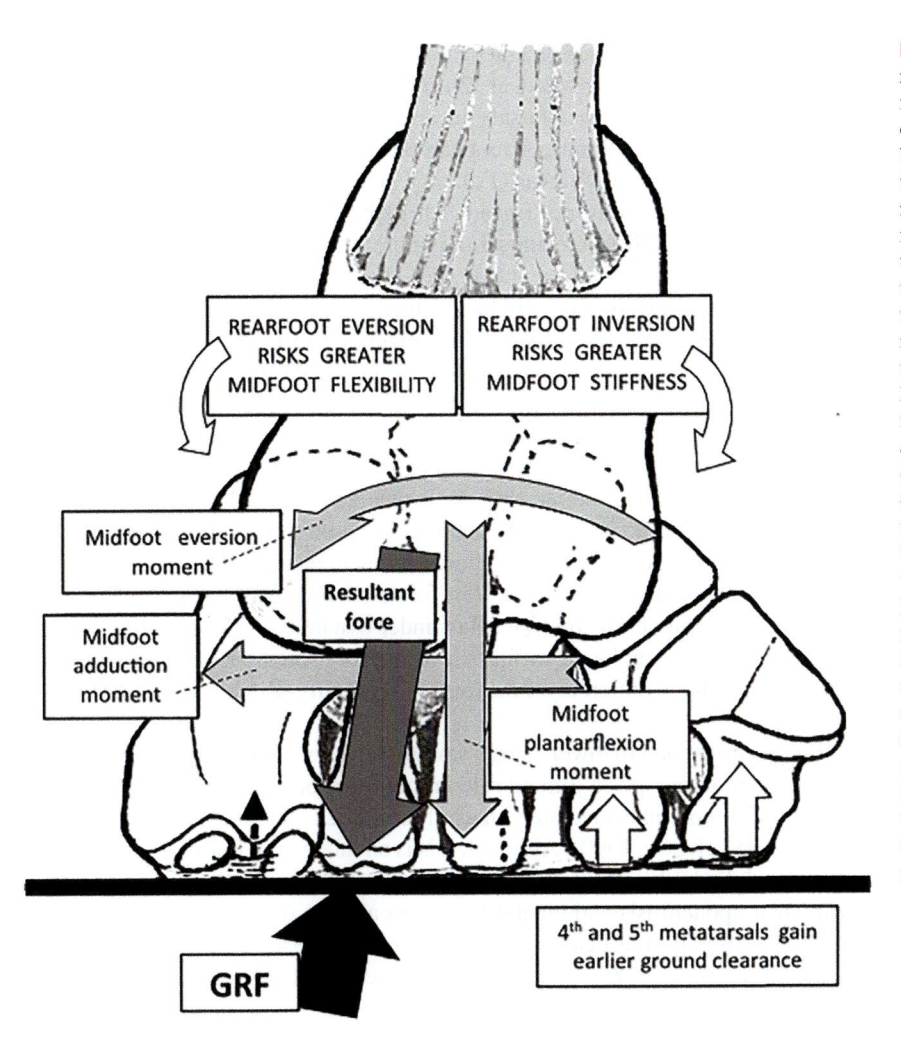

FIG. 3.2.9h Forefoot to rearfoot postural relationships across the midfoot can influence foot stiffening by 'twisting' or 'untwisting' connective tissue fibres, much as proposed by the osteoligamentous plate model. During terminal stance, the rearfoot offloads permitting more freedom of motion between forefoot and rearfoot across the midfoot. With tibialis posterior and peroneus longus between them setting up eversion, adduction, and plantarflexion moments *(grey arrows)*, the lateral metatarsals offload from ground contact while the medial metatarsals become increasingly loaded under the acceleration GRF derived from the resultant force *(black arrow opposing dark-grey arrow)* created by combined muscular activity. Which metatarsal head is the last one to remain loaded is variable (but usually the 2nd is most stable) and this is influenced by factors such as muscle strengths, soft tissue weaknesses, and local articular flexibilities. Rearfoot positions, inverted or everted to the forefoot during midstance and terminal stance, influence the amount of twist within the connective tissues, affecting midfoot and forefoot stiffness. Rearfoot inversions will tend to increase the twist as long as the midfoot and forefoot still evert, adduct, and plantarflex, increasing stiffness. Eversion of the rearfoot with these usual midfoot motions will tend to reduce connective tissue torsional stiffness. *(Permission www.healthystep.co.uk.)*

Huang et al., 1993), whereas transverse plane stiffening has been modelled to provide around 50% (Yawar et al., 2017). These findings suggest that most of the foot's stiffness comes from muscle activity and ligamentous stiffening which precedes any that is provided by the aponeurosis. Well after heel lift, during mid-terminal stance, the windlass mechanism comes into more significant play coupled to increasing digital extension angles when muscle-induced and body weight-derived forces through the forefoot are decreasing after the F3 and F5 peaks of the force–time curve.

3.2.10 The influence of foot shape and gait speed on compliance and stiffening events

Foot shape

Foot dimensions may have significant implications on mechanical advantages or disadvantages via the ability to provide levels of compliance and stiffness during gait. The passive stiffness levels direct the need for active muscular intervention to control the foot's material properties. For example, transverse radius of curvature is proportional to the square of the width and would thus be quadratically larger on the wider foot. This implies that transverse plane stiffening is harder in the wider foot with consequential reduced sagittal plane coupled stiffness in broader feet (Yawar et al., 2017).

The geometric externally visible curvature of the foot (i.e. planus or cavus) appears less important than the orientation of the articular axes of adjacent metatarsals (Yawar et al., 2017). Metatarsal torsion change has been an area of 'high evolutionary' activity in the hominin foot (Drapeau and Harmon, 2008, 2013), suggesting that the rotations within the metatarsal shafts creating transverse curvature across the forefoot have served an important role in plantigrade bipedalism and seem to predate evolution of medial longitudinal arch profile changes.

The elastic energy stored in the stretching of the distal transverse metatarsal ligaments is proportional to the angle of misalignment between the articular axis of adjacent metatarsals (Yawar et al., 2017). The foot may appear to be devoid of a transverse arch, but the preferred direction of flexion for adjacent metatarsals may be misaligned by virtue of their articular surface geometry and ligamentous layout at the tarsometatarsal joints. A foot can present as a geometrically 'flat foot' in appearance, but still have a functional transversely curved foot (Yawar et al., 2017). Such geometrically flat but functionally curved structures are found in the pectoral fins of bony ray-finned fish like the perch (Nguyen et al., 2017). The arrangement of fin rays to induce curvature equates well to metatarsal curvature in both the transverse and frontal planes, as they involve hard rays (fin bones or metatarsals) linked by soft tissues. This means that anatomical flat feet may function perfectly well despite a flat geometric shape because they still have effective functional curvature in the transverse plane of their vaults via the metatarsals, despite lacking it in the longitudinal plane. The importance of the soft tissue elements in maintaining pedal structural integrity and vault curvatures gives a highly significant role for soft tissues in the physiological health of the locomotor system.

Quick repeatable clinical measures of foot stiffness form an essential tool in the progression of diseases such as diabetes (Rogers et al., 2011; Hastings et al., 2016). However, many existing methods of foot assessment rely on shape only such as arch height index (Butler et al., 2008; McPoil et al., 2008), longitudinal arch angle (McPoil et al., 2016), and foot posture index (Redmond et al., 2006; Behling and Nigg, 2020) that do not identify changes in width within the transverse plane. This may explain the difficulty in being able to link pathological risk to such clinical assessment techniques and the variable results in linking foot type to kinematic and kinetic data (Gates et al., 2015). New forms of clinical assessment are probably warranted to establish foot flexibility. Quantifying the amount of forefoot splay under dynamic forefoot loading force, in both ankle-dorsiflexed and ankle-plantarflexed positions to replicate stiffness levels before and after heel lift, is likely to prove useful. Arch height flexibility (AHF) seems a suitable option to assess for foot flexibility (Zifchock et al., 2019; Cen et al., 2020).

Both high and low calcaneal inclination angles correspond to lower energy storage capacity within the foot (Simkin and Leichter, 1990). This may give static weightbearing radiographs a limited role in foot function assessment. Moving away from the extremes of calcaneal inclination angles to more intermediate values of around 22° markedly improves energy storage capacity in linear spring models (Simkin and Leichter, 1990). Although this is an interesting finding, it is important to point out that connective tissues demonstrate nonlinear lengthening at low and high stress loading, a behaviour which could influence results. By considering the plantar aponeurosis and plantar ligaments as only functioning to store energy within the sagittal plane without consideration to the highly significant effect that the transverse arch has on foot stiffening, may also limit the value of calcaneal inclination angles. However, such sagittal plane modelling does suggest possible mechanisms that might lead to injury during energy storage.

Gait speed

Another important consideration is gait speed. The lower limb behaves with greater stiffness during faster walking speeds (Kim and Park, 2011), and the foot is no different. Muscles, in comparison, are less able to form actin–myosin cross-bridges at faster speeds. The GRF generated in gait is dependent on mass × acceleration, whether or not the CoM is accelerating towards or away from the support surface when creating a collision. The direction of the GRF reflects the primary direction of these collisions. The GRF peaks are therefore greater with faster walking speeds. The F2 trough produces a greater drop in force the higher the CoM rises over the extending limb, which indicates CoM acceleration away from the support surface without a large impulse. At slower gait speeds, the trough tends to be shallower as GRFs are higher. Thus, at faster walking speeds compared to slow speeds, initial contact collision energy dissipation mechanisms are more important as they have to handle greater energy. However, weight transfer forces during midstance are likely to stress the midfoot more when slow walking. Faster walking speeds also require greater acceleration and therefore require higher levels of foot stiffness to improve acceleration mechanics in order to transfer ankle power into GRFs. GRFs in the middle of midstance are higher at slower walking speeds while the F1 and F3 peaks are lower. This suggests that midfoot vault-loading mechanics become more significant with slower walking speeds, increasing midfoot impulses. Midfoot impulses are higher in more flexible feet and lower in stiffer feet, reflecting the GRF interaction with foot stiffness and speed (Stolwijk et al., 2014; Hedrick et al., 2019; Cen et al., 2020). For a visual review of these concepts, see Fig. 1.2.4c, Figs 3.1.4d and 3.1.8a.

Healthy stiff and flexible feet produce different impulse loading patterns, with stiff feet having lower rearfoot and midfoot impulses in walking and running but higher forefoot impulses than more flexible feet (Cen et al., 2020). Combining this information with the effects of GRF on walking speed, different requirements and effects are likely to result from changing the walking speed in both flexible and stiff feet. Fast walking, healthy stiff feet are likely to increase their rearfoot impulses and forefoot impulses, although their midfoot impulses are likely to lower further and their forefoot impulses likely to increase the most. Flexible feet will follow a similar pattern, but their rearfoot impulses are likely to raise the most whereas their midfoot impulses will probably remain higher than those of stiff feet, while their forefoot impulses will be lower. In slow walking, healthy stiff feet are likely to experience lower rearfoot and forefoot impulses but greater midfoot impulses. However, these should not be as high as those found in slow walking flexible feet which should demonstrate proportionately lower forefoot impulses. Accelerating at a slow speed on a flexible foot is thus likely to require more muscle force to be generated for foot stiffening.

Lengthening and shortening of the foot vault link to the stiffness of the foot and the loading rate. Thus, walking at a faster pace causes the foot to increasingly stiffen its connective tissues so that the overall amount of vault lengthening changes decrease at faster gait speeds. However, the shortening from elastic recoil increases (Stolwijk et al., 2014). At slow speeds, lengthening changes increase while shortening decreases because the foot becomes less elastic and more compliant in its behaviour and thus more reliant on muscle activity for stiffness.

The preferred walking speeds of healthy individuals probably reflect the interaction of the GRF and foot stiffness properties as well as the foot's energy-dissipating and generating capacities throughout the lower limbs. Thus, complex relationships are created that finally explain the preferred walking speed. The effects of pathology-induced flexibility changes in feet are as yet not fully known. However, it is clear that age, disease, and dysfunction tend to slow gait, while diseases such as diabetes stiffen connective tissues which result in aged/diseased-induced reduced walking speeds that do not lower the tissue stresses and impulses as expected. Musculoskeletal damage may stiffen feet through arthritis-induced joint mobility loss or increased flexibility through muscle weakness and/or tendon or ligament mechanical failure. The effects on and compensations in the individual are likely to reflect the complex total limb summary of energy dissipation and stiffness relationships.

3.2.11 The foot in modulating viscous spring-damping in running

The human foot is described as having spring-like qualities in running that use vault mechanics to improve economy by delivering some 8%–17% of the mechanical energy for each step (Ker et al., 1987; Stearne et al., 2016; Kelly et al., 2018). The proportion of energy dissipated increases with running speed, suggesting that the foot functions as a viscous spring-damper (Kelly et al., 2018). Differences in impact shock frequency content between forefoot and heel contact runners indicate that different mechanisms of impact energy attenuation are at play with different foot strike patterns (Gruber et al., 2014). The impact-modulating mechanisms in heel strike running are much the same as in walking with heel plantar fat pad, tibialis anterior, shank wobbling masses, and the skeletal frame providing energy dissipation, but subjected to higher loading impact forces resulting from increased acceleration.

Running requires the lower limb to increase its flexion angles, acting as a spring–mass, with GRF at its highest in midstance and the knees considerably more burdened with energy dissipation. The increased loading rate will also cause the connective tissues associated with these structures to behave more stiffly through time-dependent properties, and thus, tissue resonance effects will be much greater as a result of higher impact forces and frequencies on stiffer joints. Forefoot contact in running is a common variant, so impact dissipation mechanisms of the foot and ankle used in forefoot running need extra consideration. As forefoot impact in walking is 'largely' abnormal, impact mechanisms that are used to dissipate energy through the foot in forefoot walking are not specifically discussed here. However, the mechanisms are the same as those available for forefoot running, but with greater compliance due to the slower loading rate. Stiffening of the foot for acceleration in relation to the spring–mass model rather than the inverted pendulum is also different during running.

The mechanisms of stance CoM translocation in the heel strike and forefoot strike spring–mass models are initially significantly different. Impact-braking energy dissipation and acceleration is still required as in walking, but under more rapid loading, the connective tissues behave more stiffly throughout. This means that the foot of a runner should behave more elastically and stiffly with less compliance than that of a walker; true of a rearfoot striker as of a forefoot runner. In all running, stiffening is easier to achieve passively than compliance because of speed-initiated tissue stiffening, but limb stiffening and compliance requires greater muscle effort because of the force–velocity relationship. However, somewhat different anatomical structures and kinematic solutions are at play in the forefoot runner compared to the rearfoot runner in dissipating impact forces (Gruber et al., 2014). Also, whether walking or running, the foot functions in a chain of lower limb events to achieve impact force dissipation, but the running foot has a more significant role in energy storage than the walking foot (Ker et al., 1987; Sasaki and Neptune, 2006).

Foot energy dissipation in running impact

The heel fat pad would appear to be potentially vulnerable to injury during high velocity heel-toe running. The connective tissue chamber walls behave more stiffly due to their time-dependent properties, so the potential ability to dissipate energy from the fluid-fat phase of the fat pad is lessened. Indeed, viscoelastic modelling indicates that an increase in the foot's contact velocity results in a rising of internal stresses within the plantar fat pad of the heel, suggesting increased risk of injury (Chen and Lee, 2015). This could be a reason why, with increasing running velocity, humans tend to start avoiding full plantigrade loading of the foot initiated by heel contact (Hasegawa et al., 2007; Hayes and Caplan, 2012; Hatala et al., 2013; Forrester and Townend, 2015). It is likely that the presence of cushioning in shoes delays the need for this changeover to forefoot running. However, some barefoot runners do persist in heel striking during running at higher speeds, particularly in habitually barefoot running populations, although their stride lengths reduce (Hatala et al., 2013). The diminished stride length likely reduces the heel strike angle to the ground so that contact occurs on a larger surface area over the plantar fat pad. Weight of the runner as well as velocity may also play a part in the shift to forefoot strike to protect the heel fat pad.

Compared to walking, energy dissipation mechanisms are required to manage higher forces during running, whatever foot contact 'type' occurs, because of higher peak GRFs resulting from higher locomotive velocities. Running soft tissue dynamics of 'wobbling masses' are more significant in impact force dissipation through damping oscillations in the first 90ms after initial contact. Wobbling mass excursions of 3mm to 4cm at 3–55Hz occur primarily in the horizontal (transverse) plane of the leg (Schmitt and Günther, 2011). In forefoot running, the lower limb's wobbling masses still have a significant role in energy dissipation, but the triceps surae–Achilles complex can also dissipate energy away from wobbling masses. As the initial heel strike transient impact GRFs are absent in forefoot running, energy dissipation is delayed and spread throughout the braking event, so that the peak stress load on these wobbling tissues is also likely to be reduced or delayed. It is worth remembering that the overall force on the body remains the same (mass x acceleration). However, the loading of the force is more evenly distributed in forefoot running away from an initial heel collision. There is reduced activity in tibialis anterior with forefoot strike (Rooney and Derrick, 2013) due to the loss of the plantarflexion moment being replaced by a dorsiflexion moment with stresses moved to the ankle plantarflexors. Thus, the combined energy dissipation of muscles such as tibialis posterior, peroneus longus, and the triceps surae–Achilles tendon during forefoot striking is increased, while reduced on the wobbling mass and tibialis anterior and removed from the heel fat pad.

In forefoot impact, the forefoot plantar fat pad and other forefoot soft tissues are loaded first. The local anatomy consists of a complex framework of skin, fascial layers, muscles, and plantar fat pads that are similar in arrangement but less highly structured than the heel fat pad (Bojsen-Møller, 1999). Little research has been attempted on investigating the role of the properties of the metatarsal plantar fat pads in impact dissipation compared to the heel's plantar fat pad. What data has been gathered demonstrates that the plantar fat pads of the forefoot in young adult runners show that elastic (Young's) modulus increases with rising impact forces with energy dissipation ratios being greatest during high impacts (Hsu et al., 2005). In healthy but aged forefeet, the elastic modulus seems to increase as does the *energy dissipation ratio* (Hsu et al., 2005). The energy dissipation ratio is effectively a measurement of hysteresis and has been proposed to demonstrate nonlinear tissue characteristics to indicate how much energy dissipation the tissues have to endure (Hsu et al., 2000). High energy dissipation indicates high tissue stresses and energy lost to elastic recoil. The data so far suggest that in young healthy adults, metatarsal fat pads have the ability to dissipate energy without too much tissue stress. However, in older adults and certainly in diabetics, high forefoot impacts may create a risk of injury through becoming stiffer, less elastic, and thinner (Hsu et al., 2005, 2007b; Chao et al., 2010) (Fig. 3.2.11a).

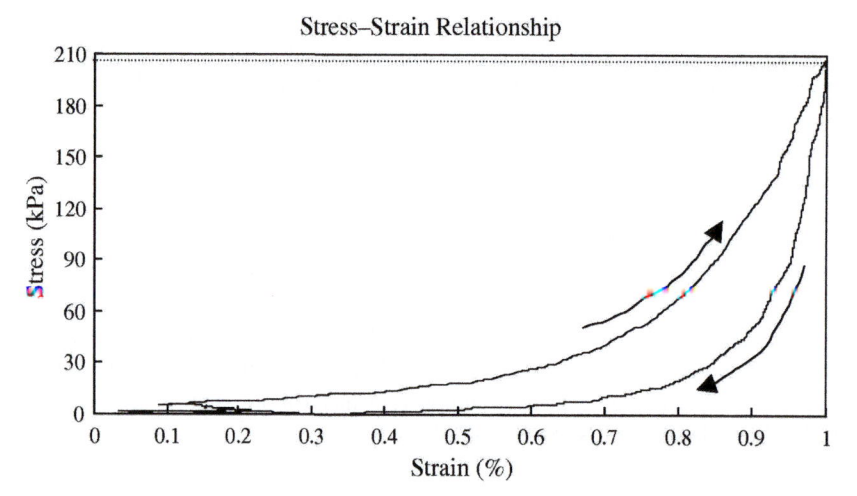

FIG. 3.2.11a The stress–strain curve of the plantar soft tissue under the left 3rd metatarsal in a 62-year-old male diabetic. The upper loading curve (↗) and lower unloading curve (✓) form an enclosed area defining the hysteresis. The energy dissipation ratio is defined as the ratio of the closed area between the loading and unloading curve to the area formed by the loading curve, the vertical line, and the X-axis. The dotted line represents the maximum loading stress of 207 kPa in this example, with a loading and unloading period lasting one second. In healthy individuals aged 30 years, pressures have been reported between 185 kPa to 198 kPa. Thus, diabetic patients endure high dissipated energy within the forefoot under the metatarsal heads. *(From Hsu, C.-C., Tsai, W.-C., Shau, Y.-W., Lee, K.-L., Hu, C.-F., 2007. Altered energy dissipation ratio of the plantar soft tissues under the metatarsal heads in patients with type 2 diabetes mellitus: A pilot study. Clin. Biomech. 22 (1), 67–73.)*

In forefoot strike patterns, the impact loading force is located anterior to the ankle's axis of rotation, with the ankle displaying a plantarflexed touchdown angle (Gruber et al., 2014). Thus, the ankle is subjected to a contact dorsiflexion moment. The triceps surae with the Achilles tendon (rather than the tibialis anterior) is now the primary muscle–tendon complex used in energy dissipation at impact. The Achilles tendon activity increases its loading time during stance to 100% during forefoot strikes (Kernozek et al., 2018). This allows the triceps surae to powerfully utilise its stretch-shortening cycle with pre-activation before impact, eccentric loading during impact braking, followed by a rapid transition to concentric-isometric contraction for acceleration (Komi et al., 1992). The mechanism utilised in energy dissipation through muscle–tendon buffering in the triceps surae is that interchange of length between tendon and muscle which has been previously outlined for tibialis anterior during heel contact gait. However, this time, muscle–tendon buffering occurs through a muscle–tendon complex with a much larger cross-sectional within an energy storage tendon. Peak tendon stress, strain, and strain rate are higher in forefoot runners than in rearfoot runners (Lyght et al., 2016; Kernozek et al., 2018). It appears that the impact dissipation stress is centred primarily on the ankle plantarflexors in forefoot running leading to a potential increased injury risk to the triceps surae and the Achilles tendon. If the plantarflexors do not take on the energy dissipation load adequately, it is possible that forefoot structures may have to increase their energy dissipation loads to assist the knee and hip mechanisms (Fig. 3.2.11b).

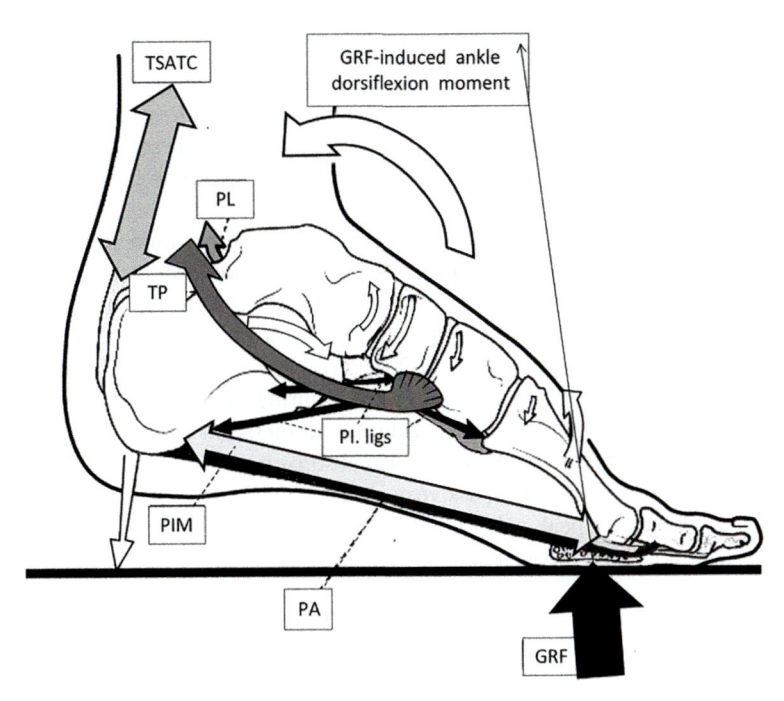

FIG. 3.2.11b Forefoot strike impact force builds more slowly than the rapid peak force of a heel strike transient. Forefoot deformation can occur over a longer percentage of stance aided by the vault's ability to deform and utilise plantar muscle tendon buffering, not available during a rearfoot contact. This allows the forefoot to have a less structured forefoot plantar fat pad. The much larger triceps surae and Achilles tendon complex (TSATC) replaces tibialis anterior as the primary energy dissipator at the ankle, utilising muscle–tendon buffering by resisting heel drop. However, other foot and ankle plantarflexors assist, utilising their own muscle tendon buffering and setting a degree of stiffness and stability across the vault to act as an elastic store of impact energy. In this role, tibialis posterior (TP) and peroneus longus (PL) play a prominent role, but the plantar ligaments (Pl. ligs), plantar intrinsic muscles (PIM), and plantar aponeurosis (PA) can all assist. High activity from TP and PL at impact provide osseous midfoot plantarflexion moments (*white arrows* within bones), resisting GRF-induced lesser tarsus and ankle dorsiflexion moments (*white curved arrows* in forefoot and at the ankle). Thus, adaptable vault mechanics for stability and stiffness are important. A small stiff base of forefoot support is not problematic when stance phase is very short, as is seen during fast running, quite unlike the situation found during toe/forefoot walking. *(Permission www.healthystep.co.uk.)*

The role of the skeletal frame in forefoot running appears not to have been investigated in any depth. Changes in the AHI as an indicator of medial longitudinal arch height do not correlate with peak forces or loading rates in forefoot running (Lees et al., 2005), suggesting that the skeletal frame does not play a prominent role in moderating forefoot loading impacts. This is likely because it must be kept stiffer and the vault profile higher so that high forefoot loading stresses can cross the midfoot to load the Achilles for impact energy dissipation. A compliant skeletal frame would likely buckle in the midfoot area, potentially injuring structures across it. Investigating the effects of forefoot running on known midfoot break subjects offers an intriguing study.

It is possible that high impact forces across the forefoot cause rapid forefoot splaying, stiffening forefoot connective tissues distally (influenced by strain rate loading) and engaging rapid deep transverse metatarsal ligament stiffening and transverse arch stiffening mechanisms that involves plantarflexor muscles across the foot. This in turn should result in longitudinal foot-stiffening mechanisms being initiated early in contact. Rapid loading rates during running lead to stiffer connective tissues (such as tendon and fascia) through their time-dependent properties, turning the foot more easily into a stiff elastic storer of energy rather than an energy dissipator at contact, although muscle–tendon buffering will do so. The initial forefoot splay may give the skeletal frame a minor role in initial energy dissipation through transverse/horizontal plane deformation as stiffening is initiated.

Stiffness in the foot for running acceleration

Kinetic energy lost at impact/braking during running is stored temporarily in the elastic structures of the lower limb (as potential energy) to be released as elastic recoil in order to save on metabolic energy (Alexander, 1984). Elastic energy storage can be used to generate vertical motions of the body's CoM without the requiring of large amounts of energy to be produced through muscle metabolism. This enables the body to bounce along on springy legs, an essential concept of the spring–mass model of running (Hamner et al., 2010). The primary parameter of this leg spring is stiffness. The human foot in forefoot running becomes an essential part of limb stiffness, capable of storing energy from strain and releasing it as quasi-elastic recoil (Ker et al., 1987). This passive mechanism of energy storage and release is important to running energetics during propulsive acceleration (McMahon, 1987) and supplements that from the rest of the lower limbs. Energy is stored primarily within the ligaments and tendons, 35% of which comes from the Achilles tendon (Ker et al., 1987; Sasaki and Neptune, 2006). However, the other soft tissue structures also play a significant role.

Muscles play a primary role in providing stiffness for stability and elastic energy storage. The primary role is fulfilled by the plantarflexors of both the ankle and the foot during walking (Ishikawa et al., 2005; Aerts et al., 2018) and running (Lai et al., 2014, 2018). With the extended active role of the ankle plantarflexors to 100% of the stance phase during forefoot loading, the significance of all the foot and ankle plantarflexors is increased in the forefoot strike position. The contribution to elastic strain energy in running of the ankle plantarflexors increases with running speed, from 53% in jogging to 74% in sprinting for soleus and 62% to 75% for gastrocnemius (Lai et al., 2014). The gastrocnemius muscle fascicles shorten more rapidly than those of soleus. This is due to the gastrocnemius crossing the knee which delays tendon recoil during stance, reflecting its ability to transfer power and work from the knee via ankle tendon stretch and through the storage of elastic strain energy (Lai et al., 2018). Thus, in running, the ankle plantarflexors take up a larger role in elastic energy storage than when walking, linking knee and ankle power together but involving foot stability. This function increases with speed.

The plantar aponeurosis also exhibits typical elastic stretch-shortening cycles with the majority of the strain being generated via arch depression, showing a similar strategy in forefoot strike running to rearfoot strike running (McDonald et al., 2016; Wager and Challis, 2016). The elastic energy stored within the lower leg and foot does not appear to be affected by the foot strike position (Wager and Challis, 2016). The plantar aponeurosis provides an energy transfer mechanism in the second half of stance phase between itself and the MTP joints (McDonald et al., 2016; Wager and Challis, 2016). The plantar aponeurosis generates power at the talonavicular joint and produces positive work during the last 45% of stance during running, increasing running efficiency through the release of elastic energy during acceleration via passive resistance and aiding vault shortening (Wager and Challis, 2016). MTP joint extension also absorbs power during late stance (Wager and Challis, 2016). How much elastic energy is released by the other passive connective tissue structures remains as yet, unknown. The talonavicular peak power occurs slightly after peak ankle power, suggesting that plantar aponeurosis power generation occurs when the Achilles power is declining, with the plantar aponeurosis elastic energy being used up by toe-off (Wager and Challis, 2016).

3.2.12 Section summary

The foot provides braking phase compliance and energy dissipation by first pre-tensioning the foot prior to impact. This creates a deformable profile that can 'give', providing controlled compliance. Energy dissipation across the foot is achieved through the plantar fat pads, muscle–tendon buffering, soft tissue resonance, and an elastically deformable skeletal frame with a vault profile that offers viscoelastic properties. During heel strike, the heel fat pad and tibialis anterior play a large part in energy dissipation, assisted by peroneus longus and tibialis posterior. In forefoot strike, the forefoot fat pads and ankle plantarflexors, particularly the triceps surae–Achilles complex, replace the heel fat pad and tibialis anterior. When faster gaits produce higher initial contact/loading forces, energy dissipation efficiency is more important with fast heel contact gaits requiring the most negative work to attenuate collision energies. This could be harder for individuals who display stiffer feet. Forefoot strikes, although changing the energy dissipation burdens and locations, reduce the need for rapid energy dissipation during fast heel contact impacts.

In stiffening the foot for acceleration, connective tissue stress-stiffening and time-dependent properties during single-limb support are utilised through vault depression as the foot increases its pronation, stiffening the foot passively. This vault depression causes increased foot length and width that induces foot stiffness. Because GRFs are higher in the middle of stance phase at slower walking speeds, the vault depression muscle activity control is more important than during faster gaits as time-dependent properties are less able to stiffen the connective tissues. Midstance stress-stiffening pre-stiffens the foot prior to heel lift, with strong evidence that this mechanism is assisted by both plantar intrinsic and plantar extrinsic muscle activities, particularly during late midstance. At the midstance-heel lift boundary, the foot should already be considerably stiffened, reducing the amount of muscle activity required to set the appropriate heel raising stiffness level.

At heel lift, the foot further increases its stiffness, led by transverse and frontal plane curvature increases initiated through soft tissue stiffening of structures such as the deep transverse metatarsal ligament. This is undoubtedly assisted through activity of the plantar extrinsic tibialis posterior and peroneus longus and the plantar intrinsic muscles such as the dorsal and plantar interossei and adductor hallucis. This transverse stiffening is coupled to longitudinal vault stiffening which includes extensive plantar intrinsic stiffening mechanisms across the midfoot, forefoot, and around the MTP joints. As the digits start to extend under the influence of heel lift and ankle and midfoot plantarflexion, the plantar aponeurosis is tightened. This initiates further stiffening of the foot, particularly during later terminal stance when other active stiffening mechanisms and ankle powers are reducing. The movement of the CoGRF medially through midfoot plantarflexion, eversion, and adduction will enhance the mechanical efficiency of acceleration during terminal stance by increasing the plantarflexion lever arm across the foot, improving the windlass mechanism's stiffening ability. The motion that permits utilisation of the medial MTP joints is known as *high gear propulsion* (Bojsen-Møller, 1979).

Running utilises all these mechanisms but the foot should always behave more stiffly and thus more elastically, returning more energy. However, the lower limb has to dissipate greater braking energies, creating a net energy deficit that must be balanced by muscle activity. Forefoot and rearfoot running utilise different energy dissipation mechanisms around the foot yet both techniques use the stiffened foot mechanisms for acceleration in much the same way.

3.3 The foot's functional units

3.3.1 Introduction

Once the foot is considered as a complete structure providing compliance and stiffening cycles during gait, it becomes helpful to approach the anatomy as a series of functional units throughout the foot. With this approach, the nature of the effects of pathology can be more easily understood. Considering the foot as a full, uninterrupted functional unit is better in understanding its mechanical role. However, to appreciate the pathomechanical consequences of failing anatomy, knowledge of the anatomical unit's functional ability from a more local perspective is helpful. Some structures are far more important to foot function than others, but each element has significance. Failure in one structure, however small, will result in a change in the tissue's stress–strain relationships. Loss of stress-loading abilities within local anatomy can increase loading elsewhere, unless the load can be functionally reduced such as by changing preferred gait speed or supporting the lost function therapeutically. This is why changes in footwear, the use of orthoses, rehabilitation, and/or rest can resolve symptoms.

Resting, healing, and strengthening to re-establish motion and stability are always the best clinical outcome. However, in situations where anatomy is permanently compromised, function must be artificially assisted. Before this can be clinically approached, the normal anatomical role of the structure must be understood. In the context of the fascial connective tissue continuum and the principles of tensegrity, the foot is a continuation of ankle function and each anatomical structure is part of the foot as a whole functional unit. Thus, problems within the foot can be managed, at least in part, through utilising complementary mechanisms within the rest of the lower limb. This underlines the need to have good lower limb functional anatomy knowledge in treating the foot and of managing the foot by indirectly treating elsewhere within the lower limb.

3.3.2 The role of cutaneous soft tissues and plantar fat pads of the foot

The cutaneous tissues' role

The cutaneous soft tissues are the foot's outer interface with the environment. In modern societies, this often means an interface being sheathed in hosiery and encased in some form of footwear contacting the support surface over the skin. The foot's skin is affected by the material it is in contact with, some of which can also damage it (Laing et al., 2015). However, footwear also protects the skin from potentially damaging object encounters and is able to redistribute and lower pressure peaks, while hosiery can reduce shear forces and decrease potentially damaging high moisture levels (Dai et al., 2006; Kirkham et al., 2014). The skin has properties capable of sustaining most biomechanical insults from the natural environment and gait if they are developed through maturation within such environments. Too often, barefoot is considered a sign of poverty, but it needs to be remembered that going barefoot is natural and, where environmentally safe to do so, should be encouraged among healthy populations. In industrial and post-industrial societies, foot tissues including skin, through having too easy a life, are often developmentally atrophied in a mechanical sense. The onset of diseases such as diabetes can then compromises such intrinsically reduced mechanical properties of the 'urban foot' even further.

The skin is a hierarchical elastin and collagen-reinforced tissue that displays nonlinear behaviour and anisotropy, with its viscoelasticity influenced by its physiological status. In the plantar foot, the skin is thicker than elsewhere because it is the interface with the external terrain. With the plantar fat pads (also known as *corpus adiposum*), the skin helps spread pressure, reducing pressure peaks from reaching harmful values. It provides for friction under contact, creating a temporary stability fixation with the ground and provides surface compliance to increase contact with the support surface. Greater surface contact increases friction and decreases peak pressures within the skin and tissues beneath. The soft tissues of the plantar foot provide a spatial filter of about 1 cm for smoothing out stresses and perturbations from terrain. This requires cutaneous tissue flexibility that helps prevent tears that would cause loss of skin as an impenetrable barrier. The area of support surface contact of the foot depends on the roughness and contour of the support surface, the shape of the foot's vault, and other foot dimensions occurring at any particular moment. Shoes and foot orthoses that contour into the plantar foot increase surface contact area and thus decrease pressures in the heel and forefoot. However, they will increase pressures across the midfoot region. This increased contouring effect of insoles in reducing forefoot and heel peak pressures should be the expected result of supporting and increasing the contacting surface area of the midfoot. This is only clinically useful if it is beneficial to reduce the pressure on the heel and forefoot and if it is safe to increase it within the midfoot. No insole can just 'decreases pressure' without increasing surface contact area or reducing the loading force via altered gait speed.

On very rough ground, the actual contact area of the foot would be smaller but for the fact that the soft tissues can deform to increase the contact area, adapting into depressions and over ridges in the supporting surface. The skin can only deform into indentations of around 1 cm. However, it can still provide some simple compliance into deeper surface indentations through its array of spring-like component structures. By extending the contact surface, the maximum stress (peak pressures) experienced over contact surfaces can be reduced. Therefore, a more prone foot posture is likely to decrease pressure peaks as the surface contour area increases. Thus, stiffer higher-vaulted feet during stance are likely to have proportionately smaller surface contact areas and higher peak forces for their size and for body weight than are lower-vaulted flexible feet.

Apart from its mechanical roles, cutaneous tissue also performs an important proprioceptive role via the cutaneous receptors. These provide for subtle motion awareness as well as information on larger joint motions beneath (Kavounoudias et al., 2001; Aimonetti et al., 2007). In the plantar foot, there are a large number of fast-adapting cutaneous receptors that provide the dynamic sensitivity necessary for balance control (Kennedy and Inglis, 2002). Cutaneous messaging informs the body of stretched and relaxed skin (Collins et al., 2000). There is the possibility that a lack of 'stimulating rough surfaces' in life causes a deterioration in the proprioceptive and exteroceptive ability of the cutaneous tissue. This opens up the possibility that appropriately designed orthoses and textured insole surfaces may be a suitable intervention in those who suffer from balance dysfunctions by stimulating the cutaneous receptors (Paton et al., 2016; Kenny et al., 2019; Ma et al., 2020).

The role of the plantar fat pads

Fat is nature's insulation of choice. On the plantar feet, it is useful in maintaining tissue temperature within the feet and protecting them from support surface temperatures. Fat has a melting point below body temperature, but the temperature below the sole of the foot is usually under 37°C and often below 20°C. Plantar fat pads contain higher proportions of unsaturated fatty acids that depress its melting point (Bojsen-Møller, 1999). Much of the role of the plantar fat pads in foot function has been discussed within the context of the heel fat pad during the foot's compliance stage of the gait cycle in dissipating impact energies (see Sections 3.2.4 and 3.2.11). It is worth repeating here that these plantar soft tissues provide an important role in influencing the load transfer to the entire body due to their viscoelastic behaviour, but particularly through their concerted interaction with the lower legs' soft tissues acting as a wobbling mass (Pain and Challis, 2001). The mechanical properties of plantar fat pads are influenced by the amount of load and their loading rate, but in producing lower leg and foot energy dissipation capacity, the heel fat pad and the soft tissues of the lower leg above are both important (Pain and Challis, 2001).

The plantar fat pads are composite materials that function with the epidermis and dermis through a biphasic fluid-like fat/adipose tissue phase and a solid connective tissue phase of the fat containment septa. These give the fat pads nonlinear viscoelastic properties (Chokhandre et al., 2012; Fontanella et al., 2012). They are capable of absorbing 30–60% of deforming compression energy (Challis et al., 2008). Under compression and shear, internal pressures inside the adipose chambers increase which is then applied to the connective tissue septa that respond by tensioning the chamber walls (see Fig. 3.2.4b). Energy is dissipated through the adipose compression and septal tension stretch, storing energy within the connective tissue walls. The energy stored within the septa on release helps reshape and reset the plantar fat pads after heel lift in readiness for the next loading cycle. Thus, this energy is lost to locomotion. Deformation on loading is

~9 mm but half this amount is reported during running (Wearing et al., 2009). This is because the plantar fat pads behave more stiffly under faster loading (Naemi et al., 2016).

It is the fluid-like (fat) content being contained within the solid components that makes the plantar fat pads near-incompressible materials (Natali et al., 2010a). The fat-filled chambers act as small pressure units that support the calcaneus, allowing it to sink into the pad on loading and thereby reduce peak pressures on the bone. By having microchambers on top of the macrochambers, the chances of tissue bulging on loading or as a result of single-chamber wall failure is reduced (Hsu et al., 2007a). The presence of high levels of elastin within each of the microchamber walls lying under the cutaneous tissue gives the microchambers higher compliance than if they only contained collagen, and they were macrochambers rather than microchambers (Naemi et al., 2016). Although particularly well developed in the heel, the same arrangement of microchambers and macrochambers is also found within the forefoot. The forefoot fat pad anatomy is complicated by the presence of the longitudinal passage of connective tissue structures that link to the toes (Bojsen-Møller, 1999). The plantar aponeurosis is able to trap areas of plantar adipose tissue in the forefoot within its superficial fibres, giving rise to forefoot fat pads positioned under the metatarsal heads (Bojsen-Møller and Lamoreux, 1979). This creates protective padding under the MTP joints particularly during the terminal stance phase of gait when the vertical GRF is directed through the metatarsals (Fig. 3.3.2).

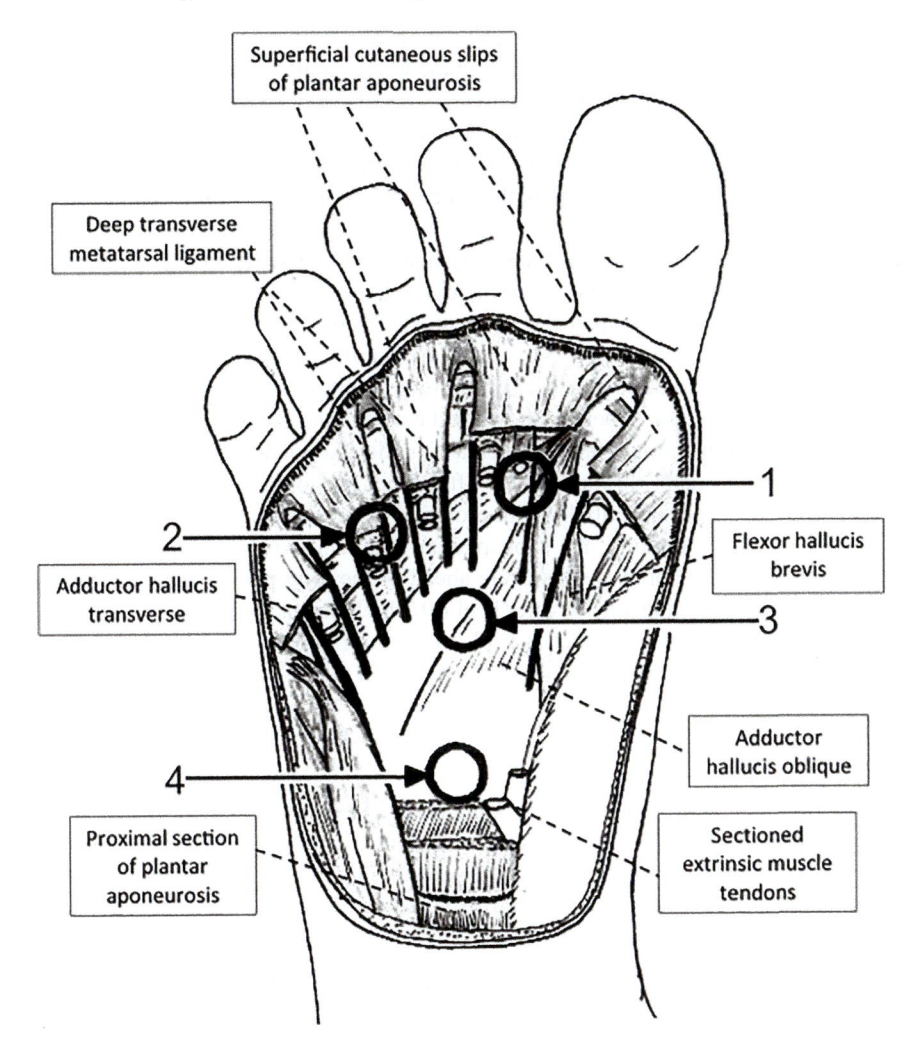

FIG. 3.3.2 A schematic of the plantar view of forefoot plantar fat pad-related anatomy in consideration to forefoot skin mobility. The central portion and proximal parts of the slips of the plantar aponeurosis are resected with the superficial muscle layer. The sagittal intermuscular septa that enclose the five metatarsal fat pads are shown as thick black lines within the proximal zone of the forefoot. The distal zone beyond these contains the superficial fibres of the plantar aponeurosis that connect into the dermis, with transverse fibres running between them. The four circles indicate sites where skin flexibility was investigated by Bojsen-Møller and Lamoreux (1979). Least mobility in the relaxed foot was found in site 2 and most at site 4, closely followed by site 3. All areas dramatically reduced their % mobility with digital extension, with most influence arising from dorsiflexion of the 1st and 2nd digits except at site 2, where extension of 2nd to 5th had most influence instead. Large digital extension ranges of 35–40° were needed to achieve a 50% reduction in skin motion. As significant digital dorsiflexion ranges are required to stiffen forefoot tissues for acceleration biomechanics, inflexible forefoot areas within footwear may prevent adequate changes in forefoot plantar fat pad biomechanics during gait. High heels may initially over-stiffen the forefoot, yet in time their frequent use may causes adaptive de-tensioning of the connective tissues that deteriorate the coupled digital extension changes to the tissue material properties. *(Adapted from Bojsen-Møller, F., Lamoreux, L., 1979. Significance of free-dorsiflexion of the toes in walking. Acta Orthop. Scand. 50 (4), 471–479.)*

Fat pads contain a rich supply of nerves and blood vessels. They also contain a high number of Pacinian corpuscles sensitive to both high-frequency shock and tissue displacement inside the fat chambers in addition to Meissner's corpuscles and other nerve end-organs (Bojsen-Møller, 1999). Together, these allow sensorimotor feedback mechanisms to adapt to ground contact impulses. The plantar pads of both the heel and forefoot have significant protective roles for loading the skeletal frame. The heel fat pad also helps protect the rest of the body from high heel impact energies, which may be compromised in some subjects reporting plantar heel pain (Wearing et al., 2009).

3.3.3 The role of the passive elastic elements of the foot

The elastic elements of the foot are the connective tissues such as the deep fascia, plantar aponeurosis, ligaments, and tendons. These structures fulfil an essential role during locomotion in preserving favourable muscle conditions, acting as an energy buffer during impacts and enhancing the power output during explosive activities (Lichtwark et al., 2007; Konow et al., 2012). They are able to store elastic energy generated during the earlier periods of stance to preserve metabolically derived energy by releasing the elastic energy during acceleration. Thus, they exchange kinematically produced kinetic elastic energy with this energy released back to affect the kinematics (Wager and Challis, 2016). Tendons are a primary site of energy storage and release with the Achilles considered to be the primary site, contributing ~30–40 J of elastic energy per step. The foot alone is thought to produce ~17 J in total (Wager and Challis, 2016). Thus, ankle plantarflexors are the primary motors of acceleration via the foot.

The foot contains many ligaments, each with a significant role to play in maintaining compliance and elasticity within the skeletal frame, while at the same time being proprioceptive. Most are mono-articular ligaments which will be discussed in relation to the anatomy of their particular joints. Some are bi-articular and multi-articular and therefore more functionally significant. They need to be considered separately in order to wholly appreciate their functional role. The medial and lateral collateral ligaments of the ankle help to stabilise the foot across the subtalar joint. As the subtalar joint only demonstrates independent flexibility when loaded by external forces (Leardini et al., 2001), the collateral ligaments provide stabilisation critical to foot function when either the ankle is loaded on the heel at a high plantarflexion angle (a rare event except in high heels) or is mildly plantarflexed and moving into low angles of dorsiflexion during heel strike, loading response, and early midstance. During late midstance, as the ankle dorsiflexes towards its maximum, ankle and subtalar motion become increasingly restricted such that other ligaments as, for example, the calcaneofibular ligament, become a more important constraint on subtalar inversion moments than the other ankle collateral ligaments that attach to the talus (Li et al., 2019).

Major foot ligaments include the long and short plantar ligaments, the spring ligament, and the deep transverse metatarsal ligament. In the sagittal plane, the plantar aponeurosis has a larger moment arm than the long, short, and spring ligaments which are therefore considered to play a smaller role than the aponeurosis itself (Wager and Challis, 2016). However, the plantar aponeurosis is most restrained and tense under toe extension and vault lengthening. Digital extension is only normally present in loading response and during terminal stance. On forefoot loading into midstance, the plantar ligaments are put under tension through foot compressive loads, being sandwiched between the BWV of the CoM and the resultant GRF. Increased foot vault 'sagging' deflection and lengthening increases under the loading force of the triceps surae–Achilles tendon complex following absolute midstance through to heel lift, further tensioning the plantar ligaments and aponeurosis. Thus, ligaments are loaded and offloaded sequentially with proximal ligaments completing their stretch-shortening elastic cycle earlier than distal ligaments. In running, the short plantar ligament is reported to complete its maximum lengthening before the spring ligament, which completes its cycle ahead of the long plantar ligament (Welte et al., 2021). The plantar aponeurosis is still under lengthening tension after heel lift which means it can still provide stiffening forces and energy dissipation across the foot vault and forefoot at heel lift, but not directly through the windlass mechanism which requires the plantar aponeurosis to start shortening (Welte et al., 2021) (Fig. 3.3.3a).

Longitudinal ligaments

Strains within the foot are spread across longitudinally and transversely orientated ligaments but will concentrate to the larger multi-articular ligaments, creating passive stretch and increasing elasticity as the vault profile depresses (Fig. 3.3.3b). Connective tissue strain rate is nonlinear, with increases in lengthening being most significant during initial loading and increased stiffness and least lengthening occurring when the loads on the ligaments are highest. This is reflected in the changes in vault profile and foot lengths (Stolwijk et al., 2014; Bjelopetrovich and Barrios, 2016; Takabayashi et al., 2020). The passive stretch is stored as potential elastic energy within the ligaments proportional to their cross-sectional areas. At heel lift, this energy can be returned to recoil the vault profile and possibly to help accelerate the foot's CoM forward. Cadaveric experiments suggest that the combined elastic energy of the plantar aponeurosis and plantar ligaments is around 17% of the mechanical energy expected during running at 4.5 m/s (Ker et al., 1987) and 8% at 2.7 m/s (Stearne et al., 2016). The difference recorded is probably the result of greater stiffness being associated with the faster loading rates of increased running speed. It has also been shown that using insoles that restrict 'arch compression' results in an increase in metabolic energy cost proportional to the elastic work lost from reducing the vault changes (Stearne et al., 2016). However, some caution to this result needs to be considered, as increased mass from the insole might also have an effect.

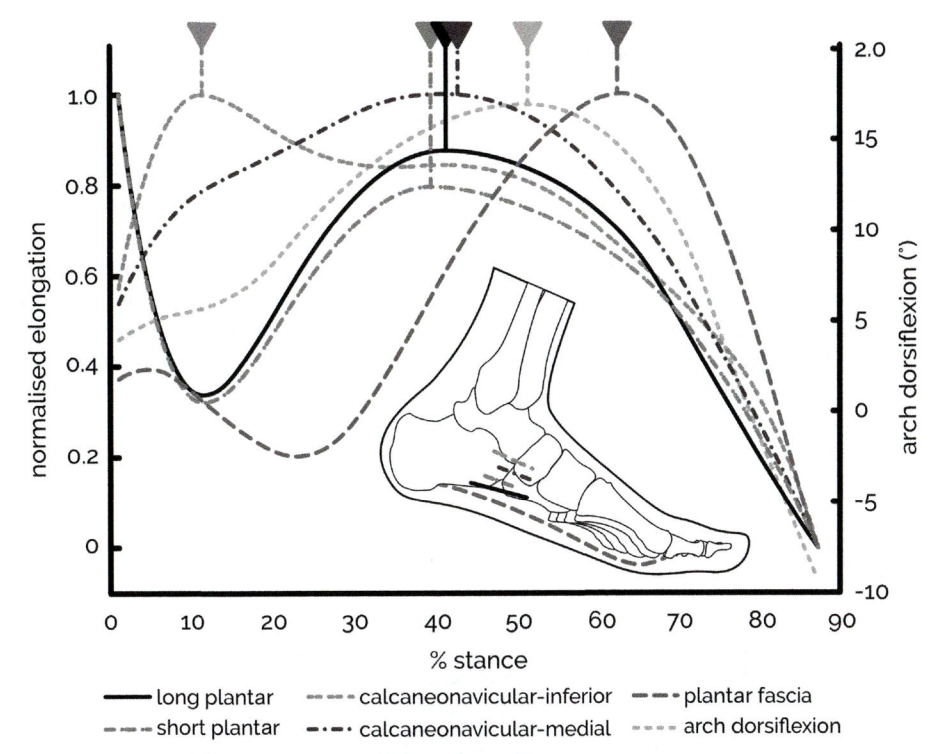

FIG. 3.3.3a During running stance phase, the more proximal ligaments complete their stretch–recoil cycles first because the peak loads and GRF move anteriorly throughout gait. Note how the long and short plantar ligaments and plantar aponeurosis initially shorten through vault rising during the windlass mechanism due to digital extension after heel strike and before forefoot loading lengthening. The plantar aponeurosis starts to lengthen last after forefoot loading and also shortens last after heel lift. It can be postulated that during walking under slower loading rates (thus greater compliance being expressed via their viscoelastic properties), the plantar ligaments and aponeurosis undergo similar cycles of length change, but with greater ranges of load-induced lengthening and smaller elastic recoil ranges of shortening during offloading. The forefoot ligaments, such as the deep transverse metatarsal ligament, are likely to complete their strain cycles late, only starting to recoil and shorten after weight transfer to the next step. *(From Welte, L., Kelly, L.A., Kessler, S.E., Lieberman, D.E., D'Andrea, S.E., Lichtwark, G.A., et al., 2021. The extensibility of the plantar fascia influences the windlass mechanism during human running. Proc. Roy. Soc. B: Biol. Sci. 288 (1943), 20202095.)*

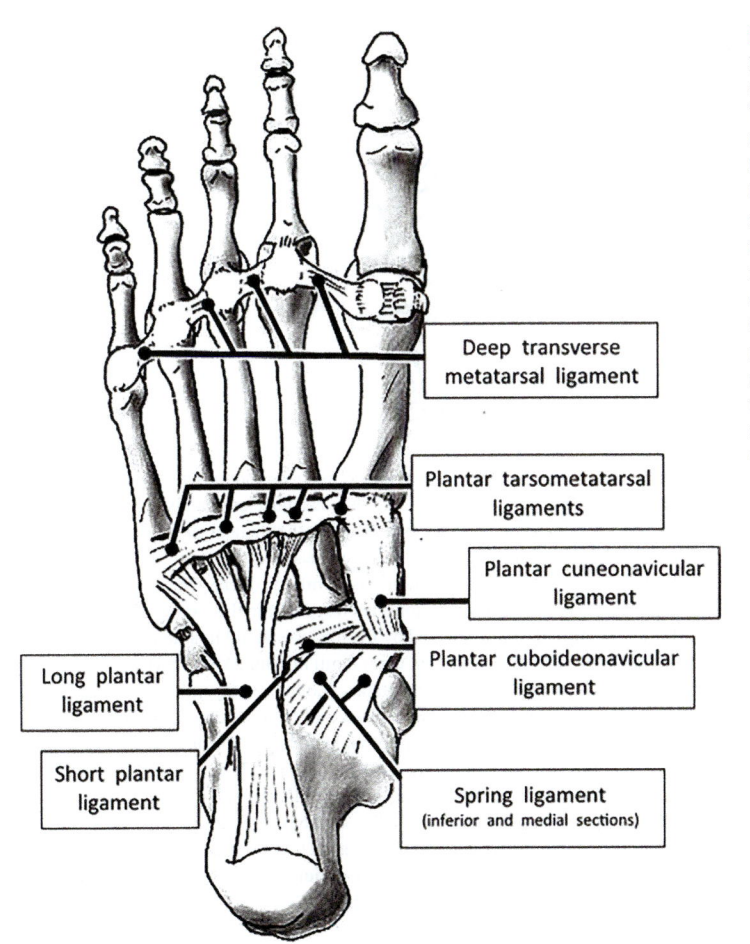

FIG. 3.3.3b A schematic of the primary ligaments of the plantar foot. Note the spring (plantar calcaneonavicular) ligament has two distinct sections: inferior (lateral) and medial. The short plantar ligament tethers the proximal lateral column together while inferiorly, the distinctly large human long plantar ligament binds the calcaneus to the tarsometatarsal region via the bases of the four lesser metatarsals. These ligaments are connected to the complex of ligaments that form an enclosing plantar tarsometatarsal network of connective tissue that links to and functions with the dorsal tarsometatarsal ligaments. Thus, the distal midfoot has a strong transverse-to-longitudinal connective tissue linkage. Finally, the deep transverse metatarsal ligament forms the anterior restraint that connects into the extensor apparatuses that enclose the metatarsal heads. Connective tissue continuity sets up multiple relationships between ligament tensions in all orthogonal planes of motion across the foot. *(Permission www.healthystep.co.uk.)*

Transverse ligaments

The deep transverse metatarsal ligament is implicated in enhanced stiffening across the whole foot (Venkadesan et al., 2017a,b; Yawar et al., 2017). The fact that the forefoot does not leave the ground until late in terminal stance phase, when the CoM is well advanced beyond the foot, presents a very limited opportunity for released energy from the transverse ligament to assist acceleration. The role of the transverse ligament is therefore considered largely, if not totally, a mechanism of stiffening to provide a stable forefoot interface with the support surface during terminal stance, concurrent with superficial soft tissue stiffness via the plantar aponeurotic effect on the plantar fat pad under digital extension. Any elastic energy stored within the deep transverse metatarsal ligament is likely used to regain the narrower unloaded forefoot width by recoiling it back into shape during swing phase. Individual frontal plane rotations and sagittal plane motions of each of the metatarsals as they load may twist and stretch linked connective tissues (Fig. 3.3.3c). This will influence transverse

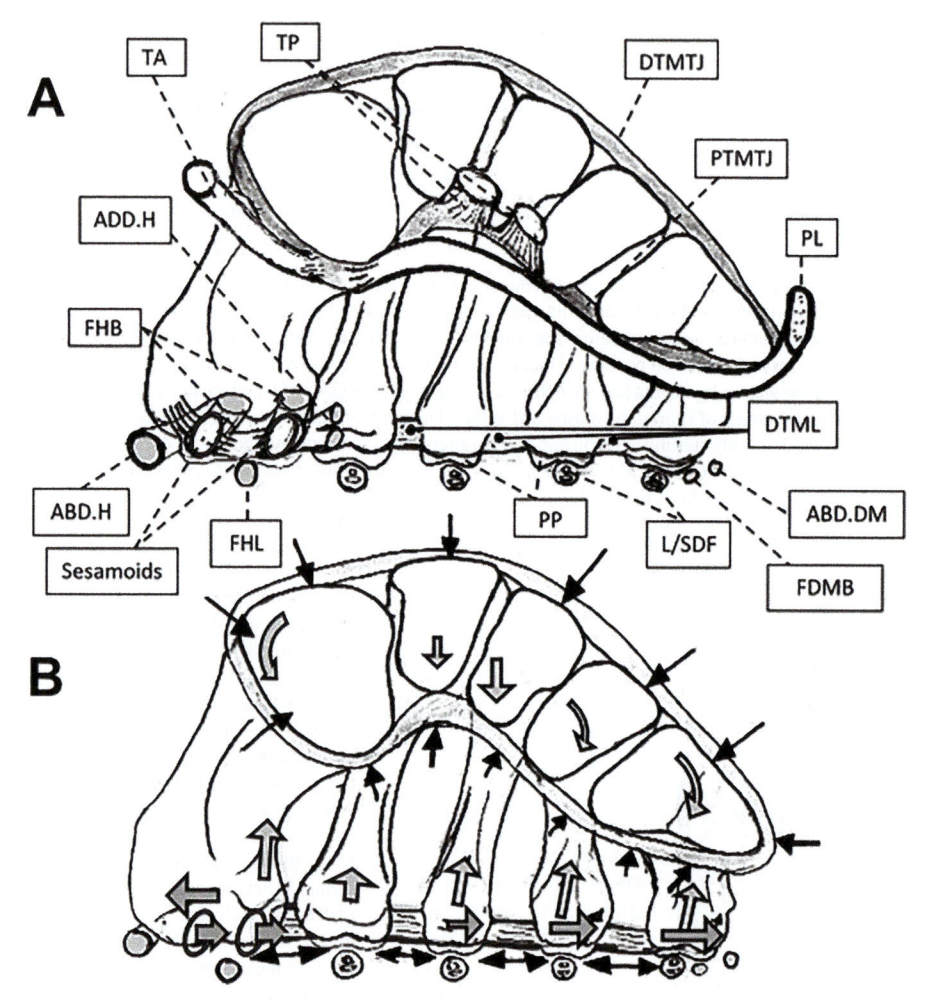

FIG. 3.3.3c The whole transverse foot-stiffening mechanism binds ligamentous tension to forefoot weightbearing loads and muscular activity in such a complex way that schematic visualisation is difficult. In (A), important anatomy is shown, except for the interosseous muscles that also play a significant part in stiffening this region. The deep transverse metatarsal ligament (DTML) form a strong distal tie-bar binding the metatarsal heads together. This ligament is connected to the fibrocartilaginous plantar plates (PP) under the metatarsal heads that aid metatarsal head rotation, with the long and short flexor tendons (L/SDF) sharing a tendon sheath beneath them. At the 1st and 5th metatarsals, the anatomy is a little different. The 1st MTP joint has additional muscles such as abductor hallucis (Abd.H) and adductor hallucis (Add.H) bound to the sesamoids enclosed within the tendon of flexor hallucis brevis (FHB), with flexor hallucis longus tendon (FHL) passing inferiorly. The sesamoids are connected by the inter-sesamoid ligament, and suspensory ligaments connect the sesamoids to the metatarsal head. The 5th MTP joint has additional muscles of abductor digiti minimi (ABD.DM) and flexor digiti minimi brevis (FDMB) but lacks the interosseus muscles of the central metatarsals. Proximally, the tarsometatarsal joints display multiple dorsal, plantar, and interosseous ligaments that create another tie-bar like a stiff, thick elastic band around the metatarsal bases and their articulations. This tie-bar has connections to the tendons of tibialis anterior (TA), tibialis posterior (TP), and peroneus longus (PL). On forefoot loading (B), the metatarsals splay and in so doing, translate horizontally away from the 2nd metatarsal head with the 1st, 4th, and 5th moving the greatest distance. The sesamoids are tensioned laterally. The metatarsal heads also dorsiflex relative to their metatarsal bases which are undergoing relative plantarflexion due to vault lowering under midfoot dorsiflexion moments. This effect is enhanced in late midstance by the ankle dorsiflexion moment. The combined effects of these motions cause individually different metatarsal rotations in the frontal plane and translations in the transverse and sagittal planes that twist and stretch to stiffen and tighten both proximal and distal ligamentous tie-bars. The results are compressive restraints on the proximal articulations vastly aided by muscular activity within the long and short digital flexors, tibialis posterior, and peroneus longus. *(Permission www.healthystep.co.uk.)*

plane stiffness beyond that of a uniplanar forefoot 'splay' mechanism in much the same way that fish bone fin rays rotate to induce fin stiffness (Nguyen et al., 2017).

The deep transverse metatarsal ligament is associated with five fibrocartilaginous plantar plates, although in the 1st MTP joint, the fibrocartilaginous plate is reduced by the presence of the sesamoids. The plantar plates and the plantar aponeurosis are intimately bound with the deep transverse metatarsal ligament as well as with the collateral ligaments of the metatarsals and the extensor apparatus of the MTP joints dorsally (Stainsby, 1997). Together, these interconnections form a longitudinal tie-bar mechanism across the forefoot, binding the distal metatarsals and MTP joints together (Stainsby, 1997).

Overall, the precise contribution of the foot's ligaments alone in the return of elastic energy remains unquantified. They seem to play a role in assisting the elastic recoil of the plantar aponeurosis and are pivotal in creating foot stiffness with muscular activity, thereby providing a semi-rigid beam to apply ankle plantarflexor forces to the forefoot after heel lift. In this capacity, they work by passive tension resulting in a compressed articular stability through the principles of biotensegrity. They also have an important sensory role that alters muscle function via 'informed' muscle spindles through the sensorimotor system, influencing posture and muscle function.

3.3.4 The role and anatomy of the plantar aponeurosis

The role of the plantar aponeurosis

The plantar aponeurosis has similarities to aponeuroses found spanning other muscles and tendons throughout the body, such as that observed between the gastrocnemius and soleus, lying superior to the Achilles. The term 'plantar aponeurosis' is better and more applicable than plantar fascia as a descriptor and is one that should be considered with reference to the different plantar aponeurotic layers and intermuscular septa that this structure is linked to. The plantar aponeurosis may contribute ~25% of total foot vault stiffness (Ker et al., 1987; Bennett et al., 1989; Huang et al., 1993). It is put under stretch tension by external forces applied to the foot vault that longitudinally lengthen it, drawing the digits into the ground. It is also put under increased tension during digital extension in loading response and terminal stance through an effect known as the windlass mechanism (Fig. 3.3.4a). At these times, its distal and proximal attachments are drawn away from each other, applying a tightening fascial stretch across the vault.

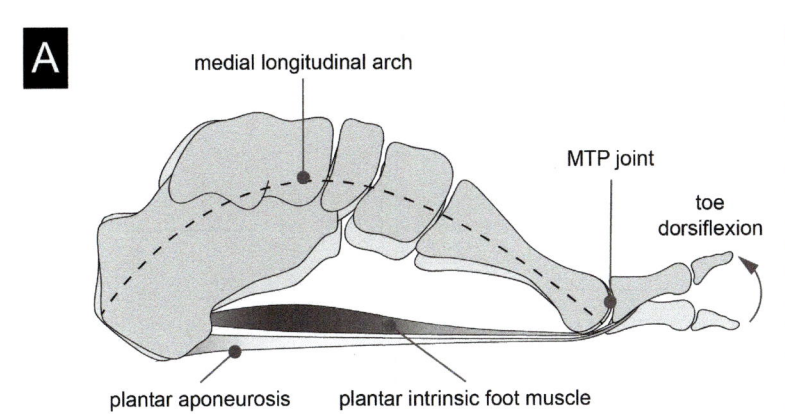

FIG. 3.3.4a A depiction of the windlass mechanism as a mechanical model that increases the radius of curvature of the medial longitudinal aspect of the foot vault (A). It has long been used to explain how the foot stiffens in terminal stance to provide a semi-stiff mechanical lever at heel lift (B). However, the aponeurosis' influence through the windlass mechanism on terminal stance stiffening is now questioned. The windlass' role is becoming to be seen as more important for forefoot loading impact dissipation, with intrinsic and extrinsic foot muscles and transverse plane stiffening being primarily responsible for initial terminal stance vault stiffness increases. *(From Sichting, F., Ebrecht, F., 2021. The rise of the longitudinal arch when sitting, standing, and walking: Contributions of the windlass mechanism. PLoS ONE 16 (4), e0249965.)*

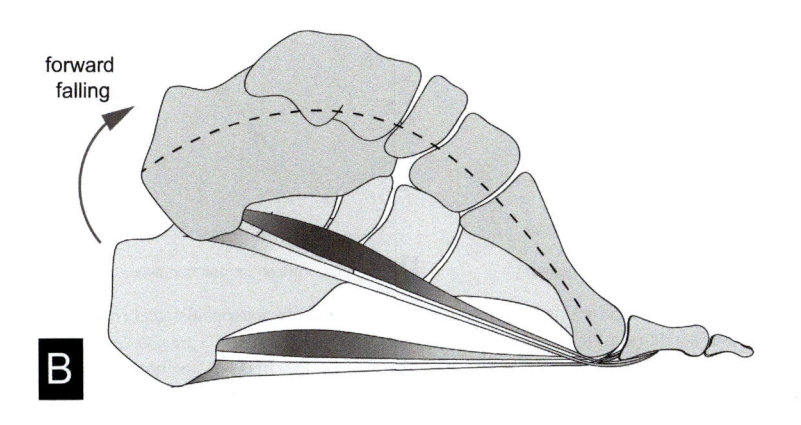

During midstance, plantar aponeurotic stretching occurs via vault flattening and it continues to stretch out in length into initial terminal stance. A little after heel lift it becomes so significantly stretch-stiffened that it starts to help shorten the foot vault under digital extension during ankle and midfoot plantarflexion. The plantar aponeurosis and vault also shorten together under the windlass effect of digital extension before forefoot loading as well as during the events of terminal stance. This is because during utilisation of the windlass mechanism, the attachment points of the plantar aponeurosis are in fact lengthening as the digits extend which tensions and stiffens it. Yet, the plantar aponeurosis is not initiating increased foot stiffness at heel lift via the windlass mechanism (Farris et al., 2020). Neither does the plantar aponeurosis work in isolation of the plantar ligaments or foot muscles in any situation. It has been calculated that the plantar ligaments' combined effect on longitudinal stiffness of the foot is either equal to or surpass that of the plantar aponeurosis alone (Venkadesan et al., 2017a). Static windlass effects, sitting and standing, poorly predict the relationship between arch dynamics and MTP joint function. This is because a decreased arch height is noted as digital extension begins at heel lift (Hunt et al., 2001; Sichting and Ebrecht, 2021). Not only this, but the windlass mechanism occurs associated with positions of ankle plantarflexion, both during loading response and after heel lift. Thus, testing the effects of the windless mechanism must be investigated within these ankle plantarflexed postures. Medial longitudinal vault height rises almost linearly with toe extension in static loading. However, it is the exponential relationship of digital extension to the rate of change in arch height dynamically that is important, for toe motion is significantly lower in sitting and standing than during walking (Sichting and Ebrecht, 2021), likely reflecting the loss of a dynamic ankle plantarflexed posture (Fig. 3.3.4b).

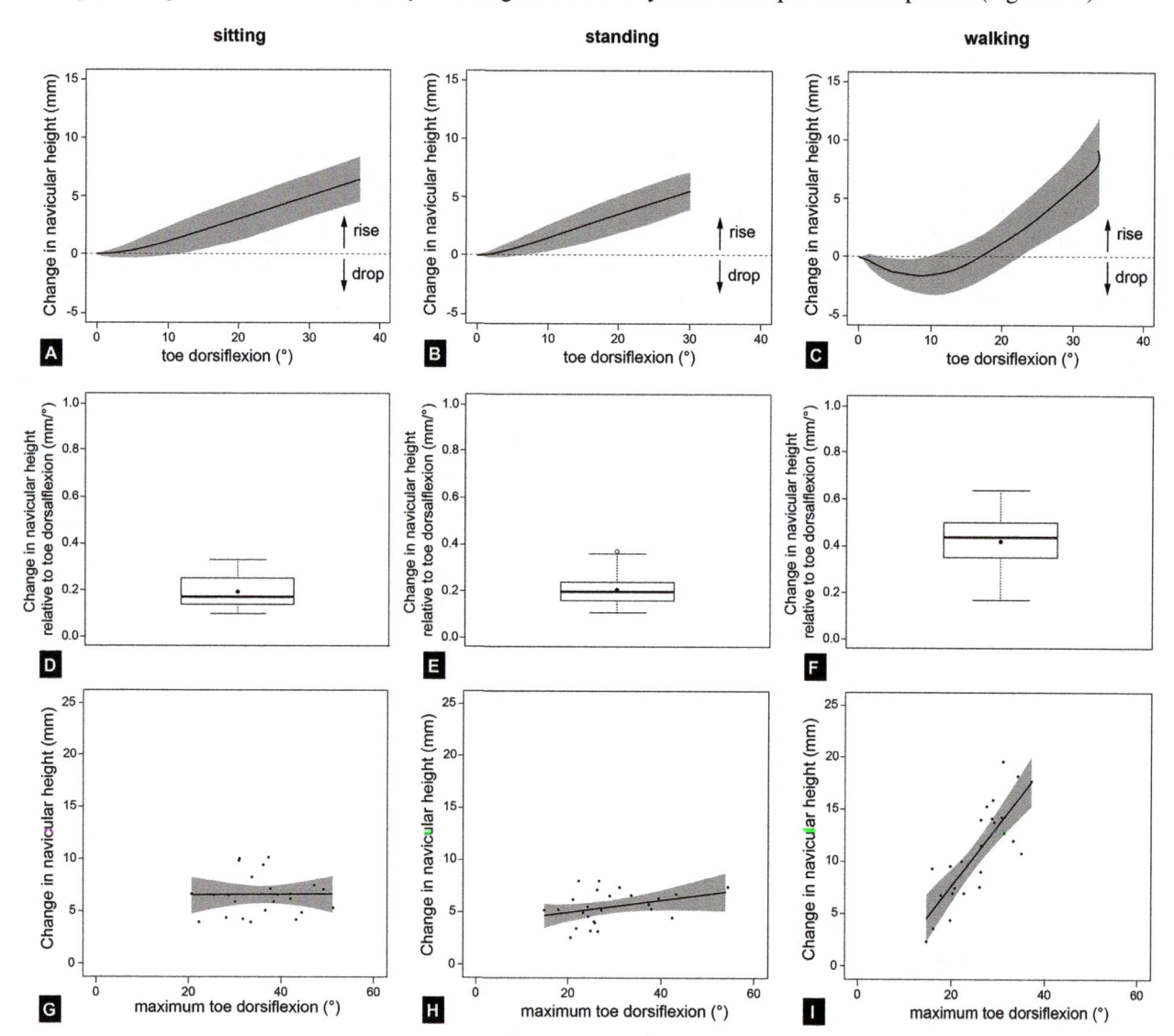

FIG. 3.3.4b Static windlass effects (sitting and standing) poorly predict the relationship between arch dynamics and metatarsophalangeal (MTP) joint motion during dynamic loading in walking. Mean changes in navicular height (mm) in relation to digital dorsiflexion (extension) (°) are shown in (A–C). Changes in navicular height relative to toe dorsiflexion (mm/°) are shown in (D–F). Black dots represent the mean value. In (G–I), the linear regression between changes in navicular height and maximum digital dorsiflexion is shown, with each circle presenting one study participant. The regression line is black, with the grey area representing the confidence interval. *(From Sichting, F., Ebrecht, F., 2021. The rise of the longitudinal arch when sitting, standing, and walking: Contributions of the windlass mechanism. PLoS ONE 16 (4), e0249965.)*

The plantar aponeurosis has complex structural anatomy with a full list of anatomical linkages that are frequently overlooked in many descriptions. As a result, it is often simplified to a 'medial longitudinal arch'-supporting structure. The plantar aponeurosis has an intimate stress–strain relationship with the Achilles tendon (Carlson et al., 2000; Erdemir et al., 2004; Cheung et al., 2006) and the two structures are continuous in early human life, although they become less so with age (Snow et al., 1995). Both structures together play a key role in transferring ankle plantarflexor power to the ground as a GRF to aid forward acceleration. When standing, both the body mass over the foot and increased triceps surae contraction-induced Achilles loading result in heightened tension within the plantar aponeurosis, with the Achilles loading demonstrating a two-fold larger straining effect than increasing mass on the body (Cheung et al., 2006). At the distal end of the plantar aponeurosis, the MTP joints also influence plantar aponeurotic strains through the windlass effect (Caravaggi et al., 2009, 2010).

Both Achilles and plantar aponeurotic strains increase with angles of dorsiflexion at the ankle (Huang et al., 2018). Therefore, increased dorsiflexion directly increases the Achilles' tensile effect on the aponeurosis (Carlson et al., 2000). This provides the plantar aponeurosis with a mechanical vault-stiffening role similar but more superficial to that of those deep plantar ligaments running longitudinally (Fig. 3.3.4c). Stiffening allows potential energy to be stored throughout the plantar aponeurosis as elastic energy, providing positive work during terminal stance and assisting in vault shortening (Wager and Challis, 2016). Thus, the aponeurosis contributes with other vault structures to enhance gait energetics. The plantar intrinsic foot muscles also play an essential role in plantar aponeurosis biomechanics, as muscle contraction has been shown to affect the length-to-width changes occurring within the aponeurosis on loading stress (Arellano et al., 2016). The effects of these stiffening mechanisms influence the force-to-velocity relationship that can be created by the ankle plantarflexors through heel lift (Takahashi et al., 2016).

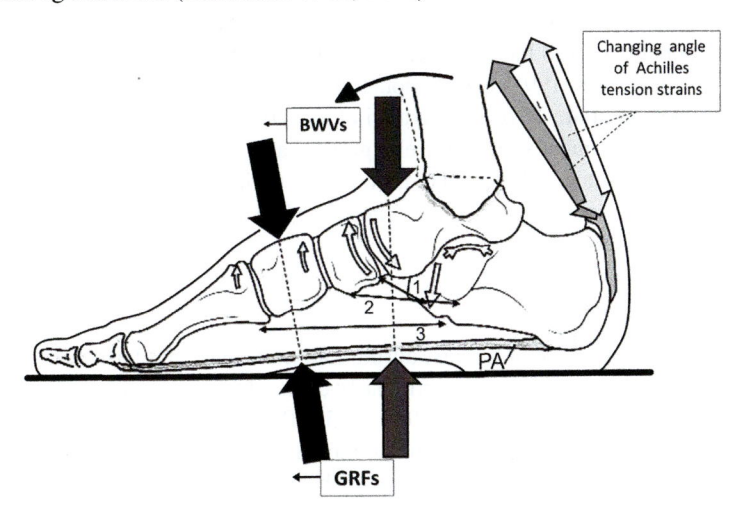

FIG. 3.3.4c Initially, during single-limb support, the vault is relatively compliant and vault spans and curvature radii increase comparatively freely due to the easier stretching of the passive tissues. From absolute midstance, the vault becomes stiffer through connective tissue stress-stiffening behaviour and increasing muscle activity under the foot. However, increasing tensions within the Achilles from triceps surae restraining the dorsiflexion moment at the ankle intensify the bending moments that cause sagging deflection across the vault. These moments centre on where the body weight vector (BWV) and ground reaction force (GRF) oppose each other, initially across the more proximal vault *(dark-grey arrows)* but becoming increasingly positioned more distally as termination of late midstance approaches. The passive restraints are the plantar ligaments (e.g. 1 = short plantar ligament; 2 = spring ligament; 3 = long plantar ligament) that increasingly stretch with the plantar aponeurosis (PA). Thus, the vault lengthens and widens as it dissipates loading energy via elastic strain. Towards the end of midstance, the BWV and GRF *(black arrows)* oppose each other more distally across the vault, permitting some elastic recoil within the more proximal ligaments such as the short plantar and spring ligaments. The more distal ligaments and plantar aponeurosis increasingly stretch as tensions and energy within the Achilles and energies crossing the vault, rise. *(Permission www.healthystep.co.uk.)*

Stance phase aponeurotic stresses

Forces within the plantar aponeurosis increase during stance, peaking during terminal stance (Erdemir et al., 2004; Gu and Li, 2012) with forces of around 538 N ± 193 N (96% body weight) and an estimated tension force of around 430 N or 77% body weight (Erdemir et al., 2004). Experimental data predict failure of the plantar aponeurosis at loads of around 916 N–1743 N (Erdemir et al., 2004). This equates to 2.15–2.80 times body weight, with failure loads being higher in males (Kitaoka et al., 1994). Both vertical and horizontal yield (widening, lengthening, and flattening) of the foot increase when the plantar fascia is cut (Arangio et al., 1998). In single-leg stance, there is a reported increase in lateral anterior shift and a reduction in vault height with increasing Achilles tendon loads. This is as a result of a plantarflexion torque on the distal calcaneus that increases plantar aponeurotic strain (Cheung et al., 2006).

Arangio et al. (1998), using mechanical modelling of the foot, found that changing vault height and including the presence or absence of the plantar aponeurosis produced some interesting results. When applying 683 N of load on the foot model investigated, it was found that an increase in the vault's profile from 20 mm to 60 mm increased the foot's stiffness with and without the presence of the plantar aponeurosis. However, the absence of the aponeurosis caused a greater change in profile upon loading the higher vault-profiled foot model than it did when loading a middle height or lower vault-profiled foot. This suggests that the plantar aponeurosis has a larger effect on changes in vault profile when under load in higher-vaulted feet than it does in lower-vaulted feet.

The complex interfaces of the plantar aponeurosis into the foot's other fascial structures give it an important linking role in both storing and releasing elastic energy within the foot (Ker et al., 1987; Natali et al., 2010b; Pavan et al., 2011; McDonald et al., 2016). Its role has primarily been considered as providing integrity to the skeletal frame (Hicks, 1954; Huang et al., 1993; Kitaoka et al., 1994; Thordarson et al., 1995; Kogler et al., 1996) and creating stiffness required for acceleration from the foot (Bojsen-Møller, 1979; Cheng et al., 2008). It is becoming clear that the aponeurosis' windlass effect is not a primary foot-stiffening mechanism in early terminal stance, a period when the intrinsic muscles and ligaments play a far more important role (Farris et al., 2020; Venkadesan et al., 2020; Williams et al., 2022). The plantar aponeurosis commences terminal stance still lengthened from midstance loading, but as vault length begins to decrease as the digits start to extend after heel lift, it should result in a similar if even slightly longer, aponeurotic length being maintained via digital extension until preswing. In running, Welte et al. (2021) noted that the aponeurosis did not start to shorten until a little after heel lift as digital extension angles were increasing. MTP joint extension enhances the aponeurosis' stiffening properties through the windlass effect as part of its initial lengthening, but more so and once the vault length starts shortening (Stolwijk et al., 2014; Griffin et al., 2015). Thus, the windlass effect only occurs coupled with ankle and midfoot plantarflexion and digital extension torques which are increasing after heel lift (Willwacher et al., 2014). The windlass effect therefore contributes to foot stiffening later in terminal stance, but not particularly at the start of heel lift. The plantar aponeurosis as a primary source of early terminal stance foot stiffness should consequently be questioned (Farris et al., 2020).

Despite this the plantar aponeurosis still has significant load-bearing roles, carrying as much as 14% of total foot load during stance phase (Kim and Voloshin, 1995) and demonstrating a stretch-shortening behaviour consistent with that of other connective tissues (McDonald et al., 2016). Caravaggi et al. (2009), in a very small study, reported that the medial slips of the aponeurosis displayed the most significant elongations and maximum tensions compared to the lateral slips. However, even within just the three subjects investigated, there was variation from this result in one of the subjects where lateral and medial slip elongation was comparable. It may therefore be suspected that mechanical aponeurosis behavioural variation within the population is common.

Loading response aponeurotic stresses

Although plantar aponeurotic loads are at their highest during late midstance and terminal stance, the aponeurosis is also 'pre-loaded' with strain during the swing phase and particularly during loading response. This is through digital extension via ankle and digital extensor muscle activity (Perry, 1992: p. 53; Byrne et al., 2007; Caravaggi et al., 2009; Wager and Challis, 2016). Both ankle dorsiflexion and digital extension increase plantar aponeurotic tension and reduce vault length (Carlson et al., 2000; Iwanuma et al., 2011a), setting the foot with some stiffness across the vault prior to impact. As stated earlier, this pre-tension, followed by forefoot loading de-tensioning through loss of digital extension after loading response, has been termed the reversed windlass effect (Stainsby, 1997; Green and Briggs, 2013), something that has since been referred to by Welte et al. (2021) as the 'inhibited forward-windlass' mechanism. In running, at initial contact, the plantar aponeurosis does not immediately elongate, but shortens through the windlass mechanism by toe extension, allowing other arch-spanning tissues to initially mitigate more of the impact load (Welte et al., 2021). Following forefoot loading, the windlass mechanism reverses and lengthening of the aponeurosis begins and continues well into gait acceleration (Welte et al., 2021). It has been noted that plantar aponeurosis pre-loading by the windlass mechanism is greater in the rearfoot strike runner than in the non-rearfoot striker, probably because of the greater reduction in ankle plantarflexion demonstrated in the former (Wager and Challis, 2016). It has been suggested that the ankle dorsiflexion angle achieved by the muscular extensors (tibialis anterior) plays a greater role in pre-tensioning the aponeurosis than do the digital extensors (Wager and Challis, 2016) (Fig. 3.3.4d).

Acceleration power in the plantar aponeurosis

Plantar aponeurosis power applied at the MTP joints follows a pattern of power generation in early stance by means of the reversed windlass flexing the digits. This is followed by power absorption during late stance through the windlass effect via digital extension acting as an energy dissipator. At the talonavicular joint, the opposite occurs with power absorption in early stance coming from its dorsiflexion motion followed by power generation in late stance (Wager and Challis, 2016) as the midfoot undergoes plantarflexion (Holowka et al., 2017). This suggests that an energy transfer is occurring between the plantar aponeurosis and the talonavicular and MTP joints; a phenomenon also noted in bi-articular muscles such as rectus

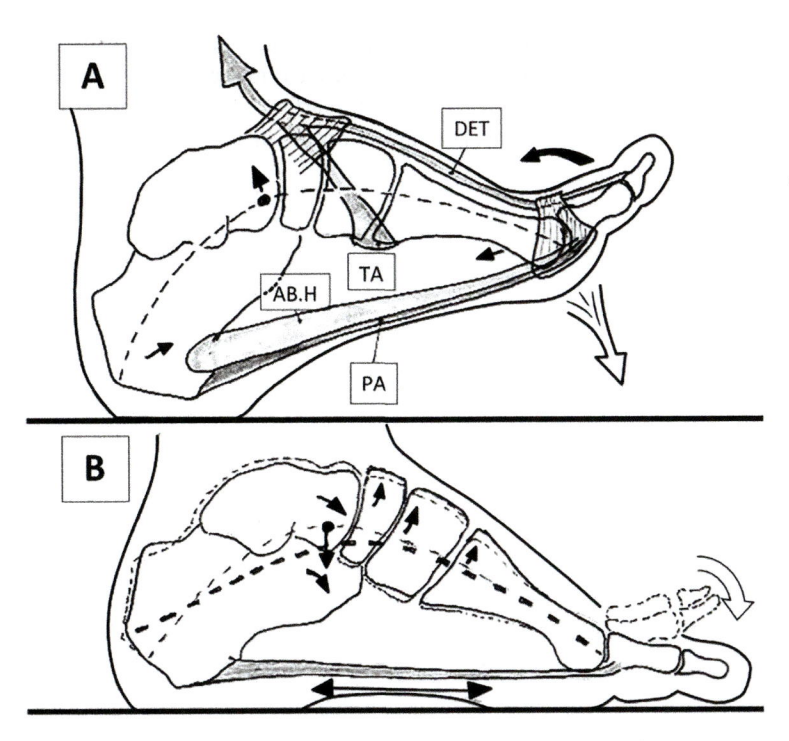

FIG. 3.3.4d The windlass mechanism may play a far more significant role in setting foot stiffness for energy dissipation (spring-like shock absorber) prior to forefoot impact than its more traditionally considered foot-stiffening role during acceleration. At and after heel strike (A), the foot is subjected to a plantarflexion moment that can be resisted by tibialis anterior (TA) and the long and short digital extensor muscles and their tendons (DET). These are restrained under the fascial extensor retinaculum at the ankle and extensor apparatuses at the MTP joints to improve their lever arm functions. The digital extensors extend the digits actively, inducing the windlass mechanism and tensioning the plantar aponeurosis (PA). This is likely aided by abductor hallucis (AB.H). Combined with TA, peroneus longus, and tibialis posterior activity (latter two muscles not shown), this decreases the longitudinal span distance and the radii of curvatures within the vault, increasing stiffness across the skeletal frame of the foot during forefoot descent. At forefoot contact (B), muscular activity reduces and the digits de-extend under passive flexion moments caused by the metatarsal heads loading under GRFs. The plantar aponeurosis relatively slackens, permitting the skeletal frame to deform more freely which increases both its radii of curvatures across the vault and its span distances. Passive ligaments and the aponeurosis can now stretch to dissipate energy with muscle–tendon buffering from muscles such as peroneus longus and tibialis posterior. The release of the windlass mechanism on forefoot loading is usually termed the 'reversed windlass'. *(Permission www.healthystep.co.uk.)*

femoris and gastrocnemius that transfer energy between their proximal and distal joints around the hip, knee, and ankle (McDonald et al., 2016). The contribution of the plantar aponeurosis to positive power in the latter part of stance is also substantially greater than the net plantar aponeurosis power purely derived from changes in its length (McDonald et al., 2016).

The ability to transfer mechanical power has two important benefits. Firstly, that of assisting acceleration through energy exchange between the MTP joints and the vault. Secondly, a reduction in the potential risk of plantar aponeurotic injury because the final power of the plantar aponeurosis contributes to the vault at acceleration and does not entirely depend on the stored elastic strain energy achieved during vault compression (McDonald et al., 2016). A link between ankle dorsiflexion, MTP joint extension, and plantar aponeurotic strain is supported (Flanigan et al., 2007; Caravaggi et al., 2009), whereas forefoot abduction and varus do not seem to affect plantar aponeurotic strains (Flanigan et al., 2007). These relationships could be used to better target and resolve plantar aponeurotic pathology.

It is estimated that the plantar aponeurosis provides around 3.1 J of elastic energy (Wager and Challis, 2016) which suggests that most of the 17 J of vault elastic energy is generated elsewhere. This seems a small quantity when compared to the Achilles' 30–40 J. However, it must be remembered that the Achilles energy is generated by the metabolic processes within muscle contraction (Fletcher and MacIntosh, 2015), whereas the plantar aponeurotic elastic energy is energetically 'cheap' (i.e. passive) to generate.

The plantar aponeurosis during running

The plantar aponeurosis demonstrates elastic stretch-shortening cycles with the majority of strain generated under vault compression forces via elastic-spring vault function associated with medial arch profile changes (McDonald et al., 2016). In the latter half of running stance phase, the plantar aponeurosis transfers energy between the MTP joints via energy absorption and the vault (via energy production) due to elastic recoil, to reduce the stain required by the plantar aponeurosis in generating positive work (McDonald et al., 2016).

During running at 2.7 m/s among individuals with foot vault postures considered as being 'normal' (using the foot posture index), the plantar aponeurosis undergoes a lengthening strain up to ~60% of stance phase, storing elastic energy. The remaining 40% of this energy is returned as elastic recoil (McDonald et al., 2016; Wager and Challis, 2016). The joint moment contributions of the plantar aponeurosis at the talonavicular and MTP joints do not seem to be different in rearfoot and non-rearfoot (midfoot/forefoot) strike runners, but the peak talonavicular work and the peak power generated are reported to occur slightly later in rearfoot runners (Wager and Challis, 2016). In forefoot running, McDonald et al. (2016) reported that the plantar aponeurosis lengthening starts at ~5% of stance, but in rearfoot strike running, it starts with forefoot loading at ~15% of stance (see Fig. 3.3.4e). Shortening of the aponeurosis occurs in acceleration only after heel lift is complete (Welte et al., 2021).

FIG. 3.3.4e Estimated plantar fascia strains during stance phase for forefoot strike (FFS) and rearfoot strike (RFS) during shod running where the foot vault is free to move. This shows that RFS runners tend to prolong their plantar aponeurotic strains, but their peak strains are lower than FFS runners, although the strains through the aponeurosis are not statistically different. *(From McDonald, K.A., Stearne, S.M., Alderson, J.A., North, I., Pires, N.J., Rubenson, J., 2016. The role of arch compression and metatarsophalangeal joint dynamics in modulating plantar fascia strain in running. PLoS ONE 11 (4), e0152602.)*

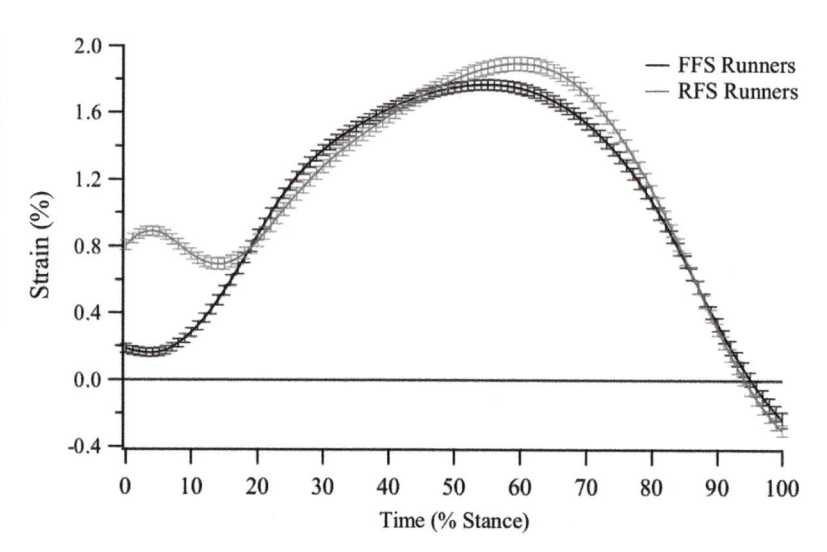

Although no statistically significant difference was noted in plantar aponeurotic energy storage in rearfoot runners compared to non-rearfoot runners, a difference was observed in the timing of the strain rate with higher strains occurring in the first 20% of stance in rearfoot running and a later higher peak strain occurring in non-rearfoot runners (Wager and Challis, 2016). There are also lower strains registered at loading but greater strain rates over the first half of stance in non-rearfoot runners (Wager and Challis, 2016) (Fig. 3.3.4f). This may relate to the findings of McDonald et al. (2016), as barefoot running and reduced heel striking seem to be associated and thus barefoot runners are more likely to forefoot strike (Lohman et al., 2011; Hall et al., 2013; Hatala et al., 2013). Plantar fascial peak strains seem to be higher and occur earlier in barefoot running conditions compared to running in footwear with MTP joint area flexibility (McDonald et al., 2016). In the latter stages of stance, elastic energy is released as the vault recoils, generating positive mechanical power. Simultaneously, energy is absorbed by extending the MTP joints (McDonald et al., 2016; Wager and Challis, 2016). McDonald et al. (2016) reported that strains within the plantar aponeurosis result in stores of energy, primarily generated by vault compression, but MTP joint extension energy loss accounts for around 11% of the energy absorbed within the lower limb during stance phase.

Anatomy of the plantar aponeurosis linked to its biomechanics

The plantar aponeurosis attaches to the plantar calcaneal tuberosity proximally. It demonstrates a heterogeneous thickness across the plantar surface, with its central portion being thickest medially along the distal line of the medial process of the plantar tubercle. The thickness varies from 0.9 mm to 4.9 mm over the calcaneus attachment, being thicker in males (Huerta and Alarcón García, 2007). It is from 1.2 mm to 4.5 mm thick at the most distal part of its proximal attachment to the calcaneus with its thickness influenced by body mass, being therefore thicker in heavier individuals (Huerta and Alarcón García, 2007). The central portion of the plantar aponeurosis runs across the inferior plantar surface of the plantar intrinsic muscles to divide into variable distinct slips at around the level of the metatarsal bases. These slips display transverse extensions between each other, keeping them partially united. The slips only become totally independent near the metatarsal heads, dividing into superficial and deep layers (Draves, 1986: p. 286).

The superficial layers run to the superficial fascia with uniting transverse extensions running between them to construct the superficial transverse metatarsal ligament. There are also slips anchoring the plantar aponeurosis to the skin that send transverse fibrous bands between them, trapping adipose tissue into chambers to create four metatarsal fat pads that encloses the digital nerves and blood vessels. The deep slips divide into medial and lateral slips to pass around the flexor tendons, forming flexor sheaths. These then join with the deep transverse metatarsal ligament and the connective tissue structures around the MTP joints. At the 1st MTP joint, the medial and lateral slips pass around the sides of the flexor hallucis longus tendon to blend with the sesamoids and fascia of flexor hallucis brevis (Draves, 1986: pp. 286–288) (Fig. 3.3.4g).

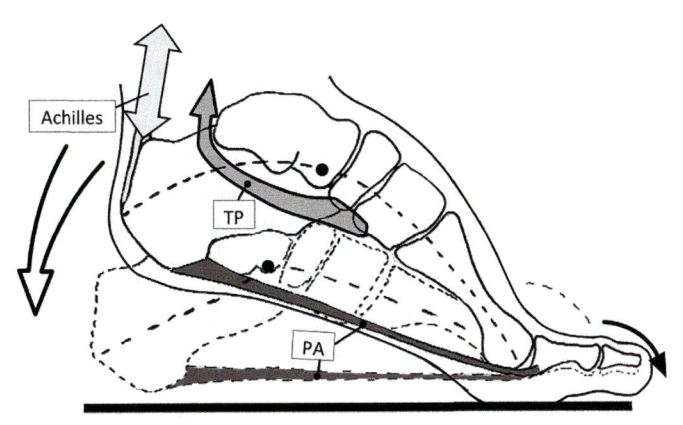

FIG. 3.3.4f Activating the windlass mechanism and reversing it during forefoot contact loading response is different to that of heel strike. Toes are extended during late swing, activating the windlass to some degree, but the ankle is set in plantarflexion for a forefoot contact. Forefoot impact sets up an ankle extension (dorsiflexion) moment, where heel descent towards the support surface is under ankle plantarflexor control. This gives the triceps surae via the Achilles a primary muscle–tendon buffering and braking role, with tibialis posterior (TP) and peroneus longus (not shown) providing more significant roles in controlling vault profile and stiffness than during heel strike. The toes will flex out of their extension angles that they have attained in terminal swing during loading response, reducing plantar aponeurosis (PA) contact tension. Tibialis anterior will not be restraining ankle plantarflexion, but does activate to a lesser degree, probably to help vault stability. Whether the heel makes ground contact after forefoot strike and how much the toes come out of relative extension will dictate how much the vault profile changes in loading response, and also how much the windlass mechanism formed during swing is reversed. As forefoot contact is associated with faster running, a stiffer foot is more desirable, so limited reversing of the windlass mechanism is an effective posture for this form of locomotion. An inability to use the reversed windlass mechanism normally to provide energy dissipation and achieve full plantigrade foot posture during walking, makes a walking forefoot contact undesirable. It may necessitate that the foot vault reduces rapidly under body mass without the heel fully weightbearing. This might help explain midfoot break commonly seen among cerebral palsy toe walkers. *(Permission www.healthystep.co.uk.)*

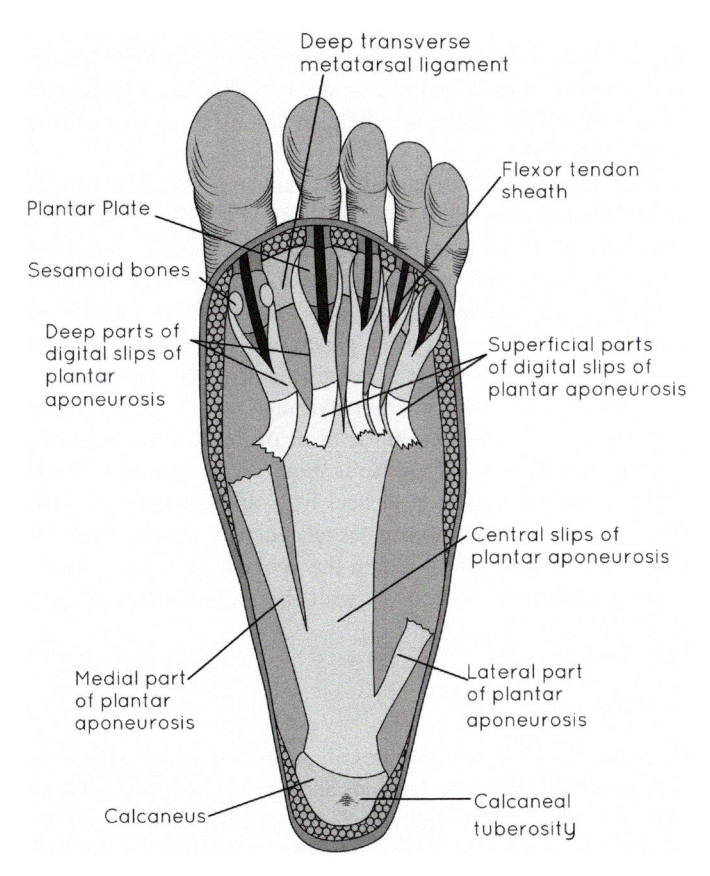

FIG. 3.3.4g The plantar aponeurosis is a structure that helps connect and bind the anatomy of the plantar aspect of the foot together. It also links to the dorsal deep fascia so that the foot works as a coherent functional unit via changing tensions occurring across it in all orthogonal planes. Here, the aponeurosis' anatomy is viewed in the horizontal plane with the cutaneous tissues, including the fat pads, removed. The superficial slips have been reflected backwards for better sight of the deeper slips binding to the metatarsophalangeal (MTP) joint anatomy. These superficial slips connect to the superficial fascia and interconnect to each other to form the superficial transverse metatarsal ligament. This arrangement of vertical and horizontal connective tissues forms walls that entrap forefoot fatty tissue, constructing the forefoot plantar fat pads and linking them to MTP joint motion. Failure of these connections makes the metatarsal heads feel more prominent. Superficial (first layer) intrinsic muscles derive attachments from the deep surface of the aponeurosis, and intermuscular compartment fascial septa run vertically and horizontally around deeper muscles and tendon sheaths, ultimately linking the aponeurosis to the periosteum and ligaments of the plantar skeletal frame. *(Permission www.healthystep.co.uk.)*

The plantar aponeurosis also has two less distinct and thinner medial and lateral sections that divide from the main central portion. The lateral part runs superficial to abductor digiti minimi, giving a proximal partial attachment for the muscle. This lateral section is thickest proximally where it runs from the plantar calcaneal tuberosity to the 5th metatarsal base. The medial section lies superficial to abductor hallucis, providing a partial attachment for the muscle. It is continuous with the flexor retinaculum, attaches firmly to the cutaneous tissue, and is continuous with the foot's dorsal fascia (*dorsalis pedis fascia*) medially and laterally. Deeply, it blends with the intermuscular septa that in turn blends to the periosteum of the bones, such as at the navicular tuberosity, medial cuneiform, and plantar surfaces of the 1st and 5th metatarsals (Draves, 1986: pp. 286–288). The deep fibres are also connected to the interosseous fascia and sundry other structures through the sagittal intermuscular septa (Bojsen-Møller and Flagstad, 1976; Hedrick, 1996). This makes the plantar aponeurosis a highly significant part of the passive functional tensegrity chain, intimately linking to the foot's vault profile longitudinally and transversely as well as to movement around the MTP joints and ankle.

Through complex fascial interconnections across the plantar foot, tension in the anterior–posterior direction of the aponeurosis will necessarily result in tightening of the intermuscular septa it is continuous with. This will result in stiffening the fascial attachments for the plantar intrinsic muscles, drawing in and tensioning together the components of the skeletal frame in all planes. The plantar aponeurosis works within a chain of events, producing stiffening through ligamentous and fascial connective tissue stress–strain behaviour and forefoot splay coupled to longitudinal stiffening. This coheres with the non-Newtonian stress-stiffening behaviour of biological viscoelastic tissue and the experimental findings on vault stiffening in vivo (Bjelopetrovich and Barrios, 2016; Takabayashi et al., 2020). This passive stress-stiffening action on the aponeurosis is likely to be enhanced by active plantar intrinsic muscle activity in a mechanism similar to that described in the gastrocnemius aponeurosis through its longitudinal and transverse plane elastic behaviour (Iwanuma et al., 2011b; Arellano et al., 2016), to which the plantar aponeurosis is structurally much akin.

Two forces govern dynamic changes occurring in the aponeurosis during muscle contraction. In low-to-intermediate levels of longitudinal force, the aponeurosis width increases in proportion to muscle fibre shortening, as muscle contraction causes radial expansion of the overlying aponeurosis (Arellano et al., 2016). As the aponeurosis is more compliant in the transverse direction (having an elastic modulus only a fifth of its longitudinal value), the strain in the aponeurosis' width is related to the amount of muscle fibre shortening strain which decreases as the muscle force increases, acting along the muscles' lines of action (Arellano et al., 2016). Essentially, this means the aponeurosis is more easily stretched longitudinally while the muscles are quiet, but on activation, the collective volume of the intrinsic muscles expands, widening the fascia, restricting longitudinal lengthening, and stiffening the aponeurosis further (Arellano et al., 2016). A link between intrinsic muscle function and plantar aponeurotic strain is thus established.

The stiffening processes during late midstance and terminal stance from muscle activity and stress–strain/strain rate fascial properties are rapidly enhanced through the later addition of the windlass effect after heel lift. This in turn increases the tension in the fascial chain that is allowing force generation in the intrinsic and extrinsic muscles to provide midfoot plantarflexion and abduction, which enhances the lever arm by shifting the CoGRF towards the longer medial metatarsals and their MTP joints (Holowka et al., 2017). These mechanisms are important supplements to transverse plane forefoot stiffening and together, increase energetics via Achilles power generation being applied to the support surface as a GRF through a stiffer forefoot. The stable GRF allows greater digital extension, which in turn enhances the stiffening effect of the plantar aponeurosis via its windlass mechanism as terminal stance phase acceleration proceeds. The lack of ability to achieve all of these events in series may lie at the heart of many musculoskeletal pathologies of the foot and lower limb derived from gait, when a stiffened stable base is required but not provided for acceleration. Failure in any part of the mechanism that delays forefoot stiffening at heel lift could cause plantar aponeurotic dysfunction and symptoms, giving rise to significant locomotive disability associated with plantar heel pain induced by injury around its proximal enthesis. Thus, the timings of peak strain within the aponeurosis and the power applied by utilising this complex fascia will both need to be considered in concert with the individual's activity if pathology within it is to be understood (Figs 3.3.4h and 3.3.4i).

3.3.5 The function and anatomy of the muscles of the foot

The intrinsic and extrinsic muscles play a substantial role in the foot's structural integrity, maintaining a functionally adaptable skeletal frame through a changing vault profile along with its soft tissue tensions and articular and osseous compressions (Bojsen-Møller, 1979; Blackwood et al., 2005; Headlee et al., 2008; Kelly et al., 2015; Okamura et al., 2018; Farris et al., 2019, 2020). Muscles, both intrinsic and extrinsic, through their tendons, moderate tensional forces loaded

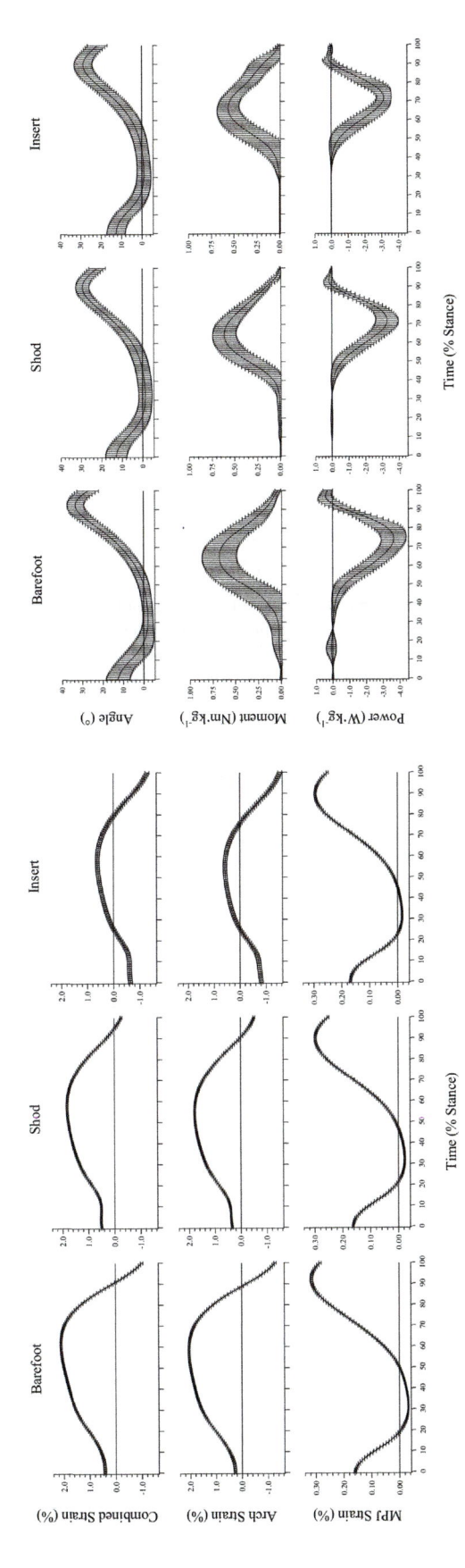

FIG. 3.3.4h The biomechanics of the plantar aponeurosis through its connections to the 1st metatarsophalangeal (MTP) joint during running have been investigated. Estimated mean plantar fascial strains during the stance phase of running are presented here (*left graphs*) as a combination of medial vault compression and MTP joint angles alone, and the effects of vault compression and MTP joint angles alone. Data are shown for barefoot, shod, and shod with custom foot orthoses (inserts). Strain values ≤0 represent lengths at which the aponeurosis is considered slack. MTP joint angles, net moment, and power traces (*right graphs*) during running stance phase under the same conditions show that the presence of shoes and custom foot orthoses significantly reduce MTP joint extension angles compared to barefoot during terminal stance. This suppresses peak moment and acceleration power. (*From McDonald, K.A., Stearne, S.M., Alderson, J.A., North, I., Pires, N.J., Rubenson, J., 2016. The role of arch compression and metatarsophalangeal joint dynamics in modulating plantar fascia strain in running. PLoS ONE 11 (4), e0152602.*)

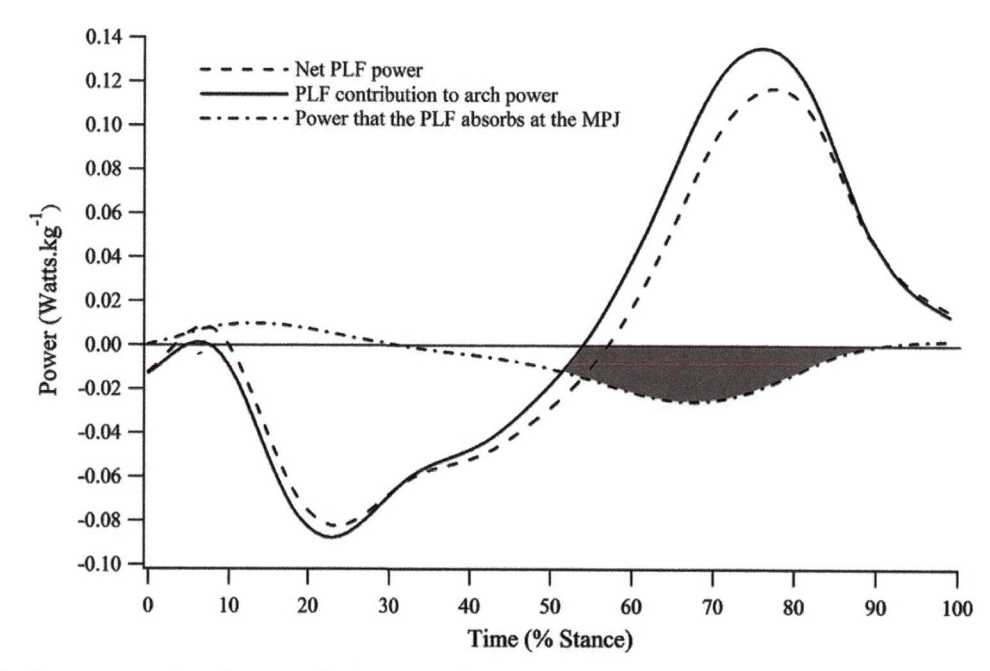

FIG. 3.3.4i The plantar aponeurosis works as part of an energy transfer mechanism, working with changes in foot vault profile and digital extension angles and moments. The aponeurosis' (plantar fascia—PLF) net power is represented by the *dashed line*, with power absorbed at the metatarsophalangeal (MTP) joints (MPJ in graph) represented by the uneven broken line. The shaded region represents the energy absorbed by the MTP joints during the acceleration phase during running gait. Similar but less dramatic effects can be expected during walking. *(From McDonald, K.A., Stearne, S.M., Alderson, J.A., North, I., Pires, N.J., Rubenson, J., 2016. The role of arch compression and metatarsophalangeal joint dynamics in modulating plantar fascia strain in running. PLoS ONE 11 (4), e0152602.)*

onto the passive support structures and this action is more pronounced when the CoM of the body is moved forward over the midfoot (Salathé and Arangio, 2002). The processes of compression and elastic recoil of the foot vault have been primarily approached by investigating the medial longitudinal arch and have long been considered to be a largely passive event. Investigation of the plantar intrinsics has proven that these muscles play a significant role in both walking and running (Kelly et al., 2015; Farris et al., 2019, 2020). However, the plantar intrinsics do not work in isolation in maintaining the appropriate posture of the skeletal frame, for not only are the passive elements at play but so too are the extrinsic foot muscles (Ferris et al., 1995; Kokubo et al., 2012). These much larger extrinsic muscles have largely been discussed in Chapter 2, Sections 2.5.8–2.5.11 in relation to ankle function. Here, they are considered more in relation to foot function.

The extrinsic dorsiflexors

The extrinsic dorsiflexors of the foot are highly active at preswing and toe-off so as to assist lifting the foot from the ground (Fig. 3.3.5a). In this action peroneus tertius, if present, aids the ground clearance by providing slight foot eversion. The extensors remain mildly active during swing phase to reduce the ankle plantarflexion angle in order to aid ground clearance. In late swing prior to heel contact, they become highly active again to 'pre-load' the plantar aponeurosis through digital/ankle extension, thus inducing the windlass mechanism's ability to stiffen the foot (Carlson et al., 2000; Caravaggi et al., 2009; Wager and Challis, 2016). This creates some 'tensional resistance' within the foot's vault (see Fig. 3.3.4d.) which can be released during forefoot loading via the reversed windlass mechanism (Stainsby, 1997; Green and Briggs, 2013). The degree of stiffness created will utilise the sensorimotor feedforward mechanisms set in expectation of the plantarflexion moment to occur with contact. These muscles continue to be highly active until forefoot loading, after which time their activity is occasionally and variably seen or absent until the next preswing (Byrne et al., 2007; Agostini et al., 2010; Di Nardo et al., 2013). The single-foot intrinsic extensor, extensor digitorum brevis, acts as an assistor to toe clearance and helps in the reversed windlass effect, having little other action but toe extension in preswing and late swing. Similar but slightly different mechanisms are at play in forefoot initial contact that significantly reduce the activity of the extensors in favour of the extrinsic plantarflexors (see Fig. 3.3.4f).

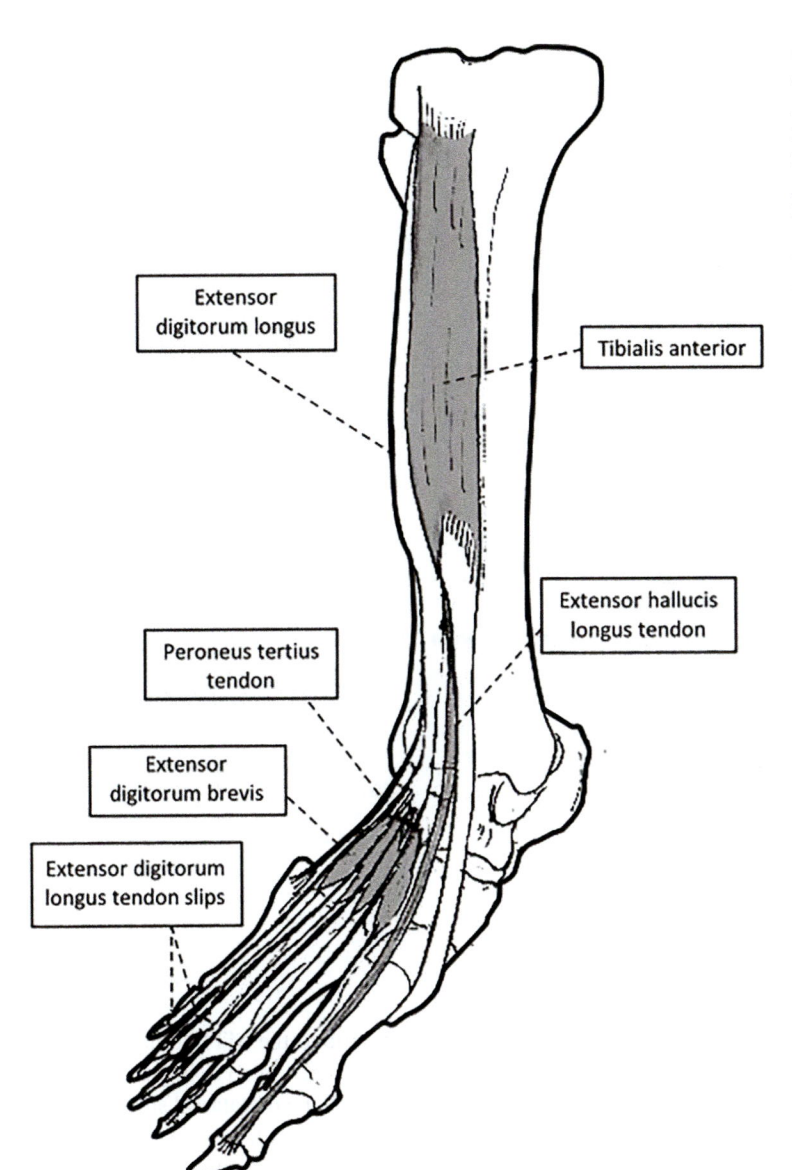

FIG. 3.3.5a The extensor muscles' vectors of the foot are dominated by forces generated by tibialis anterior (TA). Although considered an ankle dorsiflexor (or ankle plantarflexor resistor), it provides important medial and plantar restraint and articular compression to the 1st tarsometatarsal joint as well as being a very weak foot invertor. During hallux extension, it can aid in increasing vault height medially. The extensor hallucis longus (EHL) muscle belly lies beneath TA, providing some attachment for it from its fibrous sheath while its tendon attaches distally to the base of the proximal phalanx and often the MTP joint capsule medially via a medial tendon slip. It usually receives a very variable tendon slip from extensor digitorum brevis (EDB) laterally. It also provides hallux extension moments as well as some weak ankle dorsiflexion and foot inversion capacity. Laterally lie the four tendon slips of extensor digitorum longus (EDL). Its muscle belly mostly lies beneath TA but emerges laterally, with its myofascial sheath providing some attachment for TA. Each slip arises from the divisions of the EDL tendon as they emerge from the tendon sheath shared with peroneus tertius (PT) to attach first to the bases of the intermediate phalanges of the 2nd to 5th digits. Before these attachments, each slip divides into a central attachment slip that attaches to the base of each intermediate phalanx, and medial and lateral slips that run over the distal interphalangeal joint before joining together again to attach to the base of each distal phalanx. This arrangement resists tendon translation medially and laterally. Before the tendon slips pass the MTP joints, the three medial EDL slips receive variable tendon slips from EDB, as does EHL. EDB attaches proximally to and lies over the distal dorsal surface of the calcaneus and dorsal surface of the cuboid. PT, if present, lies under and often blends with EDL, attaching distally to the dorsal surface of the base (styloid process) of the 5th metatarsal. EDL and PT can be considered weak foot evertors and EDL and EDB are providers of lesser digit extension. They all acting as weak ankle dorsiflexors assisting TA. *(Permission www.healthystep.co.uk.)*

The extrinsic plantarflexors

The extrinsic plantarflexors of the foot play a critical role during stance, for they provide the primary control and power to the foot. The most significant are the triceps surae, tibialis posterior, and peroneus longus. The triceps surae dominates sagittal plane motion of the leg at the ankle and indirectly affects the vault's mechanics through the downward compression from ankle dorsiflexion during late midstance. The triceps surae is also responsible for generating the ankle power that induces heel lift and provides the large GRF on the forefoot in terminal stance. The other plantarflexors provide limited assisting plantarflexion power to the forefoot GRF, with the contribution of flexor hallucis longus to the plantarflexion torques increasing with gait speed (Péter et al., 2015).

The extrinsic plantarflexors are (near-) isometrically active in midstance (Hofmann et al., 2013). Through this activity, the extrinsic flexor tendons actively and directly influence vault kinematics (Ferris et al., 1995) through foot stiffening, midfoot and forefoot stabilising, midstance energy buffering, and energy storing within their myofascial structures. The long digital flexors, by bridging across the length of the foot vault to the distal phalanges of the toes, are able to influence motion of the vault throughout late midstance. They are also critical in providing proximally directed stability forces at the MTP joints, helping fix the fulcrums to the ground after heel lift (Fig. 3.3.5b).

FIG. 3.3.5b The plantar extrinsic flexor tendons apply large cross-sectional area muscle power from the posterior calf into the foot vault. This anatomy greatly enhances force generation capacity within the foot without, importantly, dramatically increasing its mass. In (A), the long digital flexors tendons are shown with that of peroneus longus passing beneath flexor digitorum longus. Flexor hallucis longus strongly influences medial vault stability with abductor hallucis, while flexor digitorum longus is heavily reliant on its associated plantar intrinsic muscles such as quadratus plantae for central vault control. Peroneus longus is important for medial to lateral vault control, working with tibialis posterior. The deeper lying tibialis posterior tendon is shown in (B), with the inferiorly situated flexor and peroneus tendons resected. Tibialis posterior's tendon attachment from the cuboid and central metatarsal bases and ligaments, combined with those from the medial–plantar navicular, cuneiform, and sometimes the calcaneus, provides powerful lateral to medial vault compression and vault-raising power. *(From McKeon, P.O., Hertel, J., Bramble, D., Davis, I., 2015. The foot core system: a new paradigm for understanding intrinsic foot muscle function. Br. J. Sports Med. 49 (5), 290.)*

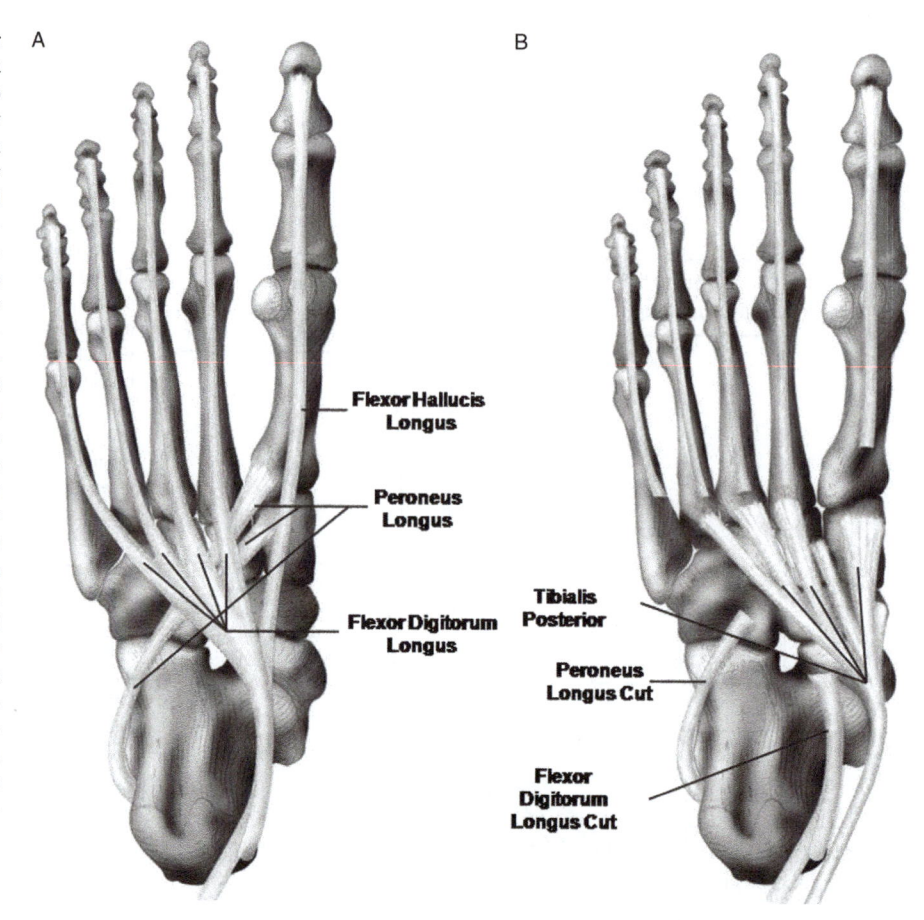

The long digital flexors play a prominent role in maintaining dynamic stabilisation around the MTP joints in concert with the intrinsic digital flexors (Ferris et al., 1995; Fiolkowski et al., 2003; Kelly et al., 2012). In so doing, they play an important part in generating tensile stresses across the vault that creates compression through the metatarsal shafts to resist bending moment deflection stresses (Ferris et al., 1995; Sharkey et al., 1995; Jacob, 2001). Although they are referred to as flexors, their primary functional role in terminal stance is to resist the passive digital dorsiflexion (or extension) that results from heel lift while stiffening the MTP joints. Increasing ankle and midfoot plantarflexion angles are intimately coupled to digital extension. Thus, the restraint of extension of the MTP joints by these muscles influences whole foot kinematics. In maintaining MTP joint stiffness, they are assisted by a complex set of passive connective tissue structures that bind the metatarsal heads and MTP joints into the deep transverse metatarsal ligament and the plantar aponeurosis (Bojsen-Møller and Lamoreux, 1979; Stainsby, 1997; Gregg et al., 2006, 2007; Dalmau-Pastor et al., 2014; Maas et al., 2016). With these complex connective tissues forming the forefoot tie-bar, the intrinsic and extrinsic flexors set up a partially controllable and variable degree of stiffness across the MTP joints (Fig. 3.3.5c).

The peroneals

The peroneus longus muscle has EMG peak activity during contact phase and particularly just before and at the midstance–terminal stance boundary (Murley et al., 2014). During late midstance–early terminal stance, both peroneus longus and brevis should move the CoGRF medially to prevent any lateral drift away from the longer medial metatarsal fulcrums of the 1st and 2nd MTP joints. Using these MTP joints is necessary to improve ankle power efficiency (Holowka et al., 2017). Peroneus brevis is likely to be a primary proximal lateral column stabiliser through the calcaneocuboid and cuboid–5th metatarsal joints, but also a lateral assistor to midfoot–forefoot eversion, helping in shifting the CoGRF medially after heel lift in conjunction with peroneus longus. Peroneus longus also compresses the lateral column together

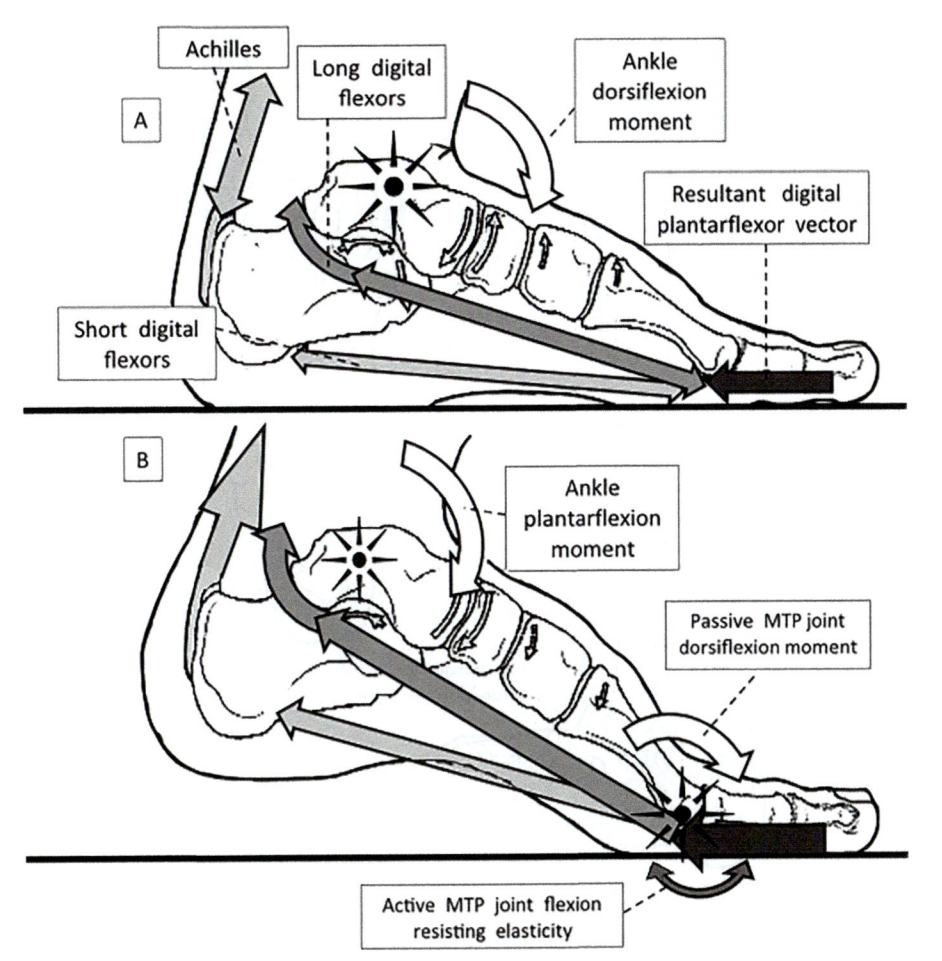

FIG. 3.3.5c The human long and short digital flexors rarely flex the toes, save when balancing on particularly uneven surfaces. Instead, their role includes helping increase vault stiffness during late midstance (A). They resist ankle dorsiflexion moments that drive midfoot sagging bending deflection by generating near-isometric contraction across the plantar vault. This action resists vault deflection and concurrently creates a proximally directed resultant force vector from the digits *(dark-grey arrow)*. This proximal vector helps limit foot lengthening and increases metatarsophalangeal (MTP) joint stiffness, creating elasticity across these joints. As a result, at heel lift (B), the passive digital extension moment will occur under tensile resistance *(represented by grey double-headed curved arrow)* derived from the digital flexion moment of the long and short flexors that provide proximal-directed compression into the MTP joints, stabilising them and stiffening local connective tissues during motion. The level of resistance set should not prevent but only restrain MTP joint dorsiflexion freedom. Excessively stiff MTP joints would lengthen the Achilles external plantarflexion moment arm for acceleration, increasing forefoot stresses. Equally problematic is unrestrained digital extension with the loss of flexion elasticity leading to excessive digital dorsiflexion and reduced forefoot stability. This would reduce the capacity to drive ankle plantarflexion power into the forefoot to create the stabilising GRF needed for acceleration. Excessively free digital dorsiflexion potentially risks MTP joint articular and supporting soft tissue pathologies. Local MTP joint anatomy has been co-opted to a locomotive extension role from structures that evolved for flexion grip. Thus, they contain intrinsic weaknesses. Digital flexor strength is extremely important for locomotion and forefoot injury protection, despite actual MTP joint flexion motion rarely being necessary for locomotion. *(Permission www.healthystep.co.uk.)*

at the calcaneocuboid joint via its tendon compression on the lateral inferior aspect of the cuboid. It has a most important role in creating and maintaining transverse vault/frontal plane stability and stabilisation. Through compression across the lesser tarsus and tarsometatarsal joints transversely, peroneus longus can couple the ability to stiffen the foot longitudinally to its transverse stiffening role with tibialis posterior (Kokubo et al., 2012; Yawar et al., 2017; Venkadesan et al., 2020). Although peroneus longus works in concert with peroneus brevis in either everting the foot or resisting inversion of the ankle, peroneus longus seems to have a far more intimate relationship with tibialis posterior with which it works antagonistically in frontal and transverse plane motions, but agonistically in stiffening and plantarflexing the midfoot (Fig. 3.3.5d).

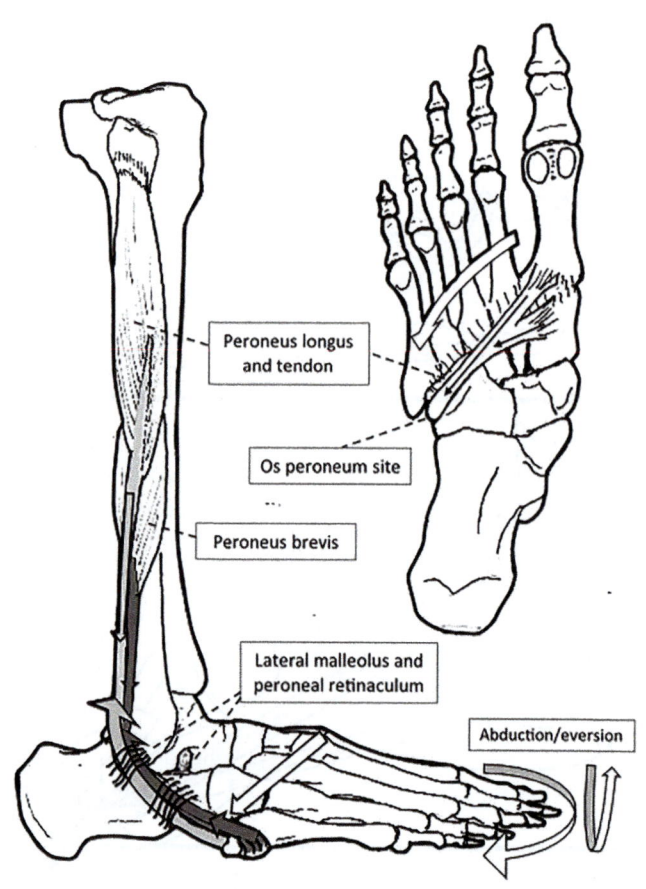

FIG. 3.3.5d A schematic to express peroneus longus' and brevis' complementary and distinctly different functions due to their commonality and variance in force vectors (A). Both muscles share a tendon sheath behind the lateral malleolus and under the superior and inferior peroneal retinaculum before parting their ways, giving them both weak ankle plantarflexion vectors. When the foot is in closed chain, both muscles set a vector that internally rotates the leg at the ankle, relatively abducting the forefoot to the leg. In open chain (including during terminal stance), they evert the foot into valgus at the ankle *(upper white arrow)* via their distal attachments. These open- and closed-chain mechanisms resist ankle inversion moments, whether induced by the foot moving below or via medial body motions above the ankle that externally rotate the leg. Peroneus brevis also utilises its distal foot attachments to compress the relatively mobile cuboid–5th metatarsal joint, stabilising the lateral column during late midstance and terminal stance. Peroneus longus, slung under the foot through the peroneal groove (B), distally attaches to the lateral side of the medial cuneiform and base of the 1st metatarsal, as well as having relationships with the tarsometatarsal region's ligaments as it runs under the long plantar ligament. Tendon slips to other metatarsal bases such as the 2nd are not uncommon. A sesamoid (os peroneum) is often present where the tendon abuts the cuboid as it passes underneath it. Peroneus longus sets up a vector that compresses the midfoot from medial to lateral, and in so doing helps raise and stiffen the foot vault with agonistic activity with tibialis posterior. However, concurrently working with peroneus brevis, the midfoot and forefoot are abducted and everted *(grey arrows)* antagonistically against tibialis posterior action. Tibialis posterior overrides the abduction moment creating midfoot adduction, but the eversion moment is essential for loading the medial forefoot for acceleration. *(Permission www.healthystep.co.uk.)*

Tibialis posterior

Like peroneus longus, tibialis posterior's peak EMG activity is around 10% of gait during loading response to forefoot contact and again at around 45%–50% of gait at the late midstance–terminal stance boundary, occurring slightly earlier with increasing walking speed (Semple et al., 2009; Murley et al., 2014). Tibialis posterior has a 'braking' role during loading response that assists the extensors in maintaining a pre-tensioned vault before forefoot loading and maintains some vault stiffness during forefoot contact. This is likely achieved by providing proximal–medial compression stability across the proximal medial column throughout its multiple midfoot attachments as well as transversely from the cuboid and 4th metatarsal base. The amount of stiffness tibialis posterior generates is achieved through its interaction with peroneus longus (Kokubo et al., 2012).

Tibialis posterior is the primary invertor of the foot (Hintermann et al., 1994; Imhauser et al., 2004). However, it is usually functioning by resisting eversion and abduction moments across the rearfoot through the orientation of its tendon under the calcaneal medial supporting prominence of the sustentaculum tali, acting against the laterally directed heel contact GRF. Tibialis posterior and its tendon also play an important part in energy dissipation and storage through loading response/early midstance vault profile changes, when its tendon lengthens while its muscle fascicles shorten (Maharaj et al., 2016) (Fig. 3.3.5e).

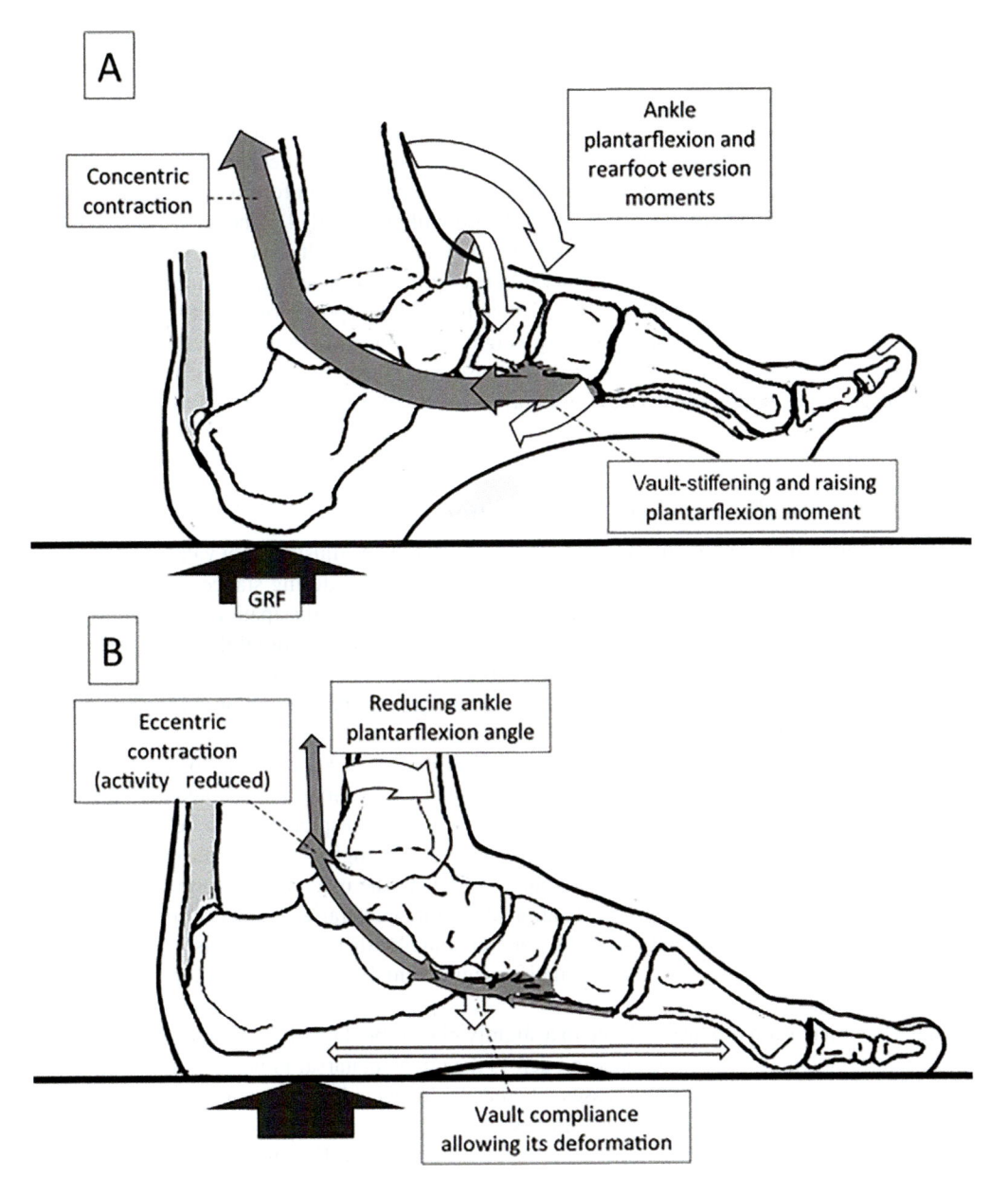

FIG. 3.3.5e Tibialis posterior has two periods of peak activity, with the second occurring during late midstance around the heel lift boundary. The first, as demonstrated here, is used to raise and stiffen the vault prior to forefoot contact via concentric contraction with peroneus longus activity (A). Its activity during this period also helps resist the eversion moment that associates with medially directed GRFs that follow a lateral–posterior heel strike, aided by tibialis anterior. At forefoot contact and into early midstance (B), its continued and yet reducing activity can control midfoot eversion and then inversion and abduction moments, as first the lateral then the medial forefoot loads. Once forefoot loading is completed, activity decreases into early midstance (although it never stops), becoming eccentric to release foot compliance for vault deformation capabilities to dissipate impact and loading energies. This allows the connective tissues to start their process of stress-stiffening and energy storage, being modified and resisted via quadratus plantae activity. The tibialis posterior tendon is now also used to store tension energy while controlling vault posture. Tibialis posterior is a 'hybrid' tendon expressing the material properties between positional and energy storage tendons. It is required to manage foot posture during single-limb support until heel lift, and then provides recoil energy released to power midfoot plantarflexion and adduction during terminal stance. *(Permission www.healthystep.co.uk.)*

Tibialis posterior's activity 'usually' reduces through midstance, rising again in late midstance and peaking before heel lift (Semple et al., 2009; Murley et al., 2014). During this phase of activity, it seems to play an important role in stiffening the foot (Kokubo et al., 2012) and moving the CoGRF anteriorly as the heel lifts (Imhauser et al., 2004). Both muscle and tendon shortening is associated with this phase which continues into terminal stance, despite decreasing muscle activity (Maharaj et al., 2016). This is quite similar to the elastic tendon behaviour of the Achilles which provides terminal stance ankle plantarflexion power, despite its activity rapidly reducing. The suggestion is that like the Achilles, tibialis posterior power during terminal stance provides work through tendon elastic recoil. In terminal stance, tibialis posterior's power assists in vault re-contouring through inducing a plantarflexion and adduction moment about the midfoot and assisting peroneus longus to move the fulcrum point to the medial metatarsals in order to achieve the best foot posture and lever arm for acceleration efficiency.

Tibialis posterior's kinematic and foot-stiffening mechanisms in conjunction with peroneus longus, through creating stability and mechanical efficiency, influence passive transverse arch-stiffening mechanics, plantar intrinsic and extrinsic activity, and the windlass mechanism. The success of tibialis posterior can therefore be indicated by all of these mechanisms together reducing the vault length seen after heel lift (Stolwijk et al., 2014). The active and passive processes that shorten the foot will allow elastic recoil from these interlinked myofascial vault structures (Ker et al., 1987; Natali et al., 2010b; Pavan et al., 2011; McDonald et al., 2016). Energy released from storage in the connective tissues is likely used to help shorten vault length as part of the vault-offloading mechanism. Tibialis posterior activity and work explains the observations of foot lengthening occurring most rapidly during weight acceptance with tendon stretching, decreasing vault length change during late midstance as tibialis posterior activity increases and peaks, and shortening at terminal stance from elastic recoil. In vault kinematics and kinetics, tibialis posterior plays a key active role, making dysfunction of this muscle a serious threat to locomotion.

Peroneus longus and tibialis posterior: One functional unit

Peroneus longus and tibialis posterior together seem to play a pivotal role in adjusting foot vault stiffness throughout stance, influencing the ability of the foot to absorb energy via increased flexibility during stance as a consequence of adjusting activity levels. These two muscles moderate stiffness, compliance, and energy dissipation within the foot most effectively by working in unison (Kokubo et al., 2012). Peroneus longus seems to control flexibility and energy dissipation while tibialis posterior activity induces stiffness (Kokubo et al., 2012). Thus, a mechanism exists whereby through moderating activity in one muscle or the other, the foot is able to manage energy dissipation and/or stiffness as required by the biomechanical consequences of gait events.

However, neither peroneus longus nor tibialis posterior have extensive attachments to the rearfoot so their activity seems to create a relationship between the ankle, midfoot, and leg in which motion is coupled between them without there being any direct control of the calcaneus or talus (there is via fascia). The suspicion has to be that control of the proximal medial column is via talonavicular and navicular–cuneiform stiffness through tibialis posterior's navicular and cuneiform attachment(s), and that through the attachments of peroneus longus to the medial cuneiform and 1st metatarsal base laterally, the medial column is stabilised distally. Stabilisation of the tarsometatarsal joints is improved by tibialis posterior's metatarsal base attachments that run medially from the tendon's lateral–distal attachments, resisting any tarsometatarsal abduction on forefoot loading. Thus, the distal vault is under a significant distolateral to medioproximal force vector as long as tibialis posterior is functional.

Peroneus longus tendon tension produces a force in a slightly plantar but laterally dominant direction around the cuboid to produce significant eversion as well as compression across the tarsometatarsal joints (Johnson and Christensen, 1999). All the articulations this muscle's tendon crosses should be expected to be compressed in a medial–distal to lateral–proximal direction arising from the cuboid 'pulley' within the peroneal groove. A 'locking' mechanism of the 1st ray of the medial column effected through the peroneus longus plantarflexing and everting the medial cuneiform and 1st metatarsal base is suggested (Johnson and Christensen, 1999). This would help stabilise the navicular anteriorly and move the medial cuneiform into abduction and eversion in relation to the navicular, allowing more freedom for the navicular to adduct and plantarflex by means of tibialis posterior activity. Tibialis posterior tensioning will adduct, plantarflex, and evert the navicular, preventing/restricting talar adduction at the talonavicular joint and will also likely restrict talar plantarflexion as a result of the navicular being pulled inferiorly, blocking talar motion anteriorly (Fig. 3.3.5f).

It is known that frontal plane motions (inversion–eversion) of the talonavicular joint are restricted in terminal stance (Leardini et al., 2007; Wolf et al., 2008) as sagittal plane motion increases as part of midfoot plantarflexion (Holowka et al., 2017). Such talonavicular and medial column kinematic interactions occurring as a consequence of the actions of tibialis posterior and peroneus longus activity influence stiffness and compliance within the foot throughout stance phase, giving a response mechanism to the terrain through the sensorimotor system. It is important to note that the combined effect of tibialis posterior and peroneus longus activity is to stiffen the foot (Kokubo, et al., 2012), and their activity usually occurs in unison (Murley et al., 2014). Thus, the foot is stiffened as a consequence of their combined synergistic activity.

Problems are likely to arise with changes in the activity relationships between these muscles. In tibialis posterior dysfunction, foot stiffness will reduce, not only through loss of tibialis posterior but also because peroneus activity in isolation increases foot flexibility. Peroneus longus dysfunction may lead to excess stiffening through unopposed tibialis posterior activity. Foot postures that change the mechanical advantage of one muscle over another will also likely influence the

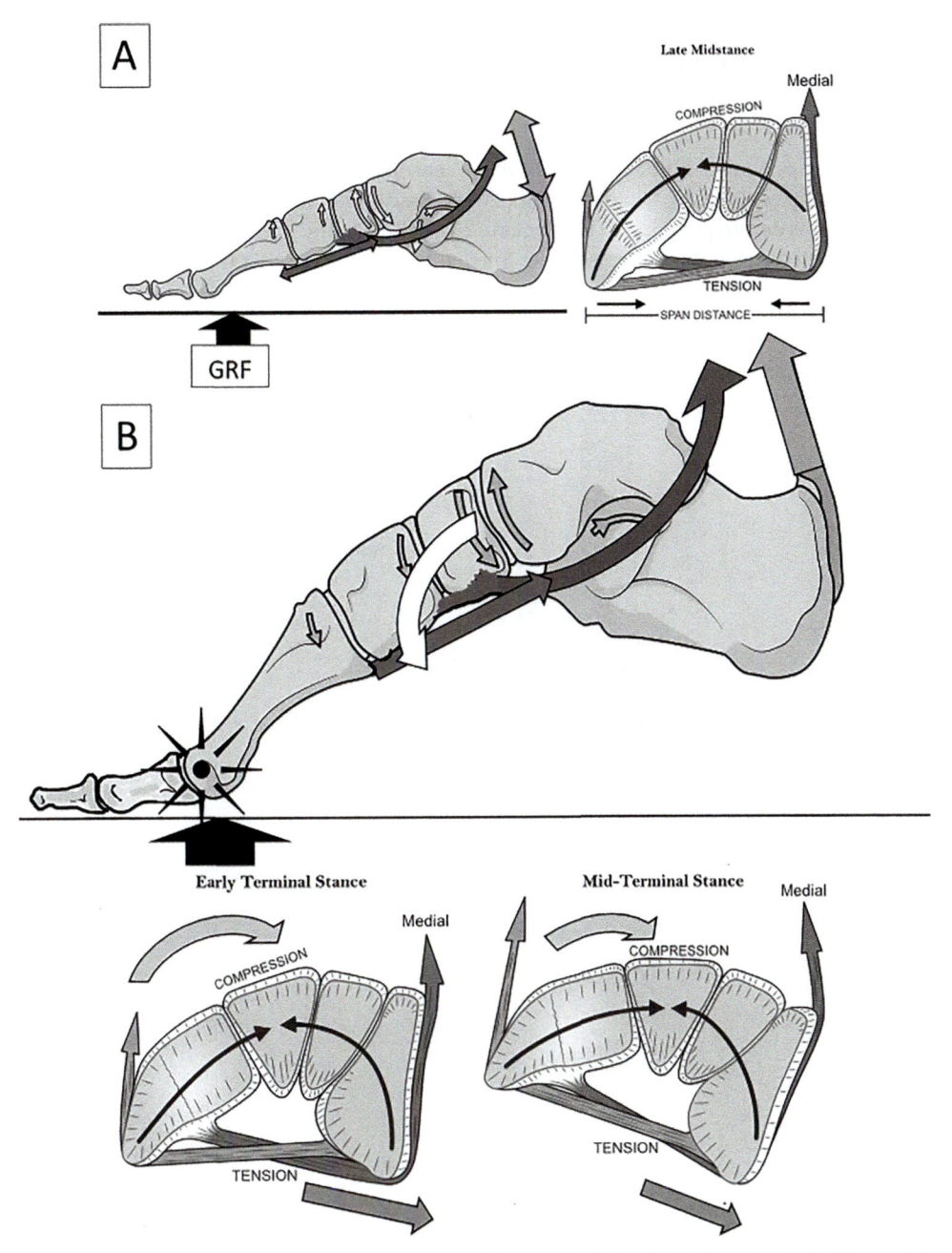

FIG. 3.3.5f Just before the heel lift boundary during late midstance (A), tibialis posterior is developing its peak activity. This combines with activity of peroneus longus to compress the midfoot together in both the longitudinal plane and transverse plane. Stiffening and starting to reduce the transverse span distance is the key component for creating a semi-rigid beam out of the foot. The proximal compression forces created by tibialis posterior's eccentric/near-isometric activity are insufficient to overcome the GRF/BWV effects of depressing the foot vault because the foot stiffens. However, increasing activity at this time continues to store energy within its tendon. Therefore, the vault profile is lowest at heel lift while the tendon has gained its maximal stored energy ready for release and the foot has become stiffened. At heel lift (B), the plantarflexion power is released from triceps surae via the Achilles' elastic recoil. This frees the energy stored within tibialis posterior's tendon with that of peroneus longus, producing midfoot plantarflexion moments *(white arrow)*. During early terminal stance, compression of the midfoot is maintained by the tibialis posterior and peroneus longus tendons, but the power of tibialis posterior generates a stronger adduction moment *(straight darker-grey arrow—lower transverse plane images)* than the peroneus longus' abduction moment. However, through its tendon compression upon the cuboid, a greater eversion moment *(curved light-grey arrow)* is applied via peroneus longus (aided by peroneus brevis) than the inversion moments provided by tibialis posterior on the navicular. Midfoot medial rotations and plantarflexions not only help twist the vault into a stiffer, higher profile, but also move the CoGRF to the medial metatarsal heads in order to set up an excellent MTP joint fulcrum point (star) and longer lever arm length for the plantarflexion power of acceleration. *(Permission www.healthystep.co.uk.)*

The plantar intrinsics

The plantar intrinsic muscles lie under the foot and are active tensors of the skeletal frame, appropriately moderating foot vault profile changes (Kelly et al., 2014; Okamura et al., 2018), assisting in forefoot stiffening during late stance, and stabilising MTP joint fulcrums during propulsion (Farris et al., 2019, 2020) (Fig. 3.3.5g). The plantar intrinsic muscles have a role in supporting the vault through their effects on the 'medial longitudinal arch' (Wong, 2007; Kelly et al., 2014, 2015)

agonistic–antagonistic relationships and activity levels of each muscle during gait. Stiff feet may require greater peroneus longus-induced stiffness reduction, while mobile feet are likely to require increased tibialis posterior stiffening activity.

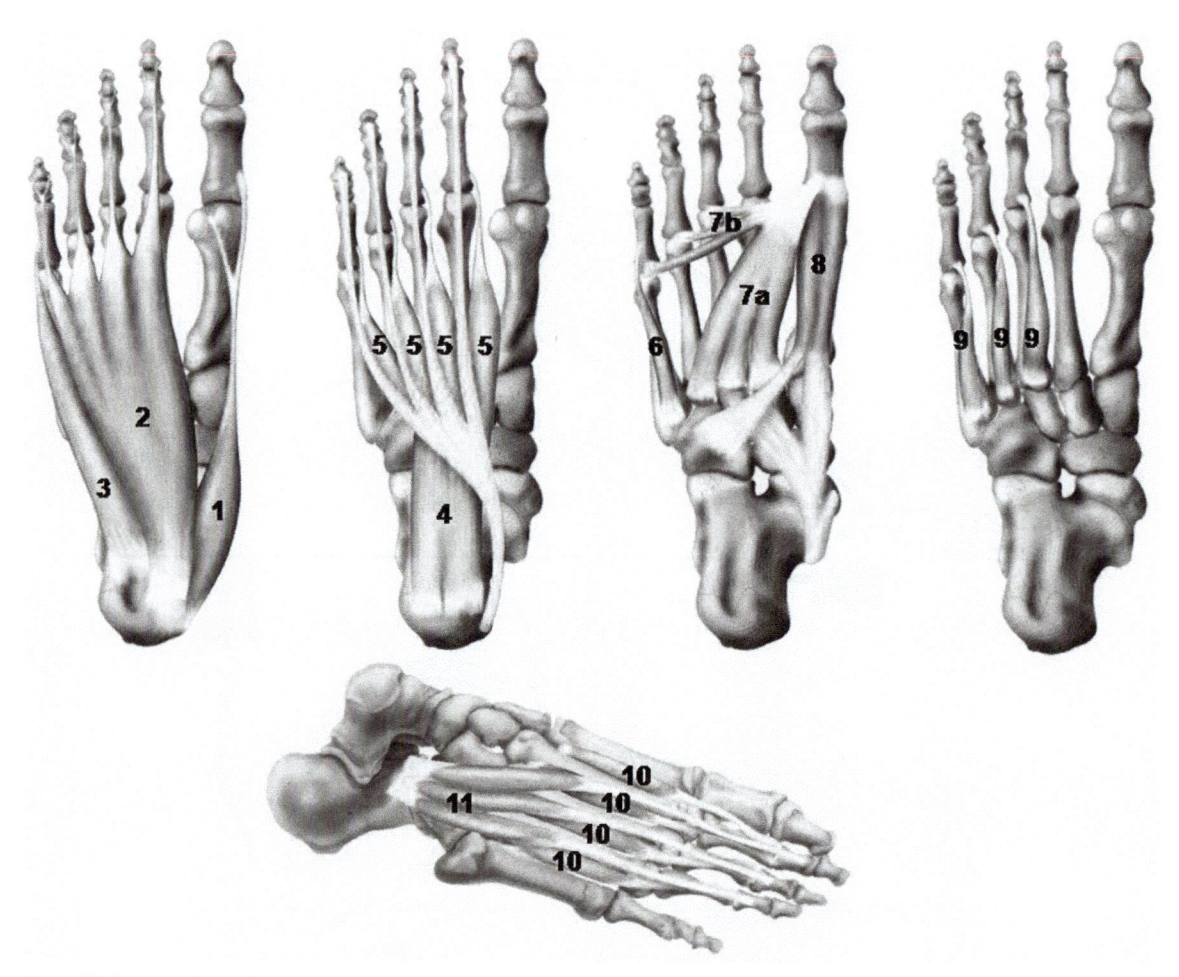

FIG. 3.3.5g Under sensorimotor input, the plantar intrinsic muscles work with the plantar extrinsic muscles to fine-tune the direction of muscular force vectors and alter compliance and stiffness levels required during locomotion. Being within the foot, their potential size is limited by the locomotive costs of adding extra mass to the foot. They are usually, albeit arguably, described as running in four layers from superficial to deep, with the interosseous muscles being within the fourth deepest layer. However, their muscular compartmental septa do not reflect these layers well. Superficially, abductor hallucis (1) is important in medial vault stability and hallux alignment, flexor digitorum brevis (2) is a part of digital extension resistance and central vault stability, while abductor digiti minimi assists lateral vault stability and 5th digit posture. These superficial muscles are the most easily studied via EMG. The second layer consists of quadratus plantae or flexor digitorum accessorius (4) that redirects flexor digitorum longus' longitudinal vector between the lesser digits and the ankle, turning itself and the long flexor into a longitudinal vault stabiliser. Adjustment of this extrinsic muscle's vector to the digits is equally true of the lumbricals (5) that arise from each of the tendon slips of flexor digitorum longus. They have the capacity to adduct and assist flexion of the lesser toes. The third layer offers flexor digiti minimi brevis (6) that attaches proximally to the 5th metatarsal's medial shaft and distally to the 5th proximal phalangeal base. It seems to offer some 5th metatarsal dorsiflexion resistance via a 5th digital flexor moment. More significant are the oblique (7a) and transverse (7b) sections of adductor hallucis that seem to have an important role in transverse distal vault stability mechanics. The proximal attachments of flexor hallucis brevis (8) to the cuboid via a ligamentous union (with often attachments to the lateral cuneiform and tibialis posterior tendon) suggest that sesamoid motion within this muscle's distal tendon slips couples longitudinal to transverse vault vectors. The plantar (9) and dorsal (10) interossei can provide vectors that resist metatarsal shaft spreading proximally while distally, they can aid in digital extension resistance. In late terminal stance, the interossei may assist digital extension with extensor digitorum brevis (11) during preswing. *(From McKeon, P.O., Hertel, J., Bramble, D., Davis, I., 2015. The foot core system: a new paradigm for understanding intrinsic foot muscle function. Br. J. Sports Med. 49 (5), 290.)*

and thereby possibly decreasing stress in the plantar aponeurosis (Wu, 2007; Okamura et al., 2018). However, the effects of the intrinsics may not be through longitudinal arch stiffening but by way of assisting in stiffening the forefoot (Farris et al., 2019). By stiffening the forefoot, they can assist the passive stiffening mechanisms suggested by Venkadesan et al. (2020) that couple transverse arch stiffening to longitudinal stiffening. They also help set the stiffness across the MTP joints. This would seem energetically logical, as assisting in transverse plane stiffening would likely be less metabolically demanding than assisting in longitudinal profile stiffening which spans a far greater distance in the foot and needs to decrease a greater radius of curvature. Anaesthetic inactivation of the plantar intrinsics during terminal stance is reported to increase stride rate and provoke increased power generation within the hip muscles, but it does not seem to influence metabolic cost of transport (Farris et al., 2019). This suggests that humans can get along without plantar intrinsic muscles although their loss shifts the energetic burden proximally.

The plantar intrinsics do not change the vault's response to initial loading in early midstance (Farris et al., 2019) as they only become significantly active during later stance from absolute midstance through to preswing (Mann and Inman, 1964; Kelly et al., 2015). Although often considered as one functional unit active together, Kelly et al. (2015) have demonstrated different activity patterns in abductor hallucis, flexor digitorum brevis, and quadratus plantae. The precise phasic activity of some of the deeper smaller intrinsics remains unclear, so caution should be maintained when considering the subtleties of intrinsic muscle activity. The three largest intrinsics are the abductor hallucis, flexor digitorum brevis, and quadratus plantae, and these all demonstrate heightened activity with increasing gait velocity and GRFs. However, they provide individually different bursts of activity during gait (Kelly et al., 2015).

Abductor hallucis is active during late swing where it may assist hallux extension prior to initial contact (Kelly et al., 2015). Its peak of activity coincides with peak vertical GRF at F3 (~heel lift), with its deactivation during terminal stance as it undergoes shortening with digital extension (Kelly et al., 2015). Quadratus plantae is also slightly active prior to heel strike through into early stance and has peak activity during midstance during running and late stance in walking (Kelly et al., 2015). Flexor digitorum brevis shows activity commencing at contact and persisting throughout stance, but deceasing in terminal stance (Kelly et al., 2015). It seems likely that the other plantar intrinsics have similarly interesting activity.

Intrinsic muscles all demonstrate slow active lengthening during midstance vault compression followed by rapid shortening on vault recoil during terminal stance (Kelly et al., 2015). Overall, it is highly likely that the intrinsics demonstrate stretch-shortening cycle properties that permit energy absorption and elastic energy storage and release, as found in other myofascial units. Using electrical stimulation of the superficial plantar intrinsics compared to anaesthetised plantar intrinsics by means of tibial nerve blocks, Okamura et al. (2018) reported that in gait during increased stimulation, vertical GRFs were lower within late stance. They proposed that the plantar intrinsics are therefore able to play a role in shock absorption. This action is likely to result from the properties of the muscle–tendon complex absorbing energy through muscle–tendon buffering. Because most of these muscles are not active at initial contact and loading response, it prevents them from being significant energy dissipators at impact. Instead, this energy dissipation occurs as vertical GRF increases during late midstance under vault compression prior to heel lift. Such energy dissipation could reduce loading stress on the plantar ligaments, MTP joint structures, and plantar aponeurosis. It has been noted that patients suffering symptoms of plantar fasciitis demonstrate less volume of the plantar intrinsic muscles (Pohl et al., 2009) and smaller cross-sectional areas (Angin et al., 2014), a finding which supports such a mechanism, as well as suggesting a treatment option.

Reduction in vertical GRFs from intrinsic energy dissipation does not seem to result in a loss of propulsive power generation within the foot, as gait velocity and progression force do not decrease in anesthetised plantar intrinsics (Okamura et al., 2018). This suggests that intrinsic muscle energy dissipation is not reducing the vault's elastic recoil. Such an indication is important because the vault is highly evolved to provide both elastic energy absorption and a stiff propulsive platform as dual key adaptations for obligate plantigrade bipedalism. It appears that the intrinsic muscles only minimally support the longitudinal vault profile and do not affect longitudinal midfoot stiffness directly (Farris et al., 2019). Navicular height can seemingly be increased by stimulating additional plantar intrinsic activity (Kelly et al., 2014), but not during loads exceeding body weight that occur during the latter half of stance phase (Caravaggi et al., 2010; Okamura et al., 2018). However, increasing plantar intrinsic activity seems to delay the onset rather than the amount of minimum navicular height (Okamura et al., 2018). Fatigue in the plantar intrinsics results in increasing amounts of navicular drop (Headlee et al., 2008), suggesting a compromise in maintaining control over the vault profile's motion. This possibly explains the finding that the foot lowers its profile and lengthens to become increasingly prone with fatigue after running (Cowley and Marsden, 2013; Fukano and Iso, 2016).

The intrinsic muscles seem to play their most significant role in stiffening the forefoot and influencing MTP joint moments during propulsion, for without their activity, the ability to generate propulsive power into the ground is impaired (Farris et al., 2019). In this role, the short flexors work with the long plantarflexors, protecting the metatarsals from bending moment stresses as well as stabilising the MTP joints (Mann and Inman, 1964; Jacob, 2001; Fiolkowski et al., 2003; Kelly et al., 2012). A consequence of intrinsic muscle failure is that ankle joint work is impaired, meaning that greater hip power is required to achieve forward progression of the support limb's CoM (Farris et al., 2019). This is a similar effect to that which is seen with a failure in triceps surae power directly, suggesting that the intrinsics aid triceps surae power application. The findings of Farris et al. (2019, 2020) indicate that the windlass mechanism of the plantar aponeurosis at heel lift is inadequate without intrinsic muscle activity that actively generates tension which are important to stiffening the MTP joints. It is likely that the concentric or near-isometric activity of the plantar intrinsic and extrinsic muscles has a significant role to play in assisting the midfoot plantarflexion moment with tibialis posterior, peroneus longus, and the long flexors. Together, these foot muscles increase the vault profile as part of foot stiffening and vault shortening during terminal stance.

The size and cross-sectional area of the plantar intrinsics may reflect the vault profile they support or may in part be responsible for the foot vault that is formed during stance. The intrinsic muscles are reported as being thicker due to a larger cross-sectional area when associated with increased body weight, with specific muscles also being influenced by the vault profile (Sakamoto and Kudo, 2020). For example, abductor hallucis, which is an important stabiliser of the medial vault, has been reported to be both thicker and thinner within different studies on flat feet (Sakamoto and Kudo, 2020). These variances may represent the differences between either a hypertrophic response to having a flatter vault profile or developing a lower vault profile through muscle atrophy. Thickness and cross-sectional area have been reported as being significantly smaller in the abductor digiti minimi muscles found in flatter foot profiles compared to normal profiles (Zhang et al., 2017; Sakamoto and Kudo, 2020), as is also the case with the oblique head of adductor hallucis (Sakamoto and Kudo, 2020). This may reflect differences in the function of differing vault curvatures.

The conclusion of recent research is that although functionally important, the role of the plantar aponeurosis and its windlass mechanism has been overestimated in stiffening the foot and is in fact assisted by multiple events, including plantar intrinsic activity (Kelly et al., 2015; Farris et al., 2019, 2020; Sichting and Ebrecht, 2021; Welte et al., 2021). The passive transverse plane-stiffening mechanisms are critical (Venkadesan et al., 2017a,b; Yawar et al., 2017), functioning as they do in concert with tibialis posterior and peroneus longus-coupled activity that together also stiffen the foot (Kokubo et al., 2012). It is likely that the plantar intrinsic muscles are an integral part of the mechanisms that couple longitudinal-to-transverse plane foot stiffening with the plantar extrinsics such as tibialis posterior, peroneus longus, and the long digital flexors all influencing the stability and stiffness of the foot vault. The ligaments straddling the vault provide further passive stiffening through their stress-stiffening and time-dependent properties. Ligaments and muscles also play a highly significant sensorimotor role, making foot stiffening, energy dissipation, and energy storage a highly responsive mechanism.

The effect of shoes on foot muscles

As the world becomes increasingly populated with urban-dwelling people, footwear designed to provide protection, stability, and comfort for the ankle-foot complex has become increasingly common. However, shoes that perform other roles such as making fashion statements or for specific occupational use are also widely used. Feet that have developed shod throughout childhood do function differently to those of the habitually unshod, with footwear generally leading to stiffer, less adaptable forefeet and larger inter-individual arch heights (Kadambande et al., 2006; D'Août et al., 2009). Shoe use is also associated with toe deformities and increased falls in the elderly which have been suggested to be a result of under-utilisation of foot muscles, reducing their strength and causing imbalances between them. It has been suggested that walking in minimalist footwear and 'flip-flop' types would interfere less with muscular function. In young individuals, when looking at the important foot and ankle plantarflexors of gastrocnemius, soleus, and flexor hallucis longus, sports shoe styles have been reported to reduce the time to peak activation variability compared to when walking in flip-flops or just socks (Péter et al., 2020). Most wearers of sports shoes also required higher muscle activity for acceleration, but not significantly so statistically; and anyway, there is always high EMG variability between individuals (Péter et al., 2020). More restrictive fashion shoes may induce quite different results, especially those involving higher heels as discussed in Chapter 6 of the companion text 'Clinical Biomechanics in Human Locomotion: Origins and Principles', Section 6.4.7. Foot strength and function may derive not only from freedom of motion but also through variability in terrain, something that the urban environment and the presence of footwear may limit the foot's exposure to. Thus, footwear may induce forefoot mechanical changes both through altered cutaneous and subcutaneous stiffness levels during each phase of gait and also through disuse muscle atrophy and force imbalances across the forefoot.

3.3.6 The role of the skeletal frame

The skeletal elements of the foot have a role in acting as a stable frame that can bear weight as a *load frame*. The soft tissues binding the frame together use biotensegrity principles to form a structure that remains stable during shape changes. Bones and ligaments form the passive elements of the biotensegrity structure, and the myofascial tissues form the active restraints, although there is some crossover as previously discussed. For the whole foot to function as a viscoelastic structure, each structural element is essential in providing the compliance and stiffening mechanisms. Thus, the skeletal parts of the load frame are not the osseous structures alone but the combined skeletal and soft tissue elements that support the bones and articulations that go to form an adaptable structural unit with an osseous core.

Ligaments, fascia, and muscle–tendon complexes need something firm they can attach to that resists significant loading deformation in order to achieve their tensional mechanical properties and create motion. In short, they need stable attachment points to pull away from or towards when under loading. If the skeletal framework was 'fixed' under a set tension, then the soft tissues would be unable to change their properties under load. It is movement via articulations of the skeletal frame under the influence of soft tissues that makes the foot's skeletal frame itself a viscoelastic structure. It is not just the overall shape of the foot that makes it so suitable for plantigrade bipedalism, it is the ability to change dimensions and curvatures and thus change stiffness and contact surface area. This allows the foot to be a mobile adaptor to surface contacts, an impact dissipator, and an elastic energy storer, yet also a unit that provides a semi-rigid beam to drive and transfer weight from one limb to the next.

Motion or stability in joints allows the soft tissues to act with variable viscoelastic properties under different loads. Such motion within foot joints allows connective tissues to tension under load, permitting elastic energy storage which assists the triceps surae in generating and transferring energy to the forefoot, thereby reducing the metabolic load on locomotive muscles. The foot can be viewed as a hydraulic damper combined with a muscular brake and a spring working in sequence. Understanding the functions and the anatomy of the skeletal frame in compliance and the processes of sequential and appropriate stiffening, helps the clinician to understand why a lack of, excessive, or inappropriate motion can be so damaging to normal kinetics and kinematics during gait. This explains how pathology is linked to the resulting tissue stress perturbations.

Although motion can occur at each articulation individually, certain joints possess a greater freedom of movement and couple their motion with some joints more significantly than they do with others. Dividing the skeletal frame into segments to examine function is helpful but risks the clinician losing site of the foot's complete function, becoming obsessed by one of its lesser anatomical functional units (e.g. the subtalar joint or the 1st MTP joint). This has certainly occurred in the past with the role of certain joints becoming the central focus of models of foot function (Lee, 2001; Petcu and Colda, 2012).

The skeletal frame is often divided into rearfoot, midfoot, and forefoot with the rearfoot and midfoot interfaces being particularly complex, possessing 20 articulations between 12 bones. Motion of the small bones such as the cuneiforms have proved challenging to capture with 3D kinematic technology, although this is now improving. The foot, however, has often been approached by separating motion into the rearfoot (or hindfoot), midfoot, and forefoot units, despite invasive in vivo studies indicating significant motion at all foot articulations (Arndt et al., 2007; Nester et al., 2007a,b; Wolf et al., 2008; Holowka et al., 2017). Before examining the joints more intimately within the rearfoot, midfoot, and forefoot, another significant functional division of the skeletal frame should be examined first: the *medial* and *lateral columns* (see Fig. 3.1.1).

3.3.7 Division of the skeletal frame: Medial and lateral columns

The medial column, proximally to distally, consists of the talus, navicular, three cuneiforms, three medial metatarsals, and perhaps should also include the medial digital bones and sesamoids as they are important to its function. The lateral column consists of the calcaneus, cuboid, 4th and 5th metatarsals, and their digital bones. The medial column is often equated to the medial longitudinal arch and the lateral column to the lateral longitudinal arch. The medial column is further subdivided into individual rays, but the *1st ray* is most often considered separately. It consists of the medial cuneiform, 1st metatarsal, and 1st tarsometatarsal joint along with its articulations with the navicular and hallux which represent the boundary (Fig. 3.3.7a).

FIG. 3.3.7a A plantar view schematic *(left)* and dorsal view *(right)* of the skeletal foot, with the talus disarticulated. The bones of the foot can be divided into a number of skeletal units as indicated, with the digits consisting of the phalanges and the forefoot consisting of the metatarsals and sesamoids. The midfoot or lesser tarsus contains the cuboid, cuneiforms, and the navicular, while the rearfoot or greater tarsus contains the talus and calcaneus. Other functional units split the bones differently. Thus, the cuboid is part of the midfoot and lateral column and the medial cuneiform is part of the midfoot, medial column, and 1st ray. All such functional divisions are human constructs that aid in understanding motion rather than representing truly independent functional units. *(Permission www.healthystep.co.uk.)*

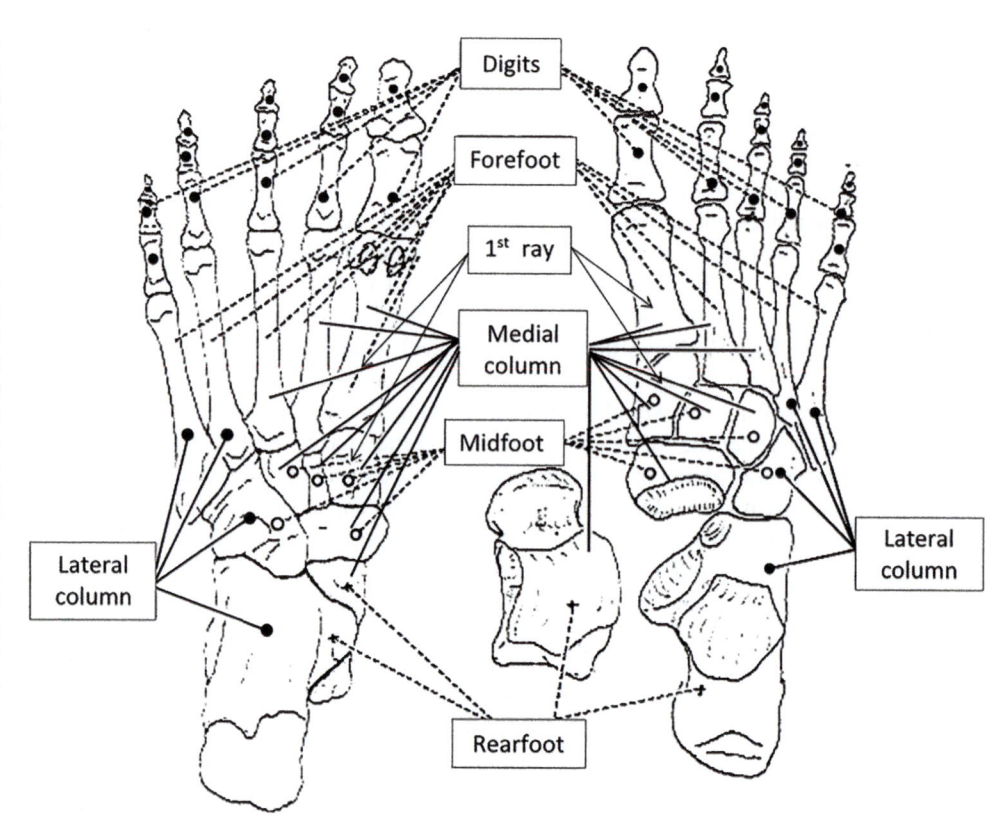

Palaeoanthropology has indicated that the 'medial longitudinal arch' and the 'lateral longitudinal arch' have different evolutionary histories (Kidd and Oxnard, 2005; Zipfel et al., 2009). This is probably because this divide within the skeletal frame underpins two discretely different motion segments that have evolved to adjust for increasing bipedalism and develop the terrestrial foot-stiffening properties required from arboreal foot function. The lateral column's longitudinal 'arch' is less compliant and less mobile and during stance, should normally change profile far less than the medial column. This is probably largely a result of a shorter span distance and less articulations. The lateral column is less capable of energy absorption and elastic energy storage as a result, but it is easier to stabilise. Earlier evolution of the lateral column suggests that lateral foot stiffness is advantageous even when terrestrial bipedalism is a more limited aspect of hominin lifestyle, being possibly important in providing better erect posture and bipedal stability.

Triangulated structures within the foot

The foot has been compared to a series of triangles in several models to explain its function. These include the tripodal model of arches to create a supporting vault from triangular support across the transverse, frontal, and sagittal planes (Kapandji, 1987: pp. 218–219). Interestingly, triangulated configurations and 'tension triangles' are also used as a means to explain tensegrity structures and are an integral part of the principles of tensegrity and biotensegrity (Scarr, 2011; Levin and Martin, 2012). Such a concomitance therefore makes triangular support mechanisms in the foot worthy of consideration. Building on the triangular theme, the foot's skeletal frame has also been compared to a triangular model consisting of two rigid, inclined triangular linkages connected by viscoelastic soft tissues, much like a *Kelvin element* (Bojsen-Møller, 1999; Vogler and Bojsen-Møller, 2000). A Kelvin element is a spring stretching instantaneously while bearing load. It is these two triangular sections, dividing the foot medially into a tibial triangle and laterally into a fibular triangle, that Bojsen-Møller (1999) termed a *load frame*.

These triangles can change shape when compressed, setting up two longitudinal triangular structures with semi-independent sagittal plane movements along the border between the medial and the lateral columns. This boundary within the skeletal frame of the foot is termed the *cleavage line* and this separates the lateral (fibular) triangular set of bones from the medial (tibial) triangular set of bones, in much the same way as is traditionally stated for the medial and lateral columns, yet slightly different as the calcaneus is a common base member to both triangular sets (Bojsen-Møller, 1999). These sets of bones form their own triangles joined at the posterior talocalcaneal facet and suggest an important anterior–posterior gliding role within the subtalar joint complex between the talus and calcaneus that gives some independence between the medial and lateral columns (Fig. 3.3.7b).

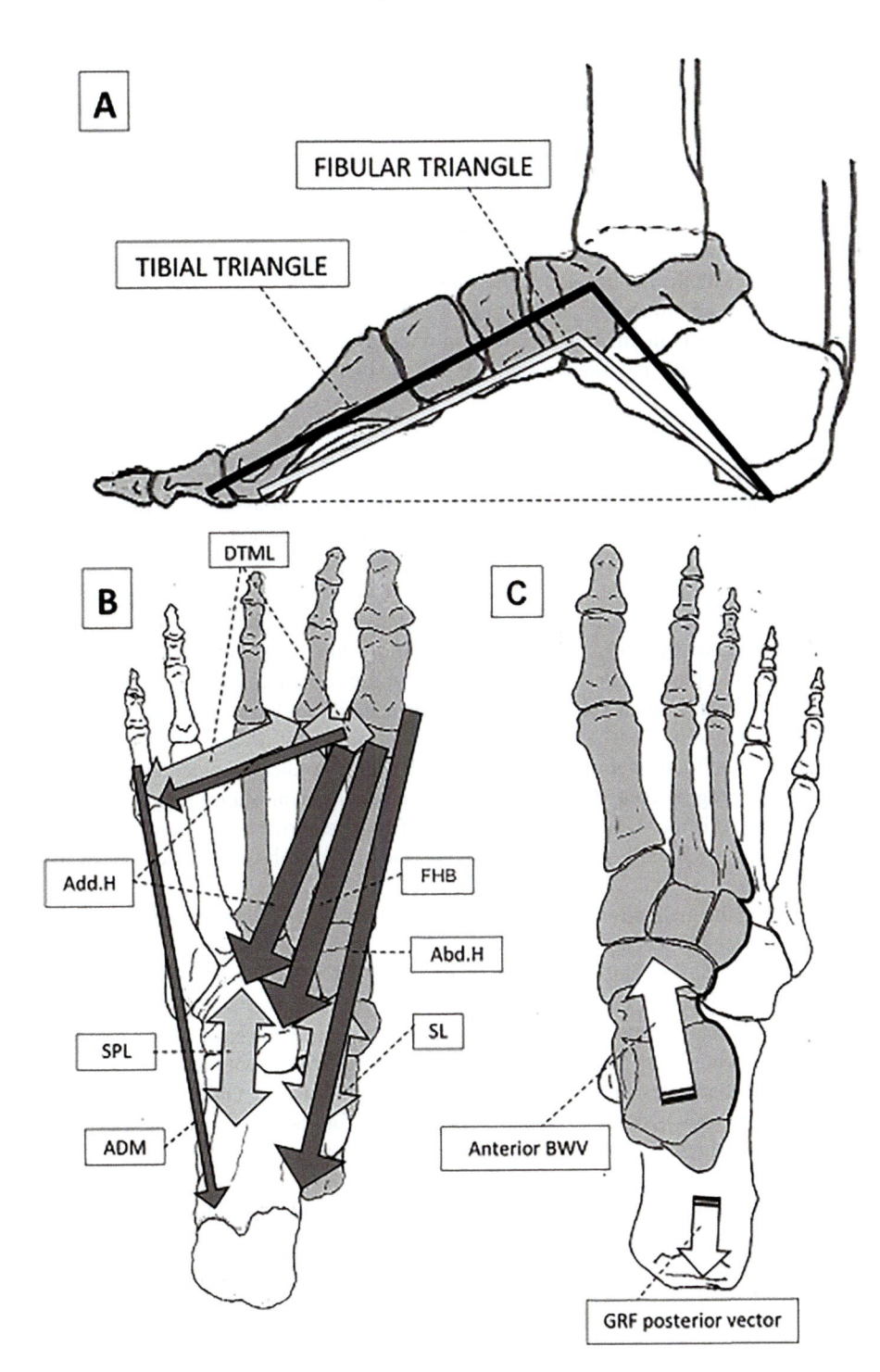

FIG. 3.3.7b The load frame consists of a tibial and fibular triangle (A), formed by the medial calcaneus and medial column and lateral calcaneus and lateral column, respectively. The floor of each triangle is constructed from the plantar soft tissue and their vectors *(dashed line)*. The angulations of the triangles are set by calcaneal inclination angles and metatarsal declination angles that change during gait. The compression members when straight or aligned can be loaded in almost pure compression. The plantar soft tissues vectors (B) consist of the short plantar (SPL), the spring (SL), and the deep transverse metatarsal (DTML) ligaments, and muscles such as abductor hallucis (Abd.H), adductor hallucis (Add.H), abductor digiti minimi (ADM), and flexor hallucis brevis (FHB). They represent tensioned members of the triangles. Soft tissues, such as the DTML, Add.H, and FHB, as well as the long plantar ligament and the plantar aponeurosis (not shown) link motion between the bones of the two triangles *(grey and white bones)*. On heel strike (C), the GRF creates a posterior-directed force against the calcaneus while the talus is driven anteriorly by the body weight vector (BWV), anteriorly translating the bones directly anterior to the talus *(grey bones)*. The soft tissues that straddle the interface between the triangles *(black articular line)* will restrain anterior shear of the medial column elements of the tibial triangle, creating elastic energy dissipation and storage. Because of differences in compliance between the tibial and fibular triangles, stance motion will usually be distinctly different and greater within the tibial triangle. *(Adapted from images by: Bojsen-Møller, F., 1999. Chapter 9: Biomechanics of the heel pad and plantar aponeurosis. In: Ranawat, C.S., Positano, R.G. (Eds.), Disorders of the Heel, Rearfoot, and Ankle, Churchill Livingstone, Philadelphia, PA, pp. 137–143.)*

The medial side of the medial triangle is occupied by abductor hallucis while the lateral side of the lateral triangle is occupied by the abductor digiti minimi, both presenting important stabilisers to the sides of these triangles longitudinally. Several large ligaments of the foot and the plantar aponeurosis span the cleavage line, including the medial portion of the bifurcate ligament of the calcaneonavicular joint, the medial distal slips of the long plantar ligament, and the deep transverse metatarsal ligament at the 3rd intermetatarsal space. Many of the plantar intrinsic muscles cross the cleavage line including the oblique and transverse adductor hallucis, the tendon slips of flexor digitorum brevis, quadratus plantae (through its long flexor tendon attachment), the tendon of peroneus longus, and the distal tendon slips of tibialis posterior. Thus, the cleavage line is bound firmly but flexibly across the transverse plane under tensions angled anteriorly to posteriorly.

The 'triangles' are therefore the result of calcaneal inclination and metatarsal declination angles forming a distinct but continuous truss medially and laterally. The calcaneal body is angled from posteroinferior to anterosuperior, rising up from the posterior calcaneal process to create the calcaneal inclination angle. The main clinical assessment of this calcaneal inclination is undertaken radiographically by taking a static stance lateral view radiograph, with interest being focused on the medial side of the medial triangle set by the calcaneal and 1st metatarsal angulations (Gentili et al., 1996). Classically, this average calcaneal inclination angle is around 18–22° (Weissman, 1989). The talar declination angle is stated to be normal around 21° and should be collinear with the medial side of the 1st metatarsal on the lateral view radiograph in stance. Thus, the 1st metatarsal angle should also be around 21° (Weissman, 1989). These lines of inclination and declination create a *calcaneal pitch angle* at the join of the inclination–declination lines, giving an apex angle to the triangle (Wearing et al., 1999). However, these angles are usually recorded in static stance and do not reflect the dynamic changing relationships.

Changes in inclination–declination angles

When weightbearing, the inclination–declination angles are associated with vault profile height assessment radiographically and loss of the collinearity of the talar–metatarsal angle is considered dysfunctional (Weissman, 1989; Gentili et al., 1996). Obviously, changes in this 'pitch' angulation are occurring throughout gait under velocity and acceleration-induced forces. A large reduction in the calcaneal pitch angle in stance has been associated with increased foot pronation and low arches (Wearing et al., 1999). Yet, failure in conforming to the average angles expected on radiographs does not necessarily relate to pathomechanics. Instead, the angulation can be modelled to explain variations in energy storage capacity in the different foot profiles. This is because extremes in these angles, both high and low, are reported to result in insufficient energy attenuation (Simkin and Leichter, 1990). The angular values need to be considered in relation to the fact that such alignments are a changing feature during gait and do not exist as fixed structural 'definitions of normality'. They may, if used in isolation, result in erroneous conclusions on foot function (Wearing et al., 1999). The calcaneus plantarflexes from ~28% of stance (Hamel et al., 2004) which will progressively decrease its inclination angle as midstance progresses to the lowest vault height at heel lift (Hunt et al., 2001). It is more likely that the amount of change during gait is more significant than the static stance angle recorded and of course, such static observation offers no indication of the transverse vault profile or its ability to function within the frontal plane.

The resulting calcaneal pitch angle formed by the inclination–declination angle in gait changes from around 161°±6° at contact to around 148°±4° during terminal stance (Kelly et al., 2015). However, rather than the actual values, it may be the timing of increases and decreases in vault profile during gait that is important. This will be of course subjected to step variables during gait, but the important factor is perhaps the ability to achieve an appropriate changing angulation of the foot that indicates the success of energy dissipation and recovery during gait appropriate for each step. There may be a percentage change, rather than a degree change, that presents a way to assess an efficient foot type to express the vault's appropriate behaviour in relation to the foot's flexibility, regardless of static vault profile. Only further research in this area will provide the answer as to whether this is possible. Nevertheless, a word of caution is necessary on these inclination–declination angles in relation to the declination angle which relies on the talar–metatarsal collinearity of the 1st metatarsal only. Each metatarsal's declination angle is different from the lesser tarsus to the metatarsal head. This means that not only do the lateral and medial columns have their own inclination–declination angles, but so does each talar–metatarsal relationship, vastly complicating the situation with regard to clinical assessment. CT data-based 3D modelling may offer solutions to this in the future.

Stressing the load frame

The loaded talus will rotate medially (adduct, plantarflex, and evert) and slide anteriorly, pushing the bones of the medial triangulated column forward. This moves the bones anterior to the talus, more anteriorly away from the lateral column bones, tensioning the ligaments and muscles that bridge the cleavage line (Bojsen-Møller, 1999). Once the muscles bridging the cleavage line become active, foot stiffness can be increased in a (hopefully) controlled manner. Initially, this allows the foot to be passively compliant under loading but increasingly resistant to any excessive sagging deflection under increasing loads (Bojsen-Møller, 1999). The talus is commonly 'cited' as having no direct muscular attachments, yet it is connected by fascial reinforcements of the crural fascia to the Achilles/soleus and both the tibia and fibula so that through tensioning the triceps surae, the fascial connections can restrain the anterior sliding of the talus (Bojsen-Møller, 1999). This would imply that as the triceps surae activates during middle and late midstance, it helps prevent further anterior talar drift. As triceps surae activity peaks at heel lift, it should be able to draw the talus posteriorly. Thus, loading forces and soft tissue mechanisms seem to exist to facilitate the positioning of the ankle axis throughout ankle dorsiflexion (talus pushed anteriorly) and plantarflexion (talus pulled posteriorly).

Bojsen-Møller (1999) proposed that motion between the medial and lateral column triangles involves three phases during stance. In contact phase, the ankle plantarflexion moment of heel contact is restrained by the action of tibialis anterior lasting until forefoot contact, during which Bojsen-Møller suggested that there is no motion between the medial and lateral triangular portions of the load frame. However, it seems likely that the posteriorly directed GRF (F4 peak) causes a posteriorly directed force through the calcaneus, while the BWV drives the talus anteriorly. It is possible that tibialis anterior activity, by means of its plantar medial attachment to the navicular and medial cuneiform, helps to restrain the anterior talar driving forces during contact phase. Extension of the digits via the reversed windlass mechanism may also provide a tensioned-induced compressive restraint on the talus through resisting the medial column translating anteriorly during loading.

Following forefoot contact, both the medial and lateral triangles are loaded, with the triangles lengthening under loading. As the tibia starts to glide anteriorly, it associates with talar anterior displacement causing the medial column to sustain greater anterior motion, thereby lengthening the medial column more than the lateral. This anterior displacement of the talus and the medial column is restricted by the soleus muscle due to its activity-inducing tension through the fascia crura on the talus, the spring ligament, the subtalar interosseous and cervical ligaments, and the medial band of the bifurcate ligament. Talar stability is assisted by any restraint of the medial column that restricts lengthening and depression of the vault. Such restraints include the medial distal slips of the long plantar ligament (and other plantar ligaments), plantar intrinsic and extrinsic muscle activity, and the deep transverse metatarsal ligament, particularly through the 3rd intermetatarsal space that links the medial and lateral members of the skeletal frame. The medial parts of the plantar aponeurosis will tension more than its lateral sections as a result. All this stress loading of the ligamentous/fascial structures should result in the storage of more elastic energy within the medial column compared to the lateral column. The lateral triangle experiences less anterior displacement, as it not coupled to the anterior displacement of the talus and is restrained by the very strong short plantar ligament and the bulk of the long plantar ligament. Any structure bridging the cleavage line between these triangles will now undergo tension from the resistive difference in anterior gliding between the two triangular members of the load frame, storing elastic energy. However, overall, motion is more restrained by the coupling of lateral column stiffness to medial column mobility (Fig. 3.3.7c).

During acceleration of terminal stance, the medial triangle is shortened more than the lateral triangle (Bojsen-Møller, 1999). Although long thought to be achieved mainly through the windlass effect, it is now clear that a number of mechanisms precede the plantar aponeurosis' role in forefoot stiffening for acceleration, including transverse arch-supporting ligament tightening that increases curvature stiffening and plantar intrinsic and extrinsic activity. Through appreciating the different motions between the medial and lateral triangular elements of the load frame, it can be seen that plantar ligamentous elastic recoil and the pulling back of the talus through soleus activity via the fascia crura also creates restraint of foot motion in terminal stance. Although somewhat simplistic, this model highlights that the medial section of the foot vault should tend to displace more than the lateral section, resulting in greater vault profile changes medially than laterally. This suggests that the medial column section of the vault possesses a greater capacity to dissipate and store energy and elastically recoil, compared to the lateral column. It also explains why more motion complexity of the medial column evolved later than the more simplistic, stable, and stiffer lateral column. This may also explain why more issues tend to present clinically that are related to structures of the medial column as there is a greater potential for anatomical structures to suffer dysfunction, not least because there are more articulations. Yet therapeutically, loss of lateral column stability and stiffness seems often more problematic to resolve.

Triangulated foot columns in summary

The triangular profiles in the foot and the division of the skeletal frame of the foot help in understanding complex articular function, although the living reality is likely more complex due to individual human anatomical differences. All foot bones

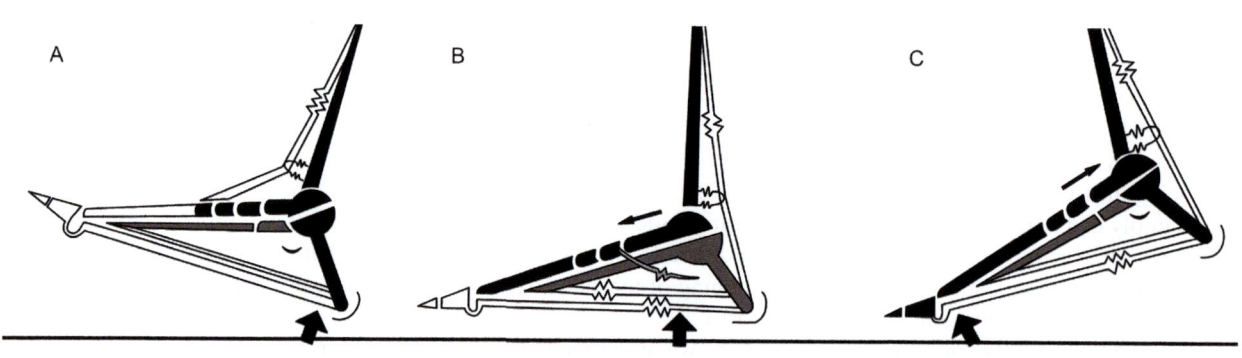

FIG. 3.3.7c A schematic of the loading frame model in its three working phases. The tibial and fibular triangles, the subtalar joint, and the tibialis anterior and soleus muscles are indicated. Compression members are demonstrated in black and grey, with tensioned members such as soleus, the plantar ligaments, plantar aponeurosis, and intrinsic and extrinsic muscles and tendons being represented in white. The digits (initially white) act more as tension members in early gait but turn increasingly into compression members under tension by terminal stance. At heel strike (A), the foot pivots over the heel rotating at the ankle, while tibialis anterior controls motion. As the hallux and lesser digits extend, they activate the windlass mechanism. This makes the digits act more as elastic tension members. The GRF applied to the calcaneus drives it relatively posteriorly to the talus which will tend to be gliding anteriorly under the BWV. The talar anterior glide is restrained by subtalar ligaments and the activation of the windlass mechanism medially. At midstance (B), the medial and lateral load frames (triangles) move differently. The calcaneus demonstrates little motion with the less mobile fibular frame *(grey)* as the tibial frame *(black)* slides anteriorly to the lateral fibular frame. Tibial loading frame displacement is restrained through plantar soft tissue tension from anterior to posterior and those tissues that cross the cleavage line between the loading frames. Increasingly, the digits provide a compression element from muscular tensions pulling them back towards the heel due to digital flexor muscle activity. During terminal stance (C), the tibial triangle frame is retracted and shortened through foot-ankle plantarflexor muscle activity and later by the plantar aponeurosis via digital extension. Plantar foot muscle activity transforms the digits into important compression members. Such models as this, proposed by Bojsen-Møller and then modified to fit new data, help in understanding the interplay of foot anatomy when setting up patterns of stiffness and compliance within gait. *(Modified from Bojsen-Møller, F., 1999. Chapter 9: Biomechanics of the heel pad and plantar aponeurosis. In: Ranawat, C.S., Positano, R.G. (Eds.), Disorders of the Heel, Rearfoot, and Ankle, Churchill Livingstone, Philadelphia, PA, pp. 137–143. (Permission www.healthystep.co.uk.)*

demonstrate significant triplanar motions between them (Wolf et al., 2008), yet this does not make more simplistic models pointless. The sliding relationship between the medial and lateral column triangles and the changes in the calcaneal pitch angle in gait can help to explain the outcomes of certain foot dysfunctions. Indeed, soft tissue-increased stresses across the cleavage line, due to the difference in motion between the two columns of the foot, may explain why there is a preponderance of digital neurofibromas (*Morton's neuroma*) in the 3rd intermetatarsal space that bridges the connection between the triangles compared to other intermetatarsal spaces (Pilavaki et al., 2004; Matthews et al., 2019).

Severe reduction of the calcaneal pitch angle can occur either as a transitory event in pedal midfoot break at heel lift or be a permanent change due to serious foot dysfunction, such as degenerative tibialis posterior tendinopathy or neuropathically induced Charcot foot. Both the transitory and the permanent loss of the triangular column relationship and the normal changing calcaneal pitch angle during gait will have serious consequences to locomotive efficiency. In the cases of an acquired pes planus resulting from permanent calcaneal pitch loss with increased mobility, a serious foot dysfunction is likely to result and be expressed as reduced energy dissipation, storage, and recoil capacity (Simkin and Leichter, 1990). Failure in the ability to create appropriately timed longitudinal foot stiffness through transverse arch coupling may result in a stiff flat foot, where all the connective tissues are stretched into their stiffest range of strain and remain largely fixed there. Alternatively, connective tissues might only fail at certain points, increasing foot mobility at affected joints until secondary osteoarthritic changes cause loss of articular motion. Loss of skeletal frame function thus leads to a loss in the ability to dissipate, store, and use elastic energy during gait in anything like an appropriate manner.

3.3.8 The rearfoot as a functional unit

Dividing the foot into functional units often finds the ankle and foot being divided into segments for the purpose of kinematic research. In such studies, these segments are assumed to be rigid between the bones of each unit. Thus, the tibia and fibula risk ending up representing two fixed bones of the lower leg in this modelling approach. The metatarsals are fixed together to become the forefoot. The navicular, three cuneiforms, and cuboid become the midfoot, and the talus and calcaneus the rearfoot. Often, the calcaneus' movements are considered to represent the rearfoot's motion. This gives a representation of how the joint axes between these segments move within each plane in relation to each other. However, it does not indicate the motions occurring between each single bone within the segments, which can be quite significant. Thus, approaching the foot as being modelled upon three rigid segments with three distinct joint axes is overly simplistic. However, for approaching and dividing foot motion, it can provide some helpful simplification (with caution) (Fig. 3.3.8a).

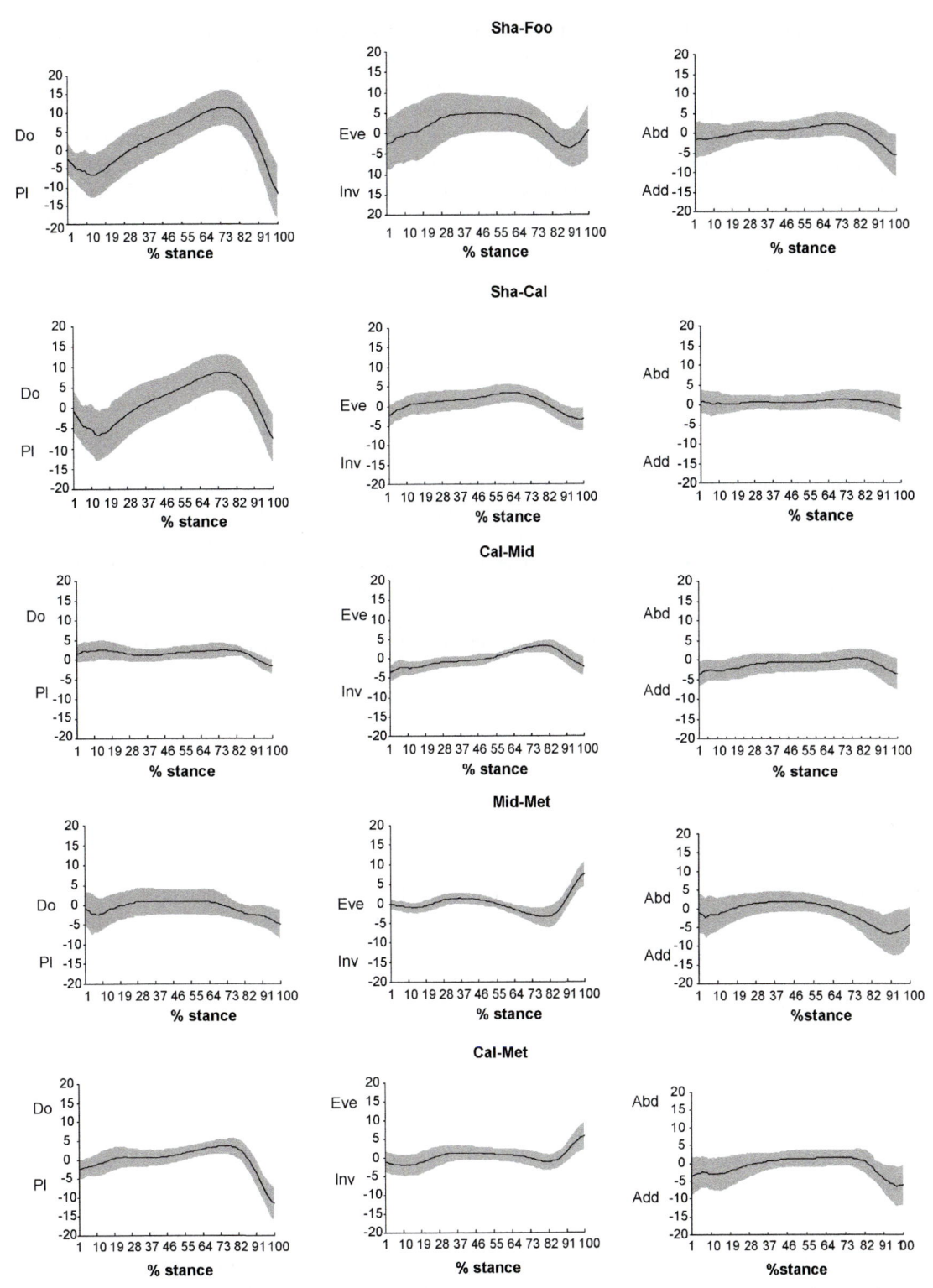

FIG. 3.3.8a Rotations of the segments over each other during the stance phase of gait with 'Sha-Foo' *(upper graphs)* being the lower leg over the whole foot at the ankle, 'Sha-Cal' (next down) being the lower leg over the calcaneus, 'Cal-Mid' *(centre graphs)* being the calcaneus rotating around the midfoot, 'Mid-Met' being the midfoot rotating in relation to the metatarsals, and 'Cal-Met' *(lower graphs)* being calcaneal rotation in relation to the metatarsals. Means are represented by solid lines and standard deviations in grey. Although such data give useful insights into gross regional movement in the feet and around the leg, the failure to isolate bones and individual articular motions means that data must be treated with caution as these are made from combined joint motions, not single articulations. *(From Leardini, A., Benedetti, M.G., Berti, L., Bettinelli, D., Nativo, R., Giannini, S., 2007. Rear-foot, mid-foot, and fore-foot motion during the stance phase of gait. Gait Posture 25 (3), 453–462.)*

The rearfoot's primary function is to assist in providing a stable structure to transfer weight from the opposite foot during the double stance phase of walking and during heel strike running. The rearfoot's posterior calcaneal tubercle is quite uneven, requiring aid from the plantar soft tissues in providing rearfoot ground contact stability. Weight transference capacity is not the rearfoot's alone, for the forefoot can also achieve this task. However, the energetics and overall stability is vastly improved if the heel and rearfoot provide initial contact during walking. The picture is not as straightforward in running. The ankle has already been discussed as a functional unit in Chapter 2, Section 2.5, but the ankle joint is also an intimate functional part of the rearfoot, and its degeneration requires articular foot compensations (Deleu et al., 2021). The other primary anatomical rearfoot articulations are the talocalcaneal and talocalcaneonavicular joints, known collectively as the subtalar joint. This 'functional' articulation can consist of either one, two, or three separate articular surfaces of variable configurations, which may or may not influence its function (Bruckner, 1987; Langdon, 2005: p. 115). Interestingly, such articular variations are also noted in chimpanzee subtalar joints (Ankel-Simons, 2007: p. 360). See Chapter 6 in the companion text 'Clinical Biomechanics in Human Locomotion: Origins and Principles', Fig. 6.3.4h for comparable images.

Rearfoot weight acceptance and stability occurs during the period of lower limb braking and energy dissipation. The long posterior tubercle of the calcaneus effectively lengthens the limb at contact, helping reduce the height drop of the heel at the end of swing. It is protected by the skin and plantar heel fat pad, with its posterior length presenting a significant lever arm for the GRF to initiate ankle joint plantarflexion. This motion brings the rest of the weightbearing surfaces of the forefoot into ground contact so as to permit completion of weight transference to the entire stance foot, while the opposite foot is offloaded as it enters preswing and early swing. To prevent forefoot slap and assist in shock attenuation, the plantarflexion moment is resisted by the foot extensors that initiate the windlass mechanism, tensioning, and therefore pre-loading the plantar aponeurosis. This helps give the skeletal frame some forefoot stability at its collision. The articular motion of the subtalar joint is likely to play some role in shock attenuation, despite it not demonstrating either free or large ranges of motion. The subtalar joint moves only under force applied externally (Leardini et al., 2001). Pain in the rearfoot and heel can be associated with subtalar joint osteoarthritis (Glanzmann and Sanhueza-Hernandez, 2007) and sometimes talocalcaneal coalitions (Bonasia et al., 2011). Symptomatic subtalar joints require treatment and so it appears that the little mobility the subtalar joint provides is significant for healthy rearfoot function, most likely during the period where it is part of the peak weightbearing structure around heel contact and throughout early midstance.

The subtalar joint complex

The subtalar joint complex is formed by the posterior talocalcaneal joint and the anterior talocalcaneonavicular joint. The talocalcaneonavicular articulation is the subtalar joint's continuation with the talonavicular joint, which is also known as the acetabulum pedis (Sarrafian, 1993). Pisani (1994) has referred to the talocalcaneonavicular joint as the coxa pedis due to its anatomy being analogically comparable to the hip joint and its rotational ability also being likened to that of the hip. This joint offers far more mobility than does the subtalar joint. The subtalar joint varies in the number of its articular surfaces, varying from one to three (posterior, middle, and anterior) in four different configurations (Viladot et al., 1984; Bruckner, 1987). These variations possibly explain the subtalar axis location variabilities reported within the literature (Leardini et al., 2001). Although the talus does not have direct muscular attachments (the only bone in the lower limb to not have any), its motion is influenced by myofascial contractions tightening its extensive fascial attachments via the talar periosteum to the fascia crura (Bojsen-Møller, 1999) and the inferior extensor retinaculum (Draves, 1986: pp. 150–152). The ankle has a close coupling relationship with the subtalar joint, such that it is difficult to discuss one without the other. As a result, it is common to discuss the ankle–subtalar complex functioning as a whole because the ankle provides a significant part of the rearfoot's motions observed in gait (Scott and Winter, 1991; Leardini et al., 2001).

The ankle offers unrestricted freedom of motion. This is unlike the subtalar joint which does not have freedom of motion but flexibility, only moving when subjected to external forces (Leardini et al., 2001). The subtalar joint is therefore described as a flexible structure with motions reflecting its ligamentous and articular configurations (Leardini et al., 2001; Hamel et al., 2004). Subtalar joint motions under external load are very limited when the ankle joint is closer to maximum dorsiflexion, with larger motions available in positions of ankle plantarflexion and when near ankle neutral or 90° (Leardini et al., 2001; Hamel et al., 2004). This means that subtalar joint motion is more likely to occur during loading response and into early midstance, when the calcaneus is subjected to GRF loading and the ankle joint is operating closer to its neutral position. By 25% of stance phase, the talus and calcaneus start to move together as one body, moving into dorsiflexion until no further significant motion at the subtalar joint occurs (Hamel et al., 2004) (Fig. 3.3.8b). Rearfoot motions observed outside loading response and early midstance are thus most likely occurring within the ankle (including frontal plane motion), and this will certainly be so after heel lift when the subtalar joint is no longer under any direct external

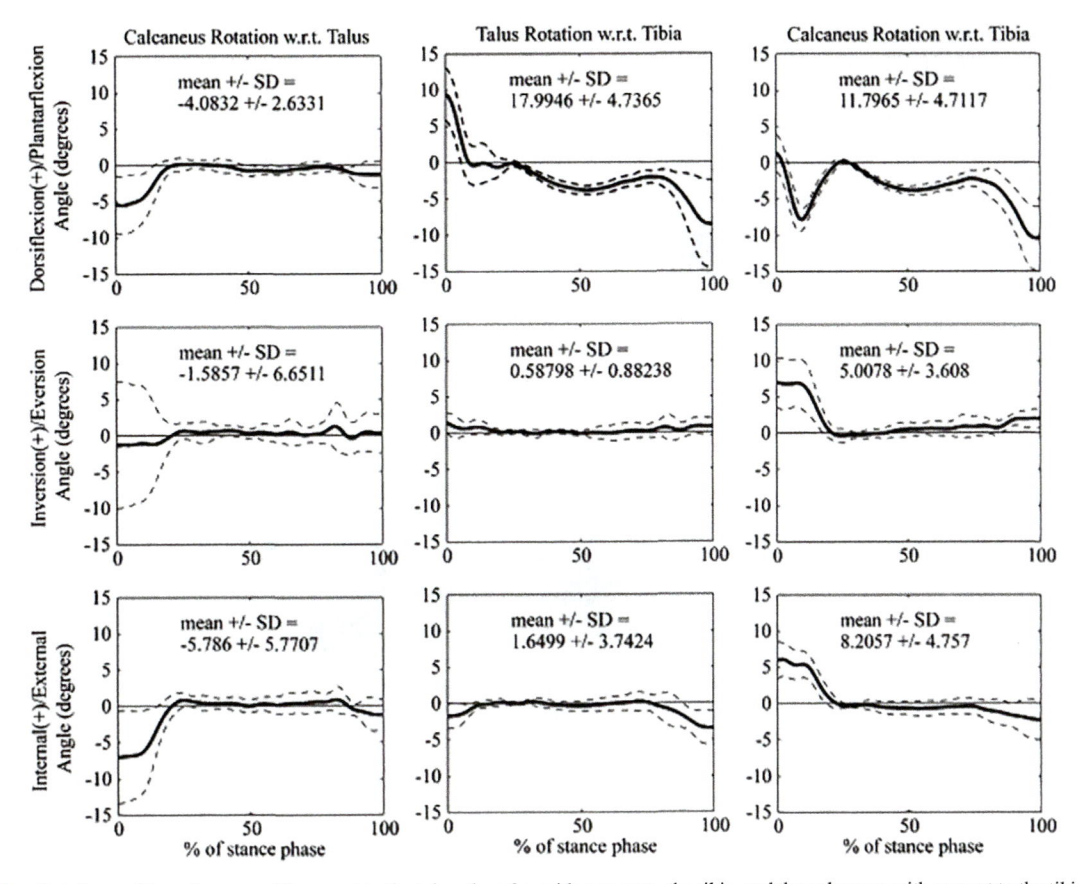

FIG. 3.3.8b Rotations of the calcaneus with respect to the talus, the talus with respect to the tibia, and the calcaneus with respect to the tibia from eight cadaveric specimens in simulated stance loading. Data derived from Euler decompositions within the different reference frames are not necessarily additive. Mean ranges of motion and their standard deviations *(dashed lines)* are shown in each plot. *(From Hamel, A.J., Sharkey, N.A., Buczek, F.L., Michelson, J., 2004. Relative motions of the tibia, talus, and calcaneus during the stance phase of gait: a cadaver study. Gait Posture 20 (2), 147–153.)*

load. The use of high heels in maintaining ankle plantarflexion under external heel loading may possibly increase subtalar motion duration during stance phase.

The articular surfaces of the subtalar joint consist of variable concave and convex ovoid surfaces interacting with each other (Sarrafian, 1993). The joint demonstrates some form closure, but stability is primarily provided by force closure, particularly via external compression forces and internal ligament tension forces (Fig. 3.3.8c). The motion generated within the subtalar joint complex is that of flexion–inversion–adduction and extension–eversion–abduction, with motion generated by the contours and orientations of the articular surfaces guided by the cervical and interosseous talocalcaneal intrinsic ligaments (Sarrafian, 1993). These ligaments permit some translation through their elasticity (Peña Fernández et al., 2020). This translation is likely to be primarily consisting of anterior–posterior gliding on loading of the joint during heel contact. Further motion guidance is achieved from the extrinsic structures of the posterior tibiocalcaneal, calcaneofibular, and deltoid ligaments (Sarrafian, 1993). The subtalar close-packed position is realised under tibiotalar vertical loading and internal or medial talar rotation. As the talus glides anteriorly on loading the BWV, the GRF is directed posteriorly in opposition onto the calcaneus. This creates maximum contact at the talonavicular joint, with the ligaments being under maximum tension and the posterior talocalcaneal joint surfaces interlocked (Sarrafian, 1993). In most of stance, the calcaneus becomes relatively fixed to the ground with soft tissue motions below, whereas the talus moves with the lower leg across the ankle. Thus, the GRF below and the rotations of the leg above provide the external forces that create subtalar joint motion (Fig. 3.3.8d).

Movements of the ankle–subtalar joint complex are fundamental to the transmission of rotations from and to the leg during the stance phase, with some authors having claimed that pure rotation occurs (Hicks, 1953; Root et al., 1966) while others asserted a more screw-like behaviour (Manter, 1941; Isman and Inman, 1969). The subtalar joint has been described as a hinge, with an oblique axis running from anterior–medial–superior to posterior–lateral–inferior at an average incline of around 42° in the sagittal plane (Manter, 1941; Hicks, 1953; Root et al., 1966; Isman and Inman, 1969). However, different

FIG. 3.3.8c An illustration of the subtalar joint. The external ligaments have been divided and the talus (B) reflected medially so its inferior articular surface is visible. The superior articular surfaces of the calcaneus (A) can be correlated to the talar articular surfaces, indicated by the dark-grey posterior facet, mid-grey middle facet, and the white anterior facet. The talonavicular facets (C) and the navicular (D) can also be visualised. The important ligamentous restraints of the subtalar joint are each labelled. *(Adapted from Peña Fernández, M., Hoxha, D., Chan, O., Mordecai, S., Blunn, G.W., Tozzi, G., et al., 2020. Centre of rotation of the human subtalar joint using weight-bearing clinical computed tomography. Sci. Rep. 10, 1035.)*

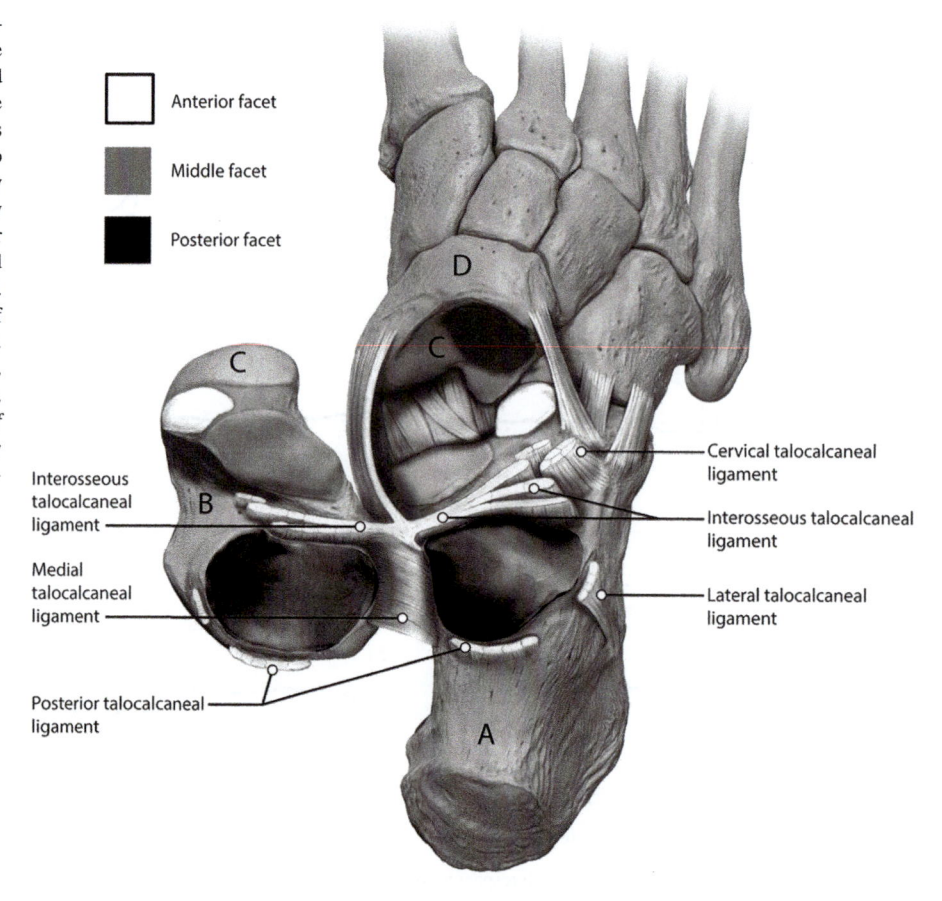

orientations in its anterior–medial deviation in the transverse plane have been recorded, from 16° (Manter, 1941) to 23° (Isman and Inman, 1969). This is suggestive of a fixed axis which is not the case (Lundberg, 1989; Lundberg and Svensson, 1993). Axis variations are reported to be large, with inclination angles reported as being between 20 and 68° and deviation angles between 4 and 47° (Leardini et al., 2001). The axis of the subtalar joint complex does not demonstrate a repeatable path due to the changing orientation of its instantaneous axes (Lundberg, 1989; Lundberg and Svensson, 1993).

The instantaneous axis location during motion in healthy weightbearing individuals is far more valuable to know clinically than the theoretical estimates gathered from cadavers. Using full weightbearing clinical computed tomography to evaluate the helical axis and the centre of rotation of the subtalar joint's inversion and eversion motion has shown consistency across healthy subjects (Peña Fernández et al., 2020). The centre of frontal plane rotation seems to be located within the middle facet of the subtalar joint with some anterior–posterior and medial–lateral translation confirmed, influenced by ligamentous elasticity (Peña Fernández et al., 2020) (Fig. 3.3.8e). Thus, it can be deduced that in individuals with restricted ligament flexibility, subtalar axis motion will be constrained, whereas in those with subtalar hypermobility or ligamentous disruption, subtalar instantaneous axis motion is likely to be greater during gait.

The calcaneus follows a unique path of motion relative to the tibia during passive flexion with motion occurring at the ankle, not the subtalar joint (Leardini et al., 1999). The subtalar joint has a single stable position in the unloaded state with no range of unresisted motion. However, load-induced motion at the subtalar joint accounts for between 70 and 90% of tibial-to-calcaneal inversion–eversion motion, with smaller contributions from the tibiotalar articulation and a minor contribution from the tibiofibular joint (Leardini et al., 2001). Two distinct axes have been found, one for inversion and one for eversion motion, with both the position and orientation of the axes being different (Leardini et al., 2001). About 8° of dorsiflexion is coupled to around 8° of pronation–supination or 11° of inversion–eversion (Hicks, 1953; Siegler et al., 1988; Lundberg, 1989; Leardini et al., 2001). The range of rotation motion in the frontal plane seems highly variable, reported to be between 6.6° and 21.2° in only eight (sixteen feet) young adult subjects (Peña Fernández et al., 2020). A mean shift in the instantaneous centre of rotation during inversion and eversion is greater in the right foot but represents no more that 30% of the total range of talar displacement, with consistently more lateral and anterior talar displacement relative to the

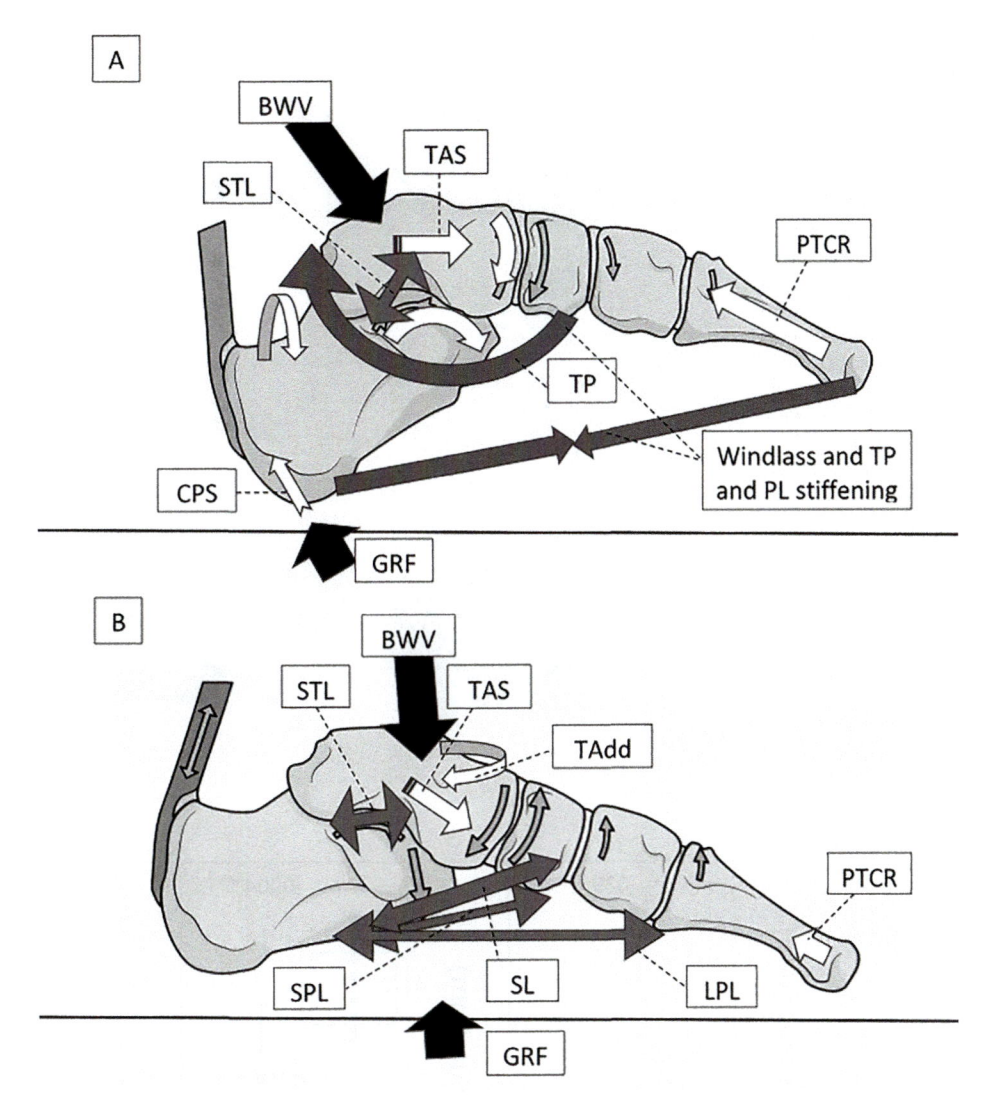

FIG. 3.3.8d The subtalar joint is subjected to two primary external force events during gait. The first occurs associated with heel contact as an impact collision (and thus not part of forefoot contact gaits). Heel strike (A) most commonly occurs slightly to the lateral aspect of the posterior heel, initiating a posterolateral GRF that drives a small eversion and a large plantarflexion moment onto the rearfoot. This creates calcaneal posterior shear (CPS) and eversion on the calcaneus through the soft tissues, while talar anterior shear (TAS) is driven by the body weight vector (BWV) loading on the talus. This is restrained by tensions developing within the subtalar ligaments (STL), namely the posterior and lateral interosseous ligaments and the cervical ligament. This talar anterior translation restraint does much to explain the macro and microfibre orientations of the subtalar ligaments. The setting of the windlass mechanism, aided by tibialis posterior (TP) and peroneus longus (not shown) via stiffening the vault, also provides posterior talar compressive restraint (PTCR) on the anterior talus via the navicular. During midstance (B), reducing ankle plantarflexion angles that become the ankle dorsiflexion motion of late midstance starts to initiate downward compression of the talus and distal calcaneus, setting up a plantarflexion moment on the rearfoot, with the anterior shear across the subtalar joint being resisted by the STL. Independent subtalar motion soon stops during later midstance. The vault depression anterior to the subtalar joint is restrained primarily by the strong short plantar ligament (SPL) laterally, the long plantar ligament (LPL) centrolaterally, and the spring (talocalcaneonavicular) ligament (SL) medially. The latter also helps restrain talar adduction (TAdd) on the calcaneus with the STL. PTCR only increases again in late midstance as digital flexor muscle activity increases, compressing the digits into the forefoot. *(Permission www.healthystep.co.uk.)*

calcaneus available in inversion than eversion (Peña Fernández et al., 2020) (Fig. 3.3.8f). This may help explain some of the higher rearfoot vulnerability to inversion sprains as opposed to those associated with eversion.

The subtalar joint is very closely linked to the talonavicular joint. This is because of its shared articulations and ligaments between the talus, calcaneus, and navicular (Savory et al., 1998; Wolf et al., 2008). The subtalar joint therefore represents two distinct joints separated by the groove of the *sinus tarsi* (Krähenbühl et al., 2017). The posterior part links

FIG. 3.3.8e A 3D rendering of subtalar joint motion in inverted, neutral, and everted positions in the left foot of one individual shown in (A) lateral, (B) superior, and (C) posterior views. Higher talar displacements occur in inversion than during eversion. The calcaneus is shown fixed to demonstrate the relative talar motion in the three configurations. Lateral sinus tarsi opening in inversion and closing in eversion can be visualised compared to the neutral position (*arrows* in A). The lateral malleolar surface of the talus approaches the sinus tarsi from inversion to eversion (*arrows* in B). The lateral talar tubercle rotates towards the sustentaculum tali from inversion to eversion (*arrows* in C). (*Adapted from Peña Fernández, M., Hoxha, D., Chan, O., Mordecai, S., Blunn, G.W., Tozzi, G., et al., 2020. Centre of rotation of the human subtalar joint using weight-bearing clinical computed tomography. Sci. Rep. 10, 1035.*)

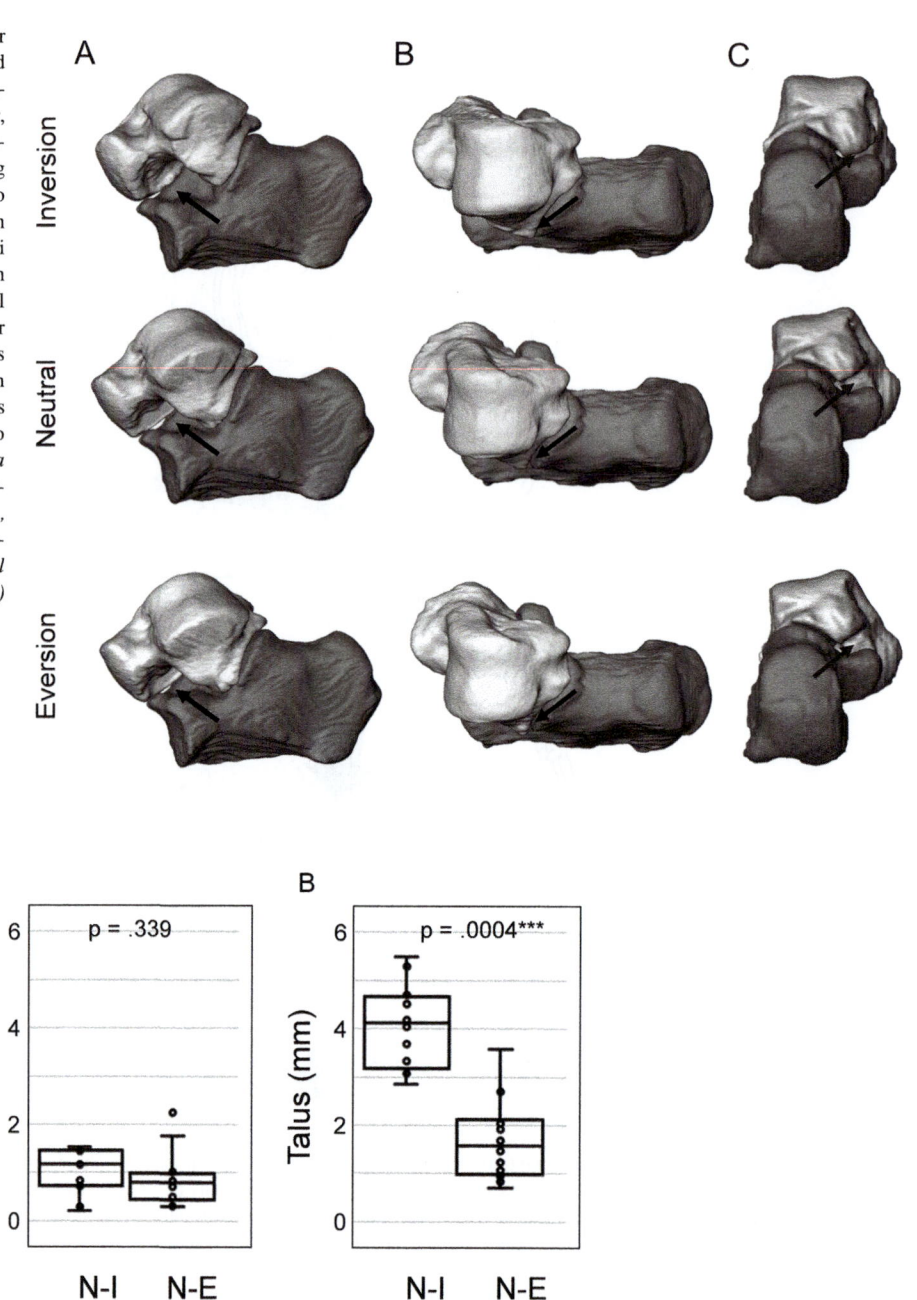

FIG. 3.3.8f Box plot distribution of the mean displacements over the entire (A) calcaneus and (B) talus from neutral to inversion (N-I) and neutral to eversion (N-E) positions. The *P* values from a two-sided Wilcoxon signed-rank test are reported in each plot. (*From Peña Fernández, M., Hoxha, D., Chan, O., Mordecai, S., Blunn, G.W., Tozzi, G., et al., 2020. Centre of rotation of the human subtalar joint using weight-bearing clinical computed tomography. Sci. Rep. 10, 1035.*)

the heel to the ankle and the anterior part links the midfoot to the rearfoot (including the ankle). The talus is the rearfoot's 'centre' and articulates with the ankle joint, subtalar joint, and the talonavicular part of the midtarsal joint. Three-fifths of the talar surface is articular, with all except its posterior process and neck enclosed within joint capsule. The talar inferior surface is totally articular, save for the deep sulcus along the talar neck that forms the roof of the sinus tarsi. This is an osseous tunnel containing fat, nerves, and blood vessels, with the sulcus calcanei forming its floor. The convex trochlear surface articulates with the ankle while the irregular concave or saddle-shaped plantar surface of the body contains the large posterior facet that occupies the entire inferior surface, transmitting forces from the leg into the talocalcaneal joint to the calcaneus. The anterior, middle, and concave talar facets, along with the navicular bone, form the distinct talocalcaneonavicular joint enclosed within a single shared joint capsule. The anteromedially projecting talar head is supported by the anterior facet of the calcaneus during calcaneal inversion and adduction, whereas the middle facet loads in calcaneal eversion and abduction.

Stability is assisted by the calcaneus' shape under the talus, with the lateral process of the talus interacting with the anterolateral process of the calcaneus and the medial tubercle with the sustentaculum tali (Bruckner, 1987). The calcaneus shares two-thirds of its superior surface with the inferior talar articular surfaces. Functionally, the subtalar joint is considered to comprise all the talocalcaneal articular surfaces, although trying to separate subtalar motion from ankle and talonavicular joint motion is problematic. The posterior subtalar joint facet sits over the middle of the calcaneal body. Inferior to this articulation, calcaneal cancellous bone radiates from the force concentration area above to spread forces across the body of the calcaneus, especially posteriorly towards its posterior third (Giddings et al., 2000).

The posterior third of the calcaneus does not display articular facets, but it does provide an enlarged posterior tubercle for the attachments of the complex soft tissues of the Kager triangle superiorly, the Achilles tendon posteriorly, and the plantar aponeurosis and the many plantar intrinsic muscles inferiorly. The length of the tubercle dictates the lever arm distance to the axis of the ankle joint (internal moment arm) (Fig. 3.3.8g). This distance is usually 6–7 cm in adults but has been shown to be variable, giving potential mechanical advantage in sprinting when it is shorter (Lee and Piazza, 2009; van Werkhoven and Piazza, 2017; Foster et al., 2021). The posterior surface of the tubercle has the Achilles attaching through a broad expansive musculotendinous junction across the middle and lower surfaces. The superior third has a smooth surface where the deep retrocalcaneal bursa sits between the calcaneal tubercle and the Achilles.

Inferiorly, the posterior tubercle develops a larger medial and smaller lateral process where the plantar aponeurosis attaches. The proximal plantar intrinsic muscle attachments lie just anterior. They include flexor digitorum brevis, quadratus plantae, part of abductor hallucis medially, and abductor digiti minimi laterally. Anterior to these attachments lies the anterior tubercle for the attachment of the short and long plantar ligaments. The short plantar ligament runs to the plantar surface of the cuboid while the thick reinforcing band of the long plantar ligament runs superficially to attach to the cuboid and the plantar lesser metatarsal bases. The long and short plantar ligaments provide strong lateral tensile restraint, compressing and stabilising the lateral column together under rearfoot plantarflexion bending moments across the rearfoot–midfoot boundary.

The spring ligament attaches to the anterior inferior surface of the sustentaculum tali to radiate over the plantar surface of the navicular, thereby helping to stabilise the proximal medial column. The sustentaculum tali projects medially from the calcaneus, with a width of 1 cm and a length of around 2.5 cm in adults, to produce a prominent shelf that articulates with the middle facet of the talus and supports the talar head at the talonavicular interface with the subtalar joint. At the end of the sustentaculum tali lies the narrow medial limit of the sulcus calcanei forming the medial termination of the sinus tarsi. These osseous structures on the calcaneus and the associated extensive soft tissue structures provide a powerful linkage proximally to the lower leg and to ankle power generation, and distally to foot-stabilising mechanisms (Fig. 3.3.8h).

Many primary anatomical features of the calcaneus such as the size of the sustentaculum tali, the extent or absence of a longitudinal groove between the medial and lateral plantar processes of the plantar posterior calcaneus, the varus angulation between the horizontal plane of the talar dome, and the posterior calcaneal tubercle are all quite variable in humans (Bonnel et al., 2013). Such variations may have implications to the success of surgical calcaneal osteotomies (Bonnel et al., 2013) as well as presenting challenges to clinical evaluation that involves assessing the heel's position through the calcaneus, such as with the Foot Posture Index (Fig. 3.3.8i).

Subtalar joint force closure

The subtalar joint's force closure comes from the numerous muscle–tendon complexes and strong passive ligament restraints that cross it. The interosseous talocalcaneal ligament occupies the sinus tarsi and consists of vertical and diagonal fibrous bands intermingled with adipose tissue, and it is angled at ~35° to the frontal plane. This ligament can be functionally considered as the central pivot of rotatory stability, equivalent to the cruciate ligaments of the knee (Bonnel et al., 2010). The interosseous talocalcaneal ligament joins the neck of the talus to the lateral edge of the calcaneus, is laterally interlinked to the extensor retinaculum, and is an anterolateral stabiliser (Bonnel et al., 2010). Moreover, with the cervical ligament, it is possibly important during loading of the foot in resisting anterior medial shear of the talus over the calcaneus. The cervical ligament or anterolateral talocalcaneal ligament also provides important stability to the joint, acting as an anterolateral subtalar 'lock' maintaining subtalar inversion resistance (Bonnel et al., 2010). Like all the subtalar ligaments, it influences the centre of rotation point within the middle facet of the subtalar joint (Peña Fernández et al., 2020). The cervical ligament, by restraining subtalar inversion, means that failure of the calcaneofibular ligament as part of the lateral ankle collateral ligament complex

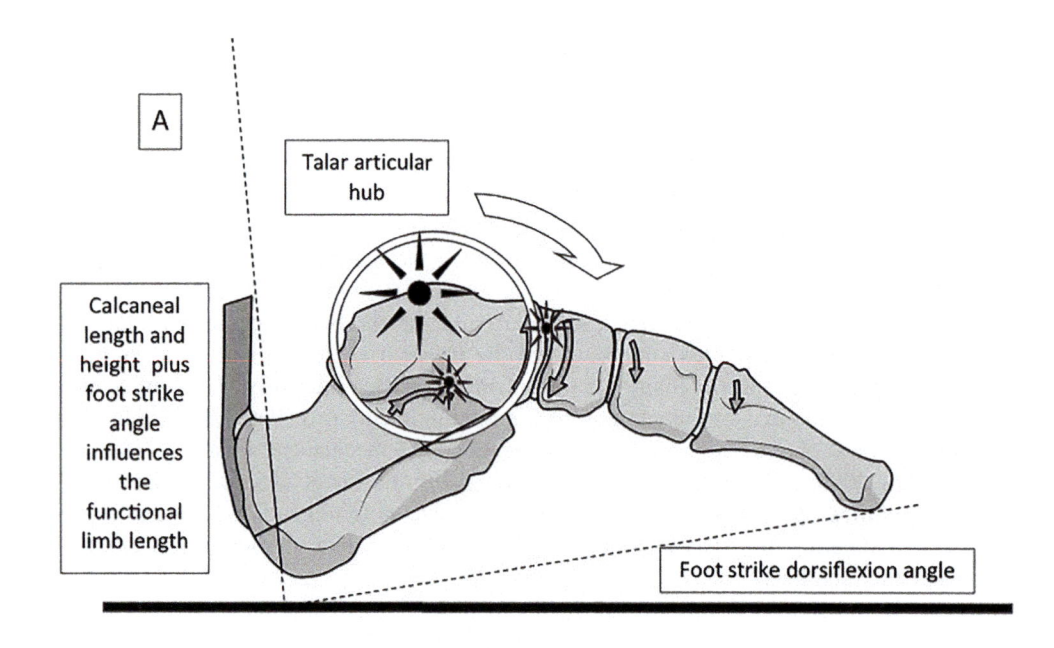

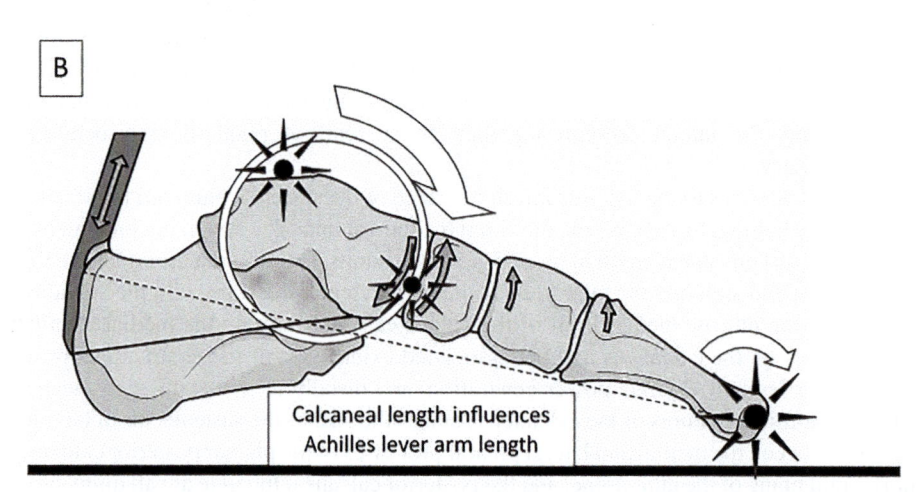

FIG. 3.3.8g The talus and the calcaneus provide distinctly different properties to the rearfoot. The talus fulfils an articular hub role, permitting correlated motions between the lower leg, calcaneus, and talonavicular joint. The calcaneus, by having a particularly long body for a short bone, uses its posterior tubercle to add functional length to the lower limb and extends both the Achilles' internal moment arm to the ankle and its external moment arm to the MTP joints. At heel contact (A), the foot's dorsiflexion angle combined with the posterior calcaneal tubercle's length increases the functional length of the lower limb, dramatically improving bipedal gait efficiency. In forefoot contact gaits, this benefit is replaced by ankle plantarflexion to use the midfoot and forefoot length to create the same limb lengthening effect. The initial loading fulcrum for foot plantarflexion is at the ankle *(large star)*, but important rotations and translations also occur at the subtalar and talonavicular joints *(smaller stars)*. These additional rearfoot joint articulations increase energy dissipation capacity and spread loading stresses across the articulations. The large calcaneus and its soft tissues provide a significant weightbearing surface through single-limb support. At the heel lift boundary (B), the posterior calcaneal tubercle length aids in the generation of ankle plantarflexion acceleration power, helping to form a long external lever arm to the MTP joint fulcrum in conjunction with the stiffened midfoot and forefoot. This longer lever arm improves mechanical efficiency but requires greater myotendinous fibre length changes (thus, slower angle changing rates) for plantarflexion motion than would a shorter tubercle. In those performing high levels of sprinting and forefoot contact gaits, the development of a long posterior calcaneal tubercle is not so beneficial. For acceleration via the MTP joint fulcrum, motion is again required between the talus, lower leg, and talonavicular joint. *(Permission www.healthystep.co.uk.)*

does not result directly in subtalar instability. However, should the cervical ligament fail, the subtalar joint and its instantaneous joint axis motion will increase. Yet, the cervical ligament's action is not yet fully appreciated and indeed, pathology of each and all the subtalar joint ligaments is not completely understood. Establishing the full roles of these ligaments is increased in complexity by all the articular facets of the talus and calcaneus being quite variable within the population, including the anterior facets with the navicular and cuboid (Viladot et al., 1984; Bruckner,

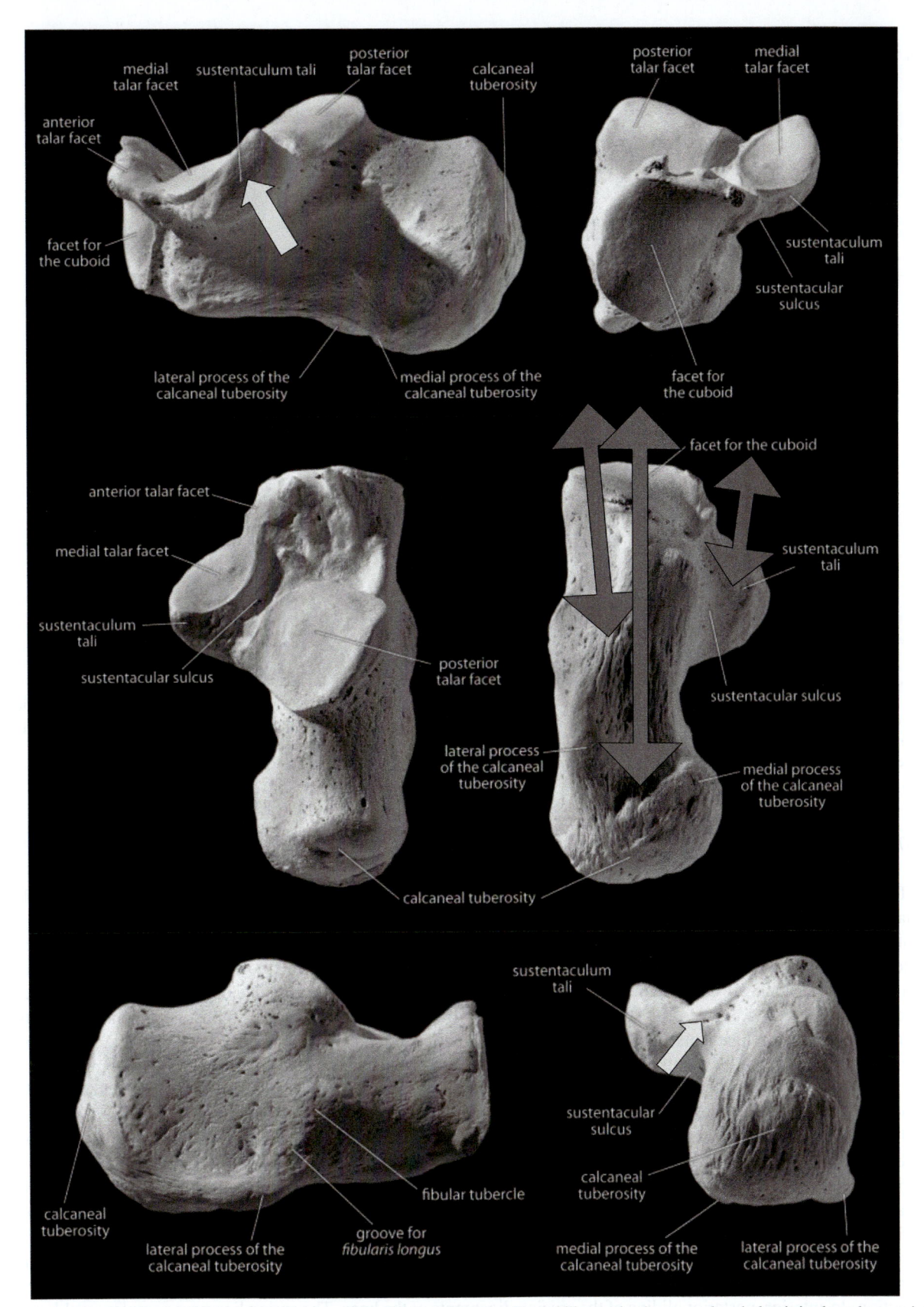

FIG. 3.3.8h The long body of the calcaneus is not only useful for influencing the Achilles tendon lever arm length, but it is also a large point of attachment for muscles and ligaments between the rearfoot, midfoot, and forefoot. These include medial to lateral: the powerful short plantar, long plantar, and spring ligaments *(grey double-pointed arrows)*. The plantar aponeurosis attaches proximally to the surface of the plantar process of the posterior tubercle, sandwiching the plantar intrinsic muscles deep to it which are attached to the plantar calcaneus. The sustentaculum tali sulcus forms an important osseous lip for the flexor hallucis longus en route to its distal attachments, allowing the tendons to apply compressive forces to the medial inferior calcaneus in a superolateral direction *(white arrow)*. The peroneal tendons are tethered to its lateral aspect by fascial retinacula and separately, they laterally compress the calcaneus above and below the fibular or peroneal tubercle. The force vectors generated under the foot via direct attachment or indirect calcaneal linkage are responsible for resisting vault sagging deflection, dissipating and storing energy, increasing tensile-induced articular compression across the vault. This creates important adaptability in foot compliance and stiffness levels. Each calcaneus reflects the lifestyle it grew under and the biomechanics it has sustained during life. *(Modified from White, T.D., Black, M.T., Folkens, P.A., 2011. Chapter 13: Foot: tarsals, metatarsals, & phalanges. In: Human Osteology, 3rd ed., Academic Press, Elsevier, San Diego, CA, pp. 271–294.)*

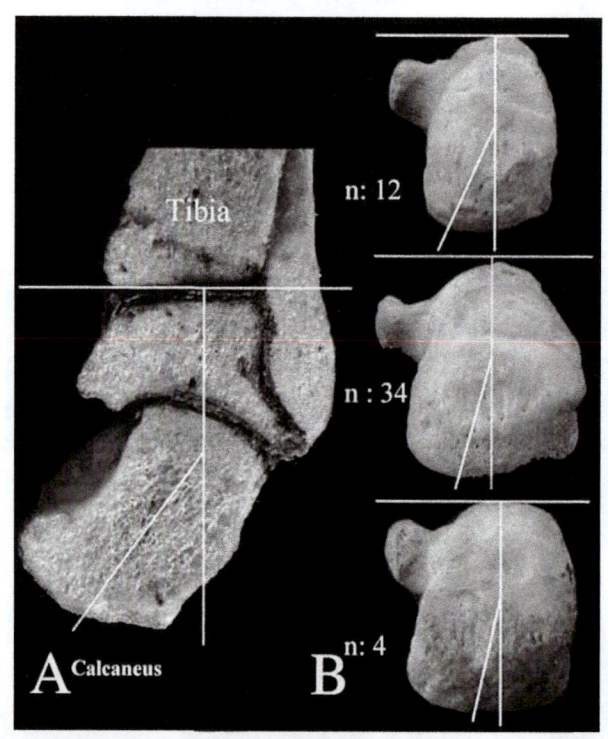

FIG. 3.3.8i (A) shows a frontal plane cross section of the talocrural and subtalar joints demonstrating a significant medial (varus) curvature within the body of the calcaneus. A posterior view of the tubercle process of three adult human specimens (B) demonstrates the calcaneus with different angles and shapes in the frontal plane. The human calcaneus is highly variable, making it an unreliable structure by which to make clinical assessments of rearfoot stance posture, predictions on subtalar positions, and expectations of foot function. *(From Bonnel, F., Teissier, P., Colombier, J.A., Toullec, E., Assi, C., 2013. Biometry of the calcaneocuboid joint: Biomechanical implications. Foot Ankle Surg. 19 (2), 70–75.)*

1987; Barbaix et al., 2000; Jung et al., 2015; Boyan et al., 2016; Prasad and Rajasekhar, 2020; Vučinić et al., 2020). See Figs 3.3.9h and 3.3.9i.

The subtalar and ankle joints and their strong interconnections to the talonavicular joint create one specialised functional rearfoot unit within the rearfoot-to-lower leg interface, with the talus at its core. The navicular therefore has a prominent duel functional unit role coupled to both rearfoot and midfoot motions. There are multiple ligaments that cross the ankle and subtalar joint to attach to the calcaneus, talus, and navicular. The coupling movements of the talus and calcaneus with the tibia and fibula are dependent on the ankle ligaments. These consist of the medial deltoid ligament (which actually consists of at least five blended thickenings, all with their own names), three relatively distinct lateral ligaments, and anterior and posterior ligaments that bind the four ankle–subtalar joint complex bones together. The most clinically infamous are the lateral ligaments, particularly the anterior talofibular ligament which is the most commonly injured structure in lateral ankle sprains. This is a common traumatic event in humans that because of the close coupling of the ankle to the rearfoot, causes as much rearfoot dysfunction as it does dysfunction of the ankle (Hertel and Corbett, 2019).

The talus and calcaneus are also bound together by the posterior, lateral, and medial talocalcaneal ligaments that reinforce the joint capsule together with the interosseous talocalcaneal and the stronger cervical ligament (Draves, 1986: p. 160). The interosseous talocalcaneal ligament is controversial in its function, with authors claiming that it tightens with subtalar eversion (Smith, 1958) and others reporting that it maintains close alignment of these bones in all positions (Cahill, 1965). It certainly limits subtalar joint motion (Bruckner, 1987). Failure of the interosseous and cervical ligaments results in abnormal rearfoot motion. It can be speculated that through their orientations, these structures may play a role in limiting anterior glide of the talus over the calcaneus during medial column loading. This function would act in a similar way to restrictions to anterior–posterior gliding of the cruciate ligaments of the knee. The spring ligament complex could achieve a similar action for the navicular. The bifurcate ligament, composed of the lateral calcaneonavicular ligament and anterior calcaneocuboid ligament, can act similarly for the navicular and cuboid, respectively.

Together, the ligaments and articular surfaces between the talus and the calcaneus are fundamental to the transmission of rotation torques from the foot to the leg during the early part of stance phase when the calcaneus and talus move independently of each other (Leardini et al., 2001). Mobility of the subtalar joint appears to be dependent on the stretching of its ligaments and the indentation of its articular cartilage surfaces, requiring the application of significant loading in order to move independently (Leardini et al., 2001). Subtalar motion only occurs when perturbations from calcaneal-tibial motion are induced by the application of external force, proximally and distally, directly on the calcaneus and talus across the subtalar joint. Such motion is resisted and then fully recovered when the force is removed through ligamentous elastic recoil (Leardini et al., 2001). As the heel contacts the ground in walking, the calcaneus is forced to rotate by the GRF into eversion, with some abduction and distal plantarflexion depressing the calcaneal inclination angle. This motion is linked to cuboid and navicular rotations, the latter via the talus (Wolf et al., 2008).

Subtalar and foot pronation and supination

The subtalar joint has long been described as providing motions of pronation and supination, but clarity on what is actually meant in the context of the functional weightbearing foot is debatable. The history of how pronation as a term became used in foot motion is complex (McDonald and Tavener, 1999; Horwood and Chockalingam, 2017). For the body or upper limb to be prone, its ventral surface must align to the ground. However, for the foot to be prone, its plantar surface must be aligned to the ground. The plantar surface is actually part of the body's dorsal surface that has been rotated under the body. The dorsal foot is part of the body's ventral surface. This means that the plantar foot is actually supine in its plantigrade position. It is far too late to reverse the terms pronation and supination of the foot, and such a move would result in unnecessary confusion. Therefore, for the foot to be pronated, the plantar surface must be prone to the support surface (Horwood and Chockalingam, 2017). Let us therefore live with this situation.

Usually, pronation of the foot is equated to the simultaneous weightbearing (closed chain) motions of calcaneal eversion, talar adduction and plantarflexion, and forefoot inversion and dorsiflexion (Gomes et al., 2019). However, this description glosses over the more subtle motions occurring across and lesser tarsus and between the metatarsals. Indeed, the midfoot and the resistance its articulations offer to torques during gait is likely to have the most effect on how prone the foot becomes on its plantar surface. Reduced resistance to midfoot torques does seem to be associated with rearfoot eversion angles and increased forefoot-rearfoot inversion and dorsiflexion motions (Gomes et al., 2019). This suggests that rearfoot and forefoot motions result from rather than actually cause changes across the midfoot, and such motions relate to the stiffness levels throughout the foot.

The plantar aspect of the foot is always prone to the ground in static and dynamic plantigrade stance in order to provide a maximum surface contact area for the foot. Pronation should perhaps only be equated to motions within the foot that are creating an increasing prone foot posture and expanding the contact surface area in static and dynamic stance. This equates to lowering the vault profile so as to increase the contact surface area of the foot with the ground. Foot pronation has been defined as:

> *Motion of the foot articulations that allow the foot to become more prone to the support surface thereby increasing ground contact surface area of the foot.*
>
> (Horwood and Chockalingam, 2017).

This description equates pronation of the foot occurring throughout the midstance phase of gait as the vault being depressed and compressed by the body's CoM passing over it during single support, with the foot most prone (vault height lowest and surface contact greatest) at the heel lift boundary at the end of midstance (Hunt et al., 2001). Through the stress–strain response of the connective tissues, the foot is stiffened under high levels of pronation. This 'proneness' is not achievable through subtalar joint motion alone which has dominant motion within the frontal plane only during early stance phase. Eversion of the subtalar joint alone is not equivalent to pronation. However, rearfoot eversion is a motion that can lead to more proneness in the proximal vault as the talus moves into coupled adduction and plantarflexion. However, plantarflexion of the distal calcaneus decreasing the calcaneal inclination angle and dorsiflexion of the midfoot expressed by navicular drop also makes the vault decrease in height, thus increasing the foot's prone posture without rearfoot eversion (Fig. 3.3.8j).

The subtalar joint needs to move in conjunction with the ankle and the joints distal to it in order to achieve motion that promotes foot 'proneness'; thus, pronation. The contribution the subtalar joint makes to weightbearing pronation is by permitting eversion and abduction through the subtalar joint complex, moving the talus and medial column anteromedially while at the same time, the distal anterior body of the calcaneus plantarflexes. This decreases the calcaneal inclination angle, lowers the vault, and increases surface contact area. The calcaneal body can take up a valgus position in relation to the ankle under loading force through pronation.

In opposition to this, forces that drive the talus into inversion and adduction cause it to posteriorly glide within the subtalar joint, giving rise to dorsiflexion of the anterior calcaneal process and increasing calcaneal inclination that can

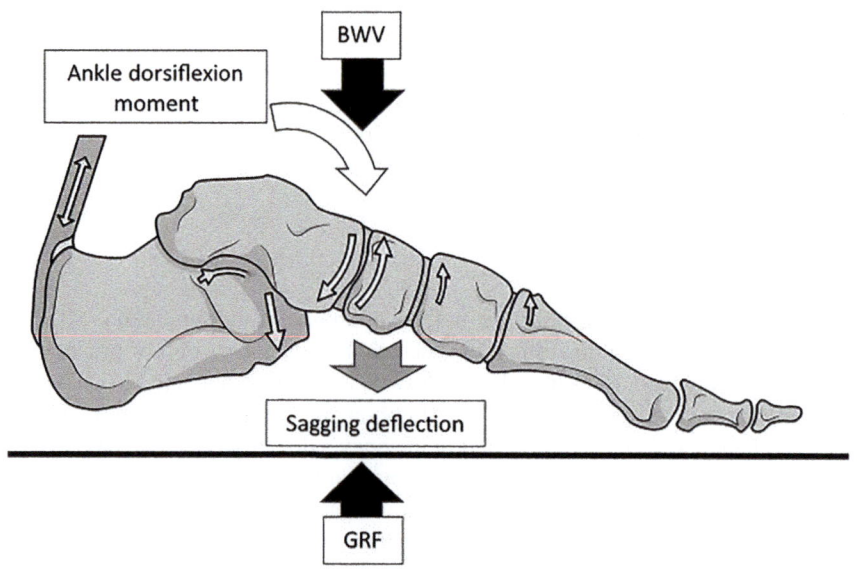

FIG. 3.3.8j   For simplicity, foot 'pronation' should be considered as articular motions that reduce foot vault profile, increasing the weightbearing plantar surface area. This occurs as a result of the vault being fully loaded when caught between the body weight vector (BWV) and the GRF and later, compressed down by the ankle dorsiflexion moment. These forces cause sagging deflection of the vault, lengthening and widening it. Pronation across the foot should initiate on forefoot loading as soft tissue tension under the plantar foot reduces via lessening of muscle activity from tibialis anterior, tibialis posterior, and peroneus longus. Pronation also results from the relaxation of plantar aponeurotic tensions through the reversed windlass mechanism. Foot pronation continues throughout midstance during most of single-limb support, reaching a peak at the heel lift boundary. However, the rate of foot pronation changes so that the higher vault at the start of midstance is pronating at a faster rate than the lower but stiffer vault of late midstance. Increasing muscle activity and rising connective tissue tensioning causes this change in the rate of pronation. Although functional units across the foot such as the rearfoot, midfoot, and forefoot can be considered as undergoing motions of pronation by contributing to vault lowering, terms such as subtalar joint pronation should be used with caution as they equate in vault-lowering terms to calcaneal plantarflexion, eversion, and abduction, and talar plantarflexion, adduction, and anterior translation. Rearfoot eversion has been the motion most commonly associated with foot pronation, but vault lowering and thus foot pronation do not require such motion. Here, the variable motions of the medial side of the medial column articulations are shown in the sagittal plane. These increase vault lowering and lengthening via proximal plantarflexion and distal dorsiflexion, but transverse plane abduction and adduction motions that widen (splay) the foot and frontal plane rotations that evert the rearfoot and invert the forefoot are all part of foot pronation. *(Permission www.healthystep.co.uk.)*

rotate the calcaneus into varus in relation to the ankle. This will reduce pronation of the foot, raise the vault, and decrease plantar contact surface area. This latter movement is usually termed supination. However, as supination means to take up a supine position (the ventral surface aligned upwards), another problem is created for foot terminology in that using this term correctly will always result in the foot being pronated and never supinated in gait. Foot supination, as the term stands, should refer to the plantar surface becoming supine meaning that the foot is only truly supine when presented upside down; that is, unless the plantar foot is considered part of the dorsal body surface when the reverse is true, but let us avoid that quandary. As the plantar foot during gait and in static stance should always be prone to the supporting surface, it would be helpful if less pronation was termed 'reduced' or 'reducing pronation' rather than 'supinated' or 'supinating', as is usually the case. Yet again, the term supination is so ingrained within the lexicon describing foot kinematics that it cannot be easily changed. Thus, 'reduced' or 'reducing pronation' becomes 'supinated' and 'supinating', respectively.

Measurements of pronation and supination within the subtalar joint or rearfoot across the research literature has produced quite different amounts of each. However, investigations have often tried to equate rearfoot motion with subtalar motion. Many studies have only report on inversion–eversion motion of the rearfoot without separating the talus from the calcaneus and the ankle from the subtalar joint (Deland et al., 1995a). Average measurements between 14.4° (Lundberg, 1989), 23.1° (Savory et al., 1998), 37.2° (Nigg et al., 1992), and 53.1° (Ball and Johnson, 1996) are recorded for the subtalar joint, with the variance explained easily through failure to separate ankle motion from subtalar joint motion. Other researchers have tried to use the change in the height of the navicular in semi-weightbearing and full weightbearing (*navicular drop* and *navicular drift*) to indicate the amount of foot pronation (Menz, 1998; Rathleff et al., 2010; Eichelberger et al., 2018). These techniques attempt to identify motion at the talocalcaneonavicular joint as part of the subtalar complex and then equate this with total foot pronation motion.

The plantarflexion or dorsiflexion of the anterior portion of the calcaneus can occur through changing its inclination angle, thus increasing or decreasing the rearfoot's 'prone' posture without necessarily creating frontal plane eversion or inversion around the ankle–subtalar complex. Both movements do not necessarily involve much motion within the subtalar joint which demonstrates only small amounts of plantarflexion–dorsiflexion, as these motions primarily occur at the ankle (Scott and Winter, 1991) and midtarsal joints. It must be remembered that subtalar motion only occurs if the calcaneus and talus are loaded by external force. Hunt et al. (2001) concluded from their studies that the greatest motion in the rearfoot occurred in the sagittal plane and the least motion in the frontal plane. They therefore concluded that rearfoot motion does not follow recognised descriptions of the traditional triplanar motions given for pronation and supination.

Foot pronation and supination should be seen as something different to the triplanar motions of pronation that are associated with the hand and forearm (see Fig. 1.1.3l in Chapter 1 of the companion text 'Clinical Biomechanics in Human Locomotion: Origins and Principles'). Simply stated, it may best be equated with degrees of 'proneness' as suggested by Horwood and Chockalingam (2017). The foot can develop a flatter foot vault profile in stance without the rearfoot significantly changing its frontal plane posture (Kanatli et al., 2006). This is not always appreciated clinically as much as it should be in that often, rearfoot eversion is automatically assumed to be occurring in the subtalar joint and is associated with a low vault profile, even if an eversion posture (rearfoot valgus to the leg) is not actually present. Small degrees of rearfoot eversion can occur at the ankle alone without the subtalar joint being involved. Foot vault height and rearfoot valgus angles must be considered separately (Kanatli et al., 2006; Behling et al., 2020).

During gait, rearfoot eversion and inversion excursions are better clinical descriptions than subtalar joint pronation and supination, not least because the origin of rearfoot motion is not always clearly identifiable to a single joint and such motions/positions in the rearfoot are visualised only in the frontal plane (rearfoot varus/valgus). Terms such as 'excessive', 'over-pronation', or 'hyper-pronation' should be restricted in use to when the foot is more prone than is necessary or desirable for the management of lower limb kinetics, or for safe kinematics during gait or other activities, and if consequentially, energetics become deleterious (Horwood and Chockalingam, 2017).

Rearfoot frontal plane motion in gait

Four asymptomatic adult patterns of rearfoot (note not subtalar alone) frontal plane motion have been identified in walking (Cornwall and McPoil, 2009). These were classified by patterns of eversion excursion as: typical, delayed, early, and prolonged (Cornwall and McPoil, 2009). 'Typical' rearfoot motions demonstrate eversion from an initially inverted position on loading the adducted lower limb, that then pass into relative or real rearfoot eversion that is sustained until around 60% of stance phase and are next followed by inversion, taking the rearfoot into inversion at 70% of stance during the terminal stance phase (see Figs 3.3.8k and 3.3.8l). This is much as had been typically described in the earlier literature on the subtalar joint (Root et al., 1977). The 'delayed' group tend to perform initial heel contact with a less inverted rearfoot, but do not evert the rearfoot until nearly 40% of stance and then follow a more similar pattern to the typical group. The 'early' group were reported to evert from a lower rearfoot degree of contact inversion but then evert rapidly before 10% of stance and reach a peak everted position to the lower limb earlier in stance at around 25%. This decreased gradually, taking the rearfoot into inversion earlier than the other groups. Finally, the 'prolonged' rearfoot evertors started in a more inverted position than is observed on average. However, they everted more gradually, not becoming everted until 35% of stance phase and then peaking in eversion at around 65% of stance to start a late inversion excursion that did not lead to rearfoot inversion until around 80% of stance, which is well into terminal stance.

It can therefore be postulated that different rearfoot joints are being utilised differently in each group starting at different lower limb adduction angles at the hip and support surface. The 'early' group is interesting in that their eversion peak occurs at 25% of stance, the moment reported as being when stance phase subtalar motion largely ceases. However, the important point with these patterns of rearfoot motion is that they were all gathered from asymptomatic healthy subjects, indicating that rearfoot frontal plane motion is not predictable in populations and that there is no single 'normal' pattern. What effects these different rearfoot eversion rates have on gait biomechanics and injury potential remains unclear, but they could simply just relate to either joint anatomical variance, joint motion dominance, or levels of stiffness within the feet.

Each pattern of rearfoot eversion allows increased sagittal plane motion across the midfoot to the forefoot (Blackwood et al., 2005). This permits vault profile depression and induces strain loading patterns into the connective tissues to absorb and store energy for acceleration recoil. As the rearfoot inverts, forefoot mobility decreases (Blackwood et al., 2005) as expected from the findings of Holowka et al. (2017) and the modelling of Venkadesan et al. (2017a,b, 2020), permitting a sequence of foot compliance, lengthening, stiffening, and plantarflexion power application as part of foot function.

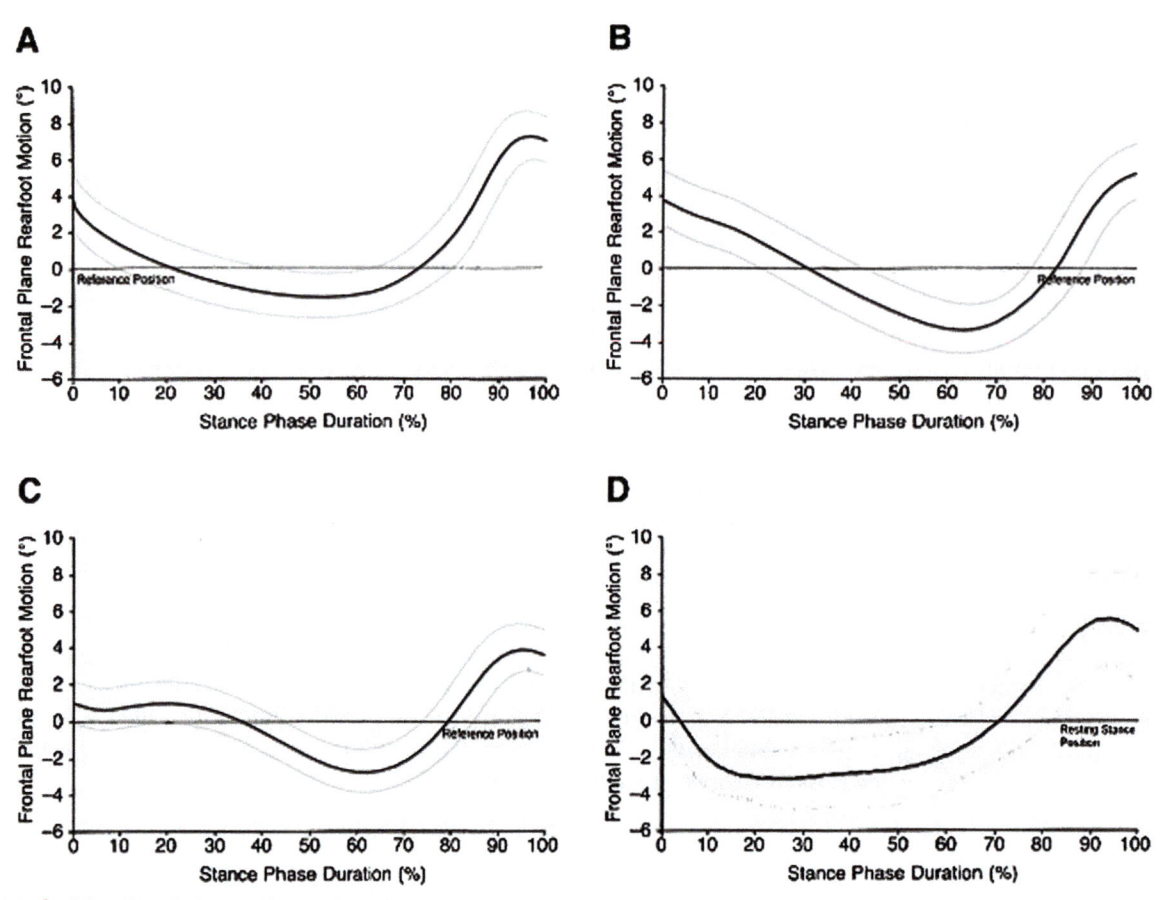

FIG. 3.3.8k Mean frontal plane rearfoot motion values for individuals with four differing rearfoot eversion patterns: (A) typical eversion; (B) prolonged eversion; (C) delayed eversion; (D) early eversion. *(Modified from Cornwall, M.W., McPoil, T.G., 2009. Classification of frontal plane rearfoot motion patterns during the stance phase of walking. J. Am. Podiatr. Med. Assoc. 99 (5), 399–405.)*

FIG. 3.3.8l Mean frontal plane rearfoot motion values compared for four different rearfoot motions during gait. *(From Cornwall, M.W., McPoil, T.G., 2009. Classification of frontal plane rearfoot motion patterns during the stance phase of walking. J. Am. Podiatr. Med. Assoc. 99 (5), 399–405.)*

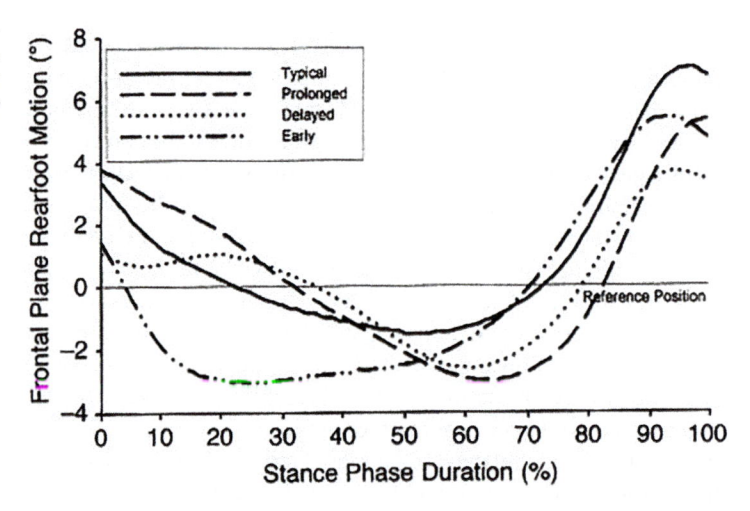

The rearfoot plays a significant role in vault motion control. This is highlighted by the findings of Cobb et al. (2009) who found that in mobile low-vaulted feet compared to more typical-vaulted feet, inversion excursion was increased (by 2.7°) and eversion excursion decreased in the rearfoot (by 1.8°), suggesting that mobile low-vaulted feet were insufficiently inverted in their rearfoot position by around 87% of stance phase. Such a rearfoot posture may be unable to easily provide the midfoot adduction and plantarflexion moments necessary to achieve a stable functional lever arm in the medial forefoot from which to accelerate. This could result in a need to continue inverting the rearfoot through the remainder of propulsion, a situation that can be addressed by either an increased eversion excursion earlier in stance or by using a similar range of excursion, but starting from a less inverted contact position of the rearfoot (Cobb et al., 2009). The use of such motion strategies may explain some of the variances noted by Cornwall and McPoil (2009) in patterns of rearfoot motion during gait, where differences in contact inversion are reduced in delayed and early invertors and increased in prolonged evertors, which also demonstrate late terminal stance inversion.

It was also noted by Cobb et al. (2009) that vault profile affected rearfoot frontal plane motion associations with the lower leg. In both typical vault profiles and mobile low vault profiles, transverse plane lower leg lateral (external) rotation occurs until around 87% of stance followed by medial (internal) leg rotation. In typical vaults, rearfoot inversion associates with lower leg external rotation and vice versa. In mobile low-vaulted feet, this association is only maintained to around 87% stance when despite continued rearfoot inversion, the leg begins to internally rotate (Cobb et al., 2009). The ability of the rearfoot to be part of a stable propulsive lever is also heavily influenced by motion distal to the rearfoot through midfoot function.

Rearfoot to forefoot coupling

Rearfoot weightbearing posture/position is reported to affect forefoot motion via the midfoot. Rearfoot eversion increases sagittal plane motion of the metatarsals when the rearfoot is in valgus (everted), with this relationship decreasing with rearfoot inversion (Blackwood et al., 2005). However, differences have not been detected for changes with forefoot inversion or eversion dependent on the rearfoot's position, despite the sagittal plane effects on motion (Blackwood et al., 2005). This sagittal plane linkage is logical in the context of vault profile changes and both compliance and stiffness within the foot. It is essential that the vault profile depresses throughout midstance. The rearfoot can provide this through plantarflexing the talus and calcaneus, thereby depressing the calcaneal inclination angle and everting the rearfoot while the talus adducts and glides anteriorly as it plantarflexes. Rearfoot eversion is provided variably within the motions of the subtalar-ankle complex rather than within the subtalar joint in isolation. Increased rearfoot sagittal and frontal plane motions that depress the vault are likely to increase relative forefoot sagittal plane motions of dorsiflexion as part of sagging deflection through the midfoot. This will cause connective tissue tightening within the forefoot and midfoot, restricting any frontal plane motion within them. Rearfoot sagittal and frontal plane motions that couple to vault raising will cause forefoot plantarflexion as well as midfoot plantarflexion.

3.3.9 The midfoot as a functional unit

The lesser tarsal joints play an important role in the compliance and stiffness cycles during gait and influence rearfoot and forefoot motion profoundly (Savory et al., 1998; Gomes et al., 2019; Magalhães et al., 2021). Undoubtedly, variable biomechanical need is why evolution has not restructured the midfoot region into either something permanently more flexible or something stiffer. By having a middle range of flexibility-stiffness adjustability, a functional compromise is achieved that can be tailored around our environment and lifestyle. The rearfoot–midfoot interface of the midtarsal (Chopart's) joint, the inter-midfoot articulations of the lesser tarsal joints, and the midfoot–forefoot interfaces of the tarsometatarsal joints are difficult to study. This is because their components are relatively small and they are intimately associated with plantar anatomy that lies deep within the foot. These multiple articulations react to loads across the entire pedal structure (Fig. 3.3.9a). Accurate study often requires intracortical bone pins for precise evaluation. Such studies tend to focus on rotations rather than on translational motions which are considered to be very small (Wolf et al., 2008). The midfoot links rearfoot motion to forefoot motion (Savory et al., 1998; Blackwood et al., 2005; Holowka et al., 2017; Gomes et al., 2019; Magalhães et al., 2021). Through the structural continuum of the soft tissues, forefoot transverse plane stiffening converts the midfoot into part of an effective propulsive lever (Blackwood et al., 2005; Holowka et al., 2017; Venkadesan et al., 2017a,b; Yawar et al., 2017; Magalhães et al., 2021).

The primary midfoot joints are the talonavicular joint and calcaneocuboid joint, and these are responsible for the function of the composite midtarsal joint. However, whether the talonavicular joint can be defined as part of an 'anterior' subtalar joint separate from a 'posterior' subtalar joint is highly debatable (Pisani, 2016). The navicular articulates with three cuneiforms distally before the midfoot ends distally at the medial three tarsometatarsal joints (cuneometatarsal joints). The cuboid straddles the whole length of the midfoot laterally to articulate with the lateral tarsometatarsal joints, but it also articulates with the lateral cuneiform. The cuneiform articulations form the inter-midfoot joints along with the cuboid and navicular. The navicular and cuneiforms are part of the medial column and the cuboid is part of the lateral column, so although the proximal articulations divide the midfoot from the rearfoot, the ligamentous interfaces between the navicular,

FIG. 3.3.9a The midfoot has boundaries at the talonavicular and calcaneocuboid joints proximally that together form the midtarsal joint *(black boundary line)*, and at the five tarsometatarsal joints distally *(white boundary line)*. Motion between the bones of the midfoot is significant, with articular motion being greater proximally on the medial side but distally on the lateral side. The cuboid stands out from the midfoot bones as being the only one that is part of the lateral column and fibular triangle. It also forms part of both the midtarsal and tarsometatarsal joint boundaries. Note the alternative names of Chopart joint and Lisfranc joint, terms that are often used within orthopaedics. *(Permission www.healthystep.co.uk.)*

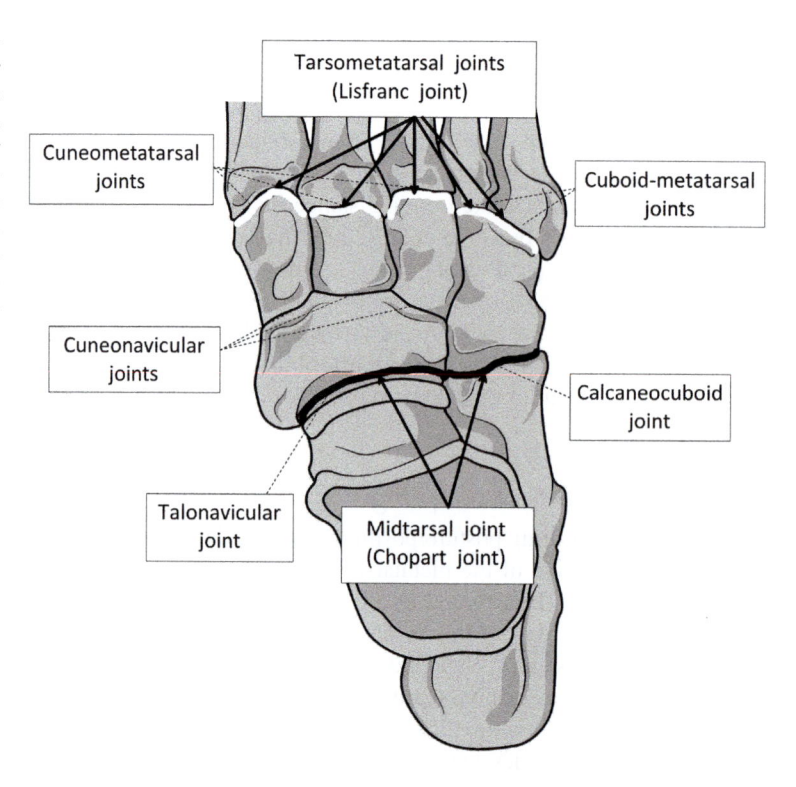

cuboid, and the cuneiform–cuboid joint represent a division separating the functional columns longitudinally throughout the foot. The medial structures in the midfoot from the navicular to the 1st metatarsal have been identified to operate as a functional unit, working and rotating together in the same direction and with the largest motions occurring proximally (Wolf et al., 2008). The talonavicular joint forms the interface between rearfoot and lower limb motions and those of the forefoot and midfoot (Fig. 3.3.9b).

A study by Savory et al. (1998) reported that fixation of joints (*arthrodesis*) in fresh frozen cadaveric feet had different effects on other foot joint motions. Fusion of the subtalar joint did not affect calcaneocuboid motion, whereas normal rotations were still present within the subtalar and talonavicular joints with calcaneonavicular fusion. Fusion of the talonavicular joint had the greatest influence on overall rearfoot motion (Savory et al., 1998) and any arthrodesis that included the talonavicular joint reduced adjacent joint motion by 2° (Astion et al., 1997). These results suggest that the talonavicular joint has a particularly significant role in influencing rearfoot and midfoot articular motions (Lundberg and Svensson, 1993; Savory et al., 1998). Due to the close coupling of motions, Wolf et al. (2008) concluded that the midfoot consists primarily of three functional distinct units with their own independent motions. The navicular–cuboid unit is rigidly bound together by a strong ligamentous relationship. The medial cuneiform–1st metatarsal joint works closely together as a unit, whereas the 5th metatarsal at the tarsometatarsal joint shows its own freedom of motion. The talonavicular and calcaneocuboid joints are thus bound closely to rearfoot motions but are not dependent upon them (Fig. 3.3.9c).

The medial midtarsal joints

The talus, navicular, and medial cuneiform are not a fixed rigid unit. The midfoot should therefore not be regarded as a single unit in gait analysis research (Wolf et al., 2008) or for that matter, in clinical appreciations of motion and pathology within the clinical situation. Relative rotations in these joints are opposite to each other resulting in negative correlations with large rotations, particularly between the talus and the navicular. The medial cuneiform rotations do not entirely follow those of the navicular, generally lagging behind them and thus resulting in a relative counterrotation (Wolf et al., 2008). This indicates that despite motion being in the same direction, distal bones seem to demonstrate smaller absolute rotations during gait than do those of the proximal joints of the medial tarsus. Talonavicular joint rotations are the largest, being more than 15° in directions of inversion–eversion and abduction–adduction, although this does show subject variability (Wolf et al., 2008). One subject in the Wolf et al. (2008) study did display extensive plantarflexion–dorsiflexion at this joint. However, such extensive plantarflexion–dorsiflexion should normally be limited by the strong spring ligament complex.

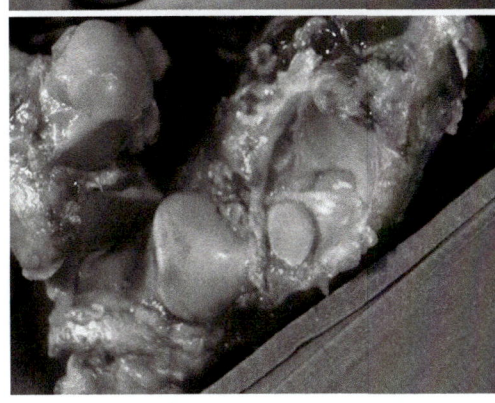

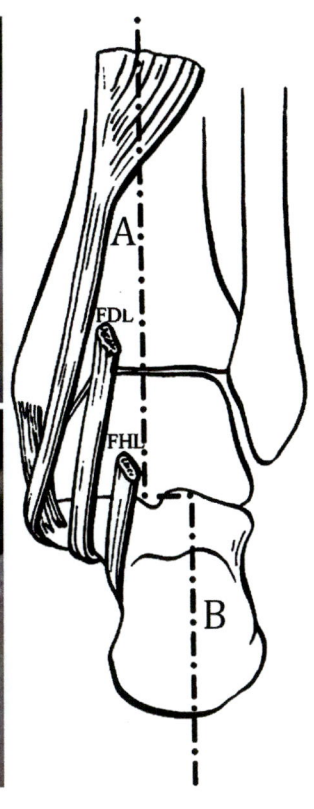

FIG. 3.3.9b The osseous elements of the left talonavicular joint *(upper left)* with the talus rotated anterolaterally and in a dissected specimen *(lower left)*, with the talus retracted laterally, show the intimate relationship between the talus and the calcaneus inferiorly and the navicular anteriorly. The talus is thus coupling ankle, subtalar, and talonavicular motions together. The right image demonstrates that viewed posteriorly, the vertical axis of the tibiotalar joint (A) misaligns medially to the talocalcaneal vertical axis (B), meaning that the subtalar and talonavicular joints are reliant on extensive soft tissue compression medially to remain stable during stance. These structures include parts of the deltoid ligament, known as the tibionavicular ligament and the tibiocalcaneal ligament and which have connections to the spring ligament complex. The flexor hallucis longus (FHL) tendon passing medially along the calcaneus and under the sustentaculum tali and tibialis posterior (superior muscle) and flexor digitorum longus (FDL) passing medially to the talus, also provide important stabilising roles on the talonavicular joint and ankle–subtalar joint complex. *(From Pisani, G., 2016. 'Coxa pedis' today. Foot Ankle Surg. 22 (2), 78–84.)*

Considering the almost ball-and-socket-like configuration of this joint, the observation of significant amounts of multiplanar motion at the talonavicular joint is not surprising.

At forefoot loading, the medial midfoot articulations demonstrate slight rotations in the frontal plane, but not in the last 20% of stance during late terminal stance as the foot stiffens (Leardini et al., 2007; Wolf et al., 2008; Holowka et al., 2017). Midfoot eversion during late stance seems to occur without significant involvement of the medial midtarsal joints. During terminal stance, the midfoot adducts and plantarflexes (Holowka et al., 2017), and the talonavicular joint likely plays the largest part in this motion. The close relationship of the navicular and medial cuneiform motion probably relates to the fact that the tibialis posterior usually has attachments to both these bones, whereas the tendons only cross the talonavicular joint without talar attachment. Only small rotations of 5° or less have been reported at the 1st cuneonavicular joint and are of a similar magnitude between the navicular and cuboid (Wolf et al., 2008). This does not mean that motion here is not insignificant in a functional context.

The cuboid joints

At the articulations between the calcaneus and cuboid and the cuboid base of the 5th metatarsal, only significant transverse plane motion has been identified. Both joints move in the same direction, although larger motions occur distally at the lateral tarsometatarsal joints during gait (Wolf et al., 2008). This is in direct contrast to the medial side of the foot where motions are smaller at the more distal articulations (Lundgren et al., 2008; Wolf et al., 2008). Significant motion occurs at both the calcaneocuboid and cuboid–5th metatarsal joints, with the cuboid-5th metatarsal joint demonstrating greater rotations than at the medial cuneiform–1st metatarsal (1st tarsometatarsal) joint. This indicates that forefoot splaying on loading (broadening-flattening) primarily derives from the lateral tarsometatarsal joints (Lundgren et al., 2008; Wolf et al., 2008).

Calcaneocuboid motion has been suggested as playing a primary role in the efficiency and energetic function of the plantar aponeurosis via the windlass effect and through use of the longer lever arm of the medial MTP joints. The calcaneocuboid joint facet is shaped as a concavoconvex surface and has been likened to an asymmetrical sector of one end of an hour-glass-shaped surface of revolution (Bojsen-Møller, 1979). The joint axis is suggested as being orientated longitudinally through the *calcanean process* or *beak* that extends posteriorly from the medial part of the cuboid, undershooting the calcaneus (Bojsen-Møller, 1979; Bonnel et al., 2013). Although variable in size and sometimes not present, the beak is articular dorsolaterally and the calcaneus is shaped to fit the proximal cuboid articular surface with both bones being held

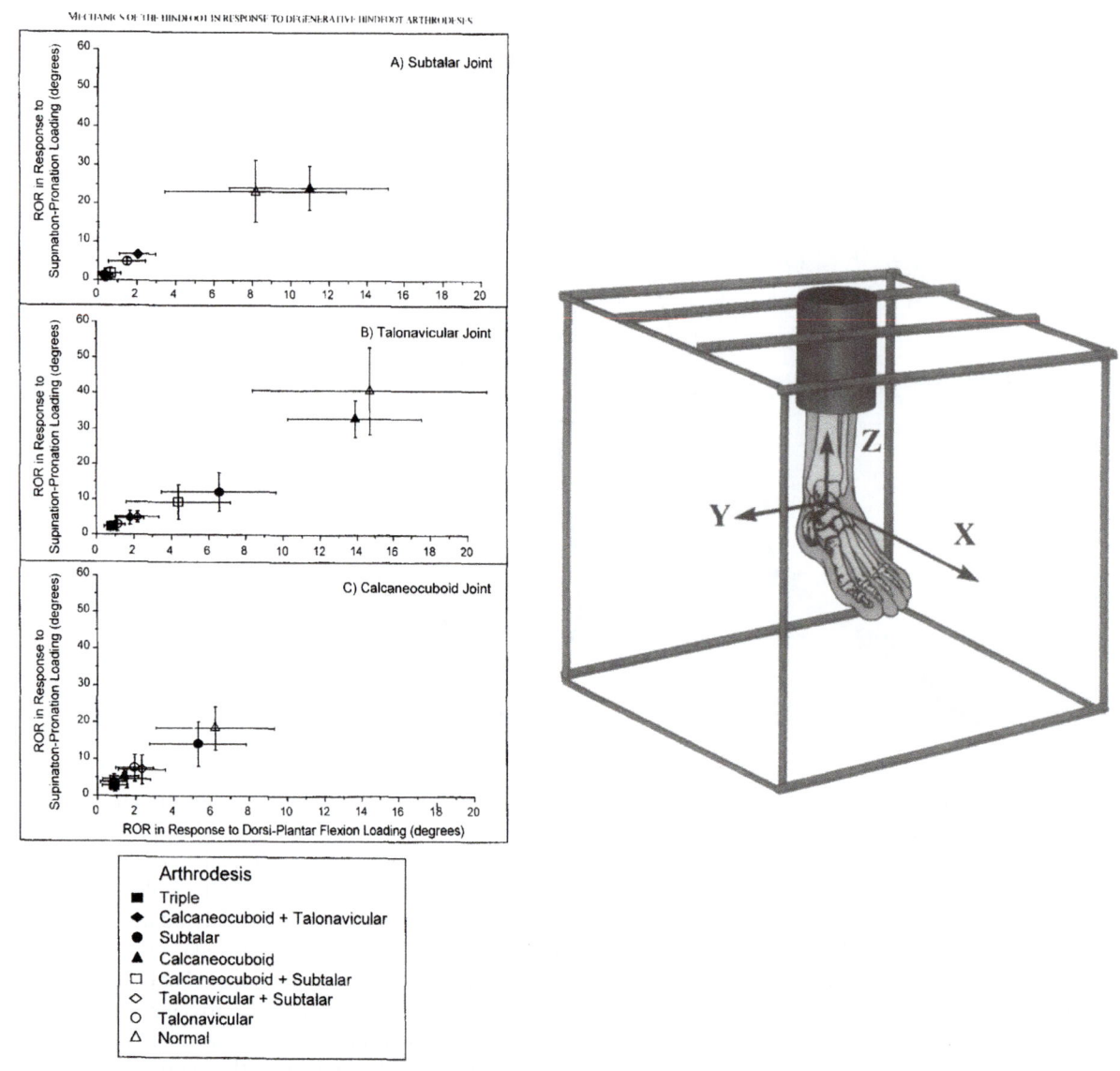

FIG. 3.3.9c Loss of motion at joints within the foot gives an indication of those joints that have most influence on internal foot kinematics. Ranges of rotations measured at the (A) subtalar joint, (B) talonavicular joint, and (C) calcaneocuboid joint are demonstrated on the left, and the orientations of the global co-ordinate axis within the test frame are shown on the right. The light-grey rectangular areas are the regions of a stable arthrodesis in each dorsiflexion–plantarflexion and supination–pronation loading direction. The area where these boxes overlap is the region of a stable joint. Along the X and Y axis of each graph is the range of rotation of the joint in response to loading dorsiflexion–plantarflexion and supination (inversion) and pronation (eversion), respectively. Measurements are expressed in degrees. *(From Savory, K.M., Wülker, N., Stukenborg, C., Alfke, D., 1998. Biomechanics of the hindfoot joints in response to degenerative hindfoot arthrodeses. Clin. Biomech. 13 (1), 62–70.)*

together by the strong plantar calcaneocuboid ligament, providing both form and force closure (Bojsen-Møller, 1979; Bonnel et al., 2013). Three types of joint surface shape have been noted for the calcaneocuboid joint. Those with an axis angle of less than 40° will have a more significant 'locking' action (Bonnel et al., 2013) (Fig. 3.3.9d).

Such an articular arrangement allows the cuboid and calcaneus to rotate around each other through a vertical axis, permitting abduction–adduction, while a second axis runs medio-superiorly through the lateral body of the calcaneus (Manter, 1941; Bojsen-Møller, 1979). On loading, the calcaneocuboid joint abducts and dorsiflexes under load as part of foot pronation until it reaches its close-packed position, blocked by the anterior dorsal overhang of the calcaneus and through ligament tightening (Bojsen-Møller, 1979; Lundgren et al., 2008; Wolf et al., 2008). Intermediate positions will allow freedom of motion. However, on adduction and plantarflexion of the midfoot, the calcaneocuboid ligaments tighten the joint again, but the joint is no longer in a close-packed position as the articular surfaces are only partially opposed (Bojsen-Møller,

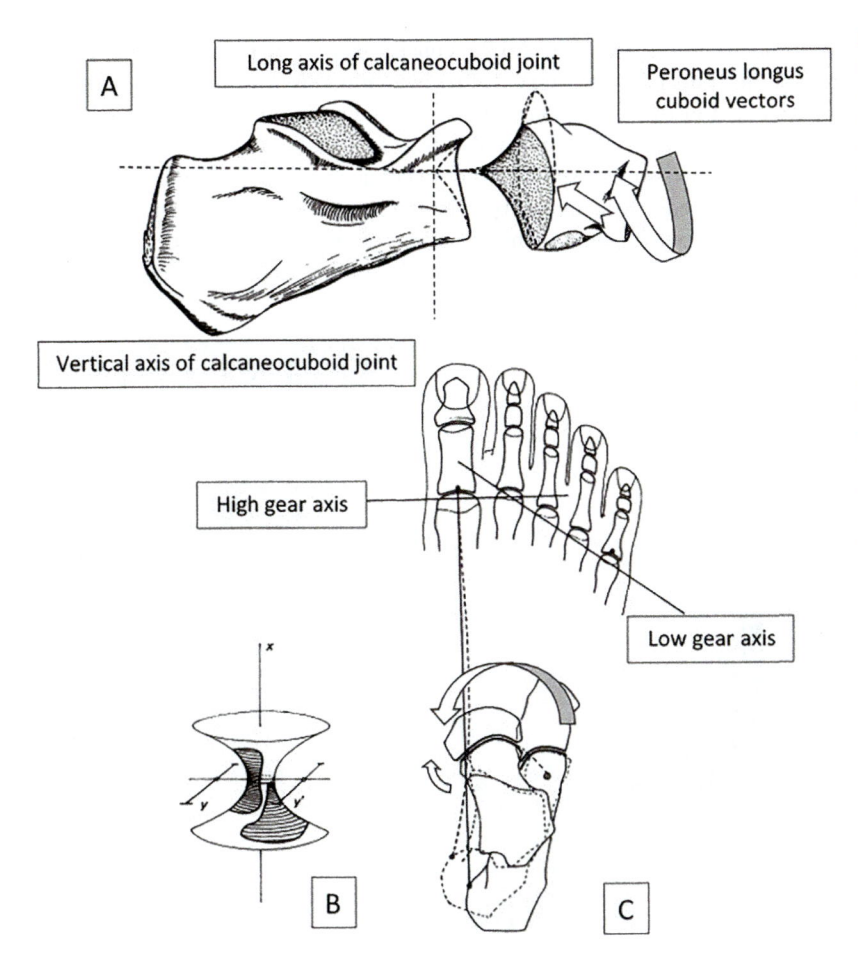

FIG. 3.3.9d A schematic of a right calcaneocuboid joint (A) with the articular surfaces shaped as asymmetrical sectors of one end of an hour-glass surface of revolution (B). The joint's instantaneous longitudinal axis is orientated through the calcaneus and cuboid, while its instantaneous vertical axis lies in the distal lateral aspect of the calcaneus. On midfoot loading during stance, the cuboid abducts around its vertical axis and dorsiflexes, increasing foot pronation via widening and flattening the lateral column. However, superior motion is usually limited because cuboid dorsiflexion is blocked by the dorsal overhang of the calcaneus at the calcaneocuboid joint and via ligament tightening. Thus, abduction and dorsiflexion bring the joint into increased close-packed form closure during midstance. The peroneal groove on the underside of the cuboid is occupied by the tendon of peroneus longus. Its activity during late midstance results in cuboid eversion under a proximally directed force. This also induces adduction around the vertical axis as well as some dorsiflexion (white arrows). The rotation and dorsiflexion of the cuboid by peroneus longus thus brings about force closure of the joint. In (C), the vertical axis passes through the lateral part of the anterior calcaneus that lies in a medial-superior direction, allowing the rearfoot to become relatively adducted *(dashed position)*, rotating around a forefoot fixed to the support surface during terminal stance. This action is part of the midfoot adduction and eversion mechanism that positions the posterior calcaneus and Achilles to align perpendicularly to the medial MTP joints as a fulcrum in order for high gear propulsion. *(Modified from Bojsen-Møller, F., 1979. Calcaneocuboid joint and stability of the longitudinal arch of the foot at high and low gear push off. J. Anat. 129 (1), 165–176.)*

1979). Stability is now provided via force closure from peroneus longus activity and ligament tightening, permitting some freedom for the calcaneocuboid joint to move into plantarflexion and adduction through the midfoot during terminal stance (Holowka et al., 2017). This lateral midfoot motion permits the foot to utilise the longer metatarsal lengths of the medial MTP joints, thereby achieving a longer Achilles external lever arm, initiating an efficient acceleration moment arm termed 'high gear' (Bojsen-Møller, 1979; Holowka et al., 2017). The torque strains accompanying these motions are very strong at the midtarsal joints and are influenced by the size and shape of the calcanean process, the decreasing radius of curvature across the joint, and the progressive tensioning of the calcaneocuboid ligaments, particularly the short plantar ligament that 'locks' the joint, stiffening the midfoot (Bonnel et al., 2013).

Failure to achieve midfoot adduction and plantarflexion at heel lift as part of vault shortening and stiffening potentially results in a drive through the shorter lateral MTP joints' fulcrum. This will result in a 'slacker' plantar aponeurosis and a shorter Achilles external moment arm for ankle plantarflexion power, both events combining to produce a less mechanically efficient 'low gear' propulsion (Bojsen-Møller, 1979). The lack of or a reduced size of the calcanean process on the cuboid (a finding observed in some individuals) or pathology within the calcaneocuboid joint itself, may alter its usual function through a loss of the stiffening capability necessary to create sufficient stability for the joint to function effectively in its different midstance and terminal stance positions. Essentially, the joint could become too mobile to stabilise in one or both positions. Alternatively, it could become too stiff to change its articular posture necessary for full gait function. Ligament failure, joint subluxation, or osteoarthritis within the calcaneocuboid joint represent the most likely functional risks. Ultimately, the ability to present effective resistance to gait torques through appropriate levels of stiffness in the midfoot influences rearfoot and forefoot motions and the ability to offer a stable foot for acceleration (Gomes et al., 2019; Magalhães et al., 2021). Cuboid stability is thus part of the passive stiffness generated throughout the midfoot.

Ligamentous reinforcements between the calcaneus and cuboid (Fig. 3.3.9e) consist of short, twisted fibres in the deep plantar *calcaneocuboid* ligament, usually termed the short plantar ligament, that is continuous with the spring ligament (Bonnel et al., 2013). This may represent a mechanism of tightening that compresses the proximal tarsal joints together, beyond the navicular and cuboid linkage, through the *bifurcate ligament* that also crosses the talonavicular joint. The short plantar ligament lies deep to the long plantar ligament, both being separated by loose connective tissue. The deeper, shorter ligament demonstrates twisted fibres found on its medial edge that are likely to be the chief elements 'locking' the calcaneocuboid joint (Bonnel et al., 2013). These fibres can be twisted tight with 'pronation' (cuboid abduction) or untwisted to loosen with adduction. The dorsal and lateral calcaneocuboid ligaments seem to have limited stabilising power.

FIG. 3.3.9e Plantar view of a left foot dissection of the deep plantar calcaneocuboid or short plantar ligament. Shorter twisted fibres are found on the medial edge of the cuboid. The short plantar ligament is medially connected and partly continuous with the plantar talocalcaneonavicular or spring ligament complex. The 5th metatarsal's styloid process (M5) is situated to help with orientation. *(From Bonnel, F., Toullec, E., Mabit, C., Tourné, Y., Sofcot, 2010. Chronic ankle instability: Biomechanics and pathomechanics of ligaments injury and associated lesions. Orthop. Traumatol. Surg. Res. 96 (4), 424–432.)*

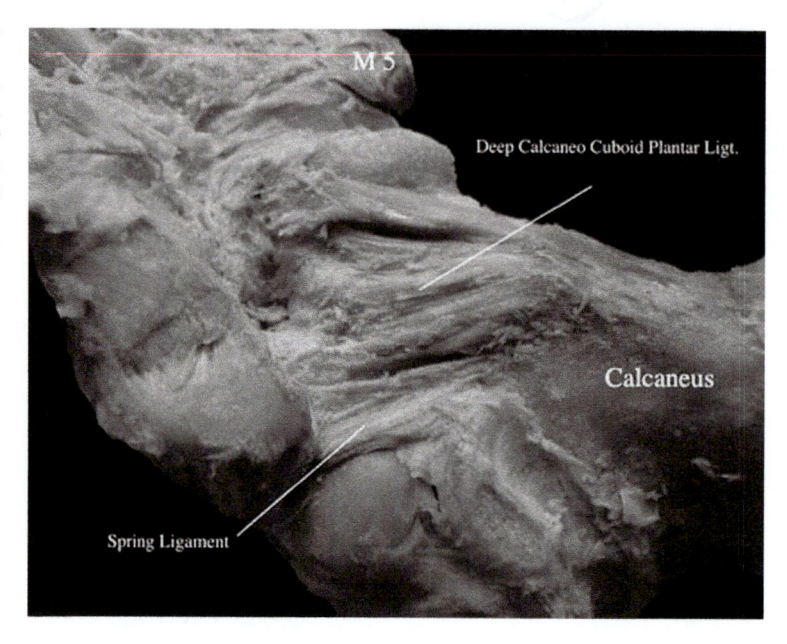

The bifurcate or *Chopart's ligament* consists of two arms shaped like a 'V' or a 'Y', passing distally from the dorsal surface of the calcaneus to the navicular medially and the cuboid laterally, where it joins with the dorsal calcaneocuboid ligament (Melão et al., 2009; Kafka et al., 2019). The medial arm is termed the lateral calcaneonavicular ligament and the lateral arm, the anterior calcaneocuboid ligament. The medial part of the bifurcate ligament mechanically couples to talonavicular joint movement, assisted by the short plantar ligament's communications with the spring ligament that run under the sustentaculum tali and the navicular.

Overall, the movements of the calcaneocuboid joint are more constrained than those of the talonavicular joint (Bonnel et al., 2013). The radius of curvature of the calcaneus' anterior articular surface is essential for calcaneocuboid joint motion, with larger radii of curvature increasing motion and smaller radii of curvature reducing motion that produce coupling limits on mobility of the talonavicular joint (Bonnel et al., 2013). Forced midfoot plantarflexion or dorsiflexion perturbations can cause failure of these ligaments, leading to Chopart's dislocations or fractures. Such injuries are difficult to manage and are associated with long-term morbidity due to the highly significant role the midfoot plays in foot function (Richter et al., 2001; Swords et al., 2008; Magalhães et al., 2021). Articular facet shapes in the proximal midfoot are variable and this may influence vulnerability to injury and/or the ability to recover foot function after injury to the ligaments of this region (Fig. 3.3.9f).

The axes of the midtarsal joint

The midtarsal joint motions are also influenced by the positions of the main joint axes of the talonavicular and calcaneocuboid joints (Bonnel et al., 2013). Two primary anatomical situations arise from the positions of the calcaneocuboid and talonavicular articular surfaces on the calcaneus and talus, respectively. Parallel or convergent axes can exist between the

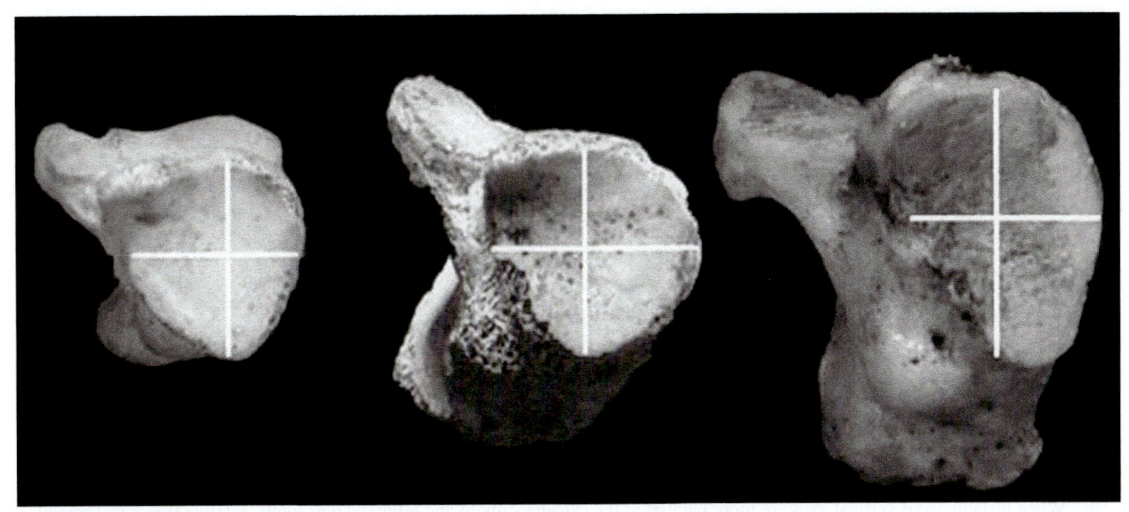

FIG. 3.3.9f Three variations demonstrated in the anterior articular facet of the human calcaneus that articulates with the cuboid. There are notable differences in length and width between vertical and horizontal axes between the specimens. Variance in motions at the calcaneocuboid articulation and across the rearfoot–midfoot boundary can be expected between individuals as a result of such anatomical variants as different articular radii of curvatures. *(From Bonnel, F., Teissier, P., Colombier, J.A., Toullec, E., Assi, C., 2013. Biometry of the calcaneocuboid joint: Biomechanical implications. Foot Ankle Surg. 19 (2), 70–75.)*

joints, directly influencing the planes of motion, with greater convergence of the axes resulting in more limited motion (Bonnel et al., 2013). These joint axes are not static but instantaneous, moving with the motions of the rearfoot and midfoot. With subtalar joint motion, the alignment of the joint axes change with calcaneal inversion and/or talar abduction bringing the axes into greater convergence, thereby restricting motion. With calcaneal eversion and/or talar adduction, the axes become more parallel, increasing midtarsal motion. Thus, osseous/articular axial alignment and the motions within the subtalar joint will together help set the level of flexibility within the region, with convergent axes providing greatest stiffness in subtalar inversion via high convergence and parallel axes in positions of eversion providing greatest flexibility (see Figs 3.3.9g and 3.3.9h).

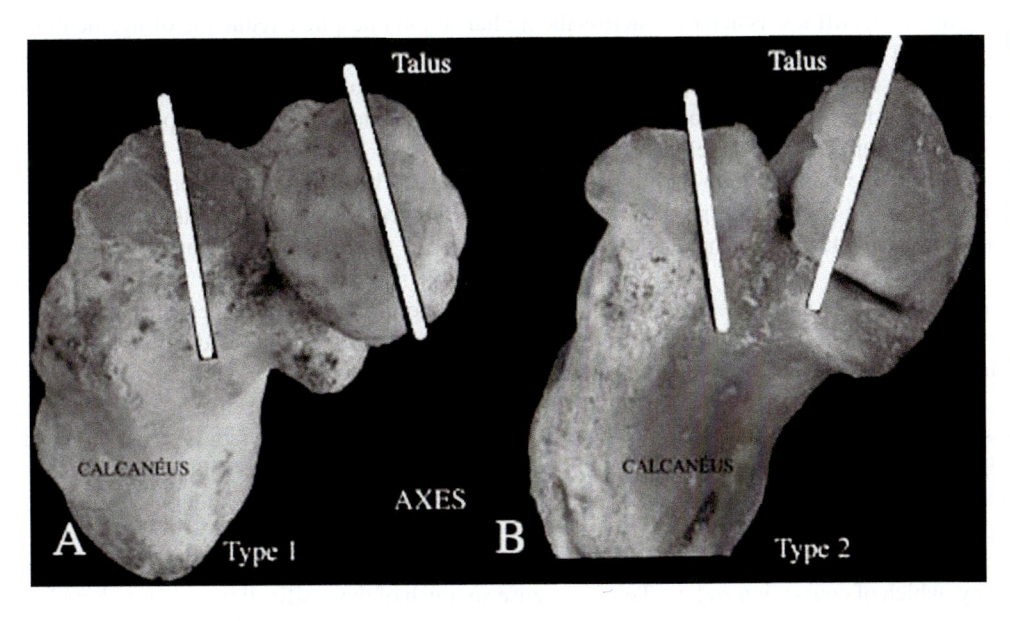

FIG. 3.3.9g The anterior view of articular surfaces of the calcaneus and talus in normal healthy human specimens showing two situations in subtalar joint maximal facet congruity. These are: a parallel axis (A) between the two facets and a convergent aligned axis (B). With the axes convergent, less motion is likely and may help explain some of the clinically observed differences in subtalar motion between individuals. *(From Bonnel, F., Toullec, E., Mabit, C., Tourné, Y., Sofcot, 2010. Chronic ankle instability: Biomechanics and pathomechanics of ligaments injury and associated lesions. Orthop. Traumatol. Surg. Res. 96 (4), 424–432.)*

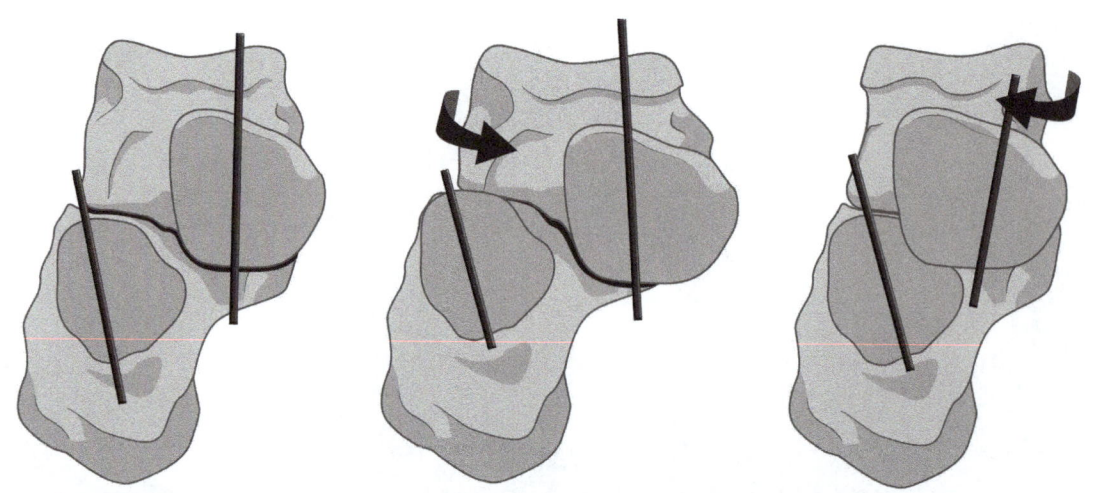

FIG. 3.3.9h When instantaneous vertical axes are more parallel at the calcaneocuboid and talonavicular joints, relative free transverse plane rotations can occur at these articulations. Axes are the result of osseous and ligament orientations supporting the articulation. At rest, the subtalar joint tends to demonstrate an individually variable offset convergence to these vertical axes, with those more parallel permitting greater proximal midfoot motion *(left)*. With subtalar joint motion, instantaneous vertical axes change their relationships. Calcaneal abduction and talar adduction across the subtalar joint tend to improve the parallel vertical alignment *(centre)*, increasing compliance and osseous mobility across the proximal vault. Such motion is associated with rearfoot contact phases of gait. Calcaneal and talar adduction *(right image)* will tend to increase convergence of the axes so that transverse motion of the cuboid and navicular will tend to bring these bones together, restricting motion. This movement tends to associate with the cuboid undergoing adduction in late stance phase, moved by peroneus longus activity bringing the cuboid into medial contact with the talar head and lateral navicular, while the navicular is blocked from abduction by the cuboid. Such motion is associated with offloading the rearfoot in late stance after heel lift. *(Permission www.healthystep.co.uk.)*

This creates an osseous restraining mechanism that works in the opposite direction to the ligaments. Thus, ligaments limit and stabilise 'pronation' or vault-flattening motion across the midtarsal joint and the orientation of the articular surface axes increase stability with vault-raising motions of 'supination', or more correctly, rearfoot inversion. This makes good biomechanical sense, as reducing the vault profile to make the foot more prone should help store more elastic energy within the connective tissues. Increasing length and 'twisting' the fibres of the ligaments will achieve this. When the rearfoot starts to invert following heel lift, the elastic energy is being returned while the foot maintains stiffness aided by osseous 'locking'. The midtarsal joints thus resist bending deflection despite ligaments shortening, able still to permit midfoot plantarflexion and adduction and to increase foot stiffness through protecting the rigid cantilever of acceleration. By changing the orientation of the axes, stiffness is achieved across the midtarsal joint despite a reduction of ligamentous tension due to a shortening of them. This stiffening is assisted by muscle tension, mainly of tibialis posterior and peroneus longus as the medial cuboid becomes 'trapped' dorsally by the tip of the anterior process of the calcaneus (if present). This stiffness is aided by the lateral cuneiform due to its proximal lifting by the peroneus longus tendon wrapping under it, helping to form a stiff proximal transverse arch (Bonnel et al., 2013).

Thus, there are five factors at play in midtarsal joint mobility–stability–stiffness. The joint surface shape of the calcaneocuboid joint, the initial anatomical alignment/misalignment of the axes (i.e. whether they be more parallel or more convergent), motion within the ankle–subtalar joints that can alter calcaneocuboid alignment, ligament integrity, and muscle power.

An individual with convergent midtarsal joint axes will have more restricted motion that becomes less restricted with calcaneal eversion and more restricted with calcaneal inversion, allowing some control of compliance and stiffening within a relatively stiff foot. In an individual with parallel axes, the same mechanisms are true. However, mobility in comparable foot positions with the former individual will always result in greater mobility across the midtarsal joint. A difference between two subjects with similarly aligned joint axes might be a result of either the congruency radius of curvature of the calcaneocuboid joint, ligamentous-connective tissue behaviour through hypermobility–hypomobility of soft tissue, or muscle strength. This underlines the difficulty of attempts in classifying feet by their vault profile alone in order to predict biomechanics during gait.

Variations in anatomy and thus axis locations are common in human populations, most likely via epigenetic expression through developmental biology, which of course, interrelate. This is because so much of the ossification and growth within the foot occurs after birth. Changes in navicular height occur from 6 to 13 years-of-age and arch height ratios change from

10 to 12 years in girls and 11 to 13 years-of-age in boys (Waseda et al., 2014). Such developmental changes would explain the differences observed in the flexibility of feet among children who partake in gymnastic-type activities, and differences seen between habitually barefoot and habitually shod populations. The type of footwear worn, combined with activity levels, will cause structures to adapt during growth to their functional environment.

The navicular joints

The talonavicular joint has considerable movement capabilities, as the articulation is ball-and-socket-like. This means that it can move in almost any direction, although transverse plane motions are usually dominant (Pisani, 2016). Manter (1941) described an oblique and a longitudinal axis for the midtarsal or transverse tarsal joint. However, undoubtedly, this modelling was flawed (Nester et al., 2001). Attempts have been made to establish a single axis for the midtarsal joint, but they have shown wide variability around an upward and medially orientated axis (Nester et al., 2001). This is not surprising as the midtarsal joint is a composite joint consisting of two semi-independent articulations that are coupled by ligaments, sharing no significant articular surface between the navicular and the cuboid. The talonavicular joint also usually shares its capsule with the anterior articular surface of the subtalar joint whereas the cuboid has no direct articular relationship with the subtalar joint's anatomy. It is no doubt better to consider the motion across this area as separate talonavicular and calcaneocuboid axes that functionally interrelate, although the navicular and cuboid do have a deep functional interrelationship (Wolf et al., 2008). The more mobile talonavicular joint as a highly asymmetrical ball-and-socket-like joint is multiplanar with an instantaneous joint axis position reflecting the particular motion occurring within it at any particular moment.

Navicular height is used clinically to infer foot posture, including among children (Aboelnasr et al., 2018). Sagittal plane motions of the navicular have been extensively utilised to appraise midfoot motion in the sagittal plane as a marker of the stability of the 'medial longitudinal arch'. They have been used for assessing or inferring foot 'pronation' (Menz, 1998), subtalar joint pronation (Hargrave et al., 2003), and talonavicular motion (Lundberg et al., 1989), clinically known as navicular drop (Lundberg et al., 1989; Rathleff et al., 2010) (Fig. 3.3.9i). Linking navicular drop to rearfoot eversion may be unwise, as a number of studies have failed to find a correlation between rearfoot frontal plane angles and medial longitudinal arch height (Kanatli et al., 2006; Behling et al., 2020). Research has confirmed significant motion occurring between the talus and navicular as well as some taking place between the navicular and medial cuneiform (Wolf et al., 2008). However, a quantifiable amount of navicular 'drop' that establishes a 'normal' has not been forthcoming (Menz, 1998; Nielsen et al., 2009; Rathleff et al., 2010). Navicular drop is probably an indication of talonavicular joint kinematics, talar plantarflexion and anterior shear, and calcaneal inclination angle on full foot loading. It is a feature primarily of talonavicular motion and surrounding osseous motion, but infers little about the rest of the foot's kinematics and kinetics (Lundberg et al., 1989), except perhaps overall foot flexibility with ranges of motion influenced by loading forces during gait. Static and dynamic measures of navicular drop should be expected to be different for the same individual, and different again between running and walking at variable gait speeds. It will also likely change with muscular fatigue during exercise (Pohl et al., 2018).

The foot's vault function needs the navicular to initially rise before forefoot contact and then fall through loading compliance, stance phase energy storage and dissipation, and during connective tissue stiffening. Navicular motion likely relates to the underlying effects of connective tissue tone (hypermobility–hypomobility), muscle tone, and vault curvatures, which are highly variable and task-specific, making a quantifiable normal amount untenable. Rathleff et al. (2010) concluded from their study that different 'types' of navicular movement may exist, with some subjects generating more variability of motion than others, all of which suggests that variability reflects healthy adaptability to daily activities such as walking and running (Pool, 1989; Goldberger et al., 2002; Georgoulis et al., 2006). Such principles of variable motion sit comfortably with sensorimotor function concepts. Attempts to simplify kinematics through such a complex structure as the foot vault (which includes the navicular surrounded by six interrelated moveable bones) may explain why using clinical measures such as navicular drop fails to establish links to increased pathological risk.

Feet classified as having a more pronated low vault foot posture when using the foot posture index tend to show small variability in navicular motion, but not in all subjects (Rathleff et al., 2010). This highlights the fact that navicular drop is controlled through all the mechanisms that influence foot stiffness. However, the tendency for less motion sits favourably with the findings of Cobb et al. (2009) that low-vaulted mobile feet tend to show slight reductions in kinematic joint excursions distal to the calcaneus compared to 'normal'-vaulted controls. Those classified as having a high vault profile demonstrate a much greater variability, with vault profiles in the middle range showing middle levels of variability. This suggests that there may be a desirable level of variability in vault motion rather than a 'correct' amount of vault motion (Rathleff et al., 2010). Interestingly, females have generally demonstrated less variability in lower limb motions than males

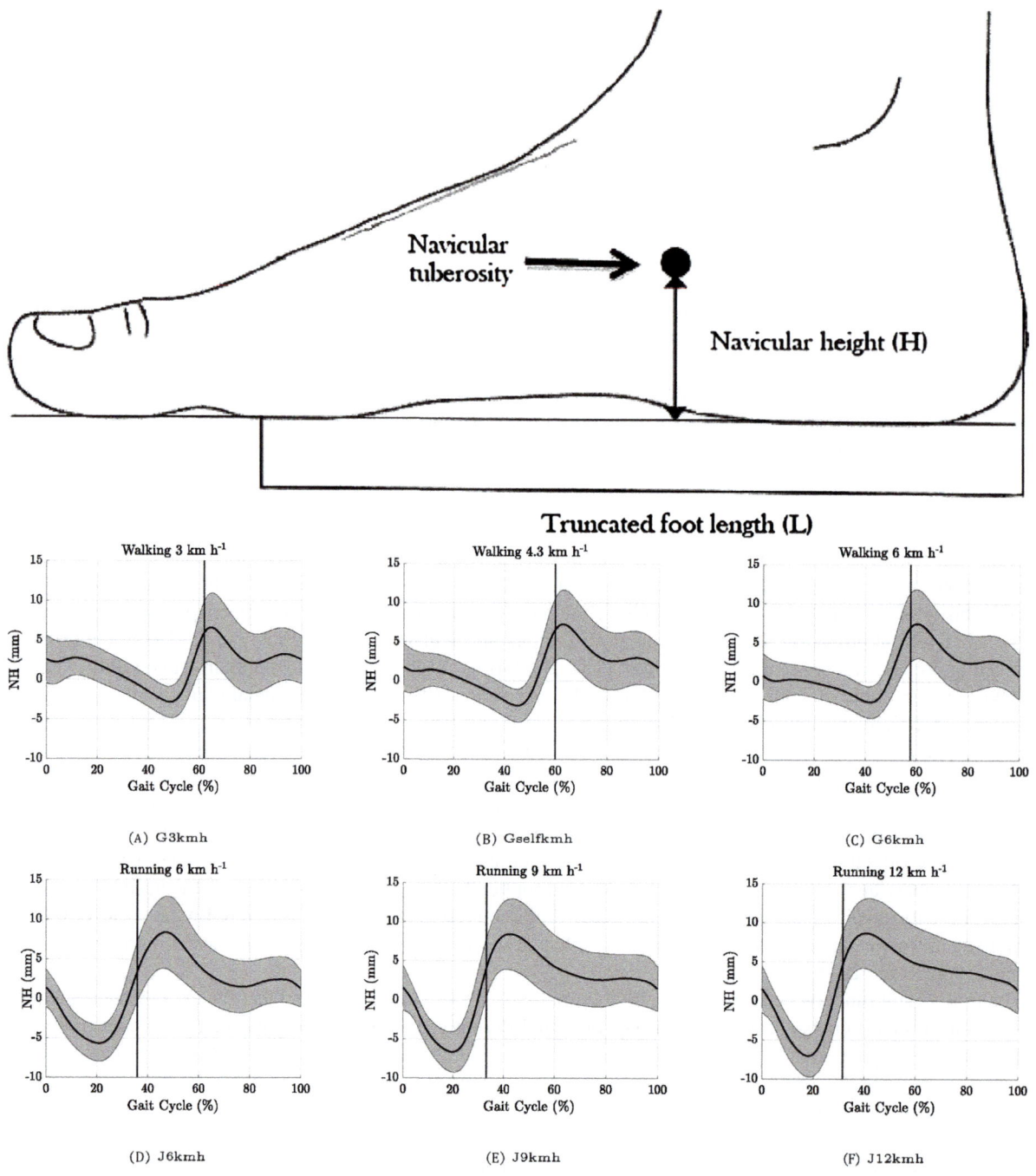

FIG. 3.3.9i Upper image demonstrates navicular height positioning, achieved by palpating the navicular tuberosity in static stance and measuring the distance to the ground. Static navicular drop is a record of the position in semi-weightbearing (usually sitting) and then measuring again in single-leg stance. Dynamic navicular drop measures require skin markers and ideally, 3D motion capture technology. Lower image shows the gait cycle time series of navicular height taken during different walking (A–C) and running (D–F) speeds, averaged among healthy 18 to 65-year-old subjects. Solid black lines indicate mean values and shaded grey areas indicate mean ± one standard deviation. Note that navicular drop peaks later during midstance but starts to decrease approaching the heel lift boundary. This is likely a result of increasing muscle activity, ligament stiffening, and the centre of the ground reaction force (CoGRF) moving further into the forefoot, allowing ligamentous recoil in connective tissues under the navicular. In running, peak navicular drop occurs around the end of braking phase at the middle of stance, but under greater loading forces, navicular drop increases. *(Upper image: Aboelnasr, E.A., Hegazy, F.A., Zaghloul, A.A., El-Talawy, H.A., Abdelazim, F.H., 2018. Validation of normalised truncated navicular height as a clinical assessment measure of static foot posture to determine flatfoot in children and adolescents: A cross sectional study. Foot 37, 85–90. Lower image: Pohl, J., Jaspers, T., Ferraro, M., Krause, F., Baur, H., Eichelberger, P., 2018. The influence of gait and speed on the dynamic navicular drop—A cross sectional study on healthy subjects. Foot 36, 67–73.)*

(Hamill et al., 1999; Pollard et al., 2005; Barrett et al., 2008; Svendsen and Madeleine, 2010), including motions within the midfoot (Rathleff et al., 2010).

Transverse plane midfoot motion

Transverse plane motion of the rearfoot and midfoot suggests a constrained mechanism across the talus, calcaneus, navicular, and cuboid so that motion tends to be similar across these articulations (Huson, 2000; Cobb et al., 2009). Thus, inversion of one of these bones link to inversion moments in the others. In hypermobile low-vaulted feet, this constraint across the joints seems to be lost, as adduction across the midfoot articulations occurs with rearfoot abduction in early midstance rather than midfoot abduction occurring together with rearfoot abduction. This suggests loss of the constraining mechanism across these articulations. The loss of the coupling of rearfoot and midfoot motion may be more significant than the amount of abduction that occurs (Cobb et al., 2009). As the primary supporting passive tensional–compressive forces stabilising the navicular are the spring ligament and the three plantar *cuneonavicular ligaments*, navicular motion variability could indicate hypermobility or insufficiency in the ligaments. Sufficient navicular depression/sagging motion increases the likelihood of presenting parallel axes across the midfoot joints, increasing midtarsal joint motion throughout gait. Foot vault drop is a necessary part of stiffening soft tissues in consequence of the timing and extent of this vault depression, but in excess, will represent a dysfunctional risk (Fig. 3.3.9j).

The great tarsal joint

One region of the midtarsal joint that remains little known functionally is the highly anatomically variable *general* or *great tarsal joint* (Fig. 3.3.9k). This consists of a shared joint capsule of the articular surfaces between the cuneiforms and the navicular, the intercuneiform articulations, the articular surface between the lateral cuneiform and cuboid, and often includes the 2nd and 3rd tarsometatarsal joints (Draves, 1986: pp. 165–167). Sometimes, the joints share their capsule with the intermetatarsal joints between the 3rd and 4th metatarsal bases and the 1st and 2nd metatarsal bases. The cuboid–navicular joint can also be part of the joint if this joint is a synovial joint rather than having only a fibrous union (Draves, 1986: pp. 165–166). There is a significant degree of sagittal plane motion reported for the 1st cuneonavicular joint, particularly of plantarflexion during 1st MTP joint extension over 20° (Phillips et al., 1996).

The cuneonavicular articular surfaces are essentially planar joints, with a slight degree of convexity on the navicular side and a concavity of the cuneiform surfaces. The joints are stabilised under the tensioned compressions from weak dorsal and relatively strong plantar ligaments, which on the medial cuneiform blend to the plantar cuneonavicular ligament and may represent a separate medial ligament (Draves, 1986: p. 166). The strongest plantar ligament of these joints is that of the medial cuneiform and navicular tuberosity, and passes to the base of the medial cuneiform. Each of these plantar cuneiform ligaments is reinforced by attachments from the tibialis posterior tendon slips which fuse to the ligaments (Draves, 1986: p. 166). It is possible that tibialis posterior dysfunction leads to increased plantar flexibility at these articulations. The two intercuneiform joints and the cuneocuboid joint are planar joints with a very slight convexity-concavity that are reinforced with dorsal plantar and strong, thick interosseous ligaments attached to roughened non-articular areas (Draves, 1986: p. 166). Dorsal ligaments run transversely but are more obliquely orientated from the cuboid. The plantar intercuneiform ligaments pass from the plantar surface transversely across each cuneiform. The plantar cuneocuboid ligament is thick, strong, and short, attached from the posterior aspect of the cuneiform to the peroneal ridge of the cuboid. All these plantar ligaments have connections to the tibialis posterior tendon slips (Fig. 3.3.9l).

The cuboid–navicular communication is either formed by an interosseous ligament in around 45%–55% of the population, a small synovial joint which communicates with the great tarsal joint, or more rarely, the talocalcaneonavicular or the calcaneocuboid joint (Draves, 1986: p. 167). If the interosseous ligament is present, it walls off the great tarsal joint. Where a planar synovial articular facet is present, the interosseous ligament is reduced or absent, permitting more gliding movement (Draves, 1986: p. 167). It can be expected that feet with an articular facet permit easier motion between the medial and lateral column cleavage line than do those with an interosseous ligament. Dorsal and plantar cuboid–navicular ligaments are present in both types of joints, with the dorsal ligament running obliquely in an anterolateral direction. However, this has a capsular thickening if the synovial facet is present. The plantar ligament runs transversely between the bones. If present, the interosseous ligament is thick and short, forming a syndesmosis (Draves, 1986: p. 167).

Soft tissue stability within the midfoot

The arrangement and strength of the ligaments associated with the tibialis posterior tendon slips found around the medial and distal midfoot make this an important region for maintaining functional proximal transverse vault stability, linked to tibialis posterior's foot-stiffening capacity. The ligaments represent structures that can dissipate, store, and return elastic

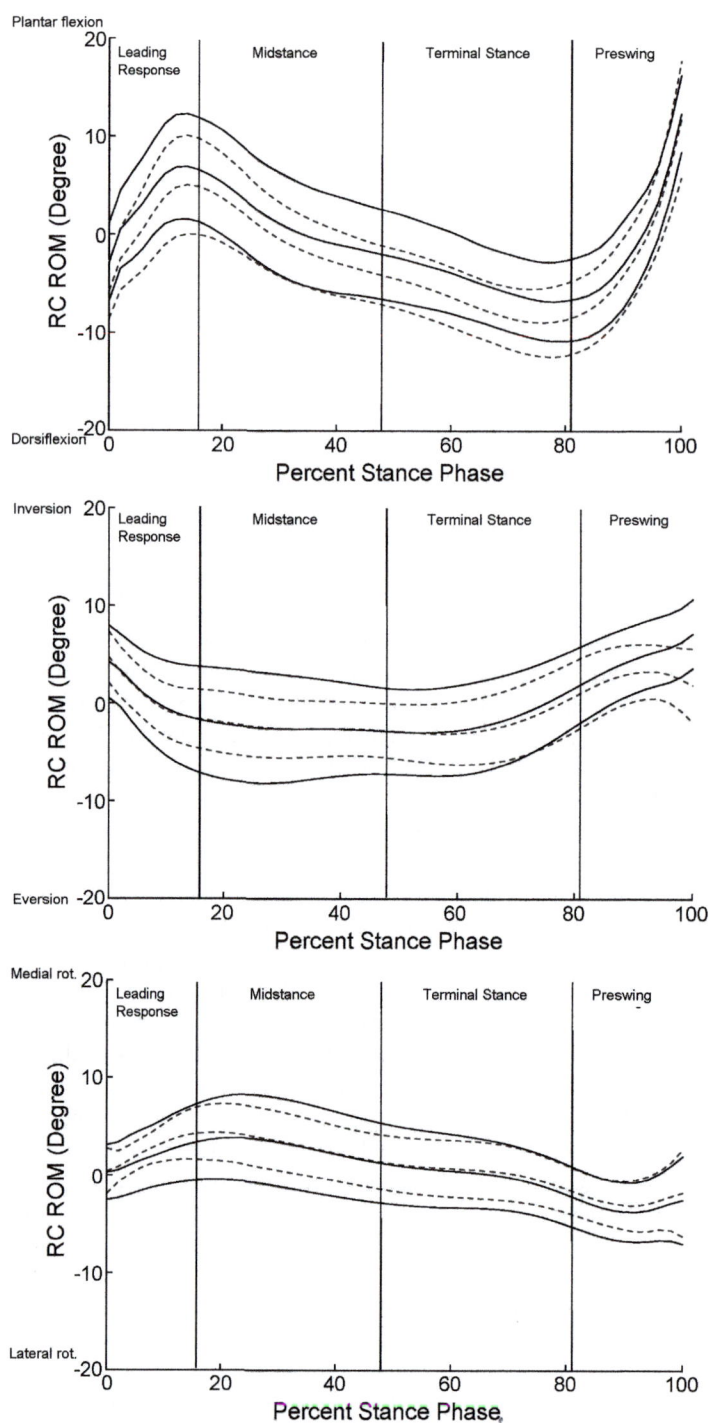

FIG. 3.3.9j Cobb et al. (2009), when comparing rearfoot complex (RC) and calcaneonavicular complex (CNC) stance phase kinematics (mean ± 1 standard deviation) of mobile low-vaulted feet *(solid lines)* to those of more typical foot postures *(dotted lines)*, found significant differences. Small inversion and eversion excursion differences between groups may be explained through differences in rearfoot frontal plane and transverse plane leg motion coupling during preswing. RC eversion excursion to the support surface did not differ significantly between groups during early stance. Both groups also demonstrate lateral (external) leg rotation to ≈87% of stance followed by medial (internal) rotation for the remainder, but in mobile low-vaulted feet, the rearfoot continues to invert as the leg begins to internally rotate, suggesting abnormal or different motion coupling. The mobile planus foot vault subjects also adduct the CNC during the majority of midstance, while typical-vaulted feet initially abduct and then gradually adduct the CNC to the end of midstance. Decreased eversion excursion during terminal stance and preswing might indicate that mobile low-vaulted feet are not sufficiently able to provide the stable forefoot base for acceleration compared to more typical-vaulted feet. *(From Cobb, S.C., Tis, L.L., Johnson, J.T., Wang, Y.T., Geil, M.D., McCarty, F.A., 2009. The effect of low-mobile foot posture on multi-segment medial foot model gait kinematics. Gait Posture 30 (3), 334–339.)*

FIG. 3.3.9k The general or great tarsal joint variably consists of a shared joint capsule enclosing most of the intertarsal joints. Demonstrated in cross section, the most common articular surfaces involved are highlighted in black outline with the joint space in grey. Two important interosseous ligaments stabilise this complex. Distally, the 1st or medial interosseous tarsometatarsal ligament (A) binds the medial cuneiform to the lateral side of the base of the 2nd metatarsal. Proximally, the cuboideonavicular interosseous ligament (B) binds the cuboid to the navicular. This ligament can be either part of a fibrous union or it bridges a synovial joint between these bones. Dorsal failure of the 2nd metatarsal base ligaments seems particularly problematic for forefoot instability, as a mobile 2nd metatarsal allows its head to significantly dorsiflex upon loading. This mobility upsets normal forefoot loading patterns and peak pressures during forefoot weightbearing, particularly from late midstance to preswing. *(Permission www.healthystep.co.uk.)*

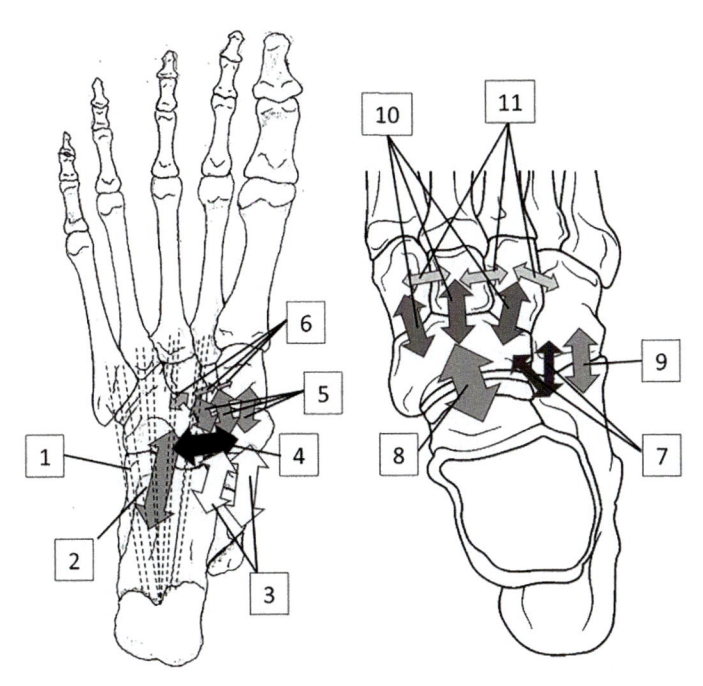

FIG. 3.3.9l A schematic to illustrate the tensile resistance of the plantar ligaments of the midfoot *(left)* and dorsal ligaments *(right)*. Plantar ligaments include the long plantar ligament (dashed outline—1), the short plantar ligaments (2), the two bands of the spring ligament (3), and the plantar cuboideonavicular ligament (4). Less often considered are the plantar cuneonavicular ligaments (5) as well as those plantar ligaments between the cuneiforms and cuboid (6). Together, they are important in resisting foot vault sagging deflection in all orthogonal planes. Dorsal midfoot ligaments are far less well known but important in preventing midfoot hogging in all orthogonal planes by assisting the plantar and interosseous ligaments in resisting frontal and transverse plane torques. They include the dorsal talonavicular ligament (7), the dorsal bifurcate ligament (8), the dorsal calcaneocuboid ligament (9), and the dorsal cuneonavicular (10), inter-cuneiform, and cuneonavicular ligaments (11). As a functional unit, these ligaments exist on a connective tissue continuum that gives the midfoot elastic capacity under twisting and stretching. This elasticity allows the whole midfoot to dissipate and store energy throughout vault motions. *(Permission www.healthystep.co.uk.)*

energy into the vault following vault depression under midstance vault bending moments. The medial side of these articulations is not supported by the strong long plantar ligament. Instead, they are more reliant on the activity of the tibialis posterior and peroneus longus muscles through their tendon slips crossing the area obliquely, laterally, and medially. This gives the medial vault more adaptability. The integrity of this area is maintained by these combined medial–lateral active and passive tensile–compressive forces of the soft tissues. Yet, the degree of translation, gliding, and the planes of motion between the bones remain largely unknown, and are likely to be subject specific. The central lesser tarsus is certainly an area that warrants more investigation.

The role of ligaments in restraining excessive motion and storing and releasing elastic energy, as well as guiding motion, is fundamental to midfoot function. Apart from the important plantar ligaments, a multitude of dorsal ligaments are running inter-articularly between the lesser tarsal bones that are thickenings in the respective joint capsules, probably serving the purpose of restraining skeletal frame lengthening and preventing intra-articular slippage/partial subluxation during midstance. They likely assist muscles in tensioning-compression across the lesser tarsus articular surfaces during terminal stance as the midfoot plantarflexes and adducts, forming the lever to apply ankle power into the ground (Holowka et al., 2017). There is a severe paucity of research on these ligaments and the intercuneiform joints they support.

Local biomechanics within the skeletal frame is important when considering the mechanics of foot function that lead to compliance, stability, and foot-stiffening mechanisms. Being small has left these joints under-researched. Yet, the functional task achieved by these 'minor' joints remains simple in that they provide a deformable but variably elastic structure that can store and release energy, complementing soft tissue mechanics. Midfoot mobility permits stretching and twist-induced stiffness within soft tissues that can behave more stiffly under higher and rapidly applied loads. Together, these material property changes in the connective tissues offset the risk of too much vault depression. In permitting controlled vault deflection to stiffen connective tissues, a more rigid structure can form a relatively stiff lever beam for the latter stages of stance proportional to the loading forces, reducing the muscular burden such that less muscle bulk is required within the foot.

The stored elastic energy within the soft tissues can be released in order to help drive power from the ankle into the ground through the forefoot, maintaining the near-zero negative-to-positive work ratio reported by Takahashi et al. (2017) and Hedrick et al. (2019). This requires the midfoot to be able to remain stable while it plantarflexes and adducts to achieve the most mechanically efficient propulsive lever arm through the MTP joints. The variability and adjustability of the mechanism that achieves this through the midfoot joints provides an adaptability to variances in terrain while at the same time preserving the best energetic efficiency for the particular motion required. These complex mechanisms exist because of the evolutionary history of the foot as an initially arboreal evolved structure. The clinician must appreciate that uniformity across different subject's vault kinematics should not be expected, but the mechanical outcomes on work performed should be. Both foot and ankle joint moments are strategies that can influence both how prone the foot becomes and the efficiency of acceleration. However, the degree of stiffness across the midfoot in absorbing and in transferring torques and energy from the lower limb muscles is critical for walking acceleration biomechanics (Magalhães et al., 2021) (Fig. 3.3.9m).

3.3.10 The tarsometatarsal joints and intermetatarsal joints as functional units

The tarsometatarsal joints or Lisfranc's joints are distally the next part of the system used in compliance and stress-stiffening mechanisms within the foot as a continuation of the midtarsal joints. They should not be considered in isolation of what is occurring posterior or anterior to them. Here, the interface is between the midfoot and forefoot as well as the short and long bones. The long metatarsal bones represent extended rigid segments longitudinally, constructing an important constituent of the lever-beam structure during the terminal stance phase of gait. However, the stability of the metatarsals, both longitudinally and transversely, is dependent on proximal and distal stability. The tarsometatarsal joints must be able to supply sufficient flexibility in order to permit vault depression and energy absorption during forefoot loading, and for passive connective tissue-stiffening mechanisms during midstance. They must also assist motion that permits plantarflexion of the midfoot during terminal stance phase. Metatarsal base articulations are rather flat, allowing only limited gliding motion under extensive ligament restraint (see Figs 3.3.10a and 3.3.10d). Having indentations into the metatarsal row transversely across the foot at the 2nd and 4th metatarsal bases improves the stability of these joints, giving both metatarsal bases unique human-derived articulations with the lateral cuneiform (Draves, 1986: p. 168; DeSilva, 2010). Only the indentation of the 2nd metatarsal is shared with African apes (Drapeau and Harmon, 2013).

Both tarsometatarsal and proximal intermetatarsal joints are reinforced by ligaments, holding the articulations firmly together under compression-tension relationships. There are a number of tarsometatarsal ligaments that include the interosseous dorsal and plantar ligaments that are assisted by the myofascial tissues that bridge these joints to provide stability. It is interesting to note that there are no tarsometatarsal interosseous ligaments at the 1st and 5th metatarsal bases (Draves,

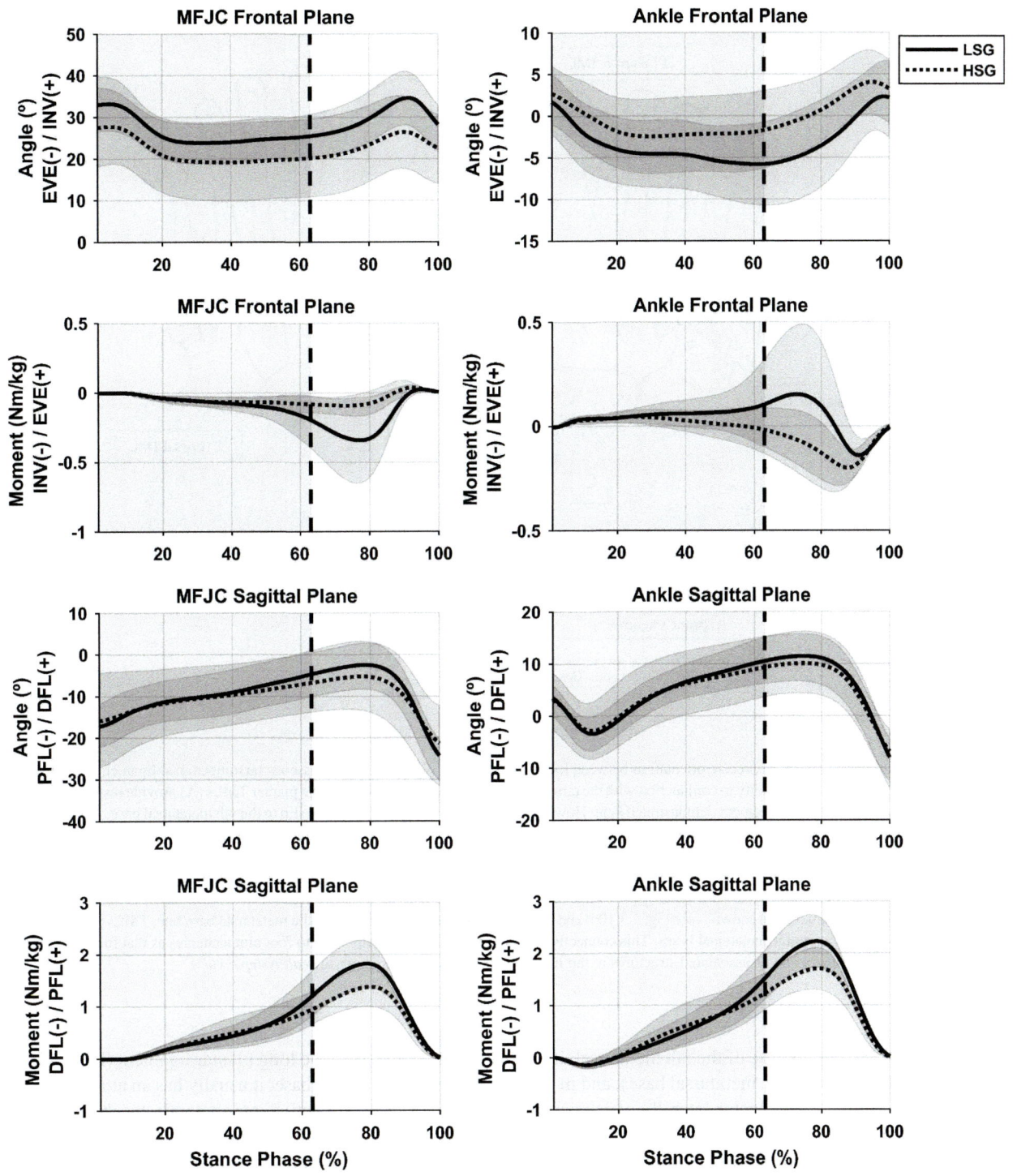

FIG. 3.3.9m The amount of stiffness achieved across the complex articulations of the midfoot is critical for foot and ankle biomechanics during terminal stance phase acceleration. More flexibility within the midfoot generates greater motion between segments and requires greater moments across segments in the sagittal plane to apply ankle plantarflexor power into the ground for acceleration. Midfoot mobility is also associated with increased frontal plane motion around the ankle as well as the midfoot. Here, mean time series for the sagittal and frontal planes are shown for less stiffness groups (LSG—solid line) and higher stiffness groups *(dashed lines)*, recording motion within the midfoot joint complex (MFJC) between the forefoot and rearfoot and the ankle joint for rearfoot to lower leg (shank) motion. PFL indicates plantarflexion, DFL dorsiflexion, EVE eversion, and INV inversion. The dashed vertical line indicates the start of terminal stance at heel lift, and shaded regions the standard deviations during loading and midstance phases from 0 to ±63%, which were not used in the analysis. *(From Magalhães, F.A., Fonseca, S.T., Araújo, V.L., Trede, R.G., Oliveira, L.M., Castor, C.G.M.E., et al., 2021. Midfoot passive stiffness affects foot and ankle kinematics and kinetics during the propulsive phase of walking. J. Biomech. 119, 110328.)*

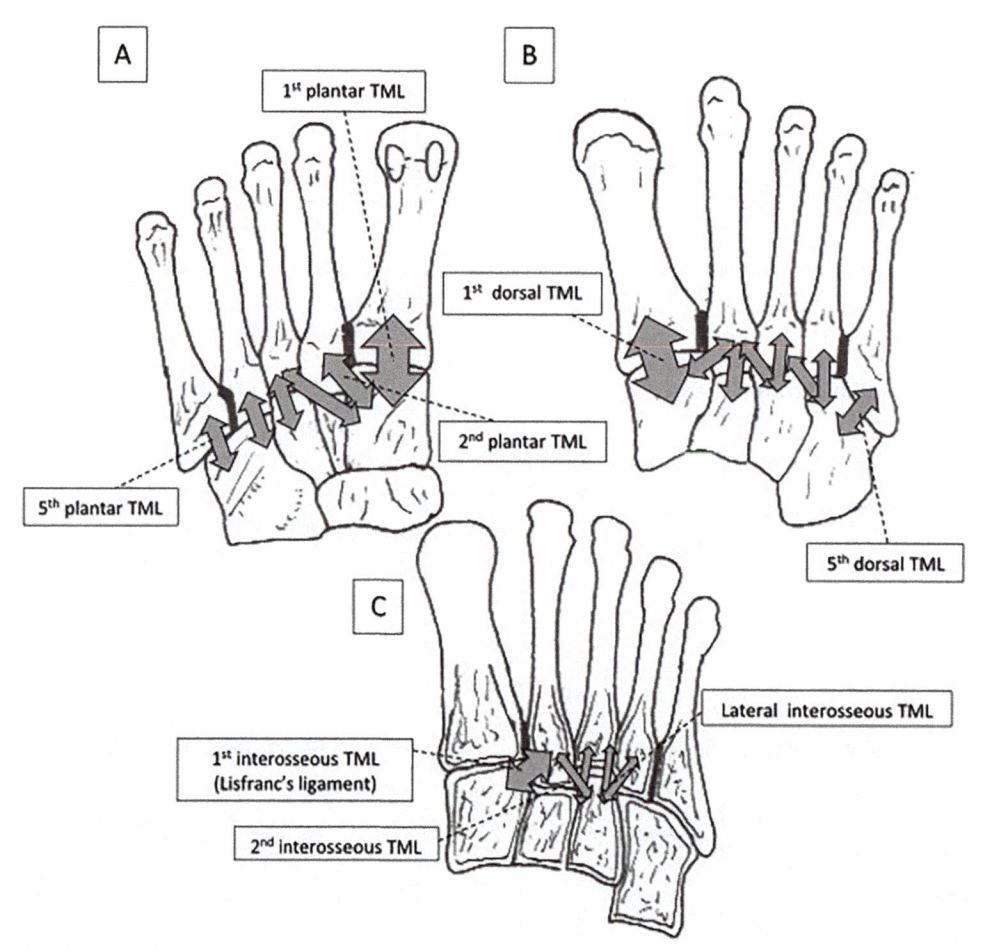

FIG. 3.3.10a Although variable in precise orientation between individuals, dorsal, plantar, and interosseous tarsometatarsal ligaments (TMLs) are an important part of tarsometatarsal stability in conjunction with the tarsal-to-metatarsal row indentations. The plantar TMLs (A) provide restraining tensional vectors, most commonly in the arrangement demonstrated here. However, the 3rd plantar TML often has a slip to the 4th metatarsal base. The dorsal TMLs (B) are more complex and more variable laterally. The interosseous TMLs (C—lesser tarsus shown in longitudinal cross section) are particularly interesting and include the important Lisfranc's ligament, the loss of which is functionally very problematic. The bases of the 1st and 5th metatarsals do not provide interosseous TML restraint. This increases mobility potential at the 1st and 5th metatarsals compared to their companions. The slips from the 2nd and lateral TMLs to the 3rd metatarsal base are variably present, possibly making some individuals more prone to 3rd metatarsal base instability. Together with the lateral Lisfranc ligament (not shown—see Fig. 3.3.10d) and aided by the osseous indentations in the metatarsal base line, TMLs constitute a distal midfoot boundary tie-bar across the metatarsal bases. This connective tissue binding is just as important to foot biomechanics as that formed by the deep transverse metatarsal ligament and its associated structures at the metatarsal heads. *(Permission www.healthystep.co.uk.)*

1986: pp. 170–171). In the case of the 5th metatarsal, it has plantar attachments to the long plantar ligament (Draves, 1986: p. 163), as do the central three metatarsal bases, and in the case of the 1st metatarsal base, it usually has an attachment from the peroneus longus tendon plantar-laterally which perhaps provides a similar yet active intermetatarsal stabilising role to the interosseous ligaments. The tarsometatarsal joints must permit some flexibility and motion during midstance to allow for vault depression by midfoot dorsiflexion. However, they must ultimately stiffen across the joints to complete the 'beam' running from the rearfoot to the MTP joints to transfer ankle power to the forefoot, while also allowing motion for terminal stance midfoot plantarflexion.

Large ranges of midfoot dorsiflexion at these joints, especially laterally, permit these joints to take an important role during heel lift in individuals with midfoot break (DeSilva, 2010). However, this ability may also compromise the transverse forefoot-stiffening mechanisms through reduced forefoot loads at heel lift that result. Reduced forefoot loads at heel lift will alter the timing and extent of transverse metatarsal ligament-stiffening biomechanics. The midfoot break also reduces the sagittal plane extension angles of the MTP joints at heel lift, which in turn couples to reduced ankle plantarflexion angles and loss of the full benefits of the windlass mechanism.

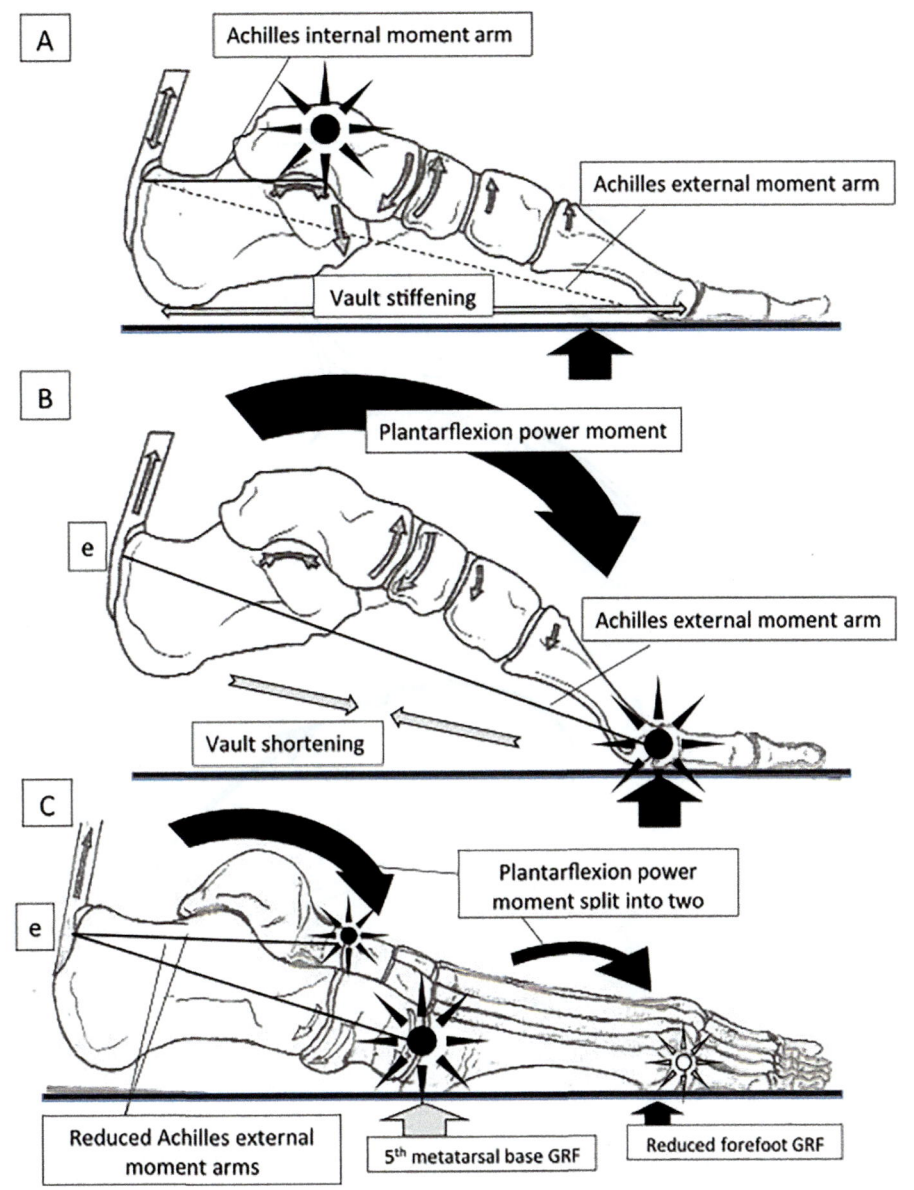

FIG. 3.3.10b Where flexibility and stability are produced within the vault matters for the formation of the plantarflexion power's lever arms for acceleration biomechanics during terminal stance. Tarsometatarsal joint stability plays an essential role in this by preventing midfoot break, particularly laterally. During late midstance (A), under increasing sagging deflection and lengthening-inducing forces from the dorsiflexion moment at the ankle fulcrum *(black star)*, the vault should increasingly stiffen through active and passive soft tissue mechanisms. This develops a semi-rigid external moment arm *(dashed black line)* from the Achilles to the MTP joints. However, during midstance, dorsiflexion is being resisted around the Achilles internal moment arm *(thin black line)*. At heel lift (B), the Achilles uses its external moment arm *(thin black line)* between its point of effort (e) to the MTP joint fulcrum *(black star)* to apply a plantarflexion power moment to form a strong CoGRF *(black arrows)* under the medial forefoot. This is only possible because the articulations between the rearfoot and forefoot are stable enough to prevent sagging deflection. Should tarsometatarsal and/or midtarsal articulations remain sufficiently flexible to permit significant sagging deflection, then plantarflexion power is dissipated through vault deformation. The two most likely joints to offer mobility within the region are the talonavicular joint and the lateral tarsometatarsal joints, as these offer most midfoot motion. High flexibility at these joints at heel lift (C—viewed laterally) alters the fulcrum positions, dramatically decreasing the Achilles external moment arm medially and laterally. Thus, these are the locations where midfoot break primarily expresses itself as articular motion. Usually, significantly increasing weight-bearing (seen as rising pressure on a pressure plate) occurs under the 5th metatarsal base at and initially following heel lift *(grey arrow)*. The plantarflexion power is thus applied through midfoot buckling until the lateral midfoot lifts from the ground when the fulcrum point then shifts to the MTP joints *(white star)*. The GRF on the forefoot is dramatically reduced because significant plantarflexion acceleration power is lost within the midfoot. As the vault is always in a semi-rigid state rather than being rigid, some motion at these joints and others within the midfoot will still occur at heel lift. Each individual step possesses a distinct level of stiffness that varies its amount and location of flexibility at the midfoot and MTP joints between steps. Thus, some steps will provide longer Achilles external moment arms and more plantarflexion power, and others will provide more energy dissipation as needed. It is the inability to utilise this variability appropriately that likely leads to gait dysfunction and pathology. *(Permission www.healthystep.co.uk.)*

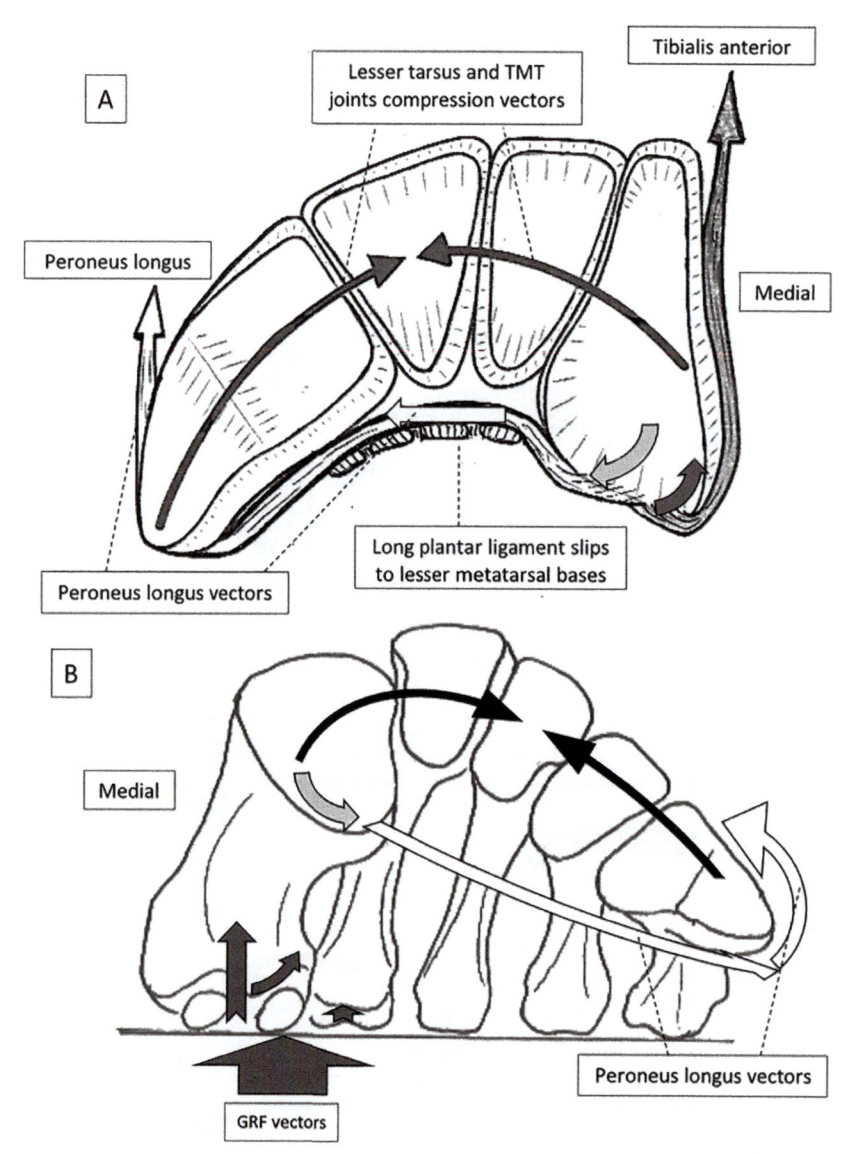

FIG. 3.3.10c Schematic of (A) the right cuboid and cuneiforms viewed from anterior and (B) the metatarsals viewed posteriorly. The tarsometatarsal (TMT) joint boundary is an important part of foot vault stability and mechanical efficiency, reflected by the complex anatomy of this region. The stability of the 1st metatarsal as part of the functional unit of the 1st ray is highly significant because the 1st MTP joint forms the medial acceleration fulcrum with the 2nd MTP joint at heel lift. Unlike the 2nd TMT joint, the 1st TMT joint is an inherently more mobile articulation. Osseous and ligamentous restraints on the 1st TMT joint are limited to a flat articular surface and 1st dorsal and plantar TMT ligaments. The joint requires stabilising forces via muscular force closure. While tibialis posterior (not shown) can stabilise the 1st cuneonavicular joint proximally via its local attachments, peroneus longus and tibialis anterior have strong plantar attachments to both the medial cuneiform and the 1st metatarsal base. This arrangement is unique to humans. Peroneus longus is held across the vault as it runs through the peroneal groove within a tendon sheath and is restrained plantarly by fibres of the long plantar ligament running to the four lesser metatarsal base attachments. Peroneus longus' attachment to the plantar posterior lateral medial cuneiform and lateral side of the 1st metatarsal base gives this muscle transverse TMT boundary compressive vectors with those of tibialis posterior *(thin black arrows)*. Peroneus longus is also generating a significant plantar-lateral and small eversion moment on the 1st TMT joint *(light-grey 1st TMT joint curved arrow)*. Tibialis anterior attaches to the medial plantar surface of the medial cuneiform and 1st metatarsal base, either as a fanning-out single tendon expansion crossing the 1st TMT joint or as two separate slips to each bone. Its force vector is one of proximal compression, dorsiflexion, and inversion of both the 1st TMT and cuneonavicular joint (dark-grey 1st TMT *joint arrow*). These muscles are both active in loading response, creating agonistic stability of the 1st ray, but only the peroneus longus forces are significant at heel lift. In single-limb support, dorsiflexion moments at the 1st TMT joint are initiated by 1st metatarsal head weightbearing GRFs that are coupled to 1st metatarsal eversion (*black arrows* in B). Note that the far more stable 2nd metatarsal will hardly move under GRFs. The influence of internal and external 1st TMT joint vectors, combined with the 1st TMT joint's angulation to the sagittal plane, may influence the development of hallux valgus deformity. *(Permission www.heathlystep.co.uk.)*

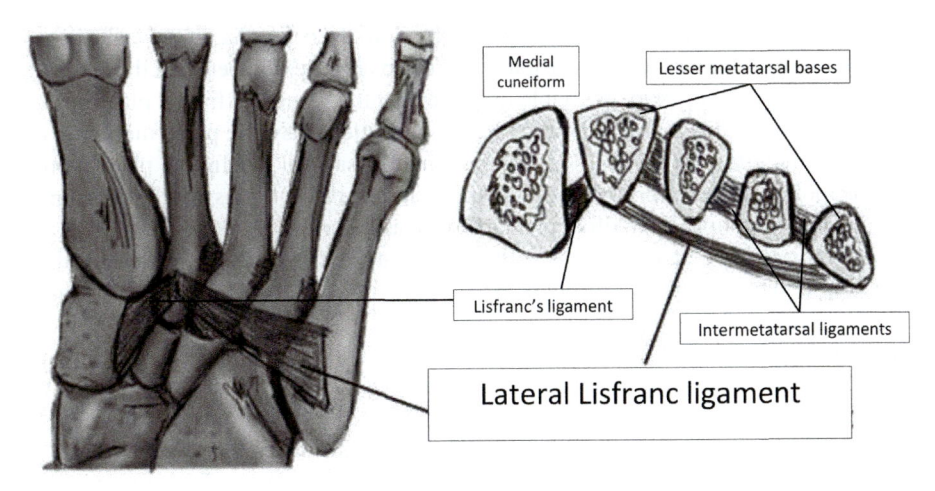

FIG. 3.3.10d Viewed plantarly *(left)* and in cross section *(right)*, the lateral Lisfranc ligament can offer a significant restraint vector of connective tissue limiting the spreading of the lateral metatarsal bases, aided by the other intermetatarsal ligaments. The separate Lisfranc's ligament running from the medial cuneiform to the 2nd metatarsal base (seen medially) does not stabilise the cuneiform to the 1st metatarsal base. Neither does the 1st cuneometatarsal joint benefit from the presence of an interosseous ligament, structures that are found stabilising the cuneiforms and the central metatarsal bases. *(Modified from Mason, L., Jayatilaka, M.L.T., Fisher, A., Fisher, L., Swanton, E., Molloy, A., 2020. Anatomy of the lateral plantar ligaments of the transverse metatarsal arch. Foot Ankle Int. 41 (1), 109–114.)*

Tarsometatarsal joint stability through limiting dorsiflexion is normally an important part of the foot's ability to become stiffened and plantarflex during propulsion (Holowka et al., 2017). It is likely that stability–stiffness induced distally across the deep transverse metatarsal ligament allows the more proximal part of the transverse vault profile to be stiffened through intrinsic muscle action (Venkadesan et al., 2017a,b; Yawar et al., 2017; Farris et al., 2019). This stiffening occurs through structures that include the interossei and the extrinsic tendon action of peroneus longus and tibialis posterior across the whole tarsal-metatarsal region, assisted later by plantar aponeurosis tightening and shortening via the windlass mechanism (Fig. 3.3.10b).

The individual synovial tarsometatarsal joints (otherwise known collectively as the functional tarsometatarsal joint or *Lisfranc's joint*) have discrete differences between them, most noticeably at the 2nd and 4th metatarsals that indent into the tarsal row (Draves, 1986: p. 168). Each articulation has a different motion reported (Lundgren et al., 2008; Wolf et al., 2008). The strong 1st or medial tarsometatarsal interosseous ligament is known as *Lisfranc's ligament*. This ligament links the medial cuneiform to the 2nd metatarsal base medially at an oblique angle and firmly compresses the 2nd metatarsal base into the tarsal row, with the 2nd interosseous ligament doing the same from the lateral cuneiform (Draves, 1986: p. 168). Each tarsometatarsal joint has a synovial planar or gliding-permitting facet running transversely in a curvilinear line (except the 2nd tarsometatarsal joint between the intermediate cuneiform and 2nd metatarsal) which sits 2–3 mm proximal to the line of the 3rd tarsometatarsal joint and around 8 mm proximal to the 1st tarsometatarsal joint (Draves, 1986: p. 168). Thus, the 2nd metatarsal base articulates with the lateral side of the medial cuneiform and the medial side of the lateral cuneiform. The recessed 2nd metatarsal base enclosure results in a restricted stiffer joint than that found in the rest of the tarsometatarsal joints. It is restrained by two oblique interosseous ligaments, two oblique and one longitudinal dorsal ligament, and one oblique and one longitudinal plantar ligament running from the three surrounding cuneiforms to the metatarsal base.

The 4th metatarsal also sits in a 3–4 mm indention of the tarsal row to the lateral cuneiform and consequently has an articulation with the lateral side of the lateral cuneiform, as well as its cuboid articulation and the medial side of the 5th metatarsal base. The tarsometatarsal articular facets of the cuboid are continuous, demarcated only by a near-vertical bony ridge of the facet surface that delineates the articular surface of the 4th metatarsal base facet from the 5th (Draves, 1986: p. 169).

The 1st ray

The medial cuneiform and the 1st metatarsal are described as forming a functional structure known as the 1st ray. It contains the 1st tarsometatarsal joint with proximal and distal boundaries at the navicular-medial cuneiform joint and the 1st MTP joint, respectively. The 1st ray also has a medial articulation at the intercuneiform joint between the medial and intermediate cuneiform and the 2nd metatarsal base (Hicks, 1953; Kelso et al., 1982). How distinct this is as an isolated functional unit is questionable. The strong Lisfranc's ligament attaches the medial cuneiform to the 2nd metatarsal, so that the 1st metatarsal seems to have greater freedom of motion with the medial cuneiform than the medial cuneiform does with the 2nd metatarsal.

Kelso et al. (1982) reported motion of the human 1st ray (medial cuneiform and 1st metatarsal together at the cuneonavicular joint) as being around a functional axis, angled 45° from the frontal and sagittal planes and slightly from the transverse plane with coupled motions of dorsiflexion–inversion and plantarflexion–eversion. Sagittal plane motion between the 1st metatarsal and the 1st cuneiform is minor and inconsistent, whereas plantarflexion between the 1st cuneiform and navicular is far more significant (Phillips et al., 1996). Variable ranges of sagittal plane and frontal plane motion have been found among subjects, demonstrating a ratio average of around 1:1 with little transverse plane motion being noted (Kelso et al., 1982; Wanivenhaus and Pretterklieber, 1989). This suggests medial sagittal and frontal plane motion is provided for by the navicular–cuneiform joint and cuneiform–metatarsal articulations, but not transverse plane motion (Kelso et al., 1982). The midfoot adduction that accompanies midfoot plantarflexion as described by Holowka et al. (2017) during terminal stance comes primarily from transverse plane motions within the talonavicular joints and calcaneocuboid joints as reported by Wolf et al. (2008). These joints also provide frontal and some sagittal plane motions complementing the joints of the 1st ray.

The 1st tarsometatarsal joint is restrained by longitudinal dorsal and plantar ligaments and may have a bursa present between the 1st and 2nd intermetatarsal bases, leading to greater mobility in some individuals. This joint has a flat, kidney-shaped articular surface on the 1st metatarsal base and a flat articular surface on the medial cuneiform limiting motion. However, in non-human primates, it is concave on the metatarsal and convex on the cuneiform and provides considerable motion (Drapeau and Harmon, 2013). The joint is also reinforced by active muscular influences from the variable attachments of tibialis anterior and peroneus longus to the plantar–medial aspects of the joint, which may prevent excessive motion at the articulation when the 1st metatarsal is initially loaded and when both of these muscles are active. In late midstance and terminal stance, the peroneus longus will create a lateral and plantarflexing stabilising force (Fig. 3.3.10c). Despite this muscular restraint or possibly because of peroneus longus activation, on weightbearing, the 1st metatarsal everts to some variable degree with ~3° having been reported on CT imaging when semi-weightbearing (Ota et al., 2019). It appears that hallux valgus deformity may have some linkage to excessive 1st metatarsal eversion on weightbearing (Watanabe et al., 2017).

Tarsometatarsal stability

Ligaments and muscles play a significant role in tarsometatarsal stabilisation, a process that involves all the plantar intrinsic and extrinsic muscles that bridge the area. This ligamentous–myofascial network forms a complex set of force vectors that can be adapted to the foot's position on the support surface. Tibialis posterior and peroneus longus, through their larger cross-sectional areas of muscle and tendon, provide significant forces arranged to create proximal and transverse compression of the tarsometatarsal region. They can resist vault depression and help restore vault profile, assisted by the flexor extrinsic and plantar intrinsic muscles such as abductor hallucis, flexor hallucis brevis, flexor digitorum brevis, and abductor digiti minimi.

The tarsometatarsal ligaments are quite complex and variable, but necessary for stability. The strongest is Lisfranc's ligament. Severance of this ligament leads to lateral destabilisation of all tarsometatarsal joints lateral to it (Draves, 1986: pp. 170–171). There is also a *lateral Lisfranc ligament* spanning between the bases of the 2nd and 5th metatarsals, on average 33.7mm in length and 4.6mm in width (Mason et al., 2020). This is located under the peroneus longus tendon and provides connection through the long plantar ligament to both the transverse and longitudinal planes of the vault (Fig. 3.3.10d). It likely has both an important stability and proprioceptive role (Mason et al., 2020). This may explain why some lateral instability-affected patients can benefit clinically by the addition of medial column support using orthoses or taping.

The 2nd interosseous ligament is the weakest and least developed of the tarsometatarsal ligaments, being absent in around 10% of the population (Draves, 1986: p. 171). The 3rd interosseous ligament is constant, but weaker than the 1st and runs obliquely from the lateral side of the intermediate cuneiform to the medial side of the 4th metatarsal base. It sometimes communicates with the lateral side of the 3rd metatarsal base (Draves, 1986: p. 171). This ligament crosses the cleavage line of the medial and lateral columns. These interosseous ligaments are aided by the intermetatarsal ligaments at the bases of the lesser metatarsals and the plantar and dorsal interossei that bind the metatarsal bases together.

The plantar ligaments are more variable, blend with the joint capsules, and are difficult to distinguish. This is particularly so where they blend with the tibialis posterior tendon slips and fibres of the long plantar ligament (Draves, 1986: p. 169). The medial plantar ligaments are the most consistent and the strongest, whereas the lateral ones are relatively weak. The dorsal ligaments are more constant.

Research indicates that the lateral tarsometatarsal joints are more mobile, permitting greater motion between the metatarsals, and increasing intermetatarsal splaying across the metatarsal heads between the 3rd, 4th, and 5th metatarsals (Lundgren et al., 2008; Wolf et al., 2008). This undoubtedly relates to weaker ligamentous tensions generated in the area (Fig. 3.3.10e). The spreading/broadening and flattening of the vault across the forefoot initiates stretching and ultimately tightening of the deep transverse metatarsal ligament, which leads to the foot stiffening through increasing the transverse plane curvature (Yawar et al., 2017; Venkadesan et al., 2020). The primary motion of the cuboid–5th metatarsal joint on loading is abduction, and greater motion occurs here than at the calcaneocuboid joint (Lundgren et al., 2008; Wolf et al., 2008). This is important for the ability in the transverse plane to dissipate energy and then stiffen on forefoot loading via intermetatarsal motion, but without destabilisation of the lateral column proximally. Abduction from the midline of the foot of both the 5th metatarsal and the 1st metatarsal needs to cause sufficient tightening of the deep transverse metatarsal ligament through lengthening on loading (energy dissipation) in order for the transverse foot-stiffening mechanism to engage.

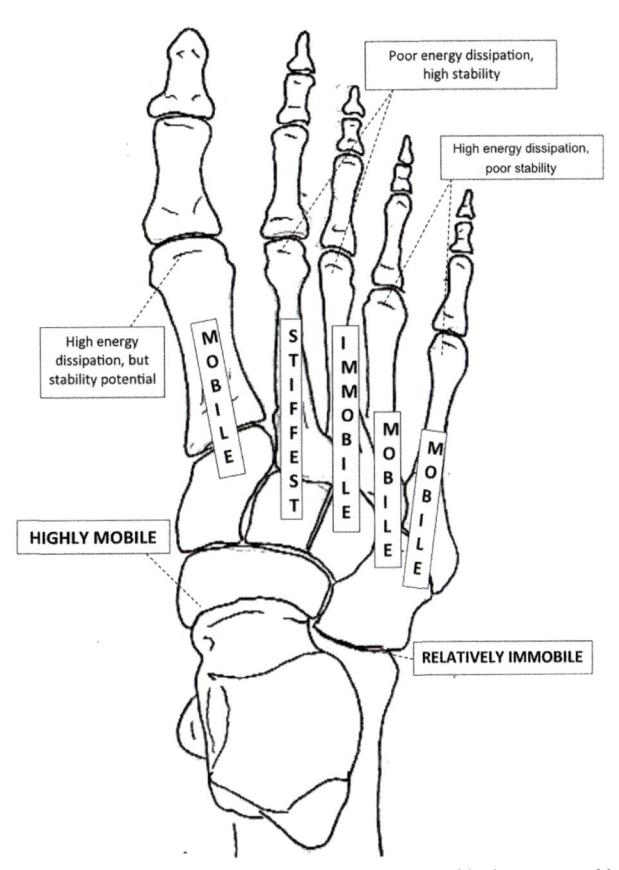

FIG. 3.3.10e   The mobility at the tarsometatarsal joints has important influences on how stable the metatarsal heads are on forefoot loading, midstance weightbearing, and terminal stance acceleration. Mobility and stability also affect how effective each metatarsal is in supplying energy dissipation from articular and osseous motion to deform and tension soft tissues across the midfoot and forefoot. High midfoot mobility is provided proximally at the talonavicular joint, but far less so at the calcaneocuboid joint. This means that the proximal vault is more stable laterally yet better at energy dissipation medially. Because the lateral tarsometatarsal joints are more mobile, forefoot loading on the lateral metatarsal heads initially provides better impact energy dissipation than will the central metatarsals with their stiffer tarsometatarsal joints. Being relatively mobile, the 1st tarsometatarsal joint can also energy-dissipate forefoot impacts, but it is less likely to do so unless the forefoot is rotated more into eversion (valgus) at loading response. During midstance and terminal stance, loading the central metatarsal heads provides a much more stable platform than loading the lateral metatarsals. The 1st metatarsal head provides an ideal acceleration platform as it presents a large weightbearing area and a thick cortical bone shaft for loading stresses. However, 1st tarsometatarsal joint mobility potential reduces its suitability as part of a stable platform. Yet if its mobility is muscularly constrained, it can provide excellent acceleration stability. Tarsometatarsal joint stability is variable between individuals, so such general trends must be compared to actual ranges of motion found during clinical examination of each patient. *(Permission www.healthystep.co.uk.)*

Should metatarsal splay stiffening fail to produce sufficient transverse forefoot stiffness for heel lift, then compensatory mechanisms will need to be activated including greater muscle activity or the permitting of further forefoot spreading before terminal stance. Lack of interosseous ligament protection at the 1st and 5th metatarsal bases makes them more vulnerable to increased splaying in individuals with poor forefoot stiffening mechanics. Such feet may require more forefoot splaying to sufficiently tighten the deep transverse metatarsal ligament to present a stable forefoot platform. However, excessive intermetatarsal spreading at the medial and lateral tarsometatarsal and intermetatarsal base joints could initiate the development of hallux valgus and/or tailor's bunions. This would create an oblique torque strain across the joint that the ligamentous structures are not aligned to resist, especially at the 1st tarsometatarsal joint where the dorsal and plantar ligaments are least obliquely orientated to resist metatarsal abduction from the foot's midline. This may couple with increased frontal plane eversion motions of the 1st metatarsal which seem to increase with hallux abducto valgus deformity (Watanabe et al., 2017; Ota et al., 2019). Some supporting evidence exists between loss of transverse vault stability at the level of the metatarsals and 5th digital varus posture in preschool children (Puszczalowska-Lizis et al., 2022), suggesting that controlling metatarsal splay is likely very important in forming the transverse vault profile of the foot and correctly positioning the digits (Fig. 3.3.10f).

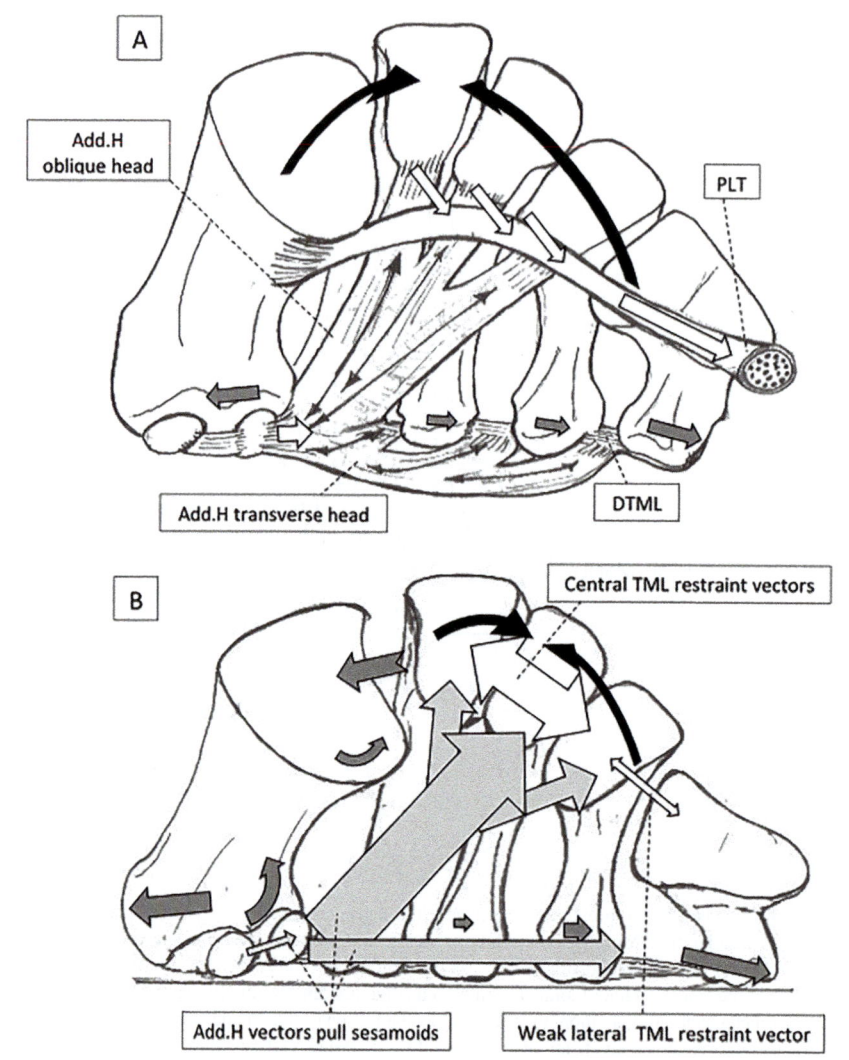

FIG. 3.3.10f Vectors across the tarsometatarsal boundary not only provide stability at their articulations via compression stability *(thin black arrows)* but also influence stability between the MTP joints, coupled transversely with the deep transverse metatarsal ligament (DTML). From forefoot loading until preswing, increasing forefoot stresses directly transfers metatarsal splay into heightened intermetatarsal ligament stiffening elasticity. As the GRF progresses further into the forefoot at the heel lift boundary and into terminal stance, muscle activity should greatly stabilise both the tarsometatarsal and MTP joints. If extrinsic muscles such as peroneus longus (PLT) or flexor hallucis longus fail to apply distal and proximal stabilising vectors to less ligament-restrained joints such as the 1st tarsometatarsal joint, then there is risk of a loss of biotensegral structural stability. This will threaten deformity of the forefoot and create digital postural instability. The schematic (A) of the primary anatomy at play shows the system balanced, with ligaments and muscles working to permit yet also limit forefoot splay while maintaining midfoot restraint of vault bending moments. The situation in (B) demonstrates that if compression vectors across all the tarsometatarsal joints are lost, then imbalances initiate. The more extensive tarsometatarsal ligament (TML) connections between the central metatarsal bases and lesser tarsus, combined with intrinsic muscle vectors such as those of adductor hallucis (Add.H), mean that these metatarsals remain relatively stable (2nd particularly). The less proximally restrained 5th metatarsal is liable to increasingly abduct, altering the line of action of the 5th MTP joint's intrinsic and extrinsic flexor vectors, thus tending to cause digit adduction and inversion (varus rotation). As the 1st metatarsal base is less bound into the lesser tarsus in the transverse plane, medial separation between the 1st and 2nd metatarsal becomes a risk, especially if the medial cuneiform's articular facet is more medially facing. GRF under the 1st MTP joint will dorsiflex and adduct the metatarsal head and evert the 1st metatarsal around its longitudinal axis, both proximally and distally, unless there is sufficient muscular restraint. As the metatarsal head slips medially and rotates, Add.H will pull the sesamoids laterally. This initially acts as a metatarsal head adduction restraint (medial drift restraint), but if sustained, the connective tissues between the metatarsal head and the sesamoids will become lengthened and lax so that the sesamoids increasingly translate laterally in relation to the MTP joint on loading, displacing MTP joint flexor muscle vectors laterally. Thus, the biomechanics of hallux valgus and tailor's bunion start to be explained via tarsometatarsal and MTP joint instability. *(Permission www.healthystep.co.uk.)*

3.3.11 The role of the metatarsal base to head orientations

The orientations between the metatarsal bases through to the metatarsal heads dictate the position and angulation of the MTP joints to their digits for directing flexion and extension moments. The torsion within the metatarsal shafts is aligned as part of the longitudinal and transverse foot vault profile, reflecting the enhanced human MTP joint dorsiflexion capability of around 74°, which is far greater than that of chimpanzees at 57° (Drapeau and Harmon, 2013). Metatarsal head positions need to create MTP joints that have transverse plane axes approximately perpendicular to the support surface, permitting smooth passive digital extension during terminal stance ankle and midfoot plantarflexion. This is particularly important for the 1st and 2nd MTP joints which present the most mechanically efficient and longest lever (Achilles external moment) arms for terminal stance acceleration power. The position of the metatarsal bases at the tarsometatarsal joints sets the angulation required for the shafts to bring the metatarsal heads perpendicular to the ground. The metatarsal shafts descend at distinctly different angles from the lesser tarsus due to their variable tarsometatarsal facet angles. Individually, these reflect the variance in the vault height in all planes. Bearing in mind that feet evolved for interaction with uneven ground, 'perfection' in MTP joint alignment was probably not as important for our ancestors than those who now near-permanently walk on 'urban' flat terrains.

The articular dorsoplantar axes of the medial two metatarsal bases with the lesser tarsus are roughly perpendicular to the ground (Drapeau and Harmon, 2013). The 1st and 2nd metatarsals therefore present little torsion throughout their shafts, reflecting the orientation of the metatarsal heads (Aiello and Dean, 1990: p. 534; Drapeau and Harmon, 2013). As the vault descends on its lateral side, the dorsoplantar axes of the bases rotate slightly, converging plantar–medially so that the metatarsals are twisted about their shafts to permit the plantar surfaces of the metatarsal heads to be more aligned to the support surface (Aiello and Dean, 1990: p. 534; Drapeau and Harmon, 2013).

The 2nd metatarsal usually presents no significant torsional rotation, averaging around 2° inverted, while the 1st has slight eversion torsion around 8° from its base (Drapeau and Harmon, 2008). On loading, the 1st metatarsal everts further, i.e. it inverts towards the shaft of the 2nd metatarsal (Kelso et al., 1982; Watanabe et al., 2017; Ota et al., 2019). This weightbearing rotation of the 1st metatarsal may influence sesamoid stability, which could be compromised in individuals with a higher 1st metatarsal eversion rotation within the shaft. The lateral three metatarsals are highly everted from the varus angulation of their bases by 18° at the 3rd, 20° at the 4th, and 16° at the 5th metatarsal (Drapeau and Harmon, 2013). This eversion rotation thus brings the metatarsal heads into approximate perpendicular alignment to a flat support surface (Fig. 3.3.11).

Metatarsal torsional rotation is not always optimal, demonstrating significant population variation (especially for the 4th and 5th metatarsals), and such wide variation found in humans suggests developmental plasticity (Drapeau and Harmon, 2013). This also raises the possibility of variation in forefoot biomechanics between individuals, which may underlie vulnerability to toe deformities and lie at the origins of pathology such as hallux valgus through increased degrees of 1st metatarsal eversion (Watanabe et al., 2017; Wagner and Wagner, 2020).

3.3.12 The metatarsophalangeal joints and digits as a functional unit

The MTP joints are fulcrums for normal heel lift. The midtarsal break, discussed in detail in Chapter 6 of the companion text 'Clinical Biomechanics in Human Locomotion: Origins and Principles', Section 6.3.8, offers an alternative fulcrum that reduces the significance of MTP joint extension and probably represents an older form of hominin heel lift technique that persists within modern populations. It works, but with energetic costs. Use of the MTP joints as a fulcrum of heel lift requires a stiffened foot vault. Dorsiflexion freedom of the digits at the MTP joints is strongly coupled to the ankle joint, with digital extension (dorsiflexion) and ankle flexion (plantarflexion) angles reflecting each other throughout the terminal stance phase (Willwacher et al., 2014). Midfoot plantarflexion undoubtedly forms part of this joint coupling. The force–velocity relationship of ankle plantarflexor power to metabolic activity and energetics reflects this relationship (Takahashi et al., 2016). Not only do humans use their MTP joints as fulcrums to pivot over, but the MTP joints are also used to apply a GRF against the support surface to drive forward from plantarflexor power-generated torques, creating forces that exceed body weight.

The digits' musculature has to provide a proximally directed force through digital base compression across the MTP joints to provide stability to the fulcrum. This should prevent any anterior shear in the metatarsal heads during heel lift. In this way, the MTP joints function like the hole (box) that the pole-vaulter places the tip of their pole into in order to initiate upward drive. If during this propulsive period the foot should lose stiffness and stability, then the desired fulcrum of the MTP joints will be lost. Thus, the MTP joints need freedom to extend, but applied through a degree of variable stiffened resistance (Fig. 3.3.12a). The best functional fulcrum, because of its length, is provided by the medial MTP joints. Adduction and eversion moments across the midfoot during late midstance and early terminal stance should position

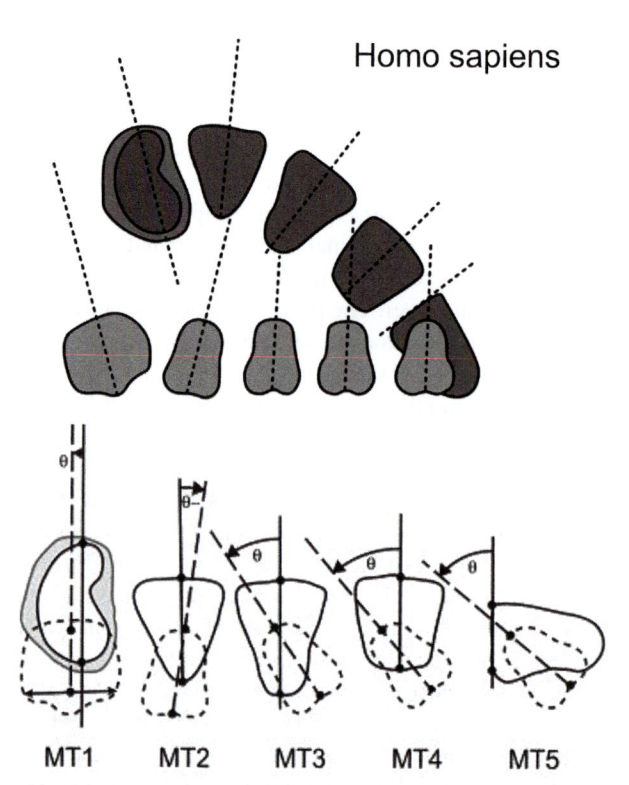

FIG. 3.3.11 Left foot average metatarsal head torsions in the modern human foot (viewed anteriorly). Upper image shows the average alignments in situ for *Homo sapiens*, with an interesting opposition of twist observed between the slightly everted 1st metatarsal head and the slightly inverted 2nd metatarsal head. In the lower image, metatarsal base bisections are represented by solid lines and the heads by *dashed lines*. Positive values represent an eversion rotation, negative values an inversion rotation, and 0° no rotation. Thus, the 1st metatarsal requires no torsion on average between the base and head, the 2nd metatarsal slight eversion to bring the head parallel to the support surface, but the lateral 3 metatarsals require increasing amounts of inversion towards the 5th metatarsal. Variations in these torsions between individuals may be important in explaining risk of digital deformity and forefoot deformity as well as other symptomatic pathologies within the forefoot. *(From Drapeau, M.S.M., Harmon, E.H., 2013. Metatarsal torsion in monkeys, apes, humans and australopiths. J. Hum. Evol. 64 (1), 93–108.)*

the foot in order for it to utilise its primary fulcrum points at the 1st and 2nd MTP joints, improving the mechanical efficiency of ankle power generation (Holowka et al., 2017). The orientation of the combined MTP joint functional axis is created by the differing functional lengths of the metatarsals involved and therefore, length relationships are an important part of gait energetics.

The forefoot soft tissues

The soft tissues of the forefoot play an essential energy management role. Like the heel, the forefoot is subjected to high forces during walking and particularly running gait, being the only pedal weightbearing surface during the terminal stance phase of gait. Thus, the forefoot receives two collision events: one as a part of an initial impact contact during loading response and then again in terminal stance. In forefoot running, it can become the only weightbearing surface throughout stance, taking the whole of the initial foot collision as well the acceleration collision. These loading events and digital dorsiflexion (extension) are also likely to have an effect on venous return from the foot, as soft tissue deformations during gait squeeze the subcutaneous venous plexus which is well developed within the forefoot (Bojsen-Møller and Lamoreux, 1979; Horwood, 2019).

Forefoot soft tissues have been described as having three transverse zones, each with a specific connective tissue frame and a specific role (Bojsen-Møller and Lamoreux, 1979). The most *distal zone* lies under the proximal phalanx of the digits and consists of lamellae of connective tissue transversely orientated and connected to the superficial fibres of the plantar aponeurosis. The *intermediate zone* lies plantar to the MTP joints and demonstrates vertical fibres forming a protective plantar fat pad cushion below each metatarsal head. The *proximal zone* consists of ten sagittal septa of

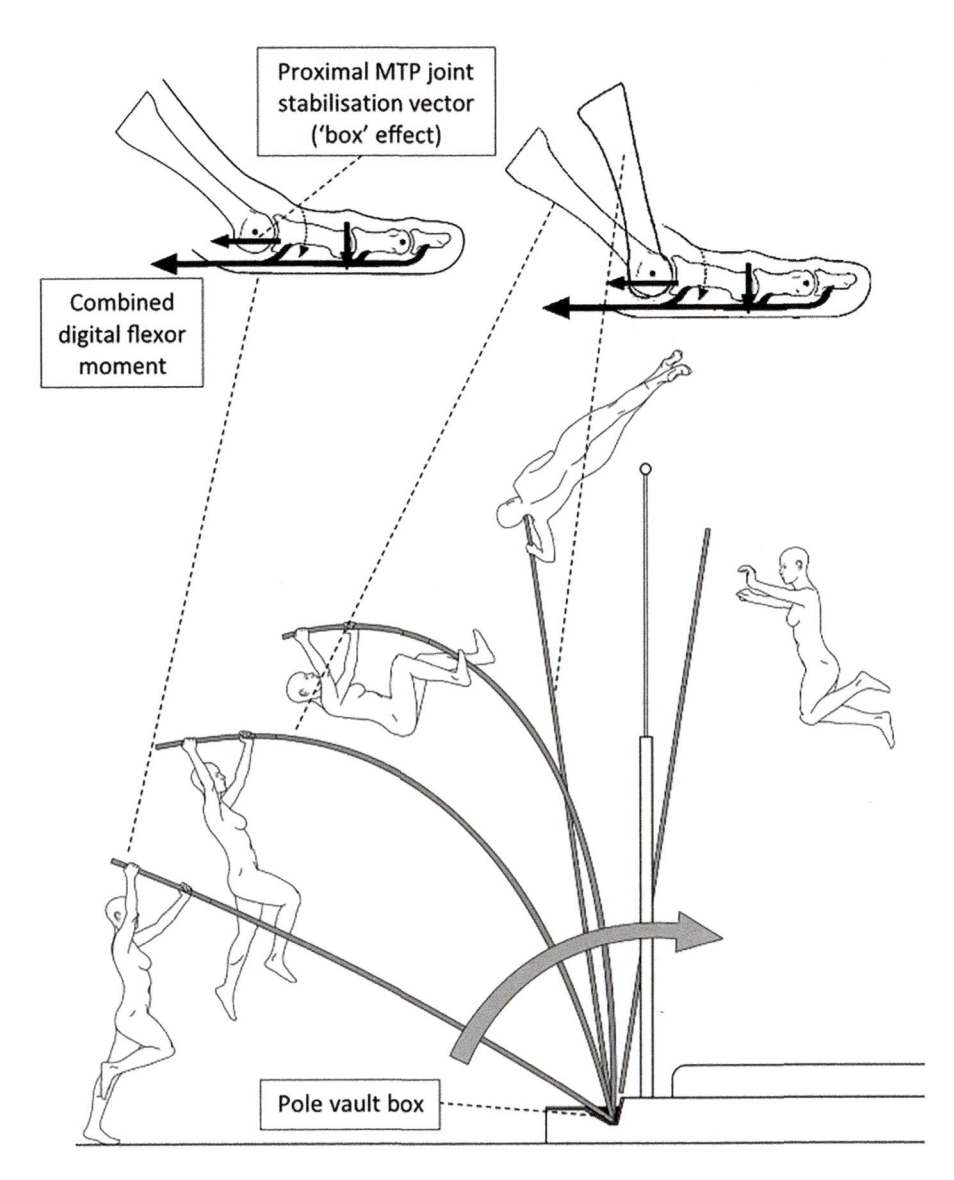

FIG. 3.3.12a Relationships between metatarsal declination angles, the position of the digits, and the forces generated by the digital flexors are appreciated by considering athletic pole vault mechanics. The pole-vaulter needs to position the pole at a declined angle within the pole vault box. This indentation fixes the pole stably into the ground so that the pole cannot displace while loading, allowing the pole and the athlete to increase their rotation over the pole's base at increasing declination angles as the pole becomes more vertical in orientation. For the metatarsal, a similar mechanism is at work. The metatarsal starts to increase its declination from the low point before heel lift, just as terminal stance initiates. The digital flexors generate a flexion moment during late midstance, creating a proximally directed force from each digit against their corresponding metatarsal head. This stabilises the MTP joints against the ground, much like the pole vault's box, giving a stable point of rotation for the metatarsal to pivot over its head as it becomes increasingly declined in orientation until metatarsal loading ceases, briefly before toe-off. Thus, metatarsal declination angles at the end of midstance matter greatly for terminal stance initiation and acceleration mechanics. *(Permission www.healthystep.co.uk.)*

anteriorly orientated lamellae running to the proximal phalanges of the digits and the deep fibres from the plantar aponeurosis also running to the proximal phalanges (Bojsen-Møller and Flagstad, 1976; Bojsen-Møller and Lamoreux, 1979). Fatty tissue is enclosed between the connective tissue septa, giving the forefoot softness and internal strength that links the skin and the proximal phalanges together (see Fig. 3.3.2). As the digits dorsiflex, the connective tissue fibres and the skin tighten up. In doing so, they transform the pliable structure into a stiffer pad that can resist shear forces (Bojsen-Møller and Lamoreux, 1979). By stiffening the soft tissues in such a manner, a more effective soft tissue platform for acceleration is created. The proximal zone is primarily loaded during loading response when it is becoming more compliant after contact through the reversed windlass mechanism, while the proximal, distal, and intermediate zones are under higher loading stress, stiffening during terminal stance under the windlass mechanism (Bojsen-Møller and Lamoreux, 1979).

Becoming compliant during forefoot loading through digital flexion motion out of extension, set during earlier loading response, allows for forefoot impact energy to be dissipated. However, during digital extension, the soft tissues stiffen to ensure that shear forces resulting from accelerations, decelerations, or rotational twisting shear are not absorbed by the skin alone but transferred into the connective tissue frame beneath (Bojsen-Møller and Lamoreux, 1979). Thus, stiffening the forefoot is important for the commencement of forefoot loading impact, but then decreases to dissipate energy via compliance-induced deformation. Shear forces through the more pliable skin across the forefoot soft tissues allow for skeletal motion above the soft tissues. Medial soft tissue pliability is influenced most through the dorsiflexion of the 1st and 2nd digits, while the lateral forefoot soft tissue is influenced most by 2nd through to 5th digital dorsiflexion. However, interestingly, like the lateral palmar surface of the hand, the lateral foot is the least pliable part at rest (Bojsen-Møller and Lamoreux, 1979).

Digital extension/dorsiflexion

Digital dorsiflexion happens twice during the gait cycle. The digits are extended actively before and during early initial contact, usually before forefoot loading (except in forefoot contact), but decrease on forefoot loading. This initial toe dorsiflexion, as part of the reversed windlass effect, gives the advantage of a stiff structure that can become more compliant as the digits move out of extension to make ground contact. Digits passively increase their extension angle throughout terminal stance. During swing, they are maintained in a slightly extended position, but reducing from the high extension angle they find themselves in at the end of terminal stance. Their extension posture is again 'fine-tuned' before initial contact (Fig. 3.3.12b).

FIG. 3.3.12b The forefoot plantar fat pad is split into three zones: proximal zone (PZ), intermediate zone (IZ), and distal zone (DZ). At initial forefoot contact, the PZ bears the greatest stresses. During and after heel strike's fat pad energy dissipation (A), the windlass mechanism is engaged via toe extension from tibialis anterior (TA) and digital extensor (DE) activity, tensioning and shortening the plantar aponeurosis (PA). Tibialis posterior (TP) and peroneus longus (not shown) aid in stiffening and raising the vault that accompanies this action. As a consequence of the windlass mechanism via connective tissue links between the MTP joints and the skin, the forefoot's plantar fat pad and skin are tensioned and stiffened. This creates greater elasticity in the solid (connective tissue) phase and decreased compressibility in the fluid (fat) phase. On forefoot contact (B), the reversed windlass mechanism resulting from the ending of TA and DE activity permits the forefoot plantar fat pad and skin to relax. This allows shear (black arrows) and compression stresses (lightning bolts) to be absorbed into the skin and connective tissues of the fat pad via deformation as part of initial foot lengthening and widening. At this stage, the intermediate zone and particularly the distal zone should not bear extensive stresses as they are anterior to the main loading surface. Toe deformity and higher-heeled shoes may alter the zone loaded at initial contact and prevent the reversed windlass from changing soft tissue properties to more compliance. (Permission www.healthystep.co.uk.)

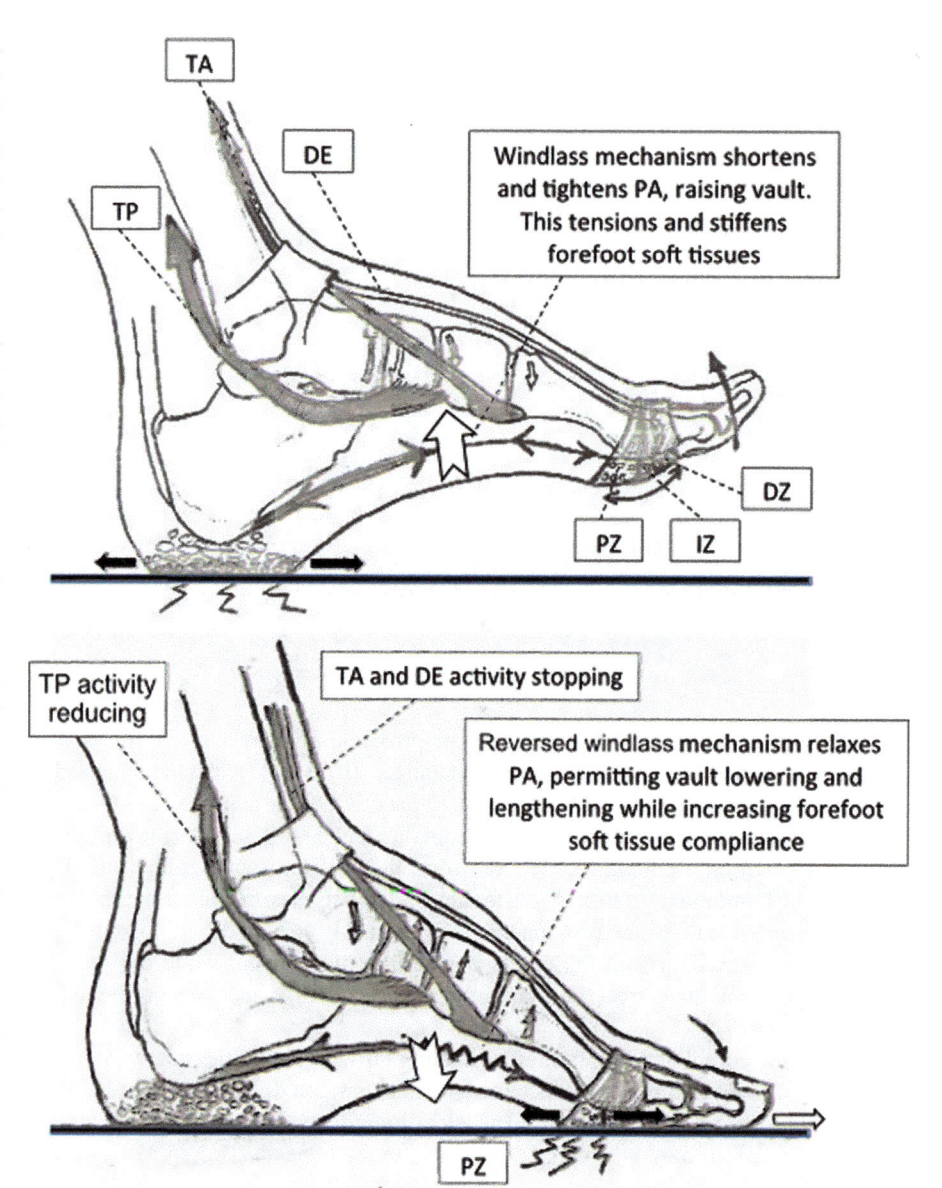

Forefoot energy dissipation occurs through the connective tissues, plantar fat pads, other cutaneous tissues, and MTP joint motion. These mechanisms are lost within the forefoot if soft tissues are not pre-stiffened by digital extension prior to forefoot contact. If not pre-stiffened, they will be too compliant to aid forefoot energy dissipation, and potentially permit excessive levels of shear within the soft tissues. This might initiate skin callus formation or ulceration in patients with tissue viability issues. Yet, an inability to bring the toes out of extension at forefoot loading will increase forefoot loading stiffness at impact, setting up high compression peak pressures and shear on a potentially pathological scale within the forefoot tissues and/or osseous structures locally. Reducing digital extension slackens the forefoot's soft tissues for absorbing energy and reducing soft tissue compression, friction, and shear. Excessive forefoot stiffness can either be caused by the toes remaining in extension or through tissue physiology changes such as are seen in diabetics. At heel strike, the hallux is dorsiflexed (extended) 20–30° in walking gaits, striking the ground around 40–120 ms after the forefoot (Bojsen-Møller and Lamoreux, 1979). It has been reported in healthy young subjects, that the 1st and 5th digits contact first, or all five digits together (Bojsen-Møller and Lamoreux, 1979). Clinically, many different patterns are noted, suggesting that these young healthy patterns can change for various reasons (Fig. 3.3.12c).

Digital dorsiflexion of 35–40° is required to change forefoot skin and underlying connective tissue stiffness by 50%. At forefoot contact, with the digits dorsiflexed 25–30°, forefoot soft tissue mobility has achieved less than half its obtainable restriction during loading of the proximal zone, giving the tissues capacity for compliance (Bojsen-Møller and Lamoreux, 1979). The proximal zone is subjected to a posteriorly directed shear during loading via the posteriorly directed GRF, pulling at its connective tissue fibres which are protected by their anteriorly sloping fibre arrangement at the proximal phalanges (Bojsen-Møller and Lamoreux, 1979).

During terminal stance, all forefoot soft tissues should become stiffened to allow force to be transferred from ankle power to the ground during acceleration from the foot. This stiffness should create significant resistance to MTP joint extension but not actually prevent motion. The appropriate stiffness level should permit a stable rotation of the foot around the digits at the MTP joints, balanced to limit energy dissipated through digital extension and increase energy stored within the connective tissues for improved mechanical efficiency. Digital extension links to degrees of ankle (midfoot) plantarflexion during terminal stance. Freedom of motion at the 1st MTP joints also influences ankle, knee, and hip kinematics during midstance (Hall and Nester, 2004). Maximal digital dorsiflexion of 55–60° is achieved during terminal stance. The distal and intermediate zones of the forefoot fat pad now become loaded. Maximal restriction of skin mobility is already reached at the distal and intermediate zones by 45–50° dorsiflexion, thus creating a firm pad for acceleration-induced shear forces (Bojsen-Møller and Lamoreux, 1979). The morphology of the connective tissue lamellae in the different zones is in keeping with the shear forces they intercept. At terminal stance, the forefoot superficial soft tissues will slide anteriorly. This is restrained by the superficial fibres of the plantar aponeurosis which insert into the distal zone and are able to transmit forces back through it to its calcaneal attachments, relieving peak stresses on the forefoot (Bojsen-Møller and Lamoreux, 1979). Terminal stance loading of the forefoot that induces calcaneal attachment loads on the plantar aponeurosis may be important to plantar fasciopathy symptoms (Fig. 3.3.12d).

The articular surface of all the metatarsal heads is continued further dorsally than is found in either human metacarpal heads or in primate metatarsal heads, due to the large range of dorsiflexion required at the toes (DeSilva et al., 2019). Digital dorsiflexion range of motion is further enhanced by the sulcus of the dorsal surface of the neck, enabling the proximal phalanx to move into full extension without dorsal joint impingement (DeSilva et al., 2019). The axis of each MTP joint is like all other joints in that it has an instantaneous axis that can move with motion and gait speed (Raychoudhury et al., 2014). Although the MTP joints are elongated from dorsal to plantar to create a sagittal axis of flexion–extension, the convex surface on the metatarsal head and the concave surface of the proximal phalanx presents the possibility of some transverse and frontal plane rotation. This is usually highly restrained by both active and passive soft tissue force vectors such as from the collateral MTP ligaments, suspensory metatarsoglenoid ligaments, and balancing forces across the MTP joints. These ligaments and the plantar MTP ligaments form part of the extension resistance forces of this joint (Fig. 3.3.12e).

The digitigrade and unguligrade phases

The terminal stance phase of plantigrade human gait is the 'forefoot-loading-alone' or *digitigrade* phase, lasting around 32% of stance. There is also an *unguligrade* phase when only the toes make contact, lasting around 14% at the end of stance phase during preswing (Bojsen-Møller and Lamoreux, 1979). 'Digitigrade' and 'unguligrade' were descriptive terms selected from the field of comparative anatomy by Bojsen-Møller and Lamoreux in order to describe late terminal stance events. During the digitigrade phase, all the toes rest on the support surface as the heel rotates around 60° over the MTP joints axes. The unguligrade phase begins after forefoot lift so that the foot rotates over an axis running through the tip of the

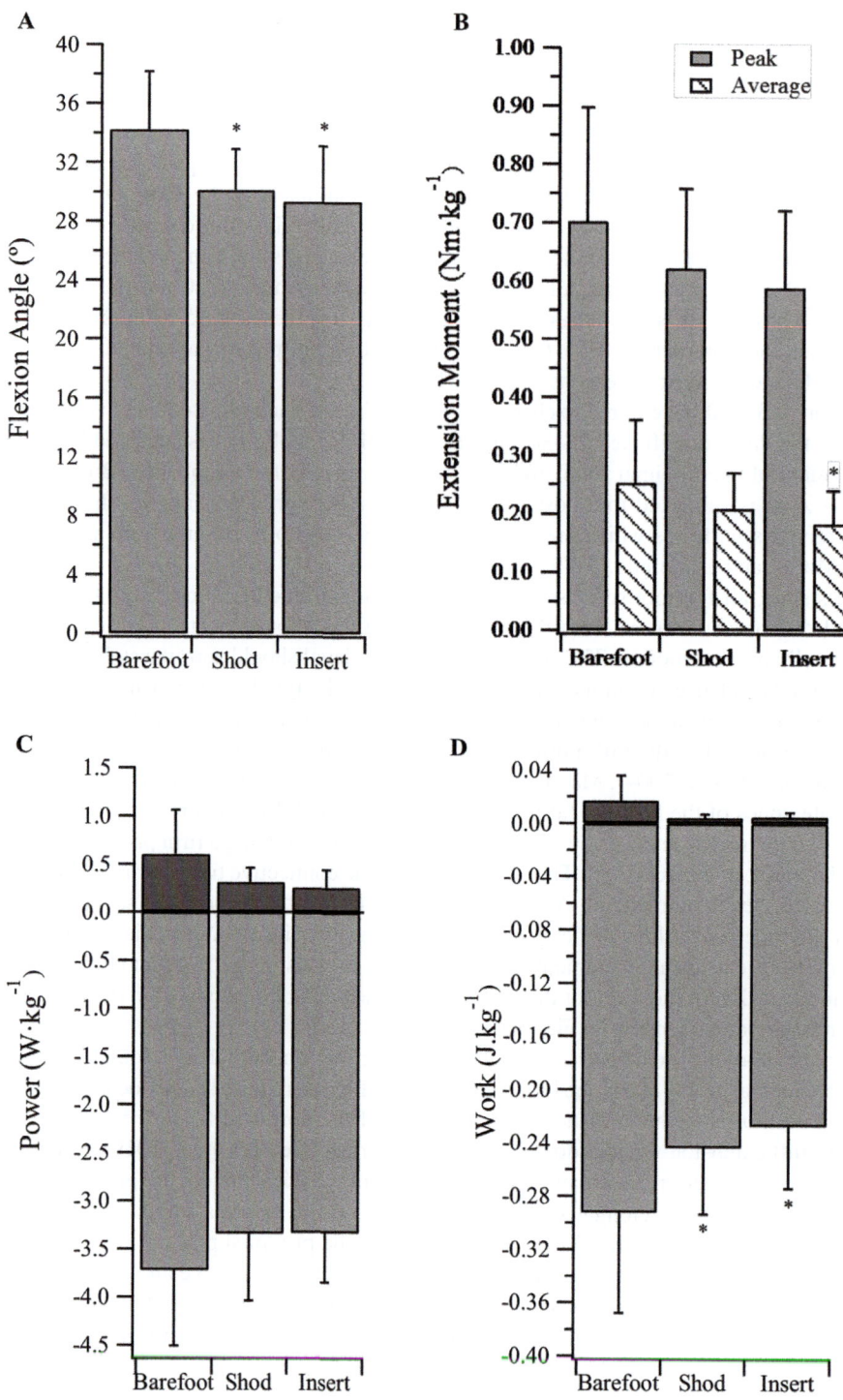

FIG. 3.3.12c Demonstration of the hallux or 1st metatarsophalangeal (MTP) joint's (A) peak flexion angle, (B) peak and average net extension moment, (C) peak positive and negative power, and (D) positive and negative work during the stance phase of running barefoot, shod, and shod with custom foot orthoses (called insert). Significant differences to barefoot results are indicated by *, and significantly different to shod by **. Foot orthoses reduced the extension moment significantly, but did not significantly alter the other parameters, although changes can be seen. The 1st MTP joint is shown to dissipate impact energy, but restricting vault lowering with an orthosis is not compensated for by increasing 1st MTP joint-derived strain through improving the extension range. *(From McDonald, K.A., Stearne, S.M., Alderson, J.A., North, I., Pires, N.J., Rubenson, J., 2016. The role of arch compression and metatarsophalangeal joint dynamics in modulating plantar fascia strain in running. PLoS ONE 11 (4), e0152602.)*

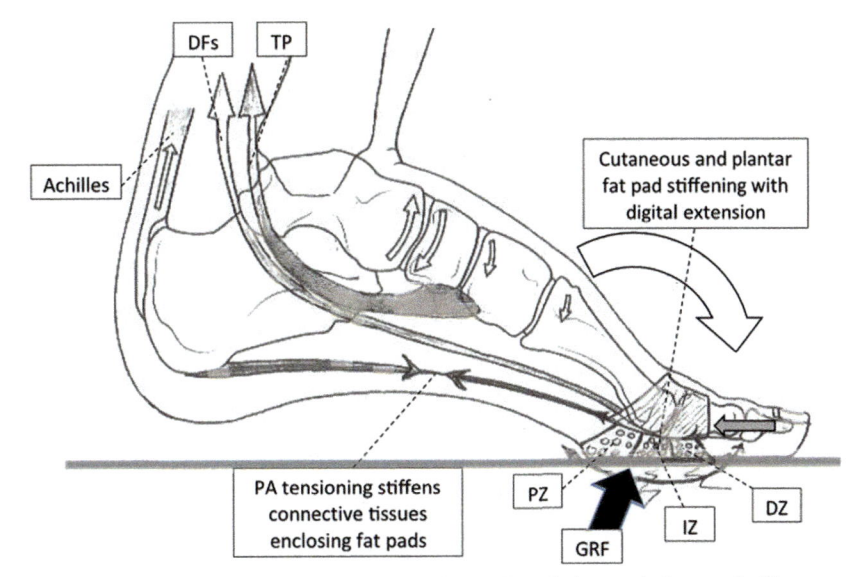

FIG. 3.3.12d The forefoot is required to provide a stable and firm acceleration platform during terminal stance. In this process, the cutaneous tissues and plantar fat pad play their part. During and after heel lift, the intermediate zone (IZ) and distal zone (DZ) sustain the primary weightbearing forces as the proximal zone (PZ) is rapidly being offloaded. The digits become increasingly dorsiflexed at the MTP joints under passive extension moments brought about by ankle and midfoot plantarflexion and the digital GRFs *(black arrow)*. Digital extension is resisted by soft tissue tensions on the plantar forefoot and within the plantar aponeurosis' (PA) heel enthesis. All the digital flexors (DFs—represented only by flexor hallucis longus in image) set up a proximally directed stabilising-compressive force vector into the MTP joints *(grey arrow)*. This tensions muscle activity and generates increased forefoot stiffness. The PA initially continues to lengthen at heel lift, which will stiffen it and its cutaneous slips enclosing the fat pads that are linked to the skin. This action stiffens the fat pad's connective tissues and helps resist anterior shear from the GRF into the skin. As terminal stance progresses, the windlass effect increases PA tensions in a proximal direction which heightens the plantar fat pad and the interconnected cutaneous stiffening mechanisms and shear resistance further. Forefoot stiffening from metatarsal splaying, muscular activity, and cutaneous interactions with the PA together provide a firm interface for the application of ankle plantarflexion power to be applied to the ground. This arises via the Achilles and is aided by tibialis posterior (TP), peroneus longus (not shown), and the long digital flexors. A stiff forefoot allows ankle plantarflexion power to create the high forefoot GRF necessary for stable weight transfer and acceleration during terminal stance. *(Permission www.healthystep.co.uk.)*

hallux and the 2nd toe, allowing the metatarsals to rotate ~90°. This is followed by loss of peak dorsiflexion of the toes at toe-off. This unguligrade period permits the rearfoot to follow a reduced curve (i.e. a straighter path) during ground clearance (Bojsen-Møller and Lamoreux, 1979). The unguligrade phase ends with the tip of the hallux being in contact with the ground. However, in most individuals, it appears that an additional axis can develop at the hallux interphalangeal joint to variable degrees (Phillips et al., 1996). In some individuals, the 1st MTP joint fulcrum is partially replaced by this more distal fulcrum, causing a hyperextension at the interphalangeal joint which may be a functional alternative to a large hallux extension or linked to dysfunction of freedom at the more proximal 1st MTP joint axis (Fig. 3.3.12f).

The 1st MTP joint sesamoids

The human 1st metatarsal is characterised by a large, mediolaterally wide metatarsal head and an articular surface that extends dorsally. The two consistent sesamoid bones of the 1st MTP joint are contained within the flexor hallucis brevis tendon and form a portion of the joint's plantar plate. Plantar plates are the fibrocartilaginous distal attachments for the plantar aponeurosis that insert firmly onto the five proximal phalanges and transversely bind to the deep transverse metatarsal ligament (Dalmau-Pastor et al., 2014; Siddle et al., 2017). These fibrocartilaginous structures are sesamoid cartilage-like and so act for the lesser MTP joints in much the same way as the sesamoid bones do for the 1st MTP joint. From the sesamoids to the proximal phalanx, the *sesamophalangeal* ligaments are situated, which are continuous with the flexor hallucis brevis tendons in much the same way as the quadriceps' tendon becomes the patellar ligament. The sesamoids are referred to as medial and lateral or tibial and fibular, respectively. They articulate on their dorsal articular surfaces with the plantar facets on the plantar aspect of the 1st metatarsal head, which is divided into two by the presence of a bone ridge or *crista*. The crista helps provide stability to the sesamoid complex by creating articular grooves for them to run in, much like the patellar groove on the femur. The sesamoids are linked together transversely by the intersesamoid ligament and are connected to the proximal phalanx through the plantar plate as well as the two sesamophalangeal ligaments continuous with the flexor hallucis brevis tendons (Anwar et al., 2005; Sims and Kurup, 2014).

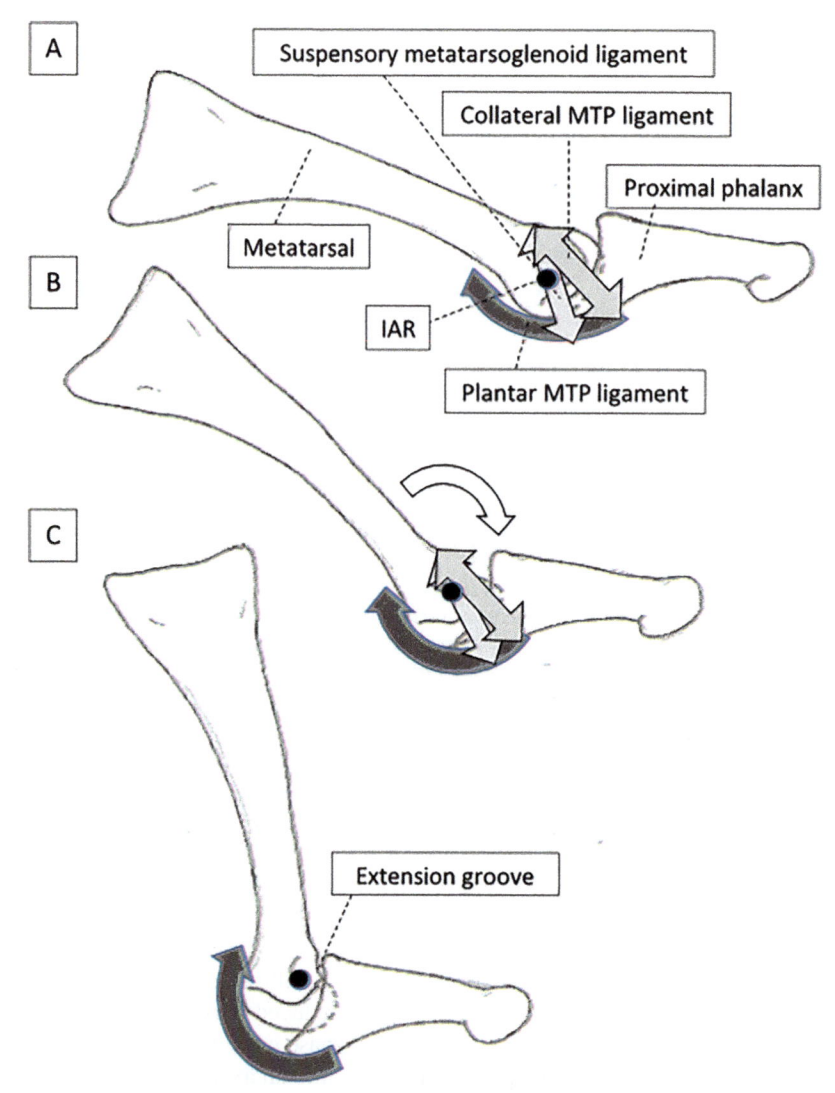

FIG. 3.3.12e The lesser metatarsophalangeal (MTP) joints consist of a sagittally elongated convex metatarsal head that articulates within a concave proximal phalangeal articular facet. These ellipsoidal joints are crudely ball-and-socket-like, potentially permitting transverse and frontal plane rotation as well as their extensive sagittal flexion–extension motions. Frontal and transverse plane movement is minimised by collateral MTP and suspensory metatarsoglenoid ligaments (often fused together as a single structure) that link to the plantar MTP ligaments. These structures bind to the deep transverse metatarsal ligament to form a ligamentous network that functions with other local connective tissues and tendons, maintaining variable compressive stabilisation of the MTP joints throughout motion. By altering the position of the instantaneous axis of rotation (IAR) superiorly during extension, the proximal phalanges can complete a smooth arc of rotation over the metatarsals' elongated facets, despite the changing metatarsal declination angles. At late midstance (A) with the MTP joint at a low angle of extension, the IAR sits relatively centrally within the metatarsal head, influenced by the resting angle of metatarsal declination under connective tissue tensions. In terminal stance (B), increasing functional metatarsal declination angles resulting from ankle and midfoot plantarflexion, move the IAR superiorly within the metatarsal head, allowing the proximal phalanx to rotate over the elongated concave surface of the metatarsal's articular facet during extension. At the end of terminal stance approaching preswing (C), the peak of digital extension may necessitate the use of the extension groove by the superior articular aspect of the proximal phalanx. Normally, MTP joint flexion only occurs out of extension, but the ability to truly flex the MTP joint on uneven terrain when barefoot is an important gripping ability that is rarely utilised on the flat urban surfaces of shod populations. *(Permission www.healthystep.co.uk.)*

The inferior sesamoid surfaces are covered by a thin layer of the flexor hallucis brevis tendon, helping to suspend the sesamoids in a sling-like complex reinforced medially and laterally by suspensory ligaments (joined to the collateral metatarsophalangeal ligaments) and linked together by the intersesamoid ligament (Fig. 3.3.12g(A)). The flexor hallucis longus tendon freely runs distally between the sesamoids passing to the base of the distal phalanx. Abductor hallucis also has a partial tendon attachment to the tibial sesamoid while the adductor hallucis tendon, via both its transverse and oblique muscle bellies, attaches to the fibular sesamoid (Fig. 3.3.12g(B)). Together, they facilitate sesamoid stability, aid hallux plantarflexion, and resist passive dorsiflexion of the 1st MTP joint by giving rise to a proximally stabilising force vector

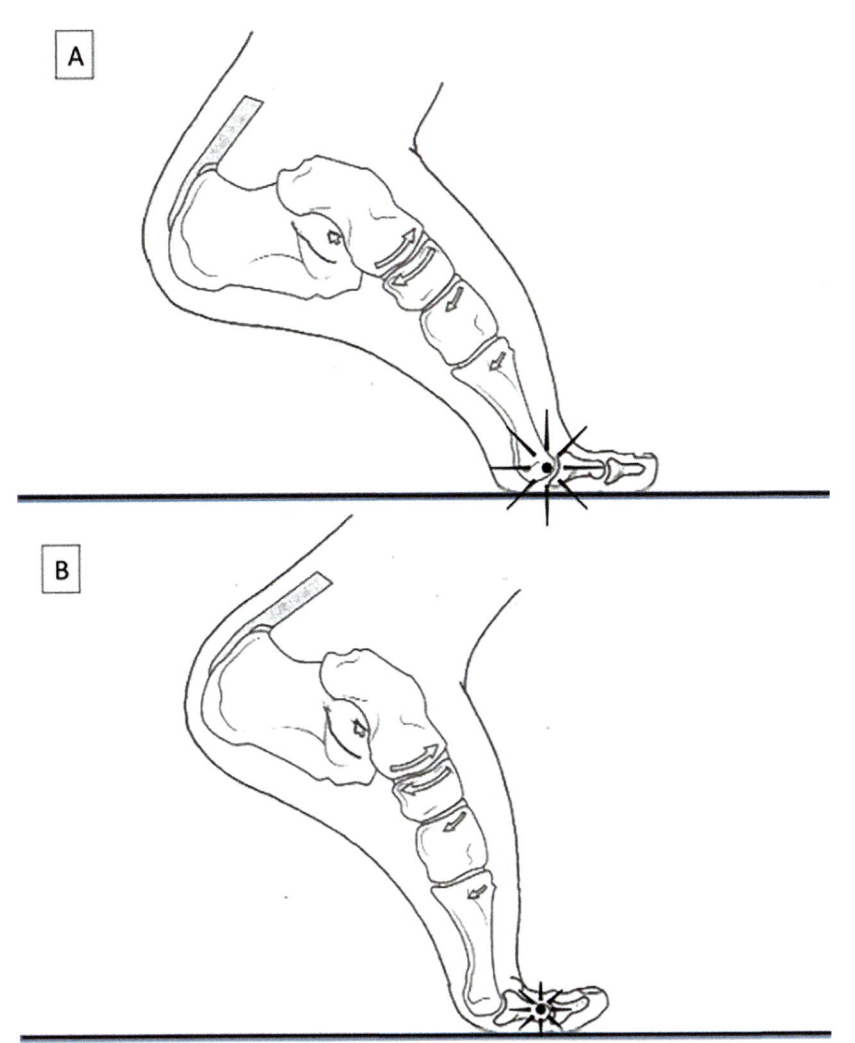

FIG. 3.3.12f The start of terminal stance ends true plantigrade posture of the foot. The foot, using the MTP joint fulcrum (star), is now becoming increasingly digitigrade (A) as the forefoot unloads before it becomes briefly unguligrade (B) around preswing. The digitigrade stage primarily occurs during the weight transfer task in walking gait, with forefoot loads moving increasingly towards the medial side but also decreasing as loads initiate at the contralateral rearfoot. The unguligrade stage occurs after the weight and CoM support phase transfer is completed. It should primarily involve loading the hallux, possibly utilising a distal fulcrum at the interphalangeal joint (IPJ) and sometimes assisted by medial lesser toe-loading. Some individuals use this hallux IPJ fulcrum as an alternative to or as compensation mechanism for any 1st and/or 2nd MTP joint fulcrum extension restrictions during gait. This tends to remodel the joint to permit increased IPJ extension (hyperextension) over time. When running, although weightbearing has not been transferred to the next foot, these stages occur after the CoM has moved anteriorly to the support foot with the unguligrade stage occurring just before the aerial phase. Forefoot running extends the percentage of digitigrade posture during the gait cycle. *(Permission www.healthystep.co.uk.)*

along the axial alignment of the 1st metatarsal shaft. During plantarflexion, the sesamoids move proximally and in dorsiflexion, distally (Anwar et al., 2005; Sims and Kurup, 2014). This enables the sesamoids to work with the plantar plate and the short flexor tendon to limit 1st MTP joint dorsiflexion, preventing proximal phalanx hyperextension.

Potentially, anything that disturbs the force vector to axial alignment of the 1st metatarsal risks dysfunction within the mechanism. This sesamoidal arrangement of anatomy applying vectors to the sesamoids has similarities to the patella, as there is also a vector dominance to the lateral side. At the 1st MTP joint, the abductor hallucis plays much the same role as the oblique section of vastus medialis. Reduced strength and function of abductor hallucis creates a similar result of lateral sesamoid tracking instability from poor medial restraint.

Tibial sesamoids are more prone to pathology than their fibular companions and acute pathologies include inflammation, fracture, and necrosis due to impingement or excessive traction (Sims and Kurup, 2014). As sesamoids can commonly consist of one or more parts, diagnostic imaging must differentiate between such normal bipartite sesamoids and a fracture. Sesamoids are also prone to dislocation and subluxation, especially in the presence of a hallux valgus deformity when the crista may become eroded (Sims and Kurup, 2014). Sesamoiditis can be caused by either cartilage abnormalities similar to that found in chondromalacia of the patella, inflammation of the surrounding soft tissue structures, or the presence of a bursa over the sesamoids resulting from soft tissue shear under the sesamoid joints if sesamoids are prominent within the soft tissues (Sims and Kurup, 2014). Osteoarthritis of the sesamoid's articulations also commonly occurs.

MTP Joint sagittal plane alignments

The MTP joints do not sit in a straight line about the metatarsal heads but extend to different lengths in the anterior–posterior direction, establishing a 'metatarsal arc' in the transverse plane (Demp, 1990, 1998) with considerable

FIG. 3.3.12g The anatomy supporting the sesamoids of the 1st MTP joint is complex. The resultant vectors set up to permit anterior–posterior gliding of the metatarsal head over the sesamoids via their metatarsosesamoid articulations throughout hallux extension and flexion. In (A), the metatarsal head is viewed anteriorly with the proximal phalanx removed and the ligaments sectioned. Frontal plane stability is improved by suspensory and collateral ligaments, with the suspensory ligaments attached to the sesamoids. The intersesamoid ligament provides a stable medial–lateral restraint to help the sesamoids move in unison, further stabilised by the crista or bony ridge on the articular surface. The sesamophalangeal ligaments are the connective tissue continuations of the tendons of flexor hallucis brevis that attach to the sesamoids. They functionally act as the distal attachments for this muscle as well as abductor and adductor hallucis to the plantar proximal phalanx. In (B), the foot is viewed from the plantar surface with vectors for adductor hallucis *(light-grey arrows)*, flexor hallucis brevis *(white arrows)*, and abductor hallucis *(dark-grey arrow)* shown in relation to the sesamoids *(black ovals)*. Despite a longer moment arm, abductor hallucis is at mechanical disadvantage to resist lateral sesamoid drift, having less extensive connections to the tibial sesamoid than to the proximal phalanx, unlike the more lateral muscles. Abductor hallucis is also required to oppose two muscles, that together containing four distinct muscle bellies that each create force vectors in a lateral direction to the sesamoids. Similarities to the anatomy and the biomechanical situation at the patella should be considered, but here the sesamoids are presented within a weightbearing surface which is further complicating function. *(Permission www.healthystep.co.uk.)*

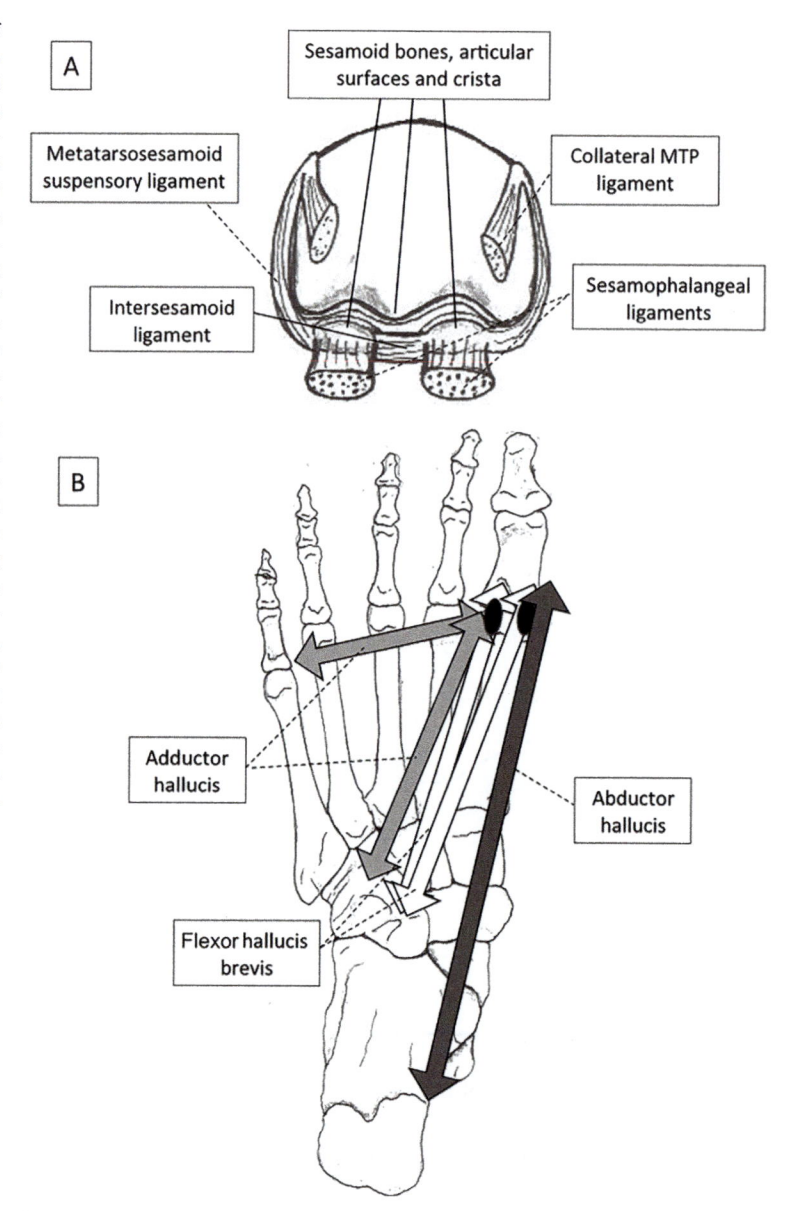

population variability in alignment/orientation (Domínguez et al., 2006; Domínguez-Maldonado et al., 2014). This sets up a curve-shaped trajectory often referred to as a *parabola*. A parabola is a set of points (the focus) in a plane from a given line (the conic section directrix). The parabola curve is created by the locus of points from the directrix. Simply put, imagine a golf ball travelling up into the air following the impact from a golf club swing. It forms an arc of travel, and by looking at the ball at multiple points during that travel, the resulting curve forms a parabola which is an open 'U' shape, symmetrical about its focus. Metatarsals have been commonly referred to as having such a parabola in the orthopaedic literature (Domínguez et al., 2006; Domínguez-Maldonado et al., 2014). However, this is incorrect because as the metatarsal arc is asymmetrical, around the 2nd metatarsal head in particular. It forms at best an ellipse or a hyperbola (Demp, 1990, 1998), although even this shape has been disputed (Robbins, 1981).

Another important consideration is that the relationship between the metatarsals is not fixed in space during gait (Venkadesan et al., 2017a,b). This makes geometric models based on static positions found on radiographs functionally of little relevance. Remember that during gait, the medial and lateral columns move anteriorly at different rates and the more mobile 1st, 4th, and 5th metatarsals can displace more on loading. The role of asymmetrical alignment in metatarsal length is probably related to the tendency for metatarsals to peak on loading at slightly different times during gait. The lateral metatarsals are often loaded first during loading response, where their increased mobility will improve energy dissipation. Getting these lateral metatarsals loaded more quickly on forefoot loading is helped by having them shorter than the more medial metatarsals. During late midstance, weight is increasingly borne by the more stable central metatarsal heads as the CoM travels towards the forefoot. After heel lift, the forefoot bears the full GRF until contralateral foot strike. The midfoot should adduct and evert in terminal stance, shifting the weightbearing surface more medially towards the medial metatarsal heads (Holowka et al., 2017). Metatarsal length influences the ankle power lever arm length (the Achilles external moment arm), so short medial metatarsals would be a disadvantage. Therefore, more distally positioned and stable medial MTP joints are mechanically advantageous.

The effect of the power generated from the ankle plantarflexors is improved by extending metatarsal lengths and is key to the mechanical efficiency of the class two lever system, inducing high gear push-off (Bojsen-Møller, 1979). The metatarsals are not and should not be loaded equally with body weight during terminal stance. The three medial metatarsals are reported to take near to 80% of terminal stance loading, with nearly 60% by the 1st and 2nd MTP joints (Hayafune et al., 1999). The lateral metatarsals do not normally function as fulcrums in late terminal stance and can therefore be shorter without energetic consequences (any reduction in material costs is energetically beneficial but shortening the external plantarflexion lever arm would not be). The 3rd metatarsal is the most likely to be loaded for longer than the other lateral metatarsals and is thus longer than the 4th and 5th. Likewise, the 4th metatarsal is longer than the 5th for similar reasons. Therefore, metatarsal lengths probably reflect in part their likely significance in extending the lever arm length to the fulcrum at acceleration, as well as their importance in forefoot loading. The use of the lateral MTP joints has been linked to the less mechanically efficient low gear propulsion (Bojsen-Møller, 1979), which is occasionally used in rough terrain walking perturbations. In certain dysfunctions such as those with restricted motion in the medial MTP joints, the lateral MTP joints may be the only fulcrum option available (Fig. 3.3.12h).

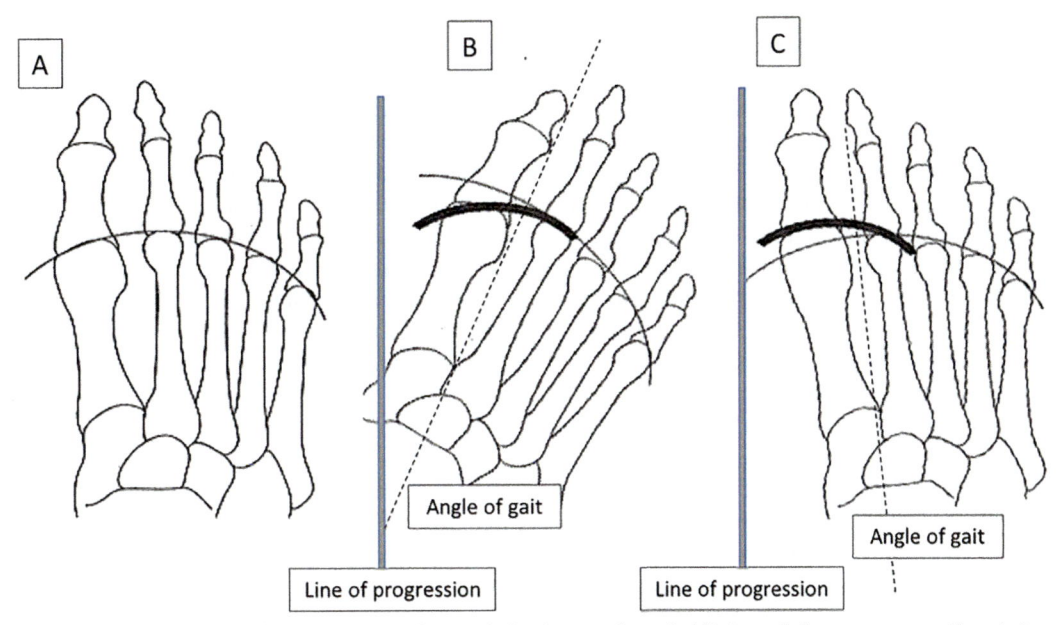

FIG. 3.3.12h Metatarsal lengths rarely if ever present a perfect parabolic shape as shown in (A). Instead, the arrangement of lengths is more an asymmetrical ellipse as demonstrated in (B) or appears as distinct and completely separate arcs (C), with the 1st and 2nd relationships usually providing their own distinct curvature *(thicker black lines)*. This likely reflects the 1st and 2nd metatarsal lengths existing as a separate mechanical relationship to the other lesser metatarsals. Their lengths will position the MTP joints to provide their high gear fulcrum axis for digital extension that they form together in the transverse and frontal plane for sagittal plane rotation. The shorter 1st metatarsal position seen in (B) better suits an abducted angle of gait *(dashed line)* to the line of progression *(thick grey line)*. In this position, the high gear axis aligns at 90° to the line of progression, ideally placed for digital extension. A longer 1st metatarsal protrusion distance (C) suits a straighter or adducted angle of gait to align the high gear axis for digital extension in the direction of the line of progression. In reality, variation between individuals in their metatarsal lengths creates a wide range of metatarsal relationships without causing issue, because they are suitable to the angle of gait the individual presents. Only where large protrusion distance differences occur between metatarsal lengths do repetitive forefoot loading patterns generate forefoot pathology as a direct result of a particular metatarsal and MTP joint bearing loads for longer or at higher peak forces than their tissue properties can sustain without injury. Poorly planned forefoot surgery that disrupts long established metatarsal relationships is a common yet wholly avoidable cause of gait problems. *(Permission www.healthystep.co.uk.)*

The relative lengths of the metatarsals may also play a significant role in creating stiffness across the forefoot, bearing in mind that the deep transverse intermetatarsal ligament is more flexible in its sections lying between the lateral metatarsals and therefore permits more transverse plane splay (Lundgren et al., 2008; Wolf et al., 2008). By loading the shorter lateral metatarsals first, they offer splay compliance, dissipating vertical loads as transverse plane (horizontal) motion which eventually stiffens the lateral metatarsal area under increased forefoot loading during midstance. Changes in uniplanar alignment and the rotations created in the metatarsals by shifting weightbearing load medially to the stiffer intermetatarsal tissues, gives rise to more stable metatarsals. This action will alter tensions within the deep transverse metatarsal ligament, further stiffening these areas for weightbearing stability. In much the same way, this is seen in the variable curvatures created by the fin rays of bony fish, which by tensing the fin membrane, create fin stiffness (Nguyen et al., 2017; Yawar et al., 2017). As is so often the case in nature, similar mechanisms are utilised across species to solve similar mechanical dilemmas. The correct positioning and motion of the metatarsal heads in gait therefore likely plays multiple roles in improving gait efficiency, while also maintaining an ability to adapt to terrain perturbations. This would help to explain why metatarsal rotations have played an important part in human metatarsal evolution.

Clinical interest in the metatarsal arc tends to lie in the magnitude of difference between the lengths of the metatarsals; in particular, how much the longest metatarsal protrudes beyond its neighbours. Most research has looked at the distance differential between the 1st and 2nd metatarsals in the sagittal plane expressed in mm of *protrusion distance*, using several different techniques (Domínguez-Maldonado et al., 2014). Static radiographic techniques as proposed by Hardy and Clapham (1951) are the most commonly used. Significant variation exists within the population between metatarsal length relationships, with females overall displaying larger protrusion distance differences than males (Domínguez-Maldonado et al., 2014). The 1st or 2nd metatarsal are the most protruded and on average, the 1st is positioned slightly longer than the 2nd (Domínguez et al., 2006; Domínguez-Maldonado et al., 2014). A long 2nd metatarsal is defined by a protrusion index of 4 mm longer than the 1st metatarsal and has been noted as a significant risk factor in plantar plate pathology (Fleischer et al., 2017). The relationships of the metatarsal heads between each other, as revealed by examination, give an indication of the alignment of the deep transverse metatarsal ligament which has an important role to play in foot stiffening in the latter stages of stance phase (Yawar et al., 2017). The deep transverse metatarsal ligament alignment across the metatarsal heads combined with the motions of abduction, eversion, and dorsiflexion at the 5th tarsometatarsal joint and 1st tarsometatarsal adduction, inversion, and dorsiflexion may be influencing the splay and torsional transverse metatarsal ligament tightening capacity of the forefoot. This splay-stiffening relationship between metatarsals may be at the heart of conditions such as tailor's bunions and hallux valgus.

Individual anatomical variations and developmental conditions such as brachymetatarsia (shortened metatarsal/metatarsals) affecting one or more metatarsals can severely disturb metatarsal length relationships. Such metatarsal alignment disruptions may reduce the ability of the foot to effectively and appropriately transfer forefoot loads and create stiffening that brings about utilisation of the most efficient fulcrum distance to the ankle plantarflexion power. Positioning metatarsal heads and MTP joints to aid mechanical efficiency in the sagittal and transverse planes is extremely important. However, positions in the frontal plane also affect loading rates, peak forces, and forefoot stiffness. A metatarsal head positioned superiorly or inferiorly to its neighbours can have detrimental effects both in how the fulcrums may move or how the forces load across each metatarsal.

If the medial metatarsal fulcrum is to be aligned to the foot's line of progression, then the relationship of the angle of gait to the transverse plane orientation of the axis of the fulcrum matters. The angle of gait should align the medial MTP joint fulcrum axis perpendicular to the line of progression. Thus, if 1st and 2nd metatarsal lengths are equal, it creates a need for the long axis of the foot to align perpendicular to the MTP joint fulcrum axis, requiring an angle of gait of around 0°. Should the foot have an abducted, externally rotated angle of gait, then the line of progression would be medial and obliquely angled to the fulcrum, inducing some angular moment outside of pure extension moment at heel lift. The larger the angle of gait, the more problematic are the out-of-plane torques on the joint. A short 1st metatarsal forming an axis with a significantly longer 2nd metatarsal would align far better to a more abducted angle of gait.

It is hoped that through developmental plasticity, the medial metatarsal length and angle of gait develop in concert under the influence of gait biomechanics, and thus reflect their requirements to each other. However, should development result in a mismatch in axis to line of progression angles, then problems in joint function and wear may develop. If gait angle is changed through injury or other dysfunctions proximally, the relationship can then be disturbed. For those who develop an in-toed gait with a negative angle of gait through internal lower limb rotations, a lateral MTP joint fulcrum may offer a better fulcrum option for the line of progression, but will result in a less energetically efficient 'low gear' forefoot acceleration.

3.3.13 MTP joints as a stabilising fulcrum

The long and short flexors, aided by quadratus plantae, the lumbricals, and the interossei create a complex mechanism of MTP joint stabilisation through their connective tissue relationships. This connective tissue network includes the extensor hood apparatuses, deep transverse metatarsal ligament, plantar plates, and plantar aponeurosis. MTP joint extension stability results from osseous alignment and soft tissue force vectors. The metatarsal declination angles should align for the forces driving through the foot to be applied appropriately through the metatarsal head and MTP joint for freedom of digital extension. Proximal soft tissue vectors are directed longitudinally across the vault under the metatarsals, changing plantar tensional bending metatarsal moments into plantar osseous compression forces. Thus, vault profile has an intimate relationship with how the forces align through the metatarsal heads. Just before heel lift, the vault should be at its flattest, the metatarsal declination angles at their lowest, the connective tissues stiffened, and muscle tensions increasing. These forces run longitudinally across the foot and through any soft tissues bridging the MTP joints. The muscular contractions and connective tissue restraints will compress the proximal phalanges into the metatarsal heads. These tensile restraints from all sources are also resisting and controlling MTP joint extension moments and rates during terminal stance (Fig. 3.3.13a).

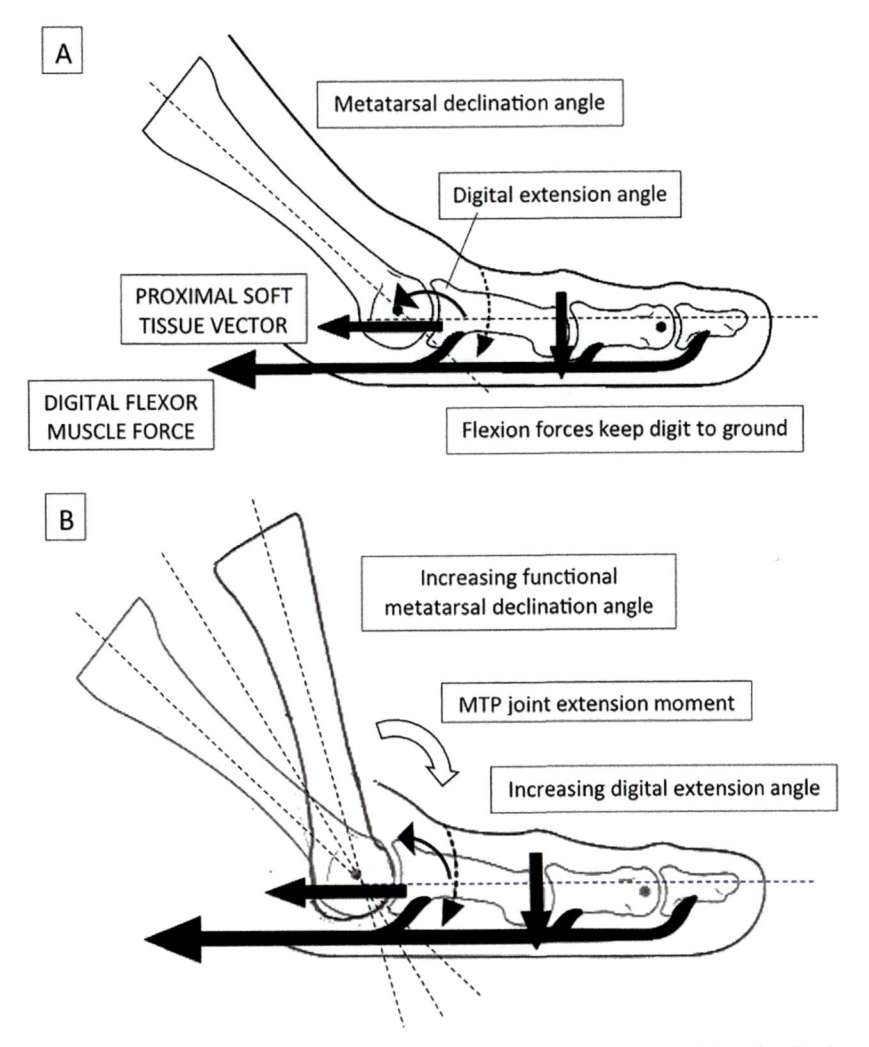

FIG. 3.3.13a The metatarsal declination angle and the proximal phalanx extension angle has a close coupled relationship, bound to the soft tissue resistance of digital extension (A). Together, the soft tissues should not only generate extension resistance but also create a proximally directed compression vector positioned along the long axis of the metatarsal via the articular surface by the proximal phalanx. Digits and metatarsals should splay together to maintain digital compression along each metatarsal's longitudinal axis. During midstance, the extension angle across each MTP joint is set by its weightbearing metatarsal declination angle before heel lift. This angle will reduce under vault sagging deflection as midstance progresses, underlying the need to set limits to vault deflection. The digits should reach their least extended position on the MTP joint just before heel lift. During terminal stance (B), the proximal stabilising compression force needs to increase (under digital flexor activity) and be maintained as the relative metatarsal declination angle increases under ankle and midfoot plantarflexion. This sets up an extension rotation through a moving instantaneous joint axis that permits the digit to ride over the metatarsal head's articular surface against resistance as the metatarsal head rotates over the flexor plate beneath it. The final peak extension angle of each digit before toe-off is set by the functional declination angles of each metatarsal at the end of terminal stance. Thus, planus feet start and likely finish with lower metatarsal declination angles using lower digital extension angles during gait than average type-vaulted feet. Cavoid-vaulted feet will tend to utilise higher digital extension angles throughout gait. Declination angles between metatarsals should create a relatively level weightbearing surface between those that remain loaded at any moment in terminal stance. It is worth remembering that the number of metatarsals loaded is decreasing during terminal stance. *(Permission www.healthystep.co.uk.)*

It is important for the next phase of heel lift that muscles increase metatarsal head compression through their activity during terminal stance, which they are known to do (Ferris et al., 1995; Kelly et al., 2015; Farris et al., 2019, 2020). At heel lift, the vault is low but starting to increase as the foot shortens (Hunt et al., 2001; Stolwijk et al., 2014), increasing the metatarsal declination angle to improve MTP joint freedom of motion. The soft tissues' proximally directed forces prevent anterior displacement of the metatarsal heads as forefoot loading forces rise. This metatarsal head 'distal-blocking' is achieved through the digital bones being pulled proximally by digital flexor muscles (including the interossei and lumbricals) and a tendinofibroaponeurotic structure known as the *extensor apparatus* (Dalmau-Pastor et al., 2014).

Dorsally, this soft tissue complex involves connections with the tendons of extensor digitorum longus and brevis, medially with the three plantar and 1st dorsal interosseous muscles and the four lumbricals, and laterally with the other three dorsal interossei. The extensor apparatus of fibroaponeurotic connective tissue wraps the digital extensors down at the MTP joint, fixing tendons to the instantaneous MTP joint axis during active toe extension (Dalmau-Pastor et al., 2014). At this level, the extensor apparatus is known as the *extensor sling* (Dalmau-Pastor et al., 2014). The digital extensor tendons run to form their rather complex attachments to the dorsal aspects of the bases of the middle and distal phalanges (Fig. 3.3.13b). However, as the digital extensors are of limited functional effect in stance, it is more the plantar structures and their relationship with the flexor tendons that become of interest.

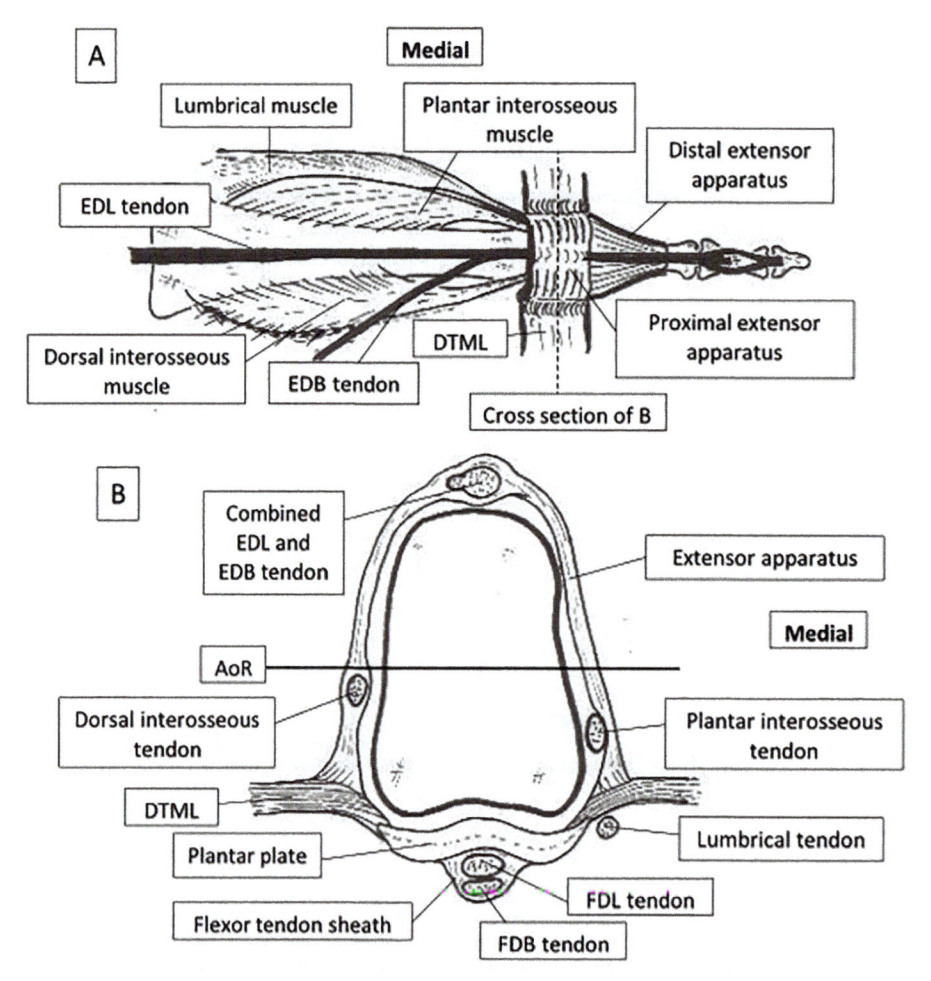

FIG. 3.3.13b A schematic of the right 3rd metatarsal viewed superiorly (A). Complex anatomy ties MTP joints together as a functional unit. Each MTP joint has subtle anatomical differences, with the 3rd and 4th being most similar. Although extensor digitorum longus (EDL) and brevis (EDB) are bound into the extensor apparatus dorsally, their role is limited to extending the digits in preswing and during swing phase for ground clearance, and at initial contact by activating the windlass mechanism for loading response. In cross section and with the metatarsal head viewed anteriorly (B), the extensor apparatus is seen connected firmly to the deep transverse metatarsal ligament (DTML). The connectivity it supplies between the DTML and the surrounding digital flexors gives the extensor apparatus high MTP joint functional importance. Both the dorsal interossei (that usually bind neighbouring metatarsal shafts together proximally via their proximal attachments) and the plantar interossei are connected to the extensor apparatus distally. The lumbrical muscles arising from the flexor digitorum longus tendon slips pass under the DTML and attach to the medial distal extensor apparatus, known as the extensor wing. The DTML links the extensor apparatus to the fibrocartilaginous plate and via this to the tendon sheath of flexor digitorum longus (FDL) and brevis (FDB). The axis of rotation (AoR) is shown but this of course is instantaneous. Thus, its location depends on metatarsal declination and digital extension angles at any moment. *(Permission www.healthystep.co.uk.)*

The plantar plates

The plantar plate (volar plate) is a fibrocartilaginous, broad, ribbon-like disc ranging from rectangular to trapezoidal in shape. It is thickest within its mid-portion, with a smooth grooved plantar surface to form a gliding plane for the flexor tendons that lie beneath (Maas et al., 2016). Its size is dependent on the metatarsal head, but averages at ∼19 mm long, ∼10 mm wide, and ∼2–5 mm thick in adults (Gregg et al., 2006, 2007). Its proximal attachment is to the metatarsal head's periosteum where it is usually thinner and more fibrous. The distal proximal phalanx attachment is thicker and stronger, having multiple interdigitations increasing its attachment surface area (Maas et al., 2016). The flexor tendon sheath attaches to the plantar plate so that the plantar plate can function like a pulley for the flexor tendons, much as the sesamoids do around the 1st MTP joint where the plantar plate is reduced in size. Thus, plantar plate function is important for digital extension.

The plantar plate is mainly composed of type I cartilage proximally, similar to that found in the menisci and the intervertebral discs. However, it is more ligamentous-like distally and demonstrates a greater concentration of fibroblasts (Gregg et al., 2006, 2007; Maas et al., 2016). Collagen fibre orientation within the plate is multi-directional, suggesting a multiplane stabilising and cushioning role, but it has a distinct layering of fibre orientation (Gregg et al., 2007). In the dorsal two-thirds, the fibres run longitudinally, suggesting a sagittal tension resistance role. The plantar one-third runs transversely and blends with the deep intermetatarsal ligament and collateral ligaments, providing a transverse and frontal plane tension resistance zone (Johnston et al., 1994; Deland et al., 1995b; Umans and Elsinger, 2001; Gregg et al., 2007). It is bound on its medial and lateral borders by the deep transverse metatarsal ligament and thus is bound into the plantar fascial slips, collateral, suspensory, and accessory MTP joint ligaments, and the extensor apparatus (Stainsby, 1997; Gregg et al., 2006, 2007; Dalmau-Pastor et al., 2014; Maas et al., 2016) (Fig. 3.3.13c).

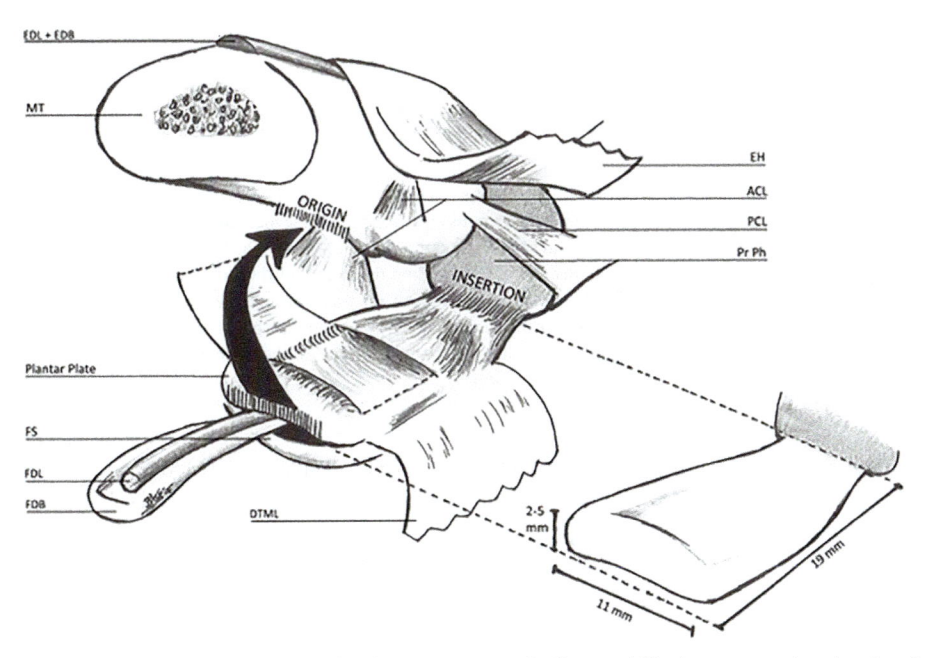

FIG. 3.3.13c A schematic of the plantar plate and its associated anatomy as part of a flexor stabilisation apparatus based on descriptive and histological data. Of note is the weaker proximal attachment to the metatarsal (origin) and much stronger distal attachment (insertion) to the plantar base of the proximal phalanx. *EDL + EDB*, extensor digitorum longus and brevis; *MT*, metatarsal; *EH*, extensor hood (extensor apparatus); *ACL*, accessory extensor ligament; *PCL*, proper collateral ligament; *Pr Ph*, proximal phalanx; *FS*, flexor sheath; *FDL + FDB*, flexor digitorum longus and brevis; *DTML*, deep transverse metatarsal ligament. *(From Maas, N.M.G., van der Grinten, M., Bramer, W.M., Kleinrensink, G.-J., 2016. Metatarsophalangeal joint stability: a systematic review on the plantar plate of the lesser toes. J. Foot Ankle Res. 9, 32.)*

The deep transverse metatarsal ligament connects to the deep slips medially and laterally emanating from the plantar aponeurosis and also to the extensor apparatus. All these interconnections form a longitudinal tie-bar mechanism across the forefoot, binding the distal metatarsals together (Stainsby, 1997). The plantar attachments of the extensor apparatus (extensor hood) are difficult to identify on dissection (Dalmau-Pastor et al., 2014). The more proximal fibres of the extensor apparatus run from superior to inferior over the extensor tendons to attach to the deep transverse metatarsal ligament and plantar plate below, on either side of the MTP joint. This is known as the *extensor sling*. A more distal triangular-shaped segment of the apparatus known as the *extensor wing* runs obliquely, with obliquely arranged fibres running from proximal-inferior to distal-superior over the dorsum of the proximal phalanx. This is continuous with the sling segment proximally (Draves, 1986: pp. 316–317; Dalmau-Pastor et al., 2014). The lumbrical muscles with their muscle bellies attached on the flexor digitorum longus tendons gain attachment distally and medially, forming the oblique border of the extensor wing after passing plantar to the deep transverse metatarsal ligament. The presence of the wing laterally remains a matter of debate (Dalmau-Pastor et al., 2014). Through the attachment to the wing, at least medially to the lumbricals, each lumbricle provides a significant flexor moment on the proximal phalanx at the MTP joint (Fig. 3.3.13d).

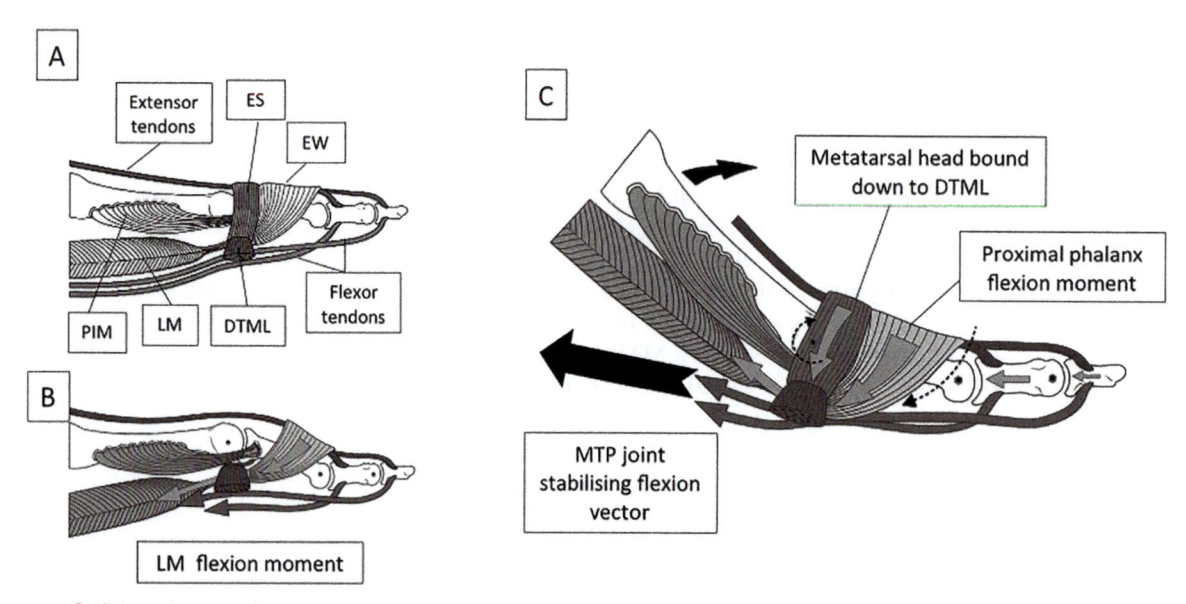

FIG. 3.3.13d Schematics of the 3rd MTP joint viewed medially. Demonstrated in (A) are the plantar interosseous muscle (PIM), the lumbrical muscle (LM), and the extensor and flexor tendons in relation to the extensor apparatus and deep transverse metatarsal ligament (DTML). The extensor apparatus is divided into proximal segments (the extensor sling—ES) and distal segments (extensor wing—EW). The lumbrical's tendon attachment to the medial wing gives the muscle a strong flexion moment (B) by passing under the DTML, aiding the long and short digital flexors. Thus, in terminal stance (C), the lumbricals add an important flexion moment to the proximal phalanx as the metatarsals rotate around the closed chain digit. This anatomy likely relates back to the grasping foot of human ancestors, appreciated most during toe flexion (B), but can now be utilised to help resist and control digital extension while providing proximal stabilisation (C). The hominin functional addition to this anatomy is likely the extensor apparatus linkage into the strong and relatively thick plantar aponeurosis. *(Permission www.healthystep.co.uk.)*

The three plantar and four dorsal interossei occupy the intermetatarsal spaces. The plantar interossei attach proximally to the medial aspect of the 3rd, 4th, and 5th metatarsals inferiorly. The larger dorsal interossei lie in each intermetatarsal space, linking each metatarsal together through their proximal attachments above the plantar muscles (Dalmau-Pastor et al., 2014). The presence of two dorsal interossei and no plantar interossei at the 2nd MTP joint may explain the more frequent 2nd digit dorsal dislocations, because they offer a less significant flexion moment as is explained more in Chapter 4, Section 4.5.7. Although the distal attachments of the interossei are much debated within the literature, they seem to occur on the phalangeal tuberosities on each side of the proximal phalanx and perhaps also to the MTP joint capsule and the distal aspect of the plantar plate (Gregg et al., 2006, 2007; Dalmau-Pastor et al., 2014; Maas et al., 2016). The dorsal interossei attach to the lateral tuberosity of the proximal phalanges, except for the 1st dorsal interosseous muscle which attaches medially. The plantar interossei attach to the medial side of their respective metatarsals. Thus, the 5th proximal phalanx has no laterally attached interossei, but has the attachments of the abductor digit minimi and flexor digiti minimi brevis tendons instead. The anatomy here is small and the differentiation of structures difficult, yet on a functional scale, the directions of the vectors of the muscles through the tissues are more important mechanically than the precise attachment details (Fig. 3.3.13e).

The long and short digital flexors pass beneath the plantar plate on their way to attach to the distal and middle phalanges, respectively. Moreover, their flexor sheath is attached to the plantar plate. The extrinsic flexors work isometrically during most of single-limb support stance phase (Hofmann et al., 2013), restraining vault lengthening in late stance and helping to maintain compression across the foot longitudinally as the vault shortens and the ankle plantarflexes. However, flexor muscle activity at the MTP joints does not account for flexion forces seen at these joints in gait, and maximum flexor muscle

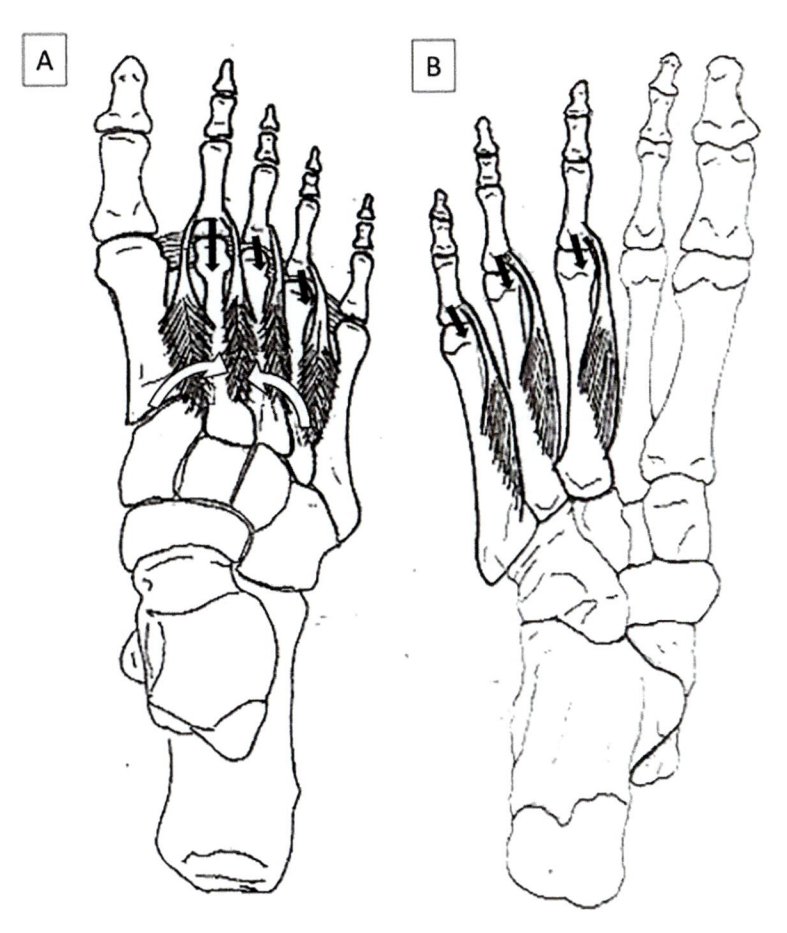

FIG. 3.3.13e A schematic of the anatomy of the interosseous muscles of the foot. Both dorsal (A) and plantar (B) interosseous muscles have their tendons passing superior to the deep transverse metatarsal ligament and are thus tendons that are relatively free to move dorsally during digital extension. Viewed from the dorsum as in (A), the four dorsal interossei, with the exception of the 1st, attach to the lateral side of the proximal phalanx with connections to the extensor apparatus. The 1st dorsal interosseous muscle attaches medially to the 2nd digit's proximal phalanx. Proximally, they attach to adjacent sides of each metatarsal, giving them the capacity to draw the metatarsals together and help dome the vault proximally when active. Distally, they assist flexion and aid the proximally directed vector that stabilises the MTP joints. With the exception of the 1st dorsal interosseous muscle, the force applied via their tendon is in a lateral–proximal direction. The plantar interossei set up a medial-proximal vector, with their flexor force likely to be slightly greater than the dorsal muscles, but without the additional benefit of direct metatarsal-to-metatarsal binding. With the lumbricals, a capacity to fine-tune the position of the proximally directed force vector via interossei in aligning digits to the long axis of the metatarsal is a useful ability, particularly when gait is on uneven support surfaces. *(Permission www.healthystep.co.uk.)*

strength is not normally used (Green and Briggs, 2013). Thus, the flexors do not seem to restrain digital extension alone. It is the flexors' relationship to the extensor sling, wing, plantar plate, and plantar aponeurosis that together stabilise the MTP joints.

The extensors only extend the mass of the digits in swing and less so the mass of the foot. They are aided by the extensor retinaculum at the ankle and extensor apparatus at the MTP joints. At initial stance, the digital extensors provide some plantarflexion moment braking but mainly stiffen the foot via the windlass mechanism, so these connective tissue structures are utilised again in initial loading response to extend the digits. It is the flexors as a part of the functional unit during terminal stance, acting throughout MTP joint extension under high forefoot loads, that are most influential to gait energetics. To stabilise MTP joint extension, the flexors require assistance from the plantar plates, extensor apparatuses, plantar aponeurosis, all the interossei, and the lumbricals. In so doing, they provide important stiffness across the forefoot and the MTP joints that provides adjustable elastic resistance. Perhaps the extensor apparatus should be replaced by the term *flexor plate mechanism*, as this would far better describe its function (Fig. 3.3.13f).

The relationship of MTP joint function to the MTP joint axis in gait

The axis of rotation for MTP joint extension lies centrally within each metatarsal head, but its relationship with phalangeal position is dependent on both the metatarsal's constantly changing declination angle and soft tissue restraint. During early terminal stance, the lumbricals and extensor wings and the interossei provide assisting plantarflexion moments to the MTP joints via the proximal phalanges because their tendons as well as the digital flexor tendons lie inferior to each MTP joint's axis. Digital extension is restrained by the plantar tissue stiffness generated by all the related flexor plate mechanism's anatomy, controlling the rate and extent of digital extension. The tensioning of the plantar plate creates a 'supporting sling' mechanism, preventing the metatarsal from driving down into the fat pad beneath while helping to compress the proximal phalanx into the metatarsal head to create a stabilising proximal vector, preventing anterior 'slip' of each metatarsal head.

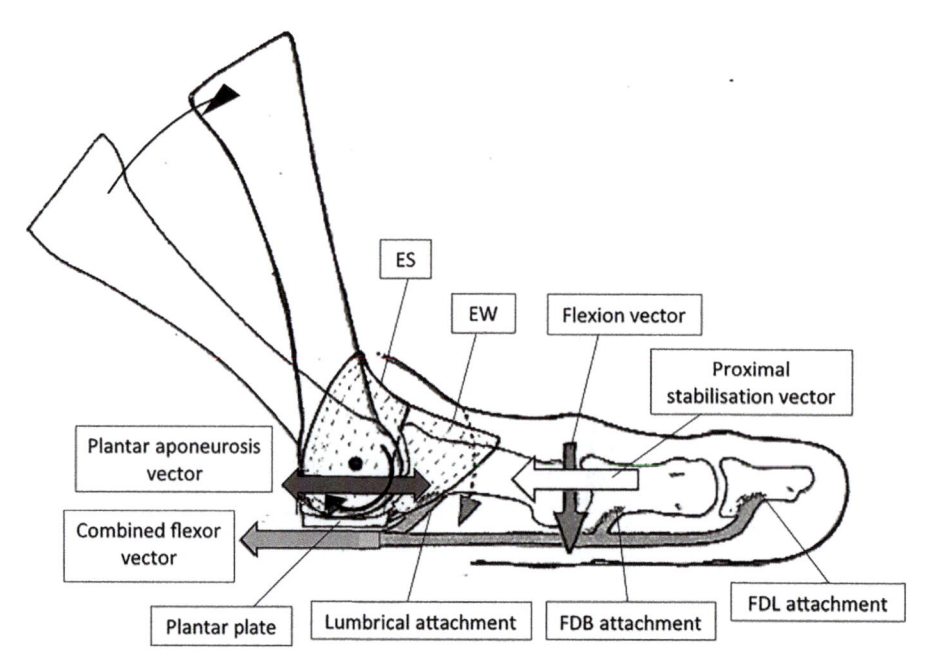

FIG. 3.3.13f A schematic of a proposed functional 'flexor plate mechanism' showing the key anatomical elements at work. The digital flexors (lumbricals, flexor digitorum brevis—FDB, and flexor digitorum longus—FDL) each add force at different locations into the digit to create a combined flexor vector that forms the bulk of proximal stabilisation forces at the MTP joint. This stabilisation vector also contains forces from the interossei (not shown), but more so in early terminal stance as their tendons displace dorsally to the joint axis (black spot) with increasing metatarsal declination angles during late terminal stance. The extensor apparatus sections of the extensor sling (ES) and extensor wing (EW) are utilised to form a plantarflexion moment on the digit, aiding the proximal stabilisation vector as it tightens and then shortens during the windlass mechanism. The lumbrical attachment to the EW also adds an extra plantarflexion moment to the proximal phalanx. The plantar plate uses its fibrocartilaginous nature to aid the rotation glide of the metatarsal head above it, while its inferior fibres bind into the stabilising deep transverse metatarsal ligament (not shown). The plantar plate is attached firmly to the proximal phalanx distally, but its proximal metatarsal head attachment represents a structural weak point. Together, all these structures provide flexion vectors on the digits that control the extension moment. The 1st and 5th MTP joints' distinct anatomy dictate that these joints have some different anatomical structures at play, but the principles are the same. *(Permission www.healthystep.co.uk.)*

Should any part of the soft tissue MTP joint stabilising mechanism fail, then the metatarsal head would no longer receive protection for the high forefoot loading stresses via plantar cushioning. Through loss of correct positioning and restraint, the digit may be too free to extend and as a consequence, either deform in its alignment to the MTP joint or impinge upon the dorsal articular surface, possibly through increased functional metatarsal declination.

In this mechanism, the deep transverse metatarsal ligament is helping through transverse vault stiffening and stabilisation of the forefoot. This is assisted by contraction of the interossei, pulling the shafts of the metatarsals together and forming an increasing transverse curvature in the vault proximally (Fig. 3.3.13g). Through the attachments to the plantar plates and the plantar aponeurosis' deep slips, the deep transverse metatarsal ligament creates a strong interconnected soft tissue anchor. It is important that stability is increased because the attachment of the plantar plate is stronger at the base of the proximal phalanx than it is at the metatarsal head (Maas et al., 2016). Thus, deep transverse metatarsal ligament-to-plantar plate attachments improve proximal tethering, thereby preventing excessive plantar plate anterior displacement from occurring with digital extension. However, this mechanism presents vulnerability in feet with particularly high metatarsal declination angles, or under weakness within the forefoot musculature (Fig. 3.3.13h).

As digital extension and metatarsal declination increase during terminal stance, the base of the proximal phalanx orientates so that the interosseous tendons become more superior to the axis of rotation, changing them to extensors. The timing of this change is important to maintain joint stability. It is likely that the dorsal interossei become extensors first followed by the plantar interossei under increasing extension angles. This is only a problem if this happens too early, possibly as a result of excessively high metatarsal declination angles seen in cavoid feet or if the remaining flexors are too weak, and/or the connective tissue restraints are compromised. In walking, the forefoot GRF should be reducing and already much reduced when the digits approach their greater extension angles through weight transference to the now weight-bearing and supporting contralateral foot. Thus, it should be safe to reduce extension resistance before preswing and allow the interosseous muscles to become extensors. As the next stage of MTP joint function is digital extension during preswing and early swing, the altered action of the interossei may help digital extension for swing (Fig. 3.3.13i).

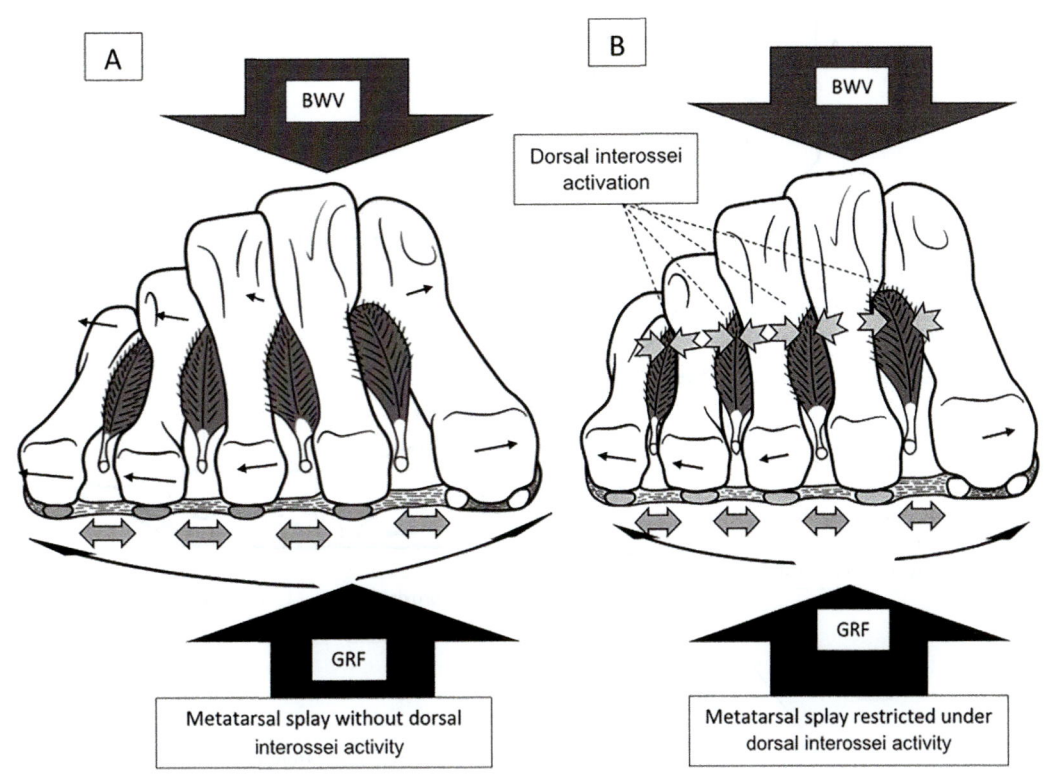

FIG. 3.3.13g Under increasing forefoot loads during late midstance and into terminal stance, derived from the body weight vector (BWV) above and the GRF below (A), metatarsal head splaying will start to increase. These transverse plane forces attempt to separate the metatarsal heads and bases from each other. Some metatarsal translation (1st, 4th, and 5th) will be greater than others, dependent on the tarsometatarsal joint ligament restraints and deep transverse metatarsal ligament (DTML) stiffness between the MTP joints. In excess, there could be disruption of the metatarsal declination angle relationships, causing some metatarsals to load excessively or deficiently or create changes in the instantaneous axis of rotation's motion within the metatarsal head during digital extension. This potentially risks impingement of the articular surfaces or overstrain of vulnerable structures such as the proximal attachment of the plantar plates (shown as dark-grey ovals within the DTML). Splay is limited anteriorly by the DTML, aided by digital flexor mechanics bound to the extensor apparatus. However, by activating the dorsal interosseous muscles (B), an active restraint is provided proximally to maintain metatarsal shaft postural relationships, which in turn further helps to protect MTP joint function. Thus, the dorsal and perhaps even the plantar interossei provide an extra level of forefoot stiffening for acceleration during gait. *(Permission www.healthystep.co.uk.)*

The mechanisms at play at the lesser MTP joints are much the same for the 1st MTP joint and indeed the 5th MTP joint, albeit their anatomy is different. At the 1st MTP joint, the plantar plate's role is largely replaced by the sesamoids and the abductor and adductor hallucis provide a similar role to lumbrical and interossei function. At the 5th MTP joint, abductor digiti minimi and flexor digiti minimi brevis (abductor quinti brevis) take on the role of the lumbricals and interossei. However, both the 1st and 2nd digits seem more prone to frontal and transverse plane deformity, so perhaps the absence of interosseous and lumbrical stability is a weakness for these MTP joints.

If the bones of the medial column do not align sufficiently to allow flexible passive hallux extension under GRF or the MTP joints permit too much extension through lack of stiffness constraint, then impingement of the superior surface of the proximal phalanx onto the superior aspect of the 1st metatarsal head's articular facet becomes a risk. This is true at any MTP joint, yet potential pathological problems at the 1st MTP joint are compounded by the risk of dysfunction of the sesamoids, including excessive articular compression at the sesamoid–metatarsal joints. Degenerative changes within the 1st MTP joint are far more frequent than all of the other MTP joints put together suggesting that despite its size, it is particularly vulnerable to digital extension biomechanical dysfunction.

3.3.14 Section summary

The foot is a complicated 'all-terrain' adaptable structure. This is reflected by the complexity of its anatomy, its variable functions, and its mechanical property behaviours during gait. It follows quite simple cycles of setting and then reducing stiffness for compliance to dissipate impact energy in order to create a braking effect, followed by stiffening mechanisms to form a stable acceleration platform. The functional units within it extensively interrelate to achieve these relatively simple

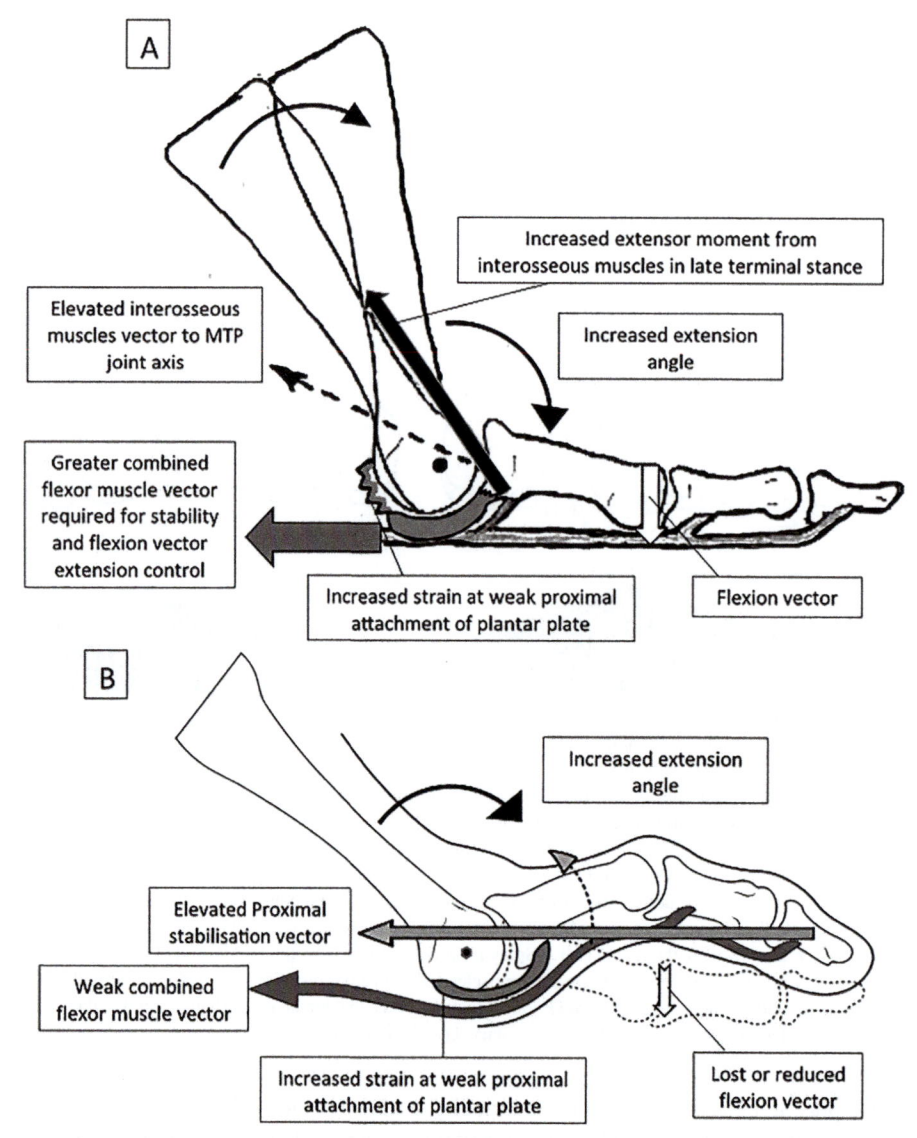

FIG. 3.3.13h The mechanics of the flexor plate mechanism are vulnerable to many dysfunctions as a result of anatomy evolving from grasping appendages. Feet possessing metatarsals with high declination angles (A), either affecting single or all metatarsals (such as seen in pes cavus) or as a consequence of environmental functional positioning (high-heeled shoes), all present particular risks. This is because the MTP joint now works through a range of motion at higher extension angles. Not only does this increase strains on the connective tissues such as the vulnerable proximal flexor plate attachment, but the moment arms of the flexors are also put at increasing mechanical disadvantage as the extension angles increase. In particular, the relative free interosseous muscle tendons can rise above the axis of rotation, turning them into proximal phalanx extensors with greater extension capability in later terminal stance. This change in vectors around the MTP joint can start to cause contraction deformity and digital buckling. Weak flexor muscular vectors can also initiate toe deformation, the type dependent on whether a single flexor or all flexors fail. Any digital buckling that results (B) can shift the proximal stabilisation vector above the axis of rotation of the MTP joint, causing an extension moment at the proximal phalanx as the functional declination angle is increasing during terminal stance. Thus, the MTP joint operates at higher extension angles, increasingly stressing the connective tissue restraints of digital extension. The proximal attachment of the flexor plate is particularly vulnerable. Thus, control of the functional declination angles of the metatarsals and extension motion of the digits are essential to maintain effective long-term MTP joint function. (*Permission www.healthystep.co.uk.*)

tasks. Function and dysfunction in one unit cannot be appreciated without consideration as to what is going on elsewhere within the foot's other functional units.

The role of each pedal functional unit is to provide appropriate levels of compliance and stiffness to match requirements during gait and activity. They must position and shape the foot so that mechanical efficiency can be maximised to the gait situations as much and as safely as possible. Each part of the foot's anatomy has some part to play in this process, differentiated by the timing in gait that each specific anatomical region loads, and the role required of the foot when each anatomical part is loaded. Injuries which seem minor can result in a significant dysfunction within the foot because of

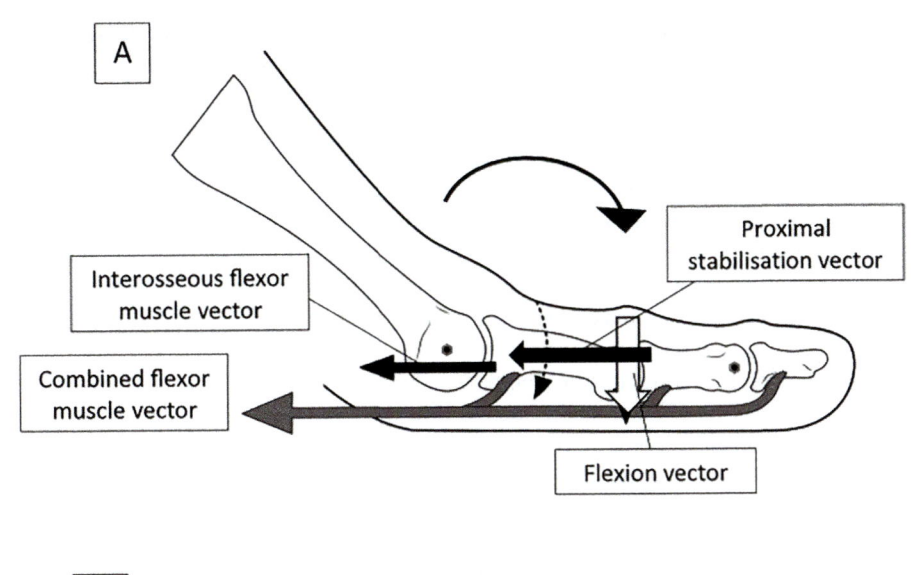

FIG. 3.3.13i The interossei provide a vector to the flexion moment of the digital flexors throughout late midstance and into early terminal stance (A), as their tendons lie below the axis of rotation of the MTP joint. Thus, they assist the proximal stabilisation vector on the metatarsal head. In late terminal stance and preswing, their tendons slip above the axis of rotation to 'transition' them into extensors (B). This alters the proximal stabilisation vector at a time when flexor muscle activity is reducing. The vector starts to tilt into a proximal-superior direction. The timing of this 'transition' to extensors for the interossei and its effect on the metatarsal head stability vector remains important to MTP joint and digital stability. The flexor-to-extensor changeover probably acts as an aid to initiating digital extension in preparation of the digital extensors becoming active during preswing. *(Permission www.healthystep.co.uk.)*

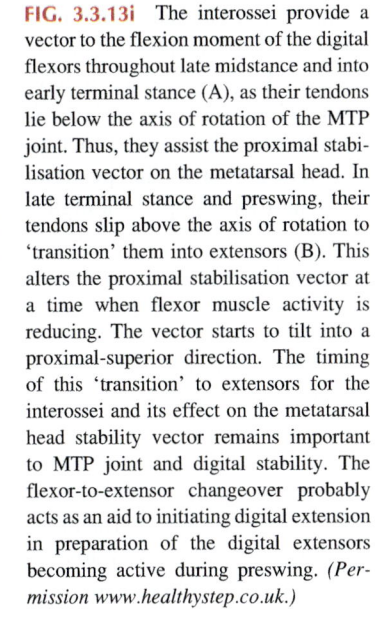

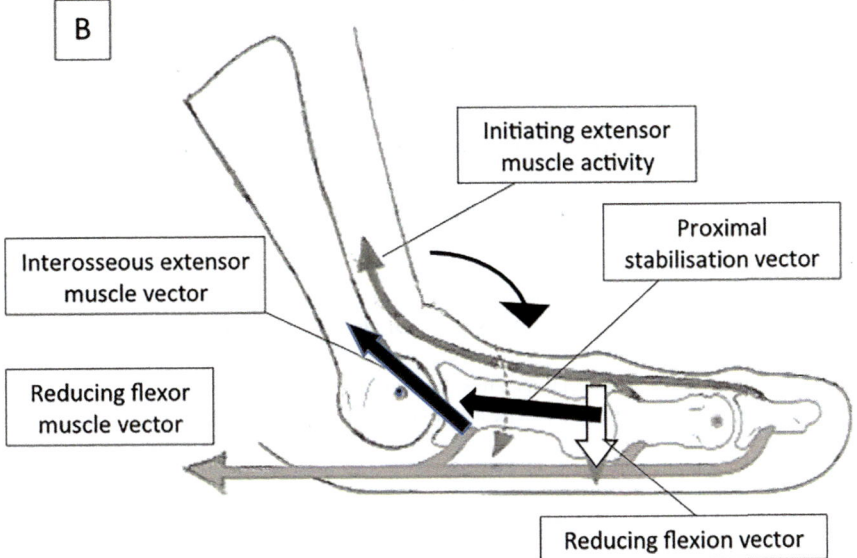

interlinked effects. Loss of the freedom of subtle joint motion at a small joint can result in a chain of restriction or compensatory excessive motion elsewhere that changes the ability of the foot to provide appropriate compliance or stiffness adequately through loss of a capability to adjust shape and properties as required. The capacity to provide stabilising force vectors around joints is essential in creating the right motion around the right joint and functional fulcrum at the right time. The passive and active mechanisms of foot stiffening play their part in setting up the most efficient fulcrum to be used. This is essential for appropriate and safe application of plantarflexion power during acceleration.

Chapter summary

Initial energy dissipation during braking is followed by allowing passage of body mass over the foot. In walking, the ankle rotates, permitting the lower limb to act as an inverted pendulum vaulting over a base of support; an action that should stiffen the foot. Finally, this stiffness across the midfoot provides an ability that helps acceleration of the CoM onto the next step. In running, braking and acceleration occurs without a lengthy vaulting stage of gradual foot stiffening, so the foot must behave more stiffly throughout stance. The foot presents a mechanical, efficiently sized structure than can act as a significantly stable base of support. It assists the lower limb in impulse management during locomotion, adds another level of safe energy dissipation and storage, and provides stiffness when required for elastic energy return. In this capacity, the human foot vault plays a fundamental role through changing its weightbearing dimensions, utilising connective tissue mechanics and intrinsic and extrinsic muscular strength, and causing tensional and shape changes across its structure. However, should the vault mechanics fail such as expressed by midfoot break at heel lift, then the foot can still operate reasonably well through its mechanical tissue properties alone. Only in vault loss combined with significant soft tissue failure is the foot unable to provide any effective safe and stable locomotion, as is seen in diabetes-induced Charcot feet or tibialis posterior and plantar ligament ruptures.

The foot also plays a highly significant role in the sensorimotor system that assists the vestibular system in postural stability. This ability is essential for such an erect and intrinsically unstable gait and posture as humans display. Impulses in gait are co-opted to create soft tissue deformations, which together with muscle contractions help return blood from the limbs as part of cardiovascular health. These mechanisms make feet so much more than a ground-to-body interface structure.

The human plantigrade foot has no other biological structure to compare it with, as it has evolved uniquely in the hominin line which now has only one species left. It is efficient and effective if healthy and well-cared-for, with regular in-life activity helping to keep it fit through biological response. However, the foot is highly vulnerable to ageing, environmental mismatch such as through footwear, and physiological illness. Being at the periphery, its tissues are often the first to express change resulting from diseases of the neurological, endocrine, and cardiovascular systems. This in turn affects the functional and material properties of the foot, resulting in changes in mechanical ability. Our current knowledge on how the foot achieves its functions and how its dysfunctions are reflected proximally into the lower limbs, indicates clearly that there is still much to learn about feet.

References

Aboelnasr, E.A., Hegazy, F.A., Zaghloul, A.A., El-Talawy, H.A., Abdelazim, F.H., 2018. Validation of normalized truncated navicular height as a clinical assessment measure of static foot posture to determine flatfoot in children and adolescents: a cross sectional study. Foot 37, 85–90.

Aerts, P., Ker, R.F., De Clercq, D., Ilsley, D.W., Alexander RMcN., 1995. The mechanical properties of the human heel pad: a paradox resolved. J. Biomech. 28 (11), 1299–1308.

Aerts, P., D'Août, K., Thorpe, S., Berillon, G., Vereecke, E., 2018. The gibbon's Achilles tendon revisited: consequences for the evolution of the great apes? Proc. R. Soc. B Biol. Sci. 285 (1880), 20180859. https://doi.org/10.1098/rspb.2018.0859.

Agostini, V., Nascimbeni, A., Gaffuri, A., Imazio, P., Benedetti, M.G., Knaflitz, M., 2010. Normative EMG activation patterns of school-age children during gait. Gait Posture 32 (3), 285–289.

Aiello, L., Dean, C., 1990. An Introduction to Human Evolutionary Anatomy. Academic Press, London.

Aimonetti, J.-M., Hospod, V., Roll, J.-P., Ribot-Ciscar, E., 2007. Cutaneous afferents provide a neuronal population vector that encodes the orientation of human ankle movements. J. Physiol. 580 (2), 649–658.

Alexander, R.M.C.N., 1984. Elastic energy stores in running vertebrates. Am. Zool. 24 (1), 85–94.

Angin, S., Crofts, G., Mickle, K.J., Nester, C.J., 2014. Ultrasound evaluation of foot muscles and plantar fascia in pes planus. Gait Posture 40 (1), 48–52.

Ankel-Simons, F., 2007. Primate Anatomy: An Introduction, third edition. Elsevier, Burlington, MA.

Anwar, R., Anjum, S.N., Nicholl, J.E., 2005. Sesamoids of the foot. Curr. Orthop. 19 (1), 40–48.

Aquino, A., Payne, C., 2000. The role of the reverse windlass mechanism in foot pathology. Australas. J. Podiatr. Med. 34 (1), 32–34.

Arangio, G.A., Chen, C., Salathé, E.P., 1998. Effect of varying arch height with and without the plantar fascia on the mechanical properties of the foot. Foot Ankle Int. 19 (10), 705–709.

Araújo, V.L., Souza, T.R., Magalhães, F.A., Santos, T.R.T., Holt, K.G., Fonseca, S.T., 2019. Effects of a foot orthosis inspired by the concept of a twisted osteoligamentous plate on the kinematics of foot-ankle complex during walking: a proof of concept. J. Biomech. 93, 118–125.

Arellano, C.J., Gidmark, N.J., Konow, N., Azizi, E., Roberts, T.J., 2016. Determinants of aponeurosis shape change during muscle contraction. J. Biomech. 49 (9), 1812–1817.

Arndt, A., Wolf, P., Liu, A., Nester, C., Stacoff, A., Jones, R., et al., 2007. Intrinsic foot kinematics measured in vivo during the stance phase of slow running. J. Biomech. 40 (12), 2672–2678.

Astion, D.J., Deland, J.T., Otis, J.C., Kenneally, S., 1997. Motion of the hindfoot after simulated arthrodesis. J. Bone Joint Surg. 79-A (2), 241–246.

Aubry, S., Risson, J.-R., Kastler, A., Barbier-Brion, B., Siliman, G., Runge, M., et al., 2013. Biomechanical properties of the calcaneal tendon *in vivo* assessed by transient shear wave elastography. Skelet. Radiol. 42 (8), 1143–1150.

Ball, P., Johnson, G.R., 1996. Technique for the measurement of hindfoot inversion and eversion and its use to study a normal population. Clin. Biomech. 11 (3), 165–169.

Barbaix, E., Van Roy, P., Clarys, J.P., 2000. Variations of anatomical elements contributing to subtalar joint stability: intrinsic risk factors for post-traumatic lateral instability of the ankle? Ergonomics 43 (10), 1718–1725.

Barrett, R., Noordegraaf, M.V., Morrison, S., 2008. Gender differences in the variability of lower extremity kinematics during treadmill locomotion. J. Mot. Behav. 40 (1), 62–70.

Bates, K.T., Collins, D., Savage, R., McClymont, J., Webster, E., Pataky, T.C., et al., 2013. The evolution of compliance in the human lateral mid-foot. Proc. R. Soc. B Biol. Sci. 280 (1769), 20131818. https://doi.org/10.1098/rspb.2013.1818.

Behling, A.-V., Nigg, B.M., 2020. Relationships between the foot posture index and static as well as dynamic rear foot and arch variables. J. Biomech. 98, 109448. https://doi.org/10.1016/j.jbiomech.2019.109448.

Behling, A.-V., Manz, S., von Tscharner, V., Nigg, B.M., 2020. Pronation or foot movement—What is important. J. Sci. Med. Sport 23 (4), 366–371.

Bennett, M.B., Ker, R.F., Alexander RMcN., 1989. Elastic strain energy storage in the feet of running monkeys. J. Zool. 217 (3), 469–475.

Beschorner, K., Cham, R., 2008. Impact of joint torques on heel acceleration at heel contact, a contributor to slips and falls. Ergonomics 51 (12), 1799–1813.

Bjelopetrovich, A., Barrios, J.A., 2016. Effects of incremental ambulatory-range loading on arch height index parameters. J. Biomech. 49 (14), 3555–3558.

Blackwood, C.B., Yuen, T.J., Sangeorzan, B.J., Ledoux, W.R., 2005. The midtarsal joint locking mechanism. Foot Ankle Int. 26 (12), 1074–1080.

Bojsen-Møller, F., 1979. Calcaneocuboid joint and stability of the longitudinal arch of the foot at high and low gear push off. J. Anat. 129 (1), 165–176.

Bojsen-Møller, F., 1999. Biomechanics of the heel pad and plantar aponeurosis. In: Ranawat, C.S., Positano, R.G. (Eds.), Disorders of the Heel, Rearfoot, and Ankle. Churchill Livingstone, Philadelphia, PA, pp. 137–143 (Chapter 9).

Bojsen-Møller, F., Flagstad, K.E., 1976. Plantar aponeurosis and internal architecture of the ball of the foot. J. Anat. 121 (3), 599–611.

Bojsen-Møller, F., Lamoreux, L., 1979. Significance of free-dorsiflexion of the toes in walking. Acta Orthop. Scand. 50 (4), 471–479.

Bonasia, D.E., Phisitkul, P., Saltzman, C.L., Barg, A., Amendola, A., 2011. Arthroscopic resection of talocalcaneal coalitions. Arthroscopy 27 (3), 430–435.

Bonnel, F., Toullec, E., Mabit, C., Tourné, Y., Sofcot., 2010. Chronic ankle instability: biomechanics and pathomechanics of ligaments injury and associated lesions. Orthop. Traumatol. Surg. Res. 96 (4), 424–432.

Bonnel, F., Teissier, P., Colombier, J.A., Toullec, E., Assi, C., 2013. Biometry of the calcaneocuboid joint: biomechanical implications. Foot Ankle Surg. 19 (2), 70–75.

Boyan, N., Ozsahin, E., Kizilkanat, E., Soames, R., Oguz, O., 2016. Morphometric measurement and types of articular facets on the talus and calcaneus in an Anatolian population. Int. J. Morphol. 34 (4), 1378–1385.

Bruckner, J., 1987. Variations in the human subtalar joint. J. Orthop. Sports Phys. Ther. 8 (10), 489–494.

Butler, R.J., Hillstrom, H., Song, J., Richards, C.J., Davis, I.S., 2008. Arch height index measurement system: establishment of reliability and normative values. J. Am. Podiatr. Med. Assoc. 98 (2), 102–106.

Byrne, C.A., O'Keeffe, D.T., Donnelly, A.E., Lyons, G.M., 2007. Effect of walking speed changes on tibialis anterior EMG during healthy gait for FES envelope design in drop foot correction. J. Electromyogr. Kinesiol. 17 (5), 605–616.

Cahill, D.R., 1965. The anatomy and function of the contents of the human tarsal sinus and canal. Anat. Rec. 153 (1), 1–17.

Cappozzo, A., Catani, F., Leardini, A., Benedetti, M.G., Della, C.U., 1996. Position and orientation in space of bones during movement: experimental artefacts. Clin. Biomech. 11 (2), 90–100.

Caravaggi, P., Pataky, T., Goulermas, J.Y., Savage, R., Crompton, R., 2009. A dynamic model of the windlass mechanism of the foot: evidence for early stance phase preloading of the plantar aponeurosis. J. Exp. Biol. 212 (15), 2491–2499.

Caravaggi, P., Pataky, T., Günther, M., Savage, R., Crompton, R., 2010. Dynamics of longitudinal arch support in relation to walking speed: contribution of the plantar aponeurosis. J. Anat. 217 (3), 254–261.

Caravaggi, P., Sforza, C., Leardini, A., Portinaro, N., Panou, A., 2018. Effect of Plano-valgus foot posture on midfoot kinematics during barefoot walking in an adolescent population. J. Foot Ankle Res. 11, 55. https://doi.org/10.1186/s13047-018-0297-7.

Carlson, R.E., Fleming, L.L., Hutton, W.C., 2000. The biomechanical relationship between the tendoachilles, plantar fascia and metatarsophalangeal joint dorsiflexion angle. Foot Ankle Int. 21 (1), 18–25.

Carson, M.C., Harrington, M.E., Thompson, N., O'Connor, J.J., Theologis, T.N., 2001. Kinematic analysis of a multi-segment foot model for research and clinical applications: a repeatability analysis. J. Biomech. 34 (10), 1299–1307.

Cavanagh, P.R., Valiant, G.A., Misevich, K.W., 1984. Biological aspects of modeling shoe/foot interaction during running. In: Frederick, E.C. (Ed.), Sports Shoes and Playing Surfaces: Biomechanical Properties. Human Kinetics, Inc., Champaign, IL, pp. 24–46 (Chapter 2).

Cen, X., Xu, D., Baker, J.S., Gu, Y., 2020. Association of arch stiffness with plantar impulse distribution during walking, running, and gait termination. Int. J. Environ. Res. Public Health 17 (6), 2090. https://doi.org/10.3390/ijerph17062090.

Challis, J.H., Murdoch, C., Winter, S.L., 2008. Mechanical properties of the human heel pad: a comparison between populations. J. Appl. Biomech. 24 (4), 377–381.

Chao, C.Y.L., Zheng, Y.-P., Huang, Y.-P., Cheing, G.L.Y., 2010. Biomechanical properties of the forefoot plantar soft tissue as measured by an optical coherence tomography-based air-jet indentation system and tissue ultrasound palpation system. Clin. Biomech. 25 (6), 594–600.

Chen, W.-M., Lee, P.V.-S., 2015. Explicit finite element modelling of heel pad mechanics in running: inclusion of body dynamics and application of physiological impact loads. Comput. Methods Biomech. Biomed. Eng. 18 (14), 1582–1595.

Cheng, H.-Y.K., Lin, C.-L., Wang, H.-W., Chou, S.-W., 2008. Finite element analysis of plantar fascia under stretch—the relative contribution of windlass mechanism and Achilles tendon force. J. Biomech. 41 (9), 1937–1944.

Cheung, J.T.-M., Zhang, M., An, K.-N., 2006. Effect of Achilles tendon loading on plantar fascia tension in the standing foot. Clin. Biomech. 21 (2), 194–203.

Chino, K., Takahashi, H., 2018. Association of gastrocnemius muscle stiffness with passive ankle joint stiffness and sex-related difference in the joint stiffness. J. Appl. Biomech. 34 (3), 169–174.

Chokhandre, S., Halloran, J.P., van den Bogert, A.J., Erdemir, A., 2012. A three-dimensional inverse finite element analysis of the heel pad. J. Biomech. Eng. 134 (3), 031002. https://doi.org/10.1115/1.4005692.

Cobb, S.C., Tis, L.L., Johnson, J.T., Wang, Y.T., Geil, M.D., McCarty, F.A., 2009. The effect of low-mobile foot posture on multi-segment medial foot model gait kinematics. Gait Posture 30 (3), 334–339.

Collins, D.F., Refshauge, K.M., Gandevia, S.C., 2000. Sensory integration in the perception of movements at the human metacarpophalangeal joint. J. Physiol. 529 (2), 505–515.

Collins, S.H., Wiggin, M.B., Sawicki, G.S., 2015. Reducing the energy cost of human walking using an unpowered exoskeleton. Nature 522 (7555), 212–215.

Cornwall, M.W., McPoil, T.G., 2009. Classification of frontal plane rearfoot motion patterns during the stance phase of walking. J. Am. Podiatr. Med. Assoc. 99 (5), 399–405.

Cowley, E., Marsden, J., 2013. The effects of prolonged running on foot posture: a repeated measures study of half marathon runners using the foot posture index and navicular height. J. Foot Ankle Res. 6, 20. https://doi.org/10.1186/1757-1146-6-20.

Croft, J.L., Bertram, J.E.A., 2020. Form in the context of function: fundamentals of an energy effective striding walk, the role of the plantigrade foot and its expected size. Am. J. Phys. Anthropol. 173 (4), 760–767.

Dai, X.-Q., Li, Y., Zhang, M., Cheung, J.T.-M., 2006. Effect of sock on biomechanical responses of foot during walking. Clin. Biomech. 21 (3), 314–321.

Dalmau-Pastor, M., Fargues, B., Alcolea, E., Martínez-Franco, N., Ruiz-Escobar, P., Vega, J., et al., 2014. Extensor apparatus of the lesser toes: anatomy with clinical implications—topical review. Foot & Ankle International. 35 (10), 957–969.

Deland, J.T., Otis, J.C., Lee, K.T., Kenneally, S.M., 1995a. Lateral column lengthening with calcaneocuboid fusion: range of motion in the triple joint complex. Foot Ankle Int. 16 (11), 729–733.

Deland, J.T., Lee, K.T., Sobel, M., DiCarlo, E.F., 1995b. Anatomy of the plantar plate and its attachments in the lesser metatarsal phalangeal joint. Foot Ankle Int. 16 (8), 480–486.

Deleu, P.-A., Naaim, A., Chèze, L., Dumas, R., Devos Bevernage, B., Goubau, L., et al., 2021. The effect of ankle and hindfoot malalignment on foot mechanics in patients suffering from post-traumatic ankle osteoarthritis. Clin. Biomech. 81, 105239.

Demp, P.H., 1990. Pathomechanical metatarsal arc: radiographic evaluation of its geometric configuration. Clin. Podiatr. Med. Surg. 7 (4), 765–776.

Demp, P.H., 1998. Geometric models that classify structural variations of the foot. J. Am. Podiatr. Med. Assoc. 88 (9), 437–441.

DeSilva, J.M., 2010. Revisiting the "midtarsal break". Am. J. Phys. Anthropol. 141 (2), 245–258.

DeSilva, J.M., Gill, S.V., 2013. Brief communication: a midtarsal (midfoot) break in the human foot. Am. J. Phys. Anthropol. 151 (3), 495–499.

DeSilva, J.M., Bonne-Annee, R., Swanson, Z., Gill, C.M., Sobel, M., Uy, J., et al., 2015. Midtarsal break variation in modern humans: functional causes, skeletal correlates, and paleontological implications. Am. J. Phys. Anthropol. 156 (4), 543–552.

DeSilva, J., McNutt, E., Benoit, J., Zipfel, B., 2019. One small step: a review of Plio-Pleistocene hominin foot evolution. Am. J. Phys. Anthropol. 168 (S67), 63–140.

Di Nardo, F., Ghetti, G., Fioretti, S., 2013. Assessment of the activation modalities of gastrocnemius lateralis and tibialis anterior during gait: a statistical analysis. J. Electromyogr. Kinesiol. 23 (6), 1428–1433.

Dixon, S., Nunns, M., House, C., Rice, H., Mostazir, M., Stiles, V., et al., 2019. Prospective study of biomechanical risk factors for second and third metatarsal stress fractures in military recruits. J. Sci. Med. Sport 22 (2), 135–139.

Doke, J., Donelan, J.M., Kuo, A.D., 2005. Mechanics and energetics of swinging the human leg. J. Exp. Biol. 208 (3), 439–445.

Domínguez, G., Munuera, P.V., Lafuente, G., 2006. Relative metatarsal protrusion in the adult: a preliminary study. J. Am. Podiatr. Med. Assoc. 96 (3), 238–244.

Domínguez-Maldonado, G., Munuera-Martinez, P.V., Castillo-López, J.M., Ramos-Ortega, J., Albornoz-Cabello, M., 2014. Normal values of metatarsal parabola arch in male and female feet. Sci. World J. 2014, 505736. https://doi.org/10.1155/2014/505736.

D'Août, K., Pataky, T.C., De Clercq, D., Aerts, P., 2009. The effects of habitual footwear use: foot shape and function in native barefoot walkers. Footwear Sci. 1 (2), 81–94.

Drapeau, M.S.M., Harmon, E.H., 2008. Metatarsal head torsion in apes, humans, and A. afarensis. [In: Abstracts of AAPA poster and podium presentations]. Am. J. Phys. Anthropol. 135 (S46), 92.

Drapeau, M.S.M., Harmon, E.H., 2013. Metatarsal torsion in monkeys, apes, humans and australopiths. J. Hum. Evol. 64 (1), 93–108.

Draves, D.J., 1986. Anatomy of the Lower Extremity. Williams & Wilkins, Baltimore, MD.

Eichelberger, P., Blasimann, A., Lutz, N., Krause, F., Baur, H., 2018. A minimal markerset for three-dimensional foot function assessment: measuring navicular drop and drift under dynamic conditions. J. Foot Ankle Res. 11, 15. https://doi.org/10.1186/s13047-018-0257-2.

Erdemir, A., Hamel, A.J., Fauth, A.R., Piazza, S.J., Sharkey, N.A., 2004. Dynamic loading of the plantar aponeurosis in walking. J. Bone Joint Surg. 86-A (3), 546–552.

Erdemir, A., Viveiros, M.L., Ulbrecht, J.S., Cavanagh, P.R., 2006. An inverse finite-element model of heel-pad indentation. J. Biomech. 39 (7), 1279–1286.

Eslami, M., Begon, M., Farahpour, N., Allard, P., 2007. Forefoot-rearfoot coupling patterns and tibial internal rotation during stance phase of barefoot versus shod running. Clin. Biomech. 22 (1), 74–80.

Farinelli, V., Hosseinzadeh, L., Palmisano, C., Frigo, C., 2019. An easily applicable method to analyse the ankle-foot power absorption and production during walking. Gait Posture 71, 56–61.

Farris, D.J., Kelly, L.A., Cresswell, A.G., Lichtwark, G.A., 2019. The functional importance of human foot muscles for bipedal locomotion. Proc. Natl. Acad. Sci. U. S. A. 116 (5), 1645–1650.

Farris, D.J., Birch, J., Kelly, L., 2020. Foot stiffening during the push-off phase of human walking is linked to active muscle contraction, and not the windlass mechanism. J. R. Soc. Interface 17 (168), 20200208. https://doi.org/10.1098/rsif.2020.0208.

Ferris, L., Sharkey, N.A., Smith, T.S., Matthews, D.K., 1995. Influence of extrinsic plantar flexors on forefoot loading during heel rise. Foot Ankle Int. 16 (8), 464–473.

Fiolkowski, P., Brunt, D., Bishop, M., Woo, R., Horodyski, M., 2003. Intrinsic pedal musculature support of the medial longitudinal arch: an electromyography study. J. Foot Ankle Surg. 42 (6), 327–333.

Fiolkowski, P., Bishop, M., Brunt, D., Williams, B., 2005. Plantar feedback contributes to the regulation of leg stiffness. Clin. Biomech. 20 (9), 952–958.

Flanigan, R.M., Nawoczenski, D.A., Chen, L., Wu, H., DiGiovanni, B.F., 2007. The influence of foot position on stretching of the plantar fascia. Foot Ankle Int. 28 (7), 815–822.

Fleischer, A.E., Klein, E.E., Ahmad, M., Shah, S., Catena, F., Weil Sr., L.S., et al., 2017. Association of abnormal metatarsal parabola with second metatarsophalangeal joint plantar plate pathology. Foot Ankle Int. 38 (3), 289–297.

Fletcher, J.R., MacIntosh, B.R., 2015. Achilles tendon strain energy in distance running: consider the muscle energy cost. J. Appl. Physiol. 118 (2), 193–199.

Fontanella, C.G., Matteoli, S., Carniel, E.L., Wilhjelm, J.E., Virga, A., Corvi, A., et al., 2012. Investigation on the load-displacement curves of a human healthy heel pad: *in vivo* compression data compared to numerical results. Med. Eng. Phys. 34 (9), 1253–1259.

Forrester, S.E., Townend, J., 2015. The effect of running velocity on footstrike angle—a curve-clustering approach. Gait Posture 41 (1), 26–32.

Foster, A.D., Block, B., Capobianco 3rd, F., Peabody, J.T., Puleo, N.A., Vegas, A., et al., 2021. Shorter heels are linked with greater elastic energy storage in the Achilles tendon. Sci. Rep. 11, 9360. https://doi.org/10.1038/s41598-021-88774-8.

Fouré, A., Cornu, C., McNair, P.J., Nordez, A., 2012. Gender differences in both active and passive parts of the plantar flexors series elastic component stiffness and geometrical parameters of the muscle-tendon complex. J. Orthop. Res. 30 (5), 707–712.

Fraser, J.J., Koldenhoven, R.M., Saliba, S.A., Hertel, J., 2017. Reliability of ankle-foot morphology, mobility, strength, and motor performance measures. Int. J. Sports Phys. Ther. 12 (7), 1134–1149.

Fukano, M., Iso, S., 2016. Changes in foot shape after long-distance running. J. Funct. Morphol. Kinesiol. 1 (1), 30–38.

Gates, L.S., Arden, N.K., McCulloch, L.A., Bowen, C.J., 2015. An evaluation of musculoskeletal foot and ankle assessment measures. Work. Papers Health Sci. 1, 11. https://www.southampton.ac.uk/assets/centresresearch/documents/wphs/LGAn%20evaluation%20of%20musculoskeletal%20foot.pdf.

Gefen, A., Megido-Ravid, M., Itzchak, Y., 2001. In vivo biomechanical behavior of the human heel pad during the stance phase of gait. J. Biomech. 34 (12), 1661–1665.

Gelber, J.R., Sinacore, D.R., Strube, M.J., Mueller, M.J., Johnson, J.E., Prior, F.W., et al., 2014. Windlass mechanism in individuals with diabetes mellitus, peripheral neuropathy, and low medial longitudinal arch height. Foot Ankle Int. 35 (8), 816–824.

Gentili, A., Masih, S., Yao, L., Seeger, L.L., 1996. Pictorial review: foot axes and angles. Br. J. Radiol. 69 (826), 968–974.

Georgoulis, A.D., Moraiti, C., Ristanis, S., Stergiou, N., 2006. A novel approach to measure variability in the anterior cruciate ligament deficient knee during walking: the use of the approximate entropy in orthopaedics. J. Clin. Monit. Comput. 20 (1), 11–18.

Ghanem, I., Massaad, A., Assi, A., Rizkallah, M., Bizdikian, A.J., El Abiad, R., et al., 2019. Understanding the foot's functional anatomy in physiological and pathological conditions: the calcaneopedal unit concept. J. Child. Orthop. 13 (2), 134–146.

Giddings, V.L., Beaupré, G.S., Whalen, R.T., Carter, D.R., 2000. Calcaneal loading during walking and running. Med. Sci. Sports Exerc. 32 (3), 627–634.

Glanzmann, M.C., Sanhueza-Hernandez, R., 2007. Arthroscopic subtalar arthrodesis for symptomatic osteoarthritis of the hindfoot: a prospective study of 41 cases. Foot Ankle Int. 28 (1), 2–7.

Goldberger, A.L., Amaral, L.A.N., Hausdorff, J.M., Ivanov, P.C., Peng, C.-K., 2002. Fractal dynamics in physiology: alterations with disease and aging. Proc. Natl. Acad. Sci. U. S. A. 99 (Suppl 1), 2466–2472.

Gomes, R.B.O., Souza, T.R., Paes, B.D.C., Magalhães, F.A., Gontijo, B.A., Fonseca, S.T., et al., 2019. Foot pronation during walking is associated to the mechanical resistance of the midfoot joint complex. Gait Posture 70, 20–23.

Graf, E., Wannop, J.W., Schlarb, H., Stefanyshyn, D., 2017. Effect of torsional stiffness on biomechanical variables of the lower extremity during running. Footwear Sci. 9 (1), 1–8.

Green, S.M., Briggs, P.J., 2013. Flexion strength of the toes in the normal foot. An evaluation using magnetic resonance imaging. Foot 23 (4), 115–119.

Gregg, J., Silberstein, M., Schneider, T., Marks, P., 2006. Sonographic and MRI evaluation of the plantar plate: a prospective study. Eur. Radiol. 16 (12), 2661–2669.

Gregg, J., Marks, P., Silberstein, M., Schneider, T., Kerr, J., 2007. Histologic anatomy of the lesser metatarsophalangeal joint plantar plate. Surg. Radiol. Anat. 29 (2), 141–147.

Greiner, T.M., Ball, K.A., 2014. The kinematics of primate midfoot flexibility. Am. J. Phys. Anthropol. 155 (4), 610–620.

Griffin, N.L., Miller, C.E., Schmitt, D., D'Août, K., 2015. Understanding the evolution of the windlass mechanism of the human foot from comparative anatomy: insights, obstacles, and future directions. Am. J. Phys. Anthropol. 156 (1), 1–10.

Grigoriadis, G., Newell, N., Carpanen, D., Christou, A., Bull, A.M.J., Masouros, S.D., 2017. Material properties of the heel fat pad across strain rates. J. Mech. Behav. Biomed. Mater. 65, 398–407.

Gruber, K., Ruder, H., Denoth, J., Schneider, K., 1998. A comparative study of impact dynamics: wobbling mass model versus rigid body models. J. Biomech. 31 (5), 439–444.

Gruber, A.H., Boyer, K.A., Derrick, T.R., Hamill, J., 2014. Impact shock frequency components and attenuation in rearfoot and forefoot running. J. Sport Health Sci. 3 (2), 113–121.

Gu, Y., Li, Z., 2012. Mechanical information of plantar fascia during normal gait. Phys. Procedia 33, 63–66.

Gwani, A.S., Adamu, A.A., Salihu, A.T., Rufai, A.A., 2017. Prediction of biological profile from foot dimensions. Could body weight and arch height affect accuracy? J. Forensic Res. 8, 3. https://doi.org/10.4172/2157-7145.1000375.

Hall, C., Nester, C.J., 2004. Sagittal plane compensations for artificially induced limitation of the first metatarsophalangeal joint: a preliminary study. J. Am. Podiatr. Med. Assoc. 94 (3), 269–274.

Hall, J.P.L., Barton, C., Jones, P.R., Morrissey, D., 2013. The biomechanical differences between barefoot and shod distance running: a systematic review and preliminary meta-analysis. Sports Med. 43 (12), 1335–1353.

Hamel, A.J., Sharkey, N.A., Buczek, F.L., Michelson, J., 2004. Relative motions of the tibia, talus, and calcaneus during the stance phase of gait: a cadaver study. Gait Posture 20 (2), 147–153.

Hamill, J., van Emmerik, R.E.A., Heiderscheit, B.C., Li, L., 1999. A dynamical systems approach to lower extremity running injuries. Clin. Biomech. 14 (5), 297–308.

Hamner, S.R., Seth, A., Delp, S.L., 2010. Muscle contributions to propulsion and support during running. J. Biomech. 43 (14), 2709–2716.

Hardy, R.H., Clapham, J.C.R., 1951. Observations on hallux valgus based on a controlled series. J. Bone Joint Surg. 33-B (3), 376–391.

Hargrave, M.D., Carcia, C.R., Gansneder, B.M., Shultz, S.J., 2003. Subtalar pronation does not influence impact forces or rate of loading during a single-leg landing. J. Athl. Train. 38 (1), 18–23.

Hasegawa, H., Yamauchi, T., Kraemer, W.J., 2007. Foot strike patterns of runners at the 15-km point during an elite-level half marathon. J. Strength Cond. Res. 21 (3), 888–893.

Hastings, M.K., Woodburn, J., Mueller, M.J., Strube, M.J., Johnson, J.E., Sinacore, D.R., 2014. Kinematics and kinetics of single-limb heel rise in diabetes related medial column foot deformity. Clin. Biomech. 29 (9), 1016–1022.

Hastings, M.K., Mueller, M.J., Woodburn, J., Strube, M.J., Commean, P., Johnson, J.E., et al., 2016. Acquired midfoot deformity and function in individuals with diabetes and peripheral neuropathy. Clin. Biomech. 32, 261–267.

Hatala, K.G., Dingwall, H.L., Wunderlich, R.E., Richmond, B.G., 2013. Variation in foot strike patterns during running among habitually barefoot populations. PLoS One 8 (1), e52548. https://doi.org/10.1371/journal.pone.0052548.

Hayafune, N., Hayafune, Y., Jacob, H.A.C., 1999. Pressure and force distribution characteristics under the normal foot during the push-off phase in gait. Foot 9 (2), 88–92.

Hayes, P., Caplan, N., 2012. Foot strike patterns and ground contact times during high-calibre middle-distance races. J. Sports Sci. 30 (12), 1275–1283.

Headlee, D.L., Leonard, J.L., Hart, J.M., Ingersoll, C.D., Hertel, J., 2008. Fatigue of the plantar intrinsic foot muscles increases navicular drop. J. Electromyogr. Kinesiol. 18 (3), 420–425.

Hedrick, M.R., 1996. The plantar aponeurosis. Foot Ankle Int. 17 (10), 646–649.

Hedrick, E.A., Stanhope, S.J., Takahashi, K.Z., 2019. The foot and ankle structures reveal emergent properties analogous to passive springs during human walking. PLoS One 14 (6), e0218047. https://doi.org/10.1371/journal.pone.0218047.

Hertel, J., Corbett, R.O., 2019. An updated model of chronic ankle instability. J. Athl. Train. 54 (6), 572–588.

Hicks, J.H., 1953. The mechanics of the foot. I: the joints. J. Anat. 87 (4), 345–357.

Hicks, J.H., 1954. The mechanics of the foot. II. The plantar aponeurosis and the arch. J. Anat. 88 (1), 25–31.

Hintermann, B., Nigg, B.M., Sommer, C., 1994. Foot movement and tension excursion: an *in vitro* study. Foot Ankle Int. 15 (7), 386–395.

Hofmann, C.L., Okita, N., Sharkey, N.A., 2013. Experimental evidence supporting isometric functioning of the extrinsic toe flexors during gait. Clin. Biomech. 28 (6), 686–691.

Holowka, N.B., Lieberman, D.E., 2018. Rethinking the evolution of the human foot: insights from experimental research. J. Exp. Biol. 221 (17), jeb174425. https://doi.org/10.1242/jep.17445.

Holowka, N.B., O'Neill, M.C., Thompson, N.E., Demes, B., 2017. Chimpanzee and human midfoot motion during bipedal walking and the evolution of the longitudinal arch of the foot. J. Hum. Evol. 104, 23–31.

Honert, E.C., Bastas, G., Zelik, K.E., 2020. Effects of toe length, foot arch length and toe joint axis on walking biomechanics. Hum. Mov. Sci. 70, 102594. https://doi.org/10.1016/j.humov.2020.102594.

Horwood, A., 2019. The biomechanical function of the foot pump in venous return from the lower extremity during the human gait cycle: an expansion of the gait model of the foot pump. Med. Hypotheses 129, 109220. https://doi.org/10.1016/j.mehy.2019.05.006.

Horwood, A.M., Chockalingam, N., 2017. Defining excessive, over, or hyper-pronation: a quandary. Foot 31, 49–55.

Hsu, T.-C., Wang, C.-L., Shau, Y.-W., Tang, F.-T., Li, K.-L., Chen, C.-Y., 2000. Altered heel-pad mechanical properties in patients with type 2 diabetes mellitus. Diabet. Med. 17 (12), 854–859.

Hsu, C.-C., Tsai, W.-C., Chen, C.P.-C., Shau, Y.-W., Wang, C.-L., Chen, M.J.-L., et al., 2005. Effects of aging on the plantar soft tissue properties under the metatarsal heads at different impact velocities. Ultrasound Med. Biol. 31 (10), 1423–1429.

Hsu, C.-C., Tsai, W.-C., Wang, C.-L., Pao, S.-H., Shau, Y.-W., Chuan, Y.-S., 2007a. Microchambers and macrochambers in heel pads: are they functionally different? J. Appl. Physiol. 102 (6), 2227–2231.

Hsu, C.-C., Tsai, W.-C., Shau, Y.-W., Lee, K.-L., Hu, C.-F., 2007b. Altered energy dissipation ratio of the plantar soft tissues under the metatarsal heads in patients with type 2 diabetes mellitus: a pilot study. Clin. Biomech. 22 (1), 67–73.

Huang, C.-K., Kitaoka, H.B., An, K.-N., Chao, E.Y.S., 1993. Biomechanical evaluation of longitudinal arch stability. Foot Ankle Int. 14 (6), 353–357.

Huang, J., Qin, K., Tang, C., Zhu, Y., Klein, C.S., Zhang, Z., et al., 2018. Assessment of passive stiffness of medial and lateral heads of gastrocnemius muscle, Achilles tendon, and plantar fascia at different ankle and knee positions using the MyotonPRO. Med. Sci. Monit. 24, 7570–7576.

Huerta, J.P., Alarcón García, J.M., 2007. Effect of gender, age and anthropometric variables on plantar fascia thickness at different locations in asymptomatic subjects. Eur. J. Radiol. 62 (3), 449–453.

Hunt, A.E., Smith, R.M., Torode, M., Keenan, A.-M., 2001. Inter-segment foot motion and ground reaction forces over the stance phase of walking. Clin. Biomech. 16 (7), 592–600.

Huson, A., 2000. Biomechanics of the tarsal mechanism. A key to the function of the normal human foot. J. Am. Podiatr. Med. Assoc. 90 (1), 12–17.

Huxel Bliven, K.C., Anderson, B.E., 2013. Core stability training for injury prevention. Sports Health Multidisc. Approach 5 (6), 514–522.

Imhauser, C.W., Siegler, S., Abidi, N.A., Frankel, D.Z., 2004. The effect of posterior tibialis tendon dysfunction on the plantar pressure characteristics and the kinematics of the arch and the hindfoot. Clin. Biomech. 19 (2), 161–169.

Ishikawa, M., Komi, P.V., Grey, M.J., Lepola, V., Bruggemann, G.-P., 2005. Muscle-tendon interaction and elastic energy usage in human walking. J. Appl. Physiol. 99 (2), 603–608.

Isman, R.E., Inman, V.T., 1969. Anthropometric studies of the human foot and ankle. Bull. Prosthet. Res. 10-11 (Spring), 97–129.

Iwanuma, S., Akagi, R., Hashizume, S., Kanehisa, H., Yanai, T., Kawakami, Y., 2011a. Triceps surae muscle–tendon unit length changes as a function of ankle joint angles and contraction levels: the effect of foot arch deformation. J. Biomech. 44 (14), 2579–2583.

Iwanuma, S., Akagi, R., Kurihara, T., Ikegawa, S., Kanehisa, H., Fukunaga, T., et al., 2011b. Longitudinal and transverse deformation of human Achilles tendon induced by isometric plantar flexion at different intensities. J. Appl. Physiol. 110 (6), 1615–1621.

Jacob, H.A.C., 2001. Forces acting in the forefoot during normal gait—an estimate. Clin. Biomech. 16 (9), 783–792.

Jahss, M.H., Michelson, J.D., Desai, P., Kaye, R., Kummer, F., Buschman, W., et al., 1992. Investigations into the fat pads of the sole of the foot: anatomy and histology. Foot Ankle 13 (5), 233–242.

Johnson, C.H., Christensen, J.C., 1999. Biomechanics of the first ray. Part I. the effects of peroneus longus function: a three-dimensional kinematic study on a cadaver model. J. Foot Ankle Surg. 38 (5), 313–321.

Johnston 3rd, R.B., Smith, J., Daniels, T., 1994. The plantar plate of the lesser toes: an anatomical study in human cadavers. Foot Ankle Int. 15 (5), 276–282.

Jung, M.-H., Choi, B.Y., Lee, J.Y., Han, C.S., Lee, J.S., Yang, Y.C., 2015. Types of subtalar joint facets. Surg. Radiol. Anat. 37 (6), 629–638.

Kadambande, S., Khurana, A., Debnath, U., Bansal, M., Hariharan, K., 2006. Comparative anthropometric analysis of shod and unshod feet. Foot 16 (4), 188–191.

Kafka, R.M., Aveytua, I.L., Choi, P.J., DiLandro, A.C., Tubbs, R.S., Loukas, M., et al., 2019. Anatomico-radiological study of the bifurcate ligament of the foot with clinical significance. Cureus 11 (1), e3847. https://doi.org/10.7759/cureus.3847.

Kanatli, U., Gözil, R., Besli, K., Yetkin, H., Bölükbasi, S., 2006. The relationship between the hindfoot angle and the medial longitudinal arch of the foot. Foot Ankle Int. 27 (8), 623–627.

Kapandji, I.A., 1987. The Physiology of the Joints. Volume 2: The Lower Limb, fifth Edition. Churchill Livingstone Inc., New York, NY.

Kavounoudias, A., Roll, R., Roll, J.-P., 2001. Foot sole and ankle muscle inputs contribute jointly to human erect posture regulation. J. Physiol. 532 (3), 869–878.

Kelly, L.A., Kuitunen, S., Racinais, S., Cresswell, A.G., 2012. Recruitment of the plantar intrinsic foot muscles with increasing postural demand. Clin. Biomech. 27 (1), 46–51.

Kelly, L.A., Cresswell, A.G., Racinais, S., Whiteley, R., Lichtwark, G., 2014. Intrinsic foot muscles have the capacity to control deformation of the longitudinal arch. J. R. Soc. Interface 11 (93), 20131188. https://doi.org/10.1098/rsif.2013.1188.

Kelly, L.A., Lichtwark, G., Cresswell, A.G., 2015. Active regulation of longitudinal arch compression and recoil during walking and running. J. R. Soc. Interface 12 (102), 20141076. https://doi.org/10.1098/rsif.2014.1076.

Kelly, L.A., Cresswell, A.G., Farris, D.J., 2018. The energetic behaviour of the human foot across a range of running speeds. Sci. Rep. 8, 10576. https://doi.org/10.1038/s41598-018-28946-1.

Kelso, S.F., Richie Jr., D.H., Cohen, I.R., Weed, J.H., Root, M., 1982. Direction and range of motion of the first ray. J. Am. Podiatry Assoc. 72 (12), 600–605.

Kennedy, P.M., Inglis, J.T., 2002. Distribution and behaviour of glabrous cutaneous receptors in the human foot sole. J. Physiol. 538 (3), 995–1002.

Kenny, R.P.W., Atkinson, G., Eaves, D.L., Martin, D., Burn, N., Dixon, J., 2019. The effects of textured materials on static balance in healthy young and older adults: a systematic review with meta-analysis. Gait Posture 71, 79–86.

Ker, R.F., Bennett, M.B., Bibby, S.R., Kester, R.C., Alexander RMcN., 1987. The spring in the arch of the human foot. Nature 325 (6100), 147–149.

Kernozek, T.W., Knaus, A., Rademaker, T., Almonroeder, T.G., 2018. The effects of habitual foot strike patterns on Achilles tendon loading in female runners. Gait Posture 66, 283–287.

Khassetarash, A., Hassannejad, R., Enders, H., Ettefagh, M.M., 2015. Damping and energy dissipation in soft tissue vibrations during running. J. Biomech. 48 (2), 204–209.

Kidd, R., Oxnard, C., 2005. Little foot and big thoughts—a re-evaluation of the Stw573 foot from Sterkfontein, South Africa. HOMO J. Compar. Hum. Biol. 55 (3), 189–212.

Kido, M., Ikoma, K., Imai, K., Tokunaga, D., Inoue, N., Kubo, T., 2013. Load response of the medial longitudinal arch in patients with flatfoot deformity: in vivo 3D study. Clin. Biomech. 28 (5), 568–573.

Kim, S., Park, S., 2011. Leg stiffness increases with speed to modulate gait frequency and propulsion energy. J. Biomech. 44 (7), 1253–1258.

Kim, W., Voloshin, A.S., 1995. Role of plantar fascia in the load bearing capacity of the human foot. J. Biomech. 28 (9), 1025–1033.

Kirkham, S., Lam, S., Nester, C., Hashmi, F., 2014. The effect of hydration on the risk of friction blister formation on the heel of the foot. Skin Res. Technol. 20 (2), 246–253.

Kitaoka, H.B., Luo, Z.P., Growney, E.S., Berglund, L.J., An, K.-N., 1994. Material properties of the plantar aponeurosis. Foot Ankle Int. 15 (10), 557–560.

Kitaoka, H.B., Luo, Z.P., An, K.-N., 1997. Effect of the posterior tibial tendon on the arch of the foot during simulated weightbearing: biomechanical analysis. Foot Ankle Int. 18 (1), 43–46.

Klingler, W., Velders, M., Hoppe, K., Pedro, M., Schleip, R., 2014. Clinical relevance of fascial tissue and dysfunctions. Curr. Pain Headache Rep. 18 (8), 439. https://doi.org/10.1007/s11916-014-0439-y.

Kogler, G.F., Solomonidis, S.E., Paul, J.P., 1996. Biomechanics of longitudinal arch support mechanisms in foot orthoses and their effect on plantar aponeurosis strain. Clin. Biomech. 11 (5), 243–252.

Kokubo, T., Hashimoto, T., Nagura, T., Nakamura, T., Suda, Y., Matsumoto, H., et al., 2012. Effect of the posterior tibial and peroneus longus on the mechanical properties of the foot arch. Foot Ankle Int. 33 (4), 320–325.

Komi, P.V., Fukashiro, S., Järvinen, M., 1992. Biomechanical loading of Achilles tendon during normal locomotion. Clin. Sports Med. 11 (3), 521–531.

Konow, N., Azizi, E., Roberts, T.J., 2012. Muscle power attenuation by tendon during energy dissipation. Proc. R. Soc. B Biol. Sci. 279 (1731), 1108–1113.
Koo, T.K., Guo, J.-Y., Cohen, J.H., Parker, K.J., 2014. Quantifying the passive stretching response of human tibialis anterior muscle using shear wave elastography. Clin. Biomech. 29 (1), 33–39.
Koshi, A., Kottalil, L., 2016. Performance comparison of through arch bridge at different arch positions. Int. J. Sci. Eng. Res. 7 (9), 515–518.
Krähenbühl, N., Horn-Lang, T., Hintermann, B., Knupp, M., 2017. The subtalar joint: a complex mechanism. EFORT Open Rev. 2 (7), 309–316.
Kuo, A.D., 2007. The six determinants of gait and the inverted pendulum analogy: a dynamic walking perspective. Hum. Mov. Sci. 26 (4), 617–656.
Kuo, A.D., Donelan, J.M., Ruina, A., 2005. Energetic consequences of walking like an inverted pendulum: step-to-step transitions. Exercise Sport Sci. Rev. 33 (2), 88–97.
Kuo, S., DeSilva, J.M., Devlin, M.J., McDonald, G., Morgan, E.F., 2013. The effect of the Achilles tendon on trabecular structure in the primate calcaneus. Anat. Rec. 296 (10), 1509–1517.
Lai, A., Schache, A.G., Lin, Y.-C., Pandy, M.G., 2014. Tendon elastic strain energy in the human ankle plantar-flexors and its role with increased running speed. J. Exp. Biol. 217 (17), 3159–3168.
Lai, A.K.M., Lichtwark, G.A., Schache, A.G., Pandy, M.G., 2018. Differences in *in vivo* muscle fascicle and tendinous tissue behavior between the ankle plantarflexors during running. Scand. J. Med. Sci. Sports 28 (7), 1828–1836.
Laing, R.M., Wilson, C.A., Dunn, L.A., Niven, B.E., 2015. Detection of fiber effects on skin health of the human foot. Text. Res. J. 85 (17), 1849–1863.
Langdon, J.H., 2005. The Human Strategy: An Evolutionary Perspective on Human Anatomy. Oxford University Press, Oxford, UK.
Leardini, A., O'Connor, J.J., 2002. A model for lever-arm length calculation of the flexor and extensor muscles at the ankle. Gait Posture 15 (3), 220–229.
Leardini, A., Benedetti, M.G., Catani, F., Simoncini, L., Giannini, S., 1999. An anatomically based protocol for the description of foot segment kinematics during gait. Clin. Biomech. 14 (8), 528–536.
Leardini, A., Stagni, R., O'Connor, J.J., 2001. Mobility of the subtalar joint in the intact ankle complex. J. Biomech. 34 (6), 805–809.
Leardini, A., Benedetti, M.G., Berti, L., Bettinelli, D., Nativo, R., Giannini, S., 2007. Rear-foot, mid-foot and fore-foot motion during the stance phase of gait. Gait Posture 25 (3), 453–462.
Leardini, A., O'Connor, J.J., Giannini, S., 2014. Biomechanics of the natural, arthritic, and replaced human ankle joint. J. Foot Ankle Res. 7, 8. https://doi.org/10.1186/1757-1146-7-8.
Ledoux, W.R., Blevins, J.J., 2007. The compressive material properties of the plantar soft tissue. J. Biomech. 40 (13), 2975–2981.
Lee, W.E., 2001. Podiatric biomechanics: An historical appraisal and discussion of the Root model as a clinical system of approach in the present context of theoretical uncertainty. Clin. Podiatr. Med. Surg. 18 (4), 555–684.
Lee, S.S.M., Piazza, S.J., 2009. Built for speed: musculoskeletal structure and sprinting ability. J. Exp. Biol. 212 (22), 3700–3707.
Lees, A., Lake, M., Klenerman, L., 2005. Shock absorption during forefoot running and its relationship to medial longitudinal arch height. Foot Ankle Int. 26 (12), 1081–1088.
Levin, S.M., Martin, D.-C., 2012. Biotensegrity: the mechanics of fascia. In: Schleip, R., Findley, T.W., Chaitow, L., Huijing, P.A. (Eds.), Fascia: The Tensional Network of the Human Body. The Science and Clinical Applications in Manual and Movement Therapy. Churchill Livingstone, Elsevier, pp. 137–142 (Part 3, Chapter 3.5).
Li, L., Gollhofer, A., Lohrer, H., Dorn-Lange, N., Bonsignore, G., Gehring, D., 2019. Function of ankle ligaments for subtalar and talocrural joint stability during an inversion movement—an *in vitro* study. J. Foot Ankle Res. 12, 16. https://doi.org/10.1186/s13047-019-0330-5.
Lichtwark, G.A., Bougoulias, K., Wilson, A.M., 2007. Muscle fascicle and series elastic element length changes along the length of the human gastrocnemius during walking and running. J. Biomech. 40 (1), 157–164.
Liu, C.L., Li, Y.P., Wang, X.Q., Zhang, Z.J., 2018. Quantifying the stiffness of Achilles tendon: intra- and inter-operator reliability and the effect of ankle joint motion. Med. Sci. Monit. 24, 4876–4881.
Lohman 3rd, E.B., Balan Sackiriyas, K.S., Swen, R.W., 2011. A comparison of the spatiotemporal parameters, kinematics, and biomechanics between shod, unshod, and minimally supported running as compared to walking. Phys. Ther. Sport 12 (4), 151–163.
Lundberg, A., 1989. Kinematics of the ankle and foot. *In vivo* roentgen stereophotogrammetry. Acta Orthop. Scand. 60 (Suppl 223), 1–26.
Lundberg, A., Svensson, O.K., 1993. The axes of rotation of the talocalcaneal and talonavicular joints. Foot 3 (2), 65–70.
Lundberg, A., Svensson, O.K., Bylund, C., Goldie, I., Selvik, G., 1989. Kinematics of the ankle/foot complex–Part 2: Pronation and supination. Foot Ankle 9 (5), 248–253.
Lundgren, P., Nester, C., Liu, A., Arndt, A., Jones, R., Stacoff, A., et al., 2008. Invasive *in vivo* measurement of rear-, mid- and forefoot motion during walking. Gait Posture 28 (1), 93–100.
Lyght, M., Nockerts, M., Kernozek, T.W., Ragan, R., 2016. Effects of foot strike and step frequency on Achilles tendon stress during running. J. Appl. Biomech. 32 (4), 365–372.
Ma, C.Z.-H., Lam, W.-K., Chang, B.-C., Lee, W.C.-C., 2020. Can insoles be used to improve static and dynamic balance of community-dwelling older adults? A systematic review on recent advances and future perspectives. J. Aging Phys. Act. 28 (6), 971–986. https://doi.org/10.1123/japa.2019-0293.
Maas, N.M.G., van der Grinten, M., Bramer, W.M., Kleinrensink, G.-J., 2016. Metatarsophalangeal joint stability: a systematic review on the plantar plate of the lesser toes. J. Foot Ankle Res. 9, 32. https://doi.org/10.1186/s13047-016-0165-2.
Magalhães, F.A., Fonseca, S.T., Araújo, V.L., Trede, R.G., Oliveira, L.M., Castor, C.G.M.E., et al., 2021. Midfoot passive stiffness affects foot and ankle kinematics and kinetics during the propulsive phase of walking. J. Biomech. 119, 110328.
Maharaj, J.N., Cresswell, A.G., Lichtwark, G.A., 2016. The mechanical function of the tibialis posterior muscle and its tendon during locomotion. J. Biomech. 49 (14), 3238–3243.
Mann, R., Inman, V.T., 1964. Phasic activity of intrinsic muscles of the foot. J. Bone Joint Surg. 46-A (3), 469–481.

Manter, J.T., 1941. Movements of the subtalar and transverse tarsal joints. Anat. Rec. 80 (4), 397–410.
Mason, L., Jayatilaka, M.L.T., Fisher, A., Fisher, L., Swanton, E., Molloy, A., 2020. Anatomy of the lateral plantar ligaments of the transverse metatarsal arch. Foot Ankle Int. 41 (1), 109–114.
Matthews, B.G., Hurn, S.E., Harding, M.P., Henry, R.A., Ware, R.S., 2019. The effectiveness of non-surgical interventions for common plantar compressive neuropathy (Morton's neuroma): a systematic review and meta-analysis. J. Foot Ankle Res. 12, 12. https://doi.org/10.1186/s13047-019-0320-7.
McDonald, S.W., Tavener, G., 1999. Pronation and supination of the foot: confused terminology. Foot 9 (1), 6–11.
McDonald, K.A., Stearne, S.M., Alderson, J.A., North, I., Pires, N.J., Rubenson, J., 2016. The role of arch compression and metatarsophalangeal joint dynamics in modulating plantar fascia strain in running. PLoS One 11 (4), e0152602. https://doi.org/10.1371/journal.pone.0152602.
McGowan, C.P., Neptune, R.R., Clark, D.J., Kautz, S.A., 2010. Modular control of human walking: adaptations to altered mechanical demands. J. Biomech. 43 (3), 412–419.
McKenzie, J., 1955. The foot as a half-dome. Br. Med. J. 1 (4921), 1068–1070.
McKeon, P.O., Hertel, J., Bramble, D., Davis, I., 2015. The foot core system: a new paradigm for understanding intrinsic foot muscle function. Br. J. Sports Med. 49 (5), 290. https://doi.org/10.1136/bjsports-2013-092690.
McMahon, T.A., 1987. The spring in the human foot. Nature 325 (6100), 108–109.
McNutt, E.J., Zipfel, B., DeSilva, J.M., 2018. The evolution of the human foot. Evol. Anthropol. Issues News Rev. 27 (5), 197–217.
McPoil, T.G., Cornwall, M.W., Vicenzino, B., Teyhen, D.S., Molloy, J.M., Christie, D.S., et al., 2008. Effect of using truncated versus total foot length to calculate the arch height ratio. Foot 18 (4), 220–227.
McPoil, T.G., Ford, J., Fundaun, J., Gallegos, C., Kinney, A., McMillan, P., et al., 2016. The use of a static measure to predict foot posture at midstance during walking. Foot 28, 47–53.
Melão, L., Canella, C., Weber, M., Negrão, P., Trudell, D., Resnick, D., 2009. Ligaments of the transverse tarsal joint complex: MRI-anatomic correlation in cadavers. Am. J. Roentgenol. 193 (3), 662–671.
Menz, H.B., 1998. Alternative techniques for the clinical assessment of foot pronation. J. Am. Podiatr. Med. Assoc. 88 (3), 119–129.
Menz, H.B., Munteanu, S.E., 2005. Validity of 3 clinical techniques for the measurement of static foot posture in older people. J. Orthop. Sports Phys. Ther. 35 (8), 479–486.
Menz, H.B., Morris, M.E., Lord, S.R., 2005. Foot and ankle characteristics associated with impaired balance and functional ability in older people. J. Gerontol. A Biol. Sci. Med. Sci. 60 (12), 1546–1552.
Morales-Orcajo, E., Souza, T.R., Bayod, J., de Las, B., Casas, E., 2017. Non-linear finite element model to assess the effect of tendon forces on the foot-ankle complex. Med. Eng. Phys. 49, 71–78.
Murley, G.S., Buldt, A.K., Trump, P.J., Wickham, J.B., 2009. Tibialis posterior EMG activity during barefoot walking in people with neutral foot posture. J. Electromyogr. Kinesiol. 19 (2), e69–e77.
Murley, G.S., Menz, H.B., Landorf, K.B., 2014. Electromyographic patterns of tibialis posterior and related muscles when walking at different speeds. Gait & Posture. 39 (4), 1080–1085.
Murphy, D., Connolly, D., Beynnon, B., 2003. Risk factors for lower extremity injury: a review of the literature. Br. J. Sports Med. 37 (1), 13–29.
Naemi, R., Behforootan, S., Chatzistergos, P., Chockalingam, N., 2016. Chapter 10: Viscoelasticity in foot-ground interaction. In: El-Amin, M.F. (Ed.), Viscoelastic and Viscoplastic Materials. IntechOpen, pp. 217–243.
Natali, A.N., Fontanella, C.G., Carniel, E.L., 2010a. Constitutive formulation and analysis of heel pad tissues mechanics. Med. Eng. Phys. 32 (5), 516–522.
Natali, A.N., Pavan, P.G., Stecco, C., 2010b. A constitutive model for the mechanical characterization of the plantar fascia. Connect. Tissue Res. 51 (5), 337–346.
Natali, A.N., Fontanella, C.G., Carniel, E.L., 2012. A numerical model for investigating the mechanics of calcaneal fat pad region. J. Mech. Behav. Biomed. Mater. 5 (1), 216–223.
Neptune, R.R., Sasaki, K., Kautz, S.A., 2008. The effect of walking speed on muscle function and mechanical energetics. Gait Posture 28 (1), 135–143.
Nester, C.J., Findlow, A., Bowker, P., 2001. Scientific approach to the axis of rotation at the midtarsal joint. J. Am. Podiatr. Med. Assoc. 91 (2), 68–73.
Nester, C.J., Liu, A.M., Ward, E., Howard, D., Cocheba, J., Derrick, T., et al., 2007a. In vitro study of foot kinematics using a dynamic walking cadaver model. J. Biomech. 40 (9), 1927–1937.
Nester, C., Jones, R.K., Liu, A., Howard, D., Lundberg, A., Arndt, A., et al., 2007b. Foot kinematics during walking measured using bone and surface mounted markers. J. Biomech. 40 (15), 3412–3423.
Nguyen, T., 2010. Effects of Curvature on the Stresses of a Curved Laminated Beams Subjected to Bending. MS dissertation. The University of Texas at Arlington. https://rc.library.uta.edu/uta-ir/handle/10106/4897.
Nguyen, K., Yu, N., Bandi, M.M., Venkadesan, M., Mandre, S., 2017. Curvature-induced stiffening of a fish fin. J. R. Soc. Interface 14 (130), 20170247. https://doi.org/10.1098/rsif.2017.0247.
Nielsen, R.G., Rathleff, M.S., Simonsen, O.H., Langberg, H., 2009. Determination of normal values for navicular drop during walking: a new model correcting for foot length and gender. J. Foot Ankle Res. 2, 12. https://doi.org/10.1186/1757-1146-2-12.
Nigg, B.M., Fisher, V., Allinger, T.L., Ronsky, J.R., Engsberg, J.R., 1992. Range of motion of the foot as a function of age. Foot Ankle 13 (6), 336–343.
Nwawka, O.K., Hayashi, D., Diaz, L.E., Goud, A.R., Arndt 3rd, W.F., Roemer, F.W., et al., 2013. Sesamoids and accessory ossicles of the foot: anatomical variability and related pathology. Insights Imaging 4 (5), 581–593.
Okamura, K., Kanai, S., Hasegawa, M., Otsuka, A., Oki, S., 2018. The effect of additional activation of the plantar intrinsic foot muscles on foot dynamics during gait. Foot 34, 1–5.
Ota, T., Nagura, T., Yamada, Y., Yamada, M., Yokoyama, Y., Ogihara, N., et al., 2019. Effect of natural full weight-bearing during standing on the rotation of the first metatarsal bone. Clin. Anat. 32 (5), 715–721.

Pain, M.T.G., Challis, J.H., 2001. The role of the heel pad and shank soft tissue during impacts: a further resolution of a paradox. J. Biomech. 34 (3), 327–333.

Papachatzis, N., Malcolm, P., Nelson, C.A., Takahashi, K.Z., 2020. Walking with added mass magnifies salient features of human foot energetics. J. Exp. Biol. 223 (12), jeb207472. https://doi.org/10.1242/jeb.207472.

Paton, J., Glasser, S., Collings, R., Marsden, J., 2016. Getting the right balance: insole design alters the static balance of people with diabetes and neuropathy. J. Foot Ankle Res. 9, 40. https://doi.org/10.1186/s13047-016-0172-3.

Pavan, P.G., Stecco, C., Darwish, S., Natali, A.N., De Caro, R., 2011. Investigation of the mechanical properties of the plantar aponeurosis. Surg. Radiol. Anat. 33 (10), 905–911.

Peña Fernández, M., Hoxha, D., Chan, O., Mordecai, S., Blunn, G.W., Tozzi, G., et al., 2020. Centre of rotation of the human subtalar joint using weight-bearing clinical computed tomography. Sci. Rep. 10, 1035. https://doi.org/10.1038/s41598-020-57912-z.

Perry, J., 1992. Gait Analysis: Normal and Pathological Function. SLACK Incorporated, Thorofare, NJ.

Petcu, D., Colda, A., 2012. Foot functioning paradigms. Proc. Roman. Acad. Ser. B 14 (3), 212–217.

Péter, A., Hegyi, A., Stenroth, L., Finni, T., Cronin, N.J., 2015. EMG and force production of the flexor hallucis longus muscle in isometric plantarflexion and the push-off phase of walking. J. Biomech. 48 (12), 3413–3419.

Péter, A., Arndt, A., Hegyi, A., Finni, T., Andersson, E., Alkjær, T., et al., 2020. Effect of footwear on intramuscular EMG activity of plantar flexor muscles in walking. J. Electromyogr. Kinesiol. 55, 102474. https://doi.org/10.1016/j.jelekin.2020.102474.

Phillips, R.D., Law, E.A., Ward, E.D., 1996. Functional motion of the medial column joints of the foot during propulsion. J. Am. Podiatr. Med. Assoc. 86 (10), 474–486.

Pilavaki, M., Chourmouzi, D., Kiziridou, A., Skordalaki, A., Zarampoukas, T., Drevelengas, A., 2004. Imaging of peripheral nerve sheath tumors with pathologic correlation: pictorial review. Eur. J. Radiol. 52 (3), 229–239.

Pini, V., Ruz, J.J., Kosaka, P.M., Malvar, O., Calleja, M., Tamayo, J., 2016. How two-dimensional bending can extraordinarily stiffen thin sheets. Sci. Rep. 6, 29627. https://doi.org/10.1038/srep29627.

Pisani, G., 1994. The coxa pedis. Eur. J. Foot Ankle Surg. 1 (2–3), 67–74.

Pisani, G., 2016. "Coxa pedis" today. Foot Ankle Surg. 22 (2), 78–84.

Pohl, M.B., Hamill, J., Davis, I.S., 2009. Biomechanical and anatomic factors associated with a history of plantar fasciitis in female runners. Clin. J. Sport Med. 19 (5), 372–376.

Pohl, J., Jaspers, T., Ferraro, M., Krause, F., Baur, H., Eichelberger, P., 2018. The influence of gait and speed on the dynamic navicular drop – a cross sectional study on healthy subjects. Foot 36, 67–73.

Pollard, C.D., Heiderscheit, B.C., van Emmerik, R.E.A., Hamill, J., 2005. Gender differences in lower extremity coupling variability during an unanticipated cutting maneuver. J. Appl. Biomech. 21 (2), 143–152.

Pontzer, H., 2017. Economy and endurance in human evolution. Curr. Biol. 27 (12), R613–R621.

Pool, R., 1989. Is it healthy to be chaotic? Science 243 (4891), 604–607.

Prang, T.C., 2016. The subtalar joint complex of *Australopithecus sediba*. J. Hum. Evol. 90, 105–119.

Prasad, S.A., Rajasekhar, S.S.S.N., 2020. Morphometric analysis of talus and calcaneus. Surg. Radiol. Anat. 41 (1), 9–24.

Puszczalowska-Lizis, E., Krawczyk, K., Omorczyk, J., 2022. Effect of longitudinal and transverse foot arch on the position of the hallux and fifth toe in preschool children in the light of regression analysis. Int. J. Environ. Res. Public Health 19 (3), 1669. https://doi.org/10.3390/ijerph19031669.

Qian, Z., Ren, L., Ren, L., 2010. A coupling analysis of the biomechanical functions of human foot complex during locomotion. J. Bionic Eng. 7 (Supplement), S150–S157.

Rathleff, M.S., Olesen, C.G., Moelgaard, C.M., Jensen, K., Madeleine, P., Olesen, J.L., 2010. Non-linear analysis of the structure of variability in midfoot kinematics. Gait Posture 31 (3), 385–390.

Raychoudhury, S., Hu, D., Ren, L., 2014. Three-dimensional kinematics of the human metatarsophalangeal joint during level walking. Front. Bioeng. Biotechnol. 2, 73. https://doi.org/10.3389/fbioe.2014.00073.

Redmond, A.C., Crosbie, J., Ouvrier, R.A., 2006. Development and validation of a novel rating system for scoring standing foot posture: the foot posture index. Clin. Biomech. 21 (1), 89–98.

Richter, M., Wippermann, B., Krettek, C., Schratt, H.E., Hufner, T., Therman, H., 2001. Fractures and fracture dislocations of the midfoot: occurrence, causes and long-term results. Foot Ankle Int. 22 (5), 392–398.

Robbins, H.M., 1981. The unified forefoot: a mathematical model in the transverse plane. J. Am. Podiatry Assoc. 71 (9), 465–471.

Roberts, T.J., Azizi, E., 2010. The series-elastic shock absorber: tendons attenuate muscle power during eccentric actions. J. Appl. Physiol. 109 (2), 396–404.

Roberts, T.J., Konow, N., 2013. How tendons buffer energy dissipation by muscle. Exerc. Sport Sci. Rev. 41 (4), 186–193.

Rogers, L.C., Frykberg, R.G., Armstrong, D.G., Boulton, A.J.M., Edmonds, M., Van, G.H., et al., 2011. The Charcot foot in diabetes. J. Am. Podiatr. Med. Assoc. 101 (5), 437–446.

Rome, K., 1998. Mechanical properties of the heel pad: current theory and review of the literature. Foot 8 (4), 179–185.

Rome, K., Webb, P., Unsworth, A., Haslock, I., 2001. Heel pad stiffness in runners with plantar heel pain. Clin. Biomech. 16 (10), 901–905.

Rome, K., Campbell, R., Flint, A., Haslock, I., 2002. Heel pad thickness—a contributing factor associated with plantar heel pain in young adults. Foot Ankle Int. 23 (2), 142–147.

Rooney, B.D., Derrick, T.R., 2013. Joint contact loading in forefoot and rearfoot strike patterns during running. J. Biomech. 46 (13), 2201–2206.

Root, M.L., Weed, J.H., Sgarlato, T.E., Bluth, D.R., 1966. Axis of motion of the subtalar joint: an anatomical study. J. Am. Podiatry Assoc. 56 (4), 149–155.

Root, M.L., Orien, W.P., Weed, J.H., 1977. Normal and Abnormal Function of the Foot: Volume II. Clinical Biomechanics Corporation, Los Angeles.

Ruina, A., Bertram, J.E.A., Srinivasan, M., 2005. A collisional model of the energetic cost of support work qualitatively explains leg sequencing in walking and galloping, pseudo-elastic leg behavior in running and the walk-to-run transition. J. Theor. Biol. 237 (2), 170–192.

Saghazadeh, M., Kitano, N., Okura, T., 2015. Gender differences of foot characteristics in older Japanese adults using a 3D foot scanner. J. Foot Ankle Res. 8, 29. https://doi.org/10.1186/s13047-015-0087-4.

Sakamoto, K., Kudo, S., 2020. Morphological characteristics of intrinsic foot muscles among flat foot and normal foot using ultrasonography. Acta Bioeng. Biomech. 22 (4), 161–166.

Salathé, E.P., Arangio, G.A., 2002. A biomechanical model of the foot: the role of muscles, tendons, and ligaments. J. Biomech. Eng. 124 (3), 281–287.

Sarrafian, S.K., 1987. Functional characteristics of the foot and plantar aponeurosis under tibiotalar loading. Foot Ankle 8 (1), 4–18.

Sarrafian, S.K., 1993. Biomechanics of the subtalar joint complex. Clin. Orthopaed. Relat. Res. 290 (May), 17–26.

Sasaki, K., Neptune, R.R., 2006. Muscle mechanical work and elastic energy utilization during walking and running near the preferred gait transition speed. Gait Posture 23 (3), 383–390.

Savory, K.M., Wülker, N., Stukenborg, C., Alfke, D., 1998. Biomechanics of the hindfoot joints in response to degenerative hindfoot arthrodeses. Clin. Biomech. 13 (1), 62–70.

Scarr, G., 2011. Helical tensegrity as a structural mechanism in human anatomy. Int. J. Osteopathic Med. 14 (1), 24–32.

Schleip, R., 2003a. Fascial plasticity—a new neurobiological explanation: Part 1. J. Bodyw. Mov. Ther. 7 (1), 11–19.

Schleip, R., 2003b. Fascial plasticity—a new neurobiological explanation: Part 2. J. Bodyw. Mov. Ther. 7 (2), 104–116.

Schleip, R., Klingler, W., Lehmann-Horn, F., 2005. Active fascial contractility: fascia may be able to contract in a smooth muscle-like manner and thereby influence musculoskeletal dynamics. Med. Hypotheses 65 (2), 273–277.

Schleip, R., Gabbiani, G., Wilke, J., Naylor, I., Hinz, B., Zorn, A., et al., 2019. Fascia is able to actively contract and may thereby influence musculoskeletal dynamics: a histochemical and mechanographic investigation. Front. Physiol. 10, 336. https://doi.org/10.3389/fphys.2019.00336.

Schmitt, S., Günther, M., 2011. Human leg impact: energy dissipation of wobbling masses. Arch. Appl. Mech. 81 (7), 887–897.

Scott, S.H., Winter, D.A., 1991. Talocrural and talocalcaneal joint kinematics and kinetics during the stance phase of walking. J. Biomech. 24 (8), 743–752.

Semple, R., Murley, G.S., Woodburn, J., Turner, D.E., 2009. Tibialis posterior in health and disease: a review of structure and function with specific reference to electromyographic studies. J. Foot Ankle Res. 2, 24. https://doi.org/10.1186/1757-1146-2-24.

Sharkey, N.A., Ferris, L., Smith, T.S., Matthews, D.K., 1995. Strain and loading of the second metatarsal during heel-lift. J. Bone Joint Surg. 77-A (7), 1050–1057.

Sharrock, C., Cropper, J., Mostad, J., Johnson, M., Malone, T., 2011. A pilot study of core stability and athletic performance: is there a relationship? Int. J. Sports Phys. Ther. 6 (2), 63–74.

Sichting, F., Ebrecht, F., 2021. The rise of the longitudinal arch when sitting, standing, and walking: contributions of the windlass mechanism. PLoS One 16 (4), e0249965. https://doi.org/10.1371/journal.pone.0249965.

Siddle, H.J., Hodgson, R.J., Hensor, E.M.A., Grainger, A.J., Redmond, A.C., Wakefield, R.J., et al., 2017. Plantar plate pathology is associated with erosive disease in the painful forefoot of patients with rheumatoid arthritis. BMC Musculoskelet. Disord. 18 (1), 308. https://doi.org/10.1186/s12891-017-1668-0.

Siegler, S., Chen, J., Schneck, C.D., 1988. The three-dimensional kinematics and flexibility characteristics of the human ankle and subtalar joints—Part I: Kinematics. J. Biomech. Eng. 110 (4), 364–373.

Simkin, A., Leichter, I., 1990. Role of the calcaneal inclination in the energy storage capacity of the human foot—a biomechanical model. Med. Biol. Eng. Comput. 28 (2), 149–152.

Sims, A.L., Kurup, H.V., 2014. Painful sesamoid of the great toe. World J. Orthopedics 5 (2), 146–150.

Smith, J.W., 1958. The ligamentous structures in the canalis and sinus tarsi. J. Anat. 92 (4), 616–620.

Snow, S.W., Bohne, W.H.O., DiCarlo, E., Chang, V.K., 1995. Anatomy of the Achilles tendon and plantar fascia in relation to the calcaneus in various age groups. Foot Ankle Int. 16 (7), 418–421.

Spears, I.R., Miller-Young, J.E., 2006. The effect of heel-pad thickness and loading protocol on measured heel-pad stiffness and a standardized protocol for inter-subject comparability. Clin. Biomech. 21 (2), 204–212.

Stainsby, G.D., 1997. Pathological anatomy and dynamic effect of the displaced plantar plate and the importance of the integrity of the plantar plate-deep transverse metatarsal ligament tie-bar. Ann. R. Coll. Surg. Engl. 79 (1), 58–68.

Stearne, S.M., McDonald, K.A., Alderson, J.A., North, I., Oxnard, C.E., Rubenson, J., 2016. The foot's arch and the energetics of human locomotion. Sci. Rep. 6, 19403. https://doi.org/10.1038/srep19403.

Stolwijk, N.M., Koenraadt, K.L.M., Louwerens, J.W.K., Grim, D., Duysens, J., Keijsers, N.L.W., 2014. Foot lengthening and shortening during gait: a parameter to investigate foot function? Gait Posture 39 (2), 773–777.

Stuber, K.J., Bruno, P., Sajko, S., Hayden, J.A., 2014. Core stability exercises for low back pain in athletes: a systematic review of the literature. Clin. J. Sport Med. 24 (6), 448–456.

Su, S., Mo, Z., Guo, J., Fan, Y., 2017. The effect of arch height and material hardness of personalized insole on correction and tissues of flatfoot. J. Healthc. Eng. 2017, 8614341. https://doi.org/10.1155/2017/8614341.

Sun, P.-C., Shih, S.-L., Chen, Y.-L., Hsu, Y.-C., Yang, R.-C., Chen, C.-S., 2012. Biomechanical analysis of foot with different foot arch heights: a finite element analysis. Comput. Methods Biomech. Biomed. Eng. 15 (6), 563–569.

Sutherland, D.H., 2001. The evolution of clinical gait analysis. Part I: kinesiological EMG. Gait Posture 14 (1), 61–70.

Svendsen, J.H., Madeleine, P., 2010. Amount and structure of force variability during short, ramp and sustained contractions in males and females. Hum. Mov. Sci. 29 (1), 35–47.

Swords, M.P., Schramski, M., Switzer, K., Nemec, S., 2008. Chopart fractures and dislocations. Foot Ankle Clin. 13 (4), 679–693.

Taş, S., Salkin, Y., 2019. An investigation of the sex-related differences in the stiffness of the Achilles tendon and gastrocnemius muscle: inter-observer reliability and inter-day repeatability and the effect of ankle joint motion. Foot 41, 44–50.

Takahashi, K.Z., Gross, M.T., van Werkhoven, H., Piazza, S.J., Sawicki, G.S., 2016. Adding stiffness to the foot modulates soleus force-velocity behaviour during human walking. Sci. Rep. 6, 29870. https://doi.org/10.1038/srep29870.

Takahashi, K.Z., Worster, K., Bruening, D.A., 2017. Energy neutral: the human foot and ankle subsections combine to produce near zero net mechanical work during walking. Sci. Rep. 7, 15404. https://doi.org/10.1038/s41598-017-15218-7.

Takabayashi, T., Edama, M., Inai, T., Nakamura, E., Kubo, M., 2020. Effect of gender and load conditions on foot arch height index and flexibility in Japanese youths. J. Foot Ankle Surg. 59 (6), 1144–1147.

Thordarson, D.B., Schmotzer, H., Chon, J., Peters, J., 1995. Dynamic support of the human longitudinal arch. A biomechanical evaluation. Clin. Orthop. Relat. Res. 316 (July), 165–172.

Tong, J.W.K., Kong, P.W., 2013. Association between foot type and lower extremity injuries: systematic literature review with meta-analysis. J. Orthop. Sports Phys. Ther. 43 (10), 700–714.

Tong, J., Lim, C.S., Goh, O.L., 2003. Technique to study the biomechanical properties of the human calcaneal heel pad. Foot 13 (2), 83–91.

Umans, H.R., Elsinger, E., 2001. The plantar plate of the lesser metatarsophalangeal joints: potential for injury and role of MR imaging. Magn. Reson. Imaging Clin. N. Am. 9 (3), 659–669.

Usherwood, J.R., Channon, A.J., Myatt, J.P., Rankin, J.W., Hubel, T.Y., 2012. The human foot and heel-sole-toe walking strategy: a mechanism enabling an inverted pendular gait with low isometric muscle force? J. R. Soc. Interface 9 (75), 2396–2402.

van Werkhoven, H., Piazza, S.J., 2017. Does foot anthropometry predict metabolic cost during running? J. Appl. Biomech. 33 (5), 317–322.

Venkadesan, M., Dias, M.A., Singh, D.K., Bandi, M.M., Mandre, S., 2017a. Stiffness of the human foot and evolution of the transverse arch. https://arxiv.org/abs/1705.10371.

Venkadesan, M., Mandre, S., Bandi, M.M., 2017b. Biological feet: evolution, mechanics and applications. In: Sharbarfi, M.A., Seyfarth, A. (Eds.), Bioinspired Legged Locomotion: Models, Concepts, Control and Applications. Butterworth-Heinemann, Elsevier Science, pp. 461–486 (Part III, Chapter 7, Section 7.1).

Venkadesan, M., Yawar, A., Eng, C.M., Dias, M.A., Singh, D.K., Tommasini, S.M., et al., 2020. Stiffness of the human foot and evolution of the transverse arch. Nature 579 (7797), 97–100.

Viladot, A., Lorenzo, J.C., Salazar, J., Rodríguez, A., 1984. The subtalar joint: embryology and morphology. Foot Ankle 5 (2), 54–66.

Vogler, H.W., Bojsen-Møller, F., 2000. Tarsal functions, movement, and stabilization mechanisms in foot, ankle, and leg performance. J. Am. Podiatr. Med. Assoc. 90 (3), 112–125.

Vučinić, I., Teofilovski-Parapid, G., Erić, M., Tubbs, S.R., Radošević, D., Jovančević, B., 2020. Morphometric analysis of the patterns of calcaneal facets for the talus in Serbian population. PLoS One 15 (10), e0240818. https://doi.org/10.1371/journal.pone.0240818.

Wager, J.C., Challis, J.H., 2016. Elastic energy within the human plantar aponeurosis contributes to arch shortening during the push-off phase of running. J. Biomech. 49 (5), 704–709.

Wagner, E., Wagner, P., 2020. Metatarsal pronation in hallux valgus deformity: a review. J. Am. Acad. Orthop. Surg. Glob. Res. Rev. 4 (6), e20.00091. https://doi.org/10.5435/JAAOSGlobal-D-20-00091.

Wanivenhaus, A., Pretterklieber, M., 1989. First tarsometatarsal joint: anatomical biomechanical study. Foot Ankle 9 (4), 153–157.

Waseda, A., Suda, Y., Inokuchi, S., Nishiwaki, Y., Toyama, Y., 2014. Standard growth of the foot arch in childhood and adolescence—derived from measurement results of 10,155 children. Foot Ankle Surg. 20 (3), 208–214.

Watanabe, K., Ikeda, Y., Suzuki, D., Teramoto, A., Kobayashi, T., Suzuki, T., et al., 2017. Three-dimensional analysis of tarsal bone response to axial loading in patients with hallux valgus and normal feet. Clin. Biomech. 42, 65–69.

Wearing, S.C., Urry, S., Perlman, P.R., Dubois, P., Smeathers, J.E., 1999. Serial measurement of calcaneal pitch during midstance. J. Am. Podiatr. Med. Assoc. 89 (4), 188–193.

Wearing, S.C., Smeathers, J.E., Yates, B., Urry, S.R., Dubois, P., 2009. Bulk compressive properties of the heel fat pad during walking: a pilot investigation in plantar heel pain. Clin. Biomech. 24 (4), 397–402.

Wearing, S.C., Hooper, S.L., Dubois, P., Smeathers, J.E., Dietze, A., 2014. Force-deformation properties of the human heel pad during barefoot walking. Med. Sci. Sports Exerc. 46 (8), 1588–1594.

Weijers, R.E., Kessels, A.G.H., Kemerink, G.J., 2005. The damping properties of the venous plexus of the heel region of the foot during simulated heel-strike. J. Biomech. 38 (12), 2423–2430.

Weissman, S.D., 1989. Biomechanically acquired foot types. In: Weissman, S.D. (Ed.), Radiology of the Foot, second Edition. Williams Wilkins, Baltimore, MD, pp. 66–90 (Section II, Chapter 5).

Welte, L., Kelly, L.A., Kessler, S.E., Lieberman, D.E., D'Andrea, S.E., Lichtwark, G.A., et al., 2021. The extensibility of the plantar fascia influences the windlass mechanism during human running. Proc. R. Soc. B Biol. Sci. 288 (1943), 20202095. https://doi.org/10.1098/rspb.2020.2095.

Williams, D.S., McClay, I.S., 2000. Measurements used to characterize the foot and the medial longitudinal arch: reliability and validity. Phys. Ther. 80 (9), 864–871.

Williams, L.R., Ridge, S.T., Johnson, A.W., Arch, E.S., Bruening, D.A., 2022. The influence of the windlass mechanism on kinematic and kinetic foot joint coupling. J. Foot Ankle Res. 15, 16. https://doi.org/10.1186/s13047-022-00520-z.

Willwacher, S., König, M., Braunstein, B., Goldmann, J.-P., Brüggemann, G.-P., 2014. The gearing function of running shoe longitudinal bending stiffness. Gait Posture 40 (3), 386–390.

Wolf, P., Stacoff, A., Liu, A., Nester, C., Arndt, A., Lundberg, A., et al., 2008. Functional units of the human foot. Gait Posture 28 (3), 434–441.

Wong, Y.S., 2007. Influence of the abductor hallucis muscle on the medial arch of the foot: a kinematic and anatomical cadaver study. Foot Ankle Int. 28 (5), 617–620.

Wu, L., 2007. Nonlinear finite element analysis for musculoskeletal biomechanics of medial and lateral plantar longitudinal arch of virtual Chinese human after plantar ligamentous structure failures. Clin. Biomech. 22 (2), 221–229.

Yawar, A., Korpas, L., Mandre, S., Venkadesan, M., 2017. Transverse contributions to the longitudinal stiffness of the human foot. https://arxiv.org/abs/1706.04610.

Zhang, X., Aeles, J., Vanwanseele, B., 2017. Comparison of foot muscle morphology and foot kinematics between recreational runners with normal feet and with asymptomatic over-pronated feet. Gait Posture 54, 290–294.

Zifchock, R.A., Davis, I., Hillstrom, H., Song, J., 2006. The effect of gender, age, and lateral dominance on arch height and arch stiffness. Foot Ankle Int. 27 (5), 367–372.

Zifchock, R., Parker, R., Wan, W., Neary, M., Song, J., Hillstrom, H., 2019. The relationship between foot arch flexibility and medial-lateral ground reaction force distribution. Gait Posture 69, 46–49.

Zipfel, B., DeSilva, J.M., Kidd, R.S., 2009. Earliest complete hominin fifth metatarsal—Implications for the evolution of the lateral column of the foot. Am. J. Phys. Anthropol. 140 (3), 532–545.

Chapter 4

Pathology through the principles of biomechanics

Chapter introduction

iological life is a biochemical–biomechanical construction. Biomechanics is not an isolated medical discipline divorced from physiology and internal medicine, playing as it does pivotal roles in all physiological systems, which includes the locomotive system. Many diseases or nutritional insufficiencies cause biochemical changes in tissues that alter mechanical properties in multiple body systems. The study and application of biomechanical principles to pathology is referred to as *pathomechanics*. The underlying principle of lower limb locomotive pathomechanics is that the applied loading stress has breached the threshold of the tissue's mechanical properties, resulting in damage and dysfunction through the biological equivalent of yield and plastic deformation. The mechanical failure of any tissue occurs through an inability to utilise each hierarchical level's mechanical properties effectively for stress management. Failure of a microstructural element in a single functional unit, if unaddressed through healing, may eventually cause failure of a whole system's biomechanical properties resulting in locomotive failure. There are many 'tissue failure roads to pathological Rome', even if some of those roads are more frequently travelled. For example, failure to adequately provide compliance and energy buffering at impact may result in too much tissue resonance within the lower leg. Failure to adequately stiffen a foot at terminal stance may increase midfoot deflection and torsional deformation through bending moments on metatarsals, thus losing energy for acceleration as a result of structural deformation and instead, sending that energy into tissues where they can 'wreak havoc'. The precise tissue that fails depends on individually developed anatomy and lifestyle-influenced tissue properties. Thus, the same kinematic event may result in different types of tissue failure at varying precise locations among individuals, even when the kinematic events consistently focus certain strains into particular areas within a functional unit.

The key to this chapter is the application of biomechanical principles in order to explain pathomechanical situations within locomotion so as to help in understanding the origins of pathology. The clinician must utilise biomechanical principles to pinpoint the functional issues and to uncover the origin of dysfunction through clinical clerking (history-taking), physical assessment, and gait analysis. It is the application of mechanical laws, informed by anatomical knowledge, that focuses the clinician on to the salient features of locomotion that require investigation.

There is much yet to learn on how pathology occurs, and yet much can be deduced from studying the research on the biological response of tissues under the influence of stresses. It must be appreciated that insufficient stress causes atrophy while moderate to high levels of stress strengthens and improves the tissue's mechanical properties. However, excessive stress or fatigue can give rise to tissue failure. Through biological response principles, the levels at which stresses become pathological are set within each individual, but with a transitory nature reflecting exercise levels and health at any particular time. The levels of loading stress acting prior to high stress or repetitive stress events are principles that have been discussed in reflection to therapeutic interventions (Mueller and Maluf, 2002). Acute, high stress traumatic events are relatively easy to understand, whereas stress related to fatigue is more complex.

Tissue stress breaches are difficult to fully explain at times. For example, it is reported that certain musculoskeletal disorders such as Achilles tendinopathy (Wezenbeek et al., 2018), and symptoms such as patellofemoral pain (Collins et al., 2019) and foot symptoms (Mølgaard et al., 2010) are more commonly reported in female sports participants than in males. Sex-related differences in postural balance ability, muscle strength, joint stability, joint mobility, and tendon and muscle stiffness have been examined to explain this, but without any consensus being achieved (Taş and Salkin, 2019). This is hardly surprising, as different females and different males are far from homogeneous morphologically, let alone the fact that all of one genetic gender do not live the same lifestyle.

It is not the aim of this chapter to discuss all conditions that can disrupt the locomotive system or to cover all the mechanisms by which a certain structure can become injured. In this chapter, an attempt is made to show how biomechanical principles can be used to understand and solve the origins and principles behind pathology. Changes in locomotive biomechanics can lead to disruption of locomotive efficiency and energetics. Examples are given within the chapter which should help in the understanding of some pathomechanical events but also act as an aid in understanding other conditions that are not discussed here.

4.1 Understanding the pathological risk

4.1.1 Introduction

Injuries can involve any combination of the fundamental elements within a functional unit that maintain segmental and overall postural stability during motion through the principles of biotensegrity. Therefore, injuries or other dysfunctions of the locomotive system which involve bone, joint, ligament, fascia, muscle–tendon, and/or the coordinating nerve supply can all lie at the origin of dysfunction. Tissue dysfunction could be due to physiological changes, genetic or developmentally induced tissue weakness, or morphology. Traumatic injures arise because the mechanical properties have been exceeded in a single event. However, they are easier to sustain when tissues are fatigued through the principles of fatigue loading that lowers the ultimate strain capabilities. Healed tissue injuries or ageing can also reduce material strength. Thus, a risk of developing pathology can become an equation that can indicate a pathological risk:

$$G\&D + EA + T + A + EF + LS = PR$$

where
 G&D = Genetics and developmental influences (tissue behaviour and anatomic traits)
 EA = Environmental Adaptation (levels of activity, lifestyle mismatch, nutritional status, etc.)
 T = Traumatic events in the past (scar tissue formation, changed mechanics, altered range of motion, etc.)
 A = Age (tissue or physiological 'age' that may not reflect chronological age)
 EF = Extrinsic factors (footwear or gait perturbation)
 LS = Loading stress: direction and quantity
 PR = Pathological risk

Internally induced or externally applied excessive stresses that produce injury have been referred to as those which exceed the *tissue threshold* (Smit and Strong, 2020). Tissue threshold is far from constant across individuals or over the loading time, remembering that stresses required to induce tissue failure in a single event tend to be higher than in failures sustained during extended periods of activity causing fatigue. Healthy tissues in young adults, especially those who are regularly loading their tissues, can be expected to sustain high stresses without fatigue and/or failure. Those of sedentary, aged, or physiologically 'unwell' populations can expect their tissues to fail at lower stresses. Children and adolescents often present with different injuries that reflect either the mechanical properties of immature anatomy, developmental abnormalities, or high levels of sporting activities that their tissues are not yet mature enough to cope with. The starting point in understanding the influence of biomechanics is through clinically 'knowing' your patient.

4.1.2 Patient clerking

The clinical approach to resolving pathomechanics begins with patient clerking. This sets in motion a process of data gathering not only of the patient's complaint, but also establishing a tissue injury risk assessment. By adding the risk factors of a patient together, an idea of a level of their tissue thresholds can be appreciated. This involves adding up health and environmental (lifestyle) risk factors, a calculation that is referred to in this text as *pathomechanical summation*. Any pathomechanical summation needs a focus, based on the diagnosis of injury or pathology. This gives context to the summation, as collecting data on features not related to the diagnosis is wasting time. Factors need to be ruled in or out during the clerking and physical assessment process in order to focus on relevant findings. Recording ranges of motion or alignment in every joint in the lower limb can prove fruitless without a pathological focus.

Each physiological system should be investigated systematically by asking the right questions and by performing the correct examinations. The physiological systems can be remembered using the mnemonic CRAGICELS which stands for **c**ardiovascular, **r**espiratory, **a**limentary, **g**enitourinary, **i**mmune system, **c**entral nervous system, **e**ndocrine, **l**ocomotive systems, and **s**kin (see Table 4.1.2a).

TABLE 4.1.2a The physiological systems that need to be considered and some of the tissue and locomotive consequences linked to these systems.

Physiological system	Potential disease examples	Clerking significance
Cardiovascular	Heart disease, vascular disease, anaemia	Oxygen and nutrient profusion of tissues
Respiratory	Asthma, chronic obstructive pulmonary diseases	Gaseous exchange of toxic carbon dioxide and essential oxygen for tissue health
Alimentary	Irritable bowel syndrome (e.g. Crohn's disease), coeliac disease	Ability to absorb nutrients and vitamins
Genitourinary	Infections of the genitourinary and alimentary system can cause Reiter's syndrome	An occasional cause of autoimmune arthritis that can induce enthesopathies
Immune system	Autoimmune diseases, rheumatoid arthritis, psoriatic arthritis, etc.	Nonmechanically originating locomotive pain
Central nervous system	Parkinson's disease, motor neurone disease, peripheral neuropathy	Disruption of gait from neurological and sensory control rather than mechanical dysfunction directly
Endocrine	Diabetes, hypothyroidism	Changes in tissue properties, such as increased stiffness or compliance. Neuropathy.
Locomotive	Osteoarthritis (degenerative joint disease), sarcopenia	Loss of joint function, change in freedom of motion at certain locations, reduced muscle power
Skin	Psoriasis, diabetes	Skin lesions or fragility, causing mechanical fragility

The clinical approach to patient clerking should follow the well-established *SOAP approach* that aids the clinician to evaluate each physiological system and form the basis for clinical notes. SOAP stands for *symptoms, objective observations, analysis,* and *plan of action*, with the plan of action being a consequence of the other three. It is the symptoms, objective observations, and assessment that are critical for pathological summation and establishing the diagnosis. Analysis needs to be cautious and meticulous, for the effects of the pathology can distract the clinician from the origin. For example, a medially unstable acquired pes planus is likely to be caused by tibialis posterior tendon dysfunction, rather than tibialis posterior pathology being the result of the acquired flat foot. Yet, a previous failure of the spring ligament or a weakness within the tibialis posterior muscle may reduce the passive and/or active ability to maintain vault stability, triggering changes in foot profile which in turn overstrains the tibialis posterior tendon. Good clinical history, clinical assessment techniques of muscle strength, and ligament integrity are essential in establishing the origins of dysfunction. This can be guided via the *SOAP approach*.

For those unfamiliar with the SOAP approach, Table 4.1.2b will prove helpful and it has been expanded to include the now more commonly used *SOAPIER approach*, guiding the clinician towards the logical systematic understanding and resolution of a patient's issue (Cowley and Lepesis, 2018a, b, c). Several different suggested meanings have been assigned to both acronyms from different sources which is extremely unhelpful. It is therefore advisable to stick to those proposed by Cowley and Lepesis for consistency. Clinical clerking is a skill that needs practice in a systematic style that works for the clinician, as not every patient is helpful in divulging the most relevant information on interview. It is important that the

TABLE 4.1.2b The SOAP and SOAPIER approach to clinical clerking.

Patient clerking	Aim	Examples & outcomes
S = SYMPTOMS COMPLAINT/ HISTORY	To establish the nature of the symptoms/ issues that the patient reports	Pain = severity (use pain scale), onset, location, duration, character, aggravating factors, ease of irritability (how easily provoked), daily (temporal) patterns of pain. Disability = limitation on activities, losses of usual activities, etc.
O = OBJECTIVE EXAMINATION/ CLINICAL & LIFESTYLE HISTORY	What the clinician sees or finds that relate to the symptoms of the patient	Joint swelling, segmental alignment issues or limb asymmetries, muscle atrophy, temperature of affected area, circulatory perfusion, etc.

Continued

TABLE 4.1.2b The SOAP and SOAPIER approach to clinical clerking—cont'd

Patient clerking	Aim	Examples & outcomes
A = ANALYSIS OF INJURED AREA/ GAIT ANALYSIS	What is found on relevant clinical examination and gait analysis that relates to the symptoms	Joint range of motion, muscle strength, symptoms provoked in active motion/resistance or passive motion, ability to perform activities including gait/ running, etc.
P = PLANNING INTERVENTION TO RESOLVE ISSUE	What the clinician intends to do to resolve the issue. Formulating a treatment plan of action	Addressing activities, training programmes, footwear advice, orthotic intervention, rehabilitation, referral on, etc.
I = Intervention	Reporting the intervention planning with the patient's consent	Should result in a patient happy to follow a pathway of intervention and knowing what action is required from clinician and patient together
E = Evaluation	Check that the patient is aware of treatment plan and need for it	A final discussion before the patient leaves to check they have absorbed the information they require
R = Re-evaluation	Review patient to confirm interventions are working	To check the problem is improving or is resolved

clinician remains focused to the task in hand, remains polite, yet at the same time be firm in gathering relevant information from the patient. Questioning is an important clinical skill in that certain questioning approaches can prompt the patient to give a suggested answer, which can be unhelpful when only the facts are required. Questions can be divided into *closed questions* such as 'Where does it hurt?', *open questions* that invite answers such as 'When does it hurt?', *searching questions* such as 'Does it hurt during any particular activity?', and *probing questions* such as 'Why do you think these shoes are the cause?'. Leading questions such as 'Does it hurt most first thing in the morning?' are best avoided, as these can result in a patient answering in a way that they think the clinician would like them to answer.

The SOAP approach should highlight serious pathology or other issues that might impede treatment, known as *clinical flags* (see Table 4.1.2c). The most important of these are the red flags that identify serious health issues requiring urgent referral for specialist investigation or other interventions rather than first receiving assessment and intervention via biomechanics. Symptoms of red flags include those that are suggestive of malignant tumours or serious infection. The other flags usually throw up more challenges to treatment options rather than revealing urgent medical needs.

TABLE 4.1.2c Information gathered through patient clerking may reveal certain issues that require urgent further investigation or mark out situations that need to be considered in treatment planning. These are known as clinical flags.

Flag colour	Indicated by	Clues to issues during clerking	Examples
RED	SERIOUS PATHOLOGY	Symptoms unrelated to activity or weightbearing, that suggest nonmechanical origin to symptoms or if symptoms are disproportionate to history	Consider systemic diseases such as inflammatory arthropathies (multiple inflamed joints), neural emergencies (drop foot gait), infections (red hot swollen), bone tumours (night pain), etc.
ORANGE	PSYCHOLOGICAL BARRIERS	Beliefs of fears that may prevent patient from accepting diagnosis or following treatment protocols	Patient convinced of their own diagnosis and of the treatment this requires and they expect
BLUE	OCCUPATIONAL or HOBBY ISSUES	Need to perform job or partake in hobbies that prevent/limit treatment options	Patient may be restricted to footwear at work, or need to complete a marathon for charity
BLACK	SOCIAL ISSUES	Family pressures, social or work issues, medical legal cases	Pressure to maintain disability benefit, avoid issues at work or school, an ongoing court case without settlement
PINK	POSITIVE ATTITUDES	Patient likely to overdo or proceed through treatment programme too aggressively	Overly keen to get better and return to activities (often athletes)

Patient questioning should establish health status and either a sole diagnosis or a few potential diagnoses that are likely to be causing their problem. This should lead to an effective and relevant clinical and diagnostic examination of the patient to confirm a suspected diagnosis rather than a blind attempt to discover what is wrong by examining everything the clinician can think of. Following assessment, the diagnosis should be as specific as possible. Knee pain, hip pain, and ankle pain are symptoms. They do not constitute a diagnosis. For example, 'plantar heel pain' is a common term that is often used as a 'diagnosis' but it is a term that alone is insufficient in promoting an understanding of the underlying pathomechanics. *Plantar fasciopathy* is a common cause of plantar heel pain in adults due to degenerative and/or hypertrophic changes at and around the attachment (enthesis) of the plantar aponeurosis to the calcaneus. Yet, many other conditions can cause plantar heel pain, each having their own distinct pathomechanical processes. To think of all plantar heel pain as being caused by overstrain of the plantar aponeurotic attachment point would be, and indeed is, erroneous. The heel is subjected to impact forces that are attenuated by several mechanisms including the heel plantar fat pad. However, the fat pad is not under significant stress at heel lift, only at heel strike. The plantar aponeurosis, on the other hand, is under some low tensional stress after initial heel contact during activation of the windlass mechanism, which is soon reversed to slacken the aponeurosis on forefoot loading. It is under far greater tensional loads and is required for energy dissipation as a consequence of foot elongation under ankle dorsiflexion moments during late midstance. It is particularly stressed at heel lift, when it transfers forefoot loading stresses to the heel attachment under initial lengthening strain before any windlass mechanism-induced shortening. Therefore, heel pain at impact is more likely to derive from the fat pad than the plantar aponeurosis. Other heel pain pathologies than these two are also available. Knowing that the plantar heel pain symptoms derives either from damage and dysfunction of the plantar fat pad or the plantar aponeurosis completely changes the diagnostic and therapeutic approach to the potential pathomechanics and what aspects of locomotion the clinician needs to assess and treat.

4.1.3 Pathomechanical summation and tissue threshold

An effective locomotive system has healthy tissues providing adequate mechanical properties to support its efficiency during all daily lifestyle activities without injury or symptoms. It should allow pain-free, mechanically efficient walking without provoking fatigue, unless it be for good reason such as an extremely long walk. Fatigue is much more likely during running. All tissues and organs are organised as pre-stressed hierarchies (systems within systems) over different size scales that adapt their structure through cellular mechanotransduction that matches their loads (Ingber, 2006). Therefore, body tissues should match their lifestyle. Modern human lifestyles can often provide minimal physiological or locomotive stresses, requiring only relatively low material properties to what humans are capable of producing. A regular long-distance runner's requirements are much higher than those of the average activity-avoiding office worker. Certain athletic activities require a more particular structural morphology, so the more elite an athlete is, the nearer they need to conform to structure that best provides for the energetics of that activity. Even small dysfunctions may affect their performance. Thus, smaller dysfunctions will tend to affect those with the most compromised tissue health, but also those operating at the highest tissue stresses for their mechanical performance, such as elite athletes. There are a number of features that are particularly important in establishing pathomechanical summation, and the significance of the findings are only relevant in combination to the activity levels and the health of the individual. In the case of several common locomotive pathologies such as medial tibial stress syndrome (MTSS), there are associations with particular repetitive motions during sporting activities, such as endurance running.

The ideal information to gather on a patient in order to build a picture of their musculoskeletal tissue threshold capability is:

- Tissue quality
- Activity
- Alignment and morphology
- Instantaneous joint axis locations at the relevant joint(s)
- Features of gait (gait determinants)

Locating the instantaneous joint axis during gait is difficult. However, feeling how a joint moves throughout its range of motion nonweightbearing or in static stance postures gives clinical insight into how a joint is likely to function and where and how the axis is likely to translate during said motion, when in gait.

Tissue threshold

Moving anatomy, including when breathing and heart pumping, requires tissues to slide relative to each other causing structural deformation. These motions that generate forces within tissues should not breach the capabilities of cells or tissues that handle them. Tissues do not manage forces uniformly because different tissues have different material properties. A functional unit only works in health and with good energetics when each tissue fulfils its mechanical role. Collision events such as initial contact impact and terminal stance acceleration represent high points of tissue stress and are tasks that are most likely to risk exceed tissue threshold in that they create the most stresses. Risk of threshold breach increases if there is weakness or there are unexpected gait perturbations, even during periods of lower force generation such as during walking's midstance. Poor locomotive energetics or prolonged activity can lead to more tissue fatigue that results in altered tissue mechanics, such as reduced energy dissipation capabilities or internal muscle contraction force generation.

Different levels of impulse, pressure, or shear force can generate pathology in different individuals because the mechanical properties of humans are not equal. This is true between individuals, and these principles are also applicable to anatomy across both left and right limbs and within different anatomical structures within each limb. Pathological summation to assess the level of tissue threshold begins with establishing the health and age of the patient. This information gives clues to the quality of the tissues involved. Limb alignments, local strengths, and joint motion quality can indicate potential 'weak' spots or force/stress concentrations within the tissues. This gives a qualitative indication of the stresses required to produce pathology within the patient. Structural and functional integrity requires appropriate stability for controllable flexibility necessary during motion. These concepts have been discussed through the principles of biotensegrity (Smit and Strong, 2020). Approaching a patient's pathology through biotensegrity principles can be helpful in understanding that the failure of any single anatomical element can result in larger-scale and more global dysfunctions. Clinicians need to ask themselves whether the dysfunction is within a passive or an active 'cable element' of tension-inducing compression stability. Alternatively, is it within the bone strut or its moveable articular interface that provides for shape changes.

Neurological control permits selection of the appropriate level of tissue rigidity, but the myofascial units must be capable of providing the structural rigidity that has been requested. The force that results from the instruction must be applied in the direction appropriate for each tissue, bearing in mind that tissues demonstrate anisotropic properties. Thus, the stress may be applied at an appropriate level but in an orientation that is not advantageous to the tissue's anisotropic arrangement.

The length of time a tissue is loaded also has profound effects through tissue's time-dependent properties. Thus, faster transitory loads are generally dealt with by stiffer/elastic tissues that can handle higher loads for short periods. Slower loading provokes a more compliant response, but these loads are under lower peak forces, otherwise potential tissue mechanics could be compromised. This is a problem for diabetics who can reduce gait speed to lower forces, yet who also express soft tissues incapable of achieving the increased compliance necessary for the shift in biomechanical forces that such reduced gait speed changes require.

In clerking, tissues can be crudely classified as capable of either sustaining very high, high, medium, or low stresses. Young adult healthy tissues should be expected to perform at the highest levels under stress, whereas older physiological dysfunctional tissues tend to perform at the lowest levels. In health, mechanical properties of tissues should only reflect the level of in-life mechanical loading the patient partakes. At a young age, tissues should quickly adjust to increased loads through the biological response. Therefore, active individuals who subject their bodies to increasing stress should develop stronger tissues, and likewise, those who are more sedentary should express lower mechanical properties. Yet, it is important to appreciate that activity levels during childhood previously will influence the rate at which the body will adapt to higher loads, and the potential peak loads that can be achieved safely. Healthy tissues are only likely to become injured if they experience sudden large mechanical loadings, i.e. in the manner of a fall or some other traumatic event, or if they encounter increased activity levels that load new peak stresses too quickly for the biological response to remodel tissues appropriately for demand. Remodelling can occur locally or throughout the whole locomotive system, depending on the tasks performed that are increasing the load.

The age of the patient also profoundly affects the mechanical properties of the tissues and functional units. Children and adolescents are not small adults. They have distinct changing gait patterns and anatomy that is healthy and quite normal for their age, yet would be abnormal in adults. Mechanical properties are different between children and adults, not just because of relative size differences but also because of a lack of structural maturity and tissue organisation that comes with tissue growth and use. As a consequence of higher levels of cartilage within them, bones tend to be less stiff during childhood. Growth plates provide areas of mechanical weakness within bones (more flexibility at soft and hard tissue interfaces) so that high stresses in children are more likely to cause avulsion fractures than in healthy mature young adults. Tissue thresholds in children reflects the age of development of the bones (and soft tissues) and the amount of stress they receive (Hamilton and Gibson, 2006; Georgiou et al., 2007; Schuett et al., 2015; Yu and Yu, 2015).

Older adults also present a different kind of tissue quality issue. Often, the bone becomes weaker than surrounding soft tissue increasing the risk of insufficiency fractures or avulsions at attachments due to bone quality issues, premature bone fatigue, or other bone pathologies (Jose et al., 2015). Changes in tissue quality with age in adults occur in a more complex manner than within children. Chronological age and tissue properties match less predictably in the older adult than they do in growing children. Older adults demonstrate some consistent changes in gait parameters, but the timings of such changes are not so clear-cut. Overall, the quality of the tissues of older adults will not only reflect their age, but also reflect the peak tissue properties that were obtained during their life as well as their present activity levels and physiological health. The older adult is far more likely to be showing physiological deterioration, even if as yet, such changes remain subclinical. Thus, physiological age is more important than actual age.

Tissue quality in pathology should be viewed both systemically as well as locally. Systemic health involves disease processes that affect the physiology of tissues widely throughout the body, such as hypothyroidism, diabetes, or cardiovascular disease. Hormonal changes can have diffuse influences on tissue mechanics as well as their ability to heal. Lack of free-flowing circulation due to heart disease, peripheral arterial disease, or congestion of tissues through venous dysfunction can cause deterioration in the affected anatomy. Other systemic diseases such as inflammatory arthropathies tend to be more selective in the tissues they affect, examples being enthesopathies associated with psoriatic arthritis or joint capsular damage and arteritis associated with rheumatoid arthritis. However, many local tissue quality issues are related to previous injury, infections, and the formation of scar tissue that induces material property deficits within or around the healed wound.

4.1.4 Activity influence on tissue threshold

Activity is crucial to understanding pathomechanics for it dictates the nature, direction, and location of peak stresses on a structure. If a patient's symptoms are only related to a specific activity, then the kinematics and kinetics of that activity need to be considered. For example, data gathered from a walking patient who is suffering from a symptom experienced during running is unlikely to provide useful biomechanical data on the injury. This is much the same as performing a running gait analysis on a sedentary diabetic patient with foot ulcers, which is also futile. Thus, the clinical biomechanist will need to apply their knowledge to an understanding of the patient's activities, be they ballroom or ballet dancing, sprinting or endurance running, or fencing or playing tennis.

Running and walking data can prove difficult enough to gather if clinical equipment like treadmills are absent or space is insufficient. If an injury is related to soccer, rugby, or golf, it may be necessary to view specific playing positions in order to construct the full picture of the injury mechanism. Slow motion video recordings captured on tablets or mobile phones by the patient's friends, coach, or relatives during sport can save the clinician considerable time in finding the pathomechanical provocation. Discussions with coaching staff can also prove highly productive.

4.1.5 Alignment and morphology in tissue threshold

The body moves in the three orthogonal planes with (most commonly) the primary aim of translation in the sagittal plane. The major joints of the lower limb that facilitate sagittal plane motion include the hip, knee, ankle, midfoot, and the MTP joints, particularly the medial ones. Aligning these joint axes in for the same sagittal plane of motion is important for the foot to be able to act analogous to a passive spring (Hedrick et al., 2019). Thus, how joints align to each other through each segment is significant to tissue stresses and power generated within gait and activity. When a knee faces anteriorly but this positioning causes the foot to become dramatically externally rotated, an energetic problem is apparent. The morphology of bones and their articulation's profiles achieved through developmental processes are highly significant to the principles of aligning joint axes perpendicular to the line of progression. Injuries that change forces across the joints from altered soft tissue vectors and alignment-altering healed bone fractures, can create new segmental orientations that are unhelpful to gait energetics.

Acetabular and femoral anteversion, coxa vara and valga, genu/tibial varum and valgum, tibial torsions, and foot posture can all influence where and how forces are applied to the lower limb across all orthogonal planes. The alignment configurations of the joints also have a significant influence on the location of the instantaneous joint axis during gait. Recording intra- and intersegmental alignment is therefore important, and can often help explain the dominance of certain moments or forces across joints. Internal and external joint torques will tend to be influenced by anteversions and tibial rotations within the bone, altering transverse plane torsion and possibly taking the joint axis out of sagittal plane alignment across the entire limb. Osseous frontal plane orientations will alter joint varus and valgus moments. Varum alignments in stance and gait will increase varus (adduction or inversion) moments and valgus alignments will increase valgus (abduction or eversion) moments at the hip, knee, and ankle, depending on the location of the angulation. A foot making initial contact on a narrower step width is likely to increase varus (adduction) moments on the ipsilateral hip and knee because the GRF becomes more medially orientated to these joints, whereas the reverse is true for valgum alignments. Therefore, limb morphology, soft tissue function, and gait kinematics interplay to form a resulting GRF that may either significantly increase, decrease, or adequately set, frontal plane forces (Fig. 4.1.5).

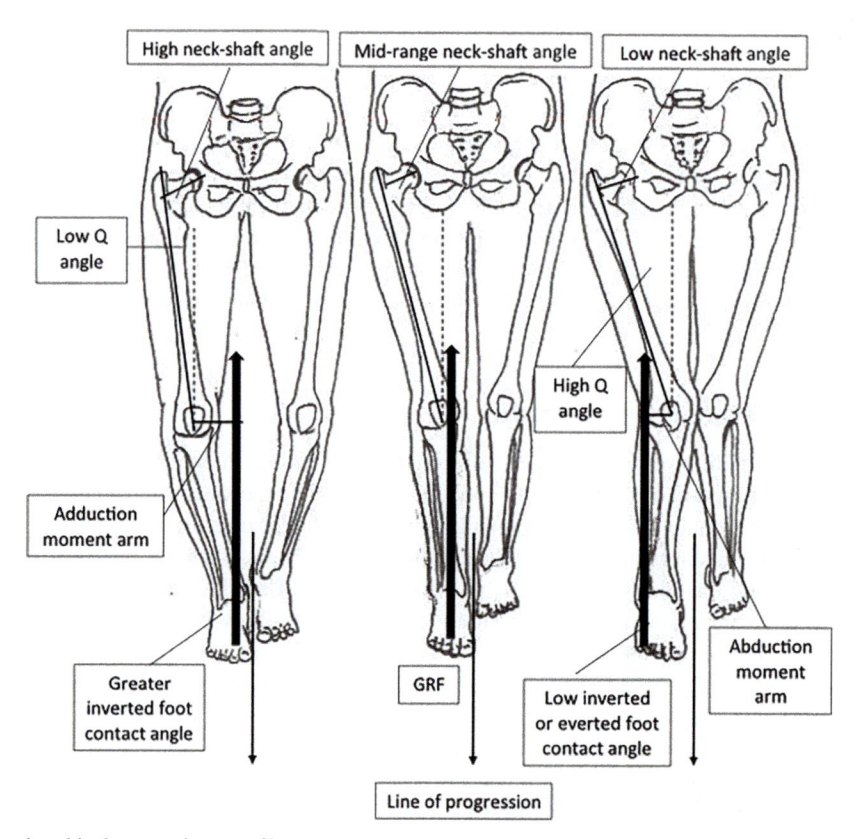

FIG. 4.1.5 Initial contact is a ubiquitous requirement of human locomotion that is not only affected by the contact area (heel or forefoot) but also by lower limb morphology of the individual. For example, a large femoral neck-shaft angle creates a coxa valgum, often coupled to genu or tibial varum (left image). This usually results in a low quadriceps (Q) angle, a narrow base of support along the line of progression (thin black arrow), and often a more inverted foot position at initial contact. The GRF (thick black arrow) from such a posture is positioned significantly medial to the knee, setting up a greater adduction (varus) moment arm from the GRF on the knee. This increases compression stresses on the medial compartment of the tibiofemoral articulation and raises tensions on the lateral soft tissues that restrain the adduction moments. A low femoral neck-shaft angle creates a coxa vara often coupled to a genu or tibial valgum (right image). This usually results in a large Q angle, a wide base of support, and a less inverted or even everted foot position at initial contact. The GRF is thus usually placed more laterally, decreasing medial knee compartment compression loads, but increasing those on lateral compartment. This also raises tensional stresses on medial soft tissues via the higher knee abduction (valgus) moments. The patellofemoral joint may be under more lateral dominant forces as a result. Although individuals that possess anatomy more between these extremes (centre image) are less likely to suffer such focused stresses within their knee and foot-ankle anatomy, they do not always avoid knee pathology. Such morphological variances away from the averages seen in populations are also not pathological conditions that need treating in themselves. An individual's anatomy may have adapted and strengthened where necessary for these particular gait biomechanics. Only if pathology is explained by excessive stresses being focused by the gait pattern, for example, medial compartment knee degeneration in a varus knee, will the frontal plane knee torques need clinical management. Where pathology does not match the expected locations of peak load during gait, changing gait patterns may not help or may even make situations worse. (*Permission www.healthystep.co.uk.*)

Vault (arch) profile has been extensively discussed in Chapter 3, Sections 3.1.4, 3.1.5, and 3.1.6, and its effects on gait are explored in Chapter 1, Section 1.4.3. It is sufficient here to remember that radius of curvature mechanics is at play, so that higher vault profiles have the potential to stiffen feet more. However, this is a mechanism that may complement and always interacts with soft tissue properties such as hypermobility–hypomobility (or something nicely in the middle), stress-stiffening properties that are affected by the extent of foot lengthening/widening, time-dependent properties from the loading rate, and the pedal muscular strength. All these mechanisms affect the foot's final (yet adaptable) instantaneous material property. High-vaulted feet that are hypomobile may not be able to adequately dissipate energy or provide shape changes necessary to improve contact, midstance, and acceleration mechanics (Honert et al., 2020). Thus, vault profile is important only when taken as a part of the foot's changing material properties and should not be used as the only determinant for initiating treatment.

4.1.6 Instantaneous joint axis location

Joint axes of rotation are not fixed but instantaneous, constantly changing with articular motion. For most joints, this changing motion occurs across the three body planes, but usually dominates in one plane more than any other. The hip, knee, and ankle joints change their axis location primarily influencing the sagittal plane rotations during motion. The subtalar joint provides more frontal plane motion than sagittal and has an axis that moves more medially within the transverse plane with eversion and more laterally with inversion. However, the subtalar joint works as part of the rearfoot with the ankle and extensively with the midfoot functional unit, with its own instantaneous axis, range of motion, and mechanics varying considerably between individuals (Jastifer and Gustafson, 2014). Changes in joint axis location result from both articular shape and soft tissue restrictions that change during motion. Appreciations of such motions of the instantaneous axis's position are important, as they should appropriately help alter the moment arms for muscles and tendons, and thus become a target for clinical interventions (Fig. 4.1.6a).

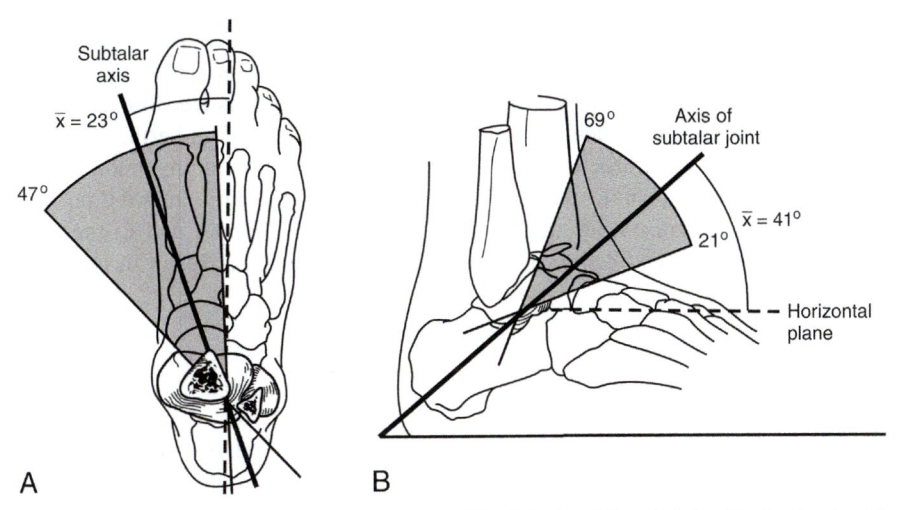

FIG. 4.1.6a Although literature providing average joint angles is widespread, its clinical usefulness is debatable, for there is wide variation and joint axis positions are instantaneous, i.e. constantly on the move during joint motion. The subtalar joint axis seems highly variable, yet its average position in the transverse plane away from the sagittal plane (A) and its inclination angle (B) are widely reported via averages gathered from research. Grey areas indicate variations reported within the literature. Clinical assessment of the individual's rearfoot motion is far more important than being aware of population averages, which may tempt clinicians into thinking that variation from these reported 'normals' represents abnormality. *(Image from Jastifer, J.R., Gustafson, P.A., 2014. The subtalar joint: biomechanics and functional representations in the literature. Foot. 24(4), 203–209.)*

Subject-specific variation is a therapeutic management complication that results from human anatomical variance in joint shape, alignment, and soft tissue restraints. The clinician should be aware of usual situations regarding joint axis motions, but it is worth assessing all of a patient's lower limb joints briefly to establish general levels of freedom and axis alignments between joints, without creating a need for collecting quantitative data on each articulation (or comparing such data to reported averages). The key is not establishing an average, but whether articular motions align flexion and extension throughout the lower limb to the line of progression. The clinician must be aware of any tendency for joint freedoms to change under different positioning of other joints, otherwise mistakes can be made in locating and establishing motion. This is more complicated within joints that are in close association, such as the ankle and subtalar joint complex or within the midfoot. Ankle dorsiflexion or plantarflexion may significantly alter motion of the subtalar joint when that joint is being loaded and assessed, while allowing motion at the ankle during testing will dramatically increase the perceived subtalar joint range of motion (Fig. 4.1.6b).

FIG. 4.1.6b A schematic that demonstrates the expected orientations of the talonavicular (T-N) and calcaneocuboid (C-C) joint axis alignments to each other when positioned within the subtalar facet maximum congruency (neutral) position (A) and when the joint is rotated into inversion (B). As discussed in Chapter 3, Section 3.3.9 (see Fig. 3.3.9g), variations in local anatomy mean that such changes within these joint alignments occur differently between individuals, so that stiffening from rearfoot to midfoot is going to involve different degrees of motion as well as varying amounts of soft tissue strength among patients. *(Image from Jastifer, J.R., Gustafson, P.A., 2014. The subtalar joint: biomechanics and functional representations in the literature. Foot. 24(4), 203–209.)*

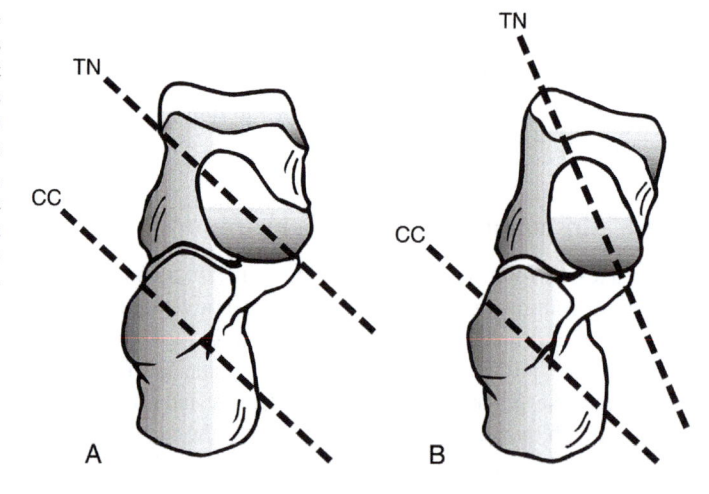

Loss of freedom of joint motion because of a minor subluxation (facet misalignment) or other blockage will alter the ability to move the joint axis, changing how forces and muscle moment arms cross the joint. This can lead to both pain and dysfunction, but not necessarily focused within the joint that loses its freedom. Such restrictions have been referred to as osteokinematic restrictions when the primary joint motion is lost, and arthrokinematic restrictions when accessory motions are lost (Hertel and Corbett, 2019). For example, following an inversion ankle sprain, reduction in ankle dorsiflexion or plantarflexion would be labelled as an osteokinematic restriction, whereas loss of talar anterior–posterior gliding would represent an arthrokinematic restriction (Hertel and Corbett, 2019). In reality, the loss of one type of motion is usually coupled to the loss of another, which can result in increased motion elsewhere to compensate.

4.1.7 Features of gait (gait determinants)

Gait determinants are features of temporal and spatial parameters and postural relationships throughout gait. Features such as step time, stance time, single support time, angle of gait, step length, stride length, and base or width of gait are all key features that indicate levels of gait efficiency. As discussed in Chapter 1, Section 1.2.6, base of gait or support (step width) influences the position of GRF forces medial or lateral across the foot and into the limb. It influences frontal plane forces throughout the lower limb while angles of gait influence the alignment across segments within the lower limb to the sagittal plane line of progression. The position of the GRF through stride length influences how well the body is stabilised due to such features as the divergence point through the placement of the GRF vector to the body's CoM's position (see Chapter 1, Section 1.2.2). These features are fundamental to the sagittal plane efficiency of gait and joint biomechanics via inter-joint postural relationships during locomotion.

Kinematic data regarding joint angles during gait, even if taken crudely from 2D video analysis, are essential in understanding pathomechanics. Knowing that a foot presented at a relatively high dorsiflexion angle to the ground during initial contact may be important to understanding anterior shin pain related to tibialis anterior overuse or the development of a calcaneal stress fracture. Both the size of the plantarflexion moment and the area of the calcaneus under initial impact are affected by the angulation of the foot to the ground during initial contact. The lack of ankle dorsiflexion during late midstance may present serious problems in maintaining the inverted pendulum mechanical efficiency. Such an ankle restriction is particularly a problem for the development of an efficient heel lift mechanisms through both restricting the freedom of the swing limb's ability to advance forward using the contralateral stance limbs anterior rotation, and also via reducing anterior displacement of the HAT segment above the stance limb. This is because the body's CoM has been unable to advance far enough ahead of the heel in readiness for contralateral foot strike before ipsilateral heel lift of the stance limb. Lost ankle dorsiflexion in midstance requires greater ankle power to facilitate acceleration through moving a larger body mass from behind the MTP joint fulcrum. This may cause injury within the triceps surae–Achilles complex through overstressing anatomy to achieve heel lift, but other tissue injury options are available to such gait dysfunction (Fig. 4.1.7a).

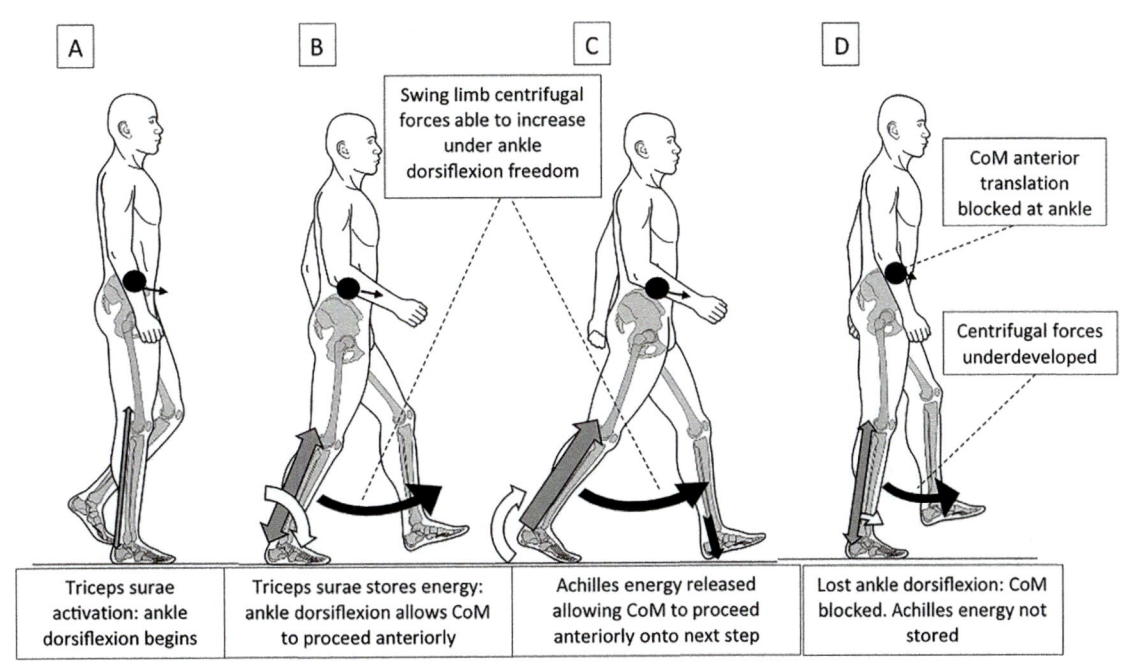

FIG. 4.1.7a Certain key events in gait are worth focusing on during gait analysis, at least until being ruled out as being related to the pathology. An example of such an event is the range of ankle dorsiflexion used during late midstance after the CoM's (black circle) high point at absolute midstance (A). It is important that the ankle provides free yet controlled ankle dorsiflexion to permit the CoM to pass anteriorly over the leg, as the swing limb passes the support limb to add centrifugal force 'drag' to the CoM's momentum during late midstance and into acceleration (B–C). This means the swing limb should advance via a reasonably lengthened stride to a position suitable to become loaded just after the support limb's heel lift. Ankle dorsiflexion also stretches the Achilles under triceps surae eccentric power, storing high levels of energy for acceleration. Thus, ankle dorsiflexion positions the CoM and the swing limb correctly, and creates the power for acceleration. Insufficient ankle dorsiflexion during late midstance (D) blocks the CoM's progression so that swing's centrifugal forces are under-developed, and thus posture will need to be altered within the trunk unless compensations in kinematics can maintain the CoM's momentum. One option to resolve this problem is an early heel lift in an attempt to maintain smooth anterior velocity of the CoM. However, this requires premature release of the Achilles' energy. Alternatively, increased midfoot dorsiflexion bending moments can compensate for loss of ankle dorsiflexion, with increased energy being moved from ankle motion into vault depression. This risks excessive plantar foot energy dissipation. Neither option, nor a mixed combination of both improves gait energetics from that of adequate and well-controlled ankle dorsiflexion motion. *(Permission www.healthystep.co.uk.)*

Another issue is limitation of MTP joint passive extension that may limit terminal stance mechanics, preventing the ankle from increasing its plantarflexion angle, without causing impingement to the dorsal aspect of the MTP joints' articular facets. MTP joint blockage of extension from osseous or soft tissue causes will interfere with ankle power used during acceleration by lengthening the external moment arm of the Achilles into the digits, which in consequence will restrict ankle and midfoot plantarflexion angles during terminal stance. MTP joint extension limitations may also result from weak ankle plantarflexors that are unable to achieve an effective ankle plantarflexion angle rapidly enough after heel lift so as to facilitate a sufficient passive MTP joint extension moment.

Increased hip flexion at initial contact will create higher loads on gluteus maximus in initially restricting and then reversing the flexion angle. Limited early midstance knee extension is associated with a more 'crouched' gait which will increase patellofemoral compression forces and limit the rise and fall of the body's CoM within midstance, reducing the efficiency of the inverted pendulum (Fig. 4.1.7b). The possibilities in alteration of gait efficiency and muscle loading that result from different combinations of joint angles are almost limitless. Therefore, the principles of efficient gait maintenance are absolutely an essential piece of the information jigsaw in the understanding of a patient's locomotive mechanics. Without any gait assessment of some form, the clinician can only guess as to what is occurring, and this is tantamount to clinical negligence. The eye alone is only providing one step up from a guess, but it is a far better alternative than never trying to assess the patient during locomotion.

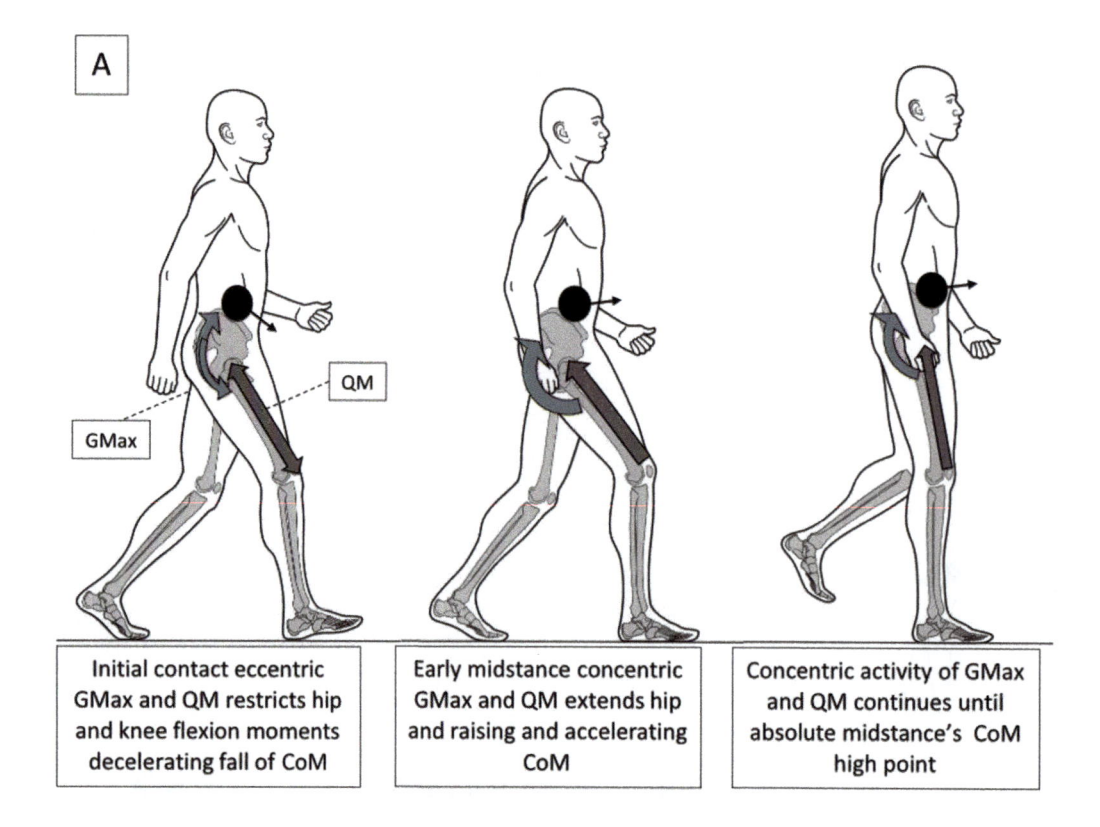

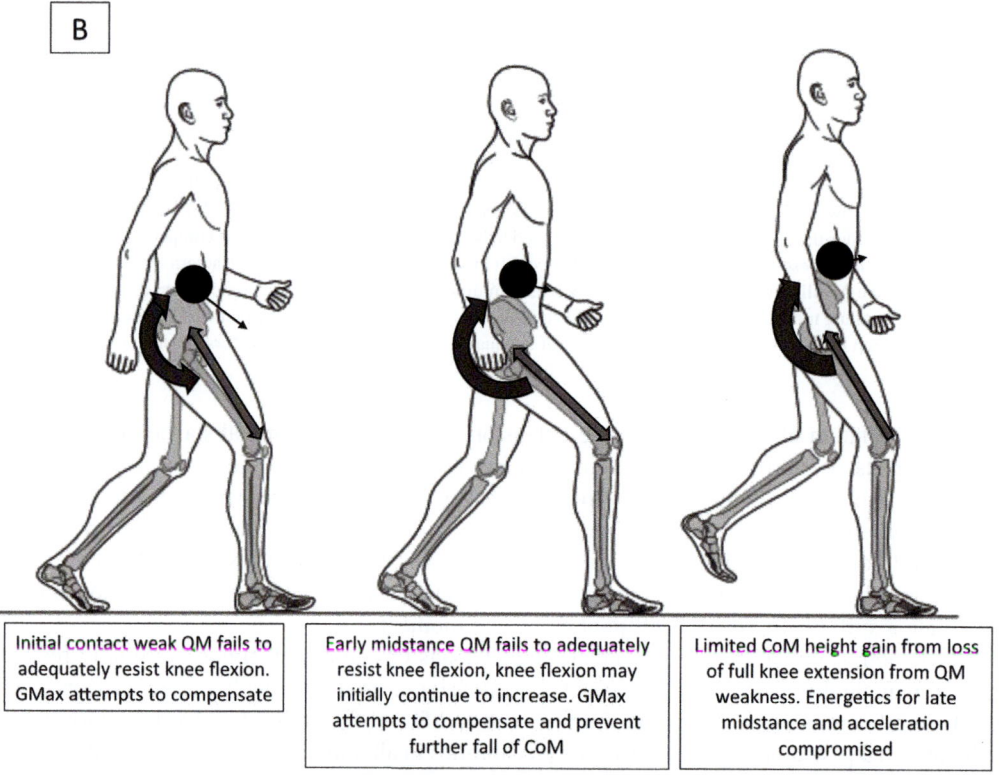

FIG. 4.1.7b Normally during walking (A), increasing hip and knee flexion angles during contact to loading response are controlled eccentrically by gluteus maximus (GMax) and the quadriceps muscles (QM). In early midstance, their concentric activity extends the lower limb to gain height for the CoM (black circle), reaching a high point at absolute midstance. Following this, potential energy from CoM height is converted into acceleration energy. If the QM show weakness (B), excessive hip flexion and/or knee flexion angles can develop during loading response, dropping the CoM too low, and this requires extending QM eccentric activity. GMax, by increasing its activity, can help to prevent further fall in the CoM's height by increasing hip flexion resistance or by extending the hip. At the end of the braking phase, greater GMax and QM activity is required to start to extend the hip and knee out of flexion during early midstance to gain height for the CoM. However, this ability is hampered by quadriceps weakness, preventing the lower limb extending to set high potential energy via CoM vertical positioning. In consequence, acceleration energetics are compromised before they initiate. Concurrent loss of GMax strength worsens the situation further, whereas GMax weakness alone causes issues for hip over-flexion which the quadriceps will need to compensate for by restricting the knee flexion range. Usually, QM are far more effective in compensating for GMax weakness than GMax for any quadriceps weakness during early stance. *(Permission www.healthystep.co.uk.)*

4.1.8 Soft tissue injuries

In locomotion, the only cutaneous tissue usually exposed to external loading stresses is the plantar skin. This is protected directly from the BWV above by the plantar fat pads of the feet during stance phase. Connective tissues are under constantly changing loads in gait via muscle activity and the external GRF. The more a muscle is given mechanical advantage by its moment arm position, the greater the efficiency of its active stability/stiffening ability, and the less metabolic muscle energy required to adapt tissue stiffness levels throughout gait. The more a muscle functions at its middle length, the more cross-bridges it can form to create muscle power. The slower the muscle contracts, the more cross-bridges it can form and the greater muscle force it can apply. These subjects of force–velocity relationships are discussed in Chapter 4 of the companion text 'Clinical Biomechanics in Human Locomotion: Origins and Principles', in Section 4.1.8, and are principles that lie at the heart of many injuries. Morphological alignment issues and changes in instantaneous joint axis location with motion influence joint mechanics that, in turn, affect the position in which a muscle has to activate. Perturbations that lead to osteokinematic or arthrokinematic restrictions influence the instantaneous joint axis and joint alignment within gait, and can result in changes in angular joint moments and force loading rates during locomotion. These alignment and joint motion issues can take muscles away from their preferred positions that usually provide them with the best mechanical advantage. Muscles operating at extreme lengths and under rapid loading rates are less able to apply appropriate stiffness and motion control. In such circumstances, passive elastic restraint takes on a greater role in stress management. One-off events are usually not problematic as long as no significant damage occurs within the myofascial tissues through a stress threshold breach. Repeated positions of compromised muscle biomechanics are likely to be more problematic and could, of course, themselves be set off by a single harmful gait perturbation that results in a change in joint kinematics, altering how a particular muscle or muscle group functions.

Once a muscle's efficiency is compromised, it has knock-on effects to the local supporting connective tissues, reflecting the ability or inability of the muscle to regulate the loading forces. If the muscle is relatively weak at the time of a perturbation, the size of the perturbation need only be very small, and then may not be identified by the patient as a significant event that they mention during clerking. In the presence of muscle and connective tissue weakness or poor sensorimotor function resulting from age or disease, the risks of breaching tissue threshold increase. In young, fit patients, significant gait perturbation events that lead to functional changes are usually remembered by the patient. If a passive connective tissue structure such as a ligament becomes compromised, then a joint may increase its freedom of motion or alter its direction of motion, which again can lead to increased demand on muscle action when in a more mechanically compromised position.

For example, increased rearfoot frontal plane eversion upon loading the foot will place the tibialis posterior muscle–tendon unit at a greater fibre length for controlling the eversion moment around the rearfoot. This means that the muscle will now be acting at a longer length and therefore disadvantaged in its ability to resist midfoot dorsiflexion through its tarsal and distal tarsometatarsal attachments, than it would be when positioned at a slightly shorter length. The speed of loading the eversion moment also affects the muscles ability to generate force via its force–velocity relationship. Changes in rearfoot kinematics could result from reactions with the terrain, lower limb and body posture above the support surface, and/or to pathology within the ankle's or foot's connective tissues. For example, previous injuries to the deltoid ligament, the spring ligament, or an interarticular ligaments of the subtalar joint during previous gait perturbation may result in increased reliance on tibialis posterior power and change the angle and speed of eversion moments applied during gait. However, similar problems could also result from degenerative changes within any of these connective tissue structures, or from weakness or degeneration of the tibialis posterior muscle or its tendon. In tendon damage or degeneration, there is the risk of reducing tibialis posterior tendon's contractile and elastic restraint used in protecting the connective tissues it works with when controlling motion within the medial rearfoot and foot vault. Thus, positional or strength dysfunction can arise from several different events and all result in further compromise in the tibialis posterior muscle's mechanical efficiency. Function will be further compromised when the foot is subjected to perturbations that increase eversion moments, or during times when gait requires increased ankle–foot stiffening through tibialis posterior activity. Decreased tibialis posterior mechanical efficiency, of whatever cause, are in consequence likely to lead to an excessively compliant foot, for tibialis posterior is an important vault stiffening element. As a result, the foot will be less able to resist vault depression in midstance or achieve midfoot stability through the proximal transverse vault at heel lift. Loss of the later ability will make terminal stance midfoot plantarflexion and adduction via tibialis posterior's combined action with peroneus longus far more difficult to achieve. Unless the eversion moment and/or tibialis posterior strength is resolved, the problem is likely to gradually intensify as peroneus longus midfoot eversion and abduction moments will start to dominate forces.

The principles of lost moment arm advantage discussed in this tibialis posterior example are applicable to any other joint moment and muscle scenario. Thus, principles can easily be reversed for peroneus longus if the eversion moment becomes an inversion moment across the rearfoot, and these principles can be applied to medial and lateral knee stability, or for that matter, flexion–extension moments throughout the lower limb. Lower limb morphologies that encourage certain angular moments more than others therefore place muscle lever arms in disadvantaged positions, increasing the risk of pathology in a coherent complementary manner. Thus, knee alignments of valgum increase abduction (valgus) knee moments, reducing the mechanical advantage of the medial knee stabilisers, risking increased tensile loading on the medial passive ligamentous and fascial restraints, and increasing lateral articular compressions. Knees in alignments of varum set up the mechanics for the reverse of this. This does not necessitate injury, for if the mechanical disadvantage is overcome due to appropriate muscle strength and connective tissue hypertrophic adaptation via the biological response, then the situation remains mechanically controlled and the tissue threshold un-breached. However, tissue property changes with acute injury, age, and/or disease may upset the biomechanical balance later in life.

Unequal vulnerability

Certain muscles are more vulnerable to disruption as a consequence of the principles of their muscle belly and tendon architecture. Pinnate muscles, through their fibre arrangements, are more likely to be operating with their muscle fibres at optimal lengths for length-tension force generation than are fusiform muscles. This enables a pinnate muscle to operate at a longer length for longer than a fusiform muscle can, before it fatigues. For a reminder of these concepts, see Chapter 4, Sections 4.1.3 and 4.1.4 within the companion text 'Clinical Biomechanics in Human Locomotion: Origins and Principles'. The nature and cross-sectional area of the particular tendons involved are also important. Energy storage tendons are more fatigue-resistant and are thus able to cope better with repeated loading events. However, they are more prone to the deleterious effects of ageing. For a reminder of these concepts, see Chapter 3, Section 3.3.9 within the companion text 'Clinical Biomechanics in Human Locomotion: Origins and Principles'. The cross-sectional area of a muscle and tendon indicates its influence on any joint torque, with larger cross-sectional areas of muscle–tendon resulting in a more significant joint torque. Not only is the size of the muscle important in considering pathomechanics, but it is also important when considering the effects on muscle–tendon length that certain joint positions make when predicting which muscle is most likely to be dysfunctional or experience dysfunction or disadvantage. Muscle–tendon units taken to long lengths do not function as powerfully, which is also true of those operating at a very short length.

Muscle and tendon injuries are clinically classified into first through to third degree and can be distinguished easily on MRI (Palmer et al., 1999). Practicalities of clinical practice often mean that clinical assessment of the injury alone may initially be called for before ordering a diagnostic image. Thus, being able to clinically quantify an injury is a useful skill. Muscle–tendon insults from first-degree strains involve a simple stretch injury, and consist of fibre disruption, interstitial oedema, and mild haemorrhage without loss of function being found on clinical assessment. Such a muscle injury may produce a feathery appearance of the muscle fascicles on MRI (Palmer et al., 1999). Mild loss of function on clinical assessment suggests second-degree strains, characterised by partial tears without retraction of tissue and haematomas that may be observed as perifascial fluid collections on MRI (Steinbach et al., 1994; Palmer et al., 1999).

Complete rupture of the myotendinous unit with complete loss of muscle function constitutes a third-degree strain. Clinically, this can be easily identified using muscle resistance testing techniques by the clinician very carefully resisting patient contraction of a muscle during motion, or the patient muscularly resisting articular motion being generated by the clinician. Rupture involves tissue retraction and its 'clumping' together (Bencardino et al., 2000). Differentiating partial ruptures from full ruptures can be clinically challenging and in some muscles such as popliteus, ruptures may be more difficult to establish than those of the triceps surae–Achilles complex or tibialis posterior. This may explain the low rate of clinical diagnoses of popliteus rupture compared to those diagnosed on MRI (Jadhav et al., 2014). Muscle complexes that provide motion through combined action, such as the quadriceps or medial hamstrings, can mask clinical presentation and make imaging necessary to confirm suspected partial or full ruptures. Contusions from direct impact usually correlate with a clear history of the point of impact.

Acute high tissue stresses can give rise to variable injury locations, depending on how and when they occurred. Tissue threshold-exceeding strains that occur due to either overuse, fatigue, or physiological degeneration tend to cause tissue injury near myotendinous junctions, tendon compression regions, or attachment points. Such locations tend to present mechanical weak points due to a locally large difference between mechanical tissue properties and via the increased risk of shear at these tissue interfaces.

4.1.9 Bone and articular injuries

Both bone and articular cartilage are structured to resist compression loading. Both structures require myofascial forces to maintain compression as the primary loading force, limit tension, and prevent shear to avert injury, while at the same time still permitting changes in bone and joint positions during motion. This is a concept most easily addressed through consideration of, and explained by, the principles of biotensegrity.

Bone injury

Bone is anisotropic and maintains a considerably higher tissue threshold in compression than in tension and shear. These properties are discussed in more detail in Chapter 3 within the companion text 'Clinical Biomechanics in Human Locomotion: Origins and Principles', Section 3.4.4. Long bones are subjected to bending stresses, whether they are positioned vertically like a column, angled like a strut, or positioned horizontally like a beam or cantilever. The deflection direction is highly influenced by their orientation. In the lower limb, only the metatarsals are subjected to high 'largely' horizontally positioned bending moments during midstance and at the heel lift boundary. Depending on the direction and orientation of the loading force, different sides of the long bone will be subjected to compression or tension. By activating the muscles on the tension side of the bone, tension force is reduced by the muscle's contractile compression force between its proximal and distal attachment. This should be achieved without inducing shear forces. Such issues are discussed in more detail in Chapter 2 with the companion text 'Clinical Biomechanics in Human Locomotion: Origins and Principles', Section 2.1.3. These are classic biotensegrity principles, with the soft tissues generating contractile tension that maintains compression on the bones throughout motion. The consequence of this interrelation between bone and soft tissue is that the high strength compression tissue threshold of bone is never breached during healthy locomotion (Fig. 4.1.9a).

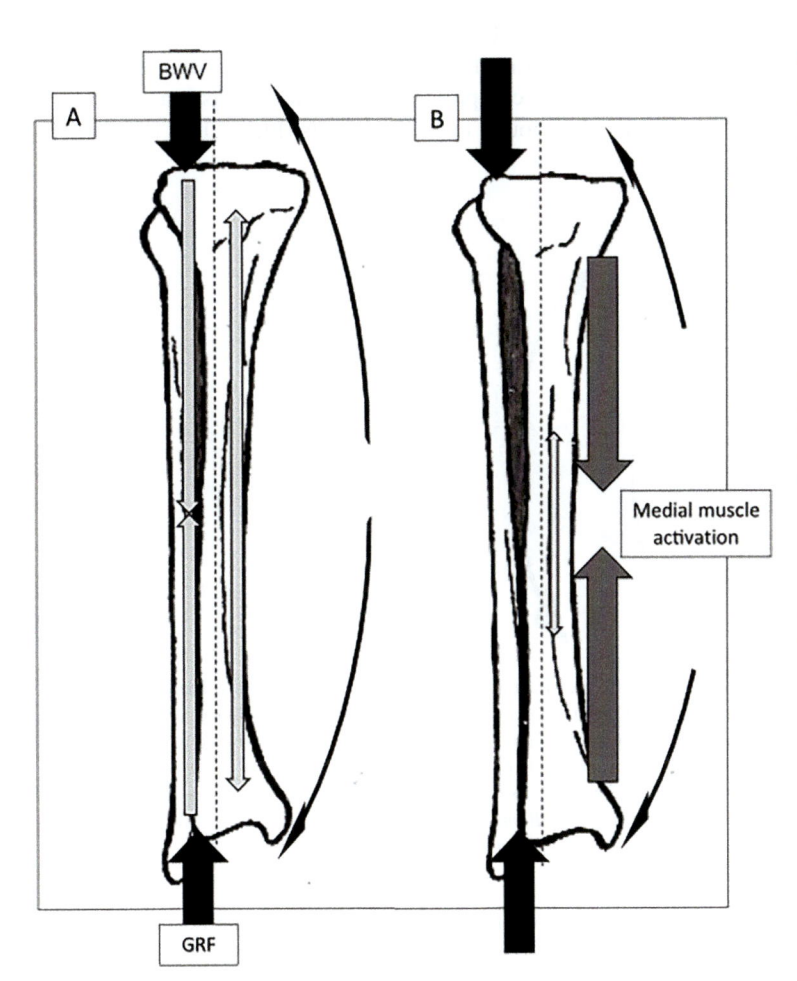

FIG. 4.1.9a Forces driven between the GRF and body weight vector (BWV) do not often align directly through a bone's long axis, but more usually focus towards one side or the other, creating a mild bending moment within the bone. Thus, bending moments across a long bone's long axis are created during locomotion. As a result, bone strain consists of both tensile (grey double-headed arrow heads directed apart) and compression elements (grey arrows directed towards each other), with the greatest strains being at the bone's surface. Muscle contraction on the side of tensile strain creates a compression force that opposes any tensile bone strain, making muscle strength an important component of osseous loading and protection. When such principles are applied to the lower leg (A), BWV and GRF vectors directed to the lateral side of the tibia and fibula (perhaps occur during an ankle eversion event) can create a laterally directed bending moment that tenses the medial tibia and compresses the lateral side of the tibia and fibula. By activating medial muscles (B) such as tibialis posterior, flexor hallucis longus, and flexor digitorum longus, some muscular compressive force (dark-grey arrows) resisting the tibial tensile strain can be created and the lateral bending force on the bone reduced. *(Permission www.healthystep.co.uk.)*

This simple mechanical protection can fail for a number of reasons. Muscles can fail to reprocess enough tension and shear into compression, and muscle weakness or fatigue are a prime reason for this. Muscle fatigue causing muscle strength loss is why this can even occur in young healthy athletic populations during extended activity. Loss of material strength lowers bone's overall compression strength, but more importantly, it lowers the more vulnerable bone tension and shear thresholds too. Even if muscles are able to reprocess tension and shear into compression, the bone may still fail if it is itself dysfunctional. Osteopenia and sarcopenia often coexist so that weak muscles fail to protect the more vulnerable bone, significantly lowering the stress tolerance levels and cyclical loading repetitions required before bone failure. The risk of fracture relates exponentially to the bone and muscle loss and the level of activity, especially if activity is suddenly increased outside of a normal sedentary lifestyle. Dancing-induced metatarsal fatigue fractures in the elderly is a known clinical phenomenon, especially after a Christmas or New Year party.

Insufficiency fractures tend to occur in older patients who have, for example, rheumatoid arthritis or an underlying metabolic bone disease such as hyperparathyroidism, osteomalacia, renal osteodystrophy, or if they have been undergoing long-term corticosteroid treatment. However, they can also occur in age-induced osteoporosis which is more common in women (Soubrier et al., 2003; Bryson et al., 2015). Osteoporotic fractures are associated with decreased bone density in areas of high levels of trabecular bone, causing insufficiency-induced compressive fractures at sites such as within the vertebral column (Geissler et al., 2015). Bisphosphonates, medically used to treat osteoporosis, suppress bone resorption by slowing the loss of trabecular bone. Long-term use, however, can lead to atypical transverse fractures in cortical bone shafts, unlike those found in high stress-induced mechanical or normal fatigue fractures (Geissler et al., 2015). This is likely due to a slowing of normal bone turnover, a process necessary for cortical bone health and the production of normal mechanical properties that bisphosphonates interfere with. Blocking this process probably results in failure of the diaphysis due to excessive stiffness through increasing collagen cross-linking, leading to advanced glycation end-products, and thus premature bone ageing (Geissler et al., 2015). The diaphysis in long-term bisphosphonate treatment tends to fail on the tension side of bone (Geissler et al., 2015). This is probably because of the effects of reducing the ultimate tensile stress of bone is a greater vulnerability when compared to reducing its ultimate compressional stress that starts at a higher level, but this may also reflect some loss of muscle strength that tends to be found in older patients (Figs 4.1.9b and 4.1.9c).

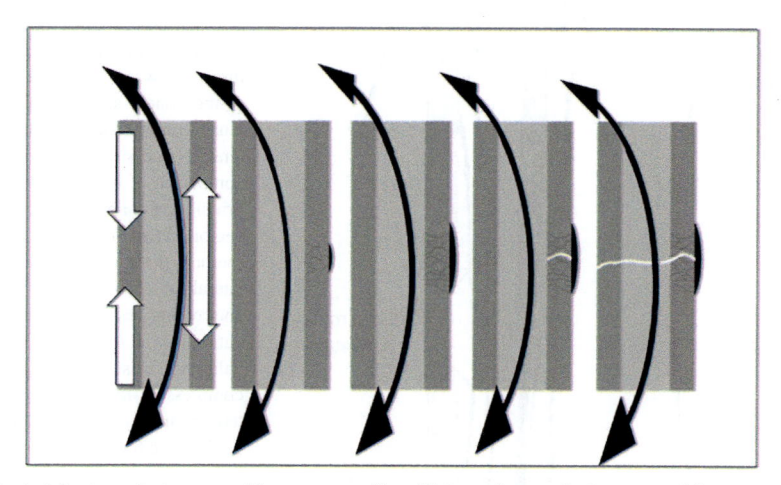

FIG. 4.1.9b A model for atypical fractures that present with symptoms of insufficiency fracture in those treated for osteoporosis. These fractures tend to occur on the tensile side of the diaphysis (double-headed white arrow), starting under the outer periosteal soft tissue layer of bone. When placed under bending moments (black double-headed arrow), one side of the long bone undergoes compression due to deflection (white arrows directed towards each other), while the other side tensions. Muscle contraction on the side undergoing tension should reduce this tension force, but when bone and muscle are in a pathological state, such muscular protective mechanisms may not reduce tension strains sufficiently on bone that is expressing lower bone tissue thresholds. *(Image from Geissler, J.R., Bajaj, D., Fritton, J.C., 2015. American Society of Biomechanics Journal of Biomechanics Award 2013: Cortical bone tissue mechanical quality and biological mechanisms possibly underlying atypical fractures. J. Biomech. 48(6), 883–894.)*

Normal trauma-induced fractures result from high stress traumatic events breaching healthy bone tissue thresholds. Injuries are predictable to the direction of the force application, where rotational shears and tensile strain usually cause mechanical failure. Pathological bone fractures do not require high stress and are not so predictable to the direction of the applied force. Osteomalacia due to low levels of vitamin D weakens both bone and muscle, compounding bone fragility by removing protective muscular torsion and shear strain protection. A version of osteomalacia referred to as *tumour-induced osteomalacia* or *oncogenic osteomalacia* is caused by decreased renal reabsorption of phosphate and low levels of active vitamin D, and is triggered by a difficult-to-locate tumour. The tumour causes diffuse bone pain, fragility fractures, and muscle weakness (Minisola et al., 2017). The bone tissue threshold may also be lowered due to local bone pathology, such as the presence of primary or secondary tumours. This can result in the generation of fractures under extremely low stresses, as abnormal tumour cells destroy normal bone architecture (Bryson et al., 2015; Kimura, 2018).

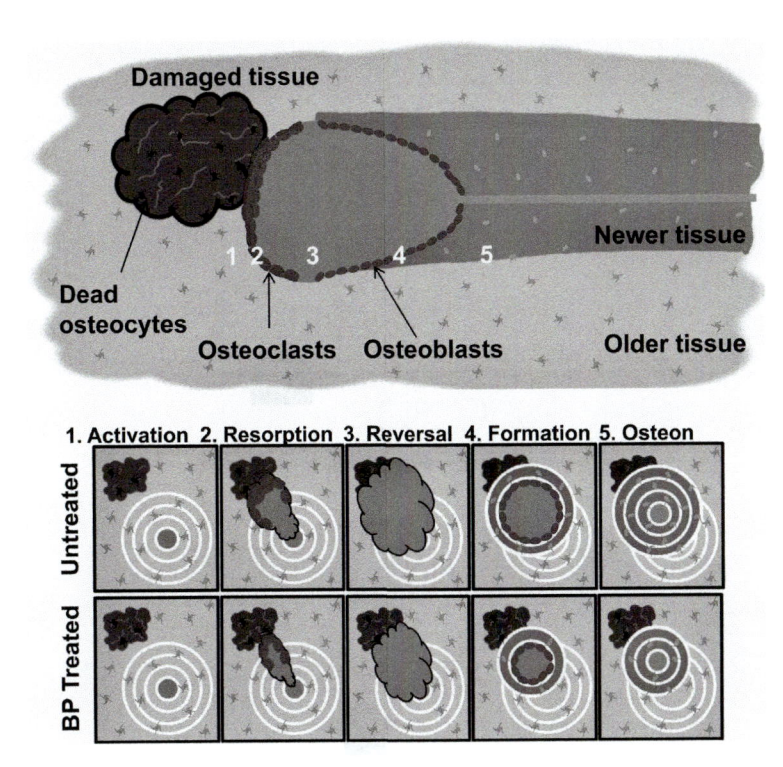

FIG. 4.1.9c Osteoclastic tunnelling remodels cortical bone (top image). This is an essential process of bone health that may explain the observation of decreased osteonal size, osteocyte lacunae density, and increased micro-crack length often found with long-term bisphosphonate-treated (BP-treated) cortical bone tissue that changes its mechanical properties (bottom image). As shown in the top image, osteoblasts normally fill in any bone resorption created by osteoclasts seeking damaged tissue. In so doing, these cells organise a central canal for the blood and nerve supply of the Haversian system. In the bottom images, 'wear and tear' bone damage in the young and healthy (untreated) is removed at a regulated rate that prevents coalescence of bone into a less unstructured mass. Long-term bisphosphonate treatment may affect bone quality via altering many of bone's basic multicellular units through interfering with normal repair and replacement processes. *(Image from Geissler, J.R., Bajaj, D., Fritton, J.C., 2015. American Society of Biomechanics Journal of Biomechanics Award 2013: Cortical bone tissue mechanical quality and biological mechanisms possibly underlying atypical fractures. J. Biomech. 48(6), 883–894.)*

Articular failure

Free-moving synovial joints need the mechanically integrated complex layers of articular cartilage, the subchondral bone beneath, the ligament guidances/restraints, and the muscles that move and protect them through stabilising compression. Thus, safe joint biomechanics requires effective connective tissue function, and any change in this relationship underlies the development of osteoarthritis (OA) (Varady and Grodzinsky, 2016). Articular cartilage is a biphasic structure that relies on its combined fluid and solid phase to provide compression resistance and impart an extremely low coefficient of friction. These concepts are discussed in detail in Chapter 3 within the companion text 'Clinical Biomechanics in Human Locomotion: Origins and Principles', Sections 3.5.3 and 3.5.4. It is worth stating again that joints suffer compression well, as long as the loading duration in one position is not sustained, otherwise its biphasic properties can be gradually lost whereupon it then has to rely only on fairly inferior solid phase cartilage mechanical properties alone.

A joint that is poorly aligned and/or stabilised can experience abnormal shear or localised concentrations of compression that can breach the tissue threshold, an example being the knee under large abduction or adduction moments. Once the cartilage's complex structure is disturbed, it loses its poroviscoelastic and loading rate (time dependent) properties (Varady and Grodzinsky, 2016). If a certain part of a joint is persistently loaded rather than spreading its compression strains over significant and changing cartilage areas, then local structural failures can occur. This leads not only to local mechanical failure, but also to failure of the total mechanical integrity of the whole cartilage on the articular surface. Cartilage has particularly poor healing properties, so avoiding cartilage damage is essential to maintaining healthy joint function.

Recurrent joint impingements, where opposing articular surfaces meet bone around the edges, are a particular problem. If bone on one side is able to compress cartilage on the other, rather than creating a cartilage–cartilage interface of materials with similar mechanical properties, then the far stiffer bone can compress into cartilage, causing trauma. Loss of the superficial zone of cartilage and via fibrillation through it, results in accelerated cartilage degradation and prevents normal transfers of compressive loads across deeper zones (Varady and Grodzinsky, 2016). It is thus important that bone aligns across joints so that forces from articular compression cannot concentrate into small areas of the superficial cartilage to initiate damage. Muscular and ligamentous positioning of the instantaneous axis of rotation and the restraint and control of the bone motions during gait is essential in preventing this from happening (Fig. 4.1.9d).

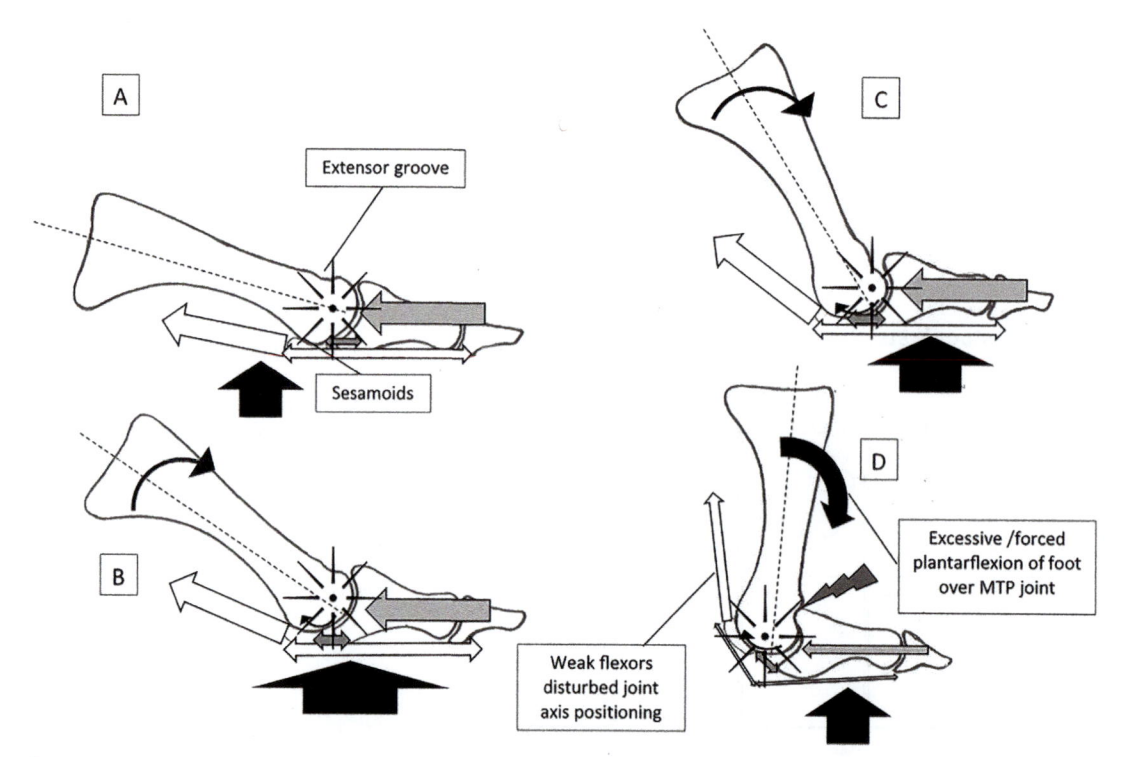

FIG. 4.1.9d A simple example of how muscle activity can prevent joint impingement is illustrated using the 1st metatarsophalangeal (MTP) joint. At the end of midstance (A), the angle of metatarsal declination (dashed line) to the support surface will be at its lowest with the GRF (vertical black arrow) still lying proximal to the 1st MTP joint. Rising hallux flexor activity via flexor hallucis longus tendon (white double-headed arrow) and the flexor hallucis brevis tendon and its sesamoid attachments (short grey doubled-headed arrow), provide a combined proximal flexor compression vector (large white-grey arrow) that pulls the proximal phalanx into the metatarsal head, stabilising it via a proximal directed vector (large grey arrow). At heel lift (B), the metatarsal head starts rotating over the sesamoids as the metatarsal declination angle increases. The flexor muscles now provide a vector of restraint against the MTP extension moment derived from ankle plantarflexion following heel lift and the rising GRF's effects on digital extension. These forces and motions help reposition the instantaneous joint axis (star) superiorly and anteriorly within the metatarsal head to allow improved rotation freedom with the concave proximal phalangeal articular surface. Muscle forces are aided by other connective tissue force vectors directly stabilising the metatarsal head as it rotates over the sesamoids, which together prevent hyperextension of the digit at the end of terminal stance before preswing (C). Failure to sufficiently provide restraint to digital extension (D) can cause hyperextension impingement along the dorsal joint line (lightning bolt). This may result from weakness within the hallux flexors' restraint or because of an altered proximal phalanx arc of rotation from a disturbed instantaneous joint axis. However, sudden exposure to a large extension force on the digit (usually from the foot going into forced ankle plantarflexion on an extended hallux) may still overcome normally adequate flexor strength, despite the axis being perfectly positioned. This is especially a risk during late terminal stance when digital flexor activity is reducing. Joint morphology, such as a shallow extension groove, may also increase some individual's susceptibility to such pathology. *(Permission wwww.healthystep.co.uk.)*

Subchondral bone mechanics forms the solid foundations of the joint and is important for articular cartilage, helping dissipate compressive loads through its 'spongy' trabecular construction. Increased stiffening of subchondral bone through sclerosis leads to loss of osseous compression protection of the cartilage via reduced subchondral flexibility under load. Subchondral sclerosis raises stiffness under the cartilage, losing any capacity of subchondral bone to dissipate energy away from the cartilage. However, too much 'give' or deformation within the subchondral bone is also an issue, for it undermines the foundations of cartilage mechanics that need to be operating over a stable base surface. Subchondral insufficiency fractures can be particularly debilitating causing the loss of normal joint contours required for spreading load and it can lead to osteonecrosis. Subchondral fractures can result from osteoporosis (Nelson et al., 2014; Jose et al., 2015; Gaudiani et al., 2020), but many are a result of acute high torque stress injuries across joints. Both events can lead to OA (joint degeneration) from changes in subchondral topography and/or levels of joint compliance or stiffness.

OA is a disease of the entire joint organ that destroys its normal mechanical properties. Beyond micro- and macro-scale biomechanics, cartilage integrity and tissue property threshold, like that of all tissues, is still dependent on normal biochemistry through physiological processes (Varady and Grodzinsky, 2016) (Fig. 4.1.9e).

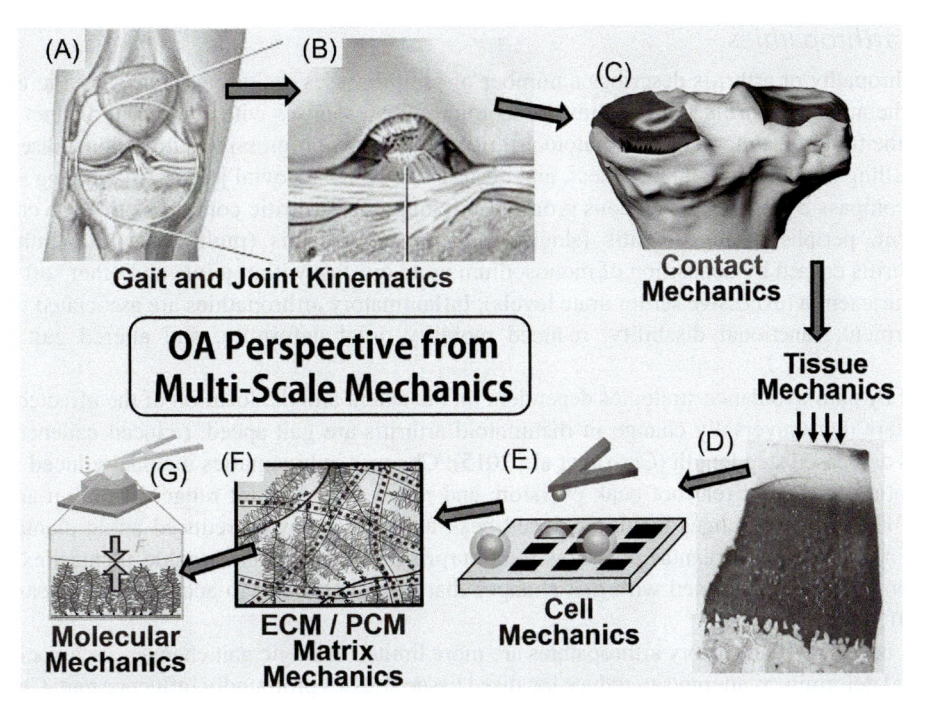

FIG. 4.1.9e Osteoarthritis (OA) pathomechanics has included traumatic effects of injuries such as anterior cruciate ligament (ACL) ruptures, but the principles of altered kinematics from loss of articular stability apply in all situations, be they traumatic or dysfunctional in origin. Thus, altered kinematics, changes in joint contact areas, and changing tissue mechanics influence cellular mechanics that cause matrix and molecular degeneration that will ultimately lead to micro and macro-scale articular dysfunction. *(Image from Varady, N.H., Grodzinsky, A.J., 2016. Osteoarthritis year in review 2015: mechanics. Osteoarthr. Cartil. 24(1), 27–35.)*

4.1.10 Disease-induced pathomechanics

Disease processes change tissue properties in different ways, so that the viscoelastic tissue properties expressed throughout gait are altered. Normal, healthy viscoelastic properties not only depend on the individual's lifestyle-induced tissues properties maintained in the present, but they also reflect those of the past, including intrauterine and childhood development. With disease and age, mechanical properties are no longer what they were when at their best. Thus, the diseased and/or aged patient is most at risk of breaching tissue threshold. Indeed, many diseases cause rapid physiological ageing of tissues, at a much rate faster than we should expect associated just with time.

Most disease processes do not afflict tissues evenly. Inflammatory arthropathies favour destruction of the connective tissues of joints and tendon sheaths (e.g. rheumatoid arthritis) or the enthesis (e.g. psoriatic arthritis). The active components of biotensegrity also take a hit from nervous system diseases such as neuropathy, and muscular atrophy and dystrophies. Sometimes both the muscles and the nerves are struck simultaneously in diseases of neuromuscular dystrophy. In diabetes, the nerves and both soft and hard tissues are damaged simultaneously. The net result of diseases in relation to their effects on tissue threshold is pretty much the same, in that they are lowered. Diseased patients are thus more easily harmed and recover their tissue function less consistently or not at all through dysfunctional healing and an abnormal biological response.

It is impossible to cover all systemic diseases here. However, the disease effects on tissue usually follow a few simple paths such as poor nutrition, poor metabolism, premature and faster rates of cell death, with excessive atrophy and tissue loss as commonly expressed by osteopenia and sarcopenia. Neuropathy tends to first alter control of the locomotive system, but then the lack of nerve stimulation is quickly followed by sarcopenia with poor active muscle strength, function, and balance because they are coupled with poor sensorimotor information from and to the muscles. Two particular groups of pathologies are worthy of some discussion here. Inflammatory arthropathies can cause considerable articular and connective tissue damage and diabetes, being as it is an ultimate disease of biochemical changes, can alter tissue biomechanics profoundly. Diabetes is a multisystem disease that causes neuropathy and poor tissue nutritional status through associated peripheral vascular disease. As a consequence of advanced glycation end-products, it can cause changes that alter the very nature of collagen mechanics. Both diabetes and inflammatory arthropathies are common, and are increasingly represented in patient populations that are arriving in biomechanics clinics.

Inflammatory arthropathies

Inflammatory arthropathy or arthritis describes a number of joint diseases including gout, psoriatic arthritis, ankylosing spondylitis, and rheumatoid arthritis. They all involve autoimmune responses with somewhat distinct differences in their presentation and the tissues they target. Rheumatoid arthritis is a chronic progressive autoimmune disease presenting with multiple joint swelling of symmetrical tenderness, and destruction of the synovial joints. Ankylosing spondylitis and psoriatic arthritis encompass a more heterogeneous group of inflammatory arthritic conditions that are characterised by vertebral involvement, peripheral oligoarthritis (single joint) or polyarthritis (multijoint), and enthesitis. Gout is an inflammatory arthritis caused by deposition of monosodium urate crystals within joints and other soft tissues and is associated with hyperuricaemia (excessive serum urate levels). Inflammatory arthropathies are associated with lower limb and foot pain, impairment, functional disability, reduced mobility, joint deformity, and altered gait strategies (Carroll et al., 2015).

Gait is altered by pain-avoidance strategies dependent on the extent and the location of the affected joints. The spatiotemporal parameters that universally change in rheumatoid arthritis are gait speed, reduced cadence, increased double support time, and decreased step length (Carroll et al., 2015). Changes in kinematics include reduced sagittal plane ankle joint range of motion, increased rearfoot peak eversion, and reduced ankle joint range of motion and angular velocity (Carroll et al., 2015). Kinetic changes include reduced peak ankle joint power, reduced ankle plantarflexion moments, and an increase in peak forefoot plantar pressures. Not surprisingly, reduction in ankle plantarflexor muscle strength has also been reported to be associated with gait changes that are often found to accompany increasing foot deformity (Carroll et al., 2015).

Although data on other inflammatory arthropathies are more limited, antalgic gait changes are associated with foot pain, enthesopathies, and deformity as attempts to reduce localised tissue stresses profoundly influence gait. Changes in spatiotemporal parameters and other kinematic and kinetic events around the ankle possibly reflect attempts to reduce stress on the Achilles tendon as is, for example, reported in the gait of patients with psoriatic arthritis (Woodburn et al., 2013). Gait data combined with good patient clerking techniques can often raise the index of suspicion with regard to the onset of these diseases before the diagnosis is known to the patient who appears to be attending with common musculoskeletal dysfunction, possible injury, and pain. Data taken on inflammatory arthropathy patients can also prove helpful in assessing treatment outcomes, or for identifying any deterioration in their condition over time through changes in gait parameters (Barkham et al., 2010).

The inflammatory arthropathy group presents with tissue thresholds lowered within specific tissues, such as joint structures and at muscular and ligamentous attachments. These will have biomechanical/biotensegrity consequences. However, without understanding the nature of these diseases and addressing their autoimmune origin, these patients cannot make progress through biomechanical interventions alone.

Diabetes mellitus

Diabetes is increasing most among those aged 65 years and over, exceeds 20% in some populations, and is expected to increase four- to five-fold by 2050 (Kirkman et al., 2012). Although type I autoimmune-induced diabetes is more associated with younger patients, clinicians need to be aware than even type II diabetes can occur in children. Diabetes is an exceptional endocrinological disease that causes multisystem ageing. Advanced glycation end-products are thought to underlie many of the biomechanical complications of diabetes, including increased skeletal fragility resulting from altered collagen stiffness through increasing the collagen fibre cross-linking within the matrix of bone (Saito and Marumo, 2015; Karim and Bouxsein, 2016; Varady and Grodzinsky, 2016). However, the primary change is an increased connective tissues stiffness and a loss of overall tissue compliance. Thus, tissue energy dissipation and safe energy storage during gait becomes an issue. Changes in soft tissue properties are a contributory factor to tissue failure, including the development of ulcers within the plantar tissues of the foot which can eventually result in amputations (Pai and Ledoux, 2012).

The viscoelastic properties of diabetic soft tissue are quite different to those of healthy soft tissues (Naemi et al., 2016a, b). There are vast differences reported in both shear and compression stiffness within diabetics, making it far harder for them to dissipate impact/collision-induced stress energies during gait, with the mechanical burden directed into the feet in particular. For example, diabetics' plantar cutaneous soft tissues have steeper linear elastic zones compared to normal plantar soft tissues (Klaesner et al., 2002; Hsu et al., 2009; Pai and Ledoux, 2010, 2012). Plantar soft tissues in diabetics are tending towards being thinner compared to age-matched nondiabetics. However, more importantly, their peak stress is significantly higher, whereas their peak strain is not significantly different (Pai and Ledoux, 2010). This may reflect stiffer plantar fat pad tissues while skin properties remain more similar to those of healthy controls (Kwak et al., 2020), rather than resulting from any actual atrophy of the plantar fat pads (Waldecker and Lehr, 2009). These changes indicate compromises in compression energy and shear strain protection properties within the plantar fat pads of the feet (Figs 4.1.10a–4.1.10c).

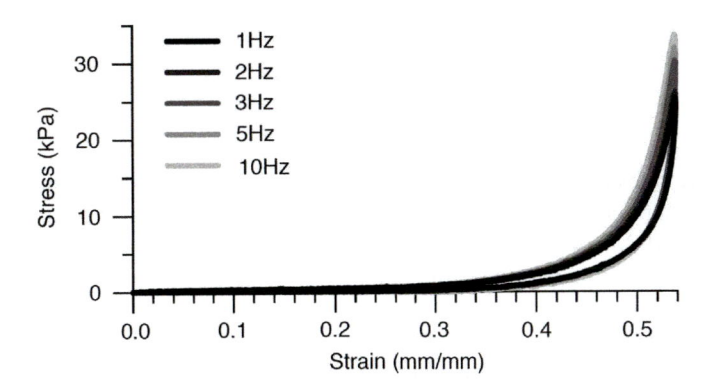

FIG. 4.1.10a Plantar soft tissue nonlinear stress–strain curves illustrate a long toe region-to-inflection point followed by a rapid increase in stiffness at higher strains. This demonstrates that plantar soft tissues increase peak stress, modulus, and energy loss with increasing frequency (Hz). Diabetic plantar tissues demonstrate much higher peak stress, slightly higher peak strains, and significantly increased modulus than do nondiabetic plantar tissues. See Figs 4.1.10b and 4.1.10c. *(Image from Pai, S., Ledoux, W.R., 2010. The compressive mechanical properties of diabetic and non-diabetic plantar soft tissue. J. Biomech. 43(9), 1754–1760.)*

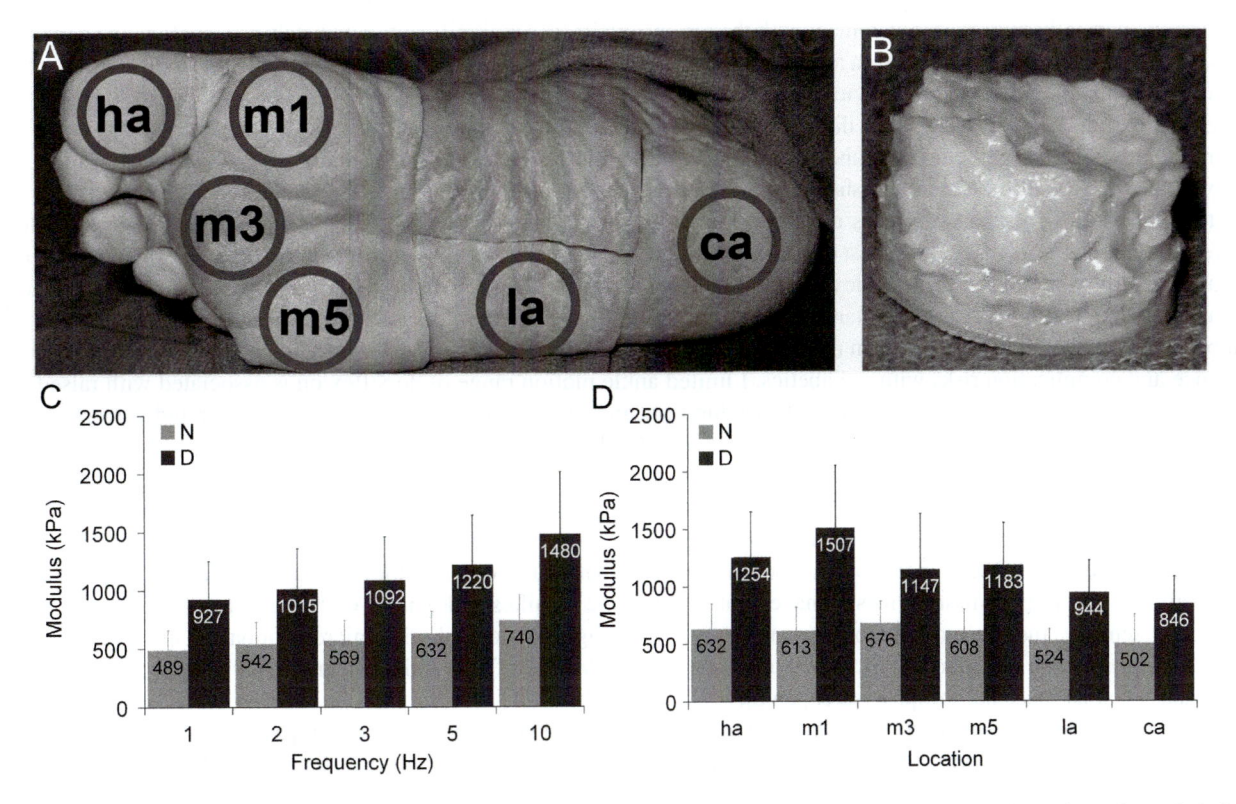

FIG. 4.1.10b Upper image (A) reveals the specimen locations used by Pai and Ledoux (2010). These were taken from cadavers for testing at the hallux (ha), 1st, 3rd, and 5th metatarsal heads (m1, m3, m5), lateral midfoot (la), and calcaneus (ca). A typical plantar tissue specimen (B) used in the study before skin removal is shown. Data from these specimens within the tables shown below, demonstrate mean modulus as a function of frequency across all locations in table C and locations across all frequencies in table D. Bars represent standard deviations. N = nondiabetic, D = diabetic. *(Image from Pai, S., Ledoux, W.R., 2010. The compressive mechanical properties of diabetic and non-diabetic plantar soft tissue. J. Biomech. 43(9), 1754–1760.)*

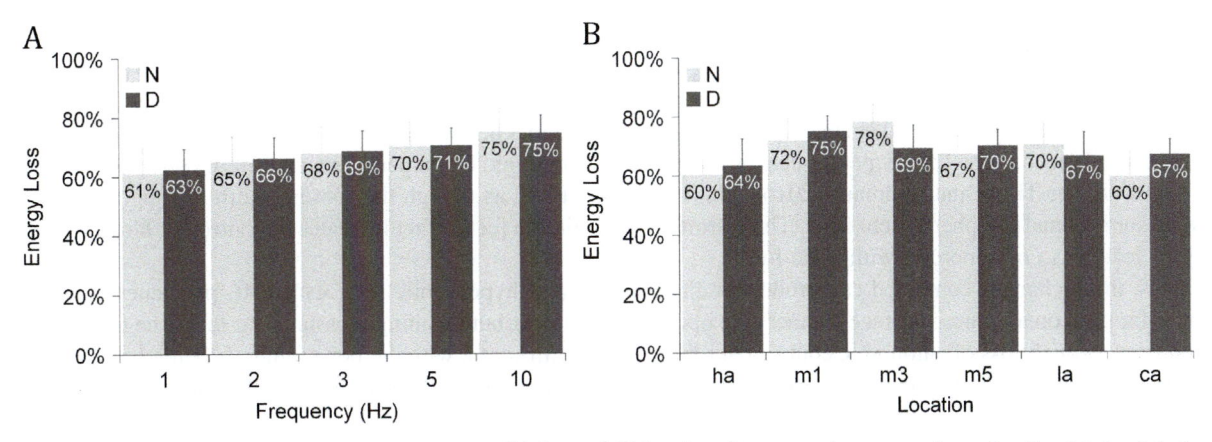

FIG. 4.1.10c Tables demonstrating normalised stress versus (A) time and (B) log time plots, comparing average shear relaxation data for diabetic = D and nondiabetic = N plantar foot soft tissue specimens. *(Image from Pai, S., Ledoux, W.R., 2012. The shear mechanical properties of diabetic and non-diabetic plantar soft tissue. J. Biomech. 45(2), 364–370.)*

Diabetics have a 20% or three-fold increase in fracture rates depending on the extent of disease and the skeletal site affected. This is despite maintaining a normal high bone mineral density to body mass index, factors that usually associate with a reduced fracture risk (Farr and Khosla, 2016; Karim and Bouxsein, 2016). Several reasons for increased fracture risk are proposed, including a heightened risk of falls, deficits in bone microarchitecture, and poor architectural bone quality (Karim and Bouxsein, 2016). Despite the fact that trabecular bone is reported to be relatively well preserved, increased cortical porosity has been reported in type II diabetics (Farr and Khosla, 2016).

Changes in diabetic gait parameters such as slower walking speeds, a wider base of gait, and prolonged double support time have been reported (Paul et al., 2009; Wrobel and Najafi, 2010; Martinelli et al., 2013; Fernando et al., 2016a). These changes probably reflect defensive strategies needed to reduce impact energies and increase stance phase stability and balance resulting from increasing soft tissue stiffness, and particularly neuropathy (Paul et al., 2009). Compliance of plantar soft tissue normally improves foot contact with the ground, so its loss is significant to gait stability let alone the concurrent loss of sensorimotor information from any neuropathy. Further kinematic protection strategies may be indicated by smaller plantarflexion, knee flexion, and pelvic obliquity movements in diabetics who have foot ulcers compared to those without (Fernando et al., 2016a). Neuropathic diabetics with plantar ulcers also demonstrate greater ranges of anterior–posterior and vertical GRFs during gait than do diabetics without neuropathic ulcers and healthy nondiabetic individuals, despite them having a slower walking speed and shorter step lengths than the latter two groups (Fernando et al., 2016a,b). Increased anterior–posterior (horizontal) forces equate to increased shear, while higher vertical forces equate to greater collision stresses. This suggests that despite best efforts in altering gait, these abnormal tissues are exposed to abnormally high tissue loads, regardless of multiple kinematic and kinetic attempts to reduce peak vertical stresses during initial contact and acceleration collisions. Weak ankle plantarflexor and dorsiflexor muscle strengths coupled with decreased ankle ranges of motion during gait may play a part in maintaining higher stresses in diabetics, as they are good predictors of gait performance and complication risks within diabetics. Limited ankle motion range of dorsiflexion is associated with raised peak plantar pressures despite having the longer stance duration, and simple stretching of calf muscles can rapidly decrease these pressure peaks (Martinelli et al., 2013; Fernando et al., 2016b; Maeshige et al., 2021).

There is a decrease in weightbearing surface area associated with a loss of foot vault mobility due to raised tissue stiffness within diabetics, which in consequence increases peak pressures and loading time on plantar tissues under direct ground contact stresses. This is a significant limb- and life-threatening risk factor resulting from acquired soft tissue hypomobility and altered (higher) foot profiles associated with diabetes. The other major risk factor is neuropathy, a factor that strongly couples to poor tissue stress management. Loss of the Achilles reflex is an early sign of peripheral neuropathy. Diabetic sensory, motor, and autonomic neuropathy are strongly associated with microangiopathy, an angiopathy that results in poor tissue nutrition and increased toxic congestion (Volmer-Thole and Lobmann, 2016). Neuropathy also leads to issues such as muscular atrophy (sarcopenia).

Atrophy of the anterior muscle group leads to increased forefoot loading pressures and impact forces from loss of tibialis anterior's deceleration braking of ankle plantarflexion moments. Atrophy of the intrinsic digital muscles, especially flexors, can give rise to toe retraction and clawing leading to shortening contraction of the plantar intrinsics despite their weakness. The digital extension associated with toe clawing or retraction increases the foot vault height profile, aided by stiffening of connective tissues, such as the plantar aponeurosis and flexor apparatus. As connective tissue stiffening is associated with nonenzymatic glycosylation end-products leading to increased tissue hypomobility (Volmer-Thole and Lobmann, 2016), diabetic feet create a combination of foot stiffening and vault 'raising' that reduces the prone foot contact area. This risk higher peak pressures and internal stresses on those tissues remaining in ground contact. These plantar tissues are intrinsically less able to provide the compliance needed to spread and dissipate stance phase gait energies via deformation. The consequences can be repetitive microtraumas, hyperkeratoses, and haematomas on the reduced surface areas under load at the heel and metatarsal heads. This is because through mechanical failure in macrostructure compliance to handle energies appropriately, strain is then expressed at a lower hierarchical lever within the tissue microstructural, causing collagen fibre failure. This situation is worsened in autonomic neuropathy due to loss of sweating and a drying out of the skin that disturbs the epidermal mechanics. The deeper tissues, through loss of autonomic blood perfusion control, increase the warming of the skin (Volmer-Thole and Lobmann, 2016). This can make it feel as though the foot is healthily supplied with blood despite some actual peripheral ischaemia. The warmth felt within the feet in fact indicates uncontrolled blood flow which may be leading to osteoporosis within the foot.

Thus, these changes combined can produce a diabetic acquired hypomobile foot, with stiff, poor energy-dissipating plantar fat pads on a reduced surface contact area upon which to distribute loading pressures. At the same time, increased blood flow through loss of autonomic control risks bone demineralisation, with the bone minerals being literally 'washed out' by the blood, creating bone fragility (Hovaguimian and Gibbons, 2011; Barwick et al., 2016; Beeve et al., 2019; Cho et al., 2020).

In Charcot foot development, a condition commonly associated with long-term autonomic and motor-sensory neuropathy, normal vault adaptability mechanics are lost along with the ability of the foot to adjust its soft tissue compliance during gait. Thus, the foot fails to adequately protect increasingly fragile foot bones and entheses. This increases the risk of tissue threshold breach through peak compression and shear strain loading on both soft tissues and bone, particularly under the ankle and midfoot dorsiflexion moments of late midstance. Diabetes-associated loss of ankle motion and surrounding muscle strengths further reduce the ability of the foot to act to protect itself. Thus, new and abnormal peak mechanical loads are applied to stiffer, more brittle tissues on a foot with less compliance and controllable mechanisms to absorb them. The foot can then be literally pulled and crushed apart via repeated exposure to the forces generated during the stance phase of gait, to form a Charcot foot.

Yet, it is worth remembering that changes in gait biomechanics notwithstanding, many diabetic ulcerations are initiated by minor skin trauma from footwear, injuring the foot while walking barefoot, or through poor nail cutting (Volmer-Thole and Lobmann, 2016). Once the skin is breached, poor healing, infections, and dysfunctional gait and tissue biomechanics maintain and compound the ulcer.

4.1.11 Principles of free-body diagrams

The pathomechanics of any situation is dependent on the individual features of lower limb morphology, tissue threshold levels, and the gait of the patient. Findings must be associated with or disassociated from the pathology. This can seem challenging to the student or novice practicing clinical biomechanics. When relating the activity of the body segment, the alignments, the instantaneous axes of rotation, and the gait determinants to the pathology, a number of critical factors are being brought together. This can cause some confusion unless a logical approach is maintained. Forces across anatomy are varied. Some will be small, others large, but between them, they create a resultant force to permit motion or achieve stability. This, in each situation, is the main stress that the anatomy has to manage without it breaching tissue threshold at any particular moment during gait.

Stance phase forces in the lower limb reflect the orientation and size of the GRF (external force), the internal tissue stresses, and the angular joint moments (internal forces) that are initiated from this body-ground interaction. They are at the same time, also influenced by the contralateral swing phase limb's biomechanics. Structures that directly oppose the resultant force or create it are thus most likely to become injured in any given loading situation. However, the failure of anatomy contributing even a small force input can change the direction of the resultant force into tissues that may be unable to cope. Failure of significant anatomy that opposes or creates the primary influence on the resultant force will obviously cause the most catastrophic changes to function, with secondary anatomy causing more subtle changes.

One of the best ways to unravel the events presented to the clinician, particularly where experience is lacking, is to use *free-body diagrams*. Free-body diagrams should bring in key mechanical events occurring during the patient's locomotion to try and bring biomechanical logic to the pathomechanics. In so doing, free-body diagrams should present a coherent therapeutic pathway to resolve the pathology (Fig. 4.1.11a). Although 3D kinematic gait analysis combined with force plates are best for presenting quality quantitative data on joint positions relative to the GRF, much can be clinically 'guesstimated' with simple 2D slow motion video analysis. Through the application of principles, even in the absence of quantifiable data, much can be understood about the potential tissue stresses occurring. Most patients undergoing biomechanical influencing treatments do not undergo expensive gait analysis testing, and yet they can still get better with the correct advice and interventions.

There are certain features that all free-body diagrams should contain: the location of the CoM, the body weight vector (BWV), and the GRF. These are central components of the stance phase of gait. Depending on the segment of interest and the pathology, other anatomic forces can be brought in. Key passive (ligament and fascia) and active (muscle) elements can be factored into the diagram. If the free-body diagram is to be constructed with the planes of motion of a joint (joints), then the suspected location of the joint axis can also be added as the fulcrum. If the pathology and forces of interest involve forces outside of a joint's axis of rotation, then these forces' moment arms around the anatomy should be considered as torques compressing or tensioning anatomy via bending moments across joints, or specific long bones. Considering the whole-body posture and the terrain or footwear is also important in free-body diagram construction (Fig. 4.1.11b).

The best starting point for understanding how to utilise free-body diagrams, especially for those new to the concept, is to consider the human 'usual' nonpathological or nondysfunctional situation, before tackling the patient's presentation. By understanding how efficient gait biomechanics operate, small variations in anatomy can be appreciated as still being able to

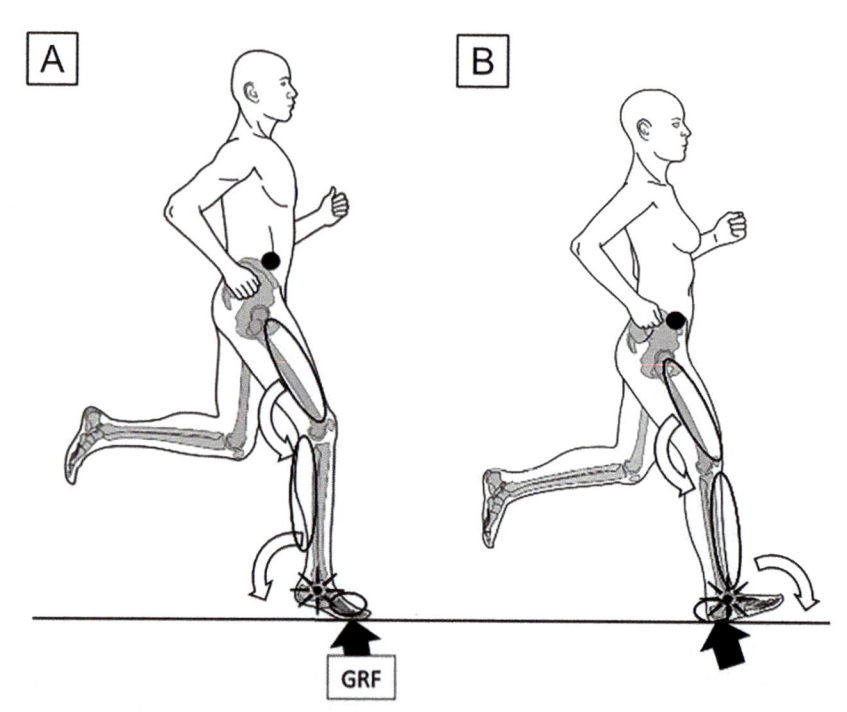

FIG. 4.1.11a Free-body diagrams can demonstrate fundamental principles to the clinician. The location of the ankle joint's instantaneous axis of rotation (star) for its sagittal plane motion and the position of the GRF (thick black arrow) and body's CoM (black spot), clearly demonstrates that impact energy dissipation by muscles around the ankle and structures within the foot are quite different in rearfoot strikes compared to those on the forefoot. Forefoot strikes (A) set up chains of resisted ankle dorsiflexion muscle activity that dissipates impact via posterior lower leg muscles. Such initial impacts present higher loading to the forefoot and Achilles. Rearfoot strike, on the other hand (B), places the initial burden on the ankle dorsiflexor muscles by resisting plantarflexion moments from GRF on the rearfoot with the heel plantar fat pad directly loaded at initial contact. In rearfoot strike, the forefoot only partakes in energy dissipation during the second separate phase of energy dissipation in later forefoot loading response. Thus, in forefoot strike runners, Achilles tendinopathy might be caused by poor muscle–tendon energy buffering at the contact phase, as well as possibly via acceleration biomechanics. However, Achilles and forefoot pathologies are unlikely to be pathomechanically linked to rearfoot strikes, moving attention immediately to the acceleration phase of gait for Achilles tendinopathy and metatarsalgias in rearfoot strikers. In contrast, injuries to the ankle dorsiflexors or heel plantar fat pad are not pathomechanically linked in forefoot strikes, but are associated risks of rearfoot strikes. Yet, regardless of strike positions, in running, the quadriceps muscles work via the knee's energy dissipation burden throughout braking which is relatively prolonged. Thus, the burden on these muscles is much increased compared to walking gait. Areas circled indicate regions of anatomy at risk of high stresses associated with particular strike position. If the pathology does not associate with locomotive biomechanics, it is likely that different phases of gait or different activities lie at the origin of the pathology. *(Permission www.healthystep.co.uk.)*

create effective mechanics, and also significant, relevant or irrelevant dysfunction can be more easily identified. Free-body diagrams of the 'usual' should present the 'ideal' human joint lever arm mechanics, angular momentums, and muscle actions in gait for comparison with the patient. At a risk of promoting misunderstanding, these ideal free-body diagrams represent a 'normal'. This is perhaps better referred to as the preferred situation.

A common mistake clinically is to consider a significant kinematic variation from normal as being the cause of pathology or dysfunction, only then to find that the affected muscle is not actually active at the period of gait when the kinematics are noted to be abnormal. For example, an unstable midstance or early terminal stance phase is unlikely to cause an injury within or be a result of tibialis anterior. By regularly applying free-body diagrams and having a ready source of normal EMG data on lower limb muscles, the clinician will find it increasingly easy to resolve each individual pathomechanical situation applicable to the patient, even when a particular problem or pathology has not been encountered by the clinician before. More dramatic traumatic events such as inversion ankle sprains, can be easily deconstructed into forces and injuries using free-body diagrams (Fig. 4.1.11c).

In the following section, the key free-body diagrams of each segment are considered and explained. This is followed by some simple examples, using free-body diagrams, of how tissue stresses can concentrate due to alignment variation and dysfunction, leading to certain pathological changes.

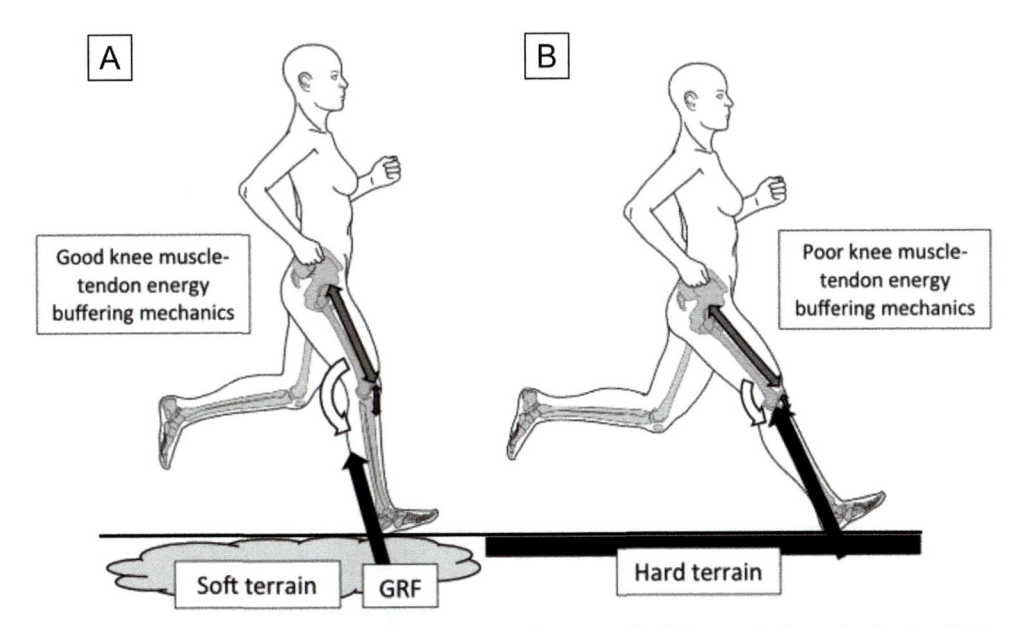

FIG. 4.1.11b Free-body diagrams can simply express biomechanical situations more clearly in a particular setting for the clinician. During loading response of running, the burden on the knee as an energy dissipator dramatically increases, making knee pathology an increased risk of running over walking. Knee shock absorption requires larger knee flexion angles and for a longer percentage of stance phase compared to walking. The quadriceps increase their activity to control the extent of knee flexion, utilising this activity to dissipate energy via muscle–tendon buffering. On soft terrains, running at slow speeds with short stride lengths (A), the knee is flexed at heel strike at a middle length for the quadriceps, aiding initiation of a flexion moment that the quadriceps can easily help to resist and provide energy-buffering under a helpful force–velocity relationship and at preferred muscle fibre length. Mechanical efficiency is therefore relatively good. Because softer ground tends to deform (forming footprints), GRF will tend to be lower on softer ground. Thus, knee impact energy dissipation mechanics are relatively easy. On harder terrain, at faster running speeds, and with a long stride length (B), quadriceps muscle requirements for GRF's force energy dissipation will be higher. If the stride length is very long, the knee will sustain impact nearer full extension, sometimes initially driving the GRF directly into the knee rather than behind it. This reduces the initial GRF flexion moment arm length to the knee. The loss or reduction of a strong knee flexion moment from the GRF initially places the quadriceps (including the patellar ligament) at a shorter initial length and the knee at a lower flexion angle from which to initiate energy-buffering flexion moments afterwards. The result can be a slight delay in initiating knee energy buffering via muscle activity before a larger requirement of total knee flexion range of motion becomes necessary to complete the braking phase. At faster speeds, muscles generate less power through their force–velocity relationship, further increasing the risk of injury if such locomotion is sustained. Despite these differences, both running styles are vulnerable to weak quadriceps, but runner (B) already has the knee disadvantaged by her running style compared to running style (A). Thus, knee injury is more likely during impact braking phase in runner (B), whereas converting runner (B) to running style (A), while also strengthening weak quadriceps, may prove beneficial. Turning the running style of (A) to that of (B) in the presence of weak quadriceps would be a potential disaster. *(Permission www.healthystep.co.uk.)*

4.1.12 Section summary

Pathomechanics is another discipline within biomechanics that has the most immediate clinical relevance. However, it is a subject underpinned by all the subjects considered in this and the companion text 'Clinical Biomechanics in Human Locomotion: Origins and Principles', and it requires the selective and considered application of many sciences by the clinician. It is the art of the clinical biomechanist to unravel the pathomechanics, but it is an art heavily underpinned by scientific knowledge. Getting the patient to provide a good history is an important starting point within the process of getting to the source of the problem.

Each patient will present with their own functional variation which is a result of their developmental processes. Being able to associate and disassociate findings is a key to success, as recording every deviation from an imagined 'normal' would be time-consuming and ultimately futile. People are functionally different. Clinicians must live with that fact. Mistaking normal variation for abnormal is a common error committed by all clinicians, but it is an error that experience will fix. A good starting point to fixing it is to realise that this can easily happen.

As each chapter within this and the companion text 'Clinical Biomechanics in Human Locomotion: Origins and Principles' has hopefully highlighted, there are key considerations within clinical biomechanics. Mechanics and biological tissues adapt in unison to gait speed/velocity. Walking mechanics is not like running mechanics, and forefoot striking and rearfoot striking that are normal variations common in running create different energy loading braking events with different stress concentrations. The lower limb, through cycles of impact braking and acceleration collisions, needs to provide the correct energy dissipation compliance and elastic stiffening mechanisms to offset the energies loaded during gait that is required for each type of locomotive event. Failure to get this load management right, risks breaching tissue threshold. Disease, age, and lifestyle adaptations influence how easily these tissue thresholds can be breached.

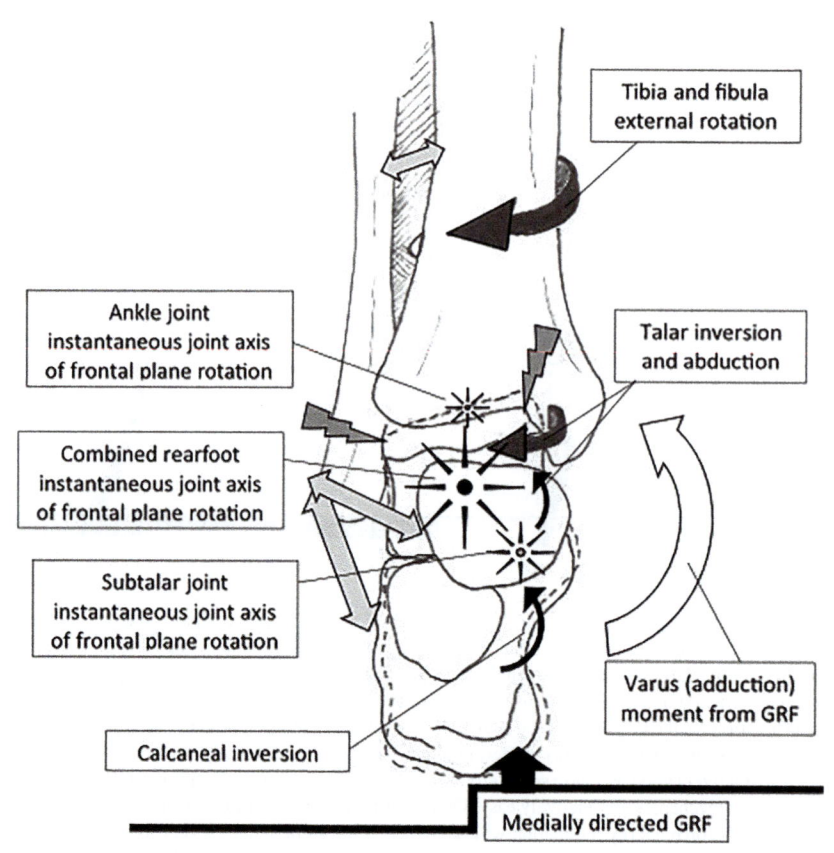

FIG. 4.1.11c Not all pathology arises from abnormal, excessive, or deficient moments acting across the primary instantaneous joint axis of sagittal plane motion. Pathomechanical torques can arise angled across the sagittal plane axis in joints such as the ankle and knee, where the joint axis has high sagittal planar dominance with freedom of motion in the frontal and transverse planes being highly restricted. The free-body diagram can be approached in the same way as handling a bending moment in a column involving points of higher flexibility (the joints within the lower limb). Frontal plane moments within joints such as the tibiofemoral joint of the knee create varus (adduction) and valgus (abduction) moments that compress the articular surface and tension-restraining ligaments. In this free-body diagram, the rearfoot joints are illustrated in anterior view. The rearfoot is particularly vulnerable to large frontal plane torques that can result from GRFs occurring only on the medial side of the rearfoot, such as occur when toppling laterally off a curb. This sets up rotation around the instantaneous joint axis lying within the ankle's *x*-axis or sagittal plane (smallest star) and that of the subtalar joint (small star) that create a combined instantaneous axis positioned between the ankle and subtalar joint (large star). The subtalar joint beneath the ankle can provide variable degrees of frontal and transverse plane protection to the ankle around its instantaneous axis, but only when it is exposed directly to external forces on the rearfoot. By inverting and adducting the calcaneus on the talus, frontal plane torques can be dissipated within the subtalar joint motions and ligament strains. Once this motion is complete under restraining subtalar ligament maximal tensions, any further inversion is expressed by talar inversion and adduction via the tibial and fibular articular surfaces, which is usually quite limited. This causes increasing articular compression superomedially and laterally (lightning strikes) and tensions the lateral ankle ligaments and the tibiofibular interosseous membrane (grey double-headed arrows). If inversion stresses cause further coupled external rotation, ligaments at both ankle and subtalar joints may fail under tension, subchondral bone and cartilage under compression, and even cortical bone may fail under torsional shear resulting in fractures. If inversion events occur when only the forefoot is loaded, the ankle will be forced to manage such perturbations as inversion or eversion without the aid of subtalar joint motion, although the midfoot may assist instead. While knee varus and valgus moments present similar mechanical concepts, they are more commonly associated with subtle long-term overload issues whereas inversion ankle sprains are one of the most common acute trauma-induced rotation torque injuries that humans suffer. *(Permission www.healthystep.co.uk.)*

4.2 Pathomechanics of the hip
4.2.1 Introduction

In this and the following sections, we consider biomechanics in relation to more specific pathology. Advice on treatment options is left to other texts. The intention here is only to understand the normal and pathomechanical situations that can arise with some conditions. This approach, in itself, should reveal the logic of certain treatment interventions.

Muscular activity within the trunk is important for providing CoM stability above the hip and correctly positioning it during locomotion. This is particularly important during initial contact, when trunk flexion moments over the hip results from a sudden posterior directed braking GRF's that drives a hip flexion moment anterior to the body's CoM. The trunk's momentum then drives it forward around the hip over a lower limb fixed in closed chain. This is a motion that must be prevented. Posterior spine extensor muscles assist in preventing this by working with the hip extensors. Indeed, the ability of abdominal and spinal muscles to adjust CoM positioning is important to lower limb function, influencing limb posture all the way into the foot. An unstable lower limb will increase CoM instability and create a need for increasing lumbopelvic muscle activity. Whether issues are arising from a descending or ascending dysfunction is important to establish and this underlines the need to approach pathology from a holistic locomotive viewpoint. Thus, the hip, being the joint between mobile limb and the stability hub of the pelvis, has a particularly tricky role to fulfil during locomotion.

The hip is the greatest provider of multiplanar freedom of motion for the lower limb around the trunk. The pelvis can align more with hip or trunk motion depending on gait speed (see Chapter 2, Section 2.2.2). Thus, the influence of, for example, sacroiliac joint mechanics on lower limb pathomechanics is dictated by preferred walking speed, as if this is slow, the pelvis will be working out of phase (antiphase) with the lower limb. The hip joint (like all other joints) should not be viewed in isolation, and the biomechanics occurring around other joints are often highly relevant to hip and pelvic dysfunction.

It is worth remembering that hip pathology can present as symptoms within the knee. Hip OA is a common source of knee symptoms referred via the saphenous nerve, so that knee pain and pain occurring below the knee does not always mean there is something intrinsically wrong with the knee. However, most hip pathology cases report groin or buttock pain (Khan et al., 2004). Multi-articular muscles can cause dysfunction and pathology at more than one joint, requiring multijoint rehabilitation to create improvement in symptoms associated with, for example, altered patellofemoral joint biomechanics occurring through altered hip abductor dysfunction (Ferber et al., 2015). Therefore, in multi-articular muscle function, free-body diagrams should be approached to include all the articulations the relevant muscle crosses. Free-body diagrams of frontal plane hip function are important for understanding patellofemoral pathomechanics, as well as for considering tibiofemoral joint positioning during stance.

Muscles that support hip function throughout the lower limb have a profound effect on the stresses placed on the osseous and articular structures at the hip (Duda et al., 1998; Heller et al., 2001). The gluteus maximus and triceps surae have a particularly strong functional relationship during gait in maintaining CoM velocity. Ankle power generation influences the requirements for hip power generation. The hip will attempt to compensate for poor ankle plantarflexor power in terminal stance. This is because both triceps surae and gluteus maximus are responsible for controlling the anterior acceleration of the body's CoM during stance. CoM acceleration is initially controlled by the gluteus maximus during loading (decelerating the CoM) and early midstance (accelerating and raising the CoM). During late midstance, triceps surae takes up the control of the CoM (decelerating it but allowing it to fall forward) and continues to do so during heel lift (accelerating the CoM and tipping it to the next step). Finally, in terminal stance, triceps surae acceleration control of what mass is left over the limb is gradually decreasing and is transferred to gluteus maximus briefly again, before swing initiates through iliacus.

4.2.2 Principles of free-body diagrams of the hip

Sagittal and frontal plane motions tend to be the focus of interest as hip motion and stability are dominated by two planes during gait. Transverse plane motions are needed for desirable alignment of the knee and ankle joints axes for sagittal plane flexion–extension of the thigh and lower leg perpendicular to the line of progression and aligned with the medial MTP joints' axis at acceleration. This ensures that the lower limb and foot can apply the ankle plantarflexion power along the line of progression. Thus, hip rotations should help position a continuation of lever arms and fulcrums at the hip, knee, ankle, and medial MTP joints at ~90° perpendicular to the line of progression of the sagittal plane. Transverse plane motion within the hip should help to stabilise this sagittal motion positioning, as well as helping to bring the support limb and foot closer to the CoM by internal rotation coupled to adduction, particularly through adductor magnus activity at loading response. However, transverse and frontal plane rotation is also used for making changes in direction. In straight walking, only relatively minor motion in the transverse plane is required between the hip and pelvis. Thus, most pathology at the hip can be explained by considering hip loads in just the sagittal and frontal planes, but free-body diagrams can be adjusted to consider alignment issues associated with transverse plane dysfunction, if it is necessary for a particular activity.

Orthogonal plane hip motion occurs simultaneously, so that sagittal and transverse plane positioning will affect frontal plane mechanics and vice versa. But certain pathologies should focus the clinician to one plane or the other as being the most likely source of pathomechanics. For example, a Trendelenburg gait (excessive hip varus tilt) can be associated with trochanteric bursitis (Speers and Bhogal, 2017). This pathology should focus attention towards the hip abductor activity in the frontal plane and receive primary clinical investigation until examinations and assessments rule out this mechanism as a possible source of the problem. Even if the origin and plane at fault is quickly identified, it is always worth considering all hip motions recorded during gait assessment in the other planes to see if a coherent linkage can be identified to other potential provocative events.

4.2.3 Sagittal plane free-body diagram of the hip

Primary motions of flexion and extension occur via the hip acting as a class three lever through proximally applied efforts on the trunk or distally applied efforts to the swing limb. This means that mechanical efficiency is compromised for increased angular motion and metabolic efficiency of muscular contraction. The critical and most difficult stage of sagittal plane hip loading during gait occurs during initial contact and throughout loading response and braking, with higher stresses again during terminal stance acceleration. This is because the hip moment arms to the GRF are at their greatest lengths during these loading and offloading events. However, during walking, loading and offloading after heel lift occurs as a double-limb stance event, reducing stresses on the hips. Normally, walking heel lift itself usually occurs still in single-limb support as it obviously does when running, making this a vulnerable moment for the hip. During walking absolute midstance, despite being in single-limb support, sagittal plane hip dysfunction is more unusual because the angular moment arms are small as the body's CoM and the GRF come to lie close to the hip's joint axis, making muscular control far easier. The GRF is also at its lowest around absolute midstance (Fig. 4.2.3a). Running is different because the hip is most flexed at the end of braking when the GRF is at its highest, CoM at its lowest, and both are close to lying directly under the hip joint axis. Swing phase hip injuries are also far less likely because the loads involved in anterior progression of the swing limb are relatively low compared to those created by the body mass movements at the hip during stance phase. Swing phase hip biomechanics should not create difficulty unless major neuromuscular or articular freedom deficits are present. However, hamstring injuries in braking terminal swing's hip and knee motions are commonly associated with running.

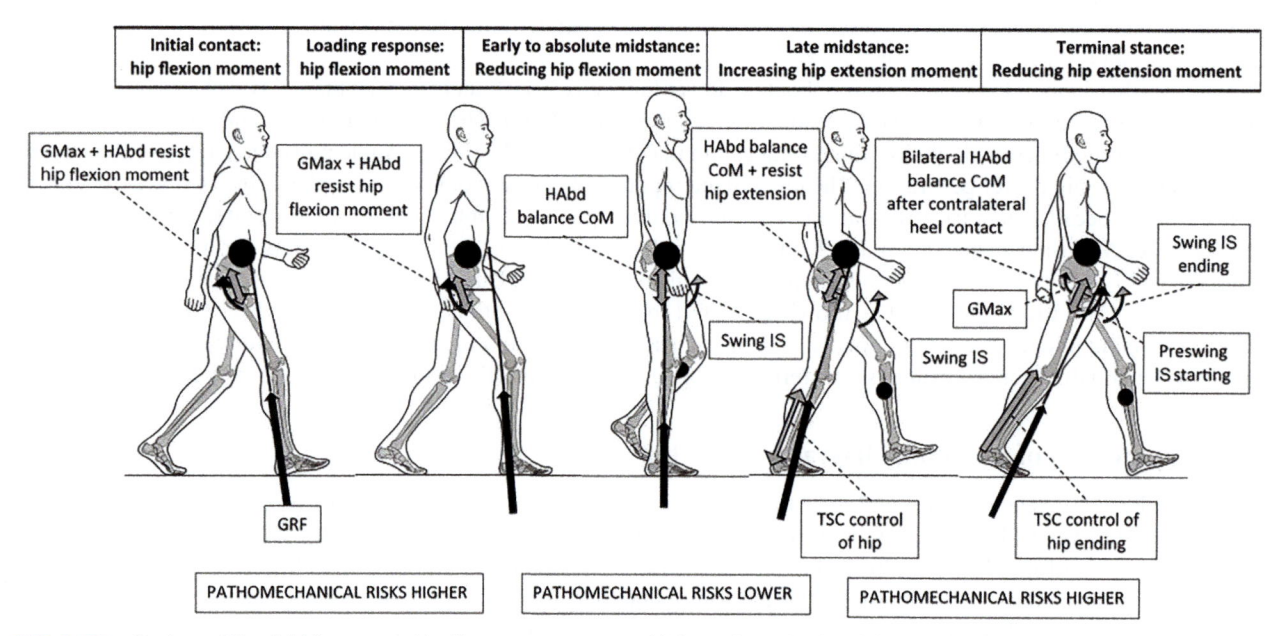

FIG. 4.2.3a During walking, initial contact and loading response presents a higher pathomechanical risk than midstance because the hip is exposed to a longer flexion moment arm from the GRF within early stance. The GRF moment arm is shortening towards absolute midstance, giving muscles greater mechanical advantage against flexion moments until the CoM (large black circle) lies directly above the support limb. In late midstance, with the hip moving increasingly into extension and the GRF moment arm lengthening behind the hip, extension motion is largely free from sagittal plane hip muscle activity. However, the CoM balance control is requiring further hip abductor activity (HAbd), despite the CoM velocity mainly being under the control of triceps surae complex (TSC). The anterior HAbd can also help resist hip extension. Hip extension becomes harder to control just before heel lift, risking hip impingement via hyperextension. In walking, most of terminal stance and loading response stress occurs under bilateral HAbd activity, stabilising the CoM during double stance phase. This makes weight transfer easier. In late terminal stance, gluteus maximus (GMax) should be briefly active before preswing as iliopsoas (IS) activates for swing phase hip flexion. IS activity in preswing helps prevent over-extension of the hip joint at the end of stance. In swing phase, although IS moves the hip through a large range of motion during swing, the CoM of the lower limb (small black circle) is relatively small compared to that of the trunk, making swing phase hip dysfunction far less likely than during stance. IS's hip flexion forces are antagonised by GMax and hamstrings, with weakness or tightness of any of these primary swing phase muscles raising the pathomechanical risk. It is worth remembering that hip and ankle motion are linked through triceps surae's influence on CoM velocity, both in ipsilateral stance and contralateral swing. *(Permission www.healthystep.co.uk.)*

By appreciating the fact that small angular moments and lower GRFs are less likely to initiate tissue threshold breaches, the midswing and the middle of midstance phases of walking gait should not be the first phases of gait investigated for hip pathology. In running, a significant hip flexion angle will persist well into stance phase under increasing vertical GRF, with

Pathology through the principles of biomechanics **Chapter | 4** 655

hip extension motion only starting after the middle of stance and continuing into acceleration. Changing from braking into acceleration during running is not divided by a weight transference phase as it is in walking. Thus, in running, gluteus maximus and the quadriceps sustain higher loads for longer and at faster loading rates, working much harder to brake hip and knee flexion. Then they are immediately put upon to start to achieve hip and knee extension moments, changing eccentric contraction immediately to concentric (Fig. 4.2.3b).

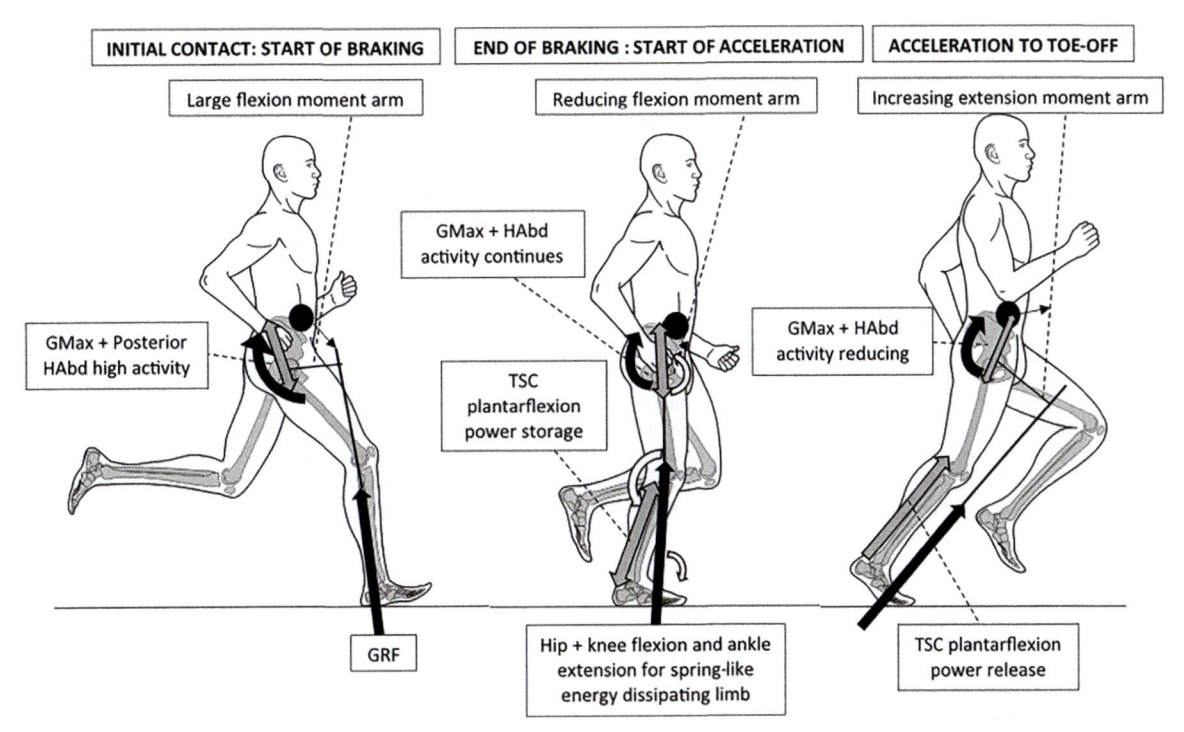

FIG. 4.2.3b The hip anatomy functioning in the sagittal plane is far more vulnerable to injury and dysfunction during running than walking. This is because running increases the acceleration effects on the forces generated, and stride lengths are often much longer, increasing the GRF flexion moment arms on the hip. Thus, gluteus maximus (GMax) is placed under greater demands to control the initial hip flexion moment. The flexion moment also continues to the middle of stance at the end of braking as the CoM (black circle) completes its fall, with the lower limb acting like a spring under compression beneath. This requires GMax to work for longer during stance under increasing vertical forces that continue to rise towards the initial acceleration phase. GMax has to help initiate the start of hip extension out of the high hip flexion angle formed at the end of loading response. The hip and knee flexion angles of running descend the CoM to a far lower level than occurs at any point during walking. Loading the limb also occurs as a single-limb support event on a more adducted limb at the hip. This necessitates greater hip abductor (HAbd) muscle activity than in walking's loading response. The very high HAbd activity only reduces in acceleration when the CoM passes anterior to the hip. Although foot strike does influence the GRF flexion moment arm, it is the stride length rather than the strike position on the foot itself that mostly influences the hip loading stresses. Once again, hip motion is also linked to ankle motion and its plantarflexion power derived from the triceps surae complex (TSC). *(Permission www.healthystep.co.uk.)*

4.2.4 The hip in loading response kinematics

During the major stress loading periods of early and late stance, hip flexion and extension moment arms around the hip reach their peak lengths, respectively. Heel strike hip moment arms are longer at initial loading because at impact, the foot is placed in its most anterior position in relation to the CoM while the hip is exposed to two (heel and forefoot) collision events as it takes on the task of bearing the weight of the trunk. Thus, high vertical and a posteriorly directed GRF occur on a long lever arm from the hip, requiring high internal muscle forces to stabilise the hip flexion moment initiated from the body's CoM lying behind the hip. This flexion moment at the hip during foot collision is used for shock attenuation via muscle–tendon buffering within the hip extensors. The longer the stride length, the greater the hip flexion moment created. Thus, increases in stride length require more gluteus maximus activity and strength in order to control the larger flexion moment, with assistance from adductor magnus, the posterior hip abductors, and also initially, the hamstrings (Fig. 4.2.4a).

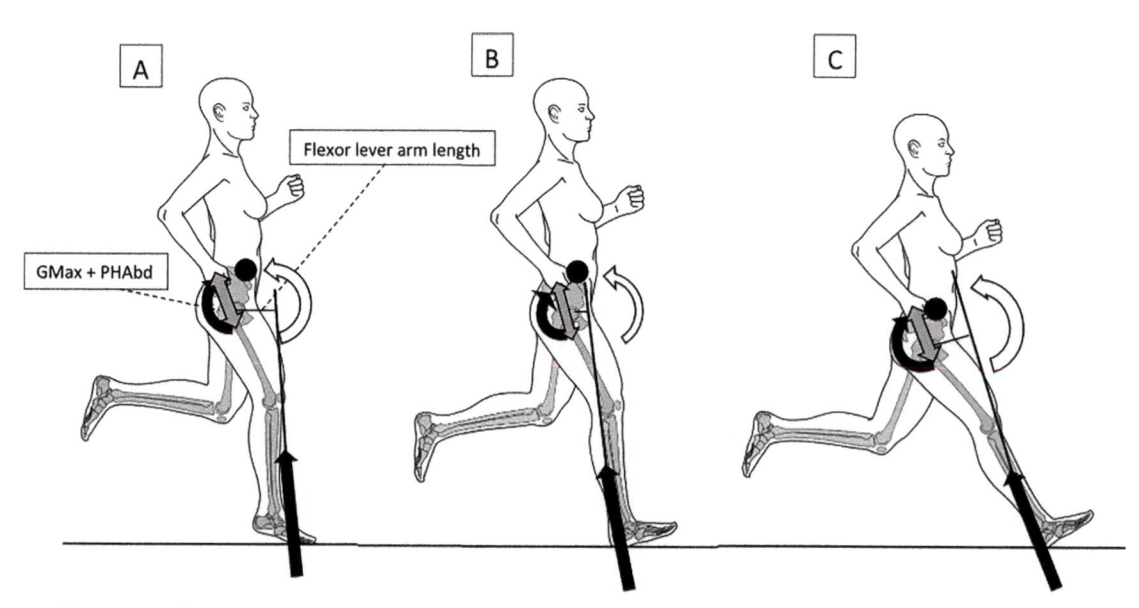

FIG. 4.2.4a The amount of hip muscle activity for gluteus maximus (GMax) and the posterior hip abductors (PHAbd) at contact is mostly affected by the stride length. Thus, the illustrated forefoot strike runner (A), landing their foot further ahead of the CoM (black circle), generates a GRF positioned more anterior to the hip joint, creating a longer flexion moment arm than a heel strike runner on a shorter stride (B). However, longest strides tend to associate with heel strikes (C), which are often attempted by runners trying to increase their running speed via taking a longer accelerated swing phase. Elite runners tend to increase their stride frequency rather than length, avoiding larger joint torques that can be increasingly difficult to manage with fatigue. Elderly and unhealthy patients with weak hip muscles also tend to reduce their stride length and gait speed in walking to reduce hip moments and forces. *(Permission www.healthystep.co.uk.)*

Late swing phase and initial contact energy dissipation at the hip

If impact energy threatens to breach tissue threshold, then slowing gait speed and reducing stride length will reduce impact forces that are initiating hip flexion. However, if energy dissipation requirement remains too high for the muscles or joint tissues to manage, then safe tissue loading may still be exceeded. A potential injury mechanism can be created by a deficit between the capacity of quadriceps and gluteus maximus to dissipate energy. Gluteus maximus, the hamstrings, iliopsoas (mainly iliacus), and the quadriceps influence swing phase hip and knee kinematics. Their combined activity positions the lower limb for initial contact. Gluteus maximus and the hamstrings activate to decelerate the momentum created through quadriceps' and iliacus' activity that have been producing lower leg swing phase acceleration and associated centrifugal forces. The gluteus maximus and hamstring activity should have a significant braking effect on the lower limb's acceleration through their eccentric or near-isometric contraction. As the posterior hip's myofascial unit lengthens through this braking action, the foot is brought into the ground under deceleration (negative acceleration), thereby reducing impact forces. The braking hip extensors can now provide energy buffering via hip flexion resistance under a reduced peak impact force that they have themselves influenced (Fig. 4.2.4b).

Initial contact involves the hip in its most flexed position during gait, the extent of this flexion angle being dependent on the stride length. The majority of hip stabilisation during the first 30% of stance is provided by the posterior sections of gluteus medius and minimus acting primarily within the frontal plane and gluteus maximus primarily acting in the sagittal plane (Anderson and Pandy, 2003). Therefore, it is the gluteus maximus that provides the most significant active restraint of hip flexion, but the posterior hip abductors also assist. Thus, the flexion-resisting primary effort of interest at the hip during initial contact and loading response is from gluteus maximus, with a secondary interest in the posterior hip abductors. The level of activity achieved influences the required hamstring and quadriceps activity and vice versa through their combined effect on positioning the lower limb and the resultant hip flexion moments. Gluteus maximus has an important energy dissipation role through muscle–tendon buffering, but the energy dissipation demand relates to stride length and gait speed. A weak gluteus maximus during late swing and loading response could increase hamstring braking and loading stresses, particularly at a longer stride length under higher impact forces, as experienced in running. Thus, such problems usually associate with running, as it creates much higher peak loading forces at higher velocities than when walking. Walking is a gait pattern that is comparatively sparing on gluteus maximus loads compared to the size of the muscle, even at faster gait speeds.

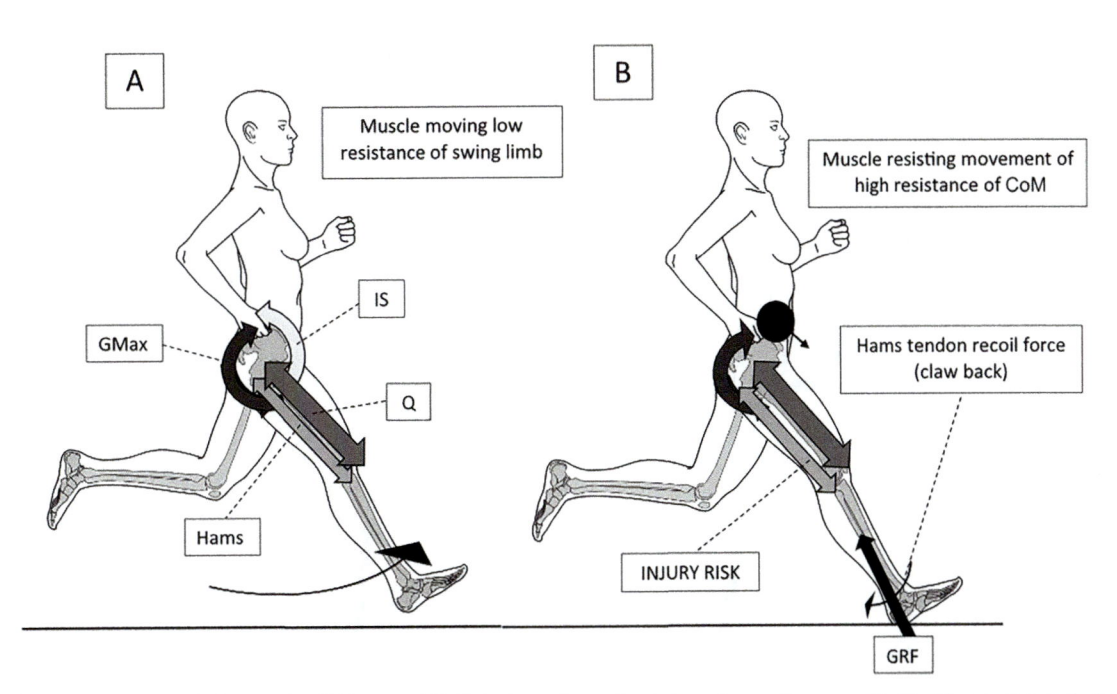

FIG. 4.2.4b Late swing phase during running (A) is potentially linked to pathomechanical events, usually as a result of braking stresses on the hamstrings (Hams). Under hip flexion acceleration from iliopsoas (IS) and rectus femoris throughout swing and via knee extension from all the quadriceps (Q) during midswing, the rotating weight of the lower limb generates centrifugal forces. The gluteus maximus (GMax) antagonises the hip moments, aided by the Hams which also resist knee extension moments. Just before heel strike, these resistance moments reach their peak, which can over-stress the Hams tendons that are acting in near-isometric contraction. At heel strike (B), Hams activity briefly continues, helping to stabilise the hip and knee as GMax and the Hams start resisting hip flexion under weightbearing load. Any 'unexpected' hip flexion and/or knee extension at this time can over-stretch the Hams. Walking rarely associates with Hams injuries because swing limb velocity is low and loading response relatively short as a percentage of gait compared to running. Running and particularly periods of sprinting with long strides, focuses stresses to the tendon of biceps femoris, making biceps femoris injuries a common athletic complaint. *(Permission www.healthystep.co.uk.)*

Despite its energy buffering role at the hip through flexion resistance, the capacity of gluteus maximus to absorb energy at contact is less than that of the quadriceps at the knee (Thompson et al., 2013). Should the quadriceps be weak, then the gluteus maximus would need to compensate for energy dissipation by increasing its activity (Thompson et al., 2013). A reduced capacity for energy dissipation at the knee should cause a patient to shorten their stride length and slow their gait speed to reduce peak impact forces. The shorter stride length reduces the hip and knee flexion GRF moment arm, reducing gluteus maximus effort required to stabilise and reduce hip flexion angles and moments, while also sparing the quadriceps. The shortened gluteus maximus moment arm, however, will also reduce its ability to dissipate energy (Fig. 4.2.4c). However, with smaller impact forces and flexion moment arms, this should not be a problem but be part of a functional benefit of reduced impact forces. Those who run with weak quadriceps are taking a significant pathomechanical risk.

Failure to make stride length adjustments in impact energy management with fatigued quadriceps, gluteus maximus, and hamstrings, opens up many pathomechanical possibilities. The obvious risk is injury to elements of the gluteus maximus and its connective tissue components and attachments. Gluteus maximus dysfunction risks loss of pelvic stabilisation at the hip, particular during initial contact and loading, shifting more of the CoM positioning and loading stresses towards the lumbopelvic anatomy. Gluteus maximus and associated posterior hip abductor weakness results in altered kinematics in all the joints gluteus maximus and its associated structures cross, such as the sacroiliac joint, the hip joint, and the knee joint via the iliotibial tract (ITT). As gluteus maximus needs to compensate for quadriceps weakness, weak quadriceps-induced increased gluteus maximus activation can increase posterior hip compression forces under raised hip extensor activity, which risks excessive restriction of hip and sacroiliac joint motion as well as more focused joint stresses. Joints must move for cartilage to remain mechanically healthy. Increased 'stiffness' or 'tightness' of gluteus maximus can thus result from attempts to increase its role as a shock absorber in compensation of a deficient quadriceps. This increased loading activation of the gluteus maximus and posterior hip abductors opens up the possibility of the lower limb becoming more externally rotated through the role of gluteus maximus as an external hip rotator. More external limb rotation and a reduced ability to restrain the anterior movement of the CoM from a reducing capability to extend the lumbopelvic region both in gait and stance, can lead to vertical forces positioning over the medial midfoot and forefoot for

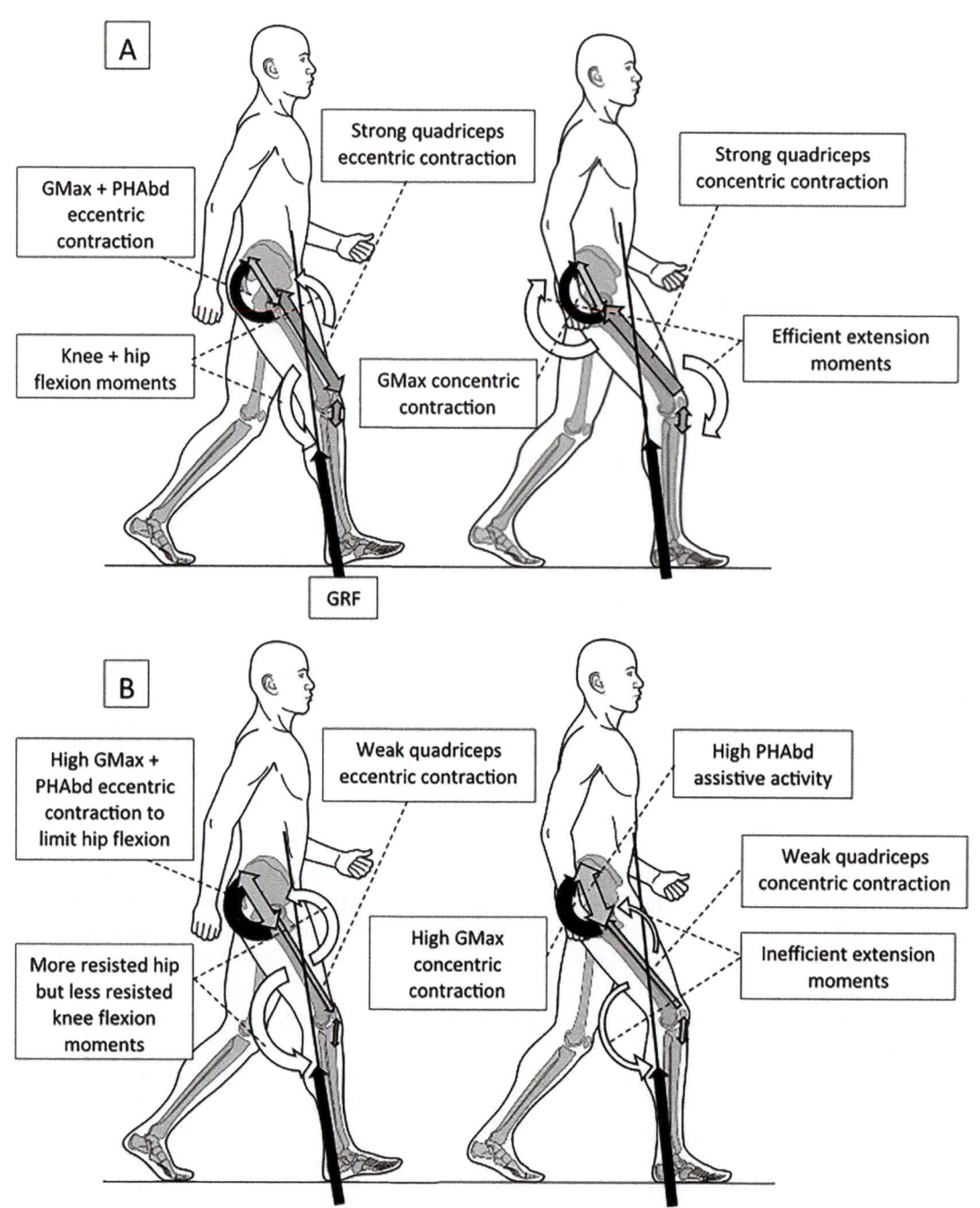

FIG. 4.2.4c In healthy walking gait (A), effective quadriceps strength sets up a limb stiffness via eccentric contraction across the hip and knee, thus acting as an energy-damping shock-absorbing spring. Although the quadriceps together cross both hip and knee, it requires gluteus maximus (GMax) and the posterior hip abductors (PHAbd) to make the hip an effective shock absorber. The hip is aided by other hip muscles such as the adductors that are fine-tuning the hip and lower limb position to the CoM's posture and setting hip stiffness appropriately with the hip abductors. Knee and hip flexion are resisted eccentrically during loading response. Once loading response is completed, concentric quadriceps, GMax, and hip abductor activity brings the limb out of knee and hip flexion. When quadriceps strength is inadequate (B), the lower limb becomes too compliant, thus allowing the knee to attain greater flexion angles before impact-induced moments are fully restrained. In these circumstances, GMax aided by the PHAbd must attempt to limit hip flexion to help restrain excess knee flexion by tending to keep the whole weakened limb more extended. Such increased gluteal activity may cause loss of normal lumbosacral motions, increase posterior joint compression, and/or increase external rotations on the lower limb. This is not mechanically efficient. If knee and hip flexion become too great in loading response, then extra concentric contraction is required from the gluteals during early midstance to start to raise the CoM. With the quadriceps weaker, the burden returns to GMax and the PHAbd working against the flexion moments operating at their longest moment arms. When repeated with each step, fatigue becomes an increasing risk, tempting muscles to only contract in a limited range of lengths to aid their ability to produce power at their preferred middle contraction lengths. Taking small strides to reduce GRF-induced flexion moment arms and keeping the whole limb more in extension is a common compensation for quadriceps weakness during walking. This posture gives more mechanical advantage to the proximal muscles against whole limb flexion moments. Running, with its longer stride lengths, increased and prolonged hip and knee flexion moments, and greater impact forces, means that even a little quadriceps weakness in runners can become a considerable injury risk. *(Permission www.healthystep.co.uk.)*

longer. This in turn could start to alter foot vault mechanics as a result of increasing torques and deflection moments across the vault, which no longer has the medial metatarsals longitudinal axis aligned to the line of progression. Thus, a descending path of dysfunction is created through altered hip biomechanics, starting from quadriceps and or gluteal muscle weakness and tightness.

Proximal muscle strength insufficiency is a greater risk factor for energy dissipation injuries in running than during walking, needing only small degrees of dysfunction to initiate significant pathologies. As gluteus maximus influences biomechanics in all planes, so its dysfunction can cause instability in frontal, transverse, and sagittal rotations. Through the gluteus maximus' and quadriceps' agonistic influence on energy dissipation, an underlying gluteus maximus and quadriceps dysfunction could be the pathomechanical route to knee pathology, such as patellofemoral pain. Keeping stride lengths short when running helps reduce the moment arms that the GRF can generate around the hip and knee, helping to make muscular control easier. This thus reduces the risk of fatigue in the hip and knee stabilisers, helping muscles to continue to offload stresses that can cause or exacerbate a number of hip and knee pathologies, for longer.

4.2.5 The hip in terminal stance

During late midstance into early terminal stance, triceps surae braking activity is resisting the body's CoM acceleration. The foot and the GRF are positioned posterior to the hip while the body's CoM lies anterior, creating an external hip extension moment. The rate and extent of hip extension is controlled by triceps surae-restraining acceleration of anterior tibial translation through ankle dorsiflexion. This hip extension creates considerable loads at the anterior superior aspect of the femoral head and neck, caused by compression within the acetabulum from the trunk's CoM falling forward. This amount of hip extension in human gait is a novel development for mammals, seemingly unique to hominin locomotion. Once the hip is sufficiently extended by the CoM's anterior displacement and the swing leg's centrifugal forces, rearfoot loading forces rapidly decrease having been moved to the forefoot. The ankle dorsiflexion that couples to the hip extension during late midstance creates energy storage within the Achilles that is released once the energy within the Achilles exceeds rearfoot loading forces. This release of ankle plantarflexion power induces the heel to lift and terminal stance to begin. This primarily triceps surae-induced power now accelerates the CoM onto the next step, but this can only happen efficiently if hip extension is adequate during midstance. Once the heel lifts, hip extension momentum should start to decrease.

In late terminal stance after heel lift, as ankle plantarflexion power starts to decrease, gluteus maximus becomes active again as the accelerator of the CoM, creating a little active hip extension to gain a little extra hip extension momentum that was lost with heel lift (Neptune et al., 2004; Neumann, 2010; DeJong et al., 2020). This action assists acceleration of the body's CoM forward onto the now rapidly loading opposite limb. The more acceleration still required at this moment, the greater the need to generate hip extension acceleration power. If ankle plantarflexor powered acceleration was sufficient to keep the CoM moving despite any reducing hip extension rates, then gluteus maximus acceleration should be minimal. In this later terminal stance acceleration role, gluteus maximus is assisted by the posterior head of adductor magnus, piriformis, and the hamstrings, although the latter's role seems to be very minor (Barrett et al., 2007). Soleus provides a similar role to gluteus maximus during terminal stance, that of accelerating the CoM forward from the trailing limb around and after heel lift. Soleus is activated earlier in midstance than the assisting gastrocnemius, but gastrocnemius also helps induce knee flexion at heel lift. Yet, both muscles' activity reduces rapidly after heel lift, requiring the CoM's acceleration role during terminal stance to derive from Achilles elastic recoil and then, as this is decreasing in power, to then pass to gluteus maximus. Gluteus maximus should only be picking up a smaller load to move forward at the point of load transfer from the triceps surae muscles and Achilles, as most body mass should have moved onto the next step, which has become the CoM's support limb. Thus, triceps surae–Achilles complex deficiency is significant to hip activity and function.

The acceleration from the ankle plantarflexion-generated power remains in the system through the Achilles' elastic properties, even after the triceps surae muscles are no longer active. This requires the Achilles to be healthy and provide its full elastic energy transfer as an elastic storage tendon. The Achilles-released plantarflexor momentum works with gluteus maximus hip extension to assist weight transfer to the next step in walking, or to assist in gaining aerial phase height during running. This muscular interaction increases velocity of hip extension and yet also limits the extent of hip extension at the end of terminal stance, together. By increasing ankle plantarflexion angles during terminal stance, the hip and knee must flex again (helped by the gastrocnemius-induced knee flexion moment), preventing excessive hip extension. Hip flexion then initiates under iliopsoas activation in preswing on a longer, yet reducing hip extension lever arm. Hip flexion increases its velocity under iliopsoas activation in preswing. The amount of hip extension associated directly with the stride length should be reducing during CoM transfer to the opposite support foot. Moments around the hip should never cause the hip to fully extend to its end range and thus risk causing high compression loading on the anterior superior femoral head and its associated structures (Fig. 4.2.5a).

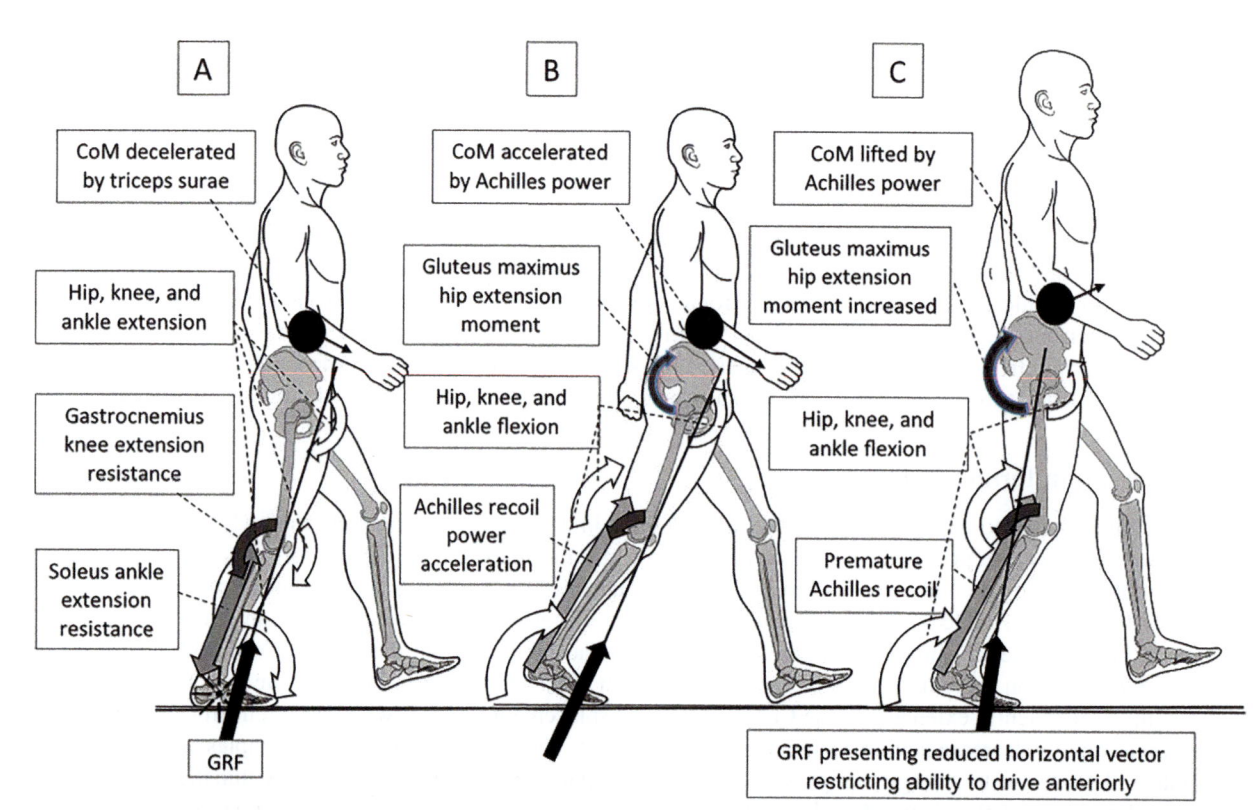

FIG. 4.2.5a From just before absolute midstance, the CoM comes under the control of mainly the triceps surae complex. Soleus primarily controls ankle dorsiflexion (extension) during late midstance (A). The GRF passes anterior to the ankle's instantaneous axis of rotation (star), allowing gravity on the CoM to exert extension moments on the ankle, hip, and knee. Gastrocnemius aids soleus in resisting ankle dorsiflexion but it also resists the knee extension moment. The hip extension moment is not directly resisted, yet restraint of the ankle couples a limitation on hip extension. This restriction on hip extension is aided by stable erect trunk posture via lumbosacral and abdominal muscles combined with activity of the hip abductors. At heel lift (B), energy stored in the Achilles via stretching under triceps surae eccentric activity is released as powerful elastic recoil, with more energy stored and released at higher gait speeds. Heel lift decelerates hip extension through the ankle plantarflexion moment flexing the knee, which applies a flexion moment to the hip, potentially losing some anterior momentum on the CoM. Thus, as the energy released from the Achilles starts to reduce, gluteus maximus often adds a little hip extension power to maintain CoM forward velocity onto the next step. The heel lift timing is important for applying appropriate CoM acceleration at the most efficient time, and helps prevent over-extension of the hip that risks excessive anterior hip joint stresses. This is a risk of a late heel lift. If the heel lifts early (C), hip extension and ankle dorsiflexion motion is prematurely lost, abruptly accelerating the CoM upwards when it should be falling forwards. By the stance limb lifting body weight too soon, energetic efficiency is reduced while Achilles stresses are increased. Thus, if loss of hip extension is the origin of premature heel lift, Achilles tendinopathy can arise from hip dysfunction. If heel lift is early or late from an ankle dysfunction, then proximal lumbopelvic and hip pathology may arise from this distal origin. *(Permission www.healthystep.co.uk.)*

In preswing at the end of terminal stance, iliopsoas activates to flex the hip out of terminal stance extension. Terminal stance and preswing forces are directed into the anterior surface of the hip joint, but should be limited by ankle plantarflexion angles that block hip extension, and iliacus activity that induces hip flexion. These anterior hip joint surface forces are mostly generated by the anterior portion of the gluteus medius and iliopsoas (Correa et al., 2010; Lewis et al., 2010) and yet, weakness within the hip flexors is associated with increased hip joint forces that risk degenerative changes within the joint (Mendis et al., 2014). The reason for this is that reduced iliopsoas activity in preswing results in increased assistor flexor activity in sartorius, rectus femoris, and tensor fascia latae, which increase the anteriorly directed hip forces beyond those of iliopsoas (Lewis et al., 2009). This increase in anterior hip joint forces probably results from an attempt to gain hip flexion momentum through less efficient effort arms than those of iliopsoas. Anterior hip joint forces reach their maximum near full hip extension during terminal stance, so that excessive hip extension resulting from a weak iliopsoas or weak ankle plantarflexors may induce hip and groin symptoms. The observed reduction in hip extension associated with the gait of individuals with hip and groin pain may be a protective mechanism for avoidance of anterior joint overload (Lewis et al., 2007, 2009, 2010), but it probably results in premature hip flexion, possibly through the assistor hip flexors that increase anterior joint compressions. In a cycle of problems, hip osteoarthritis seems to induce muscular atrophy at the hip, which possibly compounds these problems over time (Fig. 4.2.5b).

Pathology through the principles of biomechanics **Chapter | 4** 661

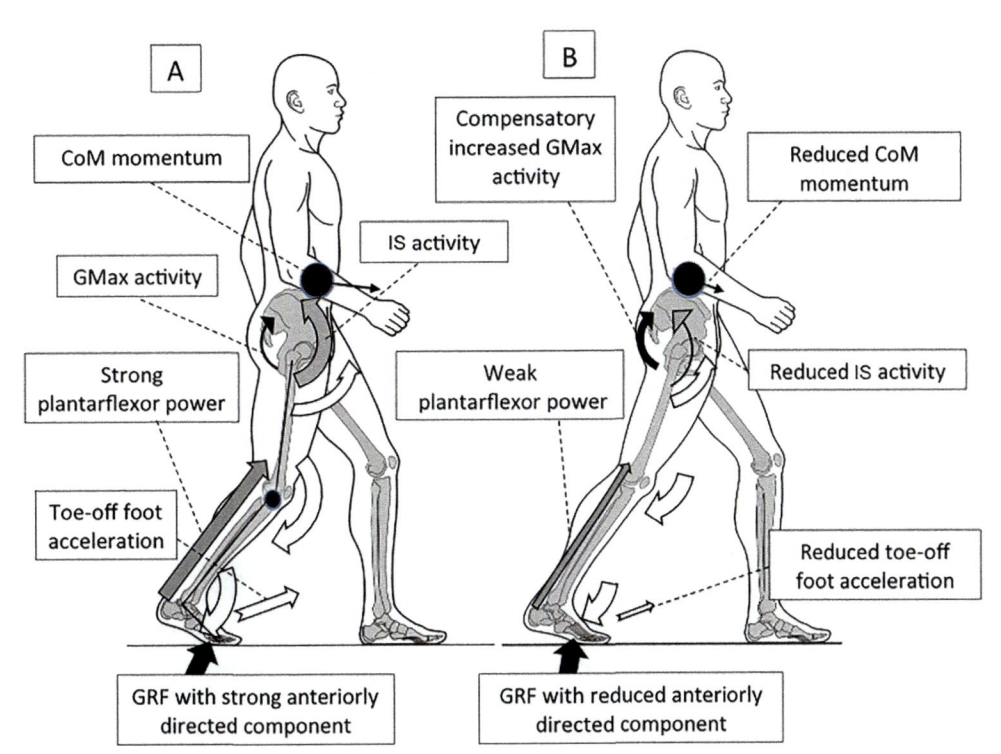

FIG. 4.2.5b A free-body diagram of healthy preswing (A) helps to explain the intricate actions in initiating hip and knee flexion by utilising ankle plantarflexion power, midfoot plantarflexion, and digital extension. When working efficiently, these motions and moments prevent the hip reaching end-range extension, risking anterior hip joint impingement while still maintaining anterior velocity of the CoM into the loading response on the contralateral limb. Release of the Achilles' stored energy provides plantarflexion power, triggering ankle plantarflexion that couples to knee flexion via fibres of gastrocnemius. This ankle plantarflexion is aided by some flexion power from peroneus longus and tibialis posterior plantarflexing the midfoot, and from digital flexor activity resisting digital extension. Any midfoot plantarflexion added to ankle plantarflexion helps to further increase the rate of knee flexion. Knee flexion reduces functional limb length, losing some momentum on the CoM, so a little gluteus maximus (GMax) activity compensates. Further hip extension is limited by the knee and ankle flexion angles, and the activation of iliopsoas (IS). This hip flexion moment is mainly derived from iliacus, aided by rectus femoris that picks up the lower limb's CoM (small black circle) when initiating the hip flexion moment that brings about toe-off. Without effective IS activity at preswing (B), the hip is far more vulnerable to hip over-extension, but multiple compensation mechanisms are available. These include increasing triceps surae activity. However, if this is not possible, the hip will need to compensate. Loss of increasing knee and ankle plantarflexion angles in terminal stance from weaker triceps surae also takes away protection against excessive anterior hip joint forces, caused by higher hip extension ranges of motion. Increasing GMax activity to maintain CoM momentum in late terminal stance may exacerbate such hip extension impingement. Thus, when examining cases of anterior hip impingement, all muscles and joints involved in protecting the anterior hip joint should be investigated, with kinematic gait data at terminal stance and preswing being most helpful in locating the origin of dysfunction. *(Permission wwww.healthystep.co.uk.)*

Thus, as long as most of the CoM is offloaded to the opposite limb efficiently through acceleration power by the ankle plantarflexors (assisted by gluteus maximus), then iliacus can start hip flexion without having the hip loaded with high levels of body CoM forces. Thus, providing a hip flexion moment to prevent anterior–superior hip compression loads breaching tissue threshold should be easy. In restricting hip extension, the degree of knee extension, foot plantarflexion, and thus gastrocnemius activity, also play a part. If the knee hyperextends in terminal stance, then the hip-to-ankle alignment will change. A reduced ankle plantarflexion angle developing after heel lift (which can follow an excessive ankle dorsiflexion moment at the late midstance-heel lift boundary) can increase the hip extension angle and the extension moment arm on the hip.

A weak soleus reducing restraint of ankle dorsiflexion in late midstance can link to excess knee extension through its altered interaction with gastrocnemius. With less restrained ankle dorsiflexion, the CoM can move anteriorly too rapidly, increasing the knee extension moment. In compensation for the weak soleus, gastrocnemius can increase its activity through its proximal attachments over the knee, but if the knee becomes fully extended, the action of gastrocnemius can change from resistor to assistor of knee extension through its changing moment arm position. In these circumstances, gastrocnemius can become a late midstance knee extensor while at the same time trying to apply more ankle plantarflexor power (resisting dorsiflexion) for decelerating the body's CoM anterior displacement before heel lift, in compensation for weakened soleus. Gluteus maximus can also compensate for the reduced ankle plantarflexor acceleration power at heel lift and thus may exacerbate the situation by activating early. This premature gluteus maximus compensation may drive excessive hip extension, further increasing the knee extension, and thereby further helping increase the potential for gastrocnemius knee extension moments. A potential delay in heel lift to initiate weight transfer to the opposite foot can then cause the iliopsoas to initiate hip flexion against a larger hip extension moment and position, requiring its increased activity (Fig. 4.2.5c). As a result, anterior–superior hip joint loads can increase at the anterior femoral head, but also potentially increase loads on the iliopsoas, gastrocnemius, posterior knee soft tissues, and the tibiofemoral articular surface anteriorly. Other injury options are also available. Thus, heel lift timing, hip extension, and knee flexion angles are critical for safe efficient late stance phase and effective swing-initiating biomechanics.

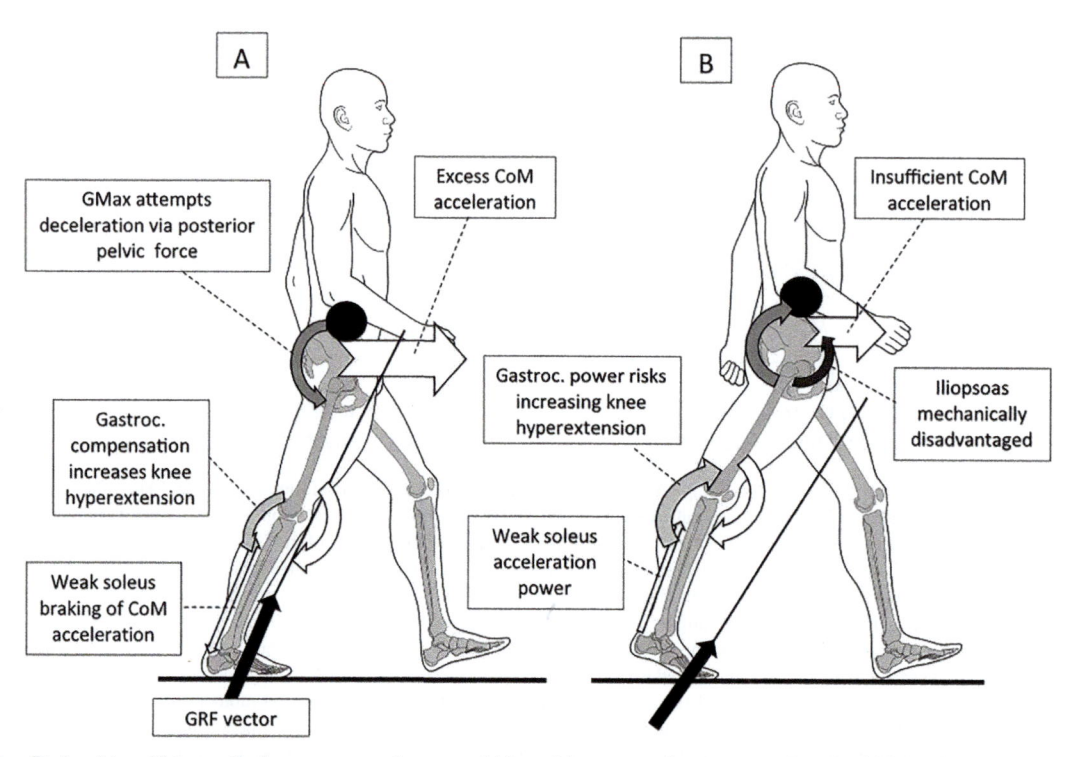

FIG. 4.2.5c During late midstance, the knee comes under external hip and knee extension moments that should be resisted by the gastrocnemius (Gastroc.), which is also assisting in restricting the rate and degree of ankle dorsiflexion with soleus. Should soleus be weak (A), then dorsiflexion deceleration can be compromised, leading to rapid ankle dorsiflexion creating excessively accelerated CoM motion that increases the external GRF moment arm for hip and knee extension. Gastroc. cannot fully compensate for soleus and risks causing knee hyperextension in this situation. Gluteus maximus (GMax) attempts to compensate for increasing CoM momentum by applying a posterior braking force on the pelvis. However, it is a poor compensator for soleus in decelerating the CoM velocity during midstance. If knee hyperextension remains unresolved at heel lift (B), gastroc. may increase its extension arm, becoming a stronger knee extensor. With a weak soleus and altered function of gastroc., plantarflexion power from the ankle to the forefoot decreases. These situations create considerable difficulty for hip function. GMax is now required to increase the acceleration moment on the CoM for terminal stance, creating greater hip extension, risking anterior impingement. A more extended knee and hip accompanying reduced CoM acceleration from an under-powered heel lift, increases the workload on iliopsoas to start the initiation of hip flexion at preswing. *(Permission www.healthystep.co.uk.)*

Terminal stance inability of the gluteus maximus to induce adequate hip extension is also significant at late terminal stance when greater acceleration off the trailing terminal stance limb is required. Therefore, the functional role of gluteus maximus in influencing gait energetics is greater in faster walking and running. Compensation for poor gluteus maximus activity can be managed through the secondary hip extensors such as the short head of adductor magnus and piriformis, or by increasing lower limb CoM acceleration via increased soleus activity. The risk of increased tissue loading would depend on which mechanisms were attempted, focusing injury risk to any one or more of the muscles and/or associated connective tissues and attachments.

As the soleus closely works with gluteus maximus in maintaining appropriate CoM acceleration during stance, soleus represents a functional solution to the lost acceleration power of the CoM in terminal stance from a weak gluteus maximus. Soleus may do so either by increasing its activity peak, by extending its duration of activity, or by keeping itself structurally stiff and tight to increase its elastic power during the acceleration phase. Stiffening by providing isometric rather than eccentric contraction, would risk positioning the ankle less dorsiflexed and the knee more flexed (through soleus and gastrocnemius attachments), tending to restrict hip extension. However, should gastrocnemius concurrently be weak, the anterior momentum of the CoM may continue to drive knee extension, hyperextending the knee on an ankle joint that is reluctant to dorsiflex.

However, triceps surae stiffening may also potentially decrease the normal anterior translation of the body's CoM via tibial rotation during midstance ankle dorsiflexion, prior to heel lift. Triceps surae myofascial stiffness that causes a block on ankle dorsiflexion, thereby preventing free anterior CoM translation, risks an earlier compensatory heel lift to maintain CoM momentum. Such a premature heel lift will result in an upward vertical displacement of the CoM when it should be falling forward, much reducing the heel lift's mechanical efficiency. This could result in soleus strains or Achilles tendon injuries which should associate with areas of the subtendon that arise from the soleus muscle fibres. As the soleus subtendon lies medially within the Achilles, excessive stress-induced failure within this subtendon's mechanics from resulting tendon tissue injury could then alter rearfoot biomechanics, further complicating the picture (Zifchock and Piazza, 2004; Lee and Piazza, 2008; Lersch et al., 2012). From this situation, it can be appreciated that through failure of hip function, it is possible that a patient could present in clinic with an Achilles tendinopathy that fails to respond to local calf rehabilitation and/or orthoses because the origin lies in hip dysfunction. Unless the hip dysfunction is addressed, the problem may persist (Fig. 4.2.5d).

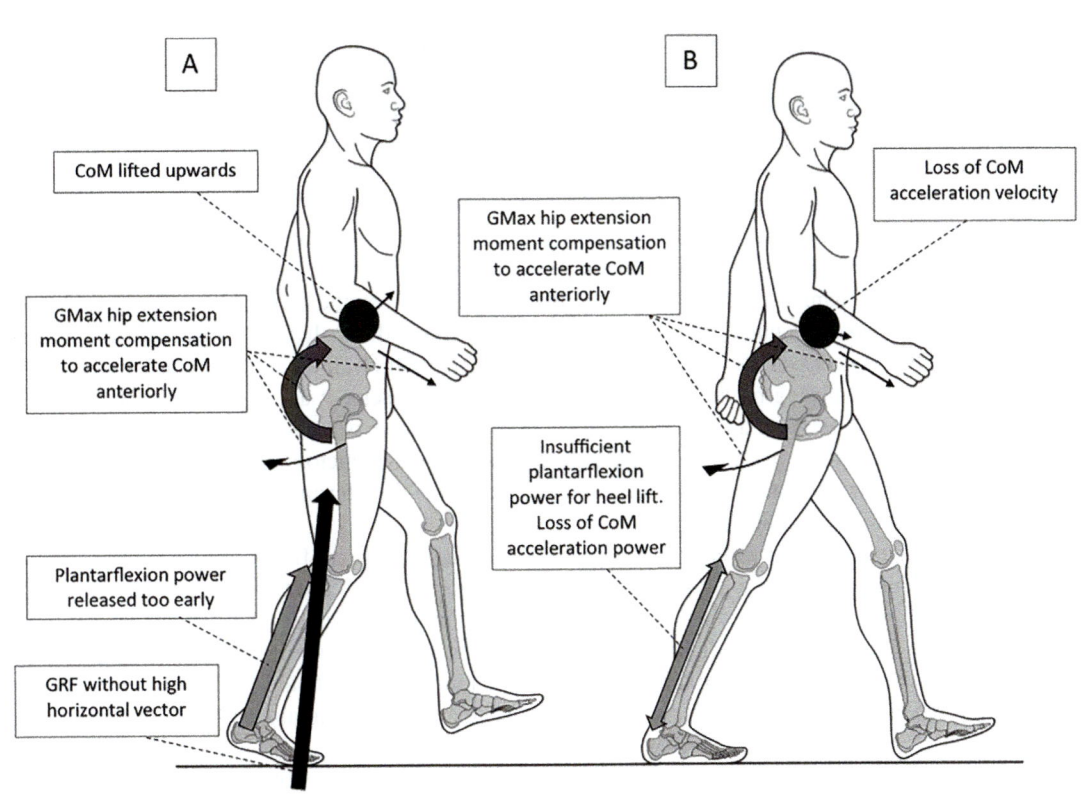

FIG. 4.2.5d Gait energetics is driven by the capacity to maintain steady CoM momentum under gentle up-and-down oscillations, which should generate GRF with strong posteriorly directed vectors towards the body during braking and anteriorly directed GRFs during acceleration. For this action, certain muscles are essential for controlling CoM velocity in the sagittal plane. Soleus, as part of the plantarflexor power and gluteus maximus (GMax) play central roles, with GMax taking the primary role from initial contact to absolute midstance around hip function, and soleus for late midstance to early terminal stance around ankle function. An interesting situation plays out during terminal stance, as GMax assists CoM acceleration onto the next step as the ankle plantarflexion power is waning from soleus. In premature heel lifts (A), the CoM is initially disrupted in its normal 'falling forward' momentum that drives an increasingly anteriorly directed GRF. Instead, the CoM initially rises upwards with a reduced horizontal element to the GRF vector. It must then fall forward from a higher height. With such heel lifts occurring with less hip extension to aid CoM acceleration via a higher horizontal force, GMax must actively power more hip extension. If this power is excessive, it could risk overextending the hip during weight transfer, risking anterior hip impingement. If heel lift is delayed (B), perhaps through soleus weakness, GMax may become active earlier to maintain CoM velocity. This increases the risk of hip hyperextension and anterior hip joint impingement, the longer the heel remains on the ground during weight transfer. The usual solution to this problem is a shorter stride, while an immediate therapeutic solution is a raised heel on or inside footwear. *(Permission www.healthystep.co.uk.)*

4.2.6 Frontal plane free-body diagrams of the hip

The role of the hip in the frontal plane during gait is to stabilise and limit the hip adduction moment (varus tilt) that occurs on the support limb side during stance phase, while facilitating continued anterior displacement of the CoM over the hip. This allows the CoM to rotate transversely and anteriorly past the hip, permitting the swing limb to advance forward for the next foot contact without causing imbalance or loss of the erect trunk posture under large frontal or transverse plane hip torques. Failure of this mechanism causes the pelvis to excessively drop downwards on the contralateral swing limb side, making ground clearance of swing more difficult. Overactivity of hip abductors causes hip abduction (valgus tilt), 'hitching' the swing leg pelvis up and increasing the ground clearance space, but at an energetic cost. This hip valgus tilt may be necessary for managing loss of swing limb shortening kinematics, such as the loss of ankle dorsiflexion capability seen in cases of foot drop. It can also result from hip abductor tightness or spasticity.

The stance hip carries the weight of the pelvis, trunk, upper limbs, the head, and the mass of the opposite limb in its swing phase during single support. The extra pendular weight of the late swing phase limb draws the CoM of the body anterior to the support limb through momentum-induced centrifugal forces from swing limb hip flexor and knee extensor muscular momentum. Thus, the anterior pull of the swing limb's mass during late swing and its medial draw on the CoM over the stance limb's hip throughout single-limb support are both significant for preparing body posture for weight transfer.

The CoM resistance moment is positioned medially to the support limb's hip axis in the frontal plane, which if uncontrolled, induces pelvic obliquity or tilt in single-limb support. The hip abductors, whose combined effort lies on the lateral side of the hip joint along the ilium, resists this varus tilt. The lever arm created in the frontal plane across the hip is that of a class one lever, which means that the CoM's distance from the hip fulcrum is critical to the effort required to counterbalance the CoM and stabilise its vertical position to avoid any significant hip adduction moment. It is also worth highlighting the fact that the hip abductors maintain the vertical displacement of the body's CoM which is important for the development of midstance potential energy from gaining CoM height during walking. Vertically, the hip abductors work as a class three lever system, as the CoM is above the hip abductors' effort arm, with the hip as a fulcrum below the proximal attachment effort of the hip abductors. With flexion and extension around the hip, the effort and resistance arms translate from anterior to posterior positions. However, the hip flexion and extension moments that the hip abductors generate, although significant, are much smaller than the hip flexors (such as iliacus) and extensors (gluteus maximus and hamstrings) because their moment arms are shorter to the hip in the sagittal plane (Fig. 4.2.6a).

Failure of the abductor mechanism is associated with a drop of the pelvis on the contralateral swing limb side, known as a Trendelenburg gait. This makes ground clearance of the swing limb more difficult, requiring greater hip and knee flexion and/or ankle extension that is necessary to functionally shorten the swing limb. Increased hip adduction moments are also linked to increased injury risk during running (Bramah et al., 2018; Ceyssens et al., 2019), suggesting a significant hip abductor effect on running energetics and fatigue resistance. In running, there is no double-limb stance during impact with loading occurring on a single limb, thereby loading only one set of hip abductors at a time throughout stance. When walking, both limb hip abductors are active on loading and offloading the limb during double-limb stance, creating periods of relative ease for each set of hip abductors.

The hip abductors in a posterior-to-anterior direction consist of the superficial fibres of gluteus maximus via the ITT, gluteus medius, gluteus minimus, and the tensor fasciae latae. Gluteus maximus and tensor fasciae latae provide only secondary abductor roles, with gluteus medius and minimus furnishing the primary hip adductor moment resistance (Rutherford and Hubley-Kozey, 2009). Initially, the more posterior parts of gluteus medius and minimus with gluteus maximus provide hip abduction stability from initial contact and loading response to absolute midstance. The other regions of the gluteal hip abductors contribute towards the end of midstance and into terminal stance (Anderson and Pandy, 2003; Semciw et al., 2013).

The orientation of the posterior segments of gluteus medius and minimus gives the muscles a more significant effort lever arm to resist hip abduction during early stance phase, assisted by gluteus maximus. These muscle sections lie posterior to the hip axis and close to the CoM while the hip is flexed. As the hip starts to extend into late midstance, the muscles lying anterior to the hip axis gain more mechanical advantage to resist the adduction moments. The anterior parts of gluteus medius, gluteus minimus, and tensor fasciae latae are now better suited to stabilise the femoral head in later mid- and terminal stance (Semciw et al., 2013). The responsibility for the hip adduction moment's resistance thus passes from posterior to anterior, helping sagittal plane facilitation of the trunk's CoM anterior displacement. The change in the segmental responsibility of gluteus medius from the posterior and middle to the anterior sections causes two bursts of EMG activity, with the anterior section's burst coming later during stance (Hof et al., 2002; Mickelborough et al., 2004; Semciw et al., 2013).

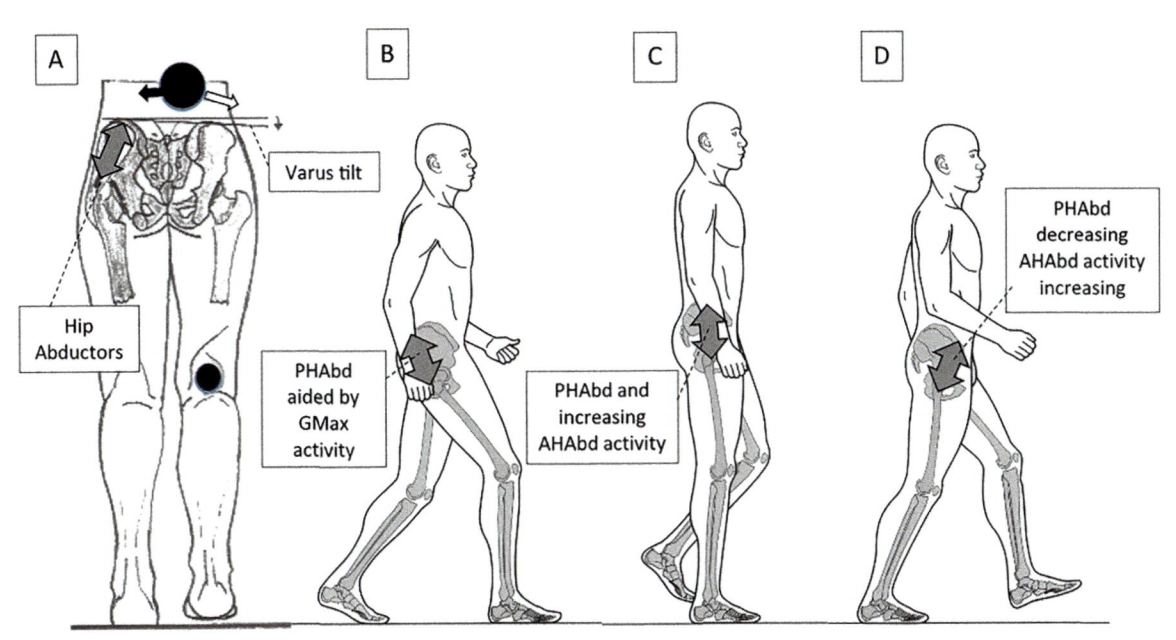

FIG. 4.2.6a The hip abductors play a fundamental role in walking gait energetics in keeping the position of the HAT CoM (large black circle) stable towards the support limb side. The CoM during walking should remain a little to the medial side of the support limb while it undergoes a stance phase cycle of a loading and braking fall, a weight-transferring lifting up and then falling forward, and a later offloading of the support limb. During this vertical oscillation, it must continue to move anteriorly at a smooth velocity appropriate to the gait speed required. The first hip abductor action is shown in (A), where the hip abductors resist the hip varus moment (black arrow opposing white arrow at CoM) created by the CoM being pulled away from the support limb as a result of gravity and the addition of the swing limb's CoM (small black circle). Varus tilt does not need to be fully corrected and indeed, allowing a little varus tilt possibly improves overall gait energetics. In early stance phase (B), the braking burden falls to the posterior hip abductors (PHAbd), consisting of the posterior and middle fibres bundles of gluteus medius, and posterior bundles of gluteus minimus aided by superficial fibres of gluteus maximus (GMax). By midstance (C), increasing anterior hip abductor activity (AHAbd) from the anterior bundles of gluteus medius, gluteus minimus, and tensor fasciae latae adds to varus tilt resistance, helping raise the CoM with hip and knee extensor activity. During late stance (D), AHAbd activity starts to peak as varus tilt control moves anterior to the hip, bringing in more power from the anterior fibres of gluteus minimus and tensor fasciae latae. GMax activity switches off at absolute midstance, yet the other posterior hip abductors activity continues throughout stance but with decreasing influence compared to anterior muscles. Thus, hip abductor activity helps explain trunk stability during the anterior translation of the CoM during stance phase. Dysfunction can be subtle, specific to individual muscle or muscle fibre bundle failure. It is therefore useful to note the starting and stopping of any hip hitch or 'Trendelenburg' hip drop during gait. *(Permission www.healthystep.co.uk.)*

The gluteus medius also generates transverse plane rotations around the hip to assist in swing phase limb advancement, by anteriorly rotating the contralateral pelvis. This is achieved through ipsilateral internal hip rotation due to its anterior attachment to the greater trochanter (Semciw et al., 2013; Uemura et al., 2018; DeJong et al., 2020). Gluteus medius and minimus also have secondary extensor and flexor roles with the posterior and middle aspects of gluteus medius and the posterior section of gluteus minimus acting as extensors, and the anterior aspects of both muscles, in conjunction with tensor fasciae latae, acting as flexors (Semciw et al., 2014). This means that when the posterior and middle sections of these muscles are active in early stance, they help resist hip flexion and reduce hip flexion angles. In late stance, when the anterior aspects of these muscles become active, they help to stabilise the hip extensor moment (Fig. 4.2.6b).

Weakness within the hip abductors is associated with significant kinematic changes around the hip with the potential to influence a wide range of injuries and dysfunctions. Significant hip adduction moment causes a Trendelenburg gait, a condition that is usually visible to the naked eye and should always be investigated. A Trendelenburg gait indicates that the frontal plane lever is compromised. Oftentimes, excessive hip adduction only occurs during running and is not noticeable in walking gait. A running Trendelenburg is harder to spot without slow motion filming. Excessive hip adduction only demonstrated when running, indicates that the hip abductor strength is inadequate when exposed to faster loading-rates and larger peak forces during single-limb support. In walking, hip abductor work is spread over both limbs during each double-limb stance phase of loading response and offloading acceleration, except at heel lift (normally a

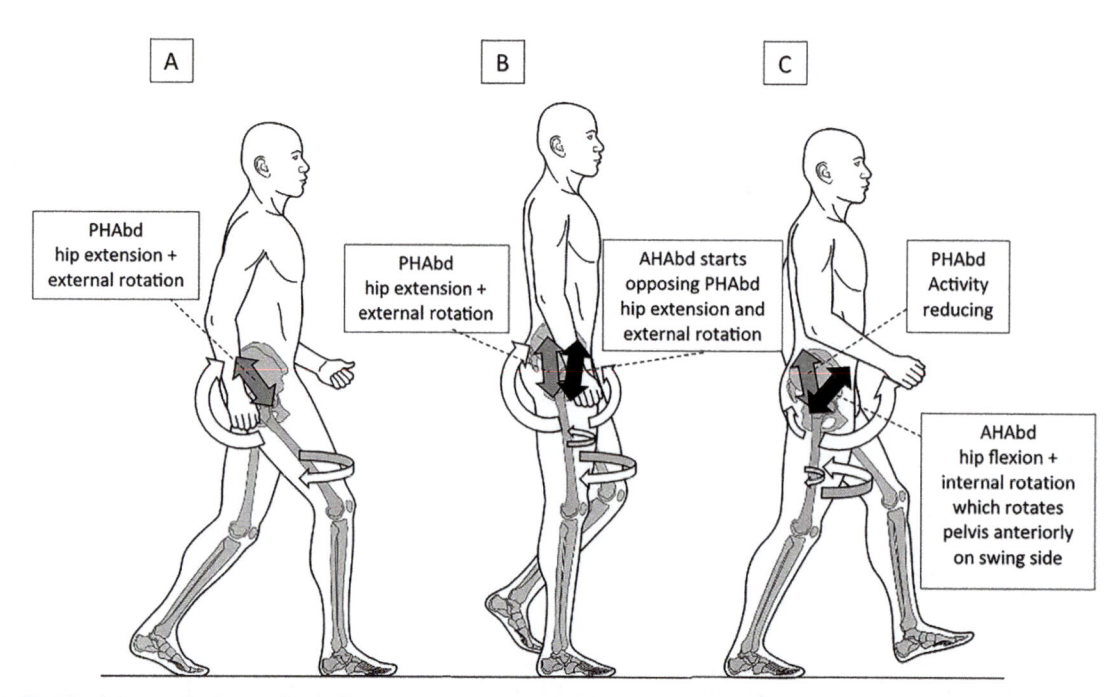

666 Clinical biomechanics in human locomotion

FIG. 4.2.6b Hip abductors also have a direct effect as extensors and flexors and as hip rotators, as well as controlling frontal plane pivoting. As a ball-and-socket joint, muscles can cause sagittal, transverse, and frontal plane motion across the hip. Although hip abductors primarily create forces within the frontal plane, their fibre bundle orientations also create forces across the other planes. The posterior hip abductors, including the superficial fibres of gluteus maximus (PHAbd—grey double-headed arrow), have their attachments posterior to the hip axis. Thus, when active, they aid hip extensor and external rotator power at the hip. This extra power is particularly important during loading response (A), a period when the hip is flexed and internally rotating, yet soon needs to start a cycle of extension and external rotation. By absolute midstance in walking (B), the anterior hip abductors (AHAbd—black double-headed arrow) that lie more anterior to the hip axis are increasing their activity, resisting the extension and external rotating power of the posterior hip abductors. During late midstance (C) and terminal stance, AHAbd power increases, allowing greater internal hip rotation that aids swing limb advancement by rotating the contralateral side of the hip anteriorly, while also helping resist hip extension moments. Thus, failure in hip abductors also alters other hip and pelvic biomechanics beyond those of the frontal plane. *(Permission www.healthystep.co.uk.)*

single-limb supporting event). Identifying which, or if all hip abductors are dysfunctional, is helpful in correct treatment planning with the level of intervention being indicated by the timing, duration, and extent of the pelvis adduction tilt. Gluteus medius provides the largest abduction moment to the hip through its large physiological cross-sectional area (Semciw, et al., 2013), thus making it highly likely that a Trendelenburg gait indicates at least some gluteus medius dysfunction.

The body's easiest mechanical resolution of the imbalance in trunk posture caused by a Trendelenburg gait is to sway the trunk to the support limb side, as this moves the CoM closer to the fulcrum. By reducing the resistance arm in such a way, the effort strength can become adequate to resolve, or at least reduce, the mechanical problem. However, the energetics of gait is compromised as it can cause the trunk to consistently sway towards the support limb and unnecessarily raise the contralateral hip during single-limb stance. By changing hip kinematics to an abduction or valgus moment, a hip hitch is created that increases frontal plane oscillations of the CoM. Minor hip abductor weakness can be recognised by identifying this hip hitch and examining hip abductor strength and flexibility. The lever arm effect of this compensatory hip abductor muscle function is to move the CoM nearer to the fulcrum which, despite the initial extra effort, results in reducing the resistance arm length through the rest of stance phase, giving the hip abductors increased mechanical advantage after an initial extra burst of activity on loading. Tightness within the hip abductors as well as increased activity can induce a hip hitch effect. However, any large gluteus medius deficit is unlikely to be resolved this way, for then the hip abductors may not have the strength to initiate this gait tactic. Indeed, the initial increased hip abduction activity necessary to achieve the effect is likely to induce earlier fatigue of these muscles in both walking and running over time. It is important here to state that the observation of an odd hip hitch or Trendelenburg among steps, especially during running, is a common and normal variant in steps to manage minor terrain, lower limb, and CoM perturbations during gait (Fig. 4.2.6c). Abnormal hip abduction angles should only be thought of as being significant if such presentations are consistently present on one or both sides.

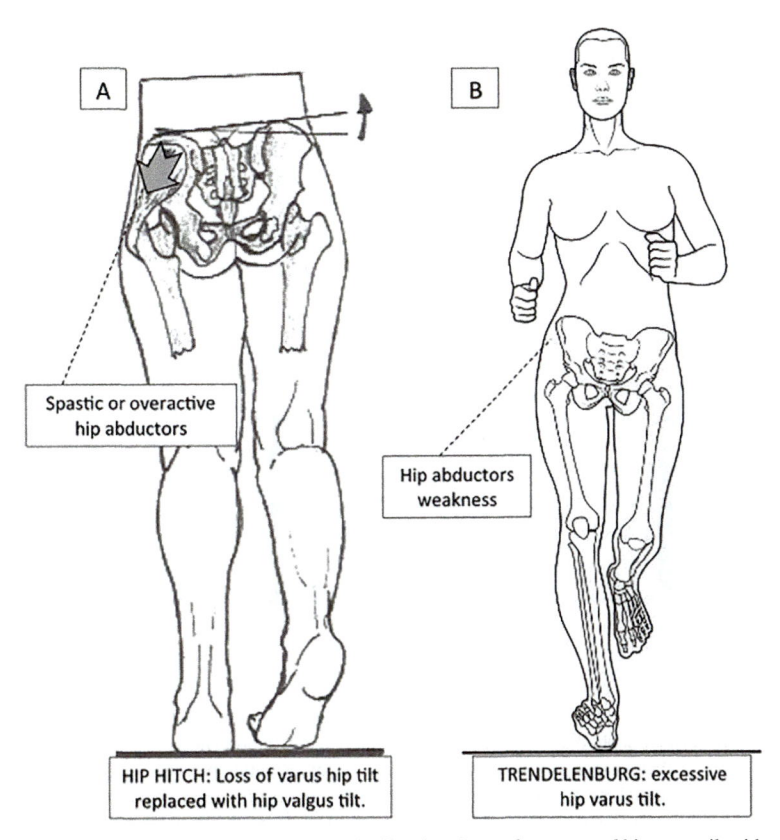

FIG. 4.2.6c Hip abductor dysfunction can be easily divided into two: dysfunction that replaces normal hip varus tilt with a valgus tilt, usually called 'hip hitch' (A), and that resulting from too much varus tilt, referred to as 'Trendelenburg gait' (B). Both problems can be caused by a number of pathologies, muscle weaknesses, or spasticities, so identifying the problem does not identify the origin. However, as single step events, such variations in hip abductor activity can just be helpful responses to gait perturbations. A hip hitch or drop must be consistent or frequent to be significant. Dysfunctions in walking gait tends to indicate more serious issues, because loading response and weight transfer occurs through a period where both limbs' hip abductors are active in CoM support through bilateral limbs being in ground contact. In running, small hip abductor dysfunction can lead to significant disruption in CoM stability, particularly as during loading response only one limb's set of hip abductors are used to help brake and stabilise the CoM over a single supporting limb, often on a relatively long moment arm from the GRF. *(Permission www.healthystep.co.uk.)*

As already mentioned, running presents a particular problem for the hip abductors because of the lack of a double-limb stance phase, requiring a large hip adduction moment to be resisted during impact, loading response, and acceleration. Another issue for the hip abductors in running is the base of support which has to be narrowed to bring the support limb and foot further underneath the body's CoM, to accommodate lack of double stance. To achieve this, the lower limb must increase adduction at the hip. This increased lower limb adduction angulation lengthens the moment arm of the hip abductors, but also increases muscle fibre length and tensions through the myofascial tissues. This means that the compressive forces between the gluteus medius and minimus muscles lying over the greater trochanter will be increased.

Anatomical bursae are strategically positioned between these muscle–tendons and the greater trochanter of the femur to prevent excessive compression and shear between bone and soft tissue. With increased thigh adduction, there is a risk that compression and friction stresses from the increased hip adduction angle will lead to greater compressive tissue loads. This situation is worsened if the runner has a low femoral head-to-shaft angle causing a coxa varum or has dysfunctional/inadequate hip abductor strength. In these situations, the adductor moment, angle, and consequential muscular demands can set up pathology such as trochanteric bursitis (Fig. 4.2.6d).

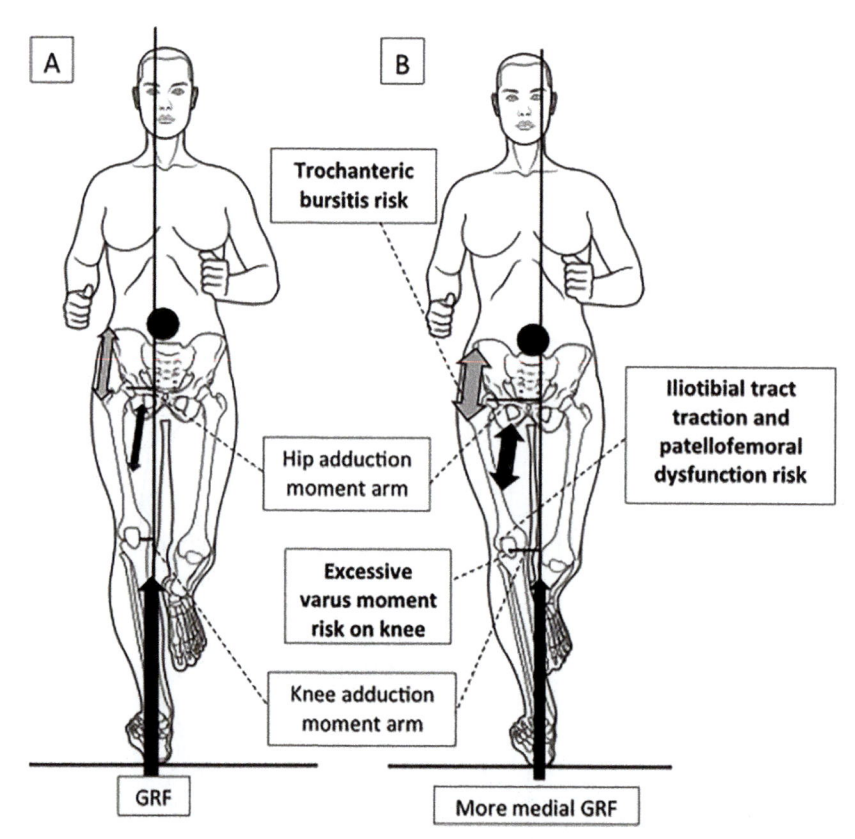

FIG. 4.2.6d Using free-body diagram of variations in the adduction angle at the hip when setting the base of support can soon indicate potential mechanical issues for the hip abductors (grey double-headed arrows). Narrower bases of support from positioning the limb further under the body due to increased hip adductor activity (black double-headed arrows) or as a result of a low femoral neck-shaft angle, puts the hip abductors at a mechanical disadvantage by moving the GRF vector (black arrow) more medially under the body. Running presents a dilemma for the hip abductors, as a gait style without double-limb contact phases during braking and acceleration requires the support limb to position more directly under the CoM (black circle). Usually (A), this means the foot still lies slightly lateral to the CoM, but more directly underneath it than would occur during walking. This posture requires the hip abductors to actively compensate for the larger hip adduction (varus) moments. However, if the foot becomes positioned increasingly medial to the hip (B), the GRF moment arm increases hip and knee varus moments. This requires larger hip abduction moments to oppose them. This can be a particular problem when the posterior hip abductors are under considerable accelerated impact forces during loading response. Inadequate posterior hip abductor strength, tight hip adductors, or the presence of a coxa varum can compound these problems. Potentially, this can set up hip or knee pathology related to excessive varus and internal rotation moments as a result of hip abductor fatigue, resulting in poor hip adduction force management and thus risking pathologies such as trochanteric bursitis. *(Permission www.healthystep.co.uk.)*

The size of the combined hip adduction moment can also generate pathology around the knee, through influencing the adduction (varus) moments on the tibiofemoral joint and/or changing the quadriceps angle to the patella, as well as affecting ITT internal tensions. It is no surprise, therefore, that larger hip adduction angles during running have been associated with a number of knee pathologies and symptoms including patellofemoral pain and ITT (band) traction syndrome (Bramah et al., 2018; Ceyssens et al., 2019).

4.2.7 Transverse plane free-body diagrams of the hip

The hip abductors (particularly the more anterior segments) also play a significant role in transverse plane rotation of the hip in which they facilitate pelvic anterior and internal rotation, helping to advance the swing limb anteriorly around the body for the next step (Al-Hayani, 2009; Semciw et al., 2013, 2014). These hip transverse plane rotations should be subtle, but may be exaggerated if hip flexion strengths on the swing side or extensor strengths on the stance side are compromised. In these cases, transverse plane hip rotations can increase in magnitude to aid stride distance. However, this is not energetically efficient as the hip abductors are not great substitutes for hip flexors and extensors. This pelvic motion is more important at faster gait speeds when the pelvis is moving more in-phase with the lower limbs (Fig. 4.2.7a).

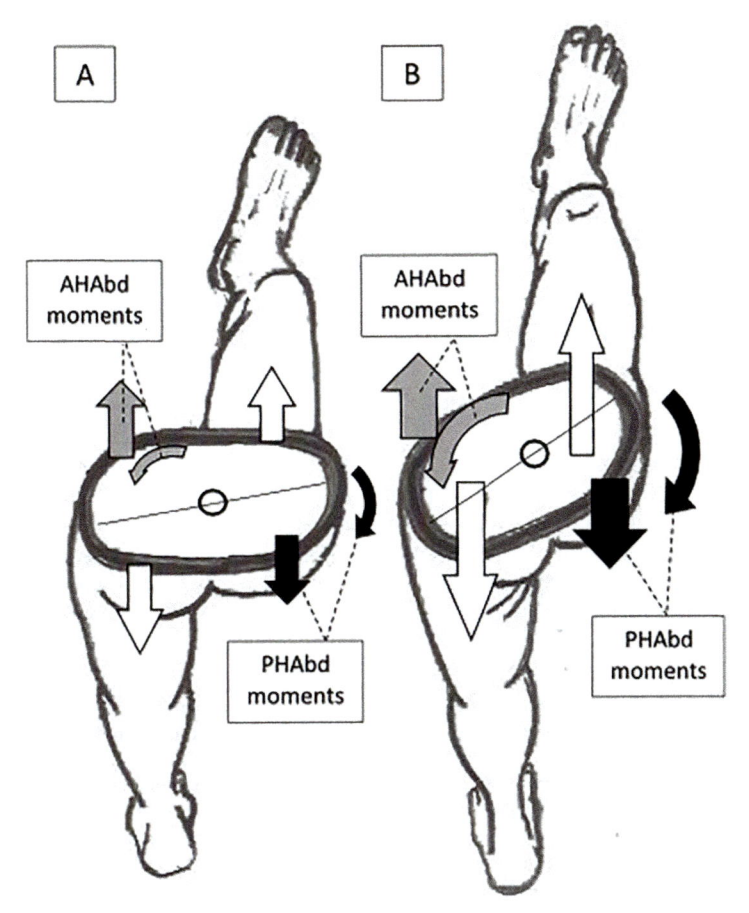

FIG. 4.2.7a As well as stabilising frontal plane torques at the hip, hip abductors are used for transverse plane hip rotation and can be used as compensations for hip flexor or extensor weakness with variable success. In slower walking with shorter strides (A), their transverse and sagittal plane activity is less pronounced. At faster gait speeds and longer strides (B), the hip abductor role in rotating the pelvis with the limbs becomes more significant. The posterior hip abductors (PHAbd) provide extension and external rotation moments on the lower limb at the hip (black arrows) during loading response that resists the internal limb rotation and hip flexion moments that accompany loading response. The anterior hip abductors (AHAbd) provide hip flexion resistance and internal hip rotation moments that pull the contralateral side of the pelvis forward during the late contralateral swing phase, helping to lengthen the stride length and improve centrifugal swing power (grey arrows). In double-limb stance when one leg is positioned posteriorly and the other anteriorly (white arrows), the antagonistic action of left and right hip abductors helps stabilise the transfer of the CoM to the anterior limb, something more difficult with a longer stride length. Thus, in faster walking, this role becomes increasingly important as double-limb stance time decreases. Hence, weakness in hip abductors is more apparent at faster walking (and running) gait speeds, as loss of their transverse plane motion and sagittal plane resistance as aids to stride length setting and weight transfer become more significant. *(Permission www.healthystep.co.uk.)*

Another common reason for abnormal transverse plane motion is high or low acetabular and femoral anteversions, which can alter the sagittal plane alignment of the lower limbs. Femoral anteversion has a strong influence on the musculoskeletal loads placed on the proximal femur, with anteversion angles of 30° increasing hip contact forces and bending moments by 28% (Heller et al., 2001). High anteversions cause increased internal limb rotation, usually with a resultant deficit in external hip rotation and vice versa with low anteversions. Anteversion issues are associated with an increased OA risk due to reduced articular surface contact during the transfer of loads across the hip joint, increasing hip instability risks (Torry et al., 2006). A free-body diagram of a patient with anteversion issues would need to be constructed with the lower limb aligned internally or externally rotated, depending on the nature of the alignment. The other primary anatomical and functional forces remain the same, even if the orientations of the forces are slightly changed. This may create functional disadvantages in certain hip muscles such as gluteus maximus and gluteus medius that provide external hip rotation action, as hip alignment variation alters the effort arms of these muscles (Fig. 4.2.7b). It is important to remember that anteversion 'normals' are age-dependent in children.

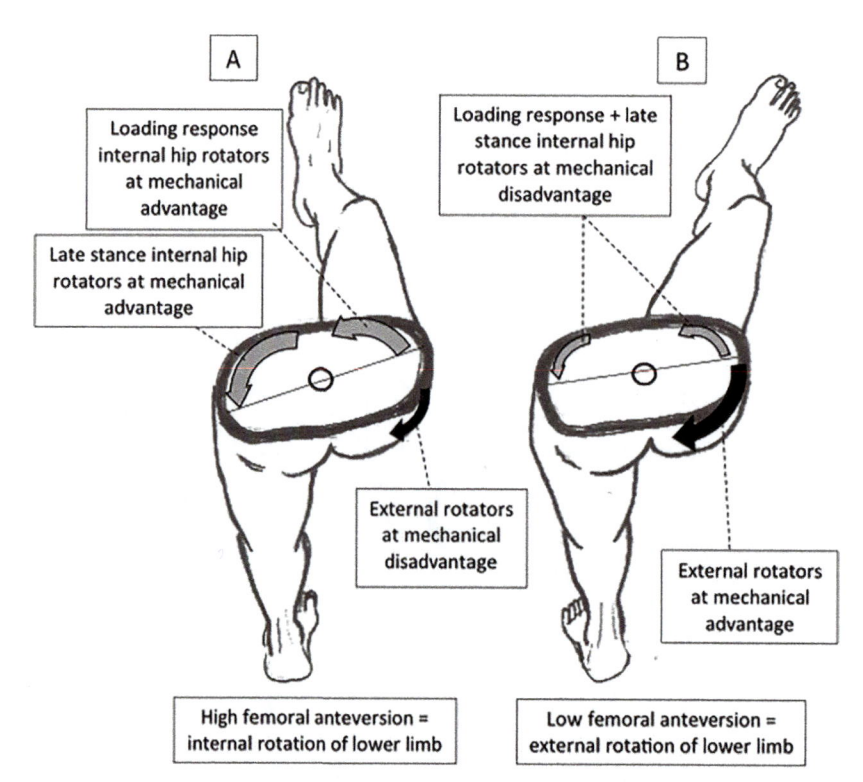

FIG. 4.2.7b Femoral anteversion can influence hip transverse plane moments via changing the lever arms for internal and external rotators of the hip. High femoral anteversion (A) causes an internal rotation of the lower limb on the femoral head and neck, making it more difficult for external rotators such as gluteus maximus and the posterior and middle fibres of gluteus medius to control internal hip torques at loading response. Internal hip rotators such as the hip adductors of adductor magnus, longus, and brevis, and pectineus at loading response have mechanical advantage. Tensor fasciae latae and the anterior fibres of gluteus minimus and medius have mechanical advantage at late midstance and terminal stance. This may cause excessive internal hip rotation at these phases of stance, threatening hip impingements. However, low anteversion angles (B) externally position the lower limb to the hip neck-shaft, giving advantage to the external rotators which may make tasks such as limb placement under the CoM at loading response and weight transfer to the opposite limb in terminal stance more difficult. Thus, changes in muscle activity may exacerbate preexisting skeletal rotational issues within the lower limbs. *(Permission www.healthystep.co.uk.)*

4.2.8 Expanding complexity of hip pathomechanics with free-body diagrams

From the examples discussed, it is clearly shown that a treatment approach to hip dysfunction is variable. A long stride length and fast gait speed in a situation of a weak gluteus maximus, linked to patellofemoral dysfunction and pathology, should be approached very differently from that associated with a weak quadriceps and a strong gluteus maximus, and a short stride length. The recording of muscle strengths on clinical examination and joint angles in gait can be utilised together through free-body diagrams to help identify the likely underlying pathomechanics, thus explaining the origin of pathology. A recurrent hamstring injury from failure in gluteus maximus braking strength is relatively easy to unravel, and a free-body diagram can illustrate that reducing the stride length and strengthening the gluteus maximus will greatly help reduce loads on the hamstrings. However, carefully constructed free-body diagrams of the hip can start to help in understanding more complex situations too.

For example, with a weak quadriceps and gluteus maximus, a long stride length and a hard running surface associated with MTSS may indicate excessive tissue resonance within the lower limb. This could be as a consequence of the loss of quadriceps/gluteus maximus energy dissipation with concurrent failure in other lower limb mechanisms to compensate. Addressing running terrain hardness, possibly through footwear 'cushioning', and by stride length reduction may not resolve the problem, if the quadriceps/gluteus maximus weakness lies at the heart of the pathomechanics and requires strength improvement.

The hip's sagittal plane loading and acceleration events during gait present particular challenges in understanding what happens during initial contact-loading response and terminal stance biomechanics. The hip plays a significant assistor role in managing impact energy dissipation with the knee. Through the quadriceps (rectus femoris) crossing both joints and with the assistor role of gluteus maximus and the hamstrings in managing knee-to-hip angles as part of a shock attenuation spring-like limb, an important functional chain is established. In terminal stance, gait models have reported that increased hip extension moments lead to increased force in the anterior hip (Lewis et al., 2007, 2010), with only a 2° increase in maximum hip extension potentially resulting in an ~25% increase in anterior hip joint force (Lewis et al., 2010). Anterior hip joint force reaches its maximum near maximal extension, at the end of terminal stance (Lewis et al., 2007, 2009). Thus, the timing and amount of activation of iliacus and the iliacus-to-gluteus maximus muscle balance related to the ankle plantarflexion power generated, become important considerations. Reduced hip extension in gait has been suggested as a possible protective mechanism associated with anterior hip pain (Lewis et al., 2010), but may be a result of compensations that can end up increasing the anterior joint forces. Increased gluteal and iliopsoas activity can reduce anterior hip joint forces (Lewis et al., 2007), and remain a rehabilitation target for the protection of the anterior hip in hip OA. However, so might the triceps surae complex.

The effect of assistor or secondary muscle action through concepts likened to Lombard's paradox should always be considered. For example, at preswing iliopsoas and gluteus maximus are CoM accelerators, but gluteus maximus is a CoM decelerator in late swing and loading response, while adductor longus changes from being a secondary hip flexor at 50° of hip flexion to being an extensor at 70° of flexion (Byrne et al., 2010). This can prove important in the context of abnormal activity or muscle weakness. Loss of gluteus maximus strength might also decrease the restraint on internal tibial torsion, because of the gluteus maximus' normal influence on the tibia via the ITT (Preece et al., 2008). Such knowledge comes with time and exploration of clinical biomechanics. Patients present clinicians with a vast range of pathologies and pathomechanical possibilities, and while most associate logically with simple local failures, many patients do not respond to treatment because the pathomechanics have not be correctly identified. The ability to think inside the realms of biomechanics is far more clinically effective over time than learning set treatment pathways for certain injuries.

The primary component when considering construction of hip free-body diagrams is the hip axis location in all orthogonal planes, which should be located at the femoral head within the acetabulum. The position and direction of the GRF vector indicated by the lower limb position and the action and strength of the principal muscles influencing hip kinematics are also all key considerations. These principal hip muscles are gluteus maximus, iliopsoas, and more surprisingly, triceps surae that between them provide the force vectors that are influencing CoM progression in the sagittal plane during stance phase around the hips. They provide the ability to resist hip flexion on loading, to extend the hip in late midstance, control extension during terminal stance, and flex the hip into preswing. Hamstrings should be added to the free-body diagram if the initial contact is of interest as they weakly assist in resisting the loading hip flexion moment but are far more important in running than in walking. They soon reduce activity and are not significant during hip extension coming out of flexion, or hip extension in preswing. Although force plates and 3D kinematic data make GRF placement more accurate, the GRF vector can be 'guesstimated' from the foot position at each stage of stance phase when just using 2D slow motion capture. Lack of costly gait analysis equipment is not an excuse for not using any gait analysis in assessing hip pathology.

For the frontal plane, the hips capacity to position the CoM for safely loading, translating, and transferring this load from one stance limb to the next primarily belongs to the hip abductors. This must be approached from the changing importance of each of these muscles and their distinct fibre bundles arranged from posterior to anterior, as the hip passes through flexion towards extension. Running's initial contact creates a particular challenge for these muscles, but if dysfunctional enough, walking CoM posture will also be disturbed, giving the potential to set up compensatory pathology from the lumbosacral region to the foot as the body and limb attempt to maintain stability of locomotion.

4.2.9 Section summary

The potential pathomechanical mechanisms that can cause tissue stress threshold breach around the hip are vast. Free-body diagrams of the hip representing the kinematics and kinetics (even without being quantified) at any particular phase of gait can provide an effective tool in understanding what is occurring at any given moment. The combined knowledge of anatomy, muscular roles, muscular activity patterns, and lever arm mechanisms around the hip can be utilised in starting to unpick the causes of pathology, both locally and also more distally within the lower limb. Seemingly small facts, such as the secondary muscle actions during certain joint motions and the 'Lombard's paradox-like' effects of changing muscle efforts across joints at different positions, can be important factors in pathomechanical problem-solving.

There are three areas of pivotal importance to hip pathomechanics. The first is the hip's role in energy dissipation, where its position at contact will influence the quadriceps' ability to dissipate energy across the hip and knee, assisted by gluteus maximus and the hamstrings. This energy dissipation event is occurring as weight is being transferred to the lower limb below the hip, so this impact dissipation must also provide stability for support of the CoM as it passes to the new stance limb. Thus, the contact hip flexion angle is of considerable importance to the lever arms and muscular lengths that these muscles are functioning on. The next important role is in providing a safe level of hip extension to aid CoM transfer over the foot in midstance and for acceleration during terminal stance for the transfer of load to the opposite lower limb. If extension is excessive, then damage to the anterior hip joint may result, but if inadequate, gait momentum and energetics will be compromised. Finally, the trunk must be held stable and erect during weight transfer across the hips and during single support phase, a function achieved via the hip abductors throughout the whole of stance phase. Through a lack of bilateral hip abductor function that can occur during double-limb stance phases of walking, running presents a particular biomechanical challenge.

4.3 Pathomechanics of the knee

4.3.1 Introduction

Unlike the hip, the knee has only one significant plane of freedom of motion, the sagittal plane, and the lever class it operates through is class three in all swing and stance phase situations. This means that mechanical efficiency is sacrificed for metabolic efficiency. Loads on the knee during swing phase are considerably less than during stance, so nearly all pathology will be associated with the stance phase limb in its closed chain weightbearing situation. The primary knee loading events in walking are during initial contact braking and terminal stance acceleration because here as in the hip, the GRF forces and CoM lie furthest from the knee's axis, creating greater moment arms for forces around the joints. Running stance phase is a stressful event for knees throughout it, both during the contact and loading response braking events, and at the start of knee extension required for acceleration. This is because of higher ranges of sagittal plane motion utilised compared to those of walking. Knees are not only subjected to sagittal torques but also significant frontal and transverse plane torques that can be important in generating pathology. Unlike the hip, the knee has poor form closure and is reliant far more heavily on its soft tissues to maintain integrity and stability via force closure. As the knee lacks free triplanar motion, any significant frontal and transverse plane torques are absorbed through soft tissue and joint deformations rather than by articular free motion.

Collisions and falls, particularly in sport, often involve high transverse and frontal plane torques across the knee which cannot be absorbed through normal sagittal joint motion. Instead, they are absorbed within soft tissues, risking injuries such as ligament tears and meniscal damage. The knee is far more vulnerable to torques applied outside its sagittal plane motion-providing joint axis, i.e. those of the frontal and transverse plane as well as anterior–posterior shears across the joint surface. Tibiofemoral joint OA is a common problem, reported to affect around 12% of the population aged over 60 years (Lawrence et al., 2008), with medial compartment OA reportedly having a rate of incidence ten times higher than OA of the lateral compartment (Felson and Radin, 1994; Dearborn et al., 1996; Pollo, 1998). This is thought to relate to the fact the human tibiofemoral joint is often subjected to knee adduction moments during gait (Sharma et al., 2010). OA is not purely driven by abnormal kinetics and kinematics, with the origin sometimes being related to physiological diseases that cause collagen and bone abnormalities. As in any joint, genetic, inflammatory, or physiological disturbance play their part in altering tissue mechanical properties (Borzì et al., 2004; Silvestri et al., 2006), but abnormal locomotive biomechanics more often plays an important part in knee pathomechanics.

4.3.2 Free-body diagrams of the knee

The knee consists of the close coupled tibiofemoral and patellofemoral joints, making them part of the central lower limb flexion–extension functional unit. The tibiofemoral joint consists of two separate articular surfaces that are intimately related. Together, the tibiofemoral and patellofemoral joints have their own distinct biomechanics in transferring quadriceps power between the pelvis, femur, and tibia. It is in the understanding of these relationships that the pathomechanics of the knee can be understood. Free-body diagrams can help in understanding the dangers of dysfunction. Symptomatic terms such as 'anterior knee pain' or 'patellofemoral pain syndrome' are not specific enough to focus on the pathomechanics, which requires greater diagnostic specificity. A diagnosis of medial patellar retinaculitis, patellofemoral ligament tear, degenerative joint disease of the patellofemoral joint (OA), medial collateral ligament (MCL) strain, or medial compartment tibiofemoral degenerative joint disease, gives the clinician a much better chance of focusing on the correct gait dysfunctional mechanics.

4.3.3 Sagittal plane free-body diagrams of the knee

As a class three lever, the knee compromises mechanical efficiency for benefits in numbers of degrees and speed of motion at reduced muscular metabolic costs. The knee, like the hip, is most vulnerable to events at initial loading and braking response and around the start of terminal stance because the internal and external forces acting across it reach their longest moment arms through the displacement between the GRF and the CoM around the knee during early and late stance. In the sagittal plane, the patellofemoral and tibiofemoral joints are largely acting in unison to manage external forces and any sagittal plane perturbations that may occur.

Swing phase and loading pathomechanics

During the swing phase, forces are much reduced within the knee as only the mass of the lower leg and foot are being moved. The effort is applied via the quadriceps through the patella and its tendinous ligament to the tibial tuberosity.

Hamstrings are vulnerable to injury in terminal swing and initial contact, as their activity peaks at this time. They are resisting the quadriceps' rapid acceleration of the lower leg at the knee in readiness for foot strike placement. This quadriceps activity is opposed by the hamstrings through their tibial and fibular attachments applying effort to resist and brake knee extension, the amount of which when coupled to hip flexion, dictates the extent of the stride length. The braking effect of the hamstrings can be sufficient to cause a posterior force, pulling the heel back into the ground and consequentially developing an anteriorly directed GRF vector at initial heel contact. This 'dragging backwards' effect, when demonstrated on a force–time curve as mentioned previously, is known as the 'claw back'. Faster heel strikes gaits are most likely to produce this effect, as this induces greater hamstring activity. Heightened hamstring activity results in increasing stretch and elastic recoil of the hamstring tendons, via the muscles' near-isometric contraction when braking swing knee extension, particularly during running (Van Hooren and Bosch, 2017a, b).

Distal effort of the large quadriceps knee extensors and flexors through the patella to the tibial tuberosity during swing is managing a relatively small mass within the lower leg, foot, and footwear (if worn). Quadriceps injury is therefore unlikely during swing phase (particularly during walking), unless a serious perturbation occurs. The main knee stress in swing is related to swing's acceleration required to achieve swing phase under a shorter duration during running. This is particularly true of sprint running activities, especially associated with fast multidirection sports such as hockey or soccer. Thus, hamstring injury is usually equated to excessive braking of knee extension just prior to, and during, initial contact (Fig. 4.3.3a).

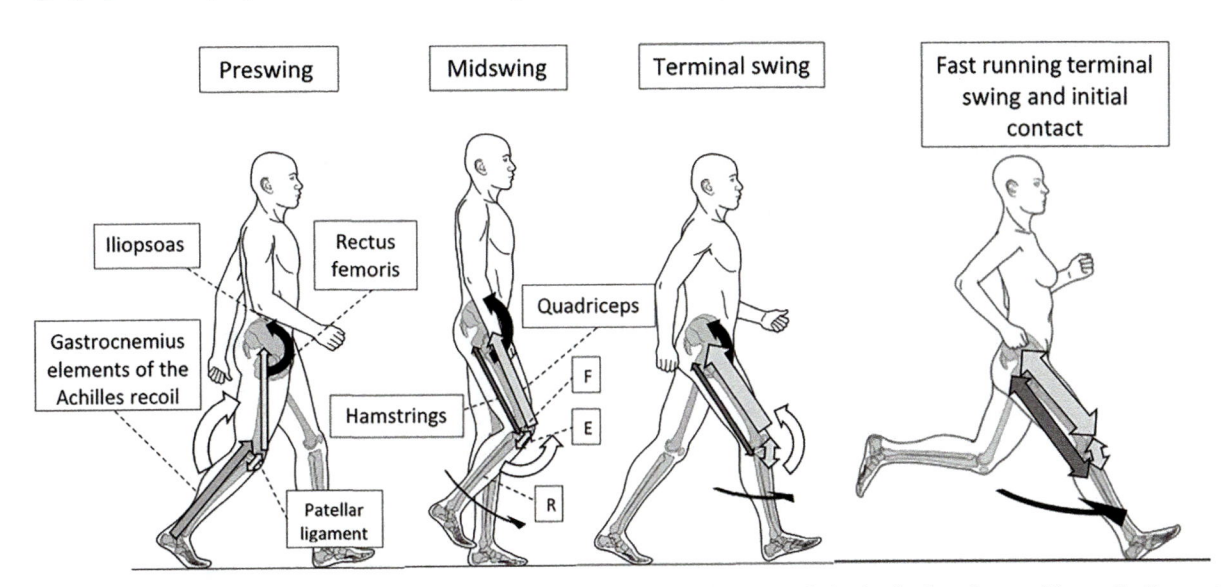

FIG. 4.3.3a The knee offers a class three lever system during swing phase with the effort (E) applied at the distal attachment of the patellar ligament, the fulcrum (F) being the knee joints axis within the distal femoral condyles, and the resistance (R) the mass of the leg and foot (as demonstrated in the midswing image). The knee joint, in swing phase, starts its motion in preswing in a 'flexion sandwich' of hip flexion forces from iliopsoas and rectus femoris and ankle plantarflexion forces from the Achilles recoil power. The Achilles' effects on the knee are primarily through the myofascial and subtendon elements arising from gastrocnemius, with its proximal attachments crossing the knee posteriorly, helping to flex it. Any loss of these moments and surrounding joint motions alters the knee flexion angle at toe-off. This may reduce ground clearance at midswing. Rectus femoris aids iliopsoas in hip flexion momentum during early swing, antagonised by increasing gluteus maximus and hamstring activity which helps maintain knee flexion. As midswing is approached, all the quadriceps become active and initiate a knee extension moment via the patellar ligament, utilising patellar superior gliding over the femur. This action dramatically increases the swing limb's anterior momentum and creates associated high centrifugal forces on the CoM. This is antagonised by the hamstrings. Hamstring braking reaches its maximum around initial contact, with the quadriceps still highly active but about to change to providing energy-dissipating eccentric contraction on ground contact under GRF-induced knee flexion. The swing limb's knee joint stresses are relatively light compared to those of the stance phase, so only substantial muscle weaknesses at the hip and knee significantly change swing biomechanics during walking. However, in fast, long-stride running, the hamstrings are often forced into a compromised longer length when braking knee extension loads. These higher loads and muscle lengths result from accelerated knee extension moments due to increased quadriceps activity. Thus, particularly with long strides, just prior and during initial contact, hamstring muscles and their tendons are vulnerable to injury. *(Permission www.healthystep.co.uk.)*

The lever arm system changes its effort point with weight transfer onto the contact foot as closed chain knee function begins. Effort is then transferred to the proximal attachment points of the quadriceps and hamstring muscles, but the lever type remains class three. In walking, the inverted pendulum function is then initiated, moving the knees out of flexion towards extension, whereas in running, the knee functions through its spring–mass function, continuing to flex as part of braking events until the start of acceleration when flexion angles start to reduce. Thus, quadriceps burden is greater and prolonged during running, for it is required to also initiate knee extension out of the high knee flexion angle at the end of braking. Weakness or imbalance among quadriceps is not well tolerated in running. On initial contact and throughout loading response, the knee plays a significant energy dissipation role, providing ~70% of energy dissipation capacity. This is achieved by using controlled active muscle–tendon buffering from the quadriceps primarily, but also a little from the hamstrings, both of which are active during initial contact for stabilising the knee.

Hamstring activity rapidly ends after initial contact, as the GRF vector (that could drive knee extension initially during impact on a longer stride heel strike) soon moves more posteriorly to the knee, leaving the quadriceps to actively manage knee flexion moments throughout the rest of loading response. In long strides, the initial GRF may lie anterior to the knee and cause an external knee extension moment that greatly increases the stress on the hamstrings, as they must now be utilised to create a protective internal flexion moment. In such situations they are braking and resisting quadriceps activity and GRF-induced knee extension moments. In more common shorter stride heel strikes or midfoot and forefoot strikes, the GRF vector is immediately driving knee flexion, increasing quadriceps energy dissipation loading and reducing hamstring stresses and their need to be active after initial contact. Reduced hamstring lengths at loading may be associated with increased patellofemoral joint reaction forces, particularly laterally, which have been reported to be significantly higher at 60° of knee flexion in squatting manoeuvres (Whyte et al., 2010).

In running, the knee flexion braking role duration as a percentage of stance is considerably increased compared to walking, persisting until the middle of stance when the knee reaches its greatest flexion angle. The quadriceps then need to initiate knee extension. In walking, the greatest flexion angle is just after completion of loading response, and after this, knee flexion angles reduce until heel lift. After absolute midstance, knee extension moments are derived from gravity on the CoM, not through further quadriceps muscle activity. Running biomechanics therefore opens the knee to far more pathomechanical risk than does walking (Fig. 4.3.3b).

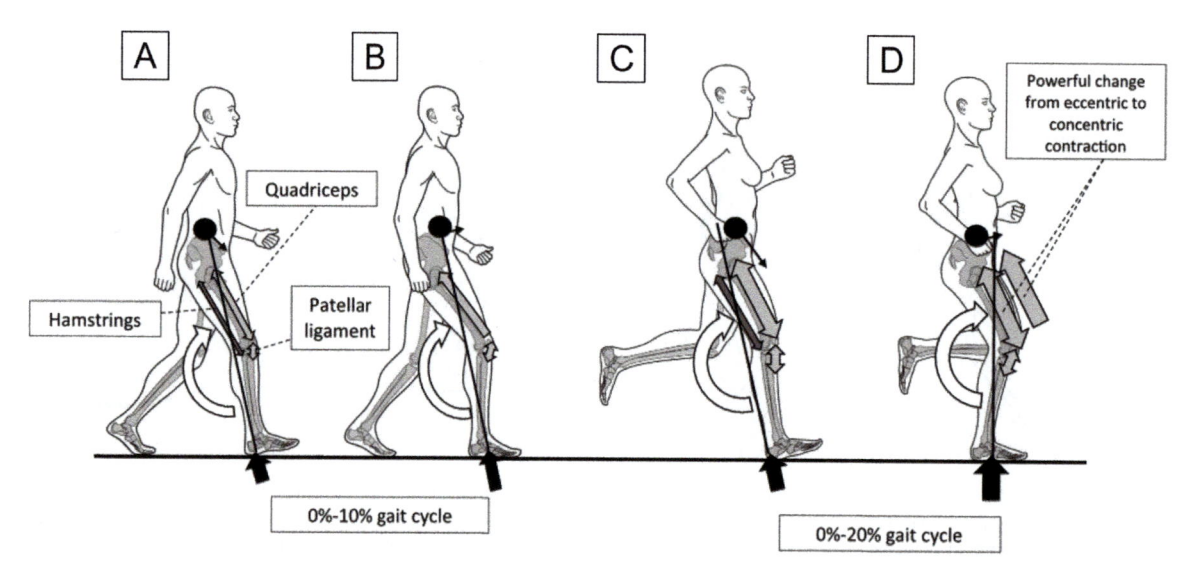

FIG. 4.3.3b The knee, utilising both tibiofemoral and patellofemoral joints, provides the body's primary shock absorption in decelerating the fall of the CoM (black circle) that is driving the GRF (black arrow). The quadriceps muscles, using their muscle–tendon buffering ability while controlling knee flexion eccentrically, absorb energy until knee extension is required. In walking (A and B), the knee shock-absorbing flexion moment (white arrow) only lasts from initial contact to the end of loading response at ~10% of the total gait cycle under relatively low acceleration forces. After loading response, the quadriceps start concentric contraction to bring the CoM up against the GRF-induced knee flexion vector. In running (C and D), shock attenuation via the quadriceps lasts to ~50% of stance phase or ~20% of the whole gait cycle, under considerably higher acceleration-induced forces from the CoM's fall. During running, the knee also enters greater ranges of flexion under higher GRFs during the middle of stance, making flexion resistance harder and for longer. After braking phase, the knee is not aided by gravity to initiate knee extension during late midstance. Therefore, acceleration is initiated by both powerful quadriceps and ankle plantarflexor activity, vastly increasing the injury risk on the quadriceps and patellofemoral anatomy during running compared to walking. (*Permission www.healthystep.co.uk.*)

Midstance and terminal stance

In early midstance during walking, the GRF vector moves nearer to the knee axis, reducing the flexion moment arm of the knee and thus offloading the quadriceps which responds by reducing activity. The quadriceps are largely inactive before absolute midstance, having completed their energy dissipation, flexion braking, and flexion-reducing knee extension roles through the first half of stance. By creating knee extension moments during early midstance, they assist the vertical height gain of the CoM for potential energy release during late stance. In these roles, the quadriceps work in cooperation with gluteus maximus at the hip. The potential energy gained from CoM height is of course released by gravity pulling the CoM back down again in late midstance, via the inverted pendular action.

Pathology through the principles of biomechanics **Chapter | 4** 675

At absolute midstance, the quadriceps are not quite finished in stance. They become active again at the end of stance after heel lift in preswing. This quadriceps activity during preswing draws the limb anteriorly and is provided by rectus femoris acting as a hip flexor. Thus, rectus femoris assists iliopsoas (iliacus) activity. This activity around toe-off is variable. It is reportedly lacking in some steps or split into distinct periods around the preswing-to-swing transition, as well as providing a more commonly reported continuous swing activity (Di Nardo and Fioretti, 2013). Variability may reflect the level of iliacus activity at the stance-to-swing transition, with rectus femoris providing assistance as and if necessary.

In running, quadriceps activity is greater than during walking, as there is no period during stance phase when GRFs and muscle moments around the knee are not significantly driving or requiring some degree of change in the flexion or extension angles. During running, the knee tends not to achieve the near-extension angles seen in walking until very late within terminal stance acceleration. Running knee motion requires the quadriceps to be changing their eccentric to concentric activity instantly at the change between braking and acceleration at the middle of stance. The middle of running stance acts as a functional changeover point, which is also the moment the lower limb is under its highest vertical GRF forces. The concentric quadriceps activity drives knee extension against GRF flexion moments in order to accelerate the CoM of the body upwards and particularly forwards, towards the aerial phase. The more horizontal the resulting GRF, the better the acceleration anteriorly. Undoubtedly, this prolonged stance quadriceps function and greater knee range of motion during running has a profound influence on the rate of patellofemoral symptoms seen in running sports compared to patellofemoral complaints associated with walking. Triceps surae power needs to be applied shortly after knee extension initiates for effective running acceleration energetics. Delayed or poor plantarflexion power from the ankle increases quadriceps burdens (Fig. 4.3.3c).

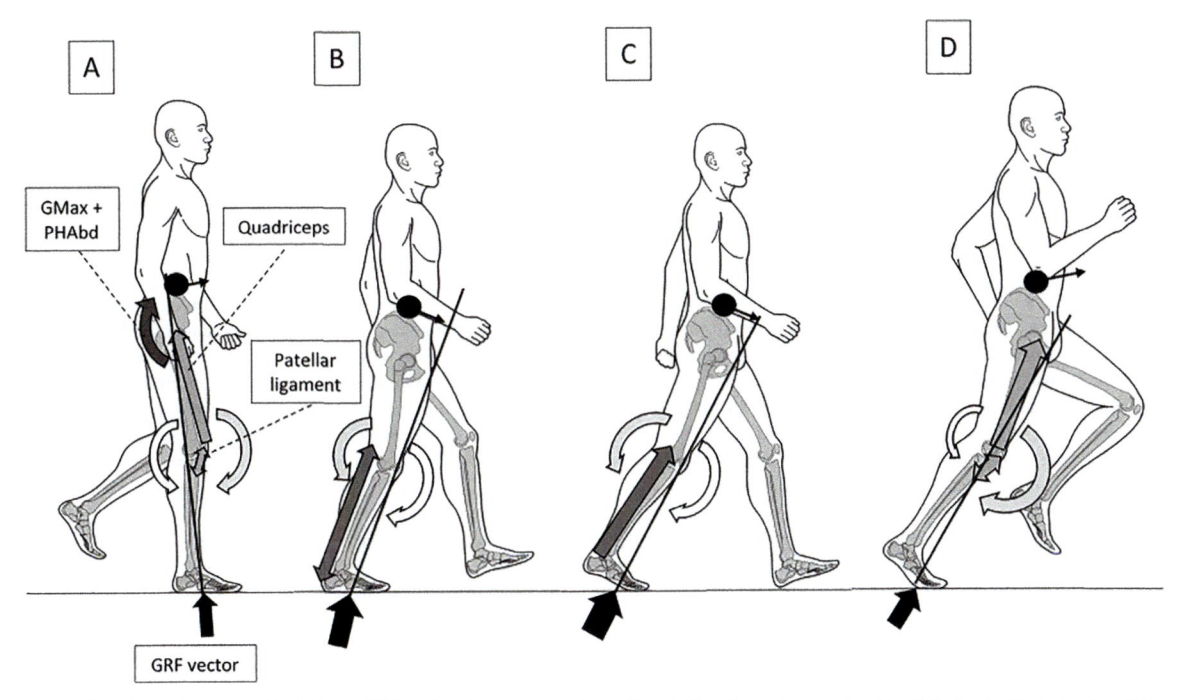

FIG. 4.3.3c Free-body diagrams of walking middle (A) and late stance phase (B and C) and running acceleration (D) clearly illustrates the difference in knee biomechanics between the different forms of locomotion. The white curved arrows show the external GRF-induced knee joint moments and the light-grey arrows, the internal muscular moments. In early midstance (A), knee extension moments derive from the quadriceps effort applied to the proximal femur that resist external knee flexion moments. Knee extension is aided by gluteus maximus (GMax) and posterior hip abductors (PHAbd) via concurrent hip extension. This combined action occurs under decreasing GRF-induced flexion moments and results in concentric muscle contraction accelerating the CoM (black circle) away from the ground. By late midstance (B), the knee is being extended by the GRF located posterior to the knee and gravity acting on the CoM lying anterior to the knee. These forces are accelerating the CoM forward under the resisted control of triceps surae. With the CoM falling anteriorly, acceleration and knee extension do not require quadriceps activity during late midstance and early terminal stance (C). As long as gastrocnemius achieves its task in preventing knee over-extension during late midstance to then help flex the knee during terminal stance, forces within the knee remain relatively low. Therefore, knee pathology in walking is usually associated with braking phase (see Fig. 4.3.3b). However, in running acceleration (D), the knee must actively extend out of a high flexion angle via strong quadriceps activity that accelerates the CoM anteriorly and upwards. This limb extension is assisted by hip and calf power that is little changed from that provided during walking, placing the increased burden on the knee. Thus, initiating running acceleration can commonly provoke knee pathology, even if loading response is still more stressful to the running knee. Running gait analysis should pay just as much attention to all stance phases in the presence of knee pathology. *(Permission www.healthystep.co.uk.)*

4.3.4 Knee muscle pathomechanics in the sagittal plane

The hamstrings consist of three individual muscles and the quadriceps four separate muscles, so individual weakness or tightness within these muscle groups as well as dysfunction of the whole functional muscle units can significantly alter torques around the knee. Weakness within the quadriceps (rectus femoris) can affect ground clearance of the foot in early swing through inadequate hip flexion, if the iliopsoas is unable to achieve the task on its own. The biggest effect of quadriceps weakness during swing is on stride length and centrifugal force production during late and terminal swing. The loss of stride length can occur through failure of the knee to extend significantly and can also be due to the loss of quadriceps-accelerated centrifugal force on lower limb extension momentum. As a consequence, the centrifugal pull on the body's CoM will also be reduced (Fig. 4.3.4a).

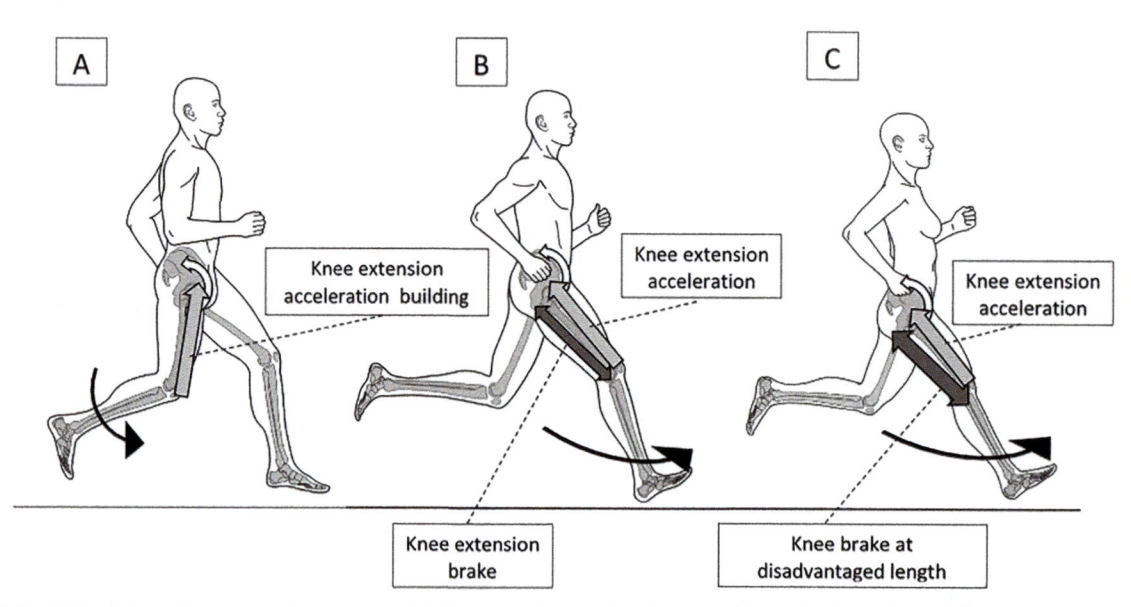

FIG. 4.3.4a With its lower forces on anatomy compared to the stance phase, swing phase knee biomechanics during walking rarely causes pathology, although knee weakness and pathology may influence overall energetics. Swing phase in running involves higher stresses and greater muscle activity than walking, while its energetics significantly influence the running speeds possible. The faster the running speed, the more influential swing biomechanics becomes to performance. At faster speeds and with fatigue, hamstring muscles become vulnerable to strains at the end of terminal swing into loading response. This is because of the need to control large centrifugal forces before impact. Diagrams of swing can help highlight the effects of dysfunction within the quadriceps on knee extension and centrifugal forces. In early running swing (A), developing rapid hip flexion and knee extension is important. While iliopsoas (white arrow) provides rapid hip flexion, the quadriceps (light-grey arrows) activate to start developing powerful knee extension, with rectus femoris assisting motion at both joints. The quadriceps extension power generated in early swing drives large centrifugal forces (black arrow) that increase the anterior draw on the CoM, dramatically improving energetics via the increased horizontal translation it brings to the CoM during late swing (B). In large stride lengths (C), centrifugal forces around the knee from quadriceps power become increasingly difficult for hamstrings (dark-grey arrows) to restrain via their near-isometric activity occurring at longer fibre lengths, risking muscular/tendon strains. Weak quadriceps risk insufficient centrifugal forces, reducing energetics and increasing the risk of running fatigue. *(Permission www.healthystep.co.uk.)*

The dangers associated with quadriceps muscular weakness are higher in running, where they are loaded for longer, but their initial braking action on impact and loading response is also absolutely necessary for walking. In loading response, quadriceps weakness can result in reduced energy dissipation during braking and failure to resist knee flexion moments. If the loading knee flexion moment is not adequately resisted, there will be excessive drop of the CoM behind the knee. The quadriceps also play a major role in impact energy dissipation, and failure of adequate energy dissipation around the knee can cause compensation in the gluteus maximus activity that accompanies initial contact (Thompson et al., 2013). This gluteus maximus compensation ability is doubly important, as increased hip and knee flexion together will seriously destabilise loading response.

Thus, should the gluteus maximus and the hip fail to adequately compensate for weak quadriceps by increasingly resisting hip flexion angles, then other lower limb energy dissipation mechanisms may be called upon to take more of the dissipation burden. Gluteus maximus, via the hip, has much lower energy dissipation capacities than the quadriceps, and other shock attenuation mechanisms are even more limited compared to knee flexion muscle–tendon buffering via the quadriceps. The other energy dissipation mechanisms having to take on GRF-induced dissipation loads are tibialis anterior, tibialis posterior, and peroneus longus muscle–tendon buffering, the plantar fat pads, soft tissue oscillations, and the reversed windlass mechanism through skeletal frame unstiffening from prior windlass-induced stiffening. These mechanisms may become overloaded and soft tissue resonance may be increased if not protected by the quadriceps, especially within the lower leg where greater soft tissue oscillations may provoke shin pains. Bone and/or articular stresses may rise as a result from an inability to adequately damp impact energy away from the joints, particularly the knee. These impact collisions leave intraarticular knee structures such as the fibrocartilaginous menisci and the hyaline cartilage vulnerable to compression and shear injuries if the quadriceps and/or any part of its compensating shock absorbing chain fail. Thus, fatigue of the quadriceps and gluteus maximus in activities such as long-distance running presents a considerable pathomechanical risk. The best option in these circumstances (besides rest from running) is to reduce the stride length to diminish both the CoM and the GRF's moment arms around the knee and also to slow gait to suppress the peaks of GRF, while rehabilitation of the quadriceps will need focusing on (Fig. 4.3.4b).

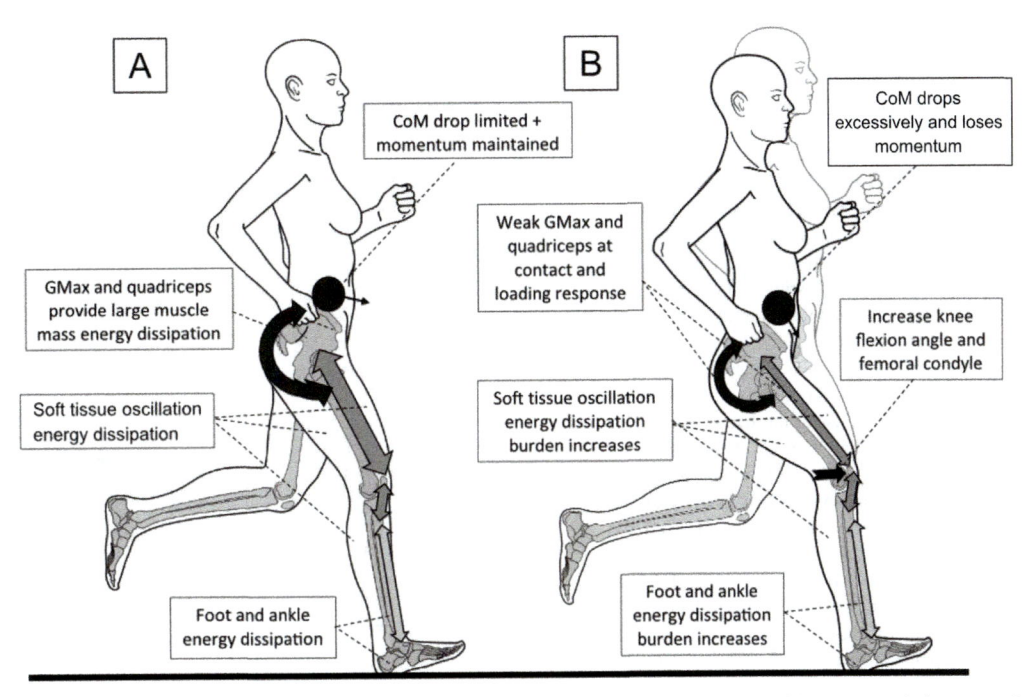

FIG. 4.3.4b As the primary shock absorber of gait, the knee must provide the ability to dissipate impact braking energy to help prevent forces breaching tissue threshold. In running (A), this energy dissipation burden increases in the knee, but less so within the hip and ankle. The quadriceps muscles and patellar ligament (dark-grey arrows) take the primary responsibility for knee energy dissipation but are greatly aided by large hip extensors like gluteus maximus (GMax—black arrows). Secondary energy dissipation mechanisms include tibialis anterior (light-grey arrows) at the ankle (or triceps surae in forefoot contact), the foot's plantar fat pads, changing foot vault compliance, and utilising soft tissue bulk oscillations within the buttocks, thigh, and calf. If there is weakness within the quadriceps, then the body is exposed to high impact energy; a situation much worsened in the presence of concurrent GMax weakness (B). In the presence of weakness in quadriceps and GMax, secondary energy dissipation mechanisms become increasingly burdened while hip and knee flexion angles can increase during loading, dropping the CoM too far and losing momentum. Increased knee flexion angles raise compression forces on the patellofemoral joint and increase shear stress within the internal knee anatomy through larger, less restrained femoral condyle anterior translation. *(Permission www.healthystep.co.uk.)*

In walking, early midstance loss of quadriceps power reduces the ability to decrease the knee flexion moment to gain limb extension height for the CoM before the middle of midstance. Thus, quadriceps weakness will reduce gait speed, reducing the F1 peak and the depth of the F2 trough. If only one or two muscles within the quadriceps are dysfunctional, then the others will experience higher loading, increasing injury risk within their myofascial units during loading response and early midstance compensations. However, possibly the biggest effect of individual quadriceps failure is in relation to the alterations and imbalances within the force vectors they set up around the patella, a subject discussed later.

4.3.5 Pathomechanics in anterior and posterior translations of the knee

Apart from the large ranges of flexion and extension within the knee, sagittal plane anterior and posterior translations also occur between the femoral and tibial articular surfaces. These anterior and posterior translations are usually coupled with internal and external rotations through the restraining properties of the anterior and posterior cruciate ligaments, respectively. The anterior cruciate ligament (ACL) is divided into anteromedial and posterolateral bundles, with the former tightening under lengthening tension loads during flexion and the posterolateral bundle tightening when moving into extension. The result of this arrangement is that anterior tibial translation and internal rotation of the femur are limited throughout motion of the tibiofemoral joints (Duthon et al., 2006; Woo et al., 2006). In midstance, with the tibia 'fixed' distally through ankle-to-foot ground contact, the ACL effectively restrains femoral posterior translation and its external rotation in relation to the tibia, as well as restraining knee valgus moments (Woo et al., 2006). This becomes most important in late midstance as the knee is under the GRF's extension moment, or during the acceleration phase of running (Fig. 4.3.5a). However, acute injuries of the ACL are associate with an external tibial twist on an internal femoral torqued event on the knee, which often causes concurrent MCL injuries through frequently associated knee valgus moments.

FIG. 4.3.5a Although acute anterior cruciate ligament (ACL) strains, tears, and ruptures are associated with high stress torques involving external tibial rotation and anterior translation (or internal femoral rotation and posterior translation), often under a valgus moment on a flexed knee, it is their effects on locomotion following injury that is more important for understanding ACL dysfunction during gait. Clinical testing associates ACL dysfunction with an anterior draw or excess motion of the tibia anteriorly, but in stance, the tibia is fixed to the ground in closed chain via the foot. There is also often concurrent medial collateral ligament dysfunction found following traumatic injuries, which destabilises any knee valgus moments during gait. During stance phase, the functional risk of ACL dysfunction is of the femur gliding too far posteriorly under extension moments during late stance, in both walking and running (A and B). Knee diagrams of normal ACL function (with the femur and patella shown in cross section) demonstrate how the femur and tibia are restrained by the ligament (C). In ACL compromise (D), the tibia can be relatively freely drawn anteriorly from its normal position (dashed outline). However, the risk in gait is for the femoral condyles to move excessively posteriorly (dashed and circle femur outline) during later stance phase when the knee is under extensor moments. In some individuals, hamstrings develop new activation patterns that improve restraint between the femur and tibia, keeping the knee more flexed. Interestingly, if this happens, the unaffected knee develops the same changes in hamstring activation. *(Permission www.healthystep.co.uk.)*

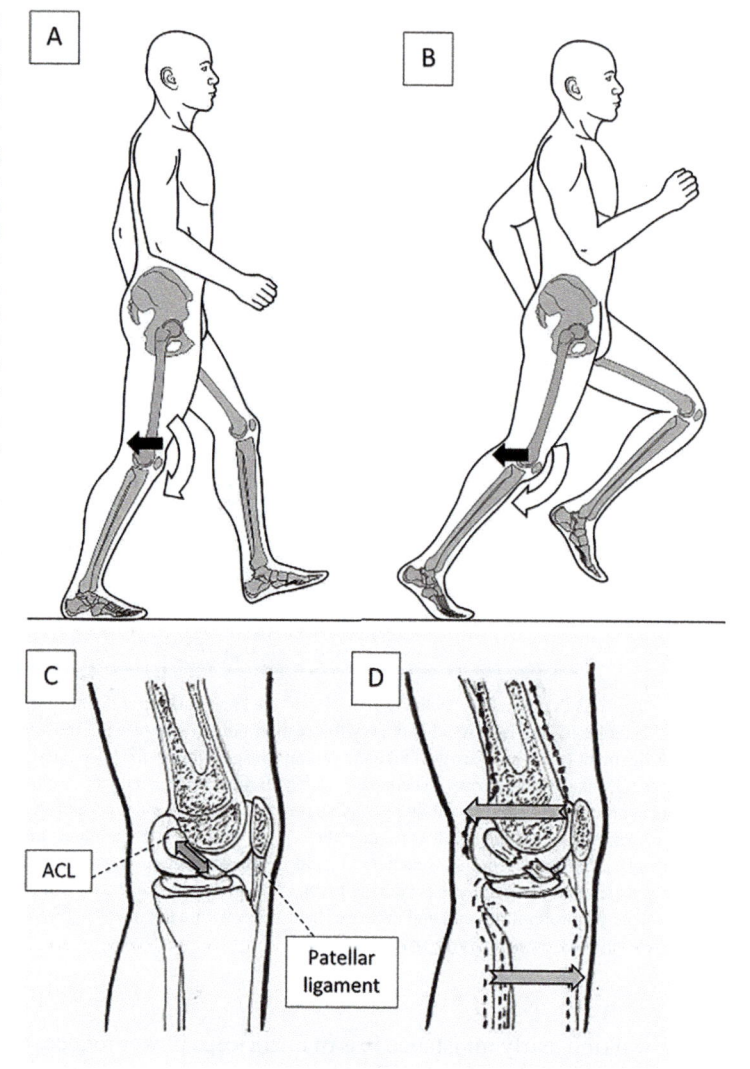

The stronger posterior cruciate ligament (PCL) restricts posterior tibial translation, particularly during flexion (Bowman and Sekiya, 2009). In single-limb support with the ankle and foot 'fixed' in closed chain, the PCL's role is to limit anterior translation of the femoral condyles. It has an anterolateral bundle that increases its tensional length with flexion to reach a mid-flexion peak and a posteromedial bundle that increases its tensional length from 60° to 120° flexion (Bowman and Sekiya, 2009). Due to fibre orientation, parts of the PCL are resisting vertical stresses, while other parts resist horizontal stresses, with increase in tensile stresses during flexion at a peak of 90°, with stresses focused on the attachment points (Ramaniraka et al., 2005).

As the knee approaches terminal extension range, the PCL's role decreases while the MCL takes on the primary role of joint stability (Bowman and Sekiya, 2009). The PCL only has a secondary role in controlling tibial external rotation and varus moments. These are roles that are mainly fulfilled by the structures of the posterolateral corner complex. The posterolateral corner complex consists of the lateral collateral ligament (LCL), popliteal tendon, and the popliteofibular ligament. Loss of the PCL's assistance affects the stabilising function of the posterolateral corner complex (Bowman and Sekiya, 2009). Therefore, failure in the PCL, particularly the anterolateral bundle, can result in increased freedom of transverse plane rotations and varus moments, increasing stresses on the posterolateral corners complex's anatomy (Fig. 4.3.5b).

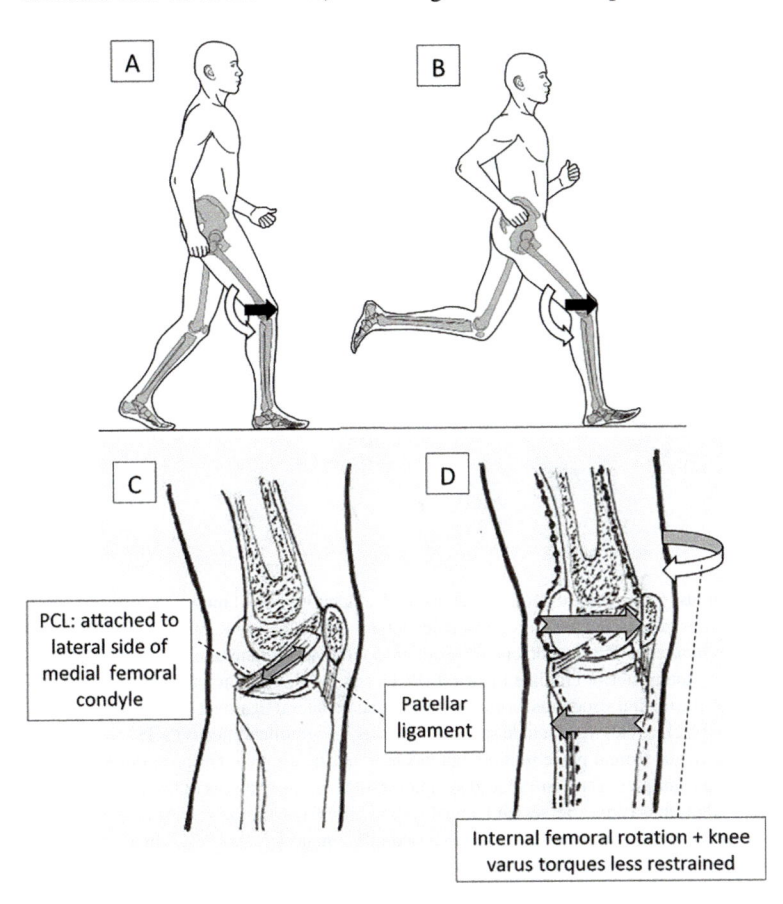

FIG. 4.3.5b The posterior cruciate ligament (PCL) is particularly important during knee flexion moments, especially under the high knee loading forces around contact and loading response, both in walking and running (A and B). The ligament helps restrain anterior femoral condyle translation over the tibia during flexion moments. A PCL free-body diagram, with the femur and patella demonstrated in cross section (C), shows the PCL's normal tensional restraint against open chain posterior tibial translation or closed chain femoral condyle anterior translation. With PCL dysfunction (D), the clinical tests reveal the tibia can position more posteriorly (solid image) to the expected normal position (represented by the dashed tibial outline). In gait, the femoral condyle can now anteriorly translate more freely, able to move to a more anterior position (solid femur image) to the tibia from that of its normal position (dashed and circle outline). Transverse plane internal femoral rotations and frontal plane varus moments may also increase, particularly if the posterolateral corner complex also becomes dysfunctional. *(Permission www.healthystep.co.uk.)*

Cruciate ligament injuries are associated with higher triplanar stresses caused by traumatic injury rather than being developed via overuse injury, although extensive degenerative joint changes can eventually cause the cruciate ligaments to fail through attachment degeneration. The loss of these structures will set up new knee joint mechanics, usually increasing anterior–posterior translation and transverse plane rotations that can lead to further pathological changes. One common consequence of cruciate ligament injury and failure is damage to the fibrocartilaginous menisci. As the menisci play a role in energy dissipation, joint stabilisation, articulation fluid flow, nutrient distribution (McDermott et al., 2008), and provide sensorimotor information (Jerosch and Prymka, 1996; Al-Dadah et al., 2011; van der Esch et al., 2013), traumatic cruciate injury can lead to joint biomechanical dysfunction on many levels, increasing the risk of articular cartilage and subchondral degenerative changes (see Fig. 4.3.6c to see how frontal moments and cruciate ligament failures can interact in pathology).

4.3.6 Frontal plane pathomechanics of the tibiofemoral joint

The tibiofemoral joint is subjected to varus (adduction) and valgus (abduction) moments which, if unrestrained, can significantly alter loading stresses between the medial and lateral articular surfaces. Varus and valgus moments across the knee result from the interaction of the GRF vector and the BWV orientations, creating either medial or lateral force vectors across the tibiofemoral joint. Osseous alignments within the lower limb, joint angles at the hip, and the width of the base of support from foot placement have considerable effects on frontal plane knee moments that can strongly link cause and effect to OA within the tibiofemoral joint (Sharma et al., 2010; Turcot et al., 2013).

Through strong soft tissue restraint, the tibiofemoral joint provides little frontal and transverse plane motion compared to those in the sagittal plane. Tensional forces medial and lateral to the knee throughout motion maintain varus and valgus moment restraints. Restraint of frontal and transverse plane motion is most restricted nearer to tibiofemoral joint extension. However, failure of the medial and lateral knee restraining mechanisms from muscles and ligaments can lead to increased valgus or varus opening/gaping across the tibiofemoral joint, respectively. Therefore, frontal plane alignment and soft tissue integrity play a significant part in the frontal plane mechanics of the knee (Fig. 4.3.6a).

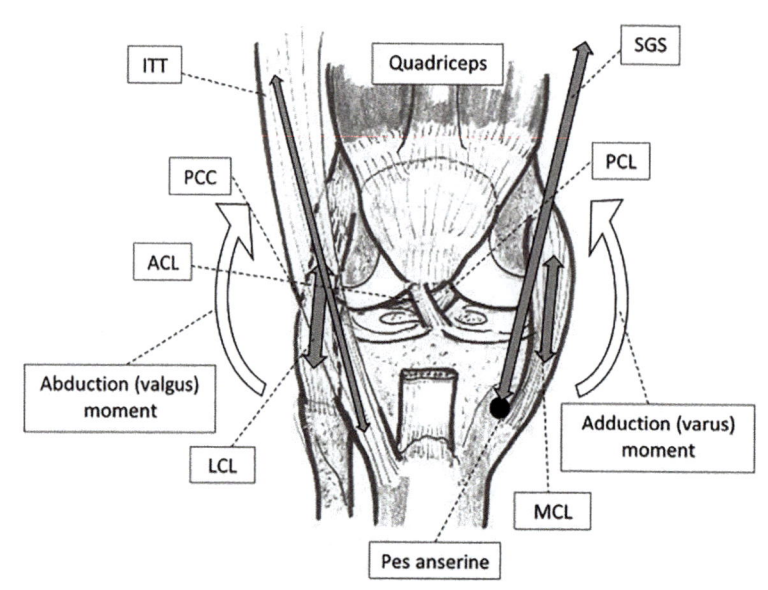

FIG. 4.3.6a A knee free-body diagram shows the primary ligamentous restraints to abduction (valgus) and adduction (varus) moments at the knee. The patellar retinaculum and ligament have been removed for better visualisation of deeper structures. If the lateral collateral ligament (LCL) becomes dysfunctional, then the knee will be compromised in resisting adduction moments. This is worsened if structures of the posterolateral corner complex (PCC) including popliteus are disrupted. Varus moments are also resisted by the posterior cruciate ligament (PCL) and the iliotibial tract (ITT) with its active muscular components proximally. Abduction moments are resisted by the medial structures, primarily the medial collateral ligament (MCL), reinforced by active contractile elements that attach to the pes anserine on the tibia (black circle) that include sartorius, gracilis, and semitendinosus (SGS—force vector shown). The anterior cruciate ligament (ACL) has some restraint of valgus frontal plane motion, but has more influence on anteroposterior translations across the knee. Quadriceps provide little assistance against frontal plane torques. Tension-induced symptoms, often at the entheses of these structures, are far more common than full or partial ruptures. Clinically induced valgus and varus stresses on manual testing provoke symptoms within the appropriate ligamentous structures, yet without demonstrating abnormality in motion if tissue strains have only caused inflammatory responses. Often, the opposite side's articular structures are also tender due to compression-induced degenerative changes from either high varus (medial compartment) or valgus (lateral compartment) moments. Observation of frontal plane motion at the knee, particularly during early stance phase, is extremely important in identifying these abnormal moments on the knee. *(Permission www.healthystep.co.uk.)*

Ideally, to keep compressive stresses within tolerance of the articular structures such as the hyaline cartilage and its bony trabecular support, moments of force in the frontal plane should lie close to the tibiofemoral joint axis. The further the GRF and BW vectors stray away from the perpendicular long axis of the femur and tibia, the longer their lever arms become for creating frontal plane moments. Such moments are acting much like bending moments across long bones, causing tension on one side and compression on the other, but are being expressed through the areas of articular flexibility. If forces vectors stray further medially, then they create varus or adduction moments, and if they stray laterally, they increase the valgus or abduction moments.

Lower limb osseous alignments can alter the orientation of the tibiofemoral joint to the BWV, CoM, and GRF during the stance phase of gait. Proximally, larger femoral neck-shaft angles of over 130° cause the femoral shaft to position more laterally to the neck which can lead to the knee becoming more laterally orientated to body, taking the foot and lower leg with it. This coxa valga alignment within the femur moves the tibiofemoral joint laterally from the CoM, so that the GRF vector generated at the foot lies very lateral to the BWV from the CoM. This will induce a valgus moment across the tibiofemoral joint and present a wider base of support. The solution for gait is to try and increase hip adduction angles to bring the lower limb under the body, but in so doing varus forces can be increased within the lower leg bones.

Coxa valga is thus usually accompanied by osseous changes within the tibia and tibula that mechanically respond to the attempt to bring the feet back under the base of support of the body's CoM. The tibia and fibula develop a medial bowing (or varus alignment), bringing the ankle towards the midline of the body. This is clinically referred to as tibial varum. Despite this postural correction of the limb, the tibiofemoral joint will remain laterally positioned to the GRF and BW vectors. The foot, becoming positioned more medially through tibial varum, moves the GRF medially to the knee, increasing the resultant varus (or adduction) moment arm at the knee. The ankle and rearfoot can be exposed to increased eversion moments, as a result of the increased tibial varum, by creating a lateral GRF to the rearfoot axis in the frontal plane (Fig. 4.3.6b(B)). Because the medial knee is usually loaded with higher stresses during stance than the lateral knee, high medial stresses during growth or bone softening diseases, such as rickets and osteomalacia, can provoke tibial varum alignment during abnormal childhood bone growth or adulthood bone maintenance processes, respectively. Seemingly spontaneous deceleration of the posteromedial growth plate of the tibial, known as *Blount's disease*, seen in infants and adolescents, can also cause pronounced medial tibial bowing that alters the position of the GRF medial to the knee, risking premature knee joint degeneration later in life (Griswold et al., 2020; Bernstein and Schoenleber, 2021).

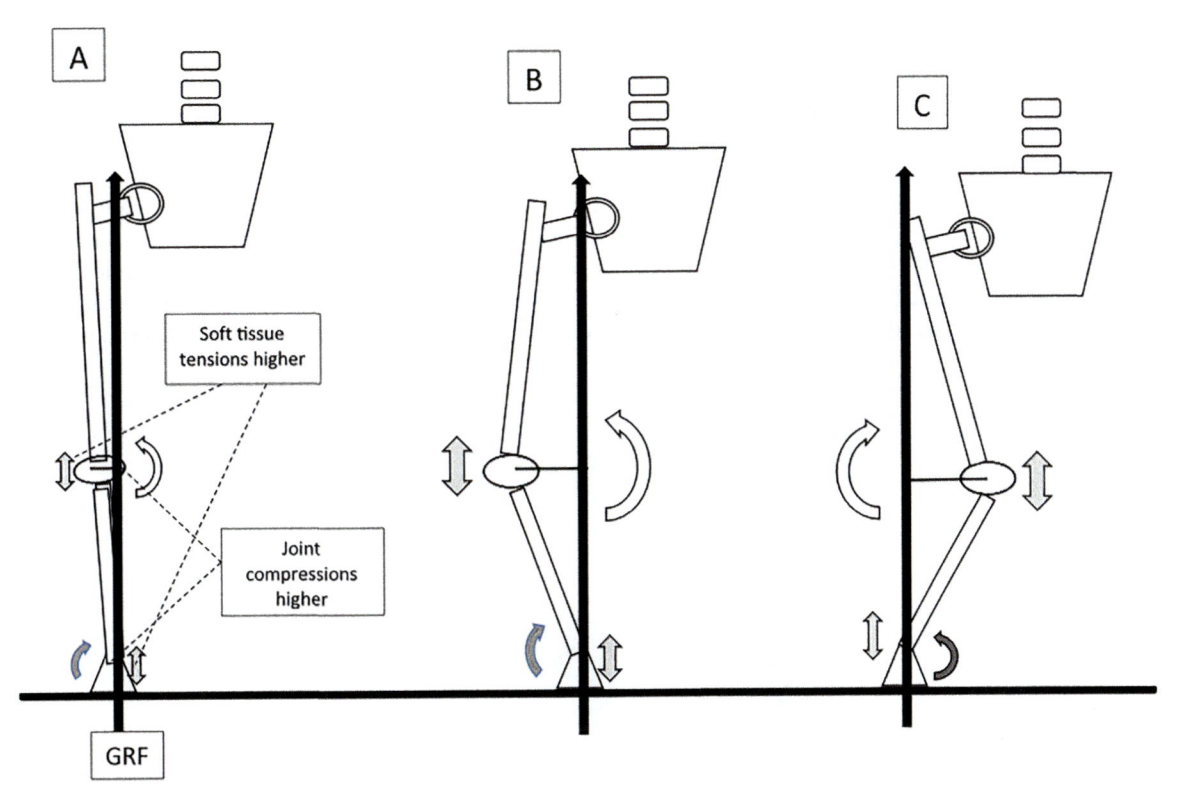

FIG. 4.3.6b Simple-styled free-body diagram of the lower limb and pelvis under the GRF positioned in the frontal plane demonstrates the concept of varus and valgus moments on the knee. In a relatively straight aligned hip and lower limb morphology (A), the small hip adduction angle positions the foot slightly medially to the knee, tending to set the initial contact GRF onto the lateral side of the heel and the medial side of the knee. This creates a small adduction (varus) moment on the knee and a small eversion (valgus) moment at the rearfoot. Lateral knee and medial rearfoot ligaments experience a little more tension within this posture compared to the medial knee and lateral rearfoot, but the medial knee and lateral ankle joints experience more compression stress. In coxa valga and genu/tibial varum alignments (B), the adduction moment arm from the GRF is lengthened, setting up greater medial knee articular compression and lateral soft tissue tension stresses, with the rearfoot also required to increase its eversion moment. In coxa vara and genu valgum alignments (C), the reverse moments are set up, with an abduction (valgus) moment created at the knee and inversion on the rearfoot and forefoot. In this situation, the lateral knee compartment articular surfaces are more likely to sustain compression and the medial soft tissues increased tensional forces. Thus, simple recording of frontal plane osseous and functional alignments during gait and static stance can give hints to the origins of medial and lateral knee symptoms. *(Permission www.healthystep.co.uk.)*

If the hip has a low tibial neck-to-shaft angle of less than 120° (the lower the more problematic), then the tibiofemoral joint can become medially orientated on a varum alignment (cox vara). This alignment increases varus moments across the tibiofemoral joint, as well increasing the femoral adduction angle to the pelvis. With this alignment, the foot will tend be become medially orientated to the base of support, even threatening to cause contact between the two limbs during swing phase. To rectify this, the base of support must be widened by hip abduction, moving the foot lateral to the knee. This can set up increased abduction forces (valgus moments) within the lower leg bones during growth and development, that can alter the orientation of the tibia near the knee and/or within its shaft, orientating the lower leg bones to align to the knee at a lateral angle. This forms a genu valgum (at the knee) or tibial valgum (within the tibial shaft). The effect of this is to realign the foot more laterally to the knee, permanently at a valgus knee angle. If the foot becomes too laterally orientated to the tibiofemoral joint, then the GRF vector is shifted to a lateral position to the knee. Positioned here, it will set up significant abduction (valgus) moments across the knee joint. An inversion moment at the rearfoot may also result from the excessively medial positioned GRFs across the foot that result (see Fig. 4.3.6b(C)).

682 Clinical biomechanics in human locomotion

Large varus and valgus moments across the knee can focus tibiofemoral frontal plane joint stresses to certain areas of the articular surfaces, increasing articular compression while also increasing tensile stresses on resistive ligaments. Varus/adduction moments increase medial compartment compression and lateral ligament tension. Valgus/abduction moments will increase lateral compartment compressions and medial ligament tensions. Only frontal plane stresses seem to increase, unless there is concurrent soft tissue structural failure. If so, then other planes of motion and moments can also increase. Concurrent loss of anterior or posterior soft tissue restraint such as the ACL and PCL will increase sagittal plane gliding under excessive compression and tensile forces.

OA associated with valgus knee alignment is dominated by central erosions on the lateral tibial plateau of the lateral compartment of the knee. This degeneration pattern is associated with less incidence of concurrent ACL damage than that seen in varus knees. This means that in general, valgus knees are associated with less anterior–posterior translation dysfunction than are varus knees (Harman et al., 1998). Varus knees are associated with central-medial tibial plateau degeneration in the medial compartment of the knee, and are more likely to express anterior–posterior translation from concurrent ACL deficiency (Vergis et al., 2002; Georgoulis et al., 2003; Moschella et al., 2006). ACL deficiency is also associated with larger degrees of varus alignment, deformity, and degeneration (Harman et al., 1998; Vasara et al., 2005; Moschella et al., 2006). The increased sagittal plane translation in the knee from ACL failure under varus compression leads to a larger area of articular degeneration, but also greater articular degeneration can lead to ACL failure (Weidow et al., 2002; Moschella et al., 2006). Thus, medial knee degeneration has great potential to set up a cycle of increasing degenerative change. Through its secondary influence on varus moments as an adjunct to the posterolateral corner complex, PCL rupture may also permit greater varus excursion through the tibiofemoral joint (Bowman and Sekiya, 2009) (Fig. 4.3.6c).

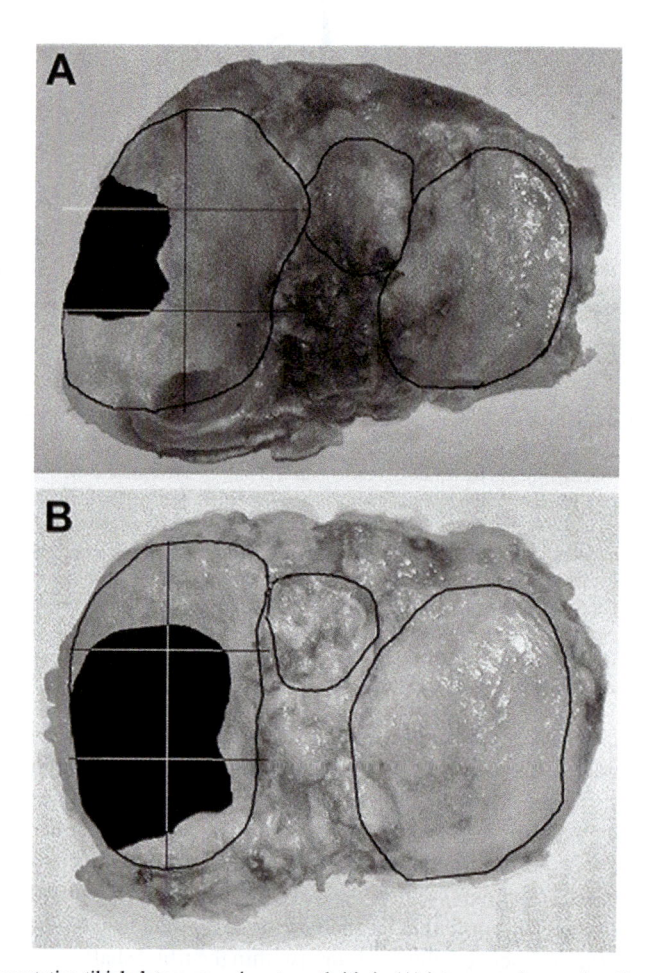

FIG. 4.3.6c Digital images of representative tibial plateau wear in osteoarthritis in (A) intact anterior cruciate ligament (ACL) and (B) with concurrent ACL failure. The loss of the ACL permits more degeneration in an anterior–posterior direction than when the anterior shear mechanics are restrained by the ACL. ACL loss expands the area of degeneration under any increased varus moment. *(Image from Moschella, D., Blasi, A., Leardini, A., Ensini, A., Catani, F., 2006. Wear patterns on tibial plateau from varus osteoarthritic knees. Clin. Biomech. 21(2), 152–158.)*

Once a tibiofemoral joint starts to degenerate, the kinematic patterns between varus or valgus alignments and their controls become substantially different, influencing the loading patterns throughout the stance phase of gait (Turcot et al., 2013). The presence of knee OA can cause further alignment disruption due to loss of joint space, which can result in changes in limb length symmetry (or previous asymmetry) and increase unicompartmental changes in varus or valgus knee alignment. This degenerative change in alignment is often an exacerbation of a preexisting postural issue, as varus alignments encourage medial compartment degeneration while valgus knees develop lateral degeneration. Mild leg length discrepancies can result from reduced articular space within the knee, with discrepancy over 1 cm known to affect the entire kinetic chain in individuals affected with knee OA, increasing sagittal loading and increasing the pelvic-to-trunk transverse plane rotations (Resende et al., 2016) (Fig. 4.3.6d).

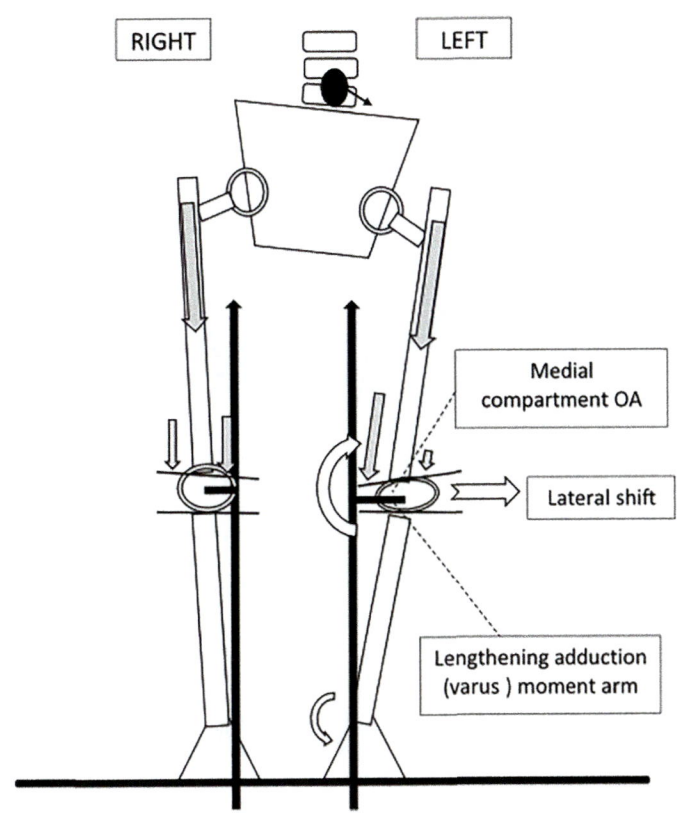

FIG. 4.3.6d A simple anterior viewed diagram to express the effects of long-term medial compartment osteoarthritis (OA) on limb length and increasing knee varus-induced degeneration. A healthy knee (as seen in the right knee) under average hip neck-shaft angles will have higher medial knee compartment pressures due to the osseous and functional alignment of the lower limb and the medial position on the GRF's moment arm. The human knee is adapted for higher varus moments through normal biological response expressed under biomechanical forces, for the knee's medial articular surfaces are larger than the lateral. However, if medial compartment articular degeneration develops, as shown in the left knee, joint space will reduce, with subchondral remodelling sclerosis occurring on the medial side. The lost joint space will result in the tibial articular surface (and possibly the femoral condyle) becoming increasingly angled into varus. This will shift the foot more medially under increased varus posturing of the limb, requiring it to use greater eversion moments to bring the plantar surface flat to the ground. By moving medially, the foot will also move the GRF more medially. This lengthens the adduction moment arm on the knee, increasing the knee varus moment and accelerating medial compartment degeneration. Any concurrent loss of normal meniscal biomechanics exacerbates peak compression forces within the knee. On loading the affected knee, a large varus moment often expresses itself through a 'lateral shift' (a deflection laterally) in the knee position, which lasts into early midstance. In time, increasing joint space loss on a more varus tilted limb creates a functional leg length discrepancy, shifting the CoM (black circle) more towards the limb with the OA knee during initial double-limb stance and static stance, unless spinal and pelvic activity can compensate. Leaving mild OA changes without therapy is not wise, as it is likely to progress as a biomechanical inevitability. Thus, early gait analysis for mild tibiofemoral symptoms may help establish risks and better direct early interventions. *(Permission www.healthystep.co.uk.)*

It is important to highlight that failure of knee mechanics in one plane is unlikely to occur in isolation. While it is important to consider the primary plane of tissue failure and the resulting tissue stresses, the clinician needs to be aware that multiple planar failure is likely to have occurred, particularly in the more advanced stages of knee pathology. The compounding issue with articular structural failure and loss of the mechanical restraints of muscles and particularly ligaments, is the concurrent loss of sensorimotor components. Sensorimotor structures inform the joint and lower limb in general to prepare for and respond to the changing kinetics and kinematics required to create stress-sparing changes in locomotion. Loss of this ability in the knee is particularly problematic, as the human knee joint is the primary energy dissipator of impact.

Pes anserine pathomechanics

A structure, which can be mistaken for pain from the patellofemoral joint or anterior medial tibiofemoral joint, is the pes anserine insertion point of sartorius, gracilis, and semitendinosus, and its associated bursa. This is known as pes anserine bursitis or tendinitis, depending on the precise diagnosis. Symptoms are located to the anterior medial knee, and seem to be far more common in females (Alvarez-Nemegyei, 2007). The largest risk factor seems to be the extent of anatomical or functional valgus knee alignment alone or in combination with MCL instability (Alvarez-Nemegyei, 2007). This suggests pathomechanics links the condition more to frontal alignment biomechanics than to sagittal plane function of the patello-femoral joint (Fig. 4.3.6e).

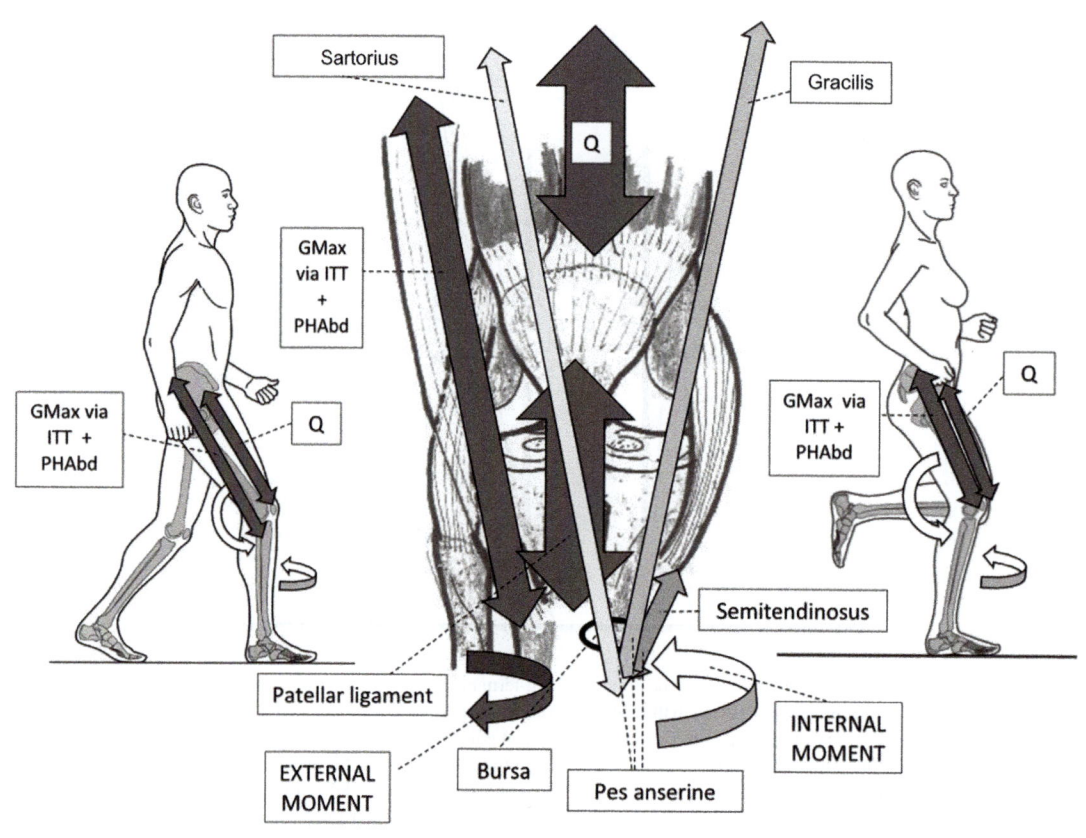

FIG. 4.3.6e Pes anserine bursitis is a fairly common, yet perhaps underdiagnosed source of anteromedial knee symptoms. It seems to be aggravated by excessive tensions within the conjoined pes anserinus tendon, derived from semitendinosus, gracilis, and sartorius above the bursae near the pes anserine attachment on the tibia. Clinically, the problem seems to associate with high ranges of knee flexion, especially in the presence of internal limb rotation during loading response when running. Stair-climbing and descent that takes the knee into higher flexion ranges also seems provocative. The vectors, coming from each part of the conjoined tendon, are different and change with degrees of knee flexion, with semitendinosus providing flexion and internal tibial rotation moments and gracilis providing knee varus and increasing internal tibial rotation at higher knee flexion ranges. Sartorius creates an internal pelvic rotation moment at the hip in a closed chain limb, and a greater knee flexion moment with increased knee and hip flexion angles. However, at angles of near-knee extension, it provides knee extension moments. Under the influence of hip adductors and GRF-induced moments, the lower limb internally rotates and flexes during loading response, requiring lower limb and hip and knee flexion moments to be resisted by gluteus maximus (GMax), the posterior hip abductors (PHAbd), iliotibial tract tensions (ITT), and quadriceps power via the patellar ligament (Q). If their combined vector is unable to resist internal and flexion torques, particularly during higher flexion angles, then a number of pathologies can be expressed including pes anserine bursitis. *(Permission www.healthystep.co.uk.)*

4.3.7 Transverse plane pathomechanics of the tibiofemoral joint

Transverse plane knee pathomechanics do not occur in isolation of the other planes, as knee structures that primarily function to provide transverse plane stability also assist in restraining motion of the other planes, often concurrently restraining knee flexion moments. Transverse plane stabilisers tend to be overlooked clinically, with the exception of the iliotibial tract (band).

Diagrams of popliteus and popliteofibular ligament pathomechanics

The popliteal muscle and popliteofibular ligament have an intimate relationship in restricting femoral and tibial transverse plane rotations. Popliteus, by causing internal tibial (external femoral) rotation, slackens the popliteofibular ligament while tibial external (internal femoral) rotations stretch popliteus and tightens the popliteofibular ligament (Ullrich et al., 2002). The popliteofibular ligament seems to affect knee motion when the lateral collateral ligament is resisting varus moments under external tibial or internal femoral rotations (Pasque et al., 2003). Popliteus is a major contributor of the stability of the posterolateral corner complex and is often overlooked as a potential source of injury clinically, yet with injuries seemingly quite common on diagnostic images (Jadhav et al., 2014).

Thus, traumatic popliteal injuries seem more common than are clinically recognised, but actual tendon attachment avulsions are rare in isolation. Popliteal injuries are most commonly found as part of concurrent structural failure of the posterolateral corner complex (Jadhav et al., 2014). Popliteus injuries are also found in combination with ACL or PCL ruptures, medial or lateral meniscal tears, and lateral collateral ligament complex tears (Jadhav et al., 2014). Popliteus is likely vulnerable to high stressed tibial external rotations or coupled internal femoral rotation during knee flexion, with the muscular belly most commonly injured and the tendon far less so (Jadhav et al., 2014) (Fig. 4.3.7a).

Free-body diagrams of iliotibial band syndrome pathomechanics

Iliotibial tract injuries are referred to as either iliotibial band syndrome, friction syndrome, or traction syndrome. It is a common complaint associated with pain around the inferior aspect of the lateral femoral condyle. It is particularly associated with runners, cyclists, rowers, military recruits, and female soccer, basketball, and hockey players (Lavine, 2010; Strauss et al., 2011). Moreover, it is considered the most common lateral knee symptom among runners and cyclists (Lavine, 2010). Its mechanical cause appears to be more associated with tensional forces 'compressing' the fascia against the lateral femoral condyle through repetitive cycles of tightening on the fascial tract, rather than with a rubbing action through free movement-induced friction from tibial or femoral rotations (Fairclough et al., 2007; Noehren et al., 2007; Ferber et al., 2010; Baker et al., 2011; Louw and Deary, 2014; Aderem and Louw, 2015). The tethering of the tract to the intermuscular septa of the lateral thigh and the thigh's fascia lata, actually prevents any free motion of the tract. Association of symptoms with an inflamed bursa is not proven, but may exist with the development of a either a cyst, bursa, or possibly inflammation of the lateral joint recess, something that is reported in some cases (Lavine, 2010). MRI findings more commonly include single changes in the soft tissues below the tract with osseous oedema and subchondral osseous erosions of the lateral femoral condyle, yet without any changes within the tract itself (Lavine, 2010).

The common clinical test associated with this condition is Ober's test undertaken to assess flexibility–distensibility of the iliotibial tract (ITT), as tightness has been thought to be linked to pathology, albeit the test's relevance is not proven (Herrington et al., 2006; Lavine, 2010), but it seems to demonstrate that there is more restriction in hip varus range in asymptomatic healthy young women than men (Gajdosik et al., 2003) (Fig. 4.3.7b). Lateral knee pain associated with the tract may actually be linked with looser rather than tighter ITTs, and to higher strain loading rates (faster lengthening) (Hamill et al., 2008). The ITT links the mechanics of the pelvis, hip, and knee directly together through its attachments to gluteus maximus and tensor fascia latae, the general deep fascia lata of the thigh, and also to its attachment to Gerdy's tubercle on the proximal tibia. The tract additionally has attachments to the lateral head of gastrocnemius and the short head of biceps femoris which also join it at Gerdy's tubercle. The ITT through its pelvic muscle attachments becomes part of the hip abductor mechanism that resists hip adduction moments, linking lateral hip and knee stability together. Thus, it helps restrain knee varus (or adduction) moments, helping create knee stability through providing up to 25% of lateral knee compartment compression force (Winby et al., 2009). The ITT distally, also assists in appropriate patellar tracking during knee extension and flexion via adding a lateral moment to the patella's soft tissues (Moorman and LaPrade, 2005).

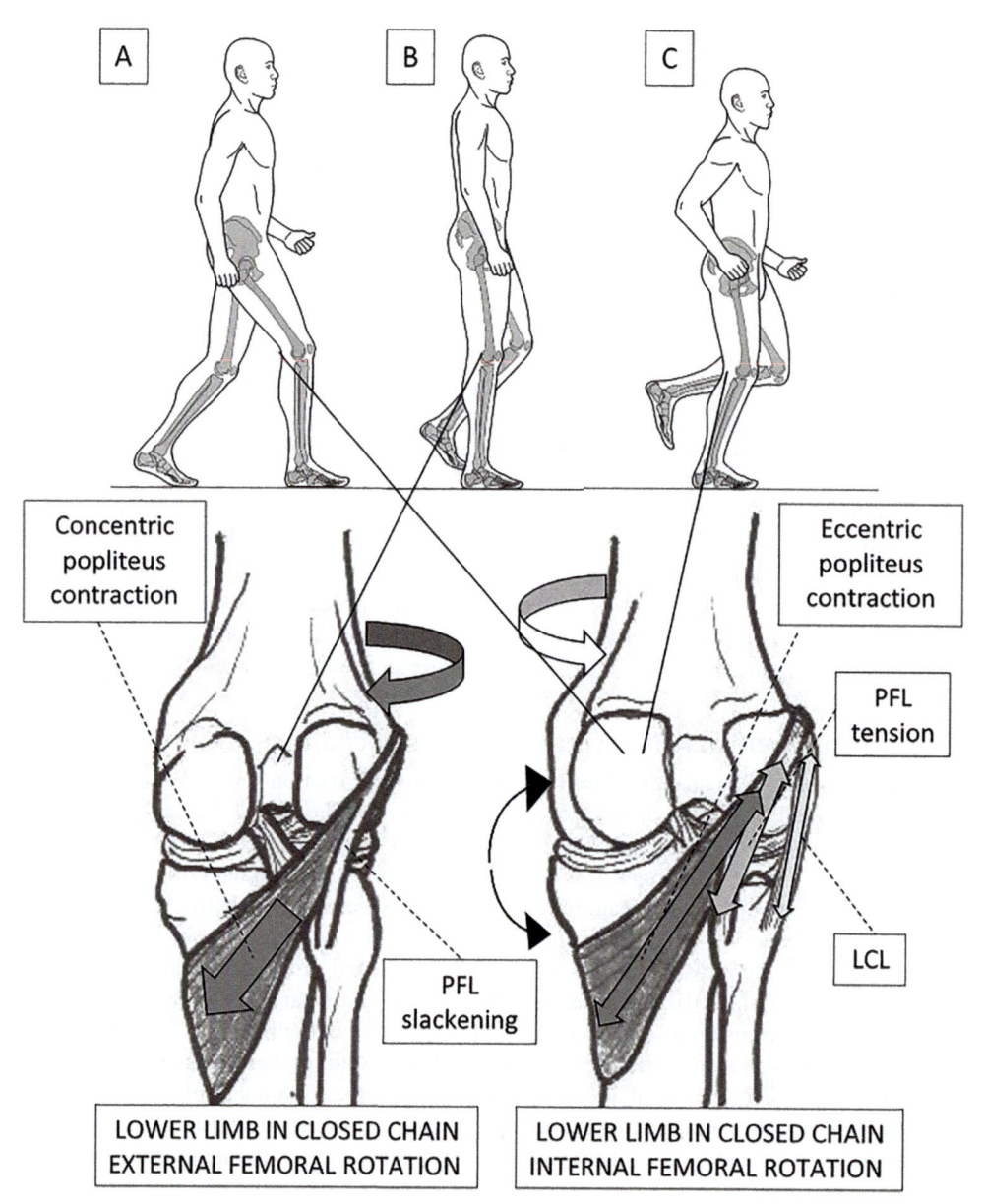

FIG. 4.3.7a Popliteus is potentially a common yet underdiagnosed cause of posterior knee pain. It is a weak knee flexor when in knee flexion angles and yet also a weak extensor at angles of near-knee extension. More importantly, popliteus internally rotates the tibia in open chain while its activity during closed chain stance phase helps it resist internal femoral rotation. With greater freedom of motion at increased flexion angles, popliteus via its proximal tendon attachment is a useful eccentric restraint on internal femoral rotations over the tibial articular plateau that occur during loading response (A). Approaching absolute midstance (B), the knee approaches near-extension and popliteus' mechanical advantage increases, aiding other muscular external rotations on the femur by its concentric contraction during this phase of gait. In running (C), most of stance phase has the knee at high flexion angles, requiring greater popliteal effort in assisting the resistance of internal femoral rotation. Popliteus appears to work agonistically with the rest of the posterolateral corner complex. Popliteofibular ligament (PFL) tightening assists in resisting internal femoral rotation and also aids the lateral collateral ligament (LCL) in resisting knee varus moments during internal femoral rotation. Popliteus' proximal attachment and tendon and the PFL are likely vulnerable during femoral internal rotations under knee flexion, to any concurrent external rotation applied to the tibia via gait perturbations, external impacts on the lower limb, or when under fierce proximal lower limb internal rotational torques. Situations that increase knee transverse, frontal, and translational freedom of motion such as cruciate ligament damage, likely increase injury risk, but trauma that damages these structures may also cause popliteal damage at the same time. This may be something overlooked during diagnostic examinations of acute knee injuries. *(Permission www.healthystep.co.uk.)*

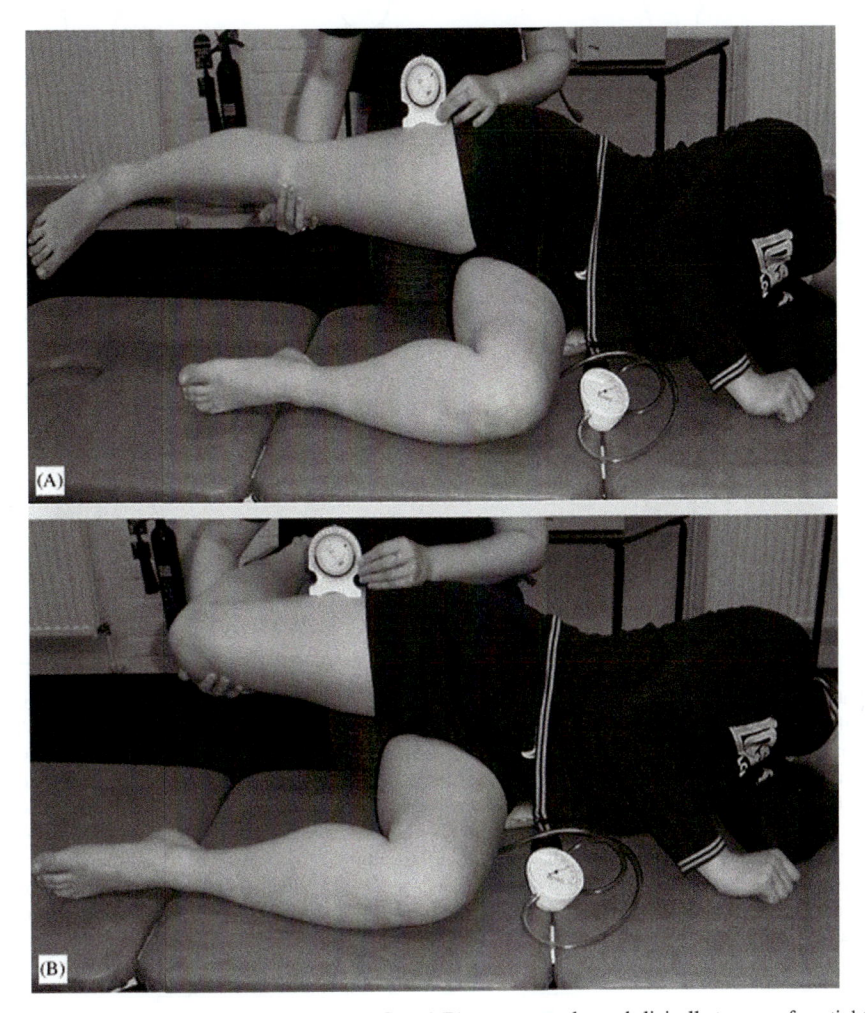

FIG. 4.3.7b The Ober test (A) and the modified Ober test with knee flexed (B) are commonly used clinically to assess for a tight iliotibial tract by seeing how far the hip can adduct when relaxed in neutral hip position. Results gathered are thought to explain iliotibial tract symptoms or patellar maltracking, but evidence for this is largely nonexistent. *(Image from Herrington, L., Rivett, N., Munro, S., 2006. The relationship between patella position and length of the iliotibial band as assessed using Ober's test. Manual Therapy 11(3), 182–186.)*

Strains on the ITT rise rapidly just after heel strike of running, probably through the narrower base of support and the larger adduction moment found in runners. Runners with a symptomatic ITT demonstrate the same strains at contact as do asymptomatic runners, but at higher strain (lengthening) rates for the remaining 90% of stance phase after initial contact (Miller et al., 2007). Yet ITT symptoms are associated to just after heel strike when the knee is around 20–30° flexed, undergoing a large loading flexion moment (Noehren et al., 2007). In this position, the ITT is most compressed on the lateral femoral condyle (Fig. 4.3.7c).

The importance of a relatively large knee flexion angle to the condition is supported by the ITT symptoms being exacerbated by downhill running, which is associated with greater flexion excursion than is experienced during level running (Noehren et al., 2007). However, a link between the knee flexion angle in runners with iliotibial band syndrome and those without has not been clearly identified (Orchard et al., 1996; Noehren et al., 2007). It is possible that the effect is only of note when accompanied by other hip or knee dysfunctions, such as hip abductor weakness or when running style becomes fatigued. Once fatigued, an increased knee flexion angle linked to heel strike has been found in iliotibial band syndrome sufferers (Miller et al., 2007). Fatigued, symptomatic, ITT runners demonstrate an increase in initial loading ITT strains, but slightly less strain around midstance and during the last 15% of stance compared to asymptomatic control runners (Miller et al., 2007). Fatigue may be a common source of weakness in the protective musculature and perhaps gives insight to rehabilitation plans aimed at the hip abductors rather than tract stretching (Fig. 4.3.7d).

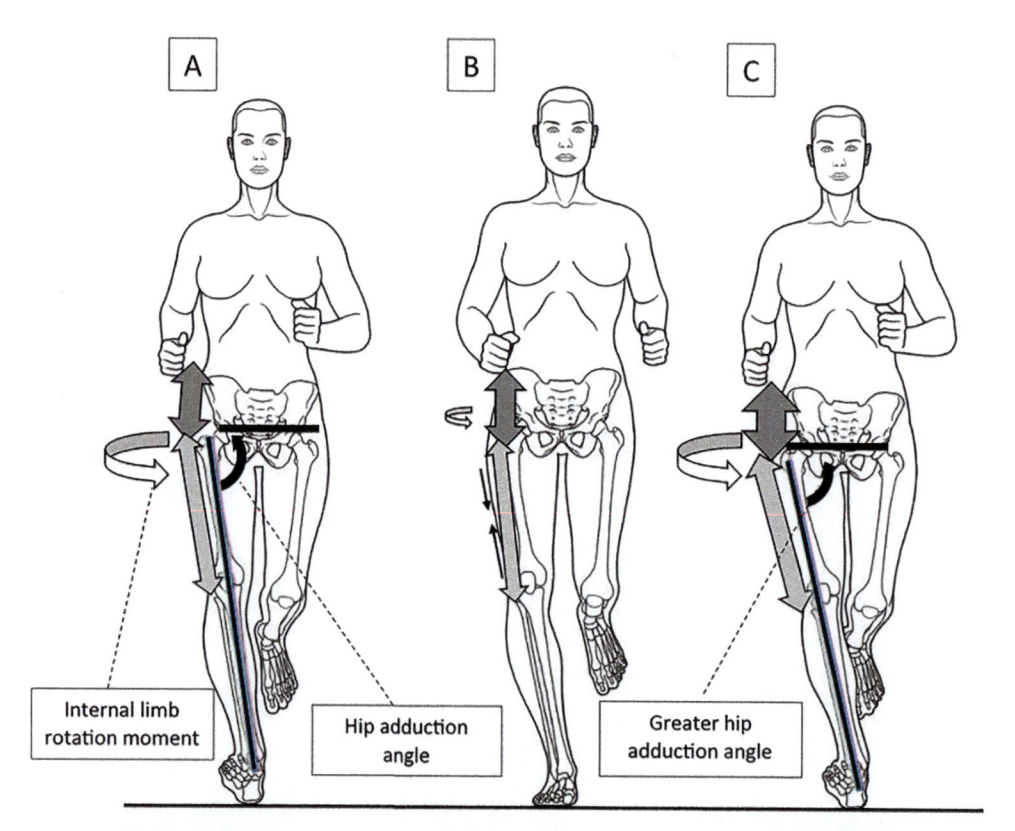

FIG. 4.3.7c From initial contact to the end of the braking phase, the iliotibial tract (ITT, light-grey double-headed arrow) is subjected to tension between the eccentrically active superficial fibres of gluteus maximus (dark-grey double-headed arrow) proximally and the tibia distally. These tensions occur during rapidly changing knee flexion angles and internal limb rotation moments. In walking, the braking period is relatively short while the knee is flexed very little, explaining the rare association of ITT symptoms with walking gait. Running, however, with its long braking phase, increasingly high knee flexion angles, and rapid internal limb rotation during the braking phase, presents far greater tensions on the ITT via increased resistance force in gluteus maximus, aided by other posterior hip abductor activation. Thus, in running, hip and knee varus moments, along with the degree of internal rotation between initial contact (A) and loading response braking (B), become the initial focus of attention for ITT symptoms. The rate of strain (indicated by thin black arrows) is reported to be greater during the middle-to-late braking phase in individuals with ITT symptoms compared to healthy controls. Those runners with more adducted limb placement are at greater risk of symptoms (C) when in the presence of functionally weak gluteal muscles or varus osseous alignments (coxa vara, tibial varum) that can disadvantage the moment arms of the hip and the knee-stabilising muscles. *(Permission www.healthystep.co.uk.)*

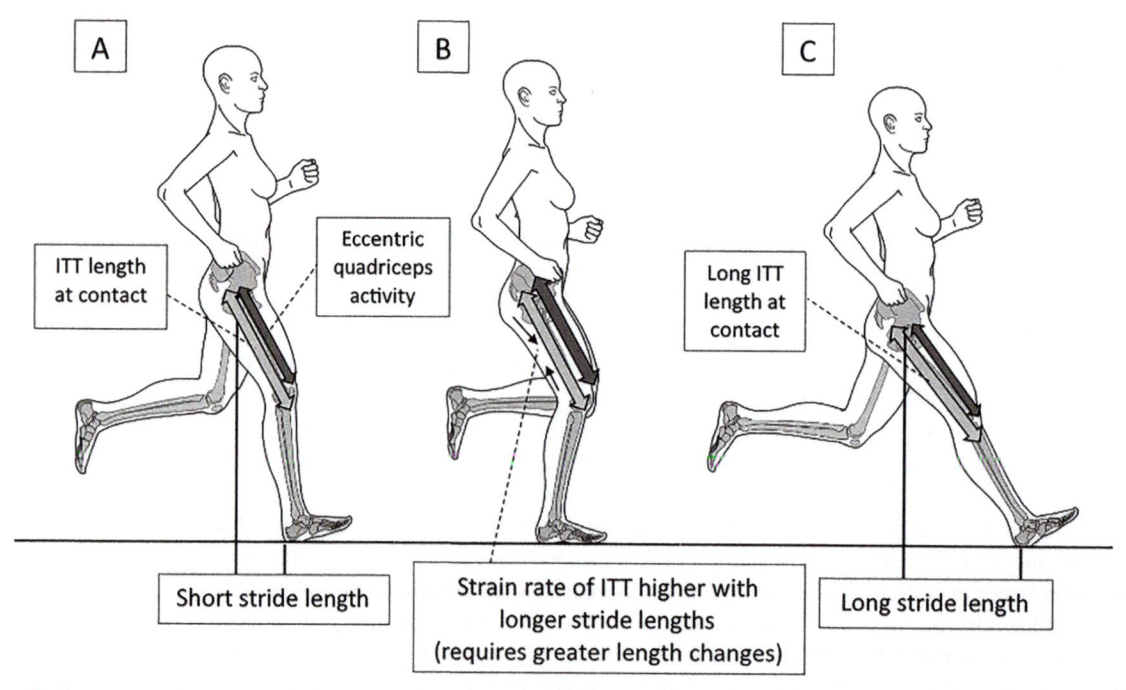

FIG. 4.3.7d Not only is the hip adduction angle, internal limb rotation, and the strengths of the muscles that resist braking hip and knee motions important to iliotibial tract (ITT) mechanics, but knee flexion angle changes are particularly significant. Under moderate running stride lengths, the knee is more flexed at initial contact (A), requiring flexion resistance over shorter angular changes. This means that the tract should experience slower strain rates (length changes) to the end of the braking phase (B), whereas strain rates will be higher following longer stride lengths. A long stride length (C) requires more extensive ranges of knee flexion, making quadriceps control more taxing and requiring a faster strain rate within the tract throughout the braking phase. Muscles under repeated higher loading rates that require greater muscle fibre length changes are also likely to fatigue earlier, potentially increasing torques and strains on passive structures such as the ITT. *(Permission www.healthystep.co.uk.)*

Because of the complex linkage, the pathomechanics of iliotibial band syndrome are likely to be multifactorial and specific only to the individual. Any situation that increases knee varus moments or large hip adduction angles, such as that found associated with hip abductor weakness, is likely to increase tension through the ITT, particularly at loading response. Higher tensions within the tract will be increasing compression over the lateral femoral condyle. Increased hip adduction angles in the stance phase of running have been widely linked to sufferers of iliotibial band syndrome (Fredericson et al., 2000; Noehren et al., 2007; Bramah et al., 2018; Ceyssens et al., 2019; Vannatta et al., 2020a). Whether this can also be related to tight or loose tracts and the rate of strain remains under debate. Through the hip linkage, *greater trochanter pain syndrome*, which was previously referred to as *trochanteric bursitis*, has also been linked to ITT dysfunction (Lavine, 2010) (Fig. 4.3.7e).

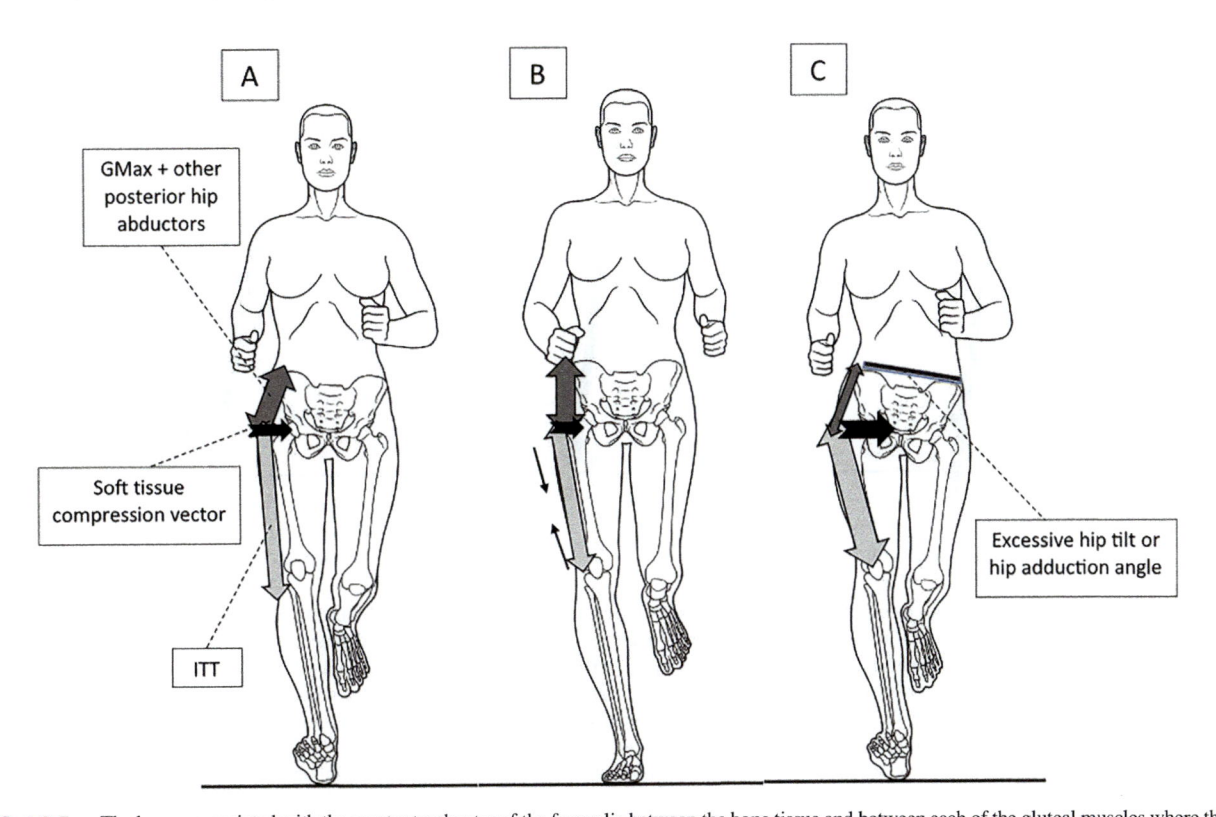

FIG. 4.3.7e The bursae associated with the greater trochanter of the femur lie between the bone tissue and between each of the gluteal muscles where they lie on top of each other, close to their trochanteric distal attachments. This not only helps reduce friction between bone and muscle but also friction between muscles under hip motions and different muscle contractions. Such frictional forces increase under larger hip adduction angles and internal hip rotations, as these postures set up an increasing tensional force between distal and proximal soft tissues. This causes soft tissue compression into the trochanteric bone (black arrow). Although trochanteric bursitis can be caused by walking hip pathomechanics, these structures are more vulnerable in running because of greater motion between the hip and knee, which the iliotibial tract (ITT) helps to stabilise via the superficial fibres of gluteus maximus (GMax) during contact (A) and braking (B). As the end of braking is approached, ITT strains and soft tissue compressions should be reducing with less internal rotation and reducing hip adduction angles. In the presence of poor control of hip adduction tilt (C), the distance between proximal gluteal origin and the ITT's attachment to Gerdy's tubercle on the tibia increases, raising tension within the tract and gluteal muscle fibres that translate into greater compression and shear through the anatomical bursae within and around the greater trochanter's anatomy. The higher the angle and the longer the knee flexion moment, the greater the changing strains within the tract that are influencing trochanteric soft tissue compressions. If highly repetitive, this increased hip adduction moment can lead to bursitis and increasing painful frictional forces on hip motion, even when just walking or using stairs. *(Permission www.healthystep.co.uk.)*

Rearfoot eversion has also been proposed as a pathomechanical factor in ITT symptoms, as a result of the coupling of tibial internal rotation to the rearfoot eversion. However, this coupling relationship is weak (Reischl et al., 1999; Pohl et al., 2007). Higher and lower incidences of rearfoot eversion compared to asymptomatic controls are reported in the literature (Messier et al., 1995; Busseuil et al., 1998; Noehren et al., 2007). After heel strike, ITT strain increases to a peak that is higher in iliotibial band syndrome runners compared to controls (Miller et al., 2007). Bearing this in mind, with the timing and coupling relationships of rearfoot eversion and knee flexion angles at initial loading that are associated with ITT pain, a rearfoot link is possible. Initial rearfoot loading occurs when increases in ITT strains are also happening, and thus, the presence or absence of large rearfoot eversion excursions may only hold a coincidental relationship. It may be that initial contact events together, rather than the rearfoot motion at contact specifically, is important. Runners with iliotibial band syndrome, when fatigued, are reported to have higher foot adduction angles, foot contact inversion angles, and internal knee rotation velocities at initial contact than controls (Miller et al., 2007). Such foot strike postures are likely to require greater rearfoot eversion, if available, to bring the medial foot flat to the support surface. Such a likelihood suggests a linked effect, rather than an actual cause of iliotibial band syndrome through 'provoking' rearfoot kinematics (Fig. 4.3.7f).

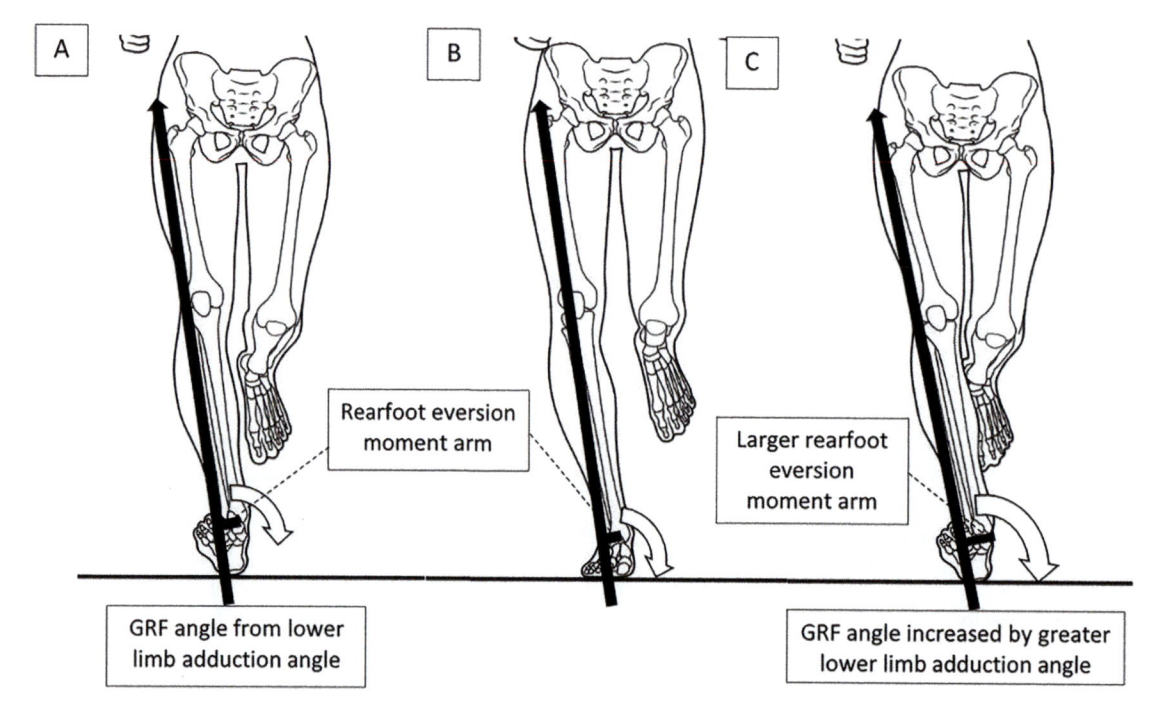

FIG. 4.3.7f The degree of rearfoot eversion, often and erroneously equated to foot pronation, has frequently been suggested as a possible cause of proximal lower limb pathological risk. Such associations have been made with patellofemoral, ITT, and generalised nonspecific knee joint symptoms, along with many other conditions. However, the amount of eversion may just be an associated consequence of the hip abduction angle. If the hip adduction angle is high, it necessitates increased rearfoot eversion loading rates and degrees to bring the foot flat to the ground from a more inverted angle set by the lower limb at initial contact. Free-body diagrams of the hip adduction angle at contact in running can demonstrate this well. At a milder adduction angle (A), the rearfoot is presented to the ground angled slightly inverted, requiring significant rearfoot eversion motion under greater GRF-induced eversion moments (white arrow) from a more lateral heel strike. This eversion is required until loading response has brought the forefoot to the ground so that the whole foot becomes 'prone' to the support surface (B). In higher hip adduction angles (C), the rearfoot contacts the ground at a higher varus attitude to the support surface, creating a distinct lateral heel contact that sets up a greater GRF-induced eversion moment on the rearfoot. Thus, the rearfoot is required to produce greater motions of eversion, often at a faster rate, to bring the foot flat to the ground the more the lower limb is adducted under the body. This means that rearfoot eversion is a capability that improves foot placement to the inverted leg, providing a safety mechanism rather than a potential pathomechanical explanation, for proximal symptoms. Other varus-creating influences such as tibial varum alignments will create a similar need. However, symptoms in rearfoot eversion resistors such as tibialis posterior may indicate that this mobile adaptation is causing pathology locally, requiring action to reduce the inverted position of the limb (via hip abductor strengthening and hip adductor flexibility) or by controlling the eversion moment that results (via medially wedged foot orthoses). *(Permission www.healthystep.co.uk.)*

Varus deformation of the knee through medial compartment OA has been suggested to be a cause of iliotibial band syndrome, due to increased tension on the fascial tract with changes in the knee varus angle (Vasilevska et al., 2009). This would suggest that genu or tibial varum may be a biomechanical risk factor for inducing ITT pathology (Fig. 4.3.7g). A link to muscle imbalances between low hamstring strength compared to quadriceps strength and genu recurvatum (hyperextension) has also been made (Lavine, 2010). This may be due to effects on the knee flexion angle and the ITT position at initial loading, or reduced braking effects from the hamstrings.

Injury to the ITT can also include rupture, seen as discontinuity and oedema on MRI (Bencardino et al., 2000). Such an injury is likely to have consequences on muscular hip adductor moment control, as well as loss of lateral knee stabilisation of varus moments and disruption of the patellofemoral tracking force vector balances.

It would appear that the pathomechanics of the ITT hinge on a number of primary functional joint positions. Knee flexion angle, hip adduction angle, and the structural and functional alignments can all affect knee varus moments. These factors, combined with loading rate stresses and fatigue, all seem to potentially play a part. This means that a

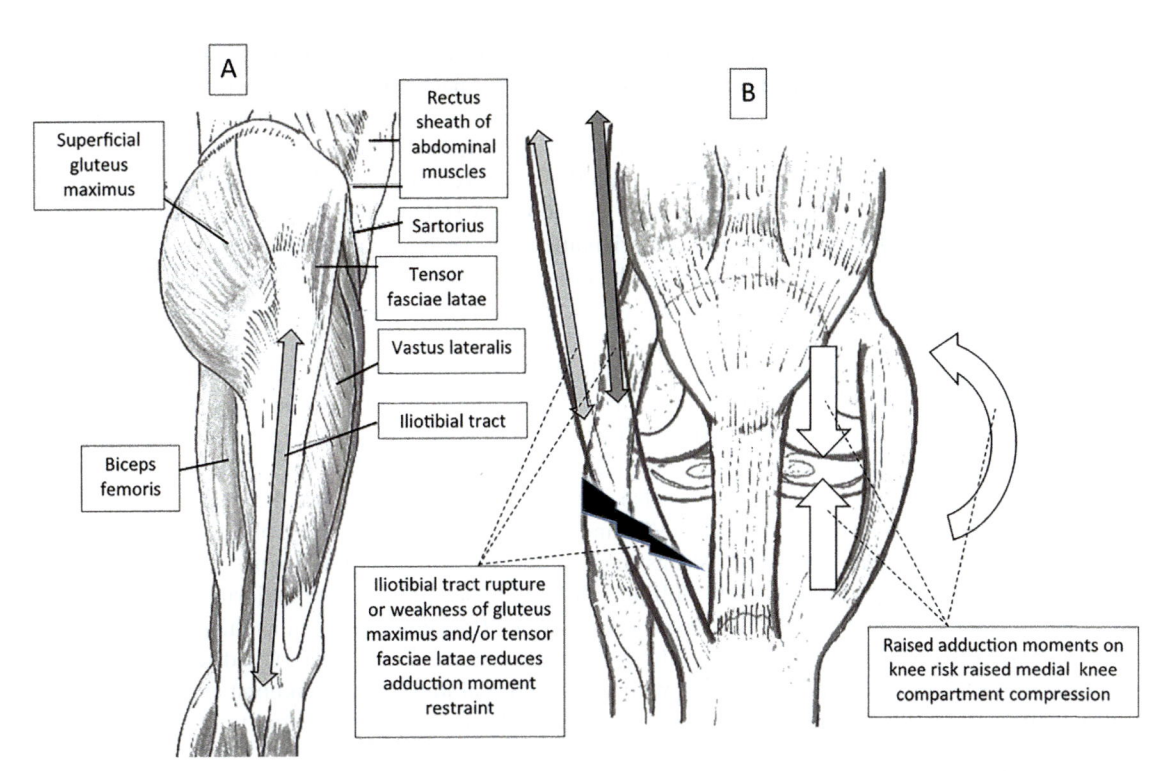

FIG. 4.3.7g The iliotibial tract provides strong lateral knee restraint (A) that is an important provider of abduction (valgus) knee moments throughout stance phase, helping to resist the usually more dominant knee adduction (varus) moments of gait. In providing its assistance to other providers of knee adduction moment restraints, such as the posterolateral corner complex and the lateral collateral ligament, the iliotibial tract protects these structures from receiving higher strains. If muscular attachments such as the superficial fibres of gluteus maximus and tensor fasciae latae are weak or the iliotibial tract tears or ruptures, then knee adduction moments may become less restrained or other adduction resistors over-stressed during gait. The result is that knee adduction moments may increase (B), risking increased articular stresses on the medial knee compartment. *(Permission www.healthystep.co.uk.)*

one-treatment-fits-all approach is probably inappropriate, a common theme in the reality of clinical biomechanics. The construction of a free-body diagram of the patient with their known features of gait may better focus the therapeutic interventions. Clinicians publishing more knee case studies with morphological, kinematic, and kinetic data during specific activities may start to create a database of important features and activity values. This may prove more helpful in the long run than trying to locate a single homogeneous risk factor. Randomised controlled trial studies of rehabilitation, orthoses, and running shoes that do not contain a morphological-specific group are probably going to be ineffectual in revealing an explanation of knee injury risks, when the pathology has many possible influencing factors that may or may not be present in each case.

4.3.8 Free-body diagrams of patellofemoral joint pathomechanics

The patella tracks in the femoral groove throughout knee extension and flexion, influencing the mechanical efficiency of the patellar ligament's ability to appropriately apply the quadriceps force to control knee flexion and extension during locomotion. The patellofemoral functional unit acts as an anterior retaining and restraining 'sling mechanism' against flexion moments. This creates considerable compressive forces between the articular surface of the patella and femur during flexion and moving back towards extension from high flexion angles. Such compressive stresses need to remain fairly evenly spread across the relatively small articular surfaces to prevent stress concentrations that can compromise cartilage properties or capsular structures. Because of the large changing joint angles at the knee, the anterior joint capsule has inward folds known as synovial plicae. They are divided into suprapatellar, medial, infrapatellar, and lateral plica. The medial plica is reported to be the one most likely to become symptomatic, as a result of inflammatory processes that cause the tissue to become inelastic, tight, thickened, fibrotic, and sometimes hyalinised (Lee et al., 2017). No specific pathomechanical explanation has been given to this process other than 'overuse' (whatever that may mean). The soft tissue force vectors around the patella are key to understanding normal and pathomechanical situations, and can themselves often be the victims of mechanical stress injury or overload. It is likely that synovial plica syndrome is just a part of overall connective tissue damage that results from poor patellofemoral loading patterns.

Patellofemoral joint motion is driven by the tibiofemoral joint's mechanics such that the two have an intimate relationship. It is unwise to assess one without recourse to the other. The mechanics of the patellofemoral joint are those of an interplay between the human's largest sesamoid bone, the anterior articular surface of the distal femur, passive fascial and ligament restraints, and active muscle force vectors (Ateshian and Hung, 2005; Senavongse and Amis, 2005; Wünschel et al., 2011; Sherman et al., 2014). Patellar stability is described as displacement induced by load or the force needed to displace the patella (Merican et al., 2009). A stable patella should glide superiorly with knee extension and inferiorly with flexion, following the orientations of the femur and tibia without significant medial or lateral displacement. Dysfunction around this joint is the most common reason for anterior knee pain, which seems to be significantly more common in athletic females (Prins and van der Wurff, 2009; Foss et al., 2014; Petersen et al., 2014).

Free-body diagrams of the patellofemoral joint should reflect the triplanar nature of the force vectors that achieve patellar stability, but these are most easily appreciated by constructing the diagram from a frontal plane anterior view. The diagram needs to reflect the primary anatomical force vectors from the quadriceps muscles associated with the quadriceps (Q) angle, from the anterior superior iliac spine to the middle of the patella (see Chapter 2, Section 2.4.6). Ligamentous and fascial restraints also need to be considered for all the medial, lateral, and vertical vectors to be accounted for. Both vastus lateralis (VL) and vastus medialis (VM) have oblique components providing combined transverse/frontal plane stability. VM has the complication of having a distinct subunit within the medial part of the muscle belly, namely vastus medialis obliquus (VMO), which provides the majority of the medially directed obliquity of the VM vector. The patella is also enveloped in fascial thickenings known as the peripatellar retinaculum, which provides passive tension resistance against the quadriceps vectors (Fig. 4.3.8a). Quadriceps strength, expressed as peak torque, is most evidenced when correlated with patellofemoral joint pain (Lankhorst et al., 2012), but the complexity of potential diagnostic and functional causes make general population correlation to a single pathomechanical factor difficult (Ireland et al., 2003; Cichanowski et al., 2007).

Sagittal plane cross-sectional free-body diagrams can also aid in the appreciation of the compression between the articular surfaces and the patellar bursae in relation to the knee's flexion angles. Tibial posterior translation, which in midstance is usually related to anterior femoral translation above the tibia, is accompanied by increased patellofemoral contact pressures (Li et al., 2004). Thus, flexion and extension angle adjustments change the joint contact areas and the extent of compression within the patellofemoral joint. The flexibility of the hamstrings also influences the patellofemoral joint reaction forces. This is particularly so laterally, and most significantly at 60° knee flexion, a finding revealed during squat tests (Whyte et al., 2010). As the hamstrings are active just before and during initial loading response, at this time, the hamstrings may also have an influence, with their significance being greatest during running as knee flexion angles and hamstring activity are greater than that found in walking (Fig. 4.3.8b).

Patellofemoral pathology includes osteochondritis or chondromalacia and eventually OA of the patellofemoral articular surface, which are conditions that relate to a decrease in the total articular contact areas, under higher peak forces. This is linked to alterations in patellar alignment such as tilt and lateral displacement. Patellar dislocation is the extreme example of lateral displacement. It often results from the presence of a shallow trochlear groove on the femur. Patellar dislocation can also be caused by trauma if severe disruption of the medial soft tissue stabilisers occurs (Fig. 4.3.8c). Patellar fractures and patellar ligament ruptures are usually linked to explosive quadriceps activities or direct trauma.

Osseous and functional alignment issues must be considered. Femoral anteversion or hip transverse plane functional rotations can affect the orientation of the patellofemoral joint to the sagittal plane line of progression of the lower limb. Situations causing the patella and knee to be externally or internally positioned change medial–lateral stabilisation dynamics, respectively. However, numerous studies have failed to show a clear correlation between patellar alignment in patients with patellofemoral pain compared to controls (Salsich and Perman, 2007). Some diagnostic imaging features on CT have been reported to be associated with patellofemoral joint pains such as increased patellar tilt with and without load-reduced contact areas, at an angle of 15° of knee flexion (Drew et al., 2016). Width of the patella combined with the amount of tibiofemoral rotation is reported to explain around 46% of variance in joint contact areas of patellofemoral joints, with width making up 31% of this variance (Salsich and Perman, 2007). This suggests that individual biomechanics of the patient as a whole is more important than looking for a single alignment issue as the origin of patellofemoral dysfunction.

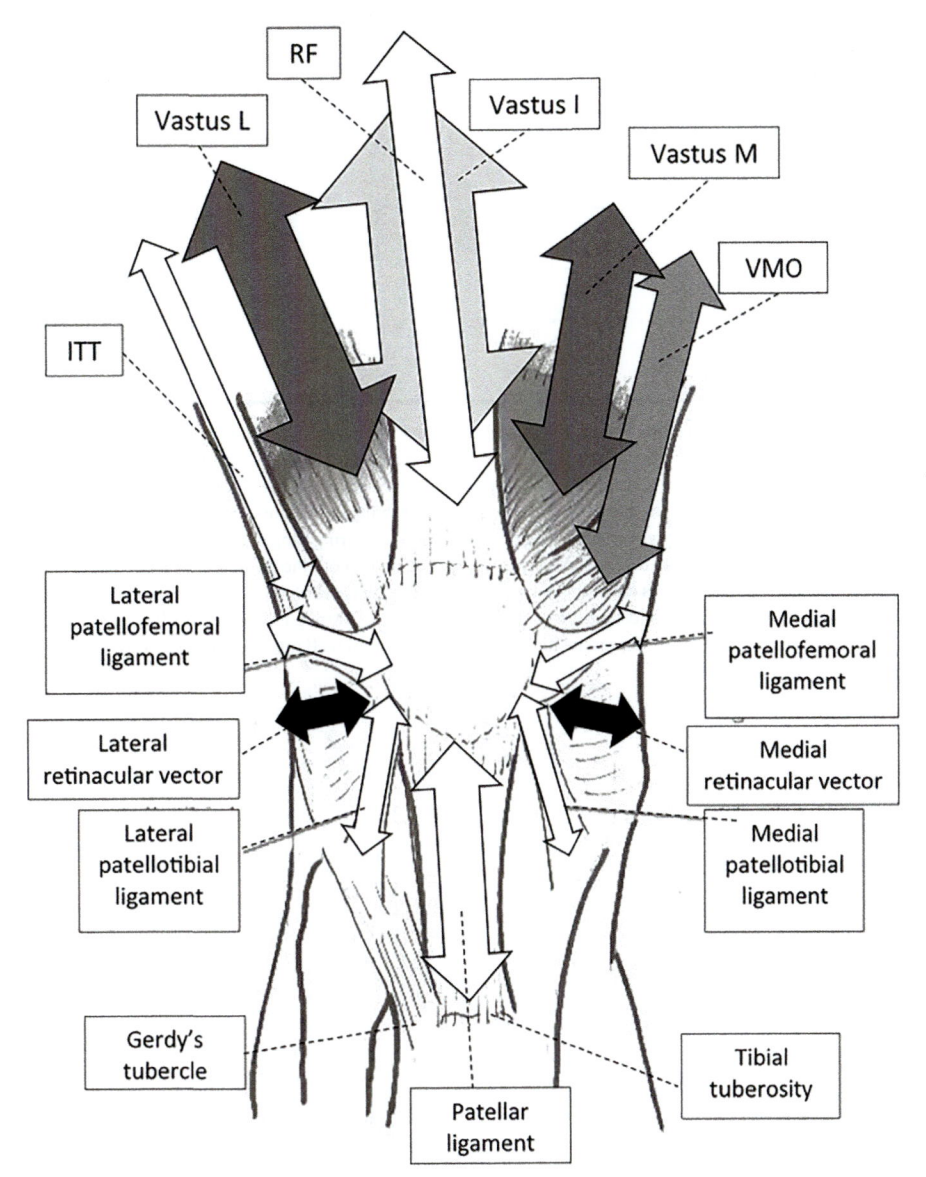

FIG. 4.3.8a When constructing a free-body diagram of the patellofemoral joint, all the vectors controlling the motion of the patella need to be considered. First, there are the muscle vectors. These consist of medial vectors from vastus medialis (Vastus M) with its longer, more vertical bundle and its distinct shorter oblique bundle (VMO). Vertical vectors are provided by the longer fibres of rectus femoris (RF) and the shorter but cross-sectionally larger fibres of vastus intermedius (Vastus I). Because of the normal femoral neck-shaft angle, these muscles provide a slightly lateralised force vector which increases with greater hip adduction angles during gait. Lateral vectors are primarily produced by vastus lateralis (Vastus L) and via the iliotibial tract (ITT) where gluteus maximus in early stance and tensor fasciae latae in late stance apply significant force to the patella and the ITT's attachment at Gerdy's tubercle. The quadriceps muscular vectors are all opposed by the patellar ligament via its proximal patellar and distal tibial tuberosity attachments. Extensive fascial reinforcement medially and laterally is provided by the patellar retinaculum. Within this structure are more organised connective tissue bundles known as the patellofemoral and patellotibial ligaments, that together set up resultant retinacular vectors on the patella. Loss or weakness of any of these vectors or changes in the hip adduction angle during gait will alter patellofemoral vectors that can provoke symptoms and pathology from recurrent patellar maltracking. By considering the anatomy and adding or subtracting vectors on a free-body diagram, the clinician can consider the origins and possible therapies for patellofemoral symptoms. *(Permission www.healthystep.co.uk.)*

Patella alta and baja

The position as well as the width of the patella may also affect pathomechanics. A superiorly positioned or 'high-riding' patella, known as *patella alta*, causes the patella to sit high within femoral trochlear groove, requiring a longer patellar ligament/tendon. As the femoral groove and its lateral ridge are important in providing some form closure to the joint, a high-positioned patella loses some lateral articular form closure, increasing the risk of lateral instability and/or dislocation

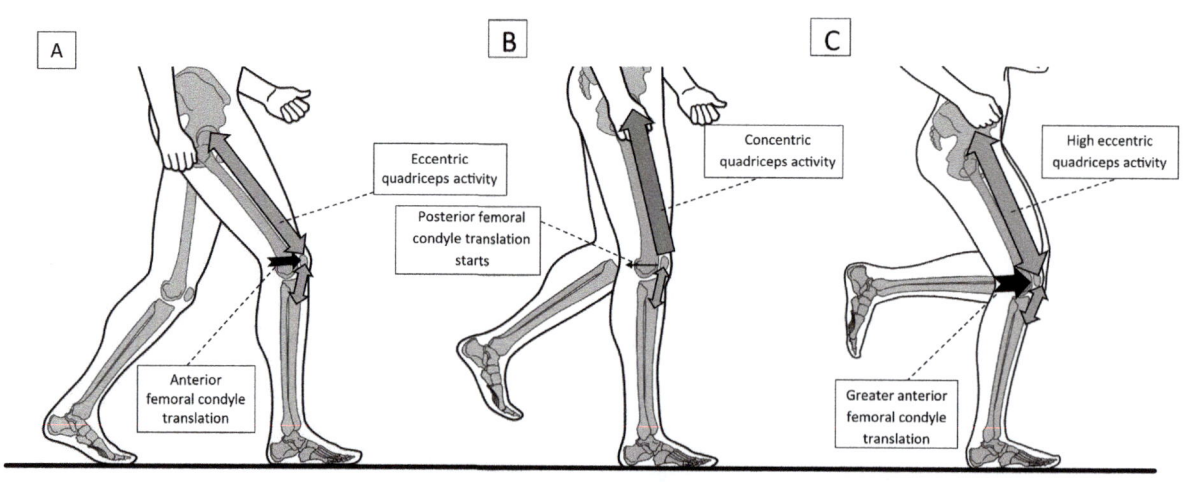

FIG. 4.3.8b The patella undergoes superior and inferior gliding over the femoral articular surface during gait, although it is femoral motion rather than that of the patella. While under eccentric quadriceps contraction during knee flexion, the patella moves towards and articulates with the inferior aspect of the femur's patellofemoral articular surface. During concentric quadriceps-induced knee extension, the reverse occurs so that at near-full extension angles, the patella sits at its highest point over the femur. In this position, mediolateral stabilisation occurs largely via the retinaculum. During swing, patellofemoral loads are not great, with the articular compression forces being at their highest during ranges of 60–90° of knee flexion. In walking stance, forces are raised, particularly at these flexion angles during loading response (A), coming under both the influence of body weight and anterior femoral condylar translation. During walking midstance (B), the knee is extending out of positions of high patellofemoral compression as the femoral condyles start to glide posteriorly, as indeed they also do throughout late midstance. However, in running (C), through the extended loading response knee flexion angles and increased anterior femoral translation, patellofemoral forces are higher and sustained until well into the acceleration phase. *(Permission www.healthystep.co.uk.)*

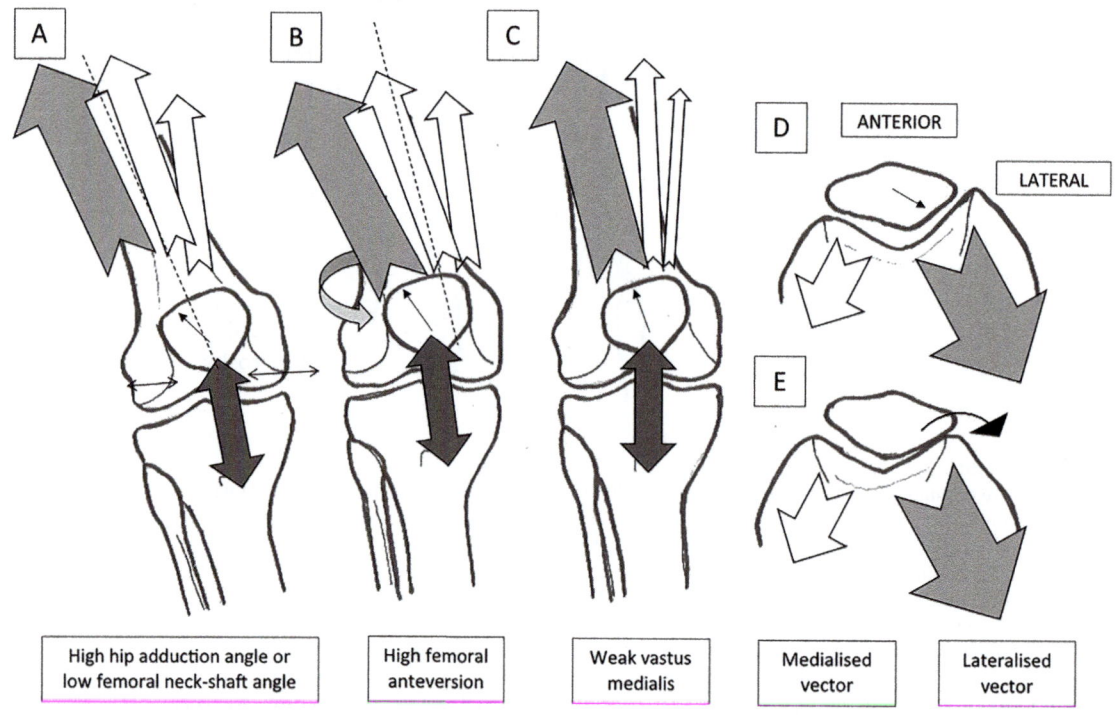

FIG. 4.3.8c Many factors influence patellar tracking within the patellofemoral joint. These primarily involve the interaction between the soft tissue vectors that set the patella's postural alignment during knee flexion and extension, with some influence from the articular shape. Osseous morphological and functional alignments can alter the muscles moment arms, which through the necessary adduction of the lower limb under the body in human bipedal posture, tend to naturally lateralise the soft tissue vectors on the patella. With increased hip adduction or low femoral neck-shaft angle (coxa vara) (A), the vectors of the iliotibial tract, vastus lateralis, vastus intermedius, and rectus femoris (large grey arrow) become more laterally positioned to the femoral long axis. These vectors are opposed by the less medially orientated vastus medialis (larger white arrow) and vastus medialis oblique (thinner white arrow). Thus, under contraction, the medial patellar retinaculum (thin black double-headed arrows) is tensioned more than the lateral. A similar situation is created by high femoral anteversion (B) which rotates the patellofemoral joint medially to lateralise the muscular moment arms. Weak vastus medialis muscle bundles (C), especially those of the oblique fibres, also cause increased lateralised resultant vectors on the patella. A large bony ridge exists on the lateral side of the patellofemoral joint, easily visualised when viewed superiorly, as demonstrated in (D). This ridge improves form closure of the joint laterally, creating a deep trochlear groove forming the patellar facet on the femur. The lateral side of the joint tends to undergo greater compression forces during knee flexion than the medial side because of vector lateralisation. The enlarged ridge helps prevent dislocation of the patella under these vectors. In some individuals, this lateral ridge is smaller creating a shallower femoral trochlear groove (E), increasing the risk that further lateralisation of the patella's soft tissue vectors could lead to subluxation or dislocation. *(Permission www.healthstep.co.uk.)*

through reduced joint contact area (Ward et al., 2007). Patella alta is common in cerebral palsy associated with a crouch gait of increased hip and knee flexion angles, resulting in increased quadriceps force (Lenhart et al., 2017). Normal gait requires up to 0.8 times body weight of force from the quadriceps, but crouched gait requires around 4.8 times up to 8 times body weight when the patella is positioned normally (Lenhart et al., 2017).

The effects of patella alta also significantly raise patellofemoral joint stresses during knee flexion (particularly during fast walking), probably as an exaggerated effect resulting from reduced joint contact area seen normally in faster walking (Ward and Powers, 2004). Although the patellar moment arm is not affected by alta, the patellar ligament and quadriceps tendon force ratios are larger than in the normal-positioned patella (Ward et al., 2005). Alta knees may experience less patellofemoral joint reaction force when trying to resist knee flexion moments in the range of between 0° and 60°, but the reduced stability and articular contact areas are the likely pathological risk of this mechanical effect (Ward et al., 2005). Having a patella alta around 1.1 cm in normal gait reduces quadriceps forces to around 0.7 times body weight but with increasingly severe crouched gaits progressively larger patella alta helps moderate the rise in quadriceps forces to only around 5.8 times body weight in severe crouched gaits (Lenhart et al., 2017). Thus, patella alta protects from high quadriceps forces, but decreases form closure of the joint, risking instability under contraction (Figs 4.3.8d and 4.3.8e).

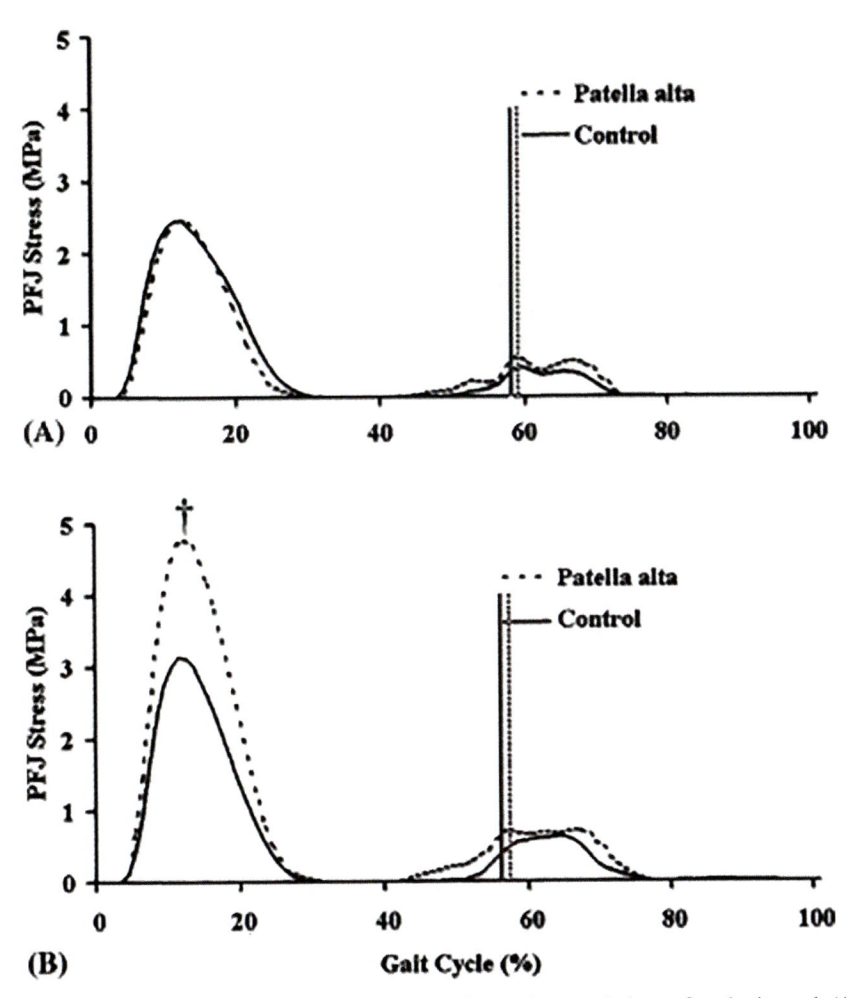

FIG. 4.3.8d Comparison of patellofemoral joint stress between patella alta and control groups during preferred gait speeds (A) and fast gait speeds (B). Vertical lines separate stance from swing phase. † indicates a significant difference in peak patellofemoral joint stress. *(Image from Ward, S.R., Powers, C. M., 2004. The influence of patella alta on patellofemoral joint stress during normal and fast walking. Clin. Biomech. 19(10), 1040–1047.)*

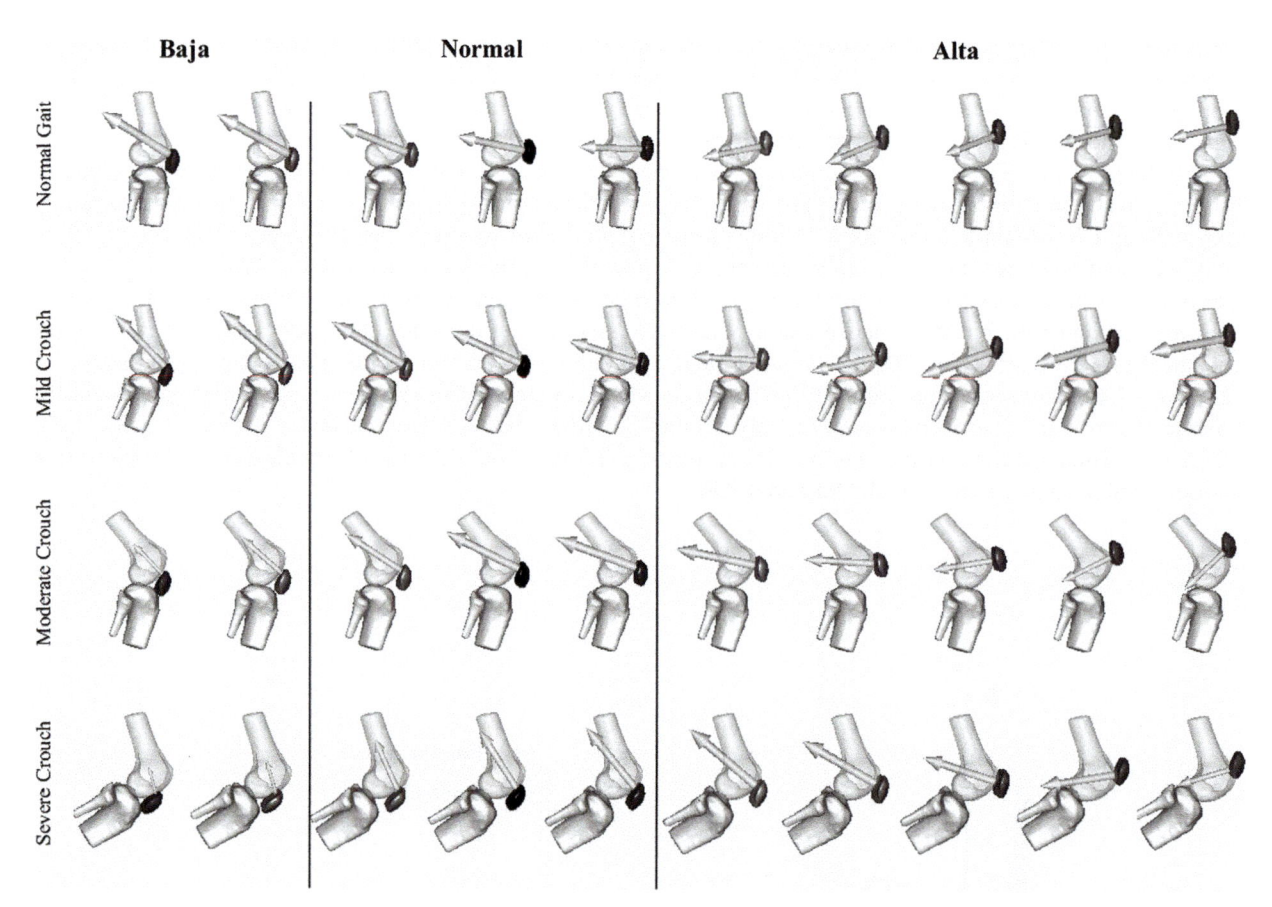

FIG. 4.3.8e A balance exists between patellofemoral compression forces and joint stability during gait. Positioning the patella lower within the trochlear groove of the femur improves joint stability but necessitates greater quadriceps forces and compression of the articular surfaces, especially within 60–90° of knee flexion. Higher patellar positions reduce the quadriceps forces, yet also reduce the form and force closure on the joint, risking instability and dislocation. The images demonstrate the interaction between patellar position and knee flexion angle to the magnitude and orientation of the patellar contact force. Arrows' size indicates the magnitude and direction of the patellar contact force upon the femur. Force vectors have been scaled for visualisation. These are constant within each row but differ between rows of crouch or knee flexion severity. The quadriceps tendon wraps over the distal femur when baja is introduced in moderate and severe crouch gait, which contributes to the apparent decrease in the patellofemoral contact area in those positions. *(Image from Lenhart, R.L., Brandon, S.C.E., Smith, C.R., Novacheck, T.F., Schwartz, M.H., Thelen, D.G., 2017. Influence of patellar position on the knee extensor mechanism in normal and crouched walking. J. Biomech. 51, 1–7.)*

Patella baja is an inferiorly positioned patella. It results in a distal position of the patella, a shorter patellar ligament, and a decreased distance between the inferior patellar pole and the tibial tuberosity. It is rare, except following surgical repair of a patellar tendon rupture, patellar fracture, total knee arthroplasty, or high tibial osteotomy (Flören et al., 2007). It is important to note that surgical high tibial osteotomies, often used for the treatment of medial compartment OA in younger patients, has been reported to alter patellar height, tilt, and rotation (Gaasbeek et al., 2007). Changes in patellar orientation following such surgery are a likely pathomechanical origin of anterior knee pain, which can sometimes be developed later.

Pathomechanics of patellofemoral pain syndrome

Symptoms of the patellofemoral joint are clinically referred to as *patellofemoral pain syndrome*. This is an unsatisfactory term lacking a clear diagnosis for symptoms that can be caused by many pathologies and thus can arise from a diversity of different pathomechanics. The source of the pain can be derived from disturbance of the articular contact areas, or from any of the many soft tissue structures and their attachment points to the patella. Patellar symptoms do not represent a homogeneous source, and this is an important point to stress for many other lower limb symptoms too. In understanding the pathomechanics, the clinician will find a more specific diagnosis helpful. By doing so, it will aid the clinician to focus therapeutic interventions. Thus, establishing the diagnosis and pathomechanics requires careful examination of patellofemoral symptoms (Petersen et al., 2014) (Fig. 4.3.8f).

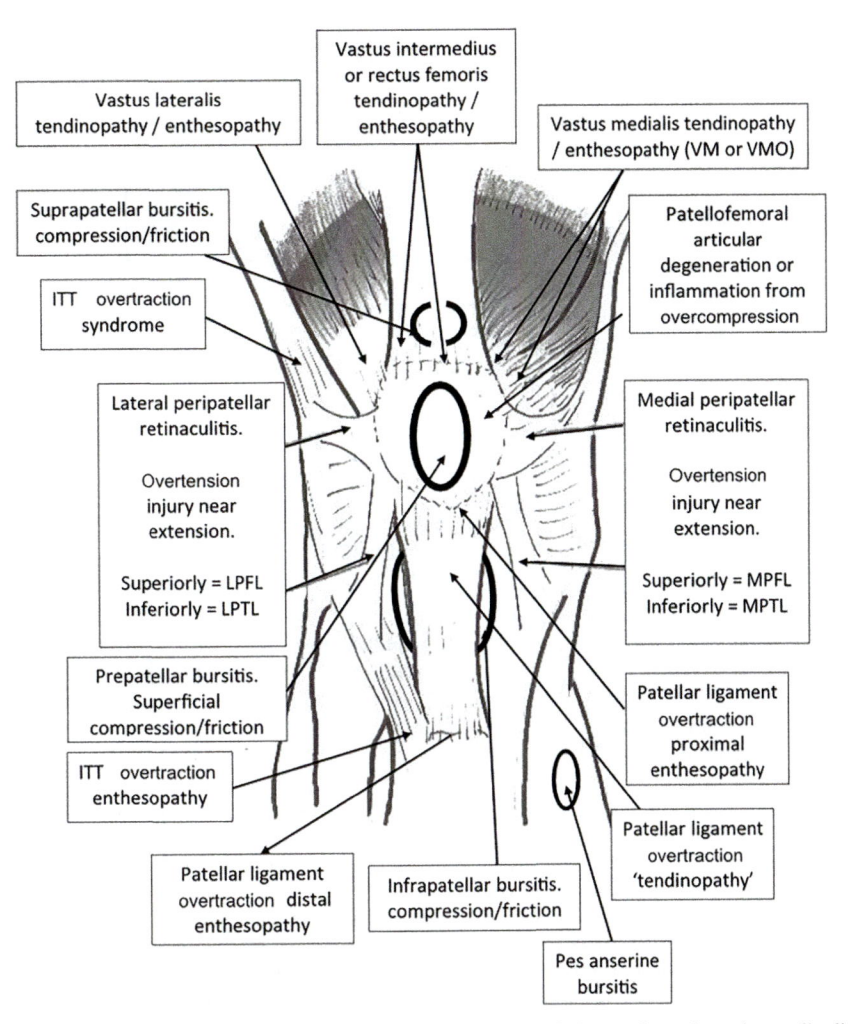

FIG. 4.3.8f Patellofemoral symptoms can arise from many anatomical structures that include muscle tendons, the patellar ligament, bursae, and connective tissue reinforcements. Terms such as anterior knee pain, patellofemoral joint syndrome, or runner's and jumper's knee are insufficient to understand the pathomechanics in any detail. Demonstrated here are the structures most commonly associated with patellofemoral (anterior) knee symptoms. LPFL and LPTL stand for lateral patellofemoral ligament and lateral patellotibial ligament, respectively, with MPFL and MPTL standing for the opposite medial connective tissue reinforcements found within the patellar retinaculum. Damage to the patellar articular surface is often termed chondromalacia patellae and is thought to relate to softening and then fissuring of the hyaline cartilage. Cartilage fibrillation and more extensive osteoarthritic changes in this joint are common and may be provoked by sustained low static compression loads on articular cartilage due to long hours of sitting in positions of high knee flexion angles. Careful clinical examination by palpation of depth, location, and stress testing the structure can clinically give a diagnosis when diagnostic images are not immediately available or appropriate. The more specific the diagnosis of the injured structures identified, the better the understanding of the biomechanical problems at play, and the more focused the intervention can become. There is little doubt that research on patellofemoral pain syndrome is of little value without a specific diagnosis being provided. *(Permission www.healthystep.co.uk.)*

Soft tissue pathology linked to patellofemoral pain includes the retinacula, bursae, individual quadriceps insertion points to the patella, and the patellar ligament itself. Symptoms are often focused either at the inferior patellar pole or the tibial tuberosity. It is common clinically to refer to pain sourced from the patellar ligament/tendon as *jumper's knee* and other patellofemoral pain syndromes sources as *runner's knee*. However, again, these are not specific pathologies. Age also influences diagnosis due to the effects of tensional forces on the immature growing bone attachments. Pain at the tibial tuberosity in adolescence from age 10 to 14 years is associated with *Osgood-Schlatter's disease* and at the inferior patellar pole with *Sinding-Larsen-Johansson syndrome* or *disease*. Both conditions result from traction strains in immature bone known as an *apophysitis*. These conditions were once reported to be more common in males, but now, rates across the sexes seem similar, probably as more teenage females take up the same sports as males (Kaneuchi et al., 2018). Onset is usually seen from early-to-middle adolescence, but is reported one year earlier on average in females, possibly reflecting their earlier bone maturation (Kaneuchi et al., 2018). Biomechanically, the same excessive or asymmetrical patellofemoral tensile factors seem to be involved as found in adults, but with the weakest point being the immature bone and fibrocartilage attachments of the patellar ligament rather than within the patellar ligament itself, or its mature enthesis (Fig. 4.3.8g).

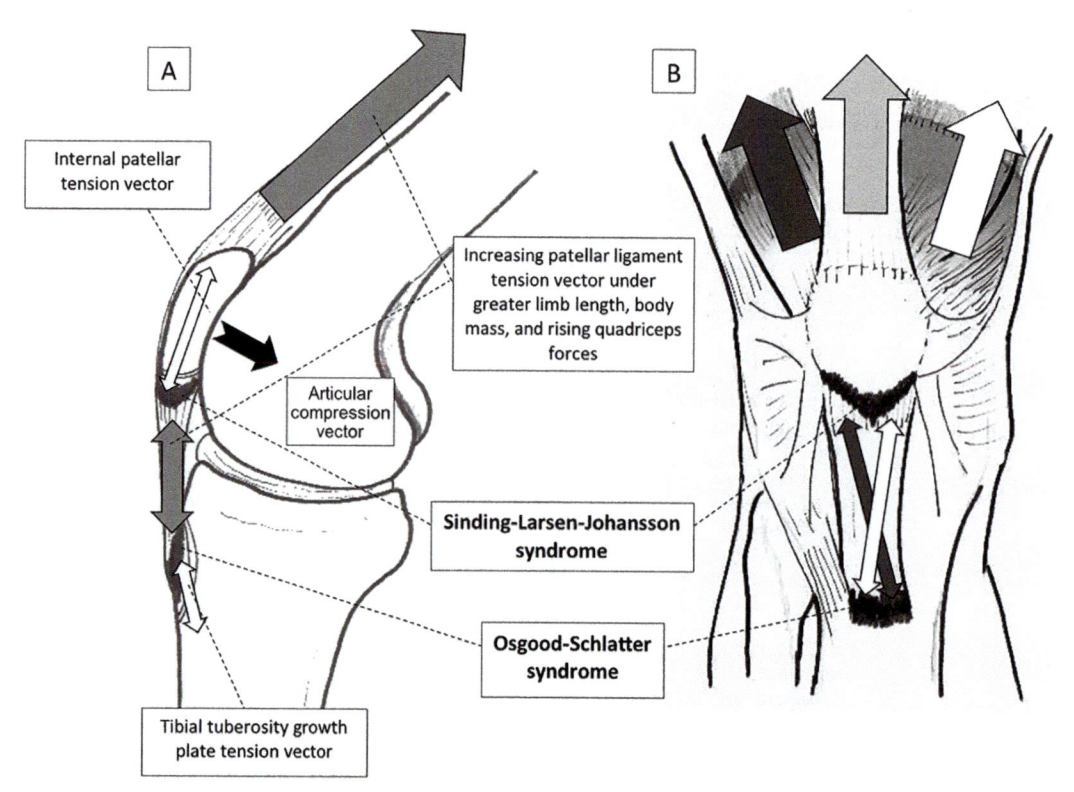

FIG. 4.3.8g Ossification of the human patella usually originates from two centres, developing during ages three to six years and finally fusing during adolescence. Fibrous rather than osseous union can occur, creating a bipartite patella which usually only presents as an incidental radiographic finding. However, symptoms are also reported with such patellae, probably for the same mechanical reasons as normal patellae. However, specific symptoms related to the immature patella and its ligament attachments are common in athletic adolescents. Whereas the connective tissue of the patellar ligament, quadriceps tendons, and the enthesopathy organ tend to present as weaker tissues in adults, the immature areas of ossification can present symptoms related to excess or unbalanced traction forces on the inferior pole of the patella or tibial tuberosity. These traction apophysitis pathologies are known as Sinding-Larsen-Johansson syndrome and Osgood Schlatter's disease or syndrome, respectively (A). In extreme cases, avulsion fractures can occur. The longer lever arms and forces from the quadriceps muscles, resulting from an increasing limb (femoral and tibial) length and rising body mass during growth coupled to raised forces during running and jumping, likely initiate the problems primarily within the sagittal plane. This is especially likely to be the case if there is vector imbalance between the soft tissues that focus stress into particular areas of the immature bone rather than more diffusely (B). If lateral vectors dominate (dark-grey) tensions across the patellar ligament running from superolateral to inferomedial, then this may be reflected in a medial location of pain within the growth plate. This would be reversed if medial vectors (white arrows) dominated. Such growth-associated patellofemoral pathology should be approached by considering and attempting to rebalance the joint vectors, as in adult patellofemoral pain. However, it must also be appreciated that bone heals poorly under tension. Rest from athletic activity for at least a while is often necessary and speeds recovery, while the underlying vector issues can start being addressed through rehabilitation under lower daily living loads. *(Permission www.healthystep.co.uk.)*

The relationship of the frontal plane geometry and the balance of strengths and weaknesses of the soft tissues around the patellofemoral joint may provide most of the patellofemoral pathomechanics. This frontal plane mechanical relationship is known as *patellar tracking*, although sagittal plane tensional forces from quadriceps tightness will also have a significant influence on tensile strains and articular compressions across the joint. Quadriceps injuries and ligament tears can cause sudden disruption in patellofemoral mechanics leading to patellofemoral pain, but in most cases, the origin of the patellar dysfunction is more insidious and often reflects lifestyle choices that cause an imbalance among the strengths of the quadriceps muscles. This *maltracking* relationship between vastus lateralis and vastus medialis has produced much research. The VMO part of vastus medialis seems particularly prone to weakness, and this weakness has been proposed to produce patellofemoral joint symptoms (Powers et al., 1996). However, it is not clear if is the only origin of patellofemoral maltracking (Petersen et al., 2014).

The patellofemoral joint is prone to lateral overpull through loss of medial force vectors or lateral osseous–articular resistance. The presence of a large quadriceps or Q-angle due to low femoral shaft-neck angles (coxa varum) may compound the issue by giving the lateral force vectors mechanical advantage (Fig. 4.3.8h). However, the evidence of Q-angle influencing patellofemoral tracking is contradictory (Petersen et al., 2014). The abnormal trochlear geometry of a shallow femoral lateral ridge reduces the lateral stability of the patellofemoral joint at around 70%, at 30° flexion, while the effect of

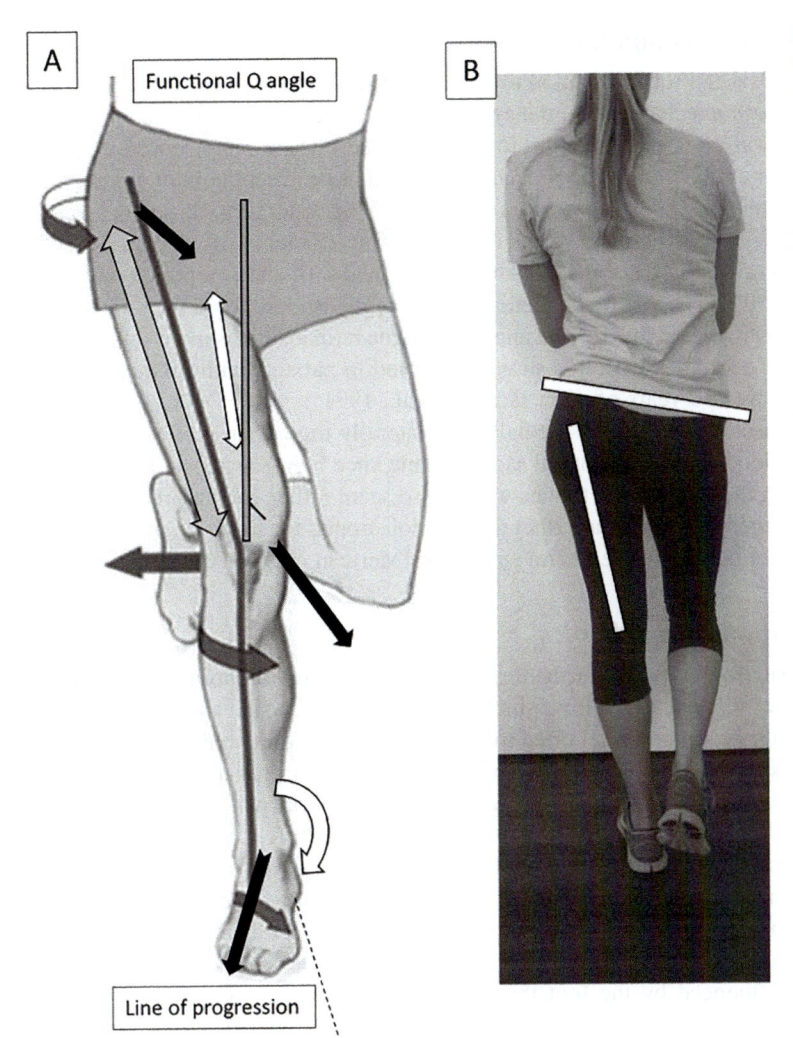

FIG. 4.3.8h Although assessing a static quadriceps (Q) angle to the patella is important, the functional Q angle under progressive degrees of knee flexion is far more significant. This dynamic angle is a result of changing vectors that control lower limb rotations in the transverse and frontal plane during knee motion (A). If the Q angles increase beyond 20° during single-limb support in an individual with patellofemoral symptoms, it suggests that hip abductors, quadriceps, and possibly foot invertors and vault stabilisers are not functioning effectively in combination. Not only will a large Q angle tend to lateralise force vectors by pulling the patella more laterally, but the hip, knee, and/or foot will also likely stray away from the line of progression (dashed line). This decreases the normal sagittal plane energetics, with joint flexion/extension occurring at the hip and knee in a differently orientated sagittal plane to each other (black arrows). Even a simple static single-limb stance test (B), to view the ability of the hip abductors to resist pelvic tilt, gives some insight into an individual's ability to control their femur and patellofemoral joint in a semidynamic alignment, although full gait assessment is obviously the better option. *(Images modified from Petersen, W., Ellermann, A., Gösele-Koppenburg, A., Best, R., Rembitzki, I.V., Brüggemann, G.-P., et al., 2014. Patellofemoral pain syndrome. Knee Surg. Sports Traumatol. Arthrosc. 22(10), 2264–2274.)*

losing VMO power causes around 30% reduction (Senavongse and Amis, 2005). For joint dislocation to occur, the poor ridge profile is possibly more significant than VMO dysfunction, but for causing increased lateral joint compression, it may still have significance. Loss of the retinaculum vectors medially causes a 49% loss of patellofemoral stabilisation in more extended positions, although the retinaculum has less influence on patellofemoral mechanics at increased knee flexion angles (Lee et al., 2001; Senavongse and Amis, 2005).

Sagittal plane position of the knee in relation to degrees of flexion has a complex interaction with joint geometry and the soft tissue force vectors. The degree of flexion in gait associated with symptoms may help in indicating the structure most likely to become overstressed. If unable to resist lateral displacement of the patella acutely, the medial soft tissues are likely to become strained, with potential damage occurring within the structure of the connective tissues as a result of tensional forces. Lateral articular compression will increase from failure of medial soft tissue restraint. Lateral articular compression and connective tissue failure associated with the VMO issues are more likely to occur at positions of greater knee flexion, than those injuries within the medial retinaculum. Recurrent patellar dislocations that are not associated with high stress traumatic events suggest a shallow trochlear ridge laterally. However, after a traumatically induced dislocation, tears to the medial restraints may result in such a significant loss of soft tissue restraint that further dislocation events become a higher risk, despite there being a normal trochlear groove.

In approaching patellofemoral force vector management

Locating the injury and having an accurate diagnosis can focus attention to the relevant pathomechanics. Treatment interventions of patellofemoral pain must reflect the pathomechanics, not the symptoms. For example, the once common surgical release achieved by cutting the lateral retinaculum (*lateral release*) to resolve the soft tissue vector imbalance is not predictable in either effect or outcomes (Lewallen et al., 1990). The procedure also fails to correct the joint compression imbalance within the patellofemoral joint (Ostermeier et al., 2007; Merican et al., 2009). Such surgery can set up a variety of other patellofemoral instability issues (Shellock et al., 1990) and vascular compromises (Sanchis-Alfonso et al., 2005). After lateral retinacular release, the patella is lateralised in flexion angles between 0° and 60°, and is reported only to be medialised at angles over this (Ostermeier et al., 2007). The loss of the medial vectors remains, despite surgery, and cutting the lateral retinaculum may only influence joint mechanics at low flexion angles when the retinaculum is more significant to patellar stabilisation. This may explain the better outcomes with this surgery reported in patients with tibiofemoral OA (Aderinto and Cobb, 2002) and in patients aged over 30 years-of-age (Jackson et al., 1991).

Cutting vastus lateralis might even up the muscle force vectors medially and laterally more than sectioning the retinaculum alone, but it is likely to cause further serious damage to normal sagittal plane knee biomechanics. Lateral release in cases where patellofemoral pathology is within the medial soft tissues would also seem rather counterproductive, creating injury where there was none before. Such surgery can create a further dysfunction around the joint, the precise effects of which will be affected by the level of the resection within the lateral structures (Merican et al., 2009).

Potential distal effects in patellofemoral pathomechanics

Tibial rotation may have a significant effect on patellofemoral biomechanics because of the patellofemoral joint soft tissue stabilisers bridging between the tibia and the upper sections of the lower limb, via the patella (Salsich and Perman, 2007). The width of the patella combined with tibiofemoral rotations is reported to have an effect on joint contact areas (Salsich and Perman, 2007). An inverse relationship of knee flexion angles and both articular contact pressures and retinacular strains exists with tibial rotation (Lee et al., 2001). At higher knee flexion angles, the effects of the peripatellar retinaculum on stability are minimised and therefore will be less affected by tibial rotations (Lee et al., 2001).

Tibiofemoral rotation in stance phase is controlled through the activity of the hip lateral (external) rotators, particularly gluteus medius (Barton et al., 2013). Dysfunction of the hip abductors may be linked to the kinematic findings of greater knee external rotation moments during moments of peak knee extension, and the greater hip adduction moments reported in running kinematics of patellofemoral pain patients (Barton et al., 2009; Bramah et al., 2018).

Tibiofemoral rotation could potentially be influenced by the foot through lower leg rotation couplings to rearfoot motion. Caution must be made when trying to correlate rearfoot frontal plane motions with tibial rotations, because of the variability and timing of the coupling of these motions (Reischl et al., 1999; Pohl et al., 2007). Foot arch height index and joint congruency angles in the static foot do not appear to correlate to patellofemoral dysfunction (Lankhorst et al., 2013). However, kinematic systematic review data give some indications that rearfoot motions could be involved, or at least linked (Barton et al., 2009; Petersen et al., 2014).

Despite the overall kinematic data correlations reported on rearfoot and patellofemoral symptoms seeming a little contradictory, a delayed peak rearfoot eversion and an increased rearfoot eversion peak have been associated with the heel strike transient during walking and running. There seems to be an overall reduction in rearfoot eversion range associated with patellofemoral pain patients when running (Barton et al., 2009). Perhaps large internal rotations and knee flexion in the lower limb associated with dynamic ankle dorsiflexion and rearfoot frontal plane motion are playing a part in some cases of patellofemoral pain, but not in all. This could be considered worthy of investigation, especially when hip external rotator function seems normal, but eversion excursion in stance phase loading is large and rapid during contact and braking, when the knee is being subjected to large flexion moments.

Weak knee extensor strength, expressed by peak torques, appears to remain the best evidenced cause of patellofemoral pain (Lankhorst et al., 2012, 2013), especially when combined with poor hip abductor/external rotator strength (Barton et al., 2009, 2013). Clinical examination of dysfunction in patellofemoral joints should start proximally and be correlated to gait analysis of the provocative event in consideration to the specific patellofemoral tissue injury. The kinematic and kinetic events in the individual patient are not best evidenced through large population studies or systematic reviews of a symptom, but through clinical examination and biomechanical data gathering. Without prospective studies and differentiation of diagnoses, identification of distinct biomechanical causes of each type of patellofemoral dysfunction remains problematic and blurred, possibly giving data that may actually be a record of the effects of having pain in the patellofemoral joint, rather than revealing its cause. Simple clinical tests on individual patients may reveal more helpful information on possible mechanisms of dysfunction (Figs 4.3.8i and 4.3.8j).

FIG. 4.3.8i A dynamic squat test may clinically reveal the development of increasing dynamic valgus angles during knee flexion, as demonstrated in this photograph. Note the abduction of the thigh and external rotation of the foot. Such a test may also provoke the anterior knee symptoms. Watch also for changes in foot posture (amount of vault-lowering pronation) and the foot angle to the line of progression during the manoeuvre (requiring the test in a barefoot state). It is worth repeating the test with the patient trying to improve hip abductor contraction during increasing hip and knee flexion angles, and also adjusting the foot posture to load the foot straighter and more laterally to see if the knee posture and/or symptoms improve. If patellofemoral pain does improve on testing under altered muscle activation and foot postures, then potential targets such as hip and quadriceps muscular rehabilitation and possibly medially supported foot orthoses may be considered as therapeutic interventions. *(Image from Petersen, W., Ellermann, A., Gösele-Koppenburg, A., Best, R., Rembitzki, I.V., Brüggemann, G.-P., et al., 2014. Patellofemoral pain syndrome. Knee Surg. Sports Traumatol. Arthrosc. 22(10), 2264–2274.)*

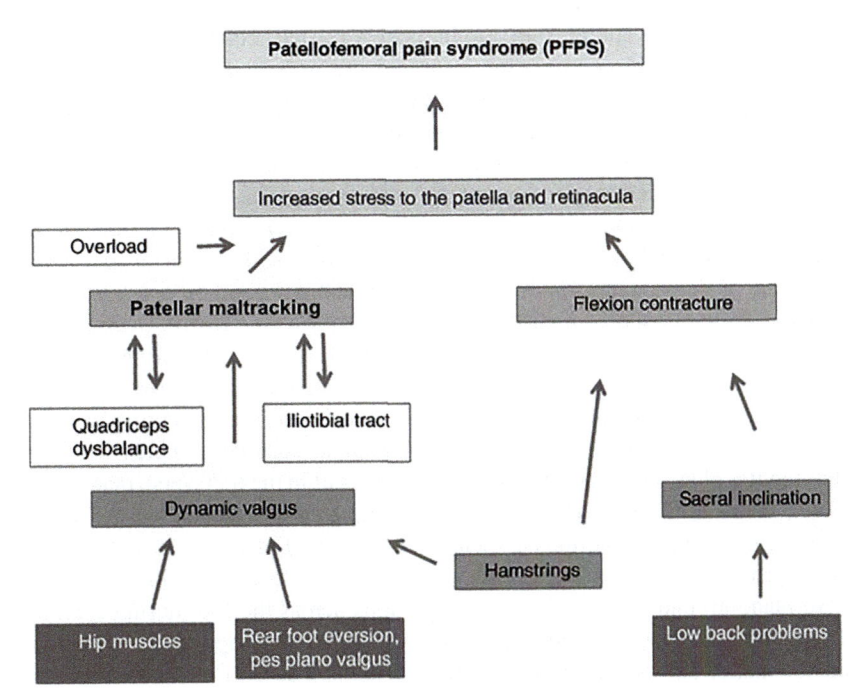

FIG. 4.3.8j The origins of patellofemoral pain are impossible to pin down to a single pathomechanical cause. Here, an algorithm for summarising potential patellofemoral pathomechanics has been developed from the published literature, which many explain the many potential pathways to pathogenesis for patellofemoral symptoms. *(Image from Petersen, W., Ellermann, A., Gösele-Koppenburg, A., Best, R., Rembitzki, I.V., Brüggemann, G.-P., et al., 2014. Patellofemoral pain syndrome. Knee Surg. Sports Traumatol. Arthrosc. 22(10), 2264–2274.)*

4.3.9 Section summary

Pathomechanics of the knee relates to the nature of the pathology, so a good, sound working diagnosis is essential to the unravelling of dysfunction. There are no such diagnoses as 'knee pain', 'medial knee pain', or 'anterior knee pain'. These are symptoms with many potential origins. Initially, knee pathomechanics should be approached by considering the mechanics of each of the functional joints separately, and then as a coupled unit. The tibiofemoral joint should also be considered through its distinct medial and lateral compartments and their anatomy. Anterior–posterior gliding dysfunction dominates sagittal plane issues but loading the knee at high or low flexion angles changes which soft tissues are under the most strain, both around the tibiofemoral joint and patellofemoral joint. Thus, injury to a specific identified soft tissue helps focus kinematic attention to the correct gait events. For example, peripatellar retinaculitis is more likely to be a diagnosis in patellofemoral dysfunction associated with the knee in more extended positions, whereas an injury to the musculotendinous units of the quadriceps is more likely to occur at increased knee flexion angles. Patellofemoral cartilage damage is also more likely to occur under compressive forces in the sagittal plane associated with high or prolonged knee flexion angles.

For OA of the tibiofemoral joints, frontal plane torques become the primary focus with medial degeneration guiding attention to the extent of varus/adduction moments on the knee, and lateral compartment degeneration focusing attention to valgus/abduction moments. Concurrent sagittal instability, excessive anterior–posterior shear, or transverse plane dysfunction will worsen the situation. When reviewing transverse and frontal plane dysfunction, the hip should also become a focus to studying kinematic events at the knee. In understanding knee pathomechanics, once a diagnosis is established, then the local functional abilities of the knee must be assessed. Following knee assessment, the stability proximally at the hip needs to be considered, before moving distally to the ankle and foot. Consider potential ascending or descending origins to pathomechanics from data and presentations found during the clinical assessment.

Although frontal plane motions of the rearfoot may not be a common significant source of knee pain, the placement of the foot to create a base of support in relation to step width and angle of gait are certainly important factors in influencing knee biomechanics and energetics during stance. Thus, a more gross overview of ankle and foot placement and stability during gait is likely to be more effective in assessing knee moments (and for that matter hip moments) than overly focusing on rearfoot frontal plane angles.

4.4 Pathomechanics of the ankle and rearfoot
4.4.1 Introduction

When considering the anatomy as part of the functional unit, the ankle was uneasily separated from the rearfoot in Chapters 2 and 3. However, when considering the pathomechanics of the ankle, it is impossible to extract the subtalar joint complex from the ankle, as the kinematics of the whole ankle–subtalar complex are too intimately related. The main consideration pertains to the subtalar joint's lack of freedom of motion replaced by a ligament-dependent flexibility induced through external forces. Essentially, for the subtalar joint to have kinematic significance, the calcaneus via the heel must be externally loaded and, most commonly, this is a result of heel contact during early stance phase. The muscles that bridge the ankle–rearfoot are therefore biomechanically influenced by the presence or lack of any heel contact.

For a forefoot runner or permanent toe-walker, all planes of motion around the rearfoot complex are supplied through the ankle alone during contact. The more plantarflexed the ankle, the more frontal and transverse plane motion can be supplied as motion within the ankle is far more restrained when within dorsiflexion angles (Leardini et al., 2001; Leardini and O'Connor, 2002). Indeed, this is true of the subtalar joint also. When the heel is loaded by external forces during heel strike to late midstance, the subtalar joint can supplement the range of motion of the rearfoot complex in all planes. However, this subtalar motion decreases with increase in angles of ankle dorsiflexion so that during late midstance its motion becomes increasingly smaller (Leardini et al., 2001). With greater rearfoot flexibility provided within ankle plantarflexion angles, high-heeled shoes that create loading of the rearfoot in plantarflexion are presenting an interesting environmental mismatch danger for increased weightbearing motion within the ankle–subtalar complex.

The primary motions of the foot in gait are sagittal plane motions followed by motions within the transverse plane, with the frontal plane providing the least (Hunt et al., 2001). Thus, the rearfoot complex is a triplanar unit that allows the foot to adapt to the support surface, but with dominance in segmental motions associated with sagittal plane ankle motion. The ankle remains responsible for most sagittal motion within the rearfoot during stance, with the forefoot and midfoot providing increasingly for frontal and transverse plane ground perturbations or adaptations across the foot during forefoot loading, midstance, and terminal stance. However, the midfoot also provides small, yet significant, sagittal plane motions that help dissipate energy and influence joint kinematics between the rearfoot and MTP joints. The MTP joints are the other main provider of sagittal plane foot motion. Subtalar joint motion is available for frontal and transverse plane roles primarily during heel loading and early midstance, when the midfoot and forefoot are less able to provide for these motions.

The intimate relationship between the rearfoot and the rest of the foot (particularly the midfoot) is highlighted by research focusing on changes within the foot after ankle surgery and injury. Using finite element analysis models, loss of ankle dorsiflexion after arthrodesis seems to cause peak plantar pressures to rise and the CoGRF to move more anteriorly increasingly rapidly within the foot during gait, compared to feet with free ankle joint motion (Wang et al., 2015). The talonavicular joint and the joints of the 1st to 3rd rays in the midfoot bear the majority of this extra loading, resulting in raised metatarsal bone stresses, probably due to the higher MTP joint extension angles and foot plantarflexion angles required at heel lift (Wang et al., 2015). With lost ankle motion, contact force and pressure decreases within the subtalar joint, but rises in the medial midtarsal joints. This suggests that arthritis in the medial midfoot is a risk after ankle arthrodesis, but any subtalar joint arthritis arising from resulting lost ankle motion probably represents preexistent osteoarthritic changes, rather than being a consequence of an ankle arthrodesis (Wang et al., 2015). These findings are supported by data on changing foot and lower leg kinematics in patients suffering from posttraumatic ankle arthritis, with planus feet tending to fare better than cavus feet (Deleu et al., 2021). More cavoid feet with an inverted arthritic rearfoot attempt to use midfoot abduction and more frontal and transverse plane lower leg motion to compensate for loss of ankle freedom, whereas the planus foot type, with more midfoot dorsiflexion being potentially available, may better compensate for the loss of ankle dorsiflexion in the sagittal plane (Deleu et al., 2021). Thus, the ankle and the foot's intimate relationship is confirmed (Fig. 4.4.1).

4.4.2 Sagittal plane free-body diagrams of the rearfoot complex

The rearfoot complex is the main provider of the foot's sagittal plane motion primary via the ankle. There is also some anterior–posterior gliding between the talus over the calcaneus at the subtalar joint, as the distal calcaneus plantarflexes. Together, this acts to lower the calcaneal inclination angle, creating some further sagittal plane motion within the rearfoot other than from the ankle. During the swing phase, the foot only pivots around the leg via the talus at the ankle. The leg also pivots around the ankle during stance but actually does so through the combined ankle–subtalar joint motion when the heel is in ground contact, primarily during early stance. The rearfoot complex provides complementary ranges of transverse and frontal plane motions that should improve the foot's ability to adapt to different terrain angulations and perturbations, while continuing to facilitate sagittal plane motion of the leg. These motions can also be used to fine-tune energy dissipation, weight transfer, and acceleration power. In walking, the ankle joint has four distinct sagittal plane lever arm functions. Three of these involve the use of class three levers, where mechanical efficiency is compromised for muscular metabolic efficiency and greater ranges of motion for less changes in muscle–tendon fibre length. Acceleration at the end of stance uses a mechanically efficient class two lever and thus avoids much of the need for higher metabolic muscular activity. Instead, heel lift utilises the released energy that has been stored within the Achilles tendon, rather than demanding concentric muscle contraction.

Swing phase

The foot is moved from its final plantarflexed position at the end of terminal stance during the preswing-swing transition towards dorsiflexion, primarily through tibialis anterior activity. The mass of the foot and the distance it moves is not great and indeed, during swing, the foot tends to remain at less than 90° angled to the lower leg rather than actually achieving a dorsiflexed position. Overall, this activity reduces the ankle plantarflexion angle by around 10° from toe-off position (Perry, 1992, p. 57). With such a small mass to move, tibialis anterior's stresses across the rearfoot complex is low during swing, with activity of the muscle generating only around 35% of maximal muscle tension forces (Byrne et al., 2007). This low stress action is unlikely to be associated with any overuse injury and foot tissue thresholds should not be breached during swing.

Swing phase dorsiflexion becomes of more clinical interest in the presence of deep peroneal nerve or L4-S2 nerve root damage, that can cause nerve-induced tibialis anterior weakness or flaccid paralysis foot drop. If foot drop is seen in gait, it must be accounted for. If this is a new development, urgent referral for spinal diagnostic imaging is required via the fastest route possible which might be via emergency departments, especially if associated with bladder and bowel disturbances when *cauda equina syndrome* must be considered. Rupture of tibialis anterior is a rare cause of foot drop in swing (the author has identified one in +30 years of practice). If a diagnosis of tibialis anterior rupture can be clinically confirmed, the situation is not urgent, but will require surgical opinion. If there is any doubt that the tibialis anterior is ruptured, then the foot drop should be considered a neurological emergency until serious nerve damage has been ruled out (Fig. 4.4.2a).

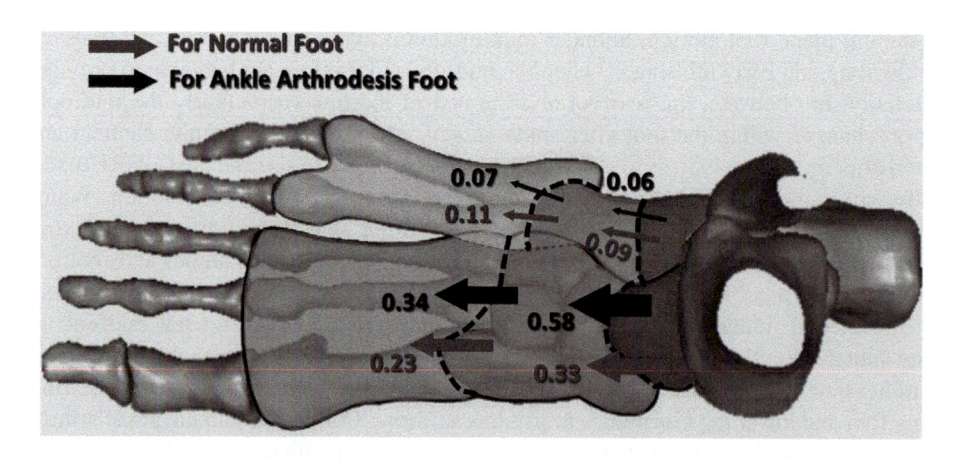

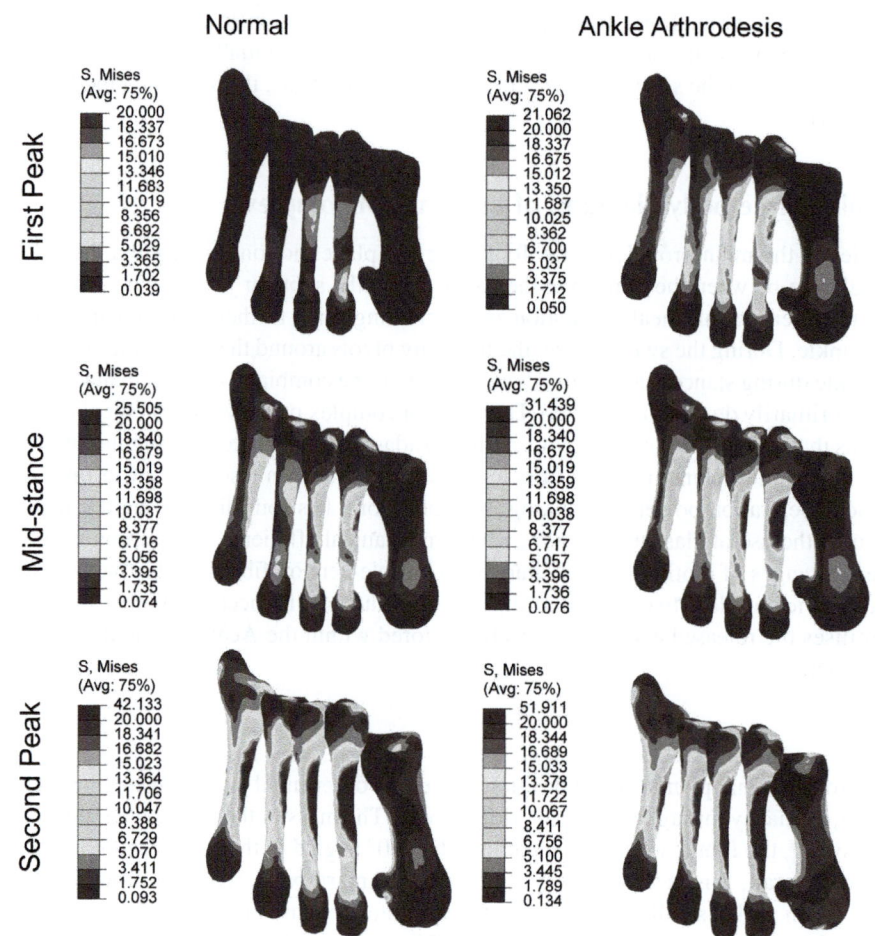

FIG. 4.4.1 Ankle arthrodesis takes away all joint motion, giving insight into the role of the ankle joint through the effects of its motion not being available. The upper image demonstrates that at loading response within a foot containing a functional ankle, the loads transferred as a multiple of body weight (grey arrows) are 0.33 times body weight at the talonavicular joint and medial lesser tarsus. They are reported at 0.23 times body weight across the medial tarsometatarsal joints. Forces are much lower at 0.09 times body weight across the calcaneocuboid joint and 0.11 at the lateral tarsometatarsal joints. When ankle joint motion is lost (black arrows), forces increase medially, becoming 0.58 times body weight at the talonavicular and medial lesser tarsal articulations, and 0.34 at the medial tarsometatarsal joints. However, they lower slightly laterally. Forces across the metatarsals (lower images) show that normally functioning ankle joint motion spares the central metatarsals from higher loads compared to cases of ankle arthrodesis during loading response and throughout midstance, helping reduce peak loads on the medial metatarsals during heel lift. These effects may exist on a continuum so that any loss of freedom in ankle motion changes the loads across the articulations of the foot. Such coupled mechanisms would help to link pathology expressed within the foot (such as talonavicular joint degeneration or second metatarsal stress fractures) to loss of ankle joint motion during locomotion. *(Images from Wang, Y., Li, Z., Wong, D.W.-C., Zhang, M., 2015. Effects of ankle arthrodesis on biomechanical performance of the entire foot. PLoS ONE 10(7), e0134340.)*

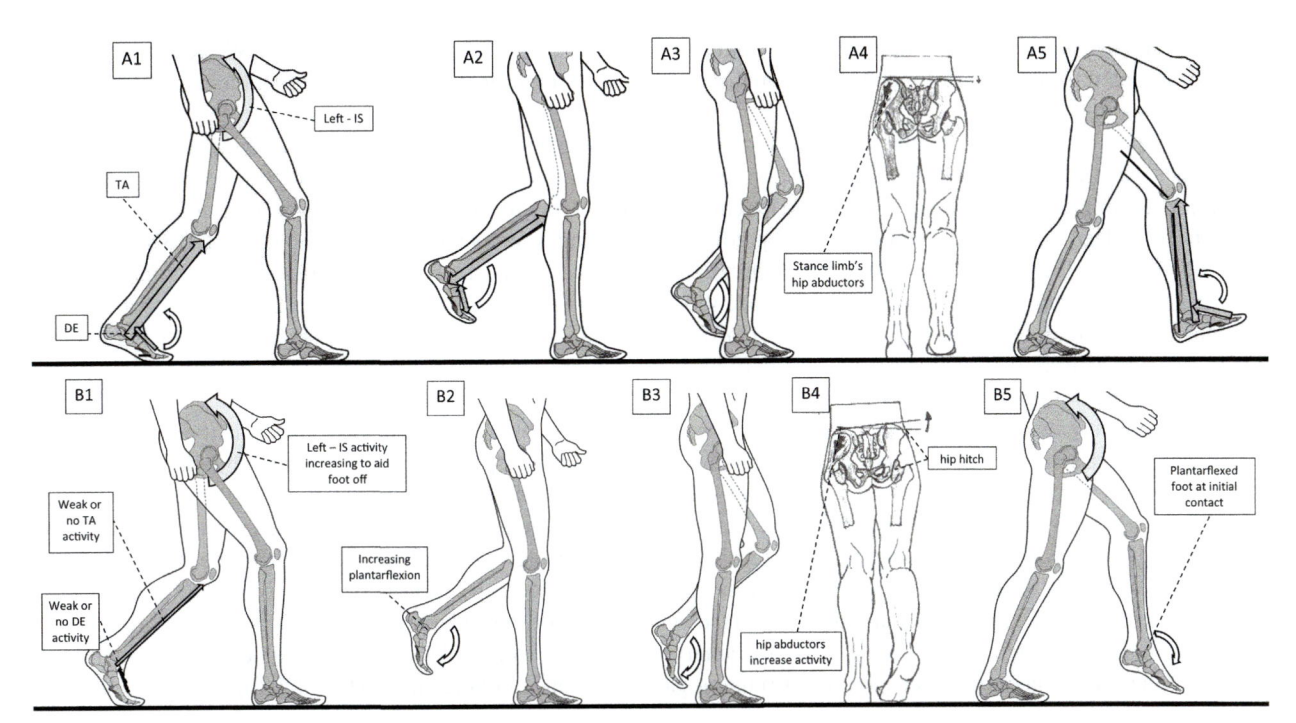

FIG. 4.4.2a Activity around the ankle is important to swing energetics, despite its range of motion appearing relatively small. Normally, (A1) the ankle dorsiflexors, led by tibialis anterior (TA) and aided by the digital extensors (DE), become active in preswing with iliopsoas (IS). As IS initiates hip flexion, TA begins reducing the ankle plantarflexion angle left at the end of acceleration, with the DE (including the extensor digitorum brevis intrinsic foot muscle) helping maintain digital extension angles from the end of terminal stance. These actions together shorten the functional limb length, aiding ground clearance during early swing (A2). The ankle extensors hold the ankle short of dorsiflexion angles, in isometric contraction. This angle is maintained throughout midswing into terminal swing (A3). By maintaining ankle in neutral or in marginal plantarflexion with knee flexion, the hip passes under the trunk with the swing limb shortened sufficiently to guarantee ground clearance as the pelvis also undergoes a small adduction motion on the stance limb's hip (A4—posterior frontal plane view). As the end of terminal swing is approached (A5), TA and DE activity increases, sometimes reducing the plantarflexion angle further and starting to extend the toes to aid in activating foot stiffness via the windlass mechanism. In anterior muscle group dysfunction (B), swing phase ground clearance is disturbed. Lifting the foot after preswing is problematic, requiring greater IS activity at the hip (B1). However, the ankle remains plantarflexed in swing which, if the DE are also affected, can actually increase its plantarflexed angle during early swing. Early swing (B2) and midswing (B3) require limb clearance compensation, usually achieved by abducting the stance limb's hip (B4) to create a hip 'hitch'. In terminal swing (B5), increased hip flexion and reduced knee extension are usually necessary to advance the swing limb in readiness for initial contact on the forefoot. Peroneal nerve neuropathy is the most common cause of foot drop although a spinal cord lesion is a dangerous source of many but tibialis anterior tendon ruptures are also recorded. The presence of concurrent loss of DE reflects the extent neuropathy and rules out TA rupture as the source of the problem. *(Permission www.healthystep.co.uk.)*

Pre and initial contact

Sagittal plane stance's contact phase motion usually begins with heel strike. The plantarflexion motion initiated at the rearfoot is primarily the responsibility of the tibialis anterior until forefoot loading is achieved. The mechanics of tibialis anterior operating in a class three lever system has already been discussed in Chapter 2, Section 2.5.9. It is a foot-braking mechanism that, excepting during fast walking and running, tends to involve low forces. Yet, its activity is also important for the braking of forefoot contact that helps absorb energy from impact through muscle–tendon buffering. This ability is lost with tibialis anterior dysfunction.

Just before heel strike and during initial contact, tibialis anterior expresses peak activity. This is to prepare for the expected plantarflexion moment generated around the ankle joint axis by initial ground contact with the posterior heel. This activity should be combined with that of tibialis posterior and peroneus longus, assisted by the long toe flexors and abductor hallucis, that together are increasing plantar aponeurotic tension and stiffening the foot. This muscle and fascia-induced stiffening can then provide the reversed windlass mechanism to aid energy dissipation and release foot vault compliance during the forefoot collision. Any part of this team of muscles failing to activate and reduce activation appropriately, potentially compromises impact energy dissipation within the foot. The loss of other muscle actions before and during forefoot loading opens up the possibility of a tibialis posterior tissue threshold breach. This is because it provides the main foot vault stiffening mechanism and must control the rate and amount of compliance that is release from the vault.

706 Clinical biomechanics in human locomotion

Failure in tibialis posterior opens up the risk of injury to other muscles active at the same time. Weakness, medially or laterally among this complex of functional anatomy risks increased rearfoot eversion or inversion moments, respectively, across the loading foot.

Tibialis anterior addresses the plantarflexion moment by resisting the velocity of forefoot flexion induced by heel strike. The longer the stride, the longer the plantarflexion moment arm. The faster the gait, the greater the acceleration to resist. Thus, fast walking or running with a long stride length is far more likely to stress tibialis anterior than slow walking with a short stride. Indeed, long stride length is a good predictor of running injuries anyway (Bramah et al., 2018), undoubtedly from the multiple effects of the increasing GRF's lever arm. The relative resistance load opposed by tibialis anterior at heel contact is small, although a heavy shoe can increase this. In walking, tibialis anterior only brakes motion that brings the forefoot to the ground, requiring ~7° of plantarflexion from the ankle that is positioned from an initial contact angle somewhere between neutral to 5° plantarflexed. This normally only requires around 45% of maximal isometric muscle strength from tibialis anterior (Byrne et al., 2007).

Serious tibialis anterior tendinopathies and myopathies are not common. However, in those exposing their tibialis anterior to a sudden increase in stress and fatigue (such as individuals initially taking up running), symptoms within the muscle–tendon unit occur with surprising frequency. Common acute examples of provoked anterior shin symptoms are encountered in situations where an individual has had to walk fast across a city in low-heeled shoes on unyielding urban surfaces, probably utilising long strides for longer periods than usual for that individual. Being a postural muscle, fatigue probably represents the biggest risk to tibialis anterior's action. However, its covering of tight deep fascia may be responsible for much of the symptoms, as any post exercise oedema in and around the muscle has little expandable space to occupy. In running heel strike, the forces initiated through greater speeds leave the anterior shin more vulnerable to pathomechanics, and this will be discussed in more detail later in this chapter under running pathomechanics in Section 4.6 (Fig. 4.4.2b).

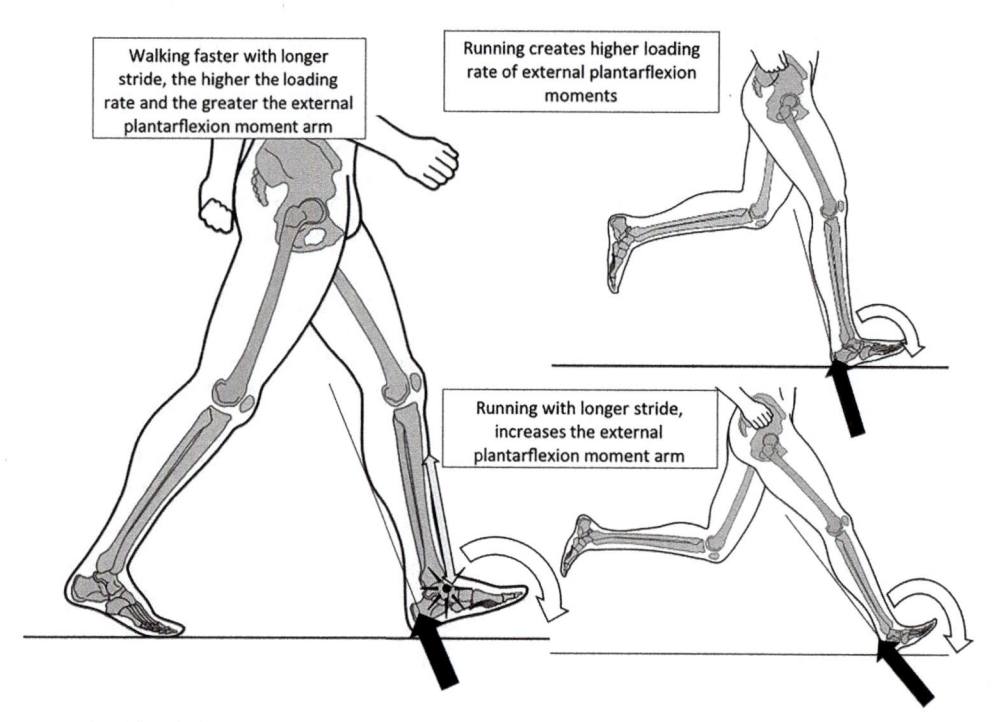

FIG. 4.4.2b Acute anterior shin pain is a common complaint noted after periods of extended fast walking with long stride lengths, especially on hard ground. This usually relates to the increased plantarflexion moment arm from the GRF to tibialis anterior and the loading rate of this moment on the muscle. Those who have recently taken up running are particularly prone as the increased velocity at heel strike induces greater ankle plantarflexion moments, significantly greater than those of walking. Thus, running sets up stresses outside the experience of normal walking strengths of tibialis anterior. Those who attempt to develop a long stride within their running pattern are particularly at risk because of the increased external moment arm from the GRF to the ankle, as a result of the increased dorsiflexion angle of the foot to the ground. *(Permission www.healthystep.co.uk.)*

Tibialis anterior is given some ankle plantarflexion braking assistance by the long and short toe extensors. The toe extensors primarily provide toe extension for the windlass mechanism's 'setting' for appropriate foot stiffness before forefoot loading in conjunction with tibialis anterior, tibialis posterior, and peroneus longus activity, as has already been discussed. In this capacity, the extensors are possibly aided by abductor hallucis. Tibialis anterior weakness risks digital extensor compensations, known as *extensor substitution*, which may cause toe deformity by repeatedly overextending the MTP joints (Fig. 4.4.2c).

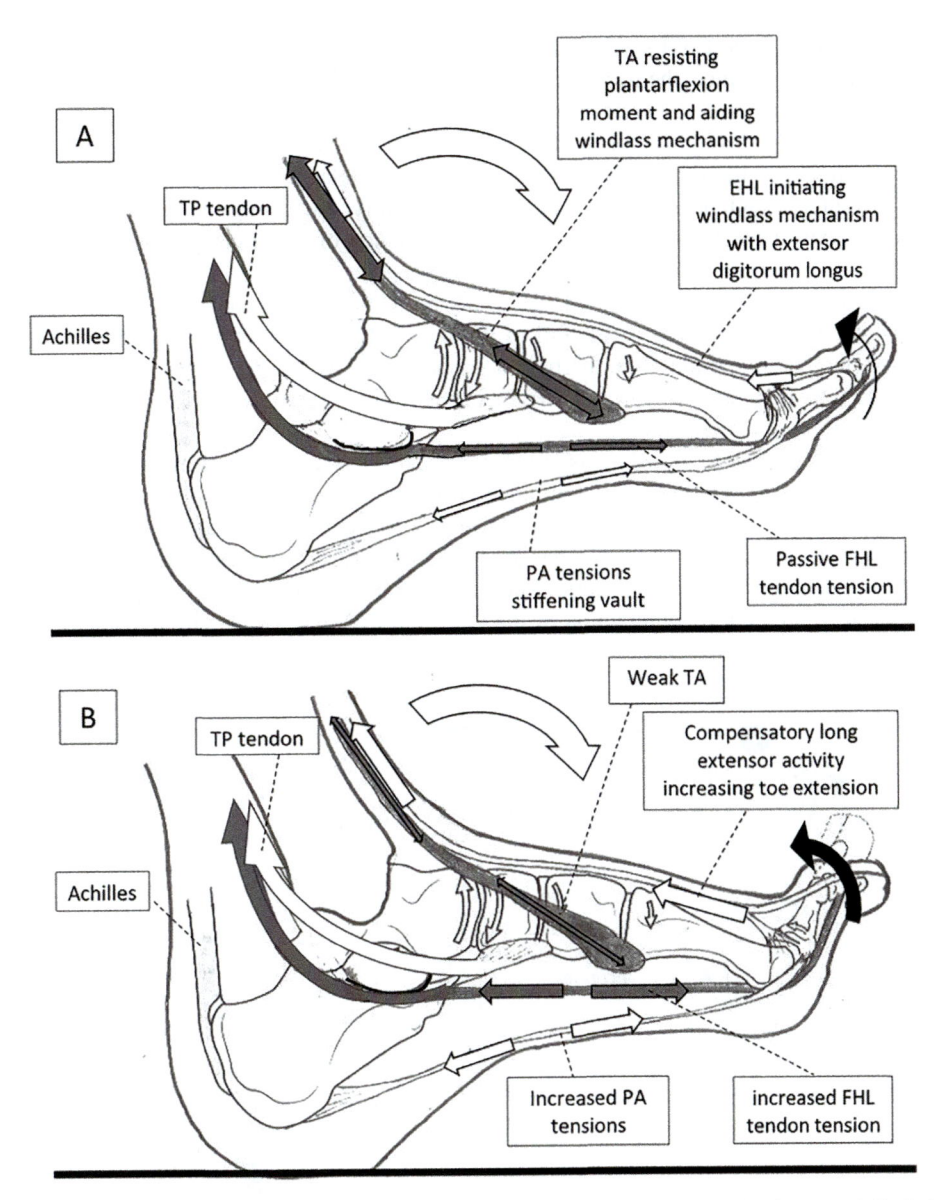

FIG. 4.4.2c The ankle extensor muscles play an important role in setting the posture for initial contact of the foot, especially during heel strike (A). In combination with stride length, the ankle angle is set by tibialis anterior (TA), but with digital extensors, such as extensor hallucis longus (EHL) assistance. Just before and during initial contact, this activity helps stiffen the foot via initiation of the windlass mechanism due to digital extension and raising of the foot vault. In this vault-stiffening action, tibialis posterior (TP) and peroneus longus (not shown) are also at play. Under the influence of the windlass mechanism, the plantar aponeurosis (PA) and digital flexor tendons such as flexor hallucis longus (FHL) tendon are tensioned. If TA is weak (B), the long extensors can increase their activity to try and set the windlass mechanism and attempt to decelerate the ankle plantarflexion moment. This requires far greater digital extension moments that are resisted by flexor tendon and PA tensions. Unresolved, long-term use of long digital extensor muscles for TA compensation can lead to retracted toe deformities. However, opposing tensions from the extensor and flexor attachments tend to result in MTP joint extension but interphalangeal (IP) joint flexions. Such long digital extensor compensations can also occur during preswing and are referred to as extensor substitution. High-vaulted cavoid feet seem more prone to such retracted toes, probably because of the higher metatarsal declination angles that they function with that tend to cause the digits to naturally operate at higher extension angles. *(Permission www.healthystep.co.uk.)*

Midstance

Once midstance begins in walking, following completion of forefoot loading response, ankle plantarflexion angles start to decrease as the proximal tibia begins its anterior displacement. The trunk moves anteriorly through hip and knee extension moments from gluteus maximus and quadriceps activity, respectively. During this phase, gluteus maximus is primarily responsible for the forward progression of the body's CoM, but as this process approaches its termination around absolute midstance, the responsibility of the CoM's forward progression changes from the hip to the ankle through triceps surae activity. At the end of early midstance, the triceps surae (particularly soleus) starts to activate in preparation for this role. However, the rearfoot complex lever arm system that limits ankle dorsiflexion does not become fully effective until after absolute midstance, as the CoM tips in front of the ankle. During early midstance, under decreasing plantarflexion angles, the ankle becomes progressively more restricted in its own and the subtalar joint's freedom of motion. By the start of late midstance, the ankle joint's instantaneous axis has advanced anteriorly, increasing the internal moment arm between the Achilles and the ankle axis. This improves the mechanical efficiency of the triceps surae to counteract anterior tibial rotation.

After absolute midstance during walking, the CoM of the body moves anterior to the ankle joint axis. This initiates dorsiflexion (extension) in the ankle joint as the fall of the inverted pendulum begins under the acceleration action of gravity. This anterior progression of the tibia, thigh, and upper body needs to be controlled and decelerated in order to give time for the swing limb (now moving towards terminal swing) to position itself in readiness for the next step and load transfer before the stance limb's heel lift.

Eccentric contraction of the triceps surae begins in a class three lever system that permits controlled progress of the CoM over the midfoot towards the forefoot, under the combined influence of gravity and an increasing CoM momentum of the swing limb's centrifugal forces. Effort is applied in this lever system via the proximal attachments of the soleus and gastrocnemius. The fulcrum lies at the ankle axis (which is moving anteriorly) and the resistance is the mass of segments above the ankle. As the gastrocnemius bridges the posterior knee, its contraction should apply a knee flexion force to resist knee extension moments that result from the interaction of the GRF and CoM vectors in late midstance (Fig. 4.4.2d).

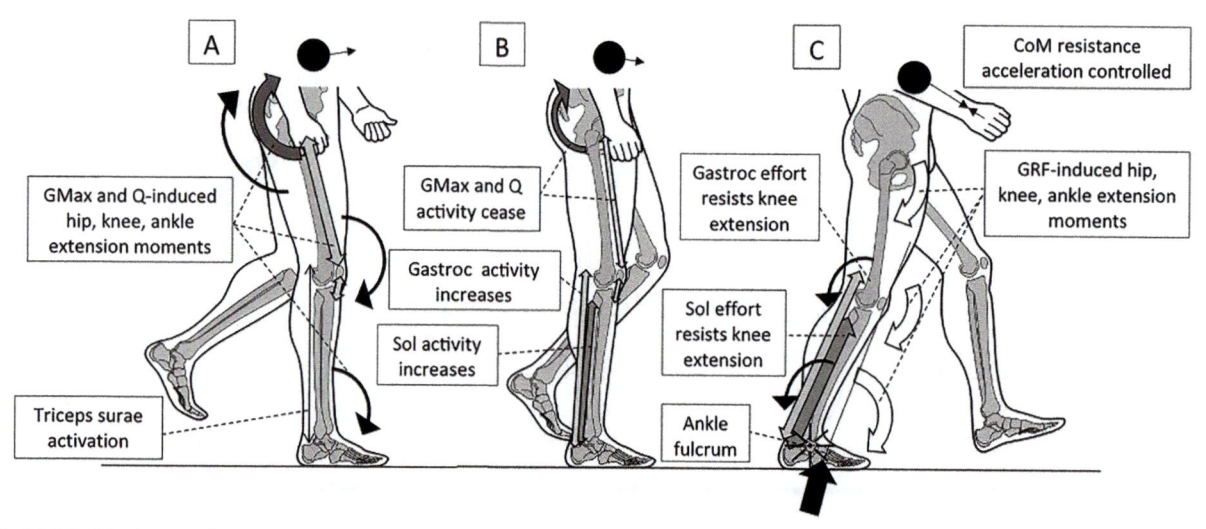

FIG. 4.4.2d Via the two-hinged inverted pendulum that the lower limb functions as in walking single-limb support, acceleration of the CoM over the limb is controlled initially by the hip and then at the ankle from absolute midstance. For this action, only slight mobility but particularly stability is required at the knee. At the end of early midstance (A), the acceleration of the CoM is still part of limb extension via gluteus maximus (GMax) at the hip and the quadriceps muscles (Q) at the knee. However, already there is increasing activity within the muscles of triceps surae (thin double-headed arrow). By absolute midstance (B), GMax and Q activity has largely ceased, with the CoM now at its highest point and the limb about to come under external forces that drive hip, knee, and ankle extension. Soleus (Sol) and gastrocnemius (Gastroc) now increase activity to act as brakes as the CoM begins its fall anteriorly. In late midstance until heel lift (C), both Sol and Gastroc are decelerating CoM acceleration. Together this sets up a class three lever than controls the CoM acting as the resistance over an ankle fulcrum, while the knee is kept stable and hip rotation controlled by hip abductor activity. However, Sol is primarily responsible for CoM acceleration control by influencing the rate of ankle dorsiflexion, while Gastroc is also responsible for preventing knee extension (providing knee stability) as it assists Sol at the ankle. Both muscles load the Achilles with tensile strain, storing energy within their respective subtendons. This means that knee and ankle angles created during gait will affect the ability of each muscle to place energy within the Achilles for acceleration. Thus, dysfunction of Sol causes different disruption in biomechanics from abnormal activity within Gastroc. *(Permission www.healthystep.co.uk.)*

The anterior movement of the ankle joint axis increases the triceps surae's lever arm and improves the mechanical efficiency, but the resistance moment arm increases at a faster rate throughout the late midstance phase as the resistance of the CoM moves anteriorly. This eccentric triceps surae activity, combined with the Achilles high energy storage properties, allows considerable energy to be stored within the myofascial-tendon unit as the Achilles is lengthened and stiffened. This stored energy can be released as kinetic energy, producing ankle power at heel lift without the need for extensive muscle contraction. Increased Achilles fibre 'twisting' and lengthening stress-stiffening are together associated with ankle dorsiflexion. This combined 'twist-stretching' action increases Achilles tendon stiffness throughout late midstance, improving the energy storage potential within the collagen fibres as long as the ankle dorsiflexion angle continues to increase. Before heel lift, the body's CoM should be anterior to the foot, positioned perfectly for mechanical efficiency at the start of terminal stance to tip the CoM to the next step of support under the action of this elastic spring.

Any failure in ankle dorsiflexion range or any triceps surae–Achilles complex dysfunction risks a compromise during midstance to position joints and the CoM correctly for energetically efficient terminal stance acceleration. If either sagittal plane facilitation of the CoM requires or causes increased vault height drop rather than free ankle dorsiflexion or if the ankle joint is blocked from achieving higher dorsiflexion angles, then the Achilles tendon stiffness, the energy storage capacity, and the joint positioning around the ankle may be disadvantaged for acceleration (Fig. 4.4.2e).

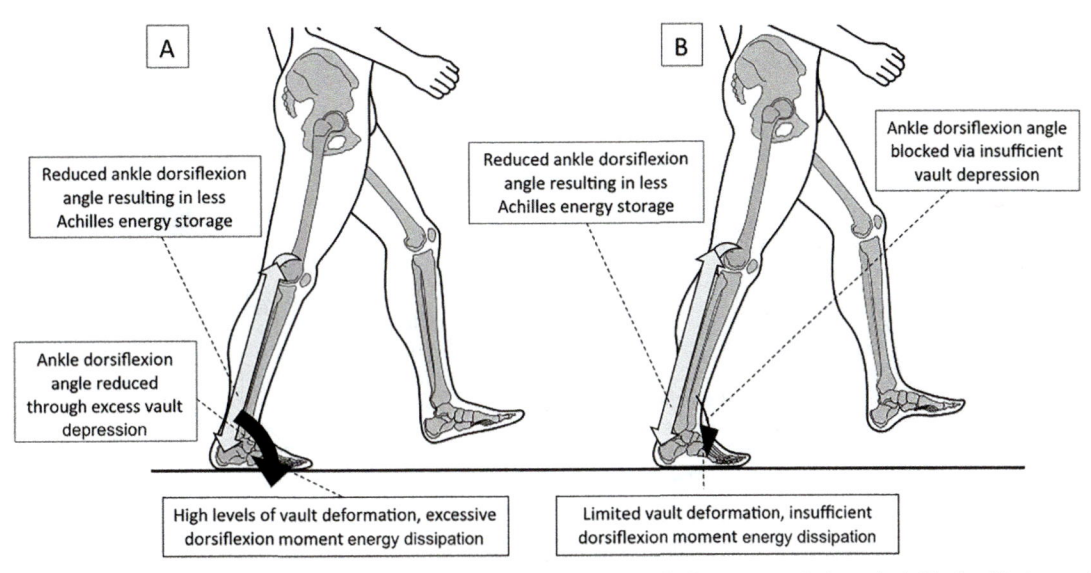

FIG. 4.4.2e The triceps surae–Achilles complex and the foot vault can both store and dissipate energy during gait. Ankle dorsiflexion angles and the amount of vault deformation work together to moderate and balance the ankle dorsiflexion moment energies to avoid tissue damage, while also improving late stance energetics. During late midstance, their synergistic relationship should generate more energy storage than dissipation for use during the initiation of acceleration. Excessively mobile foot vaults (A) associated with hypermobile planus feet or weak foot musculature can result in the dorsiflexion moment being increasingly dissipated via vault deformation. This may reduce Achilles strains as a result of a reduction in ankle dorsiflexion angle necessary before heel lift, further reducing energy storage potential before acceleration. Stiff, cavoid feet (B) can also limit ankle dorsiflexion via loss of vault lowering that normally permits some talar plantarflexion to occur at the talocalcaneonavicular joint. This can cause the ankle to reach its close-packed position prematurely, limiting the ankle dorsiflexion angle. The result is reduced energy dissipation within the vault and possibly less Achilles energy storage as a result of reduced ankle joint dorsiflexion straining and stiffening the Achilles. This can lead to an early heel lift to maintain the CoM's progression, requiring greater triceps surae muscle fibre shortening. Thus, foot vault articulations are unable to avoid high energies because of insufficient flexibility to deform and dissipate energy via controlled vault flexibility. *(Permission www.healthystep.co.uk.)*

Terminal stance

At heel lift, the triceps surae, particularly soleus, changes from a structure that decelerates the progression of the CoM to an assistor of CoM acceleration, aiding the swing limb's centrifugal force and the effects of gravity. The power it provides into terminal stance later works with the gluteus maximus to accelerate full transfer of all weight off the terminal stance foot to the next step. Ideally, heel lift should occur at approximately the point where the swing limb heel is ~1 cm from the support surface. This allows the stance foot's heel lift to quickly aid swing limb heel drop and start to accelerate weight transfer onto the contralateral foot. This heel lift should thus occur very briefly before opposite foot contact.

To achieve efficient heel lift, the ankle lever system becomes class two, with the fulcrum point moving from the ankle to the MTP joints. This means that the internal lever arm of the Achilles' effort point to the ankle's instantaneous axis and the external lever arm from the ankle's axis to the MTP joints are effectively combined together, creating a longer lever arm of plantarflexion power to the MTP joint axis. This does not mean that the internal Achilles moment arm is no longer significant. The medial MTP joints offer a longer external lever arm and thus a better efficiency for ankle plantarflexor power application than the lateral. A class two lever offers higher mechanical efficiency, but at the cost of a reduced range of motion for muscle fibre shortening and thus, increases muscular metabolic efficiency, despite the better mechanical efficiency. This should not be a problem, as only a few degrees of reduced ankle dorsiflexion are necessary for creating plantarflexion that should tip the contralateral heel towards the floor, initiating weight transfer of the body to the opposite foot. Also, heel lift is achieved primarily by passively reducing the heel loading forces to allow release of the stored Achilles tendon energy, thus requiring little if any muscular contraction for actual lifting. The elastic recoil of the Achilles coupled with the small range of motion under decreasing load spares the triceps surae from any high metabolic concentric contractions during walking acceleration (Fig. 4.4.2f).

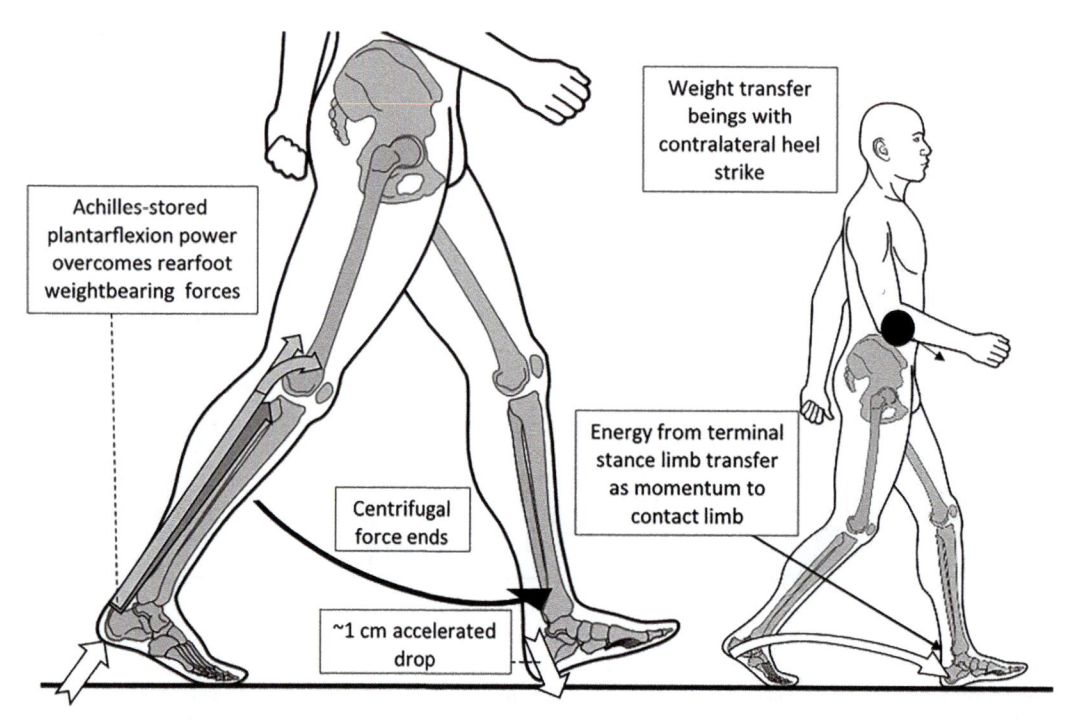

FIG. 4.4.2f Heel lift by powered ankle plantarflexion is the act of transferring acceleration energy stored within the tissues of the stance limb into HAT segmental motion, so it can be utilised in maintaining CoM velocity into the next step. This extra power should arrive just as the swing limb's centrifugal force reaches the end of its arc of swing, threatening to start a back-swing on the limb that could decelerate CoM velocity. This is why claw back (caused by a slight back-swing of the swing limb before heel contact) may be a useful way to mildly reduce vertical impact forces at heel strike. For energetics to be maximised, ankle plantarflexion should initiate in a Goldilock's period when the swing limb's foot is ~1 cm from the ground so that heel raise ends contralateral swing and starts weight transfer. For heel lift, the power stored within the Achilles should be greater than the weightbearing forces remaining on the heel as the swing limb approaches the end of its arc of swing, which is pulling the CoM forward reducing the weightbearing forces on the heel. Thus, swing limb velocity and controlled ankle dorsiflexion freedom during late midstance set the parameters for heel lift efficiency. As important is the ability to create appropriate stiffness within the foot so that plantarflexion power can be applied from the ankle to the forefoot to facilitate heel lift. A semirigid foot prevents excessive energy loss from the Achilles power via foot deformation. The ankle is thus the most important joint for the application of acceleration power, and its dysfunction will always compromise acceleration biomechanics. *(Permission www.healthystep.co.uk.)*

To maintain a low metabolic cost on the plantarflexor muscles in achieving this task, it is essential that the foot provides adequate stiffness to act as a stable beam at heel lift, otherwise energy will be lost within a compliant beam's deflection, increasing energy dissipation, and thus losing much of the plantarflexor-generated power to the forefoot. Increasing foot stiffness raises the peak GRF under the forefoot, as the power from the plantarflexors can be transferred across the stiffened foot to aid acceleration, while also increasing the plantarflexor moment effort arm (Takahashi et al., 2016). Thus, a strong stabilising force is applied to the forefoot as the rearfoot, knee, and hip flex as they move towards preswing.

For this system to work during walking, while also avoiding risks of tissue threshold breaches, heel lift should only transfer the weight of the lower limb and rearfoot that remains posterior to the MTP joints. This should occur at a time when the bulk of body mass is well head of the stance foot, falling anteriorly under gravity. It should not be required to move all the body mass. Therefore, the rest of body weight (CoM) should be tipping forward behind the terminal swing limb under gravity and centrifugal forces, just before ground contact. Heel lift should primarily help to add power to the contralateral foot's impact collision, helping initiate the inverted pendulum on the opposite limb as described through the rimless wheel model proposed by Ruina et al. (2005). Thus, high power derived from the ankle plantarflexor muscle contractions should not be required during and after heel lift (during the F3 force peak and its downslope).

Muscle power generated around the ankle in late midstance is stored within the connective tissues, particularly in the energy storage specialist, the Achilles tendon. This is released at heel lift, making heel lift primarily a passive elastic recoil event without significant metabolically induced muscle fibre length changes. The shortening of the triceps surae is thus more a result of a connective tissue elastic recoil event than a metabolic-induced muscle fibre contraction. The result should be that during walking, heel lift and terminal stance are an efficient CoM accelerator with minor active extrinsic muscle contraction required outside of a slight burst of gluteus maximus activity approaching preswing. The intrinsic foot muscles and long flexors are, however, required to maintain activity into late terminal stance to maintain a stable vault and forefoot. Once the heel has lifted, ankle plantarflexion and vertical displacement occurs through changes in hip and knee flexion posture, that is increased later via iliopsoas activity. The foot continues to plantarflex at the midfoot and ankle through the elastic recoil of tibialis posterior and peroneus longus tendons, and digital flexor activity.

4.4.3 Pathomechanics of weight transference and acceleration

The plantarflexion power that accelerates the CoM in terminal stance arises from the muscles that cross the ankle posterior to its axis. The power is built up during late midstance through increasing muscle activity that peaks just before heel lift. Triceps surae provides almost all of this power, but other contributing muscles also have important roles in stiffening the foot as previously discussed. The flexor extrinsic muscles of the digits are near-isometrically active during midstance and terminal stance (Hofmann et al., 2013; Péter et al., 2015). This action allows them to aid in the resistance of vault depression, helping to create plantar anterior–posterior compression forces across the skeletal load frame against sagging deflection bending moments developed during weight transference over the foot. This digital plantarflexor activity also aids the plantarflexion power generation at heel lift, while stabilising the MTP joints extension rotation during CoM acceleration. Tibialis posterior and peroneus longus generate their plantarflexor role through their midfoot plantarflexion moments via stiffening the foot before and during heel lift (Kokubo et al., 2012). By providing foot stiffening, these assistor rearfoot plantarflexors set the foot's semi-stiffened properties for the application of the primary triceps surae plantarflexion power. Their increased activity during late midstance also significantly resists foot vault depression, which helps prevent excessive foot deformation and energy dissipation aiding in storing energy within their tendons that can be released for ankle and midfoot plantarflexion. Thus, these muscles have important ankle–foot postural stabilising and energy storage roles.

Failure in any of these assistor muscles may not significantly reduce the ankle plantarflexor power directly, but their functional loss reduces the ability to resist bending deflection moments across the vault during midstance. This reduces the ability to protect the load frame from experiencing excessive plantar ligament tensions and dorsal articular compressions due to vault sagging during weight transference. Failure in digital flexor function will reduce MTP joint stabilisation at heel lift, as well as risking increased midfoot flexibility. Failure in tibialis posterior will not only influence forefoot loading stability, but it will also result in a serious loss of foot stiffening capability around heel lift that is likely to result in unrestrained midfoot abduction before and during heel lift. The increased and unrestrained midfoot abduction and the loss of midfoot adduction will prevent the establishment of stable medial MTP joint fulcrums. Unantagonised peroneal activity will also add midfoot eversion into the mix, increasing loads on an unstable medial foot. The loss of tibialis posterior's important midfoot plantarflexion and stiffening role will cause foot dysfunction throughout all stages of late midstance and terminal stance. This will result in a serious dissipation of plantarflexion power intended for acceleration, and loss of the more efficient medial MTP joint fulcrums (Fig. 4.4.3a).

Loss of peroneus longus function will not result in increased foot compliance as long as tibialis posterior remains intact. In fact, the foot risks excessive stiffness particularly within the medial midfoot through unantagonised tibialis posterior function. Peroneus longus loss risks 1st metatarsocuneiform joint instability and excess lateral column mobility, allowing the cuboid to drop into a more plantarflexed position during midstance, resulting in excessive lateral midfoot sagging. Loss of peroneus longus function at terminal stance will diminish the midfoot plantarflexion power and reduce or maybe lose the ability to create eversion across the midfoot via the proximal 1st ray so as to engage the medial MTP joints. A loss of this ability may force utilisation of the shorter length lateral MTP joint fulcrum, reducing energetic efficiency of CoM acceleration via low gear propulsion as a result on unrestrained tibialis posterior midfoot inversion and adduction. Indeed, frontal and transverse plane tibialis posterior activity trying to compensate for lost peroneus longus midfoot plantarflexion power may result in increased inversion moments on the midfoot that drive low gear propulsion (Fig. 4.4.3b).

Primary ankle plantarflexion power is generated by triceps surae. It plays a pivotal role in generating the GRF within the forefoot and stabilises and maintains postural stability during weight transference to the next step (Honeine et al., 2013). This action is not one that directly 'drives' the CoM forward. Rather, the ankle plantarflexion power aids CoM momentum posteriorly into the next step, while maintaining a fixed forefoot to the ground (Honeine et al., 2013). The power of the

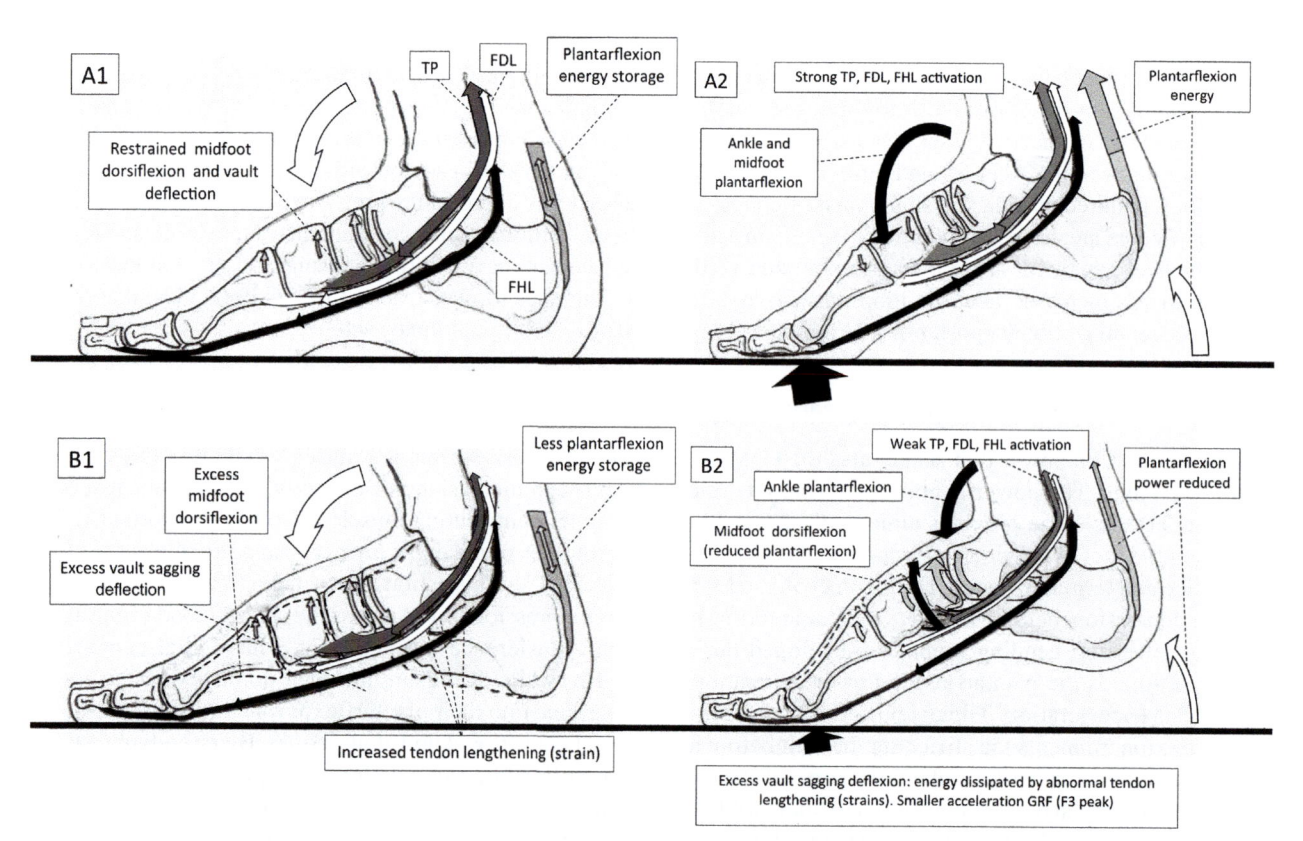

FIG. 4.4.3a For the ankle to functioning efficiently during acceleration, its relationship to forefoot biomechanics is pivotal. The application of ankle plantarflexion power requires the foot to provide a semirigid beam and stable anterior platform across a potentially highly mobile structure. If this is not provided, acceleration power will be dissipated within vault deflection and forefoot splay deformation. The plantar intrinsic muscles activity and plantar connective tissue strain-stiffening properties are important for this process, yet extrinsic muscle tendons that pass the ankle and rearfoot are also important. This includes tibialis posterior (TP –grey tendon), flexor hallucis longus (FHL—black tendon), flexor digitorum longus (FDL—white tendon), and peroneus longus and brevis (not shown). Normally, foot stiffening is well under way by late midstance (A1), with plantar extrinsic muscles raising their activity in concert with the plantar intrinsics. This loads tendons with tensile energy, stiffening them during ankle dorsiflexion and vault deformation. Their activity in late midstance limits foot vault deflection. At heel lift (A2), the energy from the Achilles is released across the stiffened semirigid midfoot beam to the metatarsals, pressing the forefoot into the ground, thus increasing forefoot GRFs that start to extend the digits. Elastic energy is also released from the tibialis posterior tendon to power medial vault-stiffening without muscle fibre contractile activity. The long toe flexor muscles remain active. If any or all of these extrinsic muscles are weak or dysfunctional, problems arise. In late midstance (B1), poor tensile stress-stiffening under the plantar foot decreases the resistance to vault deflection, allowing the foot vault to take up a lower profile (dashed skeletal outline). This permits the plantar soft tissues to dissipate rather than store energy for acceleration. If the foot at heel lift is inadequately stiffened to provide the rigid beam for plantarflexion power to transfer to the forefoot (B2), then the vault deflects into lower metatarsal declination angles (dashed skeletal outline) creating a lower forefoot impulse during acceleration as a result of lost energy. The reduced momentum on the CoM during weight transfer must be compensated for elsewhere, such as from gluteus maximus at the hip. High vault mobility can reduce the functional metatarsal declination angles during terminal stance, risking MTP joint impingements. Failure in only medial muscles, such as TP, FDL, and FHL, may cause failure of medial vault mechanics leading to medial instability during acceleration, but failure in the peroneals alone may cause lateral vault instability. *(Permission www.healthystep.co.uk.)*

triceps surae is developed during its eccentric control of CoM deceleration during weight transference through late midstance. This results in considerable elastic energy being stored within the Achilles and associated triceps surae fascial connective tissues, such as the gastrocnemius aponeurosis and intramuscular fascia. The loads generated within the musculature and the connective tissues associated with the triceps surae present a high potential injury risk. This explains the need for the highly specialised four subtendons as multifascial hierarchy, forming a tendon with areas of variable tensile response and the capacity to move past each other and yet still provide a single functional tendon with an overall large combined cross-sectional area for strength.

The triceps surae–Achilles complex's mechanical efficiency is dictated by a number of factors, including the extent and duration of ankle dorsiflexion during late midstance and the positioning of the CoM of the body at heel lift. The energy storage capacity within the Achilles tendon, assisted by other plantarflexors and foot and ankle connective tissues, requires appropriate foot stiffness during midstance. For that energy to be released effectively for acceleration requires foot and ankle stiffness at heel lift. The alignment of the foot to the long axis of the leg in the line of progression also influences

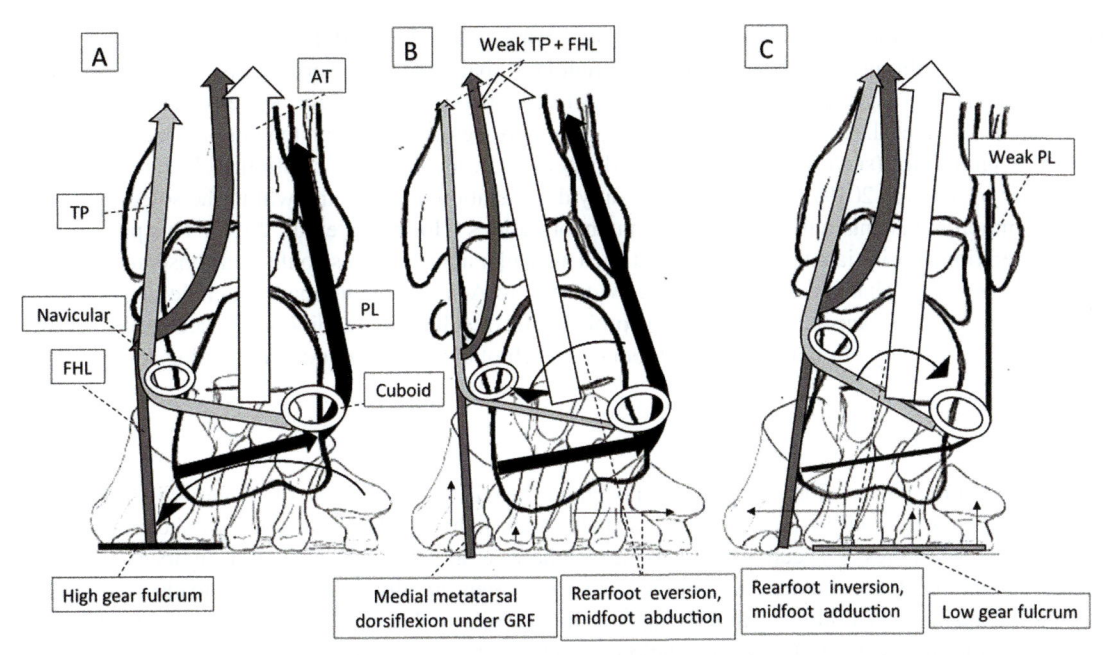

FIG. 4.4.3b The balancing of forces around the rearfoot is important for the application of plantarflexion power to the appropriate area of the forefoot for acceleration. The aim is to utilise the high gear fulcrum located between the 1st and 2nd MTP joints. This gives a longer external moment arm for Achilles power to improve the second-class lever arm for raising the stance limb's CoM. Simplified schematics help to explain such mechanics, with the midfoot removed and the cuboid and navicular represented by white ovals that are rotating in a pulley-like fashion. Ideally (A), muscle function around the rearfoot should provide a shifting of the CoGRF towards the medial fulcrum under MTP joint stabilisation. This is aided by tibialis posterior (TP—light-grey tendon vectors) providing midfoot abduction and vault plantarflexion and stabilisation via its medial cuboid and navicular attachments. Peroneus longus (PL—black tendon vectors) provides midfoot eversion and plantarflexion via its plantar medial cuneiform and 1st metatarsal base attachments. Flexor hallucis longus (FHL—dark-grey vectors) provides a moderating effect, balancing medial and lateral muscular forces around the rearfoot, while stabilising the medial vault and MTP joints with other digital flexors. Together, these muscle vectors align the Achilles tendon vector (AT—white tendon vector) to the high gear axis (black line) via a net eversion, adduction, and plantarflexion moment across the forefoot (thin black arrow). The foot and ankle can now act as a linear elastic spring. If TP is dysfunctional (B), the resultant acceleration foot motion is compromised. Metatarsals become more unstable in resisting dorsiflexion moments from the GRF as midfoot and forefoot plantarflexion moments are compromised; a situation worsened with concurrent FHL dysfunction. Although the GRF positions more medially, the high gear fulcrum is unstable and often unusable. The rearfoot (through ankle motion) will evert under the displaced eversion moment that is normally applied to the stable midfoot by PL. PL will also generate midfoot abduction moments, causing further medial instability. The AT vector is now at an everted angle when applying its plantarflexion power, risking higher medial tensional forces within it. When PL is dysfunctional (C), its eversion and plantarflexion moments across the midfoot are lost. Although tibialis posterior and FHL are usually able to compensate for the lost midfoot plantarflexion, their increased and unopposed activity risks creating midfoot inversion as well as adduction, threatening to invert the rearfoot during acceleration. This will change the AT vector angle, increasing tensions on the lateral subtendons of the Achilles. Inverting the foot in acceleration will cause the shorter external moment arm of the low gear axis (thick grey line) to be utilised, reducing the efficiency of the plantarflexion power. Note that peroneus brevis, extensor digitorum longus, and the plantar intrinsics are important components not illustrated to avoid overcomplication. *(Permission www.healthystep.co.uk.)*

how efficiently the triceps surae–Achilles complex's energy storage and release capacity acts as a passive spring (Hedrick et al., 2019). A failure in any of these complementary mechanisms can result in a significant change in tissue stresses across the triceps surae muscles and, in particular, within the Achilles tendon.

4.4.4 Achilles tendinopathy

The Achilles tendon is commonly injured despite it being the strongest and largest human tendon by cross-sectional area (Magnan et al., 2014). Approximately 6% of the population report Achilles tendon pain during their lifetime (Chimenti et al., 2017) including 56% of recreational runners (Lyght et al., 2016). The Achilles can load with up to eight times body weight (Giddings et al., 2000) and it is thought that collagen microtrauma during repetitive high loading can result in degeneration if insufficient time to repair is given before more cycles of high loading reoccur (Lyght et al., 2016). The peak incidence of Achilles tendinopathy is associated between ages 30 and 55 years when populations are usually still active, but age-related degenerative changes, such as decreased cellularity, increased glycosaminoglycan content, loss of fibre organisation, and a reduction in tendon elasticity are occurring within later adult life (Magnan et al., 2014). Achilles symptoms are usually split into insertional attachment (enthesis) pain and noninsertional or mid-portion tendon pain.

There are many factors suggested to influence Achilles tendinopathy development. Systematic review evidence links to higher BMI scores (Scott et al., 2013; de Sá et al., 2018), prior episodes of lower limb tendinopathies or fractures, the use of fluoroquinolone antibiotics (such as ofloxacin, ciprofloxacin etc.), alcohol use, creatinine clearance, and cold weather training (van der Vlist et al., 2019). The presence of fatty degeneration and oedema within the soleus and gastrocnemius muscles are common in patients with Achilles tendon abnormalities (Hoffmann et al., 2011). This possibly suggests that muscular degeneration may initiate tendinopathy, but such muscular abnormalities could be a consequence of tendinopathy rather than a cause. Biomechanical factors that are evidenced to Achilles tendinopathy include decreased forward progression during 'propulsion' (acceleration), which is probably the result of the tendinopathy, decreased isokinetic plantarflexor strength (slow velocity of contraction), and reduced rearfoot eversion excursion at loading response (van der Vlist et al., 2019).

Running risk

In running, the Achilles tendon fibres are relied upon more heavily to provide most of the lengthening of the triceps surae–Achilles complex, as well as the elastic recoil under load, rather than lengthening and shortening in concert with the triceps surae muscle fibres as occurs when walking (Ishikawa et al., 2007). Thus, risk of injury is much higher during running than in walking. When walking, the muscle fascicles shorten initially during contact loading, induced by the heel contact plantarflexion moment. Following this during late midstance, both the muscle fascicles and the tendon fibres lengthen (eccentric contraction) to decelerate CoM acceleration through ankle dorsiflexion until heel lift, when muscle and tendon fibres shorten together for acceleration during the last 70%–100% of stance (Ishikawa et al., 2007). In running, triceps surae muscle fascicles only lengthen during initial loading (0%–10% of stance). Thereafter, they shorten while the Achilles tendon performs the lengthening necessary to produce CoM deceleration of ankle dorsiflexion and the fibre shortening during elastic recoil acceleration (Ishikawa et al., 2007). This near-isometric contraction increases the energy burden and the loading forces on the Achilles, undoubtedly explaining much of running's (and sports) association with Achilles tendinopathy.

Achilles energy and vascular pathomechanics

Connective tissue energy storage during gait consists of 5%–10% being converted into heat. Failure to control and dissipate the heat energy risks tendon cell death (Magnan et al., 2014). Loss of vascular supply to a tendon (hypovascularisation or strain-induced ischaemia) may be a physiological result of stresses within tendons which in turn, makes heat loss and healing processes more difficult. Potentially, this risks a cycle where stress damage in exercise outstrips repair due to hypoxia. However, hypoxia induces certain genes to start to express protein factors that stimulate hypervascularisation, changing normally poorly vascularised tendon into becoming degenerative and hypervascularised. This hypervascularisation requires tendon matrix degradation, while a return to a more hypoxic environment promotes collagen matrix synthesis (Magnan, et al., 2014). Thus, it appears that normal tendon hypoxia is necessary to maintain healthy tendon matrix.

Chemicals such as fluoroquinolone antibiotics, that inhibit fibroblast metabolism, reduce cell proliferation, and matrix synthesis, present an increased risk of tendon degeneration (Magnan et al., 2014). With age, further degenerative changes within the tendon will lead to decreasing levels of stress tolerance, potentially provoking pathology even in low level exercise, rather than the higher, more heat-inducing stress loading cycles.

Achilles heterogenicity in pathomechanics

The large cross-sectional area reported for the Achilles is a slight misrepresentation, as it is composed of four very distinct tendon subunits that have been referred to as subtendons, sitting in a hierarchical position between fascicle and full tendon (Handsfield et al., 2017). Strain is therefore not distributed homogeneously through the Achilles tendon, but variably between the subtendons, depending on the ankle–rearfoot's kinematics (Bogaerts et al., 2016; Handsfield et al., 2017). The reason for the presence of subtendons within the Achilles is that the triceps surae, which supplies the Achilles tendon with its contractile muscular elements, is not constructed from a single muscle belly with one single nerve supply. The triceps surae consists of the soleus and two very distinct gastrocnemius muscles, i.e. medial and lateral (medialis and lateralis). The medial gastrocnemius is further divided into distinct medial and lateral fibre bundles. Thus, each of the distinct four muscular units of triceps surae has their own Achilles subtendon. These may provide subtle mechanical differences to account for the different tendon compliance needs of the gastrocnemii compared to soleus, for the gastrocnemius tendon subunits are energetically superior when they are less compliant than is soleus' (Uchida et al., 2016). These variances in compliance within the tendon subunits may be fundamental to maintaining efficient Achilles energetics (Fig. 4.4.4a).

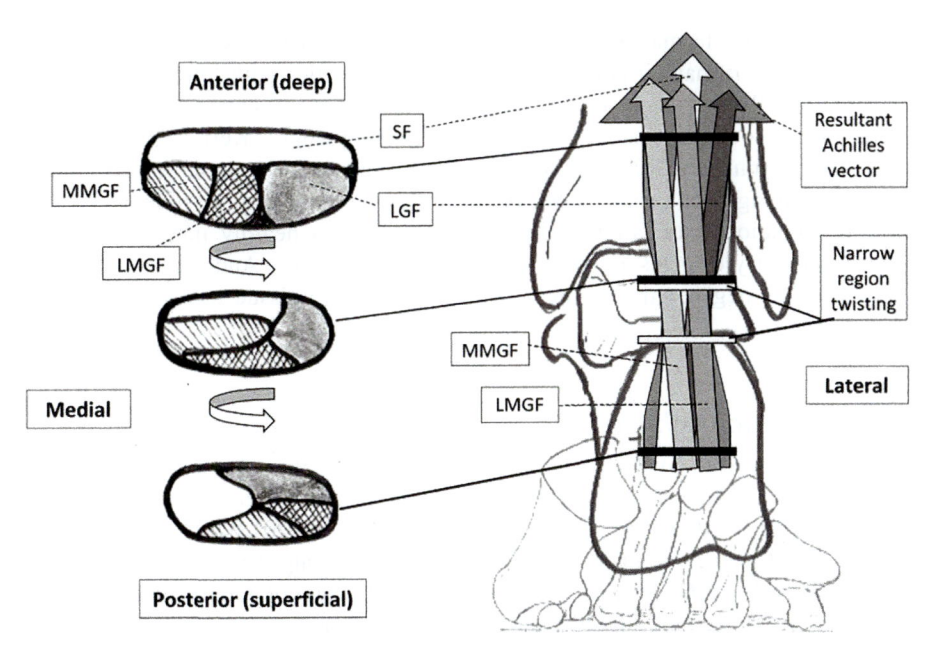

FIG. 4.4.4a The Achilles is an important anatomical structure with combined functions to store energy generated by decelerating CoM motion and then releasing it for acceleration as a powerful ankle plantarflexion moment. The Achilles is a highly specialised energy storage tendon, possessing all the benefits and disadvantages of being such a structure. Its high cross-sectional area is aided by being composed of subtendons which slowly twist in their orientation as they descend. This twisting creates a vulnerable central section with a narrower cross-sectional area. These subtendons have important differences in their mechanical loading with relatively high levels of inter-subtendon gliding possible. Their twisting arrangement increases under tension as the ankle dorsiflexes, recoiling and un-twisting during ankle plantarflexion motion. A schematic of a right Achilles demonstrates that at the level of the musculotendinous junction, soleus subtendon fibres (SF) occupy the anterior or deep aspect, with medial fibres of medial gastrocnemius (MMGF) lying posteromedially, lateral fibres of medial gastrocnemius (LMGF) posteriorly, and lateral gastrocnemius fibres (LGF) posterolaterally. This largely mirrors the position of the muscle fibres above them. By midway inferiorly down the tendon, subtendons have rotated so that SF starts to move more medially with the other subtendons following. By the insertion, the soleus-derived subtendon usually occupies the medial side of the Achilles, the LGF occupies the anterior position, and medial gastrocnemius' two subtendons occupy the posterior and posterolateral aspect. The twisting of the subtendons explains the narrowing of the Achilles at around 3–4 cm above the enthesis, something obvious when vector arrows for each subtendon are added to the diagram. Note that there is considerable variation in the size and precise positioning of each subtendon between individuals, but these trends in organisation are now well-established within the literature. Thus, location of any pathology within the Achilles is of paramount importance in understanding the precise nature of the pathomechanics and the effect it will have on the resultant Achilles vector. *(Permission www.healthystep.co.uk.)*

Despite these divisions and the considerable inter-subtendon gliding, the complete Achilles remains an extremely important single functional unit powering CoM velocity. Yet, each fascicle and each subtendon within the tendon gliding past each other allows individual stress lengthening properties within the Achilles as a whole structure. The large combined cross-sectional area of the subtendons influences the response to loading stresses, helping the Achilles resist fatigue effects such as creep under cyclical loading (Wren et al., 2003). This makes the whole much stronger than the individual parts, but the individual parts provide more adaptability. Yet, despite this complex spread of loading stresses within the Achilles and its highly structured calcaneal enthesis attachment, the Achilles, triceps surae muscles, and the gastrocnemius aponeurosis commonly develop pathology. In part, this can be explained through reduced mechanical properties resulting from age, disuse, and physiological diseases such as diabetes, which can alter levels of tendon compliance. However, putting these problems aside, there are also a number of key biomechanical events that increase the vulnerability within the Achilles structures during late midstance and terminal stance walking, and particularly during running.

Individuals vary in the morphology of their tendon structure, both in muscular anatomy (Bojsen-Møller et al., 2004; Lee and Piazza, 2009) and the degree of subtendon fascicle twisting (Edama et al., 2015; Pękala et al., 2017). This might alter the kinetics with regard to pathological risk in certain individuals. Achilles tendinopathy is prevalent across adult human lifespans and in active and sedentary people alike. It can occur in the mid-portion of the tendon, within the peritenon, the enthesis, or in any combination of these. It has been reported that these different conditions require different treatments to effectively resolve them (Cook et al., 2018). This suggests that a diverse number of mechanical origins can provoke Achilles tendinopathy, including excessive tensile, compressive, shear, or frictional forces that require different solutions to resolve them.

In health, the Achilles is built to manage high stresses and strains. Patients who develop Achilles tendinopathy in walking should be suspected of having poor tendon material properties with low tissue thresholds, likely a result of significant physiological disease, such as cardiovascular system degeneration. In running, stresses are much higher, with forefoot running styles increasing Achilles loading time (Kernozek et al., 2018) and stresses (Almonroeder et al., 2013; Nunns et al., 2013). Fatigue will alter muscular contractile and tendon compliance properties regardless of the running foot contact position. Achilles tendinopathy in runners is far more likely to be related to kinematic perturbations and fatigue, changing the tendon's healthy mechanical properties, rather than physiologically induced property changes associated with diseases and age. And yet, with the persistence of tendon pathomechanics, physiological degenerative changes within the tendon's tissue will lower the tissue threshold under stress before damage.

Approaching Achilles pathomechanics

Achilles pathomechanics will need to consider potential abnormalities of the material's property through physiological changes, before the kinematics and kinetics are then approached. It should be remembered that energy storage tendon properties are known to change with age (Godinho et al., 2017; Delabastita et al., 2018). Loss of Achilles tendon stiffness will disproportionally reduce the efficiency of gastrocnemius, possibly increasing the mechanical burden onto soleus and its Achilles subtendon (Uchida et al., 2016; Orselli et al., 2017). Reduced ranges of dorsiflexion in midstance may fail to adequately stiffen the Achilles before heel lift through insufficient twisting or stress-stiffening length changes. Excess ranges of ankle dorsiflexion could over-stiffen the Achilles, causing tendon ischaemia through either fibre over-twisting or by creating lengthening overstrain, compressing the blood vessels that penetrate to supply it. Excessive connective tissue stiffness within the muscle or tendon may also cause similar ischaemic effects, despite loss of full ranges of ankle dorsiflexion motion during late midstance.

Considering the patient's kinematic events through comparison with normal free-body diagrams can prove useful to understanding stress concentrations within the Achilles, as well as planning treatment. Gait findings can be considered in relation to the likely pathomechanics of each area of subtendon strain, i.e. medial or lateral, mid-portion, distal insertion, bursae, and/or paratenon. There is a need to reflect on the mechanical stresses and the anatomical function and positioning during free-body diagram construction in order to address issues of energy storage and release, stiffness or compliance, and compression or frictional forces against posterior calcaneal tubercle's bone.

The stresses arising in the triceps surae–Achilles complex during midstance directly relate to the angle and velocity of ankle dorsiflexion (extension) being resisted. The energy stored within eccentric restraint mechanics becomes the primary mechanism that induces heel lift. When the energy is sufficient to overcome the load remaining on the support limb's rearfoot, then the heel will lift. The active calf energy produced should account for any differences between energy stored and that required for heel lift. Obviously, some supplementary energy will be stored in the other flexor tendons such as tibialis posterior and peroneus longus as well as the ankle and foot connective tissues, which all influence the final energy equation for ankle plantarflexion power generated at heel lift. This means that energy stored during ankle dorsiflexion through all sources is essential for efficient power generation and heel lift at the ankle. Yet the greatest burden of heel lifting ankle plantarflexion power should be generated through the triceps surae–Achilles complex (Fig. 4.4.4b).

The deep (anterior) Achilles fibres undergo consistently larger displacements than do the superficial (posterior) fibres and these displacements are usually greater when under lower knee flexion angles nearer to extension (Arndt et al., 2012; Slane and Thelen, 2014, 2015; Handsfield et al., 2017). This is important to consider during late terminal stance, as there is much individual variability in the knee angle that should be near to, but not in, knee extension. Gastrocnemius plays an important role in setting this angle. The anterior Achilles fibres also undergo greater compression against the calcaneus through increasing ankle dorsiflexion angles in midstance (Lersch et al., 2012). Frontal plane torques also influence the location of strains in the Achilles (Zifchock and Piazza, 2004; Lersch et al., 2012). Generally, distal Achilles strains are greater than proximal, with proximal strains greater to the lateral side and distal strains greater medially (Lersch et al., 2012). Intratendon sliding of the subtendons increases displacement nonuniformly, with some lesser influence caused by the twisting of the Achilles that occurs with dorsiflexion, indicating that the complex internal Achilles structure strongly affects the interaction between muscle forces and tendon behaviour (Handsfield et al., 2017). These heterogeneous properties of Achilles strain set up potential mechanisms of pathomechanics when strains become concentrated by specific cyclical motions during gait (Arndt et al., 2012).

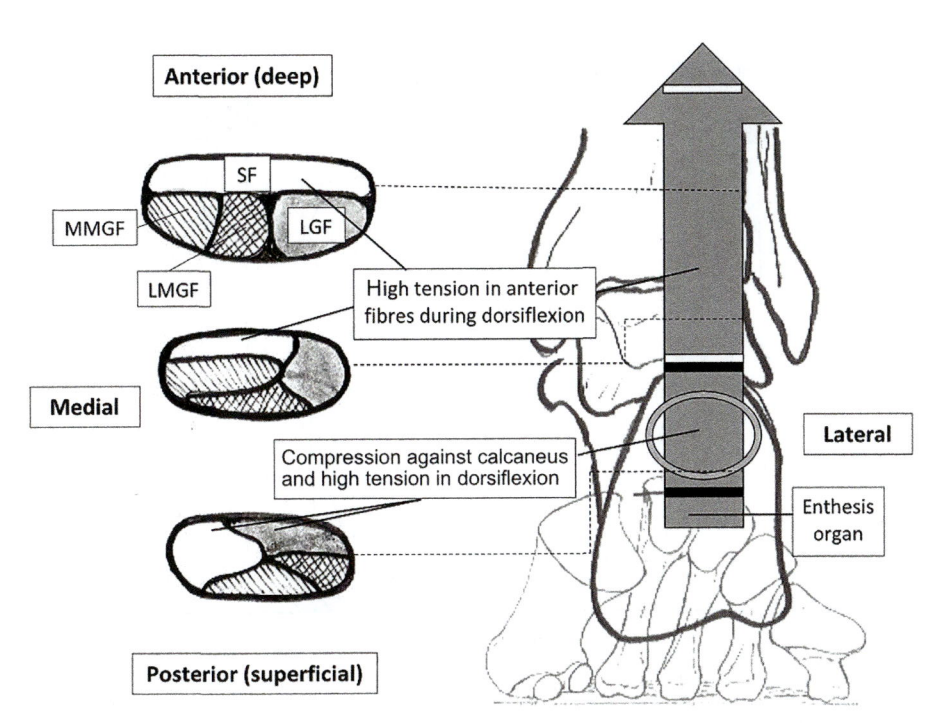

FIG. 4.4.4b A simple schematic to help demonstrate that strains within the Achilles are not consistent throughout the anatomy or over the tendon's entire length. Throughout most of the Achilles, fibre length changes are greatest anteriorly (deeper fibres) with strains focused distally, indicated within the area between the black lines. Proximally, this burden is placed on the soleus subtendon fibres (SF) but as the fibres descend, medial rotation of the subtendons moves the strain more into the fibres of lateral gastrocnemius (LGF) from SF. The distal region of the tendon overlies the posterior calcaneus which during higher ankle dorsiflexion angles, compresses the Achilles against the bone. This is thus a region of anterior fibre tension and compression. Rotation of the subtendons means that these fibres can consist of some SF medially, but primarily those of LGF more laterally. To protect this region, the retrocalcaneal bursa lies anterior to the tendon, while an area of sesamoid cartilage lies within the tendon (grey circle area) before the attachment at the enthesis organ. The more superficial medial and lateral fibres of medial gastrocnemius (MMGF and LMGF) usually avoid high tensional forces and compression, although they can become compressed by the heel counter of footwear. A subcutaneous bursa is usually present over the superficial fibres to prevent cutaneous shear and friction, but this can become inflamed with ill-fitting heel counters. The superficial tendon fibres tend to express their highest strains more proximally (between the white lines) than deep tendon fibres that express theirs distally. *(Permission www.healthystep.co.uk.)*

4.4.5 Sagittal plane rearfoot motion and Achilles pathomechanics

Problems can arise if ankle joint dorsiflexion is either restricted, occurs too rapidly, or if there is insufficient power generated within the ankle plantarflexors and related connective tissues to end dorsiflexion because the plantarflexion power is insufficient to initiate heel lift. Each change in timing and power has consequences to tissue stresses within the triceps surae, Achilles, and elsewhere. Failures in triceps surae strength and the Achilles' mechanical properties can be caused by changes in ankle motion, lower limb and trunk posture, and/or insufficient power generation. Failure in the gastrocnemii units will influence soleus mechanical demands and vice versa, creating a potentially quite complex picture as to where pathology is likely to arise, and from via what specific part of changing gait biomechanics. The gluteus maximus as a late assistor to terminal stance acceleration can also be involved in the changing events, as can any of the ankle plantarflexor assistors.

Early heel lift

Restriction of ankle joint dorsiflexion during late midstance prevents normal anterior progression of the body's CoM via anterior tibial displacement under inverted pendular mechanics. Blocking dorsiflexion causes insufficient time for centrifugal forces to fully develop for the swing foot to position itself 'suitably-anterior' for quick, efficient heel contact and weight transfer at the stance limb's heel lift. It also changes the position of the functional joint axis of the ankle which cannot move as far anteriorly, and this in turn affects the plantarflexion moment arm's length at heel lift, with a consequential need for greater muscle–tendon shortening per unit of ankle rotation (Wade et al., 2019) (Fig. 4.4.5a).

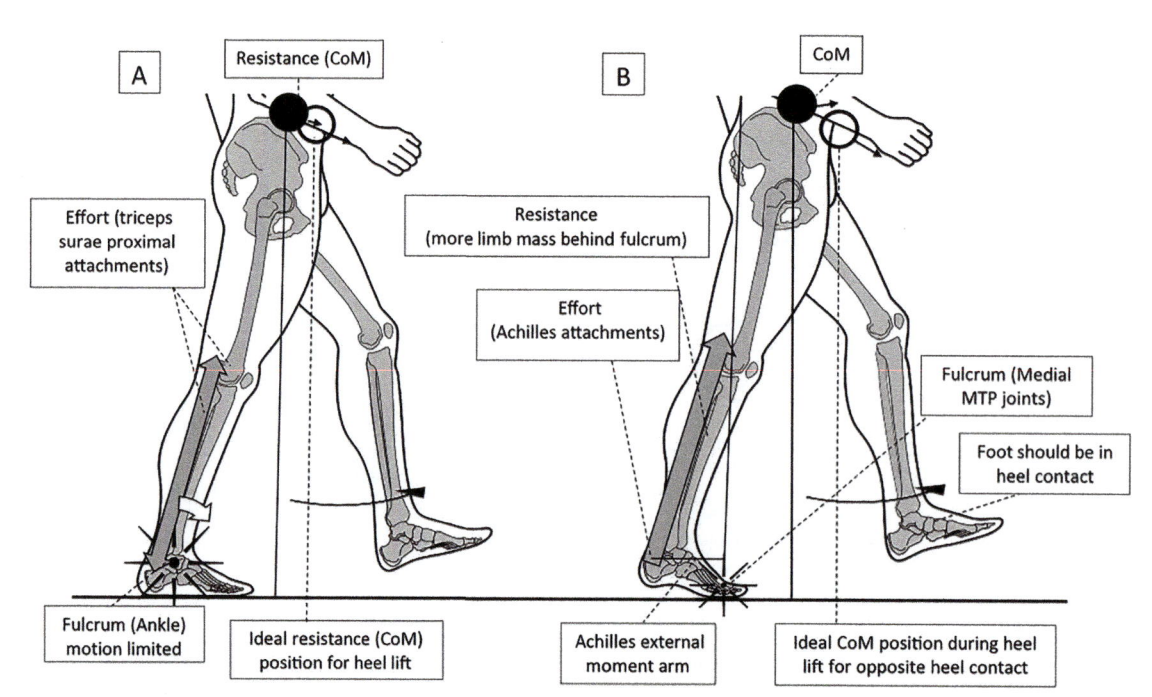

FIG. 4.4.5a A free-body diagram helps illustrate the effects of limited ankle dorsiflexion during midstance on the Achilles. The ideal location for the CoM (black ring and dashed lines) is compared to the position resulting from loss of midstance dorsiflexion (black circle and solid line). Ankle dorsiflexion restriction is common, resulting from a number of reasons including excessively stiff triceps surae myofascial and tibiofibular blockage of rotation on the anterior talar trochlear surface, which can be associated with high-vaulted cavoid feet. With loss of ankle dorsiflexion freedom during late midstance (A), the third-class lever system is unable to allow the body's CoM's (acting as the resistance) free forward progression over the foot controlled by triceps surae eccentric activity. There are two main consequences to the resulting shorter resistance arm just before heel lift. The first issue is that resistance remains further over the foot, unable to easily be drawn forward by the centrifugal forces of the swing limb. The second problem is that the fibres of the Achilles are less lengthened and so are less able to store elastic energy through tensional strain, yet the effort arm in comparison with the usually longer resistance arm is given an easier restraint on the forward motion of the resistance. In cavoid feet, the high calcaneal inclination angle may increase the compression between the anterior Achilles fibres and the posterior calcaneus. This means strains around the distal Achilles may increase (including at the enthesis) and yet decrease within the proximal Achilles. Once heel lift initiates (B), the second-class lever finds itself disadvantaged. More lower limb mass remains behind the MTP joint fulcrum, so more power is required to lift the heel. Yet with less Achilles lengthening in midstance, the power may have to come from concentric muscle contraction rather than elastic recoil power from the tendon alone. The HAT's CoM may have its anterior fall interrupted by an earlier heel lift, causing increased Achilles tensions at the start of heel lift, increasing fibre lengthening when it should be shortening. Centrifugal swing limb forces will also be less well anteriorly positioned to pull the CoM forward to offload the heel. Essentially, the resistance arm is abnormally long compared to the Achilles effort arm, and the effort power is often compromised, when it actually needs to be increased. Loads will tend to focus into the distal Achilles and enthesis organ when limited ankle dorsiflexion is associated with premature heel lift. *(Permission www.healthystep.co.uk.)*

Loss of late midstance ankle dorsiflexion will bring about a reduction in passive lengthening strain within the connective tissues and elastic energy storage capacity of the Achilles, potentially causing a greater elastic recoil energy deficit for heel lift between that stored passively and that required actively. To continue smooth anterior progression velocity of the trunk's CoM via anterior tibial translation under reduced ankle dorsiflexion conditions, heel lift may be required to occur earlier, tipping the tibia upwards and thus forward rather than rotating it anteriorly at the ankle.

Early heel lift continues anterior tibial progression but without increasing ankle dorsiflexion by initiating early ankle plantarflexion, thus 'tipping' the tibia over the foot rather than rotating around it around the ankle. This allows the CoM of the trunk to also continue to move forward but the price for this premature heel lift is that more of the body's CoM will sit behind the MTP joint fulcrum as the rearfoot plantarflexes, because the CoM will not have advanced as far forward. Loading forces on the rearfoot are likely to be still quite high when premature heel lift starts. This higher loading mass must be accelerated forward at a time when it would normally still be being decelerated. There will also be less time for the swing limb to become positioned correctly to receive body weight, so that heel lift will occur without the normal, almost instantaneous weight transfer. Thus, the forefoot will be loaded for longer, under potentially a greater plantarflexion moment created by abnormal concentric triceps surae activity. The Achilles is also likely to be less loaded with elastic energy as it is less stretched and stiffened prior to lifting, unless there is an excessive stiffness issue within the Achilles itself (as seen with diabetes). Thus, more active triceps surae muscular concentric length change is required to produce an earlier heel lift to overcome a larger CoM, under less Achilles elastic recoil.

An early heel lift should be visible on a force–time curve as an earlier F3 peak, often with a steeper upward slope than usual. This peak may be greater than the F1 peak, as more plantarflexion power is required to overcome greater resistance, sometimes necessitating triceps surae concentric contraction. The increased plantarflexor muscle activity necessary will also occur on a shorter plantarflexion lever arm at the ankle, because the ankle's instantaneous axis is unable to move as far anteriorly at the initiation of heel lift. If plantarflexor strength is adequate and the foot's loading frame stable enough and appropriately stiffened, then early heel lift can function relatively well as a locomotive technique, creating nothing more dramatic than a 'bouncier-looking' gait from greater CoM vertical displacements, although energetics will be less efficient. However, the risk of muscle strain and early fatigue will likely increase as a consequence. In the long term, this may cause earlier degenerative changes within the Achilles tendon through functional misuse via increased stresses and the reduced energetics that result.

An early heel lift on a more compliant foot from inadequate foot muscle strength and/or reduced stress-stiffened connective tissue, will risk increased plantarflexion power dissipation into the plantar soft tissues via increased vault sagging deflection at heel lift. This potentially risks foot injury as well as depowering the limb in acceleration. If the foot is excessively stiffened, then the extra energy from the acceleration plantarflexion moment may not be absorbed by the plantar soft tissues, but instead directed into the midfoot articular structures. This risks degenerative joint changes within the midfoot or a Charcot foot in a neuropathic diabetic.

The greater distance the CoM has been moved towards the fulcrum before any heel lift, the less deficit between active and passive energy requirements. If the bulk of the CoM has moved beyond the fulcrum before heel lift, then the deficit will be less. However, the more resistance that is left behind the fulcrum at early heel lift and the less the ankle can dorsiflex prior to heel lift, the greater the muscular energy input must be. The total effect in an individual will also depend on the hip, knee, and MTP joint motions available, as early heel lift requires earlier knee and hip flexion and MTP joint extension (Figs 4.4.5b and 4.4.5c).

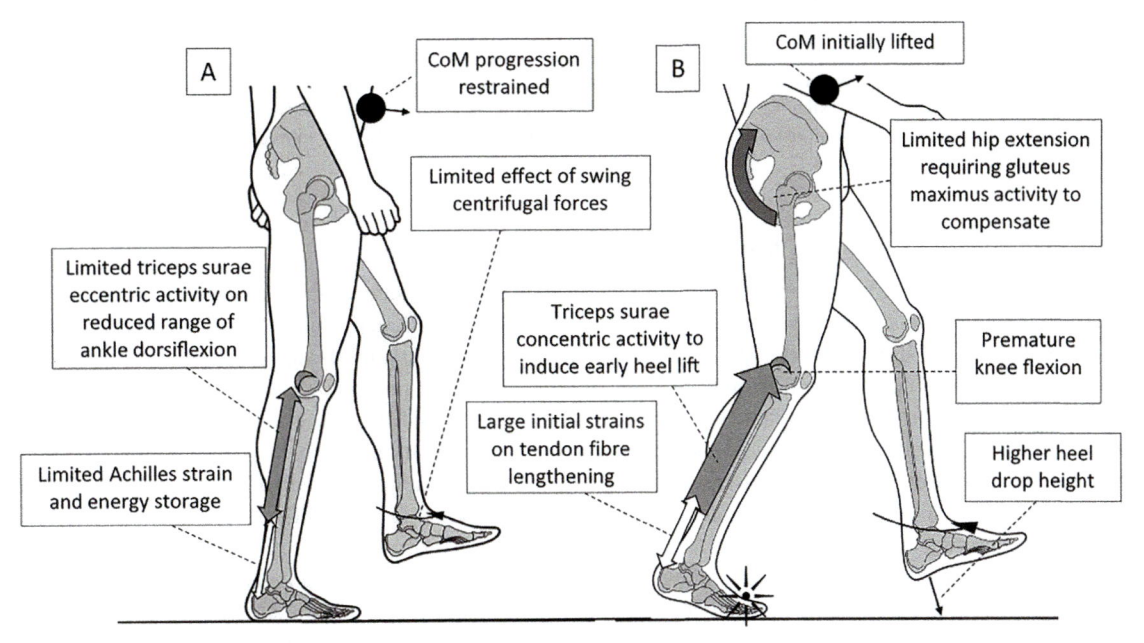

FIG. 4.4.5b A number of compromises are made in walking gait if ankle dorsiflexion range is limited during midstance (A). This can occur for a number of reasons. The key issues are loss of energy storage within the Achilles, limitation of the CoM's anterior progression through decreased CoM vaulting over the foot, and as a result, restraint on centrifugal forces from the swing limb in being able to pull the CoM forward. The preferred compensation is to increase triceps surae activity to raise energy storage within the Achilles via greater muscle fibre shortening, actively tensioning Achilles fibres to induce a premature heel lift (B). The premature heel lift results in early ankle plantarflexion, limiting hip extension and starting knee flexion moments. This makes the use of gluteus maximus acceleration assistance at the end of terminal stance more difficult. Forward acceleration of the CoM is initially upwards rather than on a downwards trajectory, requiring lifting more body mass behind the MTP joint fulcrum which reduces the mechanical efficiency of terminal stance. Once the CoM starts to fall, it drops more rapidly from a higher point, potentially increasing impact forces. This makes gait appear rather more 'bouncy' than usual. Premature heel lift risks earlier triceps surae fatigue and Achilles tendinopathy from suddenly increasing tendon strains from trying to move a larger resistance weight at heel lift via an abnormally timed triceps surae contraction. *(Permission www.healthystep.co.uk.)*

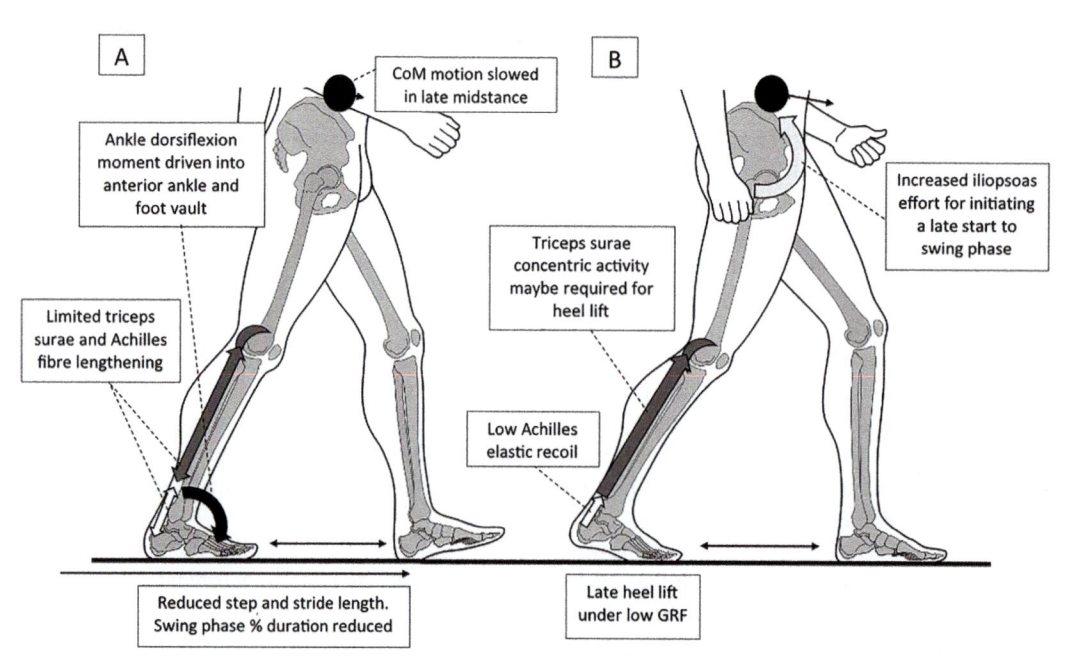

720 Clinical biomechanics in human locomotion

FIG. 4.4.5c If triceps surae strength is unable to provide an early heel lift as a compensation for restricted ankle dorsiflexion (A), late midstance CoM progression is blocked. As a consequence of lost ankle dorsiflexion freedom, the anterior ankle and midtarsal joints will sustain increasing articular compression forces under the ankle dorsiflexion moment generated by the CoM vaulting the foot. These compression forces increase upon the anterior talar articular surfaces including at the talar head at the talonavicular joint. These forces are driving plantarflexion moments into the foot as compensation to lost ankle dorsiflexion. As a result, greater sagging vault deflection moments cause increased dorsiflexion across the midfoot. This situation continues until the CoM has been transferred to the contralateral foot when it becomes the supporting limb during late double-stance phase (B). Such late heel lift after opposite limb contact can then require less ankle plantarflexion power under a much-reduced GRF remaining on the heel. This action necessitates a shortened step and stride length that can create an earlier double-limb phase and a reducing swing time. The dorsiflexion moment's energies, undissipated by ankle motion and less stored within the Achilles during late midstance, are driven into the foot vault instead. This risks midfoot degenerative joint changes and connective tissue injuries within the plantar vault. Inflexible vaults tend to stress articular and osseous structures more (including the ankle), while more flexible vaults will direct stresses towards the connective tissues. Diabetic patients with vulnerability within both bone and connective tissues are at high risk of serious damage, with concurrent neuropathy preventing defensive pain, warning of impending disaster. *(Permission www.healthystep.co.uk.)*

Articular loss of dorsiflexion in the ankle

Late midstance ankle dorsiflexion is normally limited by increasing form closure, as the joint surfaces become increasingly compressed together, and by the triceps surae–Achilles tendon becoming stiffened. Ankle dorsiflexion restrictions can arise from bone, soft tissue, and articular pathology. The talus, being wider anteriorly on its superior articular trochlear surface, is capable of causing loss of dorsiflexion. Should the wide anterior trochlear surface become essentially 'jammed' between the fibula and tibia, the talus would be unable to continue to rotate to provide full ankle dorsiflexion. The development of an anterior osteophyte on the talar neck or the tibial anterior joint line are examples of issues that will increase dorsiflexion blocking risk, but myofascial stiffness is the most common cause. The ankle joint can also become limited in dorsiflexion motion through a subtle slip anteriorly of the talus on the tibia. This can commonly result from even minor talofibular and/or subtalar ligament injuries during ankle sprains. Restrictions in the anterior-to-posterior glide or translation of the talus on the tibia are well documented to be found in association with limited dorsiflexion (Hertel and Corbett, 2019). Joints can and do, quite commonly, slightly 'misalign' their bony positions across an articulation, and as a result, lose some freedom of motion. Clinically, this minor misalignment and motion loss seems a particularly common problem in the small joints of the foot, known as an either arthrokinematic restriction or osteokinematic changes depending on whether the restriction is in a secondary joint or the original joint injured, respectively (Hertel and Corbett, 2019).

Loss of ankle motion and the development of OA is often associated with recurrent ankle sprains. This is due to altered anterior talar drift from loss of ligament stability that results in restricted dorsiflexion (Hertel and Corbett, 2019). High localised stress impingements in a joint's articular surfaces can result in cartilage failure and the development of OA. Ankle OA can cause further loss of dorsiflexion due to concurrent osteophytic proliferation of bone around the anterior joint edges, increasing impingement stresses anteriorly. Ankle joint stresses are normally predominantly along the anterior edge of the joint (Potthast et al., 2008) which increases the risk of degenerative changes that limit dorsiflexion. These anterior joint line stresses are balanced medially and laterally through a tensile bracing mechanism, controlled via rearfoot myofascial invertor–evertor balances (Potthast et al., 2008). Loss of this balance can set off articular degenerative changes through uneven ankle joint stresses, either on the medial anterior or lateral anterior aspect of the ankle.

Anterior talar displacements can thus occur as a consequence of poor muscular function as well as through traumatic events such as lateral ankle sprains or other active perturbation events, that can themselves, appear quite minor at the time. But any resulting amounts of anterior talar displacement restricts the normal talar gliding that maintains articular congruency and joint axis motion in and out of dorsiflexion. Many patients also demonstrate the displacement of the distal fibula relative to the tibia, which also results in a loss of free talar glide (Hertel and Corbett, 2019). The more plantarflexed the ankle at the original perturbation, the greater the ankle freedom of motion available at the time for potential displacement. Thus, perturbations that alter talar position are more likely to follow after injuries that occurred within ankle plantarflexion positions. However, the amount of stress applied during the injury always finally dictates the vulnerability in all ankle positions. Weak protective muscles during perturbation events, previous ligament injury, and poor sensorimotor function are all possible exacerbating factors, but most ankle sprain cases originally occur with nothing more than a little bad luck.

Clinically, ankle and midtarsal joint misalignments are referred to as *subluxations*, but rather than a true partial misalignment of a joint, they are often quite subtle bone 'shifts' in interarticular alignment that alter joint congruency and articular compressions during motion. They usually represent a loss of freedom of joint motion in certain directions, which can affect normal changes in the instantaneous joint axis location during joint movement, as well as the overall freedom and normal shape changes within the functional unit or indeed a whole structure like the foot. As minor events that can cause subluxations are not always noticed as significant events by the patient, many such misalignments may not have an obvious history of onset. Clinical assessment of the quality of joint motion is always important, and opposite limb sides should be compared in order to establish possible subtle motion dysfunction. These 'subluxations' and restrictions are targets of mobilisation and manipulation.

Other causes of ankle dorsiflexion loss

Freedom of motion at the fibula is also important to ankle motion. The fibula has flexibility under ligamentous tension to permit external rotation and slight gaping of the distal tibiofibular joint, to provide for the wider talar trochlear surface to be accommodated during dorsiflexion. In providing controlled, free tibiofibular motion, the tibiofibular interosseous ligament and anterior and posterior tibiofibular ligaments tension to act as torsional shock absorbers. This ability protects the fibula and tibia from torsional injuries arising through rearfoot transverse and frontal plane rotations. Failure of the distal tibiofibular joint to rotate and gap appropriately may also cause restriction in ankle dorsiflexion, as the anterior trochlear surface may become restricted earlier in dorsiflexion than normal. In assessing ankle function/dysfunction, the proximal as well as the distal tibiofibular joint should also be checked for free gliding motion during ankle flexion and extension, as restriction here can link to restriction of ankle motion.

Stiff, high-vault profiled feet may present a challenge to achieving adequate ankle dorsiflexion during late midstance. Most foot lengthening/widening-induced vault height changes should occur shortly after loading response with the foot continuing to lengthen and stiffen at a slower rate throughout late midstance (Stolwijk et al., 2014; Bjelopetrovich and Barrios, 2016; Takabayashi et al., 2020), with the vault profile reaching its lowest at heel lift (Hunt et al., 2001). This vault lowering allows the anterior talus to tilt into plantarflexion, coupled to a reducing calcaneal inclination angle and an anterior medial column distal displacement. This allows tibial anterior rotation to derive from some midfoot dorsiflexion as well as ankle dorsiflexion, relatively reducing the final ankle dorsiflexion angle necessary before heel lift. Vault deflection is thus a functional part of ankle dorsiflexion freedom during midstance.

A stiff foot that prevents vault deflection could be a hindrance to anterior tibial rotation, especially if it is combined with a high vault profile with an associated high calcaneal inclination angle. A stiff cavoid foot potentially limits talar and distal calcaneal plantarflexion, and thus proximal midfoot dorsiflexion. Any blocking of such increased vault sagging deflection from supplementing limited ankle joint dorsiflexion for the permitting of more anterior tibial rotation, means that the compensatory motion for lost ankle dorsiflexion is not available. When ankle and midfoot dorsiflexion are both restricted, it necessitates a need for an earlier heel lift to allow for CoM anterior translation. However, the other risk if ankle and midfoot dorsiflexion are limited, is the risk of lost energy dissipation resulting in injury. Energy dissipation is normally provided for by vault deflection during midstance that tensions the plantar soft tissues. Stiff midfoot joints that hardly move due to tense myofascial hypomobility, shift the energy burden that is normally dissipated within the plantar connective tissues into the vulnerable midfoot joint's articular cartilage.

The Achilles in reduced dorsiflexion angles of cavoid feet

A further complication is that a high calcaneal inclination angle associated with a cavoid foot can cause the superior edge of the posterior calcaneal tubercle to be angled so that it presents a more prominent edge for the Achilles to wrap over during changing tendon angles that associate with increased ankle dorsiflexion positions. If, during midstance, the inclination angle of the calcaneus does not reduce through a lack of vault lowering under ankle dorsiflexion moments, then the Achilles will become more compressed against the superior portions of the posterior calcaneal surface during late midstance. The distal Achilles will become increasingly compressed into the retrocalcaneal bursa and posterior calcaneal surface, as the enthesis organ becomes increasingly tensioned under anterior tibial rotation. This risks possible enthesopathy, bursitis, and distal tendon fibre damage. Thus, stiff cavoid feet with sufficient ankle dorsiflexion available to avoid an early heel lift, despite the loss of the assistance of adequate vault lowering, are more at risk of insertional Achilles tendinopathy. Those with insufficient muscle power to initiate an early heel lift, despite an insufficient ankle dorsiflexion angle, are also at risk of Achilles insertional tendinopathy, but also midportion tendinopathy (Fig. 4.4.5d).

FIG. 4.4.5d Failing to permit a reduction in the calcaneal inclination angle during midstance, as seen in stiff cavoid feet, can cause the ankle joint to compress into its anterior articular surfaces creating their end-range close-packed positions earlier. This close-packing limits ankle dorsiflexion freedom in late midstance, setting up a number of biomechanical disadvantages. CoM anterior progression is impeded by reduced ability to vault over the foot, in turn influencing swing limb acceleration. The loss of vault deformation changes limits internal stresses within the foot from being dissipated and also decreases overall tensional elongation within the Achilles fibres. This reduces Achilles energy storage. However, distal anterior (deep) fibres of the Achilles may become increasingly compressed against the steeply inclined posterior calcaneal surface over the periosteal fibrocartilage and increasingly tensioned at and near the enthesis. Thus, limited dorsiflexion freedom can concentrate strains within the distal tendon and around the enthesis organ rather than spreading them more evenly throughout the Achilles, risking Achilles insertitis or enthesopathy. *(Permission www.healthystep.co.uk.)*

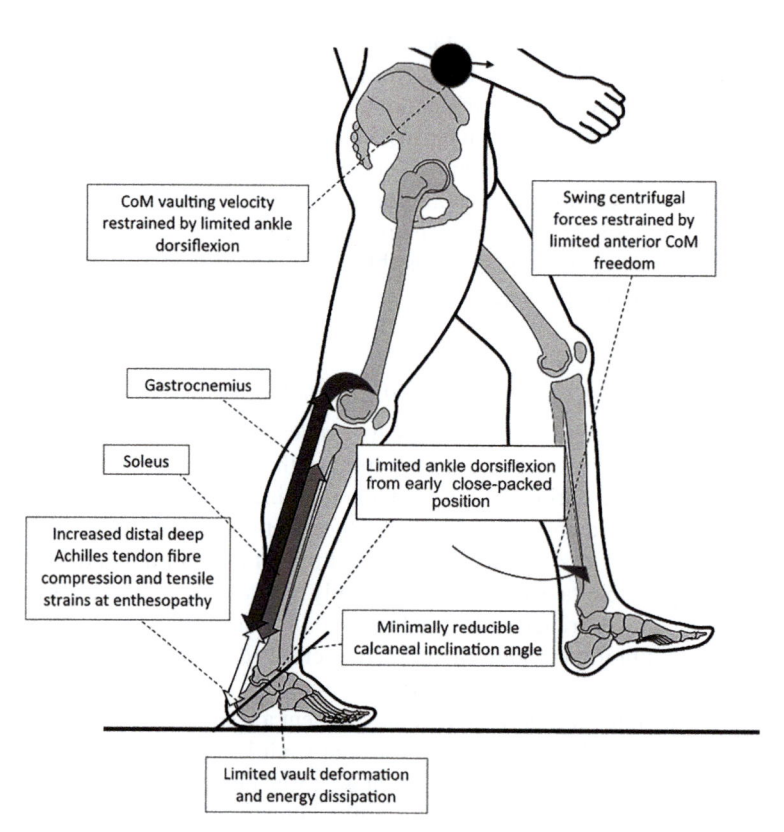

The deep areas of the distal Achilles usually undergo greater compressive strains than do the superficial tendon portions (Lersch et al., 2012; Handsfield et al., 2017), requiring protection from the posterior calcaneus by the deep retrocalcaneal bursa, periosteal fibrocartilage, and sesamoid fibrocartilage positioned around the Achilles enthesis. These structures reduce compression on the posterior calcaneus during ankle dorsiflexion. However, despite the protection, increased calcaneal inclination exacerbates the compression situation (Lersch et al., 2012), a situation worsened with failure to adequately decrease calcaneal inclination during midstance vault deflection. Thus, reduced vault compliance proximally will compound the compression forces around the Achilles attachment, risking retrocalcaneal bursitis, tendinopathy (especially around the sesamoid fibrocartilage), and enthesopathy.

As previously discussed, an early heel lift increases the muscular power-initiated tensile strain on a more compliant Achilles in an attempt to overcome the increased resistance mass from the failure to translate enough body weight anterior to the MTP joints' fulcrum. In a cavoid foot with early heel lift, this premature, high force Achilles loading possibly combines high distal tensional stress with high compression on the posterior calcaneus. Therefore, just prior to heel lift, the

Achilles insertion will be subjected to increased tensile stresses, while still undergoing high compression along the deep surface around the enthesis. This sets up ideal conditions to provoke an enthesopathy. In time, the inflammatory consequences of the stress concentrations on the enthesis may cause an increase in bone deposition through fibrocartilage ossification. The problem is then further compounded by the stiffening of the local attachment structures compared to the tendon above it, focusing stresses into the attachment area. The resulting bone and enthesis organ reaction to this situation may be to form a Haglund's deformity by increasing tissue turnover and restructuring.

The influence of muscle activity

Muscles found around the ankle provide the stresses to keep the joint stable by maintaining compression forces relatively evenly across the articular surfaces of the rearfoot (Potthast et al., 2008). Weakness in muscles is often counteracted by increased myofascial tightness which may be a cost of more sedentary lifestyles (Fatima et al., 2017), physiological diseases (Powers et al., 2016), and neurological dysfunction, as seen in cerebral palsy (Barber et al., 2011, 2012). Thus, weak and tight calf myofascial tissues may also be associated with loss of ankle dorsiflexion.

Indeed, cerebral palsy with its spasticity due to velocity-dependent stretch reflex hyperactivity often presents with an extreme limitation of any midstance ankle dorsiflexion, resulting in a lack of any plantarflexion power for acceleration (Bar-On et al., 2018). To overcome this inability to transfer the CoM over and around the ankle requires *ankle equinus* or toe-walking myofascial adaptations, which also result in high ankle plantarflexion angles during swing phase. A plantarflexed stance foot allows the CoM to pass over the foot possessing a shorter base of support consisting of just the forefoot. However, this gait technique loses all the ability to store high levels of elastic energy within the Achilles during late midstance dorsiflexion, dramatically reducing gait energetics. Stretching the ankle plantarflexors so as to improve their flexibility to achieve dorsiflexion in isolation, may be clinically counterproductive if it produces flexibility via muscle fibre length changes without muscular strength to manage ankle dorsiflexion biomechanics. Equally, lengthening the Achilles surgically may create slack connective tissues and thus ankle dorsiflexion, but without high levels of elastic recoil capability within the tendon. Ankle dorsiflexion permits CoM anterior acceleration that must be actively resisted if gait is to remain stable and elastically 'propulsive'. Calf flexibility without strength or without tendon elasticity, can be problematic even if ankle dorsiflexion freedom is available. Adding resistance training as part of strengthening to the stretching programmes of the ankle plantarflexors in cerebral palsy seems more beneficial (Blazevich et al., 2014; Kalkman et al., 2019). In less marked calf tightness seen in the general population, such a relationship may also exist without the extremes of toe-walking that may still result in loss of acceleration power and ankle flexibility.

The plantarflexors have an important compensatory effect on musculoskeletal deficits in both hip and knee extensors' and flexors' strength, whereas unfortunately plantarflexor weakness is not compensated for effectively by the hip and knee muscles during gait (Goldberg and Neptune, 2007). Therefore, although the stronger proximal muscles can help, they cannot fully resolve loss of ankle plantarflexor strength. Weakness in the plantarflexors requires functional and structural changes in order to maintain braking control of CoM progression. In weakness, stiffening the muscle–tendon unit can increase passive restraint and increase energy storage within connective tissues to counterbalance the poorer production of muscular power. By stiffening the muscle–tendon unit through limiting eccentric contraction and increasing either isometric or concentric contractions within muscle, connective tissue stretch–recoil mechanisms in the specialised energy-storage Achilles tendon offers a solution to a weakness-induced problem to power heel lift. Shortening stride length will also negate some of the ankle dorsiflexion required before the opposite swing cycle is complete. However, if an attempt is made to maintain stride length, the ankle dorsiflexion restriction from increased muscle stiffness may cause a premature heel lift on a triceps surae with limited strength.

Alternatively, heel lift delay may result from insufficient ankle power generation at the time when the swing foot is positioned correctly for ground contact. The weaker the triceps surae, the more likely this will occur, as stiffening the triceps surae to overcome strength deficits still requires significant power to initiate heel lift, so as to accelerate the CoM onto the next step. Should this heel lifting ankle power not be forthcoming, contralateral heel strike and weight transfer would have to occur without a heel lift. Heel lift can then follow after all significant weight has been transferred onto the contralateral foot, and instead, iliopsoas-induced hip flexion can add lifting of the heel to its tasks that initiate swing. If stride length is to be maintained, hip and knee extension will need to increase before heel lift. However, an alternative option is to shorten stride length, but this requires a change in the preferred walking speed to something much slower (see Fig. 4.4.5c).

The long-term consideration with the change in stride length is how well the individual adapts mechanically and physiologically to the new gait parameters. It is likely that passive energy storage through connective tissue stretch will decrease, leaving a deficit that metabolic muscular energy will need to compensate for. If the patient has comorbidities or is developing sarcopenia, then the increased muscular energy burden may not be tolerated well. As the plantarflexors are not well compensated for by other proximal lower limb muscles or other ankle plantarflexors, loss of triceps surae ankle plantarflexor strength poses a considerable mechanical challenge. It is possible that with modern lifestyles that promote gracility of structures, poor initial development in childhood, and weakening of the calf muscles through lack of use in adulthood is likely to be implicated in much locomotive dysfunction and pathology.

4.4.6 Achilles tendinopathy in forefoot strikes

While all normal walking and most distance running produces a rearfoot strike, midfoot, and forefoot strikes also occur as part of normal running techniques. In all running strike patterns, there is a rapid early triceps surae–Achilles complex muscle–tendon unit fascicle lengthening at the first 10% of contact. The fascicles then shorten throughout the rest of contact while the tendinous tissue alone lengthens before elastic recoil, after which the tendon also shortens during acceleration (Ishikawa et al., 2007). Unlike walking, where the whole myofascial unit is stress loaded, running's concentric-near-isometric muscular contraction places the burden of elastic energy storage and acceleration firmly within the Achilles tendon.

Forefoot strikes in runners generate greater Achilles tendon stresses, strains, and strain rates (Almonroeder et al., 2013; Kulmala et al., 2013; Lyght et al., 2016), and increase loading time during stance from 25% to 100% compared to rearfoot strikers (Kernozek et al., 2018). The reason for this has previously been discussed in Chapter 1, Section 1.4.9, but it is worth considering the situation again. This is not least because the situation with the Achilles acting as an energy dissipator is also true of abnormal toe-walkers, and also to events where humans land on their toes deliberately or through perturbations. The more extreme the forefoot strike, the more dramatic the initial loading stress on the triceps surae–Achilles complex is likely to be (Fig. 4.4.6a). If highly repetitive, then the triceps surae is likely to fatigue earlier requiring a change in running patterns to offload the Achilles.

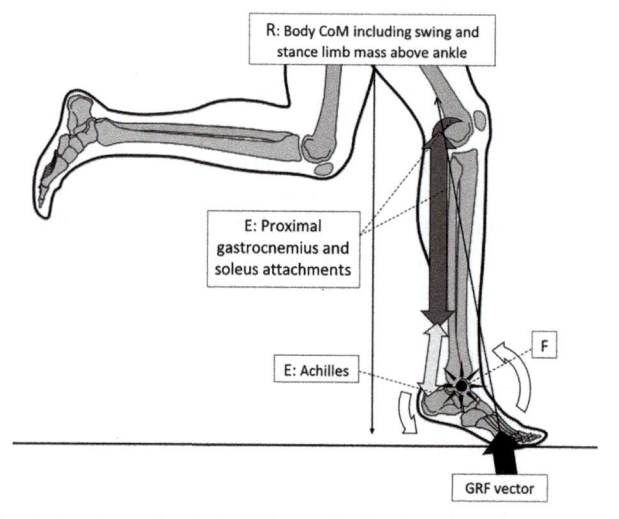

FIG. 4.4.6a A free-body diagram of forefoot running strike clearly indicates why the triceps surae and the Achilles become the myofascial structures of ankle energy dissipation. The forefoot GRF vector (black arrow) creates a dorsiflexion moment at the ankle, acting as the fulcrum (F—star). The CoM of the body above the ankle (including the stance and swing limb) becomes the resistance (R) that is driving the heel towards the ground (thin black arrow). This is resisted by the effort (E), provided by the triceps surae applying its forces to both proximal and distal attachments to limit heel drop. These forces occur through muscle and tendon fibre lengthening. Both effort points are closer to the ankle fulcrum than the resistance in the sagittal plane and particularly in the vertical direction, setting up a class three lever system. Thus, in forefoot running, the triceps surae and Achilles are subjected to prolonged high stresses from contact to acceleration, something that heel strike running can avoid. *(Permission www.healthystep.co.uk.)*

Runners with Achilles mid-portion tendinopathy demonstrate deficits in plantarflexor torques and endurance, explained by loss of soleus force-generating capacity rather than that of gastrocnemius (O'Neill et al., 2019). This may explain why forefoot runners often revert to heel striking over time during distance running and that the vast majority of elite marathon runners are rearfoot strikers (Hanley et al., 2019). Interestingly, female runners are more likely to maintain a nonrearfoot strike throughout a marathon than are males (Hanley et al., 2019), perhaps reflecting their lower transport mass.

Although the loading rate is rather different in forefoot running, habitual toe-walkers face similar issues in relation to the length of stance phase triceps surae–complex loading during locomotion. In toe-walking, ankle plantarflexors become starkly biphasic, making the triceps surae the primary muscle–tendon buffering unit during impact loading (Houx et al., 2013). This changes the triceps surae–Achilles complex into a walking braking energy dissipator as well as a midstance decelerator of the CoM in late midstance, while terminal stance ankle plantarflexion power becomes much reduced for acceleration (Matjačić et al., 2006; Houx et al., 2013). In plantarflexed ankle angles, the ankle plantarflexors find themselves at a contractile disadvantage, reducing their ability to generate force (Neptune et al., 2007). With an increased ankle plantarflexion angle throughout gait, toe-walkers tend to walk with a high CoM stance position, decreased hip motion, demonstrate half the normal knee extensor moments (Matjačić et al., 2006), and express smaller vertical displacements of their CoM than full plantigrade walkers (Usherwood et al., 2012).

Toe-walking is common in young children and can be associated with a number of neurological conditions. However, there are a number of idiopathic childhood toe-walkers that are without a known neurological cause (Pendharkar et al., 2012), as is discussed in Chapter 1, Section 1.5.6. Toe-walking may be present in 100% of steps or found to be variable and transitory, and may resolve over time without intervention. Toe-walkers change their pattern and increase the variability of ankle angles, particularly during midstance when compared to the greater consistency of normal plantigrade walkers (Pendharkar et al., 2012). Not only is knee, hip, and ankle flexion increased at contact, but many other changes also occur, including increased knee varus moments, reduced hip adduction moments, an internal foot progression angle, and less knee flexion during swing (Houx et al., 2013). In persistent cases under neuropathic pathologies such as cerebral palsy, toe-walking creates even greater clinical challenges as there tends to be more gait sway compared to the idiopathic toe-walkers. As toe-walkers grow and develop a larger body mass, toe-walking becomes more energetically problematic due to the instability generated by this distinctive gait's high CoM and potentially poor ankle stability (Fig. 4.4.6b). Persistent toe-walking in older, more experienced walkers requires urgent therapeutic interventions such as surgery, rehabilitation, specialist shoes, and/or orthosis management.

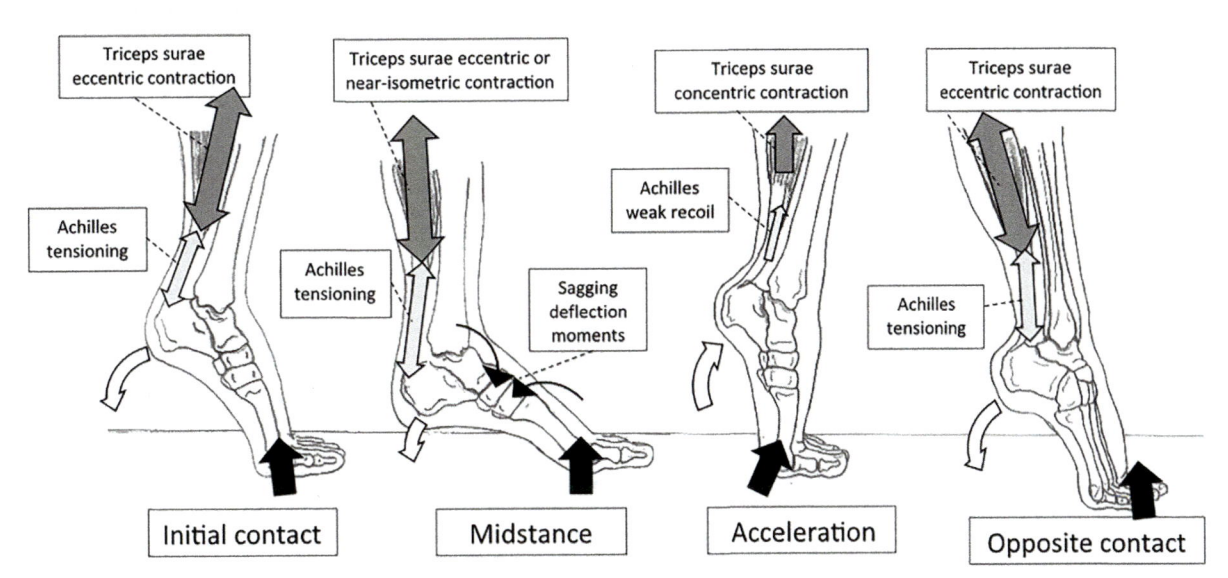

FIG. 4.4.6b Toe-walking (equinus gait) presents issues for the triceps surae and the Achilles as well as many other structures. Equinus gait varies under individual motions within the midfoot during midstance and acceleration. In all cases under ankle dorsiflexing GRF (black arrows), triceps surae becomes the provider of ankle energy dissipation at contact and maintains this throughout loading response. The Achilles is rapidly strained with tension and must sustain dual roles of energy dissipation and storage from loading response and throughout midstance. In acceleration, the Achilles can release some energy, but as the ankle is already plantarflexed, the power of elastic recoil is severely compromised. Young children in good health may not be significantly compromised in toe-walking gait when they are light in their body mass. However, under increasing body weight, the foot may become less vertical in its orientation during midstance and thus, foot anatomy as well as the Achilles may start to suffer under abnormal, more foot-horizontal loading patterns. In neurological cases such as those seen in cerebral palsy, the midfoot is often subject to high sagging deflection bending moments that result in midfoot break throughout stance and particularly during acceleration. These vary in their ability to generate forces of acceleration medially or laterally across the foot. (Permission www.healthystep.co.uk.)

4.4.7 Achilles ruptures, posterior calcaneal avulsion fractures, and Sever's disease

Sudden high-loading events on the triceps surae–Achilles complex tend to present differently to repetitive high cyclical loading strains of overuse. Achilles ruptures and avulsion fractures of the Achilles enthesis organ can present with similar symptoms, and both can result from excessive strain within the Achilles tendon independent of strain rate (Wren et al.,

2001). What separates the excessive strain causing an Achilles rupture from that causing an avulsion fracture is the bone mineral density in the posterior tubercle of the calcaneus. Wren et al. (2001) found that in human cadaveric specimens, avulsion fractures always occurred when bone density was less $0.04\,\mathrm{g\,cm^{-2}}$. However, the tendon always failed first under strain if the bone density was above $0.08\,\mathrm{g\,cm^{-2}}$ with a majority failing at the tendon when the bone density was between the two readings. It appears that tendon and ligament material properties are better maintained with age than is bone density.

Age and pathology influence which of the tissues around the enthesis are likely to fail first during loading. Under the age of 20 years, high tensional stresses are more likely to result in avulsion fractures since in this age group, the Achilles tendon's tensional strength should exceed that of immature bone under tension. From age 20 to 55 years, tendon ruptures predominate as bone has tensile properties that are higher than the Achilles (Wren et al., 2001). However, after this age, failure predominantly tends to move back to the bone, increasing avulsion fracture risk, because tendon strength is better maintained than bone strength (Wren et al., 2001). These findings sit well with clinical reports that Achilles ruptures mainly affect middle-aged men (Landvater and Renström, 1992), whereas Achilles avulsion fractures mainly affect middle-aged and elderly women (Michael and Banerjee, 1993; Levi et al., 1997).

The enthesis of the Achilles is a vulnerable location for pathology under loading stresses in late childhood and early adolescence due to skeletal maturing from immaturity, and possibly through differences in the local soft tissue anatomy. The Achilles tendon and the plantar aponeurosis are continuous below the calcaneus when young, but this continuation reduces with age (Snow et al., 1995). While this Achilles tendon–plantar aponeurosis continuation is present in life, it can more closely couple ankle motion to vault mechanics and forefoot motion across the foot. Increasing loads and developing foot vault mechanics may influence the development of conditions such as Sever's disease through over-tensioning, compressing, and shearing of the immature secondary ossification centre located at the posterior calcaneal tubercle. The posterior calcaneus secondary ossification centre remains cartilaginous until age five but then starts to ossify from its plantar aspect by age seven, starting to fuse around age twelve, and completing its union between the ages of fourteen and eighteen (Rossi et al., 2016). This maturing ossification centre can be overstressed by increasing strains within the triceps surae myofascia and/or via the plantar aponeurosis (Fig. 4.4.7). It has been suggested that Sever's disease is a stress fracture of this secondary ossification

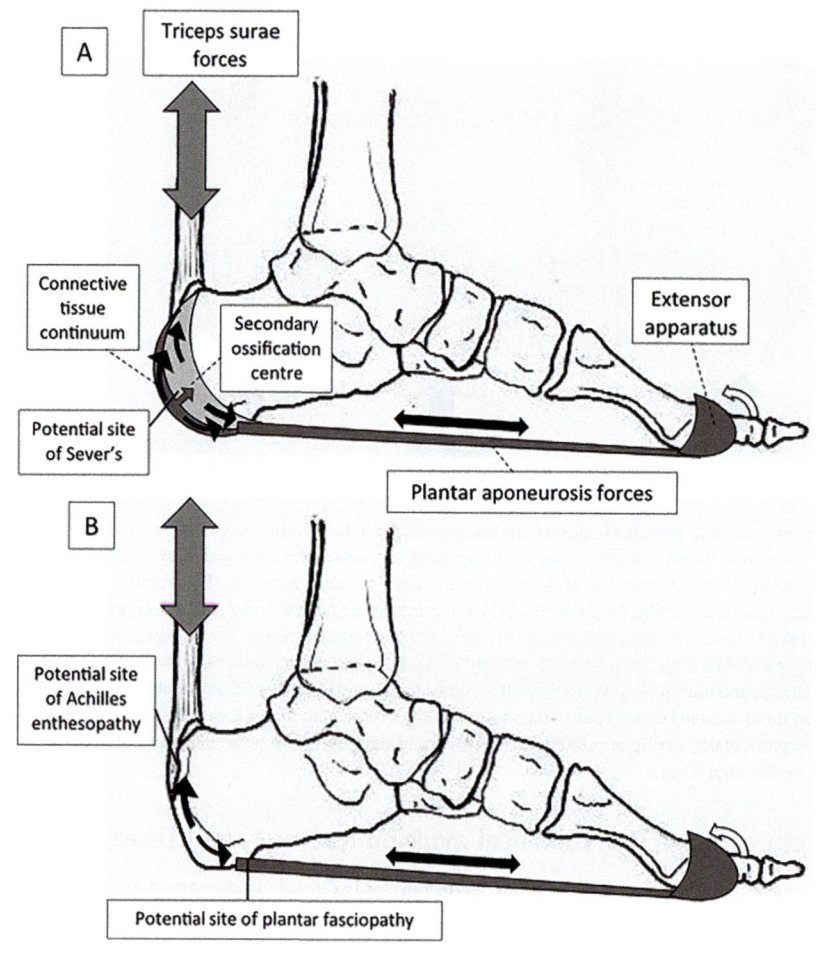

FIG. 4.4.7 See figure legend on opposite page

centre (Ogden et al., 2004), but increased density and fragmentation of the calcaneal apophysis is a common finding in asymptomatic individuals and looks very much the same as in symptomatic cases. Thus, the actual cause of the symptoms and the pathomechanics involved remains unclear (Scharfbillig et al., 2008; Perhamre et al., 2013).

4.4.8 Frontal and transverse plane rearfoot relationships in pathomechanics

The frontal and transverse plane mechanical restraints of the rearfoot limit torques within these planes so that sagittal motion can dominate. The ankle is heavily reliant on ligament restraint to hold the ankle bones together, especially at the relatively minimally mobile tibiofibular joints through the proximal and distal anteroposterior tibiofibular ligaments and syndesmosis. The ankle joint is therefore strongly braced both medially and laterally to restrict frontal and transverse plane torques, but freely permits sagittal plane motion under less ligament restraint.

The combined restriction by force and form closure greatly alters rearfoot joint freedom throughout its motion. In dorsiflexion, the ankle becomes increasingly form-closed through the shape of the talar trochlear surface within its fibular and tibial articulations. Ankle dorsiflexion-associated transverse plane torques are increasingly absorbed through the minor 'separating' motions and external rotation of the fibula on the tibia, rather than through the rotations within the articular surface. This relies heavily on ligament stiffness and recoil elasticity. Rearfoot rotation is controlled through the anterior and posterior tibiofibular ligaments and the syndesmosis, forming a stiff elastic recoiling transverse plane shock absorber. The fibula can also glide slightly superiorly with dorsiflexion, giving the tibiofibular joint a very small frontal plane torque-absorbing role with dorsiflexion and eversion. Tibiofibular motion should be considered part of ankle function (Fig. 4.4.8a).

In ankle plantarflexion, form closure is much reduced with soft tissues force closure taking over ankle stability as freedom of motion within the ankle increases. Passively, the medial and lateral ankle ligaments are the essential restraining mechanisms, not only because of their mechanical properties, but also because of their important sensorimotor information influence on the surrounding muscular activity. The passive ligaments are strongly assisted in providing rearfoot stability medially by the invertor/adductor muscles consisting of tibialis posterior, flexor digitorum longus, flexor hallucis longus, and tibialis anterior. Laterally, muscular stability is provided by the evertor/abductor muscles of peroneus longus and brevis (Fig. 4.4.8b). Failure of any of these passive or active soft tissue elements risks ankle instability, which becomes increasingly more significant as the ankle moves away from end-range dorsiflexion.

Tendon lever arm positions in relation to the axis of the rearfoot complex influence how effective these muscles can be at resisting frontal and transverse plane motions. The soft tissue structures used to control ankle joint frontal and transverse plane mechanics also cross the subtalar joint. The clinician must recognise the phases of gait when both the ankle and subtalar joint are able to move around their functional axis. When both joints move, the axis is that of the ankle–subtalar or rearfoot complex axis. Considering the ankle axis alone or combined rearfoot axis depends on whether the calcaneus is loaded or unloaded by external forces. When the rearfoot is directly unloaded, frontal and transverse plane rotations can only occur at the ankle. Also, when the ankle is dorsiflexed, both ankle and subtalar joint motion become more restrained. The ankle axis location moves anteriorly and posteriorly with dorsiflexion and plantarflexion, respectively, altering not only the internal and external ankle moment arms but also the plantarflexion moment arm from Achilles to the MTP joint fulcrum. The subtalar joint axis moves more medially with eversion and laterally with inversion. The combinations of each axis location during motion positions the instantaneous axes of rotation for the entire rearfoot.

Soft tissue rearfoot stabilisers are particularly important during contact phase and after heel lift into terminal stance. Because of greater plantarflexed ankle freedom, the more plantarflexed the position of the rearfoot during loading, the more

---

FIG. 4.4.7, CONT'D The relationship between forces from rearfoot to forefoot alters with anatomical changes that occur with age. In children (A), some superficial connective tissues from the Achilles are continuous over the plantar surface of the posterior calcaneal tubercle to the proximal plantar aponeurosis. The posterior calcaneal tubercle also has a secondary ossification site which starts to ossify in its plantar third by age seven years, with fusion beginning around age twelve years. The process is usually completed between fourteen and eighteen years. While the ossification process and the fibrous continuation exist, the ossification site and fibres will transfer forces between the Achilles and the plantar aponeurosis directly. These tissues have to mature to cope with rising forces associated with growth, increased body weight, and exercise. Shear motion and increased compressions may occur between the separate growth plate and the rest of the body of the calcaneus due to the more extensive connective tissue continuum running over the maturing bone. In excess, such strains may provoke posterior or plantar calcaneal symptoms in children, a condition known as Sever's disease. The fibrous continuation tends to decrease during the third decade of life. In mature and older adults, the loss of the connective tissue link under the calcaneus means that forces are transferred from the Achilles enthesis across the rigid calcaneus to the enthesis of the plantar aponeurosis. From here, they pass across the aponeurotic connective tissue fibres to the forefoot via connections to the extensor apparatuses and flexor plate mechanisms. This makes the Achilles and aponeurosis enthesis sites of potential stress concentration as soft and hard tissues create unions of highly different mechanical properties (B). Plantar fasciopathy (fasciitis) and Achilles insertitis are common adult pathologies of the rearfoot, whereas Sever's is the common rearfoot pathology of late childhood and early adolescence. *(Permission www.healthystep.co.uk.)*

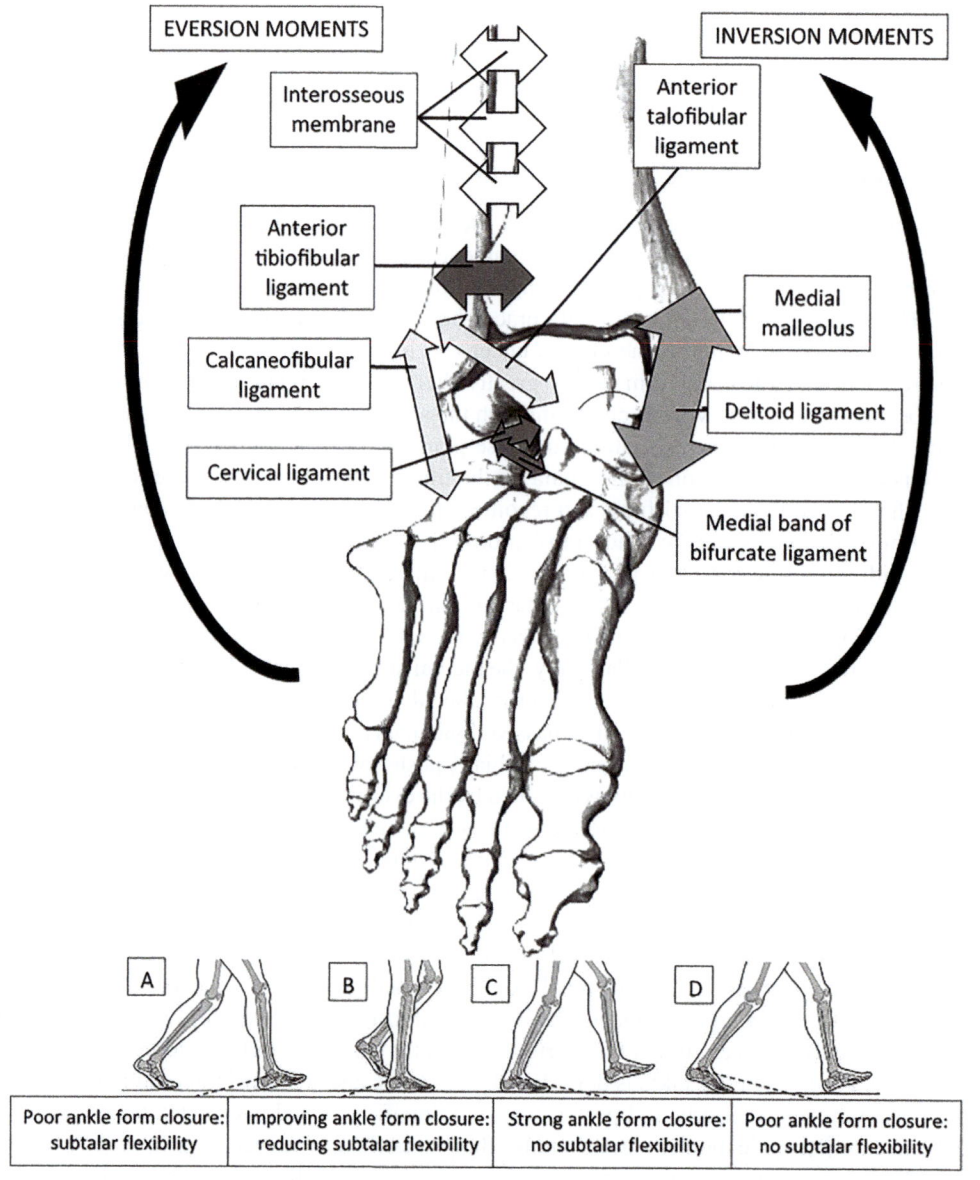

FIG. 4.4.8a Ligamentous stabilising vectors around the rearfoot vary in their significance with changes in joint flexion and extension. Sagittal rearfoot motion alters the degree of form closure of the ankle joint and flexibility of motion within the subtalar joint. The top image demonstrates the location of lateral ligament restraining vectors such as the anterior talofibular, calcaneofibular, and posterior talofibular (not shown) that resist rearfoot inversion moments. The medial deltoid ligament vectors restrain eversion. Subtalar ligament vectors of the cervical, interosseous, and medial (calcaneonavicular) band of the bifurcate ligaments restrain medio-anterior displacement of the talus and navicular in relation to the calcaneus, respectively. These are under most strain primarily during loading response and early midstance. The distal anterior tibiofibular ligament vector increases under ankle plantarflexion moments as do all the anterior ligament force vectors. The interosseous membrane and ligament vectors between the tibial and fibular shafts restrain external rotation of the talus and fibula, while also absorbing transverse plane torque energies. When considering the rearfoot during gait, initial heel contact (lower images—A) increases subtalar flexibility and via rearfoot plantarflexion, decreases ankle form closure. Thus, there is more reliance of muscular force closure to stabilise the rearfoot. By absolute midstance (B), the subtalar joint flexibility is being reduced and ankle form closure is increasing. In late midstance (C) under ankle dorsiflexion and increased muscle activity, form closure and force closure reach their peak and subtalar flexibility is minimal, reducing the risk of frontal plane ankle perturbations. In terminal stance (D), subtalar joint flexibility is absent and ankle form closure is rapidly reducing. Thus, ligaments at the ankle are more susceptible to injury during loading response and terminal stance, with subtalar ligaments being at greater risk during loading response and early midstance perturbations. However, the size and direction of the perturbation force remains the most important factor to where and how injuries develop. *(Permission www.healthystep.co.uk.)*

Pathology through the principles of biomechanics **Chapter | 4** 729

EVERSION and ABDUCTION MOMENTS

Flexor hallucis longus

INVERSION and ADDUCTION MOMENTS

Fibula

Tibia

Rearfoot inversion resistance:
Peroneus longus
Peroneus brevis
Peroneus tertius
Peroneus quartus

Rearfoot eversion resistance:
Tibialis posterior
Flexor digitorum longus
Tibialis anterior

Talus

A	B	C	D
Tibialis anterior Tibialis posterior Peroneus longus = net eversion resistance	Tibialis posterior Peroneus longus (Tibialis anterior occasionally if required) = adaptable balance	Tibialis posterior Flexor digitorum longus Flexor hallucis longus Peroneus longus Peroneus brevis = balance stability	Tibialis posterior Flexor digitorum longus Flexor hallucis longus Peroneus longus Peroneus brevis = adaptable stability

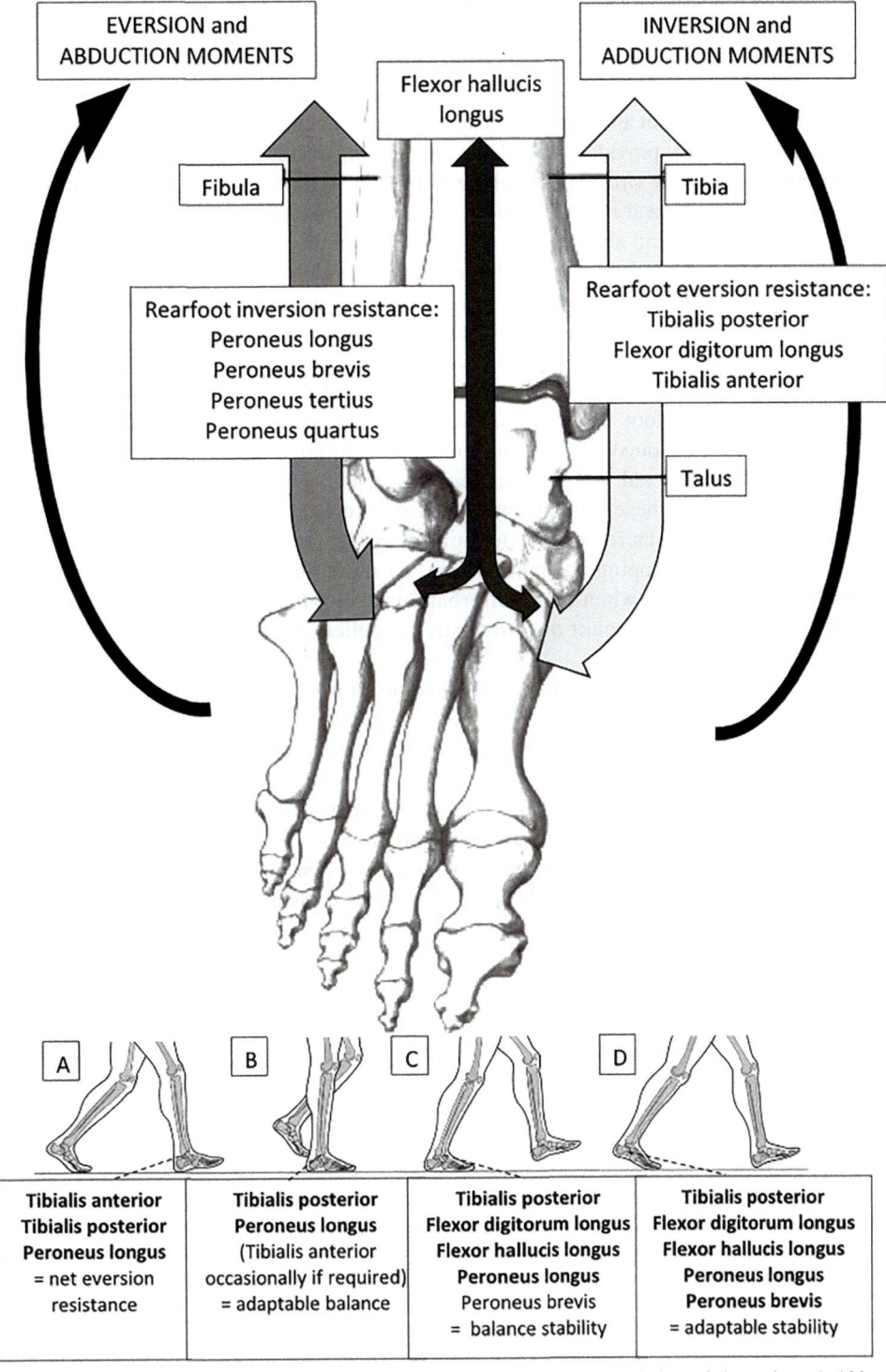

FIG. 4.4.8b Force closure around the ankle is maintained through balancing muscular force vectors during gait (upper image). Although the triceps surae via the Achilles subtendons and the digital extensors have some influence, the primary rearfoot vectors for frontal and transverse plane motion derive from tibialis posterior, flexor digitorum longus, and tibialis anterior for resisting eversion and abduction of the foot at the rearfoot while the peroneal muscles resist inversion and adduction. Flexor hallucis longus seems to play a homogenising role, helping to balance active vectors during stance phase. During heel strike loading response (lower images—A), subtalar freedom is high and ankle form closure low, with the rearfoot usually subjected to eversion and plantarflexion moments. This is restricted by tibialis anterior's and posterior's activity modulated by peroneus longus, with a variable input from other peroneals if necessary. In early midstance (B), muscular activity is reduced, yet tibialis posterior and peroneus longus remain active to provide adaptable restraints with occasional tibialis anterior activity if required. During late midstance (C), tibialis posterior and flexor digitorum longus increase inversion and adduction moments, while peroneus longus and gradually increasing peroneus brevis activity helps restrain them. Flexor hallucis longus homogenises these muscular vectors to create adaptable stability. In terminal stance (D), balanced stability in frontal and transverse plane moments across the rearfoot should be achieved and yet permit a net midfoot adduction and eversion moment during midfoot plantarflexion that increases loads towards the medial metatarsal heads. *(Permission www.healthystep.co.uk.)*

reliant the rearfoot is on the soft tissues for frontal and transverse plane stabilisation. Therefore, any gait situation that limits ankle dorsiflexion is likely to increase the significance of the active and passive frontal/transverse plane stabilisers. Thus, rearfoot strikes on longer strides and forefoot strikes are more prone to result in frontal or transverse plane perturbations. The close functional relationship of the ankle and the subtalar joint when the heel is loaded by external forces, means that instability problems around the rearfoot are better understood by approaching the pathomechanics around a single rearfoot functional unit. The ankle primarily provides sagittal plane function, with the subtalar joint providing a mechanism for reducing frontal and transverse torque stresses through perturbations in these planes, but only during heel weightbearing (Leardini et al., 2001). Without calcaneal external loading, the ankle is on its own in managing frontal and transverse plane perturbations, save that the midfoot can absorb some torques between the forefoot and the ankle.

The inability for the subtalar joint to provide motion at higher ankle plantarflexion angles explains the requirement for more frontal and transverse ankle freedom to be available with plantarflexion. However, when the ankle is plantarflexed and loaded during terminal stance, the midfoot and forefoot offer frontal and transverse plane torque absorption capacity. They also provide such motion in late midstance when the ankle and subtalar joint do not offer any significant frontal or transverse motion, being restrained within ankle dorsiflexion. Forefoot and midfoot motion can spare the rearfoot from any small frontal and transverse plane torques while the forefoot is weightbearing. Large frontal and transverse perturbations will initially be transferred proximally to the hip and knees when the rearfoot is dorsiflexed and thus stiffened and stable, risking more proximal lower limb injuries. When the ankle is plantarflexed, these frontal and transverse torques are absorbed initially within the rearfoot complex. The plantarflexed ankle's increased freedom of motion leaves it more vulnerable to frontal and transverse plane perturbations such as when stepping down an incline, on stairs, into a hole, or while wearing high heels. If the calcaneus is loaded by external force when such perturbations occur, then the subtalar joint can assist in absorbing the torques. Yet ultimately, it is the amount of torque stress applied and areas loaded that dictates the final injury risk, regardless of the rearfoot posture.

Many ankle plantarflexors are also rearfoot invertors-adductors or evertors-abductors which in unison, create a tensile frontal plane bracing mechanism around the ankle (Potthast et al., 2008). The invertors in isolation will increase medial intraarticular stresses, while the evertors in isolation cause lateral intraarticular stresses. Harmony is created by balancing stresses across the ankle joint, with flexor hallucis longus playing an important homogenising role on ankle joint stresses (Potthast et al., 2008). Potentially, dysfunction within this muscular balance will disturb the ankle joint's intraarticular stresses, with tibialis posterior having a particularly large medial influence (Potthast et al., 2008). With the ankle articular motion most restricted in maximal dorsiflexion angles and with stresses concentrated to the medial aspect of the ankle joint throughout late midstance, tibialis posterior is likely to have the most influence on intraarticular stress management with flexor hallucis longus, just before heel lift.

The primary muscles involved in the frontal and transverse plane power balances are tibialis posterior, peroneus longus, and brevis muscles. Tibialis posterior and peroneus longus are antagonists of each other's frontal and transverse plane torques across the foot, yet they work agonistically in controlling foot stiffness and midfoot plantarflexion throughout the stance phase of gait (Kokubo et al., 2012). The peroneus longus is assisted by peroneus brevis to overcome the tibialis posterior's strong inversion moment and thus together provide the eversion moment around the ankle–rearfoot and lateral column articulations before and during heel lift. This moves the CoGRF medially towards the longer fulcrum created by the 1st and 2nd metatarsals by their MTP joints. By doing so, the plantarflexion moment arm length increases for applying ankle plantarflexion power, moving the MTP joint fulcrum to align to the lower limb's line of progression. Any inversion perturbation during heel lift will be resisted by increasing activation of already active, peroneus longus and brevis. As only peroneus longus is normally active at contact and loading response, with only low or intermittent peroneus brevis activity, it alone takes the burden of resisting inversion moments during loading response. For peroneus brevis assistance in loading response to inversion perturbations, sensorimotor activation is required. However, the neurological system is often too slow to activate for this assistance. Thus, loading inversion events are more reliant on passive ligament restraint and thus more susceptible to ligament injury than are terminal stance inversion events that are more actively constrained (Fig. 4.4.8c).

Eversion moments are primarily resisted by tibialis posterior, with some assistance from the long flexors. These muscles are 'preprogrammed' to be active during initial contact and terminal stance, potentially giving

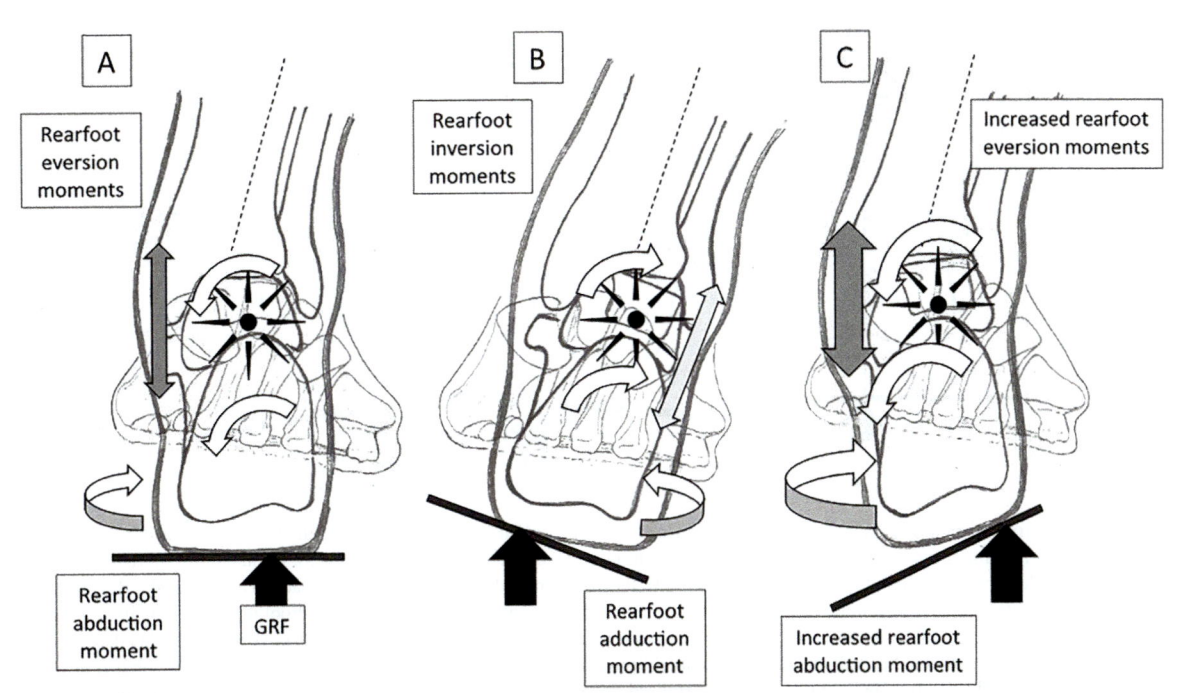

FIG. 4.4.8c The heel contact position during gait is important as it influences not only the sagittal plane moments on the rearfoot but also the frontal and transverse plane moments. On flat terrain (A), because the lower limb is normally presented inverted to the ground due to the lower limb contact adduction angle at the hip (dashed line), the lateral posterior aspect of the heel contacts first, creating a GRF across the ankle and subtalar joints that induces rearfoot eversion around a combined functional axis located between these joints (star). Such eversion moments are restrained by active and passive medial soft tissue vectors (dark-grey double-headed arrow). Structures include tibialis posterior, deltoid ligaments, and the spring ligament. This eversion moment will be influenced by the stride length (tending to increase the posterior position of the GRF on longer strides) and the adduction angle of the lower limb, which is influenced by such features as coxa vara/valga and genu/tibial varum/valgum if present, and hip abductor/adductor activity/flexibility. Longer strides tend to increase anterior translation moments on the tibia and talus from the calcaneus, putting more strain on the subtalar ligaments. On terrain angled laterally downward to the foot (B), the medial heel may contact first, moving the GRF medially to initiate an inversion moment around the combined functional axis. This angular rotation should be restrained by active and passive anatomy such as the peroneal muscles and lateral ligaments (light-grey double-headed arrow). If the terrain is angled down medially (C), the GRF may occur more laterally than on flat terrain, increasing the eversion moment across the combined axis. This requires greater medial soft tissue restraint including higher levels of muscle activation from tibialis posterior to moderate strain on the deltoid and spring ligaments. Injuries within the rearfoot occur when soft tissue-resisting vectors fail to adapt and/or when frontal plane angular moments become too large as a result of increasing GRF distance from the axis of rotation. Injuries can arise from ligament, muscle, tendon fibre tears, or from changes in peak articular compressions locations within either the ankle or subtalar joints, or within the bone. These result from frontal plane and/or associated rearfoot abduction/adduction rotation torques. *(Permission www.healthystep.co.uk.)*

the rearfoot greater eversion resistance capacity as long as muscular function is normal. However, the initial contact position of the foot is important to how the muscles react because in forefoot contact, only the ankle joint offers motion within the rearfoot (Fig. 4.4.8d). A similar situation exists for maintaining terminal stance ankle stability (Fig. 4.4.8e).

The ankle ligaments are divided into three groups. The lateral collateral ligaments consist of the anterior tibiofibular ligament, the calcaneofibular ligament, and the posterior tibiofibular ligament. The medial collateral ankle ligament is the deltoid ligament, a unified ligament with five distinct thickenings divided into a deep and superficial layer. The final group are the ligaments of the tibiofibular ligaments and syndesmosis (Golanó et al., 2010). The lateral ligaments resist inversion moments, the medial ligaments resist eversion moments. The tibiofibular syndesmotic membrane and ligaments between the tibia and fibula, also absorb transverse plane torques assisted by both the anterior and posterior positioned ankle collateral ligaments (Golanó et al., 2010).

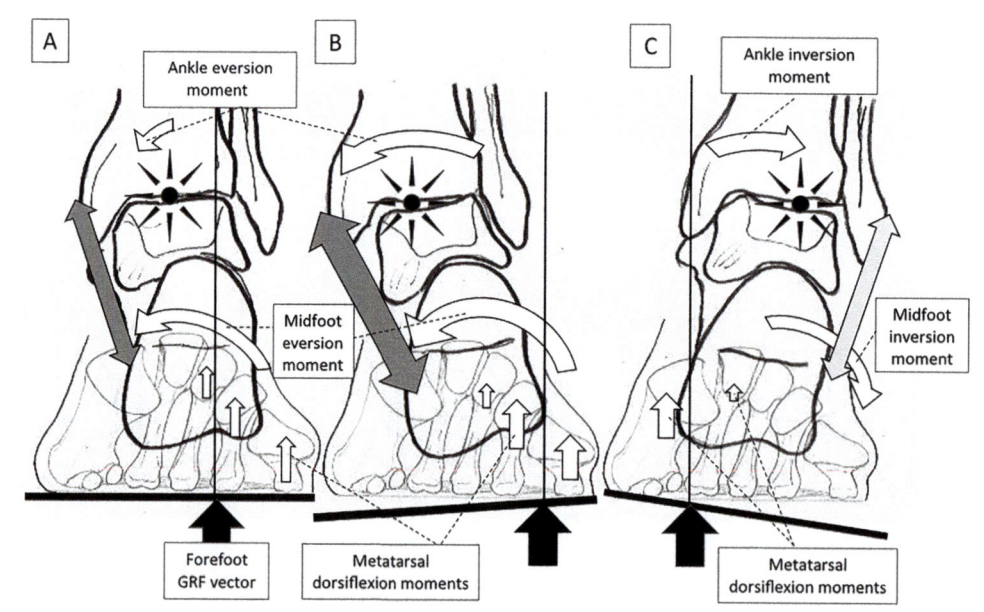

FIG. 4.4.8d During forefoot contacts, rearfoot biomechanics are altered for the subtalar joint does not normally have flexibility of motion, and the ankle is now in plantarflexion at initial contact, giving it greater freedom of motion. Thus, the axis of rotation (star) for frontal (and transverse) motion becomes located within the ankle joint alone. On flat terrain forefoot strikes (A), the lower limb adduction angle tends to cause a more lateral GRF, creating eversion and dorsiflexion moments across the midfoot and tarsometatarsal joints. If motion within the forefoot and midfoot is sufficient to dissipate the eversion moment, the rearfoot may be spared any significant frontal and transverse plane motions. Any that reach the rearfoot should be easily dissipated by medial rearfoot soft tissues (dark-grey double-headed arrow), although tibialis posterior will need to resist increased midfoot eversion and abduction moments. On terrain angled down medially (B), the GRF will be further lateralised, increasing the eversion moment that makes rearfoot involvement more likely, causing higher medial soft tissue strains and requiring greater resistive muscular activity. If the terrain is angled down laterally (C) or indeed, the limb is more abducted to the trunk, then the GRF at forefoot contact may load medial metatarsals involving the more restrained 2nd metatarsal's tarsometatarsal joint. If a resulting inversion moment is not dissipated across the forefoot and midfoot, the rearfoot may be subjected to inversion moments that will stress the lateral rearfoot soft tissues (light-grey double-headed arrow) requiring increased peroneal muscle activity. For injury, the size of the angular momentum, the soft tissue strengths, and the forefoot and midfoot mobilities become important factors to consider. *(Permission www.healthystep.co.uk.)*

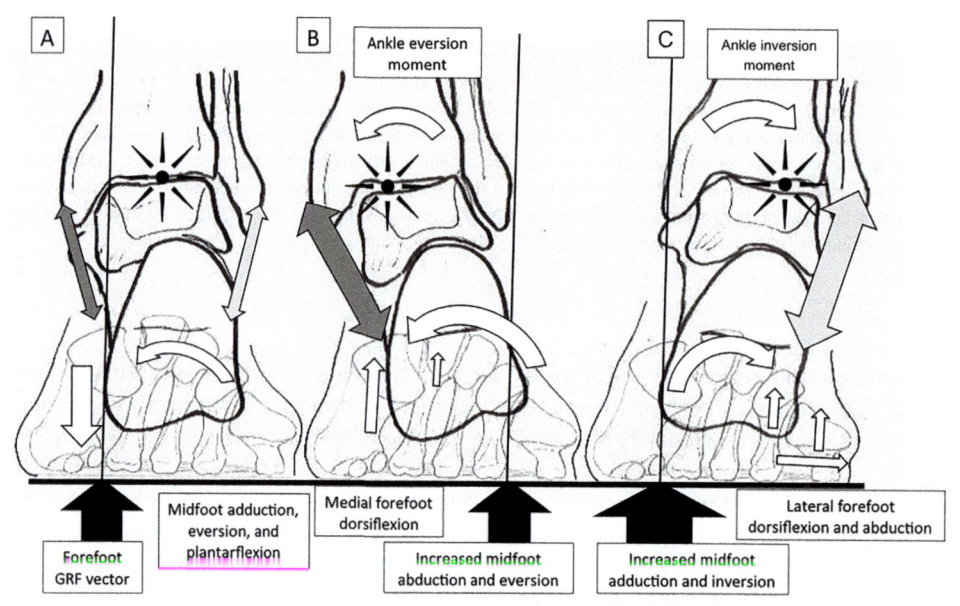

FIG. 4.4.8e After heel lift, with GRFs is positioned under the forefoot alone, rearfoot motion should only occur via rotation at the ankle joint's axis (star). The ankle has increasing mobility as its form closure reduces with rising angles of plantarflexion. The CoGRF should move towards stable medial MTP joints under midfoot peroneal eversion and plantarflexion moments, tibialis posterior adduction and plantarflexion moments, and flexor hallucis longus 1st metatarsal plantarflexion moments aided by plantar intrinsic muscle activity. This activation pattern should prevent any significant frontal plane perturbations on the rearfoot (A). If a sudden increase in lateral GRFs occurs on the forefoot (B), for example, if the ground gives way medially or if the medial forefoot is hypermobile and unstable, then the ankle may be forced to evert during terminal stance. Medial rearfoot soft tissues will be put under increased strain (dark-grey double-headed arrow) unless active medial vectors such as those from tibialis posterior and flexor digitorum longus can respond by increasing their forces of eversion resistance. If the medial forefoot GRF suddenly increases (C) and the forefoot and midfoot cannot adequately compensate, then the lateral rearfoot soft tissues may come under increased strain (light-grey double-headed arrow), requiring increased peroneal activation. The more mobile lateral metatarsals may be able to absorb some of the inversion moment through their dorsiflexion and abduction, but they are also thus sites of possible injury as a result. Higher-heeled shoes create an increased vulnerability through unusual rearfoot external forces that can permit subtalar flexibility while the ankle is in plantarflexion, increasing overall rearfoot joint mobility throughout stance phase. Stiletto heels compound the issue via reduced contact area under the footwear's heel, increasing perturbation risk. *(Permission www.healthystep.co.uk.)*

4.4.9 Ankle sprain pathomechanics

Ankles can sprain both medially by eversion moments or laterally by inversion moments across the rearfoot, with inversion sprains being far more common (Hertel and Corbett, 2019). The medial collateral (deltoid) ligaments are damaged by large eversion moments, while the lateral collaterals are usually damaged during inversion moments (Golanó et al., 2010). A degree of transverse plane torque is also associated with these injuries, with an external tibial rotation and rearfoot adduction associated with inversion sprains and the reverse being the case in an eversion sprain (Hertel and Corbett, 2019). This reflects the obliquity of the rearfoot joint complex instantaneous axes throughout motion. The combined planar motions can set up a number of injuries during frontal plane perturbations, depending on the size of each planar torque.

Eversion sprains

Eversion ankle sprains risk damage to the muscular invertors particularly tibialis posterior, and any component of the medial collateral (deltoid) ligament (Golanó et al., 2010). The medial ligaments not only support the medial ankle but, through its continuity, also supports the rearfoot and proximal midfoot via navicular and spring ligament attachments (Bonnel et al., 2010). The passive resistance of the deltoid ligament in many ways replicates the active resistance of the tibialis posterior, linking their protective functions together. Failure of either ligamentous or muscular structures will have significant mechanical consequences on the other. Eversion sprains are less common than inversion sprains but are often associated with a significant external (abduction) torque at the rearfoot. This can prove catastrophic to ankle–foot function, risking both osseous and soft tissue trauma (Fig. 4.4.9a).

Tibialis posterior dysfunction increases medial rearfoot and midfoot plantarflexion instability through loss of eversion restraint and vault-stiffening of the foot. Any part of the deltoid ligament becoming dysfunctional will change the tibialis posterior loads through its linkage with the spring ligament. Failure of the spring ligament will also affect deltoid ligament and tibialis posterior function, because of their complex interdependency (Imhauser et al., 2004).

Inversion sprains

Inversion or lateral ankle sprains are among the most prevalent musculoskeletal injuries (Golanó et al., 2010; Roos et al., 2017; Hertel and Corbett, 2019), particularly among people who take part in sports and rough terrain recreational activity (Delahunt et al., 2018). Risk factors for lateral ankle sprains have been investigated, but many cases occur due to little more than 'bad luck' because we a dealing with a tall biped that is utilising a small base of support, often moving on unstable terrains. This makes identification of predisposing factors of only some limited value. Inversion ankle sprains often occur as the foot makes initial contact on uneven ground, especially when the ankle is in a plantarflexed position. However, they can occur anytime during stance events, even in ankle neutral or dorsiflexed positions if perturbation stresses are large enough (Delahunt et al., 2006; Hertel and Corbett, 2019).

Even though there is increased vulnerability to frontal plane perturbations when the ankle is plantarflexed, it is all still a matter of the amount of torque. Landing on the forefoot, especially from a jump, stepping into a hole with the foot in a plantarflexed attitude, or entering terminal stance on an unstable surface are common factors involved in ankle sprains. The mechanism of injury is related to the frontal plane perturbation moment arm from the GRF and the position of the BWV passing down the distal leg that are setting up angular moments across the axis of the rearfoot complex (Fig. 4.4.9b).

Effects of lower limb and rearfoot morphology on frontal plane rearfoot moments

The BWV being driven through the lower limb to the rearfoot means that rearfoot complex angular momentum is a consequence of the combined effects of the lower limb alignments to the rearfoot and the direction of the GRF. This means that osseous morphologies such as tibial or knee (genu) valgum or varum influence the direction of the BWV to the rearfoot and its distance from the rearfoot axis. Tibial valgum alignments tend to increase inversion moments at the rearfoot during stance phase. Varus alignment issues tend to increase rearfoot eversion moments, but when tibial varum is particularly large, it can still leave the rearfoot more inverted in posture to the ground through insufficient rearfoot eversion compensation. Osseous and articular morphology within the rearfoot can also predispose to increasing frontal plane torques, with an inverted rearfoot or calcaneal alignment increasing the risk of exposure to large inversion moments, while a hindfoot valgus position can increase eversion torques throughout gait. Such relationships are supported by the studies of Van Gheluwe et al. (2005). However, Van Gheluwe et al only considered the effects of these lower limb alignments on the subtalar joint, which is very hard to track in vivo, thus possibly making their findings more appropriate to the whole rearfoot.

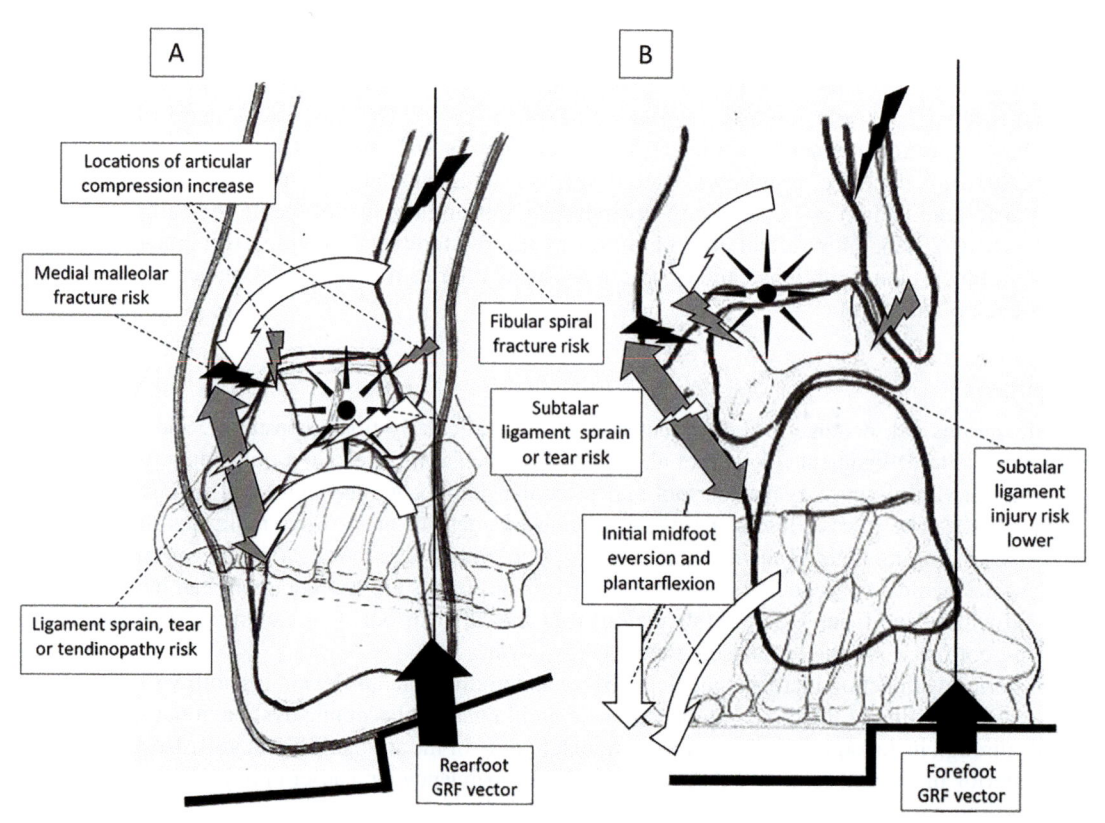

FIG. 4.4.9a Ankle injuries as a result of large lateral GRFs can occur at any time during the stance phase if perturbations within the frontal plane are large enough. The foot falling medially into a terrain undulation is a common example, as demonstrated in the images above. However, when the ankle is plantarflexing during heel contact on a medially declined surface (A), the rearfoot structures are more vulnerable. The functional rearfoot axis (star) of frontal plane rotation lies between the ankle and subtalar joint at this time. The subtalar joint is also under anterior talar translation forces over the calcaneus. Thus, the subtalar ligaments, the deltoid ligaments, and the spring ligament are all exposed to increased strains under high eversion moments during contact. They are protected by active restraint applied by medial extrinsic muscle activity and tendon compression around the medial rearfoot. High rearfoot eversion moments during heel contact evert and plantarflex the subtalar joint towards its close-packed position, increasing the risk of lateral articular compressions. Large talar eversion moments increase inferomedial and lateral ankle articular compressions and ultimately through external rotation, risk fibular spiral and inferomedial malleolar fractures. In terminal stance (B), when only the forefoot is loaded, similar injury situations exist, but the subtalar joint is unlikely to be directly involved, lowering any subtalar injury risk. Instead, the midfoot and forefoot will help absorb eversion energies initially through their greater mobility, moving the injury risk to forefoot and midfoot structures. Only if there are insufficient midfoot and forefoot eversion compensations will the rearfoot follow the eversion and abduction motion, and thus structures such as the deltoid and spring ligaments be at risk. Tibialis posterior, by bridging the rearfoot and midfoot is at particular risk during both rearfoot and forefoot contact periods, and its weakness and dysfunction is also likely to increase the effects of eversion moments on foot and ankle motion both during loading and acceleration. *(Permission www.healthystep.co.uk.)*

Rearfoot alignment can cause a dominant frontal/transverse plane displacement within the subtalar joint, a situation that can increase the displacement of the subtalar joint axis in the transverse plane. With increased supination (inversion), the subtalar joint axis is moved laterally, while with increased subtalar pronation (eversion), the axis moves medially (Leardini et al., 2001). The position of the subtalar joint under load is dependent on the internal ligament tensions and the generation of medial and lateral torques around the rearfoot, that should balance each other. This means that loss of balancing soft tissue torques between weakening invertors and stronger evertors could result in more weightbearing subtalar pronation (talar eversion, talar anterior glide, adduction, plantarflexion, and distal calcaneal plantarflexion), shifting the subtalar axis medially, and thus promoting further eversion moments around its joint axis.

Poor evertor function gives advantage to any inversion moments, and could increase the potential for subtalar motions towards supination (talar inversion, talar posterior glide, abduction, dorsiflexion, and distal calcaneal dorsiflexion) associated with moving the subtalar functional axis laterally with increasing supination. These effects on the axis location alter the moment arms of the GRF, tending to lengthen them relative to the subtalar joint axis, while decreasing the lengths of the muscular moment arms that resist them. This has consequences for combined rearfoot functional axis stability, with supination gaining mechanical advantage for the invertor muscles and medial GRF vectors through lateral axis displacement, and pronation gaining advantage for the evertor muscles and lateral GRFs through medial axis displacement (Fig. 4.4.9c).

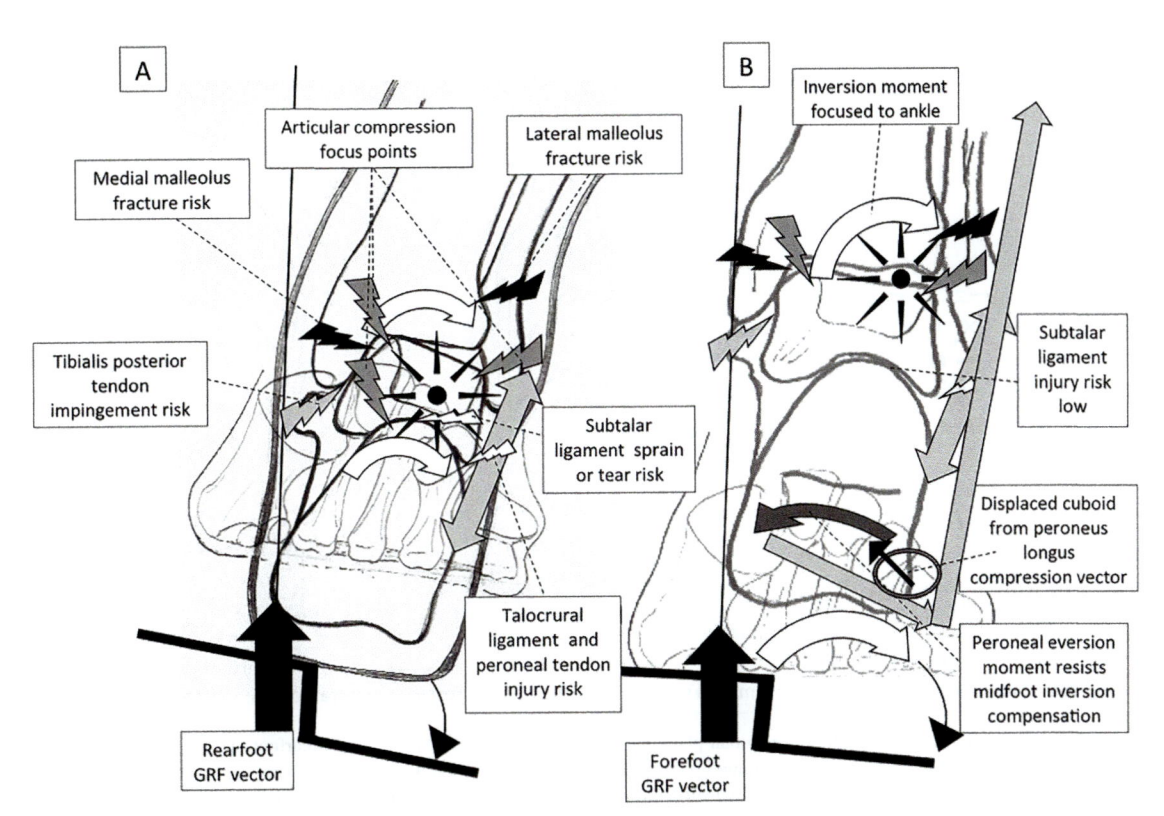

FIG. 4.4.9b The human lower limb adduction angle at the hip increases the likeliness of a rearfoot inversion perturbation when the GRF displaces medially. However, the heterogeneous nature of inversion injuries reflects the size of the inversion moment, the phase of stance when the perturbation occurs, the quality and reactive ability of the soft tissues, the specific articular anatomy of the individual, and the amount of transverse plane motion that associates with the frontal plane perturbation. When the rearfoot is exposed to large medial external moments under increasing plantarflexion angles during heel loading (A), the subtalar joint has flexibility of motion while the ankle is decreasing its form closure. The functional axis of frontal rotation (star) thus sits somewhere between the ankle and subtalar joint. Inversion moments during heel contact are dissipated and restrained by subtalar and lateral talocrural ligaments and the peroneal muscles (mainly longus). Under large inversion moments, these structures are at risk (particularly if already dysfunctional). Articular compression will increase medially as the subtalar joint moves towards its open-packed position, especially if subtalar ligament fibres tear under increasing lengthening between their attachment sites. Ankle joint articular compressions will tend to focus into the medial superior and lateral aspects, risking hyaline cartilage and subchondral bone injuries. Fractures of the medial and lateral malleoli can result from talar inversion and adduction (internal) rotation torques. Sometimes, the tibialis posterior tendon can become pinched within its inferior malleolar groove between the talus and the medial malleolus. When the rearfoot is unloaded as it is in terminal stance (B), inversion moments will be initially dissipated by midfoot inversion and dorsiflexion, but as the midfoot is usually everting, adducting, and plantarflexing at this time, midfoot inversion compensation can potentially be restricted. This leaves the ankle joint alone as the primary mechanism of inversion dissipation. Thus, ankle inversion events are most easily initiated when the forefoot is forced into inversion on a plantarflexed and unloaded rearfoot. Both peroneus longus and brevis are active during terminal stance, making injury to the tendon, musculotendinous junction, peroneal retinaculum, and muscular entheses possible. The cuboid can also displace dorsally and medially due to high tendon compression forces from a strongly activated peroneus longus tendon acting to prevent foot inversion. *(Permission www.healthystep.co.uk.)*

A model has been proposed suggesting that the position/deviation of the axis within the subtalar joint directly influences frontal plane torques across the foot and as a result, can underlie foot pathology by changing the moment arm to the frontal plane torques (Kirby, 1989, 2000, 2001). The proposed model helps explain the frontal plane effect of rearfoot torques to an axis positioned around a series of theoretical subtalar joint positions within the transverse plane. However, this model primarily relies on the effect of the GRF across the subtalar axis alone, and does not fully consider the combined effects of passive and active torques on the ankle–subtalar joint complex. Neither does the model always consider the influence of the BWV as applied through the lower limb to the subtalar joint, which also affects the direction and size of the frontal plane torque. It is the sum of all torques across the ankle–subtalar complex axis that generates the resultant torque, so that the location of the subtalar joint axis alone (whether it be medially or laterally placed) has less of an effect than the distance and the direction that each force application has on the resultant torque. This is as expected, with any angular momentum around a joint being most easily controlled through muscle function when the GRF and the BWVs lie close to the axis of rotation, and when muscles are functioning at their middle lengths, contracting slowly (Fig. 4.4.9d).

A	B	C
Rearfoot not significantly everted or inverted: soft tissue under equal stress	Rearfoot everted: medial soft tissue under higher stresses	Rearfoot inverted: lateral soft tissue under higher stresses

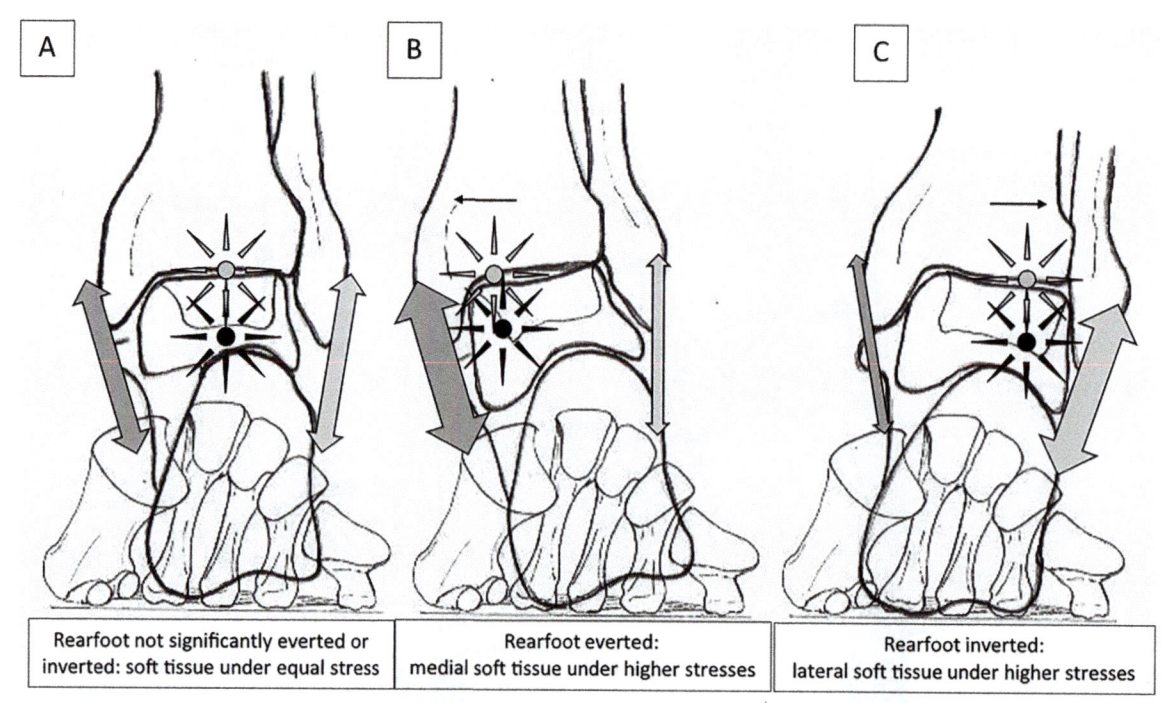

FIG. 4.4.9c The instantaneous axis of rotation for frontal plane motion within the rearfoot moves with rearfoot motion. As the ankle–subtalar joint complex is a combined functional unit with changing ranges of freedom/flexibility of motion during stance phase, the instantaneous axis lies inferiorly between the ankle and subtalar joint. This axis runs longitudinally through the talus (black star) when the subtalar joint offers motion due to the application of external force via a weightbearing calcaneus. When the calcaneus is not loaded, the rearfoot's instantaneous axis moves superiorly into the ankle joint (grey star). Ankle plantarflexion increases ankle range of motion and permits more motion within the subtalar joint when externally loaded. Ankle dorsiflexion during midstance reduces ankle joint freedom through increasing form and force closure, reducing flexibility within the loaded subtalar joint. To complicate matters further, the axis of rotation moves medially and laterally with rearfoot motion. Thus, when the rearfoot joints operate with little frontal plane angulation (A), the axis locates centrally and forces between medial soft tissues (dark-grey double-headed arrow) and lateral soft tissues (light-grey double-headed arrow) remain balanced. In this position, no mechanical advantage is given to eversion or inversion moments. When the rearfoot everts (B), the axis locates more medially providing a longer moment arm to the rearfoot axis, creating a more lateralised GRF on the foot to drive eversion. This disadvantages the resisting medial muscle forces, putting them at risk of higher stresses. When the rearfoot is inverted (C), the axis moves laterally increasing the moment arm for medial GRFs that induce inversion angular momentum. Rearfoot inversion thus tends to disadvantage lateral muscle vectors that create eversion moments, placing more stresses upon them as their own moment arm is reduced. *(Permission www.healthystep.co.uk.)*

Thus, a laterally positioned subtalar axis can still be subjected to a large eversion moment if either the GRF lies far enough laterally or the BWV moments passing through the lower leg lie far enough medially to the rearfoot axis to overcome the forces of the soft tissue restraints. By the same rules, a medial subtalar axis can still experience a rearfoot inversion moment. Such situations indicate the significance of both the torque generated by the GRF and the angle that the BWV is applied through the lower limb to the rearfoot axis. Large lower limb angles to the rearfoot can commonly occur during cutting manoeuvres in sports such as soccer, rugby, American football, and racket sports, such as tennis. They can cause inversion ankle sprains through a varus torque caused by motion and postures above the ankle, rather than motions of the foot around the ankle (Fig. 4.4.9e).

Common inversion sprain injuries

As ankle plantarflexion is the most vulnerable position for inversion ankle sprains, the anterior talofibular ligament is the most likely injured ligament. Concurrent calcaneofibular ligament injuries are generally associated with more severe inversion events (Golanó et al., 2010; Hertel and Corbett, 2019). It does not appear that the talofibular ligament offers any major mechanical inversion resistance, but the calcaneofibular ligament certainly does (Li et al., 2019). The calcaneofibular ligament influences inversion ranges of motion at the ankle mostly in ankle plantarflexion and then controls the subtalar joint during ankle dorsiflexion (Li et al., 2019) (Fig. 4.4.9f).

Inversion sprains cause the collagen fibres of the ligaments to be stretched into plastic deformation, with loss of continuity occurring in some fibres. Many or all of the fibres break with a rupture. Ligament injuries result in pain, swelling (to

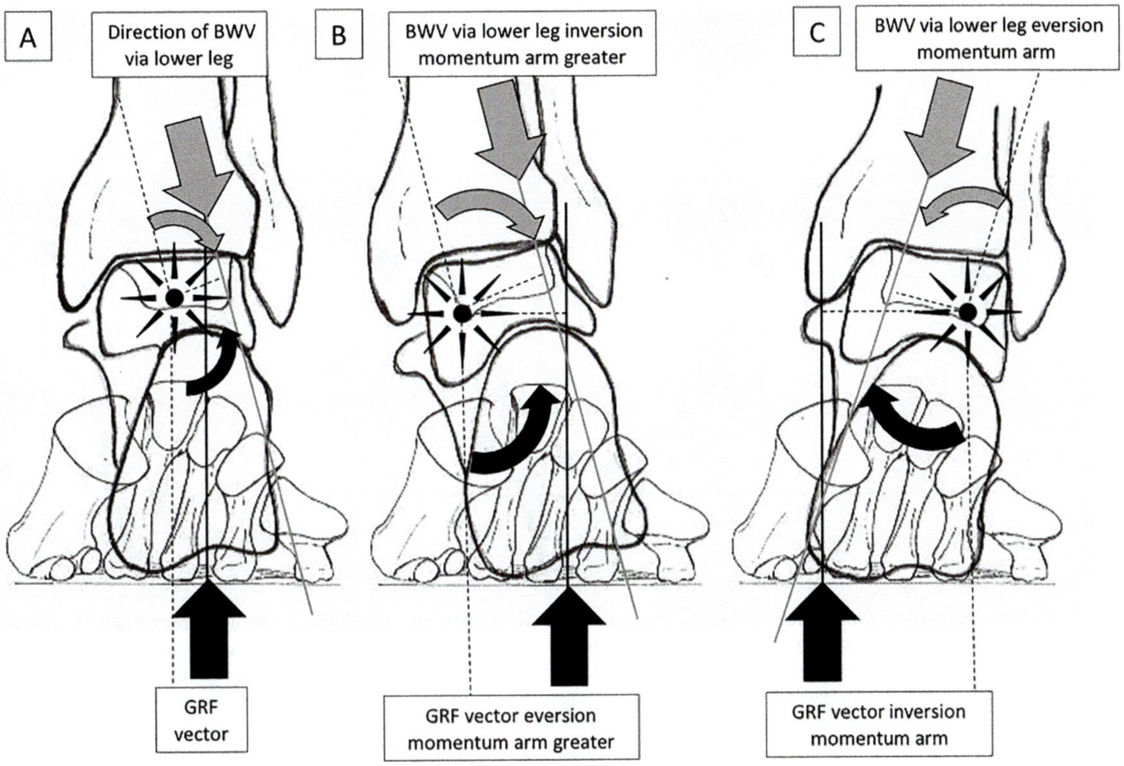

FIG. 4.4.9d Both the location and direction of the GRF and its resultant moment (black arrows) and those of the BWV (grey arrows) to the instantaneous rearfoot axis position are important for frontal plane torques. It is the sum or the angular momentum on the joint that dictates the final inversion or eversion moment that results. With the rearfoot axis (star—representing the combined ankle–subtalar complex axis) positioned centrally (A), a lateral GRF will drive an eversion angular moment proportional to the size of its force and the distance from the axis (the distance between the dashed line to GRF vector—solid black line). However, if the forces driving through the lower limb from the BWV are angled slightly laterally to the axis, then this will create an inversion moment proportional to the BWV's force and its distance from the axis (grey line). The resultant angular moment will reflect the overall resultant angular force vector. When the rearfoot is everted, the instantaneous axis moved medially (B). With the GRF lying in the same lateral position under the foot as in (A), the eversion moment it generates will be larger, as the distance from the axis is greater. However, should the BWV become more laterally orientated through the lower limb, then its inversion moment increases from the BWV's greater distance from the axis. If the rearfoot is inverted (C), moving the instantaneous axis laterally with a GRF applied medially to the foot, a potentially large moment arm for inversion is generated. Yet, the BWV also has an influence, for should that be directed medially to the axis, for example, via a tibial varum morphology or a large hip adduction angle, then the resultant rearfoot angular momentum would be more balancing and quite different to a situation where the BWV was directed laterally to the rearfoot axis. *(Permission www.healthystep.co.uk.)*

initiate natural stiffening and splinting), inflammation, and finally healing. Injury can result in ligament laxity that can be clinically identified through an increased anterior draw translation sign of the affected talus (Hertel and Corbett, 2019). Disruption of the ligament can result in sensorimotor system disturbances. This can then drive kinematic impairments that can cause an individual to deviate from successful healing pathways and towards long-term instability problems (Hertel and Corbett, 2019).

The list of potential injuries from inversion events do not stop with the lateral ankle ligaments, with other injuries often being misdiagnosed (Mansour et al., 2011). The combination of inversion and adduction of the rearfoot complex can cause fibular fracture, anterior inferior tibiofibular ligament injuries, and tibiofibular syndesmotic tears or ruptures (Hertel and Corbett, 2019). Other injuries associated with rearfoot inversion are 5th metatarsal fractures, peroneus brevis strains and enthesis avulsion fractures, subtalar joint bifurcate and cervical ligament sprains/ruptures, rearfoot osteochondral lesions, peroneal tendon injuries, peroneal retinacular lesions, damage to the superficial peroneal, tibial, and/or sural nerves, sinus tarsitis, and even tibialis posterior pathology (through medial impingement) (Mansour et al., 2011; Hertel and Corbett, 2019).

Chronic lateral instability

The consequences of lateral ankle sprains can be long-term and significant, with approximately 40% of ankle sprains proceeding to cyclical impairments, episodic events of 'giving way', and being painful or swollen at times. This is termed

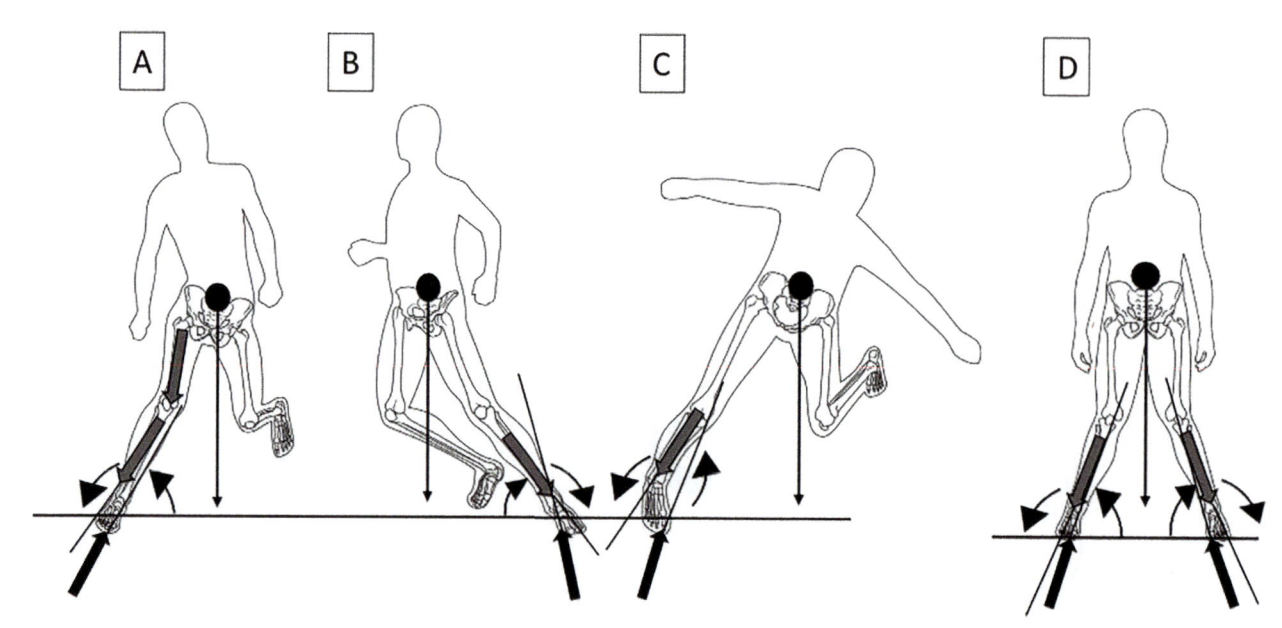

FIG. 4.4.9e For ankle injuries associated with frontal plane torques, the position of the trunk's CoM (black circle) and the rest of the lower limb's posture to the closed chain foot's position is just as important as the foot's position to the ground. Cutting manoeuvres during sport often cause inversion ankle sprains because of the interaction between the GRF vector lying medial to the rearfoot axis and the BWV being directed lateral to its instantaneous axis. As the axis moves laterally with rearfoot inversion, the GRF moment arm is potentiated. Images (A), (B), and (C) show some of the common postures seen in multidirectional sports that can provoke an inversion sprain if active restraint vectors from evertors are not adequately responsive. The cutting posture of image (A) may result in the knee sustaining a medial collateral and anterior cruciate ligament injury from valgus and internal knee torques, while also risking a lateral ankle sprain. Inversion moments usually associate with internal (adduction) torques on the rearfoot, so that injurious torques couple together. Even in static stance, more extreme coxa vara and genu/tibia valgum alignments (D) can create high rearfoot inversion moments. This derives from the GRF vector lying medial to the instantaneous rearfoot axis of the ankle–subtalar complex, while the BWV is applied laterally via the lower leg from the knee's valgus angle. What happens to such individual's ankles during dynamic gait depends on how the CoM of the trunk is positioned over the stance limb, particularly during single-limb support. An in-toed (adducted) angle of gait tends to worsen the lateral instability situation when combined with coxa vara-genu valgum postures. *(Permission www.healthystep.co.uk.)*

chronic ankle instability (Gribble et al., 2016a,b) and is defined as having a propensity to further ankle sprains for over twelve months following the initial event (Hertel and Corbett, 2019). The link of the foot to ankle biomechanics is further suggested by the findings that foot vault posture probably influences postural control, and any vault dysfunction may be significant to further ankle sprains in situations of chronic instability (Hogan et al., 2016). Thus, foot posture (pes cavus, inverted rearfoot, and valgus forefoot) and lower leg morphologies (ankle dorsiflexion limitations, tibial varum and large tibial valgum) are important considerations (Fig. 4.4.9g).

Chronic lateral ankle instability seems to result in a decreased range of rearfoot motion (Chinn et al., 2013; Herb et al., 2014). Pressure analysis reports an increased lateral centre of pressure (CoP) trajectory and increased lateral plantar pressures during jogging in subjects with chronic ankle instability (Morrison et al., 2010). Loss of dorsiflexion, that can result from an inversion sprain, further increases the risk of ankle sprains through reduction in form-closed positions that derive from greater dorsiflexion angles being achieved during gait (Hertel and Corbett, 2019). This can, in turn, lead to problems with soft tissue adaptive restriction in the triceps surae–Achilles complex as a result of neuromuscular spasm mediated by motor neurones and/or myofascial restraints (Hertel and Corbett, 2019).

Chronic ankle instability sufferers have been found to produce less eversion force from the peroneal muscles, both in neutral ankle positions and in positions of plantarflexion (Donnelly et al., 2017). Reduced peroneal reaction time to inversion perturbations from the sensorimotor deficit compound the problem (Hertel and Corbett, 2019). This increases the susceptibility of further inversion events as one of the primary inversion torque restraining mechanisms is lost. A history of chronic ankle sprains causes runners to alter their gait across all running speeds, with alterations becoming more conspicuous with higher speeds compared to healthy controls. This includes a decrease in rearfoot eversion excursions and eversion velocity (Colapietro et al., 2020), which potentially predisposes those with chronic ankle sprains to repeat injuries.

Much of the consequences of ankle sprains relate to the neurological and sensorimotor deficits that results from ligament injuries, and thus, there is a combination of mechanical and sensory deficits (Hertel, 2002; Delahunt et al., 2006, 2018; Hertel and Corbett, 2019). The alteration of sensorimotor function is thought to lie behind a delay in peroneal muscle activation creating an eversion force that leads to failure in protection from further inversion perturbations (Hertel and Corbett,

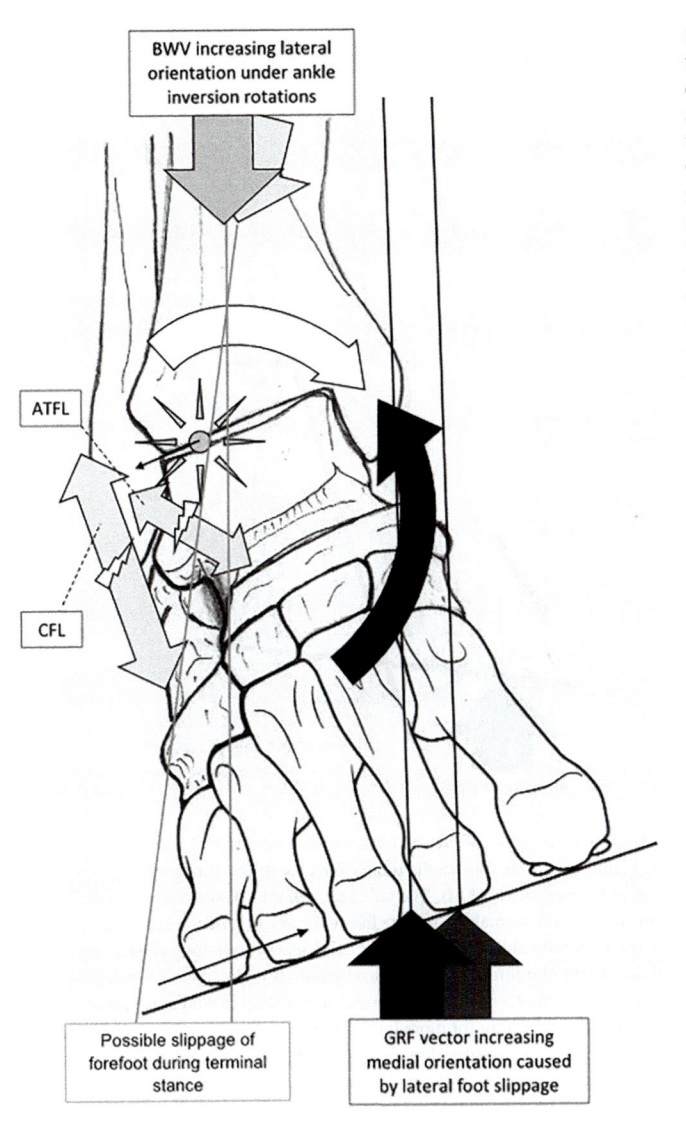

FIG. 4.4.9f On inclined surfaces, particularly if unstable or slippery, the risk of rearfoot inversion sprains within the foot positioned lower down on the slope are much higher. The downhill foot must take up an inverted rearfoot posture to maintain support surface contact, shifting the instantaneous axis laterally (star). This means both the BWV and GRF vector can lie increasingly medial to the axis, driving inversion moments that must be resisted by lateral soft tissue vectors. If the foot slips on a slope medially, causing internal and adduction rotations during terminal stance when the ankle has more motion, the possible inversion velocity and joint mobility is increased. Once the ankle starts to invert, the instantaneous axis moves further laterally. This will compound the GRF's influence on the inversion moment, helping to send the medial malleolus and medial foot towards one another. The BWV will now move laterally as the lower leg rotates, increasing inversion rotation as the GRF moves medially. During such rearfoot inversion perturbations, the anterior talofibular ligament (ATFL) and calcaneofibular ligament (CFL) are the two most frequently injured structures. The ATFL is very thin and seems more important in providing sensorimotor information than actual mechanical restraint. It is probably acting within the sensorimotor ankle hub to help keep medial and lateral muscular vectors and the body posture above balanced during foot plantarflexion manoeuvres. The CFL provides important ankle inversion restraint during plantarflexion, but when the ankle is becoming less mobile under angles of increasing dorsiflexion, its influence seems to be more to control inversion of the subtalar joint, which unlike the ATFL, it also crosses. *(Permission www.healthystep.co.uk.)*

2019). Structural changes in both the anterior talofibular ligament and the osteochondral surface have been noted in subjects with chronic ankle instabilities. Volumetric changes are also reported in the intrinsic and extrinsic foot muscles, with weakness in isometric contraction compared to controls noted in the evertors, invertors, and plantarflexors, but not in the dorsiflexors (Hertel and Corbett, 2019).

There are also possible changes in more proximal muscle function as a result of chronic ankle instability (DeJong et al., 2019) and interestingly, a prospective study has linked reduced hip muscle strength as an independent risk factor of lateral ankle sprains among youth soccer players (De Ridder et al., 2017). A fear of ankle movement can develop (*kinesiophobia*) among sufferers of chronic ankle instability (Houston et al., 2018; Hertel and Corbett, 2019). Such kinesiophobia may alter results when performing gait analysis on such individuals, giving data on the results of injury, rather than the causes of recurrent instability. This is worth noting before gathering data on chronic lateral instability patients.

4.4.10 Frontal plane Achilles pathomechanics

Achilles subtendon orientations have potentially interesting relationships as part frontal plane and transverse plane biomechanics. Sagittal plane moments localise tensile strain through the Achilles. The deep (anterior) Achilles fibres undergo more displacements than the superficial (posterior) fibres on ankle motion and are usually increased at lower knee flexion angles nearer to extension, as found during late midstance (Arndt et al., 2012; Slane and Thelen, 2014, 2015; Handsfield

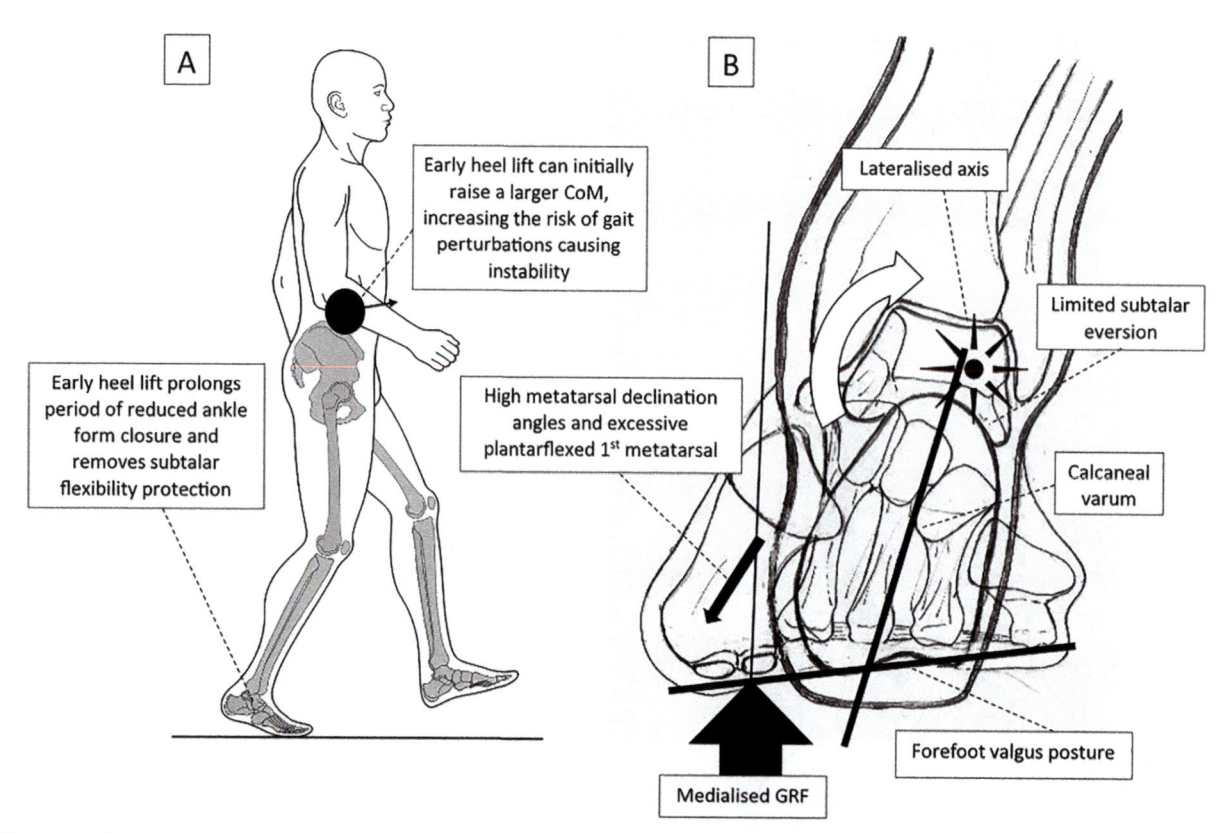

FIG. 4.4.9g Morphology, including articular surface shape and specific anatomical variances of the rearfoot, increases the risk of inversion or eversion ankle sprains between individuals. In the case of inversion ankle sprains, high, and immobile foot vaults increase rearfoot inversion risks. Limited ankle dorsiflexion during midstance (A), often associated with cavoid feet, can initiate a premature heel lift. The ankle thus plantarflexes out of its form closure for longer, under higher body weight resistance, initially raising the CoM's position, therefore making it more likely for any inversion perturbation to cause locomotive instability and injury. Those individuals with limited subtalar eversion motion and/or with inverted calcaneal morphology (B) can cause their rearfoot axis to remain more laterally positioned. Individuals with a stiff valgus (everted) midfoot and/or plantarflexed 1st metatarsal also risk developing higher GRFs on their medial forefoot, providing larger inversion moment arms to the rearfoot's instantaneous axis for initiating frontal plane rotations. This will tend to move pressures and the CoGRF towards the lateral foot via inversion, tilting the foot for more lateralised weightbearing. All or many of these issues can combine together within stiff cavoid feet, whether they are idiopathic in nature or have developed with neuropathic conditions such as Charcot–Marie–Tooth disease type 1. In the latter condition, degenerative sensory deficits and peroneal and other muscle atrophies can compound lateral instability problems. *(Permission www.healthystep.co.uk.)*

et al., 2017). With the knee near extension and the ankle dorsiflexed as seen in late midstance, the superficial distal Achilles experiences its highest strains (Lyman et al., 2004), while the deep Achilles fibres undergo their greatest compression against the calcaneus (Lersch et al., 2012). Strains on the distal Achilles will be highest just before heel lift, when the ankle is most dorsiflexed at the end of midstance. The location of the strain will depend on which muscles are most active during gait. Forces in the medial tendon are generated by the medial fibres of medial gastrocnemius and soleus, while lateral forces are higher when lateral gastrocnemius is more activated, and when all the triceps surae muscles are firing equally (Arndt et al., 1999) (Fig. 4.4.10a).

Frontal plane torques, like sagittal plane moments, influence the location of strains within the Achilles (Lersch et al., 2012). Generally, distal Achilles strains are higher than proximal strains, with proximal strains being greater laterally and distal strains higher medially (Lersch et al., 2012). As has already been stated, intratendon sliding of the subtendons increases displacement nonuniformly, with some lesser influence being caused by the twisting of the Achilles that occurs during increasing dorsiflexion angles. This changing pattern of stresses indicates that the complex internal Achilles structure strongly affects the interaction between muscle forces and tendon behaviour (Handsfield et al., 2017). These heterogeneous properties of Achilles strain set up potential mechanisms of complex pathomechanics, as strains become concentrated within specific Achilles areas during rearfoot motions in locomotion (Arndt et al., 2012).

The Achilles subtendons twist as they approach their distal attachments with some degree of individual variation (Dalmau-Pastor et al., 2014). Usually, the soleus subtendon dominates the medial enthesis, the medial fibres of the medial

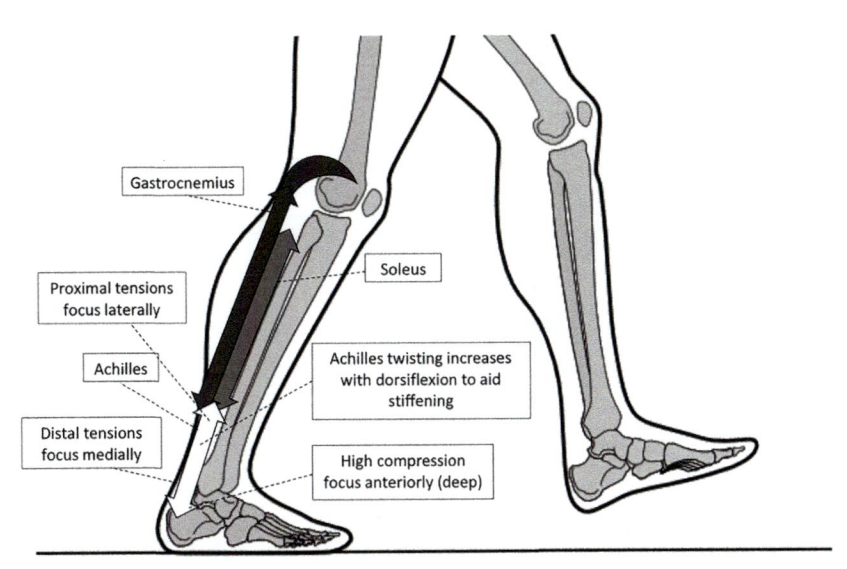

FIG. 4.4.10a The energy storing specialised Achilles tendon undergoes increasing strains during late midstance via ankle dorsiflexion, reaching a peak of stress, strain, and hence energy storage at the heel lift boundary. This process involves lengthening and twisting the Achilles tendon fibres within their subtendons, which increases overall fibre stiffness within the tendon's elasticity. With the knee near extension, Achilles stress focuses to the distal and medial fibres (primarily within the soleus subtendon) to peak just before heel lift. The deep Achilles fibres (mainly gastrocnemius lateralis and soleus subtendons) come under increasing compression against the posterior surface of the calcaneus, protected by the sesamoid cartilage within the tendon, the retrocalcaneal bursa, and the fibrocartilage overlying the calcaneus above the Achilles enthesis. Excessive tendon tension and twisting risks midportion tendinopathies while high compressions distally can provoke retrocalcaneal bursitis. More distal and insertional tendinopathies are risked by high strains of compression and tension combined. Enthesopathies and heel spurs seemingly develop under high tensional strains at the enthesis organ. With ageing, energy storage tendons are prone to mechanical property changes, and this lies at the heart of many middle-aged and older patient's Achilles pathologies. However, Achilles strains are adjusted and moderated by triceps surae muscle activity and strengths as well as rearfoot kinematics. For understanding Achilles pathomechanics, the specific location of Achilles pathology within the anatomy is very important. Treatment can then be focused on the correct tissue threshold breach within a complex of anatomies. *(Permission www.healthystep.co.uk.)*

gastrocnemius dominate the posterior or superficial part of the enthesis, the lateral fibres of medial gastrocnemius dominate the lateral posterior aspect, and the lateral gastrocnemius' fibres dominate the lateral and anterior or deepest parts, which also occupy the largest area of the enthesis (Edama et al., 2015; Pękala et al., 2017). The twisting of the Achilles means that distally, the soleus fibres of the Achilles become increasingly posterioned medially, the medial gastrocnemius subtendons more posteriorly and laterally positioned, and the lateral gastrocnemius subtendon fibres more anteriorly positioned. The frontal plane orientations at the calcaneal enthesis to the proximal force-generating muscles can change across the sub-tendons with rearfoot motion. If the triceps surae functioned as a single force generator and the calcaneal enthesis functioned as a single point for the administration of effort, then the situation would be straightforward. For when the rearfoot inverted, the point of force application would move medially to the rearfoot's instantaneous axis, increasing the inversion moment while eversion of the rearfoot would move the attachment force laterally, increasing the eversion moment arm (Zifchock and Piazza, 2004; Lee and Piazza, 2008).

However, as the Achilles is made of localised subtendons attached to muscles with subtly different activation patterns and significant freedom to slide past each other, the rearfoot frontal plane orientations become more interesting. With rearfoot inversion, the point of effort moves laterally through increasing lateral subtendon fibre tensions, so that the Achilles offers an active eversion moment arm that can resist rearfoot inversion (Zifchock and Piazza, 2004; Lee and Piazza, 2008). Rearfoot eversion, on the other hand, causes the effort point of the Achilles to reorientate medially, generating an inversion moment arm around the rearfoot frontal plane position of the joint's axis. A purely vertical (neutral) rearfoot orientation results in a pure plantarflexion moment (Zifchock and Piazza, 2004; Lee and Piazza, 2008) (Fig. 4.4.10b).

Lersch et al. (2012) have reported that mediolateral, proximodistal, and dorsoventral distributions of tendon strains within the Achilles appear more influenced by the rearfoot position than through the muscular variations in force generation within triceps surae, in most cases. This means that the rotation twist within the triceps surae–Achilles complex, combined with the muscular and tendon fascicle orientations, are of paramount importance to balancing heterogeneous tendon strains in relation to calcaneal position across the rearfoot (Lersch et al., 2012). This is logical when we consider that the primary goal of the rearfoot complex is to present a powerful plantarflexion moment from the ankle to the forefoot at heel lift. This

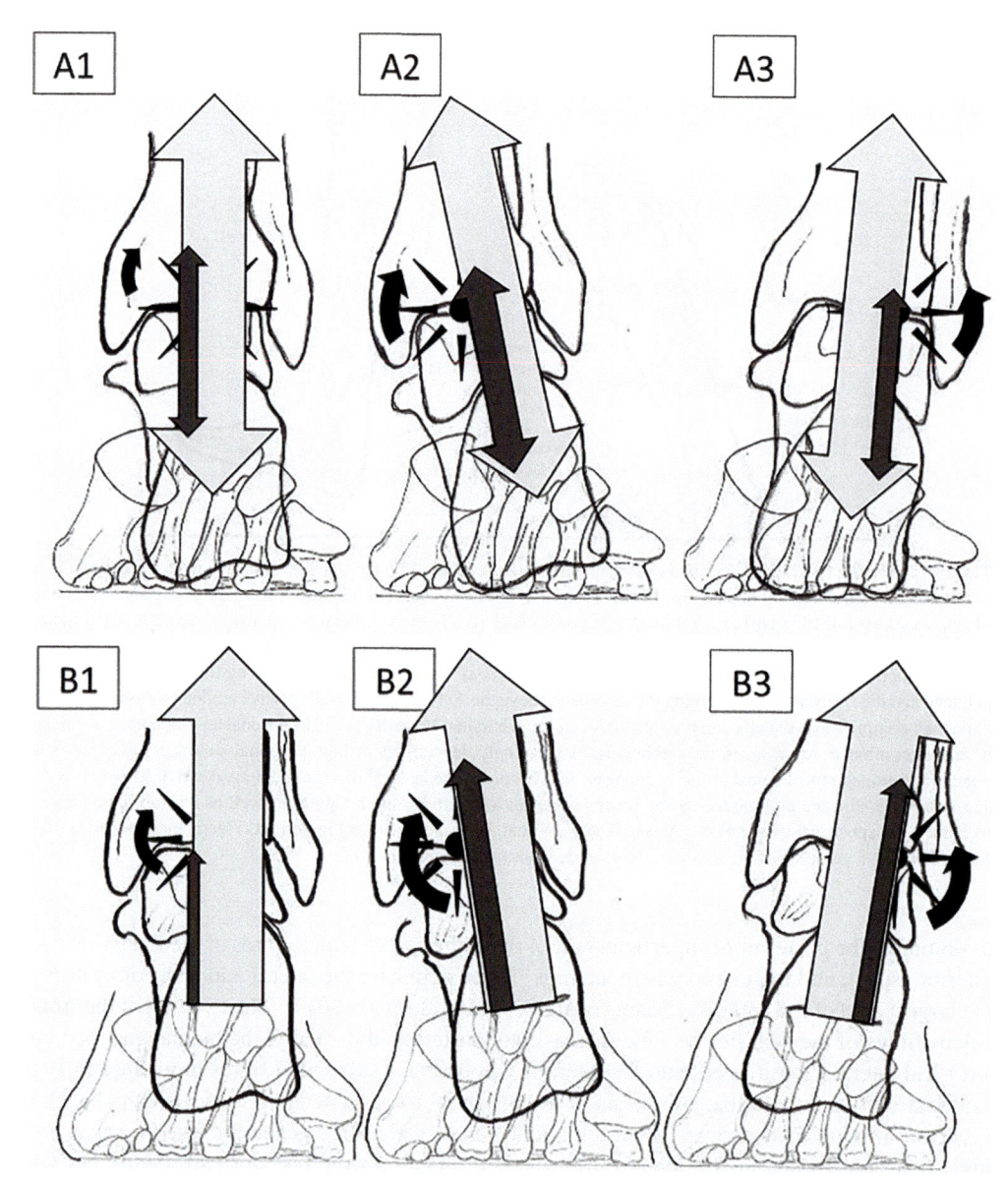

FIG. 4.4.10b Schematics to illustrate the effect of rearfoot motion on the Achilles operating just before the heel lift boundary (A) and just after heel lift (B). With the rearfoot vertical and tendon fibres lengthening at the end of midstance (A1), the Achilles stresses focus to the medial and deep distal fibres. These usually lie within the soleus subtendon medially (tensional) and the anterior part of the soleus and gastrocnemius lateralis subtendon (tensional and compressive). If the medial Achilles fibres respond by increased stiffening, they can provide a small eversion resistance moment on the rearfoot. This could be useful when the vault is at its lowest profile medially. The focus of stresses within these subtendon locations should reflect normal triceps surae and ankle–subtalar complex biomechanics, but changes within sagittal plane kinematics and triceps surae muscular strengths or fatigue may still lead to tissue damage. With rearfoot eversion in late midstance (A2), tensional forces will rise medially, likely requiring greater soleus muscular activity adaptation to moderate subtendon fibre lengthening, creating some eversion resistance. Under rearfoot inversion in late midstance (A3), distal medial stresses may reduce, but the lateral subtendon from lateral fibres of medial gastrocnemius will increase. This can provide some inversion resistance under lateral gastrocnemius action. In early terminal stance, ankle plantarflexion reduces the anterior fibre compression and as the knee flexes, stresses tend to move more proximally to the lateral Achilles fibres. However, just at the point of heel lift, anterior and distal forces remain high. With a vertical rearfoot posture (B1), Achilles tendon fibres are shortening to release their eccentrically stored energy, with generally more energy having been stored and released medially than laterally. This can help produce a very small rearfoot inversion moment as the heel lifts. Heel lift with a rearfoot valgus (everted) posture (B2) will release more energy for creating an inversion moment, which is helpful in these circumstances as the frontal plane instantaneous axis of the rearfoot will have moved medially via its everted orientation. Heel lift initiated in rearfoot varus (inverted) posture (B3) will likely reduce the medial subtendon elastic recoil, but increase that from the lateral fibres helping to create more of an eversion moment. This is a helpful correction for a lateralised rearfoot axis. However, sustained rearfoot valgus and varus postures around heel lift boundary during gait are likely to risk persistently higher stresses within medial or lateral distal subtendon and enthesis attachment points, respectively. *(Permission www.healthystep.co.uk.)*

allows the foot and ankle to work as a passive linear spring along the line of progression. Thus, any deviation within the frontal plane needs to be resisted to reduce the risk of losing this powerful action. The medial Achilles subtendon is from soleus, so that soleus action also involves an element of eversion resistance, which rises with increasing rearfoot eversion. Therefore, soleus helps resist eversion moments of the rearfoot. The lateral fibres of medial gastrocnemius and part of the lateral gastrocnemius subtendon, with their lateral attachments, would seem to have a reciprocal role in counteracting rearfoot inversion moments that increase with rearfoot inversion perturbations.

Lersch et al. (2012) have reported greatest tensile strain in the Achilles associated with rearfoot eversion, with increased strain in all frontal plane angles tested outside of rearfoot vertical. The deep lateral fibres of the Achilles tendon are also subjected to increased compression at the enthesis against the calcaneus during rearfoot eversion, while the deep medial portion undergoes greater medial tensile strain (Lersch et al., 2012). It can therefore be suspected that Achilles pathology located only on the medial side associates with rearfoot eversion moments when the triceps surae is active during gait. Achilles insertional pathology medially might also associate with rearfoot eversion. On the other hand, laterally located Achilles mid-portion tendinopathy is more likely associated with rearfoot inversion moments, as well as a laterally dominant insertitis. The clinician must remain aware that individuals vary, and that the positions of the subtendons are rotating anteromedially to posterolaterally proximally through the Achilles. Thus, how high up the tendon the pathology is located matters to which muscle's subtendon is being over strained.

As subtalar joint frontal plane motion is lost when the heel is unloaded, significant rearfoot frontal plane motion is most likely to occur in loading response and early midstance. However, as the triceps surae is most active in late midstance, attention should be directed to the rearfoot posture at late midstance. This would seem to be a primary phase of gait in which to investigate rearfoot frontal plane orientation effects on the Achilles in a patient presenting with a single side dominant Achilles tendon pathology. Of course, it is not unfeasible for the ankle to evert significantly after heel lift without subtalar motion, especially in a medially unstable foot resulting from perhaps, some degree of tibialis posterior dysfunction. However, Achilles load decreases rapidly after heel lift, and compression on the deep Achilles surface from calcaneus impingement quickly reduces under ankle plantarflexion that accompanies heel lift. The summary of the evidence so far suggests that the biomechanical events just prior to and during initial heel lift leave the Achilles most vulnerable to frontal plane perturbations, at least in walking and heel strike running. Thus, tibialis posterior's and peroneal muscle activity around heel lift also becomes highly significant.

Runners with symptomatic Achilles tendinopathy have been reported to have longer rearfoot eversion durations, but not greater excursions or velocity of eversion compared to healthy controls (Becker et al., 2017). There are indications that foot 'pronation' (hyperpronation?) associated with flatter foot vault profiles may be associated with restriction of blood flow through the Achilles, possibly via increasing the tendon twisting that may heighten with rearfoot eversion as part of pronation and ankle dorsiflexion, particularly in individuals with high subfascicle torsions (Edama et al., 2015; Karzis et al., 2017; Pękala et al., 2017). It can only be theorised at present that subtle transverse planes motions, coupled with frontal plane torques may also influence stress–strain distribution across the subtendons. If this is the case, then the effects of angle of gait may also be significant. It is the angle of gait through the foot ability to align to the lower leg's longitudinal axis that influences the ability of the ankle and Achilles to work together with the foot as a passive linear spring, improving energetics (Hedrick et al., 2019). Loss of this is likely to cause compensatory stress changes throughout the triceps surae–Achilles complex.

4.4.11 Other tendon injuries around the rearfoot

As generators of the resistive force against frontal/transverse motion across the rearfoot joints, pathology within the rearfoot invertor-adductors and the evertor-abductor muscles may result in abnormal torques as well as create them across this region. As already discussed, if motion across the rearfoot complex starts to dominate in eversion or inversion as a result of a loss of frontal plane control, then the subtalar joint axis will orientate differently, moving its and the combined rearfoot instantaneous joint axis more medially or laterally. With subtalar pronation, it will move medially and more laterally with subtalar supination (Leardini et al., 2001). The result of this is that the lever arm of the invertors-adductors to the ankle–subtalar joint axis within the frontal/transverse plane shortens with increased pronation, while increased supination shortens the evertor-abductor muscles' moment arms. This makes it difficult to determine whether a tibialis posterior dysfunction that permits the subtalar joint and rearfoot to become more everted and abducted, thus shifting the instantaneous joint axis more medially, is the cause of or a result of increased rearfoot eversion from a medially positioned rearfoot axis. Therefore, assessing the strength of the foot's evertors and invertors can help develop a clear picture of the likely origin of an increased frontal/transverse plane torque during gait that can then clinically direct intervention appropriately.

Tibialis posterior and the digital plantarflexors

As the primary active rearfoot invertor, tibialis posterior will tend to be injured with increasing eversion loading, with symptoms usually presenting in the medial ankle and/or proximal medial vault. In its rearfoot inversion or eversion-resistance role, it is assisted by the passive ligamentous structures of the deltoid ligament, the ligaments of the subtalar joint, and particularly the spring ligament via supporting the talocalcaneonavicular joint. Although the tibialis posterior has an important medial ankle stabilisation role through its complex distal attachments to the navicular and plantar surface of several lesser tarsal bones and metatarsal bases, it also has profound effects on midfoot/forefoot plantarflexion and adduction as well as controlling vault compliance and stiffness levels.

Thus, although tibialis posterior pathology is commonly located in its tendon sheath within the tendon's fibrocartilaginous region around the medial malleolus, its effects extend beyond the rearfoot into vault and foot-stiffening mechanics. Tibialis posterior dysfunction will thus be considered in more detail in the next section. This situation is also true of the long flexor muscles that can give weak inversion assistance, and often present with tendon pain and pathology behind the medial malleolus (flexor digitorum longus) or posterior to the talus (flexor hallucis longus) (Fig. 4.4.11a). Their pathomechanics and results of dysfunction tend to be more significant to the forefoot than to the rearfoot. The reason that their pathologies often present in the rearfoot is that where tendons pass over bone, there are areas of fibrocartilage within the tendon that compress against bone, presenting focal areas of tensional weakness.

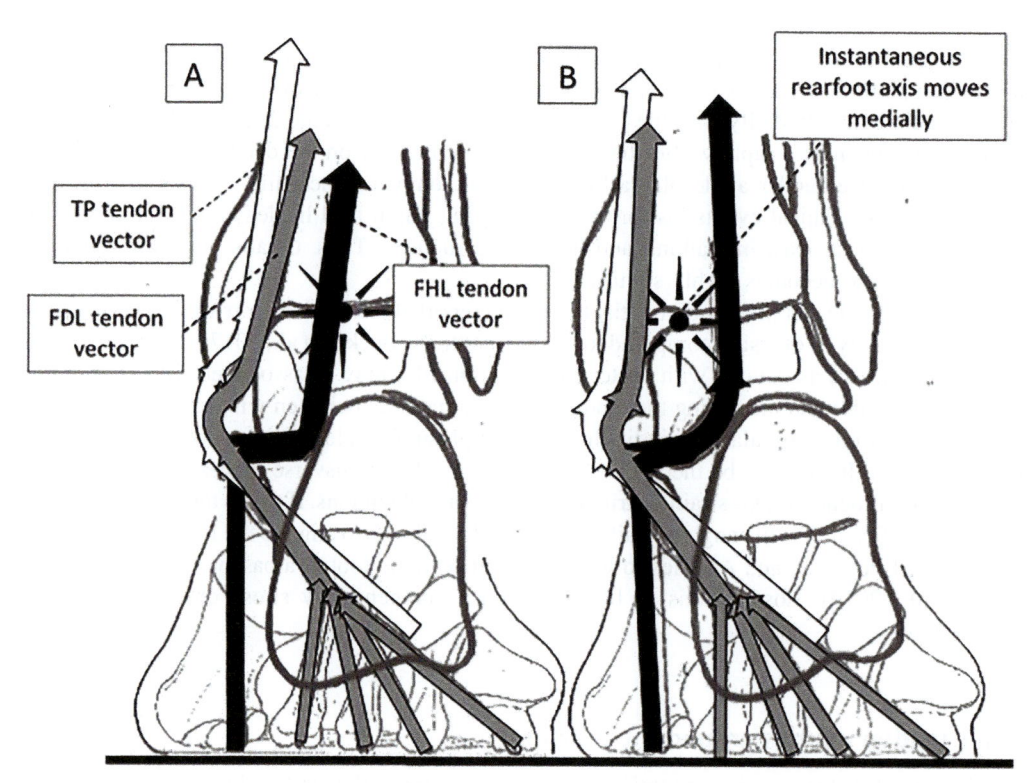

FIG. 4.4.11a Of the medial plantar extrinsic muscles, only tibialis posterior (TP) is significantly active during contact and loading response, demonstrating its first peak of biphasic activity. During late midstance, its second greater peak of activity occurs, creating important force vectors into terminal stance when all the medial plantar extrinsics play important roles in successful acceleration. With the rearfoot in a vertical posture (A), TP tendon's vector (white arrow) expressed via elastic recoil provides a strong rearfoot inversion moment around the medial malleolus from its long medial lever arm to the rearfoot axis (star) at its navicular attachment. This action is aided by its adduction and plantarflexion moment stabilising the midfoot, derived from its distal attachments. Flexor digitorum longus' (FDL) vector (dark-grey arrows), via its tendon behind and around the medial malleolus and slips to the lesser digits, can aid the TP moment. It also creates proximal metatarsal head stabilisation forces as the lesser digits extend. Flexor hallucis longus' (FHL) vector (black arrow) from its tendon running behind the talus and under the sustentaculum tali produces a more central moderating force vector under the foot, despite its tendon running to the distal phalanx of the hallux. This is because its tendon behind the talus lies close to or over the rearfoot frontal plane rotational axis. Eversion of the rearfoot occurring at the heel lift boundary (B) can cause greater strain on the TP tendon or be a consequence of TP muscle weakness or tendon dysfunction. In health, increased eversion moments will require greater TP activation aided by FDL, as the FHL can become an evertor under eversion, shifting its vector laterally to the instantaneous rearfoot axis as it drifts medially. Weakness in FDL can cause increasing TP stresses and require raised TP activity. Weakness in FHL may compromise the ability for the rearfoot to moderate moments from the medial and lateral muscles around the rearfoot. *(Permission www.healthystep.co.uk.)*

The peroneals

A significant foot functional role is also true of peroneus longus which has a complex antagonistic everting and abducting, yet agonistic midfoot plantarflexion stabilising and foot-stiffening relationship with tibialis posterior (Kokubo et al., 2012). Peroneus longus produces active eversion and abduction moments across the rearfoot and midfoot, with peroneus brevis aiding eversion power when it is also active. Peroneus brevis is considered primarily as a rearfoot-influencing muscle of terminal stance, although it also assists in calcaneocuboid and cuboid–metatarsal compression–stability through its attachment to the styloid process of the 5th metatarsal during late midstance and terminal stance. It assists peroneus longus in generating midfoot and rearfoot eversion moments during late midstance into terminal stance, which can be assisted itself by variably present accessory peroneal muscles such as *peroneus quartus* and *peroneus digiti quinti* (Yammine, 2015). Peroneus tertius should only function during swing phase.

Peroneal injuries consist of tendinitis, tenosynovitis, tendon subluxations, tendon dislocation, and tendon splits and tears (Davda et al., 2017). Chronic peroneal tears and degenerative tendinopathy are often overlooked and misdiagnosed as an underappreciated source of chronic lateral ankle pain that can occur with and without identifiable traumatic events (Dombek et al., 2003; Davda et al., 2017). Manual muscle strength testing may show little or no weakness, making identification of the problem outside of soft tissue diagnostic imaging difficult. A number of osseous and soft tissue anatomical failings have been associated with peroneal degeneration and injury including inversion ankle sprains, chronic ankle ligament laxity, and peroneal tendon subluxations. Peroneal subluxations are more likely in the presence of a flat or convex fibular groove, a low-lying or otherwise abnormal muscle belly, superior peroneal retinaculum incompetence, posterior lateral tubercle spurring, and a high-vaulted foot profile (Dombek et al., 2003; Davda et al., 2017).

A high vault foot profile, especially if combined with a hypomobile varus rearfoot, is likely to result in increased lateral foot loading during midstance, making terminal stance midfoot eversion more problematic for the peroneals, aiding tibialis posterior's mechanical advantage. A flat or convex peroneal groove as well as a low-lying peroneal muscle belly resulting from the presence of an accessory muscle, hypertrophy, or tenosynovitis, can cause overcrowding and shallowing of the retrofibular region. These factors can destabilise the lateral pulley around the posterior lateral malleolus, risking peroneal tendon subluxation (Dombek et al., 2003; van Dijk et al., 2016; Davda et al., 2017) (Fig. 4.4.11b). Peroneal tendon subluxation decreases combined plantarflexion–eversion power and anteroposterior stability across the lateral column. A sharp posterior fibular ridge may damage the peroneal tendons leading to longitudinal tendon tears that can occur during high compression–tension events around the malleolus (Karlsson and Wiger, 2002; Dombek et al., 2003; Davda et al., 2017).

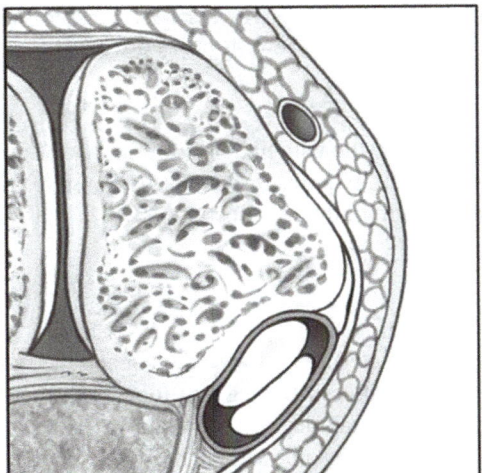

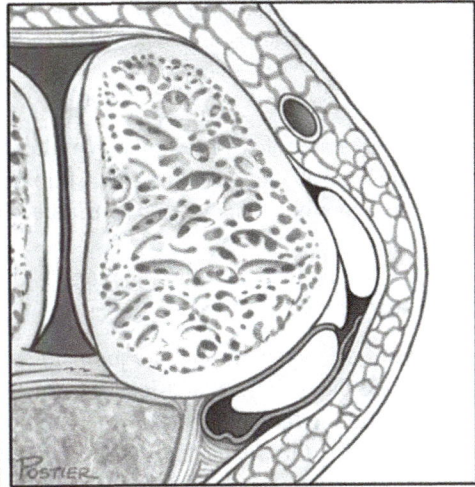

FIG. 4.4.11b A diagram of the peroneal groove and tunnel with normal concave anatomical profile (left) compared to anatomy with a convex shape that risks peroneal subluxation (right). Swelling or increased tissue mass within the groove can cause similar effects, functionally shallowing the groove. *(Image from van Dijk, P.A.D., Gianakos, A.L., Kerkhoffs, G.M.M.J., Kennedy, J.G., 2016. Return to sports and clinical outcomes in patients treated for peroneal tendon dislocation: a systematic review. Knee Surg. Sports Traumatol. Arthrosc. 24(4), 1155–1164.)*

Peroneus longus tears are reported in three distinct anatomic zones: at the lateral malleolus, at the peroneal tubercle/trochlear surface of the calcaneus, or within the cuboid notch (Dombek et al., 2003). These may represent distinct pathomechanics, with lateral malleolar injuries being associated with increased compressions around the pulley-like peroneal tubercle region. This is pathology associated with lateral rearfoot excursions (Fig. 4.4.11c). Pain associated with cuboid notch pathology seems related to midfoot eversion and plantarflexion moments. It has been reported that around 33%–37% of peroneus longus pathology is concurrent with peroneus brevis pathology (Dombek et al., 2003).

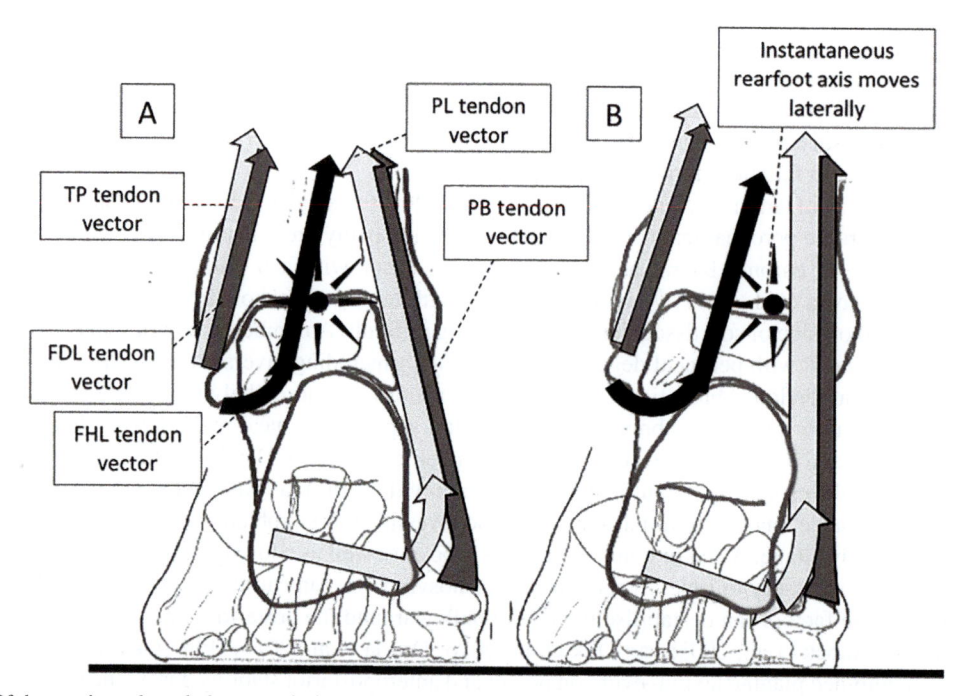

FIG. 4.4.11c Of the consistent lateral plantar extrinsics, only peroneus longus is significantly active at loading response. In terminal stance with the rearfoot in vertical posture (A), peroneus longus' (PL) tendon vector (lateral light-grey arrow) combines its forces to that of peroneus brevis' (PB—dark-grey arrow) which together provide a strong eversion moment to the rearfoot (ankle) frontal plane axis (star). Peroneus longus also adds a strong midfoot eversion, abduction, and plantarflexion moment as it wraps under and around the plantar aspect of the lateral cuboid. These lateral eversion vectors are balanced by medial vectors from tibialis posterior (TP) and flexor digitorum longus (FDL). Flexor hallucis longus (FHL) balances these forces. If the rearfoot inverts during late midstance or at heel lift (B), the peroneals must increase their activation to resist lateral instability. This is not helped by the rearfoot axis moving laterally during inversion, causing the FHL vector to move more medial to the axis and thus increasing its and the other medial extrinsics' inversion moment. *(Permission www.healthystep.co.uk.)*

Peroneus longus' important role as an agonist in midfoot plantarflexion with tibialis posterior and as an antagonist producing midfoot eversion and abduction acting against tibialis posterior, requires further discussion in relation to pathomechanics of the vault and forefoot. This will be considered further within the next section.

4.4.12 Section summary

The rearfoot complex of the ankle and subtalar joints together is the interface between the leg and the foot. It utilises triplanar motion to change the quantities and locations of flexibility between the leg and foot. Both rearfoot joints offer motion, but it is ankle motion that dominates the sagittal plane with small degrees of flexibility being provided through the subtalar joint in primarily the frontal and transverse planes, along with some sagittal motion in certain ankle positions during external rearfoot loading forces. The ankle can provide variable motion within the frontal and transverse planes, primarily in plantarflexion angles that decrease with ankle dorsiflexion during midstance. The subtalar joint only provides flexibility from external loading on the calcaneus. Both joints express decreased motion with ankle dorsiflexion. Thus, the ankle provides for frontal plane perturbations when the heel is unloaded and the forefoot loaded. The subtalar and ankle joints work together in initial heel contact loading response and early midstance to adjust posture for GRF and BWV perturbations to maintain sagittal plane progression of the leg over the foot. Anything that disrupts this relationship is likely to create significant biomechanical dysfunction within the rearfoot.

To assist the series of changing joint freedoms and motions around the rearfoot, multiple connective tissue restraints bind the leg and foot together. These primarily restrain frontal and transverse plane motions, but also create limitations on sagittal plane motion outside that necessary for normal anterior limb progression. Anterior positioned muscles act as brakes to ankle plantarflexion accelerations. Posterior muscles control leg and CoM accelerations over the foot during midstance, ending their activity with a largely elastically induced release of plantarflexion power to generate heel lift. This power drives the forefoot into the ground to create a stable platform in order to transfer weight onwards in an anterior direction. Muscles lying medially and laterally to the rearfoot control frontal and transverse plane moments so as to maintain postural stability and improve lower limb energetics, accommodating terrain variation. Again, any perturbation to this function risks pathomechanical events. Loss of any part of the function will set up compensatory mechanisms which can, in time, lead to further dysfunction and pathology as stresses become disproportionally focused into certain structures.

Through the intimate functional and anatomical relationships of the structures that cross the rearfoot into the rest of the foot, understanding rearfoot function and pathology can never be fully appreciated without recourse as to what is occurring in the midfoot and forefoot.

4.5 Pathomechanics of the foot

4.5.1 Introduction

Science can often give complex answers to questions, whereas answers that are simple are often preferred. In attempting to understand foot function, simplistic models to explain pathomechanics have been clinically very popular. However, sadly, they do not reflect the complexity of the structure that they are dealing with. The foot, particularly the midfoot, is a complicated region consisting of multiple articulations spanning from the rearfoot to the forefoot. The midfoot, through its many articulations, allows the foot to demonstrate variable compliance and stiffest levels while managing any subtle perturbations in stance through changing GRF applications. There is thus plenty of scope to develop pathology. It is not the intention to approach the midfoot through consideration to a 'normal' vault profile from which anything lower or higher represents aberrant function, but only to consider that such vault variation still demands the same fundamental requirements of being able to adapt in and out of being a more compliant structure to a stiffer structure throughout the gait cycle. Vault height influences this ability. However, so do other factors such as muscular strength and connective tissue mechanical properties as has been discussed in previous chapters in more detail. It is only by consideration of all the factors that a foot becomes identifiable as functional and healthy or dysfunctional and vulnerable to pathomechanical events.

The foot follows principles of biotensegrity in that tensional forces provide compressive stability throughout the skeletal frame across articular motions that permit the foot's shape to change. Therefore, it is important that the articulations move freely, but in a controlled manner. Soft tissues, both passive and active, provide that managed control. Failure of any functional part of the foot will cause dysfunction in another. The author often explains to patients that the foot is much like a 'Rubik's Cube', in that if a joint is slightly misaligned or otherwise restricted, then it affects the freedom to move, not only in that joint but across the whole Cube (foot). These puzzle cubes move in 3D in each of the orthogonal planes, although unlike the foot, only in one plane at a time. They are also held together through tension and compression forces, much like all musculoskeletal anatomy. Symptoms within the foot can often be linked to a restriction in a foot joint which can be resolved quickly and easily through a manual release or mobilisation technique, just as can a slightly misplaced part of a Rubik's Cube (Fig. 4.5.1).

A small articular misalignment or over-compression due to excessive soft tissue tension within a single foot joint can restrict not only local joint freedom but can also cause segmental dysfunction during gait. This underlines the importance of each part within the whole foot's structure and function; a situation that can be appreciated by considering the principles of biotensegrity. This is of course true of the whole body, but with so many anatomical structures residing in such a small area subjected to high loading stresses, the foot seems to magnify these effects. Small dysfunctions in the foot underline the principles of optimal individual energetics and loading of stresses during gait. Increasing evidence is emerging that is helping clinicians to understand the effects of arthrokinematic restrictions of the ankle and foot (Hertel and Corbett, 2019).

Equally, excessive motion or compliance resulting in loss of control can also prove problematic. The midfoot straddles the functional relationship of the rearfoot complex to the forefoot, and is pivotal in compliance–stiffness changes. For the rearfoot to perform effectively during gait, the midfoot must also perform effectively and vice versa. The midfoot also allows for motion through it and the forefoot to spare the rearfoot from perturbation torques, particularly in the frontal and transverse planes. For the forefoot to perform its roles during gait sufficiently, the midfoot must fulfil its functional role, and vice versa. The midfoot bridges the rearfoot mechanics to the forefoot and transfers body weight load and energy across it. At heel lift, the midfoot must allow the ankle plantarflexor power (energy) to pass into the forefoot. In addition, the forefoot must be stable enough to apply the ankle power generated throughout late midstance across the midfoot into a

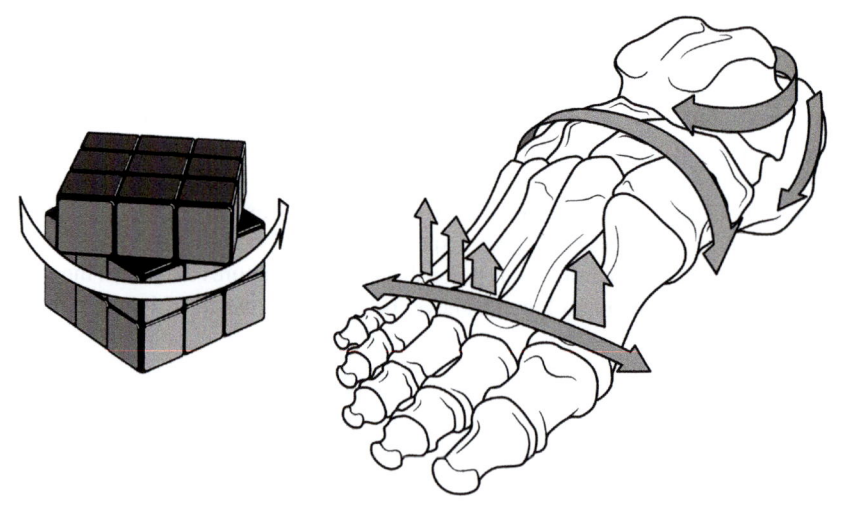

FIG. 4.5.1 The 3D puzzle of the Rubik's Cube has some similarities to the foot as it permits rotations in all three planes of orientation, as long as each of its articulations is aligned to permit freedom of motion. The many bones of the feet, restrained by ligaments and compressed by muscle contractions, demonstrate a similar relationship. If articulations are not aligned to each other to facilitate motion, restrictions at one joint interferes with the ability of the rest of the foot to move freely. A little 'shuffling' of a Rubik's Cube's segments often frees motion. Simple mobilisation and clinically applied manipulative forces can release joint restrictions within the foot that greatly enhance function and improve stress management across the whole foot. Lost motion reduces energy dissipation, focuses stresses, and reorientates instantaneous axes of motion within the foot. *(Permission www.healthystep.co.uk.)*

stabilising GRF under the forefoot to ease the transfer of load towards the next step (Takahashi et al., 2017; Venkadesan et al., 2020).

The midfoot functions in the transverse and frontal planes, but the most significant motion occurs in the sagittal plane primarily at its medial proximal articulations (Holowka et al., 2017). However, the more subtle frontal and transverse plane motions are still an important part of setting the foot's position so that it can utilise its longer ankle plantarflexion powered moment arm at heel lift, directly influencing forefoot function.

The foot fulfils three functional stance roles. The first is to provide energy dissipation in support of the lower limb's shock attenuation capability, ability, and capacity. The second role is in providing a stable base during weight transference of the body's CoM over the foot through the fulcrum of the ankle joint during ankle dorsiflexion, with the rotating tibia (moving the rest of the limb and trunk) acting at the lever beam base of motion. This is followed by a sudden change to the foot's third role, acting as a beam rotating around the MTP joint fulcrums from heel lift. In running, the shock absorbing braking action and acceleration roles occur concurrently with weight transfer. These three functions primarily occur through the sagittal plane. The other planes are far less dominant (Hunt et al., 2001) and will be considered in relation to pathology later. The simplest approach to pathology is to consider the three primary stance roles first, which equate to the three stance phases of gait, then consider if the problem might relate more to disruption within the other supporting planes.

4.5.2 Contact to loading response pathomechanics in the foot

The foot has to manage a collision impact event while adapting itself to the support surface under increasing load from weight transference, without loss of limb or trunk postural stability. This involves a number of significant events that can go wrong with potentially significant pathomechanical results.

Energy dissipation in the foot during initial contact and loading

The foot uses two primary mechanisms for energy dissipation during its impacts. The first is provided by the shock attenuation abilities of the plantar fat pads, usually those of the heel at the initial collision, then those of the forefoot during the second. The other mechanism is by using the skeletal frame as an elastic semiflexible loading structure that can deform and yet maintain stability via biotensegrity principles. The foot's shock attenuation is actually a complementary mechanism assisting the other larger shock attenuation capacities of the lower limb, in particular, the knee flexion moment via knee extensor muscle–tendon buffering. Tibialis anterior and the heel fat pad provides the energy buffering at the rearfoot during heel contacts. The triceps surae, forefoot fat pad, and the skeletal frame of the foot vault provide the shock attenuation at forefoot contacts, regardless as to whether they occur following a heel contact or during an initial forefoot strike. In the skeletal frame's ability to shock attenuate, the tibialis posterior and peroneus longus stiffening mechanism plays an important role (Kokubo et al., 2012), as also does the plantar aponeurosis through helping in setting and then reversing the windlass mechanism to initially increase and then reduce foot stiffness (Caravaggi et al., 2009; Wager and Challis, 2016). Heel fat pads are essential during heel contact, while both the plantar fat pads and skeletal frame mechanisms are important during forefoot contact. Forefoot loading offers more skeletal compliance than does rearfoot loading, as the rearfoot bones only offer two articulations, whereas the forefoot contact can utilise variable vault compliance. Thus, the heel fat pad has a greater role in the initial loading collision shock attenuation during heel contact gaits than does the forefoot fat pad, particularly as the latter is aided in energy dissipation by prior tibialis anterior braking and energy buffering of the ankle plantarflexion moments.

Plantar fat pad dysfunction

The plantar fat pads function through the composite properties of their biphasic fluid-like fat/adipose tissues and their solid connective tissue frame containment septa, working with the epidermis and dermis, providing nonlinear viscoelastic properties (Chokhandre et al., 2012; Fontanella et al., 2012). These fat pads are capable of absorbing 30%–60% of deforming energy (Challis et al., 2008). On loading, the plantar fat pad deforms under shear and compression, increasing the pressure inside the adipose chambers. This increase in internal pressure applies compression forces into the connective tissues tensioning the chamber walls. Energy is dissipated through the adipose compression and connective tissue septa's tensional stretch, storing energy within the connective tissue walls. This stored energy is lost to locomotion as the elastic recoil returns energy when the fat pads become offloaded after heel lift and during preswing within the heel and forefoot, respectively. The released energy instead reshapes and resets the plantar fat pads ready for the next loading cycle. Due to its viscoelastic properties, the faster the plantar fat pad is loaded, the stiffer it behaves (Naemi et al., 2016a).

Forefoot initial contacts can result in either a secondary impact loading of the heel fat pad or, if the heel never reaches the support surface, the forefoot fat pad becomes the only loaded fat pads throughout stance. Forefoot initial contact increases the loading time of the forefoot plantar fat pads, which will remain loaded throughout the entire length of the stance phase. In heel strike gait, the forefoot plantar fat pad is only loading at the end of contact phase during later loading response. The forefoot is then weightbearing throughout the rest of stance to be offloaded during preswing. Forefoot plantar pressures are the highest on the foot during terminal stance. Therefore, elastic recovery time of the forefoot plantar fat pads is dependent on the time the plantar fat pads are offloaded during swing phase until forefoot contact after heel strike, or only during the swing phase if forefoot strike follows swing.

In normally healthy connective tissue, loading stress on the plantar pads should not be an issue but in the presence connective tissue disease, age associated atrophy, or when connective tissue mechanical properties are fatigued, the recovery time may become a significant factor. Fat pad dysfunction and damage is unable to maintain a short realistic recovery time within their material properties. Indeed, poor soft tissue health commonly affects the plantar aspect of the metatarsal region, leading to cutaneous and plantar fat pad breakdown (Chao et al., 2010).

Any disruption within the heel or forefoot fat pads, particularly the connective tissue properties, will reduce the foot's shock attenuation capacity. Fat pad dysfunction can lead to excessive compliance or, in the case of diabetes, excessive stiffness. Increased connective tissue compliance leads to increased loading phase deformation and a reduction in the offloading elastic recoil phase (Fontanella et al., 2016). Atrophy of the connective tissue walls or septa of the plantar fat pad appears to be associated with ageing due to elastic fibre and connective tissue wall degradation, with concurrent loss of epidermis and dermis thinning which interestingly, is reported scientifically in older rats. Human plantar fat pads share common physiological functions and histological structures with those of rats (Molligan et al., 2013), for rodents and primates are closely related in an evolutionary sense. It is reported that the volume of adipose tissue in the plantar fat pads of older rats lowers to around 65% of that of the young rats, and the number of clusters of adipose tissue without enclosing surrounding septa increases along with the loss of elastic fibres in the septa walls (Molligan et al., 2013). In diabetics, the resulting abnormally stiff connective tissue chamber walls are more fragile and thus more easily damaged and torn, allowing the fat to herniate across the structure of the fat pad.

There has been a loss of compliance reported in the plantar soft tissues of humans under the hallux and at the 1st MTP joint, the 3rd and 5th metatarsal heads, and the heel with age (Kwan et al., 2010). The plantar soft tissues were not only stiffer, but also thinner at the 2nd metatarsal head with age (Chao et al., 2010). The resulting loss of energy absorbing deformation elasticity in the plantar soft tissues may be part of the mechanism for explaining increasing problems in elderly feet. Callus, that often develops on the plantar forefoot with age, possibly represents a protective mechanism that attempts to compensate for the loss of normal plantar soft tissue mechanical properties and yet has the benefit that it does not seem to influence tactile sensitivity of the feet (Holowka et al., 2019). However, barefoot populations often develop thicker and harder calluses than do shod populations (Holowka et al., 2019), suggesting perhaps that shod population skin callus may not be as effective in protecting deeper soft tissues as that seen in healthy unshod feet.

Most commonly, the loss of the plantar fat pads is an insidious process of ageing. Yet, large foot impact stresses can cause acute failure of the fat pad, sometimes associated with calcaneal fractures linked with either falls from a height or during plantar foot collisions associated with road traffic collisions (Levy et al., 1992). Corticosteroid injections (Basadonna et al., 1999) and scarring on the foot from pedal surgery or trauma may also disturb the local plantar fat pad's properties. Dysfunction of fat pad chambers causes a reduction in the energy dissipation capacity that must be accounted for by other energy dissipation mechanisms so as to prevent tissue stress thresholds from being breached elsewhere within the energy dissipation mechanisms (Fig. 4.5.2a).

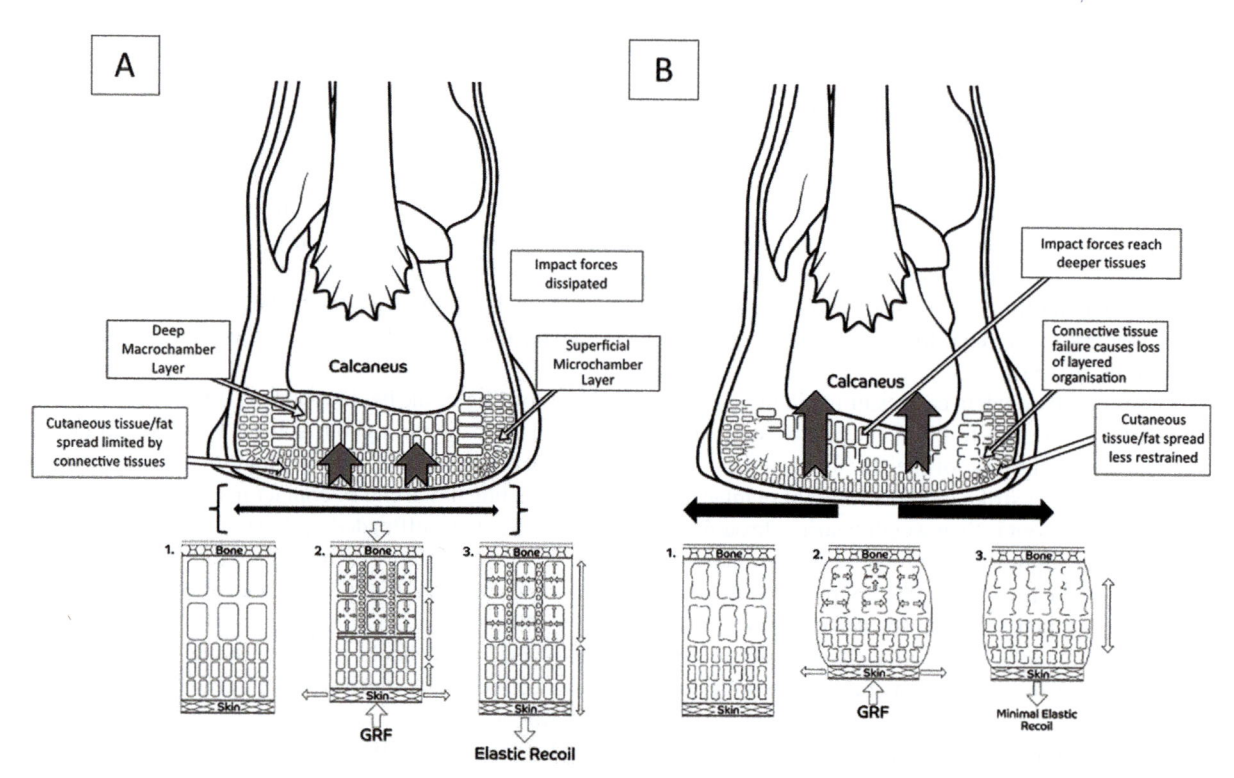

FIG. 4.5.2a The heel and forefoot plantar pads use the same combination of biphasic properties created by fat being enclosed within collagen and elastin connective tissue walls. This arrangement acts as noncompressive fluid-filled elastic chambers to dissipate impact energy. The heel fat pad is more structured through its combination of specialised layers of superficial microchambers and macrochambers (A) than the forefoot fat pad. If the connective tissue walls that compartmentalise the fat should tear and rupture, then the fat can start to displace horizontally on loading, losing the capacity to resist deformation and causing thinning of the fat pad's depth (B). Fat displacement on loading means connective tissue elastic recoil will be compromised between steps so that energy dissipation ability is reducing over sustained steps. Without fat pad energy dissipation at impact and loading response, deeper, often more delicate structures will come under increasing compression and shear stresses. Heel pain as a result of bone marrow lesions and oedema within the trabecular bone can result from heel fat pad dysfunction, even risking fatigue fracture of the calcaneus during endurance running. Similar processes within forefoot plantar fat pad dysfunction are more problematic during acceleration than impact because during walking and heel strike running, much impact energy is already dissipated by the heel before the forefoot collision. Loss of forefoot connective tissue adaptable stiffening and elasticity around the fat decreases the forefoot fat pad's ability to dissipate shear stresses during acceleration, particularly in its more distal zone. Forefoot fat pad failure risks burning metatarsalgia, plantar callus, or more seriously, metatarsal fatigue fractures with physiological bone disease, and in the presence poor vascular tissue viability, the development of soft tissue ulceration. *(Permission www.healthystep.co.uk.)*

In diabetic plantar tissues, problems are associated with abnormal stiffening due to increased collagen fibril densification from tissue glycation (Kwak et al., 2020) rather than through the loss of compliance due to atrophy of connective tissue and loss of plantar fat pad bulk (Waldecker and Lehr, 2009). The lack of energy dissipation may help to explain the findings that, despite slower walking speeds, peak vertical and horizontal (anterior–posterior) loading stresses remain high in neuropathic diabetic walking gaits compared to healthy controls (Fernando et al., 2016b). Diabetic soft tissue stiffness will decrease the tissues' capacity to dissipate energy and manage shear strains, leaving the softer skin above more vulnerable to injury (Kwak et al., 2020). Essentially, diabetics load on stiffer plantar connective tissues that are less able to deform beneficially on loading to protect other foot tissues deeper as well as superficial, to them (Fig. 4.5.2b).

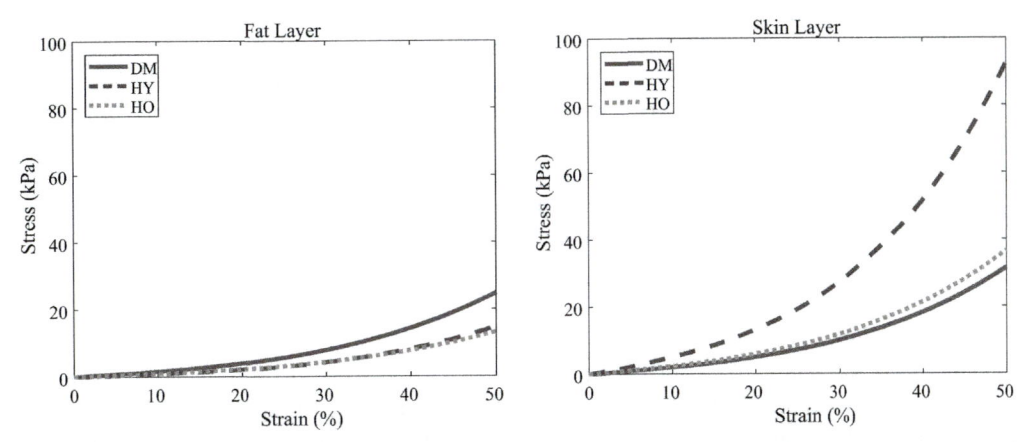

FIG. 4.5.2b Graphs demonstrating estimations of stress–strain curves of plantar fat (left) and skin (right) layers using the mean stress and strain values on groups of individuals with diabetes mellitus (DM) compared to healthy young (HY) and healthy old (HO) controls. Diabetics have stiffer plantar fat pads offering less compliance to impact, while their skin becomes a little less elastic than in healthy older adults. This combination compromises the plantar cutaneous tissues' ability to protect deeper structures and makes the skin and connective tissue more fragile under load. Although connective tissue chamber walls may remain intact, the loss of elastic compliance in the septa surrounding the fat means that normal biphasic mechanics of energy dissipation are disturbed in diabetics. However, overly stiffened connective tissues tend to become more brittle, risking rupture of chamber walls and displacement of the fat horizontally. *(Image from Kwak, Y., Kim, J., Lee, K.M., Koo, S., 2020. Increase of stiffness in plantar fat tissue in diabetic patients. J. Biomech. 107, 109857.)*

Loss of fat pad energy dissipation can present with a number of symptoms, the most common being weightbearing plantar heel pain. The symptoms are generally similar, if sometimes more indistinct, than those associated with plantar aponeurosis pathology. Symptoms are often worse on any loading of the heel, especially barefoot or in flat, thin-soled shoes. Examination will often reveal that the inferior surface of the calcaneus can be palpated easily and can be uncomfortable under pressure. Squeezing plantar soft tissue back over the plantar calcaneus and reapplying pressure will usually remove the discomfort temporarily. If the plantar fat pad of the forefoot is failing, then the plantar aspects of the metatarsal heads can be palpated more easily, and symptoms usually associated with metatarsal head pain (*metatarsalgia*) on weightbearing. This is often worse in thin-soled shoes, especially if the shoe's heel height is large which increases the forefoot loading time and digital extension angles. High-heeled footwear thus creates greater digital extension angles that stiffen the forefoots fat pad, further exposing the fat pad at the metatarsal heads to longer and higher loads within the distal zone. This can lead to the development of fibrosis, adventitious bursae, soft tissue chondroma, and/or callus development within the cutaneous tissues under the metatarsal heads (Studler et al., 2008).

Whether the loss of the plantar fat pads' energy dissipation affect areas outside the foot will depend on other shock attenuation abilities throughout the whole lower limb, but effects are usually localised to foot symptoms generated via local tissue damage.

Skeletal frame energy dissipation dysfunction

The vault profile of the skeletal frame of the foot can act as an adjustable deformable structure providing variable energy dissipation properties. The properties used to create energy dissipation are created through the elastic properties of the connective tissues acting across the skeletal struts during articular motion and the muscle–tendon buffering effects of the active muscles under motion within the skeletal frame. The more stiffly the skeletal frame is held at initial contact, the more it can relax and potentially deform over a larger range of movement on loading through connective tissue stretch and muscle slackening. Lack of deformation–motion on loading will increase the stress within the bone struts of the frame, risking articular and bone injures such as fatigue fractures. The foot also needs compliance during loading response to adapt to the support surface contours for providing a stable base during midstance. Thus, a stiff foot at initial contact that can become increasingly compliant by the end of loading response is the initial pedal functional aim, rather than presenting a stiff nondeformable or a too easily deformable structure, both of which will compromise any cushioning-energy dissipation properties.

Appropriate tensioning of the skeletal frame to form an energy damper throughout forefoot loading involves tibialis anterior, tibialis posterior, peroneus longus, and the long and short digital extensors and abductor hallucis. The digital extensors stiffen the plantar aponeurosis via the windlass mechanism and then reduce activity to release the reversed windlass mechanism (Caravaggi et al., 2009; Farris et al., 2019, 2020). In forefoot strikes, the action of tibialis anterior is replaced by triceps surae via the Achilles, and the skeletal frame has to manage an initial impact under a large ankle dorsiflexion moment rather than after a decelerated ankle plantarflexion moment. Thus, initial forefoot contact applies greater loading stresses to the foot's skeletal frame than a forefoot collision following heel strike.

Once skeletal frame loading starts, the reversed windlass effect can gradually reduce tensions across the skeletal frame utilising connective tissue slackening under tension and muscle–tendon buffering that dissipate energy. Failure in any of the muscles, both in tensioning before forefoot contact and untensioning the frame after forefoot contact, can compromise energy dissipation capabilities of the foot during later stages of loading response. In fast forefoot running, the foot is maintained in a stiffened state increasing the need for proximal energy dissipation outside the foot.

Dorsiflexor and plantarflexor loading dysfunction

Dysfunction in any part of the energy dissipation mechanisms can result in pathology. In single and in varied combinations of structure failure, variations in the kinematics and kinetics can arise. In heel strikes, failure of tibialis anterior will cause reduced or unrestrained ankle plantarflexion moments and a higher acceleration of forefoot impact, resulting in high velocity forefoot loading stresses. If the digital extensors are able to compensate, then they may be able to reduce the effect by increasing the reversed windlass tension through increased digital extension. However, the restraint of the dorsiflexion moment this mechanism supplies will not match the energy dissipation and braking effect of a healthy tibialis anterior with its relatively large cross-sectional area. The increased MTP joint extension that can result from this compensation will stiffen the forefoot fat pad (Bojsen-Møller and Lamoreux, 1979). This potentially exposes the metatarsal heads to increased bending moments upon loading, risking impact fatigue fractures and soft tissue threshold breaches because of the stiffer forefoot tissues being impacted. Long-term overuse of digital extensor muscles also risks deforming toes via extensor substitution mechanisms, causing MTP joints to 'fix' at increased extensions angles (Fig. 4.5.2c).

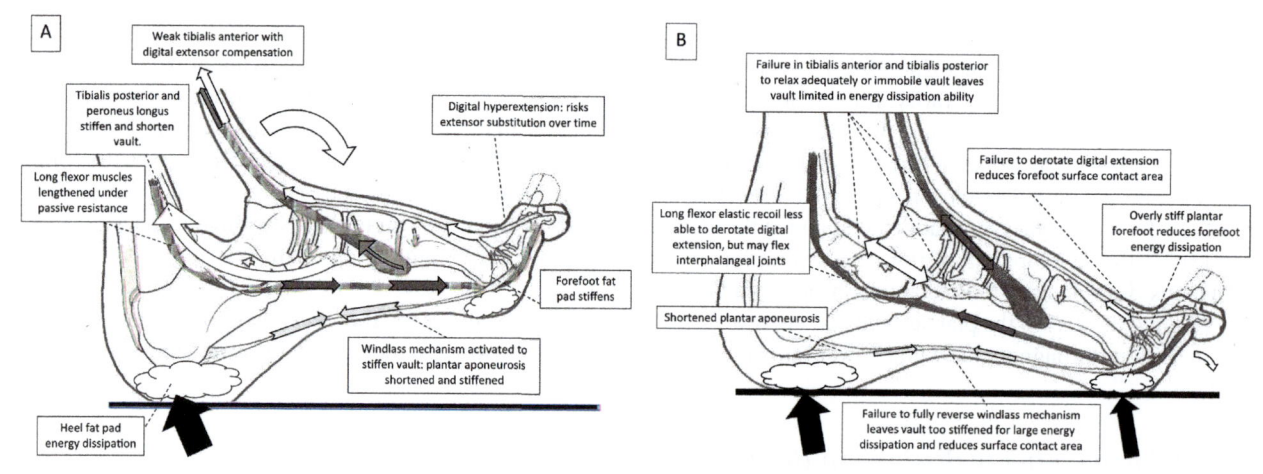

FIG. 4.5.2c At heel strike (A), the plantar fat pad provides the foot's energy dissipation role because the rearfoot initially offers little flexibility to deform. Failure of the fat pad's biphasic material properties at heel contact is thus problematic. Once the plantarflexion moment is initiated, the ankle takes on an energy dissipation role via tibialis anterior eccentric activity, aided a little by the digital extensors via them extending the toes to help set the windlass mechanism for foot stiffening. If tibialis anterior is weak, digital extensor compensation can overextend the toes, increasing forefoot plantar fat pad stiffness. Tibialis posterior and peroneus longus are also actively stiffening the vault prior to forefoot contact. Stiff cavoid feet tend to give mechanical advantage to such contact stiffening mechanisms. At forefoot loading (B), the forefoot fat pad should become more compliant as digits are relaxed and rotate out of extension under rising forefoot GRFs. Tibialis anterior and digital extensor relaxation permits reversal of the windlass mechanism, increasing vault compliance for energy dissipation. Tibialis posterior and peroneus longus activity should also reduce to aid this. Failure in all these actions decreases the foot vault's energy dissipation abilities and reduces the plantar surface contact area, increasing the plantar peak pressures. Such failures during gait are often indicated if toes remain visibly extended into midstance with little vault height reduction. Cavoid feet with higher metatarsal declination angles tend to give mechanical advantage to toe extensors and present stiffer vault mechanics at forefoot loading and early midstance, requiring greater active muscular energy dissipation compensations to avoid tissue threshold breaches. If peroneus longus is weak within cavoid feet, the vault often becomes even stiffer from unrestrained tibialis posterior activity and the foot can become more laterally unstable. *(Permission www.healthystep.co.uk.)*

Failure of triceps surae strength in forefoot strikes will cause a rapid heel drop to the support surface (Fig. 4.5.2d). Weak plantarflexors have a far more detrimental effect on CoM acceleration and momentum maintenance if large degrees of heel drop occur after forefoot contact. A large heel drop will cause a backward drag on the CoM, an effect that needs to be negated (usually by gluteus maximus) before CoM momentum can continue forward again. Individuals with very weak, flexible calf muscles resulting from neuropathy tend to develop compensatory spasticity in the ankle plantarflexors that prevent complete heel drop, as often seen in cerebral palsy sufferers. The situation in cerebral palsy, tends to be managed by a loading midfoot break that allows more of the plantar forefoot to make ground contact without heel weightbearing and by developing greater whole limb rigidity. This compensation technique loses the ankle dorsiflexion-to-knee extension coupling relationship of normal walking gait, creating very poor energetics and resulting in a situation known as *lever arms disease* (Gaston et al., 2011).

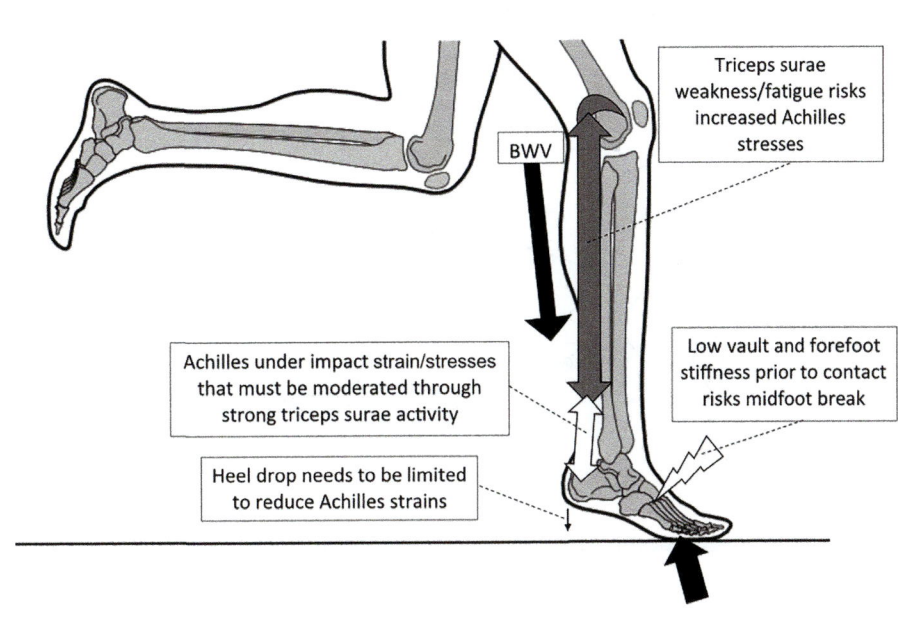

FIG. 4.5.2d Forefoot strike requires energy dissipation from the forefoot fat pad and the soft tissues of the vault, including the deep transverse metatarsal ligament. This ligament it is also used in forefoot loading energy dissipation following heel strike. However, forefoot impact is a single collision event rather than a secondary part of an impact. Thus, elasticity/stiffness within the vault and the forefoot, via extrinsic/intrinsic plantar foot muscles and the windlass mechanism, must be greater prior to contact than that required after a heel strike. Yet, the vault must be able to respond by providing and controlling some compliance during loading response. In protecting the forefoot, proximal joint motion and soft tissue energy dissipation are essential and include triceps surae deceleration of the dorsiflexion moment generated by forefoot contact on a plantarflexed ankle. Failure in foot vault stiffening may result in contact phase midfoot break (as is often seen in cerebral palsy toe-walking) with dorsiflexion energy being attenuated via the midfoot. Weakness within the triceps surae will cause a marked and accelerated heel drop which may result in a secondary heel collision. Such triceps surae weakness risks increased Achilles strains in a gait style that already prolongs the percentage loading time of this tendon, risking Achilles pathologies. *(Permission www.healthstep.co.uk.)*

Regardless of initial contact position, failure of the windlass mechanism to activate before contact will decrease foot stiffness that can be released under forefoot loading. This loss of this energy dissipation technique may require the other energy dissipation mechanisms to compensate. Failure of the reversed windlass mechanism to unextend the toes during forefoot loading will result in excess foot stiffness at loading response, leaving the metatarsal heads more exposed to loading GRF forces and plantar pressures. This will increase reliance on the forefoot plantar fat pads. Yet, with excess toe extension, the forefoot fat pads also remain stiffer. In diabetic stiffened soft tissues, common associated toe extension deformities such as hammer toes, may compromise energy dissipation further by preventing the reversed windlass mechanism, further increasing plantar fat pad stiffness. This can become a serious threat to cutaneous tissue threshold via undissipated shear and pressure forces on fragile tissues.

Tibialis posterior and peroneus longus loading dysfunction

Failure of tibialis posterior and/or peroneus longus function creates more complex effects on energy dissipation. Both muscles are also significant in influencing forefoot and digital contact areas and pressures, but not as much as the digital flexors (Ferris et al., 1995). Tibialis posterior and its tendon is known to have muscle–tendon energy buffering effects (Maharaj et al., 2016). However, tibialis posterior and peroneus longus work synergistically to control the stiffness levels within the vault of the foot (Kokubo et al., 2012). Both muscles also have frontal and transverse plane effects across the vault and influence rearfoot frontal and transverse plane stability. Tibialis posterior provides the primary foot-stiffening mechanism and eversion and abduction resistance across the foot and its vault, as first the heel and then the forefoot load. Failure in tibialis posterior function at contact phase reduces the foot's initial contact stiffness and increases forefoot compliance during loading response, with concurrent loss of eversion and abduction moment resistance at the rearfoot and across the vault. Thus, tibialis posterior dysfunction results in reduced energy dissipation capacity through failure to stiffen the foot before impact, with loss of contact compliance control and foot stability during forefoot loading. This results in less constrained rearfoot eversion and abduction moments around the midfoot complex across the vault, leading to medial instability within the loading foot. In forefoot contact, such problems start distomedially, moving proximally back into the vault, rather than from the rearfoot passing anteriorly into the vault (Fig. 4.5.2e).

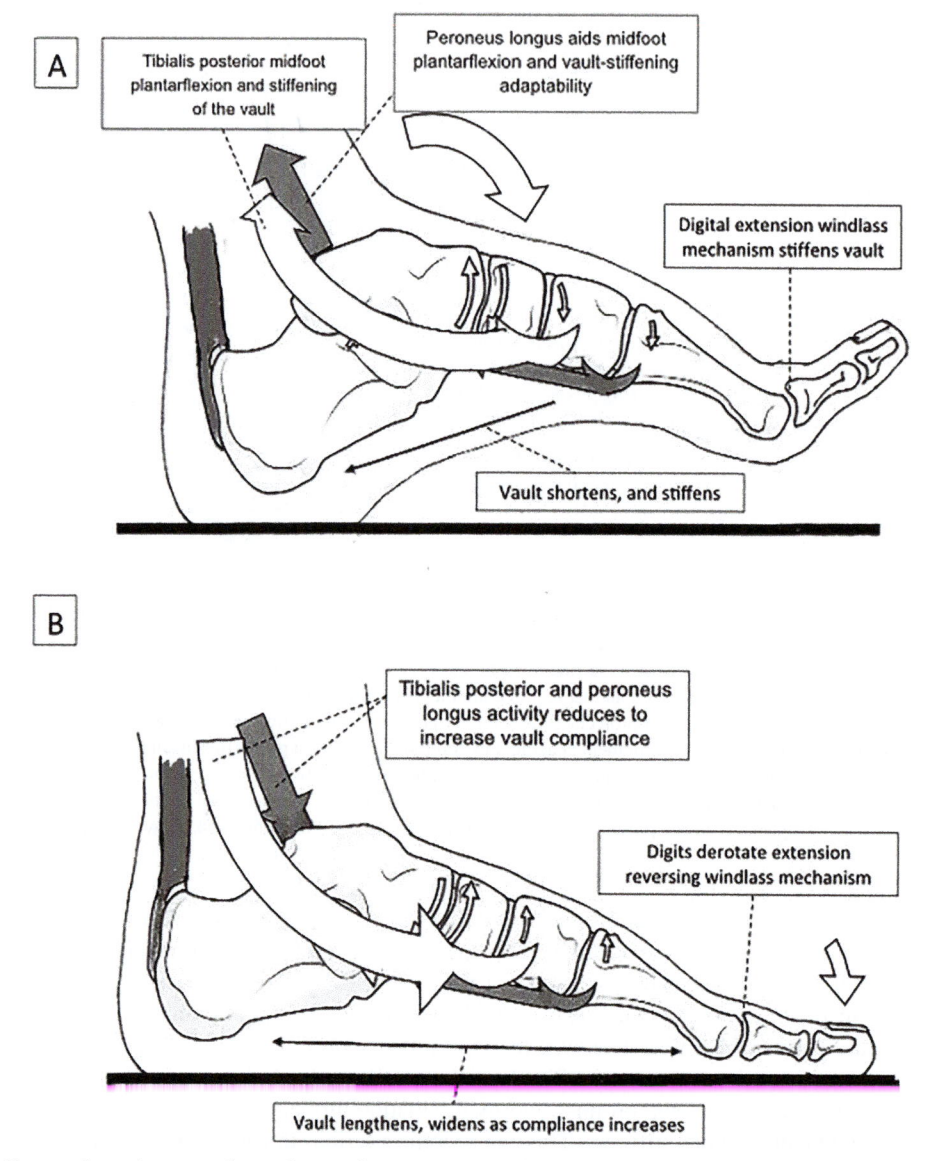

FIG. 4.5.2e Tibialis posterior and peroneus longus have an important synergistic role in active foot vault biomechanics throughout stance phase by influencing foot elasticity levels. During contact and loading response (A), these muscles initially increase vault stiffness to improve the vault's elasticity prior to forefoot contact, aided by the windlass mechanism. While tibialis posterior is primarily responsible for stiffness, peroneus longus adds adaptability by both antagonising and agonising this process. Insufficient strength of both of these muscles or the tibialis posterior weakness alone leads to a less stable, excessively compliant vault during forefoot loading. Weakness in peroneus longus alone can permit tibialis posterior to overly stiffen the vault. At forefoot contact (B), the muscles reduce activity, allowing muscle and tendon fibres to lengthen eccentrically as the windlass mechanism is also reversed. This permits freedom across the vault to increase its lowering, lengthening, and widening, a process that loads vault ligaments with strain. By raising the vault's level of compliance, the foot becomes an adaptable energy dissipator while it also increases its surface contact area as the single-limb support phase begins. This lowers peak pressures. Loss of power within these muscles at this time will result in excessive foot compliance, causing excessive deformation of the vault, reducing the ability to dissipate and store loading energies. Loss of tibialis posterior alone is particularly destabilising of the medial foot, and can lead to injuries within plantar vault ligaments, such as the spring ligament, significantly degrading lower limb gait stability further. *(Permission www.healthystep.co.uk.)*

Peroneus longus increases foot compliance in the absence of tibialis posterior function (Kokubo et al., 2012) so that a healthy functioning peroneus longus in the presence of tibialis posterior dysfunction can actively decrease foot stiffness at contact, reducing the skeletal frame's energy dissipation capacity by making it too easily deformable. Effectively, the foot starts to behave like bottomed-out foam in its inability to cushion impact. If peroneus longus is dysfunctional itself and tibialis posterior is functional, then the increase in compliance expected during the skeletal frame's loading response could be compromised and the foot remain overly stiffened. This is particularly so on the medial side with the lateral vault being less stable through loss of peroneus longus tightening its tendon under the cuboid. Under unrestrained tibialis posterior activity, foot stiffness could be too great to allow effective energy dissipation, with the vault becoming excessively stiffened to be able to absorb energy easily. With peroneus longus dysfunction, the foot will also lose restraint of any loading inversion and adduction moments across the rearfoot complex and foot vault, during loading response. This can potentially lead to lateral instability on a less stable lateral column, as well as causing energy dissipation compromises.

An overly active windlass effect can result in a late or a 'reluctance' to reverse the windlass mechanism, preventing normal vault height reduction that occurs as part of normal foot loading response compliance. This could provoke digital MTP joint hyperextension and excessive GRF-induced stress at the forefoot plantar fat pads and metatarsal heads during forefoot loading.

Pathomechanical consequences within the lower limb in relation to poor foot shock attenuation/energy dissipation will depend on whether other lower limb energy dissipation mechanisms such as knee and hip extensor muscle–tendon buffering and tissue resonance can compensate by increasing their damping activity. This compensation is more possible during slower walking when impact energies are lower than when faster walking, and particularly during running, when higher tissue energy dissipation is required. However, should the large joints' energy dissipation capacity already be compromised, loss of foot energy dissipation can add significantly to pathological risk. In the cases of tibialis posterior, tibialis anterior, and peroneus longus dysfunction, the kinematic changes often guide the therapeutic interventions, as the instabilities they cause can dominate the clinical picture. However, reflection on the chain of mechanisms proximally that provide energy dissipation through appropriate stiffening and compliance properties, should be considered in any proposed management.

Compensation mechanisms and foot types in loading

How the appropriate levels of stiffness and compliance are maintained in the lower limb for energy dissipation management is affected by an individual's foot type variant. This can be clinically approached by considering the effects of vault profile curvature and tissue flexibility. Cavoid foot vaults with stiff soft tissues are most likely to suffer from the effects of compliance reduction. The increased vault curvature stiffens the foot through curvature mechanics, particularly if other stiffening mechanism, such as soft tissues flexibility cannot compensate for lost motion. The increased metatarsal declination angle associated with cavoid feet may enhance digital extension, aiding and potentially increasing the setting of windlass mechanism stiffness prior to forefoot contact and causing compliance compromises in any inability to switch it off through reversing digital extension. An inability to permit the reversed windlass effect during forefoot loading, possibly through extensor overactivity on the windlass or tibialis posterior over activity on the vault at loading response, will keep the vault profile higher, maintaining stiffness into a period when increasing compliance should be available for dissipating the loading of full body weight under CoM braking events. Reduced muscle–tendon buffering effects through tighter muscles across the vault will exacerbate the problem and be more problematic in the presence of underlying connective tissue hypomobility or connective stiffening disease such as diabetes.

Loss of full peroneus longus activity will also exacerbate cavoid foot stiffness, as will increased tibialis posterior and digital extensor activity. Frontal plane alignments such as an inverted rearfoot profile or semirigid or fixed forefoot valgus rotations across the midfoot, which are often associated with cavoid feet, will also increase the inversion moment and decrease eversion moments during loading. This places peroneus longus at a mechanical disadvantage in having larger inversion moments to resist. This further increases the loading stress on peroneus longus which in turn, gives mechanical advantage to stiffening mechanisms, and may shift the subtalar joint axis laterally through increased rearfoot supination, worsening the situation further.

Feet that have appropriate flexibility should be able to cope with higher initial loading impact forces. However, a low planus-vaulted foot with hypermobility is more likely to suffer from decreased foot-stiffening mechanisms before impact, giving less ability to provide energy dissipation as a result of its low stiffness, and thereby acting like a failed shock absorber on a vehicle. Low metatarsal declination angles may make it harder to achieve the benefits of an efficient windlass mechanism through toe extension before forefoot loading. Thus, initial impact stiffness may be set too low to afford adequate energy dissipation through controlled deformation, rather like trying to cushion with bottomed-out foam.

This will result in a lack of 'spring resistance' (elasticity) within the system to dissipate the loading energies and will tend to increase loading rearfoot eversion excursion and midfoot dorsiflexion and abduction. Hypermobile planus feet are therefore likely more reliant on more muscular action to provide active foot stability due to a lack of passive connective tissue stiffness and/or reduced vault profile curvatures. This risks earlier foot fatigue. Unfortunately, it appears that planus feet often possess less muscle bulk and strength than that found associated with higher vault profiles (Angin et al., 2014). Excessive compliance issues can also be provoked by loss of plantar ligament integrity through loss of both mechanical restraint and sensorimotor input. Loss of the subtalar ligaments that are important in restraining talar anterior drift during loading response and early midstance, may also set up excessive anterior motion within the medial or tibial load frame and thus cause excess vault mobility (see Section 3.3.7 in Chapter 3), particularly during the earlier half of stance (Fig. 4.5.2f).

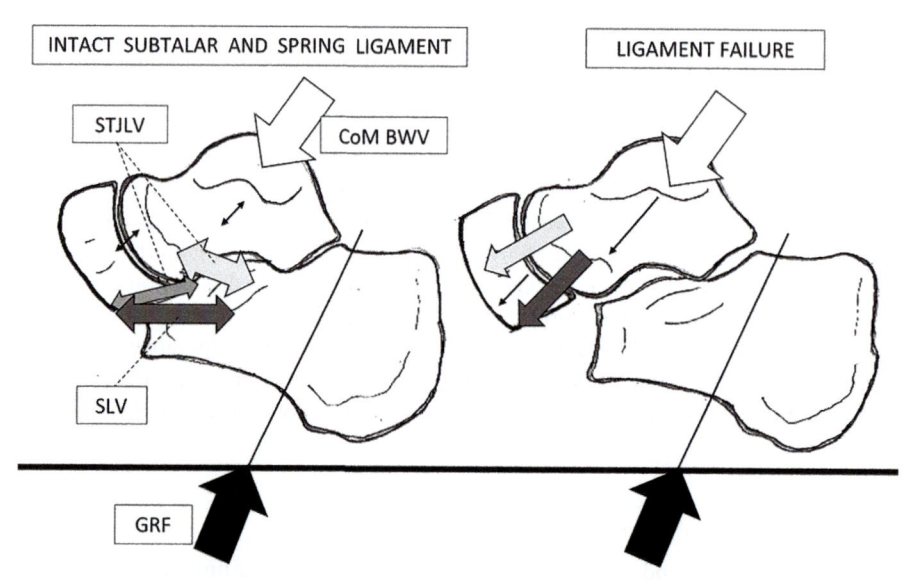

FIG. 4.5.2f The subtalar joint ligament vectors (STJLV) of the cervical ligament and interosseous talocalcaneal ligament (light-grey double-headed arrow), and calcaneonavicular branch of the bifurcate ligament (mid-grey double-headed arrow) are important in keeping the talus and navicular from excessive anterior and medial drift during loading response and early midstance. Via calcaneal and navicular attachments and a link to the deltoid ankle ligaments, the spring ligament also provides an important spring ligament vector (SLV—dark-grey double-headed arrow) along the medial proximal vault. Contact, loading response, and very early midstance expose the rearfoot to a GRF that has a posteriorly directed horizontal vector, while the CoM still drives inferiorly via an anterior and medial BWV. This creates forces that drive the talus anteroinferiorly, pushing the navicular and anterior medial bones anteriorly with it away from the calcaneus. The vector provided from the combined restraint of these ligaments prevents excessive vault drop. Failure of subtalar ligament vectors may cause the medial column (medial or tibial load frame) to excessively displace anteriorly and medially (light-grey arrow), which also risks midfoot pathomechanics. If the SLV and/or tibialis posterior tendon support are also dysfunctional, then motion will be directed more inferoanteriorly (dark-grey arrow), driving the navicular down and forwards. Such plantar-medial foot hypermobility will become a pathomechanical issue for vault function during gait. *(Permission www.healthystep.co.uk.)*

Therefore, loss of compliance-inducing factors is of more significance to pathomechanics in hypomobile, high-vaulted feet, while pathology associated with dysfunctions within the stiffness-inducing mechanisms are going to be more significant to hypermobile and/or low-vaulted feet. This may explain increased bone injuries and laterally located injuries reported in cavoid-footed runners and increased soft tissue injures in planus-footed runners (Williams et al., 2001). It follows that compliance mechanisms are under more stress in stiff feet trying to increase compliance appropriately and are thus more likely to fail, as are articular structures that are prevented from providing motion. Stiffening mechanisms are under increased stress when trying to maintain stiffness within feet expressing increased compliance, risking pathology within muscles that provide for increased stiffness.

This may explain the different loading impulse patterns seen in both heel contact running and walking between different levels of foot vault stiffness, where healthy individuals with high foot vault stiffness tend to reduce rearfoot impulses compared to those with higher vault flexibility (Cen et al., 2020). Healthy stiff feet work harder to reduce impact loading forces, while healthy flexible feet can cope with them. What happens to unhealthy stiff feet and unhealthy flexible feet with symptoms needs to be compared. Adding pathologies that influence foot stiffness levels into the research may provide interesting results that give insights into the links between pathology and the ability for the lower limb to manage impact energies.

4.5.3 Early midstance pathomechanics

In early midstance of walking, the foot should be in maximum compliance from the digital de-extending that prompts the reversed windlass effect, reducing the vault curvature and compliance via the reduction in tibialis anterior, tibialis posterior, and peroneus longus activity. The increased compliance allows the foot to freely adapt to the contours on the support surface. However, sustained high freedom for foot articulations to move within this compliance is not an advantage to later single support stability. Early midstance is thus a period of gradually reducing peak foot compliance associated with the rapid drop of the forces through the foot seen after the F1 impulse peak through to the F2 trough of the force–time curve. The lower impulse in walking is caused by the CoM rising upwards with limb extension. Reducing the residual compliance from loading response through to absolute midstance is beneficial for stabilisation of the CoM, providing a stable base for increasing its height via limb extension, and maximising the potential energetic effects gained from this height under gravitational forces.

Increasing plantar contact surface area, created initially through foot compliance, is associated with gradually raising foot stiffness that accompanies increasing foot pronation and vault depression. This explains the faster rates of initial foot lengthening under load at the start of midstance as the plantar contact area is required to increase quickly for stability. This change in foot dimensions starts to stiffen the foot, expressed through a slowing in the rate of foot lengthening, most markedly after absolute midstance during late midstance (Stolwijk et al., 2014). Thus, the initial vault drop loses a little potential height to the CoM but gains greater foot contact and stiffening adaptability. A lowering, a widened, and a stiffening vault should now help facilitate single-limb support stability with weight becoming distributed relatively equally, balanced between the forefoot and heel around absolute midstance. Thus, early midstance is associated with significant foot lengthening on a compliant foot with increase in foot stiffness under lowering loading forces, while increasing GRFs towards the F3 peak are associated with foot stiffening during late midstance for acceleration (Stolwijk et al., 2014; Bjelopetrovich and Barrios, 2016; Takabayashi et al., 2020). Both the vault and rearfoot complex's stability are increasing as the ankle reduces its plantarflexion angle, while the CoM of the body is drawn upwards and anteriorly through quadriceps and gluteal activity. At absolute midstance, the vault should be significantly lower than at the end of loading response, but significantly stiffer and more stable (Fig. 4.5.3a).

Early midstance should be a relatively low pathomechanical risk, as stresses are generally lower throughout this period of stance because of the falling vertical and horizontal forces associated with the descent of GRFs seen with the F2 trough. However, what happens here affects what happens throughout late midstance, and that in turn affects the acceleration phase of terminal stance. Compliance has to be appropriate to needs and be in a 'Goldilocks zone' that reflects the individual's vault height and soft tissue properties to produce tensile restraint for a stable acceleration from the foot. Early midstance can become dysfunctional if compliance is too great, too low, or too restricted in its ability to change from one towards the other, as it represents the start of a period of adapting from compliance to increased stiffening mechanism activations.

Activating stance muscle activity

Although EMG studies have identified peaks of activity of muscles throughout stance phase which are often mistaken to mean expected 'normal', variability within individuals between their different steps and also between individuals are noted and should not be considered abnormalities. These findings most certainly reflect changes brought about by the sensorimotor system identifying needs of postural stability and limb function relationships. It seems highly probable that variability in muscles such as tibialis posterior, tibialis anterior, and peroneus longus activity during gait reflects balancing needs from medial to lateral torques during stance that also reflect a response to the foot's changing compliance-stiffening requirements within each step. It has been noted that some individuals have less consistent activity relationships between tibialis posterior and peroneus longus in stance and variability between steps, rather than always having the same distinct peaks of activity (Murley et al., 2009, 2014). This probably reflects a need to change foot stiffness differently with each step throughout stance depending on the events occurring during each step. Thus, variance in muscle activity between steps should be seen as normal and desirable if stability and momentum are to be maintained. This is as long as gait perturbations do not result. As the foot's stiffness is influenced by gait speed, speed changes require adjustments in muscle activation to influence stiffness and compliance levels. Key stiffening and compliance muscles should still present similar activity, even if they have slightly earlier activity with increasing gait speeds, but only subtle changes should be seen across walking during consistent locomotive activity and speeds (Murley et al., 2014) (Fig. 4.5.3b).

Muscle activation variance within the population and individual step compliance–stiffness needs will change the rearfoot complex's kinematics as much as the vault profile of the foot, through its ability to make shape changing adjustments. It has been noted that asymptomatic healthy young individuals present several variable patterns of frontal plane rearfoot complex kinematics during the stance phase of gait (Cornwall and McPoil, 2009). Whether certain variants are more prone to particular pathologies is unknown. Such rearfoot variations possibly just reflect differing pedal compliance

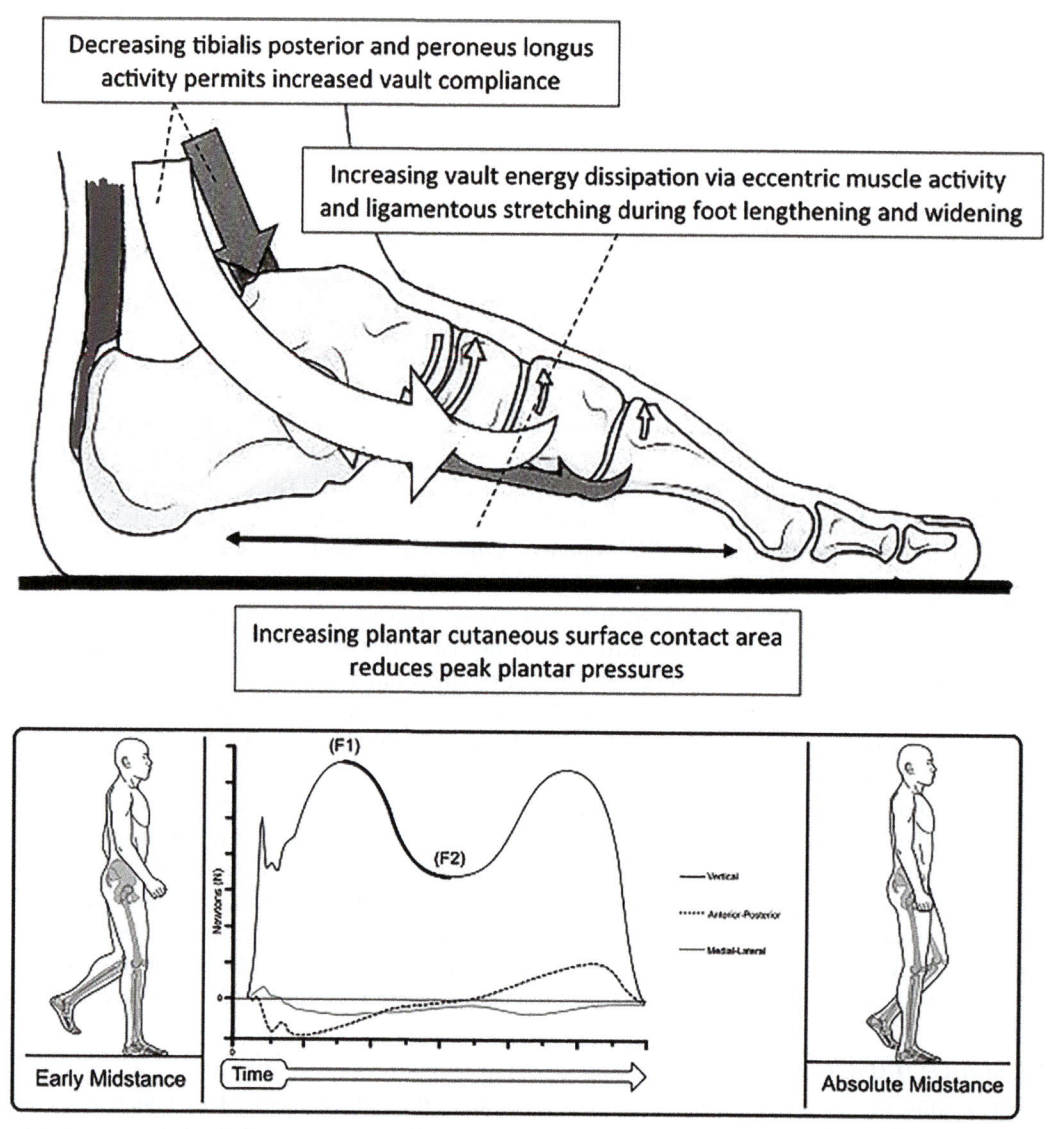

FIG. 4.5.3a Via the flexibility released through the vault as a result of reduced tibialis posterior and peroneus longus activity, the foot lengthens, widens, and lowers during early to absolute midstance. This is a period of relatively high foot compliance that rapidly increases the foot surface contact area, helping improve the base of support's stability under the lower limb and reducing peak pressures. This compliance stretches the plantar ligaments. Their presence provides passive energy dissipating elasticity and extra sensorimotor information supplementary to that of the muscle spindles. The period from the F1 peak to the F2 trough is a phase of decreasing GRF which moderates the loading stresses while the foot is more compliant. Walking slower tends to increase the forces between the F1 peak to the F2 trough, which may increase the muscular activity required to control foot deformation to protect passive structures. Weakness of tibialis posterior at this time will lead to greater foot deformation and energy dissipation, risking a breach in tissue threshold within ligaments. Abnormally stiffened connective tissues and slower preferred gait speeds associated with diabetic patients may increase midstance stresses on plantar ligaments, for the ability to stretch and dissipate energy within connective tissues may be limited. This will be especially problematic if tibialis posterior is concurrently weak. *(Permission www.healthystep.co.uk.)*

needs during stance in feet that have differing intrinsic levels of flexibility within lower limbs with differing morphologies. As the early midstance represents peak compliance and the start of stiffening, muscular activity and the resultant kinematics will significantly affect the foot's behaviour in the following stance phase periods. These variations within themselves may not be pathological, but each circumstance will set up different risks of one tissue being injured over another should fatigue, disease, or sudden external forces alter the mechanical situation in the future, or if they repetitively strain certain structures far more highly than others.

If pes planus hypermobility risks pathology due to excessive compliance, then the reverse is likely true for hypomobile cavoid feet with restricted compliance capability. The amount of foot lengthening on loading in such cavoid feet compared to others would be expected to be lower, and changes in foot stiffness under load less significant, requiring different muscle

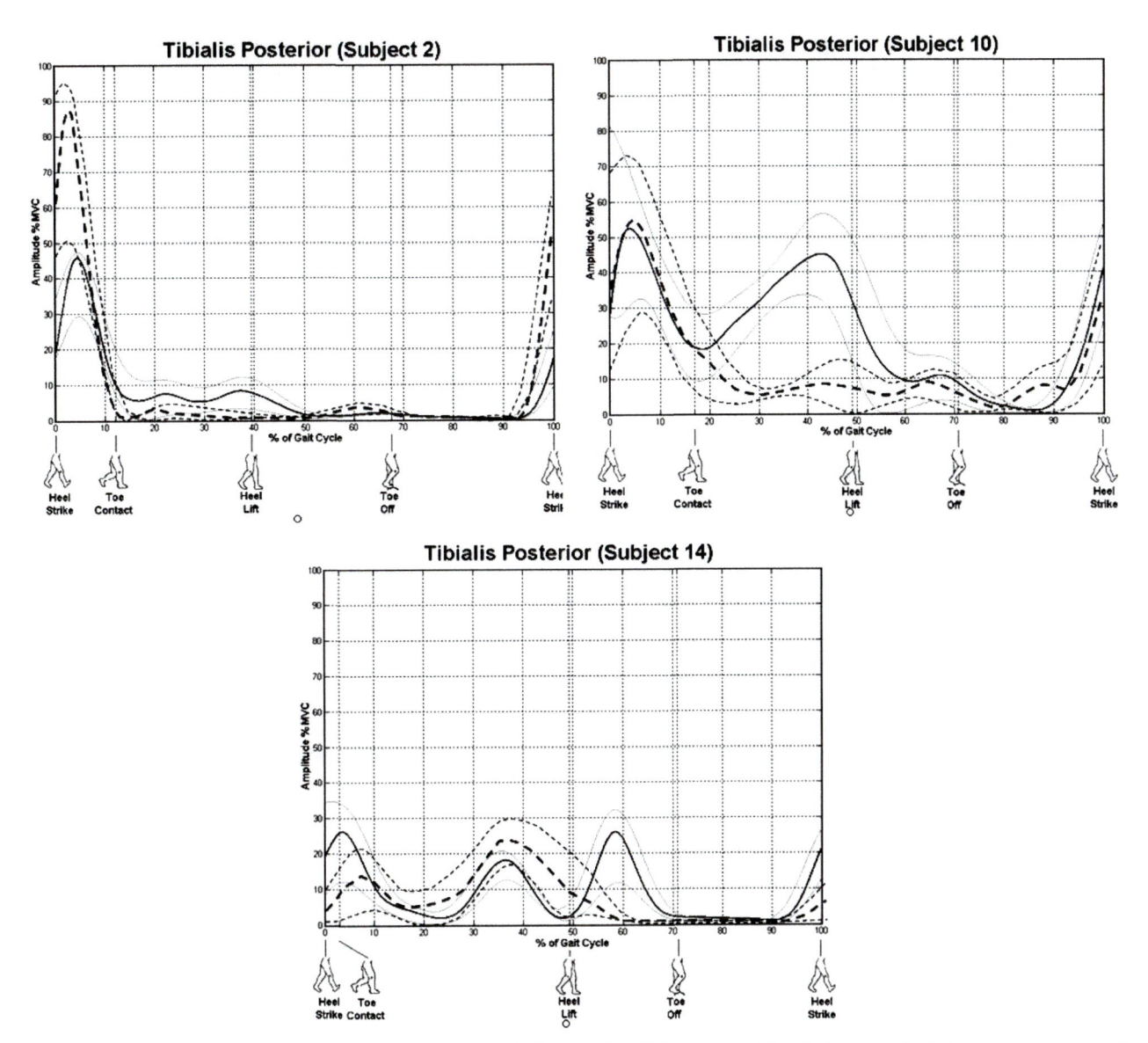

FIG. 4.5.3b The average EMG traces for tibialis posterior of three individuals from different tests, with dashed curves and solid curves representing each different test and with standard deviation lines shown above and below each solid and dashed curve. Not only do individuals vary in comparison with others, but each step also has its own muscular activity variable to other steps. This reflects the fact that there are subtle differences in posture with every step expected in healthy individuals that require fine-tuning of foot compliance-stiffening and balance patterns throughout stance. *(Image from Murley, G. S., Buldt, A.K., Trump, P.J., Wickham, J.B., 2009. Tibialis posterior EMG activity during barefoot walking in people with neutral foot posture. J. Electromyogr. Kinesiol. 19(2), e69–e77.)*

activities between the foot types. This is not because of their vault profiles alone, but because of their compliance-stiffening requirements. Tibialis posterior activity requirements during midstance would be expected to be reduced in stiffened and/or cavoid feet. Such feet could be expected to express higher peroneus longus activity in midstance (possibly of longer duration), increasing the risk of pathology within the peroneus longus or those structures related to compliance activity, than within stiffen structures such as tibialis posterior.

One particular kinematic risk of a failure within the foot to adequately lengthen and reduce the vault profile in early midstance is the effect that this may have on ankle dorsiflexion angles during late midstance, as is discussed in more detail in Section 4.4.5. Plantarflexion of the anterior calcaneus reduces the inclination angle of the calcaneus on loading, taking the talus into plantarflexion with it, while the ankle is reducing its plantarflexion angle. This manoeuvre 'opens' the anterior ankle joint, permitting further free rotation at the ankle into dorsiflexion without the talar trochlear surface becoming 'restricted' from its wider anterior surface too soon, avoiding entering its close-packed position prematurely. Failure of this calcaneal declination at early midstance associated with loss of the rapid foot lengthening increases the risk of anterior talar trochlear 'impingement' during late midstance, when the largest stance range of ankle dorsiflexion angle is required. If a hypomobile cavoid foot with a distinctly high calcaneal inclination angle is involved, then there is a high risk that anterior ankle stresses are increased and that ankle dorsiflexion becomes too restrained for efficient late midstance kinematics that transfer the CoM anteriorly, linking increased foot stiffness to decreased anterior tibial rotational freedom. This, in turn, will tend to reduce the final ankle plantarflexor power for acceleration provided by the Achilles' energy storage. It is likely that a hypermobile planus foot permits too much midfoot dorsiflexion, offering too much ankle freedom and thus requiring greater muscular restraint of tibial acceleration over the foot. This also may reduce ankle plantarflexion power, but as a result of too much energy absorption within the mobile midfoot (Fig. 4.5.3c).

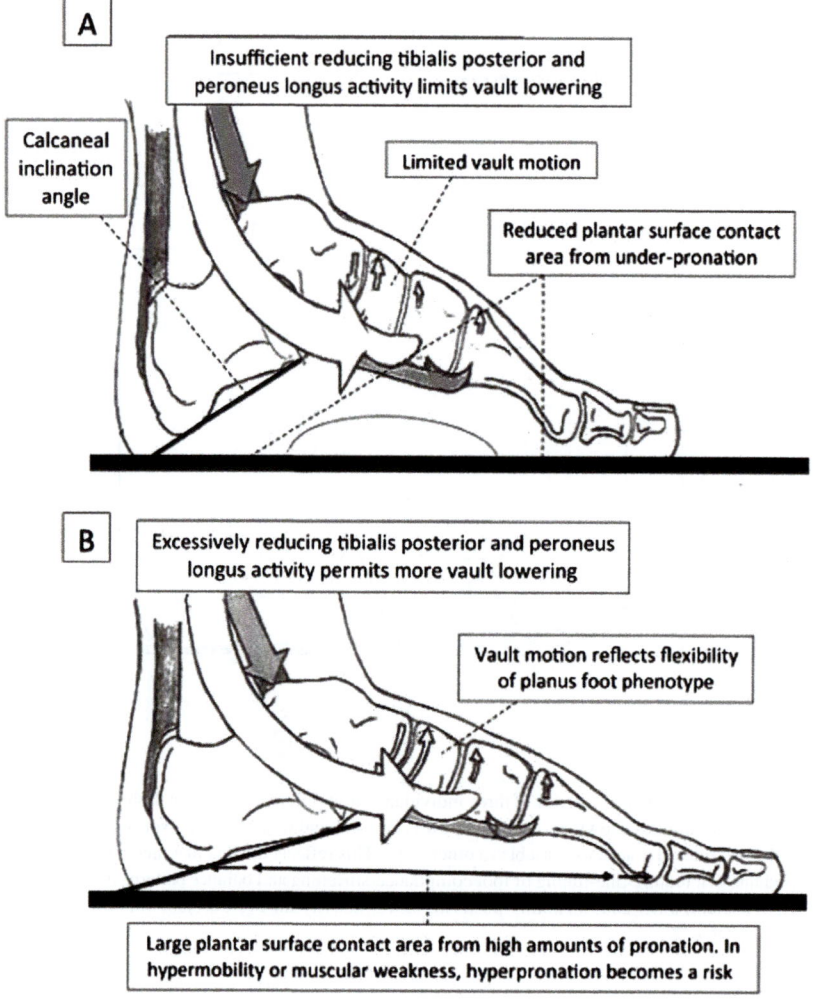

FIG. 4.5.3c Increase of the foot's dimensions at the end of loading response associated with a pronating posture should continue into early midstance and is important. These movements across the skeletal frame provide for energy dissipation, increased support contact area with the ground, and as a result, provide a reduction in peak pressures. These events are occurring as the stance limb enters single-limb support. As absolute midstance is approached, the plantar muscles begin to increase activity in readiness for decreasing compliance with the passive stiffening of the connective tissues as they move out of their linear elastic zone towards stress-stiffening. Stiff cavoid feet (A) have intrinsic mechanical advantage given over to stiffening mechanisms. This results from greater vault curvatures set from the high calcaneal inclination angle (solid black line). Such feet can struggle to provide the complementary mechanisms of tissue protection through an inability to change dimensions and increase surface contact areas. Such under or hypopronation risks a loss in energy dissipation mechanisms, exposing articular and osseous tissues to greater strains and requiring proximal joints to provide more shock attenuation. However, more mobile cavoid feet may avoid these problems. Planus feet (B) tend to give advantage to compliance. However, they may also have reduced changes within their profile if they are hypomobile, resulting in less energy dissipation capacity, and yet their low profile still provides a large surface contact area to moderate peak pressures. If the pes planus is hypermobile, early midstance can provide excessive energy dissipation via the foot's large profile changes and soft tissues strains, making restiffening of the foot for acceleration more challenging. Excess relaxation of vault-stiffening muscles or their weakness may create similar problems. *(Permission www.healthystep.co.uk.)*

Thus, early midstance is an opportunity to start to reset the foot's internal stiffness for the acceleration phase of gait. This requires freedom for the foot vault to lower, visualised through a higher rate of vault lengthening during early midstance. This allows the rearfoot complex the freedom to rotate through plantarflexion of the talus with the reducing calcaneal inclination angle, while initiating connective tissue-stiffening mechanisms.

4.5.4 Late midstance pathomechanics

After absolute midstance in walking, the rate of foot lengthening starts to reduce under increased loads associated with foot stiffness, yet without further large vault height changes (Stolwijk et al., 2014; Bjelopetrovich and Barrios, 2016; Takabayashi et al., 2020). This stress-stiffening mechanism in the foot increases rapidly from the bottom of the F2 trough of the force–time curve, increasing to the F3 peak around heel lift. At a significantly slower rate than that seen during early midstance, further lowering of the vault profile during late midstance will enhance the fall of the CoM under increasing total ankle and midfoot dorsiflexion angles. Further vault depression increases the potential weightbearing surfaces of the plantar foot through increased pronation, helping to moderate plantar pressure peaks under rising forces, particularly upon the forefoot.

Forces across the vault arise from the action of the plantarflexor moment decelerating the anterior translation and fall of the CoM. The bending moment of midfoot dorsiflexion depresses the vault. Such vault sagging deflection is resisted by the active plantarflexor muscles, fasciae, and passive ligaments that cross it. Continued foot pronation from the vault profile reducing should help prevent the talus impinging anteriorly within the ankle during continued ankle dorsiflexion as well as avoiding navicular impingement through concurrent talocalcaneonavicular joint plantarflexion. The intrinsic and extrinsic muscles throughout late midstance undergo near-isometric contraction and the connective tissue stretch-stiffening loads the plantar soft tissues with elastic potential energy, such as is reported within the tibialis posterior tendon (Maharaj et al., 2016). It is important for tibialis posterior function that the plantar ligaments are intact, otherwise its medial vault stabilisation can be compromised (Imhauser et al., 2004). This is undoubtedly true of all the vault-crossing tendons that are now under load. Failure in either connective tissue properties or muscle activity risks a soft tissue overstrain injury somewhere under the vault, rather than achieving an even spreading of soft tissue loads for efficient energy storage or dissipation.

When considering the foot's biomechanics in aged populations in relation to digital flexor activity across the vault, digital flexor tendons have been reported to reduce their capacity to store elastic energy with increasing age in pigs, sometime after maturity (Silver et al., 2003). If this change in functional capacity also occurs in human energy storage flexor tendons, then the vault's functional role may inevitably deteriorate with age.

Connective tissue stiffening during late midstance is supported by evidence that the foot vault is lowest, longest, and stiffest at heel lift (Hunt et al., 2001; Stolwijk et al., 2014; Bjelopetrovich and Barrios, 2016; Takabayashi et al., 2020). Foot-stiffening plantar intrinsic and plantar extrinsic muscles including peroneus longus and tibialis posterior, also demonstrate peak activity during late midstance, with their combined activity increasing foot stiffness (Ferris et al., 1995; Kokubo et al., 2012; Murley et al., 2014; Farris et al., 2019, 2020). Therefore, pathomechanics in the late midstance phase can be associated with either excess stiffness, vault stiffening peaking too soon, or failure to generate adequate stiffness across the foot vault before heel lift (Fig. 4.5.4a).

Too much connective tissue compliance in late midstance will require greater muscular activity to prevent potential increased connective tissue tensile strains from vault-bending moments that could damage connective tissue structures. Increased muscle activity to compensate risks early muscle fatigue and/or tendon pathology. Weakness within myofascial tissues risks increased strains among the connective tissue. These interrelationships of dysfunction and strain management reflect the nature of biotensegrity structure. The plantar tensional restrictions across the vault are necessary to moderate the vault-bending moments from focusing dorsal compression and plantar tensile strains into the skeletal frame. By restraining sagging deflection, both the dorsal compressions and plantar tensile strains into the skeletal frame of the vault are moderated, protecting bone and articular tissues. This is created by generating compression stability across the articulations that stabilise the foot. The tarsometatarsal articulations and the bones of the medial column tend to bear the greater proportions of loading stress (Halstead-Rastrick, 2013), so failure within this mechanism will tend to focus pathology into these regions. Bone marrow oedema patterns seen in the foot on MRI may follow loading patterns of gait biomechanics and increase in response to rising internal stresses, developing for example, within the medial column of feet that are more pronated (Schweitzer and White, 1996; Hirji et al., 2020) (Fig. 4.5.4b).

A Vault deflection/lengthening adequate for energy dissipation but with net energy storage from stiffening

Proximal compression vectors

B Low vault deflection inadequate for energy dissipation but high net vault energy storage

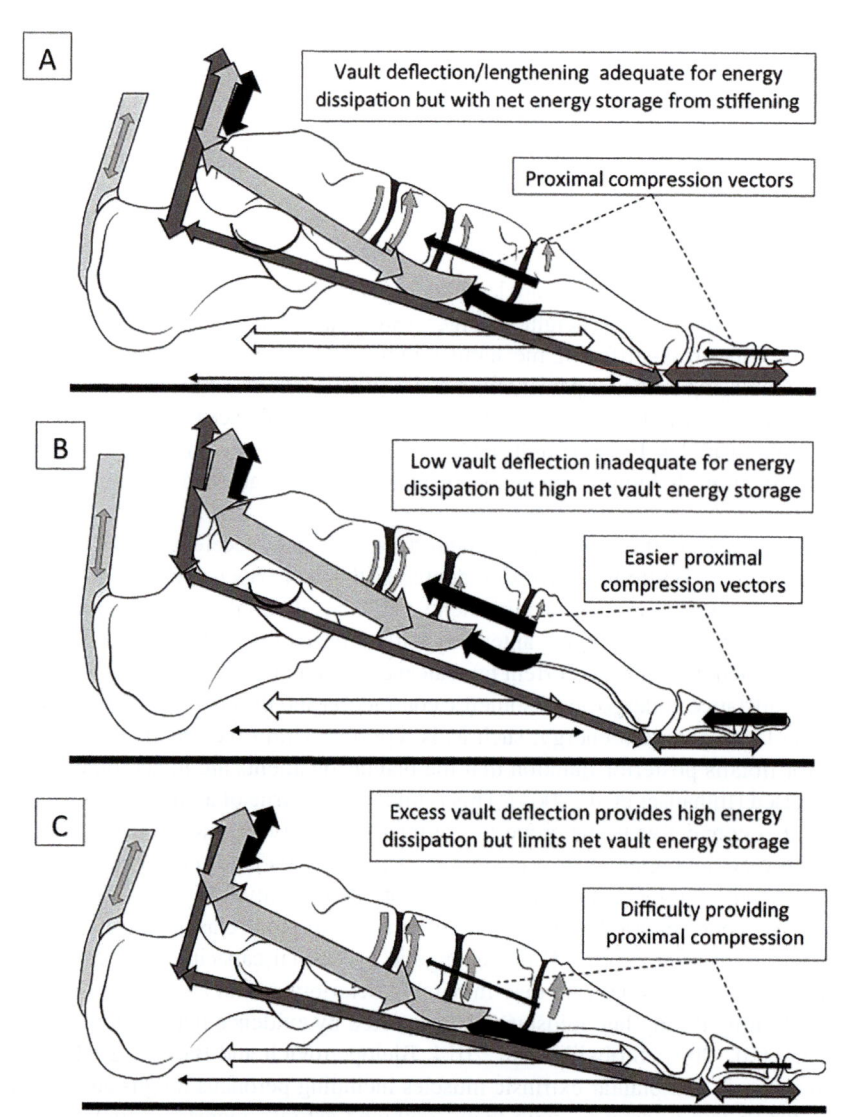

Easier proximal compression vectors

C Excess vault deflection provides high energy dissipation but limits net vault energy storage

Difficulty providing proximal compression

FIG. 4.5.4a Late midstance is a period of increasing vault and forefoot stiffness. This process is important for gait energetics, not only to help initiate heel lift by creating a semirigid beam within a cantilever and providing for a stable acceleration platform, but also to prevent excessive midfoot deflection and thus too much energy dissipation via an unrestricted ankle dorsiflexion moment. The energy stored within the foot's connective tissues before heel lift can be released to reduce the metabolic costs of acceleration. This recoil acceleration energy largely derives from the Achilles tensioning under triceps surae eccentric activity. However, other extrinsic plantarflexor tendons such as tibialis posterior (light-grey arrows), peroneus longus (black arrows), and the long digital flexors (dark-grey arrows) are still applying important compressive stability and adding energy across vault articulations via their eccentric contractions. They are also providing midfoot plantarflexion power after heel lift. Plantar intrinsic muscular activity and distal plantar ligament and plantar aponeurosis tensions (white arrows) also play an important role in providing stability for the cantilever action and for forming the forefoot acceleration platform stability. In ideally functioning feet (A), energy dissipation continues during late midstance but there is net energy storage across the vault, while rising tensional stiffening compresses the articulations together under biotensegrity principles. Stiff cavoid feet (B) offer a very stable foot beam and forefoot platform but may suffer from reduced energy storage within the Achilles and foot via both restricted ankle dorsiflexion and limited vault deformation. Forming compression across joints tends to be mechanically easy, for muscles in such feet often contract at shorter distances. Pes planus feet that demonstrate high mobility (C) are able to provide high energy dissipation for the ankle dorsiflexion moment, but their vault 'flattening' can reduce the amount of ankle dorsiflexion necessary to pass the CoM over the midfoot. Such excessive foot vault lowering will reduce Achilles and foot energy storage. Proximal articular compression stability can be compromised due to the muscles contracting at a longer length. The vault can thus lack the stiffened stability necessary for efficient heel lift. Both intrinsic and extrinsic muscles will be required to compensate by increasing their activity, risking earlier fatigue during gait, creating multiple pathology risks. As a result, for some individuals, a midfoot break may offer a better mechanical option. *(Permission www.healthystep.co.uk.)*

Legend				
0-5%	6-10%	11-15%	16-20%	21-25%

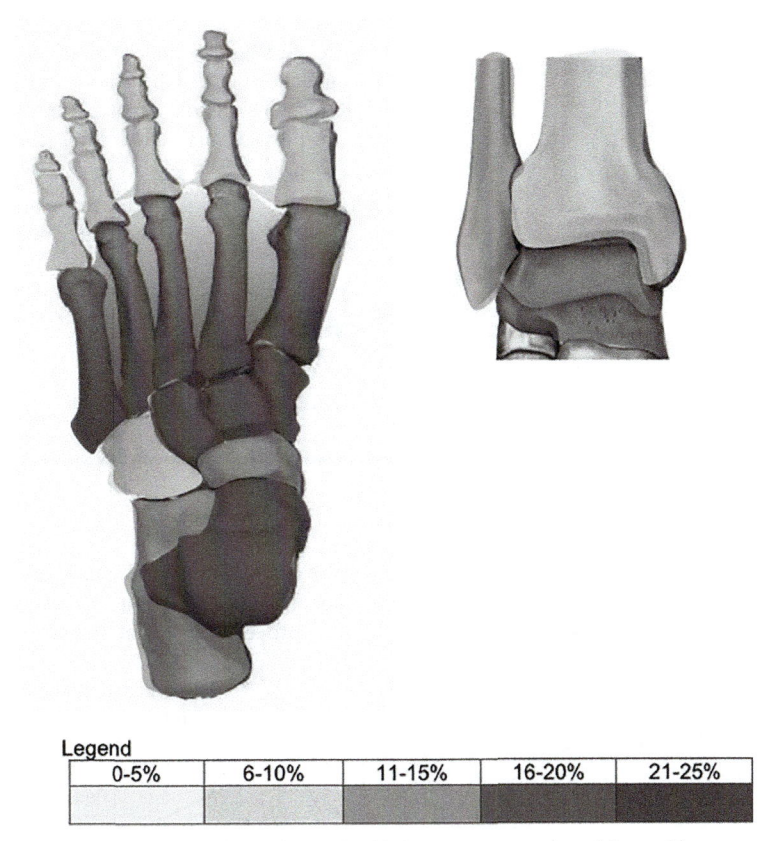

FIG. 4.5.4b A map of the pattern of bone stresses within the foot and ankle bones as a proportion of the total bones summarised from ten symptomatic and asymptomatic research studies. *(Image reproduced with permission from Halstead-Rastrick, J., 2013. Modification of midfoot bone stress with functional foot orthoses. PhD Thesis. University of Leeds School of Medicine, Leeds Institute of Rheumatic and Musculoskeletal Medicine, UK.)*

The pes planus in midstance

A pes planus with good active and passive stiffening mechanisms should be able to function energetically during midstance, as long as the mechanisms engage at appropriate times to dissipate and store energy. It will, of course, just be operating and working around a lower starting vault curvature. Hypermobile pes planus can potentially sustain greater midfoot bending deflections, resulting in increased midfoot dorsiflexion flexibility (sagging deflection) in late midstance (Caravaggi et al., 2018). This occurs because of reduced connective tissue stiffness, possibly exacerbated by an initial low vault curvature that allows excessive lengthening at initial forefoot loading compared to other feet. This means that larger impulses will tend to be absorbed within the heel and midfoot, as has been reported in such feet by Cen et al. (2020). Such hypermobile planus feet have routes to maintain gait energetics but with some associated risks. One option is to continue to try and maintain midfoot stiffness through increased muscular activity. Thus, active tension is used to substitute for less passive tension and curvature stiffness. Increased tibialis posterior combined with peroneus longus and the plantar intrinsics, will be the primary targets for this action, as they can all increase stiffness (Kokubo et al., 2012; Farris et al., 2019, 2020).

If muscles can successfully control midfoot bending dorsiflexion deflections, then there will be force–velocity benefits from the plantarflexors at heel lift. However, the risk of this is increased fatigue on sustained exercise of the plantar intrinsic and extrinsic muscles. Failure to provide adequate stiffness, especially during muscle fatigued states, risks sagging deflection overstraining the plantar connective tissues such as the spring, long, and short plantar ligaments, as well as potentially overstraining the tendons of muscles attempting to stiffen the vault.

Another option for the hypermobile pes planus is to use the midfoot bending moment for a change in mechanical advantage, by initiating midfoot break as the fulcrum of heel lift, rather than at the MTP joints. This has the benefit of reducing the metabolic demands that occur through active muscle contraction under the vault, but reduces the force–velocity effects and the lever arm length of the plantarflexors in terminal stance. This will increase metabolic demand in acceleration rather than during the deceleration of late midstance before the heel lift boundary. Should this option be developed throughout childhood and growth, then the foot's morphology and muscular development should become shaped for this mechanical need. Foot musculature should be expected to be less active, but the rest of the lower leg musculature will be required to compensate for the loss of acceleration power in terminal stance. This seems consistent with the reduced foot musculature and increased leg muscular activity reported in some pes planus feet by Angin et al. (2014) (Fig. 4.5.4c).

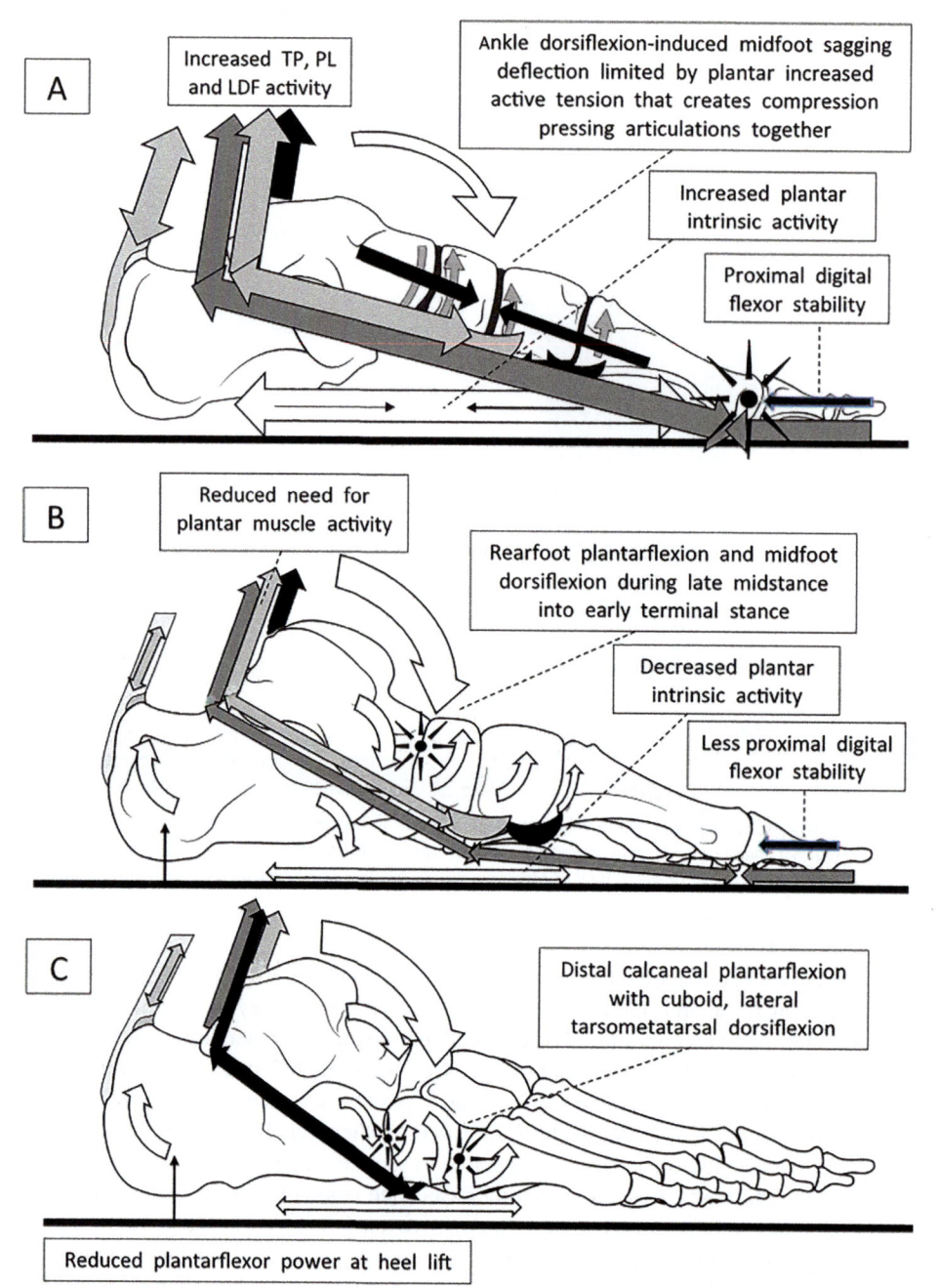

FIG. 4.5.4c The hypermobile pes planus foot type has two options in late midstance to prepare for heel lift. Both options offer potential higher injury risks than a more easily stiffened vault. By increasing activation of muscles (A), the medial MTP fulcrum can still be utilised to maintain ankle plantarflexion power mechanics. Raised activity in tibialis posterior (TP), peroneus longus (PL), and the long digital flexors (LDF) can provide midfoot dorsiflexion resistance power coupled to active vault stiffening that is adequate to still create a stiffened beam, even if it is one that is more flexible than average. This compensation mechanism will benefit from additional high activity from the plantar intrinsics, particularly abductor hallucis and the short digital flexors. The biotensegrity principles at play indicate targets for rehabilitation of such feet, for with a higher risk of muscular fatigue via compensation for associated higher connective tissue flexibility, such feet can benefit from maintaining higher muscle strengths. The other option for hypermobile planus feet is to develop a different heel lift strategy by using midfoot break (B), where plantar muscles can become far less active. However, it results in significant plantarflexor power being dissipated within midfoot flexibility. Allowing rearfoot acceleration by heel lift at a midfoot fulcrum avoids the need for generating a semistiff beam across the vault. Sagging deflection that causes midfoot dorsiflexion in midstance can be permitted to develop, exposing the vault to continued dorsiflexion (midfoot hyperpronation moments) during and after heel lift until the start of the opposite foot's support of body weight. A lateral view of the right foot (C) indicates that significant midfoot break motion occurs at the cuboid articulation as well as the talonavicular joint, and particularly at the lateral tarsometatarsal joints which provide a lateral heel lift fulcrum. Midfoot break presents a low injury risk if used since childhood as articular and ligamentous anatomy follows function during development, curving the cuboid articulations to aid motion. If midfoot break is developed later, then torques across the midfoot risk injury within anatomy that has been developed to limit midfoot dorsiflexion. *(Permission www.healthystep.co.uk.)*

Each option is a compromise for an overly compliant foot, with different pathological risks associated within each scenario. Whichever locomotive path has been chosen by the developing foot will be reflected by the individual's lifestyle and the amount of hypermobility present. The mechanical solution utilised will then influence development of the foot vault profile during childhood. Once a foot has chosen a mechanical solution to a problem, it will need to be maintained, despite changes in morphology and mass throughout growth. Thus, as long as the mechanical system is kept healthy, locomotion pathology is not an inevitability in hypermobile pes planus. However, developing a midfoot break as an adult through anatomy that does not reflect the midfoot fulcrum adaptation would be highly pathological for the midfoot articulations, especially the cuboid, plantar ligaments, and the tibialis posterior tendon. This likely explains why acquired pes planus feet are potentially so destructive, while an anatomical pes planus can be of no or little hindrance.

It must be remembered that foot-stiffening mechanisms are the sum of the parts, not just one aspect of the whole. A more hypomobile pes planus foot might cause some loss of energy dissipation ability and yet they could still offer perfectly good stiffness for terminal stance acceleration. They may offer no mechanical hindrance to the use of the medial MTP fulcrum during terminal stance, save a lower starting metatarsal declination angle at heel lift. A stiff pes planus foot may be subjected to proportionally higher midfoot bending deflection moments due to its lower vault profile, but the foot stiffness achieved may be perfectly adequate in resisting this. If anatomical development and growth has always been associated with this foot type, then the anatomy should have adapted to reflect this mechanical situation. Only with the application of a novel environment, a change in lifestyle, or a tissue disease should a stiff planus foot become a pathomechanical issue during gait.

Hypomobile cavoid feet during midstance have been largely already discussed in relation to the Achilles (Sections 4.4.4 and 4.4.10) and the functional ankle dorsiflexion angle (Section 4.4.5). The other issue with failure of the vault to adequately depress throughout late midstance relates to the increasing forefoot forces, peak pressures, and loading rates. As the vault depresses, the plantar weightbearing area across the forefoot should increase, increasing the surface area that can distribute pressure, thus helping to moderate peak pressures. In hypomobile cavoid feet, the forefoot weightbearing surface is relatively less due to the higher metatarsal declination angles and the high resistance to skeletal frame deformation under load. This risks higher peak forces on a more limited contact area, for longer. Moreover, this problem increases as the CoM moves anteriorly during late midstance and this high stress situation will persist into terminal stance. In neuropathic diabetics, these high forefoot loads are magnified by an inability of abnormally stiffened soft tissues to dissipate the loading energies, while also suffering from a loss of the ability to feel that forefoot forces are excessive. Thus, forefoot stresses are particularly significant in high-vaulted diabetic feet at the late midstance boundary and during terminal stance.

If high-vaulted feet sustain significant vault deflection during midstance, then the angulation of the midfoot articulations within the high vault can create some dorsal articular stress issues. High vaulted foot profiles with steep calcaneal inclinations and metatarsal declinations mean that sagging deflection under load increases tarsometatarsal dorsal articular compression and tensions on overly tight plantar ligaments. This risks degenerative changes dorsally within the midfoot joints and possible enthesis injuries to the plantar ligaments supporting them (Halstead-Rastrick, 2013). Two mechanisms seem to be used to protect the tarsometatarsal joints from increasing forces, that of increasing the contact surface area under higher loads and by spreading them across other tarsometatarsal joints (Lakin et al., 2001). It can be suspected that similar techniques are used across the midfoot, requiring muscle activity to adapt forces across the midfoot to keep articular and osseous stresses away from tissue threshold breaches. Thus, weak plantar muscles and connective tissues allow excess midfoot compliance that can be pathological to the midfoot articulations themselves.

4.5.5 Terminal stance phase pathomechanics

Foot stiffness developed during late midstance is essential for the beginnings of acceleration at heel lift. The foot should offer a semirigid beam-like structure to perform the role within a class two lever situation that lifts and tilts the rearfoot and the rest of the lower limb anteriorly, much like a wheelbarrow being emptied anteriorly (Fig. 4.5.5a). The power of the ankle plantarflexors sets up a force–velocity relationship that accelerates the remaining mass of the body to the next step, an event that reflects gait speed.

Foot-stiffening pathomechanics

The high level of stiffness required at heel lift is potentiated through increasing forefoot loads during late midstance, that rapidly increase at heel lift. These loads displace (splay) the metatarsals laterally and medially away from each other, widening the forefoot. The 1st metatarsal splays medially from the 2nd which is highly stable, with lateral spreading occurring between the 3rd and 4th and particularly the 4th and 5th metatarsals through motion at their cuboid articulations (Lundgren et al., 2008; Wolf et al., 2008). Because the metatarsal heads are not level lengthwise within the sagittal plane when they

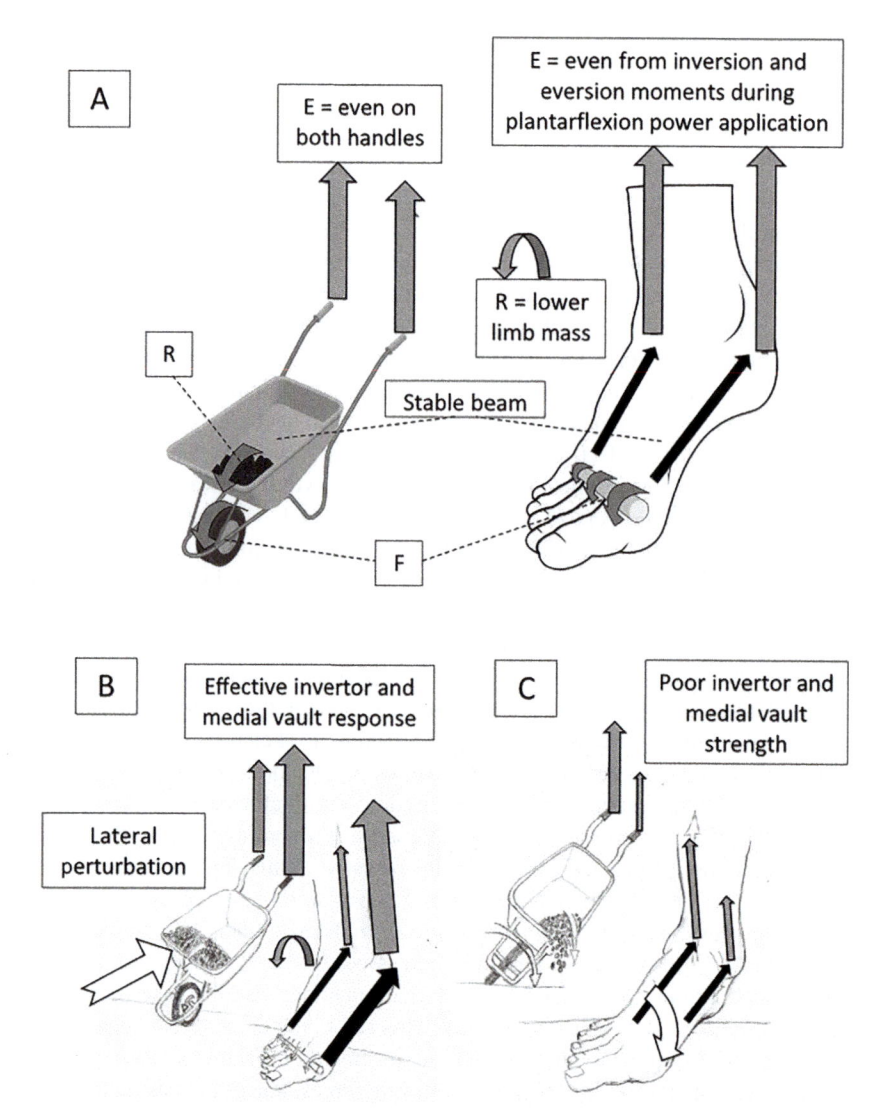

FIG. 4.5.5a The application of plantarflexion power to the foot at heel lift has much in common with a class two lever system, much like a wheelbarrow tipping out a load over a surface (A). The effort (E) must be applied evenly upwards on both handles to create a straight pivot at the wheel (fulcrum—F), so that the load (resistance—R) can be poured straight ahead out of the barrow. For the effort to be applied efficiently to move the load, the barrow must be 'solid', for any buckling of the barrow will lose energy to sagging bending deflection. The nearer the load is to the front edge of the barrow, the easier it is to tip it out, for the effort moment arm is thus lengthened in proportion to the resistance arm. Thus, for heel lift acceleration to replicate this mechanism, effort must be applied evenly from medial to lateral sides of the rearfoot to lift the heel without inversion or eversion. The further body (HAT) CoM has moved in front of the MTP joints before heel lift, the less resistance there is to be moved and the shorter the moment arm of R to F. The foot vault must remain stiffened so that as much effort as is possible can be applied to moving the resistance at the fulcrum. A little ability to 'give' within the vault helps absorb any excess energy that could be potentially tissue-damaging, particularly to the medial MTP joints used for acceleration. If a perturbation occurs during acceleration, then the ability for the foot and ankle to provide variable and adaptable medial-to-lateral vectors becomes essential in maintaining the correct direction for acceleration. A lateral eversion causing perturbation on the foot during acceleration (B) will require counterbalancing by reducing lateral muscle vectors and increased inversion moments via medial vectors which can be provided for by combinations of increased soleus, flexor digitorum longus, tibialis posterior, and/or abductor hallucis activity. Failure in strength of these muscles or the loss medial vault stability (C), will result in the plantarflexion power being accelerated too medially from the foot. Thus, medial soft tissues will disproportionately become strained under acceleration energy that risks injury. *(Permission www.healthystep.co.uk.)*

splay, they not only change their transverse plane orientations, but also their sagittal plane orientations to each other and thus their functional lengths to the line of progression. This increases the curvature between the heads in the transverse, and frontal planes, restrained by all the intermetatarsal ligaments and myofascial structures tightening between the metatarsal heads. Through stress–strain properties, the ligaments stiffen with intermetatarsal splay, increasing the transverse and frontal plane curvatures. Both greater tension and raised curvature increase vault stiffening. This transverse/frontal plane stiffening influences the foot's stiffness proximally and longitudinally (Venkadesan et al., 2017, 2020; Yawar et al., 2017).

Each metatarsal shaft has its own independent declination angles and rotations so that the central metatarsals create a transverse vault profile, highest around the 3rd metatarsal (Drapeau and Harmon, 2013). This transverse vault curve between the metatarsal shafts stiffens as the metatarsals become maximally splayed distally under increased forefoot loads following heel lift. This occurs under the peaks of force seen in the F3 and F5 peaks on a force–time curve, just before and at the weight transference to the contralateral limb, which initiates the rapid decline in terminal stance forces. Failure in the passive structures to restrain forefoot splaying and thus convert intermetatarsal motion into ligament stress-stiffening and increased curvature, risks the loss of a stable stiffened platform for acceleration and a loss in applying plantarflexor power as a stable GRF. To maintain forefoot stiffness without this passive stiffening mechanism will require other active mechanisms to compensate.

Active foot-stiffening and pathomechanics

The passive stiffness induced by increased curvature across the metatarsals at heel lift is likely potentiated through muscle activity, working with the passive mechanisms to create appropriate forefoot stiffness. Muscular stiffening is thus a potential compensation mechanism, should the passive mechanisms prove inadequate to setting the right levels of semi-stiffness during heel lift. The plantar intrinsics are highly influential in foot-stiffening at this time (Farris et al., 2019, 2020) as are the plantar extrinsics (Ferris et al., 1995; Murley et al., 2009, 2014; Kokubo et al., 2012). Tibialis posterior and peroneus longus stiffen more proximally through the midfoot transverse plane, across the cuneiforms and tarsometatarsal region as the forefoot stiffens. The medial-to-lateral pull of the peroneus longus attachments and the lateral-to-medial pull on the tibialis posterior distal attachments increase the lesser tarsus and tarsometatarsal curvature concurrently, with the increasing metatarsal curvatures. Distally, the interossei, particularly the dorsal interossei which bridge the intermetatarsal spaces, can restrain the metatarsal bases together by actively tightening the proximal forefoot. This combined muscle activity will form a semiconical structure narrowing proximally, with the metatarsal heads splaying distally. This is likely to enhance stability of the forefoot further (Fig. 4.5.5b).

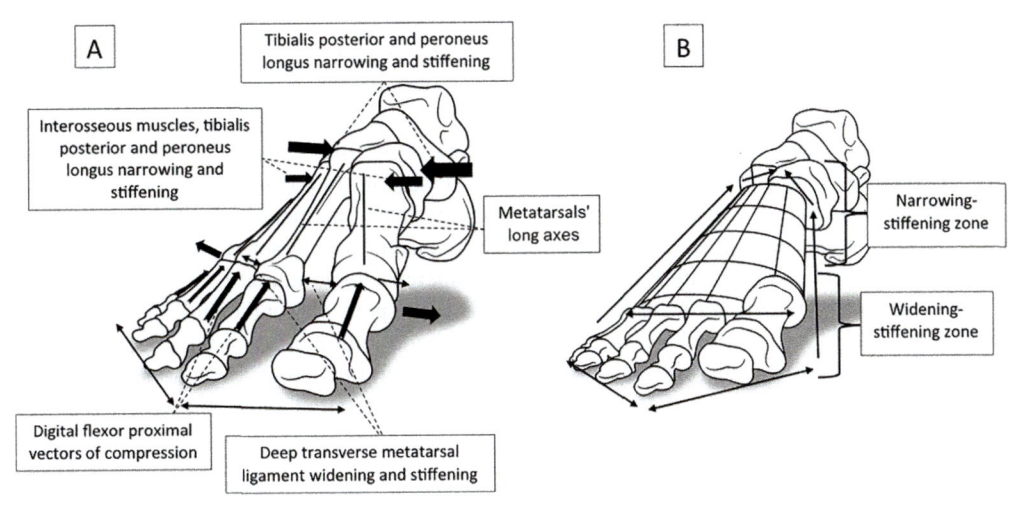

FIG. 4.5.5b Forming a stable semirigid half cone-shaped beam at the heel lift boundary provides for a cantilever that can permit efficient heel lift and provide a stable acceleration platform. The adaptable semirigid vault formed before and during heel lift means that the foot can provide for either more energy transfer or greater energy dissipation of plantarflexion power. Two key areas use biotensegrity principles to form the stiffening necessary: the forefoot and midfoot. The forefoot splay under increasing metatarsal head GRFs tightens the deep transverse metatarsal ligament. Thus, the distal forefoot widens to become stiffer. At the same time, digital flexor activity creates proximally directed forces that compress the digits into the metatarsal heads in the direction of the long axis of the metatarsals. This also helps stabilise the tarsometatarsal joints. Proximally, the interossei pull the metatarsal bases together as part of digital flexor activity, forming a narrowing stability across the midfoot. This is aided by the attachments of tibialis posterior and peroneus longus that cross the tarsometatarsal joints and who's muscular activity is stiffening the midfoot. These muscles are compressing lesser tarsal bones together from the 1st metatarsal base, medial cuneiform, and navicular, towards the cuboid. This narrows and stiffens the midfoot. Thus, a 'half cone' shape is formed that is concave plantarly, narrow proximally, and wider distally across the transverse plane of the foot (A). This transverse curving greatly enhances stiffening across the vault, so that as the rearfoot lifts under plantarflexion power, forces are applied to a broad and stable forefoot support surface with significant flexibility permitted only at the MTP joints. Midfoot eversion from the peroneal muscles and adduction from the tibialis posterior vector will move the forefoot weightbearing surface towards the stabilised medial metatarsal heads during terminal stance. These muscles also provide midfoot plantarflexion which is coupled to ankle plantarflexion. By midfoot plantarflexion, these muscles increase the plantar concavity in the sagittal plane, stiffening the foot further. Such stiffening is aided by the windlass mechanism and digital flexor muscle tightening initiated by further MTP joint extension. A geometric interpretation of this midfoot contouring at the heel lift boundary is illustrated in (B). Although highly effective in providing an all-terrain locomotive appendage, such a complex adaptable stiffening mechanism potentially offers much that can go wrong. *(Permission www.healthystep.co.uk.)*

These more proximal transverse muscular stiffening mechanisms can compensate for failure in passive forefoot load stiffening. Their failure in this function can increase strains across the passive transverse metatarsal ligament, while it is spreading. Combined active and passive failures will be particularly problematic. Thus, the transverse plane stiffening mechanisms may be deficient for many reasons, causing loss of coupled sagittal plane stiffening with loss of the ability to produce an efficient transfer of ankle power into a forefoot GRF that creates acceleration stability.

Stabilising MTP fulcrums and pathomechanics

The long and short toe flexors work with the interossei, lumbricals, abductor digiti minimi, and flexor digiti minimi brevis at the 5th MTP joint, and abductor and adductor hallucis at the 1st MTP joint. They are aided by the associated plantar plate or sesamoid apparatus, all working together to stabilise the extending MTP joints of terminal stance. Their combined activity prevents anterior translation of any particular metatarsal head by creating a proximally directed force via the proximal phalanx of the digit, while restraining digital extension. This helps compress the foot longitudinally and, in turn, aids midfoot stiffening. By doing so, stable MTP joints and soft tissue-stiffened (elastic) restraint against MTP joint extension are created (Fig. 4.5.5c). Stable digital extension permits passive engagement of the plantar aponeurosis windlass mechanism during terminal stance. This means that the plantar aponeurosis does not primarily influence foot stiffening at heel lift but likely does so shortly after through increasing digital extension angles that shorten its and the vaults length during the latter stages of terminal stance (Farris et al., 2020).

Failure in the longitudinal stiffening mechanisms due to loss of MTP joint extension stability can result from failed transverse plane stiffening, or through more local dysfunctions. Dysfunction leads to the loss of metatarsal head fulcrum stability, risking loss of the application of plantarflexion powered GRF forefoot stability for acceleration.

Tibialis posterior and peroneus longus-induced plantarflexion, foot shortening, and its influence on the pathomechanics of acceleration

With heel lift, forefoot loading induces further stiffening beyond the midtarsal joints. The ankle and midtarsal joints are thus free to plantarflex unhindered by hogging deflection caused by bending moments across foot acting as a cantilever at heel lift. The combined action of tibialis posterior and peroneus longus in stiffening the midfoot's transverse/frontal plane, assisted by the other mechanisms, allows two fundamental roles to be performed by these muscles. Both the tibialis posterior and peroneus longus have important sagittal plane and transverse plane moment arms across the tarsometatarsal joints and across the rearfoot complex, as well as having important frontal plane moment arms. This means that after heel lift, continued power through these tendons not only compresses the cuneiforms and tarsometatarsal joints for stability, but also creates plantarflexion moments at the midfoot articulations that increase longitudinal vault curvature.

It is likely that tibialis posterior and peroneus longus are primarily responsible for the midfoot plantarflexion reported in human feet during terminal stance (Holowka et al., 2017). Most single joint motion of the midfoot occurs at the talonavicular joint (Wolf et al., 2008), a joint with considerable influence on other foot joint motions (Savory et al., 1998). The assistance of peroneus brevis to peroneus longus activity during acceleration allows the midfoot to evert as it plantarflexes after heel lift, moving the CoGRF across the forefoot towards the medial MTP joint fulcrum, with tibialis posterior and peroneus longus increasing the medial 'shift' through combined midfoot adduction, eversion, and plantarflexion. This hopefully places the medial MTP joints in the correct orientation for the application of the plantarflexion power to the lower limb's line of progression. Any disruption in the mechanism compromises acceleration biomechanics.

Following heel lift, ankle plantarflexion power is quickly lost as the triceps surae, tibialis posterior, and peroneus longus are reducing activity dramatically (Fig. 4.5.5d). Thus, much of the continued terminal stance ankle plantarflexion power is brought about from tendon elastic recoil of these muscles. When recoil power is spent, the lower limb lifting action of hip flexion is achieved by iliopsoas on entering preswing. Despite decreasing muscular activity, foot length should shorten throughout terminal stance (Stolwijk et al., 2014) as a result of vault recontouring from plantarflexor elastic energy, continued intrinsic and extrinsic flexor muscle activity, and the windlass effect.

Should both tibialis posterior and peroneus longus be dysfunctional, then midfoot plantarflexion, eversion, and adduction power will be reduced or lost. If tibialis posterior alone is dysfunctional, then midfoot plantarflexion power will also be reduced and the adduction moment across the midfoot lost, possibly replaced with abduction from unresisted peroneus longus and brevis activity. The foot will also increase its terminal stance compliance, especially across the proximal transverse vault. This reduces stability at the MTP joint fulcrums, making use of the medial MTP joint fulcrums under efficient plantarflexor power very difficult (Fig. 4.5.5e). Tendinopathy of the tibialis posterior through loss of tendon elastic recoil will result in a loss of tibialis posterior power, even if the muscle itself remains strong (Maharaj et al., 2016).

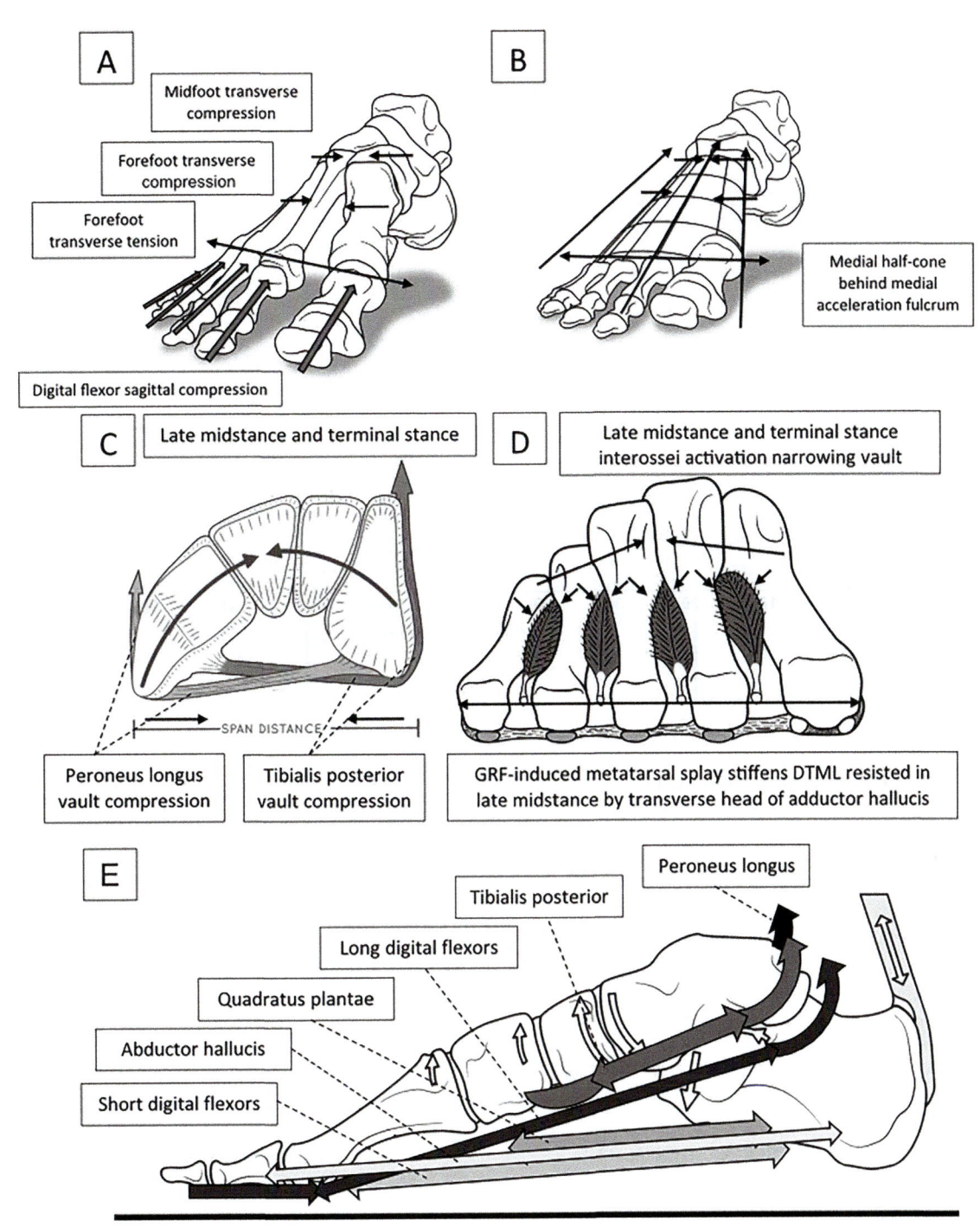

FIG. 4.5.5c The development of an increased transverse plane curvature profile in forming a stable half-cone across the midfoot to forefoot (A) is critical to the degree of stiffness possible within the foot vault's sagittal plane. This 'half-cone' profile is structured in such a way that it can still function as the foot shifts its weight to use the medial acceleration fulcrum (B), with the very stable 2nd metatarsal now forming the lateral boundary, and the stabilised 1st metatarsal creating the medial boundary. The process starts with increasing deep transverse metatarsal ligament (DTML) stiffening under forefoot loading metatarsal splay that results in distal transverse and frontal plane curvature increasing across the vault. However, muscle activity provides the power to enhance and complete the transformation from compliance to a variably semi-stiffened structure with sufficient stability to apply acceleration plantar-flexion power across the vault sagittally, without injury. Muscle failures that directly compromise transverse plane stiffening include tibialis posterior, peroneus longus, (C) and the interossei (D). Other muscle weaknesses within abductor hallucis, adductor hallucis, and the flexor hallucis muscles will also have a significant influence on overall stabilising transverse vector powers. Once transverse plane stiffening starts, it couples to longitudinal stiffening. Without adequate longitudinal stiffening, the foot's full length to the MTP joints cannot be fully utilised as a beam. Sagittal plane active stiffening across the longitudinal vault profile (E) can be compromised primarily by dysfunction within tibialis posterior, peroneus longus, the long and short digital flexors, abductor hallucis, and quadratus plantae. Muscle weakness is also problematic as muscles help protect the plantar ligaments. Plantar ligaments that provide passive stiffening and sensorimotor input are extremely important in aiding and achieving vault stiffness characteristics during each step by helping in setting muscle spindle activity, which in turn informs predictive muscle contraction. Peroneus longus failure may lead to excessive stiffening of the vault medially, but can result in excessive lateral column flexibility. *(Permission www.healthystep.co.uk.)*

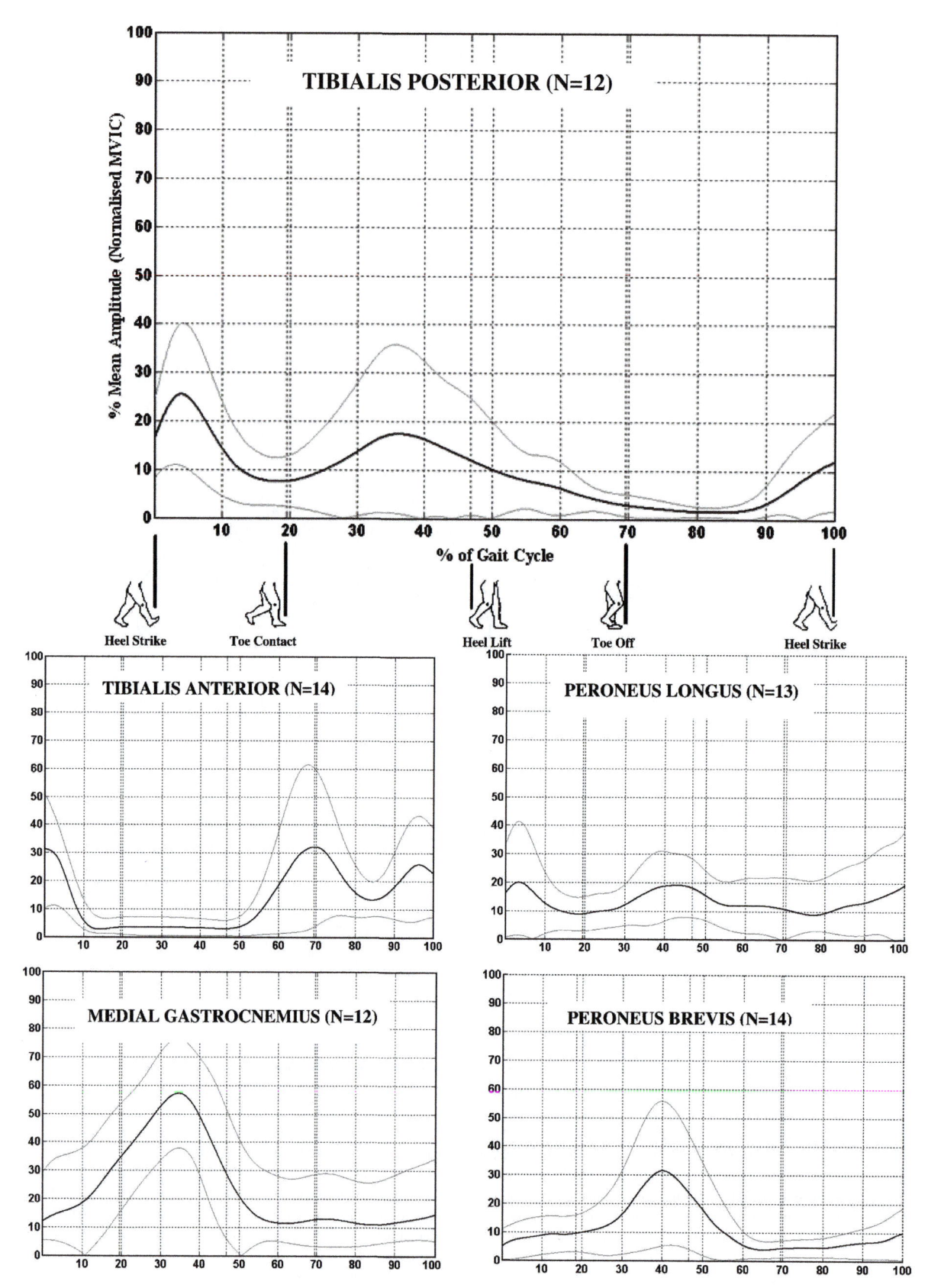

FIG. 4.5.5d The average EMG data with standard deviations above and below of tibialis posterior, tibialis anterior, peroneus longus medial gastrocnemius, and peroneus brevis showing the activation patterns around heel lift and into terminal stance. With the exception of tibialis anterior, these muscles all lose significant activity before toe-off, being most active before and during heel lift. These extrinsic muscles should be a key focus of clinical attention for understanding acceleration capacity within a foot. *(Image from Murley, G.S., Buldt, A.K., Trump, P.J., Wickham, J.B., 2009. Tibialis posterior EMG activity during barefoot walking in people with neutral foot posture. J. Electromyogr. Kinesiol. 19(2), e69–e77.)*

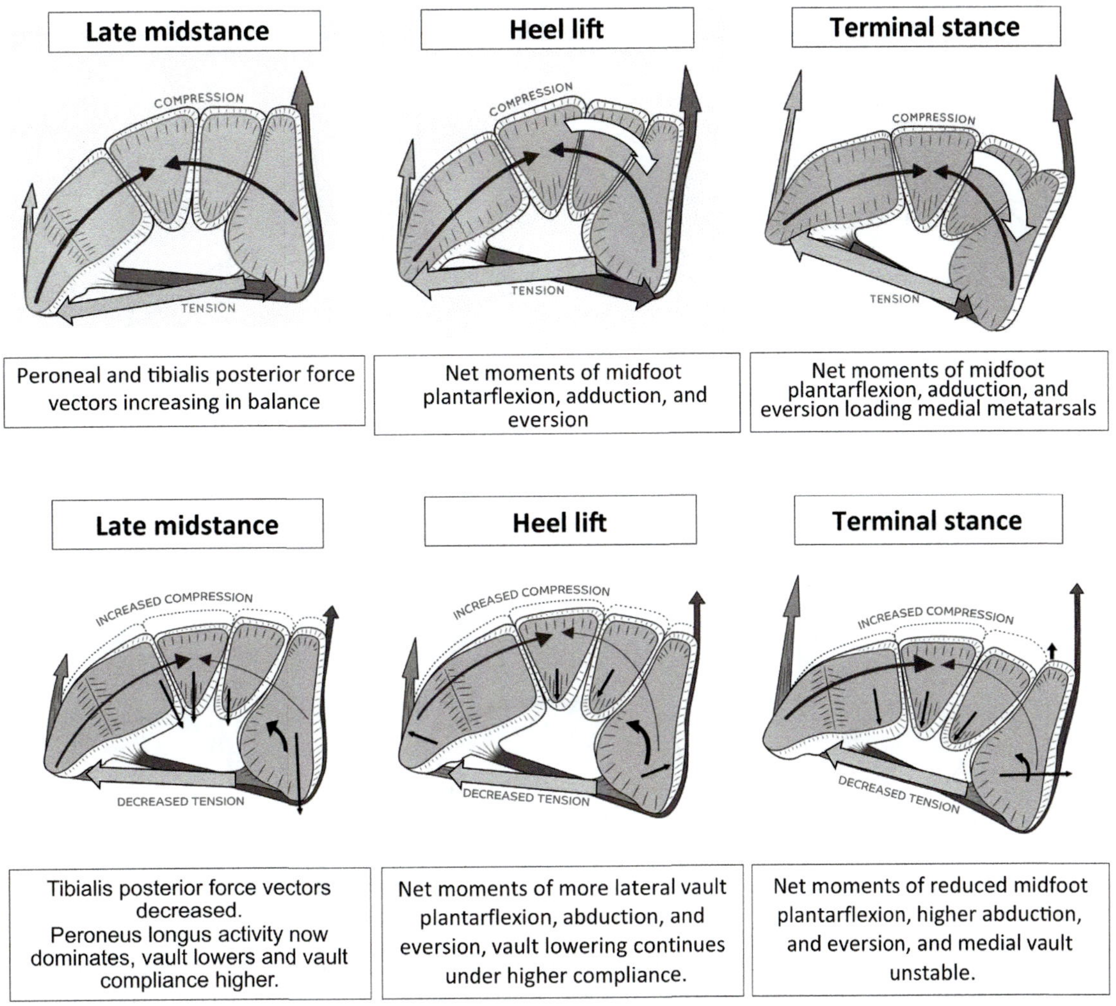

Pathology through the principles of biomechanics **Chapter | 4** 771

Late midstance	Heel lift	Terminal stance
Peroneal and tibialis posterior force vectors increasing in balance	Net moments of midfoot plantarflexion, adduction, and eversion	Net moments of midfoot plantarflexion, adduction, and eversion loading medial metatarsals

Late midstance	Heel lift	Terminal stance
Tibialis posterior force vectors decreased. Peroneus longus activity now dominates, vault lowers and vault compliance higher.	Net moments of more lateral vault plantarflexion, abduction, and eversion, vault lowering continues under higher compliance.	Net moments of reduced midfoot plantarflexion, higher abduction, and eversion, and medial vault unstable.

FIG. 4.5.5e Top images demonstrate the normal action of tibialis posterior and peroneus longus in stiffening the foot and positioning the plantarflexion power for high gear propulsion. The lower images indicate the changes resulting from tibialis posterior dysfunction. Normally, rising tibialis posterior activity works with the force vectors increasing from peroneus longus that together stiffen the midfoot in late midstance by increasing transverse vault curvature. Their increased tendon tensions compress the articulations evenly together from the tarsometatarsal joints to the ankle (biotensegrity principles). At heel lift, the adduction and plantarflexion moments from tibialis posterior and eversion and plantarflexion moments from peroneus longus (aided by peroneus brevis) are converted into motion that medially rotates and stiffens the vault towards the loading of the medial column. Much of this power now comes the tendons' elastic recoil rather than further muscular contraction. This positions the foot for use of the medial MTP joints for high geared acceleration during terminal stance. In tibialis posterior dysfunction (lower images), the ability to compress the midfoot together is compromised, permitting midfoot sagging deflection and lowering the vault excessively before heel lift. Dorsal aspect compression of midfoot articulations may now become raised, while plantar ligament tension increases from inferior articular gaping. Despite the peroneus longus tendon still being tensioned, the medial cuneiform may start to invert, lower, and adduct away from the intermediate cuneiform as the navicular behind it dorsiflexes on the talus. This destabilises the 1st ray, despite power still being applied to the medial cuneiform and the 1st metatarsal base because stability here relies on medial stability from tibialis posterior proximally. The foot working as a semirigid beam in a cantilever is not possible because the vault is too flexible to easily transfer plantarflexion power to the forefoot. With midfoot plantarflexion and adduction power reduced, the midfoot and forefoot abduct on the rearfoot under unantagonised peroneal power, making it impossible to form a stable acceleration platform medially for high gear propulsion. Peroneal eversion moments will now only further pronate and destabilise the forefoot and vault. *(Permission www.healthystep.co.uk.)*

Should the peroneus longus fail, then the foot may increase its stiffness, particularly medially. However, the midfoot plantarflexion moment will be compromised (particularly laterally), forefoot adduction will be less restrained, and the eversion moment that shifts the CoGRF medially to engage the medial MTP joint fulcrums may be reduced or lost. A normal medial MTP joint fulcrum or high gear propulsion strategy may no longer be possible. Should peroneus brevis concurrently be compromised, then the situation will be dramatically worsened, with the loss of another eversion moment. The consequence might not only be a low gear propulsion, but lateral instability during terminal stance (Fig. 4.5.5f).

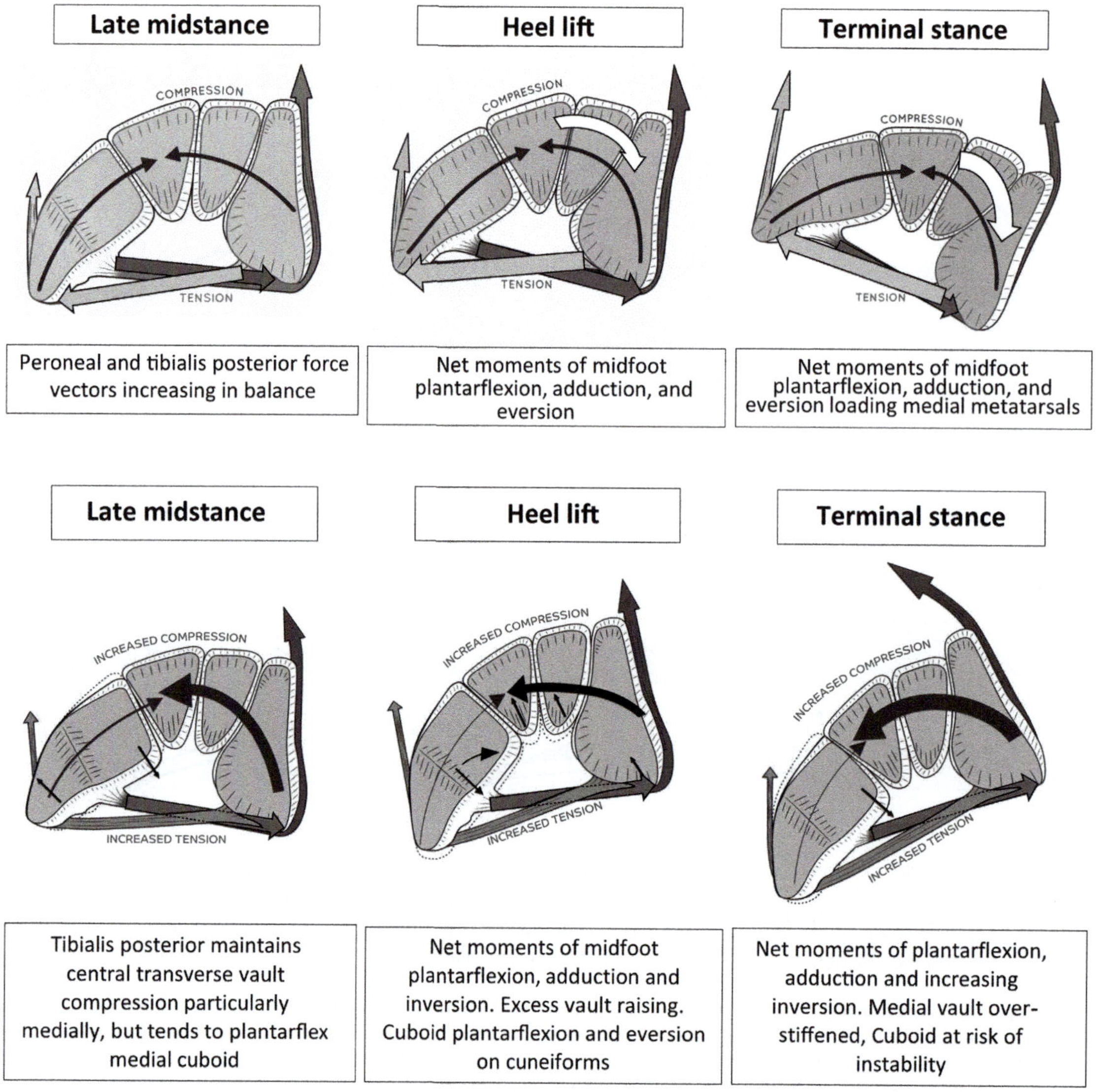

FIG. 4.5.5f Top images demonstrate the normal action of tibialis posterior and peroneus longus in stiffening the foot, helping in positioning the Achilles' plantarflexion power for high gear propulsion. Lower images illustrate peroneus longus dysfunction, showing that the midfoot can become overly stiffened, particularly medially, despite loss of medial-to-lateral tension and compression from around the cuboid. This is because tibialis posterior attaches to the medial cuboid. Without adequate peroneus longus power, unrestrained tibialis posterior's adduction moments may over-stiffen the vault and initiate a midfoot inversion force. At heel lift and during terminal stance, midfoot plantarflexion and unantagonised adduction may further increase the inversion moment. This tips the whole foot towards lateral forefoot weightbearing, making it more likely for the plantarflexion power to be directed towards the lateral MTP joints to force a low gear propulsion. Instability of the lateral column through poor cuboid plantarflexion support from the underlying peroneus longus tendon, may result in increased cuboid eversion and plantarflexion as part of lateral column compliance, further increasing lateral forefoot instability during terminal stance. *(Permission www.healthystep.co.uk.)*

Although dysfunction of the peroneal muscles is significant to all feet, the level of significance is reflected in part by the initial foot morphology. If the foot is morphologically stiffened by soft tissue hypomobility and/or from a cavus vault, then the effects are more dramatic. This would be particularly so if alignments included an inverted calcaneus or an everted (valgus) forefoot posture. These morphologies shift the GRF more laterally in midstance, and passively resist foot eversion and the medial drift of the foot's CoGRF during terminal stance. This is a subject already covered in more detail in Section 4.4.11.

The aim of terminal stance stiffness

Terminal stance stiffness is required for application of ankle plantarflexion power to induce heel lift on a stable platform of acceleration, preferably in the line of progression. Midstance foot vault lengthening and widening causes connective tissue stress-stiffening, combining with increasing plantar intrinsic and extrinsic activity to pre-stress the foot for acceleration. Rapidly increasing forefoot metatarsal splay under increasing forefoot loads at heel lift initiates transverse plane foot stiffening into an environment of increasing intrinsic and extrinsic foot muscle activity. The increasing transverse plane curvatures across the vault require passive and active tensional inputs distally for the skeletal frame to increase stiffness proximally, followed by coupled longitudinal stiffening that is eventually added to by the plantar aponeurosis when active muscular stiffening power is decreasing. Because of the influence that levels of stiffness within the midfoot has on both rearfoot-to-forefoot motion and the transfer of torques from the leg and rearfoot to the forefoot, and midfoot passive stiffness levels are all critical for acceleration biomechanics (Gomes et al., 2019; Magalhães et al., 2021).

Through terminal stance stiffness and increasing ankle plantarflexion and digital extension, the midfoot can plantarflex, adduct, and evert to allow the plantarflexion power to utilise the longer lever arm of the medial MTP joint fulcrums through digital extension. This enhances the windlass effect, improving foot stability later. The midfoot plantarflexion, adduction, and eversion coupled to the windlass effect raises the vault and shortens the foot to a higher vault profile again before swing. The ability to change shape and maintain stability is biotensegrity mechanics in action. Tensile passive and active forces achieving bone strut compression across the foot's articulations continuously through relatively rapid foot shape changes occurring throughout stance, alter compliance and stiffness properties. Failure in any part of this mechanism from a restricted articular motion to loss of a tensile element is likely to have consequences in mechanical efficiency, and may concentrate stresses in anatomy, risking tissue threshold breach. The significance of any failure will depend on the role of the structure and the nature of the foot affected. One of the most devastating dysfunctions in the foot is that of tibialis posterior failure, which is worth considering in further detail.

4.5.6 Tibialis posterior dysfunction

Tibialis posterior is a provider of medial ankle–rearfoot and vault stability that resists eversion and dorsiflexion moments around the rearfoot and talonavicular joint. This activity dissipates and stores loading energy into its tendon and linked connective tissues. It applies active muscular contractive restraint and elastic tendon recoil power into the foot during late midstance, that continues into terminal stance (Maharaj et al., 2016). Its dysfunction can affect each task of the stance phase from late swing pretensioning of the foot prior to initial contact through to late terminal stance. It is only normally inactive in stance during preswing.

Loss of its role in energy dissipation through loss of the pretension setting of the foot prior to forefoot impact and through its tendon-buffering effects during forefoot impact, will also affect medial stability of the foot during heel contact and forefoot loading (Kokubo et al., 2012; Maharaj et al., 2016). Tibialis posterior prevents forces focusing medially on the foot and helps to keep loading forces moving anteriorly through the foot during stance (Imhauser et al., 2004). Its midstance activity is an essential part of foot-stiffening and energy storage prior to heel lift (Kokubo et al., 2012; Maharaj et al., 2016). Thus, it stores and creates power that helps drive the CoGRF distally and anteriorly into terminal stance on a stiff foot (Imhauser et al., 2004; Kokubo et al., 2012; Maharaj et al., 2016). It also helps to influence forefoot and digital loading pressures in conjunction with the digital flexors and the peroneals (Ferris et al., 1995). After the triceps surae and Achilles, tibialis posterior is undoubtedly the most important muscle of the foot, having a complex relationship with many other muscles including the triceps surae. It is certainly more than a foot antipronator, as Maharaj et al. (2016) have indicated.

Tibialis posterior's energy dissipation and stiffening abilities have complex relationships with other anatomical structures where compensations for or by tibialis posterior can cause significant deterioration in stress distribution and energetics. The shape of the vault profile also influences the loading of the muscle and its tendon. A mobile flatfoot causes a medial shift in the loading of the foot prior to heel lift, probably resulting in increased loading of medial column anatomy. Excessive strain on the tibialis posterior tendon in midstance reduces tibialis posterior's ability to maintain even rates of anterior progression of the CoGRF under the foot (Imhauser et al., 2004). This may create premature high pressure under the 1st metatarsal, possibly increasing 1st tarsometatarsal joint stresses. With tibialis posterior tendon severance, the plantar foot ligaments increase their stresses, stretching the ligaments into yield, causing an acquired flat foot deformity (Imhauser et al., 2004).

When foot ligaments are intact, the tibialis posterior is an effective subtalar joint invertor (Imhauser et al., 2004) as it has a large inversion arm around the rearfoot complex through its proximal navicular attachment. Loss of tibialis posterior at heel lift increases eversion ranges by 4.14° (Hansen et al., 2001). Tibialis posterior also has significant influence on midfoot bone translations, affecting the height of navicular position and rotations in the bones along the medial side and across the transverse aspect of the vault (Imhauser et al., 2004). This action is provided by the tendon's proximal and distal attachments and the pulley action around the navicular and medial malleolus. Thus, tibialis posterior compresses the rearfoot and midfoot bones together across the sagittal, frontal, and transverse planes ready for heel lift. It also resists internal rotation of the tibia, plantarflexion of the talus, and rearfoot and navicular eversion moments, restraining vault profile sagging deflection (Imhauser et al., 2004). In this way, tibialis posterior provides the stiffening mechanism described by Kokubo et al. (2012), but to operate efficiently, the passive connective tissues must remain intact.

Tibialis posterior without ligaments

Loss of ligament restraints of the vault such as the long, short, and spring ligaments profoundly influences tibialis posterior's ability to provide spatial control of the rearfoot and midfoot. There will be loss of the ability to control talar rotation and navicular height in the sagittal plane with loss of the spring ligament (Imhauser et al., 2004). However, the tibialis posterior can still control and support the medial cuneiform's height, as these ligaments do not cross the 1st cuneiform (Imhauser et al., 2004). These ligamentous influences explain the observation by Niki et al. (2001) that reconstruction of the tibialis posterior tendon in acquired flat foot does not restore the vault profile or the rearfoot kinematics to normal, as ligament dysfunction can both result from and be caused by tibialis posterior dysfunction. Thus, treatment must address the failure of tibialis posterior differently if failure of the plantar ligaments is also suspected (Imhauser et al., 2004) (Fig. 4.5.6a).

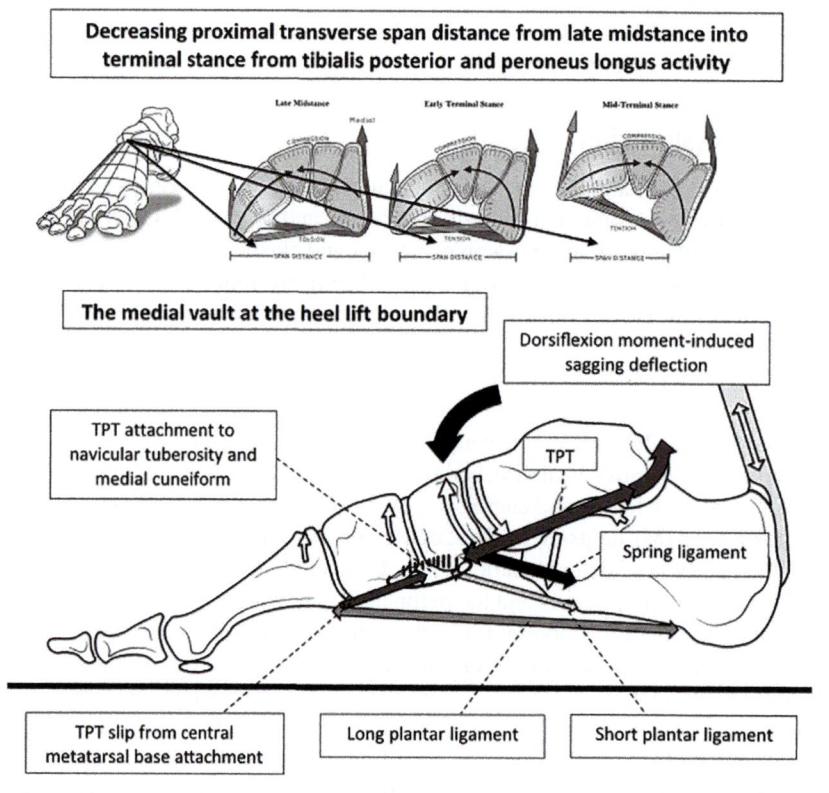

FIG. 4.5.6a Tibialis posterior and its tendon (TPT) and the plantar ligaments have an intimate relationship that helps set the increasingly stiffened, narrowed, and plantar concave-profiled proximal end of the conical-shaped vault at the heel lift boundary. Tibialis posterior resists the dorsiflexion moment's sagging deflection on the vault during late midstance, allowing the muscle to reduce activity during early midstance, but requiring increasing activity to peak at the heel lift boundary. Key plantar ligaments passively replicate the active support of tibialis posterior, and just as important, provide sensorimotor information on joint posture to help set muscle spindle activity within tibialis posterior. The spring (plantar calcaneonavicular) ligament supports the talar head, bound to the navicular, supplying information on and helping resist excess talocalcaneonavicular motion with tibialis posterior. This ligament is also bound to the calcaneus and the deltoid ligament of the ankle-rearfoot, replicating medial rearfoot support with that of the tibialis posterior tendon passing around the medial malleolus and talus. The short plantar ligament binds the plantar calcaneus to the cuboid, a bone that also has an important distal attachment site for tibialis posterior (not shown). The long plantar ligament binds the bases of the 2nd to 5th metatarsals to the calcaneus, helping in stabilising the cuboid, lesser tarsus, and tarsometatarsal joints, much as does tibialis posterior's distal attachment to the central metatarsal bases. The long plantar ligament is also attached to the peroneus longus tendon setting up an important relationship with this muscle, much as does tibialis posterior. Loss of any of these ligaments leads to mechanical and sensorimotor dysfunction of tibialis posterior and vault biomechanics. Loss of effective tibialis posterior power (usually through tendon degeneration) both at loading response, midstance, and particularly at the heel lift boundary, leads to vault failure and higher ligament stresses that can cause tissue threshold breaches that can result in plantar ligament damage. *(Permission www.healthystep.co.uk.)*

In anatomical pes planus with a strong healthy functioning tibialis posterior and intact ligaments, there is no reason to suspect an increased risk of tibialis posterior dysfunction. However, a lower vault profile with concurrent hypermobility of the passive ligament restraints is likely to result in a greater reliance of tibialis posterior to control sagittal, frontal, and transverse plane translations across the vault. This risks an increase in tibialis posterior muscle and tendon fatigue and tissue threshold breach, particularly with increasing age.

The result of combined tibialis posterior and plantar ligament failures is to increase midfoot dorsiflexion bending moments throughout midstance and terminal stance. Effects will peak around heel lift, where dorsiflexion moments derived before heel lift can be directed into the midfoot causing an increased need for vault energy dissipation. If the midfoot is unable to transfer ankle plantarflexor power to the MTP joint fulcrums, then the plantarflexion power can be directed into the midfoot deflection sagging, causing further talonavicular and tarsometatarsal dorsiflexion (Fig. 4.5.6b). This is a greater risk if ligament and tibialis posterior dysfunction are both present within a foot affected by a hypermobile lower vault profile. The consequence in time of such mechanics will be an acquired flat foot presenting with a developing midfoot break. In neuropathic diabetics, expressing peripheral motor and autonomic neuropathy, osseous weakness, and abnormal soft tissue fragility along with excessive stiffness, such unrestrained midfoot dorsiflexion moments can result in fracturing around the midfoot articulations and plantar ligament ruptures and/or avulsion leading to a Charcot foot.

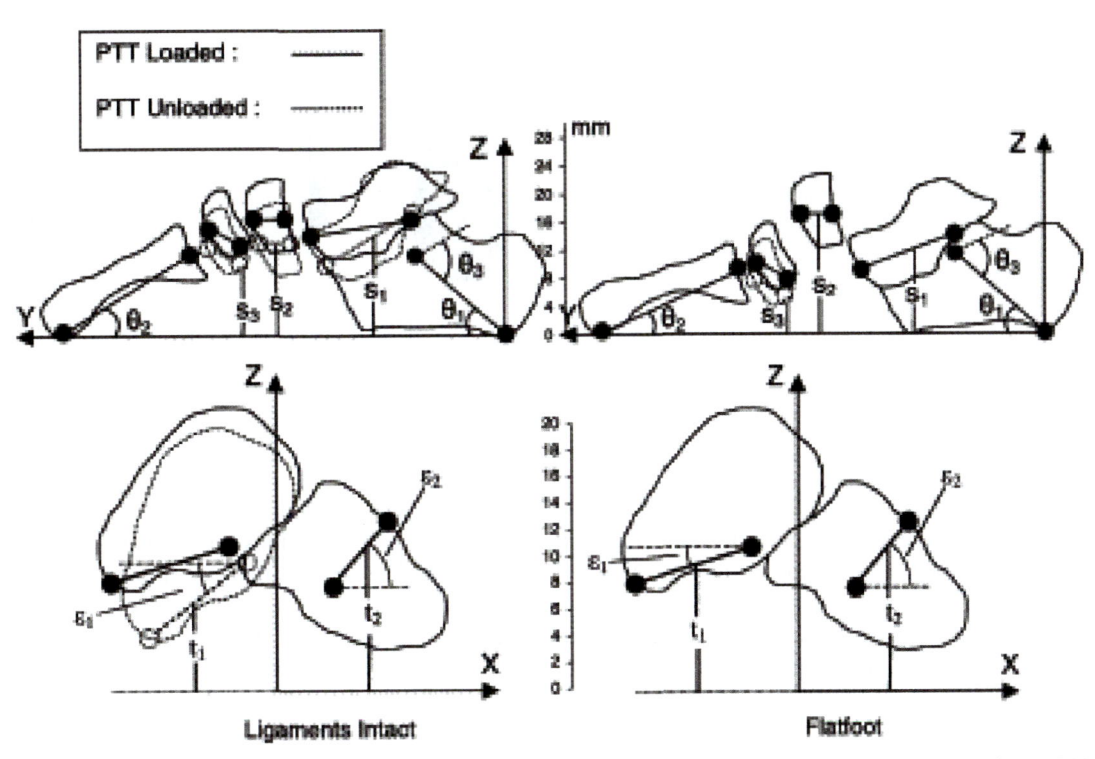

FIG. 4.5.6b Using cadaveric feet, kinematic changes under simulated gait loads are illustrated. The solid osseous outlines show the medial longitudinal vault and the transverse vault profile when the tibialis posterior tendon (PTT—posterior tibial tendon) is loaded, and the dotted outline when the tendon is unloaded. The left figures show the vault with the ligaments intact and the right figures demonstrate the vault after the long plantar ligament, the superior medial part of the spring ligament, the anterior portion of the deltoid ligament, and the plantar aponeurosis were sectioned. The normal function of tibialis posterior can be seen to be highly dependent on the passive connective tissue restraints of the midfoot and rearfoot. *(Image from Imhauser, C.W., Siegler, S., Abidi, N.A., Frankel, D.Z., 2004. The effect of posterior tibialis tendon dysfunction on the plantar pressure characteristics and the kinematics of the arch and the hindfoot. Clin. Biomech. 19(2), 161–169.)*

Tibialis posterior plays a fundamental role in assisting anterior acceleration by influencing the direction of the CoM over the foot at heel lift as part of plantarflexion ankle power generation (Imhauser et al., 2004). Failure to achieve this action through loss of tendon function or muscle weaknesses will compromise acceleration of the CoM's progression. Concurrent loss of medial plantar ligament forces results in sustained loading of the medial forefoot. Thus, tibialis posterior dysfunction compromises the action of the triceps surae–Achilles complex, directly influencing the position of the CoGRF placement anteriorly and thus affects the placement of the ankle plantarflexors' power at heel lift. Loss of medial ligamentous restraint and tibialis posterior power permits excessive medial drift of the CoGRF, increasing pressure medially on the foot. Any eversion and abduction perturbations resulting from less restrained midfoot bending moments and greater peroneal activity across the midfoot and forefoot, can alter frontal plane mechanics of the Achilles, potentially increasing medial strains within the soleus subtendon (Zifchock and Piazza, 2004; Handsfield et al., 2017). Changes in plantar pressure distribution gives hints as to what is happening in patients so affected (Fig. 4.5.6c). Thus, the management of tibialis posterior needs careful planning in the use of orthoses, rehabilitation, and/or surgery to make sure no part of the dysfunction (including concurrent ligament failure) or compensation mechanisms are not overlooked.

FIG. 4.5.6c The plantar pressure profile changes and the centre of pressure (CoP) position at simulated heel lift are demonstrated for a cadaveric specimen from the study by Imhauser et al. (2004). Unloading the tibialis posterior tendon (PTT) simulates PTT dysfunction, while sectioned plantar ligaments are presented in the flatfoot condition. The results of the cadaveric 'normal' PTT function and intact ligaments (far left) equate well with reports on normal CoP trajectories in live subjects where the CoP tends to lie under the 2nd MTP joint at heel lift. Loss of tibialis posterior power delays the anterior progression of CoP, while plantar ligament dysfunction allows the CoP to drift more medially. *(Image from Imhauser, C.W., Siegler, S., Abidi, N.A., Frankel, D.Z., 2004. The effect of posterior tibialis tendon dysfunction on the plantar pressure characteristics and the kinematics of the arch and the hindfoot. Clin. Biomech. 19(2), 161–169.)*

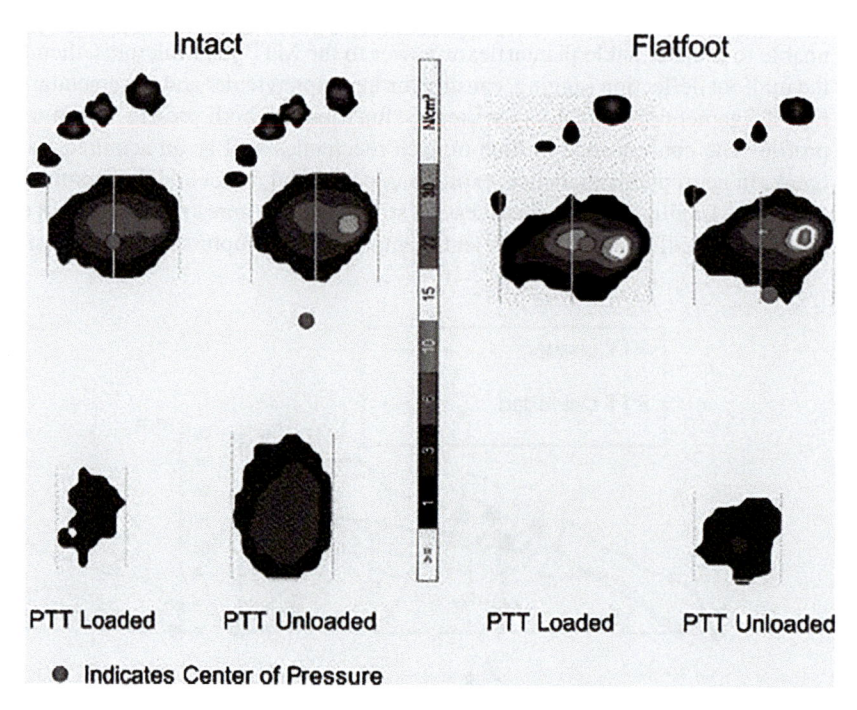

Thus, tibialis posterior dysfunction can result in severe disability and a loss of quality of life through limited locomotive capabilities, yet early intervention can reduce the requirements of invasive interventions such as surgery (Durrant et al., 2011). Ideally, any dysfunction needs urgent and prompt intervention before the situation deteriorates. Treatments must reflect all the elements necessary for improved tibialis posterior function, rather than just rearfoot complex eversion moments, and may require rehabilitation of the triceps surae and gluteus maximus to maintain terminal stance CoM acceleration to compensate for lost tibialis posterior power. A one-approach-to assessment and diagnosis is not likely to provide the best and most appropriate care (Durrant et al., 2016).

4.5.7 The MTP joint fulcrum in pathomechanics

MTP joints sustain considerable compressive and tensile forces in gait (Gregg et al., 2007). They use functional instantaneous axes that are mobile during joint motion, moving more anteriorly and superiorly with increased walking speed (Raychoudhury et al., 2014). This probably relates to increasing soft tissue stiffness associated with increasing loading rates, as discussed in Chapter 3, Section 3.1.1. Where the foot pivots to initiate heel lift matters to the relationship between the power generated through force–velocity relationship and the metabolic costs of the plantarflexor muscles, from the lever arm lengths generated (Takahashi et al., 2016; Wannop and Stefanyshyn, 2016; Oh and Park, 2017). When foot and lower limb anatomy is healthy, the mechanical 'Goldilocks' fulcrum point of gait efficiency is at the medial MTP joints. These medial MTP joints offer a longer lever arm-to-fulcrum point (Achilles external moment and plantarflexion power lever arms) for the ankle plantarflexors via the Achilles than do the lateral metatarsals (Holowka et al., 2017). The mechanically ideal fulcrum point at heel lift should be located around the 1st to the 3rd MTP joints, depending on metatarsal length-to-angle of gait relationships, with the 2nd MTP joint being most consistently significant.

The MTP joints are energy dissipators, reducing the potential force–velocity relationship of the ankle plantarflexors. However, their energy absorbing flexibility also reduces the metabolic costs of utilising the plantarflexors by shortening the length of the foot (from the tips of the digits) as a lever beam. Thus, having immobile MTP joints presents a longer lever arm, but increases the amount of muscle contraction required to achieve heel lift. This would increase the force and power of the ankle plantarflexors, but at the cost of increased metabolic demand. As an extra benefit, MTP joint passive extension, especially around the medial MTP joints, allows the windlass mechanism to provide some 'metabolically free' passive foot stability in terminal stance; something lost with immobile MTP joints. Thus, some but not too much MTP joint flexibility is ideal. Being able to adapt this flexibility offers mechanical advantages in different situations.

It has been found that the presence of an upward curvature in the toe-end of shoes that set the toe into extension before terminal stance, known as *toe springs*, can reduce the work required for walking (Sichting et al., 2020). This highlights the importance of the MTP joint extension fulcrums in human walking. It is also worth pointing out that loss of ankle plantarflexion through pathology, such as ankle OA, severely compromises stance acceleration biomechanics and increases the work required during walking. This is because of the close coupling of ankle plantarflexion with MTP joint extension angles (Fig. 4.5.7a).

Alternative fulcrums and potential pathomechanics

Individuals may present with foot mechanics and anatomy set up for an alternative heel lift through a midfoot break fulcrum rather than an MTP joint fulcrum because it is safer for them to do so. This type of heel lift requires less ankle plantarflexor power as it occurs on a shorter lever arm, requiring less muscle contraction and/or tendon elastic recoil energy to raise the heel from the ground. In chimpanzees (and possibly also in early hominins) with a shorter Achilles tendon, a midfoot fulcrum is a better metabolic option for bipedal gait. However, it reduces the plantarflexion power that can be utilised for CoM acceleration, resulting in loss of extra 'momentum power' into the next step. It also makes efficient running problematic. Thus, medial MTP joint fulcrums in modern human feet offer the best mechanical energetics. Forces occurring across the metatarsal heads in terminal stance of healthy young adult subjects in walking gait, as a percentage of body weight, are reported to be 29.1% for the 1st metatarsal head, 28.3% for the 2nd, 22.3% for the 3rd, 4.3% for the 4th, and only 3.5% for the 5th metatarsal head (Hayafune et al., 1999). This highlights the medial MTP joint dominance in the acceleration of the GRF in most humans. The pathomechanical relevance of the midfoot break is thus its detrimental effects on acceleration energetics and increased fatigue of the acceleration muscles during longer periods of locomotive activity, rather than creating foot pathology directly. The exception is if midfoot break is acquired later in life, when soft tissue structures within the vault and articular structures within the midfoot become highly vulnerable to new abnormal stresses and motion (Fig. 4.5.7b).

Lateral MTP joint fulcrums may be utilised if the lower limb posture presents the foot's longitudinal alignment internally to the line of progression, but also if medial MTP joints are unable to extend freely. They may also be used if the foot sustains an inversion moment at the heel lift boundary. In the case of an internal angle of gait, the lateral metatarsals may now present a functionally longer lever arm to the plantarflexors within the sagittal plane orientation of the lower limb joints than the medial MTP joints, which are now angled inappropriately for anterior acceleration. Loss of medial MTP joints as an acceleration fulcrum, such as can occur with 1st MTP joint degenerative OA changes, usually reduce gait energetics, but lateral fulcrums may offer a safer form of acceleration in these circumstances. However, the lateral MTP joints, particularly the 4th and 5th, are less well developed to sustain long periods of high loading stresses, due to greater mobility at their tarsometatarsal joints and their smaller cross-sectional areas. This sets up a potential tissue mismatch to forefoot loading stresses (Fig. 4.5.7c).

The terms 'high gear propulsion' and 'low gear propulsion' to describe the medial and lateral MTP joint fulcrums, respectively, are used to indicate a more or less efficient engagement of the windlass effect, linking calcaneocuboid stability around midfoot eversion to the use of the efficient high gear mechanism (Bojsen-Møller, 1979). However, the terms also suit well the concept of the better application of ankle plantarflexion power by setting the Achilles external moment arm and also aligning plantarflexion power moment arm, perpendicular to the high gear fulcrum axis at heel lift.

MTP joint freedom and pathomechanics

Apart from fulcrum positioning, the ease of freedom of motion at the MTP joints is also important. The amount of passive dorsiflexion available at each MTP joint will reflect how much and which of these foot fulcrums are easiest to use. Morphological features, such as metatarsal bone density, cross-sectional area, metatarsal declination angle, and the extensor

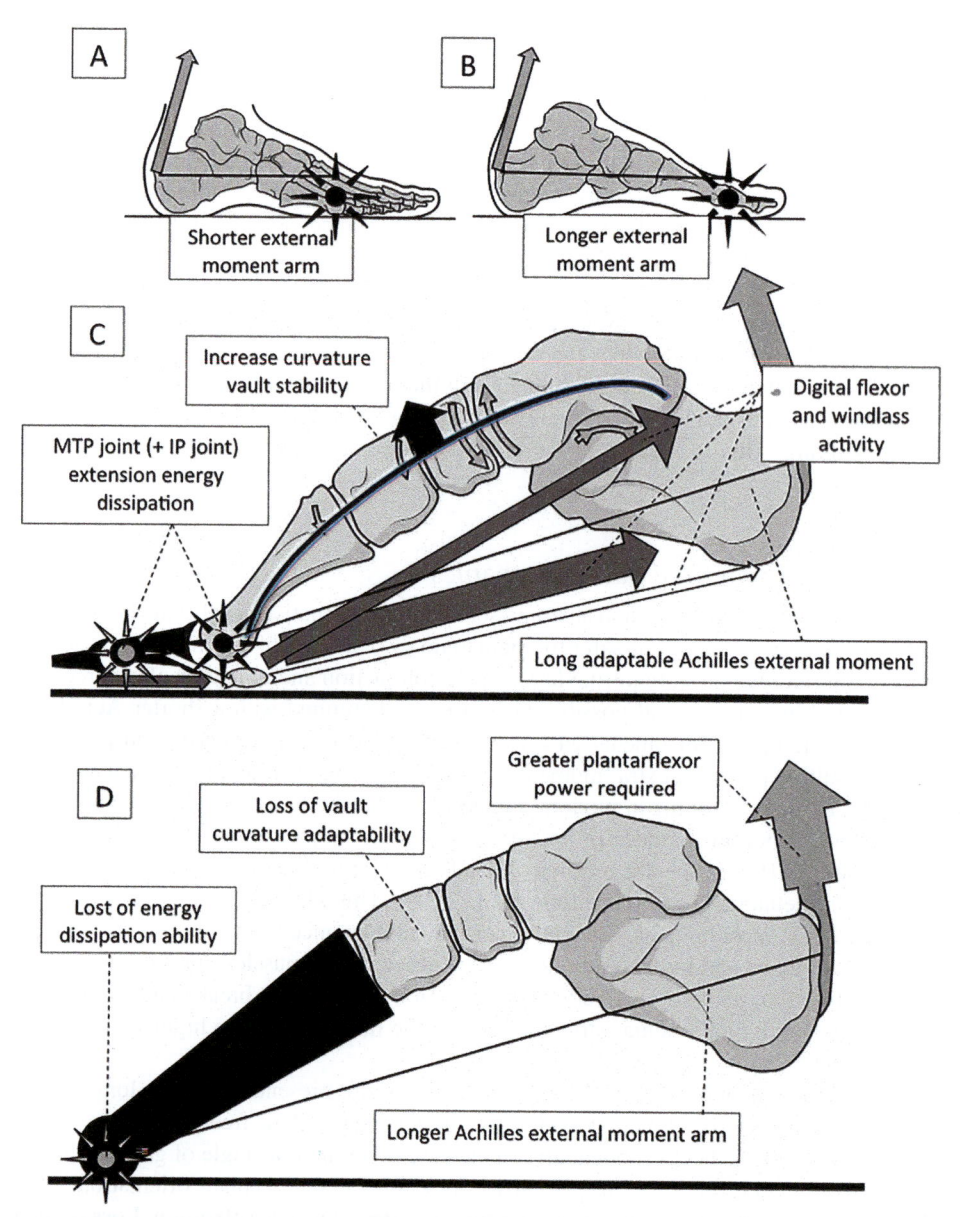

FIG. 4.5.7a The length of the external moment arm of the Achilles for applying ankle plantarflexion power for acceleration is a critical mechanical component of human gait. The lateral column (A) offers a generally more stable foot vault and would seem to offer a good lever arm for acceleration. However, its smaller vault span and shorter metatarsal lengths decrease the distance between the Achilles attachment and the lateral MTP joints than that offered by medial metatarsals. This shorter, less powerful ankle plantarflexion moment arm is termed 'low gear', reflecting the depowering of acceleration. Despite being intrinsically more flexible, the medial column (B) in health can provide a longer semirigid acceleration beam. If more energy needs to be dissipated at acceleration, then variable flexibility within the medial vault can be supplied. By utilising all stiffening mechanisms, a stable medial vault offers longer metatarsal lengths as well as a greater span, increasing the Achilles external moment arm and raising the capacity to create plantarflexion power into the forefoot. By having medial MTP joint fulcrums with freedom to extend under GRFs (C), some energy from the plantarflexion power will be dissipated. However, this energy, transferred into digital extension, helps tighten the plantar aponeurosis to initiate the windlass mechanism. The windlass helps to increase vault curvature and foot stiffness to assist the long and short flexors in stabilising the forefoot, just as the tibialis posterior and peroneus longus vault-stiffening energies are starting to wane in later terminal stance. Having secondary interphalangeal (IP) joint fulcrums gives some further flexibility for lengthening the moment arm or helping to dissipate more energies by IP joint extension, should MTP joint extension energy dissipation prove inadequate. In contrast, a rigid bone from the tarsometatarsal region to the anterior tip of the foot (D) would lengthen the plantarflexion moment arm, thus allowing more plantarflexion power to be applied to the forefoot. However, such anatomy would require greater plantarflexion power to initiate heel lift, lose the ability for any MTP joint protective energy dissipation during acceleration, and remove much of the ability to increase vault curvature during terminal stance via tensioned digital extension and the windlass mechanism. Thus, a 'Goldilocks zone' balance of plantarflexion power application and protective energy dissipation via MTP joint extension seems to exist, which may be lost with MTP joint dysfunctions. *(Permission www.healthystep.co.uk.)*

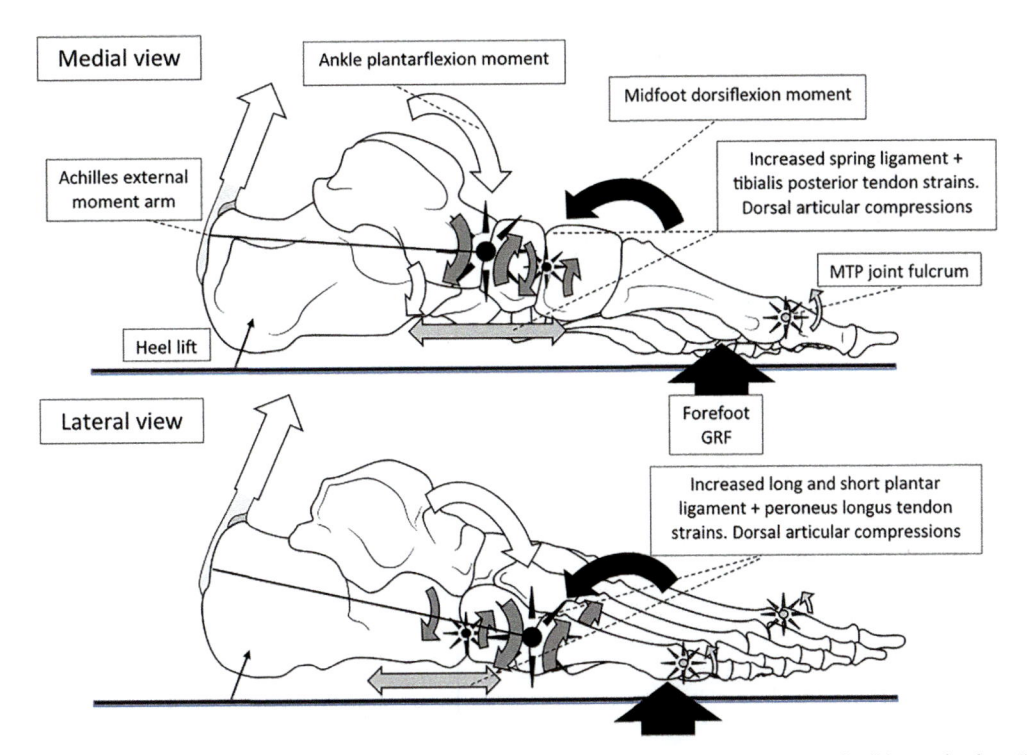

FIG. 4.5.7b Midfoot break heel lift primarily uses fulcrums (stars—larger being more significant) at the more flexible proximal medial vault articulations and the more flexible distal lateral vault articulations. The primary medial fulcrum is at the talonavicular joint, aided by navicular and cuneiform joint dorsiflexions. Laterally, the more mobile distal midfoot joints at the 4th and 5th tarsometatarsal joints act as the primary fulcrums, aided by some calcaneocuboid motion. While these articulations can occasionally provide a midfoot break in those individuals who usually use MTP joint fulcrum acceleration as an option of step adaptability, it is the persistent use of midfoot break that deteriorates foot energetics repeatedly. For some, it is the best gait option, but for others who tend to develop midfoot break later in life, it can be highly pathomechanical. Plantarflexion power in midfoot break is driven on a short Achilles external moment arm with fulcrums primarily focused to the talonavicular and the cuboid's tarsometatarsal joints. This means plantarflexion power energy dissipation derives from midfoot dorsiflexion (extension) rather than extension at the MTP joints that have evolved to freely rotate under digital flexor resistance, coupled to the flexor plate mechanism's protective restraint. Some limited digital extension usually still occurs at MTP joint fulcrums (small pale stars), dependent on the angles the toes need to reach before toe-off. Plantar ligaments that straddle the proximal, medial, and distal lateral vault are under greater tension, being utilised with the tendons of tibialis posterior and peroneus longus as high energy dissipators of the plantarflexion power. Under high midfoot dorsiflexion moments and motions at heel lift, articular structures are vulnerable unless they have developed their articular contours under midfoot break mechanics during childhood that increases articular facet curvatures during growth. When connective tissues are abnormally stiff, brittle, and inelastic, as seen within diabetic patients, energy dissipation across the midfoot during midstance can become highly destructive, causing vault collapse and in time, a rocker-bottom convex plantar profile associated with a Charcot foot. The creates an extreme form of midfoot break acceleration mechanics. *(Permission www.healthystep.co.uk.)*

grooves on the metatarsal neck, reflect the use of these joints during growth and development as well as the intrinsic developmental preprogramming of anatomy. It is likely that a change in foot fulcrum location during adulthood reflects a greater danger to the development of pathology than does a fulcrum point that an individual has developed during growth, even if the fulcrum point is not the energetic ideal.

Both foot vault profile and soft tissue restraint will influence MTP joint freedom and utilisation. Ideally, there should be appropriate stiffness for stability and restraint of free digital extension, but not so much that it interferes with the freedom to heel lift. A foot with a midfoot break will achieve a certain percentage of midfoot dorsiflexion angulation to achieve heel lift, which will be reflected in the extension angle required at the MTP joints. The more midfoot break, the less MTP joint dorsiflexion required. MTP joints usually provide all the dorsiflexion that is required to maintain a force–velocity-to-metabolic expenditure balance at heel lift. If there is limited freedom of extension available at the MTP joint, then a point further distally may be required to act as part of the extension fulcrum. The interphalangeal joints of the toes, which are not built for high extension motion normally, offer a possible solution. Such a more-distal fulcrum also leads to the force–velocity-to-metabolic expenditure of the plantarflexors being altered. The hallux interphalangeal (IP) joint often hyperextends to accommodate insufficient hallux extension at the MTP joint, but this seems to occur less frequently at other digital IP joints. Such IP joint compensations can be associated with pathology if the extension moment applied is large, creating cutaneous lesions under an IP joint, a joint that lacks extensive plantar fat padding and provides less extension resisting structures.

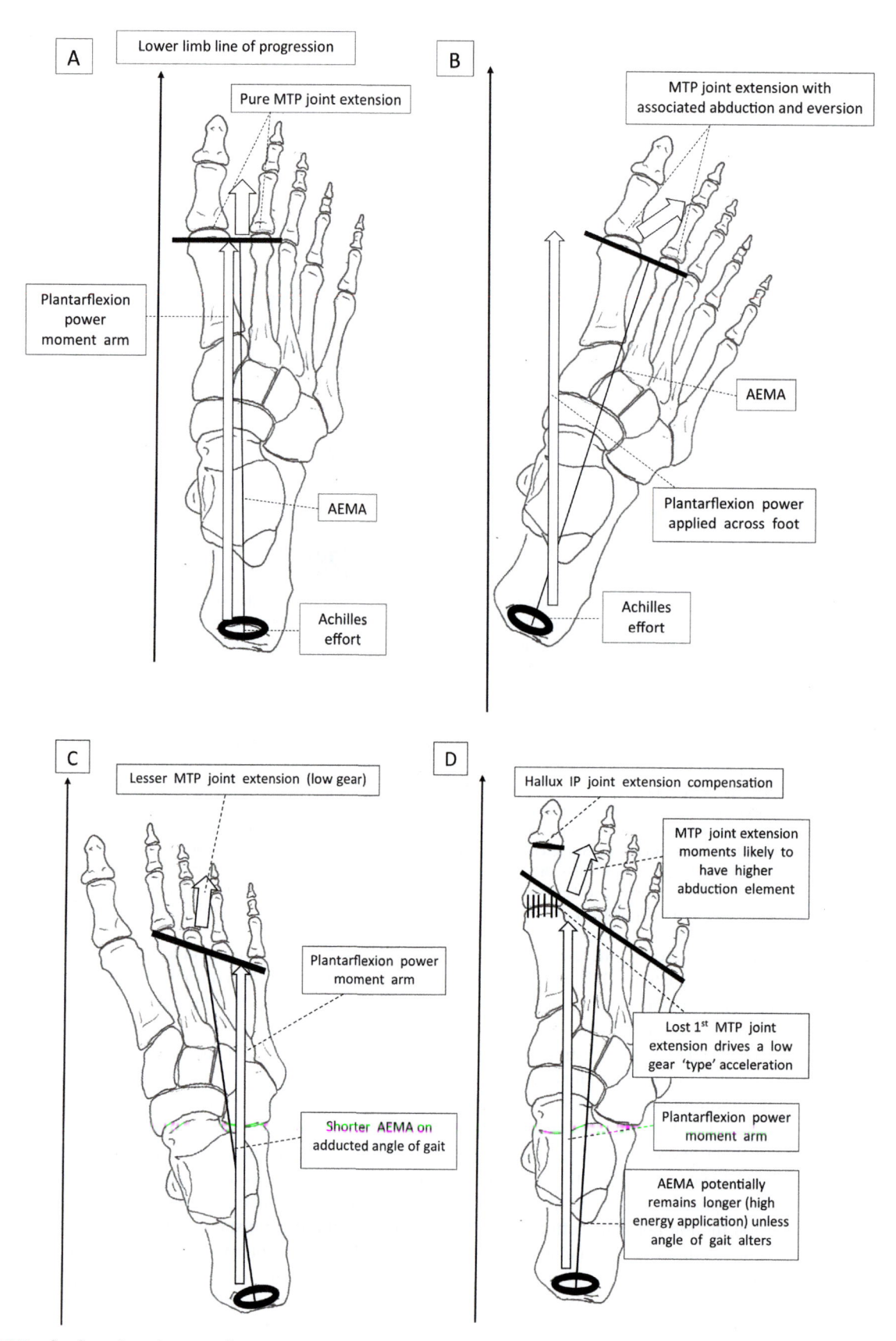

FIG. 4.5.7c See figure legend on opposite page

Declination angles and MTP joint extension

Low metatarsal declination angles at the start of extension place the proximal phalanx at a low extension angle relative to the metatarsal shaft, potentially limiting access to the extended dorsal articular aspect of the metatarsal head. The low angulation of the metatarsal may restrict the ability of the proximal phalanx to also access the extensor groove at the end of terminal stance. The low extension angle at the MTP joint at heel lift associated with low metatarsal declinations, may present a risk of dorsal impingement of the MTP joint's articular surface via the proximal phalanx colliding with the metatarsal head. Repetitive impingement can damage articular structures through localised cartilage threshold breaches. Once the biphasic properties of cartilage are compromised, OA degeneration is highly likely. However, low declinations associated with midfoot break may not create such a problem. With midtarsal break, there is a reduced need for high MTP joint extension angles, thus avoiding metatarsal head impingement (Fig. 4.5.7d).

An extension impingement situation can occur with an individual metatarsal possessing an isolated significantly low metatarsal declination, even when the other metatarsals around it are positioned normally. Such a position of the metatarsal is likely to result in later loading extension of the digit, but still risks impingement if the extension moment is large enough by the end of terminal stance. This is a finding often reported in the 1st metatarsals in a condition known as *metatarsus primus elevatus* (MPE). Although a correlation with reduced range of hallux dorsiflexion is noted, no firm correlation with 1st MTP joint OA (often called hallux rigidus) and MPE has been established, despite considerable debate (Roukis, 2005; Bouaicha et al., 2010). Radiographically, a 5 mm MPE may be a predictive factor of hallux rigidus (Bouaicha et al., 2010), suggesting that MPE is a risk factor but not an inevitable cause of 1st MTP joint OA. The delayed extension that elevation causes may reduce the extension range required from the joint, saving it from developing impingement. It is possible that MPE is a manifestation of limited hallux extension, providing a protective mechanism, rather than being a cause of OA pathology (Sanchez et al., 2018).

In a cavoid foot with exaggerated metatarsal declination angles, the MTP joints may start their motion in positions of relatively high resting dorsiflexion compared to the metatarsal shafts. The range of passive dorsiflexion required may bring the proximal phalanges to their end-ranges of motion too soon during terminal stance, despite them moving into the extensor groove. As the restricted ankle dorsiflexion in midstance is also a common problem in cavoid feet, as discussed in Section 4.5.4, a premature heel lift will also lead to a longer duration and a greater range of MTP joint extension associated with the increased ankle plantarflexion angle. A similar artificial situation can be induced with wearing high-heeled shoes, as the foot is held in ankle plantarflexion and the digits in high angles of extension, throughout stance phase (Fig. 4.5.7e).

Increased MTP joint stresses and impingements of the proximal phalanx onto the metatarsal heads have been suggested as mechanisms associated with excessive dorsiflexion that induce pathologies, such as MTP joint OA (McMaster, 1978) and Freiberg's disease at the 2nd metatarsal head (Longworth et al., 2019). These proposed models suggest pathology is

---

FIG. 4.5.7c, CONT'D The gearing of acceleration power from the Achilles at heel lift into terminal stance results from the relationship between the Achilles external moment arm (AEMA) length and the plantarflexion power moment arm it associates with (long white arrow), the medial MTP joint fulcrum position (solid black line), and the lower limb line of progression (thin black arrow). Ideally (A), plantarflexion power will be mechanically expressed at heel lift through this relationship to maximise energetics. Thus, the medial metatarsal sagittal plane instantaneous axis position for MTP joint extension is best located perpendicular to the line of progression of the lower limb. In this position, the AEMA and the plantarflexion power are applied perpendicular to the fulcrum axis along the line of progression. This will maximise sagittal plane freedom producing relatively 'straight' tensile strain on connective tissues, avoiding shears and torques within the frontal and transverse plane that could create injury. Thus, with energetics higher and injury risk lower, the term 'high gear acceleration' would seem appropriate. This arrangement can be disturbed if the angle of gait strays significantly from the line of progression. Abducted angles of gait (B) lateralise the medial metatarsal fulcrum angle so that the AEMA faces away from the line of progression. If the medial MTP joint fulcrum is still used, the lower limb extension and flexion moments must become externally rotated. A solution is to apply the plantarflexion power across the foot medially, placing inversion moments into the medial midfoot and forefoot and also by replacing the 'pure' extension torque at the MTP joints with increased levels of abduction and eversion. This will increase shear stresses on foot plantarflexors, ligaments, and articulations, particularly at the medial MTP and tarsometatarsal joints. This will increase energy dissipation and depowering acceleration. Adducted (in-toe) gait (C) rotates the medial MTP joints inwards to the line of progression, putting the lateral metatarsal axis in a better position to apply the plantarflexion power along the line of progression. However, the lateral metatarsals offer a shorter AEMA depowering acceleration which requires energy balancing compensations from other foot plantarflexors. Loss of medial MTP joint motion (D), perhaps resulting from degenerative joint disease, an inappropriately low 1st metatarsal declination angle, or 1st ray instability, can cause extension fulcrum blockage. Such issues compromise the AEMA to the acceleration fulcrum, although heel lift itself may still be possible through limited initial MTP joint extension, setting up a high plantarflexion powered heel lift that will transfer high energy into the medial MTP joints until the extension motion is blocked. Depending on the extension motion available, the IP joint may compensate, adding some more extension freedom to dissipate extension energies as terminal stance progresses. However, the foot may be forced to utilise the low gear lateral MTP joint axis once further medial MTP joint motion is lost. The relationship of metatarsal lengths and frontal plane postures in setting the MTP joint fulcrum's position also determines the most appropriate angle of gait for the application of the plantarflexion power to the line of progression. *(Permission www.healthystep.co.uk.)*

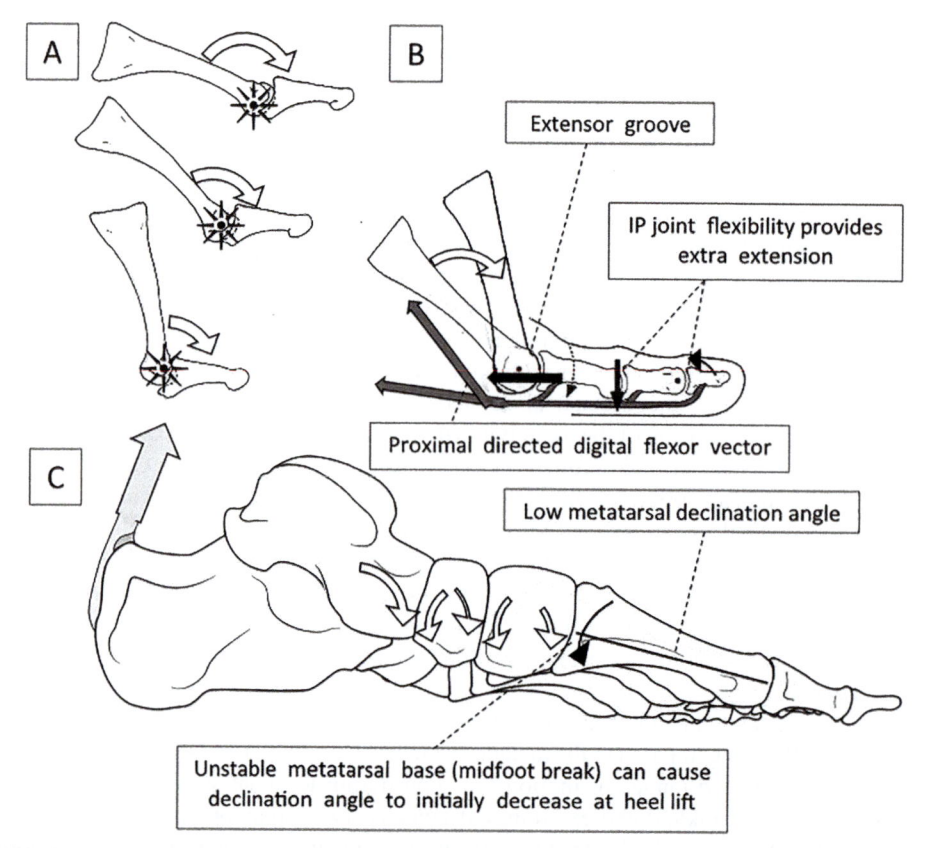

FIG. 4.5.7d For MTP joint extension to couple to ankle/midfoot plantarflexion, metatarsals need to increase their functional declination angles during terminal stance (A). By doing so, the instantaneous axis can move superiorly with the metatarsal head during extension, maintaining an arc of proximal phalanx rotation freedom. The proximally directed vector from the digital flexors is essential in achieving this freedom of motion (B), keeping the metatarsal head stable while it rotates on the plantar plate fibrocartilage or sesamoid bones. The tensions provided by flexor tendons and the flexor plate mechanism help provide extension resistance control and energy dissipation. At the end-range of extension, the presence of a significant extensor groove on the dorsal surface of the metatarsal neck reduces the risk of impingement. Interphalangeal joints can provide points of extra extension if required. Foot morphology, such as a low vault profile with a planus foot or the elevated metatarsal of metatarsus primus elevatus (MPE), presents very low declination angles at heel lift. These situations may prove inadequate to allow MTP freedom via superior instantaneous axis motion. Poor extensor groove development may also increase impingement risks. Midfoot break can actually cause the metatarsal declination angle to initially decrease at heel lift (C). *(Permission www.healthystep.co.uk.)*

resulting from hyperextension at an MTP joint arises from end-range impingement (Fig. 4.5.7f). Hypoextension of the MTP joint has also been suggested as a mechanism provoking articular damage from 1st MTP joint impingement (Dananberg, 1985, 1986). These different models equate to excessive functional metatarsal declination leading to hyperextension and thus end-range MTP joint impingement, and insufficient functional declination in hypoextension impingement, resulting from a premature encounter with the superior aspect of the metatarsal head. In cases of low declination of the metatarsal, the instantaneous axis becomes 'fixed' in its position causing the arc of rotation of the proximal phalanx to be blocked by the superior aspect of the metatarsal head's articular surface. Thus, any failure to engage the proximal phalanx into the extension groove at end-range extension presents a risk. An individual with an underdeveloped extension groove, despite an appropriate range of MTP joint extension, is also therefore vulnerable. For a surgeon tempted to alter the declination angle of a metatarsal to resolve insufficient 1st MTP joint extension, the depth or absence of the extension groove should be considered.

Metatarsal length effects on MTP fulcrums

Intermetatarsal length relationships will also affect the ease of engaging low or high gear MTP joint fulcrums for gaining appropriate levels of MTP joint extension flexibility. These concepts link metatarsal lengths to angle of gait. Metatarsal length relationships set up fulcrum axes through groups of metatarsals, tying in concepts of high and low gear propulsion

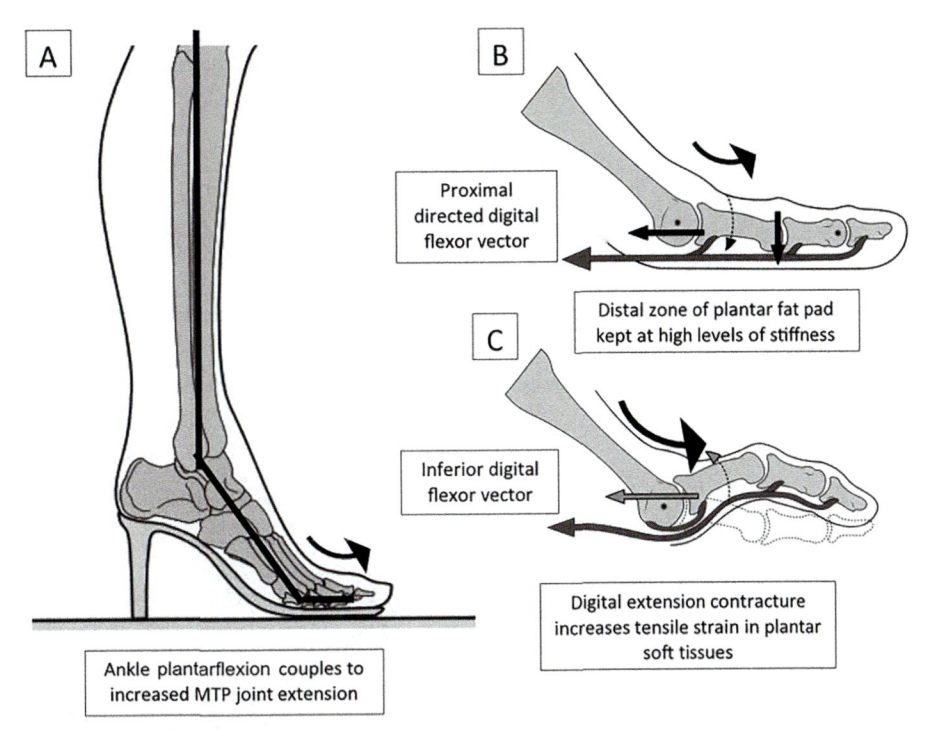

FIG. 4.5.7e High-heeled footwear create artificially high metatarsal declination angles throughout stance phase as a consequence of positioning the foot in ankle and midfoot plantarflexion. Not only does this limit energy storage within the Achilles during late midstance (unless triceps surae shortens), but the MTP joint extension that results also tensions the forefoot plantar tissues into higher levels of stiffness (A). Thus, discomfort in calf muscles and a 'burning soreness' across the forefoot should be expected with high-heeled footwear use. Acceleration will now start at a high functional metatarsal declination angle. Normally under digital extension (B), the digital flexors generate a vector that compresses the proximal phalanx to stabilise the metatarsal head as it rotates within the flexor plate mechanism (extensor apparatus) over the plantar plate. However, when digital extension is maintained throughout static stance and during dynamic stance phase via high-heeled footwear use (C), flexor muscle vectors will be divided into inferior flexor vectors (long, short flexors, and lumbricals) and also superior extensor vectors, as the interosseous muscle tendons will slip dorsally. Under sustained tensile strain, the plantar soft tissues can cause adaptation in accord with Davis' law, providing some more connective tissue compliance but also weakening MTP stability. These can cause digital deformity via increasing extension freedom. A similar forefoot situation can arise with cavoid feet, where acceleration also starts with metatarsals in higher declination angles and digits at greater extension angles. *(Permission www.healthystep.co.uk.)*

and also the preferred moment arm for the ankle plantarflexor power. Angles of gait, slightly abducted by a few degrees, open up high gear fulcrums at the 1st and 2nd MTP joints aligned to the longitudinal axis of the leg to facilitate the foot and ankle to work as a passive linear spring (Hedrick et al., 2019). This presents the preferred lever arm length for the application of the ankle plantarflexion power to the line of progression (Holowka et al., 2017; Takahashi et al., 2017). Through variation in angle of gait, the longer MTP joint fulcrum axis can be aligned to the leg's longitudinal axis position during locomotion to engage the long lever arm for the ankle plantarflexors within the line of progression, despite individual variation in the 1st and 2nd metatarsal lengths. This adaptability makes precise length relationships unnecessary for human gait energetics.

However, the protrusion distance difference between the medial metatarsals may influence how easily the fulcrum can be utilised to match the foot's gait angle and line of progression. This, in turn, can influence the proportionate amount and nature of the stress that goes through each joint. The relationship of the foot's longitudinal angle to the lower limb's line of progression and the metatarsal protrusion distance relationships all interplay to affect the efficiency of the ankle power, the force–velocity relationship of the plantarflexors' metabolic costs, and the internal MTP joint stresses, through fixing the fulcrum points' line of action. Sagittal plane ankle plantarflexion torque must be translated into 'purely' (or as much as possible) sagittal plane MTP joint extension torques to absorb energy passing across the forefoot's joints, rather than generating frontal and transverse plane torques that could damage articular structures and soft tissue supports. Small frontal and transverse plane torques should not be pathological. However, the bigger they become, the greater the risk of tissue stress threshold breach.

Abducted angles of gait move the sagittal line of progression to the medial side of the foot. Large abducted angles of gait may lead to only the 1st MTP joint becoming a practical fulcrum option, thereby concentrating most ankle power within the

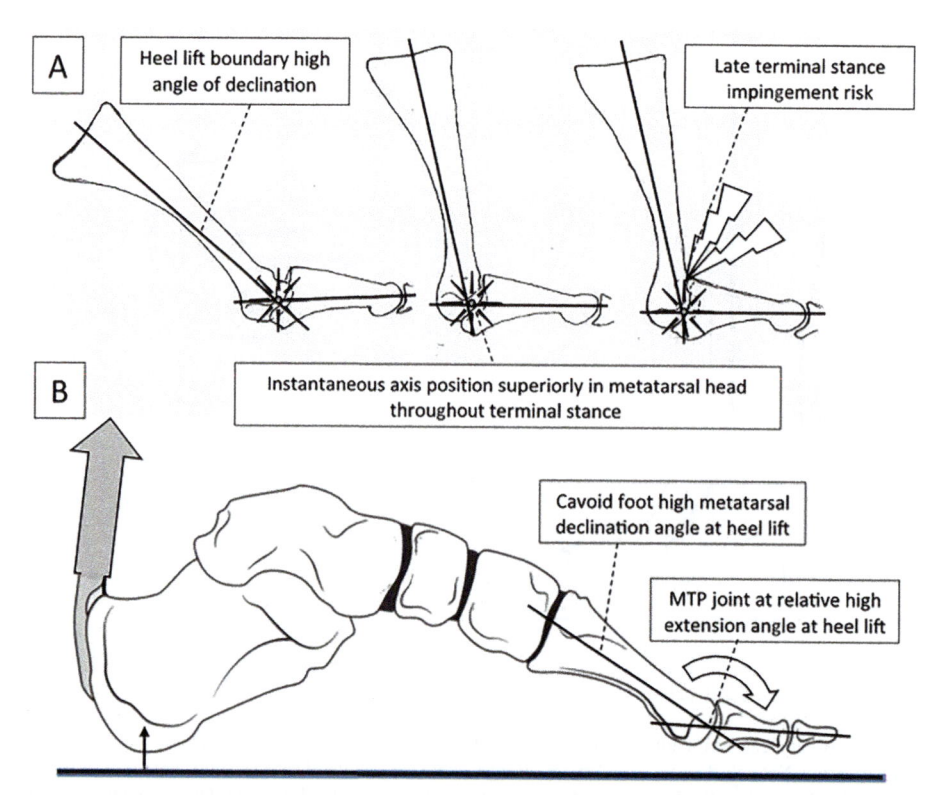

FIG. 4.5.7f Via their high metatarsal declination angles, cavoid feet start heel lift with their MTP joints at higher angles of extension. This means that the instantaneous axis of rotation is positioned more superiorly from the start (A). Thus, in cavoid feet, MTP extension moments tend to have greater freedom of motion during terminal stance. This means that any significant weakness in soft tissue flexor restraint may result in hyperextension and impingement of the dorsal articular joint surface on a mild but recurrent basis during late terminal stance. This can lead to gradual articular degeneration over time. There is a tendency for the digits of cavoid feet to develop overextension toe deformities at the MTP joints as a result of these biomechanical imbalances. The use of footwear with a heel lift will exacerbate the situation and indeed create a similar mechanism for those with more average barefoot metatarsal declination angles. *(Permission www.healthystep.co.uk.)*

1st MTP joint during terminal stance. In extreme cases, the hallux will be exposed to an abduction (external rotation) and eversion torque during extension. Adducted angles of gait will tend to move the MTP joint fulcrums more laterally, and in extreme cases, this could necessitate the use of lateral MTP joints that may then experience high loading stresses, possibly outside of their anatomical capabilities. How significant the angle of gait is to generating potentially pathological MTP joint torques, depends on the metatarsal length relationships (Fig. 4.5.7g(A) and (B)).

Most human angles of gait tend to be slightly abducted. In this position, a slight or no significant protrusion distance between the 2nd and the 1st MTP joint is advantageous to create a fulcrum axis perpendicular to the line of progression. This posture spreads MTP joint stresses and extension moments relatively evenly between these joints. A slightly longer 1st metatarsal protrusion requires a lower abduction angle of gait or straight position of the foot. This angular relationship maintains the long fulcrum distance to the ankle joint plantarflexor power, presenting an efficient MTP joint acceleration potential. Should the 1st metatarsal protrusion distance be considerably longer than the 2nd, the fulcrum line may disassociate with the foot's line of progression, and forces will tend to be disproportionately applied to the 1st MTP joint, unless the angle of gait rotates the foot more internally (more intoed). If the 2nd metatarsal protrusion distance is particularly large, then the fulcrum line may disassociate with the line of progression medially, and stresses are more likely to increase within the 2nd MTP joint. By changing the angle of gait to a more abducted position accommodates this fulcrum alignment so that disproportionate joint stresses and disruption of ankle power energetics can be compensated for. However, if the foot loses MTP joints aligned to the lower limb's line of progression to achieve safe MTP joint stresses, then all benefits in gait energetics will be lost, altering kinematics and kinetics elsewhere within the lower limb chain. Thus, anything that starts to alter stability of the lower limb to the line of progression such as hip abductor strengths and flexibility, knee rotator stability, or dysfunction within ankle-bridging muscles like tibialis posterior or peroneus longus, becomes a risk factor for altered acceleration mechanics. However, individual lesser metatarsal declination angles that mismatch to their companions on the weightbearing surface, will more likely just influence MPT joint biomechanics and local soft tissue stresses (see Fig. 4.5.7g(C)).

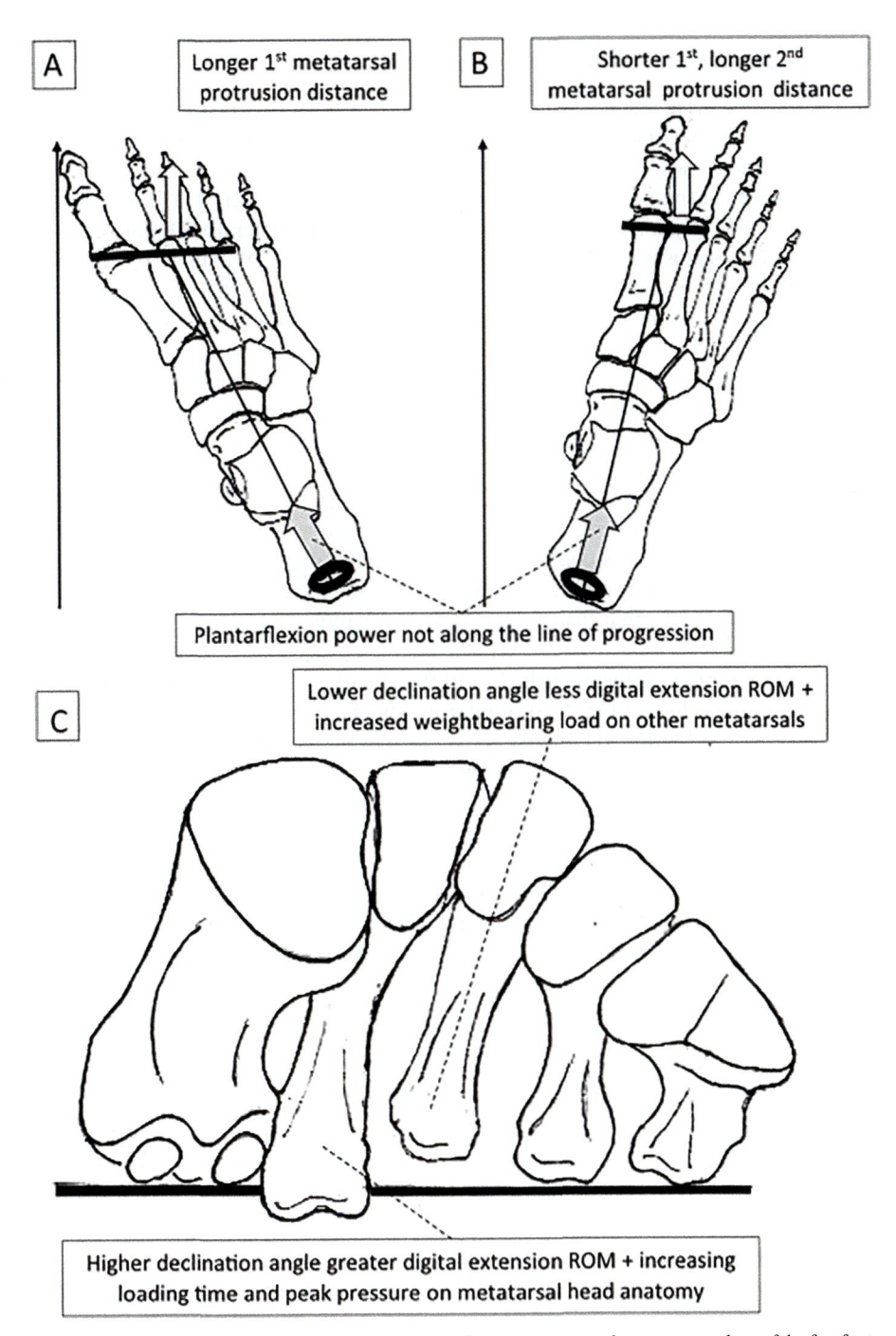

FIG. 4.5.7g Large metatarsal protrusion distances that disrupt a gentle arc of curvature across the transverse plane of the forefoot or differences between metatarsal declination angle that set MTP joints at different levels within the weightbearing frontal plane, can detrimentally influence the acceleration fulcrum mechanics across the forefoot. Significantly longer 1st than 2nd metatarsals (A) require more of an adducted gait angle to maintain the fulcrum perpendicular to the line of progression. Shorter 1st metatarsals work better on a more abducted angle of gait (B). However, if differences are large, both will upset energetics by taking the long axis of the foot away from the line of progression and the preferred direction for the Achilles external moment arm to express plantarflexion power most efficiently. Elevated metatarsals (C—3rd metatarsal) decrease the declination angle, which can cause loss of MTP joint extension within the affected MTP joint and reduce the affected metatarsal's ability to bear loads. Excessive metatarsal declinations (C—2nd metatarsal) compared to surrounding metatarsals can lead to excessive MTP extension angles at the end of terminal stance and higher tension within the structures of the flexor plate mechanism. Such metatarsal plantarflexed postures also result in longer loading times and peak pressures on the affected MTP joint during late midstance and terminal stance. *(Permission www.healthystep.co.uk.)*

4.5.8 MTP joint compliance and stiffness in pathomechanics

MTP joint extension is a passive event that is passively and actively restrained, dependent on active muscle contraction of the long and short flexors and the integrity of the connective tissue anatomy. Together, these factors set appropriate levels of stiffness at the MTP joints. In this way, MTP joint extension should not be too easy or too hard, as each situation risks pathology. Despite similar mechanisms of function between the hallux and the lesser MTP joints, degenerative joint disease is far less common in the lesser MTP joints than the 1st. In the case of the more lateral MTP joints, this may reflect the lesser loading time and lower stresses that usually occur within the joints. The 2nd MTP joint is more frequently affected by OA than the other lesser MTP joints. Its metatarsal has a very stable tarsometatarsal joint. This may increase its susceptibility to MTP joint OA and metatarsal fatigue fracture, as it is less able to move and use soft tissue to absorb loading torques and energies. However, this stability may be somewhat protective to its MTP joint extension motion compared to that of the 1st MTP joint, which like the 1st, also undergoes high weightbearing forces. Activity of digital flexors undoubtedly plays a significant role in MTP joint protection (Ferris et al., 1995). This could be a potential problem in forefoot and digital mechanics under ageing, because there is a known significant loss of strength in these muscles with age (Endo et al., 2002; Menz et al., 2006).

Toe contact forces have been investigated in the presence and absence of ankle, midfoot, and digital flexor activity. When triceps surae, peroneus longus and brevis, tibialis posterior, flexor digitorum longus, and flexor hallucis longus are all active at heel lift, each toe generates contact pressures with the support surface. However, this is dominated by the medial digits (Ferris et al., 1995). When different muscles are deactivated, changes occur not only in the digits but also in forefoot loading areas, suggesting that changes in compliance within the forefoot occur with loss of extrinsic muscle activity (Ferris et al., 1995).

Studies by Ferris et al. (1995) reported that with the loss of flexor hallucis longus, the contact areas on the hallux and the 2nd digit reduces. With concurrent loss of the flexor digitorum longus, pressures under the lesser toes and contact areas under all toes dramatically reduce. Further muscle inactivity, such as the concurrent loss of tibialis posterior, reduces the digital contact area further, with quite similar results being observed with loss of peroneal activity (Ferris et al., 1995). Additional loss of triceps surae activity causes little further change from the loss of tibialis posterior and the peroneals. These results indicate the particular importance of the long digital flexors in resisting metatarsal shaft strains compared to other extrinsic muscles (Ferris et al., 1995). However, the plantar intrinsics were not investigated by Ferris et al. (1995), yet they are known to have significant effects in forefoot stiffening across the MTP joints (Farris et al., 2019, 2020).

Other anatomical features also explain MTP joint pathomechanics. The 1st and 5th metatarsals are relatively unstable at the tarsometatarsal joints, lacking the interosseous ligaments of the other tarsometatarsal joints that restrict transverse plane translations. The 1st metatarsal has significant dorsal and plantar translations at its tarsometatarsal joint. There is also distinct sagittal plane motion at the 1st cuneonavicular joint that produces gliding plantarflexion coupled to hallux extension, when 1st MTP joint extension is over 20° (Phillips et al., 1996). This greater freedom of motion within the articulations of the 1st ray may make the 1st metatarsal more reliant on muscular stability (force closure) across its tarsometatarsal joint than other metatarsals. It also requires the sesamoid complex to be stable to prevent 1st MTP joint dysfunction. Like the 2nd MTP joint, it is part of the medial 'fulcrum of choice' throughout terminal stance, and as a result, it is subjected to the large GRFs generated by ankle plantarflexion power that stabilise the action of weight transfer from one foot to the other. Thus, its stability is important in assisting CoM acceleration, so that excessive compliance here creates particular problems.

MTP joint stiffness energetics

Loss of MTP joint extension may relate to the changes in power generation ability through altered force–velocity effects. Plantarflexor metabolism reflects altered power requirements, which will increase if the mechanical advantage of the longer fulcrum at the medial MTP joints is lost (Takahashi et al., 2016; Holowka et al., 2017). The disadvantage of not having free MTP joint motion to dissipate some of the plantarflexor energy generated at heel lift, is an increase in ankle plantarflexors' metabolic demand. Also, undissipated acceleration energies applied to MTP joints that are restricted in their motion, may eventually overload the articular structures.

Research into shoe stiffness across the level of the MTP joints may help reveal the fine balance that exists between MTP joint functional extension and lower limb biomechanics. It is proposed that a critical stiffness exists that can be defined as the *MTP joint torque*, that gives a threshold to the elastic benefit to the dorsiflexion (extension) moment at the MTP joints (Oh and Park, 2017). The consideration of elasticity (stiffness) across the MTP joints in footwear may also help give the clinician some indication as to the significance of changes in MTP joint mobility in an individual, and thus the most suitable treatment options for that individual. Appropriate levels of shoe stiffness in sports shoes, linked to foot bending stiffness, can improve performance and affect injury rates, with different requirements for different sports and a preferred stiffness for best performance (Wannop and Stefanyshyn, 2016; Oh and Park 2017). Oh and Park (2017) altered running shoe stiffness by inserting differing stiffnesses of insole into shoes, changing elasticity across the MTP joint region. They reported that stiffening within shoes that exceeded critical stiffness caused a reduction in angular impulses of the muscle–tendon at the MTP joint, and slightly increased ankle, knee, and hip joint angular moments. However, when compared to no insole, low levels of insole stiffness (using semiflexible insoles) assisted running propulsion and reduced MTP joint muscle–tendon effort and metabolic costs until a critical level of stiffness that restricted natural 1st MTP joint motion was reached (Oh and Park, 2017). With restriction of natural MTP joint motion, metabolic costs slightly increased. No significant changes were found in take-off velocity of the CoM, although horizontal forces were significantly reduced. This suggested that restriction of the natural MTP joint motion acts as a determinant of running energetics (Oh and Park, 2017) (Fig. 4.5.8a). This relationship is probably true of walking, but it is likely that less stiffness is required for walking energetics than for running.

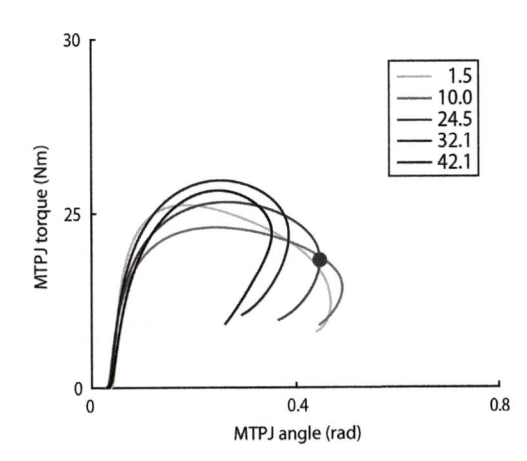

FIG. 4.5.8a Trajectories of the joint torques and angles at the MTP joints demonstrated in one representative subject from the study by Oh and Park (2017) with various insole stiffness levels. Dark lines indicate trials with insoles with stiffness levels from 1.5 to 42.1 Nm/rad (torque per unit of deflection). The trial corresponding to the insole stiffness closest to the subject's critical MTP joint stiffness is marked as a dark spot. Insoles offering semistiffness at the MTP joints seem to offer the most benefit for running. *(Image from Oh, K., Park, S., 2017. The bending stiffness of shoes is beneficial to running energetics if it does not disturb the natural MTP joint flexion. J. Biomech. 53, 127–135.)*

Thus, there are 'gearing effects' on ankle plantarflexion and MTP joint extension that links kinematics to the kinetic effects of ankle force–velocity power and MTP joint energy dissipation to the metabolic consequences of the ankle plantarflexors. The key to maintaining a 'Goldilocks zone' of terminal stance heel lift leverage is not so much the amount of dorsiflexion that occurs at the MTP joints, but the level of stiffness/resistance offered to the lever arm at the fulcrum. This will involve not only the direct joint structures of the MTP joint, but also the connective tissues forming the flexor or sesamoid complex (extensor apparatus), the plantar aponeurosis, and muscle strength and activity around the MTP joints, as well as any external influences of footwear (Fig. 4.5.8b).

Too much compliance will risk joint hyperextension and a reduction in gait acceleration energetics. Weak or damaged MTP joint flexor restraints or high metatarsal declination angles risk this outcome. Too much stiffness (elasticity) and the energetic benefits of MTP joints are lost due to increased requirements for ankle plantarflexor muscle metabolism in overcoming the MTP joint resistance, or via increasing the fulcrum point lengthening at the IP joints or beyond. Although the increased elasticity in footwear enhances forward propulsion, at excessive stiffness levels, it increases the ankle joint moment arm and decreases the forefoot GRF achieved (Willwacher et al., 2013, 2014).

Excess stiffness within the MTP joints may produce similar effects to excessive shoes stiffness under these joint, resulting from overly stiffened plantar soft tissues around the MTP joints or degenerative joint changes, reducing joint freedom and range of motion. Tight soft tissues can include overactivity of digital flexors or tightness of connective tissues such as the plantar aponeurosis. Rehabilitation may resolve these problems. OA or other osseous limitations within the MTP joints or sesamoid articulation can cause excessive MTP joint stiffness. This can have the added complication of pain which can then also cause avoidance of the affected joint fulcrum, creating new kinematic pathways that may cause their own biomechanical and metabolic consequences.

Too much compliance across the MTP joints will also have consequences to the force–velocity relationship of ankle power generation at acceleration. Thus, soft tissue weakness bridging the MTP joints represents dysfunction that reduces

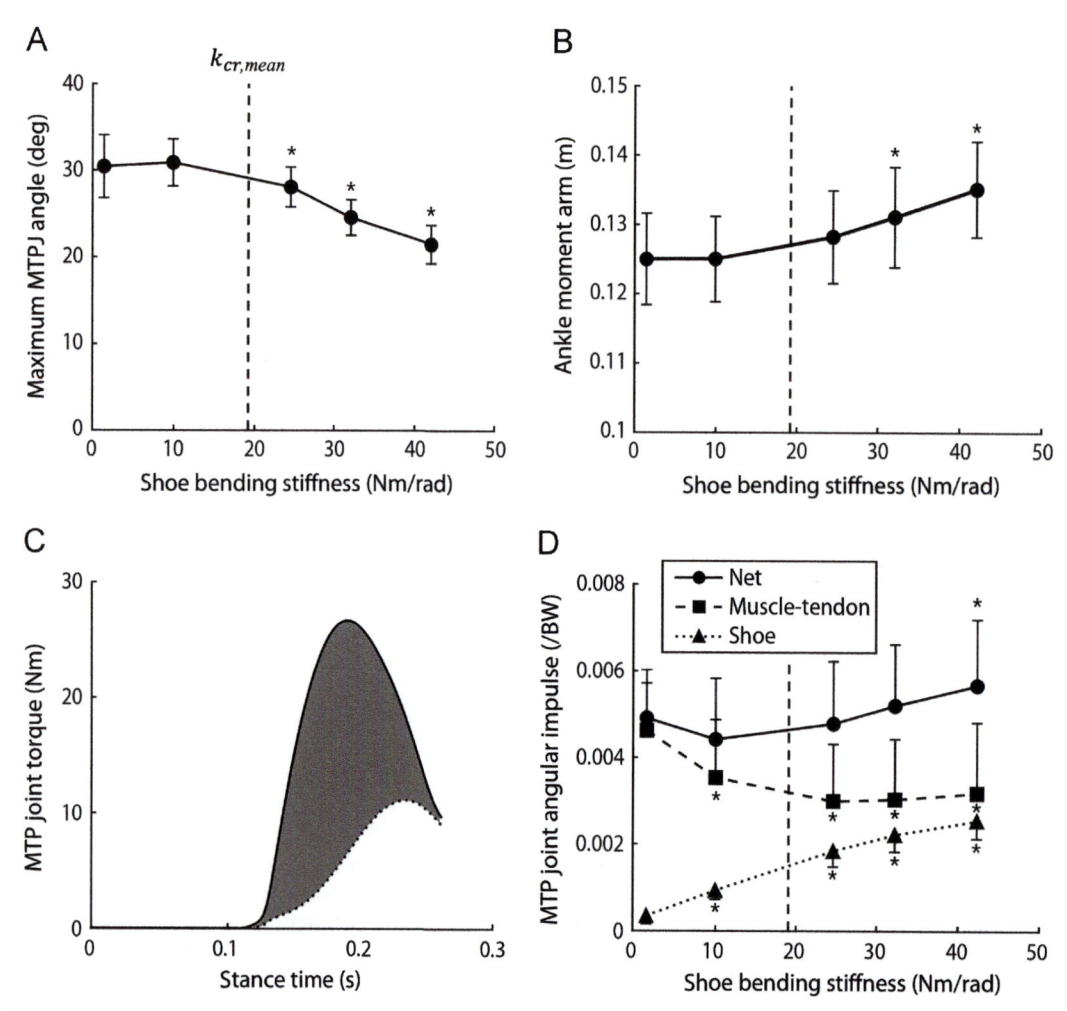

FIG. 4.5.8b Data from the research of Oh and Park (2017) demonstrates the effect of added elasticity from insole stiffness on MTP joint kinetics. Graph (A) demonstrates maximum MTP joint flexion angle as a function of the insole stiffness and (B) the corresponding moment arm of the GRF from the ankle joint. The vertical grey dashed line indicates the mean value of critical stiffness across multiple subjects tested. In (C), the net MTP joint torque (solid line) and the elastic-restoring torque exerted by the elastic insole (dotted line) is shown. The difference between the net MTP joint torque and the torque caused by the elastic insole reflects the contribution of the joint torque of the MTP muscle–tendon (MT) as shown in the shaded area. Graph (D) demonstrates the magnitude of the angular impulse of the net MTP joint torque (black circles with solid line), the angular impulse of the elastic insole (black triangles with a dotted line), and the angular impulse of the MTP joint muscle–tendon (grey squares with a dashed line). The error bar indicates the standard deviation. *(Image from Oh, K., Park, S., 2017. The bending stiffness of shoes is beneficial to running energetics if it does not disturb the natural MTP joint flexion. J. Biomech. 53, 127–135.)*

energetics and risks pathology. Unrestrained MTP joint dorsiflexion can result in end-range articular impingement, excessive soft tissue extension that can cause tears within joint capsules, or detachment of the weak proximal attachment of the digital restraining plantar plate (Gregg et al., 2007; Maas et al., 2016). For every foot, there will be a preferred level of stiffness across the MTP joints that allows controlled, yet free MTP joint extension to associate with ankle plantarflexion power that results in an energetically effective force–velocity relationship to the ankle power for acceleration over a stable forefoot. Footwear selection will thus influence acceleration biomechanics through altered forefoot stiffness levels (Fig. 4.5.8c).

These principles of changing compliance and stiffness around the MTP joint fulcrums must be considered when options such as changing shoe stiffness, using foot orthoses, or 1st MTP joint surgical arthrodesis are considered as treatment options. The clinician must seek the establishment of an 'energetic-metabolic Goldilocks zone', or a situation that is closest to it, without overloading any tissue's threshold. The MTP joint stiffness is likely to change joint impulses throughout the lower limb.

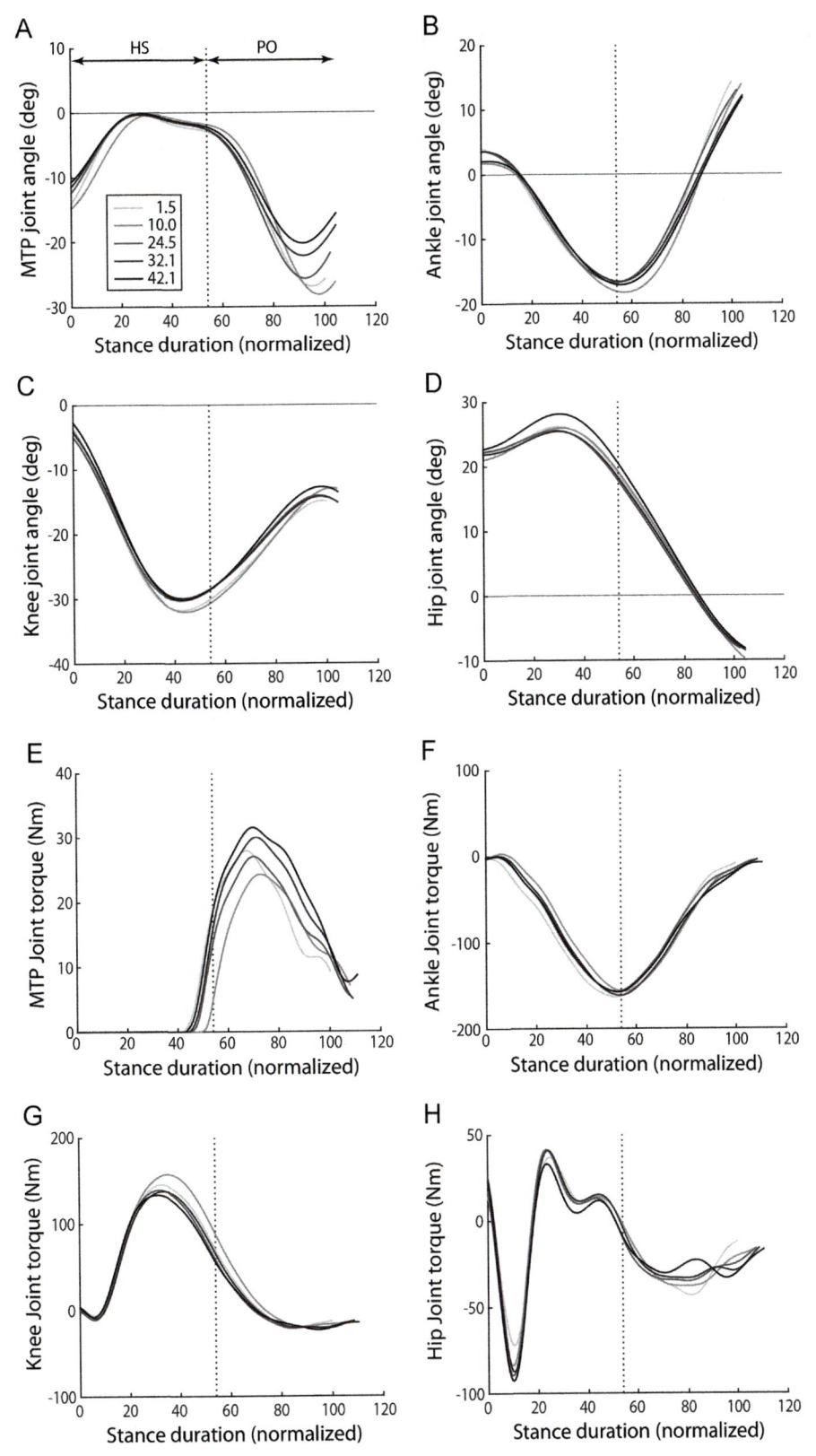

FIG. 4.5.8c Joint angle (A–D) and torque (E–H) trajectories of a representative subject from the study by Oh and Park (2017). (A) MTP joint; (B) ankle; (C) knee; and (D) hip joints. Dark lines indicate trials using insoles with stiffness ranging from 1.5 to 42.1 NM/rad. The vertical dotted line marks the initiation of the push-off (acceleration) phase defined by the anterior–posterior GRF. Torques are: (E) MTP joint; (F) ankle; (G) knee; and (H) hip joint. *(Image from Oh, K., Park, S., 2017. The bending stiffness of shoes is beneficial to running energetics if it does not disturb the natural MTP joint flexion. J. Biomech. 53, 127–135.)*

4.5.9 Soft tissue pathomechanics in the MTP joint fulcrum stability

An MTP joint's level of flexibility and stability to create appropriate stiffness for the force–velocity relationship is dependent on a complex soft tissue mechanism. Its action, as discussed previously in Chapter 3, is much like that of the hole known as a 'box' that is used in pole-vaulting to stabilise the pole that permits the athlete to vault over the pole and (hopefully) the bar. MTP joint stability is brought about through the linkage of the passive soft tissues of the forefoot through the extensor apparatuses (sling and wing expansions), deep transverse metatarsal ligament, the plantar plates, and the plantar aponeurosis. This should be considered a flexor plate mechanism that permits safe flexor power that restrains yet permits, adequate digital extension. Passive stability is assisted by the active stability components from the plantar intrinsic (Farris et al., 2019) and long digital flexor extrinsic muscles (Ferris et al., 1995). This complex soft tissue arrangement has been referred to by this author, for lecturing purposes, as the *flexor plate mechanism* for clarity from the anatomical fibrocartilaginous plantar plate alone.

The soft tissues of the MTP joint can control the rate and restrain the extent of the passive digital extension occurring through the MTP joint fulcrums. Any structure failing within the complex can result in instability that can lead to pathology and symptoms within the local anatomy, commonly referred to as *metatarsalgia*. The action of the plantar intrinsic muscles during late midstance and terminal stance, in concert with the long digital plantarflexor extrinsics, is to provide an active stabilising mechanism that engages with the passive connective tissue biomechanics to stabilise MTP joint flexibility. This mechanism is distinct to each MTP joint, as each joint has subtly different anatomy (the 3rd and 4th are the most similar), being quite distinctly different at the 1st MTP joint compared to lesser MTP joints due to the increased anatomical complexity of the sesamoid complex. The 5th MTP joint also has greater anatomical complexity through the abductor digiti minimi and flexor digiti minimi brevis muscles, but like the 4th, it is not usually loaded with high acceleration stresses. Therefore, initially, the lesser MTP joints are discussed with focus on the 2nd and 3rd MTP joint, although the fundamental principles apply to all.

Lesser MTP joint stability enables the flexor plate mechanisms to distribute forces appropriately around the MTP joints to achieve passive digital extension as the ankle and midfoot plantarflex proximally in a coupled relationship. Thus, the amount of active plantarflexion at the ankle and midfoot equate to the degrees of passive extension required at the MTP joints. Without MTP joint extension, ankle and midfoot plantarflexion will be altered and/or restricted. Many plantar intrinsic muscles have direct complex connections with the extensor apparatus, which in turn links to the deep transverse metatarsal ligament and the plantar aponeurosis. In late midstance and terminal stance, they become an interdependent functional unit (Fig. 4.5.9a).

The plantar plates, extensor apparatus, and the deep transverse metatarsal ligaments form a tie-bar mediolaterally, preventing excessive intermetatarsal splay that also provides digit stability (Stainsby, 1997; Gregg et al., 2006, 2007; Dalmau-Pastor et al., 2014; Maas et al., 2016). This anatomy is an important structural part of forefoot (and whole foot) stiffening in terminal stance (Venkadesan et al., 2017, 2020; Yawar et al., 2017). Any frontal plane torque on the metatarsal head will be resisted by either the medial or lateral sides of the deep transverse ligament tightening around the plantar plate, with eversion and inversion moments, respectively. This frontal plane torque can act as a complementary stiffening mechanism. It is possible that lack of a ligament stabiliser medially at the 1st MTP joint and a lateral stabiliser at the 5th MTP joint requires the extra muscular 'abductor' muscle forces found at these joints to provide stability (Fig. 4.5.9b). It may also explain why both of these joints are prone to frontal and transverse plane deformities, i.e. hallux abducto valgus (hallux valgus) and tailor's bunions (bunionettes), respectively. The plantar plate is largely replaced by sesamoids at the 1st MTP joint, with sesamoid articulations and a sesamoid complex that can develop a lateralised force vector dominance if abductor hallucis dysfunction occurs or the 1st metatarsal splays too far medially on loading. Indeed, it is worth noting that the tensile properties of the deep transverse metatarsal ligament between the 1st and 2nd metatarsals are inferior in the presence of hallux valgus deformity compared to normal forefeet (Abdalbary et al., 2016).

In the sagittal plane, the digital plantarflexors and the fibrocartilaginous plantar plate as part of the extensor apparatus, provide the primary plantarflexion resistance mechanism to digital extension. This mechanism's stability is dependent on the deep transverse metatarsal ligament's stiffness and stability that binds each extensor apparatus to each other. Failure in either plantarflexor muscles or plantar plate stability will tend to result in increased digital extension freedom. The proximally directed compression vector from these structures on the proximal phalanx will be reduced or lost, potentially enabling the metatarsal head to translate anteriorly on loading. The plantar plate has an intrinsic anatomical weakness at its proximal attachment where it can relatively easily rupture or avulse, causing distal migration of the plantar plate with digital extension (Maas et al., 2016). Part of the reason for this failure lies in the MTP joints' evolutionary past of being articulations of flexion for arboreal grip rather than for passive extension in bipedal gait. Similar anatomy is found within the human hand's metacarpal and digital joints to that of the foot. Wearing shoes with raised heels may create

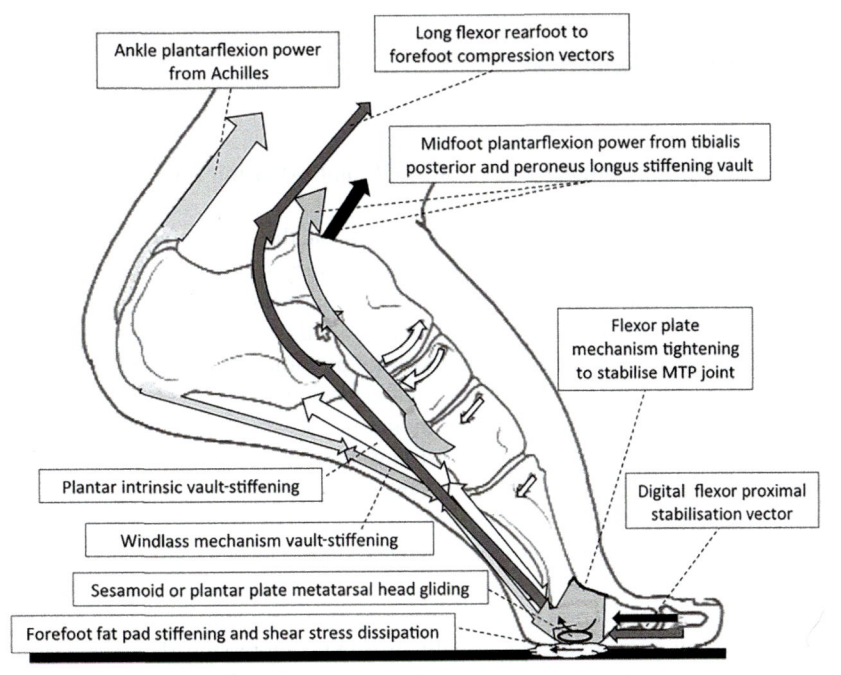

FIG. 4.5.9a The anatomy of the forefoot must work with ankle and midfoot plantarflexion to create MTP joint extension by forming a stable acceleration platform that can dissipate tensile strain and shear under high plantarflexion power acceleration. Here, key elements of terminal stance acceleration power and stability in the sagittal plane are demonstrated. The complexity of the appropriately timed interaction of soft tissues means much can go wrong. When healthy, foot tissues have a large capacity to spread loads throughout their structure via biotensegrity principles. Hence, plantar aponeurotic shortening and stiffening via the windlass mechanism, the release of power from extrinsic tendon elastic recoil, flexor plate stiffening, fat pad stiffening, and shear energy dissipation, as well as active intrinsic muscle contraction are all essential components of acceleration. However, over-stiffened and less elastic (more brittle) connective tissues resulting from end-products of glycation (either from natural ageing or diabetes) means that this capacity can be compromised. Sarcopenia or just general weakness as a result of poor developmental or lifestyle loading means that active stabilising force vectors may be underpowered and can cause early fatigue, leaving the foot too compliant to accelerate safely from. When strength and physiological health are poor, forefoot tissues may start to degenerate leading to chronic metatarsalgia. Then, a clinical hunt among the forefoot structures for the source(s) of the pathology and dysfunctions that are provoking pain is required to target therapeutic intervention. *(Permission www.healthystep.co.uk.)*

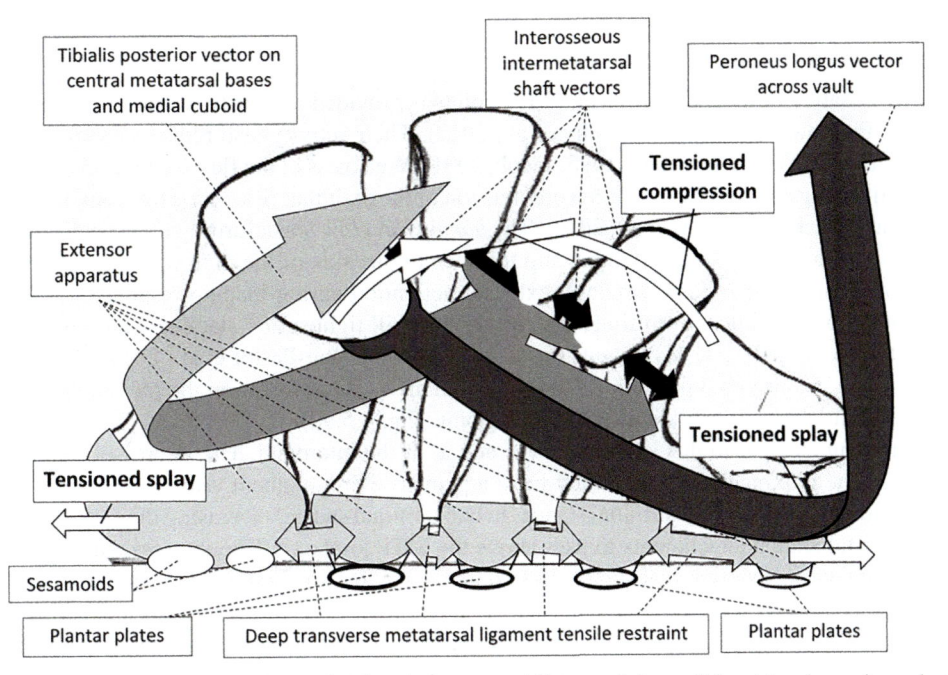

FIG. 4.5.9b In stabilising the forefoot to provide an acceleration platform, especially towards the medial metatarsals, passive and active tensional biotensegrity principles are at play. These principles create distal tension through passive forefoot splay and proximal tension-induced compression via active muscular contraction, such as that provided by interossei. The deep transverse metatarsal ligament is tensioned by metatarsal splay under increasing GRF on the forefoot, stiffening the transverse plane of the vault. This passive tensioning also binds the extensor apparatuses (including the plantar plate or sesamoids) to the MTP joints through their connective tissue linkages. Proximally, the forces released from tibialis posterior and peroneus longus tendon recoil pull the metatarsal bases together. Concurrent plantar intrinsic muscular activity, including that from the interossei, improves compression of the tarsometatarsal joints via increasing tension under the vault. This complex mechanism is able to spread stresses across these adaptable structures that are working in all three orthogonal planes. Without active muscular assistance, the deep transverse metatarsal ligament will be subjected to increased tensions and shear, which can not only disrupt forefoot stability but also start to initiate higher rotational torques at the MTP joints outside of (almost) 'pure' sagittal plane extension. Over time, connective tissue adaptations under Davis' law may leave the forefoot increasingly more compliant as a result. Re-establishing forefoot biotensegrity requires muscular strengthening to prevent tissue stress breaches or to protect the forefoot from pathology that has already developed. *(Permission www.healthystep.co.uk.)*

environmental mismatch by demanding higher ranges of MTP joint extension for longer in gait due to increases in the functional metatarsal declination angle.

Normally, as the plantar plate moves distally with digital extension, the plantar fat pad tensions via the plantar aponeurosis slips that encloses and forms it. This glides the plantar plate forward, deep to the now firmer plantar fat pad, to protect the articular surfaces of the metatarsal heads that are increasingly being exposed to weightbearing forces during extension (Bojsen-Møller and Lamoreux, 1979; Maas et al., 2016). If there is a failure of the plantar plate's proximal attachment, then increased extension freedom is likely to develop (having lost the proximal plantar plates attachment vector of restraint), with reduction in extension-resistance stiffness. This increased extension freedom creates vulnerability to change in any of the long and short toe flexors' functional lever arms. The digital flexors should exert their efforts at the base of the intermediate and terminal phalanges, and if connective tissue restraint is lessened, their activity needs to increase. Although they normally create a proximally directed stabilisation force via the proximal phalanx into the metatarsal head, they exert their effort directly through the more distal phalanges. Only the far weaker interossei, lumbricals, and plantar plate attach directly to the proximal phalanx, thus making sure the proximal phalanx aligns correctly within the MTP joint for the muscular-induced proximal force to be driven towards the metatarsal head via the distal phalanges. Should the proximal phalanx start to extend too early or too freely, then activation of the long and short digital flexors can cause interphalangeal (IP) flexion, rather than an MTP joint extension restraint and proximally directed compression. The result of this IP joint flexion is to create a 'buckling' effect that causes rather than resists, MTP joint extension. The proximal detached plantar fat pad is then often drawn anteriorly from under the metatarsal heads, taking the plantar fat pad with it via the plantar aponeurosis. Thus, the MTP joint loses the shear and compression stress dissipation protection afforded by the plantar fat pads on forefoot loading and the elastic restraint on MTP joint extension during terminal stance at the same time (Fig. 4.5.9c).

MTP joint capsule rupture is reportedly more common than plantar plate rupture (Kier and Abrahamian, 2010), but either injury will disturb the proximal flexor vector that restrains MTP joint extension. The resulting direction of the stabilising vector lost depends on which side of the joint capsule fails. If the medial capsule ruptures, then frontal and transverse torque may become unbalanced, leading to valgus rotation and abduction vectors at the MTP joint. If the lateral side fails, then varus rotation and adduction vectors may be unresisted. These can add frontal and transverse plane deformity to dorsal displacements of the digits at the MTP joints.

MTP joint muscular imbalance and lesser toe deformity

Dynamic active stabilisation of the metatarsals and MTP joints is provided primarily by the extrinsic and intrinsic flexors (Ferris et al., 1995; Fiolkowski et al., 2003; Kelly et al., 2012). Their activity also reduces bending moments across the metatarsals (Ferris et al., 1995; Sharkey et al., 1995; Jacob, 2001). Weakness of the flexors themselves are likely to instigate problems, as their stiffening restraints prevents hyperextension of the digit that risks stretching out, tearing, or rupturing the proximal plantar plate attachment (Maas et al., 2016). The *hammer* or *claw toe* deformity that can develop as a result of lost MTP joint stability can cause the extensor tendon slip to move to one side of the digit's phalanges, giving rise to further clawing dysfunction during the reversed windlass' extensor activation during loading response, rather than normal pure digital extension at the MTP joints. Given time, the soft tissues will tighten any slackness the digital deformity creates, fixing the deformity across the IP and MTP joints (Stainsby, 1997; Dalmau-Pastor et al., 2014). The increased MTP joint extension moment from the proximal phalanx can cause a plantarflexion moment on the metatarsal head, creating higher peak pressures and shear strains under the affected MTP joint (Fig. 4.5.9d).

Another confounding element to this situation is the action of the interossei. The interossei through their attachment points to the sides of the proximal phalanx usually have a plantarflexion moment arm around and below the MTP joint axis during the late midstance into early terminal stance. In late terminal stance, increasing digital extension at the proximal phalanx causes the distal tendon attachments to pass above the MTP joint axis. The interossei can then become weak but active digital extensors during late terminal stance. In so doing, toe extension is stabilised ready for toe-off. Therefore, the timing of the change from flexor to extensor and the size of the moment arms of the interossei are critical to how the interossei stabilise the MTP joint. They initially resist extension, then change to aid extension. If mistimed, they can cause increased toe extension for longer, helping to destabilise the MTP joint plantarly.

The functional MTP joint axis is reported to move more anteriorly and superiorly with faster walking speed, giving some extra time for the interossei to maintain a plantarflexion moment, resisting the greater passive extension torques related to gait speed (Raychoudhury et al., 2014). However, the change of one to the other is partly dependent on the metatarsal's functional declination angle at the start of acceleration. The steeper the angle, the earlier the interossei are likely to become extensors because the position of the functional joint axis is shifted relatively inferiorly to the base of the proximal phalanx with increasing metatarsal declination. Thus, high anatomical declination angles associated with cavoid feet or

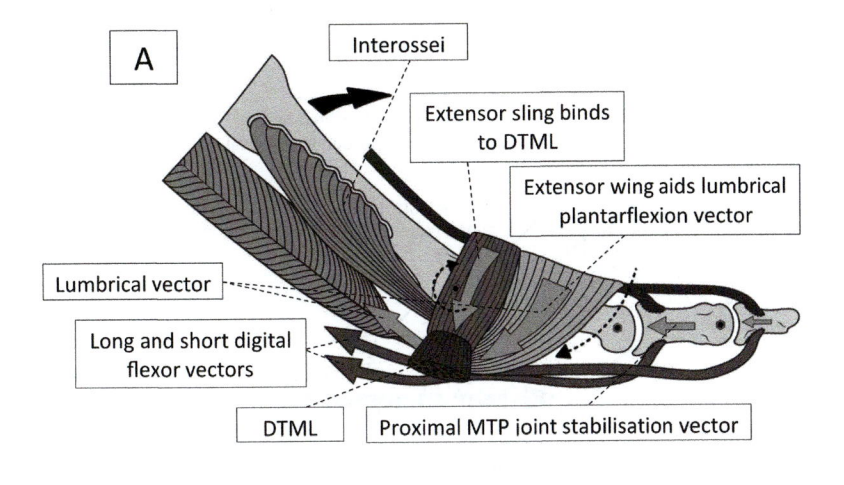

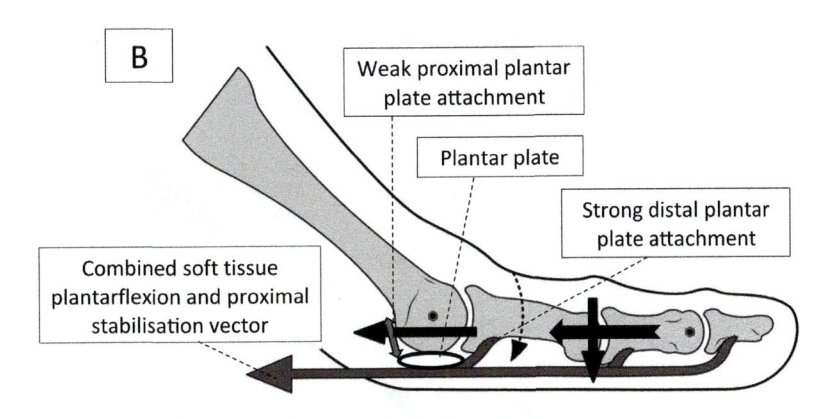

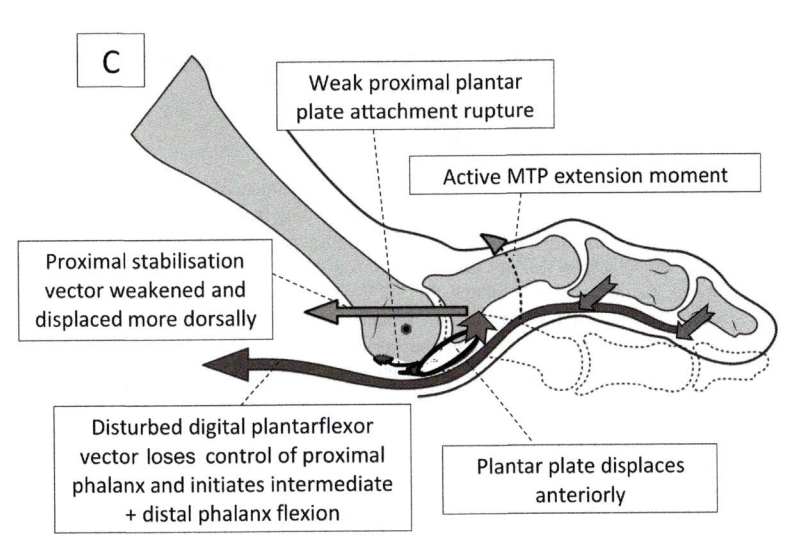

FIG. 4.5.9c Flexor plate mechanisms are formed by a complex of structures that tie into the transverse stabilising anatomy of each section of the deep transverse metatarsal ligament across the forefoot. For forefoot sagittal plane fulcrum stability, a critical balancing of forces is required. This involves the long and short digital flexors and both the lumbricals and interossei with their connections to the extensor sling and wing of the extensor apparatus (A). With the extensor apparatus stripped away, the effect of the digital flexor vector on plantar and proximal stabilisation and resistance to extension can be demonstrated (B). An important element to this mechanism is the ability for the metatarsal head to rotate over the fibrocartilaginous plantar plate (or sesamoid bones at the 1st MTP joint), creating the extension moment that sets the length of the Achilles external moment arm while dissipating some of the acceleration energies to reduce metatarsal and Achilles stresses. If the plantar plate becomes detached proximally (C), then forces that position the proximal phalanx to the metatarsal head are compromised. The plantar plate will now be drawn anterior away from the metatarsal head during MTP joint extension. Vectors from the lumbrical and interossei may be insufficient to maintain a phalangeal posture aligned distally to the metatarsal head. The long and short digital flexors may now produce IP joint flexion rather than a proximally directed MTP joint stabilisation vector. This, combined with increasingly dorsally displaced interosseous vectors, can result in MTP joint extension at the proximal phalanx, as well as via metatarsal head rotation during acceleration. Similar effects are seen with MTP joint capsular tears, but the destabilising results of this depend on the location of such connective tissue failures. *(Permission www.healthystep.co.uk.)*

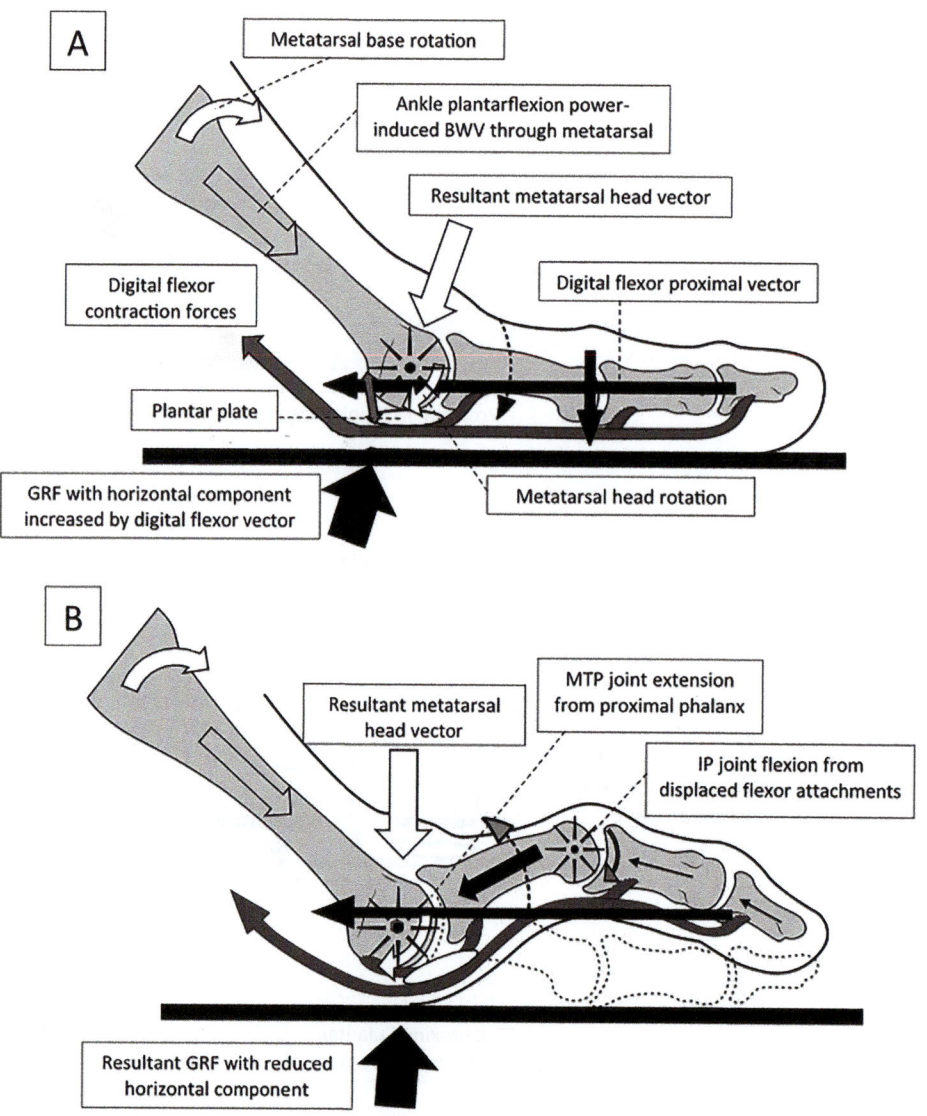

FIG. 4.5.9d The digital flexor proximal vector lying below the metatarsal head's instantaneous axis (star) provides stabilisation on the MTP joint by pulling the proximal phalanx into the metatarsal head (A). This is important for avoiding forefoot injury under the high forces of terminal stance acceleration generated by ankle plantarflexor power. A strong digital flexor contraction-generated proximal vector uses the MTP joint's connective tissues, bound to the deep transverse metatarsal ligament, to pull the digit into the support surface. This creates a stable base as the metatarsal head starts its rotation and proximal glide over the plantar plate. As the metatarsal rotates, the BWV generated from the ankle plantarflexor power directed through the foot becomes increasingly vertical, pressing the metatarsal head towards the ground. This resultant metatarsal head vertical vector is moderated by the horizontal digital flexor proximal force, so that the GRF generates a significant horizontal component under this and via the body's CoM being driven forward, reducing compression on plantar soft tissues. Sagittal plane instability of the proximal phalanx can cause the digital flexor proximal vector to displace above the instantaneous axis (B). This leads to a larger metatarsal head plantarflexion moment that in turn increases the rate of MTP joint extension. The change in the digital flexor proximal vector position tends to induce not only MTP joint extension but also IP joint flexion, reducing the horizontal component force and increasing the compression forces between the metatarsal head and the support surface. This potentially 'crushes' and increases shear strain on the plantar plate, plantar fat pad, and cutaneous tissues. The formation of corns, calluses, haematomas, and ulcers under the metatarsal heads is common, with the extent of damage reflecting the physiological health of the tissues affected. Failure of connective tissue extension restraints (plantar plate, joint capsule, flexor tendons, etc.) or loss of strength in the extrinsic and/or intrinsic muscles can all cause failure of proximal phalanx stabilisation and reduction in extension-resisting plantarflexor power at the MTP joint that provides a safety valve on energy dissipation. *(Permission www.healthystep.co.uk.)*

functional declination caused by high-heeled footwear are likely to cause earlier interossei extension moments, with concurrent loss of their extension resistance moments.

The dorsal interossei, lying within the 1st to 3rd metatarsal interspaces, have a more dorsally positioned moment arm than do the plantar interossei, because their proximal attachments are more superiorly positioned on the metatarsal shafts and their tendons approach the MTP joint more superiorly, than do the plantar interossei tendons. The plantar interossei have more inferior proximal attachments on the 3rd to 5th metatarsal shafts, so not only are the adduction moments of the dorsal interossei countered by the plantar interossei abduction moments, but the plantarflexion moments they produce remain plantarflexion moments momentarily longer during digital extension (Fig. 4.5.9e). The 2nd MTP joint is likely more vulnerable to longer interossei extension moments because it has two dorsal interossei. This situation may be compounded if the individual has a foot with an increased 2nd metatarsal declination angle, as this may create a larger active extension moment (Longworth et al., 2019).

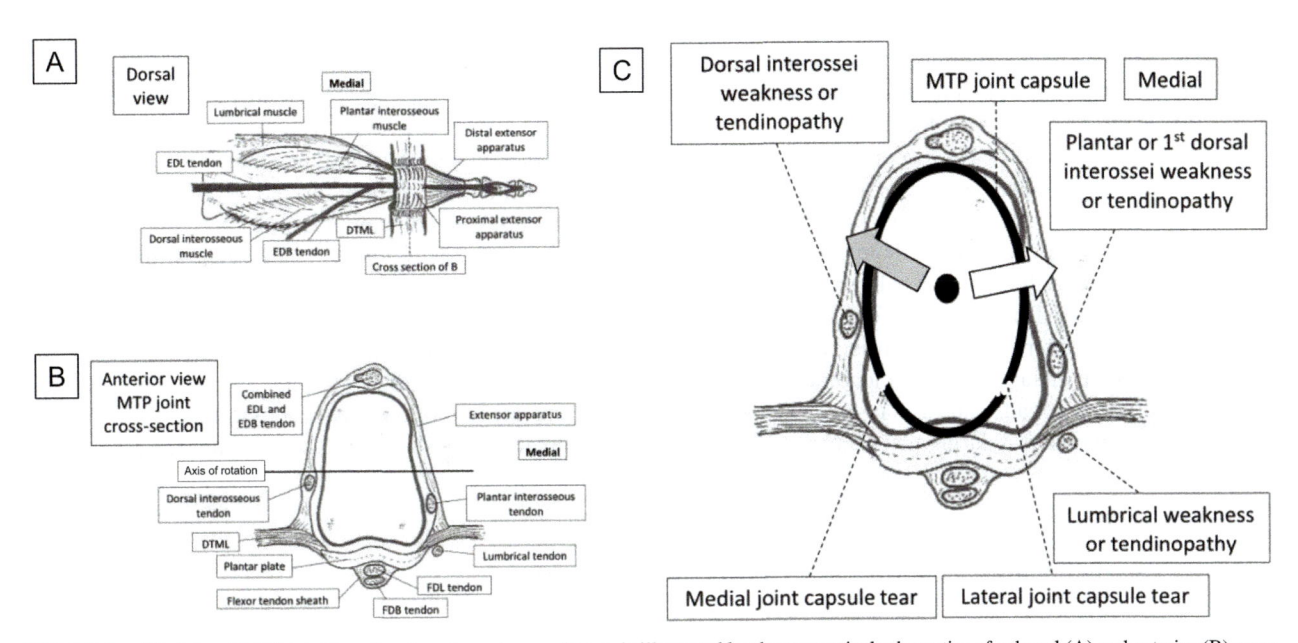

FIG. 4.5.9e Digital and MTP joint functional anatomy is complex, as is illustrated by the anatomical schematics of a dorsal (A) and anterior (B) cross-sectional view on the MTP joint. Although sagittal extension displacement of the proximal phalanx is by far the most common issue related to toe deformities, the digit is also exposed to frontal and transverse plane torques (C). When balanced, digits should only move significantly with the sagittal plane around a horizontal axis, but their concave-to-convex articular surfaces permit digital rotation around the long axis of the digit (black circle), as well as a vertical axis through the metatarsal head. The lumbricals provide plantar-medially directed vectors, so their failure tends to permit the toe to abduct, evert, and dorsiflex (light-grey arrow). Failure of the plantar interossei and the 1st dorsal interosseous muscle at the 2nd MTP joint causes a similar effect, although the dorsiflexion element is usually less pronounced, save at the 2nd MTP joint. Failure within the 2nd, 3rd, and 4th dorsal interossei results in adduction, inversion, and very slight dorsiflexion of the 2nd, 3rd, and 4th digits (white arrow). Joint capsular tears tend to occur nearer the plantar aspect because of higher extension-induced tensions. The joint capsule torn area (medially or laterally) dictates the direction the digit is likely to start to drift during its extension as a result of a reduced restraining vector on the side of the tear. *(Permission www.healthystep.co.uk.)*

Weakness in active flexor restraint is likely a common cause of excessive MTP joint extension moments through loss of stiffened restraint, thus failing to protect the passive tissues from excessive extension strains. In time, such imbalances can lead to a malalignment of the digit into hammer toes etc., with possibly frontal and transverse torques depending on the particular structures failing. The alignment of the interossei and the lateral attachments of the lumbrical-associated anatomy makes varus rotations of most lesser toes more likely. The 2nd digit with its two dorsal interossei is far more likely to create a hyperextended deformity, which can eventually lead to the proximal phalanx hyperextending the digit on the metatarsal head into subluxation and dislocation as the dorsal tissues retract without antagonistic tensional opposition from the flexors.

The four lumbrical muscles, through their attachments proximally from the tendon slips of the flexor digitorum longus, have their tendons running to the medial sides of the extensor wing under the deep transverse metatarsal ligament. They induce an adduction/varus torque as well as a significant plantarflexion moment on the proximal phalanx via the extensor wing of the extensor expansion. Imbalance of vectors around the MTP joint can give the lumbricals mechanical advantage, increasing adduction, varus, and plantarflexion moments on the toes. In lumbrical dysfunction, loss of abduction and extension resistance results in reduced adduction moments, increasing the risk of digital abduction as well as reduced extension restraint. Mechanical advantage or dysfunction of the lumbricals can occur on an individual basis, so that sometimes only one rather than all of the lesser MTP joints can be affected, but failure in quadratus plantae and the flexor digitorum longus strengths will also make these muscles dysfunctional.

Intermetatarsal pathomechanics

The control of MTP joint passive extension moments through the tendon–fascia flexor plate complex and transverse stability through the deep transverse metatarsal ligament–plantar plate interface, are important. They reduce soft tissue tensions, compressions, and shear of structures such as the intermetatarsal bursae and the plantar digital nerves that pass under the deep transverse metatarsal ligament to supply the digits. These nerves can often have unions and divisions around and beneath the deep transverse metatarsal ligament which can become compressed, tensioned, and sheared if the metatarsal splay and passive extension is not restrained by the connective tissue and muscles. If the metatarsals can splay too freely, then the deep transverse metatarsal ligament can compress the structures passing beneath it when weightbearing, a situation exacerbated in high degrees of digital extension that can also carry the plantar fat pad anteriorly with the plantar plate, while also stiffen the plantar forefoot fat pad. This can be due to the use of higher heels on footwear or can result from MTP joint muscular dysfunction and digital deformity.

Increasing intermetatarsal stresses around the deep transverse metatarsal ligament can inflame bursae or cause recurrent myelin sheath trauma, that can result in fibrosis of the myelin sheath of the nerves leading to the development of *neurofibromas*, commonly called neuromas. Both nerves and bursae can be aggravated by compression, once enlarged. In such cases, tight shoes may reveal the pathology by provoking symptoms around the MTP joints of a dull ache with radiating or burning pain through the nerves locally. The underlying pathomechanics may lie in increased tension–compression and shear stress resulting from poor forefoot stability during terminal stance. The most common site for a neurofibroma is the 3rd metatarsal interspace where it is known as a *Morton's neuroma*. This area and also the 4th interspace are sites of increased intermetatarsal splay on loading (Lundgren et al., 2008; Wolf et al., 2008), but the 3rd and 4th metatarsals receive much higher loads and display greater difference in transverse plane motion between them than occur between the 4th and 5th metatarsal during gait (Hayafune et al., 1999). This likely increases the risk of intermetatarsal pathology to the 3rd interspace over the others.

4.5.10 Toe deformities as part of MTP joint pathomechanics

As previously discussed, passive and active stability of the MTP joints is necessary to stiffen MTP extension to establish a force–velocity relationship with the ankle plantarflexors that improves forefoot stability and enhances gait energetics. Thus, slower walking increases muscle stiffening capacity across the foot at the expense of connective tissue elasticity, while faster gait increases connective tissue elasticity but reduces the ability of muscles to generative contractile stiffness and power, requiring greater metabolic activity. This relationship helps define the preferred walking speed for each individual, making the MTP joint a pivotal part of acceleration biomechanics.

Failure in calf–ankle–foot–MTP joint mechanics can alter force vectors around the digits, changing joint moments via the digits that decrease fulcrum stability. If these abnormal vectors are sustained, then they can result in digital deformations such as retracted, clawed, or hammer toes. Each deformity involves some degree of overextension of the proximal phalanx that can in time, become fixed in this position nonweightbearing. Cavoid feet, due to their high metatarsal declination angles, are associated with retracted toes caused by prolonged proximal phalanx-extension angle at the MTP joints. At heel lift, MTP joint extension start with the digits at higher extension angles to the metatarsals and position flexors such as the dorsal interossei into positions where their moment arms are more likely to produce extension moments for longer during late terminal stance. Many toe deformities are functional, being visualised only on forefoot loading during gait or just during terminal stance. Long-term dysfunction at the MTP joints can cause deformity to become fixed throughout gait, and then to become noticeable when nonweightbearing. Toe deformity reflects the functional anatomy and vector imbalance at each MTP joint. Digital plantarflexor muscle dysfunction/weakness is usually the primary driver of deformity such as hammer toes, with extensor tendon tightening often followed by extensor tendon slippage off the dorsal surface of the phalanges as a secondary developmental feature that helps to fix the deformities. The extent of hyperextension retraction and interphalangeal joint flexion depends on the different vector failures and the presence or absence of tendon tightening.

Mallet toes, with a large distal phalanx flexor deformity, likely represent long flexor tightening, while other structures remain functional. Claw toes probably reflect long and short flexor tendon tightening, with the interossei and lumbricals remaining functional. Hammer toes likely indicate long and short flexor tightening, with loss of significant interossei plantarflexor assistance and/or joint capsule or plantar plate failure. Retracted toes could indicate loss of interossei plantarflexion moments and/or capsular or plantar plate restraint without flexor tendon tightening compensation, maintaining some significant flexor activity. However, it is important to point out that most of these deformities appear on a continuum of variations (Fig. 4.5.10a).

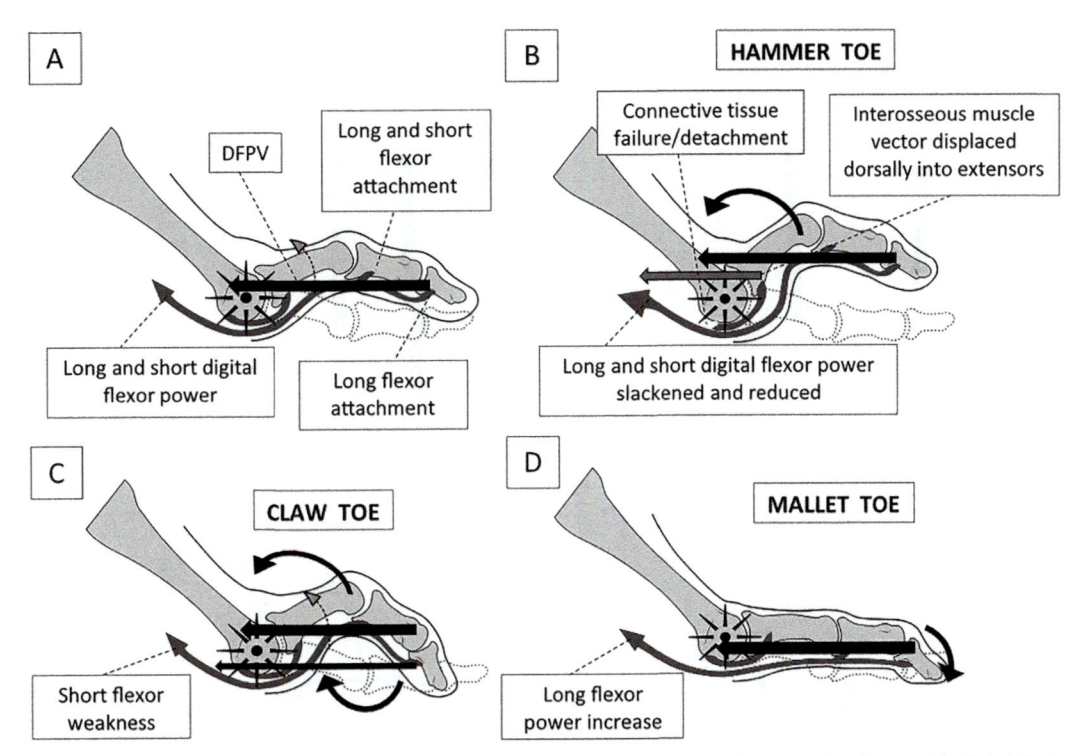

FIG. 4.5.10a The digital flexor proximal vector (DFPV—black straight arrow) is composed of all the muscles that provide digital flexion, such as the long and short flexors, lumbricals, interossei, and in the case of the hallux and 5th toe, the digital abductors and adductors. The DFPV is essential for MTP joint stabilisation and for providing controlled resistance to the rate and extent of MTP joint extension by positioning below the joint's instantaneous axis (star). This important vector of late midstance and acceleration can be disturbed by the presence of toe deformities. However, toe deformity is often the result of dorsal displacement of this vector due to connective tissue or muscle–tendon unit dysfunctions. Once the DFPV starts to displace dorsally (A), maintaining flexion at the proximal phalanx becomes difficult as the muscles and connective tissue attachments to the plantar proximal phalanx produce much smaller flexion forces than the long and short digital flexors that attach to the plantar bases of the intermediate and distal phalanx. Thus, the proximal phalanx can find itself under an increasing extension vector if this displaces above the axis, while the more distal attachments of the long and short digital flexors continue to apply flexion vectors to the interphalangeal joints. If the plantar connective tissues fail or the interosseous muscle attachments displace above the axis of rotation (B), then a hammer toe deformity can become progressive. In the worst-case scenario, the digit can sublux or dislocate dorsally from the MTP joint articulation due to unopposed extensor tendon tightening in response to slackening of flexor tendon tensions. When the long flexor is considerably stronger than the short flexor that only attaches to the intermediate phalanx, the digit can start to claw under higher distal flexion forces. Such altered forces pull the distal phalanx into the ground and buckle the toe by simultaneously flexing the intermediate and distal phalanx while extending the proximal phalanx, creating a claw toe deformity (C). This deformity can bring some of the DFPV down below the instantaneous axis, but the bulk of it still remains above the axis producing proximal phalanx extension. If the short flexor remains powerful and the connective tissues intact, proximal IP joint stability can be maintained via the long flexor attachments. However, if long flexor power increases in an attempt to increase the DFPV at an unstable MTP joint, the distal phalanx may flex at the distal IP joint alone (D), creating a mallet toe deformity. Combinations of these situations create hybrid digital deformities on a continuum. *(Permission www.healthystep.co.uk.)*

Joint instability is reported frequently at the 2nd MTP joint (Fortin and Myerson, 1995). These toes are particularly prone to MTP joint hyperextension deformities, in part due to their consistently high loading as a primary fulcrum point, but also likely because of the joint being supplied by two dorsal interossei that can become extensors earlier in gait. However, should vectors also be disturbed medially or laterally to the MTP joint, then frontal and transverse plane deformations can occur concurrently or independently of proximal phalanx hypertension. If the 1st dorsal interosseus muscle becomes dysfunctional or a medial capsular tear occurs, then the 2nd toe rotates into valgus to tilt its plantar surface towards the 3rd toe, while abducting towards the lateral toes. If the 2nd dorsal interosseus muscle or lateral capsule fails, then the 2nd toe will adduct towards the hallux and may rotate into varus, tilting its plantar surface towards the hallux. This can also occur with spasticity or tightening of the 1st dorsal interosseus, even with the lateral vectors of the 2nd toe remaining intact by overwhelming it. Lesser toe imbalances, particularly of the 2nd, are often initiated in the presence of HAV. The increased intermetatarsal angle and valgus rotation of the 1st metatarsal will disturb the proximal attachments of the 1st dorsal interosseus muscle, which usually bridges its attachments between the 1st and 2nd metatarsal shafts (Fig. 4.5.10b).

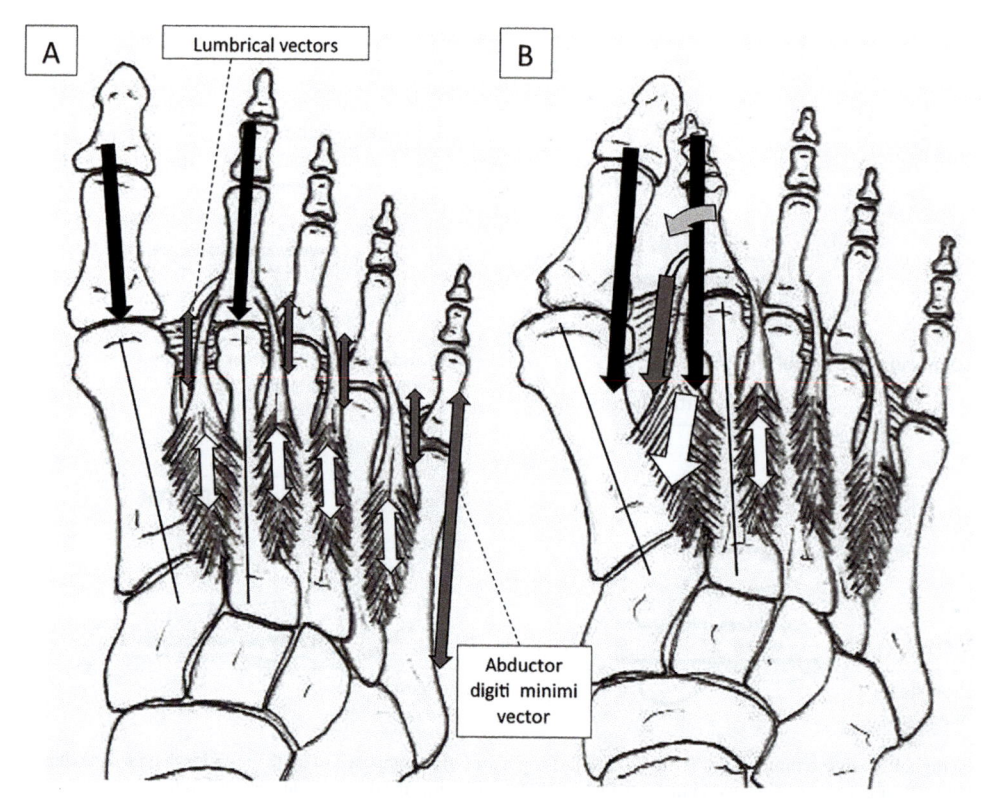

FIG. 4.5.10b Transverse and frontal plane digital stability provides a firm structure for the application of the digital flexor proximal stabilisation vectors (black arrows) on the MTP joints via proximal phalanx compression to the long axis (thin black line) of the metatarsal (A). Stability involves interosseous muscle vectors (white arrows) and those from the lumbricals (grey arrows). Only the dorsal interossei are shown, but at the 3rd, 4th, and 5th MTP joints, the plantar interossei (not shown) add medial vectors to those of the lumbricals while abductor digiti minimi adds a lateral vector at the 5th MTP joint. For more on the vector complex at the 1st MTP joint, see Figs 4.5.12a and 4.5.13c. Stability via this mechanism is vulnerable to dysfunction. Loading displacement of more mobile metatarsals, such as the 1st, 4th, and 5th, can upset this balance. In the presence of large 1st metatarsal displacement (B) via eversion rotation and medial adduction (as seen in hallux abducto valgus), the deep transverse metatarsal ligament slip between the 1st and 2nd metatarsals can become excessively tensioned/stretched, often displacing the sesamoids laterally. This metatarsal postural change moves the medial attachments of the 1st dorsal interosseous muscle medially, increasing this muscle's moment arm for 2nd digit proximal phalanx adduction. Coupled to the 1st lumbrical's adduction moment, this can easily overcome the 2nd dorsal interosseous muscle's abduction moment, causing the 2nd toe to adduct towards the hallux. The hallux moves into abduction as part of hallux valgus. Together, these digital displacements can cause the hallux to make contact with the 2nd digit. Hallux abduction on 1st metatarsal adduction causes the proximal flexor force to displace laterally to the 1st metatarsal's long axis while 2nd digit adduction causes the vector to displace medially to its metatarsal, worsening the situation for both MTP joints. In extreme cases, the 2nd toe may override the hallux, especially if lesser digital flexor strength is also poor. Although lateral vectors dominate at the 1st MTP joint, medial vectors tend to dominate at lesser MTP joints and thus lesser toe deformities of adduction are more common than abduction. *(Permission www.healthystep.co.uk.)*

The 3rd MTP joint is next most frequency affected by proximal phalanx hyperextension deformities such as hammer and retracted toes. The 4th and 5th toes, outside of deformities associated with high declination angles, are more susceptible to the development of adduction and varus deformity, because they experience more transverse plane splay through having greater mobility at their metatarsal base articulations than do the other metatarsals (Lundgren et al., 2008; Wolf et al., 2008). They do so especially in more mobile forefeet. The vectors of the lumbricals and interossei dominate towards varus, as both the plantar interossei and lumbricals have medial MTP joint distal attachments, resisted by only the dorsal interossei at the 3rd and 4th MTP joints and abductor digiti minimi at the 5th. Abductor digiti minimi can have its vectors converted to one of plantarflexion and flexor digiti minimi to one of adduction, should the 5th metatarsal splaying cause excessive abduction and valgus (eversion) rotation of the 5th metatarsal, allowing the tendons to slip plantarly and medially, respectively.

Stiffening of the forefoot via metatarsal spreading is an important component of whole foot stiffening for heel lift. The 1st, 4th, and 5th metatarsals have a greater capacity to translate in the transverse plane, as discussed in Chapter 3, Section 3.3.10. This freedom to splay occurs medially in the case of the 1st, and laterally in the case of the 4th and 5th metatarsals. In the lateral two lesser MTP joints, and to a lesser extent the 3rd, increased lateral splay will induce transverse arch stiffening, but in excess it will also increase the potential varus and adduction moments of the plantar interossei and lumbricals. In hypermobile feet, greater lateral metatarsal splay will be required to induce stiffness across the deep transverse metatarsal ligament, and thus increases the potential valgus rotation of the 4th and 5th metatarsal shafts, which in turn, increases the varus and adduction moments of the lumbricals/plantar interossei, risking adductovarus digital deformity.

Tailor's bunion

The 5th metatarsal is particularly vulnerable to metatarsal splay pathomechanics because of the ligamentous arrangement of its tarsometatarsal joint with the cuboid. The 5th metatarsal–cuboid joint lacks the interosseous ligament of the 4th and is reliant on just a dorsal and plantar ligament for passive stability. The medial side of the 5th MTP joint is linked to the deep transverse metatarsal ligament through the flexor plate complex, but there is no such restraint laterally. This is compensated for by active restraint from abductor digiti minimi and to a lesser extent, flexor digiti minimi brevis.

With increased intermetatarsal splay, the 4th and 5th metatarsals are able to spread further than the 3rd can from the 2nd metatarsal (Lundgren et al., 2008; Wolf et al., 2008). The lack of interosseous ligament stability at the 5th tarsometatarsal joint can allow further lateral displacement of the 5th metatarsal into greater abduction and valgus rotation. This can cause the abductor and flexor digiti minimi tendons, with their distal attachments to the 5th toe, to slip plantarly and may, through the varus rotation of the proximal phalanx and abduction of the metatarsal shaft, find themselves on the medial side of the 5th MTP joint. This changes their activity into 5th digit adductors and plantarflexors, rather than plantarflexors and abductors, causing a bowstringing effect that increases 5th metatarsal abduction splay and 5th digital varus and adduction rotation. This produces a 5th metatarsal and digit deformity known as a bunionette or more commonly, a 'tailor's bunion' that can lead to footwear accommodation problems as well as some lateral forefoot loading instability issues (Fig. 4.5.10c). This deformity can result in pain at the 5th metatarsal head as it becomes increasingly exposed to loading forces due to displacement from the plantar fat pad. Compression from a tight shoe will worsen the situation.

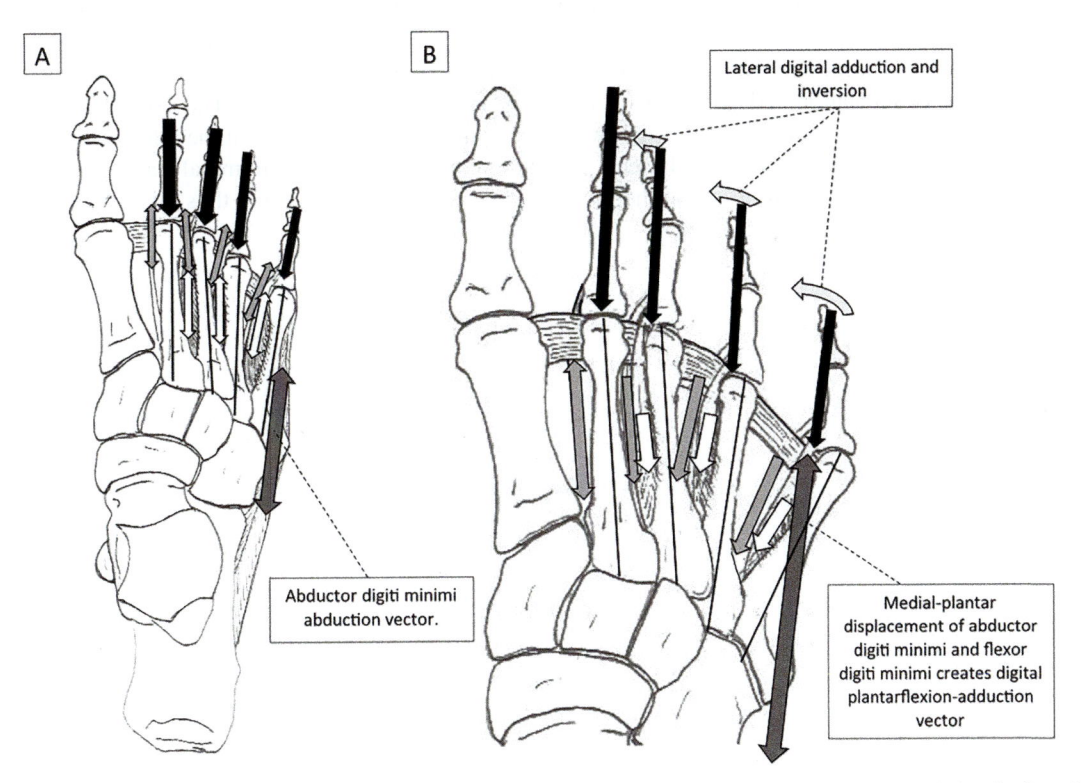

FIG. 4.5.10c Schematic of the vectors of the lateral MTP joints demonstrated with the dorsal interossei removed to aid visualisation of the plantar interosseous muscles (A). Small degrees of lateral metatarsal abduction on loading are important for forefoot splay energy dissipation that also stretch-stiffens the deep transverse metatarsal ligament. This should help stabilise the forefoot as late midstance progresses. Digital flexor proximal stabilisation vectors (black arrows) should guide and follow metatarsal motions so that the digit remains aligned to the long axis of each metatarsal (thin black line). Plantar interossei vectors (white arrows) and lumbrical vectors (grey arrows), aided by the flexor plate mechanism attachments, help make sure this relationship is maintained via their medial proximal phalanx attachments. Laterally, the dorsal interossei vector balances these medial forces (not shown). Tailor's bunion is a deformity of lateral 5th metatarsal abduction and 5th digit adduction (B). It can result from loss of general forefoot stability when it will be seen in conjunction with the development of hallux abducto valgus and other lesser toe deformities. However, it can also be due to a local loss in lateral forefoot biotensegrity, sometimes affecting the 5th metatarsal alone. This may be a result of a more laterally angled cuboid articular facet developed during growth that causes higher transverse plane flexibility at the 4th and particularly the 5th tarsometatarsal joints. Excessive metatarsal abduction displacement couples to metatarsal eversion, causing the 5th digit to develop adduction drift and some inversion rotation via rising medial muscular tensions. Digital adduction and inversion become more pronounced, the greater the metatarsal abduction. Thus, the 5th metatarsal is most prone to such deformity, but the 4th is also vulnerable and sometimes the 3rd can be affected. Thus, metatarsal abduction results in the medial displacement of the digital flexor proximal stabilisation vector on the metatarsal long axis. This combines with the forces derived from the now more medially displaced, lumbrical and plantar interossei vectors, both below and above the deep transverse metatarsal ligament, respectively. The 3rd and 4th dorsal interossei abduction moments are now disadvantaged. In the case of the 5th digit, abduction and eversion of the 5th metatarsal alters the vectors from the abductor digiti minimi and flexor digiti minimi brevis muscles, displacing them medially into becoming a flexor and a digital adductor, respectively. *(Permission www.healthystep.co.uk.)*

The angulation of the articular facet between the 5th metatarsal and cuboid may also influence the development of a tailor's bunion. If there is an increased lateral angulation of the cuboid articular surface to the 5th metatarsal facet, then there is an increased risk of 5th metatarsal transverse plane translation into abduction. However, unless prospective studies are performed, this feature remains only a theoretical pathological risk, particularly as the presence of tailor's bunion biomechanics is likely to induce facet angle changes through bone adaptation principles. This makes identification of 'a cause or an effect' difficult to differentiate in established cases unless prospective studies are undertaken in children. However, even in children, those who develop strong forefoot biotensegrity are unlikely to develop abnormal articular angles during growth through the principles of developmental tissue biomechanics, in that form follows function.

4.5.11 Pathomechanics at the 1st MTP joint

Despite a lack of homogeneity between each MTP joint's anatomy, both the extreme medial and lateral MTP joints are necessarily very different to the central three lesser MTP joints. Forming the medial and lateral borders of the forefoot, they are not enclosed within the forefoot's tie-bar of the deep transverse metatarsal ligament and its extensor apparatus/flexor plate complex reinforcements, being only linked in on one side. Unlike the 5th metatarsal, which usually only plays a small part in managing forefoot stresses and normally no part in providing a primary stabilising fulcrum for acceleration, the 1st MTP joint is a significant provider of forefoot loading and acceleration stability. In this capacity, it works closely with the more restrained, least mobile 2nd metatarsal and its own sesamoids that form a secondary set of articulations at the 1st metatarsal head. The complexity of the 1st MTP joint, coupled with its evolutionary origins in providing an opposable hallux for manipulation and grip, gives it a high pathomechanical potential. The 1st MTP joint is the most common joint affected by OA in the foot, followed by the 2nd cuneometatarsal and then the talonavicular joints, which are all joints of the medial column (Roddy and Menz, 2018). For ease of function in terminal stance, the medial column joints need a stable fulcrum point, but it is also the most mobile section of the foot and thus this situation increases the potential pathomechanical risks (Fig.4.5.11a).

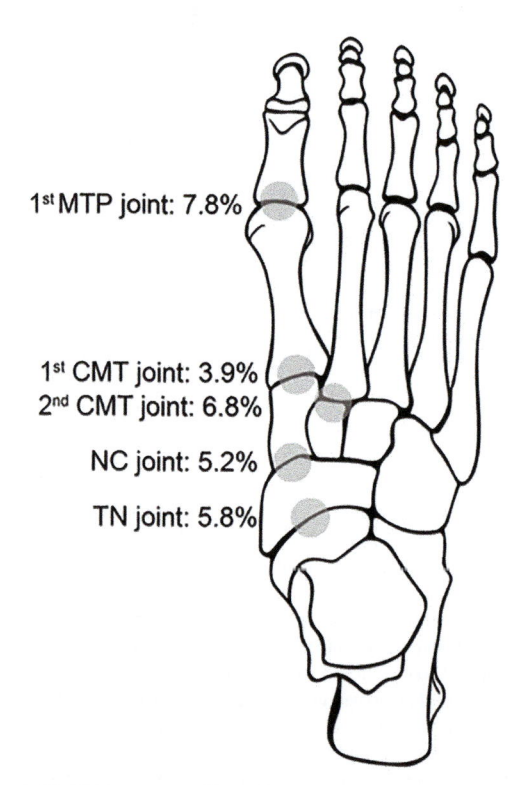

FIG. 4.5.11a Symptomatic feet with osteoarthritis (OA) seen on radiograph are reported to affect 17% of adults aged fifty years or over. Demonstrated here is the population prevalence of symptomatic OA locations seen on radiograph within the joints of the foot. *(Image from Roddy, E., Menz, H.B., 2018. Foot osteoarthritis: latest evidence and developments. Therap. Adv. Musculosk. Dis. 10(4), 91–103.)*

Extension dysfunction of the 1st MTP joint
The 1st MTP joint needs to provide compliance to extend (and also flex a little) but must not present unrestrained flexibility, so that it can allow movement, provide some energy dissipation, and maintain distal stability for the medial column. To this end, the hallux proximal phalanx is well restrained against extension by two strong attachments with the sesamoids that themselves connect to a number of MTP flexor muscles. Restriction in appropriate extension freedom during terminal stance means a loss or limitation of the preferred medial fulcrum that improves energetics at heel lift and into terminal stance. This has consequences to the potential 'Goldilocks zone' of ankle power force–velocity-to-metabolic output from the ankle plantarflexors (Takahashi et al., 2016; Wannop and Stefanyshyn, 2016; Oh and Park, 2017). Extension of the 1st MTP joint couples to motion within the 1st ray, particularly plantarflexion of the cuneonavicular joint during hallux extension over 20° (Phillips et al., 1996). This probably gives 1st MTP joint extension a significant role in maintaining medial stabilisation of the foot vault profile, that increases in height during terminal stance. Thus, if the 1st MTP cannot extend over 20°, it cannot significantly assist in medial vault stiffening during acceleration. Restriction in 1st MTP joint extension during gait also influences ankle kinematics, increasing ankle dorsiflexion, knee flexion, and hip extension during midstance, and reducing ankle plantarflexion in terminal stance (Hall and Nester, 2004). Thus, the biomechanical effects of the 1st MTP joint go beyond the foot.

Functional loss of 1st MTP joint freedom during gait, even before the development of any distinct pathology, could account for the loss of motion that has been referred to as *functional hallux limitus* (Dananberg, 1985; Van Gheluwe et al., 2006; Durrant and Chockalingam, 2009). It has been suggested that this is a source of gait dysfunction and pathology in the foot and lower limb resulting from hallux hypoextension that is initiated by a 'jamming' (impingement) of the 1st MTP joint, preventing full MTP joint extension freedom (Dananberg, 1986; Van Gheluwe et al., 2006; Durrant and Chockalingam, 2009). This is proposed to result from premature and/or high loading pressure under the 1st MTP joint in gait that reduces the 1st metatarsal declination angle through GRFs causing its elevation (Dananberg, 1986; Van Gheluwe et al., 2006; Durrant and Chockalingam, 2009). However, the elasticity (degree of stiffness) around the 1st MTP joint may be more significant to gait energetics and biomechanics than just the potential extension range of motion, or size of the GRF under the 1st MTP joint (Wannop and Stefanyshyn, 2016; Oh and Park, 2017).

Loss of end-range hallux extension required for gait may result from failure of the hallux to engage into its extensor groove at high angles of extension, for which there may be several explanations. As in lesser digital extension, the angle of declination of the metatarsal is significant, both in relation to the starting angle of extension and also the ability to dorsally glide into the extensor groove to increase extension range without articular impingement. In cases of low 1st metatarsal declination angle, the hallux starts its cycle of extension at a lower position on the metatarsal facet. If the declination angle does not increase adequately during terminal stance, then the instantaneous joint axis may be prevented from moving anterosuperiorly. The extension arc of motion around the MTP joint axis may as a result, cause the proximal phalanx to impinge on the dorsal metatarsal articular surface, rather than move into the extension groove which it cannot access.

Although functional hallux limitus (restricted hallux extension without degenerative joint changes) has been recorded in gait (Halstead and Redmond, 2006; Van Gheluwe et al., 2006; Gatt et al., 2014), the static clinical test of the Hübscher manoeuvre (Jack's test) in identifying the problem without actual gait data has proved questionable in identifying 1st MTP joint extension ranges during gait (Halstead and Redmond, 2006; Gatt et al., 2014). Indeed, this test was not even developed for this purpose (Jack, 1953). 1st MTP joint motion during gait may be correlated with the amount of foot pronation in stance phase (Gatt et al., 2014), which will decrease metatarsal declination angles associated with a loss of vault height at the end of midstance and the start of heel lift. This finding would fit in with cadaveric and live subject studies that indicate that adduction, plantarflexion, and anterior talar migration are associated with reduced 1st metatarsal declination and restricted 1st MTP joint extension mobility (D'Amico and Schuster, 1979; Roukis et al., 1996). It is reported that a negative correlation exists between weightbearing navicular drop and 1st MTP joint extension, where more navicular drop equates to less 1st MTP joint dorsiflexion motion, explaining ~33% of the variation in MTP joint maximum dorsiflexion between individuals (Paton, 2006).

There are multiple potential causes of low metatarsal declinations, such as the development of metatarsus primus elevatus alignment, or a more prone foot posture at the heel lift boundary, which is commonly associated more with pes planus foot types. Any display of 1st metatarsal dorsiflexion restraint instability on weightbearing of the medial column also decreases metatarsal declination, thereby risking degenerative changes in the 1st MTP joint (Chan and Sakellariou, 2020). Such excessive medial column/1st ray compliance can cause dorsiflexion moments at the tarsometatarsal joints at heel lift, with resulting loss of 1st MTP joint extension further reducing 1st cuneometatarsal stability during terminal stance. Excessive dorsiflexion mobility of the 1st metatarsal at its proximal joints may prevent or delay normally increasing functional metatarsal declination angles that should occur with increased ankle plantarflexion during terminal stance.

Medial column instability also risks increased dorsal compression impingements at the dorsal aspect of the any of the medial column articulations, but particularly at the 1st cuneometatarsal joint (Fig. 4.5.11b).

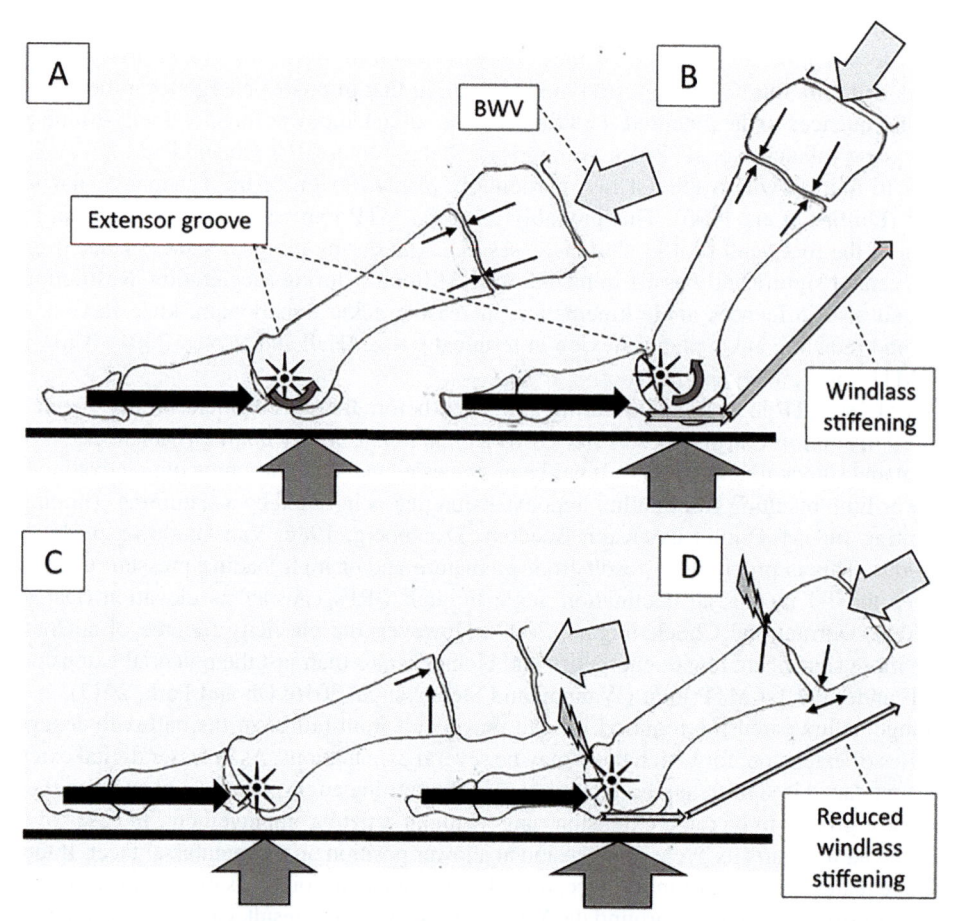

FIG. 4.5.11b Schematics of normal 1st ray kinematics that encourage 1st MTP joint extension freedom at heel lift and into terminal stance, compared to dysfunction. Normal late midstance motion involves the superior movement of the instantaneous joint axis (star), the forefoot GRF, the digital flexor proximal vector (black arrow), and the body weight vector (BWV) as demonstrated in (A). As the functional metatarsal declination angle increases after heel lift (B), the digital flexor proximal stabilisation vector compresses the metatarsal base into the medial cuneiform against the BWV, while the midfoot and ankle continue to plantarflex. This increases medial column stabilisation and thus improves 1st MTP joint stability under extension as the metatarsal head rotates over the sesamoids. Via increasing ankle and midfoot plantarflexion, the BWV becomes more vertical, compressing the cuneiform into the metatarsal base as the windlass mechanism tightens the plantar aponeurosis across the vault. These mechanisms work together so that extension improves stability, helping guide end-range proximal phalanx extension into the extensor groove. A foot with a low starting declination angle (C) takes longer for the 1st MTP joint to reach 20° extension, delaying 1st cuneonavicular plantarflexion stabilisation which under an increasing 1st MTP joint GRF, may cause the proximal 1st ray joints to glide dorsally instead. This can be restricted by strong peroneus longus, hallux flexor, and abductor hallucis activity. Any dorsal gliding at the 1st tarsometatarsal or cuneonavicular joints reduces the rate of increasing functional 1st metatarsal declination angles during rearfoot plantarflexion, further reducing the effects on the digital flexor proximal stabilisation vector in increasing the declination angle and stabilising the MTP joint. At lower MTP joint extension angles, the windlass mechanism will not activate as effectively, making it harder for the vault to self-stabilise passively. Depending on the extent of forefoot instability, sometime during later terminal stance (D), the dorsal aspect of the 1st metatarsal may impinge on the extending proximal phalanx on a less declined metatarsal head with a lower positioned instantaneous axis. In this posture, the 1st MTP joint is unable to offer its extensor grove to the proximal phalanx. If such biomechanics are recurrent, then articular damage and osteoarthritic changes, including osteophytes, will further reduce the extension range before impingement. Creating hallux IP joint hyperextension may offer some relief on blocked 1st MTP joint forces. (Permission www.healthystep.co.uk.)

The plantar 1st metatarsal–cuneiform ligament plays an important role in preventing 1st metatarsal dorsiflexion, with severance reported to increase 1st metatarsal dorsal translation by more than 5 mm on average (Mizel, 1993). Midfoot bending moments causing 1st ray dorsal instability are reported to be resisted by flexor hallucis longus (Sharkey et al., 1995; Jacob, 2001). This perhaps indicates that flexor hallucis longus activity during late midstance is important in stabilising the 1st MTP joint through proximal compression, aided by tibialis posterior and peroneus longus at the medial column's more proximal articulations. Stability and stress management of extension of the 1st MTP joint is likely to involve the extensive plantar intrinsic and extrinsic muscles that are part of vault profile stability (Fiolkowski et al., 2003; Kelly et al., 2012; Farris et al., 2019, 2020). Their function should enable the correct metatarsal declination angle to be set during heel lift and through terminal stance to facilitate 1st MTP joint extension freedom levels. Weakness of flexor hallucis longus or other vault-supporting structures may thus underlie many cases of 1st MTP joint dysfunction.

Thus, both soft tissue elasticity in the plantar tissues and metatarsal postures at heel lift are likely to be significant. Increased MTP joint stiffness could occur through sesamoid complex restrictions or soft tissue pathologies (diabetes). The result of either can alter force–velocity-to-ankle plantarflexor power relationships, leading to increased stresses around the 1st MTP joint, thus explaining the development of OA within the joint, or 1st MTP joint plantar ulceration often seen in diabetics.

Some planus feet could be more susceptible due to their low metatarsal declinations and their reduced plantar intrinsic muscle mass and strength, compared to other feet (Angin et al., 2014). High medial column loading and increased mobility associated with tibialis posterior dysfunction (Imhauser et al., 2004; Kokubo et al., 2012) could prevent free hallux extension at heel lift. However, if a hypermobile pes planus uses midfoot break, then the requirement for hallux extension at the lesser MTP joints will be reduced. Therefore, limited hallux extension in such feet may not be functionally significant to the individual's gait pattern. The situation changes if interventions such as foot orthoses limit the midfoot break, suggesting that some caution in raising the vault in such feet is required, or else increased degenerative changes within the MTP joints could result.

Long 1st metatarsals may also risk impingement of their dorsal aspects by the hallux through later initiation of their passive extension moment, compared to a slightly shorter 2nd MTP joint. This nonhomogeneous medial MTP joint extension can interfere with the medial fulcrum's position to the foot's line of progression. The change in the MTP fulcrum axis angles to the line of progression may also explain the proposed links of metatarsus primus elevatus, hallux valgus, and metatarsus adductus, which have also been proposed as being involved in 1st MTP joint OA pathomechanics (Chan and Sakellariou, 2020). Variance in the angle of gait of individuals may explain why some patients do and others do not associate 1st MTP joint OA with these deformities. For a visual review of these principles, see Fig. 4.5.7c.

1st MTP joint hyperextension pathomechanics

Thus, a 'functional' hallux limitus has been proposed as a mechanism for the development of degenerative changes in the 1st MTP joint resulting from the joint's impingement dorsally. This can be viewed as pathomechanics resulting from the loss of extension moment freedom of the 1st MTP joint. However, hyperextension of the 1st MTP joint has also been proposed as a mechanism of degeneration through dorsal joint impingement (McMaster, 1978). High metatarsal declination angles, as found in cavoid feet, have been referred to as *pseudoequinus* (Whitney and Green, 1982). Such high metatarsal declination angles can cause MTP joint hyperextension, but any excessive requirements of 1st MTP joint dorsiflexion (such as squatting on extended toes while working) may lie at the origin of some cases pathomechanics. In the presence of high 1st metatarsal declination angles, the digit starts its arc of passive extension in higher extension angles relative to the metatarsal head than that of the more average declinations. The arc of extension required throughout terminal stance may be more than the articular surface extensor groove can accommodate, despite the dorsally extended articular surface found in humans, leading to impingement of the dorsal aspect of the proximal phalanx within the extensor groove. Repetitive impingement will cyclically over-compress the articular facet during extension moments at the joint (Fig. 4.5.11c). Increasing footwear heel heights increases the functional declination angle, giving the potential effect of increasing the extension angle required at the MTP joints in order to accommodate the increased plantarflexion at the ankle caused by the shoe's heel raise, and the increased declination angles of the metatarsals that result (Sussman and D'Amico, 1984).

MTP joint articular degeneration

Once the mechanical properties of cartilage are compromised, persistent cycles of cartilage damage, initially through either insufficient or excessive joint extension freedom, will cause degeneration and OA-induced loss of motion (Chan and Sakellariou, 2020). Degenerative joint stiffening unfortunately causes increased joint stresses through raised plantarflexor power that has to try and overcome the increased joint resistance to heel lift.

Confusingly, the early stages of OA change in the 1st MTP joint have also been referred to as hallux limitus, although the term was originally adopted to describe *hallux flexus*, a deformity of excess 1st MTP joint plantarflexion with concurrent extension reduction (Camasta, 1996). This must not be confused with 'functional hallux limitus', which is referred to as a functional loss of extension in gait, not an actual loss of motion available (Dananberg, 1985, 1986). It is probably safer to refer to OA or *degenerative joint disease* (*DJD*) of the 1st MTP joint to avoid confusion, giving a clearer diagnosis. Advanced degenerative changes that relate to a severe loss of hallux extension, and sometimes also plantarflexion, are commonly known as *hallux rigidus*. This is a term that is also used for all stages of DJD of the 1st MTP joint (Chan

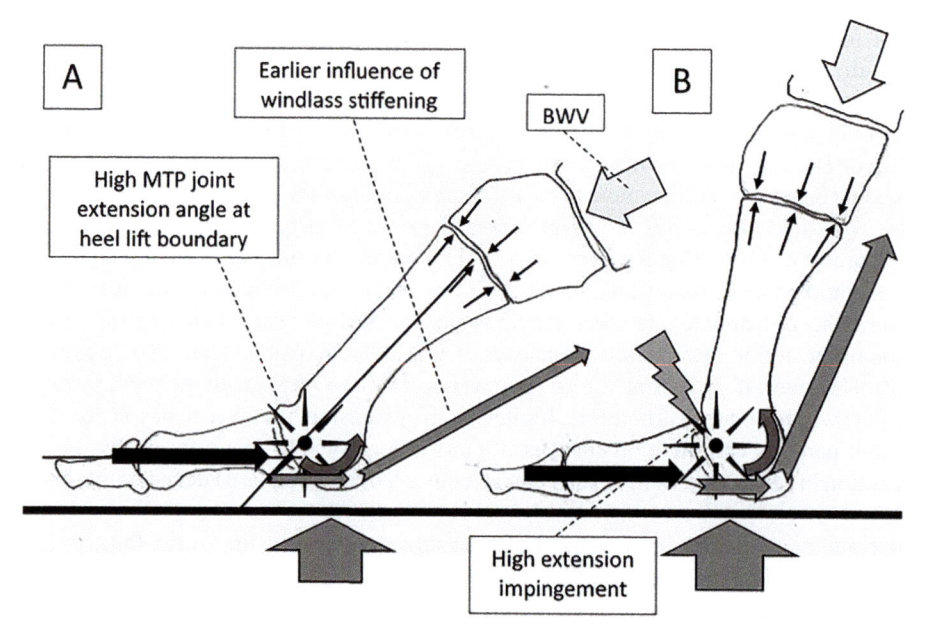

FIG. 4.5.11c High metatarsal declination angles at the start of MTP joint extension can occur for many reasons, including cavoid feet, high-heeled shoes, or uneven terrain perturbation adaptations. By starting extension with digits at high extension angles (A), the instantaneous joint axis is positioned more superiorly, tending to offer an improved freedom of extension motion. The windlass mechanism should more easily stiffen the foot, giving the plantar aponeurosis more influence on foot stiffness within cavoid foot types. Body weight vectors (BWV) directed through the medial column are verticalised more quickly so that when acting with the digital flexor proximal vector (black arrow), 1st metatarsocuneiform joint compression stability is likely to be excellent, resisting any GRF-induced tendency for any 1st ray dorsiflexion. The sesamoids may find themselves under greater peak pressures as a result of the higher earlier loading associated with the higher 1st metatarsal declination angle. During late terminal stance (B), higher starting extension angles risk the MTP joint requiring a greater degree of end-range extension, often with such extension being less resisted by plantarflexing soft tissues of the MTP joint. This risks the proximal phalanx requiring more extension than the metatarsal head and its extensor groove can offer, potentially resulting in dorsal joint impingement. See also Figs 4.5.7e and 4.5.7f. *(Permission www.healthystep.co.uk.)*

and Sakellariou, 2020). However, where rigidus starts and limitus ends is open to debate. A grading system of 1st MTP joint DJD (OA) has been classified by Coughlin and Shurnas (2003), with hallux limitus best fitting grades of between 0 and 1 (see Table 4.5.11). Another foot-specific radiographic grading system has also been proposed for foot joints (Menz et al., 2007), but as always, is subject to clinician interpretation.

TABLE 4.5.11 1st MTP joint DJD grading.

Grade	Dorsiflexion range (extension)	Radiographic findings	Clinical findings
0	40–60° (and/or 10%–20% loss compared to other side)	Normal	No pain, only stiffness and loss of motion on examination
1	30–40° (and/or 20%–50% loss compared to the other side)	Dorsal osteophytes main finding, joint space narrowing, peri-articular sclerosis, and metatarsal head flattening minimal	Mild or occasional pain and stiffness, pain at extreme dorsiflexion and/or plantarflexion on examination
2	10–30° and/or 50%–75% loss compared to other side	Dorsal, lateral, and possibly medial osteophytes flattening appearance of metatarsal head, no more than 1/4 of dorsal joint space involved on lateral view radiograph, mild to moderate joint space narrowing and sclerosis	Moderate to severe pain and stiffness that can be constant; pain occurs just before maximum dorsiflexion and plantarflexion on examination
3	<10° (and/or 75%–100% loss compared to other side) and noticeable loss of plantarflexion	As grade 2 with substantial joint space narrowing and possible peri-articular cystic changes; more than 1/4 of dorsal joint space involved on lateral view radiograph; sesamoids enlarged, cystic or irregular	Nearly constant pain and substantial stiffness of extremes of range of motion, but not during mid-range motion
4	Same as grade 3	Same as grade 3	Same as grade 3 but with definite pain in mid-range of passive motion

Adapted from Coughlin, M.J., Shurnas, P.S., 2003. Hallux rigidus: grading and long-term results of operative treatment. J. Bone Joint Surg. 85-A(11), 2072–2088.

OA of the 1st MTP joint is the most commonly reported degenerative joint change within the foot (Roddy and Menz, 2018), and may follow insidious microtraumatic events in the form of repetitive functional impingements, as already discussed. However, 1st MTP joint damage may follow a single traumatic event such as a forced hyperextension (or occasionally hyperflexion) injury that can also damage the sesamoid-plantar plate mechanism with a sudden high force MTP joint impingement. This is a traumatic event that is well known in sport as *turf toe* (Chan and Sakellariou, 2020). Interestingly, a positive family history of 1st MTP joint OA is reported in around two-thirds of cases, with 95% of these showing bilateral symptoms (Chan and Sakellariou, 2020). This may reflect the inheritance of foot types and associated pathomechanics, or lifestyle influences that increase 1st MTP joint OA such as hyperextended toe squatting postures, or something more developmental, genetic, epigenetic, or physiological in origin in MTP joint structure or cartilage properties (Fig. 4.5.11d).

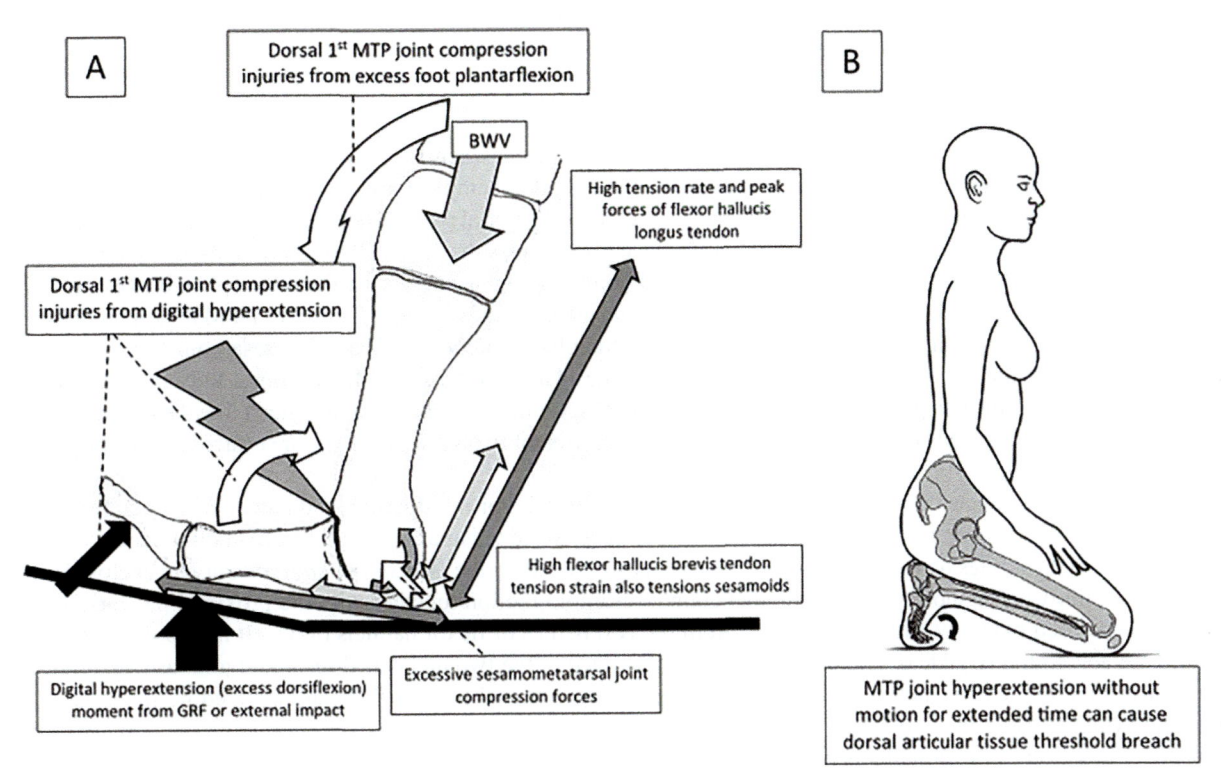

FIG. 4.5.11d Although 'turf toe' is a clinical term used both for acute forced hyperextension and hyperflexion injuries of the 1st MTP joint, hyperextension (hyperdorsiflexion) is far more common. A schematic of vectors (A) demonstrates that injuries can arise from external forces that hyperextend the digit directly, or from extension moments generated by the metatarsal from forced ankle plantarflexion impacts, or terrain perturbations that leave the hallux fixed to the ground. A range of compression and tension injuries can result, including 1st metatarsal sesamoid joint structures, hallux flexor tendons, and particularly impingement damage on the dorsal 1st MTP joint line. Dorsal aspect articular cartilage splitting, subchondral fractures, and oedema within the metatarsal head are commonly found in conjunction with flexor tendinopathies and sesamoiditis following hyperextension trauma. Initially, restricting 1st MTP joint motion to only that which is pain-free during gait, combined with gentle nonweightbearing rehabilitation motion and anti-inflammatory techniques, are essential to gradually allow healing and the repair of anatomy. This should permit remodelling of tissues for the normal high biomechanical stresses of acceleration around the 1st MTP joint. If poorly managed postinjury, permanent articular tissue damage and soft tissue scarring stiffness can lead to loss of articular biphasic biomechanics and freedom of motion, leading further to the inevitable degenerative consequences of osteoarthritis. All impact traumas to the 1st MTP joint offer similar risks, making seemingly minor injuries a potentially serious long-term risk to locomotive health. Chronic 1st MTP joint degeneration can also result from long periods of sitting/working with the hallux in static end-range extension under body weight (B). Full articular mechanical biphasic tissue properties are only available during motion, otherwise there is reliance on connective tissue properties alone which will start to fail under continual high compression. *(Permission www.healthystep.co.uk.)*

4.5.12 Pathomechanics of 1st MTP sesamoids

The sesamoids represent a different aspect of 1st MTP joint function as two separate articulations with the metatarsal head. Sesamoids remain ultimately coupled to the motion, functions, and dysfunctions of the 1st MTP joint, in a similar mechanism to the patellofemoral joint associated with tibiofemoral joint motion. The difference between the joints primarily rests on the fact that the MTP joint and the sesamoids act as part of a directly weightbearing structure (the 1st MTP joint), which means that without free MTP joint and sesamoid motion, the sesamoids can become relatively fixed hard weightbearing structures within the plantar fat pad.

Sesamoid influence of 1st MTP joint freedom

1st MTP joint extension freedom can be affected by degenerative changes at the sesamoid metatarsal articulations, which increase medial foot fulcrum stiffness. Extension at the 1st MTP joint is dependent on free sagittal plane translation of the sesamoids within its soft tissue complex, in much in the same way that the flexor plate complex facilitates appropriate lesser MTP joint extension. If the sesamoid gliding motion is restricted, so is extension of the hallux. It is not uncommon for OA changes to occur on the sesamoid's articular surfaces in conjunction with 1st MTP joint degeneration, increasing the 1st MTP joint's stiffness further. When found together, it can be clinically difficult to establish which joint's degeneration arose first, as the pathomechanics of one inevitably leads to compromise in function of the other. Two planes of motion tend to be linked to sesamoid pathology: sagittal plane flexibility for extension/flexion, and the sesamoid complex transverse plane vectors that should balance medial and lateral forces through 1st MTP joint extension. Minor frontal plane displacements can also occur, particularly with hallux abducto valgus.

Sesamoiditis

There are significant similarities in the biomechanical principles of the sesamoid joints of the hallux to that of the patella. This is not least because they both involve sesamoids within a tendon complex, arising from several muscular inputs that form multidirectional vectors stabilising a resultant force within the sagittal plane. The role of the sesamoids at the 1st MTP joint is to maintain a proximally directed tensional force applied across the 1st MTP joint throughout changing angles of extension. It is likely that the higher forces of the 1st MTP joint would make a fibrocartilaginous plantar plate alone insufficient for stability, for the plantar plate has inherent proximal attachment weakness. Thus, the 1st MTP joint requires more stable tendon-enclosed sesamoids running over grooves in the plantar metatarsal head. The sesamoids' articular surfaces facilitate gliding of the metatarsal head as it rotates over the plantar soft tissues. In so doing, they influence levels of stiffness around the joint acting with the long and short hallux flexor muscles. This is a role much like the fibrocartilaginous plantar plates of the lesser MTP joints, with the added complication of two synovial articular surfaces divided by a ridge of bone called the crista.

The two sesamoid bones under the hallux are commonly bipartite or multipartite, and can appear fractured on radiographs to the unexperienced eye. This type of formation does not seem to have any dramatic effect on function, as most are seen as incidental findings on radiographs. The sesamoids are joined by the *intersesamoid ligament* and enclosed within a shared small fibrocartilaginous plantar plate that couple sesamoid motion together with the proximal phalanx. The sesamoids are enclosed plantarly within the tendons from the two bellies of flexor hallucis brevis. It is important that the sesamoids are able to freely glide over the hyaline cartilage surface of the two metatarsal sesamoid joint articulations. These articulations are divided by the crista on the grooved plantar metatarsal head surface and covered by the intersesamoid ligament lying between the two sesamoids, binding them together. The presence of the sesamoids within their tendons improves the action of flexor hallucis brevis through changing MTP joint angles like a pulley. The sesamoids also create a pressure-reducing groove between them for the flexor hallucis longus tendon to proceed through to its attachment on the distal phalanx of the hallux. This allows the long flexor tendon within its sheath to move freely without high compression between the metatarsal head, plantar fat pad, skin, and the ground externally.

The sesamoids improve stability and flexibility at the MTP joint, without creating excessive freedom for 1st MTP joint extension (Fig. 4.5.12a). There are a number of common pathologies associated with sesamoids, which will either cause increased stiffness or excess flexibility at the 1st MTP joint. These include cartilage damage and degeneration of the metatarsosesamoid joints, bone marrow lesions within the sesamoids, fatigue (stress) fracture, acute sesamoid fractures, sesamoid avascular necrosis, plantar bursitis, and inflammation and/or fibrosis of the associated soft tissue structures of the sesamoid complex (Sims and Kurup, 2014).

Sesamoid pathomechanics with hyperextension and hypoextension

High 1st metatarsal declination angles can expose the 1st MTP joint to high ranges of hallux extensions during terminal stance, often associated with cavoid feet or high-heeled shoes. In such feet, the sesamoids can start forefoot loading with the sesamoids being more distally positioned in their articular grooves, as a result of the higher MTP joint extension angles. This can position the sesamoids into weightbearing positions under the metatarsal heads, rather than sitting more proximally within their articular grooves as they would under less hallux extension and at lower metatarsal declinations. This higher compression loading of the sesamoid-metatarsal joints may prevent the sesamoids translating sufficiently during terminal stance, in turn preventing the metatarsal head freely gliding over and around the sesamoids. This rotation of the metatarsal head over the sesamoid articulations is usually stabilised through proximally directed forces from the hallux flexors through its proximal phalanx during terminal stance 1st MTP joint extension. Thus, overly compressed sesamoids

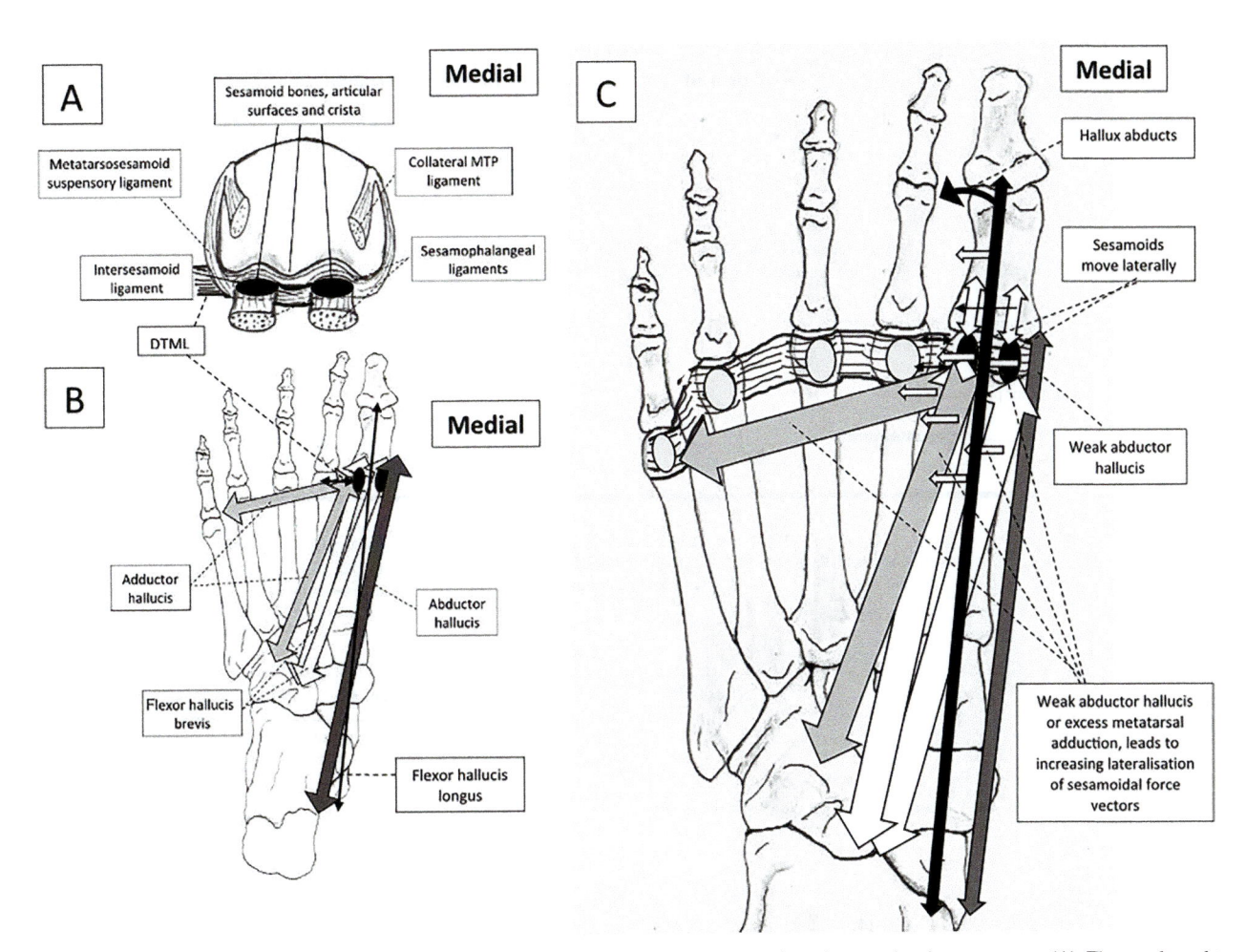

Pathology through the principles of biomechanics **Chapter | 4** 807

FIG. 4.5.12a Sesamoids (black ovals) are 'suspended' under the 1st metatarsal head by a number of connective tissue structures (A). They are bound to the extensor apparatus superiorly, the deep transverse metatarsal ligament (DTML) laterally, and the sesamophalangeal ligaments distally. Sesamoid motion is coupled to that of the proximal phalanx and the activity of flexor hallucis brevis, its tendon enveloping the sesamoids plantarly. MTP joint extension tightens the suspensory and collateral ligaments, increasing extension resistance of the flexor hallucis brevis vectors and helping to compress metatarsal and sesamoid articular surfaces together, aiding sagittal plane stability both distally and proximally at the same time. However, these tension and compression forces should not prevent gliding of the metatarsal head, allowing it to rotate with adequate freedom over the sesamoids. Sesamoid stability needs to help maintain the sagittal plane position of both the flexor hallucis brevis and flexor hallucis longus (which passes between the sesamoids plantar to the intersesamoid ligament) so that the digital flexor proximal stabilisation vector remains aligned to the long axis of the 1st metatarsal during any 1st metatarsal adduction splaying during forefoot loading. This is achieved by medial and lateral muscle vectors to the sesamoids and proximal phalanx remaining balanced (B). Thus, as the hallux hopefully adducts with the metatarsal head medially and under abductor hallucis activity, the sesamoids are pulled slightly laterally to maintain the flexor vector along the metatarsal's long axis. However, should the abductor hallucis hallux adduction moment become weak (C) or 1st metatarsal loading adduction drift become excessive, then the abductor vectors of adductor hallucis and flexor hallucis brevis, coupled to possible increased tension on the DTML, will start to draw the sesamoids laterally. This in turn moves the flexor hallucis longus tendon laterally, shifting the whole digital flexor proximal vector laterally to the long axis of the 1st metatarsal. As a result, abduction on the hallux via 'bowstringing' occurs as the hallux flexor tendons apply an increasing abduction moment across the 1st MTP joint. The lateralisation of the sesamoids that results draws the medial sesamoid into compression with the bony ridge of the crista, which can in time cause it to erode. Sesamoid dysfunction, pathology, and 1st MTP instability now become an increasing risk. It is when describing such events that the inappropriateness of the names of the muscles that supply the 1st MTP joint become obvious. *(Permission www.healthystep.co.uk.)*

can become entrapped within their articulation, so that the sesamoid complex may now increase stiffness instead of aiding joint freedom. In over-compression, sesamoid bone marrow lesions are the most likely pathologies to develop but shear motion-compensatory adventitious bursae can form within the cutaneous tissue beneath the sesamoids. Over time, OA is likely to develop when sesamoids do not glide freely during MTP joint motion. OA causes the sesamoids to become relatively 'fixed' to the metatarsal head, preventing free coupled motion with the proximal phalanx of the hallux. Without free sesamoid motion, MTP joint extension is limited, preventing the proximal phalanx from moving through an arc of motion to avoids dorsal impingement on the metatarsal head, potentially initiating 1st MTP joint degeneration (Fig. 4.5.12b).

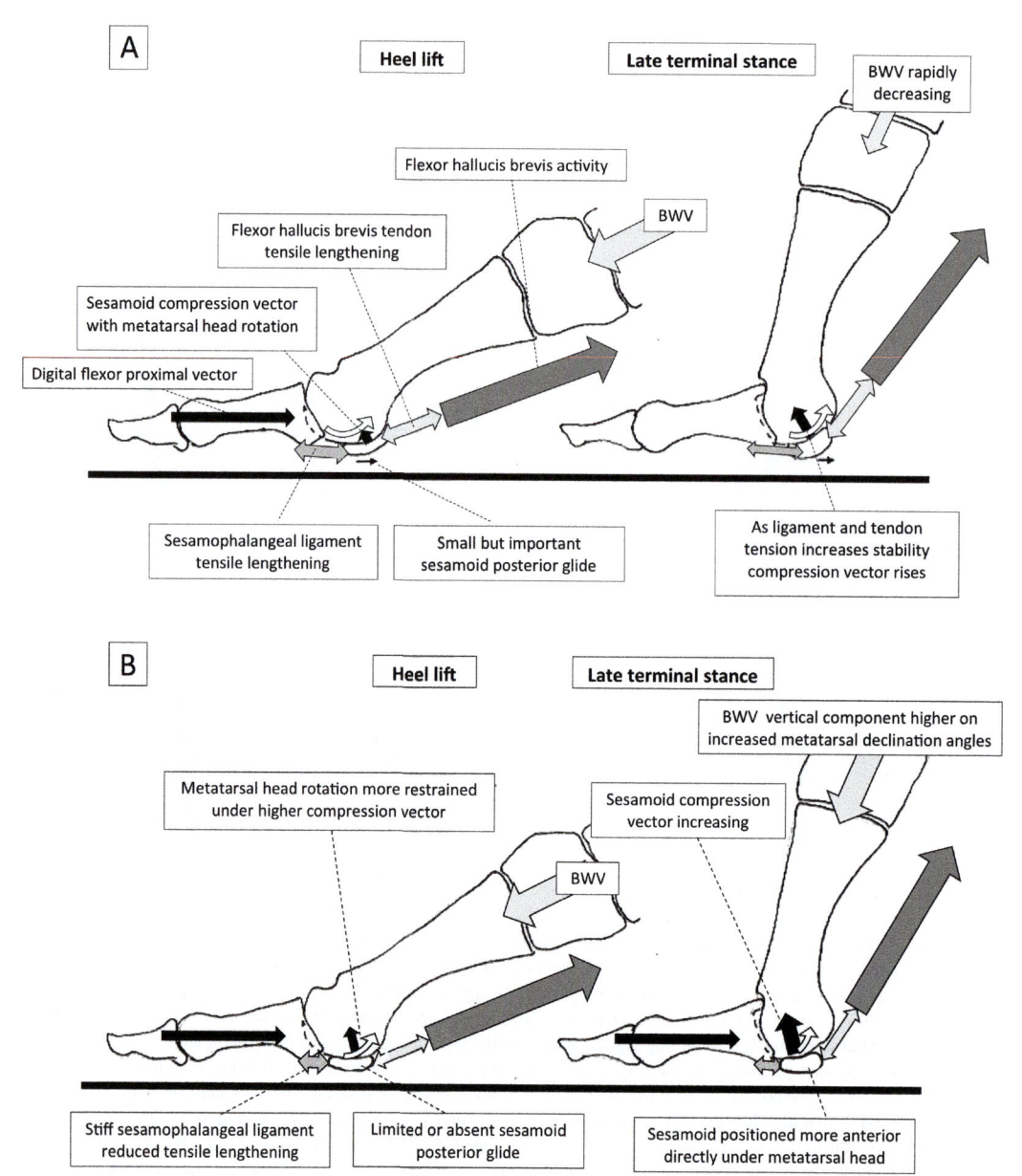

FIG. 4.5.12b During terminal stance, the increasing 1st MTP joint extension angles change the positional relationship of the metatarsal head over the sesamoids from heel lift to preswing (A). However, the sesamoids displace little from the base of the proximal phalanx. This is because they are tethered firmly by the sesamophalangeal ligaments that are distal extensions of the more flexible flexor hallucis brevis tendons lying proximally to the sesamoids. As connective tissue tensions increase, compression within the articular surfaces of the sesamoids also rise under biotensegrity principles, improving sesamoid stability by acting as a posteriorly-restraining and supporting sling during later terminal stance. Higher compression should only occur under decreasing body mass loading of the forefoot in late terminal stance, moderating the forces applied to the articular cartilage. The sesamoids should only become directly positioned under the BWV/GRF when most of the body mass has transferred to the opposite foot and forefoot GRFs are rapidly decreasing. Gait requiring higher angles of MTP joint extension will tend to position the sesamoids under the rotating metatarsal heads earlier during terminal stance, and thus place them under higher GRF earlier. This can be an issue with high metatarsal declination angles associated with cavoid feet. Total inflexibility within the sesamophalangeal ligaments (B) may position the sesamoids more anteriorly throughout terminal stance, potentially increasing the loads from the vertical component of the GRF on to the sesamoids, increasing articular compression peaks and loading times. Stiffer connective tissues are more commonly found in cavoid feet which combines increased metatarsal declination angles, higher MTP joint extension angles, and less sesamoid ligament strain deformation availability. Also, any sesamoid articular degeneration that arises will increase compression forces within the sesamoids due to loss of joint freedom. Thus, poor sesamoid positioning that restricts or loses free metatarsal head rotation can lead to oedema or fatigue fractures within the sesamoids, osteoarthritis in the sesamoid articulations, and/or ligament tears and tendinopathy within the soft tissues that enclose them. *(Permission www.healthystep.co.uk.)*

Lack of sesamoid motion can cause fatigue-induced avulsion fractures. Such sesamoid fatigue fractures can result from tendon tensions increasing without inducing actual motion, although the compression stresses that block motion can themselves risk 'crush' fatigue fractures. Overextension of the sesamoid complex in acute hyperextension events can also cause over-tension sesamoid fractures. Avascular necrosis is also a vulnerability due to the sesamoids' proximally derived arterial supply (Anwar et al., 2005), which could be disturbed easily if the sesamoids are forced to move too far distally under hyperextension of the 1st MTP joint (Fig. 4.5.12c). The tendency towards more lateral force vectors from the sesamoid complex will tend to dominate pathology to the tibial (medial) sesamoid.

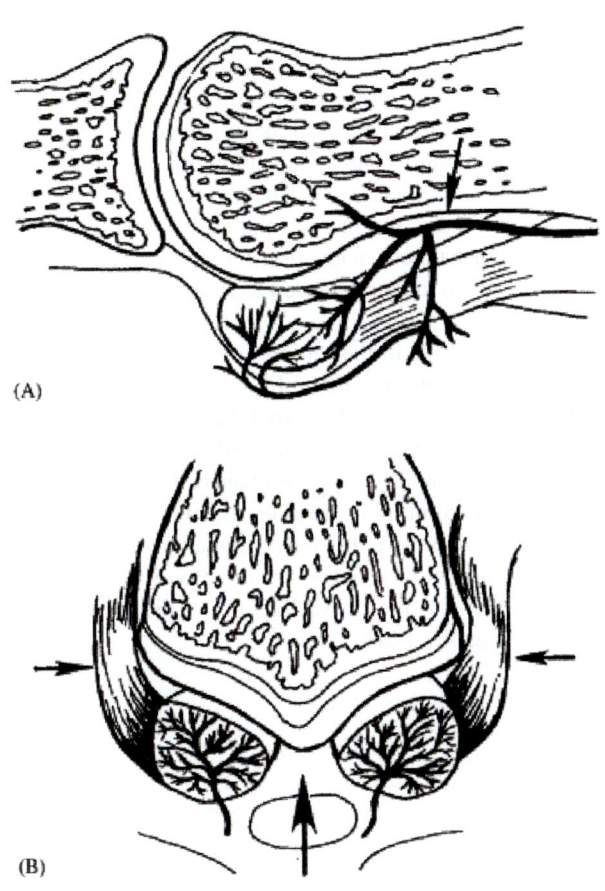

FIG. 4.5.12c A schematic of the vascular supply to the sesamoid. In the sagittal plane (A), the major vascular supply enters the sesamoid proximally. An anastomosis occurs between the distal and proximal surfaces. Viewed in the frontal plane (B), the plantar vascular supply to the sesamoid is significant. The black arrows indicate the sesamoid suspensory ligaments. The sesamoids, with their significant vascular supply, can be highly responsive metabolically to changes in biomechanical loads. In hyperextension injuries such as turf toe, the vascular supply is vulnerable to disruption which risks avascular necrosis or can cause delays in the healing response. *(Image from Sobel, M., Hashimoto, J., Arnoczky, S.P., Bohne, W.H.O., 1992. The microvasculature of the sesamoid complex: Its clinical significance. Foot Ankle 13(6), 359–363.)*

Transverse plane sesamoid dysfunction

Flexor hallucis brevis arises from proximal attachments to the cuboid, the lateral plantar area of the lateral cuneiform, and the tendon continuation of tibialis posterior, thereby linking the proximal stability of this muscle to both peroneal function and that of tibialis posterior. The muscle runs distally and medially to divide into two bellies, one passing to enclose each sesamoid as the bellies become tendinous. From the sesamoid it encloses, the tendons become more ligamentous-like to attach to the medial and lateral side of the proximal phalanx from medial and lateral sesamoid ligaments, respectively. The force vector provided is for extension resistance, but it is slightly obliquely orientated due to the muscles more lateral position to the 1st MTP joint, in much the same way that the knee extensors vastus lateralis, vastus intermedius, and rectus femoris combined vector pulls obliquely in a lateral direction on the patella. This means that the combined resultant flexion moment of the short hallux flexor is slightly dominant to the lateral side.

To achieve extension resistance, a number of secondary soft tissue stabilisers lie on the medial and lateral sides of the 1st MTP joint. Medially and laterally, the collateral MTP joint and sesamoid suspensory ligaments passively stabilise the joint.

Laterally, these ligaments link to the deep transverse metatarsal ligament. This means, once again, that the lateral side of the 1st MTP joint has potentially more restraint. The lateral sesamoid is also directly attached to the adductor hallucis which has both a transverse and oblique head, creating lateralised force vectors. The proximal attachment of the oblique adductor hallucis is to the peroneus longus tendon sheath and the plantar bases of the 2nd, 3rd, and 4th metatarsals, meaning that its proximal attachment stability is dependent on tarsometatarsal stability. The transverse head of adductor hallucis attaches to the deep transverse metatarsal ligament between the 2nd through to the 5th ligament areas. This means that lateral stability of this muscle belly will be affected by forefoot splay, with increasing spread of the lateral metatarsals potentially increasing the lateral force vector.

On the opposing medial side of the 1st MTP joint lies abductor hallucis that attaches distally to the medial side of the plantar tubercle of the proximal phalanx, providing a hallux adduction moment arm. Proximally, it attaches to the medial plantar surface of the calcaneal tuberosity, the flexor retinaculum, the plantar aponeurosis, and the intermuscular septum between itself and the flexor digitorum brevis. The hallux adduction force derived from abductor hallucis activity occurs with concurrent 1st metatarsal splay, which requires the lateralisation of the sesamoid vectors to maintain proximal digital flexor vector alignment to the long axis of the 1st metatarsal. This leaves a significant potential soft tissue vector imbalance in the sesamoid flexor complex, for it only takes fatigue or another dysfunction of abductor hallucis to lose the medial vectors to the sesamoids. Thus, if abductor hallucis' adduction force is reduced or lost it upsets the whole mechanism, as the lateralised sesamoid vector forces are no longer balanced to resist hallux adduction but instead directly pull the sesamoids laterally.

Abductor hallucis also plays a supporting role in the setting up of the reversed windlass effect, being active at initial contact as well as demonstrating peak activity during late midstance through into terminal stance. It provides forces that help to compress the medial side of the medial column of the foot. In dysfunction, not only can sesamoid imbalance result but there is also increased mobility on the medial side of the medial column. The effect on the sesamoid complex of abductor hallucis weakness will be increased lateral vector dominance. The sesamoids will pull laterally when tensioning the complex, compressing the medial sesamoid into the crista on the metatarsal head. This reduces the ability of the sesamoids to change their sagittal plane orientation to facilitate freedom and control of the 1st MTP joint extension moment. The effect is to create increased compression of the medial sesamoid on its articular surface as it is bound to the lateral sesamoid by the intersesamoid ligament that is being pulled laterally, causing the hallux to start to displace laterally into abduction relative to the 1st MTP joint (Fig. 4.5.12d).

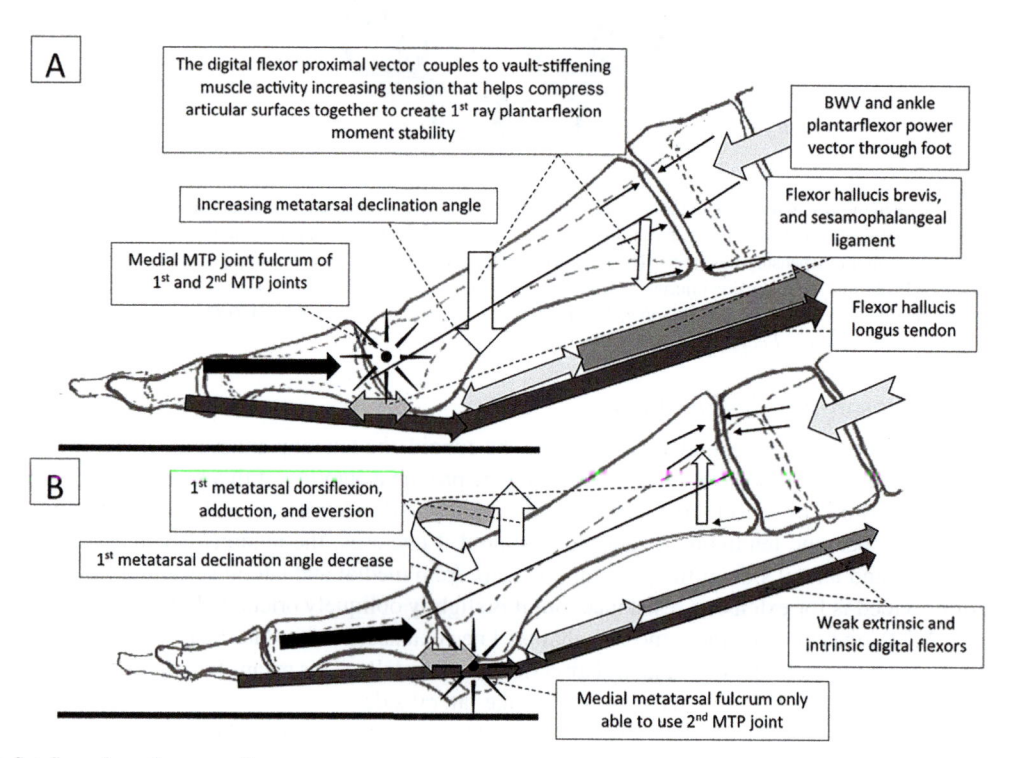

FIG. 4.5.12d See figure legend on opposite page

The tendency towards medial instability at the 1st MTP joint is associated with other common foot pathomechanics, examples being tibialis posterior dysfunction or plantar ligament failures. The imbalance between the lateral and medial force vectors around the sesamoid complex may help in the development of or be a result of, hallux abducto valgus.

4.5.13 Hallux abducto valgus

Hallux valgus, or hallux abducto valgus (HAV), is a common forefoot deformity similar to the tailor's bunion but causing a hallux deformity at the 1st metatarsal articulations. The deformity involves 1st metatarsal adduction towards the midline of the body but an abduction from the foot's midline, with the hallux moving in the opposite direction. HAV tends to develop first in one foot followed by the other, and then more usually, more extensively in the foot first affected (Young et al., 2013). While rarely necessary for diagnosis, radiological evaluation of HAV is established through angular alignments in the transverse plane between the hallux and the 1st metatarsal (hallux valgus abduction angle or hallux valgus angle), between the 1st and 2nd metatarsals (intermetatarsal angle), and the distal metatarsal articular angle. Hallux abduction angles and intermetatarsal angles define the classification of the deformity into mild, moderate, or severe (Coughlin et al., 2002; Thomas and Barrington, 2003). The severity of the deformity has been used to suggest certain surgical procedures of correction (Thomas and Barrington, 2003; Favre et al., 2010), procedures which themselves will change the biomechanics of the 1st MTP joint postoperatively, and not always favourably (Fig. 4.5.13a). Biomechanical tests indicate that reversed-L osteotomies or long plantar arm osteotomies may offer the best maintenance of stiffness in cantilever loading bending moments that are representative of those experienced in gait (Favre et al., 2010; Ravenell et al., 2011). Yet, it may be better for long-term locomotive capabilities if HAV can be conservatively managed and prevented from worsening. However, this has proved clinically challenging.

HAV is a nonhomogeneous, variably multiplane deformity, where rotation deformity is particularly significant within the transverse and frontal planes. Lateral abduction deviation in the hallux occurs with adduction and dorsiflexion of the 1st metatarsal with eversion (valgus) rotation. 1st metatarsal eversion rotation has been termed a 'pronation' deformity (Kim et al., 2015). Although the role of the deep transverse ligament lying between the 1st and 2nd metatarsals has been debated, with HAV the ligament seems to offer inferior tensile properties that are likely to disrupt the biomechanics of the sesamoid apparatus to the 2nd metatarsal head via changes in time-dependent and stress-stiffening properties (Abdalbary et al., 2016). This is likely to lead to greater displacement of the 1st metatarsal head medially on forefoot loading, but whether the changes in tensile properties are a cause or result of the deformity is as yet, unknown. Either way, the 1st metatarsal displacement explains malpositioning of the sesamoids and the perceived increased rounded head shape of the 1st metatarsal seen on radiographs (Wagner and Wagner, 2020). The sesamoids can appear orientated laterally to the 1st MTP joint, and this has long been thought to indicate that the sesamoids are subluxed or dislocated laterally. Many are, in fact, moving into valgus or eversion with the rest of the 1st MTP joint anatomy, as confirmed by CT imaging (Wagner and Wagner, 2020). Thus, in some cases, it is the 1st metatarsal rotating rather than the sesamoids moving that creates a pseudosubluxated appearance on radiographs, while other HAV cases do develop true sesamoid subluxations (Kim et al., 2015) (Fig. 4.5.13b).

FIG. 4.5.12d, CONT'D Medial column stability at the heel lift boundary is essential for acceleration energetics and fundamental to normal sesamoid function. The sesamoids are part of the anatomy that provides the distal stability necessary for the metatarsal head to rotate for 1st MTP joint extension concurrently with that of the 2nd MTP joint. Free sesamoid motion in turn improves proximal 1st metatarsal base stability via the digital flexors' proximal vector (black arrows via digits). Once the 1st ray is stable, it can work effectively with the intrinsically more stable 2nd MTP joint as a provider of the medial acceleration fulcrum (A). However, should the 1st ray remain too compliant (B) at heel lift, medial forefoot GRFs may dorsiflex the 1st metatarsal at its proximal joints, causing the midfoot to medially destabilise and invert. Rearfoot eversion may be forced to follow this drop in the medial vault height during acceleration. This may result in increased medial forefoot forces on an unstable 1st ray that will limit the ability of the 1st metatarsal head to glide and rotate proximally over the sesamoids. Without metatarsal head rotation, the sesamoids become functionally redundant and thus may become little more than hard weightbearing masses within the forefoot plantar weightbearing soft tissues. Dorsiflexion of the 1st tarsometatarsal joint is associated with variable amounts of 1st metatarsal eversion and medial adduction. Those individuals display larger degrees of 1st metatarsal adduction and elevation risk destabilisation of the medial to lateral 1st MTP joint/sesamoid vectors, and thus may only be able to use the 2nd MTP joint as an acceleration fulcrum. This instability may lead to increasing lateral tensions and greater vector advantage for the anatomy lying to the lateral aspect of the joint. For the consequences of this, see Fig. 4.5.12a. However, if the 1st ray remains stable in the sagittal plane and abductor hallucis remains strong, despite 1st metatarsal adduction, the 1st MTP joint may continue to be effective and efficient for gait, with its medial to lateral vectors remaining balanced. The stability is helped if the medial MTP joint fulcrum is well accommodated within the angle of gait and line of progression of the lower limb for the application of plantarflexor power. Such mechanics may explain why hallux abducto valgus can seriously disturb gait for some individuals, yet not for many others. *(Permission www.healthystep.co.uk.)*

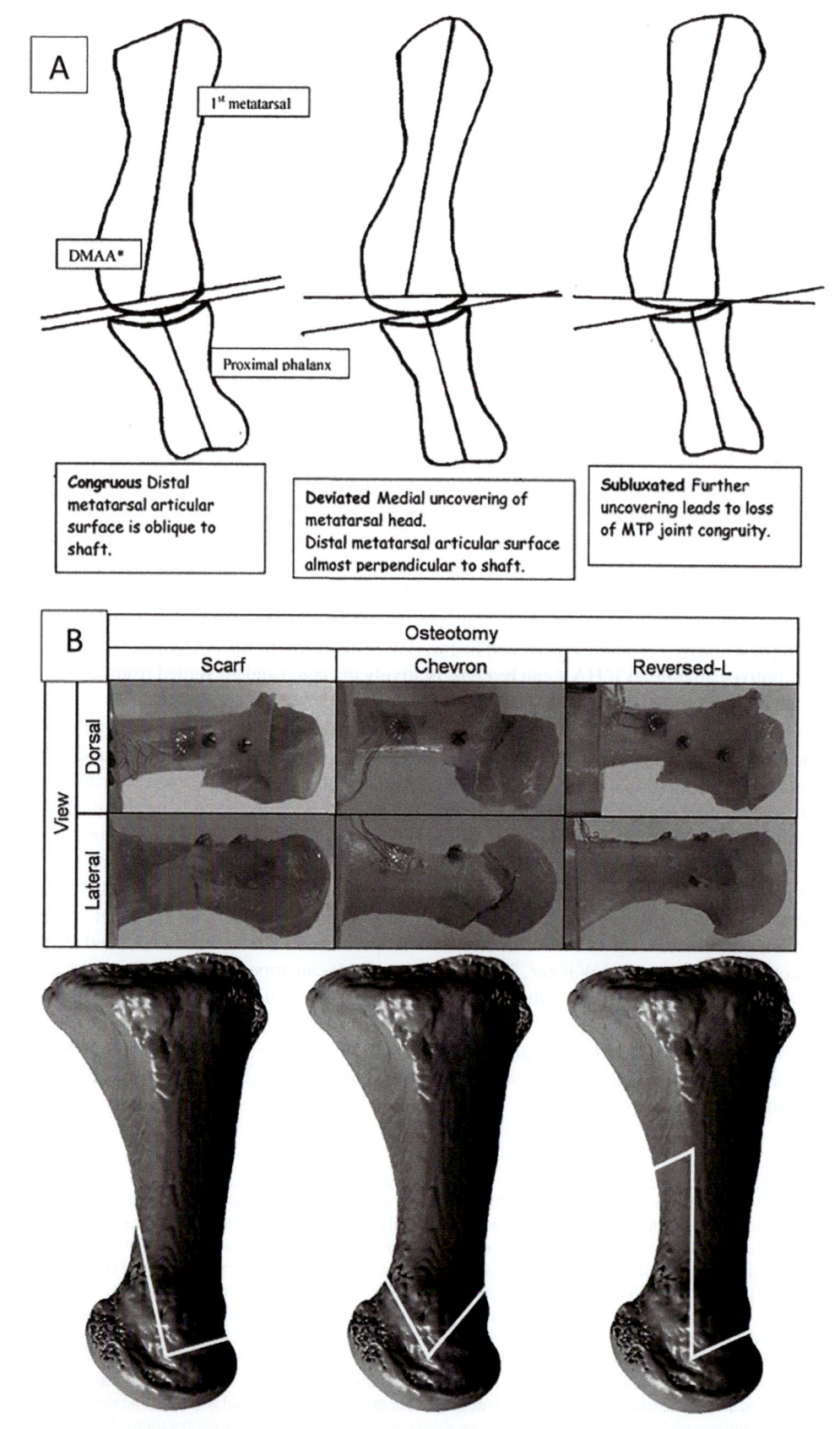

FIG. 4.5.13a HAV is a nonhomogeneous deformity, with the same degrees of hallux abduction able to develop on the metatarsal head arising from different areas of deformity. The articular surface of the 1st MTP joint can be orientated obliquely to the metatarsal shaft (distal metaphyseal articular angle—DMAA), taking the digit with it or the digit can sublux on the articular surface causing loss of articular congruency, as demonstrated in image (A). It is therefore unlikely HAV arises from only one pathomechanical origin. Generally, HAV surgery, such as distal head osteotomies (B), tends to be chosen more by the degree of deformity rather than via reflection on its origin. *(Image (A) from Thomas, S., Barrington, R., 2003. Hallux valgus. Curr. Orthop. 17 (4), 299–307. Images (B) from Favre, P., Farine, M., Snedeker, J.G., Maquieira, G.J., Espinosa, N., 2010. Biomechanical consequences of first metatarsal osteotomy in treating hallux valgus. Clin. Biomech. 25(7), 721–727.)*

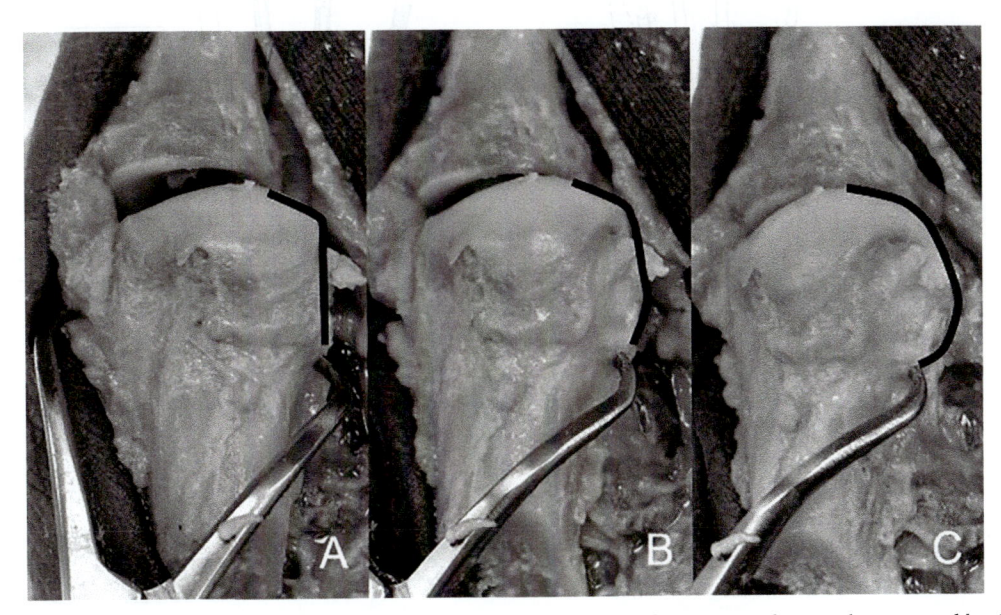

FIG. 4.5.13b A cadaveric specimen of the 1st metatarsal can demonstrate how eversion of the metatarsal causes the metatarsal head to appear more rounded when viewed dorsally. In (A), with the medial side to the left, the metatarsal is held relaxed in its anatomical position within the clamp. In (B), 15° of eversion has been applied, and finally in (C), 30° eversion has been applied. The outline of the lateral aspect of the metatarsal head becomes increasingly rounded with higher degrees of eversion. Thus, dorsal–plantar radiographs (known as anterior–posterior or A-P views) tend to capture this profile and have caused an association with HAV and a rounded metatarsal head deformity that is not really there, being just an optical illusion of rotation. *(Image from Wagner, E., Wagner, P., 2020. Metatarsal pronation in hallux valgus deformity: a review. J. Am. Acad. Orthop. Surg. Glob. Res. Rev. 4(6), e20.00091.)*

HAV is usually classified by a transverse plane angle on a dorsal view radiograph between the 1st and 2nd metatarsals of 15°. The valgus rotation of the 1st metatarsal is often called varus (Thomas and Barrington, 2003) because of the varus rotation in relation to the midline of the foot, and valgus rotation in relation to the midline of the body. The reality is that the 1st metatarsal everts and dorsiflexes, the hallux abducts and rotates into valgus from the midline of the body with the medial joint capsule and ligaments becoming 'stretched' through adaption, allowing the hallux to rotate laterally. In time, this change in hallux-to-metatarsal position takes the flexor and extensor tendon's distal attachments into lateral orientation, so that the tendons 'bowstring' to the lateral side of the 1st MTP joint as the 1st metatarsal everts. This pulls the sesamoid complex laterally, risking sesamoid subluxation. Four different scenarios are reported with HAV relative to the relationship between the metatarsal head and sesamoids: HAV with and without an everted 1st metatarsal, and with and without subluxation (Kim et al., 2015). In each situation the 1st MTP joint vectors change with HAV (Fig. 4.5.13c).

The degree of 1st metatarsal eversion rotation seems quite variable, with reported figures of between 4° and 27° (Wagner and Wagner, 2020). It is important to note that semiweightbearing increases 1st metatarsal eversion, both in individuals with and without HAV (Watanabe et al., 2017; Ota et al., 2019). This is something reported in cadaveric studies where 1st metatarsal dorsiflexion couples with eversion (inversion towards the 2nd metatarsal) (Kelso et al., 1982). This suggests that 1st metatarsal eversion is, to some degree, a normal part of 1st metatarsal weightbearing kinematics, a rotation possibly playing an important part of the foot's transverse plane stiffening mechanism through tightening by stretching and twisting the medial side of the deep transverse metatarsal ligament.

Through its complex anatomy, if the 1st metatarsal starts to evert, it takes MTP joint and sesamoid complex structures with it. As the deformity starts to become more long-standing, the deep transverse metatarsal ligament and adductor hallucis tendon maintain their lateral sesamoid attachment. This creates a constant lateral force vector with the now more lateralised positioned long and short flexor tendons, while the abductor hallucis tendon finds itself more dorsally orientated on the base of the proximal phalanx of the hallux. In time, the muscle increasingly becomes a weak hallux dorsiflexor rather than a hallux abductor. With loss of the main medial force vector at the 1st proximal phalanx and MTP joint, lateral force vectors gain mechanical advantage. They slowly stretch the medial joint capsule and pull the sesamoids laterally, potentially resulting in their dislocation (Fig. 4.5.13d). Considering that human 1st metatarsals have an average eversion rotation of 8° (Drapeau and Harmon, 2008), some individuals are likely to demonstrate greater eversion rotations than others, possibly increasing their potential HAV risk.

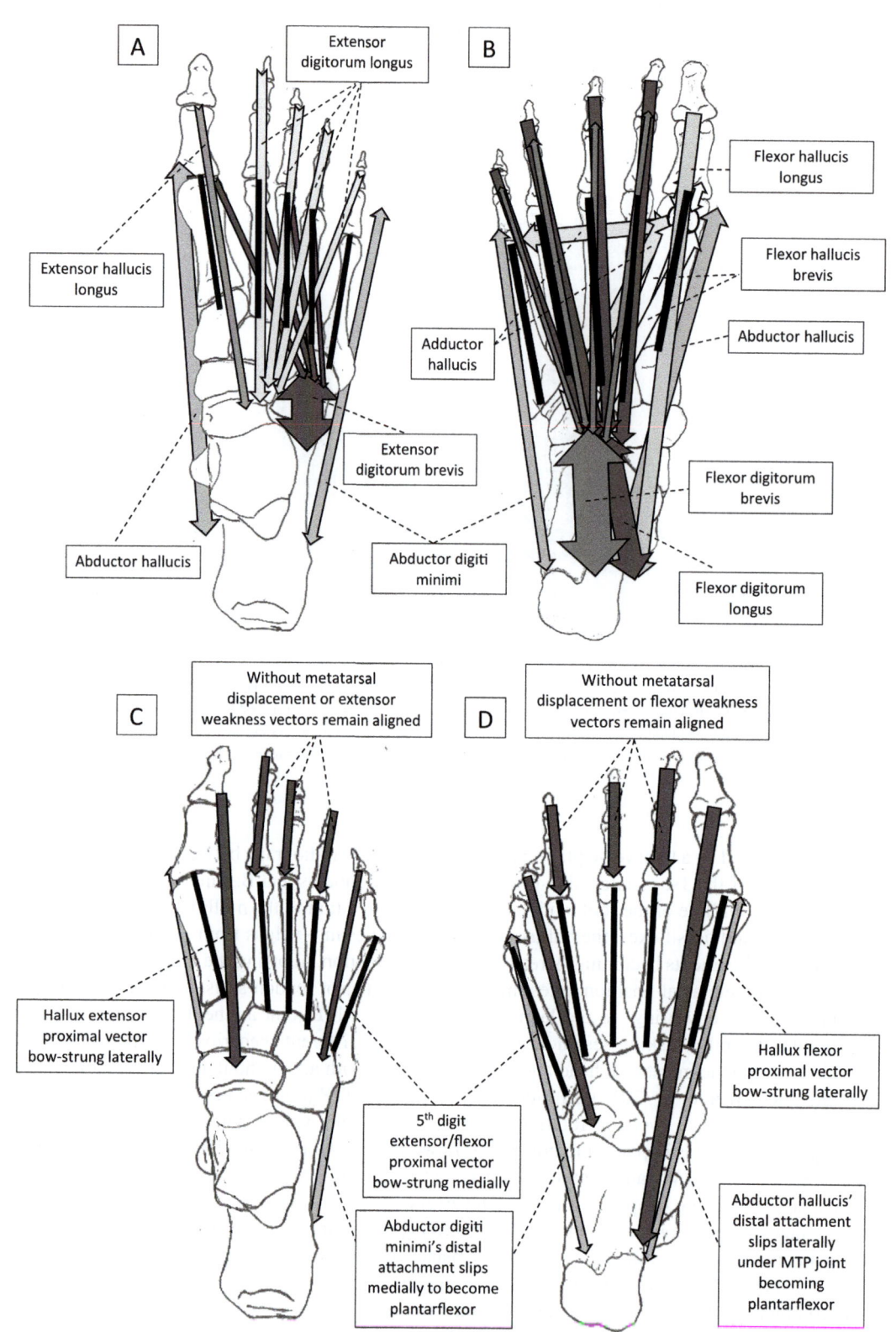

FIG. 4.5.13c Schematics of the primary muscular vectors maintaining sagittal digital posture to the long axis of the metatarsals (A dorsal view, B plantar view). Interosseous and lumbrical muscles are not shown to aid image clarity, but their small vectors are also important in aiding digital vectors to remain aligned to their metatarsals' long axis (solid black lines) across MTP joints. The abductor hallucis adduction moment on the hallux and the abductor digiti minimi abduction vector at the 5th toe are essential to aligning the 1st (via adduction) and 5th digits (via abduction) with their spreading metatarsals' long axes. By keeping digital muscular vectors central to the metatarsal long axis, MTP joint extensions should occur with minimal associated frontal or transverse plane torques. The vectors the extensor muscles generate during loading response via digital extension for setting the windlass mechanism are relatively small compared to the passive extension-resisting forces developed during late midstance for terminal stance acceleration. The flexor muscle vectors (B) are much larger and more significant for MTP joint stability, for they resist the GRF force-induced extension moment that is part of ankle and midfoot plantarflexion initiated by the plantarflexor power. The 1st and 5th MTP joints require abductor hallucis and abductor digiti minimi to help balance their vectors because these metatarsals are prone to splay more in the transverse plane on loading than the central metatarsals. For an ex-gripping appendage, this arrangement is an advantage, as strong centrally pulling flexors improve flexion grip whereas extensors only open up the digits to let go. In HAV deformity (C and D), the 1st metatarsal's long axis and the digital flexor proximal vectors become misaligned, allowing the tendons and vectors to bowstring laterally. This sets up the mechanics of increasing metatarsal and digital misalignment deformity. Long-term tissue restructuring is a consequence of Davis's law. The central metatarsals are intrinsically more stable, unless they too start excessive displacement within the transverse plane. (Permission www.healthystep.co.uk.)

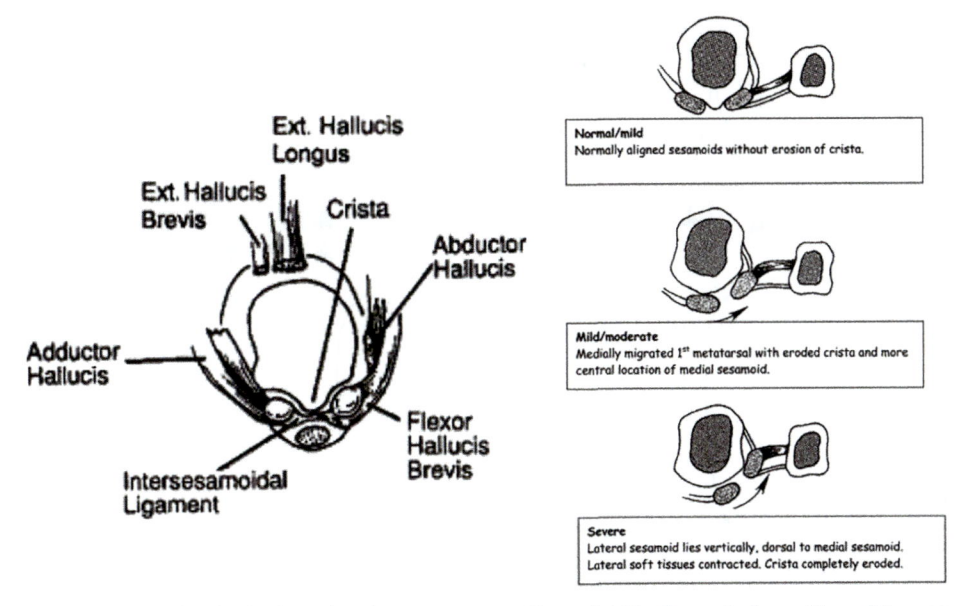

FIG. 4.5.13d The degree of HAV deformity has been thought to cause reasonably predictable changes in the positions of the anatomy as viewed in the frontal plane. However, the sesamoids may follow the metatarsal head's rotations, remaining relatively congruent to their articulations or as schematically demonstrated here, the medial sesamoid may erode the crista, allowing the sesamoids to sublux laterally. *(Images from Thomas, S., Barrington, R., 2003. Hallux valgus. Curr. Orthop. 17(4), 299–307.)*

HAV leads to a multitude of deformities and local pathologies. It has also been indicated through biopsy that the bone of the 1st metatarsal head compared to that sampled from non-HAV-affected metatarsal heads has a higher porosity, with change being proportional to the level of deformity (Lamonaca et al., 2014). This results in changes in the elastic property of the bone, permitting increasing deformability through reduction in the inorganic framework under unbalanced 1st metatarsal loading stresses (Lamonaca et al., 2014).

The size of the HAV, as far as foot function is concerned, matters. Milder HAV is set at a hallux abduction angle of between 15° and 30° and a 1st intermetatarsal angle of less than 13°, demonstrated on radiographs. Milder HAV demonstrates altered plantar pressure distribution compared to controls, but with greater deviations, larger changes occur (Martínez-Nova et al., 2010). Increased pressures are reported under the hallux and 1st MTP joint, the latter correlating with greater symptoms during activity (Martínez-Nova et al., 2010). This would support the model of increasing adduction of the 1st metatarsal, requiring hallux IP joint motion to compensate for a functional shortening of the medial fulcrum lever arm resulting from 1st metatarsal adduction, its functional shortening, and its elevation.

There could be a point where the adduction translation reduces the functional sagittal length of the 1st metatarsal so much that the 1st metatarsal is no longer a functional medial fulcrum option. This will necessitate an alternative, perhaps at the abducted hallux's IP joint. It is possible that the increasing plantar hallux pressure reported by Martínez-Nova et al. (2010) is an indication of this happening. However, these reported correlations are weak, and the higher hallux pressures are not significant, although in larger HAV deformities, more medial foot pressures persist compared to controls (Plank, 1995). This may indicate a point where HAV deformity cannot continue to maintain a medial MTP joint or hallux fulcrum (Fig. 4.5.13e).

Imbalance in stability across the forefoot can also lead to other lesser toes deformities. Although milder HAV often exists as a deformity in isolation, 2nd hammer toes, 3rd through to 5th adductovarus hammer toes, and tailor's bunions are commonly associated with more severe HAV. Although more research is certainly needed, it can be suspected that once HAV is established, transverse forefoot stress-stiffening mechanisms may become compromised, which in turn may disturb the stiffness capability to apply plantarflexor power to the forefoot effectively. However, this dysfunction is likely variable, and in milder cases, it is quite possible that forefoot stiffness remains adequate to maintain efficient and stable gait. This may explain the difference between symptomatic HAV and those without symptoms or gait dysfunction. Undoubtedly, footwear choice accommodation plays a considerable part in driving patients to seek medical assistance.

Control Group   **Hallux Valgus Group**

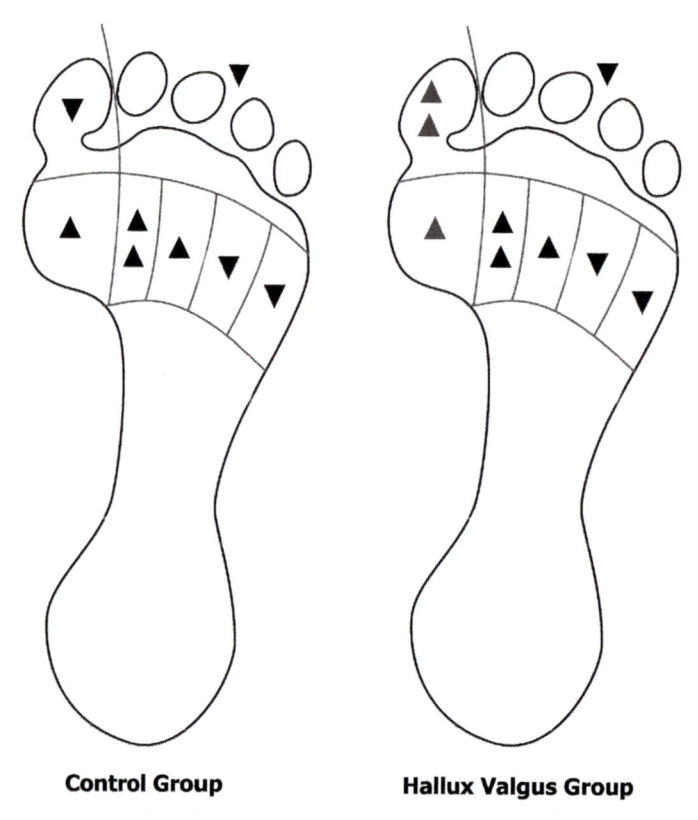

FIG. 4.5.13e A schematic of the typical trace of plantar pressures in controls (left) and HAV group (right), with lighter grey arrows under the 1st MTP joint and hallux showing areas of significant difference. *(Image from Martínez-Nova, A., Sánchez-Rodríguez, R., Pérez-Soriano, P., Llana-Belloch, S., Leal-Muro, A., Pedrera-Zamorano, J.D., 2010. Plantar pressures determinants in mild hallux valgus. Gait Posture. 32(3), 425–427.)*

The origins of HAV

Although the mechanism of deterioration seems easily explained, the origin of HAV remains elusive (Kilmartin and Wallace, 1991; Coughlin and Jones, 2007; Wagner and Wagner, 2020). It is possible that 1st metatarsal eversion and dorsiflexion coupling at the cuneometatarsal joints may be the source of instability (Wanivenhaus and Pretterklieber, 1989; Mortier et al., 2012). A correlation between 1st metatarsal eversion ranges and flatter foot profiles is also reported (Cychosz et al., 2018). Yet, there also seems to be something that lies in inheritance and lifestyle (Coughlin and Jones, 2007; Nery et al., 2013; Hollander et al., 2017). This suggests that HAV works at the interface of developmental biology and biomechanics. It does not appear to be genetic in origin, despite earlier indications of a genetic link (Piqué-Vidal et al., 2007; Hannan et al., 2013; Nery et al., 2013). An inherited link rather than a link to genes is supported by data from HAV rates in identical (monozygotic) and nonidentical (dizygotic) twins, which indicates that dizygotic twins are just as likely (or unlikely) to share HAV deformity as monozygotic twins (Munteanu et al., 2017). Yet, HAV certainly does associate with something inheritable (Piqué-Vidal et al., 2007; Hannan et al., 2013; Nery et al., 2013; Munteanu et al., 2017). Of course, lifestyles as well as genes can be inherited, but so can conditions such as ligamentous laxity (Hakim et al., 2004; Malfait et al., 2006), and foot vault types, although planus foot types do not appear to be a risk factor (Kilmartin and Wallace, 1992).

Therefore, it is possible that most HAV deformities owe their existence primarily to lifestyle choices such as the use of toe and foot-constrictive footwear (Coughlin and Jones, 2007). This is likely a result of interference in the digital flexor proximal vector direction and its power. Some footwear may influence the biomechanics of the 1st ray, 1st metatarsal, and 1st MTP joint, altering local force vectors and redirecting the proximal force from the digit away from the long axis of the metatarsal. The inherited morphological features, such as long 1st metatarsals and metatarsus primus adductus, may explain some family tendencies towards the condition, although the latter could in fact be a sign of developing HAV (Kilmartin and Wallace, 1992; McCluney and Tinley, 2006). Involvement of the roundness of shape of the 1st MTP joint articular surface was once considered a potential cause (Kilmartin and Wallace 1991; Ferrari and Malone-Lee, 2002), but eversion of the 1st metatarsal changing the view angle of the metatarsal head on radiographs seems to explain and resolve this issue (Wagner and Wagner, 2020).

HAV is reported to develop in around 12.3% of 9-year-old British school children (Kilmartin et al., 1994). Kilmartin et al. (1994) observed that of the 122 children found with HAV among 6000 school children, 16 were male and 106 were female. With gait becoming more adult-like around ages 7–9 years, we could expect to see an early group of developers of HAV among this age group, as supported by Kilmartin et al's data. Yet, if HAV developed from intrinsic biomechanical factors alone, we should expect to see males and females equally affected, at least earlier in life before changes from fashion footwear environments might influence rates of development. However, fashion differences in footwear between boys and girls are variably significant very early in shoe selection. Another possible issue might be connective tissue hypermobility, because young females are more prone to display this trait. Hypermobility is reported to have a significant genetic influence, being more common in concordant monozygotic female twins than in dizygotic twins, although environment and lifestyle are factors that also play their part (Hakim et al., 2004).

It has been suggested that males have feet more resistant to increases in intermetatarsal angles after age 11 years than in females (Gottschalk et al., 1984). Male HAV cases are reported to show strong heredity from their female line, an earlier onset, higher severity on average, and deformity mostly associated with changes in the distal metatarsal articular angle, rather than via metatarsal eversion (Nery et al., 2013). Such data suggest a stronger inheritance link in present day male cases of HAV than in those of females. Today, HAV is largely considered to be a female-dominant pathology with reported ratios of 15:1 female to male (Nery et al., 2013). This does not always seem to have been the case, as historical samples suggest that males once dominated HAV rates (Mafart, 2007).

Palaeopathological studies indicate that footwear, through the principles of environmental mismatch, may be biomechanically influencing HAV development. British skeletal remains from earlier and later medieval burials have shown HAV as being totally restricted to the late medieval period, when popularity among the rich developed a taste for pointed, constrictive footwear (Mays, 2005; Dittmar et al., 2021). In these late medieval burials, HAV prevalence is equal among men and women (Mafart, 2007). The prevalence of HAV increases in burials among men in 16th- and 17th-century France, possibly reflecting the popularity of high-heeled, stiff leather shoes among men of the time (Mafart, 2007) (Fig. 4.5.13f). This model of a stiff, constrained footwear influence fits in with the absence of HAV reported in Japan until the mid-to-late 20th century (Kato and Watanabe, 1981).

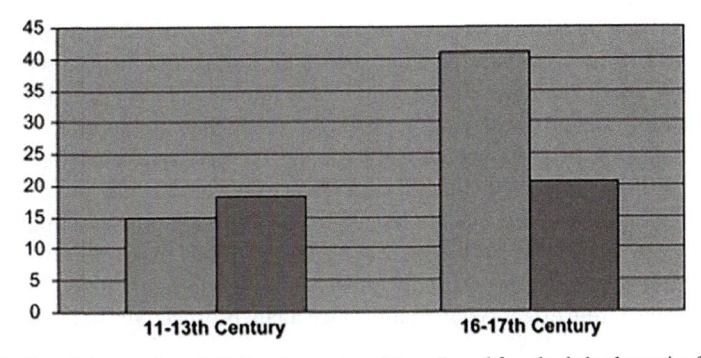

FIG. 4.5.13f The prevalence of hallux abducto valgus (HAV) is demonstrated in male and female skeletal remains from the 11th to 13th centuries and 16th to 17th centuries in a burial site at Notre-Dame-du-Bourg in Digne, France. Males indicated by light-grey and females in dark-grey blocks. While ratios are fairly even within the earlier medieval period when footwear styles of men and women were similar, the later trend for higher heels and more pointed shoes, especially among men, increased the overall prevalence and the ratio of male sufferers of HAV to female sufferers. *(Image from Mafart, B., 2007. Hallux valgus in a historical French population: paleopathological study of 605 first metatarsal bones. Joint Bone Spine 74(2), 166–170.)*

A historical change in the dominance of male to female rates of HAV and the late development of the deformity in some populations, suggests footwear style may play a prominent part in HAV pathomechanics. When comparing feet that have developed in footwear compared to those brought up barefoot, the habitual shod feet demonstrate a lower resting vault height and lower hallux abduction angles with the 1st metatarsal, which seems to indicate an inward hallux drift towards the midline of the foot with footwear use in childhood and adolescence (Hollander et al., 2017). Perhaps the inability to allow toes to splay with metatarsals through footwear constriction could be setting conditions and biomechanics around the hallux for increased HAV risk via altered digital vectors, foot strength, and forefoot adaptability. The majority of modern HAV cases studied are therefore probably due to individual environmental/lifestyle factors that influence biomechanics, such as activity levels and particularly, footwear use. Yet, it is likely that such cases are still affected to a more marginal degree by genetic and developmental factors, that can be inherited. This might explain the emergence of new lines of HAV inheritance that could develop once a population changes to a new set of environmental factors, such as enclosed restrictive footwear. Again, the reports of relatively recent population expressions of the condition in Japan seem to support this (Kato and Watanabe, 1981).

Metatarsal splay-induced transverse vault-stiffening mechanisms and MTP joint stability vectors may lie at the heart of this common foot deformity, and may help account for the interaction of inherited, lifestyle, and developmental aspects of the condition. Like the 5th metatarsal, the 1st metatarsal lacks an interosseous ligament at its tarsometatarsal joint. This increases the potential for medial translation and eversion of the 1st metatarsal to achieve deep transverse metatarsal ligament tightening. Both the hallux and the 5th digit may need to orientate more obliquely to the foot to maintain proximal stability into their respective MTP joints. This would require moving into an increasingly medial position for the hallux and moving more laterally for the 5th digit. This aids in maintaining straight lines of force vectors through the toe flexors along the long axis of the shaft of their respective metatarsals, to achieve proximal directed MTP joint stability. Footwear might prevent such digital adjustment (Fig. 4.5.13g).

Feet requiring greater forefoot widening to initiate the transverse vault-stiffening mechanism through a more flexible deep transverse metatarsal ligament, might need more space for metatarsal splay. The distal digits would require even more space to spread to align their proximally directed vectors appropriately. This might be space denied them by tighter-fitting modern footwear. Footwear might also influence the ability to induce adequate transverse tightening of the forefoot by preventing adequate metatarsal splay. Through digital compression restraint medially and laterally, the hallux and 5th digit may be unable to align to their respective metatarsal shaft to cause proximal MTP joint stability. Instead, they will angle their digital flexor vectors lateral to the 1st and medial to the 5th MTP joints. This supplies a large adduction (1st) or abduction (5th) torque on the metatarsal in the transverse plane. The use of footwear in children and adolescents is shown to alter muscle function, reduce foot width, forefoot spreading and vault height on loading during gait (Franklin et al., 2015; Hollander et al., 2017), offering yet another compounding mechanism to alter the biomechanics across the forefoot. Some evidence is emerging that decreasing the transverse vault of the foot has an influence on the severity of varus alignment of the 5th toe among children at age six (Puszczalowska-Lizis et al., 2022), evidence which may support the link between the forefoot transverse plane stability and digital alignment with footwear use (Fig. 4.5.13h).

This does not mean that loss of medial–lateral balance of force vectors of the 1st MTP joint would always result in HAV. As already discussed, other conditions related to poor 1st MTP joint motion, such as sesamoiditis, could be the only pathology that results from force vector displacement. However, metatarsus adductus has been shown to be a significant statistical factor in the development of juvenile HAV (Banks et al., 1994; Ferrari and Malone-Lee, 2003) which would be expected if 1st metatarsal adduction translations were significant to the deformity. Intermetatarsal angle increases have also been noted to occur in children before changes in the alignment of the hallux have occurred (Kilmartin et al., 1991). Hypermobility and plantar gapping of the 1st tarsometatarsal joint is also found in HAV (Coughlin and Jones, 2007), which supports a model of instability at this more proximal joint in playing a role both in the sagittal and transverse planes. This could link well into the findings of increased 1st tarsometatarsal joint and 1st metatarsal eversion associated with HAV (Mortier et al., 2012; Watanabe et al., 2017). It is possible that this is a result of developmental problems in angling the 1st tarsometatarsal joint, which starts its life more convex and ape-like in juvenile humans only reaching an adult-like flatter and more distally orientated morphology by age 6 years (Gill et al., 2015). This certainly makes this a potential footwear development influencing target during the first years of walking. Sadly, most studies involve data found in the presence of existing HAV, making the disassociating of cause from effect problematic. Once again prospective studies following lifestyle choices may be required to understand the origins of HAV.

Length of the 1st metatarsal could also have an effect through the function of the medial MTP fulcrum. A long 1st metatarsal may prevent the fulcrum axis from associating with the line of progression of the foot and lower limb. By adducting the 1st metatarsal, the functional sagittal length of the 1st MTP joint fulcrum axis could be reduced. Such a change in 1st MTP joint orientation could also prevent sagittal plane joint impingement at the 1st MTP joint, but the price could be a hallux valgus deformity risk, with the benefit of a more efficient gait pattern due to force–velocity-to-plantarflexor metabolism relationship, for ankle plantarflexion power use. To support this possibility, associations have been reported with a positive 1st metatarsal protrusion distance found in juvenile HAV (McCluney and Tinley, 2006) and with increased 1st metatarsal lengths in adult HAV (Coughlin and Jones, 2007; Munuera et al., 2008). Interestingly, this difference has been reported as being more marked in male feet (Munuera et al., 2008) (Fig. 4.5.13i).

The confounding factors of lifestyle, inheritance, and the resulting biomechanics may dictate the chance of turning risk factors into the reality of HAV. It is likely that the origins or risks lie in the evolutionary and the developmental history of each human foot, but some individual's feet are far more at risk than others. Connective tissue mobility lies on a continuum from extremely hypermobile through to very hypomobile, and represents connective tissue properties that change throughout life. In those with genetically induced connective tissue hypermobility, attempting to induce forefoot stiffness may require greater transverse translation between the medial metatarsals in order to achieve an effective medial MTP joint fulcrum and stable transverse vault. This need to achieve greater forefoot splay may destabilise the tarsometatarsal joints development, and/or perhaps, the development at such joints is compromised through the use of footwear (Fig. 4.5.13j). The quest for the causes of HAV (rather than a cause) continues.

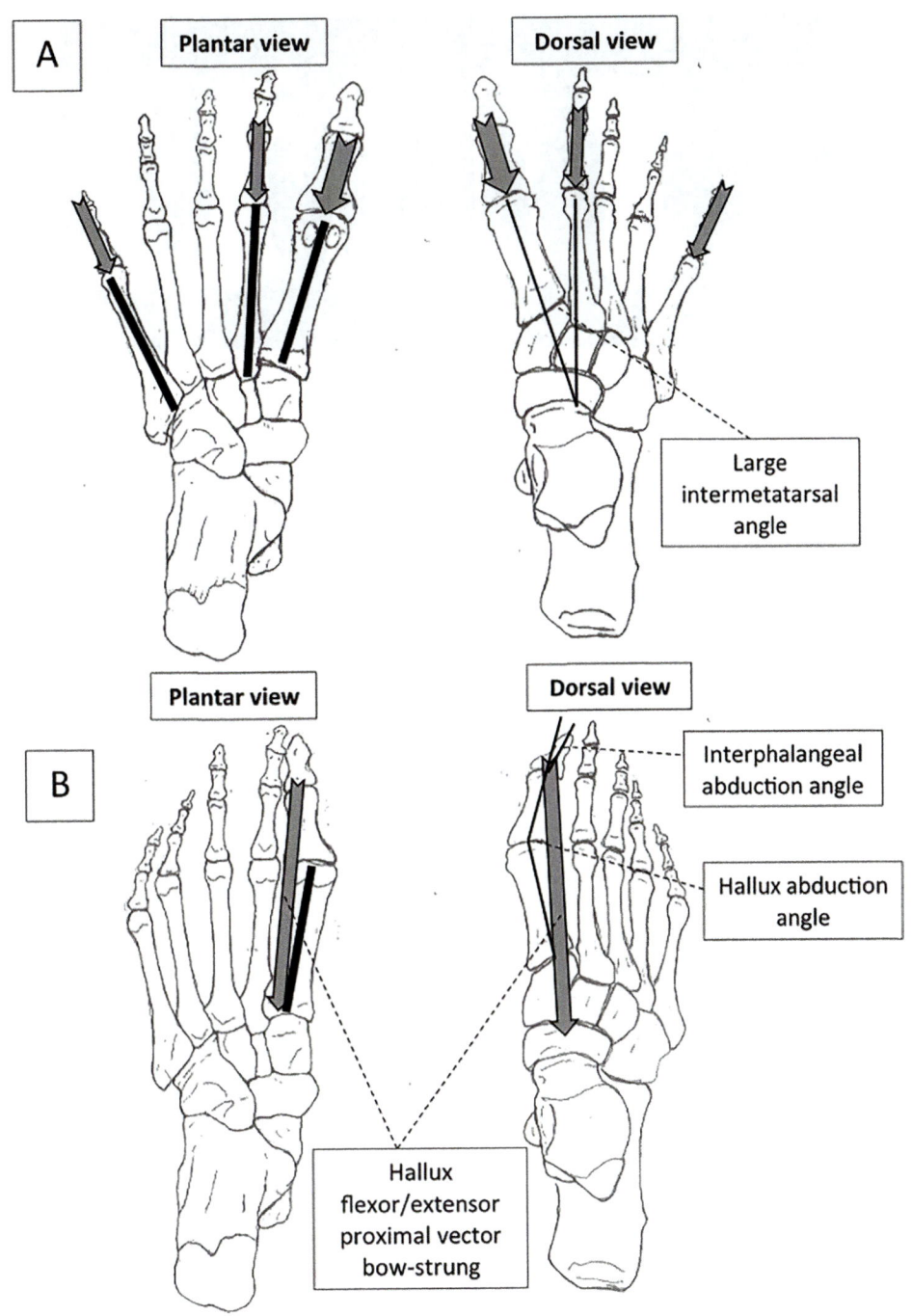

FIG. 4.5.13g HAV is a nonhomogeneous deformity with a highly variable degree of deformity. In some, deformity is large at the tarsometatarsal joints, creating sizeable intermetatarsal angles while in others, deformity is mainly at the MTP joint creating high hallux abduction angles. Deformity may also occur at the IP joint creating hallux abductus interphalangeus. Where the deformity develops likely reflects combinations of developmental and environmental mismatch factors at play. Those children that do not develop a less medially angled 1st metatarsocuneiform (tarsometatarsal) joint during the early years of growth and gait from that found at birth are likely to demonstrate more 1st metatarsal adduction angles, and thus large metatarsal splay on forefoot weightbearing (A). There will be considerable variability in this motion, reflecting the joint angle and the soft tissue anatomy and flexibility developed at this joint. Large degrees of transverse plane deviation of the 1st metatarsals required for deep transverse metatarsal ligament stiffening should not necessarily result in HAV development. As long as the digital flexor and extensor proximal vectors (grey arrows) remain directed along the line of the long axis of the metatarsal head and shaft (solid black lines), large transverse plane motions of metatarsals should only result in a wider forefoot, especially at the level of the digits. Habitually barefoot individuals usually demonstrate the widest part of the foot across the digits, whereas this is not usually the case within habitually shod populations. Where intermetatarsal angles are not particularly large (B) or indeed remain within averages, deformity may be more pronounced at the hallux abduction angle within the MTP joint and at the IP joint. Given time, the displacement of the digital flexor proximal vectors may start to change tarsometatarsal joint angles due to the biological response of bone to changing articular biomechanics. The nature of HAV therefore reflects individual biomechanics found from the tarsometatarsal joints to the digits over a lifetime. *(Permission www.healthystep.co.uk.)*

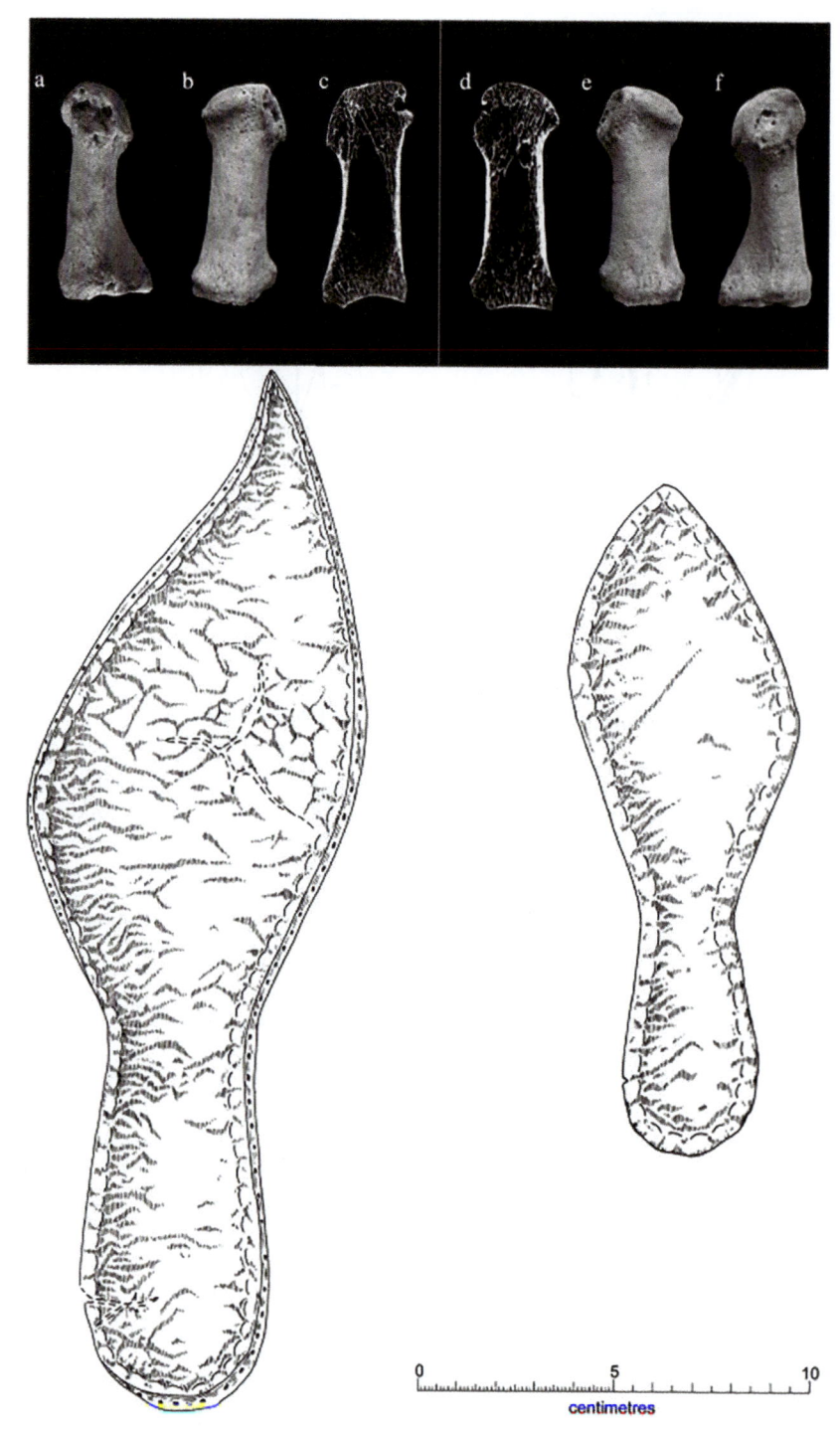

FIG. 4.5.13h Photographs (upper image) of lesions associated with hallux abducto valgus (HAV) found in adult female skeletal remains from medieval Cambridge. They demonstrate lytic cavitation at the attachment of the collateral and sesamoid suspensory ligaments in (A) and (F) on the medial aspect of the 1st metatarsal. Dorsal views of the 1st metatarsal in (B) and (E) demonstrate lateral deviation of the articular surface. CT images (C) and (D) show lytic lesions on the medial side of the metatarsal heads. The lower image consists of sketches of 14th-century leather shoes known to have been worn in Cambridge at the time the individual lived. These consist of (left) an adult's left foot turnshoe sole and (right) a child's turnshoe sole for a right foot. Skeletal remains from medieval periods before such shoe styles were developed do not show osseous signs of HAV. *(Images from Dittmar, J.M., Mitchell, P.D., Cessford, C., Inskip, S.A., Robb, J.E., 2021. Fancy shoes and painful feet: hallux valgus and fracture risk in medieval Cambridge, England. Int. J. Paleopathol. 35, 90–100.)*

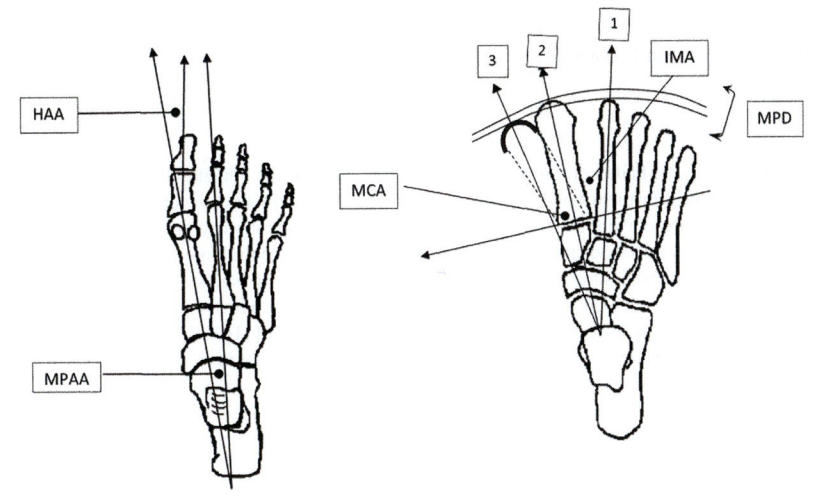

FIG. 4.5.13i A number of measurements of skeletal alignment are used to assess HAV via dorsal–plantar (A-P) view radiographs. The hallux abductus angle (HAA) records the deviation of the proximal phalanx of the hallux to the long axis of the 1st metatarsal shaft. Metatarsus (primus) adductus angle (MPAA) is determined by a bisection line running through the long axis of the 2nd metatarsal and calcaneus to the long axis of the 1st metatarsal. The metatarsocuneiform angle (MCA) records the angle between the proximal articulation at the metatarsal base to the long axis of the shaft of the 1st metatarsal. This gives some indication of the obliquity of the 1st tarsometatarsal joint, an angle that should reduce during the first six or so years of life. The intermetatarsal angle (IMA or sometimes just IM) is the angle between the long axis of the 2nd metatarsal and the 1st (between 1 and 2). With HAA, IMA is the commonest measurement recorded by clinicians. Increasing IMA and MCA are associated with increasing 1st metatarsal adduction which changes the metatarsal protrusion distance (MPD) between the 1st and 2nd metatarsals to the long axis of the foot. Increasing the IMA from positions 1–2 to that of 1–3 effectively shortens the functional length of the 1st metatarsal (thick black metatarsal head outline compared to dashed shaft outline). These measurements do not record frontal plane alignments and only indicate the primary location of the deformity at either the 1st MTP joint or tarsometatarsal joint. Such data tell us little about the origins of the deformity. In the early development of children, the osseous and articular alignment and shape changes of the 1st tarsometatarsal joint may be important, but with increasing age, instability at the tarsometatarsal joints or 1st MTP joint itself may become the focus of dysfunction that sets off the development of HAV. *(Images modified from McCluney, J.G., Tinley, P., 2006. Radiographic measurements of patients with juvenile hallux valgus compared with age-matched controls: a cohort investigation. J. Foot Ankle Surg. 45(3), 161–167.)*

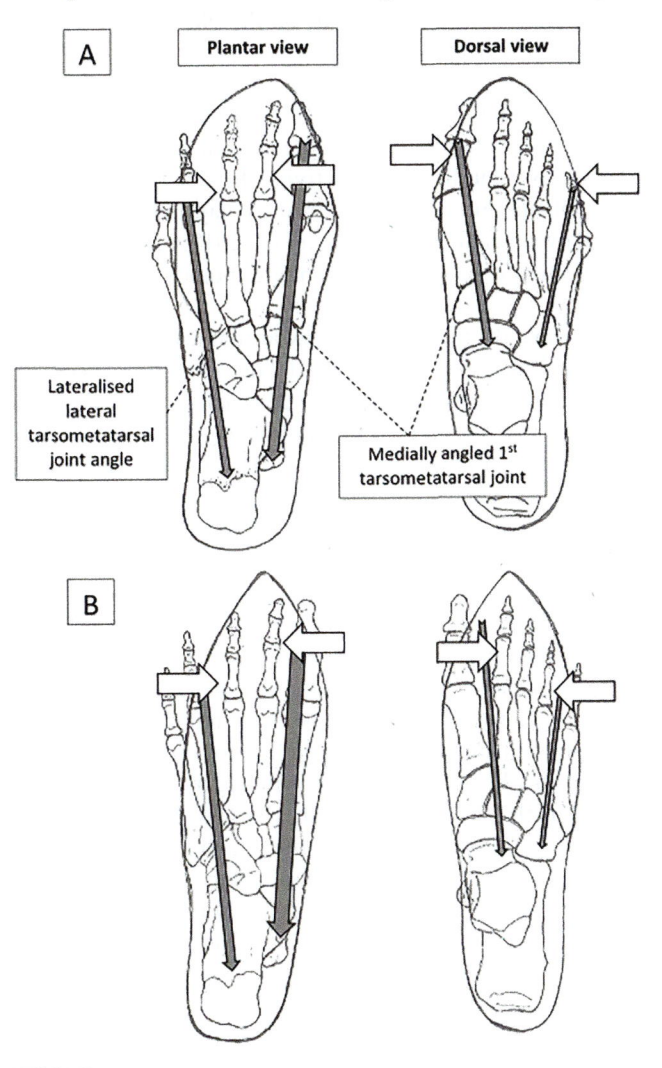

FIG. 4.5.13j See figure legend on next page

FIG. 4.5.13j, CONT'D Footwear may alter hallux proximal vectors (and often also for the 5th and 4th digits) due to restricted toe boxes that cannot accommodate digital splay, instead compressing them in the transverse plane. When the forefoot has developed a particularly wide loaded posture, such as is seen associated with high medial angulation of the 1st metatarsocuneiform joint or with lateralised lateral tarsometatarsal joints of the cuboid, even shoes that are perceived as having a relatively rounded toe box may still displace digital vectors via central foot-directed compression (A). Fashion footwear often tapers towards a point at the apex of the 2nd and 3rd toe area that can expose the MTP joints to persistent offsetting of digital vectors (B). This can occur even when intermetatarsal angles are small and aligned in a straight (rectus) manner. Children with more medially orientated 1st tarsometatarsal joints are likely to be more prone to the environmental influence of footwear, risking the development of HAV deformity in late childhood and adolescence as they require greater free digital spreading to maintain vector-to-metatarsal stability. However, given time with enough provocative footwear-induced digital vector displacements, some adults will succumb to HAV and other digital deformities. This is especially so if they also develop weakness within the muscles that provide forefoot and vault stability during late midstance and early terminal stance, and/or if footwear is also undersized. Thus, under the influence of vector offsetting, the level and degree of HAV deformity (and/or tailor's bunion) likely reflects the individual developmental and environmental biomechanics of each foot. *(Permission www.healthystep.co.uk.)*

4.5.14 Metatarsal fatigue fractures

The passive and active elements of the forefoot not only provide the stability to permit forefoot stiffening through initiating changes that increase transverse vault curvature, but also their tensile effects on joint compression and motion through biotensegrity principles are essential in limiting bending moments and torsional and shear stresses through the metatarsals. Bone being anisotropic has excellent strength under compression, but it is more vulnerable to tension-torsion shear strains. The curvature and profile of the metatarsals, which are cross-sectionally deeper in a dorsal-to-plantar direction than they are in their width in a medial-to-lateral direction, helps resist torsion within their shafts, tensile strain at their plantar surfaces, and transverse plane bending moments. To maintain high compression loading of stress, the metatarsals require soft tissue 'fixation' relative to the support surface. In particular, they require muscle contraction to convert bone tension as much as possible into compression strain. This is primarily achieved by digital flexor power creating tension-induced compressions on the plantar surface of the metatarsals, which are subjected to tension strains under sagging deflection on weightbearing.

In circumstances where the soft tissues fail to provide adequate compression across the metatarsals, increased torsional forces, bending moments, and translations in any or all orthogonal planes can result. Both loss of passive ligament tension and, particularly, muscular dysfunction due to fatigue or generalised weakness resulting from previously low activity levels, cause loss of compressive stabilisation. Inactivity also decreases tissue threshold, making injury at lower levels of stress far more likely. The plantar intrinsics and extrinsics play an important role in providing dynamic stability and stiffening of the forefoot (Fiolkowski et al., 2003; Kelly et al., 2012; Kokubo et al., 2012; Farris et al., 2019, 2020), protecting the metatarsals from the tensional effects of bending moments (Ferris et al., 1995; Sharkey et al., 1995; Jacob, 2001) (Fig. 4.5.14a).

Fatigue fractures can occur throughout the metatarsals at any anatomical location, dependent on the point of stress concentration, particularly if their bone mechanics are compromised. Such compromise could be due to osteopenia or low bone density as a result of previously low activity levels, bone diseases such as rickets/osteomalacia, osteogenesis imperfecta, dietary disorders such as anorexia and scurvy, and the effects of ageing of particularly, postmenopausal women. Females are generally more prone to stress fractures (Rizzone et al., 2017). There is also an association with reduced serum vitamin D_3 (Miller et al., 2016). Most metatarsal fatigue fractures occur in the metaphysis of the metatarsal neck. The metatarsal neck provides an area of least cross-sectional area in the diaphysis of the metatarsal, accentuated by the presence of the hominin evolved extensor groove which can concentrate sagging bending forces towards the neck region (Fig. 4.5.14b). An angled digital extension moment outside of pure sagittal motion may also increase the loading forces in the transverse and/or frontal plane, changing the stress dominance from compression to shear and torsional moments. This can alter the direction and nature of the bending forces. Such misalignment of the digital extension moment can result from flexor plate complex dysfunction.

Very high levels of activity are particularly prone to inducing metatarsal stress fractures. Foot strike position may also increase the risk of metatarsal stress fractures, as forefoot strikers are reported to generate greater 2nd metatarsal stresses during early stance, although peak stresses are comparable with rearfoot strikers throughout the whole of stance (Ellison et al., 2020). Internal bone geometry may be more significant with regard to internal forces than the external forces generated (Ellison et al., 2020). For example, the 1st metatarsal is largely exempt from fatigue fractures, with few cases reported in the literature (Minoves et al., 2011). This is probably due to its much larger cross-sectional area, particularly at the metatarsal neck. The ability of the 1st ray to translate dorsally may also help dissipate loading energy away from the bone and into articular/ligamentous tissues, altering tissue stress injury potential away from the 1st metatarsal bone to its articulations and ligaments. The sesamoid joint's extra mobility may also help dissipate loading stresses on the 1st metatarsal.

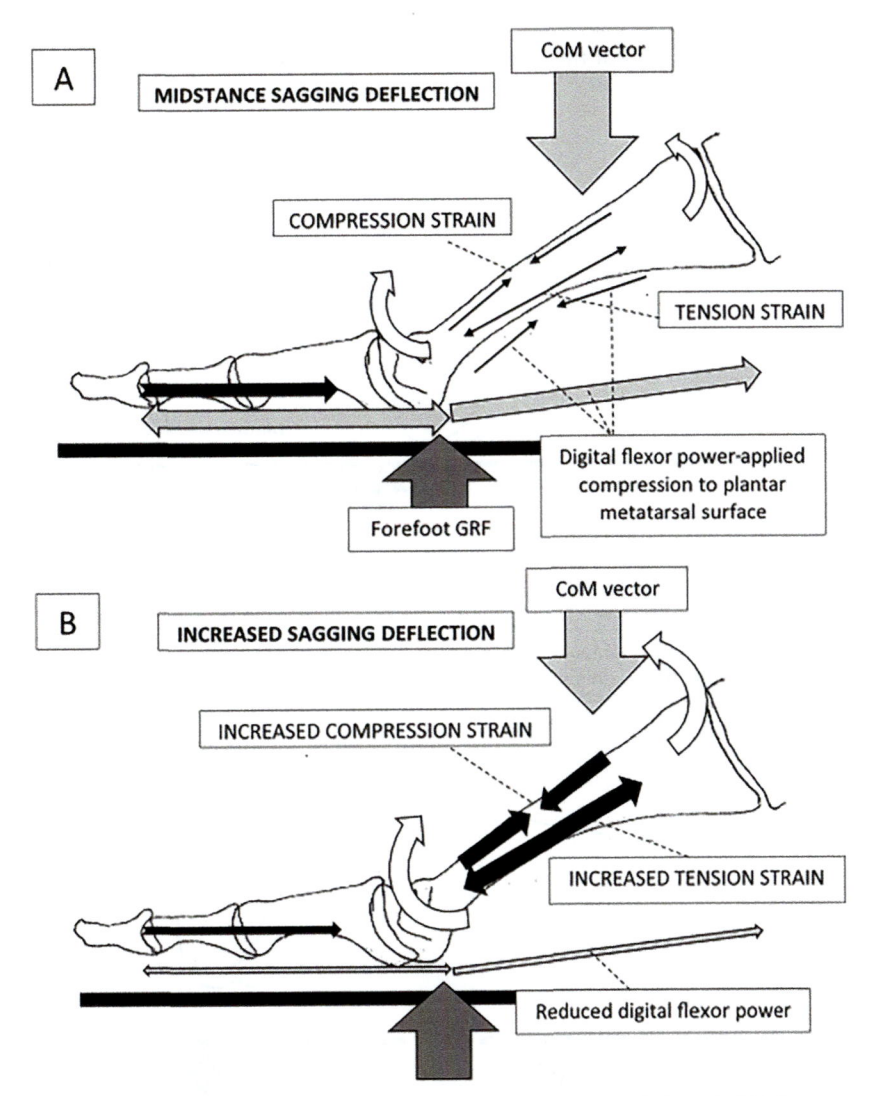

FIG. 4.5.14a The curved, dorsally convex profile of the metatarsal reflects its osseous stresses during gait. This is not an accidental relationship as developmental tissue biomechanics during growth are reflected within each adult metatarsal, such that greater weightbearing activity forms mechanically strengthened and structured metatarsals. In locomotion, metatarsals are exposed to sagging deflection during both midstance when the forefoot acts within the encastre beam and particularly as a cantilever during acceleration just after at heel lift. Thus, compression forces are concentrated to the convex dorsal surface and tensional strain towards the plantar concave surface. The plantar extrinsic and intrinsic force vectors running to and from the digits, bound into the MTP joints' connective tissues, are essential in creating tension-induced compression forces across the plantar vault to moderate osseous tensional loads from sagging. By doing so, they create the proximally directed digital flexor vectors (black arrows within digits). Factors such as low metatarsal cross-sectional areas (for physiological or mechanical reasons) and/or reduced plantar muscle strengths, risk raising osseous tensional and shear forces that could breach tissue threshold under repetition. Given time, rising bone oedema and microfractures can lead to metatarsalgia and even fatigue fracture, should tensional stresses within the affected metatarsal not be reduced. *(Permission www.healthystep.co.uk.)*

Restrictions in ankle dorsiflexion have also been associated with an increased metatarsal fatigue fracture risk, a risk thought to relate to premature heel lift and earlier and extended metatarsal head loading time (Dixon et al., 2019). This may also reflect increased resistance load (body weight) as a consequence of early engagement of the class two lever at heel lift. Restriction of ankle dorsiflexion might also increase the stress-stiffening sagging deflection moments on the soft tissues under the vault during late midstance due to increased compensatory vault depression, generating greater stresses across the forefoot through the force–velocity ankle plantarflexor power generation relationship. If the soft tissues are unable to cope with controlling and dissipating these higher strains, the metatarsals could experience greater bending moments across their shafts (Fig. 4.5.14c).

Both the 2nd metatarsal, and to a lesser extent the 3rd metatarsal, have more restricted motion on loading due to their tighter tarsometatarsal joint anatomy. These two metatarsals are far more commonly affected with fatigue fractures than

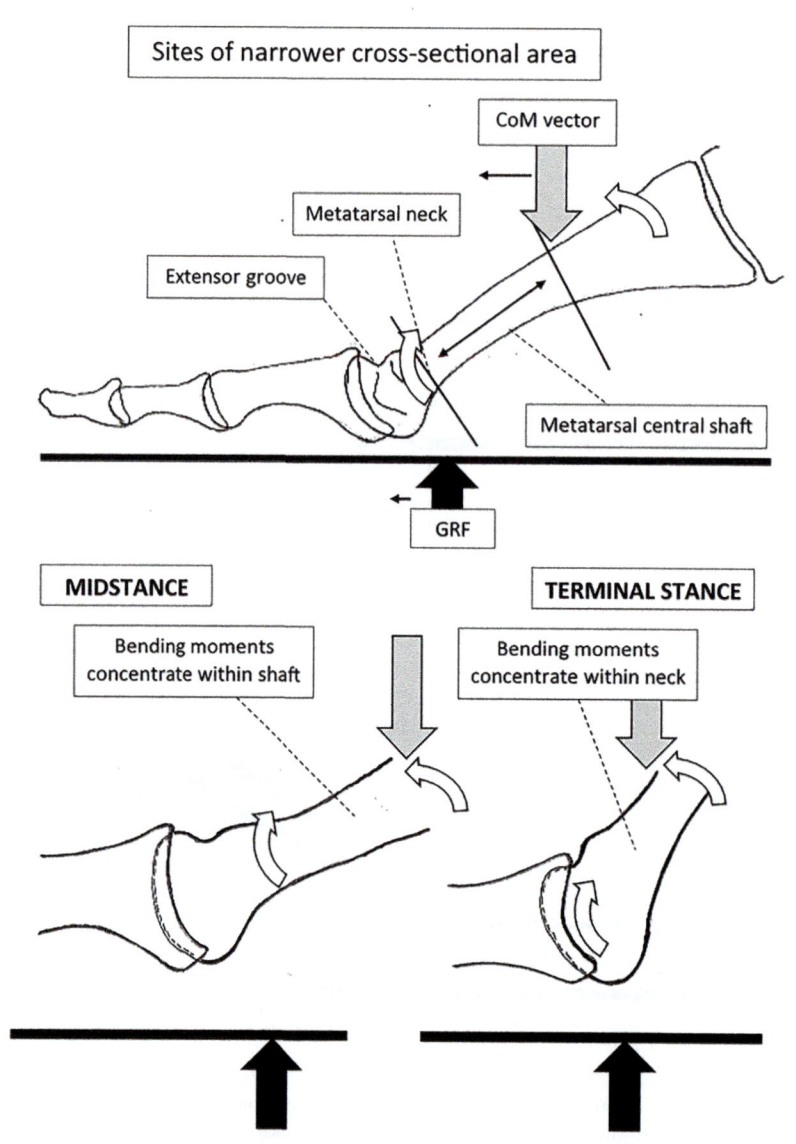

FIG. 4.5.14b The cross-sectional area is important for long bone strength. With bone physiology being equal throughout its length, the weakest areas are those with the narrowest cross-sectional areas. Thus, the 1st metatarsal is rarely afflicted with fatigue fracture as it is by far the thickest metatarsal. Mobility in surrounding joints also helps to dissipate energy, so metatarsals with more mobile tarsometatarsal joints tend to be less prone to higher peak loads than more 'fixed' metatarsals such as the 3rd and particularly the 2nd. Stiffer foot types, such as found associated with cavoid feet, tend to offer less tarsometatarsal flexibility, raising their internal bending moment deflection stresses during gait (white arrows). It is important that tensional and shear strain is minimised by muscular contraction where cross-sectional area is narrowest and the surrounding joint energy dissipation minimal. The narrower metatarsal areas are found through the central shaft (especially where narrowing initially becomes more pronounced) and particularly at the neck. The proximal and central shaft is more vulnerable during midstance and the neck more so within terminal stance, because of the interaction of the anteriorly progressing CoM vector and the GRF. The extensor groove and neck are areas where forces can concentrate during MTP extension during initial acceleration, when the forefoot sustains its highest loads. Concurrent torsional shears within the frontal and/or transverse planes between the head and shaft are also injury-provocative, which is far more likely when MTP joint and foot vault-supporting muscles are weak or fatigued. Such torques outside of the sagittal plane will exacerbate bone failure risks. (*Permission www.healthystep.co.uk.*)

other metatarsals. They represent around 60% of fatigue fractures reported in military recruits (Ross and Allsopp, 2002). The metatarsal with the most restricted freedom of motion, the 2nd metatarsal, is the location of most fatigue fractures. The 2nd metatarsal fatigue fracture is known as a *march fracture* and is associated with an increase in activity levels (Al Nazer et al., 2012), vertical loading, lower forefoot abduction angles of gait, and higher foot vault profiles compared to controls (Dixon et al., 2019). Higher vaulted feet have less soft tissue surface contact area of the foot and are reported to have increased loading of the 2nd metatarsal (Sun et al., 2012).

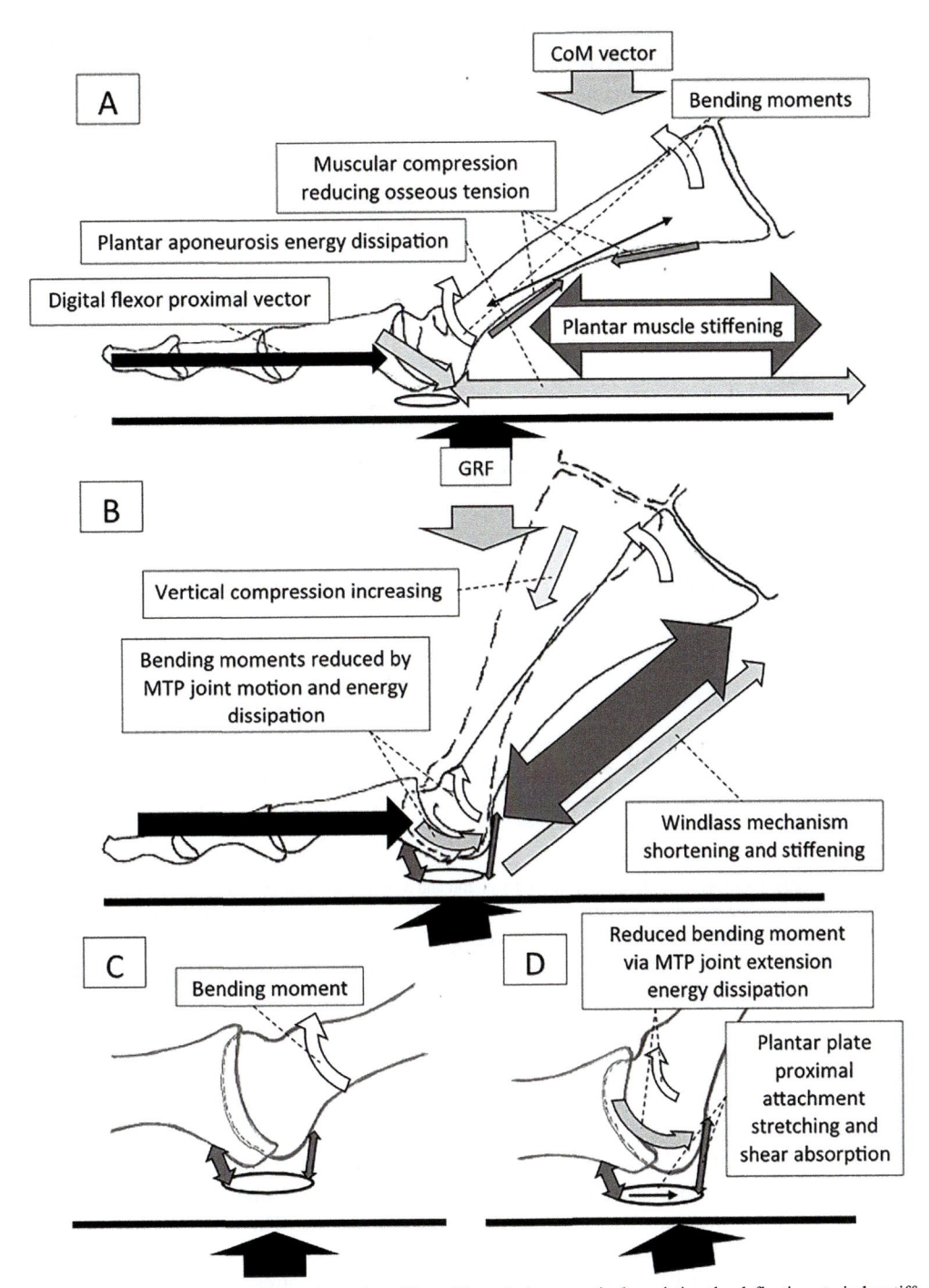

FIG. 4.5.14c Bending moments across the metatarsals can be mitigated by soft tissues actively resisting the deflection strain by stiffening the plantar aspect of the vault. This concurrently also reduces tensile strain on the plantar metatarsal surfaces via increasing contractile compression between muscular proximal and distal attachments during late midstance (A and C). Effectively, forefoot-stiffening stabilisation at this time should also prevent any significant transverse or frontal plane torques on the metatarsals and MTP joints. The maintenance of some flexibility within the vault allows the plantar soft tissues such as the plantar aponeurosis to continue to energy-dissipate, thus protecting the metatarsals from higher peak loads. At heel lift and into terminal stance (B and D), free, yet elastically resisted MTP joint motion becomes an energy dissipator, protecting the metatarsal neck area. The plantar plate is important for this role. Its stronger, stiffer, distal attachment keeps the plate fixed to the proximal phalanx while the more flexible proximal attachment stretches to allow the metatarsal head to rotate over the fibrocartilage without producing high osseous shear strains. This, combined with the other MTP joint muscle and connective tissue tensions, should primarily permit 'pure' sagittal extension, avoiding risky shear and torque stresses around the metatarsal neck. Any joint impingement or pathology resulting in loss of early MTP joint extension energy dissipation, risks fatigue fracture. In later terminal stance, weightbearing forces are rapidly reducing, while metatarsal internal stresses now consist of more vertical compression rather than bending moments resulting from the metatarsals being positioned vertically, more column-like. This should reduce the risk of osseous failure. *(Permission www.healthystep.co.uk.)*

3rd metatarsal stress fractures have been found to be associated with greater foot adduction angles in gait and a greater magnitude of forefoot loading, with a later peak metatarsal loading of pressure compared to controls (Dixon et al., 2019). They are more sensate to laterally directed loading, suggesting that individuals with 3rd metatarsals prone to stress fractures demonstrate different propulsive vertical and horizontal loading than those associated with 2nd metatarsal fractures (Dixon et al., 2019). This may reflect the greater transverse plane freedom of the 3rd metatarsal compared to the 2nd (Fig. 4.5.14d).

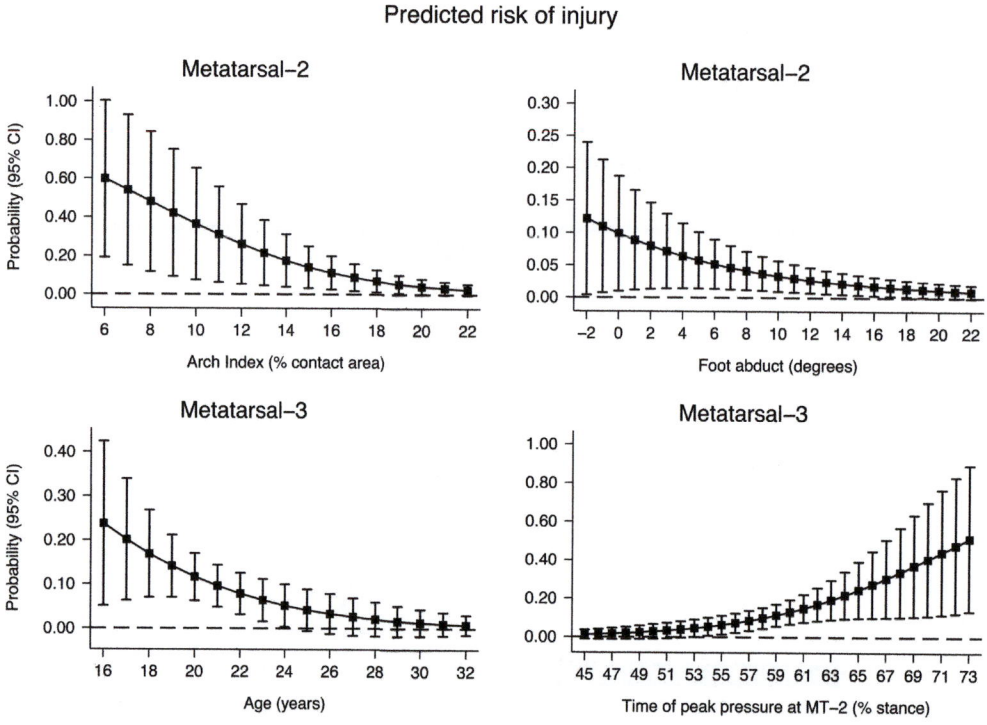

FIG. 4.5.14d Predicted probabilities for risk of fatigue fracture injury at 2nd and 3rd metatarsals in military recruits. These are related to foot vault arch index or the percentage of surface contact area of the foot, the degree of forefoot abduction, the age of the individual, and the time to peak pressure under the 2nd MTP joint as a percentage of stance phase. Less surface contact area associated with a higher foot vault is associated with 2nd metatarsal fatigue fractures. Fatigue fractures of 3rd metatarsals are reported to associate with a delay in forefoot loading and are associated with increased forefoot abduction. It seems that the less mobile highly stressed 2nd metatarsal is vulnerable to reduced mobility within the forefoot, and the more mobile but still insignificantly loaded 3rd metatarsal is associated to individuals with greater forefoot mobility. *(Image from Dixon, S., Nunns, M., House, C., Rice, H., Mostazir, M., Stiles, V., et al., 2019. Prospective study of biomechanical risk factors for second and third metatarsal stress fractures in military recruits. J. Sci. Med. Sport 22(2), 135–139.)*

Metatarsal internal stresses are known to experience high compressive stress in the physiological (safe) zone defined by the mechanostat theory in most activities except during jumping, when stress–strain moves into the mechanostat theory's overload zone (Al Nazer et al., 2012). The compression stresses that metatarsals are subjected to are possibly influenced by muscular fatigue within the foot. Arndt et al. (2002), studying barefoot walking, found an increase in compressive stress but a decrease tensile stress in 2nd metatarsals associated with muscular fatigue. This suggests that compressive-to-tensile strain relationships are controlled through foot muscle activity, helping to explain the previous data by Sharkey et al. (1995) and Jacob (2001) that flexor hallucis longus limits bending moment stresses through the medial metatarsals. It has been reported that fatigue of the muscles at the hallux increases strain on the 2nd metatarsal (Sharkey et al., 1995).

Fatigue fractures tend to occur following fairly rapid increases in exercise levels. Loads occurring gradually over time give tissues time to respond through biological response, allowing bone cross-sectional area and cortical thickness to increase and for the foot's soft tissues to strengthen, increasing their fatigue resistance. The loading dose response favours frequent loading at lower stresses to achieve these effects, which should direct the therapeutic approach to preventing fatigue fractures in rehabilitation through physical training programmes (Al Nazer et al., 2012). This principle is fundamental to the importance of coaching patients to increase loading stress through increasing their activity gradually at relatively lower levels of intensity but more frequency, to reduce fatigue risk that underlies so much of pathomechanics. Thus, the metatarsal's mobility and its supporting soft tissue strengths are important to consider in understanding a fatigue fracture's pathomechanics and recovery (Fig. 4.5.14e).

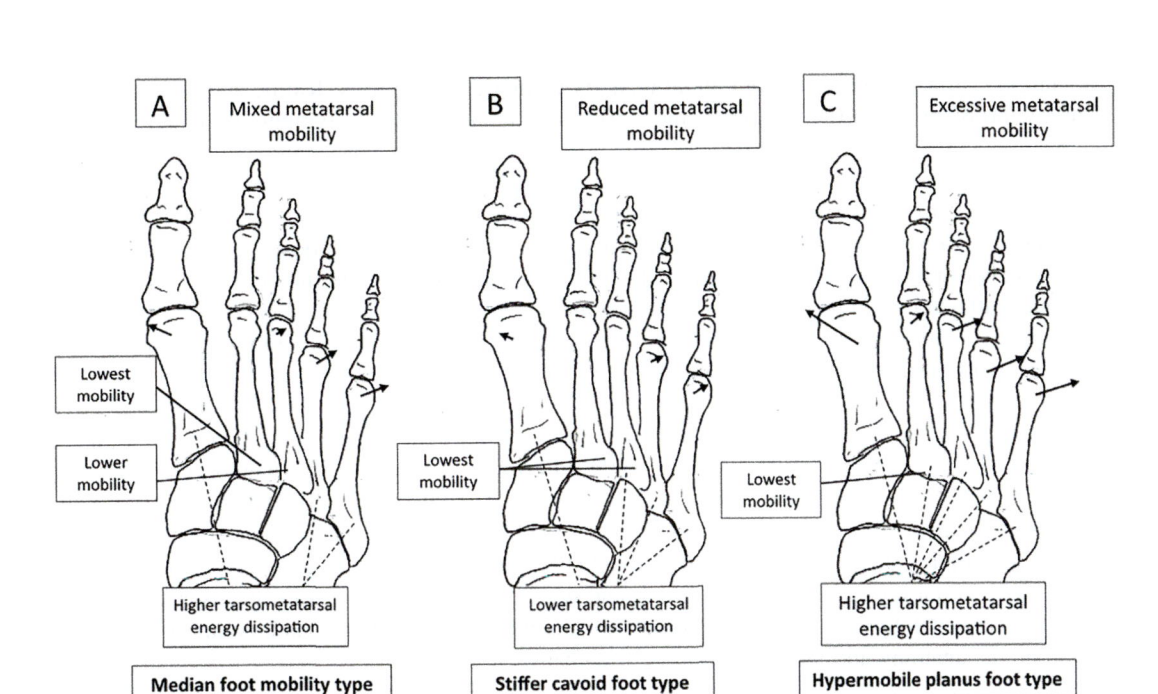

FIG. 4.5.14e On forefoot loading, metatarsals dorsiflex and splay as forefoot compliance increases under relaxation of the windlass mechanism and reducing tibialis posterior and peroneus longus activity. This is important for forefoot collision energy dissipation. The intermetatarsal space can thus continue to widen during midstance as the vault lowers, with the stiffening this induces gradually reducing the rate of intermetatarsal displacement as the deep transverse metatarsal ligament tightens. Normal anatomical differences usually dictate that the 1st, 4th, and 5th metatarsals are able to spread the most (A), meaning that their tarsometatarsal joints provide higher levels of energy dissipation. Those of the 3rd and particularly the 2nd, are held more firmly, usually exposing these metatarsals to higher forefoot loading forces than the others during forefoot contact to preswing. During midstance and terminal stance, the more rigidly held central metatarsals provide the bulk of forefoot stability, with very often the 2nd metatarsal head being the last to offload as part of the medial fulcrum of acceleration. Therefore, 2nd metatarsals are more at risk of fatigue fracture, although local joint pathologies or individual morphology in metatarsal lengths and frontal plane positioning can alter metatarsal loads from those expected. Stiffer cavoid feet (B) tend to demonstrate increased fatigue fracture risk especially to the 2nd metatarsal, because they offer less shock-absorbing joint mobility. In contrast, more mobile feet (C) may increase the risk of fatigue fracture on the 3rd metatarsal compared to other metatarsals, possibly due to greater shear and tensional torques on a still relatively highly loaded but more mobile metatarsal. However, metatarsal fatigue fracture risk within hypermobile pes planus feet is usually lower than that of stiffer feet. This is because lower metatarsal declinations and surrounding high joint mobility tends to reduce metatarsal osseous stresses during acceleration, loading them instead more towards the plantar ligaments, muscles, and articular soft tissue energy-dissipating structures of the vault. *(Permission www.healthystep.co.uk.)*

Footwear choice may also be significant. The forefoot bending stiffness of footwear may play a role in metatarsal fatigue fractures, as increased shoe stiffness moves the CoP anteriorly, modifying peak pressures across the forefoot (Stefanyshyn and Wannop, 2016). The work of Oh and Park (2017) clearly indicates that there is a preferred amount of shoe stiffness across the forefoot that should not be exceeded because of a risk of preventing normal MTP joint extension and thus increasing forefoot stresses from the ankle plantarflexors through an inability to dissipate energy through MTP joint extension. The exception to this is the careful therapeutic use of stiff forefoot rocker shoes for significant forefoot dysfunction, such as for compensating for the loss of MTP joint extension, but this must be done in the knowledge that this may increase ankle plantarflexor stresses.

In another limited study, plantar pressure distribution was shown to increase loading beneath the 2nd metatarsal in walking subjects who became fatigued with in vivo measurements that indicated increased dorsal tensional forces of the 2nd metatarsal when wearing more flexible footwear (Arndt et al., 2003). Thus, wearing very flexible shoes at times of increased foot fatigue may be associated with risks of metatarsal fatigue fracture. Increasing bending stiffness across the MTP joints may have the potential to act as a mechanism to prevent and manage metatarsal fatigue fractures, but also indicates the significance of improving intrinsic forefoot stiffening mechanisms in avoiding forefoot pathology.

4.5.15 Plantar fasciopathy and its pathomechanics

Plantar fasciopathy is pathology around the calcaneal enthesis of the plantar aponeurosis, commonly known as plantar fasciitis. Fasciopathy seems more appropriate than fasciitis because tissue thickening and degenerative changes are more commonly observed than are inflammatory changes (Monteagudo et al., 2018). Plantar fasciopathy is one of the commonest causes of foot symptoms in both athletic and nonathletic populations (Riddle and Schappert, 2004). It can become a chronic problem where the fascia becomes infiltrated by macrophages, lymphocytes, and plasma cells along with vascularisation and fibrosis, showing degenerative rather than inflammatory changes (Lemont et al., 2003). Peak incidence occurs between 40 and 60 years of age and is associated with increased body weight, the presence of a bony calcaneal spur, reduced ankle dorsiflexion, decreased 1st MTP joint dorsiflexion, and prolonged occupational standing (Irving et al., 2006). Symptoms consist of plantar heel pain associated with weightbearing after rest (poststatic dyskinesia) and pain that is worse after exercise (Irving et al., 2006). Most commonly, the site of the pain is towards the medial side of its calcaneal attachment, and it is slightly more common in women (Taunton et al., 2002; Orchard, 2012). Yet, not all plantar heel pain is plantar fasciopathy, with many other conditions including radiculopathy, plantar fat pad atrophy, plantar calcaneal neuromas, plantar bursae, subtalar OA, and calcaneal stress fractures all presenting with plantar heel pain.

Clinically, plantar fasciopathy is usually diagnosed from symptom patterns and history, and from palpation of the painful site over the plantar tubercle of the heel. Presently, ultrasound scanning is the diagnostic imaging technique of choice (Hormozi et al., 2011; Draghi et al., 2017; Boussouar, 2019; Aggarwal et al., 2020), but increased plantar aponeurotic thickness does not always associate with pain (Zhou et al., 2020). The condition is often self-limiting and symptoms can resolve with just simple conservative treatment in around 90% of patients (Monteagudo et al., 2013). However, for the remaining sufferers, a vast array of therapies, injections, and surgeries are listed for therapeutic use (Mao et al., 2019). The aetiology is rather conveniently left as 'multifactorial', for the plantar aponeurosis has a complex multifunctional role within the foot and its anatomy and mechanical properties change with age.

Pathomechanics of plantar fasciopathy

By bridging the rearfoot to the forefoot on the plantar surface, the plantar aponeurosis plays roles in midfoot energy dissipation and assists plantarflexion power transfer into the forefoot. It also has responsibilities in stabilising the forefoot through the flexor plate linkage and maintaining the integrity and position of the plantar fat pad. In terminal stance, it plays an important part in linking ankle plantarflexion angles to MTP joint extension angles, helping to maintain appropriate levels of elasticity around the MTP joint fulcrum. This binds the plantar aponeurosis into the complexity of force–velocity relationships during the acceleration phase of gait. The plantar aponeurosis provides an important 'outer' tie-bar under the foot's vault, not only longitudinally, but also obliquely through its medial and lateral slips. It also forms partial muscular attachments for the 1st layer of the plantar intrinsic muscles of the foot. Through its intimate association to the intermuscular septa (Ling and Kumar, 2008), it ultimately functions in concert with all the plantar intrinsics, extrinsics, and plantar ligaments of the foot (Fig. 4.5.15a). Thus, it is significantly tensioned from forefoot loading as part of the reversed windlass mechanism and undergoes increasing tension throughout midstance due to vault deflection and depression, but potentially experiences highest tension stresses at the start of terminal stance during heel lift, when it reaches its longest lengths under the ankle plantarflexion power of acceleration (Welte et al., 2021).

At heel lift, the vault is at its lowest and longest. It is stabilised by the stress-stiffened connective tissues, with the plantar aponeurosis undergoing further tension at the initiation of toe extension. It is likely that the plantar aponeurosis plays a significant role in setting up some passive forefoot stiffness with the plantarflexors of the digits, but only takes up a primary role in forefoot stiffening as muscle activity and elastic recoil power through the other myofascial structures starts to decrease with increase in toe extension that occurs in later terminal stance. This terminal stance foot-stiffening mechanism is achieved through the action of the windlass mechanism (Farris et al., 2020).

The complex soft tissue and bone linkages in the anatomy of the plantar aponeurosis open up a vast number of possible influences to its function and dysfunction. Thus, the plantar aponeurosis functions throughout the stance phase of gait, subject to varying tensional stresses from initial contact to the end of terminal stance. Pretensioning of the foot vault involving the plantar aponeurosis starts before initial contact as part of the preparation for the activation of windlass mechanics, which then sees reducing tension in the aponeurosis during the reversed windlass at forefoot contact (Caravaggi et al., 2009). If the mechanism fails to reverse, tensional strains on the plantar aponeurosis in loading response may remain high, rather than lowering during forefoot contact. During forefoot loading, the plantar aponeurosis is loaded with the toes extended and the ankle plantarflexed, which may potentially increase plantar aponeurotic stresses during loading response (Fig. 4.5.15b).

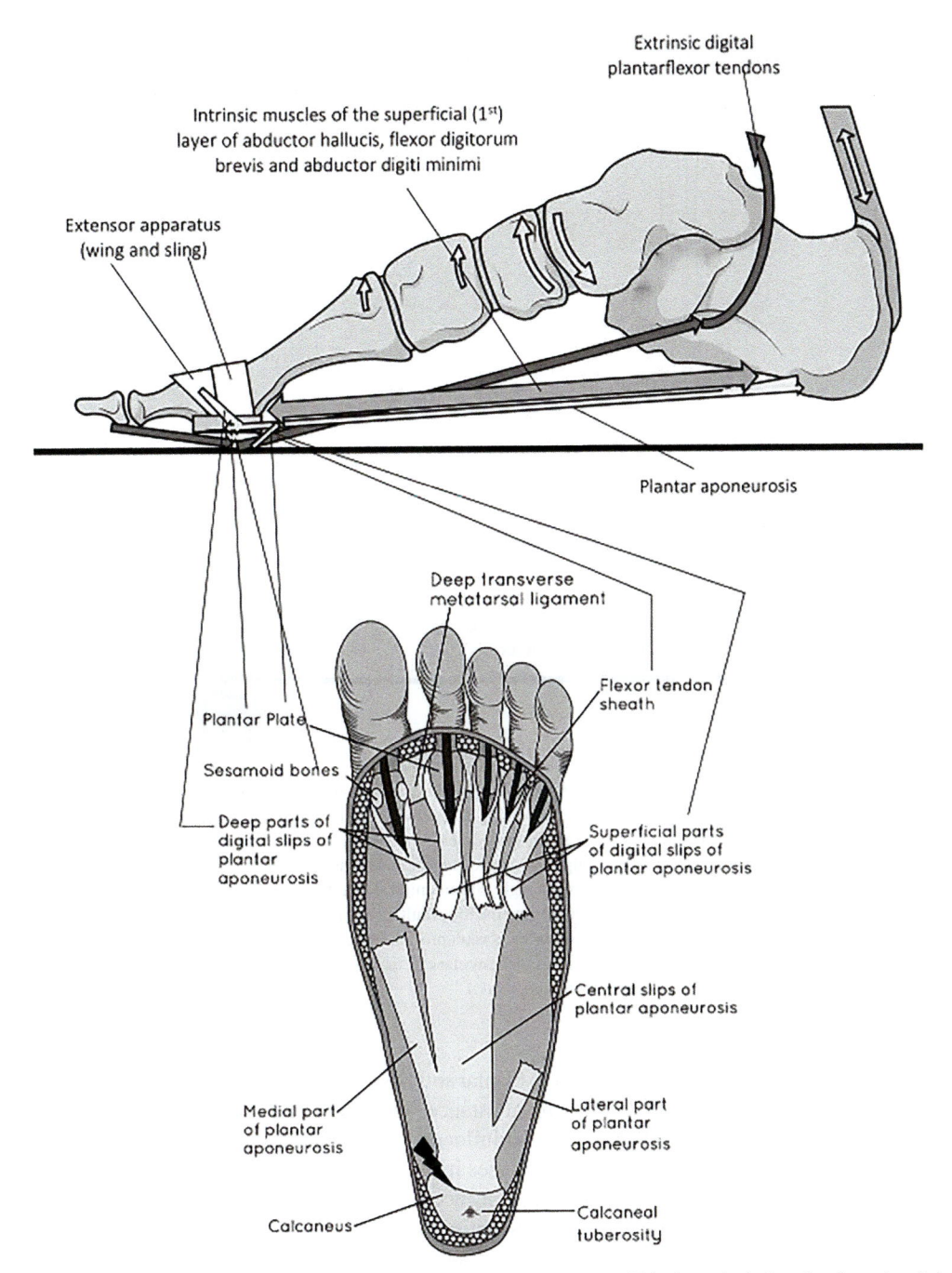

FIG. 4.5.15a The plantar aponeurosis simplified to reveal some of its anatomical linkages within the sagittal plane (top image) and also viewed on its plantar surface (lower image). This is the primary linking connective tissue structure of the plantar foot, with connections that are far more complex than shown here. The plantar aponeurosis has a direct kinetic relationship to the Achilles and soft tissue structures of the foot vault. It is involved in energy dissipation and in helping to set foot stiffness from initial contact to preswing. It undergoes a cycle of initial shortening under the windlass mechanism followed by lengthening and widening as part of foot vault compliance and energy dissipation of midstance, to only start shortening again a little after heel lift. Thus, the chance of discovering a single causative factor in the development of the common plantar heel pain-producing condition of plantar fasciopathy or fasciitis (black lightning strike shows most common site) is highly unlikely. The conditions, location, and presentations are not homogeneous between all individuals. *(Permission www.healthystep.co.uk.)*

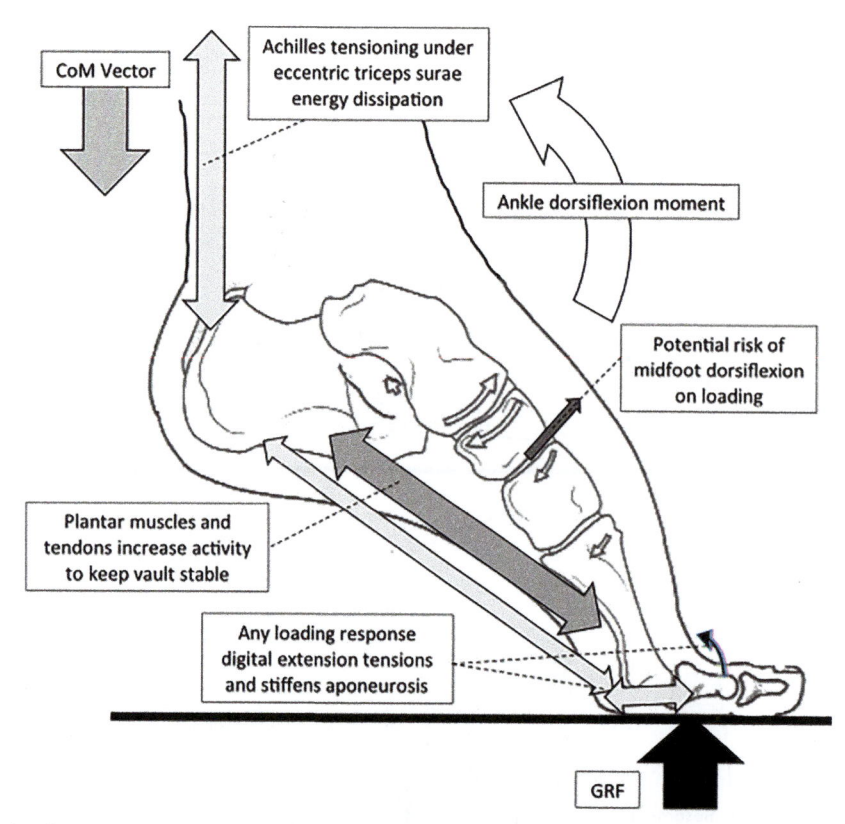

FIG. 4.5.15b With heel strike gait, the plantar aponeurosis is tensioned by the windlass mechanism prior to forefoot loading. However, the digital extension loads on the aponeurosis are small compared to those occurring at forefoot contact, when the plantar aponeurosis must then start to rapidly lengthen and energy dissipate immediately. A forefoot strike, as illustrated here, may present higher initial tensile loads on the plantar aponeurosis if the MTP joints are initially forced to extend by the GRF. Such digital extension at forefoot loading is a higher risk during 'toe strike'-type forefoot running, when the foot makes contact at a more extreme ankle plantarflexion angle. Strong midfoot plantarflexor muscles are likely to prevent injury to the plantar aponeurosis. Such muscles should be preactivated before contact, turning the foot vault via its myofascial tissues into a powerful energy dissipator by being an active, stiffer, and more elastic cantilever-like structure with a shortened vault profile. Dangers arise only if the midfoot is forced to excessively dorsiflex at forefoot strike, lengthening the vault. This is a risk of weaker/fatigued myofascial tissues. Indeed, clinically, plantar fasciopathy associated with such running kinematics seems a rare issue. *(Permission www.healthystep.co.uk.)*

The plantar aponeurosis along with the other passive plantar soft tissues plays a significant role in dissipating and storing energy by stress-stiffening under foot elongation during midstance. The stored energy can be released from the aponeurosis during the terminal stance phase as it shortens under the influence of the windlass mechanism (Hicks, 1954; Caravaggi et al., 2009; Farris et al., 2020). The plantar aponeurosis uses its tensional properties to help initially dissipate, then store and release energy that can assist in late foot-stiffening, aiding acceleration during terminal stance. These abilities make it a considerable feature in energy balance across the foot. The muscles that tension the plantar aspect of the foot are important for controlling the tensional loading of the aponeurosis just prior to and during heel lift, when the vault profile has lowered to its lowest point. Thus, as an important muscular attachment, the plantar aponeurosis plays a significant part in preventing excessive vault lengthening and widening through aiding the plantar intrinsics (Arangio et al., 1998; Arellano et al., 2016). Loss of plantar intrinsic strength will increase the aponeurotic burden in resisting bending moments through increased tension and having to store more of the tensional energy across the plantar foot.

This resistance against the bending moment links aponeurotic stresses at midstance to the ankle joint dorsiflexion moment. If dorsiflexion is insufficient in range or insufficiently resisted by the triceps surae, then greater midfoot bending moments across the vault can occur to absorb the torque created by the CoM progressing anteriorly. The aponeurosis, like the plantar ligaments, will thus absorb excessive dorsiflexion-induced energies that are not dissipated by ankle motion and the Achilles.

Active intrinsic muscles, through their fascial attachments, tension the plantar aponeurosis, helping to spread tensional strain across the network of vault ties, reducing peak strain within the foot. The plantar aponeurotic interstructural connections help to dissipate energy across the fascial network during midstance vault depression, preventing stress concentration within the aponeurosis or the deeper structures such as the plantar ligaments. This network is reinforced medially by the flexor retinaculum but interlinks to the extensor retinaculum, dorsal fascia, the extensor apparatus, and thus to the flexor plate complex of the MTP joints and the distal plantar aponeurosis. The peroneal retinaculum links laterally to the plantar aponeurosis to complete this fascial continuum enveloping the foot. Dysfunction in any part of the network has the potential for increased stress and strain somewhere within the remaining network. Its widespread anatomy makes the plantar aponeurosis particularly vulnerable, as it has so many direct links to multiple foot structures and is placed in the most vulnerable superficial plantar position (Fig. 4.5.15c).

Increasing ankle dorsiflexion motion increases Achilles energy storage tension

Increasing ankle dorsiflexion moment increases vault deflection dorsiflexing midfoot and forefoot

Plantar aponeurosis lengthens and stiffens under sagging deflection providing energy dissipation with some storage

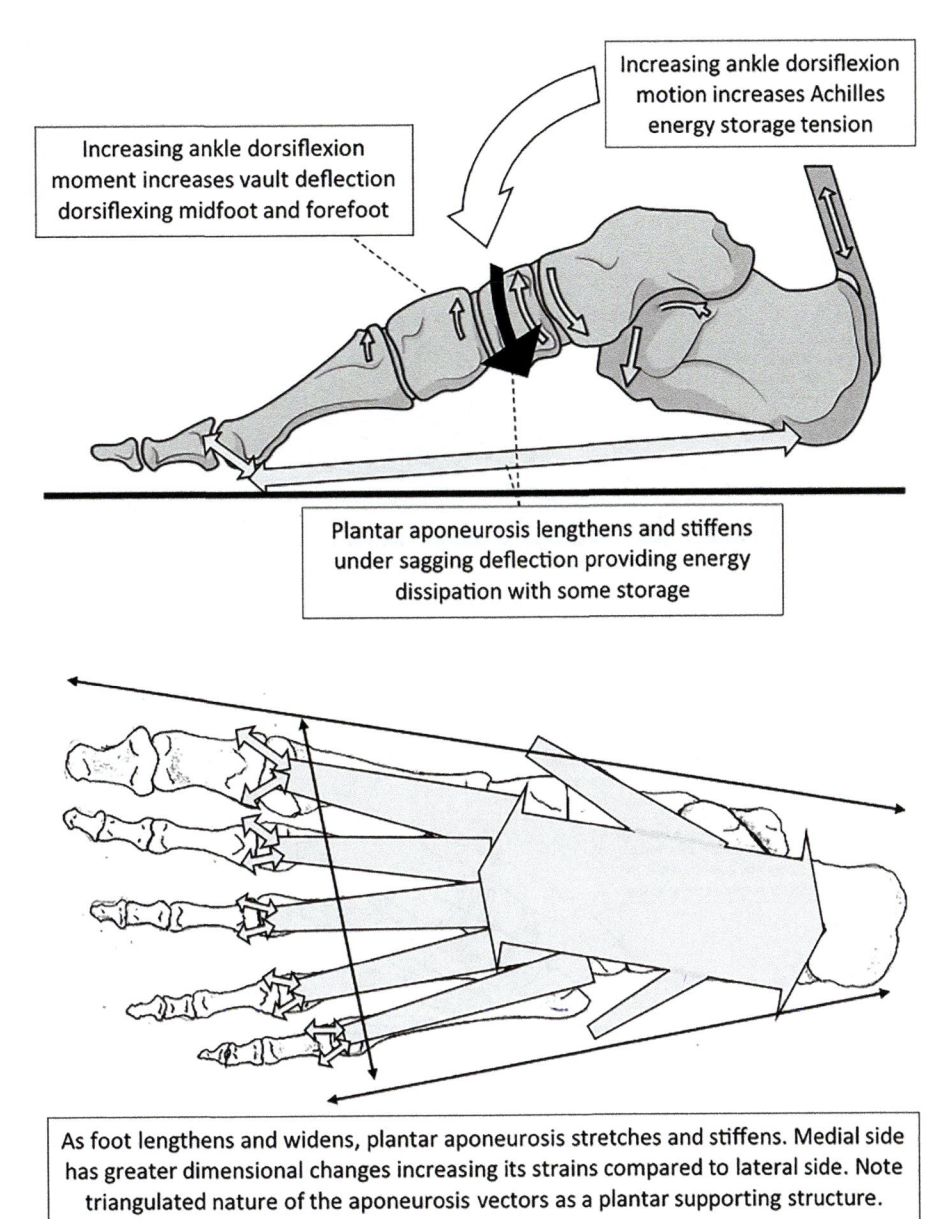

As foot lengthens and widens, plantar aponeurosis stretches and stiffens. Medial side has greater dimensional changes increasing its strains compared to lateral side. Note triangulated nature of the aponeurosis vectors as a plantar supporting structure.

FIG. 4.5.15c From forefoot loading and throughout midstance, the plantar aponeurosis plays an important part in vault energy dissipation. Its elongation coupled to that of the plantar ligaments provides some resistance to vault compliance that lengthens and widens the vault. In so stretching, the connective tissues such as the aponeurosis move towards their stress-stiffening range. In early midstance, only tibialis posterior and peroneus longus tendons and the proximal vault ligaments offer consistent vault-supporting/energy-dissipating assistance. As the GRF moves anteriorly, the proximal connective tissue cycle of stretch and elastic recoil completes. The plantar aponeurosis, by bridging the calcaneus to the MTP joints, continues to lengthen, tension, and stiffen until after heel lift. With vault profile changes greater within the medial side, it is subjected to greater strain, probably explaining why most plantar fasciopathies present on the medial side of the heel. In late midstance, stress-stiffening elongation occurs with extensive and increasing plantar muscular activity, which should protect the plantar aponeurosis from higher stresses derived from the ankle dorsiflexion moment deflecting the vault at the heel lift boundary. Thus, the coupled relationship of the Achilles to plantar aponeurotic stresses is explained. This is an action seen in all foot vault profile types from high to low, but it seems the plantar aponeurosis may play a more significant role in maintaining vault stability and energy dissipation in higher vault-profiled feet. The higher connective tissue compliance of more mobile feet will likely shift mechanical vault restraint and energy dissipation burdens towards muscles. As connective tissue often become stiffer with age and muscles often weaken, an explanation as to the increasing rate of plantar fasciopathy among active older adults is suggested. *(Permission www.healthystep.co.uk.)*

832 Clinical biomechanics in human locomotion

The Achilles and the plantar aponeurosis have an intimate stress–strain relationship (Carlson et al., 2000; Erdemir et al., 2004; Cheung et al., 2006; Zhou et al., 2020). Ankle dorsiflexion stiffens both of them (Huang et al., 2018). Both structures work to store elastic energy prior to heel lift, and then work to apply the power generated from the ankle plantarflexors to the forefoot as a stabilising GRF during terminal stance. This relationship brings the plantar aponeurosis into the force–velocity relationship of the ankle plantarflexors through foot stiffening (Takahashi et al., 2016). Thus, ankle and midfoot plantarflexion angles, MTP joint angles, MTP joint forefoot stiffness, and which fulcrums are used for acceleration are all important (and all-important) factors influencing plantar aponeurotic stresses (Fig. 4.5.15d).

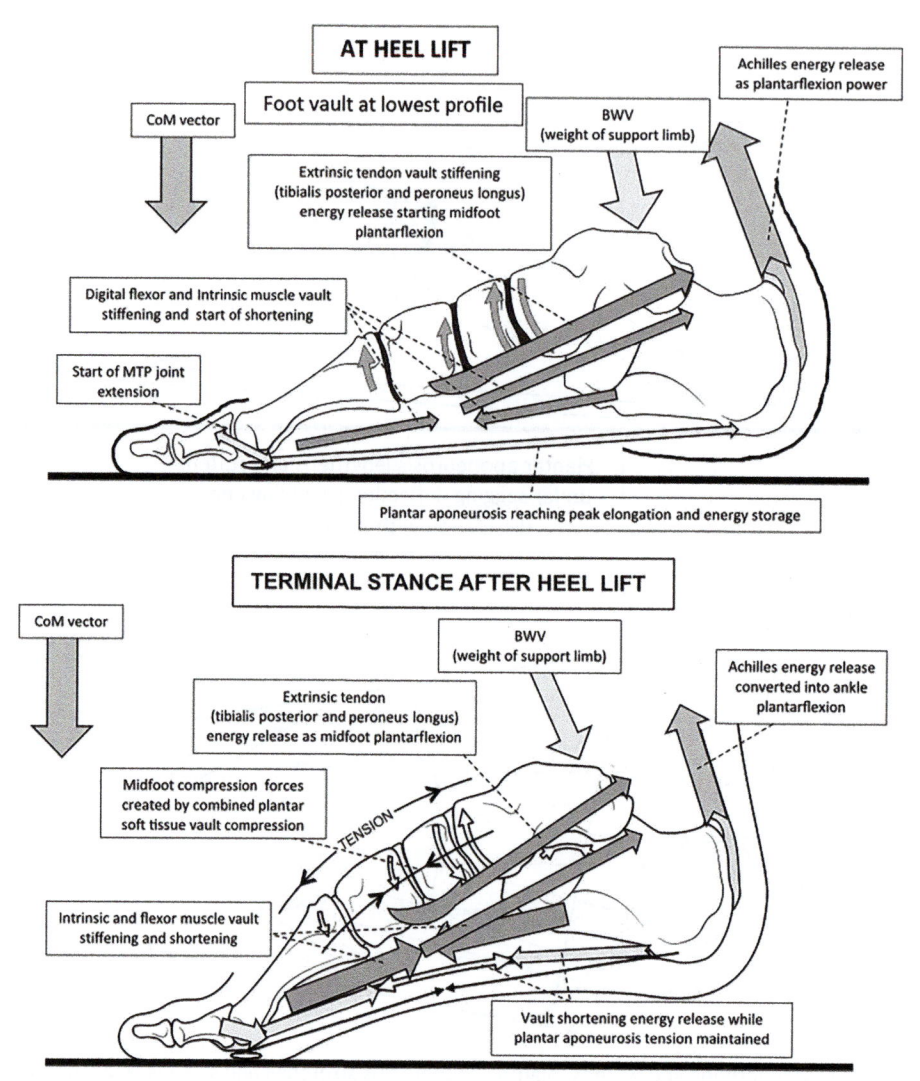

FIG. 4.5.15d The heel lift boundary seems to present a particular mechanical challenge for the plantar aponeurosis. These simplified schematics help demonstrate that the foot vault is at its lowest at heel lift, stretching the plantar aponeurosis to its longest length across the foot at this time. As the digits start to extend, further tension is applied via its digital and cutaneous forefoot slips. Thus, the plantar aponeurosis continues to lengthen into early terminal stance, dissipating energy until the vault starts to shorten. Vault raising and shortening initiates under the changing angles of ankle and midfoot plantarflexion, permitting the plantar extrinsic and intrinsic tendons to start to recoil, while the intrinsic muscles continue their isometric and concentric contraction. It is only after heel lift that the plantar aponeurosis is able to stop its energy dissipation and safely release its stored energy to help stiffen the foot via the windlass mechanism. In timing and extent of energy release, the activity of the plantar extrinsics and intrinsics is essential, with their weakness/fatigue likely to extend the lengthening duration and delay energy release. Such delay risks connective tissue stress threshold breaches. The energy placed into the vault by the tensioning Achilles under the midfoot dorsiflexion moment and the rate of Achilles energy release in initiating ankle and midfoot plantarflexion and MTP joint extension, is critical to understand the amount of energy that the vault must handle during late stance. The amount of this energy loaded into the plantar aponeurosis depends on the capacity of other passive and active plantar vault anatomies to dissipate and store energy during gait. The ability of the plantar aponeurosis to manage the energy it receives appropriately will depend on age, health, and the mechanical properties of its connective tissue, which will tend to deteriorate with age or physiological disease. *(Permission www.healthystep.co.uk.)*

Thus, any influences on these features such as intrinsic and extrinsic muscle strengths, ankle joint freedom, MTP joint freedom, and angle of gait will affect how evenly the plantar aponeurosis is loaded. The individual component muscles of the triceps surae likely have their own relationships with the aponeurosis, as it has been reported that patients with plantar fascial symptoms demonstrate increased passive medial gastrocnemius stiffness compared to asymptomatic individuals, but not in lateral gastrocnemius (Zhou et al., 2020).

As if this were not complex enough, the aponeurosis also has differing attachment points with age (Snow et al., 1995). This is likely to change the aponeurosis' mechanical relationship with the Achilles over a lifetime, as the fibrous continuations of the two structures are gradually lost throughout early adulthood (Snow et al., 1995). While fibres run from the Achilles calcaneal insertion to the plantar aponeurosis in a continuum, some stress will be kept away from the calcaneal osseous entheses of both structures, for loads can be transferred directly through the soft tissue connective tissue fibres to the forefoot. However, once these continuum fibres are lost, all distal Achilles forces and proximal plantar aponeurosis forces are concentrated into the calcaneus before transfer to the forefoot (save any force carried in other deep fasciae).

Fibre continuum, in conjunction with immature (mechanical weaker) calcaneal bone, may explain the development of Sever's disease as a cause of heel pain in the early adolescent, while loss of the continuum and calcaneal maturity changes the stressed point to the calcaneal entheses. Sever's disease is usually referred to as an over-tensioned traction of the apophysis (immature soft tissue attachment point of the Achilles) and thus an apophysitis (Fig. 4.5.15e). However, it has been argued that the condition may be a metaphyseal trabecular stress fracture, as it does not involve the apophysis ossification centre, but instead demonstrates fragmentation and sclerosis across the metaphysis between the Achilles attachment and the proximal attachment of the plantar aponeurosis (Ogden et al., 2004). Yet, this presentation also occurs

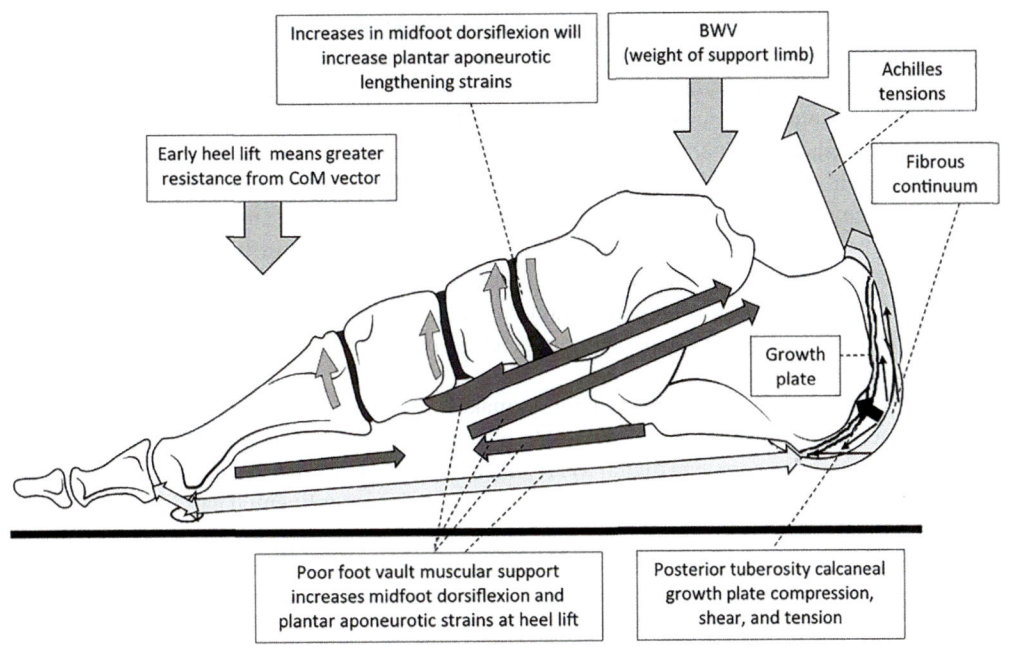

FIG. 4.5.15e It is probable that the strains that initiate Sever's disease arise from pathomechanics at the heel lift boundary applied to an immature calcaneus. This is because the foot vault is lowest, at its longest length, and at its widest width when the plantarflexion power of the Achilles is released across the foot for heel lift. The vault may be at its stiffest, but the foot only offers semirigidity at the initiation of acceleration, meaning that there is some initial midfoot dorsiflexion as the MTP joints initiate extension. The effect of this is to further lengthen the plantar aponeurosis initially, until the vault starts to shorten after heel lift. The complication for the immature heel is that a growth plate lies between the posterior calcaneal tuberosity and the body of the calcaneus. This is attached proximally to the Achilles and distally to the proximal plantar aponeurosis via connective tissue fibres that are continuous between both structures, running over the posterior tubercle. This means that rising Achilles and plantar aponeurotic tensions will compress the bone of the separate posterior tuberosity into the ossifying fibrocartilage of the growth plate. Strains between the Achilles and plantar aponeurosis may also create shear and tensional forces through the growth plate. Increasing maturation of the growth plate with the development of partial union occurs around late childhood and early adolescence, when the size of the individual is increasing, gait maturation is finalising, and often when participation in organised sport is increasing. Attempting to reduce strains and motions across the fibrous continuum from the Achilles to the plantar aponeurosis seems biomechanically appropriate. Rehabilitation to improve triceps surae flexibility and strength to help avoid an early heel lift, and by increasing plantar muscle bulk and strength to reduce midfoot dorsiflexion at the heel lift boundary, has proved clinically beneficial. Using rearfoot and vault taping, orthoses, and/or footwear to temporarily raise the heel and increase metatarsal declination in an attempt to ease and lower plantar aponeurotic tensile strains at the initiation of heel lift, seems also beneficial in relieving pain, often allowing patient exercise to continue. *(Permission www.healthystep.co.uk.)*

in asymptomatic adolescent heels (Scharfbillig et al., 2008; Perhamre et al., 2013). Perhaps Sever's disease represents an excessive shear and/or compression as well as tension through the immature bone in the longitudinal axis, made possible by the fibrous tissue continuum over the plantar surface? For more on Sever's disease, see Section 4.4.7. Sufferers often present with concurrent tight calf and hamstring myofascia. Stretching and strengthening coupled to temporary foot orthoses for pain relief, usually resolves the problem. Healing is made easier if activity levels can be reduced for a little while, for bone does not heal so well under tension; something that is true of immature bone as well as mature.

Foot vault lengthening and digital extension increases tension in the Achilles tendon (Carlson et al., 2000). Thus, plantar fasciopathy is likely to be provoked within this stress relationship, variably. At terminal stance, the plantar aponeurosis is tightened through passive digital extension as part of the windlass effect. This process is initiated just as the peak plantarflexor power is applied to the foot from the ankle plantarflexors, when the plantar aponeurosis has reached a peak of stretch at the vault's lowest profile found at heel lift. With the medial side of the vault lengthening the most and the medial MTP joints being the fulcrums of choice for terminal stance acceleration, stresses across the aponeurosis running proximally will tend to focus to the medial calcaneal attachment. Where two very mechanically different tissues meet is always a focus for stress. The enthesis organ helps to prevent tensional stresses focusing, but there are limits. How these stresses are managed by the tissues is likely reflected by the age of the anatomy, with immature bone failing first in the early adolescent (Sever's disease), and less compliant connective tissues and the enthesis organ in the mature adult (plantar fasciopathy). Obviously, foot vault profile, tissue mobility, and tissue strength will play their part in focusing stresses within each individual, with physiological age influencing tissue mechanical behaviour under load.

Although the aponeurosis is not a major part of increased foot stiffening at heel lift (Farris et al., 2020), being that the windlass is just engaging at the start of digit extension while muscle activity and connective tissue stress-stiffening will be at their peak, it seems to still play a part in stress-stiffening the foot for the remainder of terminal stance. The windlass mechanism becomes an increasing part in stabilising and maintaining levels of stiffness throughout terminal stance, linked to the flexor plate complex that stabilises the MTP joints. The flexor plate complex links the plantar aponeurosis to the deep transverse metatarsal ligament, an important structure that influences transverse vault stiffening. Thus, the long and short flexors and the other plantar intrinsics are bound to the tensions within the plantar aponeurosis and the plantar aponeurosis tensions are bound to the primary terminal stance phase stiffening mechanisms. The plantar aponeurosis is essentially a part of a tensile-stress 'sandwich'. This biomechanical sandwich is formed by the deep connective tissue ligamentous restraints of the foot vault representing the upper (superior) slice of bread and the plantar aponeurosis representing the lower (inferior) slice. This sandwich contains a plantar intrinsic muscle filling, mechanically garnished by long digital flexors, peroneus longus, and tibialis posterior tendons. It is the aponeurotic outerslice that is pulled apart the most, and the restraint of the inner ligamentous slice and muscular filling that restrain its lengthening. Thus, failure in forefoot stability and terminal stance stiffening mechanisms in both the transverse and sagittal planes hold significant potential for increasing plantar aponeurotic stresses.

In summary, the complex interlinking of anatomy makes the plantar aponeurosis a significant player in the force–velocity power generation relationship of the ankle plantarflexors. This directly influences the metabolic costs of achieving heel lift and dissipating energy via digital MTP joint extension at the medial fulcrum of the foot, not least through its role in foot-stiffening mechanics (Takahashi et al., 2016). It is therefore an influential part in fixing the 'Goldilocks zone' of heel lift/acceleration energetics. Thus, by bridging ankle plantarflexion to digital extension, the aponeurosis is subjected to considerable stress, particularly if other soft tissues involved are in any way dysfunctional. The midfoot plantarflexion and adduction that occurs in terminal stance (Holowka et al., 2017) helps moderate aponeurotic tension stress after heel lift. Failure of the deep transverse metatarsal ligament and the plantar intrinsics to create foot stiffness or tibialis posterior and peroneus longus activity to provide midfoot plantarflexion and stability, is likely to increase plantar aponeurotic tensional stress–strain loading during early terminal stance.

Pinning down the plantar aponeurosis pathomechanics

The many-faceted, synergistic stress-loading interrelation between the ankle plantarflexors, the plantar extrinsic muscles, the plantar intrinsic muscles, and the connective tissue network of the plantar foot, makes the establishment of a distinct origin and therefore a single pathomechanical cause of plantar fasciopathy challenging, to put it mildly. Even adjudged direct links between plantar fascial thickening and pain levels need to be questioned (Zhou et al., 2020). Failure due to muscle or connective tissue dysfunction within this whole complex will threaten the appropriate spreading of stress throughout the plantar vault. Potentially, this failure can occur in any stage of the stance phase, dependent on the nature of the structural failure, with the plantar aponeurosis always being a likely candidate for increased stress–strain management because it connects to so many anatomical structures and regions within the foot. Connective tissue hypermobility or hypomobility will change the loading stresses across the soft tissue network throughout the stance phase of gait. If the midfoot has increased compliance, then it will allow greater energy dissipation through fascial/connective tissue stretching, but will reduce the ability to store energies for elastic recoil during terminal stance acceleration and reduce foot-stiffened stability. This situation is likely worsened if the individual is particularly heavy compared to their foot size. Such a scenario will strain the plantar aponeurosis more during midstance, shifting the foot-stiffening burden (and possibly the injury) into the muscles.

The role of the plantar aponeurosis in vault profile changes may be greater in cavoid feet than in others (Arangio et al., 1998). This is probably because it works in a more taught state throughout midstance with less compliant energy dissipation capacity, as the vault only has limited deformation capabilities. Foot vault depression aids ankle dorsiflexion capacity for CoM progression over the foot, dissipating and storing energy while doing so. Vault compliance also permits the foot to adapt to the support surface, increasing its contact surface area. During late midstance, through increasing elongation of the foot, the plantar aponeurosis and the myofascial vault network will become increasingly taught through stress-stiffening principles of viscoelasticity. The plantar foot should reach an 'ideal' stiffness-to-energy storage ratio, just before heel lift. Cavoid feet with stiff connective tissues may provide premature stiffness without sufficient prior energy dissipation, and thus direct increased stress into the plantar aponeurosis, without significant energy being lost before safe levels are stored. This will threaten connective tissue fibre overloading at the enthesis that presents a mechanically weaker point. An early heel lift from restricted ankle and midfoot dorsiflexion will exacerbate plantar aponeurotic loading stresses for the reasons associated with prematurely lifting an increased resistance (more of the body mass), that interruptions the anterior falling of CoM's downward oscillation during late midstance.

Assessment of plantar fasciopathy pathomechanics should first focus on ankle and MTP joint mobility and strength. Through vault, ankle, and MTP joint function, there is a relationship of increasing stiffness with rising ankle plantarflexion moments in both the Achilles tendon and the plantar aponeurosis (Huang et al., 2018). Increasing tension and stiffening in the Achilles tendon during late midstance creates an increasing plantarflexion torque on the calcaneus that increases the plantar aponeurotic strain applied proximally (Cheung et al., 2006). Much of the tension energy from stresses throughout the plantar aponeurosis and vault is stored as elastic energy for the provision of additional positive work during terminal stance (Wager and Challis, 2016). This is assisted by the plantar extrinsics/intrinsics and other vault-related connective tissues.

Thus, the timing of heel lift in relation to the position of the CoM of the body, the preparedness of the opposite swing foot to become weightbearing just after the stance limb's heel lift, the MTP joint flexibility/elasticity, and the chosen fulcrum point are all important factors. So too are the application of ankle plantarflexion power to the forefoot, and the overall muscle strength around the foot and ankle. All these factors potentially affect plantar aponeurotic pathomechanics. Structural fatigue may also be related to plantar fasciopathy, linked to a higher body mass (van Leeuwen et al., 2016). However, increased mass will also tend to be associated with higher forces. Thus, strength-to-body mass ratios are also important to many different lower limb repetitive strain injuries. Plantar aponeurotic pathology is not likely to associate clearly with a single foot type or one particular dysfunction within the population, for each case will present differing pathomechanics, despite the same pathology resulting. For example, runners require differing mechanical behaviours from their plantar foot than walkers, needing to be more elastic in the former group. Yet, runners also require different soft tissue behaviours in their feet depending on the running technique. Distance track runners, who always run counterclockwise, have been reported to have different shear wave velocities (a measure of tissue stiffness) in their aponeurosis within their left foot compared to their right foot (Shiotani et al., 2021). This may be an adaptation to asymmetrical loading during training that requires more vault lowering (pronation) on the outside right foot when taking corners. Interestingly, although stiffness was higher on the right, there was no difference in aponeurotic thickness or dimensions between feet (Shiotani et al., 2021).

Whatever the origin of the increasing fascial loads, the fascia is potentially responding through biological response to increased peak stresses within the tissue. This can be achieved through increasing surrounding muscle tone or increasing aponeurotic cross-sectional area. Variable thickness is a common feature of the human plantar aponeurosis. Should inflammation and fibrosis develop around the aponeurosis enthesis, changes can result within the enthesis with parts of the fibrocartilage ossifying to form a plantar calcaneal spur. Although this does not in itself seem to provoke pain, as a calcaneal spur can occur without symptoms, it probably reflects abnormal tensional loading adaptation, hence some weak association between spur formation with plantar fasciopathy symptoms (Beeson, 2014). Yet, plantar fasciopathy can also commonly occur without spur formation. Calcaneal spurs may only reflect the amount of time that high stresses have been focused at the enthesis of the plantar aponeurosis. Histologically, intra-fascial or supra-fascial plantar calcaneal spurs present in a very similar way to posterior intra-fascial calcaneal spurs associated with the Achilles tendon, in that both types taper close to their bases with bulky tips directed to the lines of soft tissue tension (Zwirner et al., 2021). The histological evidence from the trabecular arrangement and the tip direction supports that both Achilles and plantar aponeurotic spurs result from sustained traction forces on the enthesis, being morphologically poorly structured for compression forces (Zwirner et al., 2021).

Wearing et al. (2009) noted that patients with 'plantar heel pain' demonstrated lower energy dissipation ratios (through hysteresis findings) compared to asymptomatic controls, but the lack of a clear diagnosis associated with this study limits the value of its data (Fig. 4.5.15f). However, effective 'stiffness' through viscoelastic properties is an important feature during foot loading, and this is supported by the asymmetrical aponeurotic stiffness levels reported in track runners by Shiotani et al. (2021). It is possible that chronic plantar fasciopathy leads to loss of the energy dissipation ability of the plantar aponeurosis. Aponeurosis pathology may have knock-on mechanical effects to the plantar fat pads. In the Wearing et al. (2009) study, heel pad thickness, peak stress, strain, and modulus were not reported different to the controls within the heel pain subjects, suggesting that the heel fat pad properties had remained intact (Fig. 4.5.15g).

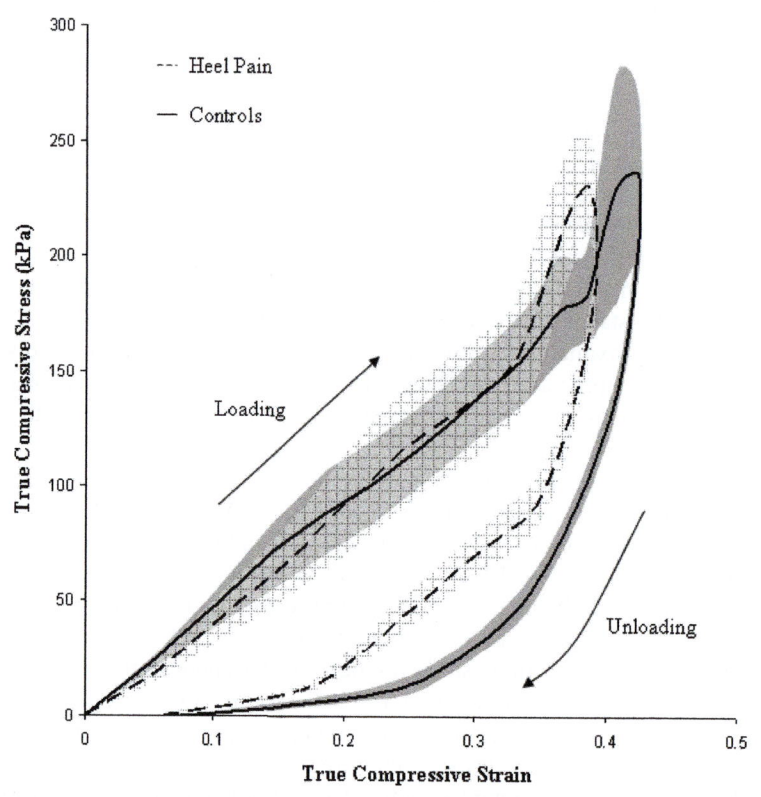

FIG. 4.5.15f A hysteresis curve demonstrating the ensemble stress–strain curves for the heel fat pad in subjects with unilateral heel pain and individually matched control subjects. That changes in heel fat pad behaviour seem to be taking place is interesting, but without a clear association with a specific diagnosis, limits the value of this information. Arrows indicate the direction of loading and unloading. *(Image from Wearing, S.C., Smeathers, J.E., Yates, B., Urry, S.R., Dubois, P., 2009. Bulk compressive properties of the heel fat pad during walking: a pilot investigation in plantar heel pain. Clin. Biomech. 24(4), 397–402.)*

Plantar fascial tears and other vault pains

Sudden high-tension loading across the vault of the foot can result in plantar aponeurotic tears, usually through unexpected foot lengthening events such as a large sagging bending moment across the forefoot, or acceleration on an unstable forefoot. These tears are usually in the mid or distal portions of the plantar aponeurosis and present as 'vault' (arch) pain, rather than plantar heel pain. They can be mistaken for *plantar fibromatosis*, a hypertrophic thickening of the plantar aponeurosis that is usually self-limiting. Plantar fibromatosis (Ledderhose disease) is a condition that is analogous to Dupuytren's contracture in the hand, and is not thought to be biomechanical in origin. The plantar intrinsics and their tendons (myopathies and tendinopathies) are also a likely source of plantar vault soft tissue symptoms, but such symptoms can be mistaken for plantar aponeurosis pathologies.

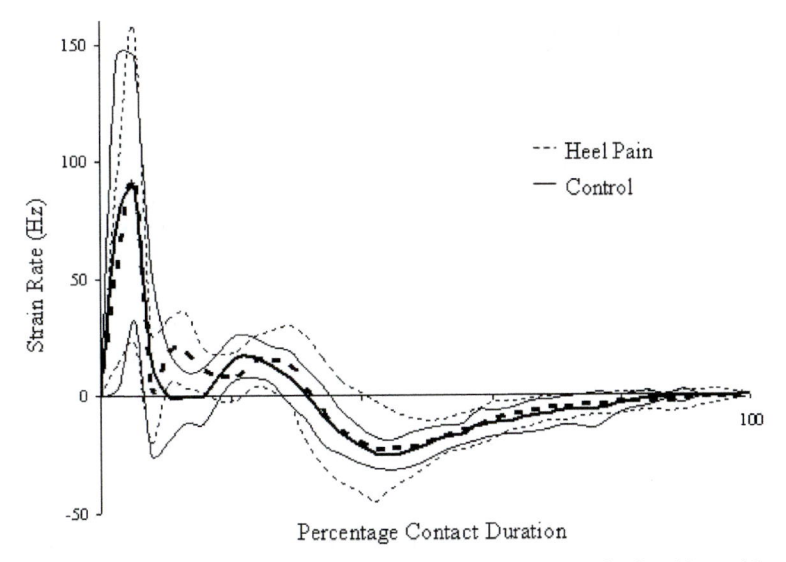

FIG. 4.5.15g An ensemble average strain rate of the heel fat pad as a function of heel contact duration in subjects with unilateral plantar heel pain and individually matched control subjects. Fine solid and dashed lines represent standard deviations for controls and symptomatic heel pads, respectively. *(Image from Wearing, S.C., Smeathers, J.E., Yates, B., Urry, S.R., Dubois, P., 2009. Bulk compressive properties of the heel fat pad during walking: a pilot investigation in plantar heel pain. Clin. Biomech. 24(4), 397–402.)*

4.5.16 Section summary

The foot is a complex structure of highly concentrated anatomy that performs complex biomechanics through changing roles. It alters its viscoelastic properties across the whole foot and across individual joints at different times. It acts as a neuromusculoskeletal interface between the lower limb and the ground, managing energies between both. In so doing, it can dissipate harmful energy and store potential energy to aid energetics and thus avoid injury. To achieve these mechanical property changes, it must alter shape and be reactive to changing environmental conditions under the foot. Through its muscular activity and sensorimotor control, it is responsible for maintaining stable posture throughout locomotion. With so many important functional roles, much can go wrong. With so many different anatomical structures, many pathologies can develop.

As always, the capability of tissues to manage daily expected in-life stresses and occasional perturbations depends on their mechanical properties, which in turn result from developmental loading throughout growth, present lifestyle, physiological health, and age. As much of physiological tissue health is reliant on vascular status, pedal tissues are particularly vulnerable as the foot is often one of the first body regions to suffer from any vascular disease process. Feet are also environmentally compromised within footwear, which in more extreme fashions can present a very novel and constraining environment for the foot to attempt to display its 'normal' kinetic properties, while having its kinematics severely altered. All these physiological and biomechanical considerations are needed in order to understand foot pathomechanics, arguably making the foot one of the hardest areas to clinically manage in clinical biomechanics.

When anatomy fails, the energies will focus within the remaining anatomy. If the rearfoot cannot manage energies sufficiently through excessive flexibility or stiffness, the rest of the lower limb, proximally, and the midfoot and forefoot, distally, has to manage the altered energetic load. If the midfoot fails, then the rearfoot and lower limb and the forefoot must adapt. Should the forefoot fail, then compensation must occur proximally. Multiple segment failure is potentially devastating for human locomotion. Treatment must be able to maintain the management of energy through the foot throughout braking and acceleration, rather than just concentrating on symptoms, which locally can often be resolved by preventing motion or loading on a tissue. If, for example, treatment for 1st MTP joint pain involves preventing motion at that site, it will result in interference with the application of energy from the plantarflexor muscles to the forefoot through coupled ankle plantarflexion and MTP joint extension. In this situation, the energy is going to focus elsewhere within the anatomy. Potentially, this could provoke a new pathology. Foot function and its pathologies will continue to be revealed by research and given time, our understanding of foot pathology will grow, as it has elsewhere in the lower limb. The fact that biomechanics will continue to play an important role in explaining foot pathology further, including presenting as yet unreported pathologies, this author has no doubt.

4.6 Pathomechanics of running
4.6.1 Introduction

Sport usually involves much higher levels of loading stress than most occupational or daily living activities produced in modern 'technological society' lifestyles. This creates a potential environmental mismatch between the development of anatomy in childhood and the sports they make take up as adults. Children exposed to sports activity from a young age are generally better protected, as their body is stimulated by repeated exposure to stress which will adapt their musculoskeletal system. However, if sport is restricted to occasional short-term high levels of exercises and repetitive manoeuvres, then there is still a risk of injury. Excessive levels of intensity or repetition of stresses in training and participation may also exceed mechanical properties of the tissues, particularly during childhood or adolescence, when tissues are immature.

Certain sports and activities can place unusual mechanical demands on the body outside of what human anatomy evolved for their locomotive system. Some dance styles, such as ballet in particular, place very high stresses on forefoot joints, while ballroom dancing via footwear constraints can alter kinematics and place stresses outside of those for which the anatomy was intended. Sports that can also involve the management of upper limb equipment at the same time as running, such as during hockey and lacrosse, place the trunk in an unusual posture during locomotion. Multidirectional sports such as soccer, American football, Australian rules football, Gaelic football, and rugby also present with unusually directed high-level stresses, such as in frequent cutting manoeuvres which can be compounded when the sport involves external impacts with other players. In cricket, fast bowlers can convert their body into a lever arm to propel a ball at high speed, requiring and generating high initial contact impacts on the lower limb. Each specialist sport has its own biomechanical demands, which need detailed study and understanding from the clinician involved in treating the players. Here, we are limited by space to considering some of the more frequent injuries related to the more common activities such as running, which is involved in most sports to some degree. Discussion is therefore restricted to repetitive overuse injuries that are rare outside of the sporting environment.

Distance running over 5 km has become an extremely popular form of exercise since the early 1970s, particularly as a way to keep fit, but also for competition. A desire to improve the runner's time over a certain distance is often more of an aim than beating the competition at events. Leisure running is a different 'biomechanical proposition' to the elite competitive runner, which often involves the biomechanics of performance overriding the risk of injury, and often requires specialised gait assessment and training technique coaching with the analysis of personal technique. These specialist biomechanical approaches are more for running coaches rather than the routine clinical practitioner. For the clinician not attached to elite level sports teams and coaching, injury prevention and resolution of injury are the key clinical biomechanical considerations, rather than performance.

The correct therapeutic approach follows from an understanding of the pathomechanics that underlie pathology. Runners seem to have a preferred movement pathway which in many ways replicates the principles of preferred walking speed. Minor disturbances to these preferred running patterns or movement pathways can be adapted with small changes in muscular activity, but unsurprisingly, large changes require large muscle adaptations that risk the loss of the preferred movement pathway (Hoitz et al., 2020). Undoubtedly, it is such large muscle adaptations to changes in running posture, altering joint torques, that are the mechanisms that lie at the origins of many running injuries, explaining why changes in distance, running style, and footwear design can provoke injury, particularly through fatigue.

4.6.2 Overuse principles

Overuse injuries are most associated with long-distance running, with rates as high as 50% of endurance runners becoming injured annually, and 25% being injured at any one time (Fields et al., 2010). The perception of 'injury' in runners may be more sensitive than in other population groups and sporting participants, such as are seen in rugby or other contact sports. The location of endurance running injuries has been reported as being around 50% at the knee, 39% in the foot, and 32% in the lower leg (shank) (van Gent et al., 2007). In the ankle, rates are reported to be around 4%–16%, injury rates at the hip of 19%, and 3%–11% in the pelvis–groin area, dependent of gender (van der Worp et al., 2015), and with differences in data being reported in running between males and females (Ceyssens et al., 2019; Vannatta et al., 2020a). The injury rate in leisure runners has changed little from the start of running's popularity in the 1970s.

Some clarity on kinematics and kinetics underlying the pathomechanics of overuse injuries is materialising from prospective research (Vannatta et al., 2020a). Prospective studies allow clinicians to look for the features of running that are likely to induce overuse injury before they are altered by the presence of injury. Most studies have looked at the biomechanics of runners who have had injury and compared them to age-matched controls who have not. The problem is whether you are identifying the features that result from injury or injury avoidance, rather than the causes. So, what can presently be derived from the research data regarding the pathomechanics of running pathology?

The actual pathomechanical principles underlying individual injuries are the same in running as they are in walking, i.e. that of stress exceeding tissue threshold. Running involves higher forces and loading rates because of the faster motions involved. Faster loading rates, in turn, tend to make and require tissues to behave more stiffly, so that faster running speeds equate to stiffer lower limbs and feet. Most running subjects exceeding their mechanical stress tolerance are otherwise usually perfectly healthy, but they have just overreached or fatigued their mechanical properties. This is quite unlike individuals with associated walking mobility-induced symptoms, where disease and age often compromise the tissue's mechanical properties. However, underlying disease must remain on the clinical radar with any patient, as sports participants can develop physiological diseases.

Certain injuries and symptoms, such as those affecting the iliotibial tract, medial tibial stress syndrome, and patellofemoral pain syndrome, are primarily running sport-associated, and often largely unknown in sedentary populations. Lower limb stress fractures in otherwise healthy individuals are commonly found among endurance sports participants and military recruits, but not in sedentary populations, where pathological bone can be expected in the presence of fatigue failure of bone. This tells us something universal about endurance sports biomechanics. In long-distance running, individuals are operating at higher intrinsic tissue stress levels without extrinsic high stress events such as those associated with contact sports. Joint alignments in relation to the GRF vectors can be very different in running compared to preferred walking speed, and can change dramatically with running techniques, speed, or fatigue. This means that concepts discussed under osseous and intersegmental alignment of anatomy, such as femoral anteversion angles, knee varum or valgus, and pes planus or pes cavus, continue to have an influence on an individual's biomechanics. However, the position of the joints in relation to each other and the size of the moment arms they create with the GRF vector can dramatically change with running technique and the onset of fatigue. Each potential postural change will alter the moment arms and the activity levels required in individual muscles around joints. The result is that certain running styles will increase stress and therefore the potential for injury in some muscles and joints, while reducing them in others.

Fatigue and energetic efficiency in pathomechanics

Running fatigue resistance is all about energetics and physical or physiological fitness. The more efficient the running posture and the more technique is boosted by the individual's morphology, the longer running can be sustained without fatigue. Once fatigue sets in, the runner will need to adjust the locomotion style to that for fatigued running to avoid injury in the muscles and tendons that as a result, are now operating below maximum capacity. This situation creates two potential problems. The first is the appropriateness of the initial running style for the runner's morphology and physical strength, and the second is, how much their running style changes when and if they become fatigued. The runner, once fatigued, needs to create kinetic and kinematic events that are now appropriate to their fatigued condition, and failure to adapt to fatigue risks injury in tissues that now demand lower and protected mechanical loads. This is something that fatigue changes should provide.

This means that each runner has their own mechanically efficient running style at a specific running speed when their muscles are at a particular level of health and fitness and are without fatigue. This will change with alterations in the running conditions, be they the terrain or the development of tissue fatigue. There is not going to be a perfect running style that suits every runner, even were terrain, running speed, and distance to be kept the same for each running subject. For each endurance runner, there will be a perfect prefatigued technique that delays the onset of a fatigued running style, and a perfectly safe fatigued running style, that must be utilised once fatigued. However, running performance in elite runners presents a different problem to injury avoidance. For high performance, there will be a most mechanically efficient way to cover a certain distance for the shortest time possible for each individual, but that might increase injury risk.

A runner's limb length is a very significant factor for performance, especially for speed where longer legs and large proximal muscles give a greater advantage. This has led to the restriction in the height of prosthetic limbs in elite bilateral limb amputee athletes. Stiffness levels across the forefoot and MTP joints are also important, and can be manipulated through forefoot flexibility within footwear. Whether an elite (or otherwise) runner can match the ideal running style by fitness training and technique alone will depend on the suitability of their underlying morphology. Some people must be made to understand that their morphology does not suit their running activity choice.

If the athlete's morphology does not permit them to achieve the best known mechanically efficient technique, then they may need to develop a technique variant that differs from the biomechanical-energetic ideal that best suits them. However, biomechanical principles do indicate why certain morphologies of human form are more prevalent in certain running distance events at the elite level. Indications are that these anatomical features are at least partly developmental, as previously discussed in relation to sprinters' Achilles internal moment arms, and asymmetrical plantar aponeurotic tissue stiffness in long-distance track runners. Biomechanical forces will affect and influence osseous and tendon structures within elite runners that directly contribute to performance.

4.6.3 Assessing the running patient's pathomechanics

Clerking a runner should be no different to any other patient except that you will also require details of their training regime and a history of their fitness levels, past injuries, and details on their running shoes. Faced with the reality that there is not a universal 'safe' running style, the clinician needs to approach the runner through the principles of biomechanics and be able to associate the injury with the history given by the running patient. Sudden switches in training programmes are a potential source of dramatic tissue stress changes which the body may not have time to adapt to via the biological response. A history of the patient's running activity both now and in the past is important because a child who runs frequently is likely to make a potentially better/safer adult runner through consequent musculoskeletal and neurological developmental processes. Sedentary children who take up running later on in life are more likely to struggle in adapting to what is in essence a significant lifestyle change. Running is a skill that is learnt subconsciously and more easily as a young child, as long as neuromusculoskeletal development is normal. An adult starting to run without previous frequent experience can potentially present clinically with greater difficulties in developing an appropriate energetic running technique, because 'learning' as an adult is generally more difficult. A history of previous running programmes with recent changes in such programmes and changes in terrain and footwear, all potentially give rise to biomechanical influences that may underpin the origin of a problem. Features, such as the running distance covered before the onset of pain and events surrounding the original onset of symptoms, are also important in indicating that provocation may only link to fatigued running. The clerking of the runner's running pattern is just as important as their medical history.

Runners need assessment of their running and ideally, this should involve both kinetic and kinematic data. Although a specific and well-equipped gait laboratory set up to assess runners is desirable, clinical limitations of space may necessitate alternatives. The undertaking of more simple gait analysis techniques with just adequate space can be sufficient, such as sending the runner outside to run with data-gathering equipment or for 2D video recording of preferred running speed. Accelerometers are a good option for gathering joint ranges of motion and joint angles. At a minimum, slow motion video is desirable so that strike position and posture can be assessed throughout the running gait cycle. Assessment by video of both the sagittal plane and frontal plane is advisable, with careful observation of any large frontal or transverse plane motion usually best visualised through filming of the frontal plane. Walking gait and static stance assessment of a runner is not sufficient (Lun et al., 2004), as joint angles and lever arms of joints can be vastly different during running than those seen in relaxed stance and walking. There is float phase in running and a lack of double support that can alter trunk and limb posture significantly from walking.

Terrain also changes joint postures, something to consider if analysis must occur on a treadmill, and particularly if performed by a runner unfamiliar with treadmill running. Thus, time for the runner to adapt to the treadmill before data collection is essential. Problems are presented by off-road cross-country runners. The best a clinician can usually do in such an encounter is to be aware of the changes that can occur in running biomechanics on uneven, undulating ground where inclined-declined surfaces randomly vary. These usually involve shorter strides and more variable gait parameters reflecting the demands of changing GRF through terrain perturbations. Fatigue presents another problem, as rarely can a clinician analyse a patient's fatigued gait unless good planning permits the runner to arrive for assessment right after a long run, or the runner can be kept running to induce fatigue before assessment is made.

Gender differences in runners

The runner's gender can be significant in understanding the kinematics and the kinetics recorded. The magnitude of forces used in muscles such as the hamstrings, gastrocnemius, and soleus are higher across the majority of the stance phase in male runners, regardless of foot strike position (Vannatta et al., 2020b). Males also demonstrate greater variability in peak gluteus maximus force and the magnitude of local peaks of activity (Vannatta et al., 2020b). However, both males and females employ their own unique lower extremity force characteristics on running (Vannatta et al., 2020b), stressing the variability of healthy running techniques generally.

Gender influences morphology, muscle, and osseous development, as the hormones released during adolescent growth influence the mechanical properties of tissues, which can then be expressed through activity levels. Thus, although the resulting morphology of the runner relies on a combination of factors, the development of muscle powers and certain osseous morphologies are more common in males than females, giving them greater physical potentials. This is not to imply that men are always better and stronger runners. Long legs and light trunk segments aid distance runners, whereas higher muscle bulk around the trunk, short posterior tubercles on the calcaneus, and longer toes are desirable for sprinters, whether they be male or female.

Running gait focus in pathomechanics

As always, in gait, certain significant biomechanical events are most important to focus on. These are initial contact/braking events and the change to and efficiency of acceleration. As motion is directed in achieving anterior displacement, the sagittal plane motions are of highest significance. A failure in providing adequate energy dissipation and acceleration, along with the development of excessive vertical, transverse, and frontal plane motion, can lie at the source of pathology, especially if sagittal plane efficiency and early fatigue onset results from muscles that either create or control these actions.

More recent systematic reviews (Vannatta et al., 2020a) and studies (Bramah et al., 2018) have suggested that certain kinematic features in gait may be associated with increased injury risk. The systematic review by Vannatta et al. (2020a) looked at thirteen prospective studies and found that gender and competition level affected injury risk, as did the amount of hip adduction motion (varus tilt), reduced ankle dorsiflexion peaks, and the amount of rearfoot eversion. Associations were often weak, inconsistent between studies, and features considered were significantly different between studies and were affected by the sample size, indicating that subject-specific issues were probably more significant than generalised kinematics across runners. The studies reviewed in Vannatta et al.'s (2020a) paper reported on subjects with patellofemoral pain, iliotibial band syndrome, Achilles tendinopathy, and knee and shin symptoms (without any specific diagnosis). The tests occurred on treadmills, soft runways, and firm runways, without shoes or running speed always being controlled for. It is likely that different injuries involve different biomechanical factors, just as different terrains, shoes, and running speeds, are known to create different running biomechanics.

Bramah et al. (2018), by looking at 72 injured runners, (also without specific diagnoses—patellofemoral pain, iliotibial band syndrome, medial tibial stress syndrome, and Achilles tendinopathy) compared to age-matched injury-free controls, found that the injured group had greater hip adduction motion (varus tilt), a more anterior trunk lean, greater knee extension, and ankle dorsiflexion at initial contact (associated with increased stride length). Furthermore, contralateral hip drop from stance limb hip adduction was found to be associated the most with injury. Walking data on metabolic costs of human locomotion indicate that certain tasks are more metabolically expensive, such as acceleration (Gottschall and Kram, 2005). It is possible that changes in metabolic demand in trunk stabilisation or braking energy dissipation may upset the metabolic balance for acceleration, inducing premature fatigue in certain muscles during running. This may at least be associated with increased hip adductor moments.

Large stride lengths have a potential to increase initial braking demand at heel strike, but also to lengthen the GRF lever arm from the knee and hip, inducing greater angular moments which requires greater muscle force production to resist them. The longer strides will also tend to place the muscles at longer initial loading lengths, reducing their contraction efficiency and power generation to control joint moments. Variance in the injuries within the groups studied is likely to involve different pathomechanics because the different structures injured have different roles in stabilising kinematics and generating kinetics. Thus, a single excessive joint motion or functional unit dysfunction cannot explain the cause of all running pathology (Fig. 4.6.3).

Trained runners seem less prone to injury and, on average, have shorter stride lengths and higher step frequencies than novices (Moore, 2016). Trained runners operating consistently at their preferred speed demonstrate a closer horizontal difference between the GRF vector and the trunk's CoM, with a lower ankle dorsiflexion angle and reduced ankle stiffness (Müller et al., 2010). Deviations in stride length that are greater than 6% are detrimental to a runner's performance (de Ruiter et al., 2014). Thus, certain key gait events should demand the most attention during assessment, starting with foot strike position, stride length, midstance hip posture, and terminal stance acceleration limb and trunk posture. The lower amount of vertical compared to horizontal, CoM motion and displacement from stance into float-aerial phase will indicate efficiency. Strike position and stride length will give indications of impact dissipation techniques and the size of the moment arms being resisted during braking. Trunk displacement and hip adduction moments indicate the effectiveness of stability and acceleration energetics. If these events do not flag up possible links to the pathology, then more subtle motions should start to be investigated.

Future research directions in understanding running pathomechanics

Prospective studies may hold the solution to a better understanding of running pathomechanics. For example, it has been reported prospectively that increased knee adduction impulse may be a risk factor for patellofemoral pain in runners (Stefanyshyn et al., 2006). But what is the diagnosis of the patellofemoral pain being investigated? Very likely, there are a number of different pathologies at play with differing pathomechanical origins. Thus, the precise location and nature

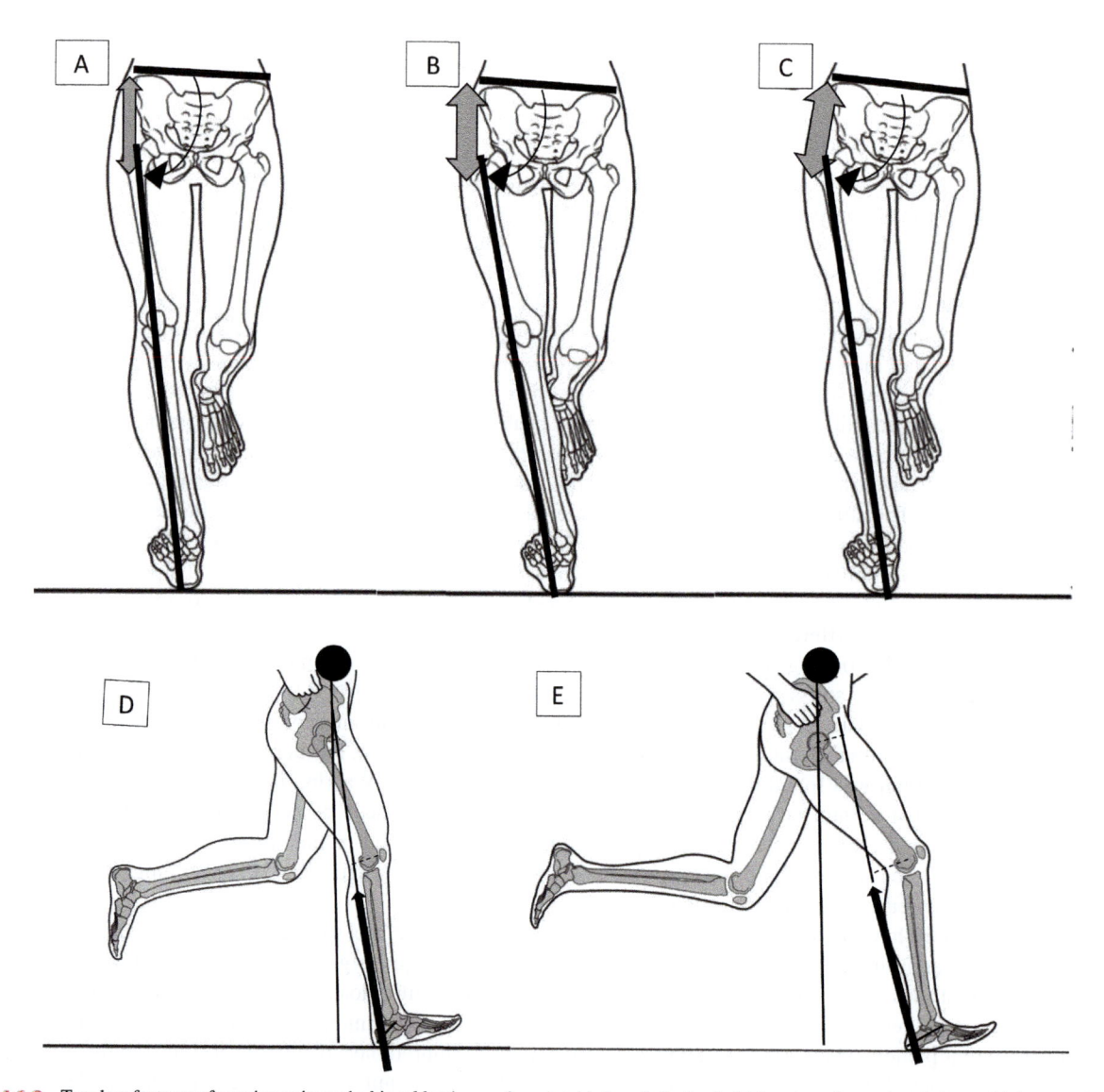

FIG. 4.6.3 Two key features of running gait are the hip adduction angle and stride length, both of which seem to have associations with increased risk of running injuries when larger. Larger hip adduction moments are reported to be associated with injury risk in prospective studies. Signs of good hip abductor function and stance phase stability are a relatively small hip adduction angle of the lower limb and an almost level pelvic posture at contact (A). The presence of a larger lower limb adduction angle (B) may risk earlier fatigue of the hip abductors which will be working on a more mechanically disadvantaged lever arm than that shown in (A). Poor hip abductor activity in loading can be identified with repetitive hip drop at initial contact and through loading response (Trendelenburg) which functionally increases the hip adduction angle (C). Stride length is also important in generating sagittal plane external GRF moment arms, with shorter strides (D) offering muscles easier moments to resist at muscle fibre lengths more appropriate for power generation against hip, knee, and ankle flexion moments, thus providing easier energy dissipation. Longer strides (E) present larger distances between the CoM and the GRF forces, driving larger hip, knee, and ankle flexion moments from contact to loading response. Elite long-distance runners are known to use shorter strides and a higher step frequency. *(Permission www.healthystep.co.uk.)*

of an injury should be known or at least suspected before treatment starts. It is undoubtedly time that specific injuries (rather than symptoms) were looked at individually on a prospective basis, with subjects being separated by the distance of the running performed, age, gender, as well as their competition and experience levels so as to see whether more coherence can be achieved in assessing the injury risk factors for each pathology. The challenge of creating homogeneity in research subjects makes research in this area difficult, and indeed, runners often change their running distance and terrain as they look for different challenges, thus complicating prospective studies.

In the present climate of paucity of knowledge in running population injury risks, the clinician is left with the practicalities of solving an individual's problem through the application of biomechanical principles applied to the data collected. This avoids having to compare a running style to a theoretical 'normal' that is proposed to not cause injury, thus allowing the clinician to concentrate on what is, rather than what should be. Through the construction of free-body diagrams coupled with appropriate anatomical and mechanical knowledge, the pathomechanics can be elucidated and the origin addressed.

4.6.4 Foot strike position free-body diagrams

Foot strike positions have been discussed in Chapter 1, Section 1.4.9, explaining how foot strike position is strongly correlated to running speed with concurrent influences from footwear. Foot strike pattern is associated with different tissue energy dissipation and energy storage management during the loading response. Failure to dissipate energy risks excessive tissue stress, regardless of strike position. Excessive energy dissipation leads to less energy storage and a larger energy deficit that needs to be made up during acceleration via increased muscle activity. Energy dissipation and generation are primarily the responsibility of the muscles via a relationship between controlled joint motion and muscle–tendon buffering, assisted by other connective tissues, tissue oscillation, and specialist structures like the plantar fat pads. The postural alignment of the lower limb and trunk at initial contact sets up lever arms of motion around the joint from the GRF vectors that direct the burden of stress management to certain muscles, preferably working at their middle lengths around short joint angular moment arms.

Thus, moment arms direct which muscles will take on the energy dissipation load as well as how hard they will need to work. This is not to suggest that foot strike position affects the actual risk of injury, a suggestion which evidence does not support (Kuhman et al., 2016; Dudley et al., 2017; Messier et al., 2018). Instead, foot strike position related to the body posture above may concentrate stresses within certain tissues, making it more likely for distinct tissues to become injured with particular strike positions. Whether these tissues can cope is another matter and is related to the specific strengths, weaknesses, and fatigue status of the runner's tissues. In heel strike gaits, tibialis anterior dissipates energy through muscle–tendon buffering (Roberts and Konow, 2013). In forefoot strike, the triceps surae, through the Achilles, receives this burden (Almonroeder et al., 2013; Nunns et al., 2013; Kernozek et al., 2018), possibly decreasing some impact load on the knee (Arendse et al., 2004) as well as tibialis anterior.

In constructing free-body diagrams of runners, clinicians should use data from the gait analysis to help visualise the effects of pathology, and then be able to logically plan a therapeutic regimen that can address the locomotive problem. The aim of the free-body diagram is to initially present a universal 'template' of concepts of the strike position involved and the muscle activities necessary for running efficiency, and then to compare this to the individual runner as a variant of these events, particularly with reference to their distinct gait kinematics and pathology. To simplify the concept for clinical application, an approach in this text is taken that the data has been gathered using simple 2D slow motion video analysis of the sagittal and frontal planes, although the principles are very easily and more effectively, applied to 3D data. When using a treadmill, the cameras will record multiple steps, but if a treadmill is felt inappropriate for the runner, then multiple 'run-pasts' of the video camera will be needed to give reasonable averages of variable joint angles. If more complex 3D motion analysis is used, then the concepts of pathomechanical risks remain the same, but more detailed quantitative data of the runner can also be assessed.

Most runners run in footwear, and a running assessment should use the runner's present footwear, initially. If the foot needs to be seen, then the clinician will need to decide whether or not the best approach is to initially have the patient run barefoot, knowing that this will be different to the motion that occurs within the shoe to some variable extent. Assessing the rest of the lower limb should be performed shod, unless assessing a barefoot runner. External shoe motion can be used in the hope that the foot largely follows the shoe (which it certainly will not, perfectly), or roughly 'guesstimate' where the foot is inside the shoe. These are the challenges of clinical biomechanics. When changing the runner's footwear during gait analysis immediately improves running posture, then the runner's footwear as a potential source of dysfunction is suggested. Needless to say, barefoot runners are far easier to assess.

Rearfoot strike pathomechanics in the sagittal plane

Sagittal plane positioning of rearfoot strike is associated with the foot angled in degrees of dorsiflexion to the support surface and a GRF vector behind the ankle axis in the sagittal plane. The angulation of the foot at heel strike is dependent on factors such as stride length and foot extensor muscle strength. The variance in heel strike angle results in biomechanical differences throughout the lower limb. Recording the foot angle is more easily done with the runner barefoot, but unless the runner normally runs barefoot, there is little point in recording the barefoot running angle. Understanding the shoe construction used by the runner, noting features such as the 'heel drop' (heel lift or height within the shoe) between the rearfoot and forefoot and the stiffness properties across the forefoot of the footwear, is useful in understanding how the foot is adapting to and influenced by the shoe. If the posterior aspect of the footwear's sole is soft, and 'collapsible', then the strike angle may increase during impact through material deformation, which during heel strike may result in slowing the loading rate of the plantarflexion impulse generated. However, it will likely increase the overall ankle dorsiflexion angle that must be managed. Should the posterior of the shoe sole be largely noncompressible, and the heel raised by the shoe's sole, then the dorsiflexion angle will be reduced. However, the hard heel lift may increase the acceleration rate of the plantarflexion moment via a hard ground impact occurring on a possibly earlier heel strike than would happen barefoot. A shoe with a posterior flare may initiate an even earlier heel strike and longer plantarflexion moment via a longer posterior moment arm (see Fig. 4.6.4a).

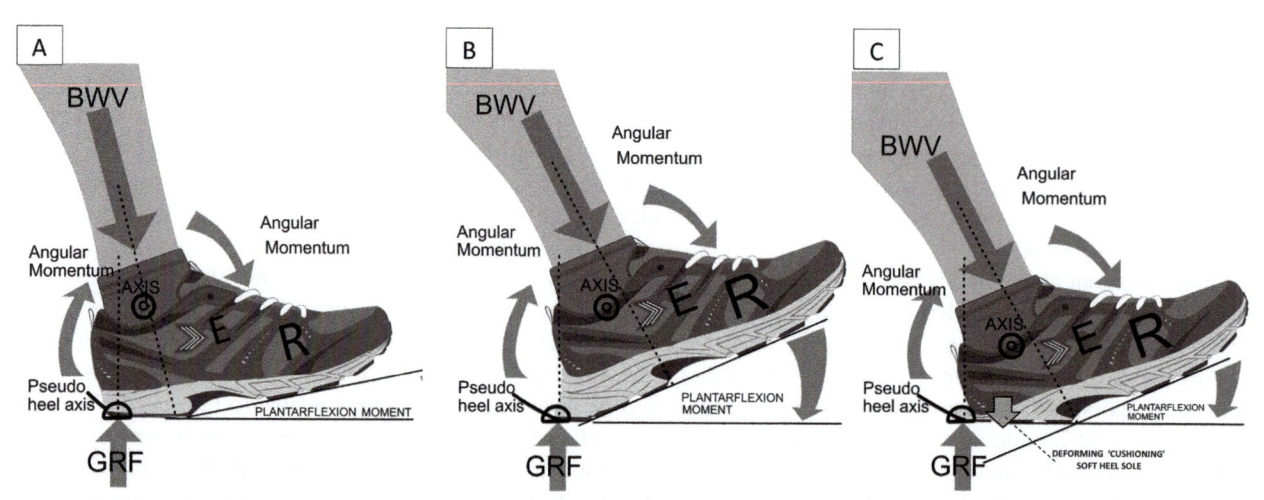

FIG. 4.6.4a Despite the fact that feet do not precisely follow the movement of running footwear (making their external surfaces poor markers for recording foot kinematics), the presence of running shoes influences running energetics and biomechanics in many ways. Next to the weight of a shoe, footwear flexibility is particularly important. How much stiffness a shoe offers and where the most flexibility is positioned across the forefoot affects the acceleration energetics, either aiding or interfering in the application of plantarflexion power or in altering MTP joint energy dissipation. At initial heel impact, as illustrated here, the design and flexibility of the heel midsole and outsole is important, affecting the size, rate, and duration of angular moments applied to ankle joint axis. Tibialis anterior decelerates the plantarflexion moment via its lever arm mechanics. The presence of footwear immediately changes the timing of initial contact, with thicker-soled shoes leading to earlier ground contacts on effectively longer limbs. This results in longer rearfoot loading times, extending the duration of heel strike transients, thereby potentially lowering peak forces. The posterior aspect of the sole of footwear effectively creates a longer heel at impact, creating a pseudo-heel contact with its own axis of rotation outside of the foot which the whole foot moves around, increasing angular momentum. The presence of the footwear sole at the heel increases the distance between the GRF and the ankle axis of rotation, creating a longer plantarflexion moment arm. This must be resisted by effort (E) from tibialis anterior applied via its distal attachment on the proximal medial vault, working within the extensor retinaculum at the ankle. Tibialis anterior is now mechanically disadvantaged compared to a barefoot heel strike, and the weight of the rest of the shoe may also increase the resistance (R) that must be decelerated over a longer distance. If the heel sole is relatively low, firm, and contoured to follow the heel profile (A), it will only increase the lever arm slightly. If stride length is also short, the footwear's effects should be minimal. A firm, posteriorly flared, thicker heel sole (B) results in an earlier heel strike, and significantly lengthens the GRF plantarflexion lever arm to the ankle joint's sagittal plane axis. If this is coupled to a long stride directing the body weight vector (BWV) more anteriorly, tibialis anterior will be placed under considerably larger stresses when attempting to brake and energy-dissipate the plantarflexion moment. This braking activity will be occurring over a longer distance created by the greater dorsiflexed foot contact angle on a longer stride. Attempting to cushion impact forces to reduce injury has led many designers to try soft deforming materials on the soles of shoes. At heel strike on a longer stride, impact forces may deform softer posterior heel soles (C), sending the heel further towards the ground. Although this deformity shortens the GRF plantarflexion moment arm, it will increase the foot dorsiflexion contact angle. Once again, despite the best efforts to 'cushion impact', such designs can inadvertently make energy dissipation harder. Similar mistakes as those discussed here can be made with ill-considered 'cushioning' or heel-lifting foot orthoses. *(Permission www.healthystep.co.uk.)*

A heel strike with the foot angled at around 6° will create a GRF vector significantly angled posteriorly against the BWV. The GRF vector lies behind the ankle joint axis, creating a strong ankle plantarflexion moment. This plantarflexion moment will match the angulation of the foot to the support surface, or more usually the foot's angulation inside the shoe. This means that a foot dorsiflexed to the support surface by 6° will create ~6° degrees of plantarflexion rotation to bring the forefoot to the ground and end the plantarflexion motion. The loading rate of the plantarflexion moment is dictated by the running speed, terrain, and shoe sole hardness under the resisting control of the ankle dorsiflexors. This task is almost completely controlled by tibialis anterior with some assistance from the long toe flexors which with abductor hallucis, are setting some of the digital extension to initiate the windlass mechanism. Tibialis posterior and peroneus longus are also engaging to 'set' an appropriate level of foot stiffness (by sensorimotor predictive activation) for forefoot loading response stability and energy dissipation. Thus, the muscles under stress and the amount of stress during braking, depends greatly on the foot strike posture (Fig. 4.6.4b).

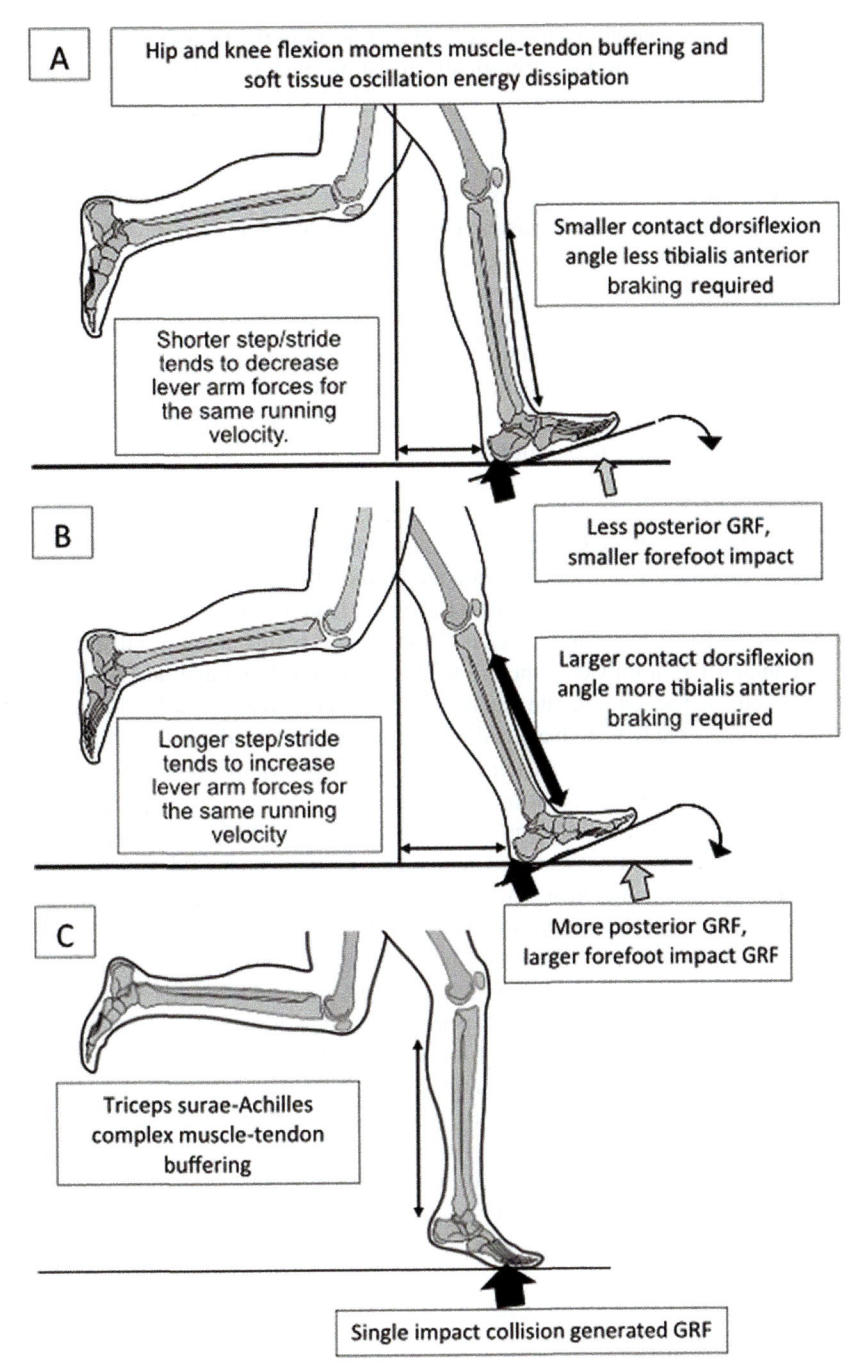

FIG. 4.6.4b The heel strike-induced forefoot impact forces of running are dependent on three critical factors as a two-collision event. The first is the speed of running and the acceleration of the limb at contact, as force is equal to mass times acceleration. This initial impact is taken by the heel fat pad, greatly aided by the proximal energy dissipation mechanisms of knee and hip flexion energy buffering and soft tissue oscillation, which continues until after forefoot impact is completed. The next factor is the ability of tibialis anterior, aided a little by the long digital extensors to slow the acceleration of the plantarflexion moment before forefoot contact. The final influence is that of the ankle dorsiflexion angle to the support surface at initial contact, which has a significant effect on tibialis anterior activity required. The plantarflexion lever arm is longer the further posterior on the heel that a heel strike occurs and the larger the ankle plantarflexion motion will be before forefoot contact. Shorter step/stride lengths (A) tend to produce lower heel strike dorsiflexion angles and reduced initial plantarflexion contact accelerations and forefoot impact forces, and also reduce proximal joint's angular momentums compared to longer step/stride lengths (B). Thus, with factors such as muscle strength and running speed being equal, rearfoot impact forces will tend to be lower on a shorter step/stride heel strike, reducing chances of tibialis anterior tendinopathies, myopathies, compartment syndromes, and forefoot pathologies such as fatigue fractures during loading response. Forefoot strikes (C) take the full collision force as a single impact event. Unable to utilise tibialis anterior braking, the triceps surae–Achilles complex becomes an important energy dissipator, risking initial contact-induced Achilles tendinopathies. Forefoot impact also relies on pretensioning the vault to create sufficient elasticity within the connective tissues of the foot to store and dissipate impact energies (with a little forefoot plantar fat pad assistance), while the proximal lower limb dissipating techniques do their job pretty much as they do during heel strikes. Indeed, safe forefoot impact in all running situations requires tibialis posterior and peroneus longus vault stiffening and the windlass mechanism to pretension the foot into an elastic shock-absorber prior to any forefoot impact, and their deficiencies also risk injury. *(Permission www.healthystep.co.uk.)*

Being aware which muscles are partaking in braking events is important, especially if they become a source of symptoms. During heel strike impact, muscles eccentrically resisting plantarflexion moments play an important role in energy dissipation through muscle–tendon buffering, along with the other energy dissipation mechanisms such as the plantar fat pad of the heel and soft tissue oscillations. If muscles are adequate in strength and fatigue resistance appropriate for mechanical need, then no issues should arise directly from the size of the plantarflexion moment arm. However, 'achy' anterior shins are extremely common in novice runners who heel strike, because the plantarflexion angle in running can be greater than walking and the loading rate is much faster. The problem can also occur during sustained fast walking with a long stride length (especially on hard surfaces) for the same reason. Usually, in novice runners, the problem will self-resolve as the tibialis anterior adapts by increasing strength, as long as the running frequency and distance is not increased (significantly) until the tissue mechanics have responded via biological response adaptations. In more persistent cases, especially when tenderness can be provoked on tibialis anterior resistance testing or by palpation, simple tibialis anterior strengthening exercises, and reduced running frequency and/or distance may be required until the problem has resolved. However, reducing the stride length of the runner will also prove beneficial, as it should reduce the foot strike angle-induced plantarflexion moment arm and decrease the duration of the ankle plantarflexion moment.

There is a direct link between the plantarflexion moment of the ankle to the dorsiflexion angle, the GRF vector's position, knee and hip angles, and stride length. As stride length increases, the foot's dorsiflexion angle to the ground usually increases, as does the hip flexion angle while the knee becomes more extended. If we consider a 13° dorsiflexion angle to the support surface compared to a 6° dorsiflexion angle, then the initial strike position will be moved more posteriorly on the heel using the greater ankle angle, and thus be further from the ankle axis. This increases the plantarflexion moment arm of the GRF vector for ankle plantarflexion. Thus, 13° of plantarflexion will then be required to complete forefoot loading of a foot angled 13° at impact, although the lower limb following along behind reduces the actual ankle plantarflexion angle that develops. The loading rate and the loading time of the plantarflexion moment will result from the dorsiflexion angle and the running speed. At high ankle dorsiflexion angles, the tibialis anterior is under higher stresses that may create inadequacies in strength that induces earlier fatigue. This may require other energy dissipators to become more active, risking injury through energy dissipation at heel impact.

Let us consider the possible consequences of failure in the muscle–tendon buffering effects of tibialis anterior. If tibialis anterior's energy dissipation and braking effect on the plantarflexion moment is limited, then other mechanisms will need to increase their dissipation role, and the precontact stiffness across the foot through the windlass mechanism will need adjusting to compensate for the change in forefoot loading rate. Thus, assistor ankle dorsiflexors, tibialis posterior, and peroneus longus will need to adjust and may potentially become increasingly loaded. The heel fat pad may experience a faster loading rate, which will cause it to behave more stiffly, possibly under greater shear strains. Foot vault stiffness may need to be changed through tibialis posterior and peroneus longus activity to adapt to changes in plantar fat pad behaviour and the likely accelerated loading response from loss of tibialis anterior braking. Changes do not stop there, as a less overall braked impact may result in a higher impact frequency causing increased lower limb tissue resonance (oscillations), requiring increased muscle damping activity which increases the energy buffering load requirements of other muscles within the lower limb.

The longer the plantarflexion moment arm generated by the higher dorsiflexion angle resulting from a longer stride length, the more likely it is that tibialis anterior mechanics will fail, either due to an initial lack of strength or from the onset of fatigue. It is worth considering the consequences in whole lower limb sagittal alignment that will accompany increased stride length. Increased stride length is likely to result in a decreased knee flexion angle and an increased hip flexion angle at impact, as the foot will increase its distance from the trunk. A longer stride length requires greater hamstring braking from a longer late swing limb knee extension, which will occur with the hamstrings at a longer length, compromising their muscle contraction capabilities.

Knee flexion at contact is the main impact energy dissipator through quadriceps resistance muscle–tendon energy buffering, with this being a bigger factor in shock absorption in running than walking. A heel strike on a very long stride will cause the GRF vector to align through the knee rather than behind it, as normally occurs. Losing the knee flexion angles at contact initially results in increased knee stiffness, as the GRF may not initiate flexion immediately to utilise quadriceps muscle–tendon buffering. The GRF vector will also tend to lie more anterior to the hip with a longer stride, increasing the hip flexion moment arm that must be resisted by gluteus maximus. In this posture, gluteus maximus will be working at a longer length, potentially reducing its contraction capacity and efficiency. The increased knee extension posture may also initially shift the energy dissipation burden from the knee to the hip, until the knee becomes sufficiently flexed under increased hamstring knee flexion, for the quadriceps to start energy buffering. A longer stride functionally lengthens the lower limb, requiring gluteus maximus, and the hamstrings to be loaded at a longer length. This reduces their initial contraction capabilities (contracting at a longer length) but also lengthens the loading time through which these muscles (particularly the hamstrings) are required to dissipate energy and move the body's CoM towards the support limb, that lies more anteriorly to it as a result of the longer stride. A shorter stride length is an obvious solution, and the more it decreases the easier it becomes for the muscles, including the quadriceps, to resist contact flexion moments. A shorter stride also decreases ankle dorsiflexion angles at contact, and the plantarflexion moment arm. Short strides also enable muscles to quickly transfer the CoM anteriorly and over the base of stability provided by the support limb and foot, helping hip abductors during loading response (Fig. 4.6.4c).

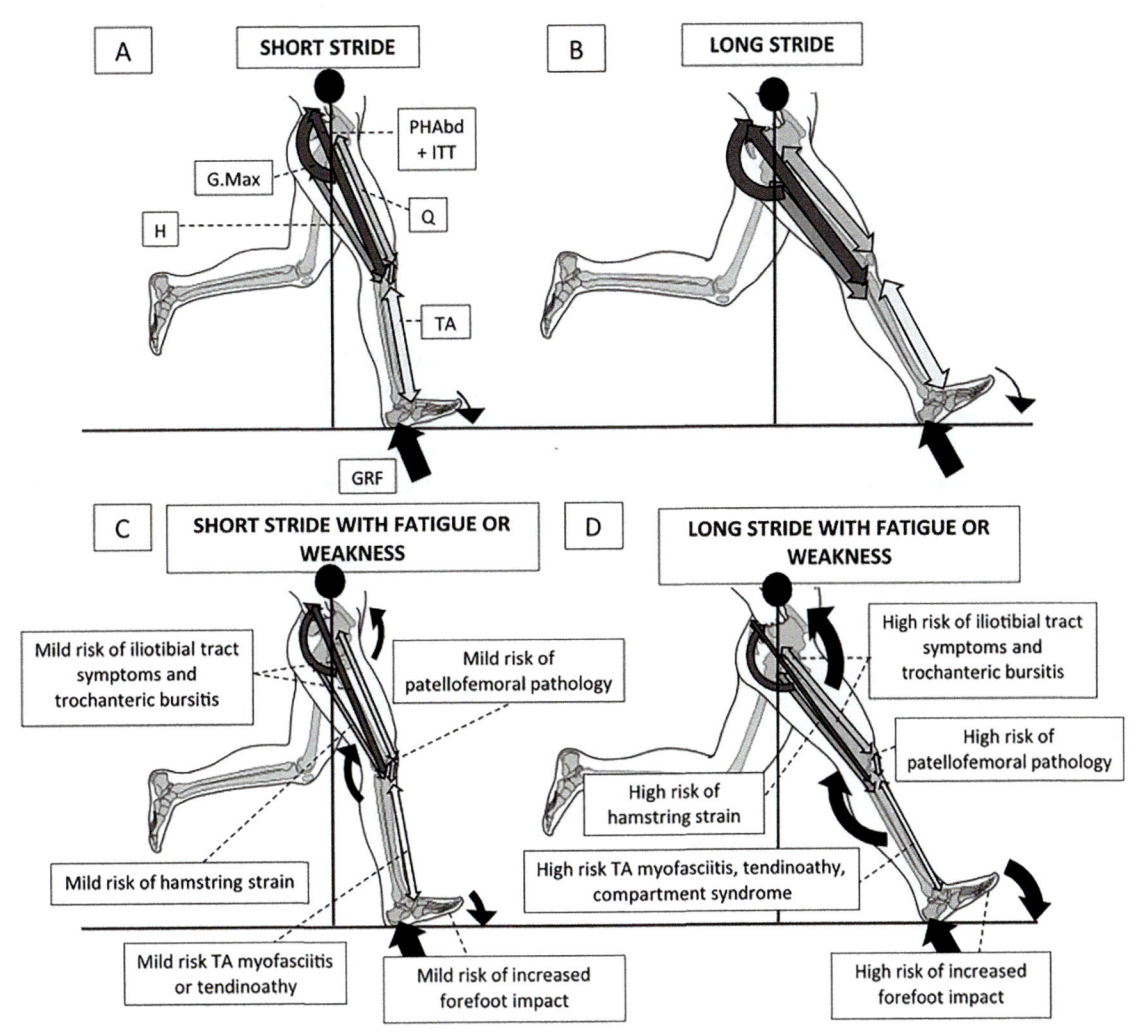

FIG. 4.6.4c Stride length matters to running biomechanics more than in walking. This is because swing limb acceleration and velocity of locomotion is greater, driving greater stresses into the musculoskeletal anatomy, such that small issues in kinematics can become the difference between safe performance and injury. Fatigue, resulting in reduced muscle strength, can cause tissues to become mechanically insufficient for safe load management. Running with a shorter stride length makes heel strike biomechanical management easier for the five muscles most relied upon for safe braking and energy dissipation. A short stride length (A) keeps the distance between the GRF and the CoM (black circle) to a more manageable length, reducing angular moments around the hip, knee, and ankle. This reduces the effort required of the hamstring muscles (H) to brake swing limb knee extension before contact. At contact, a shorter stride helps the gluteus maximus (G.Max) linked to iliotibial tract (ITT) function and the posterior hip abductors (PHAbd) activity, the quadriceps muscles (Q), and tibialis anterior (TA) to all brake flexion moments and provide energy dissipation at their respective joints. A longer stride length (B) increases the distance between the CoM and the GRF, increasing the angular moments on the joints, necessitating greater muscular effort at longer muscle fibre lengths to achieve safe and efficient braking and energy dissipation. This is likely to cause muscles to fatigue sooner. When fatigued or if muscles are inadequately prepared for running, a shorter stride length (C) is still less likely to provoke injury, for flexion angular moments are smaller. When fatigued, a longer heel strike (D) will result in a more dramatic forefoot impact from a larger plantarflexion moment on a more dorsiflexed foot angle at contact. The hamstrings are also required to apply their braking of swing limb knee extension at a longer, more dangerous length. Because the CoM is further behind the GRF and the hip more flexed, hip extensors are likely to allow increased hip flexion at contact and knee extensors may struggle to control larger knee flexion ranges of motion. The uncontrolled kinematics and poorly dissipated impact energies that can result will seek out mechanical weak points within the individual, expressing their tissue threshold breaches through pain and pathology. *(Permission www.healthystep.co.uk.)*

Midfoot and forefoot strike in sagittal plane pathomechanics

Midfoot (foot-flat initial contacts) and forefoot strikes are only normal in running. Midfoot strikes involve contact with the heel and forefoot at the same time (or close to the same time), resulting in a CoGRF vector more within the middle of the foot, closer to the axes of the ankle and other proximal lower limb joints. The foot angle with the support surface will be perpendicular or very near to perpendicular. The GRF vector lies close to or only slightly anterior to the sagittal plane ankle joint axis, so a forefoot plantarflexion moment arm is absent or very small, requiring little if any significant tibialis anterior activity. Because of the small momentum around the ankle joint, the ankle plantarflexors or dorsiflexors are under less stress, but they cannot provide as much energy dissipation either, requiring other structures to make up the loss of these muscle contributions. The knee and hip extensors, plantar fat pads, and the foot vault via the reversed windlass mechanism, will primarily dissipate the impact loading. With the short GRF moment arms and middle range muscle lengths involved in contraction, this contact posture should offer greater fatigue resistance.

Forefoot strikes involve ground contact with the foot at a plantarflexed ankle angle, setting up a GRF vector anterior to the ankle axis. This creates a significant dorsiflexion momentum arm that requires triceps surae to prevent the heel accelerating to the ground. The triceps surae takes on a significant energy dissipation role via muscle–tendon buffering action within the Achilles. Gastrocnemius may now contribute more to increased energy dissipation at the knee through its proximal attachment concurrently. The more anterior the GRF vector is to the ankle, the longer the GRF momentum arm is, and the greater the triceps surae action will need to be. Therefore, the nearer to 'toe contact' the strike position, the longer the potential lever arm may be, unless the plantarflexed ankle moves the forefoot more vertically to lie below the ankle axis. However, the greater the distance the heel lies above the forefoot, the greater the potential distance the heel has to drop under the ankle dorsiflexion moment. Often, a strong triceps surae will result in the runner not actually contacting their heel to the support surface in this position, (as is normal for sprinting) but many long-distance forefoot runners will dorsiflex their ankles until they contact the support surface with the heel. More heel drop can associate with barefoot or flat shoes than in footwear with a larger incorporated heel lift that can block full heel drop (Fig. 4.6.4d).

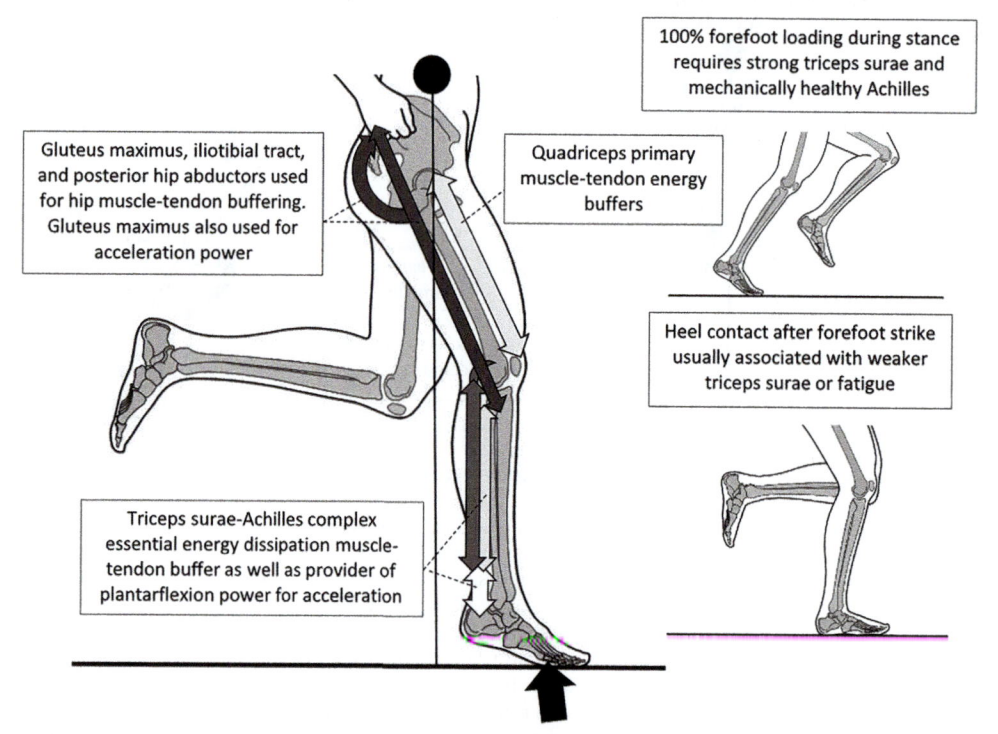

FIG. 4.6.4d The knee, via the quadriceps muscles, is the primary joint for impact energy dissipation in all forms of running, dramatically increasing its burden from walking. Forefoot running tends to keep stride lengths shorter, for it is difficult to make a long stride and maintain loading on the forefoot alone when the CoM is a long way behind the GRF. Often, short stride length forefoot strike runners will not make heel contact during stance phase, and move from forefoot contact braking to acceleration directly off the forefoot. This ability is affected by running speed and duration, with shorter distance and faster speeds producing least heel contacts. In long-distance running over time, most runners start to heel contact after forefoot strike, but many will become rearfoot strikers. This probably reflects fatigue within the triceps surae which has to remain highly active throughout stance in forefoot running. The heel fat pad is better adapted to energy dissipate adequately and safely when running pace slows with fatigue, than is the Achilles. Runners that remain on their forefeet require very strong triceps surae and mechanically healthy Achilles tendons. On running gait analysis, runners that demonstrate a heel contact immediately after forefoot strike are likely to demonstrate inadequate triceps surae strength and present Achilles tendinopathy-vulnerability which needs addressing. Those already displaying Achilles tendinopathy may do better by changing to a heel strike or foot-flat midfoot strike. *(Permission www.healthystep.co.uk.)*

Other structures under increased impact loading in forefoot strikes are the forefoot plantar fat pads and the foot vault. Foot stiffness upon loading is maintained by tibialis posterior, peroneus longus. The windlass mechanism may be partially engaged by passive digital extension from the contact GRF. Obviously, there is a potential injury implication to the triceps surae and its associated connective tissues (gastrocnemius aponeurosis and Achilles tendon) with forefoot strike. More widespread potential issues are possible should elements of the triceps surae–Achilles complex be weak and compromised or if it becomes fatigued as a result of resisting the GRF-induced dorsiflexion moment of impact, the size of which being associated with how extreme the forefoot strike is (Kernozek et al., 2018). Should the triceps surae not tolerate the increase in loads and an energy buffering role, calf and Achilles injuries may result. Triceps surae weakness or fatigue may cause a loss of heel drop control which in turn, may interrupt smooth hip and knee flexion energy buffering.

4.6.5 Frontal plane effects of foot strike positions

In the frontal plane, foot strikes can occur centrally (evenly) or to the medial or lateral sides of the foot. Soft tissue tension torques across the frontal plane at joints and osseous alignments influence the contact position of the foot prior to and during initial contact. In running, the base of support tends to narrow via increased hip adduction moving the lower limb more under the trunk to improve the base of support beneath the CoM lying over the single stance limb. Thus, running initial contacts tend to occur more towards the lateral foot, whereas in walking with contact occurring during double-limb contact, such a degree of limb adduction is not required. Running adds variable rearfoot, midfoot, and forefoot strikes into the equation as an extra complicating factor to managing larger frontal plane torques. Extreme lateral and medial strike positions influence angular moments of inversion–eversion on the foot and adduction–abduction moments at the hip and knee. Thus, recording the foot's strike side and well as the position on the foot, can prove helpful in understanding the pathomechanics when both the loading rate and duration of frontal plane motions may help explain pathology.

Lateral rearfoot strikes set up eversion moments through the rearfoot complex on a narrow base of support, increasing varus (adduction) moments at the knee and hip. Such moments might increase the knee's medial compartment articular compressions or lateral collateral ligament complex tensional strains, and/or increased varus load resistance required by the hip abductors during loading response (Figs 4.6.5a and 4.6.5b). Running lateral forefoot strikes will set up midfoot eversion moments, following initial dorsiflexion/abduction of the 4–5th metatarsals at their cuboid articulations. If these joint motions resolve the moment, the rearfoot complex is spared. However, if rearfoot eversion is also initiated by the lateral forefoot GRF, it will occur through the ankle joint alone, as the subtalar joint lacks flexibility as it is not loaded with external force during a forefoot strike. In excess, the rearfoot eversion can irritate medial ankle (deltoid) ligaments or tibialis posterior-related structures such as its tendon sheath or the tendon itself. Forefoot contacts on a narrow base of support will increase varus (adduction) moments across the knee and hip, just as in rearfoot strikes, but by utilising the forefoot the joints across the midfoot can reduce these varus effects on the knee, something that is less possible with rearfoot strikes.

Medial foot contacts in running are quite rare because of the need for a narrower base of support than during walking. As a general rule, they associate mostly with novice of runners unless such contacts develop with fatigue. Medial rearfoot strike effects are dependent on where they occur. A wide base of support may induce medial rearfoot loading, creating increased eversion moments across the rearfoot and valgus moments at the knee and hip. This requires greater tibialis posterior rearfoot eversion braking, and/or causes increased lateral articular compression within the knee and tensile strain in the medial collateral ligaments. A narrow base of support with a medial heel strike can create a GRF force that initiates rearfoot eversion torques that can either cause knee adduction or abduction torques depending on whether the GRF is medial or lateral to the knee, respectively. Thus, if medial heel strikes do occur, then can have the potentially to be particularly problematic within the lower leg and foot, but because running demands a narrower base of support, GRF external loads on the hip are usually adduction loads, regardless of the rearfoot-to-knee positions.

Forefoot medial strikes usually initially load on the 1st metatarsal, leading to dorsiflexion and eversion of the 1st metatarsal at the 1st tarsometatarsal joint, before inducing dorsiflexion and inversion of the midfoot. This will be resisted primarily by tibialis posterior and peroneus longus through their vault-stiffening mechanisms attempting to maintain forefoot stability on loading via creating a more rigid foot. If midfoot dorsiflexion and inversion fails to resolve the GRF frontal plane vector, then the rearfoot may start to evert through the ankle joint (as the subtalar joint is unloaded). The reaction of the hip and the knee will depend on where this GRF vector lies in relation to these larger joints.

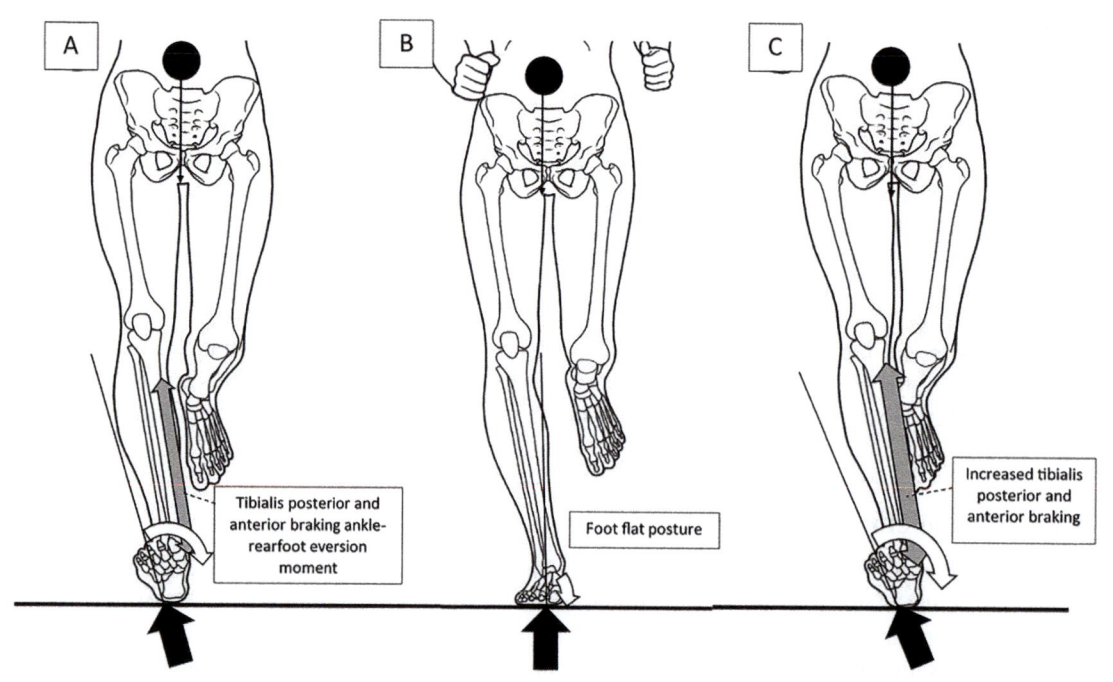

FIG. 4.6.5a Due to the need to adduct the lower limb under the trunk to bring the base of support closer to lying below the CoM, the foot tends to strike the ground more laterally when running than during walking. This is derived from the lack of a double-limb support, particularly at loading, but also during offloading acceleration. As a result, earlier hip abductor activity attempts to keep the CoM more over the single-support limb during running. Higher limb adduction angles occur regardless of the running foot contact position although how the foot responds, depends on the joints that need to bring the foot flat to the support surface. In rearfoot strikes, the lateralised initial contact GRF lies lateral and further from the instantaneous rearfoot axis for frontal plane rotation, and the medial foot is further from the flat support surface (A). The result is a greater eversion moment across the rearfoot that continues for a greater distance to place the medial foot into ground contact (B). This eversion moment must be braked by tibialis posterior, assisted by tibialis anterior, with the loading rate experienced by these myofascial structures dependent on running speed and stride length. Such events risk injury within these structures, especially when the runner is fatigued. Pathology that can commonly associate with such issues are tibialis posterior and tibialis anterior tendinopathies, anterior compartment syndrome, medial tibial stress syndrome, and tibial fatigue fractures. A more adducted strike position for the limb (C) increases the eversion moment. A less adducted angle decreases the eversion moment but places the base of support in a less advantageous position for single-limb support. Forefoot strikes face a similar safe adduction-limb position dilemma. However, the shorter stride length associated with this style of running deceases the loading rate on myofascial structures and places much of the burden of rotating the medial foot to the support surface through the midfoot, rather than the more rigid rearfoot. *(Permission www.healthystep.co.uk.)*

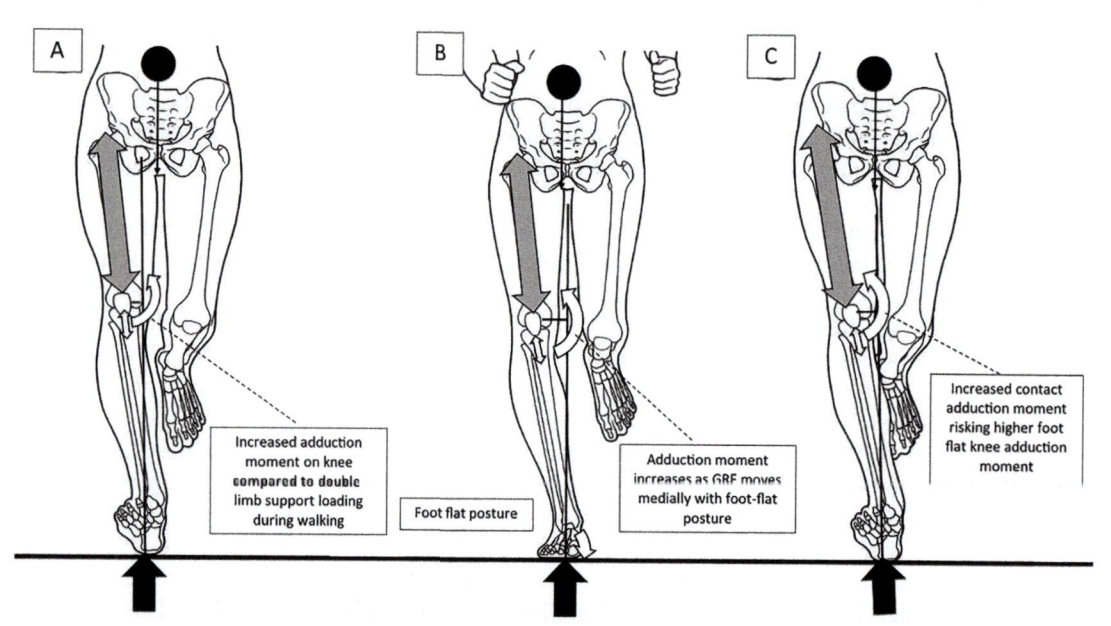

FIG. 4.6.5b Already the primary shock absorber of gait, the knee during running dramatically increases its burden under higher flexion motions until the end of braking around the middle of stance. This is immediately followed by active knee extension during acceleration, placing considerable stresses on the quadriceps–patellar tendon complex that are quite different to those of walking. This explains much of running-associated knee symptoms and pathology. The adduction angle of the lower limb also plays its part in stressing the knee. A more lateralised heel strike from the need to keep the limb more under the CoM than is necessary for walking, can move the initial GRF to a more medialised position to the knee, although laterally applied to the foot (A). The position of the vertical component of the GRF increases the adduction (varus) moment applied to the knee at contact, which will continue and often increase into the foot-flat posture (B), only decreasing during acceleration. In running styles that have a very narrow base of support (C), this situation can worsen, increasing the risk of medial compartment articular tissue threshold breaches within the tibiofemoral joint. This is particularly so should quadriceps–patellar complex muscle–tendon buffering be insufficient. Fatigue of the quadriceps–patellar complex will play its part in the development of such pathomechanics, where often articular degeneration is also concurrent within the patellofemoral joint. *(Permission www.healthystep.co.uk.)*

Transverse plane-to-frontal plane relationships in strike position and acceleration consequences

Extreme medial and lateral foot strikes in running are more common in abducted (out-toed) and abducted (in-toed) angles of gait, where the foot is abducted or adducted to the line of progression of the lower limb, respectively. Such foot alignments to the lower limb or indeed the whole lower limb's alignment to the trunk's line of progression reduce the ability of the ankle plantarflexors across the ankle and foot to act like a passive spring. For the ankle plantarflexors to perform as a passive spring during running requires the foot to be positioned perpendicular to the long axis of the lower leg as seen in walking (Hedrick et al., 2019). Loss of perpendicular trunk-to-lower limb or lower limb-to-foot alignments to the line of progression not only exposes the more extreme medial or lateral sides of the foot to initial contact, but also changes the force–velocity relationship of the ankle plantarflexors during acceleration. Such alignments are problematic to walking energetics, but are also highly significant in running energetics, which are poorer than in walking to start with. Not only is there a risk of a loss of acceleration power that is essential for running, but such loss of power through the line of progression can also change sagittal extension/flexion moments across joints into greater torques across other planes of the lower limb joints. Hips, knees, and ankles are not evolved to manage large transverse and frontal plane torques which can thus generate potentially tissue-damaging torsional shear strains (Fig. 4.6.5c).

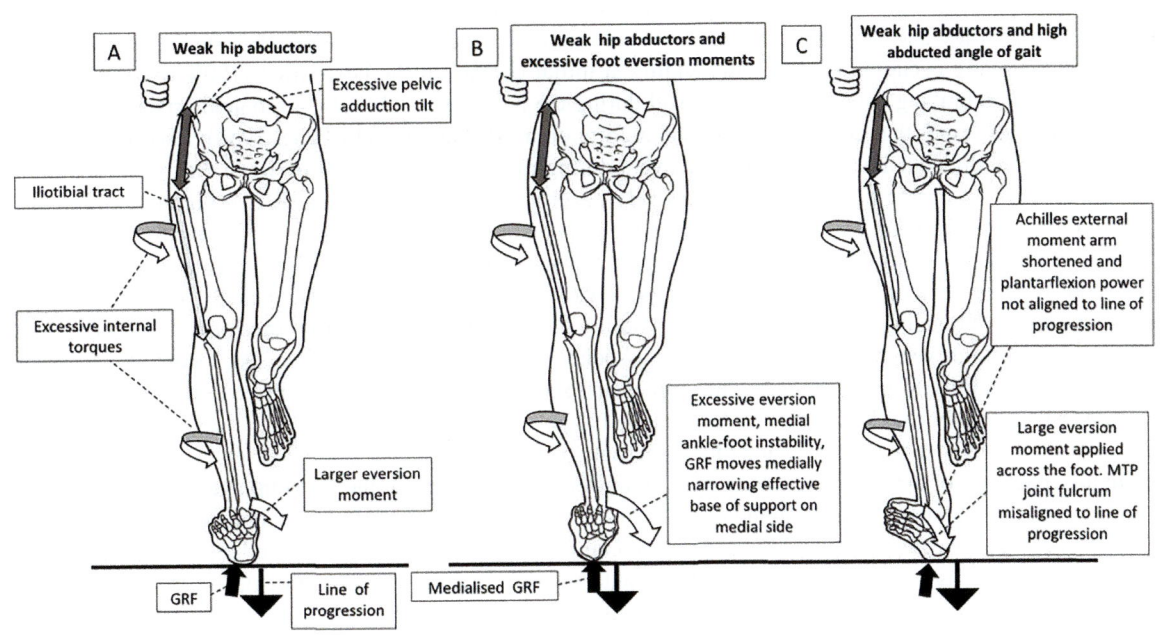

FIG. 4.6.5c Usually, the more adducted lower limb posture of running creates a lateral heel/forefoot strike, with the foot angled a few degrees abducted to the line of progression. Aligning the leg and foot to the line of progression can be complicated by proximal instability at joints such as the hip and knee when, for example, the hip abductors show insufficient strength (A). A slightly abducted foot alignment sets up a small loading eversion excursion across the rearfoot or midfoot as part of energy dissipation. This is limited by inversion vectors from the medial ankle and foot muscles and the ligaments, but also assisted by the low abduction angle of the foot to the line of progression. Thus, eversion is applied from proximal to distal, primarily along the long axis of the foot. However, poor lateral stability at the hip and knee resulting from weak hip abductors will result in increased loading–braking phase internal limb rotation that may exaggerate the eversion torque on the ankle and foot through its coupled foot and ankle abduction motion. By being slightly abducted, the foot should be positioned to align the medial MTP joints near-perpendicular to the line of progression for the ideal fulcrum position for acceleration. Proximal instability is likely to disrupt this alignment. A foot that excessively everts throughout loading response (B) will destabilise the support limb medially, preventing the foot from acting as a functionally wide base of support both medially and laterally to the CoGRF, with the CoGRF becoming medialised instead. Large eversion moments also increase strains on the medial soft tissues that restrain this motion, risking injury of deltoid ligaments, tibialis posterior and anterior, and the deep compartmental fascia of the posterior lower leg, or other anatomy. An excessively abducted foot (C) can result in a more lateral initial contact position, increasing the resulting eversion moment arm that will act more across the foot from lateral to medial rather than along the length of the foot, as seen with a less abducted angle of gait. Once again, medial soft tissues are likely to come under higher stresses as a result of the change in the angle of the moment arm. The abducted foot also presents problems for the placement of the medial fulcrum to the line of progression, unless the 1st metatarsal is short enough to maintain a perpendicular position for the fulcrum. However, regardless of this forefoot situation, the Achilles will be offset to store and then apply the plantarflexion power for acceleration, acting around a shorter Achilles external moment arm. This reduces the quality of energetics and thus risks earlier acceleration fatigue. An adducted angle of gait will create similar energetic compromises around the low gear axis of the lesser MTP joint fulcrums. *(Permission www.healthstep.co.uk.)*

4.6.6 Running acceleration pathomechanics

Running does not demonstrate a period of separate weight transmission with the CoM height increasing and then falling as is found during walking's midstance phase. In running, once the CoM completes its fall during braking at the middle of stance, it immediately changes its action to acceleration and CoM raising. Therefore, the amount of knee and hip flexion and the total fall of the CoM gives some indication as to the amount of energy that has been dissipated during braking and what will be required for acceleration. The balance that exists here is that hip and knee flexion with some ankle dorsiflexion motion (with or without initial plantarflexion) dissipates impact energies associated with CoM drop, permitting elastic connective tissue stretching through motion that stores energy (Ker et al., 1987; Farley and McMahon, 1992; Lee and Farley, 1998). However, excessive hip and knee flexion leads to excessive CoM drop without extra energy storage, and the CoM must be raised again through metabolic acceleration energy, requiring greater muscle activity under reduced energetic efficiency. Each patient thus needs to find an energetic balance between impact energy dissipation flexion (CoM drop) and the need to raise the CoM to a high point at the aerial phase through terminal stance acceleration and limb extension, which includes ankle flexion (plantarflexion).

Gravity causes a horizontal torque as the runner's CoM moves in front of the support foot during late terminal stance prior to the aerial phase, when the extensor muscles of the knee, hip, and the ankle plantarflexor power are highly active. Significant elastic recoil primarily through energy-storage tendons, regains much of the CoM's height lost during early stance (Romanov and Fletcher, 2007). Muscle activity is able to produce an adjustable stiffness in the lower limb, with the limb's 'spring' being stiffest when the CoM reaches its lowest point in the middle of running stance (McMahon and Cheng, 1990; Farley and González, 1996). A whole running stride (100%) usually has stance ending with toe-off at around ~40%, swing occupying ~60%, and ending at initial contact of the next stride (100%–0%). The first ~15% involves the collision braking event. This is followed during the next ~15%–34% by a rebound arising from the storage of energy within the tissues that control limb flexion deformation (McMahon and Cheng, 1990). Maximum speed during this 'bouncing' glancing collision with its rebound event during running is a trade-off between the magnitude of the GRF and step frequency that can be obtained at progressively faster speeds (Weyand et al., 2010) (Fig. 4.6.6a).

Acceleration phase mechanics relates to the ability to generate high horizontal moments to drive forward while raising the CoM, with efficiency being improved by creating greater horizontal than vertical CoM displacement (Romanov and Fletcher, 2007). Large vertical CoM displacement represents wasted energy and can quite easily be visually identified in runners as 'vertically bounding gaits', reflecting an inability to develop high anterior–posterior-directed GRF in the horizontal plane. Such high vertical displacements can associate with pathomechanics because locomotive energetics are poorer. Indeed, the ability to increase horizontal forces and reduce vertical displacement is seen more in elite distance runners rather than in novices (Williams and Cavanagh, 1987).

Elastic recoil through tendon stiffness and muscle activity of the hip, knee, and ankle is essential to achieve acceleration by 'springing' back out of the midstance flexions. The power required to bring the CoM up against gravity with positive horizontal acceleration is supplied from connective tissue stored energy and by muscle metabolic energy input, reflected by the amount of stretch or compression within the elastic components of anatomy (Gullstrand et al., 2009). Thus, running acceleration creates a high injury risk in the ankle plantarflexors (including the Achilles), the quadriceps, the patella and its ligament/tendon, and the gluteus maximus and its associated structures, including the iliotibial tract. Weakness or fatigue in all these units is a particular risk for running pathomechanics. Acceleration also occurs under the influence of the contralateral swing limb's centrifugal forces that are significantly greater during running than in walking. Thus, the swing limb's kinematics during stance phase acceleration are also important to note. Alterations in either limb's posture during running can dramatically change the mechanical advantage (Biewener et al., 2004), and should be the focus of attention in the transition from braking to acceleration and then throughout acceleration (Fig. 4.6.6b).

4.6.7 Pathomechanics of fascia in shin symptoms of runners

For understanding shin pain symptoms in runners, it is recommended that the information provided in Chapter 1, Sections 1.4.5 and 1.4.6 on energy dissipation techniques in human gait are considered. There are four common and distinct shin pains associated with running. Most anterior shin pain (shin splints) is related to overuse myalgia, and sometimes tendinopathy associated with the energy dissipation and muscle–tendon buffering action through plantarflexion moment restraint of the tibialis anterior muscles, as discussed in Section 4.6.4. The other three conditions are extremely rare outside of athletic/running activities, and are all linked to repetitive high impact during sporting endeavour. These are *exercise-induced compartment syndromes*, tibial and fibular *fatigue (stress) fractures*, and the less easily definable *medial tibial stress syndrome*.

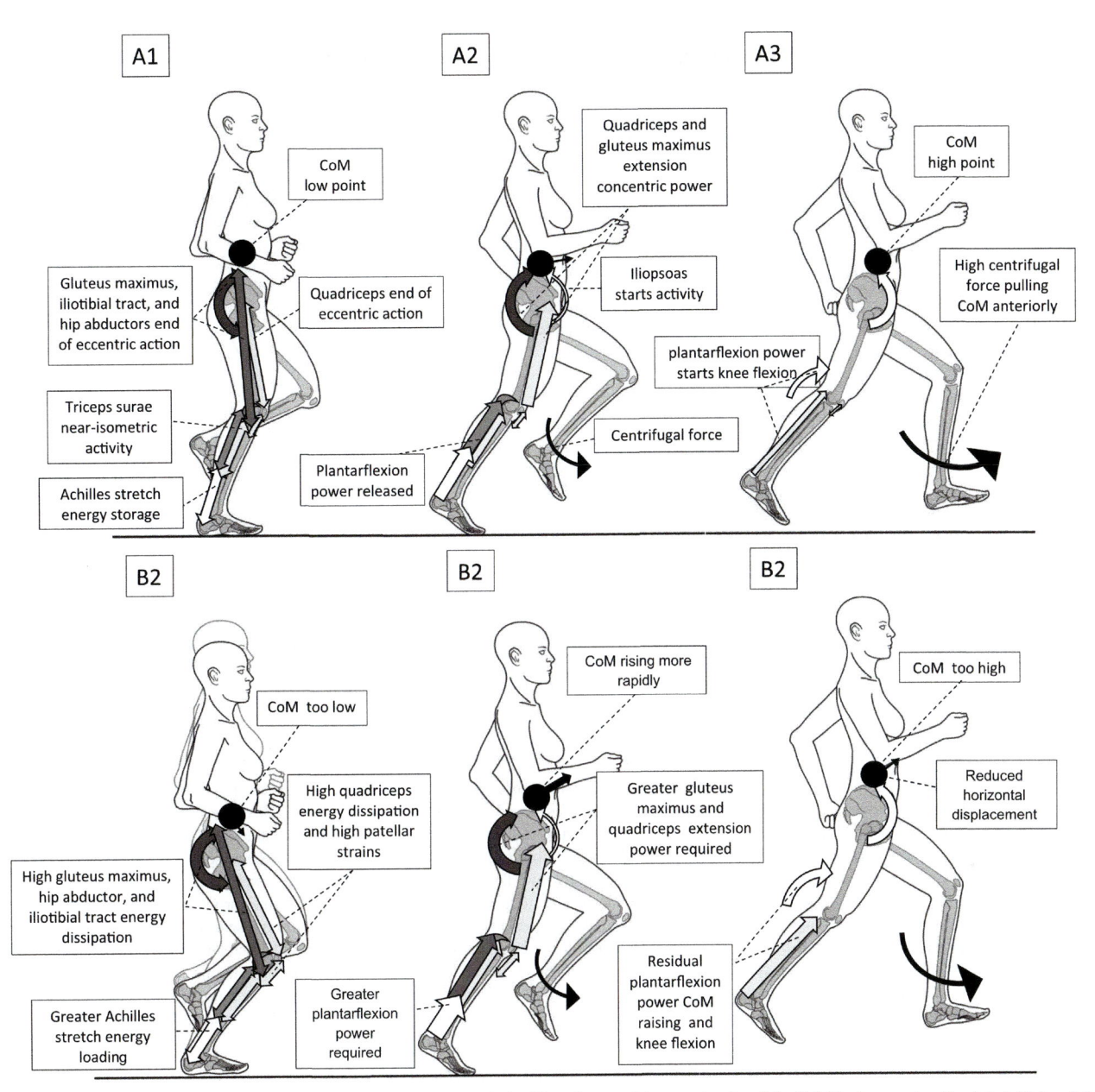

FIG. 4.6.6a The anterior fall of the CoM (black circle) throughout the braking phase and the anterior rise of the CoM during acceleration into initial aerial phase is an important part of running energetics. If the task is performed well, the risk of fatigue and hence running on weakened muscles is reduced. At the end of braking (A1), the CoM should be at its lowest point to achieve safe energy dissipation. Muscles should have completed their eccentric (or near-isometric) activity using muscle–tendon buffering to dissipate energy, but with the Achilles having stored as much energy as safely as possible for acceleration. Acceleration (A2) is initiated by concentric contraction of gluteus maximus and the quadriceps muscles to extend the hip and knee, while the plantarflexion power of the Achilles is released by heel lift and ankle plantarflexion. This combination of joint motions lengthens the limb, with as much horizontal displacement as possible. This anterior power drives the body and lower limb off the ground into the aerial phase (A3), raising the CoM to the running high point. This action is greatly aided by the terminal swing limb's centrifugal forces. If braking results in excessive limb flexion before acceleration, the CoM will drop too low for good energetics (B1). To compensate, large hip and knee concentric muscle activity and Achilles recoil power is required to initiate acceleration (B2). Equally problematic is the amount of vertical displacement required during acceleration into the aerial phase (B3), as with a more vertical displacement of the CoM, the following impact energies are going to be higher requiring more energy dissipation. Greater vertical displacement also reduces the horizontal power, slowing potential running velocity, lowering energetic potential, and increasing the risk of fatigue. *(Permission www.healthystep.co.uk.)*

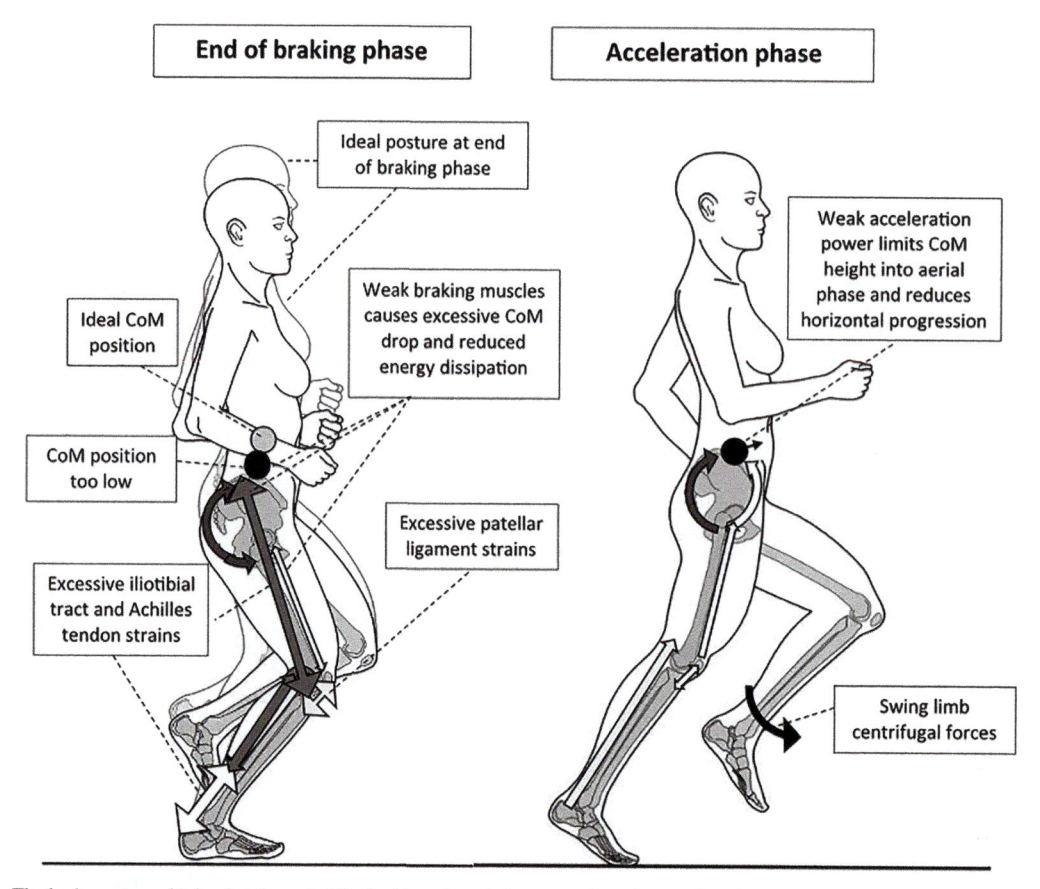

FIG. 4.6.6b The body posture obtained at the end of the braking phase is important in understanding the energy required from muscles to achieve acceleration. Although it is important to shed energy from the initial collision(s) to avoid impact-generated pathologies, excessive hip and knee joint flexion angles used for muscle–tendon buffering will cause unnecessary lowering of the CoM. An ideal posture (ghost image) should position the CoM high enough (grey circle) for easy recovery of height during acceleration phase CoM raising. If too much flexion occurs because braking muscles are too weak (solid image), the CoM (black circle) will be allowed to drop excessively low. This will require greater muscle power generation from the hip and knee extensors to draw the hip and knee out of their higher flexion angles to start to accelerate the CoM upward and anteriorly towards toe-off and the initial aerial phase. Commonly, excessive large joint flexions are associated with weaker muscles or fatigue that compromise lower limb elasticity, so that the capacity to compensate for acceleration by increased muscle power may not be possible. The result of this will be that anterior, vertical, and horizontal CoM displacement will be limited during acceleration, despite the centrifugal efforts of the swing limb. Many runners attempt to overcome the problem of weak muscles by reducing joint motion during impact. However, this in turn sets up a new pathway to pathology, as the muscle–tendon buffering ability is reduced if large joint motion becomes increasingly constrained when overall limb elasticity is lowered by fatigue. *(Permission www.healthystep.co.uk.)*

Chronic exercise-induced (exertional) compartment syndrome

Chronic exercise-induced or *exertional compartment syndrome* induces pain through exercise, muscle compartment swelling, and impaired muscle function, which may have been first mentioned by Edward Wilson, a physician who was part of Scott's Antarctic expedition of 1910–12 (Aweid et al., 2012). It is primarily associated with young, athletic, active individuals, particularly those involved in running sports and military training (Aweid et al., 2012). It is thought to arise from insufficient compliance within the fascia surrounding the muscle or as a consequence of contraction expansion and hypertrophy of the muscle, which can increase its volume by 20% within a muscular compartment (Aweid et al., 2012). The resultant compartment pressure rise from muscle volume expansion can interfere with circulation within the compartment's muscles, causing ischaemic and temporary neurological deficits during exercise. Arterial inflow into the compartment and to the muscles may only be able to occur during periods of muscle relaxation between contractions (Aweid et al., 2012).

The condition can occur in any muscular compartment. However, it is most commonly reported in the lower leg within the anterior and deep posterior muscular compartments and occasionally in both simultaneously, but rarely in all four (Aweid et al., 2012). The anterior compartment fascia is stiffer compared to that of the superficial posterior compartment (Stecco et al., 2014), a finding that possibly reflects the nature of the positional function of the tibialis anterior tendon (Lichtwark and Wilson, 2005; Birch et al., 2008; Herod et al., 2016) (Fig. 4.6.7). The use of anabolic steroids and high

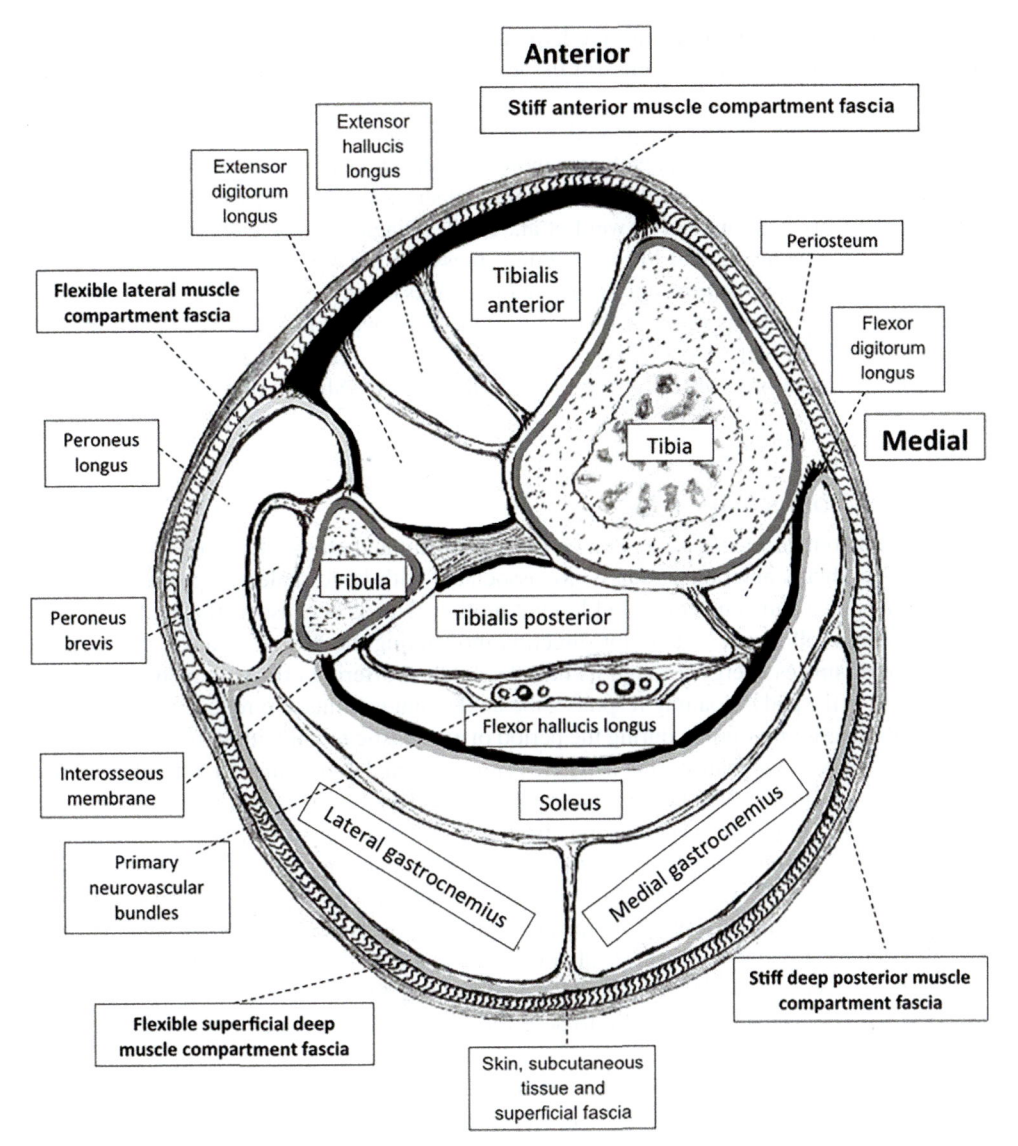

FIG. 4.6.7 A simplified schematic of the muscles in the distal third of the left lower leg (shank), demonstrating its four distinct muscular compartments. Two of the compartmental fascias are relatively stiff and inelastic compared to the others. These are the anterior and deep posterior compartments that contain tibialis anterior and the long digital extensors and tibialis posterior and the long digital flexors, respectively (stiff fascial compartments outlined in black). The lateral compartment of the peroneal muscles and the superficial posterior compartment containing soleus and gastrocnemius are far more flexible, reflecting their greater muscular length changes during locomotion (outlined in grey). Each muscle within each compartment is also surrounded by its own connective tissue intermuscular septum connected to its epimysium, and thus to the perimysium of fascicles and muscle bundles. Therefore, descending or ascending, connective tissue hierarchy links biomechanics to and from macrostructure and microstructure. The muscular fascia also binds to the periosteum of the bone, with some muscle fibres directly tethered to the periosteum. Energy dissipation by soft tissue oscillation works by the connective tissues' capacity to stretch and recoil under shock waves that displace soft tissue mass. Excessive muscular fluid within any level of connective tissue compartmentalisation will start to interfere with the connective tissue's elastic properties by increasing fascial tone stiffness. If intra-connective tissue fluid pressures are able to increase as a result of more pronounced myofascial inflammatory processes, then intercompartmental pressure may start to approach arterial blood pressures that supply muscles. This is particularly a risk under high muscle contraction forces that naturally raise intercompartmental pressures; a phenomenon used for the muscle-pumps of venous return. Muscular ischaemia produced during exercise is a threat to continued exercise and may even result in myofascial necrosis. Stiffer, less elastic compartments are more at risk. Avoiding the production of dense scar tissue within the connective tissue as a result of this pathology is critical to a full return to exercise. The correct approach to training and appropriate running techniques to avoid high impact-induced soft tissue oscillation and premature fatigue are the best approaches to avoiding the problem, and also for returning an athlete to exercise following a compartment syndrome. *(Permission www.healthystep.co.uk.)*

levels of eccentric calf exercises can induce muscle hypertrophy and decrease fascial elasticity, as can any myofascial scarring. Venous hypertension and posttraumatic soft tissue injury-induced inflammation can also increase intracompartmental pressures.

There is no universally accepted diagnostic procedure for compartment syndrome, and early diagnosis can be difficult, despite being clinically desirable. The clinical picture is more distinct for the anterior compartment, whereas the symptoms of the deep compartment must be differentiated from medial tibial stress syndrome, popliteal nerve entrapment, deep muscular myopathies, or sural nerve entrapment (Aweid et al., 2012). Diagnosis is usually achieved through intramuscular compartment pressure studies, showing that lower extremity pressures are elevated from data collected before and after exercise, a procedure that is invasive, painful, and associated with risks (Aweid et al., 2012). The evidence for this approach remains weak as a definitive diagnosis because there is considerable overlap in compartmental pressures with healthy normals (Aweid et al., 2012; Roberts and Franklyn-Miller, 2012). The results of surgical intervention are mixed and may not enable the athlete to return to full activity following fasciotomy (Slimmon et al., 2002).

The role of fascial pathomechanics

As already noted, anterior compartment fascia is much less compliant and elastic than that of the posterior compartment (Stecco et al., 2014). The tibialis anterior tendon is not associated with large scale energy storage and usually does not develop large muscle–tendon excursions during ankle joint motion. This is because as the ankle plantarflexes under heel strike GRFs, the lower limb follows in an anterior direction moderating the degree of ankle plantarflexion that actually occurs. Tibialis anterior dissipates energy through its muscle–tendon buffering effects in restraint of ankle plantarflexion following heel strike, but at the end of loading response, the ankle joint continues to reduce its ankle plantarflexion angle. This slackens the tibialis anterior myofascia during the period of any elastic recoil within the tendon, helping in dissipating and losing the tendon's energy rather than storing it. There is normally no large change in muscular volume on contraction, so its fascial attachments need to be stable rather than expansive and flexible. This is why this muscle utilises a positional tendon physiology. If the tibialis anterior becomes hypertrophied, inflamed, or increases its volume through adaptation to higher usage and larger ranges of motion, then compartment pressure may increase. More vulnerable individuals to such issues include those who are increasing their endurance running distance and/or frequency. This is particularly true of those that have previously taken part in muscle power training and sports (rugby, American football, etc.) that are heavier runners, especially if they also demonstrated a heel strike with a large foot contact dorsiflexion angle. If tibialis anterior becomes inflamed and oedematous through overuse myositis or the anterior deep fascia becomes scarred and less free-moving, then the anterior fascia has less intrinsic ability to adapt to a change in muscular or fluid volumetric increases during exercise through its inherent stiffness.

Exercise-induced leg pain is rare in the posterior superficial compartment (Aweid et al., 2012), probably because the fascia associated with these muscles is more compliant than elsewhere in the lower leg (Stecco et al., 2014). The posterior fascia crura covers the very fleshy energy storage muscle–tendon unit of the triceps surae and Achilles which in gait moves considerably within the soft tissues through the anterior translation of the tibia, with relatively large ranges of ankle dorsiflexion during midstance and plantarflexion during terminal stance. Such alterations induce large volumetric changes in the triceps surae, requiring an expansive compliant fascia for both muscle lengthening and joint change requirements. Hypertrophy of the triceps surae is unlikely to induce compartment syndrome due to the relatively high fascial compliance lying over the triceps surae. The fascia over the peroneals is similarly more flexible for the same reasons as the superficial compartment. Intrinsic high fascial compliance makes increased stiffness from scarring or other causes less likely to induce compartment syndrome than in the anterior shin.

The deep posterior fascia crura links to the intermuscular septa of the deeper posterior calf muscles (of tibialis posterior and the long digital flexors) before finally having a connective tissue link to the periosteum of the posterior osseous surfaces of the tibia, fibula, and the interosseous membrane. It also joins the superficial compartment's fascia along the posterior medial border of the tibia (Stecco et al., 2014). The deep and superficial compartments' fascia also links to the intermuscular septa of the peroneals laterally, but does not link directly to the anterior compartment. This is another compartment bound by stiffer fascia than the superficial posterior or the lateral peroneal fascia. Such binding makes it another potential candidate for compartment syndrome if tibialis posterior, flexor hallucis longus, or flexor digitorum longus become hypertrophied or inflamed and oedematous. Its point of unification with the superficial posterior fascia along the posterior medial border of the tibia represents an area bound together but of different fascial mechanical properties at a soft tissue–bone interface where stresses can be focused. This mechanical tissue area of variable zones of elasticity may be a factor in the development of medial tibial stress syndrome, as will be discussed later.

Muscular fascial compartments form part of a continuum of tube-like networks that permit sliding of macro and microanatomy within them (Dawidowicz et al., 2016), and such fascia plays a significant role in muscular force transmission (Huijing et al., 2003; Huijing and Baan, 2008). Dysfunction of the muscular compartments may have significant effects in changing the anatomical structural sliding of muscles, nerves, and blood vessels upon movement within the fascia on both a macro and microscale. Muscle dysfunction may also change the amounts of force being transmitted through the fascia to the bone during muscular contractions. As fascia also has myofibroblasts with an ability to adjust fascial stiffness and tone under autonomic nervous system control (Tozzi, 2012; Dawidowicz et al., 2015), dysfunction of the autonomic nervous system that controls the tone may also be an influence on compartment elasticity and pressures, either as a cause or as a response to pathological compartmental changes.

For understanding the origin of an exercise-induced compartment syndrome, the mechanisms that are causing hypertrophy or myositis and oedema within a muscle must be identified. Once identified, it must be addressed if the athlete is going to return to full activity, with symptom management alone unlikely to resolve issues in the long term. As muscles working at long lengths on long lever arms are more likely to become injured, the biomechanical investigations should start to look for these features, bringing stride length and foot strike positions into focus. Assessing energy dissipation techniques of the patient to see if the burden of impact is being focused into too few muscular or connective tissue structures, is also likely to prove helpful.

4.6.8 Tibial fatigue fractures

Fatigue fractures are common in running (Taunton et al., 2002; Fredericson et al., 2006; Rauh et al., 2006) and tibial fractures are one of the most commonly reported fatigue fractures of athletes (Matheson et al., 1987; Sanderlin and Raspa, 2003; Milner et al., 2006). They present as a very local source of pain on palpation, although the symptoms complained of may be more difficult for the patient to isolate to a specific locality. Furthermore, they can present similarly to medial tibial stress syndrome, but the symptoms are usually far greater in a much smaller area, on examination. A small 'hop' on the spot on the affected leg often induces pain, a test that should only be repeated once or twice at a time to avoid further damage.

Although commonly called 'stress fractures', mechanically they are bone 'fatigue' fractures. All fractures are caused by stress. Such fatigue fractures are common in the lower limbs of the athletic population with sites including the tibia, fibula, calcaneus, talus, cuboid, cuneiforms, sesamoids, and metatarsals. Metatarsal and tibial fractures are most common and are more frequent in females (Mayer et al., 2014; Welck et al., 2017). Fatigue fractures can be difficult to diagnose clinically, although it is extremely important to identify them early in order to prevent more serious escalation of the problem (Mayer et al., 2014). Exercise-induced fatigue fractures are brought on by repetitive stress events occurring at stresses below the expected mechanical failure threshold of bone. The stresses involved in 'pathological' fatigue fractures may be low and are known as insufficiency fractures. These are linked to conditions such as anorexia, rickets/osteomalacia, scurvy, osteogenesis imperfecta, and are most commonly found in postmenopausal women and the elderly (Koenig et al., 2008; Marshall et al., 2018). Anorexic athletes present an especially high risk of fatigue fracture.

Fatigue fractures result from the property of materials to fail under repeated loading below the material's ultimate tensile strength. Viscoelastic materials like bone are particularly vulnerable as their material properties actually change under repeated loading over time (fatigue response), giving them a lower tissue threshold once they become fatigued. Fatigue fractures are therefore more associated with events and activities that induce fatigue. Fatigue can cause soft tissue myopathy, tendinopathy, and ligament failure, but in bone, it causes fracture because of its brittle nature. Concurrent fatigue of the soft tissues removes protective tension-induced compression forces from bone, a protection that is usually provided by muscular contraction across it.

Two main nonmutually exclusive hypotheses explain fatigue fractures. The first model suggests that bone remodelling in response to rapid increased loading stresses is inundated with too much remodelling to maintain bone structural integrity, resulting in initial weakening before strengthening (Marshall et al., 2018). The second model suggests that as muscles fatigue during exercise, they no longer protect the bone through their energy dissipation and tensioned-compression force-providing properties, causing changes in the normal lines of stress distribution within the bones (Fig. 4.6.8a). Thus, fatigued muscle fails to moderate tensile and shear bone stresses, and therefore, they fail to reduce total bone stresses during locomotion. Such fatigue-induced loss of active soft tissue tensional elements (cables), change stresses within biotensegrity structures, leaving the bones open to increased torsional shear and tensional stresses, with reduction in protective compression.

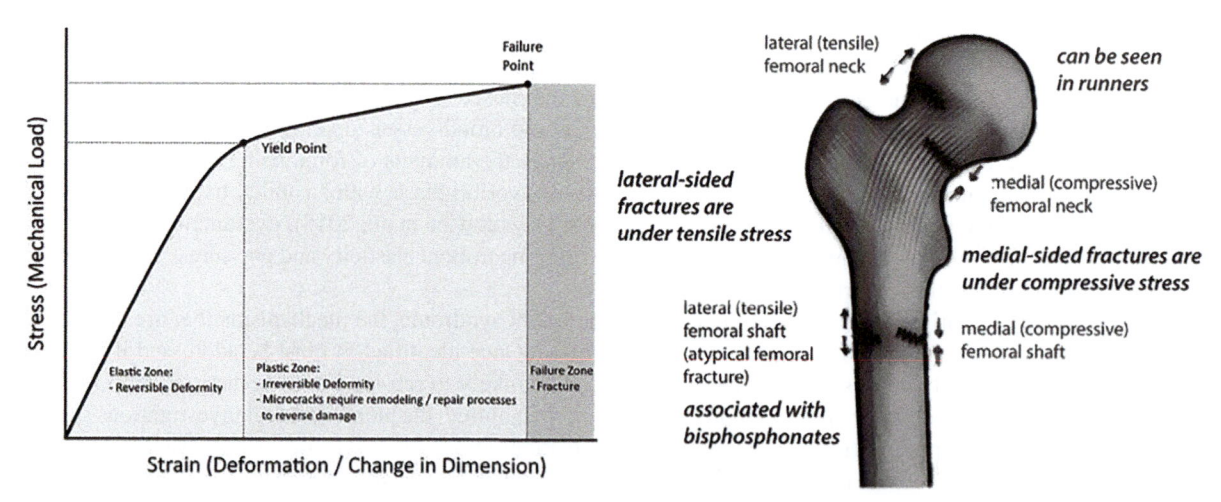

FIG. 4.6.8a A reminder of a stress–strain curve for understanding mechanical behaviour of bone tissue under loading (left image). The right image is a schematic of the locations of fatigue fractures and atypical fractures of the femur associated with runners. Note the origin of the strain in each type of fracture, with tensile strains being associated with greater difficulties in healing as bone heals best in response to compression. Because cross-sectional areas are smaller, more distal long bone fatigue fractures are more commonly encountered than those of the femur. *(Images from Marshall, R.A., Mandell, J.C., Weaver, M.J., Ferrone, M., Sodickson, A., Khurana, B., 2018. Imaging features and management of stress, atypical, and pathologic fractures. Radio-Graphics 38(7), 2173–2192.)*

Female athletes have an increased susceptibility to stress fractures that presents most commonly within metatarsals, calcanei, and tibiae (Mandell et al., 2017a, b; Rizzone et al., 2017). Part of the reason for this gender difference may be from the smaller (on average) cross-sectional areas of bone in females compared to those in males (Crossley et al., 1999; Beck et al., 2000) (Fig. 4.6.8b). Deep limb pain in female athletes should always raise fatigue fracture suspicions, including thigh pain due to the possibility of a diaphyseal femoral fatigue fracture. The female athletic triad of eating disorders (such as anorexia), menstrual irregularity, and osteoporosis makes many female athletes highly susceptible to fatigue failure during high training stresses experienced under bone insufficiency conditions (Koenig et al., 2008). However, most lower limb athletic fatigue fractures are found below the knee.

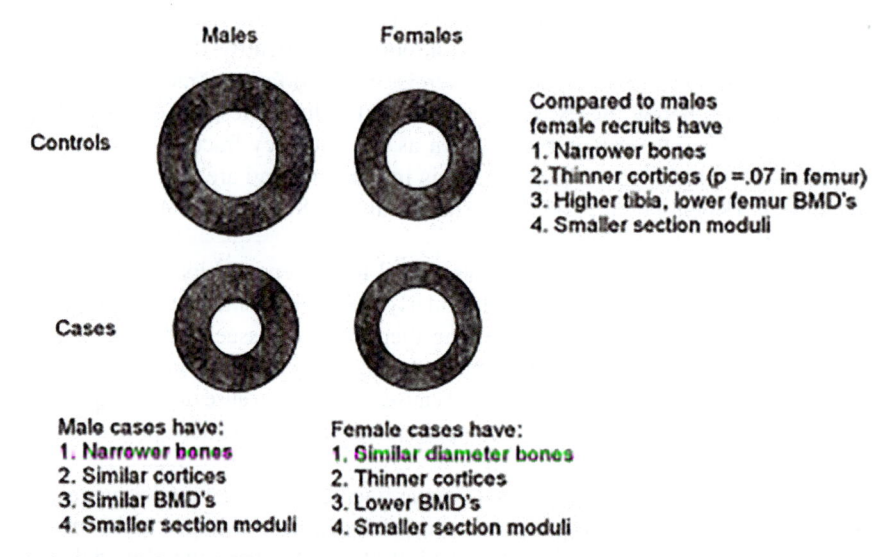

FIG. 4.6.8b A schematic depiction of observed differences in bone geometry and areal bone mineral density (BMD) between cases and controls for each gender from a study by Beck et al. (2000) on the prospective development of fatigue fractures, examining those who developed fatigue fracture presented by the inferior 'cases'. The differences have been exaggerated for clarity. Females suffering stress fractures tend to have thinner cortices and lower BMD than unaffected females, while males with stress fractures have similar cortical thickness and BMDs to unaffected males but display narrower bones. This suggests that bone gracility developed prior to exercise training is an underlying risk factor. Once again, the need for good tissue development under high biomechanical loads during childhood is essential for effective and adequate adult tissue properties. *(Image from Beck, T.J., Ruff, C.B., Shaffer, R.A., Betsinger, K., Trone, D.W., Brodine, S.K., 2000. Stress fracture in military recruits: gender differences in muscle and bone susceptibility factors. Bone 27(3), 437–444.)*

Pathomechanics of the tibial fatigue fractures

The precise aetiology of tibial fatigue fractures is unknown (Milgrom et al., 2015), but higher average and peak loading rates are suggested (Milner et al., 2006; Crowell et al., 2010; Zadpoor and Nikooyan, 2011). The tibia is subjected to both compression and tensile stresses through bending moments acting on a lower leg behaving as a column/strut. By resisting the external influence of the GRF, muscle contraction manages tensile and shear stresses, turning them as much as possible into safe internal bone compression forces. The muscle–tendon units are also essential for managing overall compression loads through energy-buffering of impact energy. Therefore, bone is loaded by both internal and external moments that need their forces appropriately directed. Indeed, the overall GRF is not reported to be different in those with tibial or metatarsal fatigue fractures compared to controls, but the vertical loading rate of the GRF is significantly faster when associated with bone failure (Zadpoor and Nikooyan, 2011). Thus, bones with smaller cross-sectional areas that are associated with a rapid loading rate seem to present an important pathomechanical focus (Fig. 4.6.8c).

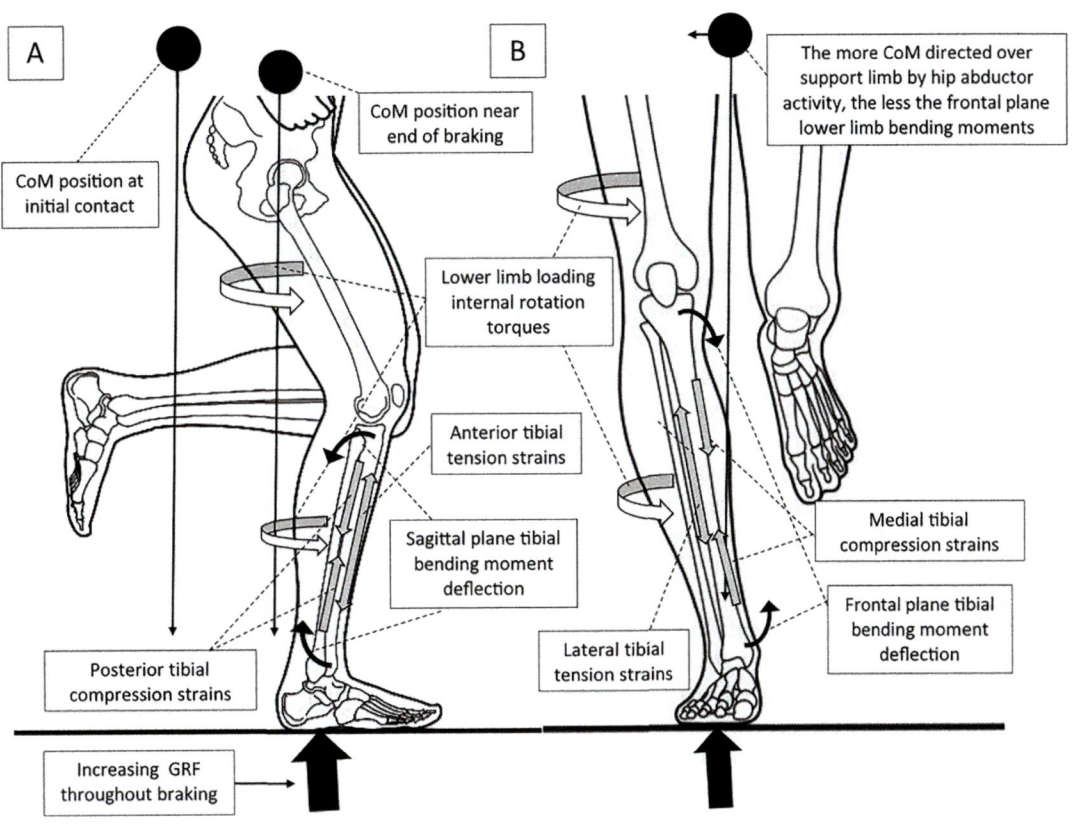

FIG. 4.6.8c Because the tibia is loaded much like a column during locomotion, bending moments and torsion are driven by displacement of the CoM as it moves from posterior to anterior during gait. In running, forces associated with impact and loading response are higher than during acceleration. In acceleration, body weight is moving rapidly anteriorly, being pulled forward by the swing limb's centrifugal forces which quickly starts offloading the tibia. Thus, peak vertical GRF is highest at the end of braking, near the acceleration boundary. However, at this stage of gait, the CoM lies over the tibia, creating nearly pure compression forces which bone tends to tolerate well. During initial contact and loading response braking, the CoM is initially offset more posteriorly to the long axis of the tibia (A), setting up a bending moment directed towards the anterior tibia, tensioning the anterior surface and compressing the posterior surface. These stress directions continue until braking completes. Higher osseous strains at this time could present tissue threshold breaches if soft tissue protection is compromised. The CoM also passes over the tibia in a medial to lateral orientation in the frontal plane (B). The CoM normally lies very slightly medial to the tibia, applying a frontal plane bending moment that causes deflection directed laterally. Thus, the lateral tibial surface is subjected to more tensile strain and the medial surface to compression stresses. This frontal plane orientation is another reason why it is important to use the hip abductors to try and keep the CoM over the support limb, as in maintains vertical compression forces closer to the long axis of the tibia. In summary, the tibia experiences higher tensile stresses on the anterior and lateral surfaces, while compression strains are higher on the posterior and medial surfaces. These compressions and tensions are moderated and controlled by muscular forces throughout the lower limb and influenced by trunk and limb positioning during the stance phase of gait. The muscle activity should limit transverse plane torques on the lower limb, as these are particularly problematic for the tibia, requiring the fibula and its articulations plus strains within the interosseous membrane and ligaments to act as torque shock absorbers. By knowing the location of any fatigue fracture, the likely damaging stresses and protective muscular dysfunctions should be suggested for investigation. *(Permission www.healthystep.co.uk.)*

The more common posteromedial border tibial fatigue fracture location is considered a 'low risk' to long-term health, as these heal better and are thus considered and evidenced to be a result of excessive compression forces. The anterior tibial surface is subjected to more tensile stress than the posterior surface, so although a little less common, the anterior cortex tibial fatigue fracture is considered higher-risk due to its tensile origin and more restricted blood supply (Mandell et al., 2017a, b). Healthy long bones have both high compressive and tensile strengths in the longitudinal axis, less so in the transverse plane and are most vulnerable to shear (Hart et al., 2017), making transverse tensions from bending moments and torsional shear the most dangerous stresses for the tibia. It is unlikely that bone is damaged easily during impact as within reasonable normal biological limits, impacts are known to be beneficial and necessary for the structural strengthening of bones (Al Nazer et al., 2012; Hart et al., 2017). However, high lower limb transverse and frontal torques during loading are a vulnerability if present, especially when a runner is fatigued.

High frequency loading with high stress loading rates, especially in torsion, is far more likely to initiate bone pathology than compression and tension in the longitudinal axial plane (Hart et al., 2017). The underlying bone morphology significantly influences the risks. The ultimate strength in 'average' human bone is approximately 193 MPa for longitudinal compression, 133 MPa for transverse compression, and 51 MPa in transverse tension, and a torsion shear strength of around 68 MPa (Hart et al., 2017), making it more likely that bone fails under tension stress rather than in compression, applied within the transverse plane causing shear. A combination of increased transverse plane torsion and shear would be particularly threatening to a bone with a low bone tissue cross-sectional area relative to its biomechanical demands, especially if concurrently expressing lower bone mineral density.

In vivo bone stresses are a challenge to quantify experimentally and are usually estimated computationally. Several excellent studies using real-life gait data and subject-specific inhomogeneous bone models based on MRI and CT data have been attempted, but have assumed that the influence of the fibula is minimal (Chen et al., 2016; Derrick et al., 2016). This is not the case, as the fibula and its ligamentous structures fulfil an important role as torque shock absorbers (Scott et al., 2007; Barsoum et al., 2011). Tibial shear strains have been recorded in running to be as high as 5533 με, taking the loaded stress into the pathological zone of modelling and thereby increasing the risk of fatigue fracture (Al Nazer et al., 2012). Military recruits make a good population to study prospectively, as exercise regimes, footwear, and interventions can be more easily controlled than in civilians. Military drill is also a unique activity that is estimated to generate high tibial shocks (Rice et al., 2018). Frontal plane force vectors have been found to be directed more medially in the midstance and late stance phases of gait in military recruits prone to tibial stress fractures, but sagittal plane GRFs were not higher than those of controls (Creaby and Dixon, 2008).

Derrick et al. (2016) looked at the forces and stresses in the distal tibia in male soldiers and modelled them within the distal tibia. This musculoskeletal model consisted of 44 muscles and subject-specific inhomogeneous bone models. It was reported that peak axial forces occurred at ~75% of stance phase and ~80% of this axial force arose from muscle contraction, the rest arising from joint reaction forces. These joint reaction forces tended to bend the tibia in a convex anterior direction, while the muscular forces bend the tibia in a concave posterior direction, with the net moment being in the posterior concave direction (Derrick et al., 2016). The resultant axial forces and bending moments produced peak tension within the anterior aspect of the tibia of around 27 MPa and peak compression within the posterior aspect of around 45 MPa (Figs 4.6.8d and 4.6.8e).

Torsional loads accounted for approximately 98% of the peak shear stressing over the cross-sectional area. Peak tensile forces occurred during weight acceptance, while peak compression force occurred at propulsion. Moreover, peak transverse stresses also occurred at propulsion (Derrick et al., 2016). As a percentage of ultimate strength, the compressive stresses recorded by Derrick et al. (2016) were only slightly greater than the tensile forces observed, being 25% of ultimate strength in compression and around 20% for tension, giving a safety factor of 4–5, respectively (Derrick et al., 2016). If those prone to stress fracture have a longer GRF moment arm to the tibia medially, then it might increase compressive loads, explaining the susceptibility of such individuals to fatigue fracture (Creaby and Dixon, 2008). This, in turn, may explain the distribution and nature of tibial stress fractures at different locations.

The locations of peak magnitudes of stress reported in studies correlate with clinical findings that the commonest location of tibial stress fractures occurs on the posteromedial tibial border on the compressive side (Al Nazer et al., 2012; Derrick et al., 2016). Slightly less common, but poorer in healing, are the anterior tibial fatigue fractures that are found on the tensile side of tibiae, where tension forces delay bone healing compared to compression that stimulates it (Boden et al., 2001; Derrick et al., 2016). As muscles create the primary forces loaded on the tibia, muscle fatigue is likely to change the stress–strain loading relationship within and across bone (Fig. 4.6.8f). It has been shown that muscle fatigue results in an increase in axial strain rates and axial tensile strains, and decreases compressive strains (Milgrom et al., 2007). Types of shoes have also been shown to affect tibial strain values and rates during walking and running, while overground tibial strain and strain rates have been reported as higher than on a treadmill's 'sprung surface' (Al Nazer et al., 2012).

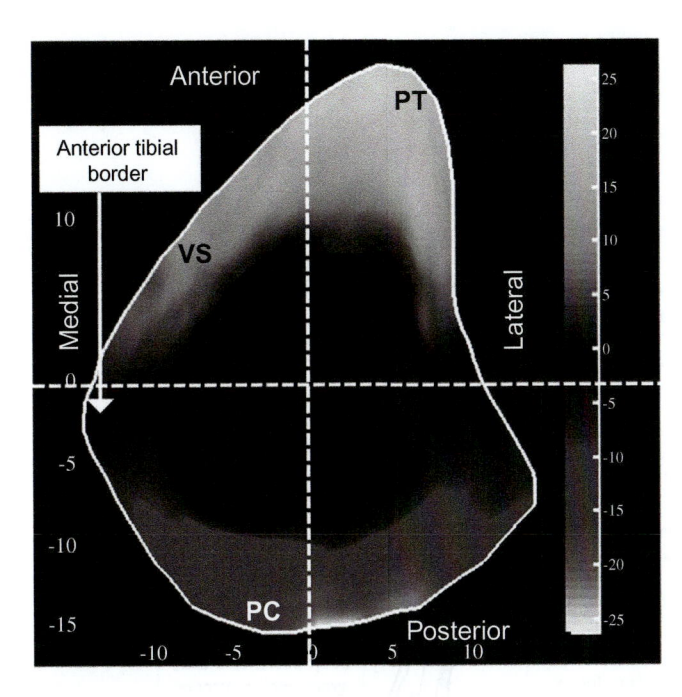

FIG. 4.6.8d A model of the locations of bending and torsional stresses in the right tibia during stance phase of running. The dashed line indicates the level of the cross-sectional model of the tibia. This proximal tibial model is of a cross section at 62% of the length from the distal end. Positive values (light-grey areas) indicate tensile stresses that primarily lie within the anterior aspect of the tibia and negative values (mid-grey areas) of compressive stress, as seen within the posterior tibia. The darkest areas indicate the least bone stress as demonstrated within the medial–posterior tibial border and lateral tibia. Note that low bone stress zones include the posteromedial border of the tibia, an area associated with medial tibial stress syndrome. 'VS' indicates the location of the vertical strain gauge used in the study model produced by Derrick et al. (2016). 'PT' indicates the typical location of peak tension and 'PC' the location of peak compression, with units in MPa. Note that neither high compression nor tension forces are usually sustained along the medial side, nor through the centre of the lateral surface. This should make direct tensional or compression lesions less likely within these areas of bone, but shear torsion remain a risk. Tibial fatigue fractures are usually generated within the posteromedial aspect associated with compressive strains, while the anterior tibial surface is susceptible to tensile strain-induced fatigue fractures. *(Image modified from Derrick, T.R., Edwards, W.B., Fellin, R.E., Seay, J.F., 2016. An integrative modeling approach for the efficient estimation of cross-sectional tibial stresses during locomotion. J. Biomech. 49(3), 429–435.)*

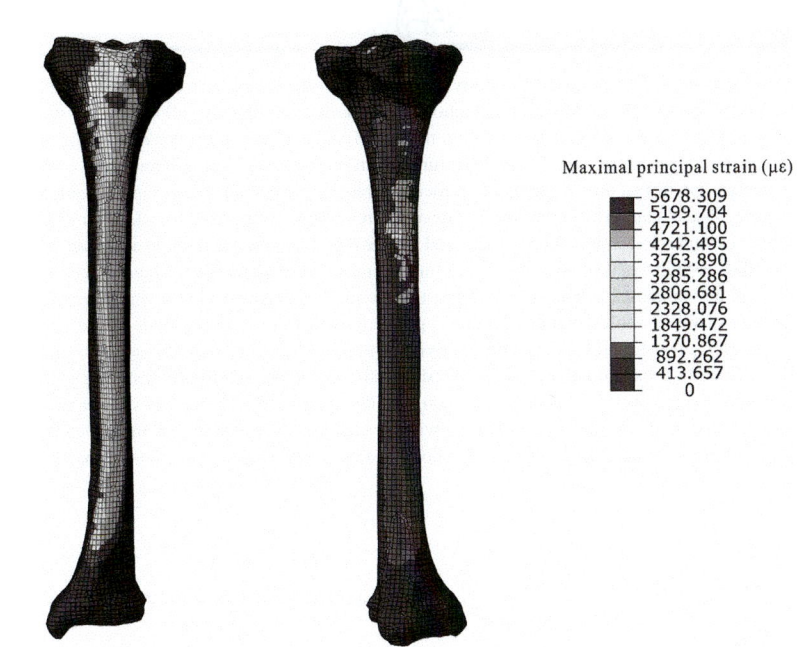

FIG. 4.6.8e Anterior and posterior views of a representative finite element model of the tibia displaying maximal principal tibial strains (με) during running. The models were loaded in bending and axial compression with tension in the anterior surface demonstrated on the left and compression on the posterior surface seen on the right. Tension strains are highest within the lighter areas along the anterior tibial surface and its lateral border, especially where it is covered by the medial edge of tibialis anterior. Compression strains are highest in the proximal third of the posterior tibia, indicated by the lighter area on the posterior right view. Note again that bone strains do not appear particularly high along the medial border of the tibia, the area most commonly associated with symptoms of medial tibial stress syndrome (see Section 4.3.9). *(Image from Chen, T.L., An, W.W., Chan, Z.Y.S., Au, I.P.H., Zhang, Z.H., Cheung, R.T.H., 2016. Immediate effects of modified landing pattern on a probabilistic tibial stress fracture model in runners. Clin. Biomech. 33, 49–54.)*

Effects of strike position on tibial stresses

Chen et al. (2016) attempted to examine the effects of tibial loading stresses through modelling based on fourteen rearfoot strike runners, to see if tibial stresses would be reduced by changing foot strike position. The study confirmed that lower impact initial loading peaks occurred in changing the strike position to the forefoot, as reported in many running forefoot strike studies, previously (Fig. 4.6.8g). However, the landing pattern switch did not change the peak tibial strains or seem to lower the risk of tibial stress fractures. Indeed, it does not appear that the size of the peak vertical GRF is a significant factor in lower limb stress fractures. However, the vertical loading rate might be significant (Zadpoor and Nikooyan, 2011).

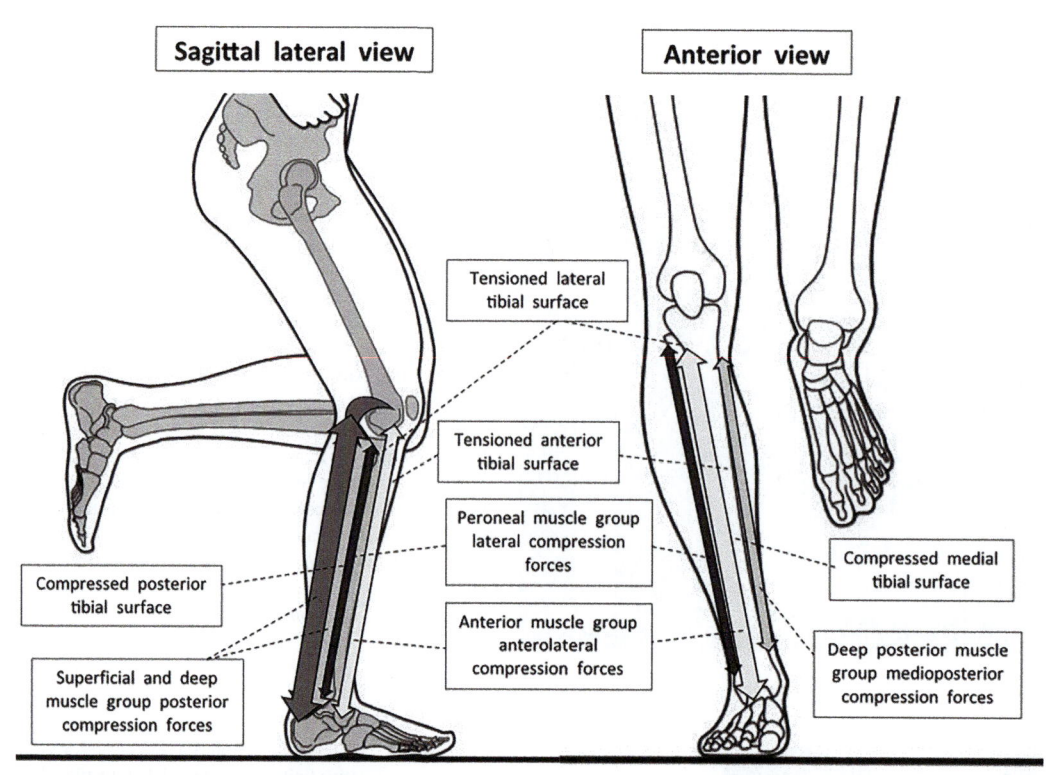

FIG. 4.6.8f In those with normal bone physiology, fatigue fractures are as much a result of failure of muscular protective systems in moderating and directing forces across the bone, as they are a result of bone material property failures. Muscle contractions are important in protecting all long bones such as the femur, fibula, and metatarsals. In the tibia, during impact and loading response, activity within the superficial and deep posterior muscle groups creates increased posterior (all muscles) and some posteromedial (for tibialis posterior and flexor digitorum longus) compression forces across areas of the tibia that are prone to higher compression forces. In running, when the superficial muscle group of all triceps surae are contracting prior to initial contact, avoiding over-compression of the bone requires controlled eccentric muscle–tendon buffering without raising compression forces from muscle contraction and body weight loading. Fortunately, long bones can tolerate high compression strains. The anterior and peroneal muscle group activation at impact and loading creates compression over areas of the tibia that are susceptible to high tensile loads. This means their increased activity produces higher protective compression forces over the lateral and anterolateral tibial surfaces that moderate tensile strains. Thus, tibialis anterior plantarflexion moment braking, activation of the windlass mechanism, and peroneus longus activity in aiding vault stiffening, all play a protective role in moderating loads on the tibia, while also helping to set other energy dissipation mechanisms. However, peroneus brevis shows little activity on initial contact, and the cross-sectional area of the anterior group compared to other muscle groups is not large. This limits the muscular capacity to resist tensile strain. The anteromedial surface of the tibia also lacks any muscle that can resist tensional strains across its surface directly. This is one reason why the anterior tibial surface is vulnerable to fatigue fractures that are more difficult to heal. Tibial morphologies also influence the degree of strain with for example, tibial varum bowing increasing medial compression and lateral tensile strains. Such morphologies increase vulnerability to tissue threshold breaches, requiring increased and responsive muscular active protection. *(Permission www.healthystep.co.uk.)*

FIG. 4.6.8g A graph to demonstrate the accumulative probability of a stress fracture with different foot strike positions from data reported by Chen et al. (2016). RFS = rearfoot strikes, MFS = midfoot strikes, and FFS = forefoot strikes. *(Image from Chen, T.L., An, W.W., Chan, Z.Y.S., Au, I.P.H., Zhang, Z.H., Cheung, R.T.H., 2016. Immediate effects of modified landing pattern on a probabilistic tibial stress fracture model in runners. Clin. Biomech. 33, 49–54.)*

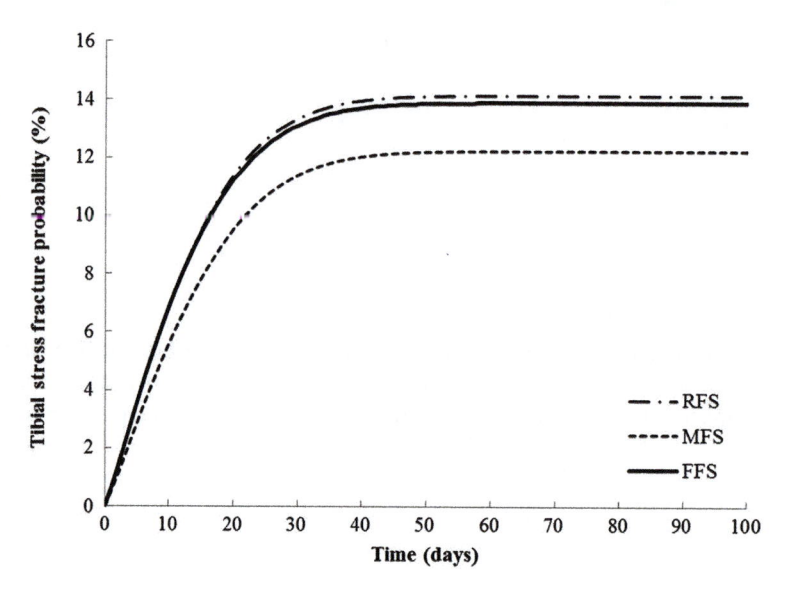

A prospective study of 1065 British Royal Marine recruits revealed several variables that increased the risk of tibial stress fractures. These were a smaller bimalleolar width (consistent with the findings of cross-sectional area), a lower body mass index, a smaller calf circumference, a greater peak heel pressure, and a lower range of tibial rotation (Nunns et al., 2016). These features all suggest a reduction in the ability to dissipate impact energies, with reported lower tibial rotation possibly indicating the loss of potential torque shock absorption through tibiofibular joint motion, stretching its long and large interosseous ligament during initial contact. This suggests that efficient motion at the tibiofibular joints is important during running. Loss of energy dissipation ability at initial impact is also supported by the finding that sufferers of tibial fatigue fractures demonstrate a faster loading rate (Zadpoor and Nikooyan, 2011). A smaller calf circumference also suggests that less soft tissue oscillation would be available distally to aid in energy dissipation via tissue resonance.

Running fatigue effects on tibial stresses

Rice et al. (2019) modelled the tibia as a hollow ellipse and applied bending moment to estimate stresses on the distal third of the tibia using beam theory, inverse dynamics, and musculoskeletal modelling using real data from runners. They reported that anterior tibial stresses increased by 15% and posterior stresses by 12% after 40 min of running compared to initially running, with changes starting to increase after only 20 min. Peak tibial accelerations may also be linked to tibial fatigue fractures. Runners with a history of tibial fatigue fracture tend to demonstrate higher peak hip adduction angles and moments than do runners with no prior history of fatigue fracture, suggesting that increased hip adduction may lead to excessive tibial torques (Yong et al., 2018).

As peak hip adduction angles are reportedly 4% higher in runners with a history of tibial fatigue fracture (Pohl et al., 2008), increased cadence by shortening stride length may reduce the risk of tibial fatigue fracture. This has been shown to lower peak tibial accelerations and peak adduction angles at the hip during treadmill running (Yong et al., 2018). Forefoot strike is also associated with lower average and peak loading rates during running (Shih et al., 2013; Boyer et al., 2014; Yong et al., 2018) and produces a small decrease in peak hip adduction angles on average (Yong et al., 2018). However, the work of Chen et al. (2016) suggests that changing to forefoot strike alone may not solve the problem, and that a fuller appreciation of all the factors including the position of the GRF in the frontal plane may be important to understanding an individual case of tibial fatigue fracture.

4.6.9 Medial tibial stress syndrome

Medial tibial stress syndrome (MTSS) presents as a diffuse pain along the edge of the medial anterior border of the tibia between the posterior and anterior surface, usually in the middle or distal one- to two-thirds, and within an area of over 5 cm (Moen et al., 2009). The condition producing pain over a distinct <5 cm medial tibial border area tends to exclude that arising from ischaemic origin or signs of tibial stress fracture (Reshef and Guelich, 2012), although the clinician should remain vigilant to these possible causes of shin pain. MTSS has been reported to occur in between 4% and 35% of military individuals and athletes (Moen et al., 2009). It is provoked by relatively high levels of activity and is not reported in sedentary populations, even though it can affect normal daily activity if the condition is allowed to progress. Symptoms are usually felt first after activity, then progress to symptoms upon initiating activity that the individual can initially run through. Symptoms gradually increase from persisting for minutes to hours, especially if activity levels are maintained or increased. The location of MTSS is shown in Fig. 4.6.9a.

This is an athletically induced problem seen primarily in long-distance runners, in other sustained running sports or training regimes, and in fast bowlers in cricket. However, it is not a common problem associated with sprint runners and it is not provoked by walking. Thus, its pathomechanics must be associated with high stress levels, but its absence in sprinters suggests that fatigue biomechanics are involved. Models have been presented to explain the pathomechanics of MTSS. One model suggests an osseous origin, the other soft tissue, but both models remain open to debate. The osseous models propose bending moments across the tibia as the cause of mechanical fatigue failure in bone, creating a periosteal or bone reaction (Reshef and Guelich, 2012), much as do the mechanisms for fatigue fracture. The other model proposes a fascia crura traction enthesopathy along the medial tibial crest, eventually causing a bone reaction under soft tissue-induced tensile stresses (Bouché and Johnson, 2007; Reshef and Guelich, 2012). There is a case to be argued for and against each model, but it is possible that both stress mechanisms could be at play in explaining this common problem, or perhaps other mechanisms that are less frequently considered. The lack of a focal point of stress concentration and pain in MTSS suggests a quite different pathomechanical event to a fatigue fracture within the tibia, indicating that something more diffuse is occurring. Fractures are essentially crack propagation issues that require concentration of forces to a localised area, something inconsistent with the diffuse symptoms of MTSS (Fig. 4.6.9b).

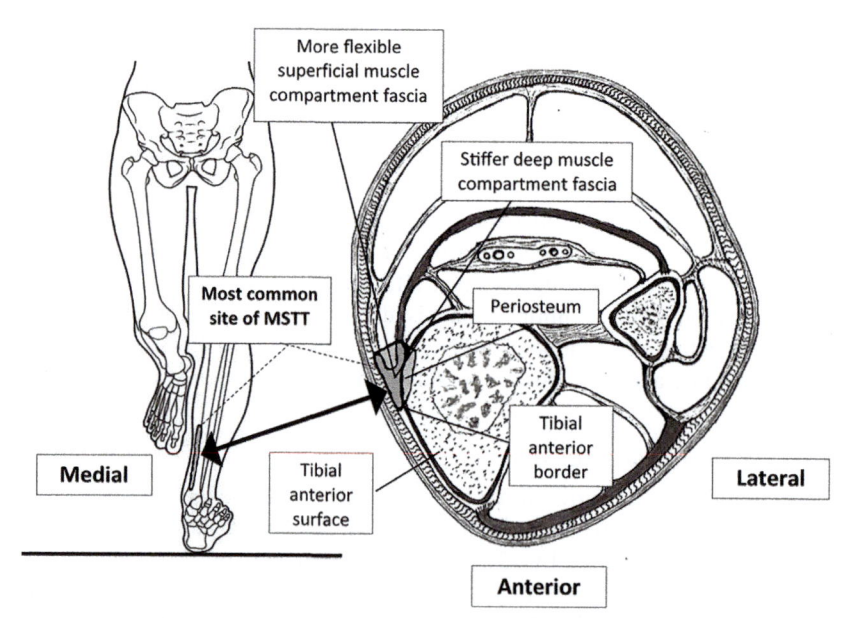

FIG. 4.6.9a A frontal plane view of the left lower limb and a cross-sectional anatomical schematic within the distal one-third of the left lower leg, demonstrating the most common location for symptoms of MTSS. For a fuller description of anatomy at this level, see Fig. 4.6.7. Pain is usually palpated along the anterior surface's medial border with the posterior surface of the tibia, usually within the distal third, but can extend almost the full length of the posteromedial bone border. This is quite unlike compartment syndrome where pain is within the muscle and often only present on exercise, and also very unlike tibial fatigue fractures that occur on the bone surfaces under high tension (anterior) or compression (posterior) forces. The laws of crack propagation dictate that fatigue fracture is a focused process that should cause pain to become very localised to a specific area, quite unlike that associated with MTSS. What is noticeable about the anatomies that come together along the medial tibial border is their quite different material properties. Such different material property 'bonding' tends to focus stresses to and within them. Under increased soft tissue oscillations as part of energy dissipation and when protective muscles are fatigued, bone may start to remodel in response to raised connective tissue stresses caused by variable resonance above them. This results from greater stretching tensions and recoil from soft tissue oscillation displacements lying superficial to it, particularly from horizontal translations. Bone responding to its surrounding connective tissue stresses would be an expected response when bone is considered as part of the connective tissue (fascial) continuum. *(Permission www.healthystep.co.uk.)*

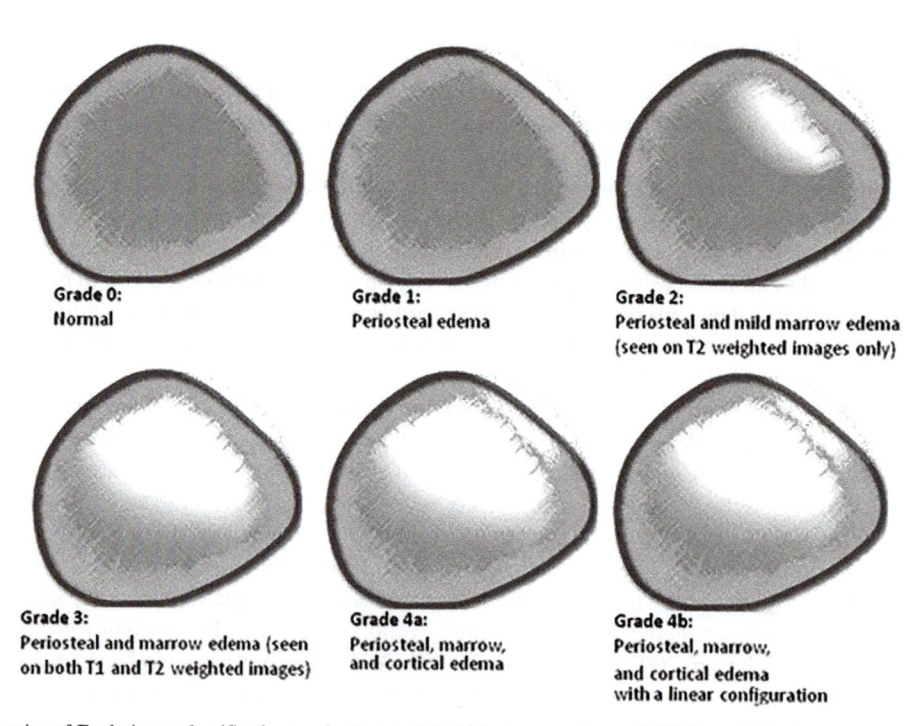

FIG. 4.6.9b Schematics of Fredericson classification grades for medial tibial stress syndrome (MTSS) on MRI. Oedema is seen without frank bone fracture. The presence of oedema (lighter areas) within the marrow and around the cortical bone long-term could start to affect the mechanical properties of the bone, threatening fatigue fracture from crack propagation if forces concentrate to one area; particularly should exercise continue without resolution of the situation. However, clinically, tibial fatigue fracture and MTSS present quite differently. *(Image from Marshall, R.A., Mandell, J.C., Weaver, M.J., Ferrone, M., Sodickson, A., Khurana, B., 2018. Imaging features and management of stress, atypical, and pathologic fractures. RadioGraphics 38(7), 2173–2192.)*

Potential causes of MTSS

Histological studies have revealed that bone resorption occurring in the area of MTSS outpaces bone formation in the tibial cortex (Moen et al., 2009), and subjects affected demonstrate lower bone mineral density than those that develop tibial stress fractures (Franklyn and Oakes, 2015). On MRI, periosteal oedema can be used as an early developing sign of the condition (Franklyn and Oakes, 2015), but periostitis has not been recorded histologically (Moen et al., 2009). Forces of compression and tension in the longitudinal axis in MTSS subjects, including those from GRF and the bending moments, are reported to be well below the tissue threshold for the tibia, being safely within the physiological zone as defined by the mechanostat theory (Al Nazer et al., 2012; Derrick et al., 2016; Hart et al., 2017). It would seem unlikely that the reported longitudinal axis stresses and strains of running are involved in damaging large areas of the posteromedial tibia (an area normally under higher compression than tensile stress, but not an area prone to high stresses generally), for there appears to be a high safety margin from recorded tibial stresses during running (Derrick et al., 2016). Yet, tissue threshold-breaching bone stresses could still be a possibility if occurring in the transverse axis, for bones' ultimate tissue strengths are lower in this plane (Fig. 4.6.9c).

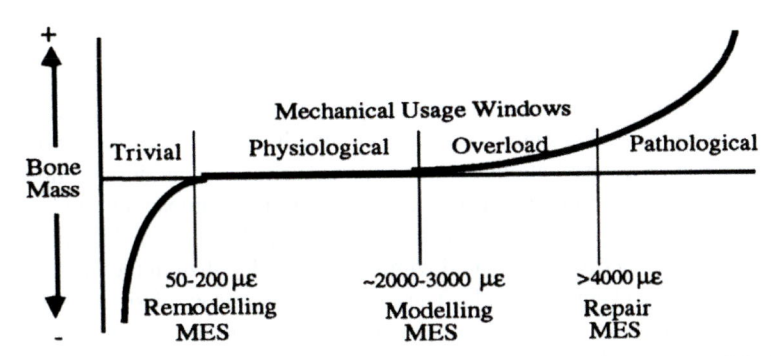

FIG. 4.6.9c Mechanostat theory relating strain magnitudes to bone response helps to explain how bone is both physiologically stimulated, and its health maintained, and yet how it also becomes damaged under high repetitive biomechanical loading. Regular loading helps set bone strength, bone mineral density, and an individual bone's cross-sectional area during growth via physiological/biological responses. However, once strain demands on tissue remodelling outstrip repair mechanisms, bone weakens. This is because bone requires time to repair and remodel to a stronger level. Maintaining exercise stresses at too high a level will reduce bone material properties, creating increasing microfractures. However, microfractures tend to focus stress through principles of crack propagation, making the development of a focused fatigue fracture more likely. The effects of fatigue on the ability of muscles to redirect strain to safer levels are undoubtedly also extremely important in strain management, but neither mechanism explains the diffuse nature of MTSS. *(Image from Forwood, M.R., Turner, C.H., 1995. Skeletal adaptations to mechanical usage: results from tibial loading studies in rats. Bone 17(4 Suppl), 197S–205S.)*

During exercise, only tibial shear strains have been reported to create strain levels within the symptomatic zone of the mechanostat theory (Al Nazer et al., 2012). Such shear strains could thus be involved, especially if MTSS is associated with individuals with an overall reduction in bone mineral density. However, other potential problems with the 'bone fatigue model' of MTSS remain. The area affected by MTSS tends to be quite large, defined as being over 5 cm, which would tend to spread the loading force out over the bone, reducing focus peak stresses. The greater the area that the symptoms cover the more the stresses should be spread across the bone. However, in MTSS, symptoms tend to heighten with increasing area. This does not usually fit with bone failure, as high osseous stresses are likely to cause microfractures which tend to concentrate forces through the principles of crack propagation. Although high bone stresses are easily implicated in the local pathology of fatigue fractures of the tibia when a runner's muscular protection is fatigued and cyclical loading has induced bone fatigue failure, the more diffuse nature of MTSS raises the possibility of other mechanical aetiological factors.

Soft tissue resonance as an aetiology of MTSS

Both stride length and stride frequency increase with running speed, with stride length and shock attenuation increasing by around 17% and 20% per m/s, respectively (Mercer et al., 2002). Impact energy dissipation through shock attenuation correlates strongly with stride length and only moderately with stride frequency across running speeds, with shock attenuation increasing linearly with running speed (Mercer et al., 2002). Shock attenuation requirements relate to increased leg peak impact accelerations. The stride length generated in running has a profound effect on running kinematic changes as well as the resulting kinetics (Mercer et al., 2002). This is implicated in running injury rates (Bramah et al., 2018), not just because of shock attenuation, but because of the changes in posture and resultant angular moments that muscles have to manage with increasing stride length, often at longer muscle lengths.

Concepts of shock attenuation in running are covered in detail in Chapter 1, Section 1.4.6, but are worth reviewing in the consideration of MTSS pathomechanics. The soft tissues of the lower limb are at the root of impact energy dampening and transmission during running (Pain and Challis, 2006; Schmitt and Günther, 2011; Riddick and Kuo, 2016). Soft tissues are able to modulate peak joint forces (Liu and Nigg 2000; Wakeling and Nigg, 2001) and dissipate mechanical energy (Riddick and Kuo, 2016). Muscles must actively offset any net energy dissipation in the body, and the amount of soft tissue deformation affects the metabolic cost of running. Impact energy not absorbed within the rearfoot is loaded proximally into the leg or distally into the midfoot and forefoot, just as energy not dissipated within the hip and thigh is loaded distally to the knee, leg, and rearfoot. In acceleration, energy not transferred into the forefoot from ankle plantarflexor power will be loaded proximally into the lower limb, ankle, or midfoot, or alternatively dissipated into the plantar musculoskeletal and cutaneous tissues of the foot. Their exist balances of energy from impact and muscle activity that must be maintained and accounted for, whatever the activity. It must be remembered that energy cannot be destroyed.

Therefore, soft tissues perform both positive work and negative work within different phases of stance, yielding substantial net negative work through energy dissipation from braking that has an increasing role at faster running speeds due to greater impacts (Riddick and Kuo, 2016). Soft tissue energy dissipation is divided into actively dissipated energy through muscles and tendons, and those tissues and mechanisms that passively dissipate energy such as plantar fat pad deformation, connective tissue stretching, and via soft tissue resonance frequencies through tissue oscillation. Soft tissues involved in oscillation include the muscle bellies and fascia, which together with the skin and fat, form wobbling masses restrained by compartmental fascia. These soft tissue oscillations or 'wobbling excursions' occur during impacts in the range of 3–55 Hz (Schmitt and Günther, 2011), and are an important mechanism in absorbing impact energy.

More energy seems to be dissipated by wobbling soft tissue masses in the horizontal (transverse plane of the leg) rather than the vertical direction, and less energy is dissipated within the lower leg than in the thigh where muscle bulk and thus wobbling mass is larger (Schmitt and Günther, 2011). The energy lost in impact dissipation within the soft tissues during running is large, and must be made up for (offset) by equal amounts of muscle activity-generated energy for acceleration. It could account for as much as ~29% of the metabolic cost of the running economy (Riddick and Kuo, 2016). These impact/braking energies applied to tissues during impact are large and potentially injurious. Thus, tissue oscillations are restrained through muscle activation, increasing the stiffness of the wobbling mass in an anticipatory manner prior to impact (Wakeling et al., 2003). By spreading energy dissipation into wobbling mass oscillations, humans avoid some of the energetic demands from muscle activity used for muscle–tendon buffering impact energy dissipation (Schmitt and Günther, 2011). By activating the muscle sufficiently, the amount of soft tissue movement possible is limited. In muscle fatigue, muscle damping of tissue vibration may significantly reduce, moving the burden of soft tissue energy dissipation into increased soft tissue oscillations. At the same time, muscular active restraint on soft tissue oscillation motion is reducing, potentially permitting greater soft tissue displacement from the underlying bone.

This combination of high impact energies and fatigued muscle possibly creates the environment for distance running-specific pathologies such as MTSS, where fatigue of active muscle in energy dissipation capabilities is surpassed and reliance on passive soft tissue wobbling mass dissipation is increased. With loss of muscle restraint on soft tissue movements, rising soft tissue oscillation motion increases the 'shock wave' effects on the fascia as it attempts to hold the soft tissues in place relative to the bone structure beneath. Where the fascia is soft and flexible such as in the superficial posterior fascia crura, extra soft tissue excursions are unlikely to create too many problems. Under the anterior fascia crura, the soft tissue bulk is not extensive, reducing the amount of soft tissue oscillations possible, plus the fascia is also stiffer. However, where large muscle mass is constrained by tight fascia, repetitive large soft tissue oscillations and excursions as part of an impact 'shock wave' could breach tissue thresholds for the fascia. An interface that consists of tissues with different mechanical behaviours would be particularly vulnerable to increased soft tissue oscillation stresses that it be should restraining. A good site for this effect would be near major fascial attachments to bone's periosteum, and sites where stiff areas and flexible areas of fascia meet covering large amounts of muscle mass. The stiffer deep posterior compartmental fascia and the more flexible superficial posterior compartmental fascia of the fascia crura join together along the medial tibial border's periosteum. This is the site of MTSS symptoms.

Large muscle bellies and other soft tissue transverse oscillating displacements between the superficial muscle group and the tibia medially, might increase tensional and shear stresses within the fascia of this region that will strain the attachments to the periosteum. This may possibly also involve some significant soft tissue motion within the deep compartment itself. Such soft tissue displacements under the fascia that are pulling away from the bone, will transversely tension–stress the superficial bone of the cortex lying beneath the periosteum that is attached to the fascia. A large muscle bulk, such as that of the human calf, positioned so low on a limb, is an unusual but a necessary anatomical feature of human gait. Its presence represents a soft tissue oscillation target at impact collisions. This inferior soft tissue mass' oscillation can result in a 'whiplash'-type effect running along the posteromedial tibial border's soft tissue-to-bone interface. In excess, this could potentially be initiating local inflammatory responses such as oedema under the periosteum and inflammation and fibrosis within the fascia, which in turn causes limitation of interfascial gliding. In time, the tensional pulling of the soft tissues could provoke the reported osseous oedema, resorption, and remodelling changes, with each part of the pathology provoking symptoms. Soft tissue oscillations will concentrate along the fascia-bone interface because these tissues are caught between the tissues that move (muscle and flexible fascia) and those that cannot freely move (bone particularly, periosteum, and stiffer fascia of the deep posterior compartment). Periosteal involvement could increase the osseous tensional stresses beneath through being part of the fascial continuum, explaining bone resorption and remodelling in the region as a response to soft tissue horizontal oscillation increases, rather than to direct tensile or compressive bone stresses causing microfractures (Fig. 4.6.9d).

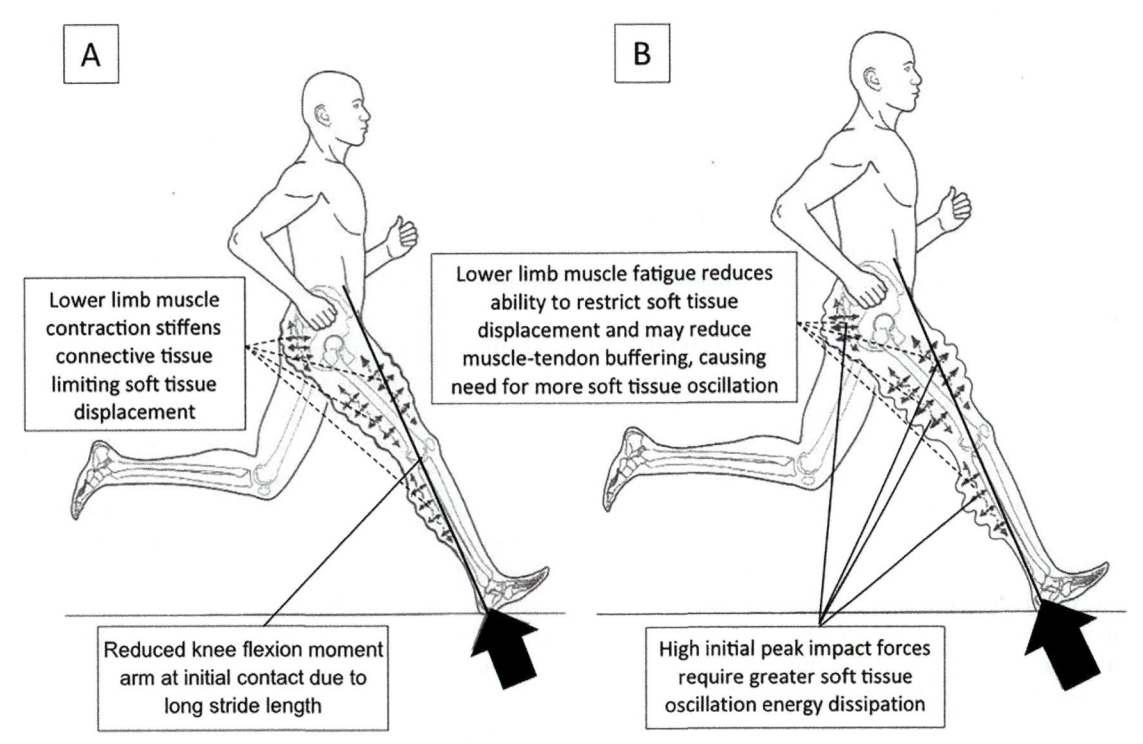

FIG. 4.6.9d Normally, soft tissue oscillations are a supplementary, although still an important part of energy dissipation at higher impacts. However, situations where knee flexion may be more limited at initial contact and initial peak impacts peaks higher, may increase tissue oscillation's role. Such situations exist where knee flexion angles are initially more minimal, such as in fast bowling in cricket or more commonly using a long stride length during running (A). This is because the GRF's angular lever arm to the knee is initially very short, reducing its knee flexion moment arm. If muscle contraction that usually suppresses soft tissue oscillations is deficient, such as due to muscular fatigue, then soft tissue oscillations may become greater with larger tissue displacement, particularly horizontally (B). If sustained, the increased tensional forces on the connective tissues holding the myofascial structures in place may cause collagen fibre damage that starts an inflammatory process, making further repetitive loading by large tissue oscillations painful. Areas where connective tissue demonstrates boundaries of different material flexibilities will tend to concentrate oscillation tensional forces. A perfect site for this to occur is at the posteromedial border of the tibia, where bone and its periosteum binds to the stiffer deep posterior compartment fascia and that of the more flexible superficial posterior compartment fascia. These posterior muscle compartments both contain relatively large muscle masses within the lower leg that can provide a distal source of large soft tissue oscillation under impact. Large distal muscle masses, so inferiorly placed on a weightbearing limb of a large terrestrial mammal, is a uniquely human trait. Its presence offers great benefits to plantigrade bipedalism, but it also presents a target for pathology if oscillation becomes excessive. *(Permission www.healthystep.co.uk.)*

If MTSS is an osseous-induced pathology, then examination of gait should be approached as in tibial stress fractures. However, the diffuse nature of the symptoms suggests something different in MTSS pathomechanics. If this is an injury resulting from excessive soft tissue resonance as a method of energy dissipation following muscle damping activity fatigue, then running events that induce fatigue of muscle damping activity are likely to lie at the origin of the pathology. Thus, again, those kinematic postures that put muscles at risk of fatigue are likely to be playing a part in the risk of developing MTSS. As the primary energy dissipation method through their action in managing joint moments is via muscle–tendon buffering, muscles at risk of initiating MTSS are likely to be either the tibialis anterior during heel strikes or the triceps surae in forefoot strikes being unable to reduce braking energy. In both strike positions, deep compartment muscles, such as tibialis posterior in particular, are working hard to set the appropriate level of foot stiffness. Thus, the fatigue resistance of lower limb soft tissue oscillation damping muscles will be important, but posterior calf muscles strength and fatigue resistance will be particularly significant because of their effects on controlling the medial fascia crura resonance locally. Once pathology has developed in the fascial-bone interface, the local fascial viscosity in the deep fascia may change, causing fibrosis or a 'densification', restricting fascial gliding properties which may need to be addressed and improved to aid in recovery (Pavan et al., 2014). This may explain why many practitioners find deep fascial manipulation clinically helpful for MTSS.

We still have much to learn in understanding running-induced pathomechanics, even in commonly encountered conditions such as MTSS.

4.6.10 Section summary

Many running injuries relate to similar pathomechanical problems as those found in walking, such as the length of muscular lever arms during loading, and the power that can be generated for acceleration, but there is a clear distinction. Most walking-provoked pathology relates either to a significant high stress-inducing traumatic events outside of normal gait patterns, or low tissue thresholds due to age-related weakness or disease. Running involves much higher forces through the consequence of greater accelerations, with mechanical tissue thresholds being lowered via fatigue events. Braking events and energy dissipation are far more significant in running than during walking, and the dissipation and transmission of energy must become more tightly controlled to avoid injury. The knee and Achilles 'pick up' much of the impact and acceleration burden of faster gait velocity during running, focusing much pathology to the knee, its muscles, and the triceps surae and Achilles.

Walking energetics are associated with the ability to allow significant rises and falls in the CoM during stance phase, with acceleration power only required to move a reducing CoM to the next foot strike following heel lift. Running involves an initial ~50% of stance phase fall in the CoM followed by an acceleration upwards and forwards through the remaining ~50% of stance that results in an aerial phase with a high CoM position. Jogging does not have an aerial phase but demonstrates all other running features. Although the requirements of the acceleration phase are similar across foot strike positions (reflecting mostly the need of speed during gait), initial contact kinematics and kinetics during braking and energy dissipation are very dependent on the foot strike position. The running CoM situation at midstance requires significant acceleration power, applied particularly as a horizontal force to drive forward on. Running also requires a rapid swing phase and a more dramatic braking of swing prior to initial contact. Acceleration requires increased triceps surae power and quadriceps activity. Thus, triceps surae activity begins much earlier in running with the Achilles experiencing an increased energy burden through prolonged stretching and elastic recoil. The quadriceps muscles are active throughout stance including acceleration, unlike during walking.

The pathomechanics of any runner warrants investigation that reflects the variation in locomotion from walking. Yet as in walking, a patient's lifestyle is not disassociated from running. The level of difference in a runner's daily lifestyle compared to their chosen sport can cause morphological mismatch between the two situations. For example, a high heel-wearing sedentary office worker taking up running is more likely to become injured than a postal worker who wears training shoes for work and walks several kilometres a day before taking up running. Those who have run extensively from childhood will also have an advantage from those who have led a sedentary childhood. Our expectations of a runner should reflect not only their lifestyle and health, but also morphology, age, and individual lifestyle factors that can heavily influence running energetic potential and injury through the tissue stresses within anatomy. The final clinical consideration of runners is that even seemingly physiologically fit patients can also develop diseases.

Chapter summary

Pathomechanics is the logical way to end the exploration of the origins and principles of clinical biomechanics of the lower limb in human locomotion. Pathomechanics is the exciting and practical part of understanding clinical practice. There is much in common with the work of a detective, with the diagnosis being the crime and the pathomechanics providing the motive and the clues to the capture of the criminal (the origin of dysfunction and injury) and the appropriate punishment, or in this case, an appropriate clinical intervention for the pathology. If there is no known diagnosis and no known pathomechanics, then providing the correct treatment will be a matter of luck. There is nothing more futile in clinical practice than the instigation of the wrong treatment programme because of the wrong diagnosis, thus trying to resolve the wrong problem.

A good detective needs to consider all the clues presented and the motive for a crime. A good clinician needs to clerk the patient for their health status, reflect on the patient's age, and in so doing, establish the risk of tissue threshold breach. The clinician also needs to locate and understand the pathology, propose a diagnosis (confirm diagnostically if appropriate), record the morphology, assess the strength and range of motions, and observe and then (if possible) record the locomotion utilised. Finally, they must reflect on all the findings. From the gathering of the origins of pathology, principles can be applied to a process that associates or disassociates the biomechanical findings to explain the pathomechanics leading to pathology.

The ability to handle the biomechanical and kinesiological data appropriately requires a knowledge of energetics, the laws of motion, tissue material properties, physiology, developmental processes, evolutionary medicine, locomotive biomechanics, and functional anatomy. Each of these subjects helps the clinician to understand the nature of the individual patient, giving context to their distinct form of locomotion. Foundational knowledge in the origins and principles of lower limb biomechanics helps to draw attention to what will be advantageous or disadvantageous to the patient's locomotion. The conclusions gathered will allow the clinician to move on to the next stage of choosing the correct therapeutic intervention or provide appropriate advice to resolve the presenting problem and maintain patient mobility. In the end, the link between the maintenance of mobility and quality of life gives purpose to clinical biomechanics.

References

Abdalbary, S.A., Elshaarawy, E.A.A., Khalid, B.E.A., 2016. Tensile properties of the deep transverse metatarsal ligament in hallux valgus: a CONSORT-Complaint article. Medicine 95 (8), e2843. https://doi.org/10.1097/MD.0000000000002843.

Aderem, J., Louw, Q.A., 2015. Biomechanical risk factors associated with iliotibial band syndrome in runners: a systematic review. BMC Musculoskelet. Disord. 16, 356. https://doi.org/10.1186/s12891-015-0808-7.

Aderinto, J., Cobb, A.G., 2002. Lateral release for patellofemoral arthritis. Arthrosc.: J. Arthrosc. Rel. Surg. 18 (4), 399–403.

Aggarwal, P., Jirankali, V., Garg, S.K., 2020. Evaluation of plantar fascia using high-resolution ultrasonography in clinically diagnosed cases of plantar fasciitis. Pol. J. Radiol. 85, e375–e380.

Al Nazer, R., Lanovaz, J., Kawalilak, C., Johnston, J.D., Kontulainen, S., 2012. Direct in vivo strain measurements in human bone—a systematic literature review. J. Biomech. 45 (1), 27–40.

Al-Dadah, O., Shepstone, L., Donell, S.T., 2011. Proprioception following partial meniscectomy in stable knees. Knee Surg. Sports Traumatol. Arthrosc. 19 (2), 207–213.

Al-Hayani, A., 2009. The functional anatomy of the hip abductors. Folia Morphol. (Warsz) 68 (2), 98–103.

Almonroeder, T., Willson, J.D., Kernozek, T.W., 2013. The effect of foot strike pattern on Achilles tendon load during running. Ann. Biomed. Eng. 41 (8), 1758–1766.

Alvarez-Nemegyei, J., 2007. Risk factors for pes anserinus tendinitis/bursitis syndrome: a case control study. J. Clin. Rheumatol. 13 (2), 63–65.

Anderson, F.C., Pandy, M.G., 2003. Individual muscle contributions to support in normal walking. Gait Posture 17 (2), 159–169.

Angin, S., Crofts, G., Mickle, K.J., Nester, C.J., 2014. Ultrasound evaluation of foot muscles and plantar fascia in pes planus. Gait Posture 40 (1), 48–52.

Anwar, R., Anjum, S.N., Nicholl, J.E., 2005. Sesamoids of the foot. Curr. Orthop. 19 (1), 40–48.

Arangio, G.A., Chen, C., Salathé, E.P., 1998. Effect of varying arch height with and without the plantar fascia on the mechanical properties of the foot. Foot Ankle Int. 19 (10), 705–709.

Arellano, C.J., Gidmark, N.J., Konow, N., Azizi, E., Roberts, T.J., 2016. Determinants of aponeurosis shape change during muscle contraction. J. Biomech. 49 (9), 1812–1817.

Arendse, R.E., Noakes, T.D., Azevedo, L.B., Romanov, N., Schwellnus, M.P., Fletcher, G., 2004. Reduced eccentric loading of the knee with the pose running method. Med. Sci. Sports Exerc. 36 (2), 272–277.

Arndt, A., Brüggemann, G.-P., Koebke, J., Segesser, B., 1999. Asymmetrical loading of the human triceps surae. I. Mediolateral force differences in the Achilles tendon. Foot Ankle Int. 20 (7), 444–449.

Arndt, A., Ekenman, I., Westblad, P., Lundberg, A., 2002. Effects of fatigue and load variation on metatarsal deformation measured in vivo during barefoot walking. J. Biomech. 35 (5), 621–628.

Arndt, A., Westblad, P., Ekenman, I., Lundberg, A., 2003. A comparison of external plantar loading and in vivo local metatarsal deformation wearing two different military boots. Gait Posture 18 (2), 20–26.

Arndt, A., Bengtsson, A.-S., Peolsson, M., Thorstensson, A., Movin, T., 2012. Non-uniform displacement within the Achilles tendon during passive ankle joint motion. Knee Surg. Sports Traumatol. Arthrosc. 20 (9), 1868–1874.

Ateshian, G.A., Hung, C.T., 2005. Patellofemoral joint biomechanics and tissue engineering. Clin. Orthop. Relat. Res. 436 (July), 81–90.

Aweid, O., Del Buono, A., Malliaras, P., Iqbal, H., Morrissey, D., Maffulli, N., et al., 2012. Systematic review and recommendations for intracompartmental pressure monitoring in diagnosing chronic exertional compartment syndrome of the leg. Clin. J. Sport Med. 22 (4), 356–370.

Baker, R.L., Souza, R.B., Fredericson, M., 2011. Iliotibial band syndrome: soft tissue and biomechanical factors in evaluation and treatment. PM&R: J. Injury Funct. Rehabil. 3 (6), 550–561.

Banks, A.S., Hsu, Y.S., Mariash, S., Zirm, R., 1994. Juvenile hallux abducto valgus association with metatarsus adductus. J. Am. Podiatr. Med. Assoc. 84 (5), 219–224.

Bar-On, L., Kalkman, B.M., Cenni, F., Schless, S.-H., Molenaers, G., Maganaris, C.N., et al., 2018. The relationship between medial gastrocnemius lengthening properties and stretch reflexes in cerebral palsy. Front. Pediatr. 6, 259. https://doi.org/10.3389/fped.2018.00259.

Barber, L., Barrett, R., Lichtwark, G., 2011. Passive muscle mechanical properties of the medial gastrocnemius in young adults with spastic cerebral palsy. J. Biomech. 44 (13), 2496–2500.

Barber, L., Barrett, R., Lichtwark, G., 2012. Medial gastrocnemius muscle fascicle active torque-length and Achilles tendon properties in young adults with spastic cerebral palsy. J. Biomech. 45 (15), 2526–2530.

Barkham, N., Coates, L.C., Keen, H., Hensor, E., Fraser, A., Redmond, A., et al., 2010. Double-blind placebo-controlled trail of etanercept in the prevention of work disability in ankylosing spondylitis. Ann. Rheum. Dis. 69 (11), 1926–1928 (Erratum in: 70(8): 1519).

Barrett, R.S., Besier, T.F., Lloyd, D.G., 2007. Individual muscle contributions to the swing phase of gait: an EMG-based forward dynamics modelling approach. Simul. Model. Pract. Theory 15 (9), 1146–1155.

Barsoum, W.K., Lee, H.H., Murray, T.G., Colbrunn, R., Klika, A.K., Butler, S., et al., 2011. Robotic testing of proximal tibio-fibular joint kinematics for measuring instability following total knee arthroplasty. J. Orthop. Res. 29 (1), 47–52.

Barton, C.J., Levinger, P., Menz, H.B., Webster, K.E., 2009. Kinematic gait characteristics associated with patellofemoral pain syndrome: a systematic review. Gait Posture 30 (4), 405–416.

Barton, C.J., Lack, S., Malliaras, P., Morrissey, D., 2013. Gluteal muscle activity and patellofemoral pain syndrome: a systematic review. Br. J. Sports Med. 47 (4), 207–214.

Barwick, A.L., Tessier, J.W., de Jonge, X.J., Chuter, V.H., 2016. Foot bone density in diabetes may be unaffected by the presence of neuropathy. J. Diabetes Complicat. 30 (6), 1087–1092.

Basadonna, P.-T., Rucco, V., Gasparini, D., Onorato, A., 1999. Plantar fat pad atrophy after corticosteroid injection for an interdigital neuroma: a case report. Am. J. Phys. Med. Rehabil. 78 (3), 283–285.

Beck, T.J., Ruff, C.B., Shaffer, R.A., Betsinger, K., Trone, D.W., Brodine, S.K., 2000. Stress fracture in military recruits: gender differences in muscle and bone susceptibility factors. Bone 27 (3), 437–444.

Becker, J., James, S., Wayner, R., Osternig, L., Chou, L.-S., 2017. Biomechanical factors associated with Achilles tendinopathy and medial tibial stress syndrome in runners. Am. J. Sports Med. 45 (11), 2614–2621.

Beeson, P., 2014. Plantar fasciopathy: revisiting the risk factors. Foot Ankle Surg. 20 (3), 160–165.

Beeve, A.T., Brazill, J.M., Scheller, E.L., 2019. Peripheral neuropathy as a component of skeletal disease in diabetes. Curr. Osteoporosis Rep. 17 (5), 256–269.

Bencardino, J.T., Rosenberg, Z.S., Brown, R.R., Hassankhani, A., Lustrin, E.S., Beltran, J., 2000. Traumatic musculotendinous injuries of the knee: diagnosis with MR imaging. RadioGraphics 20 (Suppl. 1), S103–S120.

Bernstein, J.M., Schoenleber, S.J., 2021. Adolescent Blount's disease: reconstructive considerations and approach. Oper. Tech. Orthop. 31 (2), 100875. https://doi.org/10.1016/J.OTO.2021.100875.

Biewener, A.A., Farley, C.T., Roberts, T.J., Temaner, M., 2004. Muscle mechanical advantage of human walking and running: implications for energy cost. J. Appl. Physiol. 97 (6), 2266–2274.

Birch, H.L., Worboys, S., Eissa, S., Jackson, B., Strassburg, S., Clegg, P.D., 2008. Matrix metabolism rate differs in functionally distinct tendons. Matrix Biol. 27 (3), 182–189.

Bjelopetrovich, A., Barrios, J.A., 2016. Effects of incremental ambulatory-range loading on arch height index parameters. J. Biomech. 49 (14), 3555–3558.

Blazevich, A.J., Cannavan, D., Waugh, C.M., Miller, S.C., Thorlund, J.B., Aagaard, P., et al., 2014. Range of motion, neuromechanical, and architectural adaptations to plantar flexor stretch training in humans. J. Appl. Physiol. 117 (5), 452–462.

Boden, B.P., Osbahr, D.C., Jimenez, C., 2001. Low-risk stress fractures. Am. J. Sports Med. 29 (1), 100–111.

Bogaerts, S., Desmet, H., Slagmolen, P., Peers, K., 2016. Strain mapping in the Achilles tendon—a systematic review. J. Biomech. 49 (9), 1411–1419.

Bojsen-Møller, F., 1979. Calcaneocuboid joint and stability of the longitudinal arch of the foot at high and low gear push off. J. Anat. 129 (1), 165–176.

Bojsen-Møller, F., Lamoreux, L., 1979. Significance of free-dorsiflexion of the toes in walking. Acta Orthop. Scand. 50 (4), 471–479.

Bojsen-Møller, J., Hansen, P., Aagaard, P., Svantesson, U., Kjaer, M., Magnusson, S.P., 2004. Differential displacement of the human soleus and medial gastrocnemius aponeuroses during isometric plantar flexor contractions in vivo. J. Appl. Physiol. 97 (5), 1908–1914.

Bonnel, F., Toullec, E., Mabit, C., Tourné, Y., Sofcot, 2010. Chronic ankle instability: biomechanics and pathomechanics of ligaments injury and associated lesions. Orthopaed. Traumatol. Surg. Res. 96 (4), 424–432.

Borzì, R.M., Mazzetti, I., Marcu, K.B., Facchini, A., 2004. Chemokines in cartilage degradation. Clin. Orthop. Relat. Res. 427 (Suppl), S53–S61.

Bouaicha, S., Ehrmann, C., Moor, B.K., Maquieira, G.J., Espinosa, N., 2010. Radiographic analysis of metatarsus primus elevatus and hallux rigidus. Foot Ankle Int. 31 (9), 807–814.

Bouché, R.T., Johnson, C.H., 2007. Medial tibial stress syndrome (tibial fasciitis): a proposed pathomechanical model involving fascial traction. J. Am. Podiatr. Med. Assoc. 97 (1), 31–36.

Boussouar, A., 2019. Thickness estimation, automated classification and novelty detection in ultrasound images of the plantar fascia tissues. PhD Thesis, Informatics Research Centre School of Computing, Science and Engineering, University of, Salford, UK.

Bowman Jr., K.F., Sekiya, J.K., 2009. Anatomy and biomechanics of the posterior cruciate ligament and other ligaments of the knee. Oper. Tech. Sports Med. 17 (3), 126–134.

Boyer, E.R., Rooney, B.D., Derrick, T.R., 2014. Rearfoot and midfoot or forefoot impacts in habitually shod runners. Med. Sci. Sports Exerc. 46 (7), 1384–1391.

Bramah, C., Preece, S.J., Gill, N., Herrington, L., 2018. Is there a pathological gait associated with common soft tissue running injuries? Am. J. Sports Med. 46 (12), 3023–3031.

Bryson, D.J., Wicks, L., Ashford, R.U., 2015. The investigation and management of suspected malignant pathological fractures: a review for the general orthopaedic surgeon. Injury 46 (10), 1891–1899.

Busseuil, C., Freychat, P., Guedj, E.B., Lacour, J.R., 1998. Rearfoot-forefoot orientation and traumatic risk for runners. Foot Ankle Int. 19 (1), 32–37.

Byrne, D.P., Mulhall, K.J., Baker, J.F., 2010. Anatomy & biomechanics of the hip. Open Sports Med. J. 4, 51–57.

Byrne, C.A., O'Keeffe, D.T., Donnelly, A.E., Lyons, G.M., 2007. Effect of walking speed changes on tibialis anterior EMG during healthy gait for FES envelope design in drop foot correction. J. Electromyogr. Kinesiol. 17 (5), 605–616.

Camasta, C.A., 1996. Hallux limitus and hallux rigidus. Clinical examination, radiographic findings, and natural history. Clin. Podiatr. Med. Surg. 13 (3), 423–448.

Caravaggi, P., Pataky, T., Goulermas, J.Y., Savage, R., Crompton, R., 2009. A dynamic model of the windlass mechanism of the foot: evidence for early stance phase preloading of the plantar aponeurosis. J. Exp. Biol. 212 (15), 2491–2499.

Caravaggi, P., Sforza, C., Leardini, A., Portinaro, N., Panou, A., 2018. Effect of plano-valgus foot posture on midfoot kinematics during barefoot walking in an adolescent population. J. Foot Ankle Res. 11, 55. https://doi.org/10.1186/s13047-018-0297-7.

Carlson, R.E., Fleming, L.L., Hutton, W.C., 2000. The biomechanical relationship between the tendoachilles, plantar fascia and metatarsophalangeal joint dorsiflexion angle. Foot Ankle Int. 21 (1), 18–25.

Carroll, M., Parmar, P., Dalbeth, N., Boocock, M., Rome, K., 2015. Gait characteristics associated with the foot and ankle in inflammatory arthritis: a systematic review and meta-analysis. BMC Musculoskelet. Disord. 16, 134. https://doi.org/10.1186/s12891-015-0596-0.

Cen, X., Xu, D., Baker, J.S., Gu, Y., 2020. Association of arch stiffness with plantar impulse distribution during walking, running, and gait termination. Int. J. Environ. Res. Public Health 17 (6), 2090. https://doi.org/10.3390/ijerph17062090.

Ceyssens, L., Vanelderen, R., Barton, C., Malliaras, P., Dingenen, B., 2019. Biomechanical risk factors associated with running-related injuries: a systematic review. Sports Med. 49 (7), 1095–1115.

Challis, J.H., Murdoch, C., Winter, S.L., 2008. Mechanical properties of the human heel pad: a comparison between populations. J. Appl. Biomech. 24 (4), 377–381.

Chan, O., Sakellariou, A., 2020. Hallux rigidus: a review. Orthopaed. Trauma 34 (1), 23–29.

Chao, C.Y.L., Zheng, Y.-P., Huang, Y.-P., Cheing, G.L.Y., 2010. Biomechanical properties of the forefoot plantar soft tissue as measured by an optical coherence tomography-based air-jet indentation system and tissue ultrasound palpation system. Clin. Biomech. 25 (6), 594–600.

Chen, T.L., An, W.W., Chan, Z.Y.S., Au, I.P.H., Zhang, Z.H., Cheung, R.T.H., 2016. Immediate effects of modified landing pattern on a probabilistic tibial stress fracture model in runners. Clin. Biomech. 33, 49–54.

Cheung, J.T.-M., Zhang, M., An, K.-N., 2006. Effect of Achilles tendon loading on plantar fascia tension in the standing foot. Clin. Biomech. 21 (2), 194–203.

Chimenti, R.L., Cychosz, C.C., Hall, M.M., Phisitkul, P., 2017. Current concepts review update: insertional Achilles tendinopathy. Foot Ankle Int. 38 (10), 1160–1169.

Chinn, L., Dicharry, J., Hertel, J., 2013. Ankle kinematics of individuals with chronic ankle instability while walking and jogging on a treadmill in shoes. Phys. Therapy Sport 14 (4), 232–239.

Cho, J.-H., Min, T.-H., Chun, D.-I., Won, S.-H., Park, S.Y., Kim, K., et al., 2020. Bone mineral density in diabetes mellitus foot patients for prediction of diabetic neuropathic osteoarthropathic fracture. J. Bone Metabol. 27 (3), 207–215.

Chokhandre, S., Halloran, J.P., van den Bogert, A.J., Erdemir, A., 2012. A three-dimensional inverse finite element analysis of the heel pad. J. Biomech. Eng. 134 (3), 031002. https://doi.org/10.1115/1.4005692.

Cichanowski, H.R., Schmitt, J.S., Johnson, R.J., Niemuth, P.E., 2007. Hip strength in collegiate female athletes with patellofemoral pain. Med. Sci. Sports Exerc. 39 (8), 1227–1232.

Colapietro, M., Fraser, J.J., Resch, J.E., Hertel, J., 2020. Running mechanics during 1600 meter track runs in young adults with and without chronic ankle instability. Phys. Therapy Sport 42, 16–25.

Collins, N.J., Oei, E.H.G., de Kanter, J.L., Vicenzino, B., Crossley, K.M., 2019. Prevalence of radiographic and magnetic resonance imaging features of patellofemoral osteoarthritis in young and middle-aged adults with persistent patellofemoral pain. Arthritis Care Res. 71 (8), 1068–1073.

Cook, J.L., Stasinopoulos, D., Brismée, J.-M., 2018. Insertional and mid-substance Achilles tendinopathies: eccentric training is not for everyone—updated evidence of non-surgical management. J. Manual Manipul. Therapy 26 (3), 119–122.

Cornwall, M.W., McPoil, T.G., 2009. Classification of frontal plane rearfoot motion patterns during the stance phase of walking. J. Am. Podiatr. Med. Assoc. 99 (5), 399–405.

Correa, T.A., Crossley, K.M., Kim, H.J., Pandy, M.G., 2010. Contributions of individual muscles to hip joint contact force in normal walking. J. Biomech. 43 (8), 1618–1622.

Coughlin, M.J., Jones, C.P., 2007. Hallux valgus: demographics, etiology, and radiographic assessment. Foot Ankle Int. 28 (7), 759–777.

Coughlin, M.J., Shurnas, P.S., 2003. Hallux rigidus: grading and long-term results of operative treatment. J. Bone Joint Surg. 85-A (11), 2072–2088.

Coughlin, M.J., Saltzman, C.L., Nunley 2nd, J.A., 2002. Angular measurements in the evaluation of hallux valgus deformities: a report of the ad hoc committee of the American Orthopaedic Foot & Ankle Society on angular measurements. Foot Ankle Int. 23 (1), 68–74.

Cowley, E., Lepesis, V., 2018a. The SOAPIER model in podiatric musculoskeletal assessment and management: a three-part series. Part 1. Podiatry Now 21 (8), 18–20.

Cowley, E., Lepesis, V., 2018b. The SOAPIER model in podiatric musculoskeletal assessment and management: a three-part series. Part 2. Podiatry Now 21 (9), 8–10.

Cowley, E., Lepesis, V., 2018c. The SOAPIER model in podiatric musculoskeletal assessment and management: a three-part series. Part 3. Podiatry Now 21 (10), 8–9.

Creaby, M.W., Dixon, S.J., 2008. External frontal plane loads may be associated with tibial stress fracture. Med. Sci. Sports Exerc. 40 (9), 1669–1674.

Crossley, K., Bennell, K.L., Wrigley, T., Oakes, B.W., 1999. Ground reaction forces, bone characteristics, and tibial stress fracture in male runners. Med. Sci. Sports Exerc. 31 (8), 1088–1093.

Crowell, H.P., Milner, C.E., Hamill, J., Davis, I.S., 2010. Reducing impact loading during running with the use of real-time visual feedback. J. Orthop. Sports Phys. Ther. 40 (4), 206–213.

Cychosz, C., Johnson, A., Phisitkul, P., 2018. Pronation of the first metatarsal in hallux valgus deformity: a weight bearing CT study. Foot Ankle Orthopaed. 3 (3). https://doi.org/10.1177/2473011418S00197 (AOFAS annual meeting 2018 abstracts).

Dalmau-Pastor, M., Fargues-Polo Jr., B., Casanova-Martínez Jr., D., Vega, J., Golanó, P., 2014. Anatomy of the triceps surae: a pictorial essay. Foot Ankle Clin. 19 (4), 603–635.

Dananberg, H.J., 1985. Functional hallux limitus and its effect on normal ambulation. Curr. Podiatric Med. 34 (4), 11–16.

Dananberg, H.J., 1986. Functional hallux limitus and its relationship to gait efficiency. J. Am. Podiatr. Med. Assoc. 76 (11), 648–652.

Davda, K., Malhotra, K., O'Donnell, P., Singh, D., Cullen, N., 2017. Peroneal tendon disorders. EFORT Open Rev. 2 (6), 281–292.

Dawidowicz, J., Szotek, S., Matysiak, N., Mielańczyk, Ł., Maksymowicz, K., 2015. Electron microscopy of human fascia lata: focus on telocytes. J. Cell. Mol. Med. 19 (10), 2500–2506.

Dawidowicz, J., Matysiak, N., Szotek, S., Maksymowicz, K., 2016. Telocytes of fascial structures. Adv. Exp. Med. Biol. 913, 403–424.

De Ridder, R., Witvrouw, E., Dolphens, M., Roosen, P., Van Ginckel, A., 2017. Hip strength as an intrinsic risk factor for lateral ankle sprains in youth soccer players: a 3-season prospective study. Am. J. Sports Med. 45 (2), 410–416.

de Ruiter, C.J., Verdijk, P.W.L., Werker, W., Zuidema, M.J., de Haan, A., 2014. Stride frequency in relation to oxygen consumption in experienced and novice runners. Eur. J. Sport Sci. 14 (3), 251–258.

de Sá, A., Hart, D.A., Khan, K., Scott, A., 2018. Achilles tendon structure is negatively correlated with body mass index, but not influenced by statin use: a cross-sectional study using ultrasound tissue characterization. PLoS One 13 (6), e0199645. https://doi.org/10.1371/journal.pone.0199645.

Dearborn, J.T., Eakin, C.L., Skinner, H.B., 1996. Medial compartment arthrosis of the knee. Am. J. Orthop. 25 (1), 18–26.

DeJong, A.F., Mangum, L.C., Hertel, J., 2019. Gluteus medius activity during gait is altered in individuals with chronic ankle instability: an ultrasound imaging study. Gait Posture 71, 7–13.

DeJong, A.F., Koldenhoven, R.M., Hertel, J., 2020. Cross-correlations between gluteal muscle thickness derived from ultrasound imaging and hip biomechanics during walking gait. J. Electromyogr. Kinesiol. 51, 102406. https://doi.org/10.1016/j.jelekin.2020.102406.

Delabastita, T., Bogaerts, S., Vanwanseele, B., 2018. Age-related changes in Achilles tendon stiffness and impact on functional activities: a systematic review and meta-analysis. J. Aging Phys. Act. 27 (1), 116–127.

Delahunt, E., Monaghan, K., Caulfield, B., 2006. Altered neuromuscular control and ankle joint kinematics during walking in subjects with functional instability of the ankle joint. Am. J. Sports Med. 34 (12), 1970–1976.

Delahunt, E., Bleakley, C.M., Bossard, D.S., Caulfield, B.M., Docherty, C.L., Doherty, C., et al., 2018. Clinical assessment of acute lateral ankle sprain injuries (ROAST): 2019 consensus statement and recommendations of the International Ankle Consortium. Br. J. Sports Med. 52 (20), 1304–1310.

Deleu, P.-A., Naaim, A., Chèze, L., Dumas, R., Devos Bevernage, B., Goubau, L., et al., 2021. The effect of ankle and hindfoot malalignment on foot mechanics in patients suffering from post-traumatic ankle osteoarthritis. Clin. Biomech. 81, 105239. https://doi.org/10.1016/j.clinbiomech.2020.105239.

Derrick, T.R., Edwards, W.B., Fellin, R.E., Seay, J.F., 2016. An integrative modeling approach for the efficient estimation of cross sectional tibial stresses during locomotion. J. Biomech. 49 (3), 429–435.

Di Nardo, F., Fioretti, S., 2013. Statistical analysis of surface electromyographic signal for the assessment of rectus femoris modalities of activation during gait. J. Electromyogr. Kinesiol. 23 (1), 56–61.

Dittmar, J.M., Mitchell, P.D., Cessford, C., Inskip, S.A., Robb, J.E., 2021. Fancy shoes and painful feet: hallux valgus and fracture risk in medieval Cambridge. England. Int. J. Paleopathol. 35, 90–100.

Dixon, S., Nunns, M., House, C., Rice, H., Mostazir, M., Stiles, V., et al., 2019. Prospective study of biomechanical risk factors for second and third metatarsal stress fractures in military recruits. J. Sci. Med. Sport 22 (2), 135–139.

Dombek, M.F., Lamm, B.M., Saltrick, K., Mendicino, R.W., Catanzariti, A.R., 2003. Peroneal tendon tears: a retrospective review. J. Foot Ankle Surg. 42 (5), 250–258.

Donnelly, L., Donovan, L., Hart, J.M., Hertel, J., 2017. Eversion strength and surface electromyography measures with and without chronic ankle instability measured in 2 positions. Foot Ankle Int. 38 (7), 769–778.

Draghi, F., Gitto, S., Bortolotto, C., Draghi, A.G., Belometti, G.O., 2017. Imaging of plantar fascia disorders: findings on plain radiography, ultrasound and magnetic resonance imaging. Insights Imaging 8 (1), 69–78.

D'Amico, J.C., Schuster, R.O., 1979. Motion of the first ray: clarification through investigation. J. Am. Podiatry Assoc. 69 (1), 17–23.

Drapeau, M.S.M., Harmon, E.H., 2008. Metatarsal head torsion in apes, humans, and *A. afarensis*. Am. J. Phys. Anthropol. 135 (S46), 92 (Abstracts of AAPA poster and podium presentations.).

Drapeau, M.S.M., Harmon, E.H., 2013. Metatarsal torsion in monkeys, apes, humans and australopiths. J. Hum. Evol. 64 (1), 93–108.

Drew, B.T., Redmond, A.C., Smith, T.O., Penny, F., Conaghan, P.G., 2016. Which patellofemoral joint imaging features are associated with patellofemoral pain? Systematic review and meta-analysis. Osteoarthr. Cartil. 24 (2), 224–236.

Duda, G.N., Heller, M., Albinger, J., Schulz, O., Schneider, E., Claes, L., 1998. Influence of muscle forces on femoral strain distribution. J. Biomech. 31 (9), 841–846.

Dudley, R.I., Pamukoff, D.N., Lynn, S.K., Kersey, R.D., Noffal, G.J., 2017. A prospective comparison of lower extremity kinematics and kinetics between injured and non-injured collegiate cross country runners. Hum. Mov. Sci. 52, 197–202.

Durrant, B., Chockalingam, N., 2009. Functional hallux limitus: a review. J. Am. Podiatr. Med. Assoc. 99 (3), 236–243.

Durrant, B., Chockalingam, N., Hashmi, F., 2011. Posterior tibial tendon dysfunction: a review. J. Am. Podiatr. Med. Assoc. 101 (2), 176–186.

Durrant, B., Chockalingam, N., Morriss-Roberts, C., 2016. Assessment and diagnosis of posterior tibial tendon dysfunction. Do we share the same opinions and beliefs? J. Am. Podiatr. Med. Assoc. 106 (1), 27–36.

Duthon, V.B., Barea, C., Abrassart, S., Fasel, J.H., Fritschy, D., Ménétrey, J., 2006. Anatomy of the anterior cruciate ligament. Knee Surg. Sports Traumatol. Arthrosc. 14 (3), 204–213.

Edama, M., Kubo, M., Onishi, H., Takabayashi, T., Inai, T., Yokoyama, E., et al., 2015. The twisted structure of the human Achilles tendon. Scand. J. Med. Sci. Sports 25 (5), e457–e503.

Ellison, M.A., Kenny, M., Fulford, J., Javadi, A., Rice, H.M., 2020. Incorporating subject-specific geometry to compare metatarsal stress during running with different foot strike patterns. J. Biomech. 105, 109792. https://doi.org/10.1016/j.biomech.2020.109792.

Endo, M., Ashton-Miller, J.A., Alexander, N.B., 2002. Effects of age and gender on toe flexor muscle strength. J. Gerontol. Ser. A Biol. Med. Sci. 57 (6), M392–M397.

Erdemir, A., Hamel, A.J., Fauth, A.R., Piazza, S.J., Sharkey, N.A., 2004. Dynamic loading of the plantar aponeurosis in walking. Journal of Bone and Joint. J. Bone Joint Surg. 86-A (3), 546–552.

Fairclough, J., Hayashi, K., Toumi, H., Lyons, K., Bydder, G., Phillips, N., et al., 2007. Is iliotibial band syndrome really a friction syndrome? J. Sci. Med. Sport 10 (2), 74–76. discussion 77–78.

Farley, C.T., González, O., 1996. Leg stiffness and stride frequency in human running. J. Biomech. 29 (2), 181–186.

Farley, C.T., McMahon, T.A., 1992. Energetics of walking and running: insights from simulated reduced-gravity experiments. J. Appl. Physiol. 73 (6), 2709–2712.

Farr, J.N., Khosla, S., 2016. Determinants of bone strength and quality in diabetes mellitus in humans. Bone 82, 28–34.

Farris, D.J., Kelly, L.A., Cresswell, A.G., Lichtwark, G.A., 2019. The functional importance of human foot muscles for bipedal locomotion. Proc. Natl. Acad. Sci. U. S. A. 116 (5), 1645–1650.

Farris, D.J., Birch, J., Kelly, L., 2020. Foot stiffening during the push-off phase of human walking is linked to active muscle contraction, and not the windlass mechanism. J. R. Soc. Interface 17 (168), 20200208. https://doi.org/10.1098/rsif.2020.0208.

Fatima, G., Qamar, M.M., Hassan, J.U., Basharat, A., 2017. Extended sitting can cause hamstring tightness. Saudi J. Sports Med. 17 (2), 110–114.

Favre, P., Farine, M., Snedeker, J.G., Maquieira, G.J., Espinosa, N., 2010. Biomechanical consequences of first metatarsal osteotomy in treating hallux valgus. Clin. Biomech. 25 (7), 721–727.

Felson, D.T., Radin, E.L., 1994. What causes knee osteoarthrosis: are different compartments susceptible to different risk factors? J. Rheumatol. 21 (2), 181–183.

Ferber, R., Noehren, B., Hamill, J., Davis, I.S., 2010. Competitive female runners with a history of iliotibial band syndrome demonstrate atypical hip and knee kinematics. J. Orthop. Sports Phys. Ther. 40 (2), 52–58.

Ferber, R., Bolgla, L., Earl-Boehm, J.E., Emery, C., Hamstra-Wright, K., 2015. Strengthening of the hip and core versus knee muscles for the treatment of patellofemoral pain: a multicenter randomized controlled trial. J. Athl. Train. 50 (4), 366–377.

Fernando, M.E., Crowther, R.G., Lazzarini, P.A., Sangla, K.S., Buttner, P., Golledge, J., 2016a. Gait parameters of people with diabetes-related neuropathic plantar foot ulcers. Clin. Biomech. 37, 98–107.

Fernando, M.E., Crowther, R.G., Lazzarini, P.A., Sangla, K.S., Wearing, S., Buttner, P., et al., 2016b. Plantar pressures are higher in cases with diabetic foot ulcers compared to controls despite a longer stance phase duration. BMC Endocr. Disord. 16, 51. https://doi.org/10.1186/s12902-016-0131-9.

Ferrari, J., Malone-Lee, J., 2002. The shape of the metatarsal head as a cause of hallux abductovalgus. Foot Ankle Int. 23 (3), 236–242.

Ferrari, J., Malone-Lee, J., 2003. A radiographic study of the relationship between metatarsus adductus and hallux valgus. J. Foot Ankle Surg. 42 (1), 9–14.

Ferris, L., Sharkey, N.A., Smith, T.S., Matthews, D.K., 1995. Influence of extrinsic plantar flexors on forefoot loading during heel rise. Foot Ankle Int. 16 (8), 464–473.

Fields, K.B., Sykes, J.C., Walker, K.M., Jackson, J.C., 2010. Prevention of running injuries. Curr. Sports Med. Rep. 9 (3), 176–182.

Fiolkowski, P., Brunt, D., Bishop, M., Woo, R., Horodyski, M., 2003. Intrinsic pedal musculature support of the medial longitudinal arch: an electromyography study. J. Foot Ankle Surg. 42 (6), 327–333.

Flören, M., Davis, J., Peterson, M.G.E., Laskin, R.S., 2007. A mini-midvastus capsular approach with patellar displacement decreases the prevalence of patella baja. J. Arthroplasty 22 (6 Suppl), 51–57.

Fontanella, C.G., Matteoli, S., Carniel, E.L., Wilhjelm, J.E., Virga, A., Corvi, A., et al., 2012. Investigation on the load-displacement curves of a human healthy heel pad: in vivo compression data compared to numerical results. Med. Eng. Phys. 34 (9), 1253–1259.

Fontanella, C.G., Nalesso, F., Carniel, E.L., Natali, A.N., 2016. Biomechanical behavior of plantar fat pad in healthy and degenerative foot conditions. Med. Biol. Eng. Comput. 54 (4), 653–661.

Fortin, P.T., Myerson, M.S., 1995. Second metatarsophalangeal joint instability. Foot Ankle Int. 16 (5), 306–313.

Foss, K.D.B., Myer, G.D., Magnussen, R.A., Hewett, T.E., 2014. Diagnostic differences for anterior knee pain between sexes in adolescent basketball players. J. Athletic Enhancement 3 (1), 1814. https://doi.org/10.4172/2324-9080.1000139.

Franklin, S., Grey, M.J., Heneghan, N., Bowen, L., Li, F.-X., 2015. Barefoot vs common footwear: a systematic review of the kinematic, kinetic and muscle activity differences during walking. Gait Posture 42 (3), 230–239.

Franklyn, M., Oakes, B., 2015. Aetiology and mechanisms of injury in medial tibial stress syndrome: current and future developments. World J. Orthoped. 6 (8), 577–589.

Fredericson, M., Cookingham, C.L., Chaudhari, A.M., Dowdell, B.C., Oestreicher, N., Sahrmann, S.A., 2000. Hip abductor weakness in distance runners with iliotibial band syndrome. Clin. J. Sport Med. 10 (3), 169–175.

Fredericson, M., Jennings, F., Beaulieu, C., Matheson, G.O., 2006. Stress fractures in athletes. Top. Magn. Reson. Imaging 17 (5), 309–325.

Gaasbeek, R., Welsing, R., Barink, M., Verdonschot, N., van Kampen, A., 2007. The influence of open and closed high tibial osteotomy on dynamic patellar tracking: a biomechanical study. Knee Surg. Sports Traumatol. Arthrosc. 15 (8), 978–984.

Gajdosik, R.L., Sandler, M.M., Marr, H.L., 2003. Influence of knee positions and gender on the Ober test for length of the iliotibial band. Clin. Biomech. 18 (1), 77–79.

Gaston, M.S., Rutz, E., Dreher, T., Brunner, R., 2011. Transverse plane rotation of the foot and transverse hip and pelvic kinematics in diplegic cerebral palsy. Gait Posture 34 (2), 218–221.

Gatt, A., Mifsud, T., Chockalingam, N., 2014. Severity of pronation and classification of first metatarsophalangeal joint dorsiflexion increases the validity of the Hubscher Manoeuvre for the diagnosis of functional hallux limitus. Foot 24 (2), 62–65.

Gaudiani, M.A., Samuel, L.T., Mahmood, B., Sultan, A.A., Kamath, A.F., 2020. Subchondral insufficiency fractures of the femoral head: systematic review of diagnosis, treatment and outcomes. J. Hip Preserv. Surg. 7 (1), 85–94.

Geissler, J.R., Bajaj, D., Fritton, J.C., 2015. American Society of Biomechanics Journal of Biomechanics Award 2013: cortical bone tissue mechanical quality and biological mechanisms possibly underlying atypical fractures. J. Biomech. 48 (6), 883–894.

Georgiou, G., Dimitrakopoulou, A., Siapkara, A., Kazakos, K., Provelengios, S., Dounis, E., 2007. Simultaneous bilateral tibial tubercle avulsion fracture in an adolescent: a case report and review of the literature. Knee Surg. Sports Traumatol. Arthrosc. 15 (2), 147–149.

Georgoulis, A.D., Papadonikolakis, A., Papageorgiou, C.D., Mitsou, A., Stergiou, N., 2003. Three-dimensional tibiofemoral kinematics of the anterior cruciate ligament-deficient and reconstructed knee during walking. Am. J. Sports Med. 31 (1), 75–79.

Giddings, V.L., Beaupré, G.S., Whalen, R.T., Carter, D.R., 2000. Calcaneal loading during walking and running. Med. Sci. Sports Exerc. 32 (3), 627–634.

Gill, C.M., Bredella, M.A., DeSilva, J.M., 2015. Skeletal development of hallucal tarsometatarsal joint curvature and angulation in extant apes and modern humans. J. Hum. Evol. 88, 137–145.

Godinho, M.S.C., Thorpe, C.T., Greenwald, S.E., Screen, H.R.C., 2017. Elastin is localised to the interfascicular matrix of energy storing tendons and becomes increasingly disorganised with ageing. Sci. Rep. 7, 9713. https://doi.org/10.1038/s41598-017-09995-4.

Golanó, P., Vega, J., de Leeuw, P.A.J., Malagelada, F., Manzanares, M.C., Götzens, V., et al., 2010. Anatomy of the ankle ligaments: a pictorial essay. Knee Surg. Sports Traumatol. Arthrosc. 18 (5), 557–569.

Goldberg, E.J., Neptune, R.R., 2007. Compensatory strategies during normal walking in response to muscle weakness and increased hip joint stiffness. Gait Posture 25 (3), 360–367.

Gomes, R.B.O., Souza, T.R., Paes, B.D.C., Magalhães, F.A., Gontijo, B.A., Fonseca, S.T., et al., 2019. Foot pronation during walking is associated to the mechanical resistance of the midfoot joint complex. Gait Posture 70, 20–23.

Gottschalk, F.A., Solomon, L., Beighton, P.H., 1984. The prevalence of hallux valgus in South African males. S. Afr. Med. J. 65 (18), 725–726.

Gottschall, J.S., Kram, R., 2005. Energy cost and muscular activity required for leg swing during walking. J. Appl. Physiol. 99 (1), 23–30.

Gregg, J., Silberstein, M., Schneider, T., Marks, P., 2006. Sonographic and MRI evaluation of the plantar plate: a prospective study. Eur. Radiol. 16 (12), 2661–2669.

Gregg, J., Marks, P., Silberstein, M., Schneider, T., Kerr, J., 2007. Histologic anatomy of the lesser metatarsophalangeal joint plantar plate. Surg. Radiol. Anat. 29 (2), 141–147.

Gribble, P.A., Bleakley, C.M., Caulfield, B.M., Docherty, C.L., Fourchet, F., Fong, D.T.-P., et al., 2016a. 2016 consensus statement of the International Ankle Consortium: prevalence, impact and long-term consequences of lateral ankle sprains. Br. J. Sports Med. 50 (24), 1493–1495.

Gribble, P.A., Bleakley, C.M., Caulfield, B.M., Docherty, C.L., Fourchet, F., Fong, D.T.-P., et al., 2016b. Evidence review for the 2016 International Ankle Consortium consensus statement on the prevalence, impact and long-term consequences of lateral ankle sprains. Br. J. Sports Med. 50 (24), 1496–1505.

Griswold, B.G., Shaw, K.A., Houston, H., Bertrand, S., Cearley, D., 2020. Guided growth for the treatment of infantile Blount's disease: Is it a viable option? J. Orthop. 20, 41–45.

Gullstrand, L., Halvorsen, K., Tinmark, F., Eriksson, M., Nilsson, J., 2009. Measurements of vertical displacement in running, a methodological comparison. Gait Posture 30 (1), 71–75.

Hakim, A.J., Cherkas, L.F., Grahame, R., Spector, T.D., MacGregor, A.J., 2004. The genetic epidemiology of joint hypermobility: a population study of female twins. Arthritis Rheum. 50 (8), 2640–2644.

Hall, C., Nester, C.J., 2004. Sagittal plane compensations for artificially induced limitation of the first metatarsophalangeal joint: a preliminary study. J. Am. Podiatr. Med. Assoc. 94 (3), 269–274.

Halstead, J., Redmond, A.C., 2006. Weight-bearing passive dorsiflexion of the hallux in standing is not related to hallux dorsiflexion during walking. J. Orthop. Sports Phys. Ther. 36 (8), 550–556.

Halstead-Rastrick, J., 2013. Modification of midfoot bone stress with functional foot orthoses. PhD Thesis, University of Leeds School of Medicine, Leeds Institute of Rheumatology and Musculoskeletal Medicine, UK.

Hamill, J., Miller, R., Noehren, B., Davis, I., 2008. A prospective study of iliotibial band strain in runners. Clin. Biomech. 23 (8), 1018–1025.

Hamilton, S.W., Gibson, P.H., 2006. Simultaneous bilateral avulsion fractures of the tibial tuberosity in adolescence: a case report and review of over 50 years of literature. Knee 13 (5), 404–407.

Handsfield, G.G., Inouye, J.M., Slane, L.C., Thelen, D.G., Miller, G.W., Blemker, S.S., 2017. A 3D model of the Achilles tendon to determine the mechanisms underlying nonuniform tendon displacements. J. Biomech. 51, 17–25.

Hanley, B., Bissas, A., Merlino, S., Gruber, A.H., 2019. Most marathon runners at the 2017 IAAF World Championships were rearfoot strikers, and most did not change footstrike pattern. J. Biomech. 92, 54–60.

Hannan, M.T., Menz, H.B., Jordan, J.M., Cupples, L.A., Cheng, C.-H., Hsu, Y.-H., 2013. High heritability of hallux valgus and lesser toe deformities in adult men and women. Arthritis Care Res. 65 (9), 1515–1521.

Hansen, M.L., Otis, J.C., Kenneally, S.M., Deland, J.T., 2001. A closed-loop cadaveric foot and ankle loading model. J. Biomech. 34 (4), 551–555.

Harman, M.K., Markovich, G.D., Banks, S.A., Hodge, W.A., 1998. Wear patterns on tibial plateaus from varus and valgus osteoarthritic knees. Clin. Orthop. Relat. Res. 352 (July), 149–158.

Hart, N.H., Nimphius, S., Rantalainen, T., Ireland, A., Siafarikas, A., Newton, R.U., 2017. Mechanical basis of bone strength: influence of bone material, bone structure and muscle action. J. Musculoskelet. Nueronal Interact. 17 (3), 114–139.

Hayafune, N., Hayafune, Y., Jacob, H.A.C., 1999. Pressure and force distribution characteristics under the normal foot during the push-off phase in gait. Foot 9 (2), 88–92.

Hedrick, E.A., Stanhope, S.J., Takahashi, K.Z., 2019. The foot and ankle structures reveal emergent properties analogous to passive springs during human walking. PLoS One 14 (6), e0218047. https://doi.org/10.1371/journal.pone.0218047.

Heller, M.O., Bergmann, G., Deuretzbacher, G., Claes, L., Haas, N.P., Duda, G.N., 2001. Influence of femoral anteversion on proximal femoral loading: measurement and simulation in four patients. Clin. Biomech. 16 (8), 644–649.

Herb, C.C., Chinn, L., Dicharry, J., McKeon, P.O., Hart, J.M., Hertel, J., 2014. Shank-rearfoot joint coupling with chronic ankle instability. J. Appl. Biomech. 30 (3), 366–372.

Herod, T.W., Chambers, N.C., Veres, S.P., 2016. Collagen fibrils in functionally distinct tendons have differing structural responses to tendon rupture and fatigue loading. Acta Biomater. 42, 296–307.

Herrington, L., Rivett, N., Munro, S., 2006. The relationship between patella position and length of the iliotibial band as assessed using Ober's test. Man. Ther. 11 (3), 182–186.

Hertel, J., 2002. Functional anatomy, pathomechanics, and pathophysiology of lateral ankle instability. J. Athl. Train. 37 (4), 364–375.

Hertel, J., Corbett, R.O., 2019. An updated model of chronic ankle instability. J. Athl. Train. 54 (6), 572–588.

Hicks, J.H., 1954. The mechanics of the foot. II. The plantar aponeurosis and the arch. J. Anat. 88 (1), 25–30.

Hirji, Z., Kaicker, J., Ariyanayagam, T.A., Howey, J., Choudur, H.N., 2020. Magnetic resonance imaging: marrow edema patterns in chronic foot pain. Indian J. Musculosk. Radiol. 2 (1), 26–31.

Hof, A.L., Elzinga, H., Grimmius, W., Halbertsma, J.P.K., 2002. Speed dependence of averaged EMG profiles in walking. Gait Posture 16 (1), 78–86.

Hoffmann, A., Mamisch, N., Buck, F.M., Espinosa, N., Pfirrmann, C.W.A., Zanetti, M., 2011. Oedema and fatty degeneration of the soleus and gastrocnemius muscles on MR images in patients with Achilles tendon abnormalities. Eur. Radiol. 21 (9), 1996–2003.

Hofmann, C.L., Okita, N., Sharkey, N.A., 2013. Experimental evidence supporting isometric functioning of the extrinsic toe flexors during gait. Clin. Biomech. 28 (6), 686–691.

Hogan, K.K., Powden, C.J., Hoch, M.C., 2016. The influence of foot posture on dorsiflexion range of motion and postural control in those with chronic ankle instability. Clin. Biomech. 38, 63–67.

Hoitz, F., Vienneau, J., Nigg, B.M., 2020. Influence of running shoes on muscle activity. PLoS One 15 (10), e0239852. https://doi.org/10.1371/journal.pone.0239852.

Hollander, K., de Villiers, J.E., Sehner, S., Wegscheider, K., Braumann, K.-M., Venter, R., et al., 2017. Growing-up (habitually) barefoot influences the development of foot and arch morphology in children and adolescents. Sci. Rep. 7, 8079. https://doi.org/10.1038/s41598-017-07868-4.

Holowka, N.B., Wynands, B., Drechsel, T.J., Yegian, A.K., Tobolsky, V.A., Okutoyi, P., et al., 2019. Foot callus thickness does not trade off protection for tactile sensitivity during walking. Nature 571 (7764), 261–264.

Holowka, N.B., O'Neill, M.C., Thompson, N.E., Demes, B., 2017. Chimpanzee and human midfoot motion during bipedal walking and the evolution of the longitudinal arch of the foot. J. Hum. Evol. 104, 23–31.

Honeine, J.-L., Schieppati, M., Gagey, O., Do, M.-C., 2013. The functional role of the triceps surae muscle during human locomotion. PLoS One 8 (1), e52943. https://doi.org/10.1371/journal.pone.0052943.

Honert, E.C., Bastas, G., Zelik, K.E., 2020. Effects of toe length, foot arch length and toe joint axis on walking biomechanics. Hum. Mov. Sci. 70, 102594. https://doi.org/10.1016/j.humov.2020.102594.

Hormozi, J., Lee, S., Hong, D.K., 2011. Minimal invasive percutaneous bipolar radiofrequency for plantar fasciotomy: a retrospective study. J. Foot Ankle Surg. 50 (3), 283–286.

Houston, M.N., Hoch, J.M., Hoch, M.C., 2018. College athletes with ankle sprain history exhibit greater fear-avoidance beliefs. J. Sport Rehabil. 27 (5), 419–423.

Houx, L., Lempereur, M., Rémy-Néris, O., Brochard, S., 2013. Threshold of equinus which alters biomechanical gait parameters in children. Gait Posture 38 (4), 582–589.

Hovaguimian, A., Gibbons, C.H., 2011. Diagnosis and treatment of pain in small-fiber neuropathy. Curr. Pain Headache Rep. 15 (3), 193–200.

Hsu, C.-C., Tsai, W.-C., Hsiao, T.-Y., Tseng, F.-Y., Shau, Y.-W., Wang, C.-L., et al., 2009. Diabetic effects on microchambers and macrochambers tissue properties in human heel pads. Clin. Biomech. 24 (8), 682–686.

Huang, J., Qin, K., Tang, C., Zhu, Y., Klein, C.S., Zhang, Z., et al., 2018. Assessment of passive stiffness of medial and lateral heads of gastrocnemius muscle, Achilles tendon, and plantar fascia at different ankle and knee positions using the MyotonPRO. Med. Sci. Monit. 24, 7570–7576.

Huijing, P.A., Baan, G.C., 2008. Myofascial force transmission via extramuscular pathways occurs between antagonistic muscles. Cells Tissues Organs 188 (4), 400–414.

Huijing, P.A., Mass, H., Baan, G.C., 2003. Compartmental fasciotomy and isolating a muscle from neighboring muscles interfere with myofascial force transmission within the rat anterior crural compartment. J. Morphol. 256 (3), 306–321.

Hunt, A.E., Smith, R.M., Torode, M., Keenan, A.-M., 2001. Inter-segment foot motion and ground reaction forces over the stance phase of walking. Clin. Biomech. 16 (7), 592–600.

Imhauser, C.W., Siegler, S., Abidi, N.A., Frankel, D.Z., 2004. The effect of posterior tibialis tendon dysfunction on the plantar pressure characteristics and the kinematics of the arch and the hindfoot. Clin. Biomech. 19 (2), 161–169.

Ingber, D.E., 2006. Cellular mechanotransduction: putting all the pieces together again. FASEB J. 20 (7), 811–827.

Ireland, M.L., Willson, J.D., Ballantyne, B.T., Davis, I.M., 2003. Hip strength in females with and without patellofemoral pain. J. Orthop. Sports Phys. Ther. 33 (11), 671–676.

Irving, D.B., Cook, J.L., Menz, H.B., 2006. Factors associated with chronic plantar heel pain: a systematic review. J. Sci. Med. Sport 9 (1–2), 11–22. discussion 23–24.

Ishikawa, M., Pakaslahti, J., Komi, P.V., 2007. Medial gastrocnemius muscle behavior during human running and walking. Gait Posture 25 (3), 380–384.

Jack, E.A., 1953. Naviculo-cuneiform fusion in the treatment of flat foot. J. Bone Joint Surg. 35-B (1), 75–82.

Jackson, R.W., Kunkel, S.S., Taylor, G.J., 1991. Lateral retinacular release for patellofemoral pain in the older patient. Arthrosc.: J. Arthrosc. Rel. Surg. 7 (3), 283–286.

Jacob, H.A.C., 2001. Forces acting in the forefoot during normal gait—an estimate. Clin. Biomech. 16 (9), 783–792.

Jadhav, S.P., More, S.R., Riascos, R.F., Lemos, D.F., Swischuk, L.E., 2014. Comprehensive review of the anatomy, function, and imaging of the popliteus and associated pathologic conditions. RadioGraphics 34 (2), 496–513.

Jastifer, J.R., Gustafson, P.A., 2014. The subtalar joint: biomechanics and functional representations in the literature. Foot 24 (4), 203–209.

Jerosch, J., Prymka, M., 1996. Proprioception and joint stability. Knee Surg. Sports Traumatol. Arthrosc. 4 (3), 171–179.

Jose, J., Pasquotti, G., Smith, M.K., Gupta, A., Lesniak, B.P., Kaplan, L.D., 2015. Subchondral insufficiency fractures of the knee: review of imaging findings. Acta Radiol. 56 (6), 714–719.

Kalkman, B.M., Holmes, G., Bar-On, L., Maganaris, C.N., Barton, G.J., Bass, A., et al., 2019. Resistance training combined with stretching increases tendon stiffness and is more effective than stretching alone in children with cerebral palsy: a randomized controlled trial. Front. Pediatr. 7, 333. https://doi.org/10.3389/fped.2019.00333.

Kaneuchi, Y., Otoshi, K., Hakozaki, M., Sekiguchi, M., Watanabe, K., Igari, T., et al., 2018. Bony maturity of the tibial tuberosity with regard to age and sex and its relationship to pathogenesis of Osgood-Schlatter disease: an ultrasonographic study. Orthop. J. Sports Med. 6 (1). https://doi.org/10.1177/2325967117749184.

Karim, L., Bouxsein, M.L., 2016. Effect of type 2 diabetes-related non-enzymatic glycation on bone biomechanical properties. Bone 82, 21–27.

Karlsson, J., Wiger, P., 2002. Longitudinal split of the peroneus brevis tendon and lateral ankle instability: treatment of concomitant lesions. J. Athl. Train. 37 (4), 463–466.

Karzis, K., Kalogeris, M., Mandalidis, D., Geladas, N., Karteroliotis, K., Athanasopoulos, S., 2017. The effect of foot overpronation on Achilles tendon blood supply in healthy male subjects. Scand. J. Med. Sci. Sports 27 (10), 1114–1121.

Kato, T., Watanabe, S., 1981. The etiology of hallux valgus in Japan. Clin. Orthop. Relat. Res. 157 (June), 78–81.

Kelly, L.A., Kuitunen, S., Racinais, S., Cresswell, A.G., 2012. Recruitment of the plantar intrinsic foot muscles with increasing postural demand. Clin. Biomech. 27 (1), 46–51.

Kelso, S.F., Richie Jr., D.H., Cohen, I.R., Weed, J.H., Root, M., 1982. Direction and range of motion of the first ray. J. Am. Podiatry Assoc. 72 (12), 600–605.

Ker, R.F., Bennett, M.B., Bibby, S.R., Kester, R.C., Alexander RMcN., 1987. The spring in the arch of the human foot. Nature 325 (6100), 147–149.

Kernozek, T.W., Knaus, A., Rademaker, T., Almonroeder, T.G., 2018. The effects of habitual foot strike patterns on Achilles tendon loading in female runners. Gait Posture 66, 283–287.

Khan, A.M., McLoughlin, E., Giannakas, K., Hutchinson, C., Andrew, J.G., 2004. Hip osteoarthritis: where is the pain? Ann. R. Coll. Surg. Engl. 86 (2), 119–121.

Kier, R., Abrahamian, H., Caminear, D., Eterno, R., Feldman, A., Abrahamsen, T., et al., 2010. MR arthrography of the second and third metatarsophalangeal joints for the detection of tears of the plantar plate and joint capsule. Am. J. Roentgenol. 194 (4), 1079–1081.

Kilmartin, T.E., Wallace, W.A., 1991. First metatarsal head shape in juvenile hallux abducto valgus. J. Foot Surg. 30 (5), 506–508.

Kilmartin, T.E., Wallace, W.A., 1992. The significance of pes planus in juvenile hallux valgus. Foot Ankle 13 (2), 53–56.

Kilmartin, T.E., Barrington, R.L., Wallace, W.A., 1991. Metatarsus primus varus: a statistical study. J. Bone Joint Surg. 73-B (6), 937–940.

Kilmartin, T.E., Barrington, R.L., Wallace, W.A., 1994. A controlled prospective trial of a foot orthosis for juvenile hallux valgus. J. Bone Joint Surg. 76-B (2), 210–214.

Kim, Y., Kim, J.S., Young, K.W., Naraghi, R., Cho, H.K., Lee, S.Y., 2015. A new measure of tibial sesamoid position in hallux valgus in relation to the coronal rotation of the first metatarsal in CT scans. Foot Ankle Int. 36 (8), 944–952.

Kimura, T., 2018. Multidisciplinary approach for bone metastasis: a review. Cancer 10 (6), 156. https://doi.org/10.3390/cancers10060156.

Kirby, K.A., 1989. Rotational equilibrium across the subtalar joint axis. J. Am. Podiatr. Med. Assoc. 79 (1), 1–14.

Kirby, K.A., 2000. Biomechanics of the normal and abnormal foot. J. Am. Podiatr. Med. Assoc. 90 (1), 30–34.

Kirby, K.A., 2001. Subtalar joint axis location and rotational equilibrium theory of foot function. J. Am. Podiatr. Med. Assoc. 91 (9), 465–487.

Kirkman, M.S., Briscoe, V.J., Clark, N., Florez, H., Hass, L.B., Halter, J.B., et al., 2012. Diabetes in older adults: a consensus report. J. Am. Geriatr. Soc. 60 (12), 2342–2356.

Klaesner, J.W., Hastings, M.K., Zou, D., Lewis, C., Mueller, M.J., 2002. Plantar tissue stiffness in patients with diabetes mellitus and peripheral neuropathy. Arch. Phys. Med. Rehabil. 83 (12), 1796–1801.

Koenig, S.J., Toth, A.P., Bosco, J.A., 2008. Stress fractures and stress reactions of the diaphyseal femur in collegiate athletes: an analysis of 25 cases. Am. J. Orthop. 37 (9), 476–480.

Kokubo, T., Hashimoto, T., Nagura, T., Nakamura, T., Suda, Y., Matsumoto, H., et al., 2012. Effect of the posterior tibial and peroneal longus on the mechanical properties of the foot arch. Foot Ankle Int. 33 (4), 320–325.

Kuhman, D.J., Paquette, M.R., Peel, S.A., Melcher, D.A., 2016. Comparison of ankle kinematics and ground reaction forces between prospectively injured and uninjured collegiate cross country runners. Hum. Mov. Sci. 47, 9–15.

Kulmala, J.-P., Avela, J., Pasanen, K., Parkkari, J., 2013. Forefoot strikers exhibit lower running-induced knee loading than rearfoot strikers. Med. Sci. Sports Exerc. 45 (12), 2306–2313.

Kwak, Y., Kim, J., Lee, K.M., Koo, S., 2020. Increase of stiffness in plantar fat tissue in diabetic patients. J. Biomech. 107, 109857. https://doi.org/10.1016/j.biomech.2020.109857.

Kwan, R.L.-C., Zheng, Y.-P., Cheing, G.L.-Y., 2010. The effect of aging on the biomechanical properties of plantar soft tissues. Clin. Biomech. 25 (6), 601–605.

Lakin, R.C., DeGnore, L.T., Pienkowski, D., 2001. Contact mechanics of normal tarsometatarsal joints. J. Bone Joint Surg. 83-A (4), 520–528.

Lamonaca, F., Vasile, M., Nastro, A., 2014. Hallux valgus: measurements and characterization. Measurement 57, 94–101.

Landvater, S.J., Renström, P.A., 1992. Complete Achilles tendon ruptures. Clin. Sports Med. 11 (4), 741–758.

Lankhorst, N.E., Bierma-Zeinstra, S.M.A., van Middelkoop, M., 2012. Risk factors for patellofemoral pain syndrome: a systematic review. J. Orthop. Sports Phys. Ther. 42 (2), 81–94.

Lankhorst, N.E., Bierma-Zeinstra, S.M.A., van Middelkoop, M., 2013. Factors associated with patellofemoral pain syndrome: a systematic review. Br. J. Sports Med. 47 (4), 193–206.

Lavine, R., 2010. Iliotibial band friction syndrome. Curr. Rev. Musculosk. Med. 3 (1–4), 18–22.

Lawrence, R.C., Felson, D.T., Helmick, C.G., Arnold, L.M., Choi, H., Deyo, R.A., et al., 2008. Estimates of the prevalence of arthritis and other rheumatic conditions in the United States. Part II. Arthritis Rheum. 58 (1), 26–35.

Leardini, A., O'Connor, J.J., 2002. A model for lever-arm length calculation of the flexor and extensor muscles at the ankle. Gait Posture 15 (3), 220–229.

Leardini, A., Stagni, R., O'Connor, J.J., 2001. Mobility of the subtalar joint in the intact ankle complex. J. Biomech. 34 (6), 805–809.

Lee, C.R., Farley, C.T., 1998. Determinants of the center of mass trajectory in human walking and running. J. Exp. Biol. 201 (21), 2935–2944.

Lee, S.S.M., Piazza, S.J., 2008. Inversion-eversion moment arms of gastrocnemius and tibialis anterior measured in vivo. J. Biomech. 41 (16), 3366–3370.

Lee, S.S.M., Piazza, S.J., 2009. Built for speed; Musculoskeletal structure and sprinting ability. J. Exp. Biol. 212 (22), 3700–3707.

Lee, T.Q., Yang, B.Y., Sandusky, M.D., McMahon, P.J., 2001. The effects of tibial rotation on the patellofemoral joint: assessment of the changes in in situ strain in the peripatellar retinaculum and the patellofemoral contact pressures and areas. J. Rehabil. Res. Dev. 38 (5), 463–469.

Lee, P.Y.F., Nixion, A., Chandratreya, A., Murray, J.M., 2017. Synovial plica syndrome of the knee: a commonly overlooked cause of anterior knee pain. Surg. J. 3 (1), e9–e16.

Lemont, H., Ammirati, K.M., Usen, N., 2003. Plantar fasciitis: a degenerative process (fasciosis) without inflammation. J. Am. Podiatr. Med. Assoc. 93 (3), 234–237.

Lenhart, R.L., Brandon, S.C.E., Smith, C.R., Novacheck, T.F., Schwartz, M.H., Thelen, D.G., 2017. Influence of patellar position on the knee extensor mechanism in normal and crouched walking. J. Biomech. 51, 1–7.

Lersch, C., Grötsch, A., Segesser, B., Koebke, J., Brüggemann, G.-P., Potthast, W., 2012. Influence of calcaneus angle and muscle forces on strain distribution in the human Achilles tendon. Clin. Biomech. 27 (9), 955–961.

Levi, N., Garde, L., Kofoed, H., 1997. Avulsion fracture of the calcaneus: report of a case using a new tension band technique. J. Orthop. Trauma 11 (1), 61–62.

Levy, A.S., Berkowitz, R., Franklin, P., Corbett, M., Whitelaw, G.P., 1992. Magnetic resonance imaging evaluation of calcaneal fat pads in patients with os calcis fractures. Foot Ankle 13 (2), 57–62.

Lewallen, D.G., Riegger, C.L., Myers, E.R., Hayes, W.C., 1990. Effects of retinacular release and tibial tubercle elevation in patellofemoral degenerative joint disease. J. Orthop. Res. 8 (6), 856–862.

Lewis, C.L., Sahrmann, S.A., Moran, D.W., 2007. Anterior hip joint force increases with hip extension, decreased gluteal force, or decreased iliopsoas force. J. Biomech. 40 (16), 3725–3731.

Lewis, C.L., Sahrmann, S.A., Moran, D.W., 2009. Effect of position and alteration in synergist muscle force contribution on hip forces when performing hip strengthening exercises. Clin. Biomech. 24 (1), 35–42.

Lewis, C.L., Sahrmann, S.A., Moran, D.W., 2010. Effect of hip angle on anterior hip joint force during gait. Gait Posture 32 (4), 603–607.

Li, L., Gollhofer, A., Lohrer, H., Dorn-Lange, N., Bonsignore, G., Gehring, D., 2019. Function of ankle ligaments for subtalar and talocrural joint stability during an inversion movement—an in vitro study. J. Foot Ankle Res. 12, 16. https://doi.org/10.1186/s13047-019-0330-5.

Li, G., DeFrate, L.E., Zayontz, S., Park, S.E., Gill, T.J., 2004. The effect of tibiofemoral joint kinematics on patellofemoral contact pressures under simulated muscle loads. J. Orthop. Res. 22 (4), 801–806.

Lichtwark, G.A., Wilson, A.M., 2005. In vivo mechanical properties of the human Achilles tendon during one-legged hopping. J. Exp. Biol. 208 (24), 4715–4725.

Ling, Z.X., Kumar, V.P., 2008. The myofascial compartments of the foot: a cadaver study. J. Bone Joint Surg. 90-B (8), 1114–1118.
Liu, W., Nigg, B.M., 2000. A mechanical model to determine the influence of masses and mass distribution on the impact force during running. J. Biomech. 33 (2), 219–224.
Longworth, R., Short, L., Horwood, A., 2019. Conservative treatment of Freiberg's infraction using foot orthoses: a tale of two prescriptions presented as a case study to open debate. Foot 41, 59–62.
Louw, M., Deary, C., 2014. The biomechanical variables involved in the aetiology of iliotibial band syndrome in distance runners—a systematic review of the literature. Phys. Therapy Sport 15 (1), 64–75.
Lun, V., Meeuwisse, W.H., Stergiou, P., Stefanyshyn, D., 2004. Relation between running injury and static lower limb alignment in recreational runners. Br. J. Sports Med. 38 (5), 576–580.
Lundgren, P., Nester, C., Liu, A., Arndt, A., Jones, R., Stacoff, A., et al., 2008. Invasive in vivo measurement of rear-, mid- and forefoot motion during walking. Gait Posture 28 (1), 93–100.
Lyght, M., Nockerts, M., Kernozek, T.W., Ragan, R., 2016. Effects of foot strike and step frequency on Achilles tendon stress during running. J. Appl. Biomech. 32 (4), 365–372.
Lyman, J., Weinhold, P.S., Almekinders, L.C., 2004. Strain behavior of the distal Achilles tendon: implications for insertional Achilles tendinopathy. Am. J. Sports Med. 32 (2), 457–461.
Maas, N.M.G., van der Grinten, M., Bramer, W.M., Kleinrensink, G.-J., 2016. Metatarsophalangeal joint stability: a systematic review on the plantar plate of the lesser toes. J. Foot Ankle Res. 9, 32. https://doi.org/10.1186/s13047-016-0165-2.
Maeshige, N., Uemura, M., Hirasawa, Y., Yoshikawa, Y., Moriguchi, M., Kawabe, N., et al., 2021. Immediate effects of weight-bearing calf stretching on ankle dorsiflexion range of motion and plantar pressure during gait in patients with diabetes mellitus. Int. J. Lower Extremity Wounds OnlineFirst. https://doi.org/10.1177/15347346211031318.
Mafart, B., 2007. Hallux valgus in a historical French population: paleopathological study of 605 first metatarsal bones. Joint Bone Spine 74 (2), 166–170.
Magalhães, F.A., Fonseca, S.T., Araújo, V.L., Trede, R.G., Oliveira, L.M., Castor, C.G.M.E., et al., 2021. Midfoot passive stiffness affects foot and ankle kinematics and kinetics during the propulsive phase of walking. J. Biomech. 119, 110328.
Magnan, B., Bondi, M., Pierantoni, S., Samaila, E., 2014. The pathogenesis of Achilles tendinopathy: a systematic review. Foot Ankle Surg. 20 (3), 154–159.
Maharaj, J.N., Cresswell, A.G., Lichtwark, G.A., 2016. The mechanical function of the tibialis posterior muscle and its tendon during locomotion. J. Biomech. 49 (14), 3238–3243.
Malfait, F., Hakim, A.J., De Paepe, A., Grahame, R., 2006. The genetic basis of the joint hypermobility syndromes. Rheumatology 45 (5), 502–507.
Mandell, J.C., Khurana, B., Smith, S.E., 2017a. Stress fractures of the foot and ankle. Part 1. Biomechanics of bone and principles of imaging and treatment. Skelet. Radiol. 46 (8), 1021–1029.
Mandell, J.C., Khurana, B., Smith, S.E., 2017b. Stress fractures of the foot and ankle. Part 2. Site-specific etiology, imaging, and treatment, and differential diagnosis. Skelet. Radiol. 46 (9), 1165–1186.
Mansour, R., Jibri, Z., Kamath, S., Mukherjee, K., Ostlere, S., 2011. Persistent ankle pain following a sprain: a review of imaging. Emerg. Radiol. 18 (3), 211–225.
Mao, D.W., Chandrakumara, D., Zheng, Q., Kam, C., King, K.K., C., 2019. Endoscopic plantar fasciotomy for plantar fasciitis: a systematic review and network meta-analysis of the English literature. Foot 41, 63–73.
Marshall, R.A., Mandell, J.C., Weaver, M.J., Ferrone, M., Sodickson, A., Khurana, B., 2018. Imaging features and management of stress, atypical, and pathologic fractures. RadioGraphics 38 (7), 2173–2192.
Martinelli, A.R., Mantovani, A.M., Nozabieli, A.J.L., Ferreira, D.M.A., Barela, J.A., de Camargo, M.R., et al., 2013. Muscle strength and ankle mobility for the gait parameters in diabetic neuropathies. Foot 23 (1), 17–21.
Martínez-Nova, A., Sánchez-Rodríguez, R., Pérez-Soriano, P., Llana-Belloch, S., Leal-Muro, A., Pedrera-Zamorano, J.D., 2010. Plantar pressures determinants in mild Hallux Valgus. Gait Posture 32 (3), 425–427.
Matheson, G.O., Clement, D.B., McKenzie, D.C., Taunton, J.E., Lloyd-Smith, D.R., MacIntyre, J.G., 1987. Stress fractures in athletes: a study of 320 cases. Am. J. Sports Med. 15 (1), 46–58.
Matjačić, Z., Olenšek, A., Bajd, T., 2006. Biomechanical characterization and clinical implications of artificially induced toe-walking: differences between pure soleus, pure gastrocnemius and combination of soleus and gastrocnemius contractures. J. Biomech. 39 (2), 255–266.
Mayer, S.W., Joyner, P.W., Almekinders, L.C., Parekh, S.G., 2014. Stress fractures of the foot and ankle in athletes. Sports Health: Multidiscipl. Approach 6 (6), 481–491.
Mays, S.A., 2005. Paleopathological study of hallux valgus. Am. J. Phys. Anthropol. 126 (2), 139–149.
McCluney, J.G., Tinley, P., 2006. Radiographic measurements of patients with juvenile hallux valgus compared with age-matched controls: a cohort investigation. J. Foot Ankle Surg. 45 (3), 161–167.
McDermott, I.D., Masouros, S.D., Amis, A.A., 2008. Biomechanics of the menisci of the knee. Curr. Orthop. 22 (3), 193–201.
McMahon, T.A., Cheng, G.C., 1990. The mechanics of running: How does stiffness couple with speed? J. Biomech. 23 (Suppl. 1), 65–78.
McMaster, M.J., 1978. The pathogenesis of hallux rigidus. J. Bone Joint Surg. 60-B (1), 82–87.
Mendis, M.D., Wilson, S.J., Hayes, D.A., Watts, M.C., Hides, J.A., 2014. Hip flexor muscle size, strength and recruitment pattern in patients with acetabular labral tears compared to healthy controls. Man. Ther. 19 (5), 405–410.
Menz, H.B., Zammit, G.V., Munteanu, S.E., Scott, G., 2006. Plantarflexion strength of the toes: age and gender differences and evaluation of a clinical screening test. Foot Ankle Int. 27 (12), 1103–1108.

Menz, H.B., Munteanu, S.E., Landorf, K.B., Zammit, G.V., Cicuttini, F.M., 2007. Radiographic classification of osteoarthritis in commonly affected joints of the foot. Osteoarthr. Cartil. 15 (11), 1333–1338.

Mercer, J.A., Vance, J., Hreljac, A., Hamill, J., 2002. Relationship between shock attenuation and stride length during running at different velocities. Eur. J. Appl. Physiol. 87 (4–5), 403–408.

Merican, A.M., Kondo, E., Amis, A.A., 2009. The effect on patellofemoral joint stability of selective cutting of lateral retinacular and capsular structures. J. Biomech. 42 (3), 291–296.

Messier, S.P., Edwards, D.G., Martin, D.F., Lowery, R.B., Cannon, D.W., James, M.K., et al., 1995. Etiology of iliotibial band friction syndrome in distance runners. Med. Sci. Sports Exerc. 27 (7), 951–960.

Messier, S.P., Martin, D.F., Mihalko, S.L., Ip, E., DeVita, P., Cannon, D.W., et al., 2018. A 2-year prospective cohort study of overuse running injuries: The Runners and Injury Longitudinal Study (TRAILS). Am. J. Sports Med. 46 (9), 2211–2221.

Michael, S., Banerjee, A., 1993. Apparent tendo achilles rupture in the elderly: Is routine radiography necessary? Arch. Emerg. Med. 10 (4), 336–338.

Mickelborough, J., van der Linden, M.L., Tallis, R.C., Ennos, A.R., 2004. Muscle activity during gait initiation in normal elderly people. Gait Posture 19 (1), 50–57.

Milgrom, C., Radeva-Petrova, D.R., Finestone, A., Nyska, M., Mendelson, S., Benjuya, N., et al., 2007. The effect of muscle fatigue on in vivo tibial strains. J. Biomech. 40 (4), 845–850.

Milgrom, C., Burr, D.B., Finestone, A.S., Voloshin, A., 2015. Understanding the etiology of the posteromedial tibial stress fracture. Bone 78, 11–14.

Miller, R.H., Lowry, J.L., Meardon, S.A., Gillette, J.C., 2007. Lower extremity mechanics of iliotibial band syndrome during an exhaustive run. Gait Posture 26 (3), 407–413.

Miller, J.R., Dunn, K.W., Ciliberti Jr., L.J., Patel, R.D., Swanson, B.A., 2016. Association of vitamin D with stress fractures: a retrospective cohort study. J. Foot Ankle Surg. 55 (1), 117–120.

Milner, C.E., Ferber, R., Pollard, C.D., Hamill, J., Davis, I.S., 2006. Biomechanical factors associated with tibial stress fracture in female runners. Med. Sci. Sports Exerc. 38 (2), 323–328.

Minisola, S., Peacock, M., Fukumoto, S., Cipriani, C., Pepe, J., Tella, S.H., et al., 2017. Tumour-induced osteomalacia. Nat. Rev. Dis. Primers 3, 17044. https://doi.org/10.1038/nrdp.2017.44.

Minoves, M., Ponce, A., Balius, R., Til, L., 2011. Stress fracture of the first metatarsal in a fencer: typical appearance on bone scan and pinhole imaging. Clin. Nucl. Med. 36 (10), e150–e152.

Mizel, M.S., 1993. The role of the plantar first metatarsal first cuneiform ligament in weightbearing on the first metatarsal. Foot Ankle 14 (2), 82–84.

Moen, M.H., Tol, J.L., Weir, A., Steunebrink, M., De Winter, T.C., 2009. Medial tibial stress syndrome: a critical review. Sports Med. 39 (7), 523–546.

Mølgaard, C., Lundbye-Christensen, S., Simonsen, O., 2010. High prevalence of foot problems in the Danish population: a survey of causes and associations. Foot 20 (1), 7–11.

Molligan, J., Schon, L., Zhang, Z., 2013. A stereologic study of the plantar fat pad in young and aged rats. J. Anat. 223 (5), 537–545.

Monteagudo, M., Maceira, E., Garcia-Virto, V., Canosa, R., 2013. Chronic plantar fasciitis: plantar fasciotomy versus gastrocnemius recession. Int. Orthop. 37 (9), 1845–1850.

Monteagudo, M., de Albornoz, P.M., Gutierrez, B., Tabuenca, J., Álvarez, I., 2018. Plantar fasciopathy: a current concepts review. EFORT Open Rev. 3 (8), 485–493.

Moore, I.S., 2016. Is there an economical running technique? A review of modifiable biomechanical factors affecting running economy. Sports Med. 46 (6), 793–807.

Moorman 3rd, C.T., LaPrade, R.F., 2005. Anatomy and biomechanics of the posterolateral corner of the knee. J. Knee Surg. 18 (2), 137–145.

Morrison, K.E., Hudson, D.J., Davis, I.S., Richards, J.G., Royer, T.D., Dierks, T.A., et al., 2010. Plantar pressure during running in subjects with chronic ankle instability. Foot Ankle Int. 31 (11), 994–1000.

Mortier, J.-P., Bernard, J.-L., Maestro, M., 2012. Axial rotation of the first metatarsal head in a normal population and hallux valgus patients. Orthopaed. Traumatol. Surg. Res. 98 (6), 677–683.

Moschella, D., Blasi, A., Leardini, A., Ensini, A., Catani, F., 2006. Wear patterns on tibial plateau from varus osteoarthritic knees. Clin. Biomech. 21 (2), 152–158.

Mueller, M.J., Maluf, K.S., 2002. Tissue adaptation to physical stress: a proposed "Physical Stress Theory" to guide physical therapist practice, education, and research. Phys. Ther. 82 (4), 383–403.

Müller, R., Grimmer, S., Blickhan, R., 2010. Running on uneven ground: leg adjustments by muscle pre-activation control. Hum. Mov. Sci. 29 (2), 299–310.

Munteanu, S.E., Menz, H.B., Wark, J.D., Christie, J.J., Scurrah, K.J., Bui, M., et al., 2017. Hallux valgus, by nature or nurture? A twin study. Arthritis Care Res. 69 (9), 1421–1428.

Munuera, P.V., Polo, J., Rebollo, J., 2008. Length of the first metatarsal and hallux in hallux valgus in the initial stage. Int. Orthop. 32 (4), 489–495.

Murley, G.S., Buldt, A.K., Trump, P.J., Wickham, J.B., 2009. Tibialis posterior EMG activity during barefoot walking in people with neutral foot posture. J. Electromyogr. Kinesiol. 19 (2), e69–e77.

Murley, G.S., Menz, H.B., Landorf, K.B., 2014. Electromyographic patterns of tibialis posterior and related muscles when walking at different speeds. Gait Posture 39 (4), 1080–1085.

Naemi, R., Behforootan, S., Chatzistergos, P., Chockalingam, N., 2016a. Viscoelasticity in foot-ground interaction. In: El-Amin, M.F. (Ed.), Viscoelastic and Viscoplastic Materials. IntechOpen, pp. 217–243 (Chapter 10).

Naemi, R., Chatzistergos, P., Sundar, L., Chockalingam, N., Ramachandran, A., 2016b. Differences in the mechanical characteristics of plantar soft tissue between ulcerated and non-ulcerated foot. J. Diabetes Complicat. 30 (7), 1293–1299.

Nelson, F.R., Craig, J., Francois, H., Azuh, O., Oyetakin-White, P., King, B., 2014. Subchondral insufficiency fractures and spontaneous osteonecrosis of the knee may not be related to osteoporosis. Arch. Osteoporos. 9, 194. https://doi.org/10.1007/s11657-014-0194-z.

Neptune, R.R., Zajac, F.E., Kautz, S.A., 2004. Muscle force redistributes segmental power for body progression during walking. Gait Posture 19 (2), 194–205.

Neptune, R.R., Burnfield, J.M., Mulroy, S.J., 2007. The neuromuscular demands of toe walking: a forward dynamics simulation analysis. J. Biomech. 40 (6), 1293–1300.

Nery, C., Coughlin, M.J., Baumfeld, D., Ballerini, F.J., Kobata, S., 2013. Hallux valgus in males. Part 1. Demographics, etiology, and comparative radiology. Foot Ankle Int. 34 (5), 629–635.

Neumann, D.A., 2010. Kinesiology of the hip: a focus on muscular actions. J. Orthop. Sports Phys. Ther. 40 (2), 82–94.

Niki, H., Ching, R.P., Kiser, P., Sangeorzan, B.J., 2001. The effect of posterior tibial tendon dysfunction on hindfoot kinematics. Foot Ankle Int. 22 (4), 292–300.

Noehren, B., Davis, I., Hamill, J., 2007. ASB Clinical Biomechanics Award Winner 2006: prospective study of the biomechanical factors associated with iliotibial band syndrome. Clin. Biomech. 22 (9), 951–956.

Nunns, M., House, C., Rice, H., Mostazir, M., Davey, T., Stiles, V., et al., 2016. Four biomechanical and anthropometric measures predict tibial stress fracture: a prospective study of 1065 Royal Marines. Br. J. Sports Med. 50 (19), 1206–1210.

Nunns, M., House, C., Fallowfield, J., Allsopp, A., Dixon, S., 2013. Biomechanical characteristics of barefoot footstrike modalities. J. Biomech. 46 (15), 2603–2610.

Ogden, J.A., Ganey, T.M., Hill, J.D., Jaakkola, J.I., 2004. Sever's injury: a stress fracture of the immature calcaneal metaphysis. J. Pediatr. Orthop. 24 (5), 488–492.

Oh, K., Park, S., 2017. The bending stiffness of shoes is beneficial to running energetics if it does not disturb the natural MTP joint flexion. J. Biomech. 53, 127–135.

Orchard, J., 2012. Plantar fasciitis. Br. Med. J. 345 (7878), e6603. https://doi.org/10.1136/bmj.e6603.

Orchard, J.W., Fricker, P.A., Abud, A.T., Mason, B.R., 1996. Biomechanics of iliotibial band friction syndrome in runners. Am. J. Sports Med. 24 (3), 375–379.

O'Neill, S., Barry, S., Watson, P., 2019. Plantarflexion strength and endurance deficits associated with mid-portion Achilles tendinopathy: the role of soleus. Phys. Therapy Sport 37, 69–76.

Orselli, M.I.V., Franz, J.R., Thelen, D.G., 2017. The effects of Achilles tendon compliance on triceps surae mechanics and energetics in walking. J. Biomech. 60, 227–231.

Ostermeier, S., Holst, M., Hurschler, C., Windhagen, H., Stukenborg-Colsman, C., 2007. Dynamic measurement of the patellofemoral kinematics and contact pressure after lateral retinacular release: an in vitro study. Knee Surg. Sports Traumatol. Arthrosc. 15 (5), 547–554.

Ota, T., Nagura, T., Yamada, Y., Yamada, M., Yokoyama, Y., Ogihara, N., et al., 2019. Effect of natural full weight-bearing during standing on the rotation of the first metatarsal bone. Clin. Anat. 32 (5), 715–721.

Pękala, P.A., Henry, B.M., Ochała, A., Kopacz, P., Tatoń, G., Młyniec, A., et al., 2017. The twisted structure of the Achilles tendon unraveled: a detailed quantitative and qualitative anatomical investigation. Scand. J. Med. Sci. Sports 27 (12), 1705–1715.

Pai, S., Ledoux, W.R., 2010. The compressive mechanical properties of diabetic and non-diabetic plantar soft tissue. J. Biomech. 43 (9), 1754–1760.

Pai, S., Ledoux, W.R., 2012. The shear mechanical properties of diabetic and non-diabetic plantar soft tissue. J. Biomech. 45 (2), 364–370.

Pain, M.T.G., Challis, J.H., 2006. The influence of soft tissue movement on ground reaction forces, joint torques and joint reaction forces in drop landings. J. Biomech. 39 (1), 119–124.

Palmer, W.E., Kuong, S.J., Elmadbouh, H.M., 1999. MR imaging of myotendinous strain. AJR: Am. J. Roentgenol. 173 (3), 703–709.

Pasque, C., Noyes, F.R., Gibbons, M., Levy, M., Grood, E., 2003. The role of the popliteofibular ligament and the tendon of popliteus in providing stability in the human knee. J. Bone Joint Surg. 85-B (2), 292–298.

Paton, J.S., 2006. The relationship between navicular drop and first metatarsophalangeal joint motion. J. Am. Podiatr. Med. Assoc. 96 (4), 313–317.

Paul, L., Ellis, B.M., Leese, G.P., McFadyen, A.K., McMurray, B., 2009. The effect of a cognitive or motor task on gait parameters of diabetic patients, with and without neuropathy. Diabet. Med. 26 (3), 234–239.

Pavan, P.G., Stecco, A., Stern, R., Stecco, C., 2014. Painful connections: densification versus fibrosis of fascia. Curr. Pain Headache Rep. 18 (8), 441. https://doi.org/10.1007/s11916-014-0441-4.

Pendharkar, G., Percival, P., Morgan, D., Lai, D., 2012. Automated method to distinguish toe walking strides from normal strides in the gait of idiopathic toe walking children from heel accelerometry data. Gait Posture 35 (3), 478–482.

Perhamre, S., Lazowska, D., Papageorgiou, S., Lundin, F., Klässbo, M., Norlin, R., 2013. Sever's injury: a clinical diagnosis. J. Am. Podiatr. Med. Assoc. 103 (5), 361–368.

Perry, J., 1992. Gait Analysis: Normal and Pathological Function. SLACK Incorporated, Thorofare, NJ.

Péter, A., Hegyi, A., Stenroth, L., Finni, T., Cronin, N.J., 2015. EMG and force production of the flexor hallucis longus muscle in isometric plantarflexion and the push-off phase of walking. J. Biomech. 48 (12), 3413–3419.

Petersen, W., Ellermann, A., Gösele-Koppenburg, A., Best, R., Rembitzki, I.V., Brüggemann, G.-P., et al., 2014. Patellofemoral pain syndrome. Knee Surg. Sports Traumatol. Arthrosc. 22 (10), 2264–2274.

Phillips, R.D., Law, E.A., Ward, E.D., 1996. Functional motion of the medial column joints of the foot during propulsion. J. Am. Podiatr. Med. Assoc. 86 (10), 474–486.

Piqué-Vidal, C., Solé, M.T., Antich, J., 2007. Hallux valgus inheritance: pedigree research in 350 patients with bunion deformity. J. Foot Ankle Surg. 46 (3), 149–154.

Plank, M.J., 1995. The pattern of forefoot pressure distribution in hallux valgus. Foot 5 (1), 8–14.

Pohl, M.B., Messenger, N., Buckley, J.G., 2007. Forefoot, rearfoot and shank coupling: effect of variations in speed and mode of gait. Gait Posture 25 (2), 295–302.

Pohl, M.B., Mullineaux, D.R., Milner, C.E., Hamill, J., Davis, I.S., 2008. Biomechanical predictors of retrospective tibial stress fractures in runners. J. Biomech. 41 (6), 1160–1165.

Pollo, F.E., 1998. Bracing and heel wedging for unicompartmental osteoarthritis of the knee. Am. J. Knee Surg. 11 (1), 47–50.

Potthast, W., Lersch, C., Segesser, B., Koebke, J., Brüggemann, G.-P., 2008. Intraarticular pressure distribution in the talocrural joint is related to lower leg muscle forces. Clin. Biomech. 23 (5), 632–639.

Powers, C.M., Landel, R., Perry, J., 1996. Timing and intensity of vastus muscle activity during functional activities in subjects with and without patellofemoral pain. Phys. Ther. 76 (9), 946–955. discussion 956–967.

Powers, S.K., Lynch, G.S., Murphy, K.T., Reid, M.B., Zijdewind, I., 2016. Disease-induced skeletal muscle atrophy and fatigue. Med. Sci. Sports Exerc. 48 (11), 2307–2319.

Preece, S.J., Graham-Smith, P., Nester, C.J., Howard, D., Hermens, H., Herrington, L., et al., 2008. The influence of gluteus maximus on transverse plane tibial rotation. Gait Posture 27 (4), 616–621.

Prins, M.R., van der Wurff, P., 2009. Females with patellofemoral pain syndrome have weak hip muscles: a systematic review. Aust. J. Physiother. 55 (1), 9–15.

Puszczalowska-Lizis, E., Krawczyk, K., Omorczyk, J., 2022. Effect of longitudinal and transverse foot arch on the position of the hallux and fifth toe in preschool children in the light of regression analysis. Int. J. Environ. Res. Public Health 19 (3), 1669. https://doi.org/10.3390/ijerph19031669.

Ramaniraka, N.A., Terrier, A., Theumann, N., Siegrist, O., 2005. Effects of the posterior cruciate ligament reconstruction on the biomechanics of the knee joint: a finite element analysis. Clin. Biomech. 20 (4), 434–442.

Rauh, M.J., Macera, C.A., Trone, D.W., Shaffer, R.A., Brodine, S.K., 2006. Epidemiology of stress fracture and lower-extremity overuse injury in female recruits. Med. Sci. Sports Exerc. 38 (9), 1571–1577.

Ravenell, R.A., Kihm, C.A., Lin, A.S., Garing, F.X., 2011. The offset V osteotomy versus the modified Austin with a longer plantar arm: a comparison of mechanical stability. J. Foot Ankle Surg. 50 (2), 201–206.

Raychoudhury, S., Hu, D., Ren, L., 2014. Three-dimensional kinematics of the human metatarsophalangeal joint during level walking. Front. Bioeng. Biotechnol. 2, 73. https://doi.org/10.3389/fbioe.2014.00073.

Reischl, S.F., Powers, C.M., Rao, S., Perry, J., 1999. Relationship between foot pronation and rotation of the tibia and femur during walking. Foot Ankle Int. 20 (8), 513–520.

Resende, R.A., Kirkwood, R.N., Deluzio, K.J., Morton, A.M., Fonseca, S.T., 2016. Mild leg length discrepancy affects lower limbs, pelvis and trunk biomechanics of individuals with knee osteoarthritis during gait. Clin. Biomech. 38, 1–7.

Reshef, N., Guelich, D.R., 2012. Medial tibial stress syndrome. Clin. Sports Med. 31 (2), 273–290.

Rice, H.M., Saunders, S.C., McGuire, S.J., O'Leary, T.J., Izard, R.M., 2018. Estimates of tibial shock magnitude in men and women at the start and end of a military drill training program. Mil. Med. 183 (9–10), e392–e398.

Rice, H., Weir, G., Trudeau, M.B., Meardon, S., Derrick, T., Hamill, J., 2019. Estimating tibial stress throughout the duration of a treadmill run. Med. Sci. Sports Exerc. 51 (11), 2257–2264.

Riddick, R.C., Kuo, A.D., 2016. Soft tissues store and return mechanical energy in human running. J. Biomech. 49 (3), 436–441.

Riddle, D.L., Schappert, S.M., 2004. Volume of ambulatory care visits and patterns of care for patients diagnosed with plantar fasciitis: a national study of medical doctors. Foot Ankle Int. 25 (5), 303–310.

Rizzone, K.H., Ackerman, K.E., Roos, K.G., Dompier, T.P., Kerr, Z.Y., 2017. The epidemiology of stress fractures in collegiate student-athletes, 2004–2005 through 2013–2014 academic years. J. Athl. Train. 52 (10), 966–975.

Roberts, A., Franklyn-Miller, A., 2012. The validity of the diagnostic criteria used in chronic exertional compartment syndrome: a systematic review. Scand. J. Med. Sci. Sports 22 (5), 585–595.

Roberts, T.J., Konow, N., 2013. How tendons buffer energy dissipation by muscle. Exerc. Sport Sci. Rev. 41 (4), 186–193.

Roddy, E., Menz, H.B., 2018. Foot osteoarthritis: latest evidence and developments. Therap. Adv. Musculosk. Dis. 10 (4), 91–103.

Romanov, N., Fletcher, G., 2007. Runners do not push off the ground but fall forwards via a gravitational torque. Sports Biomech. 6 (3), 434–452.

Roos, K.G., Kerr, Z.Y., Mauntel, T.C., Djoko, A., Dompier, T.P., Wikstrom, E.A., 2017. The epidemiology of lateral ligament complex ankle sprains in National Collegiate Athletic Association sports. Am. J. Sports Med. 45 (1), 201–209.

Ross, R.A., Allsopp, A., 2002. Stress fractures in Royal Marines recruits. Mil. Med. 167 (7), 560–565.

Rossi, I., Rosenberg, Z., Zember, J., 2016. Normal skeletal development and imaging pitfalls of the calcaneal apophysis: MRI features. Skelet. Radiol. 45 (4), 483–493.

Roukis, T.S., 2005. Metatarsus primus elevatus in hallux rigidus: fact or fiction? J. Am. Podiatr. Med. Assoc. 95 (3), 221–228.

Roukis, T.S., Scherer, P.R., Anderson, C.F., 1996. Position of the first ray and motion of the first metatarsophalangeal joint. J. Am. Podiatr. Med. Assoc. 86 (11), 538–546.

Ruina, A., Bertram, J.E.A., Srinivasan, M., 2005. A collisional model of the energetic cost of support work qualitatively explains leg sequencing in walking and galloping, pseudo-elastic leg behavior in running and the walk-to-run transition. J. Theor. Biol. 237 (2), 170–192.

Rutherford, D.J., Hubley-Kozey, C., 2009. Explaining the hip adduction moment variability during gait: implications for hip abductor strengthening. Clin. Biomech. 24 (3), 267–273.

Saito, M., Marumo, K., 2015. Effects of collagen crosslinking on bone material properties in health and disease. Calcif. Tissue Int. 97 (3), 242–261.

Salsich, G.B., Perman, W.H., 2007. Patellofemoral joint contact area is influenced by tibiofemoral rotation alignment in individuals who have patellofemoral pain. J. Orthop. Sports Phys. Ther. 37 (9), 521–528.

Sanchez, P.J., Grady, J.F., Lenz, R.C., Park, S.J., Ruff, J.G., 2018. Metatarsus primus elevatus resolution after first metatarsophalangeal joint arthroplasty. J. Am. Podiatr. Med. Assoc. 108 (3), 200–204.

Sanchis-Alfonso, V., Roselló-Sastre, E., Revert, F., García, A., 2005. Histologic retinacular changes associated with ischemia in painful patellofemoral malalignment. Orthopedics 28 (6), 593–599.

Sanderlin, B.W., Raspa, R.F., 2003. Common stress fractures. Am. Fam. Physician 68 (8), 1527–1532.

Savory, K.M., Wülker, N., Stukenborg, C., Alfke, D., 1998. Biomechanics of the hindfoot joints in response to degenerative hindfoot arthrodeses. Clin. Biomech. 13 (1), 62–70.

Scharfbillig, R.W., Jones, S., Scutter, S.D., 2008. Sever's disease: What does the literature really tell us? J. Am. Podiatr. Med. Assoc. 98 (3), 212–223.

Schmitt, S., Günther, M., 2011. Human leg impact: energy dissipation of wobbling masses. Arch. Appl. Mech. 81 (7), 887–897.

Schuett, D.J., Bomar, J.D., Pennock, A.T., 2015. Pelvic apophyseal avulsion fractures: a retrospective review of 228 cases. J. Pediatr. Orthop. 35 (6), 617–623.

Schweitzer, M.E., White, L.M., 1996. Does altered biomechanics cause bone marrow edema? Radiology 198 (3), 851–853.

Scott, J., Lee, H., Barsoum, W., van den Bogert, A.J., 2007. The effect of tibiofemoral loading on proximal tibiofibular joint motion. J. Anat. 211 (5), 647–653.

Scott, R.T., Hyer, C.F., Granata, A., 2013. The correlation of Achilles tendinopathy and body mass index. Foot Ankle Special. 6 (4), 283–285.

Semciw, A.I., Pizzari, T., Murley, G.S., Green, R.A., 2013. Gluteus medius: an intramuscular EMG investigation of anterior, middle and posterior segments during gait. J. Electromyogr. Kinesiol. 23 (4), 858–864.

Semciw, A.I., Green, R.A., Murley, G.S., Pizzari, T., 2014. Gluteus minimus: an intramuscular EMG investigation of anterior and posterior segments during gait. Gait Posture 39 (2), 822–826.

Senavongse, W., Amis, A.A., 2005. The effects of articular, retinacular, or muscular deficiencies on patellofemoral joint stability: a biomechanical study in vitro. J. Bone Joint Surg. 87-B (4), 577–582.

Sharkey, N.A., Ferris, L., Smith, T.S., Matthews, D.K., 1995. Strain and loading of the second metatarsal during heel-lift. J. Bone Joint Surg. 77-A (7), 1050–1057.

Sharma, L., Song, J., Dunlop, D., Felson, D., Lewis, C.E., Segal, N., et al., 2010. Varus and valgus alignment and incident and progressive knee osteoarthritis. Ann. Rheum. Dis. 69 (11), 1940–1945.

Shellock, F.G., Mink, J.H., Deutsch, A., Fox, J.M., Ferkel, R.D., 1990. Evaluation of patients with persistent symptoms after lateral retinacular release by kinematic magnetic resonance imaging of the patellofemoral joint. Arthrosc.: J. Arthrosc. Rel. Surg. 6 (3), 226–234.

Sherman, S.L., Plackis, A.C., Nuelle, C.W., 2014. Patellofemoral anatomy and biomechanics. Clin. Sports Med. 33 (3), 389–401.

Shih, Y., Lin, K.-L., Shiang, T.-Y., 2013. Is the foot striking pattern more important than barefoot or shod conditions in running? Gait Posture 38 (3), 490–494.

Shiotani, H., Yamashita, R., Mizokuchi, T., Sado, N., Naito, M., Kawakami, Y., 2021. Track distance runners exhibit bilateral differences in the plantar fascia stiffness. Sci. Rep. 11, 9260. https://doi.org/10.1038/s41598-021-88883-4.

Sichting, F., Holowka, N.B., Hansen, O.B., Lieberman, D.E., 2020. Effect of the upward curvature of toe springs on walking biomechanics in humans. Sci. Rep. 10, 14643. https://doi.org/10.1038/s41598-020-71247-9.

Silver, F.H., Freeman, J.W., Seehra, G.P., 2003. Collagen self-assembly and the development of tendon mechanical properties. J. Biomech. 36 (10), 1529–1553.

Silvestri, T., Pulsatelli, L., Dolzani, P., Frizziero, L., Facchini, A., Meliconi, R., 2006. In vivo expression of inflammatory cytokine receptors in the joint compartments of patients with arthritis. Rheumatol. Int. 26 (4), 360–368.

Sims, A.L., Kurup, H.V., 2014. Painful sesamoid of the great toe. World J. Orthoped. 5 (2), 146–150.

Slane, L.C., Thelen, D.G., 2014. Non-uniform displacements within the Achilles tendon observed during passive and eccentric loading. J. Biomech. 47 (12), 2831–2835.

Slane, L.C., Thelen, D.G., 2015. Achilles tendon displacement patterns during passive stretch and eccentric loading are altered in middle-aged adults. Med. Eng. Phys. 37 (7), 712–716.

Slimmon, D., Bennell, K., Brukner, P., Crossley, K., Bell, S.N., 2002. Long-term outcome of fasciotomy with partial fasciectomy for chronic exertional compartment syndrome of the lower leg. Am. J. Sports Med. 30 (4), 581–588.

Smit, H.J., Strong, P., 2020. Structural elements of the biomechanical system of soft tissue. Cureus 12 (4), e7895. https://doi.org/10.7759/cureus.7895.

Snow, S.W., Bohne, W.H.O., DiCarlo, E., Chang, V.K., 1995. Anatomy of the Achilles tendon and plantar fascia in relation to the calcaneus in various age groups. Foot Ankle Int. 16 (7), 418–421.

Soubrier, M., Dubost, J.-J., Boisgard, S., Sauvezie, B., Gaillard, P., Michel, J.L., et al., 2003. Insufficiency fracture. A survey of 60 cases and review of the literature. Joint Bone Spine 70 (3), 209–218.

Speers, C.J.B., Bhogal, G.S., 2017. Greater trochanteric pain syndrome: a review of diagnosis and management in general practice. Br. J. Gen. Pract. 67 (663), 479–480.

Stainsby, G.D., 1997. Pathological anatomy and dynamic effect of the displaced plantar plate and the importance of the integrity of the plantar plate-deep transverse metatarsal ligament tie-bar. Ann. R. Coll. Surg. Engl. 79 (1), 58–68.

Stecco, C., Pavan, P., Pachera, P., De Caro, R., Natali, A., 2014. Investigation of the mechanical properties of the human crural fascia and their possible clinical implications. Surg. Radiol. Anat. 36 (1), 25–32.

Stefanyshyn, D.J., Wannop, J.W., 2016. The influence of forefoot bending stiffness of footwear on athletic injury and performance. Footwear Sci. 8 (2), 51–63.

Stefanyshyn, D.J., Stergiou, P., Lun, V.M.Y., Meeuwisse, W.H., Worobets, J.T., 2006. Knee angular impulse as a predictor of patellofemoral pain in runners. Am. J. Sports Med. 34 (11), 1844–1851.

Steinbach, L.S., Fleckenstein, J.L., Mink, J.H., 1994. Magnetic resonance imaging of muscle injuries. Orthopedics 17 (11), 991–999.

Stolwijk, N.M., Koenraadt, K.L.M., Louwerens, J.W.K., Grim, D., Duysens, J., Keijsers, N.L.W., 2014. Foot lengthening and shortening during gait: a parameter to investigate foot function? Gait Posture 39 (2), 773–777.

Strauss, E.J., Kim, S., Calcei, J.G., Park, D., 2011. Iliotibial band syndrome: evaluation and management. J. Am. Acad. Orthop. Surg. 19 (12), 728–736.

Studler, U., Mengiardi, B., Bode, B., Schöttle, P.B., Pfirrmann, C.W.A., Hodler, J., et al., 2008. Fibrosis and adventitious bursae in plantar fat pad of forefoot: MR imaging findings in asymptomatic volunteers and MR imaging-histologic comparison. Radiology 246 (3), 863–870.

Sun, P.-C., Shih, S.-L., Chen, Y.-L., Hsu, Y.-C., Yang, R.-C., Chen, C.-S., 2012. Biomechanical analysis of foot with different foot arch heights: a finite element analysis. Comput. Method Biomech. Biomed. Eng. 15 (6), 563–569.

Sussman, R.E., D'Amico, J.C., 1984. The influence of the height of the heel on the first metatarsophalangeal joint. J. Am. Podiatry Assoc. 74 (10), 504–508.

Taş, S., Salkin, Y., 2019. An investigation of the sex-related differences in the stiffness of the Achilles tendon and gastrocnemius muscle: inter-observer reliability and inter-day repeatability and the effect of ankle joint motion. Foot 41, 44–50.

Takabayashi, T., Edama, M., Inai, T., Nakamura, E., Kubo, M., 2020. Effect of gender and load conditions on foot arch height index and flexibility in Japanese youths. J. Foot Ankle Surg. 59 (6), 1144–1147.

Takahashi, K.Z., Gross, M.T., van Werkhoven, H., Piazza, S.J., Sawicki, G.S., 2016. Adding stiffness to the foot modulates soleus force-velocity behaviour during human walking. Sci. Rep. 6, 29870. https://doi.org/10.1038/srep29870.

Takahashi, K.Z., Worster, K., Bruening, D.A., 2017. Energy neutral: the human foot and ankle subsections combine to produce near zero net mechanical work during walking. Sci. Rep. 7, 15404. https://doi.org/10.1038/s41598-017-15218-7.

Taunton, J.E., Ryan, M.B., Clement, D.B., McKenzie, D.C., Lloyd-Smith, D.R., Zumbo, B.D., 2002. A retrospective case-control analysis of 2002 running injuries. Br. J. Sports Med. 36 (2), 95–101.

Thomas, S., Barrington, R., 2003. Hallux valgus. Curr. Orthop. 17 (4), 299–307.

Thompson, J.A., Chaudhari, A.M.W., Schmitt, L.C., Best, T.M., Siston, R.A., 2013. Gluteus maximus and soleus compensate for simulated quadriceps atrophy and activation failure during walking. J. Biomech. 46 (13), 2165–2172.

Torry, M.R., Schenker, M.L., Martin, H.D., Hogoboom, D., Philippon, M.J., 2006. Neuromuscular hip biomechanics and pathology in the athlete. Clin. Sports Med. 25 (2), 179–197.

Tozzi, P., 2012. Selected fascial aspects of osteopathic practice. J. Bodyw. Mov. Ther. 16 (4), 503–519.

Turcot, K., Armand, S., Lübbeke, A., Fritschy, D., Hoffmeyer, P., Suvà, D., 2013. Does knee alignment influence gait in patients with severe knee osteoarthritis? Clin. Biomech. 28 (1), 34–39.

Uchida, T.K., Hicks, J.L., Dembia, C.L., Delp, S.L., 2016. Stretching your energetic budget: how tendon compliance affects the metabolic cost of running. PLoS One 11 (3), e0150378. https://doi.org/10.1371/journal.pone.0150378.

Uemura, K., Atkins, P.R., Fiorentino, N.M., Anderson, A.E., 2018. Hip rotation during standing and dynamic activities and the compensatory effect of femoral anteversion: an in vivo analysis of asymptomatic young adults using three-dimensional computed tomography models and dual fluoroscopy. Gait Posture 61, 276–281.

Ullrich, K., Krudwig, W.K., Witzel, U., 2002. Posterolateral aspect and stability of the knee joint. I. Anatomy and function of the popliteus muscle-tendon unit: an anatomical and biomechanical study. Knee Surg. Sports Traumatol. Arthrosc. 10 (2), 86–90.

Usherwood, J.R., Channon, A.J., Myatt, J.P., Rankin, J.W., Hubel, T.Y., 2012. The human foot and heel-sole-toe walking strategy: a mechanism enabling an inverted pendular gait with low isometric muscle force? J. R. Soc. Interface 9 (75), 2396–2402.

van der Esch, M., Knoop, J., Hunter, D.J., Klein, J.-P., van der Leeden, M., Knol, D.L., et al., 2013. The association between reduced knee joint proprioception and medial meniscal abnormalities using MRI in knee osteoarthritis: results from the Amsterdam osteoarthritis cohort. Osteoarthr. Cartil. 21 (5), 676–681.

van der Vlist, A.C., Breda, S.J., Oei, E.H.G., Verhaar, J.A.N., de Vos, R.-J., 2019. Clinical risk factors for Achilles tendinopathy: a systematic review. Br. J. Sports Med. 53 (21), 1352–1361.

van der Worp, M.P., ten Haaf, D.S.M., van Cingel, R., de Wijer, A., Nijhuis-van der Sanden, M.W.G., Staal, J.B., 2015. Injuries in runners; a systematic review on risk factors and sex differences. PLoS One 10 (2), e114937. https://doi.org/10.1371/journal.pone.0114937.

van Dijk, P.A.D., Gianakos, A.L., Kerkhoffs, G.M.M.J., Kennedy, J.G., 2016. Return to sports and clinical outcomes in patients treated for peroneal tendon dislocation: a systematic review. Knee Surg. Sports Traumatol. Arthrosc. 24 (4), 1155–1164.

van Gent, R.N., Siem, D., van Middelkoop, M., van Os, A.G., Bierma-Zeinstra, S.M.A., Koes, B.W., 2007. Incidence and determinants of lower extremity running injuries in long distance runners: a systematic review. Br. J. Sports Med. 41 (8), 469–480.

Van Gheluwe, B., Kirby, K.A., Hagman, F., 2005. Effects of simulated genu valgum and genu varum on ground reaction forces and subtalar joint function during gait. J. Am. Podiatr. Med. Assoc. 95 (6), 531–541.

Van Gheluwe, B., Dananberg, H.J., Hagman, F., Vanstaen, K., 2006. Effects of hallux limitus on plantar foot pressure and foot kinematics during walking. J. Am. Podiatr. Med. Assoc. 96 (5), 428–436.

Van Hooren, B., Bosch, F., 2017a. Is there really an eccentric action of the hamstrings during the swing phase of high-speed running? Part I. A critical review of the literature. J. Sports Sci. 35 (23), 2313–2321.

Van Hooren, B., Bosch, F., 2017b. Is there really an eccentric action of the hamstrings during the swing phase of high-speed running? Part II. Implications for exercise. J. Sports Sci. 35 (23), 2322–2333.

van Leeuwen, K.D.B., Rogers, J., Winzenberg, T., van Middlekoop, M., 2016. Higher body mass index is associated with plantar fasciopathy/'plantar fasciitis': systematic review and meta-analysis of various clinical and imaging risk factors. Br. J. Sports Med. 50 (16), 972–981.

Vannatta, C.N., Heinert, B.L., Kernozek, T.W., 2020a. Biomechanical risk factors for running-related injury differ by sample population: a systematic review and meta-analysis. Clin. Biomech. 75, 104991. https://doi.org/10.1016/j.clinbiomech.2020.104991.

Vannatta, C.N., Almonroeder, T.G., Kernozek, T.W., Meardon, S., 2020b. Muscle force characteristics of male and female collegiate cross-country runners during overground running. J. Sports Sci. 38 (5), 542–551.

Varady, N.H., Grodzinsky, A.J., 2016. Osteoarthritis year in review 2015: mechanics. Osteoarthr. Cartil. 24 (1), 27–35.

Vasara, A.I., Jurvelin, J.S., Peterson, L., Kiviranta, I., 2005. Arthroscopic cartilage indentation and cartilage lesions of anterior cruciate ligament-deficient knees. Am. J. Sports Med. 33 (3), 408–414.

Vasilevska, V., Szeimies, U., Stäbler, A., 2009. Magnetic resonance imaging signs of iliotibial band friction in patients with isolated medial compartment osteoarthritis of the knee. Skelet. Radiol. 38 (9), 871–875.

Venkadesan, M., Yawar, A., Eng, C.M., Dias, M.A., Singh, D.K., Tommasini, S.M., et al., 2020. Stiffness of the human foot and evolution of the transverse arch. Nature 579 (7797), 97–100.

Venkadesan, M., Mandre, S., Bandi, M.M., 2017. Biological feet: evolution, mechanics and applications. In: Sharbarfi, M.A., Seyfarth, A. (Eds.), Bioinspired Legged Locomotion: Models, Concepts, Control and Applications. Butterworth-Heinemann, Elsevier Science, pp. 461–486 (Part III, Chapter 7, Section 7.1).

Vergis, A., Hammarby, S., Gillquist, J., 2002. Fluoroscopic validation of electrogoniometrically measured femorotibial translation in healthy and ACL deficient subjects. Scand. J. Med. Sci. Sports 12 (4), 223–229.

Volmer-Thole, M., Lobmann, R., 2016. Neuropathy and diabetic foot syndrome. Int. J. Mol. Sci. 17 (6), 917. https://doi.org/10.3390/ijms17060917.

Wade, F.E., Lewis, G.S., Piazza, S.J., 2019. Estimates of Achilles tendon moment arm differ when axis of ankle rotation is derived from ankle motion. J. Biomech. 90, 71–77.

Wager, J.C., Challis, J.H., 2016. Elastic energy within the human plantar aponeurosis contributes to arch shortening during the push-off phase of running. J. Biomech. 49 (5), 704–709.

Wagner, E., Wagner, P., 2020. Metatarsal pronation in hallux valgus deformity: a review. J. Am. Acad. Orthop. Surg. Global Res. Rev. 4 (6), e20.00091. https://doi.org/10.5435/JAAOSGlobal-D-20-00091.

Wakeling, J.M., Nigg, B.M., 2001. Modification of soft tissue vibrations in the leg by muscular activity. J. Appl. Physiol. 90 (2), 412–420.

Wakeling, J.M., Liphardt, A.-M., Nigg, B.M., 2003. Muscle activity reduces soft-tissue resonance at heel-strike during walking. J. Biomech. 36 (12), 1761–1769.

Waldecker, U., Lehr, H.-A., 2009. Is there histomorphological evidence of plantar metatarsal fat pad atrophy in patients with diabetes? J. Foot Ankle Surg. 48 (6), 648–652.

Wang, Y., Li, Z., Wong, D.W.-C., Zhang, M., 2015. Effects of ankle arthrodesis on biomechanical performance of the entire foot. PLoS One 10 (7), e0134340. https://doi.org/10.1371/journal.pone.0134340.

Wanivenhaus, A., Pretterklieber, M., 1989. First tarsometatarsal joint: anatomical biomechanical study. Foot Ankle 9 (4), 153–157.

Wannop, J.W., Stefanyshyn, D.J., 2016. Editorial. Special issue: Bending stiffness: performance and injury effects. Footwear Sci. 8 (2), 49–50.

Ward, S.R., Powers, C.M., 2004. The influence of patella alta on patellofemoral joint stress during normal and fast walking. Clin. Biomech. 19 (10), 1040–1047.

Ward, S.R., Terk, M.R., Powers, C.M., 2005. Influence of patella alta on knee extensor mechanics. J. Biomech. 38 (12), 2415–2422.

Ward, S.R., Terk, M.R., Powers, C.M., 2007. Patella alta: association with patellofemoral alignment and changes in contact area during weight-bearing. J. Bone Joint Surg. 89-A (8), 1749–1755.

Watanabe, K., Ikeda, Y., Suzuki, D., Teramoto, A., Kobayashi, T., Suzuki, T., et al., 2017. Three-dimensional analysis of tarsal bone response to axial loading in patients with hallux valgus and normal feet. Clin. Biomech. 42, 65–69.

Wearing, S.C., Smeathers, J.E., Yates, B., Urry, S.R., Dubois, P., 2009. Bulk compressive properties of the heel fat pad during walking: a pilot investigation in plantar heel pain. Clin. Biomech. 24 (4), 397–402.

Weidow, J., Pak, J., Kärrholm, J., 2002. Different patterns of cartilage wear in medial and lateral gonarthrosis. Acta Orthop. Scand. 73 (3), 326–329.

Welck, M.J., Hayes, T., Pastides, P., Khan, W., Rudge, B., 2017. Stress fractures of the foot and ankle. Injury 48 (8), 1722–1726.

Welte, L., Kelly, L.A., Kessler, S.E., Lieberman, D.E., D'Andrea, S.E., Lichtwark, G.A., et al., 2021. The extensibility of the plantar fascia influences the windlass mechanism during human running. Proc. R. Soc. B Biol. Sci. 288 (1943), 20202095. https://doi.org/10.1098/rspb.2020.2095.

Weyand, P.G., Sandell, R.F., Prime, D.N.L., Bundle, M.W., 2010. The biological limits to running speed are imposed from the ground up. J. Appl. Physiol. 108 (4), 950–961.

Wezenbeek, E., De Clercq, D., Mahieu, N., Willems, T., Witvrouw, E., 2018. Activity-induced increase in Achilles tendon blood flow is age and sex dependent. Am. J. Sports Med. 46 (11), 2678–2686.

Whitney, A.K., Green, D.R., 1982. Pseudoequinus. J. Am. Podiatry Assoc. 72 (7), 365–371.

Whyte, E.F., Moran, K., Shortt, C.P., Marshall, B., 2010. The influence of reduced hamstring length on patellofemoral joint stress during squatting in healthy male adults. Gait Posture 31 (1), 47–51.

Williams, K.R., Cavanagh, P.R., 1987. Relationship between distance running mechanics, running economy, and performance. J. Appl. Physiol. 63 (3), 1236–1245.

Williams 3rd, D.S., McClay, I.S., Hamill, J., 2001. Arch structure and injury patterns in runners. Clin. Biomech. 16 (4), 341–347.

Willwacher, S., König, M., Potthast, W., Brüggemann, G.-P., 2013. Does specific footwear facilitate energy storage and return at the metatarsophalangeal joint in running? J. Appl. Biomech. 29 (5), 583–592.

Willwacher, S., König, M., Braunstein, B., Goldmann, J.-P., Brüggemann, G.-P., 2014. The gearing function of running shoe longitudinal bending stiffness. Gait Posture 40 (3), 386–390.

Winby, C.R., Lloyd, D.G., Besier, T.F., Kirk, T.B., 2009. Muscle and external load contribution to knee joint contact loads during normal gait. J. Biomech. 42 (14), 2294–2300.

Wolf, P., Stacoff, A., Liu, A., Nester, C., Arndt, A., Lundberg, A., et al., 2008. Functional units of the human foot. Gait Posture 28 (3), 434–441.

Woo, S.L.-Y., Wu, C., Dede, O., Vercillo, F., Noorani, S., 2006. Biomechanics and anterior cruciate ligament reconstruction. J. Orthop. Surg. Res. 1, 2. https://doi.org/10.1186/1749-799x-1-2.

Woodburn, J., Hyslop, E., Barn, R., McInnes, I.B., Turner, D.E., 2013. Achilles tendon biomechanics in psoriatic arthritis patients with ultrasound proven enthesitis. Scand. J. Rheumatol. 42 (4), 299–302.

Wren, T.A.L., Yerby, S.A., Beaupré, G.S., Carter, D.R., 2001. Influence of bone mineral density, age, and strain rate on the failure mode of human Achilles tendons. Clin. Biomech. 16 (6), 529–534.

Wren, T.A.L., Lindsey, D.P., Beaupré, G.S., Carter, D.R., 2003. Effects of creep and cyclic loading on the mechanical properties and failure of human Achilles tendons. Ann. Biomed. Eng. 31 (6), 710–717.

Wrobel, J.S., Najafi, B., 2010. Diabetic foot biomechanics and gait dysfunction. J. Diabetes Sci. Technol. 4 (4), 833–845.

Wünschel, M., Leichtle, U., Obloh, C., Wülker, N., Müller, O., 2011. The effect of different quadriceps loading patterns on tibiofemoral joint kinematics and patellofemoral contact pressure during simulated partial weight-bearing knee flexion. Knee Surg. Sports Traumatol. Arthrosc. 19 (7), 1099–1106.

Yammine, K., 2015. The accessory peroneal (fibular) muscles: peroneus quartus and peroneus digiti quinti. A systematic review and meta-analysis. Surg. Radiol. Anat. 37 (6), 617–627.

Yawar, A., Korpas, L., Mandre, S., Venkadesan, M., 2017. Transverse Contributions to the Longitudinal Stiffness of the Human Foot. https://arxiv.org/abs/1706.04610.

Yong, J.R., Silder, A., Montgomery, K.L., Fredericson, M., Delp, S.L., 2018. Acute changes in foot strike pattern and cadence affect running parameters associated with tibial stress fractures. J. Biomech. 76, 1–7.

Young, K.W., Park, Y.U., Kim, J.S., Jegal, H., Lee, K.T., 2013. Unilateral hallux valgus: is it true unilaterality, or does it progress to bilateral deformity? Foot Ankle Int. 34 (4), 498–503.

Yu, S.M., Yu, J.S., 2015. Calcaneal avulsion fractures: an often forgotten diagnosis. AJR: Am. J. Roentgenol. 205 (5), 1061–1067.

Zadpoor, A.A., Nikooyan, A.A., 2011. The relationship between lower-extremity stress fractures and the ground reaction force: a systematic review. Clin. Biomech. 26 (1), 23–28.

Zhou, J.-P., Yu, J.-F., Feng, Y.-N., Liu, C.-L., Su, P., Shen, S.-H., et al., 2020. Modulation in the elastic properties of gastrocnemius muscle heads in individuals with plantar fasciitis and its relationship with pain. Sci. Rep. 10, 2770. https://doi.org/10.1038/s41598-020-59715-8.

Zifchock, R.A., Piazza, S.J., 2004. Investigation of the validity of modeling the Achilles tendon as having a single insertion site. Clin. Biomech. 19 (3), 303–307.

Zwirner, J., Singh, A., Templer, F., Ondruschka, B., Hammer, N., 2021. Why heel spurs are traction spurs after all. Sci. Rep. 11, 13291. https://doi.org/10.1038/s41598-021-92664-4.

Appendix

A Further concepts in gait mechanics

This book avoids equations that are used for calculating gait speed or the costs of locomotion that can become distracting to those unlikely to need them in a clinical situation. However, the concepts of leg length on influencing gait speed and those involved in the cost of locomotion are expanded upon a little further here as they have direct effects on the preferred gait speed and the risk of fatigue within individuals that can help explain gait dysfunction and pathomechanics.

B Lower limb length and gait speed

Optimum stride length at maximum velocity has a high correlation to lower limb length, functional and anatomical. Clinically this is usually referred the 'leg length'. Stride length is approximately 2.1–2.5 times the lower limb length. Thus, the ability to adjust functional leg length via hip, knee, and ankle flexion–extension angles is important in setting the functional lower limb length. Utilizing transverse plane pelvic rotational motions during gait also increases stride length, allowing those individuals with a wider pelvis to compensate for shorter limbs. This probably explains why those with wider pelvises (a morphology more common in females) are not hindered in their ability to walk long distances with shorter lower limbs on average than those with narrower pelvises (a morphology more common in males). In those with wider pelvises but shorter lower limbs, gait energetics are maintained by increasing anterior pelvic transverse plane rotation on the stance limb during late midstance. This gives them greater ability to functionally lengthen the stride of the swing limb via their hip joints. Thus, lower limb joint angles and motions in all orthogonal planes act as important adaptable control mechanisms of stride length and gait speed via altering limb lengths at any particular moment during gait. Wider pelvises seem to provide extra options that allow easier variance in gait speed without dramatically decreasing energetic efficiency from the preferred gait speed (Wall-Scheffler and Myers, 2017).

In gait, the CoM is moving in an arc of a circle with a radius set by the leg length rotating at the ankle acting under the force of gravity that is accelerating it from the centre of the circle (the ankle joint axis during late midstance) after absolute midstance. Therefore, functional leg length and gravity determine maximum walking speed. The speed at which the CoM can be moved (V) will result from the radius derived from the stance leg length (L) and the acceleration of gravity (g) at 9.8 m/s.

Thus:

$$g = V^2/L$$

$V = $ square root of gL.

Dorsiflexing the ankle at heel strike and plantarflexing the ankle during forefoot strike and acceleration helps to functionally lengthen the lower limb at these periods during the stance phase on gait. Thus, the length of the posterior tubercle of the calcaneus and the total length of the foot and digits also have a role to play in creating the functional limb length during stance phase events.

Although lower limb length is a good predictor of maximum walking speed, it does not predict the maximal walking speed of experienced race walkers who appear to optimise their technique to make use of nonvisible flight periods of less than 40 ms (Harrison et al., 2018). Thus, experienced race walker 'run' without being noticed.

Theoretically, a longer lower limb length will produce a greater velocity for a given angular velocity, but a longer lower limb is also accompanied by greater inertia from a potentially larger mass. In sprinting, running speed does not seem to corelate directly to lower limb length. However, a longer leg in sprinting does predict a decrease in step frequency over a distance and provides a longer support time and support distance at a given speed, while also influencing lower limb joint angles (Miyashiro et al., 2019).

C Locomotive costs

Locomotor economy ($m\,kg\,J^{-1}$) can be defined as the mass-specific distance travelled per unit of energy expended expresses as the inverse cost of transport.

To calculate locomotive costs, locomotive expenditure is divided by body mass to give the mass-specific cost of locomotion:

$J\,kg^{-1}\,s^{-1}$

And then this is divided by speed: $m\,s^{-1}$

To give the cost of transport: $J\,kg^{-1}\,m^{-1}$

The energy expended to move one kilogram of body mass one meter can then be calculated.

D Locomotive power

Each footstep can generate power because every step utilises energy and energy cannot be destroyed. Indeed, if the energy of all human walking in the world could be collected and stored in batteries, the power created would do much to solve the world's energy crisis. Power is a measure of how quickly energy is transferred. The unit of power is the watt (W) which is equal to one joule per second. It represents the rate at which work is done.

$$\text{Power} = \frac{\text{Work Done}}{\text{Time Taken}}$$

$$\text{Power} = \frac{\text{Energy Expenditure (W)}}{\text{Time Taken}}$$

E Locomotive work done

Force (Newtons) for distance (metres) moved. SI unit for work is the Joule.

$$N \times m = \text{Work (J)}$$

REFERENCES

Harrison, A.J., Molloy, P.G., Furlong, L.-A.M., 2018. Does the McNeill Alexander model accurately predict maximum walking speed in novice and experienced race walkers?. J. Sport Health Sci. 7 (3), 372–377.

Miyashiro, K., Nagahara, R., Yamamoto, K., Nishijima, T., 2019. Kinematics of maximal speed sprinting with different running speed, leg length, and step characteristics. Front. Sports Active Living 1, 37. https://doi.org/10.3389/fspor.2019.00037.

Wall-Scheffler, C.M., Myers, M.J., 2017. The biomechanical and energetic advantages of a mediolaterally wide pelvis in women. Anat. Rec. 300 (4), 764–775.

Glossary: Gait and pathomechanical principles

Absolute midstance the middle of midstance in single-limb support where the CoM reaches its highest point

Accessory bone (ossicle) an inconsistent (supernumerary) bone existing as an anatomical variant

Acetabular anteversion anterior positioning of the acetabulum (socket) of the hip that internally rotates the lower limb's position. It is generally angled anteriorly a little higher in females

Acetabular retroversion posterior positioning of the acetabulum of the hip that externally rotates the lower limb's position rather than expressing the normal anterior rotation. It is associated with an increased risk of hip osteoarthritis

Achilles external moment arm the distance between the centre of rotation around the fulcrum and the site of Achilles calcaneal effort used in heel lift to apply plantarflexion power. Usually, this is the distance between the axis of rotation (fulcrum) for the medial metatarsophalangeal joints and the distal attachment of the Achilles. This moment arm changes length during ankle and foot motion

Achilles internal moment arm the distance between the ankle joint axis of rotation (fulcrum) within the sagittal plane and the point of effort from the Achilles on the calcaneus. This moment arm length changes with ankle motion

Actuator a device that produces motion by converting energy and signals going into the system. Muscles behave as actuators

Aerial phase a phase of gait where neither foot is in ground contact, only demonstrated during running. The presence of an aerial phase defines running. Also known as float phase

Agonise the act of providing motion required by concentric muscle contraction

Agonist a muscle whose contraction provides the motion required

Akinesia restricted movement. Usually associated with motor deficits including slow voluntary movements, fatigability, and the freezing of motion

Angle of sacral incidence sagittal plane angle of the sacrum which correlates with the lumbar spine curvature

Angular motion/displacement motion of a body around an axis of rotation that causes displacement. Angular forces generated are moments

Anisotropy material properties expressed differently dependent on the plane of direction a load is applied to a structure

Ankle strategy moving the whole body's CoM primarily around ankle motion, with only limited use of hip torques

Ankle–subtalar joint complex the ankle (talocrural) joint and the talocalcaneal and talocalcaneonavicular joints (subtalar joint) working together to provide interdependent ankle–rearfoot motion. Functionally inseparable during locomotion

Antagonise the act of resisting or slowing motion

Antagonist muscle that opposes the primary motion, usually by lengthening under eccentric (or near-isometric) contraction and applying less force than the concentrically contracting agonist muscle that generates the motion. These actions are common in open chain swing phase. However, muscles often act as antagonists of GRF-induced motions rather than those generated by other muscles

Anteversion (of the femur) is the angle between the projection of the femoral neck axis onto the horizontal plane and the distal condylar angle within the horizontal plane. A positive anteversion brings the head of the femur anterior to the condylar axis. In excess, it will internally rotate the lower limb. If negative (retroversion), it will increase external rotation

Apophysis normal developmental outgrowth of bone (growth plate) arising from a separate growth centre that will fuse to the rest of the bone during bone maturation

Apophysitis inflammation of the growth plate in children and adolescents

Aponeurosis an area of highly structurally organised deep fascia

Aponeurotic fasciae highly organised thickenings of fascial tissue

Arch height index a measurement of dorsal foot vault (arch) height defined as the dorsal height at one half of the foot length

Arthrokinematic restriction restriction of articular freedom in a joint providing secondary coupled motion to a primary joint

Arthropathy a joint disease of which osteoarthritis is a common type, but also includes haematological, infective, and autoimmune causes of joint degeneration and articular damage

Autonomic nervous system the nervous system that controls largely unconscious actions such as heart rate, digestion, respiration, and pupillary response. It is comprised of the sympathetic and parasympathetic nervous systems

Autophagy body system-initiated cellular destruction used to clear out damaged or old cells in order to generate new healthier ones. The term is also used for the breakdown of muscle proteins during starvation

Auxetics materials that become thicker perpendicular to the applied load

Avascular necrosis tissue death resulting from loss of blood supply

Balance the ability to maintain the centre of gravity (centre of mass) within the body's base of support during motion

Barefoot condition locomotion without footwear

Barefoot footwear/shoes footwear characterised by zero lift (drop) between heel and forefoot, with no 'arch' support, minimal or no attempt at adding cushioning material, and a very thin sole of around 3–10 mm. They only serve to protect the foot from abrasions

Bi-articulate/bi-articular having or consisting of two joints. However, the term is often used to mean that multiple joints are involved or crossed by a muscle

Bicondylar angle the angle between the diaphysis or shaft of the femur and a line perpendicular to the infracondylar plane (also known as carrying angle)

Biological response in biomechanics, this refers to the ability of tissues to change their tissue structure in response to biomechanical loads. Thus, tissues strengthen by becoming more organised under increasing repetitive loads or weaken and become less organised under decreasing loads

Biotensegrity the ability of biological structures to use tensegrity principles of compression and tension strains via tension cables and compression beams (struts) to create stability, remain strong during motion, and also to avoid/minimise shear and bending moments

Biphasic materials containing quite different properties such as fluid and solid parts, or muscles with two bursts of activity on EMG

Body weight vector the direction of application of internal (body) derived forces from elements of body mass

Bone contusion bone bruise

Bunion an adventitious bursa developed within the cutaneous tissue, commonly found over the 1st metatarsal head often associated with hallux abducto valgus (hallux valgus) deformity. The term is often used erroneously to mean hallux abducto valgus deformity

Bursa(e) a sack of synovial fluid position to reduce friction between anatomy. They can be anatomical, being formed during embryogenesis and natural growth at sites of high anatomical shear, or adventitious, forming in later life in response to new repetitive shear stresses

Calcification a process by which calcium salts build up within soft tissue causing them to harden

Callus thickened skin that develops in response to high or excessive loading. It can be described as either physiological or pathological

Cam morphotype (hip) low concavity in both the anterior and posterior aspects of the femoral neck–shaft junction which when present is more prone to developing hip osteoarthritis

Carrying angle see bicondylar angle

Central nervous system the information processing and action responding centres found within the brain and spinal cord

Centre of ground reaction force the position of the centre of the ground reaction force when averaged to a single location (often termed centre of pressure in biomechanics, but not the same as true centre of pressure)

Centre of mass the point where the surrounded mass of a system is averaged to

Centre of pressure the position of the centre of all pressures averaged. Not to be confused with the centre point of averaged ground reaction forces, also known as centre of pressure, which in this text is termed centre of ground reaction force to avoid confusion

Central pattern generators neural circuits producing rhythmic outputs in the absence of rhythmic inputs that drive motor behaviours such as walking and breathing

Chopart's joint the transverse tarsal or midtarsal joint comprising the talonavicular and calcaneocuboid articulations

Claw toe a deformity of interphalangeal joint flexion within the digits

Closed chain motion motion where distal limb segments are restrained through contact with an external object, usually the ground
Collagen the most abundant protein in humans. It is the main structural protein within the extracellular matrix. Because fibrillar collagens are the primary components of connective tissue, collagen profoundly influences tissue mechanics
Collision an instance of objects colliding together (an impact)
Concentric contraction muscle contraction produced by muscle fibre shortening
Conduction the process by which energy is transferred through a material without movement of the material. In nerves, continuous conduction is the transmission of the action potential (electrical energy) moving along the entire axon length (as seen in unmyelinated axons). Saltatory conduction allows the action potential to jump from node to node along the myelin sheath
Contralateral belonging to or occurring on the opposite side of the body
Coordinate systems a method used to assign motion to orthogonal planes using a reference frame
Corns a circumscribed build-up of hard, thick areas of skin that indent the epidermis into the dermis
Coxa pedis a term for the talocalcaneonavicular joint (the talonavicular part)
Coxa recta a straight, nonspherical shaped superoposterior femoral head or head–neck junction on a short femoral head shifted anteroinferiorly that provides mammals with an easily stabilised hip joint for walking and particularly, running
Coxa rotunda a rounded femoral head or head–neck junction usually on a long femoral head, suitable for high hip mobility: as seen in mammals that are climbers and swimmers
Coxa vara a low femoral neck–shaft angle in the frontal plane that causes the knees to become positioned closer together via a greater femoral adduction (varus) angle
Coxa valga a high femoral neck–shaft angle in the frontal plane that causes the knees to be positioned further apart through a reduced femoral adduction (varus) angle
Coxal bone the ilium, ischium, and pubis joined together. See also innominate bone
Creep deformation under sustained loading. A form of fatigue behaviour
Crural fascia see fascia cruris
Crosstalk recording information that is contaminated from other sources
Cushioning cushioning is defined as the reduction in the amplitude of the vertical ground reaction force during impact
Davis's law a model that explains how soft tissue adapts and remodels under biomechanical loads. A 'law' now being explained through the increased understanding of the interrelationship between biochemical cellular signalling and biomechanical stresses
Deflection a degree by which a structural element is displaced under load, as seen in bending a beam
Delayed-onset muscle soreness pain in muscle developed on the following day after high exercise, resulting in oedema and inflammation. Pain probably derives from the surrounding deep fascia and muscular fascia
Densification a change in viscosity caused by high concentrations of hyaluronic acid within fasciae that impedes free gliding between fascial layers
Dermatological pertaining to skin structure, function, and diseases
Dermis the 'stretchy' part of the cutaneous tissues made primarily of connective tissue that underlies the epidermis. It binds the skin to the superficial fascia
Developmental plasticity the ability of a species (genotype) to adapt anatomy during development in response to environmental pressures, creating changes that form variable phenotypes
Diabetes Diabetes mellitus as opposed to diabetes insipidus, where the second condition is always fully identified within the text
Diabetes mellitus see diabetes and diabetic
Diabetic a sufferer of diabetes mellitus
Diaphysis (plural: diaphyses) the shaft or central part of a long bone
Dizygotic twins nonidentical twins
Dorsal surface the posterior surface of the body that includes the inferior or plantar surface of the foot. The dorsal surface of the foot is its anterior surface, and thus part of the ventral surface of the body
Dorsiflexion extension motion of the ankle or foot joints
Double-limb stance (or support) when both limbs are in part of their stance phase together
Double stance see double-limb stance
Dynapenia loss of strength, power, and force generation in muscles, usually due to age but without necessarily a loss of muscle mass (sarcopenia)

Early midstance the period of single-limb stance phase between loading response to absolute midstance and after contralateral preswing, and occurring concurrently during contralateral early swing phase

Eccentric contraction muscle power generation produced under muscle fibre lengthening. A contraction of low metabolic output by high force production

Ehlers–Danlos syndrome a group of rare inherited conditions that affect connective tissues as a result of abnormal structure or function of collagen. Among the problems they cause include joint hypermobility, fragile but flexible skin, eye diseases, and cardiac and vascular fragility

Elastin the main elastic protein found within mammals

Endomysium a delicate framework of connective tissue surrounding each individual muscle fibre. It contains the vessels and nerves supplying muscle fibres and the proteoglycan matrix for ion flux and metabolic exchange necessary for muscle cell function

Energetics energy under transformation. The branch of science that deals with the properties of energy and the way in which it is redistributed in physical, chemical, or biological processes

Enthesis organ (entheses) the more anatomically specialised tendon or ligament attachments to bone

Epidermis outermost layer of the skin composed of epithelial cells. Grows from epidermal stems cells derived from its inner layer (stratum germinativum)

Epigenetics the study of how behaviour and environment can influence the activation and expression of genes, without altering DNA sequences. Effects can bridge generations

Epimysium the sheath (tube) of connective tissue that envelopes a muscle. Links to other myofascial tissues and the deep fascia

Epiphysis (plural: epiphyses) the expanded ends of long bones which ossify separately from the shaft and usually contain the articular surfaces

Epiphyseal line a dense ossification line found in mature bone indicating the previous growth plate's closure

Equinus meaning 'horse like'. Used to denote limitation of ankle dorsiflexion (extension) but has not been quantitatively defined with agreement. Also used for limited or absent midfoot dorsiflexion (forefoot equinus). In gait, the term should only be used for persistent toe walking as a result of insufficient ankle dorsiflexion to provide plantigrade foot posture

Excessive (excess) pronation see hyperpronation

External forces/moments forces or moments derived from sources external to the body. The most common are those provided via ground contact, the ground reaction force, but also include friction and air resistance

External moment arm the moment acting at a joint to overcome external moment as a sum of all the moments (internal and external), creating the net joint motion. For example, the Achilles external moment arm that creates the plantarflexion moment of the foot by extension of the metatarsophalangeal joints at heel lift via ankle rotation

Extensor apparatus (foot) the highly structured connective tissue found around the metatarsophalangeal joints. Very similar in form to the extensor apparatus of the hand

Extensor sling the proximal section of the extensor apparatus that closely attaches to the deep transverse metatarsal ligament

Extensor wing the distal segment of the extensor apparatus that has connective tissue fibres that run obliquely. Functions with the lumbrical muscles to provide proximal phalangeal stability

Exteroceptive relating to sensing external stimuli to the body

Fascia(e) collagen-based connective tissue of variable fibre organisation. The term can be used for less specialised connective tissues only, or it can be used more inclusively for all connective tissues including ligaments, tendon, and bone

Fascia cruris deep fascia that forms a sheath enveloping the leg, continuous with the fascia lata above and that of the foot below

Fascia lata deep fascia than forms an enclosing sheath over the muscles of the thigh and is continuous with the deep fascia of the leg below
(fascia cruris)

Fast-twitch fatigable muscle fibres also known as type IIB(b) or type IIX fibres that generate fast contraction via anaerobic metabolism. They are more easily fatigued compare to slow-twitch type I fibres

Fast-twitch fatigue-resistant muscle fibres also known as type IIA (a) fibres that generate moderately fast contraction, but they are more fatigue-resistant than type IIB (IIX) fibres

Fatigue physiological changes brought about through repetitive exercise. A reduction in material properties from cyclical loading, properties that are also present in all biological materials as a result of their viscoelastic nature

Fatigue fracture fracture of a material under stress below its expected material strength. A more appropriate term for clinical 'stress fractures' that occur in bone

Feedback loops/controls processing incoming information in creating a response

Feedforward loops/controls anticipatory regulation of expected changes

Femoral anteversion anterior rotation of the femoral neck to the femoral shaft within the transverse plane. Highly variable in humans. When excessive, the femoral condyles become internally rotated

Femoral retroversion posterior rotation of the femoral neck to the femoral shaft within the transverse plane. Causes the femoral condyle to become externally rotated. Rare in humans but common in great apes

Fibrillation fraying or splitting of articular (hyaline) cartilage. Also used as a term for irregular heartbeats

Fibroblast the most common cell of connective tissue. Secretes collagen matrix. Also important in healing

Fibro-adipose a fibrous reinforced fatty tissue. Found in areas of frequent higher contact pressures such as the plantar foot and digits

Fibrocartilage a more fibrous form of cartilage tissue. Often found within tendons under compression and also forms the menisci of the knee

Fibrosis thickening or scarring of tissue as a result of high levels of connective tissue repairing

Flexor plate mechanism the combined anatomy of the extensor apparatus, plantar plate, plantar aponeurosis, and intrinsic and extrinsic muscles that resist digital extension

Float phase another name for the aerial phase

Foot-flat the end of loading response when the forefoot makes contact with the ground after heel contact. When the foot first becomes fully plantigrade within a step

Foot vault the combined medial and lateral longitudinal and the transverse arches of the foot when considered as a single 3D structural arrangement that lifts the central region of the foot from the plantar surface

Force closure an ability to maintain joint stability primarily through muscular contraction

Force–velocity relationship (of muscle) determines the maximum force produced by a muscle fibre (sarcomere). Slower muscle contraction results in greater force than can be achieved at higher rates of contraction. This is because slower contraction generates greater amounts of actin-to-myosin cross-bridging. Faster contraction velocities produce a reduced peak force due to less cross-bridging

Form closure an ability to maintain joint stability via articular congruency, aided by ligament restraint without muscle contraction

Froude number is a dimensionless number used as a way of determining leg length-to-body height ratios. If total leg length is used as the characteristic length, then the theoretical maximum speed of walking is a Froude number of 1.0 with a higher value resulting in take-off with the foot missing the ground. The typical transition speed from bipedal walking to running occurs at a Froude number of ≈ 0.5

Functional hallux limitus restriction of hallux extension at the 1st metatarsophalangeal (MTP) joint, only present during gait via a failure in functional biomechanics to offer freedom of hallux extension. There is no actual osseous or soft tissue restriction on the freedom of 1st MTP joint motion

Gait cycle the complete sequence from the point of one limb's initial point of contact to the next contact with the same foot. 100% of gait

Gastroc- or gastrocnemius lateralis the lateral muscle of gastrocnemius

Gastroc- or gastrocnemius medialis the medial muscle of gastrocnemius

Glabrous skin found on the plantar and palmar surfaces, it is innervated by specialised nerve endings for detailed tactile discrimination. It is thicker than other skin, having an extra layer (lucidum) within the epidermis

Glycation the covalent attachment of a sugar to a lipid or protein. It involves a spontaneous non-enzymatic reaction of free reducing sugars with free amino groups of proteins, DNA, and lipids to form products that lead to the formation of advanced glycation end-products. These are substances that change the nature of the viscoelastic properties of biological tissues, increasing their resting stiffness

Glycation end-products the substances that result from glycation that cause increased connective tissue stiffness

Glycoproteins sugar and protein combination, often used as a binding substance between other macromolecules or cellular surfaces

Glycosaminoglycans negatively charged polysaccharide compounds found on cellular surfaces and within cellular matrix: also known as mucopolysaccharides. They exist as four types: hyaluronic acid (hyaluronan), chondroitin sulphate (sulphate), heparan sulphate, and keratin sulphate

Gout a disease resulting from defective metabolism of uric acid that causes an inflammatory arthropathy. Any joint can be affected, but small joints of the feet are common sites. Often associated with kidney disease. However, genetic and environmental factors are also at play

Ground reaction force the force generated between an object (human body) and the ground. Its forces are expressed under Newton's third law

Haemarthrosis bleeding into a joint space. Can be caused by injury or physiological disease such as haemophilia or scurvy

Haematoma a pool of blood that forms within tissue or a body space, usually as a result of ruptured blood vessels following trauma. Common under the skin and within the dermis of the foot when gait or footwear-induced forces become repetitively high and/or prolonged at the same site

Haemophilia a group of genetic diseases that cause dysfunctional clotting proteins (factors) most commonly affecting males. Haemarthrosis is common among sufferers

Hallux abducto valgus deformity of the 1st metatarsophalangeal joint and/or 1st metatarsal where the hallux becomes abducted to the midline of the body and often rotates into valgus. The 1st metatarsal often concurrently moves into adduction and eversion

Hallux flexus a deformity of excessive hallux plantarflexion and reduced extension

Hallux limitus restriction in hallux extension of the 1st MTP which may be due to osteoarthritis or other degenerative joint changes

Hallux rigidus osteoarthritis or other joint disease within the 1st MTP joint causing loss of joint motion. Usually, this is a term used for the most severe cases of loss of motion

Hallux valgus shortened version of hallux abducto valgus. Also erroneously called a bunion

Hammer toes a deformity of toes where middle toe joints (interphalangeal joints) are drawn into flexion, often concurrent with increased metatarsophalangeal joint extension

Heel contact the foot making first contact with the ground via the heel during a step. Normal first contact position during walking, but also common in running

Hip strategy a technique of moving the body as a double-segmented inverted pendulum in counterphase motion at the hip and ankle so that the HAT segment rotates forward and downward, imposing an extension on the lower limb while decreasing the moment of inertia about the ankle. This permits the ankle torque to affect a higher angular acceleration of the HAT's CoM

Hogging upward bending deflection. A negative bending moment

Homeostasis the dynamic process by which an organism maintains and controls its internal environmental stability against external perturbations

Hominins the human family, past and present

Hormones chemical messengers within the body

Hyaline cartilage articular cartilage found covering joint facets that in health has a low coefficient of friction

Hydrostatic pressure/properties the pressure exerted by a fluid within a confined space

Hyperextension a range of extension motion in excess of need or beyond the safe range of motion to avoid joint damage. Most associated with the knee joint that should only operate in degrees of flexion

Hyperflexion a range of flexion motion excess to need or beyond the safe range of motion to avoid joint damage

Hyperglycaemia excessively elevated levels of blood glucose

Hypermobility flexibility of motion greater than that necessary or usually desirable for joint stability. Flexibility is usually expressed via connective tissue mechanical properties. Can be an isolated finding (joint hypermobility syndrome or hypermobility spectrum disorder) or associated with systemic genetic connective tissue disorders such as Ehlers–Danlos syndrome, Marfan's syndrome, osteogenesis imperfecta, or Stickler syndrome

Hyperpronation pronation of the foot (increasing the prone posture of the plantar aspect of the foot) beyond that necessary for achieving safe kinematic and kinetic events during weightbearing activities

Hypomobility restricted motion. Usually refers to widespread body-wide restriction in motion which can be idiopathic in origin

Hysteresis the amount of energy lost during deformation and recoil of a material

Iliotibial band another name for iliotibial tract as used in symptomatic iliotibial band syndrome

Iliotibial tract highly organised fascial thickening of the lateral deep fascia (fascia lata) of the thigh. It runs from the lateral hip musculature to the proximal anterolateral tibia

Impact the force or energy derived from a collision (impact) event

Infarction tissue death due to inadequate blood supply from any arterial blockages, mechanical compressions, or vessel rupture (see avascular necrosis)

Initial contact the first point of contact between the lower limb and the ground. The start of the gait cycle at 0% and its completion at 100% at the next initial contact after swing phase

Innominate bone the pelvic bone formed by fusion of the ilium, ischium, and pubis (see coxal bone)

In-series elastic components also referred to as series elastic components. The contractile and the elastic components of muscles and tendons that act as a spring to store elastic energy when a muscle is loaded. Usually functioning in chains of myofascial tissue across multiple muscles

In-series stiffness the stiffness (resistance to deformation) of the contractile and elastic elements of the myofascial unit
Internal forces/moments the forces achieving the net of work of motion or stability (positive or negative) derived from within the body
Internal moment arms the distance between a muscle acting on a joint and the instantaneous joint axis (fulcrum). For example, the Achilles internal moment arm derived from the distance between the calcaneal attachment and the ankle joint axis
Ipsilateral belonging to or occurring on the same side of the body
Isometric contraction muscle contraction without muscle fibre length changes
Isotropy material property behaviours being equal in all directions
Kager's fat pad an adipose structure found between the Achilles, flexor hallucis longus, and the calcaneus. Contains extensive fascial connections to surrounding anatomy
Lacuna (plural lacunae) a cavity or depression, especially within bone. It provides the space to house living cells such as the osteocytes of bone or chondrocytes of cartilage
Late midstance the period of the walking gait cycle that begins after absolute midstance and ends with heel lift. It is associated with energy storage within the Achilles tendon during triceps surae's braking of ankle dorsiflexion that is necessary to gain heel lift and aid acceleration
Left lateral decubitus position lying on the left side of the body
Line of progression the line that runs between the footprints of left and right feet indicates the direction of the body's CoM motion during locomotion. In straight directions of locomotion, the line of progression should align with lower limb joint's flexions and extensions
Lisfranc's joint the tarsometatarsal joints
Loading rate speed in time of the application of force or pressure
Loading response the period between initial contact and the ending of braking deceleration of the CoM. In walking, it starts with heel strike and ends at forefoot loading with the foot prone (flat) and thus fully plantigrade to the support surface. This should be concurrent with contralateral toe-off. In running, it is the period from initial contact throughout braking to the start of acceleration around the middle of stance phase
Lower leg shank segment from below the knee to the ankle
Lubrication the use of substances leading to reduction in friction coefficients
Marfan's syndrome a genetic disorder of connective tissues. Tends to cause individuals to be tall and thin with long limbs and appendages and express hypermobility and sclerosis. Sufferers commonly develop mitral valve prolapse, aortic aneurysms, and respiratory disorders
Mechanoreceptors sensory receptors converting mechanical stimuli within tissues into neural impulses
Mechanostat the process by which mechanical loading influence tissue structure formation
Mechanostat model/theory a theory developed by Frost that describes the adaptation of bone tissue to its mechanical environment by a simple feedback loop. Thus, bone under higher stresses becomes denser in its tissue structure while areas under low loading become less dense and over time, can lose bone substance if loads continue to reduce
Mechanotransduction the ability for cells to respond appropriately to mechanical stress that should promote tissue strengthening and appropriate tissue construction by converting mechanical signals into biochemically induced responses
Metaphysis (plural: metaphyses) the areas of long bone adjacent to the growth plate between the shaft (diaphysis) and the expanded ends (epiphyses)
Metaphyseal zone pertaining to the metaphysis
Metatarsophalangeal (MTP) joint torque the critical level of stiffness that is required for stable MTP joint extension and stability, giving a threshold to the elastic benefit of MTP joint dorsiflexion
Metatarsus adductus a morphological alignment of the metatarsals of adduction (towards the midline of the body in the transverse plane) when compared to the rearfoot and midfoot
Metatarsus primus elevatus a 1st metatarsal with a low declination angle positioned above the other metatarsals in the frontal and sagittal planes
Midtarsal joint the talonavicular and calcaneocuboid joints as a functional unit
Minimalist footwear/shoes footwear designed to create minimal interference with natural foot motion by expressing high flexibility, by being very light, and by possessing no lift at the heel (low heel-to-toe drop) or any motion-control (stability) device
Mitochondria generators of energy-rich nucleotides and ATP. The site of cellular respiration
Moments forces of rotation

Monozygotic twins genetically identical twins
Motor unit a motor neurone and all the muscle fibres it stimulates
MTP joint torque see metatarsophalangeal joint torque
Multi-articulate/multi-articular having or consisting of multiple joints
Multiple sclerosis a disease that causes myelin sheath scarring that interferes with action potential transmission
Muscle fatigue inability to maintain contraction or tension strength within muscle as a result of an inability to produce enough ATP to meet contraction demand
Muscle spindles mechanoreceptors within muscles, also known as intrafusal fibres
Myofascia/myofascial muscle and its connective tissues/pertaining to
Myofibroblasts cells that provide a contractile element to connective tissues
Myofibrosis replacement of muscle tissue by fibrous tissue
Myonecrosis infarction (avascular necrosis) of skeletal muscle due to infection, trauma, or other loss of arteria blood flow
Myosin a superfamily of motor proteins used to form thick contractile filaments in muscle fibrils but also used for motility in virtually all eukaryotic cells. They are ATP-dependent for their action with actin
Myotendinous junction the area of union of muscle to tendon
Myotendinous unit the muscle fibres and all its connective tissues (including tendon) functioning together
Near-isometric contraction muscle contraction without muscle fibre length changes, reflecting the fact that associated connective tissue does lengthen under loading due to its viscoelastic behaviour
Necrosis tissue or organ death
Neuromuscular control the ability to produce controlled movement through coordinated muscle activity. It is an unconscious efferent motor response of a muscle to a signal regarding dynamic joint stability
Neuromuscular disease a very broad term encompassing many different conditions that impair muscle function, either directly as pathologies within voluntary (skeletal) muscles or indirectly as pathologies of the peripheral nervous system or the neuromuscular junctions
Neuromuscular junction a chemical synapse between the motor nerve and muscle fibres that allows the action potential to pass from the axon to the muscle fibre
Neurone nerve cell. The axon is the part of the nerve cell that carries that action potential
Oedema excess fluid in body cavities or within the tissues, often causing swelling
Open chain motion motion with distal limb segments free to move
Osgood-Schlatter's disease traction apophysitis of the tibial tuberosity for the patella ligament attachment
Osteitis pubis inflammation of the pubic symphysis and/or surrounding bone
Osteoarthritis inflammatory degenerative joint disease of primarily mechanical origin, better termed osteoarthrosis
Osteoarthrosis degenerative joint disease of mechanical origin causing the loss of articular cartilage. See osteoarthritis (OA)
Osteoblasts large specialised mesenchymal stem cells that work in groups to synthesise bone matrix and coordinate mineralisation of the skeleton. They are used for initial bone growth and for mature bone maintenance and remodelling
Osteoclasts large multinucleated cells that degrade bone for normal remodelling and repair processes. They can mediate bone loss in pathological situations by their increased resorptive activity
Osteocytes cells found in the lacunae of fully formed calcified bone. They are derived from osteoblasts surrounded by the products they secreted and make up 90%–95% of bone cells. Osteocytes play a regulatory role in orchestrating bone's response to immunological changes under inflammatory changes
Osteogenesis imperfecta a group of genetic diseases, often known as brittle bone disorders. Fractures occur under little or no trauma. Associated with short stature, scoliosis, joint contracture deformities, hearing loss, blue sclerae of the eyes, and respiratory issues
Osteokinematic restriction joint motion restriction in the primary joint of interest
Osteomalacia a softening and weakening of the bone due to reduced mineral density that can lead to pain and deformity within bones. Usually, it is a result of low dietary calcium or deficiency of vitamin D required for calcium uptake and utilisation. Known as rickets in individuals where bone is still immature
Osteonecrosis bone death as a result of the interruption of blood flow (infarction) to bone tissue by any cause
Osteoporosis a condition where the rate of bone creation is unable to keep pace with the normal processes of bone reabsorption. It occurs as part of bone health and maintenance resulting in reduced bone mass. Trabecular bone is most affected. It results in increased bone fragility under locomotive loads and trauma
Osteotendinous junction the area of union between tendon and bone. Also known as the enthesis organ
Overpronation see hyperpronation

Parallel elastic components the components of a muscle that provide resistive tension when a muscle is stretched passively. Titin plays a major part in this as do the muscle's associated connective tissues
Paratenon loose elastic connective tissue sleeve around certain tendons
Patellar tracking how well the patella moves within the patellofemoral joint. Poor patellar tracking is referred to as maltracking
Pathomechanical/pathomechanics the altered mechanics involved in the development or result of damage within tissues, particularly those of the musculoskeletal (locomotive) system and skin
Pathomechanical summation the adding up of lifestyle and environmental risks of developing locomotive tissue pathology to establish a high or low tissue threshold of stress capability before damage occurs
Peak pressure highest recorded pressure
Pennation (pinnation) a feather-like or multidivided arrangement seen in the fibre organisation of some muscles, pulling on a tendon at an angle (pinnation angle). See pinnation
Perimysium the connective tissue (fascial) sheath surrounding a bundle of nerve fibres
Periosteum connective tissue (fascial) membrane or sheath covering bones
Peripheral nervous system consists of all the nerves and ganglia outside the central nervous system (CNS) of the brain and spinal cord that are responsible for carrying signals to and from the CNS to the body
Perturbation an event that changes normal motions and locomotion patterns
Pes cavus a foot with a particularly high plantar vault profile. Often associated with greater stiffness than other foot profiles. Usually idiopathic and developmental but can also be associated with neurological disorders, especially if vault height has become greater over time
Pes planovalgus a low plantar vault foot profile associated with eversion of the rearfoot. Often more flexible than other foot vault profiles. Can be idiopathic and developmental in origin, can be part of connective tissue hypermobility issues, or acquired through plantar ligament ruptures and/or muscular dysfunctions
Pes planus a low-profiled plantar foot vault as seen in pes planovalgus feet, but without necessarily displaying any rearfoot eversion posture
Pes rectus more middle-ranged plantar vault foot profiles. Where pes cavus and pes planus arbitrarily begin and pes rectus starts or stops is somewhat debatable, but it is often dictated by calcaneal inclination angles and metatarsal declination angles measured on lateral view radiographs. A technique itself with known errors
Pinnation (pennation) feather-like or multidivided arrangement seen in fibre organisation of some muscles, pulling on a tendon at an angle (pennation or pinnation angle)
Plantar digital neurofibroma a benign tumour that forms from the endoneurium (connective tissue of peripheral nerves) around plantar nerves that supply the digits. Thought to derive from repetitive mechanical irritation. It can cause symptoms consistent with a local nerve entrapment
Plantar fasciopathy a common cause of heel pain involving degenerative changes within the proximal attachment of the plantar aponeurosis. Often called plantar fasciitis. In essence, it is an enthesopathy with surrounding degeneration and inflammation of the proximal plantar aponeurosis
Plantarflexion flexion motion of the ankle and foot joints
Poisson's ratio the change in a material's dimensions as a result of stress
Poroelastic properties the interacting mechanics of fluid pressures within solid materials under load
Poroviscoelasticity poroelastic properties acting within viscoelastic materials (see also viscoporoelasticity, as both terms are used)
Postural control the ability to control posture
Postural equilibrium the state of forces balanced and acting on the centre of mass/centre of gravity within minimum limits of stability from an optimal body position, both in statics and dynamics
Power work done divided by the time taken. The scientific (SI) unit is the watt (W)
Pressure a scalar; force divided by area
Preswing the period (~10%) of the gait cycle at the end of stance phase that ends with toe-off before initial swing. It involves the initiation of muscle activity required for early swing phase as the contralateral foot completes its loading response
Pronation motions that present the plantar/palmar aspect of the foot/hand to the ground. In the plantar foot, it should be equated to the foot becoming more prone to the support surface, increasing its plantar surface contact area
Proprioceptors information on self-body motion, including that from the vestibular system within the inner ear. It refers to the acquisition of stimuli originating from musculoskeletal tissue receptors sent to the CNS for processing and excludes external environmental receptors that are still part of the whole sensorimotor system

Proteoglycans proteins that are heavily glycosylated with a protein core chain attached to glycosaminoglycan chains. These are incompressible molecules found between collagen fibres within the extracellular matrix

Q angle the angle formed between the quadriceps muscles' long axis and the bisection of the patellar ligament (tendon)

Rearfoot complex the ankle and subtalar joint along with the associated distal tibial, fibular, talar, and calcaneal anatomies. Another name for the ankle–subtalar complex, which means that the talocalcaneonavicular joint should also be considered

Relaxed calcaneal stance position the position of the rearfoot (heel) when stood in a relaxed base of support in static stance

Response mechanism a response that achieves joint stability and balance following a perturbation

Reversed or reverse windlass relaxing of the windlass mechanism and thus un-tensioning the foot vault and plantar aponeurosis by bringing the digits out of extension (usually passively under GRFs). Occurs after forefoot impact during gait as part of the energy-dissipation techniques

Rheumatoid arthritis an autoimmune disease associated with largely symmetrical chronic inflammatory synovitis and polyarthritis and other significant extra-articular manifestations

Rickets much like osteomalacia but occurring within immature bones

Rocker-bottom foot a convex plantar foot profile with loss of any longitudinal plantar vault 'arch' profile. Can be associated with a rigid congenital vertical talus encountered in very young children or acquired during adulthood associated with neuropathies as part of a Charcot foot. Charcot foot is most commonly associated with diabetic-induced neuropathy, for diabetes also damages the musculoskeletal tissues' mechanical properties extensively across the foot via ischaemic and biochemical changes. In acquired cases, the talus displaces its head in a plantar direction concurrently with multiple midfoot fractures and plantar ligament damage

Rocker-bottom shoe a shoe with a convex midsole and outsole

Sacral angle the angle between the long axis of the lumbar spine and the long axis of the sacrum

Sacral slope inclination angle of the sacrum

Sagging downward bending deflection. A positive bending moment

Sarcopenia atrophy of muscle

Sciatica a pain that radiates along the path of the sciatic nerve

Sclerosis the abnormal stiffening or hardening of soft tissue, usually caused by the replacement of organ or muscle-specific tissue with abnormal levels of connective tissue

Sensorimotor mechanisms/process a series of events to achieve a specific sable posture or motion via the sensorimotor system

Sensorimotor system or somatosensory system is the neuromechanical system used to maintain functional joint stability or locomotive homeostasis during body motions. It uses integration of CNS and PNS

Sesamoid bone bone located within a tendon or tendon–ligament complex

Sesamoid fibrocartilage/cartilage organised fibrocartilage within a tendon or tendon–ligament complex as a compression-resistive structure

Septa sheet-like structures subdividing anatomy that form highly organised connective tissue walls

Septum singular form of septa

Shank lower leg segment from below the knee to the ankle

Sharpey's fibres highly organised thread-like processes of the periosteum that penetrate the tissue of the superficial lamellae of the bone, particularly at sites of connective tissue attachments such as tendon, ligament, and aponeurosis entheses. Each fibre is accompanied by an arteriole and one or more nerve fibres

Silfverskiöld test the clinical test used to distinguish myofascial tightness within gastrocnemius from that derived from soleus or the Achilles tendon directly. It is the range of dorsiflexion possible with the knee extended compared to the range possible with the knee flexed. If ankle dorsiflexion is only limited with knee extension, gastrocnemius-associated tissues may be overly stiffened. Ankle dorsiflexion ranges at the time the knee is close to full extension are more important during walking than for running as the knee remains more flexed during running's stance phase

Sinding-Larsen and Johansson syndrome (disease) traction apophysitis of the interior part of the patella

Sinus tarsi the non-articular surface groove that separates the talus from the calcaneus below

Sinus tarsitis (tarsi syndrome) pain derived from the soft tissues within or near the sinus tarsi

Single-limb support when only one limb is in its stance phase and the other in its swing phase

Single stance see single-limb support

Somatic nervous system the sensory nerve endings of the afferent nerve fibres which relay sensation from the body to the central nervous system and the motor nerves of the efferent fibres that relay motor commands back to the musculoskeletal system

Somatosensory system see sensorimotor system
Spatial summation the number of muscle fibres activated at one time
Strain change in dimensions under stress
Stance phase the period of gait when the lower limb is in contact with the ground, usually via parts or the whole foot
Step the act of moving the body mass from one limb to the next limb
Step length the distance between a lower limb's first ground contact and the next opposite lower limb's first contact point (usually via the foot)
Stress force over a cross-sectional area
Stress fracture failure in bone due to fatigue response to cyclical loading. See fatigue fracture
Stress-stiffening the ability to increase elastic modulus under increasing stress to make materials behave more stiffly, creating nonlinear elasticity
Stride a complete sequence of one step to the next step with the same foot
Stride length the distance between two sequenced steps of the same foot
Stiffness in biology, it is the degree of resistance offered by tissues in response to forces of lengthening and/or widening
Subtalar joint the talocalcaneal and talocalcaneonavicular joints as a single functional unit. It expresses flexibility rather than freedom of motion due to strong intra-articular interosseous ligaments
Subtendon a hierarchical level below tendon but a tendon unit still made up of many fascicles. Most associated with the Achilles tendon hierarchy
Supination motions that present the plantar/palmar aspect of the foot/hand away from the ground. During locomotion, supination should be equated to the foot using motion to become less prone to the support surface, decreasing its plantar surface contact area. The term should be used with caution when describing motions within the foot
Swing phase the period of gait when a lower limb (usually via the foot) has no contact with the support surface during locomotion
Sympathetic nervous system the division of the autonomic nervous system involved in producing localised rapid physiological adjustments of the 'fight or flight' response such as increasing the heart rate, widening bronchial passages, constricting blood vessels, raising blood pressures, and starting sweating
Synapse the junction between two nerve cells, consisting of the synaptic cleft that must be crossed by diffusion of a neurotransmitter substance
Synovia synovial fluid. It is an alkaline viscid transparent fluid with a low friction coefficient
Synovial fluid the fluid found within articular spaces and structures and tendon sheaths to reduce friction (see synovia)
Synovial sheaths connective tissue (fascial) tunnels that protect tendons from friction that are filled with synovial fluid
Talocrural Joint ankle joint
Tarsal tunnel syndrome a nerve entrapment of the posterior tibial nerve within the medial ankle
Temporal summation the frequency of muscular activation
Tendon sheath connective tissue 'tube' investing tendon and filled with synovial fluid to reduce friction between the tendon and other anatomy. See synovial sheaths
Tensile strength the ability of a material to resist deformation under stress (see also ultimate tensile strength)
Tissue stress threshold the level of stress necessary to induce tissue damage
Tibial plafond the horizontal area of the tibia above its distal articular surface
Tibiofemoral Joint the part of the knee joint between the femur and tibia
Time-dependent properties the faster a material is loaded, the stiffer it behaves. A classic feature of viscoelastic materials such as biological tissues constructed of collagen based connective tissue
Tissue threshold the level of stress a tissue can be loaded with without damage
Titin giant protein important in muscle properties. Connects the Z-line to the M-line within sarcomeres, aiding force transmission and providing the muscle's resting tension
Toe-off the last moment of stance phase as all digital ground contact is lost
Toe region a period of initially rapid lengthening under low stresses demonstrated by viscoelastic materials. In connective tissue, this is caused by crimp configuration of collagen fibres
Translation movement where a fixed location within a body remains parallel to itself, moving at the same velocity and acceleration
Triplanar motion motion occurring in all three orthogonal planes at the same time (sagittal, frontal (coronal), transverse)
Tropocollagen the basic structural unit of collagen, consisting of a triple helix of polypeptide chains of approximately 1000 amino acid residues each
Turf toe force impingement injury to the 1st metatarsophalangeal joint, usually through hyperextension
Ultimate tensile strength the tensile stress that starts to produce plastic deformation of a material

Vault see foot vault

Venous calf pump the muscular force that returns venous blood from the lower extremity towards the heart during contractile activity of muscles within the calf that are utilising pressure changes and the unidirectional flow of blood through valves within the veins

Venous foot pump the muscular and ground reaction forces within and on the foot that during locomotion and stance, alter internal vein pressures and utilise the unidirectional flow of blood through the venous valves of the foot. The foot's venous blood is returned to the rest of the lower limb's veins and back towards the heart

Ventral surface the anterior surface of the body. It includes the superior or dorsal surface of the foot

Viscoporoelasticity poroelastic properties acting within viscoelastic materials (see also poroviscoelasticity, as both terms are used)

Vitamin D a fat-soluble secosteroid with an endocrine mechanism of action that is responsible for absorbing calcium, magnesium, and phosphate that also has many other biological effects. The most important form is vitamin D_3 produced within the skin that is converted into a hormone for calcium absorption: a mineral essential for musculoskeletal health

Voluntary nervous system the somatic nervous system that supplies voluntary control of the body. Although this is very often acting on a subconscious level, locomotive activities such as running or walking can be overridden consciously

Voluntary control use of the voluntary nervous system

Windlass mechanism a mechanism of the foot that utilises digital extension at the metatarsophalangeal joints to tension the plantar connective tissues (primarily the plantar aponeurosis) to induce increased levels of stiffness across the foot's vault. It imitates the effect of an engineering windlass to raise the height of the foot vault. The effect is used both in loading response before forefoot contact and later again in terminal stance after heel lift

Wolff's law a model to explain how osseous tissue adapts and remodels under stresses as part of biomechanical loading. Has applications to the mechanostat model

Work the product of force multiplied by the distance over which a force acts

Index

Note: Page numbers followed by *f* indicate figures and *t* indicate tables.

A

Abdominal aponeurosis, 295, 296*f*, 298
Abductor digiti minimi, 252*f*, 477–479, 483*f*, 538, 550, 554, 768, 798
Abductor hallucis, 91–92, 147–148, 477–479, 502–503, 505, 516, 549, 601, 790, 810
Acceleration, 3–4, 37–39, 87–89, 113, 498–499, 524, 534–535, 711–713, 768–772
Acceleration stiffness, 498–499
Accelerometers, 73, 840
Acetabular anteversion, 310, 310–311*f*
Acetabular labrum, 304, 307, 315–317
Acetabular notch, 307, 315–317
Acetabulum, 281, 304, 307, 310
Acetabulum pedis. *See* Talonavicular joint
Achilles enthesis, 420–423, 421*f*, 725–726
Achilles heterogenicity, 714–716, 715*f*
Achilles rupture, 428, 725–727
Achilles tendinopathy, 743
  barefoot running, 148
  in forefoot strikes, 724–725, 724–725*f*
  incidence, 713–714
  influencing factors, 714
  pathomechanics, 716, 717*f*
  running risk, 714
Achilles tendon, 28–31, 100, 134, 166, 418–420, 419*t*, 420*f*, 496, 510, 523
  dysfunction, 49
  mechanical properties, 419*t*
  moment arm, 137–138, 395
  strain distributions, 423–425, 424*f*
  strength and stiffness, 425–426
  stress contributions, 423
  triceps surae, 29, 412
Active foot-stiffening, 767–768, 767*f*
Active forefoot stiffening, 514–515
Adolescent gait, 183, 183*f*
Aerial phase, 11–12, 37, 107
Ageing
  changing gait strategy, 191
  endurance and fatigue in walking, 191
  of gait, 188–190
  inclined/declined surfaces, 96–98
Agonistic muscles, 16
AHF. *See* Arch height flexibility (AHF)
AHI. *See* Arch height index (AHI)
Alzheimer's disease, 210
Angle of gait, 69, 71, 71*f*
Angle of sacral incidence, 281

Angular momentum, 17, 37, 81–84
Anisomelia. *See* Leg length discrepancy (LLD)
Ankle and rearfoot, 702
  Achilles ruptures, 725–727, 726–727*f*
  Achilles tendinopathy, 713–716
  finite element analysis models, 703
  free-body diagrams (*see* Free-body diagrams)
  frontal and transverse plane relationships, 727–732, 728–729*f*, 731–732*f*
  frontal plane Achilles, 739–743, 741–742*f*
  intimate relationship, 703, 704*f*
  peroneals, 745–746, 745–746*f*
  posterior calcaneal avulsion fractures, and Sever's disease, 725–727, 726–727*f*
  primary motions, 702–703
  sagittal plane motion, 717–724
  sprain, 733–739
  tibialis posterior and the digital plantarflexors, 744, 744*f*
Ankle collateral ligaments, 389
Ankle dorsiflexion, 7
  articular loss, 720–721
  causes, 721
  cavoid foot, 722–723, 722*f*
Ankle equinus, 723
Ankle, functional unit
  adaptation and pathology, 437–439, 438*f*
  ankle–subtalar complex, 380
  axis location, 395
  biomechanics and energetics, 395–396, 397*f*
  collateral ligaments, 389, 391–392*f*
  concert with muscles, 394
  extensor muscles, 407–411, 408*f*, 410*f*
  kinematics, 381–384, 381–383*f*
  leg to foot rotations, 435–437, 436*f*
  lever arm, 397
  lever systems, 398–405
  movement, 395–396
  muscle action, 405–407, 406–407*f*
  osseous topography, 385–387, 386–388*f*
  peroneals, 432–433, 432*f*
  peroneus brevis, 433
  peroneus longus, 433–435, 434*f*
  plantarflexors, 395, 396*f*, 426–435, 427–428*f*
  primary flexor muscles, 411–426
  rearfoot, 384, 392–394
  rolling rotational-sliding motion, 395–396
  rotational motion, 395–396
  sprains, 387–388

  subtalar joint, 398
  subtalar ligaments, 389–392, 393–394*f*
  tibialis posterior, 429–432, 430–431*f*, 434–435, 434*f*
  tibiofibular ligaments, 388–389, 390*f*
  triplanar axis, 397
Ankle osteoarthritis, 206
Ankle plantarflexion, 7–8, 198, 383, 383*f*
Ankle–rearfoot ligaments, 392–394
Ankle rocker, 41–42, 42*f*
Ankle sprain
  eversion, 733, 734*f*
  inversion, 733, 735*f*
Ankle–subtalar complex, 380, 559–560
Ankle torque reduction, 189
Annulus fibrosus, 278–280, 278*f*
Anterior cruciate ligament, 173, 678, 678*f*
Anterior intermeniscal ligament, 343
Anterior knee pain, 672
Antetorsion, 312
Antiphase motions, 268, 269*f*
Aponeurotic stress
  loading response, 534
  stance phase, 533–534
Apophysitis, 697, 833–834
Arches, in vault, 471–473
Arch height flexibility (AHF), 467, 468*f*, 506, 520
Arch height index (AHI), 467, 511*f*, 520
Arcuate ligament, 355, 370
Arm swing, 2–3
  in running, 128–132
  on vertical and horizontal forces, 66
  in walking, 66
Arthritis, 203–207. *See also* Inflammatory arthropathy
Articular cartilage, 116, 279–280, 309, 566–567, 643
Articular failure, 643–644, 644–645*f*
Articular motion, 256–257
Assistor muscles, in extensor energy dissipation, 504–505
Asymptomatic pes planus, 163

B

Back pain, 52–53, 191–192
Barefoot
  running, 114–116, 147–148, 843
  walking, 501–502, 826

901

Barrel vault, 475–476
Base width, 69
Beighton scale score, 171f, 172
Biceps femoris, 96, 96f, 102–103, 258, 317, 370–371
Bifurcate ligament, 482, 555, 576
Bilateral amputation, 199
Bipedal gait, human, 12
Body weight vector (BWV), 54, 54f, 58f, 499
Bone injuries, 641–642, 641–643f
Bout, 176
Braking impulse, 87–89
BWV. See Body weight vector (BWV)

C

Cadence, 72
Calcaneal inclination angle, 146–147, 168, 169f, 520, 554, 721–722
Calcaneal pitch angle, 554, 556
Calcanean process, 573–574
Calcaneocuboid joint, 165–166, 435, 571–576, 575f, 578
Calcaneopedal unit, 464–465
Calf muscle strength, 190
Cam deformity, 310
Cantilever stiffness, 483–485
Capsular ligaments, 291, 316–317
Carpal tunnel syndrome, 173–174
Cartilage, 28, 273, 275, 281
Cavoid feet, 143–144, 170, 610, 722–723
Central edge angle of Wiberg, 310
Central fovea capitis, 309
Central inferior acetabular fossa, 307
Centre of mass (CoM), 2–5, 7, 496, 520
    acceleration, 43
    curvilinear motion, 10
    early midstance, 23, 38f
    head, arms, and trunk (HAT), 8, 10, 17
    influence of stability, 49–50
    inverted pendulum-like gait, 23–26
    late midstance, 39f
    limb's, 13–15
    oscillations, 2, 50, 93
    resistance moment, 51
    stability, 51f
    transfer of midstance, 37
    up-and-down motions, 24
Centre of pressure (CoP), 51, 57, 75–77, 77f
Centre of the ground reaction forces (CoGRF), 21, 31, 51–53, 58, 66–67, 77, 496, 509
Centrifugal forces (CFs), 3–4
Cerebral palsy (CP), 201–203
Charcot foot, 167, 215, 556, 649
Chopart's joint, 517, 571
Chronic ankle instability, 737–738
Chronic exercise-induced compartment syndrome, 854–856, 855f
Claw back, 34, 63, 64f
Cleavage line, 552–553
Cloister, 475–476
Close-packing joint, 263
CoGRF. See Centre of the ground reaction forces (CoGRF)
Collagen, 501

Collision, 107
    elastic, 6
    forefoot, 594, 752, 827f
    glancing, 4–5, 6f
    heel, 494–495
    loading and off-loading, 6, 7f
    plastic, 6
    pseudo-elastic, 6, 22–23
    in sport, 672
    walking, 7–8
Concave polygon, 7
Connective tissue, 17, 47–48, 244, 265, 411, 460, 462, 465–467, 486, 496, 579, 643, 648
Contact energy dissipation, 656–659, 657f
Core stability, 253–254, 254f
Corpus adiposum, 119, 526
Counternutation torque, 267–268, 269f
Coupling, 259–260
Coxa pedis, 304, 558
Coxa saltans, 316–317
Coxa valga, 158, 311, 680–681
Cross-talk, 84, 98–99
Crouch gait, 201, 693–695
Cruciate ligament injuries, 679
Cruising, 176
Cuboid joints, 573–576
Cuboid–navicular joint, 581
Cuneiform–cuboid joint, 571–572
Cuneonavicular ligaments, 482, 581
Cushioning, 114–116
Cutaneous soft tissues, 525–526

D

Deceleration, 56, 87–89, 128, 134, 656
Deep transverse metatarsal ligament (DTML), 469, 477–479, 485f, 486, 516, 530–531, 555, 590–591, 604, 607–608, 610, 790, 796, 800
Degenerative joint disease, 803–804
Deltoid ligaments, 385f, 389, 601, 733
Diabetes mellitus, 213–215, 646–649, 647f
Digital extension/dorsiflexion, 596–597
Digital neurofibromas, 555–556
Digital plantarflexors, 744, 744f
Digitigrade and unguligrade phases, 597–599
Disease-induced pathomechanics, 645–649
    diabetes mellitus, 646–649, 647f
    inflammatory arthropathies, 646
Distance running, 138–139, 724, 838
Divergent point, 21
Domical vault, 475–476
Dorsal interossei muscles, 482, 516–517, 517f, 606, 608, 767, 794–795
Dorsalis pedis fascia, 471–472
Double-limb contact phases, 11–12
Double pendulum gait, 31–34
DTML. See Deep transverse metatarsal ligament (DTML)

E

Early heel lift, 717–719, 718f
Elastic collision, 6
Elastic modulus, 121, 343, 504, 522

Elastic springs, 4–5, 92, 472–473, 535
Electrogoniometers, 73–74
Electromyography (EMG), 181, 503
Endurance, 191
Energetic collision strategy, in gait, 4–7, 6f
Energy dissipation, 28, 37, 522
    extensor, assistor muscles in, 504–505
    forefoot, 597
    tibialis anterior in, 503–504, 504f
    wobbling mass in lower leg, 505–506
Enhanced Coleman block test, 167
Entheses, 243
Equinus gait, 186–188, 188f
Erector spinae muscle, 260–261, 293, 293f
Eversion ankle sprain, 733, 734f
Exercise-induced leg pain, 856
Exertional compartment syndrome. See Chronic exercise-induced compartment syndrome
Expanding conical vault, 475–476, 480
Extension dysfunction, 801–803, 802f
Extensor digitorum brevis, 403, 409–411, 410f, 540
Extensor muscles
    ankle, 407–411, 408f, 410f
    hip, 325–327, 326f
    knee, 366–375, 366–367f
Extensor sling, 606–608, 608f
Extensor substitution, 405, 706, 752
Extensor wing, 607–610, 608f
External abdominal oblique, 296
External eversion moment, 81
External extension moment, 81
External flexion moment, 81
External inversion moment, 81
External valgus moment, 81
External varus moment, 81
Extrinsic dorsiflexors, 540, 541f
Extrinsic ligaments, 288–289, 288–289f
Extrinsic plantarflexors, 541–542, 542f

F

Fabello-fibular ligament, 355–356, 370
Fascia lata, 317
Fascial pathomechanics, 856–857
Fast twitch fibres, 118, 254–255
Fatigue, 367–368, 839–840
    in plantar intrinsics, 549
    in running, 139–140
    in walking, 191
Fat pad, biomechanics, 502
Fatty synovial fringe, 389
Femoral anteversion, 312–313, 312f, 312t
Femoral loading patterns, 313–315, 314–315f
Femoral neck–shaft angle, 311, 311f
1st ray, 589–590
Flat foot. See Pes planus
Flexor digitorum brevis, 147–148, 477–479, 549, 563
Flexor muscles, at knee, 368–372, 369–371f
Flexor plate mechanism, 609, 610f, 612f, 790
Float phase, 11–12, 37, 107, 305
Foot, 462–463, 462f
    as beam, 468–469

core stability, 472, 472f
functional units, 525–613
  cutaneous tissues, 525–526
  intermetatarsal joints, 584–592
  metatarsal base to head orientations, 593
  metatarsophalangeal joint and digits, 593–604
  metatarsophalangeal joint as stabilising fulcrum, 604–611
  midfoot, 571–584, 572f
  muscles, 538–550
  passive elastic elements, 528–531
  plantar aponeurosis, 531–538
  plantar fat pads, 526–527, 527f
  rearfoot, 556–571
  skeletal frame, 551–556
  tarsometatarsal joints, 584–592
in gait, 494–525
mechanical constraints on role
  adaptable material properties, 490–491
  neural transmission, 489–490
  size, 489
  transverse vault profile in foot stiffening, 491
  vault adjustability, 491
multi-tied asymmetrical expanded conical vault, 474–480
muscle action in energy dissipation, 502–505
in sensorimotor system, 460
shape, 519–520
skeletal frame, 506
triangle structures, 552–554
types, 143–146
vault, 464–474
vault stiffness, 480–487
viscous spring-damping in running, 521–524
Foot drop, 16
Foot energy dissipation, 521–523
Foot function variance
  midfoot break, 165–167, 166f
  pes cavus, 167–170
  pes planus, 160–165
Foot ligaments, 528
Foot, pathomechanics, 747–748
  articular misalignment, 747
  compensation mechanisms, 755–756, 756f
  dorsiflexor, 752–753, 752–753f
  early midstance, 757–761, 758f
  energy dissipation, 748–749
  functional stance, 748
  hallux abducto valgus, 811–821, 812–816f
  late midstance, 761–765, 762–763f
  metatarsal fatigue fractures, 822–827, 823–827f
  metatarsophalangeal joint, 776–789
  midfoot functions, 748
  peroneus longus loading dysfunction, 753–755, 754f
  plantar fasciopathy, 827–836
  plantar fat pad dysfunction, 749–751, 750–751f
  plantarflexor loading dysfunction, 752–753, 752 753f
  skeletal frame, 751–752

stance muscle activity, 757–761, 759–760f
terminal stance phase, 765–773
tibialis posterior dysfunction, 753–755, 754f, 773–776
Foot Posture Index (FPI), 140
Foot shortening, 768–772
Foot's shock attenuation, 748–749
Foot stiffness, 464–465, 465f
  pathomechanics, 765–767
  for running acceleration, 524
  transverse vault profile, 491
Foot strike position
  free-body diagrams, 843–849
  frontal plane effects, 849–851, 850f
  running, 132–134, 132–133f
  transverse plane-to-frontal plane relationships, 851, 851f
Foot vault, 152, 249
  arches in, 471–473
  height in foot function, 467
  injury rates associated with, 146–147
  lengthening and shortening, 521
  mechanics, 473
  and midfoot break, 469–470
  normal, 487–489
  pes cavus, 487–489
  pes planus, 487–489
  role of, 464–474
  span distance and curvature effects, 480–487, 481f
  stiffness, 480–489
  tensional resistance, 540
  transverse arch, 473–474
  viscoelastic asymmetrical conical, 486–487
Footwear, 157, 195–196
  biomechanics in running, 147–149
  design, 149–150
  and foot vaults, 152
Force closure, 260–263, 261–262f
Force vectors, 81–84
Forefoot
  contact, 403
  fat pads, 527
  loading, 573
  rocker, 43
  running, 505
  soft tissues, 594–596
  splaying, 496
Forefoot strike, 133, 848–849
  Achilles tendinopathy, 724–725, 724f
  toe-walking, 725, 725f
Free-body diagrams
  hip
    expanding complexity, 670–671
    frontal plane, 664–668, 665–668f
    principles, 649–650, 650–652f, 653
    sagittal plane, 654–655, 654–655f
    transverse plane, 668–669, 669–670f
  of iliotibial band syndrome, 685–691, 687–691f
  of knee, 672–677
  patellofemoral joint, 691–701, 693–694f
Froude number, 158
Functional hallux limitus, 801, 803–804

Functional pes planus, 161
Functional units
  ankle, 380–439
  foot, 525–613
  hip, 304–334
  knee, 334–380
  lumbar spine and pelvis, 266–304
  soft and hard tissue, 244–265

G

Gait
  cadence, 72
  data, 49–80
  in disease, 175–178
  energetics, 2–8
  head, arms, and trunk in, 8–9, 9f
  human models, 22–34
  interpreting pressure measurement, 74–77
  joint segment motion measurement, 73–74
  locomotor segments and pelvis in, 10
  motion, 10–12
  muscle function, 80–105
  in pregnancy, 173–174
  principles, 1–49
  Rancho Los Amigos divisions, 34–48
  running, 105–156
  spatial parameters, 69–71, 69t
  speed, 72–73, 73f
  stability maintenance in stance phase, 54
  symmetry, 72
  temporal parameters, 71–72, 72t
  upright posture, challenges of, 19–22
  variability in, 77–80
  variance in, 156–200
  velocity, 72
  walking phases, 12–19
Gait analysis data
  ground reaction force (GRF), 49–56
  stability, 49–56
Gait cycle, 35
  bipedal walking, 35–37
  events of, 36t
  Rancho Los Amigos divisions, 34–43, 35f
  running, 107–109, 108f
  stance phase, 36–37, 36f
  swing phase, 36
Gait determinants, 636–638, 637–638f
Gait disturbance, 191
Gait mechanics, 887
Gait speed, 72–73, 124–125, 492, 520–521
Gait variability, 69, 77–80, 189
Gait velocity, 19, 72, 188–190, 293, 303
Gastrocnemius muscle, 18, 28–29, 81, 85, 96–98, 102–103, 255, 336, 353, 374–375, 412, 416, 524, 550, 659, 661–662
Gastrointestinal tract, 243
Generalised joint hypermobility (GJH), 170–173
Glancing collision model, 4–5, 6f
Gluteal raphe, 301
Gluteus maximus, 13–15, 80, 89, 96–98, 102–103, 143, 258, 301, 323–325, 653, 656–659

Gluteus medius, 83–85, 124, 129, 323–324, 329–330, 660
Gluteus minimus, 316–317, 323–324, 330, 331f, 664
Goniometers, 73
Gout, 646
Gracilis, 374
Gravity, 10, 111
Greater trochanter pain syndrome, 689, 689f
Great tarsal joint, 581
Ground reaction force (GRF), 3–4, 503, 520
   anterior–posterior components in walking, 63
   claw back, 34, 63, 64f
   forefoot, 463f, 469, 491–492
   heel contact, 402
   heel strike in walking and running, 382
   measurements and interpretation, 56–59
   medial–lateral components, 66, 67f
   moment arm, 654, 654f, 657
   Pedotti diagrams, 66–68, 68f
   recording, 55–56
   vectors, 23, 55–56, 496, 671
   vertical components in walking, 59–63

H

Hallux abducto valgus (HAV), 163
   causes, 818, 821–822f
   degree of deformity, 818, 819f
   measurements, 818, 821f
   origins of, 816–821
   prevalence, 817, 817f
   skeletal remains, 818, 820f
Hallux flexus, 803–804
Hallux rigidus, 803–804
Hallux valgus. *See* Hallux abducto valgus (HAV)
Head, arms, and trunk (HAT), 8–9, 9f
Heel contact, 402–403, 402f
   compliance, 499–501
   walking, 335–336, 336f
Heel fat pad, 120–121, 120f
   in compliance, 501–502, 502f
   deep macrochamber layer, 119–120
   superficial microchamber, 119–120
Heel length, 137–138
Heel lift, 29–31, 89, 90f, 193, 469–470, 488, 494, 512–513, 517
Heel pressure, 166
Heel rocker, 41, 42f
Heel strike
   ankle's axis of rotation, 503
   energy dissipation, 500
   hip moment arms, 655
   role, 6–7
   running, 57, 57f, 105, 105f
   transient, 60
   walking, 57, 57f, 114
Heel-toe walking, 494–499
High gear propulsion, 525
Hip flexor, 13–15, 25–26, 90–91, 191, 324, 664
Hip, functional units
   abductors, 329–332, 330–332f
   adductors, 327–329, 328f
   ball-and-socket joint, 304
   deep rotators, 327, 328f
   extensors, 325–327, 326f
   fascia lata, 317
   flexor, 324, 324f
   frontal plane lever, 321–322, 321–322f
   instantaneous joint axis, 317–318, 318f
   kinematics, 305
   lever arms, 319
   ligaments, 315–317, 316f
   muscle action, 323–333, 323f
   osseous topography, 307–315, 307–309f
   pathology, 333–334
   pelvic stability, 304
   sagittal plane levers, 319–320, 320f
   stabilisation, 332–333
Hip hitch, 16, 198
Hip muscle, 298–299, 299–300f
Hip, pathomechanics
   free-body diagrams (*see* Free-body diagrams)
   kinematics, 655–659, 656f
   muscular activity, 652–653
   terminal stance, 659–663, 660–663f
Hip rotation angle, 312
Hip torques, 25
Hominin
   evolution, 19
   locomotive biomechanics, 2, 4f
Homo erectus, 2
Homo floresiensis, 2
Homo naledi, 2
'Hoop stress', 344
Human bipedal gait, 12
Human gait models, 22–34
Human inverted pendular walking gait model, 23–26
Huntington's disease, 210
Hypermobile cavus foot, 472–473
Hypermobile planus foot, 472–473
Hypermobility, 251
Hyperpronation, 25–26, 160–161, 174

I

Idiopathic equinus, 188
Idiopathic pes cavus, 167–168
Iliolumbar ligaments, 281, 288–289, 289f
Iliopatellar band, 357
Iliopsoas, 300
Iliotibial band syndrome, 685–691, 687–691f
Iliotibial tract, 357–358, 357f
Impulse, 113–114
Inclination–declination angles, 554
Independent ambulation, 176
Inflammatory arthropathy, 646
In vivo bone stresses, 860
Initial contact, 35, 37
In-shoe pressure systems, 74
Instantaneous joint axis location, 635–636, 635–636f
Insufficiency fractures, 642
Intermetatarsal joints, 584–592, 586f
Intermittent claudication, 212
Internal abdominal oblique, 296–297
Internal moment, 81, 83f
Interossei muscles, 608, 609f
Interosseous ligament, 285, 287, 581, 584–586, 590, 799
Interpreting pressure, 74–77
Interpubic disc, 285–286
Interspinous ligament, 275
Intervertebral discs, 277–280, 278–279f
Intra-individual gait variability, 189
Intrinsic ligaments, 287, 287–288f
Inversion ankle sprain, 733, 735f
Inversion sprain injury, 736–737, 739f
Inverted pendular gait
   energetics of, 26–27
   plantigrade foot in, 28–29

J

Jogging, 107, 108f, 125, 868
Joint angles, changes in, 198–199
Joint congruency, 263
Joint hypermobility, 170–173
Joint hypermobility syndrome (JHS), 170
Joints
   flexion angles, 37
   force vectors and angular momentum in, 81–84
   motion measurement, 73–74
   stability, clinical implications, 264–265
   stiffness, 118–119

K

Kager's fat pad region, 419–420, 421f
Kelvin element, 552
Kinetic energy, 4–5, 111, 524
Knee adduction moments, 158
Knee extension, 15
Knee flexion, 90–91, 119
Knee, functional unit
   active stabilisation effects, 375, 376f
   adaptation and pathology, 375–380, 377f
   articular surfaces, 334
   cruciate ligaments, 346–348, 346–349f
   extensor muscle action, 366–368, 366–367f
   flexor muscles, 368–372, 369–371f
   gastrocnemius, 375
   gracilis, 374
   iliotibial tract, 357–358, 357f
   kinematic role, 335–337, 335–337f
   lateral soft tissues, 355–357, 355–356f
   latter joint, 334
   lever systems, 362–365, 363–365f
   medial epicondyle, 349–352, 351–352f
   medial muscles, 352–354, 353–354f
   meniscal role, 344–345, 344–345f
   menisci, 342–343, 342–344f
   osseous topography, 339–345, 340–342f
   patellar ligament, 368
   patellofemoral joint, 358–361, 358–362f
   popliteus muscle action, 372–373, 373f
   posterior soft tissues, 349, 350f
   sartorius, 374, 374f
   stability, 334
   tibiofemoral joint, 334
Knee osteoarthritis, 206
Knee, pathomechanics

anterior and posterior translations, 678–679, 678–679f
free-body diagrams (*see* Free-body diagrams)
frontal plane, 679–684, 680–683f
mechanical efficiency, 672
pes anserine, 684, 684f
sport, 672
transverse plane, 685–691

L

Lateral ankle instability, 394
Lateral Lisfranc ligament, 590
Lateral longitudinal arch, 471–472
Lateral vault span, 482
Late swing phase, 656–659, 657f
Late terminal stance stiffness, 517–519
Latissimus dorsi, 291–292, 292f
Laws of motion, in gait cycle, 10–11
Ledderhose disease. *See* Plantar fibromatosis
Leg length discrepancy (LLD), 191–195, 192f
    functional discrepancy, 192
        measurement, 195
        posthip replacement, 192–193
        true discrepancy, 192
        walking gait, 194, 194f
Lever arm disease, 202
Ligament failures, 377–378
Ligamentum flavum, 275
Ligamentum orbicularis, 315
Ligamentum teres, 307
Limb
    advancement, 40–41
    amputation, 196
Lisfranc's joint, 517, 589
L4–L5 lumbar spine articulation, 280–281, 280f
Load frame, stressing, 554–555
Loading and off-loading collision, 6, 7f
Locomotive costs, 888
Locomotive power, 888
Locomotive soft tissues, 244
Locomotor economy, 888
Locomotor segments in gait, 10
Long digital flexors, 542
Longitudinal ligaments, 528–529
Loose-packing joint, 263
Low back pain, 301, 302f
Lower back dysfunction, 266
Lower limb, 10, 463, 733–736
    alignment, 158–159
    amputees, 196–199
    joint angles, 13–15, 15f
    joint compensations during running, 155
    kinematics, 72–73
    length and gait speed, 887
    motion, 82, 82f
    shock attenuation in, 117–121
    soft tissue compliance and stiffening, 92–93
Lumbar spine and pelvis, functional units
    adaptation and pathology, 301–304, 302f
    extrinsic ligaments, 288–289, 288–289f
    in-phase and antiphase motion, 268–270, 269–270f
    intrinsic ligaments, 287, 287–288f
    kinematic role, 266–270, 267–269f

load distribution, 289–290, 290f
lumbar vertebrae, 273–277, 273–277f
lumbosacral joint, 273–277, 273–277f, 281, 282–283f
muscle action, 291–301
pelvic ligaments, 286–287
pelvic tilt angle, 286, 286f
sacroiliac joint, 281–286, 283–285f
4th–5th (L4–L5) articulation, 280–281, 280f
vertebral bodies and intervertebral discs, 277–280, 278–279f
Lumbar vertebrae, 273–277, 273–277f
Lumbosacral joint, 273–277, 273–277f, 281, 282–283f

M

Male and female runners, 140–143
Maximal or extreme cushioning, 94
Medial collateral ligament, 348, 352f
Medial epicondyle, 349
Medial longitudinal arch, 471–472, 471f
Medial midtarsal joints, 572–573
Medial patellofemoral ligament, 350, 361
Medial tibial stress syndrome, 863–868, 864f
    aetiology, 865–868
    causes, 865
Medial vault span, 482
Meniscal stiffness, 343
Menisci, 334
    anatomy and mechanical role, 342–343
    compressive effects, 345f
    material properties, 343
    movements, 342f
    role, 344–345
    stress–relaxation curve, 344f
    structural properties, 345
    structural variation, 343
    variable ligaments, 343
Meniscofemoral ligaments, 343, 343f
Metabolic energy, 10, 12, 127, 489
Metatarsal declination angle, 554, 595f, 604–605, 605f, 755, 781
Metatarsal fatigue fractures, 822–827, 823–827f
Metatarsalgia, 751, 790
Metatarsal torsion, 519, 593
Metatarsophalangeal (MTP) joint, 7, 29, 150–151, 179, 470, 494, 497–498, 534–535, 593–604
    articular degeneration, 803–805, 804t, 805f
    axis in gait, 609–611
    compliance, 786–789
    extension dysfunction, 801–803, 802f
    fulcrum, 604–611, 768, 776–785, 778f
    hyperextension, 803, 804f, 806–809, 808–809f
    hypoextension, 806–809, 808–809f
    intermetatarsal, 796
    muscular imbalance and lesser toe deformity, 792–795, 794–795f
    pathomechanics, 800–805, 800f
    sagittal plane alignments, 601–604
    sesamoid influence, 806
    sesamoiditis, 806, 807f
    sesamoids, 599–601

soft tissues, 790–796, 791f, 793f
stiffness energetics, 786–789, 787–789f
tailor's bunion, 799–800, 799f
toe deformities, 796–800, 797–798f
torque, 786–787
transverse plane sesamoid dysfunction, 809–811, 810–811f
Midfoot
    dorsiflexion, 8
    as functional unit, 571–584
    impulses, 520–521
    locking, 518–519
    plantarflexion, 517–518
    soft tissue stability, 581–584
    strike, 848–849
    transverse plane motion, 581
Midfoot break, 8, 42–43, 165–167, 166f
    in cerebral palsy, 202, 204f
    foot vault and, 469–470
Midsole cushioning, 150
Midsole hardness, 150
Midsole stiffness, 150–151
Midstance, 403–404, 404f
    absolute, 44–46
    early, 44–46
    knee, 674–675, 675f
    late, 44–45
    pes planus, 763–765, 764f
    of rearfoot complex, 707–709, 708–709f
Midtarsal joint, axes of, 576–579
Momentum, 113
Morton's neuroma, 796
Multiarticular muscles, 653
Multidirectional sports, 838
Multifidus, 293–294
Multi-tied asymmetrical conical vault, 480
Muscle activation and dysfunction, 98–104
Muscle activity, 243
    energetics, 81
    in energy dissipation through foot, 502–505
    force vectors and angular momentum in joints, 81–84
    influence of, 723–724
    preswing, 89–92
    in running, 121–127
    stance activation, 757–761
    stance phase, 87–89
    swing phase, 89–92
Muscle coactivation, 125–126
Muscle contraction, 12
Muscle–induced stiffness, 482
Muscle joint relationships, 257–260, 258f
Muscle's route, 243
Muscle–tendon buffering, 17, 118, 133f, 139, 198, 505
Muscle–tendon complexes, 255–256, 255f
Muscular rehabilitation programmes, 199
Muscular transverse plane stiffening, 515–517
Musculoskeletal disease, 203–208
Musculoskeletal soft tissue failures, 208
Myofascial tone. *See* Resting muscle tone

N

Navicular–cuboid unit, 572
Navicular drift, 568
Navicular drop, 568, 579
Navicular height, 579
Navicular joints, 579–581
Negative impulse, 113–114
Neural transmission, 489–490
Neurological disease, 208–211
Neuromas, 796
Neuropathic pes cavus, 167
Neutral shoes, 151–152
Newton's third law, 10
Nucleus pulposus, 278
Nutation motion, 267–268, 269f

O

Ober's test, 685, 687f
Obesity, 184
Oblique popliteal ligament, 349
Oncogenic osteomalacia, 642
Open-packed joint, 263–264
Optimal lumbopelvic stability, 272
Optimal walking speed, 158
Osgood-Schlatter's disease, 697
Osseous restraining mechanism, 578
Osseous topography
    ankle, 385–387, 386–388f
    hip, 307–315, 307–309f
Osteitis pubis, 285–286
Osteoarthritis (OA), 196–197, 206–207, 333, 375–376, 644, 645f
Osteoligamentous plate, 464–465, 464f
Overuse injuries, 838–840

P

Paediatric gait
    adolescent, 183, 183f
    development, 175–178, 176–177f
    equinus gait, 186–188, 188f
    influences on childhood, 184–185, 185f
    kinematic changes, 179–181, 179–180f
    maturation, 181–183
    pes planus, 185–186, 186–187f
Palaeoanthropology, 552
Parabola, 601–602
Parkinson's disease, 210
Passive elastic elements, of foot, 528–531
Passive stability, 51
Patella alta/baja, 693–696, 695–696f
Patellar ligament, 368
Patellar tracking, 698
Patellofemoral dysfunction, 379–380, 379f
Patellofemoral joint, 691–701, 693–694f
Patellofemoral pain syndrome, 672, 696–699, 697–699f
Pathological pes planus, 161
Pathomechanical summation, 628, 631–633
Pathomechanics, 627
    ankle and rearfoot, 702–747
    foot, 747–837
    hip, 652–671
    knee, 672–702
    patellofemoral pain syndrome, 696–699
    patient clerking, 628–631
    running, 838–868
    weight transference and acceleration, 711–713, 712–713f
Peak activity, 408
Peak pressure, 75
Peak pressure–time curve, 75
Pedal stiffness, 497–498
Pedicles, 275
Pedotti diagrams, 66–68, 68f
Pelvic step, 268
Pelvic tilt angle, 286, 286f
Pelvis
    adaptation and pathology, 301–304, 302f
    extrinsic ligaments, 288–289, 288–289f
    in gait, 10
    in-phase and antiphase motion, 268–270, 269–270f
    intrinsic ligaments, 287, 287–288f
    kinematic role, 266–270, 267–269f
    L4–L5 articulation, 280–281, 280f
    load distribution, 289–290, 290f
    lumbar vertebrae, 273–277, 273–277f
    lumbosacral joint, 273–277, 273–277f, 281, 282–283f
    motion in running, 128–132
    muscle action, 291–301
    pelvic ligaments, 286–287
    pelvic tilt angle, 286, 286f
    sacroiliac joint, 281–286, 283–285f
    vertebral bodies and intervertebral discs, 277–280, 278–279f
Peripheral vascular disease, 212–213
Peroneals
    in ankle function, 432–433, 432f
    foot, 710f, 733–736
    rearfoot, 745–746, 745–746f
Peroneus brevis, 433
Peroneus longus-induced plantarflexion, 768–772
Peroneus longus loading dysfunction, 753–755, 754f
Peroneus longus muscle, 434–435, 434f, 542–543, 544f, 546–548
Pes anserine bursitis, 684, 684f
Pes cavus, 167–170, 487–489
Pes planus, 487–489
    classification, 160–163
    effects on gait, 163–165
    paediatric gait, 185–186, 186–187f
Physiological systems, 628, 629t
Piezoelectric/piezoresistive elements, 56
Plantar aponeurosis, 505, 524, 527–528, 834–836
    acceleration power in, 534–535
    anatomy, 536–538
    biomechanics, 538, 539f
    energy transfer mechanism, 538, 540f
    loading response aponeurotic stress, 534
    role of, 531–533
    during running, 535–536
    stance phase aponeurotic stress, 533–534
    stress–strain relationship with Achilles tendon, 533
Plantar fascial tears, 836
Plantar fasciopathy, 631, 827–836
Plantar fat pad dysfunction, 526–527, 749–751, 750–751f
Plantar fibromatosis, 836
Plantarflexion torque, 25
Plantarflexor loading dysfunction, 752–753, 752–753f
Plantar heel pain, 631
Plantar intrinsic muscles, 513, 516, 548–550, 548f
Plantar plates, 607–609, 607f
Plantar pressure, 170, 176–178
Plantar vault, 474–475, 475f
Plantigrade foot, in inverted pendular gait, 28–29
Plastic collisions, 6
Popliteal fossa, 349
Popliteal line, 372
Popliteofibular ligament, 349, 350f, 355
Popliteomeniscal fascicles, 372
Popliteus muscle action, 372–373, 373f
Poroelasticity, 247
Positive impulse, 113–114
Posterior cruciate ligament, 678, 679f
Posterior oblique ligament, 350
Posterolateral corner complex, 348, 349f
Potentiometers, 73
Pregnancy, gait in, 173–174
Pressure–time integrals, 75
Preswing, 10, 36, 39, 91f
    kinematics, 91
    knee flexion in, 90–91
    muscle activity, 89–92
Pre-tensioning, of foot vault, 501
Pronated midfoot break, 167
Pronation, 81, 466–467, 490, 567–569
Propulsion, 13
Prosthesis, 196–197
Prosthetic limb, 157
Protrusion distance, 604
Proximal femoral rotation, 312
Pseudo-elastic collision, 6, 22–23
Pseudoequinus, 803
Psoas major, 300
Push-off, 13

Q

Older runners, 143
Quadratus lumborum, 294, 294–295f
Quadratus plantae, 505, 549
Quadriceps angle, 359–360
Quadriceps femoris, 366
Quasistiffness, 248

R

Rancho Los Amigos divisions
    additions and modifications, 43–48
    of gait description, 34–35, 35f
    of walking gait cycle, 35–37

Index

Rearfoot
  ankle, 384
  articulations, 558
  forefoot coupling, 571
  frontal plane motion, 569–571, 570f
  as functional unit, 556–571
  posterior calcaneal tubercle, 558
  primary function, 558
  subtalar joint complex, 558–563
  weight acceptance and stability, 558
Rearfoot eversion, 689–690, 690f
Rebound phase, 111
Rectus abdominis, 297–298, 297–298f
Relaxed calcaneal stance position (RCSP), 163, 185–186
Relaxin, 174
Resting muscle tone, 254
Retrorotation, 312
Retrotorsion, 312
Retroversion, 312
Reverse windlass, 409–411
Rheumatoid arthritis, 206, 646
Road running, 152–153
Running economy, 106, 139, 147–148
Running footwear, 152
Running, pathomechanics
  acceleration, 852, 853–854f
  assessment, 840–843
  distance running, 838
  energetic efficiency, 839–840
  fatigue, 839–840
  foot strike positions, 843–849
  gait, 841, 842f
  gender differences, 840–841
  medial tibial stress syndrome, 863–868
  overuse injuries, 838–840
  shin pain symptoms, 852–857
  sport, 838
  tibial fatigue fractures, 857–863
Running speed, 136

S

Sacroiliac (SI) joint, 260–261, 281–286, 283–285f, 303–304
Sacroiliac ligaments, 285
Sacrospinous ligaments, 286–287, 289
Sacrotuberous ligaments, 286–287, 289
Sagittal plane free-body diagrams
  of hip, 654–655, 654–655f
  of knee, 672–675
  of rearfoot complex, 703–711
Sartorius, 374, 374f
Semimembranosus muscles, 371
Semitendinosus, 371
Sesamoiditis, 601
Sesamoids
  with hyperextension and hypoextension, 806–809, 808–809f
  influence, 806
  sesamoiditis, 806, 807f
  transverse plane sesamoid dysfunction, 809–811, 810–811f
Sever's disease, 725–727
Shear forces, 595–596

Shock attenuation
  in lower limb, 117–121
  in running, 116
Shock wave, 114
Sinding-Larsen-Johansson syndrome, 697
Single-limb support, 37
Sinus tarsi, 561–563
Skeletal frame, 265
  division, 551–556
  energy dissipation dysfunction, 751–752
  medial and lateral columns, 551–556, 552f
  role, 551
  vault, in foot compliance, 506
Slipped capital femoral epiphysis, 310
Slow twitch fibres, 254–255
SOAPIER approach, 629–630, 629–630t
Soft and hard tissue
  articular motion and stability, 256–257
  close-packing, 263–265
  core stability, 253–254, 254f
  foot, 244
  form and force closure, 260–263, 261–262f
  joint motion, 257–260, 258–259f, 260t
  joint stability, 264–265
  kinematic joint, 244
  muscle–tendon complexes, 255–256, 255f
  musculoskeletal system, 244
  skeletal frame, 265
  spine, 244
  stability–mobility, 254–256
Soft tissue dysfunction, 208, 250–252, 251–252f
Soft tissue injuries, 639–640
Soft tissue rearfoot stabilisers, 727–730
Spatial filter, 490
Spring–mass model, 31–34, 524
Sprinting, 136–138
Stability shoes, 151–152
Stance knee motion/stability, 101
Stance limb, 10, 14f
Stance phase, 16–19, 18f
  aponeurotic stress, 533–534
  hip extension, 101
  muscle activity, 87–89
  muscle function and dysfunction, 100–101
  running, 107
Static windlass effects, 532, 532f
Step length, 35, 69
Step width. See Base width
Stiff planus foot, 473
Strain gauges, 56
Stress fractures, 146–147. See also Tibial fatigue fractures
Stress–strain behaviour, 492
Stretch-shortening cycles, 255–256, 255f
Stride, 35
Stride length, 35
Structural curvature, 484–485
Subject-specific variation, 635
Subtalar joint, 528, 558–563, 561f
  foot pronation and supination, 567–569
  force closure, 563–567
Subtalar ligaments, 389–392, 393–394f
Superficial fascicles, 504
Supinated midfoot break, 167

Supination, 81, 466–467, 567–569
Supported walking, 176
Supraspinous ligament, 275
Sustentaculum tali, 389
Swing initiation, 13–15
Swing limb, 10–11, 14f
Swing phase, 12–16, 14–17f, 404–405, 405f, 672–674, 673f
  muscle activity, 89–92
  muscle function and dysfunction, 102–103, 102f
  rearfoot complex, 703–704, 705f
  running, 119
Symptoms, objective observations, analysis (SOAP) approach, 629–630, 629–630t
Syndesmotic impingement, 389

T

Talocalcaneonavicular joint, 558
Talonavicular joint, 558, 572, 579
Tarsometatarsal joints, 482, 584–592, 586f, 591f
Tarsometatarsal stability, 590–592
Temporal filter, 490
Tendons, 528
  in running energetics, 127
Tendon slip, 516–517
Tensegrity, 244–248
Tensor fasciae latae, 330
Terminal stance
  of hip, 659–663, 660–662f
  of rearfoot complex, 709–711, 710f
  stiffness, 468–469, 512–519, 772–773
Terminal swing, 40
Terrain
  running, 152–155
  walking, 78–80, 78–79f
Thoracolumbar fascia, 266, 291
Tibial fatigue fractures, 857–863
Tibialis anterior, 408–409, 408f, 503–504, 504f
Tibialis posterior, 162, 162f, 429–432, 430–431f, 544–548, 545f, 744, 744f, 753–755, 754f, 773–776, 774f
Tibial plafond, 385
Tibial sesamoids, 601
Tibial stresses
  running fatigue effects, 863
  strike position, 861–863, 862f
Tibiofemoral degenerative joint disease, 378–379
Tibiofemoral joint, 679–684, 680f
  popliteal muscle, 685, 686f
  popliteofibular ligament, 685
Tibiofibular syndesmosis ligaments, 387–389
Tissue dysfunction, 628
Tissue threshold, 628, 631–633
  activity influence, 633
  alignment and morphology, 633–635, 634f
Toe deformities, 796–800
Toe springs, 777
Torques, 12, 19, 21–22, 25, 150–151, 249–250, 258, 267–268, 364–365, 490, 500–501, 574–575, 640, 672, 727, 730, 734
Total hip arthroplasty, 333–334
Total hip replacement surgery, 205–206

Transfemoral amputation, 196–197
Transverse acetabular ligament, 307
Transverse arch, of vault, 473–474
Transverse genicular ligament, 343
Transverse ligament, 315, 530–531, 530f
Transverse plane curvature, 484–486
Transverse plane free-body diagrams, of hip, 668–669, 669–670f
Transverse plane motion, of midfoot, 581
Transverse vault span, 482
Transversus abdominis, 297
Traumatic injures, 628
Treadmill running, 152–153
Triangulated foot columns, 555–556
Triceps surae–Achilles complex, 187–188, 411–412, 412f, 469–470
Triceps surae component muscles, 412–418, 413–418f
Trunk leading strategy, 191
Tumour-induced osteomalacia, 642

U

Ultra-distance running, 138
Uni-articular muscles, 258
Upright posture, challenges of, 19–22

V

Variability, in gait, 69, 77–80, 189
Variance, in gait
　ageing and aged-like gait, 188–191
　foot function, 160–170
　footwear, effects of, 195–196
　gender and morphological differences in walking, 157–159
　joint hypermobility, 170–173
　leg length discrepancy/inequality, 191–195
　lower limb amputees, 196–199
　paediatric gait, 175–188
　pregnancy, 173–174
Vastus intermedius, 84, 97f, 98–99, 101, 257–258, 258–259f, 366
Vastus medialis longus, 354, 359, 359f, 366
Vastus medialis obliquus, 257–258, 351f, 692
Vault deflection, 430f, 496, 497f, 584, 721, 765
Vault depression, 62–63, 506, 510, 510f, 524–525, 584, 711, 835
Vault span
　lateral, 482
　medial, 482
　stability mechanics, 482–484
　transverse, 482
Vault stiffness, 473–474
　soft tissue failures, 506
　span distance and curvature effects on, 480–487, 481f
Villefranche criteria, 161
Viscoelastic asymmetrical conical foot vault, 486–487

W

Walking
　arm swing in, 66
　collisions, 7–8
　double-limb contact phases, 11–12
　endurance and fatigue in, 191
　gait phases, 12–19
　gender and morphological differences, 157–159
　influence of terrain on gait analysis, 78–80, 78–79f
　initial contact in, 37
　initial contact-to-loading response, 100, 100f
　knee joint, 81
　metabolic cost, 2–4, 5f
　primary muscle function, 84–92
　and rolling, 7, 8f
　running transition, 112
　stance phase, 16–19, 18f
　swing phase, 12–16, 14–17f
　terrain, velocity, and gradient, 94–98
　treadmill declinations, 96
Weight acceptance, 37
Weightbearing
　articular surfaces, 275
　flatter foot vault profile, 250
　forefoot, 63–64, 491, 749
　heel lift, 512–513
　inclination–declination angles, 554
　rearfoot, 262, 571
Weight transference/acceleration, 37–39
Windlass effect, 409–411, 494–495, 498, 512, 532
Wobbling mass, 505–506, 522
Wobbling structures, 117

Z

Zona orbicularis. *See* Ligamentum orbicularis

9780443158605